W9-BSK-654

CONSTRUCTION

CONSTRUCTION

Principles, Materials, and Methods

SEVENTH EDITION

H. Leslie Simmons, RA, CSI

Previous Editions Developed By:
Harold B. Olin, FAIA

with contributions from
John L. Schmidt, AIA
Walter H. Lewis, FAIA

John Wiley & Sons, Inc.
New York Chichester Weinheim Brisbane Singapore Toronto

This book is printed on acid-free paper. ♾

Published simultaneously in Canada.

Library of Congress Cataloging-in-Publication Data:

Simmons, H. Leslie.
Construction : principles, materials, and methods / H. Leslie Simmons.—7th ed.
p. cm.
"Previous edition developed by Harold B. Olin, John L. Schmidt, Walter H. Lewis."
Includes bibliographical references and index.
ISBN 0-471-35640-9 (cloth : alk. paper)
1. Building. I. Olin, Harold Bennett. Construction. II. Title.
TH145.S513 2001
690—dc21 00-043581

Printed in the United States of America.

10 9 8 7 6 5

Contents

Preface ix
Acknowledgments xi
Disclaimer xiii

CHAPTER 1
GENERAL REQUIREMENTS 1

Introduction 2
Applicable *MasterFormat*™ Sections 2
1.1 Building Design 2
1.2 Construction Documents 7
1.3 Bidding and Negotiation 10
1.4 Construction Contract Administration 11
1.5 Industry Standards 12
1.6 Codes 18
1.7 Barrier-Free Design 26
1.8 The Metric System of Measurement 38
1.9 Land Surveys and Descriptions 73
1.10 Properties of Materials 77
1.11 Additional Reading 88
1.12 Acknowledgments and References 89

CHAPTER 2
SITE CONSTRUCTION 91

Introduction 92
Applicable *MasterFormat*™ Sections 92
2.1 Soils 92
2.2 Site Preparation 98
2.3 Earthwork 99
2.4 Surface- and Groundwater Problems 109
2.5 Lawns and Landscaping 112
2.6 Additional Reading 116
2.7 Acknowledgments and References 116

CHAPTER 3
CONCRETE 119

Introduction 120
Applicable *MasterFormat*™ Sections 120
3.1 History 120
3.2 Concrete Materials 123
3.3 Formwork 129
3.4 Reinforcement 134
3.5 Accessories 141
3.6 Joints 143
3.7 Mixtures and Mixing 146
3.8 Handling, Transporting, Placing, and Consolidating 155
3.9 Finishing 158
3.10 Curing and Protection 164
3.11 Concrete Foundation Systems 166
3.12 Concrete Slabs on Grade 173
3.13 Cast-in-Place Structural Concrete 180
3.14 Precast Concrete 189
3.15 Glass-Fiber-Reinforced Concrete 195
3.16 Concrete Toppings 195
3.17 Structural Insulating Roof Decks 196
3.18 Cement-Based Underlayment 197
3.19 Additional Reading 198
3.20 Acknowledgments and References 199

CHAPTER 4
MASONRY 201

Introduction 202
Applicable *MasterFormat*™ Sections 202
4.1 History 202
4.2 Mortar and Grout 203
4.3 Reinforcement, Ties, Anchors, and Flashing 211
4.4 Clay Masonry Units 214
4.5 Concrete Masonry Units 228
4.6 Unit Masonry Design 237
4.7 Unit Masonry Erection 259
4.8 Stone 276
4.9 Glass Unit Masonry 284
4.10 Properties of Selected Masonry Walls 285

4.11 Additional Reading 289
4.12 Acknowledgments and References 290

CHAPTER 5
METALS 295

Introduction 296
Applicable *MasterFormat*™ Sections 296
5.1 Iron and Steel Materials and Products 296
5.2 Iron and Steel Design and Construction 320
5.3 Miscellaneous Iron and Steel Elements 333
5.4 Aluminum 333
5.5 Other Metals 349
5.6 Additional Reading 356
5.7 Acknowledgments and References 357

CHAPTER 6
WOOD AND PLASTICS 363

Introduction 364
Applicable *MasterFormat*™ Sections 364
6.1 Properties of Wood 364
6.2 Lumber 379
6.3 Plywood and Other Panels 391
6.4 Treated Wood Foundations 406
6.5 General Framing Requirements 410
6.6 Conventional Framing and Furring 412
6.7 Other Framing Systems 447
6.8 Finish Carpentry 462
6.9 Plastic Fabrications 464
6.10 Additional Reading 474
6.11 Acknowledgments and References 474

CHAPTER 7
THERMAL AND MOISTURE PROTECTION 477

Introduction 478
Applicable *MasterFormat*™ Sections 478
7.1 Moisture Control 478
7.2 Waterproofing and Dampproofing 496
7.3 Building Insulation 499
7.4 Exterior Insulation and Finish Systems 511
7.5 Low-Slope Roofing 513
7.6 Steep Roofing 531
7.7 Siding 570
7.8 Flashing and Sheet Metal 589
7.9 Metal Roofing 594
7.10 Fireproofing 599
7.11 Joint Sealing 601
7.12 Additional Reading 607
7.13 Acknowledgments and References 609

CHAPTER 8
DOORS AND WINDOWS 617

Introduction 618
Applicable *MasterFormat*™ Sections 618
8.1 Metal Doors and Frames 618
8.2 Wood Doors 623
8.3 Aluminum Entrances and Storefronts 634
8.4 Aluminum Windows and Sliding Glass Doors 638
8.5 Wood Windows and Sliding Glass Doors 645
8.6 Storm Windows and Doors 653
8.7 Door Hardware 656
8.8 Glazing 661
8.9 Glazed Aluminum Curtain Walls 679
8.10 Additional Reading 686
8.11 Acknowledgments and References 687

CHAPTER 9
FINISHES 695

Introduction 696
Applicable *MasterFormat*™ Sections 696
9.1 Plaster Materials 696
9.2 Plaster Support Systems, Bases, and Accessories 703
9.3 Gypsum Plaster Application 709
9.4 Portland Cement Plaster Application 713
9.5 Gypsum Board Systems 719
9.6 Tile 749
9.7 Terrazzo 772
9.8 Acoustical Treatment 783
9.9 Wood Flooring 797
9.10 Stone Flooring 806
9.11 Resilient Flooring 811
9.12 Carpet 832
9.13 Resinous Flooring 846
9.14 Special Coatings 849
9.15 Paints 854

9.16 Vinyl Wall Coverings 872
9.17 Additional Reading 873
9.18 Acknowledgments and References 875

CHAPTER 10
SPECIALTIES **885**

Applicable *MasterFormat* ™ Section 886
10.1 Fire Safety 886
10.2 Fire Protection Specialties 889
10.3 Additional Reading 890
10.4 Acknowledgments and References 890

CHAPTER 11
EQUIPMENT **893**

Introduction 894
Applicable *MasterFormat* ™ Sections 894
11.1 Residential Appliances 894
11.2 Unit Kitchens 899
11.3 Additional Reading 901
11.4 Acknowledgments and References 901

CHAPTER 12
FURNISHINGS **903**

Introduction 904
Applicable *MasterFormat* ™ Sections 904
12.1 Wood Casework 904
12.2 Window Treatment 910
12.3 Additional Reading 913
12.4 Acknowledgments and References 914

CHAPTER 13
SPECIAL CONSTRUCTION **915**

Introduction 916
Applicable *MasterFormat* ™ Sections 916
13.1 Sound Control 916
13.2 Lightning Protection 964
13.3 Intrusion Prevention and Detection 965
13.4 Additional Reading 969
13.5 Acknowledgments and References 970

CHAPTER 14
CONVEYING SYSTEMS **971**

Introduction 972
Applicable *MasterFormat* ™ Sections 972
14.1 Elevators 972
14.2 Additional Reading 979
14.3 Acknowledgments and References 979

CHAPTER 15
MECHANICAL **981**

Introduction 982
Applicable *MasterFormat* ™ Sections 982
15.1 Plumbing 982
15.2 Heating, Ventilating, and Air Conditioning (HVAC) 1026
15.3 Fire Sprinkler Systems 1058
15.4 Additional Reading 1058
15.5 Acknowledgments and References 1058

CHAPTER 16
ELECTRICAL **1063**

Introduction 1064
Applicable *MasterFormat* ™ Sections 1064
16.1 Fundamentals of Electricity 1064
16.2 Electric Power Distribution 1068
16.3 Service and Distribution 1070
16.4 Lighting 1088
16.5 Communications Systems 1093
16.6 Conservation of Energy 1094
16.7 Additional Reading 1095
16.8 Acknowledgments and References 1095

APPENDIX A
DATA SOURCES **1097**

APPENDIX B
GLOSSARY **1119**

INDEX **1149**

Preface

Construction: Principles, Materials, and Methods is a widely adopted course text in nearly 200 colleges and universities offering architectural and building technology curricula. It is also becoming a standard reference source in professional offices.

This seventh edition is the result of almost 40 years of research and editorial effort costing several million dollars. During the book's first 19 years, much of this work was carried out in an ongoing membership education program initiated by the United States League of Savings Institutions in 1964. The sixth edition was the result of that research and a major program carried out during the 3 years prior to its publication by its publisher at that time, Van Nostrand Reinhold. This program consisted of completely revising the data in the previous edition to reflect the many changes that had occurred in the field and to bring that data up to date with latest industry standards. Furthermore, new data was added to introduce construction materials and methods not in general use when the previous edition was prepared. Materials and construction methods that relate to commercial construction, including high-rise buildings, were also included. Previous editions had primarily covered residential construction. The sixth edition addressed materials and methods used in both.

The seventh edition has been written to further develop the book as a modern teaching and research tool for residential, commercial, and institutional building construction. To this end, the seventh edition contains more than 1100 pages divided into 16 chapters. The organization and names of these chapters are similar to those used in *MasterFormat*™, which makes this edition easier to relate to the specifications and data filing formats currently being used in the construction industry. A list of applicable *MasterFormat*™ sections has been included as a subsection in each chapter. The numbers and titles listed there are from *MasterFormat* (1995 edition), which is published by the Construction Specifications Institute (CSI) and Construction Specifications Canada (CSC) and is used with permission from CSI, 1999. For those interested in a more in-depth explanation of *MasterFormat*™ and its use in the construction industry, contact:

Construction Specifications Institute (CSI)
99 Canal Center Plaza, Suite 300
Alexandria, VA 22314
(800) 689-2900; (703) 684-0300
CSINet URL:www.csnet.org

An attempt has been made in the seventh edition to cover every principle, material, and method used to design and construct both large and small buildings of most types. The information presented includes the background and history of the materials and systems described. In each case, materials and their manufacture are discussed first, followed by the methods of construction used to erect these materials.

Earlier editions of this book, which were directed primarily toward the home construction portion of the construction industry, included coverage, much of it extensive, of wood construction; masonry; interior and exterior finishes; heating, ventilating, and air conditioning; plumbing; electrical systems; and many other subjects as they related to small residential construction. The sixth edition included most of these as well, but supplemented them with information about materials and methods used in the construction of other building types. The construction systems covered included those of both precast and cast-in-place concrete, steel, wood, and masonry. The discussions of mechanical and electrical systems were still limited to requirements for residential construction. Other subjects covered in earlier editions were expanded to address most kinds of commercial, larger residential, and institutional buildings. As a result, the sixth edition introduced materials and systems used in most types of building construction.

The sixth edition text also covered doors and windows; glazed curtain walls; glazing; and finishes, including plaster, gypsum board, tile, terrazzo, acoustical ceilings and other acoustical treatment, wood flooring, resilient flooring, carpet, and painting and finishing.

Coverage was given to industry standards, codes, land surveys and descriptions, properties of materials, barrier-free design, metrication, sitework (including excavation, grading, shoring, sheeting, and other earthwork), and sound control in buildings.

The seventh edition covers most of the subjects addressed in the sixth edition, but supplements it with much new data. New subjects in the seventh edition include a discussion of the architect's role in building design and construction, including the development of construction documents, and the architect's responsibilities during the bidding and negotiation phase and during the construction phase of a building construction project. The sections on codes and standards have been expanded to include discussion of their effect on building design. Additional subjects include glass-fiber-reinforced concrete, concrete toppings, cement-based underlayment, glass unit masonry, miscellaneous metal fabrications, heavy timber framing, finish carpentry, wood siding, metal roofing, door hardware, stone flooring, resinous flooring, wall coverings, fire protection specialties, residential appliances, unit kitchens, elevators, and fire sprinkler systems. The discussions of mechanical systems have been greatly expanded to include information related to buildings other than residences. For example, there is coverage of the different types of HVAC systems, including heating systems (forced air, air-water, steam, water, electrical), components (boiler types, furnaces, controls, finned tube radiators, radiant heaters and panels, heat recovery systems, ducts, pipes, diffusers, grilles, and registers), and fuels (coal, gas, electricity). Cooling systems are also covered, including discussions of the refrigeration cycle, cooling components, heat pumps, direct refrigeration, and delivery systems (fan coil units, unit ventilators). The discussion of electrical systems has been expanded to include commercial lighting and cable distribution systems.

In addition, some chapters have been reorganized to make the data they contain more easily usable. For example, Chapter 5 now separates steel materials from steel products. Some material has been relocated to a chapter corresponding to its location in *MasterFormat*™.

Metric equivalents have been added to measurements throughout the book. To avoid changing copyrighted figures, however, metric measure has not generally been added to them. Instead, Figure 1.8-27 is included, which lists metric conversions that are applicable to these figures.

Each chapter now has a section called "Additional Reading" that lists sources of additional information about the subjects discussed in that chapter. To make it easier for a reader to find data, the tables and other data in Appendix A of the sixth edition have instead been placed in the various chapters where the particular subjects are discussed. To enable a reader to more easily find references for the sources of information in the book, the data in the sixth edition's Appendix B, "References," has been moved to a section entitled "Acknowledgments and References" at the end of each relevant chapter.

In Appendix A, the seventh edition contains a list of the names, addresses, telephone numbers, e-mail addresses, and Web sites (where applicable) of the organizations, associations, and agencies that contributed either directly or indirectly to the book. Appendix B is an extensive glossary of terms used in the book.

Acknowledgments

It takes the help of many people and organizations to produce a book such as this, which include all of those who prepared previous editions. First and foremost of these are the original authors, Harold B. Olin, John L. Schmidt, and Walter H. Lewis, without whom this book would not exist. The preface for the fifth edition, written by the three authors, acknowledges the contributions of several persons. Included are Senior Research Associate Christina Farnsworth, for her contribution in revising old chapters and in authoring chapters on barrier-free design, home security and safety, air plenum systems, and metrication; Richard Laya, who prepared the chapters on heat control, passive solar heating, and pole foundations; and Production Specialist Betsy Pavichevich who was responsible for the typing, layout, and preparation of graphics.

Even though the material these people prepared was, in most cases, extensively revised in the sixth edition and has been further revised in the seventh edition to bring it up to date with current materials, methods, and standards, the contribution of these people in producing the original data is to be commended.

My work on the sixth edition was aided immensely by Brenda Bertozzi, my primary research associate, who located sources and obtained permission to use their materials, and by Irene Demchyshyn, who helped me gather the research data necessary to update and expand this book.

I am indebted also to the professionals, both practicing architects and teachers, who lent their time, energy, and expertise to review the sixth edition manuscript so that it would be more accurate and more useful as a text for students of architecture and allied fields. These include Harold B. Olin, FAIA, the original primary editor of previous editions; Daniel W. Halpin, Ph.D., Professor and Head, Division of Construction Engineering and Management, Purdue University; Larry Grosse, Ph.D., Associate Professor, College of Architecture, Texas A&M University; and Terry L. Patterson, AIA, Professor, College of Architecture, University of Oklahoma, whose extensive, well-informed, and detailed comments contributed greatly to making this book as good as it could be. Professor Patterson was also the author of the sixth edition *Student Workbook* and *Instructor's Manual.*

ACKNOWLEDGMENTS FOR THE SEVENTH EDITION

To determine the needs of the readers of *Construction: Principles, Materials, and Methods*, the sixth edition was reviewed by current users, including teaching professionals and architects in practice. Their comments were of immeasurable help in developing the seventh edition. Among them were the aforementioned Harold B. Olin and Terry Patterson; and Atilla Lawrence, University of Nevada, Department of Architecture; Dana S. Mosher, Professor of Building Construction Technology, NHCTC/Manchester; Robert O. Segner, Jr., Professor of Construction Science, Department of Construction Science, College of Architecture, Texas A&M University; Khalid T. Al-Hamdouni, Rogers University School of Engineering; Maryrose McGowan, AIA, IIDA; and the late Tim Kirby, consultant to the construction industry,

I would like to gratefully acknowledge the contributions of the many manufacturers, trade and professional associations, standards-setting bodies, government agencies, periodicals, book publishers, and individuals who provided valuable research material and illustrations, as well as text reviews and comments. Without their help, this book would not have been possible. The names of most are listed in the "Acknowledgments and References" sections at the ends of the chapters.

I would also like to thank my wife, Nancy, whose help and support have been invaluable.

I am also indebted to the professionals, both practicing architects and teachers, who lent their time, energy, and expertise to review the manuscript for the seventh edition so that it would be more accurate and more useful as a text for students of architecture and allied fields and as a reference for professionals in practice. They include: Harold B. Olin, William C. Mason, Owens Community College, Toledo, OH; Linda Swoboda, University of Nebraska; John Lebduska, Hope, NJ; Harold Tepper, New Jersey Institute of Technology; Felix Rospond, West Orange, NJ; and Anthony Abbate, Florida Atlantic University.

Finally, I would like to express my thanks to the professionals at John Wiley & Sons, Inc., who lent their talents to this work. They include the acquiring editor, Amanda L. Miller, Executive Editor, and her assistant, Nyshie P. Perkinson; Jennifer Mazurkie, Associate Managing Editor, who ably guided the book through production; and Publishers' Design & Production Services, who designed and typeset the book. I would also like to thank Judith Joseph, who initially contacted me about revising the book; My sincere thanks to all those, both named and unnamed here, who participated in making this work better than it might otherwise have been.

H. Leslie Simmons, RA, CSI

Disclaimer

The information in this book was derived from data published by trade associations, standards-setting organizations, manufacturers, and government organizations and statements made to the author by their representatives; model codes; and related books and periodicals. The author and publisher have exercised their best judgment in selecting data to be presented, have reported the recommendations of the sources consulted in good faith, and have made every reasonable effort to make the data presented accurate and authoritative. However, neither the author nor the publisher warrants the accuracy or completeness of the data nor assumes liability for its fitness for a particular purpose. Users are expected to apply their own professional knowledge and experience or to consult with someone who has such knowledge and experience when using the data contained in this book. Users are also expected to consult the original sources of the data, obtain additional and updated information as needed, and seek expert advice when appropriate.

A few manufacturers and products are mentioned in this book. Such mention is intended merely to indicate the existence of such products and manufacturers, not to imply endorsement by the author or the publisher of the mentioned manufacturer or product. Nor is there any implied endorsement of other products made by a mentioned manufacturer, or any statement made by a mentioned manufacturer or associated in any way with any product, its accompanying literature, or in advertising copy.

Similarly, literature and standards produced by various manufacturers, associations, and other organizations are mentioned as well. Such mention does not imply that the mentioned item is the only available data or standard, or the best of its kind, or is the legal requirement or accepted standard in a particular locality. The author and publisher expect the reader to seek out other manufacturers, appropriate associations, and other organizations to ascertain whether they have similar literature, or will make similar data available, and to determine whether the data presented conforms with code and other legal requirements in a particular locale.

CHAPTER 1

General Requirements

Introduction

Applicable *MasterFormat*™ Sections

Building Design

Construction Documents

Bidding and Negotiation

Construction Contract Administration

Industry Standards

Codes

Barrier-Free Design

The Metric System of Measurement

Land Surveys and Descriptions

Properties of Materials

Additional Reading

Acknowledgments and References

Introduction

Many factors influence an architect's work related to building design. In addition to architectural design, an architect must be aware of and conversant in site, structural, mechanical, and electrical design. He or she must also be aware of the legal constraints, such as codes, laws, and regulations, and of the many industry standards that influence design and construction. An architect must also be knowledgeable and conversant in the production of construction documents and must understand the means and methods used in constructing buildings. He or she must understand the construction process and be able to render an architect's services during the construction phase of a building project. He or she must understand the financial constraints on building construction, and be able to design within those constraints. And in all of these, an architect must not be just a jack-of-all-trades, he or she must be a master of them all.

Chapters 2 through 16 of this book address construction materials and methods of design and construction of which an architect must be knowledgeable.

The first four parts of this chapter discuss some of the services architects provide related to a building construction project. The American Institute of Architects (AIA) has divided an architect's services into the categories *basic* and *additional*.

Basic services are those included in standard services contracts developed by AIA and included in the architect's basic fee for services.

Additional services are optional and are performed only when agreed to by the architect and the owner, with additional compensation to the architect.

Following the flow of a project from conception to the completion of the warranty period (one year after construction completion) an architect's services can be broken down into *predesign* services, *design* services, *construction* services, *postconstruction* services, and *supplemental* services.

Predesign services are additional services. They include such acts as programming, existing facilities studies, project budgeting, and site analysis.

Basic services include *design* and *construction* services. Design services (see Section 1.1) are further broken down into *schematic* design, *design development* (a further refinement of schematic design documents), and *construction documents* services (see Section 1.2).

Construction services include services performed during the *bidding and negotiation* phase (see Section 1.2) and those performed during the *construction contract administration* phase (see Section 1.4).

Postconstruction services are additional services performed after substantial completion of the building (see Section 1.4). They include such acts as maintenance and operational programming, record drawings, start-up assistance, and warranty review.

Supplemental services are additional services. They include such items as renderings, models, life cycle cost analysis, quantity surveys, graphic design, and many others.

The rest of this chapter covers some additional facets of the building design and construction process that an architect must understand to be able to carry out the architect's responsibilities in the design and construction of buildings.

Applicable *MasterFormat*™ Sections

The following *MasterFormat* Level 2 sections are applicable to this chapter.

00100 Bid Solicitation
00200 Instructions to Bidders
00300 Information Available to Bidders
00400 Bid Forms and Supplements
00490 Bidding Addenda
00500 Agreement
00600 Bonds and Certificates
00700 General Conditions
00800 Supplementary Conditions
00900 Addenda and Modifications
01100 Summary
01200 Price and Payment Procedures
01300 Administrative Requirements
01400 Quality Requirements
01500 Temporary Facilities and Controls
01600 Product Requirements
01700 Execution Requirements
01800 Facility Operation
01900 Facility Decommissioning

1.1 Building Design

An architect's first and primary contractual responsibility related to building construction is design. Building design requires training, experience, an aesthetic sense, and an understanding of certain basic principles. Among these principles are (1) the objectives good design should strive for, (2) an architect's responsibilities related to design, (3) basic building use and shape types, and (4) available construction systems and methods.

1.1.1 DESIGN OBJECTIVES

An architect's primary design objective should be to produce buildings that serve their intended purpose and that permit the activities that take place in them to proceed with appropriate dispatch and ease. They should be efficient in their use and operation. In addition, commercial buildings should be capable of producing a profit.

An architect's buildings should be of good quality construction, and should be able to be built at as low a cost as is practicable. An architect's designs should produce individual buildings that are aesthetically pleasing and that do not diminish the beauty of the natural environment around them.

1.1.2 THE BUILDING CONSTRUCTION TEAM

Many organizations and individuals must work together to produce buildings. These include owners, architects and their consultants, constructors, and supporting professions and industries.

Owners are the architect's clients. They are not necessarily the users of a building, but they begin, finance, and usually own the project.

A *design professional* is a person or organization who designs a construction project. The *prime design professional* is the design professional who is hired by the owner to lead the design team. In building construction projects, the prime professional is usually an architect. An engineer may be the prime design professional on some primarily engineering projects—for example, in the construction of bridges or in the major renovation of an existing heating, ventilating, and air conditioning system. In this section, it is assumed that the project being considered is a building and that the prime design professional is an architect. In other project types, the chores here delineated as the responsibility of the architect may fall to an engineer, as the prime design professional for a particular project.

It is ordinarily the architect's responsibility to (1) determine the legal, financial, and other constraints on project design, (2) program and design the project, (3) produce contract documents, (4) provide professional services during the bidding or negotiation phase, and (5) provide construction contract administration services. For a residence or other small building, an architect may carry out these functions alone. Larger and more complicated buildings often present design problems that are beyond the expertise of most architects. For these more complicated building construction projects, an architect functions as a member, usually as the leader, of a team of design professionals that includes structural, mechanical, civil, and electrical engineers, and interior designers, who function as consultants to the architect.

The architect and each of the architect's consultants will design, produce construction documents, and provide construction contract administration for that one portion of a building's components that fall within his or her field of expertise. The architect coordinates the activities of all team members.

The construction process often also requires input from a second group of design professionals working as consultants, either to the owner directly or to one of the team members. These other professionals include, but are not limited to, those with special knowledge about schools, hospitals, food service facilities, laboratories, industrial complexes, computer systems, communication systems, furniture, specialized equipment, and many other components. The architect usually coordinates the activities of these other professionals.

Constructors, also called *builders*, are usually a group of organizations that together erect construction projects. They consist of many types of *contractors*, including, but not limited to, *general contractors*, who oversee the work of, and usually hire, the others; and specialty, or *trade*, contractors, such as those who provide sitework, excavation, concrete, masonry, steel, carpentry, casework, moisture protection, doors, windows, finishes, specialties, equipment, and conveying, plumbing, electrical, and mechanical systems. Supporting these contractors are suppliers, who provide construction equipment, such as cranes, and product suppliers who furnish the materials, products, systems, and equipment that go into a building.

Supporting professions and industries include, but are not limited to, legal professionals; accountants; lenders and investors, who provide construction money and long-term loans that permit construction projects to be erected; insurance providers; testing and research agencies, which develop new products and test existing ones; regulators, including code and law writers and enforcers, who control health and safety issues, aesthetics, environmental issues, zoning, utilities, financial institutions, and design professionals' licensing and practices.

1.1.3 BUILDING USE TYPES

Construction projects can be identified by their use (residential, commercial [stores, office buildings, etc.], institutional [hospitals, schools, jails, etc.], industrial [manufacturing, laboratories, etc.], and nonbuilding types [bridges, towers, etc.]). From this point forward, this chapter addresses only those building construction projects for which an architect is the primary professional. In such projects, it is the architect's job to determine the design requirements specific to each use. For example, there is little resemblance between the requirements for a single-family residence and those of a hospital. There may be major differences even within a group. The are great differences, for example, between the requirements for a single-family residence and a high-rise apartment building. A local jail will probably bear little resemblance to a federal prison.

Some buildings are designed for *common use*, meaning that they have more than one use type in the same structure. Street-front stores may have residential or office spaces above them. High-rise buildings may house commercial uses on the lower floors, office uses on intermediate floors, and apartments on the upper floors.

1.1.4 BUILDING SHAPE TYPES

Buildings take many forms and shapes, depending on their use, the materials used to build them, the needs and desires of the owner, the construction budget, the building's potential operating costs, and the designer's preferences. Buildings other than single-family residences and townhouses are so varied in size and shape as to make simplification of their types difficult (Fig. 1.1-1). However, some basic types and

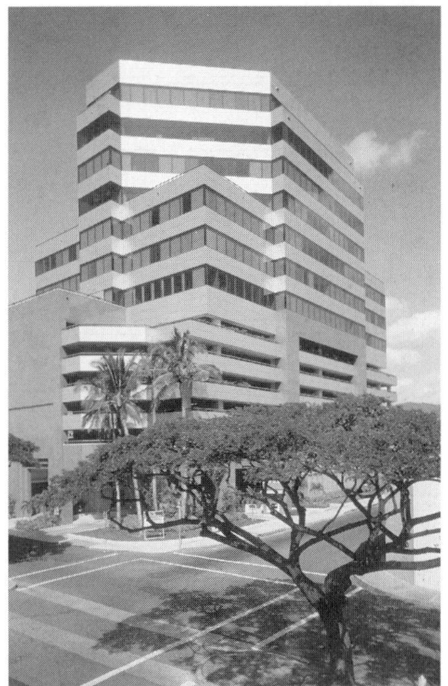

FIGURE 1.1-1 Modern buildings are seldom rectangles. (Honvest Corporation, Honolulu, Hawaii. Architect Leo A. Daly and Associates. Photo courtesy Bethlehem Steel Corporation.)

construction methods can be identified (Fig. 1.1-2).

The simplest building is a 1-story, single-span, slab-on-grade structure with a flat roof (Fig. 1.1-2a). Similar buildings with basements are also commonly built (Fig. 1.1-2b). Single-story structures with more than one structural span (Fig. 1.1-2c), in which one or more intermediate rows of walls or columns supports the roof structure, enclose more space per unit of exterior wall cladding than do smaller buildings.

Single-story buildings may also have full or partial basements. The structural systems in buildings of this type may be concrete, masonry, steel-framed, or wood-framed bearing walls with steel, concrete, or wood roof framing systems; steel, concrete, or wood interior and exterior columns with steel, concrete, or wood roof framing; or a combination of these systems. Foundations are usually poured concrete, but treated wood foundations are sometimes used (see Section 6.4). The roof of a single-story building may be either flat or any of a wide variety of shapes (Fig. 1.1-2d, 1.1-2e, and 1.1-2f). Roof decks may be of wood, steel, or concrete. Basements may have either poured concrete or reinforced masonry walls, depending on the level of the earth against the wall and the height and the hydrostatic head of adjacent underground water. Floors above basements may be steel framed with a concrete or wood floor, concrete framed with a concrete floor, steel framed with a concrete floor, wood framed with a wood floor, or a combination of these systems.

The same principles apply to multistory structures (Fig. 1.1-2g). The construction materials and structural systems in multistory structures and the height of such buildings are usually dictated by economic factors, such as land cost, but may be affected by codes and laws that restrict building height, land area coverage, or the materials that may be used. Fire codes, for example, may restrict the types of construction systems and the materials that may be left exposed. Many fire codes do not permit wood construction or the exposure of wood finishes on the exterior of buildings in certain locations.

Multistory buildings require less roof surface than single-story buildings with the same floor area. This results in a savings in the cost of roofing materials. In addition, multistory buildings require less land per unit of usable space. Because of their higher ratio of interior space to building shell area, they are also generally more energy efficient than single-story buildings. Except in rare instances, these advantages increase with the number of

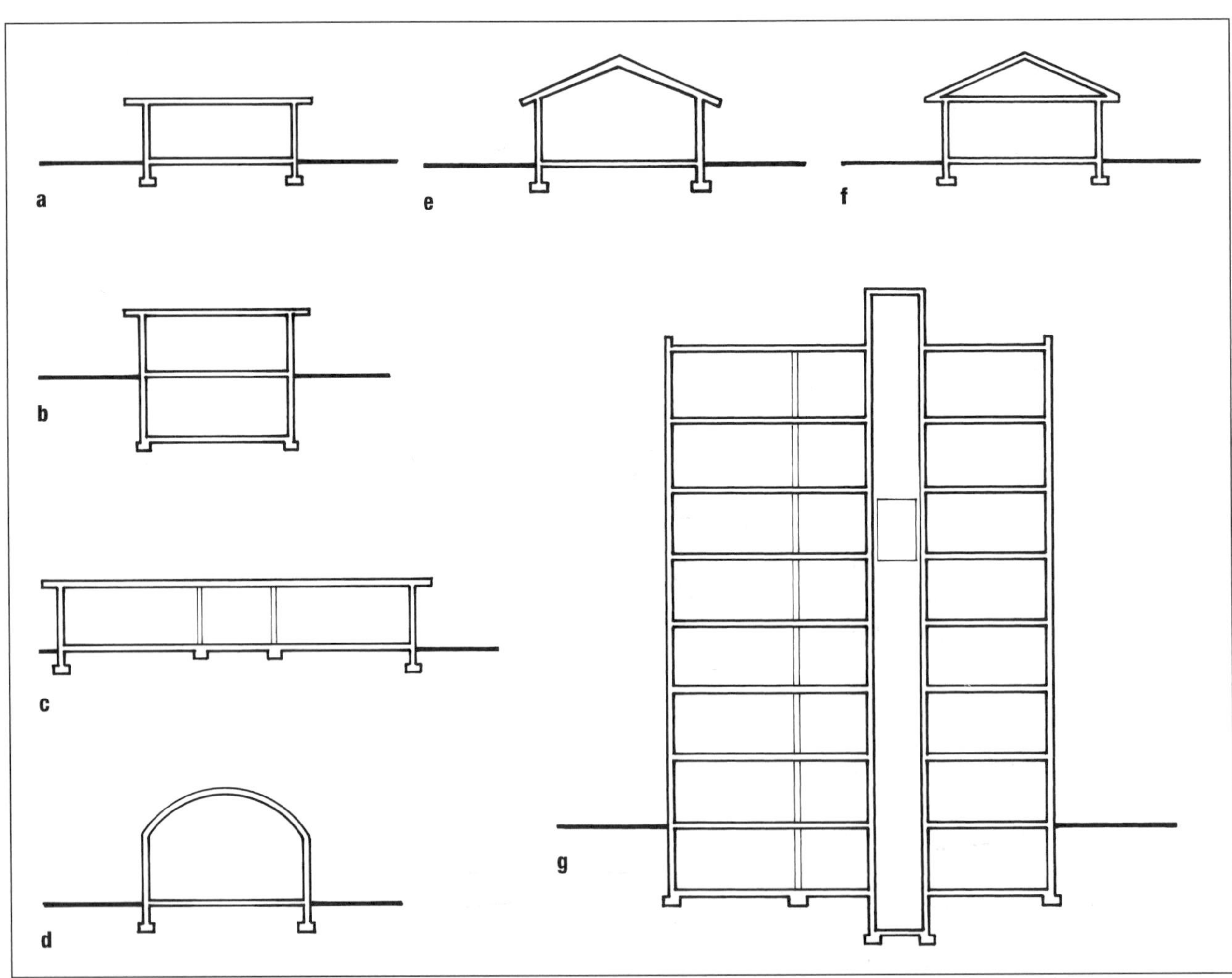

FIGURE 1.1-2 Basic building types for other than single-family residential buildings: (a) one-story; (b) one-story with basement; (c) one-story with multiple framing bays; (d, e, and f) typical roof shapes; and (g) multistory. (Drawings by HLS.)

stories. The lower costs are somewhat offset by the increased costs for maintenance of the exterior surfaces of multistory buildings, the relatively high costs of materials that can be used there, and the increased cost of construction associated with moving materials to high levels and working with them far above ground level.

Low-rise multistory buildings may be of steel or concrete construction or a combination of these. Some even have masonry bearing walls. Steel columns and concrete floors are common. Foundations are usually poured concrete spread footings, although poor soil conditions sometimes dictate the use of piles or caissons.

High-rise buildings are usually framed in steel, with thin concrete floor slabs, because concrete structures of great height have heavier and larger framing members than steel structures, which reduces the amount of usable space and increases the cost of construction. Some recent very high buildings have been designed as a series of steel shells or tubes that extend for the entire height of the building; others have been designed using the same principles as tall radio and television towers. Foundations are either poured concrete footings or pads, piles, or caissons, depending on the soil conditions and the size and load imposed on the soil by the building.

Sometimes the desire to create a statement for ego-enhancing or advertising purposes affects the size, height, and appearance of a building. For example, a corporation may wish to use its headquarters building as a symbol or may just want to own the tallest, largest, or most spectacular building in town.

Multistory buildings need elevators or escalators to make their use practicable. In addition, in most types of uses, federal accessibility laws and rules make elevators or wheelchair lifts a legal requirement in every building that is not inherently accessible to the handicapped (see Section 1.7), which, of course, includes every multistory building. The additional cost of this vertical transportation must be considered in deciding whether to construct a multistory building.

The basic building types used in single-family and townhouse construction are easier to define. Figure 1.1-3 shows some common types. Most of these types are also used for buildings other than single-family homes or townhouses, however, so they should not be thought of for only these restricted applications. The most prevalent of these is a 1-story building (Fig. 1.1-3a), because this type provides the most size and shape variations. These buildings may or may not contain a basement. Their roofs may be sloped, as shown, or flat.

One-and-one-half-story buildings, with or without basements (Fig. 1.1-3b), are sometimes used for housing. They offer more living space than single-story buildings with a minimum of additional cost. The second-floor space varies with the building size and roof slope. Light, ventilation, and view can be provided by dormers. One-and-one-half-story buildings are seldom built for other types of uses because their inherently small second-floor rooms, with their sloped ceilings, while adequate for sleeping rooms, often do not make satisfactory work spaces.

Two-story (Fig. 1.1-3c) or taller buildings, with or without basements, provide the maximum usable area at relatively low cost. Two- and three-story single-family houses and townhouses are common. These types of buildings can reduce construction cost, depending on the value of the land. When they must be accessible to the handicapped (see Section 1.7), buildings of more than 1 story require elevators, as described earlier for multistory buildings.

Bilevel buildings (Fig. 1.1-3d) are well suited for single-family houses, townhouses, or small commercial buildings in hillside locations. They provide habitable space at both grade levels when connected with full flights of stairs. In certain types of uses, accessibility restrictions may require that elevators be included. This configuration can also be used for two different occupancies, such as an apartment on one level and a small store on the other. In this case, both levels can be easily made

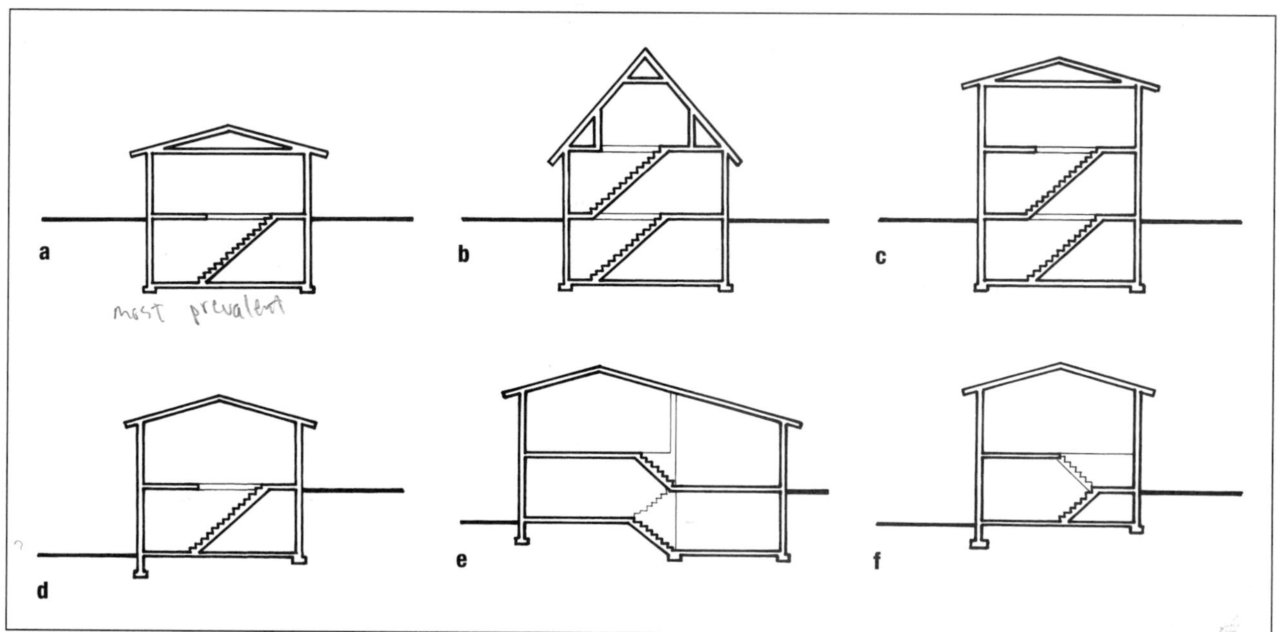

FIGURE 1.1-3 Basic single-family residential building types: (a) single-story; (b) one-and-one-half-story; (c) two-story or higher; (d) bilevel; (e) split-level; and (f) bilevel/split-entry. (Drawings by HLS.)

independently accessible to the handicapped. Roofs may be either sloped, as shown, or flat.

Split-level (Fig. 1.1-3e) and bilevel/split-entry (Fig. 1.1-3f) buildings are used mostly for single-family houses and townhouses. They are infrequently used for other purposes because of the difficulty of making them accessible to the handicapped. Split-level houses offer distinct separation of functions, either on three levels or four, including a basement. These are best suited for sloping lots. They offer numerous design possibilities, but can have awkward proportions if not carefully designed.

Bilevel/split-entry buildings are also best suited to sloping lots. They are characterized by a split foyer between two full living levels. This configuration can provide either a sunken 2-story (or more) house without a basement or a raised one-story (or more) house with a finished basement.

1.1.5 CONSTRUCTION SYSTEMS AND METHODS

The selection of construction methods and systems that produce the basic building types shown in Figures 1.1-2 and 1.1-3 is usually governed by three criteria: functional requirements, cost, and the desired appearance.

These basic criteria may require a consideration of climate, site topography, initial costs, maintenance costs, building codes, zoning ordinances, or other laws, availability of materials and labor, builder resourcefulness and size, owner taste, local custom, and other factors.

The selection of methods and systems is further complicated by the thousands of materials, products, and construction system choices available, many of which are interdependent. Sometimes the relationship of these building elements to each other will create situations in which the construction method or system is the major influence on a building's design. For example, the selection of a dome as the means to roof a coliseum may dictate that the shape of the building be circular or near-circular. Conversely, design requirements may dictate the framing system. A dome, for example, may be a poor choice for roofing a theater because of the inherent acoustical difficulties of a dome, and because a circular building may not be preferable for the kind of theater that is desired. The cost and availability of very large laminated wood (gluelam) structural elements, as compared with smaller gluelam units, may influence the width or even the shape of a church, or the way in which the gluelam units are fitted together to make a roof structure. The permissible span of the available wood decking may further influence the spacing of these gluelam units or the design of the roof structure and the placement of purlins.

Although there have been experiments with a few revolutionary construction systems since World War II, most new homes and many small commercial and institutional buildings in the United States are still built using conventional light-wood-platform framing (see Chapter 6), often with wood-truss-framed roofs. In many areas, the use of preassembled components, such as those discussed in Chapter 6, is common. In the future, advanced industrialization techniques using new materials and methods may offer new construction forms far different from those typical today.

Other basic systems in use today include wood-post-and-beam framing and wood-pole construction (see Chapter 6); masonry bearing-wall construction (see Chapter 4), sometimes with concrete floors (see Chapter 3), often supported on metal framing or bar joists; concrete-framed construction; and steel-framed construction (see Chapter 5).

In small construction, conventional wood framing still offers many advantages. As a complete construction system, it still is one of the most economical ways to build. The ease of working and fastening wood together with simple tools provides flexibility, which permits job changes without extensive reengineering. Wood framing is still the basis for most building codes and labor practices and will probably remain so for some time to come. Conventional framing is adaptable to site fabrication by the smallest builder handling each member piece by piece, as well as to off-site fabrication of individual pieces into larger preassembled components that require additional manpower or machinery for erection.

1.1.6 THE FUTURE

Further industrialization, using more and larger prefinished and prefabricated components, appears essential to help offset the rising costs of land, labor, and materials. Off-site fabrication permits maximum utilization of labor and materials under factory-controlled conditions with little loss in on-site time owing to bad weather. Efficiency may be increased with the use of power tools and machinery; volume purchasing of materials and stockpiling of finished parts is possible; greater convenience for workers and better protection for finished materials is provided; and site erection of components can usually be accomplished more economically and in less time by semiskilled or even unskilled labor.

To save costs, mechanical components for small buildings have been developed that combine a furnace, air conditioner, water heater, and electric power panel in one package. Larger mechanical components include completely furnished kitchens and bathrooms. The concept of prefinishing complete rooms has been extended to prefabricating as much as half a small building, such as a house, so that upon setting and joining two halves, an entire building is completed. Future developments may include assembling an entire building and completely finishing it prior to site placement.

Some future building construction methods will be highly sophisticated and closely integrated systems. For instance, integrated floor and ceiling systems available for use in commercial construction include structure, lighting, acoustical control, heating, cooling, and air distribution in a single system.

Components should be capable of satisfying varying design requirements, should permit simple modifications in the field in case of errors, and should be sized for ease of shipping, storage, and assembly. As component size increases, design and construction problems increase and design flexibility is lessened. The dimensions of large units are restricted to what can be transported physically and legally over the highways, and larger components usually require more manpower and larger erection equipment at the site.

Accordingly, the design, engineering, or selection of preassembled components requires judgments between size and flexibility. The most useful systems will combine the advantages of fully standardized factory-built modular units, which capitalize on the inherent savings resulting from repetitive production, and those that offer the design advantages of custom fabrication in the field.

Unfortunately, there are also certain disadvantages associated with prefabrication that have so far limited its use. For example, to be profitable, large components require a large market willing to accept a standardized design, which has not been forthcoming. In comparison, because they can be adapted to many building sizes, shapes, and designs, there is a huge market for prefabricated roof trusses, making them relatively inexpensive and readily available.

There are also potential disputes among construction trade unions and between trade unions and manufacturers about the right to do certain work. Union-member plumbers, for example, are not likely to be pleased when the plumbing piping and fixtures in a prefabricated building are installed by nonunion factory workers.

Other problems with prefabrication include consumer and builder resistance to prefabricated structures, associated with the preconceived notion that prefabricated buildings will be shabbily constructed and look like house trailers. In fact, while the construction may actually be superior to stick-built units, the appearance possibilities are somewhat limited, and design variety is difficult to achieve.

A final deterrent to prefabrication is the lack of consistency among building codes. These differences can require slight, but costly, modifications in prefabricated units to comply with the codes in different jurisdictions.

Construction practices vary widely across the country. Conventional methods and systems are not usually engineered but are based on a combination of long-established custom, rules of thumb, and arbitrary building code requirements. These practices have resulted in buildings that have usually provided satisfactory performance over the years. However, when these practices are unduly conservative, as is often the case, they foster excessive waste and therefore higher cost. New performance standards for methods and systems are constantly being established. These design criteria, based on laboratory and field testing, can materially reduce overdesign, waste, and cost. The members of the building construction industry should continue to encourage the development of criteria based on performance rather than the requirement that a specific product or system be used. They should also take steps to further the design of methods and systems based on these performance criteria. In addition, the industry needs to find the means for quicker acceptance of innovations in the marketplace as they occur.

1.2 Construction Documents

Architects and their consultants provide predesign and design services. In the predesign phase of their work, they may be engaged by the owner to provide such services as programming, existing facilities surveys, feasibility studies, and site analysis. In the design phase of their services, they provide schematic design, design development, and construction document services. Building predesign and design methods and the presentation of designs are beyond the scope of this book.

Construction documents include both drawn and written documents. Their presentations vary somewhat with different project delivery systems. This section addresses documents for a normal building project to be built under a single prime contractor agreement. In such agreements, a single general contractor is the contractor of record who signs a construction agreement with the owner. There is only a single construction agreement.

Documents will be organized differently for other project delivery systems. For example, in projects with more than one prime contractor there will be a separate owner/contractor agreement for each prime contractor. That is, there will be one agreement for a general works contractor, a separate agreement for a site works contractor, another for a mechanical contractor, still another for an electrical contractor, and so on. Such a project will require as many sets of construction documents as there are prime contractors, and each set may be presented in a slightly different order.

In a *design/build* contract, a team of design professionals (architects and engineers) and one or more constructors are employed by an owner to deliver a complete functioning building ready for occupancy and use. In such projects, drawings are usually similar to those for a single-contract project, though they may be somewhat less detailed. The written construction documents will contain some different provisions, especially in the front-end portions where legal and contractual obligations are delineated.

In a *construction management* (CM) contract, the owner employs a firm, or individual, as a construction manager to oversee the project. Many CM projects are also *fast-track* projects, in which the general contractor is brought on board during the design phase and construction documents are issued as needed for the contractor to price and bid subcontract work. The front-end documents (see Section 1.2.2.3), such as bidding requirements, contract forms, and contract conditions, will be agreed to by the owner and the contractor before the other contract documents have been completely developed. Building costs are usually established as a guaranteed maximum price (GMP) with one or more contingency amounts to take care of unforseen conditions.

In a fast-track delivery system, foundation and structural framing drawings and specifications may be issued as a package before the rest of the drawings and specifications have been completed. This *fast tracking* often permits an accelerated construction schedule, lowering the cost of construction, but these cost reductions are somewhat negated by errors that are inevitable in the fast-track method. Ultimately, when all the drawings have been completed, they will be organized in the same format as they would be for a single prime contractor project. Specifications Division 1 (see Section 1.2.2.2) will differ from those of other projects, but Divisions 2 through 16 will be the same as for other project delivery systems.

1.2.1 DRAWINGS

AIA has defined drawings as construction documents that "show in graphic and quantitative form the extent, design location, relationships, and dimensions of the work to be done. They generally contain site and building plans, elevations, sections, details, schedules, and diagrams."

The organization of drawings differs from one architectural firm to another, but there are some generalities. Sitework drawings are sometimes issued as a separate set of drawings, but often are placed at the beginning of the architectural drawings. The architectural drawings are usually placed first in a set, followed by structural, mechanical, plumbing, and electrical drawings, in that order. Drawings for other elements have been placed in various locations within a set of drawings. For example, furniture and equipment drawings may be placed immediately after the architectural drawings or may be issued as separate drawings attached at the end of the full set.

There is no universally accepted order within the disciplines. There have been numerous attempts to standardize the organization, but none has yet achieved consensus. One possible order for architectural drawings places general notes first, followed by overall building plans, building cross sections and exterior elevations, floor plan blowups, interior elevations, reflected ceiling plans, vertical circulation (elevators, stairs, escalators, etc.), exterior wall sections and details, and, finally, interior sections and details. Drawings for the other disciplines follow similar arrangements, with notes first, plans next, then elevations, details, schedules, and diagrams.

Until relatively recently, drawings were produced manually on a drawing board using a T-square, parallel bar, or drafting instrument. Standard details and notes were sometimes repeated from project to project, and some firms issued them in a separate bound volume. Drawing sheets had standardized boarders and title blocks.

An advancement to the manual drafting system included the use of overlay drafting techniques, in which a base drawing was made manually and reproduced on Mylar™ or other plastic film. The film was drilled to fit over a pinned registration bar. Additional sheets were pinned over the first to produce equipment, furniture, mechanical, and electrical drawings. The floor plan was drawn only once. This method effected savings in time and, subsequently, in document production cost and tended to produce more accurate drawings.

Recently, *computer-aided design and drafting* (CADD or simply CAD) has all but supplanted the earlier methods of drawing production. CADD systems are more expensive to implement than the earlier systems. They require purchase of expensive hardware and software and extensive employee training. Even so, they are rapidly replacing earlier methods. They are increasingly required by owners, notably federal and state government agencies. Many architectural firms are now requiring new employees to be able to use CADD programs.

Early CADD packages were used primarily to develop construction contract documents in much the same way that was used to produce manually generated drawings, but newer programs are capable of interactive functions that permit a change in one view to automatically change all other views. Change a plan, for example, and sections and elevations change automatically to reflect the plan change. Add a door in a perspective drawing, and the door is automatically added in plans and elevations. Change a wall section, and the elevations change automatically. Even more advanced programs allow a viewer to change the viewpoint of a computer-generated perspective drawing at will, allowing a user to, in effect, walk through a design, looking in every possible direction and from every possible altitude during the walk. Virtual reality systems have been demonstrated that permit a user to see a design in three dimensions and to move or resize elements by merely raising a hand. A window, for example, can be moved in a wall by a wave of the operator's hand. At some time in the future such systems will revolutionize the design of buildings and render all present methods obsolete.

1.2.2 THE PROJECT MANUAL

A *project manual* is a single volume that contains all written requirements for a building construction project.

1.2.2.1 Concept and Contents

A project manual provides a standardized format and location for all written project requirements. It includes bidding requirements, contract forms, conditions of the contract, information about alternates and allowances, and specifications. The number of possible requirements in project manuals is large, but not all are needed for every project. For example, one project may be of concrete construction and have no requirement for wood trusses. Another may be framed with wood and contain no structural steel.

1.2.2.2 The Construction Specifications Institute and *MasterFormat*

In 1948 a group of construction specifiers, frustrated with the chaotic specifications practices of that time, organized the Construction Specifications Institute (CSI) with the broad goal of improving construction specifications practices. CSI's current membership includes primarily specifiers, product manufacturer's representatives, and government agency representatives interested in specifications. Construction Specifications Canada (CSC) is a similar organization operating in Canada.

Project manuals are complex and need a numbering system to make it possible to find individual requirements. Without a standard to follow, design firms tend to invent their own particular numbering systems and methods of organizing these materials. This makes it difficult for a contractor to find a particular requirement, thus increasing the possibility of errors in estimating that affect project cost, and errors in the field that affect the completed building. For instance, in a certain project, the requirements for structural steel may be in the third section of the manual, and in another project manual similar requirements may be in the twelfth section. To unify the location of the parts of a project manual, in 1963, CSI and CSC jointly produced a publication entitled *MasterFormat*.

MasterFormat has become the industry standard for naming and numbering the data in project manuals. It has been adopted by AIA in its publications and by most federal government agencies for their construction projects. It is also the format generally used in manufacturers' product literature. It is the basis of organization for Sweet's Catalog Files, McGraw-Hill's yearly collection of manufacturer's catalogs, which are a major source of product literature in most architects' offices, and in all CSI-sponsored publications, including another major source of manufacturer's data, *The Architect's First Source for Products*. It is also the basis for data filing in many design offices. More important to our purposes here, *MasterFormat* is used on the vast majority of private-sector and government building construction projects in the United States.

Because of its almost universal use in the U.S. construction industry, students and professionals need to become familiar with *MasterFormat*. *MasterFormat* divides the work associated with construction projects into 16 *divisions*. The chapters of this book are named and numbered in the same way as *MasterFormat's* divisions. In addition, the material here is roughly equivalent to the recommended material in the corresponding divisions of

MasterFormat. The exact order of sections in *MasterFormat* tends to change slightly from edition to edition, so the subjects in this book may differ slightly in location from the order used there. Moreover, the limited size of this book precludes discussing every section listed in *MasterFormat*.

1.2.2.3 Bidding and Contracting Requirements

Bidding requirements include invitations to bid, instructions to bidders, a listing of information available to bidders, and bid and bid bond forms. They are issued before bids on a project are offered and are not considered part of the contract for construction.

Contracting requirements documents are also issued to potential bidders before they offer bids, but completed forms and other contracting requirements documents are considered part of the construction contract. Contracting requirements documents include the owner/contractor agreement, bonds (performance and payment), certificates (insurance, conformance, etc.), the conditions of the contract (general and supplementary), and addenda and modifications issued prior to acceptance of bids.

The general conditions of the contract and standard contract documents have evolved over the years from the early beginnings of AIA documents to today's complex and extensive interrelated documents. For example, a current standard form of owner/contractor agreement produced by AIA is fully compatible with AIA A201, "General Conditions of the Contract for Construction."

Engineering organizations have also formed the Engineers Joint Contract Documents Committee (EJCDC), which has developed standard contract documents for use on engineering projects.

1.2.2.4 Specifications

Specifications constitute that portion of the written requirements for a building construction project that are contained in Divisions 1 through 16 of a project manual. AIA has defined specifications as "written requirements for materials, equipment, construction systems, standards, and workmanship for the work as well as standards for the construction services required to produce the work."

Originally, specifications were just notes on the drawings. As the construction world became more complex, lawsuits proliferated, and many new materials and systems were introduced, the notes became too voluminous to put on the drawings. Architects and engineers began to type them and issue them as separate documents. Legal and procedural requirements now occupy as many pages as an entire specification set did in the 1950s.

For most construction projects today, specifications are included in the project manual and are organized according to *MasterFormat*. Within each of the 16 *MasterFormat* divisions, specifications are organized into level 2 sections, and level 3 sections. Each level represents a more detailed subdivision of the work. Sections are numbered with a five-digit system in which the first two digits are the division number and the remaining digits are the section number. For example, in Division 10, "Specialties," there is a Section 10150, "Compartments and Cubicles."

Specifications should include clear and accurate descriptions of technical requirements related to materials, products, and services. They should also state a project's requirements for quality and the prescribed use of materials and methods to produce a desired product, system, application, or finish. And they should do all this in a form that is understandable and in a format in which the requirements can be easily found and clearly understood.

Recognizing that a clear and concise format, that was the same from job to job, was essential to producing clear specifications, shortly after the introduction of *MasterFormat* CSI introduced a three-part section format and soon after that a recommended page format. In the section format, the data in each specifications section are divided into three *parts* as follows:

Part 1: General
Part 2: Products
Part 3: Execution

The content of each part is divided into articles and the articles into paragraphs.

The content and order of the data in each part have changed somewhat over the years, but the general intent is the same. Part 1, "General," includes requirements that affect the entire section and that are general in nature, such as requirements for allowances; unit prices; alternates and alternatives; submittals; quality assurance; delivery, storage, and handling; project and site conditions related to that section; sequencing; scheduling; warranties; and maintenance. Part 1 also includes definitions related to that section and cross-references to other sections.

Part 2, "Products," includes the names of acceptable manufacturers; descriptions of required materials, manufactured units, equipment, and accessories; and requirements for mixes, shop fabrication and assembly, tolerances, and source quality control, including tests and performance verifications.

Part 3, "Execution," includes requirements for examination of conditions prior to installation; preparation for installation; and erection, installation, or application, as applicable. Part 3 also includes requirements for field quality control and testing; adjustments, cleaning, and protection of the installed work; and requirements for demonstration of proper operation of items installed under the section and requirements for scheduling and coordination of the work of this section with other work.

Each part is organized into articles and paragraphs using a format as follows:

Part 1 General
1.01 Article
    a. Paragraph
    b. Paragraph
1.02 Article
    a. Paragraph
    b. Paragraph

The CSI page format addresses the margins and page arrangement, including recommended indents for each level of articles and paragraphs. Most guide specifications (see Section 1.2.2.5) are already set up in the CSI page format or in an organization resembling it.

1.2.2.5 Guide Specifications

Nonproprietary guide specifications sets encompassing all 16 *MasterFormat* divisions are commercially available today from organizations representing national associations. For example, Arcom Master Systems, sponsored by AIA, produces MASTERSPEC®; Construction Sciences Research Corporation, sponsored by CSI, produces SPECTEXT®. Commercial guides are also available from several private organizations, such as Kalin Associates and Building Design Systems, Inc., which produces an automated guide set named *SpecLink*®. Most of these sets are available in either paper or electronic form, using various methods of delivery. An advantage of using such sets is that they are updated periodically and cover a wide range of materials and systems. In addition, they are produced by organizations with no vested interest in a particular product, which may make them more reliable in some ways. Some of these sets come with detailed instructions and recommendations for selecting one product type over another. Most of them are thoroughly researched and reasonably up-to-date.

Proprietary guide specifications are available from building product manufacturers by mail and at their Web sites. Some organizations offer direct links to an extensive list of manufacturers' Web sites where guide specifications can be accessed. In addition, guide specifications sections produced by manufacturers are available from a variety of organizations in sets. These sets are often not as complete as those produced by the associations. Examples of organizations offering access to manufacturers' sites are ARCAT, Architects' First Source, and Sweet's Group.

Guide specifications are also produced by federal agencies, such as the General Services Administration (GSA) and the various military services. These are intended for use only on government projects but contain data that can be valuable to a specifier even in private-sector projects.

1.3 Bidding and Negotiation

Construction contracts are let based on dollar amounts and contract terms determined by either *bidding* or *negotiation*. *Bidding* is a process whereby a prime design professional engaged by an owner, and the design professional's consultants, prepare bidding documents and issue them to a group of constructors, who then submit the dollar amounts they will charge to build the building, under the terms of the bidding documents. The prime design consultant may be an architect or an engineer, depending on the project type. For example, for a bridge, a structural engineer may be the prime design professional. This section assumes that the project under consideration is a building and that the prime design professional is an architect.

Negotiation is a process whereby an architect engaged by an owner, and the architect's consultants, prepare negotiation documents; the owner, often with the advice of the architect, selects a potential constructor; and the architect issues the documents to the constructor, who then submits an offer to construct the project for a set number of dollars and within defined contract conditions. Then, either this offer is accepted or the offer and the terms are negotiated between the owner, with advice from the architect, and the constructor until an amicable arrangement is agreed upon.

1.3.1 BIDDING

1.3.1.1 Selecting Bidders

There are several methods for selecting bidders for a building construction project. In most government projects, all licensed contractors must, by law, be permitted to offer a bid. The only way to limit bidders is by imposing requirements for performance and payment bonds. Contractors who are unable to perform the work because of financial or other difficulties will not be able to obtain the required bonds and can be disqualified on this basis.

Some private work contracts are handled in the same way as government projects, whereby bidding is open to all legitimate constructors, but in most cases constructors requested to bid are restricted in some way. In some instances, the owner will simply develop a list of constructors it wants to bid, and exclude all others. Sometimes this list is a result of consultation between the owner and the architect, but not always.

For some projects, potential bidders are examined and *prequalified* to ascertain whether they are legitimate, reputable, licensed contractors who are large enough and financially stable enough to perform the work and are not disqualified by legal restrictions. This investigation is usually conducted by the architect, who forwards the results to the owner. The final decision of which constructors are permitted to bid lies solely with the owner. Prequalification is usually performed with the use of standard forms such as AIA's Document A305, "Contractor's Qualification Statement."

Some sophisticated owners develop and maintain lists of prequalified bidders and select from that list for each project. Government agencies are also permitted to prequalify contractors and develop a list of qualified bidders, but the failure to qualify must be based on defined restrictions, such as insufficient financial assets or company size.

1.3.1.2 Bidding Process

For government projects and for private-sector projects in which the owner has elected to use completely *open* bidding, the owner or the architect places an advertisement for bids in one or more newspapers and in other publications deemed appropriate. This advertisement lists the pertinent project information, such as its name and address, a general description of the project, the type of contract, the date, time, and place of receiving bids, bid security requirements, and other relevant information.

When the bidders have been preselected, either the owner or the architect issues an invitation to bid to all bidders on the prequalified list. The invitation to bid contains the same information as in an advertisement for bids.

1.3.1.3 Bidding Documents

When a legitimate bidder responds to an advertisement for bids or an invitation to bid and requests bidding documents, the architect issues one or more sets of bidding documents to each requester. In addition to the advertisement or invitation to bid, the bidding documents issued initially include the following:

- Instructions to bidders, which explain bidding procedures and list the requirements of the bid, such as the requirements for bonds
- Bid forms, which are developed by the owner or the architect to keep the content of all bids in the same format, so as to make interpretation easier and to ensure that bids for required alternates, allowances, and unit prices are included
- Information related to bonds, including bid bonds required to ensure that a bidder will sign an agreement based on his or her bid or forfeit the bond amount, and performance and payment bonds that ensure completion of

the signed contract and payment of the constructor's related debts

- Form of the owner/contractor agreement
- The conditions of the contract (general and supplementary)
- Drawings
- The project manual

Bidding documents also include addenda and other modifications to the previously issued documents that are issued prior to contract signing.

1.3.2 NEGOTIATION

Bid projects are often negotiated with the lowest bidder, or several of the lowest bidders, to arrive at a final cost and determine changes to the original proposal. Conversely, in a *negotiated* project, a single constructor is selected and the contract sums and terms are negotiated between the constructor and the owner, usually with advice from the architect.

1.3.2.1 Selecting Potential Constructors

An owner may select a potential constructor by firsthand knowledge of the firm or by reputation, or may select several potential constructors and reduce the list to one by a prequalification process similar to that described for bid projects.

A negotiated contract may be a straight single-prime-contractor contract or may have multiple prime contractors. It may result in a fixed-price stipulated sum, cost-plus-fee, or unit price agreement. It may result in a construction management system with a guaranteed maximum price (GMP). It may also be used for design/build agreements.

1.3.2.2 Negotiation Process

The constructor in a negotiated contract is sometimes engaged before the architect's and associated consulting engineer's design work has been completed. The advantage of this arrangement is that the constructor can be called on to issue advice related to systems and products as the design work progresses. This advice can be related to costs, availability, and interface problems that may be encountered in the field. The disadvantage is that materials and systems selections and design judgments are sometimes based on factors other than just costs and ease of construction. Care must be taken to ensure that the architect maintains control of the use, safety, and aesthetic aspects of the project.

1.3.2.3 Negotiation Documents

Negotiation documents are mostly the same as bidding documents. However, there is no advertisement, invitation to bid, or instructions to bidders. Additional requirements must sometimes be established, however, for subcontract bids. The conditions of the contract will also vary somewhat based on the delivery system to be used.

1.3.3 ARCHITECT'S RESPONSIBILITIES

In addition to the responsibilities already delineated, the architect should conduct public bid openings, unless the owner elects to do so. The architect should assist the owner in evaluating bids and negotiated contract terms. The architect should also advise the owner in regard to acceptance of bids and selection of a contractor, but the responsibility for these duties and for executing the owner/contractor agreement rests solely on the owner.

1.4 Construction Contract Administration

Some owners undervalue the potential contribution of their architects and consulting engineers during building construction and do not continue their services past the bidding and negotiation phase. Fortunately, most owners understand the advantages of these continued services and employ their design professionals until the construction has been completed. In some types of delivery systems, the design professionals' involvement during construction is essential. Obviously, in a design/build project, the design professionals are part of the design/build team and must continue with the project until completion.

Fast-track delivery systems also require the services of design professionals through the construction process. For the most part, in such projects the design and construction documents are not finished until a short time before the construction has been completed. Construction management projects are also frequently fast-track and need the services of design professionals throughout.

When engaged by owners to provide services during the construction phase of a building project, architects and engineers can provide the following services:

1. Hold preconstruction conferences to clarify procedures and establish the responsibilities of the various team members. Who the team members are will vary with the delivery system. At the least, there will be an owner, an architect, consulting engineers, and the general contractor.

2. Review and process the contractor submittals, such as shop drawings, samples, product data, mockups, test results, bonds, and record drawings.

3. Conduct periodic field observations to ensure that the work is proceeding in accordance with the contract documents, to issue field reports detailing these observations and noting the progress of the work, and to reject work found to be not in compliance with the contract documents.

4. Respond in writing to the contractor's questions concerning the intent of the contract documents.

5. Issue documents to make changes in the work. Minor changes can be made by a simple written order by the architect. Changes affecting project delivery time or costs are made by *change orders*. Change orders are sometimes initiated by the architect, who asks the contractor in writing to submit a *proposal request* for the desired changes. The contractor then responds with a proposal, indicating changes in contract time or cost associated with the requested change. In other cases, *proposal requests* are initiated by the contractor. In either case, the architect reviews the contractor's proposal and advises the owner whether to accept or reject it. Once the terms of the proposal request are acceptable to both the owner and the contractor, the architect issues a *change order*, which must then be signed by both the owner and the contractor. Signed change orders become a part of the contract for construction.

6. Attend regularly scheduled project progress meetings with representatives of the owner and the contractor.

7. Review and process the contractor's applications for payment. At the beginning of a project, the contractor submits to the architect a *schedule of values*, which is broken down into the various portions of the work. Periodically, usually once each month, the contractor submits to the architect an *application for payment*, indicating the progress of the work to date and the amount of the contract sum due that month. The application is broken down in exactly the same way as the schedule of values so that the two can be compared. The architect and his or her consultants visit the project site and compare the contractor's application with the work accomplished. The consultants report their findings to the architect, who then certifies payment by filling in, on the contractor's application for payment, the dollar amount due that month based on the architect's and consultants' observations. The amount certified may be the amount shown in the application or may be less, depending on the architect's judgment of the work actually completed. When a discrepancy is found, there is usually an attempt to have the contractor revise the proposal, but the certified amount is based ultimately on the architect's judgment, taking into account the consultants' opinions. The architect must submit written justification for payments certified in amounts less than those requested. The owner bears the final responsibility for payment and can pay more or less than is certified. Underpayment, however, can trigger a legal challenge by the contractor and should not be done without documented justification and consultation with an attorney.

8. Handle closeout procedures. When the contractor considers that the work is complete, it initiates contract closeout. To do this, the contractor submits a request for a review by the design professional pursuant to *substantial completion*. Substantial completion means that the owner can occupy the premises and use it for its intended purpose. If minor items remain to be completed, the contractor's request must include a list of these items, along with reasons for the incompleteness, and the dollar value of each incomplete item. The contractor must also submit all closeout documents required by the contract documents, including, but not limited to, warranties, maintenance agreements, operating instructions, certificates of inspection and occupancy issued by the local jurisdiction, bonds, insurance certificates, and record documents, including drawings and a project manual marked to show all changes made during construction. The contractor must also submit maintenance and replacement stock and keys, and must instruct the owner's personnel in operation of equipment as required by the contract documents.

Upon receipt of a request for a substantial completion inspection, the architect and the architect's consultants will inspect the project. If they consider the work not substantially complete, the architect will so notify the contractor with a list of observed incomplete work. The contractor must then correct the deficiencies and submit a new request for inspection. The architect and the architect's consultants will again visit the site and conduct a second inspection. When the architect and the architect's consultants deem the work to be substantially complete, and after all required closeout materials have been received, the architect will issue a *certificate of substantial completion*. If items still remain incomplete at that time, the certificate will be accompanied by a list (*punch list*) of these remaining items and a date, usually 30 days, within which the remaining work must be completed.

9. Issue a certificate of final completion. When the items on the punch list have been completed and the architect is convinced that all contract work and the contractor's other obligations under the contract have been completed, the architect will issue a certificate of final completion. The contractor may then submit a request for final payment.

10. Review and certify the contractor's final request for payment. This will complete the architect's basic services under most construction contracts.

1.5 Industry Standards

The building construction industry is made up of so many diverse interest groups that it has not been possible to develop a single comprehensive set of criteria or standards acceptable to all concerned. In addition, groups that are more intimately involved in a product or construction system are generally best qualified to establish standards for them. The result is a myriad of organizations that establish standards.

Before beginning a discussion of construction industry standards, it is necessary to clarify some terms. There is much confusion in the industry about the use of the terms *specifications*, *standards*, and *codes*. Unfortunately, the three are often erroneously used interchangeably, which can lead to some confusion. Specifications are discussed in Section 1.2, codes in Section 1.6.

In the sense that they are a detailed, precise description of a product or practice, some construction industry standards could be called *specifications*, and some are so called by their producers. Some of the manufacturers' data that designers, builders, and owners rely on are also specifications in the dictionary sense, and those data are sometimes called specifications by the manufacturers. But calling either of them *specifications* sometimes leads to confusion. Therefore, this book refers to construction industry standards as *standards* and manufacturer's data as *product descriptions* or *product literature*. Unless specifically modified in the text, the term *specification* is used in its narrow construction industry meaning, as defined in Section 1.2.

1.5.1 TYPES, OBJECTIVES, AND USES OF STANDARDS

When selecting materials or determining the suitability of materials and methods, specifiers and builders refer to a variety of *industry standards*. These are also sometimes called *reference standards*.

1.5.1.1 Types of Standards

Some standards result from the efforts of manufacturers, professionals, and tradespeople to simplify and increase the efficiency of their work or to ensure a minimum level of quality. Other standards are the work of governmental agencies and other groups interested in establishing minimum levels of safety and performance. Therefore, standards take a variety of forms depending on their source and purpose.

MATERIAL STANDARDS

Standards that define the properties of a material are called *material standards*. For example, these include standards for extruded aluminum bars, rods, shapes, and tubes, or for a particular type of steel item. They usually specify the constituents of a material, its physical properties, and its performance under stress and varying climatic conditions.

PRODUCT STANDARDS

The requirements of a specific product, such as aluminum windows, are defined by *product standards*. These often define terms, classify constituent materials, and state acceptable thicknesses, lengths, and widths. They may also spell out the acceptable methods of joining separate materials, of fabricating various parts of a product, or of assembling systems.

DESIGN STANDARDS

Design standards define the requirements for sound design using a particular material, product, or system. They are published by such organizations as the American Concrete Institute (ACI), the Architectural Woodwork Institute (AWI), and the United States Department of Housing and Urban Development (HUD).

WORKMANSHIP STANDARDS

Workmanship standards are standards for installing materials, products, and systems. The American Society for Testing and Materials (ASTM) produces many of these.

TEST METHOD STANDARDS

Test method standards spell out acceptable criteria for testing materials and systems. Again, ASTM is a prime producer of standards of this type.

1.5.1.2 Objectives of Standards

Construction standards have two basic objectives: (1) to establish levels of quality that may be recognized by a user, specifier, approver, or buyer of a material, product, or system and (2) to standardize or simplify such variables as dimensions, varieties, and other characteristics of specific products so as to minimize variations in manufacture and use.

1.5.1.3 Uses of Standards

Construction standards are used by manufacturers, specifiers, consumers, communities, and others. Standards may be used and referred to either separately or within collections, such as in the HUD *Minimum Property Standards for Housing*. Standards may also be incorporated into municipal or state building codes by inclusion or reference. Such larger works may have broader objectives than to act only as construction standards. Codes, for example, are concerned basically with minimum acceptable standards of public health, safety, and welfare; the HUD *Minimum Property Standards for Housing* establishes minimum requirements of design and construction for the insurance of mortgage loans. Standards are often incorporated by reference into construction document specifications to help establish requirements for materials, equipment, finishes, and workmanship for a particular construction project.

1.5.2 PRIVATE INDUSTRY STANDARDS

Standards are established by two kinds of private industry organizations: trade associations and standards-setting and testing agencies.

1.5.2.1 Trade Associations

Manufacturers, tradespeople, and suppliers, working through *trade associations*, prepare most private industry standards. Some trade associations are called societies or institutes.

A *trade association* is an organization of individual manufacturers or businesses engaged in the production, supply, or installation of materials or services of a similar nature. A basic function of a trade association is to promote the interests of its membership. Those interests generally are best served by the proper use of the groups' materials, products, and services. The proper use of building materials and methods is guided largely by the development of suitable levels of quality for their manufacture, use, and installation. Some of the most important activities of trade associations are directed toward research into the use and improvement of materials and methods, and toward formulating performance standards. In many instances, trade associations also sponsor programs of certification in which labels, seals, or other identifiable marks are placed on materials or products manufactured to particular standards.

STANDARDS-SETTING TRADE ASSOCIATIONS

Many trade associations engage in developing standards for materials and products their members manufacture and services they perform. The American Plywood Association (APA) is an example of a trade association active in developing standards. This association is supported by a large membership of manufacturers producing softwood plywood. The production of various types and grades of plywood requires the selection and classification of wood veneers according to strength and appearance. Producing plywood from these veneers demands careful manufacturing control of such factors as moisture content and adhesive type. The completed product, properly assembled, bonded, and finished, must conform to a body of material standards, manufacturing procedures, and performance testing standards.

Plywood manufacturers, through their own research and combined efforts within the APA, financially and technically support the research and development of criteria that the association formulates into industry standards. The members of the association then agree to produce plywood that conforms with these standards.

TRADE ASSOCIATIONS STANDARDS AS A BASE FOR OTHER STANDARDS

The standards developed by trade associations often are adopted by or used as a base from which other groups, such as the American National Standards Institute (ANSI), may develop standards. For example, industry standards may be promulgated as *ANSI standards*, which are often incorporated into the HUD *Minimum Property Standards for Housing* or building codes.

CERTIFICATION

Many trade associations and other industry groups provide assurance that established standards have been met by materials and manufactured products. Certification of quality may take the form of *grade marks*, *labels*, or *seals*. For example, the APA maintains a continuing program of product testing during and after manufacture with grade markings applied directly to the plywood. A *grade mark* is a visible statement that the appropriate APA product standard has been met.

The reliability of certifications issued by manufacturers or their associations varies. Some are excellent. Others are worthless. The most reliable certification is one issued by an *independent testing agency*.

1.5.2.2 Standards-Setting and Testing Agencies

In addition to trade associations, organizations have been established whose primary purpose is the setting of standards or the testing of materials and products to ensure that they comply with established standards. When a material or product is compliance tested by its manufacturer, its trade association, or an independent testing agency, a *testing standard* produced by one of the agencies discussed in this subsection is often used as the standard for the testing procedure. The addresses, telephone numbers, and, where applicable, Web site addresses of the agencies mentioned in this section can be found in Appendix A, "Data Sources."

AMERICAN SOCIETY FOR TESTING AND MATERIALS (ASTM)

The American Society for Testing and Materials (ASTM) is an international, privately financed, nonprofit, technical, scientific, and educational society. The objectives of the society are "the promotion of knowledge of the materials of engineering, and the standardization of specifications and methods of testing."

ASTM membership consists of individual engineers, scientists, and educators, as well as organizational members, including companies, government agencies, and universities. Technical committees formulate and recommend ASTM standards covering many types of materials; a number of administrative committees deal with publications, research, testing, consumer standards, and other activities. The society is supported mainly by membership dues, with some income from the sale of publications.

ASTM started in 1898 as the American Section of the International Society for Testing Materials. It was incorporated in 1902 and became the American Society for Testing Materials. In 1961, the name was changed to the American Society for Testing and Materials to emphasize its interest in basic information about materials.

Two general categories of information and publications are available from ASTM: (1) *ASTM standards*, which include definitions of terms, materials standards, workmanship standards, and methods of test standards used throughout the industry, and (2) data dealing with research and testing of materials, including monthly and quarterly publications and technical publications that cover symposia and collections of data.

ASTM standards are designated by the initials ASTM, followed by a code number and the year of last revision. For example, ASTM C 55-99 refers to ASTM Document C 55, "Standard Specification for Concrete Brick," as last revised in 1999.

AMERICAN NATIONAL STANDARDS INSTITUTE (ANSI)

The American National Standards Institute (ANSI) is the name adopted by the United States of America Standards Institute (USASI) in October 1969. USASI was created in 1966 by the complete reorganization of the earlier standards organization, the American Standards Association (ASA). ASA was founded during World War I to prevent duplication and waste in war production. In 1918 five leading American engineering societies, including the American Society of Mechanical Engineers (ASME), the American Society of Civil Engineers (ASCE), and ASTM, and three departments of the federal government, Commerce, War, and Navy, formed the American Engineering Standards Committee to coordinate the development of national standards. This committee was reorganized in 1928 into the American Standards Association, which was later incorporated into USASI and subsequently renamed ANSI.

More than 3000 American National Standards have been developed and approved under ANSI procedures. These standards apply in the fields of engineering, industry, safety, and consumer goods.

An American National Standard is designated by code number and date of last revision; for example, ANSI A117.1-1998, "Guidelines for Accessible and Usable Buildings and Facilities."

Unlike ASTM, ANSI does not formulate its own standards or provide testing services. Instead, one part of ANSI, composed of national trade, professional, and scientific associations, establishes and maintains procedures for the approval of standards developed by other associations, agencies, or groups, as American National Standards. In this way, a standard developed by a trade association, such as the American Architectural Manufacturers Association (AAMA), can become an American National Standard.

A second part of ANSI consists of representatives of industrial firms, labor, and government. The two parts work closely together to recommend areas of standardization deemed essential and to review standards.

ANSI is privately financed by voluntary membership dues and from the sale of the published American National Standards. These national standards are available for voluntary use and often are incorporated in regulations and codes.

UNDERWRITERS LABORATORIES, INC. (UL)

Underwriters Laboratories, Inc. (UL) is chartered as a nonprofit organization to establish, maintain, and operate laboratories for the examination and testing of devices, systems, and materials. The stated objectives of UL are (1) to determine the relation of various devices, systems, and materials to life and property and (2) to ascertain, define, and publish standards, classifications, and specifications for materials, devices, products, equipment, constructions, methods, and systems affecting hazards to life and property.

UL, formed in 1894, was originally subsidized by stock insurance companies. Before the turn of the century, as new electrical devices and products came rapidly into the market, it became necessary to test and inspect them to ensure public safety. The National Board of Fire Underwriters (now the American Insurance Association) organized and sponsored UL to meet this demand.

UL became self-supporting in about 1916. To sustain its testing program, UL contracts with a product submitter for testing, reporting, and listing of devices, systems, or materials on a time and material basis. The cost of the inspection service is provided for either by an annual fee or by service charges for labels, depending on the type of service. Materials and products carrying UL labels and certificates must meet published standards of performance and manufacture and are subjected to UL inspection during manufacture.

Although primarily interested in public safety, UL's policy is to list and label only products that perform their intended function. If a product does not perform with reasonable efficiency, even though it may be perfectly safe, it does not qualify for a UL label.

UL standards are designated by the initials UL, followed by a code number. No date of last revision is indicated by the number. For example, UL 70, "Septic Tanks, Bituminous Coated Metal," was issued in 1993.

NAHB RESEARCH FOUNDATION, INC.

The NAHB Research Foundation is a wholly owned subsidiary of the National Association of Home Builders. Its objectives are as follows:

1. To conduct and disseminate the results of research and development with respect to homes, apartments, and light commercial structures for the purpose of lowering the cost and improving the quality of buildings constructed by the U.S. homebuilding industry
2. To conduct, for itself or by contract for others, tests and investigations into and on materials, products, systems, and other matters related to the design, construction, or occupancy of homes and other buildings
3. To encourage lower construction costs and improved quality in the design and construction of residential and related structures
4. To provide a system for labeling materials and products, and to grant a seal or certificate of quality or similar device

The NAHB Research Foundation was created in 1965 as an expansion of the NAHB Research Institute, which was founded in 1952. The Research Foundation's activities have included the design and construction of a number of "research houses" that incorporated new methods and materials. The Research Institute also developed the successful TAMAP system, which marked the first time that industrial engineering techniques were used to improve productivity in the design and construction of new homes.

The Foundation has also carried out product, standards, and systems research, development, and evaluation studies for many homebuilding industry manufacturers and associations. The Foundation is thereby supported by its clients, including the National Association of Home Builders, a number of building industry manufacturers, trade associations, and other organizations. The Foundation's laboratories include a broad range of facilities that can conduct ASTM standard tests, vibration studies, acoustical measuring, and temperature-humidity control testing.

NATIONAL FIRE PROTECTION ASSOCIATION (NFPA)

The National Fire Protection Association (NFPA) was organized in 1896 to promote the science and advance the methods of fire protection. NFPA is a nonprofit educational organization that publishes and distributes various publications on fire safety, including model codes, materials standards, and recommended practices. These technical materials, aimed at minimizing losses of life and property by fire, are prepared by NFPA Technical Committees and are adopted at the annual meeting of the Association. All are published as National Fire Codes, a 12-volume compilation of NFPA's official technical material. *The National Electrical Code* is Volume 3.

THE AMERICAN SOCIETY OF HEATING, REFRIGERATING AND AIR CONDITIONING ENGINEERS (ASHRAE)

Since 1894, ASHRAE and its predecessor societies have pursued their objective of advancing the arts and sciences of heating, ventilating, and air conditioning buildings. ASHRAE conducts an extensive research program, publishes meeting transactions, and establishes standards.

ASHRAE standards are established to assist industry and the public by offering a uniform method of testing for rating purposes, by suggesting safe practices in designing and installing heating, ventilating, and air conditioning equipment and systems, and by providing other information that may serve to guide the industry. The creation of ASHRAE standards is determined by need. Conformance is voluntary.

ASHRAE standards are updated on a 5-year cycle; each title is preceded by a hyphenated number. The digits before the hyphen are the standard's numerical designation; the digits after the hyphen are the year of approval, revision, or update. For example, ASHRAE Standard 90.1-1989 "Energy Efficient Design of New Buildings Except Low-Rise Residential Buildings" describes a standard of designation 90.1, approved in 1989.

ASHRAE has developed standards not only in the traditional areas of heating, ventilating, and air conditioning equipment, but on such diverse subjects as fire safety in buildings, energy conservation, solar energy, pollution control, and ozone depletion.

1.5.3 FEDERAL GOVERNMENT STANDARDS

Many federal agencies either develop standards themselves or commission their development by other federal agencies or private-sector organizations.

1.5.3.1 Department of Commerce

Manufacturers seek to encourage product acceptance and improve their own efficiency by establishing basic levels of quality for materials and products and by coordinating dimensions, terminology, and other variables such as type and style.

Manufacturers cannot legally agree to establish unreasonable standards that might rule out the success of individual competitors, nor can they engage in price-fixing agreements. Therefore, in recognition of the desirability of certain types of industry-supported standardization, the U.S. Department of Commerce provides for the development of Product Standards.

PRODUCT STANDARDS (PS)

Product Standards are developed by manufacturers, distributors, and users in cooperation with the Office of Engineering Standards Services of the National Institute of Standards and Technology. The purpose of a Product Standard may be either (1) to establish standards of practice for sizes, dimensions, varieties, or other characteristics of a specific product or (2) to establish quality criteria, including standard methods of testing, rating, certifying, and labeling of the manufactured products.

The adoption and use of a Product Standard are voluntary. However, when reference to a Product Standard is made in contracts, labels, invoices, or advertising literature, the provisions of the standard are enforceable through usual legal channels as a part of the sales contract.

A Product Standard usually originates with the manufacturing segment of the industry. The sponsors may be manufacturers, distributors, or users of the specific product. One of these three elements of industry (the *proponent*) submits to the Office of Engineering Standards Services the necessary data to be used as the basis for developing a Product Standard. The Office, by means of assembled conferences, letter referenda, or both, assists the sponsor group in arriving at a tentative standard of practice and thereafter refers it to the other elements of the same industry for approval or for constructive criticism that will be helpful in making necessary adjustments. The regular procedure of the Office ensures continuous servicing of each Product Standard through review and revision whenever, in the opinion of the industry, changing conditions warrant such action.

A Product Standard is designated by the letters PS, followed by an identification number and the last two digits of the year of issuance or last revision. For example, PS 1-83 is the Product Standard for "Construction and Industrial Plywood." It was originally issued in 1966 as the first Product Standard. Product Standards will gradually replace the older Commercial Standards (CS) and Simplified Practice Recommendations (SPR), previously published by the Department of Commerce.

A list of Product Standards, Commercial Standards, and Simplified Practice Recommendations currently available, a price list, and ordering instructions may be obtained from the Office of Engineering Standards Services, National Institute of Standards and Technology. Also available from that office are copies of *Procedures for the Development of Voluntary Product Standards*, which explain the process through which such voluntary standards are developed.

1.5.3.2 General Services Administration (GSA)

The General Services Administration (GSA) of the United States government develops *Federal Standardization Documents*, including *Federal Specifications*, *Interim Federal Specifications*, and *Federal Standards*, through the cooperation of federal agencies and industry groups. The purpose of these documents is to standardize the variations and quality of materials and products being purchased by governmental agencies. Approximately 5600 Federal and Interim Federal Specifications have been developed by the General Services Administration.

The *Index of Federal Specifications, Standards, and Commercial Item Descriptions*, which lists those documents alphabetically by title and numerically, may be purchased from the Superintendent of Documents, U.S. Government Printing Office, and is also available at the GSA Web site.

FEDERAL SPECIFICATIONS

In the pure sense of the dictionary definition, Federal Specifications can be called specifications. According to the definitions we are using in this book, however, and according to the standard practices of the construction industry, even on projects for the government, they are actually used as standards. They are not permitted, for example, as a part of a *project manual* for a building construction project, except by reference. It is not possible to enter Federal Specifications intact into a project manual, and editing them for this purpose is neither permissible nor desirable. However, since GSA calls them specifications and this terminology is generally accepted in the building industry, we will accede to this convention.

A new Federal Specification is developed when a government procurement need arises, when a present specification becomes obsolete, or when revision is required for other reasons. Although federal specifications are gradually being replaced by industry standards, such as those promulgated by ASTM, Federal Specifications are still referenced in government guide specifications and are the only standards available for some products. In addition, Federal Specifications are still referenced by some product manufacturers even when industry standards are available. There may come a day when Federal Specifications are no longer used in the construction industry, but that day has not yet arrived.

GSA may assign the development of a particular Federal Specification to a federal agency that has specialized technical competence and has available the necessary facilities. However, nationally recognized industry, technical society, and trade association standards, such as those by ASTM, are used and adopted in Federal Specifications to the maximum extent practicable.

Federal Specifications are designated by a letter and number code. For example, FF-A-A-1812B is the Federal Specification for naphthalene insecticide. The term Federal Specification is often abbreviated as Fed. Spec. or simply FS.

1.5.3.3 Military Agencies

The federal government is one of the world's largest buyers of equipment, materials, and supplies, with annual purchases in the billions of dollars. Various departments within the Department of Defense have developed specifications covering materials, products, and services used predominately by military activities.

Military Specifications may be used by any interested civilian organization or specifier. Military Specifications are indexed by the title and code letter prefix MIL.

1.5.3.4 Department of Housing and Urban Development (HUD)

The National Housing Act, enacted by Congress in 1934 and amended from time to time, created the Federal Housing Administration (FHA) to stimulate home construction by insuring mortgage loans. The functions of this agency were transferred by Congress in 1965 to the newly created Department of Housing and Urban Development (HUD), and FHA became part of this larger cabinet-level department.

The overall purpose of HUD is to assist in the sound development of the nation's communities and metropolitan areas. Encouragement of housing production through mortgage insurance and various subsidies has been one of HUD's chief objectives. Improvement in housing quality and in land planning standards has been another HUD objective mandated by Congress.

FHA/HUD HOUSING PROGRAMS

FHA/HUD makes no loans, nor does it plan or build housing. It functions mainly as an insuring agency for mortgage loans made by private lenders, such as savings associations and commercial banks. For instance, through the Section 203(b) program, FHA/HUD encourages lenders to make loans with low down payments and long maturities on one- to four-family dwellings. The borrower pays an annual insurance premium of a small percentage of the average principle outstanding over the premium year. The Secretary of HUD sets the interest rate ceiling on FHA/HUD loans at a level required to meet market conditions. Another frequently used section, 221(d)(4), provides for mortgage insurance of new or rehabilitated low- or middle-income rental housing.

The traditional role of the FHA was transformed when the agency became the administrator of interest-rate subsidy and rent-supplement programs authorized by Congress since 1965. The Section 235 program combines insurance with interest assistance payments for owner-occupied homes. In addition to insuring the loan, FHA/HUD pays part of the interest the borrower owes the mortgage lender. Section 236 offers insurance and interest assistance for rental projects. Section 237 provides insurance on loans to borrowers with poor credit histories. Section 238 authorizes insurance for mortgage loans in high-risk situations, such as transitional urban areas, not covered by other programs.

HUD MINIMUM PROPERTY STANDARDS

Because not all housing programs authorized by Congress involve mortgage insurance, not all of them are administered by FHA/HUD. For instance, Section 8 of the Housing Act of 1974 authorizes rental subsidies for leased low-income housing. The housing may be existing or new and may be financed either conventionally or with FHA/HUD mortgage insurance.

Before 1973, FHA-insured private housing had to conform to the FHA *Minimum Property Standards* and subsidized public housing was regulated by a different set of standards. With the adoption in that year of the HUD *Minimum Property Standards* (MPS), uniform standards became applicable to all HUD housing programs.

The MPS were intended to provide a sound technical basis for the planning and design of housing under the numerous programs of the Department of Housing and Urban Development. The standards described those characteristics in a property that would provide initial and continuing utility, durability, desirability, economy of maintenance, and a safe and healthful environment.

Environmental quality was considered throughout the MPS. As a general policy, property development was required to be consistent with the national program for conservation of energy and other natural resources. Care had to be exercised to avoid air, water, land, and noise pollution and other environmental hazards.

The MPS consisted of three volumes of mandatory standards: (1) *MPS for One- and Two-Family Dwellings*, HUD 4900.1; (2) *MPS for Multifamily Housing*, HUD 4910.1; and (3) *MPS for Care-Type Housing*, HUD 4920.1. Variations and exceptions for seasonal homes intended for other than year-round occupancy were listed in HUD 4900.1. Exceptions for elderly housing were listed in HUD 4900.1 and 4910.1. A fourth volume, the MPS *Manual of Acceptable Practices*, HUD 4930.1, contained advisory and illustrative material for the three volumes of mandatory standards.

Today these documents have been withdrawn and replaced by a single document, HUD 4910.1, *Minimum Property Standards for Housing*. This document is intended to supplement the requirements of the applicable local and state building codes and the *One- and Two-Family Dwelling Code* (see Section 1.6.3.4).

MATERIALS RELEASES

The Architectural Standards Division of HUD issues *Material Releases* for specific proprietary products for which no recognized standards exist. Materials Releases relate only to the technical elements of a product. Each release describes a product and its use and is issued to HUD field offices for guidance in determining the acceptance of the product.

The absence of a release for a particular product does not preclude its use. Materials Releases are not intended to indicate approval, endorsement, or acceptance by HUD. Manufacturers of materials and products for which Materials Releases have been issued are not authorized to use them in any manner for sales promotion. Copies of Materials Releases are on file in HUD field offices but are not available for general distribution.

HUD has additional provisions for the review of special materials, products, and construction methods that it may be asked to insure. Design, materials, equipment, and construction methods other than those described in *Minimum Property Standards for Housing* are considered for use, provided complete substantiating data satisfactory to HUD are submitted. Local HUD field offices are authorized to accept variations from the standards for specific cases, subject to conditions outlined in the standards.

Variations on an area or regional basis, or variations involving a substantial number of properties on a repetitive basis, are authorized only after consideration of recommendations by the HUD field office and approval by the Architectural Standards Division. Under certain conditions some variations are established and published as Local Acceptable Standards (LAS) for a specific area.

1.5.4 EUROPEAN STANDARDS

In January 1993, 12 European countries joined economically into a *Single European Market* (SEM). In addition, the European Community (EC) is moving steadily toward the opening of borders and establishment of the free trade of ideas and goods between the member nations. These major changes in Europe will greatly affect trade and other relationships between the United States and the EC. They will also affect the U.S. building industry in many ways.

One effect that will be greatly felt is the change in European standards. As a part of the establishment of SEM, the EC member states have deemed it necessary to unify each of their existing 12 separate groups of national standards into common European standards. The European Committee for Standarization (CEN) and the Committee for European Electrotechnical Standarization (CENELEC) have been charged with publishing standards for the EC member states.

The International Organization for Standardization (IOS) is a nongovernmental organization made up of representatives from the standards institutions of 91 countries. The United States is a member, represented by ANSI, but is not very influential there. IOS approves and publishes standards produced by its members. Most of those published by IOS are European standards.

The new EC standards will be based either on reconciling the differences between existing standards in the EC states or simply by adopting the IOS standards. The United States could exert greater influence in this area by increasing its involvement in IOS, but there seems to be little movement in that direction. Although development is in progress, single European standards for construction products do not yet exist. Until they do, EC member states will continue to enforce either their own standards or interim standards.

American firms doing work in Europe and U.S. products sold in Europe today have to comply with the standards of the country in which the work is being done. Eventually they will all have to comply with the unified EC standards. In the interim, there will be a morass of conflicting and possibly overlapping standards in the various countries. Some standards will be the old national ones, some will be the new EC standards, and others will be IOS standards. Probably the best first step of anyone contemplating working in the EC would be to contact the U.S. Department of Commerce Office of European Community Affairs for advice.

What eventual effect the new EC standards will have on U.S. standards is unknown at this point. There will probably be some changes in our standards relatively soon. Down the road there may be extensive changes as the United States tries to compete economically with a unified Europe.

1.6 Codes

The planning, construction, location, and use of buildings are regulated by a variety of laws enacted by local, state, and federal governments. These statutes and ordinances include zoning, building, plumbing, electrical, and mechanical codes that are intended to protect the health, safety, and general welfare of the public. These codes incorporate many recognized construction industry standards (see Section 1.5), but they do not necessarily contain criteria that ensure efficient, convenient, or adequately equipped buildings.

A *zoning code* (see Section 1.6.2) establishes requirements for land use.

A *building code* (see Section 1.6.3) establishes requirements for the construction and occupancy of buildings. It contains standards of performance and requirements for materials, methods, and systems. It also covers structural strength, fire resistance, adequate light and ventilation, egress, occupant safety, and other considerations determined by the design, construction, alteration, and demolition of buildings.

A collection of building requirements becomes a code when it is adopted by a municipality as a public ordinance or law. Local communities may write their own codes or may legally adopt other codes, such as state building codes or one of the model codes (see Section 1.6.3.3).

1.6.1 HISTORY

Laws controlling building construction are not new. The ancient code of the Babylonian emperor Hammurabi, dating back to about 1800 B.C., is often cited as the first recorded building code. It provided severe penalties for construction practices that violated the health or safety of citizens. For example, if a building collapsed killing the occupants, its architects and builders were put to death. This ancient code, based on the idea that the strong should not injure the weak, sets the principle for today's construction regulations: The public has a right to be protected from the harmful acts of others.

Modern building regulation evolved over time, starting in the early nineteenth century with the adoption of fire laws in some large cities. These laws prohibited the construction of wooden buildings in congested areas. At about the same time, cities also began to adopt health regulations to improve the living conditions of the poor, which are the forerunners of today's minimum housing codes. Some courts, however, resisted the enactment of such laws as infringements on personal property rights.

As the validity of fire and health laws was slowly established, courts began to accept governmental control of all aspects of building construction involving the health and safety of the public. Today most states and cities, and a great many counties and towns, regulate the planning, construction, and installation of building systems through a variety of laws and ordinances.

The right to regulate building construction constitutionally rests in the states, but before 1960 few state governments exercised that right. The states usually delegated to local governments the power to regulate buildings to protect the public health, safety, and welfare. Today many states are taking an active role in building regulation. More than half the states have now adopted some form of statewide building regulation concerned with the construction of industrialized buildings, mobile homes, and conventional construction.

Federal legislation in the areas of occupational health and safety, environmental protection, pollution controls, and consumer protection, coupled with increased state legislative activity, offer evidence that the building regulatory process will become ever more complex.

1.6.2 ZONING CODES

Zoning codes are developed, interpreted, enacted, and enforced by local jurisdictions (cities, counties, townships, etc.) to achieve the following:

- Promote the general welfare by ensuring adequate light, air, and convenience of access to buildings
- Provide for safety from fire, flood, and other dangers
- Reduce congestion in the public streets
- Create a convenient, attractive, and harmonious community
- Expedite adequate police, fire, and rescue protection
- Expedite emergency evacuation
- Provide adequate public facilities (schools, parks, playgrounds, etc.)
- Reduce encroachment on historic areas
- Protect against overcrowding of land
- Prevent undue density of population
- Encourage economic development

Zoning codes affect building design by controlling where building types may be located, both within their communities and on their particular sites. Zoning codes also restrict the size of buildings that can be built on a specific site and affect the shapes of buildings. Many high-rise buildings owe their shapes to zoning setback rules that require higher floors to be farther from the property line.

A typical zoning code contains specific requirements for the type of use that is permitted in individual zones within the community. The naming of zones differs from community to community. The following are some typical zones:

R-1, Single-Family
R-2, Single-Family and Two-Family
R-3, Apartment and Multiple-Family Dwelling
RTH, Residential Town House
R-PO, Residential-Professional Offices
B-1, General Business
B-2, General Business, High Density
B-3, Shopping Centers
P-BO, Professional Business
M-L, Limited Industrial

M-1, Heavy Industrial
M-2, Light Industrial

Special districts may also be established to control requirements for special-use zones, such as historic zones; wetlands; areas where flood damage is likely; architectural districts, where design appearance is controlled; planned development districts, where special rules apply, such as allowance of increased occupant density in exchange for leaving green areas; and water runoff restricted districts established to protect lakes, rivers, and bays. Typical restrictions in a zone include requirements such as those for:

- Permitted uses
- Building height
- Minimum lot size and street frontage length
- Minimum floor area
- Required yards (distance from a building to a property line in front, rear, and sides)
- Building setback (distance from the building to lot lines at each height above the ground, which often increase in cities as the building grows taller)
- Permitted types of accessory buildings and their permitted size and location
- Off-street parking and loading requirements
- Lighting of parking and other site areas
- Visual clearance on corner lots
- Swimming pools
- Site fences and walls

1.6.3 BUILDING CODES

The construction industry is enmeshed in an extraordinary network of building codes that attempt to ensure that buildings and their environs will be safe. They generally accomplish that purpose; however, codes and code administration are criticized widely as being restrictive to building progress by retarding the acceptance of new and improved uses of materials and methods and thereby unnecessarily increasing construction costs. In some instances, the adoption of improved codes has stimulated better building practices. But the existing complex and chaotic building code situation is recognized as one of the problems facing the building industry today because of the use of descriptive-type codes (see Section 1.6.3.1), the lack of code uniformity, the multiplicity of codes, the slow response of codes to change, and inadequate performance standards in the codes (see Section 1.6.3.3).

The major building officials organizations and numerous federal, state, and local groups are constantly working to improve codes and the code regulatory system. The latest efforts center on standardizing the major model codes (see Section 1.6.3.3) into a single code for each discipline.

Some so-called codes are actually *standards* because they are voluntary and have no status in law. Only when standards are adopted by legislative bodies and incorporated into law do they become codes. In one sense, the model codes discussed in Section 1.6.3.3 are actually standards that have been adopted in many, but not all, jurisdictions as codes. Other examples of documents called codes that actually are not codes include various American National Standards Association (ANSI) and National Fire Protection Association (NFPA) codes.

Standards that are called codes by their producers often fall in the design standards category (see Section 1.5.1) but may encompass one or more of the other types.

Because some documents called codes are universally recognized as such, we will accede to that convention in this book. Therefore, publications such as ANSI A17.1 "Safety Code for Elevators and Escalators" and NFPA 101 "Life Safety Code" and "National Electrical Code" are here called codes.

1.6.3.1 The Effect of Building Codes on Design

Building codes affect building design in a number of ways. One way is by limiting the construction type and size for each building use. Figure 1.6-1 is a table from the 1999 edition of the BOCA National Building Code. Similar tables are included in all the model codes (see Section 1.6.3.3). The table shown lists 22 *Use Groups*. These use groups are defined and described in the body of the code. The table then shows height and area restrictions for each of these use groups, based on each of 10 *types of construction*. The types of construction are also defined in the text of the code. For example, suppose an architect has been commissioned to design a building that falls in group E, "Educational," that is required to have 40,000 sq. ft. (3716 m^2) of area. The table shows that if the building is Noncombustible Type 1A construction, the area and height are not limited by code. Therefore the building could be entirely on a single floor or built in 5 floors of 8000 sq. ft. (743 m^2) each.

Our example building can be also constructed of Type 2A or 2B Noncombustible, Protected Construction, but not of Type 2C, because we need 40,000 sq. ft. (3716 m^2) and 2C permits only 2 floors of a maximum 14,400 sq. ft. (1338 m^2) each, for a total of only 28,800 sq. ft. (2676 m^2). Our building could also be built of Type 3A construction, but not 3B; or of Type 4, but not Type 5.

Codes further affect building design by controlling materials and methods of construction. This effect is illustrated in Figure 1.6-2, another table from the BOCA code. In this table, structural elements are listed down the left side and types of construction across the top. These construction types are the same as those used in Figure 1.6-1. The figures at the intersection of the lines and columns are the hourly fire rating of the assemblies in the left column for each construction type listed across the top of the table. For example, the floor construction (Line 10) in a Type 1A building must have a 3-hour fire rating.

Building codes also affect building design in other ways. Building codes are either descriptive or performance oriented. A *descriptive* (sometimes called specification or prescription) type building code establishes construction requirements by reference to a particular material or method. For example, a code may require that exterior walls be built of 2 × 4 wall studs spaced 16 in. on center and covered by 1-in.-thick board sheathing applied diagonally. A builder seeking to space studs at 24 in. (610 mm) on center, which may be structurally sound and safe, or to use another type of sheathing, even one that is better than the diagonal sheathing, would be in violation of the specifically stated code requirements.

A *performance* type code does not limit the selection of methods and systems to a single type, but establishes requirements for the performance of building elements. Such a code states design and engineering criteria without reference to specific materials or methods of construction. For example, an outside wall may be required to support loads (wind, dead, live, etc.) with specific defined values and to meet or exceed stated insulating and permeability requirements. Any system performing as required by the code would be acceptable, regardless of the materials or methods used.

True performance codes are idealistically excellent, but impractical in use. To

Table 503
HEIGHT AND AREA LIMITATIONS OF BUILDINGS
Height limitations of buildings (shown in upper figure as stories and feet above grade plane)[m], and area limitations of one- or two-story buildings facing on one street or public space not less than 30 feet wide (shown in lower figure as area in square feet per floor[m]). See Note a.

Use Group		Type of construction									
		Noncombustible					Noncombustible/Combustible			Combustible	
		Type 1		Type 2			Type 3		Type 4	Type 5	
		Protected Note b		Protected		Unprotected	Protected	Unprotected	Heavy timber	Protected	Unprotected
	Note a	1A	1B	2A	2B	2C	3A	3B	4	5A	5B
A-1	Assembly, theaters	Not limited	Not limited	5 St. 65′ 19,950	3 St. 40′ 13,125	2 St. 30′ 8,400	3 St. 40′ 11,550	2 St. 30′ 8,400	3 St. 40′ 12,600	1 St. 20′ 8,925	1 St. 20′ 4,200
A-2	Assembly, nightclubs and similar uses	Not limited	Not limited 7,200	3 St. 40′ 5,700	2 St. 30′ 3,750	1 St. 20′ 2,400	2 St. 30′ 3,300	1 St. 20′ 2,400	2 St. 30′ 3,600	1 St. 20′ 2,550	1 St. 20′ 1,200
A-3	Assembly — Lecture halls, recreation centers, terminals, restaurants other than nightclubs	Not limited	Not limited	5 St. 65′ 19,950	3 St. 40′ 13,125	2 St. 30′ 8,400	3 St. 40′ 11,550	2 St. 30′ 8,400	3 St. 40′ 12,600	1 St. 20′ 8,925	1 St. 20′ 4,200
A-4	Assembly, churches — Note c	Not limited	Not limited	5 St. 65′ 34,200	3 St. 40′ 22,500	2 St. 30′ 14,400	3 St. 40′ 19,800	2 St. 30′ 14,400	3 St. 40′ 21,600	1 St. 20′ 15,300	1 St. 20′ 7,200
B	Business	Not limited	Not limited	7 St. 85′ 34,200	5 St. 65′ 22,500	3 St. 40′ 14,400	4 St. 50′ 19,800	3 St. 40′ 14,400	5 St. 65′ 21,600	3 St. 40′ 15,300	2 St. 30′ 7,200
E	Educational — Note c	Not limited	Not limited	5 St. 65′ 34,200	3 St. 40′ 22,500	2 St. 30′ 14,400	3 St. 40′ 19,800	2 St. 30′ 14,400	3 St. 40′ 21,600	1 St. 20′ 15,300 Note d	1 St. 20′ 7,200 Note d
F-1	Factory and industrial, moderate	Not limited	Not limited	6 St. 75′ 22,800	4 St. 50′ 15,000	2 St. 30′ 9,600	3 St. 40′ 13,200	2 St. 30′ 9,600	4 St. 50′ 14,400	2 St. 30′ 10,200	1 St. 20′ 4,800
F-2	Factory and industrial, low — Note h	Not limited	Not limited	7 St. 85′ 34,200	5 St. 65′ 22,500	3 St. 40′ 14,400	4 St. 50′ 19,800	3 St. 40′ 14,400	5 St. 65′ 21,600	3 St. 40′ 15,300	2 St. 30′ 7,200
H-1	High hazard, detonation hazards — Notes e, i, k, l	1 St. 20′ 16,800	1 St. 20′ 14,400	1 St. 20′ 11,400	1 St. 20′ 7,500	1 St. 20′ 4,800	1 St. 20′ 6,600	1 St. 20′ 4,800	1 St. 20′ 7,200	1 St. 20′ 5,100	Not permitted
H-2	High hazard, deflagration hazards — Notes e, i, j, l	5 St. 65′ 16,800	3 St. 40′ 14,400	3 St. 40′ 11,400	2 St. 30′ 7,500	1 St. 20′ 4,800	2 St. 30′ 6,600	1 St. 20′ 4,800	2 St. 30′ 7,200	1 St. 20′ 5,100	Not permitted
H-3	High hazard, physical hazards — Notes e, l	7 St. 85′ 33,600	7 St. 85′ 28,800	6 St. 75′ 22,800	4 St. 50′ 15,000	2 St. 30′ 9,600	3 St. 40′ 13,200	2 St. 30′ 9,600	4 St. 50′ 14,400	2 St. 30′ 10,200	1 St 20′ 4,800
H-4	High hazard, health hazards — Notes e, l	7 St. 85′ Not limited	7 St. 85′ Not limited	7 St. 85′ 34,200	5 St. 65′ 22,500	3 St. 40′ 14,400	4 St. 50′ 19,800	3 St. 40′ 14,400	5 St. 65′ 21,600	3 St. 40′ 15,300	2 St. 30′ 7,200
I-1	Institutional, residential care	Not limited	Not limited	9 St. 100′ 19,950	4 St. 50′ 13,125	3 St. 40′ 8,400	4 St. 50′ 11,550	3 St. 40′ 8,400	4 St. 50′ 12,600	3 St. 40′ 8,925	2 St. 35′ 4,200
I-2	Institutional, incapacitated	Not limited	Not limited	4 St. 50′ 17,100	2 St. 30′ 11,250	1 St. 20′ 7,200	1 St. 20′ 9,900	Not permitted	1 St. 20′ 10,800	1 St. 20′ 7,650	Not permitted
I-3	Institutional, restrained	Not limited	Not limited	4 St. 50′ 14,250	2 St. 30′ 9,375	1 St. 20′ 6,000	2 St. 30′ 8,250	1 St. 20′ 6,000	2 St. 30′ 9,000	1 St. 20′ 6,375	Not permitted
M	Mercantile	Not limited	Not limited	6 St. 75′ 22,800	4 St. 50′ 15,000	2 St. 30′ 9,600	3 St. 40′ 13,200	2 St. 30′ 9,600	4 St. 50′ 14,400	2 St. 30′ 10,200	1 St. 20′ 4,800
R-1	Residential, hotels	Not limited	Not limited	9 St. 100′ 22,800	4 St. 50′ 15,000	3 St. 40′ 9,600	4 St. 50′ 13,200	3 St. 40′ 9,600	4 St. 50′ 14,400	3 St. 40′ 10,200	2 St. 35′ 4,800
R-2	Residential, multiple-family	Not limited	Not limited	9 St. 100′ 22,800	4 St. 50′ 15,000 Note f	3 St. 40′ 9,600	4 St. 50′ 13,200 Note f	3 St. 40′ 9,600	4 St. 50′ 14,400	3 St. 40′ 10,200	2 St. 35′ 4,800
R-3	Residential, one- and two-family and multiple single-family	Not limited	Not limited	4 St. 50′ 22,800	4 St. 50′ 15,000	3 St. 40′ 9,600	4 St. 50′ 13,200	3 St. 40′ 9,600	4 St. 50′ 14,400	3 St. 40′ 10,200	2 St. 35′ 4,800
S-1	Storage, moderate	Not limited	Not limited	5 St. 65′ 19,950	4 St. 50′ 13,125	2 St. 30′ 8,400	3 St. 40′ 11,550	2 St. 30′ 8,400	4 St. 50′ 12,600	2 St. 30′ 8,925	1 St. 30′ 4,200
S-2	Storage, low — Note g	Not limited	Not limited	7 St. 85′ 34,200	5 St. 65′ 22,500	3 St. 40′ 14,400	4 St. 50′ 19,800	3 St. 40′ 14,400	5 St. 65′ 21,600	3 St. 40′ 15,300	2 St. 30′ 7,200
U	Utility, miscellaneous	Not limited	Not limited	5 St. 65′ 19,950	4 St. 50′ 13,125	2 St. 30′ 8,400	3 St. 40′ 11,550	2 St. 30′ 8,400	4 St. 50′ 12,600	2 St. 30′ 8,925	1 St. 20′ 4,200

Note a. See the following sections for general exceptions to Table 503:
Section 504.2 Allowable height increase due to automatic sprinkler system installation.
Section 506.2 Allowable area increase due to street frontage.
Section 506.3 Allowable area increase due to automatic sprinkler system installation.
Section 506.4 Allowable area reduction for multistory buildings.
Section 507.0 Unlimited area one-story buildings.
Note b. Buildings of Type 1 construction permitted to be of unlimited tabular heights and areas are not subject to special requirements that allow increased heights and areas for other types of construction (see Section 503.1.3).
Note c. For height exceptions for auditoriums in occupancies in Use Groups A-4 and E, see Section 504.3.
Note d. For height exceptions for day care centers in buildings of Type 5 construction, see Section 504.4.
Note e. For exceptions to height and area limitations for buildings with occupancies in Use Group H, see Chapter 4 governing the specific use groups.
Note f. For exceptions to height of buildings with occupancies in Use Group R-2 of Types 2B and 3A construction, see Sections 504.6 and 504.7.
Note g. For height and area exceptions for open parking structures, see Section 406.0.
Note h. For exceptions to height and area limitations for special industrial occupancies, see Section 507.1.
Note i. Occupancies in Use Groups H-1 and H-2 shall not be permitted below grade.
Note j. Rooms and areas of Use Group H-2 containing pyrophoric materials shall not be permitted in buildings of Type 3, 4 or 5 construction.
Note k. Occupancies in Use Group H-1 are required to be detached one-story buildings (see Section 707.1.1).
Note l. For exceptions to height for buildings with occupancies in Use Group H, see Section 504.5.
Note m. 1 foot = 304.8 mm; 1 square foot = 0.093 m^2.

FIGURE 1.6-1 Table 503, "Height and Area Limitations of Buildings" from 1999 *BOCA National Building Code.* Other codes may vary.

Table 602
FIRERESISTANCE RATINGS OF STRUCTURE ELEMENTS[k]

Structure element Note a		Type of construction Section 602.0									
		Noncombustible					Noncombustible/Combustible			Combustible	
		Type 1 Section 603.0		Type 2 Section 603.0			Type 3 Section 604.0		Type 4 Section 605.0	Type 5 Section 606.0	
		Protected		Protected		Unprotected	Protected	Unprotected	Heavy timber Note c	Protected	Unprotected
		1A	1B	2A	2B	2C	3A	3B	4	5A	5B
1 Exterior walls	Loadbearing	4	3	2	1	0	2	2	2	1	0
		← Not less than the fireresistance rating based on fire separation distance (see Section 705.2) →									
	Nonloadbearing										
		← Not less than the fireresistance rating based on fire separation distance (see Section 705.2) →									
2 Fire walls and party walls (Section 707.0)		4	3	2	2	2	2	2	2	2	2
		← Not less than the fireresistance rating required by Table 707.1 →									
3 Fire separation assemblies (Section 709.0)	Fire enclosure of exits (Sections 1014.11, 709.0 and Note b)	2	2	2	2	2	2	2	2	2	2
	Shafts (other than exits) and elevator hoistways (Sections 709.0, 710.0 and Note b)	2	2	2	2	2	2	2	2	1	1
	Mixed use and fire area separations (Section 313.0)	← Not less than the fireresistance rating required by Table 313.1.2 →									
	Other separation assemblies (Note i)	1	1	1	1	1	1	1	1	1	1
		← Note d →									
4 Fire partitions (Section 711.0)	Exit access corridors (Note g)	← Not less than the fireresistance rating required by Section 1011.4 →									
		← Note d →									
	Tenant spaces separations (Note f)	1	1	1	1	0	1	0	1	1	0
		← Note d →									
5 Dwelling unit and guestroom separations (Sections 711.0, 713.0 and Notes f and j)		1	1	1	1	1	1	1	1	1	1
		← Note d →									
6 Smoke barriers (Section 712.0 and Note g)		1	1	1	1	1	1	1	1	1	1
7 Other nonloadbearing partitions		0	0	0	0	0	0	0	0	0	0
		← Note d →									
8 Interior loadbearing walls, loadbearing partitions, columns, girders, trusses (other than roof trusses) and framing (Section 716.0)	Supporting more than one floor	4	3	2	1	0	1	0	see Sec. 605.0	1	0
	Supporting one floor only or a roof only	3	2	1½	1	0	1	0	see Sec. 605.0	1	0
9 Structural members supporting wall (Section 716.0 and Note g)		3	2	1½	1	0	1	0	1	1	0
		← Not less than fireresistance rating of wall supported →									
10 Floor construction including beams (Section 713.0 and Note h)		3	2	1½ Note l	1	0	1	0	see Sec. 605.0 Note c	1	0
11 Roof construction, including beams, trusses and framing, arches and roof deck (Section 715.0 and Notes e, m)	15′ or less in height to lowest member	2	1½	1	1	0	1	0	see Sec. 605.0 Note c	1	0
				← Note d →							
	More than 15′ but less than 20′ in height to lowest member	1	1	1	0	0	0	0	see Sec. 605.0	1	0
		← Note d →									
	20′ or more in height to lowest member	0	0	0	0	0	0	0	see Sec. 605.0	0	0
		← Note d →									

Note a. For fireresistance rating requirements for structural members and assemblies which support other fireresistance rated members or assemblies, see Section 716.1.

Note b. For reductions in the required fireresistance rating of exit and shaft enclosures, see Sections 1014.11 and 710.3.

Note c. For substitution of other structural materials for timber in Type 4 construction, see Section 2304.2.

Note d. For fireretardant-treated wood permitted in roof construction and nonloadbearing walls where the required fireresistance rating is 1 hour or less, see Sections 603.2 and 2310.0.

Note e. For permitted uses of heavy timber in roof construction in buildings of Types 1 and 2 construction, see Section 715.4.

Note f. For reductions in required fireresistance ratings of tenant separations and dwelling unit separations, see Sections 1011.4 and 1011.4.1.

Note g. For exceptions to the required fireresistance rating of construction supporting exit access corridor walls, tenant separation walls in covered mall buildings, and smoke barriers, see Sections 711.4 and 712.2.

Note h. For buildings having habitable or occupiable stories or basements below grade, see Section 1006.3.1.

Note i. Not less than the rating required by this code.

Note j. For Use Group R-3, see Section 310.5.

Note k. Fireresistance ratings are expressed in hours.

Note l. In buildings which are required to comply with the provisions of Section 403.3, the required fireresistance rating for floor construction, including beams, shall be 2 hours (see Section 403.3.3.1).

Note m. 1 foot = 304.8 mm.

FIGURE 1.6-2 Table 602, "Fireresistance Ratings of Structural Elements" from 1999 *BOCA National Building Code.* Other codes may vary.

enforce them, local code administrators would have to be extraordinarily competent and equipped to interpret performance criteria and evaluate proposed methods, uses, and systems. Such people and equipment are rare. A workable solution lies somewhere between descriptive and performance codes. Both types of codes should adequately provide for the acceptance of alternate methods and systems.

1.6.3.2 Building Code Enforcement

Local administration and enforcement of codes are usually done by a building inspector or engineer who has the authority to approve materials and methods that may not be directly referenced in the code. Qualified people are necessary to administer a building code properly. No matter how good a code may be, it must be enforced by someone who is experienced, informed, and objective. Builders complain about inspectors who do not understand construction and may be thus arbitrary and inconsistent. They are seldom upset by a careful and competent inspector who is consistent even though tough. A competent code enforcer knows when the letter of the code should prevail and when subjective interpretation should be made.

However, local code administrators, faced continually with difficult decisions, may well argue that their job is solely to check compliance. An often-drawn analogy is that a local policeman does not make judgments about whether a law is right or just, but is charged solely with determining compliance. Perhaps it is best that the determination of performance criteria and judgment as to whether methods and systems perform suitably must be made by technically qualified people, and not by local code administrators. Members of the model code groups (see Section 1.6.3.3) are qualified to offer this type of service.

Model code groups, which are supported by building officials themselves, have performed a great service to the building industry. But model codes have not fully solved the problems of code uniformity. The reason is, partly, that even though most communities have adopted a model code, some have adopted them with modifications. The National Association of Home Builders (NAHB) has said that many of the changes related to housing made to model codes to adapt them to local conditions come from local codes prepared 20 or more years earlier. Sometimes, progressive time- and money-saving requirements of a model code are revised so that antiquated provisions apply instead, simply to agree with local custom.

1.6.3.3 Model Codes

Problems with code development, use, and enforcement occur for several reasons.

LACK OF UNIFORMITY

Because establishing and enforcing building codes are local functions, a designer or builder who works in more than one community is often faced with a frustrating variety of requirements. A product manufacturer must win acceptance by thousands of local building code administrators, instead of concerning itself only with performance and public acceptance. Much effort has been made to unify the codes used by communities, and considerable improvement has been gained through the local adoption of model codes.

For years, code developers and building industry practitioners have been talking about developing a single national building code, and these efforts may be actually reaching a satisfactory conclusion in the ICC International Codes (see ICC Codes later in this section).

MULTIPLICITY OF CODES

If designers and builders were dealing with a single code, their problems would be greatly simplified. However, in addition to local building codes that tell them how a structure must be assembled, they also have to satisfy a number of other codes covering a variety of subjects such as plumbing, electrical wiring, traffic, utilities, health and sanitation, land planning, building occupancy, access by the handicapped, and zoning.

The lack of code uniformity and the multiplicity of additional regulations often so complicate matters that buildings are designed and built with the use of very conservative and often expensive criteria and methods. Designers and builders alike are often unable to improve their techniques because of the restrictive nature of one or more of the applicable codes.

SLOW RESPONSE TO CHANGE

Codes are often criticized for failing to recognize new materials and methods. Judging new products on the basis of performance criteria must be performed by technically qualified and equipped organizations.

This function is one in which the model code groups can exercise great leadership and provide a major stimulus to progress. The groups have responded reasonably to progressive change. However, they are subject to many pressures that deter progress, and they have sometimes been slow in acting on proposed changes. For example, it took almost 4 years for all of the major code groups to accept the NAHB's proposal to eliminate floor bridging, a proposition well documented through extensive research and testing.

INADEQUATE PERFORMANCE STANDARDS

As discussed in Section 1.5, no single group of building industry standards exists. In selecting materials or determining the suitability of materials and methods, a specifier and builder make reference to a variety of trade association certification programs, grade marks and trademarks, Underwriters Laboratories (UL) labels, and many other construction standards.

There are independent testing agencies, such as UL, and standards-setting bodies, such as ASTM, and some manufacturers have substantial commitments in research and testing facilities, but no single agency is presently recognized by all groups interested in building construction.

To help fill the need for workable codes, three major organizations of industry and professional groups and states have developed building codes that may be adopted into law for use in local communities. These codes are commonly referred to as *model codes*. The major model code groups are the Building Officials and Code Administrators International (BOCAI), the International Conference of Building Officials (ICBO), and the Southern Building Code Congress International (SBCCI). The major national model codes and their sponsors are listed in Figure 1.6-3.

Model codes have been widely accepted by local communities and now are used by more than 80% of those communities with a population of more than 10,000. The organizations that prepare model codes also provide for the continuous updating necessary to include recommendations based on industry research and the development of new materials and methods.

Other model codes that deal with specialized subjects not fully covered in the major model codes are also available. Most of these codes and their sponsors are listed in Figure 1.6-3. Many local communities have also adopted these codes.

NATIONAL BUILDING CODE

The first model building code was introduced in 1905 by the National Board of Fire Underwriters, later called the American Insurance Association (AIA). This organization was concerned primarily with public protection against fire hazards, and the original code was designed to guide communities in setting up fire safety standards. With subsequent additions, however, its *National Building Code* laid the groundwork for the development of building codes throughout the country. AIA produced its last edition of the *National Building Code* in 1976.

In 1950, BOCAI published the *Basic Building Code*. When AIA ceased publishing the *National Building Code*, BOCAI picked it up and has published it since. BOCAI also publishes the other codes listed for it in Figure 1.6-3.

STANDARD BUILDING CODE

The Southern Building Code Congress International (SBCCI) drafted the *Southern Standard Building Code* in 1945. Now renamed the *Standard Building Code*, this model code is designed to recognize the special problems, such as high winds, in the southern region of the country and is used extensively in this area. SBCCI also publishes the other codes listed for it in Figure 1.6-3.

UNIFORM BUILDING CODE

In 1927, the International Conference of Building Officials (ICBO) (initially the Pacific Coast Building Officials Conference) published its *Uniform Building Code*. Since that time, the code has gained wide acceptance, particularly in the West. ICBO also publishes the other codes listed for it in Figure 1.6-3. It copublishes the *Uniform Fire Code* and the *Uniform Fire Code Standards* with the Western Fire Chiefs Association. ICBO also jointly publishes the *Uniform Mechanical Code* with the International Association of Plumbing and Mechanical Officials.

FIGURE 1.6-3 Model Building Codes and Their Sponsoring Organizations

Code	Sponsoring Organization
Building Codes	
National Building Code	Building Officials and Code Administrators International, Inc.
Standard Building Code	Southern Building Code Congress International
Uniform Building Code	International Conference of Building Officials
One and Two Family Dwelling Code	Council of American Building Officials
Electrical Codes	
National Electrical Code	National Fire Protection Association
One and Two Family Electrical Code	National Fire Protection Association
Elevator Codes	
Safety Code for Elevators and Escalators	American National Standards Institute
Safety Code for Manlifts	American National Standards Institute
Fire Prevention Codes	
National Fire Prevention Code	Building Officials and Code Administrators International, Inc.
Uniform Fire Code	International Conference of Building Officials
Housing Codes	
Property Maintenance Code	Building Officials and Code Administrators International, Inc.
Standard Housing Code	Southern Building Code Congress International
Uniform Housing Code	International Conference of Building Officials
Plumbing Codes	
National Plumbing Code	Building Officials and Code Administrators International, Inc.
National Standard Plumbing Code	National Association of Plumbing-Heating-Cooling Contractors
Standard Plumbing Code	Southern Building Code Congress International
Uniform Plumbing Code	International Association of Plumbing and Mechanical Officials
Mechanical Codes	
National Mechanical Code	Building Officials and Code Administrators International, Inc.
Uniform Mechanical Code	International Conference of Building Officials
Standard Gas Code	Southern Building Code Congress International
Standard Mechanical Code	Southern Building Code Congress International
Miscellaneous Codes	
Boiler and Unfired Pressure Vessel Code	American Society of Mechanical Engineers
Flammable and Combustible Liquids Code	National Fire Protection Association
Safety Code for Mechanical Refrigeration	American Society of Heating, Refrigerating, & Air Conditioning Engineers, Inc.
Life Safety Code	National Fire Protection Association

NATIONAL STANDARD PLUMBING CODE

The *National Standard Plumbing Code* is sponsored by the National Association of Plumbing-Heating-Cooling Contractors (NAPHCC). The *National Standard Plumbing Code* covers the proper design, installation, and maintenance of plumbing systems according to principles of sanitation and safety, but not necessarily for efficiency, convenience, or adequacy for good service or future expansion of system use. Standards for materials and fixtures are based largely on industry standards such as Commercial Standards, American Standards, and ASTM standards.

UNIFORM PLUMBING AND UNIFORM MECHANICAL CODES

The Western Plumbing Officials Association (now the International Association of Plumbing and Mechanical Officials [IAPMO]) established the first *Uniform Plumbing Code* in 1929 and continues to publish it today. IAPMO also publishes the *Uniform Mechanical Code* jointly with ICBO. These codes cover the design, installation, and maintenance of plumbing, heating, and air conditioning systems. Materials and equipment standards are based largely on other industry standards. These codes have gained widespread use in the western states, as well as in other communities across the country.

NATIONAL ELECTRICAL CODE

The *National Electrical Code* is produced by the National Fire Protection Association (NFPA). Its purpose is to ensure the safeguarding of persons and of buildings and their contents from hazards arising from the use of electricity for light, heat, power, and other purposes. The code contains basic minimum provisions for safety which, with proper maintenance, will result in installations free from hazard, but not necessarily efficient, convenient, or adequate for good service or future expansion of electrical use. The *National Electrical Code* makes reference to many industry standards, such as ANSI/ASME A17.1, and to UL labels. It is revised periodically and is adopted by reference into many building codes and federal and state laws and regulations.

LIFE SAFETY CODE

The *Life Safety Code* began in 1912 with a pamphlet entitled *Exit Drills in Factories, Schools, Department Stores and Theaters* and has evolved over the years to cover most building types. It is published by the National Fire Protection Association (NFPA) to establish minimum requirements for a reasonable degree of safety from fire in structures. It addresses requirements for egress; design and construction that will prevent undue danger from fire, smoke, and gases; fire detection and alarm systems; and automatic sprinklers and other fire extinguishing systems. It includes requirements for building design and construction and for heating, ventilating, air conditioning, and electrical systems.

ONE- AND TWO-FAMILY DWELLING CODE

In 1971, the American Insurance Association, BOCAI, ICBO, and SBCCI adopted a consensus code entitled the *One- and Two-Family Dwelling Code*. This model code is now published by the International Code Council (ICC). It includes a compilation of data from the major model codes produced by BOCAI, ICBO, and SBCCI, and from the ICBO and IAPMO *Uniform Mechanical Code*, the IAPMO *Uniform Plumbing Code*, and the NFPA *National Electrical Code*.

The *One- and Two-Family Dwelling Code* contains requirements for building planning and construction, including requirements for heating, cooling, and plumbing. HUD requires compliance with the *One- and Two-Family Dwelling Code* in addition to its own *Minimum Property Standards for Housing* and local and state building codes for consideration for HUD/FHA loan guarantees.

Publication of this single national model code for one- and two-family dwellings constituted a significant step toward uniform minimum regulations in the housing industry, but it still has not been recognized and adopted by every state and local government.

INTERNATIONAL CODE COUNCIL (ICC)

ICC produces codes developed as a cooperative effort between BOCAI, ICBO, and SBCCI. Current ICC codes available at the time of this writing include the following:

- International Building Code (IBC)
- International Plumbing Code
- International Private Sewage Disposal Code
- International Mechanical Code
- International Fuel Gas Code
- International Zoning Code
- One- and Two-Family Dwelling Code

The intent of the *ICC International Codes* is to replace similar codes now produced by BOCAI, ICBO, and SBCCI. The first ICC codes appeared in 1997. The first edition of the *International Building Code* appeared in 2000. These codes, like all others, must be adopted by local and state jurisdictions before they have effect. If history can be relied on, this will take a long time. The code groups are therefore continuing to update most of the earlier codes. For example, there is a 1999 edition of the BOCAI *National Building Code*.

1.6.4 NATIONAL ENERGY CODES

In the United States, energy consumed in buildings amounts to about one-third of all energy consumption. Building energy consumption has been increasing rapidly for many years. The U.S. Department of Commerce's earlier estimate, that energy consumption would increase by 25% between 1984 and the year 2000, appears to have been fairly accurate. That increase represents an annual growth rate of 1.14% for housing and 1.66% for commercial use. Slowing the rate of increase through increased efficiency in building systems design can significantly reduce overall energy demand.

The *National Energy Plan* calls for substantial decreases in energy consumption through conservation. A major element of an energy conservation program is enforcement of thermal efficiency standards and insulation requirements in new and renovated buildings by state and local governments.

The National Conference of States on Building Codes and Standards (NCSBCS) in 1973 requested the National Bureau of Standards (NBS) (now the National Institute of Standards and Technology [NIST]) to develop standards that could be used by all states for energy-efficient design in new buildings. NBS completed these standards in 1974, but they were considered too complex for most state and local code enforcement officials to administer. As a result, NCSBCS requested the American Society of Heating, Refrigerating and Air Conditioning Engineers, Inc. (ASHRAE) to translate the NBS standards into enforceable language.

In 1975, ASHRAE published its Standard 90-75, "Energy Conservation in New Building Design." This standard was the first nationally recognized standard for energy-efficient design applicable to all building types. It was, however, written for use by design engineers. State and local practitioners still needed a document that could be administered through the traditional code enforcement system. In 1976, NCSBCS translated ASHRAE Standard 90-75 into the Model Code for Energy Conservation in New Building Construction. Under a grant from the Energy Research and Development Administration (now the Department of Energy), NCSBCS worked with building officials from all parts of the country who administer model codes so that the Model Code for Energy Conservation would be applicable to all geographic areas in the country and could be incorporated easily into existing codes. The NCSBCS Model Code for Energy Conservation in New Building Construction was published in January 1977.

The *Energy Policy and Conservation*

Act (EPCA) became law in 1975. The law provided for substantial grants to those states that developed and implemented statewide energy conservation plans aimed at reducing statewide energy consumption 5% by 1980. State energy conservation plans had to include at least five elements: (1) mandatory lighting efficiency standards for public buildings, (2) programs to promote the availability and use of car pools and van pools, (3) mandatory standards and policies related to procurement practices of state and local governments, (4) a traffic law that permits a right turn on red after stopping, and (5) mandatory thermal energy standards and insulation requirements for new and renovated buildings.

In 1976, Congress passed the *Energy Conservation and Production Act* (ECPA). Title III of ECPA contained the following provisions: (1) the Department of Housing and Urban Development (HUD) is directed to develop performance-oriented thermal energy efficiency standards, (2) HUD is to monitor the progress of state and local governments in implementing the standards, and (3) if a substantial amount of new housing is not constructed in conformance with the standards, HUD can recommend to the Congress that the standards be made mandatory. If Congress concurs:

1. All construction assisted by federal money will have to be consistent with the standards. This will include federal grants, loans, and loan guarantees and will apply to all savings and loan associations that are federally insured.
2. Local governments will not be required to adopt the standards into their building codes. HUD and the states will be responsible for ensuring that all proposed construction is in conformance with the standards. However, local governments that elect to adopt the standards will be reimbursed for the cost of the certification.
3. If a local government incorporates the federal standards into its codes, then no further approvals will be required. Money will be available to communities to assist them in incorporating and enforcing the federal standards.

Current requirements for insulation are addressed in Section 7.3. The currently accepted standards for energy conservation and thermal design criteria are discussed in Chapter 15.

1.6.5 CODE ADVANCEMENT

The growing complexity of building design and construction and the subsequent increase of regulatory controls have induced several organizations to improve the effectiveness of the regulatory process with better code documents, administration, and enforcement.

1.6.5.1 National Institute of Building Sciences

The Housing and Community Development Act of 1974 authorized the creation of the National Institute of Building Sciences (NIBS). NIBS was initiated by the federal government with the assistance of the National Academy of Science and the National Academy of Engineering Research Council to help improve the way building construction is regulated.

NIBS gives the United States a national center for (1) assembly, storage, and dissemination of technical data and related information on construction, (2) development and promulgation of nationally recognized performance criteria, standards, test methods, and other evaluating techniques, and (3) evaluation and prequalification of existing and new building techniques.

1.6.5.2 Council of American Building Officials

In 1972 the governing bodies of BOCAI, ICBO, and SBCCI established the Council of American Building Officials (CABO). CABO's policies are determined by a board of trustees composed of representatives of the governing bodies of the member organizations.

Examples of CABO's work include (1) advancing the *One- and Two-Family Dwelling Code* as a recognized national standard through the procedures of the American National Standards Institute (ANSI), (2) sponsoring the development of model performance standards that will complement the requirements of model codes, (3) sponsoring a national research activity that provides a single approval agency for manufacturers of building components, systems, and materials, (4) developing and adopting a model mechanical, plumbing, and fire prevention code as the standard for all three organizations, and (5) establishing ICC to implement a single group of international codes to replace all the model codes in current use.

1.6.5.3 Associated Major City Building Officials

In 1974 building code officials from the nation's 30 largest cities created an organization called Associated Major City Building Officials (AMCBO). Its purpose is to exchange ideas, experience, and information on the growing complexities of code administration and enforcement in densely populated urban areas.

AMCBO is a membership organization and the technical arm of the National League of Cities, the International City Management Association, and the National Association of Counties.

1.6.5.4 National Conference of States on Building Codes and Standards

The National Conference of States on Building Codes and Standards (NCSBCS) was formed in 1967. The organization is composed of state delegates designated by their respective governors to represent the 50 states in discussions and programs pertaining to building regulation; state and local building officials; various representatives of the building industry's design, manufacturing, and construction sectors; federal government officials; and consumers. It has affiliations with governmental agencies, such as NIST and the Consumer Product Safety Commission. Its members participate in the activities of many organizations, including CABO, BOCAI, ICBO, SBCCI, NFPA, ASHRAE, HUD, NIBS, the American Institute of Architects (AIA), ANSI, the Council of State Community Affairs Agencies (COSCAA), and ASTM.

NCSBCS was established to (1) provide a forum for states to discuss and develop solutions to code regulatory problems, (2) promote adoption and administration of uniform building codes and standards that regulate construction in and between states, (3) create an effective voice for state input on the committees of nationally recognized standards-writing organizations, (4) support comprehensive training and educational programs for code enforcement personnel, and (5) foster cooperation and encourage innovation between code administration officials and the design, manufacturing, and consumer interests affected by the code regulatory system.

1.6.5.5 International Code Council

The International Code Council (ICC) was established by BOCA, ICBO, and

SBCCI to develop international codes to replace the many various codes in use today. NFPA and ICC are reported to be working on the International Fire Code to join the growing number of ICC codes, listed earlier in this section.

1.7 Barrier-Free Design

As related to building design, a *physically handicapped person* is an individual who has a physical impairment, including impaired sensory, manual, or speaking abilities, that results in a functional limitation in gaining access to and using a building or facility.

Until recently most of the buildings, facilities, transportation systems, and other structures and spaces in which we work, play, and live (the built environment) were designed for use by average, able-bodied, young adults who walk without aid or assistance (ambulatory). People who were older, younger, smaller, taller, or who had physical handicaps were at a disadvantage in most places. Many buildings were inaccessible to them, and some were even dangerous. Fortunately, this picture is rapidly changing.

Physical conditions that make a building or facility unsafe or confusing or that prevent physically handicapped people from using them are called *architectural barriers*. *Barrier-free*, or *accessible*, design eliminates or avoids creating such barriers. Barrier-free design is frequently thought of as a way to accommodate a few special people called the *handicapped*. The wheelchair symbol, used to designate public parking spaces, toilets, telephones, and water fountains, tends to foster the belief that barrier-free design offers accessibility mainly to those in wheelchairs. Actually, barrier-free design benefits everyone, because it makes facilities safer and more convenient to use.

Barrier-free design benefits not only the physically handicapped, but also (1) children who are physically and mentally immature, (2) pregnant women who have reduced agility, stamina, or balance, (3) people who care for children and must carry them, hold their hands, or maneuver their baby carriages, (4) older people who may suffer progressive degeneration of physical, perceptual, and mental faculties, (5) those disabled by size-related disorders such as giantism, dwarfism, or obesity, (6) able-bodied people who suffer temporary illness or injury, and (7) even able-bodied people who are carrying large packages.

Some people are born with a limited or unusable physical, mental, or sensory function (*disability*), but even a normally able-bodied person can become either temporarily or permanently disabled at any moment as the result of an illness or injury. *Disabilities* include visual impairment, hearing impairment, mental or perceptual impairment, confinement to a wheelchair, and coordination disability.

Nationwide, about 22% of those between the ages of 15 and 64 and about 59% of those 65 and older are disabled to some degree. That adds up to about 20% of the total population over 15 years old, or more than 37 million disabled people. Some sources put the number of disabled people at more than 43 million. It is most likely that the percentage of disabled people will rise in the future, because many people today survive illnesses and injuries that were once fatal. In fact, the average life span has increased dramatically during this century, from about 47 years in 1900 to more than 75 years today. In addition, the U.S. population as a whole is getting older. The number of people more than 65 years old is projected to grow from a little more than 10% of the total population in 1986 to more than 20% by 2040.

This section divides barrier-free design into four broad categories: (1) safety and (2) general accessibility criteria applicable to all buildings, (3) recommendations applicable to buildings for people with specific disabilities, and (4) requirements for making buildings adaptable for future use by the handicapped.

1.7.1 THE LAW AND APPLICABLE STANDARDS

In the late 1950s, the *President's Committee on Employment of the Handicapped* joined with consumer groups, such as the *National Easter Seal Society*, and standards makers to develop standards for making buildings accessible to handicapped persons. The result was ANSI A117.1, which was first published in 1961 by the American Standards Association (now the *American National Standards Institute*). ANSI A117.1 quickly became, and remains today, a major standard for design for the handicapped. It is a voluntary standard, of course, having force only when adopted by a governing body.

In 1968, Congress passed the Architectural Barriers Act, which required that access for the handicapped be achieved in accordance with standards to be established by the General Services Administration (GSA), the Department of Housing and Urban Development (HUD), and the Department of Defense (DOD). The rules they developed established ANSI A117.1 as the generic standard of access for buildings owned or leased by the federal government.

Enforcement, however, was spotty at best. As a result, Congress passed the Rehabilitation Act of 1973, which established the *Architectural and Transportation Barriers Compliance Board* (ATBCB) to enforce the Architectural Barriers Act. The ATBCB issued a document called *ATBCB Minimum Guidelines and Requirements*.

The various states followed the federal government in passing laws requiring that buildings be made accessible to the handicapped. Unfortunately, many of them did not require compliance with ANSI A117.1 or with the *ATBCB Minimum Guidelines*, but rather developed their own standards. In an attempt to gain wider acceptance, ANSI expanded and reissued A117.1, but it still failed to gain universal acceptance. To make matters even more complicated, different federal agencies required compliance with different editions of ANSI A117.1.

In 1984, in the midst of all this confusion, GSA, HUD, DOD, and the U.S. Postal Service jointly issued the *Uniform Federal Accessibility Standard* (UFAS). Its purpose was to cause all federal agencies to follow the same technical requirements for complying with the Architectural Barriers Act. It was based on the *ATBCB Minimum Guidelines* but incorporated the basic requirements of ANSI A117.1 with some changes and additions.

As a result, various agencies of the federal government were requiring compliance with four different standards, two versions of ANSI A117.1, UFAS, and the *ATBCB Minimum Guidelines.* In some states, those four were being used in addition to the state's own standards. Some states actually require compliance with more than one standard. Even at the federal level, compliance with more than one was required by some agencies.

Gradually the two major standards have become the latest edition of ANSI A117.1, which at the time of this writing was the 1998 edition, for private sector work and the UFAS for federal government work, although the others mentioned are still required by some state and local authorities. More than half of the states have adopted some version of ANSI A117.1 or incorporated it into their own standards. All model codes and the National Fire Protection Association (NFPA) NFPA 101 *Life Safety Code* incorporate or reference some edition of ANSI A117.1.

The *Federal Fair Housing Act of 1988, Title VIII of the 1968 Civil Rights Act* prohibits housing discrimination based on a physical handicap in housing projects with four or more dwelling units. Basically, it says that everything in such dwelling units must be adaptable for accessibility by the handicapped, and all public areas serving such dwelling units must be accessible. Compliance must be in accord with the *Final Fair Housing Accessibility Guidelines* published in the Federal Register on March 6, 1991.

The *Americans with Disabilities Act of 1990* prohibits discrimination on the basis of disability in places of employment, public service (transportation facilities), or in public accommodations (restaurants, hotels, theaters, doctors' offices, retail stores, museums, libraries, parks, private schools, day care centers, and others), and in telecommunications services. The standards developed by the ATBCB to implement this law differ slightly from and expand on ANSI A117.1 and the UFAS, but may ultimately incorporate one or the other as the basis of the new standard.

The recommendations in this section come from many sources, but most of them comply with the requirements of ANSI A117.1, the *Uniform Federal Accessibility Standards*, the HUD *Minimum Property Standards for Housing*, and the HUD document *Adaptable Housing*. Keep in mind that there are some differences between these standards, and that state and local requirements may contain some other differences. In addition, these standards and the *Fair Housing Accessibility Guidelines* issued in conjunction with the *Federal Fair Housing Act of 1988* are subject to modification at any time. Therefore, when dealing with an actual project, whether for a federal or state agency or for a private sector owner, it is necessary to determine the specific requirements of the project and local, state, and federal laws and ordinances that dictate the requirements for accessibility.

Rules for providing accessibility have developed slowly and will no doubt continue to expand and change as more is understood about the needs of handicapped people. Until a single standard emerges, it is a mistake to follow any one standard alone, or any edition of a standard, before verifying the actual requirements. It is necessary also to remember that both the *Federal Fair Housing Act of 1988* and the *Americans with Disabilities Act of 1990* are civil rights laws rather than code compliance laws. Accordingly, it is best to comply with their spirit and not try to circumvent their intent. In the event of a compliance dispute, a court is likely to lean in favor of the intent of the law rather than in favor of the wording of the standards with which it requires compliance.

1.7.2 SAFETY

Many accidents in buildings can be traced to obvious causes, such as slippery floors, the lack of grab bars, or inappropriate or faulty stair railings. Accident prevention requires elimination of these obvious causes as well as other measures. Floors that are likely to get wet, such as those of approach walkways, stoops, entryways, corridors, toilet rooms, bathrooms, shower rooms, locker rooms, and kitchens, should have slip-resistant surfaces. Throw rugs and area rugs should not be used in public spaces; where they are appropriate, they should have a nonslip backing.

1.7.2.1 Doors and Sidelights

Safety glazing consisting of tempered glass, laminated glass, or plastic should be used in all glazed or glazing-insert doors and sidelights. Most building codes and regulations require such glazing.

Rounded door and jamb edges or resilient door edges minimize injury to fingers.

1.7.2.2 Stairs

To help reduce the great many accidents that occur on stairs, (1) the number of risers in a series should be at least three, because people are careless on fewer, (2) treads and risers should be uniform, (3) treads should be no less than 11 in. (280 mm) wide, measured from riser to riser, and should have a slip-resistant surface, and (4) risers should be not more than 7 in. (180 mm) high and should be closed.

NOSINGS

Nosings should project not more than 1½ in. (38 mm), and their undersides should be rounded. The radius at the leading edge of the tread should not be greater than ½ in. (13 mm).

HANDRAILS

Handrails should run continuously along both sides of a stairway, extend parallel to the floor at least 12 in. (305 mm) beyond the top and bottom of the staircase, and be free of protrusions that might snag clothing. Rails should be securely mounted at a height of 30 to 34 in. (815 to 865 mm) from the floor. If children regularly use the stairs, an extra railing should be mounted at a height of 24 in. (610 mm) from the floor. The space between balusters, if present, should be not more than 5 in. (127 mm) since wider spacings may allow a child's head to become trapped.

LANDINGS

Because long flights of stairs can be tiring, a landing should be provided for flights with more than 16 risers.

TOILET ROOMS, BATHROOMS, SHOWER ROOMS, LOCKER ROOMS

Most accidents in toilet rooms, bathrooms, shower rooms, and locker rooms involve falls caused by slipping or burns caused by scalding.

Water A person can receive a third-degree burn in just 2 seconds in 150°F (60°C) water, 6 seconds in 140°F (65.6°C) water, and 30 seconds in 130°F (54.4°C) water. Many people have limited sensitivity to heat and may not be able to adequately gauge the temperature of water in a lavatory, sink, tub, or running shower. Others may not be able to react quickly enough to a sudden surge of hot water.

To prevent scalding, the temperature of hot water should be regulated by (1) setting the temperature of the heated water to below 115°F (40.6°C), (2) providing a temperature-regulating device on tub fillers and shower heads, or (3) providing a temperature-regulating device on the hot water supply lines. The use of either individual fixture-mounted temperature controls or a temperature-regulating device on supply lines is preferred because other uses, such as dishwashing and clothes washing, require temperatures from 120°F (48.9°C) to 140°F (60°C). Fittings are available that can be set to maintain a specific water temperature either at the fixture or at the supply pipe.

In addition, exposed hot water supply and drain pipes should be wrapped with insulation to prevent contact burns.

Showers and Tubs *Nonslip strips* should be provided on the bottom surfaces of tubs and showers. Soap dishes, tub fillers, and controls should be recessed so that users cannot fall on them.

Grab Bars Firmly mounted grab bars should be placed in and near tubs and showers and adjacent to water closets to help wheelchair users transfer from the chair to the fixture and to help those with other disabilities and the elderly get up and down and maintain their balance.

The width of a grab bar gripping surface should be 1¼ to 1½ in. (32 to 38 mm). The shape of the bar should allow a natural grip. There should be no protrusions or rough surfaces to catch clothing or cause injury.

Bars should be spaced 1½ in. (38 mm) from the wall, to allow proper grasping but prevent the arm from slipping behind the bar.

1.7.2.3 Kitchens

Kitchen accidents, such as burns and falls, can be reduced by proper planning.

APPLIANCES

More accidents are related to cooktops and ranges than to any other kitchen appliance. To prevent such injury, cooktop and range *controls* should be in the front so that a cook does not have to reach over hot burners to adjust them. When dwelling residents include small children, controls should be out of sight on a horizontal surface and should either have a protective lid or require an extra movement, such as pushing, before a control can be turned to activate a burner. Electric cooktops should be equipped with a light to indicate that burners are on.

CABINETS

Reaching for cabinets above refrigerators requires standing on a stool, which presents a falling hazard. Reaching for cabinets over cooktops or stoves may cause burns. Therefore, placing cabinets over refrigerators and stoves is not desirable.

1.7.2.4 Protruding Objects

Objects, such as telephones, located between 27 and 80 in. (685 and 2030 mm) above a finished floor should not project more than 4 in. (100 mm) into a room or circulation path (see Section 1.7.3.1). Objects mounted at or below 27 in. (685 mm) may project any amount because they are detectable by a moving cane. Free-standing objects mounted between 27 and 80 in. (685 mm and 2030 mm) on posts or pylons may overhang a maximum of 12 in. (305 mm). No projection should reduce the clear width of an accessible route or maneuvering space.

When objects project into circulation spaces, a change in texture or a contrast in color of the floor surface can serve as a warning. Such warning devices should, however, be used with discretion, because overuse will diminish their effectiveness.

1.7.2.5 Lighting

Sufficient glare-free lighting should be available everywhere to meet the varying intensity needs of different tasks and to eliminate dark shadows that may conceal hazards. Changes in light intensity should be gradual, so as to give the eyes time to adjust.

The paths people follow to get from one place to another in the built environment are called circulation paths (see Section 1.7.3.1). Good, glare-free lighting in circulation paths, especially those in potentially hazardous areas such as stairways, is particularly important in the creation of a safe environment.

1.7.3 GENERAL ACCESSIBILITY

An *accessible* site, building, facility, or portion thereof is one that complies with current standards and can be approached, entered, and used by physically handicapped people.

1.7.3.1 Accessible Route

A *circulation path* is an exterior or interior way of passage from one place to another for pedestrians, including, but not limited to, walks, hallways, courtyards, stairways, and stair landings.

An *accessible route* is a continuous unobstructed circulation path connecting every accessible element and space in a facility that can be negotiated by a person with a severe disability using a wheelchair and that is also safe for and usable by people with other disabilities.

1.7.3.2 Site Planning

Barrier-free design should include the planning of accessible site facilities. Individuals with disabilities should have as much freedom of movement outside as inside.

Barrier-free buildings demand barrier-free routes of access, including (1) convenient parking, (2) accessible walks and curb ramps, (3) suitable paving, and (4) logical organization.

Direct routes should be provided into a building from public transportation, parking lots, and passenger loading zones.

PARKING

Parking spaces and passenger loading zones for people with disabilities should be located as close as possible to an accessible entrance and should be identified by the *international symbol of accessibility* (Fig. 1.7-1). If an accessible entrance cannot be seen from the parking area, signs should direct users to it.

FIGURE 1.7-1 International symbol of accessibility.

Accessible parking spaces should be at least 96 in. (2440 mm) wide and have an adjacent access aisle at least 60 in. (1525 mm) wide. Two adjacent spaces may share one access aisle.

Passenger loading zones should have an *access aisle* at least 48 in. (1220 mm) wide and 20 ft. (6100 mm) long, adjacent and parallel to the vehicle pull-up space.

Curbs between access aisles and walks should have *accessible curb cuts* and *curb ramps*, as described in a following section, "Curb Ramps."

WALKS

Walks and other access paths should be at least 36 in. (915 mm) wide for single passage and 60 in. (1525 mm) wide for passing. Paths that are less than 60 in. (1525 mm) wide should have periodically spaced areas where a pedestrian or wheelchair user can move over to allow passing. The clear *passage space* should not be reduced by protruding objects.

Walks should be smooth and nonslip, with gaps of not more than ¼ in. (6 mm) and level changes of not more than ½ in. (13 mm) (see Section 1.7.3.4). Textural changes should warn users of level changes, projecting objects, and other hazards.

CURB RAMPS

Curbs define and separate walks from vehicular traffic areas. A *curb ramp* is a short ramp cutting through a curb or built into it to remove barriers for wheelchair travel. There should be a curb ramp wherever an accessible route crosses a curb and the path changes level more than ½ in. (13 mm).

Curb ramps should be at least 36 in. (915 mm) wide and should not project into vehicular traffic (Fig. 1.7-2c), unless there is a *safety zone* on the roadway. Their slope should be 1:12 or less.

If anyone must cross perpendicular to a curb ramp, the side should be flared at a maximum slope of 1:10 (see Fig. 1.7-2a) or protected by planting strips or some other non-walking surface (see Fig. 1.7-2b).

Curb ramps should be of a different texture than the surrounding passage to warn visually handicapped users of hazards. This *textural warning* area should extend for the entire width and depth of the curb ramp.

PLANTING STRIPS

Figure 1.7-3 shows typical *planting strips*. These strips are a good way to keep protruding objects and street furniture from intruding on a clear accessible path. Planting strips can be combined with benches to provide convenient resting places for the elderly or for those with limited stamina (see Fig. 1.7-3b). Hazards should be marked with *detectable warnings*, as recommended in Section 1.7.4.1.

LIGHTING

Proper illumination is just as important outdoors as it is indoors. Weatherproof fixtures should provide a minimum of 5 footcandles (53.82 lx) of light on the

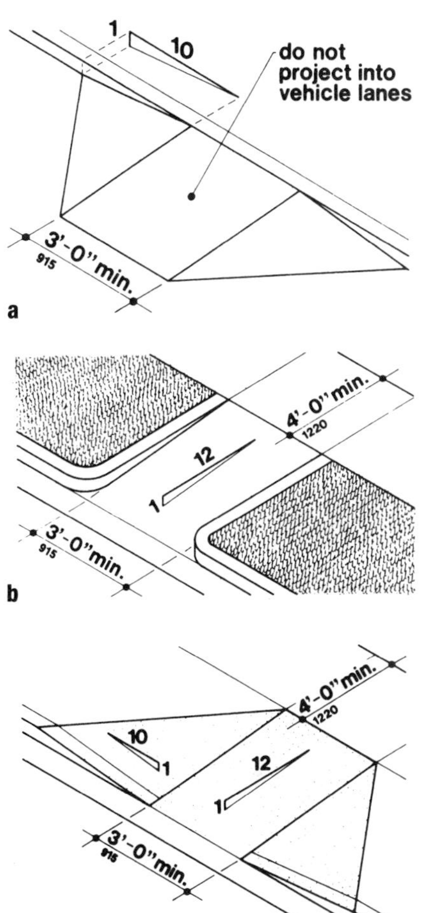

FIGURE 1.7-2 Curb ramps may either (a) start at the curb or walkway and project into the roadway or (b and c) be curb cuts.

FIGURE 1.7-3 (a) Planting strips help organize and separate street furniture from the pedestrian path. (b) Rest areas may be integrated with planting strips to provide convenient stopping places outside the circulation path.

travel surfaces at entrances, ramps, and steps. Weatherproof surface or post lights or floodlights should provide 1 footcandle (10.76 lx) and at least ½ footcandle (5.38 lx) of light along paths of travel and in parking areas.

ORGANIZATION AND ARCHITECTURAL HARMONY

Frequently used public spaces inside and outside a building should be easy to find and use. Elements that make buildings accessible, such as ramps, curb ramps, and signs, should blend harmoniously with other building elements. Careful design and site development in new construction can often reduce the need to add special elements that single out the handicapped. For example, a single-level building with appropriately sized doors and grade-level entrances would require no ramps, elevators, platform lifts, or special entrances.

Landscaping also contributes to the site (Fig. 1.7-4). Care should be taken to choose plants that do not obstruct the view or leave excessive debris in walkways. Use of poisonous and thorny plants should be avoided.

1.7.3.3 Entries and Passageways

Entries should be at grade level or made accessible by means of elevators or lifts (see Section 14.1). Doorways should have a minimum clear opening of 32 in. (815 mm) with the door open 90 degrees, measured from the face of the door to the opposite stop.

Hallways and corridors should be at least 36 in. (915 mm) wide. Where a door opens into a corridor, the corridor should be 54 in. (1370 mm) wide. Corridors where wheelchair users must pass each other or where wheelchairs must turn around should be 60 in. (1525 mm) wide.

1.7.3.4 Ramps

Passageways and walks are said to *blend to a common level* when their various portions meet in such a way that there is no abrupt rise or drop in the surface. Level changes up to ¼ in. (6 mm) high need no edge treatment; those between ¼ in. (6 mm) and ½ in. (13 mm) should be beveled with a slope no greater than 1:2. Level changes greater than ½ in. (13 mm) require a ramp in order to be accessible by wheelchair users (Fig. 1.7-5) and to provide easier passage for those who must use crutches, canes, or walkers. Ramps are the primary means by which wheelchair users enter buildings that do not have level entryways.

A *ramp* is a walking surface in an accessible space that has a *running slope* greater than 1:20. A *running slope* is the slope of a pedestrian way that is parallel to the direction of travel. Conversely, a *cross slope* is the slope of a pedestrian way that is perpendicular to the direction of travel.

FIGURE 1.7-5 Ramps can be made to blend with other building elements.

FIGURE 1.7-4 An attractively landscaped, fully accessible space provides a pleasant meeting area for everyone.

Ramp slopes should be as gradual as possible. The maximum slope should be 1:12, but more permissive ramp dimensions are sometimes allowed for existing construction where space limitations prevent the use of this preferred slope. The maximum rise for a ramp is 30 in. (760 mm); minimum clear width is 36 in. (915 mm).

Level *landings* at least as wide as the ramp itself and at least 60 in. (1525 mm) long should be provided at the top and bottom of each ramp. Landings are also necessary at intermediate levels where the *rise* of a ramp exceeds 30 in. (760 mm). If a ramp changes direction, the landing should be a minimum of 60 by 60 in. (1525 by 1525 mm). The width of the platform at a *switch-back ramp* should be at least as wide as the width of the two ramp portions plus the width of the space between them (Fig. 1.7-6).

The surfaces of ramps and landings should be *slip resistant* and should not collect water. Their edges should have curbs, walls, railings, or other protection to prevent people from falling.

Handrails should be provided on both sides of ramps that rise more than 6 in. (150 mm) or are longer than 72 in. (1830 mm). Handrails should (1) extend at least 12 in. (305 mm) beyond the top and bottom of a ramp, (2) be parallel with the ramp, (3) have a clear space of a 1½ in. (38 mm) between the handrail and any adjacent wall, and (4) be 30 to 34 in. (760 to 865 mm) above the ramp surfaces. Inside handrails on switch-back or *dogleg ramps* should be continuous.

Because many handicapped people find it easier to negotiate stairs than ramps, both should be provided in buildings with an above-ground-level entry.

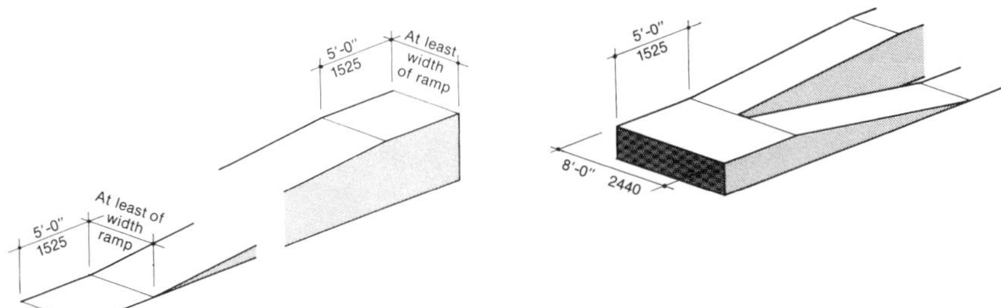

FIGURE 1.7-6 Layout and sizes of flat platforms at ramps. For clarity, railings are not shown.

1.7.3.5 Stairs

Stairs should comply with the requirements stated in Section 1.7.2. In addition, there should be *detectable warnings* (see "Texture" in Section 1.7.4.1) at least 36 in. (915 mm) wide at the top of every stair run. Stairs should have *closed risers* of uniform height, treads should be slip resistant, and nosings should not be square or abrupt.

Steps and landings should have sharp color contrasts to help people with limited sensitivity to light and dark and limited depth perception. It is also helpful if the edge of a stair contrasts with the rest of the tread.

1.7.3.6 Signage

Signage is defined as verbal, symbolic, and pictorial information presented in a graphic, two-dimensional format.

Informational and warning signs should be easy to find and to read; that is, (1) they should be positioned as close as possible and perpendicular to the path of travel, (2) the information on them should contrast with the background, and (3) they should be made of glare-free materials.

Signage should also have a consistent format or use *international symbols*.

LOCATION AND VIEWING DISTANCE

Most people can see within an angle of 30 degrees to either side of the centerline of their faces without moving their heads. People with disabilities often have limited head movement or reduced peripheral vision. Therefore, signs should be positioned as close as possible and perpendicular to the path of travel. Signs close enough to touch should be positioned for easy hand detection by a standing person, 54 to 60 in. (1372 to 1524 mm) from the floor.

The distance from which a sign will be read determines the size of the letters, numbers, and symbols used on it.

DESIGN AND PROPORTION OF LETTERS, NUMBERS, AND SYMBOLS

Letters and numbers should be of sans serif (block) type and should be sized and proportioned as recommended in Figure 1.7-7.

Characters on signs that must be read from a distance should be large enough to be read easily from the maximum viewing distance expected. Characters on signs close enough to touch should be raised at least 1⁄32 in. (0.8 mm) from their background. They should also be at least 5⁄8 in. (16 mm) tall, but no taller than 2 in. (50 mm), and have a stroke width of at least 1⁄4 in. (6 mm) in order to be easily detectable by one finger (Fig. 1.7-8).

Braille is a special raised touch alphabet for the blind that uses a cell of six dots. Unfortunately, only about 10% of those with visual impairments can read braille, but raised Arabic symbols can be read by anyone, regardless of visual handicap. To aid those who can read it, braille characters in the standard dot size and spacing may be added to the left of standard characters.

Borders, which may confuse a reader

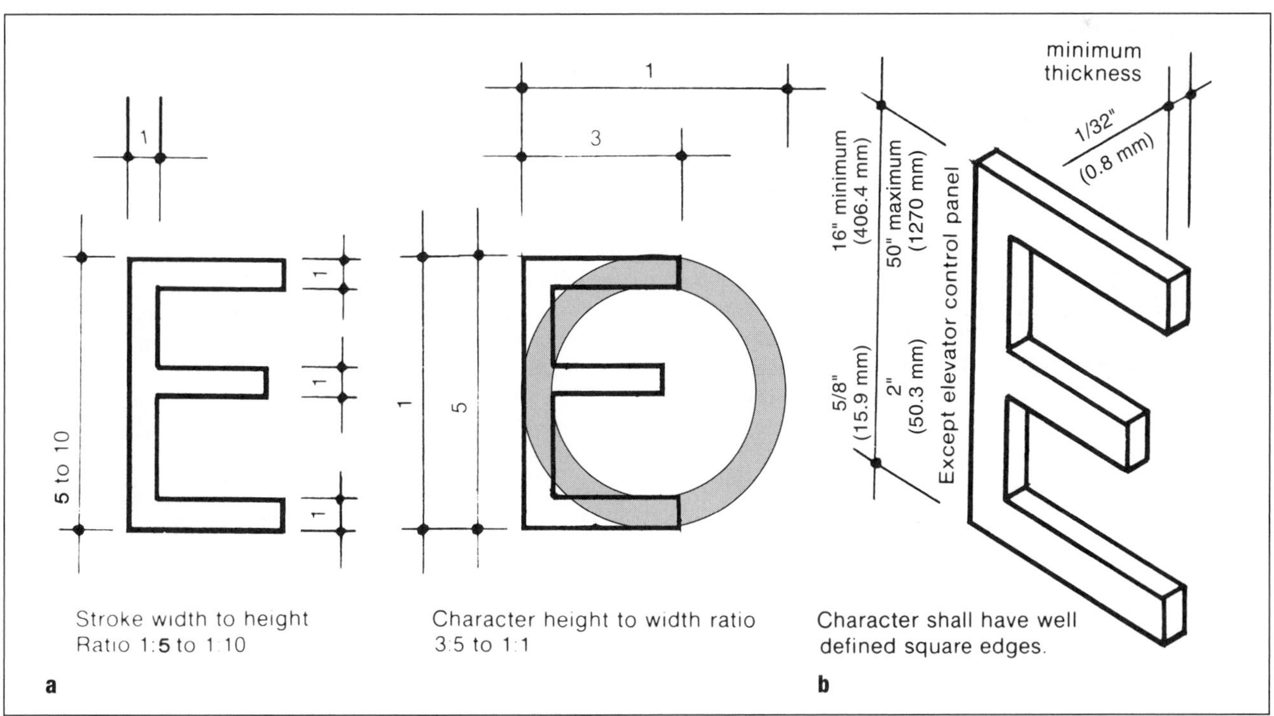

FIGURE 1.7-7 (a) Proportions of letters expressed as a ratio; (b) dimensions of letters for signs close enough to be touched.

FIGURE 1.7-8 A sign with raised lettering can be read by both those with normal sight and those with visual impairments.

with a visual impairment, are not recommended on signs.

COLOR CONTRAST

The background and characters on a sign should contrast. Dark characters may be used on a light background, but light characters on a dark background are more easily readable. Surfaces should be glare free.

1.7.3.7 Surfaces

Floor surfaces should be nonslip. When practicable, they should be carpeted to add the advantage of noise reduction, which helps those who must use hearing aids.

Carpet should (1) be securely attached, (2) have either a firm backing or no backing, and (3) have a pile height not in excess of ½ in. (13 mm) so they remain usable by those in wheelchairs. Shag carpets should be avoided. Carpets should be of *antistatic* materials or treated to be static free, because static electricity interferes with hearing aid operation. Humidifiers may be used to help control static electricity. Exposed carpet edges should have carpet trim along the entire carpet length.

1.7.3.8 Electrical

Electrical outlets should be centered at least 15 in. (380 mm), and preferably 24 in. (610 mm), above the floor.

The best height for wall switches is centered 48 in. (1220 mm) above the floor, but they should never be placed higher than 54 in. (1370 mm) above the floor. Where persons in wheelchairs will approach from a forward position, as is the case when switches are located over counters with knee spaces, switches should be centered between 36 and 40 in. (915 and 1016 mm) above the floor. In no case should switches and outlets over counters be placed more than 48 in. (1220 mm) above the floor.

Lighting should be as described in Sections 1.7.2 and 1.7.3.2.

1.7.4 SPECIFIC HANDICAPS

The kinds of specific handicaps discussed in this section include (1) *visual impairments*, (2) *hearing impairments*, (3) *mental* and *perceptual impairments*, (4) *disabilities* that require the use of a wheelchair, and (5) *coordination disabilities*.

Designs that *remove barriers* for people with one kind of disability sometimes *create barriers* for those with a different set of disabilities. For example, curb ramps (see Fig. 1.7-2) installed to permit easy street crossings for wheelchair users can cause a person with a visual impairment to wander unknowingly into automobile traffic. Both environmental needs can be met by providing a textured pavement that will act as a warning to those with visual impairments but will not impede wheelchair travel.

The recommendations in Sections 1.7.2 and 1.7.3 apply to spaces for those with specific handicaps, but the more demanding provisions discussed in this section are also necessary to accommodate specific disabilities.

1.7.4.1 Visual Impairment

Only about 10% of those with visual disabilities are totally blind. Legal blindness is usually defined as vision at or below 20/200. Some of those with visual disabilities have impaired peripheral vision; others cannot distinguish light and dark; some cannot perceive the full spectrum of colors.

Many people who have a visual impairment rely on memory instead of their eyes. They use a long cane, a guide dog, or other devices such as an *infrared detector* as aids for traveling to specific destinations. These destinations have often been explored with the help of someone with sight. Dwelling units may need little adaptation, because those with visual impairments become familiar with their surroundings.

Persons with visual impairments learn to rely to a certain extent on *tactile response* (touch). A *tactile object* is one that can be perceived using the sense of touch.

SIGNAGE

The proper type of signage is a very important means of communication for visually handicapped people. This signage should follow the rules set down in Section 1.7.3.6.

TEXTURE

A *cue* is a device that alerts a handicapped person to an upcoming condition. Cues may be *audible*, *visual*, or *textural*. A cue that consists of a standardized surface texture applied to or built into a walking surface or other element, to warn visually impaired people of a hazard in the path of travel, is called a *detectable warning*.

Detectable warnings help those with visual impairments to avoid hazards such as automobile traffic, level changes, and physical obstacles. Detectable warnings on walking surfaces include (1) exposed aggregate concrete, (2) rubber or plastic cushioned surfaces, (3) raised strips, and (4) grooves. Textures used for warning should contrast with surrounding surfaces but should be uniform in design within any building or site. Grooves should be used indoors only. Handles on doors leading to hazardous areas (such as spaces housing mechanical equipment) should be knurled or roughened as a warning.

OBSTACLES

Benches and other obstacles in the direct path of travel should be surrounded by detectable warnings to aid those with visual impairments, unless the obstacles are encircled by other indicators such as curbs (Fig. 1.7-9).

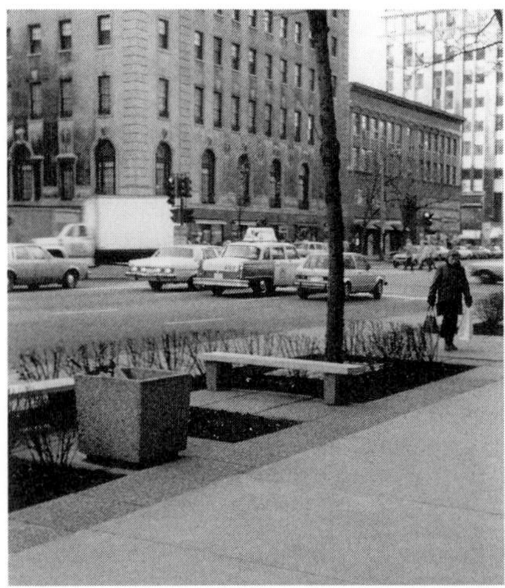

FIGURE 1.7-9 The textured surface around the planter and the bench serves as a warning to those with visual impairments.

PROJECTIONS

Many people with severe visual impairment use white canes to aid their mobility. However, the cane technique is useful only if hazards are detectable. Refer to Section 1.7.2 for requirements related to projecting objects.

SOUND

An *audible cue* is a sound or verbal alert. These are other means of helping those with visual impairments. *Audible alarms* should be incorporated into emergency warning systems. Alarm signals should be at least 15 decibels (dB) louder than the room's normal sound level, but should not exceed 120 dB.

1.7.4.2 Hearing Impairment

There are more than 13,000,000 people in the United States with partial or total hearing loss. Just as sound compensates those with visual impairments for the loss of sight, *visual cues* help those with hearing impairments to make up for their loss.

WARNING AND INDICATION LIGHTS

Emergency signals, such as fire and burglar alarms, should have *visual* as well as *audible warning systems*. These should be located so that the signal or its reflection is clearly visible and should flash with a frequency of not more than 5 cycles per second (Hz).

Elevator directional lights should be at least 2½ in. (63 mm) in the smallest dimension and should be mounted at least 72 in. (1830 mm) above the floor. Refer to Section 14.1 for additional requirements.

SOUND

Control of background noises and amplification of sound can help those with partial hearing loss to communicate.

For hearing aids to function properly, reduction of background noise, including reverberation, and control of sound frequencies are important. Use of ultrahigh-frequency sound security systems and low-cycle electric transformers should be avoided. Further information may be found in Section 13.1 and in Chapter 16.

Telephones for those with hearing impairments should be fitted with adjustable volume controls and other devices. Such telephones in public areas should be labeled as shown in Figure 1.7-10.

FIGURE 1.7-10 Sign identifying a special phone for those with hearing impairments.

1.7.4.3 Mental and Perceptual Impairment

Clear signage and good organization are especially important for those who are *mentally immature* or who have *perceptual* or *cognitive disabilities*.

If people must frequently ask for the location of a commonly used facility such as a washroom, that environment may have a barrier. Many individuals with disabilities are self-conscious and will not ask for help; if they feel uncomfortable in a facility, they often will not use it. Public facilities should be easy to locate and to utilize.

SIGNAGE

It is especially important to those with mental impairments that signs be easy to locate and to read. The basic requirements for signs are discussed in Section 1.7.3.6.

Many individuals with mental impairments who are unable to read learn to recognize information by word length. For example, *women* is a longer word than *men*. Use of the words *ladies* and *gentlemen* may confuse those familiar with the three- and five-letter words.

ORGANIZATION

The location of specific rooms within an area should be clearly identified. If corridors in different areas are similar in appearance, an effort should be made to create *landmarks* using color or other decoration. Stairways and other circulation areas should be clearly identified or in highly conspicuous locations.

1.7.4.4 Wheelchair Users

The abilities of wheelchair users differ widely. Disability may affect only the legs or may involve the entire body. Pressure sores, muscle atrophy, and blood pooling are frequent problems. Paralysis reduces the muscle tone of the affected area. Barrier-free environments encourage mobility, and the exercise gained helps prevent additional medical problems.

WHEELCHAIR ACCESS

Most adult-sized wheelchairs are of a standard size; some special wheelchairs may be larger. Wheelchairs need an accessible route from parking spaces or bus stops to the interior of a building and within the building to the final destination of the wheelchair user.

Site Access Refer to Sections 1.7.3.2 and 1.7.3.4 for requirements.

Elevators and Lifts Unless every space is accessible by level movement, or ramps or other vertical means of access are provided, elevators or platform lifts should be provided for wheelchair users. In some applications, *chair-type stair lifts* can be provided to assist persons who cannot readily climb stairs, when the stairs cannot be eliminated and elevators or platform lifts cannot be provided.

Entries and Passageways Refer to Sections 1.7.2 and 1.7.3.3 for requirements for doors, hardware, and thresholds.

Figure 1.7-11 shows the recommended minimum approach spaces necessary for convenient maneuvering of wheelchairs at doorways.

Wheelchair users frequently use the wheelchair itself to push a door open, scuffing or scratching the door. Therefore, the bottom 10 in. (255 mm) of hinged accessible doors should have a smooth, hard-surface *kickplate* (Fig. 1.7-12).

ROOM SIZE AND ARRANGEMENT

Accessible spaces for wheelchair users should be on one level and have adequate circulation and maneuvering space. Windows and doors should be located so that furniture arrangements do not obstruct wheelchair circulation.

The minimum clear floor area that will accommodate a wheelchair is 30 by 48 in. (760 by 1220 mm). The minimum diameter needed for a wheelchair user to execute a 180 degree turn is 60 in. (1525 mm).

Figure 1.7-13 shows examples of accessible bedroom plans. In an accessible dwelling, at least two bedrooms should be on the main level.

TOILET ROOMS, SHOWER ROOMS, AND BATHROOMS

Wheelchair users should be able to enter and use toilet, shower, and bathroom facilities without squeezing past obstructing fixtures or doorways.

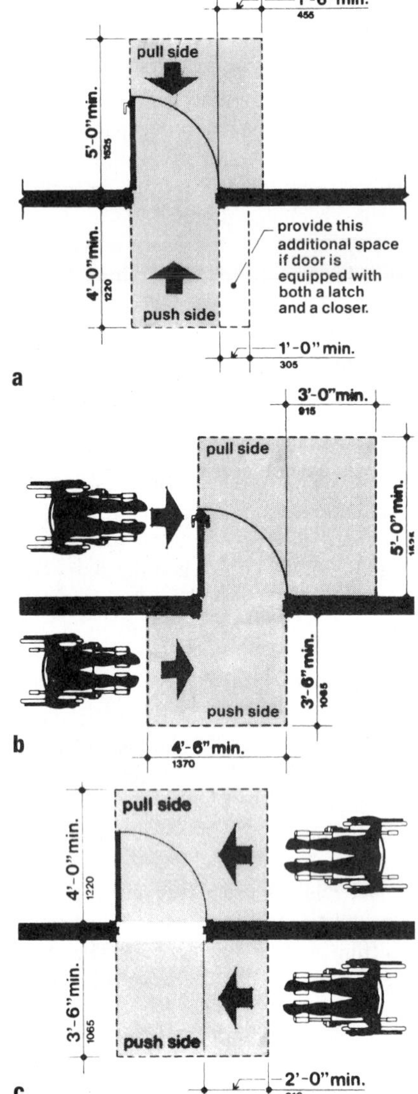

FIGURE 1.7-11 Minimum clear floor space at doors: (a) front approach, (b) hinge side approach, and (c) latch side approach.

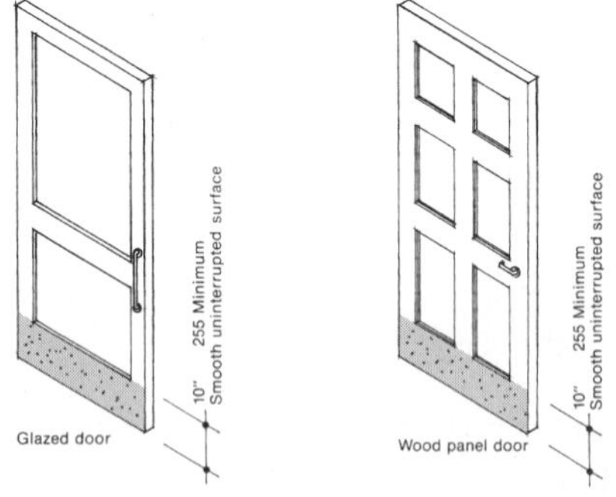

FIGURE 1.7-12 An abrasion-resistant kickplate of stainless steel or hard plastic should protect the lower 10 in. (254 mm) of doors intended for wheelchair users.

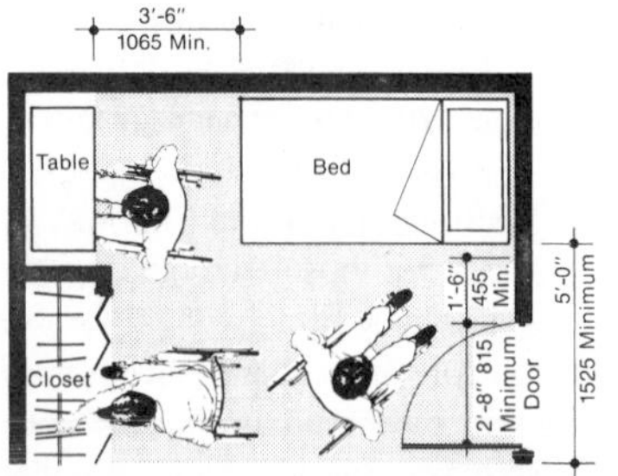

Single bed

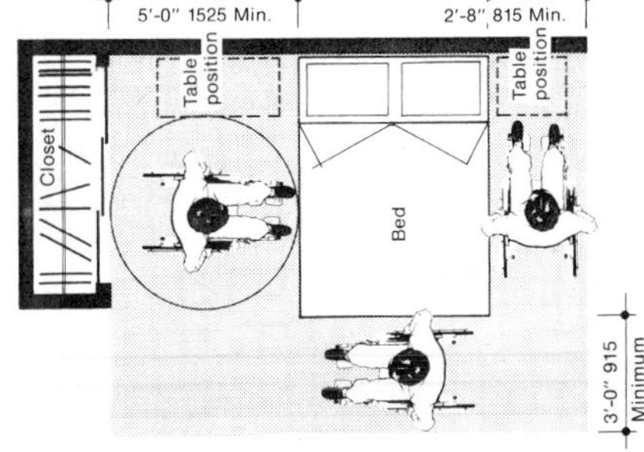

Double bed

FIGURE 1.7-13 Minimum dimensions for accessible bedrooms.

Clear Floor Space Accessible toilet rooms, shower rooms, and bathrooms should have the recommended clearances around plumbing fixtures and controls. Ideally, an accessible facility should have a clear floor area 60 in. (1525 mm) in diameter so that a wheelchair user can make a 180 degree turn. The clear floor spaces of the fixtures, controls, and turning space may overlap.

Doors Doors should have a minimum clear opening of 32 in. (815 mm). Doors should not swing into the clear floor area of a fixture or obstruct entrance to the room. If a door must swing out, it should not obstruct passage through a corridor or any other circulation space.

Water Closets Water closets in an accessible public toilet room should be 17 to 19 in. (430 to 484 mm) high measured to the top of the toilet seat. In a dwelling, the seat height should be at least 15 in. (380 mm). The bowl should be centered at least 18 in. (455 mm) from each side wall.

Lavatories Lavatories should be mounted so as to provide at least 29 in. (735 mm) of knee space to the underside of the apron. Exposed water pipes that intrude into the knee space should be padded and insulated.

Mirrors Mirrors should be mounted with the bottom edge not higher than 40 in. (1015 mm) from the floor. Medicine cabinets should be mounted with a usable shelf not higher than 44 in. (1120 mm) above the floor.

Bathtubs and Showers Depending on the location of the bathroom door, a bathtub should have a clear floor space 30 to 48 in. (760 to 1220 mm) wide, and 60 to 72 in. (1525 to 1905 mm) long. Showers should have a clear floor space at least 36 in. (915 mm) wide and 48 in. (1220 mm) long for *transfer stalls* and 60 in. (1525 mm) long for *roll-in stalls*.

Seats should be provided in both transfer shower stalls and tubs. In shower stalls the seat should extend the full depth of the stall on the wall opposite the controls.

A shower spray unit that can be used either as a fixed shower head or a handheld shower should be provided. The hose should be at least 60 in. (1525 mm) long.

Curbs in shower stalls 36 in. (915 mm) square should be not higher than 4 in. (100 mm). Stalls, such as those 30 by 60 in. (760 by 1525 mm), that allow a wheelchair to be rolled in should not have curbs.

Enclosures on bathtubs and showers should not obstruct the use of controls or the transfer from wheelchairs onto seats.

Faucets The controls on bathtubs, showers, and lavatories should be within comfortable reach of a wheelchair user. Controls such as handle and drain mechanisms should be operable with one hand and not require tight grasping or twisting.

STORAGE

A clear floor space at least 30 by 48 in. (760 by 1220 mm) should be provided near storage areas. Closet rods should be mounted not higher than 54 in. (1370 mm) above the floor.

KITCHENS

Suitable kitchen appliances, proper countertop heights, and adequate circulation space are the primary requisites of a wheelchair user's kitchen.

Clear Floor Space The minimum clear floor space between opposing base cabinets, countertops, appliances, and walls should be 40 in. (1015 mm), unless the kitchen is U-shaped. U-shaped kitchens should have 60 in. (1525 mm) clear floor

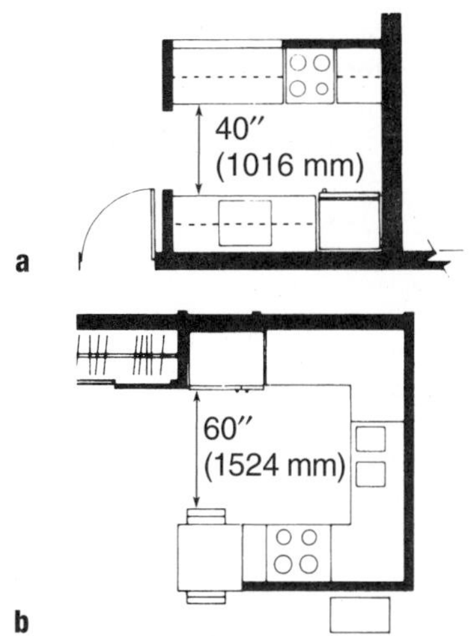

FIGURE 1.7-14 Minimum clearances between opposing cabinets in (a) parallel and (b) U-shaped kitchens.

space between opposing cabinets (Fig. 1.7-14). Clear floor space of 30 by 48 in. (760 by 1220 mm) should be provided at appliances such as a cooktop, oven, refrigerator, and dishwasher.

Controls Knobs, faucets, and appliance controls must be within comfortable reach and operable with one hand.

Counters One 30-in. (760 mm) section of counter, intended for food preparation, should be adjustable to a range of heights, including at least three positions: 28, 32, and 36 in. (710, 815, and 915 mm).

There should be no cabinets under the adjustable counter, and it should have a clear knee space 30 in. (760 mm) wide by 19 in. (485 mm) deep over a finished floor space. Counter thickness and supports should be 2 in. (50 mm) maximum over the necessary clear space. The underside of the counter should not have any sharp protrusions or rough surfaces.

Appliances Refer to Section 11.1.2 for accessibility requirements for kitchen appliances.

Cabinets Both base and wall cabinets should be reachable from a wheelchair. At least one continuous shelf in the cabinets above a work counter should not be higher than 48 in. (1220 mm). Upper cabinet doors should have pulls located as close to the bottom as possible; base cabinets should have pulls located as close to the top as possible. Because of the necessity for knee space, cabinets should not be built under counters. Instead, pantries should be used for storage. They should have storage below the 54-in. (1370 mm) height and a clear space of 30 by 48 in. (760 by 1220 mm).

Pull-out shelves in base cabinets, and lazy Susan rotating corner units are highly recommended.

Sinks Sink heights should be adjustable to 28, 32, and 36 in. (710, 815, and 915 mm). The minimum width of a counter surrounding a sink should be 30 in. (760 mm). Plumbing should be roughed in for sinks mounted at the 28-in. (710 mm) height. Counter thickness and supports should be 2 in. (50 mm) maximum over the necessary clear space. A sink bowl should be not more than 6½ in. (165 mm) deep; in a double-bowl sink only one of the bowls needs to be so shallow.

Faucets should be at the side of the sink and operated by a *single lever control.*

ELECTRICAL

The location of electrical outlets and switches should be as described in Section 1.7.3.8. Lighting requirements are in Section 1.7.2.

1.7.4.5 Coordination Disabilities

Coordination disabilities include balance, leg, upper limb, and flexibility impairments. Many people have problems with balance, stamina, and coordination, which do not confine them to a wheelchair but do limit their use of an environment. They may have difficulty walking and may need aids such as walkers or crutches, which require approximately the same amount of space as wheelchairs. Others may have difficulty bending, sitting, kneeling, or rising; still others may find it difficult to push open doors or turn knobs and faucets.

The needs of those with coordination disabilities and the needs of the elderly frequently overlap. However, younger people with disabilities do not want to be associated or housed solely with the elderly.

ENTRY AND CIRCULATION

People who use walking aids need the same amount of *circulation space* as those who use wheelchairs. Many people with coordination disabilities prefer stairs to ramps because ramps are more tiring to use. However, ramps may be beneficial to crutch and walker users who cannot negotiate stairs as easily. Ramps should be constructed according to the criteria established in Section 1.7.3.4. Wherever possible, both stairs and ramps should be provided.

SURFACES

Unevenness, raised joints, and debris can be hazardous to those with walking disabilities. Surfaces should be kept free of debris; joints and level changes should be less than ¼ in. (6 mm).

1.7.4.6 The Elderly

Accessibility recommendations for the elderly overlap those for wheelchair users and persons with coordination disabilities. Sensory losses accelerate at various ages. For example, hearing starts to diminish at 40; vision, at 50; taste, between 55 and 60; smell, after 70. The actual ages and degree of loss vary considerably.

The elderly have a higher rate of visual impairment than those in other age groups. Age impedes the eyes' ability to change focus, adjust to sudden changes in light intensity, and function at low levels of light intensity. For example, an 80-year-old may need three times as much light for comfortable reading as a 16-year-old. As it ages, the lens of the eye becomes increasingly rigid, yellow, and opaque. This process affects the ability to discern color, especially blues and greens, and to distinguish between colors such as pink and lavender. With aging, depth perception decreases and glare becomes a greater problem.

Aging also increases hearing loss, reducing the ability to hear sounds in general and high frequency sounds in particular.

The cushioning of the foot degenerates with advancing age. This loss can cause pain and contributes to the shuffling gait characteristic of many elderly people (Fig. 1.7-15).

The mobility of the elderly varies from complete ambulatory freedom to wheelchair confinement. Entries and other access facilities should be barrier-free for elderly persons, including wheelchair users. Only those criteria that add to or differ from those established for wheelchair users will be mentioned in this section.

SIGNAGE

Many elderly people suffer visual impairment, and others may be easily confused; signage and lighting should conform to the recommendations outlined in Sections 1.7.3.6 and 1.7.4.1.

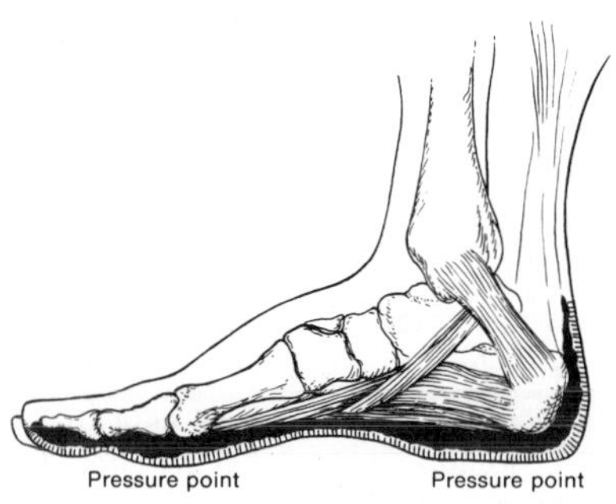

a

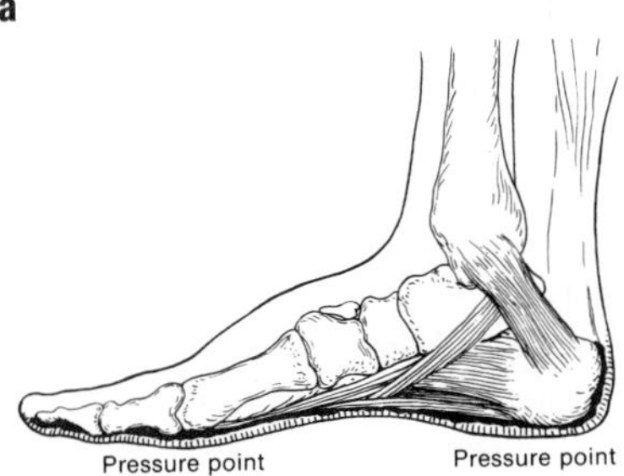

b

FIGURE 1.7-15 The foot of a younger person (a) contains more cushioning material at pressure points than that of an older person (b).

ROOM SIZE AND ARRANGEMENT

Recommended room sizes for the elderly are greater than the minimums specified for wheelchair users. This is partially because the elderly sometimes have coordination problems that make it more difficult for them to maneuver a wheelchair within the minimum spaces recommended for other wheelchair users.

The criteria for corridor widths and floor surfaces should follow those indicated in Sections 1.7.3.3 and 1.7.3.7, respectively.

COLOR AND TONE

As mentioned earlier, changes in the eye lens caused by aging often affect color perception. It may be difficult for some elderly people to distinguish between the light tints of pastel colors such as blue, lavender, and pink, and sometimes even between dark colors such as navy, brown, and charcoal. Therefore, the selection of colors for elderly persons should emphasize objects against their background. Examples include light entry–dark door jamb; light floor–dark furniture.

ELECTRICAL

Electrical outlet and switch locations and lighting requirements should be as described in Section 1.7.3.8.

BATHROOMS

Because the elderly may have to use a wheelchair at times, bathroom requirements should be the same as for wheelchair users.

KITCHENS

The requirements for kitchens used by the elderly vary. Some of the elderly are reasonably able-bodied; others use wheelchairs. Kitchens for the elderly generally should follow the recommendations for adaptable kitchens (see Section 1.7.5), thus permitting ready alterations to suit the specific needs of the occupants.

Kitchen storage should be accessible without endangering the safety of the elderly. Because not all of the elderly are wheelchair users, the use of wall cabinets is acceptable. However, these are not recommended above a refrigerator or cooktop. Pantries are an excellent means of providing safe and accessible storage. Base cabinets should have *pull-out* shelves and corner lazy Susans to make items in them more accessible.

1.7.5 ADAPTABILITY

Most people with disabilities resent being segregated into housing built specifically for the handicapped. They would prefer mainstreaming, a process which (1) equips people who have disabilities with the personal devices and adaptive skills needed to function effectively in the built environment and (2) removes physical barriers that prevent handicapped people from functioning like able-bodied individuals.

Current laws and regulations require that most types of buildings be made fully accessible to handicapped people. Other laws and regulations have been proposed that would mandate making new housing also fully accessible. When they take effect, the following discussion may become moot. Until then, however, one way to prevent the segregation of people who become disabled into housing prepared specifically for handicapped people is to build adaptable buildings. An *adaptable* building is one that can be modified to satisfy the needs of an occupant's physical limitations with ease and without excessive costs. The kinds of elements that may be modified or added in an adaptable house, for example, include, but are not limited, to kitchen counters, sinks, and grab bars.

Adaptability for wheelchair users requires the most modification and should therefore receive the greatest attention in the design process.

The recommendations in Sections 1.7.2 and 1.7.3 apply fully to adaptable buildings. In general, adaptability requires that the following design requirements be followed.

ENTRIES

When practicable, entrances should be at grade level, so that ramps or stairs will not be needed. When this is not practicable, an adaptable dwelling can be on a level accessible by a ramp or an elevator or platform lift.

DOORS

Passage doors should have a minimum clear opening of 32 in. (815 mm) with the door open 90 degrees, measured from the face of the door to the opposite stop.

Hardware Door knobs, handles, pulls, latches, and locks should have a shape that is easy to hold with one hand and does not require tight grasping, pinching, or twisting. Extensions are available that adapt a regular door knob for people with manual disabilities such as those caused by arthritis.

Thresholds Thresholds on exterior sliding glass doors should not be higher than ¾ in. (18 mm). A recessed track is preferred where suitable, such as for interior sliding doors. Thresholds on hinged passage doors should not exceed ½ in. (13 mm) in height and should be beveled. Exterior doors should have weather stripping applied to the bottom edge of the door, because weather stripping applied to thresholds can present a tripping hazard (Fig. 1.7-16).

Room Size and Arrangement A single-level dwelling is the preferred type for

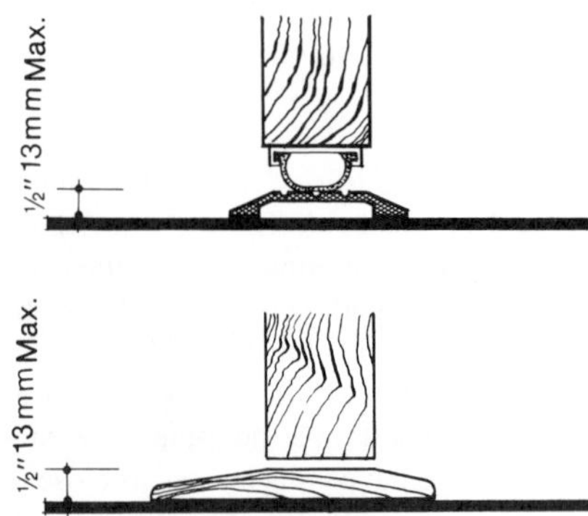

FIGURE 1.7-16 Thresholds.

adaptability. Rooms should be large enough to permit convenient furniture arrangement and wheelchair maneuvering. Most living, dining, and bedrooms already meet the requirements for adaptability. Bathrooms and kitchens require the most modification.

CORRIDORS

Hallways and corridors should be a minimum of 36 in. (915 mm) wide. If a door opens into a corridor, the passage width should be 54 in. (1370 mm); corridors where people must pass each other should be 60 in. (1525 mm) wide.

BATHROOMS

An adaptable bathroom has the appearance of a conventional bathroom but with the layout and clearances suitable for later adaptation to the needs of those with disabilities.

Clear Floor Space A typical bathroom should have a minimum width of 60 in. (1525 mm). Clear floor spaces around fixtures should be as shown in Figure 1.7-17.

Doors Bathroom doors should have a minimum clear opening of 32 in. (815 mm); doors should not swing into the clear floor space around fixtures or obstruct entrance into the bathroom. Bathroom doors can be rehung to open out in order to gain the necessary clear door space, but should not obstruct corridors or other circulation space outside the bathroom.

Lavatories Lavatories should be mounted so as to provide at least 29 in. (735 mm) clearance from the floor to the underside of the apron. The clear floor space should be at least 30 in. (760 mm) wide by 48 in. (1220 mm) deep; the lavatory basin should take no more than 19 in. (485 mm) of that depth, and pipes should not protrude into the knee space.

Water Closets Water closet seats should be at least 15 in. (380 mm) above the floor. Clear floor space and structural wall reinforcement for grab bars should be provided.

Mirrors The bottom edge of mirrors should be mounted not higher than 40 in. (1015 mm) above the floor. Medicine cabinets should be mounted with at least one usable shelf not higher than 44 in. (1120 mm) above the floor.

Bathtubs and Showers Depending on the location of the bathroom door, bathtubs and showers should have adequate clear floor space to accommodate wheelchairs.

Structural wall support and floor space should be provided so that grab bars and a seat can be readily installed for a resident with a handicap. A flexible-hose shower unit that can be used as either a fixed or hand-held head should be provided. Shower stalls 36 by 36 in. (915 by 915 mm) should have curbs not higher than 4 in. (100 mm). Stalls 30 by 60 in. (760 by 1525 mm) should not have curbs, so that a wheelchair may be rolled in; the stall floor should be pitched slightly toward the drain to help avoid drainage problems.

Grab Bars Structural reinforcement or other provisions should be provided near bathroom fixtures in the locations where grab bars would be installed.

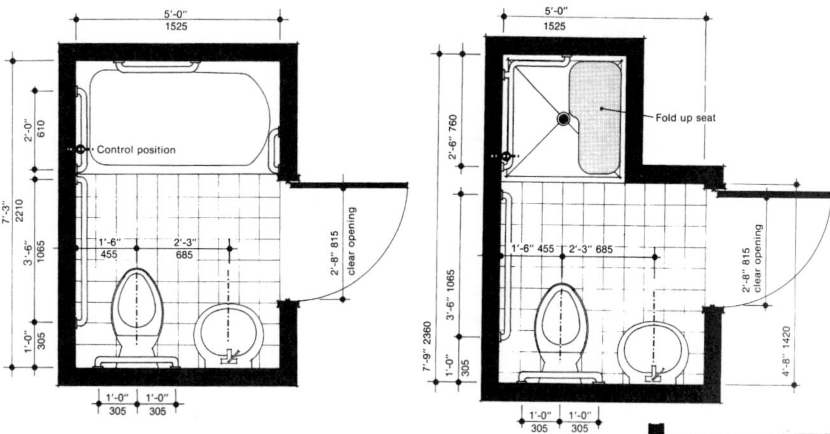

FIGURE 1.7-17 Recommended minimum dimensions for accessible bathrooms.

KITCHENS

Adjustable-height counters, sinks, and cooktops, and adequate circulation space are the key elements of an adaptable kitchen that allow a wheelchair user to freely maneuver within the space.

Clear Floor Space Clear floor space requirements are the same as those for wheelchair users (see Section 1.7.4.4).

Counters One section of counter 30 in. (760 mm) wide should be capable of being repositioned to heights of 28, 32, and 36 in. (710, 815, and 915 mm) to provide a comfortable area for food preparation. Cabinets may be mounted under the adjustable counter, but should be readily removable in order to provide the necessary knee space for a wheelchair user.

Appliances Appliances in adaptable kitchens should fulfill the accessibility requirements indicated in Section 11.1.

Cabinets Wall cabinets should have pulls located as close to the bottom as possible; base cabinets should have pulls located as close to the top as possible.

Cabinet pulls should be smooth and rounded, without any sharp edges or pointed projections. Because storage under adjustable spaces may have to be removed, pantry storage is very desirable. Other helpful devices are lazy Susans for corner cabinets, pull-out shelves in base cabinets, and cutting boards with mixing bowl cutouts.

Sinks The requirements for sinks are the same as those outlined in Section 1.7.4.4, except that cabinets may be mounted under the sink; however, they should be removable in order to provide the necessary knee space for a wheelchair user.

FLOORS

Flooring should be *nonslip* even when wet.

ELECTRICAL

Lighting should be as described in Section 1.7.2. Electrical outlet and switch locations should be as indicated in Section 1.7.3.8.

COST

When practicable, new dwellings should be made adaptable, because the cost of making an adaptable dwelling suit the needs of a wheelchair user is small in comparison with the cost of retrofitting an existing conventional house. In addition, making a house adaptable does not in any way harm its use by the able-bodied.

1.8 The Metric System of Measurement

Metrication is a term coined in Britain to describe *metric conversion*, which is the process of changing from the *customary* to the *metric* system of measurement, including the planning and coordination necessary for the change.

The *customary* system of measurement referred to here is the collection of English and other nonmetric units currently used in the United States. The *metric* system is a system originally developed in France that has been adopted by more than 90% of the nations of the world. It is based on the *metre* and six other standardized *base units*. *Metric units* already in general use include, but are not limited to, the *second*, *ampere*, *candela*, *watt*, *volt*, *ohm*, *farad*, *coulomb*, and *lumen*.

Much of the material in this section is based on two current metric standards:

1. IEEE/ASTM SI-10, which is a replacement for American Society for Testing and Materials (ASTM) E 380-89a, "Standard for Metric Practices."
2. ASTM E 621, originally a supplement to ASTM E 380, was based on the now withdrawn National Bureau of Standards (NBS) (now National Institute of Standards and Technology [NIST]) publication 938, "Recommended Practice for the Use of Metric (SI) Units in Building Design and Construction." NBS 938 was authored by world-renowned *metrication* expert Hans J. Milton while he was on loan to the NBS by the Australian government.

This section describes the history of measurement and the metric system, the properties of the *International System of Units* (SI) version of the metric system, and its applications and proper usage. It also includes tables that contain:

- Lists of units in the SI system;
- Fraction, decimal, and *metric* equivalents tables;
- Conversion tables for selected *customary* and SI measurement units. One of these contains conversions for English units that are not converted in the tables in this book.

This section also includes text and tables showing rules and recommendations for:

- Presentation of SI units and symbols;
- Presentation of numerical values in conjunction with SI;
- Application for SI quantities and associated names and symbols.

If the events in the rest of the world can be used as a guide, conversion to the metric system of measure in this country will probably change the entire measurement base of the construction industry, including guidelines, codes, standards, drawings, specifications, and associated documents. The most extensive use of metric measure in the building industry will be for linear measurement, area, volume, and mass.

Metrication disturbs some people. Manufacturers worry about the cost of retooling and introducing new product sizes. People in general worry about learning many new names and numbers.

As its critics point out, metrication will force a complete revision in the accepted units of measurement and calculation, but at the same time it will offer a logical opportunity to revise standard product sizes, performance criteria, and methods of testing. This in turn will reduce inventory, rationalize criteria, and simplify methods of testing and reporting test results.

Experience in other countries has shown that metrication takes less time and costs less than estimated. The cost of metrication occurs once; the benefits last forever. Approximately one-third of the contracts awarded to the top U.S. construction firms have been for work outside the United States. American know-how is one of the exports not usually figured into foreign trade discussions. The change to the metric system of measure will help the United States to maintain a role of technical leadership and keep it from lagging behind the metric world.

1.8.1 THE HISTORY OF MEASUREMENT

Measurement is a comparison between an unknown quantity and a known quantity. It is also a ratio of how many times a known thing can be divided into something unknown. Measurement tells how large, how many, how fast, and how much.

A unified system of measurement is a means of communication. Throughout history, various number systems and representational symbols have been used. Although there are existing remnants of systems based on the numbers 5, 12, and 20, only those systems based on the numbers 2, 10, and 60 are still in wide use. The metric system uses the base-10 (*decimal*) system of numbers.

It took many centuries to develop accurate units of measurement. The first were based on parts of the body. People used the human hand or the human foot; they also used the length of a furrow made by a plow, or the distance a man could walk in a day. Because it is harder and it takes longer to walk uphill than downhill, the Chinese even made an uphill mile shorter.

All of these forms of measurement were inexact. The size of a human hand or foot varied among individuals. Different people plowed different furrows and walked different distances in a day. There was a need for a unified standard system of weights and measures—one that everybody could agree on.

The Romans established a unified system of weights and measures. They made a bronze rod of specific length and called it a *pes*, Latin for "foot"; they made a bronze weight called a *libra*, Latin for "pound." Both the *pes* and the *libra* were kept under guard in a temple and used as standards of measurement throughout the Roman Empire. The U.S. *customary* abbreviation for pound, lb., is derived from the Latin *libra*.

After the fall of the Roman Empire, units of measurement once again became localized, and there was confusion when different regions tried to communicate numerically.

1.8.1.1 Noses, Thumbs, and Barleycorns

Because local measurements were often based on the king's body measurements, they differed from country to country and depended on the size of the current king.

The U.S. customary system uses the *yard* as a standard unit of measurement. Legend says that King Henry I of England established by royal decree the length of the yard as 36 in., which was the distance from Henry's nose to his thumb.

Another arbitrary and unrelated unit of the customary system is the system of thirteens, used to measure shoes. It is based on the size of a barleycorn, which is about one-third of an inch. Shoe sizes vary about one-third of an inch from size to size. A size 7 shoe, for example, is 1 in. smaller than a size 10.

1.8.1.2 Number Systems in Current Use

There are several numbering systems in current use around the world.

SEXAGESIMAL SYSTEM

The *sexagesimal* system is based on the number 60. First used by the Babylonians, it survives in the subdivision of time and angular measurement in the customary system. For instance, an hour has 60 minutes and a circle has 360 degrees. Both minutes and degrees can be further subdivided into 60 seconds.

BINARY SYSTEM

The *binary* system is based only on the numerals 0 and 1; it is used in electronic computers because it can represent yes/no or true/false decision logic.

FIGURE 1.8-1 Preferred and Nonpreferred Multiples, Submultiples, and Prefix Names for Some Powers of 10

Multiplication Factor			Prefix Name	Prefix Symbol	Pronunciation
Preferred					
10^{12}	or	1 000 000 000 000	tera	T	as in *terrace*
10^{9}	or	1 000 000 000	giga	G	*jig'a*
10^{6}	or	1 000 000	mega	M	as in *megaphone*
10^{3}	or	1 000	kilo	k	as in *kilowatt*
10^{-3}	or	0.001	milli	m	as in *military*
10^{-6}		0.000 001	micro	μ	as in *microphone*
10^{-9}	or	0.000 000 001	nano	n	*nan'oh*
10^{-12}	or	0.000 000 000 001	pico	p	*peek'oh*
Nonpreferred					
10^{2}	or	100	hecto	h	*heck'toe*
10^{1}	or	10	deka	da	*deck'a*
10^{-1}	or	0.1	deci	d	as in *decimal*
10^{-2}	or	0.01	centi	c	as in *sentiment*

(Copyright ASTM; reprinted with permission.)

DECIMAL SYSTEM

The *decimal* system, which is based on the number 10, is the system on which the metric system is founded. It was probably derived from the early use of human fingers to count. The Greeks and Romans first used a decimally based word structure for numbers, then adopted letter numerals for their representation. Roman numerals, which were used exclusively in Europe until about A.D. 1100, were gradually replaced by Arab numerals of Hindu-Arabic origin, which were much simpler to use.

The *Roman numeral* system consists of *quinary* (base-five) symbols: I equals 1, V equals 5, X equals 10, L equals 50, C equals 100, D equals 500, M equals 1000. In-between numbers are formed by addition to or subtraction from the seven basic symbols. When a smaller number precedes a larger, it is subtracted; when the smaller follows the larger number, it is added. Thus, 1978 is written MCMLXXVIII: C before M means 900; all other numbers retain their absolute values and are added.

The *Hindu-Arabic* system we use today expresses all numbers by combinations of 10 symbols, each having an *absolute* value and a *position* value (Fig. 1.8-1).

Decimal position notation relates each number's *position* value as a power of 10. For example, 10^1 equals 10, 10^2 equals 100, 10^3 equals 1000. Numbers to the right of the decimal are expressed in negative powers of 10. For example, 10^{-1} equals 0.1, 10^{-2} equals 0.01, and 10^{-3} equals 0.001. Each multidigit number is written as a linear combination of powers of 10. For example:

$456 = 4 \times 10^2 + 5 \times 10^1 + 6;$
$37.5 = 3 \times 10^1 + 7 + 5\ (\times 10^{-1}).$

1.8.2 THE METRIC SYSTEM

The scientific evolution of the eighteenth and nineteenth centuries, the industrial revolution of the nineteenth century, and the desire for more international trade accelerated the movement toward a universal system of weights and measures. In 1795, the French National Assembly adopted the metric system of measure.

The French metric system was based on a standard unit of length called a *metre* (not meter). The word was derived from the Greek *metron*, meaning "measure." The French metre was expressed in geodetic terms, that is, related to the size and shape of the earth. A metre was the 10-millionth part of an imaginary line running through Paris from the equator to the North Pole. Eventually, that distance was marked on a platinum/iridium bar, which became the international standard (prototype). It was kept at the International Bureau of Weights and Measures in France.

This system, as does its modern equivalent, established reference values, called *units*, for each measurable attribute of a physical phenomenon, which is called a *quantity*. A unit, then, was the reference value of a given quantity. For example, the metre was the unit for the given length (quantity) established by the prototype.

Additional units in the system included those for the quantities of *area*, *volume*, *capacity*, and *mass*. These were related decimally to the metre through the properties of water. For example, 1 *litre* represented 1/1000 of 1 cubic *metre*. When filled with water at its maximum density, a *litre* had a mass (weight) of 1000 *grams* or 1 *kilogram*.

The system also contained a set of decimal prefixes that could be attached to reference units to alter their magnitude. For example, *milli* means 1/1000 and *kilo* means 1000 times.

In 1881, the *International Electrical Congress* adopted a unit of time to produce the *centimetre-gram-second* (cgs) system. At about 1900, *metric* measurements began to be based on the *metre-kilogram-second* (MKS) system. In 1935, the *ampere* became the fourth *base unit*, resulting in the *metre-kilogram-second-ampere* (MKSA) system. The *General Conference on Weights and Measures* (CGPM), an international treaty organization, added the *kelvin* as a unit of *temperature* and *candela* as the unit of *luminous intensity*.

All three metric systems were then in use in different countries, making international trade and conversation difficult, even though many nations were using a form of the metric system of measurement.

1.8.3 THE INTERNATIONAL SYSTEM OF UNITS

In 1960, to help further unify the system of measurement internationally, the CGPM formally introduced a new metric system, called the *International System of Units*, for which the abbreviation is SI in all languages. The abbreviation represents the French *Système International d'Unités*. SI was derived from earlier decimal metric systems but supersedes them.

In response to demands by scientists and technologists for a means of precisely defining the *metre* in a way that could be accurately reproduced without having to refer to the prototype, CGPM introduced in SI a new definition based on the electromagnetic emission of a certain *krypton* atom.

1.8.3.1 Adoption of SI

SI has been adopted by every country that has changed to metric measurement since 1960. Countries that were using the cgs, MKS, or MKSA systems before 1960 are changing those units that have been superseded by the SI system. The decimal metric system is now used in more than 90% of the world's nations. Paradoxically, the United States, the first nation to introduce a system of decimal currency in 1785, and an original party of the 1875 *Treaty and International Metric Convention* that established the CGPM, is one of the last nations to adopt SI. It shares this dubious distinction with only a few small countries.

The metric system was legalized by Congress in 1866, but it was not made mandatory. Renewed interest at about the time SI appeared in the 1960s led to publication of the *ASTM Metric Practice Guide*, which in 1976 became ASTM E-380, "Standards for Metric Practice." Meanwhile, the government, responding to Public Law 90-472, signed in 1968, conducted a survey to determine the impact, desirability, feasibility, implications, and difficulties of converting the United States to the metric system of measurement.

At about the same time, ANSI established a *Metric Advisory Committee*, and in 1973, in response to the newfound interest in conversion, ANSI established an independent group called the *American National Metric Council* (ANMC) to provide assistance in the conversion to metric measurement through coordination, planning, and information services to its membership and all segments of society in the United States. Three years later, ANMC separated itself from ANSI, becoming a totally independent, private, and self-supporting national organization.

In 1975, in response to the completed study required by Public Law 90-472, Congress passed Public Law 94-168, the *Metric Conversion Act*. It stated that voluntary metrication was to be national policy and established the *U.S. Metric Board* to plan, publicize, and implement a voluntary change to the metric system based on SI, but the results of this watered-down law were predictably less than were needed to make conversion a reality. A timetable for the changeover to SI units was established by the construction industry. The target date for *metric conversion*, called M Day, was January 1, 1985, which came and went with little accomplished.

Then, with enactment of the *Omnibus Trade and Competitiveness Act of 1988*, metrication was again off and running. This law states that the metric system is the "preferred system of weights and measures for United States trade and commerce," and requires that when it is practicable and economically feasible, federal agencies use the metric system when making grants and procurements and in other business-related activities by the end of fiscal year 1992. Implementation is the responsibility of the Department of Commerce. Citing the Omnibus Trade and Competitiveness Act of 1988, the Department of Commerce announced on January 2, 1991, that metrication was no longer voluntary for federal agencies.

Accordingly, by the middle of 1991 the various federal agencies began to make plans to convert. It remains to be seen whether states and local governments will follow, but their response to other federal government initiatives indicates that they will. The private sector will not be far behind.

1.8.3.2 Structure of SI

A lifetime of using the customary system gives people a feel for the length of an inch or a foot. New associations or recognition points must be developed in order for them to be comfortable using SI (Fig. 1.8-2). This conversion in thinking happens rather more quickly than most people expect. It can be equated with the relative ease of learning a foreign language when living in the nation of its origin, as opposed to learning the same language in a classroom far removed from its general use. In addition, many SI and related units are already in our vocabulary—for example, *volt*, *ampere*, *watt*, *ohm*, *second*, and *lumen*.

The application of SI units and new

FIGURE 1.8-2 Some Metric Recognition Points

Quantity	Unit	Description
Linear measurement	1 mm	Approximate diameter of a paper clip wire
	25 mm	Vertical dimension of an ordinary U.S. postage stamp; nearest equivalent to 1 in.
	100 mm	International cigarette length; basic metric building module; nearest metric preferred dimension to 4 in. (1.6% less)
	600 mm	Height of three courses of concrete blocks including mortar joints
	2 m	Approximate height of a standard door opening (6 ft. 8 in.)
Area measurement	500 mm^2	Face area of an ordinary U.S. postage stamp [20 mm × 25 mm]
	1 m^2	1 square metre; approximate area of a shower base
Volume measurement	1 L	1 litre: 5.7% more than a U.S. quart; new soft drink bottle size
Mass [weight]	1 g	Approximate weight of a paper clip or a dollar bill; artificial sweetener package size
	5 g	mass [weight] of a nickel [5¢ coin]
	1 kg	1 kilogram: the base unit of mass in SI; mass of water in a cube with 100 mm sides (approximately 2.2 lb.)
Temperature	0°C	Freezing point of water (32°F)
	20°C	Comfortable thermostat setting in winter (68°F)
	37°C	Normal body temperature (98.6°F)
	100°C	Boiling point of water (212°F)

preferred (see Section 1.8.6) numeric values will simplify measurements in construction and reduce errors in calculations.

SI symbols have an agreed-upon form and the same meaning worldwide. There is only one recognized SI unit for each physical quantity. All units are derived from seven *base* and two *supplementary units* (Fig. 1.8-3). The entire system consists of *base units*, *supplementary units*, and *derived units*, as summarized in Figure 1.8-4.

SI is a universal, rational, coherent, and preferred measurement system that uses the base-10 (decimal) system of numbers. Relations between the units in the system contain as a numerical factor only the number 1, or "unity," because all derived units have a unity relationship to the constituent base and supplementary units. For example, one litre is equal to ⅟1000 of 1 cubic metre (0.001 m^3) and (when filled with water) has a mass of 1 *kilogram*.

BASE UNITS

SI has fewer units than the customary system. A single unit of length, the *metre* (m), and its decimally related multiples and submultiples, such as *kilometre* (km) and *millimetre* (mm), will replace a variety of customary units, including the *mile*, *furlong*, *chain*, *link*, *rod*, *fathom*, *yard*, *foot*, *hand*, *inch*, and *mil*. In addition, changing to the metre will correct the difference between the *standard foot* and the *survey foot* (which is two parts per million longer), by replacing both. SI uses the French spelling of *metre*. However, the use of the SI spelling is controversial in the United States, and the incorrect spelling *meter* may actually win out over the correct SI spelling and become the U.S. standard.

SI originally addressed the base units for only six physical quantities: *length*, *mass* (which is closely related to weight), *time*, *temperature*, *electrical current*, and *luminous intensity*. In 1971, because of the growth of thermal dynamics, nuclear physics, and electronic science, a seventh quantity, *amount of substance*, was added. Figure 1.8-5 compares the names for the base units for these seven quantities in SI and in the customary system.

The base units shown in Figure 1.8-5 are the building blocks of SI. Except for the *kilogram*, they are reproducible under controlled conditions anywhere (Fig. 1.8-6). The *kilogram* is a platinum-iridium *prototype* kept by the International Bureau of Weights and Measures in Sevres, France.

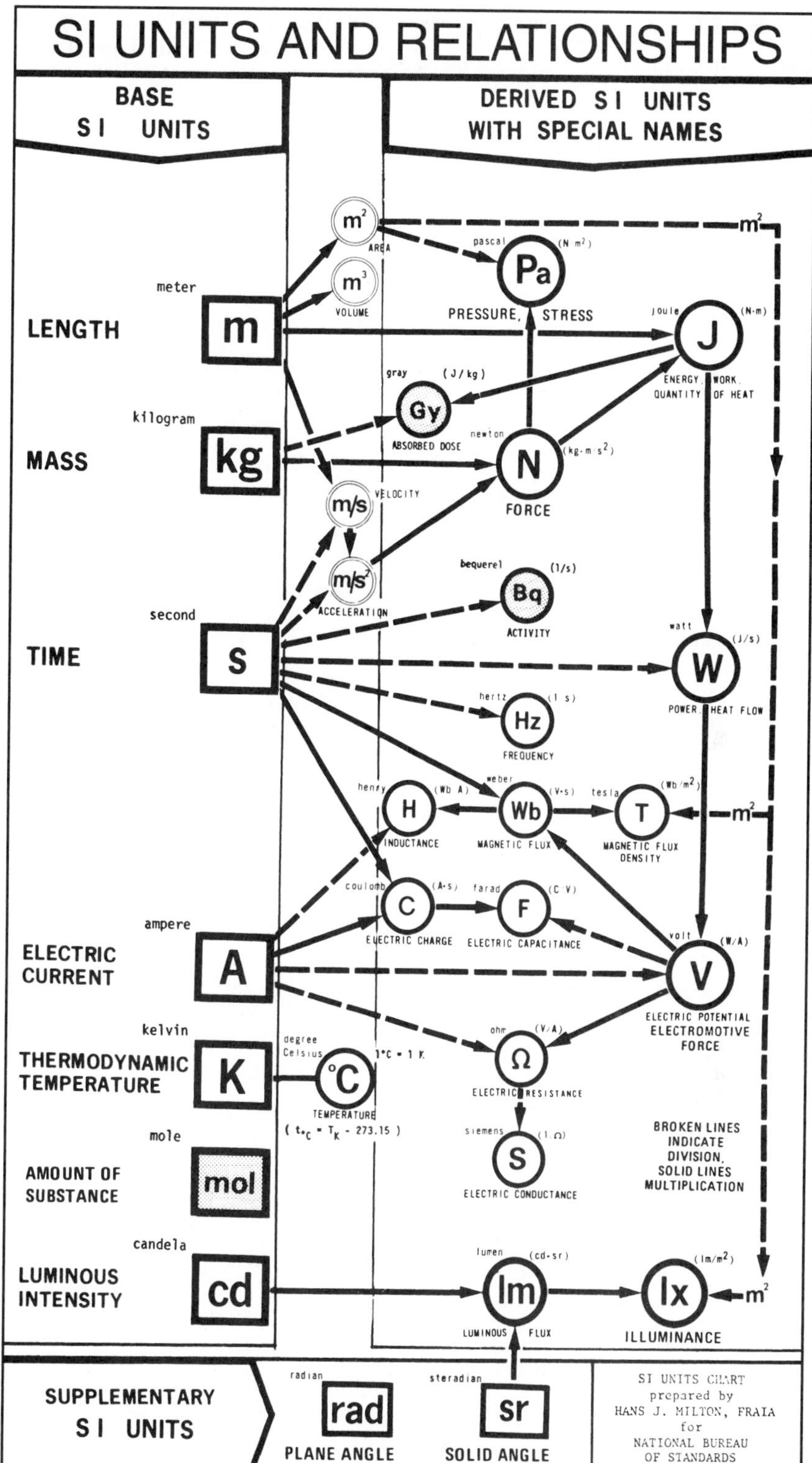

FIGURE 1.8-3 **Relationship of base, supplementary, and derived SI units.** (Copyright ASTM; reprinted with permission.)

SUPPLEMENTARY UNITS

In addition to the seven base units, SI has two supplementary units, the *radian* (plain angle) and the *steradian* (solid angle).

DERIVED UNITS

All other quantities are derived from the seven base and two supplementary units. *Derived units* are formed by combining

FIGURE 1.8-4 Units in the International System (SI). See Section 1.8 for a Detailed Discussion on the Use of the Tables

Unit Group Quantity	Unit Name	Symbol	Formula	Unit Derivation	Remarks
Base Units					
Length	metre	m			An alternative spelling is *meter*.
Mass	kilogram	kg			
Time	second	s			Already in common use.
Electric current	ampere	A			Already in common use.
Thermodynamic temperature	kelvin	K			The customary unit for temperature is the degree Celsius (°C).
Amount of substance	mole	mol			The mole has no application in construction.
Luminous intensity	candela	cd			Already in common use.
Supplementary Units					
Plane angle	radian	rad			Already in common use.
Solid angle	steradian	sr			Already in common use.
Derived Units with Special Names					
Frequency (of a periodic phenomenon)	hertz	Hz	1/s	s^{-1}	The hertz replaces *cycle per second.*
Force	newton	N	$kg{\cdot}m/s^2$	$m{\cdot}kg{\cdot}s^{-2}$	
Pressure, stress, elastic modulus	pascal	Pa	N/m^2	$m^{-1}{\cdot}kg{\cdot}s^{-2}$	
Energy, work, quantity of heat	joule	J	$N{\cdot}m$	$m^2{\cdot}kg{\cdot}s^{-2}$	
Power, radiant flux	watt	W	J/s	$m^2{\cdot}kg{\cdot}s^{-3}$	Already in common use.
Quantity of electricity, electric charge	coulomb	C	$A{\cdot}s$	$s{\cdot}A$	Already in common use.
Electric potential, potential difference, electromotive force	volt	V	J/C or W/A	$m^2{\cdot}kg{\cdot}s^{-3}{\cdot}A^{-1}$	Already in common use.
Electric capacitance	farad	F	C/V	$m^{-2}{\cdot}kg^{-1}{\cdot}s^4{\cdot}A^2$	Already in common use.
Electric resistance	ohm	Ω	V/A	$m^2{\cdot}kg{\cdot}s^{-3}{\cdot}A^{-2}$	Already in common use.
Electric conductance	siemens	S	A/V or 1/Ω	$m^{-2}{\cdot}kg^{-1}{\cdot}s^3{\cdot}A^2$	The *siemens* was formerly referred to as *mho*.
Magnetic flux	weber	Wb	$V{\cdot}s$	$m^2{\cdot}kg{\cdot}s^{-2}{\cdot}A^{-1}$	Already in common use.
Magnetic flux density	tesla	T	Wb/m^2	$kg{\cdot}s^{-2}{\cdot}A^{-1}$	Already in common use.
Electric inductance	henry	H	Wb/A	$m^2{\cdot}kg{\cdot}s^{-2}{\cdot}A^{-2}$	Already in common use.
Luminous flux	lumen	lm	$cd{\cdot}sr$	$cd{\cdot}sr$	Already in common use.
Illuminance	lux	lx	lm/m^2	$m^{-2}{\cdot}cd{\cdot}sr$	
Absorbed dose	gray	Gy	J/kg	$m^2{\cdot}s^{-2}$ (a)	(a) kg is canceled out. No application in construction.
Derived Units with Generic Names					
a. Units Expressed in Terms of One Base Unit					
Area	square metre	m^2		m^2	
Volume, capacity	cubic metre	m^3		m^3	(1 m^3 = 1000 L)
Section modulus	metre to third power		m^3	m^3	
Second moment of area	metre to fourth power		m^4	m^4	
Curvature	reciprocal (of) metre		1/m	m^{-1}	
Rotational frequency	reciprocal (of) second		1/s	s^{-1}	Revolution per second (r/s) is used in specifications for rotating machinery.
Coefficient of linear thermal expansion	reciprocal (of) kelvin		1/K	K^{-1}	
b. Units Expressed in Terms of Two or More Base Units					
Linear velocity	metre per second		m/s	$m{\cdot}s^{-1}$	
Linear acceleration	metre per second squared		m/s^2	$m{\cdot}s^{-2}$	
Kinematic viscosity	square metre per second		m^2/s	$m^2{\cdot}s^{-1}$	
Volume rate of flow	cubic metre per second		m^3/s	$m^3{\cdot}s^{-1}$	
Specific volume	cubic metre per kilogram		m^3/kg	$m^3{\cdot}kg^{-1}$	
Mass per unit length	kilogram per metre		kg/m	$m^{-1}{\cdot}kg$	
Mass per unit area	kilogram per square metre		kg/m^2	$m^{-2}{\cdot}kg$	
Density (mass per unit volume)	kilogram per cubic metre		kg/m^3	$m^{-3}{\cdot}kg$	In this SI form, mass density is conveniently 1000 times specific gravity.
Moment of inertia	kilogram metre squared		$kg{\cdot}m^2$	$m^2{\cdot}kg$	
Mass flow rate	kilogram per second		kg/s	$kg{\cdot}s^{-1}$	
Momentum	kilogram metre per second		$kg{\cdot}m/s$	$m{\cdot}kg{\cdot}s^{-1}$	

FIGURE 1.8-4 *(Continued)*

Unit Group Quantity	Unit Name	Symbol	Formula	Unit Derivation	Remarks
Derived Units with Generic Names (Continued)					
Angular momentum	kilogram metre squared per second		$kg{\cdot}m^2/s$	$m^2{\cdot}kg{\cdot}s^{-1}$	
Magnetic field strength	ampere per metre		A/m	$m^{-1}{\cdot}A$	
Current density	ampere per square metre		A/m^2	$m^{-2}{\cdot}A$	
Luminance	candela per square metre		cd/m^2	$m^{-2}{\cdot}cd$	
c. Units Expressed in Terms of Base Units and/or Derived Units with Special Names					
Moment of force, torque	newton metre		N·m	$m^2{\cdot}kg{\cdot}s^{-2}$	
Flexural rigidity	newton square metre		$n{\cdot}m^2$	$m^3{\cdot}kg{\cdot}s^{-2}$	
Force per unit length, surface tension	newton per metre		N/m	$kg{\cdot}s^{-2}$ (b)	(b) m is canceled out.
Dynamic viscosity	pascal second		Pa·s	$m^{-1}{\cdot}kg{\cdot}s^{-1}$	
Impact ductility	joule per square metre		J/m^2	$kg{\cdot}s^{-2}$ (c)	(c) m^2 is canceled out.
Combustion heat (per unit volume)	joule per cubic metre		J/m^3	$m^{-1}{\cdot}kg{\cdot}s^{-2}$	
Combustion heat (per unit mass), specific energy, specific latent heat	joule per kilogram		J/kg	$m^2{\cdot}s^{-2}$ (d)	(d) kg is canceled out.
Heat capacity, entropy	joule per kelvin		J/K	$m^2{\cdot}kg{\cdot}s^{-2}{\cdot}K^{-1}$	
Specific heat capacity, specific entropy	joule per kilogram kelvin		J/(kg·K)	$m^2{\cdot}s^{-2}{\cdot}K^{-1}$ (e)	(e) kg is canceled out.
Heat flux density, irradiance; sound intensity	watt per square metre		W/m^2	$kg{\cdot}s^{-3}$ (f)	(f) m^2 is canceled out.
Thermal conductivity	watt per metre kelvin		W/(m·K)	$m{\cdot}kg{\cdot}s^{-3}{\cdot}K^{-1}$	
Coefficient of heat transfer	watt per square metre kelvin		$W/(m^2{\cdot}K)$	$kg{\cdot}s^{-3}{\cdot}K^{-1}$ (g)	(g) m^2 is canceled out.
Thermal resistance, thermal insulance	square metre kelvin per watt		$m^2{\cdot}K/W$	$kg^{-1}{\cdot}s^3{\cdot}K$ (h)	(h) m^2 is canceled out.
Electric field strength	volt per metre		V/m	$m{\cdot}kg{\cdot}s^{-3}{\cdot}A^{-1}$	
Electric flux density	coulomb per square metre		C/m^2	$m^{-2}{\cdot}s{\cdot}A$	
Electric charge density	coulomb per cubic metre		C/m^3	$m^{-3}{\cdot}s{\cdot}A$	
Electric permittivity	farad per metre		F/m	$m^{-3}{\cdot}kg^{-1}{\cdot}s^4{\cdot}A^2$	
Electric permeability	henry per metre		H/m	$m{\cdot}kg{\cdot}s^{-2}{\cdot}A^{-2}$	
Electric resistivity	ohm metre		Ω·m	$m^3{\cdot}kg{\cdot}s^{-3}{\cdot}A^{-2}$	
Electric conductivity	siemens per metre		S/m	$m^{-3}{\cdot}kg^{-1}{\cdot}s^3{\cdot}A^2$	
Light exposure	lux second		lx·s	$m^{-2}{\cdot}s{\cdot}cd{\cdot}sr$	
Luminous efficacy	lumen per watt		lm/W	$m^{-2}{\cdot}kg^{-1}{\cdot}s^3{\cdot}cd{\cdot}sr$	
d. Units Expressed in Terms of Supplementary Units and Base and/or Derived Units					
Angular velocity	radian per second		rad/s	$s^{-1}{\cdot}rad$	
Angular acceleration	radian per second squared		rad/s^2	$m^{-2}{\cdot}rad$	
Radiant intensity	watt per steradian		W/sr	$m^2{\cdot}kg{\cdot}s^{-3}{\cdot}sr^{-1}$	
Radiance	watt per square metre steradian		$W/m^2{\cdot}sr$	$kg{\cdot}s^{-2}{\cdot}sr^{-1}$ (i)	(i) m^2 is canceled out.

Note: Metre is the preferred international spelling applied to this base unit and all its multiples and submultiples.

FIGURE 1.8-5 Customary and SI Base Units

Quantity	Customary	SI
Length	foot	metre (m)
Mass (weight)	pound	kilogram (kg)
Time	second	second (s)
Temperature	degree Fahrenheit	kelvin (K)
Electric current	ampere	ampere (A)
Luminous intensity	candela	candela (cd)
Amount of a substance		mole (mol)

base and supplementary units and other derived units according to algebraic relations linking corresponding quantities (Fig. 1.8-7). For example, the metre is a base unit, the radian is a supplementary unit, and the pascal is a derived unit.

Standard prefixes allow a change in magnitude from the subatomic scale (10^{-18}) to the astronomic scale (10^{18}) (see Fig. 1.8-1). The use of multiples of SI units that cannot be handled by using the SI prefixes is discouraged.

FIGURE 1.8-6 SI Technical Definitions

Base Units

Metre
The length equal to 1 650 763.73 wavelengths in vacuum of the radiation corresponding to the transition between the levels $2p_{10}$ and $5d_5$ of the krypton-86 atom.

Kilogram
The unit of mass, it is equal to the mass of the international prototype of the kilogram. The kilogram is the only base unit whose name for historic reasons contains a prefix.

Second
The measure of time with a duration of 9 192 631 770 periods of the radiation corresponding to the transition between the two hyperfine levels of the ground state of the cesium-133 atom.

Ampere
The measure of current flow equal to that flow that, if maintained in two straight parallel conductors, would produce between those conductors a force equal to 2×10^7 newtons per metre of length. The conductors are assumed to be of infinite length, of negligible circular cross section, and placed 1 metre apart in vacuum.

Kelvin
The unit of thermodynamic temperature equal to the fraction 1/273.16 of the thermodynamic temperature of the triple point of water.

The temperature intervals of the degree Celsius and the kelvin are identical and the scales using these two units are related by a difference of exactly 273.15 kelvins.

Mole
The amount of substance of a system that contains as many elementary entities as there are atoms in 0.012 kilogram of carbon-12. When the mole is used, the elementary entities must be specified and may be atoms, molecules, ions, electrons, other particles, or specified groups of such particles.

Candela
The luminous intensity in the perpendicular direction of a surface of 1/600 000 square metre of blackbody at the temperature of freezing platinum under a pressure of 101 325 newtons per square metre.

Supplementary Units

Radian
The plane angle between two radii of a circle that cut off on the circumference an arc equal in length to the radius. The radian is equal to $360/2\pi$ and is an angle of approximately $57° 17' 44.6''$.

Steradian
The solid angle that, having its vertex in the center of a sphere, cuts off an area of the surface of the sphere equal to that of a square with sides of length equal to the radius of the sphere. The total solid angle of a sphere is 4π steradians.

NON-SI UNITS FOR USE WITH SI

To preserve the advantages of SI as a *coherent* system, the use of units from other systems should be minimized. There are, however, acceptable but noncoherent traditional units retained for use with SI. Non-SI units accepted for general use with SI include units for volume (*litre*), mass (*metric ton*), time (*minute, hour, day, year*), plane angle (*degree of arc*), and velocity (*kilometre per hour*). Three other non-SI units are accepted for limited use. These include units for *area* (*hectare*), *energy* (*kilowatt-hour*), and *speed of rotation* (*revolution per minute*). These quantities and their associated units, along with their symbols, and their relationship to SI *units* are given in Figure 1.8-8.

ABANDONED METRIC UNITS

Many metric units in use before SI was adopted have been abandoned. SI supersedes all previous metric systems. Abandoned units include *centimetre-gram-second* (cgs) *units* such as *abampere* and *statvolt*. Decimal multiples of SI units, such as the *angstrom* (0.1 mm), that cannot be handled by using the SI *prefixes* should not be used.

SPECIAL NAMES

Except for *litre*, *metric ton*, and *hectare*, special names or multiples and submultiples of SI units should be avoided,

1.8.3.3 Recommendations for the Use of SI

SI symbols and many rules for their proper use are internationally agreed upon and should be used (Figs. 1.8-9 and 1.8-10).

FIGURE 1.8-7 Derived SI Units with Special Names

Physical Quantity	Unit	Symbol	Definition
Absorbed dose	gray	Gy	The energy imparted by ionizing radiation to a mass of matter corresponding to one joule per kilogram.
Activity	becquerel	Bq	The activity of a radionuclide having one spontaneous nuclear transition per second.
Electric capacitance	farad	F	The capacitance of a capacitor between the plates of which there appears a difference of potential of one volt when it is charged by a means of electricity equal to one coulomb.
Electric conductance	siemens	S	The electric conductance of a conductor in which a current of one ampere is produced by an electric potential difference of one volt.
Electric inductance	henry	H	The inductance of a closed circuit in which an electromotive force of one volt is produced when the electric current in the circuit varies uniformly at a rate of one ampere per second.

FIGURE 1.8-7 *(Continued)*

Physical Quantity	Unit	Symbol	Definition
Electric potential difference, electromotive force	volt	V	The unit of electric potential difference and electromotive force that is the difference of electric potential between two points of a conductor carrying a constant current of one ampere, when the power dissipated between these points is equal to one watt.
Electric resistance	ohm	Ω	The electric resistance between two points of a conductor when a constant difference of potential of one volt, applied between these two points, produces in this conductor a current of one ampere, this conductor not being the source of any electromotive force.
Energy	joule	J	The work that is done when the point of application of a force of one newton is displaced a distance of one metre in the direction of the force.
Force	newton	N	The force that, when applied to a body having a mass of one kilogram, gives it an acceleration of one metre per second squared.
Frequency	hertz	Hz	The frequency of a periodic phenomenon of which the period is one second.
Illuminance	lux	lx	The illuminance produced by a luminous flux of one lumen uniformly distributed over a surface of one square metre.
Luminous flux	lumen	lm	The luminous flux emitted in a solid angle of one steradian by a point source having a uniform intensity of one candela.
Magnetic flux	weber	Wb	The magnetic flux that, linking a circuit of one turn, produces in it an electromotive force of one volt as it is reduced to zero at a uniform rate in one second.
Magnetic flux density	tesla	T	The magnetic flux density given by a magnetic flux of one weber per square metre.
Power	watt	W	The power that gives rise to the production of energy at the rate of one joule per second.
Pressure or stress	pascal	Pa	The pressure or stress of one newton per square metre.
Quantity of electricity	coulomb	C	The quantity of electricity transported in one second by a current of one ampere.

FIGURE 1.8-8 Other Units Whose Use Is Permitted with SI

Quantity	Unit Name	Symbol	Relationship to SI Unit	Remarks
Units for General Use				
Volume	litre[a]	L[a]	$1\ L = 0.001\ m^3 = 10^6\ mm^3$	An alternative spelling is *liter*. The litre may only be used with the SI prefix milli-.
Mass	metric ton[b]	t[b]	$1\ t = 1000\ kg = (1\ Mg)$	
Time	minute	min	$1\ min = 60\ s$	
	hour	h	$1\ h = 3600\ s = (60\ min)$	
	day (mean solar)	d	$1\ d = 86{,}400\ s = (24\ h)$	
	year (calendar)	a	$1\ a = 31{,}536{,}000\ s = (365\ d)$	
Temperature interval	degree Celsius	°C	$1°C = 1\ K$	The Celsius temperature 0°C corresponds to 273.15 K exactly. ($t_{°C} = T_K - 273.15$)
Plane angle	degree (of arc)	°	$1° = 0.017453\ rad = 17.453\ mrad$	$1° = (\pi/180)\ rad$
Velocity	kilometre per hour	km/h	$1\ km/h = 0.278\ m/s$	$1\ m/s = 3.6\ km/h$
Units Accepted for Limited Use				
Area	hectare	ha	$1\ ha = 10{,}000\ m^2$	For use in land measurement.
Energy	kilowatthour	kWh	$1\ kWh = 3.6\ MJ$	For measurement of electrical energy consumption only.
Speed of rotation	revolution per minute	r/min	$1\ r/min = 1/60\ r/s = 2\pi/60\ rad/s$	To measure rotational speed in slow-moving equipment only.

[a]The international symbol for litre is the lowercase l (ell), which can be easily confused with the numeral 1. Several English-speaking countries have adopted the script ℓ as a symbol for litre to avoid any misinterpretation. The symbol L (capital ell) is recommended for U.S. use to prevent confusion.

[b]The international name for metric ton is tonne. The metric ton is equal to the megogram (Mg).

FIGURE 1.8-9 Rules and Recommendations for the Presentation of SI Units and Symbols

	Typical Examples	Remarks
General		
1. All unit names should either be denoted by correct symbols or be written in full. In the interest of simplification and to reduce the amount of writing, use unit symbols rather than fully written forms.		
2. Do not use mixtures of names and symbols.	Use: J/kg *or* joule per kilogram	Not: joule per kg Not: J/kilogram
Symbols for Unit Quantities and Prefixes		
1. SI symbols are internationally agreed upon and there is only *one* symbol for each unit quantity. Multiples and submultiples are formed by using the unit symbol and attaching a prefix symbol in front of it.	m, kg, s, A, cd, K	See also Nos. 5–7, below.
2. All unit symbols are shown in upright letters, and can be produced by a normal typewriter keyboard, with the exceptions of the symbols for the SI unit ohm and the prefix micro-, which are represented by Greek letters Ω and μ, respectively.		Exceptions: Ω, μ
3. Unit symbols are never followed by a period (full stop) except at the end of a sentence.	60 kg/m	Not: 60 kg./m.
4. Unit symbols are normally written in lowercase, except for unit names derived from a proper name, in which case the initial is capitalized. Some units have symbols consisting of two letters from a proper name, of which *only* the first letter is capitalized. (The symbol for the unit name ohm is the capital Greek letter Ω.)	m, kg, s, mol, cd, etc. A, K, N, J, W, V, etc. Pa, Hz, Wb, etc.	Exception: L
5. Prefixes for magnitudes from 10^6 to 10^{18} have capital upright letter symbols.	M, G, T, etc.	See also cl
6. Prefixes for magnitudes from 10^{-18} through 10^3 have lowercase upright letter symbols. (The symbol for 10^{-6} or micro is the lowercase Greek letter μ.)	p, n, μ, m, k, etc.	See also cl
7. Prefix symbols are directly attached to the unit symbol *without a space* between them.	mm, kW, MN, etc.	Not: m m, k W, M N
8. Do not use compound prefixes to form a multiple or submultiple of a unit (e.g., use nanometre, do not use micromillimetre *or* millimicrometre).	nm	Not: μmm or mμm
9. In the case of the base unit kilogram, prefixes are attached to the gram (e.g., milligram, not microkilogram).	mg	Not: μkg
10. Use only one prefix when forming a multiple or a submultiple of a compound unit. Normally, the prefix should be attached to a unit in the numerator. An exception to this rule is made for the base unit kilogram.	km/s, mV/m	Not: mm/μs, μV/mm Exception: MJ/kg not: kJ/g
Areas of Possible Confusion Requiring Special Care		
1. The symbols for SI units and the conventions that govern their use should be strictly followed. A number of prefix and unit symbols use the same letter, but in different form. Exercise care to present the correct symbol for each quantity.	g (gram), G (giga), k (kilo), K (kelvin), m (milli), m (metre), M (mega), n (nano), N (newton)	Others: c (centi), C (coulomb), °C (degree Celsius), s (second), S (siemens), t (metric ton), T (tesla), T (tera)
2. All prefix and unit symbols retain their prescribed form regardless of the surrounding typography. In printouts from limited character sets (telex, computer printers) special considerations apply to symbols for mega, micro, ohm, and siemens. Where confusion is likely to arise, write units in full.		
Unit Names Written Out in Full		
1. Unit names, including prefixes, are treated as common names and are *not capitalized*, except at the beginning of sentences or in titles. (The only exception is Celsius in degree Celsius, where degree is considered as the unit name and is shown in lowercase, whereas Celsius represents an adjective and is capitalized.)	metre, newton, etc.	Not: Metre, Newton Exception: degree Celsius
2. Where a prefix is attached to an SI unit to form a multiple or submultiple, the combination is written as one word. (There are three cases where the final vowel of the prefix is omitted in the combination: megohm, kilohm, and hectare.)	millimetre, kilowatt	Not: milli-metre Not: kilo watt
3. Where a compound unit is formed by multiplication of two units, the use of a space between units is preferred, but a hyphen is acceptable and in some situations more appropriate, to avoid any risk of misinterpretation.	newton metre *or* newton-metre	Not: newtonmetre
4. Where a compound unit is formed by division of the two units, this is expressed by inserting the word *per* between the numerator and the denominator.	metre per second joule per kelvin	Not: metre/second Not: joule/kelvin
5. Where the numerical value of a unit is written in full, the unit should also be written in full.	seven metres	Not: seven m

FIGURE 1.8-9 *(Continued)*

	Typical Examples	Remarks
Plurals		
1. Units written in full are subject to the normal rules of grammar. For any unit with a numerical value greater than one (1), an *s* is added to the written unit to denote the plural.	1.2 metre*s* 2.3 newton*s* 33.2 kilogram*s*	But: 0.8 metre
2. The following units have the same plural as singular when written out in full: hertz, lux, siemens.	350 kilohertz 12.5 lux	
3. Symbols never change in the plural.	2.3 N, 33.2 kg	Not: 2.3 N*s*, 33.2 kg*s*
Compound Unit Symbols—Products and Quotients		
1. The product of two units is indicated by a dot placed at midheight between the unit symbols.	kN·m, Pa·s	Not: kNm, Pas Not: kN m, Pa s See also f3 and f5
2. To express a derived unit formed by division, any one of the following methods may be used:		
a. a solidus (slash, /)	kg/m^3, W/(m·K)	See also f3 and f5
b. a horizontal line between numerator and denominator	$\frac{kg}{m^3}$; $\frac{W}{m \cdot K}$	
c. a negative index (or negative power).	kg·m^{-3}, W·m^{-1}·K^{-1}	
3. *Only one* solidus may be used in any combination.	m/s^2, m·kg/(s^3·A)	Not: m/s/s Not: m·kg/s^3/A
4. Do not use the abbreviation *p* for per in the expression of a division.	km/h	Not: kph *or* k.p.h.
5. Where the denominator is a product, this should be shown in parentheses.	W/(m^2·K)	

Note: Metre is the preferred international spelling applied to this base unit and all its multiples and submultiples.

FIGURE 1.8-10 Presentation of Numerical Values in Conjunction with SI

	Typical Examples	Remarks
Decimal Marker		
1. Whereas most European countries use the comma on the line as the decimal marker and this practice is advocated by ISO, a special exception is made for documents in the English language, which have traditionally used the point (dot) on the line, or period, as the decimal marker.		See also under g
2. The recommended decimal marker for use in the United States is the point on the line (period), and the comma should not be used. In handwritten documents the decimal marker may be shown slightly above the line for clear identification.	9.9; 15.375	Not: 9,9; 15,375
3. *Always* show a zero before the decimal point for all numbers smaller than 1.0 (one).	0.1; 0.725	Not: .1; .725
Spacing		
1. *Always* leave a gap between the numerical value associated with a symbol and the symbol of at least half a space in width. In the case of the symbol for degree Celsius this space is optional, but the degree symbol must always be attached to Celsius.	900 MHz; 200 mg; 10^6 mm^2 20°C or 20 °C	Not: 900MHz; 200mg Not: 10^{6}mm^2 Not: 20° C
2. In non-SI expressions of plane angle (°, ′, ″), do not leave a space between the numerical value and the symbol.	27°30′ (of arc)	Not: 27 ° 30 ′
3. *Always* leave a space on each side of signs for multiplication, division, addition, and subtraction.	100 mm × 100 mm; 36 MPa + 8 MPa	Not: 100 mm×100 mm Not: 36 MPa+ 8 MPa
Fractions		
1. *Avoid* common fractions in connection with SI units.	Write: 0.5 kPa	Not: ½ kPa
2. *Always* use decimal notation to express fractions of any number larger than 1.0 (one).	1.5; 16.375	Not: Not: 1½; 16⅜
3. Although the most common fractions such as half, third, quarter, and fifth will remain in speech, *always* show decimal notation in written, typed, or printed material.	0.5; 0.33; 0.25; 0.2	Not: ½; ⅓; ¼; ⅕
Powers of Units and Exponential Notation		
1. When writing unit names with a modifier *squared* or *cubed*, the following rules should be applied:		
a. In the case of area and volume, the modifier is written before the unit name as *square* and *cubic*.	cubic metre; square millimetre	Not: metre cubed; millimetre squared
b. In all other cases, the modifier is shown after the unit name as *squared, cubed, to the fourth power*, etc.	metre per second squared	Not: metre per square second (or metre per second per second)

FIGURE 1.8-10 *(Continued)*

	Typical Examples	Remarks
c. The abbreviations *sq.* for square, and *cu.* for cubic should *not* be used.		Not: sq. millimetre Not: cu. metre
2. For unit symbols with modifiers (such as square, cubic, fourth power, etc.) *always* show the superscript immediately after the symbol.	m^2; mm^3; s^4	Not: $m\ ^2$; $mm\ ^3$; $s\ ^4$
3. Show the superscript as a numeral reduced in size raised half a line space. Where a typewriter without superscript numerals is used, the full-sized numeral should be raised half a line space, provided that this does not encroach on the print in the line above.	mm^3; m/s^2	Permitted: mm^3; m/s^2
4. Where an exponent is attached to a prefixed symbol, it indicates that that multiple (or submultiple) is raised to the power expressed by the exponent.	$1\ mm^3 = (10^{-3}\ m)^3$ $= 10^{-9}\ m^3$ $1\ km^2 = (10^3\ m)^2$ $= 10^6\ m^2$	Not: $1\ mm^3$ $= 10^{-3}\ m^3$
Ratios		
1. Do not mix units in expressing a ratio of like unit quantities.	0.01 m/m 0.03 m^2/m^2	Not: 10 mm/m Not: 30,000 mm^2/m^2
2. Wherever possible, use a nonquantitative expression (ratio or percentage) to indicate measurement of slopes, deflections, etc.		Preferred: 1:100; 0.01; 1% 1:33; 0.03; 3%
Range		
1. The choice of the appropriate prefix to indicate a multiple or submultiple of an SI unit is governed by convenience to obtain numerical values within a practical range and to eliminate nonsignificant digits.		
2. In preference, use prefixes representing ternary powers of 10 (10 raised to a power that is a multiple of 3).	milli; kilo; mega	Avoid: centi; deci; deka; hecto
3. Select prefixes so that the numerical value or values occur in a common range between 0.1 and 1000.	120 kN 3.94 mm 14.5 MPa	In lieu of: 120 000 N 0.003 94 m 14 500 kPA
4. Compatibility with the general range must be a consideration; e.g., if all dimensions on a drawing are shown in millimeters (mm), a range from 1 to 99 999 (a maximum of five numerals) would be acceptable to avoid mixing of units.		Note: Drawings should include the notation *All dimensions in millimetres.*
Presentation and Tabulation of Numbers		
1. In numbers with many digits it has been common practice in the United States to separate digits into groups of three by means of commas. This practice must *not* be used with SI, to avoid confusion. It is recommended international practice to arrange digits in long numbers in groups of three from the decimal marker, with a gap of not less than half a space, and not more than a full space, separating each group.	54 375.260 55	Not: 54,375.260,55
2. For individual numbers with four digits before (or after) the decimal marker, this space is not necessary.	4500; 0.0355	
3. In all tabulations of numbers with five or more digits before and/or after the decimal marker, group digits into groups of three: e.g., 12.5255; 5735; 98 300; 0.425 75	12.525 5 5 735 98 300 0.425 75 104 047.951 25	
Use of Unprefixed Units in Calculations		
Errors in calculations involving compound units can be minimized if all prefixed units are reverted to coherent base or derived units, with numerical values expressed in powers-of-ten notation.	Preferred: 136 kJ = 136×10^3 J 20 MPa = 20×10^6 Pa 1.5 t (Mg) = 1.5×10^3 kg	Also acceptable: (or 1.36×10^5 J) (or 2×10^7 Pa)

Note: Metre is the preferred international spelling applied to this base unit and all its multiples and submultiples.

WRITING UNIT SYMBOLS

In general, unit symbols should be:

1. Printed in upright type regardless of the type style of the surrounding text;
2. Unaltered in the plural;
3. Not followed by a period except when at the end of a sentence;
4. Written in lowercase unless the unit is derived from a proper name;
5. Written with a space between the unit symbol and the numerical value.

Because a comma is used in some countries as a decimal marker, it should not be used to separate digits into groups of three. Instead, digits should be divided into groups of three by using a space at least equal to the width of the letter "i" to separate these groups. A space is not necessary in four-digit numbers. Examples are: 0.133 47, 2.141 596, 73 722, and 7372.

The word *billion* means a *thousand million* (prefix *giga*) in the United States, but a *million million* (prefix *tera*) in many other countries. Therefore, terms such as *billion* and *trillion* should not be used in technical writing.

PREFERRED RANGE OF VALUES

Whole numbers between 1 and 1000 should be used with the proper prefix wherever possible. For example, 725 m is preferred over 0.725 km or 725 000 mm. However, the prefix that adequately covers the whole quantity range should be used. For example, if 725 m is part of a group of numbers shown in *kilometres*, it should be expressed as 0.725 km.

Exponential notation, instead of prefixes, is preferred in calculations. For example, 900 mm^2 equals 0.9×10^{-3} m^2.

One measurement unit should be used throughout a set of drawings so that numerical values can be represented by numbers only. In a drawing dimensioned in millimetres, five-digit numbers are acceptable; for example, 32 845 mm.

1.8.4 SI UNITS FOR DESIGN AND CONSTRUCTION

There are two principal factors in any measurement statement, a *number*, which is always written first, and a *measurement unit*. Correct selection of units for use in building construction is essential to minimize errors and optimize coordination. Figure 1.8-11 lists SI units needed for building construction in terms of activity.

FIGURE 1.8-11 SI Units Used in Building Construction

Activity	Quantity	Unit	Symbol
Land surveying	Linear measure	kilometre, metre	km, m
	Area	square kilometre	km^2
		hectare (10 000 m^2)	ha
		square metre	m^2
Excavating	Linear measure	metre, millimetre	m, mm
	Volume	cubic metre	m^3
Concreting	Linear measure	metre, millimetre	m, mm
	Area	square metre	m^2
	Volume	cubic metre	m^3
	Temperature	degree Celsius	°C
	Water capacity	litre	L
Constituents	Mass (weight)	kilogram, gram	kg, g
Reinforcement	Cross section	square millimetre	mm^2
Trucking	Distance	kilometre	km
	Mass (weight)	metric ton (1000 kg)	t
Paving and plastering	Linear measure	metre, millimetre	m, mm
	Area	square metre	m^2
Bricklaying	Linear measure	metre, millimetre	m, mm
	Area	square metre	m^2
	Mortar-volume	cubic metre	m^3
Carpentry/joinery	Linear measure	metre, millimetre	m, mm
Steelworking	Linear measure	metre, millimetre	m, mm
	Mass (weight)	metric ton (1000 kg)	t
		kilogram, gram	kg, g
Roofing	Linear measure	metre, millimetre	m, mm
	Area	square metre	m^2
	Slope	millimetre/metre	mm/m
Painting	Linear measure	metre, millimetre	m, mm
	Area	square metre	m^2
	Capacity	litre, millilitre	L, mL
Glazing	Linear measure	metre, millimetre	m, mm
	Area	square metre	m^2
Plumbing	Linear measure	metre, millimetre	m, mm
	Mass (weight)	kilogram, gram	kg, g
	Capacity	litre	L
	Pressure	kilopascal	kPa
Drainage	Linear measure	metre, millimetre	m, mm
	Area	hectare (10 000 m^2)	ha
		square metre	m^2
	Volume	cubic metre	m^3
	Slope	millimetre/metre	mm/m
Electrical services	Linear measure	metre, millimetre	m, mm
	Frequency	hertz	Hz
	Power	watt, kilowatt	W, kW
	Energy	megajoule (1 kWh = 3.6 MJ)	MJ
	Electric current	ampere	A
	Electric potential	volt, kilovolt	V, kV
	Resistance	ohm	Ω
Mechanical services	Linear measure	metre, millimetre	m, mm
	Volume	cubic metre	m^3
	Capacity	litre	L
	Airflow	metre/second	m/s
	Volume flow	cubic metre/second	m^3/s
		litre/second	L/s
	Temperature	degree Celsius	°C
	Force	newton, kilonewton	N, kN
	Pressure	kilopascal	kPa
	Energy, work	kilojoule, megajoule	kJ, MJ

Figures 1.8-12, 1.8-13, 1.8-14, 1.8-15, 1.8-16, and 1.8-17 contain lists of recommended units for use in building design and construction in terms of quantity and application.

1.8.4.1 Linear Measurement

The preferred SI units for linear measurement in building design and construction are the millimetre (mm) and the metre (m). These have an inherently greater precision than units in the customary system, which uses decimalized fractions in sitework, but uses feet, inches, and fractions in construction work. Using the millimetre as the principal unit of construction measurement makes fractions unnecessary. One millimetre provides an accuracy of 1⁄25 in.; five digits can express any length up to 328 ft. The use of whole numbers greatly speeds calculations and helps to avoid errors; calculations can be done twice as fast with only one-fifth the rate of error (Fig. 1.8-18). In the customary system, fractions have to be reconciled to a common denominator and then added, inches have to be added and then con-

FIGURE 1.8-12 Space and Time: Geometry, Kinematics, and Periodic Phenomena

Quantity and SI Unit Symbol	Preferred Units (Symbols)	Other Acceptable Units	Unit Name	Typical Applications	Remarks
Length (m)	m		metre	**Architecture and General Engineering** Levels, overall dimensions, spans, column heights, etc., in engineering computations. **Estimating and Specification** Trenches, curbs, fences, lumber lengths, pipes and conduits; lengths of building materials generally. **Land Surveying** Boundary and cadastral surveys; survey plans; heights, geodetic surveys, contours. **Hydraulic Engineering** Pipe and channel lengths, depth of storage tanks or reservoirs, height of potentiometric head, hydraulic head, piezometric head.	Use metres on all drawings with scale ratios between 1:200 and 1:2000. Where required for purposes of accuracy. show dimensions to three decimal places.
	mm		millimetre	**Architecture and General Engineering** Spans, dimensions in buildings, dimensions of building products; depth and width of sections; displacement, settlement, deflection, elongation; slump of concrete, size of aggregate; radius of gyration, eccentricity; detailed dimensions generally; rainfall. **Estimating and Specification** Lumber cross sections; thicknesses, diameters, sheet metal gauges, fasteners; all other building product dimensions. **Hydraulic Engineering** Pipe diameters; radii of groundwater wells; height of capillary rise; precipitation, evaporation.	Use millimetres on drawings with scale ratios between 1:1 and 1:200. Avoid the use of centimetres (cm). Where cm is shown in documents, such as for snow depth, body dimensions, or carpet sizes, etc. convert to mm or m.
	km		kilometre	Distances for transportation purposes, geographical, or statistical applications in surveying; long pipes and channels.	
	μm		micrometre	Thickness of coatings (paint, galvanizing, etc.), thin sheet materials, size of find aggregate.	
Area (m^2)	m^2		square metre	**General Applications** Small land areas; area of cross section of earthworks, channels, and larger pipes; surface area of tanks and small reservoirs; areas in general. **Estimating and Specification** Site clearing; floor areas; paving, masonry construction, roofing, wall and floor finishes, plastering, paintwork, glass areas, membranes, lining materials, insulation, reinforcing mesh, formwork; areas of all building components.	($1\ m^2 = 10^6\ mm^2$) Replaces sq. ft.; sq. yd.; and square. Specify masonry construction by wall area × wall thickness.

FIGURE 1.8-12 *(Continued)*

Quantity and SI Unit Symbol	Preferred Units (Symbols)	Other Acceptable Units	Unit Name	Typical Applications	Remarks
Area (m^2) *(Continued)*					
	mm^2		square millimetre	Area of cross section for structural and other sections, bars, pipes, rolled and pressed shapes, etc.	Avoid the use of cm^2 (square centimetre) by conversion to mm^2 ($1\ cm^2 = 10^2\ mm^2 = 100\ mm^2$).
	km^2		square kilometre	Large catchment areas or land areas.	
		ha	hectare	Land areas; irrigation areas; areas on boundary and other survey plans.	($1\ ha = (10^2\ m)^2 = 10^4\ m^2 = 10\ 000\ m^2$)
Volume, capacity (m^3)	m^3		cubic metre	**General Applications** Volume, capacity (large quantities); volume of earthworks, excavations, filling, waste removal; concrete, sand, all bulk materials supplied by volume, and large quantities of lumber. **Hydraulic Engineering** Water distribution, irrigation, diversions, sewage, storage capacity, underground basins.	$1\ m^3 = 1000\ L$ As far as possible, use the cubic metre as the preferred unit of volume for all engineering purposes.
	mm^3		cubic millimetre	Volume, capacity (small quantities)	
		L	litre	Volume of fluids and containers for fluids; liquid materials, domestic water supply, consumption; volume/capacity of fuel tanks	The litre and its multiples or submultiples may be used for domestic and industrial supplies of liquids $1\ L = 1\ dm^3 = 1000\ cm^3$ $1\ mL = 1\ cm^3$
		mL	millilitre	Volume of fluids and containers for fluids (limited application only)	
		cm^3	cubic centimetre	Limited application only (small quantities)	$1\ cm^3 = 1000\ mm^3 = 10^{-6}\ m^3$
Modulus of section (m^3)	mm^3		millimetre to third power	Geometric properties of structural sections, such as plastic section modulus, elastic section modulus, etc.	
	m^3		metre to third power		
Second moment of area (m^4)	mm^4		millimetre to fourth power	Geometric properties of structural sections, such as moment of inertia of a section, torsional constant of a cross section.	
	m^4		metre to fourth power		
Plane angle (rad)	rad		radian	Generally used in calculations only to preserve coherence.	Slopes and gradients may be expressed as a ratio or as a percentage: 26.57 = 1:2 = 50% = 0.4637 rad (1 rad = 57.2958°)
	mrad		milliradian		
		(_._°)	degree (of arc)	**General Applications** Angular measurement in construction (generally using decimalized degrees); angle of rotation, torsion, shear resistance, friction, internal friction, etc. **Land Surveying** Bearings shown on boundary and cadastral survey plans; geodetic surveying	
Time, time interval (s)	s		second	Time used in methods of test; all calculations involving derived units with a time component, in order to preserve coherence.	Avoid the use of minute (min) as often as possible
		h	hour	Time used in methods of test; all calculations involving labor time, plant hire, maintenance periods, etc.	(1 h = 3600 s 1 d = 86 400 s = 86.4 ks)
		d	day		
		a	annum (year)		

FIGURE 1.8-12 *(Continued)*

Quantity and SI Unit Symbol	Preferred Units (Symbols)	Other Acceptable Units	Unit Name	Typical Applications	Remarks
Frequency (Hz)	Hz		hertz	Frequency of sound, vibration, shock; frequency of electromagnetic waves	(1 Hz = 1/s = s^{-1}) Replaces cycle(s) per second (c/s or cps)
	kHz		kilohertz		
	MHz		megahertz		
Rotational frequency, speed of rotation (s^{-1})		r/s	revolution per second	Widely used in the specification of rotational speed of machinery; use r/min (revolutions per minute) only for slow-moving machinery.	(1 r/s = 2π rad/s = 60 r/min)
Velocity, speed (m/s)	m/s		metre per second	Calculations involving rectilinear motion, velocity and speed in general; wind velocity; velocity of fluids; pipe flow velocity	(1 m/s = 3.6 km/h)
		km/h	kilometre per hour	Wind speed; speed used in transportation; speed limits	
		mm/h	millimetre per hour	Rainfall intensity	
Angular velocity (rad/s)	rad/s		radian per second	Calculations involving rotational motion	
Linear acceleration (m/s^2)	m/s^2		metre per second squared	Kinematics, and calculations of dynamic forces	Recommended value of acceleration of gravity for use in U.S.: $g_{us} = 9.8\ m/s^2$
Volume rate of flow (m^3/s)	m^3/s		cubic metre per second	Volumetric flow in general; flow in pipes, ducts, channels, rivers; irrigation spray demand	(1 m^3/s = 1000 L/s)
		m^3/h	cubic metre per hour		
		m^3/d	cubic metre per day		
		L/s	litre per second	Volumetric flow of fluids only	
		L/d	litre per day		

Note: Metre and *litre* are preferred international spellings applied to these base units and all their multiples and submultiples, such as kilometre and millilitre.

FIGURE 1.8-13 Mechanics, Statics, and Dynamics

Quantity and SI Unit Symbol	Preferred Units (Symbols)	Other Acceptable Units	Unit Name	Typical Applications	Remarks
Mass (kg)	kg		kilogram	Mass of materials in general, mass of structural elements and machinery	Use kilograms (kg) in calculations and specifications
	g		gram	Mass of samples of material for testing	Masses greater than 10^4 kg (10 000 kg) may be conveniently expressed in metric tons (t): 1 t = 10^3 kg = 1 Mg = 1000 kg
		t	metric ton	Mass of large quantities of materials, such as structural steel, reinforcement, aggregates, concrete, etc.; ratings of lifting equipment	

FIGURE 1.8-13 *(Continued)*

Quantity and SI Unit Symbol	Preferred Units (Symbols)	Other Acceptable Units	Unit Name	Typical Applications	Remarks
Mass per unit length (kg/m)	kg/m		kilogram per metre	Mass per unit length of sections, bars, and similar items of uniform cross section	Also known as Linear Density
		g/m	gram per metre	Mass per unit length of wire and similar material of uniform cross section	
Mass per unit area (kg/m^2)	kg/m^2		kilogram per square metre	Mass per unit area of slabs, plates, and similar items of uniform thickness or depth; rating for load-carrying capacities on floors (display on notices only)	Do not use in stress calculations
		g/m^2	gram per square metre	Mass per unit area of thin sheet materials, coatings, etc.	
Mass density, concentration (kg/m^3)	kg/m^3		kilogram per cubic metre	Density of materials in general; mass per unit volume of materials in a concrete mix; evaluation of masses of structures and materials	Also known as Mass per Unit Volume ($1\ kg/m^3 = 1\ g/L$) ($1\ g/m^3 = 1\ mg/L$)
		g/m^3	gram per cubic metre	Mass per unit volume (concentration) in pollution control	
		$\mu g/m^3$	microgram per cubic metre		
Momentum (kg·m/s)	kg·m/s		kilogram metre per second	Used in applied mechanics; evaluation of impact and dynamic forces	
Moment of inertia ($kg \cdot m^2$)	$kg \cdot m^2$		kilogram square metre	Rotational dynamics; evaluation of the restraining forces required for propellers, windmills, etc.	
Mass per unit time (kg/s)	kg/s		kilogram per second	Rate of transport of material on conveyors and other materials handling equipment	1 kg/s = 3.6 t/h
		t/h	metric ton per hour		
Force (N)	N		newton	Unit of force for use in calculations	$1\ N = 1\ kg \cdot m/s^2$
	kN		kilonewton	Forces in structural elements, such as columns, piles, ties, prestressing tendons, etc.; concentrated forces; axial forces; reactions; shear force; gravitational force	
Force per unit length (N/m)	N/m		newton per metre	Unit for use in calculations	
	kN/m		kilonewton per metre	Transverse force per unit length on a beam, column, etc.; force distribution in a linear direction	
Moment of force, torsional, or bending moment (N·m)	N·m		newton metre	Bending moments (in structural sections), torsional moment; overturning moment; tightening tension for high strength bolts; torque in engine drive shafts, axles, etc.	
	kN·m		kilonewton metre		
	MN·m		meganewton metre		
Pressure, stress, modulus of elasticity (Pa)	Pa		pascal	Unit for use in calculations; low differential pressure in fluids	($1\ Pa = 1\ N/m^2$)
	kPa		kilopascal	Uniformly distributed pressure (loads) on floors; soil bearing pressure; wind pressure (loads), snow loads, dead and live loads; pressure in fluids; differential pressure (e.g., in ventilating systems)	Where wind pressure, snow loads, dead and live loads are shown in kN/m^2 change units to kPa

FIGURE 1.8-13 *(Continued)*

Quantity and SI Unit Symbol	Preferred Units (Symbols)	Other Acceptable Units	Unit Name	Typical Applications	Remarks
Pressure, stress, modulus of elasticity (Pa) *(Continued)*	MPa		megapascal	Modulus of elasticity; stress (ultimate, proof, yield, permissible, calculated, etc.) in structural materials; concrete and steel strength grades	1 MPa = 1 MN/m^2 = 1 N/mm^2
	GPa		gigapascal	Modulus of elasticity in high strength materials	
	μPa		micropascal	Sound pressure (20 μPa is the reference quantity for sound pressure level)	
Compressibility (Pa^{-1})	1/Pa		reciprocal (of) pascal	Settlement analysis (coefficient of compressibility), bulk compressibility	(1/Pa = 1 m^2/N)
	1/kPa		reciprocal (of) kilopascal		
Dynamic viscosity (Pa·s)	Pa·s		pascal second	Shear stresses in fluids	(1 Pa·s = 1 $N{\cdot}s/m^2$) The centipoise (cP) = 10^{-3} Pa·s will not be used
	mPa·s		millipascal second		
Kinematic viscosity (m^2/s)	m^2/s		square metre per second		The centistokes (cSt) = 10^{-6} m^2/s will not be used 1 cSt = 1 mm^2/s
	mm^2/s		square millimetre per second	Computation of Reynolds' number, settlement analysis (coefficient of consolidation)	
Work energy (J)	J		joule	Energy absorbed in impact testing of materials; energy in general; calculations involving mechanical and electrical energy	
	kJ		kilojoule		
	MJ		megajoule		
		kWh	kilowatthour	Electrical energy applications only	1 kWh = 3.6 MJ
Impact strength (J/m^2)	J/m^2		joule per square metre	Impact strength; impact ductility	
	kJ/m^2		kilojoule per square metre		
Power (W)	W		watt	Power in general (mechanical, electrical, thermal); input/output rating, etc. of motors, engines, heating and ventilating plant and other equipment in general	
	kW		kilowatt		
	MW		megawatt	Power input/output rating, etc. of heavy power plant	
	pW		picowatt	Sound power level (1 pW is the reference quantity for sound power level)	

Note: Metre is the preferred international spelling applied to this base unit and all its multiples and submultiples.

FIGURE 1.8-14 Thermal Effects and Heat Transfer

Quantity and SI Unit Symbol	Preferred Units (Symbols)	Other Acceptable Units	Unit Name	Typical Applications	Remarks
Temperature value (K)	K		kelvin	Expression of thermodynamic temperature; calculations involving units of temperature	($t_{°C} = T_K - 273.15$)
		°C	degree Celsius	Common temperature scale for use in meteorology and general applications; ambient temperature values	Temperature values will normally be measured in °C (degree Celsius)
Temperature interval (K)	K		kelvin	Heat transfer calculations; temperature intervals in test methods; etc.	(1 K = 1°C) The use of K (kelvin) in compound units is recommended
		°C	degree Celsius		
Coefficient of linear thermal expansion (1/K)	1/K		reciprocal (of) kelvin	Expansion of materials subject to a change in temperature (generally expressed as a ratio per kelvin or degree Celsius)	
		1/°C	reciprocal (of) degree Celsius		
Heat, quantity of heat (J)	J		joule	Thermal energy calculations. Enthalpy, latent heat, sensible heat	
	kJ		kilojoule		
	MJ		megajoule		
Specific energy, specific latent heat, combustion heat (mass basis) (J/kg)	J/kg		joule per kilogram	Heat of transition; heat and energy contained in materials; combustion heat per unit mass; calorific value of fuels (mass basis); specific sensible heat; specific latent heat in psychrometric calculations	
	kJ/kg		kilojoule per kilogram		
	MJ/kg		megajoule per kilogram		
Energy density, combustion heat (volume basis) (J/m^3)	J/m^3		joule per cubic metre	Combustion heat per unit volume	
	kJ/m^3		kilojoule per cubic metre		(1 kJ/m^3 = 1 J/L)
	MJ/m^3		megajoule per cubic metre	Calorific value of fuels (volume basis)	(1 MJ/m^3 = 1 kJ/L)
Heat capacity, entropy (J/K)	J/K		joule per kelvin	Thermal behavior of materials, heat transmission calculations, entropy	
	kJ/K		kilojoule per kelvin		
Specific heat capacity, specific entropy (J/(kg·K))	J/(kg·K)		joule per kilogram kelvin	Thermal behavior of materials, heat transmission calculations	
	kJ/(kg·K)		kilojoule per kilogram kelvin		
Heat flow rate (W)	W		watt	Heat flow rate through walls, windows, etc.; heat demand	(1 W = 1 J/s)
	kW		kilowatt		
Power density, heat flux density, irradiance (W/m^2)	W/m^2		watt per square metre	Density of power or heat flow through building walls and other heat transfer surfaces; heat transmission calculations	
	kW/m^2		kilowatt per square metre		

FIGURE 1.8-14 *(Continued)*

Quantity and SI Unit Symbol	Preferred Units (Symbols)	Other Acceptable Units	Unit Name	Typical Applications	Remarks
Heat release rate (W/m³)	W/m³		watt per cubic metre	Rate of heat release per unit volume over time (for gases and liquids)	$W/m^3 = J/(m^3 \cdot s)$
	kW/m³		kilowatt per cubic metre		
Thermal conductivity (W/(m·K))	W/(m·K)		watt per metre kelvin	Estimation of thermal behavior of materials and systems; heat transmission calculations Thermal conductivity of structural and building materials in fire-resistance testing, insulation, etc.	1 W/(m·K) = 1 W/(m·°C) (k = value)
Coefficient of heat transfer (thermal conductance) (W/(m²·K))	W/(m²·K)		watt per square metre kelvin	Heat transfer calculations for buildings, building components, and equipment. Transmittance of construction elements	(U = value)
	kW/(m²·K)		kilowatt per square metre kelvin		
Thermal resistivity ((m·K)/W)	(m·K)/W		metre kelvin per watt	Heat transmission calculations (reciprocal of thermal conductivity)	
Thermal insulance (thermal resistance ((m²·K)/W)	(m²·K)/W		square metre kelvin per watt	Heat transmission calculations (reciprocal of thermal conductance)	(R = value)

Note: Metre and *litre* are the preferred international spellings applied to these units and all their multiples and submultiples.

FIGURE 1.8-15 Electricity and Magnetism

Quantity and SI Unit Symbol	Preferred Units (Symbols)	Other Acceptable Units	Unit Name	Typical Applications	Remarks
Electric current (A)	A		ampere	Maintenance rating of an electrical installation. Leakage current	
	kA		kiloampere		
	mA		milliampere		
	μA		microampere		
Magnetomotive force, magnetic potential difference (A)				Used in the calculations involved in magnetic circuits	
Magnetic field strength, magnetization (A/m)	A/m		ampere per metre	Magnetic field strength used in calculation of magnetic circuitry such as transformers, magnetic amplifiers, and general cores	(1 kA/m = 1 A/mm)
	kA/m		kiloampere per metre		
Current density (A/m²)	A/m²		ampere per square metre	Design of cross-sectional area of electrical conductor	
	kA/m²		kiloampere per square metre		
		A/mm²	ampere per square millimetre		(1 A/mm² = 1 MA/m²)

FIGURE 1.8-15 *(Continued)*

Quantity and SI Unit Symbol	Preferred Units (Symbols)	Other Acceptable Units	Unit Name	Typical Applications	Remarks
Electric charge, quantity of electricity (C)	C		coulomb	The voltage on a unit with capacitive characteristics may be related to the amount of charge present (e.g., electrostatic precipitators). Storage battery capacities	1 C = 1 A·s Do not use ampere hour: 1 A·h = 3.6 kC
	kC		kilocoulomb		
	μC		microcoulomb		
	nC		nanocoulomb		
	pC		picocoulomb		
Electric potential, potential difference, electromotive force (V)	V		volt		1 V = 1 W/A
	MV		megavolt		
	kV		kilovolt		
	mV		millivolt		
	μV		microvolt		
Electric field strength (V/m)	V/m		volt per metre	The electric field strength gives the potential gradient at points in space. This may be used to calculate or test electrical parameters such as dielectric strength	
	MV/m		megavolt per metre		
	kV/m		kilovolt per metre		
	mV/m		millivolt per metre		
	μV/m		microvolt per metre		
Active power (W)	W		watt	The useful power of an electrical circuit is expressed in watts (W). (The apparent power in an electrical circuit is expressed in volt-amperes, (V·A).)	1 W = 1 V·A
	GW		gigawatt		
	MW		megawatt		
	kW		kilowatt		
	mW		milliwatt		
	μW		microwatt		
Capacitance (F)	F		farad	Electronic components. Electrical design and performance calculators	1 F = 1 C/V
	mF		millifarad		
	μF		microfarad		
	nF		nanofarad		
	pF		picofarad		
Resistance (Ω)	Ω		ohm	The design of electrical devices with resistance, such as motors, generators, heaters, electrical distribution systems, etc.	1 Ω = 1 V/A
	GΩ		gigaohm		
	MΩ		megaohm		
	kΩ		kiloohm		
	mΩ		milliohm		
Conductance, admittance, susceptance (S)	S		siemens		The siemens (S) was formerly known as mho.
	MS		megasiemens		
	kS		kilosiemens		
	mS		millisiemens		
	μS		microsiemens		
Resistivity (Ω·m)	Ω·m		ohm metre		
	GΩ·m		gigaohm metre		
	MΩ·m		megaohm metre		

FIGURE 1.8-15 *(Continued)*

Quantity and SI Unit Symbol	Preferred Units (Symbols)	Other Acceptable Units	Unit Name	Typical Applications	Remarks
Resistivity (Ω·m) *(Continued)*	kΩ·m		kiloohm metre		
	mΩ·m		milliohm metre		
	μΩ·m		microohm metre		
	nΩ·m		nanoohm metre		
Electrical conductivity (S/m)	S/m		siemens per metre	A parameter for measuring water quality	
	MS/m		megasiemens per metre		
	kS/m		kilosiemens per metre		
	μS/m		microsiemens per metre		
Magnetic flux, flux of magnetic induction (Wb)	mWb		milliweber	Used in the calculations involved in magnetic circuits	1 Wb = 1 V·s
Magnetic flux density, magnetic induction (T)	T		tesla	Used in the calculations involved in magnetic circuits	1 T = 1 Wb/m^2
	mT		millitesla		
	μT		microtesla		
	nT		nanotesla		
Magnetic vector potential (Wb/m^2)	kWb/m^2		kiloweber per square metre	Used in the calculations involved in magnetic circuits	
Self-inductance, mutual inductance, permeance (H)	H		henry	Used in analysis and calculations involving transformers	1 H = 1 Wb/A
	mH		millihenry		
	μH		microhenry		
	nH		nanohenry		
	pH		picohenry		
Reluctance (1/H)	1/H		reciprocal of henry	Design of motors and generators	
Permeability (H/m)	H/m		henry per metre	Permeability gives the relationship between the magnetic flux density and the magnetic field strength	
	μH/m		microhenry per metre		
	nH/m		nanohenry per metre		

Note: Metre is the preferred international spelling applied to this base unit and all its multiples and submultiples.

FIGURE 1.8-16 Lighting

Quantity and SI Unit Symbol	Preferred Units (Symbols)	Other Acceptable Units	Unit Name	Typical Applications	Remarks
Luminous intensity (cd)	cd		candela		
Solid angle (sr)	sr		steradian		
Luminous flux (lm)	lm		lumen	Luminous flux of light sources, lamps, and lightbulbs	1 lm = 1 cd·sr Already in general use
	klm		kilolumen		

FIGURE 1.8-16 *(Continued)*

Quantity and SI Unit Symbol	Preferred Units (Symbols)	Other Acceptable Units	Unit Name	Typical Applications	Remarks
Quantity of light (lm·s)	lm·s		lumen second		1 lm·h = 3600 lm/s
		lm·h	lumen hour		
Luminance (cd/m^2)	cd/m^2		candela per square metre	Assessment of surface brightness; luminance of light sources, lamps, and lightbulbs; calculation of glare in lighting layouts	Replaces stilb (1 sb = 10^4 cd/m^2) and apostilb (1 apostilb = $cd/\pi m^2$)
	kcd/m^2		kilocandela per square metre		
		cd/mm^2	candela per square millimetre		
Illuminance (lx)	lx		lux	Luminous flux per unit area used in determination of illumination levels and design and evaluation of interior lighting layouts. (Outdoor daylight illumination on a horizontal plane ranges up to 100 klx.)	(a) Formerly referred to as illumination 1 lx = 1 lm/m^2 (b) Replaces (1 ph = 10^4 lx) (c) Luminous exitance is described in lm/m^2
	klx		kilolux		
Light exposure (lx·s)	lx·s		lux second		
	klx·s		kilolux second		
Luminous efficacy (lm/W)	lm/W		lumen per watt	Rating of luminous efficacy of artificial light sources	

Note: Metre is the preferred international spelling applied to this base unit and all its multiples and submultiples.

FIGURE 1.8-17 Acoustics

Quantity and SI Unit Symbol	Preferred Units (Symbols)	Other Acceptable Units	Unit Name	Typical Applications	Remarks
Wavelength (m)	m		metre	Definition of sound wave pitch	
	mm		millimetre		
Area of absorptive surface (m^2)	m^2		square metre	Calculations of room absorption	
Period, periodic time (s)	s		second	Measurement of time and reverberation time	
	ms		millisecond		
Frequency (Hz)	Hz		hertz	Frequency ranges in sound absorption calculations and sound pressure measurement	1 Hz = 1 cycle per second (cps)
	kHz		kilohertz		
Instantaneous sound pressure (Pa)	Pa		pascal	Measurement of sound pressure; reference level for sound pressure is 20 μPa, but sound pressure is shown in decibels (dB) based on a logarithmic scale Sound pressure level $L_p = 20 \log_{10} \frac{\text{actual pressure (Pa)}}{20 \times 10^{-6}\ \text{(Pa)}}$	Do not use dyne (1 dyn = 10 μPa)
	mPa		millipascal		
	μPa		micropascal		
Sound power, sound energy flux (W)	W		watt	Measurement of sound power; reference level for sound power is 1 pW Sound power level $L = 10 \log_{10} \frac{\text{actual power (W)}}{10^{-12}\ \text{(W)}}$	
	mW		milliwatt		
	μW		microwatt		
	pW		picowatt		

FIGURE 1.8-17 *(Continued)*

Quantity and SI Unit Symbol	Preferred Units (Symbols)	Other Acceptable Units	Unit Name	Typical Applications	Remarks
Sound intensity (W/m^2)	W/m^2		watt per square metre	Measurement of sound intensity; reference level for sound intensity is 1 pW/m^2	
	pW/m^2		picowatt per square metre	Sound intensity level $L_1 = 10 \log_{10} \dfrac{\text{actual intensity } (W/m^2)}{10^{-12}\ (W/m^2)}$ dB	
Specific acoustic impedance ((Pa·s)/m)	(Pa·s)/m		pascal second per metre	Sound impedance measurement	(1 (Pa·s)/m = 1 (N·s)/m^3)
Acoustic impedance, resistance ((Pa·s)/m^3)	(Pa·s)/m^3		pascal second per cubic metre	Sound impedance measurement	

Note: Metre is the preferred international spelling applied to this base unit and all its multiples and submultiples.

FIGURE 1.8-18 Comparison of Calculations in Customary Units and in SI Units

Division of a Linear Dimension into Equal Parts

Customary Units	SI Units
A detailed drawing shows a height of 8′ 10″ between two floors connected by a stair with 14 risers, but not the height of each riser. What is the target height of risers so that none of them varies by more than 1⁄16″?	A detail drawing shows a height of 2690 mm between two floors connected by a stair with 14 risers, but not the height of each riser. What is the target height of risers so that none of them varies by more than 1 mm?
8′ 10″ = 106″ 106 ÷ 14 = 7.57″ 7.57″ ≅ 7 9⁄16″ (Check) 7 9⁄16″ × 14 = 105 7⁄8″ (−1⁄8″) Target height of riser: 7 9⁄16″ with 2 risers at 7 5⁄8″ (+1⁄16″ each)	2690 ÷ 14 = 192.1 mm (Check) 192 × 14 = 2688 mm (+2 mm) Target height of riser: 192 mm with 2 risers at 193 mm each

Comment: Division of measurement in customary units requires decimalization and change to and from decimal fractions.

Area and Volume Calculations to Determine Order Quantity

Customary Units	SI Units
A flat concrete slab for an industrial structure has the following dimensions:	A flat concrete slab for an industrial structure has the following dimensions:
Length: 25′0″; width: 26′6″; thickness: 5″	Length: 7.5 m; width: 8.4 m; thickness: 125 mm (0.125 m)
Determine the order quantity of concrete to the nearest half of a cubic yard.	Determine the order quantity of concrete to the nearest half of a cubic meter.
$\dfrac{25 \times 26.5 \times 5}{12} = 276$ cu. ft. 276 ÷ 27 = 10.22 cu. yd. Order: 10 1⁄2 cu. yd.	7.5 × 8.4 × 0.125 = 7.85 m^3 Order: 8 m^3

Comment: Customary units need to be decimalized, and factors are required with inches and conversion of cubic feet to cubic yards. Metric volume is derived directly by multiplication of dimensions.

verted to feet and inches, and, finally, the feet must be added.

Sizes and dimensions should not be listed in centimetres, because (1) the order of magnitude between millimetres and centimetres is only 10 and using both units causes confusion and (2) using millimetres permits the use of whole numbers rather than fractions within appropriate tolerances for all building and product dimensions.

Unit symbols may be deleted on a drawing if a note such as "Dimensions are shown in millimetres (or metres), unless otherwise noted," appears on the drawing.

Whole numbers always indicate millimetres. A five-digit number can indicate a length up to 328 ft.; 327 ft. 10 11⁄16 inches equals 99 941 mm. A length up to 32 ft. 9 in. can be shown by a four-digit number; a length of 3 ft. 3 5⁄16 in. by a three-digit number; 1⁄25 in. is 1 millimetre.

Numbers carried to three decimal places always indicate metres; for example: 3.600, 0.300, 0.025.

Using millimetres and metres saves time and space in drawing, typing, and computer applications. It also improves clarity in drawings that have many dimensions.

The drawing scales used by architects and engineers have traditionally differed. Architects have generally used scales such as 1⁄4 in. equals 1 ft., which resulted in nondecimal scale ratios; even when engineers used seemingly metric scales such as 1 in. equals 100 ft., the ratio is nondecimal (Fig. 1.8-19). *Metrication* will eliminate these differences.

The consistent use of fewer scales has definite benefits: (1) drawings prepared on transparent media drawn to the same scale may be easily superimposed, which makes drawing faster and easier and reduces the possibility of errors and omissions, (2) drawings lend themselves more to computer-aided reduction and enlargement techniques, and (3) consistently sized drawings are easier to understand and speed construction.

FIGURE 1.8-19 Metric and Nonmetric Drawing Scales

Traditional Scales (Expressed as Ratios)		Metric Scales Preferred	Metric Scales Other
Full size	(1:1)	1:1	
Half full size 4″ = 1′0″ 3″ = 1′0″	(1:2) (1:3) (1:4)		1:2
		1:5	
2″ = 1′0″ 1½″ = 1′0″	(1:6) (1:8)		
		1:10	
1″ = 1′0″ ¾″ = 1′0″	(1:12) (1:16)		
		1:20	
½″ = 1′0″	(1:24)		
			1:25*
⅜″ = 1′0″ ¼″ = 1′0″	(1:32) (1:48)		
		1:50	
1″ = 5′0″ 3⁄16″ = 1′0″ ⅛″ = 1′0″	(1:60) (1:64) (1:96)		
		1:100	
1″ = 10′0″ 3⁄32″ = 1′0″ 1⁄16″ = 1′0″	(1:120) (1:128) (1:196)		
		1:200	

*Limited use

1.8.4.2 Area

In general, the preferred unit for area is the derived SI unit, *square metre* (m^2). Large areas can be expressed in square *kilometres* (km^2) or *hectares* (ha). *Hectare* is an alternative name for square *hectometre*, which is equal to 10 000 m^2. However, the *hectare* (ha) should be used only for land and water surface area measurements. Small areas should be expressed in *square millimetres* (mm^2). The *square centimetre* (cm^2) should be avoided to minimize confusion. Sometimes the area of buildings or products is indicated by linear dimensions (for example: 40 by 90 mm); generally, the width is indicated first and height second.

1.8.4.3 Volume and Fluid Capacity

The derived SI unit, *cubic metre* (m^3), is the preferred measurement unit for volume in construction and for large storage tank capacities.

The non-SI units *litre* (L) and *millilitre* (mL) are the preferred measurement units of *fluid capacity* (liquid volume) and gas volume. A *litre* is equal to 1⁄1000 of a cubic *metre* or 1 cubic *decimetre* (dm^3) and can be expressed as 1 L = 10^{-3} m^3; 1 L = 1 dm^3; or 1 m^3 = 1000 L.

A cubic metre contains one billion (10^9) cubic millimetres. Cubic centimetres and decimetres should be converted to the preferred cubic metres or millimetres whenever possible (see Fig. 1.8-12).

1.8.4.4 Geometric Cross-Sectional Properties

The expression of *geometric cross-sectional* properties of structural sections involves raising the unit of length to the third, fourth, or sixth power with *exponential notation* (see Fig. 1.8-1), as follows:

Modulus of section (S) is written in units of mm^3 or m^3. For example, 1 mm^3 = 0.000 000 001 = 10^{-9} m^3.

Second moment of area (I) and torsional constant (J) are written in units of mm^4 or m^4. For example, 1 mm^4 = 10^{-12} m^4.

Warping constant (C) is written in units of mm^6 or m^6. For example, 1 mm^6 = 10^{-18} m^6).

Figure 1.8-20 shows the *cross-sectional properties* of a wide flange beam 460 mm deep weighing 82 kg/m per unit of length.

FIGURE 1.8-20 Properties of a 460-mm-Deep Wide Flange Beam Weighing 82 kg/m

Property	Designation and Units
Plastic modulus	$Z_x = 1.835 \times 10^6$ mm^3 (1.835×10^{-3} m^3)
Second moment of area	$I_{x-x} = 0.371 \times 10^9$ mm^4 (0.371×10^3 m^4)
Torsional constant	$J = 0.691 \times 10^6$ mm^4 (0.691×10^{-3} m^4)
Warping constant	$C_w = 0.924 \times 10^{12}$ mm^6 (0.924×10^{-6} m^6)

1.8.4.5 Plane Angle

The preferred unit for expressing *plain angles* is the SI supplementary unit, *radian* (rad). The radian should be used in engineering calculations when possible, because it is a *coherent unit.* However, use of the customary *degree of arc* and its decimal submultiples (as in 26.250) is permissible when using the *radian* is not convenient, especially on maps and engineering drawings, and in surveying and construction operations. The use of *minutes* and *seconds* is discouraged, with the recognition that they will probably continue in use for some time.

1.8.4.6 Time Intervals

The *second* (s) is the SI base unit for time measurement. The *day* (d), *hour* (h), and *minute* (min) are non-SI alternatives. The minute should be avoided as much as possible, however, to minimize the variety of units in which time is a dimension.

Flow rates should be expressed in *cubic metres per second, litres per second,* or *cubic metres per hour*:

1 m^3/s = 1000 L/s (do not use 60 m^3/min)
1 L/s = 3.6 m^3/h (do not use 60 L/min)
1 m^3/h = 1000 L/h (do not use 1667 L/min)

The *month*, which varies in length, should not be used, unless a specific calendar month is indicated.

The *calendar year* (*a* for *annum*) represents 365 days, equal to 31 536 000 seconds.

1.8.4.7 Temperature

The *kelvin* (K) is the SI base unit of temperature. The derived SI unit, *degree Celsius* (°C), however, is acceptable and will replace the customary unit, *degree Fahrenheit* (°F), in everyday use (Fig. 1.8-21). The earlier name for Celsius, *centigrade*, is no longer used.

The temperature interval of 1 degree Celsius exactly equals 1 kelvin. A temperature expressed in degrees Celsius is equal to the temperature in kelvins less 273.15: K = –273.15°C; 0°C = 273.15 K.

The zero point of the kelvin scale is *absolute zero*, which is the point at which all molecular movement stops. Therefore, there are no negative (minus) temperatures in the kelvin scale. The lack of negative numbers makes the kelvin scale useful for scientific applications, but the large numeric values make it cumbersome for everyday use. For example, room temperature may be expressed as 68 *degrees Fahrenheit*, 20 *degrees Celsius*, or 293.15 *kelvin*; body temperature as 98.6 *degrees Fahrenheit*, 37 *degrees Celsius*, or 310.15 *kelvin*; boiling water as 212 *degrees Fahrenheit*, 100 *degrees Celsius*, or 373.15 *kelvin*.

1.8.4.8 Mass, Weight, and Force

SI differs from both the customary system and other forms of metric measurement in the use of explicit and distinctly separate units for *mass* and *force*. There is considerable confusion in the use of the term *weight* to mean either *force* or *mass*.

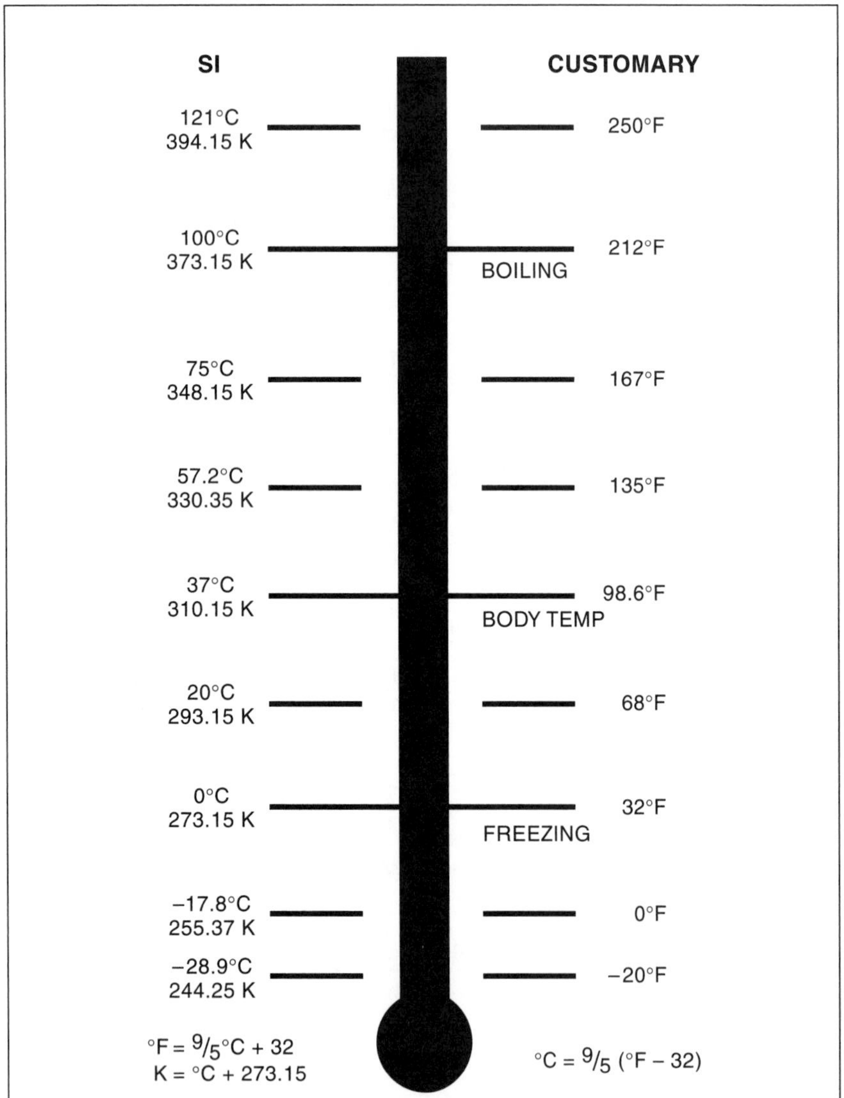

MASS

In everyday use, *weight* generally means *mass*. When one speaks of a person's weight, the quantity referred to is really *mass*.

Mass is the amount of matter there is in a body. An iron ball and a baseball of exactly the same size take up the same amount of space, but there is more matter (greater density or mass) per unit volume in the iron ball; therefore, the iron ball has more mass. *Density* is the ratio of mass to volume.

The *kilogram* (kg) is the SI base unit for mass, the amount of matter in an object that is constant and independent of gravitational attraction. Multiples of the kilogram are formed by attaching an SI prefix to *gram* (*milligram*, *megagram*). A *megagram* (Mg), which is equal to 1 000 000 grams, should be used to measure large masses formerly expressed in tons.

The name *ton* has been given to several large mass units that are widely used in commerce and technology: the *long ton* of 2240 lb., the *short ton* of 2000 lb., and the *metric ton* of 1000 kg, which is also called a *tonne*. None of these terms is applicable to SI. The term *metric ton* may be used, but it should be restricted to commercial usage and never used with *prefixes*. Use of the term *tonne* is discouraged.

FORCE

In technical papers, *weight* has often been used to mean *force of gravity*, which is a particular force related to *gravitational acceleration*. The force of gravity varies from place to place on the earth and on other planets. A body of the same mass will weigh more or have more force on the earth than on the moon (Fig. 1.8-22).

To alleviate confusion, the term *weight* should be avoided in technical practice. *Force of gravity* should denote *force*. The *newton* (N) is the derived SI unit of force. It equals mass times acceleration: F = kg × m/s^2.

Previous metric systems have used the term *kilogram-force of newton*, often dropping the suffix *force*, which caused confusion. The use of *kilogram-force* should be avoided. The *newton* also supersedes the term *kilopond* (kp). The *newton* can also be used to designate *pressure*, *stress*, *energy*, *work*, and *quantity of heat*, *power*, and many *electrical units*.

SI	CUSTOMARY
121°C 394.15 K	250°F
100°C 373.15 K	212°F BOILING
75°C 348.15 K	167°F
57.2°C 330.35 K	135°F
37°C 310.15 K	98.6°F BODY TEMP
20°C 293.15 K	68°F
0°C 273.15 K	32°F FREEZING
−17.8°C 255.37 K	0°F
−28.9°C 244.25 K	−20°F
°F = 9/5°C + 32 K = °C + 273.15	°C = 9/5 (°F – 32)

FIGURE 1.8-21 Temperature equivalents.

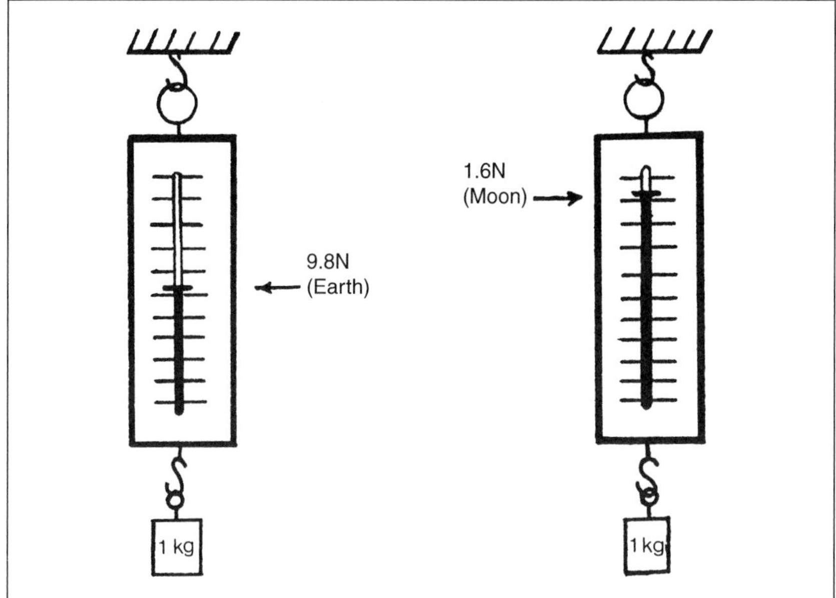

FIGURE 1.8-22 Although the mass is the same, a kg of matter weighs more than six times as much on the earth as it does on the moon.

1.8.4.9 Pressure, Stress, and Modulus of Elasticity

The derived unit *pascal* (Pa), which is the equivalent of *newtons per square metre*, is the unit for pressure, stress, and elastic modulus.

Pressure is a force applied uniformly over a surface and measured as a force per unit of area.

Stress is the effect of a force on a body, such as squeezing (compression), pulling (tension), and bending (compression plus tension). Stress is generally accompanied by *strain*, a physical change in the body's shape or dimension.

The pascal and its prefixed multiples—*megapascal* (MPa), *kilopascal* (kPa), and *gigapascal* (GPa)—replace a large number of customary units such as pounds/square inch (lb./in.^{2} or psi). It also supersedes a few non-SI metric units such as the *bar* (equal to 100 kPa) and the *millibar* (equal to 100 Pa). Old *metric gravitational units* for pressure and stress such as *kilogram-force per square centimetre* (kgf/cm^2) should not be used; neither should units such as *bar* and *torr*. Use of the *millibar* will continue in order to permit meteorologists to communicate within their profession, but even they should use the *kilopascal* (kPa) to present data to the public.

Though it may sometimes be useful to report test results in N/mm^2 (identical to MN/m^2), calculations and results should be shown in MPa or kPa.

1.8.4.10 Energy, Work, and Quantity of Heat

The *joule* (J) with its multiples is the derived SI unit of *energy*, *work*, and *quantity of heat*. It is equal to a *newton metre* (N·m) and a *watt second* (W·s). The *joule* supersedes many traditional units, such as *Btu*, *therm*, *calorie*, *kilocalorie*, and *foot pound-force*.

The *kilowatthour* (kWh) has been the customary unit of *energy consumption*; the preferred unit is the *megajoule* (MJ). Since it would be needlessly costly to recalibrate electric meters, the kWh will continue to be used in electrical applications, but should not be introduced into new areas.

The *joule* should never be used for *torque*; for this property the designated unit is *newton metre* (N·m).

For dimensional consistency in rotational dynamics, *torque* should be expressed as *newton metre per radian* (N·m/rad) and *moment of inertia* as *kilogram square metre per radian squared* (kg·m^2/rad^2).

1.8.4.11 Power and Heat Flow Rate

The *watt* (W), the derived SI unit for *power* and *heat flow rate*, is already in worldwide use as a unit for *electrical power*. It will replace some traditional units of power and *heat flow rate*: (1) *horsepower* (electric, boiler), (2) *foot pound-force per hour* (or *minute* or *second*), (3) *Btu per hour*, (4) *calorie* (or *kilocalorie*) *per minute* (or *second*), and (5) *ton of refrigeration*.

1.8.4.12 Electrical Engineering

The only changes in electrical engineering will be the renaming of the unit of conductance to *siemens* (S) instead of *mho* and the use of *hertz* (Hz) for frequency instead of *cycles per second* (cps).

1.8.4.13 Lighting Units

The *candela* (cd), the SI base unit for *luminous intensity*, is already in common use and replaces *candle* and *candlepower*.

Luminance will be expressed in *candela per square metre* (cd/m^2), which replaces *candela per square foot*, *foot lambert*, and *lambert*.

The derived SI unit *lumen* (lm) is also currently in common use. *Illuminance* is expressed in *lux* (lx) which is equal to *lumens per square metre* (lm/m^2) and replaces *lumens per square foot* and *footcandle*.

1.8.4.14 Velocity

There is no base, supplementary, or derived SI unit expressing *velocity*. The accepted non-SI unit is *kilometre per hour*.

1.8.4.15 Speed of Rotation

There is no SI unit for *speed of rotation*. The accepted non-SI unit is *revolutions per minute*.

1.8.4.16 Dimensionless Quantities

Ratios such as *relative humidity* (RH), *specific gravity* (SG), and *decibel* (dB) remain unchanged in the conversion to SI.

1.8.5 CONVERSION OF NUMERICAL VALUES

Metrication is the process of change to a new set of reference units. It involves a change in numerical values by the ratio between customary and SI reference units. This ratio is known as a *conversion factor*.

Types of conversions are (1) *exact*, (2) *soft*, (3) *hard*, and (4) *rationalization*.

1.8.5.1 Exact Conversion

An *exact*, also called *direct*, *conversion* is a change from a customary value to its pre-

FIGURE 1.8-23 Exact, Soft, and Hard Conversions of Customary Values

Customary Value	Exact Conversion	Soft Conversion	Hard Conversion	% Change for Hard Conversion
24 inches (2 ft.)	609.6 mm	610 mm	600 mm	−1.6
1 square (100 sq. ft.)	9.2903 m^2	9.3 m^2	10 m^2	+7.6
4 cu. yd.	3.058 m^3	3.1 m^3	3 m^3	−1.9
1 quart	0.946 L	0.95 L	1 L	+5.7

cise SI equivalent, generally expressed to a number of decimal places. For example, 24 in. is exactly 609.6 mm (Fig. 1.8-23).

Exact conversion is often needed in the early transitional period of a change to SI units but should be thought of as only the first step to the other types of conversions. Exact conversions will rarely, if ever, be needed in design and construction, but they are an essential step before rounding or *rationalization* of traditional values can begin. Figure 1.8-24 shows exact conversion of fractions and decimals to three places. Figure 1.8-25 shows conversion tables for the most common SI units used in building design and construction.

Conversion of quantities should be handled carefully so that accuracy is neither sacrificed nor exaggerated. *Conversion factors* and quantities should not be rounded prior to multiplication or division.

Calculated values should be rounded to the minimum number of digits that will maintain the appropriate accuracy. For example, if a metric tape measure shows values to the nearest millimetre, showing lineal values more precisely would be unnecessary if the final dimensions are to be applied using a tape measure.

For calculations using numbers of differing precision, the final answer should be no more precise than the least precise number (Fig. 1.8-26).

1.8.5.2 Soft Conversion

Soft conversion is a change in the way a product's size is described without significant change in its actual size. In general, a soft conversion is the rounding of an exact conversion to a more workable numerical value. For example, a piece of 24-in.-wide gypsum lath would become 609.6 mm wide if exact conversion were desired. A soft conversion would simply round off the number to 610 mm. Such a practice would give the appearance of metric conversion but would not result in the most ideal product width, 600 mm, which would be attained through hard conversion (see Fig. 1.8-23).

Since soft conversion is only a metric veneer, a change made only on paper, soft converted products are in production now. Some are already labeled with both customary and metric measures. Although soft conversion may be of some value in specifications, standards, codes, and other technical data, it is inappropriate in product sizing because it perpetuates customary practices and diminishes the advantages of SI.

1.8.5.3 Hard Conversion

In a *hard conversion* there is a definite change in physical size or magnitude from the customary values to preferred SI values (see Fig. 1.8-23). Such a conversion often leads to incompatibility between products made using the SI system and those earlier ones produced in customary measure. For example, in a hard conversion, the 24-in.-wide gypsum lath mentioned in Section 1.8.5.2 might actually be changed to 610 mm wide, instead of

FIGURE 1.8-24 Fraction, Decimal, and Metric Equivalents

Fractional Inch	Decimal Equivalent*	Millimetres	Fractional Inch	Decimal Equivalent*	Millimetres
1/64	0.0156	0.397	23/64	0.3594	9.128
1/32	0.0313	0.794	3/8	0.3750	9.525
3/64	0.0469	1.191	25/64	0.3906	9.922
1/16	0.0625	1.588	2/5	0.4000	10.160
5/64	0.0781	1.984	13/32	0.4063	10.319
1/12	0.0833	2.117	5/12	0.4167	10.583
3/32	0.0938	2.381	27/64	0.4219	10.716
1/10	0.1000	2.540	7/16	0.4375	11.112
7/64	0.1094	2.778	29/64	0.4531	11.509
1/8	0.1250	3.175	15/32	0.4688	11.906
9/64	0.1406	3.572	31/64	0.4844	12.303
1/6	0.1667	4.233	1/2	0.5000	12.700
11/64	0.1719	4.366	33/64	0.5156	13.097
3/16	0.1875	4.700	17/32	0.5313	13.494
1/5	0.2000	5.080	35/64	0.5469	13.891
13/64	0.2031	5.159	9/16	0.5625	14.288
7/32	0.2188	5.556	37/64	0.5781	14.684
15/64	0.2344	5.953	7/12	0.5833	14.817
1/4	0.2500	6.350	19/32	0.5938	15.081
17/64	0.2656	6.747	3/5	0.6000	15.240
9/32	0.2813	7.144	39/64	0.6094	15.478
19/64	0.2969	7.541	5/8	0.6250	15.875
3/10	0.3000	7.620	41/64	0.6406	16.272
5/16	0.3125	7.937	21/32	0.6563	16.669
1/3	0.3333	8.467	2/3	0.6667	16.933
11/32	0.3438	8.731	43/64	0.6719	17.066

FIGURE 1.8-24 *(Continued)*

Fractional Inch	Decimal Equivalent*	Millimetres	Fractional Inch	Decimal Equivalent*	Millimetres
11/16	0.6875	17.463	27/32	0.8438	21.431
7/10	0.7000	17.780	55/64	0.8594	21.828
45/64	0.7031	17.859	7/8	0.8750	22.225
23/32	0.7188	18.256	57/64	0.8906	22.622
47/64	0.7344	18.653	9/10	0.9000	22.860
3/4	0.7500	19.050	29/32	0.9063	23.019
49/64	0.7656	19.447	11/12	0.9167	23.283
25/32	0.7813	19.844	59/64	0.9219	23.416
51/64	0.7969	20.241	15/16	0.9375	23.813
4/5	0.8000	20.320	61/64	0.9531	24.209
13/16	0.8125	20.638	31/32	0.9688	24.606
53/64	0.8281	21.034	63/64	0.9844	25.003
5/6	0.8333	21.167	1″	1.000	25.400

*To convert a decimal to percentage, carry the decimal point two places to the right. Thus, 63/64, or 0.9844, equals 98.44%.

just being called that. This would be within normal tolerances and create little interface problems. More likely, however, the width would be changed to an even 600 mm wide, which is 1.6% smaller. Additional examples include the following: a 1-quart container would be replaced by a 1-*litre* container, which is a 5.7% increase in volume; a 12-ft. length would become a preferred 3600 mm, which is a 1.6% decrease in length.

Though the changes are more difficult to make, hard conversion is better than the other means of conversion because it creates a simpler SI system. Figure 1.8-23 shows the differences in converting customary values using exact, soft, and hard conversions.

Unfortunately, hard conversion is unrealistic and unacceptable in some cases, such as in changing the standard railroad gauge. The U.S. railroad gauge currently has a nonpreferred dimension of 4 ft. 8½

FIGURE 1.8-25 Conversion Factors for the Most Common Units Used in Building Design and Construction

Metric to Customary			Customary to Metric		
Length					
1 km	= 0.621 371	mile (international)	1 mile (international)	= *1.609 344*	km
	= 49.7096	chain	1 chain	= 20.1168	m
1 m	= 1.093 61	yd	1 yd	= *0.9144*	m
	= 3.280 84	ft	1 ft	= *0.3048*	m
1 mm	= 0.039 370 1	in		= *304.8*	mm
			1 in	= *25.4*	mm
			(1 U.S. survey foot	= 0.304 800 6	m)
Area					
1 km²	= 0.386 101	mile² (U.S. survey)	1 mile² (U.S. survey)	= 2.590 00	km²
1 ha	= 2.471 04	acre (U.S. survey)	1 acre (U.S. survey)	= 0.404 687	ha
1 m²	= 1.195 99	yd²		= 4046.87	m²
	= 10.7639	ft²	1 yd²	= 0.836 127	m²
1 mm²	= 0.001 550	in²	1 ft²	= 0.092 903	m²
			1 in²	= *645.16*	mm²
Volume, Modulus of Section					
1 m³	= 0.810 709 × 10⁻³	acre ft	1 acre ft	= 1233.49	m³
	= 1.307 95	yd³	1 yd³	= 0.764 555	m³
	= 35.3147	ft³	100 board ft	= 0.235 974	m³
	= 423.776	board ft	1 ft³	= 0.028 316 8	m³
1 mm³	= 61.0237 × 10⁻⁶	in³		= 28.3168	L (dm³)
			1 in³	= 16 387.1	mm³
				= 16.3871	mL (cm³)
(Fluid) Capacity					
1 L	= 0.035 314 7	ft³	1 gal (U.S. liquid)*	= 3.785 41	L
	= 0.264 172	gal (U.S.)	1 qt (U.S. liquid)	= 946.353	mL
	= 1.056 69	qt (U.S.)	1 pt (U.S. liquid)	= 473.177	mL
1 mL	= 0.061 023 7	in³	1 fl oz (U.S.)	= 29.5735	mL
	= 0.033 814	fl oz (U.S.)			
			*1 gal (U.K.) approximately 1.2 gal (U.S.)		
Second Moment of Area					
1 mm⁴	= 2.402 51 × 10⁻⁶	in⁴	1 in⁴	= 416 231	mm⁴
				= 0.416 231 × 10⁻⁶	m⁴

FIGURE 1.8-25 *(Continued)*

Metric to Customary			Customary to Metric		
Plane Angle					
1 rad	= 57° 17′ 45″	(degree)	1° (degree)	= 0.017 453 3	rad
	= 57.2958°	(degree)		= 17.4533	mrad
	= 3437.75′	(minute)	1′ (minute)	= 290.888	μrad
	= 206 265″	(second)	1″ (second)	= 4.848 14	μrad
Velocity, Speed					
1 m/s	= 3.280 84	ft/s	1 ft/s	= *0.3048*	m/s
	= 2.236 94	mile/h	1 mile/h	= *1.609 344*	km/h
1 km/h	= 0.621 371	mile/h		= *0.447 04*	m/s
Acceleration					
1 m/s^2	= 3.280 84	ft/s^2	1 ft/s^2	= *0.3048*	m/s^2
Volume Rate of Flow					
1 m^3/s	= 35.3147	ft^3/s	1 ft^3/s	= 0.028 316 8	m^3/s
	= 22.8245	million gal/d	1 ft^3/min	= 0.471 947	L/s
	= $0.810\ 709 \times 10^{-3}$	acre ft/s	1 gal/min	= 0.063 090 2	L/s
1 L/s	= 2.118 88	ft^3/min	1 gal/h	= 1.051 50	mL/s
	= 15.850 3	gal/min	1 million gal/d	= 43.8126	L/s
	= 951.022	gal/h	1 acre ft/s	= 1233.49	m^3/s
Temperature Interval					
1°C	= *1* K	= *1.8*°F	1°F	= 0.555 556	°C or K
				= 5⁄9°C	= 5⁄9 K
Equivalent Temperature Value ($t_{°C} = T_K - 273.15$)					
$t_{°C}$	= $5/9\ (t_F - 32)$		t_F	= $9/5\ t_{°C} + 32$	
Mass					
1 kg	= 2.204 62	lb (avoirdupois)	1 ton (short)*	= 0.907 185	metric ton
	= 35.2740	oz (avoirdupois)		= 907.185	kg
1 metric	= 1.102 31	ton (short, 2000 lb)	1 lb	= 0.453 592	kg
ton	= 2204.62	lb	1 oz	= 28.3495	g
1 g	= 0.035 274	oz	1 pennyweight	= 1.555 17	g
	= 0.643 015	pennyweight			
			*(1 long ton (2240 lb) = 1016.05 kg)		
Mass per Unit Length					
1 kg/m	= 0.671 969	lb/ft	1 lb/ft	= 1.488 16	kg/m
1 g/m	= 3.547 99	lb/mile	1 lb/mile	= 0.281 849	g/m
Mass per Unit Area					
1 kg/m^2	= 0.204 816	lb/ft^2	1 lb/ft^2	= 4.882 43	kg/m^2
1 g/m^2	= 0.029 494	oz/yd^2	1 oz/yd^2	= 33.9057	g/m^2
	= $3.277\ 06 \times 10^{-3}$	oz/ft^2	1 oz/ft^2	= 305.152	g/m^2
Density (Mass per Unit Volume)					
1 kg/m^3	= 0.062 428	lb/ft^3	1 lb/ft^3	= 16.0185	kg/m^3
	= 1.685 56	lb/yd^3	1 lb/yd^3	= 0.593 276	kg/m^3
1 t/m^3	= 0.842 778	ton/yd^3	1 ton/yd^3	= 1.186 55	t/m^3
Moment of Inertia					
1 $kg{\cdot}m^2$	= 23.7304	$lb{\cdot}ft^2$	1 $lb{\cdot}ft^2$	= 0.042 140 1	$kg{\cdot}m^2$
	= 3417.17	$lb{\cdot}in^2$	1 $lb{\cdot}in^2$	= 292.640	$kg{\cdot}mm^2$
Mass per Unit Time					
1 kg/s	= 2.204 62	lb/s	1 lb/s	= 0.453 592	kg/s
1 t/h	= 0.984 207	ton/h	1 ton/h	= 1.016 05	t/h
Force					
1 MN	= 112.404	tonf (ton-force)	1 tonf (ton-force)	= 8.896 44	kN
1 kN	= 0.112 404	tonf	1 kip (1000 lbf)	= 4.448 22	kN
	= 224.809	lbf (pound-force)	1 lbf (pound-force)	= 4.448 22	N
1 N	= 0.224 809				
Moment of Force, Torque					
1 N·m	= 0.737 562	lbf·ft	1 lfb·ft	= 1.355 82	N·m
	= 8.850 75	lbf·in	1 lbf·in	= 0.112 985	N·m
1 kN·m	= 0.368 781	tonf·ft	1 tonf·ft	= 2.711 64	kN·m
	= 0.737 562	kip·ft	1 kip·ft	= 1.355 82	kN·m

FIGURE 1.8-25 *(Continued)*

Metric to Customary			Customary to Metric		
Force per Unit Length					
1 N/m	= 0.068 521 8	lbf/ft	1 lbf/ft	= 14.5939	N/m
1 kN/m	= 0.034 260 9	tonf/ft	1 lbf/in	= 175.127	N/m
			1 tonf/ft	= 29.187 8	kN/m
Pressure, Stress, Modulus of Elasticity (Force per Unit Area) (1 Pa = 1 N/m^2)					
1 MPa	= 0.072 518 8	tonf/in^2	1 tonf/in^2	= 13.7895	MPa
	= 10.4427	tonf/ft^2	1 ton/ft^2	= 95.7605	kPa
	= 145.038	lbf/in^2	1 kip/in^2	= 6.894 76	MPa
1 kPa	= 20.8854	lbf/ft^2	1 lbf/in^2	= 6.894 76	kPa
			1 lbf/ft^2	= 47.8803	Pa
Work Energy, Heat (1 J = 1 N·m = 1 W·s)					
1 MJ	= 0.277 778	kWh	1 kWh (550 ft·lbf/s)	= *3.6*	MJ
1 kJ	= 0.947 817	Btu	1 Btu (Int. Table)	= 1.055 06	kJ
1 J	= 0.737 562	ft·lbf		= 1055.06	J
			1 ft·lbf	= 1.355 82	J
Power, Heat Flow Rate					
1 kW	= 1.341 02	hp (horsepower)	1 hp	= 0.745 700	kW
1 W	= 3.412 14	Btu/h		= 745.700	W
	= 0.737 562	ft·lbf/s	1 Btu/h	= 0.293 071	W
			1 ft·lbf/s	= 1.355 82	W
Heat Flux Density					
1 W/m^2	= 0.316 998	Btu/(ft^2·h)	1 Btu/(ft^2·h)	= 3.154 59	W/m^2
Coefficient of Heat Transfer					
1 W/(m^2·K)	= 0.176 100	Btu/(ft^2·h·°F)	1 Btu/(ft^2·h·°F)	= 5.678 26	W/(m^2·K)
Thermal Conductivity					
1 W/(m·K)	= 0.577 789	Btu/(ft·h·°F)	1 Btu/(ft·h·°F)	= 1.730 73	W/(m·K)
Calorific Value (Mass and Volume Basis)					
1 kJ/kg	= 0.429 923	Btu/lb	1 Btu/lb	= *2.326*	kJ/kg
(1 J/g)	= 0.429 923	Btu/lb	1 Btu/lb	= *2.326*	(J/g)
1 kJ/m^3	= 0.026 839 2	Btu/ft^3	1 Btu/ft^3	= 37.2589	kJ/m^3
Thermal Capacity (Mass and Volume Basis)					
1 kJ/(kg·K)	= 0.238 846	Btu/(lb·°F)	1 Btu/(lb·°F)	= *4.1868*	kJ/(kg·K)
1 kJ/(m^3·K)	= 0.014 910 7	Btu/(ft^3·°F)	1 Btu/(ft^3·°F)	= 67.0661	kJ/(m^3·K)
Illuminance					
1 lx (lux)	= 0.092 903	lm/ft^2 (footcandle)	1 lm/ft^2 (footcandle)	= 10.7639	lx (lux)
Luminance					
1 cd/m^2	= 0.092 903	cd/ft^2	1 cd/ft^2	= 10.7639	cd/m^2
	= 0.291 864	footlambert	1 footlambert	= 3.426 26	cd/m^2
1 kcd/m^2	= 0.314 159	lambert	1 lambert	= 3.183 01	kcd/m^2

Note: Conversion factors are taken to six significant figures, where appropriate. Values in italics are exact conversions.

(Copyright ASTM; reprinted with permission.)

in. The soft conversion is 1435 mm. Since railroad tracks must conform to the large number of existing railroad cars, the soft conversion to 1435 mm is acceptable and would cost little as compared with a hard change to 1450 mm or 1500 mm.

In general, hard conversion should be used when possible because it results in preferred values, which are easier to work with. However, the decision to make a hard conversion must always be weighed against the cost of changing the many products or components that are intended to work together.

1.8.5.4 Rationalization

Rationalization, which often occurs during the conversion from customary to SI units of measure, is the development of standards for greater product compatibility and cost-effectiveness. It results from selection of the most rational, preferred, and economical alternatives, after research into the technical implications and an appraisal of the economic impact. Rationalization usually results in changes in product sizes and often in a reduction in the number of a particular product's available sizes. For instance, during metrication, the number of reinforcing bar sizes has been reduced from 11 to 10 in Australia, from 11 to 9 in Britain, and from 11 to 8 in Canada.

Rationalization can also be applied to product standards, combining them and reducing their number.

FIGURE 1.8-26 Calculating Exact Conversion

Addition or Subtraction	
Original numbers:	153 787 400
	647 200 000
	21 000 000
	821 987 400
Rounded to the least precise and added:	
	154 000 000
	647 000 000
	21 000 000
	822 000 000

Multiplication or Division

175.5 × 3.475 = 609.8625

Round the answer to 609.9 because the least precise number, 175.5, has the decimal carried to only one place.

The objectives of rationalization are (1) to reduce the variety of products by eliminating unnecessary or uneconomical sizes, (2) to gain optimum steps in a new product range, (3) to harmonize various technical data, standards, and codes by using preferred values, and (4) to simplify practices and procedures through less complicated numerical description.

1.8.6 PREFERRED DIMENSIONS AND COORDINATION

The internationally recommended basic *building module* is 100 mm. It is small enough to provide flexibility in design, large enough to promote simplification of the number of component sizes, and acceptable worldwide. It is slightly smaller than the most common module of the customary system, 4 in. For example, the width of a wood stud is nominally 4 in.; the width of a brick and the height of a block are nominally 8 in.; the spacing of studs and the nominal length of a block are 16 in.; the width of gypsum lath and insulating sheathing is 24 in.

Preferred SI sizes are those that are preferred over others for building components and assemblies. These are usually selected multiples of the basic 100 mm building module. During the change to SI, it is important to select preferred dimensions for building products to ensure that they will work together when combined into walls, floors, and other building assemblies.

Dimensional coordination is the systematic application of preferred and related dimensions in the design of buildings and the manufacture and positioning of building components, assemblies, and elements.

Modular coordination is dimensional coordination using multiples of the international 100-mm building module.

1.8.6.1 Advantages of Preferred Dimensions

Selecting preferred building dimensions, such as floor-to-floor heights, provides an incentive to produce components of a standardized height for use in high-rise buildings, such as precast concrete stairs, ducts, shaft walls, and cladding panels.

Preferred building dimensions will also (1) simplify distributors' inventory by reducing product-size variety, (2) make it easier to combine interrelated components into a building, and (3) reduce the time and waste normally associated with assembling uncoordinated products.

In addition to reducing construction cost and time, the application of uniform preferred dimensions will simplify communications and give additional impetus to the use of computers for design and detailing.

1.8.6.2 The 100-mm Module

The 100-mm *building module* has many compelling advantages:

1. It is 1.6% smaller than the common 4-in. module of the customary system, which means that existing equipment can often be used to manufacture new metric products and that new products will often fit into existing openings.

2. The use of the *millimetre* as the working unit in building design and construction makes multimodular dimensions visible immediately. For example, 6400 mm is exactly 64 basic 100 mm modules; it is also 16 modules of 400 mm each, as opposed to sixteen 16-in. stud spaces. In customary units, the use of feet and inches hides the multiplier in modular dimensions. For example, it is not obvious that 21 ft. 4 in. is 64 whole 4-in. modules, or that 26 ft. 8 in. is a whole multiple of 16-in. stud spacings.

3. The modular square, 100 mm × 100 mm, has an area of 0.01 m^2. This relationship is useful in cost comparisons because the unit cost per modular square, in cents, has the same number as the unit cost per square metre in dollars. For example, carpet costing 30 cents a modular square would cost $30.00 a square metre.

4. The modular cube, 100 mm × 100 mm × 100 mm, represents a volume of exactly 1 litre, or 0.001 m^3. One litre of water has a specific gravity of 1 and a mass of exactly 1 kilogram, a relationship that is useful in many calculations. For example, a modular cube of concrete, with a specific gravity of 2.4, has a mass of 2.4 kilograms. The calculation of the mass of a 100-mm-wide concrete wall 2800 mm high and 7200 mm long is 28 × 72 × 2.4 = 4834.4 kg. The mass per unit length is 28 × 10 (1 metre) × 2.4 = 672 kg/m.

5. One dimension, the millimetre, expresses all building dimensions, including tolerances and limitations, which will enable much greater accuracy in construction.

1.8.7 METRIC CONVERSIONS APPLICABLE TO THIS BOOK

Previous editions of this book did not contain metric measure equivalents for the customary system units shown. Metric units have been added to this edition. However, in order to avoid altering copyrighted documents, metric units have not been added to some figures in this edition that are copyrighted and reprinted here by permission of the copyright holders. Figure 1.8-27 is a list of the customary units included in the figures for which metric units have not been indicated.

FIGURE 1.8-27 Metric Conversions for Customary Units Used in This Book and Not Converted So as Not to Alter Them as Provided by the Acknowledged Organizations

Temperature

°F	°C	°F	°C	°F	°C	°F	°C	°F	°C
-40	-40	18.6	-7.4	39	3.9	72.2	22.3	110	43.3
-30	-34.4	19	-7.2	40	4.4	73	22.8	200	93.3
-20	-28.9	20	-6.7	41	5	73.1	22.83	204.4	95.8
-10	-23.3	20.3	-6.5	43	6.1	73.5	23.1	300	148.9
0	-17.8	21	-6.1	44	6.7	74	23.3	400	204.4
1	-17.2	21.3	-5.9	44.6	7	74.3	23.5	426.7	219.3
5	-15	22	-5.6	45	7.2	74.8	23.8	500	260
6	-14.4	22.6	-5.2	46	7.8	75	23.9	600	315.6
6.3	-14.3	23	-5	47	8.3	75.6	24.2	648.9	342.7
7	-13.9	23.6	-4.7	48	8.9	76.1	24.5	800	426.7
8	-13.3	24	-4.4	49	9.4	76.9	24.9	871.1	466.2
9	-12.8	25	-3.9	50	10	77	25	1000	537.8
10	-12.2	26	-3.3	51	10.6	77.5	25.3	1093.3	589.6
10.3	-12.1	26.3	-3.2	55	12.8	78	25.6	1100	593.3
11.3	-11.5	27	-2.8	58	14.4	78.3	25.7	1200	648.9
12	-11.1	27.6	-2.4	60	15.6	79.7	26.5	1315.6	713.1
13	-10.6	28	-2.2	61	16.1	80	26.7	1500	815.6
13.6	-10.2	28.6	-1.9	62	16.7	81.2	27.3	1600	871.1
14	-10	30	-1.1	65	18.3	82	27.8	1800	982.2
15	-9.4	31	-0.6	66	18.9	83	28.3	2000	1093.3
15.3	-9.3	32	0	68	20	85	29.4	2400	1315.6
16	-8.9	33.6	0.9	69	20.6	90	32.2	2700	1482.2
16.3	-8.7	34	1.1	70	21.1	95	35	2800	1537.8
17	-8.3	35	1.7	71	21.7	100	37.8	3000	1648.9
17.6	-8	36	2.2	71.1	21.72	105	40.6	5000	2760
18	-7.8	38	3.3	72	22.2	108	42.2		

Mass

grains/lb.	kg/kg	lb.	kg	lb.	kg	lb.	kg	lb.	kg
24	0.0034	5	2.27	35	15.88	80	36.29	144	65.32
36	0.0051	6	2.72	40	18.14	81	36.74	158	71.67
40	0.0057	10	4.54	48	21.77	82	37.19	180	81.65
54	0.0077	15	6.80	50	22.68	90	40.82	192	87.09
77	0.0110	20	9.07	55	24.95	94	42.64	200	90.72
80	0.0114	21	9.53	60	27.22	100	45.36	240	108.86
108	0.0154	22.5	10.21	64	29.03	105	47.63	250	113.40
120	0.0171	24	10.89	65	29.48	110	49.90	300	136.08
160	0.0229	25	11.34	70	31.75	120	54.43	360	163.29
200	0.0286	30	13.61	72	32.66	125	56.70		
		33	14.97	75	34.02	140	63.50		

Mass per Unit Area

oz./sq. ft.	g/m²
2	610
4	1220

psi	MPa	psi	MPa	psi	MPa	psi	MPa	psi	MPa
		360	2.48	1375	9.5	2100	14.5	5000	35
		400	2.76	1400	9.6	2150	14.8	6000	41
		450	3.1	1500	10	2200	15	7000	48
		500	3.4	1575	10.9	2250	15.5	8000	55
		600	4	1600	11	2300	15.8	10,000	69
55	0.38	640	4.4	1650	11.4	2325	16	19,000	131
60	0.41	700	4.8	1675	11.5	2400	16.5	20,000	138
70	0.48	720	4.96	1700	11.7	2500	17	30,000	207
75	0.52	750	5.2	1750	12.1	2550	17.6	40,000	276
100	0.69	800	5.5	1800	12.4	2650	18.3	50,000	345
115	0.79	900	6.2	1825	12.6	2700	18.6	60,000	413
120	0.83	950	6.6	1875	12.9	2850	19.6	70,000	482
140	0.96	1000	6.9	1925	13.3	3000	21	75,000	517
160	1.1	1020	7	1950	13.4	3150	21.7	80,000	551
200	1.4	1175	8.1	1975	13.6	3300	22.7	100,000	689
225	1.6	1200	8.3	2000	13.8	3500	24	120,000	827
300	2	1250	8.6	2035	14	4000	28	140,000	965
350	2.4	1300	8.96	2050	14.1	4500	31		
320	2.2	1350	9.4						

FIGURE 1.8-27 *(Continued)*

Mass per Unit Area *(Continued)*

E	MPa
1	6890
1.2	8268
1.4	9646
1.5	10335
1.6	11024
1.7	11713
1.8	12402
1.9	13091
2	13780
2.2	15158

E	MPa
2.3	15847
2.4	16536
2.5	17225
2.6	17914

psf	kg/m²
20	97.65
25	122.06
30	146.47
35	170.88

psf	kg/m²
40	195.3
45	219.71
50	244.12

psf	MPa
100	4.79
150	7.18
200	9.58
400	19.15

psf	MPa
3000	143.64
6000	287.28

lb./square	kg/square
200	90.72
220	99.79
250	113.39
260	117.93
280	127

lb./square	kg/square
350	158.75
450	204.11

tons/sq. ft.	Mg/m²
1	9.77
2	19.53
2.5	24.61
3	29.29
5	48.82
10	97.65

Mass per Unit Length

lb./ft.	kg/m	lb./ft.	kg/m	lb./ft.	kg/m	lb./ft.	kg/m	lb./ft.	kg/m
0.376	0.5595	1.502	2.2352	2.67	3.9734	4.303	6.4036	7.65	11.384
0.668	0.9941	2.044	3.0418	3.4	5.0597	5.313	7.9066	13.6	20.239
1.043	1.5522								

Mass per Unit Volume

lb./cu. ft.	kg/m³	lb./cu. ft.	kg/m³	lb./cu. ft.	kg/m³	lb./cu. ft.	kg/m³	lb./cu. yd.	kg/m³
10	160.19	15	240.28	20	320.37	105	1681.94	520	308.5
13	208.24	18	288.33	85	1361.57	125	2002.31	564	334.61

Length

mil	mm
.35	0.0089
1	0.0254
2	0.0508
4	0.1016
6	0.1524

in.	mm
0.0141	0.36
0.0142	0.36
0.015	0.38
0.020	0.51
0.0255	0.65
0.028	0.71
0.0343	0.87
0.038	0.97
0.044	1.12
0.0449	1.14
0.050	1.27
0.056	1.42
0.0625 (1/16)	1.60
0.064	1.63
0.071	1.80
0.0821	2.09
0.085	2.16
0.098	2.49
0.102	2.56
0.113	2.87
0.125 (1/8)	3.20
0.131	3.33
0.143	3.63
0.148	3.76
0.162	4.11
0.1799	4.57
0.180	4.58

in.	mm
0.188 (3/16)	4.76
0.191	4.83
0.192	4.88
0.203	5.16
0.2031	5.16
0.207	5.19
0.219 (7/32)	5.60
0.225	5.72
0.2299	5.839
0.2300	5.842
0.239	6.07
0.244	6.073
0.2499	6.35
0.25 (1/4)	6.40
0.262	6.65
0.263	6.68
0.286	7.26
0.313 (5/16)	7.9
0.334	8.43
0.344 (11/32)	8.73
0.350	8.89
0.375 (3/8)	9.50
0.383	9.73
0.431	10.95
0.437	11.10
0.469 (15/32)	11.91
0.487	12.37
0.500 (1/2)	12.70
0.525	13.34
0.540	13.72
0.594 (19/32)	15.1
0.612	15.54
0.625 (5/8)	15.90
0.648	16.46

in.	mm
0.7	17.78
0.719 (23/32)	18.3
0.750 (3/4)	19.10
0.79	20.07
0.864	21.95
0.875 (7/8)	22.2
0.889	22.58
0.987	25.07
1	25.40
1.128	28.65
1.178	29.92
1.185	30.10
1.25 (1 1/4)	31.80
1.270	32.26
1.410	35.81
1.5 (1 1/2)	38.10
1.571	39.90
1.58	40.13
1.693	43.00
1.75 (1 3/4)	44.45
1.963	49.86
2	50.80
2.2 (2 1/5)	55.88
2.25 (2 1/4)	57.20
2.257	57.33
2.356	59.84
2.5 (2 1/2)	63.50
2.625 (2 5/8)	66.70
2.666 (2 2/3)	67.72
2.749	69.82
2.75 (2 3/4)	69.85
2.813 (2 13/16)	71.44
3	76.20
3.142	79.81

in.	mm
3.2 (3 1/5)	81.28
3.5 (3 1/2)	88.90
3.544	90.02
3.625 (3 5/8)	92.08
3.75 (3 3/4)	95.30
3.990	101.35
4	101.60
4.25 (4 1/4)	107.95
4.43	112.52
4.5 (4 1/2)	114.30
5	127.00
5.32	135.13
5.375 (5 3/8)	136.53
5.5 (5 1/2)	139.70
5.625 (5 5/8)	142.88
5.75 (5 3/4)	146.10
6	152.40
6.25 (6 1/4)	171.50
7	177.80
7.09	180.09
7.5 (7 1/2)	190.50
7.625 (7 5/8)	193.68
8	203.20
8.5 (8 1/2)	215.90
8.625 (8 5/8)	219.08
9	228.60
9.625 (9 5/8)	244.48
10	254.00
10.5 (10 1/2)	266.7
11	279.40
11.5 (11 1/2)	292.10
11.625 (11 5/8)	295.26
12	304.80
12.25 (12 1/4)	361.95

in.	mm
13.75 (13 1/4)	336.55
13	330.20
14	355.60
15	381.00
15.5 (15 1/2)	393.70
15.625 (15 5/8)	396.88
16	406.40
17	431.80
18	457.20
19	482.60
20	508.00
22	558.80
22.25 (22 1/4)	565.15
22.5 (22 1/2)	571.50
24	609.60
25.9375 (25 15/16)	658.81
30	762.00
32	812.80
32.75 (32 3/4)	831.85
34	863.60
35	889.00
36	914.40
38	965.20
39.75 (39 3/4)	1009.65
40	1016.00
48	1219.20
50	1270.00
53	1346.20
54	1371.60
71	1803.40
72	1828.80
75	1905.00
108	2743.20
288	7315.20

FIGURE 1.8-27 *(Continued)*

Length *(Continued)*

ft. & in.	m	ft. & in.	m	ft. & in.	m	ft. & in.	m	ft. & in.	m
1	0.305	5-5	1.651	7-2	2.184	9-10	2.997	24	7.315
2	0.609	5-6	1.676	7-3	2.210	9-11	3.023	25	7.62
3	0.914	5-7	1.702	7-4	2.235	9-11¾	3.042	28	8.534
4	1.219	5-8	1.727	7-5	2.261	10	3.048	30	9.144
4-1	1.245	5-9	1.753	7-6	2.286	10-2	3.099	32	9.754
4-2	1.270	5-10	1.778	7-7	2.311	10-3	3.124	33	10.058
4-3	1.295	5-11	1.803	7-8	2.337	10-9	3.277	34	10.363
4-4	1.321	6	1.829	7-9	2.362	10-10	3.302	34.6	10.546
4-5	1.346	6-1	1.854	7-10	2.388	11	3.352	35	10.668
4-6	1.371	6-2	1.880	7-11	2.413	11-3	3.429	36	10.973
4-7	1.397	6-3	1.905	7-11½	2.426	11-7	3.531	50	15.240
4-8	1.422	6-4	1.930	8	2.438	12	3.658	84	25.603
4-9	1.448	6-5	1.956	8-1	2.464	12-1	3.683	100	30.480
4-9¼	1.454	6-6	1.981	8-2	2.489	12-7	3.835	123	37.490
4-10	1.473	6-7	2.007	8-4	2.540	12-11⅝	3.953	125	38.100
4-11	1.499	6-8	2.032	8-6	2.591	13-2	4.013	150	45.720
5	1.524	6-9	2.057	8-8	2.692	13-9	4.191	156	47.548
5-1	1.549	6-10	2.083	8-10	2.718	14-11½	4.559	200	60.96
5-2	1.575	6-11	2.108	9	2.743	20	6.096	250	76.200
5-3	1.600	7	2.134	9-3	2.819	22	6.706	300	91.440
5-4	1.626	7-1	2.159	9-6	2.896				

Area

sq. in.	mm²	sq. ft.	m²	sq. ft.	m²	sq. ft.	m²	sq. ft.	m²
0.11	70.97	9	0.836	54	5.017	109	10.126	181.5	16.862
0.2	129.03	9.4	0.873	55	7.110	110	10.219	188	17.466
0.26	167.74	10	0.929	57	5.295	112	10.405	198	18.395
0.31	199.99	11	1.022	59	5.481	113	10.498	200	18.581
0.44	283.87	11.5	1.068	61	5.667	115	10.684	201	18.674
0.6	387.10	12	1.115	63	5.853	120	11.148	206.5	19.184
0.79	509.68	12.6	1.171	64	5.946	126	11.706	207	19.231
1	645.16	13	1.208	65	6.039	127	11.799	209	19.417
1.27	819.35	14	1.301	66	6.132	130	12.077	213	19.788
1.56	1006.45	15	1.394	67	6.225	132	12.263	218	20.253
2	1290.32	15.5	1.44	69	6.410	133	12.356	220	20.439
2.25	1451.61	15.7	1.459	70	6.503	136	12.635	226	20.996
4	2580.64	16	1.486	72	6.689	138	12.821	230	21.368
sq. ft.	**m²**	18.8	1.747	72.5	6.735	140	13.006	236	21.925
		19	1.765	75	6.968	141	13.099	240	22.297
1	0.093	22	2.044	79	7.339	145.5	13.517	242	22.483
1.25	0.126	25	2.323	80	7.432	146.5	13.610	251	23.319
2	0.186	25.1	2.332	84	7.804	150	13.935	254	23.597
2.25	0.209	28	2.601	85	7.897	151	14.028	254.5	23.644
3	0.279	28.3	2.629	86	7.990	153	14.214	264	24.526
3.5	0.325	31	2.880	88	8.175	154	14.307	276	25.641
4	0.372	31.4	2.917	90.5	8.408	154.5	14.354	283	26.292
5	0.465	34.6	3.214	93	8.640	157	14.586	302	28.057
5.5	0.511	37.7	3.502	94	8.733	160	14.864	311	28.893
6	0.557	38	3.530	100	9.290	163.5	15.190	314	29.172
7	0.65	43	3.995	101	9.383	170	15.794	339	31.494
7.5	0.697	44	4.088	104	9.662	173	16.072	346	32.144
8	0.743	47	4.366	106	9.848	176	16.351	377	35.024
8.5	0.79	50	4.645	106.5	9.894	180	16.723	500	46.45

Volume

gal.	L	gal.	L	gal.	L	gal.	L
0.24	0.908	10.3	38.990	14.8	56.024	16.25	61.513

FIGURE 1.8-27 *(Continued)*

Volume Rate of Flow

cu. ft./hr.	m^3/hr	cfm./in.^{2}	m^3/smm^2	cfm./in.^{2}	m^3/smm^2	cfm./in.^{2}	m^3/smm^2	cfm./in.^{2}	m^3/smm^2
1000	28.317	1	7.314E-07	7.3	5.339E-06	12	8.777E-06	20	1.463E-05
2000	56.634	3.5	2.560E-06	7.8	5.705E-06	13	9.509E-06	24	1.755E-05
3000	84.950	4.4	3.218E-06	8.4	6.144E-06	14	1.024E-05	28	2.048E-05
4000	113.267	4.7	3.438E-06	8.5	6.217E-06	15	1.097E-05	31	2.267E-05
5000	141.584	5.8	4.242E-06	9.4	6.876E-06	17	1.243E-05		
		6.8	4.974E-06	10	7.314E-06	19	1.390E-05		

Moisture Penetration

perms.	ng/(Pa·s·m^2)	perms.	ng/(Pa·s·m^2)	perms.	ng/(Pa·s·m^2)	perms.	ng/(Pa·s·m^2)	perms.	ng/(Pa·s·m^2)
0.002	0.11	0.2	11.49	1.6	91.92	5	287.26	20	1149.05
0.02	1.15	0.3	17.24	1.9	109.16	5.4	310.25	30	1723.58
0.05	2.87	0.4	22.98	2	114.91	5.5	315.99	31	1781.03
0.06	3.45	0.5	28.73	2.1	120.65	5.8	333.22	50	2872.64
0.08	4.60	0.8	45.96	2.4	137.89	9.5	545.80	85	4883.46
0.1	5.75	1	57.45	3	172.36	10	574.53	90	5170.73
0.14	8.04	1.2	68.94	3.3	189.59	11	631.98	116	6664.49
0.15	8.62	1.5	86.18	4	229.81	15	861.79	120	6894.3
0.16	9.19								

Lumber

1 × 4	25 × 100	2 × 4	50 × 100	2 × 8	50 × 200	2 × 10	50 × 250	2 × 12	50 × 300
1 × 6	25 × 150	2 × 6	50 × 150						

Metal Gauges

gauge	mm
22	.85
24	.70

Modulus of Subgrade Reaction

lb./sq. in./in.	kPa/mm
50	184.2
100	368.4
200	736.8
300	1105.2

Heat

Btu	kJ	Btu	kJ	Btu	kJ	Btu	kJ	Btu	kJ
0.8	0.84	100	105.5	240	253.2	3711	3915	40,000	42202.4
1	1.055	140	147.7	400	422	5000	5275	56,000	58083
60	63.3	160	168.8	1000	1055.1				

Thermal Conductivity (k-value)

Btu/sq. ft./°F/in./hr.	W/mK	Btu/sq. ft./°F/in./hr.	W/mK	Btu/sq. ft./°F/in./hr.	W/mK
0.5	0.865	5	8.654	300	519.219
1	1.731	6	10.384	500	865.365
2	3.461	10	17.307	1500	2596.095
2.5	4.327	100	173.073	2500	4326.825
3	5.192	150	259.610		

Thermal Resistivity (1/k)

R-value	R-value
11	4.68
19	8.09

1.9 Land Surveys and Descriptions

The transfer of land ownership and the mortgaging of property require a legal description to locate and establish the boundaries of the affected land parcel. A legal description is prepared on the basis of a *survey* of the land. A survey is the measuring and marking of land, accompanied by maps and field notes to describe the measures and marks made in the field. The lengths and directions of boundary lines are established between reference points on the ground. The reference points may be natural marks, such as rivers, or they may be stone, iron, concrete, or other artificial markers set in the field by a surveyor. Natural points, noted as permanent, and permanent markers set in the field by surveyors are called *monuments*.

Land descriptions and *surveys* have been used for centuries. It is not possible to assign the birth of the science of surveying to a particular year or even a country, but the earliest known records refer to the making of skillful measurements and calculations with respect to land. The Chinese, at an early date, and the ancient Egyptians practiced the art of surveying. In Egypt, it appears that it was necessary every spring to reestablish corners and boundary lines obliterated by floods of the Nile River.

Two general systems of survey are used in the United States: the *metes and bounds system* and the *rectangular system.*

1.9.1 METES AND BOUNDS SYSTEM

Survey by *metes and bounds* (Fig. 1.9-1) consists of beginning at a known point and *running out* the boundaries of the area by *courses* (directions) and *distances* (lengths), and fixing natural or artificial *monuments* at the corners. The *place of beginning* (P.O.B.) must be a known point that can be readily identified. The point must be established and witnessed so that it can be located again with certainty if the monument that identifies it is destroyed or removed.

Portions of the text in earlier versions of this section were taken from publications by and were used with permission of The Chicago Title and Trust Company. This edition has been extensively revised from earlier versions, and does not reflect the opinions of, and is not endorsed by The Chicago Title and Trust Company.

The metes and bounds system, one of the oldest known ways of surveying and describing land, was used in colonial days. Land being bought, claimed, or inherited would be identified by known natural points, such as trees and rivers, and artificial points, such as roads, structures, or other manmade marks. For example, a survey might start at the middle of a white pine stump, move northward to a small white oak tree, and continue with a series of measures by "chains and links" between trees, stakes, and piles of stones, until it returned to the stump where it begun.

Early metes and bounds surveys and descriptions were often based on monuments that lacked permanency. The surveying of large and irregular tracts of land without regard to uniformity and the failure of surveyors to make their survey notes a matter of public record gave rise to frequent boundary line disputes and litigation.

Present-day surveys and descriptions may refer to government survey section lines and corners (see Section 1.9.2), in addition to permanent monuments placed by surveyors.

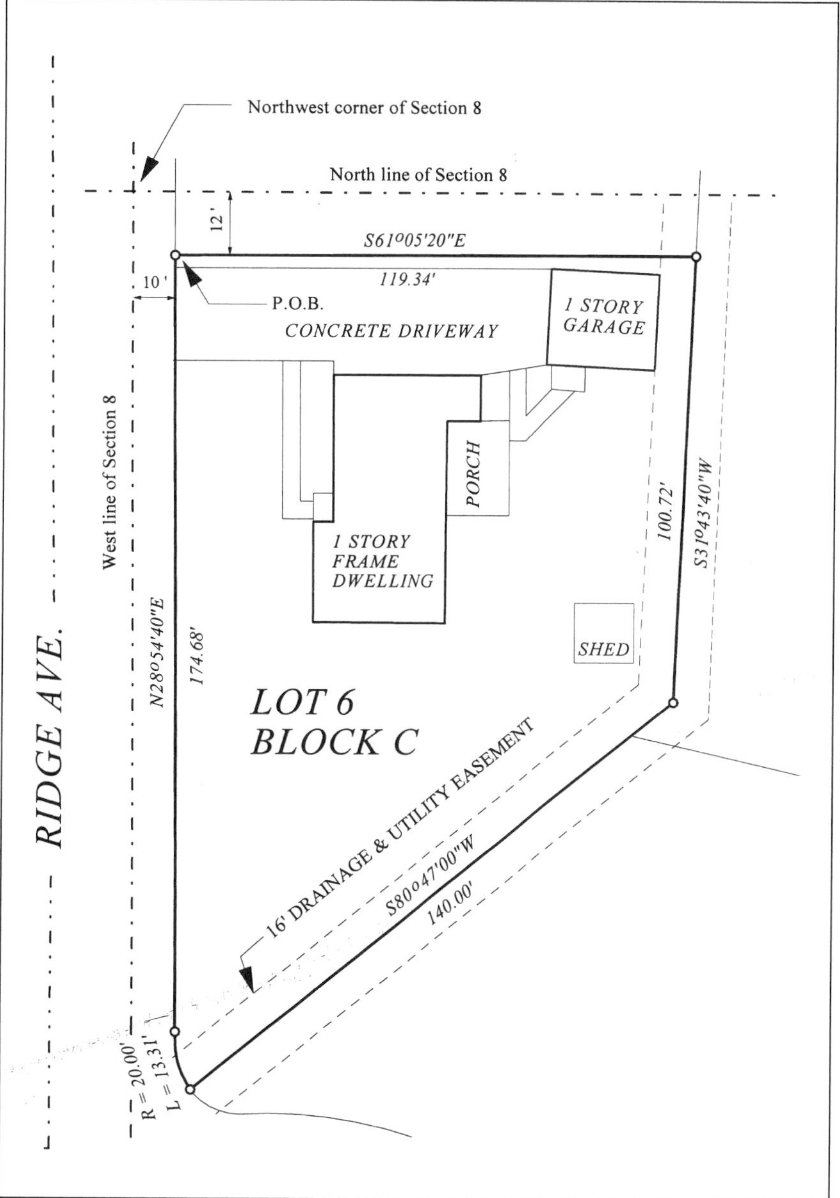

FIGURE 1.9-1 A typical metes and bounds survey plat. (Drawing by HLS.)

1.9.2 RECTANGULAR SYSTEM

After the Revolutionary War the federal government found itself with vast tracts of undeveloped and uninhabited land, with few natural characteristics suitable for use as monuments in metes and bounds descriptions. A committee headed by Thomas Jefferson evolved a new standard system of describing land so that parcels could be located readily and permanently. This system, designated the *rectangular system* (also called the *government system*) of survey, divided land into a series of rectangles. The Continental Congress adopted this system in 1785, and it is in use today in a majority of the 50 states.

1.9.2.1 Meridians and Base Lines

The world is divided into a grid of imaginary north-south lines called *meridians of longitude*, which extend from the north pole to the south pole. The meridian that passes through the Royal Observatory at Greenwich, England, is designated as 0 (zero) degrees longitude and called the *Greenwich*, or *prime*, *meridian*. Locations east or west of Greenwich are measured in degrees from the prime meridian.

The reference lines running around the globe from east to west are called *parallels of latitude*. The *equator* is designated as 0 (zero) degrees latitude. Locations that fall north or south of the equator are measured in degrees from the equator.

The rectangular system of survey is based on a series of 35 *principal meridians* and 32 *base lines*.

PRINCIPAL MERIDIANS

In establishing the rectangular system of survey, surveyors first selected a substantial landmark from which a start could be made in surveying an area of land. Usually a place was selected that could be identified readily, such as the mouth of a river, and whenever possible a monument was placed. From this point the surveyors ran a line due north and south and designated it as the principal meridian for that particular state or area. The location of the *principal meridian* (its longitude) was fixed by a reading measured in degrees, minutes, and seconds (see Glossary) west of the prime meridian.

As territories were opened and surveyed by the government, principal meridians were established for each area opened. Some of the principal meridians were referred to by number, such as the First, Second, and Third Principal Meridians. The Third Principal Meridian, for example, is the line located 89 degrees, 10 minutes, and 15 seconds west of Greenwich. It extends from the mouth of the Ohio River to the northern boundary of Illinois. Other principal meridians were named for the states, such as the Michigan Meridian, which covers the survey of that state.

BASE LINES

After the principal meridian for a particular area was fixed, a point on the meridian was selected from which a line was run at right angles, due east and west. This line was designated as the base line for the area, and its location (its latitude) was established in degrees, minutes, and seconds north of the equator.

Sometimes the base line for a new territory was established by extending the base line of an adjoining territory. For example, the base lines for both the Second and Third Principal Meridians are at 38 degrees, 28 minutes, and 20 seconds north latitude.

CORRECTION LINES

Meridians meet at the north and south poles. They are not parallel lines. An accurate measure of a midwestern area 6 miles square would show its north line to be about 50 ft. shorter than its south line. To compensate for this convergence, additional east and west lines (*correction lines*), parallel to the base line, were located at intervals of 24 miles north and south of the base line.

GUIDE MERIDIANS

After the east-west correction lines were set, *guide meridians* running due north and south at 24-mile intervals on each side of the principal meridians were established. Guide meridians extend from the base line to the first correction line, and then from correction line to correction line. The guide meridians and the correction lines form squares approximately 24 miles on each side, as shown in Figure 1.9-2.

TOWNSHIP AND RANGE LINES

The 24-mile squares were then divided into 16 smaller tracts by *township lines* and *range lines*. Township lines run east and west at 6-mile intervals. Range lines run north and south at 6-mile intervals. This cross-hatching resulted in a grid of squares, called *townships*.

Two *reference numbers* were assigned to each square: a *township number* and a *range number*. Rows of squares (*tiers*) were numbered consecutively to the north or south of the base line. Thus, each square in the first row north of the base line is called Township 1 (one) North (T.1N.); each in the second row, Township 2 (two) North (T.2N.), and so forth. Similarly, each square in the first row south of the base line is called Township 1 South; each in the second row, Township 2 South, and so on. In the same manner, rows of squares (*ranges*) were numbered consecutively to the east and west of the principal meridian. Thus, each square in the first row east of the principal meridian is called Range 1 East (R.1E.), those in the second row Range 2 East (R.2E.), and so forth, and on the west of the principal meridian, those in the first row were numbered Range 1 West, those in the second row, Range 2 West, and so on.

SECTIONS

Townships were the smallest divisions of land provided for in the 1785 act that created the rectangular system of survey. In making early surveys, the outside boundaries of the townships were surveyed and monuments were placed at every mile on the township lines.

It soon became apparent that a township 6 miles square was too large an area in which to describe and locate a given tract of land. In 1796, Congress passed an act directing that townships already surveyed be subdivided into 36 *sections*, each to be approximately 1 mile square containing about 640 acres. The sections were numbered consecutively from 1 to 36, beginning with number 1 in the northeast corner of the township and ending in the southeast corner with number 36, as shown in Figure 1.9-2b. A tract of land can be located within a square mile by giving the section number, the township number, and the range number.

FRACTIONAL SECTIONS

A *fractional section* is one that is larger or smaller than others in a township. Fractional sections may result from geographical factors, such as land area bordering on lakes. They may also occur where the surveys of separately surveyed areas do not tie together perfectly, resulting in sections that are less than 1 mile square.

In addition, because of the convergence of the meridians, each township does not form a perfect square and cannot

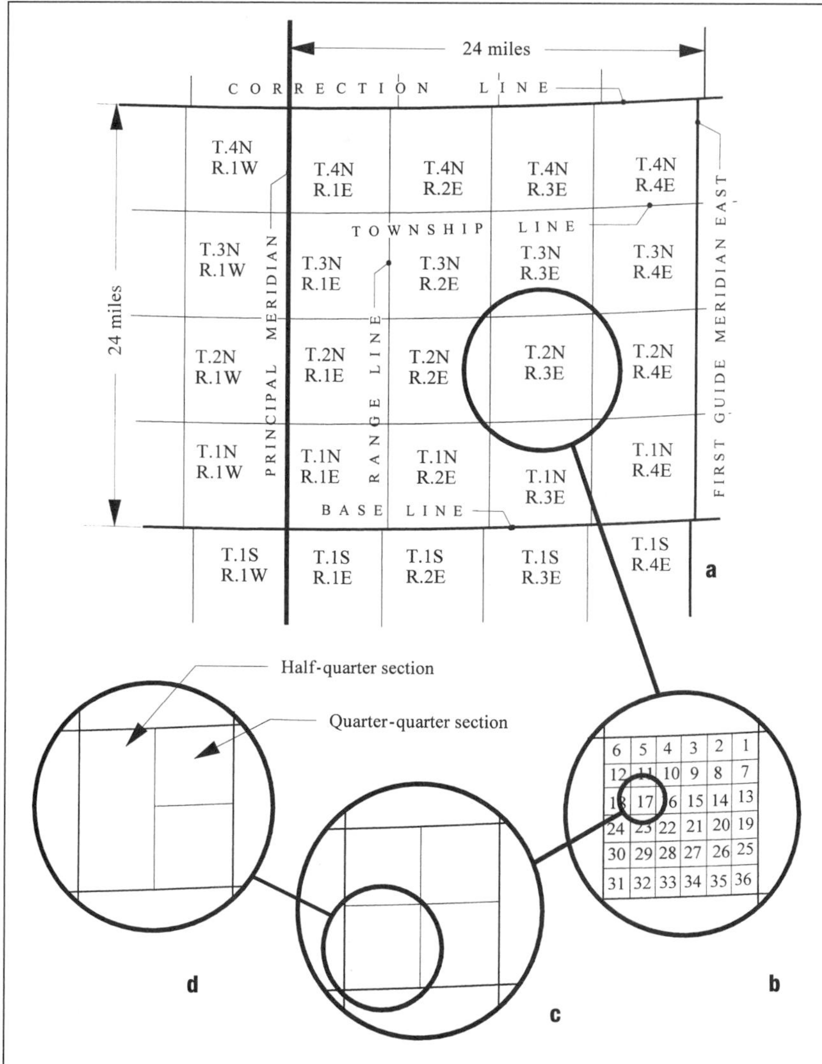

FIGURE 1.9-2 Division of land in the rectangular system: (a) A square (24 mi. on a side), shown with townships delineated. (b) A township (6 mi. on a side); this is Township 2 North (T.2N.), Range 3 East (R.3E.) from the square shown in (a), with 36 sections delineated. (c) A section (640 acres) shown with quarter section divisions; this is Section 17 from the township shown in (b). (d) A quarter section (160 acres) shown with half-quarter (80 acres) and quarter-quarter sections (40 acres) delineated. In a legal description, the quarter-quarter section in (d) would be described as "the northeast quarter of the southwest quarter of Section 17, Township 2 North, Range 3 East of the First (Second, Third, New Mexico, etc., as applicable) Principal Meridian." (Drawing by HLS.)

be surveyed to form 36 identical sections in shape and area (Fig. 1.9-2c). Therefore, excess or deficiency in south-to-north measurements is added to or deducted from the north portion of the sections in the north row in the township (Sections 1 through 6). Excess or shortage in the east-to-west measurements is added to or deducted from the west portion of the west row in the township (Sections 6, 7, 18, 19, 30, and 31). These adjusted sections are classified as *fractional sections*.

After fractional sections are divided into equal fractional parts as far as possible, remaining portions are divided into lots of approximately the same size. These *government lots* are numbered.

DIVISION OF SECTIONS

Sections are also divided into still smaller units (Fig. 1.9-2d). In 1800, Congress directed the subdivision of sections into east and west halves of approximately 320 acres each by running lines north and south through the centers of the sections. In 1805, regulations provided for *quarter sections* by running east and west lines through the centers of sections (see Fig. 1.9-2d).

Congress later provided for the further division of quarter sections into *half-quarters* and, finally, *quarter-quarters*. These subdivisions were made by running east-west lines and north-south lines through the quarter sections. A quarter-quarter section of 40 acres is the smallest statutory division of regular sections in the rectangular survey system (Fig. 1.9-2d).

1.9.3 LEGAL DESCRIPTIONS

A *legal description* identifies, locates, and defines a specific piece of land to distinguish it from all other tracts of land. Legal descriptions in the United States may be based on (1) the metes and bounds survey system, (2) the rectangular survey system, or (3) reference to recorded maps and plats. It is not uncommon to find a combination of these methods in the same description. For example, a lot in a subdivision may be described by reference to a recorded plat of a land, which in turn is described by a metes and bounds survey within the framework of the rectangular system (Fig. 1.9-3).

> All that certain lot or parcel of land with the improvements thereon and appurtenances thereto, belonging, lying, being, and situated in the City of Rightown, Virginia, being known, numbered, and designated as Lot No. six (6), in Block C, Section three (3), on a certain plat of Hunt Club View Hills, Section three (3) dated March 28, 1989, made by Simpton, Inc. Surveyors, and recorded in the Clerk's Office of the Circuit Court of the City of Rightown, Virginia, in Plat Book 2, at page 1.
>
> Said property commencing at the North West corner of Section 8, thence South along the section line 12 feet; thence East 10 feet to the point of beginning; thence South 61 degrees, 5 minutes, 20 seconds East for 119.34 feet; thence South 31 degrees, 43 minutes, 40 seconds West for 100.72 feet; thence South 80 degrees, 47 minutes West for 140 feet to a point of tangency; thence along a curve convex Northwesterly and having a radius of 20 feet; thence North 28 degrees, 54 minutes, 40 seconds East, for a distance of 174.68 feet to the point of beginning.

FIGURE 1.9-3 Legal description for the plat shown in Figure 1.9-1.

1.9.3.1 Metes and Bounds Description

A metes and bounds description consists of a series of statements describing each portion of the boundary around the parcel of land being described. Starting from a *place of beginning* (P.O.B.), the direction and distances of the property lines are *called* (stated in the description), until the perimeter has been traced around the entire property, returning to the starting point. Each leg of the perimeter is described first by direction, then by length (see Fig. 1.9-3). The survey description must "close"; that is, if the courses and distances of the description are traced in order, one must return to the starting point.

DIRECTION

Directions (*courses*) that fall along major compass directions are so named. Courses other than due north, south, east, or west (cardinal points) are called *angular courses*. The direction of an angular course is stated in degrees, minutes, and seconds as an angular deviation east or west of due north or south (see Fig. 1.9-3). (There is one exception to this convention, explained in the next section, "Prolongation.")

The direction is always written with the statement *north* or *south*, followed by the angle of eastward or westward deviation from north or south. For example, "north 30 degrees east" defines an angular course direction 30 degrees eastward of due north. A *call* "south 42 degrees 15 minutes west" defines a direction 42 degrees and 15 minutes westward of south.

PROLONGATION

Sometimes course direction is described by stating the amount of deviation from the preceding course line. An angle is measured from an extension, a *prolongation*, of the preceding line, and may be stated as "thence northerly 30 degrees to the left of an extension of the last described line."

CURVED LINES

A curved line is sometimes referred to in a description, such as: "thence south 80 degrees east 100 feet to a point of tangency; thence along a curve convex northeasterly and having a radius of 30 feet; a distance of 40 feet." The point of tangency (the point of contact between the last described straight line and the curve) is the starting place of the curve. *Convex* refers to the outside of a curve; *concave*, if used, refers to the inside of a curve. "Convex northeasterly" means that the outside of the curve lies to the northeast.

DISTANCE

Distance is usually stated in feet and decimal parts of feet, and sometimes in feet and inches or fractions of feet. Very old surveys may state distances in chains, rods, and links. Measure is made from point to point along straight lines and along the curve of a curved line.

1.9.3.2 Rectangular System Descriptions

A tract of land in the rectangular survey system can be described by reference to its principal meridian, base line, township and range numbers, section number, and, if necessary, its location within the section (see Fig. 1.9-2).

Government lots are described by identifying the lot number and the quarter section in which the lot is located.

1.9.3.3 Reference to Recorded Plat

Curved street patterns often produce irregularly shaped blocks and lots. An entire subdivision may be described within the rectangular system or by a metes and bounds description, but it may be impractical to use these systems to describe individual blocks or lots. Therefore, many states provide for the legal description of property by the reference to plats properly recorded, generally in the office of the appropriate county.

A surveyor surveys the tract and prepares a *plat*. The tract or subdivision may be divided into blocks and then into lots. Lots and blocks may be given consecutive numbers or letters to enable identification of a particular area of land within the subdivision by a lot and block number. For further identification it is common practice to give the subdivision a name (see Fig. 1.9-3).

The surveyor's plat is captioned with the legal description of the land. It shows all boundary lines, all necessary monuments and dividing lines for blocks, lots, and streets, and the numbering and dimensions of each lot.

Easements and *restrictions* may also be indicated on a plat. Upon approval by the proper authorities (such as municipal or county zoning boards, planning or building commissions), a plat is recorded generally in the appropriate government recorder's office.

When recorded, the lots in a subdivision may be easily identified by reference to lot and block numbers, the subdivision name, and the section, township, and range in which the entire subdivision boundaries are described within a rectangular survey or metes and bounds description.

1.9.3.4 Uncertainties

Mistakes, errors, and judgments in a survey may result in conflicts and uncertainties in the legal description of land. There may be differing opinions as to the precise location of the boundary lines of a tract of land, and the exact location of a point on the land can be subject to interpretation.

LOCATION OF MARKERS

Through the years, monuments may have been moved or uprooted during earth-moving projects, such as road construction, or in farming. Relocating a missing corner may require measuring from monuments in existing surveys (which may also contain uncertainties) or from other references.

The disappearance of monuments constantly causes uncertainties in surveying. Metes and bounds surveys in which a monument was a tree or a post that no longer exists may require the surveyor to dig carefully in the probable vicinity of the lost monument in an attempt to find the decayed remains of the tree or post.

Roads described in an old survey may have been abandoned and the pavement removed. Even though title companies may have unrecorded evidence of such roads, surveying judgment is required to fix a line to substitute for the missing road.

Many other uncertainties are caused by conjecture as to the location of lakes and rivers, which may have receded, dried up, or changed course and location over the years.

CONVERGENCE

The original government surveys that divided land areas into 24-mile squares each containing 16 townships, and subsequent divisions into sections and fractional sections, made a practical but not perfect effort to adjust for the convergence of meridians of longitude as they travel north. Sections were marked by lines spaced at 1-mile intervals, running parallel to the range line closest to the principal meridian. This practice causes a divergence in direction, which becomes progressively more pronounced in each successive tier of sections. Therefore, an ambiguity may result when a cardinal point is used as a course in a legal description. For instance, the call "thence north 150 feet to a point . . . ," unless marked by a monument, is usually interpreted as indicating a course that runs parallel with the east line of the section. If the section line deviates from true north, an uncertainty exists.

INACCURACIES

The measurement of distances called for in some legal descriptions is virtually impossible. For example, the distance "126.01 feet" seems precise, but it is unlikely that two surveying crews measuring this distance would observe the same measurement. A distance of .01 ft. is a little less than 1/8 in. Differences in the tautness of a measuring tape and changes in the length of the tape owing to the effect of temperature fluctuations can easily negate the accuracy of such a measurement.

In surveying a tract, a surveyor must locate the boundaries of abutting tracts as shown in earlier surveys and reconcile the perimeter lines of the tract being surveyed with them. The surveyor must also obtain legal descriptions of the abutting tracts.

Section corners and other marks in the rectangular system were sometimes placed incorrectly in early surveys, which were based on lengths such as *chains* and *links*. Surveyors made field measurements with a metal chain made of 100 links. A chain was 66 ft. long. The combined length of 80 chains was equal to 1 statute mile, or 5280 ft. Weather changes expanded and contracted these chains, and wear and use caused them to stretch. Sometimes, surveyors would count the revolutions of the wheel of the vehicle in which they rode to measure long, straight stretches. Distances were also "stepped off" by a measured stride. Even though they were far from accurate, markers placed according to these measurements are considered legal points of reference.

In addition to errors in distance, there are also deviations in the perimeter lines of a section from true east-west or north-south courses. In locating the south boundary of a tract described as the "north 10 acres of the northwest quarter" of a section in which the north line deviates from a true east and west course, some judgment is required. The south line may be considered to be parallel with the north line of the section, or it may run due east and west. Again, the surveyor must search for physical evidence, such as a stake set by an earlier surveyor or a fence indicating a line of occupation, in order to decide which line was intended.

Courses may show an appreciable error in direction. For example, a metes and bounds course shown as "south 7 degrees west" in an earlier survey may be observed to actually be "south 6 degrees west," indicating that an error could have been made by an earlier surveyor or that faulty equipment was used.

Mistakes, such as careless observations and inaccurate measurements, are random and unpredictable. Even the best surveying instruments are subject to a certain degree of error, but these errors are predictable and can be allowed for in observing measurements. Again, however, this calls for some human judgment.

1.9.3.5 Recorded Plats

When an individual lot is described by reference to a recorded subdivision plat, the plat together with all information and data shown thereon is part of the lot description (see Fig. 1.9-3). It is possible that the lot size may not be exactly as shown on the plat. The incorporation by reference is of the entire plat and all data appearing thereon. This includes not only dimensions and expressions of quantity and area, but also the notations of the courses and distances of the lines of the subdivision and of the identity and locations of the monuments marking the boundaries.

When this information is incorporated into a legal description, it is, in effect, a metes and bounds description and is subject to rules of interpretation, should any conflict or discrepancy be contained in the data of the plat. For example, monuments prevail over courses and distances, courses prevail over distances, and expressions of quantity or area are generally the least reliable.

It is possible that the overall recorded measurements of a block may not equal the recorded widths of the lots in the block. If such an error is found, the widths of some or all of the lots must be adjusted accordingly.

The systems of survey on which legal descriptions are based are as precise as is possible. There is always the possibility of uncertainties in separate surveys, and mistakes may occur in the writing and rewriting of legal descriptions. Descriptions may identify land parcels adequately, but surveys are necessary to actually locate boundary lines.

The complexities of resolving uncertainties in either land survey or legal land description are a matter of real estate law and interpretation and are beyond the scope of this book.

1.10 Properties of Materials

In this century there has been a huge increase in the number of construction products and even in the number of totally new materials used for construction purposes. The entire array of plastic materials we use today is one example. More than 300 new products continue to appear annually. Some of these are made by merely varying the ingredients in or formulation of existing materials, such as sealants, or by adding new shapes and finishes to a line of masonry units, cladding panels, and other products. A few are entirely new products, often made by combining existing materials. Glass products made by mixing other substances into molten glass are examples of new products with characteristics that did not exist in older products. In addition, many new uses for existing materials are developed annually. The introduction of glass fibers to reinforce poured-in-place concrete is an example. In order to analyze and use these innovations effectively, designers and builders need a basic understanding of the general properties of materials.

A material's suitability for a particular use depends on its ability to perform satisfactorily under the conditions to be encountered. When a material's properties are known, its performance in a particular situation can be predicted. The four general groups of properties of interest to those in the construction industry are mechanical, thermal, electrical, and chemical properties. Each depends on the composition of the material in question.

The millions of materials in our universe are all made from just three basic building blocks: *atoms*, *ions*, and *molecules*. These are bonded together by one of several basic chemical bonds. Materials can be classified according to these basic structural elements and bonds into three major categories with different properties:

1. *Ceramics and glasses*—hard, brittle, poor conductors of heat and electricity
2. *Metals*—more ductile than ceramics, good conductors of heat and electricity
3. *Molecular materials*—low melting temperatures, fair strength, poor conductors of heat and electricity

Since all construction materials fall into one of these three categories, a greater understanding of such materials can be obtained by recognizing the category in which a particular material belongs and remembering the properties of that category. For instance, glass wool and foamed polystyrene may have identical insulating capabilities, but the glass wool falls into the ceramic category and the foam into the molecular category. From that knowledge a designer can surmise, without seeing manufacturers' specifications for the materials, that the foam will have a lower melting temperature than the glass wool, because *the melting temperature of molecular materials is lower than that of ceramic materials.*

This section is intended to expand and sharpen the reader's instinct for categorizing materials. First, the basic structure of matter is discussed, then the four sets of properties that determine a material's performance are considered in some detail. For more information about the properties of a particular material, refer to the chapter in which that specific material is discussed.

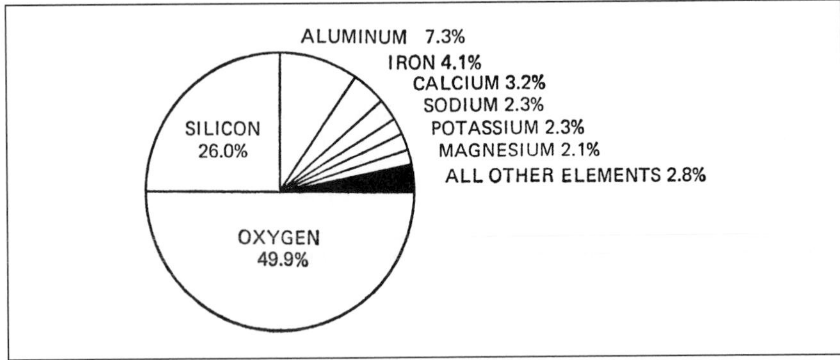

FIGURE 1.10-1 A handful of elements make up most earthly matter.

1.10.1 STRUCTURE OF MATTER

Matter is formed by the chemical bonding of atoms, ions, and molecules. The particular combination of components and bonding methods determines the final material properties.

1.10.1.1 Building Blocks of Matter

There are 102 chemical substances that cannot be subdivided into other substances. These basic materials are called *chemical elements*. Just two chemical elements, oxygen and silicon, make up more than three-quarters of all the matter on earth; just eight chemical elements constitute more than 97% of all earthly substances (Fig. 1.10-1).

A given quantity of any element can be divided into smaller and smaller quantities until a minute particle is reached. This entity, the smallest particle that retains the properties of the original material, is called an *atom*.

Ordinarily, atoms are neutral, having no electrical charge. When atoms acquire a positive or negative charge by losing or attracting negatively charged particles, they are then called *ions*. Atoms may combine into simple units called *molecules*.

ATOMS

The simplest representation of an atom looks like a miniature solar system (Fig. 1.10-2). The atom has at its core a *nucleus*, consisting of positively charged *protons* and electrically neutral *neutrons*. Orbiting about the nucleus are negatively charged *electrons*.

Because protons and neutrons are much denser than electrons, almost all of the weight of an atom resides in the nucleus. The sum of the weights of the protons and neutrons in one of its individual atoms is called the *atomic weight* of an element. In a normal electrically neutral atom, the positive charge of the protons in the nucleus is balanced by an equal negative charge in the electrons. The number of electrons in each of its atoms, which is equal to the number of protons in a single atom, is the *atomic number* of an element.

Groups of electrons moving in neighboring orbits are said to belong to the same *shell* (see Fig. 1.10-2). The shell closest to the nucleus can accommodate only 2 electrons; the second, a maximum of 8; the third, 18; and the fourth, 32. The distribution of electrons around the nucleus gives elements some of their inherent chemical properties. Figure 1.10-3 shows the chemical symbols and common names of all 102 elements, as well as their

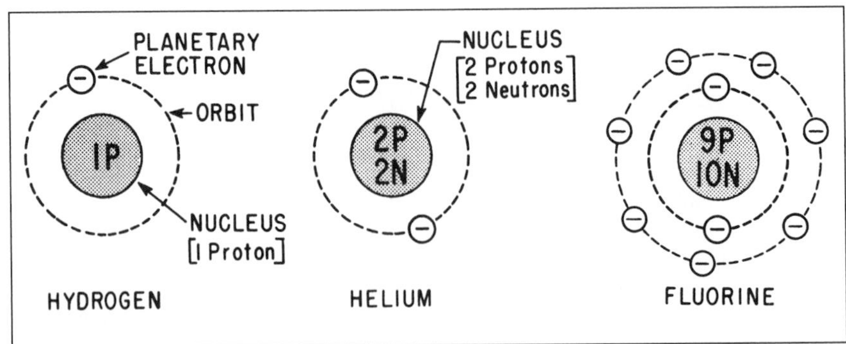

FIGURE 1.10-2 Electrons move in orbits around a nucleus. Helium's two electrons move in the same orbit, but fluorine's nine electrons move in different orbits. Groups of electrons with identical or neighboring orbits constitute a shell; hydrogen and helium have one shell, fluorine two.

FIGURE 1.10-3 The Elements: Their Atomic Numbers, Chemical Symbols, Common Names, and Shell Distributions

Atomic	Element		Number of Electrons in Shells				
No.	Symbol	Name	1	2	3	4	5
1	H	Hydrogen	1				
2	He	Helium	2				
3	Li	Lithium	2	1			
4	Be	Beryllium	2	2			
5	B	Boron	2	3			
6	C	Carbon	2	4			
7	N	Nitrogen	2	5			
8	O	Oxygen	2	6			
9	F	Fluorine	2	7			
10	Ne	Neon	2	8			
11	Na	Sodium	2	8	1		
12	Mg	Magnesium	2	8	2		
13	Al	Aluminum	2	8	3		
14	Si	Silicon	2	8	4		
15	P	Phosphorus	2	8	5		
16	S	Sulfur	2	8	6		
17	Cl	Chlorine	2	8	7		
18	A	Argon	2	8	8		
19	K	Potassium	2	8	8	1	
20	Ca	Calcium	2	8	8	2	
21	Sc	Scandium	2	8	9	2	
22	Ti	Titanium	2	8	10	2	
23	V	Vanadium	2	8	11	2	
24	Cr	Chromium	2	8	13	1	
25	Mn	Manganese	2	8	13	2	
26	Fe	Iron	2	8	14	2	
27	Co	Cobalt	2	8	15	2	
28	Ni	Nickel	2	8	16	2	
29	Cu	Copper	2	8	18	1	
30	Zn	Zinc	2	8	18	2	
31	Ga	Gallium	2	8	18	3	
32	Ge	Germanium	2	8	18	4	
33	As	Arsenic	2	8	18	5	
34	Se	Selenium	2	8	18	6	
35	Br	Bromine	2	8	18	7	
36	Kr	Krypton	2	8	18	8	
37	Rb	Rubidium	2	8	18	8	1
38	Sr	Strontium	2	8	18	8	2
39	Y	Yttrium	2	8	18	9	2
40	Zr	Zirconium	2	8	18	10	2
41	Nb	Niobium	2	8	18	12	1
42	Mo	Molybdenum	2	8	18	13	1
43	Tc	Technetium	2	8	18	14	1
44	Ru	Ruthenium	2	8	18	15	1
45	Rh	Rhodium	2	8	18	16	1
46	Pd	Palladium	2	8	18	16	1
47	Ag	Silver	2	8	18	18	1
48	Cd	Cadmium	2	8	18	18	2
49	In	Indium	2	8	18	18	3
50	Sn	Tin	2	8	18	18	4
51	Sb	Antimony	2	8	18	18	5

Atomic	Element		Number of Electrons in Shells						
No.	Symbol	Name	1	2	3	4	5	6	7
52	Te	Tellurium	2	8	18	18	6		
53	I	Iodine	2	8	18	18	7		
54	Xe	Xenon	2	8	18	18	8		
55	Cs	Cesium	2	8	18	18	8	1	
56	Ba	Barium	2	8	18	18	8	2	
57	La	Lanthanum	2	8	18	18	9	2	
58	Ce	Cerium	2	8	18	20	8	2	
59	Pr	Pra'mium	2	8	18	21	8	2	
60	Nd	Neodymium	2	8	18	21	8	2	
61	Pm	Promethium	2	8	18	23	8	2	
62	Sm	Samarium	2	8	18	24	8	2	
63	Eu	Europium	2	8	18	25	8	2	
64	Gd	Gadolinium	2	8	18	25	9	2	
65	Tb	Terbium	2	8	18	27	8	2	
66	Dy	Dysprosium	2	8	18	28	8	2	
67	Ho	Holmium	2	8	18	29	8	2	
68	Er	Erbium	2	8	18	30	8	2	
69	Tm	Thulium	2	8	18	31	8	2	
70	Yb	Ytterbium	2	8	18	32	8	2	
71	Lu	Lutetium	2	8	18	32	9	2	
72	Hf	Hafnium	2	8	18	32	10	2	
73	Ta	Tantalum	2	8	18	32	11	2	
74	W	Tungsten	2	8	18	32	12	2	
75	Re	Rhenium	2	8	18	32	13	2	
76	Os	Osmium	2	8	18	32	14	2	
77	Ir	Iridium	2	8	18	32	17	0	
78	Pt	Platinum	2	8	18	32	17	1	
79	Au	Gold	2	8	18	32	18	1	
80	Hg	Mercury	2	8	18	32	18	2	
81	Tl	Thallium	2	8	18	32	18	3	
82	Pb	Lead	2	8	18	32	18	4	
83	Bi	Bismuth	2	8	18	32	18	5	
84	Po	Polonium	2	8	18	32	18	6	
85	At	Astatine	2	8	18	32	18	7	
86	Rn	Radon	2	8	18	32	18	8	
87	Fr	Francium	2	8	18	32	18	8	1
88	Ra	Radium	2	8	18	32	18	8	2
89	Ac	Actinium	2	8	18	32	18	9	2
90	Th	Thorium	2	8	18	32	18	10	2
91	Pa	Pr'tinium	2	8	18	32	20	9	2
92	U	Uranium	2	8	18	32	21	9	2
93	Np	Neptunium	2	8	18	32	22	9	2
94	Pu	Plutonium	2	8	18	32	23	9	2
95	Am	Americium	2	8	18	32	24	9	2
96	Cm	Curium	2	8	18	32	25	9	2
97	Bk	Berkelium	2	8	18	32	26	9	2
98	Cf	Californium	2	8	18	32	27	9	2
99	Es	Einsteinium	2	8	18	32	28	9	2
100	Fm	Fermium	2	8	18	32	29	9	2
101	Md	Mendelvium	2	8	18	32	30	9	2
102	No	Nobelium	2	8	18	32	31	9	2

Period	Group IA	Group IIA	Transition Elements										Group IIIA	Group IVA	Group VA	Group VIA	Group VIIA	Group O
1	1 **H** 1.00797		Metals										Non-Metals				1 **H** 1.0079	2 **He** 4.0026
2	3 **Li** 6.939	4 **Be** 9.012											5 **B** 10.811	6 **C** 12.011	7 **N** 14.007	8 **O** 15.9994	9 **F** 18.998	10 **Ne** 20.183
3	11 **Na** 22.990	12 **Mg** 24.312	Group IIIB	Group IVB	Group VB	Group VIB	Group VIIB	Group VIII			Group IB	Group IIB	13 **Al** 26.98	14 **Si** 28.086	15 **P** 30.97	16 **S** 32.064	17 **Cl** 35.453	18 **Ar** 39.95
4	19 **K** 39.102	20 **Ca** 40.08	21 **Sc** 44.96	22 **Ti** 47.90	23 **V** 50.94	24 **Cr** 52.00	25 **Mn** 54.94	26 **Fe** 55.85	27 **Co** 58.93	28 **Ni** 58.71	29 **Cu** 63.54	30 **Zn** 65.37	31 **Ga** 69.72	32 **Ge** 72.59	33 **As** 74.92	34 **Se** 78.96	35 **Br** 79.91	36 **Kr** 83.80
5	37 **Rb** 85.47	38 **Sr** 87.62	39 **Y** 88.91	40 **Zr** 91.22	41 **Nb** 92.91	42 **Mo** 95.94	43 **Tc** 99	44 **Ru** 101.07	45 **Rh** 102.91	46 **Pd** 106.4	47 **Ag** 107.87	48 **Cd** 112.40	49 **In** 114.82	50 **Sn** 118.69	51 **Sb** 121.75	52 **Te** 127.60	53 **I** 126.90	54 **Xe** 131.30
6	55 **Cs** 132.90	56 **Ba** 137.34	57–71 **La** series*	72 **Hf** 178.49	73 **Ta** 180.95	74 **W** 183.85	75 **Re** 186.2	76 **Os** 190.2	77 **Ir** 192.2	78 **Pt** 195.1	79 **Au** 196.97	80 **Hg** 200.59	81 **Tl** 204.37	82 **Pb** 207.19	83 **Bi** 208.98	84 **Po** 210	85 **At** 210	86 **Rn** 222
7	87 **Fr** 223	88 **Ra** 226	89– **Ac** series†															

*Lanthanide Series	58 **Ce** 140.12	59 **Pr** 140.91	60 **Nd** 144.24	61 **Pm** 147	62 **Sm** 150.35	63 **Eu** 151.96	64 **Gd** 157.25	65 **Tb** 158.92	66 **Dy** 162.50	67 **Ho** 164.93	68 **Er** 167.26	69 **Tm** 168.93	70 **Yb** 173.04	71 **Lu** 174.97
†Actinide Series	90 **Th** 232.04	91 **Pa** 231	92 **U** 238.03	93 **Np** 237	94 **Pu** 239	95 **Am** 241	96 **Cm** 242	97 **Bk** 249	98 **Cf** 252	99 **Es** 254	100 **Fm** 253	101 **Md**	102 **No**	103 **Lw**

FIGURE 1.10-4 Periodic table of the elements.

atomic numbers (electron numbers) and shell distribution.

The elements also can be classified according to ascending atomic numbers and number of electron shells in an arrangement called the *periodic table* (Fig. 1.10-4). The *period number* represents the number of electron shells in each element; thus, all elements with one electron shell (H and He) are listed horizontally opposite Period 1; those with two shells, opposite Period 2; and so on up to 7. For each element, the chemical symbol is shown, with its atomic number written above and its atomic weight below.

Each vertical column of the table is labeled with a Roman numeral, and elements within a column possess similar properties. For example, copper (Cu), silver (Ag), and gold (Au) in column IB are dense, soft, and fairly chemically inert metals. The elements on the right of the table exhibit nonmetallic behavior, whereas those on the left exhibit metallic behavior to various degrees.

IONS

Under certain conditions electrons may leave an atom, resulting in a net positive charge. The atom then becomes a *positive ion*. This tendency is especially prevalent in elements of groups IA and IIA having just a few electrons in their outermost shells, inasmuch as the attraction of the positive nucleus for these remote electrons is limited. Metallic elements exhibit this tendency for positive ionization most markedly.

Certain other elements show a strong tendency to acquire excess electrons in their outer orbits and to become *negative ions*. These elements, which usually have five or more electrons in their outermost shells to begin with and are almost invariably nonmetallic, appear at the right of the periodic table (groups VIA and VIIA).

MOLECULES

A material may consist of billions of one or more kinds of atoms (*atomic substances*), or of billions of several kinds of ions (*ionic substances*). Still other substances are formed by the joining together of units, each unit consisting of a number of atoms. These units are called *molecules* and can be broken into their constituent atoms only with difficulty. A molecule of methane, for example, is composed of one carbon atom and four hydrogen atoms (Fig. 1.10-5).

Substances composed of molecules are called *molecular materials*. Water, for in-

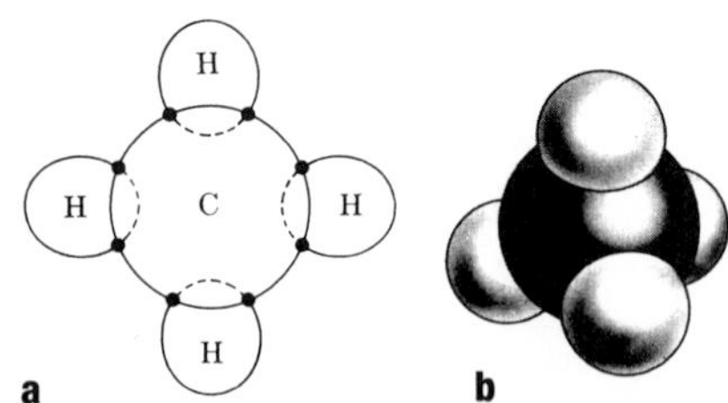

FIGURE 1.10-5 Models of a molecule of methane, CH_4: (a) two-dimensional representation; (b) a three-dimensional model.

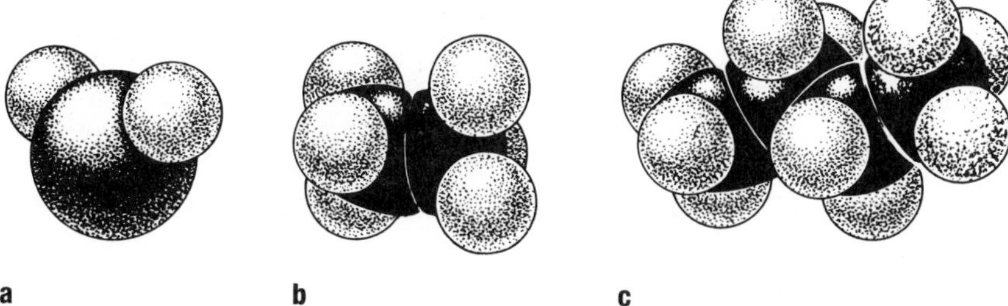

FIGURE 1.10-6 Three molecules of varying sizes: (a) water; (b) ethane; and (c) butane.

stance, is a molecular material. Each water molecule contains two hydrogen atoms and one oxygen atom and has a triangular shape. When large numbers of molecules are bonded together, the water exists as ice. As the tightness of bonding of the triangular molecules is decreased, water assumes the liquid form. When bonding is very loose, water exists as water vapor. In all three forms of the material the triangular units remain intact, and the individual hydrogen and oxygen atoms are not free to leave their respective molecules.

Molecules can vary greatly in size, as shown in Figure 1.10-6. A molecule of water contains only 3 atoms, whereas a molecule of ethane contains 8 and butane 14. Some molecules contain thousands of individual atoms; molecules of this size are called *polymers*, which form the basis of materials such as plastics (see Section 6.9).

1.10.1.2 Bonding of Matter

One of the major reasons that atoms, ions, and molecules bond together to form substances is the strong tendency in nature for outermost electron shells to contain as many pairs of electrons as possible. When a shell contains the maximum number of electron pairs, it is said to be *filled.* A filled shell results in a low energy condition on the atomic scale, meaning that the matter is relatively inert.

Because systems in nature tend to remain in the state of lowest energy, materials with filled outer shells are stable and resist change. Elements that occur naturally with filled outer shells are inert gases, such as helium (He), neon (Ne), and argon (Ar), found in group O at the far right of the periodic table. These elements refuse to combine with other elements, including each other.

Whenever the outer shell of an atom or ion of a noninert gas element is filled, the atom or ion is said to have achieved the *inert gas configuration* (IGC), in which it is content to remain because of its stability in that situation. Nature moves atoms and ions so strongly toward the IGC that these particles can be thought of as employing many tricks—including capturing electrons, surrendering electrons, and sharing electrons with their neighbors—in an effort to achieve chemical stability.

In exchanging or sharing electrons, atoms and ions interact with their neighbors, are attracted to them, and bond into matter. There are four types of bonding: *ionic*, *metallic*, *covalent*, and *secondary*.

IONIC BONDING

When atoms with almost empty outer electron shells exist in the vicinity of atoms with almost filled shells, the natural tendency of each set of atoms to achieve the IGC can be fulfilled. The atoms with few outer electrons surrender their "free" electrons and become positive ions, and atoms with many outer electrons accept these electrons to fill their outer orbits and become negative ions. A situation now exists in which positive ions and negative ions are in close proximity. The oppositely charged particles are attracted and bonded into matter by *ionic bonding*. Metals tend to have almost empty outer orbits, and nonmetals tend to have them almost filled; thus, ionic substances usually consist of elements from the left- and right-hand groups of the periodic table, bonded together.

Probably the most familiar ionically bonded substance is common table salt, NaCl. Atoms of sodium (Na) have just a single outer electron and therefore have a strong tendency to form positive (Na^+) ions. Atoms of chlorine (Cl) have seven outer electrons, and upon receiving one more electron, fill their outer orbits and achieve the IGC by becoming Cl^- ions. The two sets of ions are then bonded into NaCl (Fig. 1.10-7).

Ceramic materials, being combinations of metallic atoms and nonmetals, are bonded primarily through the ionic mechanism. Examples of this category of materials are brick, tile, portland cement, concrete, and natural stone. Their properties follow from the quality of the ionic bond: they have high melting tempera-

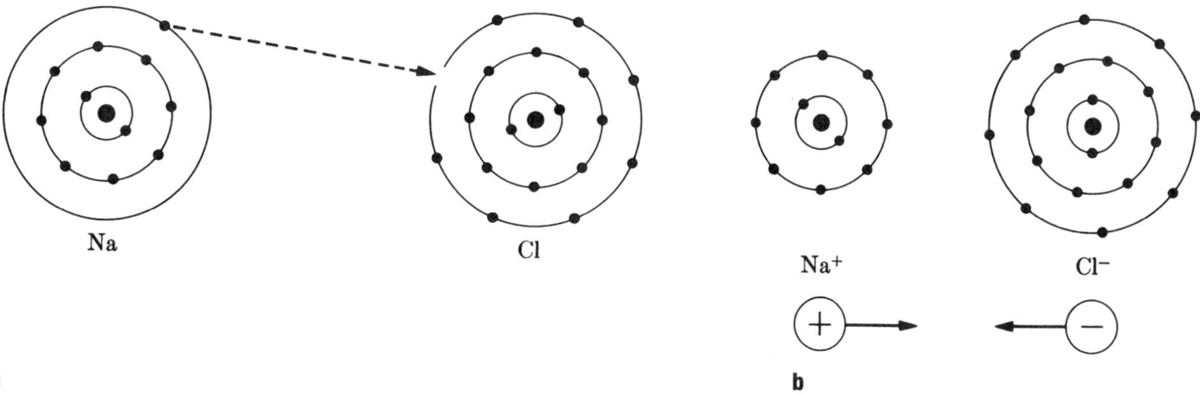

FIGURE 1.10-7 Ionic bonding in NaCl: (a) Na gives up a single electron, becoming a positive ion, and Cl accepts a single electron, becoming a negative ion; (b) positive and negative ions are then attracted to each other and bond together to become NaCl.

tures and are chemically inert because both types of ions have achieved the IGC and are tied by a strong ionic bond. Melting involves separating the bonds, and chemical attack requires separation and recombination of ions. Both processes are difficult when the original substance is stable to begin with.

Ceramic materials tend to be brittle, because the regular and extremely rigid arrangement of positive and negative ions resists shape changes. They tend to shatter, rather than be forced from one shape into another. Brittleness is generally associated with good strength in compression but low strength in tension. These materials also are poor conductors of electricity and heat, because the electrons are tightly bound into the stable ions and are not free to conduct electrical or thermal energy in response to an applied voltage or temperature.

The typical properties of ceramic materials follow logically from the nature of their constituent ions and of the ionic bond. Hundreds of new ceramic materials with enhanced properties may be developed in the future. It is difficult to predict all of their properties, but in general each of them will have a high melting temperature, high compressive strength, chemical inertness, and good thermal and electrical resistances.

METALLIC BONDING

Metallic atoms tend toward the IGC by attempting to surrender their few outer electrons, thus becoming positive ions. When large numbers of atoms are brought together, they each contribute these electrons to a mobile "sea" of electrons that circulates near the original parent atoms. The metal ions are now in a stable situation since they no longer possess their few outer electrons. They exist in a three-dimensional shape, as a metal crystal, with an electron "sea" dispersed throughout the solid (Fig. 1.10-8).

This peculiar cooperative surrendering of electrons by metal ions results in a *metallic bond*. The characteristic properties of metals follow from the nature of this bond. They are strong and have fairly high melting temperatures because the metallic bond is strong. Unlike ceramics, metals are good conductors of heat and electricity because the electrons in the "sea" are extremely mobile and free to transport thermal and electrical energy.

Metals may or may not be chemically inert. Their ability to transport electrical current makes them susceptible to chemical degradation in the form of corrosion. The tendency toward corrosion sometimes outweighs the intrinsic strength of the metallic bond.

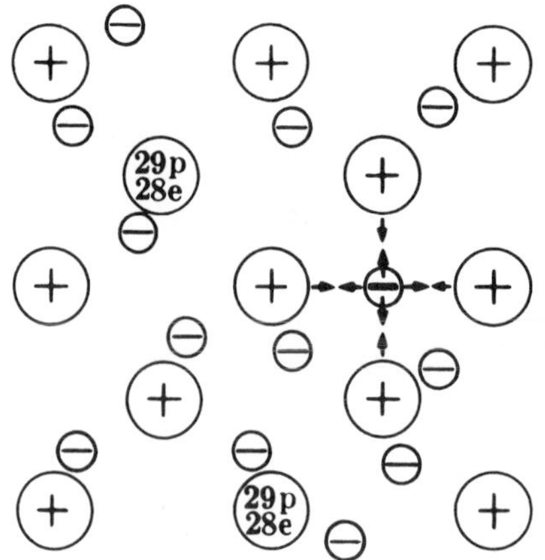

FIGURE 1.10-8 A metallic bond results from the attraction of the positive cores of the atom (copper) to the negative electrons in the surrounding "sea."

COVALENT BONDING

Many elements found in groups IVA and VA of the periodic table (see Fig. 1.10-4), such as carbon (C) and nitrogen (N), lack the strong tendency to form either positive ions or negative ions. These elements, with moderate numbers of electrons in their outer shells, reach the IGC by sharing electrons with similar elements. The process of mutual sharing of outer (*valence*) electrons by a cluster of atoms to create a stable entity is known as *covalent bonding*.

Covalent bonding is the most common mechanism whereby small numbers of atoms are bound into molecules. These molecules are then joined together by weak secondary bonds to produce molecular materials. The covalent bond is very strong, as evidenced by the fact that it is very difficult to break up a molecule into its constituent atoms.

A few substances are produced entirely by covalent bonding. Diamond is a familiar example. These substances are extremely hard and have very high melting temperatures, due to the strength of the covalent bonds. However, they seldom occur in nature, as suggested by the fact that diamonds are rarely found and then only in small amounts.

SECONDARY BONDING

To become molecular materials, atoms are bound into molecules by covalent bonding, and the molecules are then joined by means of weak *secondary bonds*. These bonds arise because positive nuclei or negative electrons in one molecule feel the weak attraction of their opposites in neighboring molecules. The weak attractions bind adjacent molecules into the substance.

Molecular materials such as wood, plastics, and bituminous products derive their properties from the behavior of the secondary bond. They have low strength and low melting temperatures as compared with metals and ceramics, because the weak secondary bond can be easily overcome by the application of force or heat. They are poor conductors of heat and electricity, because individual electrons remain tightly constrained within covalently bonded molecules.

Surprisingly, the weak secondary bond is not broken by many of the strong chemical compounds that attack metals and ceramics. Thus, molecular materials are chemically inert in a large number of environments. They are attacked by molecular solvents such as acetone, but their resistance to attack by most salts, acids, and industrial atmospheres makes them complementary to metals and ceramics.

1.10.2 PROPERTIES OF MATTER

Anyone who has been alone in a building on a quiet evening knows that materials are "alive." The creaks and groans are material responses to external stimuli such as heat, wind, and gravity loads. These stimuli fall into four categories, which define the four major properties exhibited by materials: *mechanical*, *thermal*, *electrical*, and *chemical*.

1.10.2.1 Mechanical Properties

Mechanical properties describe the response of a material to *static* (continuous) or *dynamic* (intermittent) loads.

STRESS AND STRAIN

When a force (load) of *x* pounds is applied to an object, the object *deflects* (sags) or *deforms* (changes shape) *y* inches or fractions of an inch. However, knowing the load in pounds or the deflection in inches does not provide enough information to predict the likely response.

For instance, a load of 100 lb. (45.36 kg) may induce a negligible reaction when distributed uniformly on the surface of a brick. The same load, concentrated at the point of a chisel, may shatter the brick. A deformation of 6 in. (152 mm) may be negligible when averaged over the height of an entire building, but the same deformation may have catastrophic consequences if localized. It is therefore necessary to consider forces in relation to the area over which they act and to view deformations with respect to the size of the region being deformed.

Stress (σ) is equal to the applied load (L) divided by the area (A) on which the load is acting:

$$\sigma = \frac{L}{A}$$

A 100 lb. (45.36 kg) load acting downward on the face of a nominal 4 × 8 in. (100 × 200 mm) brick exerts a stress of only 3 psi (20.68 kPa), which is well below the brick's inherent compressive strength. The same load concentrated at the point of a chisel 0.5 in. (12.7 mm) wide and sharpened to 0.02 in. (5.08 mm) thick exerts a stress of 10,000 psi (68.9 MPa), well beyond the strength of the typical brick. In engineering texts, stress is usually designated by the Greek symbol σ when the applied load is at right angles to the area in question, as in *tension* (stretching) or *compression* (squeezing).

Stress takes into account the fact that an applied load may or may not be shared by a substantial area of a body. Strain (ε) correspondingly accounts for the fact that deformation may be distributed over a large area or localized in a small region. If the original dimension of a body along a certain line is designated d, and the deformation (change in that dimension) is designated Δd, then:

$$\varepsilon = \frac{\Delta d}{d}$$

Strain is stated generally in percentage of dimension change; thus, sometimes,

$$\varepsilon = \frac{\Delta d}{d}(100)$$

Elastic Deformation An applied load, no matter how small, always generates both a stress and a strain in a solid object. A feather floating downward and landing on a steel beam will cause the beam to deflect minutely. The resulting deflection cannot be seen and the resulting strain cannot be measured even with today's sophisticated instruments, but the effect can be theoretically calculated and is really there. When the feather is removed, the strain will disappear and the beam will return to its original dimensions.

In this case, deflection is present only so long as the load is present, and its effect can be reversed by simply removing the load. This reversible type of strain is called *elastic deformation*. Many forces acting on structures cause only elastic deformation. For instance, the swaying of a tall building under the action of wind is an example of elastic deformation. In general, elastic deformation is harmless because the structure is restored to its original configuration when the force is removed.

Most building design criteria contain factors of safety to ensure that a structure will not be stressed beyond its elastic range. The most common measure of a material's *stiffness*, or ability to resist elastic deformation, is its modulus of elasticity, E.

Plastic Deformation Permanent or irreversible deformation is called *plastic deformation*. This type of deformation is essential when an aluminum ingot is formed into structural shapes. It would be most inconvenient if the shapes sprang back elastically into the shape of the original ingot. In the final structure, however, load-bearing areas are made large enough so that applied loads will not generate large enough stresses to cause permanent, plastic deformation.

THE STRESS-STRAIN TEST

Many of a material's mechanical properties can be accurately determined from its performance in a *stress-strain tensile test*, in which a material sample is stretched to fracture (breaking) and a continuous record is made of both the applied stress and resulting strain. Acceptance of parameters in a stress-strain test as illustrating mechanical behavior requires that such testing be conducted according to rigid standards. Descriptions of test procedures for materials, ranging from structural steel to plastic electrical insulation, are found in the ASTM standards.

A typical ductile metal, such as aluminum, has a stress-strain behavior as plotted in Figure 1.10-9. The behavior in

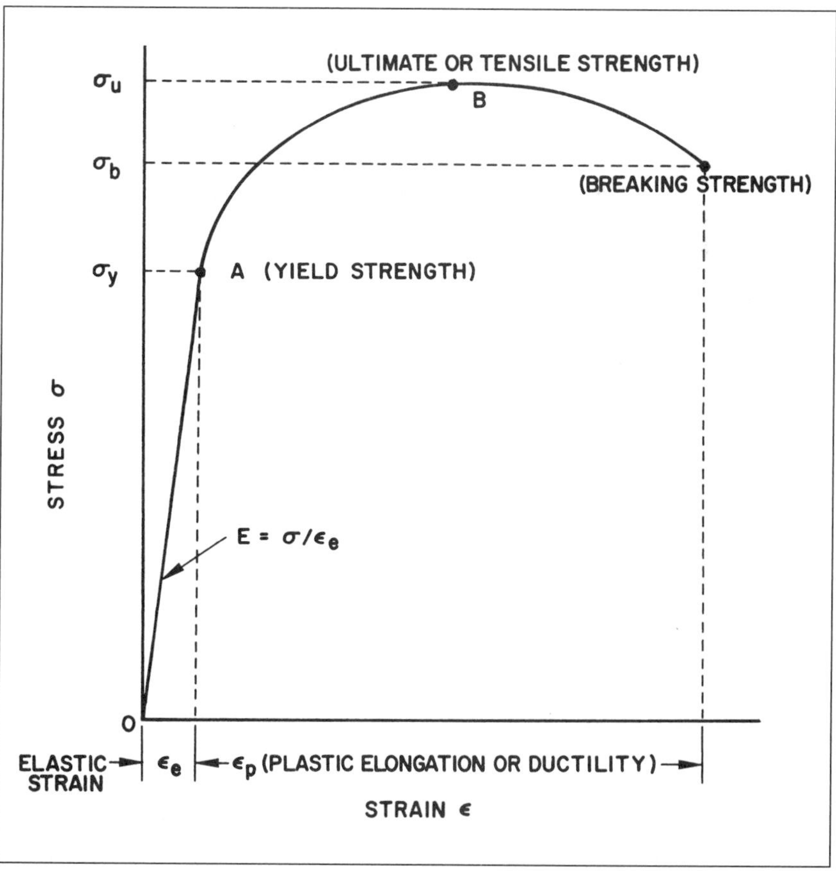

FIGURE 1.10-9 **Typical stress-strain behavior of a ductile metal.** (Courtesy Addison-Wesley Publishing Company.)

region OA is elastic and linear. If any stress between O and A is applied and then removed, the strain in the body will return to 0%. The straight line shows that for each unit of increase in stress, the strain will increase by a constant amount; in other words, strain is proportional to stress.

The proportionality constant between stress (σ) and strain (ε) is the modulus of elasticity (E) and is represented by the slope of line OA. The steeper the stress-strain curve, the stiffer the material. It follows, then, that for stresses within the elastic range:

$$E = \frac{\sigma}{\varepsilon}$$

The modulus E is an important property because it enables calculation of the elastic strain accompanying a particular stress or loading condition, and it enables calculation of deflection. Figure 1.10-10 compares the E-values of construction materials belonging to the metals, ceramics, and molecular categories. Metals are generally quite stiff, ceramics slightly less stiff, and plastics still less stiff.

When the stress rises above A, into the nonlinear portion of the curve, permanent deformation will appear in the body. If the stress is then reduced to 0, the elastic strain (ε_e) in the body will disappear, but some plastic strain (ε_p) will remain. The stress value on the curve marking the transition from elastic to plastic behavior is called the *yield point*, or *yield strength* (σ_y).

The stress-strain curves for most materials do not show clearly defined yield points. For these materials, a theoretical yield strength is calculated from the stress-strain curve. Determining the yield strength of a ductile material is useful, because it is the basis for a common structural design method. The precise definition of σ_y and detailed methods for determining it are given in ASTM standards.

If stresses greater than σ_y are applied in the test, the material behaves as shown by the AB portion of the curve, with the entire region AB representing plastic behavior. At the stress marked by point B, the material rapidly begins "necking down," as does a piece of chewing gum stretched to the breaking point. The quantity σ_u is the *ultimate* or *tensile strength.*

Beyond point B, deformation continues even with application of increasingly smaller loads for a ductile material and thus leads to the eventual fracture of the material at stress σ_b, which is the *fracture*, or *breaking*, *strength* of the material. The total plastic stretching of the material to failure is given by ε_p, the *elongation*, which is a measure of the material's ductility.

Figure 1.10-11 shows the stress-strain behavior of four typical construction materials.

Steel has a much higher E than copper and therefore deforms much less in the elastic range. The yield strength of steel is also much higher than that of copper, which indicates that it can withstand greater stresses before it undergoes permanent deformation. On the other hand, steel stretches only about one-half as much as copper, and its greater strength is obtained at the price of higher *brittleness* (less ductility). The behavior of steel and copper is typical of the metallic category of materials and is compared with that of plastics in Figure 1.10-12.

Because of the strength of a metallic bond as compared with that of a secondary bond (see Section 1.10.1.2), metals resist permanent deformation more effectively than plastics. Note in Figure 1.10-12 the range of yield strengths in the metals resulting from appropriate processing such as annealing, heat-treating, and work-hardening.

Ceramics are omitted from Figure 1.10-12 because they do not deform plastically before fracture and are therefore not generally characterized by a specific yield strength. Ceramic materials are typified by the curves for concrete shown in Figure 1.10-11. They have moderately high E values, represented by slopes greater than those for plastic but not as great as those for steel. They respond rigidly to elastic stresses, showing small deformations for even fairly large stresses. At the same time they are very brittle, showing little plastic deformation before fracture. The brittleness of ceramic materials is accentuated when they are stressed in tension, which accounts for the common use of concrete, brick, tile, and stone to resist compressive loads, but not tensile loads. The absence of an elastic-plastic transition and a maximum in the stress-strain behavior means that yield strength, ultimate strength, and breaking strength are the same for individual ceramic materials.

The polyethylene curve in Figure 1.10-11 is typical of materials in the molecular category. The linear elastic region is limited, and a small elastic stress produces a large deflection. This behavior is often a disadvantage when molecular materials are used as load-bearing elements. The molecular materials that show the greatest tendency to stretch elastically are the class of rubbers known as elastomers, which are used in rubber bands and sealants.

Under the severest conditions, metals and plastics can be stressed to ultimate strength, σ_u, before catastrophic fracture occurs, whereas ceramics can be stressed

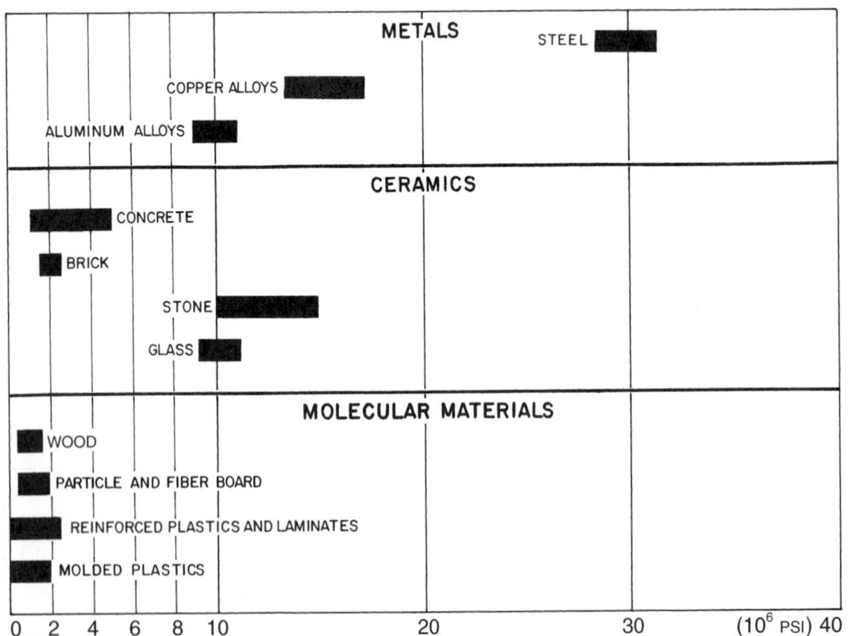

FIGURE 1.10-10 **Comparative stiffness of construction materials.** (Courtesy Addison-Wesley Publishing Company.)

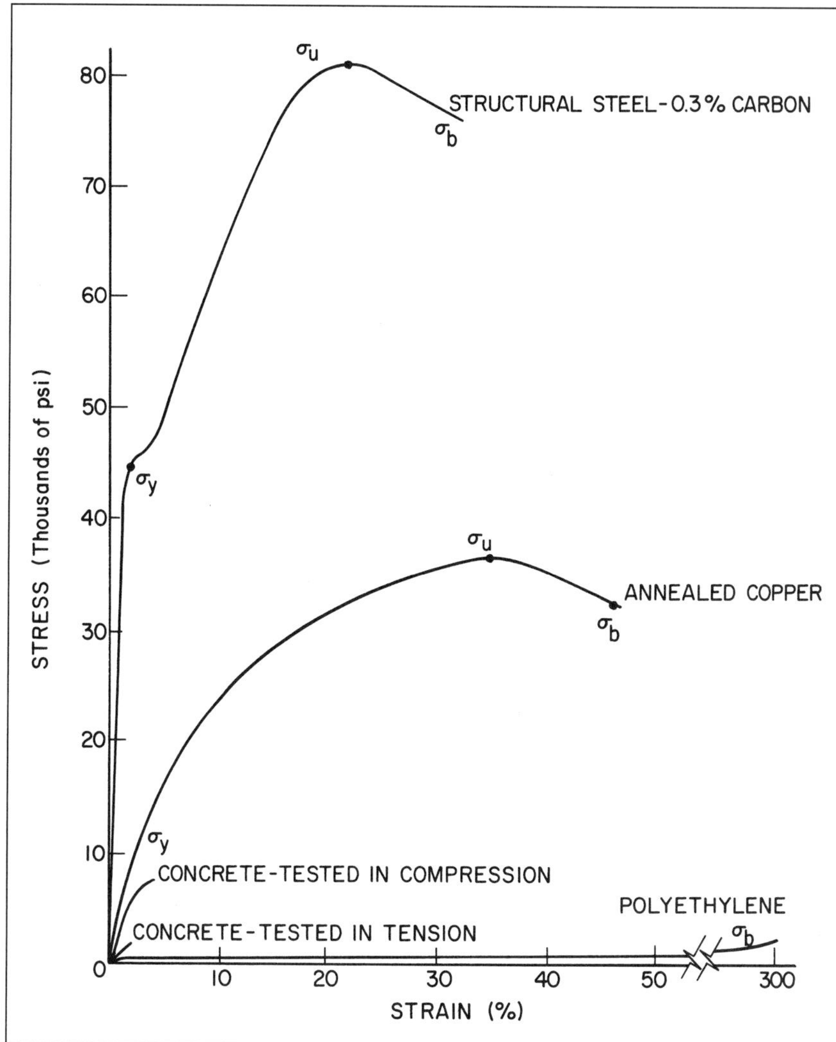

FIGURE 1.10-11 Stress-strain behavior of four construction materials. (Courtesy Addison-Wesley Publishing Company.)

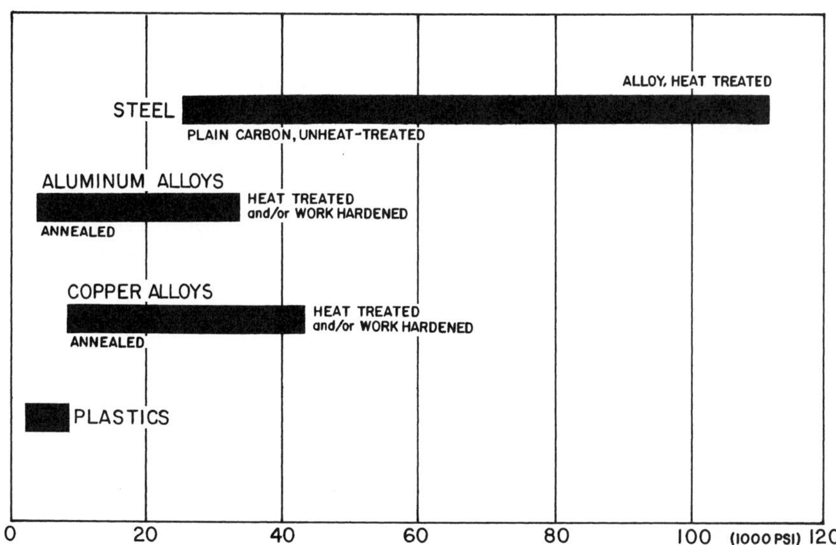

FIGURE 1.10-12 Yield strengths of metals and plastics. (Courtesy Addison-Wesley Publishing Company.)

to breaking strength, σ_b, which for them is identical to yield strength, σ_y, and ultimate strength, σ_u. Figure 1.10-13 compares the extreme limits of strength for construction materials falling into each of the three categories.

Figures 1.10-11, 1.10-12, and 1.10-13 substantiate generalizations about material properties derived from their structure. Metals are strong, elastically rigid, and fairly ductile. Ceramics are rigid, strong in compression, and brittle, especially in tension. Molecular materials have adequate mechanical properties for many construction applications, but are limited by other properties.

MISCELLANEOUS MECHANICAL PROPERTIES

Tabulated stress-strain data provide a quick means for gauging a material's mechanical performance. However, where materials are subject to mechanical loading, knowledge of other properties is necessary to predict performance.

Hardness is a measure of a material's ability to resist indentation or penetration. It is determined by various tests, described in ASTM standards, in which indenters are forced into materials under carefully prescribed conditions. In general, the harder a material is, the greater its wear and abrasion resistance. Materials with large values of E and σ_y tend to be hard and wear-resistant, so in the absence of hardness data, stress-strain data give a measure of wear resistance.

Fatigue resistance is a measure of a material's ability to withstand cyclic (repeated) stresses. When repeatedly stressed, even at stresses below yield strength, many materials will fracture without warning. Some airplane accidents have been caused by failures resulting from fatigue. The lifetimes of pumps and other mechanical devices in a building's electrical and mechanical systems depend on the fatigue behavior of the materials employed, and the use of adequate shock-absorbent mountings deserves careful consideration.

Damping capacity is a measure of a material's ability to dissipate or deaden mechanical vibration. Because sound is mechanical vibration, a material's ability to absorb sound is directly related to its damping capacity.

Impact strength or *toughness* marks a material's capacity to absorb impact without fracturing. It is defined as the total energy, from elastic deformation to fracture,

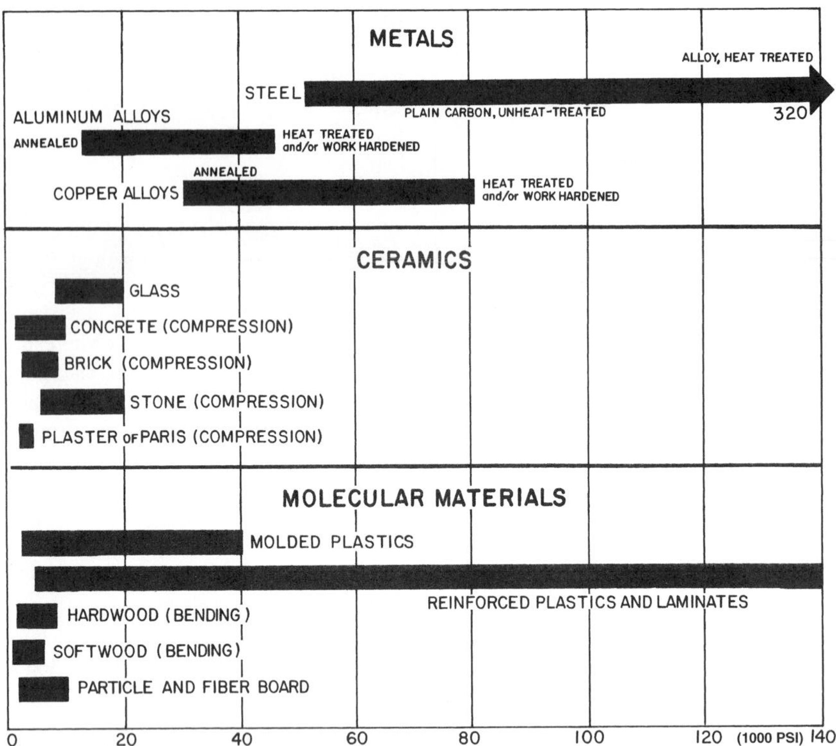

FIGURE 1.10-13 **Ultimate strengths of representative materials.** (Courtesy Addison-Wesley Publishing Company.)

that a material can absorb before breaking under impact. A material's strength and ductility affect this property. Ceramics are strong, but do not deform significantly under static or dynamic load, so they lack toughness and shatter under impact. Plastics can stretch up to several hundred percent under load, but their strength is relatively low, so they absorb little energy under impact. Metals, with their good strength and ductility, are the toughest of the common construction materials.

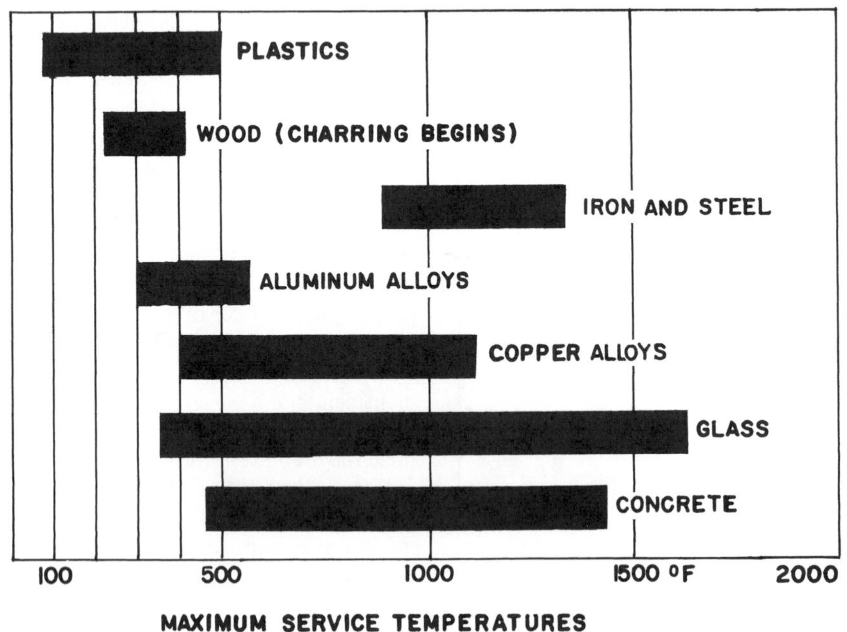

FIGURE 1.10-14 **Service temperatures of construction materials.** (Courtesy Addison-Wesley Publishing Company.)

1.10.2.2 Thermal Properties

When subjected to temperature changes, a material may change its state (solidify, melt, or vaporize), expand or contract, and conduct or reflect heat.

MELTING TEMPERATURE

As noted earlier, certain mechanical properties, such as tensile strength, give an indirect measure of other properties, such as hardness. A material's melting temperature can be similarly employed. Materials with high melting points usually retain their mechanical properties over a greater temperature range. In addition, they tend to be stronger and more chemically inert than materials with lower melting points.

Materials with strong ionic bonding and high melting temperatures, such as ceramics, tend to perform best at high temperatures. Metals perform moderately well, and molecular materials, with their weak secondary bonding, perform less well. Figure 1.10-14 compares graphically the temperature limits for continuous service of common construction materials. A more quantitative summary of the thermal properties of typical construction materials is given in Figure 1.10-15.

THERMAL CONDUCTIVITY

A building designer is concerned with a solid material's response to heat. Of special importance is the material's thermal conductivity, which is its ability to transfer heat from a region of high temperature to a region of lower temperature (see Fig. 1.10-15). Because the free electrons in metals transport heat effectively, metals typically exhibit the highest thermal conductivities; ceramics have much lower conductivities, and molecular materials the lowest.

A material's *thermal conductivity* is the number in British thermal units (Btu's) (kilojoule [kJ]) conducted in an hour through 1 sq. ft. (0.093 m^2) of 1-in. (25.4 mm)-thick material, for each 1°F (–17.2°C) of temperature differential. Conductivity is measured by placing a 1-sq.-ft. (0.093 m^2) by 1-in. (25.4 mm)-thick slab of the material to be tested in a laboratory oven with the material's "hot" face 1°F (–17.2°C) hotter than its "cold" face. The heat energy in Btu's (kJ) conducted through the material in 1 hour is the material's thermal conductivity.

FIGURE 1.10-15 Thermal Properties of Typical Construction Materials

Category	Material	Melting Temperature (°F)	Thermal Conductivity (Btu/sq. ft./°F/in./hr.)	Thermal Expansion (10^{-6} in./in./°F)
Metals	Steels (carbon or low alloy)	2700–2800	300–500	6.5
	Stainless steels	2700–2800	100–150	7
	Aluminum alloys	1100–1200	500–1500	9
	Copper alloys	1800–2000	500–2500	12.5
Ceramics	Brick	2000–5000	3–6	5
	Concrete	3000	5–10	7
	Plate glass	1500	5–6	1
Molecular materials	Wood	400 (chars)	1–2	10–30
	ABS plastic	200–300	1–2	50
	Acrylics (PMMA)	200–300	1	30–50
	Vinyls (PV)	200–300	0.5–1	30–100
	Polyethylene (PE)	200	2–5	60
	Epoxies (EP)	200–500	1–2	30
	Silicones (SI)	400–600	1	1–100

(Refer to Figure 1.8-27 for metric conversions.)

THERMAL EXPANSION

One reason that buildings seem "alive" at times is that temperature changes cause materials to expand and contract. The crackle of ductwork as a furnace starts up is an example of thermal expansion in action. *Thermal expansion* is measured by the coefficient of thermal expansion, which is stated in terms of length change per unit of length per degree of temperature change (in./in./°F [mm/mm/°C]).

Figure 1.10-15 shows that, in general, materials with low melting temperatures have high thermal expansion coefficients and vice versa. Ceramics tend to have slightly lower coefficients of thermal expansion than metals, and plastics usually have much higher coefficients than either ceramics or metals.

Structural steel has an expansion coefficient of 0.0000065 in./in./°F, which indicates that each 1-in. length of steel expands 6.5 millionths of an inch for every 1°F temperature rise. A steel bridge 1 mile long expands about 12 in. (305 mm) as its temperature is raised from 70°F (21.1°C) to 100°F (37.8°C).

Thermal expansion is a critical property in applications where several materials are joined. For example, aluminum has twice the thermal expansion of steel. Thus, if aluminum trim is fixed to a steel base, the uneven expansion of the two materials will cause internal stresses and elastic deformation in the assembly. Noises, as well as unattractive buckling effects, may result. The need for properly designed expansion joints in such cases is obvious.

1.10.2.3 Electrical Properties

The primary electrical property of interest to a building designer is *electrical conductivity*, a characteristic that is closely related to a material's *thermal conductivity*. The free electrons in metals allow heat and electrical energy to be conducted easily, so conductivity is high. Ceramics have lower electrical conductivities, and molecular materials have the lowest. This property accounts for the extensive use of molecular materials, such as plastics, in electrical insulation products.

Electrical conductivity is usually given in mho (ohm spelled backwards) units per foot of conductor length. It is the reciprocal of electrical resistivity, which is expressed in *ohms per square foot*. For example, metals have *low* electrical *resistivity* and *high* electrical *conductivity*. Figure 1.10-16 summarizes the resistivities of typical construction materials and illustrates the excellent insulating capabilities of ceramics and molecular materials.

FIGURE 1.10-16 Electrical Resistivity of Typical Construction Materials

Category	Material	Electrical Resistivity (ohm/ft.)
Metals	Steel (carbon or low alloy)	.000006
	Stainless steel	.00003
	Aluminum alloys	.000001
	Copper alloys	.0000005
Ceramics	Brick	4,000,000
	Concrete	4,000,000
	Plate glass	10,000,000,000,000
Molecular materials	ABS plastic	1,000,000,000,000,000
	Acrylics (PMMA)	100,000,000,000,000
	Vinyls (PV)	100,000,000,000
	Polyethylene (PE)	1,000,000,000,000
	Epoxies (EP)	100,000,000,000,000
	Silicones (SI)	1,000,000,000,000

1.10.2.4 Chemical Properties

Air and airborne moisture contain small amounts of active chemical compounds, which, under certain conditions, react with building materials and degrade their properties.

Ceramics resist chemical attack in most normal environments, and plastics resist attack from all but a few organic solvents. The situation is more complicated, however, in the case of metals.

Metals degrade or corrode through the transport of minute amounts of electricity from regions called *anodes* to other regions called *cathodes*. A cathode accepts electrons and remains intact, but an anode is degraded by the chemical reaction.

The essential elements in a corroding system are the anode, the cathode, and the current-carrying medium (*electrolyte*) connecting them. An electrolyte is usually a solution of water and a gas, such as carbon dioxide or sulfur dioxide. Preventing corrosion, therefore, is simply a matter of removing one of the three elements from the system. For example, most paints are molecular materials, which are poor conductors of electricity. Painting metallic surfaces not only shelters them directly from corrosive environments, but also provides a nonconductive barrier to electrical current flow.

Corrosion can occur when an entire material, or a given region in a material, is subjected to a situation conducive to anodic behavior. Such situations develop in the presence of certain impurities in the metal (acting as cathodes) and when metals with different *galvanic potentials* are placed in close proximity. Galvanic tables ranking materials according to their tendency to become anodic can be used to predict potentially corrosive situations (Fig. 1.10-17).

Materials at the top of the list have a strong tendency to become anodic, and those toward the bottom cathodic. If a material from near the top of the list, such as carbon steel, is placed in contact with a material lower on the list, such as brass, corrosion is likely to result. The steel, acting as an anode, and the brass, acting as the cathode, will cause an electric current to flow. Corrosion of the steel is inevitable.

FIGURE 1.10-17 Galvanic Series of Common Metals and Alloys

Electrolytic Tendency	Metal or Alloy
Anodic (read up)	Magnesium alloys
	Zinc
	Aluminum alloys
	Carbon steel
	Stainless steel (active)
	Lead
	Tin
	Brass
	Copper
Cathodic (read down)	Bronze
	Stainless steel (passive)
	Gold

Several points regarding galvanic reactions and corrosion problems should be remembered:

1. Intimate contact between materials widely separated in the galvanic table should be avoided.
2. Zinc can be used as a protective coating (galvanizing) on steel. Because it is higher than steel in the galvanic series, the zinc serves as a "sacrificial anode" and is intentionally permitted to corrode, thus saving the steel. As long as the anode is present, the cathode (the steel) will not corrode.
3. Tin is cathodic with respect to steel, so, as a coating over steel, it simply acts as a physical barrier to the corrosive environment. Should the tin coating be scratched or removed in spots, the exposed steel will corrode quickly. Care must be taken, for example, with *terneplate* (a tin and lead coating over steel), as with painted surfaces, to maintain a continuous film of the protective coating.

Surprisingly, aluminum and stainless steel are inherently "actively" anodic and susceptible to corrosion. The corrosion resistance normally associated with these metals develops when they are exposed to oxygen. The oxygen forms a thin, impervious, transparent oxide film that protects the underlying metal from further contact with atmospheric oxygen. The anodizing of aluminum alloys is simply an electrochemical process that produces a thicker oxide coating than forms naturally.

1.11 Additional Reading

More information about the subjects discussed in this chapter can be found in the references listed in Section 1.12 and in the following publications.

Allen, Edward. 1999. *Fundamentals of Building Construction: Materials and Methods*. 3d ed. New York: John Wiley & Sons, Inc.

Allen, Edward, and Joseph Iano. 1995. *The Architect's Studio Companion: Rules of Thumb for Preliminary Design*. 2d ed. New York: John Wiley & Sons, Inc.

Bannister, Jay M. 1991. *Building Construction Inspection: A Guide for Architects*. New York: John Wiley & Sons, Inc.

Lebovich, William L. 1993. *Design for Dignity: Studies in Accessibility*. New York: John Wiley & Sons, Inc.

Liebing, Ralph W. 1999. *Architectural Working Drawings*. 4th ed. New York: John Wiley & Sons, Inc.

Mays, Patrick C., and B. J. Novitski. 1997. *Construction Administration: An Architect's Guide to Surviving Information Overload*. New York: John Wiley & Sons, Inc.

Pachner, Edmond. 1992. *Architectural Contract Administration*. New York: John Wiley & Sons, Inc.

Snyder, James. 1998. *Architectural Construction Drawings with AutoCAD® R14*. New York: John Wiley & Sons, Inc.

Stovall, Gerald L. 1995. "Project Contract Administration." *The Construction Specifier* (5)(May):149.

Wakita, Osamu, and Richard Linde. 1994. *The Professional Practice of Architectural Working Drawings*. 2d ed. New York: John Wiley & Sons, Inc.

Yee, Rendow. 1997. *Architectural Drawing: A Visual Compendium of Types and Methods*. New York, John Wiley & Sons, Inc.

1.12 Acknowledgments and References

ACKNOWLEDGMENTS

We gratefully acknowledge the assistance of the following organizations and individuals in preparing this chapter. We are also indebted to them for permission to use their illustrations when requested and for the use of their publications as references.

Addison-Wesley Publishing Company
American National Standards Institute (ANSI)
American Society of Heating, Refrigerating and Air Conditioning Engineers (ASHRAE)
American Society for Testing and Materials (ASTM)
Bethlehem Steel Corporation
Building Officials and Code Administrators International (BOCA)
Capital Development Board of the State of Illinois
Chicago Title and Trust Company
Geoffrey Lee Farnsworth, AIA
Professor N. F. Fiore
International Conference of Building Officials (ICBO)
National Association of Home Builders (NAHB) Research Foundation
National Easter Seal Society
National Institute of Building Sciences
National Institute of Standards and Technology (NIST) of the U.S. Department of Commerce
Betsy A. Pavichevich
Plumbing-Heating-Cooling Contractors Association (PHCCA)
Standard Building Code/Southern Building Code Congress International (SBCCI)
Associate Professor Edward Steinfeld
T & S Brass and Bronze Works, Inc.
Underwriters Laboratories, Inc.
U.S. Architectural and Transportation Barriers Compliance Board (ATBCB)
U.S. Department of Housing and Urban Development (HUD)
U.S. General Services Administration, Specifications Unit
Professor Lawrence H. Van Vlack

REFERENCES

We would also like to thank the authors and publishers of the publications in the following list for their contribution to our research for this chapter.

American Institute of Architects. 1997. *A201 General Conditions of the Contract for Construction.* 1997 ed. Washington, DC: AIA.

———. *The Architects Handbook of Professional Practice.* 12th ed. Washington, DC: AIA.

American National Standards Institute. 1996. A17.1, "Safety Code for Elevators and Escalators." New York: ANSI.

———. 1998. A117.1, "Guidelines for Accessible and Usable Buildings and Facilities." New York: ANSI.

———. 1999. A156.10, "Power-Operated Pedestrian Doors." New York: ANSI.

———. 1997. A156.19, "Power-Assist and Low Energy Power-Operated Doors." New York: ANSI.

American Society for Testing and Materials (ASTM). Standards.

E 621, "Practice for the Use of Metric (SI) Units in Building Design and Construction." West Conshohocken, PA: ASTM.

E 713, "Guide for the Selection of Scales for Metric Building Drawings." West Conshohocken, PA: ASTM.

Architectural Graphic Standards (See Ramsey/Sleeper).

Arcom Master Systems. MASTERSPEC®. Basic Sections:

01020, "Allowances." Salt Lake City, UT: Arcom.

01027, "Applications for Payment." Salt Lake City, UT: Arcom.

01200, "Project Meetings." Salt Lake City, UT: Arcom.

01300, "Submittals." Salt Lake City, UT: Arcom.

01311, "Schedules and Reports." Salt Lake City, UT: Arcom.

01340, "Shop Drawings, Product Data, and Samples." Salt Lake City, UT: Arcom.

01631, "Substitutions." Salt Lake City, UT: Arcom.

01700, "Contract Closeout." Salt Lake City, UT: Arcom.

01720. "Project Record Documents." Salt Lake City, UT: Arcom.

01730. "Operation and Maintenance Data." Salt Lake City, UT: Arcom.

01740 "Warranties." Salt Lake City, UT: Arcom.

Champagne, Roger D. 1986. "Planning for Plumbing Efficiency." *The Construction Specifier* 39(6)(June):77–80.

Construction Specifications Institute. 1996. *Manual of Practice.* Alexandria, VA: CSI.

Council of American Building Officials (CABO). *One- and Two-Family Dwelling Code.* Falls Church, VA: CABO. This code is in the process of being transferred to the control of the International Code Council (ICC).

Freund, Eric C. 1990. "U.S. Construction and Practice in Europe." *The Construction Specifier* 43(9)(September): 68–82.

Goldman, Charles D. 1985. "Removing Barriers to the Handicapped." *The Construction Specifier* 38(4)(April): 40–45.

———. 1988. "Accessibility, the Evolution Continues." *The Construction Specifier* 41(2)(February):50–53.

Institute of Electrical and Electronics Engineers, Inc. (IEEE). 1997. IEEE/ASTM SI 10, *Standard for Use of the International System of Units (SI): The Modern Metric System.* New York: IEEE/ANSI.

Meier, Hans. 1989. *Construction Specifications Handbook.* 4th ed. Englewood Cliffs, NJ: Prentice-Hall.

National Institute of Standards and Technology (NIST). Special Publication 504, *Metric Dimensional Coordination—the Issues and Precedents.* Gaithersburg, MD: NIST.

———. Special Publication 530, *Metrication in Building Design.* Gaithersburg, MD: NIST.

———. Technical Note 990, *The Selection of Preferred Metric Values for Design and Construction.* Gaithersburg, MD: NIST.

Phillips, Ruth Hall, and Ronald L. Mace. 1985. "Designing and Specifying for Accessibility." *The Construction Specifier* 38(4)(April):46–57.

Raeber, John A. 1988. "Complying with Accessibility Codes." *The Construction Specifier* 41(2)(February):45–49.

———. 1991. "Redundancy in Fire Protection." *The Construction Specifier* 44(2)(February):17–19.

———. 1991. "Federal Accessibility." *The Construction Specifier* 44(8)(August):82–89.

Ramsey/Sleeper, The AIA Committee on

Architectural Graphic Standards. *Architectural Graphic Standards*. 9th ed. New York: John Wiley & Sons, Inc.

Rosen, Harold J. 1999. *Construction Specifications Writing: Principles and Procedures*. 4th ed. New York: John Wiley & Sons, Inc.

Solomon, Nancy B. 1991. "Equal Opportunity Design." *Architecture* (February):101–102.

Stubbs, Stephanie. 1991. "AIA Gives ATBCB ADA A-D-V-I-C-E." *The American Institute of Architects Memo* (May):4–7.

U.S. Department of Housing and Urban Development (HUD). 1984. *Minimum Property Standards for Housing*. Washington, DC: HUD.

———. 1989. *Adaptable Housing*, HUD-1124-PDR(1). Washington, DC: HUD.

U.S. General Services Administration (GSA). 1988. *Uniform Federal Accessibility Standards (UFAS)*. Washington, DC: GSA.

———. *Index of Federal Specifications, Standards, and Commercial Item Descriptions*. Washington, DC: Superintendent of Documents, U.S. Government Printing Office.

Van Vlack, Lawrence H. *Elements of Materials*. 2d ed. Reading, MA: Addison-Wesley.

Webb, William A. 1991. "Rights of Passage." *The Construction Specifier* 44(8)(August):91–95.

Wyatt, David J. 1998. "Successful Contract Closeout: Part 1." *The Construction Specifier* (3)(March):23.

———. 1998. "Successful Contract Closeout: Part 2." *The Construction Specifier* (4)(April):23.

CHAPTER 2

Site Construction

Introduction

Applicable *MasterFormat*™ Sections

Soils

Site Preparation

Earthwork

Surface- and Groundwater Problems

Lawns and Landscaping

Additional Reading

Acknowledgments and References

Introduction

This chapter discusses some aspects of site construction, which is as important to a building's success as any other feature. Site construction includes work to be done on the grounds, accessways, parking areas, lawns, plantings, and other site features. These have important aesthetic and site-use implications and greatly affect site and building access. Site construction also includes building location, orientation, and relationship to the finish grade. Besides their obvious aesthetic implications, a building's location and orientation and the lawns and plantings around it can alter the climate immediately at the building and somewhat control the effects of sun and wind on the building. These aspects can increase the comfort of the building's occupants and reduce not only the size of its heating and air conditioning systems, but their operating costs as well.

Site construction also includes subsurface investigation; site clearing; dewatering; excavation and backfilling for site improvements and buildings; surface- and groundwater control, including dewatering during construction and prevention of water intrusion into the completed building; and grading. These components can have important effects on construction costs, and the way in which they are done can change the final look of a building site. They can also be crucial to a completed building's structural stability. Improper excavation and backfilling can be dangerous during the work, and damaging to the completed structure later, if supporting materials are improperly disturbed or fill is not compacted as it should be. Proper grading will ensure that water drains away from a building rather than toward it, which will prevent potentially disastrous leaks and flooding.

Site construction also includes water distribution (see Section 15.1), sewage and drainage (see Sections 15.1 and 15.2), and on-site electrical transmission lines (see Section 16.2).

Site characteristics that may affect the design and construction of slabs on grade, building foundations, and site improvements include:

- Underlying soil type and properties
- Moisture conditions
- Thermal conditions
- Geographic factors that may require construction precautions to control termites or unusual conditions

Applicable *MasterFormat*™ Sections

The following *MasterFormat* Level 2 sections are applicable to this chapter.

02050 Basic Site Materials and Methods
02200 Site Preparation
02300 Earthwork
02400 Tunneling, Boring, and Jacking
02450 Foundation and Load-Bearing Elements
02600 Drainage and Containment
02700 Bases, Ballasts, Pavements, and Appurtenances
02800 Site Improvements and Amenities
02900 Planting

2.1 Soils

Some buildings rest on a continuous mass of solid crystalline *bedrock*, or a continuous sedimentary rock, but soil underlies most building sites and supports the loads imposed by most low-rise and many high-rise building foundations. Because a building's structural integrity is dependent on the material that supports its foundations, a basic understanding of the nature of soils and how they perform under loads is important.

2.1.1 CLASSIFICATION

Proper classification is a critical factor in identifying problem soils. Several soil classification systems are available, but the one most commonly used for soils associated with buildings is that developed by the American Society for Testing and Materials (ASTM). The applicable ASTM standards are D 2487 and D 2488.

In the ASTM system, soils are classed broadly as being either *granular* or *cohesive* (Fig. 2.1-1). Granular soils, such as boulders, gravel, and sand, consist of relatively large particles visible to the eye. In cohesive soils the particles are relatively small, and in some soil types, such as fine-grained clays, the particles cannot be seen even through a low-powered microscope. Silts are soils made up of particles finer than sand but coarser than clay (Fig. 2.1-2). Actual soil samples usually consist of mixtures of many soils of different particle sizes. They are described as combinations of the predominate soils, such as silty sand, clayey sand, or gravelly sand.

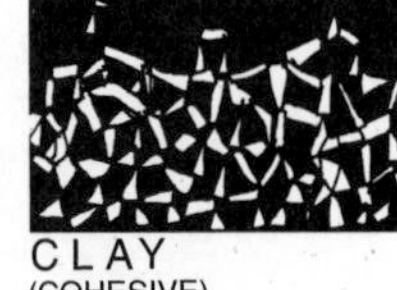

FIGURE 2.1-1 Sand grains are large and angular, and the volume of the voids is relatively small. Clay grains are minute and scaly, and the volume of the voids is quite large.

The soil classification chart in Figure 2.1-3 identifies soils and divides them into coarse-grained and fine-grained soils according to the ASTM system. Each type of soil is given a descriptive name and a letter symbol indicating its principal characteristics. As indicated by the typical names in Figure 2.1-3, field classification of soils includes a variety of different ma-

FIGURE 2.1-2 Soil Types

Soil Types	Approximate Size Limits of Soil Particles
Boulders	Larger than 3 in. (76.2 mm) diameter
Gravel	Smaller than 3 in. (76.2 mm) diameter but larger than #4 sieve
Sand	Smaller than #4 sieve[a] but larger than #200 sieve[b]
Silts	Smaller than 0.02 mm diameter but larger than 0.002 mm diameter
Clay	Smaller than 0.002 mm diameter

[a]Approximately ¼ in. (6.4 mm) in diameter.

[b]Particles less than #200 sieve are not visible to the naked eye.

terials. When soil is a combination of two types, it is described as a combination of both. For example, clayey sand describes a soil that is predominantly sand but that contains an appreciable amount of clay. A sandy clay has the properties of clay but contains an appreciable amount of sand.

2.1.2 PROPERTIES

The stability and structural properties of soil in relation to foundations are determined in large measure by the shearing resistance of the soil. Shearing resistance in most subgrade materials is the combined effect of internal friction and cohesion.

2.1.2.1 Internal Friction

The *internal friction* of a soil can be described as the resistance of the soil grains to their tendency to slide over one another.

2.1.2.2 Cohesion

The binding force that holds soil grains together is called *cohesion*. It varies with the moisture content of the soil. Cohesion may be very high in clays, but sand and silt have little if any cohesion.

2.1.2.3 Other Properties

Other soil characteristics that influence building foundations and slabs-on-grade include the following:

Classification and composition
Grain size
Moisture content
Density
Capillarity
Drainage characteristics
Stratification
Consistency
Compressibility
Climatic variations

FIGURE 2.1-3 ASTM Soil Classification System

	Major Divisions	Group Symbols	Typical Names	Presumptive Bearing Capacity, tons/sq. ft.[a]	Modulus of Subgrade reaction, *k*, lb./sq. in./in.
Coarse-grained soils more than 50% retained on #200 sieve[b]	Gravels 50% or more of coarse fraction retained on #4 sieve				
	Clean gravels	GW	Well-graded gravels and gravel-sand mixtures, little or no fines	5	300 or more
		GP	Poorly graded gravels and gravel-sand mixtures, little or no fines	5	300 or more
	Gravels with fines	GM	Silty gravels, gravel-sand-silt mixtures	2.5	200 to 300 or more
		GC	Clayey gravels, gravel-sand-clay mixtures	2	200 to 300
	Sands more than 50% of coarse fraction passes #4 sieve				
	Clean sands	SW	Well-graded sands and gravelly sands, little or no fines	3.75	200 to 300
		SP	Poorly graded sands and gravelly sands, little or no fines	3	200 to 300
	Sands with fines	SM	Silty sands, sand-silt mixtures	2	200 to 300
		SC	Clayey sands, sand-clay mixtures	2	200 to 300
Fine-grained soils 50% or more passes #200 sieve[b]	Silts and clays, liquid limit 50% or less	ML	Inorganic silts, very fine sands, rock flour, silty or clayey fine sands	1	100 to 200
		CL	Inorganic clays of low to medium plasticity, gravelly clays, sandy clays, silty clays, lean clays	1	100 to 200
		OL	Organic silts and organic silty clays of low plasticity		100 to 200
	Silts and clays, liquid limit greater than 50%	MH	Inorganic silts, micaceous or diatomaceous fine sands or silts, clastic silts	1	100 to 200
		CH	Inorganic clays of high plasticity, fat clays	1	50 to 100
		OH	Organic clays of medium to high plasticity		50 to 100
	Highly organic soils	PT	Peat, muck, and other highly organic soils		

[a]National Building Code, 1976 Edition, American Insurance Association.

[b]Based on the material passing the 3-in. (75-mm) sieve.

(Courtesy Portland Cement Association.)

DENSITY AND CONSISTENCY

The most significant property of cohesionless soils is *relative density*, which is described in such terms as *very loose, loose, medium, dense,* and *very dense.* Most coarse-grained soils are cohesionless.

The most significant property of cohesive soils is *consistency*, described in such terms as *very soft, soft, medium, stiff, very stiff,* and *hard.* Most fine-grained soils are cohesive. Purposely densifying or increasing the unit weight of a soil mass by compaction with rollers, tampers, or vibrators can improve its structural properties (see Section 2.3.7).

PLASTICITY

If a soil within some range of water content can be rolled into thin threads, it is called *plastic*. All clay minerals are plastic at some range of water content. Most very-fine-grained soils contain clay minerals and therefore are plastic. The degree of plasticity may be defined by the terms *fat* and *lean*. Lean clays are only slightly plastic because they usually contain a large proportion of silt or sand. The *plasticity index* (PI) is a value expressing the numerical difference between the *liquid limit* and the *plastic limit* of the soil as determined by test, such as ASTM D 4318. The plasticity index is equal to the liquid limit less the plastic limit ($PI = LL - PL$). The liquid limit (LL) is the amount of moisture present when the soil changes from a plastic to a liquid state. The plastic limit (PL) is the amount of moisture present when the soil changes from semisolid to plastic.

A clay soil with a plasticity index greater than 15 may expand when its moisture content increases, as would occur as a result of a rising water table. If the soil beneath a building is not consistent, as is commonly the case, expansion of the part that is of such clay while adjacent more stable materials remain stable could result in uneven heaving. This will cause cracks in the foundation and the supported structure, which can be severe enough to result in serious structural damage. Structural damage can also occur when the moisture content of such clay beneath exterior foundations is increased by groundwater seeping into the clay, while the same clay beneath interior foundations remains stable.

2.1.3 BEARING CAPACITY

Soil samples and tests to determine foundation type and size should always be made (see Section 2.1.7) where tall buildings and heavy live loads are present, which is the case for most commercial, institutional, and industrial buildings and high-rise multifamily dwellings. Such testing is also prudent where numerous single-family house foundations are being planned, such as in a subdivision.

Since the loads transmitted to soils by most small buildings are not large, the sizing of their foundations is often based on local rules of thumb, rather than on soil tests and engineering design. Building codes often list maximum bearing capacities for soils of several general classes that may be used without testing (Fig. 2.1-4). These bearing capacities normally are conservative and rarely take into account the compressibility of soils. In addition, the general descriptions of the soils are subject to different interpretations, and the use of generalized tables may lead to erroneous conclusions regarding the true load-carrying capability of the soil at a specific site. As a result, using the loads permitted by a code often results in overdesigned foundations. Soil sampling and testing are, therefore, often valuable even for individual houses and other small buildings, where the reduced costs of smaller foundations may offset the cost of soil sampling and testing.

FIGURE 2.1-4 Presumptive Bearing Values of Foundation Materials

Class of Material	Bearing Pressure, lb./sq. ft.*
1. Crystalline bedrock	12,000
2. Sedimentary rock	6,000
3. Sandy gravel or gravel	5,000
4. Sand, silty sand, clayey sand, silty gravel, and clayey gravel	3,000
5. Clay, sandy clay, silty clay, and clayey silt	2,000

*1 psf = 47.88 P.

(BOCA National Building Code/1990, copyright 1989, Building Officials and Code Administrators International, Inc. Published by arrangements with author. All rights reserved.)

2.1.4 SETTLEMENT

When the soil being loaded is dense and granular, such as coarse sand or gravel, little settlement is likely and it occurs quickly after the load is applied (Fig. 2.1-5). When the soil is a cohesive clay, settlement is likely to be greater, but occurs slowly, perhaps taking several years.

The two principal causes of settlement are (1) reduction of the volume of soil voids (spaces between soil particles containing air or water) and (2) lateral displacement of the supporting soil.

2.1.4.1 Reduction of Voids

The reduction in the volume of the voids within soil that is loaded is related to the soil's compressibility. Because sand contains large angular grains and a relatively small percentage of void spaces (see Fig. 2.1-1), foundation pressures can cause little decrease in the volume of the voids. Clay, however, has a scale-like grain structure and a relatively large percentage of voids. Therefore, decreases in the volume of the voids in clays subjected to foundation loads can be quite large.

When soil voids are filled with water, this water must be pressed out if the volume of the voids is to be decreased. Sand is permeable and offers little resistance to water passage, so that total void reduction

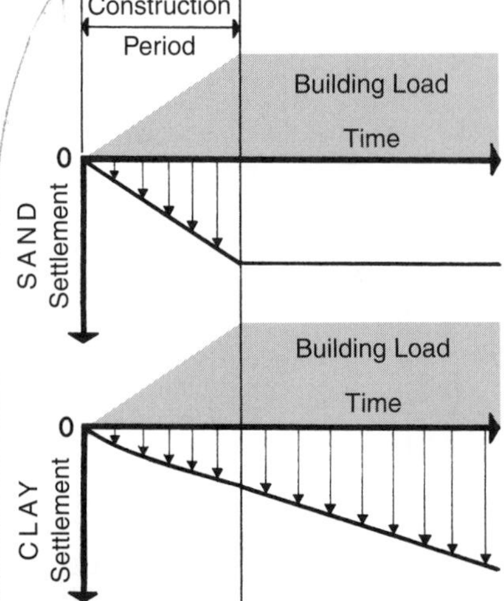

FIGURE 2.1-5 Building settlement on sand tends to be quick and slight; on clay it tends to be greater and may occur over a long period of time.

and settlement occur almost simultaneously with load application. Impermeable clays, on the other hand, require a long time for compression (*consolidation*) to take place, since water cannot pass out of them easily.

Air contained in the voids of sand offers no resistance to compression since it is expelled as soon as the load is applied. A certain amount of compression may occur instantaneously in clays if air is present in the voids, since air is compressible.

2.1.4.2 Lateral Displacement

Lateral displacement in soil is the horizontal movement of soil under heavy foundation loads. Settlement as a result of lateral displacement is not common in small buildings and generally offers no problem with small building foundation loads, unless the lateral support to the soil under the foundation areas is reduced or disturbed by adjoining excavation. Lateral displacement is more likely when heavy loads are applied and must be accounted for in foundation design for large buildings.

2.1.5 NONPROBLEM SOILS

Nonproblem soils (Fig. 2.1-6) are defined as those that are dense, coarse-grained, and properly drained. Most of these cause few problems in light buildings for slab-on-grade construction or foundations. These soils possess excellent bearing capacity because they are not subject to significant volume change. Gravels, for instance, usually occur in a dense state and are unlikely to cause any settlement problems. Usually, the only consideration is that the natural confinement of the soil not be disturbed by trenching or grading operations that would reduce the necessary lateral support.

Nonproblem soils acceptable for use in the construction of fills, backfills, and embankments for commercial, institutional, and industrial buildings and for large multifamily housing projects are generally limited to those in the GP, GM, GC, SP, and SC categories. General backfill outside of structures is generally limited to soil classified as GP, GM, GC, SP, or SM (Fig. 2.1-6). However, in either case, some other classifications may be permitted in some circumstances and subject to additional liquid limit and plastic limit requirements. The decision to permit other soil classifications should be made by a qualified and experienced soils engineer.

Controlled fills are used to bear loads, usually, but not always, beneath buildings. Soils permitted in controlled fills are generally limited to those classified as GW, GP, SW, and SP. These are often further limited to a liquid limit not to exceed 35% and a plastic limit not to exceed 12%, with a further limit on the percentage of fines that can be present. The actual requirements and the locations where controlled fills are required or permitted should be determined by a soils engineer.

The load-bearing capabilities of both cohesive and cohesionless soils can be improved by compaction. Soils beneath structures and pavements and soils adjacent to buildings are usually compacted (see Section 2.3.7).

When sand or silt is encountered, it should be determined whether it is dense, medium dense, or loose. Loose sand or silt frequently can be economically compacted to a dense state before slab construction.

Loess soils, which are unstratified loamy materials in the ML or MH groups, should be treated in the same manner as other soils in the same groups. The loads in light-building construction are generally not sufficient to cause problems with these soils unless the slab or foundation supports masonry or other heavy, concentrated loads. Heavily loaded slabs and foundations may settle unduly when loess soils become saturated; therefore, proper drainage is extremely important when these soils are encountered. Loess soils should not be permitted for bearing or adjacent to larger buildings.

FIGURE 2.1-6 Description of Nonproblem Soils

Soil Classification	Soil Description	Density or Consistency*
GW, GP	Gravels or sand-gravel mixtures, little or no fines	All densities
GM, GC	Gravels or gravel-sand mixtures, with clay or silt	Medium-dense to dense
SW, SP	Sands with little or no fines	Medium-dense to dense
SM, SC	Silty or clayey sands	Medium-dense to dense
ML	Silty or clayey fine sands	Medium-dense to dense
ML, MH	Inorganic silts	Medium-dense to dense
CL	Inorganic and silty clays of low to medium plasticity	Stiff to hard

*See Figure 2.1-3 for definition of terms.

2.1.6 PROBLEM SOILS

Problem soils include all soils that are not listed in Figure 2.1-6 as nonproblem soils. They possess less than suitable bearing characteristics for support of slabs, foundations, walks, or pavements and are usually not satisfactory even for grassy or planted areas adjacent to buildings. Special construction precautions are necessary with such soils as highly compressible or highly expansive clays, highly plastic soils, sands, silts that are very loose, and loess soils that are heavily loaded or not well drained. When a supported slab, foundation, or pavement is not sufficiently stiff or does not possess the necessary strength, detrimental effects to the supported element and even the building's superstructure may result. These effects will occur due to settlement or differential movement caused by lack of density of, or a change in the moisture content of, the supporting soil.

2.1.7 SUBSURFACE INVESTIGATION

The accuracy of soil classification and the descriptions defining density, consistency, plasticity, bearing capacity, and other soil properties depend on the skill of the individual making the observation and the extent of the controls under which the observations are made. As a result, classification by field observation is subject to uncertainties and inaccuracies. The extensiveness of soil investigation will vary with building type, local knowledge of subsurface conditions, and with conclusions drawn from preliminary investigations.

2.1.7.1 Preliminary Investigation for Small Buildings

For single-family homes and other small buildings, a preliminary investigation should be made to determine general information about the nature and origin of existing soils. This investigation should determine the need for and the necessary extent of further investigation. In some cases, it can eliminate the expense of extensive soil boring and testing. The preliminary examination usually consists of at least one test boring at the site to determine the acceptability of the soil for ground-supported slabs and foundations. Such test borings can be made with simple tools (Fig. 2.1-7) to determine soil types and the extent of each to the depth usually necessary to predict the performance of small building foundations.

Obtaining the professional advice of a soils engineer experienced in a given locality may preclude the need for even a single test boring at a single-family home site. This is especially true if the engineer's advice is coupled with an examination of other structures in the immediate area for evidence of apparent settlement or expansion of the soil, provided that similar conditions of soil, topography, and proposed construction prevail.

No additional soil investigation is required for a single-family home site if the preliminary site investigation indicates that the site has nonproblem soils consisting of firm, nonexpansive material, that no surface- or groundwater problems exist, and that fill thickness greater than 3 ft. (914 mm) will not be required.

When preliminary investigation reveals problem soils at a single-family dwelling site, the procedures, test borings, and soil investigation procedures indicated in the following paragraphs for commercial, institutional, and other building types should be followed.

FIGURE 2.1-7 **Manual and power-driven hand augers for preliminary field investigation of soils.** (Courtesy ELE International, Inc., Soiltest Products Division.)

2.1.7.2 Subsurface Investigation for Medium-Sized and Large Buildings

An engineering study should be conducted by a qualified soils engineer for every building other than a single-family dwelling. This study should be used to determine necessary design and construction procedures.

Depending on the extent of the project, this subsurface investigation may be made by digging test pits, driving sounding rods, or making test borings. Though limited in depth, test pits are effective investigative means when the footings will not go deeper than about 8 ft. (2.44 m). Load tests can be made at their bottoms, soil samples can be taken in them, and the water table can be measured easily after water fills the pit to the maximum level of the table.

Bearing capacities can be measured using standard penetration tests by driving rods using a standard hammer and counting the number of blows needed to drive the rod a given distance.

However, the most versatile method of obtaining soil test samples and determining bearing capacities is by using soil borings. Borings can be made much deeper into the soil than can either test pits or sounding rods.

DENSITY AND CONSISTENCY TESTS

The relative density of coarse-grained soils can be estimated using a standard penetration test with a split-barrel sampler and power-driven equipment (Fig. 2.1-8). This test is not always necessary for single sites for one- or two-family residences, but should be required for other types of buildings. The relative density may be estimated from the number of blows causing the sampler to penetrate 1 ft. and is classified from very loose to very dense (Fig. 2.1-9).

The consistency of clay (cohesive) soils should be determined by test from undisturbed samples using a thin-walled sampler (see Fig. 2.1-8), or the assumed characteristics may not be valid.

A direct numerical value of consistency determined by laboratory test is the load per unit of area that causes failure of an unconfined soil sample in a simple compression test. The value obtained is called the *unconfined compressive strength* (q_u) of the soil. Where CL, CH, OL, or OH soils (see Fig. 2.1-3) are encountered, this laboratory test should be performed to determine the unconfined compressive strength in order to select the type of slab to be used. The consistency of cohesive soils is given in Figure 2.1-10.

TESTS FOR OTHER PROPERTIES

Soil samples should be tested in a soils laboratory to establish their other properties. These include the soil's particle sizes, plasticity, cohesiveness, shear strength, water content, shrinkage coefficient, compressive strength, and the amount that the soil can be expected to consolidate (compress) under load.

2.1.8 EFFECTS OF WATER

Moisture resulting from either surface water from precipitation or groundwater, including the effects of seepage fields, (1) may result in volume change or reduction of the bearing capacity of a soil, (2) may impose additional requirements for site preparation with regard to grading and drainage, and (3) may necessitate special precautions such as a base course or vapor retarder (see Sections 2.3.5 and 2.4.3.2).

The amount and frequency of precipitation have primary influences on problem soils, because they may cause changes in the moisture content of soil supporting a slab or foundation. Exact values for the amount of precipitation, its seasonal variation, the amount and duration of each occurrence, and its effect on supporting soil under a slab or foundation are difficult to measure and almost impossible to predict. The most important considerations are whether precipitation or its absence will change the moisture content of the soil intended for support of slabs or foundations

a b

FIGURE 2.1-8 Test borings to determine soil properties are made with power-driven equipment and (a) a split-barrel sampler for coarse-grained soil or (b) a thin-walled sampler for fine-grained soils. (Courtesy ELE International, Inc., Soiltest Products Division.)

FIGURE 2.1-9 Relative Density of Coarse-Grained Soils

Soil Description	Rule-of-Thumb Field Guide[a]	ASTM Method No. of Blows[b]	Relative Density (%)
Very loose		0–4	0–20
Loose	Easily penetrated with ½-in. (12.7-mm) reinforcing rod pushed by hand	4–10	20–40
Medium	Easily penetrated with ½-in. (12.7-mm) reinforcing rod driven with 5-lb. (2.27-kg) hammer	10–30	40–60
Dense	Penetrated 1 in. with ½-in. (12.7-mm) reinforcing rod driven with 5-lb. (2.27-kg) hammer	30–50	60–80
Very dense	Penetrated only a few inches with ½-in. (12.7-mm) reinforcing rod driven with 5-lb. (2.27-kg) hammer	Over 50	80–100

[a]Field guide intended as an example of one of many field procedures currently in use for indicating density and is not necessarily a preferred method.

[b]Number of blows required to penetrate to a depth of 1 ft. (304.8 mm) as measured in accordance with ASTM D 1586.

during and after construction, and whether the change will have a detrimental effect on the building.

When variations in moisture content occur on a type of soil that expands as it absorbs moisture and shrinks as it loses moisture, the slab or foundation it supports is subjected in turn to cycles of uplift as the soil expands and settling as the soil shrinks. When a period of high moisture content is followed by drought, soil moisture evaporates more rapidly around the perimeter of a building, and soil moisture under the center of the building, which may be due to capillary action and migration, becomes trapped and sealed from direct exposure. Moisture can be retained under the center of a building long after extended periods of drought.

Trees and shrubbery immediately adjacent to a building's perimeter can also affect the soil moisture content by providing a shield from natural precipitation and by extracting soil moisture during the growing season. Moreover, the watering of gardens and lawns adjacent to a building can increase the underlying soil's moisture content.

When soil changes from a low to a high moisture content, a similar but opposite phenomenon develops. If prolonged periods of alternating drought and moisture occur, considerable differential moisture can develop in different areas under a building. If the soil is a type that undergoes substantial volume changes with changes in moisture, one of two conditions may result. If the building's slabs and foundations are relatively flexible, as in the case of unreinforced or only lightly reinforced slab Types I and II (see Section 3.12), they will follow the uneven contour of the supporting soil that results from uneven volume change, exposing them to distortions and possible damage to both slab or foundation and the supported superstructure (Fig. 2.1-11a).

On the other hand, if a slab or foundation is sufficiently rigid, as is the case with a properly structurally reinforced and stiffened Type III slab or concrete foundation, it will resist deformations even though the contour of the supporting soil is uneven. Higher soil pressures develop under such a slab or foundation over the plateaus, with reduced soil pressures at the valleys. The slab or foundation will be subjected to bending between these uneven contours, and the soil at the plateaus may deform owing to higher bearing pressures, distributing loads to adjacent areas (Fig. 2.1-

FIGURE 2.1-10 Consistency of Undisturbed Cohesive Soils

Soil Description	Rule-of-Thumb Field Guide[a]	Unconfined Comprehensive Strength, q_u[b] (MPa)
Very soft	Core (height = 2 × diameter) sags under own weight	0.25 (23.94)
Soft	Can be pinched in two between thumb and forefinger	0.25–0.50 (23.94–47.88)
Medium	Can be imprinted easily with fingers	0.50–1.00 (47.88–95.76)
Stiff	Can be imprinted with considerable pressure from fingers	1.00–2.00 (95.76–191.52)
Very stiff	Barely can be imprinted by pressure from fingers	2.00–4.00 (191.52–383.04)
Hard	Cannot be imprinted by fingers	4.00 (383.04)

[a]Field guide is only an indication of soil consistency. Values of q_u are given as basic values of consistency by which field classification can be verified and should not be used for design purposes without laboratory verification.

[b]q_u is unconfined compressive strength in tons/sq. ft. (not the ultimate bearing capacity of the soil for design).

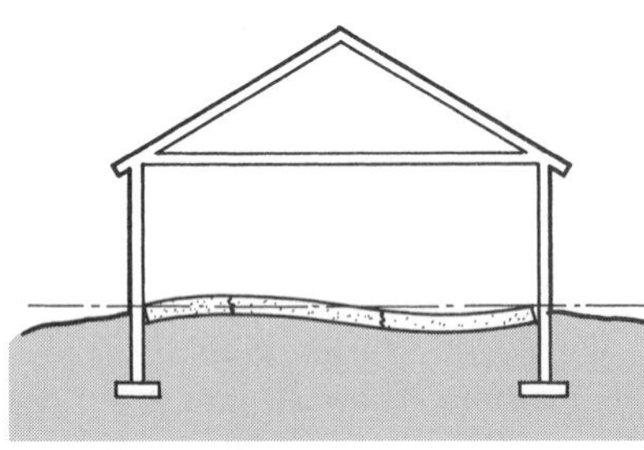

a

b

FIGURE 2.1-11 Slab distortion and soil movements (highly exaggerated): (a) a flexible slab will distort with volume changes in the supporting soil; (b) a rigid slab is designed to span uneven contours caused by volume changes in the supporting soil.

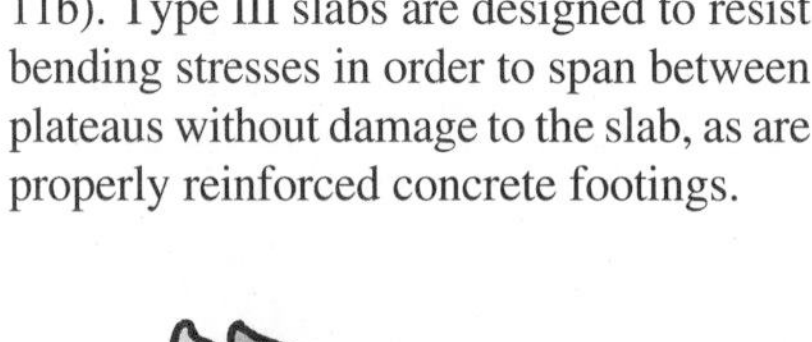

11b). Type III slabs are designed to resist bending stresses in order to span between plateaus without damage to the slab, as are properly reinforced concrete footings.

2.1.9 EFFECTS OF TEMPERATURE

The combined effects of temperature and moisture influence the depth of perimeter grade beams and foundations, as well as the selection of slab types, because some soils are subject to volume changes, called *frost heave*, that result from the formation of ice layers when water in soil freezes. Unless a Type III slab is used, or if non-heaving soils (GW, GP, SW, SP) are present and the water level is below the frost line, foundations should always be carried below the frost line (Fig. 2.1-12).

The temperature of a building slab can affect the comfort of occupants, especially during cold periods. Insulation is often introduced to help keep slabs warmer. Some buildings are partially heated by radiant heat generated by pipes carrying warm water or ducts carrying heated air built integrally with a slab-on-grade. This configuration warms the slab as well as the air above it, which can produce a more comfortable environment for occupants than if the air alone is heated.

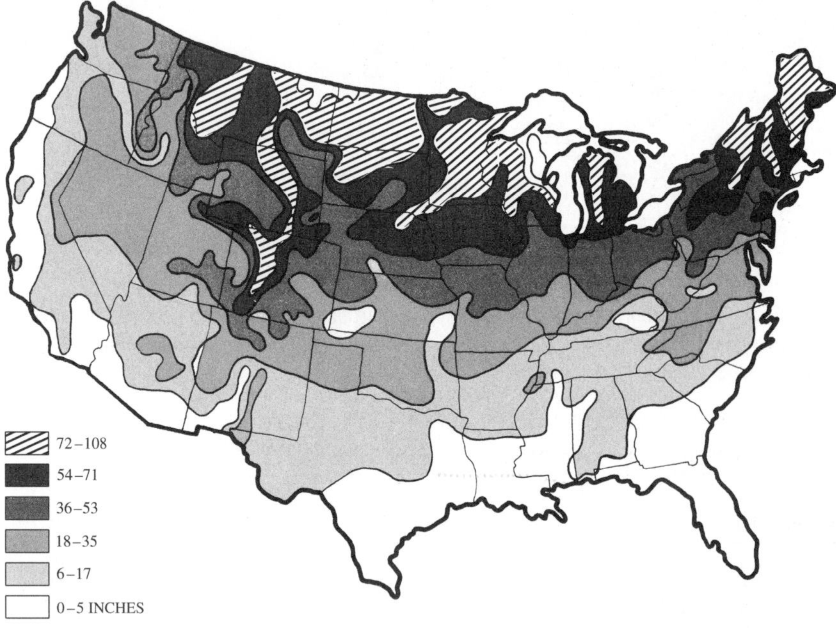

FIGURE 2.1-12 Maximum depth of frost penetration in inches. (Courtesy National Weather Service.)

2.2 Site Preparation

Site preparation includes removal from the site of existing features such as buildings, site improvements, vegetation, and topsoil that would interfere with new construction. Movable structures, such as house trailers, may be simply towed away. Houses and other light buildings are sometimes removed from the site intact.

2.2.1 DEMOLITION

A major aspect of some projects is the demolition and removal of existing buildings or sections of existing buildings that are not to be a part of the new work. When structures are to be removed as part of the general construction contract, the term *demolition* is often used to refer to those parts of such structures that are above ground. Removing below-grade portions of structures is usually thought of as excavation work (see Section 2.3.2). However, when demolition is done under a separate contract, removing basement walls and slabs and building foundations is usually a part of the separate contract.

Demolition may be either *complete* or *selective*. In complete demolition, an existing structure is demolished completely, and the debris is removed from the site. *Selective demolition* consists of removing a portion of an existing structure, usually to make way for remodeling or an addition.

Demolition is potentially hazardous to the workers involved, to the public, to adjacent property, and to the environment. Therefore, extraordinary measures should be taken to protect all of them from such danger. Most heavily populated jurisdictions have code and other legal requirements for public and property safety during demolition. There are also usually restrictions on practices that affect the health and welfare of the workers and the public or cause damage to the environment. For example, there are regulations for the handling of hazardous materials like asbestos, toxic chemicals, explosive materials, flammable materials, and radioactive waste.

Moreover, there are often extensive requirements relative to the bracing and shoring of buildings during demolition. These restrictions are usually more stringent when a building has been damaged in a fire, a storm, or an earthquake. Restrictions are likely to be extensive regarding the use of explosives.

Even when safety and environmental restrictions are not mandated by code or laws, precautions should be observed. Sprinkling with water, temporary enclosures, and other appropriate means should be used to limit the spread of dust and dirt in the air. Pedestrian and vehicular traffic should be routed around the area of danger. Rodents and insects should be controlled before and during demolition using legal and environmentally sound means so that they are not driven into adjacent structures. These means should not damage the environment.

In selective demolition there is usually a need to protect building portions that will not be demolished. Cutting and patching of existing materials is often required. Protective enclosures should be erected between areas to be demolished and areas to be let alone. People will often occupy spaces adjacent to the area in which selective demolition will take place. They should be protected from danger and from exposure to fumes, dust, dirt, loud noises, and toxic materials by whatever means are appropriate: temporary dustproof walls, sound barriers, barricades, and covered passageways.

Portions of a building to remain should similarly be protected. In addition, they should be protected from the weather by temporary walls and roofs or by weatherproof tarpaulins or plastic sheeting.

When portions of a building undergoing selective demolition remain occupied, utility service interruptions should be kept to a minimum and should be done so that the occupants are not harmed, placed in danger, or caused significant discomfort. For example, fire sprinklers should not be interrupted during times when a building is occupied. Exits should be maintained at the number and size required by code. Electric service, water, sewer, and other services should not be interrupted for extended periods while the building is occupied.

2.2.2 SITE CLEARING

Site clearing consists of removing site improvements and vegetation that will not be part of the new work. This includes removal of trees and other vegetation, clearing and grubbing, topsoil stripping, and removing above- and below-grade site improvements.

Before site clearing is started, adjacent public and private property should be protected from harm by the erection of fences and barricades. Before trees and other vegetation are removed, those to remain should be identified, marked, and protected by temporary fences or barricades. Similarly, water, electric, sewer, and other utility lines should be marked and staked so that they will not be inadvertently damaged.

Clearing and grubbing consists of removing trees, shrubs, and other vegetation, including their roots and stumps. Depressions caused by these procedures should be filled with satisfactory backfill material, as discussed in Section 2.3.4.

Topsoil should be stripped from areas where new construction is to take place. Topsoil is often stockpiled on the site and later reused in lawn and planting areas. However, topsoil should not be removed from within the farthest extent of the branches (*dripline*) of trees to remain, because doing so will damage the root structure of the trees.

Roadways, walkways, site walls, and other structures that will not be used in the new work should be demolished and the debris removed from the site.

One means of disposing of materials removed from a building site is dumping it in public or private *landfill* or *spoil* areas. Landfill areas often accept organic materials and construction debris, whereas spoil areas usually accept only earth materials, such as excavated soil and unusable topsoil. Burning on the site, which was once the most prevalent means of disposing of combustible waste, is often prohibited in heavily populated jurisdictions today. Even when it is legal, some owners will not permit burning because of environmental and safety concerns.

2.3 Earthwork

Earthwork includes:

- Excavation and filling to make way for a building
- Backfilling after the building has been built
- Grading preparatory to installing paving, walkways, lawns, and landscaping

Related site construction components include:

- Underpinning existing structures to protect them from damage during adjacent excavation
- Slope protection and erosion control
- Chemically treating the soil to protect the building against subterranean termites

2.3.1 UNDERPINNING

Underpinning is the time-consuming and expensive process of supporting the foundations of an existing building by placing new footings, pilings, or caissons beneath the existing footings. Underpinning is often needed when the excavation for a new building or an addition to an existing building will undercut existing foundations (Fig. 2.3-1). Underpinning may also be needed when an existing footing is too small to support a vertical addition to a building, when an existing undersized footing has resulted in undue settlement and damage, and in other cases when additional support is needed.

Underpinning can be accomplished in many different ways, depending on the nature of the problem. Only two of the many possibilities are covered here:

- Placing a new footing and foundation wall below the existing ones
- Using piles

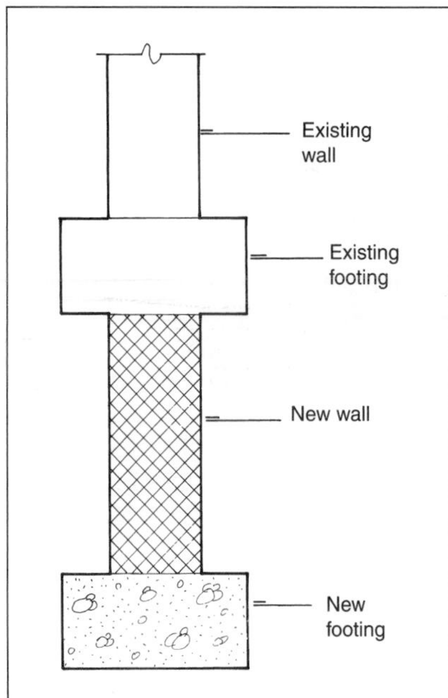

FIGURE 2.3-2 The lower-footing method of underpinning. (Drawing by HLS.)

When new building foundations are lower than those of an existing building, the existing footings can be supported by placing new footings lower than the existing ones (Fig. 2.3-2). One way to accomplish this is to place a series of *needle beams* through the existing building walls to support them until the new foundation wall and footing have been built (Fig. 2.3-3).

Another way to place a lower foundation wall and footing is to dig intermittent pits at the existing foundation down to the level of the new footing (Fig. 2.3-4). These pits must be small enough to permit the existing footing and wall to span the pit with support by temporary jacks and shoring. The footing between the pits remains supported by the original soil. After the underpinning foundation wall and footing have been placed (see Fig. 2.3-2), another section of pit is dug, until the entire wall has been underpinned.

Underpinning can also be accom-

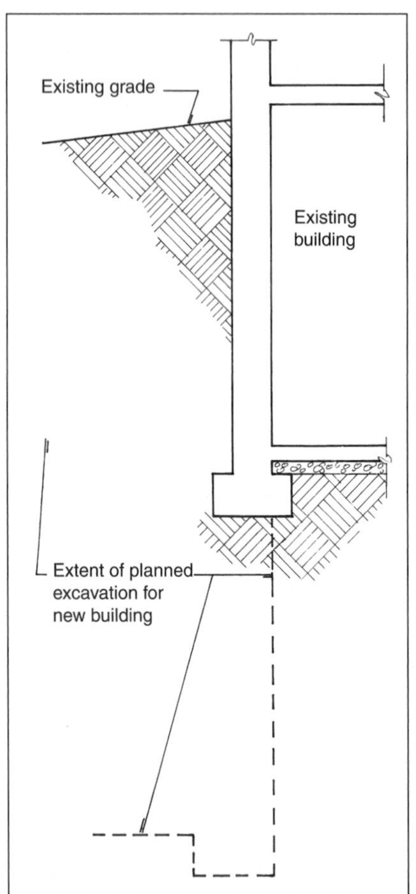

FIGURE 2.3-1 The existing footing here will most likely slide into the adjacent new excavation unless it is underpinned. (Drawing by HLS.)

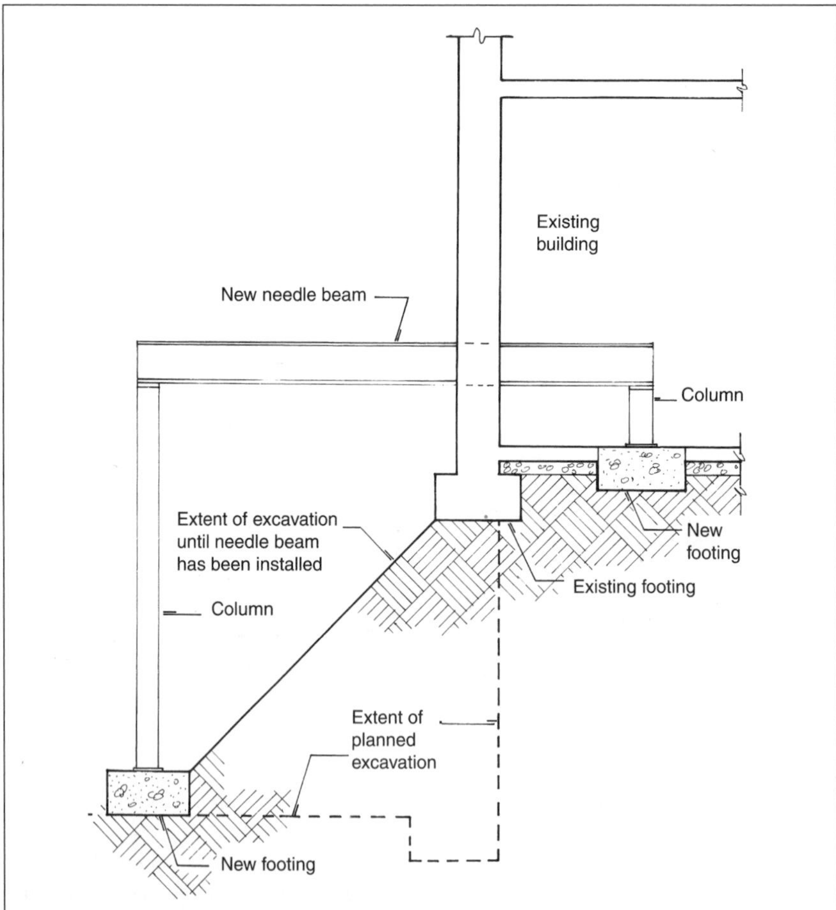

FIGURE 2.3-3 The exterior ends of needle beams may be supported at the existing grade level or on footings at the bottom of the new excavation. In either case, the soil against the existing wall must not be removed until the beams are in place. (Drawing by HLS.)

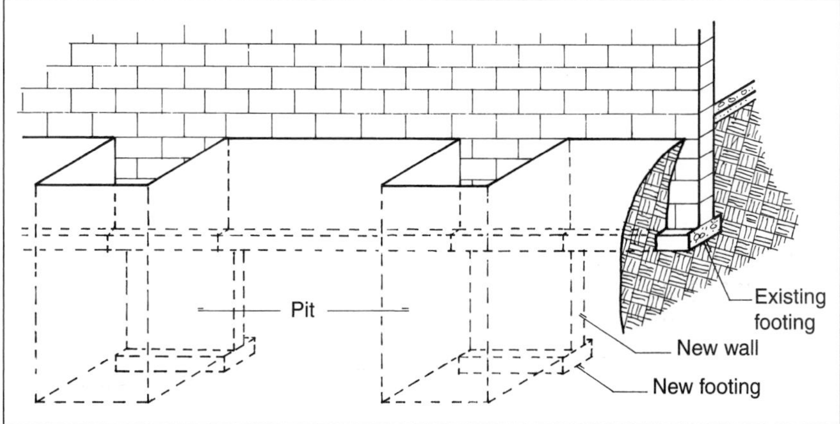

FIGURE 2.3-4 Pits dug to permit underpinning must themselves be braced and shored to prevent collapse and injury to workers. (Drawing by HLS.)

plished using piles (see Section 2.3.3). In general, the method is to place new piles adjacent to the existing footings. A section of the foundation wall is then removed at the top of each set of piles, and a concrete cap is poured in place so that it rests on the piles and penetrates the wall (Fig. 2.3-5). A proprietary system is available that substitutes a foundation support and jack assembly for the cap. This assembly rides down on a steel pipe pile and is rotated toward the existing footing until it extends beneath the footing. It is then raised hydraulically until the loads transmitted by the footing are picked up by the support assembly and transmitted to the pile. This type of assembly is particularly useful when an existing footing has settled.

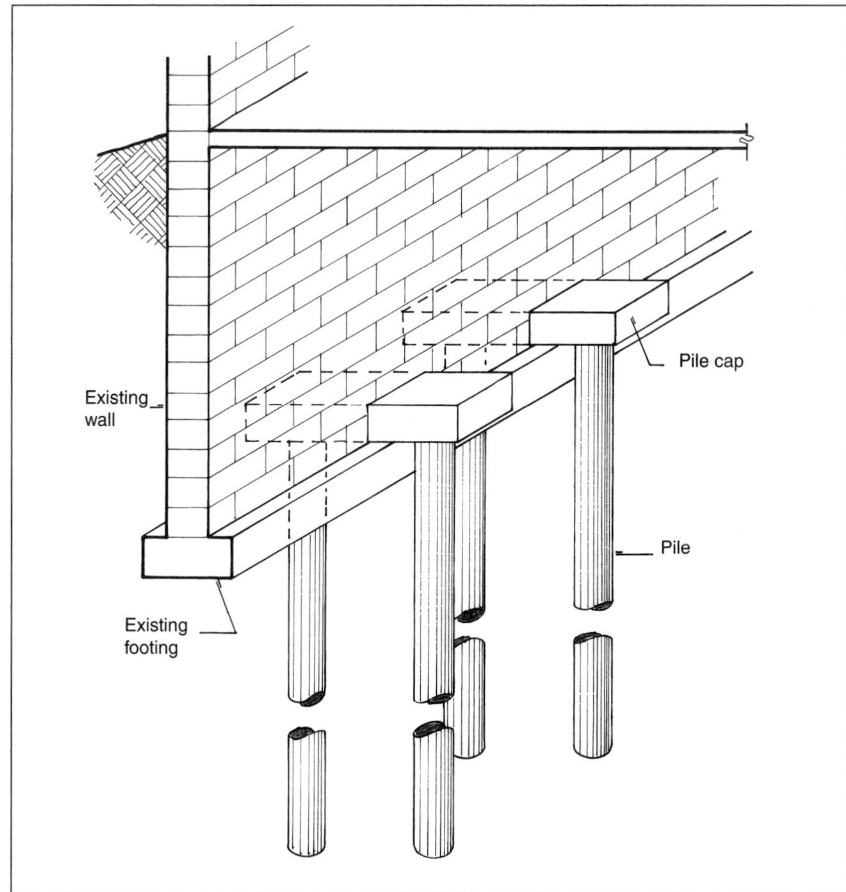

FIGURE 2.3-5 Vertical piles used as underpinning. (Drawing by HLS.)

2.3.2 EXCAVATING

Even for single-family residences and other small buildings with floor slabs on grade, excavation is necessary for trenches to contain footings and utilities. For most buildings, much more excavation than this minimum is required.

Building excavation is usually defined as the removal of materials down to the level required by the contract documents and disposal of the materials excavated.

In some small applications, excavation is done using hand tools, but most excavation requires power equipment (Fig. 2.3-6). This equipment ranges from the bulldozers, bucket loaders, trench excavators, and backhoes common in earth excavation to the rippers, power shovels, pneumatic hammers, and drop balls used to excavate in rock.

Before excavation is permitted to begin, underground utilities that are to remain should be located and staked to prevent damage to them.

The materials that underlie a site are often called *unclassified*, meaning that it is unknown whether they consist of all soil or whether rock is present. This common practice is unfair to contractors, who, when they present their bids, must gamble as to the nature of the material that will have to be excavated. Excavation of rock is far more expensive than excavation of earth. It takes more time, requires different equipment, and often necessitates blasting. Blasting is restricted in many jurisdictions to certain hours and days. In addition, the size of the explosive is sometimes restricted, requiring the use of smaller and more charges. A far fairer method would be to identify the materials to be excavated as either earth or rock.

The definitions of the terms *earth* and *rock* vary with the source. One way to define rock is to say that it includes all material that cannot be removed with a specific piece of earth-moving equipment. Earth, then, is any material that is not rock.

Excavation generally includes removal of aboveground site improvements such as pavements, walks, and site walls, as well as underground items like old foundations, piping that will not be used in the new work, wells, and cisterns.

Materials that are suitable for fill and backfill (see Section 2.3.4) should be

FIGURE 2.3-6 Many types and sizes of earth moving equipment are used in building excavation.

stockpiled for reuse. Excess and unsuitable materials should be disposed of.

Excavations are classified as *open excavations, trenches,* or *pits.* Open excavations are those that cover the large portion of a building excavation. They are usually excavated with power equipment to the level of the lowest floor slab.

Trenches and pits are dug into an open excavation or into soil that has not been otherwise excavated. Trenches for footings and utilities are usually dug by machine and finished by hand. In many cases, pits for isolated footings, sumps, and mechanical devices are dug using hand tools. Larger pits for elevators and large mechanical equipment are usually excavated by machine and finished by hand.

The bottoms of excavations should be protected against erosion, softening as a result of water saturation, and freezing. Disturbed material at the bottoms of excavations for trenches and pits should be removed. Disturbed material beneath slabs that is suitable for that location should be compacted.

2.3.3 EXCAVATION SUPPORT SYSTEMS

Except for shallow trenches and pits, the side walls of excavations, including open excavations, trenches, and pits, need to be supported until the permanent construction is in place and able to support them. Support can be obtained by simply sloping the sides of an excavation when the site is large enough to permit it. This slope should be at such an angle that the soil will not slide down it. The angle must be shallow for soils with a high percentage of sand and gravel, but may be steeper for soil that contains clay. The maximum slope should be determined by testing the actual soil to determine the angle at which it will not slide. Applicable codes and other legal restrictions may dictate the permissible slopes. As excavations become deeper, it is sometimes necessary to step the sides back (Fig. 2.3-7). Excavated materials stockpiled for use in backfill should be stored away from the edges of the excavation to reduce the height of the material adjacent to the excavation.

When the space to store excavated soil is small, or excavations are too close to property lines or other structures to properly slope the sides, other types of supports are necessary. Architects are seldom called on to design excavation support systems. This is usually left to the contractor. Therefore, the type used depends on the conditions at the site and the preferences of the contractor. An architect may, however, be required by law to review a contractor's plans for excavation support systems.

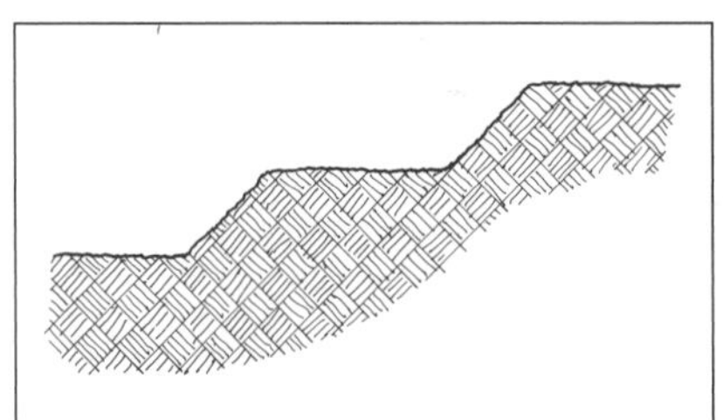

FIGURE 2.3-7 Stepping the sides of an excavation when there is room may preclude the need for shoring. (Drawing by HLS.)

On shallow excavations (12 to 15 ft. [3.66 to 4.57 m]) bracing may be done with *sheet piles.* Most of these are wood or metal (Fig. 2.3-8), but precast concrete is sometimes used. Sheet pile sections are interlocked and driven into the ground before excavation begins.

Deep excavations are usually braced with *soldier piles and lagging* (Fig. 2.3-9). Soldier piles are H-shaped steel beams, usually wide-flange sections. Lagging consists of heavy wood timbers. Soldier piles are driven into the ground before excavation starts. Lagging is placed from the bottom as the excavation is made. Lagging may be placed behind the flanges of the soldier piles, as shown in Figure 2.3-9a, or placed on the face of the soldier piles and held in place by battens that are bolted through the joints in the lagging to the flanges of the soldier piles (Fig. 2.3-9b).

On very shallow excavations, sheet piles driven to sufficient depth to act as

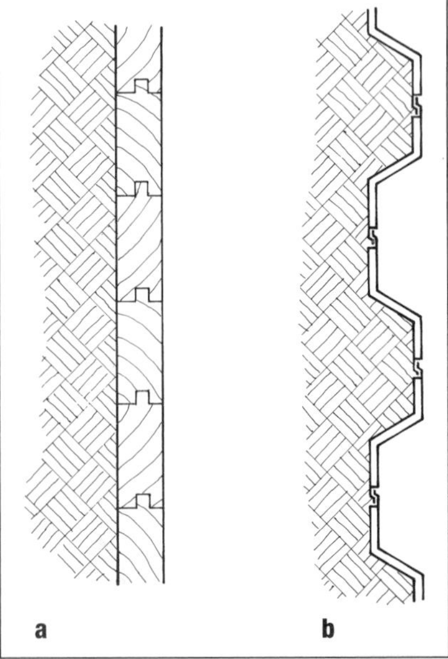

FIGURE 2.3-8 (a) Timber sheet piling and (b) one type of steel sheet piling. Other configurations and other joint types are also available. (Drawings by HLS.)

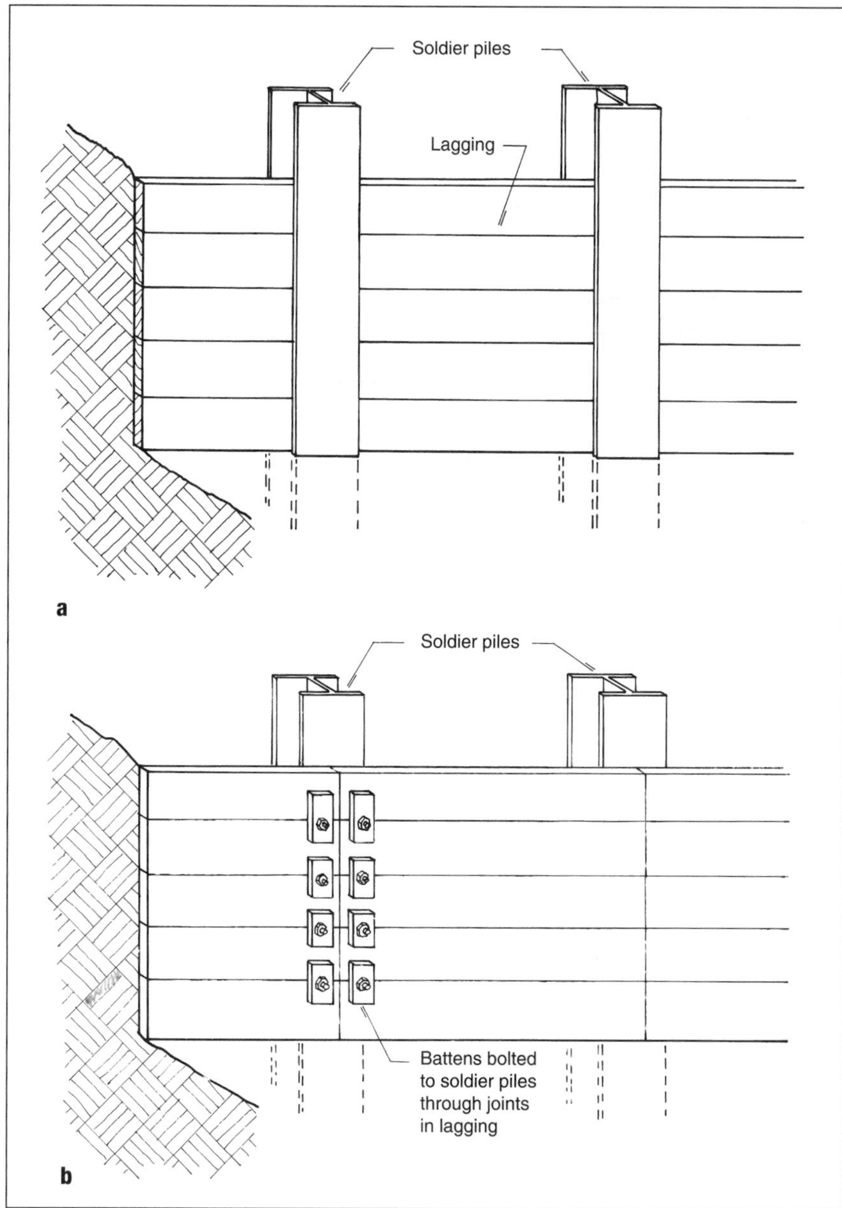

FIGURE 2.3-9 (a) Soldier piles and (b) lagging. (Drawings by HLS.)

cantilevers are sometimes strong enough to support the loads encountered without additional support. When excavations are deeper, however, both sheet and soldier piles and lagging need additional support to withstand the loads imposed by the soil and groundwater behind them. This support is provided by *shoring and bracing*. On very deep excavations, several tiers of bracing may be needed.

Sheet piles and soldier piles and lagging may be supported using *rakers*, *cross-bracing*, or *tiebacks*. All three systems are designed to support horizontal continuous *wales* that extend along the face of the sheet piles or soldier piles. Wales are usually wide-flange steel beams, but other steel shapes may be used.

The two bracing systems shown in Figure 2.3-10 have a common drawback. They both obstruct activities in the excavation. For example, rakers can be placed only after the excavation has been extended to the level of the concrete footer or heel. But the sheet piles or soldier piles and lagging can be exposed only after the bracing is in place. This means that excavation at the location of the footer or heel must be made to the level of the piles, and the rakers must be in place before the earth can be removed from the face of the sheet piles or soldier piles and lagging. The heavy dashed line labeled "Limit of excavation" in Figure 2.3-10 indicates the level of the excavation until the rakers have been placed. If more than one line of rakers is needed, the earth beneath them must be removed in stages, extending just below each level. The proper angle of slope must be maintained so that the remaining soil will stay in place and support the load imposed by the soil and groundwater on the opposite side of the sheet piling or soldier piles and lagging. In addition, excavation of the soil remaining against the sheet piles or soldier piles and lagging is difficult, because the rakers obstruct movement of excavation equipment.

Cross-bracing also obstructs the movement of equipment, but does permit excavation to continue uninterrupted from one wale level to the next lower one without slopes. In wide excavations, cross-bracing requires additional support, which is provided by driving steel columns at intermediate points across the span and installing additional wales on the columns. As the excavation is lowered, additional levels of braces can be added as needed.

The third method uses earth or rock anchors called *tiebacks*. This method does not obstruct activities in the excavation. Tiebacks are constructed by drilling a hole at each line of wales through the sheet piling or lagging into stable rock or soil. Tiebacks are not appropriate in unstable or very loose soils. An anchor rod or cables are inserted into the drilled hole, and the space around the rod or cables is filled with grout under pressure. When the grout has set, the rod or cables are *post-tensioned* (stretched) using a hydraulic jack and anchored securely to a continuous wale so that the tension is maintained. Figure 2.3-11 shows a wale made from steel channels, but other shapes, such as wide-flange steel beams, are frequently used. When the soil tends to be somewhat unstable, or a type of soil, such as sand or gravel, is present that would collapse into the hole, a steel pipe may be inserted to keep the hole open until the grout has been placed.

Cross-bracing and rakers are removed as soon as permanent construction makes them unnecessary. Tiebacks are left in place, but are not relied on for support after permanent construction is in place. Soldier piles and lagging are often removed as soon as permanent construction negates their need. However, when their removal would disturb new or existing construction, soldier piles and lagging are

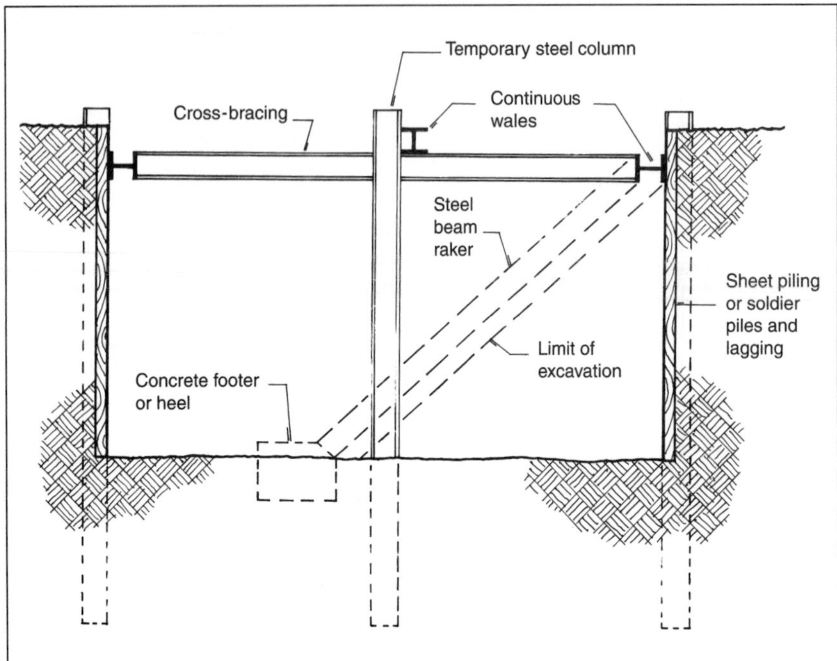

FIGURE 2.3-10 **Two foundation bracing systems. The solid lines show cross-bracing. The dashed lines represent a raker system. The two systems generally are not used together.** (Drawing by HLS.)

left in place. Sometimes they are covered with a drainage board or mat and used as the outer form for concrete walls (see Section 2.4.3.3). Sheet piling may be either cut off below ground level and left in place or removed completely, although wood sheet piling is usually removed.

2.3.4 FILL AND BACKFILL

Compacted fills are used for building sites in undulating or hilly terrain to create reasonably level ground for slab construction. *Engineered fill*, which is fill whose material selection, installation, and compacting are done according to engineering principles, is sometimes used below foundations in small buildings when the existing soil is unsuitable. In any case, the bad soil is removed and replaced with engineered fill. Soils used for fill should contain no vegetation or foreign material that would cause uneven settlement. Beneath slabs, a base course should be provided when necessary.

Topsoil and other unsuitable materials should be removed from areas beneath foundations and slabs. There should be no soil or other materials present that would cause a slab or footing to settle unevenly or excessively.

Fills should be used beneath slabs to establish the desired finished slab elevation and should be compacted to the maximum practical density. The slab bed for Types I and II ground-supported slabs (see Section 3.12) should not expand or contract due to moisture. Therefore, organic or active clays and silts of the CH, CL, OL, OH, and PT groups, as classified in ASTM D 2487, are not suitable for fills under these types of slabs. Soil groups GW, GP, SW, and SP (Fig. 2.3-12) may be used as fill material beneath slabs without a base course. These latter soil groups are also those used in engineered fill and in most fill in other locations. Groups GM and SM are also suitable under some conditions.

Before fill material is placed, *clearing and grubbing* should have been completed where the fill will be placed. Sloped surfaces should be plowed or otherwise broken up so that fill material will mix with existing material and not slip away from it.

Suitable excavated soils may be used to backfill excavations. These include soils classified in ASTM D 2487 as GW, GP, GM, SM, SW, or SP (see Fig. 2.3-12), but some jurisdictions may further limit the acceptable types. In addition, other qualifications may apply in some projects. For example, a soil that is not in the preceding list may be acceptable if its liquid limit and plastic limit are in a range that is acceptable to the project's soils engineer and its use does not violate the building code or other restrictions. When there are insufficient quantities of suitable excavated materials, acceptable soil should be brought from off-site (*borrow material*) to complete the required backfilling and grading.

Filling and backfilling materials should be placed in layers (*lifts*) of about 8 in. (200 mm) loose thickness when compaction

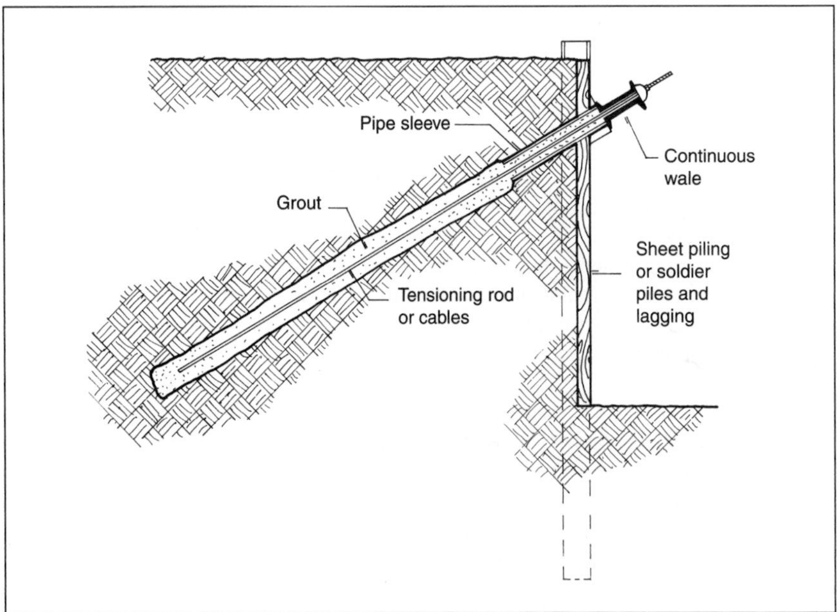

FIGURE 2.3-11 **A tieback anchoring system for excavation supports.** (Drawing by HLS.)

FIGURE 2.3-12 Requirements for Slab Beds According to Soil Classification

	Major Division	Group Symbols	Typical Names
Coarse-grained soils (More than half of material is larger than the smallest particle visible to the naked eye)	Gravels (more than half of coarse fraction is larger than ¼ in.) Clean gravels (little or no fines)	GW	Well-graded gravels, gravel-sand mixtures, little or no fines
		GP	Poorly graded gravels or gravel-sand mixtures, little or no fines
	Gravels with fines (appreciable amount of fines)	GM	Silty gravels, gravel-sand-silt mixtures
		GC	Clayey gravels, gravel-sand-clay mixtures
	Sands (more than half of coarse fraction is smaller than ¼ in.) Clean sands (little or no fines)	SW	Well-graded sands, gravelly sands, little or no fines
		SP	Poorly graded sands or gravelly sands, little or no fines
	Sands with fines (appreciable amount of fines)	SM	Silty sands, sand-silt mixtures
		SC	Clayey sands, sand-clay mixtures
Fine-grained soils (More than half of material is smaller than the smallest particle visible to the naked eye)	Silts and clays (liquid limit is less than 50)	ML	Inorganic and very fine sands, rock flour, silty or clayey fine sands or clayey silts with slight plasticity
		CL	Inorganic clays of low to medium plasticity, gravelly clays, sandy clays, silty clays, lean clays
		OL	Organic silts and organic silty soils, elastic silts
	Silts and clays (liquid limit is greater than 50)	MH	Inorganic silts, micaceous or diatomaceous fine sandy or silty soils, elastic silts
		CH	Inorganic clays of high plasticity, fat clay
		OH	Organic clays of medium high plasticity, organic silts
	Highly organic soils	PT	Peat and other highly organic soils

Generally may be used for fill without a base course

Fill that requires a base course

Not permitted for foundation fill

Note: Based on ASTM D 2487.

will be done by power equipment, and about 4 in. (100 mm) loose thickness when hand tampers will be used. When the fill or backfill material has a lower or higher than acceptable moisture content for proper compaction, it should be moistened or aerated, respectively, until it has reached the optimum moisture content. The required moisture content varies with the soil material and should be established by testing.

Backfill should not be placed until underground utilities to be covered are in place and have been tested, concrete formwork has been removed, shoring and bracing have been removed, and sheet piling to remain has been cut off below grade level. In addition, backfilling against walls should not be done until building slabs are in place and the walls have been braced either by temporary or permanent construction.

2.3.5 BASE COURSE FOR SLABS-ON-GRADE

For slab-on-grade construction, some preparation must be made to the subgrade, depending on the type of soil and other conditions encountered at the site. At the least, the area beneath the slab must be cleared and grubbed and unsuitable soils, such as topsoil, must be removed. The slab bed for ground-supported slabs should:

- Provide the necessary bearing capacity for slab support;
- Control ground moisture;
- Establish the proper slab elevation.

Fill material and undisturbed soil that are not in a suitably dense state should be properly compacted. Materials that will not support the loads from the slab should be removed and replaced with a compacted fill of suitable materials.

A slab bed may consist of undisturbed soil if it is of the proper type, density, and consistency. Soil groups GW, GP, SW, and SP are in this category (see Fig. 2.3-12). Where other types of soils form the subbase, they should be removed to the proper level and replaced with a base course to control capillary moisture rise and to provide a uniform and level bed for structural support of the slab (Fig. 2.3-13). A base course should be at least a 4-in. (100 mm)-thick compacted layer of a well-graded mixture of gravel, crushed stone, sand, or crushed slag.

2.3.6 GRADING

The rough and final grades on a building site are established by *grading*, using a combination of power equipment and hand tools. Grading is necessary where lawns, planting, pavement, walks, and building slabs will be placed.

The site around a building should be

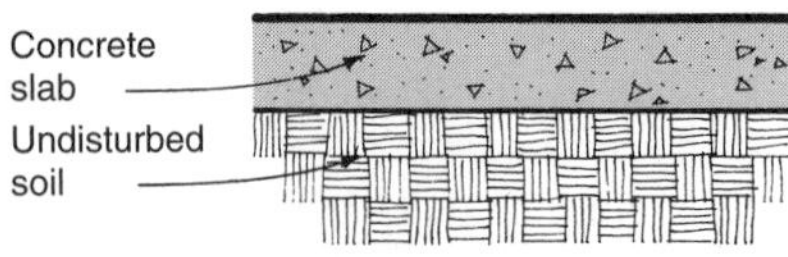

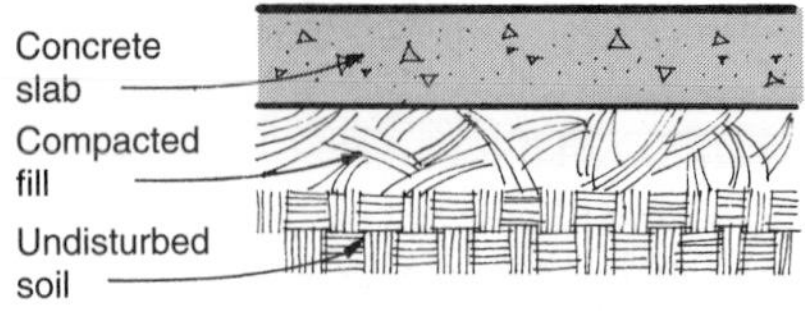

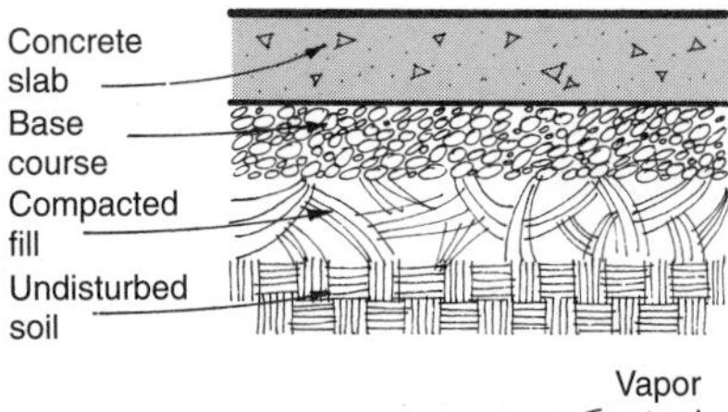

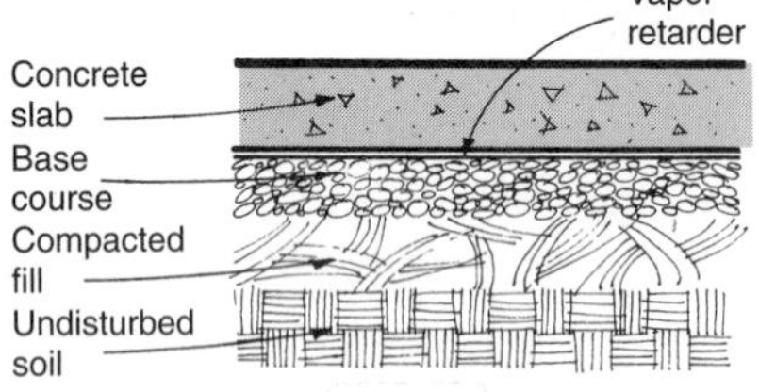

FIGURE 2.3-13 Slab beds for ground-supported slabs. A base course or a vapor retarder may be needed to control moisture.

graded to form protective slopes that will drain surface water away from the foundation so that water will not collect under the slab. Surface runoff should be directed to streets or drainage structures. Finish grade should slope downward in all directions away from a building at a minimum fall of 6 in. (150 mm) in each 10 ft. (3.05 m), which is a slope of about 5%. A percent of slope is the unit of rise (or fall) in each 100 units of run (1 ft. [305 mm] in 100 ft. [30.48 m], for example). When the area adjacent to a building is paved, a 1% slope, which is about ⅛ in. (3.2 mm) in each foot (305 mm) may be sufficient to direct water away. However, when such a paved area has less than a 5% slope, positive drainage, such as by yard drains or catch basins, should be introduced at sufficient intervals to prevent ponding caused by imperfections in the paving slope and buildup of water during heavy rainfall. Hillside sites should be graded to divert surface water around and away from structures.

Areas beneath lawns and walks should be graded to within 0.10 ft. (30.5 mm) above or below the design subgrade elevations, taking into account the thickness of sod and the walkway surface. Areas beneath pavements and building slabs should be graded to within ½ in. (12.7 mm) above or below the design subgrade elevations.

2.3.7 COMPACTING

Compacting applies energy to soil to consolidate it by compressing air voids to increase the soil's dry density. Proper compacting:

- Minimizes settling;
- Increases load-bearing characteristics;
- Increases soil stability;
- Reduces water penetration.

Proper compacting will usually prevent slab and pavement cracking caused by differential settlement in the soil. It will also reduce the settling of backfill placed against foundation and retaining walls and in other locations to prevent low spots where water will pond.

Compaction requirements are generally specified as a percentage of the maximum dry density obtainable in the test procedures in ASTM D 1557 or D 698. The relative density can be specified for fills constructed with free-draining sand and gravel soils (GW, GP, SW, SP) in accordance with ASTM D 4253 and D 4254. Varying degrees of compaction for the many different types of soils that may be encountered cannot be stated without engineering analysis. Generally, however, clean sand fill (SW, SP) should be compacted to 95% of the optimum dry density, and all other fine-grained, nonexpansive fills to 90% of the optimum dry density, in accordance with ASTM D 1557. Compacting requirements for expansive soils should be determined from engineering analysis and may require somewhat less than 90% of optimum dry density.

Using hand or power equipment (Fig. 2.3-14), compacting of fill material should be performed in 4- to 8-in. (100 to 200 mm) layers (*lifts*) of soil. Regardless of the type of soil being compacted, the thinner the lift, the better the compaction.

Structural slabs (Type IV) that receive their support at the edges and through the center by intermediate foundations require only sufficient compacting beneath the slab portion to provide temporary support of the dead load imposed by the slab until the slab develops the necessary strength to span between its foundations.

a

b

FIGURE 2.3-14 (a) Soil compacting at most building sites is done using power equipment. (b) Hand tamping is occasionally used in locations inaccessible to power-driven equipment. Hand tamping may also be used exclusively in small residential buildings and other small buildings when the soil is an easily compactible type.

Backfill adjacent to foundation walls should be uniformly compacted to at least 90% of optimum dry density unless other requirements dictate a greater degree of compacting. For example, backfill beneath walks, steps, and pavements should be compacted to at least 95% percent of optimum dry density.

2.3.8 SLOPE PROTECTION AND EROSION CONTROL

Exposed earth on a construction site should be protected from erosion, and, indeed, most jurisdictions have codes, ordi-

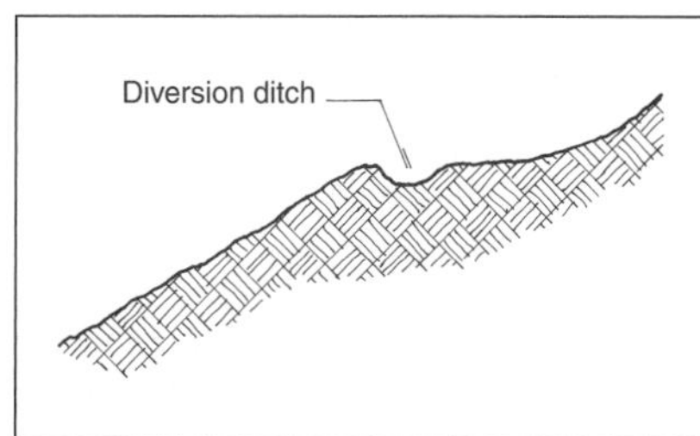

FIGURE 2.3-15 A diversion ditch at the top of a slope will slow the flow of groundwater and can be used to direct this flow to a paved outfall or into an interceptor drain that will conduct it to the lower level through pipes. (Drawing by HLS.)

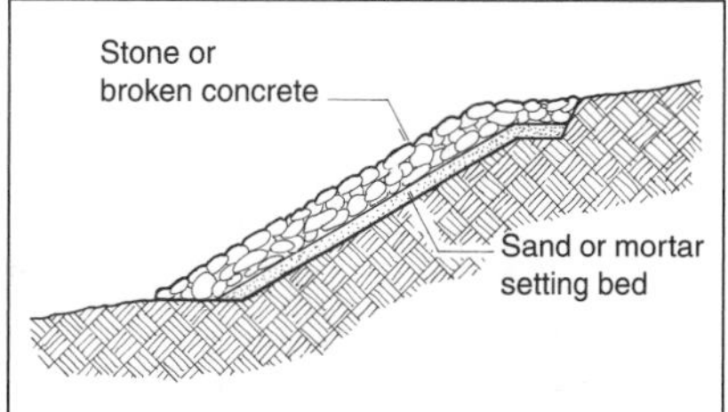

FIGURE 2.3-18 Riprap can be either temporary or permanent. It can be used to cover a slope, as shown here, or to line drainage channels or ditches. It is also used to protect roadway embankments and stream beds. (Drawing by HLS.)

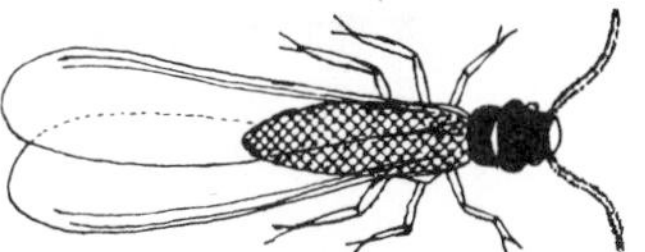

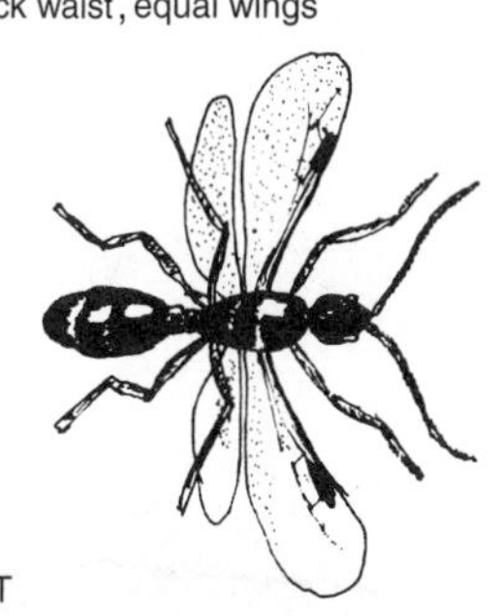

FIGURE 2.3-20 Ants are often mistaken for termites. Subterranean termites may be identified by their thick waists and wings of equal length. Ants are thin-waisted and have unequal-length wings.

nances, or rules that dictate such controls. These controls may be as simple as the staking in place of hay bales to prevent eroded materials from fouling waterways, or may be more elaborate systems with filter fabrics or riprap. In addition, providing diversion ditches (Fig. 2.3-15) and terracing of slopes (Fig. 2.3-16) may be needed to reduce the speed of water running down a slope and direct it to drainage channels protected with filter fabrics or riprap.

Filter fabrics are open woven mats of nylon or other synthetic fibers. They are staked down over a slope to permit water to flow without eroding the underlying soil (Fig. 2.3-17). Some of these products can be left in place and later covered with topsoil to receive lawns or other vegetation.

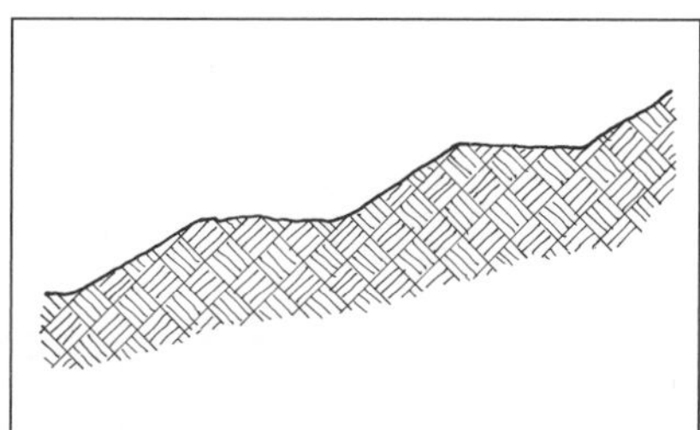

FIGURE 2.3-16 Terraces cut into a slope will reduce the velocity of groundwater and reduce erosion. (Drawing by HLS.)

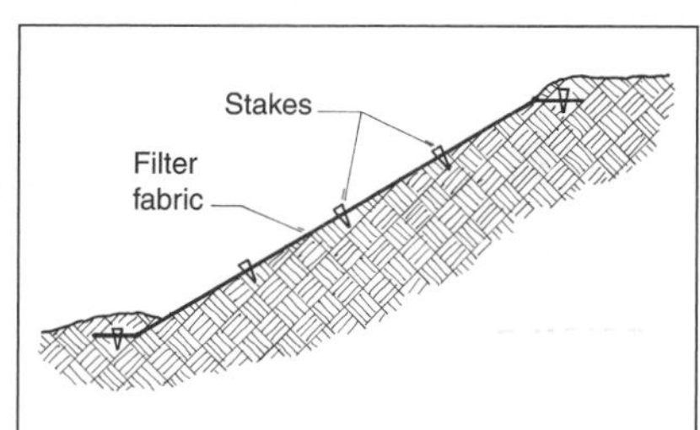

FIGURE 2.3-17 Filter fabric installed to protect a slope. (Drawing by HLS.)

Riprap is a layer of stone or broken concrete designed to form a pathway for the flow of water and prevent erosion of underlying soil (Fig. 2.3-18). Riprap consists of pieces large enough to resist their movement down the slope by the force of water. Riprap is sometimes used as a drainage channel and sometimes as protection for an entire slope.

2.3.9 TERMITE CONTROL

Two types of termites cause damage to wood in the United States, subterranean and nonsubterranean. As Figure 2.3-19 shows, the hazard is greater in areas having mild temperatures.

Subterranean termites account for about 95% of all termite damage in the United States. These termites (Fig. 2.3-20) live in nests and develop their colonies underground, but build tubes of earth through the air to reach wood and other cellulose material such as fiberboard, fabrics, and paper. Termites must have a constant source of moisture or they die. They become most numerous in moist, warm soil containing an abundant supply of food.

Subterranean termites are not carried into buildings in lumber but establish themselves by entering from ground nests after the building has been constructed, using three basic routes: (1) attacking

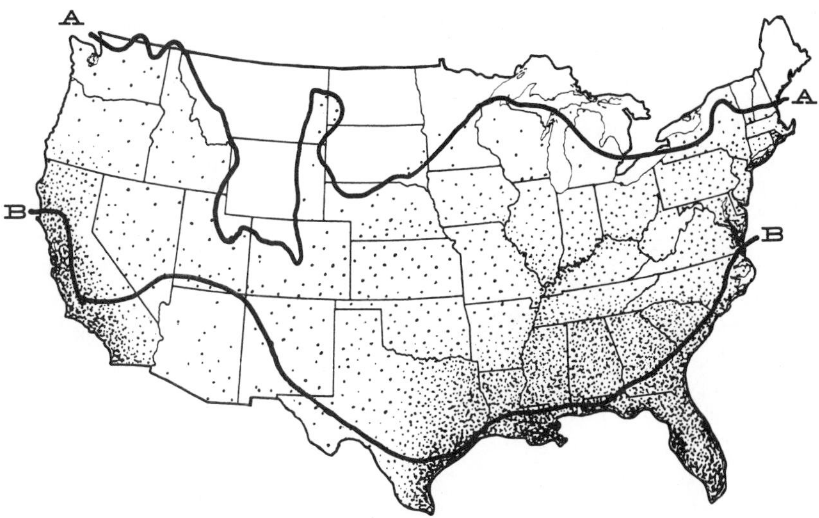

FIGURE 2.3-19 Relative hazard of termite infestation. Subterranean termites have been found as far north as line A-A. The northern limit of nonsubterranean termites is indicated by line B-B.

FIGURE 2.3-21 Subterranean termite infestation of foundation plates. Improper clearance between wood siding (removed for photo) and ground permitted termite entry.

wood in direct contact with the ground (porch stairs, trellises, and siding installed too close to grade), (2) entering through cracks and voids as small as 1⁄32 in. (0.8 mm) in slabs, piers, or foundation walls, or (3) building mud tubes over materials they cannot go through (Figs. 2.3-21, 2.3-22, and 2.3-23). During certain seasons, winged forms (see Fig. 2.3-20) swarm (fly) from the nest and, if successful, develop new colonies.

The control of subterranean termites may consist of one of the following protective methods or more than one method used in combination: (1) properly installed metal shields that isolate termites from wood or force them to build tubes around the shields, (2) chemically treated soil, which forms a barrier through which termites cannot pass to reach the construction, (3) chemically treated, pressure-impregnated wood, which is not subject to termite attack (see Section 6.1), (4) the use of poured concrete foundations, provided that no cracks greater than 1⁄64 in. (0.38 mm) are present, or (5) the use of poured reinforced concrete caps, at least 4 in. (100 mm) thick, on unit masonry foundations, provided no cracks greater than 1⁄64 in. (0.38 mm) occur.

Regardless of the protective method used, its design and construction should permit visual inspection wherever possible.

Nonsubterranean termites have been found only south of line "B-B" shown in Figure 2.3-19. Their total damage is much less than that caused by subterranean termites, but they may present a problem in the areas where they do occur. Also called *drywood termites* because of their ability to live in damp or dry wood without moisture or contact with the ground, they often are recognized by slightly compressed pellets of partially digested wood pushed through surface openings. Where this type is found, lumber should be inspected to see that it is not infested before arrival at the site. When construction is under way during the swarming season, lumber should be inspected for infestation,

Slab-on-grade construction requires special considerations in areas where subterranean termite hazard is a significant problem (see Fig. 2.3-19). Slabs are expected to crack in varying degrees, and termites can penetrate through cracks as small as 1⁄32 in. (0.8 mm), through joints between the slab and the foundation wall, through control joints, and through openings made for plumbing and conduits.

Some degree of protection of wood structural elements is provided by positive site and foundation drainage and sep-

FIGURE 2.3-22 Termites build mud tubes to permit access from ground to wood. A tube is shown here on a concrete foundation pier. Note that the tube has been built around the metal shield at the top of the pier.

FIGURE 2.3-23 Tubes shown here have been built from the wood above down to the ground. Tubes can be built up foundation walls or posts many feet from ground to wood.

aration of wood elements from the ground. Foundations and piers of hollow masonry units should have solid masonry caps in which all joints or voids are filled with mortar. Unless the foundation fill is of limited capillarity, at least a 4-in. (100 mm) base course of limited capillarity and a suitable vapor retarder should be provided under the slab.

Where local experience indicates that additional protection against termites is necessary, termite barriers should be provided. Additional protection can be afforded by using pressure-treated wood, which will resist decay as well. See Sections 6.1.5 and 6.5.2.

Soil poisoning is the most effective termite barrier. The area under and around a building should be treated, as well as all possible points of entry. The termiticides registered with the U.S. Environmental Protection Agency (EPA), and currently most commonly used, are chlorpyrifos, cypermethrin, fenvalerate, isofenphos, and permethrin. The recommended concentrations of these chemicals in treatment solutions range from 0.25% to 1%. The actual concentration, the application of the chemicals, and precautions involved in their use should be in accordance with the manufacturer's recommendations, with U.S. Department of Agriculture Home and Garden Bulletin 64, "Subterranean Termites, Their Prevention and Control in Buildings," and with applicable building codes.

The effective performance of termite shields depends on proper construction and workmanship and long-term integrity of the slab without cracking, which is unpredictable. Therefore, the effectiveness of termite shields is questionable.

2.4 Surface- and Groundwater Problems

Provisions must be made to prevent groundwater and surface water from entering a building through the walls or floors. If water penetrates basement walls, the habitable space there may become undesirable. Water in a crawl space or other enclosed space may result in decay, musty odors, mold growth, and corrosion of some metals. Water below an improperly constructed slab-on-grade can migrate through it to cause delamination of floors and can support condensation and frost accumulation at the perimeter of associated walls. In some soils, groundwater can rise high enough to fill heating ducts embedded in or buried beneath a concrete slab.

In addition to the measures discussed here, foundation walls enclosing basement spaces usually require additional protection by applications of *parging* and *dampproofing* on the exterior surface below grade (see Sections 4.4.3 and 7.2). When water under pressure (*head*) occurs above a footing, *waterproofing* may be necessary (see Section 7.2).

2.4.1 SURFACE WATER

Surface water is water that runs along the surface of the ground as a result of rain, downspout discharge, melting snow, or another source. Most surface water problems come from improper grading of the site adjacent to a building.

2.4.2 GROUNDWATER

Groundwater is water that is in the ground most of the time, although its level (*water table*) may fluctuate during dry and rainy seasons. Most groundwater problems result from locating foundations too close to the subsurface water table. Problems then occur when (1) the level of the water in the ground temporarily rises because of a seasonal increase in precipitation or (2) the water in the ground rises because of the capillarity of the soil. When the water table is so high that a foundation would be submerged constantly, either the site should be abandoned or more extensive waterproofing methods must be used to make the foundation watertight (see Section 7.2).

2.4.2.1 Water Table Rise

Problems resulting from seasonal fluctuations in precipitation that cause the water table to rise temporarily above the level of the foundation may be solved by subsurface drainage. The most effective means of resisting water infiltration through foundations is to collect the water and discharge it away from the site before it reaches the foundation.

2.4.2.2 Capillary Rise

Water will rise in certain soils in opposition to gravity as a result of capillary action. This *capillarity* occurs when the molecular attraction of water molecules to adjacent soil molecules is greater than the molecular attraction of water molecules to each other. This causes water in a small space (capillary tube), such as the voids between soil grains, to spread out and upward on the sides of the space (soil grain surfaces), forming a concave water surface (Fig. 2.4-1). The water's surface tension, which always tries to reduce the surface area, pulls the top of the water upward, following the edges. The capacity of the water to resist forces tending to pull it apart holds the tube of water together as it creeps upward.

In general, capillarity increases with a decrease in the size of the space between the grains in a soil. In fact, moisture can move upward through the voids within some clays as much as 11 ft. (3.35 m) above the water table (Fig. 2.4-2). However, because of the large spaces between soil particles, *capillarity* does not occur in coarse, granular soil types such as gravel. Therefore, one method of control is to replace soil having high capillarity with granular materials.

Water vapor will also rise through the

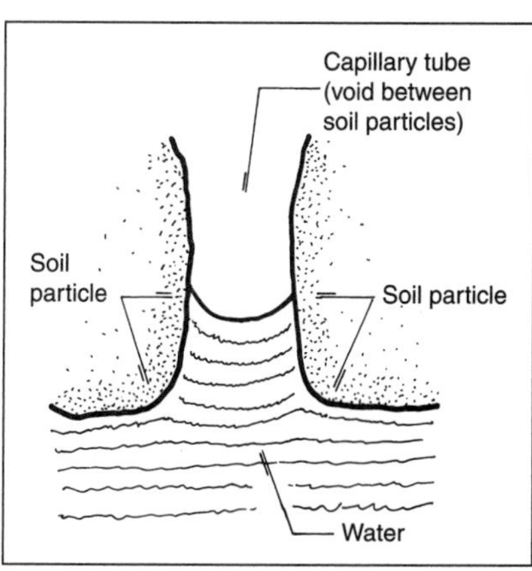

FIGURE 2.4-1 **Water rises in the voids between soil grains in some soils where the grains are tightly packed.** (Drawing by HLS.)

Capillary Rise	Soil Type	Saturation Zone
Greater Than 8′ (2438 mm)	CLAY	Greater Than 5′ (1524 mm)
Greater Than 8′ (2438 mm)	SILT	Greater Than 5′ (1524 mm)
3′ to 8′ (914 mm to 2438 mm)	FINE SAND	1′ to 5′ (305 mm to 1524 mm)
1′ to 3′ (305 mm to 914 mm)	COARSE SAND	0 to 1′ (0 to 305 mm)
0	GRAVEL	0

FIGURE 2.4-2 Water rises in soils by capillary action. Clays and silts may become fully saturated to almost 6 ft. (1800 mm) above a water table, and some water may rise more than 11 ft. (3353 mm). Note that even coarse sand may permit a rise of up to 3 ft. (914 mm). Coarse gravel prevents the rise of water by capillarity.

voids between soil particles, whether or not granular soil materials are used. This rise of water vapor can be interrupted and prevented from entering a crawl space or a building by covering the ground surface with a vapor retarder (Fig. 2.4-3). Refer to Section 7.1.3.2 for a discussion of vapor retarders in crawl spaces.

2.4.3 CONTROL OF WATER PROBLEMS

Groundwater and surface water can cause problems both during construction and in a completed building. Means of controlling surface water in a completed project by grading are discussed in Section 2.3.6. This section covers removal of groundwater both during construction and from the completed building.

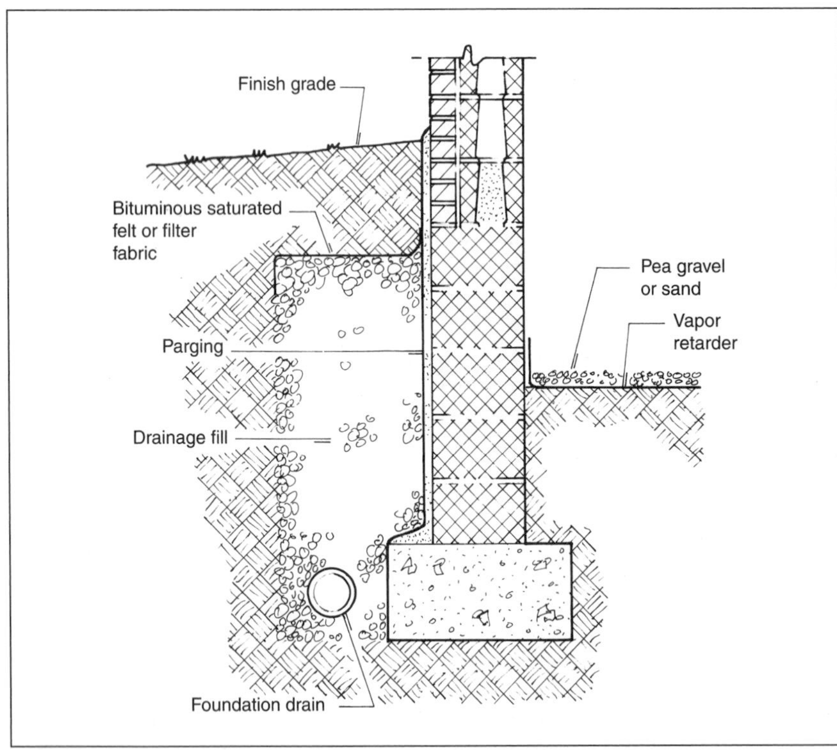

FIGURE 2.4-3 General moisture control: a sloped finished grade provides surface water runoff from foundation, a vapor retarder resists the rise of water vapor caused by capillarity, and foundation drains provide subsurface drainage. (Drawing by HLS.)

2.4.3.1 Dewatering

Dewatering is the prevention of water from entering an excavation, when possible, and the removal of water that does find its way in. Groundwater can be prevented from entering an excavation by means of berms and drainage channels or ditches.

Water that enters an excavation as rain, snow, or other precipitation can often be removed by means of portable pumps. Such water should be conducted to pump locations by depressions or ditches. Foundation trenches should not be used as drainage ditches, however, because the water will erode or otherwise damage the bearing surface at the bottom of a trench.

When the water table is higher than the bottom of an excavation, it is necessary to continuously lower the level of the water while the work is going on. The most common way of doing this is to sink a series of pipes with screened ends (*well points*) around the perimeter of the excavation to a level below the foundations. These well points may each be fitted with a pump, but more often they are interconnected by *suction lines* (pipes to one or more larger pumps that empty all of them simultaneously). In very deep excavations, two rows of well points may be needed. Well points and pumps lower the water table and keep it lower until pumping stops. The *discharge* from such wells should be conducted to storage ponds or positive outfalls that lead to a stream or storm sewer, depending on which is permitted or required in the jurisdiction of the project.

2.4.3.2 Slabs on Grade

Settlement or differential movement of slabs and foundations caused by moisture fluctuations in problem soils is discussed in Section 2.1.8. Additional provisions for moisture control are necessary for the long-term performance of concrete slab-on-grade construction. Slabs should be protected from water and water vapor to prevent damage to finish flooring materials and pipes or heating ducts embedded in the slab, and to guard against reduced effectiveness of thermal insulation. Problems may be caused by either surface water or groundwater.

The methods and extent of protection depend on:

- Slab elevation with respect to the elevation and slope of finished grade
- Elevation of groundwater table or artificial water sources, such as seepage fields
- Drainage properties of the soil or fill beneath the slab
- Type of finish flooring

Many of the moisture problems associated with slab-on-grade construction can be minimized or eliminated by proper preliminary grading, correct selection of fill or base course materials, and installation of a vapor retarder.

Slab base courses are discussed in Section 2.3.5; water control by grading in Section 2.3.6.

SLAB ELEVATION

The tops of slabs should be located at least 8 in. (200 mm) above the exterior finished grade level. The bottoms of heating ducts placed in or under a slab should be at least 2 in. (50.8 mm) above the adjacent exterior finished grade, unless the ducts are concrete, plastic, mineral-fiber-cement, or ceramic.

VAPOR RETARDER

Where no drainage or soil problems exist, and in regions where irrigation and heavy sprinkling are not done, a vapor retarder may not be necessary. However, because it is impossible to correct moisture problems after construction and because a vapor retarder can usually be installed at nominal cost during construction, a vapor retarder is recommended for all slab-on-grade construction when impermeable floor coverings will be used and when floor coverings, furnishings, or equipment must be protected from damage by moisture conditions (see Fig. 2.3-13). A vapor retarder membrane should have a permeableness of less than 0.30 perms (17.2 ng/[Pa·s·m^2]) and should resist deterioration as well as puncture from heavy traffic during construction. A vapor retarder should be installed with a 6-in. (150 mm) lap and be fitted tightly around utility and other service openings.

Vapor retarder materials suitable for slab-on-grade construction include single-layer membranes such as polyethylene sheet or multiple-layer membranes such as glass-fiber-reinforced waterproof paper with polyethylene film extrusion-coated on both sides. Polyethylene sheet used over sand or firmly compacted soil under unreinforced slabs should be at least 4 mils (0.1 mm) in nominal thickness; when used over gravel or crushed stone or under any structurally reinforced slab, it should be at least 6 mils (0.15 mm) in nominal thickness.

Except when a high-strength concrete with a lower than normal slump is used, a vapor retarder should not be placed directly beneath a slab. Instead, a 2- to 3-in. (50.8 to 76 mm) layer of sand should be placed over the vapor retarder. This permits water that bleeds from a high-slump or low-strength concrete to pass into the sand rather than migrate upward through the slab, where it will cause the slab to crack more than usual and weaken the surface, resulting in dusting and spalling.

2.4.3.3 Subdrainage Systems

Subdrainage systems include *foundation drains* and *underslab drains*. Some codes require that all buildings with basements or crawl spaces below grade have foundation drains. This is a good idea in any case, whether or not required by code, to prevent groundwater or surface water from finding its way into crawl spaces or into basements through the walls. Foundation drains are frequently used in conjunction with waterproofing (see Section 7.2). In addition, a grid of drains should be installed to prevent the buildup of hydrostatic head beneath large buildings and buildings where the water table is above the slab on grade or above the grade inside a crawl space.

Many drain pipe materials are available. Among these are nonperforated concrete drain tile, clay drain tile, cast iron soil pipe, and polyvinyl chloride pipe; porous concrete pipe; and perforated concrete, clay, polyvinyl chloride, and bituminous fiber pipe. Foundation and underslab drains are usually made from 4- or 6-in. (100 or 150 mm) pipe.

Nonperforated pipes are laid with open joints. These joints are then covered with a layer of screening to prevent fine particles from the fill above the pipe from entering the pipe through the joints. This screening material is usually heavy mesh burlap, coal-tar-saturated felt, copper mesh screen, or synthetic filter fabric.

Perforated pipes are laid with the perforations facing down and the joints closed in accordance with the pipe manufacturer's instructions.

Drainage fill material is gravel or crushed stone evenly graded so that all of it passes a 1½-in. (38.1 mm) sieve and not more than 5% passes a No. 50 sieve.

Foundation drains are placed in a bed of compacted drainage fill material around the exterior perimeter of a building's foundations (see Fig. 2.4-3). Drain pipe should be placed far enough below the floor slab to properly drain and with its lowest *invert* (bottom of pipe) at least 4 in. (100 mm) above the bottom of the footing to prevent possible washout of the soil beneath the footing. Footings should be lowered if necessary to meet these parameters.

Underslab drains are placed in trenches filled with drainage fill material (Fig. 2.4-4). Pipes are usually placed in parallel rows across the building. Depending on the distances involved, interceptor pipes may be placed perpendicular to these rows to reduce the depth of the piping required to establish a positive drainage slope.

In traditional installations, foundation drain pipes are set in a compacted 4-in. (100 mm)-thick bed of drainage fill material laid with its bottom on the undisturbed subgrade at the bottom of the footing. The pipes are then covered with drainage fill material that extends from 4 to 6 in. (100 to 150 mm) on each side of the pipe up to a level about 12 in. (305 mm) below the surface of the ground (see Fig. 2.4-3). The drainage fill should be placed in 3-in. (76 mm)-thick layers, and each layer should be compacted before the next layer is placed. The drainage fill material is then covered with a layer of No. 15 asphalt or tar-saturated felt or filter fabric to prevent fine materials from infiltrating the drainage fill. An impervious fill material (earth) is then applied to bring the grade to that

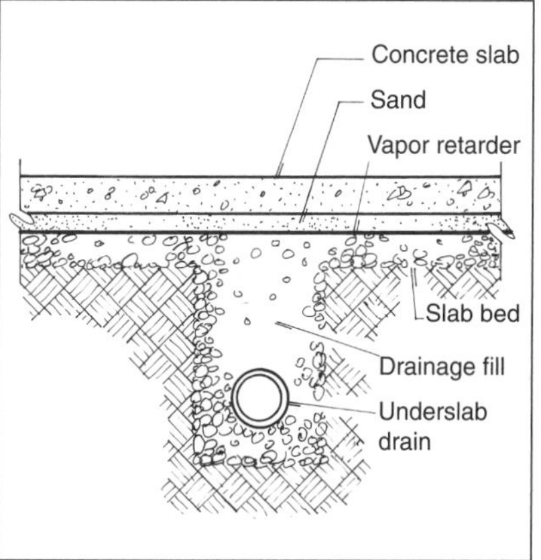

FIGURE 2.4-4 Underslab drain. (Drawing by HLS.)

required, with care taken to slope it away from the building.

Another method is used today in some installations to overcome the difficulty in a traditional installation of keeping the drainage fill material and adjacent backfill materials separated. The preceding discussion about the traditional method applies to this method as well, except that the drainage fill material is stopped 4 to 6 in. (100 to 150 mm) above the drain pipe and an *in-plane drain* is placed against the foundation wall from the bottom of the drainage fill to a level about 12 in. (305 mm) below grade (Fig. 2.4-5).

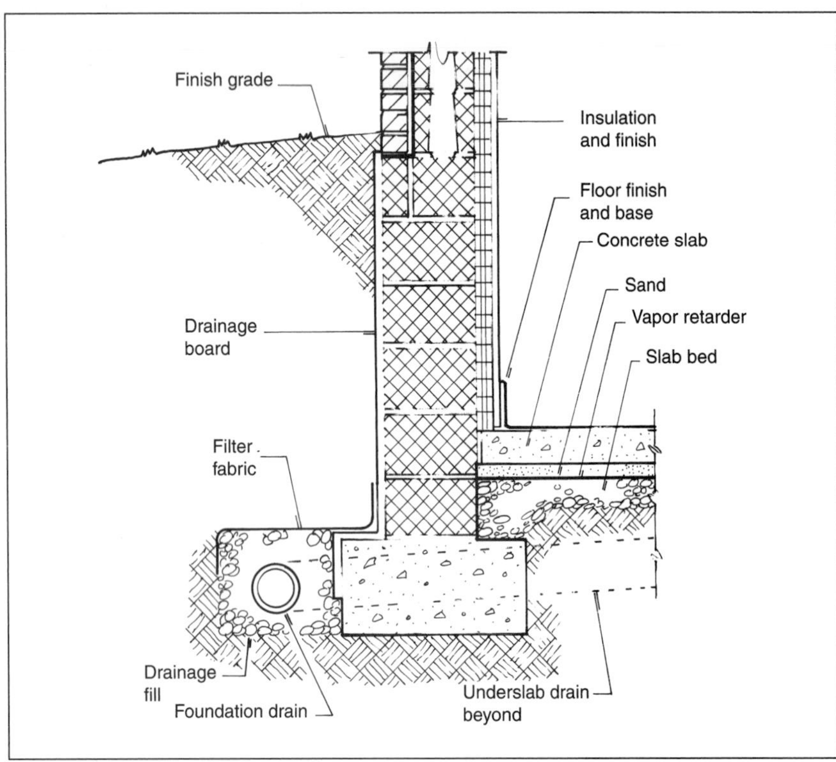

FIGURE 2.4-5 Foundation drain with drainage board, showing connection of underslab drains where they occur. (Drawing by HLS.)

In-plane drains (also called *drainage boards*) are thin boards (usually about 1 in. [25.4 mm] thick) consisting of two parts. The drain portion is made of a plastic material configured so that it forms multiple open pathways that permit water to flow freely through them. The side facing the earth is a layer of synthetic filter fabric that permits water to enter the drain but excludes fine soil. These boards can be installed with the use of adhesives or mechanical fasteners as directed by the manufacturer.

Whether a traditional or an in-plane drainage system is used, the drainage system should be tested to be sure that it properly drains before it is covered by drainage fill materials. The method of test is usually dictated by code or local practice.

Collected water may drain by gravity to a natural outfall at a lower elevation, or may be discharged by positive machine pumping from one or more collecting sumps located within the foundation area. Some codes dictate the minimum size of pits and pumps, but these may be insufficient for a particular project. Therefore, it is very important that both pit size and pump capacity be based on engineering design.

2.5 Lawns and Landscaping

Most people think first of the aesthetic appeal of lawns and landscaping, but they also serve as buffer zones between adjacent buildings and between buildings and roadways or other property. When properly placed, plants can help to keep a building cooler in summer and warmer in winter, thereby reducing dependence on heating and air conditioning equipment.

2.5.1 PLANNING REQUIREMENTS

Plantings can reduce reflection and radiation of solar energy from paved areas and adjacent buildings (Fig. 2.5-1). They can shield exterior walls from the sun, thus reducing heat transmission through them. Earth berms can also shield walls and outdoor living spaces.

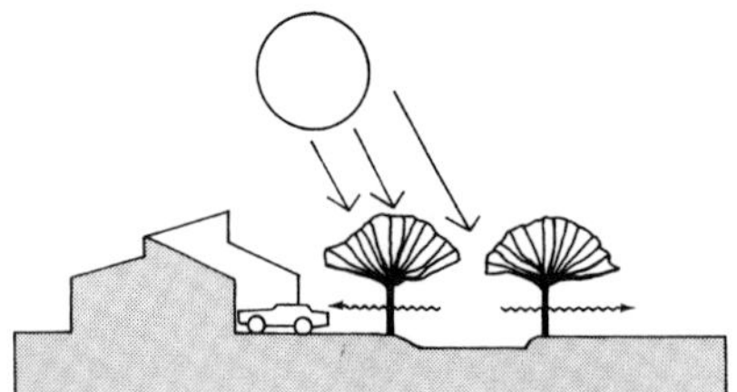

FIGURE 2.5-1 Shading paved areas prevents heat buildup and channels breezes.

Properly located *coniferous* (evergreen) trees can block prevailing winter winds. *Deciduous* (leaf-dropping) trees on south and west exposures shade walls and windows in summer but drop their leaves to allow desirable winter sun penetration (Fig. 2.5-2).

Ponds and fountains act as natural coolants to reduce air temperature around building exteriors (Fig. 2.5-3) and should not be blocked by vegetation.

Plants must be selected and placed carefully for several reasons. For example, placing deciduous trees on northern exposures can leave a building unprotected from cold winter winds. In addition, plants that hide a building can also hide an intruder. Trees should be far enough away from a building that a burglar cannot climb them and enter the second story. Bushes should be low enough that they do not obscure windows or doors. Hedges used as fences should be no more than 40 in. (1016 mm) high. Landscaping should not obscure a building from the street or parking lot. Small plants

FIGURE 2.5-2 Proper trees in the right places will reduce energy needs.

FIGURE 2.5-3 Air flowing over cool water acts as natural air conditioning. Plants should not be positioned so as to block this flow.

and floral borders can serve to define the borders between private and public spaces.

Refer to Section 15.2.2 for additional site planning recommendations related to thermal considerations.

2.5.2 TREES AND SHRUBS

Tree and shrub quality should be based on the American Nursery Landscape Association's (ANLA) *American Standard for Nursery Stock*. This publication contains a standardized system of sizing and describing plants. It covers shade and flowering trees, deciduous shrubs, coniferous evergreens, broadleaf evergreens, rose shrubs, young plants, fruit trees, bulbs, corms, and tubers. It describes and limits the dimensions of container plants, balled and burlapped plants, bareroot plants, balled and potted plants, and processed balled plants. It also contains the names of trees and shrubs that fit into each of its categories.

The U.S. Department of Agriculture (USDA) has identified 11 zones of plant hardiness in the United States. Figure 2.5-4 is an adaptation of the USDA Plant Hardiness Zone Map in USDA Agricultural Research Service Miscellaneous Publication Number 1475. USDA has also identified many plants as they relate to the Plant Hardiness Zone Map. Figure 2.5-5 contains a list of 10 of the 11 zones and examples of plants that will succeed in each of them. (Zone 11 includes areas where the average annual temperature is above 40°F (4.4°C), which are therefore essentially frost free.) Figure 2.5-6 is a list of additional plants and the zones in which they will survive. The zones listed for each plant are optimum. Listed plants may actually survive in warmer or colder

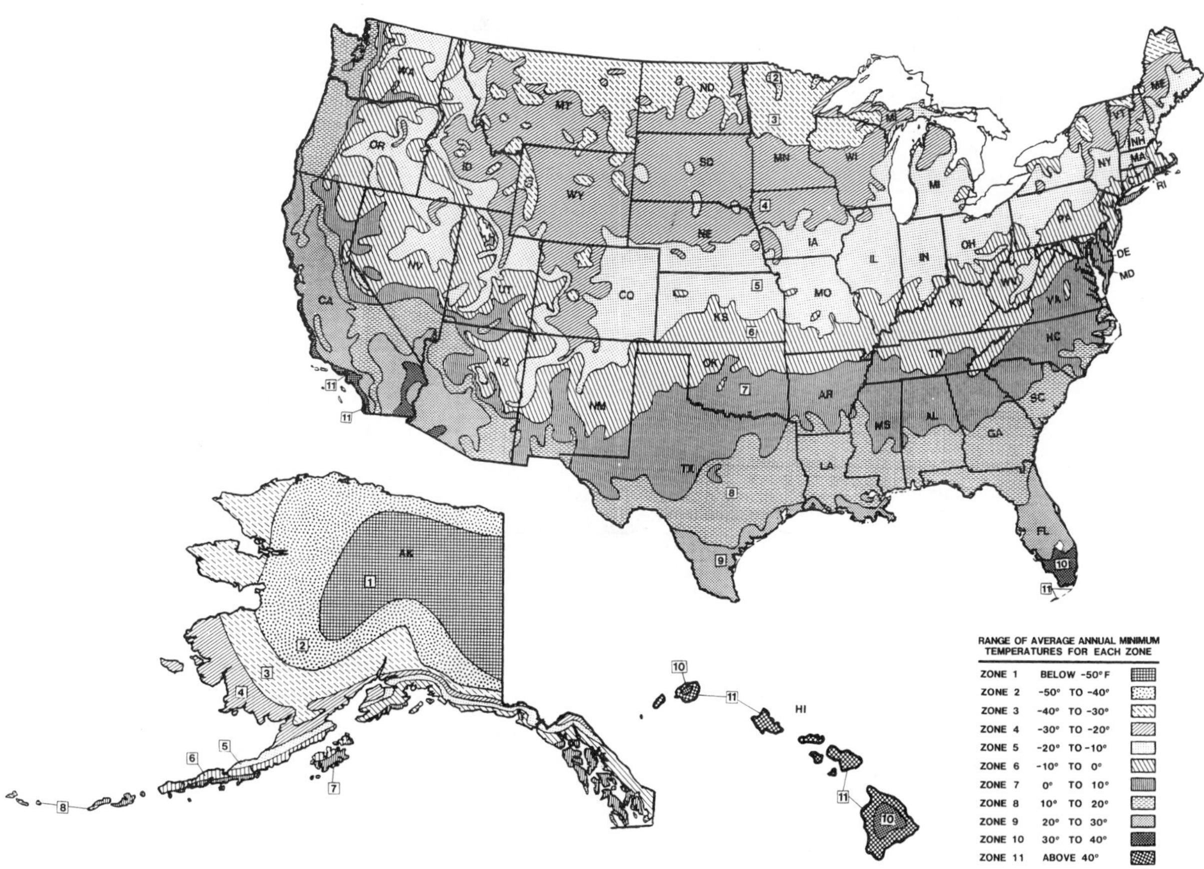

FIGURE 2.5-4 USDA plant hardiness zone map. (From U.S. Department of Agriculture, Agricultural Research Service Miscellaneous Publication 1475, "USDA Plant Hardiness Zone Map.")

FIGURE 2.5-5 Indicator Plant Examples

Zone	Botanical Name	Common Name
1 °F Below –50 °C Below –45.6	*Betula glandulosa* *Empetrum nigrum* *Populus tremuloides* *Potentilla pensylvanica* *Rhododendron lapponicum* *Salix reticulata*	Dwarf birch Crowberry Quaking aspen Pennsylvania cinquefoil Lapland rhododendron Netleaf willow
2 °F –50 to –40 °C –45.6 to –40	*Betula papyrifera* *Cornus canadensis* *Elaeagnus commutata* *Larix laricina* *Potentilla fruticosa* *Vibrunum trilobum*	Paper birch Bunchberry dogwood Silverberry Eastern larch Bush cinquefoil American cranberry bush
3 °F –40 to –30 °C –40 to –34.5	*Berberis thunbergii* *Elaeagnus angustifolia* *Junipercus communis* *Lonicera tatarica* *Malus baccata* *Thuja occidentalis*	Japanese bayberry Russian olive Common juniper Tatarian honeysuckle Siberian crabapple American arborvitae
4 °F –30 to –20 °C –34.5 to –28.9	*Acer saccharum* *Hydrangea paniculata* *Juniperus chinensis* *Ligustrum amurense* *Parthenocissus quinquefolia* *Spiraea* x *vanhouttei*	Sugar maple Panicle hydrangea Chinese juniper Amur River privet Virginia creeper Vanhoutte spirea
5 °F –20 to –10 °C –28.9 to –23.3	*Cornus florida* *Deutzia gracilis* *Ligustrum vulgare* *Parthenocissus tricuspidata* *Rosa multiflora* *Taxus cuspidata*	Flowering dogwood Slender deutzia Common privet Boston ivy Japanese rose Japanese yew
6 °F –10 to 0 °C –23.3 to –17.8	*Acer palmatum* *Buxus sempervirens* *Euonymus fortunei* *Hedera helix* *Ilex opaca* *Ligustrum ovalifolium*	Japanese maple Common box Winter creeper English ivy American holly California privet
7 °F 0 to 10 °C –17.8 to –12.3	*Acer macrophyllum* *Rhododendron* Kurume hybrids *Cedrus atlantica* *Cotoneaster microphylla* *Ilex aquifolium* *Taxus baccata*	Bigleaf maple Kurume azalea Atlas cedar Small-leaf cotoneaster English holly English yew
8 °F 10 to 20 °C –12.3 to –6.6	*Arbutus unedo* *Choisya ternata* *Olearia haastii* *Pittosporum tobira* *Prunus laurocerasus* *Viburnum tinus*	Strawberry tree Mexican orange New Zealand daisy-bush Japanese pittosporum Cherry-laurel Laurestinus
9 °F 20 to 30 °C –6.6 to –1.1	*Asparagus setaceus* *Eucalyptus globulus* *Syzygium paniculatum* *Fuchsia* hybrids *Grevillea robusta* *Schinus molle*	Asparagus fern Tasmanian blue gum Australian bush cherry Fuschia Silk-oak California pepper tree
10 °F 30 to 40 °C –1.1 to 4.4	*Bougainvillea spectabilus* *Cassia fistula* *Eucalyptus citriodora* *Ficus elastica* *Ensete ventricosum* *Roystonea regia*	Bougainvillea Golden shower Lemon eucalyptus Rubber plant Ensete Royal palm

Note: Representative persistent plants listed under the coldest zones in which they normally succeed. Such plants may serve as useful indicators of the cultural possibilities of each zone.

(From U.S. Department of Agriculture, Agricultural Research Service Miscellaneous Publication 1475, "USDA Plant Hardiness Zone Map.")

climates, but their performance will probably not be as satisfactory.

Other factors that affect plant growth and hardiness include the amount of rainfall and the soil and its amendments. Plants can be grown in areas that have less than optimum rainfall for them, but must be watered artificially in those locations.

Before selecting plants for use in a particular area, it is advisable to verify their use with a local horticulturist or nurseryman or with a state or local agricultural agent. In some states agricultural extension services are good sources for such information.

2.5.3 LAWNS

There are some 600 grass genera, of which about 40 are suitable for lawns. However, not all of these should be used in every climatic zone. For example, bluegrass (*Poa pratensis*) grows very well in many areas but is often not successful in subtropical zones. St. Augustine grass (*Stenotaphrum secundatum*) needs a warm, moist environment and will not fare well in cold, dry zones.

Some grasses are more suitable for use in shaded areas than others, but not one is really suitable for use in deep shade. Those that sometimes work in light shade include Bahia (*Paspalum notatum*), bent (*Agrostis*), carpet (*Axonopus offinis*), centipede (*Eremochloa ophiuroides*), fescue (*Festuca rubra*), rye (*Lolium perenne*), St. Augustine, and zoysia (*Zoysia matrella*). Those suitable for deeper shade include fescue, St. Augustine, and zoysia. Successful growth of a particular grass in shade depends on such other factors as the amount of moisture present or delivered artificially, drainage, soil fertility, and the type of vegetation providing the shade. All shading vegetation will draw some moisture from adjacent grass, but trees and shrubs with shallow roots may present a greater problem than deep-rooted ones. Local sources should be contacted for advice before selecting a grass for use in shaded areas.

Some species, such as bluegrass and fescue, propagate successfully from seed. Many southern species, and some northern creeping species, however, should be established from living material such as sod, plugs, sprigs, or stolens. These species include zoysia, St. Augustine, centipede, and many northern bent grasses. Such methods should also be used, regardless of the grass selected, on slopes of

FIGURE 2.5-6 Cold Hardiness Ratings for Some Additional Wood Plants

Botanical and Common Name	Zone
Abeliophyllum distichum (white forsythia)	5b
Acer platanoides (Norway maple)	4
Aesculus x *carnea* (red horsechestnut)	4
Araucaria araucana (monkeypuzzle)	7b
Arctostaphylos uva-ursi (bearberry)	2b
Aristolochia durior (Dutchman's-pipe)	4b
Aucuba japonica (Japanese aucuba)	7b
Bauhinia variegata (purple orchid tree)	9b
Berberis darwinii (Darwin barberry)	8
Betula pendula (European white birch)	3
Bouvardia "Coral" (Coral bouvardia)	9
Butia capitata (Pindo palm)	8b
Camellia reticulata (reticulata camellia)	9
Camellia sasonqua (sasanqua camellia)	7b
Carya illinoinensis "Major" (pecan)	5 (grows) 6 (fruits)
Casuarina equisetifolia (Australian pine)	9b
Ceanothus impressus (Santa Barbara ceanothus)	8
Cedrus dodara (deodar cedar)	7b
Cercis chinensis (Chinese redbug)	6b
Chamaecyparis lawsoniana (Lawson cypress)	6b
Chamaecyparis pisifera (Sawara cypress)	5
Cinnamomum camphora (camphor tree)	9
Cistus laurifolius (laurel rockrose)	7
Cistus x *purpureus* (purple rockrose)	8
Cronus alba (Tatarian dogwood)	3
Cronus kousa (Japanese dogwood)	5b
Cunninghamia lanceolata (cunninghamia)	7
Cytisus x *praecos* (Warminster broom)	6
Elaeagnus multiflora (cherry elaeagnus)	5
Elaeagnus pungens (thorny elaeagnus)	7
Eriobotrya japonica (loquat)	8
Euonymus alatus (winged euonymus)	3b
Euphorbia pulcherrima (poinsettia)	10
x *Fatshedera lizei* (botanical-wonder)	8
Forsythia ovata (early forsythia)	4b
Forsythia suspensa (weeping forsythia)	5b
Fremontodendron mexicanum (flannel bush)	9
Ginkgo biloba (ginkgo, maidenhair-tree)	5
Hibiscus rosa-sinensis (Chinese hibiscus)	9b
Hibiscus syriacus (shrub althea)	5b
Hypericum "Hidcote" (Hidcote St. Johnswort)	6
Iberis sempervirens (evergreen candytuft)	5
Ilex crenata "Convexa" (convexleaf Japanese holly)	6b
Jacaranda acutifolia (green ebony)	10
Juglans regia (English or Persian walnut)	6b
Juniperus horizontalis (creeping juniper)	3
Koelreuteria paniculata (goldenrain-tree)	6
Laburnum x *watereri* (Waterer laburnum)	5b
Lagerstroemia indica (crapemyrtle)	7
Mahonia aquifolium (Oregon hollygrape)	5b
Malus x *arnoldiana* (Arnold crabapple)	4
Melia azedarach (chinaberry)	7b
Metasequoia glyptostroboides (dawn redwood)	5b
Myrtus communis (true myrtle)	8b
Nandina domestica (heavenly bamboo)	7
Nerium oleander (oleander)	8b
Olea europaea (common olive)	9
Osmanthus heterophyllus (holly osmanthus)	7
Picea abies (Norway spruce)	3
Pieris japonica (Japanese andromeda)	6
Pinus mugo var. *mughus* (Mugo pine)	3
Pinus radiata (Monterey pine)	7b
Pinus strobus (eastern white pine)	3b
Prunus yedoensis (Yoshino cherry)	6
Rhaphiolepis indica "Rosea" (Indian hawthorn)	8
Rhododendron "America" (hybrid rhododendron)	5
Rhododendron "Loderi King George" (hybrid rhododendron)	8
Rhododendron mollis hybrids (mollis azalea)	5
Rhododendron prinophyllum (roseum) (roseshell azalea)	4
Rhododendron "Purple Splendor" (hybrid rhododendron)	7
Rhododendron southern Indian hybrids (Indian azalea)	8b
Rosa rugosa (rugosa rose)	3
Schinus terebinthifolius (Brazilian pepper-tree)	9b
Sequoia sempervirens (redwood)	8
Sequoiadendron giganteum (giant sequoia)	7
Stewartia pseudocamellia (Japanese stewartia)	6
Syringa vulgaris (common lilac)	3b
Ulmus americana (American elm)	2
Viburnum burkwoodii (Burkwood viburnum)	5b
Zelkova serrata (Japanese zelkova)	5b

(From U.S. Department of Agriculture, Agricultural Research Service Miscellaneous Publication 1475, "USDA Plant Hardiness Zone Map.")

1:6 and steeper to prevent loss of seed through erosion, and when an "instant lawn" is desired.

The name of the proper grass to use in a particular location should be obtained from a local horticulturist or nurseryman, or from a state or local agricultural agent. In general, the following grasses work well in cool, humid zones: bluegrass, both red and alta fescue, and rye. Grasses for dry areas include beach (*Ammophila*), buffalo (*Buchloe*), blue gamma (*Bouteloua*), and wheat (*Agropyron*). In some cases, Bermuda (*Cynodon dactylon*), centipede, fescue, and zoysia grasses may also work in dry zones. Bent grass and bluegrass may be used in dry zones when irrigation or sprinklers are used. Bahia,

Bermuda, carpet, centipede, St. Augustine, and zoysia work well in warm, moist zones.

2.5.4 PLANTING SOIL

Planting soil for trees, shrubs, and lawns should be topsoil that is fertile, friable, natural loam. It should not contain subsoil, clay lumps, brush, weeds, litter, stumps, roots, rocks, or any other material that would harm the plants or endanger their growth.

In addition, planting soil for trees, shrubs, and lawns must contain the nutrients each particular plant needs. In general, plants need the proper amounts of some 10 to 14 nutrients. Trees and shrubs need a pH value between 6 and 7. The numbers in the pH scale, which ranges from 0 to 14, represent the relative level of alkalinity or acidity. Seven is neutral; above is alkaline; below is acidic. The further away from 7, the more alkaline or acidic. The required pH value for grasses varies with the species. The actual nutrients needed and their percentage of the soil material vary with the plant species. In addition, some trees and shrubs require a pH lower or higher than the 6 to 7 level. It is necessary to determine the exact requirements for each particular plant from a local horticulturist or nurseryman or a state or local agricultural agent.

Soil samples taken from the site should be tested in a qualified laboratory to determine its pH level and the amounts of necessary nutrients present. Some state agricultural extension services will perform such tests. Private laboratories are also available in some places. The level of the proper nutrients can then be adjusted as necessary, using such soil amendments as humus, mulch, sawdust, manure, and fertilizer. The pH of an acidic soil can be lowered by adding lime in the form of natural dolemitic limestone. Alkaline soils can be adjusted by adding aluminum sulfate. Heavy or cohesive soils may be loosened or lightened by adding perlite or vermiculite and, in some cases, sand. Amended soils should be retested to confirm that they have the correct amounts of the proper nutrients and the correct pH value for the plant material to be used.

2.6 Additional Reading

More information about the subjects discussed in this chapter can be found in the references listed in Section 2.7 and in the following publications.

Ambrose, James. 1988. *Simplified Design of Building Foundations*. 2d ed. New York: John Wiley & Sons, Inc.

Bowles, Joseph E. 1996. *Foundation Analysis and Design*. 4th ed. New York: McGraw-Hill.

———. 1992. *Engineering Properties of Soils and Their Measurement*. 4th ed. New York: McGraw-Hill.

Cernica, John N. 1995. *Geotechnical Engineering: Foundation Design*. New York: John Wiley & Sons, Inc.

Das, B. M. 1983. *Advanced Soil Mechanics*. New York: McGraw-Hill.

Head, K. H. 1996. *Manual of Soil Laboratory Testing*. 2d ed. Vol. 1, *Soil Classification and Compaction Tests*. New York: John Wiley & Sons, Inc.

Hightshoe, Gary L. 1987. *Native Trees, Shrubs, and Vines for Urban and Rural America: A Planting Design Manual for Environmental Designers*. New York: John Wiley & Sons, Inc.

Monahan, Edward J. 1993. *Construction of Fills*. 2d ed. New York: John Wiley & Sons, Inc.

Powers, J. Patrick. 1992. *Construction Dewatering: New Methods and Applications*. 2d ed. New York: John Wiley & Sons, Inc.

Ratay, Robert T. 1996. *Handbook of Temporary Structures in Construction*. 2d ed. New York: McGraw-Hill.

Rubenstein, Harvey M. 1996. *A Guide to Site Planning and Landscape Construction*. 4th ed. New York: John Wiley & Sons, Inc.

Simonds, John O. 1997. *Landscape Architecture: A Manual of Site Planning and Design*. 3d ed. New York: McGraw-Hill.

Stitt, Fred. 1999. *Ecological Design Handbook: Sustainable Strategies for Architecture, Landscape Architecture, Interior Design, and Planning*. New York: McGraw-Hill.

Walker, Theodora D. 1991. *Site Design and Construction Detailing*. 3d ed. New York: John Wiley & Sons, Inc.

2.7 Acknowledgments and References

ACKNOWLEDGMENTS

We gratefully acknowledge the assistance of the following organizations and individuals in preparing this chapter. We are also indebted to them for permission to use their illustrations when requested and for the use of their publications as references.

American Concrete Institute (ACI)
American National Standards Institute (ANSI)
American Society for Testing and Materials (ASTM)
American Wood Preservers Bureau (AWPB)
American Wood Preservers Institute (AWPI)
Building Officials and Code Administrators International (BOCA)
ELE International, Inc.
Molded Fiberglass Concrete Forms Company
National Academy of Sciences/National Research Council
National Association of Home Builders (NAHB) Research Foundation
National Forest Products Association
National Institute of Building Sciences (NIBS)
National Institute of Standards and Technology (NIST) of the U.S. Department of Commerce
Portland Cement Association (PCA)
Underwriters Laboratories, Inc. (UL)
U.S. Department of Agriculture

U.S. Department of Housing and Urban Development (HUD)

U.S. Weather Service

REFERENCES

We would also like to thank the authors and publishers of the publications in the following list for their contributions to our research for this chapter.

American Concrete Institute-International. 1996. ACI 302.1R-96, "Guide for Concrete Floor and Slab Construction." Farmington Hills, MI: ACI-Int.

———. 1980. ACI 543R-80, "Recommendations for Design, Manufacture, and Installation of Concrete Piles." Farmington Hills, MI: ACI-Int.

———. 1998. ACI 336.1R-98, "Reference Specification for the Construction of Drilled Piers." Farmington Hills, MI: ACI-Int.

American Nursery Landscape Association (ANLA). *American Standard for Nursery Stock.* Washington, DC: ANLA.

American Society of Heating, Refrigerating and Air Conditioning Engineers. 1997. *ASHRAE Handbook; 1997 Fundamentals.* Atlanta, GA: ASHRAE.

American Society for Testing and Materials (ASTM). Standards:

A 74, "Specification for Cast Iron Soil Pipe and Fittings." West Conshohocken, PA: ASTM.

A 252, "Standard Specification for Welded and Seamless Steel Pipe Piles." West Conshohocken, PA: ASTM.

C 4, "Standard Specification for Clay Drain Tile and Perforated Clay Drain Tile." West Conshohocken, PA: ASTM.

C 412, "Standard Specification for Concrete Drain Tile." West Conshohocken, PA: ASTM.

C 444, "Standard Specification for Perforated Concrete Pipe." West Conshohocken, PA: ASTM.

C 654, "Standard Specification for Porous Concrete Pipe." West Conshohocken, PA: ASTM.

C 700, "Standard Specification for Vitrified Clay Pipe, Extra Strength, Standard Strength, and Perforated." West Conshohocken, PA: ASTM.

D 427, "Test Method for Shrinkage Factors of Soils by the Mercury Method." West Conshohocken, PA: ASTM.

D 698, "Test Method for Laboratory Compaction Characteristics of Soil Using Standard Effort (12,400 ft-lbf/ft)." West Conshohocken, PA: ASTM.

D 1143, "Standard Test Method for Piles Under Static Axial Compressive Load." West Conshohocken, PA: ASTM.

D1557, "Test Method for Laboratory Compaction Characteristics of Soil Using Modified Effort (56,000 ft-lbf/ft3(2,700 kN-m/m3))." West Conshohocken, PA: ASTM.

D 1586, "Standard Test Method for Penetration Test and Split-Barrel Sampling of Soils." West Conshohocken, PA: ASTM.

D 1883, "Standard Test Method for CBR (California Bearing Ratio) of Laboratory Compacted Soils." West Conshohocken, PA: ASTM.

D 2487, "Standard Classification of Soils for Engineering Purposes (Unified Soil Classification System)." West Conshohocken, PA: ASTM.

D 2488, "Standard Practice for Description and Identification of Soils (Visual-Manual Procedures)." West Conshohocken, PA: ASTM.

D 2729, "Specification for Poly(Vinyl Chloride) (PVC) Sewer Pipe and Fittings." West Conshohocken, PA: ASTM.

D 4253, "Standard Test Methods for Maximum Index Density of Soils Using a Vibratory Table." West Conshohocken, PA: ASTM.

D 4254, "Standard Test Method for Minimum Index Density and Unit Weight of Soils and Calculation of Relative Density." West Conshohocken, PA: ASTM.

D 4318, "Standard Test Method for Liquid Limit, Plastic Limit, and Plasticity Index of Soils." West Conshohocken, PA: ASTM.

F 405, "Standard Specification for Corrugated Polyethylene Tube and Fittings." West Conshohocken, PA: ASTM.

American Wood Preservers Association. *Book of Standards.* Seattle, WA: American Wood Preservers Association.

Architectural Graphic Standards. See Ramsey/Sleeper.

Arcom Master Systems. MASTERSPEC®. Basic Sections:

02080, "Utility Materials." 6/96 ed. Salt Lake City, UT: Arcom.

02230, "Site Clearing." 5/97 ed. Salt Lake City, UT: Arcom.

02240, "Dewatering." 5/96 ed. Salt Lake City, UT: Arcom.

02260, "Excavation Support and Protection." 5/96 ed. Salt Lake City, UT: Arcom.

02282, "Termite Control." 2/92 ed. Salt Lake City, UT: Arcom.

02300, "Earthwork." 5/97 ed. Salt Lake City, UT: Arcom.

02455, "Driven Piles." 2/97 ed. Salt Lake City, UT: Arcom.

02456, "Concrete-Filled Steel Piles." 2/97 ed. Salt Lake City, UT: Arcom.

02457, "Prestressed Concrete Piles." 2/97 ed. Salt Lake City, UT: Arcom.

02458, "Steel H Piles." 2/97 ed. Salt Lake City, UT: Arcom.

02459, "Timber Piles." 2/97 ed. Salt Lake City, UT: Arcom.

02466, "Drilled Piles." 8/96 ed. Salt Lake City, UT: Arcom.

02711, "Foundation Drainage Systems." 3/94 ed. Salt Lake City, UT: Arcom.

02900, "Landscaping." 11/93 ed. Salt Lake City, UT: Arcom.

02930, "Lawns and Grasses." 11/93 ed. Salt Lake City, UT: Arcom.

02955, "Trees and Shrubs." 11/93 ed. Salt Lake City, UT: Arcom.

02956, "Ground Cover and Plants." 11/93 ed. Salt Lake City, UT: Arcom.

02900, "Landscape Work." 8/88 ed. Salt Lake City, UT: Arcom.

Briscoe, Frank. 1991. "Wood-Destroying Insects." *Old-House Journal* 19(2) (March/April):34–39.

Building Officials and Code Administrators International, Inc. 1999. *National Building Code.* Country Club Hills, IL: BOCA.

Council of American Building Officials. 1983. *One- and Two-Family Building Code.* Falls Church, VA: CABO (in the process of being transferred to the control of the International Code Council [ICC]).

National Association of Plumbing-Heating-Cooling Contractors. 1990. *National Standard Plumbing Code.* Falls Church, VA: National Association of Plumbing-Heating-Cooling Contractors.

National Institute of Building Sciences (NIBS). 1993. *Wood Protection Guidelines, Protecting Wood from Decay Fungi and Termites.* Washington, DC: NIBS.

Portland Cement Association (PCA). *Concrete Floors on Ground.* Skokie, IL: PCA.

Ramsey/Sleeper, The AIA Committee on Architectural Graphic Standards. 1988. *Architectural Graphic Standards.* 8th ed. New York: John Wiley & Sons, Inc.

Underground Space Center. 1988. *Building Foundation Design Handbook.* Minneapolis, MN: Underground Space Center.

United States Department of Agriculture, Agricultural Research Service. Miscellaneous Publication Number 1475, *USDA Plant Hardiness Zone Map.* Washington, DC: USDA.

United States Department of Agriculture, Forest Service. Home and Garden Bulletin 64, *Subterranean Termites, Their Prevention and Control in Buildings.* Washington, DC: USDA.

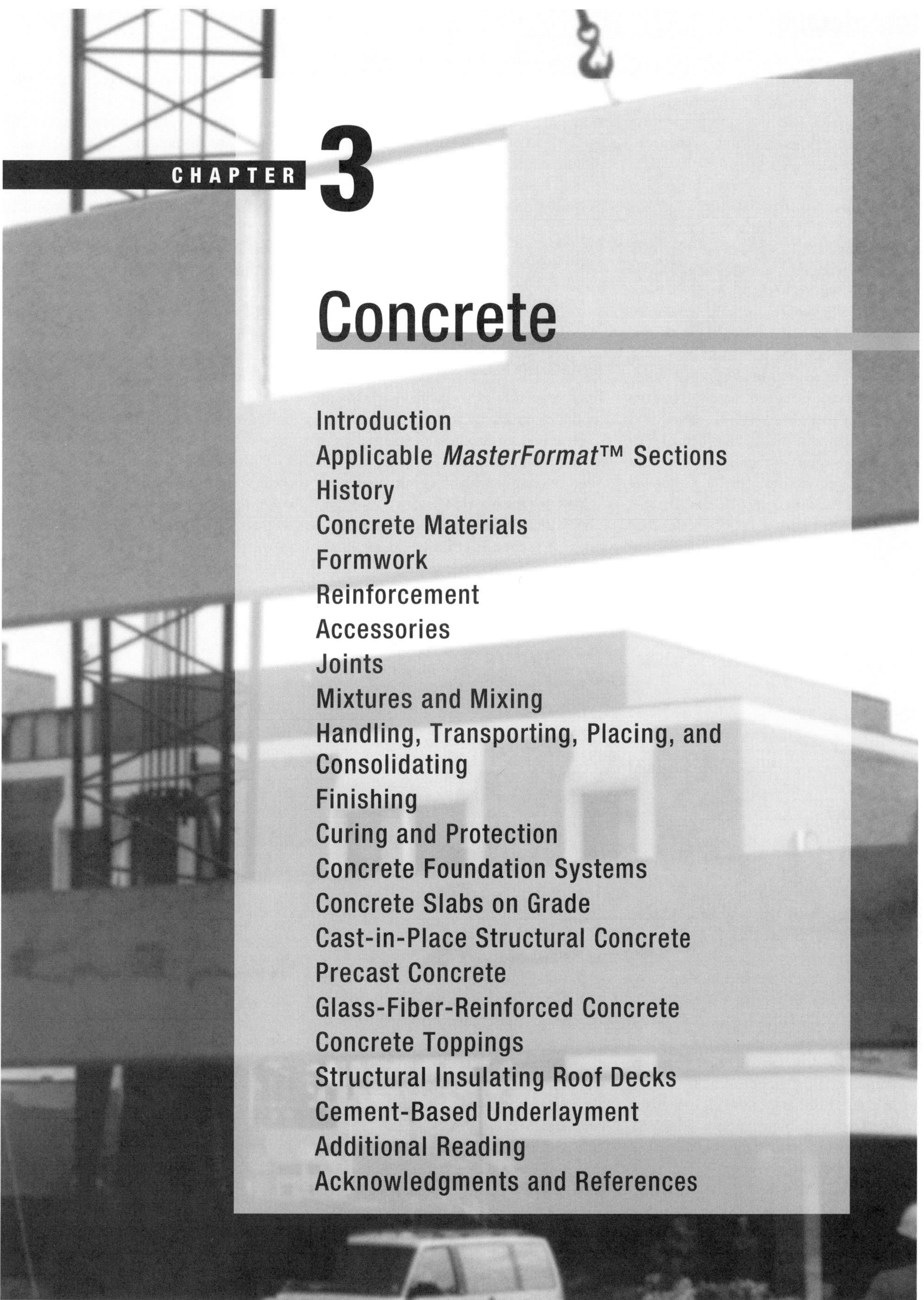

CHAPTER 3

Concrete

Introduction
Applicable *MasterFormat*™ Sections
History
Concrete Materials
Formwork
Reinforcement
Accessories
Joints
Mixtures and Mixing
Handling, Transporting, Placing, and Consolidating
Finishing
Curing and Protection
Concrete Foundation Systems
Concrete Slabs on Grade
Cast-in-Place Structural Concrete
Precast Concrete
Glass-Fiber-Reinforced Concrete
Concrete Toppings
Structural Insulating Roof Decks
Cement-Based Underlayment
Additional Reading
Acknowledgments and References

Introduction

Concrete is used in almost every type and size of architectural and engineering structure. Its major uses are for foundation walls and footings, building superstructures, floor slabs, stairs, sidewalks, paving, curbs and gutters, streets, highways, bridges, and for site structures such as storm water inlets and piping, septic tanks, and drainage swales. The evidence is that the use of concrete is increasing. This is partly due to the inherent fire resistance of concrete, which eliminates the expensive fireproofing required on steel structures. Ease of erection has increased the use of precast concrete units in all types of buildings. Concrete is an old material, but the technology associated with it continues to evolve. Fiber reinforcement is being specified increasingly to toughen concrete slabs and help control cracking, thus reducing the need for control joints. The use of glass-fiber-reinforced concrete (GFRC) for exterior cladding, column covers, and decorative panels is also increasing. Super plasticizers are being used more and more to increase the fluidity of concrete, which makes pumping it to its final location more practicable.

The term *concrete of a suitable quality* and similar phrases used in this book refer to concrete that will perform satisfactorily in its intended use. To do so, it must possess:

- Strength to carry superimposed loads
- Sufficient watertightness to prevent water penetration
- Durability to resist wear and weather
- Workability, to ensure proper handling, placing, finishing, and curing

These properties can be achieved only with good materials and careful workmanship.

Required surface treatments may range from smooth-troweled surfaces for finished floors to highly decorative geometric patterns for use on terraces, walls, and other surfaces. The plastic quality of concrete offers almost unlimited design possibilities of form, pattern, and texture.

The in-place cost of concrete is three to six times the cost of its ingredients. The in-place cost as well as the long-term cost of maintenance is dependent largely on proper care in handling, placing, finishing, and curing. Skillful selection and combining of ingredients in the mix are necessary to achieve the most economical concrete. The same factors also influence concrete quality. No concrete mix is ever better than the workmanship it receives in the field.

To produce concrete of suitable quality, sound materials must be carefully selected, proportioned, and combined into the correct mixture with a consistency that can be worked into place to provide the required surface finish. The placed concrete must then be properly cured and protected.

Unless quality control measures are observed, undesirable properties may occur in the hardened concrete, especially on the wearing surface. These undesirable properties may lead to a soft or *dusting* surface, permeable concrete, cracking, or poor durability.

Applicable *MasterFormat*™ Sections

The following *MasterFormat* Level 2 sections are applicable to this chapter.

03050 Basic Concrete Materials and Methods
03100 Concrete Forms and Accessories
03200 Concrete Reinforcement
03300 Cast-in-Place Concrete
03400 Precast Concrete
03500 Cementitious Decks and Underlayment
03600 Grouts

3.1 History

The Romans were using a form of concrete made with lime, broken stones, and sand to construct temples and other buildings as early as the third century B.C. At first, they left the surface of concrete rough and finished it with a form of stucco. Soon they began to embed small stones in the surface to produce a decorative finish. Eventually, they incorporated broken terra-cotta roof tiles, embedding these at the surface with their smooth sides outward. This innovation eventually led to the manufacture of clay bricks.

Roman architecture was altered forever by the accidental discovery near Mt. Vesuvius of a native natural cement called *puzzolana.* The Romans combined this silica- and alumina-bearing material with crushed limestone to produce a slow-drying cement that was much stronger, harder, and more adhesive than the cements they had been making using lime. In addition, it was *hydraulic,* meaning that it would set both in air and under water. The name, puzzolana, now spelled *pozzolan,* is used today to describe a variety of siliceous or siliceous and aluminous materials used as a substitute for a part of the portland cement in some modern portland cement concrete.

Better concrete soon led the Romans to use more of this material and to build larger structures, especially arches, vaults, and domes. Concrete then, as now, had no permanent form of its own before it set and needed some sort of form or mold to hold it in the desired shape. The Romans used wooden forms, often supported by temporary supports (*centering*). To conserve materials and labor, they invented reusable wood forms and centering that could be used repeatedly. They also used brick or stone to form the outer layers of a structure and filled the voids between the outer layers with concrete. They built impressive domed and vaulted structures using unreinforced concrete to house temples, baths, and other large spaces, some with spans as large as 150 ft. (46 m).

With the collapse of the Roman Empire, concrete technology largely fell into disuse and remained virtually unknown until the rebirth of learning during the Renaissance. In the fifteenth century, the 10-book series, *De Architectura*, written by Roman architect and engineer Marcus

Vitruvius Pollio in the first century B.C., became a popular topic of study.

It was not until near the end of the eighteenth century, however, that research related to concrete technology resumed, and not until 1824 that the essential ingredient in modern concrete was discovered. In that year a brick mason named Joseph Aspdin, living in Leeds, England, patented a new type of cement. He made it by treating, heating (*calcining*), and cooling raw materials into small rocklike shapes called *clinker*, which consisted essentially of calcium silicates, and crushing them into a fine powder. Aspdin called his discovery *portland cement*, because of its resemblance to a gray rock found on the Isle of Portland in the English Channel. Aspdin's cement was an overnight success. However, even though ample raw materials were present in the United States, manufacture of portland cement did not begin in this country until 1872.

Concrete has many uses on its own, but modern concrete buildings would not be possible using concrete alone. Concrete has high strength in compression, but little tensile strength. Even though other forces are also involved, the primary forces acting on a beam or unsupported slab are those that tend to push it together (compression) at the top of the beam or slab and those that pull it apart (tension) at the bottom. Therefore, a concrete slab or beam of any appreciable length will pull apart at the bottom and collapse, especially if a weight (load) is applied to it (Fig. 3.1-1a). This problem can be solved by placing in the concrete unit a material, such as iron or steel, that is strong in tension. Such a *reinforced concrete* unit is capable of safely spanning large distances (Fig. 3.1-1b).

In the 1850s, the race was on to develop a practical method of embedding iron and steel in concrete shapes. Early experimentation was not only with beams, but also with boats and other devices. In 1857, a French gardener named F. Joseph Monier patented a system for producing reinforced concrete. He used it to produce flowerpots, water tanks, and garden bridges. The techniques were quickly adapted to building construction. In 1857, the first application of reinforced concrete in the United States occurred in Port Chester, New York. By 1900, the use of reinforced concrete in buildings was widespread.

A reinforced concrete unit's strength can be increased by reducing tensile stresses in the concrete. This is done by stretching the reinforcing steel, which in turn places the concrete in compression before it is loaded. The scientific basis for modern *prestressing* of concrete was developed in the late 1920s by a French engineer named Freyssinet for use in long-span bridge construction. Prestressing is used extensively in precast concrete units. Prestressed precast concrete requires the use of a high-strength concrete and special curing methods developed in this century.

In addition to more conventional structures built with columns, beams, and slabs, concrete is effective in erecting thin shell structures, some with very large unsupported spans. Because concrete is placed as a fluid, many configurations of such structures are possible. Figure 3.1-2 depicts some of the more common shapes.

Load

a

Load

Reinforcing rod

b

FIGURE 3.1-1 Dashed lines show beam shape (exaggerated) after load is applied. (a) Unreinforced beam bottom will separate, leading to collapse. (b) Reinforced beam also deflects, but magnitude is less. Concrete cracks at bottom, but does not collapse. (Drawing by HLS.)

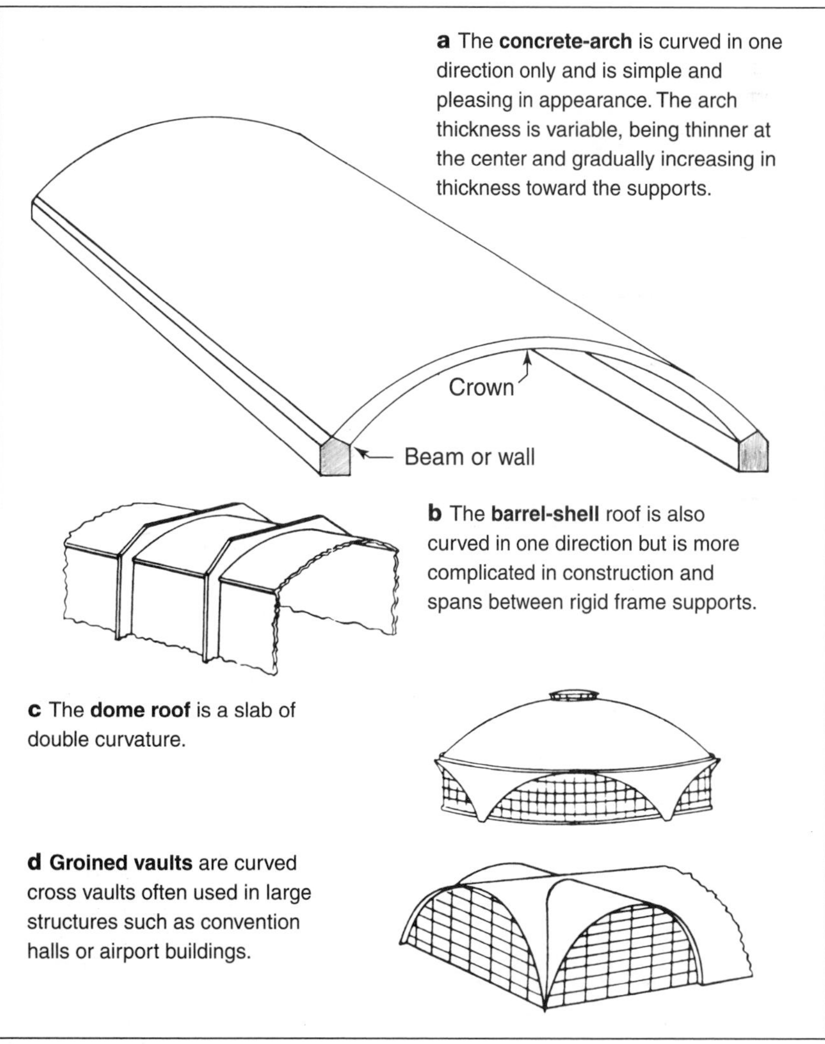

a The **concrete-arch** is curved in one direction only and is simple and pleasing in appearance. The arch thickness is variable, being thinner at the center and gradually increasing in thickness toward the supports.
Crown
Beam or wall
b The **barrel-shell** roof is also curved in one direction but is more complicated in construction and spans between rigid frame supports.
c The **dome roof** is a slab of double curvature.
d **Groined vaults** are curved cross vaults often used in large structures such as convention halls or airport buildings.

FIGURE 3.1-2 (a) Concrete arch; (b) barrel shell; (c) dome; (d) groined vaults. (Courtesy Concrete Reinforcing Steel Institute.)

Concrete has been used by such renowned architects as Frank Lloyd Wright, LeCorbusier, Pier Luigi Nervi, Felix Candela, Eero Saarinen, Paul Rudolph, and I. M. Pei. The sketches in Figures 3.1-3 through 3.1-6 depict some of their modern concrete masterpieces. Contemporary architects continue to use concrete today, in both high-rise and smaller buildings (Figs. 3.1-7, 3.1-8, and 3.1-9).

Advances in concrete technology continue at a rapid pace with the relatively recent development of:

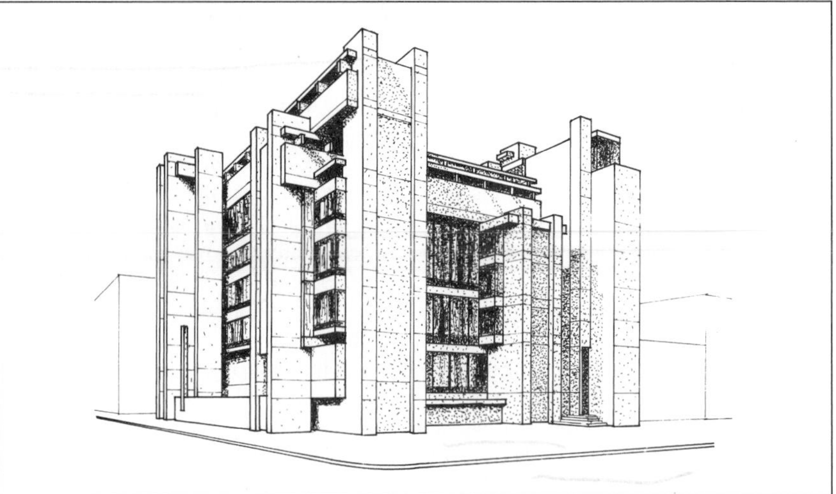

FIGURE 3.1-6 **Art and Architecture Building, Yale University, New Haven, Connecticut.** (Architect: Paul Rudolph; drawing by HLS.)

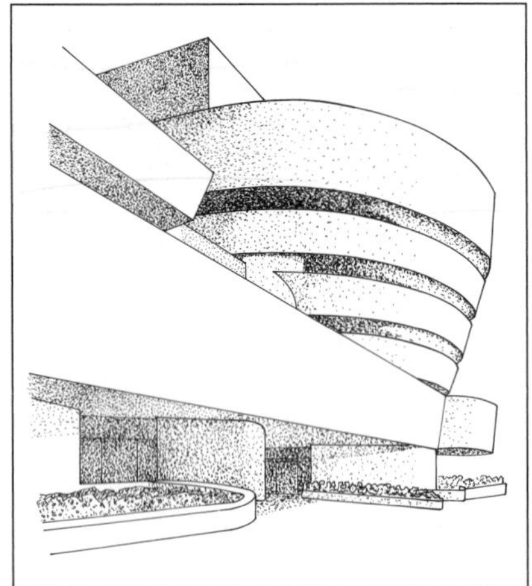

FIGURE 3.1-3 **The Guggenheim Museum, New York.** (Architect: Frank Lloyd Wright; drawing by HLS.)

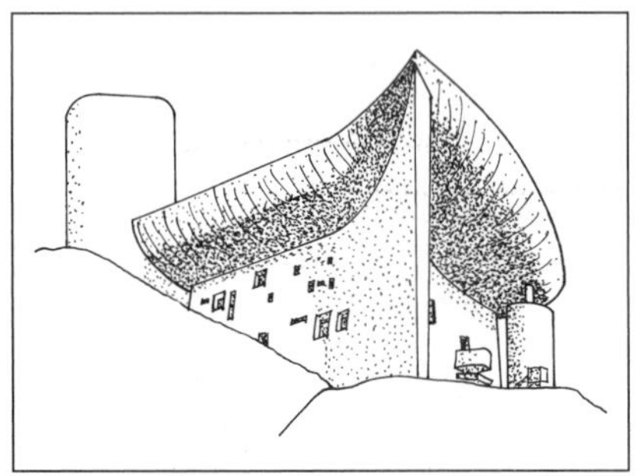

FIGURE 3.1-4 **The chapel of Notre Dame du Haut at Ronchamp, France.** (Architect: LeCorbusier; drawing by HLS.)

- Plastic and metal reusable forms, some of which impart textured finishes to concrete
- *Flying forms* that permit rapid form erection and removal
- Advanced chemical formulas for portland cement that produce different cement formulations (high early strength, air-entrained, etc.) for different uses
- Admixtures to control setting time, ease of delivery, and other characteristics of concrete mixes
- High-strength steel strands that make prestressing possible
- Steel and plastic fiber reinforcement that makes slab surfaces stronger and reduces the need for joints
- Strong and light glass-fiber-reinforced concrete building cladding
- Machine delivery devices such as transit mixing trucks, batch mixing plants, and concrete pumps
- Machine placing and finishing devices, such as internal and form vibrators, motor-driven screeds, and mechanical riding trowels

Currently developing technology will, in the future, produce concrete that will:

- Be corrosion resistant, for use in bridges, parking garages, roads, and radiant floor heating systems;
- Conduct electricity. Currently, concrete acts as an insulator. Future concrete that conducts electricity may have far-reaching uses, including electrical transmission tower support structures and bridge decks that will dissipate lightning strikes.

FIGURE 3.1-7 **One Magnificent Mile, Chicago, Illinois.** (Architect: Skidmore Owings and Merrill; photo courtesy Portland Cement Association.)

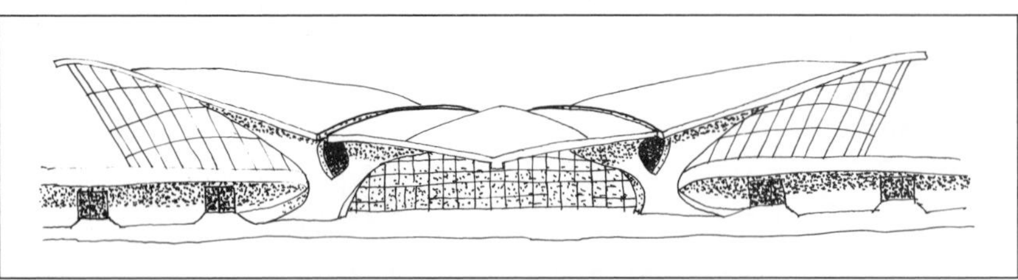

FIGURE 3.1-5 **TWA Terminal, John F. Kennedy Airport, New York.** (Architect: Eero Saarinen; drawing by HLS.)

FIGURE 3.1-8 Commissioners of Public Works Administrative Offices, Charleston, South Carolina. (Architect: Lucus Stubbs Pascullis Powell and Penny, Ltd; photo courtesy Portland Cement Association.)

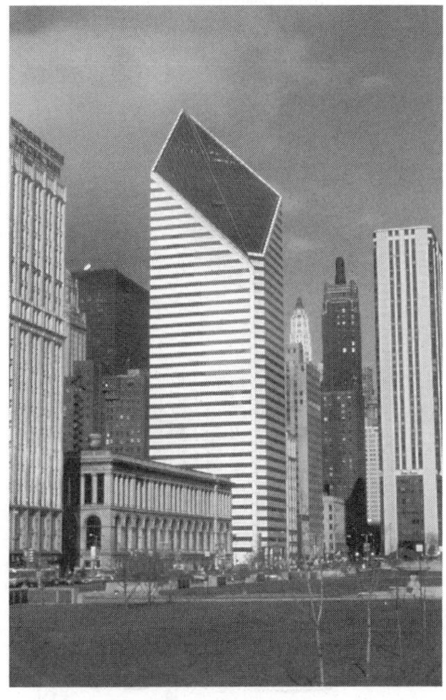

FIGURE 3.1-9 Associates Center, Chicago, Illinois. (Architects: A. Epstein and Sons; photo courtesy Portland Cement Association.)

3.2 Concrete Materials

Concrete is made by combining four materials into a mixture of basically two parts: aggregate and portland cement paste. The paste, composed of portland cement and water, binds the fine aggregate (sand) and coarse aggregate (gravel or crushed stone) into a rocklike mass when the paste hardens. Concrete hardening (*setting*) results from a chemical reaction (*hydration*) between portland cement and water. The portland cement and fine and coarse aggregates used should conform to the requirements of the standards of the American Society for Testing and Materials (ASTM) mentioned in this section and summarized in Figure 3.2-1.

Concrete is often reinforced with steel mesh or rods to increase its tensile strength (see Section 3.4).

FIGURE 3.2-1 Concrete Materials

Material	Standard
Portland cement	
Types I, II, III, IV, V	ASTM C 150
Types IA, IIA, IIIA (air entraining)	ASTM C 150
Pozzolan and blast furnace	ASTM C 595
Ready-mixed concrete	ASTM C 94
Aggregates	
Normal weight	ASTM C 33
Lightweight	ASTM C 330
Admixtures	
Air-entraining agents	ASTM C 260
Accelerators: calcium chloride	ASTM D 98

(Courtesy Portland Cement Association.)

3.2.1 PORTLAND CEMENT

Portland cement is a finely pulverized material consisting principally of compounds of lime, silica, alumina, and iron. It is manufactured from selected materials under closely controlled processes (Fig. 3.2-2). The process starts by mixing limestone or marl with such other ingredients as clay, shale, or blast furnace slag in proper proportions and then burning this mixture in a rotary kiln at a temperature of approximately 2700°F (1482°C) to fuse it into *clinker* (rocklike shapes of varying size averaging ¾ in. [19 mm] in diameter). The clinker is cooled and then pulverized together with a small amount of gypsum, which slows the setting time. The finished product, portland cement, is so fine that nearly all of it will pass through a sieve with 40,000 openings to the square inch (62 openings per mm^2).

When portland cement is mixed with water, a paste is formed which first *sets* (becomes stiff) and then hardens into a solid mass. The setting and hardening (*hydration*) are brought about by chemical reactions between the portland cement and water. The reaction is not simply one of evaporation or drying.

3.2.1.1 Types of Portland Cements

Portland cements are made to meet different physical and chemical requirements for specific purposes. ASTM C 150 describes eight portland cement types: Type I, Normal; Type IA, Normal, Air-entraining; Type II, Moderate; Type IIA, Moderate, Air-entraining; Type III, High Early Strength; Type IIIA, High Early Strength, Air-entraining; Type IV, Low Heat of Hydration; and Type V, Sulfate Resisting. These types are summarized in Figure 3.2-3.

TYPE I, NORMAL

Type I is a general-purpose portland cement suitable for most uses when the special properties of other types are not required. Type I portland cement should

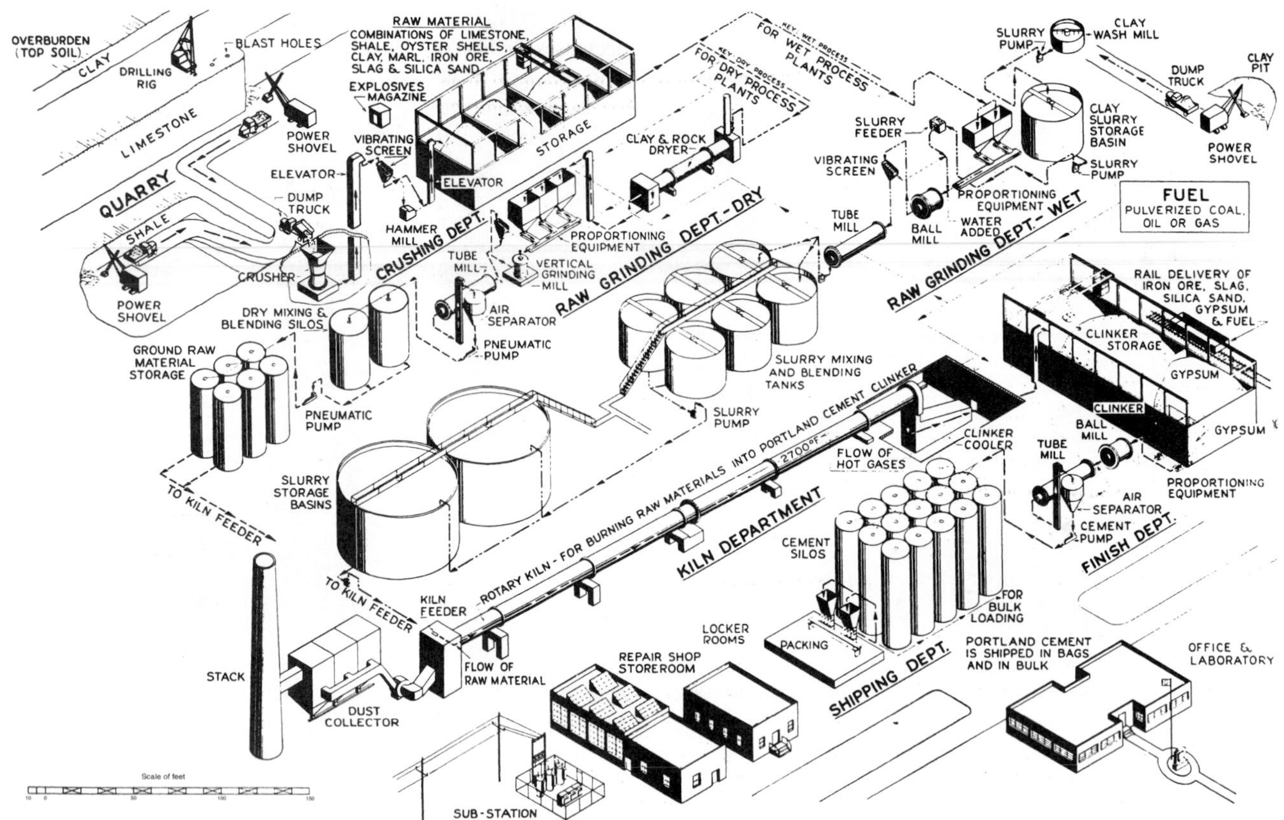

FIGURE 3.2-2 **Steps in the manufacture of portland cement, showing both dry and wet processes.** (Courtesy Portland Cement Association.)

not be used where concrete will be in contact with high-sulfate soils or will be subject to high temperatures during hydration.

TYPE II, MODERATE

Type II portland cement is used where precaution against moderate sulfate attack is important, as in drainage structures where sulfate concentrations in the groundwater are higher than normal. Type II portland cement generates less heat of hydration and cures at a slower rate than Type I. This moderate heat of hydration reduces temperature rise, which is especially important when concrete is placed in warm weather in structures of considerable mass, such as in large piers or heavy retaining walls.

TYPE III, HIGH EARLY STRENGTH

Type III portland cement is used when high strengths are desired at a very early time, usually a week or less. It is used:

- When early form removal is desirable;
- When concrete must be put into service quickly;
- In cold weather to reduce the period required for protection against low temperatures;
- When its use will secure high early strengths more satisfactorily or more economically than the use of richer mixes of Type I portland cement.

TYPE IV, LOW HEAT

Type IV is a special portland cement for use where the amount and rate of heat generated during hydration must be kept to a minimum. The development of strength is also at a slower rate. Type IV is intended for use in large masses of concrete, such as in large gravity dams, where temperature rise resulting from heat generated during hardening is a critical factor.

TYPE V, SULFATE RESISTING

Type V is a special portland cement intended for use only in construction exposed to severe sulfate action, such as in some western states having soils and waters of high alkali content. It has a slower rate of strength gain than normal portland cement.

AIR-ENTRAINING PORTLAND CEMENTS

Types IA, IIA, and IIIA correspond in composition to Types I, II, and III, respectively. They produce concrete that is resistant to severe frost action and to scaling

FIGURE 3.2-3 **Portland Cement Types and Uses**

ASTM C 150	Use	ASTM C 175
Type I	In general construction when special properties of other types are not required	Type IA*
Type II	In general construction where moderate heat of hydration is required	Type IIA*
Type III	When high early strength is required	Type IIIA*
Type IV	When low heat of hydration is required	
Type V	When high sulfate resistance is required	

*Air-entraining portland cement.

(Courtesy Portland Cement Association.)

caused by chemicals used for snow and ice removal. In these portland cements, small quantities of air-entraining additions have been incorporated by grinding them in along with clinker during manufacture. These additions should conform to the requirements of ASTM C 226. Concrete made with these portland cements contains tiny, well-distributed, and completely separated air bubbles. Air-entraining portland cements cost no more than the corresponding non-air-entraining types in ASTM C 150.

3.2.1.2 White Portland Cement

White portland cement is manufactured to conform to ASTM C 150, although it is not specifically mentioned there. It is used primarily in precast wall and facing panels, terrazzo, stucco, cement paint, tile grout, and decorative concrete. It is recommended wherever white or colored concrete or mortar is desired. It is made of selected raw materials containing negligible amounts of iron and manganese oxide, the substances that give concrete its gray color.

3.2.1.3 Portland Blast-Furnace Slag Cement

In portland blast-furnace slag cement, granulated and selected blast-furnace slag, which is obtained by rapidly chilling or quenching molten slag in water, steam, or air, is added to and ground along with portland cement. This type of portland cement should meet the requirements of ASTM C 595, Type IS and Type IS-A, Air-entraining, which requires that the slag constituent be between 25% and 65% of the total weight of the portland cement.

Portland blast-furnace slag cement is usually available only in areas where it can be obtained economically, such as near blast-furnace sites.

3.2.1.4 Portland-Pozzolan Cement

In portland-pozzolan cement, pozzolan, which consists of siliceous (or siliceous and aluminous) material, is blended with ground portland cement. Typical pozzolan materials are fly ash, volcanic ash, and calcined shale and clay. This type of portland cement should meet the requirements of ASTM Standard C 595, Type IP and Type IP-A, Air-entraining, which requires that the pozzolan constituent be between 15% and 40% of the total weight of the portland cement.

Concrete with pozzolan has a higher cured strength than concrete containing only portland cement, aggregate, and water. Pozzolan also decreases the heat of hydration, which is important in mass concrete structures. Pozzolans should not be used, however, where high early strength is a requirement, because they tend to slow the curing process.

3.2.2 WATER

Mixing water should be clean and free from oil, alkali, and acid. In general, water that is fit to drink is suitable for use in concrete. However, water with excessive quantities of sulfates should be avoided even when it is potable.

3.2.3 AGGREGATES

Aggregates constitute 60% to 75% of the volume of concrete (Fig. 3.2-4). Therefore, both the cost and quality of a concrete are affected by the kind of aggregates used in it. Aggregates should be obtained from reliable dealers whose materials are of a quality known to make suitable concrete. They should comply with the requirements of ASTM C 33, which places limits on the allowable amounts of damaging substances and specifies requirements for grading, strength, and soundness of normal-weight aggregate.

The most commonly used aggregates—sand, crushed stone, and air-cooled blast-furnace slag—produce normal-weight concrete weighing between 135 and 160 lb./ft.^{3} (2163 and 2563 kg/m^3). Aggregates of expanded shale, clay, slate, and slag are used to produce structural lightweight concrete weighing from 85 to 115 lb./ft.^{3} (1362 to 1842 kg/m^3). Other lightweight materials, such as pumice, scoria, perlite, vermiculite, and diatomite, are used to produce insulating concretes ranging in weight from 15 to 90 lb./ft.^{3} (240 to 1442 kg/m^3) (Fig. 3.2-5).

Both fine and coarse aggregates should be uniformly graded from their finest particles up to their largest. They should also be clean and free of loam, clay, and vegetable matter, because these foreign particles prevent portland cement paste from properly binding aggregate particles together. Concrete containing these objectionable materials will be porous, have low resistance to weathering, and have low strength. It can also develop surface defects such as popouts.

In addition, the most economical concrete results when both the fine and coarse aggregates are uniformly graded, because this requires the least amount of portland cement paste to surround the aggregate and fill all the spaces between the particles (Fig. 3.2-6). Less paste means a reduction in shrinkage, because most shrinkage occurs in the paste.

3.2.3.1 Fine Aggregate

Fine aggregate consists of sand or another suitable fine material. A good concrete sand will contain particles varying uniformly in size from very fine up to ¼ in. (6.4 mm) in diameter (Fig. 3.2-7). In well-graded sand the finer particles help fill the spaces between the larger particles. Sufficient fine particles are necessary for good workability and smooth surfaces, but an excess of fines requires an increase in the amount of portland cement–water paste to achieve a desired strength, thus in-

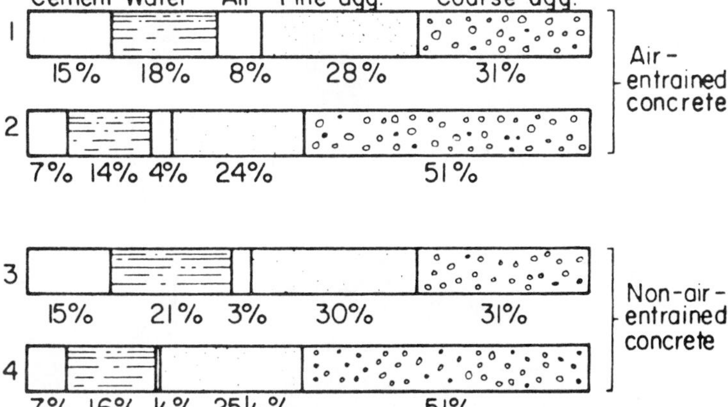

FIGURE 3.2-4 Range in proportions of materials used in concrete, by absolute volume. Bars 1 and 3 represent rich mixes with small aggregate. Bars 2 and 4 represent lean mixes with large aggregate. (Courtesy Portland Cement Association.)

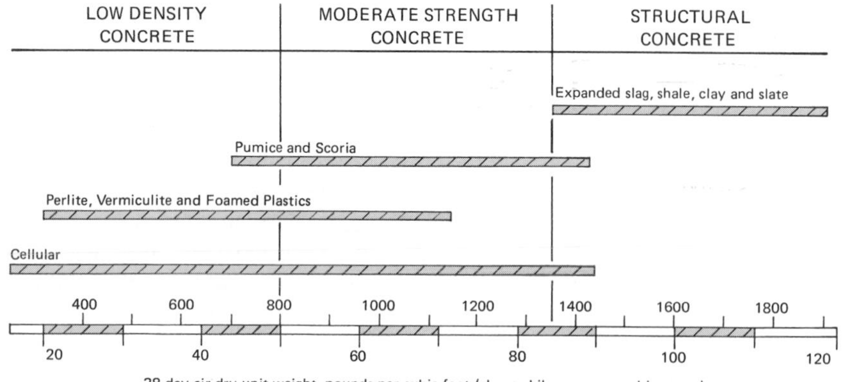

FIGURE 3.2-5 Concrete made with lightweight aggregate. (Courtesy Portland Cement Association.)

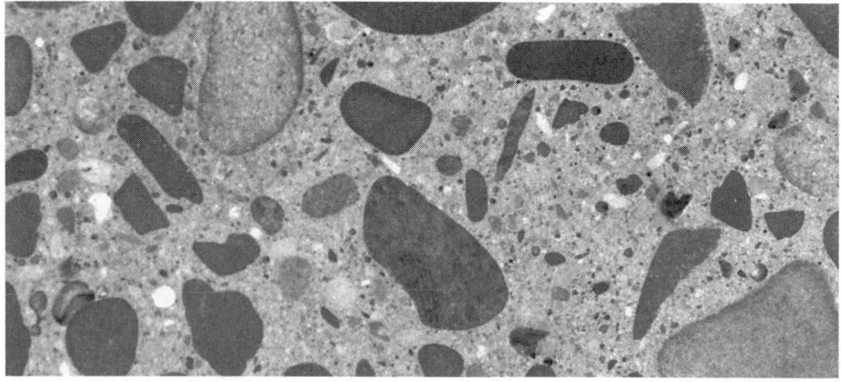

FIGURE 3.2-6 Cross section of concrete showing cement paste completely surrounding each aggregate particle and filling the spaces between particles. The quality of its paste largely controls the quality of a concrete. (Courtesy Portland Cement Association.)

creasing cost. Failure to increase the portland cement content when there is an excess of fines in the aggregate will lower the strength of a concrete. A poor concrete sand is shown in Figure 3.2-8.

Figure 3.2-9 shows a typical grading of fine aggregates for light concrete slabs on grade.

3.2.3.2 Coarse Aggregate

Coarse aggregate consists of gravel, crushed stone, or other suitable materials larger than ¼ in. (6.4 mm) in diameter (Fig. 3.2-10). Coarse aggregate that is sound, hard, and durable is best suited for making concrete. Material that is soft or flaky or wears away rapidly is unsatisfactory.

In most concrete, normal-weight coarse aggregate sizes are limited to those between ⅜ in. (9.5 mm) and 1½ in. (38.1 mm). Thin toppings and slabs may be limited to ⅜-in. (9.5 mm) aggregate. Coarse aggregate as large as 6 in. (152.4 mm) in diameter is often found in massive concrete structures, such as dams. Lightweight aggregate is usually limited to a maximum of 1 in. (25.4 mm). In walls, the largest piece of any type of coarse aggregate should never be more than one-fifth the overall wall thickness. The maximum size in slabs should be about one-third the slab thickness. The largest

FIGURE 3.2-7 Well-graded sand before and after separation into its various sizes. The particles vary from dust to about ¼ in. (6.4 mm) in diameter. (Courtesy Portland Cement Association.)

FIGURE 3.2-8 A poor concrete sand, before and after being separated into four sizes. This sand lacks particles larger than 1/16 in. (1.6 mm). More cement is needed with fine sand. (Courtesy Portland Cement Association.)

FIGURE 3.2-9 Gradings of Fine Aggregate for Concrete Slabs

	Percent of Aggregate Passing Sieve	
Sieve Size	**Normal Weight**	**Lightweight**
⅜ in. (10 mm)	100	100
No. 4	95–100	85–100
No. 8	80–90	—
No. 16	50–75	40–80
No. 30	30–50	30–65
No. 50	10–20	10–35
No. 100	2–5	0–5

(Courtesy Portland Cement Association.)

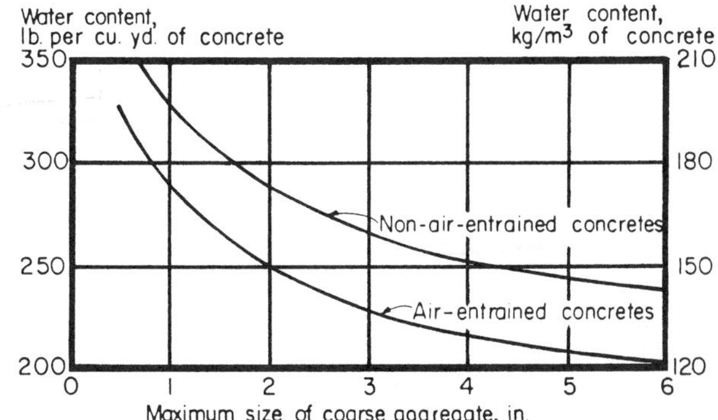

FIGURE 3.2-11 The effect of aggregate size on required water content. (Courtesy Portland Cement Association.)

a

b

FIGURE 3.2-10 (a) A well-graded coarse aggregate for concrete. Notice how the small pieces fit between the larger ones. (b) The same material separated, from left to right, into ¼- to ⅜-in. (6.4 to 9.5 mm); ¾-in. (19.1 mm); and 1½-in. (38.1 mm) sizes. (Courtesy Portland Cement Association.)

piece of aggregate should never be larger than three-quarters of the width of the narrowest space through which the concrete will be required to pass during placing. This is usually the space between the reinforcing bars or between the bars and the forms.

Coarse aggregate is well graded when particles range uniformly from ⅜ in. (9.5 mm) up to the largest size that may be used on the kind of work to be done. Using the maximum allowable particle size usually results in less drying shrinkage and more economical concrete.

The size of coarse aggregate used in concrete has a bearing on economy. More water is usually required for small aggregates than for large. The water required for a slump of 3 to 4 in. (76 to 100 mm) is shown in Figure 3.2-11 for a wide range of coarse aggregate sizes. For a given water-cement ratio (see Section 3.2.1), the amount of portland cement required decreases as the maximum size of coarse aggregate increases. However, the increased cost of obtaining and handling aggregates larger than 2 in. (50.8 mm) may offset the savings in portland cement.

3.2.4 ADMIXTURES

Admixtures are materials other than portland cement, aggregates, and water that are added to concrete either immediately before or during its mixing to alter the properties of the concrete in a variety of ways. For example, they can be used to:

- Improve workability;
- Reduce separation of coarse and fine aggregates caused by the settling out of the heavier coarse aggregate;
- Entrain air;
- Accelerate or retard setting and hardening.

Concrete should be workable, finishable, strong, durable, watertight, and wear resistant. The following factors should be considered in deciding whether to use admixtures in a concrete mix:

- Changing aggregate gradation, portland cement type or amount, or mix proportions may be a better, surer, and more economical way to achieve the desired objectives than will incorporating an admixture.
- Many admixtures affect more than one property of concrete, sometimes hurting desirable properties.
- The effect of some admixtures is changed by mix wetness and richness, aggregate gradation, and the character and length of mixing.
- Some admixtures will not react the same with all portland cements, even of the same type.

Accordingly, admixtures should be used with caution since specific effects that will result from their use can seldom be predicted accurately. In addition, two or more admixtures should not be used in the same concrete mix unless their combined effect is well known and will not produce an undesirable condition.

ASTM C 494 classifies some chemical admixtures by function, as follows:

Type A: Water-Reducing
Type B: Retarding
Type C: Accelerating
Type D: Water-Reducing and Retarding
Type E: Water-Reducing and Accelerating
Type F: High-Range Water-Reducing
Type G: High-Range Water-Reducing and Retarding

Of these, the most frequently used are air-entraining agents, water-reducing agents, retarders, and accelerators.

3.2.4.1 Water-Reducing Admixtures

Water-reducing admixtures, such as hydroxylated carboxylic acid, should be used in almost all concrete. They permit a lower water content, improve workability, and increase the efficiency of the portland cement in a mix, which lowers a concrete's cost relative to its performance.

High-range water-reducing admixtures (*super plasticizers*) are used mostly in concrete that is to be pumped. They produce a mix that flows easily with no increase in its water content.

3.2.4.2 Retarders

Admixtures that have a retarding effect on the set of portland cement overcome the accelerating effect that temperature has on setting during hot weather and in large masses of concrete, and delay the early stiffening of concrete placed under difficult conditions. Retarder solutions are also sometimes applied directly to the surface of concrete to retard the set of a surface layer of mortar so that it can be readily removed by brushing, thus exposing the aggregate and producing textured surface effects.

Because most retarders also act as water reducers, they are frequently called *water-reducing retarders*. Retarders may also entrain some air in concrete. Many chemicals have a retarding influence on the normal setting time of portland cement. Some of these are variable in action, retarding the set of some portland cements and accelerating the set of others. Unless experience with a retarder has determined the extent of its effect on the setting time and other properties of a concrete, its use as an admixture should not be attempted without technical advice or advance experimentation with the portland cement and other concreting materials involved to determine its effect on the properties of the concrete.

3.2.4.3 Accelerators

Accelerators increase the rate of early strength development in concrete to:

- Reduce the time before finishing operations can be started;
- Permit earlier removal of forms and screeds;
- Reduce the period required for curing in certain types of work;
- Advance the time when a structure can be placed in service;
- Partially compensate for the slow gain in strength of a concrete even with proper protection during cold weather;
- Reduce the period of protection required for initial and final set in making emergency repairs and other work.

Under most conditions, commonly used accelerators cause an increase in the drying shrinkage of concrete. In many cases, it must be decided whether to use an admixture, increase the portland cement content, use high-early-strength portland cement, provide greater protection or a longer curing period, or to use a combination of these.

Calcium chloride is used to accelerate the time of set and to increase the rate of strength gain. It should meet the requirements of ASTM D 98. The amount used should not exceed 2% by weight of the portland cement used. Greater amounts may cause rapid stiffening, increase drying shrinkage, and corrode reinforcement. Calcium chloride should always be added in solution as part of the mixing water to ensure uniform distribution. If it is added in dry form, not all of the dry particles may completely dissolve during mixing. These undissolved lumps can cause popouts or dark spots in hardened concrete. Calcium chloride should never be used as an antifreeze. To lower the freezing point of concrete appreciably would require using so much calcium chloride that the concrete would be ruined. Instead, protective measures should be taken to prevent the concrete from freezing (see Section 3.10.3.1).

Calcium chloride or admixtures containing soluble chlorides should not be used in:

- Prestressed concrete, because of possible corrosion hazards;
- Concrete containing embedded aluminum, such as conduit, because serious corrosion can result, especially if the aluminum is in contact with embedded steel and the concrete is in a humid environment;
- Concrete exposed to soils or water containing sulfates;
- Floor slabs intended to receive dry-shake metallic finishes;
- Hot weather generally.

3.2.4.4 Air-Entraining Agents

Air-entrained concrete can be made either with admixtures added to the mix or with air-entraining portland cement as listed in ASTM C 150, which already contains the admixtures. Air-entraining admixtures produce many microscopic stable air bubbles in concrete. These materials should conform with the requirements of ASTM C 260, which includes specifications and methods of testing. Such admixtures will, at little or no additional cost, improve a concrete's workability and durability and produce a hardened concrete that is resistant to severe frost action and the effects of salt applied for snow and ice removal. Common air-entrainment admixtures include wood resin, sulfonated hydrocarbons, and fatty resinous acids. Because careful control is essential in producing air-entrained concrete, its use should be restricted to plant or transit-mixed batches.

Properly proportioned air-entrained concrete has better workability and contains less water per cubic yard (m^3) than non-air-entrained concrete of the same slump (see Section 3.7.3.1 for a discussion of slump tests). This results in a more solid, weather-resistant, blemish-free surface. Air-entrained concrete can be handled and placed with less segregation of materials and less tendency to bleed (see Section 3.9.1.1). These properties indirectly aid in promoting durability by increasing the uniformity of the concrete.

In concrete made with air-entraining portland cement or admixtures containing air-entraining agents, the quantity of water in the mix must be adjusted to maintain the desired slump. Such mix designs and adjustments should be made only by a qualified technician.

Because of its proven record, both in the laboratory and in the field, and because it increases workability and durability, all concrete would benefit from entrained air regardless of exposure conditions. Generally, however, air entrainment is specified only for concrete that is exposed to cycles of freezing and thawing because, except in very lean and harsh mixtures, air entrainment reduces both compressive and flexural strengths by 2% to 6% for each 1% of air added. However, when freezing and thawing will be encountered, and when watertightness is necessary, air entrainment is essential. Whether by use of air-entraining portland cement or admixture, a proper percentage of entrained air in a concrete mix should be obtained, depending on the maximum size of aggregate, as shown in Figure 3.2-12.

3.2.4.5 Synthetic Fiber Reinforcement

Synthetic fiber reinforcement is glass, steel, or polypropylene fibers with a nominal fiber length of 2 in. (50 mm). It is sometimes added to concrete slabs to help prevent plastic and shrinkage cracking and can be quite effective when properly used. Fiber reinforcement is not a substitute for wire mesh, however, and should not be used at all where service temperatures will exceed 300°F (150°C). Concentrations above 0.1% by volume are not cost-effective. The effective standard for synthetic fiber reinforcement is ASTM C 1116.

FIGURE 3.2-12 Recommended Air-Entrainment Amounts for Various Maximum Coarse Aggregate Sizes

Nominal Maximum Size Coarse Aggregate	Total Air Content Percentage by Volume of Concrete
⅜ in.	6 to 10
½ in.	5 to 9
¾ in.	4 to 8
1 in.	3½ to 6½
1½ in.	3 to 6

(Courtesy Portland Cement Association.)

3.3 Formwork

Because concrete is unable to maintain a particular shape before it sets, it must be placed in a form or mold. In building construction, this form or mold is called *formwork*. Architects usually select the form liners and inserts often used to impart special finishes or patterns to concrete. Architects and structural engineers, working together, also select special forms for ribbed slabs. But the design of the structural portions of formwork and necessary external supports should be, and usually is, the responsibility of the building contractor. Codes require certain minimum provisions for formwork, and a particular circumstance may cause a structural engineer to place minimum limits on form design. The minimum location and number of slab form supports and shoring may be dictated, for example. Unfortunately, in some jurisdictions the architect is legally prevented from disclaiming responsibility for form design.

The discussion here is general in nature. Additional requirements for formwork are mentioned in the sections of this chapter where a particular type of concrete unit is discussed. For example, there are discussions about forming slabs, including lift-slab forming, in Section 3.13.4 and some additional requirements for precast concrete formwork in Section 3.14.

3.3.1 MATERIALS

Forms are made from wood, metal, and plastic. They can be divided into two classes: temporary forms and permanent forms.

3.3.1.1 Temporary Forms

Most formwork is temporary, for two reasons: (1) permanent forms would detract from the desired finish, and (2) it is more economical to reuse formwork.

Most temporary concrete forms are made from wood. The parts of the form in contact with the concrete are often high-grade lumber or plastic-overlaid plywood. Wood structural forms are also used to support metal and plastic pans and domes and metal- and plastic-form textured liners. Temporary forms for pan slabs and ribbed slabs (Section 3.13.4.1) are either steel or plastic.

Round column forms made of metal, molded glass-fiber-reinforced plastic, and multilayered paper or fiber are available. These are often plastic coated on the interior to improve the smoothness of the cast column and to eliminate the lines imparted by the spiral winding of the form.

An almost unlimited number of textures and patterns can be imparted to concrete using metal or plastic *form liners*. Most of these are proprietary items, but custom designs are also possible.

3.3.1.2 Permanent Forms

Permanent corrugated-steel forms are used to make supported concrete slabs. A supported slab rests on beams or columns, rather than on earth (slab on grade). Permanent steel forms are made from corrugated sheet steel.

3.3.2 DESIGN AND CONSTRUCTION

The potential failure of formwork represents a great hazard to people working around and on it and to the property adjacent to or supported by it. Therefore, formwork must be designed and built using the same skills and practices that would be applied to a permanent structure. It must be strong enough to withstand the great pressures imposed by heavy, wet concrete and the additional stresses that result from placing and consolidating the concrete. In addition, formwork must be rigid enough to maintain the desired shape under these stresses. It must also remain tight, so that cement paste and water are not lost from the concrete during hydration. Such loss will affect the quality of the cured concrete.

Forms must be accessible so that concrete can be easily placed, moved in the form, and consolidated. They must also be designed so that they can be easily removed without damaging the concrete or the form. For example, there must be no undercuts, intrusions, or sharp edges left in the concrete, because these will either be broken off or will tear the forms apart as the forms are removed. Corners should be rounded off or beveled by inserting wood, rubber, or plastic chamfering strips in the forms.

Form surfaces must impart the desired degree of smoothness and texture to the concrete. Form surfaces that will be in contact with concrete should be coated with a wax, oil, or plastic compound form-release (parting) agent to prevent the concrete from sticking to them. Where the concrete will be exposed to view, the parting agent should be nonstaining.

The following discussions assume that forms will be custom-built on the site. However, there are available today prefabricated proprietary wood, steel, and aluminum systems for forming footings, columns, and walls (Fig. 3.3-1). These systems are similar in concept to the custom-built systems described here, although they may use different materials for the various components. For example, a column form may use angle-shaped steel

FIGURE 3.3-1 Prefabricated formwork. (Courtesy the Burke Company.)

FIGURE 3.3-2 Prefabricated formwork, some with bracing, some without. (Courtesy the Burke Company.)

clamps for yokes instead of wood members or steel bands to hold the form together; footing forms may use corrugated steel sides rather than boards or plywood; wall forms may be framed with aluminum or steel studs instead of wood studs; and wales and braces may be metal as well, or may not be necessary at all (Fig. 3.3-2). When these types of systems are used, it is necessary to follow the manufacturer's instructions with regard to erection, safety, and removal.

3.3.2.1 Forming Footings

Footings should be formed, unless they are in very stable soil and have been cut cleanly and accurately to the shape of the footing, which is a rare condition. Continuous footing may be formed using plywood, as shown in Figure 3.3-3, or using wood boards. Small footing forms can be braced using stakes and spreaders as shown. On small footings, steel rods are sometimes used instead of stakes.

Large footings are framed in a similar manner, except that plywood or multiple boards are used for the side forms and these are braced by vertical boards (*cleats*) and continuous horizontal framing (*wales*) (Fig. 3.3-4). Tie wires are passed through the form from side to side and fastened to the cleats. Stakes may be used to prevent the formwork from spreading. Steel straps or lumber spreaders are sometimes necessary across the top of larger forms to prevent the side walls from spreading apart when the concrete is placed.

3.3.2.2 Forming Walls

Custom forms for walls are built at the site using plywood supported by studs, tied together by continuous wales, and braced diagonally to the ground (Fig. 3.3-5).

To erect a wall form, the inner liner and its framing are installed first and braced back to the ground. The reinforcing steel is placed next. Then the other liner and its framing are placed and braced (Fig. 3.3-6). As the formwork is erected, the wales are tied together using form ties to keep the form from spreading. There are two types of ties: *snap ties* and *screw ties* (Fig. 3.3-7).

Snap ties are inserted through holes in the forms as the forms are erected. A washer on the tie rod at the interior side of each form surface maintains the width of the wall and prevents concrete from leak-

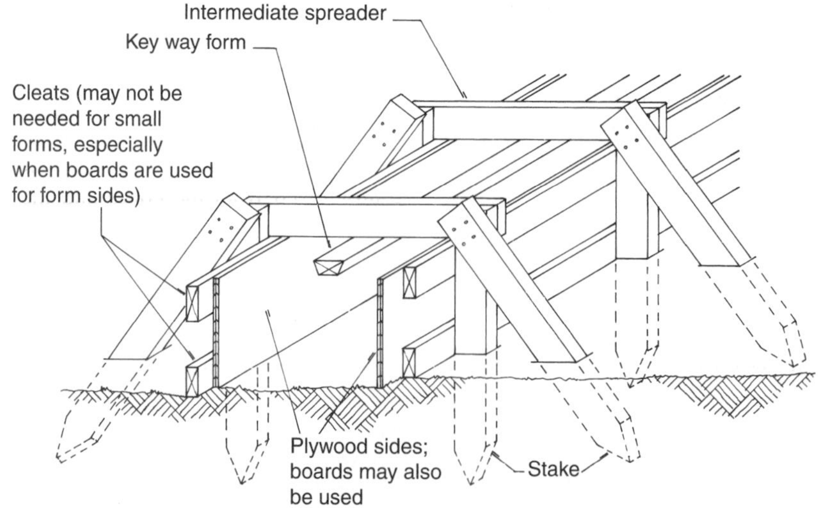

FIGURE 3.3-3 A typical footing form. (Drawing by HLS.)

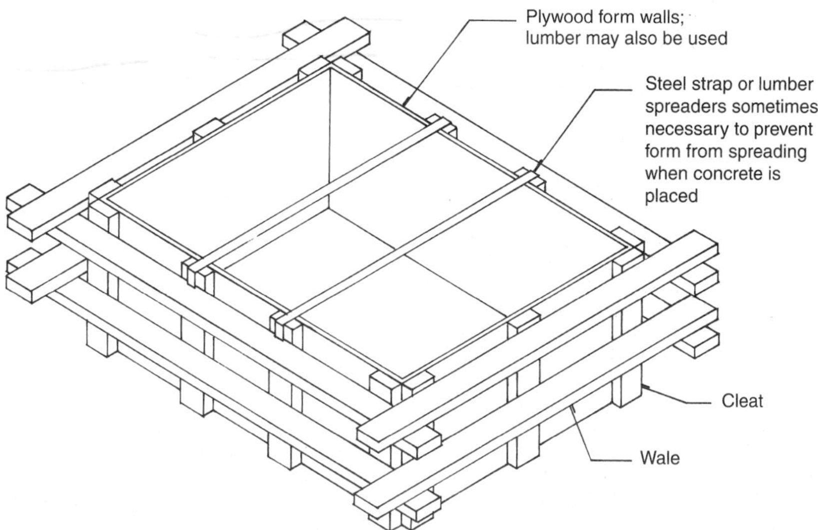

FIGURE 3.3-4 One of several methods commonly used to form large column footings. (Drawing by HLS.)

FIGURE 3.3-5 Typical wall forms. (Courtesy Portland Cement Association.)

ing out through the hole. The ends are then clamped at the wales with slotted tapered wedges. The rod is deformed just inside each washer, so that when the forms have been removed the rod may be twisted until it breaks at the weakened spot. This process leaves a ragged hole in the concrete, which can be filled with mortar.

The cones on *screw ties* have interior threads. These ties are inserted into the forms with the cones against the inner lining of the form. Threaded rods are then passed through holes in the form and screwed into the cones at each wall face. The cones prevent concrete from leaking out through the holes. The threaded rods are then locked into place at the wales by slotted tapered wedges or plates. When the forms are removed, the threaded rods are unscrewed from the cones and the cones are unscrewed from the tie, leaving a clean cone-shaped hole in the concrete, which can be left open or filled with mortar or a sealant.

3.3.2.3 Forming Columns

Columns require a liner, bracing, and yokes, straps, or clamps to hold the form together (Fig. 3.3-8). The straps shown may be replaced by wood yokes held together with threaded rods and bolts, or by a proprietary system using two overlapping angle brackets (Fig. 3.3-9). Larger columns may require additional rows of bracing inside the strapping or brackets.

3.3.2.4 Forming Beams and Girders

Beams and girders are formed with wood or plywood. One way to form light beams is with braces, kickers, and spreaders, as shown in Figure 3.3-10. The bottoms of these forms are supported on the shoring system. A deep beam or girder form is built with ties and diagonal braces, much like a wall form. The bottom is supported on continuous wood cleats turned with the long edges horizontal (as shown) or vertical, depending on the size of the beam or girder. Forms for large beams and girders may also require additional bracing.

3.3.2.5 Forming Floors

Both slabs on grade and supported floors require forming.

SLABS ON GRADE

The edges of and openings through slabs on grade require forming. This is usually done using wood or metal forms staked to the ground, similar to the wall footing form shown in Figure 3.3-3.

SUPPORTED FLOORS

In general, beams and girders are formed first, then the slab is formed. Flat slabs are formed with plywood supported on joists. Pans and domes are supported on a wood frame of joists and plywood called *centering* (Fig. 3.3-11). Alternatively, the plywood beneath pans or domes is omitted and wide boards are placed across the joists at the location of each rib. The forms for beams and floors are supported by a system of temporary wood or metal beams or joists. These are in turn supported by adjustable-length steel columns called *shores*, wood posts, or a series of adjustable ladderlike steel assemblages called *staging*. The carrier shown in the figure is a part of this form support system. The entire support system for forms is called *shoring*.

3.3.2.6 Flying Forms

Flying forms are made by building a large section of form, as described earlier for centering, and supporting the entire section on deep steel trusses. This assembly can then be lifted by crane and positioned in the next location without dismantling it, thus saving cost and speeding construction.

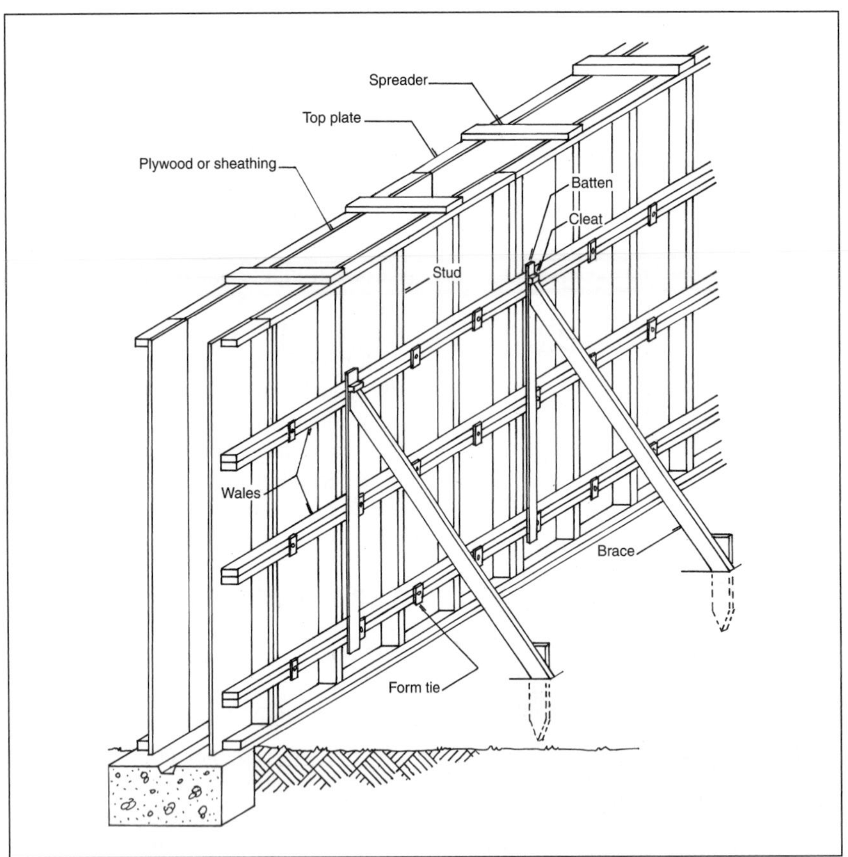

FIGURE 3.3-6 Detail of one type of wall form. (Drawing by HLS.)

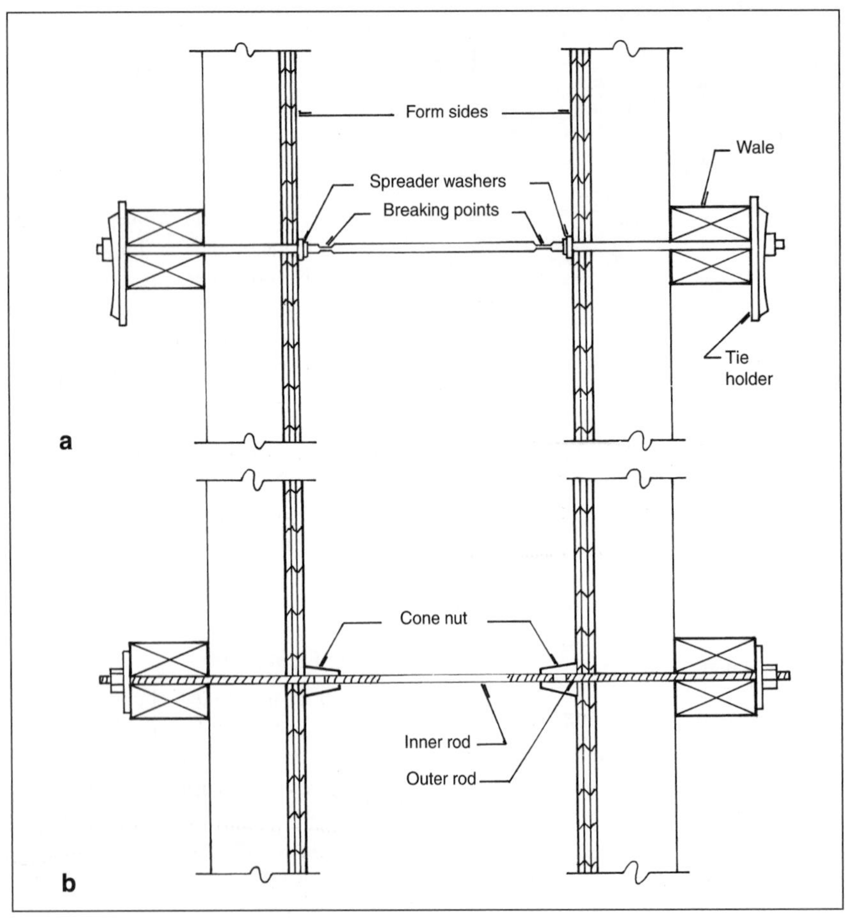

FIGURE 3.3-7 Wall form ties: (a) a snap tie and (b) a screw tie. (Drawing by HLS.)

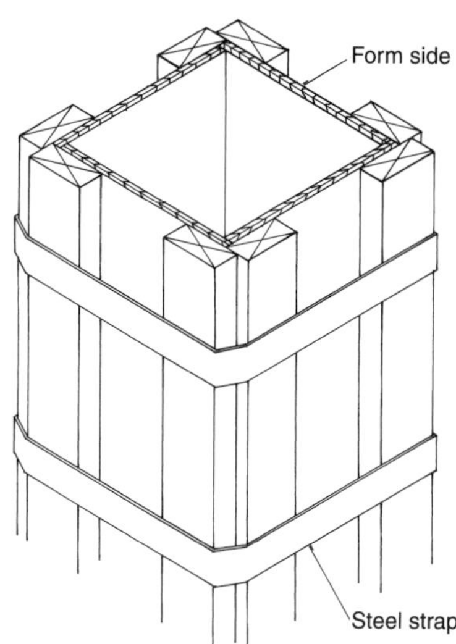

FIGURE 3.3-8 A column form with steel straps. (Drawing by HLS.)

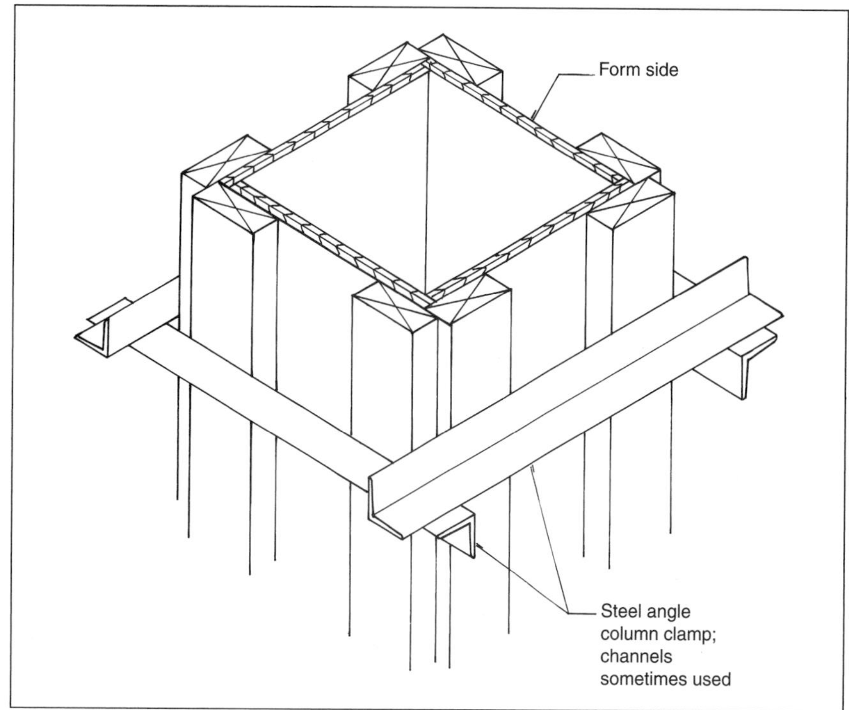

FIGURE 3.3-9 A column form with overlapping angle brackets. (Drawing by HLS.)

3.3.2.7 Slip Forming

Slip forming is a method of continuously moving a form for vertical structures, such as elevator or stair shafts, upward on jacks as new concrete is placed on top of the old.

3.3.3 FORM REMOVAL

Leaving forms in place as long as possible helps in the curing process and protects freshly placed concrete from damage (see Section 3.10.2.1). However, it is sometimes desirable to remove forms from beams and slabs as soon as it can be done safely. For example, patching and repairing of formed surfaces should be done as early as possible, which requires removal of the forms. It is also often necessary to remove forms quickly to permit their reuse.

In any case, forms should not be removed until the concrete has attained sufficient strength to ensure structural stability and to safely carry both the dead load and construction loads that may be imposed on it. Concrete should be hard enough that its surfaces will not be injured in any way when reasonable care is used in removing forms.

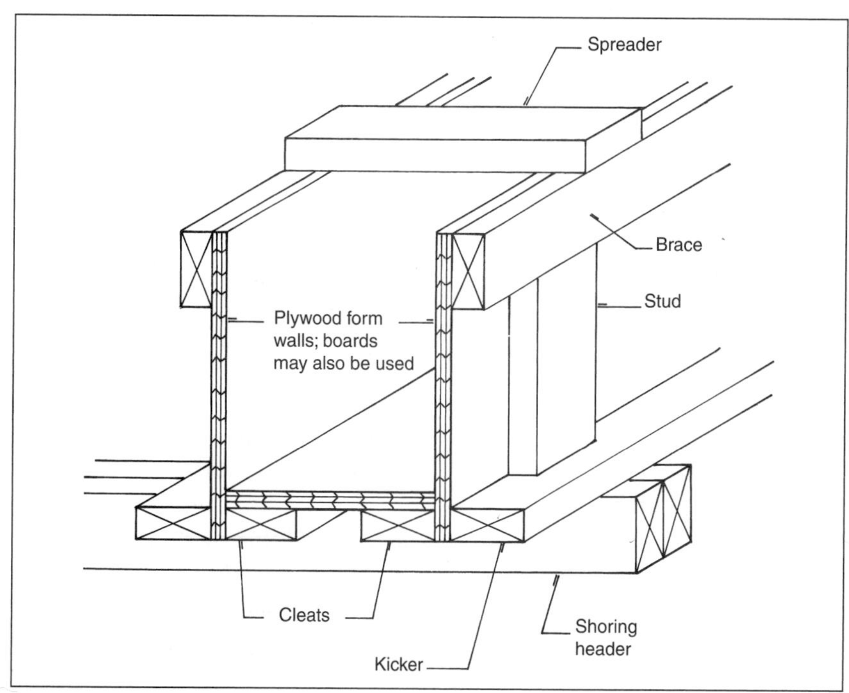

FIGURE 3.3-10 A typical beam form. Heavier beams may need wall ties or diagonal braces back to the header. Additional framing members will also be necessary to support floor slab forms. Floor slab form framing will usually eliminate the need for spreaders. (Drawing by HLS.)

When their forms are removed, slabs, beams, and girders, although strong enough to support themselves, will often not be sufficiently strong for several more weeks to support construction loads that will be imposed on them. This is especially true when additional concrete will be placed on higher floors, whose forms will be at least partially supported by the concrete from which the forms have just been removed. So that work may continue uninterrupted, wood posts or adjustable metal columns called *reshores* are placed beneath slabs and beams from which the forms have recently been removed (see Fig. 3.13-12). These reshores are then left in place until the beams and slabs are able to support the imposed loads.

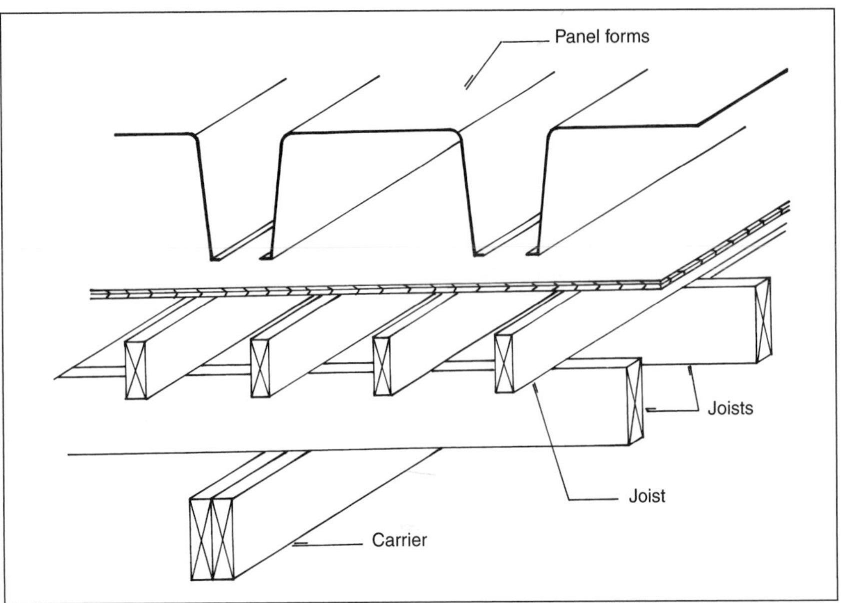

FIGURE 3.3-11 **Typical centering for a supported floor system.** (Drawing by HLS.)

3.4 Reinforcement

Until *reinforced concrete* was developed, concrete could only be used to resist forces that tended to compress it (compressive loads) (Fig. 3.4-1). Concrete is strong in resisting *compressive* (pushing) loads, but its ability to resist *tensile* (pulling) forces is practically nonexistent. All slabs, beams, and girders, and even some columns and walls, are subject to both compressive loads (*compression*) and tensile loads (*tension*) caused by bending (see Fig. 3.1-1), drying shrinkage, temperature changes, and vertical and horizontal shear. The net result of vertical and horizontal shear forces is called *diagonal tension*. Tensile forces that result from bending will cause an unreinforced concrete beam or slab to crack along the bottom and collapse (see Fig. 3.1-1a). Diagonal tension will cause a crack at the support, which will act at an approximate 45-degree angle (Fig. 3.4-2). Diagonal tension at a column or other support is called *punching shear*.

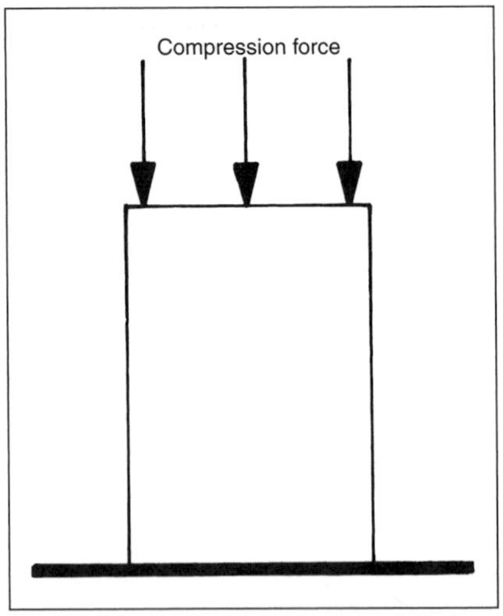

Compression force

FIGURE 3.4-1 **Concrete is best at supporting compression loads.** (Drawing by HLS.)

The requirements in this section apply in principle to all reinforced concrete. Specific requirements for reinforcing steel in different types of slabs, beams, girders, walls, and columns are addressed in other sections of this chapter.

3.4.1 PURPOSE OF CONCRETE REINFORCEMENT

The concept of *reinforced concrete* is simple. Concrete resists compression forces; steel bars resist tensile forces and shear. When both are placed in the same structural member (slab, wall, beam, girder, or column) the steel will resist the tensile, bending, and shear forces that would otherwise tear the concrete apart, and the concrete will resist applied compression forces that would bend the steel.

The theory is simple; the practice is not. The location and amount of steel in a concrete structural component are critical and vary with the type of component, the loads that will be applied, whether the unit is prestressed or not, and whether the unit is cast in place or precast. For example, the reinforcement required in a simple (single-span) beam varies considerably from the reinforcement in a beam that is continuous across several spans.

Other complications also occur. For instance, because steel is also strong in compression, it is sometimes used to resist a portion of the applied compression loads. This often occurs in columns, and in beams and girders when limiting their depth is important.

In addition, steel is used in concrete even when it is not required to resist tensile and shear forces. For example, slabs on grade simply pass loads through them to the underlying base. Structural cracks in slabs on grade are due to a failure in the base and not in the concrete or its lack of reinforcement. However, slabs on grade do tend to crack as a result of drying shrinkage and when they are subjected to widely varying temperatures. Therefore, a layer of light welded wire fabric is often added to them. This is not structural reinforcement, and should not be counted on to help a slab span depressions in the slab base. While it will not prevent cracking, it will hold cracks together so they are less unsightly. Some practitioners believe that welded wire fabric is not necessary in

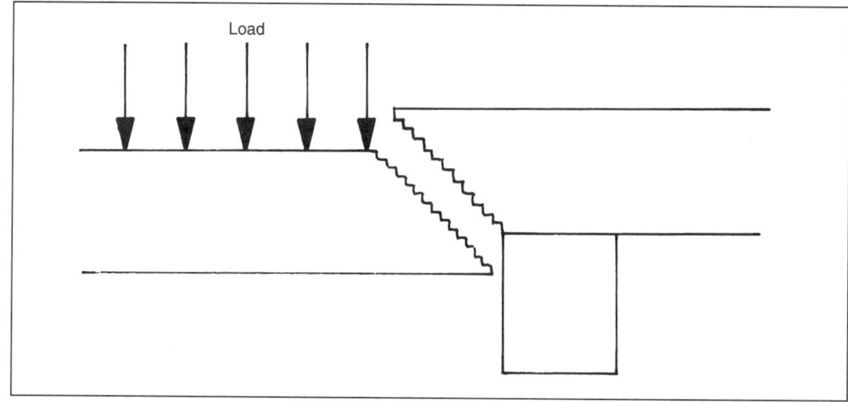

FIGURE 3.4-2 Diagonal tension (shear) failure. (Drawing by HLS.)

slabs on grade and routinely omit it. The Portland Cement Association's *Concrete Floors on Ground* suggests that welded wire fabric may not be necessary when the slab is uniformly supported and joints are located close together. The spacing of joints that would permit deletion of fabric is not specified.

Reinforcing bars are also needed in other types of slabs to resist cracks caused by drying shrinkage and temperature changes. This steel is called *temperature steel.* When structural reinforcement occurs, it acts to resist these stresses, and temperature steel is not required. When structural reinforcement is placed in one direction only, temperature steel is required in the perpendicular direction (see Section 3.13.4).

The use of concrete and steel together is made possible by the chemical compatibility of the two materials. Concrete bonds very strongly to steel, and this bond is not broken by thermal movement because the coefficients of expansion of steel and concrete are close to the same. The alkalis in concrete protect steel from corroding.

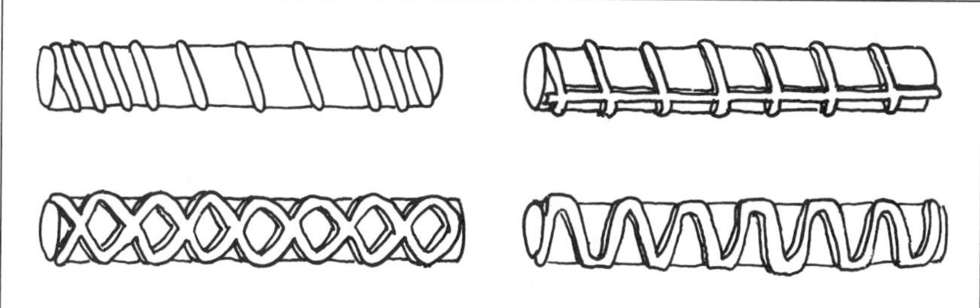

FIGURE 3.4-3 Typical deformed steel bars. (Drawing by HLS.)

3.4.2 MATERIALS

Steel reinforcement is either *structural reinforcement* or *temperature reinforcement.* Structural reinforcement resists primary tensile, shear, and compression forces. It consists mainly of steel bars, but welded wire fabric is also used in structural slabs and walls and in some precast concrete elements.

Temperature reinforcement holds together cracks that occur from drying shrinkage and from expansion and contraction due to temperature variations. It consists of light welded wire fabric in slabs on grade and of small bars in supported slabs. Synthetic fiber reinforcement is also sometimes added to concrete slabs to help prevent plastic and shrinkage cracking (see Section 3.2.4.5).

3.4.2.1 Reinforcing Bars

Reinforcing bars are either plain or *deformed* (Fig. 3.4-3). Deformations im-

FIGURE 3.4-4 Deformed Bar Designation Numbers, Nominal Weights, Nominal Dimensions, and Deformation Requirements

		Nominal Dimensions[a]			Deformation Requirements, in.		
Bar Designation No.[b]	Nominal Weight, lb./ft.	Diameter, in.	Cross-Sectional Area, sq. in.	Perimeter, in.	Maximum Average Spacing	Minimum Average Height	Maximum Gap (Chord of 12½% of Nominal Perimeter)
3	0.376	0.375	0.11	1.178	0.262	0.015	0.143
4	0.668	0.500	0.20	1.571	0.350	0.020	0.191
5	1.043	0.625	0.31	1.963	0.437	0.028	0.239
6	1.502	0.750	0.44	2.356	0.525	0.038	0.286
7	2.044	0.875	0.60	2.749	0.612	0.044	0.334
8	2.670	1.000	0.79	3.142	0.700	0.050	0.383
9	3.400	1.128	1.00	3.544	0.790	0.056	0.431
10	4.303	1.270	1.27	3.990	0.889	0.064	0.487
11	5.313	1.410	1.56	4.430	0.987	0.071	0.540
14	7.65	1.693	2.25	5.32	1.185	0.085	0.648
18	13.60	2.257	4.00	7.09	1.58	0.102	0.864

[a]The nominal dimensions of a deformed bar are equivalent to those of a plain round bar having the same weight per foot as the deformed bar.
[b]Bar numbers are based on the number of eighths of an inch included in the nominal diameter of the bars.

prove the bond between steel and concrete. They are made in 11 standard sizes (Fig. 3.4-4) and in several grades (Fig. 3.4-5). They should comply with ASTM A 615, A 616, A 617, or A 706. Most concrete reinforcement for building construction is made from Grade 40 or Grade 60, deformed new billet or ingot steel bars, conforming to ASTM A 615.

Reinforcing bars typically have numbers and letters stamped on them to identify the producing mill, bar size, type of steel, and grade (Fig. 3.4-6).

3.4.2.2 Welded Wire Fabric

Welded wire fabric is made by welding either smooth or deformed reinforcing bars into a square- or rectangular-shaped mesh. Welded wire fabric is designated by the size and spacing of the reinforcing bars (*wires*) used to make it. An example of the designation system is WWF 6 × 12–W16 × W8, which denotes welded wire fabric (WWF) with a 6-in. (152 mm) spacing between longitudinal wires and a 12-in. (305 mm) spacing between transverse wires, with smooth W16 longitudinal wires and smooth W8 transverse wires. The W represents smooth wire. The letter D substituted for the W would represent deformed wires. The number following the W (or D) is the cross-sectional area of the wire in hundredths of an inch. Figure 3.4-7 shows WWF styles designated by the Wire Reinforcement Institute as common stock styles. Figure 3.4-8 shows the nominal wire diameters and weights of steel associated with size numbers. The normal welded wire fabric used in light slabs on grade is WWF 6 × 6–W1.4 × W1.4.

Lighter gauges of welded wire fabric are available in rolls; heavier gauges in sheets (see Fig. 3.4-7).

3.4.2.3 Accessories

Accessories are used to support structural steel and to make splices in it. The devices shown in Figure 3.4-9 are used to support reinforcing steel in forms. Welded wire fabric in slabs on grade should be supported on similar devices having flat plates (*sand plates*) at their legs to prevent the legs from sinking into the slab base. The common practice of supporting reinforcement in slabs on grade using pieces of concrete or concrete bricks is discouraged because concrete often does not adhere to such materials. This produces cracks, which permit water to reach and rust the reinforcement, destroying its effectiveness. The plastic caps shown on the feet of the support devices in the figure are used when the underside of the slab will be exposed, and in other locations to prevent the supports from rusting and transferring this rust to the reinforcement.

In beams, girders, walls, and slabs, when bar space is not at a premium, splices are usually made by lapping the bars a prescribed number of bar diameters and tying them together with wire to keep them in place until the concrete has set. Sometimes, in columns when space is limited, bars are butted end to end and connected using special devices made for the purpose or are welded together.

3.4.3 DESIGN AND PLACEMENT

The primary standards for reinforcing steel design, fabrication, and installation is the Concrete Reinforcing Steel Institute publication *Manual of Standard Practice* and the American Welding Society publications D1.4, "Structural Welding Code-Reinforcing Steel," and "Connections in Reinforced Concrete Construction."

Reinforcing steel is usually fabricated in a mill or shop following shop drawings prepared by the fabricator from the structural engineer's drawings and approved by the structural engineer and the architect. The fabricated steel is bundled, tagged as to its location in the building, and delivered to the construction site. Details of *bends* and *hooks* are standard (Figs. 3.4-10 and 3.4-11). Bundles are delivered to their place in the building by

FIGURE 3.4-5 ASTM Standards Covering Various Bars by Grade

Material	ASTM Standard	Grade	Yield Strength, psi (MPa)	Bar Sizes Available
Billet or ingot steel	A 615	40	40,000 (275.6)	3–6
	A 615	60	60,000 (413.4)	3–11, 14, 18
	A 615	75	75,000 (516.75)	11, 14, 18
Rail steel	A 616	50	50,000 (344.5)	3–11
	A 616	60	60,000 (413.4)	3–11
Axle steel	A 617	40	40,000 (275.6)	3–11
	A 617	60	60,000 (413.4)	3–11

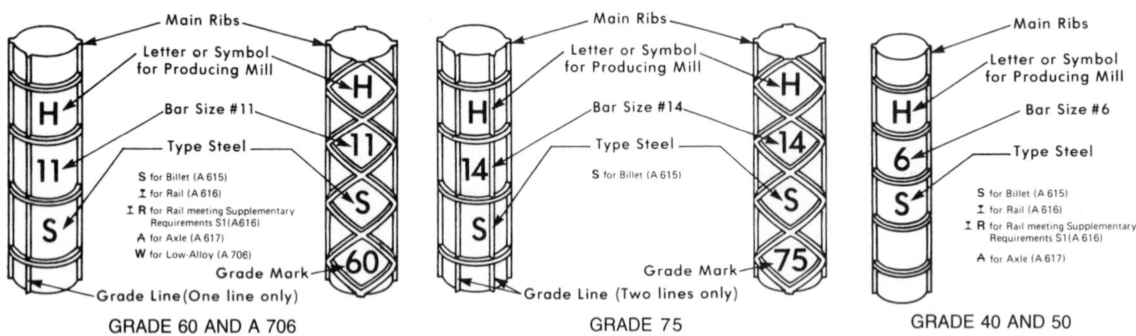

FIGURE 3.4-6 Standard markings on steel bars. (Courtesy Concrete Reinforcing Steel Institute.)

FIGURE 3.4-7 Common Stock Styles of Welded Wire Fabric

Style Designation		Steel Area, sq. in./ft. (mm²/m²)		Approximate Weight, lb./100 sq. ft. (kg/m²)
New Designation (by W-number)	Old Designation (by Steel Wire Gauge)	Longitudinal	Transverse	
Rolls				
6 × 6–W1.4 × W1.4	6 × 6–10 × 10	.03 (208.33)	.03 (208.33)	21 (1.03)
6 × 6–W2 × W2	6 × 6–8 × 8[a]	.04 (277.78)	.04 (277.78)	29 (1.42)
6 × 6–W2.9 × W2.9	6 × 6–6 × 6	.06 (416.67)	.06 (416.67)	42 (2.05)
6 × 6–W4 × W4	6 × 6–4 × 4	.08 (555.56)	.08 (555.56)	58 (2.83)
4 × 4–W1.4 × W1.4	4 × 4–10 × 10	.04 (277.78)	.04 (277.78)	31 (1.51)
4 × 4–W2 × W2	4 × 4–8 × 8[a]	.06 (416.67)	.06 (416.67)	43 (2.10)
4 × 4–W2.9 × W2.9	4 × 4–6 × 6	.09 (625)	.09 (625)	62 (3.03)
4 × 4–W4 × W4	4 × 4–4 × 4	.12 (833.33)	.12 (833.33)	86 (4.20)
Sheets				
6 × 6–W2.9 × W2.9	6 × 6–6 × 6	.06 (416.67)	.06 (416.67)	42 (2.05)
6 × 6–W4 × W4	6 × 6–4 × 4	.08 (555.56)	.08 (555.56)	58 (2.83)
6 × 6–W5.5 × W5.5	6 × 6–2 × 2[b]	.11 (763.89)	.11 (763.89)	80 (3.11)
4 × 4–W4 × W4	4 × 4–4 × 4	.12 (833.33)	.12 (833.33)	86 (4.20)

[a]Exact W-number size for 8-gauge is W2.1.
[b]Exact W-number size for 2-gauge is W5.4.
(Courtesy Concrete Reinforcing Steel Institute.)

hand or crane. The reinforcement is then wired in place to maintain its location and alignment until the concrete has set. In some cases, sections of reinforcement, such as those for a column, are preassembled and delivered to the site in one piece. Such assemblies may also be made up at the site on the ground and lifted into place by crane.

Steel should be accurately positioned, secured to prevent its displacement during the placing of concrete, and cleaned of rust and foreign matter before concrete is placed against it. Steel should be protected from possible harm by fire and corrosion by covering it with at least the thickness of concrete shown in Figure 3.4-12.

Welded wire fabric should be installed in the largest practicable length. Adjoining pieces should be lapped at least one full mesh and tied together with wire to prevent movement during concrete placement.

FIGURE 3.4-8 Sectional Area and Weight of Welded Wire Fabric

Wire Size Number[a]		Nominal Diameter,[b] in.	Nominal Weight, lb./lin. ft.	Area of Width for Various Spacings, sq. in./ft. — Center-to-Center Spacing						
Plain	Deformed			2 in.	3 in.	4 in.	6 in.	8 in.	10 in.	12 in.
W20	D20	0.505	.680	1.20	.80	.60	.40	.30	.24	.20
W18	D18	0.479	.612	1.08	.72	.54	.36	.27	.216	.18
W16	D16	0.451	.544	.96	.64	.48	.32	.24	.192	.16
W14	D14	0.422	.476	.84	.56	.42	.28	.21	.168	.14
W12	D12	0.391	.408	.72	.48	.36	.24	.18	.144	.12
W11	D11	0.374	.374	.66	.44	.33	.22	.165	.132	.11
W10.5		0.366	.357	.63	.42	.315	.21	.157	.126	.105
W10	D10	0.357	.340	.60	.40	.30	.20	.15	.12	.10
W9.5		0.348	.323	.57	.38	.285	.19	.142	.114	.095
W9	D9	0.338	.306	.54	.36	.27	.18	.135	.108	.09
W8.5		0.329	.289	.51	.34	.255	.17	.127	.102	.085
W8	D8	0.319	.272	.48	.32	.24	.16	.12	.096	.08
W7.5		0.309	.255	.45	.30	.225	.15	.112	.09	.075
W7	D7	0.299	.238	.42	.28	.21	.14	.105	.084	.07
W6.5		0.288	.221	.39	.26	.195	.13	.097	.078	.065
W6	D6	0.276	.204	.36	.24	.18	.12	.09	.072	.06
W5.5		0.265	.187	.33	.22	.165	.11	.082	.066	.055
W5	D5	0.252	.170	.30	.20	.15	.10	.075	.06	.05
W4.5		0.239	.153	.27	.18	.135	.09	.067	.054	.045
W4	D4	0.226	.136	.24	.16	.12	.08	.06	.048	.04
W3.5		0.211	.119	.21	.14	.105	.07	.052	.042	.035
W3		0.195	.102	.18	.12	.09	.06	.045	.036	.03
W2.9		0.192	.099	.174	.116	.087	.058	.043	.035	.029
W2.5		0.178	.085	.15	.10	.075	.05	.037	.03	.025
W2.1		0.162	.070	.126	.084	.063	.042	.031	.025	.021
W2		0.160	.068	.12	.08	.06	.04	.03	.024	.02
W1.5		0.138	.051	.09	.06	.045	.03	.022	.018	.015
W1.4		0.134	.048	.084	.056	.042	.028	.021	.017	.014

Note: The above listing of smooth and deformed wire sizes represents wires normally selected to manufacture welded wire fabric styles to specific areas of reinforcement. Wire sizes other than those listed here, including larger sizes, may be available if the quantity required is sufficient to justify manufacture.
[a]The number following the prefix W or the prefix D identifies the cross-sectional area of the wire in hundredths of a square inch.
[b]The nominal diameter of a deformed wire is equivalent to the diameter of a smooth wire having the same weight per foot as the deformed wire.
(Courtesy Concrete Reinforcing Steel Institute.)

SYMBOL	BAR SUPPORT ILLUSTRATION	BAR SUPPORT ILLUSTRATION PLASTIC CAPPED OR DIPPED	TYPE OF SUPPORT	TYPICAL SIZES
SB	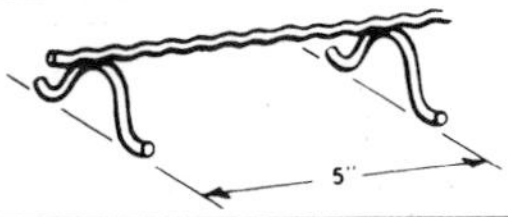 5"	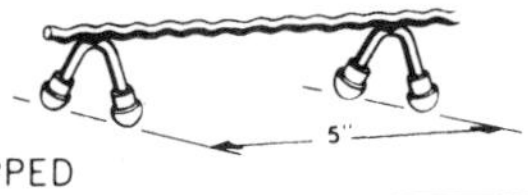 CAPPED 5"	Slab Bolster	¾, 1, 1½, and 2 inch heights in 5 ft. and 10 ft. lengths
SBU*	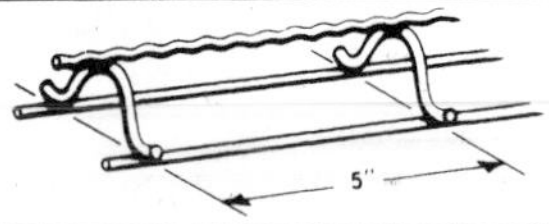 5"		Slab Bolster Upper	Same as SB
BB	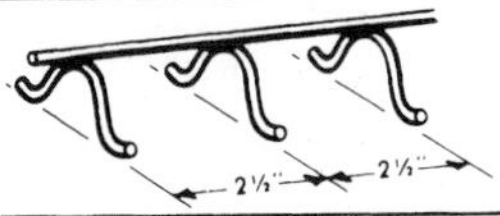 2½" 2½"	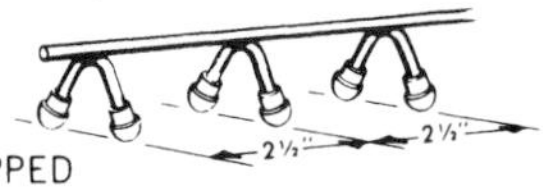 CAPPED 2½" 2½"	Beam Bolster	1, 1½, 2, over 2" to 5" heights in increments of ¼" in lengths of 5 ft.
BBU*	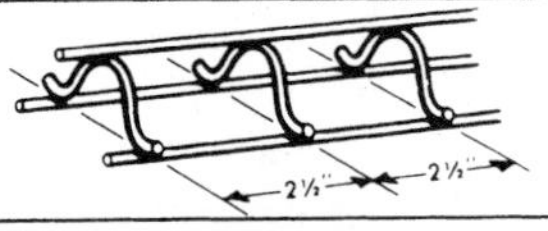 2½" 2½"		Beam Bolster Upper	Same as BB
BC	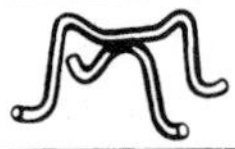	DIPPED 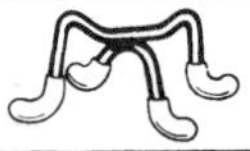	Individual Bar Chair	¾, 1, 1½, and 1¾" heights
JC	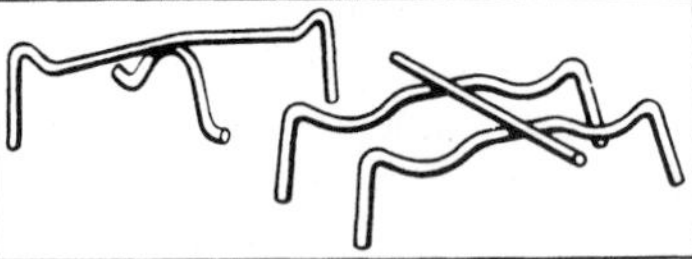	 DIPPED DIPPED	Joist Chair	4, 5, and 6 inch widths and ¾, 1 and 1½ inch heights
HC	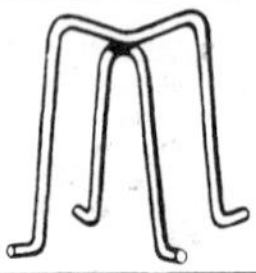	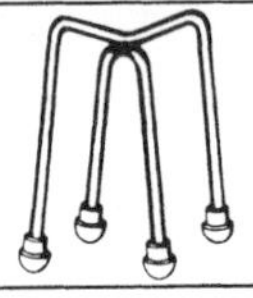 CAPPED	Individual High Chair	2 to 15 inch heights in increments of ¼ inch
HCM*	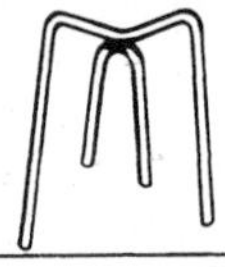		High Chair for Metal Deck	2 to 15 inch heights in increments of ¼ in.
CHC	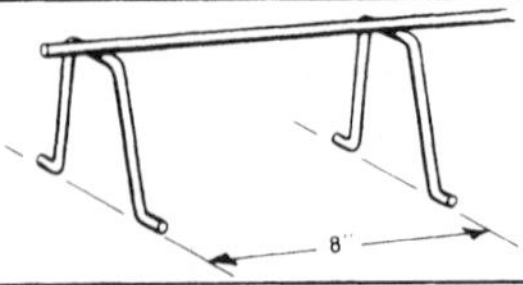 8"	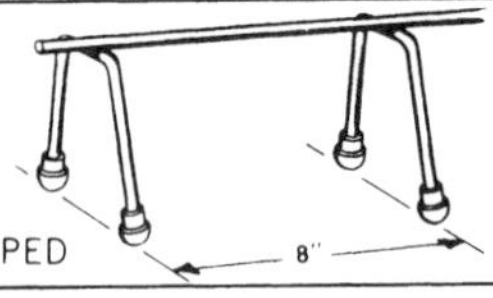 CAPPED 8"	Continuous High Chair	Same as HC in 5 foot and 10 foot lengths
CHCU*	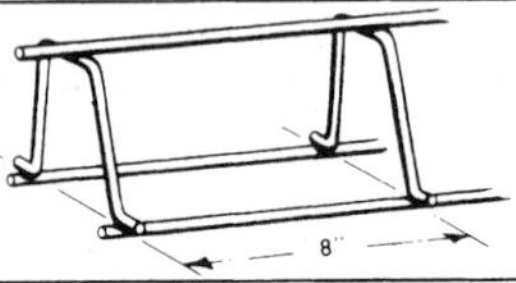 8"		Continuous High Chair Upper	Same as CHC
CHCM*	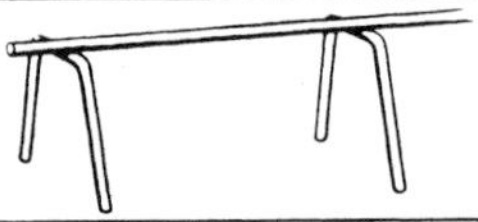		Continuous High Chair for Metal Deck	Up to 5 inch heights in increments of ¼ in.
JCU**	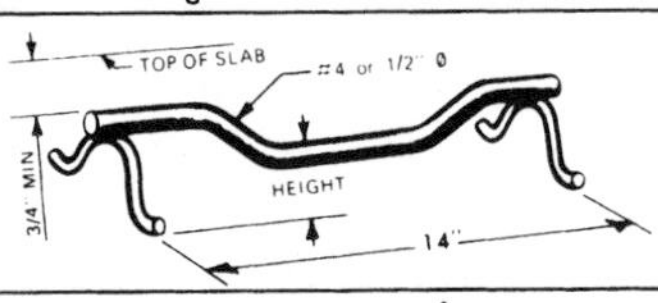 TOP OF SLAB #4 or 1/2" ⌀ HEIGHT 3/4" MIN 14"	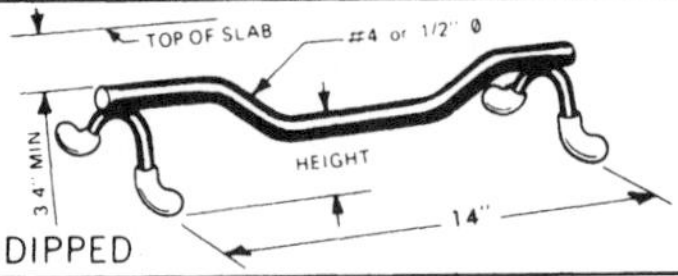 TOP OF SLAB #4 or 1/2" ⌀ HEIGHT 3/4" MIN 14" DIPPED	Joist Chair Upper	14" Span Heights −1" through +3½" vary in ¼" increments
CS	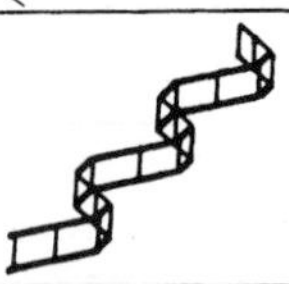		Continuous Support	1½" to 12" in increments of ¼" in lengths of 6′-8″

*Usually available in Class 3 only, except on special order.
**Usually available in Class 3 only, with upturned or end bearing legs.

FIGURE 3.4-9 Typical types and sizes of wire bar supports. (Courtesy Concrete Reinforcing Steel Institute.)

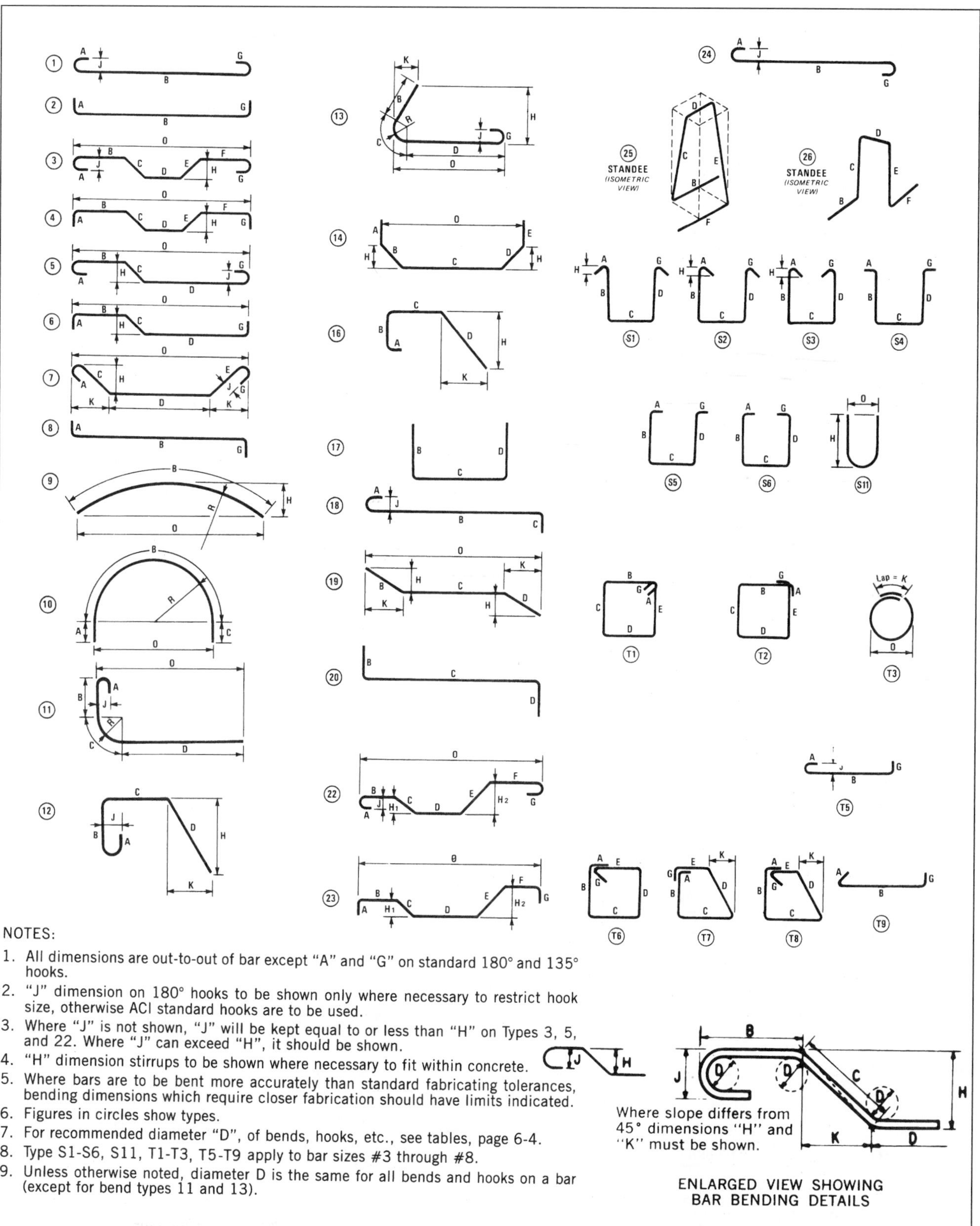

STANDEE (ISOMETRIC VIEW)
NOTES:
1. All dimensions are out-to-out of bar except "A" and "G" on standard 180° and 135° hooks.
2. "J" dimension on 180° hooks to be shown only where necessary to restrict hook size, otherwise ACI standard hooks are to be used.
3. Where "J" is not shown, "J" will be kept equal to or less than "H" on Types 3, 5, and 22. Where "J" can exceed "H", it should be shown.
4. "H" dimension stirrups to be shown where necessary to fit within concrete.
5. Where bars are to be bent more accurately than standard fabricating tolerances, bending dimensions which require closer fabrication should have limits indicated.
6. Figures in circles show types.
7. For recommended diameter "D", of bends, hooks, etc., see tables, page 6-4.
8. Type S1-S6, S11, T1-T3, T5-T9 apply to bar sizes #3 through #8.
9. Unless otherwise noted, diameter D is the same for all bends and hooks on a bar (except for bend types 11 and 13).
Where slope differs from 45° dimensions "H" and "K" must be shown.
ENLARGED VIEW SHOWING BAR BENDING DETAILS

FIGURE 3.4-10 **Typical bar bends.** (Courtesy Concrete Reinforcing Steel Institute.)

All specific sizes recommended by CRSI below meet minimum requirements of ACI 318

RECOMMENDED END HOOKS
All Grades

D=Finished bend diameter

Bar Size	D	180° HOOKS A or G	180° HOOKS J	90° HOOKS A or G
# 3	2¼	5	3	6
# 4	3	6	4	8
# 5	3¾	7	5	10
# 6	4½	8	6	1-0
# 7	5¼	10	7	1-2
# 8	6	11	8	1-4
# 9	9½	1-3	11¾	1-7
#10	10¾	1-5	1-1¼	1-10
#11	12	1-7	1-2¾	2-0
#14	18¼	2-3	1-9¾	2-7
#18	24	3-0	2-4½	3-5

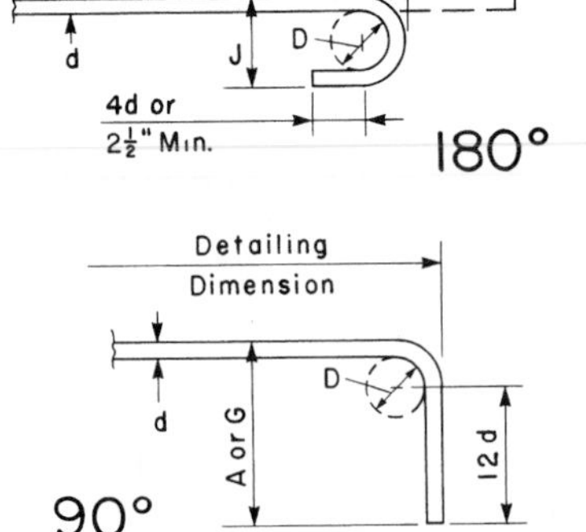

Detailing
Dimension
D
d
A or G
12d
90°

STIRRUP AND TIE HOOKS

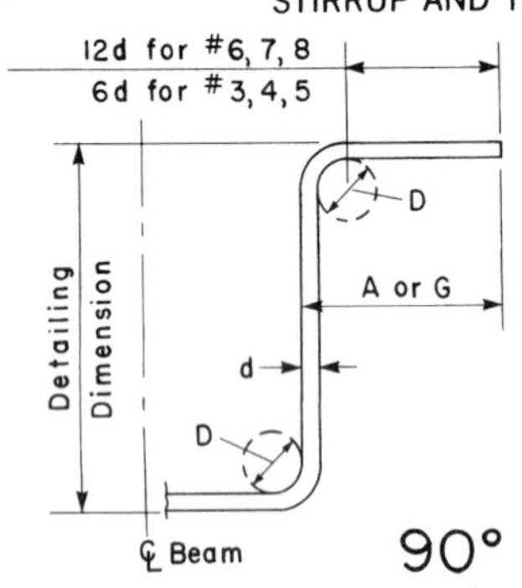

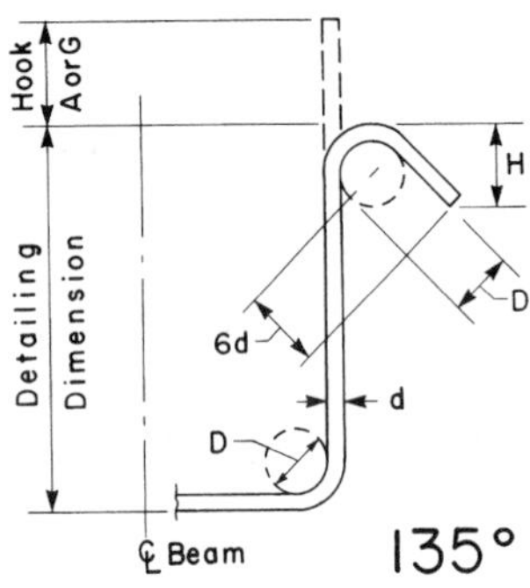

STIRRUPS
(TIES SIMILAR)

135° SEISMIC STIRRUP/TIE HOOKS

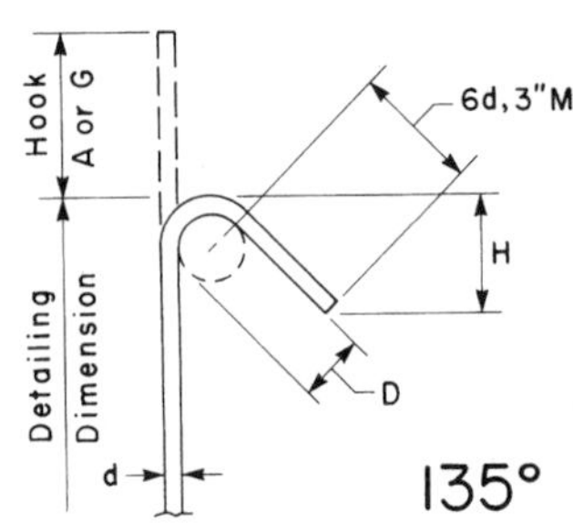

STIRRUP AND TIE HOOK DIMENSIONS
Grades 40-50-60

Bar Size	D (in.)	90° Hook: Hook A or G	135° Hook: Hook A or G	135° Hook: H Approx.
#3	1½	4	4	2½
#4	2	4½	4½	3
#5	2½	6	5½	3¾
#6	4½	1-0	8	4½
#7	5¼	1-2	9	5¼
#8	6	1-4	10½	6

135° SEISMIC STIRRUP/TIE HOOK DIMENSIONS
Grades 40-50-60

Bar Size	D (in.)	135° Hook: Hook A or G	135° Hook: H Approx.
#3	1½	4¼	3
#4	2	4½	3
#5	2½	5½	3¾
#6	4½	8	4½
#7	5¼	9	5¼
#8	6	10½	6

NOTES:

1. 180° hook J dimension (sizes #10, #11, #14 and #18), and A or G dimension (#14 and #18) have been revised to reflect test research using ASTM/ACI bend test criteria as a minimum.
2. Tables for Stirrup and Tie Hook dimensions have been expanded to include sizes #6, #7, and #8 to reflect current design practices.

FIGURE 3.4-11 Standard hooks. (Courtesy Concrete Reinforcing Steel Institute.)

FIGURE 3.4-12 Recommended Minimum Concrete Cover for Reinforcing Steel

Location and Use	Minimum Cover
Against ground without forms	3 in.
Exposed to weather or ground but placed in forms	
Greater than ⅝-in.-diameter bars	2 in.
Less than ⅝-in.-diameter bars	1½ in.
Slabs and walls (no exposure)	¾ in.
Beams, girders, columns (no exposure)	1½ in.

(Courtesy Portland Cement Association.)

3.5 Accessories

Many types of accessories are used in concrete construction. These include forming accessories (see Section 3.3); reinforcement accessories (see Section 3.4); anchors and fasteners; sleeves; joint inserts; reglets; and waterstops.

3.5.1 ANCHORS AND FASTENERS

Anchors and fasteners are used to anchor other materials, such as wood, stone, and masonry, to concrete. They range from simple threaded anchor bolts to elaborate adjustable inserts. Figures 3.5-1 through 3.5-6 show a few of the many types of anchors and fasteners available for use in concrete.

Figure 3.5-1 shows a simple anchor bolt. The bent portion turns out into the concrete to hold the bolt in place. Anchor bolts are used for fastening precast concrete, metal, and wood construction to concrete. Examples are precast concrete columns, steel column plates, and wood plates. The diameter and length of such bolts vary, depending on the loads that will be applied and the construction materials they will anchor.

Figure 3.5-2 shows two of many available concrete insert configurations. Figure 3.5-2a is a *shelf-angle insert* designed to receive a special bolt. Figure 3.5-2b is an *internally threaded insert*, which receives a standard bolt. In each case, the insert is fastened to the inside face of the forms before concrete is placed. The shelf-angle insert shown is one of many available configurations, all of which work in essentially the same way. An anchor bolt with a special head designed to fit the insert is placed into the cross slot on the insert and dropped into the vertical slot. The device to be supported is slipped over the bolt and the washer and nut are installed and tightened, pulling the anchor head against the inside of the insert. With a threaded insert, a bolt is passed through the item to be supported into the threaded insert. Then a washer and nut are installed and the nut is tightened.

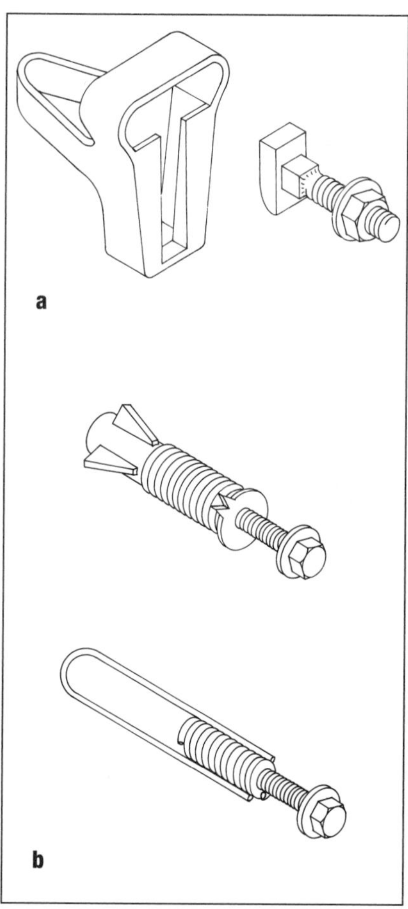

FIGURE 3.5-2 Concrete inserts are installed in formwork so that they finish flush with the concrete: (a) shelf-angle insert and anchor; (b) two types of light-duty internally threaded inserts. (Drawing by HLS.)

Figure 3.5-3 shows a dovetail slot and two of the many types of anchors that can be used in such a slot. These are used to anchor masonry and stone to concrete. The slot is formed from galvanized steel, copper, or stainless steel. The anchors may be galvanized steel or stainless steel. Dovetail slots are available with a felt or plastic-foam filler to prevent intrusion of concrete and other materials, or with no filler. Dovetail slots are nailed to the inside of the form before the concrete is placed. The wedge-shaped ends of the anchors are slipped into the slot vertically and twisted to lock them in place.

Figure 3.5-4 shows two types of expansion anchors. Figure 3.5-4a is representative of the many types of heavy-duty expansion anchors available. They all work in essentially the same way. A hole is drilled in the concrete after it has set. The anchor is placed in the hole. The bolt

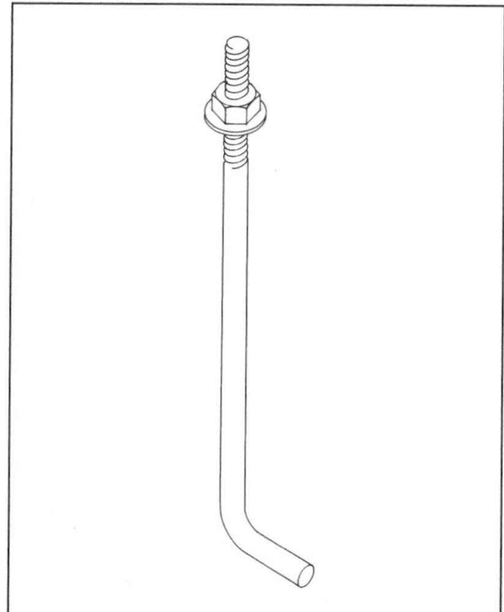

FIGURE 3.5-1 Threaded anchor bolt with nut and washer. (Drawing by HLS.)

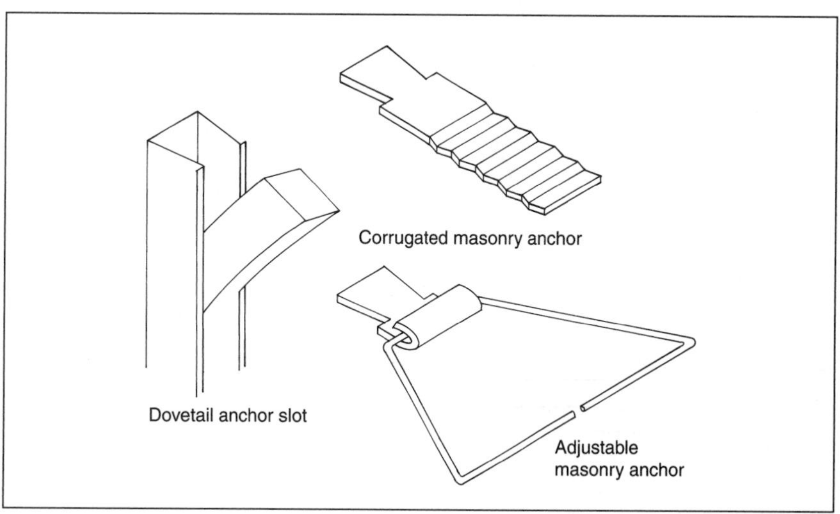

FIGURE 3.5-3 Dovetail anchor slots are designed to receive anchors for stone and masonry. (Drawing by HLS.)

furnished with the anchor is passed through the item to be supported and into the anchor shell. As the bolt is tightened, the expanding portion of the anchor is forced apart, wedging the anchor securely into the drilled hole.

Figure 3.5-4b shows one of many types of light-duty expansion anchors. The one shown has a *drive pin*; others have bolts or screws. Light-duty expansion anchors are placed in drilled holes in set concrete. A pin, screw, or bolt is passed through the item to be supported and into the anchor. A drive pin is driven with a hammer. Screws are tightened with a screwdriver; bolts with a wrench. As the pin, screw, or bolt is forced into the anchor, the anchor expands, locking itself into the drilled hole.

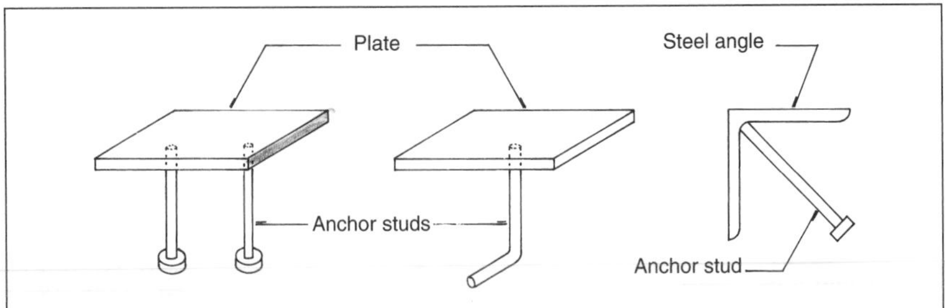

FIGURE 3.5-6 Various types of anchor plates and angles. (Drawing by HLS.)

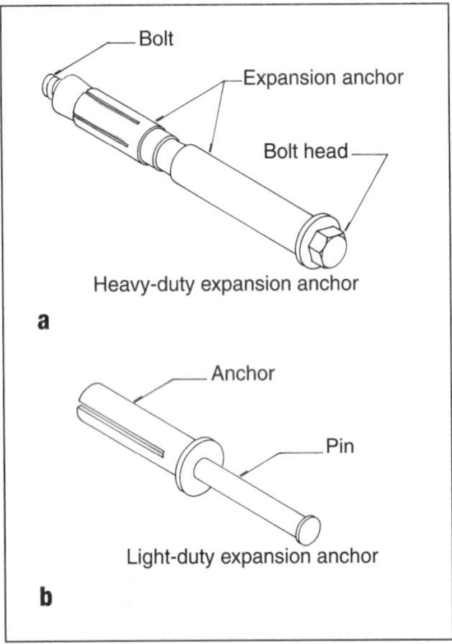

FIGURE 3.5-4 Expansion anchors. (Drawing by HLS.)

Figure 3.5-5 shows two types of powder-actuated fasteners. These are driven into concrete by an exploding charge of gunpowder in a gun specially designed for the purpose. Figure 3.5-5a is a drive pin that is driven through the item to be supported into the concrete. Figure 3:5-5b is a stud, which has threads so that a supported item can be held in place with washers and nuts.

Figure 3.5-6 shows several configurations of anchor plates and angle clips used in concrete construction. These devices are fastened to the forms before concrete is placed, with the anchors shown toward the concrete. Bolts may be welded to plate and angle clips to support or anchor other construction, or other steel shapes may be welded to them.

Other anchor types not shown include hammer- and powder-driven nails and screws.

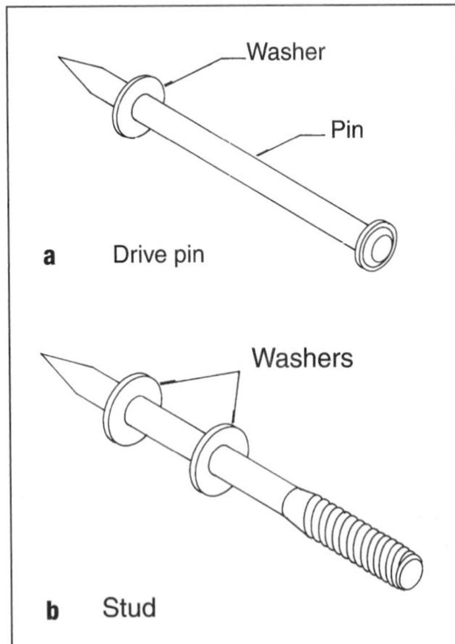

FIGURE 3.5-5 Powder-actuated fasteners. (Drawing by HLS.)

3.5.2 SLEEVES

Metal, cardboard, and plastic sleeves are used to produce holes in concrete for the insertion of pipes, railings, and other items that pass through or are set into concrete. Openings for banks of several pipes, conduits, and the like are often formed with a wood box set into the form. Openings for single pipes are usually formed with galvanized steel pipes or cardboard or plastic tubes. When railing standards or other items that must be wedged in tightly for support are placed in sleeves, paper and concrete materials are removed; metal sleeves are left in place (Fig. 3.5-7).

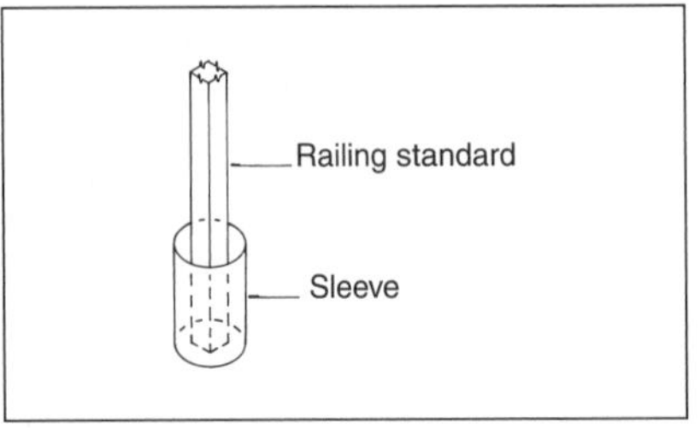

FIGURE 3.5-7 A sleeve for use in concrete. (Drawing by HLS.)

3.5.3 JOINT INSERTS

Joint inserts are used to form control joints in slabs. Figure 3.5-8 shows a typical tongue-and-groove control-joint insert for use in slabs on grade. These inserts are made of either plastic or galvanized steel. Similar inserts are used in supported slabs.

Similar inserts are also used to form keyways in concrete to permit bonding to adjacent concrete or masonry (Fig. 3.5-9).

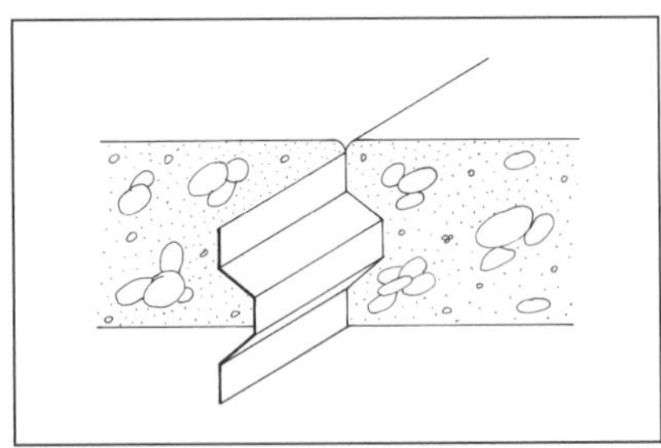

FIGURE 3.5-8 Control-joint insert. (Drawing by HLS.)

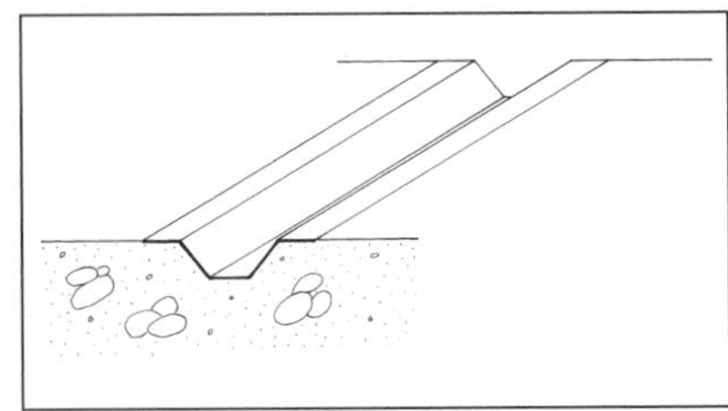

FIGURE 3.5-9 Keyway insert. (Drawing by HLS.)

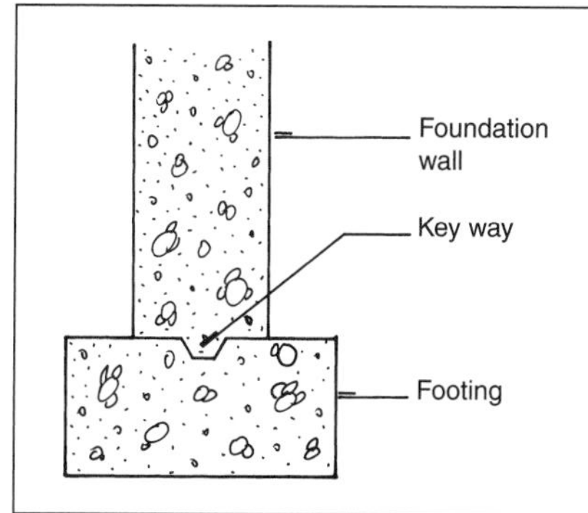

FIGURE 3.5-10 Keyway used with a concrete wall. (Drawing by HLS.)

Keyways are often placed in the tops of footings to key in the foundation walls, whether the wall is concrete or masonry (Fig. 3.5-10).

Plastic strips are used as inserts to create control joints. These may be just flat strips or flanged units that remain in place. Some of these strips have removable tops to permit installation of sealant on top of the insert (Fig. 3.5-11).

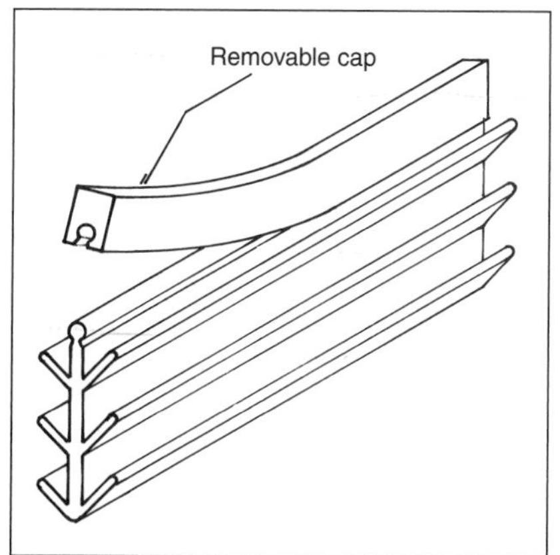

FIGURE 3.5-11 Control joint insert with removable plastic cap. (Drawing by HLS.)

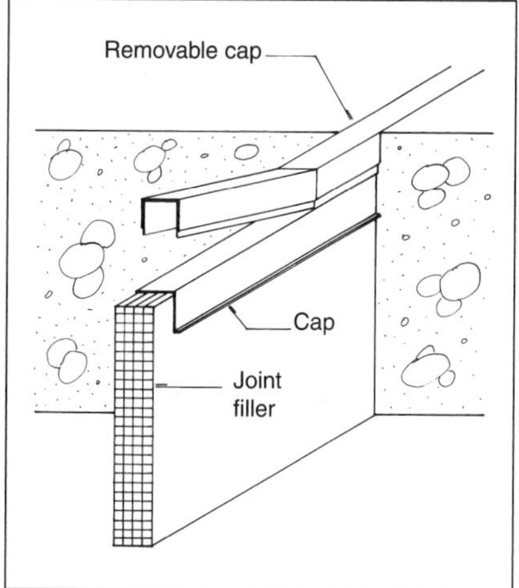

FIGURE 3.5-12 Removable cap for joint filler. (Drawing by HLS.)

Plastic caps are sometimes used over fibrous joint fillers to permit easier application of sealant in the joint (Fig. 3.5-12).

3.5.4 REGLETS

Reglets are devices that form slots in concrete for the insertion of flashings. Figure 3.5-13 shows one reglet configuration. Others are available, but the general characteristics of all reglets are similar. They consist essentially of a slot into which flashing is inserted. They are available in galvanized steel, copper, stainless steel, and plastic.

3.5.5 WATERSTOPS

Waterstops are rubber or vinyl inserts designed to be placed in concrete joints to prevent water from penetrating the joints. There are many available configurations. A few of the more commonly used waterstops are shown in Figure 3.5-14.

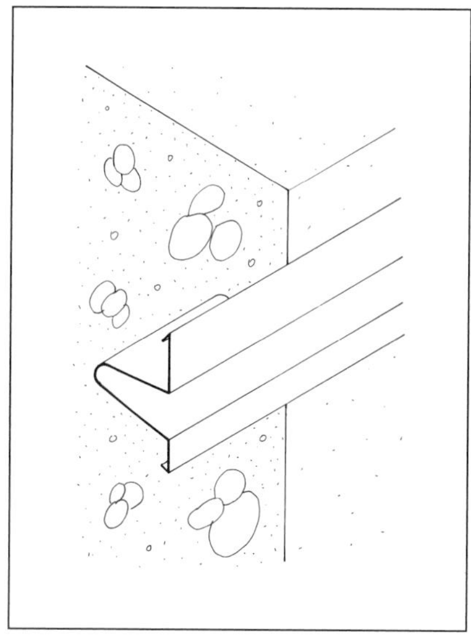

FIGURE 3.5-13 A reglet. (Drawing by HLS.)

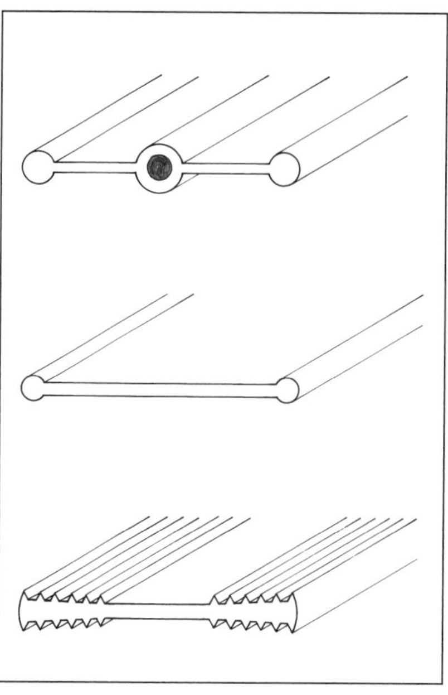

FIGURE 3.5-14 Waterstops. (Drawing by HLS.)

3.6 Joints

It is not possible to entirely prevent cracks in concrete, but good jointing practice will help to reduce their number and encourage them to occur in more acceptable locations. Joints in concrete work permit movement and volume changes; the handling, placing, and finishing of conveniently sized areas; and the separation of independent elements. The types of joints required in concrete include control joints, construction joints, isolation and separation joints, and building expansion joints.

The amount of thermal movement in concrete can be calculated by using its approximate coefficient of expansion (and contraction) of 0.0000055 in./in./°F (0.00014 mm/mm/17.77°C). For example, a 100-ft. (30.5 m) length of unrestrained concrete would expand 0.66 in. (16.76 mm) in length for each 100°F (37.8°C) rise in temperature. If the temperature were to drop 100°F (37.8°C), the same amount of contraction would occur. Concrete would be exposed to a 100°F (37.8°C) rise in temperature if it were placed in the spring when the temperature was 50°F (10°C) and then in the summer the temperature rose to 150°F (65.6°C).

However, concrete shortens about 0.72 in. (18.29 mm) per 100 ft. (30.5 m) while drying from its saturated condition at placing to an average hardened state, with a moisture content equilibrium with air at 50% relative humidity. This shrinkage slightly exceeds the expansion caused by such an extreme increase in temperature as 100°F (37.8°C). Furthermore, in most outdoor applications concrete reaches its maximum moisture content (and moisture expansion) during the season of low temperature (and thermal contraction). Thus, the volume changes resulting from moisture and temperature variations frequently offset each other. Therefore, there is no need for expansion joints in most concrete associated with buildings. The common reference to *expansion* is misleading, because it implies that an increase in size after placement must be allowed for. There may be rare instances, in very large structures and in highways or other large paved surfaces, where extensive surfaces of concrete are subject to large temperature variations and sufficient expansion may occur to justify expansion joints. In most instances, however, concrete is at its greatest mass when it is placed.

Nevertheless, some practitioners in warm regions insist that expansion joints are needed and include them as a safety factor, especially in concrete paving. Where they are used, expansion joints are similar to the isolation joints discussed in Section 3.6.3.

3.6.1 CONTROL JOINTS

As concrete sets or hardens, its excess mixing water is lost through evaporation and hydration. This initial water loss produces shrinkage that is larger than subsequent increases in the size of the hardened mass resulting from an increase in temperature or moisture content.

Shrinkage as water dries from large areas of freshly placed concrete results in jagged irregularly spaced cracks. By anticipating shrinkage and installing *control joints* to limit areas and control where cracking occurs, concrete work can be made more attractive, serviceable, and relatively free from unsightly random cracking. These joints provide a break or a reduction in slab thickness and thus create a weakened section that encourages cracking to occur at that location. When concrete shrinks, the cracks in these joints will open slightly, reducing the number of irregular and unsightly random cracks.

The maximum spacing between control joints depends on the concrete's thickness and shrinkage potential, the curing environment, and the absence or presence of reinforcement.

FIGURE 3.6-1 Jointing using a jointing tool. (Courtesy Portland Cement Association.)

Control joints in sidewalks should be spaced at intervals generally equal to the width of the slab (Fig. 3.6-1), but should not be more than 6 ft. (1800 mm) apart. Joints in driveways should be spaced generally equal to the width of the slab, but never more than 20 ft. (6000 mm) apart.

In large slab areas that do not contain structural reinforcement (Types I and II, as described in Section 3.12.2.1) control joints should divide the slab into approximately square panels, with a maximum spacing of 20 ft. (6000 mm). Actual spacings depend on slab thickness (Fig. 3.6-2c). A rule of thumb for plain concrete slabs is that joint spacing in feet should not exceed two slab thicknesses in inches for unreinforced concrete made with coarse ¾-in. (19 mm) maximum aggregate. Thus, a 4-in. (100 mm)-thick plain slab would require control joints at intervals not to exceed 8 ft. (2400 mm). Additional control joints should be provided at slab intersections (see Fig. 3.12-4).

Control joints are unnecessary in a slab that is to receive a floor finish, such as resilient flooring or carpeting, because the flooring will cover any unsightly cracks that result from shrinkage. Control joints are usually unnecessary in structurally reinforced slabs (Types III and IV, as described in Section 3.12.2).

Control joints can be made by sawing a groove in hardened, but not yet fully cured, concrete with a power saw or by installing a keyed joint (Fig. 3.6-2a and b). Sawed joints are approximately ⅛-in. (3.2 mm)-wide grooves cut in the concrete to a depth equal to one-fourth of the total slab thickness, but not less than ¾ in. (19 mm), and at least equal to the maximum size of the aggregate. Tooled control joints should be of a similar minimum depth. However, grooves cut only for decorative purposes may be shallower.

It is good practice to use a straight board as a guide when making a groove in a concrete slab (see Fig. 3.6-1), because tooled joints should be straight and perpendicular to the edge of the slab. The same care must be taken in running joints as in edging, for a tooled joint can add to or detract from the appearance of the finished slab.

On large concrete flat surfaces it may be more convenient to cut joints with an electric or gasoline-driven power saw fitted with an abrasive or diamond blade. Joints can be cut as soon as the concrete surface is firm enough not to be torn or damaged by the blade (within 4 to 12 hours) and before random shrinkage cracks form in the slab.

3.6.2 CONSTRUCTION JOINTS

Construction joints are necessary in large construction work because it is not practicable to handle, place, and finish a large area in one operation. When construction joints are installed, concrete can be conveniently and practically placed in several operations with no loss in the appearance or performance of the complete job. Although they may double as control or isolation joints, construction joints must allow no vertical movement in the completed floor and are thus often keyed joints, as shown in Figure 3.6-2b.

Plastic or rubber waterstops are inserted in construction joints below grade and in other locations to prevent water penetration through a concrete structural element (Fig. 3.6-3).

3.6.3 ISOLATION AND SEPARATION JOINTS

Isolation and separation joints are often necessary to separate concrete sections and prevent the bonding of one concrete section with another, or to separate a concrete section from another material or structural part so that one can move independently from the other. *Isolation joints are sometimes called expansion joints.*

Isolation and separation joints are usually formed by installing ⅛-in. (3.2 mm)-thick, or slightly thicker, asphalt-

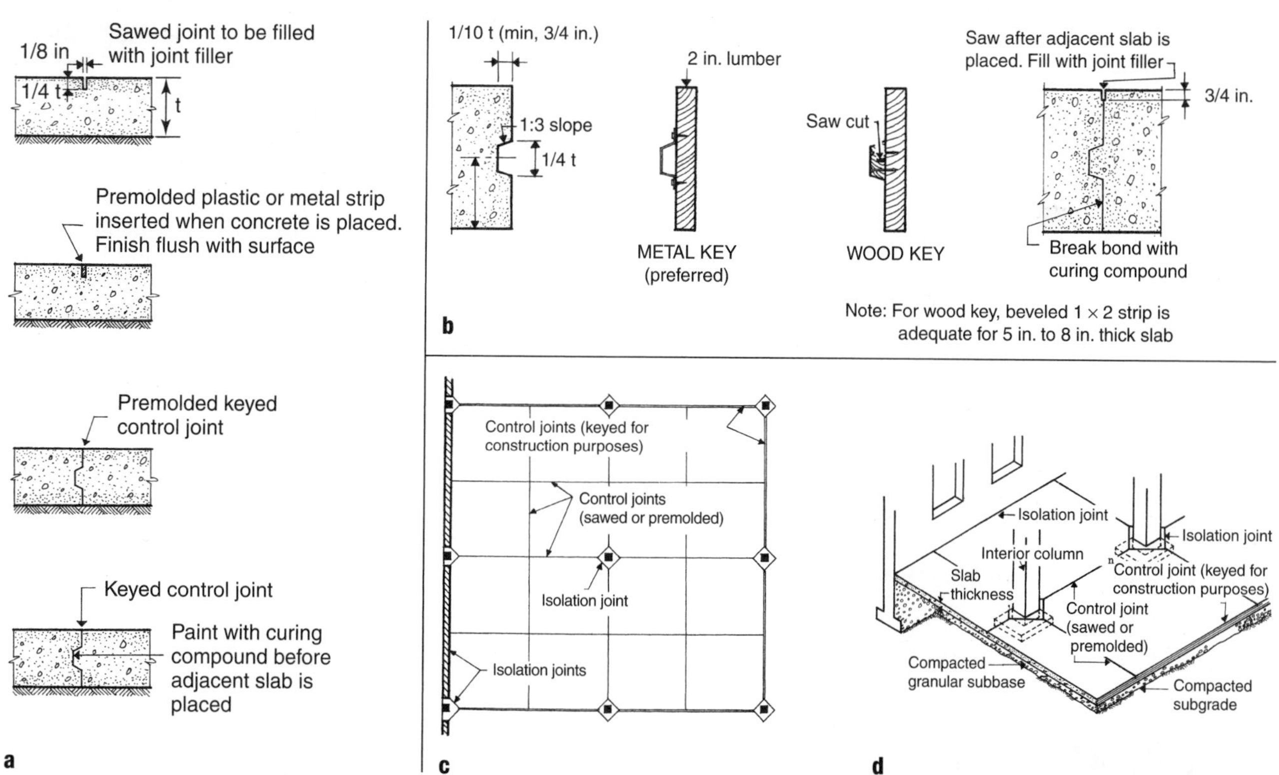

FIGURE 3.6-2 (a) Several types of control joints for slabs on grade. (b) Keyed joint details for slabs on grade. Using too large a key may cause a crack parallel to the joint; if the key is too small, it may shear off, resulting in loss of transfer capacity. A horizontal saw cut permits the wood key strip to swell slightly without damaging the concrete and allows for easy removal of the strip. The joint should be filled with joint filler. (c) Isolation joint for interior column. Column is boxed out so that the corners meet the control joints. This eliminates cracks radiating from the column corners. Circular fiberboard forms can be used instead of square wood forms. (d) Components of a slab on grade. (All courtesy Portland Cement Association.)

impregnated fiber sheets in floors at columns, footings, the junctures between floors and walls, and anywhere else that adjacent surfaces are required to move independently. Figure 3.6-2c shows how isolation joints are formed around a column. Similar consideration should be given around the perimeter of a slab, as shown in Figure 3.6-2d. Figure 3.6-4 shows an isolation joint at a column.

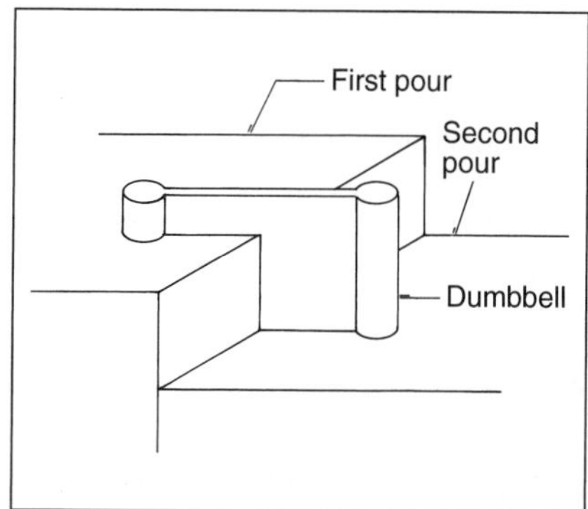

FIGURE 3.6-3 Waterstop in a control joint. (Drawing by HLS.)

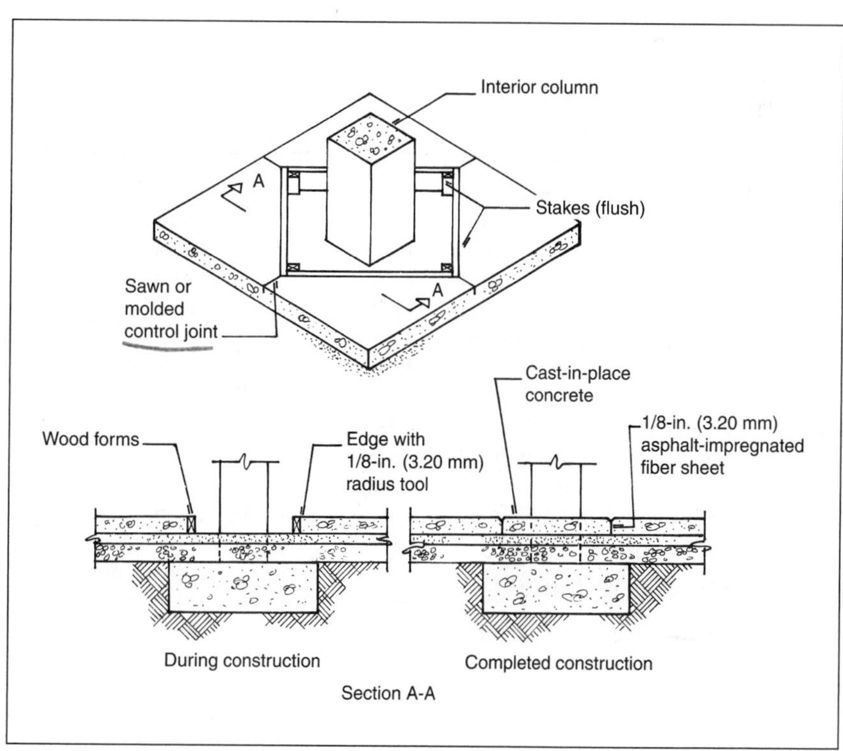

FIGURE 3.6-4 A construction joint used as an isolation joint. (Drawing by HLS.)

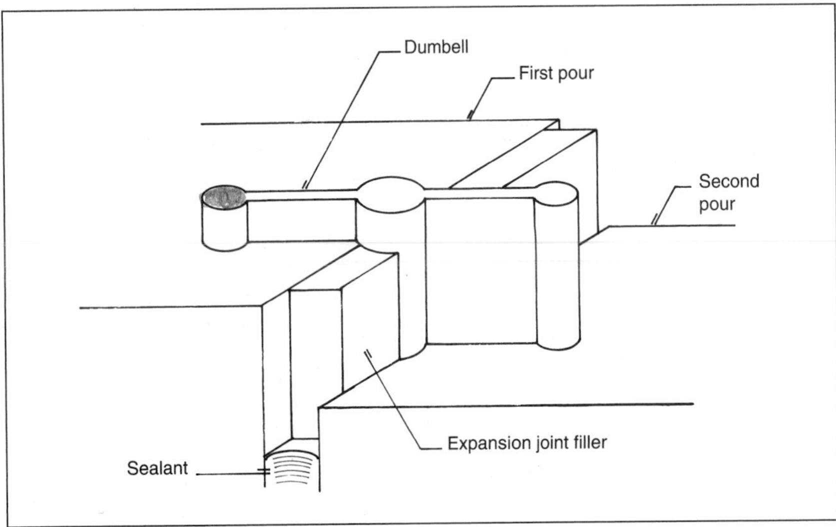

FIGURE 3.6-5 **Waterstop in a building expansion joint.** (Drawing by HLS.)

3.6.4 BUILDING EXPANSION JOINTS

While expansion of the concrete itself is seldom a problem, expansion of an entire structure often requires the insertion of *building expansion joints* that pass through both structural elements and finishes. Concrete that crosses a building expansion joint must be interrupted. Various types of joints have been developed to cover such joints.

In slabs and in walls, where water is likely to pass through a building expansion joint, waterstops are placed in the concrete (Fig. 3.6-5).

3.7 Mixtures and Mixing

The mixture of its ingredients will determine whether the consistency, workability, and finishability of plastic concrete will be in a desirable range. It will also determine whether the essential properties of hardened concrete (strength, durability, watertightness, and wear resistance) are obtained, and whether the concrete will be economical.

In properly made concrete, every aggregate particle, no matter how large or small, is surrounded completely by portland cement and water paste, and the spaces between aggregate particles are all filled completely with paste. This is illustrated in Figure 3.2-6, which shows a section cut through hardened concrete. Aggregates are considered inert materials, and the paste is the cementing medium that binds the aggregate particles into a solid mass. Therefore, the quality of a concrete is greatly dependent on the quality of its paste, and it is important that the paste itself have the required strength, durability, and watertightness.

3.7.1 WATER-CEMENT RATIO

A concrete mixture's water-cement ratio is the ratio by weight of water to portland cement in the mixture. This means pound (kg) of water per pound (kg) of cement (or ounce [g] per ounce [g]). A water-cement ratio of 0.31 could be thought of as 0.31 pounds (kg) of water for each pound (kg) of cement. The units (pounds, ounces, kg) are not used, because they cancel out mathematically (pound/pound = 1). The cementing or binding properties of a cement paste are due to a chemical reaction between the cement and water (hydration) that requires time and favorable temperature and moisture conditions. The reaction takes place very rapidly at first and then more slowly for a long time under favorable conditions. Only a relatively small amount of water, about 3½ gal. (13.25 L) per 94 lb. (42.64 kg) of portland cement, is actually required to complete the chemical reaction. This amount is equivalent to a water-cement ratio of 0.31. But a water-cement ratio ranging from 0.35 to 0.80 is usually maintained for the sake of workability and placeability. Another advantage of using more water is that more aggregate can be used, with resulting economy. All the water that is not required for hydration eventually evaporates, leaving the small holes that account for the porosity of concrete.

For successful results, a proper proportion of water to portland cement is essential. If there is a high water-cement ratio, meaning that too much water has been added, a portland cement paste becomes thin or diluted and will be weak and porous when it hardens. Using too little water may result in a mix that cannot be properly placed and finished. Portland cement paste made with the correct amount of water has strong binding qualities and is watertight and durable when cured. If its portland cement paste and aggregates are strong and durable, a concrete will be strong and durable. The strength, durability, freeze-thaw resistance, watertightness, and wear resistance of a concrete are largely controlled by using a sufficiently low ratio of water to portland cement. Additional control can be gained by adding an adequate amount of entrained air.

The water-cement ratio selected should be the lowest value required to meet design considerations such as durability, strength, and impermeability. For instance, concrete that will be exposed to a combination of wet-dry and freeze-thaw cycling and de-icing chemicals, requires the following for durability: (1) a low water-cement ratio; (2) air entrainment; (3) suitable materials; (4) adequate curing; and (5) good construction practices.

Figure 3.7-1 is a guide for selecting the water-cement ratios for various exposure conditions.

3.7.1.1 Strength

The compressive strength required of a particular concrete is dependent on several factors. One is the exposure condition (see Fig. 3.7-1); another is the applied load. Strengths as low as 2500 psi (17

FIGURE 3.7-1 Recommended Concrete Mixtures for Various Exposure Conditions

Concrete Construction Element	Exposure Condition[a]	Minimum Compressive Strength at 28 days,[b] Plant or Transit, psi	Practical Water-Cement Ratio by Weight: Non-Air-Entrained Concrete	Practical Water-Cement Ratio by Weight: Air-Entrained Concrete	Normal Maximum Coarse Aggregate Size,[e] in.	Minimum Cement[f] Content, lb./cu. yd.	Air-Entrainment by Volume, %	Slump, in.
Foundations, basement walls, and slabs not exposed to weather	Severe					564		5-in. maximum for hand methods of strike-off and consolidation; 3-in. maximum for mechanical strike-off and consolidation
	Moderate	3000[c]	See footnote c	0.55	1		See footnote f	
	Negligible					520		
Foundations, basement walls, exterior walls, and other concrete work exposed to weather	Severe	3500[c]	See footnote c	0.45		564	6–8	
	Moderate	3000	0.58	0.50	1	520	5–7	
	Negligible	2500	0.67	0.55			See footnote f	
Concrete exposed to sulfate attack	Severe	5000[c]	See footnote d	See footnote d		564	6–8	
	Moderate				1	520	5–7	
	Negligible	5000					See footnote f	
Driveways, garage floors, walks, porches, patios, and stairs exposed to weather	Severe	4000[c]	See footnote c	0.45		564	6–8	
	Moderate	3000[c]		0.50	1	520	5–7	
	Negligible	2500	0.67	0.55			See footnote f	

[a]See Figure 3.7-8 for severity of exposure, regional weathering areas.

[b]With most materials, the water-cement ratios shown will provide average strengths greater than required.

[c]Use air-entrained concrete only.

[d]Proportions should be established by the trial batch method.

[e]Maximum size of coarse aggregate should not be larger than one-fifth the narrowest dimension between forms, nor larger than three-quarters the minimum clear spacing between reinforcing bars.

[f]2% to 3% air entrainment, to improve cohesiveness and reduce bleeding, is recommended but not essential.

(Courtesy Portland Cement Association.)

MPa) may be sufficient in some small building foundations; columns in very tall buildings may employ concrete with a strength of 19,000 psi (1312 MPa). Even higher-strength concretes may be used in the future when technology permits their development.

There is a direct link between the strength of concrete and its water-cement ratio. In fact, for given materials and handling conditions, strength is determined primarily by the water-cement ratio, as long as the mixture is plastic and workable. The age-strength relationships for Types I and III portland cements are shown in Figure 3.7-2.

For a given water-cement ratio, the strength at a certain age is practically fixed, if the mixture is plastic and workable, the aggregates are strong, clean, and sound, and the concrete is properly cured. More water will result in less strength and less water will result in greater strength.

After concrete has been placed, its strength continues to increase while moisture and optimum temperatures are present to promote hydration, and as long as the evaporation of its water (*drying*) is prevented. When concrete is permitted to dry, the water that would cause hydration is gone and the chemical reactions between the portland cement and water cease. This will result in lower strengths than would occur if the concrete were kept moist. Therefore, it is desirable to keep concrete constantly moist as long as possible after placing (Fig. 3.7-3).

The temperature at which concrete is mixed and cured affects the rate of the chemical reactions between the portland cement and water. Seventy-five°F (23.9°C) is considered the normal curing temperature. Most specifiers require that the range be between 50° and 85°F (10° and 29.4°C). Some will permit temperatures as high as 90°F (32°C). At higher temperatures, concrete increases in strength faster than normal during the first few days of curing, but much slower than normal during later periods (Fig. 3.7-4). The final strength of concrete cured at temperatures above 90°F (32°C) is lower than that of concrete cured at between 50° and 90°F (10° and 32°C). Concrete made and cured at 45°F (7.2°C) will gain strength slowly during curing and perhaps never reach the strength of concrete cured at between 50° and 90°F (10° and 32°C). As the temperature in concrete rises, the slump decreases. Adding water to maintain a desirable slump decreases the concrete's strength and adversely affects its other properties.

3.7.1.2 Durability

There is a direct link between a concrete's durability and its water-cement ratio. The most destructive natural weathering force is freezing and thawing while concrete is wet or moist, because of the expansion of water as it changes to ice. A portland cement paste made with a low water-cement ratio will be much more resistant to damage by freezing and thawing than one with a higher water-cement ratio. This is demonstrated by the concrete cubes in Figure 3.7-5, which have been subjected to 70 cycles of freezing and thawing while sat-

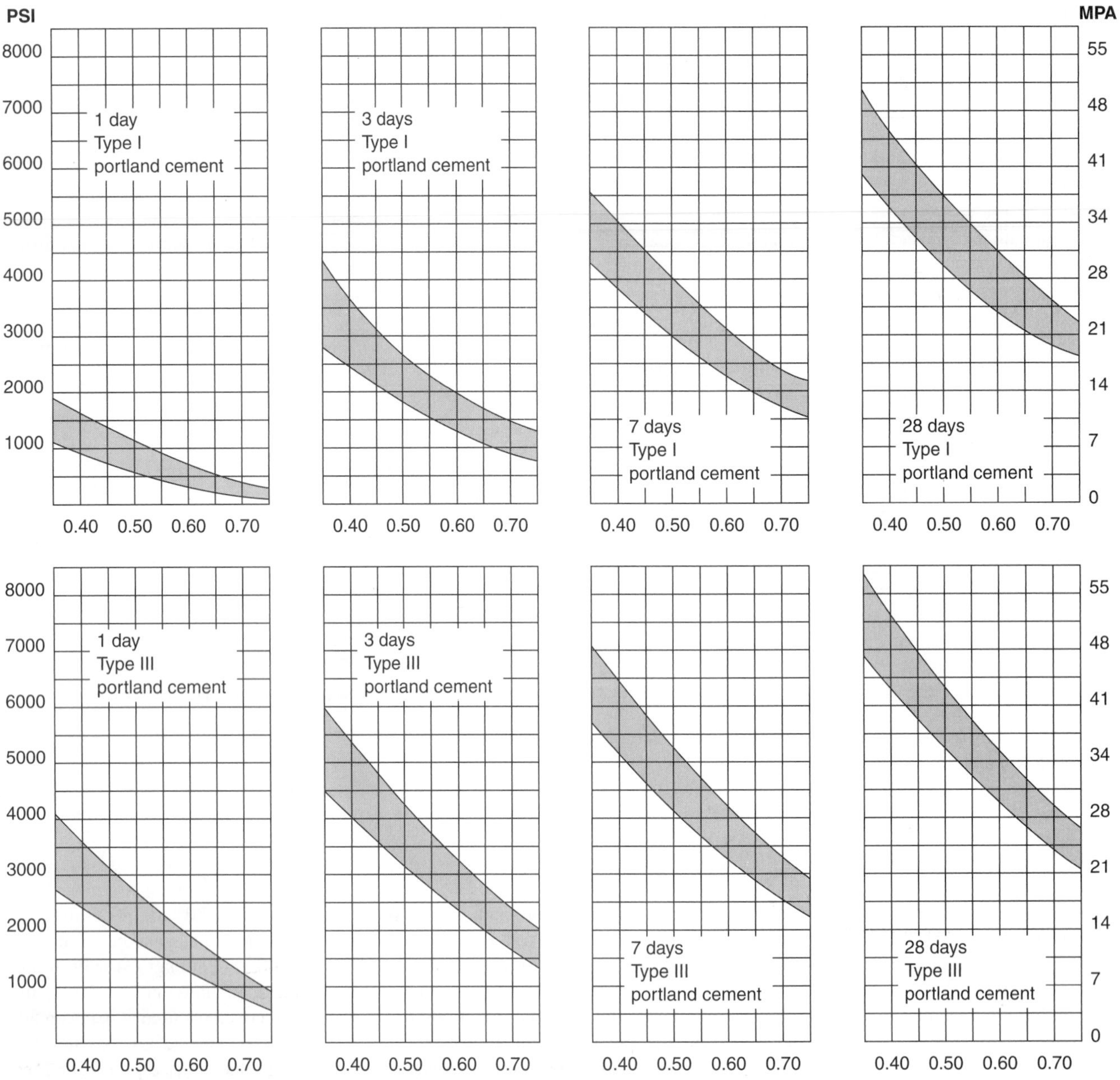

FIGURE 3.7-2 **The relationship between age and compressive strength for Types I and III portland cement concretes. A majority of the results of tests for compressive strength made by several laboratories, using a variety of materials complying with ASTM standards, fall within the shaded areas.** (Drawing by HLS.)

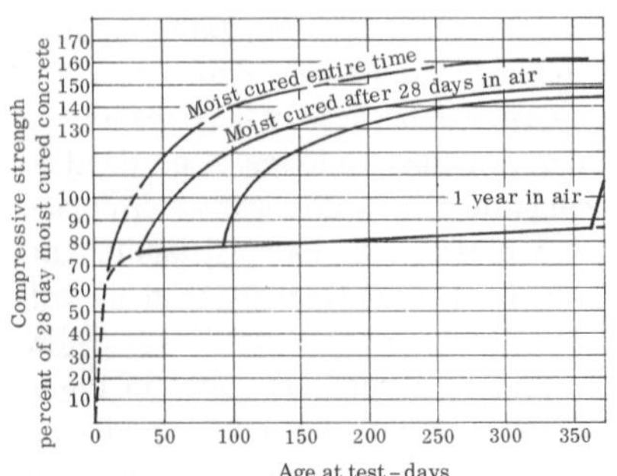

FIGURE 3.7-3 **The strength of a concrete continues to increase as long as moisture is present to promote hydration of the cement. Notice that a resumption of moist curing after a period of drying also increases the strength. However, the test specimens used in this study were small as compared with most concrete members. Because resaturation is difficult in the field, moist curing should be continuous.** (Courtesy Portland Cement Association.)

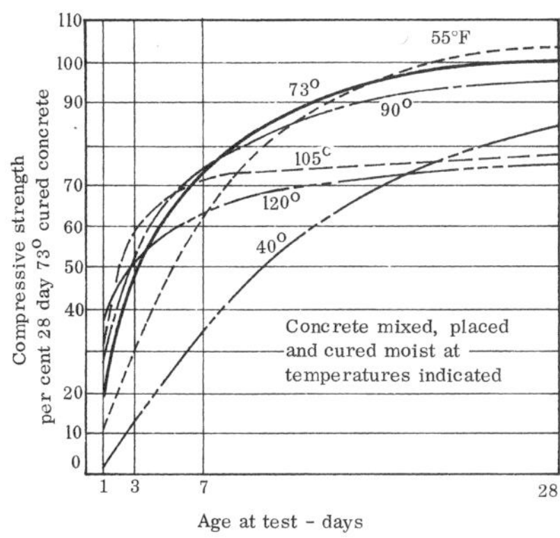

FIGURE 3.7-4 The effect of temperature on the strength of a concrete. Other characteristics are also affected. (Courtesy Portland Cement Association.)

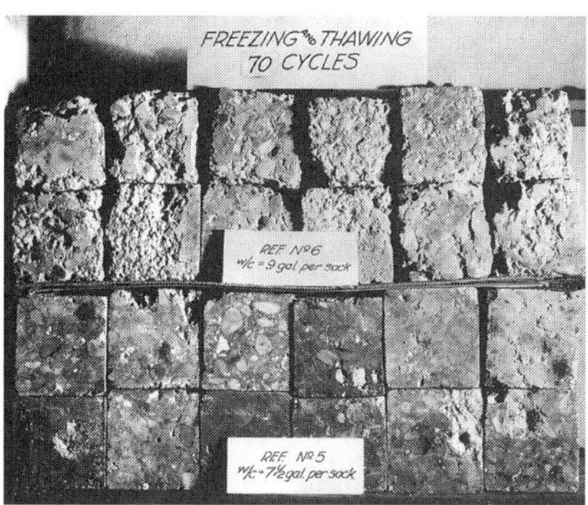

FIGURE 3.7-5 The durability of a concrete exposed to freezing and thawing is affected by the quality of its paste. The specimens at the bottom of this photograph were made with a lower water-cement ratio (0.64) than those in the upper part (0.80) and were, therefore, more resistant to damage. The same aggregate was used in all the specimens. (Courtesy Portland Cement Association.)

urated with water. Those in the upper portion, which were made with a water-cement ratio of 0.80, show much more disintegration than those in the lower portion, which were made with a water-cement ratio of 0.64. Furthermore, introducing entrained air into concrete will significantly increase its resistance to freezing and thawing damage and the application of de-icer salts.

3.7.1.3 Watertightness

For a concrete to be watertight and impervious, its portland cement paste must be formulated to produce this result. As with strength and durability, there is a direct link between a concrete's watertightness and its water-cement ratio.

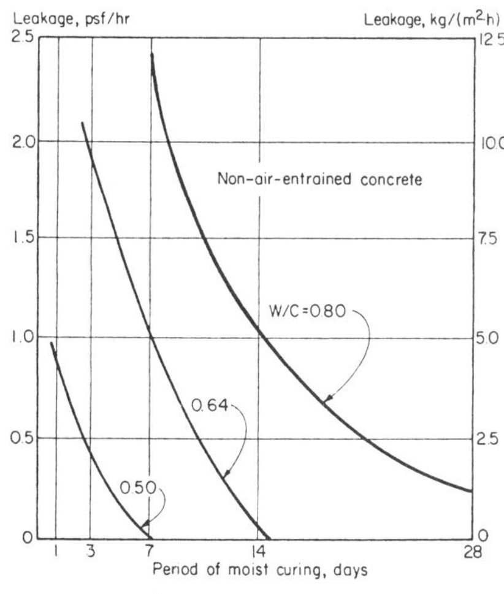

FIGURE 3.7-6 The effect of water-cement ratio on watertightness. Leakage is reduced as the water-cement ratio is decreased and the curing period is increased. Specimens were 1 × 6 in. (25 × 150 mm) mortar discs. Pressure was 20 psi (0.14 MPa). Values are 48-hour averages. (Courtesy Portland Cement Association.)

Tests show that a portland cement concrete's permeability depends on the amount of mixing water in its water paste and the extent to which the chemical reactions between the portland cement and water have progressed. The results of subjecting discs made with portland cement mortar, fine aggregate, and water to 20 psi (0.14 MPa) water pressure are shown in Figure 3.7-6. In these tests, mortar cured moist for 7 days had no leakage when made with a water-cement ratio of 0.50, but there was considerable leakage with mortars made with the higher water-cement ratios. Moreover, in each case, leakage became less as the curing period was increased. Mortar discs with a water-cement ratio of 0.80 leaked, even when moist-cured for a month.

Air entrainment improves watertightness by allowing reduction of the water-cement ratio. To be watertight, concrete must also be free from cracks and honeycombing.

As a result of testing and experience in the field, definite recommendations can be made regarding the maximum amount of mixing water that should be used for various construction applications (see Fig. 3.7-1).

3.7.2 SELECTING A CONCRETE MIX

The best proportions for a concrete mix are those that result in the most economical combination of portland cement, aggregate, and other materials that will produce concrete of the required durability, abrasion resistance, and strength, and that can be placed and finished properly. For slab construction, the properties of the plastic concrete that affect placing and finishing are as important as the properties of the hardened concrete (abrasion resistance, durability, and strength). The plastic properties will greatly affect the quality of the top 1/16 to 1/8 in. (1.6 to 3.2 mm) of the slab surface. However, because these properties are not easily measured, there is a tendency to emphasize the importance of more readily determined properties such as compressive strength.

Recommendations for slump, strength, maximum coarse aggregate size, and minimum portland cement content required for each size aggregate are given in Figure 3.7-7. The portland cement content of a mix should be properly proportioned to produce concrete that can be easily placed and finished and will, in addition, possess adequate strength and durability. Mixtures with low portland cement content produce harsh concretes and slabs that bleed excess water to the surface, are difficult to finish, and have poor surface characteristics. Mixtures with excessive portland cement content can result in increased drying shrinkage.

When concrete mixtures are not established by tests before construction work begins, the concrete mixtures shown in Figure 3.7-1 may be used. It is first necessary to determine the exposure condition of the concrete from the weathering areas shown in Figure 3.7-8. Areas shown in the severe exposure range are subject to many freeze-thaw cycles per year and to de-icer chemicals; moderate exposure areas are subject to few freeze-thaw cycles and no de-icer chemicals; negligible exposure areas have no freeze-thaw cycles, and de-icer chemicals are not used.

When the exposure condition of a concrete construction element is known, it is possible to use Figure 3.7-1 to determine the recommended minimum compressive strength. It is also possible to determine the maximum permissible water-cement ratio necessary to obtain the specified minimum compressive strength. The minimum portland cement content is included in addition to the lowest maximum water-cement ratio. Minimum portland cement requirements ensure satisfactory finishability, improved water resistance, and suitable appearance of vertical surfaces.

FIGURE 3.7-7 Recommended Mixtures for Light Slabs on Grade[a]

Maximum Size of Coarse Aggregate,[b] in. (mm)	Minimum Cement Content, lb./cu. yd. (kg/m³)	Maximum Water/Cement by Weight: Non-Air-Entrained Concrete	Maximum Water/Cement by Weight: Air-Entrained Concrete	Slump, in. (mm)	Probable Compressive Strength,[c] psi (MPa)
1½ (38.1)	470 (278.84)				
1 (25.4)	520 (308.10)				
¾ (19.1)	540 (320.37)	0.51	0.40	2–4 (50.8–101.6)	3500 (24)
½ (12.7)	590 (350.03)				
⅜ (9.5)	610 (361.90)				

[a]For both air-entrained and non-air-entrained concretes.

[b]Normal-weight aggregate; different mixes may be required for lightweight aggregate concretes.

[c]Strength refers to approximate compressive strength of cylinders made and tested according to applicable ASTM standards at 28 days that have been continuously moist cured.

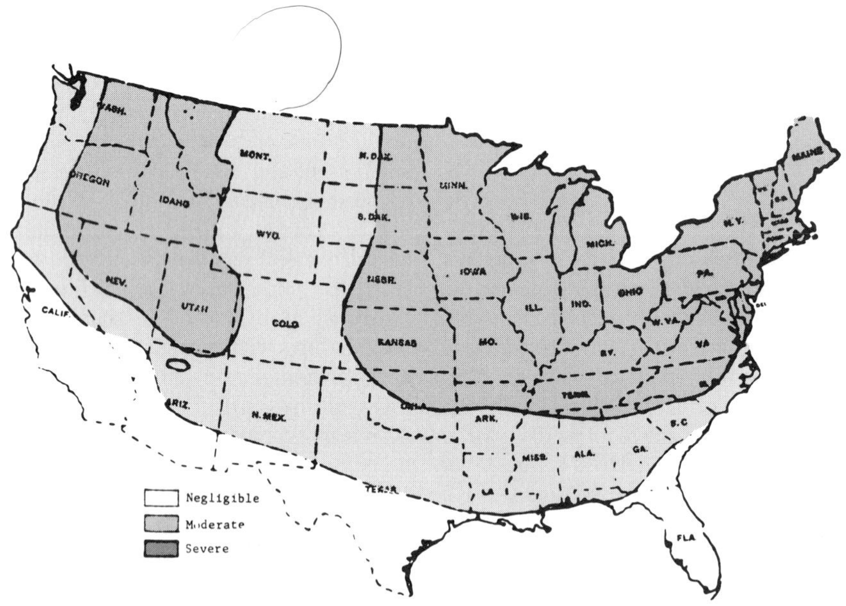

FIGURE 3.7-8 Regional weathering areas. Areas are based upon ASTM C 62 weathering index map adjusted to conform with an experience survey by HUD and the Portland Cement Association. Local conditions may be more or less severe than indicated by regional classification.

The percentage of air entrainment for various maximum aggregate sizes is also given; slumps should be in the ranges shown.

An example showing how to use Figure 3.7-1 follows. The proposed project is in the Chicago area, which is in a severe weathering region (see Fig. 3.7-8). According to Figure 3.7-1, the minimum compressive strength required for basement walls that will be exposed to weather is 3500 psi (24 MPa). Reading across the table shows a practical water-cement ratio of 0.45, and 6% to 8% air-entrained concrete. The maximum portland cement content likely to be required to achieve that strength at the recommended air-entrainment level is 564 lb./yd.³ (334.6 kg/m³).

Even though 2500 psi (17 MPa) concrete may satisfactorily support typical vertical and lateral loading in a residential basement wall, this lower-strength concrete would be less watertight than the concrete recommended here. A higher-strength concrete provides a margin of safety against the possibility of foundation cracking and leaking.

High-strength concrete, as may occur in tall building columns and in mass concrete for gravity dam structures, is produced by including additives, such as silica fume, fly ash, and superplasticizers, in the mix.

3.7.2.1 Allowance for Water in the Aggregate

When selecting the proper amount of water to add to a concrete mix to produce the desired water-cement ratio, it is necessary to allow for the water contained in most fine aggregates. A simple test for determining whether sand is *damp*, *wet*, or *very wet* is to press some together in your hand. If the sand falls apart after your hand is opened, it is damp (Fig. 3.7-9a); if it forms a ball that holds its shape, it is wet (Fig. 3.7-9b); if the sand sparkles and leaves water on your hand, it is very wet (Fig. 3.7-9c).

The most practical procedure is to make trial mixes. On large jobs where many materials or unusual materials may be under consideration, or where strength is particularly important, the trials may be full-size batches checked in the field. The first trial mix may be selected on the basis of experience or established relationships such as those given in Figure 3.7-1. This table has been developed from experience and data from several sources and indicates the suggested proportions of all the various ingredients that should result in good workability.

3.7.3 CONSISTENCY OF THE MIX

With a given amount of portland cement paste, more aggregate is used in stiff mixes than in more fluid mixes; consequently, stiff mixes are more economical

a

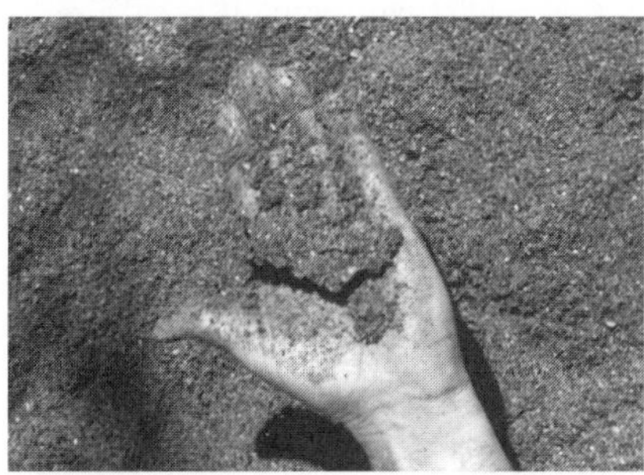

b

c

FIGURE 3.7-9 (a) Damp sand; (b) wet sand; (c) very wet sand. (Courtesy Portland Cement Association.)

in cost of materials. Stiff mixes require more labor in placing, however, and when a mixture is too stiff, the additional cost of placing may offset savings in materials. Mixes should always be of a consistency and workability that will permit the concrete to be worked into the angles and corners of forms and around reinforcement without permitting the materials to segregate or excess free water to collect on the surface. Thin members and heavily reinforced members require more fluid mixes than large members or flat surfaces containing little reinforcing.

A *plastic* concrete is one that is readily molded and yet will change its form only slowly if the mold is removed. Mixes of plastic consistency are required for most concrete work. Concrete of such consistency does not crumble but flows sluggishly without segregation. Thus, neither very stiff, crumbly mixes nor very fluid, watery mixes are of plastic consistency (Fig. 3.7-10).

The ease or difficulty of placing concrete in a particular location is referred to as *workability*. A stiff but plastic mix with large aggregate would be workable in a large, open form but not in a thin wall with closely spaced reinforcement.

Excessive slump and consequent bleeding and segregation are primary causes of poor concrete performance. For slabs, the slump may vary from 2 to 4 in. (50 to 100 mm) (see Fig. 3.7-7), which will produce slabs that can be easily placed and properly consolidated, yet will have a consistency not conducive to bleeding or segregation during placing and finishing. Concrete for slabs where appearance is important, and concrete for ramps and sloping surfaces, should have a slump that does not exceed 3 in. (76 mm). Different batches of concrete placed in the same slab, ramp, or sloping surface should have approximately the same slump. When lightweight concrete is placed at slumps in excess of 3 in. (76 mm), the coarser lightweight aggregate may rise to the surface and excessive bleeding may also occur.

Reinforced foundation systems may have slumps as low as 1 in. (25.4 mm), but should not exceed 3 in. (76 mm). Concrete that will be struck off and consolidated by hand may have slumps of up to 5 in. (127 mm). The slump of concrete that will be struck off and consolidated with mechanical devices should be limited to 3 in. (76 mm). Concrete that contains high-range water-reducing admixtures may have slumps up to 8 in. (203 mm) after the admixture has been added. But the slump before the admixture is added must be between 2 and 3 in. (50 and 76 mm). Slumps for other concrete should be not less than 1 in. (25.4 mm) nor more than 4 in. (100 mm).

3.7.3.1 Slump Test

A slump test conforming with the requirements of ASTM C 143 may be used as a rough measure of the consistency of concrete. This test is not a measure of workability, and it should not be used to compare mixes of entirely different proportions or containing different kinds of

a

b

c

FIGURE 3.7-10 (a) A concrete mixture with insufficient cement-sand mortar to fill the spaces between the coarse aggregate. This mixture will be difficult to handle and place and will result in honeycombed surfaces and porous concrete. (b) A concrete mixture with the correct amount of cement-sand mortar. With light troweling, the spaces between aggregate are filled with mortar. Note the appearance on the edge of the pile. This is a good workable mixture that will give the maximum yield of concrete. (c) A concrete mixture with too much cement-sand mortar. This mixture is plastic and workable and will produce smooth surfaces, but the yield of concrete will be low and uneconomical and the concrete will probably be porous. (Courtesy Portland Cement Association.)

aggregates. Any change in slump on the job indicates that changes have been made in grading or proportioning the aggregate, or in water content. The mix should be corrected immediately to get the proper consistency by adjusting amounts and proportions of sand and coarse aggregate, with care taken not to change the water-cement ratio.

In a slump test, a test specimen is made in a mold called a *slump cone*, which is made of 16-gauge (1.6 mm) galvanized metal to the shape shown in Figure 3.7-11a. The base and top are open. The mold is provided with foot pieces and handles, as shown.

To make a slump test, a concrete sample is taken just before its batch is placed in the forms. A slump cone is placed on a flat surface, such as a smooth plank or concrete slab, is held firmly in place by the tester standing on the foot pieces, and is filled with concrete to about one-third of its volume. Then the concrete is puddled with 25 strokes of a ⅝-in. (15.9 mm)-diameter rod about 24 in. (610 mm) long and bullet pointed at the lower end (Fig. 3.7-11b). The filling is completed in two more layers, each about one-third of the volume of the cone. Each new layer is rodded 25 times, with the rod penetrating into the underlying layer on each stroke. After the top layer has been rodded, it is struck off with a trowel so that the mold is filled completely. The mold is removed immediately after it has been struck off, by gently raising it vertically.

The slump of the concrete is measured, as shown in Figure 3.7-11c, immediately after the cone is removed. If the top of the slump pile is 5 in. (127 mm) below the top of the cone, the slump for this concrete is 5 in. (127 mm).

3.7.3.2 Compression Test

Compression tests are made in accordance with ASTM C 31 to determine whether a concrete has the specified compressive strength. In these tests, concrete specimens are tested at 7 days and at 28 days after they are taken to determine the concrete's rate of strength gain.

Standard procedures provide for curing the specimens either in the laboratory or in the field. Laboratory curing gives an indication of the potential quality of the concrete when properly cured under ideal conditions. Field-cured specimens may give a more accurate interpretation of actual strength in the structure or slab, under the less than ideal curing conditions at a job site, but they offer no explanation as to whether lack of strength is due to error in proportioning, poor materials, or unfavorable curing conditions. Sometimes both methods are used, especially when the weather is unfavorable, in order to interpret the tests properly. When differences occur, the laboratory test results prevail.

Compressive strength test samples are taken at three or more regular intervals during discharge of a concrete batch. Samples should not be taken at the beginning or end of discharge. The batch of concrete thus sampled should be noted as to its location in the work, the air temperature, and unusual conditions at the time. The usual number of test specimens in a sample is four; one is tested at 7 days, two at 28 days, and the fourth is held in reserve in case additional testing is necessary. If both laboratory and field tests are required, duplicate samples are taken.

A compressive test specimen is placed in a watertight cylindrical mold to prevent water loss. Standard cylindrical molds, like those shown in Figure 3.7-12, are 3 in. (76 mm) in diameter by 6 in. (152 mm) long, 4 in. (101.6 mm) in diameter by 8 in. (203 mm) long, and 6 in. (152.4 mm) in diameter by 12 in. (304.8 mm) long. The 6-in. by 12-in. (152.4 by 304.8 mm) cylinders are generally used when the coarse aggregate does not exceed 2 in. (50.8 mm) in nominal size. A mold is filled in three equal layers. Each layer is puddled with 25 strokes of a ⅝-in. (15.9 mm)-diameter steel rod about 24 in. (610 mm) long with a bullet-pointed tip. Reinforcing bars and other tools should not be used as puddle rods. After the top layer has been rodded, the surface of the concrete is struck off with a trowel and covered to prevent evaporation.

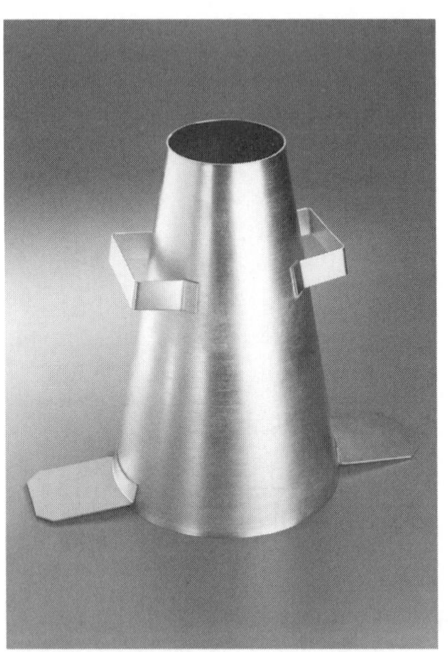

a

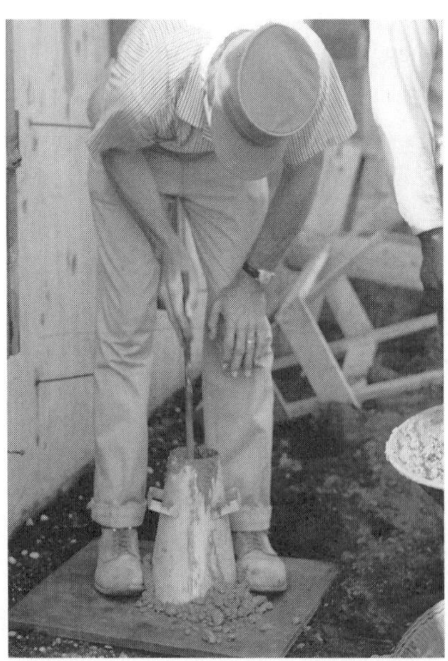

b

c

FIGURE 3.7-11 (a) Slump cone (mold) for slump test. (Courtesy ELE International, Inc., Soil Test Products Division.) **(b) Rodding the concrete in a slump cone to ensure complete filling.** (Copyright ASTM, reprinted with permission.) **(c) Measuring slump by the depression below a rod laid across the cone. This amount of slump indicates a medium-wet mixture.** (Copyright ASTM, reprinted with permission.)

FIGURE 3.7-12 Standard plastic test cylinders. Waxed cardboard cylinders with metal bottoms are also available. (Courtesy ELE International, Inc., Soil Test Products Division.)

The cured specimens are crushed in a laboratory testing machine to determine their strength (Fig. 3.7-13). If the strength of the specimens is too low, test cores should be cut from the concrete structural unit containing the suspect concrete. These should then be tested for compression. If this test indicates unsatisfactory strength, the concrete should be removed and new concrete of adequate strength should be placed.

3.7.4 MATERIALS MEASUREMENT

Producing uniform batches of concrete of proper proportions and consistency requires that the ingredients be measured accurately. The effects of the water-cement ratio on concrete quality make it necessary to measure all the ingredients, including the water. A troublesome factor is the effect of the varying amounts of moisture nearly always present in aggregate, particularly in natural sand. The amount of free moisture introduced in the mixture with the aggregates must be determined and allowance made for it if accurate control is to be obtained.

3.7.4.1 Measuring Portland Cement

On small projects where sacked portland cement is used, the batches of concrete should be of such size that only full sacks are used. If partial sacks are used, they should be weighed for each batch. It is not satisfactory to divide sacks of portland cement on the basis of volume.

Bulk (unsacked) portland cement, which is used today for most concrete, should be measured by weight for each batch.

3.7.4.2 Measuring Water

Dependable and accurate means for measuring mixing water are essential. Portable mixers should be equipped with water tanks and measuring devices that are accurate when properly operated. The measuring device most generally used operates on the principle of the siphon. The tank is filled, and the desired amount of water is siphoned off. When the water level reaches the point at which the bottom of the siphon is set, the water is shut off automatically.

3.7.4.3 Measuring Aggregates

Measuring aggregates by volume cannot be depended on except under the most careful supervision. A small amount of moisture in fine aggregate, which is nearly always present, causes the aggregate to bulk up (increase in volume), as indicated in Figure 3.7-14. The amount of bulking varies with the amount of moisture present and the grading; fine sands bulk more than coarse sands for a given amount of moisture, because a given quantity of smaller particles presents more surface to which water can adhere than does the same amount of aggregate with larger particles. Note in Figure 3.7-14 that a fine sand with a 5% moisture content increases in volume almost 40% over its dry volume. Sand often contains almost enough moisture to produce the maximum bulking. Small changes in moisture can cause large changes in the amount of bulking, until all particles have been surrounded with water. After that, the addition of

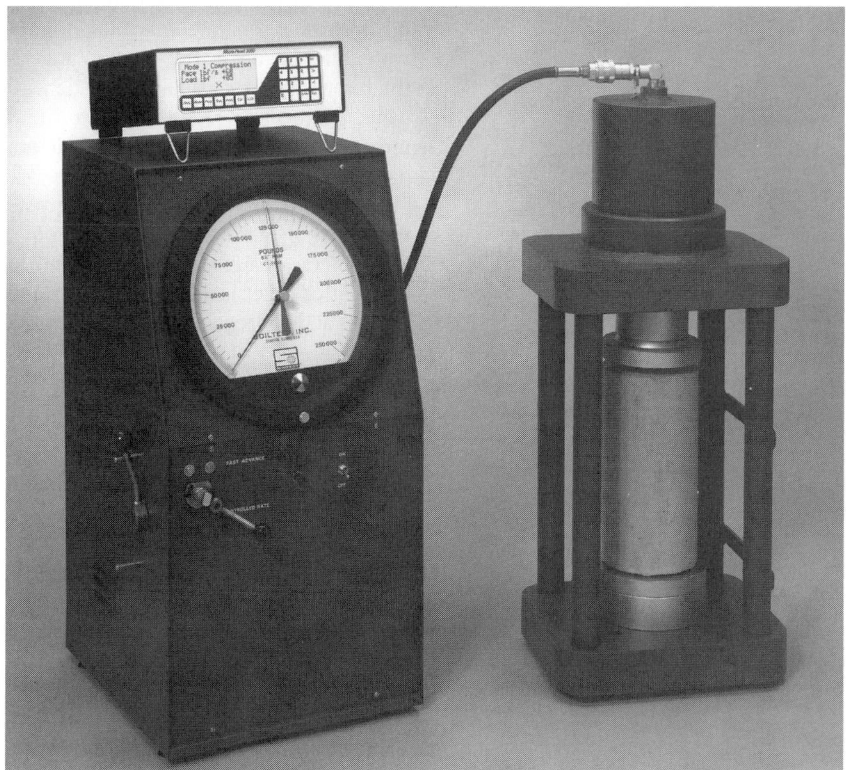

FIGURE 3.7-13 A concrete compression tester. (Courtesy ELE International, Inc., Soil Test Products Division.)

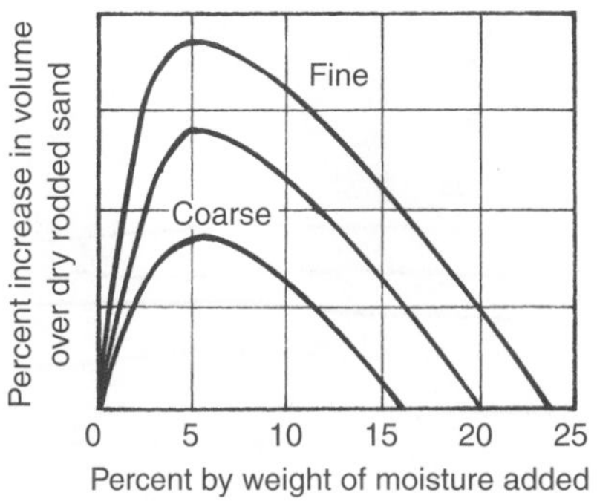

FIGURE 3.7-14 Bulking in fine aggregate caused by surface moisture. (Courtesy Portland Cement Association.)

water causes a lesser degree of increased bulking. Therefore, the relationship between moisture and bulk is not linear. For these reasons, aggregates should always be measured by weight.

To reduce segregation of aggregates and to produce uniformity from batch to batch, it is desirable to weigh the coarse aggregate in two or more sizes, especially if the maximum size exceeds 1 in. (25.4 mm). Generally, the ratio of the maximum size particle to the minimum size for coarse aggregate separations should not exceed 2:1 for materials larger than 1 in. (25.4 mm), and should not exceed 3:1 for finer materials. Therefore, 1½-in. (38 mm) aggregate would be separated into two batches, one with ¼- to ¾-in. (6.4 to 19.1 mm)-sized aggregate and the other with ¾- to 1½-in. (19.1 to 38 mm) sizes.

3.7.5 MIXING

The ingredients should be in the proper condition when they are charged into a mixer. Aggregates should have the correct moisture content, and portland cement should be free of lumps. If lumps are present that cannot be pulverized between the thumb and finger, the portland cement should not be used.

Concrete should be mixed thoroughly until its appearance, including its color, is uniform and its ingredients are uniformly distributed. The time required for thorough mixing depends on several factors. Specifications usually require a minimum of 1 minute mixing for mixtures up to 1 cu. yd. (0.765 m^3) capacity with an increase of 15 seconds mixing for each ½ cu. yd. (0.38 m^3), or fraction thereof, of additional capacity. In general, longer mixing gives more uniform results. The mixing period should be measured from the time all solid materials are in the mixer drum, provided that all of the water is added before a quarter of the mixing time has elapsed. Many mixers are equipped with timing devices, some of which can be set for a given mixing time and locked so that the batch cannot be discharged until the designated time has elapsed.

Mixers should not be loaded above their rated capacity and should be operated at approximately the speeds for which they are designed. If the blades of the mixer become worn or coated with hardened concrete, the mixing action will be less efficient. Badly worn blades should be replaced, and hardened concrete should be removed before each run of concrete.

Under usual operating conditions, up to about 10% of the mixing water should be placed in the drum before the aggregates and portland cement are added. Water should then be added uniformly with the dry materials, leaving about 10% to be added after all other materials are in the drum. When heated water is used during cold weather, this order of charging may require some modification to prevent flash setting of the portland cement. In this case, addition of the portland cement should be delayed until most of the aggregate and water have intermixed.

The concrete should be discharged completely from a mixer before the mixer is recharged.

FIGURE 3.7-15 **Loading trucks at a ready-mix plant.** (Courtesy Portland Cement Association.)

3.7.5.1 Ready-Mixed Concrete

Most concrete used in building construction originates in a central ready-mix plant (Fig. 3.7-15). Ready-mixed concrete should comply with ASTM C 94, which requires, among other things, that the hauling be done in agitator trucks or transit mixer trucks operated at agitator speed.

In some ready-mixed operations, the materials are dry-batched at the central plant and then mixed en route to the job in truck mixers. In others, the concrete is mixed in a stationary mixer at the central plant only enough to intermingle the ingredients, generally about a half minute. The mixing is completed in a truck mixer en route to the job.

Truck mixers consist essentially of a mixer with a separate water tank and a water measuring device mounted on a truck chassis. They usually are made with capacities of 1 to 5 cu. yd. (0.76 to 3.8 m^3). Agitator trucks are similar but do not have separate water tanks.

ASTM C 94 requires that concrete must be delivered and discharged from a truck mixer or agitator truck within 1½ hours after water is added to the portland cement and aggregate.

3.7.5.2 Remixing Concrete

The initial set of concrete does not usually take place until 2 or 3 hours after the portland cement is mixed with water, but fresh concrete that is left standing tends to dry out and stiffen somewhat before the portland cement sets. Such concrete may be remixed and used within 1½ hours of the time water was added to the mix, if when remixed it becomes sufficiently plastic to be compacted in the forms. However, adding water to make the mixture more workable (*retempering*) should never be permitted, because it lowers the quality just as would adding a larger amount of water in the original mixing.

3.8 Handling, Transporting, Placing, and Consolidating

Each step in handling, transporting, placing, and consolidating concrete should be controlled carefully to maintain uniformity within each batch and from batch to batch so that the complete structure has uniform quality throughout. It is essential to avoid separation of coarse aggregate from mortar and of water from the other ingredients. Equipment for conveying concrete from mixer to place of final deposit should do so without material loss or separation, so as to provide a continuous flow of concrete at the point of delivery. Segregation at the point of discharge from the mixer can be avoided by providing a down pipe at the end of the chute so that the concrete will drop vertically into the center of the receiving bucket, hopper, or form.

Concrete is handled and transported by:

- Chutes
- Push buggies
- Buckets handled by cranes
- Pumping through a pipeline
- Pneumatically forcing through a hose (shotcrete)

The methods of handling and transporting concrete and the equipment used should not place restrictions on the consistency of the concrete, which should be governed instead by the placing conditions. If placing conditions permit a stiff mix, the delivery equipment used should be designed and arranged to facilitate handling and transporting such a mix.

3.8.1 PREPARATION

Before concrete is placed, the surfaces to receive the concrete should be properly prepared, forms should be erected, and reinforcement should be installed.

3.8.1.1 Requirements Common to All Concrete

Concrete should not be placed on a frozen subgrade. Snow, ice, and debris must be removed from within forms and excavations and from slab beds before concrete is placed.

When concrete is to be placed on rock, loose material should be removed and cut faces should be nearly vertical or horizontal rather than sloping.

Chlorides should not be used in concrete in which dissimilar metals are embedded, such as reinforcing steel and heating ducts.

Reinforcing steel should be clean and free of loose rust and mill scale at the time concrete is placed. Hardened mortar should be removed from the steel.

3.8.1.2 Slabs on Grade

Subgrades should be trimmed to specified elevations and uniformly compacted before concrete is placed. Except when a vapor retarder, waterproof membrane, or separator has been installed immediately below a slab, the subgrade should be moistened (Fig. 3.8-1) to prevent too rapid extraction of water from the concrete. However, at the time of placing, there should be no mud, soft spots, or free water standing where concrete will be placed.

The Portland Cement Association (PCA) recommends that a vapor retarder be placed under every concrete slab on grade where an impervious floor finish will be installed and in every other location where the passage of water vapor through a slab on grade is undesirable. However, a vapor retarder placed in contact with the concrete will cause excess moisture in the concrete to bleed to the surface rather than downward, increasing the number of capillaries in the concrete and contributing to a weakening of the surface, more cracking, and higher permeability of the slab. Therefore, PCA recommends that a vapor retarder not be placed directly beneath a slab, but rather be covered by a 3-in. (76 mm)-thick layer of granular, self-draining material, such as sand. Refer to Section 2.3 for a detailed discussion.

FIGURE 3.8-1 Wetting the subgrade before concrete is placed. (Courtesy Portland Cement Association.)

When a vapor retarder is not used beneath a slab on grade, a separator of building paper or other material that will withstand handling and construction traffic is sometimes installed. Such separators are usually installed between the base course or fill and the slab to prevent the fines or paste of plastic concrete from seeping down into the base course or fill. However, if such a separator is complete and impervious, it should not be placed directly in contact with the slab. If it is, it may act in the same way a vapor retarder often acts and contribute to warping (curling) of the slab and exacerbate drying shrinkage cracking.

Steel reinforcing, ductwork, heating pipes, and other items embedded in a slab should be set and supported at the proper elevation before concrete is placed.

3.8.1.3 Formed Concrete

Forms should be tight, adequately braced, and constructed of material that will impart the desired texture to the finished concrete. Before concrete is placed in them, forms should be cleaned to remove dirt, nails, sawdust, wood chips, and other foreign materials. To facilitate form removal, forms should be treated with a release agent such as oil or lacquer. For architectural concrete, the agent should be a nonstaining lacquer or emulsified stearate. Wood forms should be moistened before concrete is placed, or they will absorb water from the concrete and swell.

3.8.2 PLACING

Concrete should be placed as nearly as practicable in its final position to avoid segregation resulting from rehandling or flowing. It should not be placed in large quantities at a given point and allowed to run, nor be worked over a long distance in the form. This practice results in segregation, because the mortar tends to flow out ahead of the coarser material. It also results in sloping work planes and potential weak bonding between successive concrete layers. In general, concrete should be placed in horizontal layers of uniform thickness, with each layer thoroughly compacted before the next is placed.

FIGURE 3.8-2 Placing concrete by crane-delivered bucket. (Courtesy Portland Cement Association.)

Layers should be 6 to 24 in. (150 to 610 mm) thick; thinner layers may be used in narrow and heavily reinforced forms. Each layer must be placed before the previous one stiffens.

3.8.2.1 Placing Systems

Concrete is either delivered to the site by truck or, in small projects, mixed at the site, and moved to its placement location by:

- Dumping directly from a truck's chute;
- Buggy;
- Dumping from a truck's chute into a large bucket, which is then lifted by a crane and deposited in its proper location (Fig. 3.8-2); or
- Dumping from a truck into a concrete pump, which delivers the mix to its final location through a hose. Pumps are capable of delivering concrete over long distances and up many floors (Fig. 3.8-3).

FIGURE 3.8-3 Placing concrete by pump hose. (Courtesy Portland Cement Association.)

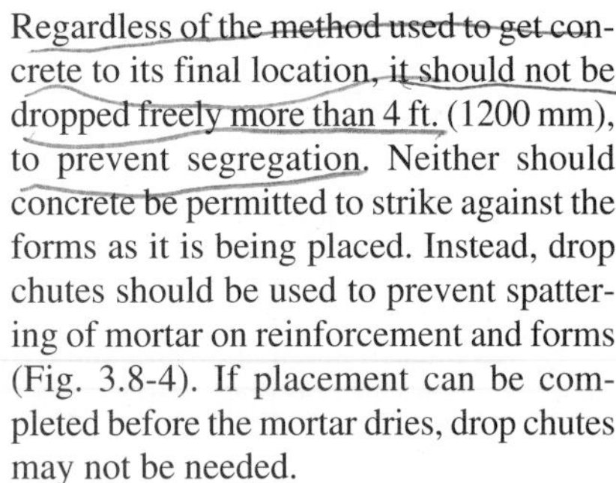

Regardless of the method used to get concrete to its final location, it should not be dropped freely more than 4 ft. (1200 mm), to prevent segregation. Neither should concrete be permitted to strike against the forms as it is being placed. Instead, drop chutes should be used to prevent spattering of mortar on reinforcement and forms (Fig. 3.8-4). If placement can be completed before the mortar dries, drop chutes may not be needed.

3.8.2.2 Order of Placing

To prevent water from collecting at the ends of forms, in corners of forms, or along form faces, the order of placing should be as follows: on walls, start at ends and work toward center; on slabs, start around the perimeter.

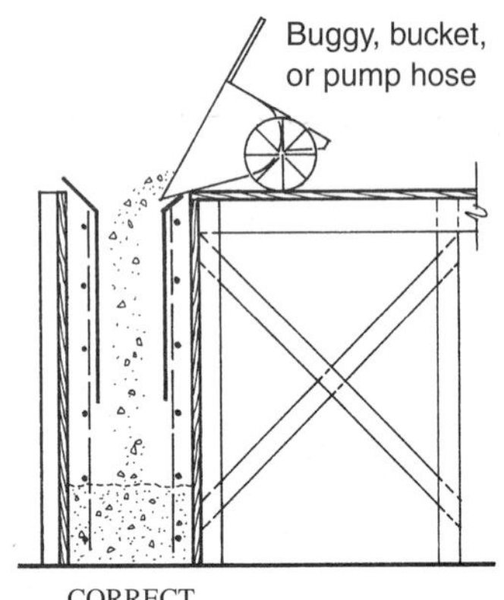

CORRECT

Separation is avoided by discharging concrete into hopper feeding into drop chute. This arrangement also keeps forms and steel clean until concrete covers them.

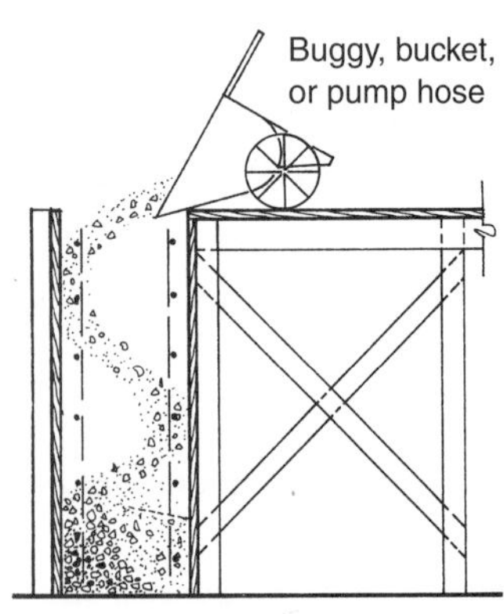

INCORRECT

Permitting concrete from chute or buggy to strike against form and ricochet on bars and form faces causes separation and honeycomb at the bottom.

FIGURE 3.8-4 Placing concrete in the top of a narrow form. (Courtesy Portland Cement Association.)

On flat surfaces, such as slabs, the placing of concrete should be started at the far end of the work so that each batch will be dumped against previously placed concrete, not away from it (Fig. 3.8-5). Concrete should not be dumped in separate piles and then leveled and worked together. If stone pockets occur, some of the excess large particles should be removed and distributed to areas where there is more mortar present to surround them.

In walls, the first batches should be placed at either end of the section; the placing should then progress toward the center. The same procedure should be used for each layer.

3.8.2.3 Placing on Hardened Concrete

To secure good bond and a watertight joint, when fresh concrete is placed on hardened concrete, the hardened concrete should be level, rough, clean, and moist, and some aggregate particles should be exposed. *Laitance* (soft layers of mortar) should be removed from the top surface of hardened concrete. An appreciable thickness of laitance indicates that segregation and bleeding have occurred, or that too much water was used in the concrete. Laitance can be prevented by using a stiffer mix or finer material, particularly in the upper part of the lift.

On floor slabs to be built in two courses, the top of the lower course may be broomed, just before it sets, with a stiff fiber or steel broom. The surface should be level, scored, and free of laitance. It must then be protected and thoroughly cleaned just before the second course is placed.

When new concrete is to be placed on old concrete, the old concrete must be roughened and cleaned thoroughly. In most cases, it is necessary to remove the entire surface to expose a new surface satisfactory for bonding.

Hardened concrete should be moistened thoroughly before new concrete is placed on it. Where concrete has dried out, it is necessary to saturate it for several days. There should be no pools of water, however, when the new concrete is placed.

Where concrete is to be placed on hardened concrete or brick, a layer of mortar should be placed on the substrate to provide a cushion against which the new concrete can be placed; this prevents stone pockets and ensures a tight joint. This mortar should be made with the same content as the concrete and should have a slump of about 6 in. (150 mm). It should be placed to a thickness of ½ to 1 in. (13 to 25 mm) and should be well worked into the irregularities of the hard surface. In two-course floor construction, a coat of portland cement and water paste of the consistency of thick paint should be brushed onto the hard surface just before the second course is placed.

3.8.2.4 Shotcrete

Shotcrete is pneumatically placed concrete, used primarily for swimming pools and other in-ground and aboveground free-form structures and for repairing damaged concrete. Its major advantage is that it does not require formwork even when placed on vertical surfaces. Shotcrete concrete is forced through a hose by compressed air onto earth, rock, existing concrete, or another layer of shotcrete.

Shotcrete is reinforced with bars and mesh, and sometimes steel fiber reinforcement, to minimize cracking. Finishes include the natural finish left by the placement gun, a screeded finish left by a sharp-edged cutting screed, a flash-coat finish produced by spraying on a final coat of shotcrete made with fine aggregate, and a float and trowel finish, which cannot be used on thin slabs.

3.8.3 CONSOLIDATING

Concrete should be compacted by a method appropriate to the material and its location to:

- Eliminate stone pockets and large air bubbles;
- Consolidate each layer with that previously placed;
- Completely embed reinforcing and fixtures;
- Bring just enough fine material to the faces and top surfaces to produce the desired finish.

3.8.3.1 Spading and Puddling

Medium- to high-slump (3 to 8 in. [76 to 200 mm]) concrete should be compacted and worked into place by spading or puddling. Spades or sticks, long enough to reach the bottoms of the forms and thin enough to pass between the reinforcing steel and forms, should be used.

3.8.3.2 Vibrating

Low- to medium-slump (1 to 6 in. [25 to 150 mm]) concrete should be compacted using mechanical vibrators either within the concrete (internal vibrators) or on the top of the concrete or on the forms (external vibrators). Relatively thin slabs that are not heavily reinforced may be vibrated by mechanical beam or truss screeds. Thick slabs, slabs with structural reinforcement, and thicker concrete members, like footings, walls, columns, beams, and girders, should be vibrated with internal vibrators (Fig. 3.8-6).

a

CORRECT

Concrete should be dumped into the face of previously placed concrete.

INCORRECT

Dumping concrete away from previously placed concrete causes separation.

b

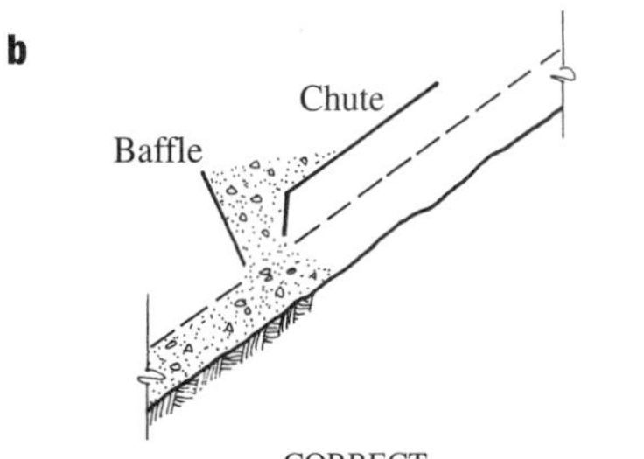

CORRECT

A baffle and drop at end of chute will avoid separation and concrete remains on slope.

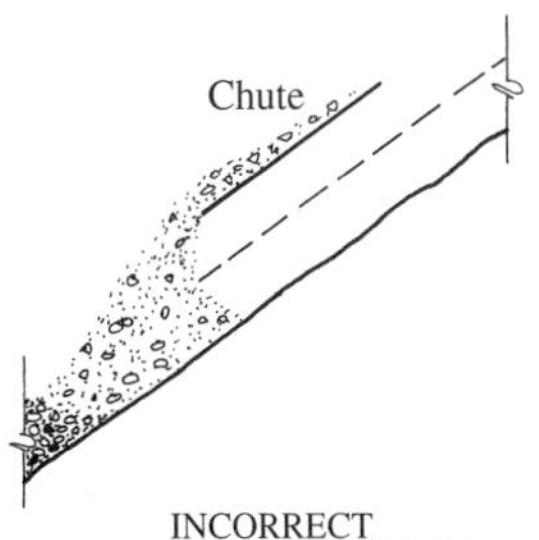

INCORRECT

Discharging concrete from free end chute onto a slope causes separation of rock which goes to bottom of slope. Velocity tends to carry concrete down the slope.

FIGURE 3.8-5 (a) Placing slab concrete; (b) placing concrete on a sloping surface. (Courtesy Portland Cement Association.)

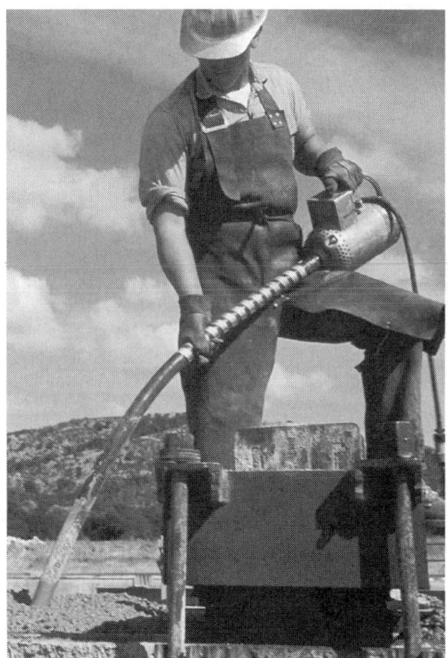

FIGURE 3.8-6 **Using an internal vibrator.** (Courtesy Portland Cement Association.)

a

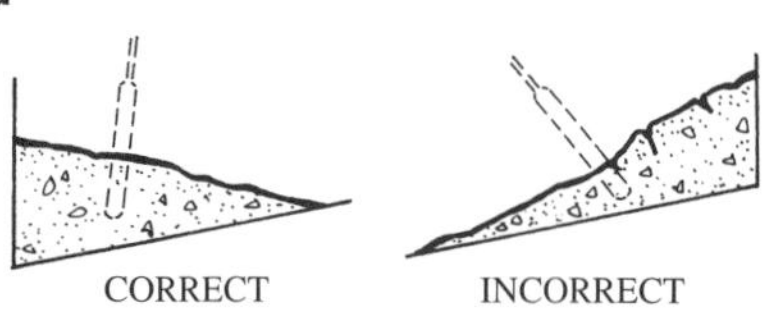

Correct: Start placing at bottom of slope so that compaction is increased by weight of newly added concrete. Vibration consolidates the concrete.

Incorrect: When placing is begun at top of slope, the upper concrete tends to pull apart, especially when vibrated below because this starts flow and removes support from the concrete above.

b

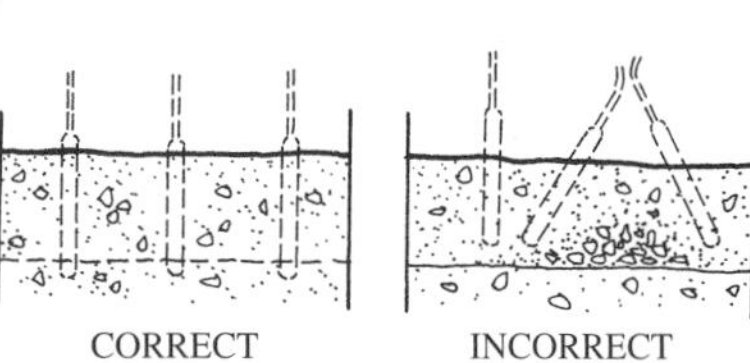

Correct: Start placing at bottom of slope so that compaction is increased by weight of newly added concrete. Vibration consolidates the concrete.

Incorrect: Haphazard random penetration of the vibrator at all angles and spacings without sufficient depth will not assure intimate combination of the two layers.

FIGURE 3.8-7 **(a) Consolidation of concrete on slopes; (b) consolidation of concrete in multiple lifts.** (Courtesy Portland Cement Association.)

When concrete is vibrated, the internal friction between the coarse aggregate particles is temporarily eliminated and the concrete behaves as a liquid; it settles in the forms under the action of gravity and the entrapped air bubbles rise more easily to the surface. Friction is reestablished as soon as vibration stops. Therefore, mixes of lower water content or leaner mixes for a given water content can be used. If less water is used per unit volume of portland cement, the concrete will be of better quality; if less portland cement per unit volume of water is used, the concrete will be more economical. Vibrators should not be used to push or move concrete laterally over long distances within a form. The concrete should be deposited as near its final position as possible. It should be distributed in layers and then vibrated (Fig. 3.8-7). Some hand spading or puddling may be necessary along with the vibration to secure smooth surfaces and reduce pitting on formed surfaces.

Precautions should be taken not to overvibrate to the point that segregation results, especially when the concrete has a higher slump than is necessary for vibration. Precaution and judgment must be used to be sure that complete consolidation is secured without segregation and that no areas are missed. Sufficient vibration is usually indicated by a line of mortar along the forms and by the submerging of the coarse aggregate particles in the mortar.

3.9 Finishing

Concrete may be finished in several ways, depending on its location, the concrete material used, and the aesthetic effect desired. In this section, general finishing procedures are summarized first, followed by techniques to produce special surface finishes.

3.9.1 FINISHING STANDARD-WEIGHT CONCRETE SLABS

The finishing of standard-weight concrete slabs proceeds through several steps in a defined order: screeding, leveling, edging, jointing, floating, troweling, and surface texturing.

3.9.1.1 Bleeding

Generally, the dry materials used in making quality concrete are heavier than water. Thus, shortly after placement, they will have a tendency to settle to the bottom and displace the mixing water to the surface, which is called *bleeding*. This occurs more readily when the concrete is not air entrained.

If any operation is performed on the surface while bleed water is present, serious scaling, dusting, or crazing of the hardened concrete can result. This point cannot be overemphasized and is the basic rule for successful finishing of concrete surfaces. Section 3.9.6 contains a more detailed discussion of scaling, crazing, and dusting.

Placing, screeding, consolidating, and leveling of placed concrete must be performed before bleeding takes place. Therefore, concrete should not be placed faster than it can be spread, struck off, and leveled, and should not be allowed to remain in wheelbarrows, buggies, or buckets any longer than is absolutely necessary. It should be dumped and spread as soon as possible and immediately struck off to proper grade and leveled.

3.9.1.2 Screeding

The surface of newly placed concrete is struck off (*screeded*) by moving a straightedge back and forth with a sawlike motion across the top of the forms and screeds. If mechanical vibrating equipment is used in the striking-off process, the need for leveling may be eliminated. Whether boards, vibratory screeds, or roller screeds are used, a small amount of concrete should always be kept ahead of the straightedge to fill in low spots and maintain a plane surface (Fig. 3.9-1).

Of all the placing and finishing operations, screeding the surface to a predetermined grade has the greatest effect on surface tolerances. Concrete should be struck off immediately after it is placed, and screed stakes used to establish sur-

FIGURE 3.9-1 Striking off (screeding). (Courtesy Portland Cement Association.)

face elevation should be removed as the work progresses so that workers do not have to walk back into areas that have already been struck off.

3.9.1.3 Leveling

Leveling is the bringing of a concrete surface to true grade with enough mortar to produce the desired finish. After a concrete slab has been screeded, it should immediately be smoothed with a *darby* to level raised spots and fill depressions (Fig. 3.9-2) left after screeding. Long-handled floats, called *bull floats*, of either wood or metal are sometimes used instead of darbies to smooth and level concrete surfaces (Fig. 3.9-3). Because it is hard to produce surfaces in plane near the edges of a slab using bull floats, darbies are sometimes needed after the bull floating has been done.

Leveling is sometimes called "darbying or bull floating," but this terminology requires that both terms be used or that the architect dictate which is to be used, which is inappropriate. In its literature the PCA uses the broader term *leveling*.

The purpose of leveling is to eliminate the ridges and voids left by screeding. In addition, it should slightly embed the

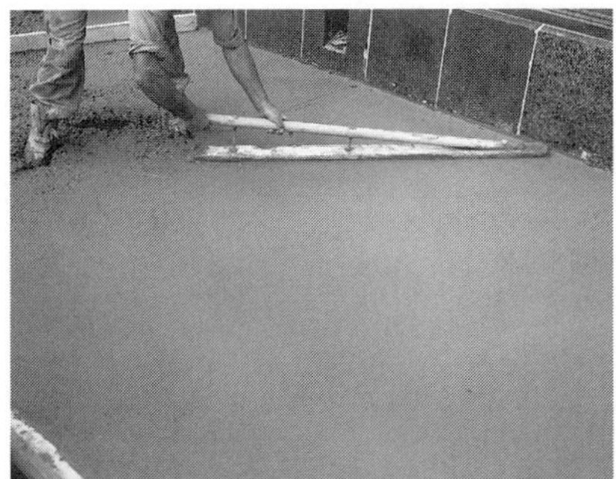

FIGURE 3.9-2 Leveling using a darby. (Courtesy Portland Cement Association.)

FIGURE 3.9-3 Leveling using a bull float. (Courtesy Portland Cement Association.)

coarse aggregate, thus preparing the surface for subsequent finishing operations of edging, jointing, floating, and troweling.

A slight stiffening of the concrete is necessary after leveling before further finishing operations are started. No subsequent operations should be performed until the concrete will sustain foot pressure with only about ¼-in. (6 mm) indentation.

3.9.1.4 Edging

When all bleed water and water sheen have left the surface and the concrete has started to stiffen, other finishing operations such as *edging* may be started (Fig. 3.9-4). Edging rounds off the formed edge of a slab to prevent chipping or other damage. An *edger* should be run back and forth until a finished edge is produced. All coarse aggregate particles should be covered, and the edger should not leave too deep an impression in the slab. If it does, the indentation may be difficult to remove with subsequent finishing operations.

Usually, edging is not required for most interior slabs on grade and is more commonly performed on sidewalks, driveways, and steps. An edger should not

FIGURE 3.9-4 Edging. (Courtesy Portland Cement Association.)

be used when a slab is to be finished with resilient flooring requiring a smooth, level subfloor. Edges at construction joints in the slab may be ground lightly with a silicon carbide stone to remove irregularities after forms are stripped and before the adjacent slab is placed.

3.9.1.5 Jointing

Except when joints will be later sawed, immediately following or during edging, premolded inserts are placed in concrete slabs to control cracking in the concrete as a result of shrinkage. Refer to Section 3.6 for a detailed discussion about joints.

3.9.1.6 Floating

After edging and hand-jointing operations, a slab should be floated. Many variables, such as concrete temperature, air temperature, relative humidity, wind, and other factors, affect the process and make it difficult to set a definite time to begin floating. This knowledge comes only through job experience. In general, floating may be done when the water sheen has disappeared and the concrete will support the weight of the finisher.

The purpose of floating is to:

- Embed large aggregate just beneath the surface;
- Remove slight imperfections, humps, and voids to produce a level or plane surface;
- Consolidate mortar at the surface in preparation for other finishing operations;
- Open the surface to permit excess moisture to escape.

Aluminum or magnesium floats should be used, especially on air-entrained concrete. This type of metal float greatly reduces the amount of work to be done by the finisher because the float slides more readily over the concrete surface, has a good floating action, and forms a smoother surface texture than a wood float. A wood float tends to stick to and "tear" the concrete surface.

The marks left by edgers and jointers should be removed by floating (Figs. 3.9-5 and 3.9-6) unless such marks are desired for decorative purposes, in which case, the edger or jointer should be rerun after the floating operation.

3.9.1.7 Troweling

Troweling is done on slabs that are to be left exposed or to receive thin finishes,

FIGURE 3.9-5 **Hand floating.** (Courtesy Portland Cement Association.)

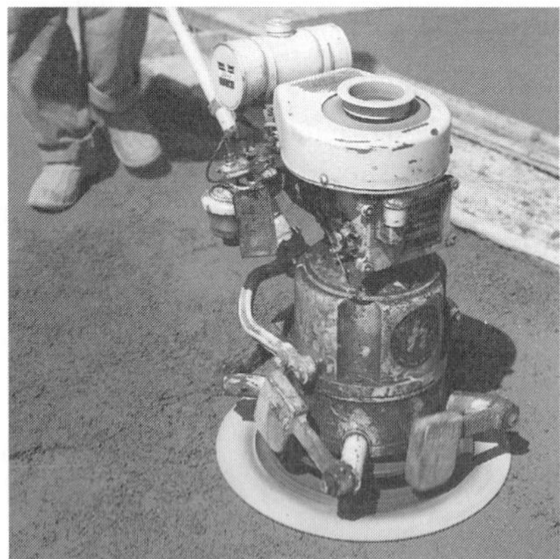

FIGURE 3.9-6 **Power floating.** (Courtesy Portland Cement Association.)

such as resilient flooring, carpet, tile, or paint. When troweling is required, the surface should be steel-troweled immediately after floating (Figs. 3.9-7 and 3.9-8). It is customary for a cement mason using hand tools to float and then steel-trowel an area before moving the knee boards. If necessary, tooled joints and edges should be rerun before and after troweling to maintain true lines, proper depths, and uniformity and to remove kinks.

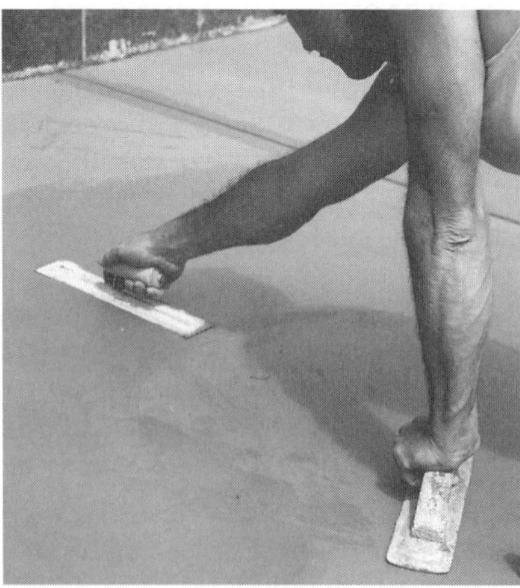

FIGURE 3.9-7 **Hand troweling.** (Courtesy Portland Cement Association.)

FIGURE 3.9-8 **Power troweling. Also available are self-propelled power trowels on which the operator rides in a seat.** (Courtesy Portland Cement Association.)

The purpose of troweling is to produce a smooth, hard surface. For the first troweling, whether by power or by hand, the trowel blade must be kept as flat against the surface as possible. If tilted or pitched at too great an angle, an objectionable *washboard* or *chatter* surface will result. For first troweling, a new trowel is not recommended. An older trowel that has been broken in can be worked quite flat without the edge digging into the concrete. The smoothness of a surface can be improved by timely additional trowelings. There should be a lapse of time between successive trowelings to permit the concrete to increase its set. As the surface stiffens, each successive troweling should be made by a smaller-sized trowel tipped at a progressively higher angle so that sufficient pressure can be applied for proper finishing.

For exposed slabs, additional troweling increases the compaction of fines at the surface, giving greater density and better wear resistance. A second troweling is recommended even if the slab is to be finished with resilient flooring, because it results in closer surface tolerances and a better surface to receive the flooring.

3.9.1.8 Broom Finish

Steel-troweled concrete surfaces are very smooth and can become slippery when wet. They can be slightly roughened to produce a nonslip surface by *brushing* or *brooming* them. A brushed surface is made by drawing a broom over the surface after steel troweling. Notice in Figure 3.9-9 that the broom is being drawn right over the edge joint, but the joint is not being marred. This indicates that the concrete surface has been steel-troweled properly and is hard enough. Other types of surface finishes are discussed in Section 3.9.4.

FIGURE 3.9-9 **Broom finishing.** (Courtesy Portland Cement Association.)

3.9.2 FINISHING AIR-ENTRAINED CONCRETE SLABS

The microscopic air bubbles in air-entrained concrete tend to hold the ingredients, including water, in suspension. This type of concrete requires less mixing water and still has good workability with the same slump. Since there is less water and it is held in suspension, little or no bleeding occurs. There is, therefore, no need to wait for the evaporation of free water from the surface, and floating and troweling can and should be started as soon as the slab can support the finisher and equipment. Many horizontal surface defects and failures are caused by performing finishing operations while bleed water or excess surface moisture is present. Therefore, better results are generally accomplished with air-entrained concrete.

As with other concrete, if floating is done by hand, an aluminum or magnesium float should be used. A wood float drags and greatly increases the amount of work necessary to accomplish the same result. If floating is done with power equipment, the only major difference between the finishing procedures for air-entrained concrete and those for other concrete is that floating may be started sooner on air-entrained concrete.

3.9.3 FINISHING LIGHTWEIGHT STRUCTURAL CONCRETE SLABS

Finishing operations for slabs made with lightweight structural concrete containing coarse aggregates of expanded clay, shale, or slag vary somewhat from those used on slabs made with normal-weight concrete. When the surface of the concrete is worked, there is a tendency for the coarse lightweight aggregate, rather than the mortar, to rise to the surface. Lightweight concrete can be easily finished if the following precautions are observed: (1) the mix should be properly proportioned and should not be over- or undersanded in an attempt to meet unit weight requirements, and (2) finishing should not be started too early and the concrete should not be overworked or overvibrated. A well-proportioned mix can generally be placed, struck off, leveled, and floated with less effort than is necessary for normal-weight concrete. Excessive leveling and floating are principal causes of finishing problems, because the heavier mortar is driven down and the coarse aggregate is brought to the surface.

3.9.4 SPECIAL SLAB FINISHES

Due to the plastic quality of concrete, many surface finishes can be applied. The surface may be scored or tooled with a jointer in decorative and geometric patterns. Some of the more common special finishes are discussed here.

3.9.4.1 Exposed Aggregate

An exposed aggregate surface is often chosen for any area where a special textural effect is desired (Fig. 3.9-10). If the surface is ground and polished, it is suitable especially for such places as entrances, interior terraces, and courtyards.

Selection of aggregates is so important that test panels should be made before the job is started. Colorful gravel aggregate that is quite uniform in gradation and in sizes ranging from ½ to ¾ in. (13 to 19 mm) is recommended. Flat, sliver-shaped particles or aggregates less than ½ in. (13 mm) in diameter should be avoided, because they become dislodged during exposing operations. Exposing the aggregate used in ordinary concrete is generally unsatisfactory, as this will not necessarily reveal a high percentage of coarse aggregates.

A 5½- to 6-sack concrete with a maximum 3 in. (76 mm) slump should be used. Immediately after the slab has been screeded and leveled, the selected aggregate should be scattered by hand and evenly distributed so that the entire surface is completely covered (Fig. 3.9-11a). The initial embedding of the aggregate is usually done by patting with a darby or the flat side of a 2 × 4 (50 × 100) (Fig. 3.9-11b). After the aggregate is quite thoroughly embedded and as soon as the concrete will support the weight of a mason on kneeboards, the surface should be hand-floated with a float or darby (Fig. 3.9-11c). This operation should be performed so thoroughly that all aggregate is embedded entirely just beneath the surface. The grout should completely surround and slightly cover all aggregate, leaving no holes or openings in the surface.

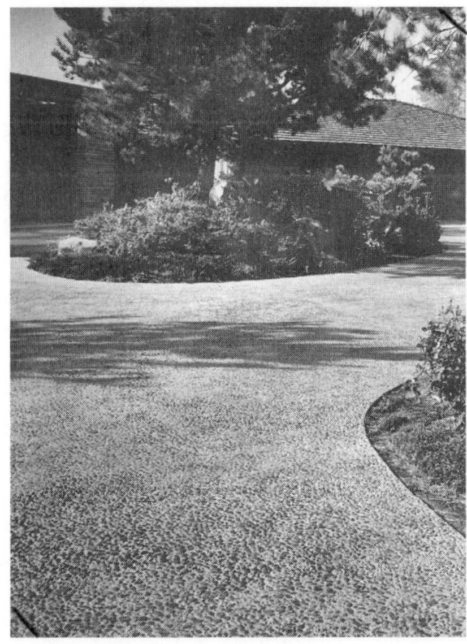

FIGURE 3.9-10 Exposed-aggregate surface. (Courtesy Portland Cement Association.)

Shortly after this floating, a reliable retarder may be sprayed or brushed over the surface following the manufacturer's recommendations. On small jobs a retarder may not be necessary. Retarders generally are used on large jobs for better control of exposing operations. Where a retarder has been used, exposing the aggregate is usually done some hours later by brushing and hosing with water. However, the manufacturer's recommendations should be followed closely.

Whether or not a retarder has been used, the proper time for exposing aggregate is quite critical. It should be done as soon as the grout covering the aggregate can be removed by simultaneously brushing and hosing (Fig. 3.9-11d), without overexposing or dislodging the aggregate. If during exposing it is necessary for masons to move about on the surface, kneeboards should be used gently. If possible,

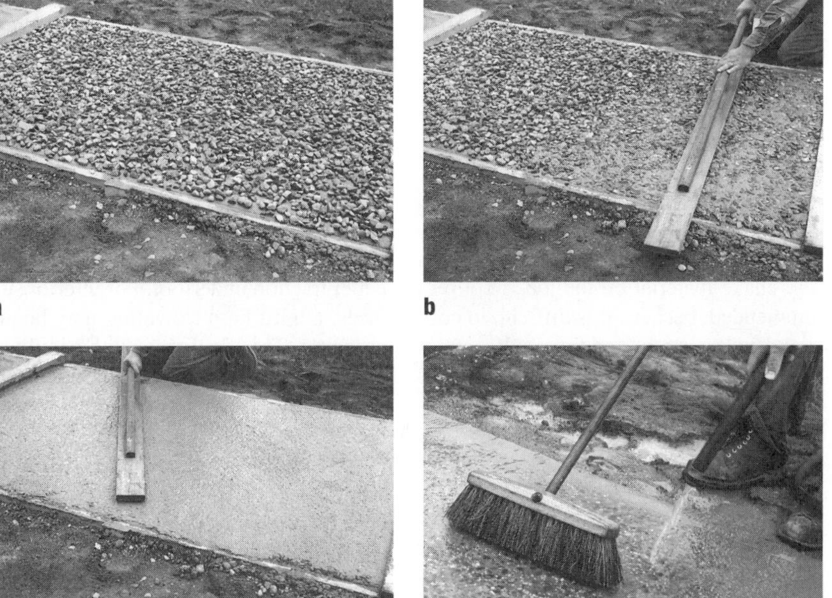

FIGURE 3.9-11 Producing an exposed-aggregate surface: (a) decorative aggregate is distributed to cover the entire surface; (b) initial embedding of aggregate; (c) floating the surface to completely cover aggregate with grout; (d) exposing aggregate is done by simultaneously brushing and hosing with water. (Courtesy Portland Cement Association.)

however, such movement should be avoided because of the risk of breaking aggregate bond.

For interior areas and where a smooth surface is desired, no retarder is used and exposure of the aggregate is accomplished entirely by grinding. This may be followed by polishing, which will produce a surface similar to terrazzo.

Because the aggregate completely covers the surface, tooled joints in this type of work are impractical. Decorative or control joints are accomplished best by sawing. Control joints should be cut 4 to 12 hours after the slab is placed and should be at least one-fifth the depth of the slab. A small-radius edger should be used before and immediately after the aggregate has been embedded to provide a more decorative edge to the slab. Another method of providing control joints is to install permanent strips of redwood before placing concrete.

Another method of placement requires spreading a 1-in. (25.4 mm)-thick or thicker top course containing the special aggregate.

As in all concrete work, exposed aggregate slabs should be cured thoroughly.

3.9.4.2 Dry-Shake Color Surface

Dry-shake colored-concrete surfaces are used for interior areas, terraces, decorative walks, driveways, and other areas where a colored surface is desired. They are made by applying any of a number of proprietary dry-shake materials. Their basic ingredients are mineral oxide pigment, white portland cement, and specially graded silica sand or fine aggregate. Other pigments are inappropriate and should not be used. Selecting the ingredients, proportioning, and mixing of a dry-shake material on the job is not recommended, because it is difficult to control the proportions and the results cannot be predicted with accuracy. Field-mixed colors are likely to produce a mottled, uneven coloring.

After concrete has been struck off and leveled and free water and excess moisture have evaporated from its surface, the surface should be floated either by power or by hand. If by hand, a magnesium or aluminum float should be used. Preliminary floating should be done before the dry-shake material is applied to bring up enough moisture for combining with the dry-shake material. Floating also removes ridges and depressions that may cause variations in color intensity. Immediately following floating, the dry-shake material is shaken evenly by hand over the surface, following the dry-shake material manufacturer's instructions. If the wrong amount of the dry-shake material is applied in one spot, nonuniformity in color and possibly surface peeling will result.

Unless the manufacturer recommends otherwise, the first application of a colored dry-shake should use about two-thirds of the total amount needed. Amounts are usually specified in pounds per square foot (kg/m^2) of surface. In a few minutes, this dry material will absorb some moisture from the plastic concrete and should then be thoroughly floated into the surface, preferably with a power float. Immediately following, the balance of the dry-shake material should be distributed evenly over the surface, thoroughly floated again, and made part of the surface, taking care that a uniform color is obtained.

Tooled edges and joints should be run both before and again after the dry-shake application.

Shortly after final floating, the surface should be power-troweled. If done by hand, troweling should immediately follow final floating. After the first troweling, whether by power or by hand, there should be a lapse of time to allow the concrete to increase its set. The length of elapsed time should depend on the temperature and humidity present. Concrete may be troweled a second time to improve the texture and to produce a denser, harder surface.

For exterior surfaces, two trowelings are usually enough. Then a fine, soft-bristled pushbroom may be drawn over the surface to produce a roughened texture for better traction underfoot. For interior surfaces, a third hard troweling may be appropriate. This final troweling should be done by hand to eliminate washboard or trowel marks and to produce a smooth, dense, hard-wearing surface.

Dry-shake colored concrete, like with other types of freshly placed concrete, must be cured thoroughly. After thorough curing and surface drying, interior surfaces may be given at least two coats of special concrete floor wax, which is available from various reliable manufacturers. It contains the same mineral oxide pigment used in the dry-shake. Care should be taken to avoid staining, such as by dirt or foot traffic, during the curing or drying period and before waxing.

3.9.4.3 Swirl Design

A swirl design, nonskid surface texture can be produced on a slab with a magnesium or aluminum float or a steel finishing trowel. When a float is used, the finish is called a swirl float finish; when a trowel is used, a swirl trowel finish.

After a concrete surface has been struck off, leveled, floated, and steel-troweled, it is ready to be given either a swirl float or swirl trowel finish. A float should be worked flat on the surface in a semicircular or fanlike motion. Pressure applied on the float with this motion will give a rough-textured swirl design, as shown in Figure 3.9-12a.

With the same motion, using a steel trowel held flat, a cement mason can obtain a finer-textured swirl design on a concrete surface (Fig. 3.9-12b). Moist curing of the slab is the final operation.

3.9.4.4 Keystone Finish

A keystone finish is a travertine-like texture. It can be used on a terrace, garden

a

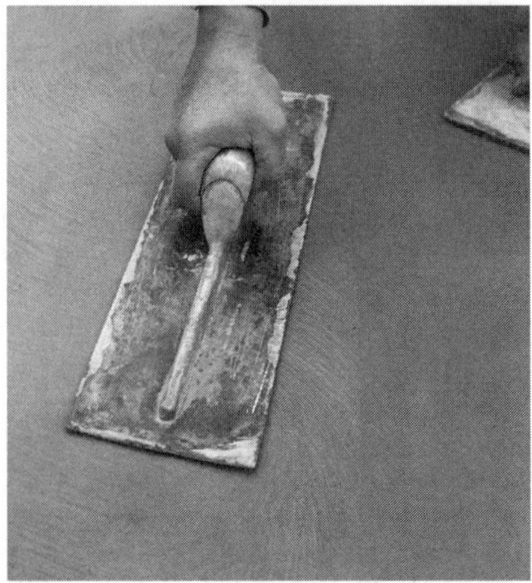

b

FIGURE 3.9-12 Producing a swirl design on a concrete surface: (a) using an aluminum float; (b) using a steel finishing trowel. (Courtesy Portland Cement Association.)

walk, driveway, deck around a swimming pool, and in any other location where an unusually decorative flat concrete surface is desired. It should not be used where it will be subjected to freezing weather.

After a concrete slab has been screeded, leveled, and edged in the usual manner, it is broomed with a stiff-bristled broom to ensure that the finish (mortar) coat will bond.

A mortar coat is made by mixing one sack of white portland cement and 2 cu. ft. (0.57 m^3) of sand with about ½ lb. (0.23 kg) of a color pigment. Yellow is usually selected to tint the mortar coat, but any mineral oxide color may be used. Care must be taken to keep the proportions exactly the same for all batches. Enough water is added to make a soupy mixture with the consistency of thick paint.

Mortar-coat material is placed in pails and thrown vigorously on the slab with a stiff-bristle brush, called a *dash brush*, to produce an uneven surface with ridges and depressions. The ridges should be about ¼ to ½ in. (6.5 to 13 mm) high.

When the surface is hard enough to permit the use of kneeboards, it is troweled with a steel trowel to flatten the ridges, leaving it smooth in the high spots and rough or coarse-grained in the low spots. Many interesting textures can be produced, depending on the amount of troweling (Figs. 3.9-13a and 3.9-13b). A textured slab can then be scored into random geometrical designs before it cures.

a

b

FIGURE 3.9-13 **(a) Keystone or travertine finish; (b) random-scored keystone finish.** (Courtesy Portland Cement Association.)

3.9.5 FINISHING FORMED SURFACES

The way in which formed surfaces are finished depends on their exposure and the aesthetic effect the designer is trying to achieve. There are several classifications used in the industry to describe the various finishes.

3.9.5.1 Rough Form Finish

A *rough form finish* is left by the formwork in locations that will be concealed from view in the completed building. The surface is improved only by the repair of tie holes and defective areas, removal of the fins left by concrete leaking out through joints in the forms, and chipping off or grinding down of projections larger than ¼ in. (6.4 mm) high.

3.9.5.2 Smooth Form Finish

A *smooth form finish* is used in two instances. The first is when form liners are used to impart a pattern or texture to the concrete. The second is for concrete that will receive paint, waterproofing, dampproofing, veneer plaster, or another thin coating or covering material. After the forms are removed, fins and projections are removed completely and imperfections are repaired. The surface is made smooth or of the texture imparted by the forms to adjacent areas, as applicable. Tie holes may be patched or left alone, depending on the designer's intent. Sometimes tie holes are used as a decorative element. *Screw ties* with cones are usually used when tie holes will not be filled (see Section 3.3.2.2).

3.9.5.3 Smooth Rubbed Finish

A *smooth rubbed finish* is applied when a smooth form or form liner is used. This finish is applied by moistening the new concrete within one day after the forms have been removed and rubbing it with an abrasive, such as a carborundum brick. Cement grout is not used during this finishing process.

3.9.5.4 Grout-Cleaned Finish

A *grout-cleaned finish* is also used on concrete that has been placed against a smooth form or form liner. It consists of applying to the surface a thick paint-consistency grout of 1 part portland cement and 1½ parts fine sand, mixed with equal parts of an acrylic or styrene butadiene bonding material and water. A blend of standard (gray) and white portland cement is used to achieve the color and tone desired by the designer. After the grout has been applied, the finish is completed by rubbing it with burlap.

3.9.6 SURFACE DEFECTS

There are many causes for concrete surface defects. Some of the major defects, their causes, and the construction techniques that should be used to prevent them are described here. Many defects result from improper curing or errors in mixing or finishing.

3.9.6.1 Scaling

Scaling is the breaking away of the hardened concrete surface of a slab to a depth of about ⅙ to 3/16 in. (1.6 to 4.8 mm). It usually occurs at an early age of the slab.

Scaling of a slab may occur if it is subjected to cycles of freezing and thawing soon after the slab has been placed. A favorable temperature must be maintained long enough to prevent injury (see Section 3.10.3.1). Cycles of freezing and thawing and applications of de-icing salts on non-air-entrained concrete can also cause scaling. This is why air-entrained concrete is recommended for all severe exposure conditions.

Performing a finishing operation while free excess water or bleed water is on the surface causes segregation of the surface fines (sand and portland cement) and also brings a thin layer of neat portland cement, clay, and silt to the surface, leaving another layer of nearly clean, washed sand that is not bonded to the concrete under it. To prevent scaling from this cause, water should be allowed to evaporate from the surface or be forced to evaporate by fans or blower-type heaters, or it should be removed by dragging a rubber garden hose over the surface before finishing operations begin.

3.9.6.2 Crazing

Crazing is the occurrence of numerous fine cracks in the surface of a newly hardened slab due to surface shrinkage. These cracks form an overall pattern similar in appearance to a crushed eggshell.

Crazing can be the result of rapid surface drying, usually caused by either high air temperatures, hot sun, drying winds, or a combination of these. It can be lessened by curing with water, because water will maintain or lower the concrete surface temperature (see Section 3.10.3.3).

Premature floating and troweling when there is an excess amount of moisture at the surface, or while the concrete is still too plastic, brings an excess amount of fines and moisture to the surface. A rapid loss of this moisture will cause shrinkage at the surface, which may result in crazing. Floating and troweling should be delayed until excess moisture has evaporated from the surface and the concrete has started its initial set. Excess moisture can be avoided by reducing the slump and using air-entrained concrete.

Overuse of a vibrating screed, darby, or bull float may also contribute to crazing by working an excess of mortar to the surface, which tends to cause additional surface shrinkage.

3.9.6.3 Dusting

Dusting is the appearance of a powdery material on the surface of a newly hardened concrete slab.

An excess of harmful fines (clay or silt) in a concrete mix with the sand and portland cement at the surface can result in dusting. This condition emphasizes the need to use clean and well-graded coarse and fine aggregates.

Premature troweling and floating mix excess surface water with surface fines, weakening the portland cement paste. Troweling should be delayed until all free water or excess moisture has disappeared and the concrete has started its initial set.

When carbon dioxide, as may be emitted from open portable fossil-fuel heaters and gasoline engines, comes in contact with the surface of plastic concrete, a reaction takes place that impairs proper hydration. Such fumes should be vented to the outside and sufficient fresh air ventilation provided.

Condensation sometimes occurs on a concrete surface before floating and troweling have been completed, usually in the spring and fall when materials have become cold due to low night temperatures. If possible, this condition should be anticipated and the concrete should be heated, or at least hot water should be used for mixing. If this is impossible, blower-type heaters should be used to lower the humidity directly over the slab, and fans should be used to increase air circulation. If heaters and fans are not available, windows and doors should be opened. When condensation is present, floating and initial troweling should be held to a minimum and the concrete surface should not be given a second troweling.

Neat portland cement and mixtures of portland cement and fine sand should never be used as a dry-shake. Because condensation may occur for several hours while concrete is beginning to harden, emergency measures may have to be taken in order to finish a slab. A well-mixed dry mixture of 1 part portland cement and 1 part well-graded concrete sand may be evenly and lightly distributed over a concrete surface, if it is followed at once by troweling. There should be no second troweling, because additional condensation may take place after the first troweling.

Winter-protection heaters may lower the relative humidity around concrete excessively and inhibit proper hydration of the portland cement. Water jackets should be placed on heaters to increase the relative humidity by evaporation, and moist curing methods should be employed. Heaters should be moved periodically so that no area will be subjected to an extreme or harmful amount of heat.

Proper curing for the correct length of time is essential. Concrete that is not cured properly will often be weak and its surface will be easily worn by foot traffic.

3.10 Curing and Protection

Concrete curing is an important construction operation that is often neglected. Even when concrete has been properly mixed, carefully placed, and correctly finished, a poor job will result if proper curing techniques are not followed.

3.10.1 PURPOSE OF CURING

The strength and watertightness of concrete improve with age as long as conditions are favorable for continued *hydration* of its portland cement. Other qualities, such as weathering and resistance to freezing and thawing, are affected similarly. Improvement is rapid early and continues more slowly for an indefinite period as long as moisture and favorable temperatures are present.

Fresh concrete contains more than enough water for hydration, but under most job conditions much of this water is lost by evaporation unless certain precautions are taken. Hydration proceeds at a much slower rate as the temperature drops until it stops altogether at about 14°F (–10°C). However, at about 40°F (4.4°C), hydration becomes too slow for concrete work to proceed without special provisions to heat the ingredients so that the concrete will be at least 50°F (10°C) when it is placed. Therefore, concrete should be protected so that moisture is not lost during the early stages of hardening, and it should be kept at a temperature that will promote hydration and also protect against injury from subsequent construction activities.

3.10.2 CURING METHODS

Concrete can be kept moist in a number of ways, including leaving its forms in place, sprinkling, ponding, and using moisture-retention covers or a seal coat that is applied as liquid and then hardens to form a thin membrane.

Curing methods that use water, such as sprinkling, ponding, and using wet coverings, are the most effective means and should be used whenever practicable. Although not usually as effective as curing with wet coverings, moisture retention membranes are widely used because of their convenience. In any case, the most important consideration is that the surface of the concrete be kept uniformly wet or moist. Partial drying of any part of the surface of a slab can result in crazing and cracking. Curing compounds will affect the bonding properties of adhesives and should not be used on slabs that are to receive resilient flooring or other adhesive-applied finishes unless they are known to be compatible with the adhesive. Curing compounds are also not recommended for slabs that will be given a further surface treatment.

3.10.2.1 Leaving Forms in Place

Leaving forms in place is of great assistance in retaining moisture. In hot, dry weather wood forms will dry out and should be kept moist by sprinkling. In all cases, exposed surfaces must be protected from moisture loss.

3.10.2.2 Sprinkling

When concrete is kept moist by sprinkling, the drying of surfaces between water applications must be prevented. Alternate cycles of wetting are conducive to surface crazing or cracking. A fine spray of water applied continuously provides a more constant supply of moisture and is better than copious applications of water with periods of drying between.

3.10.2.3 Ponding

Ponding is sometimes used on flat surfaces, such as pavements, sidewalks, and floors. A small dam of earth or other water-retaining material is placed around the perimeter of a surface, and the enclosed area is kept flooded with water. Ponding gives a more constant condition than does sprinkling.

3.10.2.4 Curing Covers

Moisture-retaining covers such as wet sand, burlap, polyethylene film, or other materials are sometimes used. Care should be taken to cover the entire concrete surface, including exposed sides of members and the sides of pavements and sidewalks where the forms have been removed. Covers should be kept constantly moist enough to provide a film of moisture on the concrete surface.

WATERTIGHT COVERS

Watertight paper is used on floors and other horizontal areas. It should be nonstaining and strong enough to withstand wind and the abrasive action of workers walking over it. Seams should be overlapped several inches and covered with glued tape. Polyethylene films often are used as watertight curing covers.

SEALING COMPOUNDS

Sealing curing compounds are available in black, red, colorless, and white pigmented coatings. Some of them can be applied in one coat, but two coats will give better results. The application should be made immediately after the concrete has been finished. If there is a delay, the concrete should be kept moist until the application is made. In extremely hot weather it is advisable to cover a slab with water for 12 hours before using curing compounds. On formed surfaces, such as beams and columns, the forms should be removed, the concrete sprayed lightly with water, and then a sealing compound applied. Where the membrane must be protected against traffic, it should be covered with at least 1 in. (25 mm) of sand or earth or by some other means. Such a protective cover should be placed not sooner than 24 hours after the sealing compound.

3.10.3 CURING TEMPERATURES

Temperature affects the rate of chemical reactions between portland cement and water. Consequently, temperature affects the rate at which concrete hardens, as well as its strength and other properties.

The temperatures and time periods in the following discussions are approximations to introduce the subject. The actual required temperatures and times vary with concrete thickness and air temperature. Refer to American Concrete Institute (ACI) publications ACI 305R and ACI 306 for more specific requirements.

3.10.3.1 Cold-Weather Construction

In cold weather it is often necessary to heat a concrete's ingredients, and to either cover fresh concrete or provide a heated enclosure, or both. Hydration of portland cement causes some heat to be generated, and if this heat is retained, it raises the temperature of the concrete. In cold weather, concrete should have a temperature at the time of placing between 50°F and 70°F (10°C and 21°C). In no case should its ingredients be heated to the point where the temperature of fresh concrete is above 70°F (21°C). If it is, built-in thermal stresses will result.

Therefore, when the temperature of the surrounding air is below 40°F, the water and aggregate used in a concrete should be heated so that the temperature of the mixed concrete at the time of placing is between 50°F and 70°F (10°C and 21°C). The temperature of the air adjacent to the placed concrete should then be kept at between 50°F and 70°F (10°C and 21°C) during the curing period. This may require covering the concrete with an insulating material like straw, surrounding the concrete with a heated enclosure, or both, until sufficient time has passed for the concrete to remain undamaged by the prevailing conditions. This time varies with the thickness of the concrete, the air temperature, and other conditions, such as high winds, that may be present. The normal range is above 70°F (21°C) for 3 days or above 50°F (10°C) for 5 days when normal portland cement (Type I) is used; or above 70°F (21°C) for 2 days or 50°F (10°C) for 3 days when high-early-strength portland cement (Type III) is used. This range may vary, however, if high winds occur, and will have to be increased if the concrete has not properly cured within the specified times.

Fresh concrete should not be placed on a frozen subgrade. When such a subgrade thaws, there is likely to be uneven settlement and cracking of the concrete member it supports. The insides of forms and reinforcing steel and embedded fixtures should be free of ice at the time concrete is placed. A thin layer of warm concrete should not be placed on cold, hardened concrete, for the thick upper layer will shrink as it cools and the lower layer will expand as it warms. Bond failure will result.

No dependence should be placed on salt or other chemicals for the prevention of freezing.

3.10.3.2 Mild-Weather Construction

In relatively mild weather, that is, when the temperature is generally above 40°F to 45°F (4.4°C to 7.2°C), with only short periods below this range, heating only the mixing water will usually provide the desired temperatures in the concrete. The mixing water should be heated, or other steps taken to raise the temperature of the concrete to between 50°F and 70°F (10°C and 21°C) at the time of placing.

In-place concrete should be kept at a favorable temperature long enough to avoid injury by exposure to the atmospheric temperature. In general, the air surrounding placed concrete made with normal portland cement (Type I) should be kept at 70°F (21°C) or above for the first 3 days, or above 50°F (10°C) for the first 5 days. When high-early-strength concrete (Type III) is used, air temperatures should be kept at 70°F (21°C) for 2 days or at 50°F (10°C) for 3 days.

In mild weather, a covering of tarpaulins may be enough protection. A layer of straw covered with tarpaulins will protect against severer conditions, but an enclosure of tarpaulins or other watertight material with artificial heat may be necessary to provide proper protection in many cases. To prevent rapid drying of concrete, especially near heating elements, it may be necessary to elevate these elements and protect concrete near them with sand kept wet constantly.

3.10.3.3 Hot-Weather Construction

In extremely hot weather, extra care is required to avoid high temperatures in newly placed concrete and to prevent its rapid drying. Keeping aggregate stockpiles moistened with cool water will help to lower the temperature. In very hot weather, mixing water should be chilled by refrigeration or by using ice as part of the mixing water. The ice should be melted by the time the concrete leaves the mixer.

A subgrade on which concrete will be placed should be saturated some time in advance, and then sprinkled just before the placing operation, so that it will not absorb water from the mix. Wood forms should be wetted thoroughly if they have not been treated otherwise. There should be no delay in placing concrete, and it should be struck off and leveled at once. Temperature covers, such as burlap kept constantly wet, should be placed over fresh concrete immediately after it has been leveled. During final finishing, only a small section immediately ahead of the finishers should be uncovered. The finished surface should then be re-covered and the cover should be kept wet. A delay in finishing air-entrained concrete in hot weather usually leads to the formation of a surface that is difficult to finish without leaving ridges.

Fresh concrete should be shaded as soon as possible after it has been finished, and moist curing should be started as quickly as can be done without marring the surface. Concrete surfaces should be kept constantly wet during the curing period.

FIGURE 3.10-1 Length of Curing

Type of Portland Cement	Curing Time, days	Concrete Surface Temperature, °F (°C)
Type I	5	70 (21) or higher
	7	50–70 (10–21)
Type II	3	50 (10)
Other than Type I or Type II	7	70 (21)
	14	50–70 (10–21)

3.10.4 LENGTH OF CURING

Sufficient curing time should always be allowed for concrete to develop adequate strength before it is loaded. Curing should be started as soon as it is possible to apply a curing medium without damaging the surface and should continue as long as is practicable. Some curing times and surface temperatures for normal mild-weather conditions are summarized in Figure 3.10-1. However, during hot weather, curing should continue for at least 7 days, regardless of the type of concrete or the temperature. In mild weather, the curing time can sometimes be reduced to 2 days when high-early-strength portland cement (Type III) is used. Even this time may have to be increased, though, in very hot weather.

In any case, concrete temperature should not be allowed to fall below 50°F (10°C) at any time during curing. Rapid cooling may induce thermal cracking, especially during the first 12 to 24 hours.

3.11 Concrete Foundation Systems

A building's foundations form its *substructure* and support its *superstructure*, transmitting loads to the earth. They also resist *lateral* (horizontal) loads from the ground and the superstructure and provide anchorage for the superstructure against uplift (Fig. 3.11-1).

A foundation system is composed of load-carrying elements, which may include footings, caissons, piles, grade beams, walls, piers, columns, pilasters (Fig. 3.11-2), and the turned-down portion of a combination footing and floor system. Foundation systems are constructed of concrete, masonry, steel, wood, and combinations of these materials. Some common variations are shown in Figure 3.11-3.

Foundations take three basic forms: basements, crawl spaces, and slabs on grade. Figure 3.11-3a,b shows a basement and a crawl space with spread footings, concrete or masonry walls, and steel-joist floor framing. The floor may also be

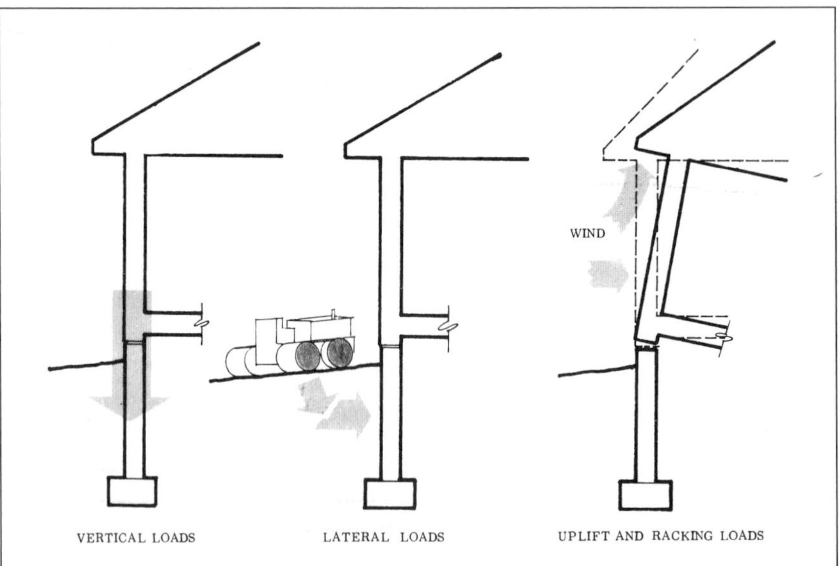

FIGURE 3.11-1 Foundation walls must (1) resist vertical loads from the weight of the building and its occupants; (2) resist lateral loads from the soil, water, and other superimposed loads, such as earth-moving equipment and earthquakes; and (3) anchor the building to its foundation.

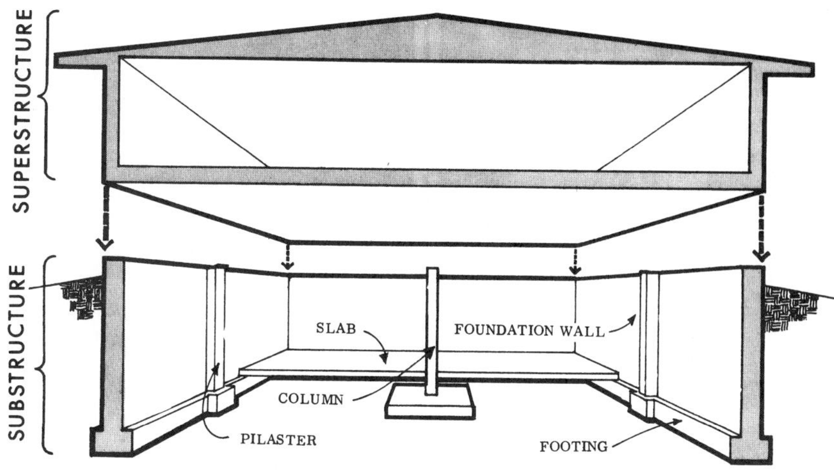

FIGURE 3.11-2 The substructure (foundations) of a building.

framed in wood, cast-in-place concrete, or precast concrete. Figure 3.11-3c,d shows a similar basement or crawl space supported on piles and caissons (the floor framing has been omitted for clarity). Basements and crawl spaces provide underfloor space for easier installation of mechanical, electrical, and plumbing systems.

Slabs on grade may turn down at the edges to form combination slab-and-foundation systems, as shown in Figure 3.11-3e, or rest on foundation walls and footings, as in Figure 3.11-3f. Slabs may also be supported on footings or grade beams resting on piles (Fig. 3.11-3g). Floating foundations (see Section 3.11.2.1) are also combination slab-and-foundation systems.

This section discusses concrete foundation systems. Other relevant chapters include:

- Chapter 2—soil materials and earthwork needed to support foundations, surface- and groundwater control, thermal control, and termite protection
- Chapter 4—masonry foundation walls, columns, piers, and pilasters
- Chapter 6—wood foundations, additional precautions necessary to prevent termite damage, and foundation anchorage
- Chapter 7—waterproofing, dampproofing, and foundation wall thermal insulation

Spread foundations, such as footings and mat and raft foundations, resting on the soil receive loads from foundation walls, piers, pilasters, and columns and must be able to distribute them directly to the soil without exceeding its allowable bearing capacity. Piles and caissons receive loads from foundation walls, isolated columns, or grade beams and transfer them to the soil or into bedrock.

Basements are habitable spaces that lie mostly below the adjacent finished grade. Foundations form the substructure for buildings whether or not there is a basement. Buildings without basements include those with a crawl space or a slab on grade. Some such buildings have a limited basement space, but many do not.

Many of the recommendations in this section are based on the American National Standards Institute's (ANSI) publication A56.1, "Building Code Requirements for Excavation and Foundations" and the Underground Space Center's *Building Foundation Design Handbook.*

3.11.1 DESIGN

A foundation system should be economical and should carry all expected loads with a reasonable factor of safety. The design of foundation elements is based on assumptions as to the kinds of loads the foundation must carry.

Loads associated with foundation design are either *vertical* or *lateral* (horizontal), *static* (nonmoving) or *dynamic* (moving), and *concentrated* at a point or *uniform* along a surface. These loads may be further divided into dead loads, live loads, wind loads, and other loads.

3.11.1.1 Dead Loads

Dead loads act constantly. They include the weight of the superstructure and of the foundation itself. Dead loads are calculated by summing the total weight of the materials used in the construction. Dead loads are usually assumed to be uniformly distributed over the length and thickness of a foundation wall.

3.11.1.2 Live Loads

Live loads are variable, periodic, and sometimes moving. They include, but are not limited to, the weight of snow or water on the roof, furniture, equipment, and people. The type and magnitude of live loads acting on a structure vary according to locality, occupancy, and other factors. Most building codes specify minimum live loads that may be used in sizing foundation walls, footings, and other elements of the substructure.

3.11.1.3 Wind Loads

Wind loads result from the intermittent effect of wind on the superstructure. They may act downward, laterally at any angle, or upward.

3.11.1.4 Other Loads

Some loads that affect foundations are difficult to categorize as purely live, dead, or wind loads. Sometimes they incorporate all three. They are usually calculated and accounted for in the design separately. Among these are lateral pressures exerted on foundation walls by the adjacent soil and water in that soil; multidirectional loads imparted by seismic disturbances (earthquakes); thrust loads transmitted into the foundations from domes, rigid frames, vaults, and arches; uplift caused by expansion due to freezing and subsequent *frost heave*; and uplift resulting from the tendency of a building with its foundations below the water table to float (*buoyancy*).

3.11.1.5 Settlement

Refer to Section 2.1.4 for a more detailed discussion of the causes and effects of settlement in different types of soils.

A building's substructure must distribute building loads so that settlement will be either negligible or uniform under all parts of the building. *Differential* (uneven) settlement under different parts of a foundation may cause substantial problems. Uneven settlement can occur when a foundation is out of plumb or not level, or when loads are not balanced on the substructure, and can cause such damage as cracks in foundation walls and in finished walls and ceilings, sloping floors, open joints in woodwork, and functional annoyances, such as binding doors and windows. When differential settlement is extreme, partial failure of the structural integrity of the building can result.

In time, all foundations built on soil will settle to some extent. The extent of settling depends on the compressibility of the underlying soil and the loads placed on it. Differential settlement can be controlled by designing footings so they will transmit an equal load per unit of area to the soil.

The bearing capacity and compressibility of soils can be calculated by site and laboratory tests on actual soil samples (see Section 2.1.7). With a knowledge of bearing capacity and the magnitude of superimposed loads, designers can use engineering analysis to size foundations and predict settlement. Although engineering analysis is used for commercial, institutional, industrial, and multifamily housing structures, this procedure is seldom followed in designing foundations for single-family dwellings or small buildings of other types. Instead, their footings are usually sized using rules of thumb based on established practices and on the performance of other foundations in their locality. Even in these types of buildings, however, it is necessary to proportion footing sizes for equal settlements and to balance applied concentrated loads, such as will occur at a chimney or column.

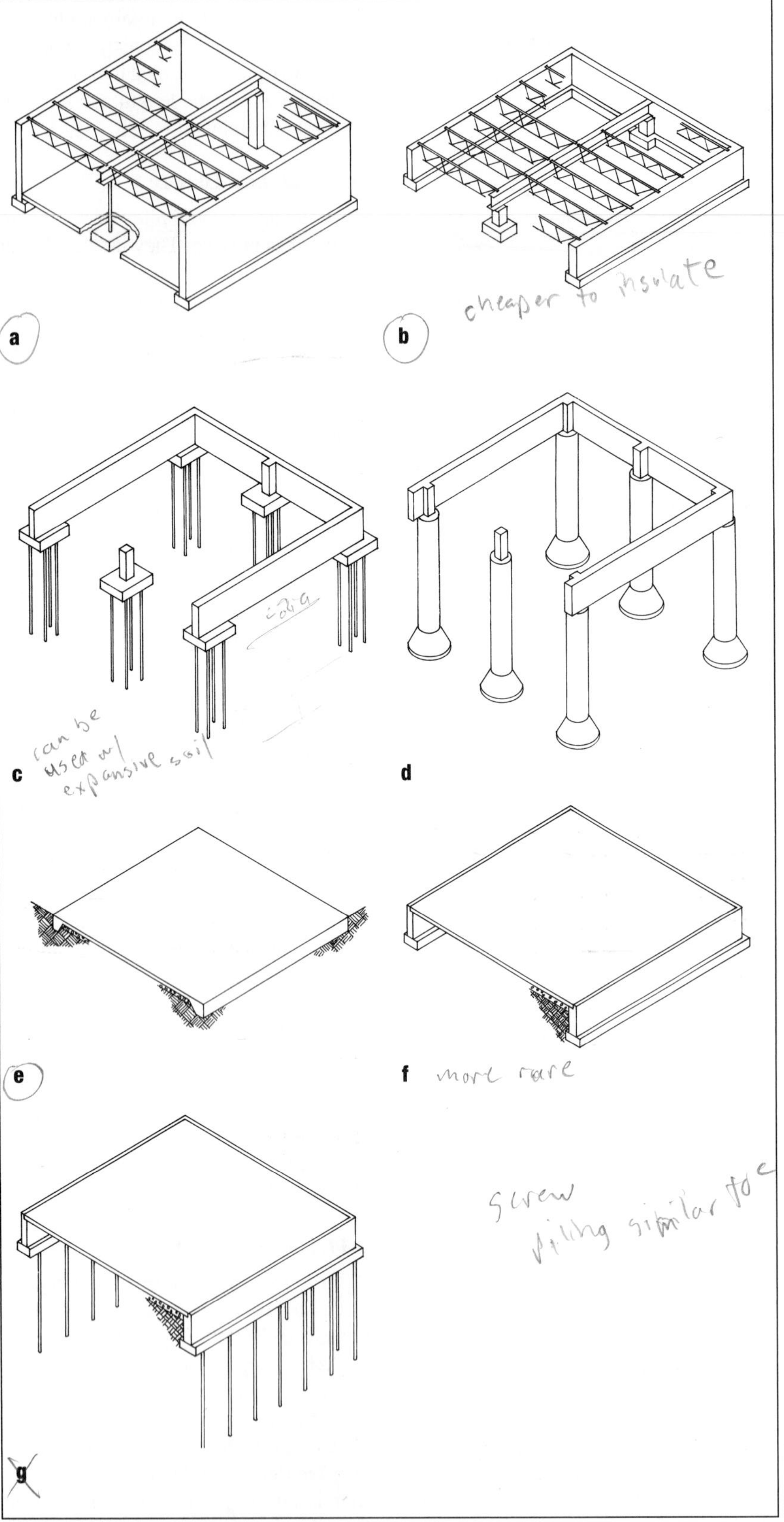

FIGURE 3.11-3 Basic foundation forms: (a) basement and (b) crawl space, both with spread footings, concrete or masonry walls, and steel-joist floor framing; (c) basement and (d) crawl space, both supported on piles and caissons; (e) slab on grade that turns down at the edges to form a combination slab-and-foundation system; (f) slab on grade that rests on foundation walls and footings; and (g) slab supported on footings. (Drawing by HLS.)

3.11.2 FOUNDATION TYPES

Foundations are of two general types: (1) *spread foundations*, which distribute a building's loads directly to a sufficient area of soil to obtain adequate bearing capacity, and (2) *pile* and *caisson* foundations, which transmit a building's loads through soils that have inadequate bearing capacity for spread footings, to deeper layers of soil or rock that have adequate bearing value.

Most foundations in single-family dwellings and other 1- and 2-story buildings are of the spread type. Many buildings of three or more floors are also supported on spread foundations. Piles and caissons are used when the soil's bearing capacity is insufficient to support the loads imposed on the soil by a building.

3.11.2.1 Spread Foundations

Spread foundations support transmitting elements, such as walls, pilasters, columns, piers, and grade beams, on an abruptly enlarged base called a *footing, mat foundation*, or *raft foundation*. These in turn spread the load directly to the supporting soil.

FOOTINGS

Except on bedrock, or in the extremely rare instances when the bearing capacity of the soil is inadequate to support a foundation wall or pier alone, a footing is required to spread the loads on the supporting soil. Foundation walls built of masonry units bonded with mortar should have footings, regardless of the soil's bearing capacity. Refer to Section 2.1.3 for a discussion of the bearing capacities of various soil types.

Footings are classified according to the way in which they are loaded, shaped, and constructed. Several types are shown in Figure 3.11-4. Unreinforced, or plain, footings carry light loads. Reinforced footings contain embedded steel to strengthen them in tension and in shear. Any footing type can be either plain or reinforced.

Footing Material Footings have been made from many materials, including wood, stone, brick, and concrete masonry units. Most footings today, however, are monolithic cast-in-place concrete. Concrete for unreinforced footings in foundations should have a minimum compressive strength of 2000 psi (14 MPa) at 28 days. Higher strengths may be required when determined by engineering analysis. Sufficient time should be allowed for the strength of the concrete to develop before subjecting a footing to loads. In cold weather, concrete should be placed as soon after excavation as possible and must be protected from freezing until it cures (see Section 3.10.3).

When a large excavation has been carried deeper than desired, the depth should be refilled with concrete rather than with earth fill material. If it is necessary to excavate beneath a footing for a sewer or water line or other small opening, the excavation may be backfilled with compacted stone or gravel or with concrete.

Continuous Footings Continuous footings support foundation walls of either concrete or masonry (Fig. 3.11-4a). They may also be used to support a row of several metal or wood columns or masonry piers to minimize differential settlement. When a single footing supports more than one column or masonry pier, it is called a *combined footing* (Fig. 3.11-4d).

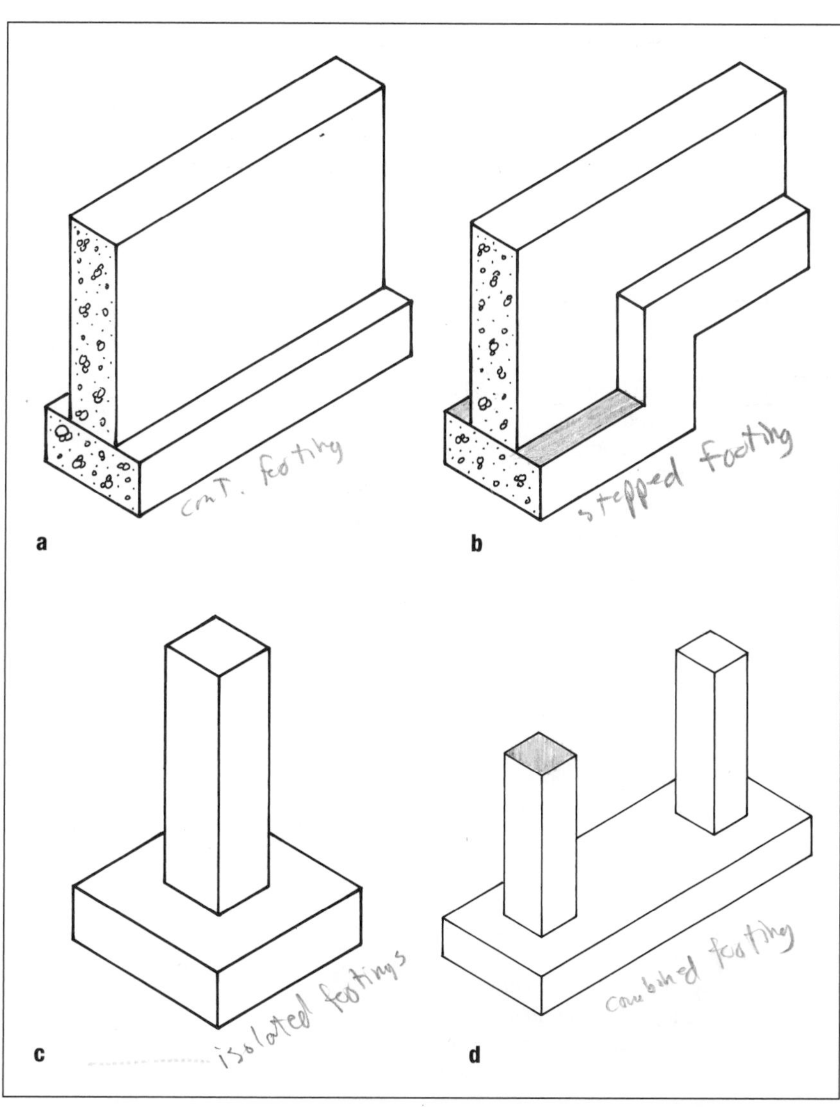

FIGURE 3.11-4 **Various types of footings spread building loads over soil area. All of the types shown can be reinforced.** (Drawing by HLS.)

Stepped Footings Stepped footings change levels in stages to accommodate a sloping grade (Fig. 3.11-4b). Typically, the number of steps necessary to traverse a slope is minimized by making them large, both vertically and horizontally. A few large steps are generally less expensive to form than many small ones.

The vertical part of the step in a stepped footing should be of the same thickness and width as the horizontal footing and should be placed at the same time. If more than one step is necessary, the horizontal distance between steps should not be less than 2 ft. (610 mm) and the vertical step should not exceed three-quarters of the horizontal distance between steps (see Fig. 3.11-4b).

Isolated Footings Isolated footings are independent footings that receive the loads of free-standing columns or piers (Fig. 3.11-4c). Masonry piers are bonded to a footing with mortar. Steel and precast concrete columns are usually anchored to a footing with embedded bolts. Wood columns are frequently held in position on a footing by protruding steel pins embedded in the concrete.

Footing Size Footings should be sized for existing soil conditions and the magnitude of dead and live loads. The usual rule-of-thumb size for normal residential footings resting on soil of average bearing capacity (2000 psf [9765 kg/m^2]) is that the width of the footing should equal twice the thickness of the wall, and the depth of the footing should equal the wall

FIGURE 3.11-5 Dimensions for Continuous Wall Footings

Number of Floors		Frame Construction t″	Frame Construction p″	Masonry or Veneer Construction t″	Masonry or Veneer Construction p″
1	Basement	6″ (150 mm)	2″ (50 mm)	6″ (150 mm)	3″ (75 mm)
1	No Basement	6″ (150 mm)	3″ (75 mm)	6″ (150 mm)	4″ (100 mm)
2	Basement	6″ (150 mm)	4″ (100 mm)	8″ (200 mm)	5″ (125 mm)
2	No Basement	6″ (150 mm)	3″ (75 mm)	6″ (150 mm)	4″ (100 mm)

Based on average soil bearing value = 2000 P.S.F. (9.76 kg/m^2).

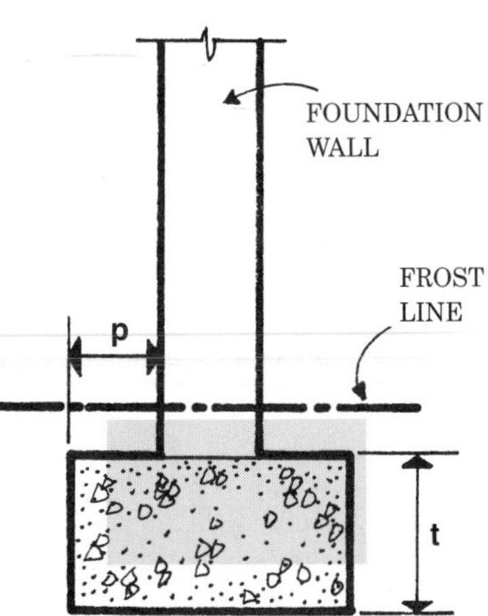

p = Footing projection
t = Footing thickness

thickness. In other words, given a wall thickness *t*, the footing width will be 2*t* and its depth will be *t*. In the absence of specific building code requirements or engineering analysis, the footing sizes given in Figure 3.11-5 have been used successfully. However, their applicability to a particular project should be verified in every instance. In no case should the sizes be smaller than those shown in Figure 3.11-5. These sizes are based on a conventionally loaded wall footing and on soil of average bearing value (approximately 2000 psf [9765 kg/m^2] or greater). Footings should not be less than 6 in. (150 mm) thick and, unless transversely reinforced, their thickness should not be less than 1½ times the footing projection.

Footings for piers, columns, and masonry chimneys should be at least 8 in. (200 mm) thick and, unless reinforced, not less than 1½ times the footing projection. In addition, the minimum footings for interior piers or columns should be 24 × 24 in. (610 × 610 mm).

Footing Location In most cases, footings should rest on undisturbed earth. Footings for small buildings are occasionally supported on *engineered fill* (see Section 2.3.4) when other types of foundations are inappropriate or would be more expensive than placing spread footings on engineered fill. In these cases, the footings must be designed by engineering analysis for these conditions. In addition, more than usual care must be taken in compacting and testing the compacted engineered fill to ensure that the necessary compaction has been achieved.

Footings should not be placed on frozen ground. As frozen soil thaws, excess water can transform the soil into liquid mud that is unable to support loads. Footings should be located far enough below ground level to protect them from possible frost heaving caused by the freezing of soil moisture (see Section 2.1.9).

In areas where soils expand due to moisture changes, footings should be deep enough to be unaffected by seasonal moisture changes or should be specifically engineered for these conditions.

Reinforced Footings In reinforced footings, steel bars are placed transversely (perpendicular to the wall) when loads must be spread over wider areas. When its projection exceeds two-thirds of its thickness, a footing should be reinforced transversely with steel as determined by engineering analysis.

Footings are reinforced longitudinally when they must bridge weak spots, such as excavations for service or sewer connections, and when the soil has low bearing capacity or varies in compressibility.

MAT AND RAFT FOUNDATIONS

Mat and raft foundations (Fig. 3.11-6) are used over soils of low bearing capacity when other foundations would be inadequate. They are made of concrete and heavily reinforced with steel, so that the entire foundation will act as a unit.

A *mat* foundation is a thickened slab that supports loads and transmits them as a single structural unit over the entire slab and soil surface area. Mat foundations are used when the foundations for a building become so large, usually due to poor soil, that it is more economical to join them into a single unit than to cast them each separately.

A mat foundation becomes a *raft* or *floating* foundation when it is combined with basement walls and placed beneath a building in such a way that the weight of the soil removed to make the basement equals the weight of the building's superstructure and substructure combined. In such a structure, settlement is minimum even on low-strength soils, because the building load on the soil is no more than the weight of the excavated soil.

3.11.2.2 Piles and Caissons

Piles and caissons are used as foundation support for buildings when the soil is not capable of supporting the loads that will be imposed by a building on spread foundations.

DRIVEN PILES

Driven piles are columnlike units that transmit loads through poor soil to rock or lower levels of soil that have adequate

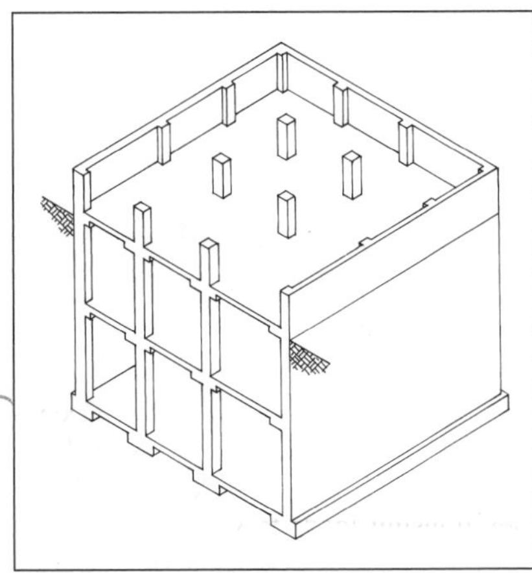

FIGURE 3.11-6 Mat and raft foundations. (Drawing by HLS.)

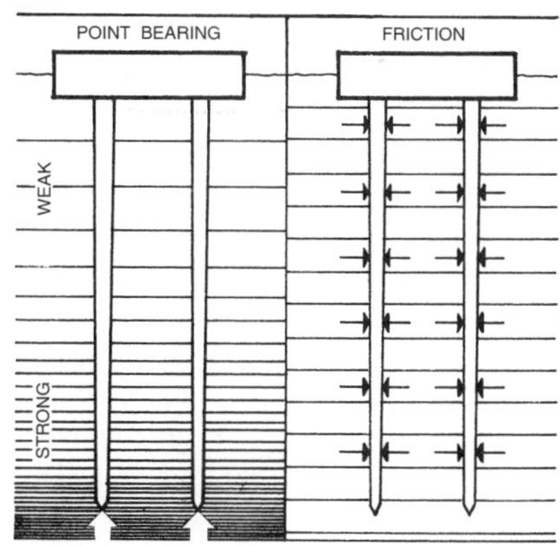

FIGURE 3.11-7 Point bearing piles reach to strong soils or rock. Friction piles depend on friction between the piles and the soil for their support.

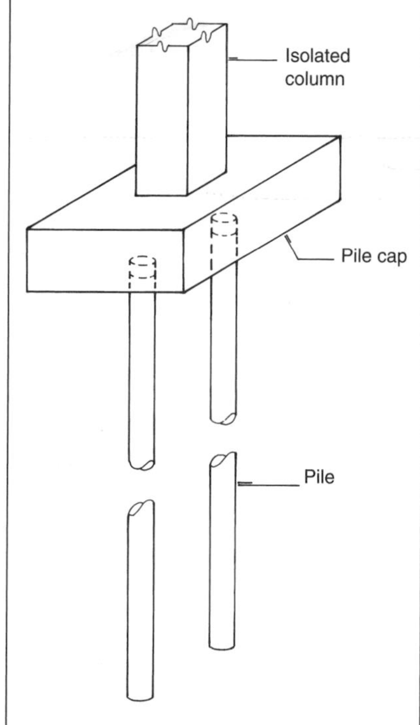

FIGURE 3.11-8 A cluster of two piles with an isolated pile cap. Similar caps are also used when the supported element is a grade beam instead of a column. Caps may be either square, rectangular, or multisided, depending on the number of piles and the size of the cap. (Drawing by HLS.)

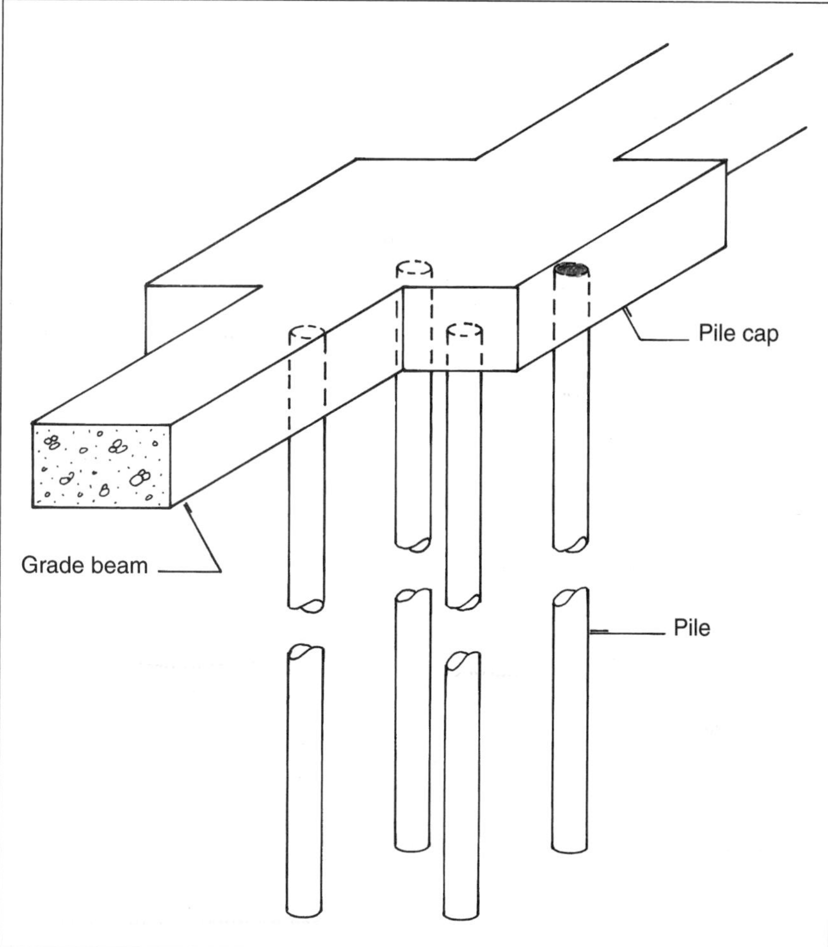

FIGURE 3.11-9 Pile caps formed integrally with a grade beam are sometimes used when the load-distributing element is a bearing wall. (Drawing by HLS.)

bearing capacity (Fig. 3.11-7). In some regions, short piles are called *piers*. Piles serve the same purpose as footings, in that they transmit loads to subbase strata capable of carrying the load. Piles are usually placed in clusters of two or more, spaced from 30 to 48 in. (762 to 1220 mm) on centers. This arrangement permits them to act together and produces a higher load-carrying capacity than can be achieved with isolated piles. The load capacities mentioned in this section are based on the piles being distributed in clusters. These load capacities are also the maximum permitted. Optimum load capacities are somewhat lower in each instance. Piles receive building loads from isolated columns and from grade beams by means of reinforced concrete pile caps (Fig. 3.11-8). Pile caps are sometimes simply a widened section of a grade beam (Fig. 3.11-9).

Pile foundations are either (1) *point bearing* types, which transmit loads to lower, stronger soil or rock through their points, or (2) *friction* types, which develop the necessary bearing capacity through surface friction between the pile and the ground (see Fig. 3.11-7).

Piles are driven with heavy *hammers* in large machines called *pile drivers*. *Drop hammers* are simply raised and dropped on a pile by force of gravity. *Differential-acting steam hammers* are rammed into the top of a pile by steam pressure or compressed air. Modern vibratory hammers and diesel-driven hammers are also being used today.

Friction piles are driven to a predetermined depth or resistance based on soil boring analysis and field tests. To verify the design analysis, test piles are usually driven and loaded before the rest of the piles are driven. Point-bearing piles are driven until additional blows of the hammer produce very little movement in the pile (*refusal*).

Piles may be made of wood, concrete, steel, or a combination (composite) of these. Timber piles have been used for at least 2000 years and probably longer. Some below-water timber piles beneath bridges in Europe are known to have remained in continuous service for more than 1000 years. They are, however, suit-

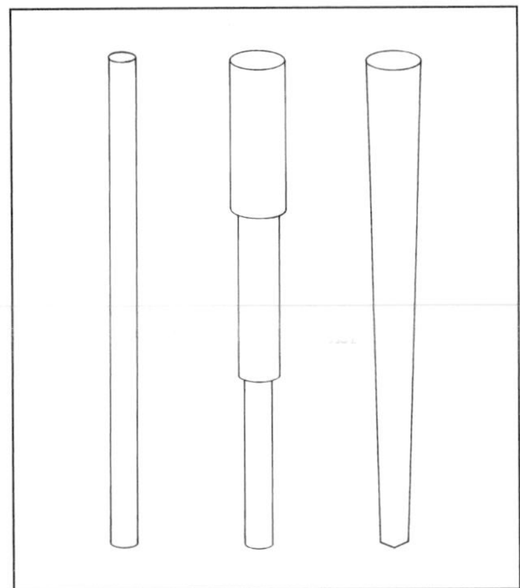

FIGURE 3.11-10 Steel pile shapes. (Drawing by HLS.)

able only for relatively light loads (40 tons [390,594 kg/m^2] maximum) and must be preservative-treated when they will extend above the water table. They are also limited in length to the effective height of the tree from which they are cut (45 to 65 ft. [13.7 to 19.8 m]), because they cannot be spliced.

Steel piles may be either H-shaped sections or pipes (Fig. 3.11-10). H-shaped piles are heavy, wide-flange sections varying in size from 8 to 14 in. (203 to 356 mm) in both depth and flange width. They can carry loads of 50 to 200 tons (488,243 to 1,952,972 kg/m^2) each and may be as much as 150 ft. (45⁄72 m) long. To produce such lengths, sections are welded together as they are being driven.

Steel-pipe piles are later filled with concrete. They are available in several types and shapes (Fig. 3.11-11). They may be heavy-walled types that can be driven directly or thinner-walled types that require a tight-fitting, heavy inner lining (*mandrel*) that is withdrawn before the concrete is placed. They may be smooth-walled or corrugated, and round or fluted. Sizes range from 8 to 24 in. (203 to 610 mm) in diameter, and maximum load capacity ranges from 75 to 200 tons (732,365 to 1,952,972 kg/m^2), depending on the type. Pipes with closed ends may be driven to as much as 120 ft. (36.6 m). Other pipe piles are limited to about 80 ft. (24.4 m).

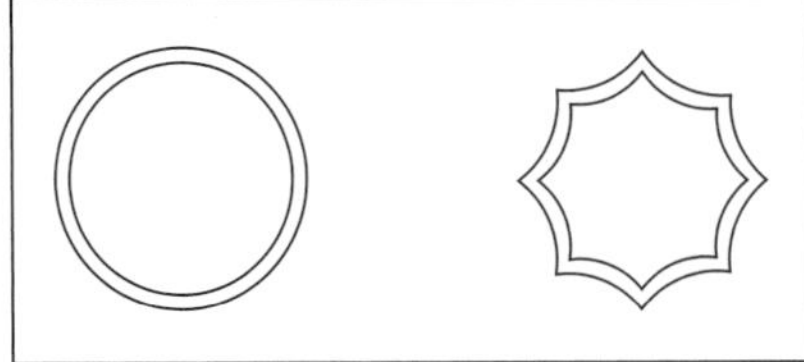

FIGURE 3.11-11 Round and fluted pile shapes. Fluted steel piles are usually tapered. (Drawing by HLS.)

Precast concrete piles are either solid concrete or open cylinders that are later filled solid with concrete. They may be square, round, or octagonal in cross section. All precast concrete piles are reinforced. Most are also prestressed. Solid precast concrete piles with simple reinforcement can be up to 80 ft. (24.38 m) long. This length increases to 150 ft. (45.72 m) for cylinder piles, and up to 200 ft. (60.96 m) when they are prestressed. Maximum capacity ranges from 100 tons (976,486 kg/m^2) for simple reinforced piles to as much as 500 tons (4,882,430 kg/m^2) for prestressed cylinder piles. Sizes vary from 10 to 54 in. (254 to 1372 mm) across.

Composite piles are constructed in several configurations. Some have timber or concrete-filled steel-pipe lower sections and concrete-filled shell with mandrel upper sections. Others have H-shaped steel lower sections and precast concrete upper sections. Composite piles sometimes permit less expensive applications. For example, the combination of timber and concrete-filled steel shell permits the use of relatively inexpensive timber piles at much lower depths than is ordinarily possible (up to 150 ft. [45.72 m]). A disadvantage of using composite piles is that their load-carrying potential is limited to that of the lower of the two elements. For example, a composite pile of timber and concrete-filled steel shell is limited to 40 tons (390,594 kg/m^2), instead of the 75 to 80 tons (732,365 to 781,189 kg/m^2) the concrete-filled steel piles can carry.

CAISSONS

Caissons are concrete columns placed in auger-drilled or excavated holes. They serve the same purpose as *piles*. While some caissons, sometimes called piers, are only 6 to 12 in. (152 to 305 mm) in diameter, most are much larger and are capable of carrying much heavier loads. They can be as much as 6 ft. (1.8 m) in diameter and carry 3500 tons (34,177,010 kg/m^2) or more. Caissons extend through unsatisfactory soil to a firm soil-bearing stratum or to bedrock. As a caisson hole is drilled or dug, a steel casing is lowered into it to keep the hole from caving in. This casing is raised and removed as the concrete is placed. Small, relatively short and lightly loaded caissons, sometimes called piers, are frequently poured into cardboard or composition casings that are not removed as the concrete is placed.

Caissons are generally one of four types: (1) *rock caissons* simply rest directly on solid rock; (2) *high-capacity rock caissons* rest in a socket cut into rock, thus supporting the load both by resting on the rock and by friction against the sides of the pocket—these are sometimes called *socket caissons*; (3) *hardpan* or *clay caissons* rest on a soil-bearing stratum; and (4) *friction caissons* are supported by the friction acting on the sides of the caisson, much as friction piles are supported—in fact, they are sometimes called *cast-in-place piles*.

Hardpan or clay caissons either have straight shafts or have an enlarged base called a *bell* (Fig. 3.11-12). Bells are produced by hand excavation or with a special device (*belling bucket*) on the auger.

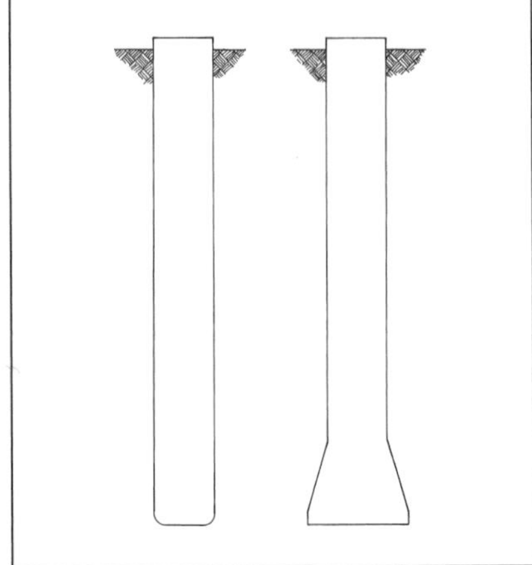

FIGURE 3.11-12 Typical caisson configurations. (Drawing by HLS.)

3.12 Concrete Slabs on Grade

The suitability of a particular slab-on-grade construction is generally influenced by the geographic location of the building, the building site, and the construction of the superstructure. Specific influences on the type and design of a slab include site preparation, topography, soil properties, climate, the type and support of the superstructure, and comfort factors such as thermal and moisture control. Producing satisfactory slabs requires properly designing the slabs, correctly selecting the concrete materials and mixtures, and carefully placing, finishing, and curing the concrete.

In this section recommendations are given for selecting and constructing concrete slabs on grade. If care is exercised to control concrete quality and if the necessary precautions dictated by the characteristics of the site, superstructure, and geographic location are taken, slabs on grade are suitable for floors and combination foundation-and-floor systems in all parts of the United States. Unless specifically modified in this section, the provisions of Section 3.2 are applicable to concrete slabs on grade.

These recommendations are based on:

- PCA's "Concrete Floors on Ground"
- American Concrete Institute (ACI) International's ACI 302
- Building Research Advisory Board, National Academy of Sciences's *Criteria for Selection and Design for Residential Slabs on Ground*
- Underground Space Center's *Building Foundation Design Handbook*

3.12.1 DESIGN

Slabs on grade may be either independent floors or combination foundation-and-floor systems. In independent floors, the slab is separate from the building's load-carrying elements, such as walls, columns, and footings, which transmit their loads directly to the supporting soil (Fig. 3.12-1a). In combination foundation-and-floor systems, the slab is an integral part of the foundation and carries superstructure loads and transmits them to the soil (Fig. 3.12-1b). Concrete foundation elements such as walls and footings, which are common to many types of foundation systems, are discussed in Section 3.11. Masonry walls, columns, piers, and pilasters are covered in Chapter 4.

Slab-on-grade construction requirements are determined largely by site characteristics, such as drainage and soil type, and the methods used to build the slab. Slabs may be ground supported or supported structurally. They may be unreinforced (*plain*), contain welded wire fabric reinforcement, or be structurally reinforced with welded wire fabric and steel-reinforced bars. Four types of slabs are described in the following section.

Slab construction depends on the following factors:

- Slab type
- Site preparation, including preparation of the slab bed, grading, and backfilling
- Precautions to control ground and surface moisture
- Thermal control when required

The performance of slabs is dependent on the quality of concrete materials and workmanship.

3.12.2 SLAB TYPES

Concrete slabs for floors and combination foundation-and-floor systems can be classified according to construction, as either:

- Ground-supported (slabs that rest directly on a slab bed consisting of undisturbed soil, compacted fill, or a base course); or
- Structurally supported (slabs supported independently of the ground by walls, piers, grade beams, or piles).

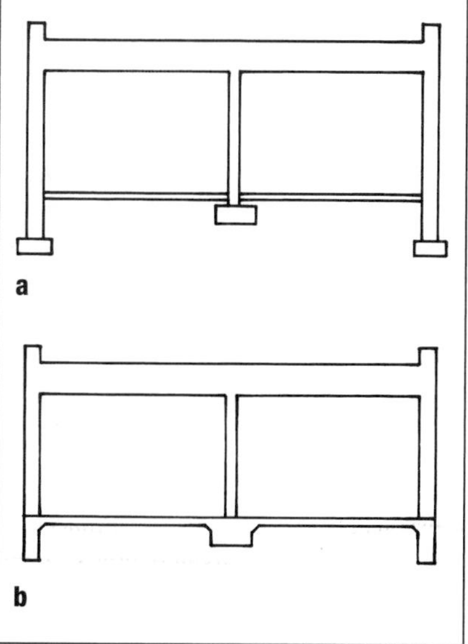

FIGURE 3.12-1 (a) Ground-supported floor slab built independently of foundation walls; (b) ground-supported slab built integrally with foundation-grade beam. (Drawing by HLS.)

The appropriateness of a concrete slab floor or combination foundation-and-floor system should be based on soil properties, climate, functional and aesthetic requirements, and economy.

3.12.2.1 Ground-Supported Slabs

Ground-supported slabs can be designated as Type I, II, or III.

TYPE I SLABS

A Type I slab is mostly unreinforced and is separated from the elements that support the superstructure (Fig. 3.12-2) so that it carries no superstructure loads. Superstructure loads are supported on foundations independent of the slab. A Type I slab has a nominal thickness of at least 4 in. (100 mm) and is cast directly on a dense or compacted slab bed. It is unreinforced but may contain welded wire fabric or bars in localized areas. Type I slabs should not be placed on a slab bed that can develop significant changes in volume over time. Differential movement in the slab bed may produce enough tensile stress in the concrete to crack the slab. Random cracking due to drying shrinkage is controlled by:

- Isolation joints
- Construction joints
- Limiting the area between control joints placed to induce cracking at preselected locations
- Fiber reinforcing

The controlling factors in the suitable performance of a Type I slab are:

- Selection of the correct slab bed
- Concrete quality
- Slab location
- Control joint spacing

Because a Type I slab is unreinforced, except for specific localized areas, it possesses only compressive strength and cannot tolerate appreciable amounts of tension or warping. Such a slab may crack from shrinkage during drying, but when used under the conditions recommended in the following paragraphs, cracks that do occur should not open excessively or

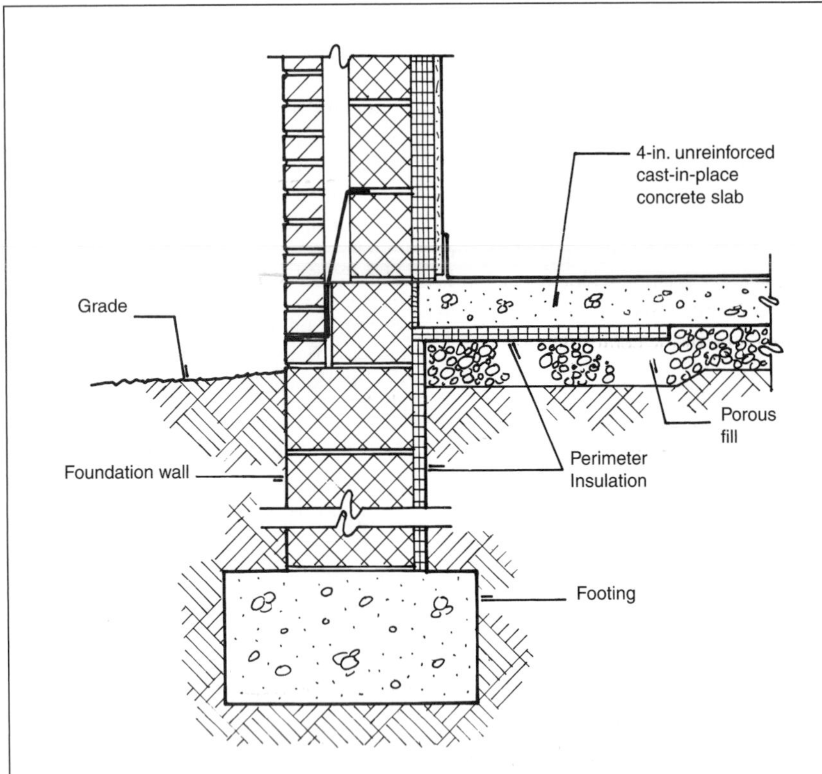

FIGURE 3.12-2 **Type I slab.** (Drawing by HLS.)

prove detrimental to the serviceability of the slab.

Type I slabs should be placed only on well-drained and properly graded coarse-grained soils (Fig. 3.12-3). Such soils are not affected significantly by climate or moisture content changes and should develop no appreciable volume change that would be detrimental to the slab. The soil must be capable of supporting a Type I slab and be uniformly and adequately compacted to ensure that warping and tensile stresses that could contribute to cracking are not induced in the slab.

Type I slabs can be applied successfully to:

- Gravelly soils of any density (Types GW, GP)
- Sandy soils (with or without silts) and clays (Types GM, GC, SW, SP, SM, SC)
- Silts (Types ML, MH), provided they are dense or medium dense
- Loose sandy and silty soils if the soil can be compacted to a dense or medium-dense state to its entire depth before slab construction (see Fig. 3.12-3)

Refer to Section 2.1.1 for a discussion of soil types.

Type I slabs should be rectangular or square in shape (Fig. 3.12-4a). To control shrinkage cracking during drying and subsequent thermal volume changes, no dimension of a slab perimeter should exceed 32 ft. (9.75 m). However, if a slab is to be left exposed, control joints spaced 15 to 20 ft. (4.57 to 6.1 m) apart are recommended (Fig. 3.12-4b).

Irregularities in shape, such as when a nonrectangular T- or L-shape is required, need control joints to divide the slab into squares or rectangles (see Fig. 3.12-4b). If control joints are not provided, diagonal cracks are likely to occur where the slab shape changes.

Irregularities in thickness that reduce a slab to less than 4 in. (100 mm) in nominal thickness should not be permitted. Stresses caused by drying shrinkage or thermal change tend to cause cracking at these locations. When the upper surface of a slab must be lowered, such as to permit installation of ceramic tile or another floor finish, the uniform slab thickness should

FIGURE 3.12-3 **Recommended Slab Bed for Slabs on Grade**

Slab Type	Soil Type	Density	Plasticity[a]	Consistency, q_u/w[b]	Climatic Rating
TYPE I	GW, GP	All	—	—	All
	GM, GC, SW, SP, SM, SC, ML, MH	Dense or medium dense[c]			
TYPE II	GW, GP, GM, GC, SW, SP, SM, SC, ML, MH	All	—	—	All
	CL, CH,	—	PI < 15	$q_u/w \geq 7.5$	All
	OL, OH	—	PI > 15	$q_u/w \geq 7.5$	$C_w \geq 45$
TYPE III	CL, CH,	—	PI > 15	$q_u/w \geq 7.5$	$C_w < 45$
	OL, OH	—	—	$7.5 > q_u/w > 1.5$	All
TYPE IV	CL, CH, OL, OH	—	—	$q_u/w < 1.5$	All
	PT	—	—	All	

[a]PI = Plasticity Index, determined in accordance with ASTM D 4318.

[b]q_u/w = unconfined compressive strength/ton/sq. ft. of undisturbed sample determined in accordance with ASTM D 2166.

[c]May be loose if properly compacted to its entire depth.

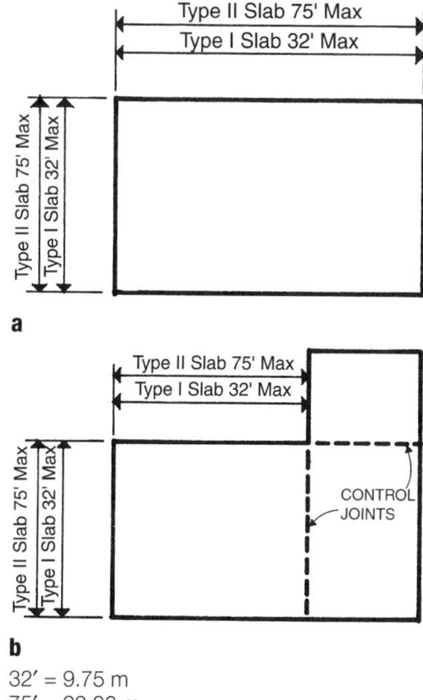

32′ = 9.75 m
75′ = 22.86 m

FIGURE 3.12-4 Maximum dimensions for Types I and II slabs.

be maintained by lowering the underside of the slab beginning at least 24 in. (610 mm) away from the slab depression (Fig. 3.12-5a). In addition, when the vertical displacement is greater than 1½ in. (38 mm), WWF 6 × 6–W2.9 × W2.9 welded wire fabric should be placed in the slab extending 25 in. (635 mm) on either side of the point of displacement (Fig. 3.12-5b).

Heating coils or pipes should not be embedded in Type I slabs because thermal stresses will be induced that the slab is unable to resist without cracking. Heating ducts may be embedded in Type I slabs if WWF 6 × 6–W2.9 × W2.9 welded wire fabric is placed to extend 19 in. (482 mm) on either side of the centerline of the duct or the slab edge, whichever is closer. Refer to Chapter 15 for further discussion about embedding heating elements in Type I slabs, including additional reinforcement needed.

Superstructure loads should be supported directly on foundations independent of the slab, but Type I slabs can accommodate partition loads up to 500 lb./lin. ft. (744 kg/m). Because a Type I slab is unreinforced, it cannot carry greater loads without the possibility of cracking. Where this loading must be exceeded, reinforcement or special construction should be provided to minimize cracking that would result. Methods of supporting partitions and concentrated loads are illustrated in Figure 3.12-6.

Openings in Type I slabs should be kept to a minimum because they can introduce stresses that are not uniform. When an opening 12 in. (305 mm) or wider must be provided for a column or other penetration, a slab should be reinforced with 25-in. (635 mm)-wide WWF 6 × 6–W1.4 × W1.4 welded wire fabric around the opening (Fig. 3.12-7).

TYPE II SLABS

A Type II slab is also separated from the elements that support the superstructure (Fig. 3.12-8), but contains welded wire fabric or reinforcing bars in localized areas. The controlling factors in the suitable performance of a Type II slab include:

- Correct slab bed preparation
- Concrete quality
- Correct size and placement of the welded wire fabric reinforcement

Type II slabs are cast on a dense or compacted bed and contain welded wire fabric to hold drying shrinkage and thermal-movement cracks together once they form. This reinforcement does not prevent cracking or add to the load-carrying capacity of a slab. Grading and loading are similar to those for Type I slabs. As with Type I slabs, the design of the slab is not influenced by the type of superstructure that is supported directly on independent foundations.

Because they contain welded wire fabric, Type II slabs can be cast onto some soils that are unsuitable for Type I slabs. However, Type II slabs cannot be expected to resist differential movement of the soil, because the amount of reinforcement provided is based on the premise that the entire slab will remain uniformly supported by the underlying soil.

Type II slabs can be constructed on all soils suitable for Type I slabs under all climatic conditions, and these soils can be of any density, including loose (see Fig. 3.12-3). In addition, Type II slabs can be applied to certain fine-grained soils (CL, OL, CH, OH) (see Section 2.1.1) under certain climatic conditions if the unconfined compressive strength and plasticity index falls within the limits shown in Figure 3.12-3.

Type II slabs must be at least 4 in. (100 mm) in uniform thickness, and no dimension of the slab perimeter should exceed 75 ft. (22.86 m) (see Fig. 3.12-4).

The same recommendations apply as for Type I slabs with respect to irregularities in thickness (see Fig. 3.12-5) and shape (see Fig. 3.12-4), with the added requirement that reinforcement be continuous across control joints or that keyed

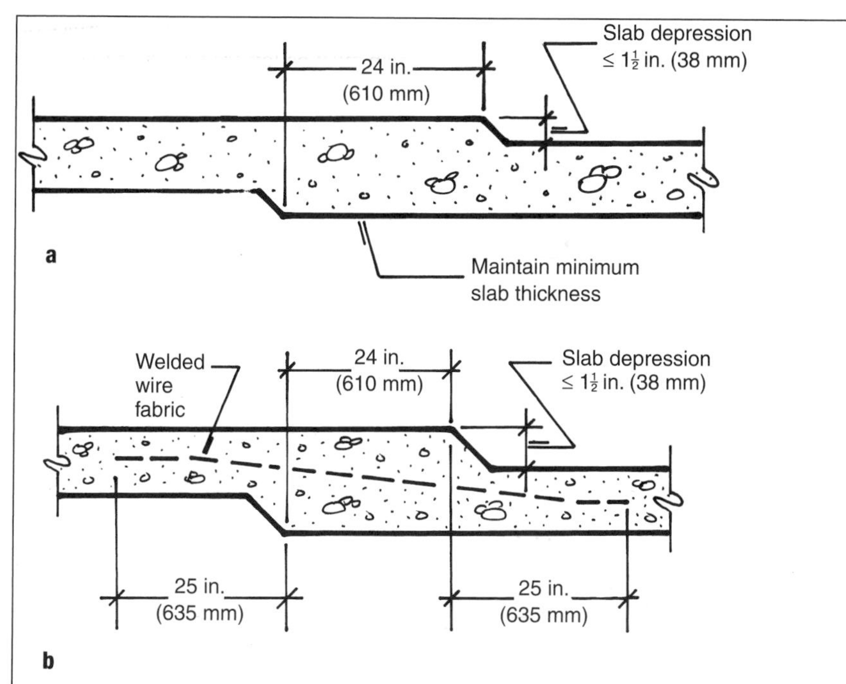

FIGURE 3.12-5 **(a) Method of maintaining minimum slab thickness; (b) location of welded wire fabric reinforcement.** (Drawing by HLS.)

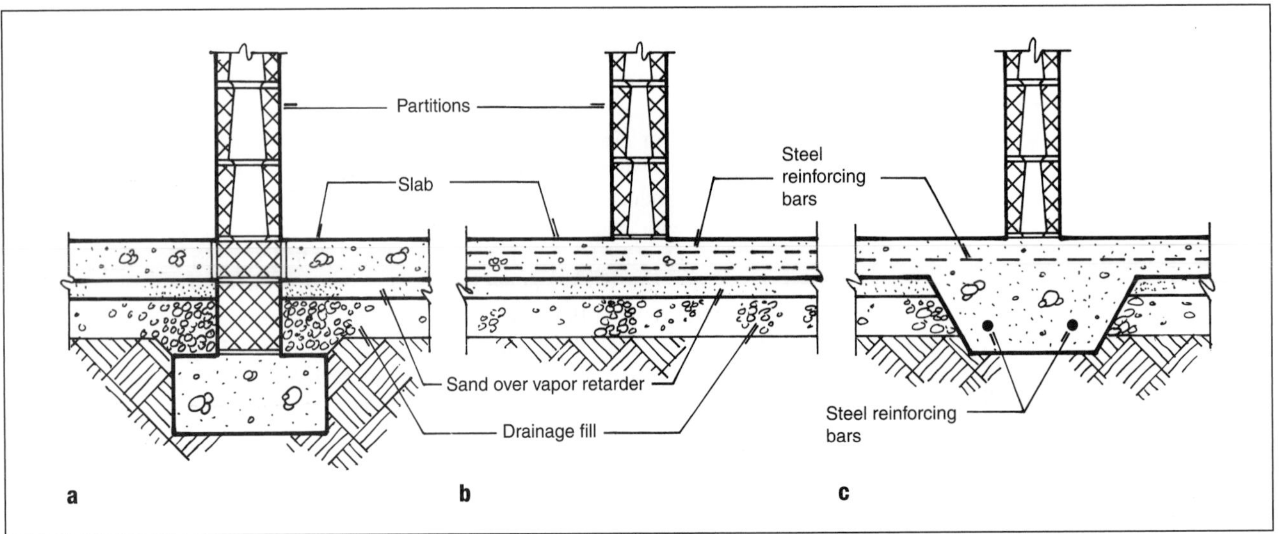

FIGURE 3.12-6 **(a) In Type I and II slabs, a heavy partition or concentrated load should be carried on a separate footing. (b, c) Alternative methods that may be used when the slab is designed using engineering methods and structural reinforcement.** (Drawing by HLS.)

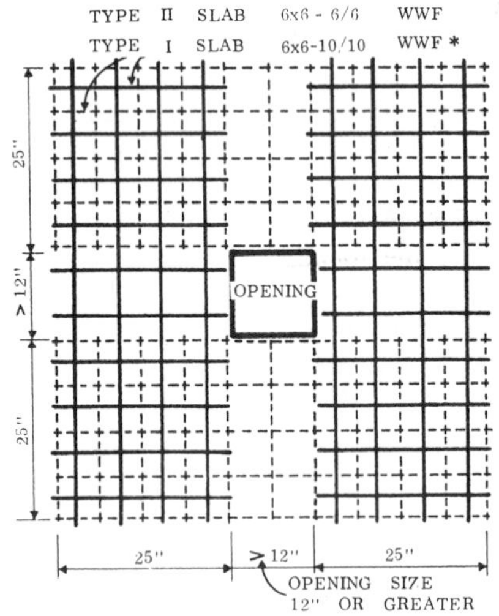

*In addition to normal slab reinforcement.
12″ = 305 mm
25″ = 635 mm

FIGURE 3.12-7 **Location of reinforcing around a hole in a Type I or II slab.**

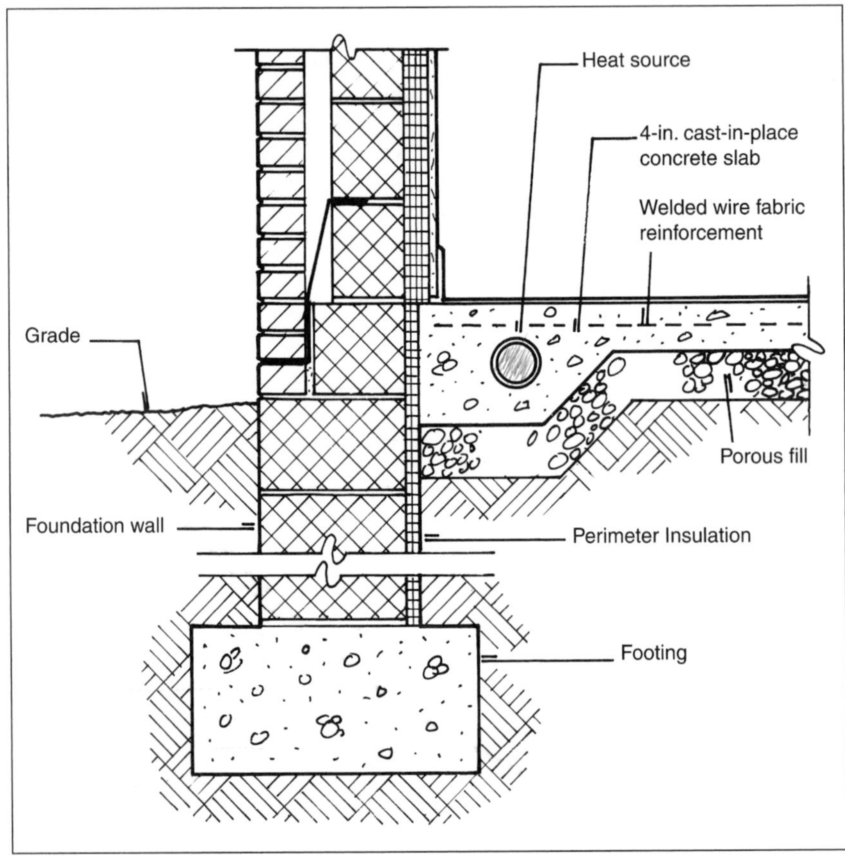

FIGURE 3.12-8 **Type II slab.** (Drawing by HLS.)

joints be provided to eliminate the possibility of vertical displacement between adjacent sections.

Welded wire fabric reinforcement in Type II slabs provides control of crack size only. Cracks that are expected to occur should be held tightly closed by the tensile reinforcement and should not be objectionable. To control crack size, the slab should be provided with the minimum welded wire fabric reinforcement recommended in Figure 3.12-9. Reinforcement should be placed in the center of the slab, midway between top and bottom, and should be supported while the concrete is being placed, preferably on metal chairs designed for the purpose. Edges should be lapped at least one mesh space. The common practice of laying the mesh on the bed and pulling it up with hooks (*hooking*) after the concrete has been placed is not recommended, because accurate and level placement of the mesh within the slab using this method is almost impossible. The practice of laying

FIGURE 3.12-9 Minimum Fabric Reinforcement for Type II Slabs

Maximum Slab Dimension, ft. (m)	Wire Spacing, in. (mm)	Wire Gauge, number
Up to 45 (13.72)	6 × 6 (150 × 150)	W10 × W10
45–60 (13.72–18.29)	6 × 6 (150 × 150)	W8 × W8
60–75 (18.29–22.86)	6 × 6 (150 × 150)	W6 × W6

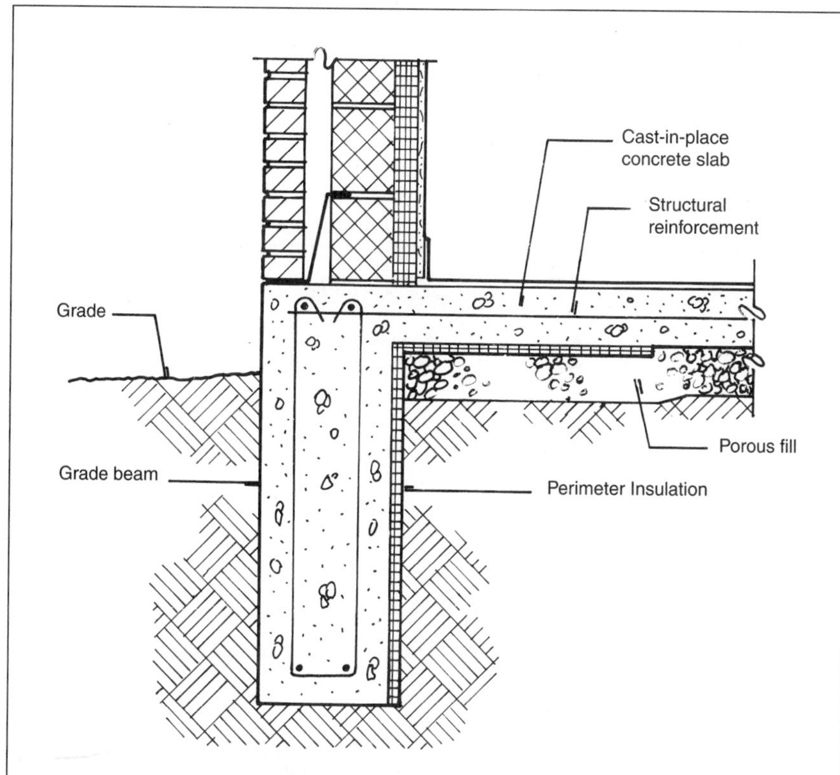

FIGURE 3.12-10 Type III slab. (Drawing by HLS.)

the mesh on top of a poured slab and embedding it by having workers walk on it is also discouraged for the same reason.

Because the reinforcement in Type II slabs will hold together cracks that form due to temperature changes, heating pipes or coils and heating ducts can be embedded in them. However, the construction recommendations given in Chapter 15 should be followed.

Type II slabs are expected to deflect with slight soil movement; therefore, superstructure loads should be supported independently on foundations. However, like Type I slabs, Type II slabs can accommodate partition loads of up to 500 lb./lin. ft. (744 kg/m) and equivalent concentrated loads. An additional layer of reinforcement can be provided to extend at least 25 in. (635 mm) on either side of the partition or concentrated load to distribute added stresses (see Fig. 3.12-6). Heavy concentrated loads, such as masonry walls, should be supported on independent or structurally reinforced footings (see Fig. 3.12-6a and c).

The recommendations for Type I slabs relative to openings apply also to Type II slabs, except that the additional layer of reinforcing placed around the opening should be 6 × 6–W2.9 × W2.9 welded wire fabric. The additional reinforcing around the opening will prevent the concentration of stresses that are likely to crack the slab at these points (see Fig. 3.12-7).

TYPE III SLABS

Type III slabs are combination foundation-and-floor systems (Fig. 3.12-10) that are structurally reinforced to make them usable over problem soils that undergo substantial volume changes with time and climate. The use of foundations with spread footings is not advisable on such soils. Type III slabs receive superstructure loads and transmit them to the foundation soil under the entire slab area. Figure 3.12-11 shows an example of a Type III slab.

Depending on its sensitivity to differential settlement, the superstructure imposes limits on the maximum tolerable deflection in a Type III slab. Because the superstructure is supported on the slab itself, walls and other load-bearing superstructure elements will deform when the slab deforms. Different materials tolerate deformations of differing intensities before developing undesirable defects such as cracking and distortions. For example, wood frame construction can generally sustain more deformation than masonry construction before warping and cracking create mechanical and aesthetic problems. Comparatively brittle cementitious materials such as concrete, stucco, plaster, and mortar can tolerate only small deformations before cracking develops.

The objective of a Type III slab is to cause the foundation-and-floor system and the superstructure to act as one monolithic unit that restricts differential movements in both slab and superstructure. To ensure that the slab and the superstructure actually act in this manner, the slab must be designed with the necessary rigidity and strength. Therefore, in addition to requiring concrete quality control, Type III slabs should be designed by a professional engineer in accordance with good engineering practice to provide the proper dimensions for stiffness and the correct amount and size of steel reinforcement for strength. The Building Research Advisory Board, National Academy of Sciences's *Criteria for Selection and Design for Residential Slabs on Ground* is applicable.

Because a Type III slab is structurally reinforced, it can be used over certain problem soils such as highly compressible, highly expansive, and highly plastic soils. The unconfined compressive strength, plasticity index, and climatic ratings for soils on which Type III slabs should be used are given in Figure 3.12-3.

Highly compressible soils include those that are high in organic content (PT), such as peat, which are so compressible that they will not support ground-supported slabs.

Highly compressible clays are soils

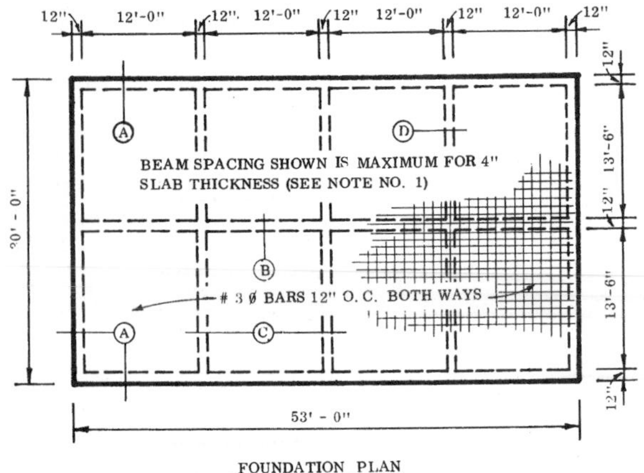

NOTES:

1. Panel size may be increased by a maximum of 15'-0" both ways and with same reinforcement by increasing slab thickness to 5", or by using a 4" slab with #3 ∅ bars 10" o.c. both ways.

2. All beams shall extend minimum of 12" into undisturbed soil.

3. If garage is to be built in lieu of carport, sec. D, the perimeter beams will be of the same size and same reinforcement as main house beams.

4. Bar laps or splice shall be a min. of 24 bar diameters.

5. This design is for 1-story construction only.

6. Foundation concrete shall be minimum 2500 psi and shall be cured by an acceptable and positive method.

7. Slab reinforcement shall be accurately supported in upper one-third of slab.

8. Bottom bars in perimeter beams shall extend around corners a min. of 24 bar diameters.

9. Porch slabs beams and reinforcement shall be same as carports.

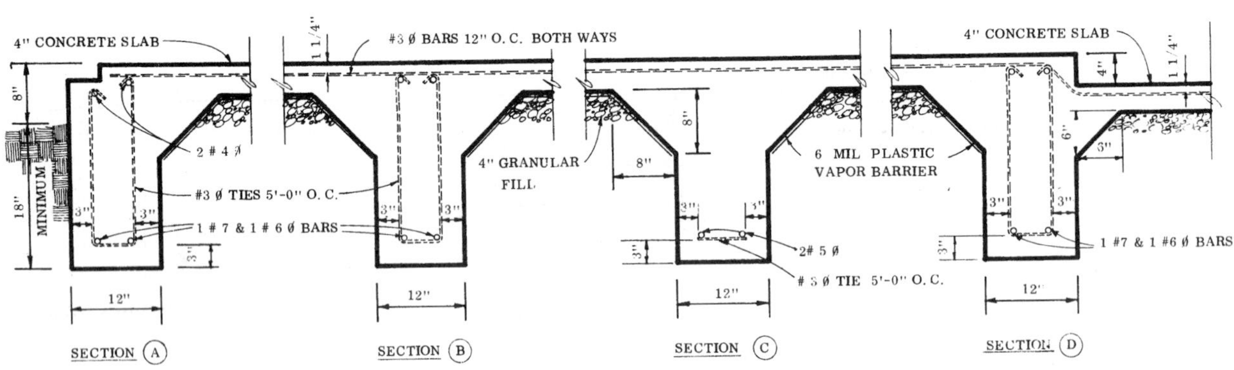

5′-0″ = 1.5 m, 12′-0″ = 3.6 m, 15′-0″ = 4.5 m, 30′-0″ = 9.1 m, 53′-0″ = 16.2 m
1¼″ = 29 mm, 3″ = 76 mm, 4″ = 100 mm, 5″ = 125 mm, 6″ = 150 mm, 10″ = 250 mm, 12″ = 300 mm, 18″ = 450 mm

FIGURE 3.12-11 **An example of Type III slab construction. This design should not be used until reviewed by a professional engineer.**

with sufficient moisture content to have a soft or very soft consistency or high organic content. Excessive settlement may result even under relatively light residential loading or from the weight of fill material.

Loose fine sands and silts can cause serious problems beneath a slab. Vibration of loose sands and silts may cause a decrease in the volume of soil voids with resultant settlement, depending on the amplitude and frequency of the vibration. When it is not practical to compact very loose sand and silt to a suitably dense state for ground-supported slabs, a Type IV slab should be used.

Highly expansive and plastic soils are problem soils because they are greatly affected by climatic conditions. Any soil with a plasticity index greater than 15 may cause problems by expanding under exposure to severe climatic conditions. However, under favorable or intermediate conditions this material may be considered for Type II slabs.

A soil with a plasticity index of 30 or higher will react vigorously to changes in moisture content and will present the most difficult design problems because of the great volume change that can be expected. Loads associated with small buildings are not usually sufficient to control such expansion by loading. In any case, when such soils are encountered, professional advice should be sought.

Loose soils can become problem soils when they are heavily loaded and become saturated. When slab construction is desirable over these soils and the slab must be heavily loaded but problems of groundwater or surface water exist, a Type III slab should be selected.

3.12.2.2 Structurally Supported Slabs

Type IV slabs are structurally reinforced and can be used over very poor soils, such as highly expansive clay, which are extremely sensitive to moisture, and soils of negligible bearing capacity or that are high in organic material content, and where Type I, II, or III slabs would not be suitable. Type IV slabs require professional design based on engineering analysis.

Like Type III slabs, Type IV slabs receive and transmit superstructure loads to the soil, but unlike Type III slabs, they do not depend on soil for support. The slab bed functions as a form and is used only for temporary support of the slab until it develops sufficient strength to be supported on beams, piers, or intermediate foundation walls. These foundation elements in turn are carried by piles or footings that distribute the loads to stable ground below the level of the slab (Fig. 3.12-12).

Type IV slabs can be used over very poor soils (see Fig. 3.12-3) that cannot be treated economically or practically to receive ground-supported slabs. These soils may include expansive clays, highly organic soils, and very loose sands and silts. Foundation elements such as grade beams should be designed so that swelling of the soils beneath the slab will not lift them off the foundations or pull the foundations out of the ground.

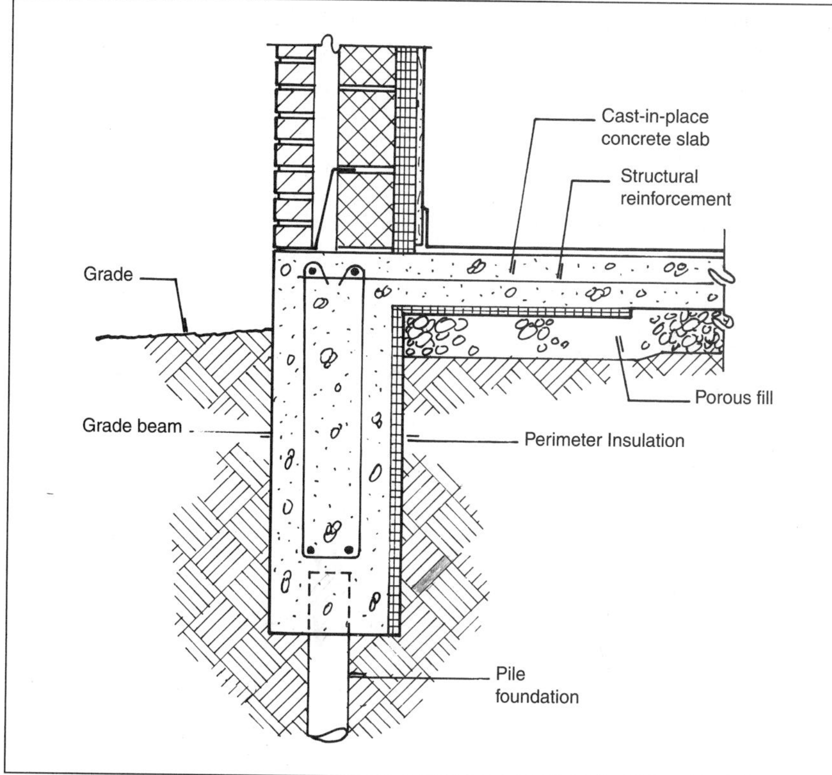

FIGURE 3.12-12 Type IV slab. (Drawing by HLS.)

A method used in some areas to prevent soil swelling from lifting grade beams is to cast the beam between its supports (piles, caissons, grade beams, etc.) on rectangular cardboard tube forms. These forms are folded from a flat sheet with a diagonal across the center. They are about 6 in. (150 mm) thick. When the earth swells, it crushes the cardboard instead of lifting the beam.

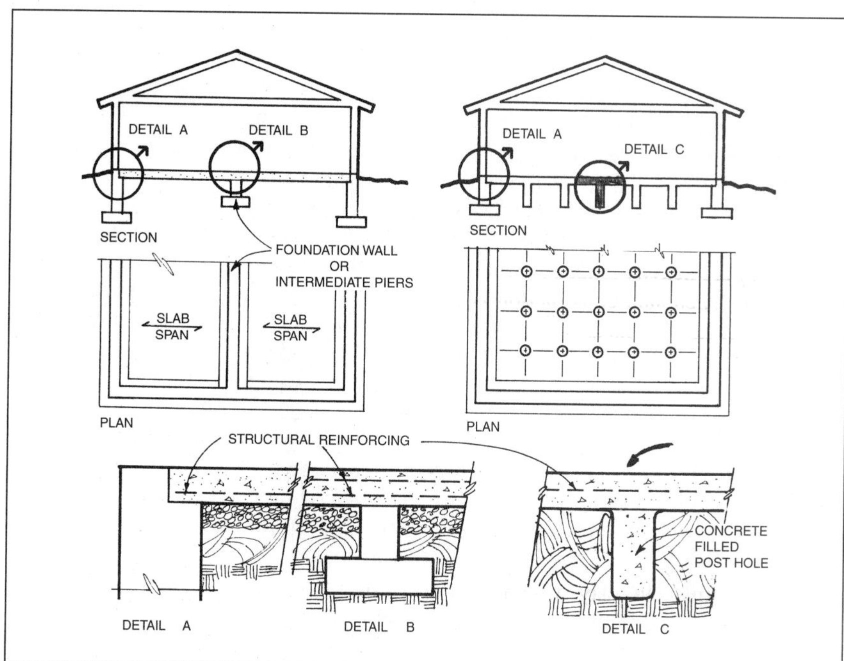

FIGURE 3.12-13 Construction details for a variation of a Type IV slab.

In local practice, variations of Type IV slabs are sometimes used over unstable soils that may not necessarily be considered extreme problem soils (Fig. 3.12-13). Foundation piles, piers, or footings should be designed to penetrate to a depth that provides adequate support below unstable soil.

3.12.3 MOISTURE CONTROL

Refer to Section 2.4 for a discussion of the control of groundwater and surface water, including provisions for slab beds and vapor retarders directly under slabs on grade.

3.12.4 THERMAL CONTROL

For the most comfortable conditions and to conserve energy, slab perimeters should be insulated in all but the warmest climate regions. Whenever walls are insulated, slabs should be insulated. Slab insulation is discussed in Section 7.3.

Slabs are sometimes used to deliver heat throughout buildings. Slabs heated by embedded pipes or containing warm air ducts and slabs in buildings for which summer cooling is to be provided are discussed in Chapter 15.

3.12.5 SLAB FLATNESS AND LEVELNESS

Flatness is a measure of the degree to which the surface of a slab deviates from a plane. Slab *levelness* is a measure of the degree to which a slab deviates from horizontal. A slab can be flat (having no depressions or rises in its surface) and still not be level. For example, a slab that is flat within the allowable tolerances may have one corner that is higher than the others or one side that is lower than the opposite side. Conversely, a level slab may have depressions, rises, or pockets in its surface.

Until recently, slab surface tolerances were specified as a dimension between ⅛ in. (3.2 mm) and ½ in. (305 mm) in every 10 ft. (3050 mm), depending on the activities that would take place where the slab was located and the finishes that would be applied. Historically, these deviations were measured with a 10-ft. (305 mm)-long straight board, with its long edge vertical, placed across the slab in each primary direction and diagonally. The slab was to be within the specified tolerance dimension of the board along its entire length. This method was inaccurate, especially in measuring levelness.

A more accurate method is called the *F-curve* or *F-number* system. F-numbers are designated F_F for flatness and F_L for levelness. The numbers have no units. The method for determining a slab's F-numbers is defined in ASTM E 1155. This system depends on marking multiple lines on a slab and measuring the deviation at 12-in. (305 mm) intervals along these lines. F-numbers are obtained mathematically from these measurements using formulas contained in ASTM E 1155.

Several devices are used to determine the deviations. One is a 10- or 12-ft. (3050 or 3658 mm)-long highway straightedge laid along the lines. The deviations are then measured down from the straightedge. This method produces an accurate reading when done properly but is time-consuming. Laser levels, optical levels, floor profilometers, and inclinometers give quick and accurate readings.

The degree of flatness and levelness increases as the F-numbers increase. Slabs requiring a closer tolerance must have higher F-numbers. The required F-numbers for a particular slab depend on the activities that will take place on it and the finish that will be applied. In general, the required F_F is 15 and the required F_L is 13 for a slab with a scratch finish, as is permitted on slabs that are to receive concrete floor toppings, mortar setting beds for tile or terrazzo, and other bonded cementitious finishes. Slabs with a float finish that are to receive membrane waterproofing, elastic roofing, or sand-bed terrazzo should have an F_F of 18 and an F_L of 15. Trowel-finish slabs that are to receive resilient flooring, carpet, paint, and other thin-film finishes should have an F_F of 20 and an F_L of 17.

F_F-numbers for slabs in buildings range from 15 to 100; F_L-numbers from 13 to 50. The higher numbers are required when an extraordinary degree of flatness and levelness are needed, such as in skating rinks.

3.13 Cast-in-Place Structural Concrete

The cast-in-place structural concrete discussed in this section includes walls, columns, piers, pilasters, beams, girders, and slabs. The discussion here related to the use and placement of reinforcing steel should be considered introductory only. The actual size and placement of bars, ties, stirrups, and accessories should be determined from structural analysis by a structural engineer. In addition, forming, installing reinforcement, and placing concrete are the responsibility of the contractor. The methods described here may not be followed on a particular project and may, in fact, be inappropriate under certain circumstances. For example, the procedures described here do not apply to mass concrete structures.

3.13.1 WALLS

Foundation walls, which rest on footings, must be able to support vertical loads from the superstructure, resist lateral (horizontal) loads from the ground and the superstructure, and provide anchorage for the superstructure against uplift. Wall construction should be delayed until the footing is sufficiently cured to support the loads. This time will vary depending on the air temperature and the magnitude of the loads to be supported.

In addition to their structural functions, foundation walls must be durable, resist free water and water vapor penetration, provide a barrier against fire, and control air and heat flow. A foundation wall that encloses a habitable space must present a suitable appearance or be able to accept finishes and must accommodate windows, doors, and other openings.

Reinforced concrete and concrete masonry are used extensively for foundation walls. Foundation walls for most commercial, institutional, and industrial buildings are designed according to engineering principles based on theoretical analysis of the forces and stresses involved. Foundations for single- and two-family homes and other small buildings are more commonly designed using minimum requirements based on past experience and observation of foundations for such buildings or on similar requirements listed in the applicable building code. Such requirements are generally empirical, particularly in designing basement walls to resist lateral earth pressures. The usual minimum thickness for concrete foundation walls for small buildings is 10 in. (254 mm) for concrete walls and 12 in. (305 mm) for concrete masonry walls. However, these thicknesses should not be used until they have been verified as acceptable under local conditions. The type of soil and the height of the water table affect the lateral loads acting on foundation walls and must be taken into account.

Figure 3.13-1 shows a typical concrete foundation wall. The key shown on the top of the footing is sometimes omitted. The procedure for building such a wall is as follows: The footing is formed. Reinforcement, including dowels, is installed, and the concrete is placed and screeded. When the concrete in the footing has cured sufficiently (about one day), one side of the wall forms is erected and coated with a form-release agent. Vertical and horizontal reinforcing steel members are placed in the wall and wired to the dowels and to each other. Figure 3.13-1 shows two layers of steel, but walls carrying lighter loads often require only one layer. The reinforcing ties shown are necessary only for walls that bear heavy vertical loads.

After the steel has been placed and inspected, wall ties are inserted through holes in the side of the form that is in place. The remaining side of the formwork is coated with form-release agent and slipped into place over the wall ties. When screw-type wall ties are used, the threaded sections are screwed into the cones. The wales and braces are then installed, and the wall ties are anchored to the wales. The formwork is inspected to ensure that it has been installed properly and that it is in the proper position, tight, and securely braced.

Foreign material is removed from the inside of the forms, the level to which the concrete will be placed is marked on the forms using surveying instruments, and concrete of the proper consistency is placed. Placing of the concrete is done by dumping it from a bucket or by hose from a concrete pump. In a low wall, the concrete may be placed directly from a transit-mix truck's chute. As the concrete is

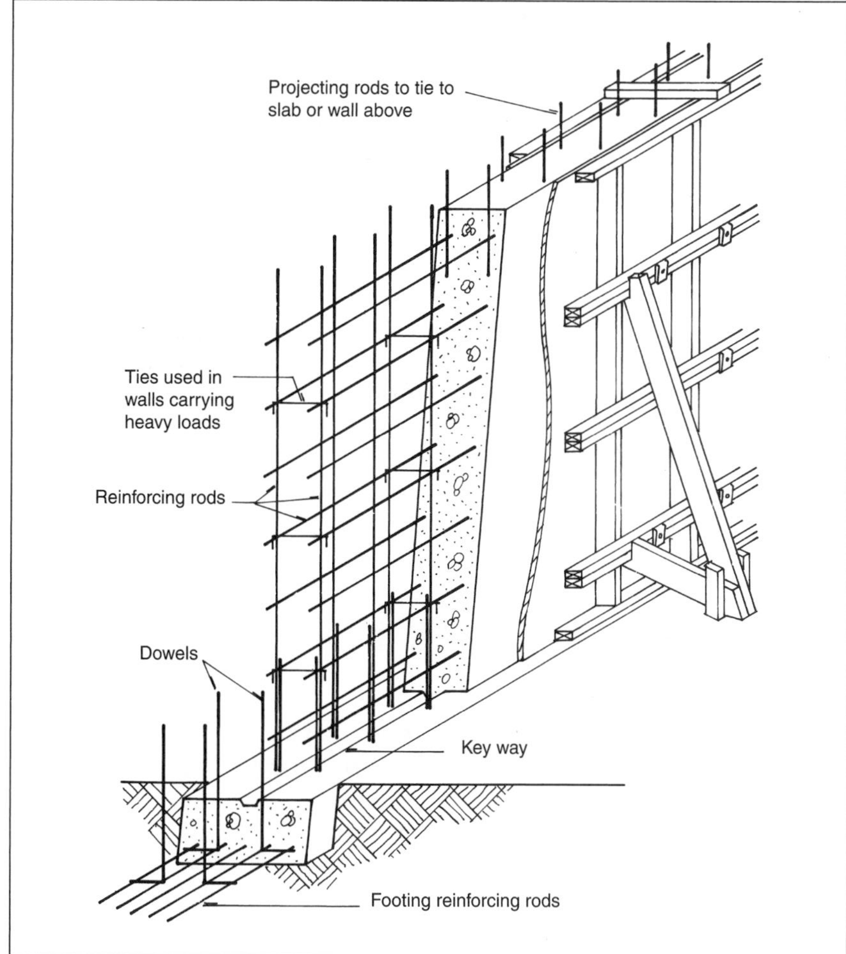

FIGURE 3.13-1 **Typical concrete foundation wall.** (Drawing by HLS.)

placed, it is consolidated using internal vibrators. When it reaches the line on the forms, the concrete is smoothed using hand floats. The top is then covered with canvas or sheet plastic to prevent drying, and the concrete is permitted to cure.

Several days later, the forms are removed and the exposed wall ties are twisted off (unscrewed in the case of screw ties) and finishing is begun.

Most cast-in-place concrete walls higher in a building are built in essentially the same way as foundation walls. Their vertical steel is tied to the steel projecting from below, rather than to dowels. Exposed-to-view joints between levels of concrete in walls should be treated in some manner, such as with a grooved joint or other positive articulation. This is needed to prevent an unsightly condition that may occur because of slight differences in the concrete or because of the difficulty of obtaining clean horizontal joints at such locations.

3.13.2 COLUMNS, PIERS, AND PILASTERS

Columns, piers, and pilasters act in compression to transmit loads to footings. Columns and piers are freestanding; pilasters are built integrally with the wall and may also function as stiffeners to provide necessary lateral support to a wall. Masonry piers and pilasters are discussed in Chapter 4.

Column, pier, and pilaster construction should be delayed until the footing is sufficiently cured to support the loads. This time will vary, depending on the air temperature and the magnitude of the loads to be supported.

Figure 3.13-2 shows the basic types of reinforcements in a concrete column. Piers and pilasters are similar. Several parts of this drawing have been stylized for reasons of clarity. For example, single vertical bars are shown. When more steel is required, bundles containing several individual bars are often used in each location. Only four vertical bar locations are shown, but additional bar locations are often necessary. Eight bars is common, with one falling halfway between each of the corner bars. Sometimes many more locations are required. In the figure, vertical steel is shown in a rectangular pattern, but round patterns are also used, and either may be used in either square or round columns. The reinforcement shown is that for a tied column. When the steel is in a circular pattern, circular ties are often used, but more expensive spiral ties are also sometimes used (Fig. 3.13-3) when the structural engineer deems their use justified. Spiral ties are shipped to the job site coiled like a spring. They are stretched and wired in place at the job.

The process for building a concrete column, pier, or pilaster is similar to that described for a wall. There are, however, some exceptions:

- Dowels are set to line up with the vertical steel.
- Reinforcement ties serve as horizontal reinforcement. The spacing of reinforcement ties is not necessarily the same throughout the height of a column, being doubled where splices occur and increased in number where loads so dictate. Their spacing and size are selected by the structural engineer.
- The reinforcement is often wired together in the horizontal position into a *column cage*, which is then hoisted into location by a crane.
- The reinforcement is placed and wired to the dowels before any formwork is installed. When space is limited within a column, bars may be spliced in line rather than being lapped.
- Form ties are not usually needed.

3.13.3 BEAMS AND GIRDERS

There are several types of beams and girders. *Grade beams* are located approximately at grade level. In contrast to a continuous footing, which is supported uniformly along its length, a grade beam receives building loads and transfers them to either spread, pile, or caisson foundations. Grade beams can take many forms. The examples in Figure 3.13-4 include (a) a reinforced concrete beam placed integrally with the concrete slab and supported on piles and (b) a precast reinforced wall carried on footings and supported laterally by concrete or masonry piers.

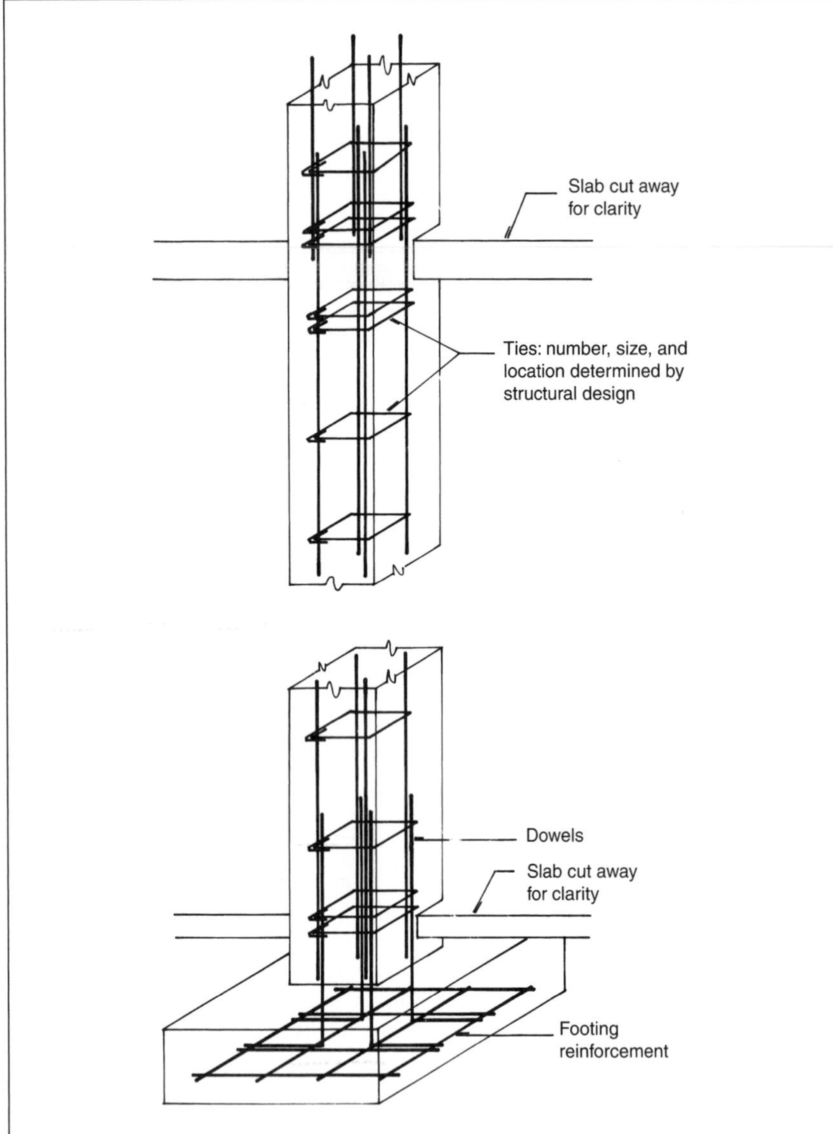

FIGURE 3.13-2 **Typical tied column reinforcement.** (Drawing by HLS.)

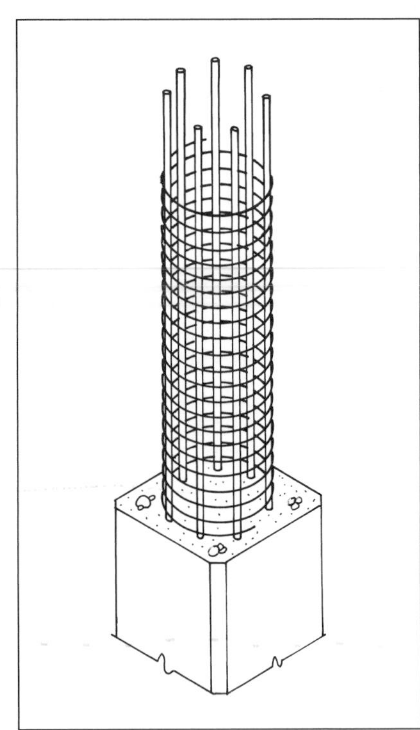

FIGURE 3.13-3 **Typical spiral column reinforcement.** (Drawing by HLS.)

Superstructure beams are located above grade level, where they serve to support floors and roofs. Beams are of three general types: simple beams, continuous beams, and beams that are part of a slab. Beams usually carry only uniform loads that act over the entire length of the beam.

Girders too may be simple or continuous, and they also often carry a uniform load, but in addition they carry beams. Large transfer girders sometimes also carry columns.

3.13.3.1 Simple Beams

Simple beams (Fig. 3.13-5) span only one bay from support to support. They are not often found in large concrete-framed buildings, but an understanding of them will help in understanding the more complex beams and girders usually found in actual construction. As discussed earlier, when a uniform load, including its own weight, is applied to a simple beam, compression stresses are imposed in the top portion of the beam and tensile stresses occur in the lower portion (see Fig. 3.1-1a). Reinforcing steel is added near the bottom of the beam to absorb the tensile stresses (see Fig. 3.1-1b). This reinforcement is supported at the proper height in the form by chairs or bolsters (see Fig. 3.4-9).

Tensile stresses in a simple beam are concentrated at the bottom of the beam in the center of its span and are strongest at the center. Near the ends of the beam, the stresses move diagonally upward, where they tend to cause the beam to crack diagonally. Although weaker than the stresses at the center of the beam, these diagonal stress are still sufficient to cause damage. *Stirrups* are added and wire is tied to the bottom reinforcing bars near the supports to absorb this diagonal stress (Fig. 3.13-6). No stirrups are needed near the center of a simple beam. Stirrups may be U-shaped (U-*stirrups*), as shown in the main figure, or closed (*closed stirrup-ties*), as shown in Figure 3.13-6a, or capped (two-piece) ties, as shown in Figure 3.13-6b. Light reinforcing rods are placed near the top of a beam with stirrups to hold the stirrups in place until the concrete has set. When heavy loads are present and the depth of a beam is limited by factors other than structural considerations, compression steel may be added to the top of a beam. Then this compression steel is used to hold the stirrups in place.

Reinforcing steel dissipates the forces it absorbs into the concrete by means of its bond with the concrete until the stresses have been neutralized. However, when the reinforcement reaches the end of a beam, it typically has some remaining stress, but there is no more concrete. This remaining stress is dissipated by bending each bar into a U shape called a *hook*.

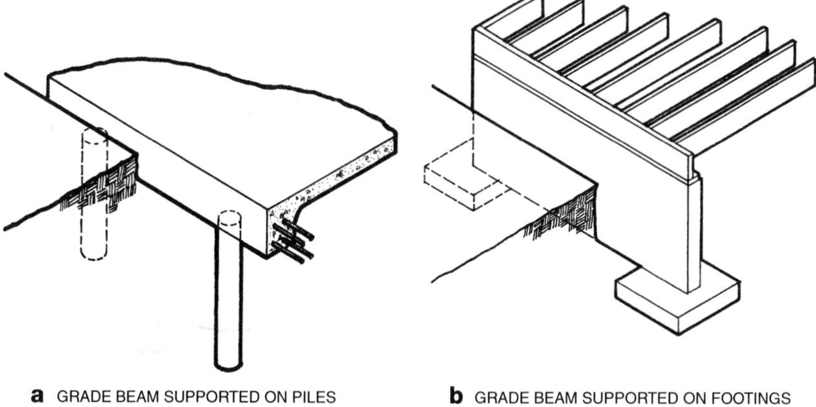

FIGURE 3.13-4 Grade beams: (a) reinforced concrete beam and (b) precast reinforced wall.

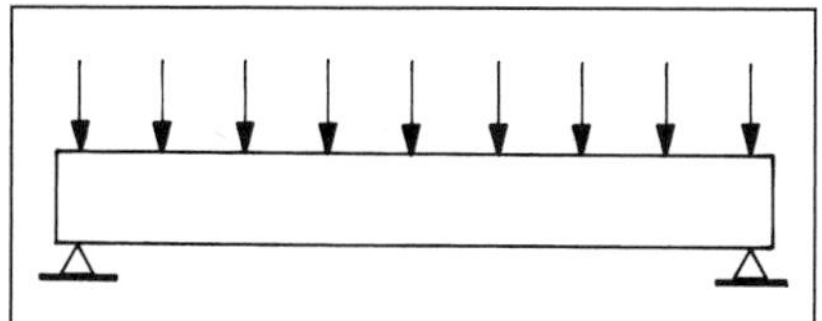

FIGURE 3.13-5 Simple beam. (Drawing by HLS.)

Beams are formed, reinforcement installed, and concrete placed in the beams in a sequence similar to that used in walls. The major differences are that the steel reinforcement in beams is supported on chairs or bolsters resting on the bottom of the formwork, and wall ties are usually not required.

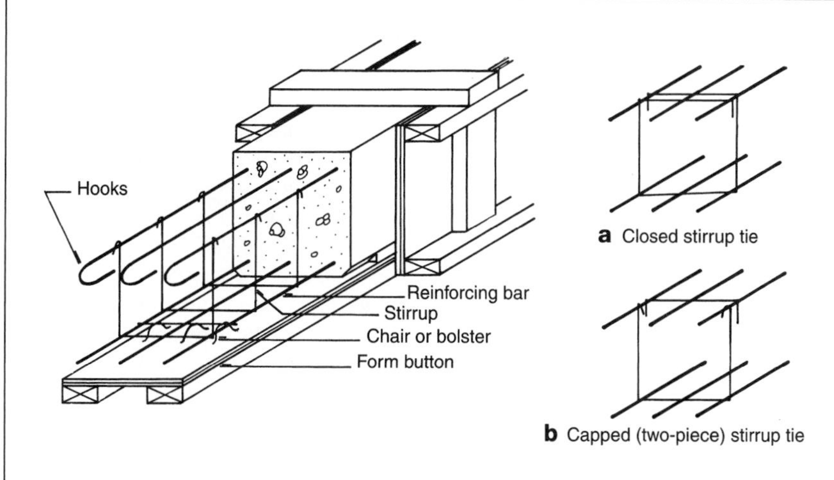

FIGURE 3.13-6 Reinforcement and forming for a typical beam. (Drawing by HLS.)

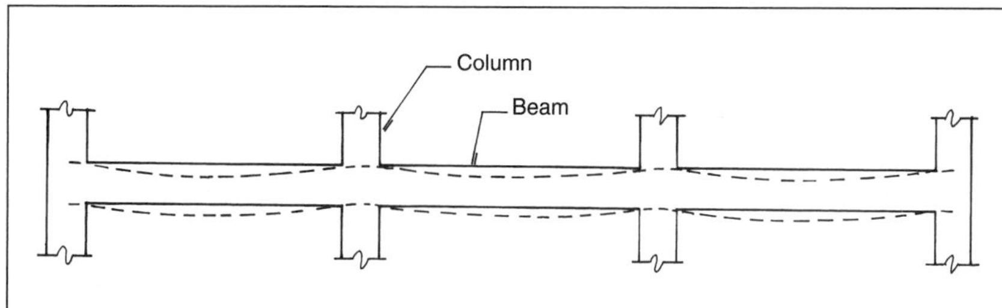

FIGURE 3.13-7 A continuous beam. (Drawing by HLS.)

3.13.3.2 Continuous Beams

A *continuous beam* crosses more than one span (Fig. 3.13-7). The dashed lines in the figure show a greatly exaggerated view of the bending that tends to occur in such a beam. Reinforcing steel is added to resist this bending. These beams are formed and steel and concrete are placed in them as described for simple beams. Some ways in which continuous beams are different from simple beams follow:

- The reinforcement in a simple beam tends to be the same throughout. In a continuous beam, more cross-sectional area of reinforcement is required at the bottom of the beam near the center of each bay than is required to pass through the columns. This usually means fewer bottom bars at the columns.
- Top steel is required in a continuous beam at every column to resist the tendency of the beam to bend upward at those locations.
- Stirrups are required at each end of a continuous beam and, in addition, in the beam on both sides of each intermediate column.
- Right-angle bends, with the steel turning downward into the columns at the ends of a continuous beam, may be used in lieu of hooks when there is sufficient room for them.

3.13.3.3 Beams That Are Part of Floor Slabs

Beams are often part of a concrete slab (Fig. 3.13-8). They do not act like simple beams or even continuous beams, but rather have some unique characteristics. For example, they act as part of the slab, rather than as separate elements, and they are affected by forces acting in the slab. Therefore, they are discussed in Section 3.13.4.

3.13.3.4 Girders

Girders are essentially large beams, and the principles discussed in Sections 3.13.3.1 and 3.13.3.2 apply to them as well. Their steel reinforcement must be specifically designed by a structural engineer, but they usually have both top and bottom steel reinforcement and heavier bars than do beams. Additional steel is required where concentrated loads are carried.

3.13.3.5 Post-Tensioning

Tensile forces near the bottoms of loaded beams tend to pull the concrete apart,

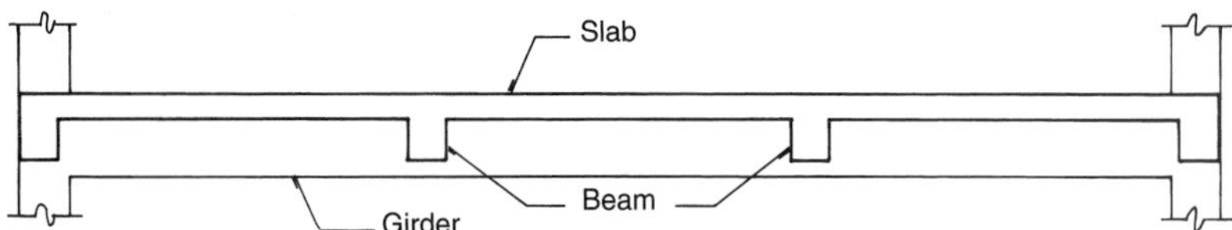

FIGURE 3.13-8 **Beam and floor slab working together.** (Drawing by HLS.)

making it crack (see Fig. 3.1-1). Even when reinforcing steel is added, these cracks are often large enough to be seen. Such cracks can be eliminated altogether by placing the concrete in compression before an external load is applied. If enough compression is applied and held there, when a load is applied, the amount of compression will be lessened but will never disappear completely. Compressing the concrete in a beam will also permit it to carry the same loads with less concrete, making it lighter and less expensive.

Compression loads can be added to a beam by stretching the steel and locking it to the concrete at the ends. As the steel tries to return to its normal length, the resulting stress is transferred to the concrete. There are two procedures used to produce this result. *Prestressing* is done before concrete is placed around the steel. This process is best done under controlled conditions in a shop or factory and is very expensive to do anywhere else. Therefore, it is used almost exclusively in precast concrete work, as described in Section 3.14.

A process called *post-tensioning* is used to add permanent tension to the reinforcing steel in cast-in-place concrete and subsequently to place the concrete under compression. Essentially, it is done using *tendons*, which are bundles of high-strength, cold-drawn steel wire strands or steel bars. The tendons are coated with oil or placed in a steel tube to prevent bonding between the concrete and the tendons. The concrete is placed and permitted to cure. Generally, the tendons are fastened to one end of the beam and stretched from the other end using hydraulic jacks. When the desired amount of tension has been achieved, the jack end of the tendon is secured to the concrete, usually with a bearing plate of some sort. When the strands are very long, they must be stretched from both ends simultaneously to ensure uniform tensioning.

Tendons can be placed level, as conventional reinforcing bars are placed. Post-tensioned beams are more efficient, however, if the tendons are placed in a shape that approximates the lines of tensile force in the beam. This is near the bottom of the beam in the center and sloping upward toward the top of the beam at the ends in a V-shape.

Post-tensioned beams tend to shorten due to elastic compression, shrinkage, and creep. These movements must be accommodated in the building design and construction. Adjacent elements that would be affected by these changes should be built after the post-tensioning has been completed, or they should be isolated from the post-tensioned unit. In addition, the amount of tension to be applied can sometimes be adjusted slightly upward to at least partially compensate for these movements.

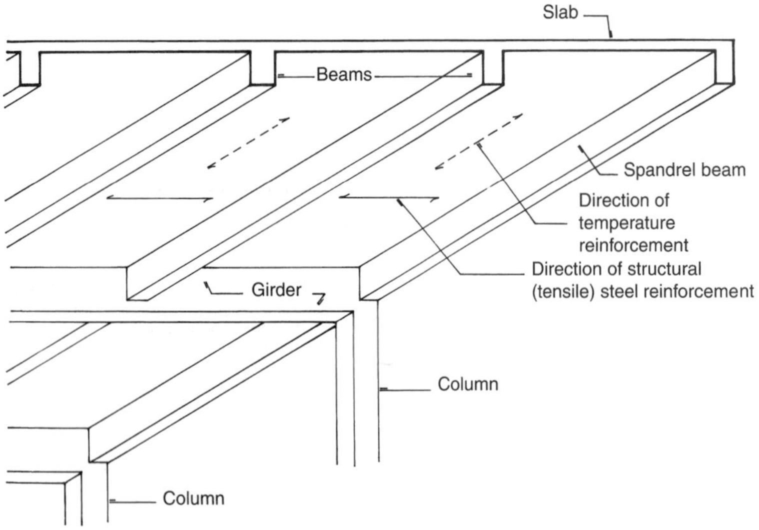

3.13.4 STRUCTURAL SLABS

The term *structural slabs* in this book refers to slabs that are not cast against the earth (slabs on grade). Structural slabs fall into one of two categories: one-way slabs and two-way slabs.

3.13.4.1 One-Way Slabs

One-way slabs span in one direction from support to support. There are two types of one-way slabs: *solid* and *ribbed* or *joist* slabs (Figs. 3.13-9 and 3.13-10).

SOLID SLABS

The solid slab shown in Figure 3.13-9 is cut to show its shape. Such slabs usually have beams, girders, or bearing walls on each of their sides (Fig. 3.13-11).

The first step in constructing a one-way solid slab is to form the supporting walls and columns, place the concrete in them, and permit it to cure sufficiently to support the loads that will be imposed. The second step is to build the forms for the beams and girders, and then the forms for the slabs. The forms should be supported on temporary supports and shores of sufficient strength and stability to resist the large loads that will be imposed by the forms and the concrete. When all the forms have been built, the side of the forms that will be in contact with concrete should be coated with form-release agent and the steel should be placed in them, following the shop drawings.

Structural reinforcement in a one-way solid slab is similar to that described in Section 3.13.3.2 for continuous beams, generally with bottom bars and top bars over the beams and girders. The structural reinforcement in one-way slabs is placed parallel to the span (see Fig. 3.13-9). Temperature steel, which usually consists of No. 3 bars, is placed perpendicular to the structural bars. Temperature steel is not needed parallel to the span, because the structural bars there will serve to resist shrinkage and temperature cracks in that direction.

When the steel is in place and has been inspected, concrete may be placed into the

FIGURE 3.13-9 A one-way solid slab. (Drawing by HLS.)

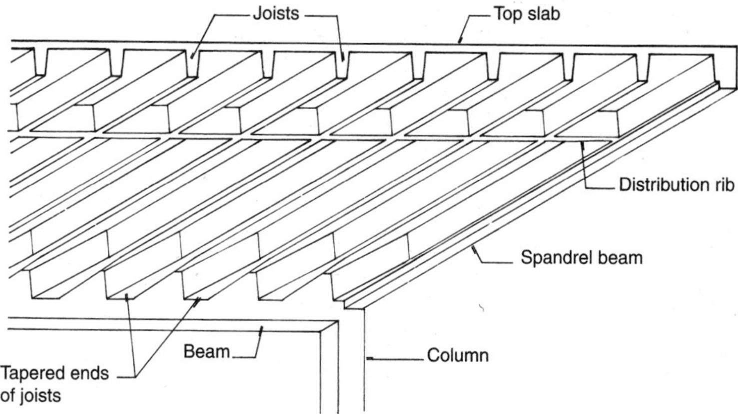

FIGURE 3.13-10 A one-way ribbed slab. (Drawing by HLS.)

forms and consolidated. The slab should then be cured and finished as described in Section 3.9. When the slab, beams, and girders are sufficiently strong to support themselves, the forms should be removed and reshores placed (Fig. 3.13-12). *Reshoring* consists of vertical metal or wood props designed to help the new concrete support construction loads for several additional weeks until it has cured sufficiently to stand alone. After reshoring has been done, formed surfaces should be finished. If there is another floor above the one just completed, formwork should be moved up to it and the process repeated.

RIBBED SLABS

One-way solid slabs become uneconomical when slab spans are long, because a slab must become thick and heavy to support the loads imposed and the steel must be placed lower in the slab. Eventually, the weight of the slab itself becomes a problem. The solution then is to eliminate part of the concrete. This is possible, because much of the concrete near the bottom of a deep slab is not normally needed to resist loads. Most of the stress there is tensile, which is resisted by the steel. The solution then is to remove part of this nonworking concrete; one way is by forming the slab using metal or plastic *pans* (Fig. 3.13-13), which produces the effect of a thick slab using less concrete.

Standard pans come in two widths, 20 in. (508 mm) and 30 in. (762 mm). They range in depth from 8 in. (203 mm) to 20 in. (508 mm). Pans are tapered to permit their easy removal after the concrete has set. Tapered ends are used to increase the amount of concrete at the beams.

The ribs formed between pans are called *joists*. Joist width can be varied by widening the spacing of the pans. Ribs are supported on beams. When these beams are kept at the same depth as the pans, they are called *joist bands*. Joists are often interrupted at midspan by a beam known as a *distribution rib*, which helps to distribute concentrated loads to several joists.

Ribbed one-way slabs are constructed by first building the supporting elements such as columns and walls and permitting them to cure sufficiently to support the loads that will be imposed. Then the supporting structure (centering) is built for the pans as described in Section 3.3.2.5. The centering decking forms the bottom of the joists. The pans should be placed on this centering and coated with a form-release agent. The steel should then be placed.

After the steel has been installed and inspected, the concrete should be placed and the slab surface cured and finished as described in Section 3.9. When the ribbed one-way slab has set sufficiently to support its weight, the centering should be removed and the pans dropped from the concrete and moved to the next forming area. The formed surfaces should then be finished.

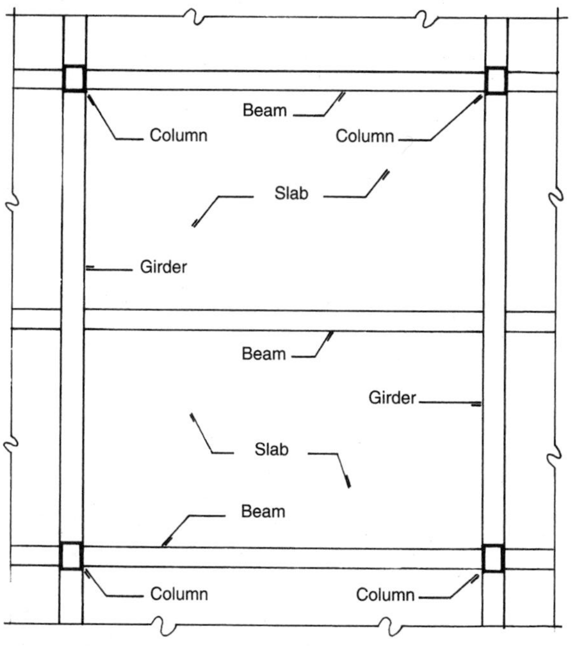

3.13.4.2 Two-Way Slabs

Two-way slabs span in two perpendicular directions at the same time. They are generally selected when the columns in a building can be located to create square or almost square bays. In this case, they are usually less expensive than comparable one-way slabs. It is possible to construct a two-way slab that is supported on all four sides by beams, similar to the one-way solid slabs described in Section 3.13.4.1, but such slabs are seldom used. Two-way slabs without beams are easier to form and thinner when the beam depth is taken into account.

There are essentially two types of two-way slabs in general use, *flat plate* and *waffle* slabs (Figs. 3.13-14 and 3.13-15). Two-way flat plate slabs are used on buildings with relatively short spans. When spans become longer, the same thing happens with two-way slabs that happens with one-way slabs: the slabs get thicker until they become uneconomical. The solution is similar. The two-way

FIGURE 3.13-11 Solid one-way slab framing. (Drawing by HLS.)

FIGURE 3.13-12 Formwork is under construction on the top floor. On the next-to-the-top floor, the formwork is still in place. The formwork has been removed from the lower floor and reshoring has been placed. (Courtesy Portland Cement Association.)

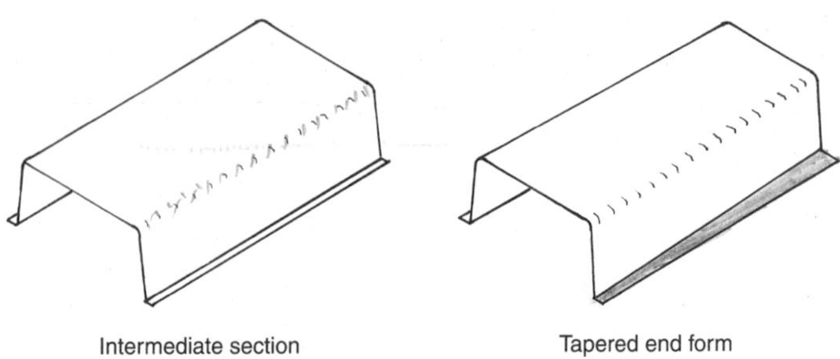

FIGURE 3.13-13 Typical one-way ribbed-slab pan. (Drawing by HLS.)

equivalent of a pan slab is a *waffle slab*, which has ribs in both directions (Fig. 3.13-15).

3.13.4.3 Flat Plate Slabs

Flat plate slabs that carry light loads, such as those in office buildings, hotels, apartment buildings, and dormitories, usually have no thickened areas (see Fig. 3.13-14). Flat plate slabs in more heavily loaded storage and industrial buildings often have dropped panels at the columns to resist the heavy shear loads associated with heavier loading (Fig. 3.13-16).

The first step in constructing a two-way flat plate slab is to form the supporting walls and columns, place the concrete in them, and permit it to cure sufficiently to support the loads that will be imposed. The second step is to build the forms for the dropped panels, when they occur, and then the forms for the slabs. The forms should be supported on temporary supports and shores of sufficient strength and stability to resist the large loads that will be imposed by the forms and the concrete. When all the forms have been built, the side of the forms that will be in contact with concrete should be coated with a form-release agent and the steel should be placed in them, following the shop drawings.

Structural reinforcement in each of the two perpendicular directions in a two-way flat slab varies, depending on its location

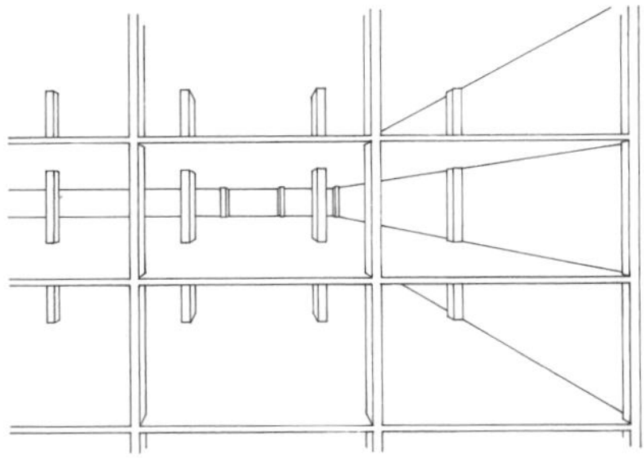

FIGURE 3.13-14 Two-way flat-plate slab. (Drawing by HLS.)

FIGURE 3.13-15 Waffle slab. (Courtesy Portland Cement Association.)

relative to the columns. The slab is divided into zones, one between the columns called the *middle strips* and another called the *column strips*, which lie along the column rows in both directions (Fig. 3.13-17). The least amount of reinforcement is placed in the middle strips where the loads are lightest. Additional steel and closer spacing of bars occurs in the column strips to accommodate the higher bending stresses there. Steel is located near the top of the beam in the column strips where upward bending stresses are present. Temperature steel is not needed in two-way slabs, because there is already steel in both directions. There is usually no additional steel in column caps when they are used. Rather, the additional concrete is relied on to resist the additional shear stress.

When the steel is in place and has been inspected, concrete may be placed into the forms and consolidated. The slab should then be cured and finished as described in Section 3.9. When the slab is sufficiently strong to support itself, the forms should be removed and reshores placed. The formed surfaces should then be finished. If there is another floor above the one just completed, formwork should be moved up and the process repeated.

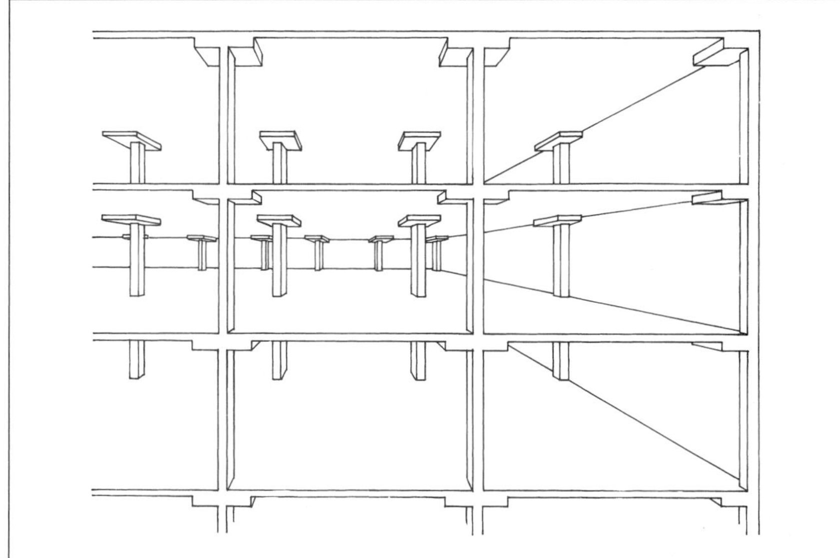

FIGURE 3.13-16 Two-way flat-plate slab with dropped panels. (Drawing by HLS.)

A variation in two-way flat slab construction that eliminates most of the formwork is *lift-slab* construction. This is a process in which the slabs for the floors of a building occurring above the ground floor are cast on the ground floor in a stack and are then lifted into place with a system of hydraulic jacks. The slabs have cast-in sleeves that ride up the columns. When the lowest slab of a stack has reached its proper level, the sleeves are welded to the columns and the jacks are used to pull the remaining slabs up the column to the next level. When the last slab has been welded into place, the jacks are removed.

3.13.4.4 Waffle Slabs

Two-way waffle slabs are formed using metal or plastic pans called *domes* (Fig. 3.13-18). The ribs formed between domes are called *joists*. There are two basic sizes of standard domes. The smaller of the two forms joists 5 in. (127 mm) wide at the bottom on 24-in. (610 mm) centers and from 8 to 14 in. (203 to 356 mm) deep. The second standard dome size produces joists 6 in. (152 mm) wide at the bottom on 36-in. (914 mm) centers and from 8 to 20 in. (203 to 508 mm) deep. Larger sizes are also available. Domes are tapered to permit their easy removal after the concrete has set. Some domes have compressed-air fittings to permit even easier removal.

Waffle slabs are most efficient when allowed to cantilever at the perimeter. When a cantilever is not desirable, a perimeter beam (*slab band*) must be placed there (see Fig. 3.13-19). In addition, column *heads* are created around the columns to accommodate the shear stresses there (see Fig. 3.13-19). Slab bands and heads are created by omitting domes.

Waffle slabs are constructed by first building the supporting elements, such as columns and walls, and permitting them to cure sufficiently to support the loads that will be imposed. Then the supporting structure (*centering*) is built for the domes as described in Section 3.3.2.5. The domes should be placed on this centering and coated with a form-release agent. The steel should then be placed.

After the steel has been installed and inspected, the concrete should be placed and the slab surface cured and finished as described in Section 3.9. When the waffle slab has set sufficiently to support its weight, the centering should be removed and the domes dropped from the concrete and moved to the next forming area. The formed surfaces should then be finished.

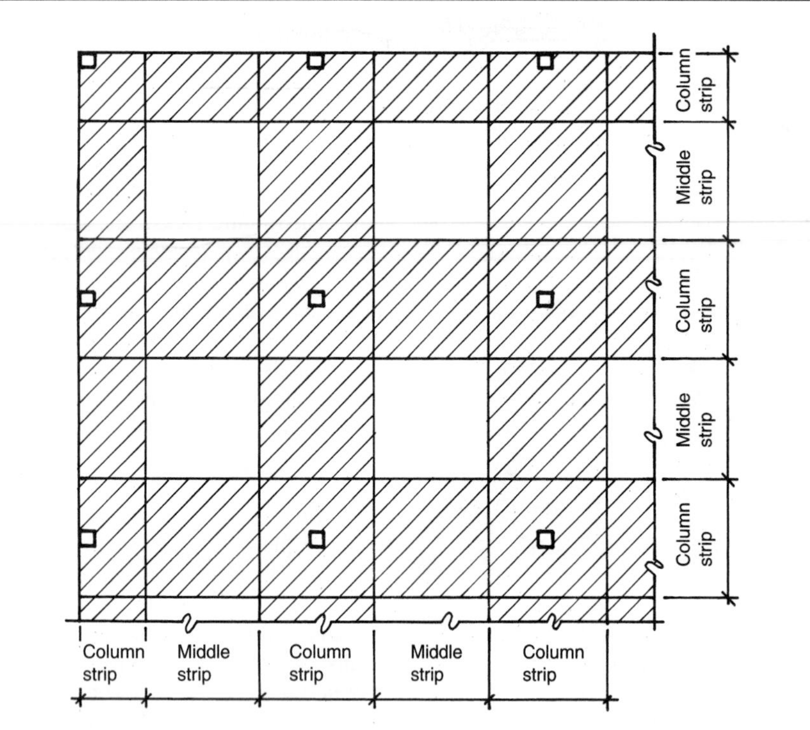

FIGURE 3.13-17 Plan showing locations of middle strips and column strips in a two-way flat slab. (Drawing by HLS.)

3.13.4.5 Slabs on Permanent Forms

In steel structures, concrete slabs are often placed on a permanent corrugated steel sheet decking (Fig. 3.13-20). The steel decking acts like a form until the concrete sets. Such slabs are usually reinforced with welded wire fabric. This produces a finished lightweight deck. Such slabs are finished and cured as described in Section 3.9.

Heavier-gauge steel decking is also used with a concrete fill to produce *composite decking* (see Chapter 5).

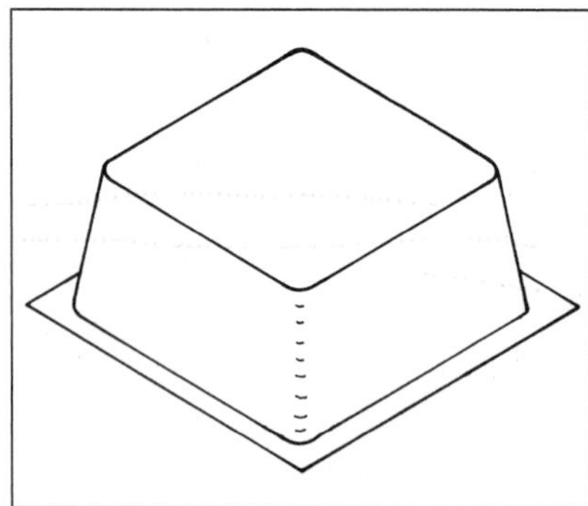

FIGURE 3.13-18 A waffle-slab dome. (Drawing by HLS.)

3.13.4.6 Slab Flatness

The principles concerning flatness requirements for slabs on grade discussed in Section 3.12.5 also apply to structural slabs.

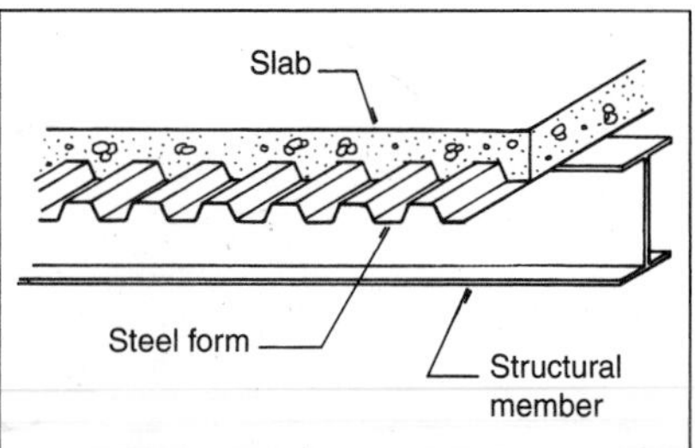

FIGURE 3.13-20 A concrete slab on a permanent steel form. (Drawing by HLS.)

3.13.4.7 Post-Tensioning

A process similar to that described in Section 3.13.3.5 for post-tensioning beams is sometimes used to post-tension flat-plate concrete slabs. Post-tensioning concrete slabs reduces their thickness, which results in smaller floor-to-floor heights and less cladding of the building. The amount of steel required can be as little as one-third of that needed for a conventional concrete slab.

3.13.5 CONCRETE STAIRS

Concrete stairs are of two types. Each run of the first type consists of an inclined reinforced slab with integral risers and horizontal landings at the top and bottom (Fig. 3.13-21). The bottoms, sides, and risers are formed. The risers usually slope outward at the top. All concrete in such a

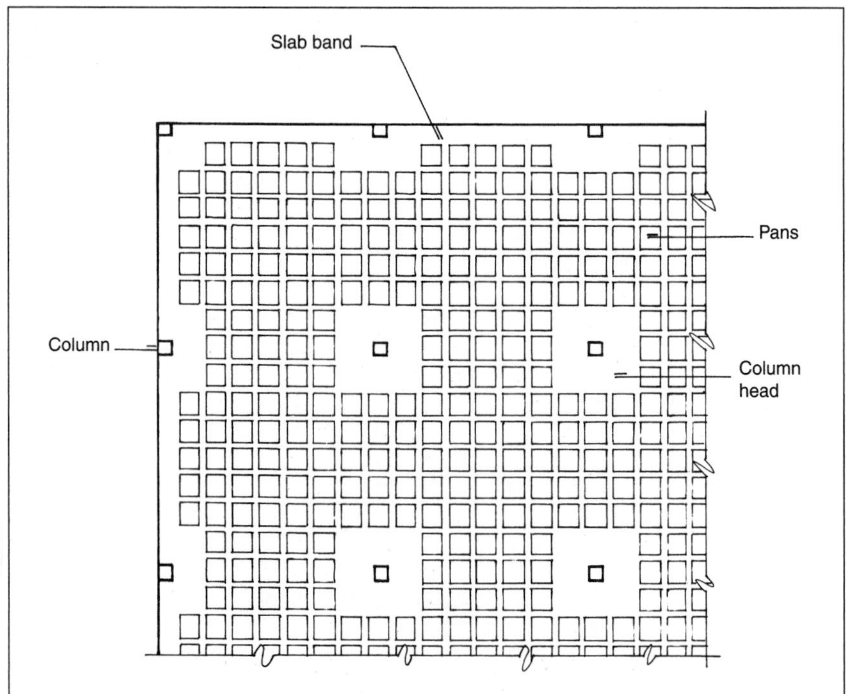

FIGURE 3.13-19 Pan slab in plan showing slab bands and column heads. (Drawing by HLS.)

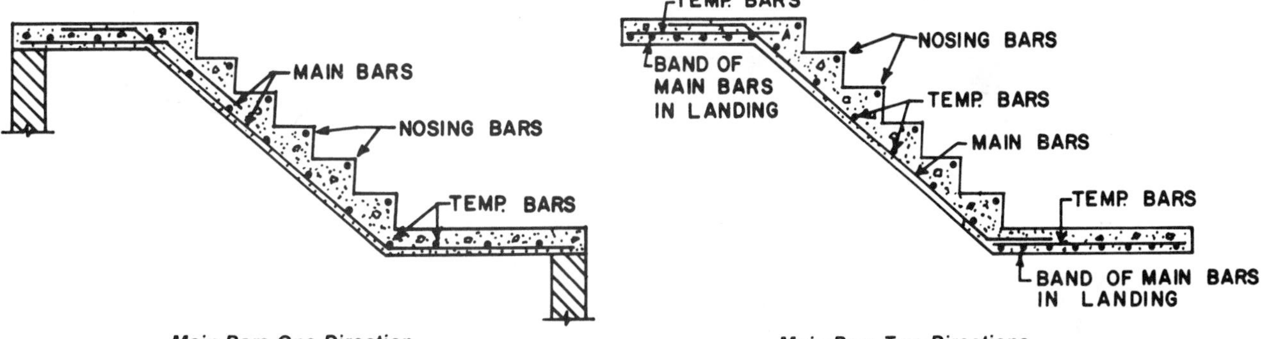

FIGURE 3.13-21 **Typical stair reinforcement.** (Courtesy Concrete Reinforcing Steel Institute.)

stair is placed simultaneously. This type of stair usually occurs in concrete-framed buildings. Similar stairs are also used at entrances and site locations with the underneath supported by earth rather than by formwork.

The second type of concrete stair consists of steel pans supported on steel stringers. These pans are filled with concrete. This type of stair is most often used in steel-framed buildings.

3.14 Precast Concrete

Precast concrete can be either site cast or plant cast.

3.14.1 SITE-CAST UNITS

Concrete structural elements are sometimes precast at a job site and lifted or tilted into place by construction cranes. Any precast element can be cast in the field on a horizontal surface such as the ground or a concrete slab and lifted into place in the building. Job-cast units are not subject to the size limitations imposed on plant-cast units. Therefore, it is sometimes advantageous to precast large elements, such as transfer beams, on the ground. Site-cast precast units are subject to the same disadvantages as cast-in-place concrete relative to curing time and adverse weather conditions.

FIGURE 3.14-1 **A tilt-up panel being lifted into place.** (Courtesy the Burke Company.)

A common element precast in the field is the *tilt-up* slab. To save time and money by eliminating most of the usual formwork, concrete wall sections are cast on a poured-in-place concrete slab and tilted or lifted into place (Fig. 3.14-1). The greatest use of tilt-up construction has been in single-story industrial buildings, but the method has also been used in other types of buildings.

Some additional reinforcement is necessary in field-precast elements to permit attachment of lines from a crane and to resist the stresses generated by lifting the units into place.

Joining sections of tilt-up panels is accomplished in several ways. The space between panels may be filled with a poured-in-place concrete pilaster or column. Panels may be joined using welded-together steel angles cast into the tilt-up panels. Tilt-up panels may be supported by separate concrete columns, by concrete columns cast as a part of the tilt-up panels, or by structural steel columns. Panels may themselves be bearing walls, supporting a roof structure without columns.

3.14.2 PLANT-CAST UNITS

The cost of forming and placing concrete in the field and the need to cope with un-

FIGURE 3.14-2 A precast concrete beam being erected by crane. (Photo courtesy Precast/Prestressed Concrete Institute.)

FIGURE 3.14-3 A precast concrete structure being erected in the snow. (Photo courtesy Precast/Prestressed Concrete Institute.)

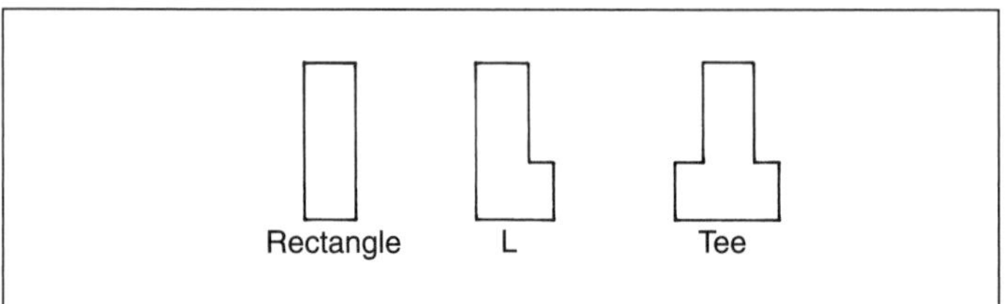

FIGURE 3.14-4 Typical precast beam shapes. (Drawing by HLS.)

favorable weather conditions while doing so have led to *plant-cast* precast concrete elements that are produced in a factory environment. Units are cast at ground level in the plant, shipped to the construction site, and set in place using cranes in much the same way that structural steel is erected (Fig. 3.14-2).

Precasting concrete elements in a plant reduces their cost and makes it easier to control their quality. For example, concrete in beds in a plant can be easily consolidated by mechanical vibration, which results in fewer voids and better surfaces than can be achieved in concrete vibrated in the field. The ease of *prestressing* structural units in an industrial plant allows the units to be made smaller. It is also possible in a plant to cure units using steam. Steam-cured precast concrete units made with high-early-strength portland cement can be removed from the casting bed within 24 hours after the concrete has been placed.

Plant-cast precast concrete units, which are heavy and bulky, must be shipped from plant to job site by truck. Legal limitations on the size of trucks and the use of roads limit the size of precast units that can be shipped, which in turn influences the types of units that can be effectively precast in a plant. For these and other reasons, plant-cast precast concrete units are usually modular and relatively small. For example, it is impracticable to precast several bays of a two-way structural slab in a plant. Entire walls and even rooms have been produced, but most plant-cast precast concrete is made in relatively small modular units in a few standard shapes.

Once plant-cast precast concrete units arrive at a job site, they can be erected quickly in most kinds of weather (Fig. 3.14-3). There is no need to wait for them to cure or to provide the moisture conditions required to cure cast-in-place concrete.

Precast concrete units can be divided into two basic categories: *structural* precast concrete units and *architectural* precast concrete units.

3.14.3 STRUCTURAL PRECAST CONCRETE

Structural precast concrete units are so called because they form part of the structural system of the building and carry loads other than their own weight.

Precast/Prestressed Concrete Institute (PCI) publication MNL-116S establishes standard practices for the manufacture of structural precast concrete elements. PCI also administers a plant certification program.

3.14.3.1 Types of Units

Structural precast concrete elements include walls, beams, girders, columns, and floor and roof systems.

WALLS

Structural precast walls are occasionally used, but most precast wall panels are architectural concrete. Therefore, this category is discussed in Section 3.14.4.

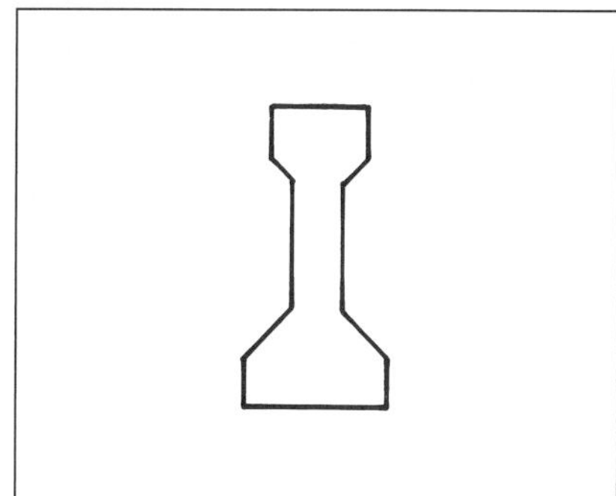

FIGURE 3.14-5 **AASHTO bridge girder shape.** (Drawing by HLS.)

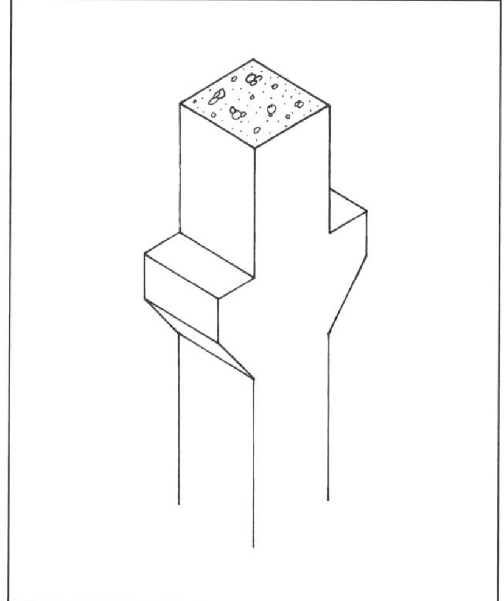

FIGURE 3.14-6 **Typical precast concrete haunched column.** (Drawing by HLS.)

BEAMS, GIRDERS, AND COLUMNS

The usual shapes for precast beams are rectangular, L-shaped, and inverted-T-shaped (Fig. 3.14-4). The projections serve as support ledges for floor and roof systems.

Girders are usually L-shaped or inverted-T-shaped, but heavier girders may be the American Association of State Highway and Transportation Officials (AASHTO) shape shown in Figure 3.14-5. These girders were originally used on bridges, but work as well in buildings.

Precast columns are usually either square or rectangular. While most structural precast concrete is post-tensioned, columns often have conventional reinforcement. Columns may be either single-story units or cast in multistory sections with corbels to support beams and girders (Figs. 3.14-6 and 3.14-7).

FLOOR AND ROOF ELEMENTS

Floor and roof elements rest on bearing walls, beams, or girders and are either left exposed to become the floor or roof slab or are covered with a concrete topping (see Section 3.16).

Floor and roof elements fall into one of two general categories: flat slab or T-shaped. Either type may be made from either normal-weight or lightweight concrete. A balance must be drawn in this selection between the decreased weight of lightweight concrete and its greater cost.

FIGURE 3.14-7 **Concrete-framed building under construction, showing two-story columns and several types of beam haunches.** (Photo courtesy Precast/Prestressed Concrete Institute.)

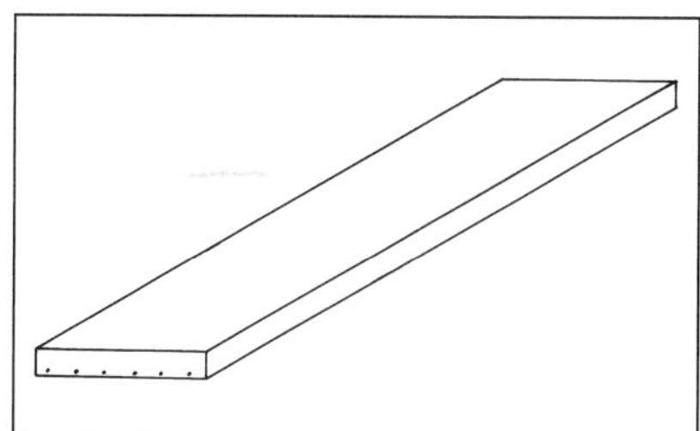

FIGURE 3.14-8 **Solid flat precast concrete slab.** (Drawing by HLS.)

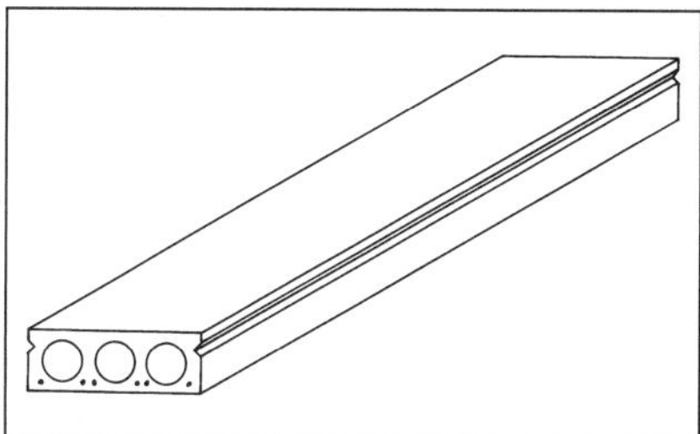

FIGURE 3.14-9 **Cored flat precast concrete slab.** (Drawing by HLS.)

Flat-Slab Precast Elements Flat-slab precast elements may be either solid or hollow-core. *Solid flat-slab units* are simple flat concrete slabs (Fig. 3.14-8). They are used where spans are about 20 ft. (6.1 m) maximum, and maximum headroom is needed.

Hollow-core flat-slab units are cast with round voids running their entire length (Fig. 3.14-9). They are appropriate for intermediate spans (about 50 ft. [15.2 m] maximum) where solid flat units would be too thick and heavy. The principle is the same as that discussed in Section 3.13.4 as justification for selecting pan slabs instead of flat-plate slabs. The cores eliminate the nonworking concrete, lightening the slab section. The steel is contained in the top and bottom of the cored slab.

Both solid and hollow-core slabs are relatively thin, which saves on floor-to-floor height. They can also be painted or given a thin textured coating and left exposed as a finished ceiling.

T-shaped Precast Elements When spans are longer than can be accommodated by cored slabs, the cores are eliminated, the slab thinned, and stems added to give stiffness to the system. The result is T-shaped floor and roof units. When the units have a single stem, they are called *single tees*; two-stemmed elements are called *double tees* (Fig. 3.14-10).

Double tees can span as much as 100

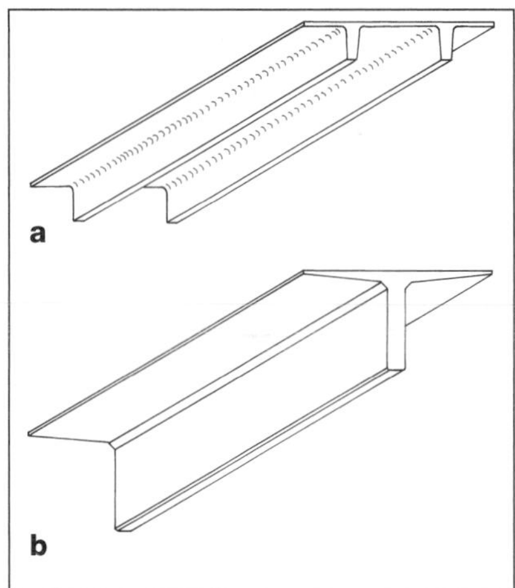

FIGURE 3.14-10 **Stemmed precast concrete elements: (a) double tee; (b) single tee.** (Drawing by HLS.)

FIGURE 3.14-11 **Installing a precast hollow-core flat-slab section.** (Photo courtesy Precast/Prestressed Concrete Institute.)

ft. (30.5 m). They can also be used on shorter spans where solid slabs and cored slabs will also work. Where headroom is not a problem and their appearance is desirable, double tees may be appropriate on short spans.

Single tees have the disadvantage of having to be braced during construction because they are subject to tipping until they are finally fixed in place. They are, therefore, seldom used except on long spans. Their normal spans of between 60 and 150 ft. (18.3 and 45.7 m) require depths from 36 to 60 in. (914 to 1500 mm), which is appropriate in few buildings.

Tees usually have steel weld plates cast into their edges so that adjacent units can be welded together to develop a diaphragm action in the entire system. This process helps to reduce differential deflection in adjacent units.

Topping Precast concrete floor and roof elements are usually finished rough and covered with a 2- to 4-in. (50.8 to 102 mm)-thick topping, as described in Section 3.16.

3.14.3.2 Manufacture

Most structural precast concrete elements are manufactured in a plant on casting beds that range from 200 to 400 ft. (61 to 122 m) in length. However, some are cast in molds. The process is essentially as follows.

Prestressing steel is placed in a casting bed and tensioned using hydraulic jacks. Bulkheads are placed in the bed to separate individual units. Bulkheads are sometimes omitted for solid and cored slabs, which are then sawed apart later. Core forms and opening forms are placed in the beds. Concrete is placed in the beds, consolidated by vibration, and leveled. The top surface is then finished, as appropriate, with floats or trowels. Steam or radiant heat is then used to cure the unit. In some plants, some units, especially flat-slab units, are wet-cured for 7 days in lieu of steam curing.

Test cylinders are taken at the same time the concrete is placed and cured. When the units are steam-cured, about 24 hours after the concrete is placed, the cylinders are tested to verify the strength of the concrete. When it is found to be satisfactory, the prestressing strands are cut and the units are separated and removed from the casting bed.

3.14.3.3 Erection

Precast structural concrete elements are handled and erected in much the same way as structural steel. Units are hoisted into place by crane (Fig. 3.14-11). Some units are held in place by gravity, others by the welding together of inserts in the elements.

CONNECTIONS

Precast concrete columns usually have a cast-in steel plate at the bottom. This plate is fastened to anchor bolts set into a concrete footing (Fig. 3.14-12). Column sections are joined using similar plates and bolts (Fig. 3.14-13).

Beams and girders are anchored to columns in a variety of ways. A typical

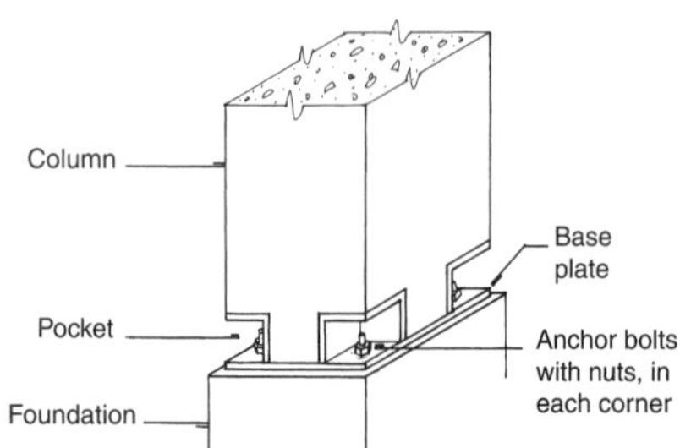

FIGURE 3.14-12 **A typical precast concrete column base at footing. Pockets are filled with nonshrink grout after anchor nuts are set.** (Drawing by HLS.)

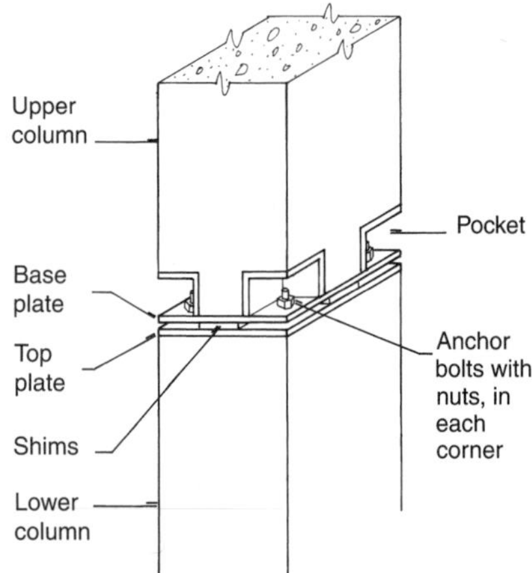

FIGURE 3.14-13 **A typical bolted column-to-column connection. Pockets and spaces around shims are packed solid with nonshrink grout after anchor nuts are set.** (Drawing by HLS.)

FIGURE 3.14-14 Tees resting on a beam and on masonry. (Photo courtesy Precast/Prestressed Concrete Institute.)

FIGURE 3.14-15 Hollow-core flat slab resting on a beam. (Photo courtesy Precast/Prestressed Concrete Institute.)

system is the use of corbels to support beams and girders (see Fig. 3.14-6).

Tees usually rest on L- or T-shaped beams or girders (Fig. 3.14-14). Tees are often, but not always, given a concrete topping (see Section 3.16).

Flat and cored slabs usually rest on beams or girders (Fig. 3.14-15). Such slabs are usually given a concrete topping (see Section 3.16).

Bearing pads are usually inserted between concrete elements and other concrete elements supported by them so that the elements may move to accommodate thermal changes and other structure movement. Bearing pads are usually synthetic rubber for beams, girders, and tees, and plastic for slabs.

Precast structural elements must be held together to resist seismic and wind loads. This is sometimes accomplished by installing reinforcing rods in a topping or by welding together plates cast into the elements. However, the stems of tees are never fastened solidly to beams or columns because doing so would prevent them from expanding and contracting freely.

3.14.4 ARCHITECTURAL PRECAST CONCRETE

Architectural precast concrete units include panels and other members used as part of a building's exterior cladding. They are usually not load bearing.

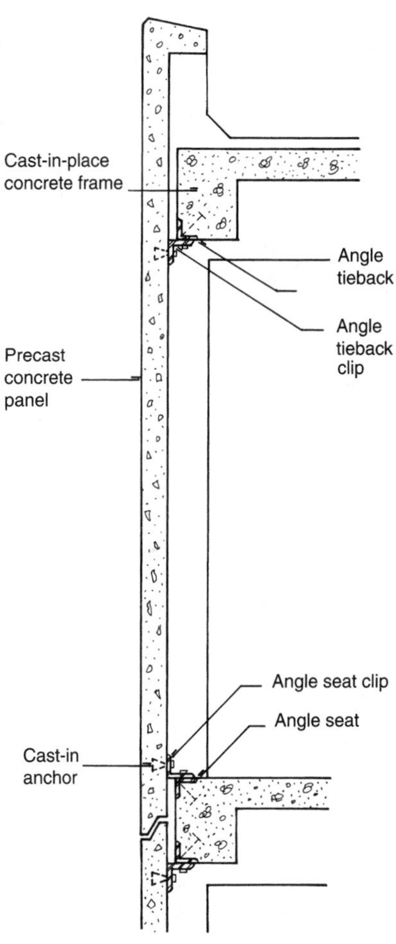

FIGURE 3.14-16 Architectural precast cladding panel supported by a cast-in-place concrete structural frame, showing a typical method of supporting and anchoring panels. Finishes, flashing, caulking, roofing, and thermal and safing insulation have been omitted for clarity. Seat and tieback clips usually have slotted bolt holes to permit adjustment and are shimmed as necessary and bolted to panel and to seat and tieback angles. Plates are sometimes used in place of seat and tieback angles. (Drawing by HLS.)

The PCI publication MNL-117 establishes standard practices for the manufacture of architectural precast concrete elements. PCI also administers a plant certification program.

Most architectural precast concrete panels are made from standard-weight concrete and reinforced conventionally. However, glass-fiber-reinforced concrete panels, which are enjoying increased use today, require no conventional steel reinforcement (see Section 3.15). These panels are lighter than conventional panels and can have steel studs cast into them, thus providing a cavity for insulation.

FIGURE 3.14-17 A light precast panel being installed. (Photo courtesy Precast/Prestressed Concrete Institute.)

FIGURE 3.14-18 A heavyweight cored precast wall panel being installed. These panels are designed to carry the weight of the panels above. (Photo courtesy Precast/Prestressed Concrete Institute.)

Architectural precast concrete panels are available in many finishes, ranging from flat, as-cast finishes to ornate articulation made using form liners. Exposed aggregate, bush-hammered, and abrasive-blasted finishes are possible, as are panels faced with granite, marble, brick, and other materials. Some units have insulation cast into them.

Many panel configurations are possible, from flat panels to those with deep shadow lines and recesses. Architectural precast concrete column covers are also available.

Architectural precast panels are usually installed by crane and fastened in place with cast-in angles and plates (Figs. 3.14-16, 3.14-17, and 3.14-18). Joints are usually treated with sealants.

3.14.5 PRESTRESSING AND POST-TENSIONING

Most structural precast concrete units and some architectural precast concrete units are prestressed. *Prestressing* is done in precast concrete for the same reasons post-tensioning is done in cast-in-place beams and slabs (see Section 3.13.3.5). The process is also similar to the post-tensioning described in Section 3.13.3.5, except that the tendons are stretched between *abutments* before the concrete is placed around them. When the concrete has cured, the tension is released from the steel, which recoils, transferring the stresses in it to the concrete in the form of compression. This load transfer is done through the bond between the steel and the concrete. Bearing plates are not required.

Usually, precast units are cast end to end over continuous tendons. When the concrete has cured, the tendons are cut at both ends of each concrete element. This reduces the need for a separate set of costly abutments for each precast element.

3.15 Glass-Fiber-Reinforced Concrete

Glass-fiber-reinforced concrete (GFRC) is a plant-cast composite product, based on portland cement reinforced with randomly dispersed glass fibers in lieu of reinforcing steel. It is used to manufacture components of exterior wall-cladding systems, such as panels, column covers, window wall units, and ancillary components of curtain wall systems, such as spandrels, mullions, fascia panels, and soffits. It is also used for mansards, sunscreens, and interior panels and column covers. GFRC units are not load-bearing building components.

GFRC weighs about 80% less than conventional precast concrete. This light weight makes it ideal for rehabilitation projects where increased weight is a problem. In new buildings, this light weight can result in significant cost savings in the structure as compared with other forms of cladding, such as conventional precast concrete.

GFRC elements should comply with the requirements of PCI standards MNL 117 and MNL 128. A manufacturer should be a member of PCI or participate in its plant certification program.

3.15.1 DESIGN FLEXIBILITY

GFRC is capable of almost unlimited design flexibility, in terms of the shapes possible. For example, deep reveals, complex curves, and even 90-degree corners are possible. Panels may be flat, fluted, sculptured, or grained to look like wood. This product is ideal for producing vertical and horizontal sunshades and deeply recessed windows, which aid in reducing solar heat gain on the glass and thereby reduce the cost of mechanical cooling systems and their operation. GFRC can be placed in rubber molds made from existing building elements to produce elements that exactly match an existing configuration.

A wide range of colors is possible, produced by colored cement in the matrix and colored aggregates. Exposed aggregate faces are also possible. Face mixes can be added that make a GFRC panel's appearance indistinguishable from conventional precast concrete, granite, or limestone, and it can be made to resemble metal panels.

GFRC can be faced with natural stone, veneer-thickness brick, tile, and architectural terra-cotta. These materials are separated from the GFRC with a bond breaker to ensure that their differential thermal movements do not damage the GFRC. However, some sources advise against using GFRC as a backup for clay masonry because of past failures possibly attributable to such use.

3.15.2 MANUFACTURE

The matrix for GFRC panels is a mixture of portland cement, aggregates (mostly sand), glass fibers, water, and a water reducer or super plasticizer. Coloring agents, air-entraining agents, fly ash, and natural pozzolan are also used by some fabricators. The glass fibers constitute between 4% and 5% of the matrix.

Most panels are manufactured by spraying the matrix on a mold. In small ornamental elements, the matrix is sometimes packed into a mold by hand.

Except on small panels, thickened sections or stiffeners are usually added to the back of a GFRC panel. Stiffeners may be steel studs or structural steel tubes or may be formed from the GFRC concrete material by spraying it over rib forms, such as polystyrene foam strips. The stiffeners, especially studs, provide a space for insulation, making high U-values possible and helping to decrease heating and cooling loads. They also provide space for electrical, mechanical, and telephone conduits.

GFRC will expand and contract up to ⅛ in. (3.2 mm) in 10 ft. (3000 mm) due to moisture and thermal effects. This movement must be accounted for in joint widths. Joints in GFRC panels are finished with polysulfide, polyurethane, or silicone sealants.

3.15.3 INSTALLATION

Quality control during installation is important if GFRC panel systems are to resist water penetration. Joints should provide for a double layer of sealant.

Anchors should be flexible to prevent them from applying stress to the GFRC. Fastening is accomplished with bolts or welding.

Windows and doors should be attached to the panel frames, not to the GFRC itself. If windows or doors would prohibit the GFRC from moving freely, use a separate framing system for them. Do not permit windows to bear on the GFRC material or to be supported by it.

Alignment of panel faces may result in the backing studs being slightly out of plane or not plumb. It is therefore necessary to provide horizontal furring members attached to the embedded stiffener studs for attachment of interior finishes. However, the cost of shimming and installing such furring may be more expensive than installing a secondary row of studs to receive finishes.

3.16 Concrete Toppings

Concrete toppings are thin slabs of concrete placed on top of other concrete, waterproofing, or insulation. They are of two basic types: *bonded* and *unbonded.*

Most toppings are composed of portland cement, coarse aggregate, and fine aggregate mixed with water. Sometimes a plasticizing agent is used instead of the water or in addition to it. The aggregates may be either normal-weight or heavy-duty types. The latter are especially hard aggregates, such as granite, quartz, or corundum, which are used when abrasive resistance is needed. Aggregates are limited to a ½-in. (13 mm)-diameter maximum size, which is generally smaller than those used in normal concrete. However, aggregates in 3-in. (76 mm)-thick toppings may be somewhat larger. Aggregates may be of selected colors if desired. Sometimes iron particles are added to the mix to improve surface wear and increase the compressive strength of the concrete.

Toppings may be machine- or ready-mixed. They are placed, cured, and finished as described for slabs in Sections 3.8 and 3.9. Toppings that contain special selected aggregate are sometimes given a ground surface finish using grinding machines to expose the aggregate.

3.16.1 BONDED TOPPINGS

A *bonded topping* is a layer of concrete placed over an underlying concrete slab or precast concrete structural element (see Section 3.13). Bonded toppings may be placed at the same time as the structural slab or later.

Bonded toppings are used for several purposes. Over monolithic concrete slabs, they can create a special finish, such as an abrasive finish, where it would be impracticable, too expensive, or otherwise undesirable to include the aggregate necessary to produce this finish in the structural concrete. Unreinforced bonded toppings are also used over monolithic concrete in heavy industrial buildings where the movement of equipment may damage the finished floor. The toppings are placed after the equipment has been placed. Bonded toppings over monolithic concrete slabs are usually 1 to 1½ in. (25 to 38 mm) thick and contain neither reinforcing steel nor welded wire fabric.

Bonded toppings are also used to make a level and smooth surface over structural precast concrete elements and to make a composite structural unit with them (see Section 3.14). Such toppings often contain a layer of welded wire fabric or steel reinforcing bars to help tie together a series of precast elements. Toppings over precast concrete elements are usually 2 to 4 in. (50 to 100 mm) thick.

3.16.2 UNBONDED TOPPINGS

Unbonded toppings are thin concrete slabs used to produce wearing surfaces over sheet membrane waterproofing or insulation (see Chapter 7). They are usually 3 in. (76 mm) thick, but are sometimes thicker. They always contain a layer of welded wire fabric.

3.16.3 SLAB FLATNESS

The principles concerning flatness requirements for slabs on grade discussed in Section 3.12.5 also apply to toppings.

3.17 Structural Insulating Roof Decks

Two kinds of structural insulating roof decks are addressed here: poured-in-place gypsum concrete decks and prefabricated board decks. The material, gypsum, is described in Section 9.1.

3.17.1 POURED-IN-PLACE GYPSUM CONCRETE ROOF DECKS

A poured-in-place gypsum concrete roof deck is a reinforced gypsum concrete slab placed on permanent formboards that may or may not be supported by subpurlins.

3.17.1.1 Materials

GYPSUM CONCRETE

Gypsum concrete mix is a factory-controlled mixture of gypsum and wood chips or mineral aggregate. It requires only the addition of water at the job site. A gypsum concrete slab, which weighs 35 to 55 lb./cu. ft. (561 to 881 kg/m^3), depending on the type and amount of aggregate, is poured to an average thickness of 2 to 3½ in. (51 to 89 mm). Gypsum concrete slabs are reinforced with galvanized steel, welded wire mesh, or woven wire mesh to provide at least 0.26 sq. in. (168 mm^2) of cross-sectional area of steel in each foot (305 mm) of slab width (Fig. 3.17-1).

FORMBOARDS

Formboards are nonstructural components of a gypsum concrete roof deck and are usually not considered in the load-bearing capacity of the slab. They serve as a form and remain in place to provide a functional underside of the deck. Various formboards are available to provide fire resistance, light reflection, insulation, sound absorption, and economy.

Gypsum formboards, which are also called *roof boards*, are surfaced on the exposed face and longitudinal edges with glass-fiber sheets or with calendered manila paper specially treated to resist fungus growth. The back face is surfaced with gray liner paper as is used on regular gypsum wallboard (see Section 9.5) or with glass-fiber sheets. Gypsum formboards are usually ½ or ⅝ in. (12.7 or 15.9 mm) thick, 32 in. (812.8 mm) wide, and either 8 or 10 ft. (2438 or 3048 mm) long, although other sizes may be available.

Wood-fiber, glass-fiber, and mineral-fiber boards and various types of precast planks are also used as forms for gypsum concrete. Their widths generally coincide with the standard 24-, 32-, and 48-in. (610, 313, and 1219 mm) nominal subpurlin spacing.

SUBPURLINS

Steel subpurlins (rails, bulb tees, or other tee sections) are welded transversely to the primary framing to support formboards and roof slab. They span between main purlins, beams, or joists. Subpurlins also anchor the deck against uplift forces, restrict deck movement caused by temperature changes, and provide lateral bracing for the main roof purlins. Subpurlins, which vary in size, weight, and shape, are selected on the basis of the span between the main purlins or joists supporting them and the required safe load-bearing capacity of the deck.

When subpurlins are not used, the formboards rest directly on purlins, beams, or joists (Fig. 3.17-2).

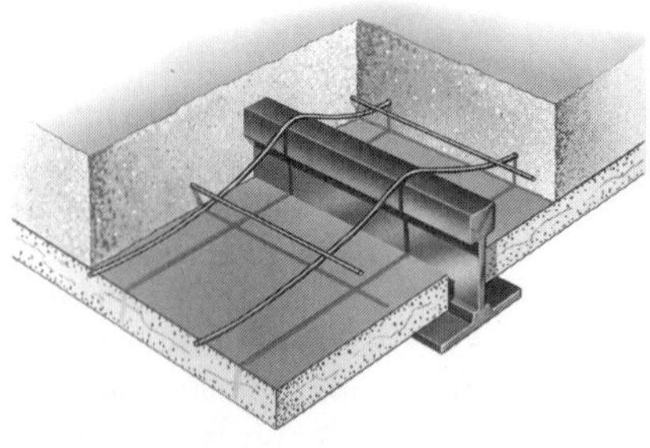
a

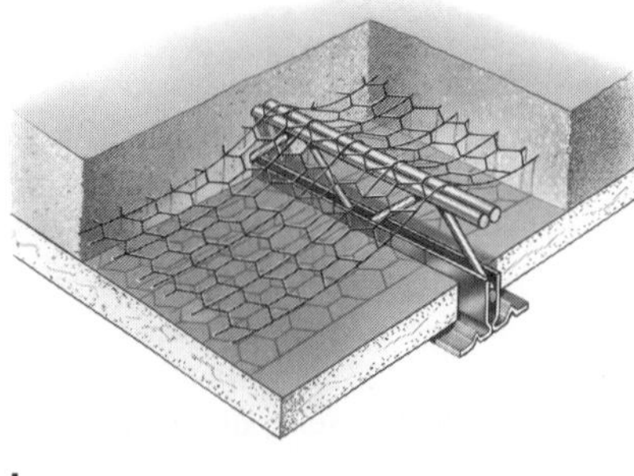
b

FIGURE 3.17-1 Gypsum concrete roof construction: (a) with subpurlins; (b) without subpurlins. (Courtesy Gypsum Association.)

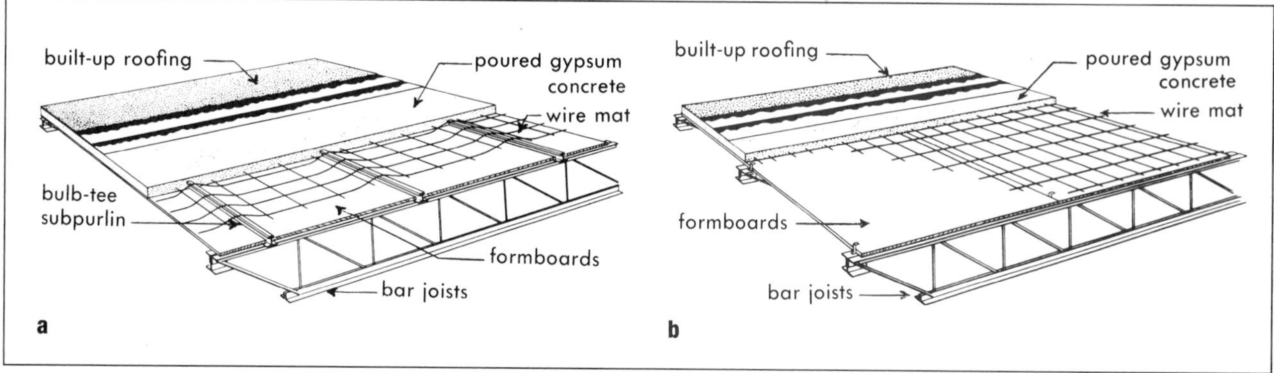

FIGURE 3.17-2 Subpurlins and reinforcing: (a) bulb tee and welded wire; (b) trussed tee and woven wire. (Courtesy Gypsum Association.)

3.17.1.2 Use and Construction

Gypsum concrete decks are used when noncombustible roof construction is required and other types are less practicable or more costly. They can be installed on virtually any roof shape, size, or configuration, even on those that are warped, sawtoothed, curved, or sloped. The primary framing can be wood, steel, or concrete. The light weight of gypsum concrete, as compared with portland cement concrete, often allows the use of a lighter structural frame. Gypsum concrete sets rapidly, and the roof slab may be used as a working surface within 30 minutes after pouring.

Decks of gypsum concrete poured over gypsum- or mineral-fiber formboards are classified as noncombustible. When combustible wood-fiber formboards are used, there is generally an increase in insurance premiums. Gypsum concrete decks, 2 and 2½ in. (50.8 and 63.5 mm) thick on gypsum formboards without a suspended ceiling below, qualify for 1- and 2-hour fire resistance ratings, respectively. The materials used in gypsum concrete should conform with the requirements of ASTM C 317; installation with ASTM C 956.

3.17.2 PREFABRICATED STRUCTURAL INSULATING ROOF DECKS

Prefabricated structural insulating roof deck boards are designed to provide a structural deck as well as thermal insulation and interior ceiling finish. They are used primarily on flat or sloping roofs with exposed beams. The underside of the decking is factory-finished in various designs or textures, some with acoustical value.

Two types of structural insulating roof decks are discussed here: lightweight aggregate boards and wood-fiber boards. These boards are available in a variety of thicknesses and sizes, in either butt-joint or tongue-and-groove edge profiles (Fig. 3.17-3). Because the underside of the deck is also the interior finish, care must be taken not to damage the surface during erection. In areas with high interior humidities, such as bathrooms or laundries, adequate roof insulation and ventilation should be provided to prevent condensation.

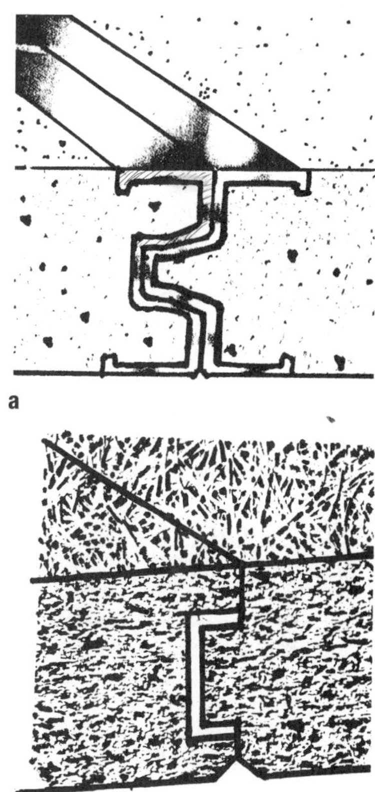

FIGURE 3.17-3 Insulating deck: (a) lightweight perlite aggregate; (b) wood fiber.

3.18 Cement-Based Underlayment

An underlayment is a material placed beneath floor finishing materials, such as tile, resilient flooring, carpets, and liquid-applied flooring, to level and stabilize the substrate to receive floor finishes. Underlayments may be cement-based or board types. This section addresses two cement-based types: gypsum cement underlayment and cement-based, polymer-modified, self-leveling underlayment. Other underlayment types are addressed in the sections covering the finish flooring materials to be applied over them.

3.18.1 GYPSUM CEMENTITIOUS UNDERLAYMENT

Gypsum-based floor underlayments have been available for more than 50 years. They were originally developed to replace conventional underlayments. They are used mostly in multifamily and commercial buildings. They have two major advantages over cement-based underlayments: they cost less, and they are lighter in weight.

Gypsum-based underlayments have two major disadvantages as compared with cement-based underlayments. They are not as strong, and they are more susceptible to moisture damage.

Early gypsum-based underlayments were limited in their use by a very low compressive strength (100 psi [6.9 MPa]). Today's formulations are capable of strengths up to 2500 psi (17.2 MPa), which makes them useful in more applications, but they are still low in strength as compared with cement-based underlayments, which have a strength of 4100 to 5500 psi (28 to 38 MPa). The lower compressive strengths make the use of gypsum-based underlayments questionable where rolling loads or heavy concentrated loads will be present. They are also more susceptible to damage when the bond between applied floor coverings is greater than the internal strength of the underlayment.

Gypsum-based underlayment manufacturers do not recommend its use in locations where it will be subjected to moisture damage. Therefore, its use is questionable over cast-in-place concrete, which will continue to emit moisture vapor for a long time after it is placed. It should not be used where the adhesives used to apply finishes over it are water-based. Water from floor cleanings can also deteriorate a gypsum-based underlayment. It should not be used as an exposed floor or on the exterior, such as on balconies. Its use below ground is limited to those situations where water intrusion is impossible. Where a gypsum-based underlayment will be exposed to water, such as in kitchens and baths, it must be covered by an impervious material, such as vinyl sheet flooring.

The advice of the flooring products manufacturer should be sought, however, regarding whether gypsum-based underlayment is acceptable beneath its products. Installation adhesives must be compatible with the underlayment, and the bond strength of the adhesive must be low. Resilient flooring manufacturers generally do not recommend the use of gypsum-based underlayments beneath their products.

3.18.1.1 Products

Gypsum-based underlayments are proprietary products made with gypsum cement, sand aggregate, and water. They are self-leveling. There are several types available, and selection is based on project requirements and cost. Some gypsum-based underlayments are of relatively low strength, ranging from 1200 to 1500 psi (8.3 to 10.3 MPa). Others range from 1600 to 2500 psi (11 to 17.2 MPa). Their point-loading resistance is similarly different, ranging from 550 to 2000 lb. (250 to 910 kg) on a 1-in. (25.4 mm) disc.

3.18.1.2 Installation

Gypsum-based underlayment is often installed over wood-based substrates, such as plywood, oriented strand board, or wafer board, but may be installed over concrete that has been allowed to cure for the time recommended by the underlayment manufacturer and has been sealed with a primer recommended by the underlayment manufacturer. It can also be installed over extruded or expanded polystyrene foam and over sound-deadening pads.

The deflection in the substrates beneath gypsum-based underlayment must be limited to L/360. Greater deflection will cause the underlayment to crack.

Installation of a gypsum-based underlayment must be done in accordance with its manufacturer's recommendations. These will vary from manufacturer to manufacturer and from product to product. Probably the worst and most frequent problem is overwatering the mix, which lowers the underlayment's compressive strength and ability to resist point loads.

Another major problem occurs when finish flooring is installed before the underlayment has had adequate time to cure. Normally, a ¾-in. (19 mm)-thick underlayment will be ready to receive finish flooring in 5 to 7 days, but the actual time required will vary, depending on the ambient temperature and humidity. The underlayment should be tested as recommended by its manufacturer for its moisture content before flooring application.

3.18.2 CEMENT-BASED UNDERLAYMENTS

Cement-based underlayments are self-leveling polymer-modified hydraulic cement products. They are primarily used to level and resurface existing cast-in-place and precast concrete, terrazzo, and tile substrates. They can also be used over vinyl asbestos tile, extruded or expanded polystyrene foam, and wood. They are not a finish floor, nor are they structural in nature. They can serve as a substrate for wood, tile, carpet, and resilient finish flooring.

3.18.2.1 Products

Cement-based underlayments are proprietary products composed of a binder of portland cement or hydraulic or blended cement, as defined in ASTM C 219, washed-gravel or sand aggregate, and water. Some manufacturers also recommend that some chemical additives be included in cement-based underlayment that will be used over certain substrates, such as cutback adhesives, wood, or metal.

3.18.2.3 Installation

Cement-based underlayments are generally installed in thicknesses varying from ⅛ to 6 in. (3 to 150 mm). Different manufacturers have different maximum thickness and filler requirements. Without a filler (sand) added, the limits vary from ½ in. (12 mm) to 1½ in. (38 mm). Most manufacturers recommend that a filler be used at all thicknesses and require it when the thickness increases beyond the ½-in. (12 mm) to 1½-in. (38 mm) maximums. Some manufacturers recommend the use of several layers of unfilled material to make up increased thicknesses.

The requirements for placing and curing concrete discussed elsewhere in this chapter apply to cement-based underlayments.

3.19 Additional Reading

More information about the subjects discussed in this chapter can be found in the references listed in Section 3.20 and in the following publications.

Ambrose, James. 1996. *Simplified Design of Concrete Structures*. 7th ed. New York: John Wiley & Sons, Inc.

Butt, Thomas K. 1994. "Concrete Slab Floor Flatness and Levelness Tolerances." *The Construction Specifier* 11(November):30.

Dobrowolski, Joseph A. 1998. *Concrete*

Construction Handbook. New York: McGraw-Hill.
Ferguson, Phil M., John E. Breen, and James O. Jirsa. 1988. *Reinforced Concrete Fundamentals*. 5th ed. New York: John Wiley & Sons, Inc.
Gencarelli, Frank. 1992. "Fiber for Secondary Reinforcement." *The Construction Specifier* 12(December):123.
Gerwick, Ben C. 1997. *Construction of Prestressed Concrete Structures*. 2d Ed. New York: John Wiley & Sons, Inc.
Hime, William G. 1995. "Avoiding Gypsum Concrete Failures." *The Construction Specifier* 7(July):38.
King, Stephen M. 1995. "Shrinkage-Compensating Concrete Floors." *The Construction Specifier* 7(July):30.
Lin, Tung Yen, and Ned H. Burns. 1981. *Design of Prestressed Concrete Structures*. 3d ed. New York: John Wiley & Sons, Inc.
Mailvaganam, Noel P. 1993. "Concreting with Chemical Admixtures." *The Construction Specifier* 7(July):106.
McCormac, Jack C. 1998. *Design of Reinforced Concrete*. 4th ed. New York: John Wiley & Sons, Inc.
McDougle, Edwin A. 1995. "GFRC Comes of Age." *The Construction Specifier* 12(December):46.
Michaud, Pete. 1996. "Design Guidelines for Concrete Anchoring." *The Construction Specifier* 12(December):38.
Nawy, Edward G. 2000. *Reinforced Concrete: A Fundamental Approach*, 4th ed. Upper Saddle River, NJ: Prentice-Hall.
Nicastro, David H. 1994. "Cracking in GFRC." *The Construction Specifier* 9(September):160.
Nilson Arthur H. 1987. *Design of Prestressed Concrete*. 2d ed. New York: John Wiley & Sons, Inc.
———. 1997. *Design of Concrete Structures*. New York: McGraw-Hill.
Peurifoy, R. L. 1996. *Formwork for Concrete Structures*. New York: McGraw-Hill.
Raeber, John A. 1991. "Architectural Precast Concrete Finishes." *The Construction Specifier* 12(December): 29.
———. 1994. "Glass-Fiber Reinforced Concrete." *The Construction Specifier* 10(October):21.
Robinson, Ralph C. 1997. "Architectural Precast Concrete Connections." *The Construction Specifier* 12(December): 32.
Socha, Dennis, Edwin Jakacki, and William Rector. 1997. "Installation Guidelines for Full-Thickness Gypsum Underlayments." *The Construction Specifier* 9(September):73.
Wang, Chu-Kia, and Charles G. Salmon. 1998. *Reinforced Concrete Design*. 6th ed. New York: John Wiley & Sons, Inc.

3.20 Acknowledgments and References

ACKNOWLEDGMENTS

We gratefully acknowledge the assistance of the following organizations and individuals in preparing this chapter. We are also indebted to them for permission to use their illustrations when requested and for the use of their publications as references.

American Concrete Institute
American Society for Testing and Materials
Building Research Board of the National Academy of Sciences (NAS)
Burke Company
Ceco Concrete Corporation
Concrete Reinforcing Steel Institute
ELE International, Inc.
Gypsum Association
Martin Fireproofing Corporation; Martin Fireproofing Georgia, Inc.
Maxxon Corporation
National Association of Home Builders (NAHB)
Portland Cement Association (PCA)
Prestressed Concrete Institute (PCI)
U.S. Department of Housing and Urban Development (HUD)
Wire Reinforcement Institute

REFERENCES

We would also like to thank the authors and publishers of the publications in the following list for their contribution to our research for this chapter.

ACI International. 1999. *Manual of Concrete Practice*. Farmington Hills, MI: ACI International.
———. 1991. ACI 211.1: *Standard Practice for Selecting Proportions for Normal, Heavy-weight, and Mass Concrete*. Farmington Hills, MI: ACI International.
———. 1991. ACI 212.3R: *Chemical Admixtures for Concrete*. Farmington Hills, MI: ACI International.
———. 1996. ACI 301: *Specifications for Structural Concrete for Buildings*. Farmington Hills, MI: ACI International.
———. 1996. ACI 302.1R: *Guide for Concrete Floor and Slab Construction*. Farmington Hills, MI: ACI International.
———. 1991. ACI 305R: *Hot Weather Concreting*. Farmington Hills, MI: ACI International.
———. 1990. ACI 306.1: *Standard Specification for Cold Weather Concreting*. Farmington Hills, MI: ACI International.
———. 1996. ACI 309R: *Guide for Consolidation of Concrete*. Farmington Hills, MI: ACI International.
———. 1999. ACI 318: *Building Code Requirements for Structural Concrete, and Commentary*. Farmington Hills, MI: ACI International.
———. 1984. ACI 332R: *Guide to Residential Cast-in-Place Concrete Construction*. Farmington Hills, MI: ACI International.
———. 1994. ACI 347R: *Guide to Formwork for Concrete*. Farmington Hills, MI: ACI International.
Adams, J. T. 1983. *The Complete Concrete, Masonry, and Brick Handbook*. New York: Van Nostrand Reinhold.
American National Standards Institute (ANSI). ANSI A56.1, *Building Code Requirements for Excavations and Foundations*. New York: ANSI.
American Society for Testing and Materials (ASTM). Standards:
A 82, "Standard Specification for Steel Wire, Plain, for Concrete Reinforcement." West Conshohocken, PA: ASTM.
A 185, "Standard Specification for Steel Welded Wire, Fabric, Plain, for Concrete Reinforcement." West Conshohocken, PA: ASTM.
A 416, "Standard Specification for Steel Strand, Uncoated Seven-Wire for Prestressed Concrete." West Conshohocken, PA: ASTM.
A 497, "Standard Specification for Steel Welded Wire Fabric, Deformed, for Concrete Reinforcement." West Conshohocken, PA: ASTM.
A 615, "Standard Specification for De-

formed and Plain Billet-Steel Bars for Concrete Reinforcement." West Conshohocken, PA: ASTM.

A 616, "Standard Specification for Rail-Steel Deformed and Plain Bars for Concrete Reinforcement." West Conshohocken, PA: ASTM.

A 617, "Standard Specification for Axle-Steel Deformed and Plain Bars for Concrete Reinforcement." West Conshohocken, PA: ASTM.

A 706/A 706M, "Standard Specification for Low-Alloy Steel Deformed and Plain Bars for Concrete Reinforcement." West Conshohocken, PA: ASTM.

A 767, "Standard Specification for Zinc-Coated (Galvanized) Steel Bars for Concrete Reinforcement." West Conshohocken, PA: ASTM.

A 775, "Standard Specification for Epoxy-Coated Reinforcing Steel Bars." West Conshohocken, PA: ASTM.

C 31, "Standard Practice for Making and Curing Concrete Test Specimens in the Field." West Conshohocken, PA: ASTM.

C 33, "Standard Specification for Concrete Aggregate." West Conshohocken, PA: ASTM.

C 39, "Standard Test Method for Compressive Strength of Cylindrical Concrete Specimens." West Conshohocken, PA: ASTM.

C 94, "Standard Specification for Ready-Mixed Concrete." West Conshohocken, PA: ASTM.

C 143, "Standard Test Method for Slump of Hydraulic Cement Concrete." West Conshohocken, PA: ASTM.

C 150, "Standard Specification for Portland Cement." West Conshohocken, PA: ASTM.

C 219, "Standard Terminology Relating to Hydraulic Cement." West Conshohocken, PA: ASTM.

C 226, "Standard Specification for Air-Entraining Additions for Use in the Manufacture of Air-Entraining Hydraulic Cement." West Conshohocken, PA: ASTM.

C 260, "Standard Specification for Air-Entraining Admixtures for Concrete." West Conshohocken, PA: ASTM.

C 309, "Standard Specification for Liquid Membrane-Forming Compounds for Curing Concrete." West Conshohocken, PA: ASTM.

C 317, "Standard Specification for Gypsum Concrete." West Conshohocken, PA: ASTM.

C 330, "Standard Specification for Lightweight Aggregate for Structural Concrete." West Conshohocken, PA: ASTM.

C 494, "Standard Specification for Chemical Admixtures for Concrete." West Conshohocken, PA: ASTM.

C 595, "Standard Specification for Blended Hydraulic Cements." West Conshohocken, PA: ASTM.

C 618, "Standard Specification for Coal Fly Ash and Raw or Calcined Natural Pozzolan for Use as a Mineral Admixture in Concrete." West Conshohocken, PA: ASTM.

C 956, "Standard Specification for Installation of Cast-in-Place Reinforced Gypsum Concrete." West Conshohocken, PA: ASTM.

C 1116, "Fiber-Reinforced Concrete and Shotcrete." West Conshohocken, PA: ASTM.

D 98, "Standard Specification for Calcium Chloride." West Conshohocken, PA: ASTM.

D 2166, "Standard Test Method for Unconfined Compressive Strength of Cohesive Soil." West Conshohocken, PA: ASTM.

D 4318, " Standard Test Method for Liquid Limit, Plastic Limit, and Plasticity Index of Soils." West Conshohocken, PA: ASTM.

E 1155, "Standard Test Method for Determining FF." West Conshohocken, PA: ASTM.

Arcom Master Systems. MASTERSPEC®. Basic Sections:

03120, "Architectural Cast-in-Place Concrete Formwork." 8/91 ed. Salt Lake City, UT: Arcom.

03300, "Cast-in-place Concrete." 8/92 ed. Salt Lake City, UT: Arcom.

03350, "Concrete Toppings." 5/91 ed. Salt Lake City, UT: Arcom.

03355, "Special Concrete Finishes." 8/94 ed. Salt Lake City, UT: Arcom.

03361, "Shotcrete." 5/95 ed. Salt Lake City, UT: Arcom.

03410, "Structural Precast Concrete—Plant Cast." 2/94 ed. Salt Lake City, UT: Arcom.

03450, "Architectural Precast Concrete—Plant Cast." 8/94 ed. Salt Lake City, UT: Arcom.

03455, "Glass-Fiber Reinforced Concrete—Plant Cast. 5/92 ed. Salt Lake City, UT: Arcom.

03470, "Tilt-Up Precast Concrete." 5/91 ed. Salt Lake City, UT: Arcom.

03520, "Insulating Concrete Decks." 2/95 ed. Salt Lake City, UT: Arcom.

03531, "Cementitious Wood-Fiber Planks." 5/93 ed. Salt Lake City, UT: Arcom.

03542, "Cement-Based Underlayment." 8/97 ed. Salt Lake City, UT: Arcom.

Concrete Reinforcing Steel Institute. 1997. *CRSI Manual of Standard Practice*. Schaumburg, IL: CRSI.

———. 1997. *Placing Reinforcing Bars*. Schaumburg, IL: CRSI.

De Cristoforo, R. J. 1975. *Concrete and Masonry*. Reston, VA: Reston Publishing Company.

Parker, Harry, and John D. Hauf. 1976. *Simplified Design of Reinforced Concrete*. 4th ed. New York: John Wiley & Sons, Inc.

Portland Cement Association (PCA). "Moisture Migration—Concrete Slab-on-Ground Construction." Skokie, IL: PCA.

———. *Design and Control of Concrete Mixtures*. Skokie, IL: PCA.

———. *Concrete Floors on Ground*. Skokie, IL: PCA.

———. *Concrete Slab Surface Defects: Causes, Prevention, Repair*. Skokie, IL: PCA.

———. *Design and Control of Concrete Mixtures*. 13th ed. Skokie, IL: PCA.

———. *Batching, Mixing, Transporting and Handling Concrete*. Skokie, IL: PCA.

———. *Resurfacing Concrete Floors*. Skokie, IL: PCA.

Precast/Prestressed Concrete Institute (PCI). MNL 116S, *Manual for Quality Control for Plants and Production of Architectural Precast Concrete Products*. Chicago, IL: PCI.

———. MNL 117, *Manual for Quality Control for Plants and Production of Architectural Precast Products*. Chicago, IL: PCI.

———. MNL 120, *PCI Design Handbook—Precast and Prestressed Concrete*. Chicago, IL: PCI.

———. MNL 127, *Recommended Practice for Erection of Precast Concrete*. Chicago, IL: PCI.

———. MNL 128, *Recommended Practice for Glass Fiber Reinforced Concrete Panels*. Chicago, IL: PCI.

———. *Architectural Precast Concrete*. Chicago, IL: PCI.

Pruter, Walter F. 1987. "The Advantages of Perlite Concrete Roof Deck Systems." *The Construction Specifier* 40(11) (November): 52–55.

Ramsey/Sleeper, The AIA Committee on Architectural Graphic Standards. *Architectural Graphic Standards*. 8th ed. New York: John Wiley & Sons, Inc.

Underground Space Center. 1988. *Building Foundation Design Handbook*. Minneapolis, MN: Underground Space Center.

Wilson, Forrest. 1984. *Building Materials Evaluation Handbook*. New York: Van Nostrand Reinhold.

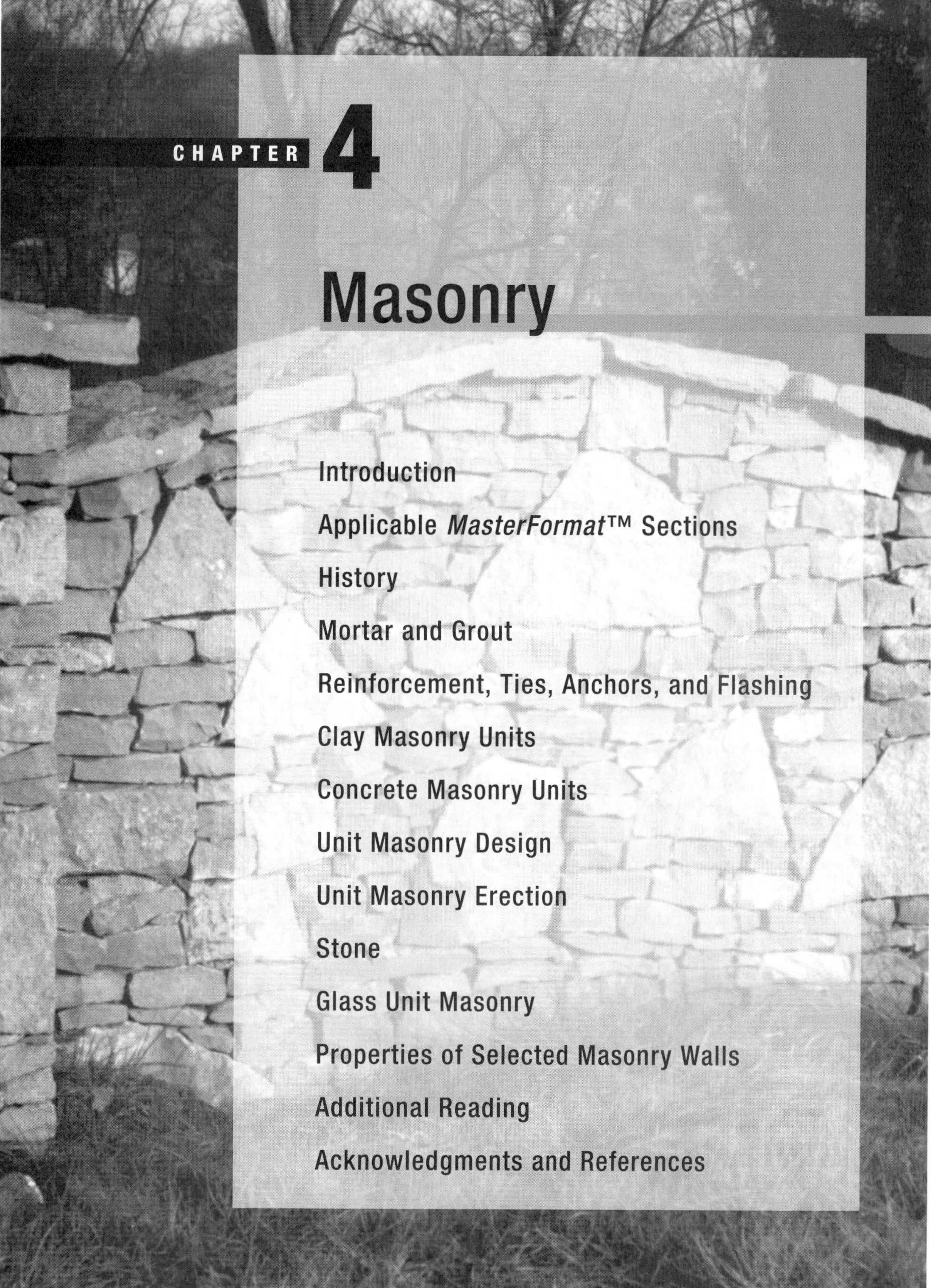

CHAPTER 4

Masonry

Introduction

Applicable *MasterFormat*™ Sections

History

Mortar and Grout

Reinforcement, Ties, Anchors, and Flashing

Clay Masonry Units

Concrete Masonry Units

Unit Masonry Design

Unit Masonry Erection

Stone

Glass Unit Masonry

Properties of Selected Masonry Walls

Additional Reading

Acknowledgments and References

Introduction

Stone, clay masonry units, and concrete masonry units are all used extensively in construction today. Brick, which is one of the oldest manufactured building materials, is an example of clay masonry. Concrete masonry units include concrete block, a relatively new material. To take advantage of their compressive strength, durability, and fireproof character, these materials are used primarily in wall assemblies, often appearing in conjunction. In addition, all three are also used as decorative screening. Stone and clay masonry units are also suitable for floor surfacing and exterior paving. Lightweight aggregates have made lighter concrete masonry units possible and permitted the fabrication of larger masonry panels than earlier technology permitted. The development of high-bond-strength mortars has made prefabricated brick masonry panels practicable. Continuing improvement of sealants, adhesives, and masonry units will ensure continued use of these time-proven masonry materials.

To serve their intended function, masonry walls must perform as they were designed to do. Exterior walls, for example, must possess at least three essential functional properties. They must be:

- Strong enough to carry applied loads;
- Watertight enough to prevent groundwater and storm water penetration;
- Durable so that they will not quickly erode due to weather or other atmospheric conditions;
- Resistant to heat, sound, and fire transmission when these properties are required.

Aesthetic considerations often influence the selection of masonry units that will be exposed to view. Appearance is affected by the following:

- Color of both the units and the mortar
- Texture of the units, glazes applied to the units, and that created by the method of laying the units
- Pattern produced by the type of bond and the unit

Assuming that the design has been done correctly, the procedures for producing satisfactory masonry assemblies can be summarized as follows:

- Materials must be sound.
- Materials must be properly handled, stored, protected, and prepared.
- The mortar type must be the one best suited for assembly use and exposure.
- Mortar constituents must be proportioned and mixed correctly.
- Units must be laid with full head and bed joints.
- Joints must be tooled properly.
- Masonry must be protected from damage by the weather during construction.
- Water must be prevented from penetrating to the inside of the assembly after it has been built, by properly located and installed flashing and weep holes.

This chapter describes materials and procedures that will help to produce masonry of suitable quality.

Applicable *MasterFormat*™ Sections

The following *MasterFormat* Level 2 sections are applicable to this chapter.

04050 Basic Masonry Materials and Methods
04200 Masonry Units
04400 Stone
04500 Refractories
04600 Corrosion-Resistant Masonry
04700 Simulated Masonry
04800 Masonry Assemblies

4.1 History

Masonry structures today are made from brick and block, which are together called masonry units, or from stone. Masonry structures predate written history. The earliest structures were huts made from unshaped native field stones piled upon one another without mortar or other materials in the joints. Where stone was not available, sod or dried mud served the same purpose. Later, clay and silt were mixed with water and formed by hand into bricks. The spaces between these bricks were sometimes packed with mud to keep out the wind and rain and to make it easier to build level walls with irregular bricks. Later still, it was discovered that clay bricks placed in or adjacent to a fire became harder and more weather resistant. The Romans used this knowledge to build kilns to produce burned-clay roofing tiles and, eventually, burned-clay bricks.

The Mesopotamians built stone and sun-dried brick buildings four thousand years before the birth of Christ. A thousand years later, the Egyptians began building temples and pyramids of cut stone. By laboriously cutting the stone using bronze tools, they made all pieces fit closely together.

Early stone buildings were limited in that openings and column spacings could only be as large as could be spanned with a single stone. Roofs were made of wood. The Babylonians built small arches over windows and other small openings, but the Romans were the first to build arches large enough to support bridges and large buildings.

When it became possible to make stone-working tools from iron, the art of stone building developed to a high order. The Greeks refined the process to produce fine details in stone. The Romans built, for the first time, buildings with large open spaces. They used arches and vaults to build temples, forum buildings, theaters, amphitheaters, baths, aqueducts, and homes.

The building of brick and stone structures without joint materials changed when the Etruscans developed a lime mortar that could be used to fill gaps between masonry units. Later, the Romans discovered how to make a hydraulic (that which will set under water) cement by burning and grinding a type of volcanic rock. This led to stronger and more watertight stone and brick structures. The same discovery led to the expanded use of concrete.

The sun-baked bricks used extensively by ancient peoples began to disappear in sophisticated cultures after the Romans invented kilns. Clay could then be baked hard enough that its resistance to the elements expanded dramatically.

The fall of the Roman Empire was followed by the emergence of the semi-oriental Byzantine Empire, centered in Constantinople (Istanbul). The Byzantine Empire flourished for a thousand years until it was overrun in 1453 by the Ottoman Turks. Byzantine architecture combined Roman arch forms with other shapes and added detail and color. The development of the *pendative* permitted the construction of a stone dome over a square space. The Romans had been limited to circular supporting shapes, such as in the Pantheon. A pendative is a spherical triangle whose lower point rests on a pier and whose curved upper edge, in combination with three other pendatives, forms a circle that can support a dome. Using this device, they built large stone domes over square buildings and were able to construct buildings with several domes. The Byzantines also used broad arches and massive buttresses to add aisles and galleries to their buildings.

Beginning in the eleventh century, a 200-year-long revival called the Romanesque period saw the beginnings of forms of stone structures that would be refined in later periods. These included ribbed vaults, membered piers, and pier buttresses.

The Gothic period, which followed, saw the development of flying buttresses that could withstand the thrust from large vaulted and domed structures. These produced even larger and more complex structures and, when combined with vaults and piers, produced what has been called a "cage of stone." Walls could be replaced with windows, leading to the development of ornate stained glass windows, especially in the cathedrals of western Europe.

Beginning in the fourteenth century, the Renaissance (Age of Discovery) stirred an interest in the neoclassical. This was a period of great strides in the arts, humanities, and science.

In the late eighteenth and early nineteenth centuries (the "age of revivals") architects actually measured Roman and Greek ruins, Gothic buildings, and other early styles and reproduced them as new buildings. Yet during this period, little was done to further the science of building.

In the late eighteenth century, the industrial revolution ushered in the modern era. Machines began to replace much of the handwork necessary to quarry and cut stone and to mold and fire bricks. Stone units became more uniform in shape and size. Bricks became more consistent in color, strength, and size.

Blocks made from cement and various aggregates appeared after the development of portland cement and portland cement concrete in the late eighteenth and early nineteenth centuries. They were cheaper and lighter than cut stone and larger than bricks, which reduced the time needed to lay them. They were more versatile than concrete in that they could be easily placed in small quantities and did not require formwork.

Until the development of the theory of elasticity in the nineteenth century, unit masonry and stone construction were based solely on experience. After that, masonry structures could be built using rational design based on calculated stresses.

Until the late nineteenth century, masonry and stone were the materials of choice throughout the world for constructing buildings, bridges, and viaducts. The advent of cast iron and steel structural elements quickly replaced masonry and stone as the major structural support elements in large buildings. Reinforced concrete had the same effect in smaller buildings and in bridges and other arched structures. The building of large unit-masonry and stone arches reached its climax at about the turn of the twentieth century; then these materials were replaced by reinforced concrete. Concrete was less labor intensive, easier to form and place, and could be made to span distances three to four times those possible with stone.

Although they have been largely replaced by steel and concrete as primary load-carrying elements in larger buildings, unit masonry and stone remain in great use today for cladding, partitions, and flooring. They are especially valuable where fire and weather resistance is required. Their varied appearance is used to create a variety of aesthetic effects. They are adaptable to current standards and methods for insulating buildings to make them more comfortable and to increase their ability to conserve energy. Modern masonry cavity walls resist water penetration and provide space to house insulation. Masonry partitions are fire resistant and help to restrict sound penetration. New mortars and construction adhesives reduce the labor needed to place masonry walls, and some walls are built from prefabricated masonry panels and held together with adhesives.

Further advances are certain. In spite of innovations in metal and concrete construction techniques and materials, masonry and stone construction are likely to be with us for a very long time to come.

4.2 Mortar and Grout

Mortar is a combination of one or more cementitious materials (portland cement, lime, or masonry cement), a clean, well-graded aggregate, such as sand, and enough water to give the mixture a plastic, workable quality. The term *grout* refers to several different types of construction products, depending on the material and its use. Here we are concerned with only two types of grouts. One type is used to fill voids in both reinforced and unreinforced masonry. This type is made from the same ingredients as mortar, but is produced in a pourable consistency. The various ingredients must be proportioned properly to give the mortar or grout a good balance of essential properties. The second type of grout is a tile-setting grout used in filling the joints in unit masonry or stone paving and flooring. In this section, the term *grout* refers to the material used to fill voids unless specifically identified as the type used in paving and flooring joints.

Mortar is placed in the joints between individual masonry or stone units in a wall or other building element to cushion the units and to provide a level setting bed for them. It also seals the spaces between the units, compensates for size variations in the units, and provides aesthetic quality by creating shadow lines and color effects. Grout is poured into voids between masonry unit wythes and into the cores in hollow masonry units to produce solid filled and reinforced masonry walls. Mortar and grout also grip metal ties and reinforcement, helping them to act integrally with the other parts of a wall. Tile-setting-type grout is used to fill the joints between brick pavers.

4.2.1 MORTAR AND GROUT PROPERTIES

Because mortar and grout are placed while plastic and then harden, they must have two sets of properties: (1) those present when they are in their plastic state and (2) those that result after they have hardened. Proper plastic properties and hardened properties are both necessary for a mortar or grout to be suitable for use in building construction. Both sets of properties affect a finished wall's strength, durability, and watertightness.

The basic constituents of most modern masonry mortars and grouts are portland cement, lime, sand, and water. Masonry cement also contains other ingredients (see Section 4.2.2.4). Sometimes additives are used in portland cement–lime mortars and grouts, even though some additives are harmful to the hardened mortar or grout. Because both the plastic and hardened properties of a mortar or grout depend largely on its ingredients, these should conform to the appropriate standards of the American Society for Testing and Materials (ASTM) and must be properly proportioned. Because a mortar's or grout's properties also depend on the proportion of its ingredients, different mortar types (see Section 4.2.4.1) and grout mixes have different properties.

4.2.1.1 Plastic Properties

The properties of mortars and grouts while they are still plastic include *workability*, *water retention*, *initial flow*, and *flow after suction*.

WORKABILITY

A *workable* mortar is uniform, cohesive, and of a consistency that makes it usable by a mason. A mortar is workable when particles in the mix do not segregate and when it spreads easily, holds the weight of the units, makes alignment easy, clings to the vertical surfaces of masonry units, and easily extrudes from the mortar joints without dropping or smearing.

Water retention, flow, resistance to segregation, and other factors affect a mortar's workability. These in turn are affected by the properties of the mortar ingredients. This complex relationship makes quantitative estimates of workability difficult. There is, in fact, no standard laboratory test for measuring it.

Grout is designed to pour and therefore must flow freely without being too soupy. It must also be uniform.

WATER RETENTION

Water retention in a mortar or grout prevents rapid loss of water and a resultant loss of plasticity when the mortar or grout contacts a masonry unit with a high absorption rate. A high degree of water retention also prevents a mortar from *bleeding* when it comes in contact with a masonry unit that has a low absorption rate. *Bleeding* is a process in which water leaves the mortar and is deposited in a thin layer between the masonry unit and the mortar. When this happens, the unit is said to *float*. Such floating materially reduces bond.

The water retention of a mortar is the ratio of a plastic characteristic called *flow* immediately after *mixing* to the *flow* of the same mortar after *suction*. The flow of a mortar immediately after mixing (initial flow) is measured by a simple laboratory test (Fig. 4.2-1). A truncated cone of mortar 4 in. (100 mm) in diameter is placed on a testing apparatus called a *flow table*. The flow table is then mechanically raised and dropped 25 times in 15 seconds through a height of ½ in. (12.7 mm). This movement causes the mortar to flow outward, increasing the diameter of the cone. The flow is the ratio of the final diameter to the original diameter of the cone, expressed as a percentage. For example, if the diameter of the mortar mass is 8 in. (200 mm) after the test, the flow is 100%. ASTM C 270 requires an initial flow of 110%, plus or minus 5%, for specimens tested for water retention or compressive strength.

A *flow-after-suction* test is then used to predict the theoretical flow of a mortar after it has lost water to an absorbent masonry unit. In this test, a sample of the same mortar previously tested for initial flow is placed over an absorbent filter in a perforated metal dish mounted over a vacuum device. Operation of the vacuum device removes some water from the mortar. This mortar is then removed from the suction device and tested for flow in the same way as in the first test. The laboratory value of water retention is the ratio of flow after suction to initial flow, expressed as a percentage. The formula is $(A/B) \times 100$, where A is the flow after suction and B is the flow immediately after mixing. ASTM C 270 requires a water retention of not less than 75%.

FIGURE 4.2-1 The laboratory test for measuring flow. (Courtesy Portland Cement Association.)

4.2.1.2 Hardened Properties

Among the properties of hardened mortar are *bond strength*, *durability*, *compressive strength*, *volume change*, and *appearance*. The most important of these are bond strength and compressive strength.

BOND STRENGTH

Bond strength is perhaps the most important property of hardened mortar and an important characteristic of hardened grout, especially of grout used in conjunction with the type of terra-cotta called ceramic veneer (see Section 4.7.7). Some of the variables that can affect bond are addressed in this section.

A mortar's or grout's other properties often determine its hardened bond strength.

Properties that may affect bond include the type and amount of cementitious materials and the amount of water in the mix (see Sections 4.2.4.3 and 4.2.5), the water retention characteristics of the mortar or grout (see Section 4.2.1.1), the entrained air content of the mortar or grout, and the mortar's or grout's compressive strength.

The amount of water in a mortar mix and its water retention affect its flow (see Section 4.2.1.1). The more water, the greater the flow. A mortar's tensile bond strength is a mechanical function rather than a chemical one. The mortar flows into pores in the masonry unit and interlocks with them. Therefore, bond strength always increases as the mortar's flow increases because wetter mortar flows more readily into these pores. Adding lime to portland cement mortar increases its workability and therefore facilitates the interlocking (bonding) process. Too little water can also affect bond (see Section 4.2.6).

When the air content of a mortar or grout exceeds 12%, there is an accompanying decrease in bond.

The type of masonry unit in contact with a mortar or grout can affect its bond strength. Masonry unit characteristics that can affect bond include surface texture, the suction of clay masonry units (see Section 4.7.1.2), and the moisture content of concrete masonry units (see Section 4.4.2.3).

Workmanship can also affect the bond strength of a mortar. For example, movement of a masonry unit after it has been laid can cause loss of bond, as can water on the surface of a clay masonry unit or a delay in laying masonry after the mortar has been placed. The air temperature and relative humidity during a mortar's curing period can also affect its bond strength.

DURABILITY

The *durability* of mortar is measured principally by its ability to resist water penetration. Water entering through a masonry joint into the adjacent masonry can cause damage to the mortar and the adjacent masonry in several ways. Repeated cycles of freezing and thawing of such water under natural weather conditions will cause the surface of the mortar or masonry unit to pop off (spall). Water entering a mortar will dissolve salts in the mortar. These salts will then leach out into or onto the surface of adjacent masonry. Upon drying, these salts will crystallize. Salts that crystallize on the exterior face of a wall (efflorescence) are unsightly but do little harm. However, when the crystallization occurs within a wall (subflorescence or cryptoflorescence), considerable damage can result. Subsequent cycles of wetting and drying cause larger and larger crystalline deposits to form until the deposits grow large enough to cause the surface of the mortar or masonry units to spall.

An increase in entrained air content may increase the durability of mortar, but will substantially decrease its bond strength and adversely affect other desirable properties. Because portland cement–lime mortar is sufficiently durable without air entrainment, admixtures should not be used to increase its entrained air content. Masonry cement mortars tend to be less durable, so air entrainment admixtures are used in them.

COMPRESSIVE STRENGTH

The compressive strength of a mortar or grout depends largely on the amount of portland cement in its mix, rising when the cement content is increased. Conversely, a larger flow (see Section 4.2.1.1) brought about by a rise in the water content of a mortar will decrease its compressive strength.

Since there are few reports of structural distress or failure due to low compressive strength problems, it is not necessary to use mortars of greater than moderate strength in normal construction. Bond strength, workability, and water retention are more important and are generally given a higher priority when mortar is being selected. Conversely, using mortar with a too high strength can cause significant problems, especially in repointing older masonry. A high-strength mortar used in an existing joint may be stronger and less flexible than the existing mortar or even the masonry units. Normal thermal and other movements can cause the stronger mortar to tear away from the weaker mortar, causing cracks that may leak. These movements may even cause the masonry units to chip or crack. Therefore, mortar strength should never be higher than necessary for the prevailing conditions, and in existing walls new mortar should match the existing mortar in composition and materials.

VOLUME CHANGE

Some people believe that mortar shrinkage can be extensive and can cause leaky walls. Actually, the maximum possible shrinkage in a properly tooled mortar joint containing sound mortar that has a good balance of all desirable properties is so small that any resultant crack could not be seen with the naked eye.

WATERTIGHTNESS

When a wall leaks significantly and there are no major holes in it, the leaks are usually through fine cracks between the mortar and the masonry or stone units, especially in the vertical joints. The watertightness of the masonry units themselves, or of the mortars commonly used today, is seldom a factor in wall leaks.

RATE OF HARDENING

The rate of hardening of mortar or grout is the speed at which it develops a resistance to indentation and crushing. Too rapid hardening may interfere with a mason's use of a mortar or the pouring of a grout. Too slow hardening may impede the progress of the work or may subject a mortar or grout to early damage from frost action during winter. A well-defined, consistent rate of hardening allows a mason to tool joints at the same degree of hardness and thus obtain uniform joint color.

MORTAR COLOR

Color is an inherent characteristic of hardened mortar. The materials needed to produce colored mortar are discussed in Section 4.2.3.4. Regardless of what the color is, color uniformity affects the overall appearance of masonry. There are several factors other than the materials used that can affect the uniformity of color and shade of a hardened mortar. These include atmospheric conditions, the moisture content of the masonry units, the admixtures used, the control exercised over the mixture, and the timing of the mason's tooling of the joints.

Controlling the mix from batch to batch and day to day is necessary if a mortar is to have the same color throughout. Careful measurement of materials and thorough mixing are important in providing uniformity.

The timing of joint tooling influences the shade of a hardened mortar. If the mason tools the joint when the mortar is relatively hard, a darker color results than if the mason tools the joint when the mortar is relatively soft. To ensure a uniform mortar color in the finished wall, the mason should tool all the joints at like degrees of mortar hardness (Fig. 4.2-2).

FIGURE 4.2-2 **Joints should be tooled when the mortar is at a like degree of hardness to prevent variations in color.** (Courtesy Portland Cement Association.)

4.2.2 KINDS OF MORTAR AND GROUT

Four kinds of mortars and two kinds of grouts are described in this section. The mortars are sand-lime, portland cement, portland cement–lime, and masonry cement. All of the discussions about mortar elsewhere in this chapter refer to either portland cement–lime or masonry cement mortars, and they are the only ones covered in ASTM C 270. Lime mortars and portland cement mortars are seldom used today except in small, special cases, such as in patching a like mortar.

The two kinds of grouts discussed here are portland cement grout and portland cement–lime grout. Lime mortars are not suitable for grout. Masonry cement is usually not permitted in grout.

4.2.2.1 Sand-Lime Mortars

Until relatively recently, all mortar was a mixture of lime, sand, and water. The lime was the cementing material. Sand-lime mortars harden at a slow, variable rate, develop low compressive strength, and have poor durability when subjected to freeze-thaw cycles. Therefore, the joints in many old masonry installations that were built with sand-lime mortar have deteriorated until there is almost no binder left. The sand can be raked from the joint with a finger.

However, sand-lime mortars have high workability and high water retention and they are more flexible than portland cement mortars. Because lime hardens only when it is in contact with air, complete hardening of a lime mortar occurs very slowly over a long period of time. This is a disadvantage in terms of project progress. The advantage of this slow hardening is that even when cracks occur they often "heal" spontaneously during the hardening process, which tends to help keep water penetration to a minimum.

Sand-lime mortar is seldom used today except in renovation work where an existing masonry was laid using a similar material.

4.2.2.2 Portland Cement Mortars and Grouts

Portland cement mortars and grouts (portland cement, sand, and water) harden quickly at a consistent rate and develop high compressive strengths. They also have good durability when subjected to freeze-thaw cycles, but their workability and water retention are low.

Portland cement grouts are used in the installation of the kind of architectural terra-cotta called ceramic veneer (see Section 4.7.7). They may also be used in other types of masonry, but are not permitted in engineered unit masonry walls.

4.2.2.3 Portland Cement–Lime Mortars and Grouts

To combine the advantages of portland cement and lime and to compensate for their disadvantages, and because their mixture produces the most predictable of mortars and grouts, portland cement–lime mortars and grouts (portland cement, lime, sand, and water) are widely used. In portland cement–lime mortar and grout, the portland cement contributes durability, high early strength, a consistent rate of hardening, and high compressive strength. Lime adds workability, water retention, and plasticity. Both contribute bond strength. Sand acts as a filler, making the mix economical, and adds strength. Water is the mixing vehicle. It creates plastic workability and initiates the cementing action.

4.2.2.4 Masonry Cement Mortars

Masonry cement mortars (masonry cement, sand, and water) are made with proprietary mortar mixes called *masonry cement*. Masonry cement manufacturers combine in one bag such ingredients as portland cement, portland blast-furnace slag cement, portland-pozzolan cement, natural cement, and slag cement. Masonry cement may also contain hydrated lime, limestone, chalk, calcareous shell, talc, or slag, and an air-entraining agent or other admixtures.

The principal advantages of using masonry cement mortar are that it is readily available, convenient to mix, and produces a mortar with generally good workability. The first of these advantages is somewhat reduced, however, by the availability of ready-mixed portland cement–lime mortars, which resemble ready-mixed concrete in that the ingredients are batched and mixed at a central location and delivered to a project site ready to use.

The principal disadvantage of using masonry cement is that the standard for it, ASTM C 91, does not limit the ingredients or their proportions or control the types or quantities of admixtures a producer can include in it. Therefore, the performance of masonry cement is relatively unpredictable as compared with portland cement–lime mortars made in compliance with ASTM C 270, which does limit the ingredients and admixtures in mortar. While it is true that laboratory test results furnished by individual manufacturers for specific masonry cements can serve as a guide to performance, masonry cements are not recommended for use in reinforced masonry or in masonry that may have to withstand lateral forces or heavy loads.

4.2.3 MORTAR AND GROUT MATERIALS

ASTM C 270, which is the currently recognized standard for mortar for unit masonry work and stone, lists the required standards controlling the ingredients for mortar. ASTM C 476 similarly lists the controlling standards for grout ingredients for use in engineered masonry. The listed standards are based on laboratory tests, field use, and many years of experience.

To comply with ASTM C 270 and C 476, mortar and grout materials must conform to the individual ASTM standards given there. Many of the numbers of those standards are repeated in this section. Some of them are:

- C 5—Quicklime.
- C 91—Masonry cement.
- C 144—Aggregates (sand) for use in mortar.
- C 150—Portland cement, Type I or II. Type III may also be used for cold-weather construction. Air-entraining portland cement is also covered, but should not be used in mortars.
- C 207—Hydrated lime (Type S).
- C 404—Aggregates for use in grout.

Some mortars and grouts, such as those used in terra-cotta, for example, may not be required by a code or other governmental regulation, or recommended by an industry standard, to comply with ASTM standards. Nevertheless, it is prudent to require that their ingredients comply with the individual ASTM standards listed in ASTM C 270 and C 476.

4.2.3.1 Cementitious Materials

Cementitious materials used in mortar and grout include portland cement, masonry cement, quicklime, and hydrated lime.

PORTLAND CEMENT

ASTM C 150 covers three portland cement types recommended for use in mortar and grout:

- Type I: Used when no special properties are required
- Type II: Used when moderate sulfate resistance or moderate heat of hydration is desired
- Type III: Used when high early strength is needed, such as in cold weather

ASTM C 150 also covers several air-entraining types of cements, but these should not be used in masonry mortar or grout because the amount of air entrained is often difficult to predict. Excess entrained air in a mortar reduces the bond between the mortar and the masonry units laid with it.

MASONRY CEMENT

Masonry cements should conform to the requirements of ASTM C 91. They should not be used in grout. Where their use is permitted in mortar, more satisfactory results may be obtained by requiring the use of a specific manufacturer's product that has proven to be acceptable for the use intended. As an alternative, masonry cements may be specified by listing the acceptable properties and limiting the ingredients. Admixtures should also be limited to those that are permitted in mortars by ASTM C 270.

Masonry cements are available with color pigments added at the factory. The advantage of using them when colored mortar is desired is that it is much easier to obtain uniformity of color when the pigments are added to the mix in the plant, rather than at the job site. Masonry cement is often selected for architectural projects for this very reason. Care must be taken, however, to ensure that the colored masonry cement selected will produce a mortar with the other properties required.

The air content of masonry cement is permitted by ASTM C 91 to be as high as 22%. When masonry cement is allowed, however, its air content should be limited to 12%, since higher amounts often result in loss of bond strength.

QUICKLIME

Quicklime, which is essentially calcium oxide, CaO, should conform to ASTM C 5. It must be carefully mixed with water (slaked) and stored for as long as 2 weeks before use and have the consistency of putty. Though used extensively in the past, quicklime is used in that form today only in preservation work to match an existing mortar.

HYDRATED LIME

Hydrated lime must conform with ASTM C 207. Hydrated lime is quicklime that has been slaked before packaging, converting the calcium oxide into calcium hydroxide, $Ca(OH)_2$. Hydrated lime, which can be mixed and used without delay, is much more convenient than quicklime. Hydrated lime is available in both S and N types, but only Type S is recommended for masonry mortar and grout, because ASTM C 207 does not control the amount of unhydrated oxides in Type N or its plasticity. Air-entraining hydrated lime is also available, but it is not suitable for masonry mortars or grouts.

4.2.3.2 Aggregates

The only aggregate suitable for use in masonry mortars is material that complies with ASTM C 144. The standard for aggregates for use in grout is ASTM C 404. Either natural or manufactured aggregates may be used, but both must be clean, sound, and well graded. Manufactured aggregate may be ground stone such as marble, granite, or other sound stone of the desired color. The aggregate used has an important effect on the workability and durability of a mortar or grout.

AGGREGATE GRADATION FOR MASONRY- AND STONE-SETTING MORTAR

For the best workability and greatest economy, aggregate for general use should be graded uniformly to include all particle sizes from very fine up to ¼ in. (6.4 mm). Aggregates with less than 5% very fine particles produce harsh, hard-to-work mortars that need additional cement or lime to become workable. Aggregates deficient in large particles generally result in weaker mortars. Nevertheless, aggregates for joints narrower than ¼ in. (6.4 mm) wide should have 95% of their particles passing a No. 16 sieve and 100% passing a No. 8 sieve.

Many commercially available sands do not comply with the gradation limits set in ASTM C 144 and should not be used in masonry mortar. Fortunately, the gradation of many commercial sands can be altered easily and inexpensively by adding fine or coarse sands as appropriate.

AGGREGATE GRADATION FOR STONE JOINT-POINTING MORTAR

Aggregates for stone joint-pointing mortar should be very fine with no granules retained on a No. 16 sieve.

AGGREGATE GRADATION FOR GROUT

Grout for use in engineered unit masonry is divided in ASTM C 476 into two categories, fine and coarse. The aggregate controlled by ASTM C 404 is similarly graded, and there are several subgrades within each category. The gradation of the aggregate for a particular grout is determined by the size of the grouted space and the height of the pour.

The aggregate recommended for use in grout associated with the kind of terracotta called ceramic veneer (see Section 4.4.4.3) is pea gravel. This pea gravel is usually defined as aggregate that is uniformly graded from ¼ in. (6.4 mm) size down to that which can pass through a No. 8 sieve. It should comply with the other requirements of ASTM C 404.

4.2.3.3 Water

The water used in mortar and grout should be clean and free from deleterious amounts of acids, alkalis, and organic materials. Water that is fit to drink is generally suitable for use in mortar or grout unless the water contains minerals that would be harmful to the mortar.

4.2.3.4 Materials for Colored Mortar

Mortar may be colored by either controlling the color of the ingredients or by adding pigments to the mix.

COLORED AGGREGATE MORTAR

The preferred method of obtaining colored mortar is by using colored aggregate and cement. This method will usually give permanent color and will not weaken the mortar. Either white or gray portland cement may be used. The aggregates may be white or brown sand or ground granite, marble, limestone, or other stone. For example, white sand, ground limestone, or ground marble may be used with white portland cement and lime to produce white joints. The disadvantage of this method, of course, is the limit on the number of possible mortar colors.

COLORED PIGMENTED MORTAR

The second method available for producing colored mortar is to add pigments to the mix. Masonry cement is available with the pigments already added. Pigments must be added in the field for colored portland cement–lime mortars.

Mortar pigments must be fine enough to disperse throughout the mix, must be capable of imparting the desired color when used in permissible quantities, and must not react with other ingredients to the detriment of the mortar. It is usually possible to meet these requirements using metallic oxides, such as iron, manganese, and chromium oxides. Carbon black and ultramarine blue have also been used successfully. Organic pigments should be avoided, as should colors containing Prussian blue, cadmium lithopone, and zinc and lead chromates.

The minimum quantity of pigment that will produce the desired results should be used, because an excess may seriously impair strength and durability. The American Concrete Institute's (ACI) 530.1/ASCE 6/TMS 602 limits the amount of mineral oxide pigments to 10% by weight of the cement in portland cement–lime mortar and 5% of the weight of masonry cement in masonry cement mortar. It limits carbon black to not more than 2% of the weight of the cement in portland cement–mortar and 1% of the weight of the masonry cement in masonry cement mortar. It permits no other pigments. Therefore, if black mortar is desired, the best pigment to use is black iron oxide.

For best results, pigments should be premixed with portland cement in large, controlled quantities. Premixing large quantities will ensure a more uniform color than can be obtained by mixing smaller batches at the project site.

4.2.3.5 Admixtures

Some admixtures are permissible in mortar and grout. For example, color pigments are an admixture. Other available admixtures, some of which are permitted in some mortars and grouts, include air-entraining agents, water-repellent agents, accelerators, retarders, and antifreeze compounds. ASTM C 270 mentions the use of calcium chloride as an accelerator, but ACI 530.1/ASCE 6/TMS 602 specifically prohibits the use of admixtures that contain chlorides in mortar and grout that are used in engineered masonry. Nonchloride accelerating admixtures are available, but if these are necessary, specific products should be selected that are known to give satisfactory results without damage to the properties of the mortar or grout.

ASTM C 476 prohibits using any kind of antifreeze compound in grout used in engineered masonry and permits other admixtures only with specific approval from the purchaser of the grout. All admixtures should be laboratory tested to ensure that they will not be harmful to a mortar or grout.

Except for latex additives and colored pigments, additives should not be used in mortar that will come into contact with stone. Some additives will discolor or otherwise harm stone.

4.2.4 MORTAR TYPES, USES, AND PROPORTIONS

Mortar types are defined in ASTM C 270.

4.2.4.1 Mortar Types

The four mortar types described in ASTM C 270 (M, S, N, and O) are those recommended for use with unit masonry and stone. Of these, only Types M, S, and N are permitted in engineered unit masonry. The four types are as follows.

TYPE M

Type M mortar is a high-strength mortar that has somewhat greater durability than other mortar types. It is specifically recommended for use in reinforced masonry and where high compressive strength is necessary. A typical use is in a reinforced masonry wall that is below grade, such as a foundation or retaining wall. It is also used extensively for walks, sewers, and manholes and to set stone in paving installations.

TYPE S

Type S mortar is a medium-high-strength mortar recommended for use where bond and lateral strength are more important than high compressive strength. Tensile bond strength between brick and Type S mortar approaches the maximum obtainable with cement-lime mortars. Type S mortar is recommended for use in reinforced masonry and in unreinforced masonry where maximum flexural strength is required. It is also used in setting beds for granite, marble, and slate in other than paving installations.

TYPE N

Type N mortar is a medium-high-strength mortar recommended for general use in exposed masonry above grade where high compressive and lateral strengths are not required. It is specifically recommended for chimneys, parapet walls, and exterior walls subject to severe exposure. It is also used to set limestone other than in paving installations.

TYPE O

Type O mortar is a low-strength mortar suitable only for general interior use in non-load-bearing masonry and for tuck pointing. It should never be used where it will be subjected to freezing conditions.

4.2.4.2 Mortar Uses

No single mortar type is best for all purposes. The basic rule to follow in selecting a mortar type is to never select one that is stronger in compression than it needs to be to satisfy the structural requirements of the wall.

Selection of the proper mortar type for each unit masonry use should be based on the masonry type and its location in the building (Fig. 4.2-3). No mortar type rates highest in all desirable properties. Adjustments in the mix to improve one property are often made at the expense of others. For this reason, the properties of each mortar type should be evaluated and the type chosen that will best satisfy the end-use requirements. Figure 4.2-3 shows the recommended mortar types for six common construction applications. Additional information about selection of mortar types can be found in the Brick Institute of America's (BIA) Technical Notes, specifically Technical Notes No. 8, 8A, and 8B.

In addition to the general uses suggested in Section 4.2.4.1, some specific cases should be mentioned. The following paragraphs suggest the mortar types usually selected for the specific conditions indicated. Mortar types for brick paving and flooring and for stone are addressed in Sections 4.2.8, 4.2.9, and 4.2.10.

CAVITY WALLS

In cavity walls where wind velocities will exceed 80 miles per hour, Type S mortars should be used. In locations where winds of lower velocity are expected, either Type S or Type N may be used.

FACING TILE

For facing tile, either M, S, N, or O types are usually acceptable. ASTM C 144 allows the use of sand particle sizes as large as ¼ in. (6.4 mm). However, facing tile is often laid with ¼ in. (6.4 mm) wide joints. In that case, the aggregate used must pass through a No. 16 sieve, which allows aggregates to pass that are slightly less than 1/16 in. (1.6 mm) in size.

DIRT-RESISTING MORTAR

For dirt resistance, and when medium resistance to staining is needed, aluminum tristearate, calcium stearate, or ammonium stearate is sometimes added to a masonry mortar or the mortar in the form of terra-cotta called ceramic veneer (see Section 4.7.7). When this is done, the amount should be limited to that amount dictated by code, if applicable, or the amount recommended by the masonry unit manufacturer. In the absence of a code restriction or manufacturer's recommendation, the amount should be limited to not more than 3% of the weight of the portland cement in the mortar.

Where maximum dirt resistance is desired, as for facing tile, for example, the mortar should be raked out to a depth of about ⅜ in. (9.5 mm) and the joints pointed with a mortar consisting of 1 part portland cement, ⅛ part lime, and 2 parts graded fine (80 mesh) sand, by volume, to which has been added aluminum tristearate, calcium stearate, or ammonium stearate in an amount equal to about 2% of the weight of the portland cement in the mortar.

Admixtures should not be used in mortar for engineered masonry unless specifically allowed by ACI 530/ASCE 5/TMS 402, ACI 530.1/ASCE 6/TMS 602, and the applicable building code.

TUCK-POINTING MORTAR

Tuck pointing is the refilling of an existing joint from which some of the mortar has been removed. Tuck-pointing mortar must be *prehydrated.* To prehydrate a mortar, the ingredients are thoroughly mixed without water, then mixed again, adding only enough water to produce a damp, unworkable compound that will retain its form when pressed into a ball. After 1 to 2 hours, enough water is added to make the mortar workable; the resulting mortar should be used immediately.

For best results, tuck-pointing mortar should have the same ingredients and mixture as the original mortar. When the original mortar type is unknown, prehydrated Type N mortar is often used.

TWO DIFFERENT MORTARS IN THE SAME JOINT

Sometimes it is helpful to place two different mortars in the same joint. These mortars should be of the same type, however, and masonry cement and portland cement–lime mortars should not be used together. Two different mortars may be installed, for example, where white, black, or colored mortar is desired. One option, of course, is to use the same mortar throughout. However, white, black, and colored mortars are usually more expensive than natural colored mortars, and their properties are not as consistent or as predictable as those of mortars that do not contain pigments. An alternative is to lay the unit masonry in a normal, natural colored mortar. After the units have been laid, rake the joints back to a depth of about ½ in. (12.7 mm), and then point them with the desired white, black, or colored mortar.

4.2.4.3 Mortar Proportions

The ingredient proportions required by ASTM C 270 for the mortar types listed in Section 4.2.4.1 are shown in Figure 4.2-4.

The mortar used in the type of terra-cotta called adhesion-type ceramic veneer is usually required to have a mixture consisting of 1 part portland cement, ½ part lime putty or hydrated lime, and 4 parts sand, by volume.

4.2.5 GROUT TYPES, USES, AND PROPORTIONS

In ASTM C 476 grout is classified as either fine or coarse. The only difference between the two is the gradation of the aggregate. They both contain 1 part portland cement, 0 to 1/10 part hydrated lime or lime putty, and damp, loose fine aggregate in an amount between 2¼ and 3 times the sum of volumes of the cementitious materials. In addition, coarse grout contains coarse aggregate in an amount equal to between 1 and 2 times the sum of the volumes of the cementitious materials. Both grout types are used to fill reinforced masonry walls, tie beams, lintels, and other reinforced masonry items.

Grout for use in anchored-type ceramic veneer is usually required to have 1 part

FIGURE 4.2-3 Recommended Mortar Types for Various Construction Applications

Construction Application	Recommended Minimum ASTM Mortar Types	Order of Relative Importance of Principal Properties		
		Plasticity[a]	Compressive Strength	Weather Resistance
Foundations, basements, walls, isolated piers[b]	M, S[c]	3	2	1
Exterior walls	S, N	2	3	1
Solid masonry unit veneer over wood frame	N	2	3	1
Interior walls, load bearing	S, N	1	2	3
Interior partitions, non-load bearing	N, O	1	—	—
Reinforced masonry (columns, pilasters, walls, beams)	M, S[c]	3	1	2

[a]Adequate workability and a minimum water retention (flow after suction of 70%) assumed for all mortars.
[b]Also any masonry wall subject to unusual lateral loads for earthquakes, hurricanes, etc.
[c]Only portland cement-lime Type S and M mortars.
(Courtesy Portland Cement Association.)

FIGURE 4.2-4 Mortar Proportions by Volume[a]

Type[b]	Portland Cement	Hydrated Lime or Lime Putty	Masonry Cement	Maximum Damp Loose Aggregate[c]	Minimum Compressive Strength, 2-in. Cubes at 28 Days, psi (MPa)
M	1	¼	—	3	2500 (17)
	or				
	1	—	1	6	2500 (17)
S	1	½	—	4½	1800 (12)
	or				
	½	—	1	4½	1800 (12)
N	1	1	—	6	750 (5)
	or				
	—	—	1	3	750 (5)
O	1	2	—	9	350 (2.4)

[a]The weight of 1 cu. ft. (0.028 m^3)of materials used is considered to be: portland cement, 94 lb. (42.64 kg), 1 bag; hydrated lime, 40 lb. (18.14 kg); lime putty, 80 lb. (36.27 kg); dry sand, 80 lb. (36.27 kg); masonry cement, weight printed on bag.

[b]Shaded figures show proportions for portland cement-lime mortar. Mortars made with masonry cement are unshaded.

[c]The damp loose aggregate should be not less than 2¼ times, nor more than 3 times, the sum of the cementitious material used.

portland cement, 1 part sand, and 5 parts pea gravel.

4.2.6 MORTAR AND GROUT MIXING

Thorough mixing is important to the development of desirable properties in a mortar or grout. Because some properties begin to deteriorate soon after a mortar or grout is mixed, no more should be mixed than can be placed before hydration (curing) begins.

Water is essential to the development of bond strength in mortar (see Section 4.2.1.2). If a mortar contains too little water, the bond will be weak and spotty. Therefore, a mortar should be mixed with the maximum amount of water possible, so long as it does not become so soupy that it will not stay where placed, extrude from the joint under the weight of the masonry units, or otherwise become unusable by a mason.

Grout should be mixed with sufficient water to produce a slump of between 8 and 11 in. (200 and 280 mm). Retempering should not be necessary. Grout should be used within 1½ hours after water is first added to the mix.

ASTM C 270 and C 476 give the proportions of the ingredients in mortars and grouts in terms of parts by volume. Ingredients may be measured in the field either by volume or by weight. In either case the proportions of the ingredients must be accurately controlled if a predictable mortar or grout is to be produced. Both ASTM C 270 and C 476 contain estimated weights per cubic foot of mortar ingredients.

Whether volume or weight measure is employed, containers of known measure must be used. Cement is usually added directly from the bag as full or half bags, depending on the amount of mortar or grout needed. The materials must also be in the condition required by ASTM C 270 and C 476. For example, moisture in sand causes it to bulk or fluff up, which can make a cubic foot of wet sand contain a different amount of sand than assumed by the design tables in ASTM C 270 and C 476. ASTM C 270 and C 476 therefore both require that sand be in a damp, loose condition.

Oversanding of mortar happens often because the sand is measured by shovel rather than in a container having a known capacity. Oversanded mortar is harsh and unworkable, provides a weak bond with masonry units, and tends to erode. The problem is that the amount of sand a shovel will hold depends on the size of the shovel and the fatigue of the person holding it. Therefore, the measurement of sand solely by shovel should be prohibited. BIA suggests several accurate methods for measuring sand. Refer to BIA Technical Note 8B for specifics.

4.2.6.1 Machine Mixing

Except on very small projects, mortar should always be mixed in a mechanical batch mixer designed for the purpose. Grout should always be mixed in a mechanical batch mixer regardless of the size of the project. The size of the mixer needed will depend on the amount of mortar to be mixed.

When machine mixing is used, mortar and grout ingredients should be batched in quantities small enough to mix properly. Water should be added slowly until the proper consistency has been reached. After all the ingredients have been added, the mortar or grout should be mixed a minimum of 3 to a maximum of 5 minutes. The mixer drum should be completely empty before it is recharged with the next batch.

4.2.6.2 Hand Mixing

When mortar must be mixed by hand, the sand should first be spread in the mortar box, the cement and other cementitious materials should be spread on top of the sand, and the ingredients should then be mixed well with a hoe from both ends of the box. Next, about three-quarters of the required water should be added and mixed until the cementitious ingredients and aggregates are uniformly damp. Water should then be added in small amounts, and mixing should be continued until the mortar reaches a suitable workability. The batch should be allowed to stand for about 5 minutes and remixed thoroughly with the hoe without additional water.

4.2.6.3 Cold-Weather Mixing

The temperature of a mortar or grout when placed should be between 70° and 120°F (21.1° and 48.9°C) when the outside temperature is below 40°F (4.4°C). Higher mortar temperatures may result in fast hardening, which can make it impossible to lay masonry units with that mortar. Even when the units can be laid, the mortar may have reduced compressive and bond strengths.

Heating the mixing water is one of the easiest methods of raising the temperature of a mortar or grout. Mixing water should not be heated above 160°F (71.1°C), however, because a flash set can occur when very hot water comes in contact with cement.

In freezing weather, the moisture that is almost always present in masonry sand will freeze. Therefore, sand must be heated, slowly and carefully to prevent scorching, to remove the ice.

The use of calcium chloride and other antifreeze admixtures to lower the freezing point of mortar or grout is not permitted by most modern codes and standards and many building codes. BIA also recommends against their use. The quantity of such materials necessary to lower the freezing point of mortar appreciably is so large that mortar strength and other desirable properties would be seriously impaired. Such materials not only hasten the corrosion of metal anchors and reinforcement, they weaken mortar, contribute to efflorescence, and may even cause the masonry to spall when the chemicals dissolve in water and then recrystallize.

For additional information concerning cold-weather masonry construction, refer to BIA Technical Note No. 1.

4.2.7 RETEMPERING

When mortar is not used immediately after it is mixed, as is often the case, some of its water may evaporate. Adding water to such mortar, which is called *retempering*, is sometimes prohibited because it is thought to have detrimental effects. While it is true that retempering slightly reduces a mortar's compressive strength, failing to retemper mortar that needs it may result in a materially lowered bond strength. Therefore, retempering should be permitted to replace water lost by evaporation, provided that the mortar has not begun to set (Fig. 4.2-5). Regardless of whether it has been retempered or not, mortar should be placed within 2½ hours after its initial mixing, and no mortar should be used after it has begun to set.

A change in the time between the spreading of a mortar and the placing of masonry units in it will affect the flow of the mortar, especially if the masonry units have a high suction. The highest bond strength is achieved when this time interval is minimal. Varying this interval to any large extent will result in a wall with varying tensile bond strengths.

4.2.8 BRICK PAVING AND FLOORING MORTAR AND GROUT

Thick-set portland cement mortar beds for brick paving and flooring should be a mixture of portland cement, hydrated lime, and fine aggregate. This material is mixed with water, but a latex additive often replaces some or all of the water to provide better adhesion between the brick and the setting bed and to produce a more flexible setting bed than can be accomplished using water alone. Setting bed mortar should comply with ASTM C 270, either Type S or Type N; Type S is preferable in most instances for interior use, and Type M for exterior use.

Thin-set brick flooring is set in a bed of either dry-set portland cement mortar conforming with American National Standards Institute ANSI A118.1 or latex portland cement mortar conforming with ANSI A118.4.

FIGURE 4.2-5 Retempering of mortar. (Courtesy Portland Cement Association.)

Grout for brick pavers should be either a mixture of portland cement and aggregate mixed with water or with a latex additive; a dry-set grout complying with ANSI A118.6; or a portland cement grout complying with ANSI A118.6.

4.2.9 STONE FLOORING MORTAR AND GROUT

Stone flooring is installed using either thin-set or thick-set beds in accordance with ANSI standards. Thin stone tiles are, however, usually set using thin-set methods.

Setting beds for stone flooring are usually portland cement mortar that complies with ANSI A108.1. Grout may be either sand–portland cement, commercial portland cement, dry-set, or latex portland cement type.

4.2.10 STONE PAVING

Stone paving bed mortar should be either Type M portland cement–lime mortar conforming with ASTM Standard C 270, or a mixture of 1 part portland cement and 3 parts sand mixed with water. Part or all of the water is sometimes replaced with a latex admixture to make the mortar bond more readily to the stone, to make the setting bed somewhat more flexible, and to reduce water penetration through the mortar. Adding a latex admixture increases the setting time of a bed and may not be advisable in all cases, especially if the additive also contains a retarder.

Pointing mortar for stone paving should be a mixture of 1 part portland cement and 2 parts aggregate. The aggregate may be either natural or ground white sand or ground marble, granite, or other stone to produce the desired color.

4.3 Reinforcement, Ties, Anchors, and Flashing

Masonry and stone walls require various types of reinforcement, ties, anchors, and flashing, depending on the masonry or stone materials in the wall and its configuration and location.

4.3.1 JOINT REINFORCEMENT, TIES, AND ANCHORS FOR UNIT MASONRY

Many types of joint reinforcements, ties, and anchors are used in unit masonry construction. Some of them are shown in Figure 4.3-1. The use of these items is discussed and illustrated in Sections 4.7.2.2 and 4.7.2.3.

Metal joint reinforcement, ties, and anchoring devices in masonry or stone walls that are exposed to salt-laden water will tend to corrode. Four inches or more of masonry cover may offer some protection, but all such devices should be of copper-coated steel, hot-dip galvanized (zinc-coated) steel, or stainless steel. Even though stainless steel is more expensive than the other two, it offers the highest degree of safety and is essential in stone construction and preferable in unit masonry construction as well. The uses, types, and relative ASTM standards for masonry joint reinforcement, anchors, and ties are summarized in Figure 4.3-2.

The requirements for some types of joint reinforcements, ties, and anchors in unit masonry walls are as follows.

Joint reinforcement. Joint reinforce-

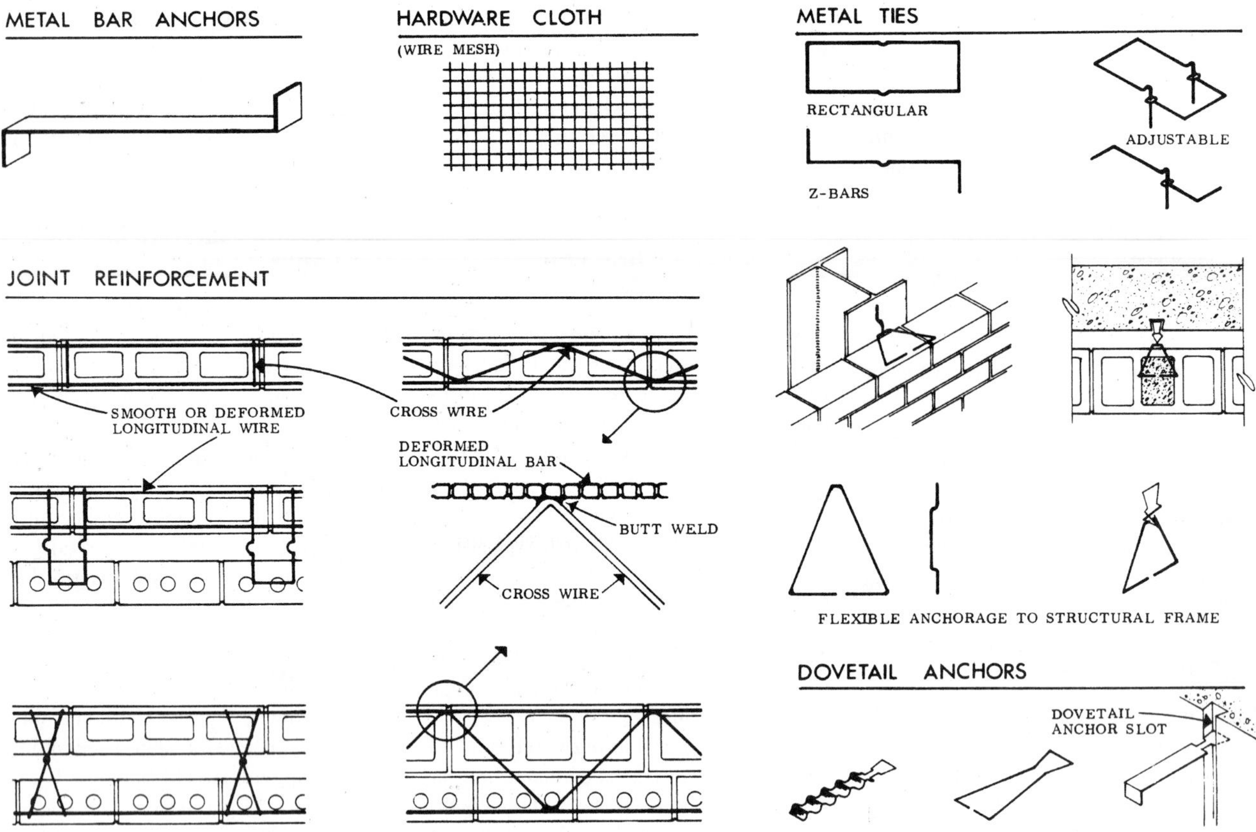

FIGURE 4.3-1 Joint reinforcement, ties, and anchoring devices.

ment for use in horizontal courses of interior masonry and other masonry protected from the weather should be fabricated from zinc-coated cold-drawn steel wire (ASTM A 82) or Type 304 stainless steel wire (ASTM A 580). Joint reinforcement for exterior walls should be fabricated from Type 304 stainless steel (ASTM A 580). Joint reinforcement should consist of two smooth or deformed longitudinal wires 9 gauge or larger, and the weld connected with 12-gauge wires. The distance between weld contacts of the cross-wires should not exceed 6 in. (150 mm) for smooth longitudinal wire or 16 in. (406 mm) for deformed longitudinal wire. The width of the joint reinforcement should be such that it can be embedded in a mortar joint with its wires not closer than ⅝ in. (16 mm) to the masonry face surface. The configuration of joint reinforcement may be either ladder design, truss design, or tab design (see Fig. 4.3-1).

Wire mesh ties. Either wire mesh ties or hardware cloth are used to anchor non-load-bearing partitions to intersecting masonry. They should be made from at least 15-gauge zinc- or copper-coated steel wire, have a ½-in. (12.7 mm) mesh, and be at least 12 in. (305 mm) long. Their width should be 1 in. (25.4 mm) less than the wall thickness.

Cavity wall ties. Ties for use in exterior cavity walls should be 3⁄16 in. (4.8 mm) diameter, Type 304 stainless steel wire (ASTM A 580) formed either in a rectangular shape not less than 4 in. (100 mm) wide (for use with hollow units having vertical cells), or in a Z shape with 2-in. (50.8 mm) legs (for use with solid units or hollow units with horizontal cells).

Ties should be long enough to permit their ends to be embedded in the outer face shell mortar beds of hollow units or the center of mortar beds of solid units.

Joint reinforcement comprising 9-gauge or larger cross-wires may be used instead of metal ties, provided that the cross-wires are spaced to render strength and stiffness per square foot of wall area equivalent to that of 3⁄16-in. (4.8 mm)-diameter Z ties used at the maximum spacing.

Faced wall ties. Ties used in faced walls should meet the same requirements given for cavity wall ties.

Veneered wall ties, wood studs. The ties usually used for brick veneer over wood studs are 22-gauge, 6-in. (152.4 mm)-long, ⅞-in. (22.2 mm)-wide corrugated galvanized steel. These are usually bent up and nailed through the sheathing into the underlying studs.

Veneered wall ties, metal studs. Ties for masonry veneer over metal studs should be triangular, flexible, stainless steel wire ties similar to those used to anchor masonry to steel structural members. The fixed portion of the ties should be screwed through the sheathing to the studs. The flexible triangular portion of the tie should then be threaded through the support and built into the masonry as it is erected.

Dovetail anchors. Dovetail anchors for use with embedded slots or inserts in concrete should be 16-gauge, zinc-coated sheet steel, ⅞ in. (22.2 mm) wide, and long enough to permit the ends of the tie to be embedded in the outer face-shell mortar beds of hollow units or the center of mortar beds in solid units.

Rigid steel anchors. Rigid steel anchors for anchorage of load-bearing walls and fire walls should be 1¼-in. (31.8 mm) by ¼-in. (6.4 mm) steel with the ends turned up not less than 2 in. (50.8 mm). They should be not less than 16 in. (406.4 mm) long for 8-in. (200 mm)-thick walls and not less than 24 in. (610 mm) long for 12-in. (305 mm)-thick walls.

FIGURE 4.3-2 Joint Reinforcement, Ties, and Anchoring Device Materials and Uses

Use	Material	ASTM Standard
Interior wall joint reinforcement	Copper-coated wire Zinc-coated wire	B 227 A 116
Exterior wall joint reinforcement	Stainless steel wire	A 580, Type 304
Reinforced masonry reinforcement	Billet steel bars Epoxy-coated billet steel bars Deformed wire Welded wire fabric Plain Deformed	A 615 A 615 and A 775 A 496 A 185 A 497
Interior wall anchors and ties	Galvanized carbon steel wire Galvanized steel sheet	A 82 A 366 and A 153
Exterior wall anchors and ties	Stainless steel wire Galvanized steel sheet Stainless steel sheet	A 580, Type 304 A 366 and A 153 A 167, Type 304
Rigid anchors	Heavy-thickness galvanized steel sheet Plates and bars Steel Painted Galvanized Stainless steel	A 635 and A 653 A 36 2 coats epoxy paint A 153 A 666, Type 304

4.3.2 STRUCTURAL REINFORCEMENT FOR UNIT MASONRY WALLS

Structural reinforcement in unit masonry walls consists mainly of steel bars and welded wire mesh.

4.3.3 REINFORCEMENT, TIES, ANCHORS, AND SUPPORT SYSTEMS FOR STONE

Various types of reinforcements, ties, anchors, and grid support systems are used in stone construction, depending on the type of stone and its use. All such devices that come in contact with stone should be fabricated from ANSI Type 304 stainless steel (ASTM A 666). Devices and support components that are not in direct contact with stone may be fabricated from hot-dip galvanized steel (ASTM A 36 for materials, ASTM A 123 for galvanizing). Some of the various types of devices used are shown in Figure 4.3-3.

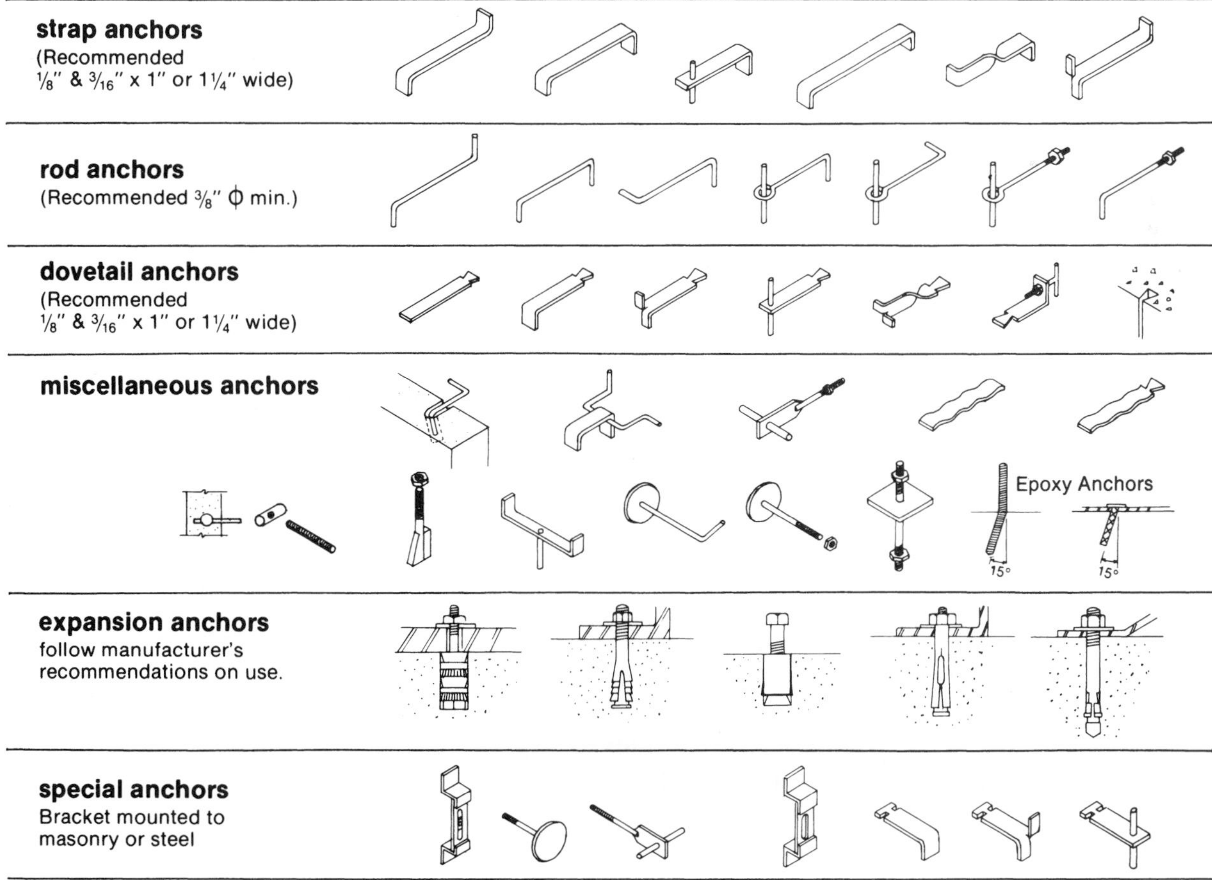

FIGURE 4.3-3 Stone anchors. (Courtesy Indiana Limestone Institute.)

FIGURE 4.3-4 Through-wall flashing in a concrete masonry wall and a rope-wick weep hole.

4.3.4 FLASHING

Flashing is installed at vulnerable locations in masonry to help prevent water from entering and to collect water that has entered the masonry and divert it through weep holes to the exterior (Fig. 4.3-4).

Flashings are formed from sheet metal, bituminous fabric membranes, plastic sheet materials, or a combination of these materials. Flashing materials are discussed in Section 7.8.

Because replacement costs can exceed the initial installed costs, it is prudent to select a durable flashing material for the original installation. Recommendations for locating and installing flashing are discussed in Section 4.7.2.8.

4.4 Clay Masonry Units

Clay masonry units are burned clay units used in buildings. They include brick, hollow clay tile, and architectural terra-cotta. They do not include ceramic tile, flue linings, drain tile, or other ceramic products. Ceramic tile and related materials are discussed in Section 9.6.

Solid clay masonry units, such as brick, and hollow clay masonry units, such as tile, are composed of burned clay, shale, fire clay, or a mixture of these, formed into the desired shape while plastic and then burned in a kiln.

Terra-cotta is Italian for "cooked earth." It is composed of an aged clay to which admixtures and sand or pulverized previously fired clay, called grog, are added. It is extruded or formed into the desired shape by hand packing and then fired at a high temperature to produce a material that is harder and more compact than most brick.

Both the raw materials used in making clay masonry and the way in which it is manufactured have a marked influence on the properties of the units in a finished wall. Therefore, a general understanding of both materials and methods is essential to the selection and use of brick and tile.

4.4.1 RAW MATERIALS

Clays used in making clay masonry units must have enough plasticity so that they can be shaped or molded when mixed with water. They must also have sufficient tensile strength to maintain their shape after they have been formed. And their particles must fuse when they are subjected to high temperatures in a kiln. The three principal forms of clays used to manufacture clay masonry units have similar chemical compositions but different physical characteristics. They are:

- *Surface clays:* sedimentary clays, found near the surface of the earth.
- *Shales:* clays that have been subjected to high pressures until they have hardened almost to the form of slate.
- *Fire clays:* clays found deeper in the ground than the other types and that have refractory qualities, making them resistant to high temperatures and useful in such applications as fire brick. They usually have fewer impurities and more uniform chemical and physical properties than surface clays or shales.

All are compounds of silica and alumina with varying amounts of metallic oxides and other impurities. The metallic oxides act as fluxes, which promote fusion at lower temperatures than would be possible without them. They also influence the color of finished units.

There are slight differences in the properties of any two finished clay masonry units, regardless of how hard their manufacturer tried to minimize variations. An attempt is made to standardize the properties of units by mixing clays from different locations in a pit, mixing clays from different pits, and making compensations during the manufacturing processes, but none of these is completely successful.

4.4.2 MANUFACTURE OF CLAY MASONRY UNITS

In general, the manufacturing process (Fig. 4.4-1) consists of mining (also called *winning*), preparing, and storing the raw materials; and forming, drying, glazing, burning, cooling, drawing, and storing the finished units.

4.4.2.1 Mining (Winning), Preparation, and Storage

In a process called *winning*, raw clay is taken from either open pits or underground mines. Raw clay that does not contain large lumps or stones is immediately passed through an inclined vibrating screen to control its particle size, and then stored.

The clay is crushed to break up lumps, and stones are removed. Then the clay is pulverized and mixed by large grinding wheels that revolve in a circular pan. Each of these wheels weighs between 4 and 8 tons (3629 to 7257 kg). The pulverized clay is then screened and stored for later use.

The clay from a single pit is often placed in several storage areas to permit later blending. Blending produces more uniformity in raw clay, helps to control the color of finished units, and permits some control over the suitability of the raw material for manufacturing a given type of unit.

4.4.2.2 Forming

The first step in the forming process, *tempering*, produces a homogeneous, plastic mass ready for molding. Tempering is usually achieved by adding water to the clay in a mixing chamber called a *pug mill*, which contains one or two revolving shafts with blades (Fig. 4.4-2a). Three methods are used to form clay masonry units: the stiff-mud, the soft-mud, and the dry-press processes.

STIFF-MUD PROCESS

The stiff-mud process is the most commonly used method for forming brick and

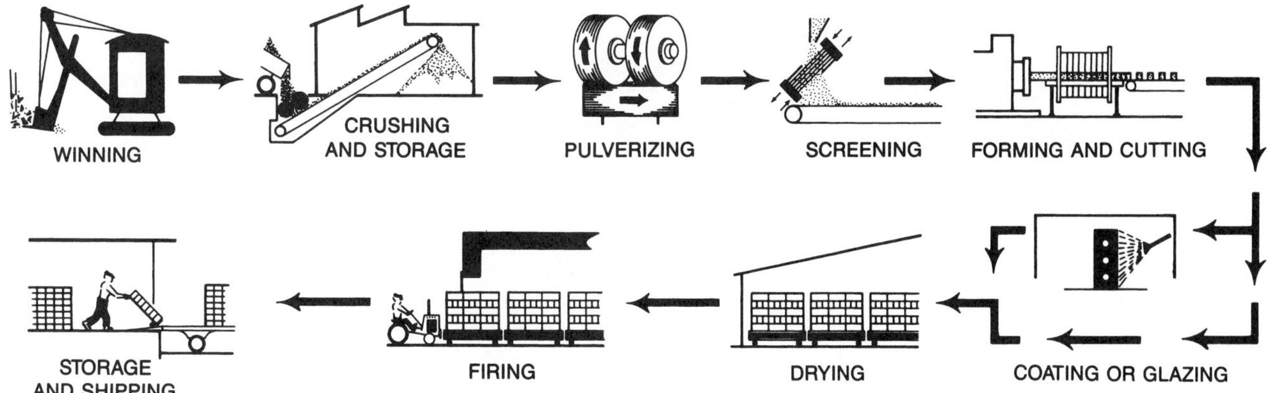

FIGURE 4.4-1 Manufacturing flowchart. (Courtesy Brick Institute of America.)

FIGURE 4.4-2 (a) Mixing powdered clay with water. (b) Clay emerging from a de-airing machine. (c) Mixture being extruded through a die that molds it into a long, flat column. (Courtesy Brick Institute of America.)

FIGURE 4.4-3 **Forming and cutting a clay column in the stiff-mud process.** (Courtesy Brick Institute of America.)

FIGURE 4.4-4 **Bricks on a belt. Core holes of various sizes are cut through the body of bricks as they are extruded from the die to make them lighter and enable them to burn more uniformly.** (Courtesy Brick Institute of America.)

clay tile. All structural clay tiles and many bricks are made this way. A similar method is used for extruded terra-cotta. In the stiff-mud process, clay is mixed with sufficient water (12% to 15% by weight) to produce the desired degree of plasticity. After thorough mixing (*pugging*), the tempered clay goes through a de-airing machine in which a partial vacuum is maintained. De-airing removes air, thus eliminating holes and bubbles, which improves the clay's plasticity and workability and results in greater strength in finished units (Fig. 4.4-2b).

After de-airing, the clay is forced (extruded) through a die, which produces a clay column of the desired shape (Fig. 4.4-2c). Here, various textures may be applied. The column then passes through an automatic cutter (Fig. 4.4-3). Die sizes and the spacing of cutter wires must be calculated to compensate for normal shrinkage during drying and burning.

After they are cut from the column, individual units proceed from the cutting table onto a continuous moving belt, where they are inspected. Acceptable units are removed from the belt as it passes and placed on dryer cars (Fig. 4.4-4). Imperfect units are returned to the pug mill.

SOFT-MUD PROCESS

The soft-mud process is used only for producing brick. It is particularly suitable for clays that contain too much natural water (20% to 30% by weight) to be satisfactorily extruded in the stiff-mud process. In the soft-mud process, raw clay is mixed with water and the mixture is then formed into units in molds. To prevent this clay from sticking, the molds are lubricated with sand and water or just water. When sand is used, the bricks are called *sandstruck* (Fig. 4.4-5h); if water is used, they are called *waterstruck* (see Fig. 4.4-5g).

A method similar to the soft-mud process for forming brick has been used to make molded terra-cotta shapes, except that a high water content is not a prerequisite.

DRY-PRESS PROCESS

The dry-press process is used with clay that has very low plasticity. This clay is mixed with a minimum of water (up to 10% by weight) and then formed in steel molds under pressure of between 500 and 1500 psi (211 and 1054.6 Mg/m^2).

4.4.2.3 Drying

When wet clay masonry units come from molding or cutting machines, they contain 7% to 30% moisture, depending on the forming method. Before brick or clay tile units are burned, most of this water is evaporated in dryer kilns at temperatures ranging from 100°F (37.8°C) to 300°F (148.9°C). Drying times vary from 24 to 48 hours, depending on the type of clay used in the units. During this time, heat and humidity must be regulated to avoid excessive cracking of the units.

Terra-cotta units must be thoroughly dried before they are fired. Unless the units are bone dry at the time of firing, they will crack or explode.

4.4.2.4 Ceramic Glazing

Ceramic glazing is the process used to produce the finish on glazed facing tile,

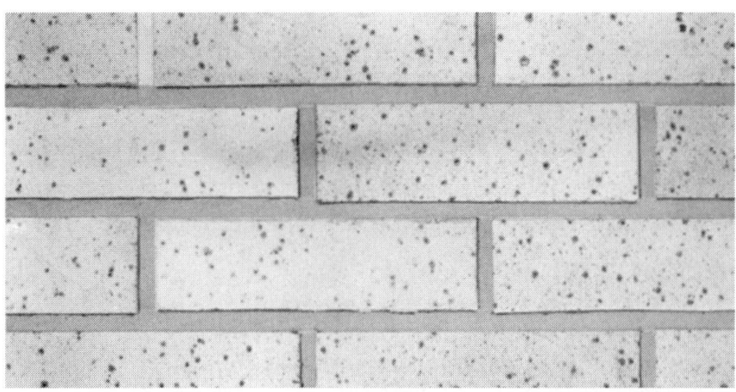

a. Smooth

b. Matte-vertical markings

c. Matte-horizontal markings

d. Rugs

e. Barks

f. Sandmold

g. Waterstruck

h. Sandstruck

FIGURE 4.4-5 Typical brick textures. These may not be available in all locations. (Courtesy Brick Institute of America.)

glazed facing brick, and glazed terra-cotta. It is a specialized, controlled procedure in which glazes are sprayed onto masonry units and then fired. Spray glazes are composed of mineral ingredients that fuse into a glasslike coating at a given temperature. They are available in almost any color. There are two basic variations of the ceramic glazing process: *high-fired glazing* and *low-fired glazing*.

HIGH-FIRED GLAZING

In the high-fired process, the glazes are sprayed on the masonry units sometimes before and sometimes after drying. The units are then kiln-fired at normal firing temperatures.

LOW-FIRED GLAZING

Some glazed colors cannot be produced at high temperatures. When these are desired, low-fired glazing is used. In this process, the glazes are applied after the units have been burned to maturity and cooled. Then the units with the glazes applied are fired again at a lower temperature.

4.4.2.5 Burning and Cooling

Burning is the process of baking or firing dried clay masonry units in ovens called *kilns* to produce finished units. *Cooling* is the process of returning the masonry units to atmospheric temperature. The two chief types of kilns used are *periodic kilns* and *tunnel kilns* (Figs. 4.4-6a and 4.4-6b). Both types are fueled by natural gas, oil, or coal.

Burning of brick and clay tile units requires 2 to 5 days, depending on the kiln used and other variables. The burning time for terra-cotta may be a little shorter.

During burning and cooling, the temperatures in a kiln must be raised and lowered at controlled rates. To permit careful control of temperatures and their rates of change, kilns are equipped with devices to permit constant monitoring and regulating of their interior temperatures.

The burning process begins when dried masonry units are loaded into kilns. In periodic kilns (see Fig. 4.4-6a), the units are placed in the kiln in a pattern that permits free circulation of hot gases. In tunnel kilns (see Fig. 4.4-6b) the units are placed on special cars that pass through various temperature zones as they traverse the tunnel.

Burning may be divided into six stages: the evaporation of free water (water-smoking); dehydration; oxidation; vitrification; flashing; and cooling. The first four are associated with rising temperatures in the kiln. The actual temperatures used in the burning process will vary somewhat depending on the type of clay. In general, for brick and clay tile units water-smoking takes place at about 400°F (204.4°C), dehydration between 300°F (148.9°C) and 1800°F (982°C), oxidation between 1000°F (537.8°C) and 1800°F (982°C), and vitrification between 1600°F (871°C) and 2400°F (1315.6°C).

To produce different colors and shades of colors, masonry units are sometimes subjected to a reducing atmosphere (one that contains insufficient oxygen for complete combustion) near the end of the burning process. The colors and shades produced by this so-called flashing will vary with the type of clay used in the units.

Cooling begins after the temperature in a kiln has reached the maximum needed to properly burn the units in it. The rate of cooling must be controlled, because it has a direct effect on the color of the units, and because excessively rapid cooling will cause the masonry units to crack or check. In tunnel kilns, cooling seldom requires more than 48 hours, but in periodic kilns, proper cooling usually requires at

a

b

FIGURE 4.4-6 Until recent years bricks were fired in (a) periodic (beehive) kilns. Today most bricks are fired in (b) tunnel kilns. (Courtesy Brick Institute of America.)

FIGURE 4.4-7 Unloading a kiln. (Courtesy Brick Institute of America.)

least 48 hours and sometimes as much as 72 hours.

4.4.2.6 Drawing

Drawing is the process of unloading a kiln after cooling (Fig. 4.4-7). During drawing, units are sorted and graded. They are then taken to a storage yard or loaded directly onto rail cars or trucks for delivery.

4.4.3 PROPERTIES OF CLAY MASONRY UNITS

The properties of finished units depend on the characteristics of the raw materials and the effects of the manufacturing process.

4.4.3.1 Compressive Strength

The clay, the method of manufacturing, and the degree of burning affect compressive strength. With some exceptions, the plastic clays used in the stiff-mud process have higher compressive strengths when burned than do the clays used in the soft-mud or dry-press processes. For a given clay and method of manufacture, higher compressive strengths are associated with higher burning temperatures. The compressive strength of brick varies from 1500 psi (1054.6 Mg/m^2) to more than 20,000 psi (14 061 Mg/m^2), due mainly to the wide variation in the properties of the clays used.

4.4.3.2 Absorption

The degree of water absorption of a clay masonry unit also depends on the clay, manufacturing method, and temperature of burning. Plastic clays and higher burning temperatures generally produce units having lower absorption. Generally, the stiff-mud process produces units with lower absorption than either the soft-mud or dry-press process. A brick's initial rate of absorption, which is called *suction*, results from pores or small openings in the burned clay that act as capillaries to draw or suck water into the unit. Suction has little bearing on the transmission of free water through a brick, but it has an important effect on the bond between brick and mortar. Maximum bond strength and minimum water penetration are obtained when the suction of the brick at the time of laying does not exceed 20 grams (0.7 oz.) of water per minute. Bricks having suction of more than 20 grams per minute should be wetted before they are laid to reduce the suction. Section 4.7.1.2 contains more details about acceptable degrees of absorption and describes a field test that will help determine suction.

4.4.3.3 Durability

The durability of clay masonry units is affected by pore size within the units and by the amount and completeness of fusion during burning. In general, higher burning temperatures produce harder clay masonry units.

A major weathering action that affects burned clay masonry units is alternate freezing and thawing in the presence of moisture. The ability of clay masonry units to withstand freeze-thaw cycles without disintegration is important in every cold climate, but is especially so where the annual average precipitation exceeds 20 in. (508 mm) of water and alternate freezing and thawing is common. Units produced from the same raw materials and by the same method of manufacture may have different resistances to damage by freeze-thaw cycles, depending on their compressive strengths and absorption. Units with high compressive strength and units with low absorption are usually the more resistant to damage during freeze-thaw cycles.

A second factor that affects the durability of clay masonry units is their soluble salts content. Such salts will dissolve when water is present and crystallize either on the surface of the masonry units or within them. Surface deposits (efflorescence), while unsightly, are not damaging. Internal crystallization (cryptoflorescence), however, will build up until the surface of the masonry units spalls. For this reason, brick that will be exposed to water in any form should be tested and rated *not effloresced* in accordance with ASTM C 67.

4.4.3.4 Natural Color

Burned clays produce a wide range of colors, from pearl grays and creams through buff, and in a descending scale of reds to purple, maroon, and gunmetal black. The clay's chemical composition, the burning temperatures, and the method of burning all control color. Terra-cotta clays produce units that vary widely in color from red to brown to white.

Clays used in masonry units contain varying degrees of many metallic oxides. Of these, iron oxides probably have the greatest effect on color. Clay containing iron in almost any form will burn red when exposed to an oxidizing fire because the iron combines with the oxygen to form ferrous oxide, which is the same as red rust. When clay containing iron is burned in a reducing atmosphere (*flashed*), it will take on a purple cast.

Lighter colors are produced when a clay is underburned. This is the method used to produce salmon-colored brick from a clay that would ordinarily burn red. Underburned clay masonry is softer, more absorptive, and has a lower compressive strength than units that are burned completely.

Overburning clay produces *clinker brick*, which is dark red to black when the clays used are those that produce red units when burned normally, and dark speckled brown when the clay used would produce buff units when burned normally. Clinker bricks are ordinarily used for aesthetic reasons and for their charm.

With few exceptions, the natural colors of clay masonry units are mixtures of shades rather than pure colors. Therefore, the accepted practice in specifying color is to require that the color of the delivered units match an approved sample panel made up of several units representative of the color range expected.

By using applied coatings, such as glazes or nonlustrous finishes, and by introducing chemicals into the kiln that vaporize and combine with the clay, color effects can be produced in almost the complete range of the spectrum.

4.4.3.5 Texture

Textures result from the dies or molds used in forming masonry units. The pressure exerted by a steel die normally produces smooth units. Many textures may be applied, however, by mounting at the outlet end of a die an attachment that cuts, scratches, rolls, brushes, or otherwise roughens the surface of the uncut clay column as it leaves the die. The possible textures range from fine through medium to coarse. Some typical textures are smooth, matte with vertical or horizontal markings, rugs, barks, sandmold, waterstruck, and sandstruck (see Fig. 4.4-5).

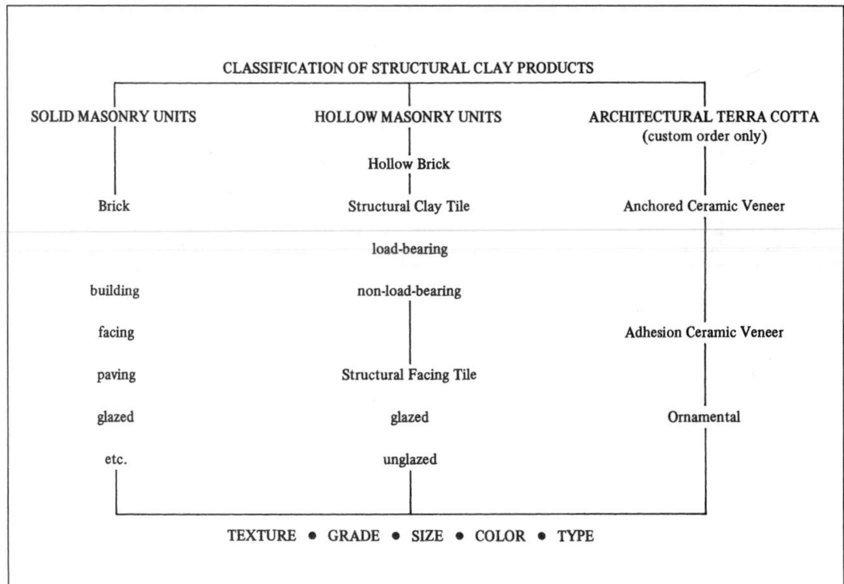

FIGURE 4.4-8 Classification of clay masonry units.

4.4.3.6 Size Variation

The clays used to manufacture brick and clay tile shrink 4.5% to 15% during drying and burning. Terra-cotta clay shrinks 6% to 7%. In general, this shrinkage can be compensated for by increasing die size and adjusting the spacing of cutter wires so that the units will approximate the desired final size after they are burned. Molds are simply made larger to compensate for shrinkage.

A more serious problem exists when units to be used together do not shrink the same amount. Units shrink at differing rates for several reasons. For example, units made with clays having different chemical compositions may not be the same size. In spite of the manufacturer's attempts to limit it, this kind of variation will always occur to at least some extent. It is not possible to have every unit made from exactly the same clay. Large differences in size can be prevented, however, by careful monitoring to ensure that all units to be used together are made from similar clays.

The amount of shrinkage during burning increases with higher temperatures, which are necessary to produce darker shades of color. As a result, when a wide range of color tone is desired, some variation in size between the dark and the light units is inevitable. Size differences can be reduced by selecting masonry units of consistent color throughout.

Manufacturers try to control shrinkage to obtain units of uniform size, but variations in raw materials and kiln temperatures make absolute uniformity impossible. Therefore, specifications should spell out permissible variations in size.

4.4.4 CLASSIFICATIONS OF CLAY MASONRY UNITS

Clay masonry units are classified (Fig. 4.4-8) as solid masonry units, hollow masonry units, and architectural terra-cotta. The many individual products are qualified further as to size, grade, type, color, and texture.

Clay masonry units should conform to the current standards of the American Society for Testing and Materials (ASTM) as follows:

Building (common) brick	ASTM C 62
Facing brick, unglazed	ASTM C 216
Facing brick, ceramic glazed	ASTM C 126
Hollow brick	ASTM C 652
Structural clay tile, load-bearing	ASTM C 34
Structural clay tile, non-load-bearing	ASTM C 56
Structural clay facing tile, unglazed facing tile	ASTM C 212
Structural clay facing tile, ceramic glazed	ASTM C 126
Architectural terra-cotta	None

FIGURE 4.4-9 Standard Nomenclature for Brick Sizes[a]

Unit Designation	Nominal Dimensions, in. w	h	l	Joint Thickness,[b] in.	Specified Dimensions,[c] in. w	h	l	Vertical Coursing
Modular Brick Sizes								
Modular	4	2⅔	8	⅜	3⅝	2¼	7⅝	3C = 8 in.
				½	3½	2¼	7½	
Engineer modular	4	3⅕	8	⅜	3⅝	2¾	7⅝	5C = 16 in.
				½	3½	2³⁄₁₆	7½	
Closure modular	4	4	8	⅜	3⅝	3⅝	7⅝	1C = 4 in.
				½	3½	3½	7½	
Roman	4	2	12	⅜	3⅝	1⅝	11⅝	2C = 4 in.
				½	3½	2¼	7½	
Norman	4	2⅔	12	⅜	3⅝	2¼	11⅝	3C = 8 in.
				½	3½	2¼	11½	
Engineer Norman	4	3⅕	12	⅜	3⅝	2¾	11⅝	5C = 16 in.
				½	3½	2¹³⁄₁₆	11½	
Utility	4	4	12	⅜	3⅝	3⅝	11⅝	1C = 4 in.
				½	3½	3½	11½	
Nonmodular Brick Sizes								
Standard				⅜	3⅝	2¼	8	3C = 8 in.
				½	3½	2¼	8	
Engineer standard				⅜	3⅝	2¾	8	5C = 16 in.
				½	3½	2¹³⁄₁₆	8	
Closure standard				⅜	3⅝	3⅝	8	1C = 4 in.
				½	3½	3½	8	
King				⅜	3	2¾	9⅝	5C = 16 in.
					3	2⅝	9⅝	
Queen				⅜	3	2¾	8	5C = 16 in.

[a]1 in. = 25.4 mm; 1 ft. = 0.3 m.
[b]Common joint sizes used with length and width dimensions. Joint thicknesses of bed joints vary based on vertical coursing and specified unit height.
[c]Specified dimensions may vary within this range from manufacturer to manufacturer.
(Courtesy Brick Institute of America.)

FIGURE 4.4-10 Other Brick Sizes[a]

Nominal Dimensions, in. w	h	l	Joint Thickness,[b] in.	Specified Dimensions,[c] in. w	h	l	Vertical Coursing
Modular Brick Sizes							
4	6	8	⅜	3⅝	5⅝	7⅝	2C = 12 in.
			½	3½	5½	7½	
4	8	8	⅜	3⅝	7⅝	7⅝	1C = 8 in.
			½	3½	7½	7½	
6	3⅕	12	⅜	5⅝	2¾	11⅝	5C = 16 in.
			½	5½	2¹³⁄₁₆	11½	
6	4	12	⅜	5⅝	3⅝	11⅝	1C = 4 in.
			½	5½	3½	11½	
8	4	12	⅜	7⅝	3⅝	11⅝	1C = 4 in.
			½	7½	3½	11½	
8	4	16	⅜	7⅝	3⅝	15⅝	1C = 4 in.
			½	7½	3½	15½	
Nonmodular Brick Sizes							
			⅜	3	2¾	8⅝	5C = 16 in.
				3	2⅝	8⅝	

[a]1 in. = 25.4 mm; 1 ft. = 0.3 m.
[b]Common joint sizes used with length and width dimensions. Joint thicknesses of bed joints vary based on vertical coursing and specified unit height.
[c]Specified dimensions may vary within this range from manufacturer to manufacturer.
(Courtesy Brick Institute of America.)

Pedestrian and light-traffic paving brick — ASTM C 902

ASTM standards do not fix the required size, color, or texture of masonry units for a specific use, but rather cover a range of sizes and a wide variety of colors and textures. When construction documents require that masonry units be of a certain type and grade as defined in an ASTM standard, it is usually necessary to add specific requirements related to the units' size, color, and texture and to the quality of workmanship required and the specific finished appearance desired. This can be done by requesting that these variables match approved samples.

4.4.4.1 Solid Masonry Units

The term *brick* usually refers to a solid masonry unit. Cored bricks, including those with large cores and those with as many as 21 small "pencil" cores, are considered solid masonry units unless the core area exceeds 25% of the unit's total cross-sectional area. Cores are introduced in brick to promote even drying during burning and to reduce their weight. Brick has been produced in many shapes and sizes, including those listed in Figures 4.4-9 and 4.4-10. Most brick, however, is produced in modular sizes having nominal dimensions based on a 4-in. (100 mm) module.

The term *modular coordination* refers to a system of components sized to fit together easily with a minimum of cutting and fitting in the field. Coordinated assemblies often finish with an overall dimension in even units without fractions. Designing masonry units with modular coordination in mind tends to increase productivity during construction and can result in significant cost savings. Masonry units so designed are referred to as *modular units*. Common designations and sizes of brick are shown in Figure 4.4-11. Not many manufacturers produce all the sizes shown, so it is important to determine during the design phase which sizes are available in the locality of the project.

The nominal dimensions of a modular brick are equal to the manufactured (actual) dimensions plus the thickness of the mortar joint for which the unit is designed. In general, joint thicknesses in brick are either ⅜ or ½ in. (9.5 or 12.7 mm).

The manufactured height for modular brick that is designed to be laid with three courses in every 8 in. (203 mm) of height is about 2¼ in. (57.2 mm). This is to en-

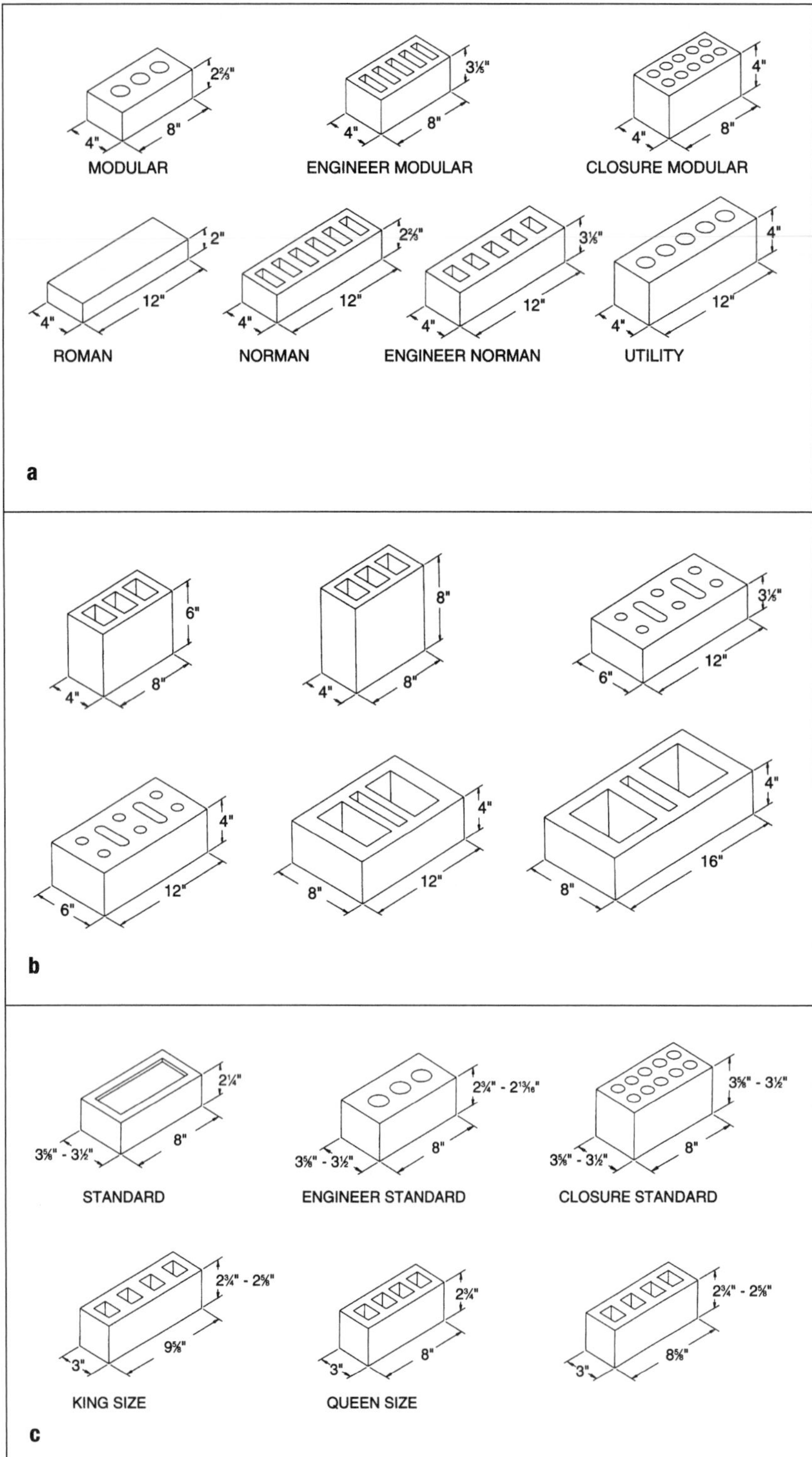

FIGURE 4.4-11 Designations for brick: (a) and (b) modular brick sizes (dimensions are nominal); (c) nonmodular brick sizes (dimensions shown are actual size). (Courtesy Brick Institute of America.)

sure that the vertical coursing in an addition built with modular brick can be made to match the coursing in an existing building constructed with either modular or nonmodular brick. The manufacturers who produce modular brick have agreed to keep the manufactured face height 2¼ in. (57.2 mm), even when the length and thickness of their units are modular. They consider as minimal the differences in mortar bed thickness necessary to maintain the modular coursing of three courses in 8 in. (203 mm).

The brick unit dimensions in all the figures in this chapter are shown with the thickness first, followed by the face dimensions (height and length), a practice that has become the industry standard. Specifications and drawings related to brick should follow this standard order of listing brick dimensions. For example, if the nominal dimensions of a Roman brick are 4 × 2 × 12 in. (100 × 50 × 300 mm) and the units are designed to be laid with ⅜-in. (9.5 mm) mortar joints, their actual size will be 3⅝ × 1⅝ × 11⅝-in. (92.1 × 41.3 × 295.3 mm), with allowance, of course, for permissible size variations in manufacture.

4.4.4.2 Hollow Masonry Units

Hollow masonry units include hollow brick and hollow clay tile. The many types of hollow clay tiles are categorized as either structural clay tile or structural facing tile. In all hollow masonry units the cores, cells, and hollow spaces exceed 25% of the gross cross-sectional area of the unit. Tile designed to be laid with its cells in a horizontal plane is called *side construction* or *horizontal cell tile*; that designed to be laid with its cells placed vertically is called *end construction* or *vertical cell tile*. The size, number, shape, and thickness of cells will vary greatly depending on the manufacturer.

Hollow brick is defined by ASTM C 652 as a hollow clay masonry unit whose net cross-sectional area in every plane parallel to the bearing surface is not less than 60% of the gross cross-sectional area measured in the same plane. Hollow brick units are cored between 25% (the upper limit for solid brick units) and 40%. At the higher limit, they are only 60% solid.

Generally, hollow brick can be used where solid brick is used; it is the same size and looks the same on the outside as solid brick. However, solid brick design standards are not applicable to hollow brick construction. Instead, hollow brick structures should be individually designed according to accepted engineering practice.

Technological improvements in masonry construction and lower production, transportation, and installation costs have given hollow brick advantages over solid

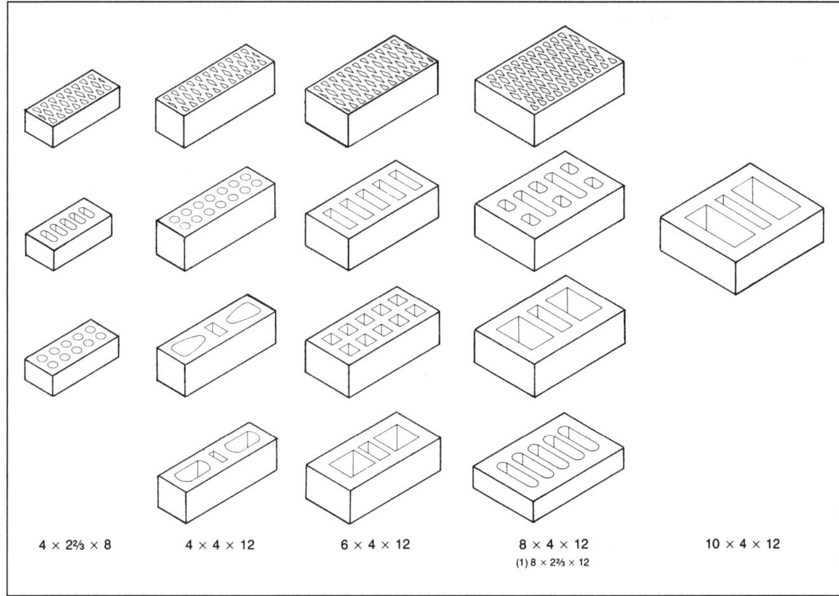

FIGURE 4.4-12 **Typical hollow brick units.** (Courtesy Brick Institute of America.)

brick in both speed of erection and economy of construction. Unfortunately, hollow bricks are not readily available in some parts of the country.

Engineered brick masonry design and construction have created a trend toward larger brick units. Typical hollow brick units are shown in Figure 4.4-12. Their greater core area makes hollow bricks lighter and permits easier handling. Larger sizes permit faster, more economical construction. For example, three units, each 8 in. (200 mm) × 4 in. (100 mm) × 12 in. (300 mm), produce 1 sq. ft. (0.093 m^2) of 8-in. (200 mm)-thick wall, while 13 standard-size brick units would be required.

Structural clay tile is available in both load-bearing and non-load-bearing units. A few of the available shapes and sizes are shown in Figure 4.4-13.

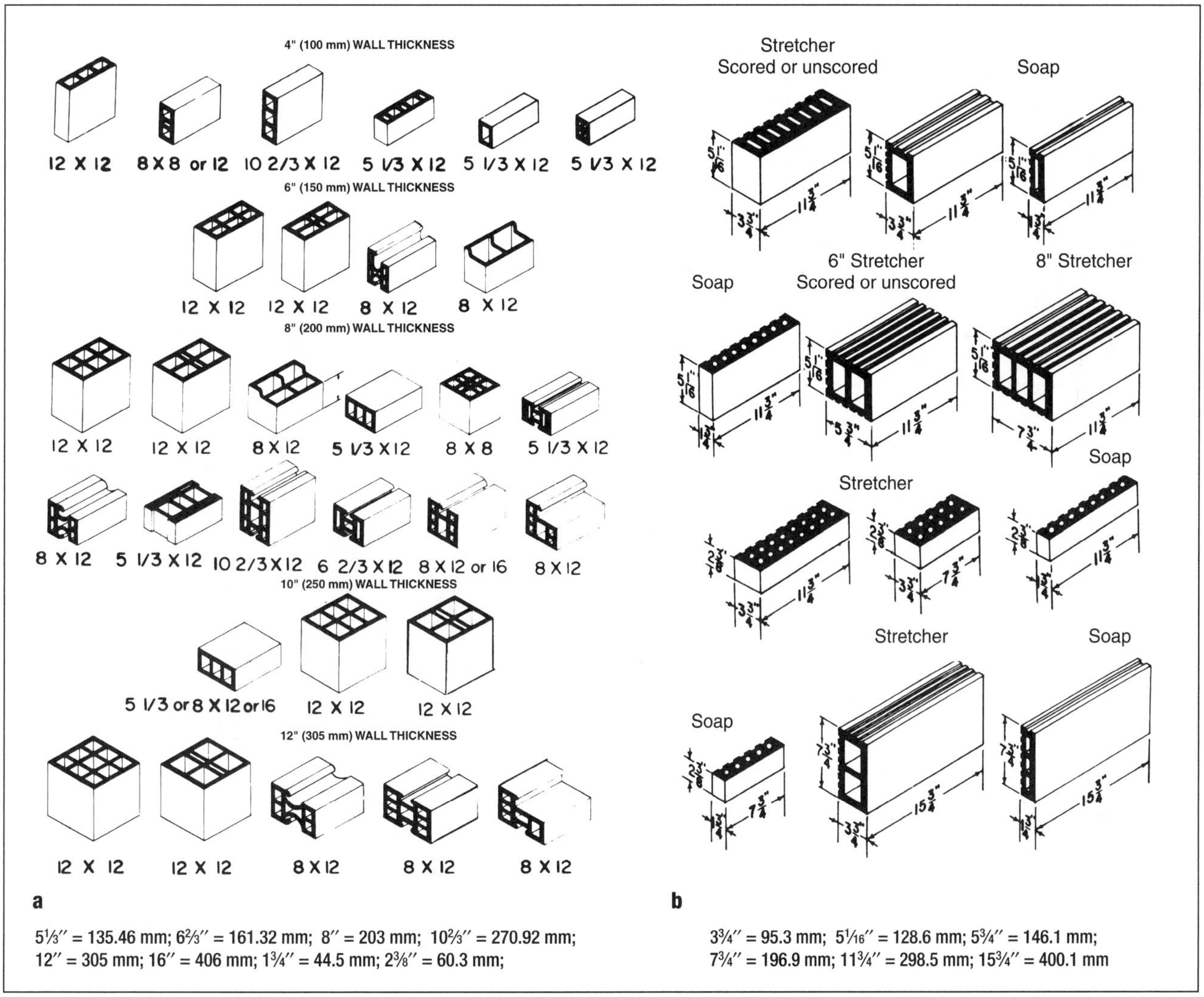

a

5⅓″ = 135.46 mm; 6⅔″ = 161.32 mm; 8″ = 203 mm; 10⅔″ = 270.92 mm; 12″ = 305 mm; 16″ = 406 mm; 1¾″ = 44.5 mm; 2⅜″ = 60.3 mm;

b

3¾″ = 95.3 mm; 5 1/16″ = 128.6 mm; 5¾″ = 146.1 mm; 7¾″ = 196.9 mm; 11¾″ = 298.5 mm; 15¾″ = 400.1 mm

FIGURE 4.4-13 (a) Structural clay tile. (b) Structural facing tile.

FIGURE 4.4-14 Nominal Modular Sizes of Structural Facing Tile*

Face Dimension in Wall		
Thickness	**Height**	**Length**
2, 4, 6, and 8 in. (50, 100, 150, and 200 mm)	4 in. (100 mm)	8 and 12 in. (200 and 300 mm)
2, 4, 6, and 8 in. (50, 100, 150, and 200 mm)	5⅓ in. (135 mm)	8 and 12 in. (200 and 300 mm)
2, 4, 6, and 8 in. (50, 100, 150, and 200 mm)	6 in. (150 mm)	12 in. (300 mm)
2, 4, 6, and 8 in. (50, 100, 150, and 200 mm)	8 in. (200 mm)	12 and 16 in. (300 and 400 mm)

*Nominal sizes include the thickness of the mortar joint for all dimensions.

Structural facing tile is available either glazed or unglazed. Glazed tile is made from a high-grade, light-burning fire clay to which a ceramic glaze has been applied. During burning, the glaze fuses into a glasslike coating on the unit. Unglazed tile is produced from light- or dark-burning clays and shales and may have either a smooth or rough textural finish. The nominal modular sizes of structural facing tile are shown in Figure 4.4-14. Figure 4.4-13 shows a few common and special shapes currently produced.

4.4.4.3 Architectural Terra-Cotta

Terra-cotta has been used to make sewer pipes, fireproofing for metal structural elements, floor vaulting, and flue liners, but it is chiefly used today for copings and decorative cladding. It is made in practically every color possible and in a variety of shapes and forms. The surface may be flat, grooved, or raised, and either smooth or textured. It may be glazed or natural.

One form of terra-cotta used for exterior decoration is called *ceramic veneer*. Made in a process similar to that used for extruded brick, it is limited to items and shapes that can be extruded. There are two types. The first, called *adhesion-type*, is designed to be installed using mortar as one would install ceramic tile. The maximum thickness of this type is 1⅛ in. (28.6 mm). The maximum face size is 540 sq. in. (0.348 m^2), which usually results in pieces that are about 18 by 30 in. (457.2 by 762 mm) or 20 by 27 in. (508 by 686 mm).

The second type of ceramic veneer is called *anchored-type* of ceramic veneer. It is designed to be held in place by anchors tied to reinforcing rods installed in a thick bed of grout. This type of ceramic veneer can be made in larger panels than can the adhesion type. The panels have ribs or scoring on their backs, making their total thickness 2 to 2½ in. (50.8 to 63.5 mm).

The second form of terra-cotta in general use today is tile. The term *tile* is used here to designate those terra-cotta shapes that are produced by casting. Earlier terra-cotta tile often had an egg-crate structure on the back side to lighten the units. Some modern tiles have vertical voids similar to those in concrete block, and some are flat and ribbed on the back, similar to ceramic tile. Tiles may be surfacing units with repetitive patterns, or they may be decorative elements with letters or numbers engraved in them. They may also be sculpted.

4.4.5 STANDARDS CONTROLLING CLAY MASONRY UNITS

Standards developed by ASTM for the many types and grades of brick and clay tile are widely used to control the quality of clay masonry units. Every architect's specification and every masonry contractor's order for brick, hollow brick, structural clay tile, and structural facing tile should include a requirement that the units delivered comply with the applicable ASTM standard. The standard should be specifically named.

ASTM standards classify clay masonry units in types, grades, and classes, based on the weathering index (Fig. 4.4-15), the appearance of the completed wall, and other factors. When ASTM standards are used to control masonry quality, it is essential that the grade, type, and class of the units be spelled out.

4.4.5.1 Weathering Index

The weathering index is a measure of the potential effect of weathering on clay masonry units. The various regions shown in Figure 4.4-15 were established by multiplying the average annual number of freezing cycle days by the annual winter rainfall.

A freezing cycle day is a day when the air temperature passes above or below 32°F (0°C). The average annual number

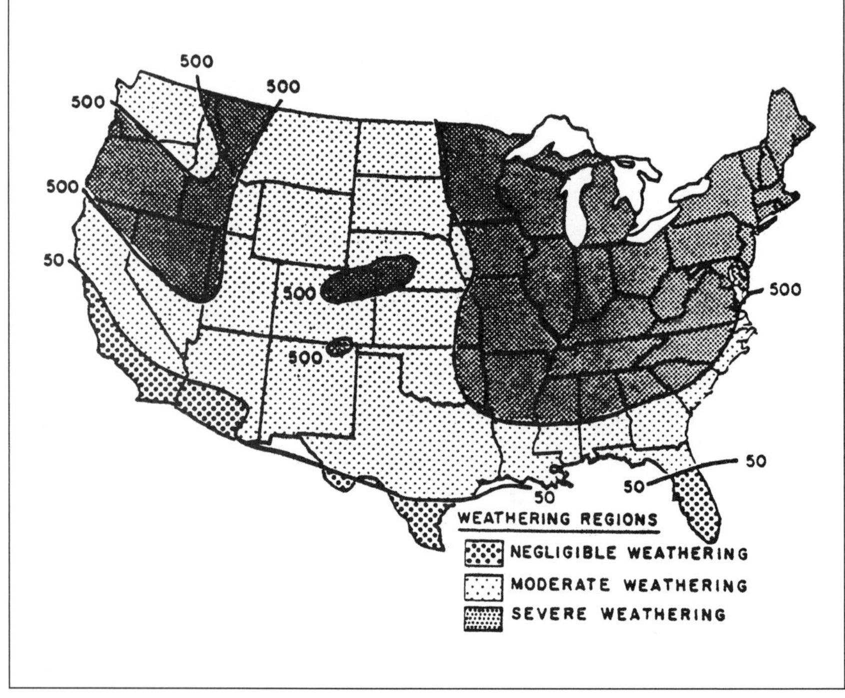

FIGURE 4.4-15 United States weathering index. (Copyright ASTM; reprinted with permission.)

FIGURE 4.4-16 Physical Requirements for Solid Clay Masonry Units by Grade

Designation	Minimum Compressive Strength (brick flatwise), Gross Area, psi		Maximum Water Absorption by 5-Hour Boiling		Maximum Saturation Coefficient*	
	Average of 5 Bricks	Individual	Average of 5 Bricks	Individual	Average of 5 Bricks	Individual
Grade SW	3000	2500	17.0%	20.0%	0.78	0.80
Grade MW	2500	2200	22.0%	25.0%	0.88	0.90
Grade NW	1500	1250	No limit	No limit	No limit	No limit

*The saturation coefficient is the ratio of absorption by 24-hour submersion in cold water to that after 5-hour submersion in boiling water.

of freezing cycle days equals the difference between the mean number of days when the minimum temperature is 32°F (0°C) or below and the mean number of days when the maximum temperature is 32°F (0°C) or below. Winter rainfall is the sum, in inches, of the mean monthly precipitation occurring between the first and last killing frosts in the fall and spring.

A severe weathering region has an index greater than 500; a moderate region, 50 to 500; and a negligible region, less than 50. Figure 4.4-15 shows regions where clay masonry units will be subjected to severe, moderate, and negligible weathering.

Masonry in contact with the earth is considered to be in a severe weathering condition, regardless of the climate.

4.4.5.2 Summary of Physical Characteristics and Uses

A summary of the physical requirements and normal uses for various types of clay masonry units follows. The accompanying figures show more detail and guides to selecting the proper types, classes, and grades of units based on their use.

Building (common) brick. ASTM C 62 covers three grades, SW (severe weathering), MW (moderate weathering), and NW (negligible weathering), based on resistance to weathering (see Fig. 4.4-15). The physical requirements for these grades are given in Figure 4.4-16, the uses in Figure 4.4-17.

Facing brick. ASTM C 216 covers two grades, SW and MW, based on resistance to weathering, and three types, FBX, FBS, and FBA, based on the appearance of the finished wall. FBX units are allowed small dimension tolerances, small distortion tolerances, little chippage, and small color range differences from unit to unit. In fact, they require such a high precision and uniformity that they are seldom specified. FBS units require less precision and uniformity, and FBA units require even less than FBS units. The most common are FBS units. The physical requirements for facing brick are the same as for building brick grades SW and MW (see Fig. 4.4-16). The uses of facing brick based on both weathering and appearance are shown in Figure 4.4-18.

Hollow brick. ASTM C 652 covers two grades, SW and MW, based on resistance to weathering, and four types, HBS, HBX, HBA, and HBB, based on the appearance of the finished wall. The physical requirements for hollow brick are the same as for building brick grades SW and MW (see Fig. 4.4-16). The uses of hollow brick are shown in Figure 4.4-19.

Structural clay tile, load-bearing. ASTM C 34 covers two grades, LBX and

FIGURE 4.4-17 Uses of Building (Common) Brick[a] (ASTM C 62) by Grade

Grade[b] (based on weathering index)	Uses
SW (severe weathering)	Foundations or other structures below grade or in contact with earth, especially when subjected to freezing temperatures
MW (moderate weathering)	Exterior walls and other exposed vertical masonry surfaces above grade
NW (negligible weathering)	All interior masonry or as backup

[a]Units with core areas less than 25% of unit's gross cross-sectional area.
[b]See Fig. 4.4-15.

FIGURE 4.4-18 Uses of Facing Brick[a] (ASTM C 216) by Grade and Type

Grade[b] (based on weathering index)	Uses
SW (severe weathering)	Masonry in contact with earth, especially when subject to freezing temperatures
MW (moderate weathering)	Exterior walls and other exposed masonry above grade
Type (based on appearance of finished wall)	**Uses**
FBX	Masonry where high degree of mechanical perfection, minimum variation in color range, and minimum variation in size are required
FBS	Where wider color ranges and greater size variations are permissible than specified for FBX
FBA	Where architectural effects are desired resulting from nonuniformity in size, color, and texture of units

[a]Units with core areas less than 25% of unit's total cross-sectional area.
[b]See Fig. 4.4-15.

FIGURE 4.4-19 Uses of Hollow Brick[a] (ASTM C 652) by Grade and Type

Grade[b] (based on weathering index)	Uses
SW (severe weathering)	Where a high and uniform degree of resistance to frost action and disintegration by weathering is desired and where brick may be frozen when permeated with water
MW (moderate weathering)	Where a moderate and somewhat nonuniform degree of resistance to frost action is permissible and where brick is unlikely to be permeated with water when exposed to below-freezing temperatures
Type (based on appearance of finished wall)	**Uses**
HBS	Exposed exterior and interior walls where wider color ranges and greater variations in size are permitted than specified for HBX
HBX	Exposed exterior and interior walls where narrow color range and minimum permissible variation in size are required
HBA	Where architectural effects are desired resulting from nonuniformity in size, color, and texture of units
HBB	Where color and texture are not a consideration and greater size variation is permitted than specified for HBX

[a]Units with core area greater than 25% of unit's total cross-sectional area.

[b]See Fig. 4.4-15.

FIGURE 4.4-20 Physical Requirements for Load-Bearing Structural Clay Tile

	Maximum Water Absorption[a] by 1-Hour Boiling, %		Minimum Compressive Strength (based on gross area),[b] psi			
			End Construction Tile		Side Construction Tile	
Grade	Average of 5 Bricks	Individual	Average of 5 Bricks	Individual	Average of 5 Bricks	Individual
LBX	16	19	1400	1000	700	500
LB	25	28	1000	700	700	500

[a]The range in percentage absorption for tile delivered to any one job should be not more than 12.

[b]Gross area of a unit should be determined by multiplying the horizontal face dimension of the unit, as placed in the wall, by its thickness.

(Copyright ASTM; reprinted with permission.)

FIGURE 4.4-21 Uses of Load-Bearing Structural Clay Tile[a] (ASTM C 34) by Grade

Grade[b] (based on weathering index)	Uses
LBX	Masonry exposed to weathering and direct application of stucco
LB	Masonry not exposed to frost action or earth; may be used in exposed masonry if protected with a facing of 3 in. (76 mm) or more of other masonry

[a]Units with core area greater than 25% of unit's total cross-sectional area.

[b]See Fig. 4.4-17.

LB, based on compressive strength and resistance to weathering. The physical requirements for load-bearing structural clay tile are given in Figure 4.4-20, its uses in Figure 4.4-21.

Structural clay tile, unglazed facing tile. ASTM C 212 covers two classes, standard and special duty, based on the thickness of the face shells, and two types, FTX and FTS, based on the appearance of the finished wall. The physical requirements for unglazed structural clay facing tile are given in Figure 4.4-22, its uses in Figure 4.4-23.

Facing brick and structural clay facing tile, ceramic glazed. ASTM C 126 covers ceramic glazed facing brick and other solid units, as well as facing tile having a finish consisting of a ceramic color glaze or a clear glaze. It includes two grades, S (select), for use with comparatively narrow mortar joints, and SS (select sized or ground edge), for use where variation of face dimension must be very small, and two types, I (single-faced units), for use where only one finished face will be exposed, and II (two-faced units), for use where two opposite finished faces will be exposed (Fig. 4.4-24). The physical requirements include minimum compressive strength, tests of finish (imperviousness, chemical resistance, and crazing), limitations on distortion, and limitations on variation in dimensions.

Architectural terra-cotta. There is no ASTM standard for terra-cotta. Terra-cotta units of every type are custom prod-

FIGURE 4.4-22 Physical Requirements for Unglazed Structural Clay Facing Tile

	Maximum Water Absorption			
	By 24-Hour Submersion in Cold Water, %		By 1-Hour Boiling, %	
Type	**Average**	**Individual**	**Average**	**Individual**
FTX	7	9	9	11
FTS	13	16	16	19
	Compressive Strength Based on Gross Area			
	End Construction Tile		**Side Construction Tile**	
Class	**Minimum Average of 5 Tests, psi**	**Individual Minimum psi**	**Minimum Average of 5 Tests, psi**	**Individual Minimum psi**
Standard	1400	1000	700	500
Special duty	2500	2000	1200	1000

FIGURE 4.4-23 Uses of Unglazed Structural Clay Facing Tile* (ASTM C 212) Based on Type

Type	Uses
FTX	Exposed masonry: smooth face, low absorption, and resistance to staining. Has high degree of mechanical perfection, minimal variation in color and face dimensions
FTS	May be smooth or rough textured, with moderate absorption and variation in dimensions, medium color range, and minor surface finish defects

*Units with core area less than 25% of unit's total cross-sectional area.

FIGURE 4.4-24 Uses of Ceramic Glazed Facing Tile and Brick* (ASTM C 126) Based on Grade and Type

Grade	Uses
S (select)	Masonry with narrow mortar joints (¼ in./6.4 mm)
SS (select sized or ground edge)	Masonry where face dimension variation must be very small
Type	**Uses**
I (single-face units)	Masonry where only one finished face will be exposed
II (two-faced units)	Masonry where two opposite finished faces will be exposed

*Units with core area greater than 25% of unit's total cross-sectional area.

ucts. Many stock dies are available, but in general terra-cotta sizes and shapes are produced to meet specific project requirements.

Pedestrian and light-traffic paving brick. ASTM C 902 covers paving units designed to support pedestrian and light vehicular traffic in patios, walkways, floors, plazas, and driveways. It lists three classes, SX, MX, and NX, based on the severity of weather exposure, and three types, I, II, and III, based on the severity of traffic and exposure. The physical requirements for the classes are given in Figure 4.4-25, its uses in Figure 4.4-26.

FIGURE 4.4-25 Physical Requirements of Pedestrian and Light-Traffic Paving Brick[a] (ASTM C 902)

	Minimum Compressive Strength (brick flatwise), Gross Area, psi		Maximum Cold Water Absorption		Maximum Saturation Coefficient[b]	
Designation Class	**Average of 5 Bricks**	**Individual**	**Average of 5 Bricks**	**Individual**	**Average of 5 Bricks**	**Individual**
SX	8000	7000	8%	11%	0.78	0.80
MX	3000	2500	14%	17%	No limit	No limit
NX	3000	2500	No limit	No limit	No limit	No limit

[a]Minimum modulus of rupture values should be considered by the purchaser for uses of brick where support or loading may be severe.

[b]The saturation coefficient is the ratio of absorption by 24-hour submersion in room temperature water to that after 5-hour submersion in boiling water.

FIGURE 4.4-26 Uses of Pedestrian and Light-Traffic Paving Brick (ASTM C 902) Based on Class and Type

Class (based on weathering index)	Uses
SX	Where brick may be frozen while saturated with water
MX	Where resistance to freezing is not a factor
NX	Interior use when an effective sealer, wax, or other suitable surface coating will be applied
Type (based on traffic)	**Uses**
I	Where brick will be exposed to extensive abrasion, such as in driveways and entranceways to public or commercial buildings
II	Where brick will be exposed to intermediate traffic, such as floors in restaurants, stores, and exterior walkways
III	Where brick will be exposed to low traffic, such as floors or patios in single-family houses

Paving brick is also classified in ASTM 902 by use in the following applications:

- PS: general-use classification for brick to be used with mortar- and grout-filled joints or in patterns, such as running bond, laid without mortar where a close tolerance in the units is necessary
- PX: for units to be laid without mortar in patterns that require a high degree of dimensional accuracy
- PA: for units with special appearance characteristics where dimensional tolerances are not restricted

4.5 Concrete Masonry Units

Concrete masonry units are molded concrete units used in building construction as an integral part of the structure, as facing for or filler panels between structural elements, and to construct partitions. Concrete masonry units include concrete brick, hollow concrete block, slump block, split-face block, and other special units. They do not include concrete splash blocks, manhole blocks, paving units, sills, or similar concrete products, or cast stone.

Solid and hollow masonry units are made from a relatively dry mixture of portland cement (and sometimes other cementitious materials), aggregates, and water. Admixtures are also often used. The ingredients are thoroughly mixed, then molded into the desired shapes by compaction and vibration, both by machine. They are then cured under controlled conditions of moisture and temperature. After a period of aging, during which the required strength, moisture content, and other desired properties are attained, the units are ready for use. Because the raw materials and manufacturing methods used both influence the strength and other properties of concrete masonry units in a finished wall, a general understanding of both will help in the selection and use of such units.

4.5.1 RAW MATERIALS

The raw materials used to manufacture concrete masonry units include cementitious materials, water, aggregates, and various admixtures.

4.5.1.1 Cementitious Materials

The primary cementitious material used in concrete masonry units is portland cement. Most of the portland cement used is normal portland cement without an air-entraining admixture. To a lesser extent, high-early-strength portland cement and both normal and high-early-strength portland cements containing an air-entraining admixture are also used. Units formed with high-early-strength cement have greater strength during the early stages of setting and curing than their counterparts made with normal portland cement. They are also less likely to break during handling and delivery.

Portland blast-furnace slag cements, fly ash, silica flour, and other pozzolanic or hydraulic materials may also be used as part of the cementing medium. Pozzolanic materials, such as fly ash, have little or no cementitious value in themselves, but combine at ordinary temperatures with lime liberated from portland cement during hydration to form cementing compounds. This reaction is slow, which contributes to strength development after several months. Materials such as silica flour react similarly at the elevated temperatures used in high-pressure (autoclave) curing. Both fly ash and silica flour are used as a portion of the cementing material with portland cement. The correct proportioning of these materials depends on the curing temperatures and type of aggregate used.

Although its use is not common, natural cement can be used satisfactorily to replace up to 20% of the portland cement in concrete masonry units.

The types of cementitious materials used in concrete masonry units and the applicable ASTM standards covering them are shown in Figure 4.5-1.

4.5.1.2 Aggregates

Aggregates constitute about 90% of the weight of a concrete masonry unit and therefore have an important effect on the desirable properties as well as the cost of the finished units. In general, local availability is the primary determinant in the use of any one type of aggregate. The applicable ASTM standards for aggregates are shown in Figure 4.5-1.

FIGURE 4.5-1 Standards Covering the Raw Materials Used in Concrete Masonry Units

Cementitious Materials	
Portland cement	ASTM C 150
Portland blast-furnace cement	ASTM C 595
Fly ash	ASTM C 618
Hydrated lime	ASTM C 207 — Type S
Siliceous and pozzolanic	No ASTM standards. Siliceous and pozzolanic materials should be shown by test or experience not to be detrimental to the durability of the concrete.
Aggregates	
Normal weight	ASTM C 33
Lightweight	ASTM C 331
Admixtures	
Air-entraining agents	ASTM C 260
Accelerators and calcium chloride	ASTM D 98
Coloring pigments, integral water repellents, etc.	No ASTM standards. These admixtures should be previously established as suitable for use in concrete or be shown by test or experience not to be detrimental to the durability.

Aggregates are classified according to their weight as either dense or lightweight. Dense aggregates, which are also called normal-weight aggregates, include sand, gravel, crushed limestone, and air-cooled slag. Lightweight aggregates include expanded shale, clay, slag, coal cinders, pumice, and scoria. The effect of aggregate type on weight, strength, and other characteristics of concrete block is shown in Figure 4.5-2. Desirable aggregate properties include:

- Toughness, hardness, and strength to resist impact, abrasion, and loading
- Durability to resist freezing and thawing and expansion and contraction resulting from moisture and temperature changes
- Uniform gradation of fine and coarse aggregate sizes to produce an economical, workable mixture and uniform appearance—the maximum size of aggregate should not exceed one-third of the smallest section of a block's shell
- Cleanliness and lack of foreign particles that might impair strength or cause surface imperfections

4.5.1.3 Admixtures

Only a few of the numerous admixtures marketed for use in concrete mixes have been found beneficial or desirable in concrete masonry units. The admixtures used include air-entraining agents, accelerators, coloring pigments, integral water repellents, and a few others. The applicable ASTM standards are shown in Figure

FIGURE 4.5-2 Properties of Concrete Masonry Units Based on the Aggregates Used

Aggregate (graded; ⅜ in. (9.5 mm) to 0) Type	Density (air-dry), lb./cu. ft. (kg/m³)	Weight, lb./cu. ft. (kg/m³) of Concrete	Weight, 8-in. × 8-in. × 16-in. (200 mm × 200 mm × 400 mm) Unit	Compressive Strength (gross area), psi* (MPa)	Water Absorption, lb./cu. ft. (kg/m³) of Concrete	Thermal Expansion Coefficient (per °F) × 10^{-6}
Normal weight						
Sand and gravel	130–145 (2082.41–2242.59)	135 (2162.5)	40 (18.14)	1200–1800 (8.3–12.4)	7–10 (112.13–160.19)	5.0
Limestone	120–140 (1922.22–2242.59)	135 (2162.5)	40 (18.14)	1100–1800 (7.6–12.4)	8–12 (128.15–192.22)	5.0
Air-cooled slag	100–125 (1601.85–2002.31)	120 (1922.22)	35 (15.88)	1100–1500 (7.6–10.3)	9–13 (144.17–208.24)	4.6
Lightweight						
Expanded shale	75–90 (1201.39–1441.67)	85 (1361.57)	25 (11.34)	1000–1500 (7–10.3)	12–15 (192.22–240.29)	4.5
Expanded slag	80–105 (1281.48–1681.94)	95 (1521.76)	28 (12.70)	700–1200 (4.8–8.3)	12–16 (192.22–256.3)	4.0
Cinders	80–105 (1281.48–1681.94)	95 (1521.76)	28 (12.70)	700–1000 (4.8–7)	12–18 (192.22–288.33)	4.5
Pumice	60–85 (961.11–1361.57)	75 (1201.39)	22 (9.98)	700–900 (4.8–6.2)	13–19 (208.24–304.35)	4.0
Scoria	75–100 (1201.39–1601.85)	95 (1521.76)	28 (12.70)	700–1200 (4.8–8.3)	12–16 (192.22–256.3)	4.0

*Multiply these values by 1.80 to obtain approximate corresponding values of strength of the concrete (strength of unit on net area).

4.5-1. Admixtures should be investigated individually to determine whether they are appropriate and to ensure that they will perform satisfactorily under the individual manufacturing plant conditions.

Air-entraining agents increase the plasticity and workability of a concrete mixture and distribute minute air bubbles uniformly throughout the hardened concrete, which increases its ability to withstand frost action. The addition of an air-entraining admixture to the mix for concrete masonry units usually permits greater compaction of the mix and produces a denser unit. In addition, the appearance of the units is improved, a closer reproduction of the contours of the mold is obtained, and breakage of freshly molded units is reduced.

Water-repelling agents, such as metallic stearates, are used to a limited extent. They are somewhat effective in reducing the rate of absorption and capillarity of the units.

Accelerators, the most common type being calcium chloride, are used by some producers to speed up the hardening of units during cold weather operations.

Workability agents, many of which are of the air-entraining type, are sometimes used to correct the harshness of concrete mixes made with poorly graded aggregates.

4.5.2 MANUFACTURE

Block plants vary in size, layout, equipment, type of automation, and manufacturing details, but their basic operations are essentially the same. They include the steps shown in Figure 4.5-3 and discussed in the following paragraphs.

4.5.2.1 Receiving Raw Materials

Materials are delivered in bulk to a plant by truck or rail and elevated into overhead steel storage bins above the batching and mixing area (Fig. 4.5-4). At some plants, cement is delivered in special tank trucks or railroad cars and forced by compressed air through pipes directly into the overhead cement storage bin. Different aggregates are handled and stored separately to permit proper blending and proportioning.

FIGURE 4.5-4 Placing materials into an overhead storage bin above a batching and mixing machine.

4.5.2.2 Batching and Mixing

Materials in the concrete mix are proportioned by weight with a *weigh batcher*. The weigh batcher is placed under the storage bins, and the correct amount of each material for a batch is discharged into the batcher. The weigh batcher is then positioned over the mixer and the materials are discharged into the mixer, where they are mixed with a controlled quantity of water for a period of 6 to 8 minutes (Fig. 4.5-5). An electronic water meter senses the water contained in the aggregate and adjusts the amount of mixing water. Admixtures in liquid form may be added as part of the mixing water.

FIGURE 4.5-5 A weigh batcher positioned over a mixer.

4.5.2.3 Molding Units

The mixed batch is then discharged into a hopper above the molding machine. Controlled quantities of the mix are fed into the machine for molding into desired shapes and sizes. Concrete is consolidated in the mold by a combination of vibration and pressure (Fig. 4.5-6). Typically, three 8 × 8 × 16 in. (200 × 200 × 400 mm) con-

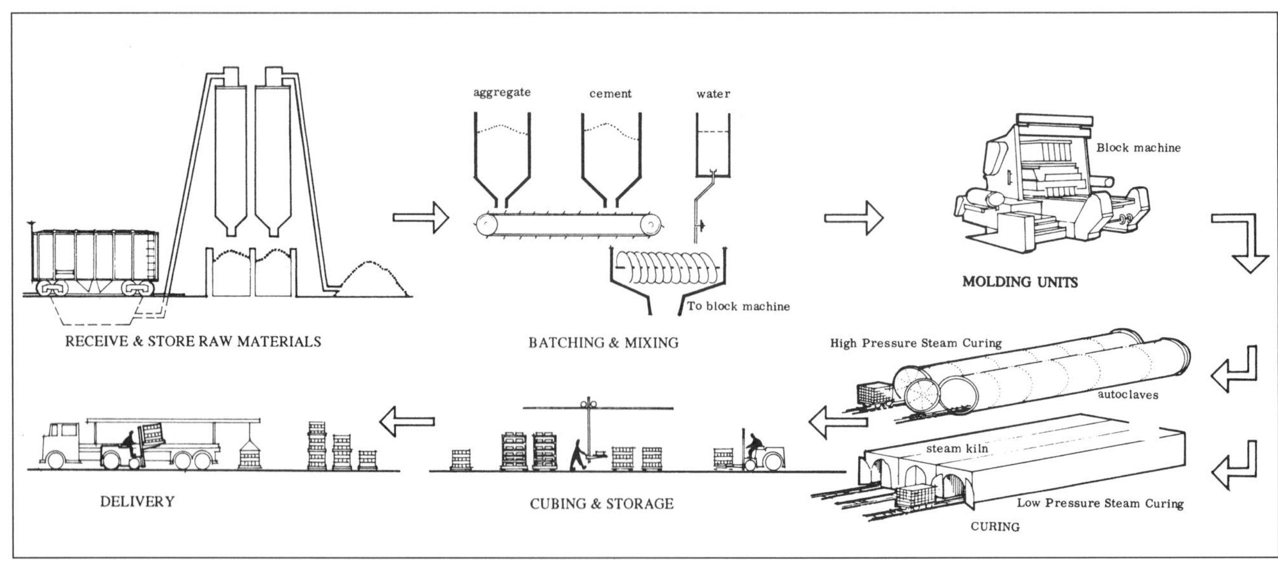

FIGURE 4.5-3 Flowchart for manufacturing concrete masonry units.

FIGURE 4.5-6 A molding machine.

FIGURE 4.5-8 An autoclave.

FIGURE 4.5-9 Cured concrete masonry units being assembled into a cube.

crete blocks, or their equivalent in volume of concrete, are molded per machine cycle at a speed of about 5 cycles per minute.

The freshly molded (*green*) concrete units emerge from the machines on steel pallets and are placed on a steel curing rack. This operation, called *off bearing*, may be done manually (Fig. 4.5-7) or with an automated rack loader. At this stage the units are sufficiently dry to hold their shape outside the mold, but are fragile and can be broken by hand pressure.

4.5.2.4 Curing

Occasionally, where the climate is favorable, units are moist cured at normal temperatures of 70° to 100°F (21.1° to 37.8°C). Most units, however, are cured using either low-pressure or high-pressure steam.

LOW-PRESSURE STEAM CURING

Steam curing at atmospheric pressure accelerates the hardening process. This *low-pressure steam curing*, which is done in a kiln, accounts for most concrete masonry units produced in the United States. The main benefit of steam curing is the rapid strength gain in the units, which allows them to be placed in inventory within hours after they are molded. Two to 4 days after molding, the compressive strength of low-pressure steam-cured block will be 90% or more of the final ultimate strength. Moist-cured block cured at normal temperatures, on the other hand, develop strength slowly and have less than 40% of their ultimate strength 4 days after molding. Low-pressure steam curing produces units of lighter color than is usually obtained with moist curing.

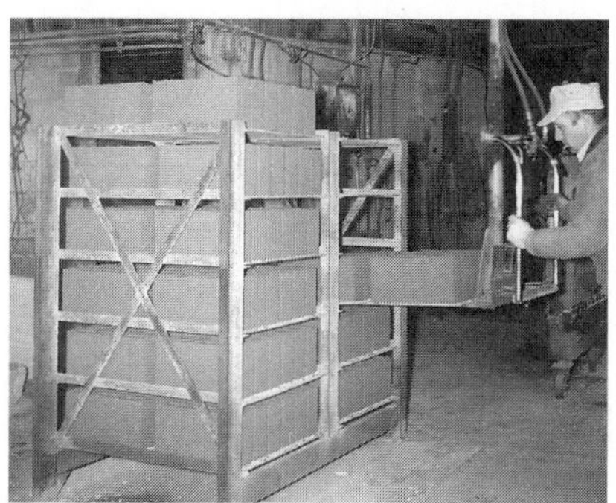

FIGURE 4.5-7 Freshly molded concrete units being placed on a steel curing rack.

HIGH-PRESSURE STEAM CURING

Units cured by application of saturated steam in an autoclave (Fig. 4.5-8) under a pressure between 125 and 150 psi (460 and 553 kPa/mm^2) are called autoclaved units. Autoclaved units, which were widely available in the past, are not made in many parts of the country today and are available only at a high price as a result of shipping costs.

The principal benefits of high-pressure steam curing are high early strength and greater stabilization against volume changes caused by varying moisture conditions. Within 24 hours after molding, usable units can be produced that have a permanent compressive strength equal to that of moist-cured units only after they have been cured for a full 28 days.

4.5.2.5 Cubing and Storage

When curing has been completed, the curing racks are conveyed to a *cubing station*, where the units are assembled into cubes, usually consisting of six layers of 15 to 18 blocks, 8 × 8 × 16 in. (200 × 200 × 400 mm), or an equivalent number of other sized units (Fig. 4.5-9). Cubes are stacked so that two low openings are left to accommodate the forks of a forklift. They are then secured together with steel bands and transported to the storage area, where they are placed in inventory in stacks three or four cubes high. Depending on the curing methods used, the units remain in inventory anywhere from a few days to several weeks before they are released for use.

4.5.2.6 Delivery

When units are ready to be released, they are loaded onto flat-bed trucks by the cube, using a forklift or other loading device (Fig. 4.5-10a). At a construction project's site, cubes of units are picked up one or two at a time by a mechanical unloader and set at convenient locations for use (Fig. 4.5-10b).

a

b

FIGURE 4.5-10 (a) Units being loaded on a flat-bed truck at the manufacturing plant. (b) A truck equipped with a mechanical device for unloading units at a construction site.

4.5.3 PHYSICAL PROPERTIES

The physical properties of concrete masonry units are determined largely by the properties of the hardened cement paste and the aggregate. In general, much of the technical knowledge relating to concrete is applicable to concrete masonry units. However, differences do occur as a result of variations in mix composition and consistency, the consolidation method used, textural requirements, method of curing, and other factors.

Concrete masonry units usually have a substantially lower cement factor than structural concrete. Between 2.75 and 4 sacks of cement are used per cubic yard in masonry units, as compared with 4 and 7 bags in concrete. The water-cement ratio is much lower too, running between 2 and 4 gallons (7.57 and 15.14 L) per sack of cement in concrete masonry and between 5 and 7 gallons (18.93 and 26.5 L) per sack in concrete. The aggregate in concrete masonry is graded finer than that used in concrete, with the largest size seldom exceeding ⅜ in. (9.5 mm). Concrete masonry units are more often made with porous or lightweight materials than is concrete. Concrete masonry units contain a relatively large volume of interparticle void spaces (spaces between aggregate particles that are not filled with cement paste). They are also cured at higher temperatures than is cast-in-place structural concrete.

Typical ranges in compressive strength, water absorption, density, and other physical properties of commercial-grade hollow concrete masonry units are shown in Figure 4.5-2.

4.5.3.1 Compressive Strength

The strength of concrete masonry units is difficult to predict from mix data, because a simple water-cement ratio is not valid for harsh, dry mixes, and because each type of aggregate exhibits different characteristics during mixing. Unlike cast-in-place concrete, in concrete masonry units the highest strengths are obtained from the wettest mixes. However, the maximum amount of mixing water is not usually added, in order to cut down on breakage of freshly molded units during handling before curing. Dry mixes do not compact as well as wet mixes and therefore have lower strengths. Through experimentation, each manufacturing plant develops mixes and techniques to produce units of optimum strength compatible with other desirable properties.

The compressive strengths shown in Figure 4.5-2 are based on the gross bearing area of the concrete masonry unit, including core spaces. The compressive strength of the unit based on net area, excluding core space, is about 1.8 times the values shown. Solid face and veneer units, such as concrete brick and split block, when made with sand and gravel or crushed stone aggregate, will generally exceed 3000 psi (21 MPa) in compression.

As shown in Figure 4.5-2, units made with lightweight aggregates have lower compressive strengths than units made with sand and gravel, limestone, or air-cooled slag. Major influences on compressive strength are:

- Type and gradation of aggregate
- Type and amount of cementitious material
- Degree of compaction attained in molding the units
- Moisture content and temperature of the units at the time of testing

4.5.3.2 Tensile Strength, Flexural Strength, and Modulus of Elasticity

The tensile strength, flexural strength, and modulus of elasticity of concrete masonry units vary in proportion to their compressive strength. Tensile strength ranges from 7% to 10% of the compressive strength, flexural strength from 15% to 20% of compressive strength, and the modulus of elasticity from 300 to 1200 times the compressive strength.

4.5.3.3 Water Absorption

Absorption tests are used to determine the density of a concrete. Absorption is a measure of the number of pounds (kilograms) of water in each cubic foot (cubic meter) of concrete. In hollow concrete masonry units this measure varies over a wide range, from as little as 4 or 5 lb./cu. ft. (64.1 to 80.1 kg/m^3) for the heaviest sand and gravel units to as much as 20 lb./cu. ft. (320.4 kg/m^3) for the very porous, lightweight aggregate types. Solid units made with sand and gravel normally have less than 7 lb./cu. ft. (112 kg/m^3) absorption. The absorption of the aggregate, which may vary from 1% to 5% of the dry weight of aggregate for dense aggregates to as much as 30% or more for the lightest aggregate, has a strong influence on the absorption of the units.

The porosity and pore structure of a concrete influence other properties, such as permeability, thermal conductivity, and sound absorption, but the influence is not always predictable, especially in comparing units made with different types of aggregates. High water absorption is not a desirable or purposely incorporated property, but it is accepted as a natural consequence when properties such as lightness in weight, higher sound absorption, or thermal insulation are desired.

A high initial rate of absorption (*suction rate*) results when concrete contains a large portion of relatively large interconnected pores and voids. Units with a high suction rate combined with high absorption will have high permeability to water, air, and sound and may also have less resistance to frost action. On the other hand, unconnected air-filled pores present in lightweight aggregate and in air-entrained cement paste impart the advantages of porosity yet minimize permeability to water, air, and sound.

An abnormally high suction rate adversely affects the structural bond of mortar to units, but this can be minimized by using mortars with high water retention properties.

Concrete masonry intended for use in exterior walls and that will not be painted should have low absorption, and mortar joints should be tooled for weathertightness.

4.5.3.4 Volume Changes

Concrete masonry units undergo small dimensional changes due to changes in temperature, changes in moisture content, and a chemical reaction called *carbonation*. Of these factors, only moisture changes can be conveniently controlled appreciably. This can be done through preshrinking the units and adequately drying them before they are used.

Carbonation causes irreversible shrinkage in masonry units as a result of a chemical reaction within the concrete that occurs when it absorbs carbon dioxide from the air. Although few test data exist, it appears that the magnitude of this change under certain conditions and over extended periods may approximate that caused by changes in moisture content.

Concrete masonry units expand when they are heated and contract upon cooling. These volume changes are fully reversible. Units return to their original shape when the temperature returns to its original level. The coefficient of thermal expansion for concrete masonry units depends largely on the coefficient of the aggregate, as it constitutes about 90% of the concrete volume. Coefficients of thermal expansion for concrete masonry units are shown in Figure 4.5-2.

Moisture changes cause concrete masonry units to expand when wetted and contract when dried. Moisture volume changes that occur when units are new may not be fully reversible because of concrete's tendency to assume a permanent contraction during the first few cycles of wetting and drying. Later moisture volume changes are usually fully reversible, however.

An important contributor to the development of cracks in concrete masonry walls is the volume change in the units from original shrinkage during drying. If the units are laid up in a wall before they have become dimensionally stable and have been allowed to shrink, tensile stresses will develop where the wall is restrained from shrinking, and cracking can be expected. Drying shrinkage can be greatly reduced with proper curing and drying so that when units are placed, their moisture content is in equilibrium with the surrounding air—that is, dried down to the average air-dry condition to which the finished wall will be exposed. For a given aggregate type, autoclave curing will sometimes reduce drying by as much as 50%, as compared with that of units cured with low-pressure steam or cured at normal temperatures.

The amount of residual shrinkage, or movement, a concrete unit will undergo because of moisture change after placement is the product of its shrinkage potential and its moisture loss after placement. In other words, residual shrinkage = shrinkage potential × moisture loss. Shrinkage potential depends on manufacturing technique, raw materials used, age of unit, and other factors. Moisture loss depends on the moisture content of the units when laid and the relative humidity and temperature of the drying environment. Although measurable in the laboratory, the small magnitudes make it impractical to measure shrinkage on a routine basis as a means of demonstrating compliance with specifications. Instead, shrinkage is controlled by limiting the moisture content of the units at the time of delivery.

Figure 4.5-11 shows the correlation between moisture content and drying environment, expressed in terms of relative humidity. Masonry units complying with the limits for moisture content shown will have a tolerable residual shrinkage, and

FIGURE 4.5-11 Maximum Moisture Content of Type I Units

Linear Shrinkage Classification,[a] %	Maximum Moisture Content, % of Total Absorption (average 5 units) — Average Annual Relative Humidity,[b] %		
	Over 75	**75–50**	**Under 50**
Up to 0.03	45	40	35
0.03 to 0.045	40	35	30
0.045 to 0.065	35	30	25

[a]As determined by ASTM C 426 test for drying shrinkage of concrete within previous 12 months.

[b]As reported by the U.S. Weather Service nearest source of manufacture.

FIGURE 4.5-12 Linear Shrinkage of Concrete Masonry Units

Aggregate Type	Steam-Curing Method	Average Linear Shrinkage,* %
Normal weight	High pressure	0.019
	Low pressure	0.027
Lightweight	High pressure	0.023
	Low pressure	0.042
Pumice	High pressure	0.039
	Low pressure	0.063

*As determined by test, ASTM C 426.

cracking of the wall will be minimized. The linear shrinkage classifications shown depend on aggregate type and curing method. Figure 4.5-12 shows the average values of linear shrinkage. The term *normal weight* refers to units made with sand, gravel, limestone, and other dense aggregates. *Lightweight* refers to all common lightweight aggregate types except pumice. Aggregate classifications are not necessarily exclusive and may overlap since many lightweight blocks contain minor amounts of normal-weight aggregate.

4.5.3.5 Surface Texture

Texture may be greatly varied (Fig. 4.5-13) to satisfy aesthetic requirements or to suit a desired physical requirement, such as a coarse texture to provide mechanical bond for direct application of plaster. Normal-weight and lightweight aggregate units can be given many different surface textures by controlling (1) the gradation of the aggregates, (2) the amount of water used, and (3) the degree of compaction at the time of molding. In addition, the surface of a unit may be ground to produce the exact texture desired (Fig. 4.5-14), and the face mold can also be changed to produce numerous contours in the face shell of the unit (Fig. 4.5-15).

Units with open surface textures offer some value in absorbing sound. Exposed concrete masonry walls built with ordinary commercial units will absorb between 18% and 68% of the incident sound. The textures shown in Figure 4.5-13 illustrate this range. Surface finishes, such as paint, tend to close the surface pores and thereby reduce the absorption significantly.

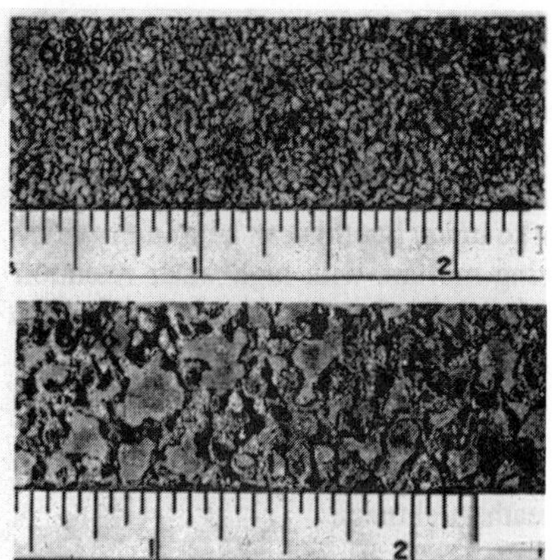

FIGURE 4.5-13 The surface texture of a concrete masonry unit can affect its insulating value and sound absorption characteristics. As shown here, surface textures may range from fine to coarse.

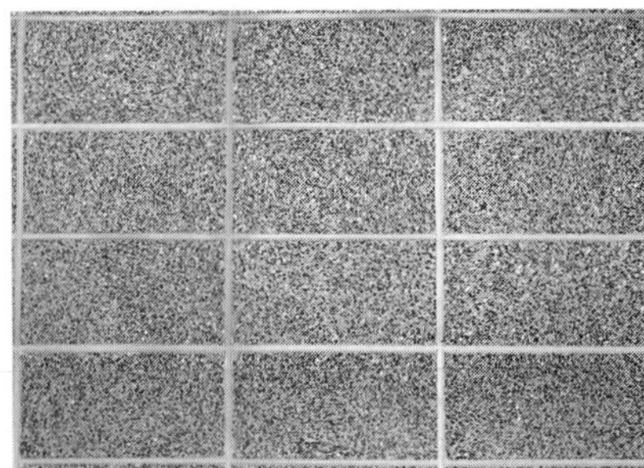

FIGURE 4.5-14 Ground-face units laid in stack bond.

4.5.3.6 Color

The usual aggregates and portland cements used in commercial production produce units with a range of color from off-white to gunmetal gray, and from light tan to brownish shades. A wider color range, including blues, browns, buffs, greens, reds, and slate grays, can be obtained using pure mineral oxide pigments singly or in combination, mixed integrally with the concrete before it is molded.

4.5.4 CLASSIFICATION OF CONCRETE MASONRY UNITS

Concrete masonry units can be classified as concrete brick, consisting of solid units (building brick and slump brick); concrete block, consisting of both solid units (load-bearing and non-load-bearing) and hollow units (load-bearing and non-load-bearing); and special units (split-faced block, faced block, and decorative block).

4.5.4.1 Concrete Brick

This category includes concrete building brick and slump brick.

FIGURE 4.5-15 The decorative concrete blocks shown are called *Shadowal.* They may be laid in other patterns to produce a variety of effects.

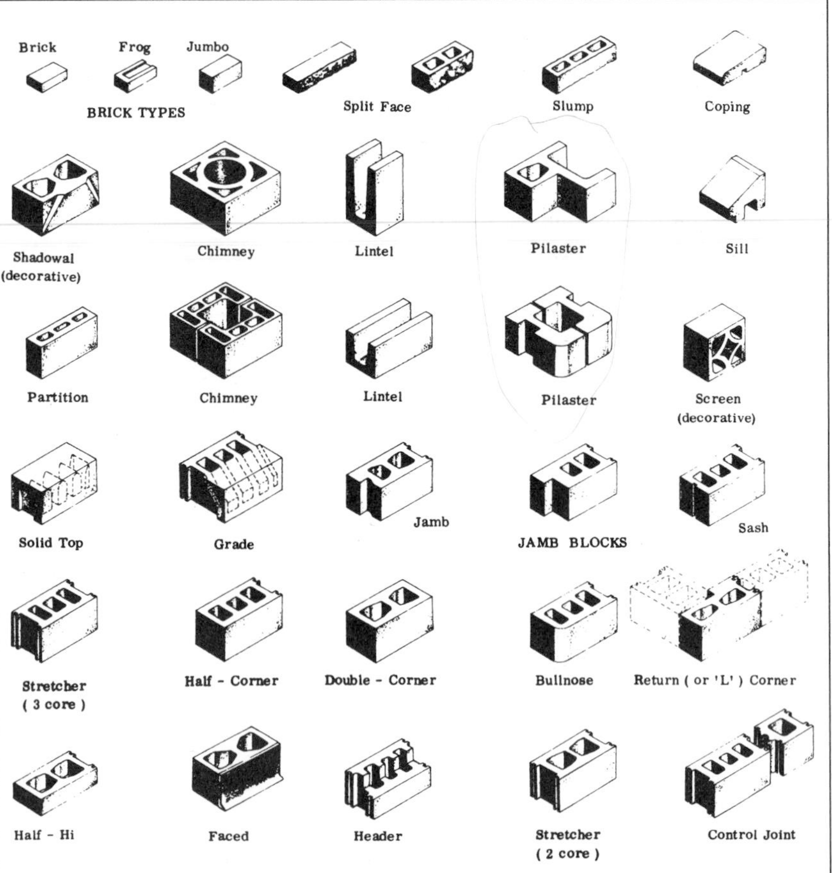

FIGURE 4.5-16 Typical shapes of concrete masonry units.

BUILDING BRICK

Building bricks are either completely solid or have a shallow depression called a *frog* (Fig. 4.5-16). The frog's purpose is to reduce weight and provide for better mechanical bond when these units are laid in mortar. Concrete building bricks are modular units sized to finish 4 in. (101.6 mm) in width and 8 in. (203.2 mm) in length when laid with ³⁄₈-in. (9.5 mm)-thick mortar joints. The thickness varies, but in most units is such that by slightly varying the bed joints a mason can lay the units with three bricks and three mortar joints in every 8 in. (203.2 mm) of wall height. Some manufacturers produce oversized jumbo (see Fig. 4.5-16) and double-brick units.

SLUMP BRICK

The concrete mixture used in making slump brick has a consistency that causes the units to sag or *slump* when they are removed from the molds, resulting in irregularly faced units that vary in height, surface texture, and appearance (Fig. 4.5-17). The slump may occur from the effects of gravity alone or may be enhanced by a mechanical device that presses down on the units after they have been removed from the molds. Slump brick units are used when their special and unusual appearance is desired.

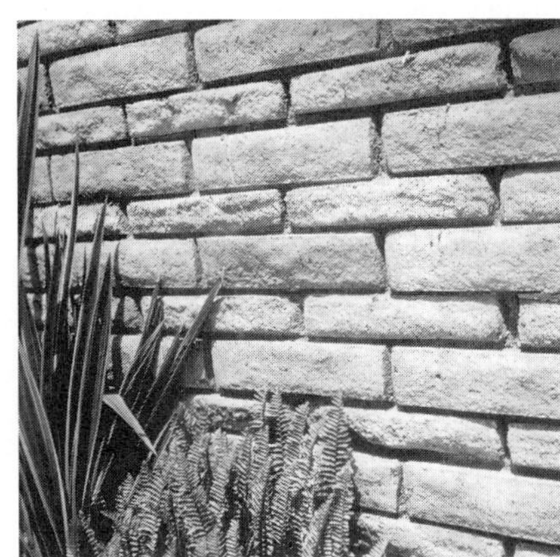

FIGURE 4.5-17 Slump blocks and slump bricks can produce bold, irregular wall textures.

4.5.4.2 Concrete Block

Concrete blocks are either solid load-bearing units, hollow load-bearing units, or hollow non-load-bearing units. Regardless of the grade or type of the unit (see Section 4.5.5) or whether it is load-bearing or non-load-bearing, a solid unit has a core area less than 25% of total cross-sectional area of the unit, while a hollow unit has a core area greater than 25% of the total cross-sectional area of the unit.

Some typical shapes of concrete blocks are shown in Figure 4.5-16, but it is seldom possible to obtain all of them in a given locality. Most producers also make half-length and half-height units for use with the full-sized units they manufacture. Concrete blocks are made in modular sizes for use with ⅜-in. (9.5 mm)-wide head and bed joints. Therefore, an 8 × 8 × 16 in. (200 × 200 × 400 mm) block is actually only 7⅝ × 7⅝ × 15⅝ in. (193.7 × 193.7 × 396.9 mm). The typical dimensions and component parts of a three-core corner unit are shown in Figure 4.5-18. Face shells and webs increase in thickness from the bottom of the unit to the top to permit easy removal of the core form during manufacture and to provide greater bedding area for mortar in the bed joints. The end flanges of most units have ¾-in. (19.1 mm)-wide by 3/16-in. (4.8 mm)-deep mortar grooves to increase transverse strength in a wall laid up with them and to provide a vertical mortar key, which helps create a weathertight joint.

Figure 4.5-18 shows a three-core unit, but units are also made with two cores. These offer some advantages over the three-core design. The principal advantage is the resulting graduated increase of the face shell thickness at the center web, which increases the strength of the unit in tension and reduces the tendency of units to crack due to drying shrinkage and temperature changes. This is particularly true when units are laid in running bond, because the head joints of the course above and below a block occur over the thickened center web rather than over an open core, as is the case with three-core blocks. Two-core blocks have three webs instead of four, which reduces heat conduction from 3% to 4%. They are also about 4 lb. (1.8 kg) lighter in weight than the same sized three-core block. And, finally, they have more vertical space to accommodate reinforcement and conduit or other utilities.

The uses of some of the available types of concrete blocks follow.

Corner blocks have one flush end for use in pilasters and piers and on exposed corners.

Bullnose blocks have one or more small radius-rounded corners. They are used instead of square-edged corner units where a rounded corner is desirable.

Jamb or sash blocks are used to facilitate the installation of windows, louvers, and other opening fillers.

Solid top blocks have solid tops for use as a bearing surface in the finishing course of a foundation wall.

Header blocks have a recess to receive the header unit in a masonry bonded wall.

Lintel blocks are U- or W-shaped. They are used to construct horizontal lintels or beams.

Partition blocks are usually 4 or 6 in. thick. They are used to build non-load-bearing partitions and walls.

Return (L) corner blocks are used in 6-in. (150 mm)-, 10-in. (250 mm)-, and 12-in. (300 mm)-thick walls at corners to maintain a consistent horizontal coursing with the appearance of full-length units and half-length units without cutting of the corner units.

Control joint blocks of various types are used to construct vertical shear-type control joints.

4.5.4.3 Special Units

Some special units (see Fig. 4.5-16) are stock units, and others are custom made. Some of the types available follow.

Split-face blocks are solid or hollow units that are produced in parallel pairs and fractured (split) lengthwise by machine after they have hardened to produce two units, each with a rough surface texture on one face (Fig. 4.5-19). Color variations are obtained by introducing mineral colors and by using aggregates of different gradation and color in the concrete mixture. The units are laid in a wall with the fractured faces exposed. Units are available in random sizes, which are often larger than brick sizes, so that broken joint patterns can be used.

Slump blocks are similar to slump brick but larger (see Fig. 4.5-17).

Faced blocks are units with ceramic glazed, plastic, polished, or ground faces. These facings are applied in separate operations on cured concrete masonry units. One type of glazed unit uses a glazing material that is a thermosetting, resinous binder combined with glass silica and color pigments or colored granules. This mixture, in a semiliquid state, is cast onto each block in individual molds. The units are then heat treated to set the glaze and bind it integrally into the pores of the concrete unit.

Decorative blocks are manufactured in many different types and configurations. Some have beveled face shell recesses. Others have surfaces with more than one plane (see Fig. 4.5-15). And still others have clear rectangular or curved holes through them. Different effects can be created by using the same block design in different pattern bonds. Standard units can also be used to produce decorative effects.

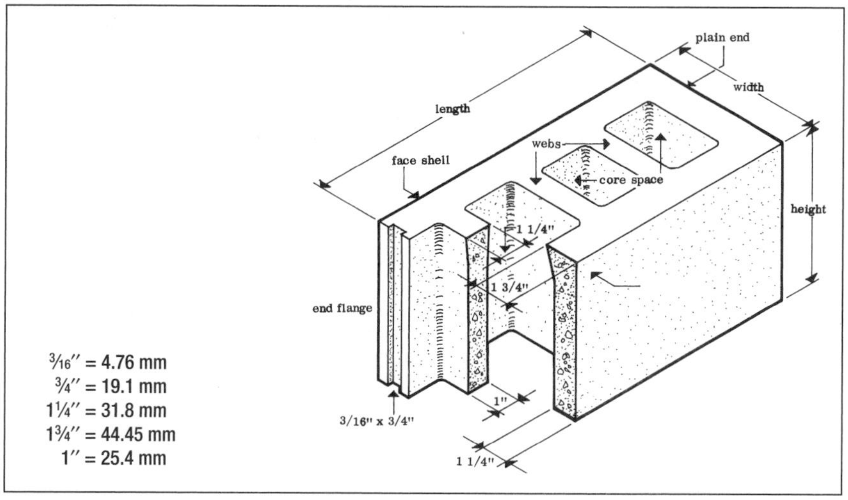

3/16″ = 4.76 mm
¾″ = 19.1 mm
1¼″ = 31.8 mm
1¾″ = 44.45 mm
1″ = 25.4 mm

FIGURE 4.5-18 A typical 8 in. (200 mm) × 8 in. (200 mm) × 16 in. (400 mm) concrete corner block.

FIGURE 4.5-19 Split-face blocks have a rough and natural stonelike texture.

Chimney blocks are designed for use with a flue lining in chimneys.

Pilaster blocks are used in plain or reinforced pilasters and columns.

4.5.5 STANDARDS CONTROLLING CONCRETE MASONRY UNITS

The four types of concrete masonry units that are controlled by ASTM standards are listed with the applicable standards in Figure 4.5-20.

4.5.5.1 Grades

Concrete masonry units intended for load-bearing construction are classified in ASTM standards as Grade N and Grade S. Grade N units are suitable for general use. They may be used in exterior walls, whether below or above grade, that may be exposed to moisture or weather, and for interior walls and backup.

Grade S units are suitable for use only where they will not be exposed to the weather or other moisture. They may not be used below grade. They may be used in exterior walls only when a weather-protective coating is applied over them. They may be unprotected in walls that are not exposed to the weather or other moisture.

4.5.5.2 Types

Concrete building brick and both load-bearing and non-load-bearing concrete block can be manufactured with a controlled moisture content. *Moisture-controlled* units are designated by ASTM as Type I units. They are available in both grades N and S. The two designations are written together (N-I or S-I). The moisture content of Type I units is regulated during manufacture to conform with the limits established by the applicable ASTM standard. Type II units as designated in the ASTM standards are not restricted to a definite moisture content.

The maximum moisture contents for Type I units, based on various average annual relative humidities and the linear shrinkage classification of the masonry units, are shown in Figure 4.5-11. The selection should be based on the local humidity conditions as reported by the U.S. Weather Service nearest the place where the masonry units are to be manufactured and the shrinkage classification of the particular masonry units used.

Units with limited moisture content should be used, in most cases, to minimize in-place shrinkage of masonry units and the wall cracks that result from it. As shown in Figure 4.5-11, when the construction is located in an area of low relative humidity and a low equilibrium moisture condition is expected, the maximum moisture content at the time of use should be lower than when the construction is located in a less arid region. A lower maximum moisture content is also required when the shrinkage potential of the masonry unit is greater. The shrinkage classification for different units can be determined from Figure 4.5-12.

In addition to regulating moisture content, ASTM standards also establish requirements for the strength and water absorption of masonry units (see Fig. 4.5-20). The effect of water absorption and other properties on masonry construction is described in Section 4.5.3.

FIGURE 4.5-20 Strength and Water Absorption Requirements for Concrete Masonry Units

		Minimum Compressive Strength on Average Gross Area, psi		Maximum Water Absorption for Units of Different Densities, lb./cu. ft.			
				Lightweight Concrete		Medium-Weight Concrete	Normal-Weight Concrete
Unit Type and Standard	Grade*	Average of 3 Units	Individual Unit	Less than 85 lb./cu. ft.	Less than 105 lb./cu. ft.	105 to 125 lb./cu. ft.	More than 125 lb./cu. ft.
Hollow load-bearing units, ASTM C 90	N-I, N-II	1000	800	—	18	15	13
	S-I, S-II	700	600	20	—	—	—
Solid load-bearing units, ASTM C 145	N-I, N-II	1800	1500	—	18	15	13
	S-I, S-II	1200	1000	20	—	—	—
Concrete building brick, ASTM C 55	N-I, N-II	3500	3000	15	15	13	10
	S-I, S-II	2500	2000	18	18	15	13
Hollow non-load-bearing block, ASTM C 129	—	350	300	—	—	—	—

*Grades S-I and S-II are limited to walls not exposed to the weather and to exterior above-grade walls protected with weather-resistant coatings. Grades N-I and N-II may be used below grade as well as above; protective coatings are recommended below grade and where required above grade.

4.6 Unit Masonry Design

Masonry walls provide excellent structural performance and offer long-term, low-maintenance finishes. They are one basic element of buildings of all kinds, from enormous skyscrapers to single-family dwellings. They are used as part of the structural bearing system in small and large structures and as part of the enclosing and finishing system of many large and small steel-, concrete-, and wood-framed buildings. To be successful, masonry wall design, selection, and construction must take into consideration the required functional performance level, loading conditions, allowable costs, exposure, type of occupancy, and desired finished appearance. These considerations are often influenced by the systems and finishes used in other construction elements.

To function adequately, a masonry wall must:

- Support the applied *dead load* (the weight of the wall and of the other building elements it carries, such as floors and roofs);
- Resist forces applied directly to the wall, such as those that result from wind and earthquakes;
- Support *live loads* transmitted into the wall by other building elements that are supported by the wall, such as floors and roofs;
- Be weathertight to adequately control the flow of heat, moisture, air, and water vapor;
- Produce satisfactory levels of sound and visual privacy;
- Have an appropriate fire resistance;
- Accommodate heating, air conditioning, electrical, and plumbing equipment;
- Be suitable for the application of various finish materials;
- Be economical;
- Permit installation of doors, windows, louvers, and other opening closers.

When requirements other than the above exist, the materials, design, and construction methods described in this chapter may be inadequate.

Wall systems must safely support superimposed vertical loads and resist horizontal and racking loads due to lateral forces, such as wind and earthquake, and, where they are load-bearing, transmit all loads to the foundation. Masonry walls on small structures are rarely specifically engineered. Ordinarily, building codes and rules of thumb based on established practices determine the materials and methods used in their construction.

The costs of a masonry wall include in-place costs, maintenance costs, and building operating costs associated with the masonry. *In-place costs* are the sum of the costs of materials, labor, and overhead required to erect a wall. The selection of a particular wall type will affect other costs, such as the cost of installing insulation, vapor retarders, and siding, other finishes, and for providing openings for windows and doors. The time and costs required to enclose a building and provide its heating, air conditioning, electrical, and plumbing systems may also be influenced by the wall system selected. The costs of maintaining the masonry and the costs of maintaining the heating, ventilating, and air conditioning systems that are attributable to a wall system, including heat loss or gain through the construction, have an important influence on the ultimate cost of a wall and should not be overlooked if long-term economy is to be achieved.

The aesthetics of a building are influenced by the location, size, and type of doors, windows, and other openings that form a part of the building's walls. Moreover, the wall construction should be compatible with the roof and floor systems used in the construction. Masonry walls are usually left exposed on their exterior sides, but they may be finished with stucco or other coatings. Interior surfaces can be finished with applications such as plaster or gypsum board or can be left unfinished.

Many of the industry's currently recognized minimum requirements for masonry wall design are based on experience and observation of the performance of masonry assemblies over many years. It naturally follows that building codes and published standards often require minimum wall thicknesses and maximum lateral support spacings, instead of demanding that walls be designed using theoretical analysis of the forces and stresses involved.

Most of the recommendations for the design of masonry walls in this section are based on the provisions of ACI 530, ASCE 5, TMS 402 and ACI 530.1, ASCE 6, TMS 602, which have been developed jointly by the American Concrete Institute (ACI) and the American Society of Civil Engineers (ASCE), and by the Masonry Society (TMS). These documents have been adopted by reference in the Standard Building Code published by the Southern Building Code Congress International and in the BOCA National Building Code published by the Building Officials and Code Administrators International, Inc. These standards consolidate and supersede several masonry codes that were the standards used for many years, but which are no longer valid.

Section 4.2 of this book explains how to choose the correct mortar type for the use and exposure of a wall. Sections 4.4 and 4.5 explain how to select the right masonry units. Section 4.7 illustrates the techniques of proper masonry installation. This section (4.6) describes the various types of masonry wall assemblies and their properties when built, and relates many details essential to satisfactory design. Together, the sections of this chapter explain how to produce masonry construction of suitable quality.

The need for insulating masonry walls and the materials and application methods generally used for that insulation are discussed in Section 7.3.

4.6.1 WALL TYPES

Masonry walls may be classified as solid walls, cavity walls, veneered walls, and reinforced walls (Fig. 4.6-1). Each type may be constructed of either solid or hollow masonry units. Solid and cavity walls are built to be either load-bearing or non-load-bearing; veneered walls are non-load-bearing; reinforced walls are load-bearing. Figure 4.6-2 shows one load-bearing condition; Figure 4.6-3 shows another. No wall type is suitable for all exposures or loading conditions, nor is any one type appropriate for all localities or types of occupancies. The final selection of a wall type should be made only after weighing the various requirements of function, economy, and appearance.

4.6.1.1 Solid Walls

A solid wall is built up of units laid close together, with all joints between units filled solidly with mortar. Solid walls may

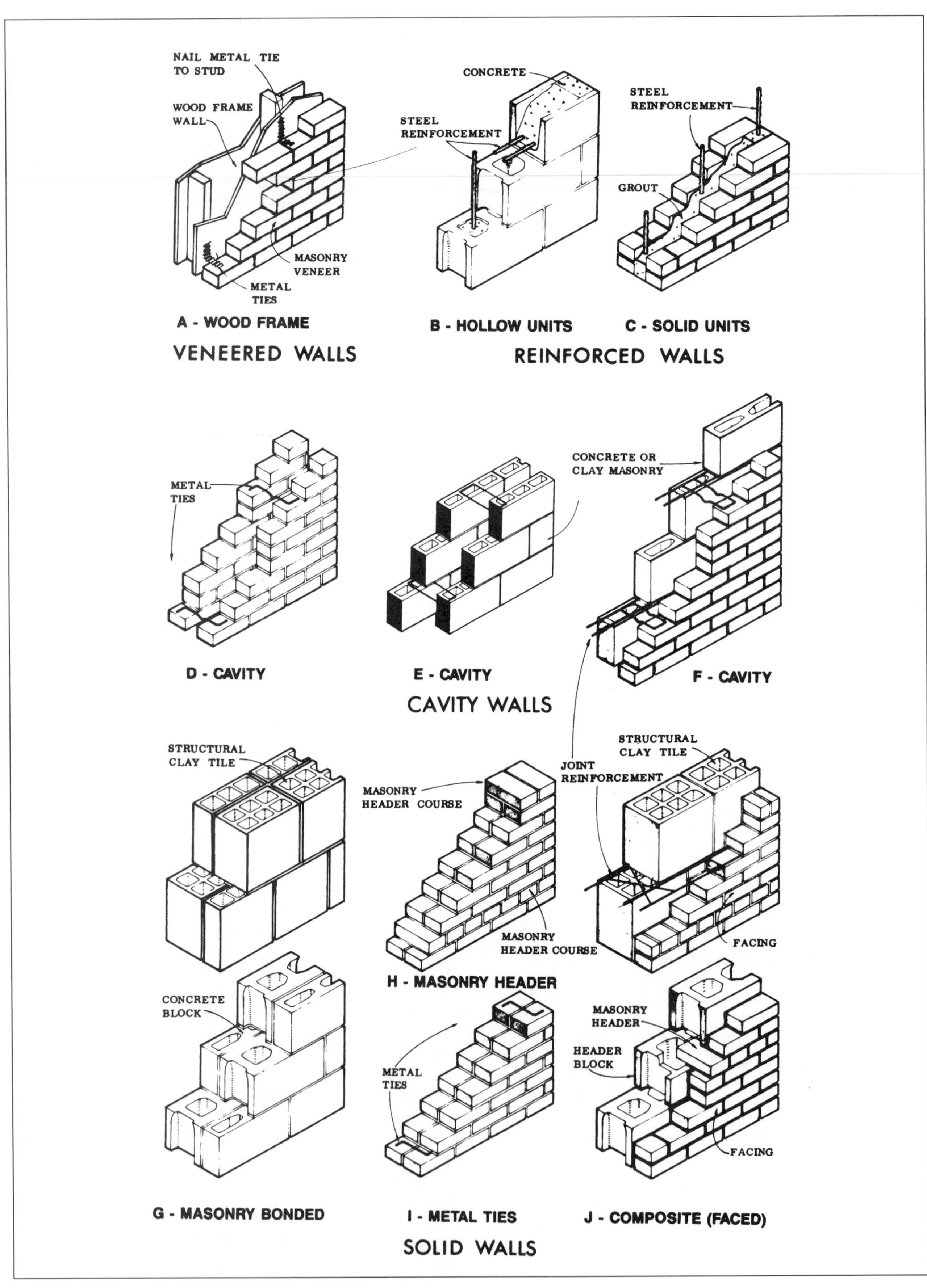

FIGURE 4.6-1 Masonry wall types.

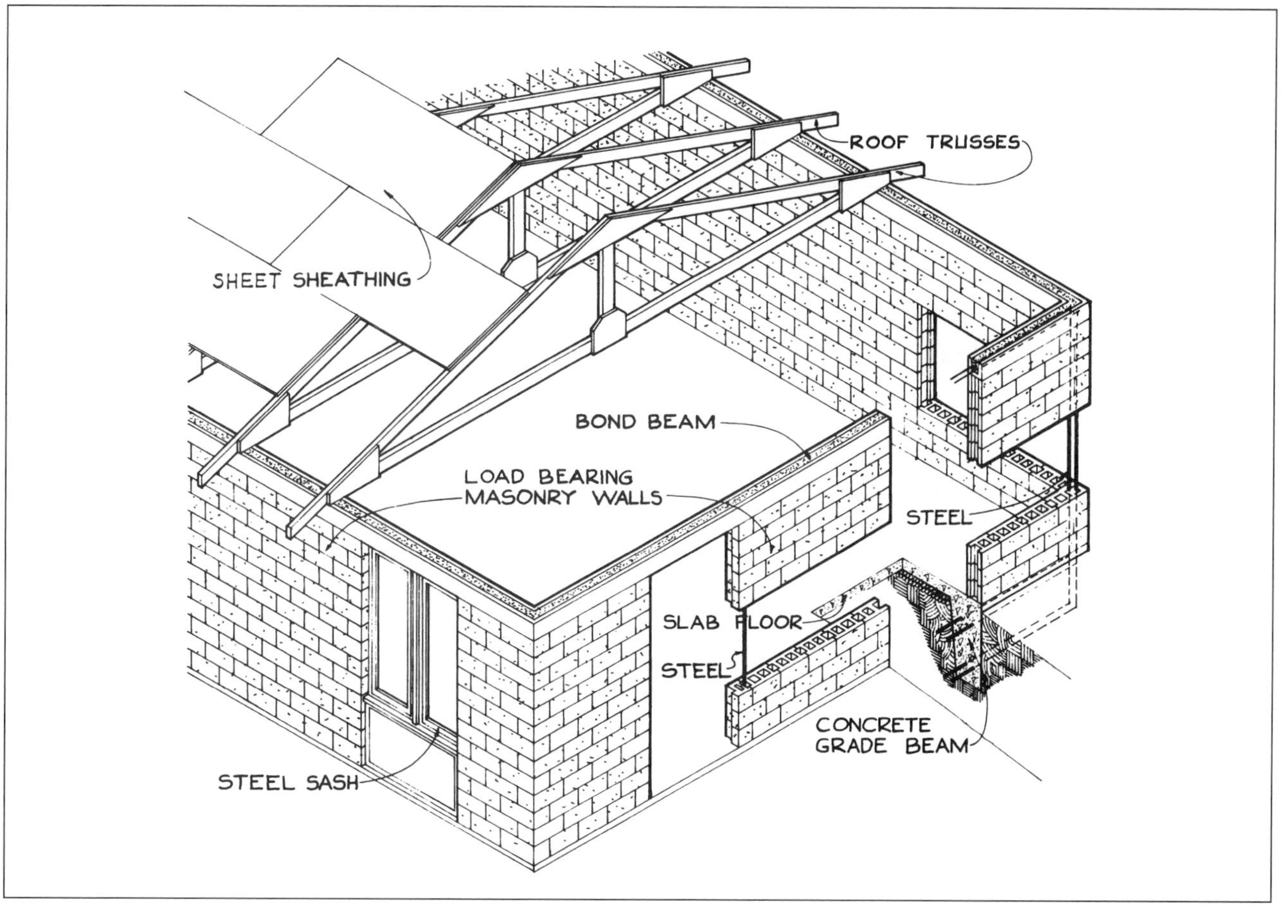

FIGURE 4.6-2 Many types of clay and concrete unit masonry load-bearing walls are possible. This drawing shows concrete masonry walls that have been reinforced to resist lateral forces.

FIGURE 4.6-3 This building has load-bearing masonry walls. The floor framing is supported directly on the masonry.

be built of either solid or hollow masonry units (or both in combination) in any required thickness and are used for either load-bearing or non-load-bearing construction. The structural bond of solid walls is provided by metal ties, masonry headers, or joint reinforcement, as shown in Figure 4.6-1. Walls that utilize masonry headers or have units overlapping in alternate courses (see Fig. 4.6-1G) are called masonry bonded. These are used in areas of slight weather exposure, for economical foundation walls, and for exterior load-bearing walls. There is a possibility that masonry headers may break due to wall movement and may not, therefore, provide a bond as good as has previously been supposed.

Solid walls may consist of solid units of brick or concrete brick and block; hollow units of concrete block, hollow brick, or structural clay tile; or may be a composite (faced) construction consisting of facing and backup units of different materials bonded so that both facing and backup are load bearing.

Typical details of 8-in. (200 mm) solid brick walls and composite walls are shown in Figure 4.6-1H, 4.6-1I, and 4.6-1J. Solid and composite walls bonded with masonry headers are less resistant to rain penetration than metal-tied walls and should be used only in areas where they will not be exposed to wind and rain (Fig. 4.6-4) (see also Section 4.6.3). The addition of furring to the inner wall reduces

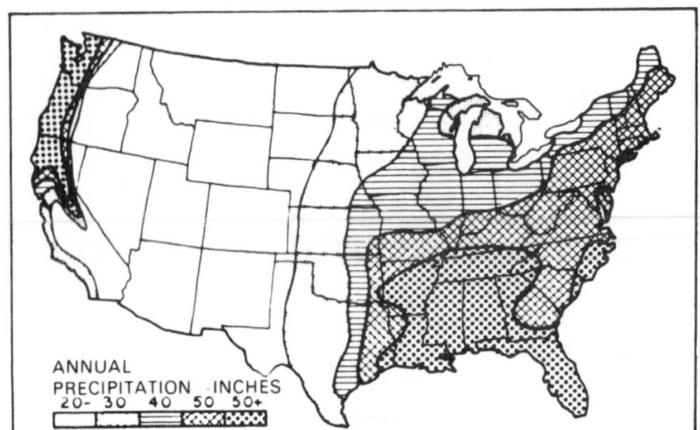

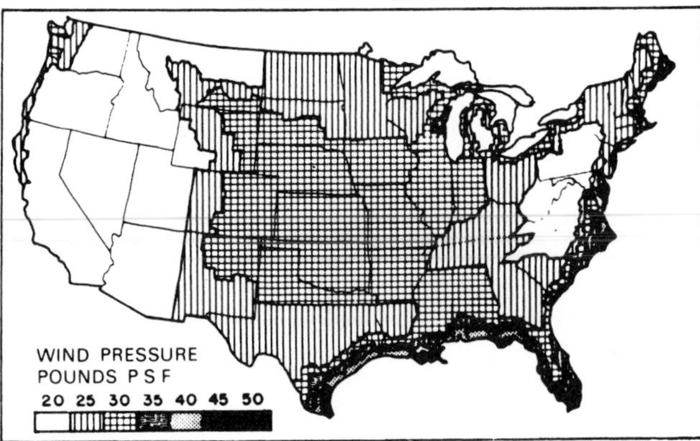

FIGURE 4.6-4 **The weather exposure of masonry units and assemblies is based on precipitation and wind pressure. These maps show the annual precipitation rates and wind pressures that may be expected in the United States.** (Courtesy Brick Institute of America.)

through-the-wall moisture penetration and permits use in areas of severe exposure. In areas subjected to hurricanes or wind-driven rains, the exterior surface of concrete block walls is usually stuccoed to resist rain penetration.

Nationally recognized building codes permit the use of exterior load-bearing 6-in. (150 mm)-thick masonry walls for 1-story, single-family dwellings where the wall height does not exceed 9 ft. (2743 mm) to the eaves or 15 ft. (4572 mm) to the peak of the gable. Figure 4.6-5 is a typical wall section using SCR brick. SCR brick is a solid unit designed to form a nominal 6-in. (150 mm)-thick wall laid up in running bond with full ½-in. (12.7 mm)-thick bed and head joints. The length and width of the SCR brick make it possible to turn corners and maintain the half-bond. An air space, formed by 2 × 2 (50 × 50) wood furring strips is flashed at the base of the wall to provide a barrier to moisture penetration. The furred space also permits installation of electrical wiring and outlets and provides room for wall insulation if it is required. Similar construction using solid or hollow concrete masonry units may also be used.

Where solid masonry walls are used, insulation and mechanical equipment are often installed within furred spaces on the interior side of the wall. On below-grade basement and foundation walls, insulation is often placed on the exterior side of the wall.

4.6.1.2 Cavity Walls

A cavity wall consists of two wythes of either solid or hollow masonry, each of which is at least 4 in. (100 mm) thick, separated by a continuous air space not less than 2 in. (50.8 mm) or more than 4½ in. (114 mm) wide, and bonded together with metal ties or joint reinforcement (see Figs. 4.6-1D, E, and F, and 4.6-6). The exterior wythe may be brick, hollow brick, or concrete masonry units. The interior wythe may be brick, hollow brick, structural clay tile, or solid or hollow concrete masonry units (Fig. 4.6-7). The exterior wythe is usually a nominal 4 in. (100 mm) thickness. The interior wythe may be 4, 6, or 8 in. (100, 150, or 200 mm), depending on the height and length of the wall and the loads to be carried. The nominal overall thickness will be 10, 12, or 14 in. (250, 305, or 356 mm) when the air space is

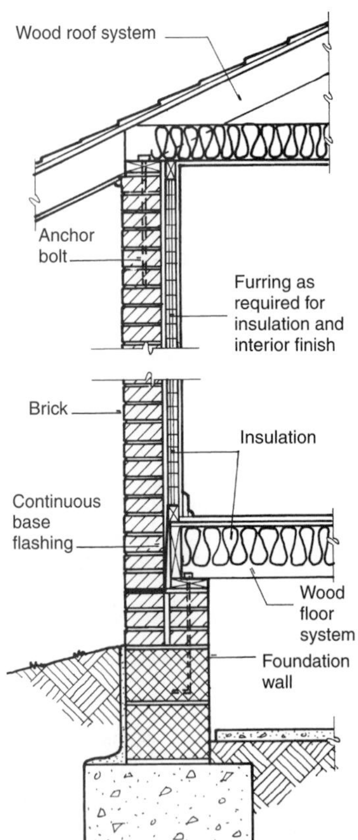

FIGURE 4.6-5 **Section through a nominal 6-in.-thick load-bearing wall.** (Drawing by HLS.)

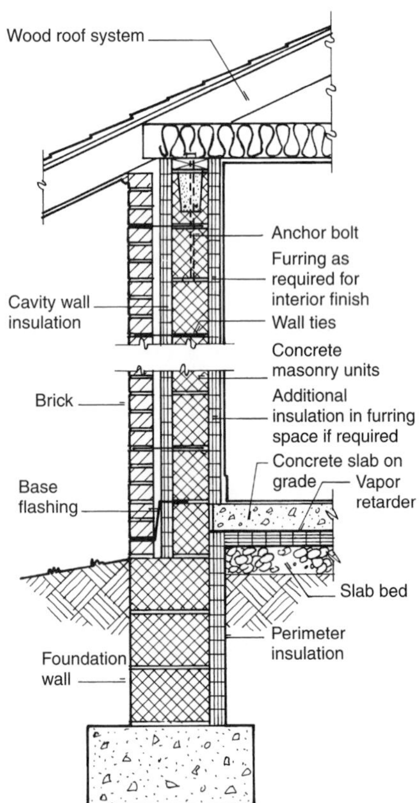

FIGURE 4.6-6 **Section through a nominal 12-in.-thick load-bearing cavity wall.** (Drawing by HLS.)

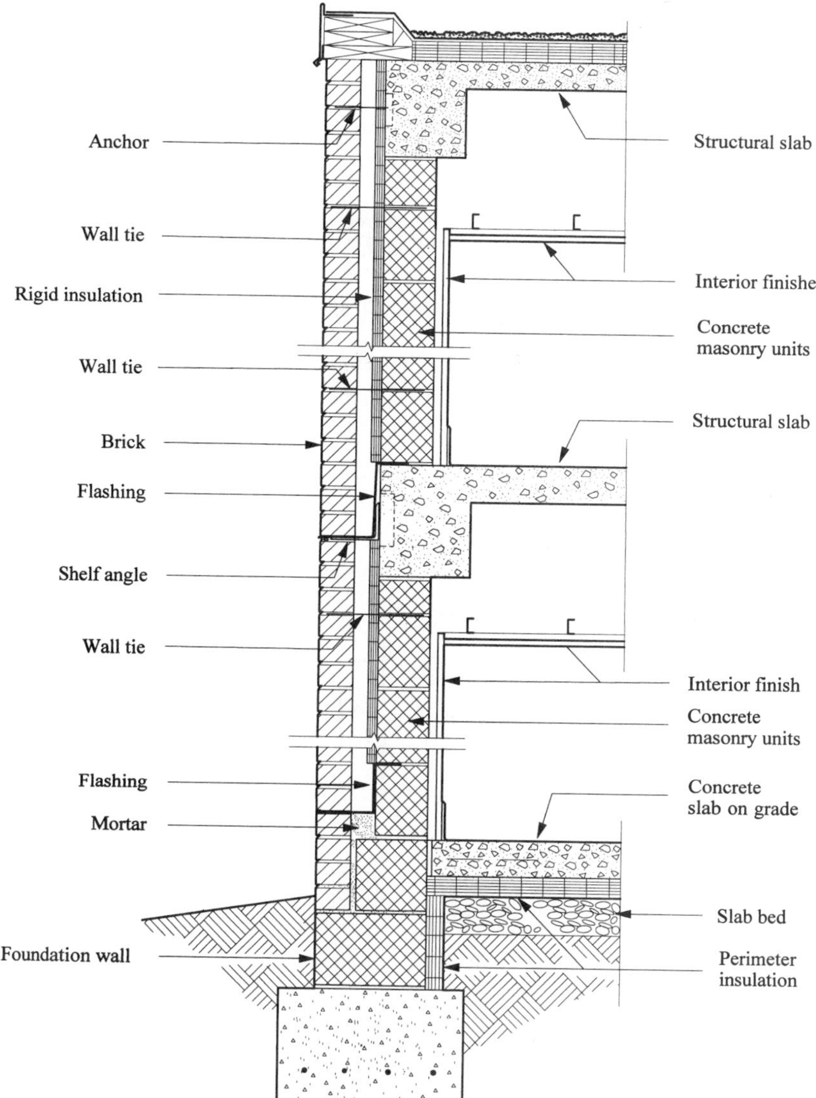

FIGURE 4.6-7 **Section through a non-load-bearing cavity wall on a concrete-framed building.** (Drawing by HLS.)

FIGURE 4.6-9 **The cavity in this wall is being used to conceal plumbing pipes and electric wiring runs.**

maintained at a nominal 2 in. (50.8 mm). The cavity may be left as all air space or insulated in a variety of ways.

A cavity offers two advantages in areas of severe exposure. The continuous air space provides insulation value to the wall and permits insulation to be installed within the wall to further reduce heat transfer (Fig. 4.6-8) (see also Section 7.3). It also acts as a barrier to moisture, eliminating the need for furring because significant rain penetration to the interior is practically impossible if proper flashing and weep holes are installed. A cavity may also be used to conceal mechanical and electrical utilities (Fig. 4.6-9).

For a cavity wall to be effective, certain precautions must be observed during its construction. The cavity must be kept free of mortar droppings that could form a bridge allowing moisture to penetrate to the interior face. Two methods for doing this are shown in Figure 4.6-10. Flashings and weep holes must be provided to drain the cavity of water that passes through the exterior wythe. Metal cavity wall ties, designed to resist both tension and compression, should be used to bond the exterior and interior wythes and should be solidly embedded in mortar so that the two wythes will act together to resist wind and other lateral forces. These ties may be separate items or may be part of continuous joint reinforcement.

FIGURE 4.6-8 **Cavity wall with rigid insulation in the cavity.** (Courtesy DOW USA.)

A vapor retarder is unnecessary in a cavity wall when the cavity is insulated with fill materials that will not retain excessive moisture, such as water-repellent vermiculite or silicon-treated perlite, or with rigid board materials, like foam glass and foamed plastics, that are at least 1 in. (25.4 mm) less in thickness than the cavity and are installed next to the inner wythe.

In cavity walls, insulation and mechanical equipment are often installed between thicknesses of masonry or within furred spaces on the interior side of the wall.

4.6.1.3 Veneered Walls

Installing masonry units as a facing material (veneer), without using their load-bearing properties, is common in all building types. Figure 4.6-11 shows typical brick veneer over wood studs and metal studs in a single-story wood-framed structure. Figure 4.6-12 shows a two-story steel-framed building with a cavity wall consisting of brick veneer over insulated metal studs. Similar construction is also used in higher buildings and those framed with a concrete structural system. The facing of a veneered wall is attached but not bonded to the studs and does not act structurally with the load-bearing portion of the wall.

In masonry veneer construction, wood

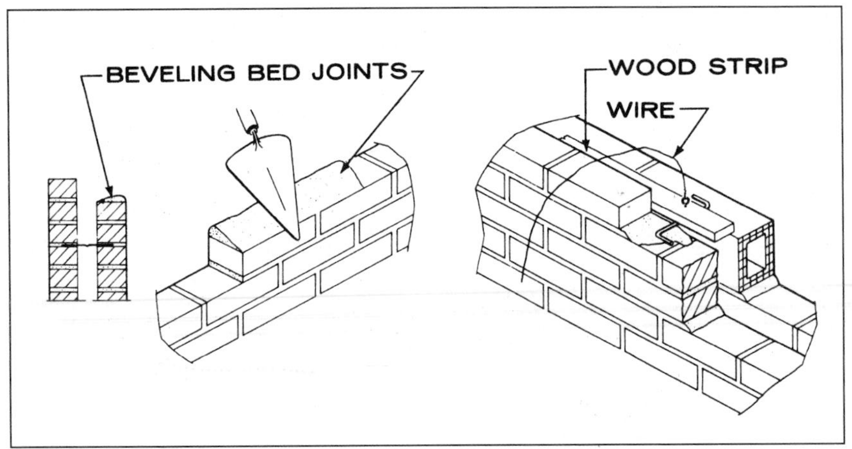

FIGURE 4.6-10 This figure demonstrates two methods of keeping mortar out of a cavity. The two drawings on the left show the beveling of a mortar bed to prevent mortar from being squeezed into the cavity when the next unit is placed on it. The drawing on the right shows how a wood strip can be placed on each tier of metal ties or continuous joint reinforcement to catch mortar droppings. The wood strip is raised as the work progresses. (Courtesy Brick Institute of America.)

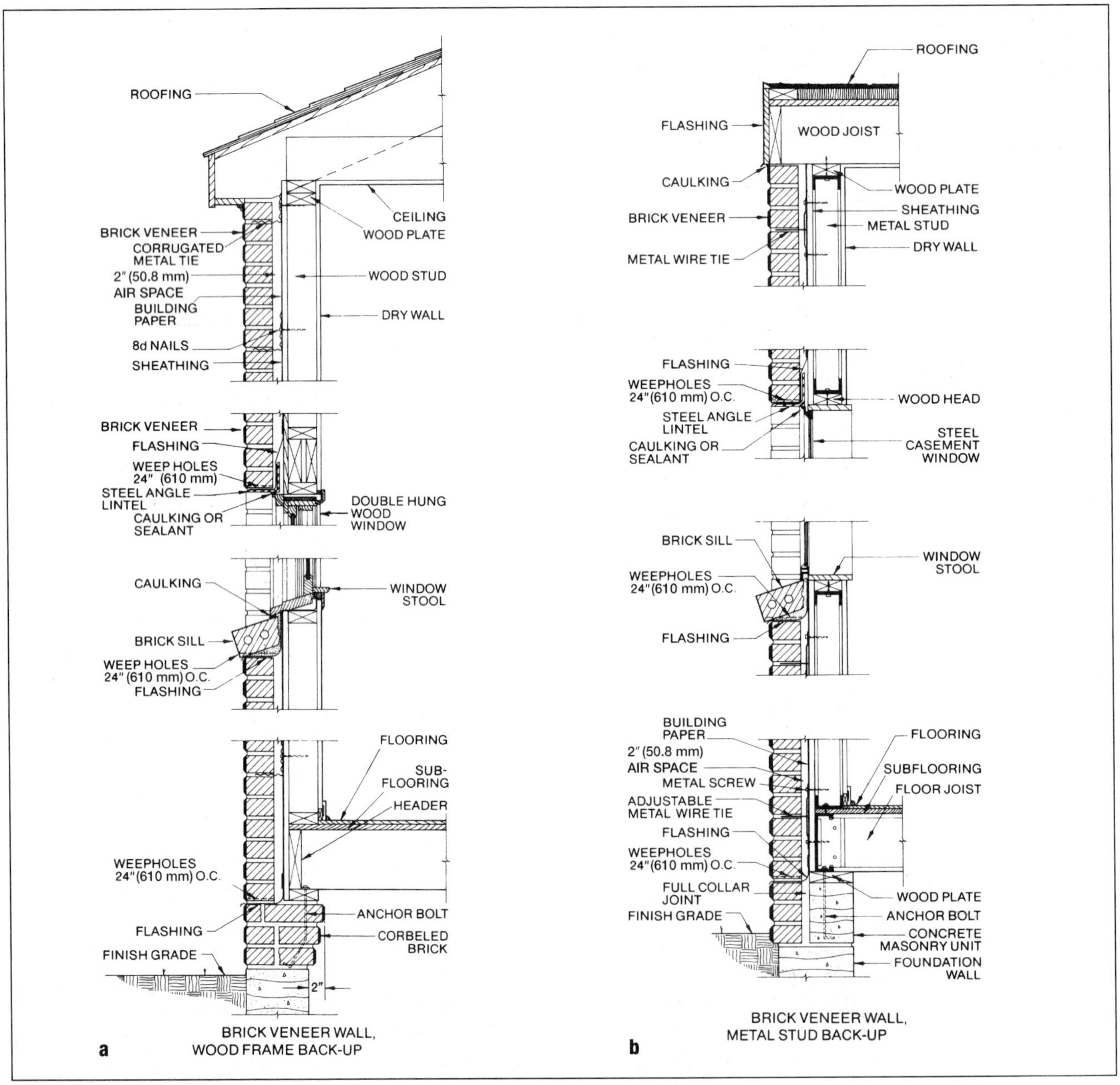

ROOFING
BRICK VENEER
CORRUGATED METAL TIE
2" (50.8 mm) AIR SPACE
BUILDING PAPER
8d NAILS
SHEATHING
CEILING
WOOD PLATE
WOOD STUD
DRY WALL
BRICK VENEER
FLASHING
WEEP HOLES 24" (610 mm)
STEEL ANGLE LINTEL
CAULKING OR SEALANT
DOUBLE HUNG WOOD WINDOW
CAULKING
WINDOW STOOL
BRICK SILL
WEEP HOLES 24" (610 mm) O.C.
FLASHING
FLOORING
SUB-FLOORING
HEADER
WEEPHOLES 24"(610 mm) O.C.
ANCHOR BOLT
FLASHING
CORBELED BRICK
FINISH GRADE
2"

a BRICK VENEER WALL, WOOD FRAME BACK-UP

ROOFING
FLASHING
WOOD JOIST
CAULKING
WOOD PLATE
SHEATHING
BRICK VENEER
METAL STUD
DRY WALL
METAL WIRE TIE
FLASHING
WEEPHOLES 24"(610 mm) O.C.
WOOD HEAD
STEEL ANGLE LINTEL
CAULKING OR SEALANT
STEEL CASEMENT WINDOW
BRICK SILL
WINDOW STOOL
WEEPHOLES 24"(610 mm) O.C.
FLASHING
BUILDING PAPER
FLOORING
2" (50.8 mm) AIR SPACE
SUBFLOORING
METAL SCREW
FLOOR JOIST
ADJUSTABLE METAL WIRE TIE
FLASHING
WEEPHOLES 24"(610 mm) O.C.
FULL COLLAR JOINT
WOOD PLATE
FINISH GRADE
ANCHOR BOLT
CONCRETE MASONRY UNIT
FOUNDATION WALL

b BRICK VENEER WALL, METAL STUD BACK-UP

FIGURE 4.6-11 Sections through brick veneer walls. Section (a) is through a wood stud wall. Section (b) is through a metal stud wall. Insulation has been omitted from the drawings for clarity.

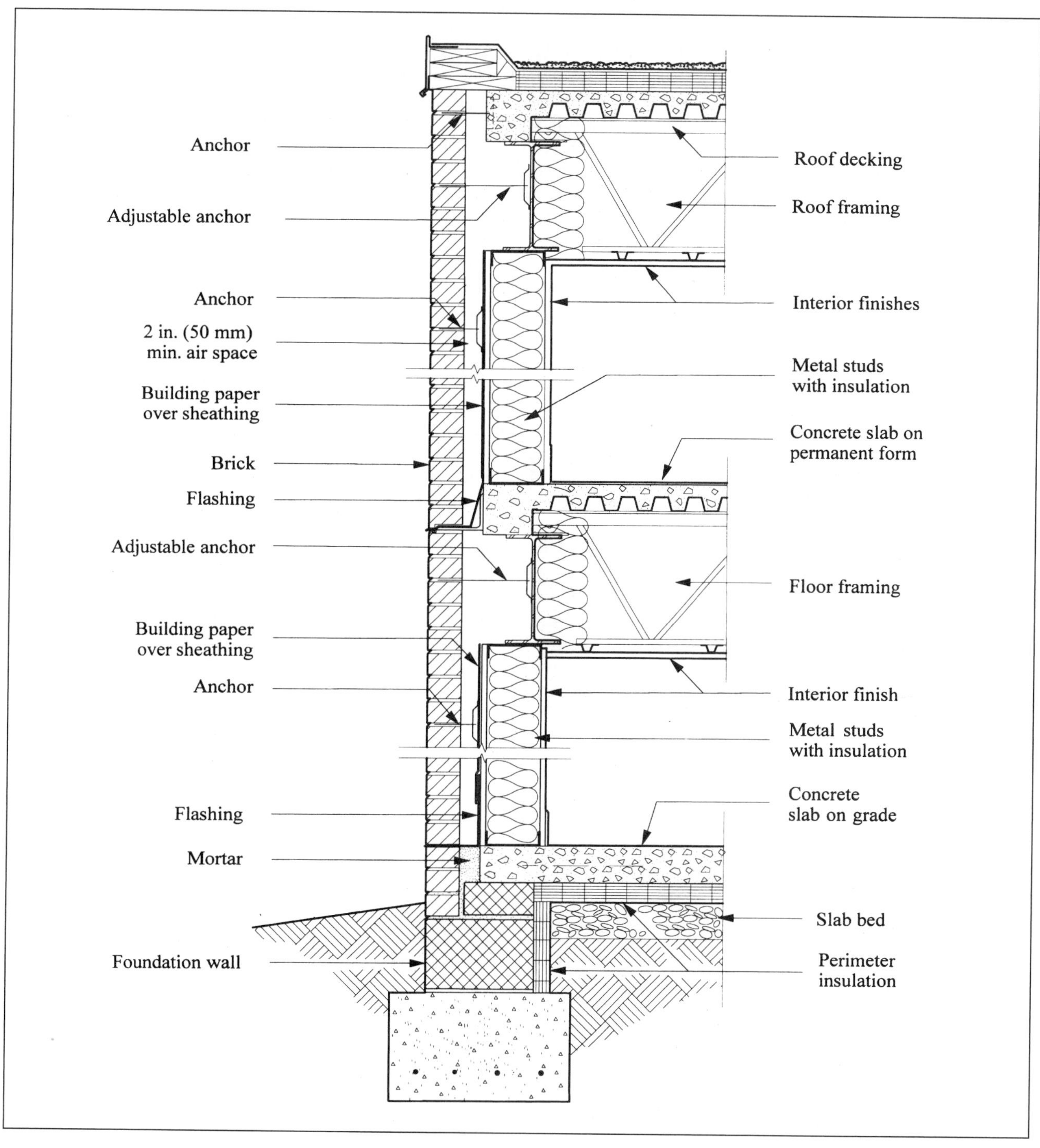

FIGURE 4.6-12 **Section through a brick veneer wall on a steel-framed building.** (Drawing by HLS.)

or metal framing carries the vertical loads while the masonry performs mostly as an exterior finish surface and provides a weather barrier. The space between the studs is typically used to contain insulation and to conceal heating, air conditioning, electrical, and plumbing lines.

Key factors in constructing brick veneer walls include correct selection of masonry materials, adequate anchorage to the backup wall, and proper installation of flashing and weep holes.

Since the veneer is isolated from the backup wall by an air space and exposed to the exterior, severe weathering (SW) grade facing brick should be used. Since a masonry veneer carries only its own weight, Type N mortar should be used in it, consistent with the principle that mortar be of the lowest strength that is compatible with structural requirements.

Refer to Section 4.7.2.2 for a description of the bonding together of brick veneer walls and a short discussion about the controversy surrounding the use of brick veneer with metal studs.

Flashing and weep holes to provide resistance to rain penetration should be located at the bottom of the wall and at all openings.

4.6.1.4 Reinforced Walls

Following the principles used in reinforced concrete, steel reinforcement is sometimes placed in a masonry wall and embedded in grout to give the wall greatly increased resistance to forces that produce tensile, shear, and compressive stresses. The chief uses of reinforced masonry are as walls, columns, lintels, and bond beams (Fig. 4.6-13).

There are two ways to reinforce masonry walls. In the first method, which is sometimes called *reinforced hollow masonry*, the reinforcing and grout are placed in the cores of concrete masonry units (see Fig. 4.6-1B) or hollow brick. In this method, concrete masonry units are laid with full face, shell bed joints. The head joints are filled with mortar to a distance in from the face at least equal to the face shell thickness. Cross webs that are adjacent to cores that will be filled with grout should be filled solidly with mortar. Hollow brick should be laid with full head joints, but the cores must be left open in the bed joints. Both concrete masonry units and hollow brick must be laid and lapped with the cores lined up vertically to permit installation of reinforcing steel and proper filling with grout.

In the second method, which is called *reinforced solid masonry* or *reinforced grouted masonry*, the reinforcement and grout are placed in the space between two wythes of masonry (see Fig. 4.6-1C). In this method the two masonry wythes are laid with completely filled bed and head joints. Cores and joints adjacent to spaces to be filled with grout should also be filled with mortar.

In both methods, care must be taken to prevent mortar droppings from entering the spaces to be filled with grout and to remove mortar fins from those spaces. Sometimes cleanouts are placed in walls to be grouted so that foreign materials can be removed.

Reinforcing bars are placed vertically in walls and tied to horizontal bars. Continuous joint reinforcement is usually also placed in reinforced masonry walls.

The kind of grout used varies with the size of the spaces to be filled. When the least clear dimension of the space is $2\frac{1}{2}$ in. (63.5 mm) or less, fine grout (see Section 4.2.5) should be used. Coarse grout should be used when the spaces to be filled are larger. The height of pour of fine grout should be limited to 12 in. (305 mm), while coarse grout can be placed in layers as much as 4 ft. (1219 mm) thick. Fine grout must be agitated or puddled to ensure that it fills the grout space completely. Coarse grout should be vibrated.

Refer to BIA's *Technical Notes on Brick Construction*, Nos. 17 through 17L and 41, for additional information about reinforced masonry. Reinforced masonry design should be done in accordance with the requirements of ACI 530, ASCE 5, TMS 402, and ACI 530.1, ASCE 6, TMS 602.

Insulation and mechanical equipment are usually installed in furred spaces on the interior side of reinforced masonry walls. Insulation is often placed on the exterior side of below-grade basement and foundation walls.

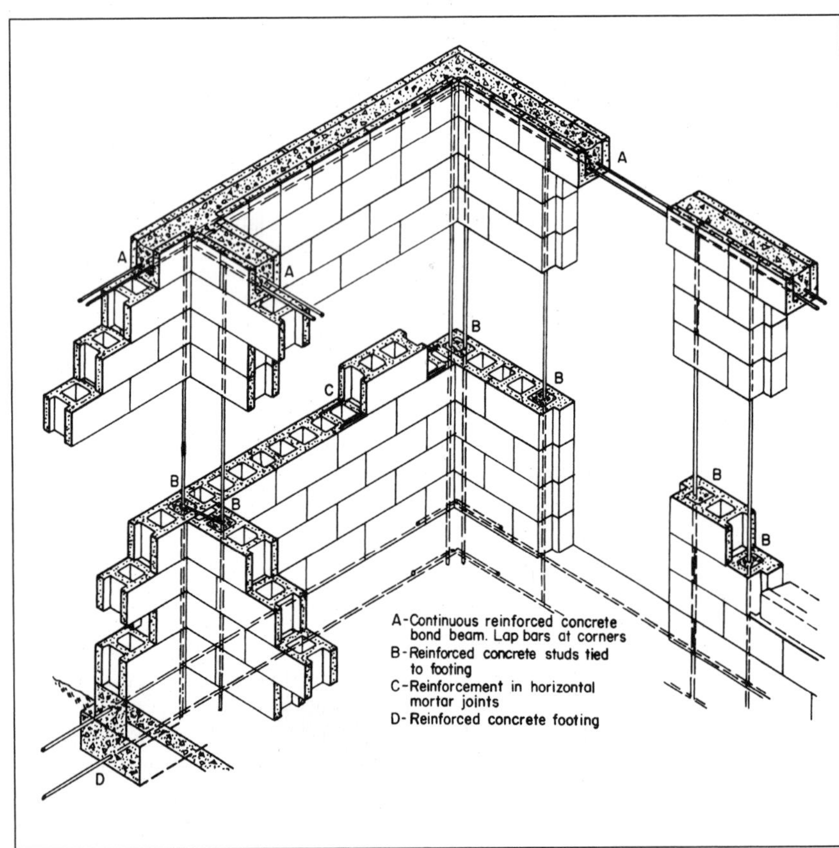

FIGURE 4.6-13 Typical methods of reinforcing concrete unit masonry walls.

4.6.2 PATTERN BONDS

The pattern formed by the masonry units and mortar joints is called pattern bond. Pattern bond may result from the type of structural bond used or may be purely decorative. Structural bond is the method by which individual masonry units are interlocked together, using masonry headers or metal ties, to cause the entire wall assembly to act as a single structural unit.

4.6.2.1 Brick Pattern Bonds

Traditional brick bond patterns still commonly used today are described briefly below and are illustrated in Figure 4.6-14.

RUNNING BOND

Running bond is the simplest of the basic pattern bonds in that there are no headers. Metal ties are used for structural bond. Running bond is used largely in cavity and veneer walls of solid units and often in concrete masonry unit and clay tile walls where masonry bonding may be accomplished by wider stretcher units (see Fig. 4.6-1G).

COMMON (AMERICAN) BOND

Common bond is similar to running bond, except that there is a course of full-length headers at every fifth, sixth, or seventh course, depending on the structural bonding requirements. In some historic buildings, header courses were placed at even

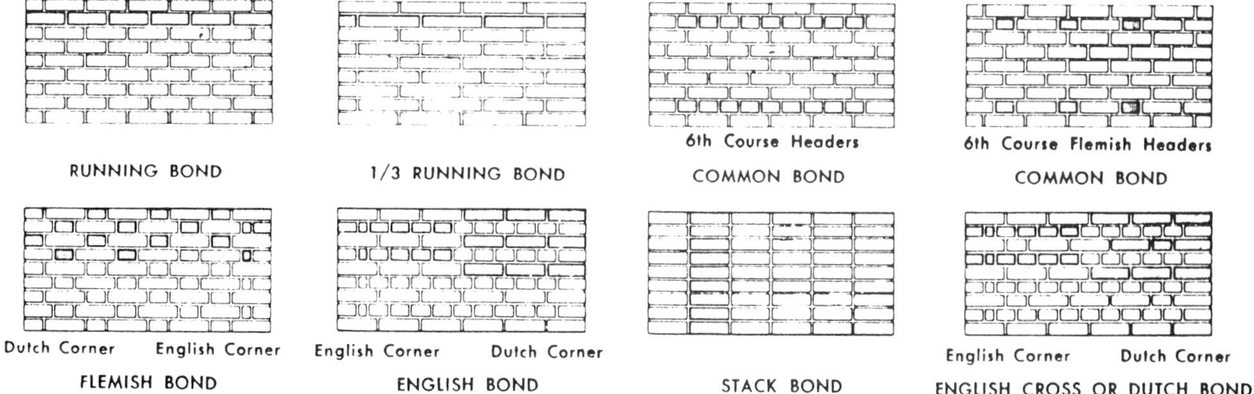

FIGURE 4.6-14 **Traditional brick pattern bonds.** (Courtesy Brick Institute of America.)

wider spacings. It is important in laying out every bond pattern that the corners be started correctly. For common bond, a three-quarter brick starts each header course at the corner.

FLEMISH BOND

Flemish bond has alternate stretchers and headers in each course. Where the headers are not used for structural bonding, half-bricks, called *blind headers*, may be used. There are two methods of starting the corners in a Flemish bond pattern. In the *Dutch* corner, a three-quarter brick starts each course. In the *English* corner, 2-in. (50.8 mm) or quarter-brick closures are used.

ENGLISH BOND

In the English bond pattern, alternate courses are composed of headers and stretchers. The joints between stretchers in all courses line up vertically. Blind headers are used, except in structural bonding courses.

ENGLISH CROSS OR DUTCH BOND

The English cross or Dutch bond pattern is a variation of English bond. Vertical joints between the stretchers in alternate courses do not line up vertically, but center on the stretchers in the courses above and below.

STACK BOND

In a stack bond pattern, all the vertical joints are aligned. Since units do not overlap, this pattern is usually bonded to the backing with rigid steel ties. In large wall areas and in load-bearing construction, such walls should be reinforced with continuous joint reinforcement or steel rods placed in the horizontal mortar joints. If the vertical mortar joints are to align, the masonry units must be dimensionally accurate or prematched units must be carefully selected for size.

CONTEMPORARY BONDS

Many traditional pattern bonds have been modified by projecting and recessing units or by leaving out units to form perforated walls and screens. A few contemporary brick bonds are shown in Figure 4.6-15. The effect of a particular pattern bond can be altered by varying the pattern formed by variations in the color and texture of individual units, by changing the joint types used, and by recessing or projecting individual units, courses, or wythes.

The basket weave and diagonal bond patterns shown for concrete masonry units in Figure 4.6-16 are also used in brick paving.

4.6.2.2 Concrete Masonry Unit Pattern Bonds

A number of bond patterns currently used with concrete masonry units are shown in Figure 4.6-16. Those more commonly used are the running bond patterns in which the units overlap the units in the next course by one-half unit and the vertical joints in every second row are in alignment. Although some of the patterns shown are decorative, they have all been tested and shown to be suitable for load-bearing construction.

4.6.3 MOISTURE CONTROL

Refer to Section 7.1 for a general discussion of moisture control in buildings and some specific requirements for moisture control in foundations and basements with masonry walls. The principal sources of moisture in masonry are rain or melting

FIGURE 4.6-15 **Some other brick pattern bonds.** (Courtesy Brick Institute of America.)

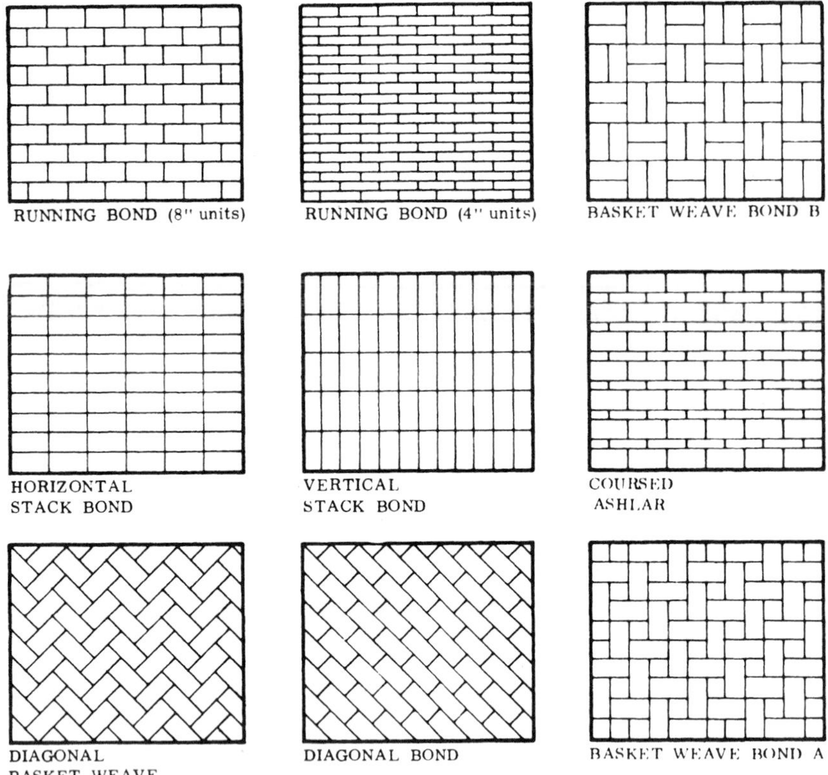

FIGURE 4.6-16 Concrete masonry unit pattern bonds.

snow penetration, capillary action from contact with free water or moisture in the ground, and condensation of vapor within the masonry. Other possible sources are faulty drainage and leaking plumbing systems.

Preventing moisture from entering a masonry wall is important to prevent the formation of efflorescence and cryptoflorescence and to prevent freeze-thaw cycle damage, both as discussed earlier (Sections 4.2.1.2 and 4.4.5.1). Excluding moisture from walls is also necessary to prevent corrosion of reinforcement and ties and the resultant damage to the wall and other building components.

Methods used to control rain and melting snow penetration into masonry walls include providing:

- Adequate flashing
- Proper tooling of mortar joints
- Parging and dampproofing
- Painting
- The filling of joints between masonry and doors, windows, and other openings and penetrations with sealants
- A sufficient slope or wash to readily drain horizontal surfaces, such as those at sills and copings
- An overhang on sills, copings, and the like, and including drips to keep rain and melted snow away from the walls
- Adequate gutters and downspouts

Methods used to control groundwater penetration into masonry foundation, basement, and retaining walls include parging and dampproofing (see Sections 4.7.3 and 7.2), waterproofing (see Section 7.2), and sloping the exterior grade to drain water away from the walls.

Water vapor that may condense within uninsulated masonry can often be controlled by installing a vapor retarder on the warm side of the wall (see Section 7.1). A water-emulsion asphalt paint applied to the wall's surface is one method used to provide such a retarder. Masonry insulated with proper types of insulation may not require an additional vapor retarder.

4.6.4 CRACK CONTROL

Differential movements within masonry walls and conditions that prohibit necessary movement can produce stresses that may cause serious cracking unless precautions are taken to minimize them. Conditions that can generate stress concentrations large enough to cause cracking include:

- Expansion or contraction due to temperature changes
- Expansion or contraction due to changes in moisture content of the units, especially of concrete masonry units
- Contraction in concrete masonry units due to chemical reactions called carbonation (see Section 4.5.1.3)
- Structural movements caused by unequal foundation settlement or by the concentration of applied loads
- Concentration of stress that develops where doors, windows, and other openings occur
- Members built into the wall that locally restrain or prevent the wall from moving, as at the junction of floors, roofs, columns, and intersecting walls

The potential for cracking in masonry walls is exacerbated by the following:

- Dissimilar materials, as often occur in composite walls, expand and contract at different rates (metal lintels, windows, and door frames have substantially greater coefficients of expansion than does masonry, which causes relatively large differential movements).
- Exterior walls having a cold exterior surface and a warm interior surface tend to warp.
- The degree of restraint to movement imposed by roofs, floors, and intersecting walls, and the effect of openings are difficult to estimate.

Expansion and control joints are built into masonry walls to reduce cracking caused by movement. The location of expansion and control joints cannot be established with mathematical accuracy because the many variables involved may act singly or in combination in a given case. Therefore, these joints are usually located using experience and observation of existing buildings. The recommendations that follow have evolved in just that way.

4.6.4.1 Expansion Joints for Clay Masonry

An unrestrained clay masonry wall 100 ft. (30.48 m) long will tend to expand ⅜ in. (9.5 mm) in response to a temperature increase of 100°F (37.8°C). If this expansion is cumulative, it could cause the wall to extend beyond its foundation (Fig. 4.6-17a). When the wall contracts, its tensile

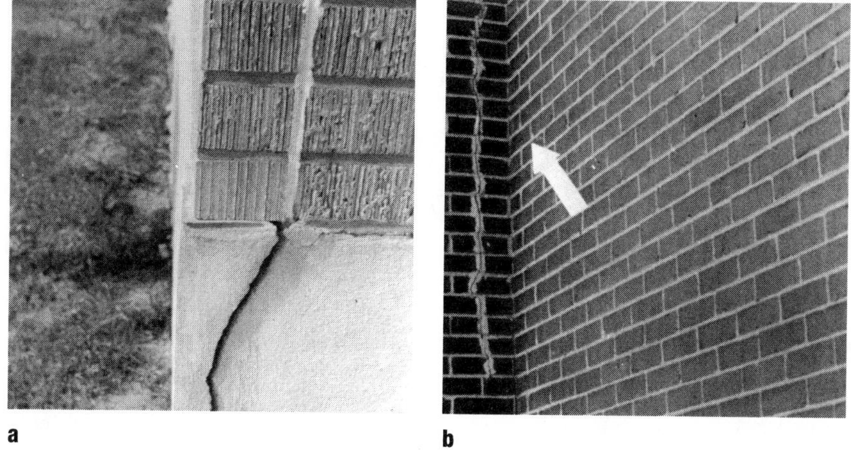

FIGURE 4.6-17 Cracks in masonry caused by movement. The crack in (a) is in a foundation wall. The one in (b) is in a building wall. (Courtesy Brick Institute of America.)

strength is not sufficient to overcome frictional resistance and cracks will appear. Cumulative expansion causes failure in walls at offsets where the masonry in the offset at right angles to a long wall is placed in bending and shear by the expansion of the long wall (Fig. 4.6-17b).

Expansion joints are installed to provide a complete separation through the structure. Figure 4.6-18 shows several of the many methods used to build these joints. Expansion joints are expensive, may be difficult to maintain, and should be avoided when it is possible to do so without having the walls crack.

In general, expansion joints should be located at offsets when the wall expanding into an offset is 50 ft. (15.24 m) or more in length and at junctions of walls in L-, T-, and U-shaped buildings. Recommended locations are shown in Figure 4.6-18. The table in Figure 4.6-19 gives recommended spacings.

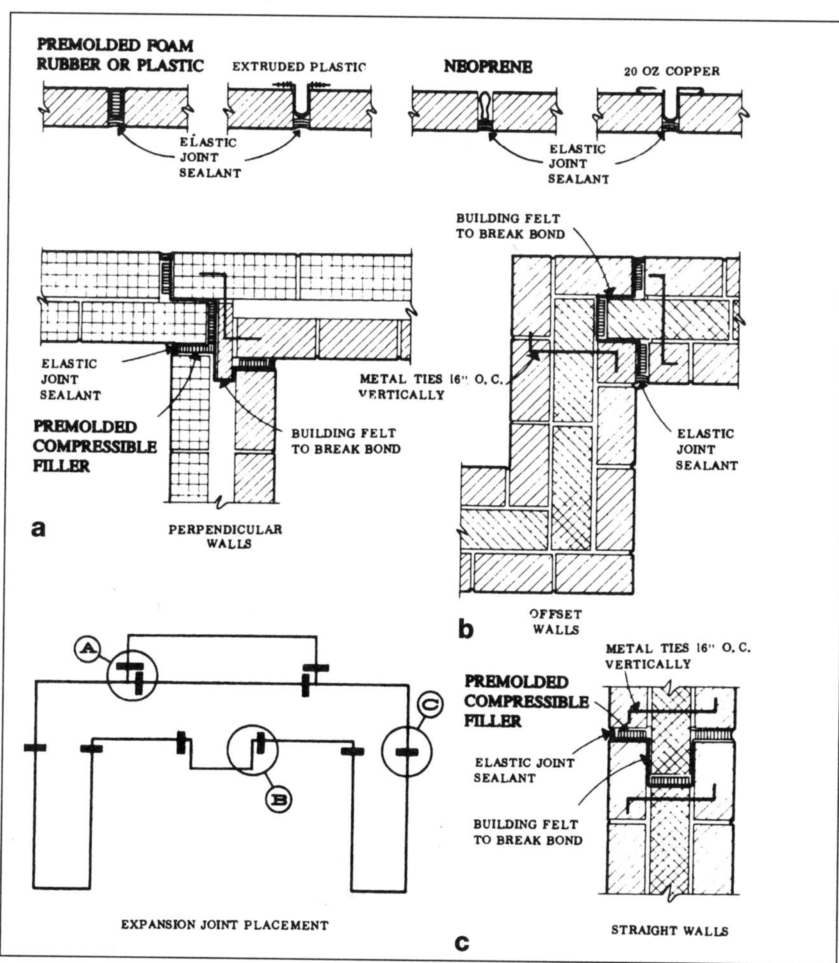

FIGURE 4.6-18 Expansion joint design and location in masonry walls. (Courtesy Brick Institute of America.)

4.6.4.2 Control Joints for Concrete Masonry

Most initial movement in a concrete masonry wall is due to contraction from drying shrinkage. As a new masonry wall tries to contract, tensile forces occur that must be resisted by the tensile strength of the wall. The problem is made worse when a foundation, roof, or other restraining element offers resistance to the normal shrinkage of the masonry. A crack will occur when the forces that occur exceed the tensile strength of either the masonry units or the mortar.

Cracking due to shrinkage can be controlled by one or more of the following methods:

- Using moisture-controlled (Type I) concrete masonry units manufactured according to ASTM standards
- Installing horizontal steel reinforcement to increase the wall's tensile resistance to cracking and to minimize the width of cracks
- Building in control joints located to accommodate wall movements

Sometimes more than one method must be used in the same wall.

Control joints are continuous vertical joints built into concrete masonry walls as they are laid up, to minimize shrinkage cracks and to control cracks at locations where a concentration of stresses or points of weakness is expected. Properly built control joints should prevent slight wall movements from cracking the wall, be themselves sealed against the weather, and stabilize the wall laterally across the joint by means of a shear key.

LOCATION AND SPACING

The layout of control joints will depend on the following variables:

- Wall length and height;
- Wall type;
- Moisture content of units at time of laying;
- Joint reinforcement spacing;
- Use of bond beams;
- Architectural details;
- Local experience as to the need for control joints in the locality where the building is being built. In the absence of specific local data or practical experience, the locations and spacings of control joints shown in Figures 4.6-20 and 4.6-21 have been used successfully. Other spacings and locations may be necessary, however, in a particular project, especially if there are unusual circumstances.

Control joints should be placed in both load-bearing and non-load-bearing walls at changes in wall height or thickness, at the junction of walls with columns and pilasters, where openings occur, at wall intersections, and at return angles in L-, T-, and U-shaped buildings.

In composite walls, where concrete masonry is used as a backup for other materials, control joints should be extended through the facing when the wall is laid with masonry bonding. Control joints do not have to extend through the facing material when it is bonded to the backup with metal ties.

When stucco or plaster is applied directly to masonry units, control joints should extend through the stucco or plaster.

Because the temperature- and moisture-volume change of concrete masonry below grade in foundation walls is usually quite low, and because it is difficult to construct a watertight joint in masonry below grade, control joints are not recommended for basement or foundation walls. The movements at a control joint in above-grade masonry may cause cracks to develop in the foundation wall below. To minimize this possibility, a continuous bond beam can be provided at the top of the foundation walls. Then the control joints in the above-grade wall can be stopped at the bond beam with no damage to the wall below it.

FIGURE 4.6-19 Recommended Spacing of Expansion Joints

Outside Temperature Ranges[a]	Maximum Length of Wall[b]			
	Unheated or Insulated		Heated, Not Insulated	
	Solid, ft.	Openings, ft.[c]	Solid, ft.	Openings, ft.[c]
100 and over	200	100	250	125
Less than 100	250	125	300	150

[a]The range from the lowest average temperature to the highest in degrees F.

[b]For insulated cavity or veneer walls, distances are approximately one-half the lengths shown.

[c]Opening: 20% or more of wall area.

(Courtesy Brick Institute of America.)

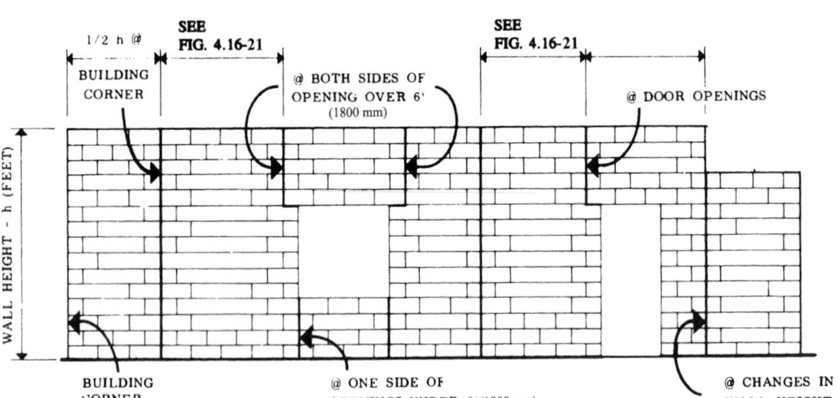

FIGURE 4.6-20 Typical control joint locations in concrete masonry unit walls and partitions.

FIGURE 4.6-21 Maximum Spacings of Control Joints in Unreinforced Masonry[a]

Panel Dimensions[b]	Vertical Spacing of Joint Reinforcements[c]			
	None	24 in. (600 mm)	16 in. (400 mm)	8 in. (200 mm)
Ratio of panel, length to height (L/H)[d]	2	2½	3	4
Panel length (L) not to exceed	40 ft. (12.9 m)	45 ft. (13.72 m)	50 ft. (15.24 m)	60 ft. (18.29 m)

[a]Moisture controlled, Type I concrete masonry units assumed. If nonmoisture control units, reduce joint spacing by 50%; if wall is solid grouted, spacing should be 33% less.

[b]A panel is a wall element in one plane lying between (a) corners or wall ends, (b) control joints, or (c) a control joint and end wall.

[c]Continuous metal ties with a minimum of two No. 9 gauge longitudinal wires.

[d]Maximum panel length (L) is applicable regardless of height (H).

CONSTRUCTING A CONTROL JOINT

Control joints should be laid up in mortar just like any other vertical joint. If the control joint is to be exposed to the weather or to view, a recess should be provided to receive a joint sealant by raking the mortar out to a depth of about ¾ in. (19 mm) after it has become quite stiff (Fig. 4.6-22). Sometimes a pressure gun and a thin,

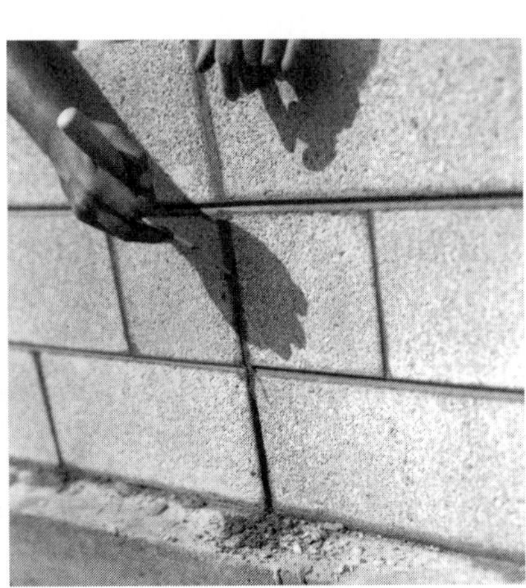

FIGURE 4.6-22 The mortar in this control joint has been raked out to a depth of about ¾ in. preparatory to applying a sealant. (Courtesy Portland Cement Association.)

FIGURE 4.6-23 This control joint is being sealed with a knife-grade caulking compound. (Courtesy Portland Cement Association.)

flat caulking trowel are used to force a knife-grade caulking compound into the joint (Fig. 4.6-23). In other cases, the methods discussed in Section 7.11 are used to apply an elastomeric sealant. To keep them as unnoticeable as possible, control joints should be built plumb and be of the same thickness as the other mortar joints.

One type of control joint can be built with regular stretcher block. Building paper or roofing felt is inserted in the end core of full- and half-length block and extended the full height of the control joint. The core is then filled with mortar for lateral support (Fig. 4.6-24). The paper, or felt, cut to convenient lengths and wide enough to extend across the joint, prevents the mortar from bonding on one side of the joint, thus permitting the control joint to function (Fig. 4.6-25). Control joint blocks with tongue-and-groove-shaped ends, which are available in many areas, provide lateral support to the wall sections on each side of the control joint (Fig. 4.6-26). These control joint blocks are made in both full- and half-length units (Fig. 4.6-27).

FIGURE 4.6-24 This control joint is being built by inserting building paper in the end core of the block. Roofing felt may also have been used. (Courtesy Portland Cement Association.)

FIGURE 4.6-25 A control joint with building paper or roofing felt and a mortar-filled core. (Courtesy Portland Cement Association.)

Because control joints can be unsightly and require maintenance, they should be avoided when possible. Methods used to increase the tensile strength of a wall and to improve the dimensional stability of the masonry units used in it will reduce the number of control joints required, as will designing to eliminate the conditions that dictate using them.

Other methods are also available to increase the crack resistance of walls. For example, two-core blocks with thickened fillets and three-core blocks with thickened center cores have higher tensile strengths than standard units. Their use can increase the crack resistance of a wall about 50% and 25%, respectively, over standard three-core blocks. Another example is the inclusion of continuous joint reinforcement in a wall. As Figure 4.6-21 shows, when joint reinforcement is introduced, the spacing of control joints may be increased.

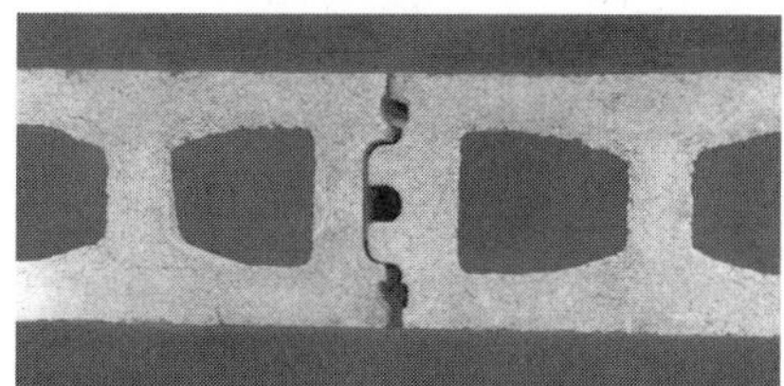

FIGURE 4.6-26 Special control joint units provide lateral stability by means of tongue-and-grove-shaped ends. (Courtesy Portland Cement Association.)

FIGURE 4.6-27 Laying a half-length control joint unit. (Courtesy Portland Cement Association.)

4.6.5 STRUCTURAL DESIGN OF MASONRY WALLS

Masonry walls are used both as bearing walls to support floor or roof structures and as non-load-bearing walls and partitions. The same basic principles apply to the design of masonry walls regardless of their use. Masonry should always be supported on other masonry, on concrete, or on steel. Wood girders and other types of wood framing should never be used to support masonry.

4.6.5.1 Compressive Strength

The strength at the point of failure, which is called the *ultimate compressive strength*, of masonry walls is closely related to the compressive strength of the masonry units themselves and of the mortar. The quality of workmanship, thickness of mortar joints, regularity of the units' bearing surfaces, and workability of the mortar are also important.

High compressive strength is developed by installing high-strength units, using proper workmanship so that every joint is completely filled with mortar. The strengths used for designing purposes, which are called allowable compressive stresses, recommended for various wall types, unit strengths, and mortar types are given in Figure 4.6-28.

The principal loads producing compressive stresses in a wall are the weight of the wall itself and the weights of associated floors and roofs that are transferred into the wall, which are collectively called the *dead load*; and the weights of occupants, furniture, and equipment, which are collectively called the *live load*. Both dead and live loads can be determined with reasonable accuracy, and once they have been determined, the resultant compressive stresses can be computed. Such stresses should not exceed the allowable stresses given in Figure 4.6-28 for the type of wall, kinds of units, and type of mortar being used. When composite walls or

FIGURE 4.6-28 Allowable Compressive Stresses in Unit Masonry

Construction; Compressive Strength of Unit, Gross Area, psi	Allowable Compressive Stresses,[a] Gross Cross-Sectional Area, psi Type M or S Mortar	Type N Mortar
Solid masonry of brick and other solid units of clay or shale; sand-lime or concrete bricks		
8000 or greater	350	300
4500	225	200
2500	160	140
1500	115	100
Grouted masonry of clay or shale; sand-lime or concrete		
4500 or greater	225	200
2500	160	140
1500	115	100
Solid masonry of solid concrete masonry units		
3000 or greater	225	200
2000	160	140
1200	115	100
Masonry of hollow load-bearing units		
2000 or greater	140	120
1500	115	100
1000	75	70
700	60	55
Hollow walls (noncomposite masonry bonded[b])		
Solid units		
2500 or greater	160	140
1500	115	100
Hollow units	75	70
Stone ashlar masonry		
Granite	720	640
Limestone or marble	450	400
Sandstone or cast stone	360	320
Rubble stone masonry		
Coarse, rough, or random	120	100

[a]Linear interpolation for determining allowable stresses for masonry units having compressive strengths that are intermediate between those given in the table is permitted.

[b]Where floor and roof loads are carried upon one wythe, the gross cross-sectional area is that of the wythe under load; if both wythes are loaded, the gross cross-sectional area is that of the wall minus the area of the cavity between the wythes. Walls bonded with metal ties shall be considered as noncomposite walls unless collar joints are filled with mortar or grout.

(Courtesy American Concrete Institute.)

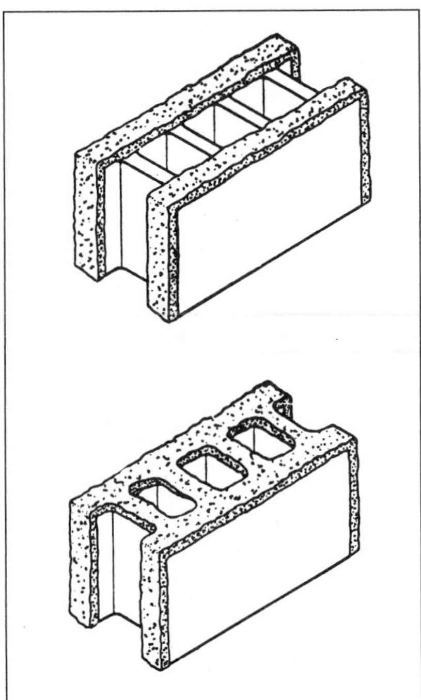

FIGURE 4.6-29 The top drawing shows face-shell mortar bedding; the bottom, full mortar bedding. (Courtesy American Concrete Institute.)

other structural masonry elements are composed of different kinds or types of units or mortars, the maximum compressive stress should not exceed the allowable stress for the weakest of the combinations of either the masonry unit(s) or the mortar type(s) of which the wall is composed.

Fortunately, typical construction today does not require most masonry walls to carry large vertical loads. For example, the superimposed load on the foundation wall of a typical 1-story residence rarely exceeds 15 psi (0.1 MPa). About 75% of the bricks produced today have compressive strengths of more than 4500 psi (31 MPa); concrete block, more than 1000 psi (7 MPa). Typical mortars used today have compressive strengths in the range of 750 psi (5 MPa) to 2500 psi (17 MPa). Therefore, it is evident that most masonry walls for 1-story buildings have far more compressive strength than they need.

Tests have established the relationship between the compressive strength of walls and the masonry units in them. The compressive strength of walls constructed of hollow concrete masonry units is about 42% of the compressive strength of the units when they are laid with face-shell mortar bedding, and 53% when they are laid with full mortar bedding (Fig. 4.6-29). So a wall built with full mortar bedding and concrete masonry units with compressive strengths of 1000 psi (7 MPa) would have a compressive wall strength of 0.53 × 1000, or 530 psi (3.7 MPa). Using an allowable compressive stress of 70 psi (0.48 MPa) (from Fig. 4.6-28), a wall having a strength of 530 psi (3.7 MPa) will have a factor of safety of 530/70 (3.7/0.48), or about 7½. A factor of safety of 4 is ample for masonry walls.

4.6.5.2 Transverse Strength

The transverse strength of a wall is a measure of its *lateral stability*, which is its resistance to such lateral forces as earth pressures, as occur in foundation, basement, and retaining walls; wind pressures; and earthquake forces. Transverse loads induce tensile stresses in masonry that must be resisted. The lateral stability of masonry walls depends on the tensile strength of the bond between mortar and masonry units. Actions that increase tensile bond strength also increase transverse strength. These include using Type S mortar mixed with the maximum amount of water consistent with workability, maintaining the suction of clay masonry units below 20 grams per minute, per 30 sq. in.

($0.02\ m^2$) (a simple field test to determine the suction of brick is described in Section 4.7.1.2), and making sure that both head and bed joints are completely full of mortar.

The lateral stability of a masonry wall is also increased by vertical loads, which produce compressive stresses that offset tensile stresses caused by transverse loads. The addition of reinforcing steel also greatly increases the lateral strength of a masonry wall.

4.6.5.3 Effect of Flashing on Wall Strength

The effect of a flashing on wall strength depends on where mortar is placed and the bond of the mortar to both the flashing and the masonry. Through-wall flashing, such as base and head flashing, does not affect a wall's compressive strength but will reduce bending and shearing strengths. If flashing is placed directly on masonry with no mortar beneath it, the flexural strength at that point is 0. Fortunately, through-wall flashing usually occurs where bending resistance is least significant.

Test data suggest that when mortar is placed immediately above and below copper flashing, flexural strength is 30% to 70% of a comparable unflashed wall. There are no sufficient data that permit a similar generalization for shearing strength under similar conditions. Assuming that there is no mortar under a base flashing, a wall's shear resistance, its ability to withstand lateral forces, depends on friction. The wall still retains some shearing resistance because the coefficient of friction between the flashing and the masonry is in the order of 0.25 to 0.50.

4.6.5.4 Lateral Support of Walls

Masonry walls require lateral support both while they are under construction and after they have been completed.

LATERAL SUPPORT DURING CONSTRUCTION

Masonry walls are seldom designed to be freestanding, without support from columns, piers, or cross walls. Wind pressures can create four times as much bending stress in a freestanding wall as they do in a wall that is supported by other structural parts of the building. This stress often occurs at the bottom of the wall where flashing and lack of bond in fresh mortar decrease the wall's strength to resist tensile wind forces (see Section 4.6.5.3).

Bracing should be installed to resist wind pressure. Figure 4.6-30 shows commonly accepted limits on wall heights based on various peak wind velocities. Where bracing is used, the heights shown are the safe heights above the bracing. For example, to withstand 50 mph (80.47 km/h) wind gusts, the freestanding, unbraced height of a 10-in. (250 mm)-thick wall weighing 67 psi (0.46 MPa) or less should not be more than 7½ ft. (2286 mm) tall. The curves in the figure assume that the mortar has no tensile strength and that the walls are freestanding, unreinforced, ungrouted concrete masonry. In cavity walls, the thickness is assumed to be two-thirds of the sum of the thickness of the two wythes. Table values may not be adequate under some circumstances.

Required bracing should be left in place until permanent bracing, such as that afforded by piers, cross walls, floors, and roof structures, have been placed and permanently brace the wall.

LATERAL SUPPORT OF COMPLETED WALLS

Masonry walls must be permanently laterally supported or braced at certain intervals by either vertical or horizontal supports. When the limiting distance is length, vertical support may be by columns, piers, pilasters, buttresses, or cross walls. When the limiting distance is height, horizontal supports may be floors, beams, or roofs. The spacing of these supports is primarily dependent on the wall thickness. Possible spacing of supports and other lateral support requirements are summarized in Figure 4.6-31. Codes often require different spacings than those shown and should be consulted before limits are imposed. In walls of different classes of units or mortars, the ratio of wall height or length to wall thickness should not exceed that allowed for the weakest of the combinations of units and mortars in the member. Sufficient bonding and anchorage must be provided to transfer the loads from the wall to its supports.

If vertical supports are provided, there is no limit on the height of walls between floors or between floor and roof. When horizontal supports are provided, there is no limit on the length of a wall, and lateral supports in the form of columns, pilasters, or cross walls are not required.

Neither the ratio of the unsupported wall height to wall thickness nor the ratio of unsupported wall length to wall thickness should be permitted to exceed the values given in Figure 4.6-31 for various wall types.

4.6.5.5 Wall Thicknesses, Heights, and Lengths

Of course, it is possible to design every wall by theoretical analysis of the forces involved, and such an analysis is imperative when unusual conditions occur or when a code requires that it be done. How-

in.		mm
4	=	100
6	=	150
8	=	200
10	=	250
12	=	300

psf		kg/m²
80	=	390.6
67	=	327.1
54	=	263.7
44	=	214.8
31	=	146.5

ft.		m
16	=	4.88
15	=	4.57
14	=	4.27
13	=	3.96
12	=	3.66
11	=	3.35

ft.		m
10	=	3.05
9	=	2.74
8	=	2.44
7	=	2.13
6	=	1.83
5	=	1.52
4	=	1.22
3	=	0.91
2	=	0.61
1	=	0.31

mph		km(h)
30	=	48.28
40	=	64.37
50	=	80.47
60	=	96.56
70	=	112.65
80	=	128.75
90	=	144.84

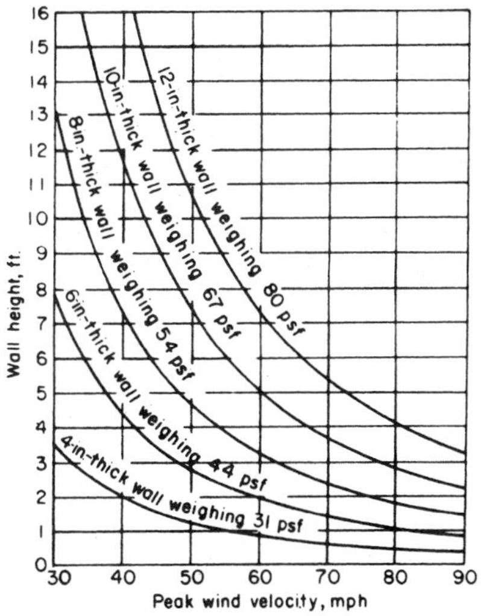

FIGURE 4.6-30 Maximum unsupported height of nonreinforced, ungrouted concrete masonry walls during construction.

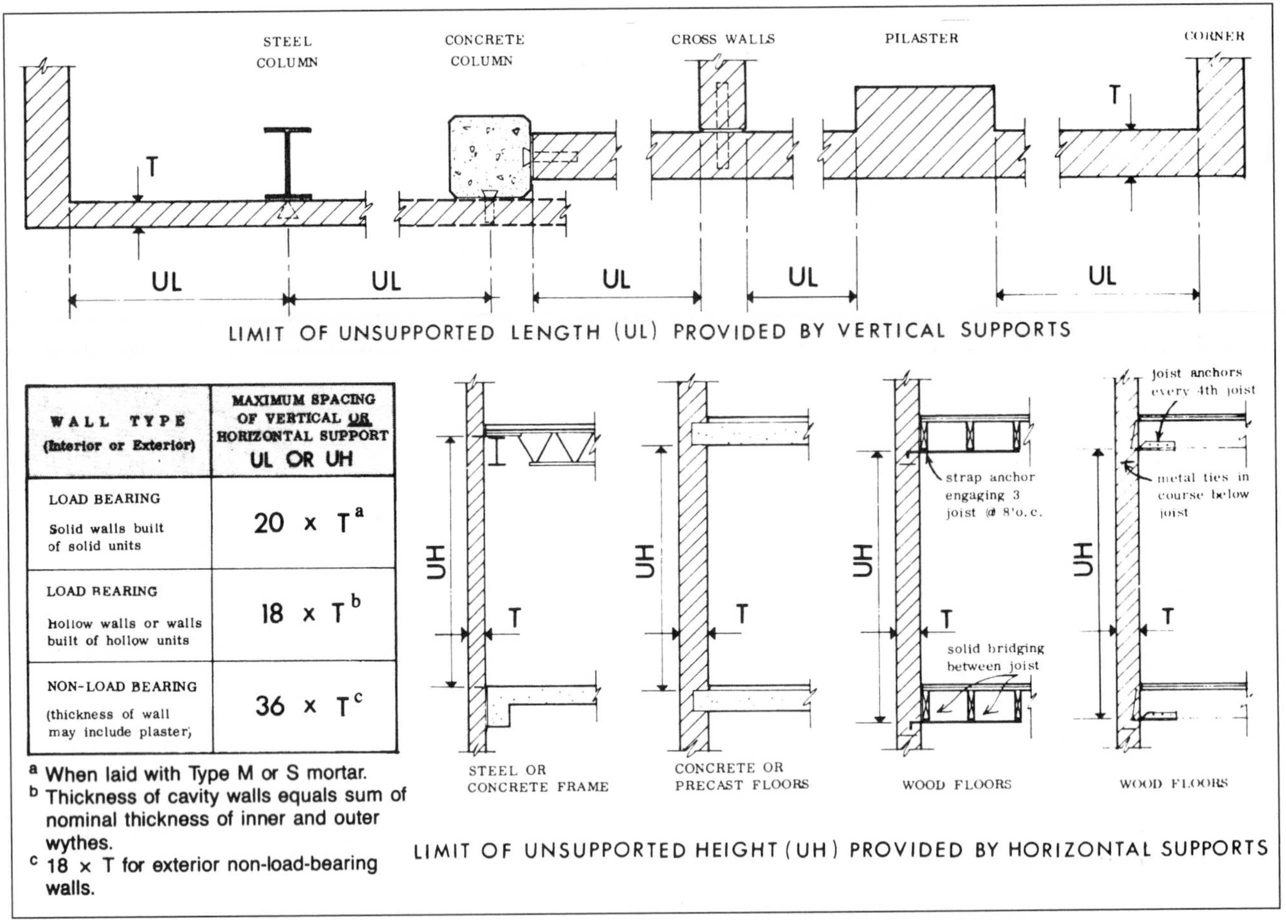

WALL TYPE (Interior or Exterior)	MAXIMUM SPACING OF VERTICAL OR HORIZONTAL SUPPORT UL OR UH
LOAD BEARING Solid walls built of solid units	20 x T[a]
LOAD BEARING Hollow walls or walls built of hollow units	18 x T[b]
NON-LOAD BEARING (thickness of wall may include plaster)	36 x T[c]

[a] When laid with Type M or S mortar.
[b] Thickness of cavity walls equals sum of nominal thickness of inner and outer wythes.
[c] 18 x T for exterior non-load-bearing walls.

FIGURE 4.6-31 Lateral support requirements for completed masonry walls.

ever, theoretical analysis is seldom needed or required by code for houses or for small buildings of any type. Therefore, the design of masonry walls in most small projects is based on empirical requirements, as are the suggestions in this section.

In general, walls should not vary in thickness between their lateral supports. An exception is made, however, in residential buildings not more than 2 stories in height, where vertical recesses may be built into 8-in. (200 mm)-thick walls if such a recess is not more than 4 in. (100 mm) deep and does not occupy more than 4 sq. ft. (2581 mm^2) of the wall's area.

FOUNDATION WALLS

Unreinforced concrete masonry is used extensively for foundation walls, especially in single-family dwellings and other single-story buildings. Experience and observation have resulted in the development of minimum requirements governing their design. The requirements are empirical, particularly in designing basement walls to resist lateral earth pressures. Therefore, foundation design for small buildings is rarely based on theoretical analysis of the forces and stresses involved. In the absence of specific building code requirements or individual engineering analysis, the wall thicknesses shown in Figure 4.6-32 have been used when the height of unbalanced fill against a wall did not exceed the limits given in the figure and no other unusual circumstances existed. However, the figure does not take into account the presence of hydrostatic pressure against the wall due to groundwater. In addition, some practitioners believe that masonry foundation walls should not be less than 12 in. (305 mm) thick, regardless of the height of earth against them.

OTHER WALLS

This section includes empirical design data for walls *other* than single-family, residential foundation walls. The data here should be verified with applicable code requirements and local experience in every case.

Load-Bearing Walls Load-bearing walls in residential buildings 3 stories or less in height may be 8 in. (200 mm) thick when not more than 35 ft. (10.67 m) high, if the roof is designed to impart no horizontal thrust.

Except where earthquake design is required, exterior walls in 1-story residential buildings may be 6 in. (150 mm) thick when they are not more than 9 ft. (2743 mm) high, the height to the peak of the gable does not exceed 15 ft. (4572 mm), and girders and concentrated loads are supported on not less than 8 × 12 in. (200 × 305 mm) integral piers or pilasters.

Interior bearing walls in dwellings not more than 1½ stories or 20 ft. (6.1 m) high may be 6 in. (150 mm) thick.

Load-bearing walls more than 35 ft. (10.67 m) high should be at least 12 in. (305 mm) thick for the top 35 ft. (10.67 m) of their height, and should increase 4 in. (100 mm) in thickness for each successive 35 ft. (10.67 m).

Cavity walls built of solid masonry units and more than 25 ft. (7.6 m) high

FIGURE 4.6-32 Foundation Wall Thicknesses Based on Wall Types and Height of Unbalanced Fill

Foundation Wall Construction	Maximum Height of Unbalanced Fill H (feet)	Minimum Wall Thickness T (inches) Frame Construction	Masonry or Veneer
Hollow Masonry	3′ (900 mm)	8″ (200 mm)	8″ (200 mm)
	5′ (1500 mm)	8″ (200 mm)	8″ (200 mm)
	7′ (2000 mm)	12″ (300 mm)	10″ (250 mm)
Solid Masonry	3′ (900 mm)	6″ (150 mm)	8″ (200 mm)
	5′ (1500 mm)	8″ (200 mm)	8″ (200 mm)
	7′ (2000 mm)	10″ (250 mm)	8″ (200 mm)

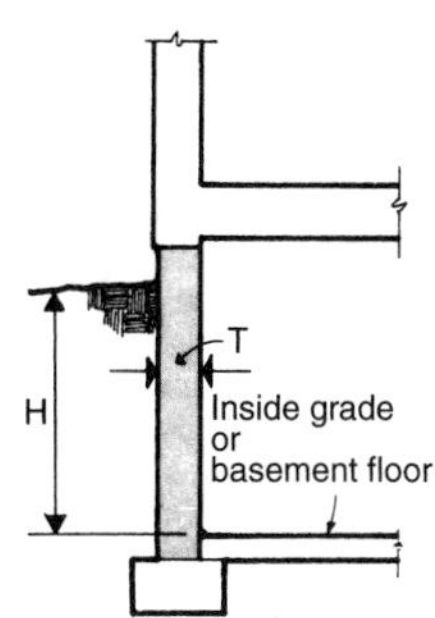

above the support of the wall may be 10 in. (250 mm) thick. The facing and backing of cavity walls should each be at least 4 in. (100 mm) thick, and the cavity should not be less than 2 in. (50.8 mm) or more than 4½ in. (114 mm) wide.

Non-Load-Bearing Walls Non-load-bearing walls on the exterior may be 4 in. (100 mm) less in thickness than is required for bearing walls but not less than 8 in. (200 mm) in thickness, except where 6-in. (150 mm)-thick walls are permitted in 1-story residences.

4.6.6 LINTELS AND ARCHES

The design of a structural member to support masonry over an opening should not be taken lightly. It is true that rule-of-thumb methods often result in overdesign. It is also true that arbitrary selection of steel shapes or reinforcing steel can result in inadequate sizing, causing structural cracks that lead to leaks and an unsightly appearance. Underdesigned supports may also fail completely, with disastrous results. Therefore, both masonry lintels and arches should be designed in accordance with engineering practice.

The dead weight of a masonry wall that must be supported over an opening can safely be assumed as the weight of a triangular section whose height is one-half the clear span of the opening (Fig. 4.6-33). Arching action of the masonry above the top of the opening may be counted on to support the additional weight. A concentrated load over an opening may be distributed along a wall length equal to the base of a triangle whose sides make an angle of 30 degrees with the vertical from the point of application of the load. The horizontal thrust resulting from an arching action must be resisted by sufficient mass in the adjoining wall.

Historically, various forms of arches have been used to span openings. Figure 4.6-34 shows some typical brick arch types. Arches built in contact with adjacent arches form a barrel vault. Domes may be thought of as arches rotated about their vertical centerlines.

Brick arch construction requires extensive labor, especially when the bricks are ground into wedge shapes with an abrasive stone to produce the curve. Most arches today are made using tapered joints between rectangular bricks to save the labor cost of producing wedge-shaped bricks. These are called *rough arches.* Most arches are built in solid walls, because they are difficult to construct and even more difficult to flash properly in cavity walls and veneer walls. The arch must extend across the cavity to present a uniform appearance across the bottom of the arch. This not only blocks the cavity, it also ties the inner and outer wythes together, producing difficulties due to differential movement between the wythes.

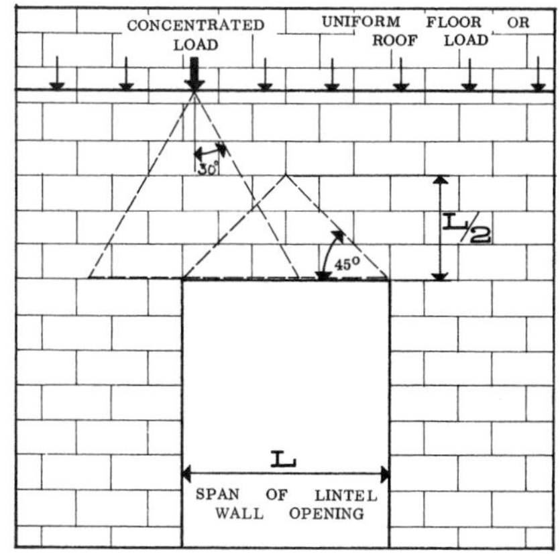

FIGURE 4.6-33 Assumptions for determining the loads imposed on lintels in masonry walls.

In North America, where labor costs are high, arches have for the most part been replaced by lintels. The most common lintel for supporting masonry over openings in concrete unit masonry is a reinforced concrete masonry lintel. Such lintels may be rectangular concrete units made either in the shop or in the field from U-shaped lintel block, reinforcing steel, and grout (Fig. 4.6-35).

The most common method of supporting masonry over exterior openings is by steel angles or structural shapes (see Fig. 4.6-35). Using galvanized steel lintels will reduce the possibility of rusting, but will slightly increase the initial costs of the lintels. Stainless steel lintels would eliminate maintenance and the possibility of failure resulting from corrosion altogether and, even though the initial costs are higher, may well be worth the difference.

When the steel reinforcement can be completely protected from the elements, reinforced masonry lintels have certain advantages over steel lintels. Their maintenance costs are lower, because the steel does not require periodic painting. Even the initial costs are lower, because less steel is required. However, if the wall around the lintel permits water to penetrate to the reinforcement, it will rust.

Lintels should bear on the wall not less than 4 in. (100 mm) at each end and should be of sufficient stiffness to carry the superimposed load without deflection of more than 1⁄360 of the clear span.

4.6.7 PIERS AND PILASTERS

Piers are freestanding columns; pilasters are columns that are bonded to and are built as integral parts of a wall. Both carry

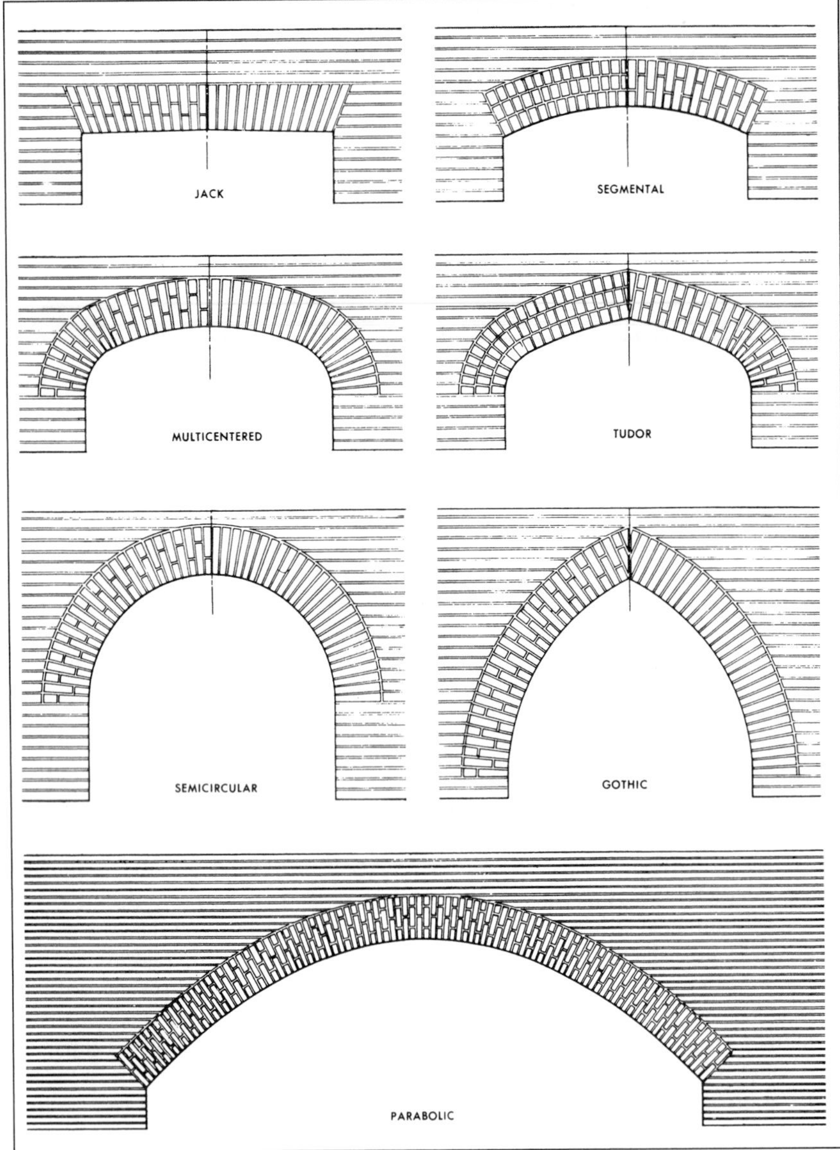

FIGURE 4.6-34 Typical brick arches. (Courtesy Brick Institute of America.)

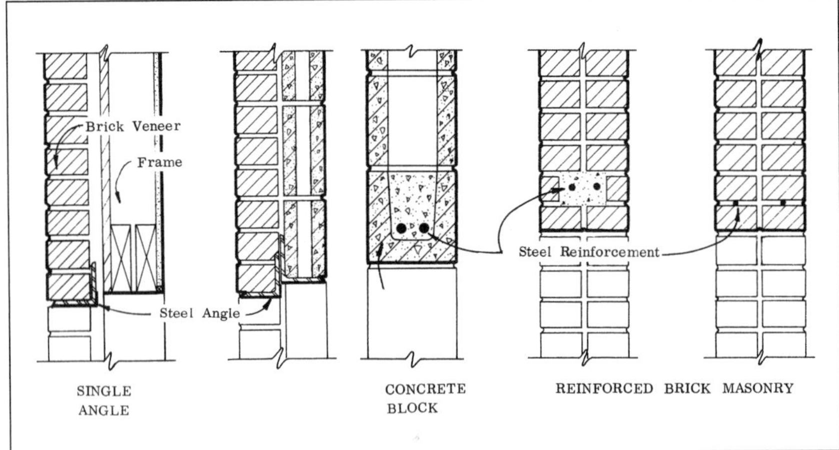

FIGURE 4.6-35 Typical lintels for clay and concrete masonry.

concentrated vertical loads. In addition, pilasters often provide necessary lateral support to walls.

Pier size is usually governed by code and is often a ratio of the height of the pier to its least cross-sectional dimension (h/t). The minimum pier section for solid units is usually 8 × 12 in. (200 × 305 mm) and for hollow units is usually 8 × 16 in. (200 × 400 mm).

The unsupported height of a solid masonry pier should not exceed 10 times its least cross-sectional dimension. The same is true of piers constructed with hollow masonry units, provided that the cells are filled with Type M or Type S mortar (see Section 4.2.4.1). Piers constructed with unfilled hollow masonry units may have unsupported heights not exceeding 4 times their least dimension.

4.6.8 FIRE RESISTANCE

The fire resistance of walls, floors, and partitions is usually based on ASTM E 119. The standard fire test does not measure the fire hazard in terms of actual performance in a real fire. Rather, it gives a comparison of the measure of performance between assemblies tested under similar conditions.

The test consists of exposing one side of a wall to a fire of controlled intensity measured by the standard time-temperature curve (Fig. 4.6-36). Immediately after firing, the hot face of the wall is subjected to a blast of water from a fire hose. The wall must withstand the fire and the hose stream without permitting flame or gases to pass through it, and the heat transmitted through the wall must be limited to the extent that there is less than a 250°F (121°C) rise in temperature. A load-bearing wall must also be able to resist its design stresses without failing. Within 72 hours after the tests have been completed,

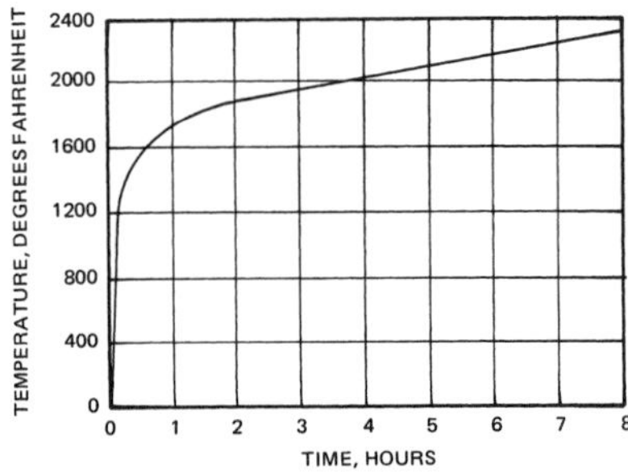

FIGURE 4.6-36 Standard time-to-temperature curve. (ASTM E 119 Fire Test). (Copyright ASTM; reprinted with permission.)

FIGURE 4.6-37 Ultimate Fire-Resistance Periods for Load-Bearing Clay and Shale Brick Walls

		Ultimate Fire Resistance Period, hours				
		Incombustible Members Framed into Wall or No Framed-in Members			Combustible Members Framed into Wall	
Nominal Wall Thick-ness, in.	Wall Type	No Plaster	Plaster[a] on One Side	Plaster[a] on Two Sides	No Plaster	Plaster on Exposed Side[a]
4	Solid	1¼	1¾	2½	—	—
8	Solid	5	6	7	2	2½
12	Solid[b]	10	10	12	8	9
12	Solid[c]	12	13	15	—	—
10	Cavity	5	6	7	2	2½

[a]To achieve these ratings, each plastered wall face must have at least ½ in., 1:3 gypsum-sand plaster.

[b]Based on load failure.

[c]Based on temperature rise (for non-load-bearing walls).

(Courtesy Brick Institute of America.)

bearing walls must safely sustain loads twice those applied during the test. Such tests provide the basis for estimating the performance of selected construction types.

Although some walls have ultimate fire-resistance test periods of more than 10 hours, they are often rated lower than their ultimate fire-resistance test periods because most building codes require ratings in multiples of 1 hour with a 4-hour maximum. Figure 4.6-37 gives ultimate fire-resistance periods for solid brick load-bearing walls. Figure 4.6-38 lists the equivalent thickness of concrete masonry units required for different aggregate types to give the fire ratings found in model building codes (see Section 1.6).

4.6.9 FIREPLACES AND CHIMNEYS

The design and construction of an efficient, functional fireplace and chimney requires adherence to some basic rules concerning dimensions and the placement of the various component parts. A correctly designed fireplace will:

- Ensure proper fuel combustion;
- Deliver smoke and other combustion products up the chimney;
- Radiate the maximum amount of heat into the room;
- Afford simplicity and fire safety in construction.

The first two of these objectives are closely related and depend mainly on the shape and dimensions of the combustion chamber, the proper location of the fireplace throat and the smoke shelf, and the ratio of the flue area to the area of the fireplace opening. The third objective depends on the dimensions of the combustion chamber, and the fourth depends on the size and shape of the masonry units and their ability to withstand high temperatures without warping, cracking, or deterioration. Typical details of fireplace construction are shown in Figure 4.6-39.

4.6.9.1 Separation from Other Construction

Fireplace walls should be separated from combustible construction as follows:

1. *Framing members:* 2-in. (50.8 mm) air space. This air space should be fire-stopped at floor level with an extension of ceiling finish, strips of fire-resistant board, or other noncombustible material.
2. *Subfloor and flooring:* ¾-in. (19 mm) air space. Trim in contact with the back of the fireplace should be of noncombustible material or should be fire-retardant wood.
3. *Wall sheathing:* 2-in. (50.8 mm) air space.
4. *Furring strips:* 1-in. (25.4 mm) air space, except that furring may be applied directly to masonry at sides or corners of fireplace.

Combustible materials should not be placed within 8 in. (200 mm) of the top or side edges of a fireplace opening.

When the floor construction or finish is combustible, an incombustible hearth should be installed. This hearth should extend at least 16 in. (400 mm) in front and

FIGURE 4.6-38 Estimated Fire-Resistance Ratings for Concrete Masonry Unit Walls of Equivalent Thickness[a]

	Members Framed into Wall or Partition						
	None or Noncombustible, in. (mm)[b]						
Concrete Masonry Units	4 hours	3 hours	2 hours	1½ hours	1 hour	¾ hour	½ hour
Expanded slag or pumice aggregates	4.7 (119.38)	4.0 (101.6)	3.2 (81.28)	2.7 (68.58)	2.1 (53.34)	1.9 (48.26)	1.5 (38.1)
Expanded clay or shale aggregates	5.7 (144.78)	4.8 (121.92)	3.8 (96.52)	3.3 (83.82)	2.6 (66.04)	2.2 (55.88)	1.8 (45.72)
Limestone, cinders, or unexpanded slag aggregates	5.9 (149.86)	5.0 (127)	4.0 (101.6)	3.4 (86.36)	2.7 (68.58)	2.3 (58.42)	1.9 (48.26)
Calcareous gravel aggregates	6.2 (157.48)	5.3 (134.62)	4.2 (106.68)	3.6 (91.44)	2.8 (71.12)	2.4 (60.96)	2.0 (50.8)
Siliceous gravel aggregates	6.7 (170.18)	5.7 (144.78)	4.5 (114.3)	3.8 (96.52)	3.0 (76.2)	2.6 (66.04)	2.1 (53.34)

[a]Equivalent thickness is the solid thickness that would be obtained if the same amount of concrete contained in a hollow unit were recast without core holes.

[b]Where combustible members are framed into the wall, the wall must be of such thickness or be so constructed that the thickness of solid material between the end of each member and the opposite face of the wall, or between members set in from opposite sides, will not be less than 93% of the thickness shown in this table.

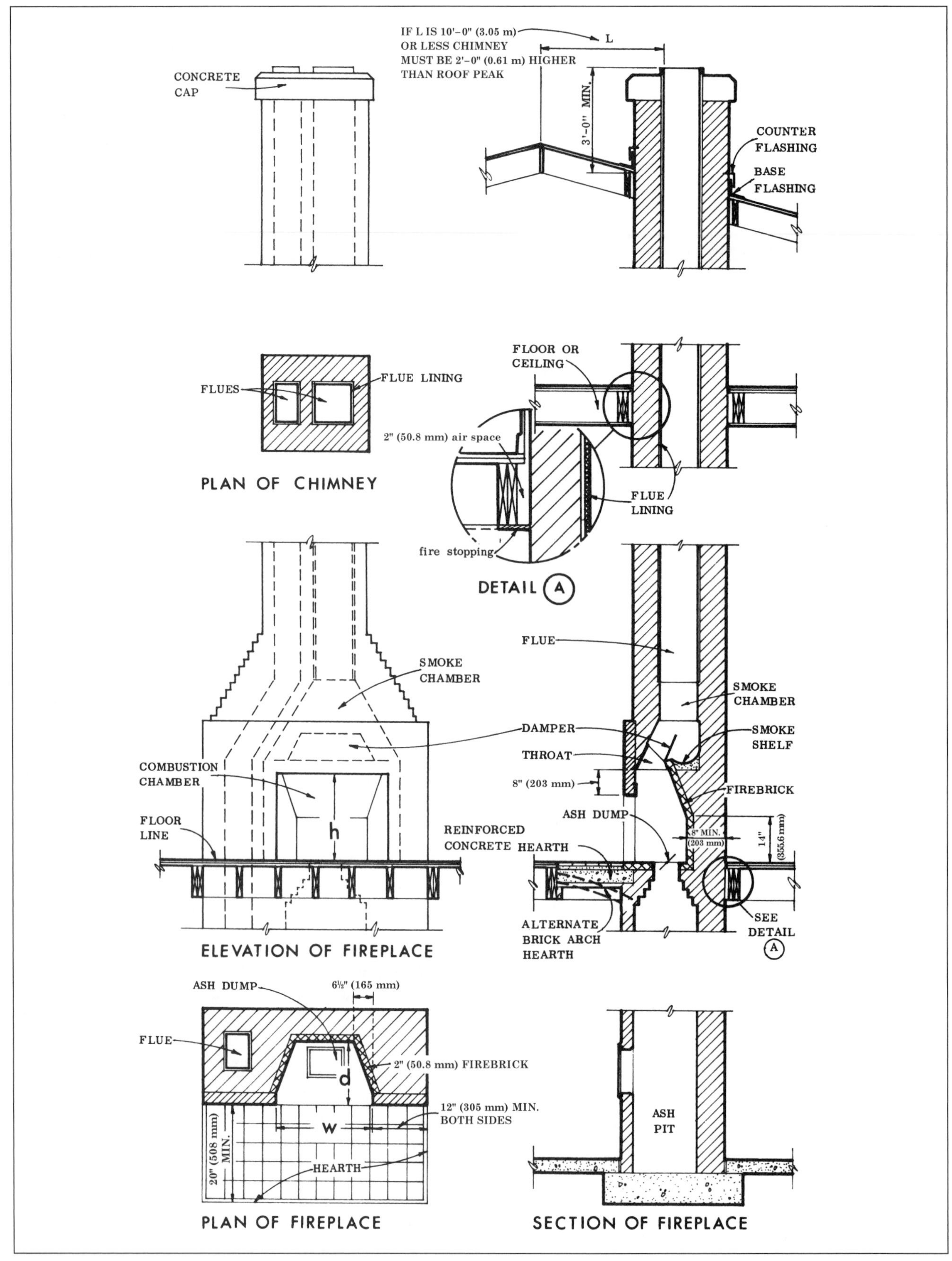

FIGURE 4.6-39 Typical fireplace and chimney construction details. Hearths are often cantilevered concrete slabs rather than the arch-supported or floor-framing-supported structures shown. Modern codes and regulations often require fresh air intakes near a fireplace to provide combustion air.

8 in. (200 mm) beyond each side of the fireplace opening when the opening is less than 6 sq. ft. (0.56 m^2) in area, and 20 in. (508 mm) in front and 12 in. (305 mm) to the sides when the opening is larger.

Wood framing and furring members should not be placed closer than 2 in. (50.8 mm) from the walls of a chimney. The space between chimneys and floor members should be filled with an incombustible material providing a firestop to prevent the transfer of fire from below through the space.

4.6.9.2 Size and Shape

The size of a fireplace depends not only on aesthetic considerations, but also on the size of the room to be heated. For a room with 300 sq. ft. (27.87 m^2) of floor area, a fireplace with an opening 30 to 36 in. (762 to 914 mm) wide is sufficient. Approximate dimensions for sizing a fireplace are given in Figure 4.6-40.

COMBUSTION CHAMBER

The shape of the combustion chamber influences both the draft and the amount of heat that will be radiated into the room. A slope at the back throws the flame forward and leads the gases, with increasing velocity, through the throat. Sloping the back and sides also helps radiate the maximum amount of heat into the room. The dimensions, wall angles, and slopes shown in Figures 4.6-39 and 4.6-40 may be varied slightly to correspond with brick coursing, but no major changes should be made in the dimensions shown.

The combustion chamber, unless it is a preformed metal one, should be lined with firebrick at least 2 in. (50.8 mm) thick laid in fireclay mortar. The back and end walls, including the thickness of the firebrick, should be at least 8 in. (200 mm) thick to support the weight of the chimney above.

THROAT

Because of its effect on draft, a fireplace's throat must be carefully designed. It should be not less than 8 in. (200 mm) above the highest point of the fireplace opening. The sloping back should extend to the same height and support the back of the damper.

A metal damper, preferably of a design in which the valve plate opens upward and toward the back, should be placed in the throat. It should extend the full width of the fireplace opening. When open, the damper's valve plate must form a barrier against downdrafts and deflect them upward with the ascending column of smoke. When the fireplace is not in use, the damper should be kept closed to prevent heat loss from the room and to keep dirt out of the flue.

SMOKE SHELF AND CHAMBER

The position of the smoke, or downdraft, shelf is established by the location of the throat. The smoke shelf should be directly under the bottom of the flue and extend horizontally the full width of the throat. The space above the shelf is called the smoke chamber. The back wall of the chimney should be built straight, and the other three sides should be sloped uniformly toward the center to meet the bottom of the flue lining.

Metal linings are available for smoke chambers to give the chamber its proper form. They provide smooth surfaces and simplify masonry installation.

FLUE

It is desirable to obtain relatively high velocities of flue gases and smoke through the throat and flue. The velocity is affected by both the area of the flue and the height of the chimney. The proper sizes of fireplace flues are given in Figure 4.6-40.

Each fireplace should have a separate flue with no other openings or connections. A flue lining should be supported on at least three sides by a ledge of protecting brick that finishes flush with the inside of the lining.

HEARTH

A widely used practice in hearth construction is to form a cantilevered rein-

FIGURE 4.6-40 Fireplace Types and Dimensions

Type	Opening Height h″ (mm)	Hearth Size w″ × d″ (mm)	Flue Size (standard modular)
Single Face	29 (736.6)	30 × 16 (762 × 406.4)	12 × 12 (304.8 × 304.8)
	29 (736.6)	36 × 16 (914.4 × 406.4)	12 × 12 (304.8 × 304.8)
	29 (736.6)	40 × 16 (1016 × 406.4)	12 × 16 (304.8 × 406.4)
	32 (812.8)	48 × 18 (1219.2 × 461.5)	16 × 16 (406.4 × 406.4)
Two Face (adjacent)	26 (666.7)	32 × 16 (812.8 × 406.4)	12 × 16 (304.8 × 406.4)
	29 (736.6)	40 × 16 (1016 × 406.4)	16 × 16 (406.4 × 406.4)
	29 (736.6)	48 × 20 (1219.2 × 508)	16 × 16 (406.4 × 406.4)
Two Face (opposite)	29 (736.6)	32 × 28 (812.8 × 711.2)	16 × 16 (406.4 × 406.4)
	29 (736.6)	36 × 28 (914.4 × 711.2)	16 × 20 (406.4 × 508)
	29 (736.6)	40 × 28 (1016 × 711.2)	16 × 20 (406.4 × 508)
Three Face (2 long, 1 short)	27 (685.8)	36 × 32 (914.4 × 812.8)	20 × 20 (508 × 508)
	27 (685.8)	36 × 36 (914.4 × 914.4)	20 × 20 (508 × 508)
	27 (685.8)	44 × 40 (1117.6 × 1016)	20 × 20 (508 × 508)

forced concrete underhearth, as shown in Figure 4.6-39. An alternative, more widely used in the past than today, is to construct an arch of brick from the chimney base to the floor construction. The arch is then filled to a level surface and receives the finished hearth (see Fig. 4.6-39).

EXTERIOR AIR SUPPLY

Some codes now require that each fireplace be provided with an exterior air intake to ensure proper combustion. This air intake should permit air to reach the fireplace either directly from the exterior or from an enclosed space open to the exterior, such as a crawl space or attic. The air intake should open into the side wall of the fireplace chamber or within 24 in. (406 mm) of the fireplace opening on or near the floor.

4.6.9.3 Fireplace Types

There are several types of masonry fireplaces being used today, but the basic principles involved in their design and construction are the same. The single-faced fireplace is the most common and the oldest variety, and therefore most of the standard design information is based on this type.

Multifaced fireplaces, if used properly, may be effective and attractive. They present some problems as to draft and opening size, which usually must be solved on an individual basis. Certain standard sizes that have been found to work satisfactorily under most conditions are shown in Figure 4.6-40.

Most fireplaces are not energy efficient, but can be made more so by the following practices:

1. Provide a source of outside air for fireplace combustion so that heated air is not drawn out of the house.
2. Adjust the damper to suit the type of fire. For example, a slow-burning log may need only a 1-in. (25.4 mm) or 2-in. (50.8 mm) damper opening to provide adequate combustion without smoking.
3. Provide a glass screen with an adjustable damper, so that the damper in the throat of the fireplace can be left open while the entrance of combustion air is controlled by the damper in the glass screen. When the fireplace is not in use, the glass screen can be closed, preventing heated inside air from being wasted.

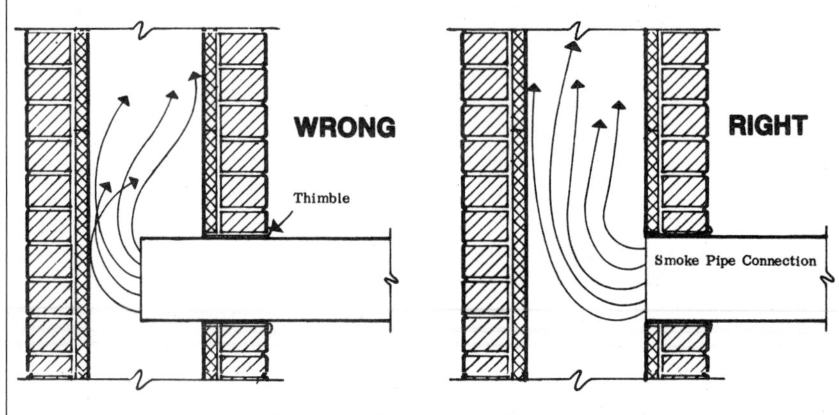

FIGURE 4.6-41 Smoke pipe connections to chimney.

4.6.9.4 Chimneys

Because of the large mass and weight of a chimney, it is quite possible that its foundation will have to support a greater load per square foot than the surrounding foundations. Therefore, special attention must be given to a chimney's footing and the rest of the foundation's design so that the unit load (psi) on the soil under all portions of the building's foundation will be nearly the same. If not, there may be uneven settling of the foundation, which may result in masonry cracking or other damage to the building.

The thickness of the masonry in the base of a chimney that contains only flues may be 4 in. (100 mm) thick, but if a fireplace is included it must be at least 8 in. (200 mm) thick. The ash pit in the base should be fitted with a tight cast-iron door and a frame of the same material anchored securely approximately 16 in. (406 mm) above the floor. If there is no basement, the door should be located one or two courses above the grade line. A flue lining is not required 8 in. (200 mm) or more below the point where the lowest smoke pipe thimble or flue ring enters the flue.

Each flue must be built as a separate unit, entirely free from other flues or openings. Joints in flue walls should be completely filled with mortar. In masonry chimneys with walls less than 8 in. (200 mm) thick, liners should be separated from the chimney wall and the space between the liner and masonry should be left unfilled; only enough mortar should be used to make a good joint and hold the liners in position. Exposed joints inside a flue should be struck smooth. The interior surface should not be plastered.

Although unlined fireplace chimneys with masonry walls at least 8 in. (200 mm) thick may perform adequately, the use of a fireclay flue liner is recommended. A flue lining provides a smooth interior surface that permits more efficient operation of the flue.

The tops of chimneys without flue linings may have to be rebuilt every few years because of the disintegrating effect of smoke and gases on the mortar. When rebuilding is necessary, a chimney must be taken down to a point where the mortar joints are solid and a new top built.

Each flue lining section should be set in place before the brick has been laid to the top of the lining section below, and then the brick should be built up around the just-placed lining to a level below the next section. Offsets and bends should be made by equally mitering both ends of abutting sections of lining. This prevents reduction of the flue area, so that the same effective area is maintained for the entire height of a flue.

Smoke pipe connections should enter the side of a flue at a thimble or flue ring, which should be built as the work progresses. A metal smoke pipe should not extend beyond the inside face of the flue (Fig. 4.6-41). The top of a smoke pipe should not be less than 18 in. (457 mm) below the ceiling, and no wood or combustible materials should be placed within 6 in. (150 mm) of a thimble.

A flue lining should project at least 4 in. (100 mm) above the top course or capping of the chimney and should be surrounded with portland cement mortar at least 2 in. (50.8 mm) thick, finished with a straight or concave slope to direct the air currents upward at the top of the flue. This wash also serves to drain water from the top of the chimney and should be constructed with a drip to keep the walls dry and clean.

Unless the applicable code has more stringent requirements, chimneys should extend at least 2 ft. (609 mm) above every part of the roof that lies within 10 ft. (3048 mm) of the chimney and at least 3 ft. (914 mm) above the roof at the point of penetration.

After a chimney has been completed and the masonry cured, a test of each flue should be conducted. The test is made by building a smudge fire at the top of the flue and, when smoke is pouring freely from the top, covering the top tightly. Any escaping smoke indicates openings that must be repaired. Because such repair is usually difficult and expensive, it is far more satisfactory and economical to see that construction is properly executed as the work progresses.

Smoke passages and chimney flues should be kept clean. The most efficient method of cleaning a chimney flue is with a weighted brush or bundle of rags lowered and raised from the top. An accumulation of soot and resins may cause a chimney fire whose sparks may ignite the roof. A chimney fire may also damage the chimney, which may in turn permit the passage of more fire and cause significant damage to the building.

4.7 Unit Masonry Erection

Selection of masonry materials is covered in Sections 4.2, 4.3, 4.4, and 4.5; design of masonry assemblies, in Section 4.6. This section addresses masonry unit and mortar preparation, unit laying, mortar bedding and jointing, joint finishing, and flashing installation. All these must be done correctly if a completed masonry assembly is to be watertight, strong, and durable and present a satisfactory appearance.

4.7.1 PREPARATION OF MATERIALS

Protection during shipping, handling, and storage of masonry and associated materials and accessories is essential to prevent damage to them. Proper preparation is necessary before masonry and related materials and accessories are placed in a masonry assembly.

4.7.1.1 Storage and Protection of Materials

All materials that will be used in masonry assemblies should be stored under cover, off the ground, in a dry place, and protected from water and freezing. They should also be kept free of soil, ice, frost, and other contaminants. If they become contaminated during storage or handling, the contaminants must be removed before the materials are used.

It is common to place clay and concrete masonry units and cementitious mortar materials outdoors on skids or platforms and to cover them during inclement weather with tarpaulins, clear plastic film, or another suitable material. It is better to store these material indoors when possible.

Structural clay tile units, either glazed or unglazed, and terra-cotta units should be covered at all times. It is best to store these units indoors.

Sand and other aggregates are often not protected as carefully as they should be, which results in their becoming soaked with water or contaminated with soil or other foreign materials. Foreign materials, excess water, and ice must be removed before aggregates are used. If different aggregates become mixed together during storage, it is necessary to grade them again, which is difficult, or to discard them, which is expensive.

Metal ties, anchors, and reinforcement must be kept dry and protected from contact with soil and away from conditions that would harm them. Before they are placed, these items must be free of loose rust and other coatings that would destroy or reduce bond.

4.7.1.2 Clay Masonry

The *suction* (initial rate of absorption) of a brick is the amount of water in grams absorbed by 30 sq. in. (0.02 m^2) of surface when the brick is placed in water to a depth of ⅛ in. (3.2 mm) for 1 minute, as tested in accordance with ASTM C 67 (Fig. 4.7-1). In almost every case, mortar bonds best to brick whose suction is 5 to 20 grams of water.

Bricks having a suction in excess of 20 grams of water per minute should be wetted sufficiently so that the suction when

FIGURE 4.7-1 Apparatus for testing the suction of a brick. (Courtesy Brick Institute of America.)

the brick is laid does not exceed that amount. If high-suction bricks are not wetted, they will absorb water from the mortar before a bond has developed. Sprinkling brick to wet it will not produce the desired results. Rather, a hose stream should be played on each brick pile until water runs from all sides. Wetting should be done several hours before the units are laid, and preferably the previous day. The surfaces of wetted brick must be allowed to dry before the brick is laid. Water on the surface of a brick will cause it to *float* on the mortar bed and prevent it from bonding with the mortar.

Conversely, if bricks with a low suction are wetted, or if other bricks are wetted too much, they may become saturated. Saturated brick will not absorb enough moisture from the mortar and will float, which usually will not permit the formation of a sufficient bond between the brick and the mortar. When brick suction exceeds 60 grams of water at the time of laying, the bond may be extremely poor regardless of the quality of workmanship or mortar type used.

An effective on-the-job method of determining whether a brick should be wetted before it is laid is to sprinkle drops of water on its flat side. If these drops are absorbed completely in less than 1 minute, the brick should be wetted. A more accurate method is to draw a circle on the flat side of a brick using a wax pencil and a quarter as a guide. With a medicine dropper, apply 20 drops of water to the surface of the brick inside the circle. If the water is completely absorbed in less than 1½ minutes, the brick should be wetted before it is laid.

4.7.1.3 Concrete Unit Masonry

ACI 530.1, ASCE 6, TMS 402 specifically prohibit the wetting of concrete masonry units. Therefore, care must be taken to keep these units dry on the job. They should be stockpiled on planks or other supports, free from contact with the ground, and covered for protection against wetting. Concrete units should never be wetted before or during laying in a wall.

The suction test described in Section 4.7.1.2 is not applicable to concrete masonry units. A concrete masonry unit with a moisture content less than 20% of the total amount it can absorb can draw water out of mortar on hot, dry days, but concrete masonry unit suction is usually insignificant because the units generally contain sufficient moisture to retard suction effects.

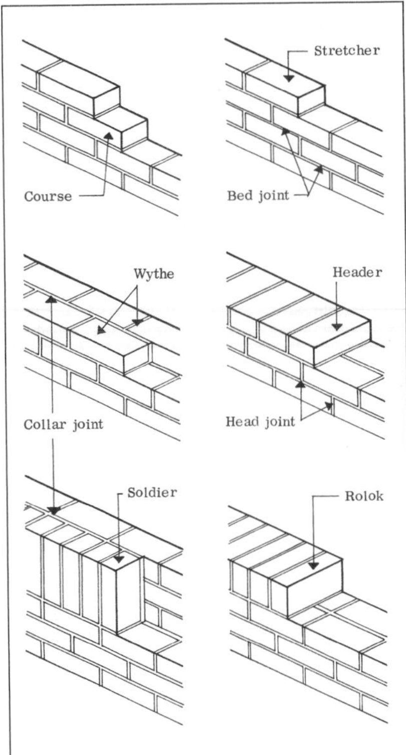

FIGURE 4.7-2 The names of the various positions in which masonry units may be placed in walls, and the names given to various joint locations.

4.7.1.4 Mortar

The strength and resistance of masonry walls to water penetration are dependent on the completeness and strength of the bond between the mortar and the masonry units. Mortar characteristics that affect bond strength are discussed in Section 4.2.1. Mortar preparation includes mixing the mortar and maintaining the proper water ratio until the mortar is used. Mixing is discussed in Section 4.2.6; retempering to control water content in Section 4.2.7. Mortar application techniques that affect bond strength and water penetration resistance are discussed in Section 4.7.2.5.

4.7.2 INSTALLATION OF SOLID AND HOLLOW MASONRY UNITS

The installation of masonry units requires laying the units, bonding them together, and anchoring them to the structure. It also includes installing continuous joint reinforcement, building in bearings for supported structural members, mortar bedding and jointing, backparging, joint finishing, and installing flashing and weep holes.

Some specific terms used to describe the positions of masonry units in a wall are shown in Figure 4.7-2.

4.7.2.1 Laying Masonry Units

Many general techniques are applicable to the laying of both solid and hollow masonry units, including clay brick, hollow clay brick, structural clay tile, concrete brick, and concrete block.

Masonry should be laid *plumb* (exactly vertical) and *true* (accurately placed) to lines. Each unit should be adjusted to its final position while the mortar is still soft and plastic. A mason should avoid overplumbing, tapping, pounding on, or otherwise attempting to realign units after they have been set in position (Figs. 4.7-3a and 4.7-3b). Such movements will often break the bond between the unit and the mortar. When the bond has been bro-

a

b

FIGURE 4.7-3 Moving masonry units after they have been laid can result in cracks. Even hammering a brick back in line, as shown in (a), can break the bond. Cracks caused by realigning units often occur at corners, as shown in (b). (Courtesy Brick Institute of America.)

FIGURE 4.7-4 Using a story pole. (Courtesy Portland Cement Association.)

FIGURE 4.7-6 Aligning the faces of a course of blocks and leveling the units. (Courtesy Portland Cement Association.)

FIGURE 4.7-7 Aligning units at a corner and checking for plumb. (Courtesy Portland Cement Association.)

ken in this manner, it usually remains broken regardless of a mason's attempts to reestablish it. When units must be adjusted after the mortar has started to harden, the units and the mortar should be removed and the units should be laid again using new mortar.

The highest bond strength is achieved when the time interval between the placing of a mortar bed and the laying of a masonry unit into it is kept to a minimum. In addition, a change in the time interval between the spreading of a mortar and the placing of masonry units into that mortar will affect the flow of the mortar (see Section 4.2.1.1), especially if the masonry units have a high suction (see Section 4.7.1.2). Varying this interval to any large extent will result in a wall with varying tensile bond strengths.

Concrete masonry units and some other hollow masonry units have thicker face shells on one side than on the other, and should be laid with the thicker face shell up to provide a larger mortar bedding area.

Anchors, accessories, flashings, and other items required to be built into masonry should be built in as the work progresses.

FIGURE 4.7-5 Spacing units at a corner without mortar before laying them. (Courtesy Portland Cement Association.)

When fresh masonry joins masonry that is partially or totally set, the exposed surface of the set masonry should be cleaned, and in the case of clay masonry wetted, so as to obtain the best possible bond with the new work.

A board marked with the courses (*story pole* or *course pole*) provides an accurate method of finding the top of the masonry for each course (Fig. 4.7-4). The same story pole should be used throughout the work.

STARTING THE WORK

After locating the corners, a mason often spaces out the units dry (without mortar) to determine the extent to which they must be cut or have their joint sizes varied to accomplish accurate horizontal coursing (Fig. 4.7-5).

A mortar bed is then spread and the units bedded in it (see Section 4.7.2.5). The corner unit is laid first and carefully placed in its correct position.

After several units have been laid, a mason's level is used as a straightedge to verify that they are in correct alignment at the proper grade and are plumb. Corrections are made by tapping the units with a trowel handle (Fig. 4.7-6). The first course of masonry should be laid carefully to ensure that it is properly aligned, leveled, and plumbed. This will assist the mason in laying succeeding courses and in building a straight, true wall.

LAYING UP THE CORNERS

After the first course has been laid, the corners of the wall are built before the rest of the wall is laid. The corners are started by laying up several courses higher than the center of the wall. Each course is stepped back by one-half unit. As each course is laid, it is checked with a level for alignment and plumb (Fig. 4.7-7) and for being level (Fig. 4.7-8). At the same time, each course is checked carefully with a level or straightedge to make certain that the faces of the units are all in the same plane so that the wall will be true and straight (Fig. 4.7-9).

FIGURE 4.7-8 Checking to see whether a unit is level. (Courtesy Portland Cement Association.)

LAYING BETWEEN THE CORNERS

Before filling in the wall between the corners, the mason stretches a line from corner to corner for each course, then lays the top outside edge of the units to this line. In laying concrete masonry units to a mason's line, the way a mason handles and grips the units is important. Tipping the units slightly toward the mason will permit seeing the upper edge of the course below, thus enabling the placement of the block's lower edge directly over the course below (Fig. 4.7-10). By rolling the block slightly to a vertical position and shoving it against the adjacent block, the mason can lay it to the line with minimum adjustment.

FIGURE 4.7-9 Checking to see whether units are in line. (Courtesy Portland Cement Association.)

FIGURE 4.7-10 A mason positions units between the corners by laying them to a line. (Courtesy Portland Cement Association.)

As stated earlier, adjustments to final position must be made while the mortar is soft and plastic. By tapping lightly with a trowel handle, a mason can level and adjust each unit to the mason's line. The use of a level between corners is necessary only to check the alignment of each unit with the face of the wall.

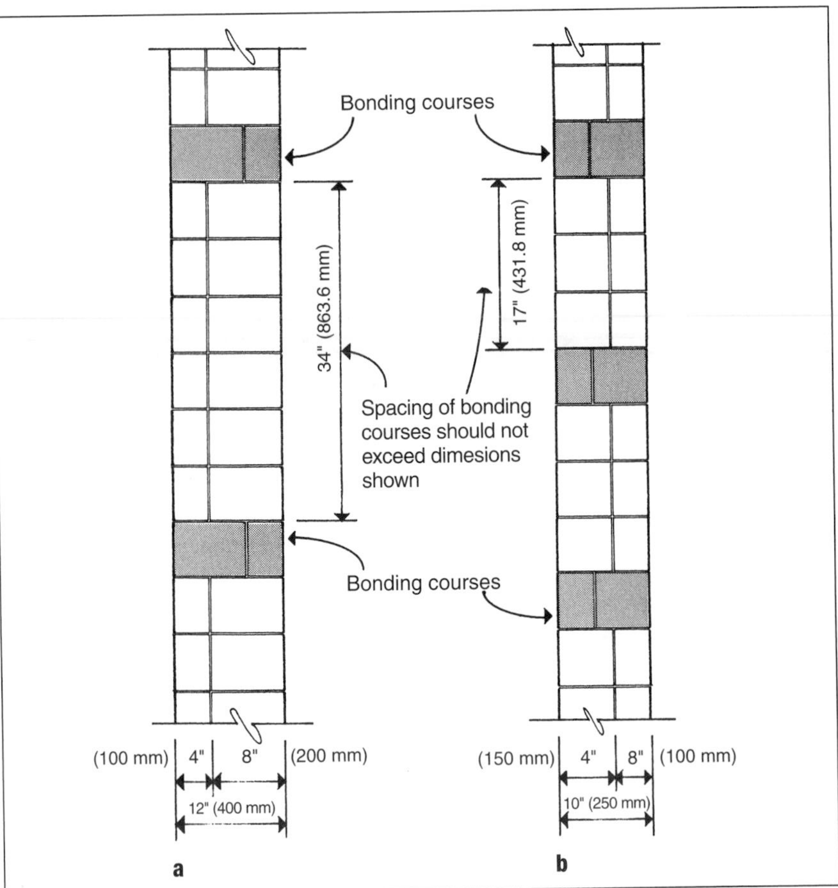

FIGURE 4.7-11 The overlapping stretcher method of bonding.

4.7.2.2 Bonding and Anchoring

The units in a single-wythe masonry wall are bonded together by the wall's pattern bond.

Two methods are generally used to structurally bond together the several wythes of wall assemblies that have more than one wythe of masonry. One method (*masonry bonded*) is accomplished by overlapping and interlocking the masonry units. The second (*metal tied*), which is the preferred method in most cases, bonds the masonry together using metal ties embedded in the mortar joints.

Masonry must also be anchored to other structural elements when it relies on them for support. Most such anchoring is done using metal anchors.

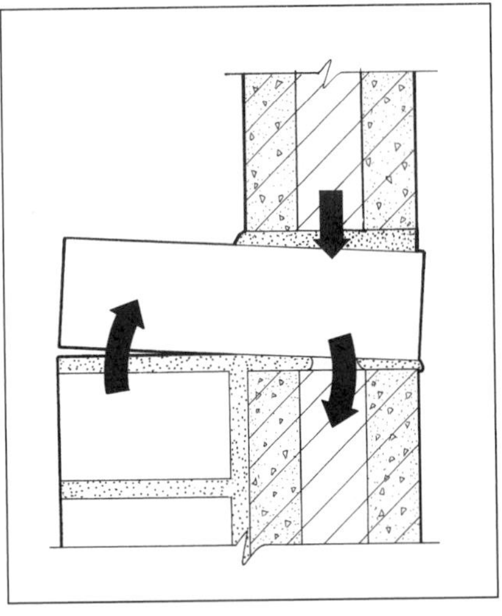

a b

FIGURE 4.7-12 (a) The effect of a heavy backup unit on a masonry header. (b) A mason laying a header course in an 8-in. (200 mm) solid brick and block wall. In this case, the header courses occur in every seventh brick course. (Courtesy Portland Cement Association.)

MASONRY BONDED SINGLE-WYTHE WALLS

The units in a single-wythe brick wall are bonded together by building the wall in the running bond pattern shown in Figure 4.6-14. Other bond patterns can be used in single-wythe walls, of course, but the headers must be cut to a length that matches the runner's thickness. This method is unsatisfactory when both sides of the wall are exposed, because the finish of the cut units cannot be made to match the face texture of the brick. When a stack bond pattern is used, it is usually necessary to provide continuous joint reinforcement (see Section 4.7.2.3).

FIGURE 4.7-13 A metal bar anchor being installed to tie a bearing wall to an intersecting wall. A metal lath was used to support the mortar filling the core where the turned-down end of the bar is being inserted. (Courtesy Portland Cement Association.)

FIGURE 4.7-14 The core into which the upturned end of the metal tie bar shown in Figure 4.7-13 projects is here being filled with mortar. (Courtesy Portland Cement Association.)

The units in a single-wythe concrete unit masonry wall are bonded together by building the walls in one of the patterns shown in Figure 4.6-16. Stack bond patterns require the use of continuous joint reinforcement (see Section 4.7.2.3).

Corners and intersections in a single-wythe brick or concrete unit masonry nonbearing wall or partition can be bonded by laying at least 50% of the units in an overlapping pattern. Each overlapping unit should bear at least 3 in. (76.2 mm) on the unit below.

Masonry bonded load-bearing single-wythe brick and concrete unit masonry walls may be bonded at intersections by overlapping 50% of the units, with alternate units bearing at least 3 in. (76.2 mm) on the unit below. This method is not recommended by some experts, however, except in earthquake zones. In other areas, they prefer that such walls be masonry bonded only at corners, and be built up separately at other intersections, and tied together there with metal connectors, as discussed later in this section.

FIGURE 4.7-15 A strip of galvanized hardware cloth being used to tie a non-load-bearing wall to an intersecting wall. (Courtesy Portland Cement Association.)

FIGURE 4.7-16 This hardware cloth strip was left protruding from the completed wall at the right and is now being bedded into a joint in an intersecting wall. (Courtesy Portland Cement Association.)

MASONRY BONDED MULTIPLE-WYTHE WALLS

The bonding of the units in a masonry bonded multiple-wythe wall is based on variations of the two traditional methods of bonding: (1) English bond, consisting of alternating courses of headers and stretchers, and (2) Flemish bond, consisting of alternating headers and stretchers in every course (see Fig. 4.6-14). In this regard, common bond is but a variation of English bond. English and common bonds are seldom used today in this country in contemporary construction. Flemish bond sometimes appears in contemporary churches. Of course, any of the three may be needed to match existing construction.

Overlapping stretchers, laid parallel with the length of the wall, develop longitudinal bonding strength; headers laid across the width of the wall bond the wall transversely. Some types of masonry bonded walls are shown in Figure 4.6-1, using brick (4.6-1H), overlapping stretchers (4.6-1G), brick and block, or tile backup (4.6-1J).

Masonry bonded brick walls should be built so that not less than 4% of the wall's surface is composed of headers. The headers should extend not less than 4 in. (100 mm) into the backing and should be not more than 24 in. (610 mm) apart either vertically or horizontally. This means that there must be at least one header course in every seven brick courses. In multiple-wythe tile or block walls, structural bonding is achieved by overlapping stretcher units rather than by using header units (see Fig. 4.6-1G). The stretcher units may be bonded at vertical intervals not exceeding 34 in. (864 mm) when the bonding course can overlap one-half of the unit below and the lap is at least 4 in. (100 mm) (Fig. 4.7-11a). When these conditions cannot be met, the bond may be achieved by overlapping headers that are at least 50% greater in thickness than the units below, and planning the bond courses at vertical intervals of 17 in. (432 mm) (Fig. 4.7-11b).

Unfortunately, water can penetrate a multiple-wythe masonry bonded wall fairly easily, for several reasons:

1. It is sometimes difficult to obtain full mortar joints between brick headers. This permits water to find its way to the inside face of the wall along the headers, which are continuous through the full wall thickness.

2. It is difficult to bond brick and larger, heavier backup units without impairing the bond between the headers and the mortar. If the backup unit is set on the header before the mortar has had time to stiffen, the header will settle unevenly and a crack may develop (Fig. 4.7-12).

3. Perhaps the most important reason that water may find its way through a masonry bonded wall is that there are differences in thermal, moisture, and other movements between different facing and backup materials. The differences are greatest between brick and concrete masonry units where movement of the concrete masonry unit backup, which shrinks as it dries, and the brick, which progressively expands over time, creates an eccentric load on the brick headers. This tends to rupture the bond between the headers and the mortar at the external face. The flexibility inherent in metal ties negates this disadvantage.

METAL-TIED, SINGLE-WYTHE WALLS

The units in the field of metal-tied, single-wythe walls are not bonded using metal ties. Metal ties are used, however, where such walls abut other walls.

When a single-wythe, load-bearing brick or concrete masonry unit wall is not masonry bonded to the walls it abuts, except at the corners, it should terminate at the faces of these other walls in control joints. For lateral support, bearing walls should be tied to walls they intersect with rigid steel tie bars ¼ in. (6.4 mm) thick, 1¼ in. (31.8 mm) wide, and 28 in. (711 mm) long, with a 2-in. (50.8 mm) right angle bend on each end (Fig. 4.7-13). Tie bars should be placed not more than 4 ft. (1219 mm) apart vertically. The bends at the ends of the tie bars should be embedded in cores filled with mortar (Fig. 4.7-14).

Single-wythe non-load-bearing masonry unit walls or partitions that are 6 in. (152 mm) or less in thickness may be tied to walls they intersect using strips of metal lath or ¼-in. (6.4 mm) mesh, galvanized hardware cloth placed across the joint between the two walls (Fig. 4.7-15). These metal strips should be placed in alternate courses (not more than 16 in. [406 mm] apart vertically) in the wall. When one wall is constructed first, the metal strips should be built into this wall and later embedded into the mortar joint in the second wall as it is built (Fig. 4.7-16). Corrugated metal tie anchors may be used in the same way. Where the two walls meet, a control joint should be provided if the joint is exposed to view.

METAL-TIED, MULTIPLE-WYTHE WALLS

Both multiple-wythe solid and cavity walls can be structurally bonded using metal ties (see Fig. 4.6-1). For exterior walls, especially where weather resistance is important or the physical characteristics of the facing and backing materials are different, bonding with metal ties is better than the masonry bonded method. This is because metal-tied walls have greater resistance to moisture penetration and the ties allow slight differential movement between the facing and backing units, which may help to relieve stress and reduce cracking.

The usual requirements for metal-tie bonding are for 3/16-in. (4.8 mm)-diameter steel ties, or metal ties of equivalent stiffness and strength, to be spaced so that at least one tie occurs in every 4½ sq. ft. (0.42 m^2) of wall area. Except where different spacings are required by code or local practice, and unless another spacing is recommended later in this section for a specific wall type or condition, in solid, composite, and faced walls the maximum vertical distance between ties should not exceed 18 in. (457 mm) and the maximum horizontal distance should not exceed 36 in. (914 mm) (Fig. 4.7-17). Ties in alternate courses should be staggered (see Fig. 4.7-17). Ties should be embedded in horizontal joints of the facing and backing. Additional bonding ties should be provided at openings. These should be spaced not more than 3 ft. (914 mm) apart around the perimeter and within 12 in. (305 mm) of the openings and, in addition, on both sides of expansion and control joints.

General and specific recommendations for some types of walls and conditions are summarized below. Notice that some conditions require masonry bonding or a combination of masonry bonding and metal ties.

Reinforced Masonry Walls The units in a single-wythe hollow reinforced masonry wall are bonded together by their pattern bond (see Figs. 4.6-14 and 4.6-16) and by the reinforcing steel and the grout in their cores (see Section 4.6.1.4).

The units in the separate wythes of reinforced solid masonry walls are bonded together by their pattern bond (see Figs. 4.6-14 and 4.6-16). The wythes are structurally bonded together by the reinforcing steel and grout in the cavity (*collar joint*) between the wythes of the wall (see Section 4.6.1.4).

Faced Walls The facing should be bonded to the backing with metal ties.

Cavity Walls Cavity walls should be bonded with metal ties. Their spacings should not exceed 24 in. (610 mm) vertically or 36 in. (914 mm) horizontally.

Brick Veneer Walls In wood frame construction (see Fig. 4.6-11a) brick veneer is often anchored to the backup wall by nailing corrosion-resistant corrugated metal ties through the sheathing to the studs. This type of construction has been used for many years, but recently some experts have begun to question its longevity. Failure may result from rusting of the ties due to water reaching them from leaks or from condensation due to high humidity. Professional guidance should be sought before electing to use this method of construction, and local practice should be followed.

There is a great deal of controversy about whether brick veneer should be used over metal studs (see Fig. 4.6-11b). There are more questions about using this combination on tall structures than about its use on 1- and 2-story residences and other small buildings. Its supporters claim that brick veneer over metal studs will give satisfactory long-term results if the proper materials are selected and are applied correctly. Detractors claim that brick veneer applied over metal studs will eventually crack, regardless of the materials used or the configuration of the wall. This will allow water to enter the wall and corrode the metal supports, resulting in veneer failure.

The key elements in the controversy are the required stiffness of the studs, the types of anchors to use, and whether any

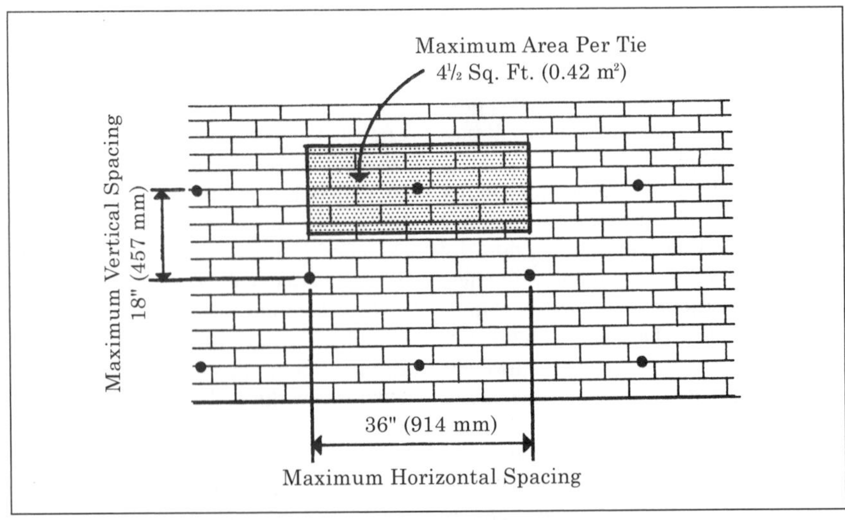

FIGURE 4.7-17 Wall tie locations in multiple-wythe walls.

available anchoring system will perform satisfactorily in the long run. The most authoritative current advice about the methods needed to construct such a brick veneered wall are offered by the BIA. However, before deciding to use brick veneer over metal studs it would be wise to contact industry sources, such as BIA, the Portland Cement Association (PCA), the Metal Lath/Steel Framing Association, and the manufacturers of the products that will be used, and obtain their latest recommendations concerning this controversial use of brick. It is also necessary to determine if this type of construction is permitted by the applicable codes and the owner. There has been some movement by some federal agencies and states to ban its use altogether.

ANCHORAGE

Masonry assemblies must be securely anchored to elements that are used for its vertical and horizontal support. These include steel and concrete columns and beams, concrete walls, concrete and masonry bond beams, and floors, other walls, and roofs. This anchoring is usually done with flexible metal anchors that permit at least some differential movement.

Exterior walls facing or abutting concrete members should be anchored by dovetailed anchors inserted in slots set into the concrete (see Fig. 4.3-1). These anchors should be spaced not more than 18 in. (457 mm) vertically and 24 in. (610 mm) horizontally.

4.7.2.3 Continuous Joint Reinforcement

Continuous joint reinforcement is often used today to control cracks in masonry walls (see Figs. 4.3-1, 4.6-1, and 4.6-9). When joint reinforcement is used, it often replaces and does the same job usually done by metal wall ties. In wythes laid in a stacked bond pattern, continuous joint reinforcement is necessary to bond the units together. Continuous joint reinforcement is also sometimes used in lieu of horizontal reinforcement in reinforced masonry walls.

Continuous joint reinforcement should be installed in every masonry wall and partition that is more than 20 ft. (6.1 m) long, and is usually installed in every wall regardless of length. Using continuous joint reinforcement not only reduces the amount of cracking associated with thermal movements, it also permits a greater spacing of control joints (see Fig. 4.6-22).

Joint reinforcement should be embedded in horizontal mortar joints so that it is at least ⅝ in. (15.9 mm) from the masonry surface and should be continuous except through control and expansion joints. Joint reinforcement sections should be lapped at least 6 in. (152 mm), and the lap should contain at least one cross-wire of each section.

Continuous joint reinforcement should be installed in most cases at not more than 16 in. (406 mm) on center vertically. In reinforced walls, continuous reinforcement is sometimes placed at closer spacings. In parapet walls and foundation walls, reinforcement is often spaced every 8 in. (203 mm). Local codes sometimes govern this placement. To compensate for loss of reinforcement at wall openings, joint reinforcement should be installed in the first and second bed joints immediately above and below openings and should extend at least 24 in. (610 mm) beyond the ends of sills and lintels.

4.7.2.4 Masonry Support for Structural Members

The bearing course of masonry under structural members such as floor joists, concrete slabs, beams, and lintels should be at least 8-in. (203 mm)-thick solid masonry or hollow masonry with the cells filled solidly with mortar or fine grout.

4.7.2.5 Mortar Bedding and Jointing

Laboratory tests and field observations of masonry indicate that good masonry construction demands that the joints be completely filled with mortar as the units are laid. Partially filling joints results in leaky walls, reduces the strength of masonry to between 50% and 60%, and contributes to disintegration and cracking when water penetrates the wall and expands as it freezes.

BRICK

Clay and concrete brick and other solid masonry units should generally be embedded and jointed in similar ways. Every head, bed, and collar joint in them should be filled solidly with mortar.

Mortar joints in contemporary brick walls usually vary from ¼ in. (6.4 mm) to ½ in. (12.7 mm) in thickness, depending on the size of the brick and the bonding. Thicker joints are sometimes used, however.

Bed Joints The mortar in bed joints in brick should be spread to a uniform thickness and furrowed only slightly (Fig. 4.7-18). As the brick is laid, excess mortar will fill the furrow and ensure full bed joints. If the bed is too thin or the furrow is too deep, there will not be enough mortar to fill the furrow completely, and the resulting void will enable water, entering an opening in the mortar or between the mortar and the brick, to penetrate the wall (Fig. 4.7-19).

To ensure a good bond, mortar should not be spread too far ahead of laying block or brick, or it will stiffen and lose its plasticity (Fig. 4.7-20). This is especially true in hot weather when absorbent bricks are being laid. One way to determine if bond is developing is to remove a freshly laid brick from the wall (Fig. 4.7-21). If the mortar is soft and plastic, it will stick to the brick placed on top of the bed as well as to the brick on which the bed is spread, and adequate bond will be attained.

FIGURE 4.7-18 In brick, the mortar furrow in the bed joint should be shallow. (Courtesy Brick Institute of America.)

FIGURE 4.7-19 The furrow in the bed joint shown here is too deep. (Courtesy Brick Institute of America.)

FIGURE 4.7-20 Mortar should be spread only a few units ahead of the masonry. (Courtesy Brick Institute of America.)

FIGURE 4.7-21 **This mason has removed a newly laid brick to determine whether adequate bond is being achieved.** (Courtesy Brick Institute of America.)

FIGURE 4.7-22 **Loading a brick before laying it.** (Courtesy Brick Institute of America.)

Head Joints Head joints are more vulnerable to water penetration than bed joints because the weight of the wall above tends to close cracks in bed joints. Therefore, every head joint in both the facing brick and backup units should be completely filled with mortar. One method is to spread the mortar on the end of a brick before it is laid (Fig. 4.7-22). When the brick is shoved into place, mortar should squeeze out at the sides and top, indicating that the head joint is completely filled (Fig. 4.7-23). An alternate method is to throw plenty of mortar on the end of the brick already in place (Fig. 4.7-24), giving the desired result (see Fig. 4.7-23) when the next brick is shoved into place. A third method is to place a full trowel of mortar on the wall (Fig. 4.7-25) and shove the next brick into this deep mortar bed, which will squeeze mortar out at the sides and top of the joint (Fig. 4.7-26).

Dabs of mortar spotted on both corners of a brick (Fig. 4.7-27) will not completely fill the head joints (Fig. 4.7-28). Attempting to fill the joints from above after the brick is placed (Fig. 4.7-29), which is called *slushing*, cannot be relied upon to fill voids left in head joints (Fig. 4.7-30).

Cross joints in header courses must

FIGURE 4.7-23 **Mortar oozing out indicates that the joint has been filled completely.** (Courtesy Brick Institute of America.)

FIGURE 4.7-24 **Buttering the end of a brick that has already been laid.** (Courtesy Brick Institute of America.)

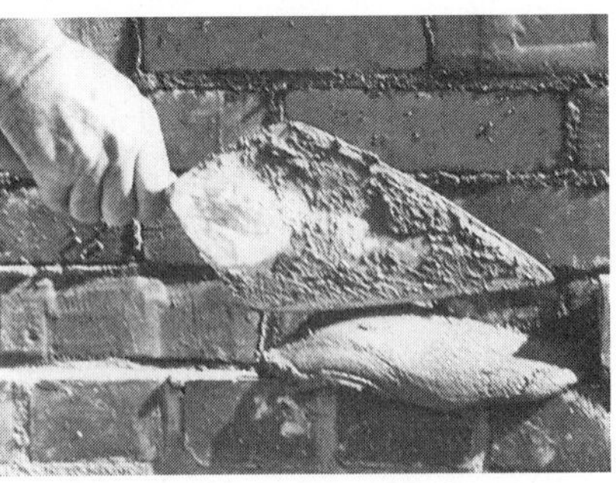

FIGURE 4.7-25 **Throwing a full trowel of mortar on an already installed brick to receive the next unit.** (Courtesy Brick Institute of America.)

FIGURE 4.7-26 **A brick being shoved into the mortar that was thrown on in Figure 4.7-25.** (Courtesy Brick Institute of America.)

FIGURE 4.7-27 **Spotting the corners of a brick. This will produce unsatisfactory head joints and must not be permitted.** (Courtesy Brick Institute of America.)

FIGURE 4.7-28 **The result of spotting the corners as shown in Figure 4.4-27. Note that the head joint is not filled completely.** (Courtesy Brick Institute of America.)

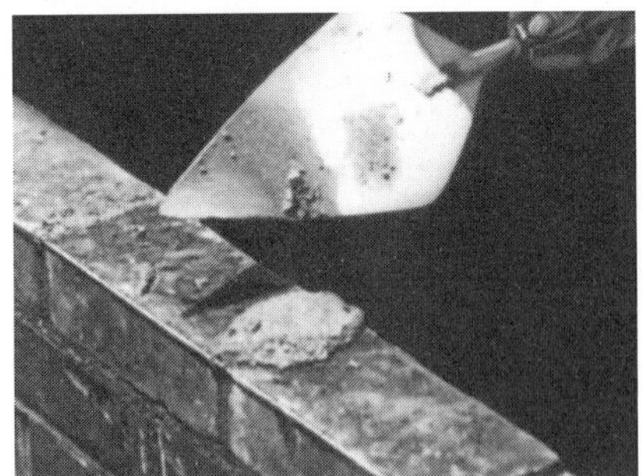

FIGURE 4.7-29 **Attempting to fill a head joint by slushing.** (Courtesy Brick Institute of America.)

FIGURE 4.7-30 **A freshly laid brick has been removed after slushing (see Fig. 4.7-29). Note that the head joint was not adequately filled.** (Courtesy Brick Institute of America.)

FIGURE 4.7-31 Filling a cross joint in a header course with mortar. The side of a header brick is buttered (top) and shoved into place (bottom). (Courtesy Brick Institute of America.)

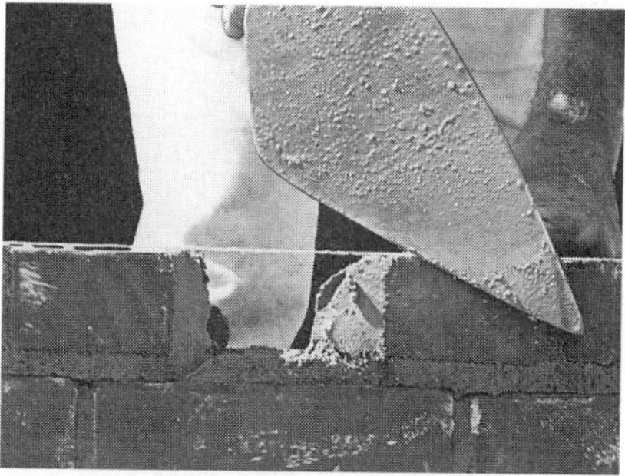

FIGURE 4.7-32 Placing mortar on a brick that is already in place. (Courtesy Brick Institute of America.)

FIGURE 4.7-33 A closer unit being buttered. (Courtesy Brick Institute of America.)

FIGURE 4.7-35 A furrowed bed joint and three blocks with their face shells buttered prior to laying. Note the full mortar bedding. (Courtesy Portland Cement Association.)

FIGURE 4.7-36 Full mortar bedding. (Courtesy Portland Cement Association.)

also be completely filled with mortar. This is accomplished by spreading mortar over the entire side of the header brick before it is placed and then shoving it into place to fill all voids (Fig. 4.7-31).

Closers The last brick in a course is called a *closer*. Before laying a closer brick, plenty of mortar should be placed both on the brick already in place (Fig. 4.7-32) and on both sides or ends, as applicable, of the closer brick (Fig. 4.7-33). The closer should then be laid without disturbing the brick already in place (Fig. 4.7-34).

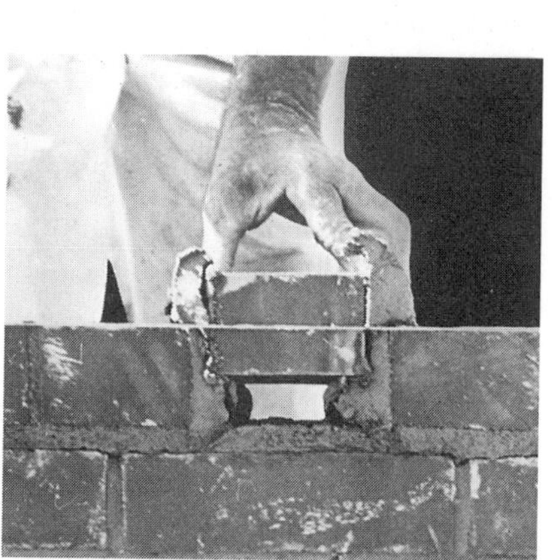

FIGURE 4.7-34 The unit in Figure 4.7-33 is here being placed in the wall. (Courtesy Brick Institute of America.)

CONCRETE MASONRY UNITS

Mortar joints for concrete masonry units should be ⅜ in. (9.5 mm) thick.

Bed Joints For the first course of both solid and hollow units, a full, thick mortar bed should be spread on the foundation or other support under the entire unit including the cells, and furrowed with a trowel to ensure that there will be enough mortar along the bottom edge of the face shells and webs of the block to completely fill the joint (Fig. 4.7-35).

In columns, piers, pilasters, and where adjacent cells will be filled with grout, the face shells and webs should be set in a full and complete bed of mortar. This is called *full mortar bedding* (Fig. 4.7-36).

Solid units should be laid in full mortar beds regardless of their use or location.

In all other concrete unit masonry work, mortar is usually applied only to the horizontal face shells of the units. This is called *face-shell mortar bedding* (Fig. 4.7-37).

Head Joints In vertical joints in concrete block, only the ends of the face shells are bedded in mortar. This bedding should, however, be complete and solid for the entire depth of the face shell.

Mortar for vertical joints in block can be applied to the vertical face shells of either the block to be placed or of the block previously laid. Some masons butter both edges to ensure filled joints. By placing three or four blocks on end, a mason can butter their vertical face shells in one operation (see Fig. 4.7-35). Each block is then brought to its final position and pushed downward into the mortar bed and against the previously laid blocks, thus producing full vertical mortar joints (Fig. 4.7-38).

Closers The last unit in a course is called a closer. Before laying closer units, plenty of mortar should be placed both on the unit already in place and on both ends of the closer unit. The closer should then be laid without disturbing the unit already in place (Fig. 4.7-39).

STRUCTURAL CLAY TILE AND OTHER HOLLOW MASONRY UNITS

In general, the requirements for hollow concrete masonry units apply as well to hollow brick, structural clay tile, and other

FIGURE 4.7-37 Face-shell mortar bedding. (Courtesy Portland Cement Association.)

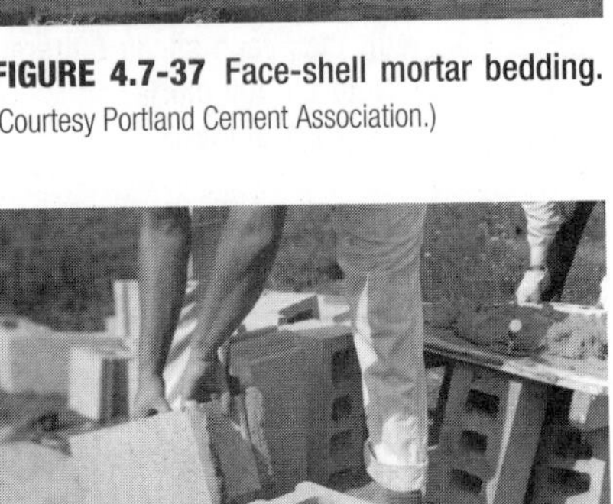

FIGURE 4.7-38 Positioning a block before setting it in place. Note the full vertical joints between the blocks that are already in place. (Courtesy Portland Cement Association.)

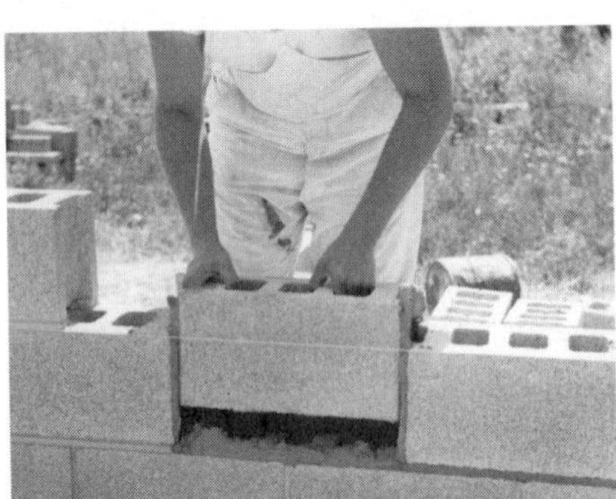

FIGURE 4.7-39 Setting a closer block. (Courtesy Portland Cement Association.)

FIGURE 4.7-40 Backparging a collar joint. (Courtesy Portland Cement Association.)

FIGURE 4.7-41 Mortar joints that were cut flush before they hardened. (Courtesy Portland Cement Association.)

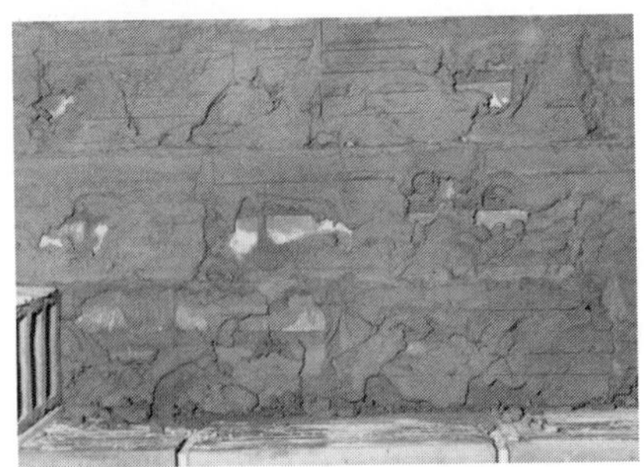

FIGURE 4.7-42 A freshly laid wythe of masonry has been removed to show the spotty and incomplete backparging that resulted from slushing a collar joint. (Courtesy Portland Cement Association.)

hollow masonry units. There are several exceptions.

Vertical-cell structural clay tile units should be laid with divided (not full) head joints and when exposed to the weather, care should be taken to prevent continuous mortar joints through the wall.

Horizontal-cell structural clay tile units that are more than 4 in. (100 mm) thick should be laid with divided bed joints. Head joints of horizontal-cell units should be placed on both sides of the tile and sufficient mortar used so that excess mortar will be squeezed out of the joints as the units are placed in position.

4.7.2.6 Backparging

The vertical, longitudinal joints (*collar joints*) between the wythes of multiwythe masonry walls should be completely filled by parging either the face of the backing or the back of the facing, to help prevent water leakage through the wall (Fig. 4.7-40). Mortar extruding into a collar joint should be cut flush before it hardens. Parging placed over hardened mortar will not bond to it and may result in a leaky wall (Fig. 4.7-41). Filling collar joints from above (slushing) will not fill the joint completely and should not be permitted (Fig. 4.7-42). For best results, parging should be done as the walls are laid up. Careful workmanship is necessary, because the pressure required to spread a parge coat of mortar on newly laid units can break the bond between the units and their mortar joints, producing cracks that may contribute to future leakage.

4.7.2.7 Joint Finishing

When masonry walls leak, the water often enters through improperly sealed cracks and voids around openings, through flashings between the wall and a roof, or through cracks where the wall intersects with other materials or construction. Even when water actually penetrates a wall, the leak is almost never through the masonry units themselves. Most of the time, water enters a masonry wall through cracks and voids between the masonry units and the mortar in the joints. Many such leaks can be prevented by properly finishing the joints. Mortar joint finishes fall loosely into three types, troweled joints, tooled joints, and sealant-filled joints.

Regardless of the joint finishing type used, care should be taken to prevent smearing mortar or sealants onto the finished surface of masonry units that will be exposed to view. Hardened, embedded mortar often cannot be removed. Mortar droppings that adhere to a wall should be allowed to dry before their removal is attempted. Removal should be done using a trowel so that the mortar will not become embedded into the face of the unit.

TROWELED JOINTS

In a troweled joint, excess mortar is simply cut off (*struck*) with a trowel and no further finishing is used. Joints should be struck flush on surfaces that will be parged, plastered, stuccoed, or covered with other masonry.

The best troweled joint is a weathered joint (Fig. 4.7-43, Column 3), which will shed water. Unfortunately, producing this type of joint does not force enough mortar against the brick above to form a watertight joint. Therefore, it should not be used where exposed to the weather.

A flush joint (see Fig. 4.7-43, Column 4) produces an uncompacted joint with a hairline crack where the mortar is pulled

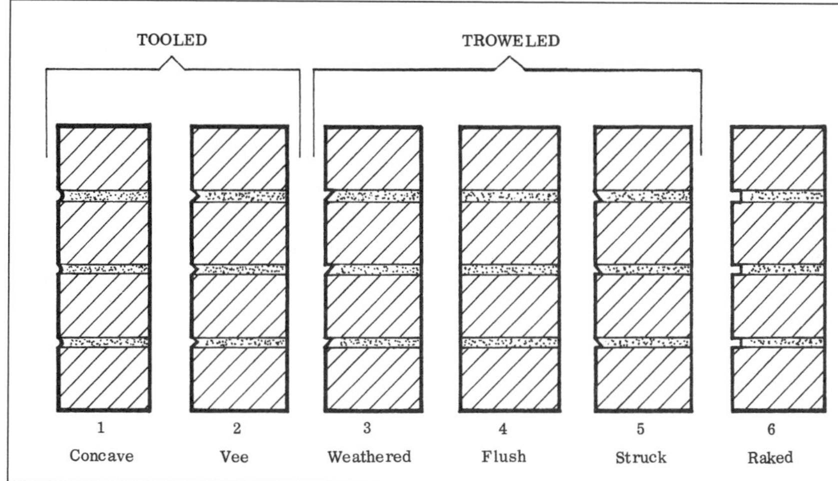

FIGURE 4.7-43 Types of mortar joints.

away from the masonry by the cutting action of a trowel (Fig. 4.7-44).

A struck joint (see Fig. 4.7-43, Column 5) and its kin, a raked joint (Column 6), which is made by removing the surface of the mortar while it is still soft, both leave a shelf on which water can collect. Raked joints are popular with some architects in spite of their drawbacks, however, because they produce crisp, dark joint lines.

TOOLED JOINTS

Because tooled joints provide excellent protection against water penetration they should be used in every wall that is exposed to the weather, except where sealant-filled joints are required. Joints in below-grade walls that will not be parged should also be tooled. For example, slab-on-grade and crawl space foundations need not be parged when moisture penetration is not a problem, and if they are not, then those joints should be tooled. Most exposed interior joints are also tooled to give them a finished appearance, even though water penetration is not usually a problem on the interior.

The most common tooled joints used today are either concave (see Fig. 4.7-43, Column 1) or V-shaped (Column 2). Many other tooled joint shapes have been used historically, however, and may be encountered in renovation work. To make a tooled joint, a special tool is used to compress and shape the mortar. The tooling operation compacts the mortar and forces it tightly against the masonry on each side of the joint (Fig. 4.7-45). Tooling should be done as soon as a section of the wall has been laid and the mortar has become "thumbprint" hard. Every joint should be tooled at the same degree of mortar hardness so that the joints will all be the same color and uniform in appearance with clean, sharp lines. The jointer for tooling joints should be slightly larger than the joint so that complete contact is made along the edges of the masonry unit, thus compressing and sealing the surface of the mortar, and upturned on the ends to prevent gouging the mortar (Figs. 4.7-46a and 4.7-46b).

SEALANT-FILLED JOINTS

Joints between masonry and the perimeter of door and window frames, louvers, and other penetrations, and joints where masonry abuts other materials should be sealant-filled joints. Expansion and control joints are also usually sealant filled. Where joints to be sealant filled contain mortar or grout, they should be cleaned out to a uniform depth of at least ¾ in. (19.1 mm) and then filled solidly with an elastic sealant compound forced into place with a pressure gun. Sealant-filled joints should not be less than ¼ in. (6.4 mm) wide. Sealant materials and methods of installation are discussed and illustrated in Section 7.11.

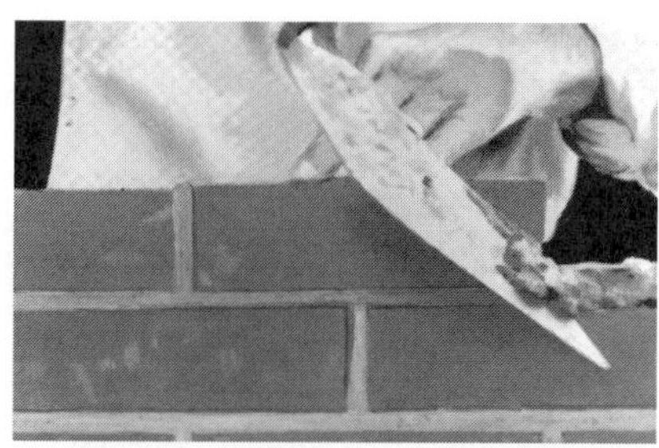

FIGURE 4.7-44 A mortar joint being cut flush. (Courtesy Brick Institute of America.)

4.7.2.8 Flashing

The general principles, materials, and methods used to install flashings are discussed in detail in Section 7.8. Wall flashing in masonry is included here, however, because, unlike many other flashing installations, it is usually installed by the mason as the masonry is erected, and its installation is generally thought of in the industry as part of the masonry work. Because most of the recommendations for flashing in Section 7.8 apply as well to the flashing discussed here, those requirements are not repeated here.

Except in areas of very slight exposure to precipitation (see Fig. 4.6-4) where

FIGURE 4.7-45 A mortar joint before and after tooling. (Courtesy Brick Institute of America.)

a

b

FIGURE 4.7-46 Here the joints are being tooled (a) in a concrete masonry unit wall with a concave jointer (courtesy Portland Cement Association) and (b) in a brick wall with a V-jointer (courtesy Brick Institute of America).

local experience recommends omitting it, flashing should be installed at the following vulnerable locations (Figs. 4.7-47 and 4.7-48):

1. At the base of a wall
2. Under the sills of wall openings
3. Over the heads of wall openings
4. At floor lines
5. Beneath copings
6. Where masonry penetrates a roof surface

PURPOSE OF FLASHING

Masonry wall flashing (Fig. 4.7-49) is installed to collect moisture that may penetrate the wall and divert it through weep holes to the exterior. The amount of exposure of masonry to rain, which varies greatly in the United States, is a basic consideration when designing masonry walls to resist moisture penetration. Exposure may be defined roughly in terms of the resultant wind pressures and precipitation maps shown in Figure 4.6-4 as either severe, moderate, or slight. *Severe* exposure indicates an annual precipitation of 30 in. (762 mm) or more, with a wind pressure of 30 psf (146.5 kg/m^2) or higher. *Moderate* exposure means an annual precipitation of 30 in. (762 mm) or more and a wind pressure of 20 to 25 psf (97.65 to 122.1 kg/m^2). *Slight* exposure means an annual precipitation of less than 30 in. (762 mm) and a wind pressure of 20 to 25 psf (97.65 to 122.1 kg/m^2), or an annual precipitation of less than 20 in. (483 mm).

The following recommendations apply to areas of severe exposure, such as the Gulf Coast, Atlantic Seaboard, and Great Lakes region, and to areas of moderate exposure, which are typical in the Midwest. Where exposure is slight, completely concealed flashing such as lintel or sill flashing, is often eliminated, and in some areas even exposed flashing is reduced to a minimum.

FLASHING LOCATIONS

The vulnerable locations in masonry where wall flashing is necessary are described below (see Figs. 4.7-47 and 4.7-48). Flashing should be provided under horizontal masonry surfaces such as sills and copings, at intersections of masonry walls with horizontal surfaces such as roofs and chimneys, over the heads of openings such as doors and windows, and frequently at floor lines, depending on the type of construction.

Base flashing and damp checks are used at the base of a wall. Base flashing diverts to the exterior any moisture that enters the wall above the location of the flashing. A damp check stops the upward capillary travel of ground moisture. If installed properly, metal shields used for termite protection of wood joists (see Sections 2.3.9 and 6.5.2) may double for these purposes.

Sill flashing should be placed under and behind every sill. The ends of sill flashing should extend beyond the sides of an opening and turn up at least 1 in. (25.4 mm) into the wall. Sills should slope and project from the wall to drain water away from the building and to prevent staining. When the undersides of sills do not slope, a drip notch should be provided or the flashing may be extended and bent down to form a drip.

Head flashing should be placed over all openings except those completely protected by overhanging projections. At steel lintels the flashing should be placed under and behind the facing material, with its outer edge bent down over the lintel to form a drip.

FLASHING INSTALLATION

When flashing is to be laid on or against masonry, the surface of the masonry must be smooth and free from projections that would puncture the flashing material or otherwise destroy its effectiveness. Through-wall flashing should be placed on a thin mortar bed, and another thin mortar bed should be placed on top of the flashing to receive and bond the next masonry course. Metal flashing should be deformed to help lock the mortar in place.

Flashing seams must be overlapped and thoroughly bonded or sealed to prevent water penetration. Although most sheet metal flashing materials can be soldered, lockslip joints are required at intervals to permit thermal expansion and contraction. Many plastics can be permanently and effectively joined by heat or an appropriate adhesive. The elastic pliability of plastic flashings eliminates the need for expansion and contraction seams.

4.7.2.9 Weep Holes

The water collected by flashing must be drained to the outside. Concealed flashings in tooled mortar joints are not self-draining without weep holes. Actually, such flashing acts as a trap to collect water and, by concentrating it in one place, can do more harm than good. Therefore, except at windowsill flashing where little water collects, weep holes should be provided in the head joint immediately above all flashing.

Weep holes may be formed in several ways, and although no one method provides better weep holes than the others, some have a few disadvantages. Weep holes may be formed by simply omitting the mortar in the head joints at the desired intervals, or by inserting oiled rods or pins and then removing them when the mortar is ready for tooling. Weep holes may also be made by inserting plastic or metal tubes, or metal or plastic vents, in the head joints and leaving them in place. Another way is to insert short lengths of sash cord and leave them in place to disintegrate later. Using these methods, weep holes should be placed at 24-in. (610 mm) intervals horizontally. Unfortunately, these methods all produce open weep holes, which provide access for insects and, especially when placed over lintel flashing, may contribute to wall staining. Care must be taken to prevent open weep holes from being clogged with mortar droppings, dirt, or other foreign materials.

Closed weep holes can be made by inserting wicks in the head joints (see Fig. 4.3-4). These wicks are made of ¼-in. (6.4 mm) glass fiber rope or a similar inorganic material. Another way is to insert ½-in. (12.7 mm) glass fiber insulation in open head joints. Since they do not drain as well as open joints, closed weep holes should be spaced every 16 in. (406 mm). Care must be taken to ensure that mortar droppings do not block the holes behind the wick or insulation, and that the materials in the holes do not themselves become clogged with mortar or dirt.

4.7.3 PARGING AND DAMPPROOFING

When the groundwater table and subsoil conditions do not cause water to build up hydrostatic pressure against masonry foundation and basement walls below grade, but moisture is present in the soil, penetration of that moisture can usually be controlled by parging and dampproofing the wall. Dampproofing is a coating applied to resist moisture penetration resulting from capillary action (see Section 7.2). Where hydrostatic pressures may develop, concrete foundation walls are normally used rather than masonry and water penetration is controlled by waterproofing. Waterproofing is a continuous water-

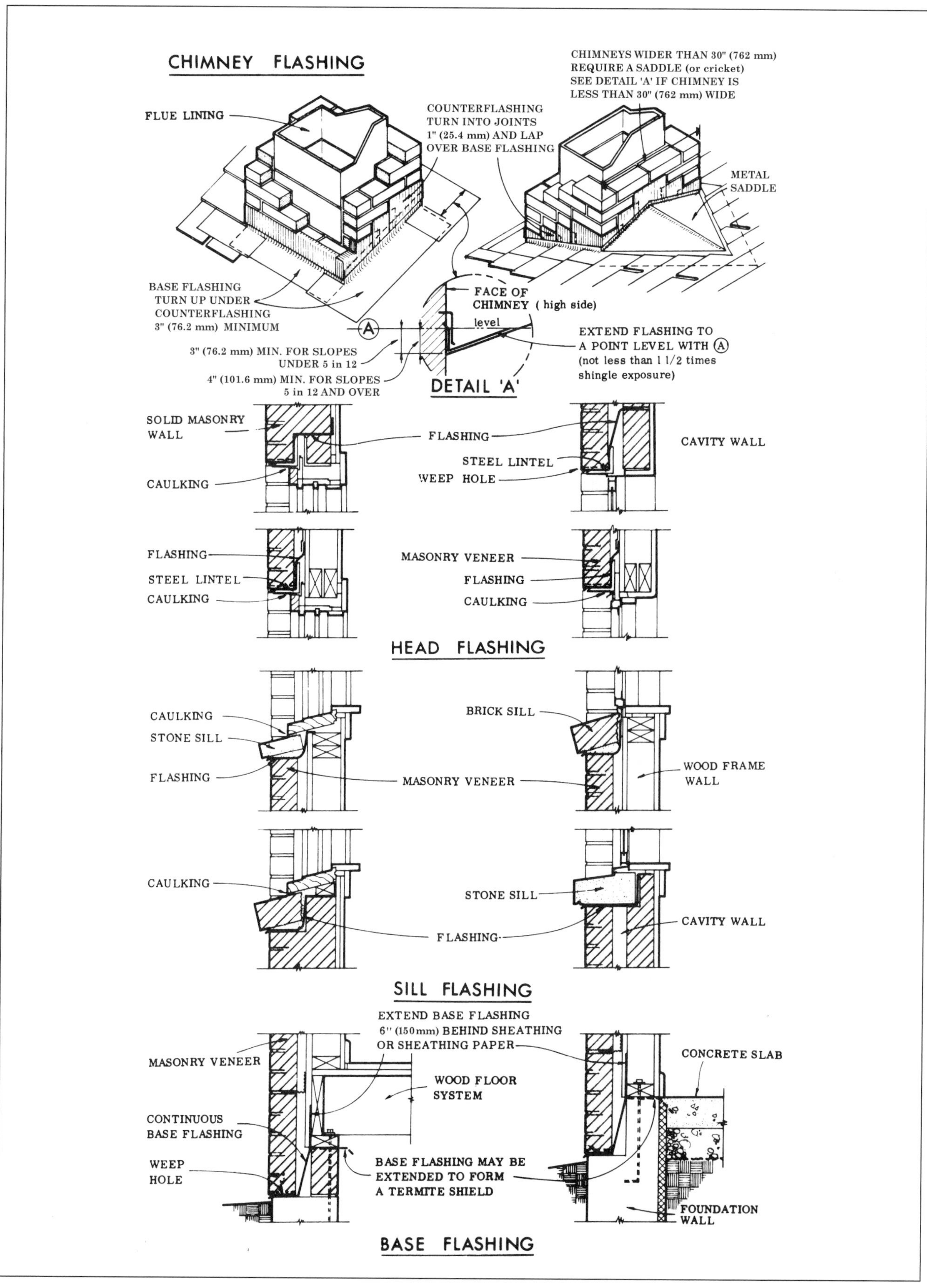

CHIMNEY FLASHING
FLUE LINING
COUNTERFLASHING TURN INTO JOINTS 1" (25.4 mm) AND LAP OVER BASE FLASHING
CHIMNEYS WIDER THAN 30" (762 mm) REQUIRE A SADDLE (or cricket) SEE DETAIL 'A' IF CHIMNEY IS LESS THAN 30" (762 mm) WIDE
METAL SADDLE
BASE FLASHING TURN UP UNDER COUNTERFLASHING 3" (76.2 mm) MINIMUM
FACE OF CHIMNEY (high side)
level
A
3" (76.2 mm) MIN. FOR SLOPES UNDER 5 in 12
4" (101.6 mm) MIN. FOR SLOPES 5 in 12 AND OVER
EXTEND FLASHING TO A POINT LEVEL WITH A (not less than 1 1/2 times shingle exposure)
DETAIL 'A'
SOLID MASONRY WALL
CAULKING
FLASHING
STEEL LINTEL
WEEP HOLE
CAVITY WALL
FLASHING
STEEL LINTEL
CAULKING
MASONRY VENEER
FLASHING
CAULKING
HEAD FLASHING
CAULKING
STONE SILL
FLASHING
BRICK SILL
MASONRY VENEER
WOOD FRAME WALL
CAULKING
STONE SILL
FLASHING
CAVITY WALL
SILL FLASHING
EXTEND BASE FLASHING 6" (150mm) BEHIND SHEATHING OR SHEATHING PAPER
MASONRY VENEER
WOOD FLOOR SYSTEM
CONTINUOUS BASE FLASHING
WEEP HOLE
BASE FLASHING MAY BE EXTENDED TO FORM A TERMITE SHIELD
CONCRETE SLAB
FOUNDATION WALL
BASE FLASHING

FIGURE 4.7-47 Details showing flashing installations in masonry assemblies.

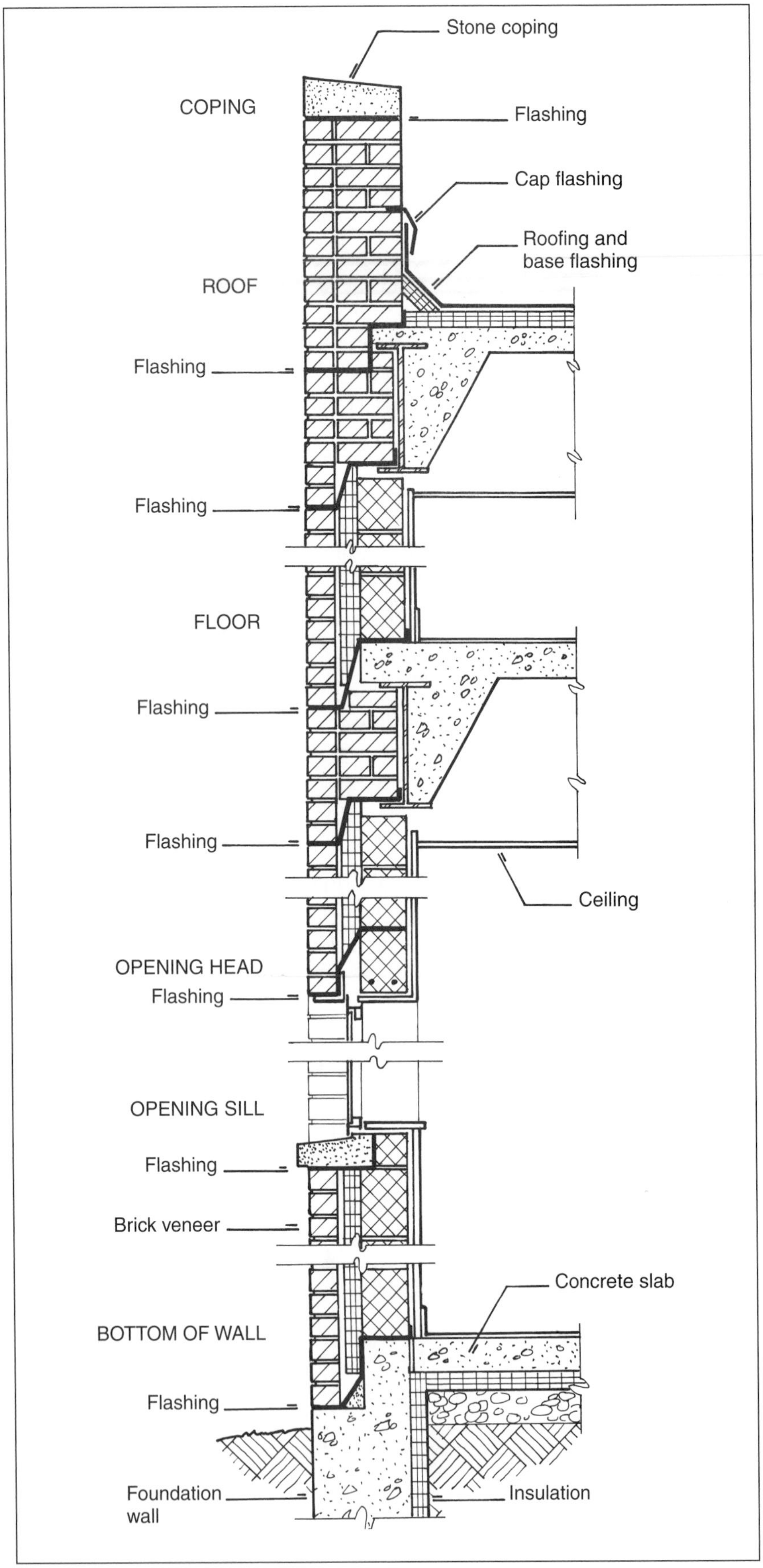

Stone coping

COPING

Flashing

Cap flashing

Roofing and
base flashing

ROOF

Flashing

Flashing

FLOOR

Flashing

Flashing

Ceiling

OPENING HEAD

Flashing

OPENING SILL

Flashing

Brick veneer

Concrete slab

BOTTOM OF WALL

Flashing

Foundation
wall

Insulation

FIGURE 4.7-48 **Wall flashing locations in masonry construction.** (Drawing by HLS.)

FIGURE 4.7-49 **Through-wall flashing.** (Courtesy Portland Cement Association.)

tight membrane capable of resisting water under pressure (see Section 7.2).

Parging consists of Type M mortar or portland cement plaster applied to the exterior face of a below-grade wall and troweled to a smooth, dense surface (Fig. 4.7-50a). Portland cement mortar applied above grade is called stucco (see Section 9.4). Parging should be thickened and turned out onto the footing to form a cove (see Fig. 4.7-50b) and should extend upward from the footing to a line about 6 in. (152 mm) above finished grade. Its top should be beveled to form a wash.

Parging should be applied in two coats. The first coat should be roughened when partially set, hardened for 24 hours, then moistened before the second coat is applied. Parging should be moist cured for at least 48 hours and dry before dampproofing is applied over it. In single-family residential buildings, parging is sometimes applied in only one ¼-in. (6.4 mm)-thick coat and is often applied in two ¼-in. (6.4 mm)-thick coats to a completed thickness of ½ in. (12.7 mm). In larger projects, two ⅜-in. (9.5 mm)-thick coats are usually applied for a total thickness of ¾ in. (19.1 mm). The latter method produces a better job and should be used even on small projects. A single ¼-in. (6.4 mm)-thick coat is much too thin to form a credible parging. Even ½ in. (12.7 mm) is too thin; the additional cost of the extra ¼ in. (6.4 mm) of parging is small when compared with the benefits. One problem with thin coats of parging is that walls are seldom straight and never smooth. The parging tends to average the projections and depressions. Over projections in a wall, a thin coat of parging tends to get even thinner. Very thin coats may flake off the wall when they are subjected to wall movement. When pieces of parging fall off, the covering dampproofing goes with them.

Bituminous dampproofing products (see Section 7.2) may be applied either over parging or directly on the masonry to increase resistance to moisture penetration. Parging is sometimes omitted when the wall to be dampproofed is smooth and the soil is well drained and dry.

Dampproofing of interior masonry walls is also discussed in Section 7.2.

4.7.4 BRICK PAVING AND FLOORING

Brick *paving* is installed on exteriors, where it is generally exposed to weather. Brick *flooring* is installed in interiors, where it is not exposed to weather. However, even in an interior, a brick traffic surface is exposed to severe wear and must therefore be highly durable. Because a unit's resistance to freeze-thaw cycles is a good measure of its durability, units with this characteristic are often chosen even for indoor use and for use outdoors even where severe weather is not a problem.

Brick paving and flooring are laid in one of many possible patterns. Figure 4.7-51 shows some of the more popular patterns.

Brick paving and flooring are classified as either *mortared* or *mortarless*. Most brick paving is installed in a mortarless system. Conversely, most brick flooring is set in mortar because of the difficulties of sealing unfilled joints or joints filled with sand to prevent liquids from entering them when the floor is cleaned.

4.7.4.1 Mortared Paving and Flooring

Mortared brick paving and flooring have mortar or grout in the joints and are laid in a mortar setting bed, either over a concrete slab on grade or on a suspended diaphragm such as a structural concrete, steel, or wood floor or roof assembly.

Most mortared brick paving and flooring are laid in a thick-set latex-modified portland cement mortar setting bed with its joints grouted using a portland cement grout mixed with a latex additive. Thick-set beds may be bonded to a slab on grade, but should be placed over a *cleavage membrane* on slabs subject to bending or deflection, so that these movements do not disrupt the flooring.

In residential construction, and in some small buildings of other types, brick flooring is sometimes laid in a thin-set mortar bed either over a concrete slab or on a properly supported wood subfloor. A

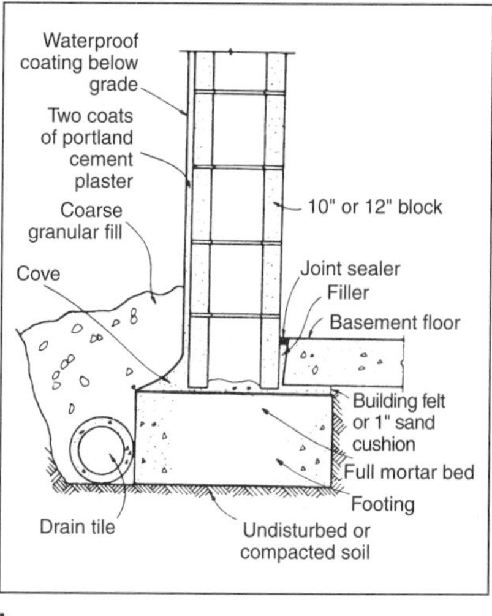

a b

FIGURE 4.7-50 (a) Applying parging; (b) a wall section showing parging and dampproofing. Note that the parging turns out onto the footing to form a cove. (Courtesy Portland Cement Association.)

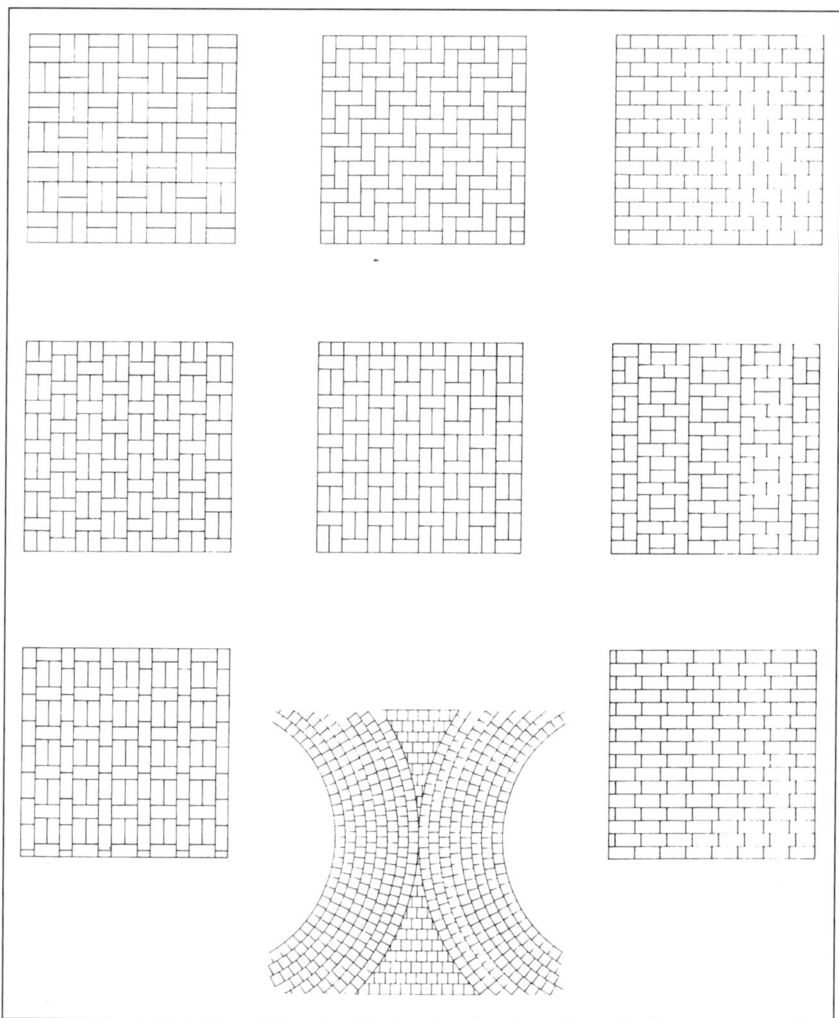

FIGURE 4.7-51 Brick flooring patterns. Patterns may be laid parallel with the borders or on a diagonal. (Courtesy Brick Institute of America.)

thin-set bed is lighter and requires less floor-to-floor height than a thick-set bed, but the substrate must be sufficiently level and properly sloped to permit the thinner application. In addition, curing compounds cannot be used on a slab that will receive brick flooring set in a thin-set mortar bed.

4.7.4.2 Mortarless Paving and Flooring

Mortarless brick paving or flooring can be laid on a rigid concrete slab, on an asphaltic concrete paving base, or over a suspended diaphragm structure. A separation membrane or thin layer of sand is often used over concrete. A thin asphalt setting bed is needed over an asphalt base. On a suspended diaphragm, a roofing or waterproofing membrane, with or without insulation, may occur beneath the pavers on a roof or plaza deck. Most mortarless paving and flooring, however, is set in a leveling bed of sand, gravel, crushed stone, or a mixture of sand and cement. This bed can be supported directly on earth or on a rigid base, such as a concrete slab or asphalt concrete paving.

4.7.4.3 Joints

Joints in mortared brick pavers and flooring are usually left open and filled with mortar or latex-modified grout.

Joints in mortarless brick paving or flooring may have open joints, or the bricks may be placed with hand-tight closed joints. Open joints are filled with latex-modified cement grout. Hand-tight joints are filled by sweeping a dry mixture of portland cement and sand into them and fogging the surface with water.

4.7.5 PROTECTION DURING CONSTRUCTION

Masonry under construction must be protected from rain and snow and, in cold weather, from freezing. Methods of providing this protection are discussed in this section. Lateral support of walls during construction is covered in Section 4.6.5.4.

4.7.5.1 Protection from Rain and Snow

Unprotected, partially finished masonry that becomes saturated with water during a rain- or snowstorm may take months to dry out, and during this drying period efflorescence may appear on the surface (see Section 4.7.6.2).

FIGURE 4.7-52 One way to protect the exposed top of a masonry wall under construction. (Courtesy Portland Cement Association.)

At the end of each day or shutdown period when there is a threat of precipitation, partially completed masonry should be protected by covering it with securely weighted-down canvas or other strong waterproof membrane. This covering should overhang at least 2 ft. (610 mm) on each side of a wall and be securely anchored (Fig. 4.7-52).

4.7.5.2 Cold-Weather Construction

As the air temperature falls, mason productivity and masonry strength are affected. The chemical reaction between portland cement and water (*hydration*) slows significantly at temperatures below 40°F (4.4°C). Below 32°F (0°C) hydration proceeds only when sufficient liquid water is available. Hydration ceases altogether at about 14°F (–10°C). Therefore, masonry should not be laid when the air temperature is 40°F (4.4°C) and falling or lower unless the proper precautions are taken.

Procedures for mortar mixing in cold weather are addressed in Section 4.2.6.3. Figure 4.7-53 summarizes the other precautions required in cold weather. The basic principle is that at air temperatures below 40°F (4.4°C), masonry materials must be heated to achieve satisfactory masonry assemblies. Masonry construction may continue even when temperatures drop below freezing, but only when the proper measures are taken to make the materials act as they would at temperatures above 40°F (4.4°C). For example, heated mortar materials have performance characteristics identical to those they have in the normal temperature range.

It is also very important that mortar and masonry units not be laid on frozen

FIGURE 4.7-53 Recommended Procedures for Masonry Construction in Cold Weather[a]

Work Day Temperature	Construction Requirement[b]	Recommended Protection[b]
All temperatures	Normal masonry procedures.	Masonry materials should be received, stored, and protected in ways that prevent water damage. Cover tops of newly laid masonry with plastic or canvas to prevent water damage.
40° to 32°F (4.4 to 0°C)	Heat mixing water to produce mortar temperatures between 40° and 120°F (4.4 and 48.9°C). Never place masonry on frozen, snowy, or ice-covered surfaces.	Cover walls and materials with plastic or canvas to prevent wetting and freezing.
32° to 25°F (0 to –3.9°C)	Heat mixing water and sand to produce mortar temperatures between 40°F and 120°F (4.4 and 48.9°C).	Provide windbreaks for wind velocities over 15 mph (24.14 km/h). Cover walls to prevent freezing. Maintain masonry above 32°F (0°C) for 24 hours using auxiliary heat for insulating blankets.
25° to 20°F (–3.9 to –6.7°C)	Mortar on boards should be maintained above 40°F (4.4°C). Masonry units below 20°F (–6.7°C) must be heated above 20°F (–6.7°C).	
20°F and below (–6.7°C)	Heat mixing water and sand to produce mortar temperatures between 40° and 120°F (4.4 and 48.9°C).	Provide enclosures and supply heat to maintain masonry enclosure above 32°F (0°C) for 24 hours.

[a]For unenclosed construction sites. These recommendations do not apply where the complete work area, including materials storage space, is enclosed and heated.

[b]Each requirement is in addition to the one above it.

materials and that the masonry units be at a temperature of at least 20°F (–6.7°C) and contain no visible frost or ice at the time of laying. Stored masonry units must be protected and warmed if necessary. When wetting of brick is necessary (see Section 4.7.1.2) in cold weather, the water used must be warmed.

Masonry must be protected from precipitation and winds in excess of 15 mph (24 km/h) and maintained at above freezing temperatures during construction and for at least 24 hours after construction. Even when temperatures are above freezing, if they are below 40°F (4.4°C) completed masonry must be kept covered and protected from rain and snow for at least 24 hours after construction.

4.7.6 POINTING AND CLEANING MASONRY

Exposed masonry should be protected against staining by covering it, and excess mortar should be wiped off the surface as the work progresses. When the masonry work has been otherwise completed, the joints should be pointed and the entire masonry should be carefully cleaned. Large particles of excess mortar should be removed using a putty knife or chisel, and the masonry should be cleaned of dirt, dust, efflorescence, stains, and other foreign materials.

4.7.6.1 Pointing

Except for weep holes, holes in exposed masonry should be *pointed* (filled), and defective joints should be cut out and repointed with fresh mortar.

Holes left by nails or line pins should be pointed (filled) immediately while the adjacent mortar is green enough to allow an adequate bond with the patch. Otherwise, water entering through nail holes may penetrate the wall (Fig. 4.7-54).

FIGURE 4.7-54 Nail and line-pin holes should be filled as soon as possible. (Courtesy Brick Institute of America.)

4.7.6.2 Cleaning Clay Masonry

Unglazed brick and tile should be cleaned to remove general dirt, grime, and stains. Cleaning should not be attempted until at least 48 hours after completion.

GENERAL CLEANING

Unglazed brick and tile on which no efflorescence or other stain appears should first be cleaned with stiff brushes and clean water. Soil not removed by that cleaning should be scrubbed off using stiff brushes, clean water, and soap powder, or another mild solution recommended by the manufacturer of the masonry unit.

Only after such mild cleaning has been unsuccessful should stronger measures be used. With the concurrence of the masonry unit manufacturer, stubborn dirt may be removed with a mild solution of acid similar to that suggested in the paragraph entitled "Mortar Stains" later in this section. The surface should be thoroughly wetted with clear water and then scrubbed with the acid, followed immediately by a thorough rinsing with clear water.

The exposed surfaces of glazed brick, glazed structural clay tile, and terra-cotta should be washed with detergent powder and warm water, applied with a scrubbing brush, and then rinsed thoroughly with clear water. Metal cleaning tools and brushes, acid solutions, and abrasive materials should not be used.

Regardless of the cleaning solution being used, the cleaning should begin at the top and proceed downward. The wall should be wetted with water before a cleaning solution is applied, and adjacent construction should be protected. Stiff fiber brushes should be used. Only a small area should be cleaned at one time. The cleaning should then be followed immediately by a thorough rinsing with clean water.

EFFLORESCENCE AND OTHER STAINS

Large amounts of soluble salts in mortar and masonry units can cause efflorescence or lime run to appear on the face of the masonry if excessive water penetrates the masonry. The most common form of efflorescence is a white powdery deposit of alkali salts and carbonates (Fig. 4.7-55), but not all white deposits on masonry are efflorescence, and salts of other colors sometimes appear as well. For example, a white stain may be *lime run*, as described later, or a silicate deposit called *scumming* or *white scum*. Other stains on brick masonry may be so-called green stain caused by vanadium salts or brown, tan, or gray manganese dioxide salts. These water-soluble salts are originally present in either the mortar or the masonry units. They are carried to the surface by movement of water that has entered the wall. The dissolved salts are left on the face of the masonry when the water evaporates.

White efflorescence will often be washed away through normal weathering, especially if it is the result of water in a

FIGURE 4.7-55 A bad case of efflorescence. (Courtesy Brick Institute of America.)

wall that dried after completion. When necessary, it can usually be removed by dry brushing alone. If that measure fails, a cleaning with water and stiff brushes will usually do the job. In severe cases, a solution of 1 part muriatic acid to 12 parts water may be used.

White scum is very difficult and sometimes impossible to remove. Green stain is also difficult to remove. Magnesium stains, while not difficult to remove, do require special techniques. All such stains should be removed in the manner and using the materials recommended by the manufacturer of the masonry units that are affected. Different methods may be needed for light-colored masonry than for darker masonry units. In the absence of specific recommendations regarding removal of stains from brick masonry, the BIA's Technical Notes 20 and 23 offer excellent guidance.

The condition called lime run, which is actually a deposit of carbonate, presents itself as a vertical white crusty area that appears to have flowed from a hole in a mortar joint. It is often mistaken for efflorescence. Despite its name, lime run is not associated with the lime in a mortar, but is thought to be a deposit of one or more calcium compounds that have been dissolved in a water stream that moved though the joint and then reacted with carbon dioxide in the air. Lime run can be removed using a weak solution of hydrochloric acid. The condition will return, however, unless the water source is removed.

MORTAR STAINS

When possible, mortar stains should be removed from brick by scrubbing with detergent and warm water. Harsher cleaning methods should not be used on buff, gray, or other light-colored bricks without the concurrence of their manufacturer. Unless specifically not recommended by the manufacturer, when milder methods have failed, mortar may be removed from darker-colored bricks using a mild acid solution.

When use of acid is unavoidable, a solution of hydrochloric acid and water (*muriatic acid*) is often used. Muriatic acid should be further diluted until it is no stronger than 1 part muriatic acid to 9 parts water. Even this dilute a solution should not be used until the mortar in the wall to be cleaned is well hardened and at least 7 days old. Acid solutions should always be tested in an inconspicuous location on a wall to ensure against adverse effects.

Lumps of mortar should be removed from terra-cotta using sharpened wood paddles. Acid should never be used on terra-cotta.

4.7.6.3 Cleaning Concrete Unit Masonry

Because hardened embedded mortar smears can never be removed from it, and because concrete masonry should not be cleaned with an acid wash, mortar should never be smeared on the surface of an exposed unit. Mortar droppings should be allowed to dry before removal is attempted. Trying to remove wet mortar will probably result in the mortar's being smeared into the surface of the unit, from which it cannot be removed. However, most dry and hard mortar can be removed with a trowel. A final brushing will then remove practically all the remaining mortar (Fig. 4.7-56).

FIGURE 4.7-56 Final brushing. (Courtesy Portland Cement Association.)

4.7.7 INSTALLATION OF ARCHITECTURAL TERRA-COTTA

Historically, architectural terra-cotta was tied to a building's structure with metal anchors. Then it was backfilled to make it more fireproof and to protect the concealed anchors.

Modern-day terra-cotta is applied in two ways. The first method is used to install adhesion-type ceramic veneer and terra-cotta tile (see Section 4.4.4.3). It is similar to the method used to install ceramic tile. In this method the wall is dampened and the terra-cotta units are soaked in water for 1 hour. Then a brush coat of portland cement and water is applied to the back of the units. Mortar (see Section 4.2) is then applied to both the units and the wall, and the units are pressed into the mortar. The units are set plumb and level with even ¼-in. (6.4 mm)-wide mortar joints. The units are supported by wood wedges placed in the joints until the mortar hardens. The joints are then tooled or raked back ½ in. (12.7 mm) deep, pointed, and then tooled.

The second method of installing terra-cotta is used to install anchored-type ceramic veneer. In this method the ceramic veneer may be mounted with or without vertical pencil rods, depending on the design and size of the units and the design of the supporting elements. Pencil rods are ¼-in. (6.4 mm)-diameter steel reinforcing rods. When they are used, they are mounted vertically and held in place by passing them through the loops of dowel anchors attached to the supporting construction. The ceramic veneer units are set in place with wood wedges in the joints to support their weight. Wire anchors are installed in slots in the units and tied to the pencil rods. When rods are not used, the anchors are tied to other anchors set into or attached to the supporting construction. After one horizontal course has been laid, the space between the units and the supporting construction is filled with the grout described in Section 4.2. The installation is completed by removing the wood wedges, pointing the joints, and cleaning the units.

4.8 Stone

Stone played an important part in the lives of people long before recorded history. Undoubtedly, early humans used naturally occurring stones as both tools and weapons. About half a million years ago, our predecessors began to flake chips off flint and other stone to make more useful implements and weapons. Early dwellings were, in many instances, natural stone caves. Where these were not available, shelter was made from wood and animal skins, which were relatively portable, a necessity for nomadic peoples. With the development of farming came more permanent settlements and the associated need for permanent dwellings. Where loose stones were available, which was often the case as land was cleared for farming, they were piled up to form crude stone huts.

Shaped stone was not used extensively until iron stoneworking tools were developed. Before the Etruscans introduced lime mortar, cut stone buildings were made by accurately cutting and fitting stones together with no joint filler, which was time-consuming and tedious work. The use of mortar joints reduced the need for extreme accuracy in stone cutting and speeded the construction process. When the Romans developed hydraulic mortar, much stronger stone structures could be made and much longer spans built.

Stone was gradually replaced, to a large extent in smaller and residential buildings, by lighter and easier to handle brick and, later, concrete masonry units. However, stone remained in extensive use in decorative and ceremonial buildings, such as cathedrals and large public buildings.

In the eighteenth and nineteenth centuries, lighter steel and concrete frameworks virtually replaced mass stone as the structural support element in buildings. Stone structures cannot compete with these newer materials in terms of economy or speed of construction. The large structures possible with these materials simply cannot be built in stone.

However, stone remains in great use as an important modern building material, but it is used mostly as cladding, trim, and paving. Although it is still sometimes used in mass, most stone today is hung on concrete or steel structural framing. Massive stone structures and stone bridges and aqueducts are things of the past.

4.8.1 MATERIALS

The types of natural stones used most often today—limestone, marble, granite, and slate—are addressed here. Other types of stones are also used, however. These include sandstone, bluestone, soapstone, and many locally available stones.

4.8.1.1 Limestone

Limestone is sedimentary rock composed of either calcium carbonate (calcite) or dolomite, which is a mixture of calcium and magnesium, or a mixture of calcite and dolomite. It is classified as either *oolitic* or *dolomitic*. Oolitic limestone is quarried mostly in Indiana and dolomitic mostly in Minnesota, but both are also found in other locations. Dolomitic limestone is more crystalline and stronger than oolitic limestone, and some will take a polish whereas oolitic limestone will not. Finishes for oolitic limestone include smooth finishes, and textured and coarse finishes such as plucked, split-face, tooled, chat-sawed, and rock-face. Finishes for dolomitic limestone include honed, sandblasted, and polished smooth. Sometimes different finishes are used together.

Limestone ranges in color from a light creamy buff to a brownish buff or from silver gray to bluish gray, while some is green. Stone primarily of one of these colors may contain various spots and streaks of the others. Variegated limestone contains both buff and gray colors, either in the same stone or in adjacent stones. Limestone ranges in texture from very fine grained to coarse grained with pit holes, reedy formations, honeycomb, iron spots, and travertine-like formations.

Limestone is not suitable for use in acidic environments or where chemicals will be used to melt ice and snow, nor should it be cleaned with harsh chemical cleaners. Acids and de-icing and cleaning chemicals will damage the stone.

The industry standard for limestone is ASTM C 568.

4.8.1.2 Marble

Marble is metamorphic stone composed principally of calcite and dolomite (limestone) that has been recrystallized over a long time by heat and pressure. Marble is classified in four categories, I Calcite, II Dolomite, III Serpentine, and IV Travertine. Marble comes in so many colors and with such a variety of veining that it is necessary to specify it by its source or by requiring that the material supplied match a sample furnished by the architect.

Many types of stones are classified as marbles. Among them are onyx, alabaster, some gypsums, and some marble and granite mixtures. The marble category known as *travertine* is a favorite of many architects due to its distinctive appearance, distinguished by its color and its irregularly shaped surface pores. Like other marbles, travertine began as limestone, which through geological shifting descended in the earth. Heated by the earth's inner core, water in underground *aquifers* rises as steam and hot pressurized water to form hot mud baths, such as "Old Faithful." This rising hot water dissolves the limestone and brings with it granules from below, forming mud beds on the surface. In time, the mud beds cool and crystallize into solid travertine.

A requirement of marble is that it be capable of taking a polish, and interior marble is often so finished. However, highly polished finishes are inappropriate on exteriors, where weathering will dull them, and on traffic surfaces. Typical finishes used on exterior surfaces are honed, sand-blasted, abrasive, axed, hammered, rock-faced, rough-sawn, and tooled.

4.8.1.3 Granite

Granite is granular igneous rock consisting mostly of quartz or feldspar. It is hard, strong, durable, and almost impervious to water. True granite ranges in color from pink to light or dark gray. Other igneous stones that do not strictly meet the chemical and mineralogical requirements for granite are nevertheless sold as granite. These come in black, brown, green, and buff. Because there are so many possible colors and grain characteristics in granite, the best way to specify it is either by its specific source or as a match for samples furnished by the architect.

Common granite finishes include mirrorlike polished, honed, thermal, split-faced, and jet-honed. It can be used under severe weathering conditions and in contact with the ground.

4.8.1.4 Slate

Slate is a microcrystalline metamorphic rock formed originally from clay. It consists of thin plates that can be easily split into sheets. It is used primarily for paving and roofing. Slate comes in red, purple, green, blue, and black. Normal finishes are *natural cleft* (which is the face left by splitting the slate), honed, and *sand-rubbed*, in which the texture left by splitting is completely removed to a smooth surface.

4.8.2 MANUFACTURE

Stone for building is obtained by *quarrying* and prepared for use by *milling*. Quarrying is simply the removal of the stone from the ground. Stone that will be used for rubble (see Fig. 4.8-2) may be blasted out of the ground or broken off using picks and crowbars. Most stone, however,

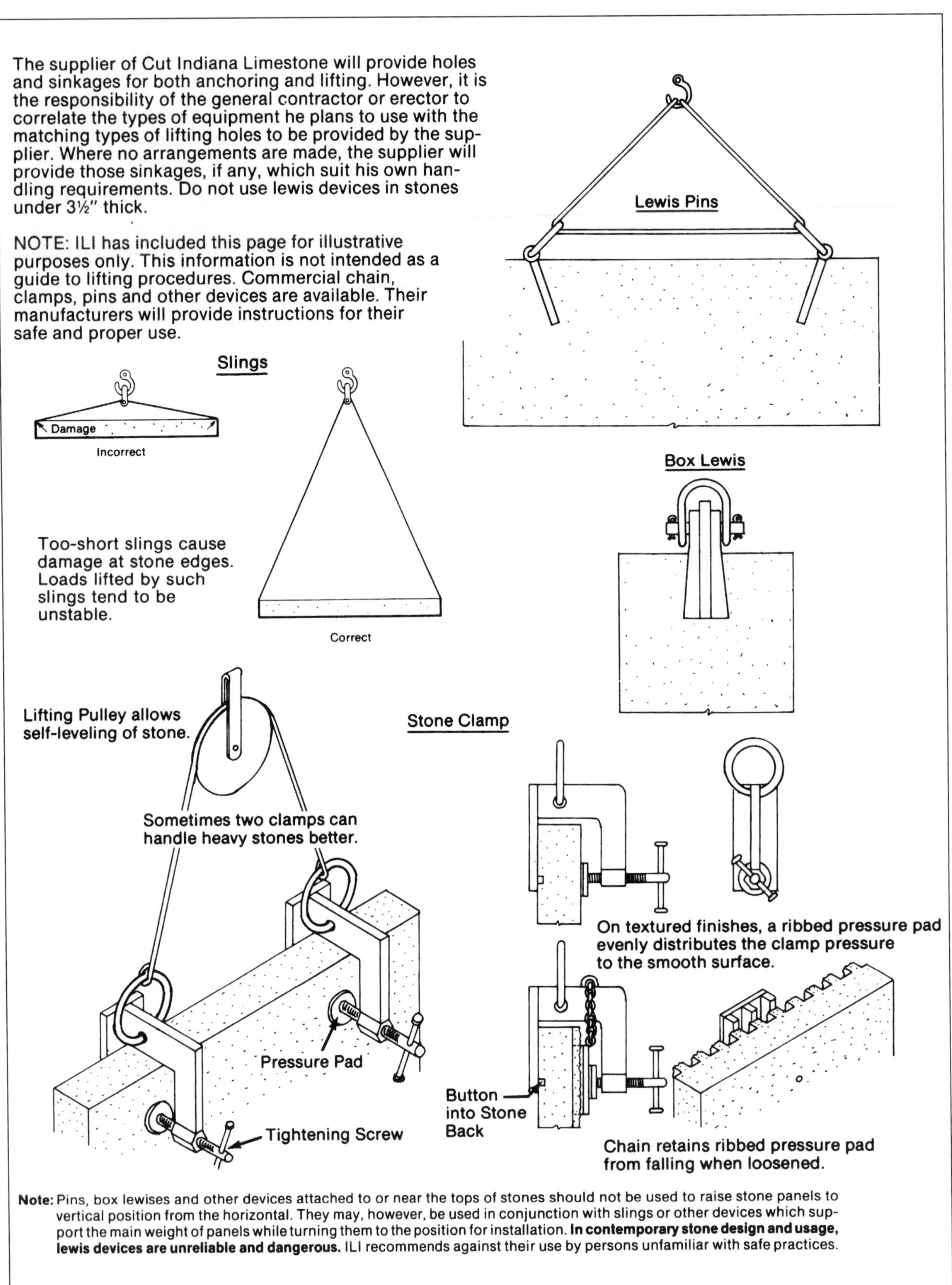

The supplier of Cut Indiana Limestone will provide holes and sinkages for both anchoring and lifting. However, it is the responsibility of the general contractor or erector to correlate the types of equipment he plans to use with the matching types of lifting holes to be provided by the supplier. Where no arrangements are made, the supplier will provide those sinkages, if any, which suit his own handling requirements. Do not use lewis devices in stones under 3½" thick.
NOTE: ILI has included this page for illustrative purposes only. This information is not intended as a guide to lifting procedures. Commercial chain, clamps, pins and other devices are available. Their manufacturers will provide instructions for their safe and proper use.
Lewis Pins
Slings
Damage
Incorrect
Too-short slings cause damage at stone edges. Loads lifted by such slings tend to be unstable.
Correct
Box Lewis
Lifting Pulley allows self-leveling of stone.
Stone Clamp
Sometimes two clamps can handle heavy stones better.
On textured finishes, a ribbed pressure pad evenly distributes the clamp pressure to the smooth surface.
Pressure Pad
Button into Stone Back
Tightening Screw
Chain retains ribbed pressure pad from falling when loosened.
Note: Pins, box lewises and other devices attached to or near the tops of stones should not be used to raise stone panels to vertical position from the horizontal. They may, however, be used in conjunction with slings or other devices which support the main weight of panels while turning them to the position for installation. In contemporary stone design and usage, lewis devices are unreliable and dangerous. ILI recommends against their use by persons unfamiliar with safe practices.

FIGURE 4.8-1 Some methods of lifting large stones. Other devices are also used. (*Indiana Limestone Handbook*, 19th ed., p. 37, reprinted with permission of Indiana Limestone Institute.)

FIGURE 4.8-2 A limestone rubble wall. (Photo courtesy Indiana Limestone Institute.)

is cut from a quarry using wire saws that are drawn through the rock to separate it into large blocks.

In the milling process, quarry blocks are sawn into smaller blocks or slabs in a shop using a different type of wire saw. The wires that actually cut the stone are continuously lubricated by a slurry of water and silicone carbide or another abrasive.

Most stone blocks and slabs are too heavy to be moved and placed by hand. Cranes and other mechanical lifting devices are used extensively in the quarry, at the mill, and at the building site. Individual stones must have the proper holes and sinkages to permit them to be lifted without damage (Fig. 4.8-1) and to be placed without interference or damage to mortar or other bed joints.

Natural stone contains moisture (quarry sap) when quarried. When the moisture dries and the stone stabilizes, it is said to be *seasoned*. Stone is often cut and used before seasoning is complete, but until it is seasoned, which can take as long as a year, stone will not reach its final color and should not be coated with water repellents of any kind.

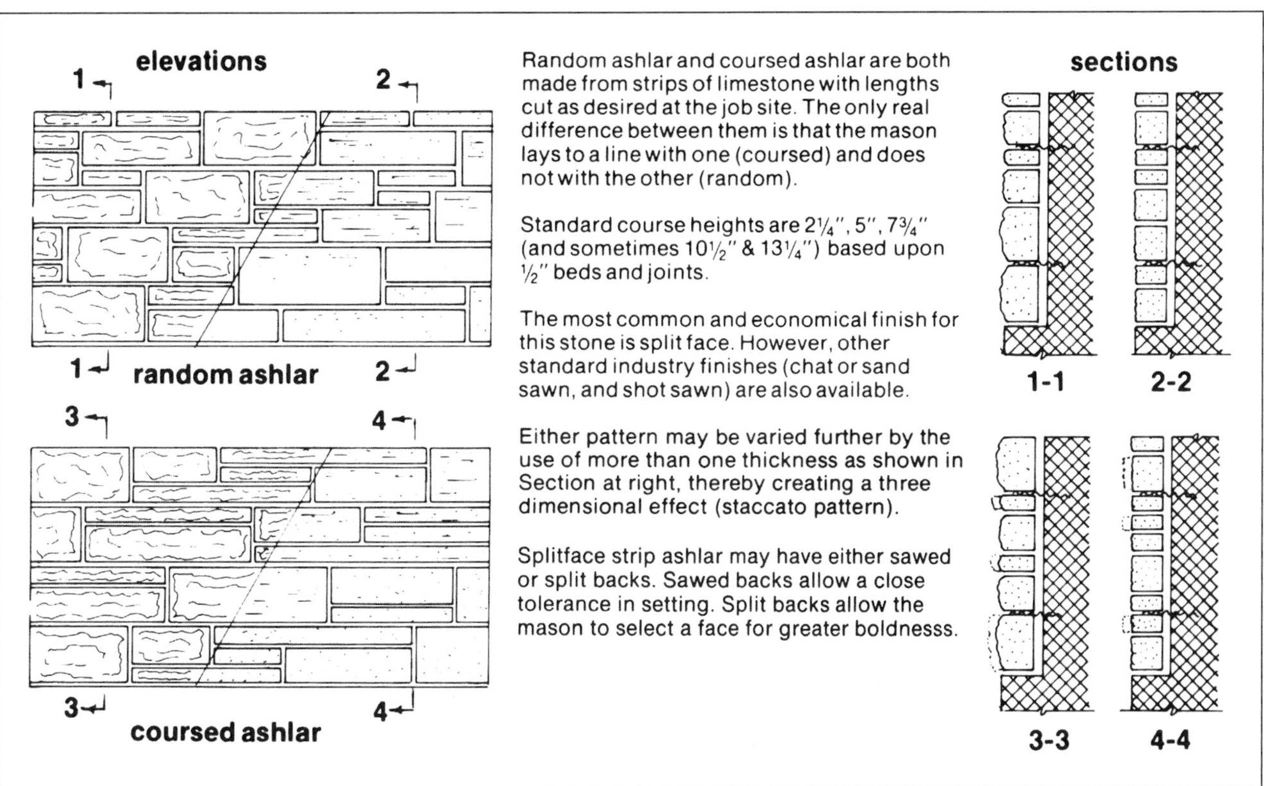

elevations
1
2
random ashlar
3
4
coursed ashlar
Random ashlar and coursed ashlar are both made from strips of limestone with lengths cut as desired at the job site. The only real difference between them is that the mason lays to a line with one (coursed) and does not with the other (random).
Standard course heights are 2¼", 5", 7¾" (and sometimes 10½" & 13¼") based upon ½" beds and joints.
The most common and economical finish for this stone is split face. However, other standard industry finishes (chat or sand sawn, and shot sawn) are also available.
Either pattern may be varied further by the use of more than one thickness as shown in Section at right, thereby creating a three dimensional effect (staccato pattern).
Splitface strip ashlar may have either sawed or split backs. Sawed backs allow a close tolerance in setting. Split backs allow the mason to select a face for greater boldnesss.
sections
1-1
2-2
3-3
4-4

FIGURE 4.8-3 Ashlar stone patterns. (Courtesy Indiana Limestone Institute.)

4.8.3 FINISHES

Cut stones are finished by grinding, polishing, or other treatment to produce the desired finish. Final cutting is often done using a diamond saw. Cylindrical shapes are turned on a lathe. Articulation is sculpted by machine or by hand, depending on the final configuration. Each stone is marked with a unique code so that its correct location in the building can be ascertained when the stone is delivered to the construction site.

Finishes applied on stone include the following:

Chat-sawed: Chat-sawed surfaces are coarse pebbled surfaces made by using a coarse abrasive during sawing.

Flamed: A flamed (thermal) finish results from the application of a direct flame at high temperatures (a blowtorch). This usually is seen in granites and some limestones. Most stones cannot withstand such treatment.

Hand-tooled: A hand-tooled surface is any surface made using hand tools.

Honed: Stones having a honed finish are smooth, like polished stones, but are nonreflective. Honed finishes are achieved by rubbing by hand or machine using abrasives.

Plucked: A plucked surface has a rough finish made by rough-planing the surface and breaking out (plucking) small particles.

Polished: Polished stone presents a mirrorlike surface. Only crystallized stone can take a polish. Essentially, at a microscopic level, polishing is the putting of a facet on each crystal, much as a jeweler puts a facet on a diamond. The result is a surface that allows light to reflect in and out of the stone, enhancing the visible light and color and giving the appearance of depth. A polished finish does not affect the porosity of the stone.

Rock-face: A rock-face surface is coarse and irregular. The stone is sawed on the top and bottom, and the face is exposed by splitting and is then dressed by machine.

Rubbed or Fine-rubbed: Rubbed finishes are achieved by rubbing with abrasives, such as sand or grit. A rubbed finish has occasional light trails or scratches. A fine-rubbed finish is smooth, but has no sheen.

Rusticated: A rusticated stone face has a deep joint effect, cut by machine and sometimes finished by hand.

Sandblasted: Sandblasted stone has a smooth matte surface.

Sawcut: A sawcut finish is simply the rough sawcut surface of the stone with no other treatment applied. Sawcut material is seldom used where it will be visible in a completed building, but tumbled marble is often made from sawcut material; if not tumbled quite enough, it will still show circular saw marks.

Sculptured: A sculptured finish is a reproduction of a sculptor's model carved in the stone by hand.

Split-face: A split-face finish is a rough, irregular surface made by sawing the stone top and bottom and splitting it.

4.8.4 DESIGN AND CONSTRUCTION

Stone may be laid in mortar beds and supported by the stone below, or be supported by metal. It may also be adhered to backing panels of concrete or metal. It may be used to build solid walls, arches, and vaults; as a facing over masonry; or as cladding. As cladding, it may be supported by the structure or mounted on a metal framework.

4.8.4.1 Traditional Stone Setting

Stone structures may be built in the traditional method using mortar beds either as a complete wall or as a facing for masonry. Finished joints may be either mortar or sealant filled. Traditional mortar bed setting methods are used today primarily to build stone retaining walls, planters, and the like; for installing stone copings and trim in masonry walls; and in restoration work on existing stone structures.

Stone laid with traditional mortar bed methods is placed either as *rubble* or as *ashlar*. Rubble consists of natural stone shapes that may be laid in either random or coursed patterns (Fig. 4.8-2).

Ashlar is cut stone with squared corners. It may be laid in either random or coursed patterns (Fig. 4.8-3). Large blocks with square corners are called *dimension stone*.

The methods used are first to place a setting bed, then set the stone in this bed. The joints are raked back. When the bed has set, the joints are either filled with mortar and tooled or filled with a sealant.

4.8.4.2 Mechanical Fastening

Stone cladding may be set on a field-erected grid system of stone anchors, fasteners, and metal struts and braces (Figs. 4.8-4, 4.8-5, and 4.8-6). Very thin stone cladding may also be installed as panels in aluminum frames. Thin stone cladding panels may be applied as a facing on concrete panels. They may also be preassembled into panels on a prefabricated metal support system, which is then lifted into place and mechanically attached to metal frames or to the structure. Panels may have insulation sprayed on their interior surfaces. Joints in mechanically fastened stone cladding are usually sealant filled.

Stone cladding may be of almost any thickness, but the trend today is toward using stone that is 2 in. (50.8 mm) thick or less. The performance requirements for stone cladding are determined in much the same way as those for aluminum curtain walls or concrete cladding. Thin stone is not likely to perform as well as thicker stone. It is more subject to water penetration; to damage due to chemical weathering because of its thinness; to damage resulting from relatively light impact forces, such as those from wind-driven rain or sleet; and to bowing from internal stresses that are less likely to cause damage in thicker stone. Therefore, greater factors of safety must be used when selecting and installing thin stone veneer. However, the savings resulting from reduction in cost of the stone material, the ease of erecting the lighter stone, and the reduction of the load on the structure often make using thin stone cladding worthwhile.

4.8.4.3 Paving and Flooring

Stone is used both as exterior paving and as flooring. Granite is particularly suited for exterior paving. The abrasive resistance of stone used as paving or flooring is important, as is the slip resistance of its finish.

The usual method of placing stone paving is to set the stone in a thick portland cement mortar bed having a latex additive. Joints are pointed with grout or filled with sealant.

Stone flooring may also be set in a thick-set bed, but thin stone tile flooring is often set in a thin-set method similar to that used to install ceramic tile (see Section 9.10.)

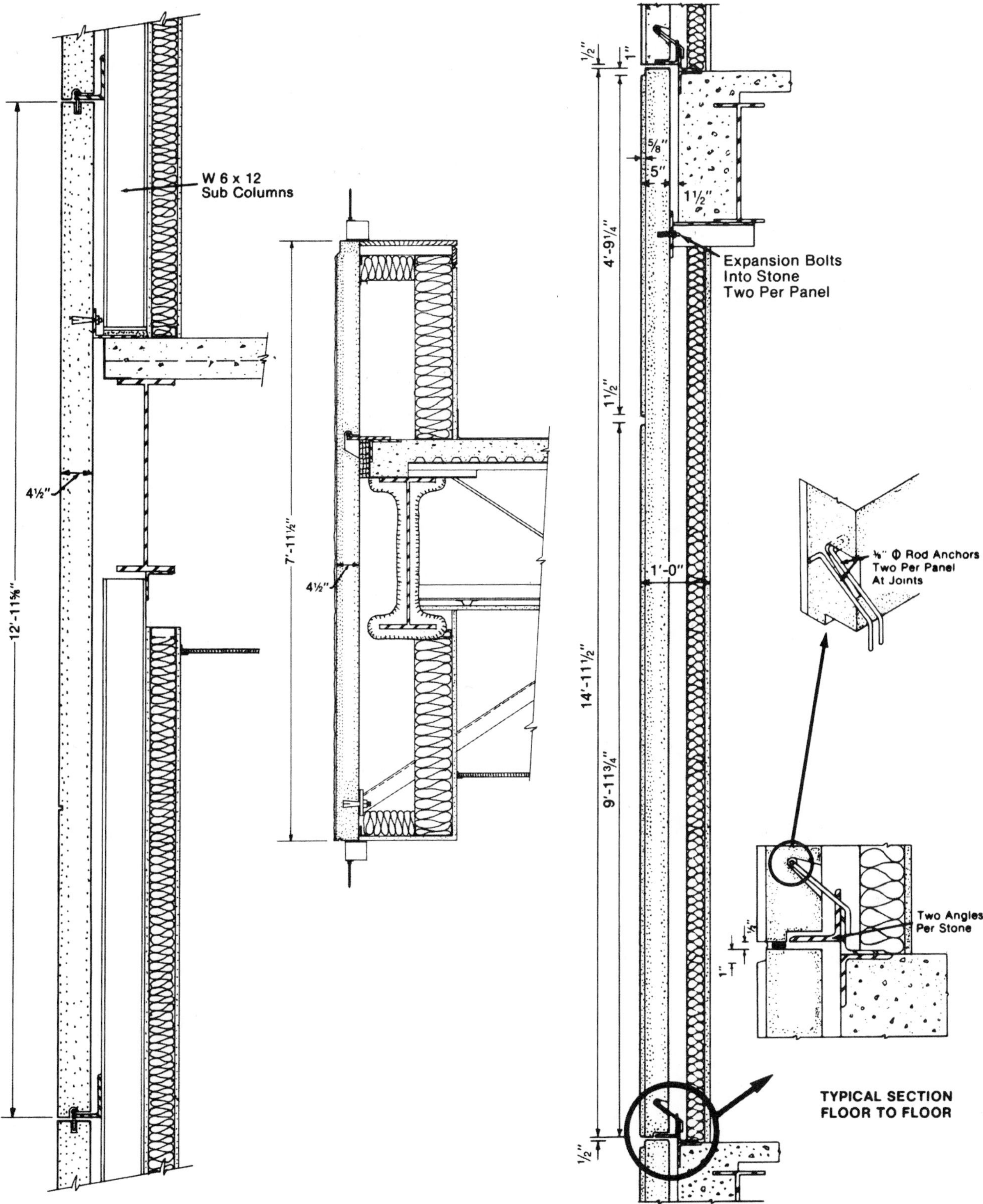

FIGURE 4.8-4 Multistory stone cladding anchored to a steel frame without masonry backup. Three different connection systems are shown. (Drawings courtesy Indiana Limestone Institute.)

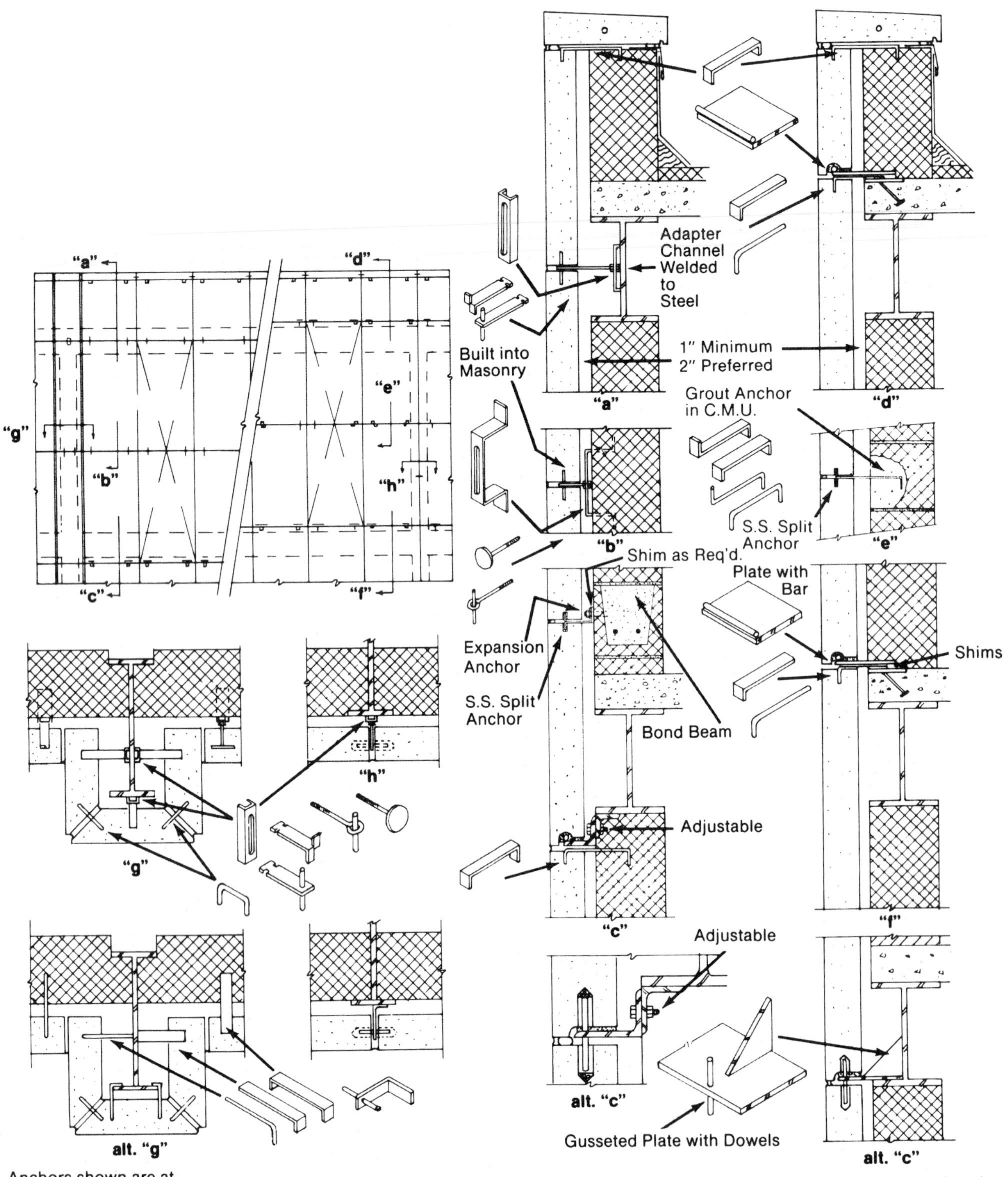

FIGURE 4.8-5 Some of the many possible ways to anchor multistory stone cladding with masonry backup to a steel structural system. The number of types of anchors in a single building should be the fewest practicable. Panels should be supported at least at each floor level by a positive structural element (angle, plate). (Drawings courtesy Indiana Limestone Institute.)

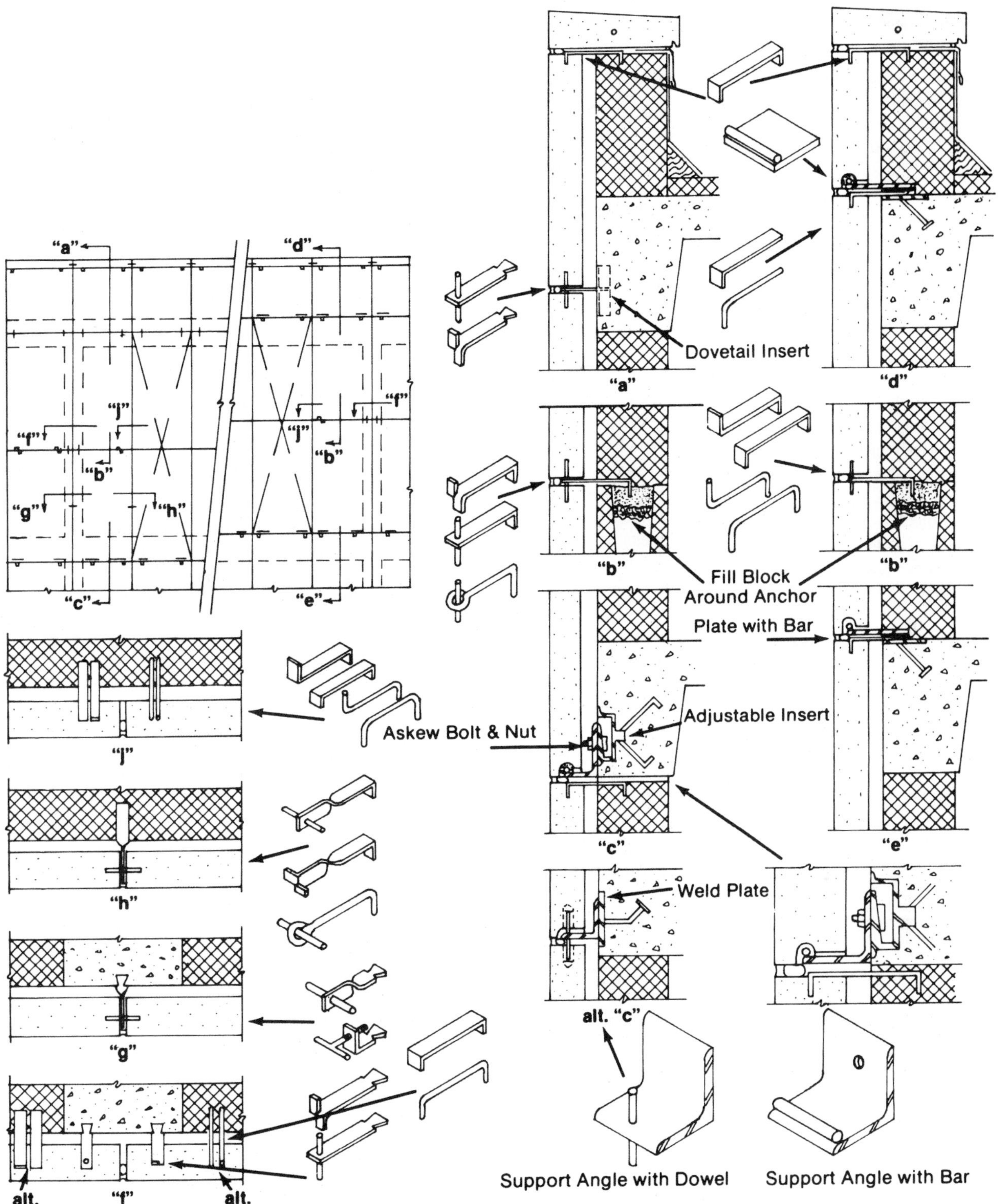

FIGURE 4.8-6 Anchoring multistory stone cladding with masonry backup to a concrete frame is similar to anchoring it to a steel frame (see Figure 4.8-5), and the same principles apply. (Drawings courtesy Indiana Limestone Institute.)

4.9 Glass Unit Masonry

Glass block, which enjoyed great popularity in the mid-1900s and then almost disappeared from the market, has returned to favor. Glass blocks are available in many sizes and shapes. They are made with clear, textured, reflective, and heat-absorbing glass and may be either solid or hollow. They can be installed in the traditional way, using mortar in the joints, in systems using aluminum frames, or in frameless systems held together with sealant.

Glass blocks are used in both interior and exterior locations. They may be placed vertically in the conventional manner, or horizontally in luminous floors and skylights. They provide a partially transparent or translucent surface, depending on the block selected.

Glass block units are not themselves fire rated, but some building codes permit their use in fire-rated assemblies in certain limited cases. For example, they can be used in 45-minute assemblies in some locations when not exceeding 15% of the surface area. They can be used in 90-minute-rated masonry walls, with certain exceptions and restrictions. It is necessary to verify their use in fire-rated assemblies with code requirements, but they should not be summarily excluded from such use without checking.

Glass units can be used to build both straight and curved walls.

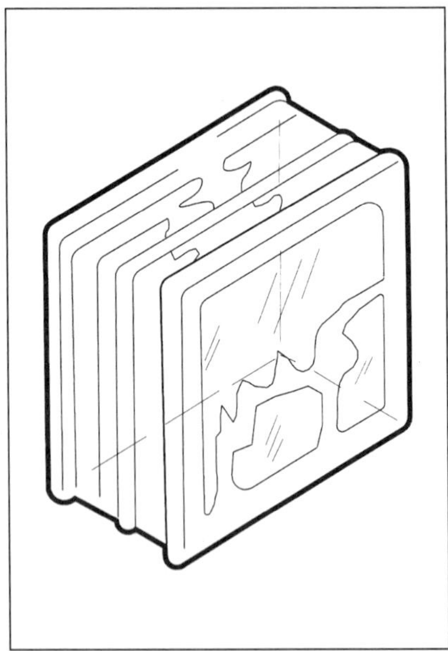

FIGURE 4.9-1 A typical glass block unit. (Drawing by HLS.)

4.9.1 MATERIALS

4.9.1.1 Glass Block

Glass blocks (Fig. 4.9-1) are available as hollow units or as solid blocks. Hollow units are made by joining two molded units together, with a partial vacuum between them. The face thickness of the blocks varies. A wide variety of patterns are available. Patterns are usually, but not always, formed on the inner surfaces of hollow units, with the outer surfaces kept flat and smooth. Decorative units may have ceramic faces fused onto the glass block.

Solar reflective units are coated on one or both outer faces with a reflective oxide.

Both single-cavity and double-cavity blocks are available. Double cavities are formed by inserting a glass fiber insert between the two halves of the block, forming two parallel air spaces. Ambient heat transmission control is good in single-cavity hollow blocks and is further enhanced in double-cavity blocks.

4.9.1.2 Mortar

Mortar for glass block installation is Type S portland cement–lime mortar in accordance with ASTM C 270. For use in exterior panels, a waterproofing admixture is included. Colored pigment mortars or colored aggregates are sometimes used.

4.9.1.3 Accessories

Accessories for mortar installations include panel reinforcement (Fig. 4.9-2a), panel anchors (Fig. 4.9-2b), expansion strips, spacers, and aluminum- or steel-channel perimeter chasers (Fig. 4.9-2c). These are usually proprietary products of glass block manufacturers or are recommended by glass block manufacturers for use with their products. Sealants and backers for such installations are those recommended by glass block manufacturers.

Accessories for framed systems are provided by or recommended by the systems' manufacturers. These include separators, aluminum framing systems, and plastic foam tape.

4.9.2 INSTALLATION

The following are general recommendations. The various manufacturers' instructions vary, and they must be followed when installing glass block.

4.9.2.1 Installation with Mortar Joints

The traditional method of installing glass block assemblies is to lay the blocks with both horizontal and vertical mortar joints between the blocks (Fig. 4.9-3). Units may be anchored to other construction at the head and jambs using sheet metal channels as shown in Figure 4.9-3, but blocks may also be anchored at jambs using either:

- Panel anchors, consisting of perforated steel strips (see Fig. 4.9-2b)
- Dovetail wire ties into dovetail slots in concrete
- Trapezoidal-shaped ties welded to steel columns

Blocks may also be anchored at the head using bolts extending down into the joints from above. This method is most often used in interior applications.

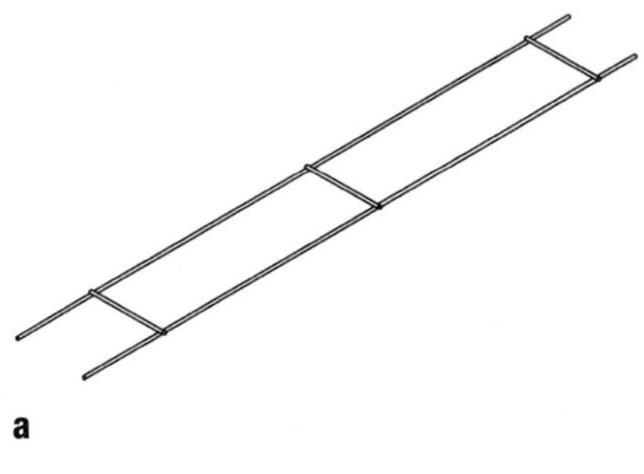

a

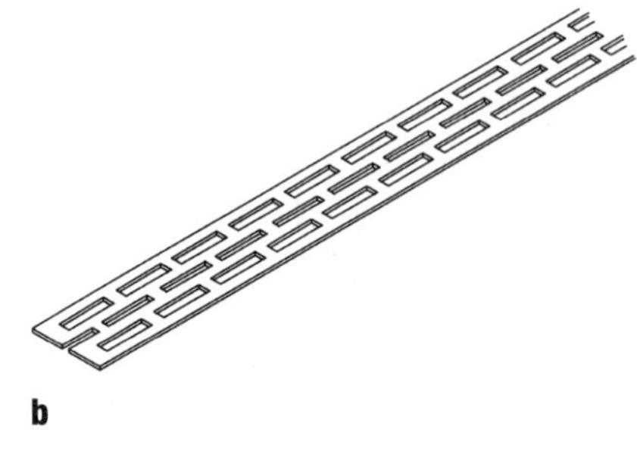

b

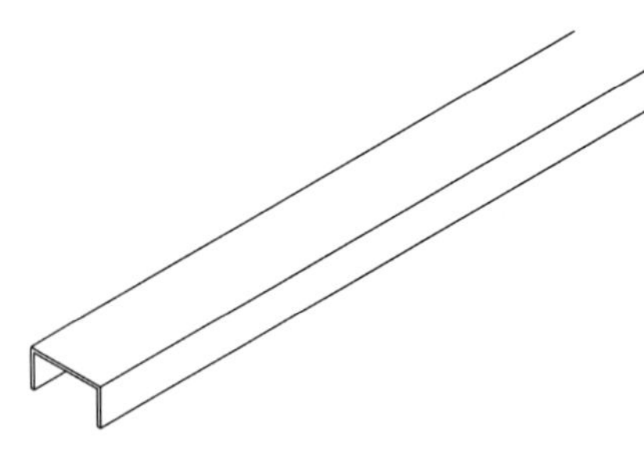

c

FIGURE 4.9-2 Glass block accessories: (a) joint reinforcement; (b) panel anchor; (c) metal channels. (Drawing by HLS.)

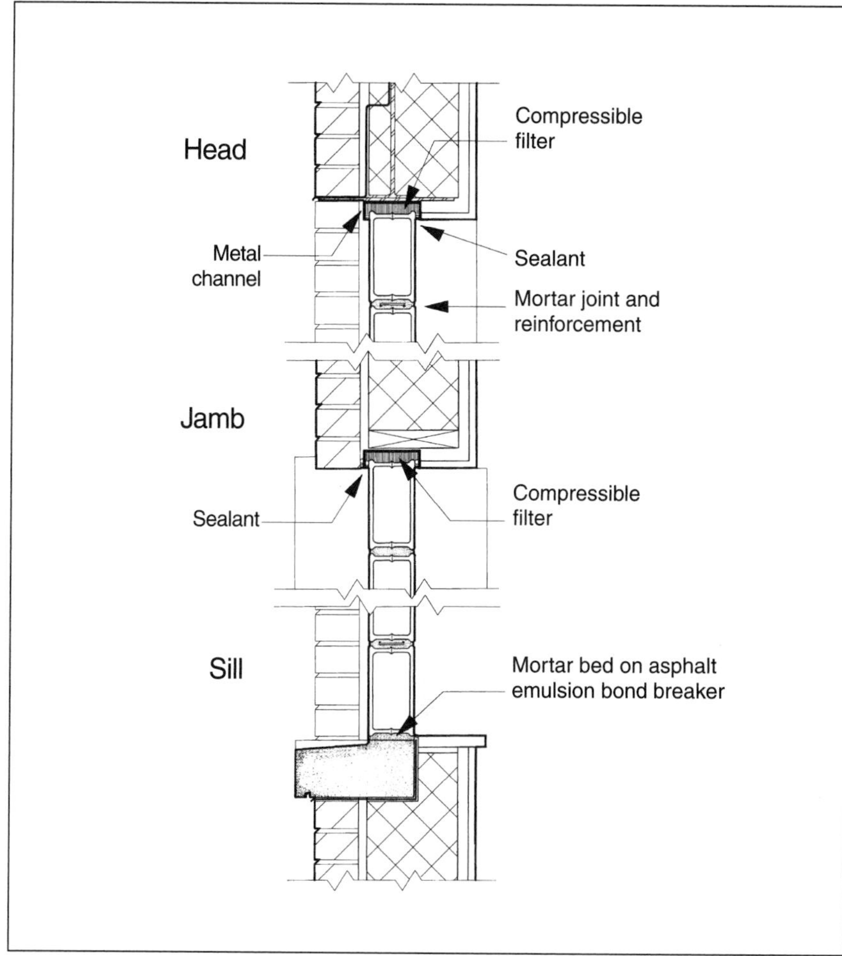

FIGURE 4.9-3 Typical exterior wall glass block installation with mortar joints and metal channels. (Drawing by HLS.)

Horizontal panel reinforcement, consisting of parallel wires joined by crosswires in a ladder fashion, is installed in the joints at about 16-in. (406 mm) intervals, depending on unit size and thickness. A compressible glass fiber or plastic foam expansion strip is installed between the glass blocks and adjacent construction to permit expansion in the blocks.

General recommendations for mortar joints with glass block are similar to those for other types of masonry. Some, but not all, glass block manufacturers recommend that joints be raked out and pointed with mortar. All manufacturers recommend that joints at adjacent materials be finished with a sealant. Some manufacturers recommend that all joints be finished with a sealant.

Joint width is usually ¼ in. (6.4 mm), but ⅜-in. (9.5 mm) joints are sometimes used.

4.9.2.2 Other Installation Methods

Framed and *frameless* glass block systems should be installed in accordance with their manufacturer's recommendations. In framed (grid) systems, the glass block is set into an aluminum grid designed and sized to accept the block. The grid encases each block. A plastic foam filler is placed between the block and the aluminum grid members to permit expansion and contraction. Joints are then sealed with a silicone sealant.

A frameless system has an overall frame like a window frame. There is no frame between the individual glass blocks as there is in a framed system. Blocks are set into the overall frame. A glass block sealant is applied to the vertical edges of each block, and a vertical separator is installed. When a course of blocks has been installed, a horizontal separator is placed across it to act as a separator and as joint reinforcement. This procedure is repeated until the entire panel is in place. Sealant is applied as the components are installed, producing a watertight structure.

4.10 Properties of Selected Masonry Walls

The figures in this section contain data that permit comparison of several masonry wall properties. The figures show various wall assemblies and thicknesses and for each, give the wall's fire rating, measured in hours; sound-transmission class (STC), measured in decibels; heat transmission, measured by the wall's U-value in Btu/sq. ft. hr. °F (K-value in W/m^2 h °C); and weight, measured in psf (kg/m^2).

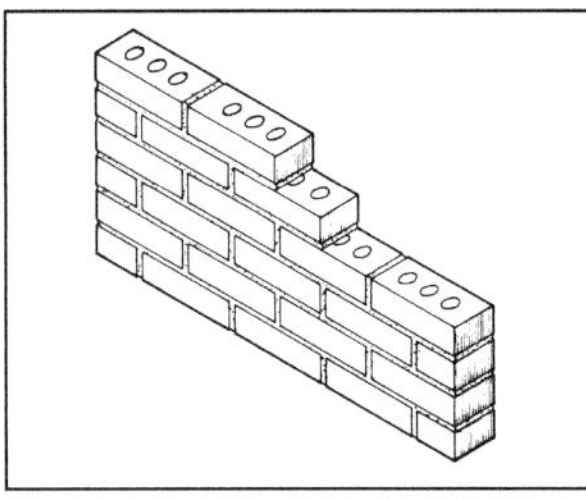

FACE BRICK (2¼ in. × 8 in.)

Fire Rating (hr.):	1
Sound Transmission Class (dB):	45
U-value (Btu/sq. ft. hr., °F):	.76
Wall Weight (lb./sq. ft.):	38

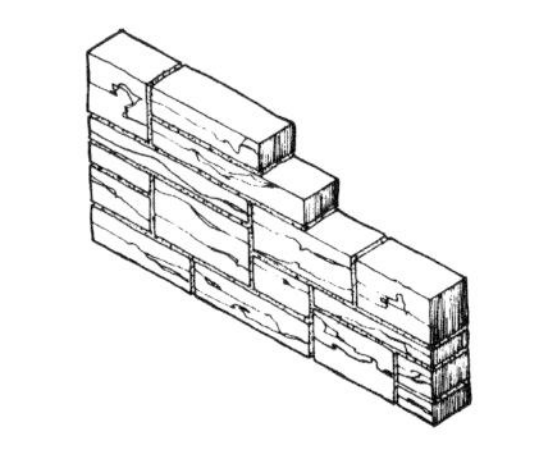

LANNON STONE

Fire Rating (hr.):	1
Sound Transmission Class (dB):	46
U-value (Btu/sq. ft. hr., °F):	.68
Wall Weight (lb./sq. ft.):	50

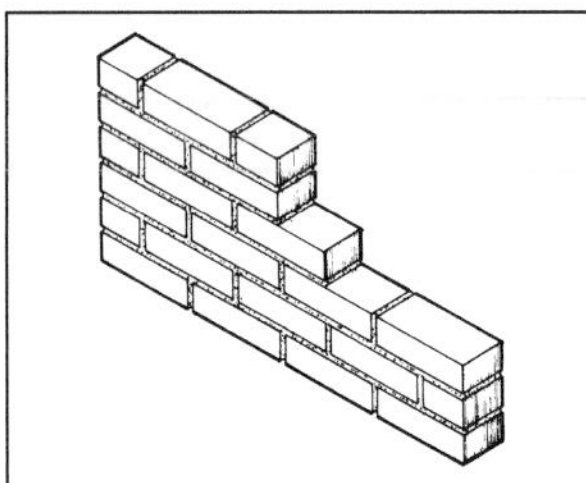

COMMON BRICK (2⅛ in. × 8 in.)

Fire Rating (hr.):	1
Sound Transmission Class (dB):	34
U-value (Btu/sq. ft. hr., °F):	.76
Wall Weight (lb./sq. ft.):	38

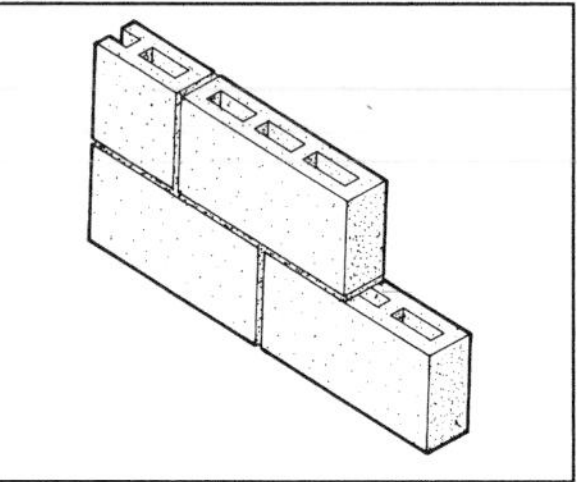

LIGHTWEIGHT CONCRETE BLOCK

Fire Rating (hr.):	1
Sound Transmission Class (dB):	40
U-value (Btu/sq. ft. hr., °F):	43
Wall Weight (lb./sq. ft.):	24

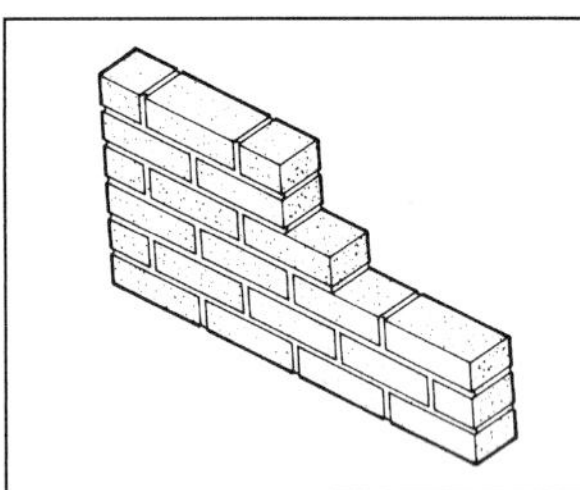

CONCRETE BRICK (2¼ in. × 7⅝ in.)

Fire Rating (hr.):	1
Sound Transmission Class (dB):	37
U-value (Btu/sq. ft. hr., °F):	.67
Wall Weight (lb./sq. ft.):	35

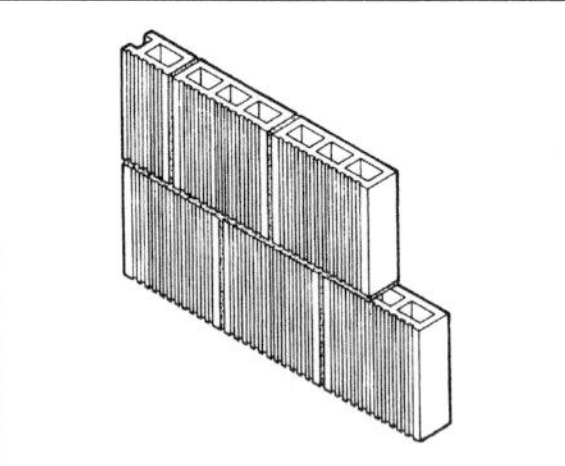

CLAY PARTITION TILE (12 in. × 12 in. Face Size) (Trowel Cut Joint)

Fire Rating (hr.):	1
Sound Transmission Class (dB):	28
U-value (Btu/sq. ft. hr., °F):	.55
Wall Weight (lb./sq. ft.):	21

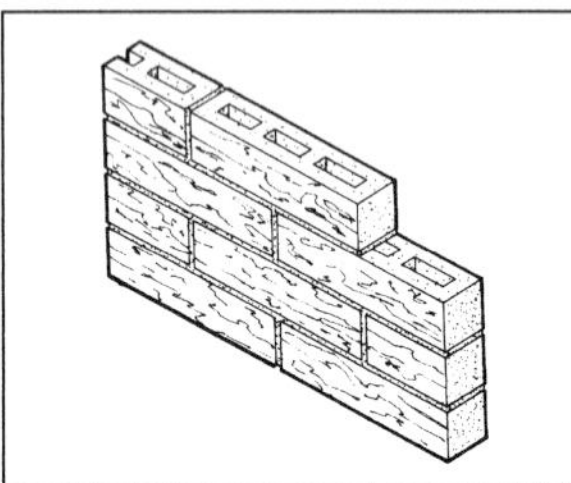

HEAVYWEIGHT CONCRETE "SLUMP" BLOCK (4 in. × 16 in. Face Size)

Fire Rating (hr.):	1
Sound Transmission Class (dB):	36
U-value (Btu/sq. ft. hr., °F):	.46
Wall Weight (lb./sq. ft.):	33

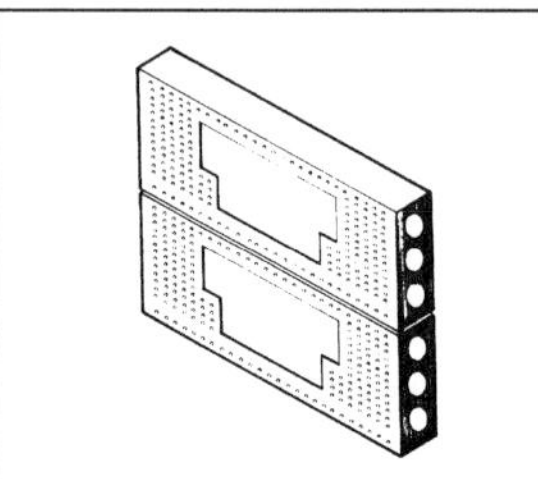

GYPSUM PARTITION TILE (12 in. × 30 in. Face Size) (Trowel Cut Joint)

Fire Rating (hr.):	4
Sound Transmission Class (dB):	42
U-value (Btu/sq. ft. hr., °F):	N.A.
Wall Weight (lb./sq. ft.):	26

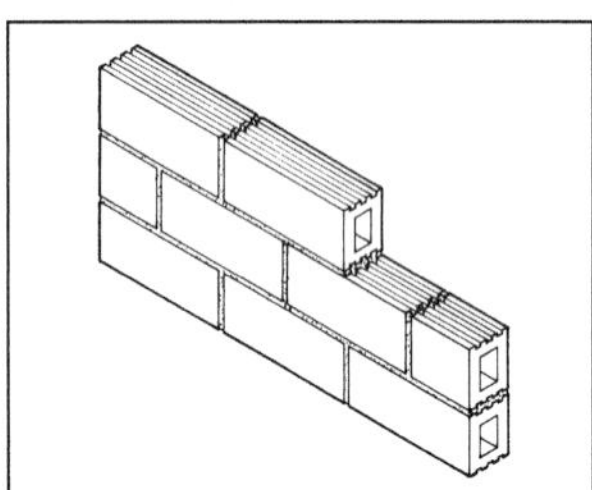

GLAZED TILE (6 TC Series and 6 TCD Series—Tooled 2 Sides)

Fire Rating (hr.):	–1
Sound Transmission Class (dB):	32
U-value (Btu/sq. ft. hr., °F):	.40
Wall Weight (lb./sq. ft.):	27

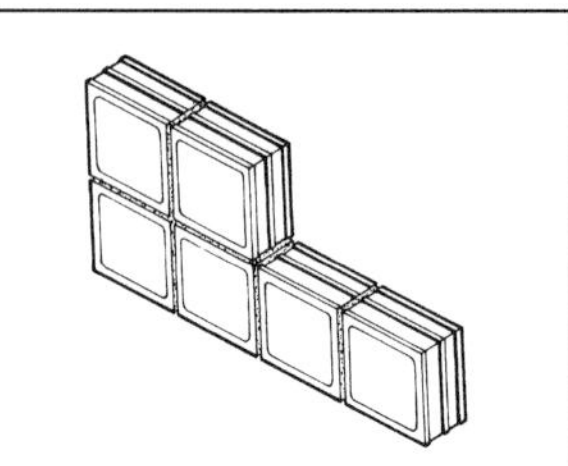

GLASS BLOCK* (8 in. × 8 in.) (Tooled 2 Sides)

Fire Rating (hr.):	–
Sound Transmission Class (dB):	27
U-value (Btu/sq. ft. hr., °F):	.56
Wall Weight (lb./sq. ft.):	20

*6 in. × 6 in. and 12 in. × 12 in. Glass Block are also available.

in.		mm
2⅛	=	53.98
2¼	=	57.15
4	=	100
6	=	150
8	=	200
12	=	300
16	=	400

U-value Btu/ft.^{2} × h.		k-value W/m^2 × k
0.40	=	2.27
0.43	=	2.44
0.46	=	2.61
0.55	=	3.12
0.56	=	3.18
0.67	=	3.80

U-value Btu/ft.^{2} × h.		k-value W/m^2 × k
0.68	=	3.86
0.76	=	4.32
lb./sq. ft.		**kg/m^2**
20	=	97.65
21	=	102.53
24	=	117.17

lb./sq. ft.		kg/m^2
26	=	126.94
27	=	131.82
33	=	161.12
35	=	170.88
38	=	185.53
50	=	244.12

FIGURE 4.10-1 Properties of 4-in. (100 mm) walls.

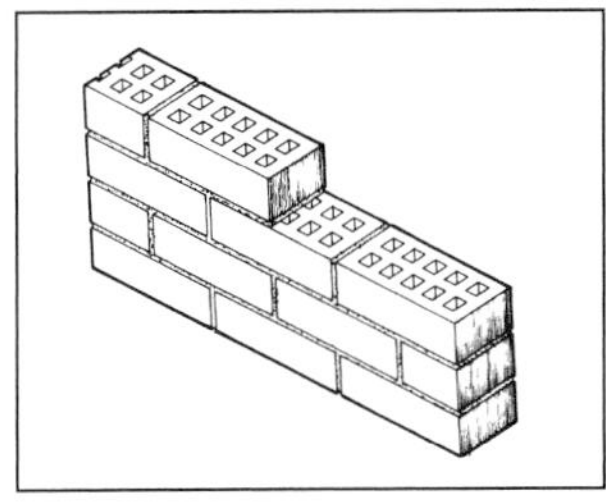

BRICK, THRU WALL UNITS
(3½ in. × 11½ in. Face Size)

Fire Rating (hr.):	2+
Sound Transmission Class (dB):	41
U-value (Btu/sq. ft. hr., °F):	.66
Wall Weight (lb./sq. ft.):	42

LIGHTWEIGHT CONCRETE BLOCK

Fire Rating (hr.):	1
Sound Transmission Class (dB):	45
U-value (Btu/sq. ft. hr., °F):	.38
Wall Weight (lb./sq. ft.):	26

in.		mm
3½	=	88.9
6	=	150
11½	=	292.1

U-value Btu/ft.^{2} × h.		k-value W/m^2 × k
0.38	=	2.16
0.66	=	3.75

lb./sq. ft.		kg/m^2
26	=	126.94
42	=	205.06

FIGURE 4.10-2 Properties of 6-in. (150 mm) walls.

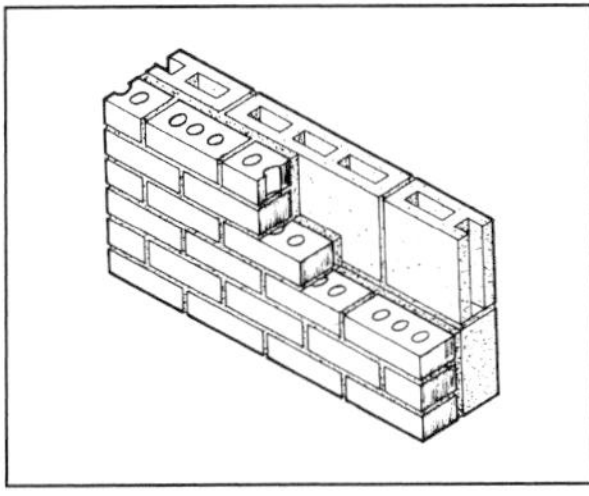

4 IN. FACE BRICK AND 4 IN. LIGHTWEIGHT CONCRETE BLOCK

Fire Rating (hr.):	3+
Sound Transmission Class (dB):	48
U-value (Btu/sq. ft. hr., °F):	.33
Wall Weight (lb./sq. ft.):	53

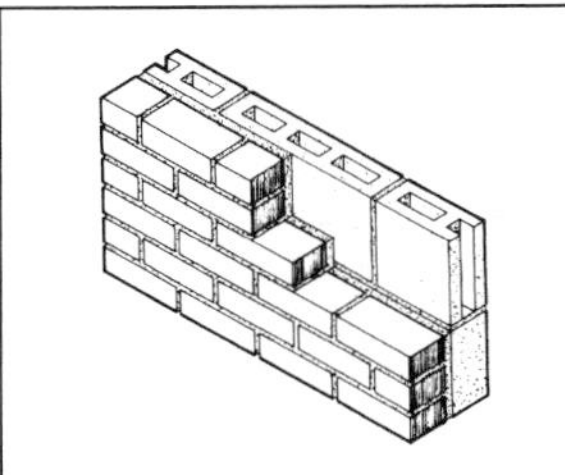

4 IN. COMMON BRICK AND 4 IN. LIGHTWEIGHT CONCRETE BLOCK

Fire Rating (hr.):	3
Sound Transmission Class (dB):	48
U-value (Btu/sq. ft. hr., °F):	.33
Wall Weight (lb./sq. ft.):	53

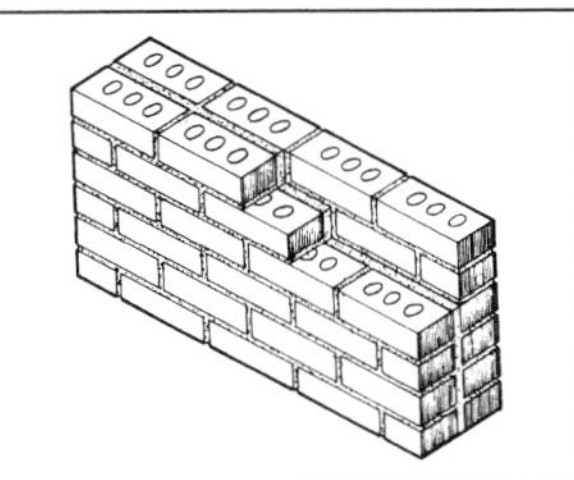

4 IN. FACE BRICK AND 4 IN. FACE BRICK (Tooled 2 Sides)

Fire Rating (hr.):	4
Sound Transmission Class (dB):	49
U-value (Btu/sq. ft. hr., °F):	.54
Wall Weight (lb./sq. ft.):	62

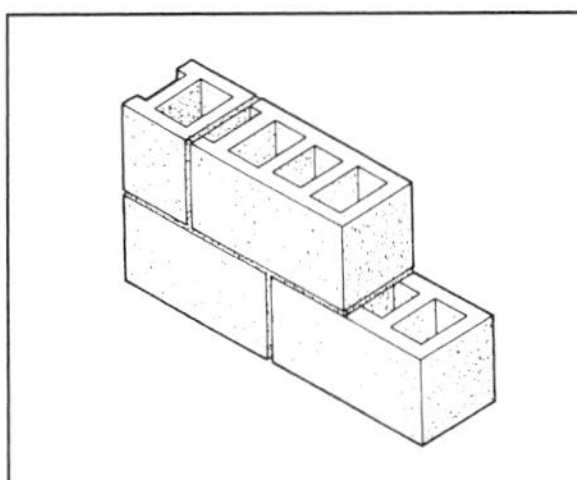

LIGHTWEIGHT CONCRETE BLOCK (Tooled 2 Sides)

Fire Rating (hr.):	3
Sound Transmission Class (dB):	46
U-value (Btu/sq. ft. hr., °F):	.35
Wall Weight (lb./sq. ft.):	32

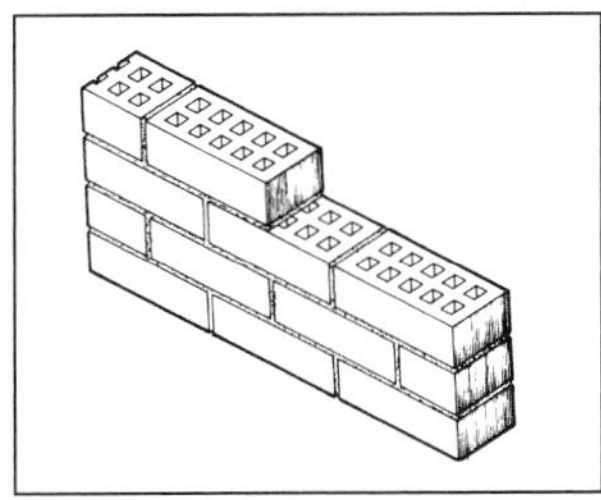

BRICK, THRU WALL UNITS
(3½ in. × 11½ in. Face Size)

Fire Rating (hr.):	3
Sound Transmission Class (dB):	50
U-value (Btu/sq. ft. hr., °F):	.47
Wall Weight (lb./sq. ft.):	61

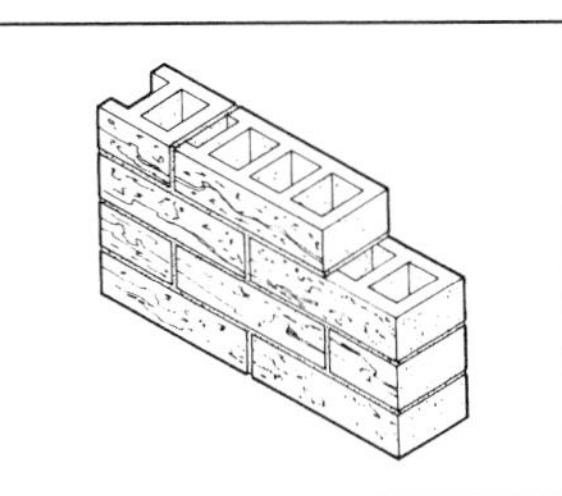

HEAVYWEIGHT CONCRETE "SLUMP" BLOCK (4 in. × 16 in. Face Size)

Fire Rating (hr.):	3
Sound Transmission Class (dB):	44
U-value (Btu/sq. ft. hr., °F):	.38
Wall Weight (lb./sq. ft.):	45

in.		mm
3½	=	88.9
4	=	100
16	=	400

U-value Btu/ft.^{2} × h.		k-value W/m^2 × k
0.33	=	1.87
0.35	=	1.99
0.38	=	2.16
0.47	=	2.67
0.54	=	3.07

lb./sq. ft.		kg/m^2
32	=	156.24
45	=	219.70
53	=	258.77
61	=	279.83
62	=	302.71

FIGURE 4.10-3 Properties of 8-in. (200 mm) walls.

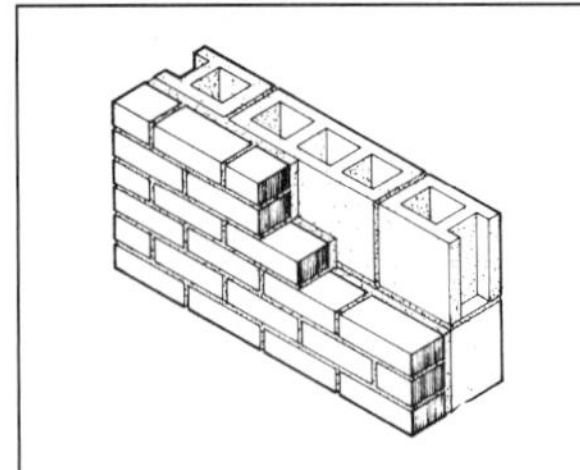

4 IN. FACE BRICK AND 6 IN. LIGHTWEIGHT CONCRETE BLOCK

Fire Rating (hr.):	4
Sound Transmission Class (dB):	50
U-value (Btu/sq. ft. hr., °F):	.32
Wall Weight (lb./sq. ft.):	62

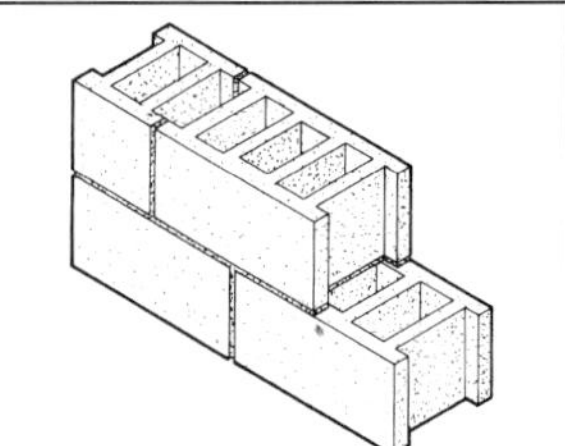

LIGHTWEIGHT CONCRETE BLOCK (Tooled 2 Sides)

Fire Rating (hr.):	3
Sound Transmission Class (dB):	48
U-value (Btu/sq. ft. hr., °F):	.33
Wall Weight (lb./sq. ft.):	39

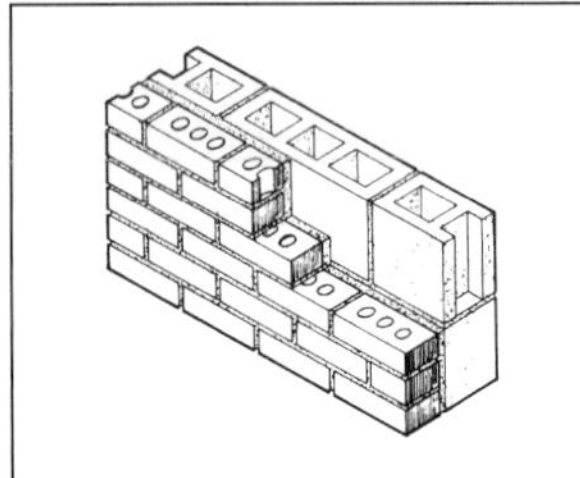

4 IN. COMMON BRICK AND 6 IN. LIGHTWEIGHT CONCRETE BLOCK

Fire Rating (hr.):	4
Sound Transmission Class (dB):	50
U-value (Btu/sq. ft. hr., °F):	.32
Wall Weight (lb./sq. ft.):	62

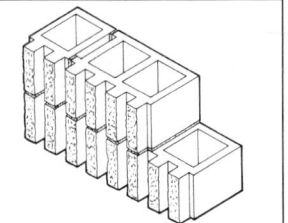

FLUTED (4 FLUTES) HEAVYWEIGHT CONCRETE BLOCK (Tooled 2 Sides)

Fire Rating (hr.):	3
Sound Transmission Class (dB):	48
U-value (Btu/sq. ft. hr., °F):	.33
Wall Weight (lb./sq. ft.):	39

in.		mm
4	=	100
6	=	150
10	=	250

U-value Btu/ft.^{2} × h.		k-value W/m^2 × k
0.32	=	1.82
0.33	=	1.87
0.49	=	2.78

lb./sq. ft.		kg/m^2
39	=	190.41
61	=	297.83
62	=	302.71

FIGURE 4.10-4 Properties of 10-in. (250 mm) walls.

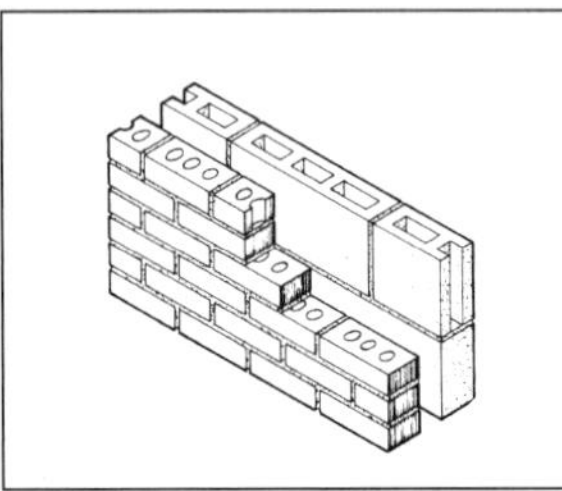

4 IN. FACE BRICK, 2 IN. AIR SPACE, AND 4 IN. LIGHTWEIGHT CONCRETE BLOCK

Fire Rating (hr.):	3
Sound Transmission Class (dB):	48
U-value (Btu/sq. ft. hr., °F):	.27
Wall Weight (lb./sq. ft.):	53

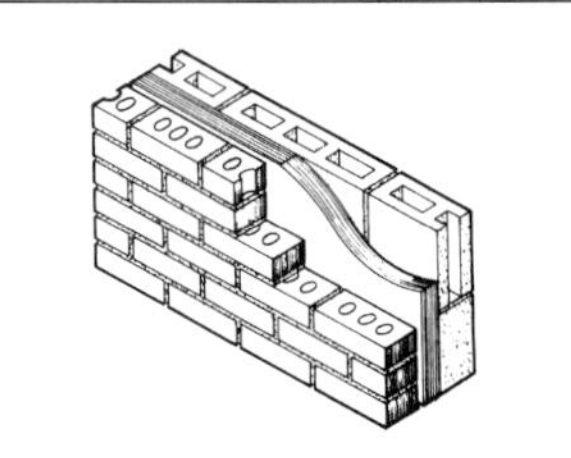

4 IN. FACE BRICK, 2 IN. URETHANE INSULATION, AND 4 IN. LIGHTWEIGHT CONCRETE BLOCK

Fire Rating (hr.):	3
Sound Transmission Class (dB):	49
U-value (Btu/sq. ft. hr., °F):	.05
Wall Weight (lb./sq. ft.):	54

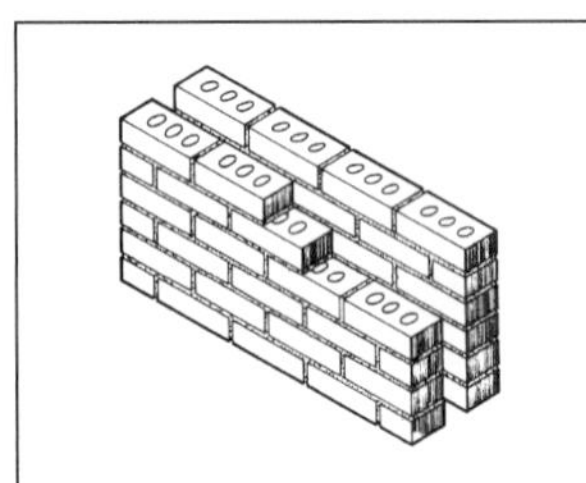

4 IN. FACE BRICK, 2 IN. AIR SPACE, AND 4 IN. FACE BRICK

Fire Rating (hr.):	4
Sound Transmission Class (dB):	51
U-value (Btu/sq. ft. hr., °F):	.37
Wall Weight (lb./sq. ft.):	64

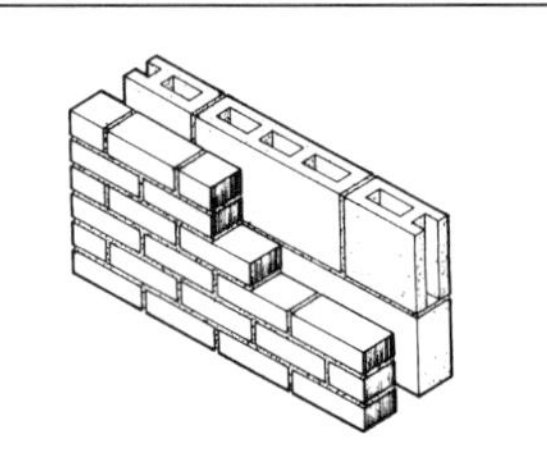

4 IN. COMMON BRICK, 2 IN. AIR SPACE, AND 4 IN. LIGHTWEIGHT CONCRETE BLOCK

Fire Rating (hr.):	3
Sound Transmission Class (dB):	48
U-value (Btu/sq. ft. hr., °F):	.27
Wall Weight (lb./sq. ft.):	53

in.		mm
2	=	50.8
4	=	100
10	=	250

U-value Btu/ft.^{2} × h.		k-value W/m^2 × k
0.05	=	0.28
0.27	=	1.53
0.37	=	2.10

lb./sq. ft.		kg/m^2
53	=	258.77
54	=	263.65
64	=	312.48

FIGURE 4.10-5 Properties of 10-in. (250 mm) cavity walls.

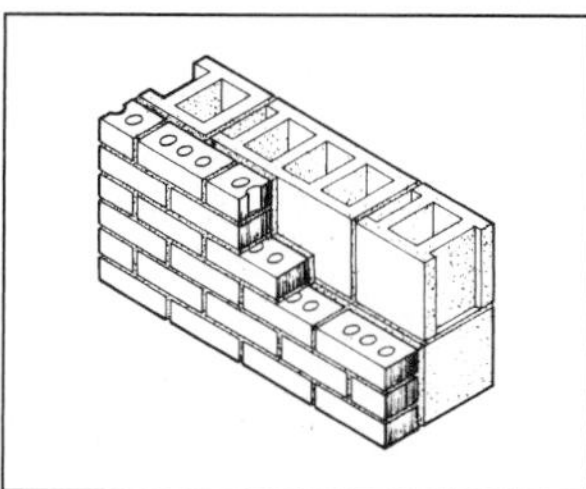

4 IN. FACE BRICK AND 8 IN. LIGHTWEIGHT CONCRETE BLOCK	
Fire Rating (hr.):	4
Sound Transmission Class (dB):	50+
U-value (Btu/sq. ft. hr., °F):	.29
Wall Weight (lb./sq. ft.):	72

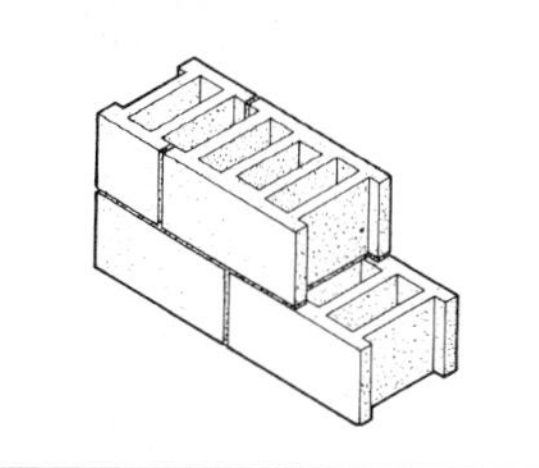

LIGHTWEIGHT CONCRETE BLOCK	
Fire Rating (hr.):	3–4
Sound Transmission Class (dB):	49
U-value (Btu/sq. ft. hr., °F):	.31
Wall Weight (lb./sq. ft.):	47

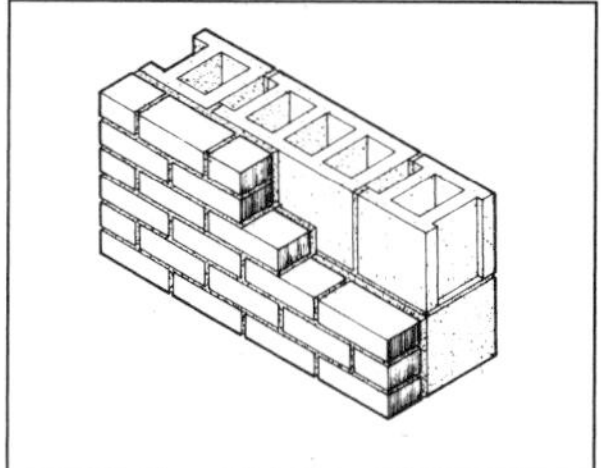

4 IN. COMMON BRICK AND 8 IN. LIGHTWEIGHT CONCRETE BLOCK	
Fire Rating (hr.):	4
Sound Transmission Class (dB):	50+
U-value (Btu/sq. ft. hr., °F):	.29
Wall Weight (lb./sq. ft.):	72

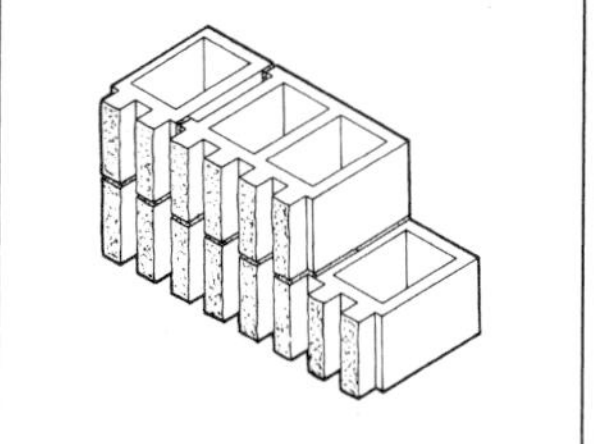

FLUTED (4 FLUTES) HEAVYWEIGHT CONCRETE BLOCK	
Fire Rating (hr.):	3–4
Sound Transmission Class (dB):	54
U-value (Btu/sq. ft. hr., °F):	.47
Wall Weight (lb./sq. ft.):	73

in.		mm
4	=	100
8	=	200
12	=	300

U-value Btu/ft.^{2} × h.		k-value W/m^2 × k
0.29	=	1.65
0.31	=	1.76
0.47	=	2.67

lb./sq. ft.		kg/m^2
47	=	229.47
72	=	351.53
73	=	356.42

FIGURE 4.10-6 Properties of 12-in. (300 mm) walls.

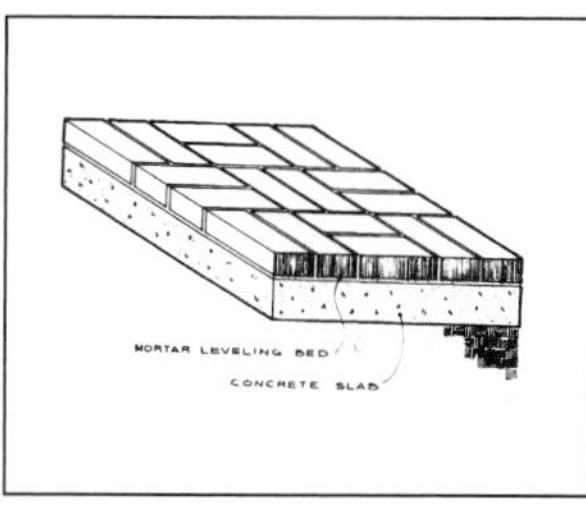

MORTAR SET—BED AND ALL JOINTS (Type "M" Mortar)

Weight: 29 lb./sq. ft.^{2} (141.59 kg/m^2)

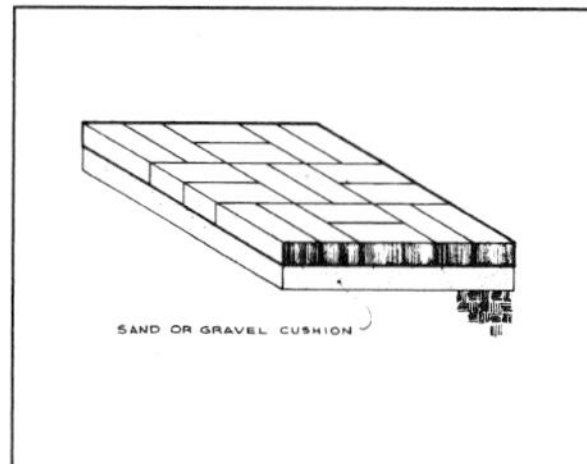

DRY SET AND BUTTED (Sand Swept Into Joints)

Weight: 27 lb./sq. ft.^{2} (131.82 kg/m^2)

FIGURE 4.10-7 Properties of brick paving.

4.11 Additional Reading

More information about the subjects discussed in this chapter can be found in the references listed in Section 4.12, and in the following publications.

Ambrose, James. 1997. *Simplified Design of Masonry Structures*. New York: John Wiley & Sons, Inc.

Cash, Kevin B., and Thomas A. Schwartz. 1994. "Stone Repair Options." *The Construction Specifier* 7 (July):72.

Chacon, Mark A. 1999. *Architectural Stone: Fabrication, Installation, and Selection*. New York: John Wiley & Sons, Inc.

Doe, Bruce R., and Susan I. Sherwood. 1985. "Acid Rain and Dimension Stone: A Dangerous Combination?" *The Construction Specifier* 2 (February):46.

Heyman, Jacques. 1997. *The Stone Skeleton: Structural Engineering of Masonry Architecture*. New York: Cambridge University Press.

Hooker, Kenneth A. 1995. "Stone Anchoring." *The Construction Specifier* 9 (September):49.

Hunter, David L. 1996. *Masonry Construction*. Upper Saddle River, NJ: Prentice-Hall.

Lore, Gordon. 1993. "New Stones on the Market." *The Construction Specifier* 5 (May):154.

Nicastro, David H. 1995. "Stone Stains." *The Construction Specifier* 11 (November):96.

Pritsker, Ellen. 1985. "Coloring the Skyline with Granite." *The Construction Specifier* 2 (February):40.

Roessler, David. 1990. "The Trinity Building: Restoring Granite and Limestone." *The Construction Specifier* 12 (December):80.

R. S. Means Company. 1998. *Means Concrete & Masonry Cost Data*. Kingston, MA: R. S. Means Company.

Sovinski, Robert W. 1999. *Brick in the Landscape: A Practical Guide to Specification and Design*. New York: John Wiley & Sons, Inc.

Weaver, Martin E. 1992. "Building and Monumental Stones of the United States." *The Construction Specifier* 3 (March):29.

Weiss, Norman R. 1987. "Restoring Architectural Stone." *The Construction Specifier* 7 (July):60.

4.12 Acknowledgments and References

ACKNOWLEDGMENTS

We gratefully acknowledge the assistance of the following organizations and individuals in preparing this chapter. We are also indebted to them for permission to use their illustrations when requested and for the use of their publications as references.

A.A. Wire Products Company
American Concrete Institute
American National Standards Institute
American Slate Company
American Society for Testing and Materials (ASTM)
Ancor Granite Tile Inc.
Athenian Marble
Brick Institute of America (BIA)
Buckingham-Virginia Slate Corporation Burlington Natstone Inc.
Building Officials and Code Administrators International (BOCA)
Building Research Board of the National Academy of Sciences (NAS)
Camara Slate, Inc.
Cold Spring Granite Co.
Dow Corning Corporation
Facing Tile Institute
Georgia Marble Company
Greek Trade Commission
Indiana Limestone Co., Inc.
Indiana Limestone Institute of America (ILI)
Marble Institute of America (MIA)
National Building Granite Quarries Association, Inc. (NBGQA)
National Concrete Masonry Association (NCMA)
Portland Cement Association (PCA)
Rock of Ages Corporation
Solnhofen Natural Stone, Inc.
Structural Slate Company
U.S. Department of Housing and Urban Development (HUD)

REFERENCES

We would also like to thank the authors and publishers of the publications in the following list for their contribution to our research for this chapter.

Adams, J. T. 1983. *The Complete Concrete, Masonry, and Brick Handbook.* New York: Van Nostrand Reinhold.

American Concrete Institute, American Society of Civil Engineers, and Masonry Society. 1999. ACI 530, ASCE 5, TMS 402, *Building Code Requirements for Masonry Structures.* Farmington Hills, MI: ACI, ASCE, TMS.

———. 1988. ACI 530.1, ASCE 6, TMS 602, *Specifications for Masonry Structures.* Farmington Hills, MI: ACI, ASCE, TMS.

———. 1999. ACI 530.1, ASCE 5, TMS 402, *Commentary on Building Code Requirements for Masonry Structures.* Farmington Hills, MI: ACI, ASCE, TMS.

———. 1999. ACI 530.1, ASCE 6, TMS 602, *Commentary on Specifications for Masonry Structures.* Farmington Hills, MI: ACI, ASCE, TMS.

American National Standards Institute. 1999. ANSI A108.1, "Installation of Ceramic Tile." New York: ANSI.

———. 1992. ANSI A118.1, "Specifications for Dry-Set Portland Cement Mortar." New York: ANSI.

———. 1992. ANSI A118.4, "Specifications for Latex Portland Cement Mortar." New York: ANSI.

———. 1992. ANSI A118.6, "Specifications for Ceramic Tile Grouts." New York: ANSI.

American Society for Testing and Materials (ASTM). 1998. *Annual Book of ASTM Standards.* Vol. 4.01, *Construction (Cement, Lime, Gypsum).* West Conshohocken, PA: ASTM.

American Society for Testing and Materials (ASTM) Standards:

A 36/A 36M, "Standard Specification for Carbon Structural Steel." West Conshohocken, PA: ASTM.

A 82, "Standard Specification for Steel Wire, Plain, for Concrete Reinforcement." West Conshohocken, PA: ASTM.

A 116, "Standard Specification for Zinc-Coated (Galvanized) Steel Woven Wire Fence Fabric." West Conshohocken, PA: ASTM.

A 123/A 123M, "Standard Specification for Zinc (Hot-Dip Galvanized) Coatings on Iron and Steel Products." West Conshohocken, PA: ASTM.

A 153/A 153M, "Standard Specification for Zinc Coating (Hot-Dip) on Iron and Steel Hardware." West Conshohocken, PA: ASTM.

A 167, "Standard Specification for Stainless and Heat-Resisting Chromium-Nickel Steel Plate, Sheet, and Strip." West Conshohocken, PA: ASTM.

A 185, "Standard Specification for Steel Welded Wire Fabric, Plain, for Concrete Reinforcement." West Conshohocken, PA: ASTM.

A 366, "Standard Specification for Steel, Sheet, Carbon, Cold-Rolled, Commercial Quality." West Conshohocken, PA: ASTM.

A 496, "Standard Specification for Steel Wire, Deformed, for Concrete Reinforcement." West Conshohocken, PA: ASTM.

A 497, "Standard Specification for Welded Wire Fabric, Deformed, for Concrete Reinforcement." West Conshohocken, PA: ASTM.

A 510, "Standard Specification for General Requirements for Wire Rods and Coarse Round Wire, Carbon Steel." West Conshohocken, PA: ASTM.

A 580/A 580M, "Standard Specification for Stainless Steel and Heat-Resistant Steel Wire." West Conshohocken, PA: ASTM.

A 615/A 615M, "Standard Specification for Deformed and Plain Billet-Steel Bars for Concrete Reinforcement." West Conshohocken, PA: ASTM.

A 635/A 635M, "Standard Specification for Steel, Sheet and Strip, Heavy-Thickness Coils, Carbon, Hot-Rolled." West Conshohocken, PA: ASTM.

A 653/A 653M, "Standard Specification for Steel Sheet, Zinc-Coated (Galvanized) or Zinc-Iron Alloy-Coated (Galvannealed) by the Hot-Dip Process." West Conshohocken, PA: ASTM.

A 666, "Standard Specification for Annealed or Cold-Rolled Austenitic Stainless Steel Sheet, Strip, Plate, and Flat Bar." West Conshohocken, PA: ASTM.

A 775, "Standard Specification for Epoxy-Coated Reinforcing Steel Bars." West Conshohocken, PA: ASTM.

B 101, "Standard Specification for Lead Coated Copper Sheet and Strip for Building Construction." West Conshohocken, PA: ASTM.

B 152, "Standard Specification for Copper Sheet, Strip, Plate, and Rolled Bar." West Conshohocken, PA: ASTM.

B 227, "Standard Specification for Hard-Drawn Copper-Clad Steel Wire." West Conshohocken, PA: ASTM.

C 5, "Standard Specification for Quicklime for Structural Purposes." West Conshohocken, PA: ASTM.

C 33, "Standard Specifications for Concrete Aggregates." West Conshohocken, PA: ASTM.
C 34, "Standard Specification for Structural Clay Load-Bearing Wall Tile." West Conshohocken, PA: ASTM.
C 55, "Standard Specification for Concrete Brick." West Conshohocken, PA: ASTM.
C 56, "Standard Specification for Structural Clay Non-Load-Bearing Tile." West Conshohocken, PA: ASTM.
C 62, "Standard Specification for Building Brick (Solid Masonry Units Made from Clay or Shale)." West Conshohocken, PA: ASTM.
C 67, "Standard Test Method for Sampling and Testing Brick and Structural Clay Tile." West Conshohocken, PA: ASTM.
C 90, "Standard Specification for Hollow Load-Bearing Concrete Masonry Units." West Conshohocken, PA: ASTM.
C 91, "Standard Specifications for Masonry Cement." West Conshohocken, PA: ASTM..
C 97, "Standard Test Method for Absorption and Bulk Specific Gravity of Dimension Stone." West Conshohocken, PA: ASTM.
C 99, "Standard Test Method for Modulus of Rupture of Dimension Stone." West Conshohocken, PA: ASTM.
C 119, "Standard Terminology Relating to Dimension Stone." West Conshohocken, PA: ASTM.
C 126, "Standard Specification for Ceramic Glazed Structural Clay Facing Tile, Facing Brick, and Solid Masonry Units." West Conshohocken, PA: ASTM.
C 129, "Standard Specification for Nonloadbearing Concrete Masonry Units." West Conshohocken, PA: ASTM.
C 144, "Standard Specification for Aggregate for Masonry Mortar." West Conshohocken, PA: ASTM.
C 150, "Standard Specification for Portland Cement." West Conshohocken, PA: ASTM.
C 170, Test Method for Compressive Strength of Dimension Stone." West Conshohocken, PA: ASTM.
C 206, "Standard Specification for Finishing Hydrated Lime." West Conshohocken, PA: ASTM.
C 207, "Standard Specifications for Hydrated Lime for Masonry Purposes." West Conshohocken, PA: ASTM.
C 212, "Standard Specification for Structural Clay Facing Tile." West Conshohocken, PA: ASTM.
C 216, "Standard Specification for Facing Brick (Solid Masonry Units Made from Clay or Shale)." West Conshohocken, PA: ASTM.
C 241, "Standard Test Method for Abrasion Resistance of Stone Subject to Foot Traffic." West Conshohocken, PA: ASTM.
C 260, "Standard Specification for Air-Entraining Admixtures for Concrete." West Conshohocken, PA: ASTM.
C 270, "Standard Specification for Mortar for Unit Masonry." West Conshohocken, PA: ASTM.
C 331, "Standard Specification for Lightweight Aggregate for Concrete Masonry." West Conshohocken, PA: ASTM.
C 404, "Standard Specification for Aggregate for Masonry Grout." West Conshohocken, PA: ASTM.
C 426, "Standard Test Method for Linear Drying Shrinkage of Concrete Masonry Units." West Conshohocken, PA: ASTM.
C 476, "Standard Specification for Grout for Masonry." West Conshohocken, PA: ASTM.
C 503, "Standard Specification for Marble Dimension Stone (Exterior)." West Conshohocken, PA: ASTM.
C 568, "Standard Specification for Limestone Dimension Stone." West Conshohocken, PA: ASTM.
C 595, "Standard Specification for Blended Hydraulic Cements." West Conshohocken, PA: ASTM.
C 615, "Standard Specification for Granite Dimension Stone." West Conshohocken, PA: ASTM.
C 618, "Standard Specification for Coal Fly Ash and Raw or Calcined Natural Pozzolan for Use as a Mineral Admixture in Portland Cement Concrete." West Conshohocken, PA: ASTM.
C 629, "Standard Specification for Slate Dimension Stone." West Conshohocken, PA: ASTM.
C 645, "Standard Specifications for Nonstructural Steel Framing Members." West Conshohocken, PA: ASTM.
C 652, "Standard Specification for Hollow Brick (Hollow Masonry Units Made from Clay or Shale)." West Conshohocken, PA: ASTM.
C 880, "Standard Test Method for Flexural Strength of Dimension Stone." West Conshohocken, PA: ASTM.
C 902, "Standard Specification for Pedestrian and Light Traffic Paving Brick." West Conshohocken, PA: ASTM.
C 955, "Standard Specifications for Load-Bearing (Transverse and Axial) Steel Studs, Runners (Track), and Bracing or Bridging, for Screw Application of Gypsum Panel Products and Metal Plaster Bases." West Conshohocken, PA: ASTM.
C 1007, "Standard Specification for Installation of Load-Bearing (Transverse and Axial) Steel Studs and Accessories." West Conshohocken, PA: ASTM.
C 1242, "Standard Guide for Design, Selection, and Installation of Exterior Dimension Stone Anchors and Anchoring Systems." West Conshohocken, PA: ASTM.
D 98, "Standard Specification for Calcium Chloride." West Conshohocken, PA: ASTM.
E 90, "Standard Test Method for Laboratory Measurement of Airborne-Sound Transmission Loss of Building Partitions and Elements." West Conshohocken, PA: ASTM.
E 96, "Standard Test Methods for Water Vapor Transmission of Materials." West Conshohocken, PA: ASTM.
E 119, "Standard Methods for Fire Tests of Building Construction and Materials." West Conshohocken, PA: ASTM.
E 413, "Standard Classification for Rating Sound Insulation." West Conshohocken, PA: ASTM.
E 514, "Standard Test Method for Penetration and Leakage Through Masonry." West Conshohocken, PA: ASTM.
Architectural Graphic Standards. See Ramsey/Sleeper.
Architectural Technology. 1984. "Rediscovering Terra-Cotta: A Glowing Recommendation." *Architectural Technology* (spring):12–13.
———. 1986. "Technical Tips: Slope the Sills." *Architectural Technology* (May/June): 66.
———. 1986. "Technical Tips: Tension Bars Save Masonry." *Architectural Technology* (May/June): 66–67.
Arcom Master Systems. *MasterSpec*, Basic Sections:
04200, "Unit Masonry," 5/94 ed. Salt Lake City, UT: Arcom.
04405, "Dimension Stone Cladding," 8/95 ed. Salt Lake City, UT: Arcom.
04410, "Stone Masonry Veneer," 11/95 ed. Salt Lake City, UT: Arcom.
04815, "Glass Unit Masonry Assem-

blies," 2/96 ed. Salt Lake City, UT: Arcom.

04901, "Clay Masonry Restoration and Cleaning," 8/96 ed. Salt Lake City, UT: Arcom.

04902, "Stone Masonry Restoration and Cleaning," 8/96 ed. Salt Lake City, UT: Arcom.

07620, "Sheet Metal Flashing and Trim," 8/94 ed. Salt Lake City, UT: Arcom.

Arumala, Joseph O., and Russell H. Brown. 1982. *Performance Evaluation of Brick Veneer with Steel Stud Backup*. Clemson, SC: Clemson University Department of Civil Engineering.

Beall, Christine. 1997. *Masonry Design and Detailing*. New York: McGraw-Hill.

———. 1988. "Controlling Moisture Movement in Masonry Walls." *The Construction Specifier* 41(8)(June): 36–47.

———. 1988. "The Types of Masonry Mortar." *The Magazine of Masonry Construction* 1(4):164–166.

———. 1989. "Weatherproof Masonry Walls." *Architecture* (March):133–136.

———. 1989. "The Many Differences Between Concrete and Masonry Technology." *Architecture* (April):123–124.

———. 1989. "Preventing Problems with Thin Brick Veneers." *The Construction Specifier* 42(9)(September):68–78.

———. 1991. "The ACI/ASCE Masonry Specification." *The Construction Specifier* 44(1)(January):36–42.

Belles, Donald. 1987. "Preventing Flame Spread on Exterior Walls." *Exteriors* 5(2)(summer):28–38.

Bock, Gordon. 1989. "Inspecting Chimneys." *The Old-House Journal* 17(2)(March/April):34–37.

Borchelt, J., ed. 1982. *STP 778—Masonry: Materials, Properties, and Performance*. West Conshohocken, PA: ASTM.

Brick Institute of America. 1980. *Compilation of Veneer/Metal Stud Problems*. Reston, VA: Brick Institute of America.

———. *Technical Notes on Brick Construction*.

No. 1. Rev. Mar. 1992. "All-Weather Construction." Reston, VA: BIA.

No. 2. Rev. Jan./Feb. 1975 (reissued Sept. 1998). "Glossary of Terms Relating to Brick Masonry." Reston, VA: BIA.

No. 3. Rev. Oct. 1995. "Overview of Building Code Requirements for Masonry Structures ACI 530/ASCE 5/TMS 402 and Specifications for Masonry Structures ACI 530.1/ASCE 6/TMS 602." Reston, VA: BIA.

No. 3A. Dec. 1992. "Brick Masonry Material Properties." Reston, VA: BIA.

No. 3B. May 1993. "Brick Masonry Section Properties." Reston, VA: BIA.

No. 7. Rev. Feb. 1985. "Water Resistance of Brick Masonry—Design and Detailing, Part I." Reston, VA: BIA.

No. 7A. Rev. Mar. 1985. "Water Resistance of Brick Masonry Materials, Part II." Reston, VA: BIA.

No. 7B. Rev. Apr. 1985. "Water Resistance of Brick Masonry—Construction and Workmanship, Part III." Reston, VA: BIA.

No. 7C. Feb. 1965 (reissued Sept. 1988). "Moisture Control in Brick and Tile Walls: Condensation." Reston, VA: BIA.

No. 7D. May 1988. "Moisture Resistance of Brick Masonry Walls—Condensation Analysis." Reston, VA: BIA.

No. 7F. Feb. 1986. "Moisture Resistance of Brick Masonry—Maintenance." Reston, VA: BIA.

No. 8. Rev. Aug. 1995. "Portland Cement-Lime Mortars for Brick Masonry." Reston, VA: BIA.

No. 8A. Rev. Sept. 1988. "Standards Specifications for Portland Cement-Lime Mortar for Brick Masonry." Reston, VA: BIA.

No. 8B. July/Aug. 1976. "Mortar for Brick Masonry—Selection and Controls." Reston, VA: BIA.

No. 9. Rev. Mar. 1986. "Manufacturing, Classification, and Selection of Brick: Manufacturing—Part I." Reston, VA: BIA.

No. 9B. Jan. 1989. "Manufacturing, Classification, and Selection of Brick: Selection—Part III." Reston, VA: BIA.

No. 10A. Rev. June/July 1973 (reissued July 1986). "Modular Brick Masonry." Reston, VA: BIA.

No. 10B. June 1993. "Brick Sizes and Related Information." Reston, VA: BIA.

No. 11. Rev. Dec. 1971. "Guide Specifications for Brick Masonry, Part I." Reston, VA: BIA.

No. 11A. Rev. June 1978. "Guide Specifications for Brick Masonry, Part II." Reston, VA: BIA.

No. 11B. Rev. Feb. 1972. "Guide Specifications for Brick Masonry, Part III." Reston, VA: BIA.

No. 11C. Rev. July 1972. "Guide Specifications for Brick Masonry, Part IV." Reston, VA: BIA.

No. 11D. Aug. 1972. "Guide Specifications for Brick Masonry, Part IV Continued." Reston, VA: BIA.

No. 11E. Rev. Sept. 1991. "Guide Specifications for Brick Masonry, Part V." Reston, VA: BIA.

No. 13. May 1962 (reissued Mar. 1982). "Ceramic Glazed Brick Facing for Exterior Walls." Reston, VA: BIA.

No. 14. Rev. Sept. 1992. "Brick Floors and Pavements, Part I—Design and Detailing." Reston, VA: BIA.

No. 14A. Rev. II. Jan. 1993. "Brick Floors and Pavements, Part II—Materials and Installation." Reston, VA: BIA.

No. 14B. Nov./Dec. 1975 (reissued Sept. 1998). "Brick Floors and Pavements, Part III." Reston, VA: BIA.

No. 15. Rev. May 1988. "Salvaged Brick." Reston, VA: BIA.

No. 16. Rev. Oct. 1974 (reissued May 1987). "Fire Resistance." Reston, VA: BIA.

No. 16B. June 1991 (reissued Aug. 1991). "Calculated Fire Resistance." Reston, VA: BIA.

No. 17. Nov.1962. "Reinforced Brick Masonry, Part I." Reston, VA: BIA.

No. 17A. Rev. June 1964 (reissued May 1987). "Reinforced Brick Masonry, Flexural Design, Part II." Reston, VA: BIA.

No. 17B. Feb. 1963 (reissued Dec. 1984). "Reinforced Brick Masonry, Axial Design, Part III." Reston, VA: BIA.

No. 17C. Mar. 1963 (reissued May 1986). "Reinforced Brick Masonry—Inspectors' Guide, Part IV." Reston, VA: BIA.

No. 17D. Rev. Mar. 1972 (reissued May 1988). "High-Lift Grouted Reinforced Brick Masonry." Reston, VA: BIA.

No. 17H. Nov./Dec. 1964 (reissued July 1986). "Reinforced Brick and Tile Lintels." Reston, VA: BIA.

No. 17J. June 1967 (reissued June 1987). "Design Tables for Reinforced Brick Masonry Flexural Members." Reston, VA: BIA.

No. 17L. Rev. Feb./Mar. 1973 (reissued Sept. 1988). "Four-Inch RBM Curtain and Panel Walls." Reston, VA: BIA.

No. 17M. July 1968 (reissued Sept. 1988). "Reinforced Brick Masonry Girders—Examples." Reston, VA: BIA.

No. 18. Rev. Jan. 1991. "Volume Changes and Effect of Movement—Part I." Reston, VA: BIA.

No. 18A. Rev. Dec. 1991. "Design and Detailing of Movement Joints—Part II." Reston, VA: BIA.

No. 19. Rev. Jan. 1993. "Residential Fireplace Design." Reston, VA: BIA.

No. 19A. Rev. May 1980. "Residential Fireplaces, Details and Construction." Reston, VA: BIA.

No. 19B. June 1980 (reissued Jan. 1988). "Residential Chimneys, Design and Construction." Reston, VA: BIA.

No. 19C. Rev. Apr. 1988. "Contemporary Brick Masonry Fireplaces." Reston, VA: BIA.

No. 19D. Jan. 1983 (reissued June 1987). "Brick Masonry Fireplaces, Part I, Russian Style Heaters." Reston, VA: BIA.

No. 19E. 1983 (reissued Sept. 1988). "Brick Masonry Fireplaces, Part II, Fountain and Contemporary Style Heaters." Reston, VA: BIA.

No. 20. Rev II. Nov. 1990. "Cleaning Brick Masonry." Reston, VA: BIA.

No. 21. Rev. Jan./Feb. 1977 (reissued May 1987). "Brick Masonry Cavity Walls." Reston, VA: BIA.

No. 21A. Rev. May/June 1977 (reissued July 1986). "Brick Masonry Cavity Walls: Insulated." Reston, VA: BIA.

No. 21B. Jan./Feb. 1978 (reissued Jan. 1987). "Brick Masonry Cavity Walls: Detailing." Reston, VA: BIA.

No. 21C. Rev. Oct. 1989. "Brick Masonry Cavity Walls: Construction." Reston, VA: BIA.

No. 23. Rev. May 1985. "Efflorescence, Causes and Mechanisms: Part I." Reston, VA: BIA.

No. 23A. Rev. June 1985. "Efflorescence, Prevention and Control: Part II." Reston, VA: BIA.

No. 24. Rev. Feb. 1970 (reissued Dec. 1986). "The Contemporary Bearing Wall." Reston, VA: BIA.

No. 24A. Rev. May 1970 (reissued May 1985). "Building Code Requirements for the Contemporary Bearing Wall." Reston, VA: BIA.

No. 24B. Rev. July/Aug. 1970 (reissued May 1985). "Design Examples of Contemporary Bearing Walls." Reston, VA: BIA.

No. 24C. Rev. Sept./Oct. 1970 (reissued May 1988). "The Contemporary Bearing Wall—Introduction to Shear Wall Design." Reston, VA: BIA.

No. 24D. Rev. Dec. 1970 (reissued Dec. 1987). "The Contemporary Bearing Wall—Example of Shear Wall Design." Reston, VA: BIA.

No. 24E. Rev. Nov. 1970 (reissued May 1987). "The Contemporary Bearing Wall—Design Tables for Columns and Walls." Reston, VA: BIA.

No. 24F. Rev. Nov./Dec. 1974 (reissued Sept. 1988). "The Contemporary Bearing Wall—Construction." Reston, VA: BIA.

No. 24G. Dec. 1968 (reissued Feb. 1987). "Contemporary Bearing Wall—Detailing." Reston, VA: BIA.

No. 24H. Rev. Jan. 1969 (reissued Feb. 1987). "The Contemporary Bearing Walls—Wall Types and Properties." Reston, VA: BIA.

No. 26. Rev. Sept. 1994. "Single Wythe Bearing Walls." Reston, VA: BIA.

No. 27. Rev. Aug. 1994. "Brick Masonry Rain Screen Walls." Reston, VA: BIA.

No. 28. Rev. Aug. 1991. "Brick Veneer, Wood Frame Construction." Reston, VA: BIA.

No. 28A. Sept./Oct. 1978 (reissued Sept. 1988). "Brick Veneer, Existing Construction." Reston, VA: BIA.

No. 28B. Rev. II. Feb. 1987. "Brick Veneer, Steel Stud Panel Walls." Reston, VA: BIA.

No. 28C. Jan. 1986. "Thin Brick Veneer, Introduction." Reston, VA: BIA.

No. 29. Rev. July 1994. "Brick in Landscape Architecture—Pedestrian Application." Reston, VA: BIA.

No. 29A. Rev. Nov. 1968 (reissued Sept. 1988). "Brick in Landscape Architecture—Garden Walls." Reston, VA: BIA.

No. 29B. Apr. 1967 (reissued May 1988). "Brick in Landscape Architecture—Miscellaneous Applications." Reston, VA: BIA.

No. 30. July 1967 (reissued Sept. 1988). "Bonds and Patterns in Brickwork." Reston, VA: BIA.

No. 31. Rev. Jan. 1995. "Brick Masonry Arches." Reston, VA: BIA.

No. 31A. Oct. 1967 (reissued July 1986). "Structural Design of Brick Masonry Arches." Reston, VA: BIA.

No. 31B. Rev. Nov./Dec. 1981 (reissued May 1987). "Structural Steel Lintels." Reston, VA: BIA.

No. 36. Rev. July/Aug. 1981 (reissued Jan. 1988). "Brick Masonry Details: Sills and Soffits." Reston, VA: BIA.

No. 36A. Rev. Sept./Oct. 1981 (reissued Jan. 1988). "Brick Masonry Details, Caps and Copings, Corbels and Racking." Reston, VA: BIA.

No. 40. Oct./Nov. 1973 (reissued Jan. 1987). "Prefabricated Brick Masonry—Introduction." Reston, VA: BIA.

No. 41. Rev. Feb. 1996. "Hollow Brick Masonry." Reston, VA: BIA.

No. 42. Nov. 1991. "Empirical Design of Brick Masonry." Reston, VA: BIA.

No. 44A. May 1986. "Fasteners for Brick Masonry." Reston, VA: BIA.

No. 44B. Mar. 1987 (reissued Sept. 1988). "Wall Ties for Brick Masonry." Reston, VA: BIA.

Brown, R. W. 1984. *Residential Foundations: Design, Behavior, and Repair.* New York: Van Nostrand Reinhold.

Catani, Mario J. 1985. "Protection of Embedded Steel in Masonry." *The Construction Specifier* 38(1)(January): 62–68.

———. 1987. "Connecting Masonry with Ties and Anchors." *The Construction Specifier* 40(8)(August):108–113.

———. 1989. "Setting New Standards for Masonry." *The Construction Specifier* 42(9)(September):108–114.

Catani, Mario J., and A. Rhett Whitlock. "Coping with Wide Cavities." *The Construction Specifier* 39(8)(August):34–42.

Chin, Ian R., John P. Stecich, and Bernard Erlin. 1986. "Design of Thin Stone Veneers for Buildings." *Building Stone Magazine* (May/June): 50–61.

Clifton, J. R., ed. 1986. *STP 935—Cleaning Stone and Masonry*. West Conshohocken, PA: ASTM.

Council of American Building Officials. 1983. *One- and Two-Family Dwelling Code*. Falls Church, VA. CABO (in process of being transferred to the control of the International Code Council (ICC).

De Cristoforo, R. J. 1975. *Concrete and Masonry*. Reston, VA: Reston Publishing Company, Inc.

Gabby, Brent A. 1990. "About Face." *Architecture* (April):113–114, 141–143.

———. 1991. "Brick Pavers: A Commentary on ASTM C 902." *The Construction Specifier* 44(1)(January): 29–31.

Gordon, Douglas E. 1985. "Brick Basics." *Architectural Technology* (fall):17–24. *See also* April 1986, "Feedback," pp. 8–9.

Grimm, Clayford T. 1985. "Clay Brick/Concrete Masonry Wall Design Checklist." *The Construction Specifier* 38(8)(August):67–69.

———. 1989. "What They Say About Brick Veneer over Steel Studs." *The Construction Specifier* 42(9)(September):76–77.

Grogan, John C. 1983. "Water Permeance in Brick Masonry." *The Construction Specifier* 36(12)(December):36–38.

Grogan, John C., and John T. Conway, eds. 1985. *Masonry Research Application and Problems (STP 871)*. West Conshohocken, PA: ASTM.

Gross, James G., Robert D. Dikkers, and John C. Grogan. 1969. *Recommended Practice for Engineered Brick Masonry*. Reston, VA: Brick Institute of America.

Harris, Harry A., ed. 1988. *STP 992—Masonry: Materials, Design, Construction, and Maintenance*. West Conshohocken, PA: ASTM.

Hayes, Arthur J. 1989. "Keeping Abreast of Masonry Technology." *The Construction Specifier* 42(9)(September):45–46.

Hunderman, Harry J., and Deborah Slaton. 1989. "Terra-Cotta: Analysis and Repair." *The Construction Specifier* 42(7)(July):50–57.

Indiana Limestone Institute of America, Inc. (ILI). *The Contractors Handbook on Indiana Limestone*. 6th ed. Bedford, IN: ILI.

———. *The Finishing Touch*. Bedford, IN: ILI.

———. *Indiana Limestone Handbook*. 19th ed. Bedford, IN: ILI.

———. *The Patton Glossary of Building Stone and Masonry Terms*. Bedford, IN: ILI.

———. *Specifications for Indiana Limestone*. Bedford, IN: ILI.

———. *ILI Technote on Safety Factors*. Bedford, IN: ILI.

Kaminetzky, Dov. 1989. "Investigating Masonry Failures." *The Construction Specifier* 42(1)(January):42–47.

———. 1983. "Matching Brick." *The Construction Specifier* 36(12)(December):46–51.

Laska, Walter. 1991. "Portland Lime Mortar, Understanding ASTM C 270." *The Construction Specifier* 44(1)(January):71–73.

Lucas, James J. 1983. "Causes and Cures for Deteriorating Masonry." *The Construction Specifier* 36(3)(March):60–70.

Marble Institute of America, Inc. *Dimension Stone—Design Manual IV*. Farmington, MI: Marble Institute of America.

Metal Lath/Steel Framing Association. 1985. *Lightweight Steel Framing Systems Manual*. 2d ed. Chicago, IL: Metal Lath/Steel Framing Association.

———. 1986. *Technical Information File*. Chicago, IL: Metal Lath/Steel Framing Association.

———. *Technical Bulletin No. 18: Fire Rated Metal Lath/Steel Stud Exterior*. Chicago, IL: Metal Lath/Steel Framing Association.

National Building Granite Quarries Association. *Specifications for Architectural Granite*. Barre, VT: National Building Granite Quarries Association.

National Concrete Masonry Association (NCMA). TEK 12-2A, "The Structural Role of Joint Reinforcement in Concrete Masonry." Herndon, VA: NCMA.

———. TEK 1-3A, "Building Code Requirements for Masonry Structures." Herndon, VA: NCMA.

———. *Architectural and Engineering Concrete Masonry Details for Building Construction*. McLean, VA: NCMA.

Nicastro, David H. 1989. "Common Problems in Masonry Detailing." *The Construction Specifier* 42(9)(September): 80–83.

Nunn, Mark A. 1989. "Exterior Brick Pavements Guide." *The Construction Specifier* 42(9)(September):100–107.

Pilling, Doug. 1980. "Brick Walks." *The Old-House Journal* 18(7)(July):73, 78–79.

Portland Cement Association (PCA). *Water Penetration Tests of Masonry Walls*. Skokie, IL: PCA.

Raeber, John A. 1989. "Composite Plaster and Masonry Construction: A Closer Look." *The Construction Specifier* 42(9)(September):31–32.

———. 1991. "Brick Veneer in Earthquake Country." *The Construction Specifier* 41(1)(January):23–24.

Ramsey/Sleeper, The AIA Committee on Architectural Graphic Standards. *Architectural Graphic Standards*. 8th ed. New York: John Wiley & Sons, Inc.

Randall, Frank A., and William C. Panarese. 1976. *Concrete Masonry Handbook for Architects, Engineers, and Builders*. 4th ed. Skokie, IL: Portland Cement Association.

Reynolds, Phillip. 1983. "Specifying Non-Industrial Exterior Brick Paving." *The Construction Specifier* 36(12)(December):39–45.

Richter, Robert J. 1991. "Masonry Failures, A Look at Trade Practices." *The Construction Specifier* 44(1)(January): 44–51.

Rhault, Robert. 1991. "Masonry Cleaning." *The Construction Specifier* 44(1)(January):53–61.

Rock of Ages Corporation. 1992. "The Rock of Ages Story." Barre VT: Rock of Ages Corporation.

Sheppard, Walter Lee. 1981. "Obtaining Sound Chemically Resistant Masonry Construction." *The Construction Specifier* 34(12)(December):20–26.

Silver, Larry. 1991. "Precast Concrete Masonry Lintels." *The Construction Specifier* 44(1)(January):66–69.

Sivinski, Valerie A. 1985. "Masonry Maintenance." *Architectural Technology* (fall):25–26.

Smith, Baird M. 1984. *Moisture Problems in Historic Masonry Walls: Diagnosis and Treatment*. Technical Report. Washington, DC: Technical Preservation Services, U.S. Department of the Interior.

Specialty Steel Industry of North America. *Stainless Steel for Wall Ties, Stone Anchors and Masonry Fastening Systems*. Washington, DC: SSINA.

Szoke, Stephen S., and Gerald J. Carrier. 1986. "Avoiding Cracks in Brickwork." *The Construction Specifier* 39(8)(August):44–56.

Szoke, Stephen S., and Hugh C. MacDonald. 1989. "Combining Masonry and Brick." *Architecture* (January): 103–106.

Tiller, deTeel Patterson. 1979. *The Preservation of Historic Glazed Architectural Terra-Cotta*. Preservation Brief #7. Washington, DC: U.S. Department of the Interior, National Park Service, Preservation Assistance Division, Technical Preservation Services.

Trimble, Brian E., and J. Gregg Borchelt. "Jack Arches in Masonry Construction." *The Construction Specifier* 44(1)(January):62–65.

Underground Space Center. 1988. *Building Foundation Design Handbook*. Minneapolis, MN: Underground Space Center.

Wilson, Forrest. 1985. "Building Diagnostics." *Architectural Technology* (winter):22–41.

Wintz, Joseph A., III, and Alan H. Yorkdale. 1983. "Brick Veneer Panel and Curtain Wall Systems—A Designer's Guide." *The Construction Specifier* 36(12)(December):24–39.

CHAPTER 5

Metals

Introduction

Applicable *MasterFormat*™ Sections

Iron and Steel Materials and Products

Iron and Steel Design and Construction

Miscellaneous Iron and Steel Elements

Aluminum

Other Metals

Additional Reading

Acknowledgments and References

Introduction

Metals have been an essential part of human development since our beginnings. The materials for very early stone structures were cut and shaped laboriously, using tools made of hard stone and flint. The first known metals were those found in nature, such as gold and silver. Later, other metals, probably discovered accidentally, were obtained by melting ores. The first relatively hard metals produced were mostly copper alloys, such as brass and bronze. These metals were vastly superior to flint and stone for some purposes. They are still used today in building construction. Copper is used in roofing, flashing, piping, wiring, and in the manufacture of equipment. Copper alloys are used to manufacture doors and their frames, windows, door and window hardware, cladding panels, plumbing fixtures, and for many decorative purposes.

The knowledge that metals could be obtained by heating ores naturally led to a search for other metals. The discovery of iron came fairly early because the process of making it was relatively easy. Iron and its modern derivative, steel, are used extensively in buildings today.

Aluminum, which was not produced until the nineteenth century, is another common metal in most of today's buildings. It is used to make cladding panels, storefronts, curtain walls, doors and their frames, windows, siding, roofing, flashing, wiring, piping, railings, hardware, and many other products.

It is impossible to construct a modern building containing no metal products. Products made from iron, steel, aluminum, copper, bronze, lead, alloys of these metals, and other metals are contained in numerous building components, from those forming the underlying structure to those in the shell. Modern mechanical and electrical systems cannot function without metals. Masonry structures contain metal reinforcements, anchors, and fasteners. Concrete structures are reinforced with steel. Wood framing is held together with metal nails, lag bolts, and screws. Finishes are hung on metal furring and framing. This chapter discusses some of the metal products used in buildings today. Covering them all in detail would take an entire book larger than this one.

Applicable *MasterFormat*™ Sections

The following *MasterFormat*™ Level 2 sections are applicable to this chapter.

05050 Basic Metal Materials and Methods
05100 Structural Metal Framing
05200 Metal Joists
05300 Metal Deck
05400 Cold-Formed Metal Framing
05500 Metal Fabrications
05600 Hydraulic Fabrications
05700 Ornamental Metal
05800 Expansion Control

5.1 Iron and Steel Materials and Products

The chief ingredient of the metals known as iron and steel is the chemical element iron (*ferrum*), from which iron-based alloys derive the generic name *ferrous metals*. Although about 5% of the earth's crust is composed of iron, it is rarely found in its metallic form in nature. It is almost always combined with oxygen (iron oxide) and sulfur silicon and is mixed with other minerals. It is found in rock, gravel, sand, clay, and mud. In this form it is mined as iron ore, which is the chief source of iron. Iron- and steelmaking involve extracting iron from ore, combining it with carbon and other elements, and then forming it into shapes.

Pure iron is inherently soft, ductile, easily shaped, and relatively weak. However, *wrought iron*, which is nearly pure, is strong enough for many construction purposes. In fact, it has been used for many bridges and buildings and as the structural frame for the Eiffel Tower. The many irons and steels used in modern construction and manufacture get their strength and hardness mainly from the element carbon. Most steels have the best combination of properties when carbon is present in amounts less than 1.2%. Ferrous metals containing substantially larger amounts of carbon become so hard and brittle that they cannot be readily shaped by either hot or cold working methods. These metals are formed by being cast into molds and are referred to as *cast iron*.

5.1.1 HISTORY

Iron was discovered at different times in different parts of the world. Iron implements that are more than 4000 years old have been found in Egypt; others in China are more than 2700 years old. By the time of Rameses II, who reigned from 1292 to 1225 B.C., iron was being widely used in Egypt. In the following thousand years, the use of iron spread throughout most of the known world. Iron was not used extensively in western Europe, however, until about 100 B.C.

Fairly high quality steel was made in ancient Syria and in Toledo, in what is now Spain, but the methods used were lost, and the steelmaking processes used during the period between the beginning of the Dark Ages and the 1730s could not control the impurities and carbon content sufficiently to produce metal with the properties of modern steel. The so-called blister steel they produced was nothing more than carburized wrought iron. Benjamin Huntsman changed that in the 1730s with his invention of the crucible process, which permits the melting of blister steel into a homogeneous mass, thus producing steel in the modern sense.

Huntsman's crucible process, which produced steel of very high quality, was quite labor-intensive and yielded only small quantities of steel with each charging. In addition, modern rolling, drawing, and heat-treating processes that improve the mechanical properties of steel were unknown. Therefore, the earliest ferrous metals used in construction were essentially wrought iron and cast iron, shaped by handworking or by casting hot metal in molds.

The first successful North American ironworks was built in Saugus, Massachusetts, in the late 1640s. Its earliest products were cast-iron cooking pots and wrought-iron nails. Nails were so expensive that old buildings were often burned so that the nails could be recovered.

The first high-tonnage steelmaking process was discovered almost simultaneously by an American, William Kelly, and an Englishman, Sir Henry Bessemer. The Civil War interrupted Kelly's work and slowed the development of commercial steelmaking here until after 1864. This gave Bessemer time to develop the process and make it practical. Therefore, the process and the furnace used in it were given his name. In the Bessemer (pneumatic) process, air was forced through molten iron contained in a pear-shaped vessel. As the air rose through the iron, it burned out impurities. The addition of manganese caused chemical changes, which converted the iron to steel.

After 1864, the Bessemer process and the discovery of abundant iron ore deposits in the Lake Superior region made the United States a world leader in steel production. The modern revolution in construction, industry, commerce, transportation, and communications is due in large part to the economical production of iron and steel.

The term *steel* covers a variety of ferrous metals of different chemical compositions and properties. These properties are affected by the alloying elements included, the production processes used, and the methods of manufacture.

5.1.2 RAW MATERIALS

In this section, the term *raw material* is used to designate all materials used to make iron and steel up to the point of producing metal ingots used for product manufacture.

The general steps in the mining and processing of raw materials, smelting them into pig iron, and refining that into steel ingots, are illustrated in Figure 5.1-1. Iron- and steelmaking starts with the separation of the metal iron from its oxide, which is the way iron appears in iron ores. This is done by heating iron ore while it is in contact with coke, a form of carbon. The carbon then combines with both the oxygen present and the iron in the ore, forming a volatile gas and a carbon-rich crude iron called pig iron. The term *pig iron* comes from the shape of the trenches into which the molten iron from early furnaces ran. The iron was poured into a main trench from which it ran along into smaller connected trenches, where it hardened into ingots. These smaller trenches, lined up against the main trench, looked to some like baby pigs lined up against their mother for suckling.

The primary raw materials used to make pig iron are iron ore, fuel (coke), air, flux, and refractories. In steelmaking, pig iron and metal scrap replace iron ore as the primary source of iron, and electricity may replace fossil fuel as a source of heat. In some processes, gaseous oxygen supplements or completely replaces ordinary combustion air. The goal in steelmaking is to remove some or most of the carbon and other impurities from pig iron.

5.1.2.1 Mining and Processing

The primary raw materials—iron ores, fuels, fluxes, and refractories—are often

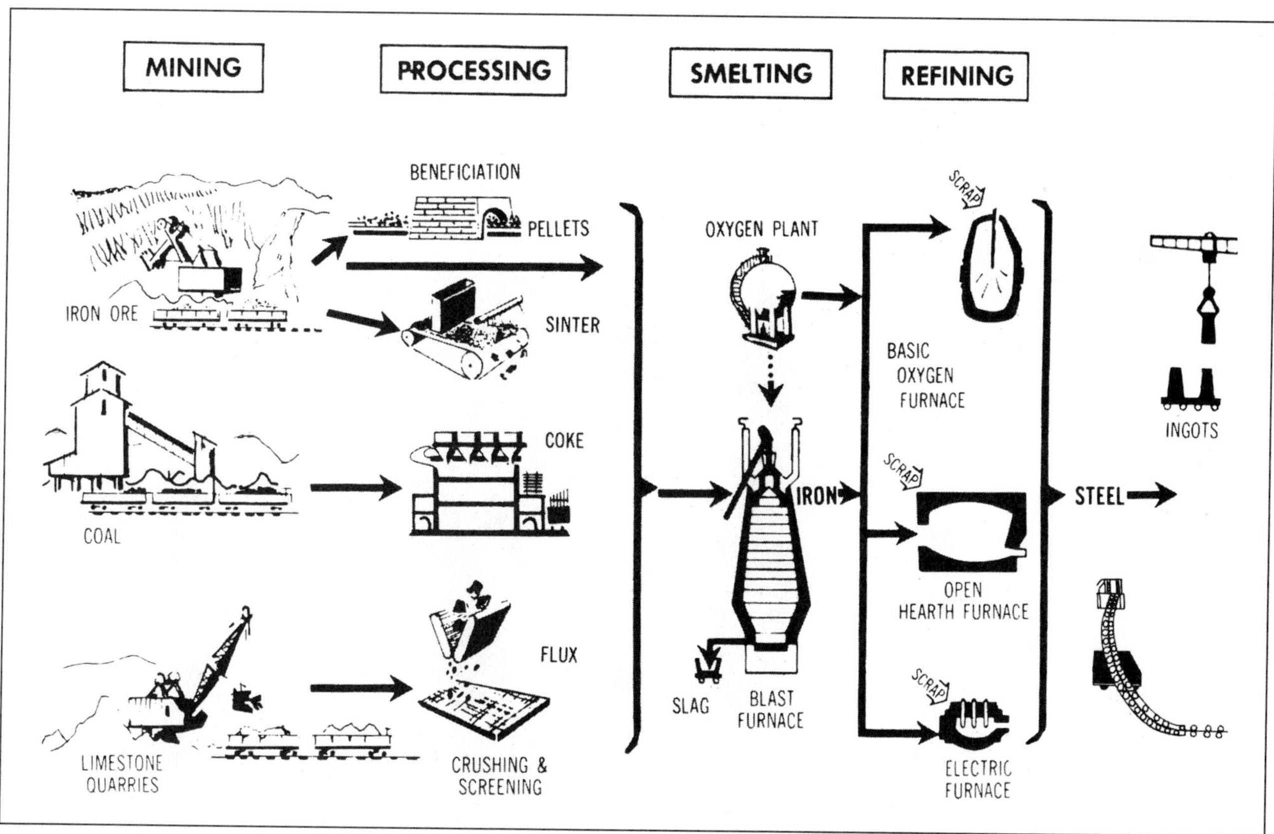

FIGURE 5.1-1 Flowchart of operations in the production of steel ingot. (Courtesy American Iron and Steel Institute.)

mined in areas that are far from iron- and steelmaking plants. To reduce shipping costs, raw materials are sometimes processed near the mines to increase their yield and reduce their bulk. They are then shipped by rail and water (Fig. 5.1-2) to steel mills or plants for direct use or further processing. Mills and plants are typically located along navigable water routes, which permit low-cost transportation of bulk materials and provide an abundant supply of the fresh water that is essential in steelmaking.

IRON ORES

Iron ore is mined by either the underground or the open pit method (Fig. 5.1-3). The iron ore of the Lake Superior region dominates steel production; other deposits are mined in Alabama, New Jersey, New York, California, and a few other states. Iron ore is also imported, with Canada providing more than half of all imports. The most common iron minerals are hematite (Fe_2O_3), siderite ($FeCo_3$), and pyrite (FeS), which have iron contents ranging as high as 70%. Limonite may be considered as magnetite plus water; taconite and jasper are names referring to a wide variety of iron-bearing rocks with iron contents ranging between 22% and 30%.

Most ore mined in the Lake Superior region is taconite, which lies close to the surface but is so hard that superheated flames must be used to drill holes for explosive charges. After it has been blasted out of the ground, taconite is processed by *beneficiation*, a treatment that increases its iron content per ton of raw material. Ores with iron contents near 50% are usually shipped as-is. Ores like taconite, with a 23% to 25% iron content, can be improved to nearly 66% iron content by beneficiation. Richer ores and beneficiated taconite can be further improved by direct reduction to an iron content of more than 90%.

Beneficiation The beneficiation of ores generally consists of (1) grinding and concentrating them to increase the iron content and (2) *agglomeration*, which increases the ore's particles to a size suitable for blast furnace use. Two forms of agglomeration are *pelletizing*, used for jasper and taconite, and *sintering*, used to recycle iron ore dust.

Pelletizing by both magnetic separation and concentration is usually done near mine sites to avoid transporting large tonnages of non-iron-bearing minerals. In one pelletizing process, magnetic taconite is crushed into a fine powder with the consistency of flour. The powder is passed through a magnetic separator to remove the iron particles. The iron powder then goes to a balling drum, where it is mixed with coal dust and a binder material to form pellets with 60% iron content. The pellets, each about 1 in. in diameter, are baked to a hard finish in a kiln or furnace.

In another pelletizing process, high-grade nonmagnetic ores are crushed and then concentrated in flotation cells. Flotation cells contain liquids in which the heavier iron-bearing particles sink to the bottom of the cell while the lighter particles remain suspended and are skimmed away. After filtering, the iron-bearing particles are formed into pellets and baked hard.

Sintering is a process by which iron ore dust, reclaimed from blast furnace gases and other manufacturing processes, is fused with coke and fluxes into a clinker-like product containing 52% to 65% iron.

Direct Reduction Direct reduction includes various processes in which ore is concentrated to high iron content that is comparable to that of molten iron from a blast furnace. Some ores are made into pellets, others into briquettes. Direct-reduced ores with over 90% iron contents can bypass traditional blast furnace processing and be charged directly into electric steelmaking and basic oxygen furnaces.

FUELS

Of the three main fuels used in the iron and steel industry—coal, oil, and natural gas—coal supplies more than 65% of the industry's heat and energy requirements.

The chief fuel used in a blast furnace is coke, which is produced in coking ovens from selected types of bituminous coal (Fig. 5.1-4). Most coal of coking

FIGURE 5.1-2 Ore and limestone are often delivered to a steel mill (background) by boat. (Courtesy Republic Steel Corporation.)

FIGURE 5.1-3 Typical terraced effect of open pit ore mining. (Courtesy American Iron and Steel Institute.)

FIGURE 5.1-4 Hot coke is forced from one of a battery of slot-shaped ovens into a quenching car for cooling with water. (Courtesy Bethlehem Steel Corporation.)

quality is mined in West Virginia, Pennsylvania, Kentucky, and Alabama. In addition to furnishing heat and carbon, coke also acts as a reducing element to separate iron from its oxide in the ores to make pig iron.

FLUXES

Fluxes are minerals that have an affinity for the impurities in iron ore or pig iron. They combine with these impurities and separate them from the molten metal in a furnace by forming a liquid slag. Some fluxes are used mainly to separate the impurities, others to make the slag more fluid, thus floating it to the surface of the metal sooner. Depending on the affinity of a flux to combine chemically with certain impurities, it is classified as either *basic* or *acid*. Limestone and dolomite are examples of basic fluxes; sand, gravel, and quartz rock are acid fluxes. Fluorspar is a neutral flux used mainly to make the slag more fluid.

REFRACTORIES

Refractories are nonmetallic materials with superior heat and abrasion resistance used for linings of steelmaking furnaces, flues, and vessels. Like fluxes, refractories may be *acid* or *basic*, depending on their chemical reaction with impurities. Theoretically, acid and basic refractories must be used with like fluxes to minimize chemical reaction between them, but in practice this rule is not always followed.

A common acid refractory material is *ganister*, a form of quartzite rock obtained from quarries and mines in Pennsylvania, Wisconsin, Alabama, Utah, and California. The most common basic refractory material is *magnesia*, which is obtained from the mineral magnesite and from sea water (Fig. 5.1-5).

FIGURE 5.1-5 Magnesia is extracted from sea water or brines for use as a basic refractory in steel furnaces. (Courtesy American Iron and Steel Institute.)

ALLOY ORES

The chemistry of steel requires many elements for alloying purposes, such as manganese, silicon, nickel, chromium, and molybdenum. Although the United States is a major steel producer, it depends on imports to furnish many of these elements. Most alloys are imported as ores and processed into useful forms called *ferroalloys*.

Manganese All steel contains manganese, both as a scavenger to remove oxygen from molten metal and as an alloying element to improve hardness and wear resistance. An average of 12 lb. (5.44 kg) of manganese goes into each ton (907 kg) of steel.

Chromium Added in amounts of 1% to 5%, chromium provides stainless steels with resistance to heat, rust, and corrosion. To other alloy steels and cast iron, it provides hardness and wear resistance. Chromium is also important as a coating metal.

Nickel Cryogenic steels (those produced at low temperatures) use nickel to improve hardness. Steels containing nickel are especially suitable for case hardening, the process of surface hardening by absorption of carbon at high temperatures.

Silicon In steelmaking, silicon serves as a deoxidizer and as an alloying element. When used in large percentages, silicon enables cast irons to withstand highly corrosive acids.

Tungsten Tungsten provides heat resistance and is used in tools used for high-speed cutting of steel and to make tungsten carbide, an abrasion-resistant material almost as hard as diamonds. It is also used as a welded-on deposit on parts exposed to extreme wear.

Molybdenum One of the more versatile alloying elements, molybdenum increases the hardness and corrosion resistance of steels.

Vanadium Vanadium is often used in conjunction with other alloying elements: with chromium in steel springs, with manganese in special plates and other structural forms, and with molybdenum in certain high-temperature steels. Small amounts (often measured in hundredths of a percent) provide the desired effect in construction steels.

Boron Known widely as a cleansing agent, boron is used as a hardening agent in steel. Minute quantities are added to obtain the desired effect, but these must be protected by other deoxidizers or must be added to a completely deoxidized molten metal bath.

AIR AND OXYGEN

Air and oxygen can be considered as much a raw material as the specially prepared iron ore, coke, and lime consumed in steel production. Air must be forced into furnaces, heated or cooled, and cleaned to control air pollution. Oxygen is a product of special plants that also produce nitrogen and argon from ordinary air. It takes about 4 tons (3630 kg) of air to make a ton (907 kg) of iron.

WATER

Each ton (907 kg) of raw steel produced requires 35,000 gallons (132,489 L) of water, mostly to cool closed systems. Often, this water must be cleaned both before and after use.

5.1.2.2 Iron Making

Smelting is a metallurgical operation by which metal ore is heated to separate the metal in it from impurities with which the metal may be chemically combined or physically mixed. The separation of chemically combined iron and oxygen is called *reduction*. Pig iron is made in a *blast furnace* by smelting, which extracts the iron by reduction of its oxide and by physically separating it from other impurities.

BLAST FURNACE

A *blast furnace* is a cylindrical structure, approximately 120 ft. (36.58 m) tall, lined with refractory brick and encased in a steel shell (Fig. 5.1-6). A *charge* consisting of ore, coke, and limestone is loaded at the top, and a blast of hot air is introduced near the bottom. A charge consists of about 1½ tons (1361 kg) of ore, ¾ ton (680 kg) of coke, and ¼ ton (227 kg) of limestone for each ton (907 kg) of iron to be produced. Oxygen in the air combines with carbon in the hot coke, causing combustion and producing heat. The charge is smelted by the heat, and, as the gases of

Iron ore, coke and limestone are the principal materials used in making iron.
IRON ORE STOCKPILE
STOCK HOUSE
receives raw materials which are carried up the inclined ramp (skip hoist) and charged into the blast furnace.
STOVES
pre-heat air for the blast furnace.
BLAST FURNACE
DUST CATCHER
MOLTEN IRON
LADLE CAR
MOLTEN SLAG
for disposal.

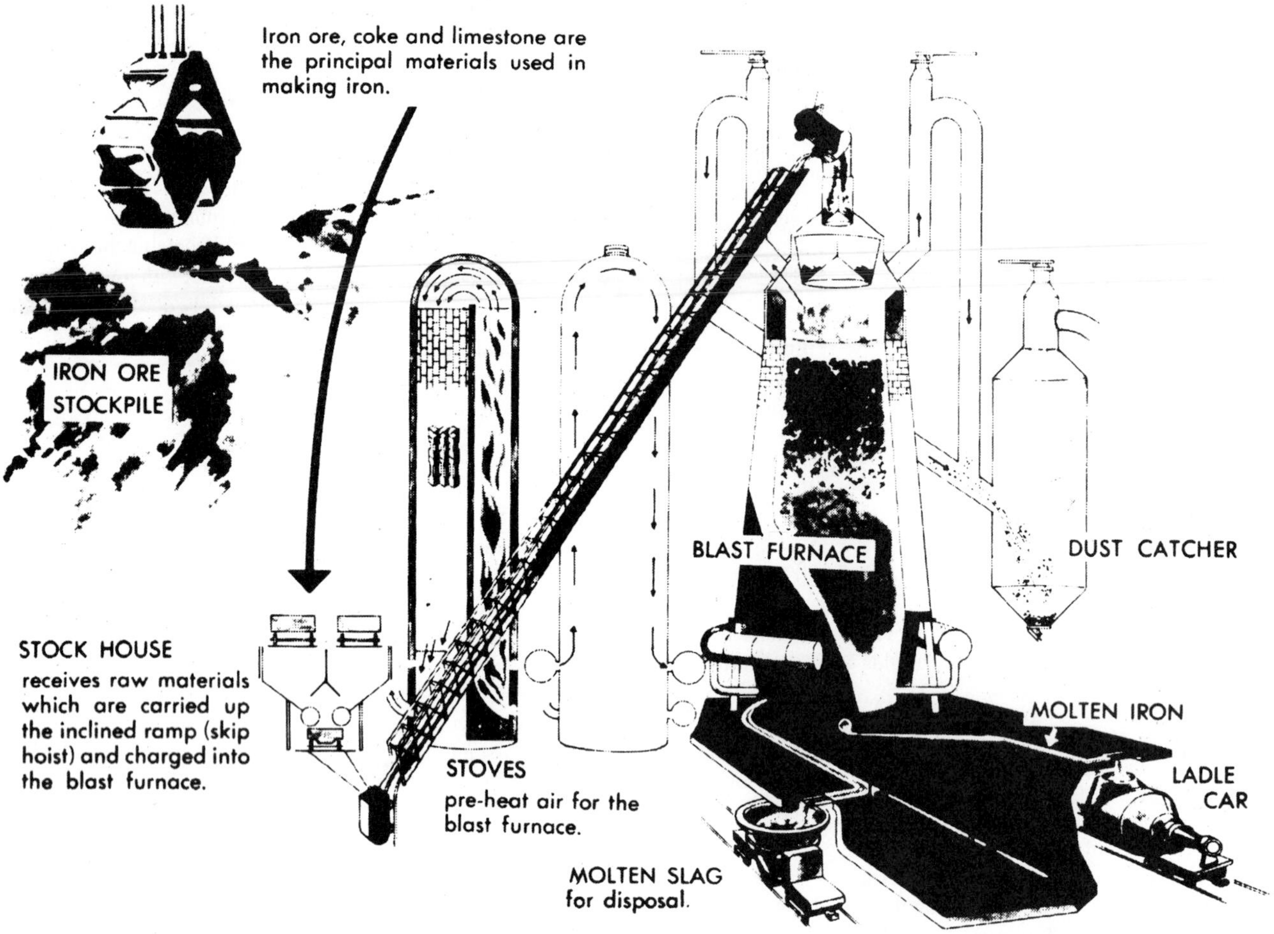

FIGURE 5.1-6 **Diagrammatic illustration of a blast furnace and its auxiliaries.** (Courtesy American Iron and Steel Institute.)

combustion pass through the hot ore, another chemical reaction takes place that frees the iron from its oxide. The hot combustion gases are cleaned in dust catchers and recycled through brick checkerworks (stoves) to heat the incoming air blast (Fig. 5.1-7). A blast furnace is operated 24 hours a day.

The impurities in the ore and the flux combine with oxygen to form a floating slag 4 to 5 ft. (1220 to 1500 mm) deep, which is tapped from the furnace at about 2-hour intervals. The molten crude iron, which contains about 93% iron and 7% extraneous elements, is tapped at a lower level at about 3-hour intervals. This is then conveyed in ladle cars to *mixers* (holding furnaces), where the liquid iron is mixed to equalize its chemical composition and kept hot pending further refinement.

Most of this molten iron is sent to open hearth or basic oxygen steelmaking facilities. Some goes to a casting machine where it solidifies into molds, known as *pigs* (ingots).

Most iron from pig casting (pig iron) is made to rigid specifications, which vary widely according to end use. Pig iron that is intended to be sold as an end product is called *merchant* pig iron.

End products made from pig iron vary from soil pipe to fancy dutch ovens. A merchant pig iron producer may have several hundred piles of different grades of iron and may blend a dozen or more grades of iron ore to arrive at a desired chemical composition. While 30,000 cu. ft. (850 m^3) of solids weighing just under 1000 tons (907,184 kg) are being processed in a furnace, the chemical composition of the mix is controlled so that the contents of carbon, silicon, sulfur, phosphorus, and manganese are maintained to within fractions of 1% of the desired amounts.

FOUNDRY

A small percentage of pig iron is remelted in *foundries* and formed into special shapes by casting in sand or loam molds. This process is called *iron founding*. Since no major change in chemical composition takes place when pig iron is cast, it is most often melted together with scrap metal and alloying elements before it is cast in order to impart desirable properties. The

FIGURE 5.1-7 **Overall view of a blast furnace and its auxiliaries. A battery of cylindrical checkerworks appears at center left.** (Courtesy American Iron and Steel Institute.)

carbon content of cast iron therefore is somewhat lower than that of pig iron, ranging between 1.5% and 4%.

The types of cast iron produced are *gray*, *white*, *malleable*, *chilled*, *alloyed*, and *nodular*. Only the first three are important in construction, and only these will be discussed here.

Gray and White Cast Iron When molten cast iron solidifies, carbon in the iron may remain chemically combined in the form of iron carbide or may separate out as graphite. *White cast iron* contains carbon mainly in the chemically combined form (iron carbide), which makes this metal hard and nonductile. In *gray cast iron*, the carbon is chiefly in the form of graphite flakes, which impart the characteristic gray fracture and promote machinability and resistance to wear.

Malleable Cast Iron White cast iron may be made somewhat softer and more ductile, though not strictly malleable, by *annealing* (controlled heating and cooling). The carbon in the iron carbide is converted to a more dispersed graphite form, which provides greater ductility than the flaky graphite of gray iron or the carbide of white iron.

Direct Reduction There are direct-reduction processes reported to be capable of bypassing the blast furnace. These include processes using solid reductants, processes using gaseous reductants, and direct steel processes. However, none of these processes is really ready to replace efficient blast furnaces.

5.1.2.3 Steelmaking

Steelmaking requires lowering the carbon content of pig iron, controlling impurities in the pig iron and in other raw materials, and adding alloying elements to obtain desired properties.

Carbon is beneficial in steel when it is contained within certain ranges, which vary with the type of steel. Carbon in quantities up to about 1.2% of the total material increases hardness and strength, but this increase is less when the carbon content is between 0.5% and 1.2% of the total. Because brittleness increases and degree of ductility decreases as the carbon content increases, lower contents are preferable in most steels.

Uncontrolled or excessive amounts of carbon, phosphorus, sulfur, manganese, and silicon are classed as impurities. Pig iron usually contains these elements as impurities originating from the raw materials of the blast furnace. Up to certain limits or within specified ranges, the effects of these elements on steel may be beneficial, and steelmaking processes attempt to regulate their contents within these limits or ranges. Phosphorus and sulfur impart some desirable properties, but must be limited to less than 0.05% in most steels due to their adverse effects on *malleability* (ability to be shaped). Silicon and manganese are strengthening elements desired in ordinary steels in the range of 0.01% to 0.35% and 0.20% to 2.00%, respectively, and may be added when raw materials contain insufficient quantities. All of these elements are present in controlled amounts in most steels. When they occur naturally, they are called *residual* elements; when deliberately added, they are called *alloying* elements.

Steelmaking processes generally start with hot pig iron delivered in ladles by overhead cranes to the steel furnace, where scrap metal and fluxes are added. After the melting operation, the hot steel is tapped from the furnace into other ladles and moved to awaiting molds for casting into ingots or to a strand casting machine. The chief steelmaking processes are *open hearth*, *electric*, and *pneumatic* (Fig. 5.1-8).

OPEN HEARTH PROCESS

The majority of open hearth furnaces are of the basic type, capable of controlling the phosphorus present in pig iron. Both

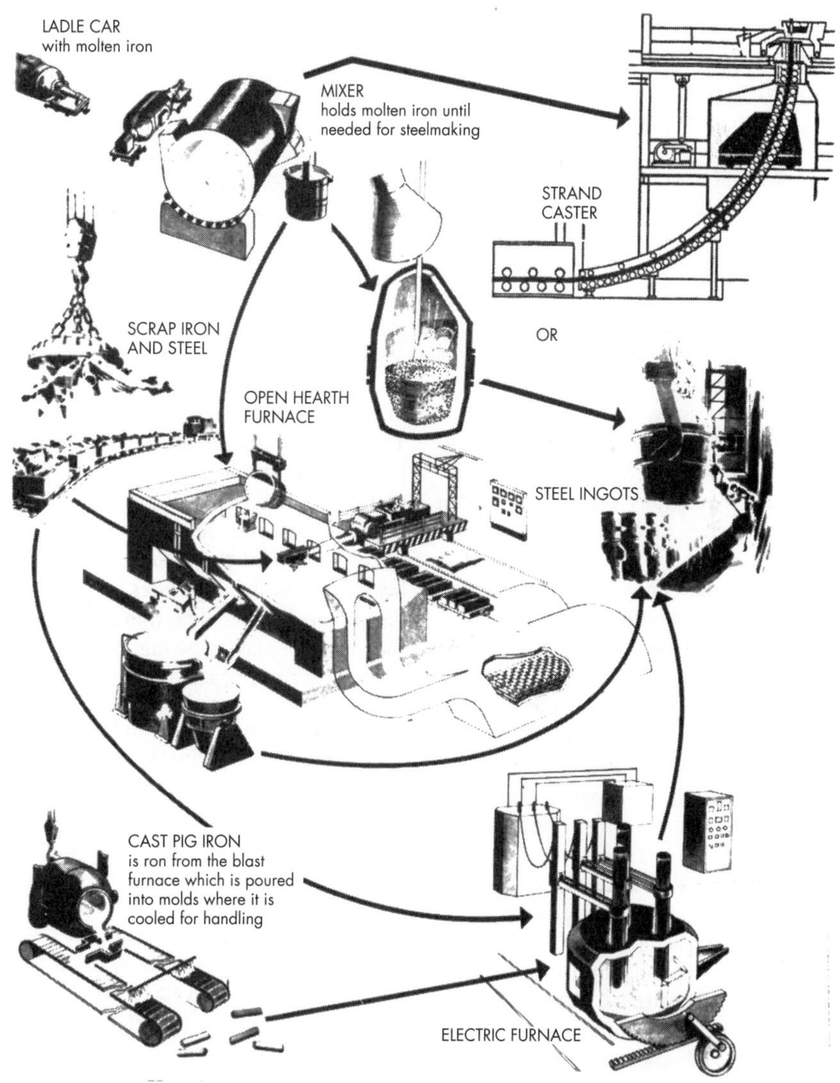

FIGURE 5.1-8 **Hot pig iron may be converted to steel in one of the steelmaking furnaces shown or in induction furnaces or basic oxygen furnaces.** (Courtesy American Iron and Steel Institute.)

FIGURE 5.1-9 An open hearth furnace. Hot pig iron is being poured from a ladle. A charging machine (foreground) adds scrap metal and limestone flux. Open hearth furnaces are seldom used today. (Courtesy American Iron and Steel Institute.)

ordinary carbon grades and high-grade alloy steels are usually produced by this process.

The name *open hearth* is applied because the saucer-shaped hearth, or floor of the furnace, is exposed to the sweep of the flames. The furnace itself is a masonry structure lined with refractory materials arranged on two levels: the upper level for charging the raw materials, the lower for tapping the molten steel and slag by gravity. Furnaces may hold 50 to 600 tons (45.36 to 544.3 Mg) and produce 25 to 50 tons (22.68 to 45.36 Mg) of steel per hour. The charge generally consists of molten pig iron, scrap metal, and fluxes (Fig. 5.1-9).

The fuel (gas, oil, tar, or a combination of these, mixed with hot air) is blown through ports at the ends of the furnace to produce combustion and heat over the hearth. As in a blast furnace, the hot gases of combustion are recycled through checkerworks, which preheat the incoming air. In addition, gaseous oxygen is usually introduced to speed the melting operation.

Refinement of the metal results when the impurities and carbon in the pig iron and the oxygen and fluxes in the mix react to each other chemically to form a slag that floats to the surface. The molten steel flows through a hole in the bottom of the hearth into ladles. When desired, the carbon content may be increased by adding pig iron in the furnace or coke or coal in the ladle. Oxidizable *ferroalloys* (iron-based compounds rich in desired alloying elements) may also be added in the ladle.

ELECTRIC PROCESSES

Most high-grade steels, such as alloy, stainless, and heat-resisting steel, are produced in electric furnaces capable of developing the high temperatures and reducing conditions necessary for melting these metals. Here electrical resistance and arc radiation, rather than combustion, are used to produce heat, and so air or oxygen is not necessary for combustion. High-purity oxygen is introduced at appropriate times to oxidize impurities, which results in better control of the oxidation process and a smaller loss of expensive alloying elements.

The charge usually consists of scrap metal, mill scale, fluxes, and, often, alloying elements. Most electric furnaces are of the basic type and produce low-phosphorus steel. Many grades of basic steel can be produced at a rate of 30 to 40 tons (27.2 to 36.3 Mg) per hour in an electric arc furnace, which may hold up to 300 tons (272.2 Mg). Special grades involving valuable alloying elements are generally produced in smaller quantities, often in an electric induction furnace.

Electric Arc Furnace The body of an electric arc furnace is a circular steel shell resembling a huge kettle, lined with refractory materials and mounted so that it may be tilted to pour off the molten steel. Three or more cylindrical electrodes project into the furnace from above. When high-voltage current is passed through the electrodes, an electric arc is produced between them and the charge, generating heat. Many of the impurities are oxidized during the melting operation and float to the surface as slag, allowing the steel to be poured off from beneath it (Fig. 5.1-10).

Induction Furnace An induction furnace consists of a magnesia pot, insulated by other refractory materials and surrounded by windings of copper tubing (Fig. 5.1-11). When high-voltage alternating electric current is passed through the copper tubing around the pot, an induction current is generated in the metal charge of the furnace. The resistance of the metal to the induced current creates the heat for the melting operation.

PNEUMATIC PROCESS

Steelmaking operations that use high-speed oxygen or air to oxidize impurities in the charge without fuel are called *pneumatic processes*. Their chief advantages are shortened heat time and high output. With furnace capacities of 300 tons (272.2 Mg) and tap-to-tap heat cycles of 45 minutes, it is possible to produce steel at a rate 6 to 10 times that of open hearth and electric arc furnaces. The pneumatic process, which takes place in a *basic oxygen furnace*, is the predominant method of producing steel in the United States.

In a basic oxygen furnace, which is a pear-shaped steel vessel lined with refractory materials (Fig. 5.1-12), a jet of high-purity oxygen is directed onto molten metal. The charge, consisting of molten pig iron, some scrap, and fluxes, is added from the top. Larger amounts of hot pig iron and lesser amounts of scrap metal are used in an oxygen furnace than in other steel furnaces.

No fuel is required in this process, because the heat of the large mass of molten pig iron is sufficient to start a chemical reaction between the oxygen, the carbon, and the other impurities. Oxidation, in turn, produces sufficient heat to continue the process without the introduction of fuel.

If the as-tapped metal has a lower carbon or manganese content than is required, *ferromanganese* (a manganese-rich ferroalloy containing carbon) may be added in the ladle. The carbon gives added strength and hardness, and the manganese, in addition to strength, ensures proper ductility at elevated temperatures by combining with excess sulfur.

Steel Casting No matter which steelmaking process is used, steel must be formed into a specific shape before further processing. To this end, it is either poured into *ingot molds* or put through a *strand-casting machine*.

Ingots Raw steel from a furnace is typically *teemed* (poured) from a ladle into cast-iron ingot molds, which may be more than 8 ft. (2400 mm) high and 3 ft. (915 mm) in diameter. These molds are coated on the inside with a releasing agent that prevents splashed metal from marring the ingot's surface quality. Often, stripped ingots are moved to furnaces called soaking pits while the outside is solid but the interior is still molten.

Strand Casting Strand casting produces a continuous ribbon of steel rather than separate molds. It can result in semi-

FIGURE 5.1-10 An electric arc furnace. (Courtesy Lukens Inc.)

FIGURE 5.1-11 A 1500-ton (1360.78-Mg) channel induction furnace. (Courtesy Ajax Magnethermic Corporation.)

FIGURE 5.1-12 A basic oxygen furnace. A ladle is pouring hot pig iron into the mouth of a tilted furnace. (Courtesy American Iron and Steel Institute.)

finished solid products, thus bypassing many of the steps necessary to produce ingots. At some point the continuous ribbon of steel is cut into desired lengths (Fig. 5.1-13).

5.1.3 PROPERTIES

The properties of various irons and steels are closely related to their chemical composition and grain structure, that is, the size and arrangement of the crystalline microscopic particles making up the metal. In the initial crude refining processes, the most important factors influencing grain structure are the carbon and its chemical reaction with iron as it cools. Further refining processes, alloying elements, heat treatments, and hot and cold working operations also affect grain structure. In ordinary steels, carbon and the other residual elements are important. In higher grades of steel, alloying elements contribute special properties to the metal.

The essential physical property that distinguishes steel from iron is its malleability as it comes from the furnace and is initially cast. Other unique properties of steel are its ability to resist high stresses by deforming without breaking (toughness) and its ability to be hardened by hot and cold working. The effects of the several residual elements originating in the raw materials and of the added alloying elements are outlined in Figure 5.1-14.

5.1.3.1 Cast Iron

Only slight variations in properties or in the grain structure of the metal occur in cast iron as a result of the reaction between its carbon and its iron. Most ordinary white and gray cast irons, therefore, like pig iron, are hard and brittle and have relatively high compressive strength and low tensile strength. However, with the proper combination of selected pig irons, scrap metals, and alloying elements, cast iron can be made with increased hardness, toughness, and corrosion and wear resistance. Casting also permits the formation of complex and massive shapes not readily produced by machining or rolling.

Drainage, vent, and waste pipe, ornamental railings, lamp posts, handrail brackets, stair treads, manhole covers, and concrete inserts for anchors are familiar cast-iron products. In sanitary ware involving enameling and baking at elevated temperatures, cast iron's resistance to warping and cracking at high heat is an important advantage.

Malleable cast iron has improved breakage resistance as a result of increased toughness and ductility. It is also easier to weld than gray or white cast iron. It is used extensively for small articles of door hardware, concrete inserts, brackets, flanges, and anchors. Special *alloy cast irons*, containing between 1% and 5% alloying elements, and *high-strength cast irons*, which contain molybdenum, chromium, and nickel, are also produced for special uses. *Nodular cast iron*, which is produced in this country by adding magnesium, has greater strength, ductility, toughness, machinability, and corrosion resistance than other cast irons.

5.1.3.2 Carbon Steels

Carbon steels are those in which the residual elements, such as carbon, man-

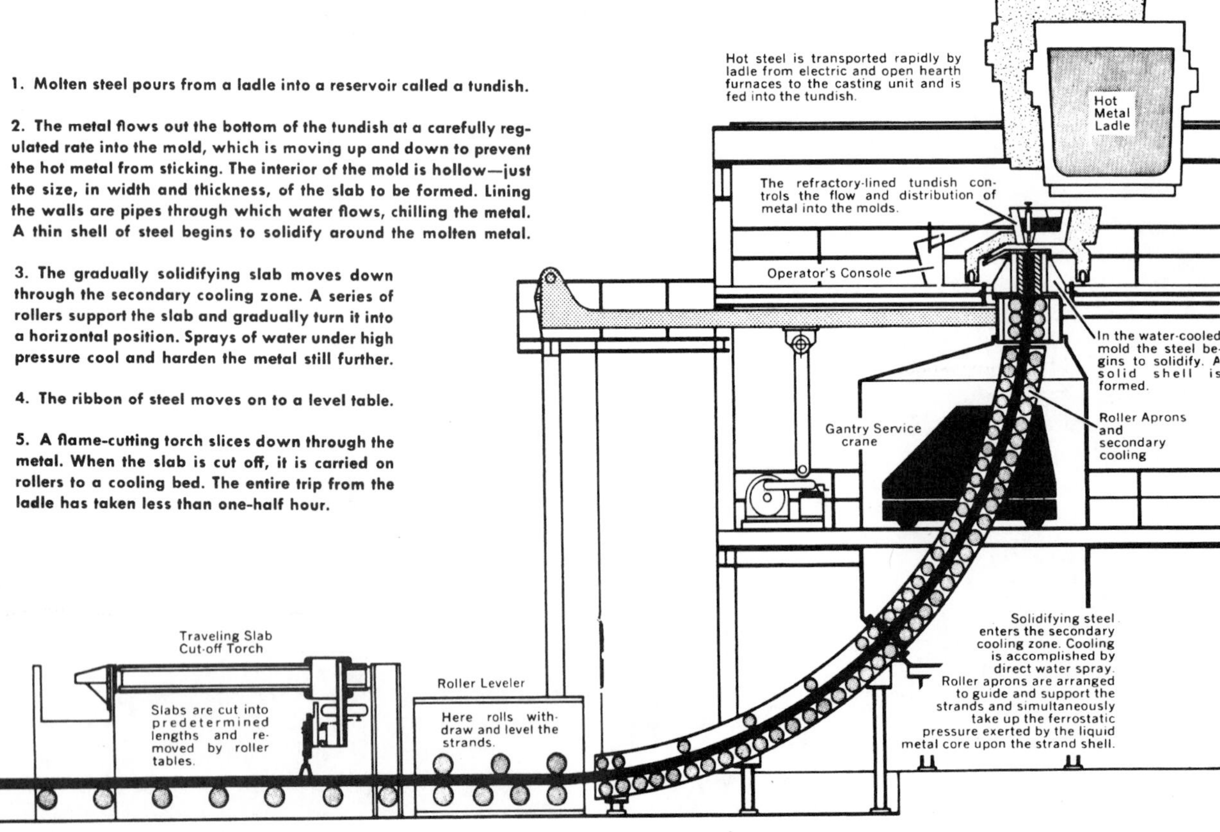

1. Molten steel pours from a ladle into a reservoir called a tundish.
2. The metal flows out the bottom of the tundish at a carefully regulated rate into the mold, which is moving up and down to prevent the hot metal from sticking. The interior of the mold is hollow—just the size, in width and thickness, of the slab to be formed. Lining the walls are pipes through which water flows, chilling the metal. A thin shell of steel begins to solidify around the molten metal.
3. The gradually solidifying slab moves down through the secondary cooling zone. A series of rollers support the slab and gradually turn it into a horizontal position. Sprays of water under high pressure cool and harden the metal still further.
4. The ribbon of steel moves on to a level table.
5. A flame-cutting torch slices down through the metal. When the slab is cut off, it is carried on rollers to a cooling bed. The entire trip from the ladle has taken less than one-half hour.
Hot steel is transported rapidly by ladle from electric and open hearth furnaces to the casting unit and is fed into the tundish.
Hot Metal Ladle
The refractory-lined tundish controls the flow and distribution of metal into the molds.
Operator's Console
In the water-cooled mold the steel begins to solidify. A solid shell is formed.
Gantry Service crane
Roller Aprons and secondary cooling
Solidifying steel enters the secondary cooling zone. Cooling is accomplished by direct water spray. Roller aprons are arranged to guide and support the strands and simultaneously take up the ferrostatic pressure exerted by the liquid metal core upon the strand shell.
Traveling Slab Cut-off Torch
Slabs are cut into predetermined lengths and removed by roller tables.
Roller Leveler
Here rolls withdraw and level the strands.
Slab Run Out Table

FIGURE 5.1-13 Strand, or continuous, casting. (Courtesy American Iron and Steel Institute.)

FIGURE 5.1-14 Effects of Alloying Elements

Element	Common Content	Effects
Carbon	Up to 0.90%	Increases hardness, tensile strength, and responsiveness to heat treatment with corresponding increases in strength and hardness
	Over 0.90%	Increases hardness and brittleness; over 1.2% causes loss of malleability
Manganese	0.50% to 2.0%	Imparts strength and responsiveness to heat treatment; promotes hardness, uniformity of internal grain structure
Silicon	Up to 2.50%	Same general effects as manganese
Sulfur	Up to 0.050%	Maintained below this content to retain malleability at high temperatures, which is reduced with increased content
	0.05% to 3.0%	Improves machinability
Phosphorus	Up to 0.05%	Increases strength and corrosion resistance, but is maintained below this content to retain malleability and weldability at room temperature
Aluminum	Variable	Promotes small grain size and uniformity of internal grain structure in the as-cast metal or during heat treatment
Copper	Up to 0.25%	Increases strength and corrosion resistance
Lead	0.15% to 0.35%	Improves machinability without detrimental effect on mechanical properties
Chromium	0.50% to 1.50%	In alloy steels, increases responsiveness to heat treatment and hardenability
	4.0% to 12%	In heat-resisting steels, causes retention of mechanical properties at high temperatures
	Over 12%	Increases corrosion resistance and hardness
Nickel	1.0% to 4.0%	In alloy steels, increases strength, toughness, and impact resistance
	Up to 27.0%	In stainless steels, improves performance at elevated temperatures and prevents work hardening
Molybdenum	0.10% to 0.40%	In alloy steels, increases toughness and hardenability
	Up to 4.0%	In stainless steels, increases corrosion resistance and strength retention at high temperatures
Tungsten	17% to 20%	In tool steels, promotes hardness at high cutting temperatures; in stainless steels, smaller amounts ensure strength retention at high temperatures
Vanadium	0.15% to 0.20%	Promotes small grain size and uniformity of internal grain structure in the as-cast metal or during heat treatment; improves resistance to thermal fatigue shock
Tellurium	Up to .05%	Improves machinability when added to leaded steels
Titanium	Variable	Prevents loss of effective chromium through carbide precipitation in 18/8 stainless steels
Cobalt	17.0% to 36.0%	Increases magnetic properties of alloy steels; in smaller amounts, promotes strength at high temperatures in heat-resisting steels

(Courtesy American Iron and Steel Institute.)

ganese, phosphorus, sulfur, and silicon are controlled, but in which no alloying elements are added to achieve special properties. They contain up to 1.2% carbon, and other elements are controlled within specified limits or ranges. More than 90% of the steel manufactured into finished products is carbon steel.

CLASSIFICATION AND USES

In the past, steels have been classified according to carbon content as follows:

Soft steel: 0.20% maximum
Mild steel: 0.15% to 0.25%
Medium steel: 0.25% to 0.45%
Hard steel: 0.45% to 0.85%
Spring steel: 0.85% to 1.15%

These broad classifications have become less useful as the need for precisely compounded steel types has increased. Today, the American Iron and Steel Institute (AISI) and American Society for Testing and Materials (ASTM) designations are more commonly used in manufacturing and construction.

AISI Designations The AISI designation system assigns code numbers that indicate the chemical composition and the steelmaking process of various types of steel. For carbon steels, each four-digit code number specifies the permissible maximums or ranges for residual elements, including carbon, manganese, phosphorus, sulfur, and, sometimes, silicon. Some types of alloy steels are designated by five numerals.

The first two digits in the AISI system indicate the type of steel (carbon or alloy) (Fig. 5.1-15). The last two digits of the four-numeral series and the last three digits of the five-numeral series indicate the approximate mean of the carbon range. For example, in series designation 1035, 35 represents a carbon range of 0.32% to 0.38%. It is necessary, however, to deviate from this system and to interpolate numbers in the case of some carbon ranges, and for variations in manganese, sulfur, or other elements with the same carbon range.

ASTM Designations Structural steels are commonly specified according to ASTM standards. The carbon steels typically used in building construction conform to ASTM Standards A 6, A 36, A 53, A 82, A 185, A 283, A 497, A 500, A 501, A 615, A 663, A 675, A 767, and A 966.

FIGURE 5.1-15 AISI Designations

Series Designation*	Type and Approximate Percentages of Identifying Elements
	Alloy Steels
13xx	Manganese 1.75
40xx	Molybdenum 0.20 or 0.25; or molybdenum 0.25 and sulfur 0.042
41xx	Chromium 0.50, 0.80, or 0.95; molybdenum 0.12, 0.20, or 0.30
43xx	Nickel 1.83, chromium 0.50 or 0.80, molybdenum 0.05
46xx	Nickel 0.85 or 1.83, molybdenum 0.20 or 0.25
47xx	Nickel 1.05, chromium 0.45, molybdenum 0.20 or 0.35
48xx	Nickel 3.50, molybdenum 0.25
51xx	Chromium 0.80, 0.88, 0.93, 0.95, or 1.00
51xxx	Chromium 1.03
52xxx	Chromium 1.45
61xx	Chromium 0.60 or 0.95, vanadium 0.13 or min. 0.15
86xx	Nickel 0.55, chromium 0.50, molybdenum 0.20
87xx	Nickel 0.55, chromium 0.50, molybdenum 0.25
88xx	Nickel 0.55, chromium 0.50, molybdenum 0.35
92xx	Silicon 2.00; or silicon 1.40 and chromium 0.70
50Bxx	Chromium 0.28 or 0.50
51Bxx	Chromium 0.80
81Bxx	Nickel 0.30, chromium 0.45, molybdenum 0.12
94Bxx	Nickel 0.45, chromium 0.40, molybdenum 0.12
	Carbon Steels
10xx	Nonresulfurized, manganese 1.00% maximum
11xx	Resulfurized
12xx	Rephosphorized and resulfurized
15xx	Nonresulfurized, manganese maximum over 1.00%

*B denotes boron steel.

(Courtesy American Iron and Steel Institute.)

The steel most used for construction purposes is that complying with ASTM A 36.

ASTM standards indicate certain partial chemical compositions and specified minimums for strength and ductility. However, they regulate only those chemical elements that directly affect the fabrication and erection of steel, making it possible to produce proprietary steels of different chemical compositions conforming with the required ASTM performance standards.

Unified Numbering System Alloy and carbon steels have been assigned a designation in the Unified Numbering System (UNS) for Metals and Alloys. The UNS was established in 1975 by ASTM and the Society of Automotive Engineers (SAE). These designations appear in ASTM E 527 and SAE Standard J1086.

A UNS number consists of a single-letter prefix followed by five digits. The prefix *E* designates steels made by the basic electric furnace process with special practices; the absence of a prefix indicates open hearth steels. The prefix *M* designates a series of merchant quality steels, *G* indicates standard alloy or carbon steels, and *H* indicates standard hardenability steels. The first four digits correspond to AISI, ASTM, or SAE steel designations, and the last digit usually relates to an additional chemical requirement such as lead or boron. The use of 6 as the last digit designates steels that are made by the basic electric furnace process with special practices. An example of the UNS is "Series Designation E52100," which is the same steel as the 1.45% chromium steel (52) shown in Figure 5.1-15, made by the basic electric furnace process (E), which has a carbon content between 0.98% and 1.10% (100).

Common Uses Carbon steels are used for most products in the building industry, from structural shapes, concrete reinforcing bars, sheets, plates, and pipes to the smallest items of builders' hardware. Carbon steels generally possess adequate strength, hardness, stiffness, malleability, and weldability, but poor resistance to corrosion.

Occasionally, copper and other elements may be added to improve atmospheric corrosion resistance, and by varying the chemical compositions, heat treatments, and subsequent hot and cold working operations, other desirable properties can be enhanced.

5.1.3.3 Alloy Steels

Alloy steel is so classified when the content of alloying elements exceeds certain limits or when a definite range of alloying elements is specified. However, the term *alloy steel* does not cover stainless steel, tool steel, specialty steel, or some other classifications, even though these steels contain specified amounts of alloying elements.

CLASSIFICATION AND USES

Alloy steels are generally grouped by the element or combination of elements used to obtain desired properties. Manganese-molybdenum steels, for example, are known for hardenability and resistance to fatigue; silicon-manganese steels for high relative strength and shock resistance; nickel steels for toughness and resistance to impact and corrosion.

AISI Designations The AISI has developed a numerical designation system for identifying alloy steels. These steels are designated by a four-digit code, from 1200 to 9800, according to chemical composition (see Fig. 5.1-15). There are specific allowable ranges for carbon, manganese, silicon, nickel, chromium, and molybdenum; limits for sulfur and phosphorus are also usually specified. The prefix *E* is used to designate electric furnace steel; the absence of a prefix indicates open hearth steels. The suffix *H* is sometimes used to identify steels of controlled hardenability; *B* or *BV* between the second and third digits indicates boron or boron-vanadium alloys, respectively.

These steels are grouped in series according to the most important alloying element(s), identified by the first two digits; the last two digits indicate the approximate carbon content. For example, AISI type 4023 steel indicates a molybdenum alloy steel (40) with an average carbon content of 0.23% (23), made in an open hearth furnace (no *E* prefix).

Common Uses Alloy steels have properties essentially similar to or exceeding those of the carbon steels. In general, particular properties are improved by certain elements. For example, a high silicon content gives steel excellent magnetic permeability and low core loss, making it useful in motors, generators, and transformers; cobalt improves magnetic properties for such uses as permanent magnets in electrical apparatus and sealing gaskets in modern refrigerators; nickel imparts great toughness, which is useful in rock drilling and air hammer equipment.

Because of their higher cost, alloy steels are not often used in construction materials. However, a particular group known as *heat-treated constructional alloy steels* have been rolled in a limited number of structural shapes and used successfully in large structures where unusually large loads or high temperatures are encountered. These steels are capable of developing very high strengths (ultimate strength up to 135,000 psi [930 MPa], yield point up to 100,000 psi [689 MPa]) and show promise of increased structural use.

5.1.3.4 High-Strength Low-Alloy Steels

High-strength low-alloy steels constitute a group with chemical compositions specially developed to impart better mechanical properties and greater resistance to atmospheric corrosion than are obtainable from conventional carbon structural steels.

CLASSIFICATION AND USES

AISI classifies high-strength low-alloy steels according to chemical composition. ASTM publishes recommended practices and standards for them.

Structural Steels Among the ASTM standards for high-strength, low-alloy structural steels are A 242, A 572, and A 588.

Sheet Steel The most widely produced and ordered high-strength, low-alloy sheet steel products are:

1. *Conventional high-strength low-alloy steel* (ASTM A 607), which is produced to various strength levels where atmospheric corrosion resistance and maximum formability are not a requirement.
2. *Improved atmospheric corrosion resistance high-strength, low-alloy sheet steel* (ASTM A 606), which is produced to 45 or 50 kips per square inch (ksi) (310.3 or 344.7 MPa) minimum yield points (1 kip equals 1000 lb. [453.6 kg]). The types of steel with enhanced corrosion resistance include Type 2 with a corrosion resistance twice that of plain carbon steel, and Type 4 with a corrosion resistance at least four times greater.
3. *Improved formability high-strength low-alloy steel* (ASTM A 715), which is produced with special forming properties to various strength levels where maximum formability is required. This type is generally furnished for making parts too difficult for the fabricating properties of the other two types.

In general, the minimum yield points for these steels vary within a range of 45 to 80 ksi (310.3 to 551.6 MPa) and their minimum tensile strength varies from 55 to 90 ksi (379 to 620.5 MPa).

Steel Bars The most frequently specified high-strength, low-alloy steel bars have a minimum yield point ranging from 42 to 80 ksi (289.4 to 551.5 MPa) and a minimum tensile strength that varies between 60 and 95 ksi (413.7 and 655 MPa).

COMMON USES

High-strength low-alloy steels are commonly used where high strength in relation to weight is important. The steels listed in the previous paragraphs are often used in bridges and in high-rise building construction. These and other high-strength steels are also used extensively in various transportation vehicles, heavy construction equipment, industrial containers, and wherever resistance to corrosion and impact damage is important.

Some proprietary low-alloy steels can be made with increased corrosion resistance when intended for exposed architectural use. These steels, sometimes called *weathering steels*, form natural rust-colored, self-healing oxide coatings that inhibit further corrosion when left unpainted. This rusting may not stop, however, if certain pollutants are present in the atmosphere. These steels are used increasingly in commercial and industrial construction, but are not used extensively in residential work.

Unfortunately, the rust-colored oxide coating that forms to protect weathering steels will wash off if water runs across it. Special detailing is necessary to prevent this runoff from staining adjacent materials.

5.1.3.5 Stainless and Heat-Resisting Steels

The outstanding characteristics of the two groups of steels referred to as *stainless* and *heat-resisting* are implied in their names. Stainless steels possess excellent corrosion resistance at varying temperature ranges. Heat-resisting steels retain their essential physical and mechanical properties at elevated temperatures. Stainless and heat-resisting steels account for about 1% of total steel production.

Chromium is the alloying element mainly responsible for corrosion-resisting and heat-resisting properties, although other elements, such as nickel, manganese, and molybdenum, also contribute to these properties. The stainless qualities of steel are derived from a self-healing chromium oxide that forms a transparent skin and prevents further oxidation. Steels with less than 4% chromium are regarded as alloy steels. Heat-resisting steels normally contain 4% to 12% chromium, and stainless steels more than 12%.

CLASSIFICATION AND USES

Stainless and heat-resisting steels are classified according to chromium content and internal grain structure into three major groups: *martensitic, ferritic,* and *austenitic*, and a minor group under the martensitic group, *precipitation hardening*.

Martensitic Steels Heat-resisting and stainless steels with a chromium content between 4% and 18% are known as *martensitic*. These may be hardened by heat treatment and are therefore sometimes referred to as hardenable chromium alloys. They also have adequate cold-forming characteristics and are satisfactory for hot working or forging, but must be slowly cooled or annealed after forging to prevent cracking. They are easily welded but require subsequent annealing or tempering to restore ductility.

Ferritic Steel Stainless steels that contain between 12% and 27% chromium are called *ferritic*. These steels do not respond readily to heat treatment and are sometimes referred to as nonhardenable chromium alloys, However, their hardness may be increased somewhat by cold working. Ductility may be improved by hot or cold working followed by annealing. These steels are not welded as easily as the martensitic steels, but have greater corrosion resistance. Fair ductibility after annealing allows product fabrication by a variety of processes such as cold forming, spinning, and light drawing. Ferritic steels possess low strength at elevated temperatures and a low coefficient of thermal expansion.

Austenitic Steels The chromium content of austenitic steels ranges between 16% and 26%. They also contain between 3.5% and 22% nickel. These steels are characterized by a greater ability to harden as a result of cold working than the ferritic types. Austenitic steels include both nickel-chromium and nickel-chromium-manganese groups of alloys. Both groups provide superior resistance to corrosion and are suited to many fabrication techniques. They have high ductility, which is important for deep drawing and forging operations, are easily welded, and have good oxidation resistance at elevated temperatures. A relatively high coefficient of thermal expansion is characteristic of these steels, which must be considered in the design of high-temperature equipment. They are also more susceptible to corrosive attack by deoxidizing sulfur gases than other stainless steels.

Precipitation Hardening Steels Precipitation hardening steels are iron-chromium-nickel alloys with additional elements, which are hardenable by solution treating and aging. These steels combine high strength, hardness, and corrosion resistance.

Industry Designations Heat-resisting and stainless steels are identified by Specialty Steel Industry of North America (SSINA), ASTM, and SAE designations. SSINA designations follow the old AISI system in which heat-resisting and stainless steels are grouped into five series bearing number designations from 200 to 500.

The six-digit UNS system mentioned earlier applies as well to stainless steel. Stainless steels are identified in the UNS with an *S* followed by five digits.

Common Uses There are many applications for stainless steels and a wide variety of alloys from which to choose.

Martensitic chromium steels are suitable for applications requiring high strength, hardness, and resistance to abrasion, such as in steam and gas turbine parts, bearings, and cutlery. SSINA Type 410 (basic type) is one of the most widely used alloys in this category.

Ferritic chromium steels are used mostly for automotive trim, applications involving nitric acid, high-temperature service requiring resistance to scaling, and uses that call for low thermal expansion. SSINA Types 405 and 430 steels are the most economical and widely used alloys in this category. Type 430 is sometimes used in building construction for gutters and downspouts, architectural trim, sills, and column covers. It is also used for kitchen applications such as range hoods, range tops, and appliance fronts.

Austenitic nickel-chromium steels are the most numerous and popular stainless steels. Steels in this category are sometimes referred to as 18-8 alloys, because several of the most common types (302, 303, and 304) contain approximately 18% chromium and 8% nickel. Among these, Types 302 (the basic 18-8 grade) and 304 are used most often for such building applications as fascias, curtain walls, storefronts, doors and windows, column covers, and railings. Their relative chemical inertness and finish-retention qualities also make them useful in food preparation and surgical equipment, hospital clean rooms, laboratories, and for such familiar kitchen applications as sinks, countertops, and appliances.

In addition to a variety of automotive and transportation equipment uses, Type 301 steel is sometimes used for structural members because it work hardens rapidly to high strength and toughness. Because it has good cold-forming properties, it is used for making roof drainage products.

Type 316 is the most resistant to attack by salt spray and corrosive industrial fumes, and is therefore used in seacoast and industrial areas where protection from such exposures is essential.

Austenitic nickel-chromium-manganese steels Types 201 and 202 are similar to their nickel-chromium counterparts, Types 301 and 302, and may be used interchangeably with them. They are used in automotive and food storage equipment, and for flatware, cooking utensils, kitchen sinks, and countertops.

5.1.4 MANUFACTURE

Steel products are manufactured by casting in foundries and by forging, extruding, rolling, and cold drawing in mills. The products of a foundry are referred to as cast products; all others are wrought products. Wrought products account for more than 90% of the total steel production, and by far the largest part of that is produced by hot rolling and cold finishing.

Steel mill products are regarded as semifinished when in the form of blooms, billets, and slabs, and as finished when in a form suitable for further fabrication into components or for use directly in construction. Figure 5.1-16 shows the variety of finished mill products produced at steel mills.

5.1.4.1 Cast-Steel Products

Castings are generally made by pouring molten steel into sand molds. Before it is cast, pig iron may be further refined and melted with scrap metal and ferroalloys at the foundry to obtain desired casting and mechanical properties.

Casting is used in preference to other forming operations when desired shapes are of a size or complexity not readily obtained by rolling or machining. Complex and massive shapes possessing great strength and impact resistance at high and low temperatures are usually cast (Fig. 5.1-17). Other cast-steel products used in construction are concrete inserts, anchors, and items of hardware. Iron and steel castings contribute many millions of tons (kg) to a broad range of uses in a typical year. Most of these are in gray and malleable cast iron, and some are in steel. Of the cast-steel products, approximately two-thirds are carbon steels; the rest are alloy and special steels.

The transportation industry relies on castings for many engine and car body parts, as does the excavating and mining equipment industry. Castings are also used extensively in many parts of steel-making furnaces and rolling mills.

5.1.4.2 Wrought-Steel Products

A few steel products that are complex and have an irregular cross section are made by *forging*. Linear products of constant cross section, such as bars, tubes, and

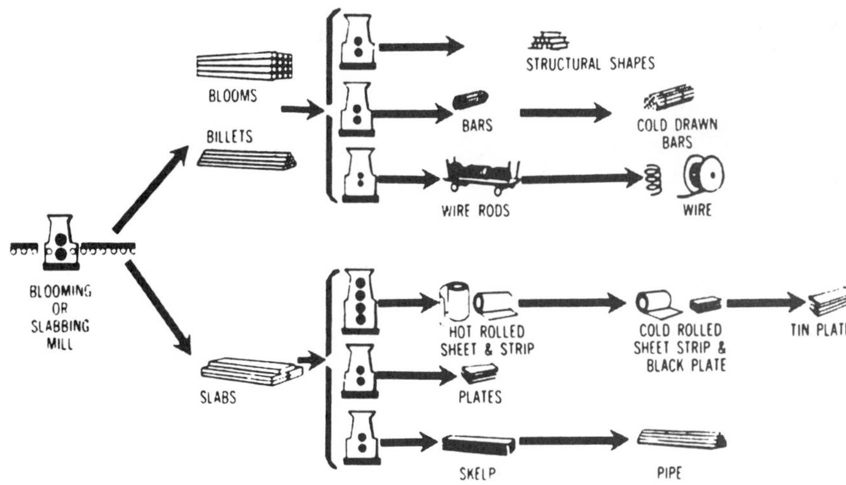

FIGURE 5.1-16 **Operations in the manufacturing sequence for steel products. Blooms, billets, slabs, bars, plates, and pipes are shown as stacks of individual units.** (Courtesy American Iron and Steel Institute.)

rods, and complex shapes may be made by *extruding*. Less complex and larger linear forms, and sheets, strips, rods, bars, and most structural shapes used in construction, are produced by *hot rolling* and *cold finishing* or by hot rolling without cold finishing.

FORGING

Forging is a method of forming hot metal into desired shapes by pressing it between heat-resistant dies. When metal is forged, toughness, strength, and ductility increase significantly along the lines of flow (Fig. 5.1-18), which is an important advantage of this forming technique. The metal to be forged is preheated to between 2200°F (1204°C) and 2400°F (1315.6°C) in special furnaces, and in some operations the dies are heated as well. The two major types of forging are *open die* and *closed die*.

FIGURE 5.1-17 **Steel castings may range in weight from a few ounces to several hundred tons, as in this bridge saddle.** (Courtesy Bethlehem Steel Corporation.)

Open-Die Forging In open-die forging, a large press squeezes (rather than strikes) steel between two flat surfaces. Temperature must be carefully controlled, which requires frequent reheating of the steel between shaping actions. Some hydraulic open-die forging presses may accept ingots weighing several hundred tons (Fig. 5.1-19).

Closed-Die Forging In closed-die forging, a hammer pounds a section of steel between carved dies until it reaches the desired shape. Closed-die forging most often uses steam hammers. Drop hammers develop forging force by the fall of a heavy weight or ram. In double-action hammers the speed of the falling weight is usually given additional impetus.

EXTRUDING

Semifinished shapes may be converted into lengths of uniform cross section by *extruding*. In this process an advancing ram (Fig. 5.1-20) forces preheated, plastic metal through a tough, heat-resistant die of the desired profile (Fig. 5.1-21). Presses vary in size from a small 700-ton (635 Mg), 2½-in. (63.5 mm) maximum circle die press designed for fast extrusion speeds of up to 25 ft./sec. (7.62 m/s), to 12,000-ton (1088.6 Mg) installations capable of extruding sections up to 21 in. (533.4 mm) in diameter at slower speeds. Stainless steel is generally extruded in shapes up to 6½ in. (165.1 mm) in diameter.

Extruding can produce more complex sections with better surface characteristics than can be produced by rolling. It can be used to shape certain alloys without undesirable residual effects (such as excessive hardness) and is more economical than other forming methods for small quantities. However, most steel extrusions are limited to 35 ft. (10.67 m) in length and 21 in. (533.4 mm) in diameter.

HOT ROLLING AND COLD FINISHING

By far the largest proportion of wrought-steel products are manufactured by *hot rolling* and *cold finishing*.

Hot rolling is used to make semifinished shapes as well as some finished products. Hot rolling may be done as part of the continuous casting process on a strand caster or by using hot ingots made by traditional steelmaking processes.

In a hot-rolling process, hot steel passes through a system of rolls, which gradually imparts rectangular bloom,

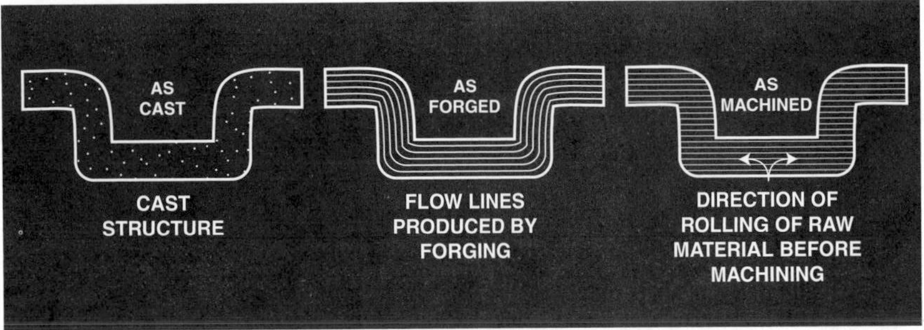

FIGURE 5.1-18 Continuity of flow lines in forged metal structure results in greater strength and toughness. (Courtesy American Iron and Steel Institute.)

FIGURE 5.1-19 A hydraulic press, which may exert a force of up to 18,000 tons (16,329.3 Mg), is particularly suitable for forging high-strength steels. (Courtesy Kaiser Aluminum and Chemical Corporation.)

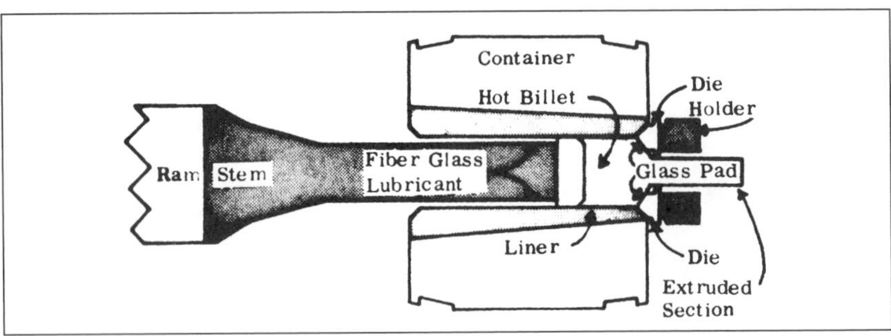

FIGURE 5.1-20 Hot steel is forced through a lubricated flow extrusion press. (Courtesy American Iron and Steel Institute.)

FIGURE 5.1-21 Bars and tubing can be extruded into a variety of open and hollow shapes. (Courtesy American Iron and Steel Institute.)

billet, or slab shapes. These are cooled, cleaned of surface irregularities, inspected, and reheated for further rolling. Semifinished shapes may be hot rolled directly into finished hot-rolled products such as structural shapes, plates, sheets, strip, and bars; or hot rolling may be used as an intermediate step before cold finishing, as in the manufacture of bar, sheet, strip, and wire. The hot rolling of finished products is sometimes referred to as *hot finishing*.

Hot-rolled products intended for cold finishing are *descaled* (cleaned of surface oxide scale), usually by *pickling*, which involves passing the steel through sulfuric or hydrochloric acid, followed by rinsing with hot and cold water, steam drying, and, usually, oiling.

Cold finishing consists of cold rolling, cold reduction, and cold drawing of previously hot-rolled, descaled shapes. *Cold rolling* involves passing metal at room temperature through sets of rolls to impart the desired shape, finish, and mechanical properties. *Cold reduction* is a form of cold rolling that drastically reduces the thickness of a flat product (sheet, strip, blackplate, etc.) and improves the strength, surface finish, and flatness. Cold reduction often hardens the metal excessively, so that it must be annealed to soften it and further *temper* (skin) rolled to obtain the proper strength and stiffness and the desired surface texture.

Cold drawing is used to make smaller or more complex bar shapes from hot-rolled annealed bars by pulling the bars through a hard, abrasion-resistant die. Wire is similarly cold drawn from wire rods. This operation results in improved machinability and strength, smoother finish, and greater dimensional accuracy.

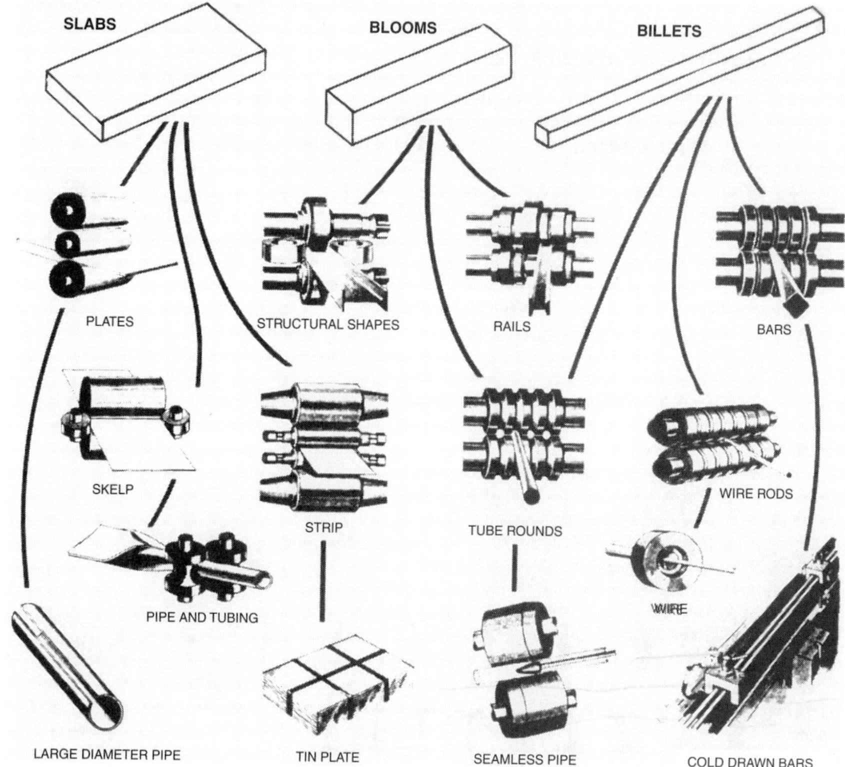

FIGURE 5.1-22 Intermediate mill processes used to convert semifinished steel shapes into finished mill products. (Courtesy American Iron and Steel Institute.)

Slabs, Blooms, and Billets Slabs, blooms, and billets are semifinished shapes from which other products are made. *Slabs* are flat rectangular shapes with a width more than twice the thickness, which is usually less than 10 in. (254 mm). *Blooms* are rectangular shapes generally larger than 36 sq. in. (23,226 mm²). *Billets* are less than 36 sq. in. (23,226 mm²), more nearly square in cross section, and longer.

The type of semifinished shape rolled depends on the intended final use. Slabs are commonly used in the rolling of flat products such as plates, sheets, strip, and skelp; blooms in the rolling of rails, tube rounds, and structural shapes; billets for making bars, tube rounds, and wire rods (Fig. 5.1-22). Skelp, tube rounds, and wire rods are also semifinished shapes. Skelp is used to make welded pipe, tube rounds for piercing into seamless pipe, and wire rods for drawing into wire.

Sheet and Strip Much of the steel shipped annually is the product of continuous sheet mills. The definition of these and other flat-rolled products is complex and varies, depending on whether they are stainless or carbon steel. Generally, the dimensions of sheet and strip lie between those of bars and plates (Fig. 5.1-23). For carbon steels, the method of rolling (whether hot or cold) is also a factor.

Hot-rolled sheet and strip are manufactured from semifinished slabs in sheet and strip mills. Preheated slabs are squeezed and shaped by progressive hot rolling until the final desired thickness and width are obtained. Sheets and strip emerge from the roller tables at speeds up to 3500 ft./min. (1066.8 m/min) and are wound into coils for shipment or for transfer to other departments for further working.

FIGURE 5.1-23 General Classification of Flat-Rolled Steel Products by Size[a]

Carbon Steel, Hot Rolled

	Width					
Thickness	**To 3½ in. incl.**	**Over 3½ in. to 6 in. incl.**	**Over 6 in. to 8 in. incl.**	**Over 8 in. to 12 in. incl.**	**Over 12 in. to 48 in. incl.**	**Over 48 in.**
0.2300 in. & thicker	Bar	Bar	Bar	Plate	Plate	Plate
0.2299 in. to 0.2031 in.	Bar	Bar	Strip	Strip	Sheet	Plate
0.2030 in. to 0.1800 in.	Strip	Strip	Strip	Strip	Sheet	Plate
0.1799 in. to 0.0449 in.	Strip	Strip	Strip	Strip	Sheet	Sheet
0.0448 in. to 0.0344 in.	Strip	Strip				
0.0343 in. to 0.0255 in.	Strip					

Carbon Steel, Cold Rolled

	Width						
Thickness	**To ¹⁵⁄₃₂ in. incl.**		**Over ¹⁵⁄₃₂ in. to 12 in. incl.**		**Over 12 in. to 23¹⁵⁄₁₆ in. incl.**		**Over 23¹⁵⁄₁₆ in.**
0.2500 in. & thicker	Bar		Bar		Strip[c]	Sheet[e]	Sheet
0.2499 in. to 0.0142 in.	Flat Wire[b]	Strip	Strip	Sheet[d]	Strip[c]	Sheet[e]	Sheet
0.0141 in. & thinner	Flat Wire[b]	Strip	Strip		Strip	Blackplate[e]	Blackplate

FIGURE 5.1-23 *(Continued)*

Stainless Steel, Hot and Cold Rolled				
	Width			
Thickness	**¼ in. to ⅜ in.**	**⅜ in. to 10 in.**	**10 in. to 24 in.**	**24 in. & Over**
3/16 in. & over	Bar[f]	Bar	Plate	Plate
⅛ in. to 3/16 in.	Flat Wire[g]	Bar[g]	Strip	Sheet[h]
Up to ⅛ in.	Flat Wire	Strip	Strip	Sheet[h]

[a]Dimensional ranges for flat-rolled alloy steels are customarily slightly different from those for carbon or stainless steel.

[b]When the material has rolled or prepared edges.

[c]When special temper, edge, finish, or single-strand rolling is specified.

[d]Cut lengths or sheet coils slit from wider coils (with resulting No. 3 edges), in thicknesses 0.0142 in. to 0.0821 in. inclusive, and widths 2 in. to 12 in. carbon 0.20% maximum.

[e]When no special temper, edge, or finish (other than Dull or Luster) is specified.

[f]Only hot rolled.

[g]May also be rolled as strip.

[h]In polished finishes Nos. 3, 4, 6, 7 and 8.

(Courtesy American Iron and Steel Institute.) (Refer to Figure 1.8-27 for metric conversions.)

Modern cold-reducing mills may receive hot-rolled steel about as thick as a half dollar and three-quarters of a mile (1207 m) long; two minutes later, the steel will be the thickness of two playing cards and more than two miles (3220 m) long.

Cold-reduced sheets and strip are made by cold reducing previously hot-rolled and pickled flat products. Substantial amounts of sheet and strip are cold rolled to meet consumer requirements for thickness, surface finish, or mechanical properties (see Fig. 5.1-26). Cold-reduced products may also be annealed and temper rolled or zinc or tin coated.

Blackplate is the classification for cold-reduced products thinner than .0141 in. (0.3556 mm) and wider than 12 in. (305 mm) (see Fig. 5.1-23).

FIGURE 5.1-24 **A hot plate is rolled back and forth in a reversing mill until the desired thickness is achieved.** (Courtesy American Iron and Steel Institute.)

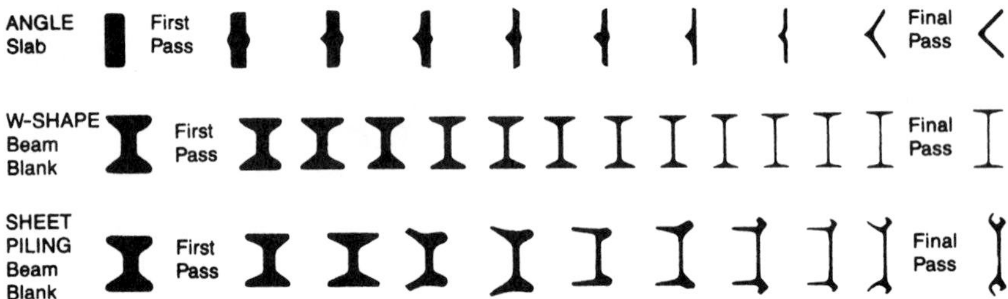

FIGURE 5.1-25 **Successive steps in rolling structural shapes. Hot blanks are roughly shaped by one set of rolls, then other rolls impart the final shape.** (Courtesy American Iron and Steel Institute.)

Plates Flat-rolled finished products thicker than sheet and strip, produced from slabs or slab ingots by hot rolling, are called plates (see Figs. 5.1-23 and 5.1-24). The two principal classes of plates are *sheared* and *universal*, depending on the type of mill on which they are rolled. Sheared plate mills have only horizontal rolls and produce plates with uneven edges and ends, which must later be flame cut or sheared straight to the desired size. Universal plate mills have vertical rolls that are capable of producing smooth, straight edges so that only the ends must be trimmed. Lighter plates may also be rolled on some sheet and strip mills from slabs produced in blooming or slabbing mills.

Structural Shapes Structural members, such as W-shape (wide flange) sections, angles, channels, tees, zees, and piling, are hot rolled from blooms or billets. As many as 40 passes through grooved rolls may be required to produce a desired shape (Fig. 5.1-25). The first sets of rolls drastically reduce the cross section and

FIGURE 5.1-26 Some Common Structural Steel Designations

Shape	Designation	Example (as used on drawings)*	
Wide flange	W	W24 × 62	24 is depth, 62 is weight
Angle	L	L5 × 3 × ¼	5 and 3 are leg lengths ¼ is leg thickness
Channel	C	C10 × 15.3	10 is depth, 15.3 is weight
Structural tee	WT	WT12 × 52	12 is depth, 52 is weight

*Dimensions are in inches; weights are in pounds per linear foot.

FIGURE 5.1-27 A Few of the Many Available Wide-Flange (W) Shapes*

	Area	Depth		Web			Flange				Distance			Nominal Wt. per Ft.	Compact Section Criteria						Elastic Properties						Torsional constant	Plastic Modulus	
				Thickness			Width		Thickness												Axis X-X			Axis Y-Y					
	A	d		t_w		$t_w/2$	b_f		t_f		T	k	k_1		$b_f/2t_f$	F'_y	d/t_w	F''_y	r_T	d/A_f	I	X	r	I	S	r	J	Z_x	Z_y
Designation	in.^{2}	in.		in.		in.	in.		in.		in.	in.	in.	lb.		ksi		ksi	in.		in.^{4}	in.^{3}	in.	in.^{4}	in.^{3}	in.	in.^{4}	in.^{3}	in.^{3}
W18 × 119	35.1	18.97	19	0.655	5/8	5/16	11.265	11 1/4	1.060	1 1/16	15 1/2	1 3/4	15/16	119	5.3	—	29.0	—	3.02	1.59	2190	231	7.90	253	44.9	2.69	10.6	261	69.1
× 106	31.1	18.73	18 3/4	0.590	9/16	5/16	11.200	11 1/4	0.940	15/16	15 1/2	1 5/8	15/16	106	6.0	—	31.7	—	3.00	1.78	1910	204	7.84	220	39.4	2.66	7.48	230	60.5
× 97	28.5	18.59	18 5/8	0.535	9/16	5/16	11.145	11 1/8	0.870	7/8	15 1/2	1 9/16	7/8	97	6.4	—	34.7	54.7	2.99	1.92	1750	188	7.82	201	36.1	2.65	5.86	211	55.3
× 86	25.3	18.39	18 3/8	0.480	1/2	1/4	11.090	11 1/8	0.770	3/4	15 1/2	1 7/16	7/8	86	7.2	—	38.3	45.0	2.97	2.15	1530	166	7.77	175	31.6	2.63	4.10	186	48.4
× 76	22.3	18.21	18 1/4	0.425	7/16	1/4	11.035	11	0.680	11/16	15 1/2	1 3/8	13/16	76	8.1	64.2	42.8	36.0	2.95	2.43	1330	146	7.73	152	27.6	2.61	2.83	163	42.2
W18 × 71	20.8	18.47	18 1/2	0.495	1/2	1/4	7.635	7 5/8	0.810	13/16	15 1/2	1 1/2	7/8	71	4.7	—	37.3	47.4	1.98	2.99	1170	127	7.50	60.3	15.8	1.70	3.48	145	24.7
× 65	19.1	18.35	18 3/8	0.450	7/16	1/4	7.590	7 5/8	0.750	3/4	15 1/2	1 7/16	7/8	65	5.1	—	40.8	39.7	1.97	3.22	1070	117	7.49	54.8	14.4	1.69	2.73	133	22.5
× 60	17.6	18.24	18 1/4	0.415	7/16	1/4	7.555	7 1/2	0.695	11/16	15 1/2	1 3/8	13/16	60	5.4	—	44.0	34.2	1.96	3.47	984	108	7.47	50.1	13.3	1.69	2.17	123	20.6
× 55	16.2	18.11	18 1/8	0.390	3/8	3/16	7.530	7 1/2	0.630	5/8	15 1/2	1 5/16	13/16	55	6.0	—	46.4	30.6	1.95	3.82	890	98.3	7.41	44.9	11.9	1.67	1.66	112	18.5
× 50	14.7	17.99	18	0.355	3/8	3/16	7.495	7 1/2	0.570	9/16	15 1/2	1 1/4	13/16	50	6.6	—	50.7	25.7	1.94	4.21	800	88.9	7.38	40.1	10.7	1.65	1.24	101	16.6
W18 × 46	13.5	18.06	18	0.360	3/8	3/16	6.060	6	0.605	5/8	15 1/2	1 1/4	13/16	46	5.0	—	50.2	26.2	1.54	4.93	712	78.8	7.25	22.5	7.43	1.92	1.22	90.7	11.7
× 40	11.8	17.90	17 7/8	0.315	5/16	3/16	6.015	6	0.525	1/2	15 1/2	1 3/16	13/16	40	5.7	—	56.8	20.5	1.52	5.67	612	68.4	7.21	19.1	6.35	1.27	0.81	78.4	9.95
× 35	10.3	17.70	17 3/4	0.300	5/16	3/16	6.000	6	0.425	7/16	15 1/2	1 1/8	3/4	35	7.1	—	59.0	19.0	1.49	6.94	510	57.6	7.04	15.3	5.12	1.22	0.51	66.5	8.06
W16 × 100	29.4	16.97	17	0.585	9/16	5/16	10.425	10 3/8	0.985	1	13 5/8	1 11/16	15/16	100	5.3	—	29.0	—	2.81	1.65	1490	175	7.10	186	35.7	2.51	7.73	198	54.9
× 89	26.2	16.75	16 3/4	0.525	1/2	1/4	10.365	10 3/8	0.875	7/8	13 5/8	1 9/16	7/8	89	5.0	—	31.9	64.9	2.79	1.85	1300	155	7.05	163	31.4	2.49	5.45	175	48.1
× 77	22.6	16.52	16 1/2	0.455	7/16	1/4	10.295	10 1/4	0.760	3/4	13 5/8	1 7/16	7/8	77	6.8	—	36.3	50.1	2.77	2.11	1110	134	7.00	138	26.9	2.47	3.57	150	41.1
× 67	19.7	16.33	16 3/8	0.395	3/8	3/16	10.235	10 1/4	0.665	11/16	13 5/8	1 3/8	13/16	67	7.7	—	41.3	38.6	2.75	2.40	954	117	6.96	119	23.2	2.46	2.39	130	35.5
W16 × 57	16.8	16.43	16 3/8	0.430	7/16	1/4	7.120	7 1/8	0.715	11/16	13 5/8	1 3/8	7/8	57	5.0	—	38.2	45.2	1.86	3.23	758	92.2	6.72	43.1	12.1	1.60	2.22	105	18.9
× 50	14.7	16.26	16 1/4	0.380	3/8	3/16	7.070	7 1/8	0.630	5/8	13 5/8	1 5/16	13/16	50	5.6	—	42.8	36.1	1.84	3.65	659	81.0	6.68	37.2	10.5	1.59	1.52	92.0	16.3
× 45	13.3	16.13	16 1/8	0.345	3/8	3/16	7.035	7	0.565	9/16	13 5/8	1 1/4	13/16	45	6.2	—	46.8	30.2	1.83	4.06	586	72.7	6.65	32.8	9.34	1.57	1.11	82.3	14.5
× 40	11.8	16.01	16	0.305	5/16	3/16	6.995	7	0.505	1/2	13 5/8	1 3/16	13/16	40	6.9	—	52.5	24.0	1.82	4.53	518	64.7	6.63	28.9	8.25	1.57	0.79	72.9	12.7
× 36	10.6	15.86	15 7/8	0.295	5/16	3/16	6.985	7	0.430	7/16	13 5/8	1 1/8	3/4	36	8.1	64.0	53.8	22.9	1.79	5.28	448	56.5	6.51	24.5	7.00	1.52	0.54	64.0	10.8
W16 × 31	9.12	15.88	15 7/8	0.275	1/4	1/8	5.525	5 1/2	0.440	7/16	13 5/8	1 1/8	3/4	31	6.3	—	57.7	19.8	1.39	6.53	375	47.2	6.41	12.4	4.49	1.17	0.46	54.0	7.03
× 26	7.68	15.69	15 3/4	0.250	1/4	1/8	5.500	5 1/2	0.345	3/8	13 5/8	1 1/16	3/4	26	8.0	—	62.8	16.8	1.36	8.27	301	38.4	6.26	9.59	3.49	1.12	0.26	44.2	5.48

*The shapes listed here are only a few of the many available. This table shows every detail needed by a designer.

(Courtesy American Institute of Steel Construction, Inc.)

FIGURE 5.1-28 Some of the Available Structural Steel Angles*

Size and Thickness	k	Weight per Foot	Area	Axis X-X I	Axis X-X S	Axis X-X r	Axis X-X y	Axis Y-Y I	Axis Y-Y S	Axis Y-Y r	Axis Y-Y x	Axis Z-Z r	
in.	in.	lb.	in.^{2}	in.^{4}	in.^{3}	in.	in.	in.^{4}	in.^{3}	in.	in.	in.	Tan α
L9 × 4 × 5/8	1 1/8	26.3	7.73	64.9	11.5	2.90	3.36	8.32	2.65	1.04	0.858	.847	0.216
9/16	1 1/16	23.8	7.00	59.1	10.4	2.91	3.33	7.63	2.41	1.04	0.834	.850	0.218
1/2	1	21.3	6.25	53.2	9.34	2.92	3.31	6.92	2.17	1.05	0.810	.854	0.220
L8 × 8 × 1 1/8	1 3/4	56.9	16.7	98.0	17.5	2.42	2.41	98.0	17.5	2.42	2.41	1.56	1.000
1	1 5/8	51.0	15.0	89.0	15.8	2.44	2.37	89.0	15.8	2.44	2.37	1.56	1.000
7/8	1 1/2	45.0	13.2	79.6	14.0	2.45	2.32	79.6	14.0	2.45	2.32	1.57	1.000
3/4	1 3/8	38.9	11.4	69.7	12.2	2.47	2.28	69.7	12.2	2.47	2.28	1.58	1.000
5/8	1 1/4	32.7	9.61	59.4	10.3	2.49	2.23	59.4	10.3	2.49	2.23	1.58	1.000
9/16	1 3/16	29.6	8.68	54.1	9.34	2.50	2.21	54.1	9.34	2.50	2.21	1.59	1.000
1/2	1 1/8	26.4	7.75	48.6	8.36	2.50	2.19	48.6	8.36	2.50	2.19	1.59	1.000
L8 × 6 × 1	1 1/2	44.2	13.0	80.8	15.1	2.49	2.65	38.8	8.92	1.73	1.65	1.28	0.543
7/8	1 3/8	39.1	11.5	72.3	13.4	2.51	2.61	34.9	7.94	1.74	1.61	1.28	0.547
3/4	1 1/4	33.8	9.94	63.4	11.7	2.53	2.56	30.7	6.92	1.76	1.56	1.29	0.551
5/8	1 1/8	28.5	8.36	54.1	9.87	2.54	2.52	26.3	5.88	1.77	1.52	1.29	0.554
9/16	1 1/16	25.7	7.56	49.3	8.95	2.55	2.50	24.0	5.34	1.78	1.50	1.30	0.556
1/2	1	23.0	6.75	44.3	8.02	2.56	2.47	21.7	4.79	1.79	1.47	1.30	0.558
7/16	15/16	20.2	5.93	39.2	7.07	2.57	2.45	19.3	4.23	1.80	1.45	1.31	0.560
L8 × 4 × 1	1 1/2	37.4	11.0	69.6	14.1	2.52	3.05	11.6	3.94	1.03	1.05	0.846	0.247
3/4	1 1/4	28.7	8.44	54.9	10.9	2.55	2.95	9.36	3.07	1.05	0.953	0.852	0.258
9/16	1 1/16	21.9	6.43	42.8	8.35	2.58	2.88	7.43	2.38	1.07	0.882	0.861	0.265
1/2	1	19.6	5.75	38.5	7.49	2.59	2.86	6.74	2.15	1.08	0.859	0.865	0.267
L7 × 4 × 3/4	1 1/4	26.2	7.69	37.8	8.42	2.22	2.51	9.05	3.03	1.09	1.01	0.860	0.324
5/8	1 1/8	22.1	6.48	32.4	7.14	2.24	2.46	7.84	2.58	1.10	0.963	0.865	0.329
1/2	1	17.9	5.25	26.7	5.81	2.25	2.42	6.53	2.12	1.11	0.917	0.872	0.335
3/8	7/8	13.6	3.98	20.6	4.44	2.27	2.37	5.10	1.63	1.13	0.870	0.880	0.340
L6 × 6 × 1	1 1/2	37.4	11.0	35.5	8.57	1.80	1.86	35.5	8.57	1.80	1.86	1.17	1.000
7/8	1 3/8	33.1	9.73	31.9	7.63	1.81	1.82	31.9	7.63	1.81	1.82	1.17	1.000
3/4	1 1/4	28.7	8.44	28.2	6.66	1.83	1.78	28.2	6.66	1.83	1.78	1.17	1.000
5/8	1 1/8	24.2	7.11	24.2	5.66	1.84	1.73	24.2	5.66	1.84	1.73	1.18	1.000
9/16	1 1/16	21.9	6.43	22.1	5.14	1.85	1.71	22.1	5.14	1.85	1.71	1.18	1.000
1/2	1	19.6	5.75	19.9	4.61	1.86	1.68	19.9	4.61	1.86	1.68	1.18	1.000
7/16	15/16	17.2	5.06	17.7	4.08	1.87	1.66	17.7	4.08	1.87	1.66	1.19	1.000
3/8	7/8	14.9	4.36	15.4	3.53	1.88	1.64	15.4	3.53	1.88	1.64	1.19	1.000
5/16	13/16	12.4	3.65	13.0	2.97	1.89	1.62	13.0	2.97	1.89	1.62	1.20	1.000
L6 × 4 × 7/8	1 3/8	27.2	7.98	27.7	7.15	1.86	2.12	9.75	3.39	1.11	1.12	0.875	0.421
3/4	1 1/4	23.6	6.94	24.5	6.25	1.88	2.08	8.68	2.97	1.12	1.08	0.860	0.428
5/8	1 1/8	20.0	5.86	21.1	5.31	1.90	2.03	7.52	2.54	1.13	1.03	0.864	0.435
9/16	1 1/16	18.1	5.31	19.3	4.83	1.90	2.01	6.91	2.31	1.14	1.01	0.866	0.438
1/2	1	16.2	4.75	17.4	4.33	1.91	1.99	6.27	2.08	1.15	0.987	0.870	0.440
7/16	15/16	14.3	4.18	15.5	3.83	1.92	1.96	5.60	1.85	1.16	0.964	0.873	0.443
3/8	7/8	12.3	3.61	13.5	3.32	1.93	1.94	4.90	1.60	1.17	0.941	0.877	0.446
5/16	13/16	10.3	3.03	11.4	2.79	1.94	1.92	4.18	1.35	1.17	0.918	0.882	0.448
L6 × 3 1/2 × 1/2	1	15.3	4.50	16.6	4.24	1.92	2.08	4.25	1.59	0.972	0.833	0.759	0.344
3/8	7/8	11.7	3.42	12.9	3.24	1.94	2.04	3.34	1.23	0.988	0.787	0.767	0.350
5/16	13/16	9.8	2.87	10.9	2.73	1.95	2.01	2.85	1.04	0.996	0.763	0.772	0.352
L5 × 5 × 7/8	1 3/8	27.2	7.98	17.8	5.17	1.49	1.57	17.8	5.17	1.49	1.57	0.973	1.000
3/4	1 1/4	23.6	6.94	15.7	4.53	1.51	1.52	15.7	4.53	1.51	1.52	0.975	1.000
5/8	1 1/8	20.0	5.86	13.6	3.86	1.52	1.48	13.6	3.86	1.52	1.48	0.978	1.000
1/2	1	16.2	4.75	11.3	3.16	1.54	1.43	11.3	3.16	1.54	1.43	0.983	1.000
7/16	15/16	14.3	4.18	10.0	2.79	1.55	1.41	10.0	2.79	1.55	1.41	0.986	1.000
3/8	7/8	12.3	3.61	8.74	2.42	1.56	1.39	8.74	2.42	1.56	1.39	0.990	1.000
5/16	13/16	10.3	3.03	7.42	2.04	1.57	1.37	7.42	2.04	1.57	1.37	0.994	1.000

*Angles in shaded rows may not be readily available. Availability is subject to rolling accumulation and geographical location, and should be checked with material suppliers.

(Courtesy American Institute of Steel Construction, Inc.)

impart the approximate shape; the finishing rolls gradually form the final shape and, for deep sections such as channels and angles, bend the legs into position. Wide-flange shapes are rolled on universal mills that have vertical rolls capable of shaping the vertical surfaces. Different weights of a shape are produced by varying the spacing of the rollers, wider spacings producing heavier sections.

Structural shapes are identified by standard designations (Fig. 5.1-26). Wide-flange shapes are available in a great many sizes, a few of which are shown in Figure 5.1-27. Those used for beams and girders tend to be relatively tall with narrow flanges; those for columns and piles tend to be more square in shape.

Angles are available with legs of either equal length or unequal length (Fig. 5.1-28). Among their many uses are as structural lintels over masonry openings, supports for precast concrete and stone cladding elements, floor opening framing, concrete slab opening edging, structure cross bracing, brackets used to connect structural beams to girders and columns, components of trusses, and supports for finishes furring.

Channels have many of the same applications as angles. They are also used as light beams, wall framing, and purlins. Tees are also used as light beams and in such applications as purlins in roof framing and as support for cementitious and plank decking.

Each structural shape is marked at the mill with its shape designation (*W, S, L, C, WT*), mill identification, and a code that identifies the batch of steel from which it was produced. This marking provides a convenient means of tracing a shape back to its source in the event of its failure. A certificate showing the chemical analysis of the batch of steel accompanies steel shapes from the mill to the fabricator to display compliance with the required specifications.

Bars Bars may be round, square, hexagonal, or multifaceted long shapes, generally thicker than wire. They are produced by hot rolling and cold drawing. In hot rolling, between 10 and 15 roll passes, taking less than 2 minutes, are required to change a square hot billet to a bar. Bars more than ¾ in. (19.1 mm) in diameter are usually sheared into lengths 16 to 20 ft. (4.88 to 6.1 m) long. Smaller bars are typically wound in coils for ease of handling and shipping. Carbon steel bars may

FIGURE 5.1-29 Irregular as well as round and rectangular bars can be easily cold drawn. (Courtesy American Iron and Steel Institute.)

be galvanized or treated with other corrosion-resistant coatings. Flat rectangular shapes (flats), which are narrower than sheet and strip, are also known as bars (see Fig. 5.1-23) and may be either hot or cold rolled.

Hot-rolled bars may be cold drawn to further reduce their cross section and produce complex shapes (Fig. 5.1-29) of closer dimensional tolerances and smoother finishes. Improved mechanical properties, such as greater tensile strength and hardness, also result from cold drawing. The degree of reduction depends on the type of metal used and the desired mechanical properties of the end product.

Wire Cold-drawn products round, square, or multifaceted in cross section and

FIGURE 5.1-30 Fine wire can be made by drawing rods through successively smaller die openings. (Courtesy American Iron and Steel Institute.)

smaller than bars are called wire. Round wire is drawn from 0.005 in. (0.127 mm) up to 1 in. (25.4 mm) in diameter. Flat wire is a cold-rolled product, rectangular in shape and generally narrower than bars (see Fig. 5.1-23).

Semifinished wire rods of low carbon steel may be drawn directly into wire by *angle draft drawing*. For finer wire sizes or harder steels, the rods must be softened by annealing before the continuous *multiple draft drawing* (Fig. 5.1-30) through a series of dies. Wire may also be annealed after drawing to make extremely soft *annealed wire*.

Cold-drawn wire of carbon steel can be produced with very high tensile strengths ranging to more than 500,000 psi. (351,535 Mg/m^2). Corrosion resistance can be achieved by galvanizing or coating with other corrosion-resistant metals.

Tubular Products Tubular products are long, hollow metal products of round, oval, square, rectangular, or multifaceted cross section. Tubular products can be either pipe or tubing, the terms being applied somewhat interchangeably. Round pipe in a wide range of sizes and square or rectangular structural tubing are widely employed in building construction. Tubular products that are mechanically worked and produced on *welded* and *seamless* mills are referred to as *wrought pipe* to distinguish them from cast pipe. Wrought tubular products may be marketed, after being milled as either hot or cold finished, as *black pipe* or as *galvanized pipe* when hot dipped in zinc.

Welded pipe is made chiefly by the butt weld and the electric weld processes. In the *butt weld* process, a strip of preheated, semifinished skelp is continuously formed into a tubular shape between concave rolls, and the edges are joined by mechanical pressure during rolling (Fig. 5.1-31a). Pipe sizes between ⅛ in. (3.2 mm) and 4 in. (101.6 mm) are commonly produced by this process.

The majority of steel pipe in this country is produced by *electric weld* processes. Preheating of the metal is not required, and localized heat necessary for joining the edges is generated during the welding process itself. Resistance welding is used to make *electric-resistance welded* (ERW) tubular products up to 20 in. (508 mm) in diameter. In this process, a strip of cold metal is formed into tubular shape by rolling, and the edges are joined by heat and pressure (Fig. 5.1-31b). Welding heat

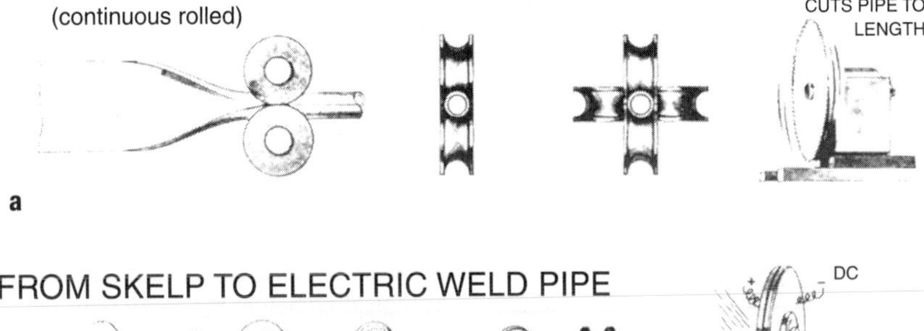

FROM SKELP TO ELECTRIC WELD PIPE

DC

WATER

CUTTING TOOLS

b

FIGURE 5.1-31 **(a) The butt-weld process can produce pipes up to 4 in. (101.6 mm) in diameter. (b) The electric-weld process can produce pipes up to 8 in. (203.2 mm) in diameter.** (Courtesy American Iron and Steel Institute.)

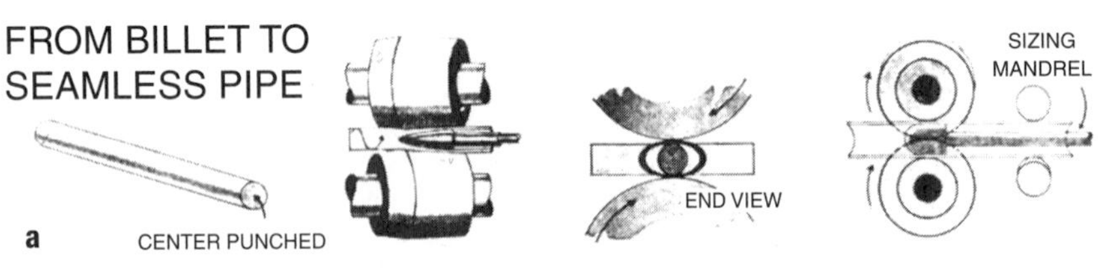

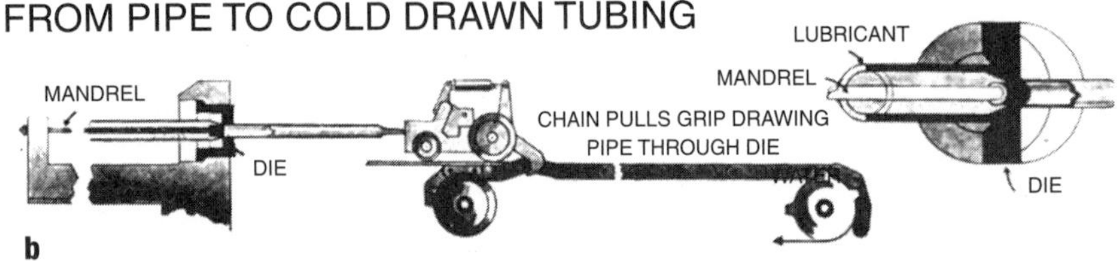

FIGURE 5.1-32 **(a) Most seamless carbon steel pipe is made by hot piercing and extruding. (b) Stainless and alloy steel tubing is cold drawn.** (Courtesy American Iron and Steel Institute.)

is generated in the seam by the resistance of the metal to an electric current introduced by a wheel-shaped roller acting as the electrode.

For larger-diameter and thick-walled pipe, one of several electric welding processes may be used. In most such processes, steel strip or plates are die formed cold by several successive forming operations, consisting of crimping (bending the edges), U forming, and O forming. The last forming operations result in a nearly round shape with edges touching and ready for welding. Submerged arc welding and metal inert-gas-shielded arc (MIG) welding are often used for this purpose. The final shape and diameter are obtained by expanding the welded shell with hydraulic power against a retaining jacket.

Seamless pipe and tubing are made by hot piercing and hot extrusion. Piercing is accomplished by continuously feeding a hot billet between pairs of tapered rolls that are skewed to apply uneven pressure, which causes a pilot hole to form in the center of the billet as it rolls between them. A piercing point enlarges the pilot hole, and a sizing mandrel shapes the final opening size and wall thickness (Fig. 5.1-32a). Most seamless products are made by hot working, although some seamless tubing is made by cold drawing a prepierced billet through a die over an internal mandrel (Fig. 5.1-32b). Stainless and alloy steel tubing up to 11 in. (279.4 mm) in diameter are often made in this manner. Hot-piercing methods can produce seamless pipe up to about 26 in. (660.4 mm) in diameter.

Hot extrusion produces a tubular shape by forcing hot predrilled billets through a die over an internal mandrel. This method is used primarily to make small sizes of stainless or high-alloy pipe for cold reducing. It is also the most economical method of producing complex tubular shapes in small quantities.

5.1.4.3 Fabrication

Most steels can be fabricated by hot forming, cold forming, and machining. Some forming methods may be the same as those previously described; however, fabrication as used here implies reworking the products of the mills and foundries in a fabricating plant. Cold-forming techniques are by far the more common ones used in the manufacture of products for building construction.

HOT FORMING

Hot-forming operations are used on metals too thick or too hard to be formed cold. These operations include (1) forging (described earlier), (2) high-temperature forming (electroforming), which uses heat generated by the resistance of the workpiece to an electric current passed through it by the forming roll, and (3) high-energy-rate forming, a method of hot (sometimes cold) shaping of materials by the sudden impact of a small explosive charge, a shockwave generated by an electric spark in a fluid, or sudden release of a compressed gas through a system of valves.

COLD FORMING

Cold-forming operations are performed on most flat-rolled and tubular products by shaping, bending, or drawing the metal beyond its yield point so that a permanent set is achieved. The most common cold-forming methods are *roller*, *stretch*, *shear*, and *brake forming*, and *deep drawing*.

Roller Forming A variety of shapes (Fig. 5.1-33) are formed by passing sheet and strip through a series of roller dies that gradually impart the desired shape (Fig. 5.1-34). Many types of corrugated and ribbed decking and wall panels are made in this manner.

Stretch Forming In stretch forming, light-gauge flat products are shaped over form blocks into shallow curved shapes by pressure. The steps in stretch forming a particular shape are illustrated in Figure 5.1-35.

FIGURE 5.1-33 **Sheet and strip may be roller formed into a variety of shapes.** (Courtesy American Iron and Steel Institute.)

FIGURE 5.1-34 **Wall and roofing panels are typically roller formed from galvanized or stainless steel sheets.** (Courtesy American Iron and Steel Institute.)

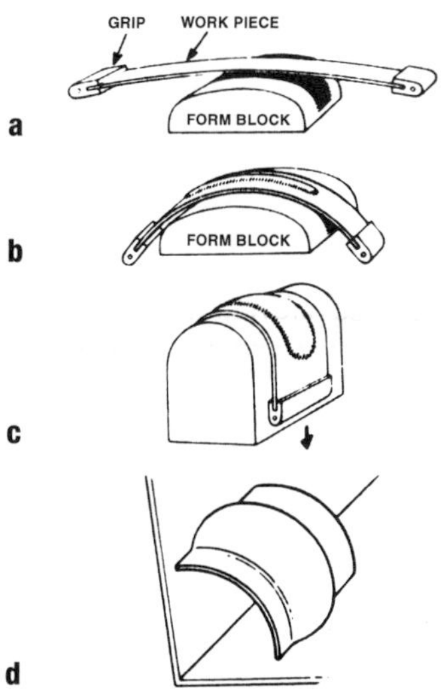

FIGURE 5.1-35 **In stretch forming, an oversized piece of sheet metal is gripped (a) and bent over a form block (b) to produce the desired contour (c) and is then trimmed to the final shape (d).** (Courtesy American Iron and Steel Institute.)

Shear Forming In shear forming, flat round blanks are formed into curved surfaces such as cylinders, domes, and cones by turning the workpiece on a mandrel and shaping it over a block with external pressure. *Spinning* is a type of shear forming in which the deformation and reduction in original blank thickness is not quite as severe (Fig. 5.1-36). Tank heads, television tube cones, and stainless steel railing base flanges and caps are examples.

Brake Forming *Brake forming* is a common method of fabricating sheet metal components for construction. In this method, sheet steel is shaped on a press brake by squeezing it between a punch and a die block in one or several consecutive operations (Fig. 5.1-37). Heating and air conditioning ducts, steel roofing products, and many of the sections used in architectural stainless steel storefront construction are fabricated by this method.

Deep Drawing Flat, thin-gauge blanks are often formed in a draw press between an operating punch and a stationary die. Many household utensils, such as pots and pans, and commercial vessels and containers are made this way. Stainless steel can be drawn to various degrees, but the austenitic group lends itself most readily to deep-drawing operations.

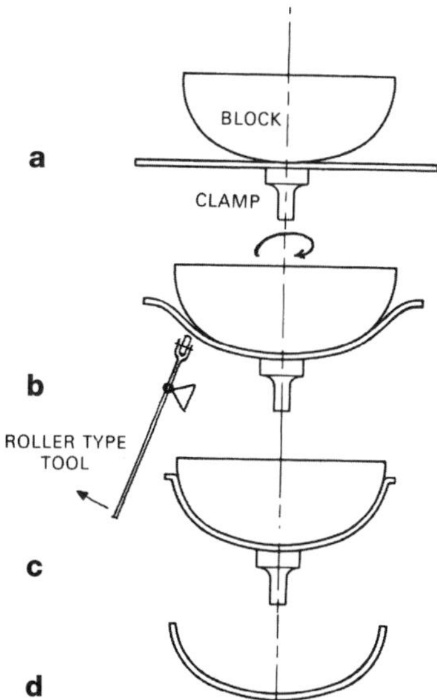

FIGURE 5.1-36 **In spinning, a round blank is clamped on a form block (a) and spun while pressure is applied (b), gradually imparting the shape of the form block (c), then is trimmed to the desired profile (d).** (Courtesy American Iron and Steel Institute.)

FIGURE 5.1-37 **Both carbon steel and stainless steel can be bent sharply in a press brake to form various products.** (Courtesy American Iron and Steel Institute.)

MACHINING

Most steels can be milled, sawed, drilled, punched, sheared, tapped, and reamed with appropriate equipment. Tool steels containing tungsten, molybdenum, and chromium that have been subjected to tempering and other heat treatments are used for this purpose. Sulfur and lead improve the machining properties of carbon steels and are added to steels requiring extensive machining. Among the stainless steels, the ferritic alloys are the easiest to machine.

5.1.4.4 Protective Coatings and Mechanical Finishes

Ordinary cast iron, most carbon steels, and some alloy steels oxidize under ordinary atmospheric conditions. Elevated temperatures and moisture accelerate this oxidation, which is known as atmospheric corrosion. The resulting surface scale (rust) deteriorates the metal surface and, if not prevented, progresses until the metal reverts to the oxide state found in nature. The chief function of protective coatings used on iron and carbon steel is to inhibit corrosion.

Not all ferrous metals require equal protection against progressive corrosion. Copper-bearing steels, some cast irons, and some high-strength low-alloy steels acquire excellent corrosion resistance from a tight surface oxide that inhibits further corrosion. Sometimes the appearance of the natural patina is objectionable, and finish coatings are applied on these steels for decorative purposes. Decorative and protective coatings are used less often on stainless steels because of their superior corrosion resistance and attractive natural finishes.

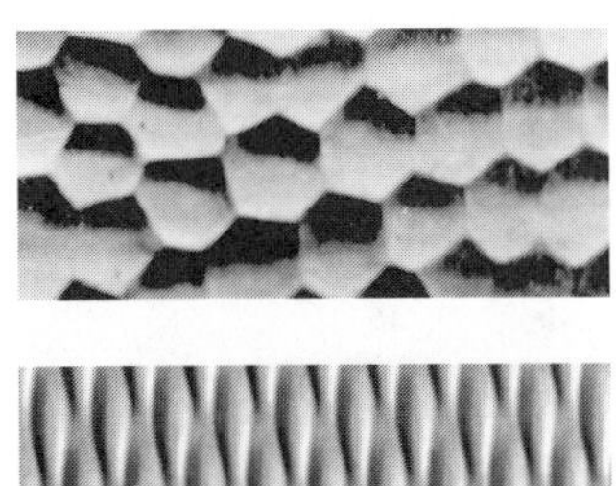
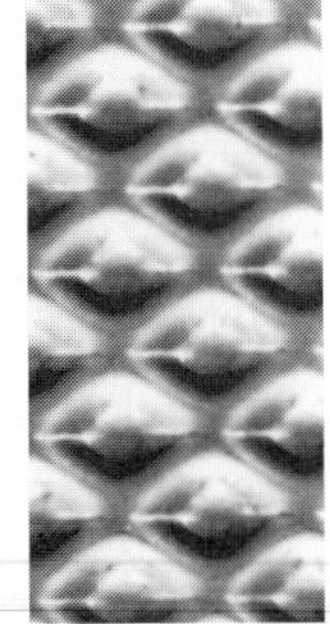
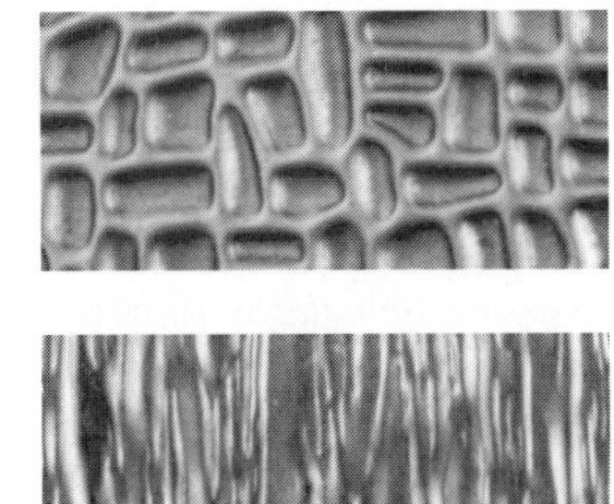
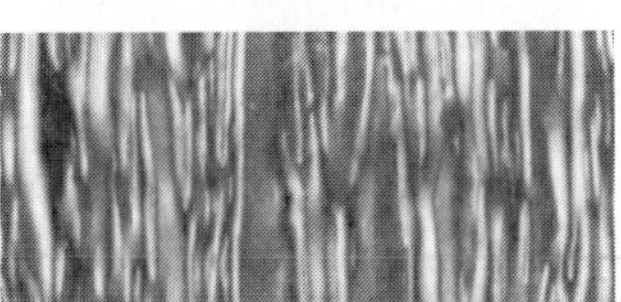

FIGURE 5.1-38 **Steel sheets can be stiffened and decorated with patterned finishes.** (Photographs furnished by Rigidized Metals Corp.)

Finishes for iron and steel may be classified as (1) *mechanical*, including the as-rolled and as-drawn natural mill finish and those imparted by further grinding, polishing, or patterning, (2) *chemical*, generally consisting of cleaning or preparatory operations for further finishing, and (3) *organic and inorganic coatings* such as metallic, vitreous, laminated, and painted finishes. Organic coatings commonly sprayed on or brush applied in the field are addressed in Chapter 9.

MECHANICAL FINISHES

The *hot-rolled mill finish* (black, as-rolled finish) on carbon steel is characterized by a mill scale and rust powder that is sometimes acceptable as a base for organic coatings applied by brush. For spray painting, this rust powder must be removed by sandblasting or wire brushing.

Cold-finished surfaces are usually smoother and contain little or no surface corrosion, but must be degreased of oil coatings and sometimes roughened to provide a good base for organic coatings. Steel products that will not be coated may be sent to a *temper mill*, which rolls the steel to a desired flatness and surface quality. Cold-reduced sheet and strip have a natural mill finish acceptable for many uses, and as a base for subsequent galvanizing, tinning, or terne coating. *Black pipe* has a natural dark finish often left uncoated. Stainless steel sheet is produced in a variety of patterned finishes (Fig. 5.1-38) imparted by roll forming for greater stiffness and decorative effect and to produce differential light reflection that emphasizes waviness of sheet surfaces in thin flat products.

Stainless Steel Sheet and Strip Because stainless steels are inherently corrosion resistant and can be made strong and rigid in thin sheets, these products are usually not protected by coatings. They are most often either left with their natural cold-rolled finish, such as Nos. 1, 2D, and 2B sheet finishes, or given a polished finish, such as Nos. 3, 4, 6, 7, and 8 (Fig. 5.1-39). An exception is terne-coated stainless steel, which is used in roofing.

Stainless Bars, Pipe, and Tubing Stainless bars, pipes, and tubing are manufactured in several neutral and polished finishes similar to those of sheets and strip described in the preceding paragraph. In industrial applications, they may be used in one of the natural finishes; for architectural work where uniformity of appearance is important, polished finishes are generally preferred.

CHEMICAL FINISHES

Chemical finishes are used chiefly as cleaning and conditioning steps for other coating applications. *Pickling*, which removes oxide scale, and other operations for removing lubricants in product manu-

FIGURE 5.1-39 Common Stainless Sheet Steel Mechanical Finishes

	Designation	Description	Uses
Natural Finishes (unpolished)	No. 1 Finish	Rough dull surface resulting from hot rolling, annealing, and descaling	Generally available on heavy gauge sheets; not used in architectural applications
	No. 2D Finish (No. 1 Strip Finish)	Smooth dull surface resulting from cold rolling, annealing, and descaling	Some architectural applications requiring low luster, as in roofing and drainage products; suitable for cold forming and further polishing
	No. 2B Finish (No. 2 Strip Finish)	Bright, smooth cold-rolled surface produced either by highly polished rolls or bright annealing	General-purpose natural finish for transportation equipment; limited use in curtain walls and storefronts; used for further polishing
Polished Finishes	No. 3 Finish	A surface produced by grinding and polishing with slightly finer abrasives than for finish No. 4	General-purpose natural finish for transportation equipment; limited use in curtain walls and storefronts; used for further polishing
	No. 4 Finish	Bright, polished surface produced by grinding and polishing either at the mill or fabricator plant	Most common for architectural uses; directional grit line permits blending of fabricated or field joints with mill-polished surface
	No. 6 Finish	Dull matte finish produced by brushing No. 4 finish	Architectural applications where softer, less reflective finish is desired
	No. 7 Finish	A highly reflective finish produced by further polishing and buffing	Chiefly in architectural applications
	No. 8 Finish	Mirrorlike finish produced by polishing with buffing rouges until surface is free of grit lines	Infrequent in architectural applications

(Courtesy American Iron and Steel Institute.)

FIGURE 5.1-40 A continuous hot-dip galvanizing unit. (Courtesy American Iron and Steel Institute.)

facture are examples. *Conversion coatings* are sometimes used as decorative finishes on carbon and stainless steels.

Carbon Steel Conversion Coatings Conversion coatings are mostly chemical treatments (as with an acid phosphate solution) that *convert* the chemical nature of the surface film to improve its bonding properties with paint and other applied coatings. In one decorative finish, textured steel is zinc plated and dyed using a chromate conversion coating, followed by buffing. Buffing causes the bright zinc to stand out in relief, highlighted by the color coating in the recesses.

Stainless Conversion Coatings Surface blackening, a common conversion coating, on stainless steel produces a decorative finish ranging from a bluish hue to dark brown or dead black. It is produced by encouraging the growth of a thin oxide, either by controlled heat treatment or dipping in a hot chemical oxidizing bath.

ORGANIC AND INORGANIC COATINGS

Carbon steels are most often finished for protective and decorative purposes with organic and inorganic coatings. These may be (1) metallic, applied by hot dipping, electroplating, or other methods, (2) vitreous, fused-on glassy materials such as porcelain enamel, or (3) laminated, involving the adhesive application of inert plastic films.

Metallic Coatings Metallic coatings provide protection against corrosive elements, either by (1) acting as a *sacrificial* metal (one that is purposely permitted to corrode), as in the case of zinc coatings, or (2) being relatively inert and corrosion resistant (thus protecting the underlying base metal), as in a nickel or chromium coating. The metals most often used for protecting carbon steels are zinc, tin, terne metal, aluminum, cadmium, chromium, and nickel. These coating metals are most commonly applied by hot dipping, electroplating, metallizing, or cladding. Galvanizing (coating with zinc) is usually done by hot dipping or electroplating. Aluminum coatings are generally applied by hot dipping. Tin plating is coating with tin, either by hot dipping or electroplating. Bonderizing on galvanized sheets and galvannealing are light metallic coatings with a suitable surface for paint coatings.

Hot dipping is a process in which steel is immersed in a molten bath of the coating metal (Fig. 5.1-40). The most prevalent method is continuous hot-dip galvanizing, which accounts for more than 10 times as much galvanized steel as the electrolytic process. The hot-dip process, which results in heavier coatings, is the chief method of galvanizing tubular and flat-rolled products. Hot-dipped galvanized sheet and strip can be made with zinc coatings of designation G90 (.90 oz./sq. ft. [0.275 kg/m^2]), which are used frequently in architectural applications. Heavier coatings of approximately 2 oz./sq. ft. (0.61 kg/m^2) are made for severe corrosive applications. These coating weights are the combined weights on both surfaces.

Hot-dipped *aluminized* sheets are usually coated with an aluminum-silicon (5% to 11%) coating known as Type 1. Aluminum-silicon coatings provide excellent resistance to atmospheric and high-temperature (up to 1250°F [676.67°C]) corrosion and are extensively used in the automotive industry.

Electroplating employs an electric current and an electrolytic solution to deposit metallic coatings on steel or iron (Fig. 5.1-41). Tin, zinc, and cadmium are commonly applied by this method, as are nickel and chromium either directly or over copper. Electroplating is the chief method of making modern tinplate and the second most important process for making galvanized products. Lighter coatings of

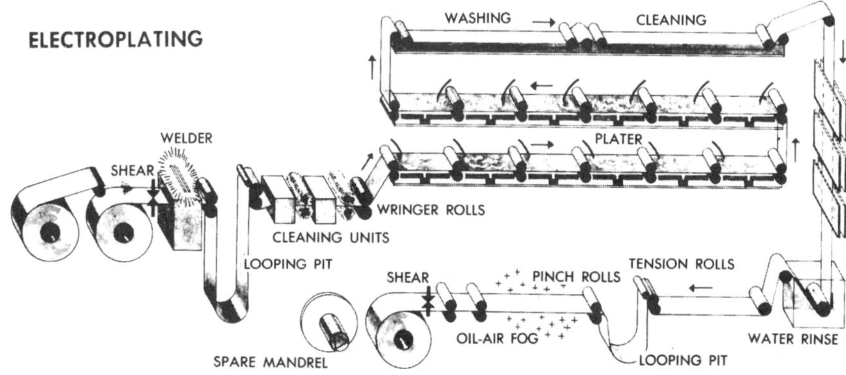

FIGURE 5.1-41 Steps in the production of tinplate and galvanized products by the electrolytic process. The looping pits shown are necessary to ensure that the proper tension is applied to the metal throughout the process. (Courtesy American Iron and Steel Institute.)

0.10 oz./sq. ft. (0.031 kg/m^2) (for minimum corrosion resistance) and greater variations in coating thicknesses are possible with electroplating. The lighter coatings are normally recommended on surfaces to be painted or to be used in mildly corrosive exposures.

Metallizing consists of spraying molten metal against the surface to be coated. It is used extensively for applying zinc and aluminum coatings. The coating adheres to the base metal by a combination of mechanical interlocking and metallurgical bonding. Metallizing is the chief method of metallic coating practical for field use.

Cladding produces bimetallic "sandwich" products consisting usually of a carbon or low-alloy steel core, covered with a thin sheet layer of coating metal. Stainless steel and aluminum coatings are applied by hot rolling under controlled, nonoxidizing conditions. Copper may also be used for cladding steel plates by hot dipping or electroplating.

Zinc-based paint pretreatments are common on steel to improve the bond of subsequent paint finishes.

Bonderizing is the most common process for producing a suitable paint base on fabricated products. Zinc-coated products are dipped in a hot phosphate solution, resulting in a thin (0.02 to 0.03 oz./sq. ft. [0.006 to 0.009 kg/m^2]) crystalline surface film of zinc phosphate.

Vitreous Coatings Although there are several types of vitreous coatings used on metal, including glass-on-steel linings applied to hot water tanks and process piping, *porcelain enamel* is the type most common in architectural applications. Common uses are porcelain enamel on cast iron or steel in bathtubs, sinks, and other sanitary ware. Porcelain-enameled building panels are used in commercial and institutional buildings.

Porcelain enamels have similar inert abrasion- and corrosion-resistant properties similar to glass. They can be produced in a great variety of lightfast colors and textures. Porcelain enamel is applied in one or two coats to sheets that have been prepared by etching. In the two-coat system, the first enamel coat contains adherence promoters, such as cobalt or nickel oxides, and is applied by spraying, dipping, or flow coating, followed by firing at high temperatures. The second (and sometimes third) coat contains the coloring elements and is applied and fired in a similar manner. Coating thickness ranges between 4 and 20 mils (0.1 and 0.51 mm), though durability is not necessarily a function of thickness; specifically, where flexure is a possibility, thinner or fewer coatings may be preferred.

Laminated Coatings Laminated coatings are thin (2 to 12 mils [0.05 to 0.3 mm] thick), tough plastic films applied with thermosetting (heat-hardening) adhesives to steel products. The most common films used are polyvinyl chloride (PVC) and polyvinyl fluoride (PVF), both of which possess good abrasion resistance and sufficient tenacity, toughness, and flexibility to withstand severe forming operations. They have excellent resistance to weathering, good color retention, and resistance to attack by a variety of chemical agents. Although color selections are more limited here than in porcelain enamels, the potential color variations are substantial.

5.2 Iron and Steel Design and Construction

Many low-rise and medium-rise buildings and all very tall buildings are built with structural steel frames. In such construction, structural steel shapes are erected and fastened together using various methods to form a rigid framework from which are hung floors and exterior walls. Floors are supported either on steel structural members or on steel joists supported by structural framing.

5.2.1 JOINTS AND CONNECTIONS

Steel and iron structural shapes can be joined with *mechanical fastenings*, *mechanically formed joints*, *welding*, and *adhesive bonding*. Similar joints and connections are also used in iron and steel product fabrication.

5.2.1.1 Mechanical Fastenings

Mechanical fastenings include a variety of permanent and semipermanent types. They are used to connect metal to metal and metal to other building materials. Permanent mechanical fastening types (welding, rivets, etc.) involve permanent deformation, so the components may be disassembled only by the destruction or cutting of the fastener or the component. Semipermanent fastenings (screws, bolts, and nails) allow disassembly without permanent damage to either the component or the fastener.

Bolting and welding are the most significant joining methods for steel construction. Riveting is almost never employed for that purpose today and is used only occasionally in shop-fabricated steel structural components. One of the disadvantages of riveting is the high noise level of riveting machines.

RIVETING

Riveting involves the joining of components by means of a rivet with a prefabricated head on one end, and deformation of the shank to form a head on the other end. Structural rivets are generally deformed by a pneumatic or hydraulic riveter while red hot. As the rivet cools it shortens in length, drawing the components together. The resulting compressive and frictional forces between the contact surfaces are important in resisting tensile and shear stresses developed in structural joints.

BOLTING

Bolting is a semipermanent mechanical method of fastening with nuts, bolts, and washers. Bolting is easier to accomplish in the field than welding, especially in locations where access is difficult and when adverse weather conditions exist. Bolted connections work best when it is necessary for them to resist only shearing forces. When moment forces also occur, bolted joints often become complex and bulky as compared with welded joints. Bolts are sometimes used to hold joints together for welding.

Specified tensile values must be developed in bolts by tightening them with calibrated torque wrenches, by turning the nuts a specified number of turns from a snug-tight position, or using a special power tool that locks onto the tip of the bolt and breaks it off when the nut is sufficiently tight, thus preventing further tightening. Proper tightening is important to develop frictional resistance in a joint

FIGURE 5.2-1 Air-impact wrenches speed high-strength bolting of steel connections in modern buildings. (Courtesy American Iron and Steel Institute.)

to counteract shearing forces. Keeping the threaded part of a bolt out of the structural parts being joined is also essential in developing the full strength of the connection.

Nuts and bolts are made in sizes ranging up to 4 in. (102 mm) in diameter and with several types of heads (square, round, hexagonal, countersunk, etc.). Most bolts have square or hexagonal heads, as do most nuts, which allows the use of wrenches. Although bolts are available in both carbon and alloy steels, using high-strength bolts for most major structural applications is becoming more prevalent (Fig. 5.2-1). Ordinary carbon steel bolts are used in securing many components in small building construction, such as sill plates to foundations, and beams to columns and to each other.

STUD WELDING

In stud welding, which is a semipermanent method of fastening, one end of a stud is arc welded by a gun-shaped unit to a steel component, and a nut is applied to the stud's threaded end (Fig. 5.2-2). Studs are used most often when nonstructural materials must be attached to steel, as in the application of architectural finishes to a structural framework. Stud welding is also used in composite structural design as a means of transferring stresses between structural steel and reinforced concrete to produce combined resistance to stresses. Studs are also available unthreaded or internally threaded, in a variety of shapes such as bent and rectangular eyebolts and j-bolts.

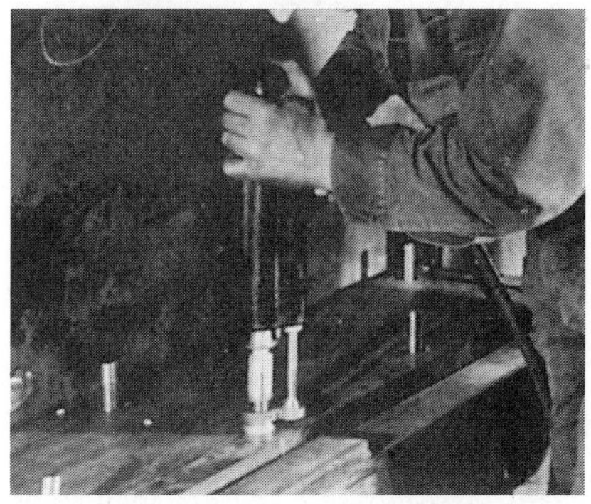

FIGURE 5.2-2 Special stud welding equipment permits fast and efficient joining of many materials to steel. (Courtesy TRW Nelson Co.)

POWDER-ACTUATED FASTENINGS

Powder-actuated fastenings are permanent or semipermanent fastenings that employ a gun powered by a small explosive charge to drive a pin into hard and tough materials. In permanent fastenings, a specially hardened drive pin is used to connect components in a fashion similar to nailing. Often, however, a pointed stud, threaded on one end, is driven into one of the components and assembly is completed with a nut. The availability of inexpensive guns and a variety of stud sizes and shapes makes this a useful fastening procedure in all types of construction (Fig. 5.2-3).

THREADED FASTENERS AND NAILS

The most familiar semipermanent fastening devices are screws, bolts, and threaded and grooved nails. These are manufactured from carbon, alloy, and stainless steels in many shapes and sizes. Their uses include ordinary household applications, assemblies for product fabrication, and the joining of structural trim components in construction.

EXPANSION BOLTS

Expansion bolts are semipermanent fasteners generally used to join materials to concrete and masonry surfaces. They are also referred to as cinch anchors, expansion anchors, and by many proprietary names. These devices consist of two or more units that develop holding power in concrete or masonry through wedge action and friction with the walls of a pilot hole by expansion. This action is produced by striking or turning the threaded machine bolt insert. Expansion bolts act in a manner similar to screws used with lead, fiber, or plastic shields, but usually develop much greater withdrawal strengths (Fig. 5.2-4).

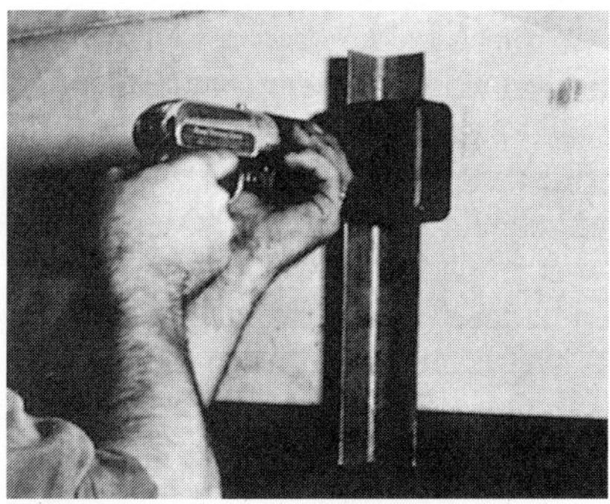

FIGURE 5.2-3 Powder-actuated fasteners are often used to secure steel members to concrete and masonry surfaces.

ADHESIVE-HELD FASTENERS

Adhesive-held fasteners are designed to be held in place by the action of an adhesive. A device consisting of two glass containers, each holding one component of a two-part adhesive, is inserted into a hole drilled in concrete. The desired fastener is driven into the hole, breaking both parts of the glass container and permitting the components of the adhesive to mix and begin curing. When the adhesive has fully cured, the fastener is held securely in place.

MECHANICALLY FORMED JOINTS

Joints in thin sheets are often made without the use of fasteners. Instead, a part of the material itself is used to make the connection. Two common methods are staking and lock-seam joining.

Staking is a method of joining flat materials with fold-over tabs inserted through

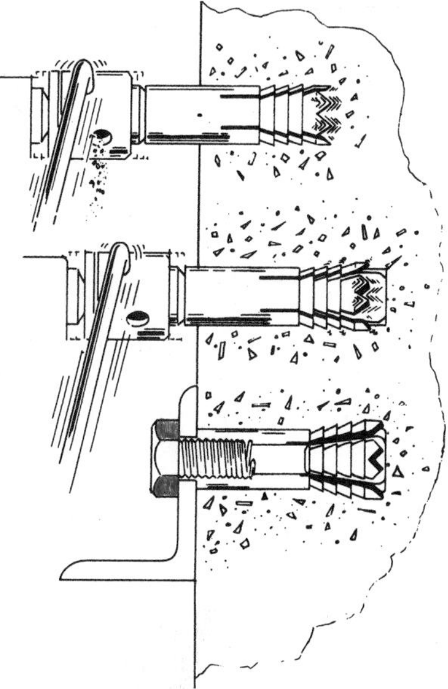

FIGURE 5.2-4 Self-drilling expansion anchors speed the joining of steel to concrete. (Courtesy ITW Ramset/Red Head.)

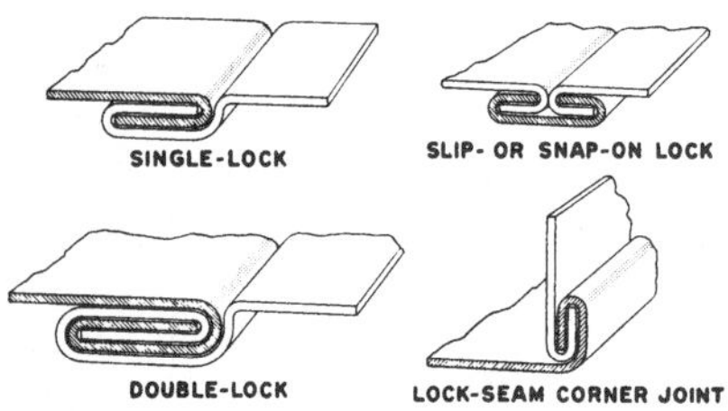

FIGURE 5.2-5 Typical methods of lock-seam joining of sheet metal components in product manufacture. (Courtesy American Iron and Steel Institute.)

matching slots in the elements to be joined. It is often used in product fabrication.

Lock-seam joining is a mechanical method of joining sheet metals in both shop and field by bending and interlocking the elements (Fig. 5.2-5). This is a common method of fabricating air ducts for heating and air conditioning, as well as roofing and roof drainage components.

5.2.1.2 Welding, Brazing, and Soldering

Welding, brazing, and soldering are methods of joining two or more metal components. Welded joints are made by melting the edges of the components to be joined and adding filler metal if required. Brazing and soldering generally employ nonferrous filler metals that have melting points below those of the base metals and do not require melting of the base metal to effect a bond. When the filler metal has a melting point below 800°F (426.7°C), the joining method is considered soldering; if above, it is brazing. Brazed connections can be as strong as welded joints, depending on service requirements. The brazing filler metal flows into a joint of any configuration by capillary attraction. Soldering and brazing are most often used in product manufacture; welding is used both in manufacturing and in structural applications.

Steel product manufacturers employ many different welding, brazing, and soldering processes. The following discussion, however, is limited to common techniques used in building construction.

Structural joints in building frames may be either welded or bolted. The decision as to whether to use welding or bolting to form a particular joint rests with the structural designer. Welding is probably best for most shop joints because control of the surrounding environment is easier than it is in the field and welding is usually more economical in the shop than bolting. For example, computer-driven robots can sometimes be used for shop welding, reducing labor costs. When joints must resist both shear and moment forces, welding usually produces cleaner and less complicated joints.

The thickness and number of welds in a particular joint is decided by the structural designer, based on the forces at the joint.

WELDING TECHNIQUES

Most welding related to building construction can be classified in two main processes: *arc* and *resistance*, both of which are *electric* welding processes. *Arc welding* depends on the heat of an electric arc established between an electrode and the components to be welded. It is most commonly used in the field for connecting steel structural components. Steel decking is also welded in place using arc welding.

Welding is an exacting process requiring care and skill. Welders should therefore be well trained and periodically tested to ensure their competence. In addition, each weld should be inspected to ensure that it was properly made. These inspections may be done visually for minor conditions, but for structural joints more sophisticated inspection methods are usually employed. These include radiographic, ultrasonic, magnetic particle, and dye-penetrant testing, each of which will detect hidden flaws and voids within welds. The decision as to which type of testing is appropriate rests with the structural designer.

Resistance welding is generally a shop process in which the components are butted together and heat is generated in the seam by the resistance of the metal to the passage of an electric current. An electrode in contact with the component introduces the current. Welded wire fabric is usually produced using resistance welding.

Oxygen and nitrogen in the air have a detrimental effect on hot metal in a weld area. Most arc processes *shield* the weld area from atmospheric contamination with inert gases or granular fusible materials. *Shielded metal arc welding* (also known as *manual metal arc* or *coated stick electrode welding*) is the most common arc process (Fig. 5.2-6a). This process uses a consumable metal stick electrode that is gradually melted by the arc and deposits filler metal in the weld. A chemical coating on the electrode releases an inert gas, which forms a shielding envelope around the weld area. This process depends on manual replacement of the stick electrode as it is consumed.

a

b

c

FIGURE 5.2-6 (a) Shielded metal arc welding. (b) Metal inert-gas-shielded welding (MIG). (c) Submerged arc welding. (Courtesy American Iron and Steel Institute.)

In *semiautomatic* processes using self-advancing wire electrodes, a wire comes off a spool at a predetermined rate and goes through a hand-held welding gun. The two chief processes of this type are *metal inert-gas-shielded arc* (MIG), or *gas-shielded metal arc* welding (Fig. 5.2-6b), and *submerged arc* welding (Fig. 5.2-6c). In MIG, an inert gas or mixture of gases (argon, helium, a combination of

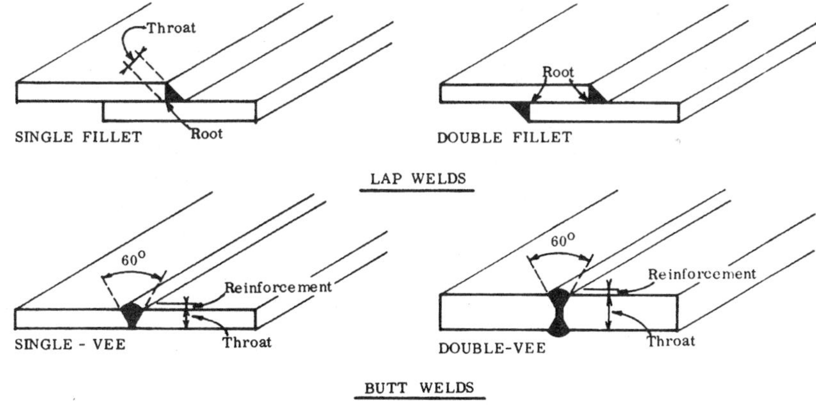

FIGURE 5.2-7 Welded joints common in structural assemblies. (Courtesy American Iron and Steel Institute.)

both, or carbon dioxide) from an external supply is directed through a gun on the weld area to shield it from atmospheric contaminants. In submerged arc welding, protection is achieved by burying the arc and immediate weld area under a blanket of fusible granular material. Arc heat melts the material to form a layer of protective slag over the molten metal. The slag and loose material are removed after welding has been completed. In MIG, filler metal is provided by a consumable electrode; in submerged arc welding, either a consumable electrode or a separate filler rod may be used.

The most common welded joint types are shown in Figure 5.2-7. Exposed welds are often ground smooth if the welds will permit it. The welding processes used with various types of steel are outlined in Figure 5.2-8. Symbols are used on drawings to denote the types of welds required (Fig. 5.2-9).

Spot welding is a type of resistance welding in which the components are joined by a series of spot welds, rather than a continuous line of weldment as in other techniques. It is produced by the resistance heating of a small area between two cylindrical electrodes under pressure.

Seam welding is actually a series of overlapping spot welds produced under pressure by a pair of disc electrodes, as in welded pipe manufacture.

FIGURE 5.2-8 Welding Processes Used to Join Various Types of Steel

Type of Steel	Metal-Shielded Arc	Submerged Arc	Metal-Inert Gas-Shielded (MIG) Arc	Spot, Seam, and Projection
Low Carbon				
Sheets	Common	Common	Common	Common
Plates and bars	Common	Common	Common	Occasional
Medium Carbon*				
Sheets	Common	Common	Common	Common
Plates and bars	Common	Common	Common	Rare
High Carbon*				
Sheets	Common	Occasional	Occasional	Common
Plates and bars	Common	Occasional	Occasional	Not used
Low Alloy*				
Sheets	Common	Common	Common	Occasional
Plates and bars	Common	Common	Common	Rare
Stainless*				
Sheets	Common	Occasional	Common	Common
Plates and bars	Common	Common	Common	Rare

*Preheating and/or post-heating (stress relieving) structures may be required in the weld zones, to prevent embrittlement and excessive residual stresses on certain carbon and low alloy steels, and to stabilize in order to prevent corrosion on stainless steels.

(Courtesy American Iron and Steel Institute.)

Projection welding is also similar to spot welding, but makes use of preapplied projective embossments on one of the components to localize the heat and effect a weld in the desired spot.

EFFECTS OF WELDING ON STEEL

Because heat is an essential element of all welding techniques and metals respond to heat exposure in varying degrees, some steels are more weldable than others. The degree of weldability is related to the sensitivity of the metal to heat, as reflected in the soundness of the weld and the degree of undesirable side effects produced by the operation, including:

- Cracks in the base and weld metals
- Decreases in the base metal's physical properties (strength, ductility, corrosion resistance, etc.)
- Finished surface oxidation

The most important element affecting the weldability of steel is its carbon content. Low-carbon steels (less than 0.15%) are the most weldable; mild steels (0.15% to 0.30% carbon) are next in weldability and can be welded by most processes up to a section thickness of ½ in. (12.7 mm). Thicker sections require greater input of heat to effect a weld, and this may have detrimental results unless proper procedures and care are used. All ASTM structural steels have a carbon content of less than 0.35% and, therefore, can be welded.

Although all stainless steels can be welded by some method, austenitic steels are the most weldable. Almost any commercial welding method can be used to join these steels without problems of soundness and porosity, or loss in toughness or ductility. However, the high rate of thermal expansion in austenitic steels must be considered in order to control distortion. Another hazard is the reduction of corrosion resistance in the weld area, because some of the chromium becomes ineffective through reaction with carbon. This often occurs when stainless steels are exposed to high welding temperatures over extended periods. Such an effect can be minimized by limiting the heat input, promoting heat dissipation, and selecting low-heat-resistance welding methods. Annealing of the finished assembly followed by fast cooling also restores corrosion resistance.

5.2.1.3 Adhesive Bonding

Many decorative and protective laminated coatings are applied to metal base materi-

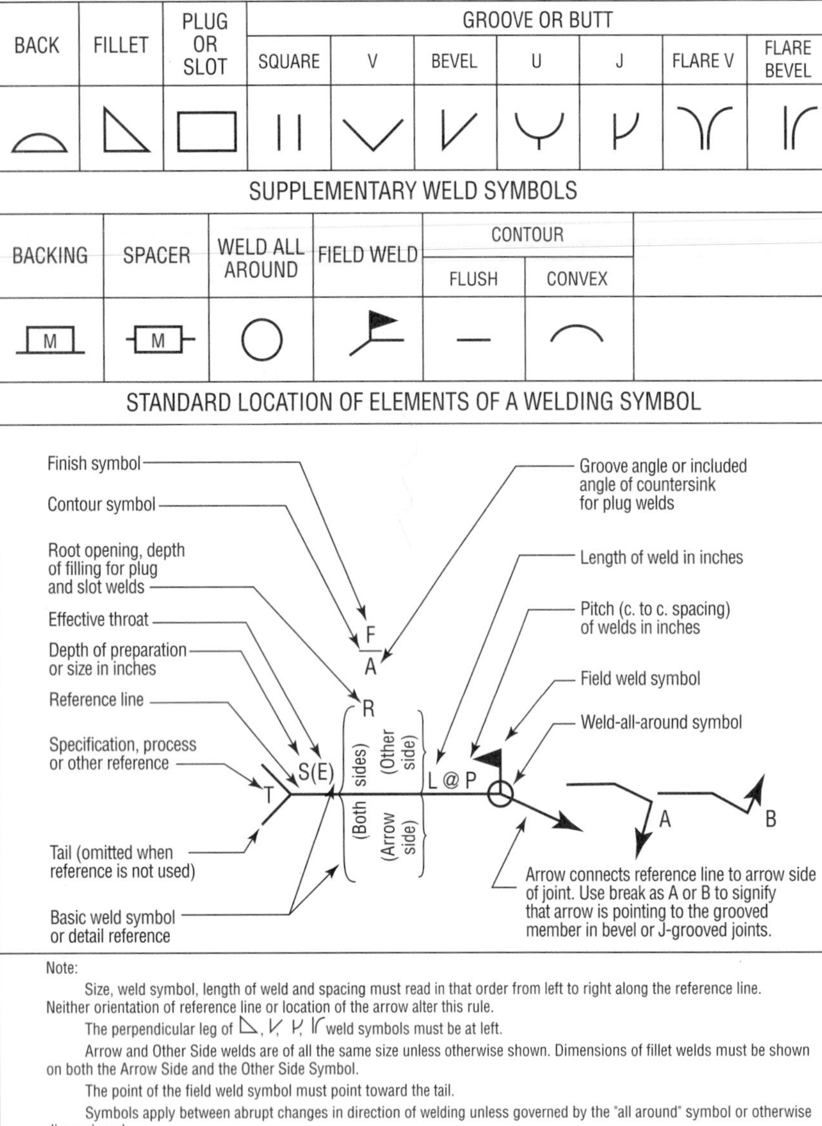

FIGURE 5.2-9 Standard welding symbols. (Courtesy American Institute of Steel Construction, Inc.)

als using adhesives. Adhesive bonding is also used to secure thin carbon, galvanized, and stainless steel sheets and foils to nonferrous materials that provide the required rigidity, flatness, strength, or insulating qualities.

Carbon steel surfacing adds abrasion and indentation resistance to softer core materials. Steel-clad honeycomb doors and steel bonded to plywood panels for column and radiator covers are common examples; galvanized or vitreous enamel-coated steel sheets, bonded with rigid urethane foam to form sandwich panels, are often used in building construction.

Stainless steel sheets also are bonded to rigid or insulating materials in the manufacture of many products with corrosion-resistant finishes. Very thin stainless steel foil is laminated to kraft paper for use in packaging where additional strength and impermeability to vapor are important considerations.

The chief advantages of adhesive bonding are uniform distribution of stresses, absence of surface fasteners, and economy of mass production. Ease of joining small parts not accessible for welding, and low weight in relation to strength make this a valuable structural joining method in the aircraft industry. However, the high initial cost of specialized equipment, low peel and creep (flow under stress) strengths, sensitivity to certain temperatures (characteristic of low-cost adhesives) at present limit their use in building construction mainly to the manufacture of panelized products.

The most common adhesives used in the lamination of steel to other materials are epoxies, phenolic resins, rubber polymers modified by resins, polyvinyl acetates, and chemically cured neoprenes. The most promising adhesives for laminating steel are the polyurethanes, which, with epoxies and phenolics, withstand considerable loads at a greater temperature range and come closest to the concept of "structural" adhesives. These adhesives often possess high peel strength, chemical inertness, resistance to wetting action, and shear and tensile strengths greater than the material bonded. Under certain conditions they may be used for joining structural members in building construction.

Adhesives are available in a variety of forms such as pressure-sensitive tapes, films, liquids, pastes, and solids. They can be applied manually with ordinary building tools or with specialized mass-production equipment that spray, roll, or flow the adhesives onto component surfaces. Some acquire permanent mechanical properties when heated (thermosetting adhesives); others set at moderate temperatures but can be softened repeatedly at high temperatures (thermoplastics). Still other adhesives depend on the chemical reaction between two agents to acquire their permanent characteristics.

Although many adhesives can perform well under a given set of service conditions, there is no universal adhesive and each must be selected to meet the anticipated operating requirements.

5.2.2 STEEL STRUCTURES AND CONSTRUCTION SYSTEMS

Conventional steel-framed structures are built from hot-rolled beams, columns, and girders, combined with bar joists and steel decking. Sometimes these components are combined into complete preengineered buildings that are shipped to the site broken down into units and fastened together in the field to form a complete building

FIGURE 5.2-10 Large-span atrium spaceframe supporting a tension fabric cladding. (Hyatt Regency Hotel, Burlingame, California; architect, Hornberger Worstell and Associates; photograph courtesy Mero Structures, Inc.)

frame and shell. Small structures and parts of larger structures carrying light loads are sometimes erected using lightweight cold-formed framing.

Long spans are achieved in steel using several different types of systems. Conventional framing can be stretched using plate girders and castellated beams. Roofs may be framed with long-span trusses. Special areas, such as arenas, stadium roofs, and airline and train terminals, may be spanned with domes, space frames (Fig. 5.2-10), or steel cable systems.

5.2.2.1 Design

Steel building frames may be designed by either architects or structural engineers, but architects seldom design frames for large buildings or space frames, steel-cable roof systems, or trusses.

THE PROCESS

The designer produces a set of working drawings (Fig. 5.2-11) and specifications detailing the type of steel, basic requirements for fasteners, the location and sizes of the basic elements of the frame (columns, girders, beams, etc.), the basic dimensions of the frame (column centers, floor heights, etc.), and unusual erection details.

Using the designer's drawings and specifications as a guide, a fabricator, employed by a building contractor or by a structural steel subcontractor, produces a set of shop drawings showing the same basic information as on the designer's drawings, but adding information about joints, member lengths, and all other details necessary to fabricate and erect the frame. The shop drawings also show details of methods and bolts necessary to anchor the frame to the foundations.

These shop drawings are then sent by way of the contractor to the designer for review. If revisions are necessary, the designer marks the drawings and returns them to the contractor for revision and resubmittal. This process continues until the designer finds the shop drawings acceptable. The designer then places a stamp on the drawings, stating that the drawings are

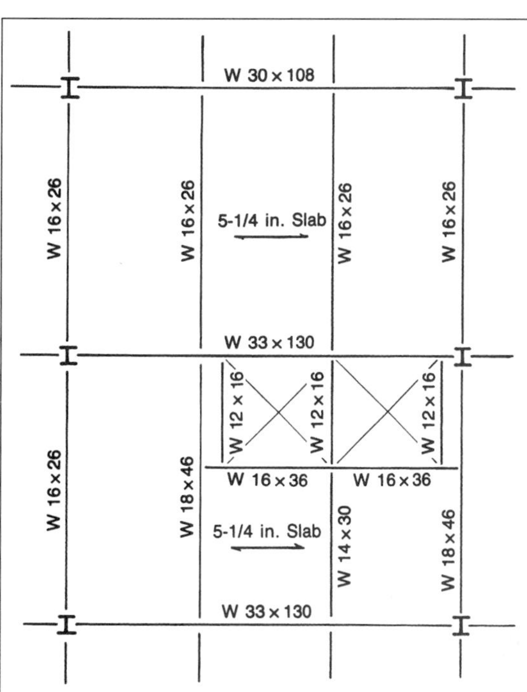

FIGURE 5.2-11 Part of a designer's steel-framing plan. Squares with Xs are elevator shafts. Column sizes and types are usually shown in tables. (Drawing by HLS.)

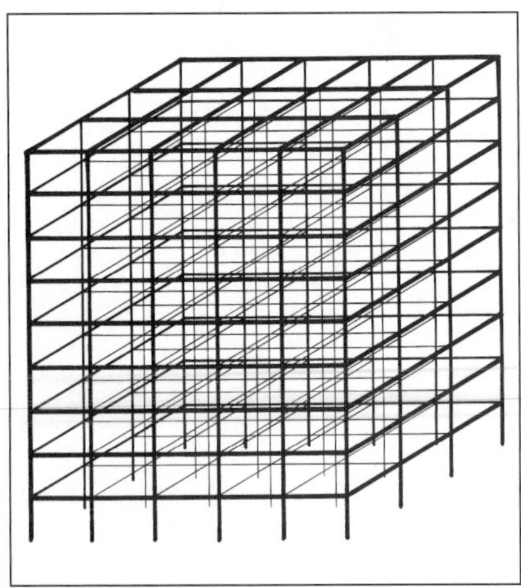

FIGURE 5.2-12 **Typical steel-framing grid.** (Drawing by HLS.)

acceptable, and returns them to the contractor for return to the fabricator.

CONVENTIONAL STEEL FRAMES

The structural frame for a conventional steel-framed building is composed of columns, girders, and beams fastened together to form a three-dimensional grid (Fig. 5.2-12).

Under normal conditions, the joints in conventional structural steel building grids are subjected to both shear and moment forces. Unless some means is employed to resist these forces, the frame will be inherently unstable and subject to wracking and ultimate collapse. The American Institute of Steel Construction (AISC) lists three methods for solving these inherent difficulties as types of steel-frame construction. In Type 1, connections must be made sufficiently rigid to resist all forces that will occur without significant change in the angle between framing members (beams, girders, columns). In Type 2, diagonal bracing or shear panels are introduced to resist moment forces. Thus, most joints need resist only shear forces. Type 3 is a semirigid system in which the joints are sufficiently rigid to resist the applicable moment forces, but not as rigid as those required in Type 1.

Joints that are required to resist only shear are usually just bolted (Fig. 5.2-13). Joints required to resist both moment and shear forces can be fastened using bolts

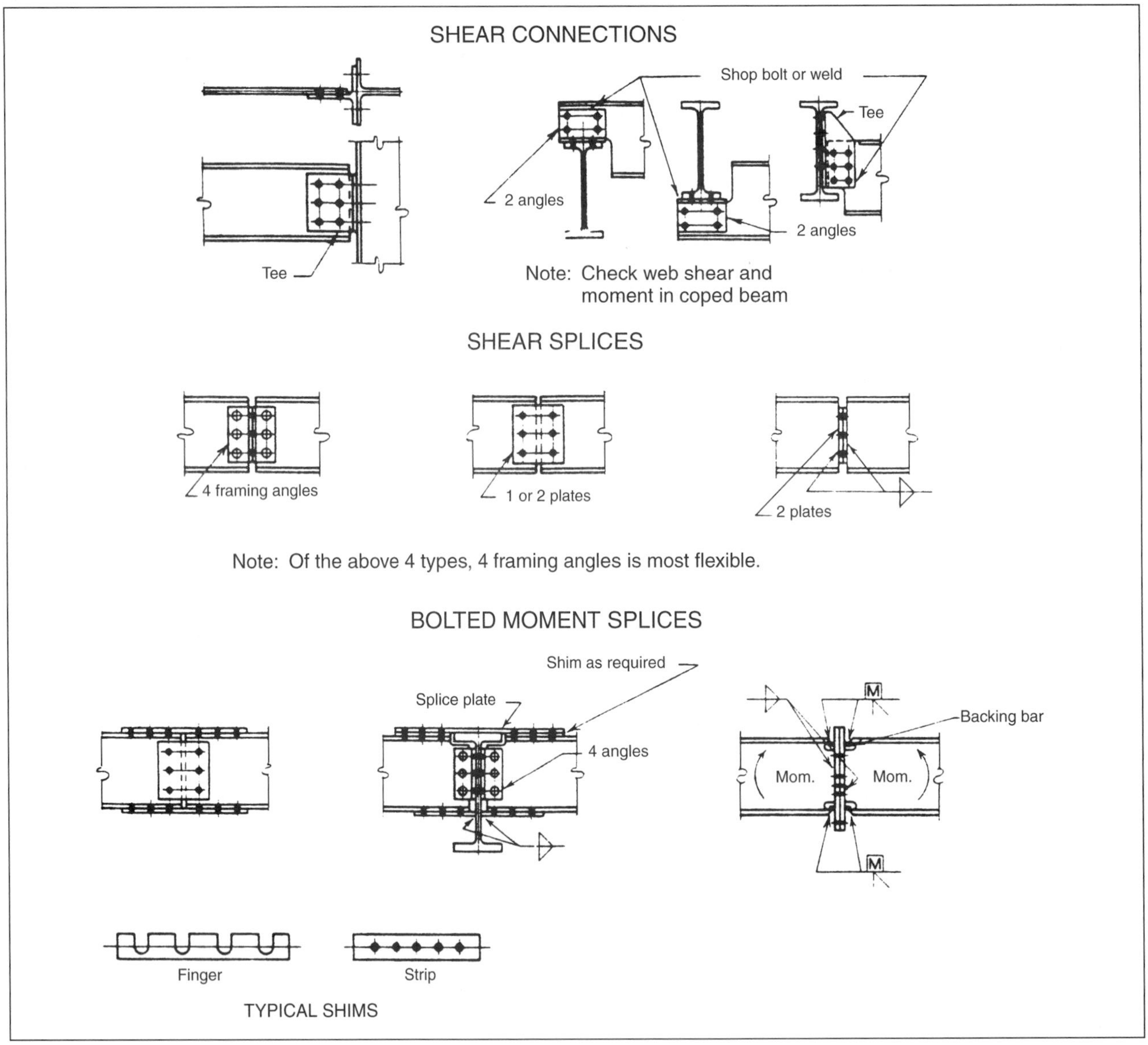

SHEAR CONNECTIONS

Shop bolt or weld

Tee

2 angles

Tee

2 angles

Note: Check web shear and moment in coped beam

SHEAR SPLICES

4 framing angles

1 or 2 plates

2 plates

Note: Of the above 4 types, 4 framing angles is most flexible.

BOLTED MOMENT SPLICES

Shim as required

Splice plate

4 angles

M

Backing bar

Mom.

Mom.

M

Finger

Strip

TYPICAL SHIMS

FIGURE 5.2-13 **Details of shear connections.** (Courtesy American Institute of Steel Construction, Inc.)

alone, but such joints are both complicated and expensive. Therefore, joints resisting both shear and moment forces are either welded or, more frequently, both bolted and welded (Fig. 5.2-14). Such joints are more expensive to produce than those designed to resist only shear forces. For this reason, an attempt is usually made, especially on large buildings, to reduce the number of moment joints in a steel frame.

Such a reduction in the number of moment joints in a building is often accomplished by making part of the frame rigid and tying the rest of the frame to the rigid portion by means of a rigid floor system, such as steel decking with a concrete topping. Such a floor system must be de-

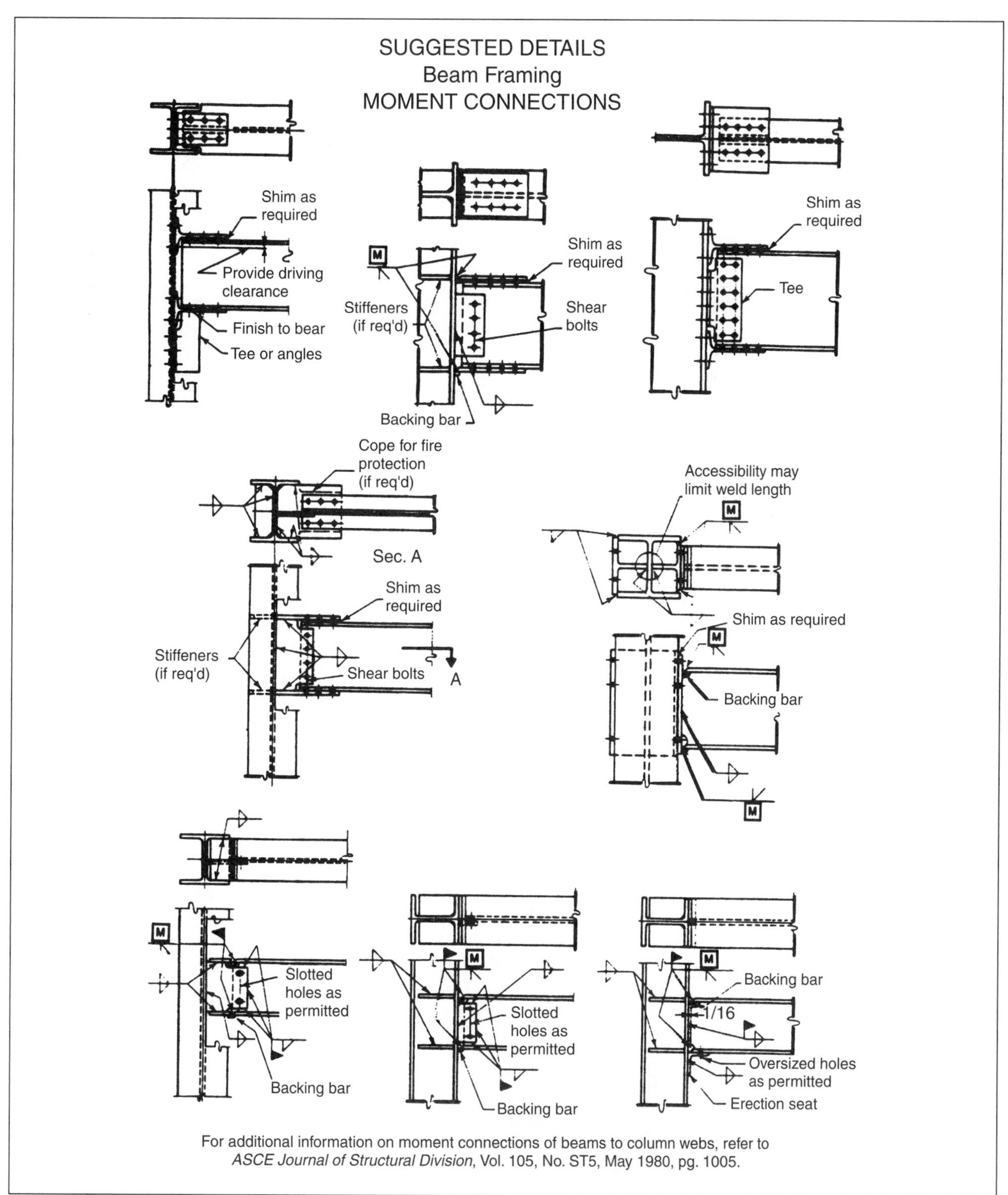

FIGURE 5.2-14 Details of moment connections. (Courtesy American Institute of Steel Construction, Inc.)

signed structurally as a diaphragm, meaning that it acts as a rigid unit to distribute horizontal forces acting on the building to the rigid portion of the frame that is, in turn, capable of resisting those forces and transmitting them harmlessly to the ground.

The rigid portion of a building frame may be either in the perimeter or in the core. The core is that portion of a building that houses elevator and stair towers and service facilities, such as toilets. The amount of the frame that must be made rigid and its location are functions of the design and are the responsibility of the structural designer.

The three basic means of creating a rigid core or perimeter of a steel building frame are diagonal bracing, shear panels, and moment connections. Most of the rest of the frame can usually be fastened with shear-only connections, although there are some exceptions. For example, cantilever beams usually require moment connections.

Diagonal bracing consists simply of steel members jointing opposite corners of a structural bay (Fig. 5.2-15) and thus forming rigid triangles within the bay. Because the triangular structure itself cannot distort, moment joints are not necessary to attach the bracing to the frame.

Shear panels often consist of rigid concrete walls, but are sometimes constructed of braced steel framing. In either case, they must resist the forces imposed on them without distorting.

Moment connections can be used to form a rigid portion of the frame with the same result as that obtained with diagonal bracing or shear panels. While structural design and safety must be considered, the final decision about the method used to stabilize a steel-framed building is often based on which method is the least expensive.

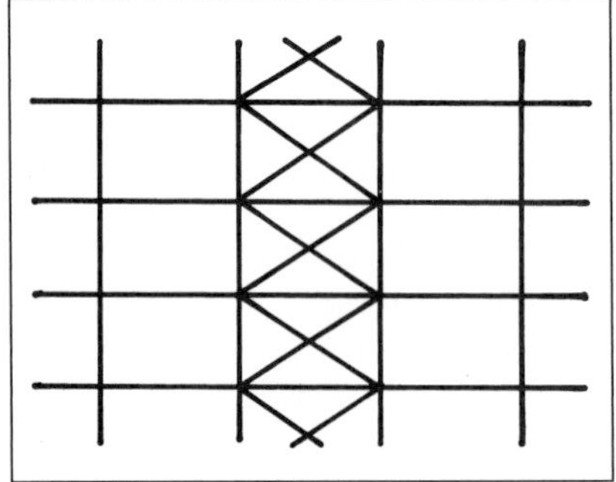

FIGURE 5.2-15 Diagonal bracing. (Drawing by HLS.)

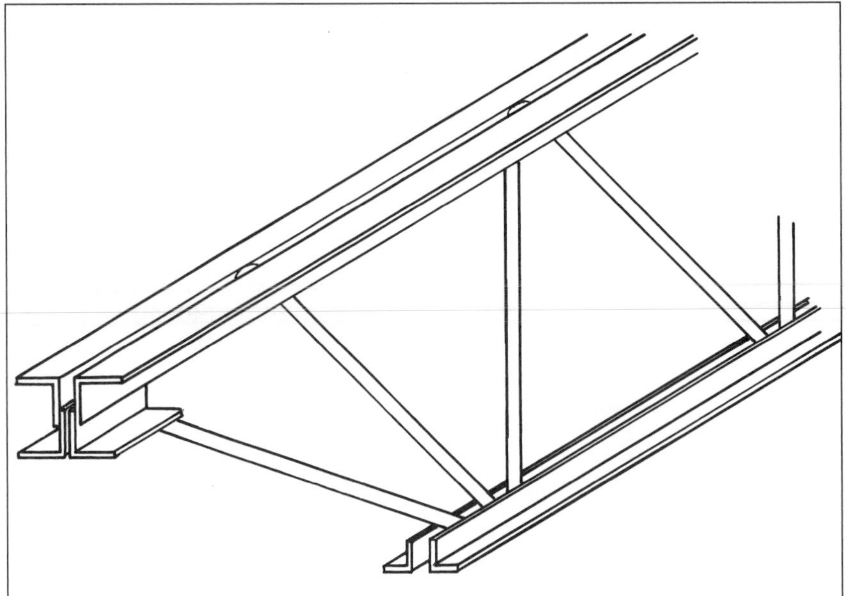

FIGURE 5.2-16 Bar joist. (Drawing by HLS.)

STEEL JOISTS AND JOIST GIRDERS

Standard open-web steel joists (Fig. 5.2-16) are used for floor and roof framing mostly in masonry or steel-framed buildings (Fig. 5.2-17). Joist girders are open-web trusses designed for longer spans and as primary framing members in lieu of beams. Both are usually built from high-strength steel to standard sizes. Selection is made by a structural designer from tables produced by the Steel Joists Institute (SJI).

Joists are classified as K-Series, LH-Series, and DLH-Series. K-Series joists range in depth from 8 to 30 in. (203.2 to 762 mm) in 2-in. (50.8 mm) increments. They are designed to span up to 60 ft. (18.29 m). Except where their ends abut over steel beams where bolts and welding can be used to provide additional support, K-Series joists must be supported at least 2½ in. (63.5 mm) on steel bearing plates.

FIGURE 5.2-17 Bar joist being placed on joist girders. (Courtesy of the Vulcraft Division of Nucor Corporation.)

LH-Series (longspan) joists are available in depths from 18 to 48 in. (457.2 to 1219.2 mm). The deeper ones can span up to 96 ft. (29.26 m). LH-Series joists must be supported by at least 4 in. (101.6 mm) of steel beam or 6 in. (152.4 mm) of steel bearing plate when resting on concrete or masonry.

DLH-Series (deep longspan) joists range in depth from 56 to 72 in. (1422.4 to 1828.8 mm). Some can span as much as 144 ft. (43.89 m). Bearing requirements for DLH-Series joists are the same as those for LH-Series joists.

Joists are usually spaced about 2 ft. (609.6 mm) apart, but spacings as wide as 10 ft. (3048 mm) sometimes occur when the floor or roof decking can accommodate such wide spacings. Joists usually require bridging to prevent them from twisting or overturning. This can be accomplished by horizontal angles or rods running across the bottom chords or by diagonal steel bracing from joist to joist. Bridging is usually welded in place.

ROOF AND FLOOR DECKING

Many types of products are used as roof decks on steel-framed buildings. Exam-

FIGURE 5.2-18 **Steel deck being placed on steel joists and joist girders.** (Courtesy of the Vulcraft Division of Nucor Corporation.)

ples include the poured-in-place gypsum concrete decks and the lightweight aggregate board and wood fiber board decks described in Chapter 3. A major type of roof decking used on steel-framed buildings is corrugated steel decking, either with or without a concrete topping (Fig. 5.2-18).

Concrete slabs are often used as floor decking in steel-framed buildings. Sometimes such slabs are formed using temporary formwork, as described in Chapter 3. More frequently, floor slabs are placed over corrugated steel decking.

Many different configurations of corrugated steel decking are available, depending on the use and span. *The Design Manual for Composite Decks, Form Decks, and Roof Decks* of the Steel Deck Institute (SDI) spells out the minimum requirements for each type.

Steel decking is usually fastened in place by welding through it to the supporting steel. In thin metal decking, and where greater strength is required, welding washers are used. Occasionally, decking is fastened in place with mechanical fasteners. When a deck must act as a diaphragm, its panels are sometimes fastened together at the laps between panels with screws or welds. The method of attachment to be used and whether deck laps need fastening are the responsibility of the structural designer. The important consideration is to ensure that the deck and its fastenings will adequately resist all forces that may act on them.

Roof Decking Roof decking is made from 22-, 20-, 18-, and 16-gauge (0.85, 1.0, 1.3, 1.6 mm) steel in depths of 1½ in. (38.1 mm) to 7½ in. (190.5 mm) (Fig. 5.2-19). It can span from 4 ft. 6 in. (1371.6 mm) for the lightest decking to as much as 30 ft. (9.14 m) for heavy, deep, cellular decking. Rib openings in 1½-in. (38.1 mm) roof decking are either 1 in. (25.45 mm) wide (*narrow rib*), 1¾ in. (44.45 mm) wide (*intermediate rib*), or 2½ in. (63.5 mm) wide (*wide rib*). Corrugated sheet steel roof decks are either *open flute* (Fig. 5.2-19a) or *cellular* (Fig. 5.2-19b), meaning that they have a sheet steel plate affixed to the bottom of the decking. *Acoustical roof decking* is also available. Some of it is of the open-flute type and some is of the cellular type. The open-flute type has openings in the vertical webs. Openings in the cellular type are in the bottom plate. The concealed spaces in acoustical roof decking are filled with sound-absorptive matter, usually glass-fiber insulation.

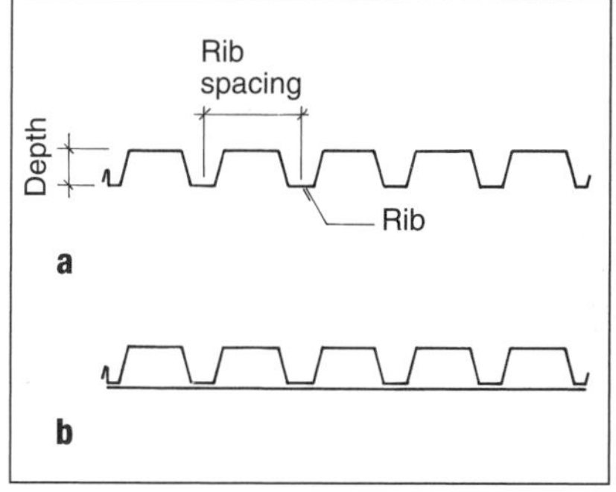

FIGURE 5.2-19 **Roof deck: (a) open flute type; (b) cellular type.** (Drawing by HLS.)

Floor Decking Corrugated steel floor decking is available in three different types: *form deck*, *composite deck*, and *composite cellular deck*.

Form deck is used as a permanent form for a concrete slab (Fig. 5.2-20a). It varies in thickness from ½ in. (12.7 mm) to 2 in. (50.8 mm). Spans range from 1 to 2 ft. (304.8 to 610 mm) for light ½-in. (12.7mm) form decking, to as much as 12 ft. (3.66 m) for heavy 2-in. (50.8 mm)-thick decking with 6-in. (152.4 mm) rib spacings. Form deck adds no strength to the slab. It simply reduces the amount of work involved in forming and eliminates the need to strip the forms from a slab.

Composite decking is designed not only to support a concrete slab while it is being placed, but also to bond with it, providing tensile reinforcement for the composite structure (Fig. 5.2-20b). The resulting longer slab spans and greater strength produce a more economical structure than a noncomposite construction. Sometimes the composite nature of the system is further enhanced by welding

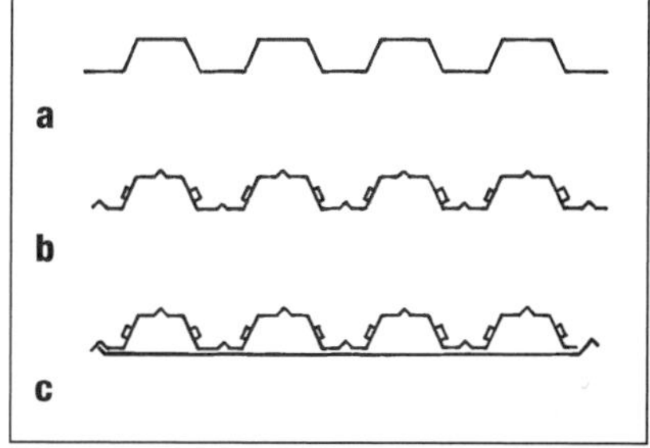

FIGURE 5.2-20 **Floor decks: (a) Noncomposite form deck. (b) Composite deck. The indentations and projections act to produce a bond with the concrete slab as it is placed. The configuration of these indentations and projections varies from manufacturer to manufacturer. Some composite decking has steel rods welded across the top of the flutes. (c) Composite cellular deck. Many configurations are available.** (Drawing by HLS.)

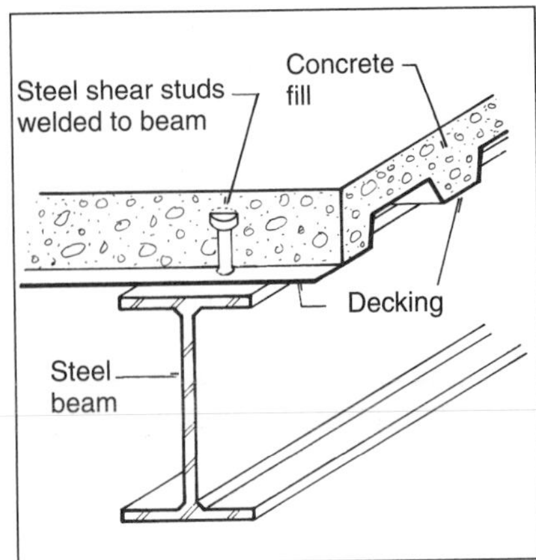

FIGURE 5.2-21 Composite construction shear studs. (Drawing by HLS.)

shear connectors to the tops of supporting beams, thus bonding the concrete slab to the steel structural members as well as to the decking (Figs. 5.2-21 and 5.2-22).

Cellular composite decking works in essentially the same way as noncellular composite decking (Fig. 5.2-20c). In addition, cellular decking can be designed to house electrical wiring or to provide acoustical treatment in the same way that cellular roof decking does.

STEEL TRUSSES

Steel trusses are manufactured from steel structural shapes to produce units that are lighter in weight than comparable solid steel construction, such as welded steel plate girders. Most trusses form roof structures rather than floors.

The structures known as *space frames* are actually three-dimensional trusses, designed to span in two directions simultaneously.

LIGHTWEIGHT STEEL FRAMING

Section 9.5 addresses steel studs used primarily for non-load-bearing partitions. Heavier *cold-formed* steel studs, joists, and accessories, capable of carrying loads other than themselves and their finishes, are also available. These provide a nonflammable and dimensionally stable structural system for light loads. Although compatible with conventional stud or bearing wall construction, lightweight steel framing also introduces the possibility of longer spans and large-scale prefabrication (Fig. 5.2-23).

PREENGINEERED BUILDINGS

Preengineered buildings are primarily used as industrial facilities, but may also be used as commercial buildings. They typically combine rolled steel sections, open-web joists or joints-girders, and sheet steel roofing and siding,

5.2.2.2 Fabrication

After receipt of shop drawings bearing the designer's stamp of acceptance, a fabricator produces full-size templets of wood or cardboard to assist in the positioning of bolt holes and cutouts, and begins the fabrication process.

Fabrication consists of measuring and cutting beams, columns, girders, base plates, and other components to the proper length and configuration; punching or boring holes for bolts; welding on stiffeners, connection devices such as angles, and smaller base plates; beveling of joints for welding; flattening and squaring of bearing surfaces for tight fit; cambering (curving slightly upward) of beams and girders where the design requires it; and assembly of large components such as plate girders, trusses, and built-up columns.

5.2.2.3 Erection

The various components of most steel-framed structures are fabricated in the shop, delivered to the construction site as individual components (beams, girders, joists, decking, etc.), and fastened together in the field (Figs. 5.2-24 and 5.2-25).

The procedures and equipment used in the erection of steel-framed buildings are the responsibility of the *erector*. The actual work is done by *ironworkers* (sometimes called *steelworkers*) (Fig. 5.2-26).

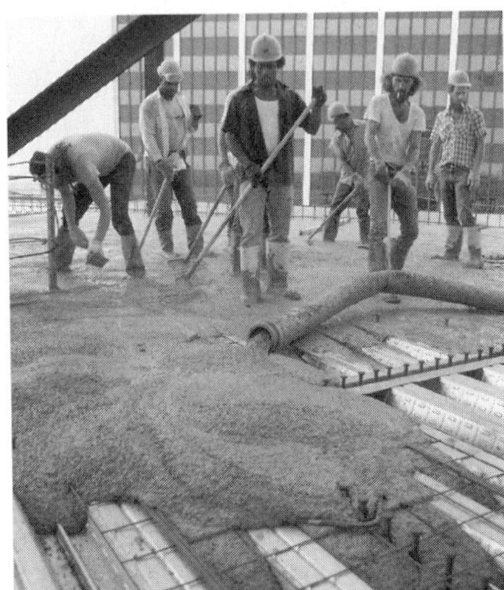

FIGURE 5.2-22 Concrete being placed on a composite steel deck. (Courtesy of the Vulcraft Division of Nucor Corporation.)

FIGURE 5.2-23 Prefabricated steel stud wall assembly can often be lifted in place by one or two workers. (Courtesy American Iron and Steel Institute.)

FIGURE 5.2-24 An ironworker high on a steel frame building. (Courtesy Bethlehem Steel Corporation.)

FIGURE 5.2-25 A complicated steel frame under construction. (Courtesy Bethlehem Steel Corporation.)

a

b

FIGURE 5.2-26 (a) Ironworkers wrestling a steel column into place. (b) Ironworkers welding steel members together. (Courtesy Bethlehem Steel Corporation.)

The designer should require that layout and positioning of the steel, positioning of anchor bolts, and establishing and checking of bearing levels and other elevations be done by a licensed surveyor. The erection should be accomplished in accordance with the requirements and within the tolerances established by AISC, unless the structural designer specifically specifies or approves deviations from AISC requirements.

Structural steel is placed in position in the field with cranes or other lifting devices. On fairly low buildings, ground-mounted cranes are used to erect the entire structure (Fig. 5.2-27).

On taller structures, either guy derricks (Fig. 5.2-28) or climbing cranes (Fig. 5.2-

FIGURE 5.2-27 Steel being erected using a ground-mounted crane. (First Indiana Plaza, Indianapolis, Indiana; architect, CSO Architects, Inc.; design architect, 3DI International; photograph courtesy of the Vulcraft Division of Nucor Corporation.)

FIGURE 5.2-28 Guy derricks are lifted onto the structure by a ground-mounted crane when the building has reached the practical limit of the ground-mounted crane. A guy derrick rests on the structure and is stabilized by guys fastened to the structure. Components of guy derricks are moved upward to a new level as the frame progresses. To move a guy derrick upward, the boom is disconnected from the mast, temporarily supported on the structure, and guyed to stabilize it. The mast is then lifted by the boom onto a higher level, where it is supported on the structure and stabilized. The crane is then reassembled. (Courtesy Bethlehem Steel Corporation.)

29), both of which are mounted on the building, are employed.

The general process usually followed in erecting a steel structure begins with the placement of anchor bolts and leveling plates and the placing of grout beneath the leveling plates by the general contractor. Leveling plates are often omitted, however, especially for larger columns.

Each of the lowest-level columns is then transported to its proper position and lowered by crane so that the anchor bolts pass through the corresponding holes in the base plate. On multistoried buildings, the lowest (first *tier*) columns are usually 2 stories high.

After the first-tier columns have been bolted in place and are plumb, the beams and girders for the first two floors are placed and fastened in place (Fig. 5.2-30).

The next tier columns, which are also usually 2 stories in height, are erected and the beams and girders for the next 2 floors are installed (Fig. 5.2-31). This process continues until the entire structure has been completed (Fig. 5.2-32).

FIGURE 5.2-29 A climbing crane rises through an opening in the structure, such as an elevator shaft. Additional sections are added at the top of the crane as the structure grows. (Courtesy Bethlehem Steel Corporation.)

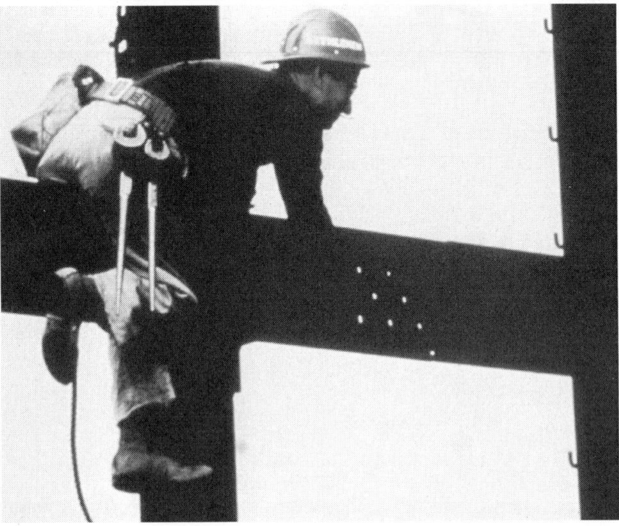

FIGURE 5.2-31 An ironworker guiding a beam into place. (Courtesy Bethlehem Steel Corporation.)

FIGURE 5.2-30 Installing beams. (Courtesy Bethlehem Steel Corporation.)

FIGURE 5.2-32 Erecting a portion of a sloping steel structure. (Courtesy Bethlehem Steel Corporation.)

5.3 Miscellaneous Iron and Steel Elements

In addition to the use of steel and iron in structural systems, ferrous metals are used in many other building products. Those mentioned in this section are among the most important.

5.3.1 SITEWORK

Ferrous metals used in sitework applications include fencing components, barricades, drains, manhole covers, guardrails, corrugated pipe for drainage, and many others.

5.3.2 ARCHITECTURAL ITEMS AND SYSTEMS

Ferrous metal components are used in virtually every architectural building system.

5.3.2.1 Roofing, Siding, Cladding, and Flashing

Both structural and nonstructural ferrous standing-seam metal roofing (see Section 7.9) have a long history of use in several styles. These products offer a broad choice of finishes and coatings for durability, texture, and color.

Siding, in traditional clapboard designs as well as in vertical and other patterns, is an increasingly popular steel product (see Section 7.7).

Wall panel systems (cladding), often in combination with lightweight steel framing, are used in many types of buildings today.

Stainless steel is used for flashing, and carbon steel sheets are used in composite flashing (see Section 7.8).

5.3.2.2 Doors and Windows

Many types and styles of exterior and interior doors and various steel windows are available. Steel doors (see Section 8.1) provide fire safety, security, durability, and dimensional stability. Modern exterior swinging steel doors contain an insulating core surrounded by steel skins and sides. Designs range from smooth flush faces to variations with glazing, moldings, textured surfaces, and vinyl laminations.

Steel-framed windows offer a complete range of designs for residential, commercial, and institutional uses.

5.3.2.3 Interior Finishes

Steel products used in interior finishes include non-load-bearing studs and furring components (see Section 9.5) and acoustical ceilings (see Section 9.8).

Non-load-bearing studs and furring members provide noncombustible, dimensionally stable framing and furring for gypsum wallboard, plaster, and other interior surfaces.

Steel acoustical ceilings are found in many types of buildings, sometimes as an integral part of a steel deck floor and ceiling system.

5.3.2.4 Ornamental Iron Elements

Steel and iron are used to make ornamental building components, such as railings and handrails, panels, and decorative elements.

5.3.2.5 Other Architectural Uses

There are many other steel products used in buildings, including, but not limited to:

- Steel pan (concrete-filled) stairs
- Grating-tread and solid-tread stairs
- Spiral stairs
- Ladders
- Prefabricated fireplaces and flues
- Laboratory, kitchen, and other cabinets
- Unit kitchens
- Rainware (flashing, gutters, and downspouts)
- Roof accessories (skylights, stacks, ventilators, roof hatches, etc.)
- Gratings
- Floor plates
- Sidewalk hatches
- Access doors
- Loose lintels
- Various castings

5.3.3 PLUMBING SYSTEMS; HEATING, VENTILATING, AND AIR CONDITIONING SYSTEMS; AND ELECTRICAL SYSTEMS

Iron and steel are used extensively in plumbing fixtures and piping, in hot water and steam piping, in radiators, valves, and accessories, and in heating and cooling ductwork.

Plumbing ware includes both cast-iron and sheet steel products, such as tubs, shower and tub enclosures, toilets, bidets, and sinks. These are often finished in porcelain enamel. Stainless steel is widely used for kitchen, laboratory, and bar sinks. Sheet steel ductwork, galvanized for corrosion resistance, is part of most heating and cooling systems that employ moving air. Such systems generally have fans, vents, louvers, and other sheet steel accessories. (See Chapter 15 for additional discussion.)

Electrical systems (see Chapter 16) include many steel and iron elements, such as panel and junction boxes, conduit, outlet boxes and outlets, among others.

5.4 Aluminum

Aluminum was first detected in 1807 as an ingredient of common clay, but it was not until 1825 that the first metallic aluminum was produced, and not until 1845 was aluminum powder successfully transformed into solid particles. In 1852, due to costly production methods, pure aluminum sold for $545 per pound and was used only for the finest of jewelry and tableware.

The 1886 discovery that metallic aluminum could be produced by dissolving alumina (aluminum oxide) in molten cryolite and then passing an electric current through the solution gave birth to the modern aluminum industry.

Today, building and other construction make up a large market for aluminum in the United States. Products that use large amounts include windows, doors, storefronts, building cladding systems, screens, siding, and mobile homes.

A unique combination of properties makes aluminum one of the most versatile engineering and construction materials. It is light in weight, yet some of its alloys have strengths greater than that of structural steel. It has high resistance to corrosion and will not rust red to stain adjacent surfaces or discolor products.

It has no toxic action, but possesses high electrical and thermal conductivities and high reflectivity to both heat and light. Aluminum can be cast or worked into a large variety of shapes and accepts many surface finishes. These characteristics give aluminum its versatility, and usually, two or more of them come into play.

One of aluminum's most important attributes is that it functions as designed and maintains its finish over extended periods. For example, an aluminum cap installed on the top of the Washington Monument in 1884 remains in good condition. A great many large and small buildings have been clad with aluminum components, and millions of homes and other small buildings have been sided with aluminum during the past 65 years.

As with other materials, however, aluminum's properties must be understood to ensure its proper use. This section discusses the production of aluminum and aluminum alloys, their various properties, and the methods by which aluminum products are manufactured. Individual aluminum products are discussed in Chapters 7, 8, and 9.

5.4.1 RAW MATERIAL

Aluminum forms 8% of the earth's crust, making it the most plentiful metal, but it is never found free in nature. Unlike gold and silver, aluminum is always combined chemically with other elements. Although nearly all common rocks and clays contain some aluminum, it has not been historically practical to extract aluminum unless the raw ore contained at least 45% aluminum oxide. This percentage may change from time to time, however, as economic conditions vary.

Such ores are called *bauxites*, after the town of Les Baux in France, where one of the first bauxite ore deposits was found. Most of the world's bauxite deposits are outside the United States, making it necessary for this country to import most of its supply. About 95% of the bauxite mined in the United States comes from Arkansas and the rest from small, lower-grade deposits in Georgia and Alabama.

Work is now under way in the United States to develop methods of using domestic ores such as kaolin and alunite as replacements for imported bauxite. The U.S. Bureau of Mines, which is coordinating the project, believes the technology exists to extract aluminum commercially from these domestic ores.

5.4.1.1 Mining

Strip or open pit mining methods are generally used to obtain aluminum ore, which is crushed, washed, screened, ground, and then dried in kilns (large revolving steel drums) at temperatures up to 250°F (121.11°C). The general steps in the mining and refining of aluminum are shown in Figure 5.4-1.

5.4.1.2 Refining

Commercial bauxite refining is done by the Bayer process, which separates aluminum oxide (alumina) from unwanted minerals.

Dried, ground bauxite is mixed with a solution of caustic soda (sodium hydroxide), which dissolves the alumina to form sodium aluminate. The silica in the bauxite reacts and precipitates out of solution. Iron oxide and other impurities are not affected chemically and, being solids, settle out.

This *green liquor* is now highly supersaturated sodium aluminate. Previously prepared hydrated alumina crystals are added to the solution. These form larger crystals, which gradually settle out. After being washed to remove remaining traces of impurities, the concentrated aluminum hydrate crystals are roasted at temperatures of more than 2000°F (1093.3°C), which drives off the water. The resulting alumina is a fine white powder, similar to sugar in appearance and consistency.

5.4.1.3 Reduction

Reduction is the electrolytic separation of aluminum from its oxide, alumina. The general reduction procedure is described in Figure 5.4-1.

This process employs a carbon-lined vessel (pot) containing molten cryolite (sodium aluminum fluoride) in which alumina is dissolved. Metallic aluminum is separated from the alumina by passing an electric current through this solution at a temperature of about 1775°F (963.33°C).

Molten aluminum collects at the bottom of the pot and is tapped or siphoned off into large ladles. It is then poured into molds to form ingots or transferred to holding furnaces for alloying. Aluminum, as it comes from the pots, is about 99.5% pure. By additional refining, super-purity aluminum, which is about 99.99% pure, can be produced.

5.4.1.4 Recycling

Recycled aluminum, metal that has already had a useful life in a product and is used over again, is playing an increasingly important role (see Fig. 5.4-1). Recycling is important because it saves 95% of the energy required to make aluminum from ore. Ingots that come from recycling are called *secondary ingots*.

Old aluminum windows, doors, and siding are among the many products that are being brought to scrap dealers for recycling.

5.4.1.5 Aluminum Alloys

In high-purity form, aluminum has a tensile strength of about 7000 psi (4921.5 Mg/m^2) and is relatively soft and ductile. Most applications, however, require greater strength and other properties achieved by the addition of other elements to produce various aluminum *alloys*.

Aluminum *alloys* can be divided into two general categories: those whose strength characteristics may be improved by heat treatment (*heat-treatable* alloys), and those whose strength may be improved by mechanical methods of cold working (*non-heat-treatable* alloys).

HEAT-TREATABLE ALLOYS

The initial strength of heat-treatable alloys is improved by the addition of alloying elements such as copper, magnesium, zinc, and silicon. Since these elements (singly or in various combinations) show increasingly solid solubility in aluminum with increasing temperature, it is possible to subject them to thermal treatments that will impart further strengthening.

The first step, called *solution heat treatment*, is an elevated-temperature process. This is followed by rapid quenching, usually in water, which momentarily "freezes" the structure and for a short time renders the alloy very workable. It is at this stage that some fabricators retain this more workable structure by storing the alloys at below-freezing temperatures until they are ready to be formed. At room or elevated temperatures the alloys are not stable after quenching, but after a period of several days at room temperature, they are considerably stronger. This is called *natural aging* or *room temperature precipitation*. Some alloys, particularly those containing magnesium and silicon or magnesium and zinc, continue to age-harden for long periods of time at room temperature.

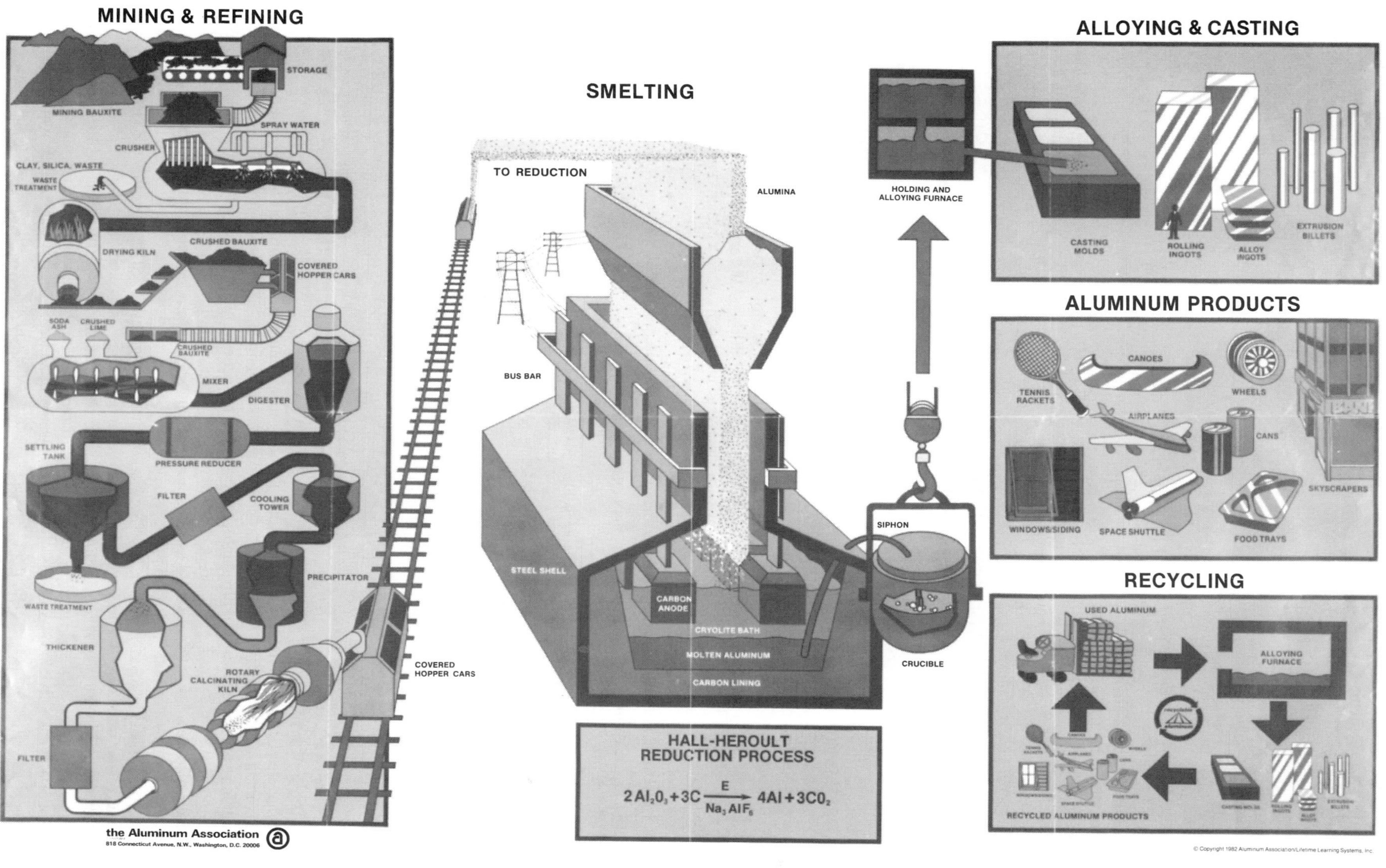

MINING & REFINING
STORAGE
MINING BAUXITE
SPRAY WATER
CRUSHER
CLAY, SILICA, WASTE
WASTE TREATMENT
DRYING KILN
CRUSHED BAUXITE
COVERED HOPPER CARS
SODA ASH
CRUSHED LIME
CRUSHED BAUXITE
MIXER
DIGESTER
SETTLING TANK
PRESSURE REDUCER
FILTER
COOLING TOWER
PRECIPITATOR
WASTE TREATMENT
THICKENER
ROTARY CALCINATING KILN
FILTER
the Aluminum Association
818 Connecticut Avenue, N.W., Washington, D.C. 20006
COVERED HOPPER CARS
SMELTING
TO REDUCTION
ALUMINA
BUS BAR
STEEL SHELL
CARBON ANODE
CRYOLITE BATH
MOLTEN ALUMINUM
CARBON LINING
HALL-HEROULT REDUCTION PROCESS
$2Al_2O_3 + 3C \xrightarrow[Na_3AlF_6]{E} 4Al + 3CO_2$
SIPHON
CRUCIBLE
HOLDING AND ALLOYING FURNACE
ALLOYING & CASTING
CASTING MOLDS
ROLLING INGOTS
ALLOY INGOTS
EXTRUSION BILLETS
ALUMINUM PRODUCTS
TENNIS RACKETS
CANOES
WHEELS
AIRPLANES
CANS
SKYSCRAPERS
WINDOWS/SIDING
SPACE SHUTTLE
FOOD TRAYS
RECYCLING
USED ALUMINUM
ALLOYING FURNACE
RECYCLED ALUMINUM PRODUCTS
© Copyright 1982 Aluminum Association/Lifetime Learning Systems, Inc.

FIGURE 5.4-1 The mining, refining, and reduction of aluminum. (Courtesy Aluminum Association.)

By heating an alloy at a controlled temperature, even further strengthening is possible and properties are stabilized. This process is called *artificial aging* or *precipitation hardening*.

The highest strengths are obtained by the proper combination of solution heat treatment, quenching, cold working, and artificial aging.

NON-HEAT-TREATABLE ALLOYS

Non-heat-treatable (*common*) aluminum alloys have alloying elements that do not show an increase in strength with heat treatment. The initial strength of non-heat-treatable alloys depends on the hardening effects of elements such as manganese, silicon, iron, and magnesium, singly or in various combinations. These alloys can be strengthened by *strain hardening*, *cold rolling*, and other mechanical working.

ANNEALING

Annealing is a thermal treatment that is applicable to both heat-treatable and non-heat-treatable alloys and to cast and wrought products. It is done by elevating the temperature of the metal to between 600°F (315.56°C) and 800°F (426.7°C) and then cooling it slowly. The purpose of this treatment is to relieve internal stresses and to return the alloy to its softest and most ductile condition. Annealing is usually performed to condition the metal for severe forming operations. Subsequent working processes increase the mechanical properties of non-heat-treatable alloys. The mechanical properties of heat-treatable alloys may be increased by appropriate thermal treatments after forming operations have been completed.

CLADDING

To increase the corrosion-resistant properties of some products or to provide for special surface appearance or preparation for additional surface treatment, an aluminum alloy or other metal coating may be metallurgically bonded to an aluminum product. Most cladding is done on sheet and plate products, but some tube, rod, and wire products are also clad.

Aluminum clad with high-purity aluminum or a non-heat-treatable alloy to improve its surface corrosion resistance is designated by its four-digit alloy number (see the following paragraph) preceded by the word *Alclad*. For example, Alclad 3004 is a common sheet and plate material used for building applications.

ALLOY DESIGNATION SYSTEMS

Aluminum products fall into two general groups: *wrought* products and *cast* products.

Alloy and temper are designated by numerical systems for both cast and wrought products. Figure 5.4-2 shows some typical uses for various alloys.

Temper Designations A notation or temper designation is used following the alloy (Fig. 5.4-3). The letter *H* is used for non-heat-treatable and the letter *T* for heat-treatable alloys. Some temper designations apply only to wrought products, others to cast products, but most apply to both. Additional digits after the *T* or *H* (H1, H2, T1, T2, etc.) indicate variations in the basic treatment.

Wrought Alloys Almost all aluminum products used in construction are made from wrought alloys, which are designated by a four-digit index system (Fig. 5.4-4). The first digit of the number indicates an alloy group or series.

The 1000 Series designates aluminum of 99% or higher purity. These alloys have excellent corrosion resistance, high thermal and electrical conductivity, low mechanical properties, and excellent workability. A moderate increase in strength may be obtained by strain hardening. This group includes alloys (1235) used in electrical and chemical fields and in aluminum foils for insulation and vapor retarders; alloy 1100 is frequently used for sheet metal work.

In the 2000 Series, copper is the principal alloying element. These alloys require heat treatment to obtain optimum properties, and in the heat-treated condition their mechanical properties are similar to and may exceed those of mild steel.

Sometimes artificial aging is employed to further increase strength. The alloys in the 2000 Series do not have as good corrosion resistance as most other aluminum alloys and in sheet form are usually clad with a high-purity alloy or a magnesium-silicon alloy of the 6000 Series. Alloy 2024 is used for fasteners; alloy 2014 for structural shapes.

The 3000 Series alloys have manganese as the major alloying element. They are usually non-heat-treatable. Because only a limited percentage of manganese (up to about 1.5%) can be effectively added to aluminum, it is used as a major element in only a few alloys.

Alloy 3003 is a general-purpose alloy used for moderate-strength applications requiring good workability. In sheet form, it is used for air ducts, acoustical ceiling pans, awnings, corrugated decks, garage doors, flashings, gutters, siding, shingles, and other products.

The major alloying element in the 4000 Series is silicon, which substantially lowers the melting point of the alloy without producing brittleness. Most alloys in this series are non-heat-treatable. Those containing appreciable amounts of silicon become dark gray when anodic oxide finishes are applied and hence are used for architectural applications.

The 5000 Series alloys use magnesium as the major alloying element. Magnesium is one of the most effective and widely used alloying elements, and when used as the major alloying element (or with manganese) the result is a moderate-to-high-strength, non-heat-treatable alloy. Magnesium is more effective than manganese as a hardener and can be added in considerably greater quantities. These alloys have especially good resistance to corrosion in marine atmospheres and sea water. Shade screening, venetian blinds, and weather stripping are often made from alloy 5052.

The 6000 Series alloys are heat-treatable alloys containing silicon and magnesium. Although not as strong as most alloys of the 2000 or 7000 Series, they possess good formability and corrosion resistance and are of medium strength. They may be formed in the solution of heat-treated temper (T4 temper) and then reach higher strengths by artificial aging (T6 temper). This series is the one most commonly used in products for residential construction. For example, 6061 is used in structural shapes, nails, and bars; 6063 in window and door frames, hardware, louvers, tubing, and curtain walls.

The 7000 Series alloys have zinc as the major alloying element, which, when combined with a smaller percentage of magnesium, results in heat-treatable alloys of very high strength. Usually, other elements such as copper and chromium are also added in small quantities. Among the highest-strength alloys available is 7075. It is often used in aircraft structures and other high-stressed parts.

Experimental alloys are also designated with the groups described here, but carry the prefix *X*, which is dropped when the alloy becomes standard.

Cast alloy designations use a four-digit

FIGURE 5.4-2 Examples of Aluminum Alloys Used for Some Building Applications

Application	Alloy Number	Product
Acoustic ceiling	3003-H14	Sheet
Air duct	3003-H14, 1100-H16, Duct stock	Sheet
Awning	3003-H14	Sheet
Builder's hardware	5050-0 B443.0, 356.0 6063-T5	Sheet Casting Extrusion
Curtain wall	3003-H14, Alclad 4043, Alclad 1235 B443.0 6063-T5, 4043-F	Sheet Casting Extrusion
Deck (corrugated)	3003-H14	Sheet
Door frame	6063-T5, 6063-T6	Extrusion
Fascia plate	6063-T42 3003-H14	Extrusion Sheet
Flashing	3003-0	Sheet
Flue lining	1100-H16	Sheet
Garage door	3003-H14	Sheet
Grating	6061-T6	Bar
Gravel stop	6063-T5, 6063-T42	Extrusion
Grille	B443.0 3003-H14 6063-T5	Sand casting Sheet Extrusion
Gutter	3003-H14	Sheet
Hood	3003-H14	Sheet
Insulation	1235	Foil
Kick plate	3003-H18	Sheet
Letters	B443.0, 514.0 3003-H14	Sand casting Sheet
Louver	3003-H14 6063-T5	Sheet Extrusion
Mullion	3003-H14 6063-T5, 6063-T6	Sheet Extrusion
Nails	6061-T913	Wire
Railing	6063-T5, 6061-T83	Extrusion
Relief panel (sculptural)	B443.0	Casting
Roofing	Special roofing alloys	Sheet
Screen	Alclad 5056-H392	Wire
Screws	2024-T4, 2017-T4	Screw machine stock
Shade screening	5052-H38	Sheet
Shingle	3003-H14	Sheet
Siding	3003-H14	Sheet
Structural shape	2014-T6, 6061-T6 6061-T6, 2014-T6	Extrusion Rolled shape
Termite shield	3003-H14	Sheet
Terrazzo strip	3003-H14	Sheet
Threshold	6063-T5	Extrusion
Vapor barrier	1235-0	Foil
Venetian blind	5052-H18	Sheet
Weatherstrip	5052-H38	Sheet
Window	6063-T5, 6063-T6, 6061-T6	Extrusion

system similar to that for wrought alloys, shown in Figure 5.4-4.

5.4.2 PROPERTIES

Aluminum has a specific gravity of about 2.70, a density of about 170 lb./cu. ft. (1.71 kg/m^3); and weighs about 0.1 lb./cu. in. (276.8 mg/m^3), as compared with 0.28 lb./cu. in. (775.04 mg/m^3) for iron and 0.32 lb./cu. in. (885.76 mg/m^3) for copper.

Pure aluminum melts at about 1200°F (648.89°C). Melting points for various alloys range from 900°F (482.2°C) to 1250°F (676.67°C).

Aluminum and aluminum alloys have relatively high coefficients of thermal expansion. For structural calculation, the coefficient of expansion is assumed to be 13×10^{-6}. This means, for example, that an unrestrained piece of aluminum 18 ft. (5.49 m) long will expand or contract 0.259 in. (6.5786 mm) (about ¼ in. [6.4 mm]) in a 100°F (37.8°C) temperature change.

The two common metals used as electric conductors are copper and aluminum. The electrical conductivity of EC grade (electric-conductor) aluminum is about 62% that of the International Annealed Copper Standard. However, with a specific gravity of less than one-third that of copper, 1 lb. (0.45 kg) of aluminum will conduct twice as much electricity as 1 lb. (0.45 kg) of copper. Unfortunately, aluminum is stiffer than copper; therefore, connections are more difficult to make, resulting in inadequate and unsafe connections.

The high thermal conductivity of aluminum makes it well fitted for use wherever a transfer of thermal energy from one medium to another is desired. However, high thermal conductivity is generally not required in most building material applications and, in fact, is often a problem in such components as window frames and in storefronts and curtain wall sections where *thermal breaks* (other nonconducting materials) are often introduced to reduce the flow of heat through the aluminum.

Aluminum is an excellent reflector of radiant energy through the entire range of wavelengths, from ultraviolet through the visible spectrum to infrared (heat) waves. Aluminum's light reflectivity of more than 80% accounts for its wide use in lighting fixtures, and its high heat reflectivity and low emissivity have led to its use as reflective insulation and improved

FIGURE 5.4-3 Aluminum Association's Temper Designations

Temper Designations		
-F	As fabricated	
-O	Annealed	
-H	Strain hardened (wrought only)	
	-H1	Plus one or more digits[a]: strain hardened only
	-H2	Plus one or more digits[a]: strain hardened and partially annealed
	-H3	Plus one or more digits[a]: strain hardened, then stabilized
	-H4	Plus one or more digits[a]: strain hardened and lacquered or painted
-W	Solution heat treated—unstable temper	
-T	Thermally treated to produce stable tempers other than -F, -O, or -H	
	-T1	Cooled from an elevated temperature shaping process and naturally aged to a substantially stable condition
	-T2	Cooled from an elevated temperature shaping process, cold worked, and naturally aged to a substantially stable condition
	-T3	Solution heat treated, cold worked, and naturally aged to a substantially stable condition
	-T4	Solution heat treated and naturally aged to a substantially stable condition
	-T5	Cooled from an elevated temperature shaping process, then artificially aged
	-T6	Solution heat treated, then artificially aged
	-T7	Solution heat treated and stabilized
	-T8	Solution heat treated, cold worked, then artificially aged
	-T9	Solution heat treated, artificially aged, then cold worked
	-T10	Cooled from an elevated temperature shaping process, cold worked, then artificially aged

[a]Second digit indicates final degree of strain hardening, i.e., 2 is ¼ hard, 6 is ¾ hard, 8 is full hard.
The temper letter *H* is used for non-heat-treatable alloys, *T* for heat-treatable alloys.

performance in such applications as paint, roofing, and ductwork.

Aluminum's nonmagnetic properties make it useful for electrical shielding purposes in electrical equipment, and its nonsparking characteristics make it an excellent material for use around inflammable or explosive substances.

The ease with which aluminum may be fabricated is one of its most important properties. The metal can be cast, rolled to many thicknesses (down to foil thinner than paper), extruded, stamped, drawn, spun, roll formed, hammered, or forged; it can be turned, milled, bored, or machined at maximum speeds.

Almost any method of joining is applicable to aluminum, and the wide variety of mechanical aluminum fasteners simplifies the assembly of products.

FIGURE 5.4-4 The Aluminum Association's Four-Digit Number System for Designating Wrought Aluminum Alloys

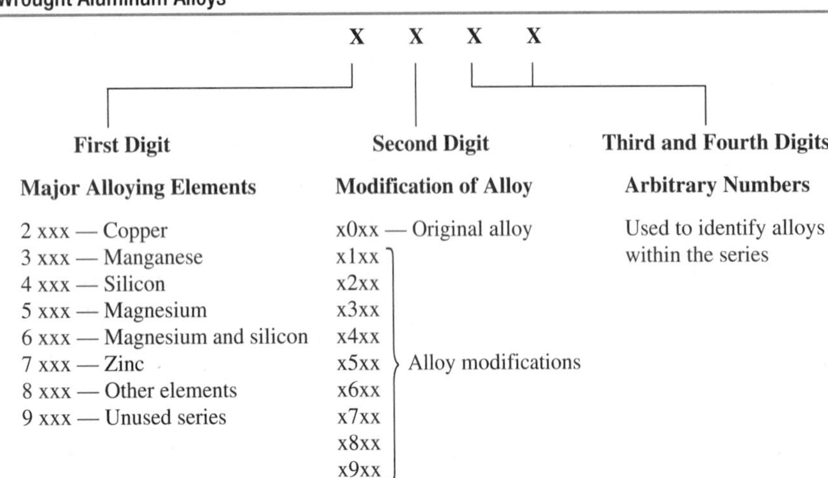

5.4.2.1 Strength

Although pure aluminum has a tensile strength of only about 7000 psi (4921.5 Mg/m^2), by working the metal, as by cold rolling, its strength can be nearly doubled. Much larger increases in strength can be obtained by *alloying* aluminum with small percentages of one or more other metals such as manganese, silicon, copper, magnesium, or zinc. Like pure aluminum, the alloys are also made stronger by cold working, and heat-treatable alloys are further strengthened and hardened by heat treatments so that tensile strengths approaching 100,000 psi (703,070 Mg/m^2) are possible (Fig. 5.4-5). However, such high-strength aluminums are seldom used in building construction products.

Strength decreases at elevated temperatures, although some alloys retain good strength at temperatures from 400°F (204.4°C) to 500°F (260°C). At subzero temperatures strength increases without loss of ductility, so aluminum is a particularly useful metal for low-temperature applications.

5.4.2.2 Corrosion Resistance

Aluminum's high resistance to corrosion is due to the very thin inert surface film of aluminum oxide that forms rapidly and naturally in air. After it reaches a thickness of about a ten-millionth of an inch, this film effectively halts further atmospheric oxidation of the metal and thus protects the surface. If this natural oxide film is broken, as by a scratch, a new protective film forms immediately. The oxide film increases in thickness with temperature and remains protective to the underlying aluminum even at the melting point.

It may be necessary to protect aluminum in certain severely corrosive environments by protective coatings such as organic paints, by cladding, or by increasing the thickness and effectiveness of the oxide film by anodizing (see Section 5.4.3.4).

WEATHERING

The extent of corrosion due to weathering depends on the type and extent of contamination of the surrounding atmosphere. More corrosive elements are present in industrial and marine areas than in rural areas.

In tests and observations (in part under the direction of the National Institute of Standards and Technology) made over a 10-year period in different atmospheric

TENSILE AND YIELD STRENGTHS—ksi
(Minimum) For .063 Sheet
In Various Alloy and Temper Combinations

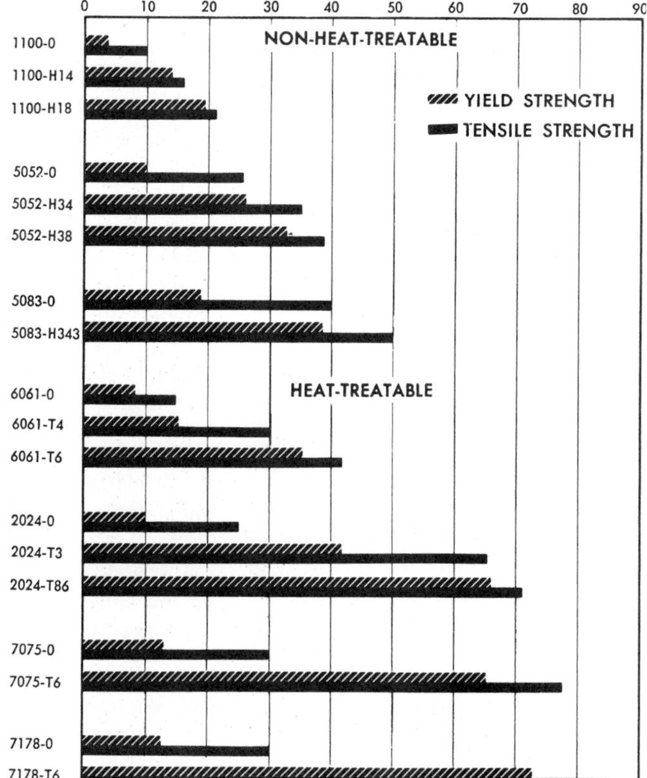

ksi		Mg/m²
10	=	7030.7
20	=	14,061.4
30	=	21,092.1
40	=	28,122.8
50	=	35,153.5
60	=	42,184.2
70	=	49,214.9
80	=	56,245.6
90	=	63,276.3

FIGURE 5.4-5 Mechanical properties of representative aluminum alloys.

FIGURE 5.4-6 Electrolytic Solution Potentials*

Metal or Alloy	Potential in Volts
Magnesium	–1.73
Zinc	–1.00
520.0-T4 alloy (cast)	–0.92
7072 (wrought)	–0.96
514.0 alloy (cast)	–0.87
Pure aluminum and 5052-0, 1100-0, 1100-H18, 3003-H18, 6061-T6 alloys (wrought) and 43 alloy (cast)	–0.83
Cadmium	–0.82
356.0-T4 alloy (cast)	–0.81
2017-T4, 2024-T4 alloys (wrought)	–0.68
Mild steel	–0.58
Lead	–0.55
Tin	–0.49
Brass (60–40)	–0.28
Copper	–0.20
Stainless steel (18–8)	–0.15

*Measured in a 5.85% sodium chloride solution containing 0.3% hydrogen peroxide. The values vary somewhat depending on the particular lot of material and on the surface preparation employed. Coupling metals with dissimilar electrolytic potentials in an electrolyte results in galvanic corrosion of the metal with the highest negative potential. Degree of attack generally increases as potential difference increases.

conditions on alloys 1100, 3003, and 3004, the depth of penetration of atmospheric corrosion was less than one mil ($^1/_{1000}$ in. [0.0254 mm]) in rural areas, about 4 mils (0.1016 mm) in industrial areas, and up to 6 mils (0.1524 mm) in seacoast locations.

GALVANIC CORROSION

When two dissimilar metals are connected by a solution that conducts electricity (an electrolyte), *galvanic corrosion* may occur. As in a battery, an electric current is created, and one metal corrodes away while the other is being plated. Galvanic attack can occur when moisture condenses from the air and contaminating elements act as an electrolyte. The threat of such corrosion is greatest in industrial areas because the atmospheric moisture is more contaminated, and in coastal areas because of the salt and moisture in the air.

The severity of galvanic attack varies, depending on how far apart the two metals are in the galvanic table (Fig. 5.4-6). For example, coupling mild steel to aluminum would produce less corrosion than would coupling copper to aluminum under identical exposure conditions. The metal that has a higher voltage potential is sacrificial and corrodes. Thus, zinc provides cathodic protection to aluminum and aluminum alloys. That is, when zinc and aluminum are in the presence of an electrolyte, the zinc usually corrodes. Thus, zinc-plated steel parts are commonly used with aluminum to prevent the harmful effect of galvanic action on the aluminum.

Cadmium-plated steel in contact with aluminum alloys is also satisfactory. Since there is no appreciable difference in the electrolytic potential of these two metals, practically no galvanic action occurs.

Aluminum is compatible with some stainless steels, chromium, zinc, and small areas of white bronze. Where permanent contact with other metals cannot be avoided, the risk of galvanic corrosion can be greatly reduced by painting the other metal and the aluminum at the contact area with zinc chromate, followed by one coat of a nonleaded paint, such as aluminum paint, or of a heavy-bodied bituminous paint. When the dissimilar metal cannot be painted, the aluminum may be given the insulating treatment. As an alternative, a strip of plastic or a similar insulator may be placed between the two metals.

In severely corrosive atmospheres and in high moisture areas, the edges of a dissimilar metal joint may be sealed with compatible building mastic or caulking.

A dangerous source of electrolytic attack may be water drainage from metals such as copper. Therefore, aluminum gutters on buildings with copper flashings must be protected by painting either the copper flashing or both the copper and the aluminum.

DIRECT CHEMICAL ATTACK

Some chemicals will dissolve metals on contact. For example, solutions of strong alkalis, sulfuric acid, hydrochloric acid, carbonates, and fluorides will attack aluminum because they can dissolve or penetrate its protective oxide coating. Some chemical attack may also occur when aluminum is placed in contact with wet alkaline materials such as mortar, concrete, and plaster. Therefore, surface protection of aluminum products is often necessary during construction.

Where aluminum is buried in concrete, it is usually not necessary to paint the parts. However, if chlorides are present from additives or other sources in steel-reinforced concrete, the aluminum parts should be coated with bituminous paint or another suitable coating.

Aluminum in contact with concrete, masonry, or other absorbent materials under even intermittently wet conditions should be protected with a coating of bituminous paint, zinc chromate primer, or a separating layer of plastic or other gasketing material. Creosote and tar coatings should not be used, however, because of their acid contents. When wet concrete or another alkaline material may splash against aluminum during construction, the aluminum surfaces should be protected. A coating of clear acrylic-type lacquer or a suitable strippable coating may be used when maintenance of appearance is important.

Although aluminum is inert when buried in many types of soil, some soils and the presence of stray electrical currents in the ground may cause corrosion. Therefore, buried aluminum pipe may have to be coated or otherwise protected.

5.4.3 MANUFACTURE

Wrought mill products make up the greater part of aluminum products used in the building industry. They include plate, sheet, foil, bar, rod, wire, extruded sections, structural shapes, tube, pipe, and forging stock. The mechanical properties of such materials are developed by alloying, hot working, cold working, heat treatment, or combinations of these processes.

Cast products include sand castings, permanent mold castings, and die castings. These are produced in a foundry by pouring molten metal into molds of a desired shape. The mechanical properties of some castings may be improved by heat treatment.

The general categories of wrought and cast products are outlined in Figure 5.4-7.

5.4.3.1 Wrought Products

Wrought mill products are those shaped by hot and cold rolling, cold drawing, extruding, and forging.

HOT AND COLD ROLLING

Hot rolling is performed in mills, in rolling machines, through which an ingot, heated to soften it, is passed until it is reduced to the desired shape. Hot rolling can be continued until a thickness of about 0.125 in. (3.175 mm) is obtained. Then, further thinning is usually accomplished by cold rolling. Cold-rolled material has a better finish and, in the case of non-heat-treatable alloys, increased strength and hardness.

Plate Plate has a minimum thickness of 0.250 in. (6.40 mm). To make it, a slab of aluminum is hot rolled in a reversing mill (acting like a rolling pin) until its thickness has been reduced from 1 to 3 in. (25.4 to 76.2 mm). Further rolling operations are then performed in a hot mill.

Sheet Sheet is material with a thickness between 0.006 and 0.249 in. (0.1524 and 6.3246 mm). It is rolled in a hot mill to

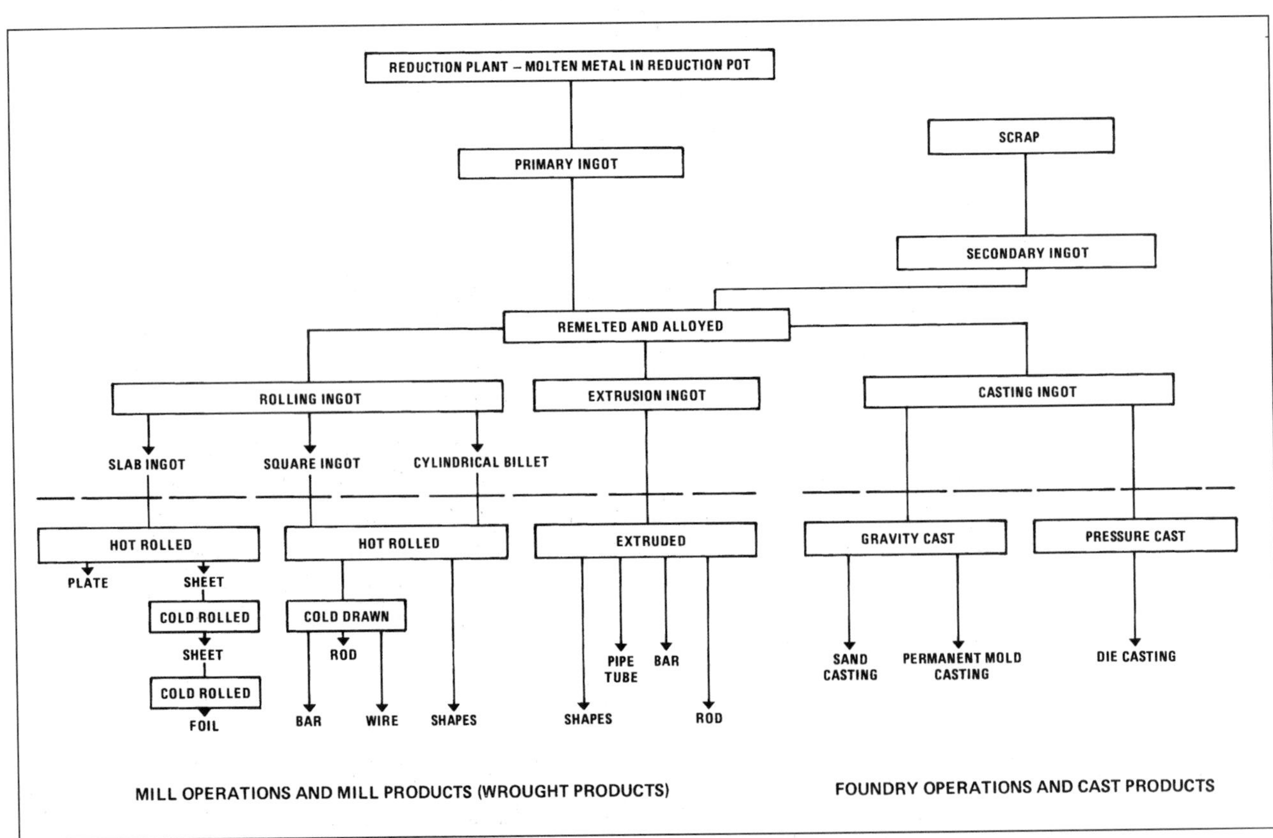

FIGURE 5.4-7 Classification of aluminum-mill operations used to manufacture wrought products and foundry operations used to make cast products.

about 0.125 in. (3.2 mm) thick, then cold rolled for additional thinning and to give it more strength and a better finish.

Coiled sheet is rolled to its finish gauge and width in a continuous strip up to several hundred feet in length and wound into coils. *Flat sheet* is sheared to proper lengths from coiled sheet. It may be stretcher leveled for improved flatness.

Brazing sheet is a special type of sheet that has a cladding that melts at a lower temperature than the core. Assemblies of parts made from this sheet can be brazed together by subjecting them to a temperature of about 1000°F (537.78°C), which does not melt the core alloy but only the cladding, joining the surfaces.

Aluminum foil has a thickness of less than 0.006 in. (0.1524 mm). It is made by rolling sheet in cold-rolling mills and is usually supplied in coil form. Its most common applications in the building industry are as thermal insulation and vapor retarders.

Structural Shapes Although most structural aluminum shapes are extruded, some are hot rolled by a series of specially shaped rolls. Complicated structural shapes require many roller passes to prevent the undue straining that is possible when too much force is applied in a single pass. Because they require considerable metal and a high degree of machining and finishing to get the desired contour, rolls are expensive. Therefore, lower tooling costs often make extruded structural aluminum shapes less expensive than rolled ones. Finishing operations include roller or stretch straightening, and, when required, the rolled shapes may be drawn through dies to improve surface finish and to obtain closer dimensional tolerances.

Standard rolled structural shapes, usually supplied in 2014 and 6061 alloys, include angles, channels, and beams.

Hot rolling is used to produce rod and bar shapes. Cold-finished rod and bar are produced by hot rolling to a size larger than specified and then cold drawing through a die for a better finish and closer dimensional tolerances. Bar is commercially available as square (round or square-edged), square-edge rectangular, and hexagonal.

Rod, bar, and other sections suitable for further shaping by impact or pressure in a die are called forging stock. Forging rod is nearly always rolled, but forging bar may be rolled or extruded, and forging shapes are usually extruded.

Redraw rod is typically ⅜ in. (9.5 mm) in diameter. This rod is suitable for further drawing to produce wire.

COLD DRAWING

Drawn products are formed by pulling (drawing) metal through a die. The purpose of cold drawing is to obtain close dimensional tolerances and a better surface finish.

Wire, which must be kept within close tolerances, is cold drawn by pulling redraw rod (⅜ in. [9.5 mm] diameter) through a series of progressively smaller dies.

Hollow tube sections are drawn in a process similar to drawing bar and rod, except that the tube is threaded over a mandrel (shaper) positioned in the center of the die. When the tube is drawn, close tolerances both inside and out are maintained.

EXTRUDING

Extruding is a process in which a heated aluminum ingot is pushed by hydraulic pressure through a die opening. The extruded aluminum assumes the same size and cross section as the die opening. Extruded products include shapes of varying cross section, rod, bar, and seamless tube. Aluminum window sash, storefront sections, and expansion joint covers are examples of extruded shapes.

Shapes that cannot be extruded as a single section because of dimensional limitations or complexity of cross section can be obtained by fitting (mating) more than one section together.

The extruding process provides great flexibility of design shape. The economical manufacture of practically any shape is possible due to the relatively low cost of dies. Close tolerances in manufacture can be obtained, and small parts of desired shapes can be sliced from a longer extrusion, often minimizing or eliminating machining. Extrusions are straighter, dimensionally more accurate, and have a better surface finish than rolled shapes.

5.4.3.2 Cast Products

Cast products are manufactured in a foundry. They include sand castings, permanent mold castings, and die castings.

SAND CASTINGS

Molds made of sand are lowest in cost and permit maximum flexibility in design changes, although labor costs are higher than for other castings. Sand castings require heavier section thicknesses than other methods, and tolerances are not as close. They are used for spandrel panels, decorative panels, lighting standard bases, and general fittings.

PERMANENT MOLD CASTINGS

Permanent mold casting employs metal molds with a metal core. It is used where a large number of castings from the same mold are required. If a sand core is used, the casting is called a semipermanent mold casting. Molten metal is poured into the mold by gravity.

DIE CASTINGS

Die castings are permanent mold castings in which the metal is forced into the mold under pressure. Die castings have smoother surfaces than permanent mold castings and permit greater repetitive use of the molds. Die casting is generally used for the mass production of small items such as handles, latches, and other door and window fittings.

The main advantages of permanent mold and die castings over sand castings are higher strength, better surface appearance, and closer dimensional tolerances.

5.4.3.3 Fabrication

Fabrication includes forming and machining operations performed on aluminum products and castings after they leave the mill or foundry.

FORMING

Aluminum, one of the most workable of common metals, can be fabricated into a variety of shapes by bending, stretching, rolling, drawing, and forging.

Formability varies greatly, depending on the aluminum alloy and temper.

Bending Temper and thickness greatly affect the minimum bend radius that can be used in forming. Certain aluminum alloys can be cold formed to a sharp 90-degree bend; others must be formed on stipulated radii.

Stretch Forming Sheets are stretched to flatten them or to form them into special shapes. The most suitable alloys for stretch forming are those with high levels of elongation (ability to stretch without fracturing). The advantage of stretch forming is that there is hardly any tendency of the material to assume its origi-

FIGURE 5.4-8 A series of cold rolls form corrugated decking from embossed sheet. (Courtesy Kaiser Aluminum and Chemical Corporation.)

FIGURE 5.4-9 A large, 8000-ton (7257.47 Mg) hydraulic press produces forgings for use as architectural panels. (Courtesy Kaiser Aluminum and Chemical Corporation.)

nal shape after forming stresses are removed.

Roll Forming Products are roll formed by passing strips through a series of roller dies (Fig. 5.4-8). Each successive pair of rolls shapes the metal slightly—only a small deformation is made by any single pair of rolls. No straightening is required since roll-formed products are ordinarily straight and free from distortion.

Forging High-strength aluminum parts are often produced by forging (Fig. 5.4-9). Most forgings are made by hammering or pressing heated aluminum between closed steel dies. Press forging forces the metal into a die cavity of the desired shape under mechanical or hydraulic pressure. Hammer forging produces the same results by repeated blows from a drop hammer. Aluminum forgings may be heat treated after fabrication.

Since forging refines and provides desirable directional characteristics to the grain structure of the metal, improving strength and fatigue resistance, it is used to make various types of door hardware where strength is important.

MACHINING

The machining characteristics of aluminum and its alloys, especially of wrought products, are excellent. Sawing, drilling, tapping, and threading methods vary little from those employed for other metals.

5.4.3.4 Finishes

The surfaces of many aluminum products require no further finishing. The natural oxide film makes protective coatings unnecessary under ordinary conditions. However, finishes are frequently applied in order to (1) change or improve the appearance for decorative purposes, (2) improve its abrasion or corrosion resistance, or (3) provide a base for subsequent surface treatment such as painting.

Natural finishes and other common surface treatments are briefly described in this section and summarized in Figure 5.4-10.

UNCONTROLLED NATURAL FINISHES

Hot-rolled products show a certain amount of discoloration or darkening, while cold-finished products have a whiter, brighter surface. Extruded products have an intermediate appearance, nearer that of cold-finished items, and show traces of longitudinal striations (*die lines*) caused by the extrusion process. Castings vary little in their surface appearance, except that die castings and permanent mold castings are smoother than sand castings.

CONTROLLED NATURAL FINISHES

Natural finishes of cold-rolled sheet may be controlled by the degree of polishing given to the rollers of the rolling mill, as follows:

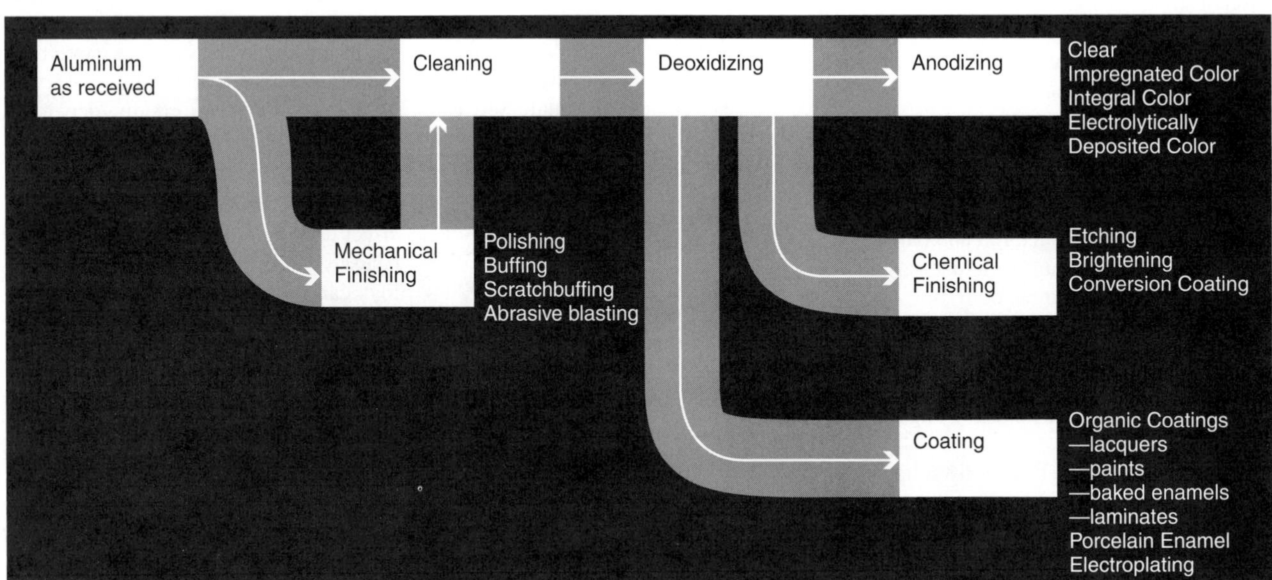

FIGURE 5.4-10 Principal aluminum finishing processes.

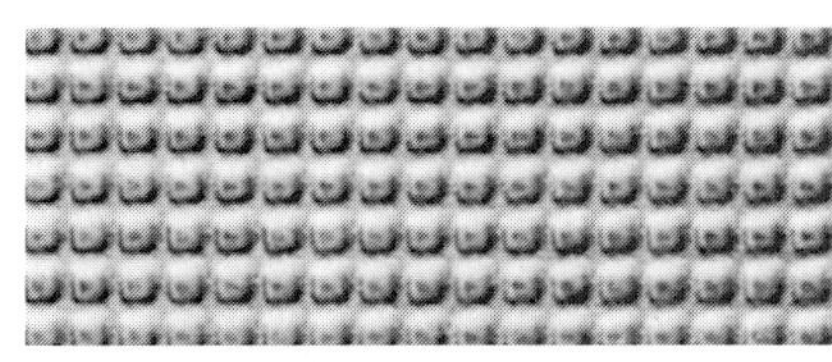

FIGURE 5.4-11 Aluminum sheet and plate can be embossed to produce a variety of surface textures.

- *Mill finish*—uncontrolled, slightly variable, not entirely free from stains or oil
- *One-side-bright mill finish*—moderate degree of brightness on one side uncontrolled on reverse side, possibly with a dull, nonuniform appearance
- *Standard bright finish*—relatively bright, uniform appearance on both sides; somewhat less lustrous than standard one-side-bright sheet
- *Standard one-side-bright sheet*—uniformly bright and lustrous surface on one side; uncontrolled on reverse side, possibly with a dull nonuniform appearance

Embossed sheet is obtained by passing a mill finish sheet through a pair of rolls with a matched design or between a design roll and a plain roll. Embossed surfaces increase the stiffness and tend to hide wear and scratches. Patterns include wood grain, leather grain, fluted, stucco, diamond, notched, ribbed, bark, and pebble designs (Fig. 5.4-11).

MECHANICAL FINISHES

Aluminum surfaces may be mechanically treated by:

- *Grinding* with a dry abrasive wheel, primarily for removing surface roughness from castings.
- *Buffing* with a fine abrasive, resulting in a highly lustrous surface.
- *Polishing* with finer abrasive than used in buffing and with a very light pressure, resulting in a mirrorlike surface.
- *Scratch brushing* with power-driven rotating wire brushes to give coarse or smooth-lined surface textures, depending on the size of the wire bristles used.
- *Shot blasting* and *sandblasting* with various combinations of pressure, steel shot, and sand to produce matte and coarse surfaces. Rough textures may be given a lacquer coat or may be anodized to minimize the collection of dust.
- *Satin finishing*, accomplished with fine wire bristles, very fine abrasives, or a muslin or felt wheel to produce a soft, smooth sheet.
- *Spin finishing*, by pressing an oily abrasive cloth against the aluminum while it is revolved rapidly. Subsequent applications of fine stainless steel wool and fine-mesh emery give a bright, silvery finish with a pattern of concentric circles.
- *Highlighting*, which provides a two-tone effect for relief surfaces. Elevated areas are masked; recessed areas are sandblasted. Then the masking is reversed, and the elevated areas are given a highlight effect by painting or polishing.

The Aluminum Association designations for various mechanical finishes are shown in Figure 5.4-12.

CHEMICAL FINISHES

Many methods of treating aluminum surfaces employ the reaction of various chemicals on the metal. Some of the most common are discussed here. The Aluminum Association designations for various chemical finishes are shown in Figure 5.4-13.

Etch Finish A decorative, silvery white finish, also known as *frosted* finish, may be produced by etching the metal with caustic soda (sodium hydroxide) or another chemical. Several methods permit partial or *design* etching to produce patterns of varying finish and colored background.

Conversion Coatings Although chemical conversion coatings are generally used on aluminum to prepare the surface for painting, some types may also be used as a final finish.

Since the natural oxide film on aluminum surfaces may not provide a suitable bond for paints and other organic coatings and laminates, its chemical nature is often converted (altered) to improve adhesion. *Conversion films*, which are also called *conversion coatings*, are generally applied by use of phosphate or chromate solutions, many of which are proprietary. One of the simplest methods of surface preparation for painting is to etch the surface with a phosphoric acid solution, thus providing a good mechanical bond.

FIGURE 5.4-12 The Aluminum Association's Designations for Various Mechanical Finishes*

As Fabricated	Buffed	Directional Textured	Nondirectional Textured
M10—Unspecified	M20—Unspecified	M30—Unspecified	M40—Unspecified
M11—Specular as fabricated	**M21—Smooth specular**	M31—Fine satin	M41—Extra fine matte
M12—Nonspecular as fabricated	M22—Specular	**M32—Medium satin**	**M42—Fine matte**
M1x—Other	M2x—Other	M33—Coarse satin	**M43—Medium matte**
		M34—Hand rubbed	**M44—Coarse matte**
		M35—Brushed	M45—Fine shot blast
		M3x—Other	**M46—Medium shot blast**
			M47—Coarse shot blast
			M4x—Other

*Finishes printed in **boldface type** are those most frequently used for architectural work.

(Reprinted from *Metal Finishes Manual for Architectural and Metal Products* with permission of the National Association of Architectural Metal Manufacturers (NAAMM), Chicago, Illinois.)

FIGURE 5.4-13 The Aluminum Association's Designations for Various Chemical Finishes*

Nonetched Cleaned	Etched	Brightened	Conversion Coatings
C10—Unspecified	C20—Unspecified	C30—Unspecified	C40—Unspecified
C11—Degreased	C21—Fine matte	C31—Highly specular	C41—Acid chromate-fluoride
C12—Chemically cleaned	**C22—Medium matte**	C32—Diffuse bright	C42—Acid chromate-fluoride-phosphate
C1x—Other	C23—Coarse matte	C3x—Other	C43—Alkaline chromate
	C2x—Other		C44—Nonchromate
			C45—Nonrinsed chromate
			C4x—Other

*Finishes printed in **boldface type** are those most frequently used for architectural work.

(Reprinted from *Metal Finishes Manual for Architectural and Metal Products* with permission of the National Association of Architectural Metal Manufacturers (NAAMM), Chicago, Illinois.)

Chemical Oxidizing Chemical oxide films have improved resistance to corrosion. As a base for painting, they provide additional protection in the event that the finish is scratched or broken. They are more economical and easier to apply than anodic processes (described in a following paragraph) that also produce oxide films, but are thinner and softer than anodic finishes. Chemical oxide coatings may be dyed, but anodic films are preferred if a superior color quality is desired.

Zincating In the zincating process, aluminum is immersed in a zincate bath providing a very thin, firmly adherent coating of zinc. This base is often used for aluminum that is to be electroplated.

Chemical Brightening A mirror-bright surface finish can be provided on aluminum by immersing the parts in a hot chemical solution, followed by subsequent rinsing. Depending on the alloy and its mechanical preparation, reflectivity values up to 85% may be obtained.

ANODIZED FINISHES

Anodized electrolytic oxide coatings are readily applied to aluminum alloys and are widely used. Thinner anodic flash films may be used as a prepaint treatment.

The distinction between a coating and a film is thickness. A film is less than 0.1 mil (0.0254 mm); any thicker covering is referred to as a coating.

Ordinarily, clear anodic coatings on architectural products are preceded by either a chemical etch, some form of mechanical finish, or both. Etching is the most economical and common prefinish, imparting a frosted appearance that helps to conceal variations in the mill finish and die lines on extrusions. Mechanical finishes also leave a pleasing texture, which is retained after anodizing.

If neither etching nor mechanical finishing is specified, the aluminum must still be chemically cleaned before it is anodized to ensure satisfactory results. Contaminants deposited on the metal surfaces during fabrication processes may affect the anodic coating and should be completely removed before anodizing.

Anodizing consists of immersing a product in an acid solution (*electrolyte*) and passing an electric current, usually of the direct type, between the metal and the solution, with the aluminum serving as the anode. This results in the controlled formation of a relatively thin but durable oxide or coating on the surface of the metal.

In general, such coatings are many times thicker than the natural oxide film and may be transparent, translucent, or opaque, depending on the alloys and the electrolyte used. Anodic coatings do not change the surface texture of the metal, but they greatly increase resistance to corrosion and provide increased resistance to abrasion.

Several anodizing processes are commonly used. The principal differences between these processes are the type of solution used for the electrolyte, the voltage and current required, and the temperature of the bath.

The *sulfuric acid process*, because of its relatively low cost, is most widely used for architectural anodizing. It produces comparatively thick, transparent, and absorptive coatings that are suitable for receiving coloring dyes and pigments (see the following paragraphs). Thin films may be produced by this process, which are excellent pretreatments for paint, enamel, and lacquer coatings.

Color Anodizing There are three methods of color anodizing in common use: sulfuric acid anodizing and dyeing, integral-color anodizing, and sulfuric acid anodizing and electrolytically deposited color (Fig. 5.4-14).

In *sulfuric acid anodizing and conventional dyeing*, the work is rinsed after it leaves the anodizing bath and is then dipped in a bath containing either an organic dye or an inorganic pigment. To set the dye, the work is dipped in a nickel acetate solution and is sealed (usually in boiling deionized water) to produce a clear film that provides minimum interferences with, and optimum lightfastness of, the color.

A wide range of colors is possible with this process, but not all of them are satisfactory for outdoor service. As indicated in Figure 5.4-14, the color is concentrated in the top third of the anodic film; therefore, it is necessary to adjust dyeing conditions (time, temperature, dye concentration, and pH) to get similar colors with different film thicknesses.

In *integral-color anodizing*, the coloring material is spread uniformly through the anodic film. Integral-color anodic films may be produced by one of two methods or by a combination of both. The first method makes use of an alloy with constituents that color the anodic film as it grows. The second method uses an electrolyte that yields a colored precipitate during anodizing, again coloring the anodic film as it grows.

A combination of these two methods increases the range of possible colors. The colors produced by integral-color anodizing are very lightfast, but, as with sulfuric acid anodizing and electrolytically deposited color, the depth of color varies with the anodic film thickness. In addition, the kind of alloy is a major variable; some colors are restricted to specific alloys.

Sulfuric acid anodizing and electrolytically deposited color concentrate the color at the base of the anodic film (see

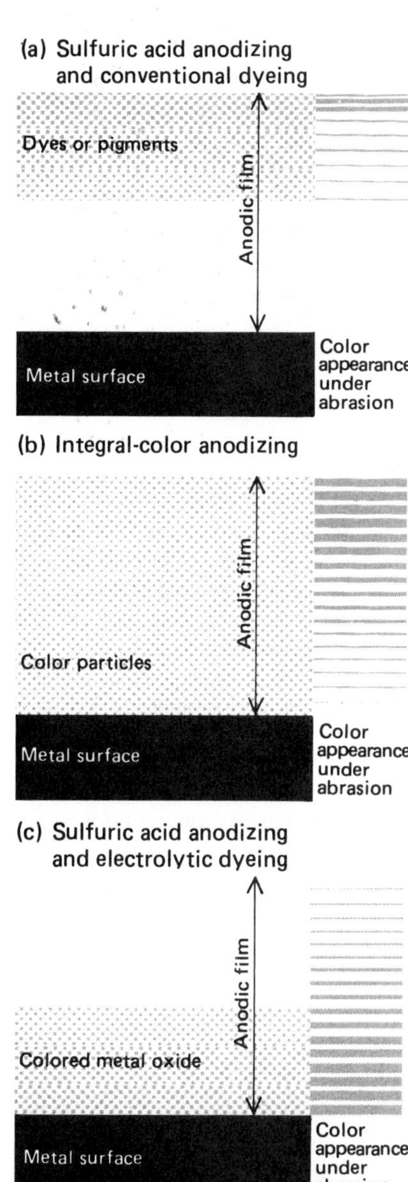

FIGURE 5.4-14 Distribution of coloring elements in the anodic film of the three main color anodizing processes.

Fig. 5.4-14). Consequently, the color receives maximum protection from sunlight and weathering.

In most applications, the pores in the anodic film are sealed off to increase resistance to staining and corrosion and to improve color durability. Because most applications require a clean film, immersion of the work in boiling deionized water with a pH of 5.5 to 6.5 is the usual method of sealing. When maximum corrosion resistance is required, potassium dichromate is added to the water.

The thickness of anodic films and coatings range from less than 0.1 mil (0.0254 mm), in the case of primer films for receiving organic finishes, to 5.0 mils (0.127 mm) for some of the engineering hardcoats.

For architectural applications, the thickness and weight of coatings are of prime importance; no coating thinner than 0.4 mil (0.1016 mm) should be used even for interior purposes. Coatings for exterior parts that will be subject to contaminated atmospheres, or will not receive regular maintenance care, should have a thickness of at least 0.7 mil (0.01778 mm) and a minimum coating weight of 27 mg/sq. in. (27 mg/645.16 mm^2).

Classification of Anodic Coatings The Aluminum Association identifies the various anodic finishes under four general types: General Anodic Coatings, Protective and Decorative Coatings, Architectural Class I Coatings, and Architectural Class II Coatings. The first two classifications apply to general industrial work. Therefore, only the last two are discussed here. The Aluminum Association designations for anodic coatings are shown in Figure 5.4-15.

Architectural Class I coatings are more than 0.7 mil (0.01778 mm) thick, weigh more than 27 mg/sq. in. (27 mg/645.16 mm^2), and include the hardest and most durable anodic coatings available. They are suitable for exterior items that receive no regular maintenance care and for interior architectural items subject to normal wear.

Architectural Class II coatings range in thickness from 0.4 to 0.7 mil (0.1016 to 0.01778 mm), with corresponding weights of from 17 mg/sq. in. (17 mg/645.16 mm^2) to 27 mg/sq. in. (27mg/645.16 mm^2). These are appropriate for interior items not subject to excessive wear or abrasion, and for exterior items such as storefronts and entrances that are regularly cleaned and maintained.

PAINT COATINGS

Field-applied paints are discussed in Section 9.15. The discussion here relates to coatings applied in the shop or factory.

A substantial volume of aluminum is painted for both decorative and protective purposes. Several important applications are siding, cladding panels, and doors.

The trend in painting aluminum is toward continuous processing that integrates pretreatment, coating, and curing into a single operation. *Roller coating* enables the painting of sheet at line speeds of up to 600 ft./min. (182.88 m/min).

First, the bare coiled sheet is passed through a series of cleaning and surface preparation baths to provide a surface to which paint will strongly adhere; then it is roller coated with an organic paint; finally, it is passed through an oven where the paint is baked hard. The resulting finish has an excellent combination of hardness, ductility, adhesion, and gloss and can undergo further fabrication without harm.

Castings and extrusions are commonly sprayed. Electrostatic spraying, in which electrically charged coating particles are

FIGURE 5.4-15 The Aluminum Association's Designations for Anodic Finishes[a]

General	Protective and Decorative	Architectural Class II	Architectural Class I
A10— Unspecified anodic coating	A21[b]— Clear (natural)	**A31— Clear (natural)**	**A41—Clear (natural)**
A11— Preparation for other applied coatings	A22[b]— Integral color	**A32— Integral color**	**A42—Integral color**
A12— Chromic acid anodic coating	A23[b]— Impregnated color	**A33— Impregnated color**	**A43—Impregnated color**
A13— Hard coating	A24— Electrolytically deposited color	**A34— Electrolytically deposited color**	**A44—Electrolytically deposited color**
A1x— Other	A2x— Other	A3x—Other	A4x—Other

[a]Finishes printed in **boldface type** are those most frequently used for architectural work.

[b]Third digit (1, 2, or 3) added to designate min. thickness in 1/10 mil.

(Reprinted from *Metal Finishes Manual for Architectural and Metal Products* with permission of the National Association of Architectural Metal Manufacturers (NAAMM), Chicago, Illinois.)

attracted to the surface by maintaining it at the opposite charge, provides a more uniform finish, with less waste, than conventional spraying.

VITREOUS COATINGS

Porcelain enamel is a vitreous inorganic coating bonded to metal by fusion at high temperatures. A porcelain enamel coating significantly improves the existing properties of aluminum, increasing durability and corrosion resistance. Practically any texture and lightfast color are available.

ELECTROPLATING

Chromium and copper plating are the most common and easiest to apply electroplated finishes for aluminum. Chromium is readily plated on aluminum either after zincating, or after copper, brass, or nickel plating. Lustrous chrome plate is easy to obtain by buffing. Chromium plates are generally applied for decorative effect or resistance to abrasion.

Aluminum may be copper plated after a zincate treatment or by other methods, including anodizing in phosphoric acid. Copper may also be electroplated after first precleaning and zincating the aluminum. Copper plates on aluminum are used where soldering operations are required, for certain electrical requirements, for a base for chromium or silver plating, and for decorative items.

Silver plating decreases electrical resistance and is sometimes used for electrical bus-bar connections. Tin and brass plates are especially good as a base for soldering. Zinc and cadmium plates provide corrosion protection. Chemically treated nickel plates yield a black decorative finish.

ELECTROPOLISHING

High-purity aluminum may be treated by electropolishing to obtain a light reflectivity of about 85%. The metal surface is anodically smoothed to a highly polished surface, suitable for use in aluminum light fixture reflectors.

Electropolished reflecting surfaces are usually given a final anodizing treatment for surface protection. Final reflectivity depends on the purity of the aluminum, the thickness of the oxide film, and the particular procedure employed.

LAMINATED COATINGS

For special use and design applications, many other materials may be laminated to aluminum surfaces. For example, polyvinyl chloride and vinyl films are used to provide nonmetallic color and long-lasting finish for exterior applications such as aluminum siding. The film is adhesive bonded to aluminum surfaces that have been suitably prepared, usually by application of a chemical conversion coating.

TEMPORARY PROTECTIVE COATINGS

Waxes and lacquers applied to aluminum surfaces provide some weather protection and make cleaning easier. Aluminum so treated can usually be kept clean by wiping with a damp cloth or rinsing with clear water.

Lacquer Clear lacquer, applied by a fabricator or contractor, provides the most common type of temporary surface protection for aluminum. Methacrylate lacquers, which are commonly used, are resistant to alkalis such as lime mortar, plaster, and concrete and provide some protection against abrasion. Under normal conditions, a methacrylate coating lasts for several years and weathers by chalking rather than by unsightly peeling. Methacrylate lacquer should not be used on color-anodized aluminum, however, because lacquer never fully adheres to an anodized coating. Where adhesion is incomplete, the lacquer will appear white. On clear anodizing, this effect is largely unnoticeable. On a color-anodized surface, however, the white spots produce a blotchy, unappealing effect.

Wax Liquid waxes or wax-based cleaners offer some protection against atmospheric weathering and retard grime accumulation on aluminum surfaces. However, after repeated waxings, surfaces may tend to yellow. If necessary, the wax may be removed with a solvent or mild abrasive cleaner before rewaxing.

Strippable Coatings Strippable coatings are rather expensive coatings that are usually applied by the fabricator to protect surfaces from the normal hazards encountered during shipping and construction. They are ideal for finer finishes that may be scratched or damaged more easily. Strippable coatings should be left in place until the final cleanup operation. Aluminum with such coatings should not be stored in places exposed to sunlight, because the coatings may tend to become tacky and difficult to remove. If they cannot be easily removed, a solvent should be used.

Tapes For protection against abrasion or staining on smaller areas, various types of tape may be used. They may be applied either by the fabricator or by the contractor.

5.4.3.5 Joints and Connections

Practically every method of joining and connecting is used in assembling aluminum products, including mechanical fastenings; mechanically formed connections; welding, brazing, and soldering; and bonding with adhesives.

MECHANICAL FASTENINGS

A great variety of permanent and semipermanent fasteners are used to join aluminum to itself and other materials. Permanent fastening requires destruction of the fastener or the product itself for disassembly. Semipermanent fastening includes methods, such as screwing and bolting, that permit nondestructive disassembly.

Riveting Riveting is a method of permanent mechanical fastening in which a headed pin (rivet) inserted through the pieces being joined is mechanically deformed so that a head is formed on both sides of the work, tightly gripping the pieces together. Riveting is probably the most widely used aluminum fastening method (particularly in the aircraft industry) and makes use of a great variety of rivet shapes, sizes, and assembling techniques.

Aluminum rivets are typically used in joining aluminum for neater appearance and uniform finish, avoidance of galvanic corrosion, low cost, and light weight.

Aluminum rivets are available in all standard designs and sizes with the usual round, brazier, flat, and countersunk heads.

Rivets can be made in practically any alloy. The following alloys meet most needs: 1100, 2024, 2117, 5056, and 6053. Alloys 1100 (commercially pure aluminum) and 5056 are non-heat-treatable. The others are heat-treatable. In general, rivets to be used with parts of a given alloy should match that alloy in mechanical and physical properties.

Although standard rivets meet requirements, many special types are available, including tubular and other designs for blind riveting (where the work is accessible from only one side).

Stitching Joining of various metals to other metals and other materials by wire sewing is called *stitching*. Aluminum's relatively low density and softness make it very suitable for stitching.

Stitching requires no backing material, prepunching, or drilling. The work need not be clamped nor the metal cleaned. Inspection is simple, as it can be determined visually whether a stitch has penetrated the work and is properly clinched. Stitching is used in assembling aluminum garage doors, on sheet metal (such as in ductwork and blower installations), and in joining aluminum to rubber, fabrics, and other materials.

Threaded Fasteners Threaded fasteners are semipermanent fasteners that are widely used in assembling aluminum products. They include a multitude of screw and bolt types. Whatever fastener is used, however, it should be aluminum or another metal compatible with aluminum, such as stainless steel.

Screws for use with aluminum are usually made from alloy 2024-T4 and are manufactured in practically any head type, size, and standard thread (Fig. 5.4-16).

Bolts, nuts, and washers are usually made from alloy 2024-T4, 6061-T6, or 7075-T73. They are manufactured in practically any head style (typically hexagonal), size, and standard machine thread.

Because of their freedom from corrosion and staining, aluminum nails are popular fasteners for roofing and siding made of aluminum or of nonmetallic materials. These nails are manufactured from alloys 6061 and 5056 which, in this form, have tensile strengths in excess of 60,000 psi (42,184.2 Mg/m^2).

Special Fasteners Many other special fasteners and rivets are available for blind fastening and other connections. Blind fasteners include pop rivets, explosive rivets, and others where only one side of the work is accessible. Other special fasteners, such as speed nuts and quick-release fasteners, are also commonly used.

MECHANICALLY FORMED JOINTS

Connections obtained by forming or cutting the parts to be joined in such manner that they interlock are called mechanically formed joints. Those used for aluminum work do not differ basically from such joints in other metals. Where high strength is of primary importance, other joining methods are usually preferred. But the building industry offers many applications, such as roof work, ducts, gutters, downspouts, and railings, where mechanically formed joints are quite satisfactory (Fig. 5.4-17).

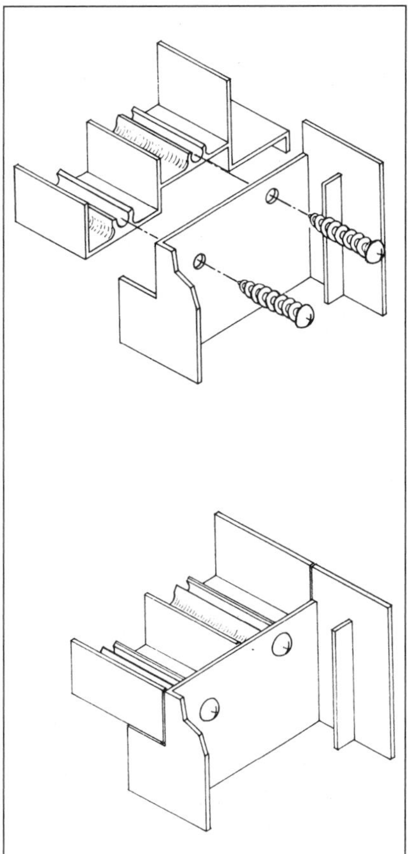

FIGURE 5.4-16 Extruded aluminum window and door frame sections are often connected using screw fasteners. A common method is to thread the screws into continuous slots.

Other mechanically formed joints are made by *staking*. A staked joint is commonly used in assembling extruded aluminum members. Tabs or prongs formed at the end of one member are inserted through mating holes in the member to be joined and are twisted, bent over, or spun down (Fig. 5.4-18).

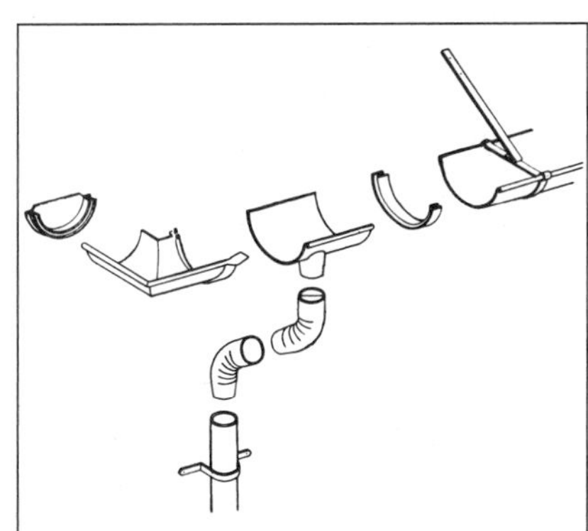

FIGURE 5.4-17 Mechanically formed slip joints are used to assemble aluminum gutters and downspouts.

WELDING, BRAZING, AND SOLDERING

Welding, brazing, and soldering are methods of forming a metallurgical bond between two or more members of an assembly. The selection of a specific process for a particular job depends on many variables, including conditions encountered in service, shape and thickness of parts, volume of production, and the alloys employed.

The choice of filler metal alloy for welding depends on joint design, service conditions, surface finish, welding process, the alloy or alloy combinations to be joined, and the desired appearance. Therefore, filler alloys vary considerably and must be selected for each particular circumstance.

There are many welding processes suitable for use with aluminum alloys. Some of the more important processes are discussed here.

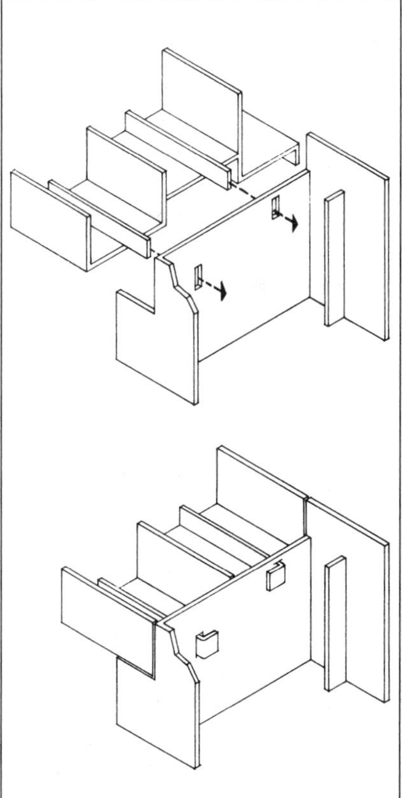

FIGURE 5.4-18 Staked joints are often used in assembling aluminum frame members of windows and doors.

Gas Welding The earliest welding process used on aluminum was gas welding. In gas welding, oxygen, acetylene, oxyhydrogen, or another gas flame is the heat source. The gas flame melts the parent metal, and a flux-coated filler rod is used to complete the weld. The disadvantages of gas welding include:

- Necessity to remove flux residue, a potential source of corrosion or weld defects in multipass welds
- Distortion, which is greater than in arc welding
- A wide heat-affected zone
- Slow welding speeds

These conditions generally result in higher costs, as compared with gas-shielded arc methods.

Aluminum from about 1/32-in. (0.794-mm) to 1-in. (25.4-mm) thick may be gas welded. Heavier material is seldom gas welded, because heat dissipation is so rapid that it is difficult to apply sufficient heat to melt the metal with a torch.

Gas welding finds its greatest use today where few or occasional welds are required. It is also the obvious choice in the field when electric power lines or portable generators for arc welding are not accessible.

Arc Welding Most forms of arc welding depend on the production of heat from high-voltage electric flow (arc) from an electrode to the parent metal.

Inert-Gas-Shielded Arc Welding Of all the fusion welding processes, the inert-gas-shielded arc processes are by far the most important and most widely used. In the tungsten inert-gas (TIG) arc process, an arc is drawn between the work and a nonconsumable tungsten electrode; the arc is shielded by a flow of inert argon or helium. Filler metal is added manually.

In metal inert-gas (MIG) arc welding, the arc in inert gas is drawn between an automatically fed filler wire and the work. Fluxes, used to dissolve the aluminum oxide surface film on the metal in other welding processes, are not required for these because of the action of the arc and the inert gas shield.

These processes are applicable to all welding positions. Thinner gauges are commonly welded by TIG. MIG is particularly appropriate for automatic welding; TIG is also appropriate there when filler metal is not required. Since welding speeds are high, the effect of heat input on strength is minimized.

Resistance Welding In resistance welding, an electric current is conducted through the parts to be joined. Heat is generated at the juncture of the two parts as a result of resistance to the passage of heavy current. When pressure is applied to the heated metal, fusion occurs. Resistance welding has practically no effect on the alloy temper because the weld is made so quickly that the adjoining area does not heat appreciably.

Flash welding is a type of resistance welding used to butt weld shapes such as tubes, extruded shapes, rods, bars, and wire. A typical process is the miter joining of aluminum window sections. In general, resistance welding is used where technically practical and where mass production is involved.

Spot and seam welding are resistance welding methods especially important in fabricating high-strength heat-treated sheet alloys, which can be so joined with practically no loss in strength. Spot welding is widely used to replace riveting, joining sheets at intervals as required. *Seam welding* is merely spot welding with the spots spaced so closely that they overlap, producing an airtight joint where desired.

Brazing Brazing is a welding process that uses a filler metal that is nonferrous, or an alloy with a melting point higher than 800°F (426.7°C) but lower than that of the materials being joined.

In brazing, the molten metal is called *brazing filler metal*; in soldering it is called *solder*.

Brazing is employed to join millions of aluminum parts each year, ranging in thickness from foil to heavy plate and castings. It is used for joints that because of their complexity of contour, limited access, or great ratio of length to part size, are unsuitable for welding.

Filler alloys for brazing aluminum include such aluminum-silicon alloys as 4047 and 4145. They have liquidus temperatures much closer to the solidus temperature of the base metal than filler alloys for brazing most other metals. Consequently, temperature control is critical, and only those aluminum alloys with relatively high melting points are brazeable. These include alloys 1050, 1100, 3003, 6063, and the self-aging foundry alloy Alcan 432.

Although alloy 6061 can be brazed, the melting range of this alloy is very close to that of alloy 4047, the most common brazing filler alloy. Alloys with high magnesium content are not considered brazeable because of poor wetting by the filler alloys. Flux is essential to remove the oxide film.

Aluminum may be brazed by several methods, differing mainly in the way that heat is applied to the work. These include torch brazing, furnace brazing, and salt bath brazing. *Torch brazing* is employed where the joints are accessible both for heating with the torch flame and for feeding filler metal into the joint.

In *furnace brazing*, the entire work assembly is held in a furnace for 1 to 5 minutes. Due to the capillary action of a brazing alloy, the furnace brazing process is particularly useful for inaccessible joints.

In the *salt bath process* (also known as *dip* brazing), the work is submerged in molten salt, which performs as the heating medium and as the flux. Time in the salt bath varies from 30 to 180 seconds.

Whatever the process, it is essential that surfaces be thoroughly cleaned before they are brazed. For 1100 and 3003 alloys, degreasing is usually sufficient; for other alloys, an etching treatment is recommended.

Under some conditions, brazed joints develop strengths comparable to those of welded joints, make more economical use of filler metal, can be made faster and have a neater appearance (requiring little finishing work, if any). Brazing makes possible the joining of extremely thin sections. In general, the corrosion resistance of brazed joints is comparable to that of welded joints.

Soldering Filler metals (*solders*) with melting points below 800°F (426.7°C) are used in soldering.

Difficulties in soldering aluminum are caused by the great affinity of aluminum for oxygen and the lack of bond between solder and the natural oxide film present on aluminum. Ordinarily, a suitable flux is applied to dissolve the oxide and prevent it from reforming, or otherwise reacting, so that the solder may bond directly to the underlying aluminum. Another way to get below the oxide film is to mechanically abrade it down through an overlaying layer of molten solder, using a soldering iron, a wire brush, or another tool.

When soldering aluminum to other metals, the aluminum surface should be *tinned* (precoated with solder) and then joined to the other metal with solder and flux. If any difficulty is encountered, the other metal surface should also be tinned first.

The essentials of soldering aluminum are to remove the surface oxide and to use a filler metal that will not encourage galvanic corrosion. Therefore, appropriate fluxes, solders with higher melting points, pure zinc, and zinc-aluminum (95-5) are used where corrosion is a factor. Where corrosion resistance is less important, zinc-tin and lead-tin alloy solders with lower melting points are used.

In corrosive atmospheres, lacquer coatings may be used to improve the corrosion resistance of soldered aluminum.

ADHESIVE BONDING

Adhesive bonding utilizes adhesives for bonding aluminum to aluminum or other materials. Design possibilities are limited only by extremes of stress, temperature, corrosive environment, and cost. A special consideration in designing joints for the use of adhesives is that peeling stresses should be kept as low as possible. The pretreatment of aluminum used in adhesive bonding is concerned mainly with cleaning the surface. Where the adhesive bond must develop the greatest strengths, a chromic–sulfuric acid dip is often used as a chemical cleaning method.

Adhesives may be applied by any of several techniques. Roller coating is particularly effective for high-speed application of a thin coat of adhesive, as in the production of laminates. Various building materials are adhesively laminated with aluminum foil to increase their thermal insulation properties.

For more limited areas, adhesives may be applied by brush, blade, or spray. Certain adhesives are adapted to mass production practice by manufacturing them in film form. The film is cut into sheets, ribbons, or special shapes that follow the contours of specific joint areas and are assembled with the metal parts before curing.

Curing processes for adhesives vary considerably in type and form. The length of curing time ranges from 3 seconds to more than 76 hours, depending on the type of adhesives and whether heat and pressure are applied to the assembled joint. Either heat, pressure, or both are used to obtain the highest bond strengths.

There is no universal adhesive. The nature of the material joined and the ultimate conditions of service determine the type of adhesive to be used. Therefore, there are many proprietary adhesives developed for varying bonding processes. Natural and synthetic rubbers, polyvinyl acetates, phenolic blends, and epoxies are among the commonly used adhesive bases.

5.5 Other Metals

Other metals are also used in buildings. Chief among these are copper and its alloys, brass and bronze. Zinc, lead, chromium, nickel, tin, and gold are also used.

5.5.1 COPPER AND ITS ALLOYS

Copper is probably the world's second most widely used metal, following the various alloys of iron, including steel. Copper occurs in nature in its metallic state, which probably explains its use as early as 5000 B.C. Today it is obtained mainly by refining the ores cuprite, which is copper oxide; malachite, which is basic copper carbonate; or chalcocite or chalcopyrite, both of which are copper sulfide. The process is a complicated one, requiring roasting, smelting, and converting the ore and then refining the resulting metal by electrolysis.

Newly refined copper is about 99.95% pure. It has a reddish yellow color and is soft, malleable, ductile, relatively strong, and a very good conductor of electricity. Copper does not corrode in dry air but will oxidize when heated. In normal atmospheric conditions, it oxidizes slowly, forming a thin surface coating of verdigris, which is similar to the ore malachite.

Copper has many uses, including making copper sulfate for printing, electroplating, and purifying public water supplies. It is also used in tanning leather, in manufacturing pigments, insecticides, and dyes for cotton, and even as an astringent. Copper is also used extensively in making coins, cooking utensils, art objects, wire, and various sheet metal items such as gutters, downspouts, and roofing. One of the primary uses of copper is in the making of various alloys, including the two most interesting to those in the building industry, brass and bronze.

5.5.1.1 The Copper Alloys

There are more than 1100 recognized alloys of copper. Of these, more than half are alloys of copper and zinc or copper and tin. The first has traditionally been known as *brass*; the second as *bronze*. Today, most copper alloys used for other than utilitarian purposes in building and related construction are called bronze, even though the many that contain zinc are really brass.

Bronze was almost certainly the first industrial alloy. In some places in the world, copper and tin ore are found naturally intermixed, which probably explains the appearance of bronze as early as 3500 B.C. Brass followed much later, but even it has been in use since about 50 B.C.

ALLOY DESIGNATIONS

The Copper Development Association (CDA) has a numbering system for copper alloys. CDA is also the administrator of the Unified Numbering System (UNS), which was developed jointly by ASTM and the Society of Automotive Engineers (SAE). The CDA system consists of three digits. The UNS adds a letter *C* prefix to the three digits and a 00 suffix. The alloy 280 (*Muntz metal*) in the CDA system is C28000 in the UNS.

COMMON ALLOYS

The CDA places the most common copper alloys used in the construction industry into three groups. The A group contains copper, including alloy 110, which is 99.9% copper, and alloy 122, which is 0.02% phosphorus.

The B group contains the common brasses and extruded *architectural bronze*. It includes alloy 220 (*commercial brass*), which is 10% zinc; alloy 230 (*red brass*), which is 15% zinc; alloy 260 (*cartridge brass*), which is 30% zinc; alloy 280 (*Muntz metal*), which is 40% zinc; and alloy 385 (*architectural bronze*), which is 40% zinc and 3% lead.

The C group contains the so-called *white bronzes*. It includes alloy 651 (*low silicon bronze*), which has between 1.5%

and 3% silicon; alloy 655 (*high silicon bronze*), which has 3% silicon; alloy 745 (*nickel silver*), which has 25% zinc and 10% nickel; and alloy 796 (*leaded nickel silver*), which has 42% zinc, 10% nickel, 2% manganese, and 1% lead.

Casting alloys are special alloys made specifically for casting. They fall in the 800 and 900 Series of CDA alloy designations.

Other alloys are used also, of course, such as when color matching is required. The CDA booklet *Copper Brass Bronze Design Handbook, Architectural Applications* includes a chart showing which alloys match others in color. When a nonstructural metal item contains several different shapes or types of components, such as sheets, pipes, extrusions, fasteners, castings, and so forth, the different parts are most likely made from different alloys, even though the colors match.

5.5.1.2 Copper Alloy Products Production

Many copper alloy products are used in producing nonstructural metal items for use in buildings and related construction. Figure 5.5-1 is a partial list of such products, showing one possible series of color-matched alloys that may be used.

FORMING METHODS

The production of final copper alloy items for use in buildings starts with the forming or fabricating of copper alloy materials into useful products, such as those listed in Figure 5.5-1. There are four basic methods generally used to form these products: casting, rolling, extruding, and drawing.

Casting Intricate shapes are often made using the casting method. Castings may later be drilled, milled, or otherwise machined to produce necessary openings and close fits with adjacent materials.

Rolling Many copper alloy items used in buildings and related construction are at some point in their development run through a hot-rolling mill. Many of them are then passed through a cold-rolling process to make their surfaces brighter and smoother. Items produced by the rolling process include plates, sheets, shapes, rods, bars, and forging stock.

Extruding Many copper alloy products used in buildings and related construction are produced by extrusion. Extruded items include structural shapes, rods, bars, seamless tubes and pipes, and the kinds of special shapes often used in nonstructural metal items.

Extrusions are superior in virtually every way to rolled products. They are straighter, more accurate in dimension, and smoother. They are, however, limited to diagonal cross sections of 6 in. (152.4 mm) and a thickness of about ⅛ in. (3.2 mm).

Drawing Tubes and wire can be made by drawing copper alloy through a die, rather than pushing it, as in the extrusion process.

FABRICATION

Some rolled, extruded, and drawn products are used exactly as they come from the mill to fabricate items used in buildings. Others require only additional finishing. Many other components, however, are made by fabricating these products into usable shapes. The methods employed include additional roll forming, forging, bending, break forming, explosive forming, and spinning. Not all methods are applicable to all alloys. Refer to the CDA publication *Copper Brass Bronze Design Handbook, Architectural Applications* for an indication of the alloys that can be formed by each method.

Roll Forming The rolled products mentioned earlier are passed through an additional roll-forming process to make the final product. Roll forming at this stage consists of passing sheets through a series of shaped rollers to produce the desired shapes.

Forging Forging is done using one of four methods. The first two methods require closed steel dies. In the first method, *hammering*, the copper alloy is hammered into the die cavity by repeated blows of a *drop hammer*.

The second method, *pressing*, involves forcing the metal into the die cavity under hydraulic pressure.

In the third method, *hydroforming*, a punch or formed steel die is forced under hydraulic pressure against the metal, which is backed up by rubber.

In the fourth method, *stamping*, the metal is bent, shaped, cut, indented, embossed, coined, or formed by means of a press or hammer that is faced with a shaped die.

Bending Many copper alloy sheet metal and plate components used in buildings, such as door frames and the like, and many roofing-related copper items are shaped by bending.

Brake Forming Brake forming is probably the most used method for shaping copper alloy sheets used in buildings. Many copper alloy components used in nonstructural metal items, in shapes such as door frames and the like, and in many roofing-related copper alloy items are shaped by brake forming.

In essence, brake forming consists of squeezing sheet metal between a punch and a die block. Several consecutive operations are usually required to make a final shape.

Explosive Forming High-energy-rate forming is done by forcing metal against a single die, using a shock wave generated by detonation of small explosive charges.

Spinning A type of shear forming, *spinning*, is done by clamping the sheet metal to a form block and spinning the block and the metal against a roller, which grad-

FIGURE 5.5-1 Copper Alloy Products That Are Compatible in Color[a]

Product	ASTM Standard	CDA Alloy Number and Name
Extrusions	B 455	385, architectural bronze
Sheets, plates, and bars	B 370	280, Muntz metal
Round tubing	B 135	230, red brass
Castings	B 271	857
Sand castings	B 584	857
Wire and rods	—[b]	280, Muntz metal

[a]The alloys listed represent only a single group of materials that generally match in color and are by no means the only alloys used for the products listed. Refer to the CDA publication *Copper Brass Bronze Design Handbook, Architectural Applications* for an indication of other alloys used for the purposes listed.

[b]No current universally recognized standard for alloy 280 wire rods.

ually forces the metal to fit the shape of the form block. Spin forming is done using either power or hand tools.

5.5.1.3 Finishes

In its *Metal Finishes Manual for Architectural and Metal Products*, the National Association of Architectural Metal Manufacturers (NAAMM) specifies the requirements for three types of basic finishes for copper alloys: *mechanical finishes*, *chemical finishes*, and *coatings* (Fig. 5.5-2).

MECHANICAL FINISHES

Mechanical finishes include those left by the manufacturing process and those created by grinding, polishing, sandblasting, and rolling. NAAMM designates the types of mechanical finishes for use on copper alloys as: *as-fabricated*, *buffed*, *directional textured*, *nondirectional textured*, and *patterned*. Within these types are many finishes (Fig. 5.5-3). Each of these finishes is given a designation that consists of the letter *M* (mechanical) followed by a two-digit number. In the first four types, the first digit corresponds to the four types of finish. The second digit designates the specific finish within that type. Thus, M21 is a smooth specular buffed finish, and M31 is a fine satin, directional textured finish.

The fifth type, *patterned*, is designated M4*x* with the *x* to be specified. The *x* usually denotes a proprietary finish.

As-Fabricated Finish An *as-fabricated finish* is simply the natural finish imparted by the casting, rolling, or extruding

FIGURE 5.5-2 Summary of Standard Designations for Copper Alloy Finishes*

Mechanical Finishes (M)

As Fabricated	Buffed	Directional Textured	Nondirectional Textured
M10—Unspecified	M20—Unspecified	M30—Unspecified	M40—Unspecified
M11—Specular as fabricated	M21—Smooth specular	M31—Fine satin	M41—(Unassigned)
M12—Matte finish as fabricated	M22—Specular	**M32—Medium satin**	**M42—Fine matte**
M1x—Other (to be specified)	M2x—Other (to be specified)	**M33—Coarse satin**	M43—Medium matte
		M34—Hand rubbed	M44—Coarse matte
		M35—Brushed	M45—Fine shot blast
		M36—Uniform	M46—Medium shot blast
		M3x—Other (to be specified)	M47—Coarse shot blast
			M4x—Other (to be specified)

Chemical Finishes (C)

Nonetched Cleaned	Conversion Coatings	
C10—Unspecified	**C50—Ammonium chloride**	(patina)
C11—Degreased	**C51—Cuprous chloride–hydrochloric acid**	(patina)
C12—Cleaned	**C52—Ammonium sulfate**	(patina)
C1x—Other (to be specified)	C53—Carbonate	(patina)
	C54—Oxide	(statuary)
	C55—Sulfide	(statuary)
	C56—Selenide	(statuary)
	C5x—Other (to be specified)	

Coatings, Clear Organic (O)

Air Dry (General Architectural Work)	Thermoset (Hardware)	Chemical Cure
60—Unspecified	70—Unspecified	80—Unspecified
6x—Other (to be specified)	7x—Other (to be specified)	8x—Other (to be specified)

Coatings, Laminated (L)

	Thermoset (Hardware)
	L90—Unspecified
	L91—Clear polyvinyl fluoride
	L9x—Other (to be specified)

Coatings, Vitreous and Metallic

Since the use of these finishes in architectural work is rather infrequent, it is recommended that they be specified in full, rather than being identified by number.

Coatings, Oils and Waxes

These applied coatings are primarily used for maintenance purposes on site. Because of the broad range of materials in common use, it is recommended that, where desired, such coatings be specified in full.

*Those printed in **boldface type** are most frequently used for architectural work.

(Reprinted from *Metal Finishes Manual for Architectural and Metal Products* with permission of the National Association of Architectural Metal Manufacturers (NAAMM), Chicago, Illinois.)

FIGURE 5.5-3 Standard Designations for Mechanical Finishes

Type of Finish	Designation	Description	Examples of Method of Finishing
As fabricated	M10	Unspecified	Optional with finisher
	M11	Specular as fabricated	Cold rolling with polished steel rolls
	M12	Matte finish as fabricated	Cold rolling followed by annealing; hot rolling, extruding, casting
	M1x	Other	To be specified
Buffer	M20	Unspecified	Optional with finisher
	M21	Smooth specular	Cutting with aluminum oxide or silicon carbide compounds, starting with relatively coarse grits and finishing with 320 grit, using a peripheral wheel speed of 6000 fpm; followed by buffing with aluminum oxide buffing compounds using a peripheral wheel speed of 7000 fpm
	M22	Specular	Cutting with compounds as for the M21 finish, followed by a final light buffing
	M2x	Other	To be specified
Directional textured	M30	Unspecified	Optional with finisher
	M31	Fine satin	Wheel or belt polishing with aluminum oxide or silicon carbide abrasives of 180–240 grit, using a peripheral speed of 6000 fpm
	M32	Medium satin	Wheel or belt polishing with aluminum oxide or silicon carbide abrasives of 120–180 grit, using a peripheral wheel speed of 6000 fpm
	M33	Coarse satin	Wheel or belt polishing with aluminum oxide or silicon carbide abrasives of 80–120 grit, using a peripheral wheel speed of 6000 fpm
	M34	Hand rubbed	Hand rubbing with #0 stainless steel wool and solvent, #0 pumice and solvent, nonwoven abrasive mesh pad or Turkish oil and emery
	M35	Brushed	Brushing with rotary stainless steel, brass or nickel silver wire wheel; coarseness of finish controlled by diameter and speed of wheel and pressure exerted Wheel polishing with aluminum oxide or silicon carbide abrasive compounds of 50–150 grit on packed or loose muslin buffs at peripheral speeds of from 3500–5600 fpm
	M36	Uniform	Wheel or belt polishing in a single pass with aluminum oxide or silicon carbide abrasives of 60–80 grit, using a peripheral speed of 6000 fpm
	M3x	Other	To be specified
Nondirectional textured	M40	Unspecified	Optional with finisher
	M41	(number unassigned)	
	M42	Fine matte	Air blast with #100–#200 mesh silica sand or aluminum oxide; air pressure 30–90 lb.; gun 12 in. from work at an angle of 60–90 degrees
	M43	Medium matte	Air blast with #40–#80 mesh silica sand or aluminum oxide; air pressure 30–90 lb.; gun 12 in. from work at an angle of 60–90 degrees
	M44	Coarse matte	Air blast with #20 mesh silica sand or aluminum oxide; air pressure 30–90 lb.; gun 12 in. from work at an angle or 60–90 degrees
	M45	Fine shot blast	Air blast with S-70 metal shot
	M46	Medium shot blast	Air blast with S-230 metal shot
	M47	Coarse shot blast	Air blast with S-550 metal shot
	M4x	Other	To be specified

(Reprinted from *Metal Finishes Manual for Architectural and Metal Products* with permission of the National Association of Architectural Metal Manufacturers (NAAMM), Chicago, Illinois.)

process. It is the finish the copper alloy has when it leaves the initial manufacturing stage; it is not the final finish after the material has been fabricated into a metal item. The term *as-fabricated finish* corresponds to the term *mill finish* used for iron and steel.

Most copper alloys with an *as-fabricated* finish will display some imperfections, which must be taken into account when deciding whether to use them without further finishing. Castings, for example, have a rough, matte finish. Sand castings are rougher than die castings. Hot-rolled copper alloy items will have a dull surface, may be darker than cast, cold-rolled, or extruded copper alloy items, and will usually have some random discolorations as well. Extrusions are usually the same color as cold-rolled material, but are often left with parallel striations (*die lines*). Most imperfections tend to become worse in appearance when the metal is further formed and assembled into metal items.

Cold-rolled copper alloy items usually come closer to the final desired appearance than those produced by the other methods. The process permits some control by varying the amount of polishing used on the rollers. The smoother the rollers are, the closer the cold-rolled surface will be to a desirable finish. Even cold-rolled copper alloy items, however, may have stains, a coat of oil, or both.

NAAMM divides as-fabricated finishes on copper alloys into four classes (see Fig. 5.5-3). An *unspecified* as-fabricated finish is the natural finish imparted by casting, hot rolling, cold rolling with unpolished rollers, or extruding.

A *specular* as-fabricated finish is a mirrorlike finish that is produced by cold rolling with polished rollers. The specular finish may be applied to one or both sides. Castings, forgings, and extrusions cannot be given a specular finish.

A *matte* as-fabricated finish is a dull

finish achieved by annealing copper alloy items that have been produced by hot rolling, cold rolling with unpolished rolls, casting, or extruding.

The *other* category is reserved for special finishes that do not fit into one of the other three categories.

Buffed Finishes Copper alloy finishes created by a process of grinding, polishing, and buffing are called buffed finishes. There are four standard buffed finishes for copper alloys (see Fig. 5.5-3). An *unspecified* buffed finish is optional with the copper alloy finisher.

A *smooth specular* buffed finish, which is the brightest and most lustrous mechanical finish that can be developed on copper alloy items, is produced by successively grinding, polishing, and buffing.

A *specular* buffed finish is produced by the same methods as a smooth specular buffed finish, but is not as smooth.

The *other* category is reserved for special finishes that do not fit into one of the other three categories.

Directional Textured Finishes The standard designation system recognizes eight finish types in the directional textured finish category (see Fig. 5.5-3). An *unspecified* directional textured finish is optional with the copper alloy finisher.

The three *satin* finishes represent different degrees of fineness. They are all produced by wheel or belt polishing. *Spin finishing*, in which an abrasive cloth is held against the metal and rotated, followed by stainless steel wool or emery polishing, produces a bright finish with a fine pattern of concentric circles.

A *hand-rubbed* finish is an expensive finish type that is produced by rubbing the metal with a fine brass brush or a nonwoven abrasive mesh pad, abrasive cloths, or stainless steel wool. It is seldom used for general finishing, but is often used for final touch-up on the other satin finishes.

A *brushed* finish is a directional finish characterized by fine parallel scratches that are produced by stainless steel, nickel steel, or brass wire brushes, sander heads, impregnated plastic discs, and other means.

A *uniform* finish is produced by a single pass of a No. 80 grit belt.

The *other* category is reserved for special finishes that do not fit into one of the other seven categories.

Nondirectional Textured Finishes Finishes that are designated as nondirectional textured finishes are produced by abrasive blasting (see Fig. 5.5-3). Most of these finishes are *matte* in appearance with various degrees of roughness. They are not widely used on metal items for building construction. Their most extensive use is on castings. They should not be used on thin metal (¼ in. [6.4 mm] or less) because they tend to distort the metal.

An *unspecified* nondirectional textured finish is optional with the finisher.

The categories with *matte* in their names are produced by blasting with silica sand or aluminum oxide. Those with *shot blast* in their names are produced by steel shot blasting.

The *other* category is reserved for special finishes that do not fit into one of the other eight categories.

Combinations of finishes are sometimes used on surfaces that have relief; the incised areas are blasted with abrasives while the high areas are masked. Then the raised areas are polished.

Patterned Finishes Thin copper alloy sheets can be given a patterned finish by rolling *as-fabricated* sheet between rollers that have been shaped to the desired design. There is no specific standard category for patterned finishes.

CHEMICAL FINISHES

Chemicals are used on copper alloys for two primary purposes: to clean the copper alloys of foreign matter and to change the metal's color and provide a final finish. There are some other chemical treatments used on copper alloys, but they are not extensively used on building construction items. The two chemical finishes we are interested in are called *nonetch cleaning* and *conversion coatings*.

Chemical finishes for use on copper alloys are designated in a way similar to that for mechanical finishes, except that the letter is a *C* (Fig. 5.5-4). The first digit corresponds to the types, the second digit to the specific finish. Thus, C11 is a degreased cleaning, and C51 is a cuprous chloride–hydrochloride acid patina.

Conversion coatings are used to change the color of a copper alloy and provide a final finish for the metal. As a copper alloy ages, it oxidizes, changing its color and producing a coating that not only changes the appearance of the material, but protects it from further oxidization. Conversion coatings are not always completely successful in duplicating, by accelerated chemical means, the natural appearance and protective coating that result from aging. The coatings produced are oxides or sulfides of the metal.

Two basic types of conversion coatings are commonly used: those that produce a *patina* (verde antique) finish and those that produce an oxidized finish known as *statuary bronze*. Figure 5.5-4 lists the seven types of materials used to produce the more common conversion coatings. *Patinas*, which are produced using acid chlorides, are more difficult to control than *statuary bronze* finishes. Patinas often have variations in color, especially over large surface areas, and sometimes fail to adhere to the metal. They are also likely to stain adjacent materials.

Oxidized (*statuary*) finishes are somewhat more stable than patinas, and their color is easier to control. They come in three tones: *light*, *medium*, and *dark*. A range of color should be expected, however, even within a single tone group.

COATINGS

Coatings are similarly denoted. The letter *O* is used for clear organic coatings, and *L* is used for laminated coatings (see Fig. 5.5-2).

THE COMPLETE SYSTEM

All the appropriate designations are used together when describing a particular finish. For example, the finish M31-M34-O7*x* has a fine satin (M31), hand rubbed (M34) directional textured mechanical finish, and a clear thermoset organic coating (O7*x*). To make the designation complete, the organic coating, represented by the *x*, must be completely described, including its characteristics and manufacturer.

When an *x* appears in a designation, it is necessary to follow that designation with a description, usually of a proprietary finish.

5.5.2 ZINC

Zinc is a bluish-white metal that is plastic and highly corrosion resistant. It is not decorative, however, and is usually given an applied finish, usually paint and other organic coatings.

Nonstructural metal items whose strength is not paramount are sometimes made from zinc. Most zinc is used, however, as an alloying metal or as a coating for other metals. Refer to Section 5.1.4.4 for a discussion about zinc coatings, such as galvanizing on iron and steel.

FIGURE 5.5-4 Standard Designations for Chemical Finishes

Type of Finish	Designation	Description	Examples of Method of Finishing
Nonetched cleaned	C10	Unspecified	Optional with finisher.
	C11	Degreased	Treatment with organic solvent.
	C12	Chemically cleaned	Use of inhibited chemical cleaner.
	C1x	Other	To be specified.
Conversion coatings	C50	Ammonium chloride (*patina*)	Saturated solution of commercial sal ammoniac, spray or brush applied. Repeated applications may be required.
	C51	Cuprous chloride Hydrochloric acid (*patina*)	In 500 ml of warm water, dissolve 164 g cuprous chloride crystals, 117 ml hydrochloric acid, 69 ml glacial acetic acid, 80 g ammonium chloride, 11 g arsenic trioxide. Dilute to 1 liter. Apply to spray, brush, or stippling. Repeated applications may be required. Avoid use of aluminum containers.
	C52	Ammonium sulfate (*patina*)	Dissolve in 1 liter of warm water, 111 g ammonium sulfate, 3.5 g copper sulfate, 1.6 ml concentrated ammonia. Spray apply. Six–eight applications may be required under high humidity conditions.
	C53	Carbonate (*patina*)	Various formulations utilizing copper carbonate as the major constituent.
	C54	Oxide (*statuary*)	Principal formulations utilize aqueous solutions of copper sulfates and copper nitrates at temperatures of from 85°C to boiling; or permanganate solutions at temperatures of from 80°C to boiling, using immersion periods of from 30 sec. to 5 min.
	C55	Sulfide (*statuary*)	2%–10% aqueous solutions of ammonium sulfide, potassium sulfide, or sodium sulfide. Solutions swabbed or brushed on. Repeated application increases depth of color.
	C56	Selenide (*statuary*)	Principally proprietary formulations. Because the solutions are toxic, user preparation should be avoided. Follow manufacturer's directions for use without deviation.
	C5x	Other	To be specified.

(Reprinted from *Metal Finishes Manual for Architectural and Metal Products* with permission of the National Association of Architectural Metal Manufacturers (NAAMM), Chicago, Illinois.)

Another major use of zinc is as an alloying metal. It is alloyed with copper to make brass and with aluminum to make high-strength, heat-treatable alloys. It is also added to magnesium-aluminum alloys to increase their ductility.

Zinc is also a pigment in some paints, particularly in primer paints for galvanized surfaces and in paints for rusted steel surfaces.

5.5.3 LEAD

The heavy metal lead is soft and weak due to its ductility. Its properties, however, especially its ductility, resistance to corrosion, and density, which make it able to stop x-ray penetration, cause it to be used extensively in the construction industry.

Lead can be easily rolled into sheets and extruded to make pipes. It is used extensively in the chemical industry because of its resistance to corrosion by many chemicals, including sulfuric acid. Its chief uses in buildings are as radiation protection, as sound barriers and baffles, and as a coating for other metals to provide corrosion protection. Sheet lead is also used in roofing and associated rainware, but lead-coated copper is a more common roofing material.

Some paints contain lead, especially primer coatings on steel. This use may diminish in the future because of the concern about lead poisoning, especially of children, but lead paints will be found on existing metals for many years even if their use is eventually banned altogether.

5.5.4 CHROMIUM

Chromium, which is very hard and highly corrosion resistant, is used mostly as a coating on steel or other metals and as an alloying metal for steel. All stainless steel, for example, contains chromium.

Chromium is so hard that it resists polishing. Therefore, to produce the familiar highly polished chrome-plated surfaces, such as those on plumbing fixtures, it must be deposited over a highly polished steel surface, usually with an intermediate layer of nickel.

5.5.5 NICKEL

The primary use of nickel in buildings is as an alloying metal. *Monel metal*, for example, is an alloy of nickel with 31% copper. *Nickel silver* is an alloy of nickel and brass. Perhaps half the nickel produced in this country ends up in steel alloys.

At one time, nickel was a popular finish coating for steel. Today it is more likely to be used as an undercoating over steel for the more popular chromium.

5.5.6 TIN

Probably the best-known use of tin in the building industry today is as an alloying metal with copper to make bronze. Tin is also used as a coating for steel and iron. In earlier days, much roofing was made from tinplate. Alloys of tin and other metals, such as zinc, are also used for coatings on steel. Solder contains tin.

Tin is used less than zinc as a protective coating for steel, because tin is much rarer than zinc, especially in North America, and because tin does not protect steel from corrosion as effectively as zinc does.

5.5.7 GOLD

Gold is seldom used in the building industry, except to gild decorative items. It is worth mentioning here, however, because there are many, especially older buildings on which gold leaf was used extensively. Gilding with gold is not quite as expensive as it may appear, because of the metal's great malleability. Gold leaf is only 1/250,000 in. (0.0001016 mm) thick. At that thickness, 1 oz. (28.3495 g) of gold will cover 120 sq. ft. (11.15 m^2) of surface.

5.5.8 METAL-COATED METAL

Some metals are plated or coated with another metal. The base metal may be either ferrous or nonferrous, but the coating or plating metal is usually nonferrous. In most cases, the two metals are bonded together but remain separate metals and do not form an alloy. In an *alloy*, the constituent materials are melted together to form a single material. For example, the various types of steel are all alloys of iron and other materials. Bronze is an alloy of copper.

Finishes are applied to metals for two purposes: to protect the metal and to decorate it. Some metallic finishes, such as galvanizing and aluminum plating on steel, are primarily intended only to protect the metal and do not serve to decorate it, although they may be left as the final finish in inconspicuous locations. When decoration is desired, these kinds of metallic finish are usually given a finish coat of another material. Other metallic finishes, such as chromium plating, both decorate and protect the underlying metal.

Metal coatings for steel are discussed in Section 5.1.4.4.

Metal coatings on nonferrous metals are used mostly to give the underlying metal a desired finish. Some, however, also protect the underlying metal. Chromium is much harder and more scratch resistant, for example, than either aluminum or copper, which it is often used to coat. Some metallic coatings protect other materials from the coated metals. Lead-coated copper, for example, prevents water runoff from the copper from staining, corroding, or otherwise damaging adjacent materials.

5.5.8.1 Methods

There are six basic methods of coating metal with another metal: hot dipping, electroplating, spraying (metallizing), cladding, alloying (cementation), and fusion welding.

In the *hot-dip* process, the underlying metal is coated with a second metal by being immersed in a molten bath of the second metal. Zinc, aluminum, tin, and terne are applied to steel, for example, by the hot-dip process.

Zinc, cadmium, aluminum, and nickel are often deposited on steel, aluminum, and copper alloy by the *electroplating* process. Chromium and copper are similarly deposited over nickel that has been deposited on steel or aluminum. Chromium is also deposited over nickel that has been deposited over copper or steel.

Most metal coatings may be applied to steel and nonferrous metals by spraying, but probably the most frequent use of that method is to apply zinc and aluminum to steel. This is the only method, other than touch-up galvanizing repair by brush, that may be effectively used in field applications.

Steel is covered with copper, stainless steel, and aluminum by the *cladding* process. This process, which may include dipping, rolling, electrolytic deposit, or electrowelding, differs depending on the metal that forms the cladding, but in all cases the cladding metal and the steel remain as separate materials.

Cementation is the forming of an alloy between a coating metal and the coated metal.

Fusion welding is, as it sounds, the fusing of a cladding metal to the base metal by welding.

The most common methods used in the building industry to coat steel with other metals are the *hot-dip* and the *electroplating* processes. *Galvanizing* (zinc coating) and *aluminizing* (aluminum coating) are the most common coatings used on steel.

The most common method used to coat nonferrous metals with other metal for use in the building industry is electroplating, although some dipping is also used.

5.5.8.2 Types

There are many types of metal-coated metal used in building construction. In addition to those discussed in Sections 5.1 and 5.4 and earlier in this section, other common types include those covered in the following paragraphs.

CADMIUM-PLATED STEEL

Steel fasteners and hardware items are sometimes electroplated with cadmium to provide an electrolytic separation between the steel and other materials to prevent galvanic corrosion. Since cadmium is toxic, cadmium-plated steel items should not be used where they will come in contact with people—especially children—or food. Cadmium-plated steel should comply with the requirements of ASTM B 766.

CHROMIUM-PLATED METAL

Chromium is used to plate steel, aluminum, and copper. Iron, carbon steel, and stainless steel are all chromium plated to produce products used in buildings. Applicable standards include ASTM B 177, B 254, B 320, and B 650.

Chromium is plated onto aluminum after zincating (see "Zinc-Plated Aluminum" below) or plating the aluminum with copper, brass, or nickel.

Copper can best be plated with chromium after a coating of nickel has been applied. Chromium-plated copper has many uses in the construction industry, especially for hardware, toilet accessories, and fasteners.

NICKEL-PLATED METAL

Nickel plating is seldom used alone in the building industry. It is sometimes used on steel as an intermediate coat beneath chromium or copper plating. Nickel plating on copper is used primarily as a base for chromium plating.

BRASS-PLATED STEEL

Brass-plated steel is sometimes used for door hardware and similar items.

COPPER-COATED STEEL

Copper-coated steel sheets are sometimes used in flashing and other sheet metal applications.

COPPER-PLATED ALUMINUM

Copper plating may be used as a base for chromium plating or as a finish.

TERNEPLATE

Terne is an alloy of lead and tin, usually in a ratio of 4 parts lead to 1 part tin. Terneplate, which is the "terne" used in older buildings, is sheet iron or steel coated with a layer of terne. Today's terne-coated steel (TCS) is copper-bearing steel or stainless steel coated with terne. Terneplate and TCS are used mostly in roofing, but may occasionally be found in other nonstructural metal items.

LEAD-COATED COPPER

Sheets of lead-coated copper are used extensively in roofing and flashing applications and in other nonstructural metal items, especially when water running across bare copper would be harmful to adjacent materials. Lead-coated copper is produced by electroplating or hot dipping. It may be on both sides of the sheet or on one side only. The standard for lead-coated copper is ASTM B 101, which lists two types and two classes. Type I denotes application by dipping. A Type II coating is electrodeposited. Class A Standard requires a lead coating weight of 6 to 7½ lb./100 sq. ft. (2.722 to 3.402 kg/9.29 m^2) of surface. Class B Heavy requires 10 to 15 lb./100 sq. ft. (4.536 to 6.804 kg/9.29 m^2). The weights given above are the weight of lead on each side of the sheet. Thus, the total weight of lead on a Class A Standard sheet with lead on both sides, which is the way B 101 lists it, is from 12 to 15 lb./100 sq. ft. (5.443 to 6.804 kg/9.29 m^2) of sheet.

ZINC-PLATED ALUMINUM

In a process called zincating, aluminum plate is immersed in a zincate bath to coat it with a thin film of zinc. Zincating is generally used to prepare aluminum for electroplating. The standard for zincating on aluminum is ASTM B 253.

GOLD PLATING (GILDING)

Gilding with gold is an ancient art, at least 4000 years old, that is still practiced today. Gold plating is the application of gold leaf to wood, plaster, metal, glass, or another material to simulate the look of solid gold. It does not include the application of gold paint. There are some imitation golds (called Dutch leaf) that are used to simulate real gold leaf, but that material tarnishes (oxidizes) even when protected by varnish. Pure gold needs no protection, since gold, for all practical purposes, does not oxidize. Even extremely thin high-carat gold leaf, when properly maintained and protected, will withstand most climates and pollution levels for between 30 and 90 years.

The purity of gold is measured by its weight in carats. Pure gold weighs 24 carats. Golds of lesser carat weight are alloys that include silver, copper, or both. Gold leaf can weigh as much as 23½ carats, but most is lighter because pure gold does not wear as well as its lower-carat alloys.

Gold leaf is available today in four forms. *Loose leaf* comes in loose sheets for use on interior surfaces only. *Patent leaf* comes in books of sheets adhered to backing paper. It is the form used for most architectural gilding, especially on exteriors. *Surface leaf* is used where close inspection is not critical, since it is made from scrap gold and has flaws and discolorations. Finally, *glass leaf* is used on signs and on glass. Among these forms there are many qualities and grades.

There are essentially three methods of gilding in general use today, although there are others. The three are *oil* (mordant) *gilding*, *water gilding*, and *glass gilding*.

Different gilders have their own ways of using each method. In essence, however, both the *oil* and *water* methods are started by applying several coats of a material known as *gesso*, which is a mixture of calcium carbonate (*gilders' whiting*) and glue. In oil gilding, the gesso is then coated with shellac or marine varnish. The gold leaf is next laid into a coat of oil sizing, which acts as an adhesive. In water gilding, the gold leaf is laid over the gesso in a wet mixture of alcohol and water called *gilders' liquor*. If the gold leaf is to be burnished, a layer of a special clay, called *burnishing clay*, is laid just ahead of the gold. In glass gilding, an *oil size* is sometimes used, but usually the gold is held in place by a *size* that is a mixture of gelatin and water.

5.6 Additional Reading

More information about the subjects discussed in this chapter can be found in the references listed in Section 5.7 and in the following publications.

Allen, Edward. 1999. *Fundamentals of Building Construction: Materials and Methods*. 3d ed. New York: John Wiley & Sons, Inc.

Ambrose, James. 1994. *Design of Building Trusses*. New York: John Wiley & Sons, Inc.

———. 1995. *Simplified Design of Building Structures*. 3d ed. New York: John Wiley & Sons, Inc.

———. 1997. *Simplified Design of Steel Structures*. 7th ed. New York: John Wiley & Sons, Inc.

Amon, Rene, David Fanella, Bruce Knobloch, and Atanu Mazumder. 1994. "Making Connections." *The Construction Specifier* (4)(April):120.

Case, John, Lord A. H. Chiver, and Carl T. F. Ross. 1999. *Strength of Materials and Structures*. 4th ed. New York: John Wiley & Sons, Inc.

Englekirk, Robert. 1994. *Steel Structures: Controlling Behavior Through Design*. New York: John Wiley & Sons, Inc.

Newman, Alexander. 1997. *Metal Building Systems: Design and Specifications*. New York: McGraw-Hill.

Nicastro, David H., and Kimball J. Beasley. 1995. "Tower Distress." *The Construction Specifier* (10)(October):96.

Scharff, Robert. 1996. *Residential Steel Framing Handbook*. New York: McGraw-Hill.

Smith, J. C. 1996. *Structural Steel Design: LRFD Approach*. 2d ed. New York: John Wiley & Sons, Inc.

Underwood, James R., and Michele Chiuini. 1998. *Structural Design: A Practical Guide for Architects*. New York: John Wiley & Sons, Inc.

Yu, Wei-Wen. 1991. *Cold-Formed Steel Design*. 2d ed. New York: John Wiley & Sons, Inc.

Zahner, L. William. 1995. *Architectural Metals: A Guide to Selection, Specification, and Performance*. New York: John Wiley & Sons, Inc.

5.7 Acknowledgments and References

ACKNOWLEDGMENTS

We gratefully acknowledge the assistance of the following organizations and individuals in preparing this chapter. We are also indebted to them for permission to use their illustrations when requested and for the use of their publications as references.

ALCOA Building Products
Allegheny Ludlum Steel Corporation
Aluminum Association (AA)
American Architectural Manufacturers Association (AAMA)
American Institute of Steel Construction (AISC)
American Iron and Steel Institute (AISI)
American Society for Testing and Materials (ASTM)
American Welding Society (AWS)
Bethlehem Steel Corporation
Ceco Corporation
Copper Development Association (CDA)
Inland Steel Company
ITM Ramset
Jones and Laughlin Corporation
Kaiser Aluminum and Chemical Corporation
Lukens, Incorporated
Modern Metals (magazine)
National Association of Architectural Metal Manufacturers (NAAMM)
Phillips Industries, Inc.
Reynolds Metals Company
Rheem Manufacturing Company
Rigidized Steel Corporation
Steel Deck Institute (SDI)
Steel Joist Institute (SJI)
TRW Nelson Company
Underwriters Laboratories, Inc. (UL)
United States Steel Corporation
Youngstown Sheet and Tube Company

REFERENCES

We would also like to thank the authors and publishers of the publications in the following list for their contribution to our research for this chapter.

Abate, Kenneth. 1989. "Metal Coatings, Fighting the Elements with Superior Paint Systems, Part I." *Metal Architecture* 5(6)(June):10, 69.

———. 1989. "Metal Coatings, Fighting the Elements with Superior Paint Systems, Part II." *Metal Architecture* 5(7)(July):40–41.

American Architectural Manufacturers Association (AAMA). 604.2-1977, "Voluntary Specifications for Residential Color Anodic Finishes." Schaumburg, IL: AAMA.

———. 606.1-76, "Voluntary Guide Specifications and Inspection Methods for Integral Color Anodic Finishes for Architectural Aluminum." Schaumburg, IL: AAMA.

———. 607.1-77, "Voluntary Guide Specifications and Inspection Methods for Clear Anodic Finishes for Architectural Aluminum." Schaumburg, IL: AAMA.

———. 608.1-77, "Voluntary Guide Specifications and Inspection Methods for Electrolytically Deposited Color Anodic Finishes for Architectural Aluminum." Schaumburg, IL: AAMA.

———. 609-93, "Voluntary Guide Specifications for Cleaning and Maintenance of Architectural Anodized Aluminum." Schaumburg, IL: AAMA.

———. 610.1-79, "Voluntary Guide Specifications for Cleaning and Maintenance of Painted Aluminum Extrusions and Curtain Wall Panels." Schaumburg, IL: AAMA.

———. 2603-98. "Voluntary Specification, Performance Requirements and Test Procedures for Pigmented Organic Coatings on Aluminum Extrusions and Panels." Schaumburg, IL: AAMA.

———. 2604-98, "Voluntary Specification, Performance Requirements and Test Procedures for High Performance Organic Coatings on Aluminum Extrusions and Panels." Schaumburg, IL: AAMA.

———. 2605-98, "Voluntary Specification, Performance Requirements and Test Procedures for Superior Performing Organic Coatings on Aluminum Extrusions and Panels." Schaumburg, IL: AAMA.

———. *Care and Handling of Architectural Aluminum from Shop to Site.* Curtain Wall Manual, Volume 10 (CW-10). Schaumburg, IL: AAMA.

American Institute of Steel Construction (AISC). *Manual of Steel Construction.* 9th ed. Chicago, IL: AISC.

American Iron and Steel Institute (AISI). 1993. SG-930, "Design and Fabrication of Cold-formed Steel Structures." Washington, DC: AISI.

———. 1996. SG-973, *Cold-formed Steel Design Manual.* Washington, DC: AISI.

———. 1996. SG-971, "Specification and Commentary for the Design of Cold-formed Steel Structural Members." 1996 ed. Washington, DC: AISI.

American Society for Testing and Materials (ASTM) Standards:

A 6/A 6M, "Standard Specification for General Requirements for Rolled Structural Steel Bars, Plates, Shapes, and Sheet Piling." West Conshohocken, PA: ASTM.

A 27/A 27M, " Standard Specification for Steel Castings, Carbon, for General Application." West Conshohocken, PA: ASTM.

A 36/A 36M, "Standard Specification for Carbon Structural Steel." West Conshohocken, PA: ASTM.

A 47/A 47M, "Standard Specification for Ferritic Malleable Iron Castings." West Conshohocken, PA: ASTM.

A 48, "Specification for Gray Iron Castings." West Conshohocken, PA: ASTM.

A 53/A 53M, "Standard Specification for Pipe, Steel, Black and Hot-Dipped, Zinc-Coated, Welded and Seamless." West Conshohocken, PA: ASTM.

A 82, "Standard Specification for Steel Wire, Plain, for Concrete Reinforcement." West Conshohocken, PA: ASTM.

A 108, "Standard Specification for Steel Bars, Carbon, Cold-Finished, Standard Quality." West Conshohocken, PA: ASTM.

A 121, "Specification for Zinc-Coated (Galvanized) Steel Barbed Wire." West Conshohocken, PA: ASTM.

A 123/A 123M, "Standard Specification for Zinc (Hot-Dip Galvanized) Coatings on Iron and Steel Products." West Conshohocken, PA: ASTM.

A 153/A 153M, "Standard Specification for Zinc Coating (Hot-Dip) on Iron and Steel Hardware." West Conshohocken, PA: ASTM.

A 167, "Standard Specification for Stainless and Heat-Resisting Chromium-Nickel Steel Plate, Sheet, and Strip." West Conshohocken, PA: ASTM.

A 185, "Specification for Steel Welded Wire, Fabric, Plain, for Concrete Reinforcement." West Conshohocken, PA: ASTM.

A 242/A 242M, "Standard Specification for High-Strength Low-Alloy Structural Steel." West Conshohocken, PA: ASTM.

A 269, "Standard Specification for Seamless and Welded Austenitic Stainless Steel Tubing for General Service." West Conshohocken, PA: ASTM.

A 276, "Standard Specification for Stainless Steel Bars and Shapes." West Conshohocken, PA: ASTM.

A 283/A 283M, "Standard Specification for Low and Intermediate Tensile Strength Carbon Steel Plates." West Conshohocken, PA: ASTM.

A 307, "Standard Specification for Carbon Steel Bolts and Studs, 60 000 PSI Tensile Strength." West Conshohocken, PA: ASTM.

A 312/A 312M, "Standard Specification for Seamless and Welded Austenitic Stainless Steel Pipes." West Conshohocken, PA: ASTM.

A 325 "Standard Specification for Structural Bolts, Steel, Heat Treated, 120/105 ksi Minimum Tensile Strength." West Conshohocken, PA: ASTM.

A 366/A 366M, "Standard Specification for Steel, Sheet, Carbon, Cold-Rolled, Commercial Quality." West Conshohocken, PA: ASTM.

A 424, "Standard Specification for Steel Sheet for Porcelain Enameling." West Conshohocken, PA: ASTM.

A 463/A 463M, "Standard Specification for Steel Sheet, Aluminum-Coated, by the Hot-Dip Process." West Conshohocken, PA: ASTM.

A 490, "Standard Specification for Heat-Treated Steel Structural Bolts, 150 ksi Minimum Tensile Strength." West Conshohocken, PA: ASTM.

A 497, "Standard Specification for Steel Welded Wire Fabric, Deformed, for Concrete Reinforcement." West Conshohocken, PA: ASTM.

A 500, "Standard Specification for Cold-Formed Welded and Seamless Carbon Steel Structural Tubing in Rounds and Shapes." West Conshohocken, PA: ASTM.

A 501, "Standard Specification for Hot-Formed Welded and Seamless Carbon Steel Structural Tubing." West Conshohocken, PA: ASTM.

A 510, "Standard Specification for General Requirements for Wire Rods and Coarse Round Wire, Carbon Steel." West Conshohocken, PA: ASTM.

A 536, "Standard Specification for Ductile Iron Castings." West Conshohocken, PA: ASTM.

A 554, "Standard Specification for Welded Stainless Steel Mechanical Tubing." West Conshohocken, PA: ASTM.

A 568/A 568M, "Standard Specification for Steel, Sheet, Carbon, and High-Strength, Low-Alloy, Hot-Rolled and Cold-Rolled, General Requirements for." West Conshohocken, PA: ASTM.

A 569, "Standard Specification for Steel, Carbon (0.15 Maximum, Percent), Hot-Rolled Sheet and Strip, Commercial Quality." West Conshohocken, PA: ASTM.

A 570/A 570M, "Standard Specification for Steel, Sheet and Strip, Carbon, Hot-Rolled, Structural Quality." West Conshohocken, PA: ASTM.

A 572/A 572M, "Standard Specification for High-Strength Low-Alloy Columbium-Vanadium Structural Steel." West Conshohocken, PA: ASTM.

A 575, "Standard Specifications for Steel Bars, Carbon, Merchant Quality, M-Grades." West Conshohocken, PA: ASTM.

A 588/A 588M, "Standard Specification for High-Strength Low-Alloy Structural Steel with 50 ksi [345 MPa] Minimum Yield Point to 4 in. [100 mm] Thick." West Conshohocken, PA: ASTM.

A 591/A 591M, "Standard Specification for Steel Sheet, Electrolytic Zinc-Coated, for Light Coating Mass Applications." West Conshohocken, PA: ASTM.

A 606-98, "Standard Specification for Steel, Sheet and Strip, High-Strength, Low-Alloy, Hot-Rolled and Cold-Rolled, with Improved Atmospheric Corrosion Resistance." West Conshohocken, PA: ASTM.

A 607, "Standard Specifications for Steel, Sheet and Strip, High-Strength, Low-Alloy Columbium or Vanadium, or Both, Hot-Rolled and Cold-Rolled." West Conshohocken, PA: ASTM.

A 611, "Standard Specification for Structural Steel (SS), Sheet, Carbon, Cold-Rolled." West Conshohocken, PA: ASTM.

A 615/A 615M, "Standard Specification for Deformed and Plain Billet-Steel Bars for Concrete Reinforcement." West Conshohocken, PA: ASTM.

A 641, "Standard Specifications for Zinc-Coated (Galvanized) Carbon Steel Wire." West Conshohocken, PA: ASTM.

A 653/A 653M, "Standard Specification for Steel Sheet, Zinc-Coated (Galvanized) or Zinc-Iron Alloy-Coated (Galvannealed) by the Hot-Dip Process." West Conshohocken, PA: ASTM.

A 663/A 663M, "Standard Specification for Steel Bars, Carbon, Merchant Quality, Mechanical Properties." West Conshohocken, PA: ASTM.

A 675/A 675M, "Standard Specification for Steel Bars, Carbon, Hot-Wrought, Special Quality, Mechanical Properties." West Conshohocken, PA: ASTM.

A 715, "Standard Specification for Steel Sheet and Strip, High-Strength, Low-Alloy, Hot-Rolled, and Steel Sheet, Cold-Rolled, High-Strength, Low-Alloy, with Improved Formability." West Conshohocken, PA: ASTM.

A 743/A 743M, "Standard Specifications for Castings, Iron-Chromium-Nickel, and Nickel-Base, Corrosion-Resistant, for General Application." West Conshohocken, PA: ASTM.

A 744/A 744M, "Standard Specification for Castings, Iron-Chromium-Nickel, Corrosion-Resistant, for Severe Service." West Conshohocken, PA: ASTM.

A 767/A 767M, "Standard Specification for Zinc-Coated (Galvanized) Steel Bars for Concrete Reinforcement." West Conshohocken, PA: ASTM.

A 780, "Standard Practice for Repair of Damaged and Uncoated Areas of Hot-Dip Galvanized Coatings." West Conshohocken, PA: ASTM.

A 786/786M, "Standard Specifications for Rolled Steel Floor Plates." West Conshohocken, PA: ASTM.

A 792/A 792M, "Standard Specification for Steel Sheet, 55% Aluminum-Zinc Alloy-Coated by the Hot-Dip Process." West Conshohocken, PA: ASTM.

A 875/A 875M, "Standard Specification for Steel Sheet, Zinc-5% Aluminum Alloy-Coated by the Hot-Dip Process." West Conshohocken, PA: ASTM.

A 996/A 996M, "Standard Specification for Rail-Steel and Axle-Steel Deformed Bars for Concrete Reinforcement." West Conshohocken, PA: ASTM.

B 26/B 26M, "Standard Specification for Aluminum-Alloy Sand Castings." West Conshohocken, PA: ASTM.

B 32, "Standard Specification for Solder Metal." West Conshohocken, PA: ASTM.

B 101, "Standard Specification for Lead-Coated Copper Sheet and Strip for Building Construction." West Conshohocken, PA: ASTM.

B 108, "Standard Specification for Aluminum-Alloy Permanent Mold Castings." West Conshohocken, PA: ASTM.

B 177, "Standard Guide for Chromium Electroplating on Steel for Engineering Use." West Conshohocken, PA: ASTM.

B 209, "Standard Specification for Aluminum and Aluminum Alloy Sheet and Plate." West Conshohocken, PA: ASTM.

B 210, "Standard Specification for Aluminum-Alloy Drawn Seamless Tubes." West Conshohocken, PA: ASTM.

B 211, "Standard Specification for Aluminum and Aluminum-Alloy Bar, Rod, and Wire." West Conshohocken, PA: ASTM.

B 221, "Standard Specification for Aluminum and Aluminum-Alloy Extruded Bars, Rods, Wire, Profiles, and Tubes." West Conshohocken, PA: ASTM.

B 247, "Standard Specification for Aluminum and Aluminum-Alloy Die Forgings, Hand Forgings, and Rolled Ring Forgings." West Conshohocken, PA: ASTM.

B 253, "Standard Guide for Preparation of Aluminum Alloys for Electroplating." West Conshohocken, PA: ASTM.

B 254, "Standard Practice for Preparation of and Electroplating on Stainless Steel." West Conshohocken, PA: ASTM.

B 320, "Standard Practice for Preparation of Iron Castings for Electroplating." West Conshohocken, PA: ASTM.

B 429, "Standard Specification for Aluminum-Alloy Extruded Structural Pipe and Tube." West Conshohocken, PA: ASTM.

B 483, "Standard Specification for Aluminum and Aluminum-Alloy Extruded Drawn Tubes for General Purpose Applications." West Conshohocken, PA: ASTM.

B 650, "Standard Specification for Electrodeposited Engineering Chromium Coatings on Ferrous Substrates." West Conshohocken, PA: ASTM.

B 766, "Standard Specification for Electrodeposited Coatings of Cadmium." West Conshohocken, PA: ASTM.

C 282, "Standard Test Method for Acid Resistance of Porcelain Enamels (Citric Acid Spot Test)." West Conshohocken, PA: ASTM.

C 283, "Standard Test Method for Resistance of Porcelain Enamel Utensils to Boiling Acid." West Conshohocken, PA: ASTM.

C 286, "Standard Terminology Relating to Porcelain Enamel and Ceramic-Metal Systems." West Conshohocken, PA: ASTM.

C 346, "Standard Test Method for 45-deg Specular Gloss of Ceramic Materials." West Conshohocken, PA: ASTM.

C 448, "Standard Test Method for Abrasion Resistance of Porcelain Enamel." West Conshohocken, PA: ASTM.

C 538, "Standard Test Method for Color Retention of Red, Orange, and Yellow Porcelain Enamels." West Conshohocken, PA: ASTM.

C 645, "Standard Specification for Nonstructural Steel Framing Members." West Conshohocken, PA: ASTM.

C 703, "Standard Test Method for Spalling Resistance of Porcelain Enameled Aluminum." West Conshohocken, PA: ASTM.

C 955, "Standard Specification for Load-Bearing (Transverse and Axial) Steel Studs, Runners (Track), and Bracing or Bridging for Screw Application of Gypsum Board and Metal Plaster Bases." West Conshohocken, PA: ASTM.

C 1007, "Standard Specifications for Installation of Load-Bearing (Transverse and Axial) Steel Studs and Accessories." West Conshohocken, PA: ASTM.

D 1730, "Standard Practices for Preparation of Aluminum and Aluminum-Alloy Surfaces for Painting." West Conshohocken, PA: ASTM.

D 1731, "Standard Practices for Preparation of Hot-Dip Aluminum Surfaces for Painting." West Conshohocken, PA: ASTM.

D 2092, "Standard Guide for Preparation of Zinc-Coated (Galvanized) Steel Surfaces for Painting." West Conshohocken, PA: ASTM.

D 3794, "Standard Practice for Testing Coil Coatings." West Conshohocken, PA: ASTM.

E 94, "Standard Guide for Radiographic Testing." West Conshohocken, PA: ASTM.

E 142, "Standard Method of Controlling Quality of Radiographic Testing." West Conshohocken, PA: ASTM.

E 165, "Standard Test Method for Liquid Penetrant Examination." West Conshohocken, PA: ASTM.

E 527, "Standard Practice for Numbering Metals and Alloys (UNS)." West Conshohocken, PA: ASTM.

E 709, "Standard Guide for Magnetic Particle Examination." West Conshohocken, PA: ASTM.

American Welding Society. AWS D1.1, *Structural Welding Code—Steel*. Miami, FL: AWS.

———. AWS D1.2, *Structural Welding Code—Aluminum*. Miami, FL: AWS.

Appleman, Bernard R. 1989. "Coatings for Steel Structures." *The Construction Specifier* 42(3)(March): 88–93.

Architectural Graphic Standards. See Ramsey/Sleeper.

Arcom Master Systems. MASTERSPEC®. Basic Sections:

05120, "Structural Steel." Salt Lake City, UT: Arcom.

05210, "Steel Deck." Salt Lake City, UT: Arcom.

05400, "Cold Formed Metal Framing." Salt Lake City, UT: Arcom.

05500, "Metal Fabrications." Salt Lake City, UT: Arcom.

05511, "Metal Stairs." Salt Lake City, UT: Arcom.

05521, "Pipe and Tube Railings." Salt Lake City, UT: Arcom.

05580, "Formed Metal Fabrications." Salt Lake City, UT: Arcom.

05700, "Ornamental Metal." Salt Lake City, UT: Arcom.

ASM International. 1990. *ASM Handbook*. Vol. 01, *Properties and Selection: Irons, Steels, and High-Performance Alloys*. Materials Park, OH: ASM International.

———. 1989. *ASM Handbook*. Vol. 17, *Nondestructive Evaluation and Quality Control*. Materials Park, OH: ASM International.

———. 1994. *ASM Handbook*. Vol. 05, *Surface Engineering*. Materials Park, OH: ASM International.

Bocchi, Greg. 1986. "Powder Coatings: A New Technology Takes Off." *The Construction Specifier* 39(9)(September):102–05.

———. 1988. "Powder Coatings Making Inroads in Metal Construction Market." *Metal Architecture* 4(4)(April): 8–9.

Brick Institute of America (BIA). 1980. *Compilation of Veneer/Metal Stud Problems*. Reston, VA: BIA.

———. *Technical Notes on Brick Con-*

struction. No. 28B. Rev. II. Feb. 1987. "Brick Veneer, Steel Stud Panel Walls." Reston, VA: BIA.

Building Design and Construction. 1988. "Focus on Metals in Building Construction." *Building Design and Construction* 29(6)(June):67.

Copper Development Association (CDA). 4104-1779, *Architectural Applications*. Greenwich, CT: CDA.

———. 4115-1929, *Copper in Architecture—Handbook*. Greenwich, CT: CDA.

———. 401/0R, *Sheet Copper Applications*. Greenwich, CT: CDA.

Council of American Building Officials (CABO). *One- and Two-Family Dwelling Code*. Falls Church, VA: CABO. This code is in the process of being transferred to the control of the International Code Council (ICC). (See Section 1.6.3.)

Factory Mutual Research Organization (FM). 1999. *Factory Mutual Research Approval Guide*. Norwood, MA: FM.

———. 1998. "Wind Loads to Roof Systems and Roof Deck Securement," *Loss Prevention Data Sheet No. 1-28*. Norwood, MA: FM.

Howell, J. Scott. 1987. "Architectural Cast Iron: Design and Restoration." *The Construction Specifier* 40(7) (July):70–74.

Johnson, Stephen. 1988. "Improvement in Domestic Steel Quality Translating into Better Metal Paneling." *Metal Architecture* 4(9)(September):5, 85.

Koller, Alice. 1981. "Hot Dip Galvanizing: How and When to Use It." *The Construction Specifier* 34(8)(September):47–51.

LaQue, F. E., and H. R. Copson, eds. 1963. *Corrosion Resistance of Metals and Alloys*. 2d ed. New York: Van Nostrand Reinhold.

Metal Architecture. 1987. "Building Repainting System Duplicates Coil Coatings in Appearance, Life Expectancy." *Metal Architecture* 3(11)(November):24.

———. 1987. "Guide to Metal Wall and Roof Panels." *Metal Architecture* 3(12)(December):20, 22–23, 28–31, 38.

———. 1988. "Guide to Pre-Insulated Panels." *Metal Architecture* 4(4) (April):40–41, 43–46.

———. 1988. "1988 Guide to Architectural Coil Coatings." *Metal Architecture* 4(11)(November):22–23.

———. 1988. "Selecting Fasteners: Manufacturer's Advice." *Metal Architecture* 4(12)(December):26–32.

Modern Metals. 1989. "Presses on Roll Form Lines: How to Choose the Best." *Modern Metals* 45(1)(February):10–20.

———. 1989. "Finishing Forum: Coil Anodized Sheet Survives with Flying Colors." *Modern Metals* 45(1)(February):22–28.

Moit, Dan. 1988. "Coatings for Metals: Preplan the Selection." *Metal Architecture* 4(9)(September):8.

Munger, Charles G. 1984. *Corrosion Prevention by Protective Coatings*. Houston, TX: National Association of Corrosion Engineers.

National Association of Architectural Metal Manufacturers (NAAMM). 1988. *Metal Finishes Manual*. Chicago, IL: NAAMM.

———. 1992. *Code of Standard Practice for the Architectural Metal Industry*. Chicago, IL: NAAMM.

———. 1992. *Metal Stairs Manual*. Chicago, IL: NAAMM.

———. 1995. *Pipe Railing Manual*. Chicago, IL: NAAMM.

Petersen, Maurice R. 1984. "Finishes on Metals: A View from the Field—Part 1." *The Construction Specifier* 37(12) (December):36–39.

———. 1985. "Finishes on Metals: A View from the Field—Part 2." *The Construction Specifier* 38(1)(January): 70–73.

Porcelain Enamel Institute (PEI). PEI-101, "Design & Fabrication of Metal of Porcelain Enamel." Nashville, TN: PEI.

———. PEI-1001 (ALS-100) "Specification for Architectural Porcelain Enamel." Nashville, TN: PEI.

———. "Color Guide for Architectural Porcelain Enamel." Nashville, TN: PEI.

———. "Weatherability of Porcelain Enamel." Nashville, TN: PEI.

Ramsey/Sleeper, The AIA Committee on Architectural Graphic Standards. 1998. *Architectural Graphic Standards*. 9th ed. New York: John Wiley & Sons, Inc.

Sagan, Vincent E. 1998. "Steel Stud Wall System Essentials." *The Construction Specifier* (5)(May):70.

Sheet Metal and Air Conditioning Contractors National Association (SMACNA). *Architectural Sheet Metal Manual*. 5th ed. Chantilly, VA: SMACNA.

Sivinski, Valerie A. 1986. "Preserving Historic Materials: Ferrous Metals." *Architecture* (November): 108–109.

Society for Protective Coatings (SSPC). 1995. *Good Paint Practice*. Vol. 1 of *Steel Structures Painting Manual*. 3rd ed. Pittsburgh, PA: SSPC.

———. *Systems and Specifications*. 1995. Vol. 2 of *Steel Structures Painting Manual*. 7th ed. Pittsburgh, PA: SSPC.

Society of Automotive Engineers (SAE). 1995. Document No. J1086, *Numbering Metals and Alloy*. Warrendale, PA: SAE.

Specialty Steel Industry of North America (SSINA). *Standard Practices for Stainless Steel Roofing, Flashing, and Copings*. Washington, DC: SSINA.

———. *Finishes for Stainless Steel*. Washington, DC: SSINA.

———. *Specifications for Stainless Steel*. Washington, DC: SSINA.

———. *Stainless Steel Fasteners*. Washington, DC: SSINA.

———. *Design Guidelines for the Selection and Use of Stainless Steel*. Washington, DC: SSINA.

Steel Deck Institute (SDI). *Design Manual for Composite Decks, Form Decks and Roof Decks*. Fox River Grove, IL: SDI.

Steel Joists Institute (SJI). 1994. *Standard Specifications and Load Tables*. 40th ed. Myrtle Beach, SC: SJI.

———. 1987. "Handling and Erection of Steel Joists and Joists Girders." *SJI Technical Digest No. 9*. Myrtle Beach, SC: SJI.

Underwriters Laboratories, Inc. (UL). *Fire Resistance Directory*. Northbrook, IL: UL.

———. *Building Materials Directory*. Northbrook, IL: UL.

U.S. Department of the Army, Corps of Engineers Guide Specifications, Military Construction:

1999. Section 05090, "Welding, Structural." Washington, DC: Office of the Chief of Engineers, Department of the Army.

1999. Section 05120, "Structural Steel." Washington, DC: Office of the Chief of Engineers, Department of the Army.

1997. Section 05210, "Steel Joists." Washington, DC: Office of the Chief of Engineers, Department of the Army.

1999. Section 05300, "Steel Decking." Washington, DC: Office of the Chief of Engineers, Department of the Army.

1999. Section 05400, "Cold Formed Steel

Framing." Washington, DC: Office of the Chief of Engineers, Department of the Army.

1997. Section 05500, "Miscellaneous Metal." Washington, DC: Office of the Chief of Engineers, Department of the Army.

Weaver, Martin E. 1989. "Fighting Rust, Part I: A Backgrounder." *The Construction Specifier* 42(5)(May):143–145.

———. 1988. "Acid Rain." *The Construction Specifier* 41(7)(July):54–62.

———. 1989. "Fighting Rust, Part II: Remedies." *The Construction Specifier* 42(6)(June):129–130.

CHAPTER 6

Wood and Plastics

Introduction

Applicable *MasterFormat*™ Sections

Properties of Wood

Lumber

Plywood and Other Panels

Treated Wood Foundations

General Framing Requirements

Conventional Framing and Furring

Other Framing Systems

Finish Carpentry

Plastic Fabrications

Additional Reading

Acknowledgments and References

Introduction

Wood is a renewable resource. Wise forest management assures a permanent supply of timber to meet foreseeable future needs. Foresters can grow trees faster than nature can without their assistance. As a result of reforestation and other conservation practices, the yearly growth of timber nationwide exceeds that harvested or lost to fire or disease.

Wood is highly prized for its decorative character, which is derived from a variety of grain, figure, texture, hue, and other natural markings. These properties are important in lumber, plywood, and laminated timber, as well as in other wood products, such as flooring, windows, doors, and millwork.

Applicable *MasterFormat*™ Sections

The following *MasterFormat* Level 2 sections are applicable to this chapter.

06050 Basic Wood and Plastic Materials and Methods
06100 Rough Carpentry
06200 Finish Carpentry
06400 Architectural Woodwork
06500 Structural Plastics
06600 Plastic Fabrications

6.1 Properties of Wood

The wide range of properties in wood makes it suitable for use in many structural, nonstructural, and decorative products and in hundreds of end uses. It has high strength relative to its weight in compression, tension, bending, and resistance to impact. It can be worked easily to desired shapes with simple tools, is highly durable when simple precautions are observed, and has excellent insulating qualities. The general properties of wood that set it apart from concrete, metals, plastics, and other materials should be understood so that wood and wood products are used properly to perform satisfactorily.

6.1.1 IMPORTANT TREE SPECIES

Figure 6.1-1 shows important commercial species of wood and the areas in which they grow. The division of species into two classes, *softwood* and *hardwood*, is a botanical difference and is not always descriptive of the softness or hardness of the wood. Because of fundamental differences in microstructure and properties, softwoods and hardwoods differ in use, size standards, and method of grading.

6.1.1.1 Softwoods

About 75% of the lumber produced in this country comes from softwood trees. This lumber is used as structural and framing lumber, sheathing, roofing, subflooring, exterior siding, flooring, trim, and interior paneling. Softwoods are called *conifers* or *coniferous* because most species bear cones. With few exceptions, softwoods have scalelike or needlelike leaves and are evergreen (Fig. 6.1-2).

6.1.1.2 Hardwoods

Lumber produced from hardwood trees is used as flooring, interior paneling, and for building cabinets and furniture. Hardwood trees, which are broad leaved, are called

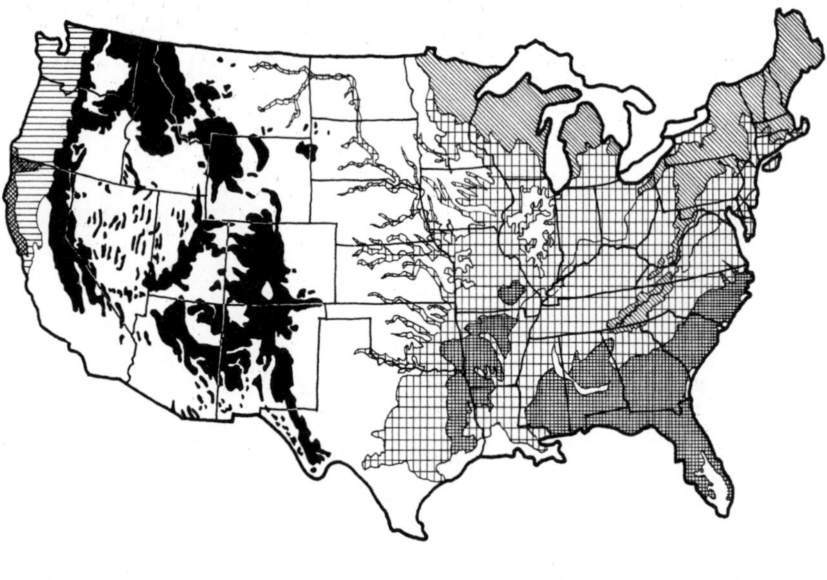

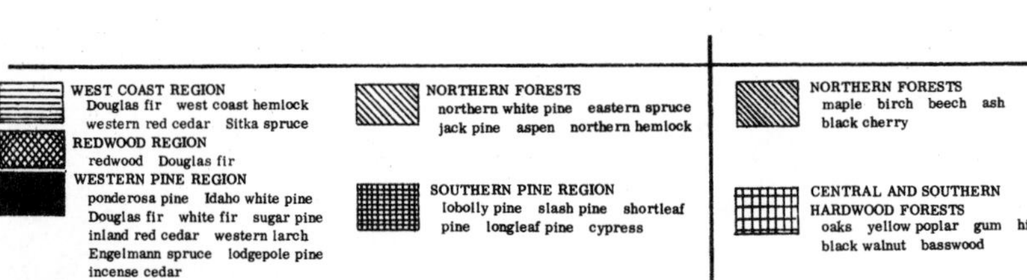

FIGURE 6.1-1 Areas of timber growth in the United States.

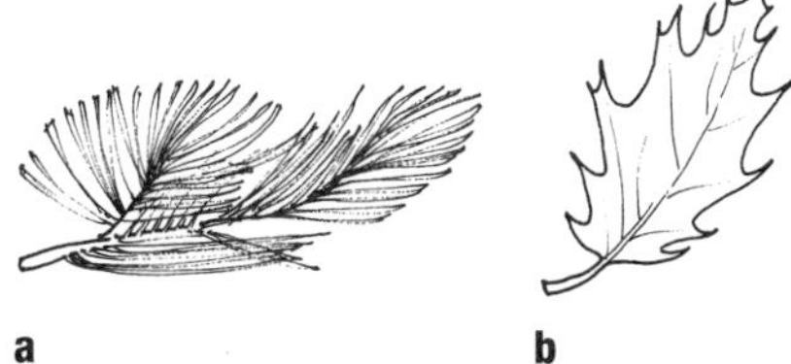

FIGURE 6.1-2 (a) Typical growth from a coniferous (softwood) tree. (b) A representative leaf from a deciduous (hardwood) tree.

deciduous because they shed their leaves at the end of each growing season, except in warmer climates (see Fig. 6.1-2).

6.1.2 WOOD GROWTH

A tree grows by the production of new cells in the *cambium* layer, a 3-to-5-cell-thick zone of active cell growth just beneath the inner bark (Fig. 6.1-3). As the cells mature, the inner layer of the cambium becomes sapwood and the outer layer becomes inner bark. The inner bark changes to outer bark as the cells become inactive. A wood cell does not develop further after it is formed and matures. All tree growth, consisting of increases in diameter and height, results entirely from new cell formation.

Each year completely new layers of wood cells envelop the preceding year's growth beneath the bark over the entire tree. Each year's growth can be seen as an annual growth ring consisting of a light-colored layer and a darker layer. In many species growth is more active during the spring, and cells formed then (*springwood*) tend to be large and thin walled as compared with cells formed later in the year when growth is slower (*summerwood*). The smaller, thick-walled cells of summerwood are denser and darker than springwood (Figs. 6.1-3 and 6.1-4). Because of its density, summerwood is both stronger and harder than springwood. In species such as Douglas fir and southern yellow pine, the amount of summerwood can be used to estimate the wood's strength and density.

The wood between the *pith* at the center of a tree and the *cambium* is divided into light-colored *sapwood*, whose cells are active in storing food and carrying sap from the roots up to the leaves, and generally darker *heartwood*, whose cells are inactive in the growth process but provide structural support for the tree. Generally speaking, sapwood and heartwood are of comparable strength. The gradual change from sapwood to heartwood involves the depositing of chemicals in the cellular structure of the heartwood, sometimes filling the cell cavities. In some species these chemicals are toxic, giving heartwood greater resistance to wood-destroying fungi and insects.

Most wood cells run parallel to the height of the tree and are called *fibers* because of their needlelike shape. However, running transversely from the *pith* to the *bark* are cells called *rays*, which convey food across the grain (Fig. 6.1-4). It is these rays that produce the decorative grain effects seen in lumber and plywood cut from species such as oak, where the rays are conspicuous.

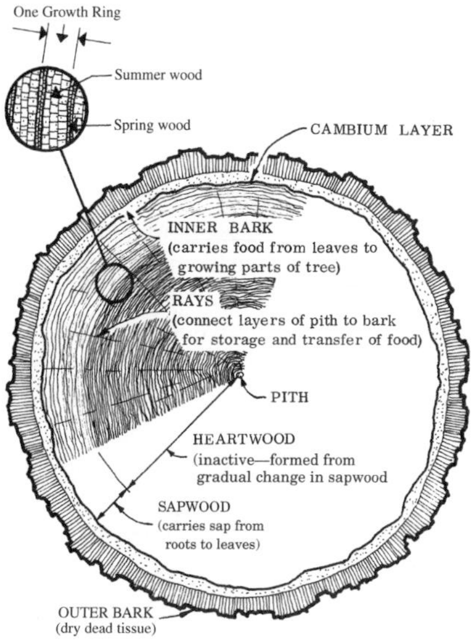

6.1.2.1 Growth Characteristics

Certain growth characteristics affect the grade and use of wood. These include *knots*, *shakes*, and *pitch pockets* (Fig. 6.1-5).

KNOTS

Knots are common growth characteristics that are formed when branches of a tree, which originate in the center at the pith, become enclosed within the wood during subsequent growth. If a branch dies and later falls off, the dead stubs are incorporated in the wood, become overgrown with new wood, and appear as knots when the wood is cut into lumber. Knots are detrimental to the structural properties of wood.

SHAKES

A *shake* is a grain separation occurring between annual growth rings, running along the grain parallel to the height of the tree (see Fig. 6.1-5). Shakes seldom develop in lumber unless they are present in the tree before it is felled.

PITCH POCKETS

A *pitch pocket* is a small, well-defined grain separation that may contain solid or liquid resin (see Fig. 6.1-5). Pitch pockets may be found in Douglas fir, western larch, pine, and spruce.

6.1.3 PHYSICAL PROPERTIES

The properties of wood are determined by its physical and chemical composition. Most characteristics of wood are related to its cellular structure.

FIGURE 6.1-3 Cross section of a living tree, showing active and inactive cell layers.

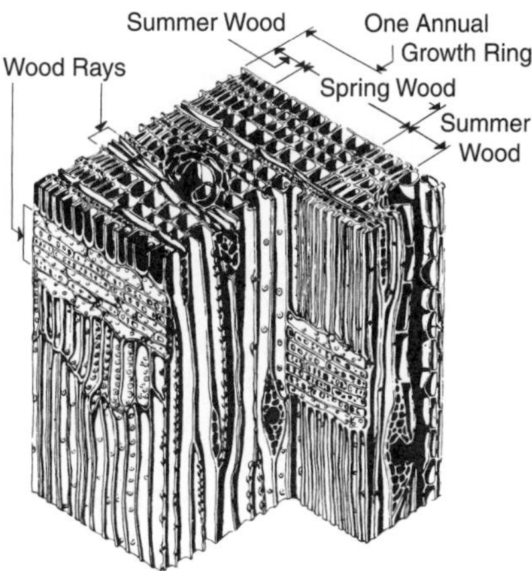

FIGURE 6.1-4 The cell structure of wood. This minute cube represents a block of wood about 1/32 in. (0.794 mm) on a side. Its top surface is parallel with the end surface of a log. (Reprinted from Forest Products Laboratory, U.S. Forest Service, USDA publication. Not subject to copyright in the United States.)

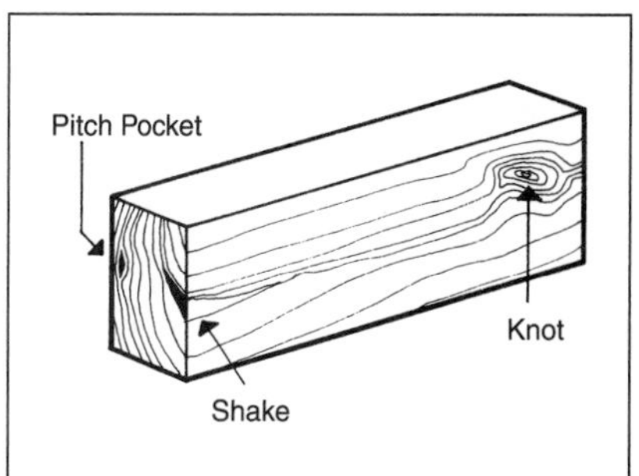

FIGURE 6.1-5 Growth characteristics of wood.

6.1.3.1 Chemical Composition

Wood cells, or fibers, are primarily *cellulose* cemented together with *lignin*. The wood structure is about 70% cellulose, between 12% and 28% lignin, and up to 1% ash-forming materials. These constituents give wood its hygroscopic properties, its susceptibility to decay, and its strength. The bond between individual fibers is so strong that when tested in tension they commonly tear apart rather than separate. The rest of wood, although not part of its structure, consists of extractives that give different species distinctive characteristics such as color, odor, and natural resistance to decay.

It is possible to dissolve the lignin in wood chips using chemicals, thus freeing the cellulose fibers. By further processing, these fibers can then be turned into a *pulp* from which paper and paperboard products are made. It is also possible to chemically convert cellulose so that it may be used to make textiles (such as rayon), plastics, and other products that depend on cellulose derivatives.

6.1.3.2 Hygroscopic Properties

Wood is *hygroscopic*, meaning that it expands when it absorbs moisture and shrinks when it dries or loses moisture. This property affects the end use of wood. Although the wet (green) condition is normal for wood throughout its life as a tree, most products made of wood require that it be used in a dry condition; therefore, *seasoning* by drying to an acceptable *moisture content* is necessary.

MOISTURE CONTENT

The moisture content of wood is the weight of water it contains, expressed as a percentage of the weight of the wood when oven dry. The weight of the water in wet wood can be twice that in wood that is oven dry. The oven-drying method, used to determine moisture content in the laboratory, is shown in Figure 6.1-6. For field use, a moisture meter is more practical: it gives an instantaneous reading by measuring the resistance to current flow between two pins driven into the wood.

In living trees the amount of moisture varies widely between different species, among individual trees of the same species, among different parts of a tree, and between sapwood and heartwood. Many softwoods have a large proportion of moisture in the sapwood and far less in the heartwood, while most hardwoods have about the same moisture content in both sapwood and heartwood. The extreme limits of moisture content in green softwoods can be shown by comparing the moisture content of the heartwood of Douglas fir and southern pine, which may be as low as 30%, to the moisture content of the sapwood of cedars and redwoods, which may be as high as 200%.

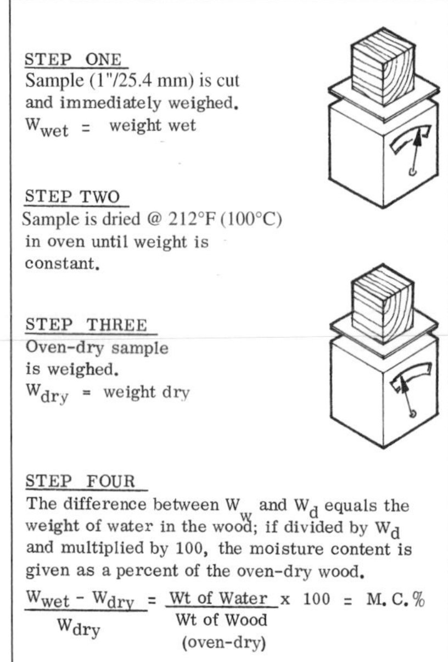

STEP ONE
Sample (1"/25.4 mm) is cut and immediately weighed.
Wwet = weight wet
STEP TWO
Sample is dried @ 212°F (100°C) in oven until weight is constant.
STEP THREE
Oven-dry sample is weighed.
Wdry = weight dry
STEP FOUR
The difference between Ww and Wd equals the weight of water in the wood; if divided by Wd and multiplied by 100, the moisture content is given as a percent of the oven-dry wood.
Wwet - Wdry / Wdry = Wt of Water / Wt of Wood (oven-dry) x 100 = M.C.%

FIGURE 6.1-6 Oven-drying method of measuring the moisture content of wood.

FIGURE 6.1-7 Moisture content and wood-cell shrinkage.

FIBER SATURATION POINT

Moisture in green wood is present in two forms: in the cell cavities as free water and within the cell fibers as absorbed water (Fig. 6.1-7). When wood dries, its cell fibers give off their absorbed water only after all the free water is gone and the adjacent cell cavities are empty. The point at which the fibers are still fully saturated, but the cell cavities are empty, is called the *fiber saturation point*. In most species this occurs at about 30% moisture content. The significance of this condition is that it represents the point at which shrinkage begins. Even lumber cut with a green moisture content as high as 200% can dry to the fiber saturation point (30% moisture content) with no shrinkage of the wood. Only when the cell fibers begin to give off their absorbed water and start to constrict does the wood shrink.

Therefore, all of the shrinkage wood can experience takes place between its fiber saturation point and a theoretical moisture content of 0% (oven-dry condition). Within this range, shrinkage is proportional to moisture loss. Once wood has reached a 30% moisture content or below that level, for every 1% loss or gain in moisture content, it shrinks or swells, respectively, about 1/30 of the total expansion or contraction. For example, at 15% moisture content wood will have experienced half of its total possible shrinkage. However, wood in service almost never reaches a 0% moisture content because of the influence of water vapor in the surrounding atmosphere. Therefore, the total possible shrinkage is far less important than the probable shrinkage under ordinary conditions.

EQUILIBRIUM MOISTURE CONTENT

After a tree is felled and cut into lumber, its moisture content begins to drop as moisture in the wood is lost to the surrounding air. The wood will then continue to give off or take on moisture until the moisture within the wood has reached a point of equilibrium with the moisture in the air. The moisture content of the wood at this point is called the *equilibrium moisture content*. The relation of atmospheric temperature and humidity to the equilibrium moisture content of wood is given in Figure 6.1-8, which shows, for example, that wood kept constantly in air at 70°F (21.1°C) with a 60% relative humidity eventually will balance at a moisture content of about 11%. Since temperature and

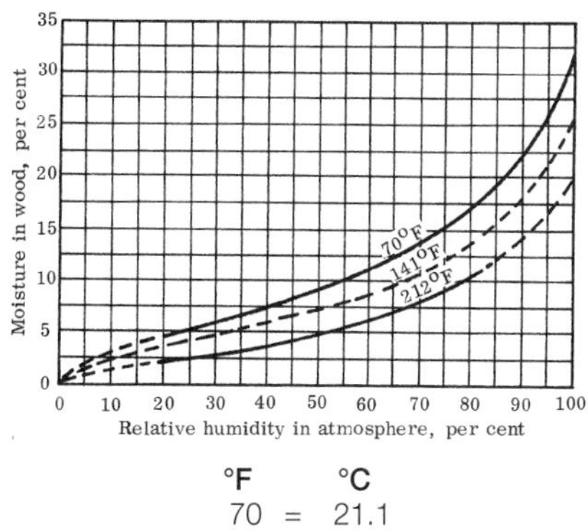

°F		°C
70	=	21.1
141	=	60.6
212	=	100.0

FIGURE 6.1-8 Equilibrium moisture content for various temperatures and relative humidities.

humidity are not constant in service, wood is subject to variations in moisture content as it seeks an equilibrium with the surrounding air. These variations tend to be gradual and seasonal. In addition, the equilibrium moisture content of wood is usually less on the interior of heated buildings than in unheated buildings or in an exterior exposure.

Knowing the equilibrium moisture content in a particular location is significant because it permits one to predict the moisture content a wood will attain in service there. To ensure that wood will experience only minor dimensional changes, it should be fabricated and installed at a moisture content as close as possible to the equilibrium moisture content it will attain in use (Fig. 6.1-9). This is especially critical when dealing with strip flooring, wood doors, and items of millwork, where a proper fit between adjoining wood elements is important. Since the equilibrium moisture content of interior wood items in most of the country varies between 6% and 12%, these products are usually fabricated with a moisture content in this range. However, since wood may lose or absorb moisture during shipment, it may end up at the site with a moisture content substantially higher or lower than the local equilibrium moisture content. Therefore, it is advisable to condition such products by storing them in the space in which they will be installed for several days before installation.

Exterior trim, siding, and board sheathing are subject to greater variations in atmospheric humidity and temperature and, therefore, also in equilibrium moisture content. In addition, a given wood's equilibrium moisture content is usually higher on the exterior than it is indoors and may differ from season to season within the same area, as well as from region to region (see Fig. 6.1-9). However, the conditions of installation and service are not as demanding for exterior elements as they are for interior ones, so these are normally fabricated with a moisture content between 12% and 15%, which is considered adequate for exterior use in all locations and for all purposes.

Framing lumber is surfaced either with a moisture content higher than 19% (grade marked S-GRN, meaning surfaced green) or at a maximum moisture content of 19% (grade marked S-DRY). Framing lumber may be surfaced at a moisture content of 15% or lower (grade marked MC 15). These grade marks should be those of an agency approved by the Board of Review of the American Lumber Standards Committee.

To increase the chances that no individual piece exceeds the maximum moisture content required by a particular grade stamp, lumber is seasoned to an average moisture content several percentage points lower than the maximum. Therefore, S-DRY lumber is dried to an average moisture content of 15%, and MC 15 lumber is dried to an average moisture content of 12%. Since the equilibrium moisture content of wood in most of the country is less than 15%, lumber that has been milled while green may, after air drying, actually have a much lower moisture content. In arid regions, framing lumber should have a maximum moisture content of 15% when it is installed on the interior of a building. In every other location, including damp regions, lumber for interior use should have a moisture content of 19% or lower (see Fig. 6.1-9).

In arid regions, either S-GRN or S-DRY lumber may achieve the desired maximum 15% moisture content if it is permitted to air dry at the yard or site, because the lumber loses moisture quickly under dry atmospheric conditions. Specifying MC 15 merely provides the visual assurance from a grade mark that this moisture content has been achieved at the mill. In other parts of the country, where air drying is not so rapid, it is best to specify dry lumber (19% maximum moisture content) for all 2-in. (50 mm)-thick or thicker framing lumber.

Large quantities of lumber less than 2 in. (50 mm) thick are used in residential construction. Boards and dimension lumber 2 in. (50 mm) thick or thinner should be purchased as either S-DRY or MC 15 and should be so identified by an appropriate grade mark.

In most species, boards and 2-in. (50 mm)-thick dimension lumber are graded and sold either dry or green. Southern pine lumber that has been graded according to Southern Pine Inspection Bureau (SPIB) rules is usually surfaced at 15% maximum moisture content and carries the KD or KD 15 grade mark. Lumber 3 in. (75 mm) and 4 in. (100 mm) thick is not readily available in dry grades in many locations, because most mills fabricate these sizes from green wood. However, such lumber, even when bearing an S-GRN grade mark, often has an acceptable moisture content because it has been air dried at the yard.

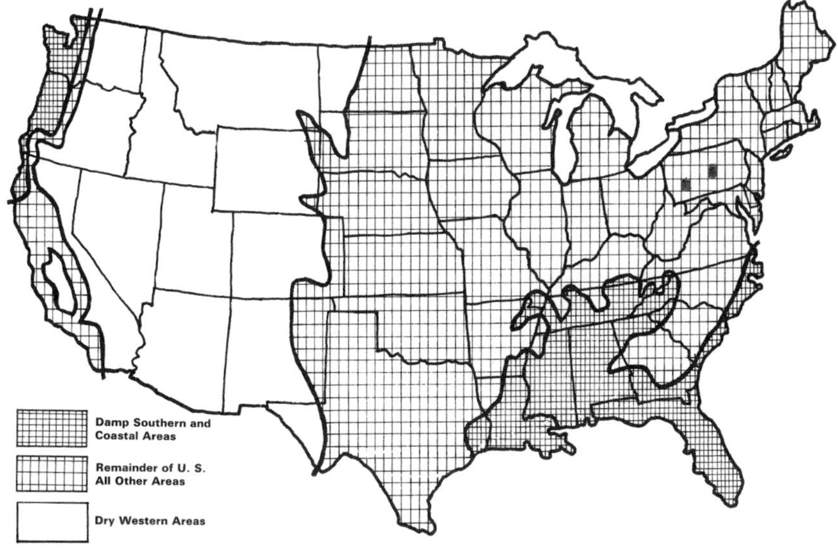

FIGURE 6.1-9 Atmospheric humidity regions of the United States.

Larger pieces, 3 in. (75 mm) and 4 in. (100 mm) thick, should have a 19% maximum moisture content and, in the absence of a grade mark, should be checked to ensure that the moisture content is at that level.

Moisture contents should be checked in accordance with the procedures outlined in American Society for Testing and Materials (ASTM) standards D 4442 and D 4444.

EFFECTS OF MOISTURE CONTENT ON PROPERTIES

As stated earlier, size variations in wood due to shrinkage occur only with changes in moisture content below the fiber saturation point. Shrinkage or swelling parallel to the grain (longitudinally with the height of the tree) is practically negligible and has little significance in construction applications. Across the grain, however, wood will shrink appreciably in both width and thickness. Shrinkage is greatest in the direction tangent to the annual growth rings and is about one-half to two-thirds as much across these rings (radial). Figure 6.1-10 shows the combined effects of tangential and radial shrinkage for various shapes due to drying from the green condition.

In general, hardwoods shrink more than softwoods, and heavier species more than lighter ones. Members with a large cross section do not shrink as much proportionately, because drying is not simultaneous in the inner and outer parts of the piece. Since the inner part dries more slowly than the outer portion, the wood near the surface is prevented from shrinking normally. The outer layers become set and tend to keep the inner portions from shrinking as they dry out. A 6 × 6 (150 × 150) or larger softwood structural lumber member will shrink about 1/32 in. (0.79 mm) per in. (25.4 mm) of face width as it dries from the green state and equalizes to its surroundings.

Softwood structural lumber 2 in. (50 mm) and 3 in. (75 mm) thick is usually partially or fully seasoned when marketed. The average amount of shrinkage in service for material of these sizes usually does not exceed 1/32 in. (0.79 mm) per in. (25.4 mm) of face width (Fig. 6.1-11).

The strength of wood increases as its moisture content decreases. This increase in strength is caused by (1) strengthening and stiffening of the cell fibers as they dry out and (2) increase in the compactness of the wood in a given volume. Wood dried from green to a 5% moisture content may add 2.5% to 20% to its density (weight of wood per unit of volume) and, in small pieces, end-crushing strength and bending strength may easily be doubled or, in some woods, even tripled. The increase in strength in small, clear specimens is much greater than in large timbers because the increase is offset by the influence of defects that develop in large members during seasoning.

Not all strength properties are equally affected by decreases in moisture content. For example, crushing and bending strength increase greatly, but some, such as stiffness, increase only moderately. Still others, like shock resistance, may show a very slight decrease. This is because shock resistance and toughness depend on pliability as well as strength, and, while drier wood is stronger, it does not bend as far before failure as does green wood. The increase in various strength properties due to a change from the fiber saturation point at 30% moisture content down to a moisture content of 19% is shown in Figure 6.1-12.

The decay resistance of wood is assured at a moisture content below 20%. In fact, the most effective and practical method of preventing decay is to control the moisture content of wood. Optimum conditions for decay occur when moisture content is about 25%. Wood that is installed and maintained below a moisture content of 20% will not decay. Most wood in use under protected conditions is below this level. If the moisture content will exceed 20% due to conditions of use or ex-

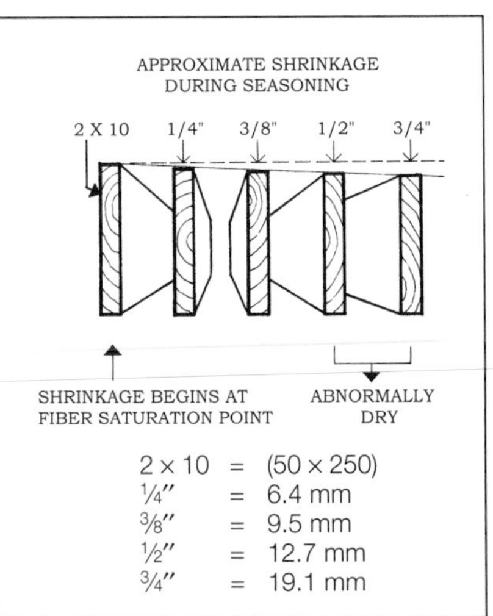

FIGURE 6.1-11 Approximate shrinkage during seasoning of a 2 × 10 (50 × 250) from green to a theoretical 0% moisture content.

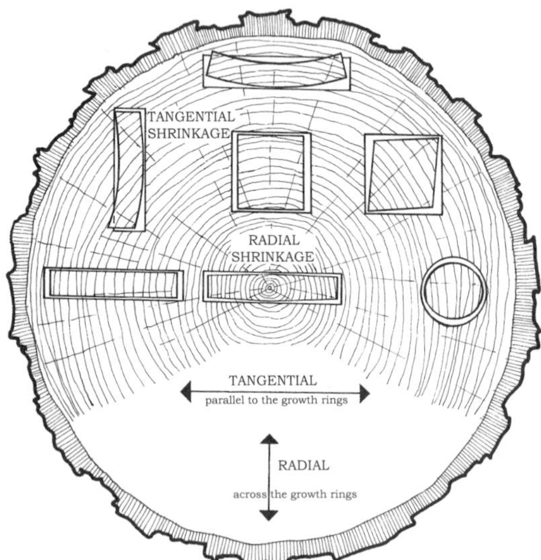

FIGURE 6.1-10 Characteristic shrinkage and distortion of members as affected by the direction of the annual rings.

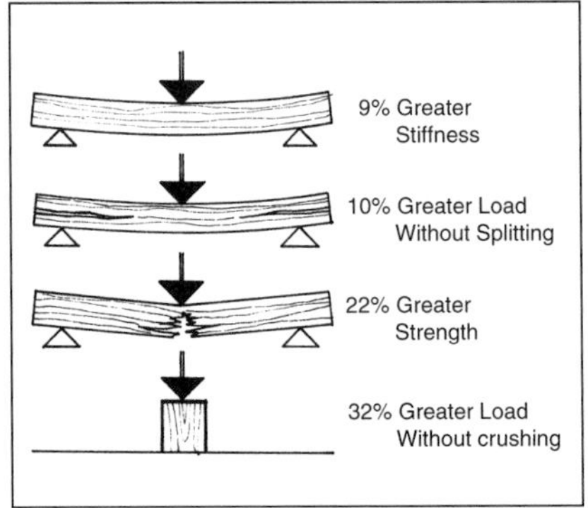

FIGURE 6.1-12 Approximate increase in wood structural properties with a decrease in moisture content from 30% to 19%.

posure, the wood should be treated with toxic chemicals to prevent decay (see Section 6.1.5) or the naturally durable heartwood of certain species may be used (see Section 6.1.3.6).

Paintability varies slightly with moisture content. The caveat that wet wood should not be painted refers to free water present on the surface and not to the moisture content of the wood. Wood painted at 16% to 20% moisture content holds paint slightly longer than wood painted at 10% moisture content.

Although the time required for paint to dry (harden) depends greatly on the paint, intensity of sunlight, temperature, and relative humidity, drying is retarded when the moisture content of a wood substrate is excessive. Therefore, wood to be painted should have a moisture content of 20% or less.

Many wood products depend extensively on gluing. The moisture content of the wood at the time of gluing affects both the strength of the bond and, later, the checking and warping of glued members. In general, glue bond improves at reduced moisture contents. Wood glues adhere well to wood with moisture contents up to 15% and even higher, but when large changes in moisture content occur after gluing, shrinking or swelling stresses may weaken both the wood and the glued joints. The moisture content of wood products at the time of gluing should be increased by the moisture in the glue, and the total should approximate the average moisture content that the glued member will attain in service.

Lumber used for interior millwork can be satisfactorily glued at a 5% to 6% moisture content; exterior millwork at 10% to 12% moisture content. Veneers used to make plywood must be dried to 2% to 6% moisture content at the time of gluing.

6.1.3.3 Specific Gravity

Specific gravity is a ratio comparison of the weight of a material to water. It is obtained by dividing the weight of the material by the weight of an equal volume of water. Water has a specific gravity of 1. Wood fiber, with a specific gravity of about 1.5, is heavier than water. However, dry wood of most species floats in water because a proportion of the volume is occupied by air-filled cell cavities. The specific gravity of most species when oven dry ranges from about 0.36 to 0.70 (Fig. 6.1-13). Variation among species in the size of cells and in the thickness of cell walls affects the amount of solid wood substance present and, therefore, the specific gravity. Specific gravity is an approximate measure of solid wood substance and a general index of strength properties, although specific gravity values may be affected somewhat by gums, resins, and extractives, which contribute little to strength. The relationship of specific gravity to wood strength is evident in the practice of assigning higher basic stress values to lumber such as Douglas fir and southern yellow pine designated as *dense*.

FIGURE 6.1-13 Average Specific Gravities of Some Wood Species

Species	Specific Gravity*
Aspen	0.39
Balsam fir	0.36
Beech–birch–hickory	0.71
Coast Sitka spruce	0.39
Cottonwood	0.41
Douglas fir–larch	0.50
Douglas fir–larch (north)	0.49
Douglas fir (south)	0.46
Eastern hemlock	0.41
Eastern hemlock–tamarack	0.41
Eastern hemlock–tamarack (north)	0.47
Eastern softwoods	0.36
Eastern spruce	0.41
Eastern white pine	0.36
Hem–fir	0.43
Hem–fir (north)	0.46
Mixed maple	0.55
Mixed oak	0.68
Mixed southern pine	0.51
Mountain hemlock	0.47
Northern pine	0.42
Northern red oak	0.68
Northern species	0.35
Northern white cedar	0.31
Ponderosa pine	0.43
Red maple	0.58
Red oak	0.67
Red pine	0.44
Redwood, close grained	0.44
Redwood, open grained	0.37
Sitka spruce	0.43
Southern pine	0.55
Spruce–pine–fir	0.42
Spruce–pine–fir (south)	0.36
Western cedars	0.36
Western cedars (north)	0.35
Western hemlock	0.47
Western hemlock (north)	0.46
Western white pine	0.40
Western woods	0.36
White oak	0.73
Yellow poplar	0.43

*The specific gravities are based on weight and volume when oven dry.

POROSITY OF WOOD

Void space in wood occurs as (1) coarse, microscopic capillaries within the basic cell system and (2) submicroscopic capillaries within the cell walls, which are at a maximum when the wood is saturated with moisture, as in the green or freshly cut condition. A single cubic foot of green wood has about 15 sq. ft. (1.394 m^2) of internal cell wall surface area and about 22,000 sq. ft. (2043.87 m^2) (about half the surface area of a football field) of internal submicroscopic capillary surface area in the cell walls (see Fig. 6.1-4).

This enormous internal surface area gives wood many of its unique properties, including:

1. Ease of impregnation with toxic chemicals for decay and insect protection.
2. Ease of impregnation with moisture repellents for slowing down moisture exchange to minimize shrinkage.
3. Excellent insulating properties.
4. Shrinking and swelling in response to variations in moisture content.
5. Excellent adherence of paint films and adhesives. Some synthetic resins adhere with such tenacity that they resist many exposure conditions.

6.1.3.4 Structural Properties

Unlike crystalline materials such as steel and concrete, wood is fibrous. The length of fibers varies from about 1/25 in. (1.016 mm) in hardwoods to between 1/8 in. (3.2 mm) and 1/3 in. (8.47 mm) in softwoods. The strength of wood, however, does not depend on the length of the fibers but on the thickness of their walls and on the direction of the fibers relative to applied loads. The fibers in wood are oriented with their length, essentially parallel to the vertical axis of the tree, and the strength of wood parallel to these fibers (parallel to the grain) is quite different from its strength perpendicular to the fibers (perpendicular to the grain).

External forces, called *loads* (such as wind, gravity, and those caused by such natural phenomena as earthquakes), produce internal resisting forces in structural members, known as *stresses. Tensile stresses* result from stretching forces; *compressive stresses* from squeezing forces. When a member is loaded so that the force is acting more or less perpendicular to its length, as is the case in a

FIGURE 6.1-14 Base Design Values for Visually Graded Dimension Lumber

Species and commercial grade	Size classification	Bending F_b	Tension parallel to grain F_t	Shear parallel to grain F_v	Compression perpendicular to grain $F_{c\perp}$	Compression parallel to grain F_c	Modulus of Elasticity E	Grading Rules Agency
		Design values in pounds per square inch (psi)						
ASPEN								
Select Structural		875	500	60	265	725	1,100,000	
No. 1	2″–4″ thick	625	375	60	265	600	1,100,000	
No. 2		600	350	60	265	450	1,000,000	NELMA
No. 3	2″ & wider	350	200	60	265	275	900,000	NSLB
Stud		475	275	60	265	300	900,000	WWPA
Construction	2″–4″ thick	700	400	60	265	625	900,000	
Standard		375	225	60	265	475	900,000	
Utility	2″–4″ wide	175	100	60	265	300	800,000	
BEECH-BIRCH-HICKORY								
Select Structural		1450	850	100	715	1200	1,700,000	
No. 1	2″–4″ thick	1050	600	100	715	950	1,600,000	
No. 2		1000	600	100	715	750	1,500,000	
No. 3	2″ & wider	575	350	100	715	425	1,300,000	NELMA
Stud		775	450	100	715	475	1,300,000	
Construction	2″–4″ thick	1150	675	100	715	1000	1,400,000	
Standard		650	375	100	715	775	1,300,000	
Utility	2″–4″ wide	300	175	100	715	500	1,200,000	
COTTONWOOD								
Select Structural		875	525	65	320	775	1,200,000	
No. 1	2″–4″ thick	625	375	65	320	625	1,200,000	
No. 2		625	350	65	320	475	1,100,000	
No. 3	2″ & wider	350	200	65	320	275	1,000,000	NSLB
Stud		475	275	65	320	300	1,000,000	
Construction	2″–4″ thick	700	400	65	320	650	1,000,000	
Standard		400	225	65	320	500	900,000	
Utility	2″–4″ wide	175	100	65	320	325	900,000	
DOUGLAS FIR-LARCH								
Select Structural		1500	1000	95	625	1700	1,900,000	
No. 1 & Btr	2″–4″ thick	1200	800	95	625	1550	1,800,000	
No. 1		1000	675	95	625	1500	1,700,000	
No. 2	2″ & wider	900	575	95	625	1350	1,600,000	WCLIB
No. 3		525	325	95	625	775	1,400,000	WWPA
Stud		700	450	95	625	850	1,400,000	
Construction	2″–4″ thick	1000	650	95	625	1650	1,500,000	
Standard		575	375	95	625	1400	1,400,000	
Utility	2″–4″ wide	275	175	95	625	900	1,300,000	
DOUGLAS FIR-LARCH (NORTH)								
Select Structural	2″–4′ thick	1350	825	95	625	1900	1,900,000	
No. 1/No. 2		850	500	95	625	1400	1,600,000	
No. 3	2″ & wider	475	300	95	625	825	1,400,000	NLGA
Stud		650	400	95	625	900	1,400,000	
Construction	2″–4″ thick	950	575	95	625	1800	1,500,000	
Standard		525	325	95	625	1450	1,400,000	
Utility	2″–4″ wide	250	150	95	625	950	1,300,000	
DOUGLAS FIR-SOUTH								
Select Structural		1350	900	90	520	1600	1,400,000	
No. 1	2″–4″ thick	925	600	90	520	1450	1,300,000	
No. 2		850	525	90	520	1350	1,200,000	
No. 3	2″ & wider	500	300	90	520	775	1,100,000	WWPA

FIGURE 6.1-14 *(Continued)*

Species and commercial grade	Size classification	Bending F_b	Tension parallel to grain F_t	Shear parallel to grain F_v	Compression perpendicular to grain $F_{c\perp}$	Compression parallel to grain F_c	Modulus of Elasticity E	Grading Rules Agency
		Design values in pounds per square inch (psi)						
DOUGLAS FIR-SOUTH *(Continued)*								
Stud		675	425	90	520	850	1,100,000	
Construction	2″–4″ thick	975	600	90	520	1650	1,200,000	
Standard		550	350	90	520	1400	1,100,000	
Utility	2″–4″ wide	250	150	90	520	900	1,000,000	
EASTERN HEMLOCK-BALSAM FIR								
Select Structural		1250	575	70	335	1200	1,200,000	
No. 1	2″–4″ thick	775	350	70	335	1000	1,100,000	
No. 2		575	275	70	335	825	1,100,000	
No. 3	2″ & wider	350	150	70	335	475	900,000	NELMA
Stud		450	200	70	335	525	900,000	NSLB
Construction	2″–4″ thick	675	300	70	335	1050	1,000,000	
Standard		375	175	70	335	850	900,000	
Utility	2′–4″ wide	175	75	70	335	550	800,000	
EASTERN HEMLOCK-TAMARACK								
Select Structural		1250	575	85	555	1200	1,200,000	
No. 1	2″–4″ thick	775	350	85	555	1000	1,100,000	
No. 2		575	275	85	555	825	1,100,000	
No. 3	2″ & wider	350	150	85	555	475	900,000	NELMA
Stud		450	200	85	555	525	900,000	NSLB
Construction	2″–4″ thick	675	300	85	555	1050	1,000,000	
Standard		375	175	85	555	850	900,000	
Utility	2″–4″ wide	1785	75	85	555	550	800,000	
EASTERN SOFTWOODS								
Select Structural		1250	575	70	335	1200	1,200,000	
No. 1	2″–4″ thick	775	350	70	335	1000	1,100,000	
No. 2		575	275	70	335	825	1,100,000	
No. 3	2″ & wider	350	150	70	335	475	900,000	
Stud		450	200	70	335	525	900,000	NELMA
Construction	2″–4″ thick	675	300	70	335	1050	1,000,000	NSLB
Standard		375	175	70	335	850	900,000	
Utility	2″–4″ wide	175	75	70	335	550	800,000	

(All species except Southern Pine) (Tabulated design values are for normal load duration and dry service conditions. See NDS 2.3 for a comprehensive description of design value adjustment factors.)

(Courtesy of American Forest & Paper Association, Washington, DC.)

beam, the resulting stress is called *bending*. A simple span member in bending develops compressive stresses in the upper part and tensile stresses in the lower part of its cross section.

The strength values of wood are determined by testing small, clear specimens in bending, compression, shear, and so forth, and assigning basic stresses (basic strength values) to different species. Because small, clear pieces are seldom used in construction, lower working stresses (also called allowable unit stresses) must be derived to take into account the effect of knots, slope-of-grain, checks, and other characteristics that reduce the strength of a member to a value less than that for clear wood.

Accordingly, the maximum allowable unit stress (*working stress*) that should be permitted to exist on a grade of lumber is determined by the effect of the maximum size and most unfavorable position of any strength-reducing characteristics permitted in the grade. The working stresses of various species of dimension lumber are shown in tables in the *NDS Supplement: Design Values for Wood Construction®* to the *National Design Specifications (NDS®) for Wood Construction®* of the American Forest & Paper Association (AF&PA). Figure 6.1-14 is part of a table from the 1997 edition of the *NDS Supplement* for visually graded dimension lumber. Figure 6.1-15 is the *NDS Supplement* table for mechanically graded dimension lumber.

The *NDS Supplement* states that "bending design values, F_b, for dimension lumber 2″ to 4″ thick, shall be multiplied by the repetitive member factor, $C_r = 1.15$, when such members are used as joists, truss chords, rafters, studs, planks, decking, or similar members, which are in contact or spaced not more than 24 inches on center, and are not less than 3 in number and are joined by floor, roof or other load distributing elements adequate to support the design load."

The tabular design values must also be adjusted by a wet service factor, C_M, when the wood is used with a moisture content

FIGURE 6.1-15 Design Values for Mechanically Graded Dimension Lumber

Species and commercial grade	Size classification	Bending F_b	Tension parallel to grain F_t	Compression parallel to grain F_c	Modulus of Elasticity E	Grading Rules Agency
		Design values in pounds per square inch (psi)				
MACHINE STRESS RATED (MSR) LUMBER						
900f–1.0E		900	350	1050	1,000,000	WCLIB, WWPA
1200f–1.2E		1200	600	1400	1,200,000	NLGA, WCLIB, WWPA
1250f–1.4E		1250	800	1475	1,400,000	WCLIB
1350f–1.3E		1350	750	1600	1,300,000	NLGA, WCLIB, WWPA
1400f–1.2E		1400	800	1600	1,200,000	NLGA
1450f–1.3E		1450	800	1625	1,300,000	NLGA, WCLIB, WWPA
1500f–1.3E		1500	900	1650	1,300,000	WWPA
1500f–1.4E		1500	900	1650	1,400,000	NLGA, WCLIB, WWPA
1600f–1.4E		1600	950	1675	1,400,000	NLGA
1650f–1.3E		1650	1020	1700	1,300,000	NLGA, WWPA
1650f–1.5E		1650	1020	1700	1,500,000	NLGA, SPIB, WCLIB, WWPA
1650f–1.6E		1650	1175	1700	1,600,000	WCLIB, WWPA
1700f–1.6E		1700	1175	1725	1,600,000	WCLIB
1750f–2.0E		1750	1125	1725	2,000,000	WCLIB
1800f–1.5E		1800	1300	1750	1,500,000	NLGA, WWPA
1800f–1.6E		1800	1175	1750	1,600,000	NLGA, SPIB, WCLIB, WWPA
1950f–1.5E	2″ & less in thickness	1950	1375	1800	1,800,000	SPIB, WWPA
1950f–1.7E		1950	1375	1800	1,700,000	NLGA, SPIB, WCLIB, WWPA
2000f–1.6E		2000	1300	1825	1,600,000	NLGA
2100f–1.8E	2″ & wider	2100	1575	1875	1,800,000	NLGA, SPIB, WCLIB, WWPA
2250f–1.7E		2250	1750	1925	1,700,000	NLGA, WWPA
2250f–1.8E		2250	1750	1925	1,800,000	NLGA, WCLIB, WWPA
2250f–1.9E		2250	1750	1925	1,900,000	NLGA, SPIB, WCLIB, WWPA
2400f–1.8E		2400	1925	1975	1,800,000	NLGA, WWPA
2400f–2.0E		2400	1925	1975	2,000,000	NLGA, SPIB, WCLIB, WWPA
2500f–2.2E		2500	1750	2000	2,200,000	WCLIB
2550f–2.1E		2550	2050	2025	2,100,000	NLGA, SPIB, WCLIB, WWPA
2700f–2.0E		2700	1800	2100	2,000,000	WCLIB, WWPA
2700f–2.2E		2700	2150	2100	2,200,000	NLGA, SPIB, WCLIB, WWPA
2850f–2.3E		2850	2300	2150	2,300,000	NLGA, SPIB, WCLIB, WWPA
3000f–2.4E		3000	2400	2200	2,400,000	NLGA, SPIB
MACHINE EVALUATED LUMBER (MEL)						
M-5		900	500	1050	1,100,000	SPIB
M-6		1100	600	1300	1,000,000	SPIB
M-7		1200	650	1400	1,100,000	SPIB
M-8		1300	700	1500	1,800,000	SPIB
M-9		1400	800	1600	1,400,000	SPIB
M-10		1400	800	1600	1,200,000	NLGA, SPIB
M-11		1550	850	1675	1,500,000	NLGA, SPIB
M-12		1600	850	1675	1,600,000	NLGA, SPIB
M-13		1600	950	1675	1,400,000	NLGA, SPIB
M-14		1800	1000	1750	1,700,000	NLGA, SPIB
M-15		1800	1100	1750	1,500,000	NLGA, SPIB
M-16		1800	1300	1750	1,500,000	SPIB
M-17	2″ & less in thickness	1950	1300	2050	1,700,000	SPIB
M-18		2000	1200	1825	1,800,000	NLGA, SPIB
M-19	2″ & wider	2000	1300	1825	1,600,000	NLGA, SPIB
M-20		2000	1600	2100	1,900,000	SPIB
M-21		2300	1400	1950	1,900,000	NLGA, SPIB
M-22		2350	1500	1950	1,700,000	NLGA, SPIB
M-23		2400	1900	1975	1,800,000	NLGA, SPIB
M-24		2700	1800	2100	1,900,000	NLGA, SPIB
M-25		2750	2000	2100	2,200,000	NLGA, SPIB

FIGURE 6.1-15 *(Continued)*

Species and commercial grade	Size classification	Bending F_b	Tension parallel to grain F_t	Compression parallel to grain F_c	Modulus of Elasticity E	Grading Rules Agency
		Design values in pounds per square inch (psi)				
MACHINE EVALUATED LUMBER (MEL) *(Continued)*						
M-26		2800	1800	2150	2,000,000	NLGA, SPIB
M-27		3000	2000	2400	2,100,000	SPIB
M-28		2200	1600	1900	1,700,000	SPIB
M-29		1550	850	1650	1,700,000	SPIB

1. **LUMBER DIMENSIONS.** Tabulated design values are applicable to lumber that will be used under dry conditions such as in most covered structures. For 2″ to 4″ thick lumber the DRY dressed sizes shall be used (see Table 1A) regardless of the moisture content at the time of manufacture or use. In calculating design values, the natural gain in strength and stiffness that occurs as lumber dries has been taken into consideration as well as the reduction in size that occurs when unseasoned lumber shrinks. The gain in load carrying capacity due to increased strength and stiffness resulting from drying more than offsets the design effect of size reductions due to shrinkage.
2. **SHEAR PARALLEL TO GRAIN, F_v, COMPRESSION PERPENDICULAR TO GRAIN, $F_{c\perp}$, AND SPECIFIC GRAVITY, G.** Design values for shear parallel to grain, F_v, and compression perpendicular to grain, $F_{c\perp}$, are identical to the design values given in Tables 4A and 4B for No. 2 visually graded lumber of the appropriate species. If "F_v" or "$F_{c\perp}$" values are shown on the grade stamp and differ from the values shown in Tables 4A and 4B, the values shown on the grade stamp shall be used for design (examples include Spruce-Pine-Fir MSR or MEL grades with "E" of 2,000,000 psi and higher where $F_{c\perp}$ = 615 psi, and Southern Pine MSR or MEL grades with "E" of 1,900,000 psi and higher where $F_{c\perp}$ = 805 psi and F_v = 100 psi).
   In some cases, grades of MSR or MEL lumber may have assigned specific gravity, G, values different from overall species values shown in several NDS tables. "G" values for these grades are sometimes determined on a mill-specific basis and may be included on the grade stamp; alternatively, where a grade of a given species group can be shown to have consistently higher specific gravity at all producing mills, the higher "G" value is sometimes assigned generically to that grade (examples include Engelmann Spruce-Lodgepole Pine MSR for 1650f and higher grades where G = 0.46, and Spruce-Pine-Fir MSR or MEL grades with "E" of 2,000,000 psi and higher where G = 0.50, and Southern Pine MSR or MEL grades with "E" of 1,900,000 psi and higher where G = 0.57). These higher than overall species "G" values can be used in formula provided by the grading rules to obtain higher "F_v" and "$F_{c\perp}$" values.
3. **MODULUS OF ELASTICITY, E, AND TENSION PARALLEL TO GRAIN, F_t.** For any given bending design value, F_b, the modulus of elasticity, E, and tension parallel to grain, F_t, design value may vary depending upon species, timber source or other variables. The "E" and "F_t" values included in the "F_b–E" grade designations in Table 4C are those usually associated with each "F_b" level. Grade stamps may show higher and lower values if machine rating indicates the assignment is appropriate. When the "E" or "F_t" values shown on a grade stamp differs from the values in Table 4C, the values shown on the grade stamp shall be used for design. The tabulated "F_b" and "F_c" values associated with the designated "F_b" value shall be used for design.
4. **COMPRESSION PARALLEL TO GRAIN, F_c.** This grade requires "F_c" qualification and quality control.

(Tabulated design values are for normal load duration and dry service conditions, unless specified otherwise. See NDS 2.3 for a comprehensive description of design value adjustment factors.)

(Courtesy of American Forest & Paper Association, Washington, DC.)

greater than 19%. Required adjustments are:

F_b = 0.85 (when F_b adjusted by the size factor is ≤1150 psi, C_M is 1)
F_t = 1.0
F_v = 0.97
$F_{c\perp}$ = 0.67
F_c = 0.8 (when F_b adjusted by the size factor is ≤750 psi, C_M is 1)
E = 0.9

The tabular F_b, F_t, and F_c values must also be adjusted by a size factor, C_F. The C_F multipliers for F_b vary with the depth and width of the member. The multipliers for select, structural No. 1 and better, and Nos. 1, 2, and 3 grades are all the same. For example, for a 2 × 8 (50 × 200) member of one of these grades, the F_b multiplier is 1.2; for a 4 × 8 (100 × 200) member, F_t is 1.3. F_t and F_c, however, vary with only the grade and member depth. For example, for an 8-in. (200 mm)-deep member F_t is 1.2 and F_c is 1.05 whether the thickness is 2 in. (50 mm), 3 in. (75 mm), or 4 in. (100 mm). There are other multipliers for stud grade, construction and standard grade, and utility grade members. The *NDS Supplement* contains a table that lists each of the required multipliers.

In addition, when members are used flat, the tabulated bending values, after adjustment for the size factor, must also be multiplied by an adjustment factor when members are used flat rather than with their long edges vertical. These factors vary, based on member size. For example, the multiplier for a 2 × 8 (50 × 200) loaded on the 8 in. (200 mm) side is 1.15, and for a 2 × 10 (50 × 250) loaded on the 10 in. (250 mm) side is 1.2.

Allowable shear values can be increased when the length of splits and shakes is less than that contemplated in the tables. The adjustment factors necessary to calculate these changes in tabular values are explained, and the formulas for them given, in the *NDS Supplement.*

The allowable unit stresses shown in Figures 6.1-14 and 6.1-15, and adjusted using size, repetitive member, flat use, wet service, and shear stress factors, are based on what NDS calls *normal load duration*. The NDS defines this condition as fully stressing a member to its allowable design value by the application of the full design load for a cumulative duration of approximately 10 years, or applying 90% of the full design load continuously throughout the remaining life of the structure. Values for normal loading are generally used for designing members to carry dead loads (such as floor or roof construction) and live loads imposed by occupants, equipment, and furnishings. For loads of brief duration, such as those produced by snow, earthquake, or wind, higher working stresses are used than required for normal loads (Fig. 6.1-16).

BENDING

Wood develops high fiber strength in bending. When a wood beam, loaded as in Figure 6.1-17a, deflects, bending stresses are produced in the fibers of the member. If the load is great enough to produce stresses that exceed the fiber strength of the wood, the beam will fail by breaking

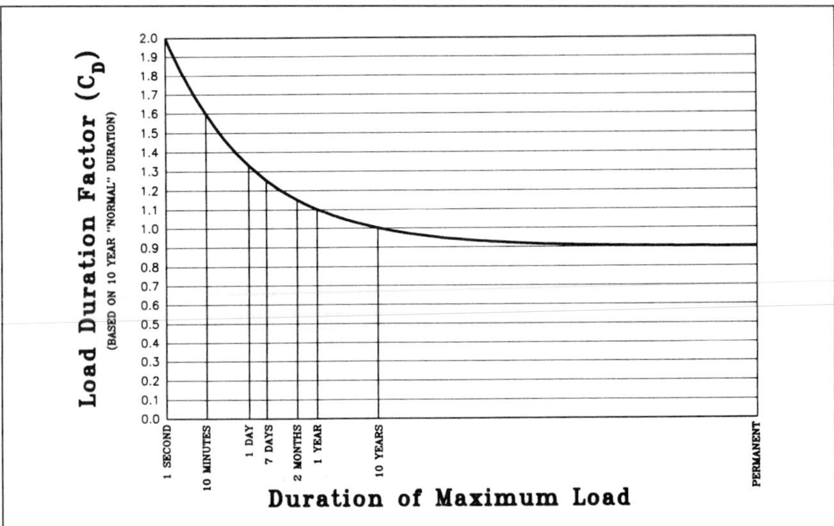

FIGURE 6.1-16 **Load duration factors. The graph plots duration of load on the horizontal axis and load duration factors on the vertical axis. The tabular allowable unit stresses in the *NDS Supplement* are multiplied by the load duration factor determined using this graph. The duration factors plotted are for load duration other than the normal duration of 10 years, which has a load duration factor of 1.** (Courtesy American Forest & Paper Association.)

(Fig. 6.1-17b). The stresses produced in the fibers are greater in magnitude the farther they are from the central axis. If a fiber is located twice as far from the central axis as another fiber, it will have twice the stress. Therefore, in a beam, bending stresses are maximum in the outermost fibers (called the *extreme fiber*) at the top and bottom of the beam. In beams subjected to bending, the fiber stresses in bending should be held below the working stress values (F_b) for the grade of wood used (see Fig. 6.1-14).

HORIZONTAL SHEAR

A beam also tends to fail by dropping down between supports (see Fig. 6.1-17c) as a result of vertical shear, the tendency for one part of the beam to move vertically with respect to an adjacent part. Wood beams are more likely to fail due to the tendency of the top fibers in the beam to move horizontally with respect to the bottom fibers. This produces stresses in fibers called *horizontal shear* (see Fig. 6.1-17d). Horizontal shearing stresses are largest at each end of a beam along the central (neutral) axis. In beams the fiber stress in horizontal shear should be held below the working stress values (F_v) for the grade of wood used (see Fig. 6.1-14).

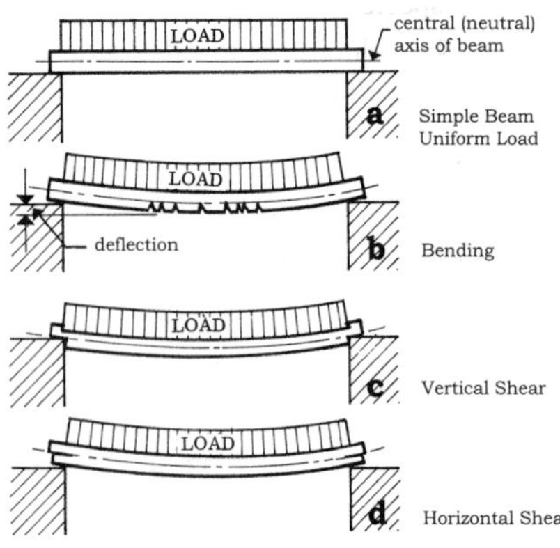

FIGURE 6.1-17 **Stresses produced in a beam subjected to bending.**

MODULUS OF ELASTICITY

The stiffness of a material is measured by its modulus of elasticity (E) (see Figs. 6.1-14 and 6.1-15). This measure of wood's stiffness determines a beam's resistance to deflection. Modulus of elasticity is also used in computing the load on long columns, because such columns depend on stiffness to resist buckling.

Mechanical grading techniques based on modulus of elasticity permit greater utilization of wood's strength properties (see Section 6.2.4.3).

TENSION

Theoretically, tensile strength parallel to the grain is the strongest property of wood, but in practice this is greatly affected by knots, splits, and checks. Since the allowable unit stress in *tension parallel to grain* (F_t) is a value applied to the entire cross section (less holes) of a piece of wood, it must reflect an average internal resistance to loading that is much more affected by wood characteristics than by a value applied to a narrow band of the cross section, such as the extreme fiber in bending stress (F_b). For most species, the allowable F_t values are approximately two-thirds of F_b values. Wood is weak in *tension perpendicular to the grain*, which permits the fibers to be pulled apart under heavy loading. Therefore, wood should not be loaded so as to create this type of stress. Allowable values for tension perpendicular to the grain usually are not tabulated.

COMPRESSION

The allowable unit stress in compression parallel to the grain is 2 to 5 times greater than it is perpendicular to the grain (Fig. 6.1-18). There are only minor differences in the strength properties in the two directions perpendicular to the grain (radially and tangentially). The values for compressive strength perpendicular to the grain ($F_{c\perp}$), and parallel to the grain (F_c) are shown in Figures 6.1-14 and 6.1-15.

6.1.3.5 Gluing Properties

The gluing properties of the woods most widely used for glued products are shown in Figure 6.1-19. These classifications are based on the average quality of typical joints of wood when glued with animal, casein, starch, urea resin, and resorcinol resin glues. A joint is considered to be glued satisfactorily when its strength is approximately equal to the strength of the wood.

Whether it will be easy or difficult to obtain a satisfactory joint depends on the density of the wood used, its structure, the presence of extractives or infiltrated materials in the wood, and the kind of glue. In general, hardwoods are more difficult to glue than softwoods, and heartwood is more difficult than sapwood. Several

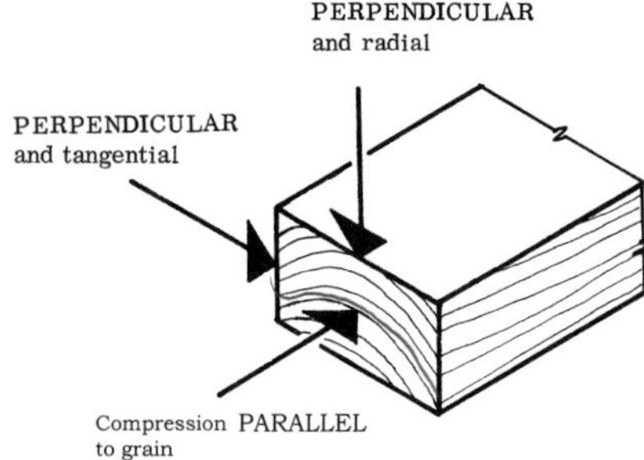

FIGURE 6.1-18 **Compression parallel and perpendicular to the grain, both radially and tangentially.**

FIGURE 6.1-19 Gluing Properties of Some Wood Species

Glueability*	Hardwoods	Softwoods
Excellent	Aspen Chestnut Cottonwood Black willow Yellow poplar	Bald cypress Western red cedar White fir Western larch Redwood Sitka spruce
Good	Alder Basswood Butternut Rock elm Hackberry	Eastern red cedar Douglas fir West Coast hemlock Northern white pine Southern yellow pine Ponderosa pine
Satisfactory	White ash Black cherry Soft maple Oak Pecan Sycamore Tupelo Black walnut	Alaska cedar
Poor	Beech Birch Hickory Hard maple	

*Ratings based on the ability of the wood to be glued under a wide range of conditions and with a variety of glues.

species vary considerably in their gluing characteristics with different glues (see Fig. 6.1-19).

6.1.3.6 Nonstructural Properties

Wood's many unique qualities result in its selection as a material for many nonstructural uses. These qualities include decorative characteristics, weathering qualities, thermal expansion and conductivity, and natural decay resistance.

DECORATIVE CHARACTERISTICS

Factors influencing the decorative value of wood include color, figure patterns, texture, and luster; ability to take filler, stains, bleaches, and other finishes; and the method of cutting or sawing. The terms *figure*, *grain*, and *texture* are often incorrectly used interchangeably.

Figure is the pattern or design created more by the abnormal than by the normal growth of a tree (wavy or swirly) and is incorrectly used synonymously with grain.

Grain is often used in referring to the annual rings (fine grain and coarse grain), but it is also used to indicate the direction of the fibers (straight grain, vertical grain). Finishers refer to the relative size of the pores (open grained, closed grained), which determines the need for a filler.

The term *texture*, although often used synonymously with *grain*, usually refers more to the finer structure and the visual dimension of depth in the wood than to the annual growth rings.

In plain-sawn boards and rotary-cut veneer, the annual rings often form ellipses and parabolas, making striking figures. In quarter-sawn boards and sliced veneers, the growth rings form stripes, and, in species such as oak, the rays add conspicuous figure. Examples of plain- and quarter-sawn boards are shown in Figure 6.1-20. In open-grained hardwoods (such as oak, mahogany, and walnut) and sliced veneers, the appearance can be varied greatly by using fillers of different colors, and in softwoods the annual growth layers can be accentuated by applying stain.

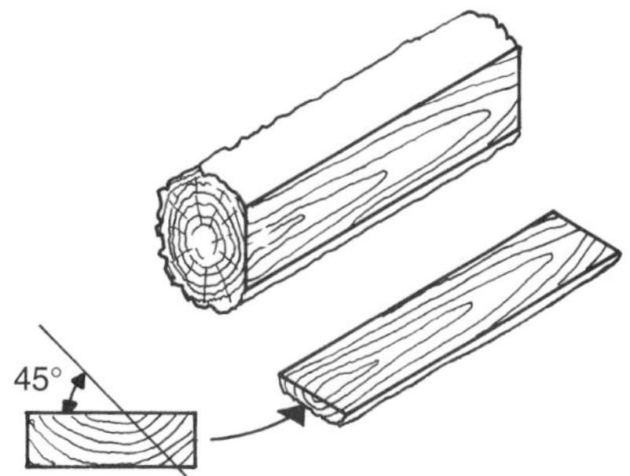

Lumber cut tangent to the annual rings is:
Plain sawn for hardwoods
Flat sawn for softwoods

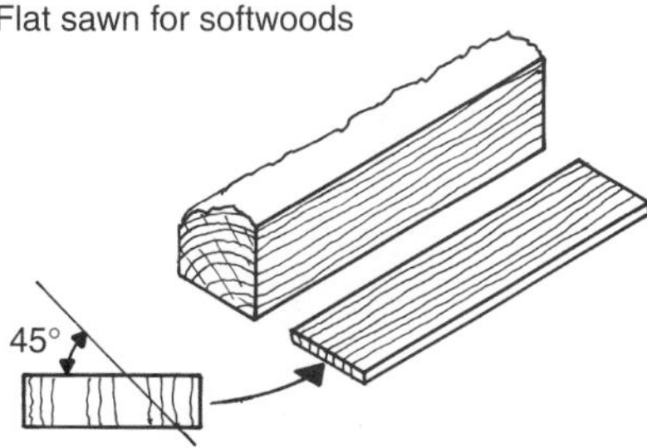

Lumber cut radially to the annual rings and parallel to the ray is:
Quarter-sawn for hardwoods
Edge grained or vertical grained for softwoods

FIGURE 6.1-20 The two basic ways to cut lumber from a log. (Drawing by HLS.)

WEATHERING

Weathering of exposed boards with no protective coating may cause:

1. Change in color (cedar and cypress weather to a light gray with a silvery sheen; Douglas fir and redwood to dark gray with no sheen).
2. Roughening and checking of the surface except in cedar, cypress, and redwood, on which weather checks are inconspicuous.
3. Warping tendencies (a slight tendency in cedar, cypress, and redwood; a distinct tendency in most of the other softwoods; and a pronounced tendency in most hardwoods). Some resistance to warping can be obtained by keeping the width of an exposed, unpainted board to not more than 8 times its thickness.

THERMAL EXPANSION

Like most materials, wood expands when heated and contracts when cooled. Except in special applications, such as buildings kept dry but subject to a wide temperature range, the thermal expansion and contraction of wood is ignored, because it is much smaller than the swelling and shrinkage due to moisture content variations occurring under normal conditions of exposure.

THERMAL CONDUCTIVITY

The thermal conductivity of wood is influenced by several of its properties, the most important of which are specific gravity and moisture content. Lighter-weight woods are better insulators, and insulation value increases as the moisture content decreases. Because of its good insulating characteristics, wood fiber is a principal constituent in insulating fiberboards.

NATURAL DECAY RESISTANCE

Natural toxic substances deposited in the cell structure of some species make their heartwood resistant to decay and termite attack. Decay-resistant species are bald cypress (tidewater red), cedars, redwood, black locust, and black walnut. Termite-resistant species are redwood, bald cypress (tidewater red), and eastern red cedar.

Naturally durable heartwood of the above species may be used in all conditions of exposure, except for members embedded in the ground to support permanent structures. For maximum durability, 100% heartwood of resistant species selected during grading should be used. For such species as redwood and cedar, several grading authorities, including the Redwood Inspection Service, the West Coast Lumber Inspection Bureau, and the Western Wood Products Association, have foundation grades selected from 100% heartwood with a high toxic extractive

FIGURE 6.1-21 White pocket (white speck) is permitted in certain lumber grades and is serviceable and acceptable for many uses. (Courtesy West Coast Lumber Inspection Bureau.)

content. Lumber with natural decay and termite resistance should be marked with the grade mark of an inspection agency authorized by the American Lumber Standards Committee. Where wood is embedded in the ground for the support of permanent structures, it should be pressure treated, as discussed in Section 6.1.5.

6.1.4 FUNGUS AND INSECT HAZARD

When lumber is used under continuously dry conditions and is adequately protected from attack by insects and fungi, preservative treatment is unnecessary. Preventive measures against these hazards during the manufacturing and merchandising stages are well established and widely practiced by reputable lumber producers. Serious decay or insect problems in buildings result most often from poor construction techniques.

6.1.4.1 Decay

Fungi (microscopic plants) cause decay, molds, and stains. Fungus growth can develop in wood only under the following conditions:

1. There must be an adequate supply of organic material, which for some fungi is the wood itself.
2. Temperatures must range from 41°F (5°C) to 104°F (40°C).
3. There must be sufficient oxygen, which is always present unless the wood is completely below the groundwater line or completely submerged in water.
4. The wood's moisture content must exceed 20%.

DRY ROT

The term *dry rot* is an often-used misnomer, because the fungi that cause decay must have access to water in order to function. Decay will not occur in wood that is maintained below a 20% moisture content, which is typical of air-dried wood. Therefore, even wood that is dry when decay is detected must have been wet earlier. Most decay in wood occurs when its moisture content is above the fiber saturation point (30% moisture content). But wood that is continuously water soaked will not decay either, because insufficient oxygen will be present.

In the later stages of decay a wood-destroying fungus seriously reduces the strength of wood. Therefore, most grading rules limit the amount of decay that may be present in structural lumber.

WHITE POCKET

Also known as white speck, *white pocket* is caused in living softwood trees by a relatively harmless fungus that ceases to develop when a log is cut into lumber and put into normal use. This fungus is as harmless as other natural characteristics normally found in lumber, such as knots, shake, or wane. It will not spread from one piece to another or even within an individual piece. Lumber with white pocket is serviceable, and certain amounts of this fungus are accepted by grading authorities in all but the higher grades of lumber (Fig. 6.1-21). However, white pocket does reduce the strength of lumber, and, more important, nail holding values are greatly reduced in lumber having white pocket.

6.1.4.2 Molds and Stains

Molds and some stains are caused by fungi that do not destroy the infected wood. Stock with such stains and molds is practically unimpaired except in appearance.

Powdery surface molds vary from white or light colors to black. These discolorations are largely superficial and often are easily brushed off or removed by surfacing the wood.

Stains appear as specks, spots, streaks, or patches that vary most commonly from bluish to bluish black or brown. These discolorations usually cannot be removed by brushing or surfacing because they extend into the sapwood.

6.1.4.3 Insect Damage

Holes caused by insects may occur in standing trees and in either unseasoned or seasoned lumber. These are generally classified as pinholes or grub holes. Such damage is taken into account when lumber is graded, but the strength of a piece of wood damaged by insects after it has been installed cannot be estimated visually. Therefore, when strength is an important factor, every installed piece found containing insect holes should be removed and replaced by a new member.

Many precautions taken to prevent decay are also effective against various insects that attack and destroy wood.

Subterranean termites account for about 95% of all termite damage in the United States and a significant portion of the total insect damage. Nonsubterranean termites (called dry-wood termites) occur only in the southern part of the country and the lower east and west coastal areas, but they are a hazard where they do exist. Refer to Section 2.3 for a discussion of the location and control of termites, to Section 6.1.5 for information about the preservative treatment of wood to prevent termite damage, to Section 6.5.2 for ways to separate wood from the ground and the installation of termite barriers, and to Sections 6.4 and 6.7.6 for some specific requirements for termite treatment of wood foundations and wood pole construction.

6.1.5 PRESERVATIVE TREATMENT

Preservative treatments protect wood against decay and insect attack. Pressure treatments, in which preservative chemicals are applied under pressure to obtain maximum penetration, afford the greatest protection for lumber and plywood against decay or termite attack. Nonpressure treatments should be restricted for use only for protection of exterior millwork not in contact with the ground.

Wood preservatives may be divided into three major classes: (1) water-borne preservatives, (2) oil-borne preservatives, and (3) creosote and solutions containing creosote.

In all areas of the country, lumber and wood products used in contact with the ground or embedded in the ground should be preservative pressure treated.

In areas of severe termite infestation (see Section 2.3) or with a high decay hazard, such as in damp coastal areas (see Fig. 6.1-9), lumber and wood products should be preservative pressure treated for use in the following locations:

1. Members in contact with masonry or concrete
2. Members at grade, below grade, or less than 8 in. (200 mm) above grade
3. Members in a crawl space (floor joists less than 18 in. [460 mm] above interior grade and wood girders less than 12 in. [305 mm] above interior grade)
4. Members used in conjunction with roofing or flashing
5. Wood siding closer than 6 in. (150 mm) to exterior finish grade

Other wood members and products may require preservative treatment where dictated by local experience. Soil poisoning (see Section 2.3), termite shields (see Section 6.5), monolithic concrete slab construction, and other construction procedures can minimize the likelihood of termite damage.

Except for structural members embedded in the ground and used as part of the foundation of permanent buildings, and except in other locations where the applicable code requires the use of pressure-treated wood, heartwood grades of naturally resistant species can be used in place of pressure-treated lumber. Some acceptable grades and species of naturally resistant dimension lumber are California redwood—Select Heart, Construction Heart, or Foundation 2 Grade; tidewater red cypress—Heart Common; and Eastern red cedar—Foundation Grade.

Exterior millwork items such as doors, window units, and casing trim, which are subjected to intermittent wetting and drying, should be preservative treated by nonpressure methods (see Section 6.1.5.4) unless more effective treatment for protection against termites is indicated by local conditions.

Job-cut ends of preservative treated lumber should be brush coated with not less than a 3% solution of an approved preservative, usually the same one used in the factory process.

6.1.5.1 Water-Borne Preservatives

Water-borne preservatives are most often used to pressure treat lumber and plywood for construction because they leave the wood clean, odorless, and easy to paint, and because they present less of an environmental hazard than other preservatives. Two types have proven to be very durable and will protect wood to be used in contact with the ground, supporting permanent building structures. These are ammoniacal copper arsenate (ACA) and chromated copper arsenate (CCA). The latter is available in three different formulations, which are considered to be equivalent in performance. Water-borne preservatives are relatively odorless and are paintable, provided the wood has been dried to the moisture content required for untreated wood. These preservatives will not cause discoloration in interior finishes unless a wet-process finish, such as plaster, is applied to treated wood without a waterproof backing paper.

6.1.5.2 Oil-Borne Preservatives

Lumber treated with oil-borne preservatives can be used in any installation that does not place it in contact with salt water, except where its use is not permitted by code, law, or other restriction. Less lumber treated with oil-borne preservatives is used each year, and its use may eventually be banned altogether due to environmental concerns. Such lumber is often prohibited for use where it will come in contact with people, especially children.

The oil-borne preservative Penta (pentachlorophenol), which had for years been the most widely used oil-borne preservative, is highly toxic to both fungi and insects, but is insoluble in water and thus permanent. It is this permanency that has caused such great concern about its use. Because it is almost impossible to break down and because of its high toxicity, Penta has been designated a toxic environmental hazard and is no longer used in the building construction industry.

6.1.5.3 Creosote

Creosote is a widely used preservative for some applications, but its use is also being prohibited in some jurisdictions where it is likely to come in contact with people. Creosote and solutions containing it are used where protection against wood-destroying organisms is of first importance, where painting is not required, and where its odor is not objectionable. Wood may be protected in severe exposures with high retentions and in less severe exposures with lower retentions.

Creosote–coal tar mixtures are the most widely accepted preservatives for marine and saltwater installations and are ideally suited for treating pilings used for shore dwellings.

Creosote-petroleum mixtures are used where economy is of first importance, but they should not be used for marine installations.

6.1.5.4 Treatment Methods

There are two methods of preservative treatment: (1) pressure processes and (2) nonpressure processes.

PRESSURE PROCESSES

There are two pressure preserving processes in general use: the *full-cell process* and the *empty-cell process*. Both are dependable means of ensuring uniform penetration and distribution of preservative. Timber to be treated by either of these pressure processes is loaded on tram cars and rolled into a long steel cylinder called a *retort*, which is then sealed.

The *full-cell process* is used for treating with creosote, creosote mixtures, water-borne preservatives, and fire-retarding chemicals (see Section 6.1.6). It provides a high retention of chemicals, which is required for wood to be installed in a marine environment, for example.

In the full-cell process much of the air is removed from the cylinder and, consequently, from the wood. Then the cylinder is filled with preservative under various combinations of temperature and pressure to force the preservative deep into the wood. In this process, not only is the

FIGURE 6.1-22 Frequently Used Preservative Treatment Standards*

AWPA Standards	
C1	Preservative treatment by pressure process
C2	Lumber, timbers, bridge ties, and mine ties, pressure treatment
C9	Plywood, pressure treatment
C15	Wood for commercial-residential construction, pressure treatment
C22	Lumber and plywood for permanent wood foundations, pressure process
C23	Round poles and posts used in building construction—preservative treatment by pressure processes

*The listed standards and procedures are the more frequently used ones. Standards and procedures related to Penta (pentachlorphenol) have been omitted from the list. Refer to AWPA literature for the applicable standards and procedures if Penta preservatives are to be used. Note that these are not permitted in some jurisdictions.

preservative absorbed by the cell walls, but the cell cavities are also filled. After the required penetration and retention are achieved, the preservative charge is removed. This process leaves the maximum amount of preservative in the wood, but when used with creosote or its solutions, will not leave as clean a surface as do the other processes. In addition, especially when water-borne preservatives are used, the wood is left wet and must be air or kiln dried before shipment.

The *empty-cell process* is primarily used for treatment with oil-borne preservatives. In this process the wood is subjected to high air pressures before the preservative is introduced. When the pressure is released, the expanding entrapped air expels the excess preservative from the wood cell cavities, thus reducing the net retention of preservative. This method provides treatment of the cell walls but leaves the cell cavities empty. Thus, it yields a drier product, gives deeper and more uniform penetration, and permits maximum penetration for a given weight.

NONPRESSURE PROCESSES

Wood treated with nonpressure processes should not be used in contact with the ground or under severe conditions of decay or termite attack, because the preservative is not sufficiently impregnated in the wood to afford adequate protection. Nonpressure treatment can be effective for exterior millwork such as precut door, window, and trim components, where the wood is generally dry but may be exposed to moisture or wetting intermittently.

Nonpressure-treated lumber should be seasoned to a maximum moisture content of 19%, or lower if necessary for the intended use. Surfaces that are cut or milled after treatment should be treated again with a brush application of preservative before final assembly or installation. *Vacuum* and *brief dip* nonpressure treatments usually provide clean, paintable surfaces.

In the *vacuum process*, a vacuum or partial vacuum is introduced to exhaust air from the cells and pores of the wood. Atmospheric pressure is then reintroduced to force the preservative into the wood.

The *brief dip* or 3-minute immersion process consists of completely submerging the wood in an open tank of the preservative solution for a period of at least 3 minutes.

6.1.5.5 Standards and Certification of Quality

Standards have been established for the wood preserving industry, and certification seals are available for treated products.

PRESSURE-TREATED WOOD

The American Wood Preservers Association (AWPA) is a technical and research organization serving the wood preserving industry. AWPA develops and promulgates requirements for preservative formulations, methods of treatment, and minimum chemical retentions, which are the recognized industry standards for the pressure-preservative treatment of wood. The most commonly used AWPA standards are shown in Figure 6.1-22.

A related organization, the American Wood Preservers Institute (AWPI), performs the promotional and educational functions for the industry.

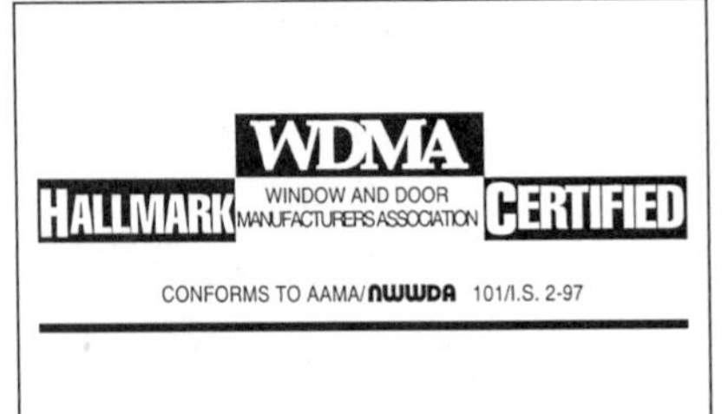

FIGURE 6.1-23 Window and Door Manufacturers Association seal. (Courtesy Window and Door Manufacturers Association.)

NONPRESSURE-TREATED WOOD

The Window and Door Manufacturers Association (WDMA) administers a testing and plant inspection program, resulting in certification and issuance of its seal (Fig. 6.1-23). The WDMA seal ensures that a product conforms to WDMA IS-4. WDMA issues a list of products that comply with IS-4.

6.1.6 FIRE-RETARDANT TREATMENT

The natural fire resistance of wood can be greatly enhanced by its impregnation with inorganic salts that react chemically at temperatures below the ignition point of wood. This reaction causes the combustible vapors normally generated in the wood to break down into nonflammable water and carbon dioxide.

Fire-retardant-treated wood has been marketed since World War II. Today's treatment methods and chemicals provide a uniformity that was earlier unattainable. Products tested and approved by the Underwriters' Laboratories, Inc. (UL) and which bear its label are recognized by model building code authorities and federal agencies as suitable for structural use in fire-resistive and noncombustible construction.

After treatment, plywood and lumber, 2 in. (50 mm) nominal or less in thickness, should be dried to a moisture content of 19% or less. Designers may want to require a lower moisture content for fire-retardant-treated wood to be used for millwork, cabinets, office paneling, and other special uses. Fire-retardant-treated wood is generally not suitable for transparent finishes. Most of it is painted when exposed to view.

Except where exterior-grade treatment complying with rain-testing requirements has been employed, fire-retardant-treated wood should not be used where it will be exposed directly to the weather. The presence of fire-retardant chemicals in treated wood may prohibit the use of some wood preservatives, such as CCA.

6.2 Lumber

Lumber is the designation given to a large number of products produced in a sawmill and planing mill. It is manufactured by sawing, edging, and trimming logs into square or rectangular pieces of wood. Veneers, barrel staves, and hammer handles, however, are not lumber. *Yard lumber* includes (1) *boards* used for sheathing, siding, flooring, paneling, trim, and patterned millwork, (2) *dimension* stock used for light framing, joists, planks, and roof and floor decking, and (3) *timbers* used for posts, beams, and stringers. Lumber varies widely in quality and therefore must be divided into commercial lumber grades. Each grade covers a narrow range of quality to provide a basis for selecting lumber for each of its many uses.

6.2.1 MANUFACTURING PROCESS

The manufacture of lumber begins in the forest. After trees are felled, the logs are transported to a lumber mill for processing. Following manufacture by a sawmill, a wholesaler locates lumber and resells it to wood-using plants and retail lumberyards, which distribute about 75% of all lumber to the ultimate user. Many retail yards and some wholesalers maintain a planing mill and other equipment to satisfy individual user requirements.

6.2.1.1 Logging Practice

Modern logging methods make it possible to transport most of the trunk of a tree from a forest to a mill. Portable power saws are used for felling trees and cutting them into logs. Depending on the terrain and other variable factors, logs are removed from the forests by trucks, tractors, trains, or by fast, powerful cable logging systems. The logs are grouped at a landing, where mobile or other types of loaders place them onto trucks for the haul to the mill (Figs. 6.2-1 and 6.2-2).

6.2.1.2 Sawmill Practice

Operations vary from mill to mill, depending on the rate of output, the type of machinery, and the end product desired, but, in the main, the pattern of operations is as follows.

Automatic machinery handles much of the sawmill process (Fig. 6.2-3), but plants still depend heavily on the judgment of skilled workers at each step during manufacture.

In some mills, logs are cut to a shorter length. In some cases, low-quality parts are removed. In other cases, logs are cut to fit the mill's machinery or to produce lumber of certain lengths. Usually, logs are sent first to a debarker, where their bark is removed by peeling, scraping, or blasting with high-pressure water jets. This reveals the shape of each log, its knots, and its irregularities. The log is then placed on a carriage and fed into a *headsaw*, which may be either a circular saw or a band saw. The main saw mechanism, called a *headrig*, is the point where a log is broken down into rough cants, timbers, planks, and boards. Some mills employ, as the head saw, a gang saw with as many as 40 straight saws. Others use a gang saw in combination with a band or circular head saw to break large cants rapidly into boards and dimension lumber.

The next step is the *edger*, a series of

FIGURE 6.2-1 **Logs are loaded aboard trailer trucks for transport to a sawmill.** (Courtesy Weyerhaeuser.)

FIGURE 6.2-2 **At a sawmill, a log unloader can remove an entire load from a truck in one bite.** (Courtesy Weyerhaeuser.)

FIGURE 6.2-3 A sawmill operator. (Courtesy Weyerhaeuser.)

small saws that rip the lumber into the desired width and remove the rounded edge. The edger also rips wide pieces into narrow widths. Edged lumber moves onto the trimmer, where the saws trim off the rough ends and defects to produce the most desirable lengths.

The final step in the sawmill is called a *green chain*, where the edged and trimmed pieces are sorted visually according to grade, species, and size (Figs. 6.2-4 and 6.2-5).

From the green chain, lumber may be sent to a seasoning yard or dry kiln (Fig. 6.2-6) or to remanufacturing plants for resawing, dressing, ripping, planing, or other treatment.

FIGURE 6.2-4 Operator making sure boards do not pile up. He is also tossing culls. (Courtesy Weyerhaeuser.)

6.2.1.3 Seasoning Practice

Lumber seasoning is the process of reducing the moisture content of the wood. Kiln drying and air drying are the methods used. In either case, the advantages to be gained over unseasoned, or green, lumber include increased strength, reduction in shrinkage and checking and warping in service, less susceptibility to fungus attack in storage, and improvement in the capacity to receive pressure-preservative treatment. Kiln drying can further reduce the moisture content to almost any desired value, can result in even greater reduction in weight, hastens the drying time over that required in air seasoning, and kills stain, decay fungi, and insects in the wood. In the case of hardwoods, air drying in outside yards is often combined with kiln drying. Following seasoning, lumber is sent to a planing mill for sizing and surfacing, after which it is graded, tallied, and prepared for shipment.

Under the *American Softwood Lumber Standard* (Product Standard PS 20), mills can fabricate dimension lumber in either the green or seasoned state, provided that green lumber is surfaced to slightly larger sizes to compensate for eventual shrinkage (Fig. 6.2-7). PS 20 also requires that lumber up to, but not including, 5 in. (127 mm) thick be grade marked S-DRY if it was surfaced at a moisture content of 19% or less; if it was surfaced at a higher moisture content, it must be marked S-GRN.

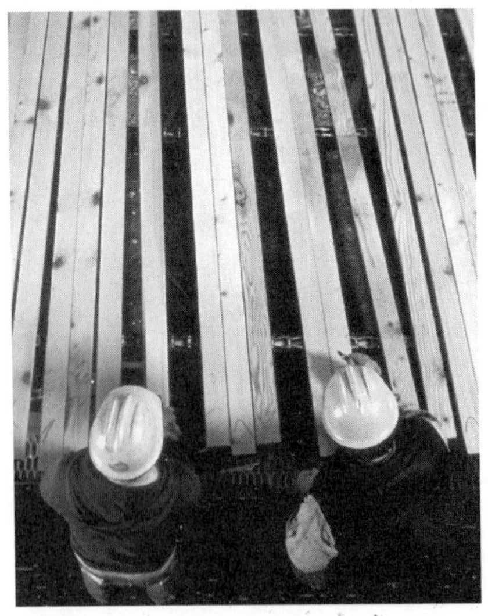

FIGURE 6.2-5 Lumber grading. (Courtesy Weyerhaeuser.)

FIGURE 6.2-6 Stacks of lumber being loaded into a restraint dryer. (Courtesy Weyerhaeuser.)

FIGURE 6.2-7 American Standard Dressed Sizes of Green and Dry Boards, Dimension Lumber, and Timbers

	Thicknesses					Face Widths				
		Minimum Dressed					Minimum Dressed			
	Nominal	Dry		Green		Nominal	Dry		Green	
Item	in.	in.	mm	in.	mm	in.	in.	mm	in.	mm
						2	1½	38	1 9/16	40
						3	2½	63	2 9/16	65
						4	3½	89	3 9/16	90
						5	4½	114	4⅝	117
	1	¾	19	25/32	20	6	5½	140	5⅝	143
						7	6½	165	6⅝	168
Boards*	1¼	1	25	1 1/32	26	8	7¼	184	7½	190
						9	8¼	210	8½	216
	1½	1¼	32	1 9/32	33	10	9¼	235	9½	241
						11	10¼	260	10½	267
						12	11¼	286	11½	292
						14	13¼	337	13½	343
						16	15¼	387	15½	394
						2	1½	38	1 9/16	40
						3	2½	63	2 9/16	65
						4	3½	89	3 9/16	90
	2	1½	38	1 9/16	40	5	4½	114	4⅝	117
Dimension	2½	2	51	2 1/16	52	6	5½	140	5⅝	143
	3	2½	63	2 9/16	65	8	7¼	184	7½	190
	3½	3	76	3 1/16	78	10	9¼	235	9½	241
						12	11¼	286	11½	292
						14	13¼	337	13½	343
						16	15¼	387	15½	394
						2	1½	38	1 9/16	40
						3	2½	63	2 9/16	65
						4	3½	89	3 9/16	90
						5	4½	114	4⅝	117
Dimension	4	3½	89	3 9/16	90	6	5½	140	5⅝	143
	4½	4	102	4 1/16	103	8	7¼	184	7½	190
						10	9¼	235	9½	241
						12	11¼	286	11½	292
						14	13¼	337	13½	343
						16	15¼	387	15½	394
Timbers	5 and thicker	½ off	13 off	½ off	13 off	5 and wider	½ off	13 off	½ off	13 off

*Boards less than the minimum thickness for nominal 1 in. but ⅝ in. (16 mm) or greater thickness dry [11/16 in. (17 mm) green] shall be regarded as American Standard Lumber, but such boards shall be marked to show the size and condition of seasoning at the time of dressing. They shall also be distinguished from nominal 1-in. boards on invoices and certificates.

(*Source:* Table 3, Proposed Voluntary Product Standard PS 20-94, "American Softwood Lumber Standard, NIST, page 12. Not subject to copyright in the United States.) (Refer to Figure 1.8-27 for metric conversions.)

Although not required by the Standard, lumber milled at a 15% moisture content is identified in most species as MC 15; in grading Southern pine, the term KD 15 (kiln-dried) is used in place of MC 15.

KILN-DRIED LUMBER

In a dry kiln, heat, humidity, and air circulation are controlled to reduce the moisture content to any practical percentage. With few exceptions, the appearance grades of lumber intended for exterior and interior finish are kiln dried, and some mills also kiln dry dimension and other yard grades. Specifying kiln-dried lumber does not necessarily ensure that it is dry, since drying may be terminated in the kiln at any time. The moisture content desired should always accompany the words *kiln dried* unless the moisture content established by the grading rules is suitable for the use intended. This is usually 6% to 12% in the finish grades of softwoods and hardwoods, and 15% to 19% in the common yard grades of softwoods. Kiln drying takes a number of days.

AIR-DRIED LUMBER

In the air-dried method, lumber dries simply by evaporation when exposed to the atmosphere, either outdoors or in unheated storage over a period of several months. Ordinarily, dimension and lower grades of softwood lumber used for framing are sent to a yard for air drying, but are sometimes shipped unseasoned (green). Structural timbers are not usually held long enough to be seasoned, although some drying may take place between their sawing and the time of shipment.

The moisture content of thoroughly air-dried lumber averages between 12% and 15% in the United States, but varies widely from as low as 6% in summer in the arid Southwest to 24% in winter in the Pacific Northwest. The various grading rules governing softwood lumber establish the maximum moisture content for each species of seasoned lumber, whether it is air or kiln dried.

Hardwoods are commonly classified according to grade and size at the time of sawing, with all stock then sent to the air-drying yard. After air drying, the wood may be kiln dried at the mill or shipped to a remanufacturing plant where it is kiln dried before being made into flooring, interior finish, cabinetwork, or other finished products.

UNSEASONED LUMBER

Lumber more than 2 in. (50 mm) thick is rarely seasoned at the mill. Lumber 3 in. (75 mm) and 4 in. (100 mm) thick requires longer periods of air drying to achieve a 19% moisture content. This thicker lumber is usually fabricated green and is not often available in a retail lumber yard with an S-DRY grade mark. However, such lumber frequently achieves an acceptable moisture content of 19% or less after air drying at the lumber yard. In the absence of a grade mark, the wood can simply be measured. If a representative sample of pieces are close to the thickness and width requirements of PS 20 for dry lumber of that size (see Fig. 6.2-7), the lumber is likely to have an acceptable moisture content. An electric moisture meter, of course, provides a more reliable estimate of moisture content.

6.2.1.4 The Planing Mill

As previously stated, lumber can be surfaced (planed) either unseasoned or seasoned. Except for some framing lumber, most lumber is surfaced after seasoning because the planing removes some of the imperfections caused by shrinkage during seasoning.

Surfacing is done at a planing mill, where the lumber is planed and worked to patterns. Equipment varies from a few machines for surfacing and resawing in some mills to complete fabricating operations in others. Dried, rough-sawed products are converted into boards and dimension, flooring, ceiling, partition, siding, casing, and other finished stock.

After going through a planing mill, unseasoned lumber is returned to be seasoned; seasoned lumber may be removed to storage sheds to await shipment, but most of it is usually moved directly to cars and loaded for immediate shipment to a wholesale or retail yard or to a woodworking or industrial plant (Fig. 6.2-8).

6.2.1.5 Reforestation

The lumber industry has had to adapt production to its source of supply—trees. Raw material requirements must be geared to the volume of timber that the forests can grow. This adaptation has been accomplished by more wood research and greater utilization, coupled with the application of forest conservation principles such as tree farming.

There are four approaches generally used for timber harvesting and management: *clear-cutting*, *selective cutting*, *shelter-wood cutting*, and *seed-tree cutting*. Clear-cutting means harvesting all the trees in a given area of forest at one time. The lumber industry prefers the clear-cutting practice for many species because it is more efficient and results in quicker regeneration. Conservationists, on the other hand, charge that this method is unsightly, causes erosion, and destroys wildlife habitats.

Clear-cutting is advisable only when there is a great deal of rainfall. Without a relatively high rainfall, seedlings will burn up as they have no relief from the hot sun. Many West Coast forests have the amount of rainfall needed for clear-cutting. Clear-cutting is also done regardless of rainfall amounts when a disease is discovered in a forest or is moving toward the forest from adjacent forests. The wood is then said to have been *saved* from the disease, which would soon destroy the living trees. Clear-cutting is also done when part of a forest has become very ragged from selective cutting. Foresters are now able to grow good trees faster than can nature unassisted by man. As soon as a stand of trees is clear-cut, forestry crews start planting new trees in the same area as soon as possible. Unfortunately, this replanting must be done by hand, which is expensive.

Entire forests are not usually clear-cut all at once. Lumber companies instead lease a limited number of acres of a government-owned forest to cut or cut a limited number of acres of their own land at a time. Some people call this *patch clear-cutting* because it sometimes resembles a checkerboard or patch quilt pattern.

Selective cutting implies taking only the older, slower-growing, and defective trees, leaving behind enough trees for natural reseeding, wildlife shelter, and erosion control. However, regardless of the method of cutting and reseeding, seedlings require 60 years or more before they produce harvestable trees. Unfortunately, selective cutting is often done too frequently and is sometimes too extensive for forest self-perpetuation. Therefore, selective-cut forests tend to eventually run out of harvestable wood and the trees that would soon produce it.

Shelter-wood cutting and *seed-tree cutting* are processes in which some trees are left standing while others are har-

FIGURE 6.2-8 **Lumber waiting to be loaded aboard ship.** (Courtesy Weyerhaeuser.)

vested. The difference between the two is that in shelter-wood cutting 10 to 20 trees are left per acre, whereas in seed-tree cutting only about 4 or 5 trees are left per acre. Shelter-wood cutting is used when there is insufficient rainfall for clear-cutting, when there are trees that cannot tolerate continuous direct sun, and when the subject tree species has a shallow root system, as does spruce, and therefore cannot withstand significant wind. The trees left standing reseed the cut area, eliminating the need for reseeding by hand. The 10 to 20 trees left per acre in shelter-wood cutting also provide some shade and wind protection for the seedlings. The fewer trees left by seed-tree cutting do not afford this protection in any significant way.

6.2.2 MARKETING AND MEASURE

PS 20 states that lumber should be "tallied board measure," which is defined as using the board foot as the unit of measure, and most lumber is marketed and sold that way. A board foot is the equivalent of a piece 1 in. thick and 1 ft. square. In metric measure, 100 board feet is equivalent to 0.236 m^3. Some products such as moldings, which have been machined further, are sold by the lineal foot (running foot), which is equivalent to 0.2048 metre. Wood shingles are packaged in bundles and sold according to the surface area they are intended to cover. Posts, poles, and pilings are sold by the piece or the lineal foot (metre).

6.2.2.1 Sizes

Lumber is bought and sold by its *nominal* dimensions. The nominal size of lumber is its rough unfinished size, sold commercially as 1 × 6 (50 × 150), 2 × 4 (50 × 100), 4 × 8 (100 × 200), and so on (Fig. 6.2-9). The dressed size is smaller than the nominal dimensions as a result of seasoning and surfacing. Thus, a 2 × 4 (50 × 100) is actually 1½ in. (38.1 mm) × 3½ in. (88.9 mm). Members up to 6 in. (150 mm) wide lose ½ in. (12.7 mm); wider members lose ¾ in. (19.1 mm). Wood in buildings constructed before 1973 will be larger. A 2 × 4 (50 × 100), for example, will be 1⅝ in. (41.275 mm) by 3⅝ in. (92.08 mm). Even older buildings may have full-sized members. A 2 × 4 (50 × 100) may actually be 2 in. (50.8 mm) by 4 in. (101.6 mm). In worked lumber (Fig. 6.2-10) the face is that portion of the piece that will be exposed to view after it has been installed.

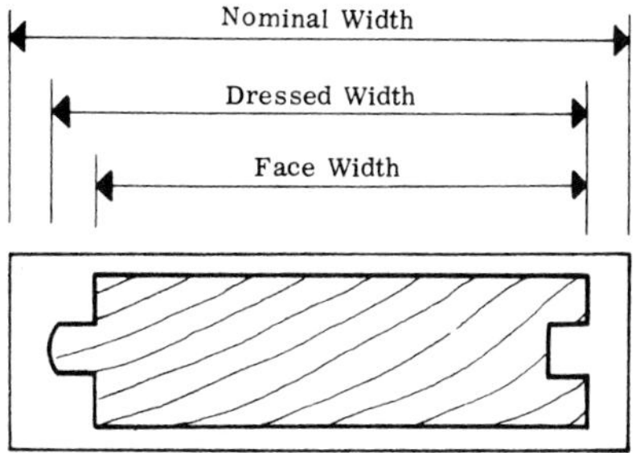

Nominal Size of lumber is its unfinished size, known commercially as 1×6, 2×4, etc. Seasoning and planing bring it to its *Dressed Size*. The *Face Size* is that portion of piece exposed to view when in place.

FIGURE 6.2-9 Nominal, dressed, and face sizes of worked lumber.

Under the provisions of PS 20-70, the dressed size of dimension lumber less than 5 in. (125 mm) thick is related to the moisture content of the wood at the time of manufacture. Therefore, lumber fabricated at a moisture content of more than 19% is called green and must be dressed to slightly larger dimensions (see Fig. 6.2-9) than lumber fabricated at a moisture content of 19% or less, which is called dry. The objective of this requirement is to ensure that both green and dry lumber achieve approximately the same actual dimensions (and strength) after air drying in service.

Softwood lumber is usually sold in multiples of 2-ft. (610 mm) lengths, ranging from 6 ft. (1.829 m) to 24 ft. (7.32 m), but many mills and lumberyards will precut members to exact dimensions to meet the demands of preassembled component construction and other special customer requirements. Hardwood lumber is usually sold by odd- and even-foot lengths or in random lengths for remanufacture into other products.

6.2.2.2 Board Measure

As stated earlier, *board measure* is used to indicate quantities of lumber. A board foot is the amount of lumber in a piece with nominal dimensions of 1 in. thick, 1 ft. wide, and 1 ft. long. This is the equivalent of 0.00236 m^3. Therefore, a 1-in. (25 mm)-thick board that is 1 ft. (0.3048 m) wide and 10 ft. (3.048 m) long has 10 board feet (0.0236 m^3) of lumber in it. The actual dimensions of this board are about ¾ in. (19.1 mm) by 11¼ in. (285.75 mm) by 10 ft. (3.048 m). To calculate the board feet in a piece of lumber, multiply the thickness in inches by the width in feet by the length in feet, using all nominal dimensions. Since the width of a board is usually expressed in inches, it is necessary to divide by 12 to convert inches to feet (see example, Fig. 6.2-11). If the nominal thickness is less than 1 in., the number of board feet equals the width in feet multiplied by the length in feet. To convert board feet to m^3, multiply by 0.002,359,74.

6.2.3 LUMBER QUALITY

As cut from a log, lumber varies widely in quality because of either natural growth characteristics or the manufacturing process. The product of a log is divided into commercial lumber grades, each having a relatively narrow range of quality. The visual grading of lumber should be done in accordance with American Society for Testing and Materials (ASTM) Standard D 245 and provisions of PS 20.

In terms of quality, or number of undesirable growth characteristics, lumber can be classified as common or select. Common grades are suitable for general construction purposes. Select grades, as the name implies, are selected to provide

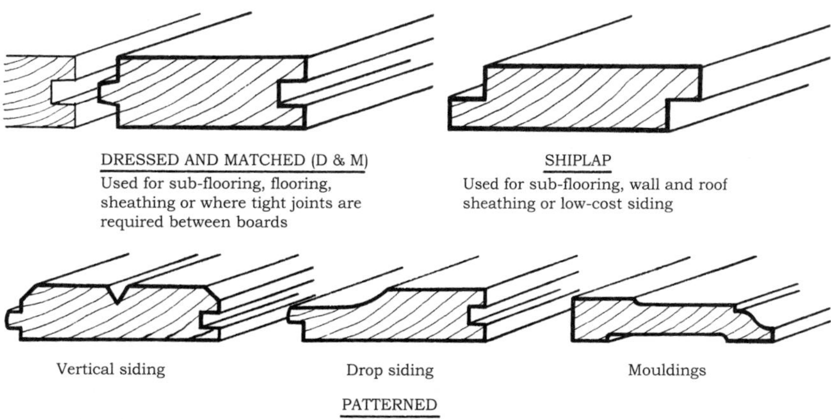

FIGURE 6.2-10 Typical cross sections of worked lumber.

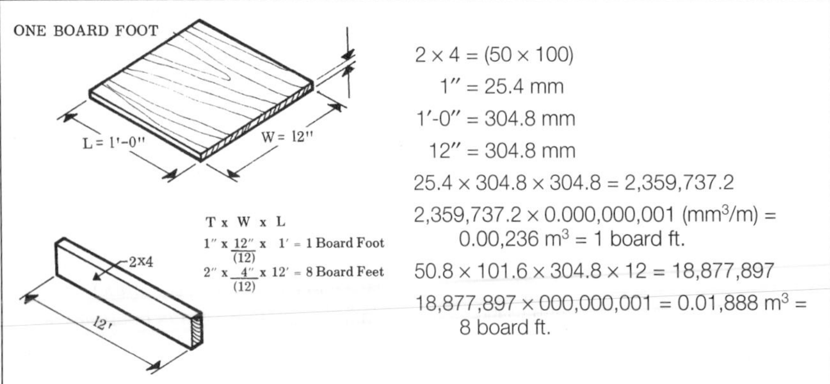

FIGURE 6.2-11 Board foot measurement.

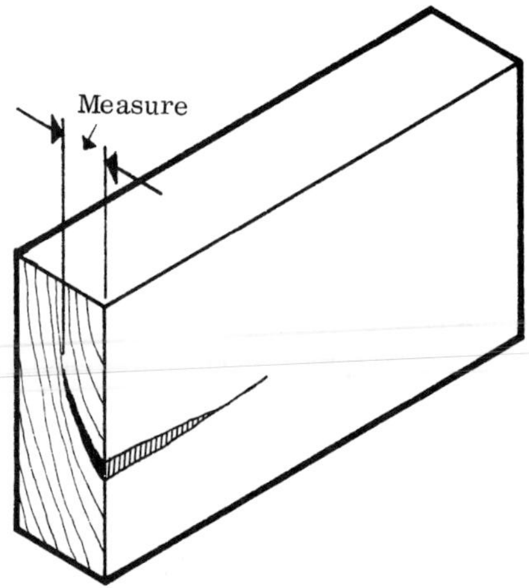

FIGURE 6.2-13 A shake is a lengthwise separation between, not across, the annual rings.

a surface suitable for painting or natural finishes. The bulk of yard lumber is graded as common lumber, with only boards (lumber up to 1½ in. (38.10 mm) thick and 2 in. (50 mm) wide or wider) available in select grades. Lumber more than 1½ in. (38.10 mm) thick also may be graded as select, but is used for fabrication into millwork and generally is not stocked as yard lumber. The word *select* in *Select Structural* and *Select Decking* does not identify a select grade. Timbers and dimension lumber grades are classified only as common grades.

Quality-reducing characteristics are faults that detract from the appearance, strength, or general utility of a piece in the use for which it is intended. The definition of such faults is often difficult in grading, because a slight or small defect in *common* lumber is often a medium or large defect in *select* (finish) lumber. In select lumber the defects are light, small, or serious depending on the size of the piece or as they come into combination with other defects. Most of the natural and manufacturing characteristics having an effect on the quality of lumber as it is graded are briefly described in the following section.

6.2.3.1 Natural Characteristics

Among the characteristics that affect the appearance, grading, or use of wood are knots, shakes, pitch and pitch pockets, decay, molds and stains, and insect damage. Reference should be made to the grading rules covering each species for the specific limitations of these characteristics in a given grade.

KNOTS

Contrary to common notion, lumber containing knots is not seriously defective for ordinary use. Knots may be undesirable because of appearance in select grades, but they are characteristic of all common grades (Fig. 6.2-12). Knots are restricted by size and location in grading because they do depreciate the strength value of wood used for structural purposes. Strength reduction is a result of the combined effects of the local cross grain, produced by knots interrupting the direction of the grain, plus checking that may develop in and around knots during drying. Strength reduction is not caused by an inherent inferiority of the material comprising the knots. Therefore, knots in lumber are more accurately defined as strength-reducing characteristics than as defects.

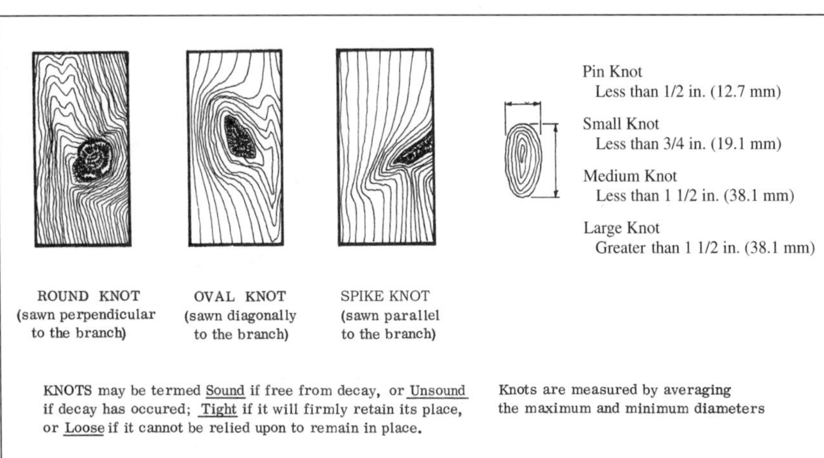

FIGURE 6.2-12 Knots are limbs embedded in a tree that have been cut through during lumber manufacture.

SHAKES

The effect of *shakes* (Fig. 6.2-13) on wood is largely in appearance, as they only slightly affect the compressive and tensile strength of members. Shakes do reduce the shear resistance of wood used for beams, however, so limitations are imposed during grading.

PITCH POCKETS

Whether pitch pockets affect strength depends on their number, size, and location in a piece. Their effect on strength has often been overestimated. Certain restrictions are imposed in the lumber grading rules.

DECAY

The presence of decomposed or destroyed wood as a result of decay is restricted by grading rules in structural grades of lumber to specific proportions of the cross section.

MOLDS AND STAINS

Molds and stains that cannot be removed by surfacing are quality-reducing factors only for appearance grades.

INSECT DAMAGE

The effect on strength of pinholes, grub holes, and other insect damage is taken into account during grading. The strength of such lumber is acceptable when properly graded. Such insect damage is of primary concern in the appearance grades.

6.2.3.2 Manufacturing Characteristics

Some characteristics that affect the appearance, grading, or use of lumber are the result of the manufacturing process. Manufacturing faults may be present due to sawing practice or finishing operations at either the sawmill or the planing mill, or they may be caused during seasoning.

EFFECTS OF SAWING

Theoretically, lumber is cut from a log in one of two distinct ways (see Fig. 6.1-20): (1) tangential to the annual growth rings, producing lumber called *plain-sawn* in hardwoods and *flat-grained* or *slash-grained* in softwoods, or (2) on a radial to the rings (parallel to the rays), producing lumber called *quarter-sawn* in hardwoods and *edge-grained* or *vertical-grained* in softwoods. Not all lumber can be cut to fit these definitions exactly. In commercial practice, a piece sawn so that the growth rings when viewed from the end of the piece form an angle of 45 degrees or more with the wide face is classified as quarter-sawn, and when the rings form an angle of less than 45 degrees, it is classified as plain sawn.

For many uses, including structural applications, plain-sawn and quarter-sawn lumber are equally satisfactory, but each has some advantages. As compared with *quarter-sawn* lumber, *plain-sawn* lumber:

- Is less expensive to produce because less labor and waste are involved;
- Shows some figure patterns involving growth rings and other figure sources more conspicuously;
- Is less subject to shrinking and swelling in thickness;
- Has round or oval knots that usually have a lesser weakening effect and affect surface appearance less than the spike knots that are more common in quarter-sawed lumber;
- Has shakes and pitch pockets that extend through fewer pieces.

Conversely, *quarter-sawn* lumber:

- Is more costly to produce;
- Displays some figure patterns peculiar to the radial surface more conspicuously;
- Shrinks and swells less in width;
- Tends to wear more evenly because the radial surface is more uniform;
- Holds paint better on radial surfaces in some species;
- Works smoother for items run to patterns or profiles;
- Twists and cups less and has less tendency to develop surface checks and splits in seasoning and in use.

A *cross grain* results when the fibers in a piece of wood are not parallel to the edges. One type of cross grain, called *diagonal grain,* occurs when a log is sawn at an angle not parallel with the bark. If the center of the log is lined up with the center of the headrig carriage, the first few cuts do not cut the full length of the log, because most logs are tapered, and cross grain may result. Cross grain may cause warpage, end shrinkage, rough faces, torn or lifted grain, and weak lumber.

Like knots, cross grain has a depreciating influence on the strength of lumber due to the deviation in grain alignment. But unlike knots, which are of a highly localized nature, cross grain may involve an entire piece. The various grading rules control cross grain in structural grades by a measurement of the slope-of-grain. Figure 6.2-14 shows how this is measured.

EFFECTS OF SURFACING

Occasionally, imperfections in lumber are created during surfacing or finishing. Depending on their size and location in a piece, the imperfections or blemishes resulting from manufacture defined in the following paragraphs are limited by the grading rules so that they will have no more effect on the utility of the piece than other characteristics permitted in the same grade.

Torn grain is a roughened area caused by a machine tearing out bits of the wood during dressing.

Skip is an area on a piece that is not surfaced smoothly as it passes through a planer.

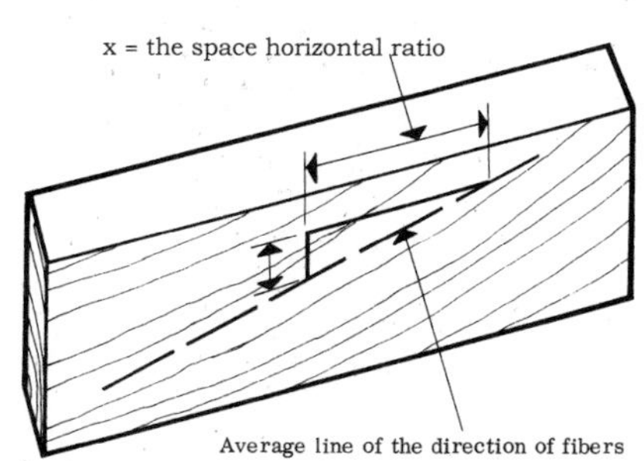

FIGURE 6.2-14 The slope of the grain affects strength.

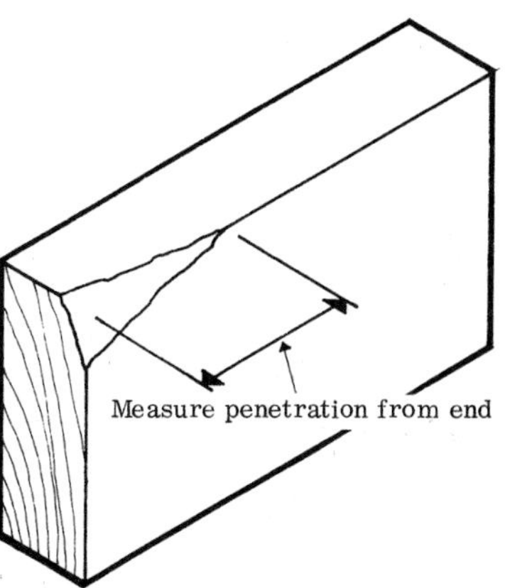

FIGURE 6.2-15 Wane.

Mismatched lumber results when adjoining pieces in worked or patterned lumber do not fit tightly or will not align in the same plane.

Wane is the presence of bark or the absence of wood on the edge or corner of a piece. It is restricted in grading rules because of limitations on bearing area, nailing edge, and appearance, rather than because it has any direct effect on strength (Fig. 6.2-15).

Machine burn is the darkening of the wood due to overheating by machine knives or rolls when pieces are stopped in a machine.

EFFECTS OF SEASONING

The lowering in quality of lumber due to seasoning, which is called *de-grade*, is usually a result of unequal shrinkage, which includes surface and end checking, loosening of knots, and various types of warp. Seasoning de-grade can be largely eliminated by proper practice in either kiln drying or air drying at the point of manufacture, in retail yard storage, and at the construction site. Lumber dried too rapidly will check and split. If it is dried too slowly and the temperature is favorable for fungal growth, stain or decay may develop.

Grading rules specify the amount of seasoning de-grade permitted in the various grades, some of which are described and illustrated in the following paragraphs.

Warp is a deviation from a true or plane surface and includes bow, crook, twist, cup, or combinations of these (Fig. 6.2-16). Warp is measured at the point of greatest deviation from a straight line.

Checking is separation across the

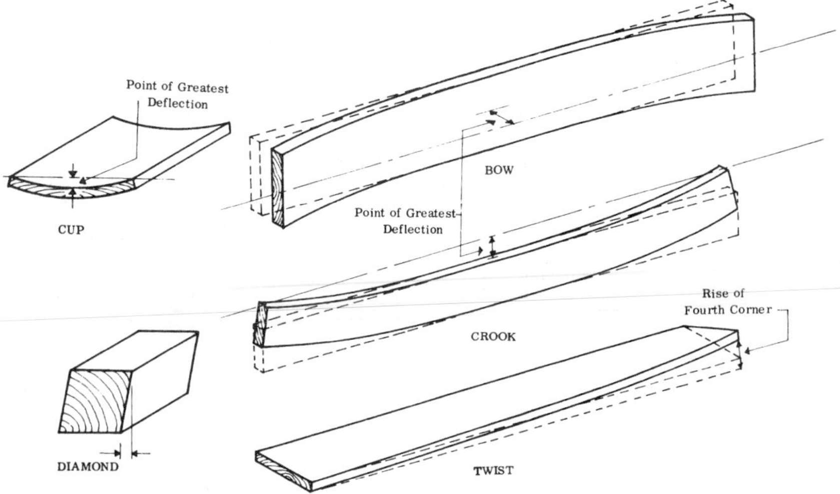

FIGURE 6.2-16 Warp.

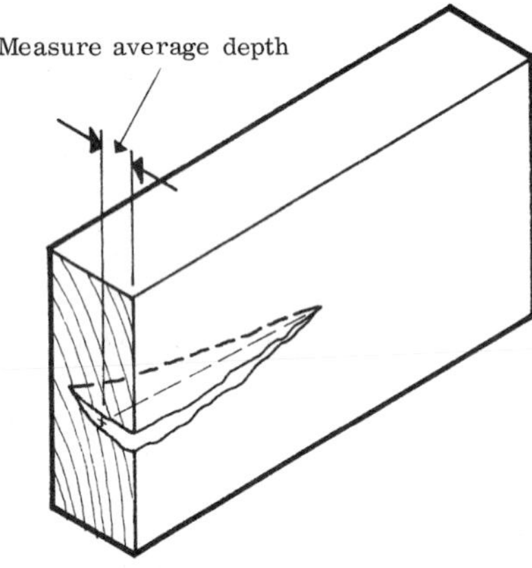

FIGURE 6.2-17 Check.

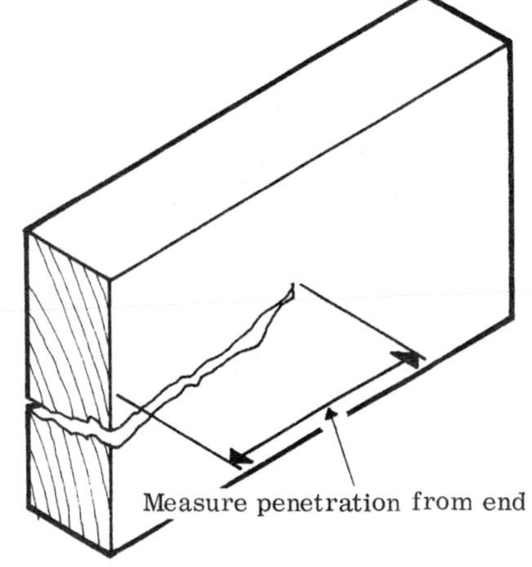

FIGURE 6.2-18 Split.

growth rings, occurring as a surface check if present on only one surface or as a through check if it extends entirely through the piece. Checking is measured by depth and length (Fig. 6.2-17). Checks seldom affect strength but do affect appearance, especially in select grades of softwoods where large grain separations are objectionable. In structural lumber, checks have little, if any, effect on tension or compression members. Checks may reduce shear resistance, and limitations in the grading rules are based on this reduction.

Splits are lengthwise grain separations extending through a piece from one surface to the other. They are measured as the penetration of the split from the end of the piece and parallel to the edge of the piece (Fig. 6.2-18).

6.2.4 SOFTWOOD LUMBER

In the early days, sawmilling was entirely a local industry without need for definite standards of size or quality; manufacturers produced lumber to local customer requirements. As the lumber industry and transportation systems expanded, however, markets were no longer local. Lumber of various species originating in widely separated localities and produced by different mills could not be sold and used successfully without common standards of quality.

Around 1900, several lumber manufacturers' associations were formed, largely to formulate grading rules for the various species manufactured by the producers in their regions. Although these grading rules had much in common, they differed in detail. Industry standardization conferences were begun in 1919, and by 1924 countrywide softwood lumber standards had emerged. The consensus document, known as the *American Lumber Standard*, was revised several times and stayed in force for many years as *Simplified Practice Recommendation SPR 16-53*. This standard underwent a major revision in 1965 and was adopted and published in 1970 as the National Bureau of Standards voluntary Product Standard PS 20-70, *American Softwood Lumber Standard*. Under this standard, for the first time, dressed lumber sizes were related to moisture content in order to ensure reliability of actual in-service dimensions and strengths. PS 20 was amended in 1986 and again in 1994 and stands today as the industry standard for softwood lumber.

A major advance of PS 20-70 was the establishment of an American Lumber Standards Committee and, under it, a National Grading Rules Committee. The function of the latter is to establish and maintain the National Grading Rule for *dimension* lumber. Its members include representatives from all the major grading associations and model code groups and many other organizations representative of the industry, designers, and consumers.

Dimension lumber, as defined in PS 20, is "lumber from 2 in. (50 mm) to, but not including, 5 in. (125 mm) in nominal thickness, and 2 in. (50 mm) or more in nominal width. Dimension may be classified as framing, joists, planks, rafters, studs, small timbers, etc."

Under the National Grading Rule, all grading authorities apply the same grade names and descriptions to dimension lumber of similar quality, regardless of species. However, the Rule does not restrict grading authorities from including additional grades of special products in their own grading rules. For example, the Southern Pine Inspection Bureau grades, and its members market, many grades not covered by the National Grading Rule. In addition, the National Grading Rule does not cover decking lumber, so grade names for this lumber are not the same for all associations. Nevertheless, PS 20 and the National Grading Rule considerably simplify grade recognition and the assignment of working stresses for softwood lumber.

6.2.4.1 Major Producing Regions

Most softwood used in construction is produced in three major regions. The timber species, approximate yearly volume of production, the association that represents producers, and the grading authorities that establish and supervise the lumber grading practices for the different species are shown for each region in Figure 6.2-19.

FIGURE 6.2-19 Major Softwood-Producing Regions of the United States

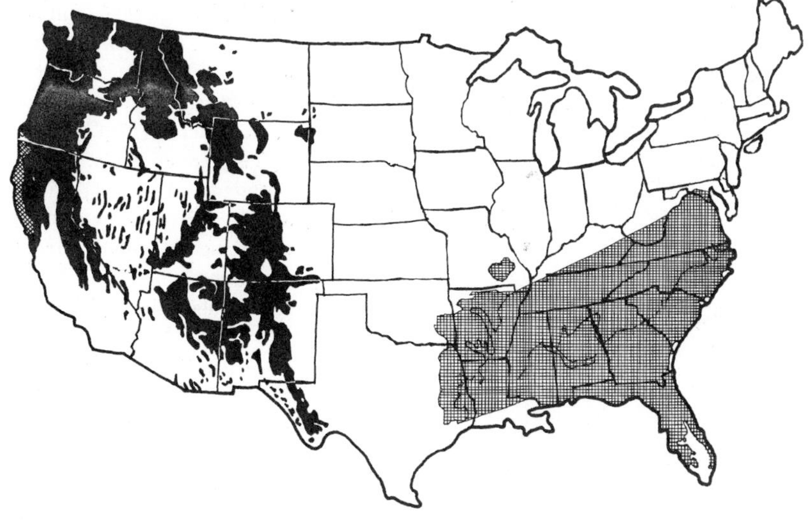

Key	Region	Species	Approximate Yearly Production	Lumber Association	Grading Authority
■	**Western Wood Region**	Douglas fir Ponderosa pine Western red cedar Incense cedar Western hemlock White fir Engelmann spruce Sitka spruce Western larch Lodgepole pine Idaho white pine Sugar pine	16.9 billion board feet	Western Wood Products Association	Western Wood Products Association West Coast Lumber Inspection Bureau
▒	**Redwood**	Redwood Douglas fir	2.7 billion board feet	California Redwood Association	Redwood Inspection Service
▦	**Southern Pine Region**	Longleaf pine Shortleaf pine	6.7 billion board feet	Southern Forest Products Association	Southern Pine Inspection Bureau

WESTERN WOODS REGION

The western woods region is the largest of the producing areas. It includes Oregon, Washington, California, Idaho, Arizona, New Mexico, Texas, Colorado, Montana, South Dakota, Nevada, Utah, and Wyoming. This region produces lumber cut from ponderosa pine, Douglas fir, Engelmann spruce, Idaho white pine, incense cedar, lodgepole pine, sugar pine, Sitka spruce, western hemlock, western larch, western red cedar, and white fir.

REDWOOD REGION

The redwood region is the smallest in area, encompassing but 12 northern California counties. About half of the total production there consists of lumber cut from redwood; the rest comes from other softwood species.

SOUTHERN PINE REGION

The southern pine region covers the entire Southeast, from the eastern part of Texas to Virginia, with some production in Maryland and Missouri. This region produces lumber cut from shortleaf, loblolly, longleaf, slash, Virginia, and pond pine.

6.2.4.2 Standards of Quality

PS 20-70 does not contain detailed requirements for selecting or specifying lumber. It does, however, establish the basic principles, provisions, and procedures for developing commercial grading rules for various species.

The commercial grading rules for each region not only conform to PS 20 and the National Grading Rule, but also reflect the properties of individual species and recognize the specific qualities of lumber desired by users. The grading rules are detailed requirements used to facilitate the manufacture, specifying, sale, and use of lumber. Their basic aim is to separate into marketable and usable classifications the wide range of lumber qualities obtained from a log. Grading is based mainly on quality-reducing characteristics that affect the appearance, strength, durability, or utility of wood.

Commercial grading rules are formulated and maintained by grading authorities, each of which covers the species and individual products manufactured in a given producing region. These authorities publish grading rules for the species produced in their regions and maintain staffs of qualified lumber inspectors who supervise grading at the mills.

Mills that manufacture lumber according to PS 20 usually stamp each piece of dimension lumber with an appropriate grade mark. Examples of official grade marks of the four chief grading authorities are shown in Figure 6.2-20. American grading authorities are approved by the Board of Review of the American Lumber Standards Committee, operating under procedures established by PS 20. The Board of Review reviews the adequacy, competency, and reliability of the U.S. agencies it has approved. The Board of Review has approved agencies in Canada on the same basis. Therefore, Canadian lumber that has been graded and stamped by an authorized agency is compatible with PS 20 and U.S. standard engineering practice.

6.2.4.3 Classifications

Softwood lumber is classified by use, manufacture, size, and grade.

USE CLASSIFICATIONS

The use classifications of softwood lumber are yard lumber, structural lumber, and factory and shop lumber.

Yard lumber consists of grades, sizes, and patterns generally intended for ordinary construction and general building purposes.

Structural lumber is lumber that is 2 in. (50 mm) or more in nominal thickness and is intended for use where working stresses are required. Structural lumber is sometimes called *stress grade* lumber. Working stresses (allowable unit stresses) for

FIGURE 6.2-20 Typical Grade Marks for Dimension Lumber.

West Coast Lumber Inspection Bureau
Construction grade light framing
Douglas fir maximun 19% m.c.
Mill No. 10

12 SEL DECK
WWP ® MC 15 INC CDR

Western Wood Products Association
Select grade decking
Incense cedar maximum 15% m.c.
Mill No. 12

SPIB® No. 1
KD19 7

Southern Pine Inspection Bureau
No. 1 grade light framing
Kiln dried less than 19% m.c.
Mill No., 7

Redwood Inspection Service
Stud grade light framing
Redwood over 19% m.c.
Mill No. 50

(Courtesy West Coast Lumber Inspection Bureau, Western Wood Products Association, Southern Pine Inspection Bureau, and Redwood Inspection Service, Novato, CA, a Division of the California Redwood Association.)

several important commercial species are given in Figure 6.1-14. Complete tabulations for commercially available species and grades are published by the American Forest & Paper Association as *NDS Supplement* to its *National Design Specification for Wood Construction* (NDS). NDS is the industry-recognized standard of proper engineering practice for structural wood design.

Factory and shop lumber is intended primarily for further manufacture into windows, doors, millwork, and similar items. Therefore, the grade of a piece is based on the amount of usable clear wood that can be cut from it.

MANUFACTURING CLASSIFICATIONS

The extent to which lumber is processed is covered in this classification, which includes *rough* lumber, *dressed* (surfaced) lumber, and *worked* lumber.

Rough lumber is lumber that has not been dressed (surfaced) but is sawed, edged, and trimmed at least to the extent of showing saw marks in the wood on four longitudinal surfaces of each piece for its overall length.

Dressed (surfaced) lumber is dressed by a surfacing machine to attain a smooth surface and uniform size. It is designated as follows:

S1S—surfaced on one side
S2S—surfaced on two sides
S1E—surfaced on one edge
S2E—surfaced on two edges
S4S—surfaced on both sides and both edges

These designations are linked to describe pieces finished in other ways. For example, a piece finished on one side and two edges would be designated as S1S2E.

Worked lumber, in addition to being dressed, is matched, shiplapped, or patterned. *Matched lumber* has been worked with a tongue on one edge and a groove on the opposite edge to provide a close tongue-and-groove joint by fitting two pieces together. For end matching, the tongue and groove are worked in the ends (see Fig. 6.2-10).

Shiplapped lumber has been rabbeted on both edges of each piece to provide a close-lapped joint by fitting two pieces together (see Fig. 6.2-10).

Patterned lumber is shaped to a pattern or to a molded form in addition to being dressed, matched, or shiplapped and includes siding and other millwork.

SIZE CLASSIFICATION

Softwood lumber is subdivided into three major categories by size: nominal size, rough size, and dressed size.

Nominal Size The nominal size classification is further subdivided into three categories: boards, dimension lumber, and timbers. Sizes here are given in nominal dimensions.

Boards are pieces less than 2 in. (50 mm) thick and 2 in. (50 mm) or more wide. They include sheathing, subflooring, roofing, finish, trim, siding, and paneling. Boards less than 6 in. (150 mm) wide may be classified as strips.

Dimension lumber, as defined earlier, is used principally for the structural elements of wood-framed buildings. It makes up most common lumber.

Timbers are lumber pieces 5 in. (125 mm) or more in the least dimension. These may be further classified as beams, stringers, posts, caps, sills, girders, purlins, and other members. Lumber thicker than 5 in. (125 mm) in either face is seldom used in ordinary construction and, when stocked by a retail yard, is frequently in a green, rough (undressed) condition. When using heavy timbers, shrinkage of the assembly should be expected, and the design of the structure should be planned to minimize the effects of this shrinkage. Laminated timber, made by gluing up smaller pieces that can readily be kiln dried, should be used when seasoned timbers are mandatory.

Most dimension and timber lumber is stress graded, with working stresses assigned to each grade. Such lumber, intended for load-bearing conditions, also is called structural.

GRADE CLASSIFICATIONS

Another method of classifying softwood lumber is to grade it as *Select*, *Finish*, or *Common*. *Select* and *Finish* grades have good appearance and are used where a natural or high-quality paint finish is required. They are used for the highest quality of interior and exterior finish, trim, moldings, paneling, flooring, and siding. In most species the two highest grades, A and B, are combined as *B and Better* (B&B or B&Btr). Grades C&Btr, C Select, and D Select are more economical and serve many finishing purposes where cost savings are more important than perfect appearance. *Finish* grades include *Superior*, *Prime*, and *E* (Fig. 6.2-21). *Common* grades are suitable for general construction and utility purposes.

The various grades of softwood framing lumber are grouped according to size and intended use into several categories, as follows:

Light framing is lumber 2 in. (50 mm) to 4 in. (100 mm) thick and 2 in. (50 mm) to 4 in. (100 mm) wide and consists of

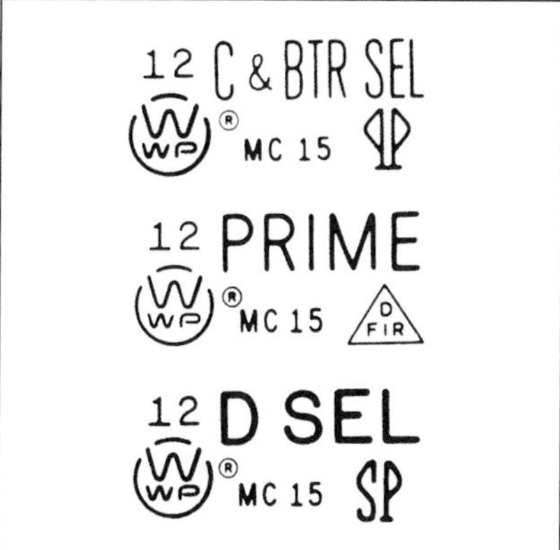

FIGURE 6.2-21 Grade stamps for finish and select grades. (Courtesy Western Wood Products Association.)

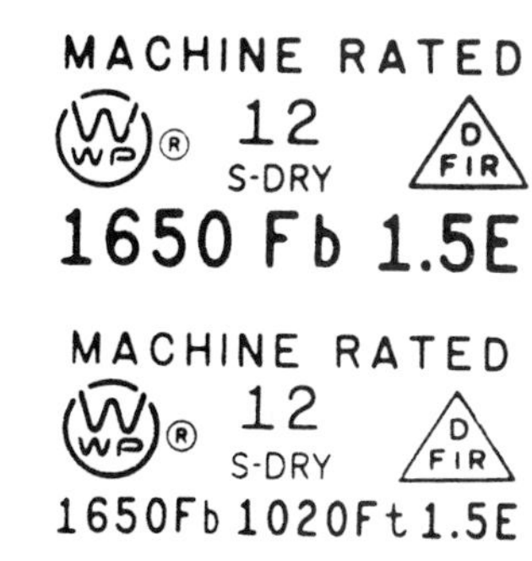

FIGURE 6.2-22 Machine stress-rated (MSR) lumber is automatically stamped with an identifying grade mark. (Courtesy Western Wood Products Association.)

several categories of products. Ordinary light-framing grades (Construction, Standard, and Utility) are intended for uses not requiring high strength, such as general framing, plates, and blocking.

Structural light-framing grades (Select Structural, No. 1, No. 2, and No. 3) are designed to fit engineering applications where higher bending strength ratios are needed, such as in trusses and in box beams.

Structural joists and planks are lumber of rectangular cross section 2 in. (50 mm) to 4 in. (100 mm) thick and 5 in. (125 mm) wide or wider, graded with respect to strength in bending when loaded either on a narrow face of a joist or on the wide face of a plank. The grades are the same as in structural light framing: Select Structural, No. 1, No. 2, and No. 3.

Studs grade members are suitable for stud use in both non-load-bearing and load-bearing walls. Studs (2 in. [50 mm] to 4 in. [100 mm] thick, 2 in. [50 mm] to 4 in. [100 mm] wide) are made in lengths 10 ft. (3048 mm) and shorter, primarily for use in walls. Strength in bending is comparable to No. 3 structural light-framing lumber.

Decking is lumber 2 in. (50 mm) to 4 in. (100 mm) thick, 4 in. (100 mm) wide and wider, intended for exposed view as planking in ceilings and roofs. The moisture content is usually limited to 15%, and the lumber is worked to a tongue-and-groove pattern. It may be ordered with V-shaped or rounded edges and with striated or grooved patterns on the face. Typical grades are Select and Commercial.

Timbers are classified as either stress rated or non-stress rated. The grades of non-stress-rated timbers are Square Edge and Sound (SE&S), No. 1, No. 2, and No. 3. Stress-rated grades are Select Structural SR, No. 1SR, and No. 2SR. Timbers that are nearly square (that is, their width does not exceed their thickness by more than 2 in. [50 mm]) are generally intended for use as posts and have slightly lower allowable bending strengths than rectangular timbers, which are intended for use mainly as beams.

Most lumber is graded visually by trained inspectors. In stress-grade lumber, limitations are imposed on slope-of-grain and the size, number, and placement of knots. Some structural grades in some species (for example, Douglas fir and southern pine) have additional grade requirements based on density and growth-ring count. Although most dimension lumber is graded visually, some lumber 2 in. (50 mm) or less in thickness is now being mechanically stress rated at the sawmill, allowing manufacturers to standardize the load-bearing capacity of each piece. This more refined method of selection permits assignment of higher stresses for much of the lumber in a grade that had been subjected to visual inspection.

In mechanical rating, lumber is fed through a machine that subjects each piece to bending and measures its modulus of elasticity, E. The machine then electronically computes the stress grade, taking into account the interacting effect of knots, slope-of-grain, growth rate, density, and moisture content. As the piece leaves the machine it is automatically stamped MACHINE RATED, with an assigned fiber stress in bending (F_b on the grade mark) and corresponding modulus of elasticity (E) rating (Fig. 6.2-22).

The close relationship between E and F_b has led to the establishment of categories of stress grades for lumber 2 in. (50 mm) or less in thickness (Fig. 6.2-23) by the Southern Pine Inspection Bureau (see also Fig. 6.1-15). Machine stress-rated lumber is also visually graded to limit certain characteristics (checks, splits, wane, warp).

6.2.4.4 Size and Grade Standardization

Historically, lumber grade names and grade requirements have been developed independently on a regional basis and, even within regions, have varied among some species. Methods of establishing stress values for stress-grade lumber have also differed among lumber-producing regions, and changes in structural grading are continuing as the newer methods of machine rating are refined.

Under PS 20, a National Grading Rule Committee (NGRC) was established for the purpose of developing uniform, countrywide grade requirements for dimension lumber. The functions of the NGRC, according to PS 20, were to "establish, maintain, and make fully and fairly available grade strength ratios, nomenclature, and descriptions of grades for dimension lumber conforming to American Lumber Standard." The requirements developed by the NGRC are known as the National Grading Rule for dimension lumber and constitute a part of all grading authority grading rules.

Individual grading rules are much more extensive and detailed than the National Grading Rule. Bending strength ratios are standard relationships between the basic stresses of an ideal, clear-wood, piece of lumber free of strength-reducing characteristics (knots, sloping grain, splits) and the stresses developed in a particular grade of lumber in that species.

Since lumber shrinks in direct proportion to moisture loss in drying, from a green condition of about 30% moisture content to the moisture content it ultimately attains in service, it is possible to establish, with reasonable tolerances, separate sizes for green and dry-surfaced lumber so that both types achieve essentially the same size in service and provide equal strength and stiffness for the same spans and loading conditions.

The *NDS Supplement* takes into account the slight oversizing of green lum-

FIGURE 6.2-23 Machine Stress-Rated Lumber, Allowable Design Values (in psi)

Extreme Fiber Stress in Bending, F_b[a]	f-E Classification	Modulus of Elasticity, E	Tension Parallel to Grain, F_t	Compression Parallel to Grain, $F_{c\perp}$
1200	1200f-1.2E	1.2	600	
1350	1350f-1.3E	1.3	750	1600
1400	1400f-1.2E	1.2	800	1600
1500	1500f-1.3E	1.3	900	1650
	1500f-1.4E	1.4		
1600	1600f-1.4E	1.4	950	1675
1650	1650f-1.4E	1.4	1020	1700
	1650f-1.5E	1.5		
1800[b]	1800f-1.5E	1.5	1300	1750
1800	1800f-1.6E	1.6	1175	1750
1950	1950f-1.5E	1.5	1375	1800
	1950f-1.7E	1.7		
2000	2000f-1.6E	1.6	1300	1825
2100	2100f-1.8E	1.8	1575	1875
2250	2250f-1.6E	1.6	1750	1925
	2250f-1.7E	1.7		
	2250f-1.9E	1.9		
2400	2400f-1.7E	1.7	1925	1975
	2400f-1.8E	1.8		
	2400f-2.0E	2.0		
2550	2550f-2.1E	2.1	2050	2025
2700	2700f-2.2E	2.2	2150	2100
2850	2850f-2.3E	2.3	2300	2150
3000	3000f-2.4E	2.4	2400	2200
3150	3150f-2.5E	2.5	2500	2250
3300	3300f-2.6E	2.6	2650	2325

Horizontal shear F_v – 90 psi
Compression perpendicular to grain $F_{c\perp}$ – 565 psi

(Courtesy Southern Pine Inspection Bureau.)

[a]The tabulated Extreme Fiber in Bending values, F_b, are applicable to lumber loaded on edge When loaded flatwise, these values may be increased by multiplying by the following factors:

Nominal Width (in.)	4	6	8	10	12	14
Factor	1.1	1.15	1.15	1.2	1.2	1.2

[b]Grade requires F_t certification and quality control.

ber and allows the same values for green and dry surfaced lumber when both are used at a moisture content of 19% or less. Lower values are assigned to lumber that will be used under exposed conditions where the moisture content is likely to exceed 19% for extended periods.

6.2.5 HARDWOOD

Hardwood accounts for about 25% of the total board foot production of lumber. It comes from many species and has uses ranging from small specialty items to the heaviest structural work. The predominant hardwood species are the oaks and walnut, which constitute more than 50% of total hardwood lumber production.

6.2.5.1 Areas of Growth

Most of the commercial hardwood species in the United States grow east of the Great Plains (see Fig. 6.1-1). The hardwoods of the west, which grow principally in California, Oregon, and Washington, amount to about 2% of the total hardwood produced.

6.2.5.2 Standards of Quality

The National Hardwood Lumber Association (NHLA) publishes size and grading rules for sawn hardwood lumber. These rules, however, are used within the lumber industry to describe hardwood lumber used to manufacture other products and are not normally used to specify hardwood used in building construction. In the construction industry, hardwood lumber is made to special order for the specific requirements of the purchaser, and the grade descriptions and characteristics are those that are agreed upon by the buyer and seller.

The grading rules set down by the Architectural Woodwork Institute in its *Architectural Woodwork Quality Standards, Guide Specifications, and Quality Certification Program* can be used to control hardwood lumber for cabinets, paneling, shelving, miscellaneous ornamental items, stairs and handrails, millwork, doors, windows, and casework.

6.2.5.3 Hardwood Dimension

A large proportion of the hardwood lumber produced at sawmills is used in paneling, furniture, flooring, and other finished or semifinished products. *Dimension stock*, or *hardwood dimension*, produced by hardwood sawmills is shipped to furniture, tool, piano, cabinet, and similar factories.

Hardwood *dimension* lumber is usually kiln dried and processed to a point where it will deliver maximum utility to the user. It is manufactured from rough boards and flitches to the specific requirements of a particular plant or industry, in specified thicknesses, widths, and lengths. It may be solid or glued-up and is classified as rough dimension, surfaced dimension, semifabricated dimension, or completely fabricated dimension.

Rough hardwood dimension consists of planks sawn and ripped to certain sizes.

Surfaced and *semifabricated* are rough dimension lumbers, carried one or more steps further, which may include one or more of several operations such as surfacing, molding, tenoning, drum sanding, equalizing, trimming, or mitering. However, such operations do not make the product a completely fabricated one ready for assembly.

Completely fabricated is lumber ready for assembly into finished products. Hardwood parts are often fabricated in hardwood dimension plants adjoining the sawmill site, and many hardwood flooring plants are operated in conjunction with hardwood sawmills.

The manufacture of hardwood lumber and its by-products is one of the more complex branches of the lumber industry because of the variety of uses for hardwood and the many species involved.

6.3 Plywood and Other Panels

Plywood is an engineered panel composed either of several wood veneer sheets placed at right angles to one another, or of veneer sheets placed across a lumber, mineral, or oriented strand board core. These components are then bonded together under high pressure using either a waterproof or water-resistant adhesive. Placing thin veneer sheets in this manner tends to balance the positive and negative aspects of wood's strength properties, improves its dimensional stability, and increases its resistance to both checking and splitting.

Both decorative and construction grades of plywood are manufactured. Construction plywood is used wherever strength is an important factor but occasionally for paneling too; decorative plywood (chiefly hardwood) is used for wall paneling, cabinetwork, and furniture.

FIGURE 6.3-1 **At a lath, a sharp steel blade peels a block into thin layers.** (Courtesy American Plywood Association.)

6.3.1 CONSTRUCTION PLYWOOD

Construction plywood is manufactured primarily of softwoods, but hardwood species are also employed. However, a plywood panel need not be composed entirely of the same species. Strength and stiffness requirements for inner plies are less demanding, and these are frequently of a different species than the outer plies. Structural strength or stiffness is not significantly changed, because the outer faces of a plywood panel provide most of its strength and stiffness lengthwise, which is the normal direction to apply a plywood panel when it is used structurally.

6.3.1.1 Manufacture

Manufacturing begins by debarking a tree, cutting it into peeler logs, and cutting these into shorter log-shaped *blocks*. These blocks are placed in a giant lathe and rotated against a long knife that peels the wood into long, continuous veneer sheets, mainly 1⁄10 in. (2.54 mm) to 3⁄16 in. (4.76 mm) thick (Fig. 6.3-1). Veneer cut in this manner is called *rotary cut*. Veneer primarily intended for decorative plywood is sometimes *sliced* rather than peeled from the block (see Fig. 6.3-2), but an insignificant amount of softwood plywood is manufactured of sliced veneer.

The veneer is next conveyed to clippers, cut into desired widths, then run

FIGURE 6.3-2 **A flitch is held in clamps and moved against a knife.**

FIGURE 6.3-3 **A dryer oven reduces the moisture content of veneer.** (Courtesy American Plywood Association.)

These laid-up panels are stacked in contact with each other and inserted into a hydraulic press, where pressure of about 150 psi (105.46 Mg/m^2) is exerted to compress the veneer and glue together to make plywood panels. The panels are then separated, and each is placed into a separate shelf of a final pressing machine (Fig. 6.3-4). The panels are not in contact with each other. The machine applies heat at about 300°F (148.9°C) to each panel and presses up from the bottom, closing the shelves against each other. This lasts from 10 to 15 seconds. The bottom of the machine then lowers and tilts the shelves. The panels slide out one at a time onto an elevated moving table that collects the panels, requiring no contact by human hands. After removal from the press, these panels are trimmed to exact size, run through large drum sanders if they are of a sanded grade, and are finally checked for grade and appearance, emerging as finished plywood ready for use.

through long drying ovens that reduce its moisture content to between 2% and 6% to provide the best possible glue bond and panel stability (Fig. 6.3-3).

The 4 ft. (1200 mm) × 8 ft. (2400 mm) veneers are then graded and placed on a long 4-ft. (1200 mm)-wide conveyer, where they pass under a device that drops glue in lines or sprays glue downward onto the veneers as they move along. A parallel conveyer above this lower system delivers other veneers onto the top of the glued veneers. The interior veneers, which have their grain running perpendicular to the face veneers, are not 4 ft. (1200 mm) × 8 ft. (2400 mm) as are the face veneers. Because they are made from poorer-quality veneers and because of their grain orientation, they are 4 ft. (1200 mm) measured parallel to their grain and some dimension less than 4 ft. (1200 mm) in the other direction, which is the direction of their movement. Consequently, it takes several of these smaller veneers side by side to match the length of the 8-ft. (2400 mm) veneers moving down the conveyor. Workers standing beside the conveyor quickly position the inner veneers dropping on the line from the upper conveyor into a 4 ft. (1200 mm) × 8 ft. (2400 mm) format. Once these inner veneers have been positioned, they pass under a glue-dropping machine. The last veneer to drop from the upper conveyor onto the stack is another 4 ft. (1200 mm) × 8 ft. (2400 mm) face veneer.

FIGURE 6.3-4 **Laid-up panels are placed in a press that presses veneers into plywood panels.** (Courtesy American Plywood Association.)

6.3.1.2 Standards of Quality

Construction plywood, except for a few specialty panels, should conform with the requirements of the latest version of "U.S. Product Standard PS 1 for Construction and Industrial Plywood." PS 1 contains the recognized grading rule for construction plywood and the accepted bond, panel construction, dimensional tolerance, and workmanship requirements for maintaining a high level of quality. It also includes procedures for testing, grading, labeling, and identifying plywood. Conformance with the requirements of PS 1 is indicated by the grade trademark or other suitable designation of a qualified testing and certifying agency performing the functions outlined in the Product Standard.

About 85% of all construction plywood is produced by mills whose owners belong to the American Plywood Association (APA). This organization maintains a quality control program to assure that plywood carrying an official grade trademark conforms with PS 1. Typical APA grade trademarks are shown in Figure 6.3-5.

Figure 6.3-6, which is excerpted from a more extensive APA table, can be used to help make a selection of plywood for general use. The grade of plywood needed can be determined by examining the data in the "Description and Use" column. It is then necessary to select either a thickness or a span rating and, for appearance grades, the desired species group. When engineering design is needed, such as for stress skin panels or unusual loading conditions, a designer should refer to the more extensive design data produced by APA.

Plywood is graded and marked according to species, exposure durability, veneer grade, panel grade, and application.

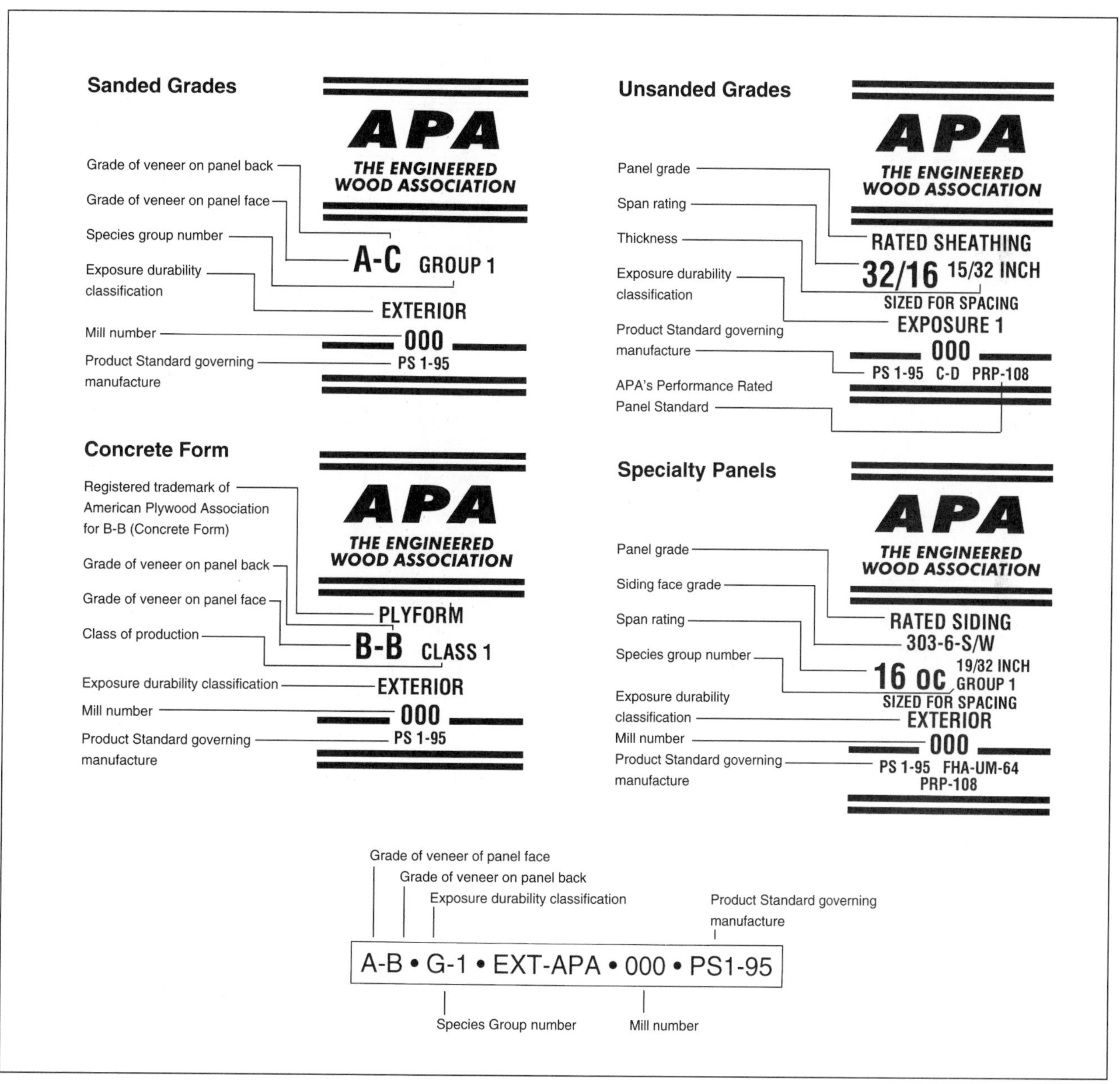

FIGURE 6.3-5 Typical grade trademarks of the American Plywood Association. (Courtesy American Plywood Association.) (Refer to Figure 1.8-27 for metric conversions.)

FIGURE 6.3-6 Guide to the Use of Plywood Panels

Interior or Protected Applications

Plywood Grade	Description and Use	Typical Trademarks	Veneer Grade: Face	Veneer Grade: Back	Veneer Grade: Inner	Common Thicknesses
APA-Rated Sheathing Exp 1 or 2[a]	Unsanded sheathing grade for wall, roof, subflooring, and industrial applications such as pallets and for engineering design, with proper stresses. Manufactured with intermediate and exterior glue.[b] For permanent exposure to weather or moisture only. Exterior type plywood is suitable.	APA THE ENGINEERED WOOD ASSOCIATION RATED SHEATHING 32/16 15/32 INCH SIZED FOR SPACING EXPOSURE 1 000 PS 1-95 C-D PRP-108	C	D	D	5/16, 3/8, 15/32, 1/2, 19/32, 5/8, 23/32, 3/4
APA Structural I Rated Sheathing Exp 1[a] **or APA Structural II**[c] **Rated Sheathing Exp 1**	Plywood grades to use where strength properties are of maximum importance, such as plywood-lumber components. Made with exterior glue only. STRUCTURAL I is made from all Group 1 woods. STRUCTURAL II allows Group 3 woods.	APA THE ENGINEERED WOOD ASSOCIATION RATED SHEATHING STRUCTURAL I 24/0 3/8 INCH SIZED FOR SPACING EXPOSURE 1 000 PS 1-95 C-D PRP-108	C	D	D	5/16, 3/8, 15/32, 1/2, 19/32, 5/8, 23/32, 3/4
APA Rated Sturd-I-Floor Exp 1 or 2[a]	For combination subfloor-underlayment. Provides smooth surface for application of carpet. Possesses high concentrated and impact load resistance during construction and occupancy. Manufactured with intermediate and exterior glue. Touch-sanded.[d] Available with tongue and groove.[e]	APA THE ENGINEERED WOOD ASSOCIATION RATED STURD-I-FLOOR 20 oc 19/32 INCH SIZED FOR SPACING T&G NET WIDTH 47-1/2 EXPOSURE 1 000 PS 1-95 UNDERLAYMENT PRP-108	C plugged	D	C and D	19/32, 5/8, 23/32, 3/4, 1 1/8 (2-4-1)
APA Underlayment Exp 1, 2, or INT	For underlayment under carpet. Available with exterior glue. Touch-sanded. Available with tongue and groove.[e]	APA THE ENGINEERED WOOD ASSOCIATION UNDERLAYMENT GROUP 1 EXPOSURE 1 000 PS 1-95	C plugged	D	C and D	1/2, 19/32, 5/8, 23/32, 3/4
APA C-D Plugged Exp 1, 2, or INT	For built-ins, wall and ceiling tile backing, NOT for underlayment. Available with exterior glue. Touch-sanded.[e]	APA THE ENGINEERED WOOD ASSOCIATION C-D PLUGGED GROUP 2 EXPOSURE 1 000 PS 1-95	C plugged	D	D	1/2, 19/32, 5/8, 23/32, 3/4
APA Appearance Grades Exp 1, 2, or INT	Generally applied where a high quality surface is required, Includes APA N-N, N-A, N-B, N-D, A-A, A-B, A-D, B-B, and B-D INT grades.[e]	APA THE ENGINEERED WOOD ASSOCIATION A-D GROUP 1 EXPOSURE 1 000 PS 1-95	B or better	D or better	C and D	1/4, 11/32, 3/8, 15/32, 1/2, 19/32, 5/8, 23/32, 3/4

(Continues)

FIGURE 6.3-6 *(Continued)*

Exterior Applications

Plywood Grade	Description and Use	Typical Trademarks	Veneer Grade			Common Thicknesses
			Face	Back	Inner	
APA-Rated Sheathing Ext[a]	Unsanded sheathing grade with waterproof glue bond for wall, roof, subfloor, and industrial applications such as pallet bins.	APA THE ENGINEERED WOOD ASSOCIATION RATED SHEATHING 48/24 23/32 INCH SIZED FOR SPACING EXTERIOR 000 PS 1-95 C-C PRP-108	C	C	C	5/16, 3/8, 15/32, 1/2, 19/32, 5/8, 23/32, 3/4
APA Structural I Rated Sheathing Ext[a] **or APA Structural II**[c] **Rated Sheathing Ext**	"Structural" is a modifier for this unsanded sheathing grade. For engineered applications in construction and industry where full exterior-type panels are required. STRUCTURAL I is made from Group 1 woods only.	APA THE ENGINEERED WOOD ASSOCIATION RATED SHEATHING STRUCTURAL I 24/0 3/8 INCH SIZED FOR SPACING EXTERIOR 000 PS 1-95 C-C PRP-108	C	C	C	5/16, 3/8, 15/32, 1/2, 19/32, 5/8, 23/32, 3/4
APA Rated Sturd-I-Floor Ext[a]	For combination subfloor-underlayment where severe moisture conditions may be present, as in balcony decks. Possesses high concentrated and impact load resistance during construction and occupancy. Touch-sanded.[d] Available with tongue and groove.[e]	APA THE ENGINEERED WOOD ASSOCIATION RATED STURD-I-FLOOR 20 oc 19/32 INCH SIZED FOR SPACING EXTERIOR 000 PS 1-95 C-C PLUGGED PRP-108	C plugged	C	C	19/32, 5/8, 23/32, 3/4
APA Underlayment Ext and APA C-C Plugged Ext	Underlayment for floor where severe moisture conditions may exist. Also for controlled atmosphere rooms and many industrial applications. Touch-sanded. Available with tongue and groove.[e]	APA THE ENGINEERED WOOD ASSOCIATION C-C PLUGGED GROUP 2 EXTERIOR 000 PS 1-95	C plugged	C	C	1/2, 19/32, 5/8, 23/32, 3/4
APA B-B Plyform Class I or II[c]	Concrete-form grade with high reuse factor. Sanded both sides, mill-oiled unless otherwise specified. Available in HDO. For refined design information on this special-use panel see *APA Design/Construction Guide: Concrete Forming*, Form No. V345. Design using values from this specification will result in a conservative design.[e]	APA THE ENGINEERED WOOD ASSOCIATION PLYFORM B-B CLASS 1 EXTERIOR 000 PS 1-95	B	B	C	19/32, 5/8, 23/32, 3/4
APA Marine Ext	Superior Exterior-type plywood made only with Douglas fir or western larch. Special solid-core construction. Available with MDO or HDO face. Ideal for boat hull construction.	MARINE • A-A • EXT- APA • 000 • PS 1-95	A or B	A or B	B	1/4, 3/8, 1/2, 5/8, 3/4
APA Appearance Grades Ext	Generally applied where a high quality surface is required. Includes APA A-A, A-B, A-C, B-B, B-C, HDO, and MDO Ext.[e]	APA THE ENGINEERED WOOD ASSOCIATION A-C GROUP 1 EXTERIOR 000 PS 1-95	B or better	C or better	C	1/4, 11/32, 3/8, 15/32, 1/2, 19/32, 5/8, 23/32, 3/4

[a]Properties and stresses apply only to APA-rated Sturd-I-Floor and APA-rated Sheathing manufactured entirely with veneers.

[b]When exterior glue is specified, i.e., Exposure 1, stress level 2 (S-2) should be used.

[c]Check local suppliers for availability before specifying Structural II and Plyform Class II grades, as they are rarely manufactured.

[d]APA rated Sturd-I-Floor 2-4-1 may be produced unsanded.

[e]May be available as Structural I.

(Courtesy American Plywood Association.) (Refer to Figure 1.8-27 for metric conversions.)

6.3.1.3 Wood Species

PS 1 classifies about 70 species of wood for use in construction and industrial plywood. These are divided into five groups according to their strength, stiffness, and other properties (see Fig. 6.3-7), with Group 1 being the strongest and stiffest. The group number shown in the grade stamp (see Fig. 6.3-5) refers to the species used in the face and back veneer plies.

6.3.1.4 Exposure Durability

Two basic types of construction plywood are produced to comply with the requirements of PS 1. These are *exterior type* and *interior type*, which refer to the exposure durability of the panel. Within each type there are several classifications based on the types of veneers and adhesives used.

EXTERIOR TYPE (WATERPROOF)

The veneers used in exterior panels must be C grade or better. Exterior plywood panels are designed to retain their glue bond under one of the three levels of exposure designated by the exposure durability classifications:

- *Exterior* panels maintain their integrity when repeatedly wetted and dried or otherwise subjected to the weather in a permanent exterior exposure to weather or moisture.
- *Exposure 1* panels may be exposed for a long time during construction, but finally protected. They have the same adhesives as are used in exterior type panels, but are not designed for permanent exposure.
- *Exposure 2* panels are used where exposure time will be limited before the panels are protected.

INTERIOR TYPE (MOISTURE RESISTANT)

Interior type plywood is made with D grade veneer or better. There are three levels of adhesive durability within the interior type classification.

- *Moisture-resistant* plywood is bonded with interior glue and is intended for interior applications where it may be temporarily exposed to the elements.
- Plywood bonded with intermediate glue has a lower moisture resistance than that bonded with exterior glue, but greater than that bonded with interior glue. Intermediate glues have high levels of resistance to bacteria, mold, and moisture. They are suitable for use in protected construction where moderate delays in providing protection may be expected or conditions of high humidity and water leakage may exist.
- Plywood bonded with exterior glue is intended for protected construction where maximum performance is required for protection against moisture exposure resulting from long construction delays or other conditions of similar severity.

6.3.1.5 Grades

The term *grade* refers both to veneer grade and panel grade. The veneer grade is a definition of the quality of the face veneer. The panel grade is identified in terms of the veneers used, by a name identifying the panel's intended use, or both.

VENEER GRADES

Each type of plywood panel is manufactured with several grades of face veneers. A veneer's grade designation is based on the species group and the quality of the veneer used on the face and back. Panels are built up of veneers glued together at right angles to each other, using one or more of five veneer grades ranging from N (highest) through A, B, C, and D (lowest). Veneer may be repaired to raise its quality to a higher grade. Repairs consisting of patches or plugs, usually oval or

FIGURE 6.3-7 Classification of Species

Group 1	Group 2		Group 3	Group 4	Group 5
Apitong[a,b]	Cedar, Port Orford	Mengkulang[a]	Alder, red	Aspen	Basswood
Beech, American	Cypress	Meranti, red[a,d]	Birch, paper	Bigtooth	Poplar, balsam
Birch	Douglas fir 2[c]	Mersawa[a]	Cedar, Alaska	Quaking	
Sweet	Fir	Pine	Fir, subalpine	Cativo	
Yellow	Balsam	Pond	Hemlock, eastern	Cedar	
Douglas fir 1[c]	California red	Red	Maple, bigleaf	Incense	
Kapur[a]	Grand	Virginia	Pine	Western red	
Keruing[a,b]	Noble	Western white	Jack	Cottonwood	
Larch, western	Pacific silver	Spruce	Lodgepole	Eastern	
Maple, sugar	White	Black	Ponderosa	Black (western popular)	
Pine	Hemlock, western	Red	Spruce	Pine	
Caribbean	Lauan	Sitka	Redwood	Eastern white	
Ocote	Almon	Sweetgum	Spruce	Sugar	
Pine, southern	Bagtikan	Tamarack	Engelmann		
Loblolly	Mayapis	Yellow poplar	White		
Longleaf	Red lauan				
Shortleaf	Tangile				
Slash	White lauan				
Tanoak	Maple, black				

[a]Each of these names represents a trade group of woods consisting of several closely related species.

[b]Species from the genus *Dipterocarpus* are marked collectively: Apitong if originating in the Philippines; Keruing if originating in Malaysia or Indonesia.

[c]Douglas fir from trees grown in the states of Washington, Oregon, California, Idaho, Montana, Wyoming, and the Canadian Provinces of Alberta and British Columbia shall be classed as Douglas fir No. 1. Douglas fir from trees grown in the states of Nevada, Utah, Colorado, Arizona, and New Mexico shall be classed as Douglas fir No. 2.

[d]Red Meranti shall be limited to species having a specific gravity of 0.41 or more based on green volume and oven-dry weight.

(Table 1, Voluntary Product Standard PS 1-83, "Construction and Industrial Plywood," National Institute of Standards and Technology, May 1984, page 7. Not subject to copyright in the United States.)

circular, along with long, narrow shims, fill voids in the veneer sheet or replace characteristics such as knots and pitch pockets. There are also free-form synthetic (plastic) patches. These are put into a panel after the veneers have been bonded together. A flaw is routed out, the synthetic patch is squirted into the void from a hand-held device resembling a glue gun. The synthetic material is then cured and sanded. In the grade stamp designation for Rated Siding in Figure 6.3-5, the 303-6-S/W designation means the following: 303 is the number code for a particular line of siding. The S stands for synthetic patches. The W stands for wood patches. The 6 is the number of patches permitted in the face. In this case, six synthetic or wood patches are permissible.

Veneer grades are illustrated in Figure 6.3-8 and described in the following paragraphs.

N Grade: A specialized veneer available on special order, used for natural finish. It is select, all heartwood or all sapwood, and free of open defects with only a few small, well-matched repairs (not illustrated). N grade veneer is of such high quality and so expensive that it is rarely specified.

A Grade: A smooth paint grade permitting a number of neatly made repairs, plugs, or patches with certain restrictions. This is the grade usually specified for high-quality work.

B Grade: A solid surface veneer, except for specified minor characteristics. It may have a considerable number of neatly made repairs and is suitable for concrete form work where a smooth surface is desired.

C Plugged Grade: A special improved C Grade intended for underlayment for floors and combination subfloor underlayment (not illustrated).

C Grade: The lowest quality of veneer permitted in exterior type panels. It is commonly used as the face of sheathing grade panels. Permitted defects include tight knots up to 1½ in. (38.1 mm), 1-in. (25.4 mm) knotholes, occasional 1½-in. (38.1 mm) knotholes under certain conditions, splits up to ½ in. (12.7 mm) tapering to a point, and other minor defects.

D Grade: The lowest quality of veneer permitted in interior type panels. This grade permits 2½-in. (63.5 mm) knotholes, occasional 3-in. (76.2 mm) knotholes under certain conditions, pitch pockets up to 2½ in. (63.5 mm) wide, and tapering splits up to 1 in. (25.4 mm).

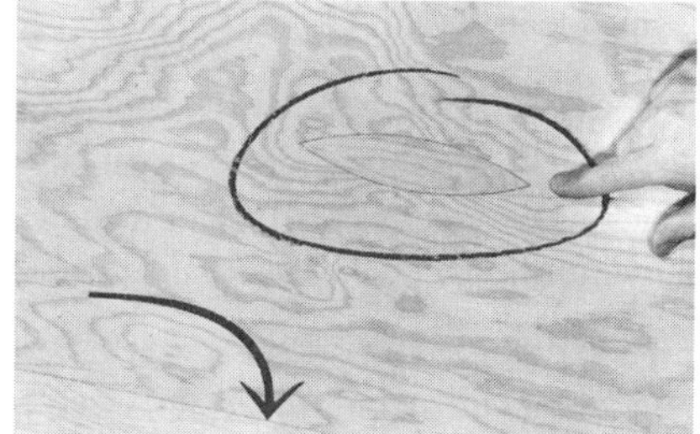

"A" Grade

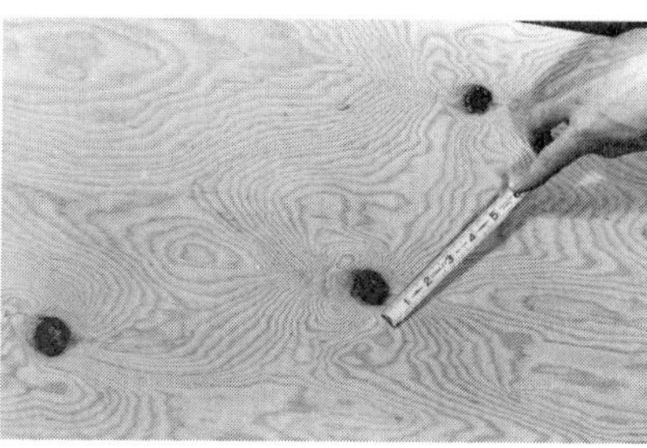

"C" Grade

"B" Grade

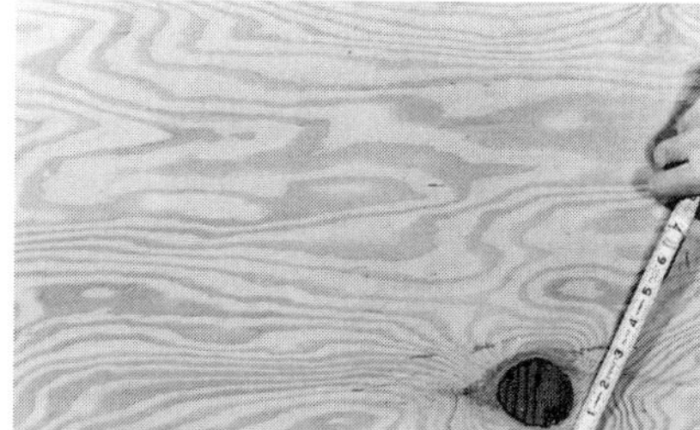

"D" Grade

FIGURE 6.3-8 Plywood veneer grades. (Courtesy APA, the Engineered Wood Association.)

PANEL GRADES

The grade of a panel is indicated by two letters, such as A-D, describing a panel with an A grade front face veneer and a D grade back face veneer. The inner plies for an interior type panel are D or better, and those for an exterior type panel are C or better. Unless a decorative or other surface is specified, panels in most grades are run through large drum sanders where they are sanded smooth on both sides for appearance and balanced construction.

6.3.1.6 Performance-Rated Panels

Performance-rated panels are also sometimes called *structural panels* or *structural-use panels*. These are panels designed for use where design properties, such as tension, compression, shear, cross-panel flexural properties, and nail bearing may be significant. Included in this group are panels designed for use as floor, wall, or roof sheathing; for subflooring, underlayment, or both; for siding; for construction of stress-skin panels, box beams, and other structural elements; and for other structural uses. This category includes most of the plywood used in building construction. It also includes oriented strand board (see Section 6.3.3) and Com-ply, which is a product made with alternating plies of wood veneer and reconstituted wood.

Performance-rated panels possess stiffness, strength, nail-holding ability, durability, dimensional stability, lateral bracing ability, and resistance to puncture and impact. Some of them must also provide a suitable base for a covering material, such as shingles, built-up roofing, siding, or flooring.

In small buildings with conventional framing, sheathing, and subflooring, no special engineering design is required if recognized standards and construction details are followed. Correct use of plywood involves selecting the appropriate panel type, grade, and span rating for the spans and loads to be carried, in addition to using proper nailing.

Performance-rated panels are identified according to their application, the quality of the face and back veneer, or both (see Fig. 6.3-6). When necessary, they are also identified as to their structural grade (Structural I or Structural II). They are classified in one of the three exterior exposure durability classifications discussed earlier (Exterior, Exposure 1, or Exposure 2).

Performance-rated panels for use as sheathing, subflooring, underlayment, or siding are assigned a *span rating* to set the support spacing required for each. This rating is included as part of the grade trademark (see Fig. 6.3-5) as a single number or as two numbers separated by a slash. When a single number appears, that is the required maximum spacing of the supports in inches when used as a flooring material. When two numbers separated by a slash appear, the two numbers denote different spacings depending on the use of the panel. For example, a panel may be used as either wall or roof sheathing, but the support spacing is not the same in the two different uses. The first number denotes the maximum spacing in inches when the panel is used as a roof decking. The second number is the maximum spacing in inches when the panel is used for flooring. Wall sheathing has its own single-number system, which denotes the maximum span in inches for the supporting studs when sheathing spans in its weak direction.

Some performance-rated panels are left unsanded for greater stiffness and strength, as well as economy. However, some panels, such as underlayment, are either sanded or touch sanded to bring adjacent panels to a uniform thickness, and some, such as siding, are sanded for appearance.

6.3.1.7 Specialty Plywoods

Special manufacturing processes and treatments produce a number of variations from standard panels, each designed to meet specific use requirements.

OVERLAID PLYWOOD

Overlaid plywood panels are high-grade panels of *exterior* plywood to which cellulose-fiber sheets containing prescribed amounts of resin have been bonded to one or both faces. Most overlaid plywood has an overlay on both sides. The resin is usually phenolic, but other resins may also be used if they meet the requirements stated in PS 1. The overlay sheets are bonded to the veneers with adhesives under pressure. Overlaid plywood is classified as either *high density* (HDO), *medium density* (MDO), or *special overlay* panels.

High-Density Overlay The face veneer quality of HDO plywood may be either A-A or B-B. The inner plies are C-plugged veneer. Concrete form overlay HDO has B-B face and back veneers. The overlay sheets on HDO are impregnated with at least 45% resin solids based on the volatile-free weight of fiber and resin exclusive of glueline. The total thickness of a resin-impregnated overlay sheet is 0.012 in. (0.305 mm) before pressing, and the total weight is at least 60 lb. (27.22 kg) per 1000 sq. ft. (92.903 m^2) or 0.293 kg/m^2.

HDO has a hard, smooth, chemically resistant, durable surface that requires no further finishing by paint or varnish. It is available in several basic colors, but also provides an excellent paint base and can be painted if color variation is preferred. HDO is especially suitable and widely used in concrete form work for architectural finishes, highway signs, and in other severe exposures.

Medium-Density Overlay The face veneer quality of MDO plywood is B-B. The inner plies are C veneer. The overlay sheets on MDO must contain at least 17% resin solids for a beater-loaded sheet and 22% resin solids for an impregnated sheet, both based on the volatile-free weight of fiber and resin exclusive of glueline. The total thickness of a resin-impregnated overlay sheet is 0.012 in. (0.305 mm) before pressing, and the total weight is at least 58 lb. (26.31 kg) per 1000 sq. ft. (92.903 m^2) or 0.28 kg/m^2.

MDO is intended to receive high-quality paint finishes and provides an opaque, smooth, nonglossy surface that generally blanks out the grain, although some evidence of the underlying grain is permitted. A number of mills manufacture special MDO precut siding, frequently mill-primed, in widths up to 16 in. (406.4 mm).

Special Overlays Overlaid plywood surfacing panels that do not fit the definition of either HDO or MDO are called *special overlays*. These must meet the glue bond requirements for overlaid plywood, and the panels must be of the exterior type.

TEXTURED SIDING PANELS

Textured plywood siding panels are used for exterior siding, gable ends, and fencing, and as forms for special effects in concrete. Refer to Section 7.7 for additional information.

PANELING

Most paneling plywoods are manufactured from hardwood species (see Section 6.3.2.1). Some softwood species, such as California redwood, knotty pine, and western red cedar, are used for interior paneling. Other manufacturers assemble softwood veneers into paneling plywood with surface textures such as embossing, striations, or relief grain. Some manufacturers produce softwood panels having printed overlays simulating hardwood grains.

TONGUE-AND-GROOVE PANELS

Plywood in ½-in. (12.7 mm), ⅝-in. (15.9 mm), and ¾-in. (19.1 mm) thicknesses, with tongue-and-groove joints, is produced by a number of mills in underlayment grades and occasionally other grades. Tongue-and-groove panels are used to eliminate the need for blocking at joints. One performance-rated panel is available in thickness from 19/32 in. (15.08 mm) up to 1 3/32 in. (27.78 mm) for use as a combination subfloor and underlayment, for use over joist spacings varying from 16 in. (410 mm) to as wide as 48 in. (1220 mm), based on the panels being continuous over at least two spans. These types of panels are tongued and grooved (Fig. 6.3-9) to eliminate the need to block the joints.

CONCRETE FORM PANELS

The face veneers for plywood panels designed for concrete form work are not less than B grade. The inner plies are not less than C grade. These panels are usually sanded on both sides and oiled at the mill. They are classified as either Class I or Class II according to the combination of wood species groups used in faces, crossbands, and center veneer plies. Class I panels require the use of stronger and stiffer species groups in faces and crossbands than do Class II panels. Concrete form panels are designed for reuse. HDO panels are also sometimes used as concrete forms.

MARINE PLYWOOD

Marine plywood panels are exterior grade plywood of grade A-A, A-B, B-B, HDO,

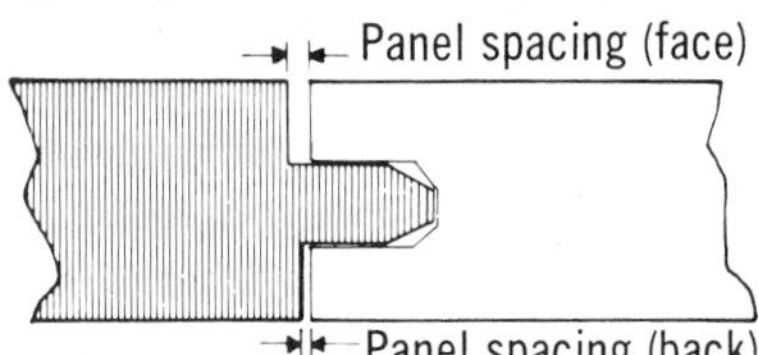

FIGURE 6.3-9 Tongue-and-groove panels. (Photograph courtesy APA, the Engineered Wood Association.)

or MDO, but with certain additional requirements. For example, the only wood species permitted are Douglas fir 1 and western larch; the number of repairs in the face veneers is more limited than normal; inner plies must be B grade or better; wood repairs must be glued with an exterior adhesive; and there are additional restrictions as well.

6.3.1.8 Sizes

Plywood is most readily available in panels 4 ft. (1220 mm) wide by 8 ft. (2400 mm) long. Some mills also produce 5-ft. (1500 mm)-wide panels and lengths up to 12 ft. (3600 mm), but these are not always stocked. Beyond these limits widths up to 10 ft. (3000 mm) and any reasonable length (which has been more than 50 ft. [15.24 mm]) can be made by scarf jointing or finger jointing panels together. Scarf- or finger-jointed construction consists of gluing panels either end-to-end or side-to-side continuously along a tapered or fingered cut to form a single panel (Fig. 6.3-10).

Plywood panels are permitted by U.S. Product Standard PS 1 to be up to 1⁄16 in. (1.6 mm) smaller but may be no larger in horizontal dimension than the specified size. Nominal available thicknesses and thickness tolerances for plywood panels are shown in Figure 6.3-11.

FIGURE 6.3-11 Plywood Panel Thicknesses and Tolerances According to U.S. Product Standard PS 1

Plywood Type	Normally Available Nominal Thicknesses, in. (mm)	Tolerances (plus or minus), in. (mm)
Unsanded panels	5⁄16 (7.9), 3⁄8 (9.5), 15⁄32 (11.9), 1⁄2 (12.7), 19⁄32 (15.1), 5⁄8 (15.9), 23⁄32 (18.3), 3⁄4 (19.1)	1⁄32 (0.79)
	7⁄8 (22.2), 1 (25.4), 1 1⁄8 (28.6)	5% of specified thickness
Sanded panels	1⁄4 (6.4), 9⁄32 (7.14), 11⁄32 (8.73), 3⁄8 (9.5), 15⁄32 (11.9), 1⁄2 (12.7), 19⁄32 (15.1), 5⁄8 (15.9), 23⁄32 (18.3), 3⁄4 (19.1)	1⁄64 (0.38)
	7⁄8 (22.2), 1 (25.4), 1 1⁄8 (28.6)	3% of specified thickness
Touch-sanded panels	1⁄2 (12.7), 19⁄32 (15.1), 5⁄8 (15.9), 23⁄32 (18.3), 3⁄4 (19.1)	1⁄32 (0.79)
	1 1⁄8 (28.6)	5% of specified thickness
Marine plywood	1⁄4 (6.4), 3⁄8 (9.5), 1⁄2 (12.7), 5⁄8 (15.9), 3⁄4 (19.1)	1⁄64 (0.38)
Overlaid plywood	11⁄32 (8.73), 3⁄8 (9.5), 15⁄32 (11.9), 1⁄2 (12.7), 19⁄32 (15.1), 5⁄8 (15.9), 23⁄32 (18.3), 3⁄4 (19.1)	1⁄32 (0.79)

Plywood panels always have an odd number of *layers* for balanced construction, the minimum number being three. A layer may consist of a single thickness of veneer, or it may be a lamination of two veneers with their grain running in the same direction. Each single thickness of veneer is called a *ply*. The minimum number of plies for various thicknesses of plywood is given in Figure 6.3-12. A diagram of how the layers are typically arranged is given in Figure 6.3-13.

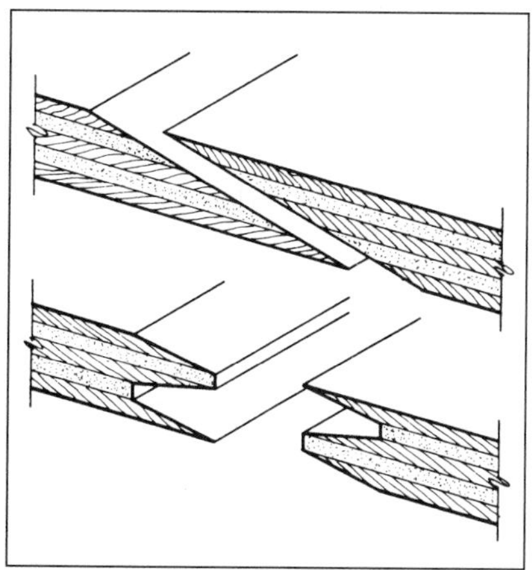

FIGURE 6.3-10 Panels can be scarf-jointed (top) or finger-jointed (bottom).

6.3.1.9 Physical and Mechanical Properties

Aside from the characteristics of the veneer used, the most important properties of plywood derive from cross-laminated construction. Plywood utilizes the strength and dimensional stability of wood along the grain by placing alternate layers at right angles. This cross lamination provides resistance to splitting, improved dimensional stability, warp resistance, and other properties, as discussed in the following paragraphs.

DURABILITY

In order to comply with the PS 1 requirements for exterior type plywood, the glueline used to bond the veneers must be as durable as the wood itself. Theoretically, exterior panels may be exposed outdoors without paint or other treatment, without affecting the durability of the glue bond or the life of the panels. In practice, exposing the edges of any plywood to weather is not a good idea. The glueline at the edge of a plywood panel is the part of the panel most vulnerable to damage by water and weather. Usually, therefore, plywood, like other wood products, is painted or stained when exposed outdoors to protect it against weathering and checking. Preservative treatments are available for plywood that is to be used in areas of high decay or termite hazard.

BENDING STRENGTH AND STIFFNESS

The cross lamination of layers in plywood panels capitalizes on the more desirable physical and mechanical properties of wood. The strength of wood parallel to the grain is many times greater than that perpendicular to the grain. Cross-laminated construction provides longitudinal grain crosswise and improves significantly on solid wood's crosswise strength and stiffness. However, plywood panels are both stiffer and stronger along the face grain than across the face grain and should be applied with the face grain perpendicular to their supports for maximum resistance to bending.

RACKING RESISTANCE

Plywood functions in its own plane as a structural bracing material when used on a wall, roof, or floor. Although 5⁄16 in. (7.94 mm) is the minimum thickness for sheathing, even 1⁄4-in. (6.4 mm)-thick plywood on a standard wood stud-framed wall provides twice the stiffness and more than twice the strength furnished by 1-in. (25 mm)-thick diagonal boards, which has been long recognized as adequate. The racking strength of plywood and other

FIGURE 6.3-12 Panel Constructions

Panel Grades	Finished Panel Nominal Thickness Range, in. (mm)	Minimum Number of Plies	Minimum Number of Layers
Exterior			
Marine	Through ⅜ (9.5)	3	3
Special exterior	Over ⅜ (12.7), through ¾ (19.1)	5	5
B-B concrete form	Over ¾ (19.1)	7	7
High-density overlay			
High-density concrete form overlay			
Interior			
N-N, N-A, N-B, N-D, A-A, A-B, A-D, B-B, B-D			
Structural I (C-D, C-D plugged, and underlayment)	Through ⅜ (9.5)	3	3
Structural II (C-D, C-D plugged, and underlayment)	Over ⅜ (12.7), through ½ (15.9)	4	3
	Over ½ (15.9), through ⅞ (22.2)	5	5
	Over ⅞ (22.2)	6	5
Exterior			
A-A, A-B, A-C, B-B, B-C			
Structural I and Structural II (C-C and C-C plugged)			
Medium density and special overlays			
Interior (including grades with exterior glue)			
Underlayment	Through ½ (12.7)	3	3
	Over ½ (12.7), through ¾ (19.1)	4	3
Exterior	Over ¾ (19.1)	5	5
C-C plugged			
Interior (including grades with exterior glue)			
C-D	Through ⅝ (15.9)	3	3
C-D plugged	Over ⅝ (15.9), through ¾ (19.1)	4	3
	Over ¾ (19.1)	5	5
Exterior			
C-C			

(Table 4, Voluntary Product Standard PS 1-83, "Construction and Industrial Plywood," National Institute of Standards and Technology, May 1984, page 17. Not subject to copyright in the United States.)

sheathing materials is further discussed later in this chapter.

Two properties give plywood superior bracing qualities: (1) high shear strength in all directions for loads applied perpendicular to panel faces and (2) relatively high nail-bearing strength, allowing nails to be driven as close as ¼ in. (6.4 mm) from the edge of a panel. Splitting is virtually impossible lengthwise or crosswise, because no cleavage plane exists.

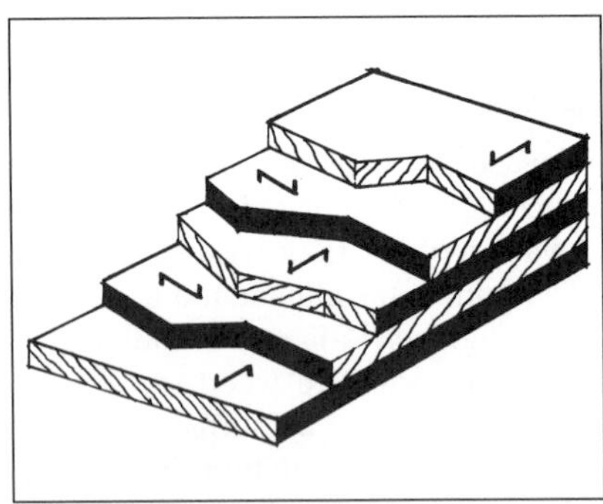

FIGURE 6.3-13 Each layer is perpendicular to adjacent layers for strength and stiffness.

IMPACT AND CONCENTRATED LOAD RESISTANCE

Stresses resulting from impact or concentrated loading are distributed over a wide area by the cross veneers in a plywood panel. This is an important property in sheathing, subflooring, siding, and wall paneling.

DIMENSIONAL STABILITY

Like solid wood, plywood is subjected to dimensional changes due to variations in moisture content, temperature, and other causes.

HYGROSCOPIC EXPANSION AND CONTRACTION

Expansion due to absorption and contraction due to loss of water in solid wood are generally insignificant parallel to the grain, even with substantial changes in moisture content. Perpendicular to the grain, dimensional changes can be considerable (see Section 6.1.3.2). In plywood, any tendency for veneers to expand or contract crosswise is greatly minimized by one or more adjacent veneers running lengthwise. The total expansion both across the width and along the length of a 4 × 8 ft. (1200 × 2400 mm) panel in service under normal variations in moisture content (8% to 14%) should average between 0.03 in. (1/32 in. [0.79 mm]) and 0.05 in. (1/20 in. [1.27 mm]).

THERMAL EXPANSION

Expansion and contraction due to thermal changes in solid wood are about 10 times greater across the grain as they are parallel to the grain, where changes are quite small. In plywood thermal expansion along the face is somewhat greater than in plain wood because of the influence of the crossbands. In an 8-ft. (2400 mm)-long Douglas fir panel with five equal plies, the expansion caused by a 120°F (48.89°C) rise in temperature should average approximately 0.021 in. (1/50 in. [0.53 mm]). This is about 40% more than for plain Douglas fir wood, but still insignificant for most uses. Across the grain, expansion due to a 120°F (48.89°C) rise is approximately 0.029 in. (0.74 mm), or only about one-sixth that for plain wood.

WARP

Bow, cup, or twist may occur in a plywood panel if it is subjected to unequal absorption of moisture along its edges, different exposure to moisture on opposite faces, or unbalanced painting or coating of its faces. However, plywood that is flat at the time of manufacture will remain so unless conditions of storage or use subject it to uneven changes in moisture content.

CHECKING

Checking may appear on any wood surface, plywood included, in the form of hairline cracks or even slightly open splits. Under severe conditions of moisture and dryness, such as experienced by an unpainted, unprotected plywood panel exposed to the weather, checks may in time become open cracks, penetrating virtually the full thickness of the face. With moisture pickup, the exposed veneer surface expands while the undersurface glued to the crossband is held in a relatively fixed position. The expanding outer fibers push against one another and acquire a *compression set*. As the fibers dry out and shrink, they pull away from each other, creating checks of different intensity.

Remedies to control checking within acceptable limits include protection from dampness or moisture during transit and storage, protection during use at the job site, and the use of protective coatings, such as conventional paint, and water-repellent dips to minimize moisture absorption and loss.

WORKABILITY

Plywood is easily worked, cut to various sizes and shapes, and may be bent to fairly sharp radii (Fig. 6.3-14).

FIGURE 6.3-14 Minimum Plywood Bending Radii

Panel Thickness, in.	Bending Radii for Panel Bend in Direction, ft. Across Grain	Parallel to Grain
1/4	2	5
5/16	2	6
11/32 and 3/8	3	8
15/32 and 1/2	6	12
19/32 and 5/8	8	16
23/32 and 3/4	12	20

(Courtesy American Plywood Association.) (Refer to Figure 1.8-27 for metric conversions.)

FIRE AND DECAY RESISTANCE

A high degree of resistance to either fire or decay can be imparted to plywood through pressure-impregnation treatment similar to that used for lumber (see Section 6.1.6). Only plywood bonded with exterior glue is treated, because the adhesive used in it is not affected by pressure impregnation of fire-retardant chemicals. Fire-retardant treatments have little effect on the rate of heat transfer through a panel, but virtually eliminate fuel contribution to a fire and prevent the wood from burning by causing it to char.

Plywood used indoors may be protected by the application of a fire-retardant paint to its surface. When exposed to high temperatures, such paints blister or bubble, forming air pockets that insulate the plywood, retard combustion, and greatly reduce flame spread along the surface.

6.3.2 DECORATIVE PLYWOOD

The division of plywood products into two categories, (1) *construction and industrial* and (2) *decorative*, is based on the two governing standards: the previously discussed PS 1 and the Hardwood Plywood and Veneer Association (HPVA) standard HP-1.

Construction and industrial plywood covered by PS 1 is made mostly of softwood species, although a few hardwood species are permitted. Decorative plywood covered by HP-1 consists entirely of hardwood-faced panels. The interior layers of these panels are often of softwood or particleboard, however, to help lower their cost and to conserve the rarer hardwood species.

Decorative plywood panels are most often used as wall paneling, in cabinet and furniture work, as thin face skins in door manufacture, and as face veneer for block flooring.

The species of wood used for the face veneers of panels identifies the hardwood plywood panels. For example, birch plywood has a face veneer and, probably, a back veneer of birch.

6.3.2.1 Manufacture

Differences between the manufacture and appearance of hardwood and softwood plywood, as well as differences among hardwood plywoods, may result from the: (1) tree species, (2) part of the tree used, (3) cutting methods used to produce the veneer, (4) methods used to match the veneers, and (5) type of core construction.

TREE SPECIES

More than 99,000 species of hardwoods have been classified. About 240 of these are available in commercial quantities in the United States, and about 50 are widely used for hardwood plywood. These include lauan, walnut, birch, gum, oak, maple, ash, cherry, elm, beech, African mahogany, Honduras mahogany, butternut, pecan, cottonwood, sycamore, hackberry, teak, and rosewood. The most popular at a given time varies with the public's changing preferences. Some typical imported and domestic hardwood veneer species and their characteristics are listed in Figure 6.3-15.

PARTS OF THE TREE

The portion of the tree used greatly affects hardwood plywood figure patterns. Most plywood comes from relatively straight portions of the tree, but some of the more interesting and valuable patterns are cut from a crotch, a burl, or a stump. Veneers cut from a crotch have a plume-like design. A burl produces veneers with a pattern of swirls. A stump yields rippling patterns with rich contrasts (Fig. 6.3-16).

VENEER TYPES

The manner in which veneers are cut is important in producing various visual effects. Two logs of the same species that are cut differently will have entirely different visual characteristics even though their veneer colors are similar. Five principal methods of cutting veneers are used (Fig. 6.3-17).

Rotary Cut To produce a rotary cut, a log is mounted centrally in a lathe and turned against a blade, producing exceptionally wide veneer as compared with that cut by other methods. Because this cut follows the log's annular growth rings, the resulting pattern has a bold, variegated ripple figure. This is the type of cut used for most softwood veneers.

Flat Slicing Flat slicing results in a variegated wavy figure. To produce it, a half log, or flitch, is mounted with the heart

FIGURE 6.3-15 Visual Characteristics of Representative Hardwood Species

Commercial Name	Color	Type of Figure
Ash, American	White to light brown	Medium open grain
Basswood	White to pale brown	
Beech	White to reddish brown	
Birch	White to light reddish brown	Curly grained, figured flat cut, plain rotary
Butternut	Pale brown	Leafy grain
Cherry, American	Light to dark reddish brown	Plain to rich mottle
Cottonwood	White to light grayish brown	
Elm	Light brownish red	Strong
Gum, black and tupelo	White, grayish white Greenish to grayish black	
Gum, red	Pink to reddish brown	Medium to highly figured
Hackberry	Pale yellow, yellowish gray to light brown	
Lauan	Pale grayish, yellowish brown to reddish brown	
Mahogany, African	Pink to reddish brown	Plain stripe to highly figured
Mahogany, tropical American	Pink to gold brown	Straight to rich mottle
Mahogany, crotch and swirls	Pink to reddish brown	Moon and feather crotch, plain and figured swirl
Maple	White to tan	Plain, curly burls
Oak, red	Pink tan to ochre	Plain to flake
Oak, white	Gray tan to ochre	Plain to flake
Rosewood, Brazilian	Pink brown and violet	Wide range figure
Rosewood, East Indian	Purple to straw	Striped and figured
Teak	Light tan, dark brown	Plain, ripple, mottle, stripe
Walnut, American	Soft gray brown	Typical figure or stripe

(Used with permission of the Hardwood Plywood & Veneer Association.)

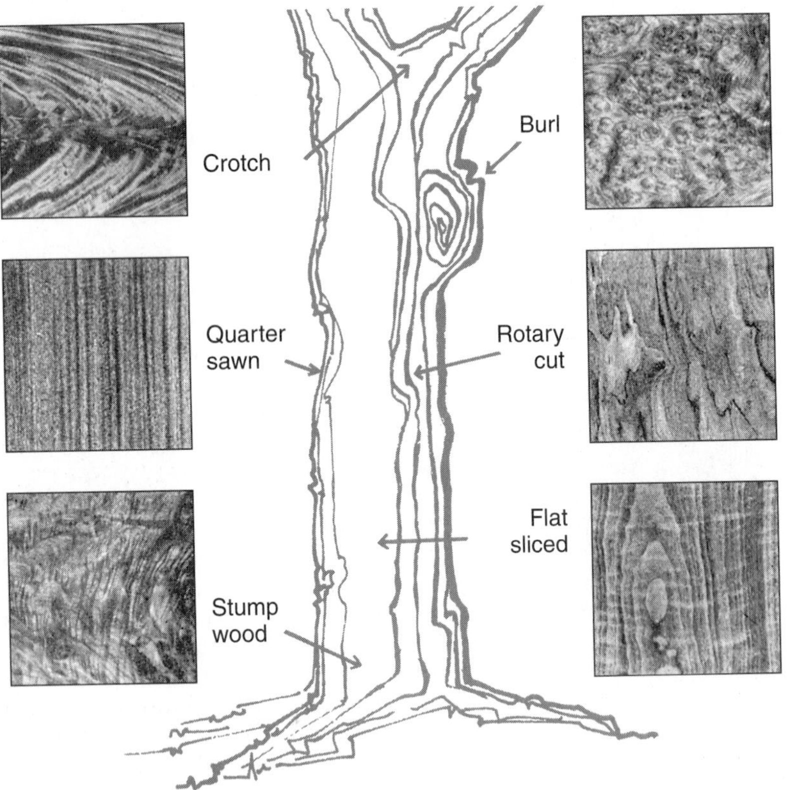

FIGURE 6.3-16 Portions of a tree from which various types of figure patterns are obtained. Some species produce several figure types. (Used with permission of the Hardwood Plywood & Veneer Association and of the Fine Hardwood Veneer Association.)

side flat against the guide plate of a slicer, and the slicing is done parallel to a line through the center of the log (see Fig. 6.3-2).

Quarter Slicing Quarter-sliced veneer has a series of stripes, straight in some woods, varied in others. To produce this type of cut, a quarter log (flitch) is mounted on a guide plate so that the growth rings of the log strike the knife at approximately right angles.

Half-Round Slicing Half-round sliced veneer shows modified characteristics of both rotary and plain-sliced veneers. This method is a variation of rotary cutting in which segments of a log (flitches) are mounted off-center in a lathe. This results in a cut slightly across the annual growth rings.

Rift-Cut Veneer Various species of oak and some other species are sometimes cut using a rift cut. In these woods the ray cells that radiate from the center of the log are conspicuous, and a rift or comb grain effect can be obtained by cutting perpendicularly to these rays either on a lathe or on a slicer.

VENEER MATCHING

The methods used to match face veneers also produce different visual characteristics. Basic and special matching methods are shown in Figure 6.3-18.

Book Matching Any type of veneer can be book matched. In book matching every other sheet is turned over like the leaves of a book. Thus, the back of one veneer meets the front of the adjacent veneer, producing a matching joint design. Book matching is also called edge matching.

Slip Matching Quarter-sliced veneers are often slip matched, but this method may also be used for other types of veneer cuts. In this type of match, adjacent sheets are joined side by side, without turning, to repeat the flitch figure. Slip-matched veneers are common in *premium*, *good*, and *specialty* hardwood plywood grades and are often used in doors and matched, ungrooved wall panels.

Random Matching The joining of veneers with the intention of creating a casual unmatched effect is called random matching. In this type of matching, veneers from several logs may be used in

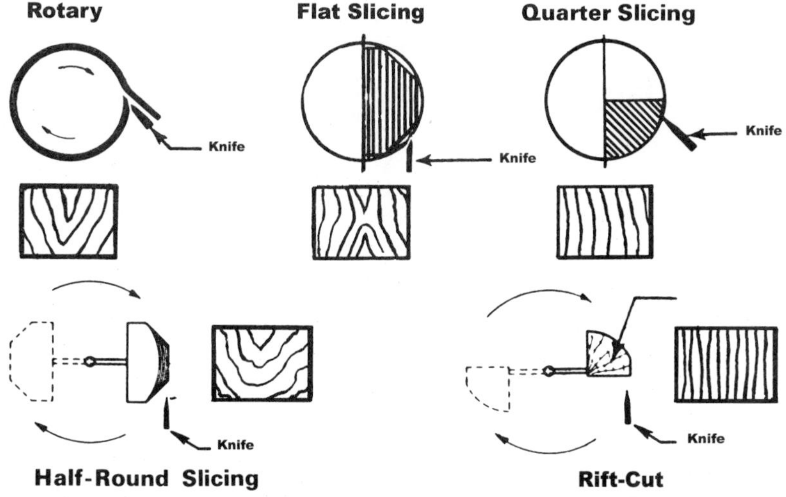

FIGURE 6.3-17 Visual characteristics of veneers depend on method of manufacture.

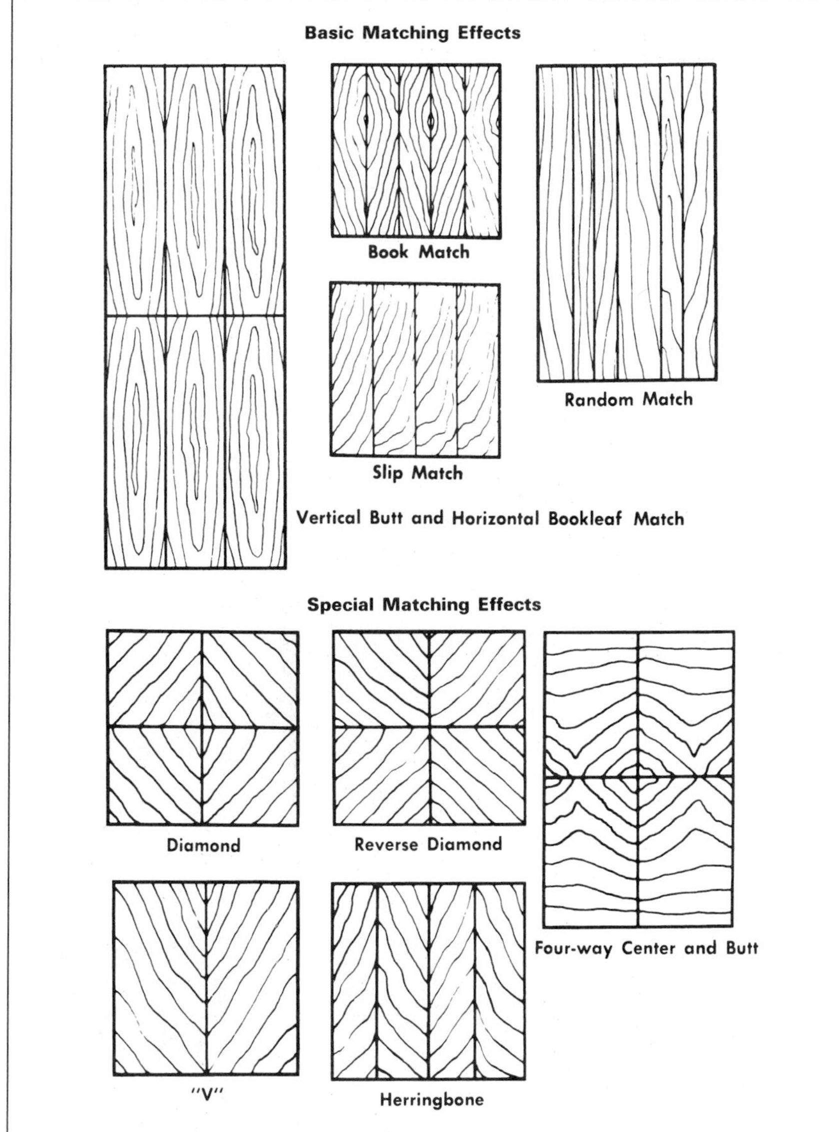

FIGURE 6.3-18 Basic and special matching effects for wood veneers.

the manufacture of a single panel. Random matching is generally used in prefinished V-grooved wall panels.

Vertical Butt and Horizontal Bookleaf Matching Flitches whose length does not permit their fabrication into a panel of the desired height can be matched using the vertical butt and horizontal bookleaf method. In this type of matching, the veneer may be matched horizontally as well as vertically. This type of matching is often used in custom wall paneling and in furniture.

Four-Way Center and Butt Matching Several *special* matching effects are used in furniture manufacture and in custom wall paneling. One of these, four-way center and butt matching, is ordinarily applied to crotch or stump veneers to reveal the figure. Occasionally, flat-sliced veneers are matched in this manner when panel length requirements exceed the length of available veneers.

CORE CONSTRUCTION

The innermost ply in a hardwood plywood panel is called the *core*. The plies between the core and the face plies are called *crossbands*. There are four basic types of cores used in hardwood plywood: *veneer*, *lumber*, *particleboard*, and *mineral*. When panels are exposed on one surface only, back and face plies are of species and thickness to balance the face veneers.

Veneer Core The veneer core method is same as that for softwood plywood. In hardwood plywood, three, five, or more plies are laid with the grain direction of adjacent plies at right angles to each other. The total number of plies is always odd so that the panels are balanced.

Lumber Core Some panels have a core of sawed lumber to which crossband plies and face veneers are glued (Fig. 6.3-19). Lumber core construction is often used in cabinetwork and furniture construction in which exposed edges or edge treatment such as doweled, splined, or dovetailed joints are desired, or where butt hinges are to be used. A lumber core is made of narrow wood strips edge-glued together. Panels with hardwood bands that match the face applied on the edges may be specially ordered.

Particleboard Core Plywood for use as table, desk, and cabinet tops often has a particleboard core. This is an aggregate

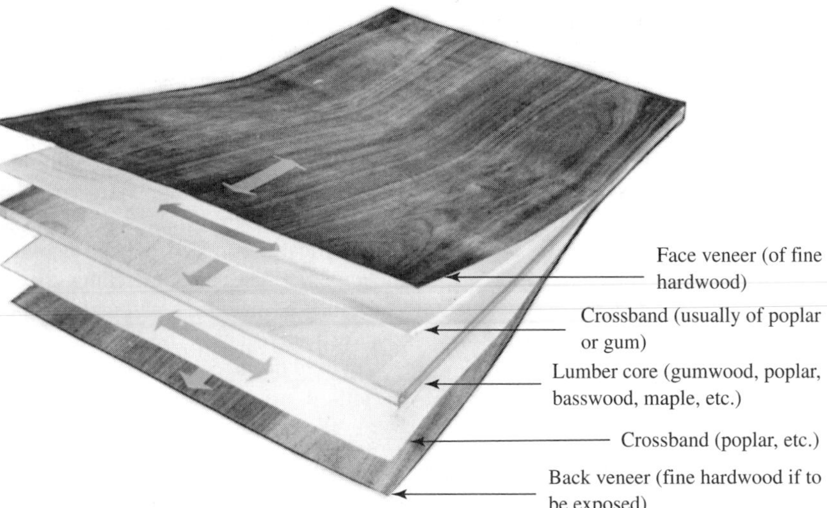

FIGURE 6.3-19 Typical 5-ply lumber core panel. Each ply (indicated by arrows) runs at right angles to adjacent plies. (Used with permission of the Hardwood Plywood & Veneer Association and of the Fine Hardwood Veneer Association.)

FIGURE 6.3-20 Standard Thicknesses of Veneer-Core Hardwood Plywood*

Plies	Thickness, in.
3	⅛, ³⁄₁₆, ¼
5	⁵⁄₁₆, ⅜, ½
5 and 7	⅝
5, 7, and 9	¾

(Used with permission of the Hardwood Plywood & Veneer Association.) (Refer to Figure 1.8-27 for metric conversions.)

*Hardwood plywood is square within ¹⁄₁₆ in. measured on short dimension. Particleboard and MDF-core hardwood plywood panels are typically 3- and sometimes 5-ply in thicknesses up to ¾ in. and thicker.

of wood particles (also called chipboard, chipcore, particle board, etc.) fused under heat and pressure. The face veneers are usually glued directly to this core, although crossbanding is sometimes used.

Mineral Core Fire-resistant panels are made with mineral cores. In this construction, the face veneers are bonded to a core of hard, noncombustible material.

SIZES

Hardwood plywood is most commonly sold in 4 × 8 ft. (1200 × 2400 mm) panels, although panels are custom made up to 40 ft. (12.19 m) long and up to 7 ft. (2.134 m) wide. Stock panels may also be purchased in lengths of 10 ft. (3048 mm) and 12 ft. (3658 mm).

Hardwood plywood is manufactured with an uneven number of plies in thicknesses ranging from ⅛ in. (3.2 mm) up to as much as 3 in. (76.2 mm) in custom panels. Common thicknesses in inches include ¼ in. (6.4 mm), ⅜ in. (9.5 mm), ½ in. (12.7 mm), and ¾ in. (19.1 mm) veneer core and ¾ in. (19.1 mm) in lumber and particleboard core. Figure 6.3-20 shows common thicknesses for varying numbers of plies.

PATTERN AND FIGURE CHARACTERISTICS OF VENEER

Figure refers to the highlights of the so-called crossfire running at right angles to grain direction. The grain character and direction are described by using the word *pattern*. Those in the industry may describe figure by saying it "has a great deal of crossfire," or "has a straight or broken stripe" or "is highly figured." Figure 6.3-21 shows some common veneer figures and the terms often used to describe them.

6.3.2.2 General Classifications

Hardwood plywood falls generally into two categories: (1) *specialty* plywoods, which are not typically or necessarily manufactured to meet industry standards, and (2) *standard* plywoods, which are produced to meet HPVA HP-1 (Fig. 6.3-22).

SPECIALTY PLYWOOD

There are no industry-recognized standards covering *specialty plywoods*. Species,

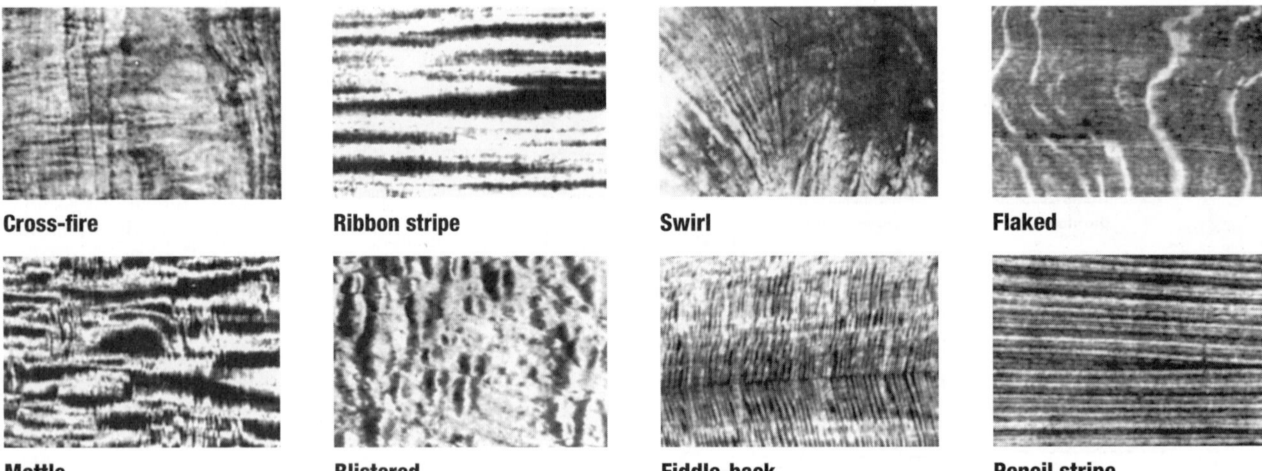

FIGURE 6.3-21 Commonly used figure descriptions. (Used with permission of the Hardwood Plywood & Veneer Association and of the Fine Hardwood Veneer Association.)

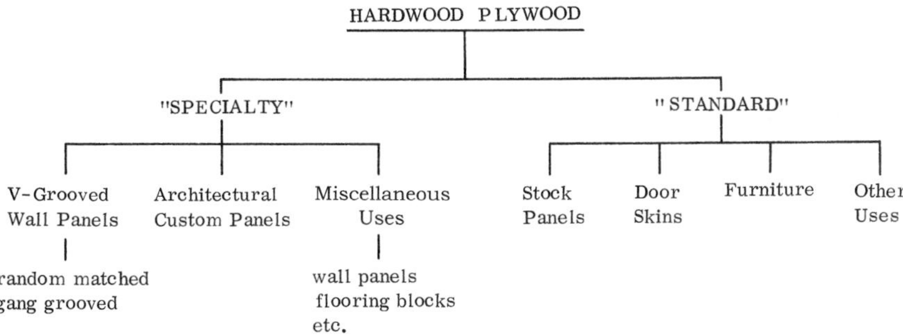

FIGURE 6.3-22 Classification of hardwood plywood.

type, match, core construction, and other requirements are as agreed upon by the seller and the buyer. Generally, panels containing the least amount of blemishes are sold at a higher price than those with large knots and filled knotholes, splits, and other surface defects. Often, the intention is that color and grain vary.

Each manufacturer develops its own grade terminology. However, such terminology is rarely by number or letter but by an arbitrary descriptive system. For example, *aristocratic* and *homestead* are names selected to appeal to different tastes rather than to establish degrees of quality. The aristocratic panels might contain few blemishes and very few filled places, while the homestead panels would contain more knots, large filled knotholes, and more character. There may be little, if any, price differential between the so-called aristocratic and homestead panels.

V-Grooved Wall Panels V-grooved panels, either natural or prefinished, account for much of the hardwood plywood used in residential construction. There are two types of V-grooved panels: random matched and gang grooved.

In random-matched panels the individual veneer sheets, of which there may be four to nine in a 4 × 8 ft. (1200 × 2400 mm) panel, are selected to simulate lumber planking. The V-grooving at the joints of the veneer sheets produces a mismatched panel with veneer sheets dissimilar in grain and color.

Gang-grooved panels are produced by passing one- or multipiece faces under a machine with grooving knives set at spaced intervals. When multipiece face veneers are used, no attempt is made to groove on the joints of the veneer. In gang-grooved panels, the spacing of grooves is identical from one panel to the next.

Architectural Custom Panels Panels made to special order are called architectural custom panels. The method is to select the desired hardwood face veneer from flitch samples before the plywood is manufactured. The selected flitch is then fabricated at a plywood plant into panels to meet the custom specification. The only limitations of architectural custom panels are those imposed by the patterns that are available in the length and width of the selected veneer.

Miscellaneous Uses Increasing amounts of hardwood plywood are produced for many uses, too numerous to discuss individually here. Among these are wall panels of varying widths less than 4 ft. (1200 mm), and flooring products such as laminated block.

STANDARD PLYWOOD

Plywood that conforms with the requirements of HPVA HP-1 is called *standard* hardwood plywood. It is manufactured primarily for use as stock panels, door skins, and furniture manufacture. The HPVA maintains a quality control program which ensures that plywood carrying its official grade trademark conforms with HP-1 (Fig. 6.3-23).

Stock panels, as the name implies, are fabricated and stocked by a plywood manufacturer and sold through local distribution warehouses to a number of different consumers. The character and quality of face veneers are dependent on market conditions, and buyer selection is made from the available stock at the time of purchase. Panels are typically 4 × 8 ft. (1200 × 2400 mm) of varying thicknesses, such as ¼ in. (6.4 mm), ½ in. (12.7 mm), and ¾ in. (19.1 mm), and may be put to a variety of uses.

Door-skin plywood comprises a large percentage of hardwood veneer and is used as door skins to face solid and hollow-core doors (see Section 8.2).

The furniture industry and other wood-product manufacturers make extensive use of hardwood plywood. Outside the construction industry, hardwood plywood finds almost endless uses in automobile, ship, boat, aircraft, and railroad car construction, and in sporting goods, musical instruments, caskets, and many other items.

STANDARD TYPES

The word *type* is used here to indicate a panel's durability, that is, the ability of the plies to stay together under various conditions of exposure. Three basic types of hardwood plywood are produced to conform with the requirements of HPVA HP-1.

Type I (exterior) hardwood plywood is made with waterproof adhesive and is used where it might come in contact with

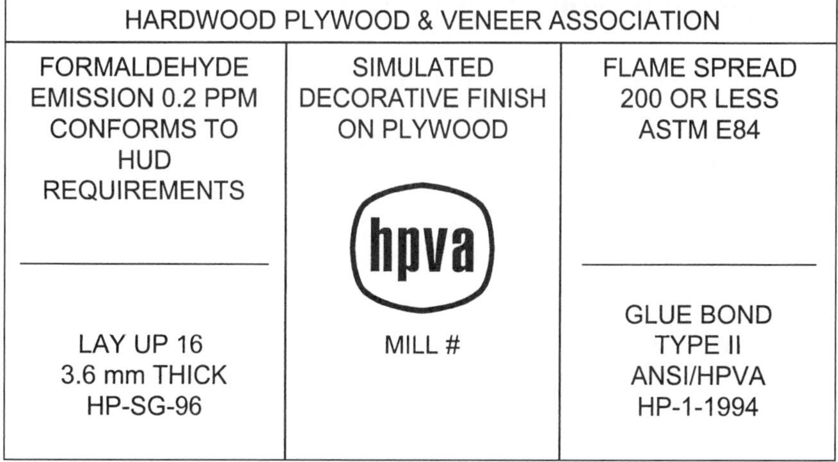

FIGURE 6.3-23 Grade stamp of the Hardwood Plywood and Veneer Association. (Used with permission of the Hardwood Plywood & Veneer Association.)

water or must withstand full weather exposure. The adhesives used are usually resins of phenol, resorcinol, phenolresorcinol, melamine, or melamine urea. Veneer core construction is preferred for Type I plywood.

The *technical type* is made with the same adhesive requirements as Type I but varies in thickness and arrangement of plies to provide approximately equal tensile and compressive strength in length and width.

Type II (interior) plywood is made with water-resistant adhesive (urea resin) and is used where it would not be subjected to contact with water, although it should retain practically all of its strength if occasionally subjected to thorough wetting and drying. Most hardwood plywood is Type II.

STANDARD GRADES

A plywood panel's *grade* designates the quality of the face, back, and inner plies. Hardwood veneer grades used under HP-1 are *face*, *specialty*, *back*, and *inner ply*.

Face Grades Face veneers may be made from more than one piece. With most species, multipiece faces must be book matched or slip matched. The quality of these veneers is high: only a few small burls, occasional pin knots, slight color streaks, and inconspicuous small patches are permitted. No other defects, such as splits, shakes, worm holes, or decay, are allowed. Face grades include AA, A, B, C, D, and E, in descending order of quality.

Specialty Grade Speciality grade (SP) includes face veneers that are unlike those in the face grades. Examples include wormy chestnut, birdseye maple, and English brown oak, all of which have unusual decorative features.

Back Grades Back-grade veneers are labeled 1, 2, 3, and 4, with 1 being the most restrictive. These grades are used on surfaces that will not be exposed in the finished work.

Inner Ply Grades There are four grades used to identify the inner plies of hardwood plywood panels, J, K, L, and M. J is the most restrictive. M is reserved for plies not directly adjacent to facing plies.

6.3.3 NONVENEER PANELS

Particleboard panels are manufactured from small fragments of wood bound together with an adhesive. These types of products have been variously called *flakeboard*, *waferboard*, and *particleboard*. Today most of them are called *oriented strand board* (OSB). Most structural nonveneer wood panels produced today are OSB, in which panels are made using strands or wafers of wood mostly oriented in the long direction of the panel. Several individual panels are laid with the direction of their strands or wafers perpendicular to each adjacent sheet and bonded together to form a final panel. This cross lamination makes such panels strong and stable for the same reasons that plywood has those characteristics.

In general, particleboard should comply with the requirements of the American National Standards Institute's (ANSI) A 208.1, but some such products are actually extruded. Particleboard may be either fire-retardant treated or left untreated.

In some panel products, particleboard panels are used in combination with wood panels to produce a board. The particleboard sheets are used as crossbands with a wood core and wood face veneers. The panels are bonded in the same manner as standard plywood. Some such products are made in a one-step pressing operation in which the voids in the veneers are filled by the wood particles in the particleboard layers.

6.3.4 FIBERBOARD

Wood- or cane-fiber panels are used primarily for sheathing wood frame walls. Panels are coated or impregnated with asphalt, or given some other treatment to provide water resistance. Occasional wetting and drying during construction will not damage this material. Refer to Section 6.6.5 for additional data about fiberboard panels.

6.4 Treated Wood Foundations

Permanent wood foundation system principles were originally developed cooperatively by the American Wood Preservers Institute (AWPI), the U.S. Forest Service, and the National Forest Products Association (NFPA) (now the American Forest & Paper Association [AF&PA]). Literature on permanent wood foundation systems is currently published by the Southern Pine Council (SPC), a promotional body coordinated and supported by the members of the Southern Forest Products Association (SFPA) and the Southeastern Lumber Manufacturers Association (SLMA).

Most local and all major model building code authorities recognize, and the U.S. Department of Housing and Urban Development/Federal Housing Administration (HUD/FHA) and the Farmers Home Administration accept, permanent wood foundation systems on the basis of the NFPA's Technical Note No. 7, "The Permanent Wood Foundation System: Basic Requirements." The text of this section follows the January 1987 editions of Technical Note No. 7 and the SPC publication *Permanent Wood Foundations: Design and Construction Guide.*

Specific design and construction recommendations for a permanent wood foundation system are based on studies conducted by the National Association of Home Builders (NAHB) Research Foundation and by APA—the Engineered Wood Association.

The system provides a practical foundation for light-frame buildings with the following advantages over more conventional foundation systems:

- A reduction in construction delays due to inclement weather
- Savings in cost
- Less erection time
- Creation of more comfortable below-grade living area

The system requires good quality control to prevent seemingly minor errors from resulting in serious problems later. For example, without such quality control, a worker may substitute a piece of untreated lumber where treated lumber is required or careless backfilling may damage the protective film layer and expose the wood to below-grade earth.

6.4.1 GENERAL DESCRIPTION

A permanent wood foundation system is fabricated of pressure-treated lumber and pressure-treated plywood. The structural elements of the substructure are wood. Foundation wall sections of nominal 2-in. (50 mm) lumber and plywood cladding may be factory fabricated or constructed at the job site.

Either conventional concrete footings or wood footing plates may be used. Wood footing plates are nominal 2-in. (50 mm) or thicker pressure-treated wood planks resting on a 4-in. (100 mm) or thicker bed of fine gravel or crushed stone.

The walls are of wood stud construction with a plywood facing on the earth side. Where a basement occurs, a finish on the interior of the wall is optional. When habitable space is enclosed, the wall is insulated.

The system can be used for both basement and crawl space construction. The below-grade portions of exterior basement walls are covered with a polyethylene film bonded to the plywood at the top. Joints are lapped and sealed with adhesive. Concrete basement floors are poured over a gravel base.

Typical details of wood foundation systems for basement and crawl space construction are shown in Figures 6.4-1 and 6.4-2. There are several permitted variations, however, so details different from those shown may be required.

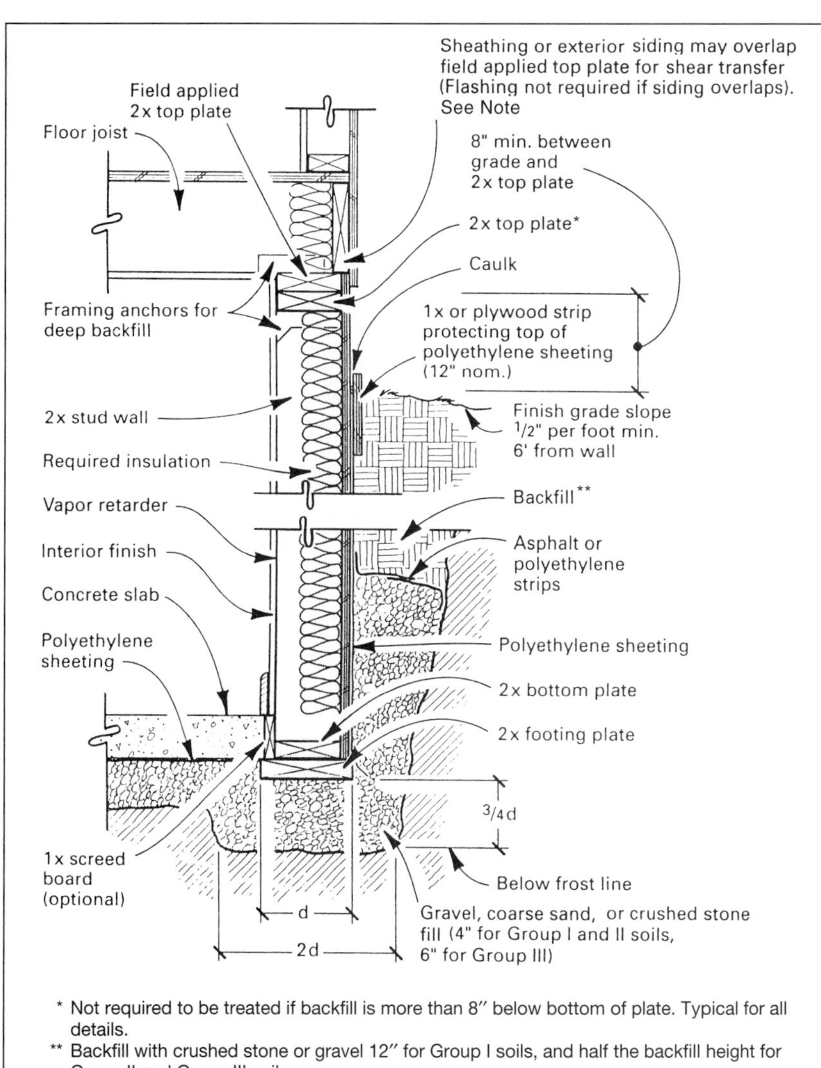

* Not required to be treated if backfill is more than 8″ below bottom of plate. Typical for all details.

** Backfill with crushed stone or gravel 12″ for Group I soils, and half the backfill height for Group II and Group III soils.

Note: For daylight basement foundations, use double header joists (stagger end joints) or splice header joist for continuity on uphill and daylight sides of building.

FIGURE 6.4-1 Basement foundation. (Courtesy Southern Forest Products Association.)

6.4.2 SITE PREPARATION

For basement construction, a house site should be excavated and leveled to the required elevation. Utility lines located below the basement floor should then be installed, and provision should be made for foundation drainage in accordance with local requirements.

When habitable space is planned below grade, unless the soil is very porous, a sump is required to ensure adequate drainage (Fig. 6.4-3). A sump can be drained by gravity or by pump to a storm sewer or drainage swale.

After grading and installation of plumbing lines and the drainage system, the basement site should be covered with a layer of gravel. For crawl space construction, trenches should be dug to the required depth and gravel placed as a base for the footing plates.

6.4.3 FOOTINGS

The width of footing plates is determined by loads on the foundation wall and the bearing capacities of the gravel base and soil. When brick veneer is used, larger footing plates are required to provide support for the veneer. Brick height should not exceed 8 ft. (2400 mm). To reduce the amount of brick required, such veneer is typically supported below grade on a wood stud knee wall that rests on and is nailed to the treated wood footing.

Where basement construction is involved, footings under exterior foundations should be placed on a strip of polyethylene film laid directly on the gravel base.

A load-bearing wall may be used as center support under the first floor (Fig. 6.4-4). If a girder support is used instead, the footings under the foundation ending walls and under the posts should be designed for concentrated loads.

Footing plates must be located below the frost line.

6.4.4 FOUNDATION WALLS

Detailed structural requirements, including member spacings and sizes, for the foundation wall of typical 1- and 2-story wood frame houses are shown in *Permanent Wood Foundations: Design and Construction Guide.*

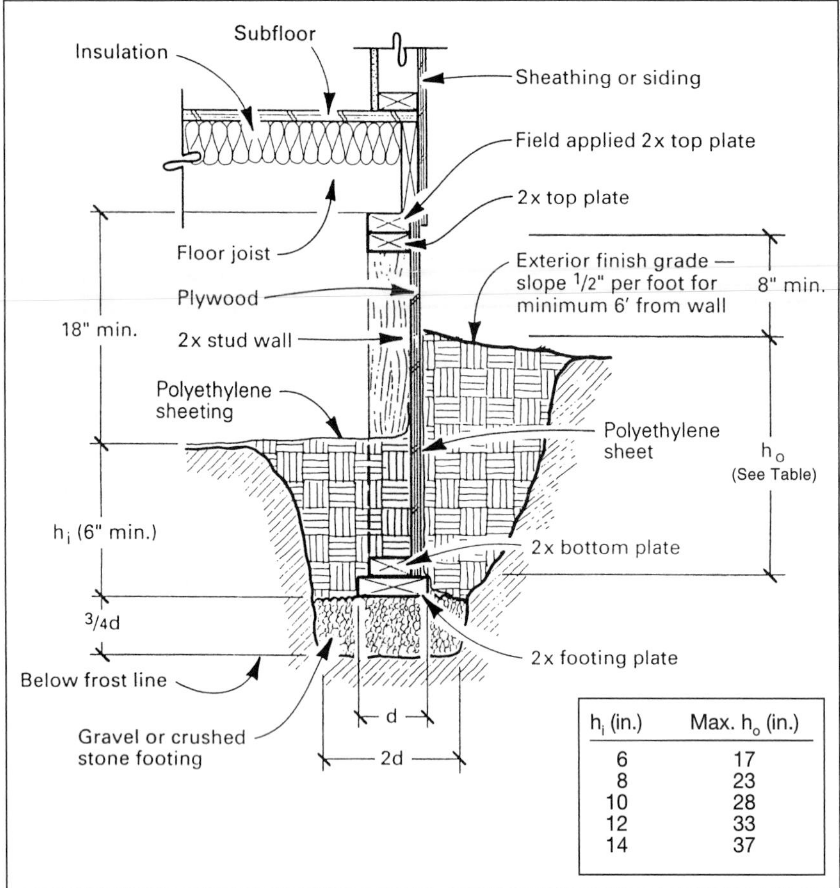

h_i (in.)	Max. h_o (in.)
6	17
8	23
10	28
12	33
14	37

FIGURE 6.4-2 Crawl space foundation. (Courtesy Southern Forest Products Association.)

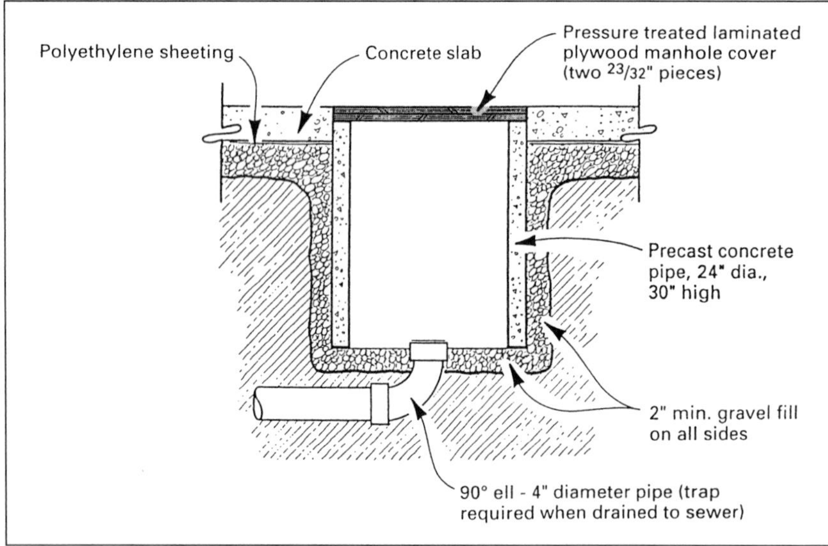

FIGURE 6.4-3 Sump for poorly drained soils. (Courtesy Southern Forest Products Association.)

6.4.4.1 Design Method

Lumber framing members in exterior walls enclosing a basement are designed to resist the lateral pressure of fill and the vertical forces resulting from live and dead loads on the structure. The plywood sheathing is designed to resist the maximum inward soil pressures at the bottom of the wall.

Foundation walls for crawl space construction are designed to resist only vertical loads when the difference between the outside grade and ground level in the crawl space is 12 in. (300 mm) or less. Framing is covered with ½-in. (12.7 mm) plywood to keep out vermin and provide additional stability.

6.4.4.2 Construction

Foundation walls are built over a previously laid wood footing. Common practice is to prefabricate a foundation wall in modular sections. Components are fastened together using corrosion-resistant nails. Treated foundation wall studs are end nailed to treated top and bottom plates. Treated plywood panels, ½ in. (12.7 mm) thick or thicker, are attached to studs. Usually, panels are applied with their face grains vertical and parallel to the studs. Vertical joints between panels within wall sections should be located over studs.

Foundation wall sections should be nailed to the wood footing and to adjacent sections. Wall panel joints should not occur over joints in the footing. An upper top plate should be face nailed to the lower top plate of the foundation wall to tie the wall sections together.

An alternate method of attaching the upper top plate is to fabricate the foundation wall sections with the edge of the plywood wall panels extending ¾ in. (19.1 mm) above the top of the treated top plate. The field-applied top plate is then face nailed to the top plate of the wall section and secured to the overlapping edge of the plywood.

Joints in the field-applied top plate should be staggered with respect to joints in the treated wood plate of the wall panels. At corners and wall intersections, the field-applied upper plate of one wall should be extended across the treated plate of the intersecting wall to tie the building together. The first floor is then installed, following standard platform frame construction practices.

Plywood panel joints in the foundation wall should provide the minimum space recommended by the manufacturer of the sealant compound selected. In both basement and crawl space construction, the exterior side of foundation walls should be covered below grade with a 6-mil (0.1524 mm) polyethylene sheet to act as a vapor retarder. This film should be bonded to the plywood and should have its joints lapped and sealed (see Figs. 6.4-1 and 6.4-2).

The top edge of the polyethylene

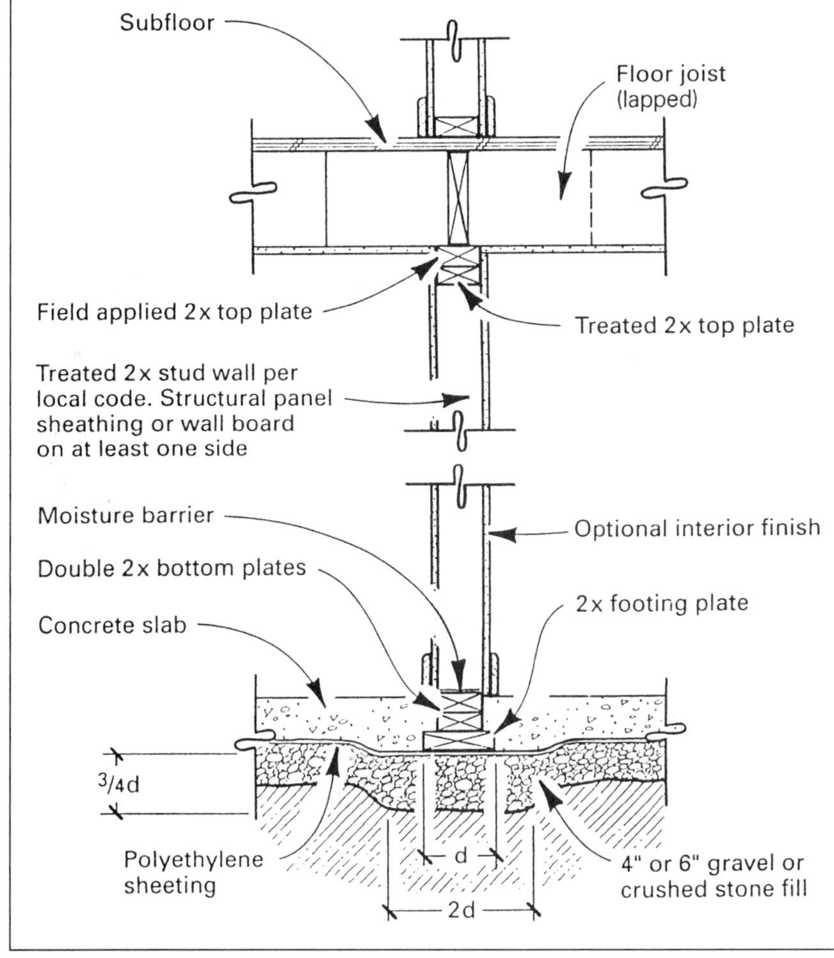

FIGURE 6.4-4 **Basement bearing partition.** (Courtesy Southern Forest Products Association.)

sheeting should be adhered to the plywood wall with adhesive. Film areas near grade level can be protected from mechanical damage and exposure using treated plywood, brick, stucco, or other covering appropriate to the architectural design. Where plywood is used, the top edge of a 12-in. (305 mm)-wide panel of treated material is attached to the wall several inches above the finish grade level. The top inside edge of the panel should be caulked its full length before the panel is fastened to the wall.

Where appropriate, suitable insulation should be installed within the wall stud space. Where habitable space occurs, an interior vapor retarder and wall finish are usually applied to the stud framing.

6.4.5 FLOORS

For basement construction, a 6-mil (0.1524 mm) polyethylene film is applied over the gravel bed, and a concrete slab at least 3 in. (75 mm) thick is placed over the film. Placement of the film and the slab can be delayed until after the building is under roof. Backfill should not be placed against the foundation walls, however, until after the concrete slab is in place and set and the top of the wall is adequately braced. Where the height of backfill exceeds 4 ft. (1200 mm), gravel should be used for the lower portion.

For crawl space construction, a 6-mil (0.1524 mm) polyethylene film ground cover should be used. Film edges should be lapped and sealed.

The first-floor framing on a wood foundation is installed according to conventional platform framing methods. Floor joists should be suspended at least 18 in. (457 mm) above the ground.

6.4.6 CENTER SUPPORT

Framing members in load-bearing walls acting as center support for a first-floor joist system are similar to exterior foundation walls (see Fig. 6.4-4), except that only vertical loads need be considered during design.

6.4.6.1 Load-Bearing Wall

In basement construction, load-bearing interior walls should be made with pressure-treated bottom plates and, in areas of low termite hazard, with untreated studs and top plates. A double bottom plate is used to provide clearance between the top of the concrete slab floor and the untreated vertical framing in the wall (see Fig. 6.4-4). In areas of high termite hazard, pressure-treated studs should be used. The interior wall framing can be covered on both sides with gypsum wallboard or other interior finish as desired.

Interior load-bearing walls in crawl space construction should be made of treated framing and covered on one side with ½-in. (12.7 mm)-thick treated plywood.

6.4.6.2 Post-and-Girder Support

Where a post-and-girder center support system is used, the posts should be supported on blocks of sufficient size to ensure that the vertical concentrated loads will not exceed the bearing capacity of the gravel base and soil. Foundation walls supporting a center girder should also be designed to carry and distribute the concentrated load from the girder to a wood footing and gravel base, so that the allowable gravel and soil bearing capacities are not exceeded. The design of a suitable post-and-girder support system requires engineering calculations.

6.4.7 LUMBER AND PLYWOOD PRESERVATIVE TREATMENT

Lumber and plywood used in exterior foundation walls below an upper top plate and wood footings, or bottom plates placed in contact with concrete or the ground, should be pressure treated. Treatment should be in accordance with American Wood Preservers Association (AWPA) Standard C22, "Lumber and Plywood for Permanent Wood Foundations, Pressure Process."

Treated lumber and plywood should bear the appropriate AWPB Quality Mark or that of an acceptable independent inspection agency that maintains continuing control, testing, and inspection of the quality of the product. Treated lumber should be dried to a moisture content not exceeding 19% after treatment. To the extent practicable, fabricating, cutting, boring, and trimming of lumber and plywood should be done before treatment. Lumber or plywood cut after treatment should have material exposed by the cutting brush-coated with a 3% solution of the same preservative used in the original treatment.

6.5 General Framing Requirements

A variety of methods and systems are used in the construction of wood-framed buildings across the country. Local climatic conditions, economic considerations, and aesthetic preferences may influence the selection of the system to be used. This section includes some of the requirements common to the various systems. Sections 6.6 and 6.7 address the more specific aspects of various framing systems.

6.5.1 ANCHORAGE REQUIREMENTS

The superstructure of a wood-framed building should be anchored to the foundation to provide resistance to lateral forces, primarily from wind. Wind pressures produce compressive, tensile (uplift), shearing, and racking loads on a foundation (Fig. 6.5-1).

6.5.1.1 Overturning, Translation, and Rotation

Overturning forces may produce critical stresses in foundations that support light superstructures subjected to high wind pressures. Connections anchoring a building's superstructure to its foundation and the foundation itself must have adequate strength in tension to resist these forces.

When the ratio of exposed building surface to the width of the building is large, high wind forces tend to cause translation (lateral movement) of the superstructure. When the wind forces are not symmetrical, building rotation can occur. Forces tending to cause translation and rotation are transferred to a foundation by the first floor acting as a diaphragm. Connections anchoring the superstructure must develop enough shearing strength to transmit these forces to the foundation. When a first floor does not have enough rigidity to provide diaphragm action, high shear and bending stresses may be produced in the foundation walls. In this case, consideration should be given to the use of shear walls, pilasters, or bond beams to stiffen the walls.

6.5.1.2 Anchorage Recommendations

Superstructures should be anchored to both masonry and concrete foundations as applicable. These anchoring requirements are covered in the model codes and in the Council of American Building Officials' (CABO) *One- and Two-Family Dwelling Code*. In general, the following requirements apply, but each should be verified with the requirements in the applicable code.

The sill plates of a wood-joist floor system should be anchored to masonry walls with ½-in. (12.7 mm) bolts, extending at least 15 in. (381 mm) into the filled cores of masonry units. Anchor bolts should be spaced not more than 6 ft. (1830 mm) apart, with one bolt not more than 12 in. (305 mm) from each end of the sill plate.

For concrete walls, sill plates should be anchored with ½-in. (12.7 mm) bolts embedded 7 in. (178 mm), with the same maximum spacing as for masonry walls. Hardened steel studs driven by powder-actuated tools also may be used, except when earthquake design is required. Their maximum spacing should not exceed that required by code.

6.5.2 DECAY AND TERMITE CONTROL

The design and construction of foundations that support buildings using wood structural floor, wall, and roof systems must include precautions for the protection of the wood from the effects of decay and attack by subterranean termites (see Sections 2.3, 6.1.5, and 7.1).

Suitable protection of wood structural elements is provided by (1) positive site and building drainage, (2) treatment of the soil with pesticides, (3) separation of wood elements from the ground, (4) preservative treatment of the wood, and (5) ventilation and condensation control in enclosed spaces.

6.5.2.1 Separation of Wood from the Ground

Figure 6.5-2 shows recommended distances for the separation of wood from the ground. These distances will (1) maintain wood at a safe moisture content for decay

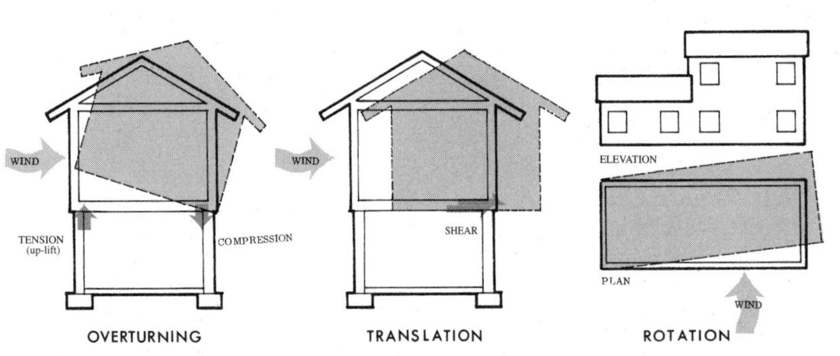

FIGURE 6.5-1 Wind forces tend to lift or overturn light superstructures, move them laterally, or rotate them on their foundations.

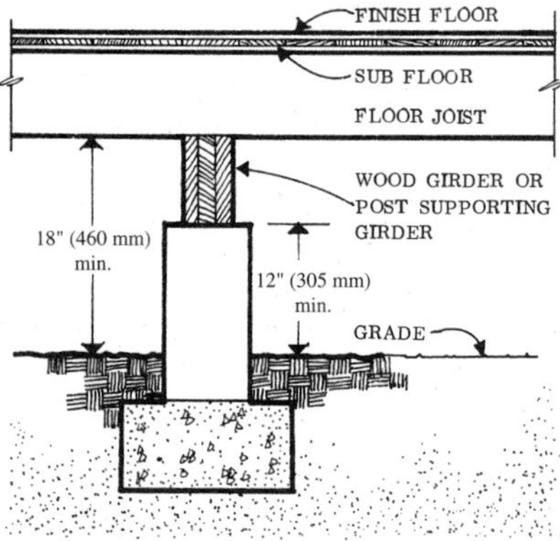

FIGURE 6.5-2 Minimum separation of wood from the ground.

protection, (2) provide a termite barrier, and (3) facilitate periodic visual inspection. When it is impracticable to comply with the clearances recommended, all heartwood of a naturally durable species or pressure-treated lumber should be used. However, when termites are known to exist, access space always should be provided for periodic visual inspection. Greater clearances may be required where mechanical equipment is located in a crawl space.

6.5.2.2 Termite Barriers

A *termite barrier* is any material that cannot be penetrated by termites and that drives the insects into the open where they can be detected and eliminated. Where local experience indicates that additional protection against termites is required, one or more of the following barriers is recommended, depending on the degree of hazard: (1) chemical soil treatment (see Section 2.3), (2) preservative-treated lumber for all floor framing up to and including the subfloor (see Section 6.1.5), or (3) properly installed termite shields.

Additional precautions include using poured concrete foundations in which no cracks greater than 1/64 in. (0.381 mm) are present, or poured reinforced concrete caps at least 4 in. (100 mm) thick on unit masonry foundations. Such caps should have no cracks greater than 1/64 in. (0.381 mm).

6.5.3 NAILING PRACTICES

Of primary consideration in the construction of a light-wood-framed building is the method used to fasten the various wood members together. These connections are most commonly made with nails, but in certain cases metal straps, lag screws, bolts, staples, and adhesives may be used.

Proper fastening of frame members and covering materials provides the rigidity and strength to resist windstorms, earthquakes, and other natural disasters. Adequate nailing also is important for the suitable performance of wood parts. In general, nails give stronger joints when driven into the side grain (perpendicular to wood fibers) than into the end grain of wood. For instance, the allowable lateral (shear) loads for nails driven into the side grain are 50% higher than those allowed when the nails are driven into the end grain. Whenever possible, connections should be designed in such a way that nails are loaded in shear, exerting a sideways pull on the nail, rather than in direct withdrawal, pulling straight out. Connections never should be designed in a way that would load an end-nailed joint in direct withdrawal (Fig. 6.5-3).

The various codes summarize in tables the required nailing practices for the framing of a conventional wood frame building. Penny designations and actual sizes of common wire nails are shown in Figure 6.5-4.

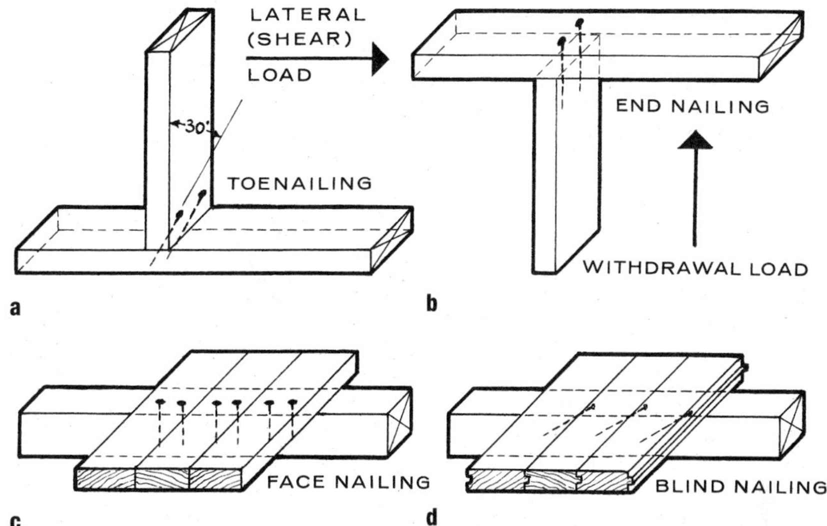

FIGURE 6.5-3 Common nailing methods: (a) slant nailing at a 30-degree angle to the face or edge of the attached piece (toe nailing), (b) nailing into the end grain of the supporting piece (end nailing), (c) nailing into side grain through the wide dimension of the attached piece (face nailing), and (d) toe nailing through the edge of a plank so as to conceal the nail head (blind nailing).

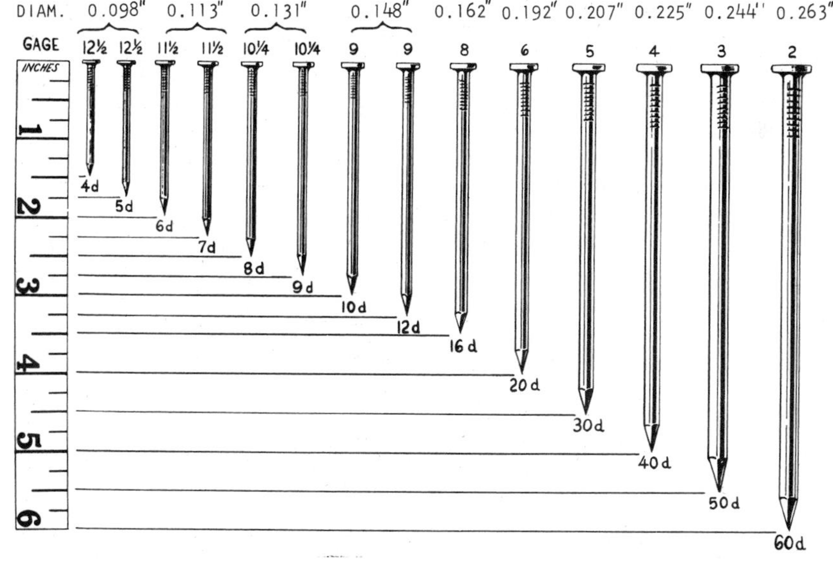

FIGURE 6.5-4 Standard sizes of common wire nails. Thicker nails in sizes 10d to 60d are known as spikes. (Refer to Figure 1.8-27 for metric conversions.)

6.6 Conventional Framing and Furring

Modern, light wood framing is an outgrowth of the *balloon frame* that originated around the middle of the nineteenth century in Chicago. It was a revolutionary development that drastically changed the direction of the housing industry and contributed substantially to the colonization of the West. Historians have claimed that if not for the balloon frame, Chicago and San Francisco could not have grown from villages to great cities in a single year. Houses and other wood frame structures were built with a rapidity possible only with this revolutionary type of construction.

Before the balloon frame evolved, frame buildings were built of heavy timbers, mortised and tenoned together and pinned with hardwood dowels (see Section 6.7.3). The principle of balloon framing involves the repetitive use of slender structural members called *studs* for walls and *joists* for roofs and floors, all tied together with simple nailed connections. This system changed house construction from a highly skilled craft to a relatively simple mechanical procedure that could be learned quickly by most handymen familiar with hand tools. For that reason, perhaps, it was regarded with suspicion and acquired its derisive name on the assumption that such a house would blow away like a balloon the first time a strong wind hit it.

The development of the balloon frame is closely associated with the level of industrialization reached in the United States in the middle of the nineteenth century. The evolution of sawmill technology and the ability to mass-produce cut nails cheaply are credited with the impetus of this revolutionary idea. When the manufacture of cut nails was started at the turn of the century, wrought nails cost 25 cents a pound. With the introduction of more sophisticated machinery, the price was reduced to 8 cents in 1820, to 5 cents in 1833, and to a mere 3 cents a pound by 1842.

Over the years, balloon framing was modified to permit the use of shorter (and cheaper) wall studs, spanning from floor to floor rather than from foundation to roof as in the original system. This new approach made it possible to erect a floor platform first, then to nail together a full-length wall on a horizontal surface, more conveniently and faster than in an upright position. This modified framing system evolved as wood framing moved west with the early settlers and came to be known as *Western* or *platform framing*. Most new wood buildings today are platform framed.

The search for economy in labor and materials is a continuing process. As the cost of labor became a more significant factor in total housing cost, plank-and-beam construction (see Section 6.7.2) and other floor and roof assemblies evolved, utilizing greater spacing of fewer and larger structural members. Such systems were more compatible with post-and-beam wall framing (see Section 6.7.1) and, as with the skeleton structure of steel and concrete buildings, enabled designers to treat the space between posts as a curtain wall, intended mainly to keep out the weather rather than to resist loads. The use of prefabricated trusses (see section 6.7.5) for roof support permitted fewer, and sometimes no, interior walls.

In spite of innovations in light wood framing, the essential construction principles inherent in the balloon frame persist. The nail still remains an essential connective element in light wood-framing systems, and the exterior covering between framing members, the sheathing, is relied on to provide structural continuity to an assembly of light members.

Much of the material in this and related sections on wood framing is based on data in the U.S. Department of Agriculture's Agriculture Handbook No. 73, *Wood-Frame House Construction*; the publications of the NFPA; the publications of the APA; and the other publications listed in Section 6.11, noted in the text, or both.

The superstructure of a wood-framed building is made up of floor and ceiling assemblies (Section 6.6.1), including their subflooring and decking (Section 6.6.4), walls (Section 6.6.2), and roof and ceiling assemblies (Section 6.6.3), including their sheathing (Section 6.6.5). Some components of the superstructure may be preassembled (Section 6.6.6). The superstructure may be conventionally framed (Section 6.6) or may be constructed using other framing systems (Section 6.7). The superstructure may be supported directly on foundation walls (Chapters 3 and 4; Section 6.4) or may rest on posts, beams, and girders (Section 6.6.1.2) that in turn are supported by the foundations.

6.6.1 FLOORS AND FLOOR AND CEILING ASSEMBLIES

The floor framing in a conventional wood-framed building consists of posts, beams (or girders), sill plates, and joists. Floor sheathing (subflooring) ties the various framing elements together and stabilizes the joists to prevent twisting and buckling. When these elements are assembled properly on a foundation, they form a level, anchored platform for the erection of partitions, additional floors, and the roof. Posts and center beams, which support the inside ends of joists, are sometimes replaced with a wood frame or masonry wall when the basement area is divided into rooms. Wood-framed buildings may also be constructed over a crawl space with floor framing similar to that used over a basement. In a slab-on-ground construction, wood floor framing is limited to the upper floors.

In multilevel buildings, the combination of each intermediate floor and the ceiling systems below it is usually recognized as a single basic element of the superstructure. A combined floor and ceiling system is called a *floor and ceiling assembly*. The design, selection, and construction of a floor and ceiling assembly is based on (1) the required level of functional performance, (2) costs, and (3) the desired finished appearance. These considerations often are influenced by the systems and finishes used in other basic elements of the construction.

A floor and ceiling assembly should have:

1. Adequate strength to support the materials used in the floor and ceiling assembly and other elements in the building that it supports (dead load) and loads imposed by the occupants and furnishings (live loads)
2. Provisions for lateral support of walls
3. Satisfactory resistance to the transmission of airborne and structure-borne sound
4. Suitable fire resistance
5. Suitability for the application of finish materials
6. Adaptability to economical methods of assembly and erection

7. Space to accommodate heating, air conditioning, electrical, and plumbing equipment
8. Control of heat loss and the flow of water vapor

Floor and ceiling assemblies must support superimposed loads safely and without excessive deflection or vibration. Design dead and live loads are usually established by building codes which require structural members to be sized so that an allowable fiber stress in bending is not exceeded when the assembly is fully loaded. Codes do not always limit deflection, because it seldom has a bearing on structural failure or safety. Deflection limitations should be established in floor and ceiling assembly design, however, to reduce deflection to acceptable limits, to assure the integrity of applied finish materials, and to limit vibration to acceptable levels of human perception. Deflection, rather than bending strength, is generally the controlling factor in the structural design of wood floor and ceiling assemblies.

Depending on the design of the foundation system, a first-floor system may be required to provide lateral support to foundation walls, and floor and ceiling assemblies may also be required to provide lateral support for other elements of the superstructure against loads caused by wind or earthquake.

The ultimate costs of a floor and ceiling assembly include in-place costs, maintenance costs, and operating costs. In-place costs are the sum of the costs of materials, labor, and overhead required to assemble and install a floor and ceiling assembly. The selection of the system may affect other costs, such as the cost of applying floor and ceiling finishes, constructing interior partitions, and providing space to accommodate heating, air conditioning, electrical, and plumbing equipment.

Maintenance and operating costs attributable to a floor and ceiling assembly, such as heat loss or gain, have an important influence on the ultimate cost of the assembly and should not be overlooked if long-term economy is to be achieved.

The makeup of a floor and ceiling assembly is influenced by the desired finish applications. The finish materials selected may:

1. Influence the spacing and span of structural members;
2. Determine the need for floor underlayment or wall or ceiling furring when varying structural component spacing is impracticable, as is often the case, or when such variation would violate code requirements;
3. Dictate methods of construction.

Prefinished floor or ceiling surfaces may not require the application of an additional finish after the system has been placed. However, the floor should provide a working platform for finishing operations during the construction period. Therefore, a floor and ceiling assembly is often selected that has a subfloor or sheathing application that is later provided with finished surfaces.

Wood-framed floor and ceiling assemblies include wood floor joists or floor trusses (discussed in this section), plank-and-beam construction (Section 6.7.2), heavy timber construction (Section 6.7.3), and panelized construction (Section 6.6.6). Concrete slab-on-grade floor construction is discussed in Section 3.12.

6.6.1.1 Lumber Materials

Framing lumber should be of the proper grade for the intended use, graded by a recognized grading authority. The moisture content of framing lumber is important in the performance of finishes. It should be as near as possible to the equilibrium moisture content (see Section 6.2) for the particular locality and should not exceed 19% when the finish material is installed. However, a 15% or lower moisture content is better in all regions of the country and essential in arid regions, especially when a finish such as gypsum board will be applied directly to the framing. Dimension lumber generally can be obtained at a moisture content of 19% or less when S-DRY is specified, or at 15% or less when MC 15 is specified. In arid parts of the country, lumber of higher moisture contents can be specified if the lumber is allowed to approach its equilibrium moisture content before interior finishes and trim, such as baseboard, shoe mold, and door casings, are applied.

In light wood-framed buildings, wood studs are typical framing members for walls and partitions; wood joists and trusses are the most common ceiling and roof framing members. These framing members are generally spaced 16 in. (406 mm) or 24 in. (610 mm) on centers, which permits direct application of most finishes, such as gypsum board and plaster. However, furring (Section 6.6.7) may be required when member spacing exceeds the maximum recommended for the particular finish material or when special performance, such as increased sound isolation, is desired.

The National Grading Rule for dimension lumber, established in accordance with Product Standard PS 20 (see Section 6.2), provides a series of uniform grade designations applicable to all species. Joists and planks (2 in. [50 mm] to 4 in. [100 mm] thick and 5 in. [130 mm] wide or wider) are graded as Select Structural, No. 1, No. 2, and No. 3 grades. All grades are suitable for floor joists if they have adequate strength characteristics for the intended spacing, span, and loading conditions (see Section 6.6.1.3). Light framing (2 in. [50 mm] to 4 in. [100 mm] thick and 2 in. [50 mm] to 4 in. [100 mm] wide) is sorted into as many as seven grades. These grades, in order of decreasing strength, are: Select Structural, No. 1, No. 2, Construction, No. 3, Standard, and Utility. An eighth, Stud grade, is similar to No. 3 grade but is cut to 10-ft. (3050 mm) or shorter lengths, is straighter, and has a better nailing surface. The lower light-framing grades should be used for sills and plates rather than for studs and posts.

6.6.1.2 Posts, Beams, and Girders

Wood or steel posts are generally used in a basement to support wood girders or steel beams. Masonry piers can also be used in basements but are most commonly employed in buildings with crawl spaces.

Round steel posts, known as pipe columns or lally columns (steel pipes filled with concrete), can be used to support both wood girders and steel beams. Columns should be provided with a steel bearing plate at each end. Secure anchoring of a girder or beam to a post or column is essential (Fig. 6.6-1).

Wood posts for freestanding use should be of solid or built-up lumber and not less than the equivalent cross-sectional area of a solid 6 × 6 (150 × 150). When combined with a frame wall, posts may be 4 × 6s (100 × 150s) to conform to stud depth. Loads may dictate other sizes. Spaced columns may also be used. Refer to *Wood Structural Design Data* by the AF&PA for design requirements for wood columns. Wood posts should be squared at both ends and securely fastened to the girder (Fig. 6.6-2a). The bottom of a post should rest on and be pinned to a concrete pedestal 2 to 3 in. (50 to 75 mm) above the

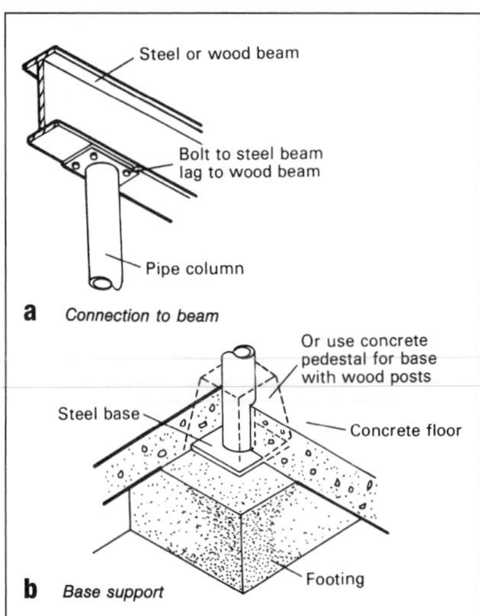

FIGURE 6.6-1 **(a) Steel top plate provides a means of anchoring a beam; (b) base plate distributes the load and anchors the steel post to the footing.** (Fig. 29, *Wood Frame House Construction*, United States Department of Agriculture, page 41. Not subject to copyright in the United States.)

finish floor (Fig. 6.6-2b). It is good practice to treat the bottom of a post with a decay preservative or to use a vapor-resistant covering such as heavy roll roofing over the concrete pedestal. A sheet metal covering is preferred in areas subject to termite infestation.

Both wood girders and steel beams are used in present-day wood-framed construction. Standard steel S-shaped and W-shaped (wide flange) beams are the most commonly used steel beam shapes. Wood girders may be either solid or built up (Fig. 6.6-3). A built-up beam is preferred because it can be made up from drier dimension lumber and is more dimensionally stable. Solid lumber or commercially available glue-laminated girders may be desirable when a beam will be exposed in a finished room.

The use of the term *beam*, in connection with steel members, and *girder*, with wood members, is based on common usage, rather than a technical difference in meaning. In engineering terminology a beam is any member with loads applied perpendicular to its direction of span, and a girder is a principal beam, frequently supporting other beams. In this broader sense, joists, girders, headers, and trimmers are types of beams. Wood joists are seldom called beams, but both built-up and solid wood members carrying joists are frequently referred to as beams or girders.

A built-up girder (see Fig. 6.6-3) is usually made up of two or more pieces of 2-in. (50 mm) lumber nailed together, with the ends of the pieces joined over a supporting post. Building codes dictate the required type, location, size, and number of fasteners. For example, a code may require that a two-piece girder be face nailed from one side with 10d nails, two nails at the end of each piece and others driven near top and bottom in two rows 16 in. (406 mm) on centers, with the nails staggered. A three-piece girder might be required to be face nailed from each side with 20d nails, two nails near each end of each piece and others in two rows driven 32 in. (810 mm) on centers staggered.

Examples of member sizes and number of members in built-up wood girders supporting floor loads for various spans are shown in Figure 6.6-4. The table indicates required lumber sizes for various conditions of loading and framing, assuming a minimum fiber stress in bending (F_b of 1000 psi [703.07 Mg/m^2] and 1500 psi [1054.6 Mg/m^2]). Species and grades meeting these requirements can be selected from tables in the *NDS Supplement*. These tables should be verified with the requirements of the applicable building code, which may vary from the table data. Solid beams with an equivalent cross section are usually permitted in lieu of built-up beams.

The ends of wood girders are generally required to bear at least 4 in. (100 mm) on foundation walls and pilasters. When the lumber is not preservative treated, a ½-in. (12.7 mm) airspace is often required at each end and at each side of wood girders

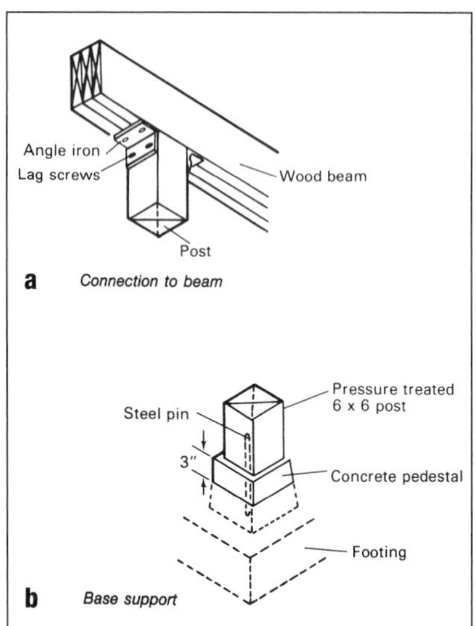

FIGURE 6.6-2 **(a) A wood post should be connected to a girder with metal angles of framing anchors; (b) the base of a wood post should be pinned and protected against decay and termites.** (Fig. 30, *Wood Frame House Construction*, United States Department of Agriculture, page 41. Not subject to copyright in the United States.)

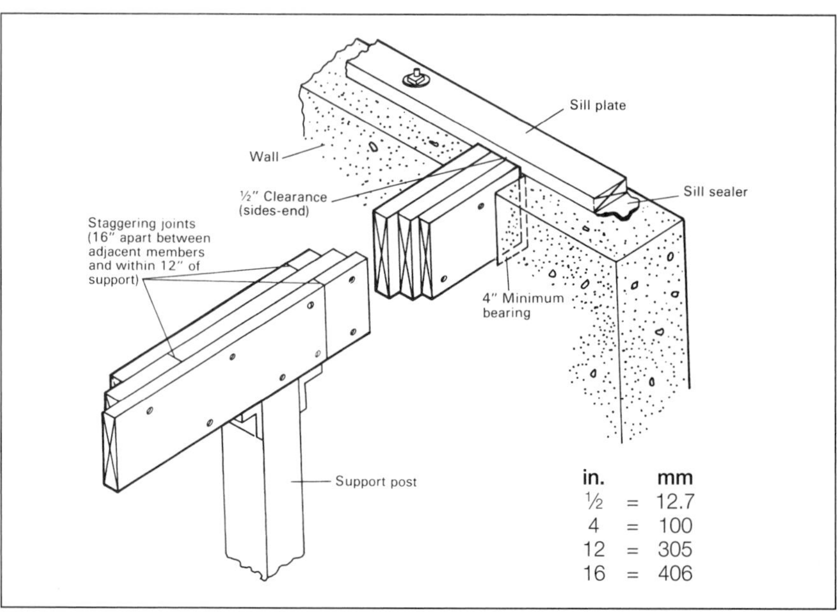

FIGURE 6.6-3 **At least one member of a built-up wood girder should be continuous across a supporting post. In termite-infested areas, a beam pocket should be lined with metal to form a termite shield.** (Fig. 31, *Wood Frame House Construction*, United States Department of Agriculture, page 42. Not subject to copyright in the United States.)

FIGURE 6.6-4 Maximum Allowable Spans for Built-Up Girders Supporting One or Two Floors

Maximum Allowable Spans ("S") for Girders Supporting One Floor[a]

Lumber		House Widths ("L")[b]								
Fb, psi	**Size**	**20 ft.**	**22 ft.**	**24 ft.**	**26 ft.**	**28 ft.**	**30 ft.**	**32 ft.**	**34 ft.**	**36 ft.**
1000	(2) 2 in. × 6 in.	4 ft.-5 in.	4 ft.-0 in.	—	—	—	—	—	—	—
	(3) 2 in. × 6 in.	5 ft.-6 in.	5 ft.-3 in.	5 ft.-0 in.	4 ft.-10 in.	4 ft.-8 in.	4 ft.-5 in.	4 ft.-2 in.	—	—
	(2) 2 in. × 8 in.	5 ft.-10 in.	5 ft.-3 in.	4 ft.-10 in.	4 ft.-5 in.	4 ft.-2 in.	—	—	—	—
	(3) 2 in. × 8 in.	7 ft.-3 in.	6 ft.-11 in.	6 ft.-7 in.	6 ft.-4 in.	6 ft.-2 in.	5 ft.-10 in.	5 ft.-5 in.	5 ft.-1 in.	4 ft.-10 in.
	(2) 2 in. × 10 in.	7 ft.-5 in.	6 ft.-9 in.	6 ft.-2 in.	5 ft.-8 in.	5 ft.-3 in.	4 ft.-11 in.	4 ft.-8 in.	4 ft.-4 in.	4 ft.-1 in.
	(3) 2 in. × 10 in.	9 ft.-3 in.	8 ft.-10 in.	8 ft.-5 in.	8 ft.-1 in.	7 ft.-10 in.	7 ft.-5 in.	6 ft.-11 in.	6 ft.-6 in.	6 ft.-2 in.
	(2) 2 in. × 12 in.	9 ft.-0 in.	8 ft.-2 in.	7 ft.-6 in.	6 ft.-11 in.	6 ft.-5 in.	6 ft.-0 in.	5 ft.-8 in.	5 ft.-4 in.	5 ft.-0 in.
	(3) 2 in. × 12 in.	11 ft.-3 in.	10 ft.-9 in.	10 ft.-3 in.	9 ft.-10 in.	9 ft.-6 in.	9 ft.-0 in.	8 ft.-5 in.	7 ft.-11 in.	7 ft.-6 in.
1500	(2) 2 in. × 6 in.	5 ft.-3 in.	4 ft.-10 in.	4 ft.-5 in.	4 ft.-1 in.	—	—	—	—	—
	(3) 2 in. × 6 in.	6 ft.-9 in.	6 ft.-5 in.	6 ft.-2 in.	5 ft.-11 in.	5 ft.-8 in.	5 ft.-3 in.	4 ft.-11 in.	4 ft.-8 in.	4 ft.-5 in.
	(2) 2 in. × 8 in.	7 ft.-0 in.	6 ft.-4 in.	5 ft.-10 in.	5 ft.-4 in.	5 ft.-0 in.	4 ft.-8 in.	4 ft.-4 in.	4 ft.-1 in.	—
	(3) 2 in. × 8 in.	8 ft.-11 in.	8 ft.-6 in.	8 ft.-1 in.	7 ft.-9 in.	7 ft.-5 in.	7 ft.-0 in.	6 ft.-6 in.	6 ft.-2 in.	5 ft.-10 in.
	(2) 2 in. × 10 in.	8 ft.-11 in.	8 ft.-1 in.	7 ft.-5 in.	6 ft.-10 in.	6 ft.-4 in.	5 ft.-11 in.	5 ft.-7 in.	5 ft.-3 in.	4 ft.-11 in.
	(3) 2 in. × 10 in.	11 ft.-4 in.	10 ft.-10 in.	10 ft.-4 in.	9 ft.-11 in.	9 ft.-6 in.	8 ft.-11 in.	8 ft.-4 in.	7 ft.-10 in.	7 ft.-5 in.
	(2) 2 in. × 12 in.	10 ft.-10 in.	9 ft.-10 in.	9 ft.-0 in.	8 ft.-4 in.	7 ft.-9 in.	7 ft.-2 in.	6 ft.-9 in.	6 ft.-4 in.	6 ft.-0 in.
	(3) 2 in. × 12 in.	13 ft.-9 in.	13 ft.-2 in.	12 ft.-7 in.	12 ft.-1 in.	11 ft.-7 in.	10 ft.-10 in.	10 ft.-2 in.	9 ft.-6 in.	9 ft.-0 in.

LL + DL = 50psf
"L/2±1 "L/2±1
Girder
"L"
"S" "S" "S"
Load diagram applies also to 2-st. house below

Maximum Allowable Spans ("S") for Girders Supporting Two Floors[a]

Lumber		House Widths ("L")[b]								
Fb, psi	**Size**	**20 ft.**	**22 ft.**	**24 ft.**	**26 ft.**	**28 ft.**	**30 ft.**	**32 ft.**	**34 ft.**	**36 ft.**
1000	(2) 2 in. × 8 in.	4 ft.-7 in.	4 ft.-2 in.	—	—	—	—	—	—	—
	(3) 2 in. × 12 in.	4 ft.-9 in.	4 ft.-4 in.	4 ft.-0 in.	—	—	—	—	—	—
	(3) 2 in. × 10 in.	5 ft.-10 in.	5 ft.-4 in.	4 ft.-11 in.	4 ft.-7 in.	4 ft.-3 in.	4 ft.-0 in.	—	—	—
	(3) 2 in. × 12 in.	7 ft.-1 in.	6 ft.-6 in.	6 ft.-0 in.	5 ft.-6 in.	5 ft.-2 in.	4 ft.-10 in.	4 ft.-6 in.	4 ft.-3 in.	4 ft.-0 in.
1500	(3) 2 in. × 6 in.	4 ft.-2 in.	—	—	—	—	—	—	—	—
	(2) 2 in. × 10 in.	4 ft.-8 in.	4 ft.-3 in.	—	—	—	—	—	—	—
	(3) 2 in. × 8 in.	5 ft.-6 in.	5 ft.-0 in.	4 ft.-7 in.	4 ft.-3 in.	—	—	—	—	—
	(2) 2 in. × 12 in.	5 ft.-8 in.	5 ft.-2 in.	4 ft.-9 in.	4 ft.-5 in.	4 ft.-1 in.	—	—	—	—
	(3) 2 in. × 10 in.	7 ft.-0 in.	6 ft.-5 in.	7 ft.-6 in.	5 ft.-6 in.	5 ft.-1 in.	4 ft.-9 in.	4 ft.-6 in.	4 ft.-3 in.	4 ft.-0 in.
	(3) 2 in. × 12 in.	8 ft.-6 in.	7 ft.-9 in.	7 ft.-2 in.	6 ft.-8 in.	6 ft.-2 in.	5 ft.-9 in.	5 ft.-5 in.	5 ft.-1 in.	4 ft.-10 in.,

LL + DL = 40psf
Wall load = 50plf
LL+DL = 50psf
"L/2±1 "L/2±1
Girder
"L"

[a]Maximum spans are based on 30 psf live load for second floor and 40 psf live load for first floor: 10 psf dead load for either floor: 50 psf dead load for partition above girder.

[b]Spans given may be interpolated on a straight line basis for house widths intermediate between those shown.

framing into a masonry or concrete wall (see Fig. 6.6-3). In termite-infested areas, these pockets should be lined with sheet metal, which should extend on the inside about 3½ in. (90 mm) to form a termite shield. A girder should be accurately aligned (true) and level. Wood shims should not be used to level girders or beams. Where adjustment is required, slate, metal, or another durable noncompressible material should be used to support the girder ends. Shim space under a beam or girder should be grouted solid.

6.6.1.3 Girder-Joist Connection

The simplest method of floor framing is one in which joists bear directly on a wood girder or steel beam. Then, the top of the beam coincides with the top of the anchored sill plate. This method is used when there is adequate headroom and a dropped girder is not objectionable. It is simple to frame and has the advantage of providing ample joist cavities for ductwork and electrical and mechanical equipment.

WOOD GIRDER OR BEAM

For more uniform shrinkage and to provide greater headroom, joists can be framed into a wood girder with joist hangers or ledger strips (Fig. 6.6-5). A ledger strip should be at least 2 × 2 (50 × 50) in size and should be face nailed to the girder with three 16d nails at each point where a joist frames into the girder.

Depending on the sizes of the joists and girders, joists may be supported on a ledger strip. In the rare case where individual boards rather than plywood are used for subflooring, continuity across the supporting beam or girder must be provided by elements other than the subfloor. A continuous horizontal tie between exterior walls can be obtained by nailing notched joists together (Fig. 6.6-5a), or a connecting scab at each pair of joists can provide this continuity and also serve as a nailing area for the subfloor (Fig. 6.6-5b). A steel strap can be used to tie joists together when the tops of a beam and the joists are level (Fig. 6.6-5c). When notched joists or scabs ride over the top of a girder, it is important that a small space be allowed above the beam to provide for shrinkage of the joists. Sometimes a ledger strip is positioned higher on a girder and the joists are notched so that the joist shoulders bear on the ledger strip and the joist bottoms are level with the girder. In such situations the notch depth should not exceed one-quarter of the depth of the joist.

When a space is required for heating ducts in a partition supported on a girder, a spaced wood girder is sometimes necessary (Fig. 6.6-6). Solid blocking is used at intervals between the two members. A single post support for a spaced girder usually requires a bolster (wide capital) of sufficient width to support the spaced girders.

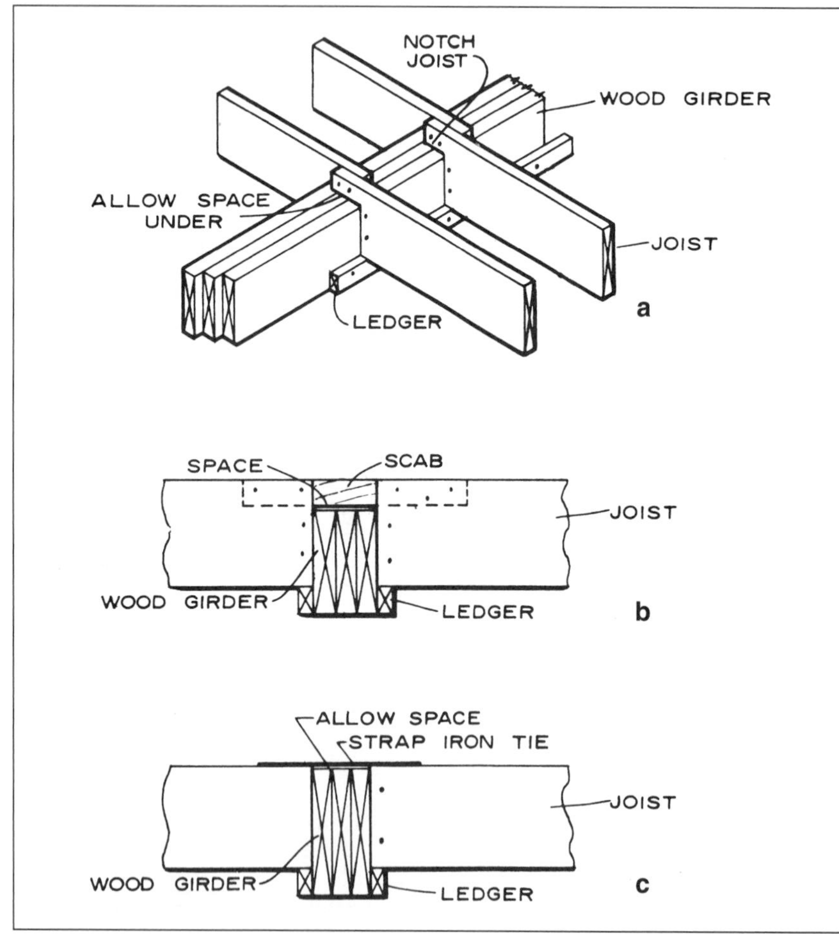

FIGURE 6.6-5 Typical joist-girder connection showing how joists can be tied together when wood board subfloor is used.

STEEL BEAMS

Joists may be supported by or framed into a steel beam (Fig. 6.6-7). In general, connections between joists and a beam should provide (1) lateral stability for the beam, (2) resistance against twisting and overturning of the joists, (3) structural continuity and diaphragm action to resist lateral pressures against the foundation walls, and (4) anchorage for the entire floor assembly. This is accomplished by bolting a 2 × 4 (50 × 100) or 2 × 6 (50 × 150) plate to the upper or lower beam flange, on which the joists are supported (Fig. 6.6-7a). Each joist should be toe nailed to the wood plate and should be bolted at 24 in. (610 mm) on centers, through either the beam's web or its flange.

When it is not feasible to use a nailing plate, special clips can be nailed to the sides of the joists to develop the kind of connection described earlier. In Figure 6.6-7b, the nailing plate has been omitted, and a metal strap tie provides continuity at each joist and anchors the floor assembly; the 2 × 4 (50 × 100) blocking fitted between the joists keeps them from twisting or overturning. Since strap ties only prevent the joists from pulling apart, the joists should be installed so the ends are in contact with the beam web to provide compressive resistance as well.

6.6.1.4 Sill Plate Construction

The sill plate over a foundation wall should be securely anchored (see Section 6.5.1.2) and should be separated from the finish grade by at least 8 in. (200 mm). In Florida and the other Gulf States, the sill

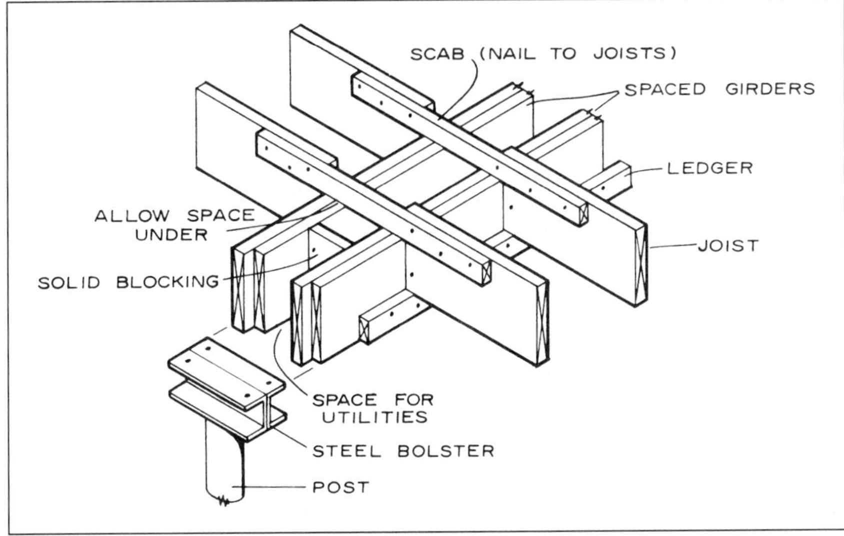

FIGURE 6.6-6 Spaced girder detail showing connections to steel post and floor joists.

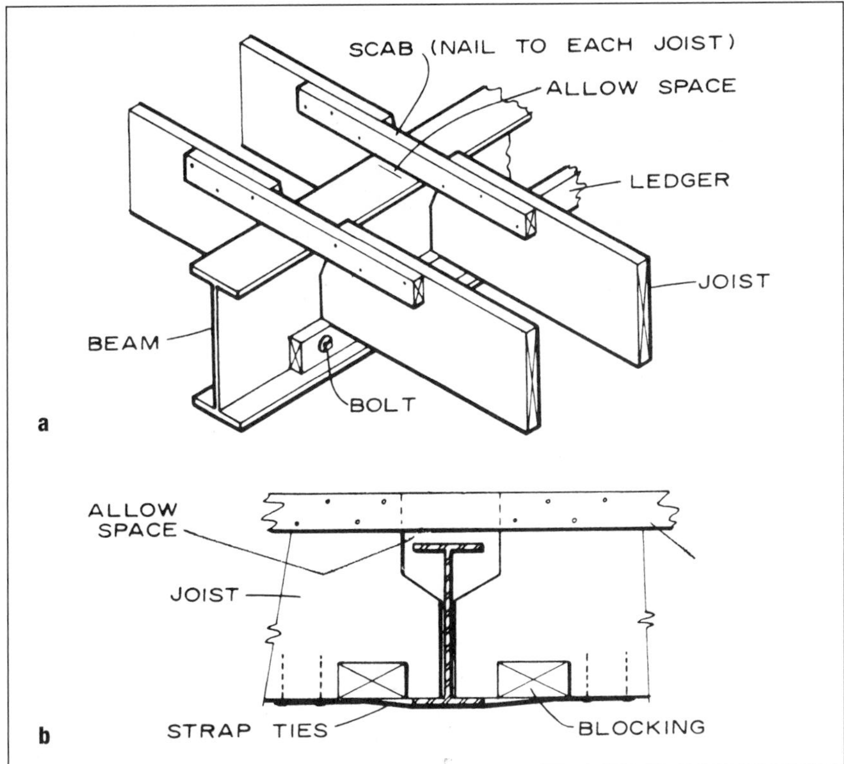

FIGURE 6.6-7 Wood joists can be supported (a) on wood ledger or (b) directly on lower steel beam flange.

plate, and all wood members within 18 in. (450 mm) of grade, should be preservative treated against decay and termites. In areas of high humidity but where termite infestation is not a hazard, inherently decay-resistant grades of redwood, cypress, or red cedar can be used.

Because of its adaptability to the use of tilt-up and component construction methods, and because it provides a stable working platform during construction, platform framing (Fig. 6.6-8) provides great flexibility for the wide variety of possible design requirements in modern light wood-framed buildings.

A *box sill assembly* is commonly used in platform framing (Fig. 6.6-9). It consists generally of a 2 (nominal 2 in. [50 mm] thick of some width) *band joist* and a 2 × 4 (50 × 100) or 2 × 6 (50 × 150) *sill plate*. The plate should be wide enough to provide a full base for the band joist plus at least a 1½-in. (38 mm)-wide bearing ledge for the floor joists. When the band joist is perpendicular to the floor joists, it is also called a *header*; when parallel, a *stringer*. To differentiate between the two conditions, this terminology has been used in this text.

Each floor joist should be toe nailed to the sill plate with three 8d or two 10d nails; a header should be end nailed to each joist with two 16d nails. If suitable rigid wall sheathing is not nailed to the sill plate, the header should be toe nailed to the sill plate with 8d nails 16 in. (406 mm) on centers. A stringer should similarly be toe nailed to the sill plate with 8d nails at 16 in. (406 mm) on centers. A sill plate should be anchored to the foundation wall over a resilient sill sealer or bed of grout. A termite shield should be used in areas subject to termite infestation.

6.6.1.5 Floor Joists

A wood joist floor is part of the floor and ceiling assembly portion of a building's superstructure, which also includes the walls and a roof and ceiling assembly. A joist floor is essentially the same whether the walls are wood framed, solid masonry, or concrete. It also makes no difference whether wood-framed walls have brick veneer or some other exterior finish.

Platform-frame constructions typically include wood floor joists of nominal 2 × (50 ×) lumber, spaced 12 in. (305 mm), 16 (406 mm), or 24 in. (610 mm) on centers. Contemporary floor joists, however, are frequently composite truss type units made up of plywood webs and lumber top and bottom chords (Fig. 6.6-10a). Longer spans are sometimes achieved using floor trusses made with top and bottom chords and diagonal wood framing members arranged in a framework of triangles in much the same way as in flat roof trusses (see Section 6.7.5). Subflooring of wood boards or plywood is applied over floor joists and floor trusses to form a structural floor and working platform (Figs. 6.6-10b and 6.6-11).

Generally, joists span from the exterior walls to center load-bearing walls or beams. Interior load-bearing walls may restrict the flexibility of room planning, however. For relatively narrow buildings

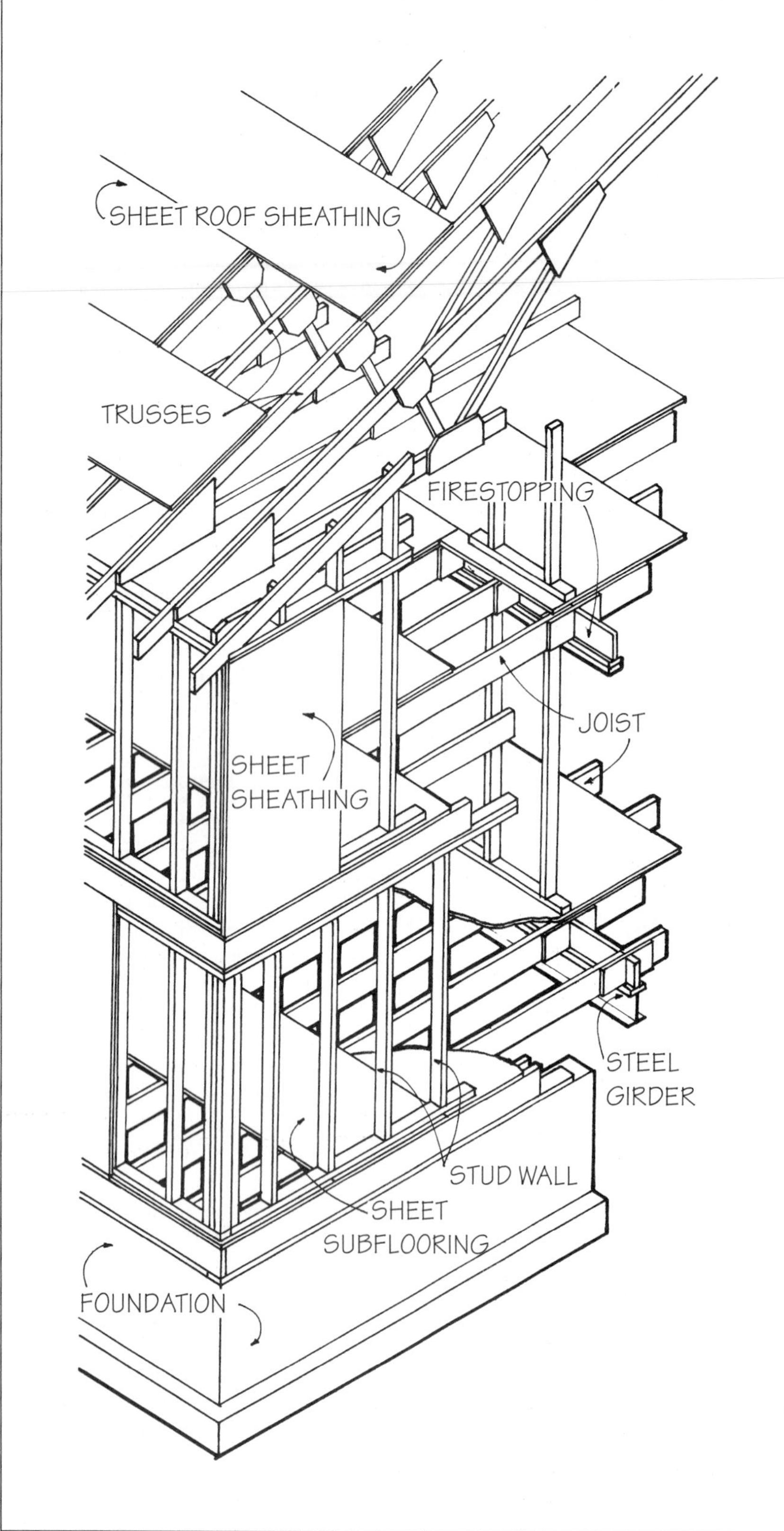

FIGURE 6.6-8 Customary platform framing. Characterized by floor joists and story-height studs spaced 16 in. (406 mm) or 24 in. (610 mm) on center; roof trusses spaced up to 4 ft. (1220 mm) on center; panel sheathing, such as plywood, gypsum, or fiberboard; and elimination of unnecessary floor bridging and corner bracing. The floor construction forms a work platform, permitting efficient tilt-up methods; adapts easily to panelized construction.

(about 16 ft. [4.88 m] or less) joists may span from exterior wall to exterior wall without a center support.

Joist floor systems provide space for heating, air conditioning, electrical, and plumbing equipment and devices, and are easily insulated to minimize heat loss and improve sound transmission characteristics.

Some types of floor finishes (such as hardwood flooring) may be installed directly over wood or plywood subflooring; other finishes (such as resilient flooring) may require an underlayment. A variety of ceiling finishes may be applied, usually directly to the underside of the joists.

MATERIALS

Floor joist lumber is selected primarily to meet strength and stiffness requirements, which depend on the dead and live loads to be carried. For average spans encountered in residential construction, stiffness requirements are more critical than strength requirements. In other words, typical joist sizes are selected on the basis of their stiffness (resistance to deflection) rather than strength (ability to carry loads without failure). Stiffness requirements mandate an arbitrary maximum deflection under loads: $\frac{1}{360}$ of the joist span, in inches. For instance, the maximum allowable deflection for a 15 ft. joist span would be 180 in. (15 × 12) divided by 360 ($\frac{180}{360}$), or ½ in. (12.7 mm).

Other desirable qualities for joist lumber are good nail-holding ability and freedom from warp. Wood floor joists are usually 2-in. (50 mm)-thick nominal lumber, 6 in. (150 mm) to 12 in. (305 mm) deep. The size depends on loading, length of span, joist spacing, and the species and grade of lumber used.

Floor joists for simple frame buildings and ordinary construction can be selected by referring to *Span Tables for Joists and Rafters*, published by the American Forest & Paper Association (AF&PA). The table there for floor joists for 40 psf (195.3 kg/m^2) live load and 10 psf (48.8 kg/m^2) dead load is reproduced in Figure 6.6-12. These tables indicate the minimum modulus of elasticity, E, and extreme fiber strength in bending, F_b, required for various spans, lumber sizes, and conditions of loading. *Span Tables for Joists and Rafters* also includes tables for 30 psf (146.47 kg/m^2), 50 psf (244.12 kg/m^2), and 60 psf (292.95 kg/m^2) live loads and for a 20 psf (97.65 kg/m^2) dead load. Suitable species and grades meeting the indicated strength

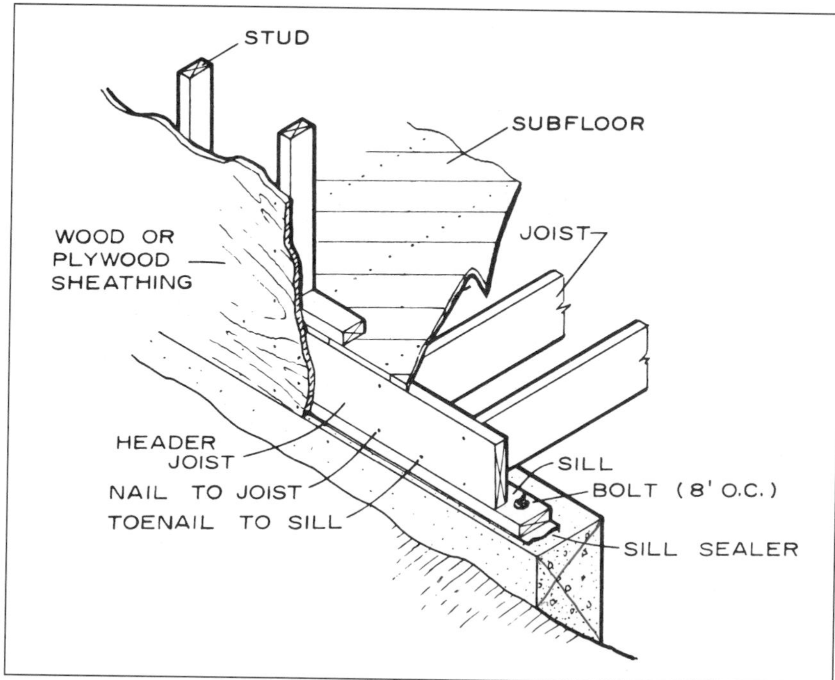

FIGURE 6.6-9 Box sill assembly. The studs rest on a sole plate. The subflooring may be boards as shown, but is more likely to be a sheet material.

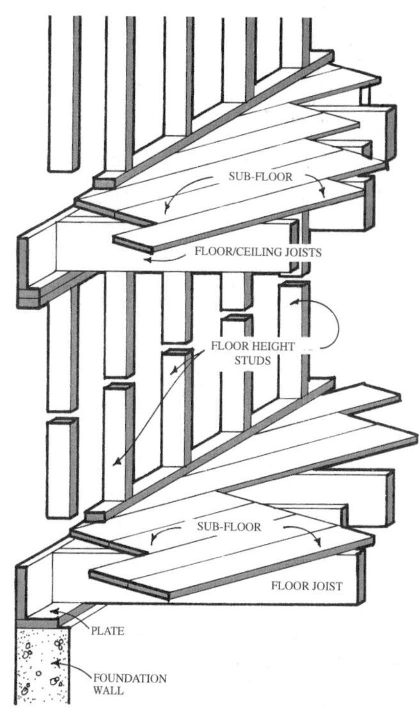

FIGURE 6.6-11 Wood joist floor/ceiling construction. Subflooring may be boards as shown, but is more likely to be a sheet material.

requirements can be selected from the tables of allowable unit stresses in the *NDS Supplement*.

In these tables, the allowable span is the clear distance between supports. Tabulated spans are applicable to green or dry lumber, provided it is used in covered structures, where it will normally achieve a moisture content of 19% or less.

For an example of floor joist selection, assume a required span of 12 ft. 9 in. (3886 mm), a live load of 40 psf (195.3 kg/m^2) and joists spaced 16 in. (406 mm) on centers. Figure 6.6-12 shows that a 2 × 8 (50 × 200) having an E value of 1,600,000 psi (1,124,912 Mg/m^2) (1.6 in the table) and an F_b value of 1255 psi (882.35 Mg/m^2) would have an allowable span of 12 ft. 10 in. (3911.6 mm). Suitable species and grades can be selected from the *NDS Supplement*. In our example, Figure 6.1-14 shows that a visually graded Douglas Fir-Larch No.1 & Btr grade has an F_b of 1200 psi (843.68 Mg/m^2) and an E of 1,800,000 psi (1,265,526 Mg/m^2). The *NDS Supplement* permits increasing the allowable stress shown in the table by a repetitive member factor (C_r), a wet service factor (C_M), a flat use factor (C_{fu}), a size factor (C_F), and a shear stress factor (C_H). These allowable increases are shown on pages 25 and 26 of the 1997 *NDS Supplement*. Of these, only the repetitive member and size factors are applicable. The F_b can be increased by a factor of 1.15 because of the repetitive member factor (C_r), producing an allowable F_b of 1380 psi (970.24 Mg/m^2). A further increase

a

b

FIGURE 6.6-10 (a) Truss type wood joists. (b) A plywood subfloor over wood joists. (Courtesy American Plywood Association.)

FIGURE 6.6-12 Maximum Allowable Spans for Floor Joists with L/360 Deflection Limits[a,b]

Joist Size, in.	Spacing, in.	Modulus of Elasticity, E, in 1,000,000 psi																
		0.8	**0.9**	**1.0**	**1.1**	**1.2**	**1.3**	**1.4**	**1.5**	**1.6**	**1.7**	**1.8**	**1.9**	**2.0**	**2.1**	**2.2**	**2.3**	**2.4**
2 × 6	12.0	8-6	8-10	9-2	9-6	9-9	10-0	10-3	10-6	10-9	10-11	11-2	11-4	11-7	11-9	11-11	12-1	12-3
	16.0	7-9	8-0	8-4	8-7	8-10	9-1	9-4	9-6	9-9	9-11	10-2	10-4	10-6	10-8	10-10	11-0	11-2
	19.2	7-3	7-7	7-10	8-1	8-4	8-7	8-9	9-0	9-2	9-4	9-6	9-8	9-10	10-0	10-2	10-4	10-6
	24.0	6-9	7-0	7-3	7-6	7-9	7-11	8-2	8-4	8-6	8-8	8-10	9-0	9-2	9-4	9-6	9-7	9-9
2 × 8	12.0	11-3	11-8	12-1	12-6	12-10	13-2	13-6	13-10	14-2	14-5	14-8	15-0	15-3	15-6	15-9	15-11	16-2
	16.0	10-2	10-7	11-0	11-4	11-8	12-0	12-3	12-7	12-10	13-1	13-4	13-7	13-10	14-1	14-3	14-6	14-8
	19.2	9-7	10-0	10-4	10-8	11-0	11-3	11-7	11-10	12-1	12-4	12-7	12-10	13-0	13-3	13-5	13-8	13-10
	24.0	8-11	9-3	9-7	9-11	10-2	10-6	10-9	11-0	11-3	11-5	11-8	11-11	12-1	12-3	12-6	12-8	12-10
2 × 10	12.0	14-4	14-11	15-5	15-11	16-5	16-10	17-3	17-8	18-0	18-5	18-9	19-1	19-5	19-9	20-1	20-4	20-8
	16.0	13-0	13-6	14-0	14-6	14-11	15-3	15-8	16-0	16-5	16-9	17-0	17-4	17-8	17-11	18-3	18-6	18-9
	19.2	12-3	12-9	13-2	13-7	14-0	14-5	14-9	15-1	15-5	15-9	16-0	16-4	16-7	16-11	17-2	17-5	17-8
	24.0	11-4	11-10	12-3	12-8	13-0	13-4	13-8	14-0	14-4	14-7	14-11	15-2	15-5	15-8	15-11	16-2	16-5
2 × 12	12.0	17-5	18-1	18-9	19-4	19-11	20-6	21-0	21-6	21-11	22-5	22-10	23-3	23-7	24-0	24-5	24-9	25-1
	16.0	15-10	16-5	17-0	17.7	18-1	18-7	19-1	19-6	19-11	20-4	20-9	21-1	21-6	21-10	22-2	22-6	22-10
	19.2	14-11	15-6	16-0	16-7	17-0	17-6	17-11	18-4	18-9	19-2	19-6	19-10	20-2	20-6	20-10	21-2	21-6
	24.0	13-10	14-4	14-11	15-4	15-10	16-3	16-8	17-0	17-5	17-9	18-1	18-5	18-9	19-1	19-4	19-8	19-11
F_b	12.0	718	777	833	888	941	993	1043	1092	1140	1187	1233	1278	1323	1367	1410	1452	1494
F_b	16.0	790	855	917	977	1036	1093	1148	1202	1255	1306	1357	1407	1456	1504	1551	1598	1644
F_b	19.2	840	909	975	1039	1101	1161	1220	1277	1333	1388	1442	1495	1547	1598	1649	1698	1747
F_b	24.0	905	979	1050	1119	1186	1251	1314	1376	1436	1496	1554	1611	1667	1722	1776	1829	1882

[a]The required bending design value, F_b, in pounds per square inch (psi) is shown at the bottom of the table and is applicable to all lumber sizes shown. Spans are shown in feet-inches and are limited to 26 ft. and less. Check sources of supply for availability of lumber in lengths greater than 20 ft.

[b]Design Criteria: Deflections—for 40 pounds per square foot (psf) live load. Limited to span in inches divided by 360. Strength—live load of 40 psf plus dead load of 10 psf determines the required bending design value.

(Courtesy American Forest & Paper Association.)

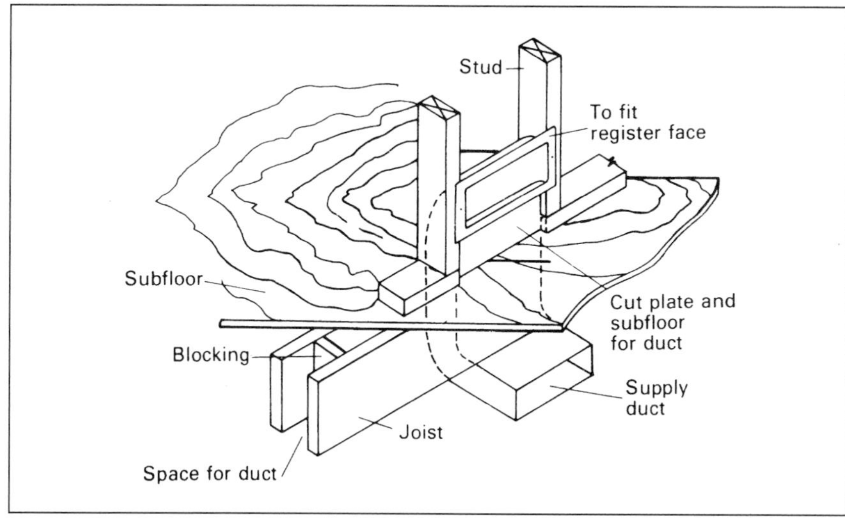

FIGURE 6.6-13 Spaced joist construction under a parallel bearing partition. (Fig. 43, *Wood Frame House Construction*, United States Department of Agriculture, page 52. Not subject to copyright in the United States.)

nailed across the top, until the subflooring is applied.

It is important when designing a wood floor system to equalize shrinkage of the framing at the exterior walls and at the center support. By using approximately the same total depth of cross-grain wood at the center girder as at the exterior wall, differential shrinkage can be minimized. This reduces plaster cracks, prevents sticking doors, and avoids other inconveniences caused by uneven shrinkage. For instance, if there is a total of 12 in. (305 mm) of cross grain wood at the foundation wall (including joists and sill plate), this should be balanced with an equal amount of cross-grain wood at the center girder. To avoid uneven shrinkage, a steel beam center support is generally preferred. If a wood girder is used, MC 15 lumber permitted by the size factor is 1.2. Therefore, the F_b can be increased to 1656 psi (1164.28 Mg/m^2). Figure 6.1-15 shows a machine-stress-rated lumber, 1700*f*-1.6*E*, of WCLIB species that has an F_b of 1700 psi (1195.22 Mg/m^2) and an E of 1,600,000 psi (1,124,912 Mg/m^2).

AF&PA advises that the tabulated size factor increases apply only to Douglas Fir-Larch, Douglas Fir-Larch (North), and Douglas Fir South. The tables already take this factor into account for Southern Pine.

Joists' grades should be selected for the specific design conditions. Where several combinations of joist size and spacing satisfy the given requirements, a deeper joist at greater spacing is generally preferred, provided, of course, that the subfloor and the ceiling finish are selected to span the greater spacing. Using deeper joists usually requires less lumber, less labor, and results in a stiffer floor and ceiling assembly.

Joist Installation Floor and ceiling assemblies intended for the direct application of gypsum board or any other board finish should have all joists evenly spaced, with bottom faces aligned in a level plane.

After the sill plates have been anchored to the foundation walls, joists are laid out over the sill plates and center supports. The floor joists are spaced uniformly at 12 in. (305 mm), 16 (406 mm), or 24 in. (610 mm) on centers, as determined by the design requirements. Extra joists are added where required to support bearing partitions (Fig. 6.6-13) or to frame floor openings (Fig. 6.6-14). Joists are braced temporarily with wood strips

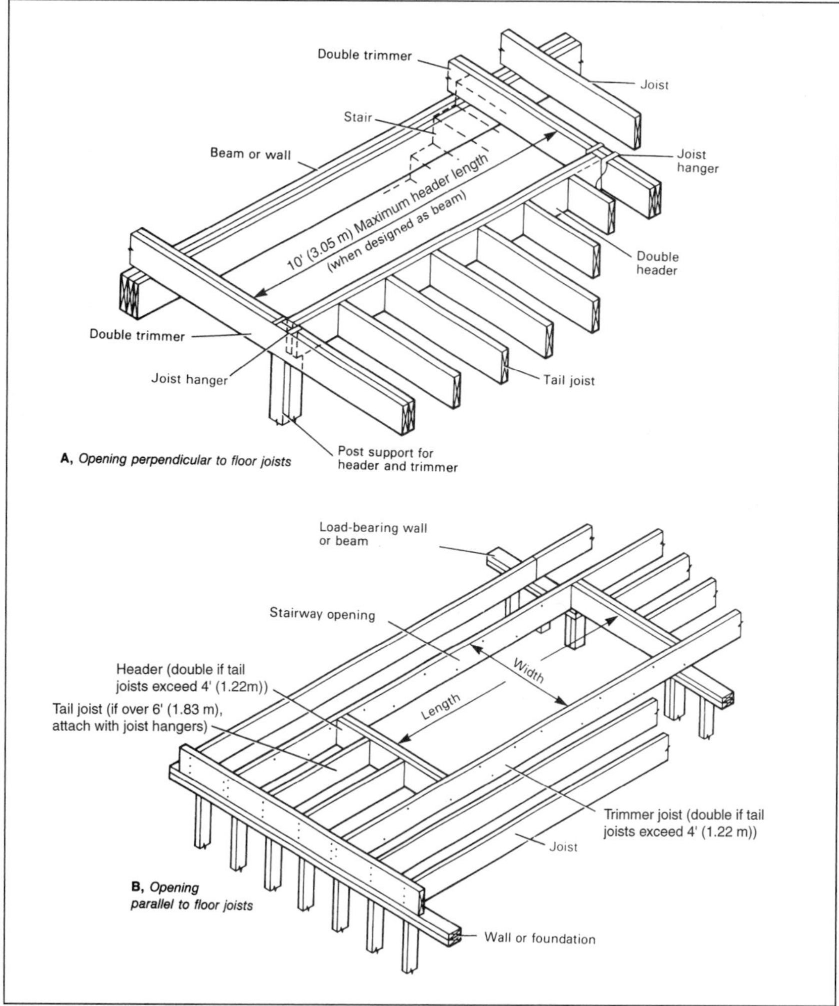

FIGURE 6.6-14 Framing for a large floor opening (a) with the long dimension of the opening perpendicular to the joists, and (b) with the long dimension of the opening parallel with the joists. (Fig. 48, *Wood Frame House Construction*, United States Department of Agriculture, page 57. Not subject to copyright in the United States.)

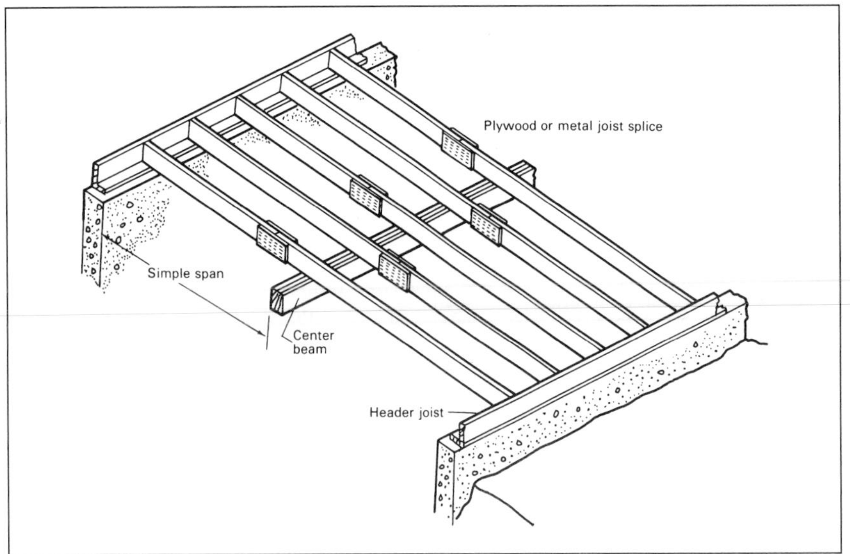

FIGURE 6.6-15 In-line joist system framed with pairs of overhanging joists and supported joists. (Fig. 35, *Wood Frame House Construction*, United States Department of Agriculture, page 46. Not subject to copyright in the United States.)

should be specified, or the joists should be supported on ledger strips rather than over the top of the girder.

Excessively bowed or crooked joists should not be used, but joists with a slight crown may be used if they are installed with the crown up. After the subfloor and normal floor loads are applied, such a joist will tend to straighten out. The largest edge knots should also be placed on top, since the upper side of the joist is in compression and the knots will have a lesser effect on strength and deflection.

Header joists should be fastened by nailing into the end of each joist with two 16d nails. In addition, the header joists and the stringer joists (parallel to the exterior walls) should be toe nailed to the sill with 8d nails spaced 16 in. (406 mm) on centers, unless rigid wall sheathing covers the sill plate and is nailed to it with 8d nails 8 in. (200 mm) on centers.

Each floor joist should be toe nailed to the sill and center beam or girder with two 10d or three 8d nails; joists should also be nailed to each other with at least three 16d nails when they lap over the center beam. The lap should be not less than 4 in. (100 mm), and the overhang of either joist beyond the edge of the girder or beam should not exceed 12 in. (305 mm). The limit on overhangs is imposed in order to reduce the possibility of lifting the subfloor resting on the overhang when a heavy concentrated load is placed on the joist near its midspan. If a nominal 2-in. (50 mm) scab is used across butt-ended joists, it should be nailed to each joist with at least three 16d nails at each side of the joint.

For greater lumber economy, an in-line joist system is sometimes used in floor and roof framing when center supports are present. This system normally allows the use of joists one size smaller than in conventional floor framing. Briefly, the system consists of joists of uneven lengths with the long, overhanging joist cantilevered over the center support and connected to the supported joist with a metal connector or plywood splice plate (Fig. 6.6-15). In typical single-family construction, the overhang varies between about 1 ft. 10 in. (558.8 mm) and 2 ft. 10 in. (863.6 mm). For Douglas fir and southern pine No. 1 lumber, total spans (between foundation walls) of up to 22 ft. (6.71 m) are possible with 2 × 6s (50 × 150s), spans up to 30 ft. (9.14 m) with 2 × 8s (50 × 200s), and spans up to 34 ft. (10.36 m) with 2 × 10 (50 × 250) joists. Other innovative floor systems that depend on the special attachment of a plywood subfloor are discussed in Section 6.6.4.

Floor joists under parallel bearing partitions should be doubled; if access into the partition is required for air ducts, the double joists should be spaced and solid blocking should be used between them as in spaced wood girder construction (Fig. 6.6-16 [see also Figs. 6.6-6 and 6.6-13]).

6.6.1.6 Bridging

Cross-bridging between joists, which has been used in house construction for many years and which is still required by some building codes, has been demonstrated by the National Association of Home Builders Research Foundation and several other independent laboratories to be virtually useless under most circumstances. In fact, even when bridging is tight fitting and well installed, there is no significant improvement in a floor's ability to transfer vertical loads after the subfloor and finish flooring have been installed. Therefore, in most wood-framed buildings, bridging between floor joists is not nec-

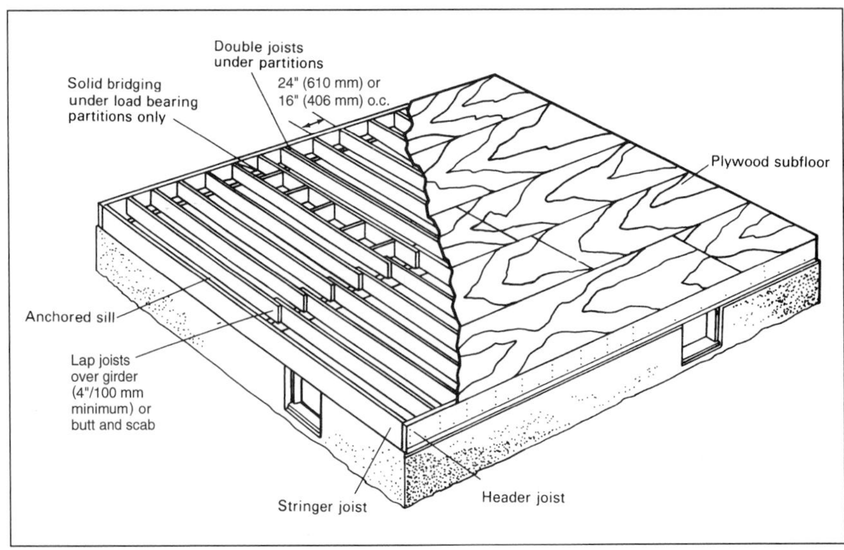

FIGURE 6.6-16 Typical platform-framed floor construction. (Fig. 34, *Wood Frame House Construction*, United States Department of Agriculture, page 46. Not subject to copyright in the United States.)

essary. Properly nailed subflooring and the nailing of the ends of joists to band joists or headers is sufficient to keep the joists aligned and to prevent them from twisting.

However, when the depth-to-thickness ratio of joists exceeds 6, or when joist twisting has been experienced, bridging is often added at midspan or at intervals of not more than 8 ft. (2440 mm). This bridging may consist of solid wood blocks, or of either metal struts or 1 × 3 (25 × 75) or larger wood members installed in an X fashion, with one end at the top of a joist and the opposite end at the bottom of the adjacent joist.

In addition, either cross-bridging or solid blocking using 2 × 4 (50 × 100) or larger lumber installed vertically between joists may improve the diaphragm action of a floor, and may increase its resistance to lateral (horizontal) loads imparted to it by foundation walls. Therefore, bridging may be required in situations where engineering design so indicates.

Solid bridging between joists should be used to provide a more rigid base for nonbearing partitions that happen to fall between joists (see Fig. 6.6-16). It also provides a fire-stop against lateral flame spread. Solid blocking should be well fitted and securely nailed to the joists. Load-bearing partitions running parallel to the joist direction should be supported by doubled joists, either by placing the partition over a normally spaced joist that has been doubled, or by providing an additional pair of joists under the partition.

6.6.1.7 Floor Openings

Large openings are often necessary in floors for ducts, chimneys, and stairs (see Fig. 6.6-14). Where joists have been cut away, the remaining *tail joists* are framed into single or double cross joists, called *headers*. The headers in turn are framed into abutting full-length joists, known as *trimmers*.

A single header can be used if the dimension of the opening along the header is less than 4 ft. (1220 mm); a single trimmer can be used to support a single header when the opening is near the end of the joist span and less than 4 ft. (1220 mm) long. Tail joists shorter than 6 ft. (1800 mm) should be connected to headers with three 16d end nails and two 10d toe nails. Longer tail joists should be connected with framing anchors or supported on a ledger strip as in joist-girder connections. Headers should be connected to trimmers with three 16d end nails and two 10d toe nails.

When framing openings larger than 4 ft. (1220 mm) by 4 ft. (1220 mm), both the trimmers and headers should be doubled. Connections between headers and trimmers should be made with framing anchors, joist hangers, or ledger strips. Tail joists more than 6 ft. (1800 mm) long should be similarly attached to headers; shorter tail joists can be attached as recommended earlier for smaller openings. The nailing sequence for double headers is shown in Figure 6.6-17; the opening is first framed out using single headers and trimmers; then the second joists are applied and face nailed to the first.

The largest floor openings are usually for stairs; for straight runs and typical floor heights, they are usually about 10 ft. (3000 mm) long and 3 ft. (900 mm) to 4 ft. (1220 mm) wide. The long dimension of stairway openings may be either parallel or at right angles to the joists. However, it is much easier to frame a stairway opening when its length is parallel to the joists (see Fig. 6.6-14a).

When the length of a stair opening is perpendicular to the joists, a long doubled header is required (see Fig. 6.6-14b). A header under these conditions without a supporting wall beneath it is usually limited to a 10-ft. (3000 mm) length and should be designed as a beam. A load-bearing wall under all or part of this opening substantially simplifies the framing, as the joists then bear on the top plate of a wall rather than on a header.

6.6.1.8 Floor Projections

The framing for floor projections beyond the supporting wall below, such as for a bay window or balcony, generally consists of extended floor joists, if they run perpendicular to the supporting wall (Fig. 6.6-18a). This extension normally should not exceed 24 in. (610 mm) unless it is specifically structurally designed for a greater projection. The joists forming each side of such a bay should be doubled. Nailing procedures should conform to those used in framing stair openings. The subflooring should be extended flush with the outer framing member.

In framing the roof of such a projection, rafters are often supported on a header over the bay area, in the plane of the supporting wall below. The header thus supports the roof load and transmits it to the foundation wall below, so that the bay wall carries less load.

Projections parallel to the direction of the floor joists should generally be limited to small areas and extensions of not more than 24 in. (610 mm) (Fig. 6.6-18b). In this construction, the stringers should frame into doubled joists. Joist hangers or a ledger strip can be used to frame joists into the double header, as in a joist-girder connection.

6.6.1.9 Framing for Mechanical and Electrical Equipment

It is frequently necessary to notch, drill, or cut framing members to accommodate heating, plumbing, and electrical equip-

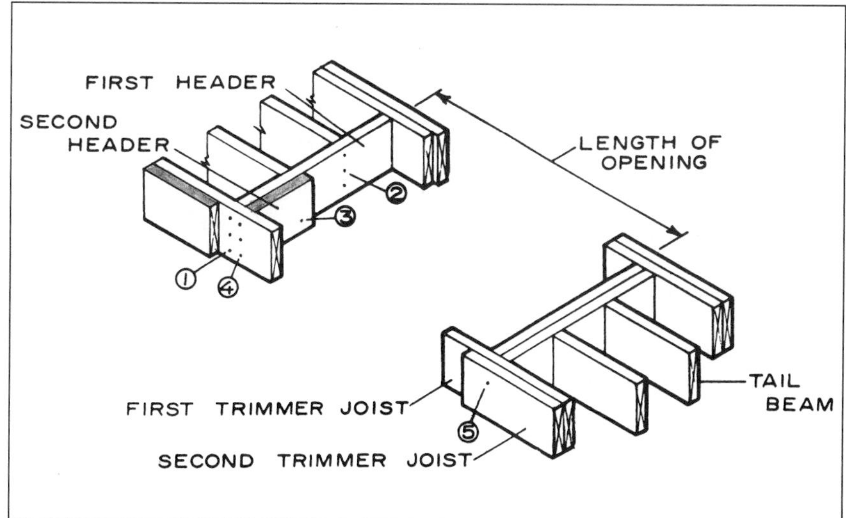

FIGURE 6.6-17 Nailing sequence for floor openings: (1) nailing trimmer to first header, (2) nailing header to tail beams, (3) nailing headers together, (4) nailing trimmer to second header, and (5) nailing trimmers together.

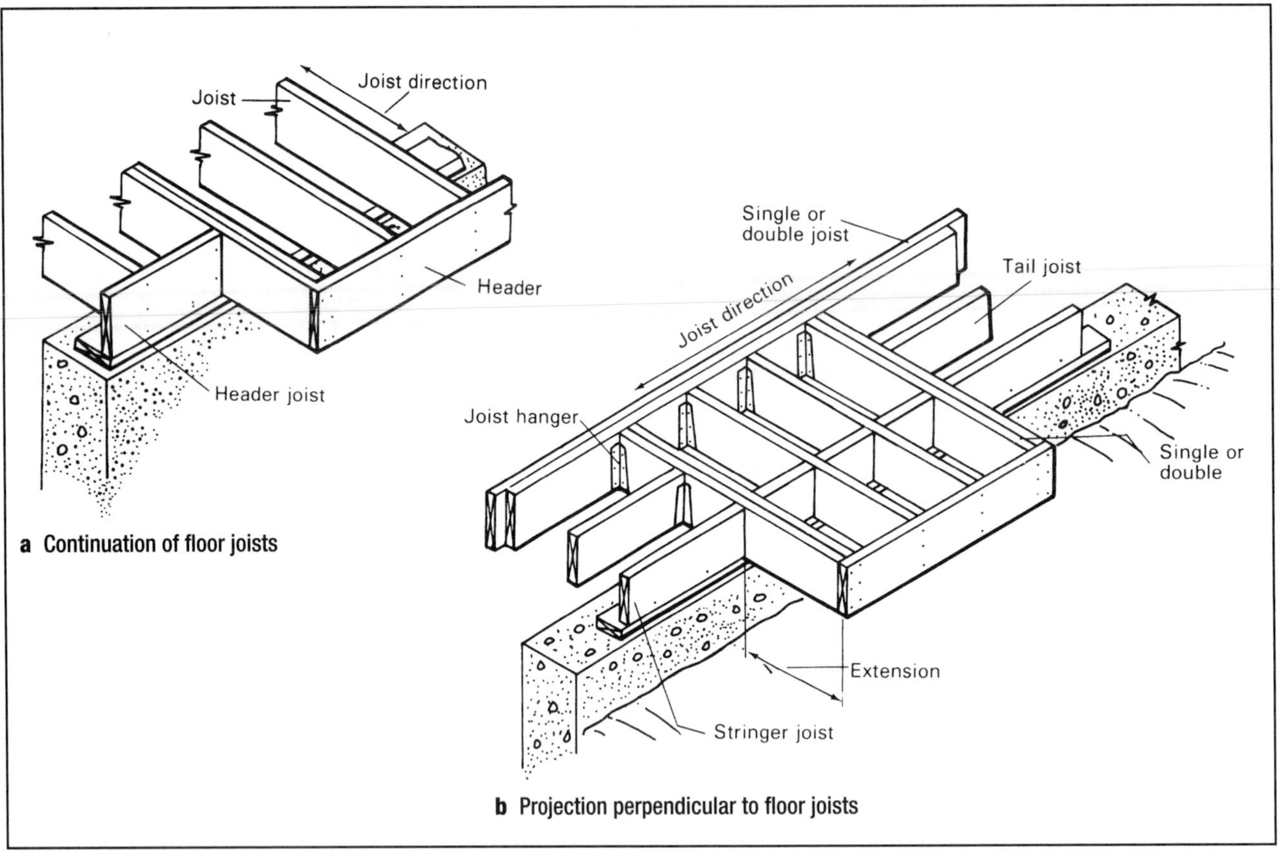

FIGURE 6.6-18 **Framing for wall projections: (a) when joists are perpendicular to wall, (b) when joists are parallel with wall.** (Fig. 39, *Wood Frame House Construction*, United States Department of Agriculture, page 51. Not subject to copyright in the United States.)

ment. Preconstruction planning can minimize the amount of cutting required. If mechanical and electrical distribution plans are worked out in advance, many time- and cost-saving adjustments in the framing can be made.

Floor joists should be notched or drilled in a way that will minimize the effect on their structural integrity. The top or bottom of a joist should be notched only in the end one-third of the span and to a depth of not more than one-sixth of the joist depth. Holes may be bored in joists, provided they are not more than 2 in. (50 mm) in diameter and not less than 2½ in. (63.5 mm) from either edge (Fig. 6.6-19). Where notching or drilling to a greater extent is necessary, the weakened joist should be reinforced with scabs on both sides or by adding another joist. If reinforcing is not possible because of conflict with the mechanical equipment, the joist should be cut away and the area treated as a floor opening, with tail beams supported on headers.

When a plumbing wall adjacent to a bathroom also serves as a load-bearing partition, the supporting double joists should be spaced and blocked (Fig. 6.6-20) as recommended for a spaced girder. The joist nearest the front apron of a tub should also be doubled to prevent excessive deflection when the tub is full of water.

6.6.2 WALL ASSEMBLIES

Wall assemblies are a basic element of a framed building's superstructure. In their design, selection, and construction, it is necessary to consider (1) the required level of functional performance, (2) costs, and (3) the desired finished appearance. These considerations often are influenced by the finishes used and other basic elements of the building.

The functional performance requirements of a wall assembly include:

1. Adequate strength to support the wall and other building elements it supports (dead loads) and live loads, such as those created by wind or earthquakes, and those transmitted into the wall from supported construction, such as floors.
2. Weathertightness (control of the flow of heat, moisture, air, and water vapor)
3. Satisfactory levels of sound and visual privacy

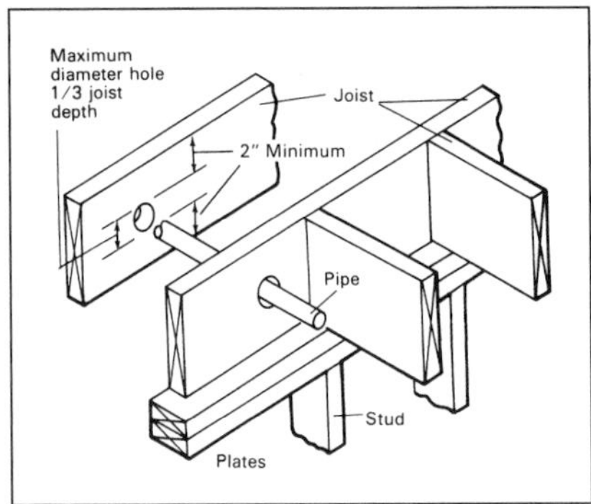

FIGURE 6.6-19 **Recommended limits for boring of holes in load-bearing floor joists.** (Fig. 31, *Wood Frame House Construction*, United States Department of Agriculture, page 52. Not subject to copyright in the United States.)

4. Suitable fire resistance
5. The ability to accommodate heating, air conditioning, electrical, and plumbing equipment
6. Suitability for the application of various finish materials
7. Adaptability to economical methods of assembly and erection
8. Provision for the installation of doors, windows, and other openings

When other requirements exist, the conventional constructions described in this section may not be adequate. Structural walls must safely support superimposed vertical loads and resist horizontal and racking loads due to lateral forces, such as wind and earthquake, and transmit all loads to the foundation. Rarely are walls specifically engineered. Ordinarily, building codes or rules of thumb based on established practices determine materials and methods of construction.

The ultimate costs of a wall assembly include in-place costs, maintenance costs, and operating costs. In-place costs include the costs of materials, labor, and overhead required to assemble and install the system. The selection of a wall assembly may affect other costs, such as those for installing insulation and a vapor retarder, applying siding or other finishes, and providing for window and door openings. The necessary time and costs to enclose the building and the costs of installing heating, air conditioning, electrical, and plumbing equipment also may be influenced by the type of assembly. Maintenance costs and operating costs attributable to a wall assembly (such as those to minimize heat loss or gain through the construction) have an important influence on the ultimate cost of the wall and should not be overlooked if long-term economy is to be achieved.

Occasionally some of the structural elements in a wood-framed building are left exposed, but wood-framed wall assemblies are usually covered with applied materials to provide the desired finished appearance and performance. The selection of finish materials may:

1. Influence the spacing and unsupported span of primary structural members;
2. Determine the need for wall sheathing and other finish bases;
3. Dictate the kind, type, and physical properties of the subsurface material;
4. Dictate methods of construction.

A building's appearance is influenced by the location, size, and type of doors, windows, and other openings that form part of its wall assemblies. In addition, wall assemblies must be compatible with the roof and floor assemblies used in the construction.

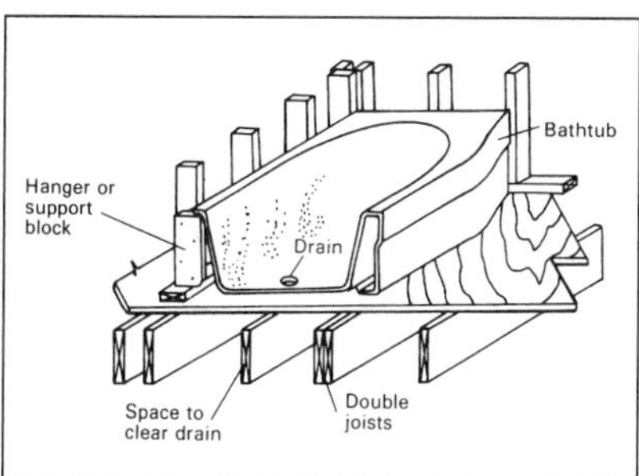

FIGURE 6.6-20 Recommended framing to minimize floor deflection under tub. (Fig. 31, *Wood Frame House Construction*, United States Department of Agriculture, page 42. Not subject to copyright in the United States.)

Typical wall assemblies include conventional stud wall construction (discussed in this section), panelized construction (Section 6.6.6), masonry construction (Chapter 4), and post-and-beam construction (Section 6.7.1).

6.6.2.1 Materials

Wall framing lumber should be of the proper grade for the intended use, graded by a recognized grading authority. Its moisture content is important in the performance of applied finishes, and should be as near as possible to the equilibrium moisture content for the particular area. In no case should it be more than 15% when finishes are applied.

6.6.2.2 Stud Walls and Partitions

In light wood-framed construction, wood studs are the typical framing members for walls and partitions; wood joists and trusses are the most common ceiling and roof framing members. These framing members are generally spaced 16 in. (407 mm) or 24 in. (610 mm) on centers, depending on the unsupported height of the member and the number of floor and roof structures supported. Member sizes and spacings are often dictated by the building code. The usual spacings permit direct application of most finishes, such as gypsum board and plaster. However, furring may be required when member spacing exceeds the maximum recommended for the particular finish material or when special performance, such as increased sound isolation, is desired.

The following discussion of conventional stud wall construction refers to stick-built erection methods, where carpenters on the job handle individual studs rather than prefabricated panels.

Buildings built with platform-frame construction typically use wood stud wall assemblies (Fig. 6.6-21). Walls may be sheathed with wood boards, plywood, fiber, or gypsum boards, which provide rigidity, form a weather barrier, and may be necessary as a nailing base to receive exterior finishes. Some types of exterior and interior finishes can be applied di-

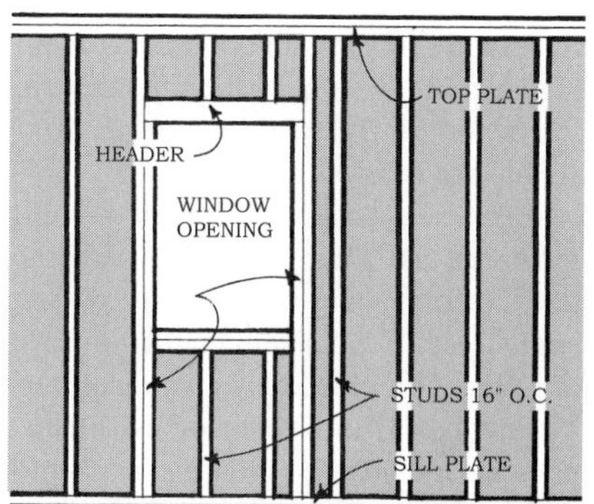

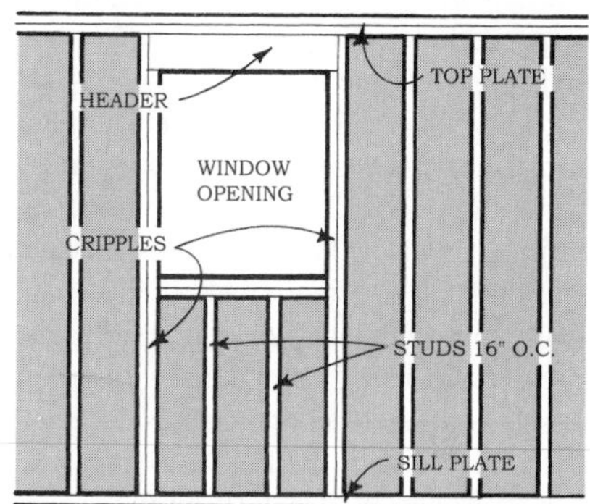

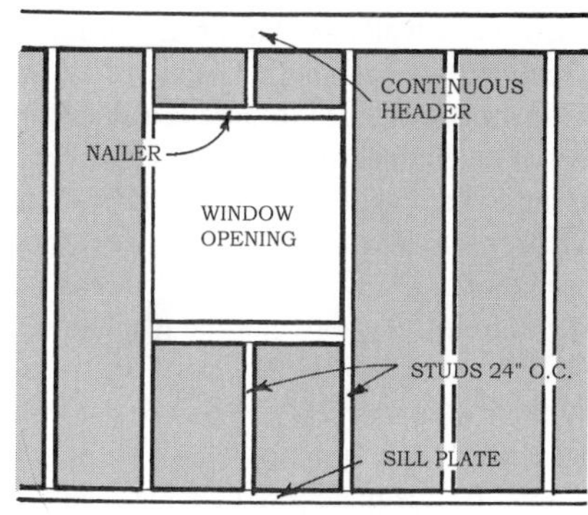

16″ = 406 mm
24″ = 610 mm

FIGURE 6.6-21 Typical wood stud construction, showing framing members. Structural wall sheathing or diagonal bracing, which is required to resist wind and other stresses, is not shown here.

rectly to studs; other finishes must be applied over sheathing or other bases.

In load-bearing partitions, studs should be of 2 × 4s (50 × 100s) or 2 × 6s (50 × 150s), usually spaced at 16 in. (406 mm) or 24 in. (610 mm) on centers, depending on the loads to be encountered. In non-load-bearing or staggered-stud partitions, 2 × 3 (50 × 75) or 2 × 4 (50 × 100) studs may be used, spaced 16 in. (406 mm) on centers when openings are present or cabinets and other elements are supported on them, and 24 in. (610 mm) on centers when no openings are included and no additional loads are imposed on them. Backup members should be provided at interior corners for support and as a nailing base for gypsum board and other finish materials. Warped or crooked studs should not be used.

Because of the existing body of building codes and other regulations and practices, stud spacing is rarely specifically engineered. The conventional 16-in. (406 mm) and 24-in. (610 mm) spacing of studs has evolved from years of established practice and is based more on accommodating finish materials applied to the framework than on actual engineering for imposed loads. For example, in one-story light wood-frame construction, 2 × 4 (50 × 100) studs could be placed at 6-ft. (1800 mm) intervals and adequately support the generally imposed roof loads. But few exterior or interior finish materials can span such intervals economically. The 16-in. (406 mm) and 24-in. (610 mm) spacings permit the use of standard widths and lengths of panel sheathing and siding, and therefore remain the convention in use. Stud spacings of 12 in. (305 mm), although rare in low-rise buildings, are sometimes used when walls are tall or lateral loads are to be superimposed. The use of 12-in. (305 mm) spacings is a structural decision based on engineering. There is currently a trend toward using 2 × 6 (50 × 150) studs at 24 in. (610 mm) on center for exterior walls to take advantage of the increased depth to install more thermal insulation.

The suitable performance of most types of siding depends in part on the wall structure that supports it. Wall studs should be adequately sized and spaced to carry superimposed loads. Conservatively, structural load requirements are normally met using:

1. 2×4 (50×100) studs at a maximum spacing of 24 in. (610 mm) on center when a wall supports only a ceiling and roof;
2. 2 × 4 (50 × 100) studs at a maximum spacing of 16 in. (406 mm) on centers when the wall supports a floor in addition to the roof;
3. 2 × 6 (50 × 150) studs at a maximum spacing of 16 in. (406 mm) on centers when the wall supports two floors in addition to the roof.

Codes permit the use of smaller members and wider spacings under certain conditions. Often, the spacing of exterior wall studs is dependent on the type and thickness of sheathing rather than on structural load requirements. Stud spacings for sheathing support are discussed in Section 6.6.5.

In stud systems, roof and other vertically imposed loads are carried by the studs. Sheathing and applied finishes resist lateral forces but are not assumed to share in carrying vertical loads.

Wood stud assemblies lend themselves to on-site or off-site fabrication. Pre-framed panel wall assemblies are wood stud assemblies fabricated in sections away from the job site. Stud wall assemblies provide flexibility in design and erection and require only the use of simple hand tools and fastenings. Changes can be made easily during construction, particularly with on-site fabrication. Space between studs provides for insulation and heating, air conditioning, electrical, and plumbing equipment. Studs may be erected individually in the traditional manner, but most stud walls today are framed on the subfloor with sheathing and siding applied and windows and doors installed. The assembly then is tilted into place.

Wood stud walls may be framed, erected, and sheathed quickly, and the building can be enclosed quickly. Interior finishing operations and mechanical installations may then take place under cover, giving maximum flexibility to job scheduling.

Wood stud wall systems are economical and are widely used. They provide excellent structural performance, make use of relatively inexpensive basic materials, permit the application of many finishes, and are adaptable to practically any method of on- or off-site fabrication (see Section 6.6.6).

In masonry veneer construction (see Chapter 4), wood framing assumes all vertical loads and the masonry performs essentially as an exterior finish surface and provides a weather barrier for the assembly. The space between the studs is typically used for the installation of insulation and heating, air conditioning, electrical, and plumbing lines.

6.6.3 ROOF AND CEILING ASSEMBLIES

In many cases, especially in wood-framed buildings, the ceiling construction is an integral part of the roof construction. When it is, the entire system is called a *roof and ceiling assembly*. Such assemblies are a basic element of a building's superstructure.

A light wood-framed building's roof and ceiling assemblies include joist and rafter construction (discussed in this section), plank-and-beam construction (see Section 6.7.2), and panelized construction (see Section 6.6.6).

In the design, selection, and construction of a roof and ceiling assembly, it is necessary to consider (1) the required level of functional performance, (2) costs, and (3) the desired finished appearance. These considerations are often influenced by the systems and finishes used in other basic elements of the construction.

Roof and ceiling assemblies are required to:

1. Have adequate strength to support the roof and ceiling assembly (dead loads) and live loads (such as from snow, ice, and wind);
2. Control the flow of heat, water, air, and water vapor;
3. Provide satisfactory levels of sound privacy;
4. Have suitable fire resistance;
5. Be suitable for the application of finish materials;
6. Be adaptable to economical methods of assembly and erection;
7. Accommodate heating, air conditioning, electrical, and plumbing equipment.

The ultimate costs of a roof and ceiling assembly include in-place costs, maintenance costs, and operating costs. In-place costs are the sum of the costs of materials, labor, and overhead required to assemble and install the assembly. The type of assembly selected may affect other costs, such as those for installing insulation, applying roofing finishes, and for providing interior wall support; it may also affect the time needed to enclose the building. Maintenance costs and operating costs related to a roof and ceiling assembly, such as those attributable to heat loss or gain through the construction, have an important influence on the ultimate cost of a building and should not be overlooked if long-term economy is to be achieved.

The form of a roof is determined by the roof and ceiling assembly selected, but usually, the assembly is covered with applied materials that provide the desired finished appearance and performance. The finish materials selected may:

1. Affect the spacing, span, and slope of structural members;
2. Determine the need for roof sheathing or underlayment;
3. Dictate methods of construction.

A roof and ceiling assembly should be compatible with the wall assembly and the floor and ceiling assemblies used in the construction.

In light wood-framed building construction, wood joists and rafter construction (discussed in this section) and trusses (Section 6.7.5) are the most common ceiling and roof framing members. These framing members are generally spaced 16 in. (406 mm) or 24 in. (610 mm) on centers, depending on the loads to be carried, which permits direct application of most finishes, such as gypsum board and plaster. However, furring may be required when member spacing exceeds the maximum recommended for the particular finish material or when special performance, such as increased sound isolation, is desired.

Ceiling assemblies intended for the direct application of gypsum board or any other board finish should have all joists evenly spaced with bottom faces aligned in a level plane. Joists with a slight crown should be installed with the crown up. Slightly crooked or bowed joists can be straightened and leveled by nailing stringers approximately at midspan (Fig. 6.6-22) (see also Section 6.7.5). Excessively bowed or crooked joists should not be used.

6.6.3.1 Materials

Framing lumber should be of the proper grade for the intended use, graded by a recognized grading authority. The moisture content of framing lumber should be as near as possible to the equilibrium moisture content for the particular area and should not exceed 15% at the time the finish material is installed.

6.6.3.2 Joist-and-Rafter Roofs

Joist-and-rafter assemblies consist of nominal 2-in. (50 mm) lumber ceiling joists and roof rafters, usually spaced 16 in. (406 mm) or 24 in. (610 mm) on centers. Typical construction for joist-and-rafter roofs is shown in Figure 6.6-23. Joists span roughly one-half the width of the building from the exterior walls to a center load-bearing wall or beam. The rafters are supported on exterior walls and bear against each other or a ridge board at the high point of the roof. When loaded, the ridge tends to lower and the rafters exert a thrusting force that tries to spread the outside walls. This thrust is resisted by the ceiling joists acting as ties from wall to wall.

Roof and ceiling framing should be of the proper grade for the intended use, graded by a recognized grading authority. Its moisture content should not be more than 15% at the time finishes are applied.

Joists and rafters for simple frame buildings and ordinary construction can be selected by referring to *Span Tables for Joists and Rafters*, published by AF&PA. These tables indicate the minimum modulus of elasticity, E, and extreme fiber strength in bending, F_b, required for various spans, lumber sizes, and conditions of loading. Suitable species and grades meeting the indicated strength requirements can be selected from the tables of Allowable Unit Stresses in the *NDS Supplement* (see Fig. 6.1-14).

A table of maximum allowable spans for ceiling joists carrying a drywall ceiling and a 20-lb. attic live load is included as Figure 6.6-24. *Span Tables for Joists and Rafters* also contains a table for ceiling joists with a ceiling but not attic storage. A table for rafters carrying a drywall ceiling, a 40 psf (195.3 kg/m^2) live load, and a 10 psf (48.8 kg/m^2) dead load is included as Figure 6.6-25. *Span Tables for Joists and Rafters* also includes tables for rafters carrying live loads of 20 psf (97.65 kg/m^2), 30 psf (146.47 kg/m^2), and 50 psf (244.12 kg/m^2) and dead loads of 15 psf (73.24 kg/m^2) and 20 psf (97.65 kg/m^2) for rafters carrying a drywall ceiling, and for rafters carrying no ceiling.

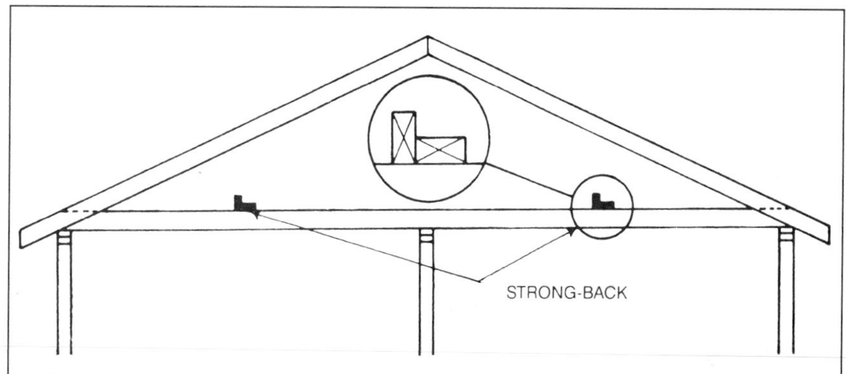

FIGURE 6.6-22 Joist spacing can be regulated with 2 × 4 (50 × 100) stringers at midpoints and can be leveled with 2 × 6 (50 × 150) stringers set on edge. (Courtesy Gypsum Association.)

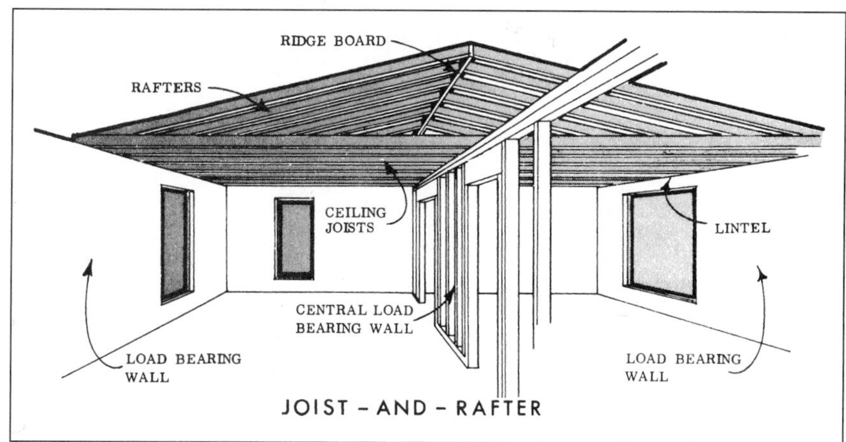

FIGURE 6.6-23 Wood joist-and-rafter roof and ceiling assembly.

FIGURE 6.6-24 Maximum Allowable Spans for Ceiling Joists with L/240 Deflection Limits[a,b]

Joist Size, in.	Spacing, in.	Modulus of Elasticity, E, in 1,000,000 psi																
		0.8	0.9	1.0	1.1	1.2	1.3	1.4	1.5	1.6	1.7	1.8	1.9	2.0	2.1	2.2	2.3	2.4
2 × 4	12.0	7-10	8-1	8-5	8-8	8-11	9-2	9-5	9-8	9-10	10-0	10-3	10-5	10-7	10-9	10-11	11-1	11-3
	16.0	7-1	7-5	7-8	7-11	8-1	8-4	8-7	8-9	8-11	9-1	9-4	9-6	9-8	9-9	9-11	10-1	10-3
	19.2	6-8	6-11	7-2	7-5	7-8	7-10	8-1	8-3	8-5	8-7	8-9	8-11	9-1	9-3	9-4	9-6	9-8
	24.0	6-2	6-5	6-8	6-11	7-1	7-3	7-6	7-8	7-10	8-0	8-1	8-3	8-5	8-7	8-8	8-10	8-11
2 × 6	12.0	12-3	12-9	13-3	13-8	14-1	14-5	14-9	15-2	15-6	15-9	16-1	16-4	16-8	16-11	17-2	17-5	17-8
	16.0	11-2	11-7	12-0	12-5	12-9	13-1	13-5	13-9	14-1	14-4	14-7	14-11	15-2	15-5	15-7	15-10	16-1
	19.2	10-6	10-11	11-4	11-8	12-0	12-4	12-8	12-11	13-3	13-6	13-9	14-0	14-3	14-6	14-8	14-11	15-2
	24.0	9-9	10-2	10-6	10-10	11-2	11-5	11-9	12-0	12-3	12-6	12-9	13-0	13-3	13-5	13-8	13-10	14-1
2 × 8	12.0	16-2	16-10	17-5	18-0	18-6	19-0	19-6	19-11	20-5	20-10	21-2	21-7	21-11	22-4	22-8	23-0	23-4
	16.0	14-8	15-3	15-10	16-4	16-10	17-3	17-9	18-1	18-6	18-11	19-3	19-7	19-11	20-3	20-7	20-11	21-2
	19.2	13-10	14-5	14-11	15-5	15-10	16-3	16-8	17-1	17-5	17-9	18-1	18-5	18-9	19-1	19-5	19-8	19-11
	24.0	12-10	13-4	13-10	14-3	14-8	15-1	15-6	15-10	16-2	16-6	16-10	17-2	17-5	17-9	18-0	18-3	18-6
2 × 10	12.0	20-8	21-6	22-3	22-11	23-8	24-3	24-10	25-5	26-0								
	16.0	18-9	19-6	20-2	20-10	21-6	22-1	22-7	23-1	23-8	24-1	24-7	25-0	25-5	25-10			
	19.2	17-8	18-4	19-0	19-7	20-2	20-9	21-3	21-9	22-3	22-8	23-1	23-7	23-11	24-4	24-9	25-1	25-5
	24.0	16-5	17-0	17-8	18-3	18-9	19-3	19-9	20-2	20-8	21-1	21-6	21-10	22-3	22-7	22-11	23-4	23-8
F_b	12.0	896	969	1040	1108	1174	1239	1302	1363	1423	1481	1539	1595	1651	1706	1759	1812	1864
F_b	16.0	986	1067	1145	1220	1293	1364	1433	1500	1566	1631	1694	1756	1817	1877	1936	1995	2052
F_b	19.2	1048	1134	1216	1296	1374	1449	1522	1594	1664	1733	1800	1866	1931	1995	2058	2120	2181
F_b	24.0	1129	1221	1310	1396	1480	1561	1640	1717	1793	1866	1939	2010	2080	2149	2217	2283	2349

[a]The required bending design value, F_b, in pounds per square inch (psi) is shown at the bottom of the table and is applicable to all lumber sizes shown. Spans are shown in feet-inches and are limited to 26 ft. and less. Check sources of supply for availability of lumber in lengths greater than 20 ft.

[b]Design Criteria: Deflections—for 20 pounds per square foot (psf) live load. Limited to span in inches divided by 240. Strength—live load of 40 psf plus dead load of 10 psf determines the required bending design value.

(Courtesy American Forest & Paper Association.)

FIGURE 6.6-25 Maximum Allowable Spans for Rafters with L/240 Deflection Limits

Rafter Size, in.	Spacing, in.	Bending Design Value, F_b, psi																					
		300	400	500	600	700	800	900	1000	1100	1200	1300	1400	1500	1600	1700	1800	1900	2000	2100	2200	2300	2400
2 × 6	12.0	5-6	6-4	7-1	7-9	8-5	9-0	9-6	10-0	10-6	11-0	11-5	11-11	12-4	12-8	13-1	13-6	13-10	14-2				
	16.0	4-9	5-6	6-2	6-9	7-3	7-9	8-3	8-8	9-1	9-6	9-11	10-3	10-8	11-0	11-4	11-8	12-0	12-4	12-7	12-11		
	19.2	4-4	5-0	5-7	6-2	6-8	7-1	7-6	7-11	8-4	8-8	9-1	9-5	9-9	10-0	10-4	10-8	10-11	11-3	11-6	11-9	12-0	12-4
	24.0	3-11	4-6	5-0	5-6	5-11	6-4	6-9	7-1	7-5	7-9	8-1	8-5	8-8	9-0	9-3	9-6	9-9	10-0	10-3	10-6	10-9	11-0
2 × 8	12.0	7-3	8-4	9-4	10-3	11-1	11-10	12-7	13-3	13-11	14-6	15-1	15-8	16-3	16-9	17-3	17-9	18-3	18-9				
	16.0	6-3	7-3	8-1	8-11	9-7	10-3	10-10	11-6	12-0	12-7	13-1	13-7	14-0	14-6	14-11	15-5	15-10	16-3	16-7	17-0		
	19.2	5-9	6-7	7-5	8-1	8-9	9-4	9-11	10-6	11-0	11-6	11-11	12-5	12-10	13-3	13-8	14-0	14-5	14-10	15-2	15-6	15-10	16-3
	24.0	5-2	5-11	6-7	7-3	7-10	8-4	8-11	9-4	9-10	10-3	10-8	11-1	11-6	11-10	12-2	12-7	12-11	13-3	13-7	13-11	14-2	14-6
2 × 10	12.0	9-3	10-8	11-11	13-1	14-2	15-1	16-0	16-11	17-9	18-6	19-3	20-0	20-8	21-4	22-0	22-8	23-3	23-11				
	16.0	8-0	9-3	10-4	11-4	12-3	13-1	13-10	14-8	15-4	16-0	16-8	17-4	17-11	18-6	19-1	19-7	20-2	20-8	21-2	21-8		
	19.2	7-4	8-5	9-5	10-4	11-2	11-11	12-8	13-4	14-0	14-8	15-3	15-10	16-4	16-11	17-5	17-11	18-5	18-11	19-4	19-10	20-3	20-8
	24.0	6-6	7-7	8-5	9-3	10-0	10-8	11-4	11-11	12-6	13-1	13-7	14-2	14-8	15-1	15-7	16-0	16-6	16-11	17-4	17-9	18-1	18-6
2 × 12	12.0	11-3	13-0	14-6	15-11	17-2	18-4	19-6	20-6	21-7	22-6	23-5	24-4	25-2	26-0								
	16.0	9-9	11-3	12-7	13-9	14-11	15-11	16-10	17-9	18-8	19-6	20-3	21-1	21-9	22-6	23-2	23-10	24-6	25-2	25-9			
	19.2	8-11	10-3	11-6	12-7	13-7	14-6	15-5	16-3	17-0	17-9	18-6	19-3	19-11	20-6	21-2	21-9	22-5	23-0	23-6	24-1	24-8	25-2
	24.0	7-11	9-2	10-3	11-3	12-2	13-0	13-9	14-6	15-3	15-11	16-7	17-2	17-9	18-4	18-11	19-6	20-0	20-6	21-1	21-7	22-0	22-6
E	12.0	0.14	0.22	0.31	0.41	0.51	0.63	0.75	0.88	1.01	1.15	1.30	1.45	1.61	1.77	1.94	2.12	2.30	2.48				
E	16.0	0.12	0.19	0.27	0.35	0.44	0.54	0.65	0.76	0.88	1.00	1.12	1.26	1.39	1.54	1.68	1.83	1.99	2.15	2.31	2.48		
E	19.2	0.11	0.18	0.24	0.32	0.41	0.50	0.59	0.69	0.80	0.91	1.03	1.15	1.27	1.40	1.54	1.67	1.81	1.96	2.11	2.26	2.42	2.58
E	24.0	0.10	0.16	0.22	0.29	0.36	0.44	0.53	0.62	0.71	0.81	0.92	1.03	1.14	1.25	1.37	1.50	1.62	1.75	1.89	2.02	2.16	2.30

The required elasticity, *E*, in 1,000,000 pounds per square inch (psi) is shown at the bottom of the table, is limited to 2.6 million psi and less, and is applicable to all lumber sizes shown. Spans are shown in feet-inches and are limited to 26 ft. and less. Check sources of supply for availability of lumber in lengths greater than 20 ft.

Design Criteria: Strength—live load of 40 pounds per square foot (psf) plus dead load of 10 psf determines the required bending design value. Deflection—For 40 psf live load. Limited to span in inches divided by 240.

(Courtesy American Forest & Paper Association.)

In each case, the allowable span is the clear distance between supports. For rafters, the span is measured along the horizontal projection (Fig. 6.6-26). Tabulated spans are applicable to green or dry lumber, provided it is used in covered structures where it will normally achieve a moisture content of 19% or less.

As an example of rafter selection, assume a horizontal projection span of 15 ft. (4.57 m) in a rafter carrying a drywall ceiling, a live load of 40 psf (195.3 kg/m^2), a 10 psf (48.8 kg/m^2) dead load, with rafters spaced 16 in. (406 mm) on centers. Figure 6.6-25 shows that a 2 × 10 (50 × 250) having an F_b value of 1100 psi (773.4 kg/m^2) and an E value of 880,000 psi (618,702 kg/m^2) would have an allowable horizontal projection of 15 ft. 4 in. (4673.5 mm). Suitable species and grades can be selected from the tables in the *NDS Supplement*. For our example, Figure 6.1-14 shows that a visually graded No. 2 Douglas Fir-Larch has an F_b of 900 psi (632.76 kg/m^2) and an E of 1,600,000 psi (1,124,912 Mg/m^2). The F_b can be increased by a factor of 1.15 because of the repetitive member factor (C_r), producing an allowable F_b of 1035 psi. (727.68 Mg/m^2). A further increase permitted by the size factor is 1.2. Therefore, the F_b can be increased to 1242 psi (873.21 Mg/m^2). Figure 6.1-15 shows that machine-stress-rated lumber, 1250f-1.4E, of WCLIB species has an F_b of 1250 psi (878.84 Mg/m^2) and an E of 1,400,000 psi (984,298 Mg/m^2).

Joist-and-rafter construction generally requires longer erection time than truss construction and may expose materials longer to possible weather hazards.

Interior load-bearing walls somewhat restrict the flexibility of room planning. However, for relatively short spans of 16 ft. (4880 mm) or less, load-bearing interior walls or beams generally are not required. If a building's plan includes a number of wings of different spans, joist-and-rafter framing may be more economical than truss construction.

Joist-and-rafter assemblies provide overhead space for the installation of insulation and mechanical equipment; in addition, with adequate rafter slopes, economical attic storage or expansion space can result.

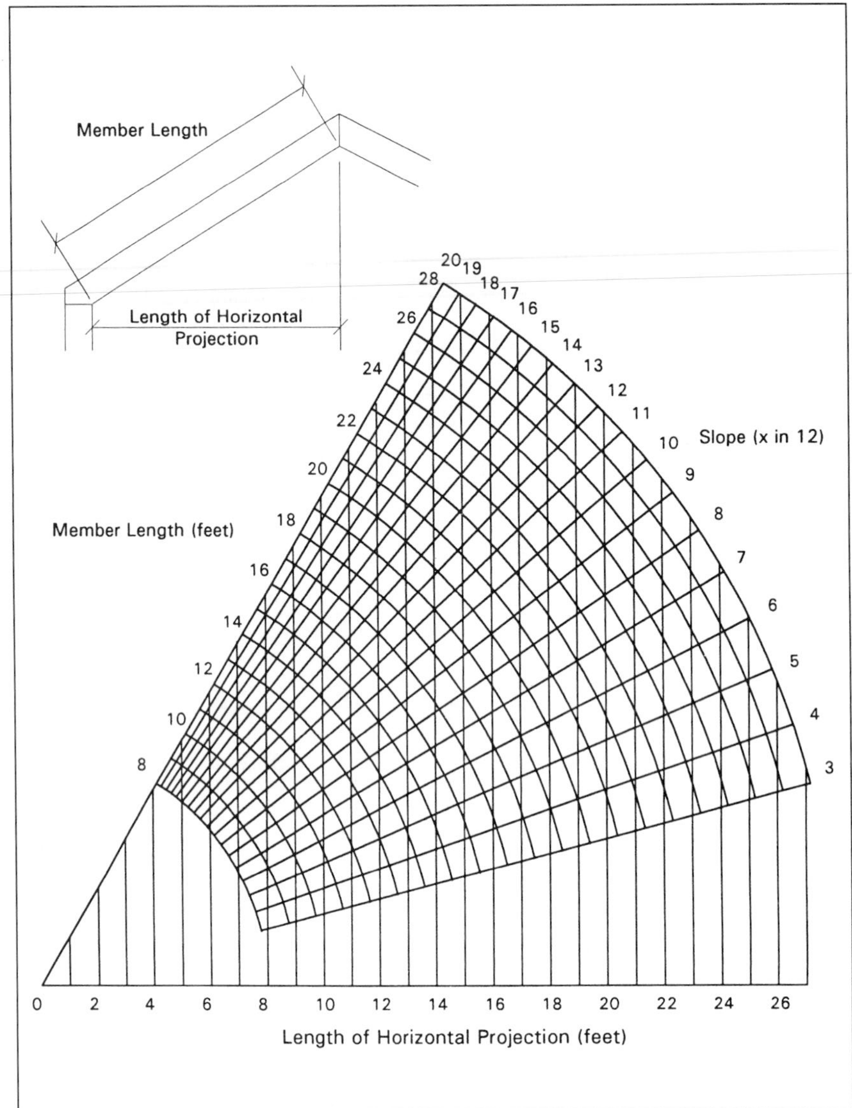

FIGURE 6.6-26 **Conversion diagram for rafters.** (Courtesy American Forest & Paper Association.)

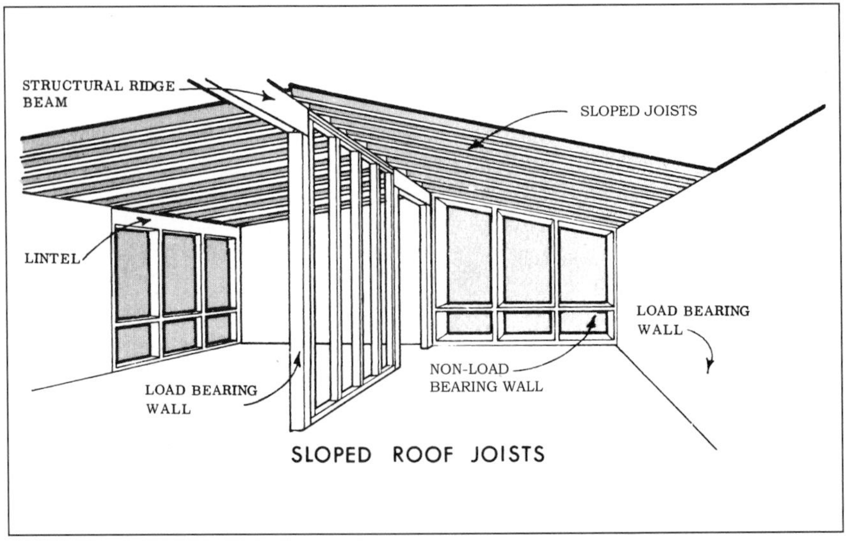

FIGURE 6.6-27 Wood joist roof and ceiling assembly.

6.6.3.3 Joist Roofs

In joist roof and ceiling assemblies, roof joists alone provide the function of both ceiling joists and roof rafters (Fig. 6.6-27).

These joist members are generally of nominal 2-in. (50 mm) dimension lumber spaced 12 in. (305 mm), 16 in. (406 mm), or 24 in. (610 mm) on centers, depending on the spans and the loads to be carried. Joists may have a clear span between exterior walls, or they may be centrally supported on load-bearing partitions or beams. When the roof is sloped rather than flat, the construction may also be called a *ridge-supported rafter assembly*, which often is used to achieve an interior sloped ceiling effect. Roof sheathing materials may be attached directly to the top of the joists, and ceiling finishes applied directly to the underside. Joists and rafters may be selected from the tables in *Span Tables for Joists and Rafters* by AF&PA.

Ductwork can be accommodated in the joist space if it is run parallel to the joists. When it is necessary to run ductwork perpendicular to the joists, it is common to use dropped-ceiling areas over hallways. Wiring can be concealed in the joist space, and the system is relatively easy to insulate. Interior and exterior finishes are applied in the same manner as for truss and joist-and-rafter roofs.

6.6.4 SUBFLOORING AND PLYWOOD UNDERLAYMENT

Subflooring is required beneath most kinds of finish flooring. An additional underlayment is also needed for some types of flooring.

6.6.4.1 Subflooring

Subflooring serves as a working platform and as a base for floor finishes. The two main types of subfloor used as a base for flooring in light wood-frame construction are wood and concrete. Concrete floors are discussed in Chapter 3. Additional requirements for subfloors are discussed in the sections where the various floor finishes are covered.

Tile, terrazzo, resilient flooring, and carpet may be installed over a double-layer base composed of a separate subfloor and an underlayment, or may be a single-layer construction consisting of a combination subfloor/underlayment material specifically designed for the purpose. Wood flooring is usually installed over a single-layer subfloor.

Whether it is covered with underlayment or not, wood subflooring should be firmly fastened and structurally adequate to carry the intended loads without excessive deflection. The maximum deflection of such subflooring under full dead and live loads after the finished flooring has been installed should not exceed $\frac{1}{360}$ of any span.

Figure 6.6-28 shows wood joist floor construction suitable for the application of resilient flooring, carpet, and ceramic tile. The only acceptable wood material base for terrazzo is plywood. Wood subflooring for wood floors may be boards or plywood.

Suspended wood floors above grade are usually sufficiently removed from excessive external moisture sources, but floors over crawl spaces at or near grade may require the crawl space to be ventilated and the ground inside the crawl space to be covered with a suitable vapor retarder (see Section 2.4).

BOARD SUBFLOORING

Board subflooring is permissible without underlayment beneath structural flooring such as wood strip flooring, beneath tile set in a mortar bed, and beneath terrazzo with an underbed. However, underlayment is recommended over board subflooring beneath thin-set tile, thin-set terrazzo, resilient flooring, and carpet.

Conventional board subflooring consists of nominal 1-in. (25 mm)-thick boards 4 in. (100 mm) to 8 in. (200 mm) wide. These boards may be applied either diagonally or at right angles to the joists. When boards are placed at right angles to the joists, wood strip or plank flooring must be laid at right angles to the boards. Diagonal board installation is preferred with wood finish flooring because it permits the finish flooring to be laid either parallel or at right angles to the joists. Boards may have square, tongue-and-groove (T&G), or shiplap edges. Except for tongue-and-grooved boards, board ends should rest over the joists. Subfloor boards less than 8 in. (200-mm) wide should be nailed to each joist with two 8d nails; those 8 in. (200 mm) wide or wider with three 8d nails. Where desired, 7d annularly threaded nails can be substituted for the 8d smooth shank nails.

The moisture content of lumber boards used for subflooring at the time of installation should be not more than 15%.

PLYWOOD SUBFLOORING

Plywood is a suitable subflooring material for tile, terrazzo, wood flooring, resilient flooring, and carpet. Additional details are covered in sections where the various types of flooring are addressed. A single-layer subfloor without underlayment may be used beneath structural flooring such as wood strip flooring, beneath tile with a thick-set mortar bed, beneath terrazzo with an underbed, and

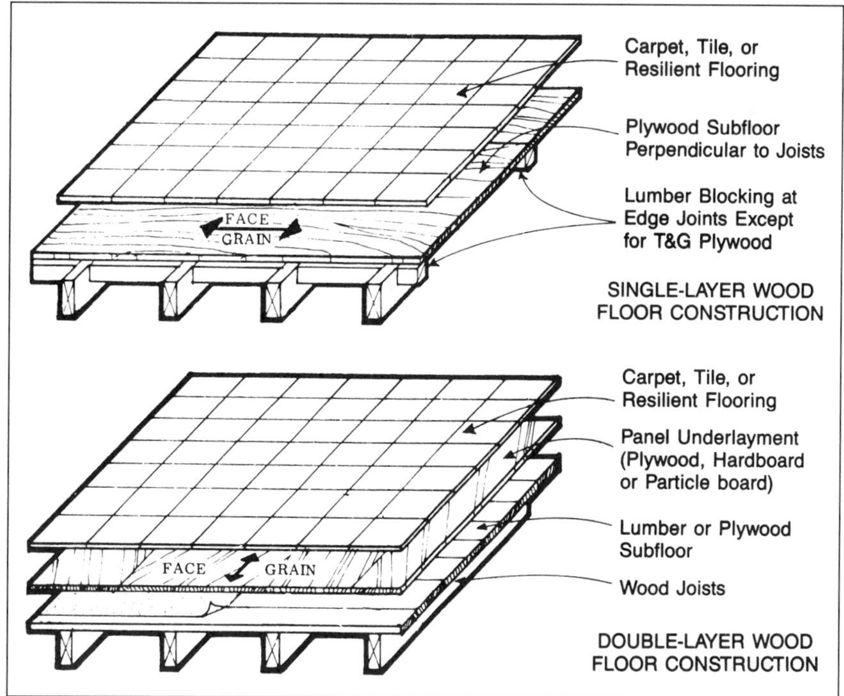

FIGURE 6.6-28 Typical subflooring construction for the application of carpet, tile, or resilient flooring.

beneath carpet with a pad. Plywood subflooring requires an underlayment for receipt of thin-set tile, thin-set terrazzo, and resilient flooring.

A conventional plywood subfloor should be APA-rated sheathing either Exposure 1 or 2; Exterior; Structural I, Exposure 1; or Structural I, Exterior, depending on its exposure to moisture. The more moisture-resistant grades should be selected when a hazard of prolonged exposure to moisture exists, such as in bathrooms, in floors adjacent to plumbing fixtures, or when the subfloor may be exposed to the weather for long periods during construction. The thickness varies from ½ in. (12.7 mm) to ¾ in. (119.1 mm), depending on the span rating, type of finish floor, and spacing of the supporting joists. Plywood subfloor panels to be covered with underlayment may be unsanded.

Plywood subflooring should be installed with the long dimension perpendicular to the joists, with end joints centered over the joists (see Fig. 6.6-28). Panels should be staggered so that the ends of adjacent panels do not fall on the same support. The longitudinal edges should be supported on 2 × 3 (50 × 75) or heavier wood blocking. The blocking may be eliminated, however, when tongue-and-groove plywood is used (Fig. 6.6-29). Supports should be spaced and panels should be nailed as shown in Figure 6.6-30.

Subfloor panels should be spaced approximately ⅛ in. (3.2 mm) apart at end and edge joints, and placed about ⅛ in. (3.2 mm) from adjacent vertical surfaces, such as walls and pipes, to allow for expansion caused by changes in humidity. Nailing should start in the middle of the panel and proceed outward to the edges, with nail heads set flush or slightly countersunk. Perimeter nails should be located not more than ⅜ in. (9.5 mm) from the edges. Panel joints should be sanded smooth and flush before finish flooring is installed. It is important to protect single-layer floors from damage during construction. However, the difficulty of furnishing adequate protection during construction may warrant the use of unsanded or touch-sanded plywood and then power sanding the entire floor area just before the flooring is installed.

FIGURE 6.6-29 Tongue-and-groove-edged APA-rated Sturd-I-Floor with the proper span rating can be installed on spacings of up to 48 in. (1220 mm), and tongue-and-groove-edged APA-rated sheathing with the proper span rating can be installed on spans of up to 24 in. (610 mm), both without edge blocking, but should be protected during construction. (Courtesy American Plywood Association.)

6.6.4.2 Plywood Underlayment

An underlayment is required over both wood board and plywood subflooring beneath thin-set tile, thin-set terrazzo, and resilient flooring. Using an underlayment is often desirable even when not ab-

FIGURE 6.6-30 APA Panel Subflooring (APA-Rated Sheathing)

Panel Span Rating	Minimum Panel Thickness, in.	Maximum Span, in.	Nail Size[a] and Type	Maximum Nail Spacing, in. — Supported Panel Edges	Maximum Nail Spacing, in. — Intermediate Supports
24/16	7/16	16	6d common	6	12
32/16	15/32	16[b]	8d common[c]	6	12
40/20	19/32	20[b,d]	8d common	6	12
48/24	23/32	24	8d common	6	12
60/32	7/8	32	3d common	6	12

[a]Other code-approved fasteners may be used.

[b]Span may be 24 in. if ¾-in. wood strip flooring is installed at right angles to joists.

[c]6d common nail permitted if panel is ½ in. or thinner.

[d]Span may be 24 in. if a minimum 1½ in. of lightweight concrete is applied over panels.

(Courtesy American Plywood Association.) (Refer to Figure 1.8-27 for metric conversions.)

solutely required for the type of finish flooring, because the subfloor, which provides a convenient working platform during most of the construction period, can be easily damaged, and installation of an underlayment can be scheduled more closely to the installation of the flooring.

Underlayment beneath resilient flooring is often either Underlayment or C-C Plugged panel grades marked *sanded face*. Other grades may also be used, however, when acceptable to the flooring manufacturer. Thin underlayment for other purposes may be APA Underlayment Interior, Exposure 1; C-C Plugged Exterior; or APA C-C Plugged Exterior. Underlayment thicker than 19⁄32 in. (15.08 mm) may be APA-rated Sturd-I-Floor Exposure 1 or 2, or APA-rated Sturd-I-Floor, Exterior. The more moisture-resistant grades should be used in conditions of greater moisture, such as encountered in bathrooms and laundry areas, and beneath tile and terrazzo.

Underlayment panels should be installed with the finished face up, and when over plywood subflooring, with the long edge across the supports. When laid over lumber board subflooring, the face grain of plywood underlayment should be laid perpendicular to the boards (see Fig. 6.6-28). The ends and edges should be staggered and should not fall within 2 in. (50 mm) of the joints in the subflooring. However, it is not necessary to stagger the joints in underlayment beneath carpet with a pad. In general, joints between panels should be left open 1⁄32 in. (0.79 mm) to permit some movement. However, wider spacings between panels may be required by some flooring manufacturers for certain applications, and their recommendations should be followed. Joints between underlayment and other material, such as adjacent vertical surfaces and pipes, should be left open 1⁄8 in. (3.2 mm) beneath resilient flooring and carpet and 1⁄4 in. (6.4 mm) beneath tile and terrazzo, unless the manufacturers of these products recommend different spacings. Surfaces of adjacent panels should be level within 1⁄32 in. (0.79 mm).

Underlayment should be of sufficient strength to withstand loads to be encountered. In general, panels beneath tile should not be less than 5⁄8 in. (15.9 mm), except that 1⁄2 in. (12.7 mm) is sometimes acceptable in small residential applications; beneath terrazzo not less than 3⁄8 in. (9.5 mm).

Panels 1⁄2 in. (12.7 mm) thick or thinner should be nailed with 3d ring-shank nails at 6 in. (150 mm) on centers along the edges and at 8 in. (200 mm) on centers each way throughout the rest of each panel. Thicker panels should be nailed using 4d nails at 6 in. (150 mm) on centers along the edges and at 12 in. (305 mm) on centers throughout the rest of each panel. Fastener length should be the same as the total thickness of the subflooring and the underlayment.

6.6.4.3 Combination Subfloor and Underlayment

APA-rated Sturd-I-Floor is a proprietary *combination subfloor and underlayment* plywood panel. It is designed specifically for use as a single-layer subfloor beneath carpet with a pad, but is also usable under other flooring types and may be used even when a separate underlayment is required. It is available in thicknesses from 19⁄32 in. (15.08 mm) to 1 1⁄8 in. (28.58 mm) and with span ratings from 16 to 48. It is available in exposure durability classifications Exterior, Exposure 1, and Exposure 2.

The support system for APA-rated Sturd-I-Floor may be a conventional floor framing system with the framing members (joists or floor trusses) at 16 in. (406 mm), 20 in. (508 mm), or 24 in. (610 mm) on centers (Fig. 6.6-31); a joist system with the joists spaced 32 in. (760 mm) on centers (Fig. 6.6-32); or a girder system with the girders at 48 in. (1220 mm) on centers (Fig. 6.6-33).

Sturd-I-Floor panels are manufactured either as conventional plywood, as a composite, or as nonveneer panels. Panels are available both square edged and tongued and grooved. The face ply can be either sanded or touch sanded; the latter is used when field sanding is intended at the end of construction, just before the floor finish is installed.

The installation of APA-rated Sturd-I-Floor panels is similar to that for other plywood subfloors: face plies perpendicular to supports; end joints staggered; panels continuous over at least two supports; panel ends resting on the supports; end and edge joints spaced 1⁄8 in. (3.2 mm);

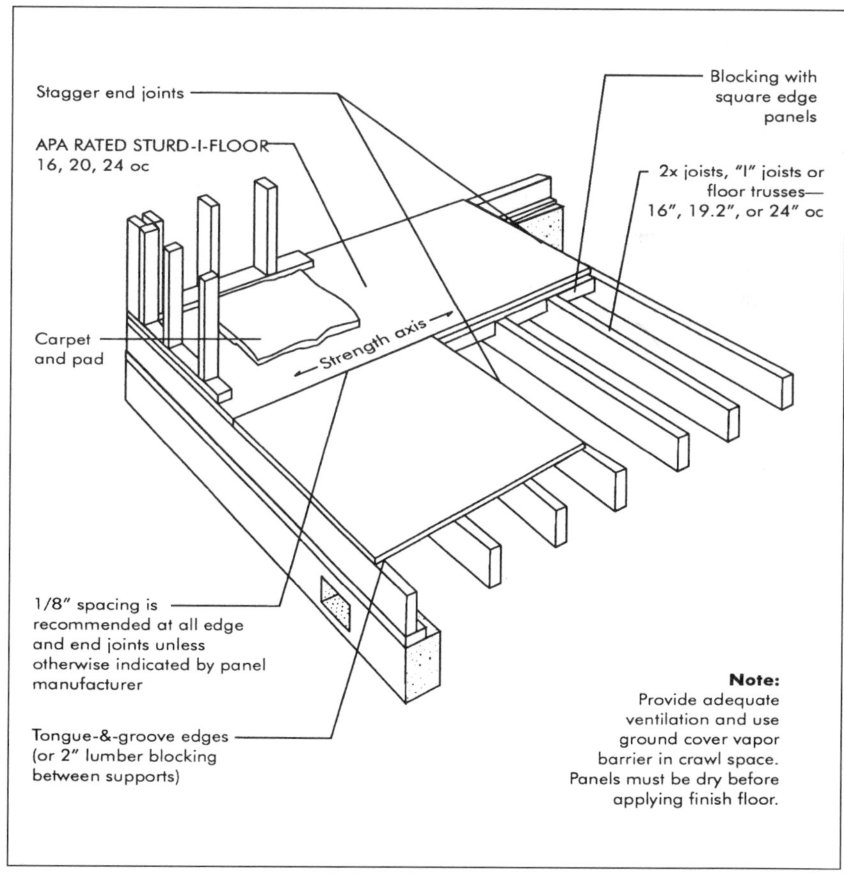

FIGURE 6.6-31 APA-rated Sturd-I-Floor, 16 in. (406 mm), 20 in. (560 mm), and 24 in. (610 mm) oc. (Courtesy American Plywood Association.) (Refer to Figure 1.8-27 for metric conversions.)

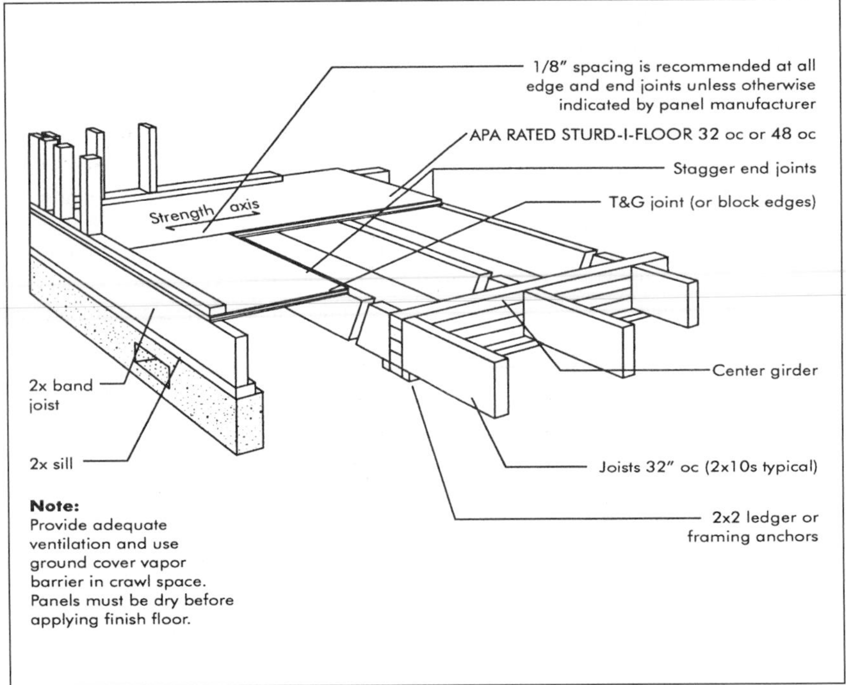

FIGURE 6.6-32 APA-rated Sturd-I-Floor, 32 in. (813 mm) and 48 in. (1220 mm) oc (over supports 32 in. [813 mm] on center). (Courtesy American Plywood Association.) (Refer to Figure 1.8-27 for metric conversions.)

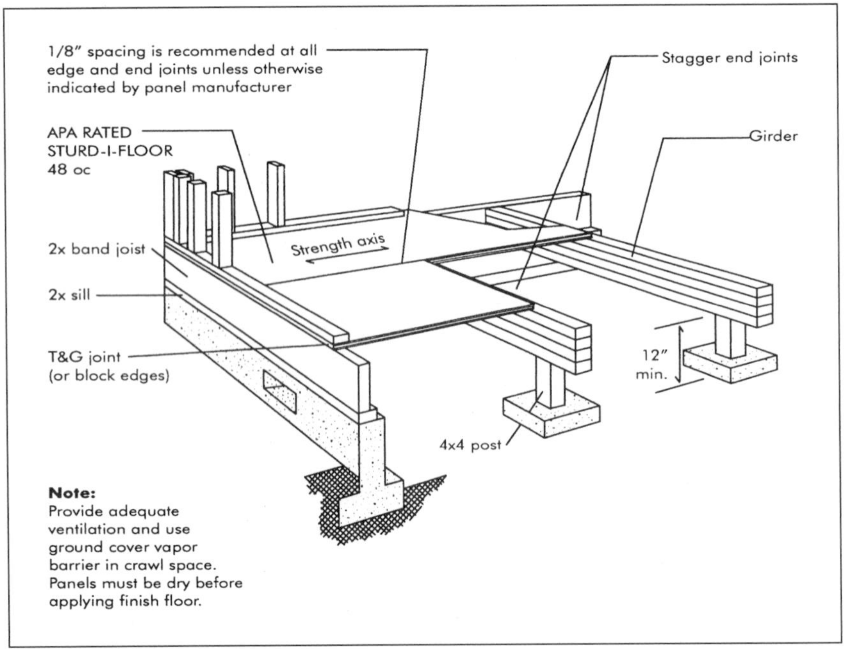

FIGURE 6.6-33 APA-rated Sturd-I-Floor, 48 in. (1220 mm) oc (over supports 48 in. [1220 mm] on center). (Courtesy American Plywood Association.) (Refer to Figure 1.8-27 for metric conversions.)

solid 2-in. (50 mm) nominal blocking beneath unsupported panel edges of square-edged panels (blocking is not required at tongue-and-groove joints).

A glued installation is recommended for Sturd-I-Floor (see Section 6.6.4.3), but this material can also be installed with nailing only. Unless the code requires a different spacing, the spacings should be in accordance with Figure 6.6-34.

6.6.4.4 Glued Floor System

In the APA glued floor system the flooring and the joists are glued together with such strength and stiffness that they act as a series of integral T-beams. In addition, the gluing virtually eliminates the squeaks which can result from shrinking lumber (Fig. 6.6-35). Under some circumstances using the APA glued floor systems will permit the use of smaller joists or longer spans for the same size joists than will a conventional nailed system. The glued system also has better creep resistance. The APA glued subfloor system can be installed quickly using readily available materials and techniques.

The subflooring in the APA glued floor system should be either APA-rated sheathing with a plywood overlay or structural flooring, such as a wood strip floor, or a one-piece combination subfloor and underlayment of APA-rated Sturd-I-Floor. The subflooring will vary in thickness from ½ in. (12.7 mm) to 1⅛ in. (28.58 mm), depending on the joist spacing and the finish flooring material to be supported. Tongue-and-groove material is recommended. If square-edged material is used, it is necessary to provide solid blocking under the panel edges between joists. The plywood is glued to this blocking. The basic joist framing is the same as for a conventionally nailed subfloor.

When the finish floor is a resilient non-textile material, a separate underlayment is required even over Sturd-I-Floor. This underlayment may be either veneer-faced Sturd-I-Floor with a sanded face or another underlayment (see Section 6.6.4.3).

The plywood subflooring panels in a glued system should be placed with the long dimension perpendicular to the joists, with end joints staggered and spaced ⅛ in. (3.2 mm) at both ends and edges. Before each panel is set in place, a bead of glue is applied to the supports with a caulking gun; for extra stiffness, the edge groove should also be glued. The panel is then secured with ring- or screw-shank nails. These nails should be 6d for ¾ in. (19.1 mm) and thinner panels and 8d for thicker panels. Nail spacing varies from 6 in. (150 mm) to 12 in. (305 mm) along panel ends and edges and in the field of the panels at each support, depending on the joist spacing and the panel thickness (see Fig. 6.6-34).

More detailed information on the glued floor system, including joist span tables, nailing sizes and spacings, application sequence, and recommended adhesives, is

FIGURE 6.6-34 APA-Rated Sturd-I-Floor[a]

Span Rating (Maximum Joist Spacing), in.	Minimum Panel Thickness,[b] in.	Fastening: Glue-Nailed[c]			Fastening: Nailed Only		
			Maximum Spacing, in.			Maximum Spacing, in.	
		Nail Size and Type	Supported Panel Edges	Intermediate Supports	Nail Size and Type	Supported Panel Edges	Intermediate Supports
16	19⁄32	6d ring- or screw-shank[d]	12	12	6d ring- or screw-shank	6	12
20	19⁄32	6d ring- or screw-shank[d]	12	12	6d ring- or screw-shank	6	12
24	23⁄32	6d ring- or screw-shank[d]	12	12	6d ring- or screw-shank	6	12
	7⁄8	8d ring- or screw-shank[d]	6	12	8d ring- or screw-shank	6	12
32	7⁄8	8d ring- or screw-shank[d]	6	12	8d ring- or screw-shank	6	12
48	1 3⁄32	8d ring- or screw-shank[e]	6	[f]	8d ring- or screw-shank[e]	6	[f]

[a]Special conditions may impose heavy traffic and concentrated loads that require construction in excess of the minimums shown.

[b]Panels in a given thickness may be manufactured in more than one Span Rating. Panels with a Span Rating greater than the actual joist spacing may be substituted for panels of the same thickness with a Span Rating matching the actual joist spacing. For example, 19⁄32-in.-thick Sturd-I-Floor 20 on centers (oc) may be substituted for 19⁄32-in.-thick Sturd-I-Floor 16 oc over joists 16 in. oc.

[c]Use only adhesives conforming to APA Specification AFG-01, applied in accordance with the manufacturer's recommendations. If OSB panels with sealed surfaces and edges are to be used, use only solvent-based glues; check with panel manufacturer.

[d]8d common nails may be substituted if ring- or screw-shank nails are not available.

[e]10d common nails may be substituted with 1 1⁄8-in. panels if supports are well seasoned.

[f]Space nails maximum 6 in. for 48-in. spans and 12 in. for 32-in. spans.

(Courtesy American Plywood Association.) (Refer to Figure 1.8-27 for metric conversions.)

contained in the booklet *APA Residential and Commercial Design/Construction Guide*.

6.6.5 WALL AND ROOF SHEATHING

Most exterior wall finishes and roofing materials applied over wood-framed construction require sheathing, and some require sheathing paper. Wall sheathing may be lumber, plywood, oriented strand board, or a nonwood material such as fiberboard or gypsum board. Roof sheathing is usually lumber, plywood, or oriented strand board.

Wood shakes and shingles can be attached to wood nailing strips fastened to the wall studs through a sheathing material. These nailing strips are discussed in the sections where the various finish materials are covered.

6.6.5.1 Wall Sheathing and Air Infiltration Barriers

Wall sheathing is applied to framing members for one or more of the following reasons:

1. When properly fastened, it is believed to help a structural frame resist racking (being forced out of shape or plumb). However, tests by the Division of Building Research, National Research Council of Canada, indicate that interior wall finishes, like plaster and gypsum board, provide about four times as much resistance to racking as lumber sheathing applied horizontally with corner bracing. Therefore, the National Building Code of Canada does not require sheathing except when the exterior finish materials require it as a nailing base.
2. It contributes insulation, which reduces heat transmission and resists infiltration of wind and moisture.
3. It functions in many cases as a nailing base for exterior finish materials.

When wall sheathing is used as a nailing base, it should be smooth and securely attached to its supports. Shingles may be applied over nailable solid sheathing consisting of wood boards, plywood, or nailable fiberboard. They may also be applied to wood nailing strips over almost any type of sheathing. Wood shakes and shingles can be applied over spaced sheathing. Plywood siding can be installed either directly over the framing or over plywood sheathing. The types of sheathing required and additional details concerning it are discussed in the sections covering the various exterior wall-finishing materials.

LUMBER WALL SHEATHING

Lumber sheathing may be solid, spaced, or nailing strips, depending on the type of finish siding to be applied.

Lumber sheathing boards beneath wall shingles should be seasoned to a moisture content between 12% and 18%. They should be free from loose knots and large knotholes. Warped boards that would prevent the siding from lying flat should not be used.

Sheathing boards up to 8 in. (200 mm) wide are face nailed at each framing member with two 7d annularly threaded nails or two 8d common wire nails. Boards 10 in. (250 mm) and 12 in. (305 mm) wide require three nails at each support. Sup-

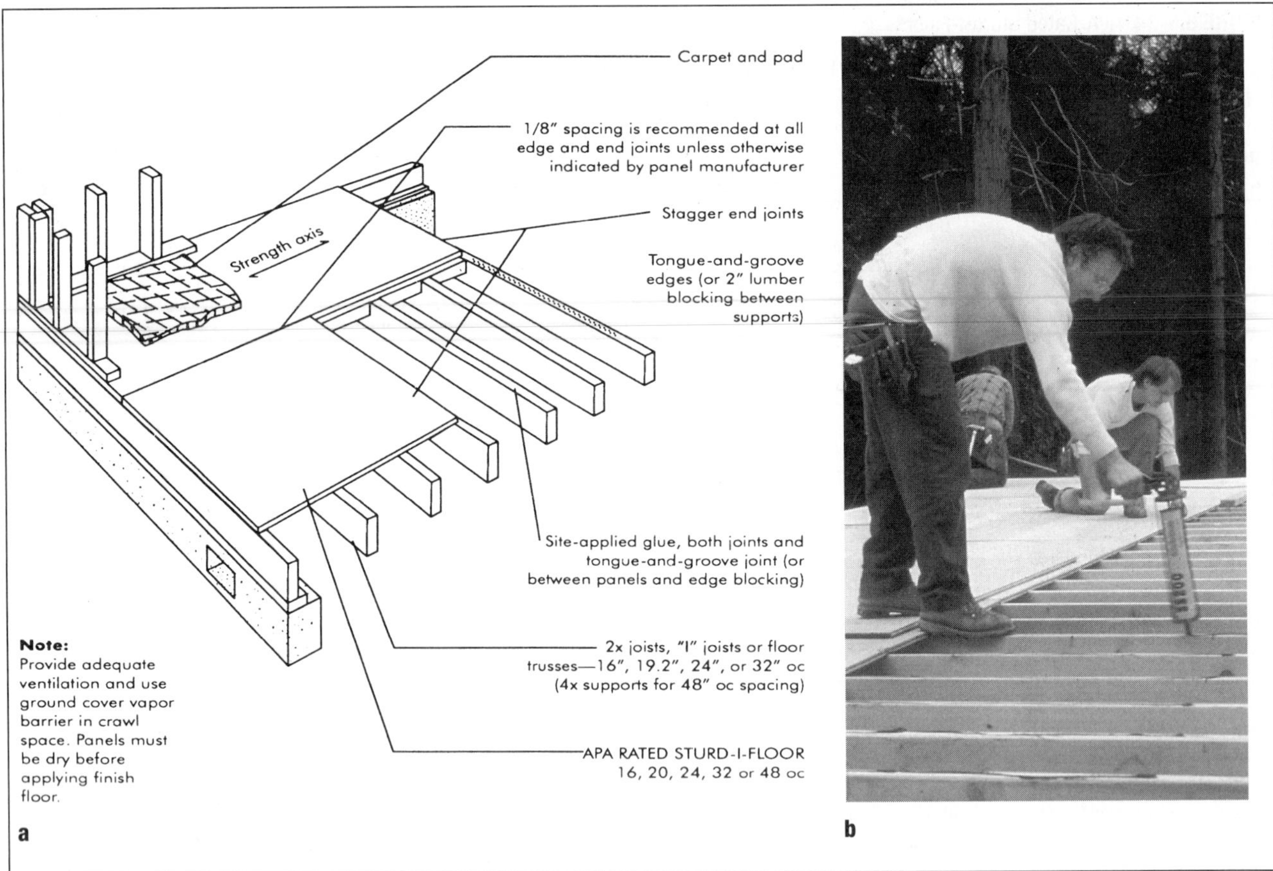

FIGURE 6.6-35 (a) APA glued floor system; (b) applying adhesive. (Courtesy American Plywood Association.) (Refer to Figure 1.8-27 for metric conversions.)

ports for nominal 1-in. (25 mm) lumber sheathing should not be more than 24 in. (610 mm) on centers.

Solid Sheathing Boards for *solid lumber sheathing* should be of nominal 1-in. (25 mm) thickness. Boards used for solid sheathing beneath mineral fiber cement shingles should be not more than 8 in. (200 mm) in nominal width and should have shiplapped or tongue-and-groove edges. Boards used for sheathing beneath wood shakes and shingles should be not more than 10 in. (250 mm) in nominal width. These may be square edged, shiplapped, or tongued and grooved.

When lumber sheathing boards are placed horizontally, structural corner bracing is required. Therefore, solid sheathing boards are usually placed diagonally, except where the finish requires horizontal placement (see Chapter 7).

Spaced Sheathing For the application of wood shingles and shakes, wood spaced sheathing should be at least 1 × 3 (25 × 75), 1 × 4 (25 × 100), or 1 × 6 (25 × 150) boards, as required for the finish material (see Chapter 7).

PLYWOOD WALL SHEATHING

Plywood sheathing should be APA-rated sheathing of Exposure 1 or 2; Exterior; Structural I, Exposure 1; or Structural I, Exterior. It should have the appropriate span rating according to the stud spacing. The span rating may be either 16 or 24 for a 16-in. (406 mm) stud spacing, but must be 24 for a 24-in. (610 mm) stud spacing (Fig. 6.6-36). These materials may include particleboard, as discussed in Section 6.3.

Nailable plywood sheathing is sheathing to which a finish material, such as siding, will be directly nailed. However, sheathing on brick-veneered and stuccoed walls should be installed in the same manner. Such sheathing may be placed with its long edge either horizontal or vertical, except that three-ply panels at studs spaced

FIGURE 6.6-36 APA Panel Wall Sheathing Material

			Nail Spacing, in.	
Panel Span Rating	**Maximum Stud Spacing, in.**	**Nail Size[a,b]**	**Supported Panel Edges**	**Intermediate Supports**
12/0, 16/0, 20/0, or Wall-16 oc	16	6d for panels ½ in. thick or less; 8d for thicker panels	6	12
24/0, 24/16, 32/16, or Wall-24 oc	24			

[a]Common, smooth, annular spiral thread, or galvanized box.
[b]Other code-approved fasteners may be used.
(Courtesy American Plywood Association.) (Refer to Figure 1.8-27 for metric conversions.)

24 in. (610 mm) on centers must be placed with the long edge across the studs, and panels ⅜ in. (9.5 mm) thick or less at studs spaced 16 in. (406 mm) on centers must be placed with the long edge across the studs. Some codes require the application of nailers (back blocking) at joints in plywood sheathing that do not fall over supports (Fig. 6.6-37). Because greater-than-normal stiffness is required in sheathing beneath stucco, panels there are often applied with the long side across the studs, and blocking is usually installed at all panel edges, whether or not required by code.

Plywood wall sheathing ½ in. (12.7 mm) thick or less should be applied with 6d common, smooth shank, annular, spiral-thread, or galvanized box nails. Thicker panels should be installed with similar 8d nails. Nails should be spaced at 6 in. (150 mm) centers along supported panel edges and at 12 in. (305 mm) centers at intermediate supports.

PLYWOOD CORNER BRACING

When nonwood sheathing is used that does not provide corner bracing, such as fiberboard and gypsum board, the corners of a building must be braced. This bracing may be plywood sheathing (Fig. 6.6-38). Such sheathing should be one of the materials mentioned earlier for plywood wall sheathing. The thickness should match the adjacent sheathing material.

NONLUMBER SHEATHING

Nonlumber sheathing materials in common use include fiberboard, gypsum board, and insulating board sheathing.

When fiberboard, gypsum board, insulating board, or other sheathing materials other than wood or plywood are recommended by the manufacturer for use as wall sheathing under a type of siding, the sheathing should be installed and attached to its supports according to its manufacturer's recommendations. The recommendations of the manufacturer supersede the suggestions in this section for installing and fastening such products.

Fiberboard, gypsum board, and insulating sheathing are not adequate nailing bases. When these sheathing materials are used, siding should be applied by nailing through the sheathing and into framing members or furring strips. Nor do these materials provide adequate frame bracing. When they are used, another form of bracing must be provided. One way to brace such a frame is using ¹⁵⁄₃₂-in. (11.91 mm) or ½-in. (12.7 mm) APA-rated sheathing plywood at the corners (see Fig. 6.6-38).

Fiberboard Sheathing Fiberboard sheathing is a homogeneous cellulosic material with square edges. It is available in ½-in. (12.7 mm) and ²⁵⁄₃₂-in. (19.84 mm) thicknesses, in 4-ft. (1220 mm) widths, and in 8-ft. (2440 mm) and 9-ft. (2700 mm) lengths. It should comply with the requirements of American National Standards Institute (ANSI) A 194.1, Type IV.

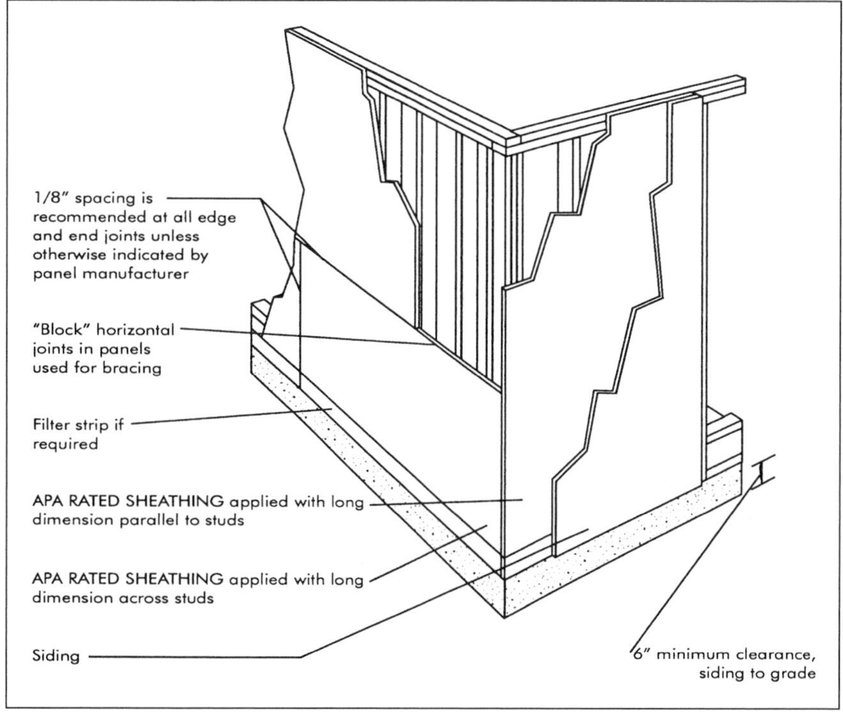

FIGURE 6.6-37 APA panel wall sheathing installation. (Courtesy American Plywood Association.) (Refer to Figure 1.8-27 for metric conversions.)

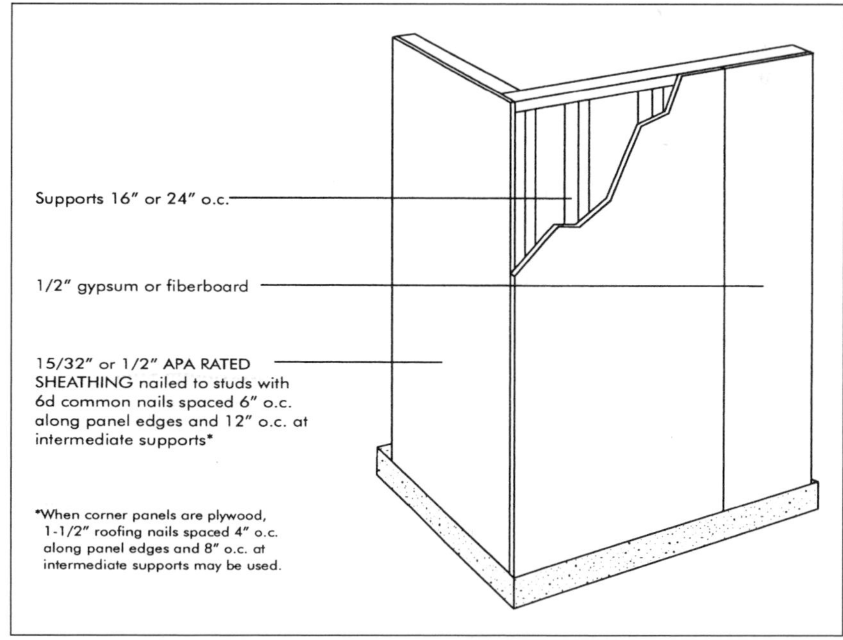

FIGURE 6.6-38 APA panel corner bracing. (Courtesy American Plywood Association.) (Refer to Figure 1.8-27 for metric conversions.)

Two densities are available: Class 1, Regular Density, and Class 2, Intermediate Density. Class 3, High Density, or nail-base fiberboard, is no longer made. Class 1 requires the addition of frame corner bracing. Class 2 may or may not require corner bracing, depending on the building's structural requirements.

Studs for fiberboard sheathing should be placed 16 in. (406 mm) on centers. Nailers must be provided, and panels must be nailed along all edges.

Unless otherwise recommended by its manufacturer, fiberboard sheathing should be applied with its long dimension parallel to the framing members (vertically) with sufficient bearing for satisfactory nailing along all edges. About ⅛ in. (3.2 mm) should be left open between the edges and ends of adjoining boards.

Fiberboard sheathing should be installed according to its manufacturer's instructions. Usually, these boards are nailed or stapled at 3 in. (75 mm) on centers along the edges and at 6 in. (150 mm) on centers at intermediate supports. Fasteners for ½-in. (12.7 mm) sheathing should be 11-gauge (3.9 mm) galvanized roofing nails, 6d common galvanized nails, or 16-gauge (1.6 mm) galvanized staples. Fasteners for 25/32-in. (19.84 mm) material should be correspondingly larger.

Gypsum Board Sheathing Gypsum sheathing may be used beneath wood, vinyl, composition, metal, and plywood siding, and beneath masonry veneer, stucco, and exterior insulation and finish systems. It is available in ½-in. (12.7 mm) regular and ⅝-in. (15.9 mm) Type X (fire-resistant) boards. The ½-in. (12.7 mm) boards come in 2-ft. (610 mm) × 8-ft. (2440 mm) and 4-ft. (1220 mm) × 8-ft. (2440 mm) sizes. The ⅝-in. (15.9 mm)-thick boards are 4 ft. (1220 mm) × 8 ft. (2440 mm). The 4-ft. (1220 mm) × 8-ft. (2440 mm) panels have square edges; the 2-ft. (610 mm)-wide panels have tongue-and-groove edges. All sizes may have either a plain or a water-resistant core. Their faces, backs, and edges are covered with either paper or a glass-fiber mat. Gypsum sheathing boards should comply with ASTM C-79.

The maximum framing spacing for gypsum sheathing is 24 in. (610 mm) on centers. Sheathing of the 2-ft. (610 mm) by 4-ft. (1220 mm) size should be installed with the long edge across the framing (horizontal) and with the tongue edge up. The ends should occur over framing members. The 4-ft. (1220 mm) × 8-ft. (2440 mm) sheathing boards should be installed with the long edges vertical, and with each edge occurring over a framing member. A continuous nailer should be installed when the ends of vertically applied boards do not fall over a sill or plate.

Gypsum sheathing may be fastened in place using nails for the ½-in. (12.7 mm) and ⅝-in. (15.9 mm) thicknesses. Staples may also be used for the ½-in. (12.7 mm) thickness. Nails should be 11 gauge (3.9 mm) with a 7/16-in. (11.11 mm) head and 1½ in. (38.1 mm) long for ½-in. (12.7 mm)-thick boards, and 1¾ in. (44.45 mm) long for ⅝-in. (15.9 mm)-thick boards. Staples should be 15 gauge (1.75 mm) or 16 gauge (1.6 mm), 7/16 in. (11.11 mm) or ½ in. (12.7 mm) wide and 1½ in. (63.5 mm) long. Fasteners should be placed not closer than ⅜ in. (9.5 mm) from the edge or end of a board. Boards may also be applied using screws, and often are, especially when the wall framing consists of metal studs. The size and spacing of drywall screws should be in accordance with the board manufacturer's recommendations.

Fasteners should be located in 4-ft. (1220 mm) × 8-ft. (2440 mm) panels at not more than 4 in. (100 mm) on centers along the edges and ends, and not less than 8 in. (200 mm) on centers along each intermediate support. Fasteners in 2-ft. (610 mm) by 4-ft. (1220 mm) tongue-and-grove boards should be located where recommended by the manufacturer of the board, but no fewer than four fasteners should be used in each 2-ft. (610 mm) width at each framing member when diagonal bracing is used, and no fewer than 7 fasteners in each 2-ft. (610 mm) width when bracing is not used.

Codes vary on the use of diagonal bracing with gypsum board sheathing. Some do not require bracing under some circumstances. Some permit the gypsum sheathing to be counted as bracing under some conditions. Others require bracing, but do not permit gypsum sheathing to serve in that capacity. In these cases, plywood panels, let-in wood, or metal corner bracing may be used.

Insulated Sheathing Several types of insulated sheathing are available. The common insulating materials they contain include polystyrene and isocyanurate. Some are plain boards without coverings. Others are covered with reflective foil or other materials that increase their thermal effect, and, in some cases, the boards themselves provide air and moisture barriers, or can be made to so function by taping the joints, thus eliminating the need for a separate layer of material to provide these functions. It is necessary, however, to take special precautions, such as installing vapor-relief strips, to ensure that moisture does not become trapped in the stud cavities behind insulated sheathing.

Most insulated sheathing products are combustible and cannot be left exposed. Therefore, their use is limited by codes to wall assemblies that protect them. They are often used, for example, in exterior insulation and finish systems (see Section 7.4). Another limitation of these products is that they are not usable or do not perform well under some kinds of finishing materials. The manufacturer of the siding or other finishing material to be used should be consulted and its advice followed regarding the specific use of insulated sheathing products.

The installation and fastening of insulated sheathing boards varies from product to product, and they should be installed according to their manufacturers' instructions.

AIR INFILTRATION BARRIERS

Air infiltration barriers reduce the number of air changes in a building and help create a more energy-efficient interior environment. They also protect sheathing against deterioration and resist the further penetration of moisture that finds its way through the siding or other finish material. Air infiltration barriers in common use over wood-framed walls today include asphalt-saturated organic felt (building paper) and more modern products such as polyethylene sheets and woven polyolefin sheet.

These materials may be applied directly to studs to resist the infiltration of air and moisture when sheathing is not used or when spaced sheathing is used. They may also be applied to the studs behind water-resistant non-nailable sheathing to make locating the studs easier for nailing finishing materials. They may also be installed over the sheathing. While not strictly required with plywood sheathing or other water-resistant sheathing materials, polyethylene and polyolefin materials may be installed as an air infiltration barrier only. Local codes may require a vapor retarder beneath masonry veneer, and may or may not permit the more modern plastic materials to be used in this application. Building paper is required beneath stucco. An air infiltration barrier is not appropriate in exterior insulation and finish systems.

Sheathing paper should be a material such as No. 15 or No. 30 asphalt-saturated felt, which has low vapor resistance (see Chapter 7). Materials such as coated felts or laminated waterproof papers which have high vapor resistance and act as vapor retarders should not be used. Such materials may permit moisture or frost to accumulate as a result of moisture condensation between the sheathing paper and the surface of the wall sheathing. Coal tar–saturated felts should not be used because they may stain siding.

When sheathing paper is used, it should be applied over the entire wall as soon as possible with at least a 4-in. (100 mm) top lap at horizontal joints and a 4-in. (100 mm) side lap at end joints. It should be lapped 6 in. (150 mm) from both sides around corners. Only sufficient fasteners need be used to hold the sheathing paper securely in place until the finish materials are applied. Wall finish materials should not be applied over wet sheathing paper.

Installation, fastening, and sealing of air infiltration barriers should be done in accordance with the recommendations of their manufacturer. Fastening is done using staples, nails, or adhesives. Overlaps should be sealed. These products degrade in ultraviolet light and therefore should not be left exposed for more than one month.

6.6.5.2 Roof Sheathing

Suitable performance of roofing is dependent on the entire roof structure. Roof joists, rafters, and other supports should be adequately sized and spaced to carry the necessary superimposed loads over the required spans (see Sections 6.6.3, 6.7.2, and 6.7.3).

Many types of roofing require a bituminous sheet underlayment for proper performance. These underlayments are discussed in the sections where the various roofing types are covered.

Roof sheathing should be smooth, securely attached to its supports, and provide a base to receive roofing nails and fasteners. Some roofing types are applied over solid sheathing composed of wood boards or plywood. Other types are applied over spaced sheathing. Refer to Chapter 7 for an indication of the sheathing type required for each roofing type.

LUMBER SHEATHING

Solid sheathing boards should be seasoned to a maximum moisture content of 15%. Boards should be tongued and grooved (T&G), in a nominal 1-in. (25 mm) thickness and not more than 6 in. (150 mm) in nominal width, to minimize shrinkage. Boards should be tightly matched and securely face and edge nailed with two 7d annularly threaded or 8d common wire nails at each framing member. Maximum spacing of supports for nominal 1-in. (25 mm) lumber boards is 24 in. (610 mm) on centers.

Spaced sheathing of 1 × 3 (25 × 75), 1 × 4 (25 × 100), or 1 × 6 (25 × 150) boards may be used in blizzard-free areas (generally where the outside design temperature is warmer than 0°F [–17.8°C]) to receive the application of wood shingles or wood shakes (see Chapter 7). The center-to-center spacing of these boards should be equal to the exposure of the shingle or shake, but not more than 10 in. (250 mm), and the openings between boards should not be wider than the individual boards. Seasoning and face nailing of spaced boards should be the same as described for solid wood board sheathing.

PLYWOOD SHEATHING

Plywood sheathing panels should be APA-rated sheathing of one of the following grades: Exposure 1 or 2; Exterior; Structural I, Exposure 1; or Structural I, Exterior. Panels exposed to the exterior should be Exterior grade. Plywood sheathing should have a span rating, panel thickness, support spacing, and edge support appropriate to the load supported (Fig. 6.6-39). Panel edges may be square or tongue-and-grooved.

Plywood sheathing should be applied with its long sides perpendicular to the framing and with a ⅛-in. (3.2 mm) gap between panels to permit thermal movement. Square-edged boards should be held level and supported with back blocking or panel clips. The number and spacing of panel clips depend on the support spacing; one between each pair of supports that are 24 in. (610 mm) apart, and two in 48-in. (1220 mm)-wide spaces is normal.

Plywood sheathing up to ½ in. (12.7 mm) thick should be fastened in place with 6d nails; from ½ in. (12.7 mm) to 1 in. (25.4 mm) thick with 8d nails. These nails may be either common smooth-shank or deformed-shank nails. Panels 1⅛ in. (28.58 mm) thick should be nailed with 8d ring- or screw-shank nails or 10d common smooth-shank nails. Nails should be spaced at 6 in. on centers around panel edges and 12 in. (305 mm) on centers at intermediate supports, except that in panels over support spacings of 48 in. (1220 mm) or more, all nails should be spaced at 6 in. (150 mm) on centers.

NONWOOD SHEATHING

When nonwood sheathing materials are recommended by their manufacturers for use as roof sheathing, the sheathing manufacturer should specify the type of fastening required to attach sheathing to supports. Such materials should not be used without the concurrence of the roofing manufacturer.

6.6.6 PREASSEMBLED COMPONENTS

Preassembled components are parts of a building's framing, floor and ceiling assemblies, wall assemblies, or roof and ceiling assemblies that are manufactured in a location other than that in which they will finally be used. They may range from small wall panels to complete structures.

Elements used in preassembled components include units made from plywood and lumber, including stressed-skin panels, sandwich panels, preframed panels, rigid frames, box beams, and various curved units. Plywood is also used to construct roof trusses and truss joists, which can be considered as preassembled components, but are discussed in Section 6.7.3.

Panelized assemblies for floors, walls, and roofs consist of normal framing systems that are built as a unit and then erected, one or more of the components listed in the preceding paragraph, or both.

6.6.6.1 Elements Used in Preassembled Components

The development of off-site, shop-fabricated building components has led to the production of such elements made of dimension lumber and plywood. These integral units may consist of plywood panels either nailed to precut lumber or fastened with glue, using either pressure- or nailed-glued techniques under controlled shop conditions. Full-pressure gluing, which requires rather expensive equipment and careful control, produces excellent results (Fig. 6.6-40). Nail gluing, where nails furnish the pressure required

FIGURE 6.6-39 Recommended Uniform Floor Live Loads for APA-Rated Sheathing and APA-Rated Sturd-I-Floor with Long Dimension Perpendicular to Supports

Sturd-I-Floor Span Rating	Sheathing Span Rating	Maximum Span, in.	Allowable Live Loads, psf* Joist Spacing, in. 12	16	20	24	32	40	48
16 oc	24/16, 32/16	16		100					
20 oc	40/20	20		150	100				
24 oc	48/24	24				100			
32 oc	60/32	32					100		
48 oc		48						100	55

*10 psf dead load assumed. Live load deflection limit is f/360.

(Courtesy American Plywood Association.) (Refer to Figure 1.8-27 for metric conversions.)

for an adequate glue bond, also requires careful and precise techniques but less equipment. Nail gluing leads to satisfactory, predictable results in the laboratory but has not always been successful in the field. This is due to poor gluing techniques, unfavorable temperatures, use of lumber not properly dried or surfaced, insufficient nailing for proper pressure, and dried-out glue lines. However, with proper supervision and control, field gluing can be successful.

Aside from the advantages—less erection time and reduced in-place costs—that may result from the use of almost any preassembled components, plywood components derive added benefits from the plywood itself. Plywood's tensile and compressive strength, panel-shear strength, split resistance, and dimensional stability enable it to function structurally as part of the component instead of serving just as a base for covering materials.

FIGURE 6.6-40 Inserting insulation and foil vapor retarder in a stressed-skin panel. (Courtesy American Plywood Association.)

Finishes that can be applied to a conventional assembly can be applied to a panelized floor and ceiling, wall, or roof and ceiling assembly. Finishes can be field applied or may be an integral, prefinished part of the panelized assembly (Fig. 6.6-41).

STRESSED-SKIN AND SANDWICH PANELS

Stressed-skin panels are panels with stressed covers that consist of plywood sheets glued to the top and, in most cases, the bottom faces of longitudinal framing members so that the assembly will act integrally in resisting bending stresses (Fig. 6.6-42). The plywood faces carry the compressive and tensile stresses induced by bending, and the framing members principally carry shear stresses. Stressing the skin allows a reduction in framing member size. For example, wall and partition studs, which normally are 2 × 4s (50 × 100s), can be reduced to 1 × 3s (25 × 75s) or 2 × 3s (50 × 75s) with stressed-skin design. The plywood must be bonded firmly with glue to the frame in order to transfer applied shearing stresses (Fig. 6.6-43). Stressed-skin panels may be used as walls, floors, and roofs (Fig. 6.6-44) (see also Section 6.6.6.2).

Sandwich panels follow the general principles of stressed-skin panel design, but the faces are glued to and separated by weaker, lightweight core material such as resin-impregnated paper honeycomb or a plastic foam core designed for this use (Fig. 6.6-45). Sandwich panels possess many of the advantages of stressed-skin panels but are lighter in weight, needing no inner panel framing except, perhaps, around the perimeter. The cores provide a uniform, stabilizing support for the strong, relatively thin faces. The cores usually have little strength in flexure but have adequate compressive strength for normal floor panels. Some cores, particularly plastic foams, also provide excellent insulation.

Although panelized construction provides many of the advantages of component fabrication, the incorporation of heating, air conditioning, electrical, and plumbing equipment requires careful design of the system. Designs have been made to incorporate mechanical and electrical equipment into stressed skin and sandwich panels. These require special handling and erection procedures in the field and may not be acceptable to local code authorities. Therefore, they have had limited success to date.

In addition, many local building codes specifically require 2 × 4 (50 × 100) studs spaced at 16 in. (406 mm) or 24 in. (610 mm) on centers, which would negate the advantage of using stressed-skin or sandwich panels unless an exception could be obtained. Some codes, however, permit the acceptance of any system that performs as well as the system specified in the code. In this case, a properly engineered and constructed panelized system may be accepted.

PREFRAMED PANELS

Preframed panels consist of fabricated, precut lumber and plywood that are nailed or mechanically fastened together. They may be used for walls, floors, or roofs (Fig. 6.6-46).

RIGID FRAMES

Rigid frames are specialized construction elements consisting of arch-shaped frames, usually comprising four pieces of straight lumber (two for posts and two for rafters). Plywood gussets, rigidly nailed or glued, connect adjacent members, transmit stresses across the joints, and consolidate the frames into integral units.

BOX BEAMS

Box beams are hollow structural units consisting of two or more vertical plywood webs that are attached, usually by gluing or nailing, to lumber flanges. Typical box beam cross sections are shown in Figure 6.6-47. As in a steel beam, the flanges carry the bending forces and the plywood webs transmit the shear. In this manner, each of the materials is employed more effectively. At intervals along the beam, stiffeners may be inserted between flanges and attached to the webs to distribute concentrated loads and resist web buckling.

Box beams have been used successfully for many purposes, including rather spectacular 48-in. (1220 mm)-deep girders for factory roof systems and other members spanning up to 120 ft. (36.58 m), as well as in many conventional applications (see Fig. 6.6-48).

Although either nail gluing or nailing alone may be satisfactory, pressure gluing with accurate controls is the preferred method to give maximum stiffness and strength to box beams.

a

b

c

FIGURE 6.6-41 (a) Sandwich panel with foamed plastic core; (b) an assembly panel made of a foamed plastic core sandwich panel; (c) an entire house structure made of sandwich panels. (Courtesy American Plywood Association.)

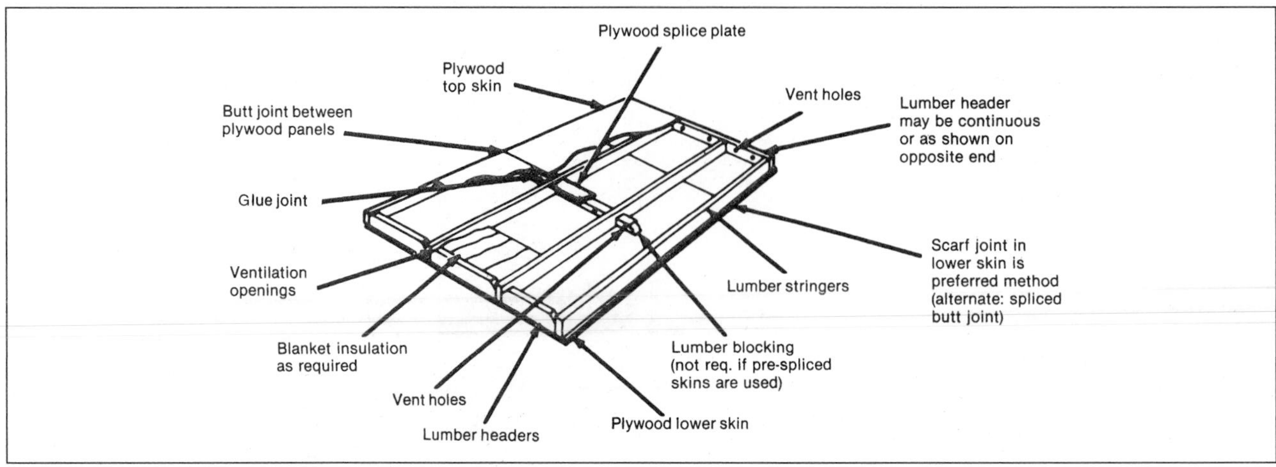

FIGURE 6.6-42 Typical two-faced stressed-skin panel. (Courtesy American Plywood Association.)

a

b

FIGURE 6.6-43 (a) Stressed-skin panels with plywood surfaces glued to a 2-in. (50 mm) lumber frame. After the glue has been placed and the plywood positioned, the entire panel is treated with heat and pressure in a press; (b) stressed-skin panel being lowered into place by crane. (Courtesy American Plywood Association.)

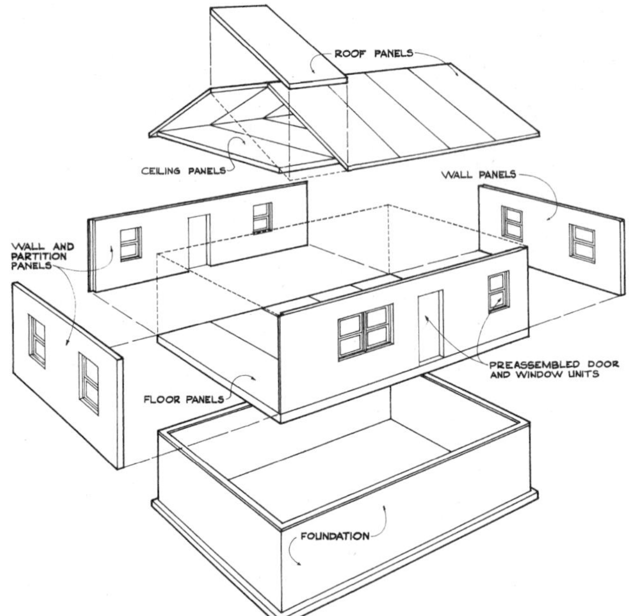

FIGURE 6.6-44 Typical uses for stressed-skin panels.

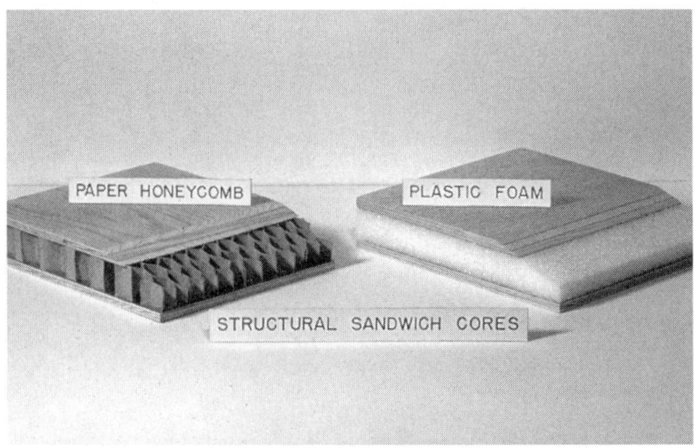

FIGURE 6.6-45 Sandwich panel with (left) honeycomb core and (right) foam core. (Courtesy American Plywood Association.)

a

b

FIGURE 6.6-46 **Preframed panels being placed (a) by crane and (b) by hand.** (Courtesy American Plywood Association.)

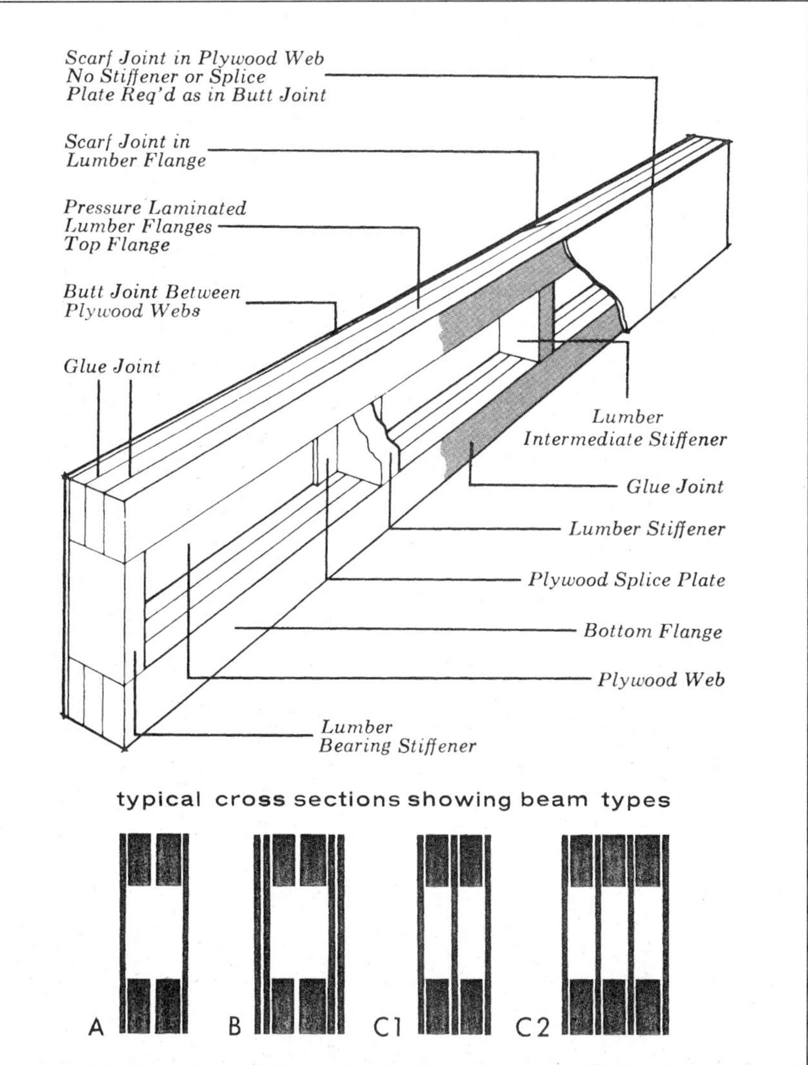

FIGURE 6.6-47 **Typical plywood box beams.** (Courtesy American Plywood Association.)

FIGURE 6.6-48 **A plywood box beam.** (Courtesy American Plywood Association.)

CURVED AND MISCELLANEOUS ROOF UNITS

Curved and other roof units permit construction of unique design shapes. Curved panels consist of built-up panels with plywood faces glued top and bottom to various core materials. These usually form an arc of a circle, as shown in Figure 6.6-49. Other design shapes, some of which are shown in Figure 6.6-50, can also be produced.

6.6.6.2 Panelized Assemblies

Floor and ceiling assemblies, wall assemblies, and roof and ceiling assemblies can

Typical Panel Using Curved Plywood Ribs

Preferred method: Scarf joint in plywood skin
Alternate: Spliced butt joint where permitted by design.
Plywood upper skin
Glue joint
Set out edge member to form tongue of tongue & groove joint
Insulation (optional)
Set in edge member to form groove
Depth
Headers
Length
Width
Lumber ribs may be used if desired
Plywood laminated ribs (laminated & curved prior to panel assembly)
Lower plywood skin

Typical Panel Using Solid Plywood Core

Scarf joint in plywood
Plywood upper skin
Glue joint
Plywood core set in to form groove
Tight butt joint in plywood core unless scarf joint required
Length
Plywood core set out to form tongue of tongue & groove joint
Width
Plywood core
Lower plywood skin

Typical Panel Using Lightweight Sandwich Core

Upper plywood skin
Edge members & headers may be plywood laminated sections
Scarf joint in plywood
Lumber edge member set out to form tongue
Length
Depth
Lumber edge member set in to form groove of tongue & groove joint
Resin-impregnated paper honeycomb core
Glue
Width
Lower plywood skin
Lumber headers
Scarf joint in plywood

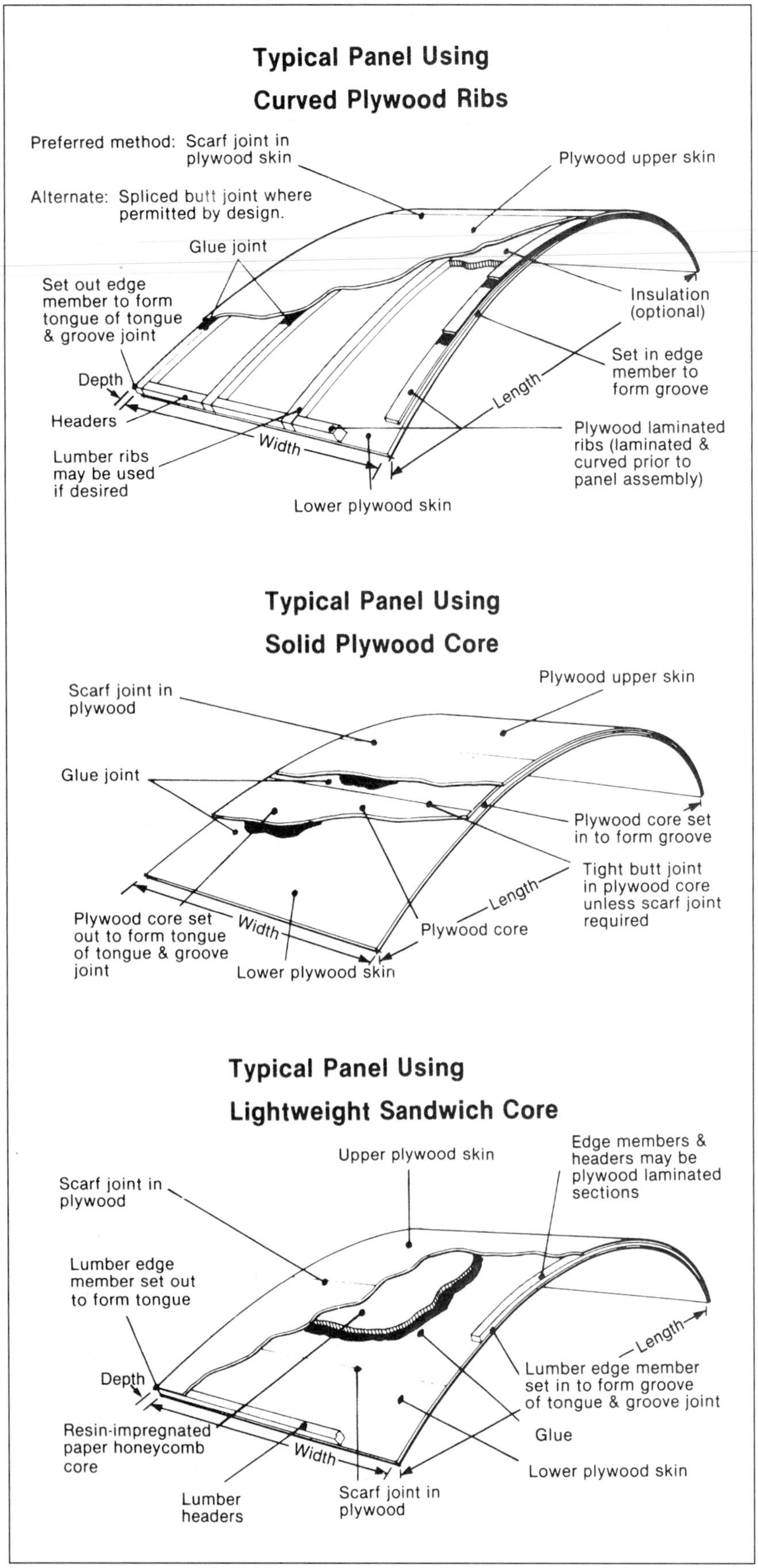

FIGURE 6.6-49 Curved stressed-skin panels. (Courtesy American Plywood Association.)

be manufactured and brought to the erection site as panelized units.

PANELIZED FLOOR AND CEILING ASSEMBLIES

Panelized floor and ceiling assemblies take many forms. One common assembly uses 1⅛-in. (28.58 mm)-thick APA-rated Sturd-I-Floor combination subfloor and underlayment plywood panels spanning between beams spaced 48 in. (1220 mm) on centers (see Fig. 6.6-33). Other systems consist of preframed panels in which the members are precut and to which subflooring is applied to form a preassembled component (Figs. 6.6-51 and 6.6-52). Frame members of nominal 2-in. (50 mm) *dimension* lumber function as floor joists. Panels may be fabricated on- or off-site in practically any size, with insulation and vapor retarder installed.

Panelized floor and ceiling assemblies may also utilize stressed-skin panels or sandwich panels (see Section 6.6.6.1). Panelized floor assemblies can be used in a variety of ways: panels may span between transverse or longitudinal beams or be centrally supported by bearing walls or beams, as with conventional floor and ceiling assemblies.

FIGURE 6.6-50 Plywood stressed-skin construction permits the use of many geometric shapes. (Courtesy American Plywood Association.)

PANELIZED WALL ASSEMBLIES

Panelized wall assemblies may consist of preframed panels using precut wood studs to which sheathing has been applied (Figs. 6.6-53 and 6.6-54). Fabrication may include the installation of insulation, air infiltration barriers, interior and exterior finishes, windows, and doors (Figs. 6.6-55 and 6.6-56), and panels may be fabricated off-site in any size that is conveniently handled.

Panelized wall assemblies may also include stressed-skin and sandwich panels (see Section 6.6.6.1).

PANELIZED ROOF AND CEILING ASSEMBLIES

Panelized roof and ceiling assemblies may consist of preframed, precut, and sheathed panels (Fig. 6.6-57). Nominal 2-in. (50 mm) *dimension* frame members of the panels function as rafters or ceiling joists. Panels may be fabricated on- or off-site in any size that can be conveniently handled.

Panelized roof and ceiling assemblies may also include stressed-skin or sandwich panel construction (see Section 6.6.6.1). Panelized roof and ceiling assemblies may be used in a variety of ways to form sloped or flat roofs and ceilings (Fig. 6.6-58). Panels may span between transverse or longitudinal beams or can be centrally supported by bearing walls or beams as in joist roofs.

Membrane or shingle roofing finishes may be applied to panelized systems. Interior finishes may be field applied or may be an integral, prefinished part of the panel.

6.6.7 FURRING

Wood furring strips are used with wood frame and masonry or concrete construction (Fig. 6.6-59) where incombustibility, decay resistance, or termites are not a consideration. When the framing is of irreg-

FIGURE 6.6-51 This preframed floor panel consists of plywood sheathing attached to 2 × 4 (50 × 100) stiffeners that span 4 ft. (1220 mm) from beam to beam. (Courtesy American Plywood Association.)

FIGURE 6.6-52 Preframed floor panel with top and bottom plywood surfaces applied to 2 × 8 (50 × 200) frame members. (Courtesy American Plywood Association.)

FIGURE 6.6-53 Panels may be built off-site in sizes that can be conveniently handled by two workers. (Courtesy American Plywood Association.)

FIGURE 6.6-54 Preframed panels 4 ft. (1220 mm) wide being erected. Studs are spaced 24 in. (610 mm) on centers. The space at the top of the panels is to receive a continuous header. (Courtesy American Plywood Association.)

FIGURE 6.6-55 Large wall panel with sheathing applied and windows installed can be fabricated off-site. (Courtesy American Plywood Association.)

a

b

FIGURE 6.6-56 Off-site fabrication may also include (a) panels with exterior finish and windows installed, (b) panels with siding, windows, and insulation installed. (Courtesy American Plywood Association.)

FIGURE 6.6-57 This preframed roof panel has been assembled on the ground, partially sheathed, and lifted into position. (Courtesy American Plywood Association.)

FIGURE 6.6-58 **Vaulted stressed-skin panels can create unusual roof and ceiling forms.** (Courtesy American Plywood Association.)

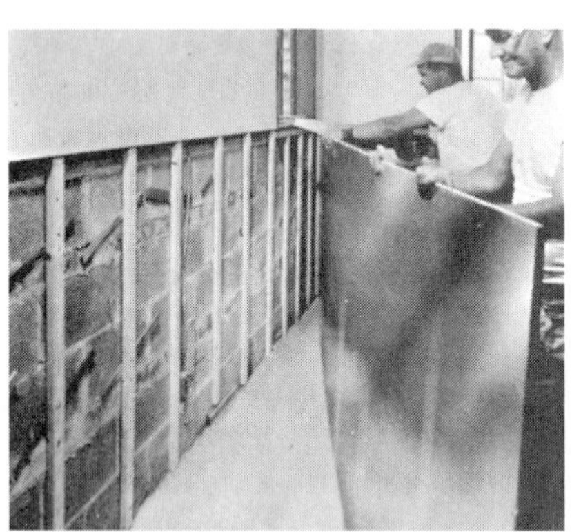

FIGURE 6.6-59 **Wood furring on masonry is shimmed to provide a plumbed, true, firm base for gypsum wallboard attachment.** (Courtesy American Gypsum Association.)

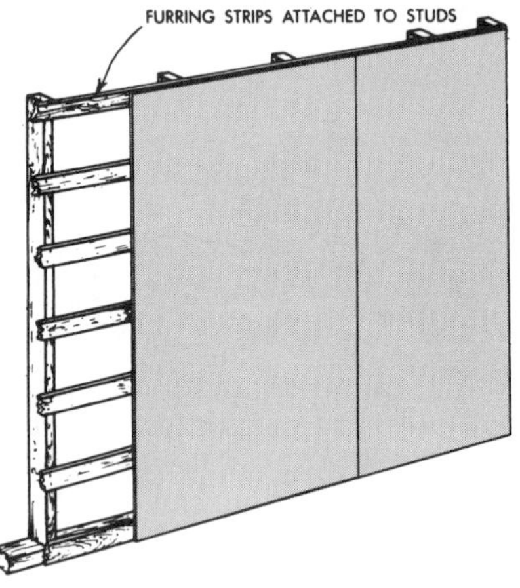

FIGURE 6.6-60 **Wood cross furring supporting gypsum board.** (Courtesy American Gypsum Association.)

ular spacing or the spacing is too great for the intended wall finish material, cross furring should be applied perpendicular to framing members (Fig. 6.6-60). Recommendations for wood cross furring are given in Figure 6.6-61.

Cross furring of wood smaller than 2 × 2s (50 × 50s) is not sufficiently stiff to permit the nailing of gypsum wallboard or gypsum plaster lath without excessive hammer rebound, which may result in loose board attachment. Thinner cross furring is acceptable for screw-attached board because little or no impact is involved when screws are driven. Thinner furring may also be acceptable if the board is applied with stud adhesive and nailing is limited to locations where the furring is fully supported by underlying framing.

When the primary purpose of furring is to increase the assembly thickness to conceal piping or to accept additional thermal insulation, 1 × 2 (25 × 50) or larger parallel furring can be applied directly over the primary framing, nailed 16 in. (406 mm) on centers with nails providing the penetration recommended for cross furring (see Fig. 6.6-61).

Furring strips over masonry or concrete can be secured with cut nails, helically threaded concrete nails, or powder-actuated fasteners of appropriate size. Maximum fastener spacing should be 16 in. (406 mm) on centers for 1 × 2 (25 × 50) strips and 24 in. (610 mm) on centers for 2 × 2 (50 × 50) or larger strips.

FIGURE 6.6-61 Minimum Requirements for Wood Cross Furring*

	Support Spacing			Nails	
Location	**in.**	**mm**	**Furring**	**Spacing**	**Penetration**
Ceilings	16 o.c.	406 o.c.	2 in. × 2 in.	2 at each support	⅞ in. (22 mm) (threaded)
	24 o.c.	610 o.c.	2 in. × 2 in.		1¼ in. (32 mm) (smooth shank)
Walls	16 o.c.	406 o.c.	1 in. × 2 in.	1 at each support	¾ in. (19 mm) (threaded or smooth shank)
	24 o.c.	610 o.c.	1 in. × 3 in.		

*Wallboard screw attached; if nailed, 2 in. × 2 in. furring should be used in all cases.

(Courtesy American Gypsum Association.)

6.7 Other Framing Systems

Other wood-framing systems include post-and-beam, plank-and-beam, heavy timber, laminated timber, roof trusses, and wood-pole construction.

6.7.1 POST AND BEAM

Post-and-beam systems (Fig. 6.7-1) consist of:

- Posts that support a series of beams
- The supported beams, which carry the loads of floors, roofs, and ceilings
- Non-load-bearing walls located between the posts (Fig. 6.7-2)

In a post-and-beam system the posts typically are 4 × 4s (100 × 100s) or are built up of several 2 by 4s (50 × 100s) and are spaced 4 ft. (1220 mm) to 8 ft. (2400 mm) apart. Wall areas between posts are framed, sheathed, and finished similarly to other wood wall systems to serve as weather barriers, provide resistance to racking, and give lateral support to the posts. Resistance to lateral forces is achieved with sheathing materials or diagonal bracing. Since these wall areas do not carry vertical loads, framing members may be placed horizontally to receive vertical exterior siding, and areas between posts can easily accommodate large windows or doors. Glass panels are often set (stopped) into the posts. No headers are required, but the placement of windows and doors is determined by the spacing and location of the posts.

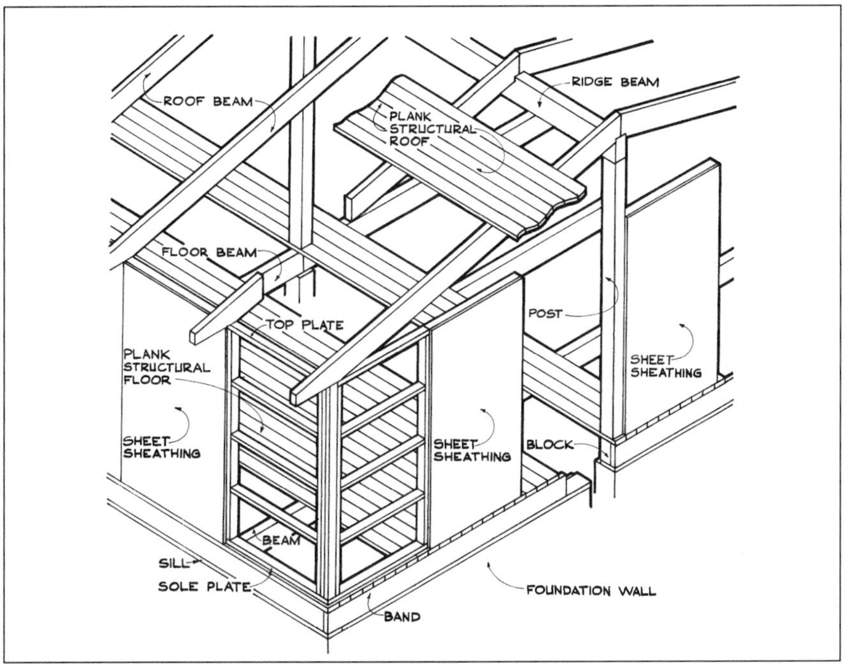

Post-and-beam systems are often integrated with plank-and-beam roof and ceiling assemblies. Planks and beams are often left exposed to form a finished ceiling, and posts may be partially exposed on the interior and exterior surfaces as part of the finished wall. In such designs, the expression of the basic structural system largely determines the interior and exterior appearance and character.

FIGURE 6.7-1 Post-and-beam framing.

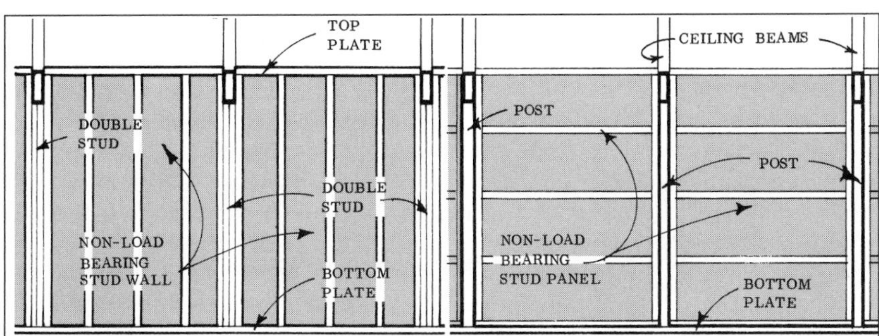

FIGURE 6.7-2 Wall construction for use with post-and-beam framing.

6.7.2 PLANK AND BEAM

Plank-and-beam construction is similar to wood joist construction, except that the subfloor is 2 in. (50 mm) or more in nominal thickness and structural members are larger and spaced farther apart. Underlayment is recommended over plank subfloors.

When heavier structural members are used for framing, support spacing can be increased, and heavier 2-in. (50 mm)-thick boards, called planks, are generally used. A 2-in. (50 mm)-thick member laid flat and stressed in bending across its 2-in. (50 mm) dimension is referred to as a *plank*, and hence this type of subfloor also is called *planking*. Planking is usually tongue-and-grooved to facilitate the transfer of concentrated loads from one plank to another. In lumber grading, such tongue-and-groove material is called *decking*.

The meanings of the terms *beam*, *plank*, and *decking* used in lumber grading do not completely agree with general construction terminology. In light wood-framed construction, 4-in. (100 mm)-thick framing members are called *beams*; yet in lumber grading these are classed as *joists* and *planks*, and the term *beams and stringers* is restricted to rectangular members 5 in. (127 mm) thick or thicker. In lumber grading, *decking* means lumber 2 in. (50 mm) to 4 in. (100 mm) thick, 4 in. (100 mm) wide or wider, with a maximum moisture content of 15% or 19%, and with tongue-and-groove edges. However, in common usage, both installations with this special product and ordinary surfaced four-sides lumber joists and planks are referred to as *planking*, or as *decking* when used in a roof assembly, which is called a *roof deck*. For instance, a traditional plank-and-beam framing system uses planking, which is really decking, and beams, which are really joists, according to lumber grading terminology.

6.7.2.1 Plank-and-Beam Floors

Plank-and-beam floor assemblies are characterized by the use of nominal 2 × 6 (50 × 150) or 2 × 8 (50 × 200) tongue-and-groove (T&G) or splined *planks* spanning between beams generally spaced 4 ft. (1220 mm) to 8 ft. (2440 mm) apart (Fig. 6.7-3). The planks serve as subflooring, as a working platform, and to transmit floor loads to fewer but larger members than occur in conventional wood joist floor systems.

Plank-and-beam floor and ceiling assemblies are often integrated with plank-and-beam roof and ceiling assemblies and post-and-beam wall systems. In such design, the basic structural system largely determines the interior appearance and character.

Conventional joist framing requires a maximum support spacing of 24 in. (610 mm) on centers. Using APA-rated Sturd-I-Floor combination subfloor and underlayment can push this spacing up to 48 in. (1220 mm) on centers. But plank-and-beam framing permits support spacings of 6 ft. (1800 mm) or more.

Tongue-and-groove or splined plank subfloors have a finished surface on the underside, which can be left exposed to provide a finished ceiling. Using such a system, however, does not provide the usual cavities for mechanical and electrical distribution equipment, and, in the absence of a separate ceiling finish, relatively poor sound isolation, particularly from impact noise. Therefore, plank floors are most suitable over unfinished spaces or where impact sound isolation from above is not an important consideration. In light wood-frame construction, plank-and-beam construction is generally more adaptable to roof assemblies or to floor systems in 1-story construction. When deep girders are not needed beneath plank floors, they permit a lower floor-to-grade elevation, resulting in a lower building height.

Floor finishes such as strip wood flooring may be applied directly over such planks, but resilient flooring and adhesively applied ceramic tile or thin-set terrazzo require an underlayment. Planks can be used as a base for carpeting with suitable cushioning (padding), if the subfloor surface is reasonably level and smooth. Heavy cushioning and deep pile carpeting, with or without underlayment, also considerably improves impact sound isolation and is recommended where sound transmission is a factor.

Plank subfloors can be built with tongue and groove (T&G) or surfaced-four-sides (S4S) lumber. T&G lumber is graded in most species as Select Decking and Commercial Decking, both of which are suitable for subfloors. Decking is manufactured 2-in. (50 mm) to 4-in. (100 mm) in nominal thickness and in widths ranging from 4 in. (100 mm) to 12 in. (305 mm). The maximum width for subfloors should not exceed 8 in. (200 mm) at a 19% moisture content, but wider boards can be used when their moisture content is limited to 15%. Thinner (2-in. [50 mm]) planks are generally made with a single tongue and groove, while 3-in. (75 mm) and 4-in. (100 mm)-thick planks have two. Planks with grooves on both edges, for use with splines, also are available. Usually one face is of better quality and is intended for exposed view, such as in a paneled ceiling. This face may also be worked to a special pattern with V-joints, rabbets, or rounded edges.

S4S lumber suitable for planking is the same as for joists and is graded as Select Structural No. 1, No. 2, or No. 3.

A plank-and-beam system is essentially a skeleton framework. Planks are intended to support moderate, uniformly distributed loads. These are carried to the beams, which in turn transmit their loads to posts or walls supported on foundations. Foundations for plank-and-beam framing are similar to those used for other wood-framing systems (Fig. 6.7-4). With beams (and sometimes posts) spaced 6 ft. (1800 mm) or more on centers, this system is well adapted to pier and crawl space foundations.

Posts should be of adequate size to carry the load and large enough to provide full bearing for the ends of beams. In general, posts should be at least 4 × 4 (100 × 100), but 4 × 6 (100 × 150) posts are preferred, particularly where beams join or where spaced beams are used. Where the ends of beams abut over a post, the 6-in. (150 mm) dimension should be parallel to the beams (Fig. 6.7-5).

Suitable beam sizes vary with grades, species, span, loading, and spacing. Beams may be solid lumber, glue-laminated members, or built up of several joists securely nailed to each other or to spacer blocks between them. Wood beams supporting plank floors or roofs should not exceed the maximum allowable spans tabulated in AF&PA publication *Plank and Beam Framing for Residential Buildings*.

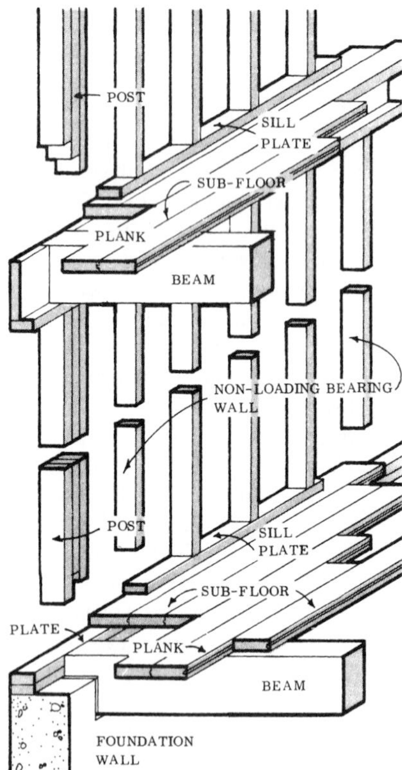

FIGURE 6.7-3 Plank-and-beam construction.

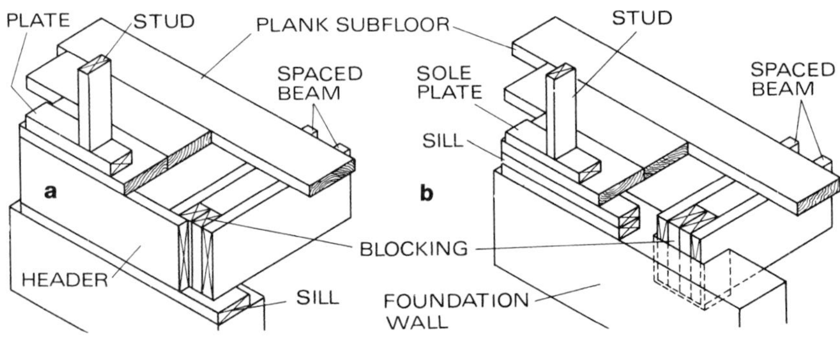

FIGURE 6.7-4 Spaced beam (or girder) for plank floors can be (a) supported on a foundation sill or (b) framed into a wall pocket. (Courtesy American Forest & Paper Association.)

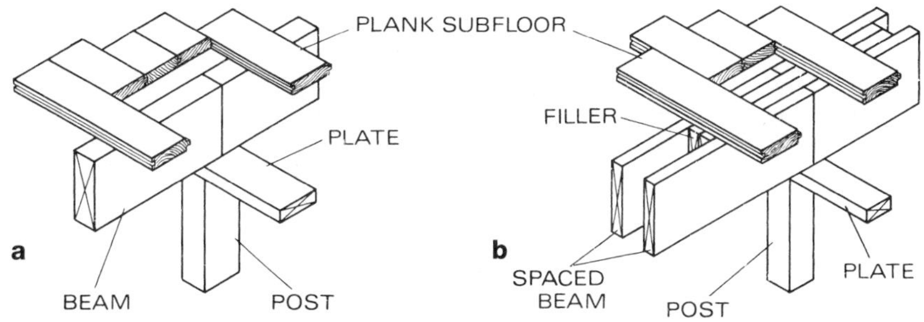

FIGURE 6.7-5 Intermediate support for plank floors can be provided by a wood post under (a) a solid beam or (b) a spaced beam. (Courtesy American Forest & Paper Association.)

Figure 6.7-6 is an example, showing a table for beams supporting a 20 psf (97.65 kg/m^2) live load with an allowable deflection of L/240. *Plank and Beam Framing for Residential Buildings* also includes tables for live loads of 30 psf (146.47 kg/m^2) and 40 psf (195.3 kg/m^2) and allowable deflections of L/300 and L/360. To use these tables, first determine the span, the live loads to be supported, and the deflection limitation. Then select from the tables the proper size of beam with the corresponding required values for *f* and *E*. The beam used should be of a grade and species that meet or exceed the selected values.

Beams should be securely fastened to posts with framing anchors or clip angles. Details of upper-floor beams framing into an exterior frame wall are illustrated in Figure 6.7-7.

Greater advantage can be taken of the strength and stiffness of plank subflooring material by making the planks continuous over more than one span. For example, using the same span and uniform load in each case, a plank that is continuous over two spans is nearly 2½ times as stiff as a plank that extends over a single span (Fig. 6.7-8).

A nominal 2-in. (50 mm)-thick plank should not exceed the maximum allowable spans shown in the tables in *Plank and Beam Framing for Residential Buildings* (Fig. 6.7-9). The types of plank

FIGURE 6.7-6 Required *f* and *E* Values for Floor and Roof Beams

Span of Beam, ft.	Nominal Size of Beam	Minimum *f* and *E* (psi) for Beams Spaced: 6 ft.-0 in.		7 ft.-0 in.		8 ft.-0 in.	
		f	*E*	*f*	*E*	*f*	*E*
10	2-3 × 6	1070	780000	1250	910000	1430	1040000
	1-3 × 8	1235	680000	1440	794000	1645	906000
	2-2 × 8	1030	570000	1200	665000	1370	760000
	1-4 × 8	880	485000	1030	566000	1175	646000
	3-2 × 8	685	380000	800	443000	915	506000
	2-3 × 8	615	340000	720	397000	820	453000
	2-2 × 10	630	273000	735	219000	840	364000
11	2-3 × 6	1295	1037000	1510	1210000	1730	1382000
	1-3 × 8	1490	905000	1740	1056000	1990	1206000
	2-2 × 8	1245	754000	1450	880000	1660	1005000
	1-4 × 8	1065	647000	1245	755000	1420	862000
	3-2 × 8	830	503000	970	587000	1105	670000
	2-3 × 8	745	453000	870	529000	995	604000
	2-2 × 10	765	363000	890	424000	1020	484000
12	2-3 × 6	1545	1346000	1800	1571000	2060	1794000
	1-3 × 8	1775	1175000	2070	1371000	2370	1566000
	2-2 × 8	1480	980000	1725	1144000	1970	1306000
	1-4 × 8	1270	840000	1480	980000	1690	1120000
	3-2 × 8	985	653000	1150	762000	1315	870000
	2-3 × 8	890	588000	1035	686000	1185	784000
	1-6 × 8	755	483000	880	564000	1005	644000
	2-2 × 10	910	472000	1060	551000	1210	629000
	1-3 × 10	1090	566000	1275	660000	1455	754000
13	2-3 × 6	1815	1711000	2110	1997000	2415	2281000
	1-3 × 8	2085	1494000	2430	1743000	2780	1991000
	2-2 × 8	1740	1245000	2025	1453000	2315	1660000
	1-4 × 8	1490	1067000	1735	1245000	1985	1422000
	3-2 × 8	1160	830000	1350	969000	1545	1106000
	2-3 × 8	1045	747000	1215	872000	1390	996000
	1-6 × 8	885	614000	1040	716000	1185	818000
	2-2 × 10	1070	600000	1245	700000	1420	800000
	1-3 × 10	1280	719000	1495	839000	1710	958000
14	2-2 × 8	2015	1555000	2350	1815000	2685	2073000
	3-2 × 8	1340	1037000	1570	1210000	1790	1382000
	2-3 × 8	1210	933000	1410	1089000	1610	1244000
	1-6 × 8	1025	766000	1200	894000	1370	1021000
	1-3 × 10	1485	899000	1730	1049000	1980	1198000
	2-2 × 10	1235	749000	1445	874000	1650	998000
	1-4 × 10	1060	642000	1240	749000	1415	856000
	3-2 × 10	825	499000	965	582000	1100	665000
	2-3 × 10	740	449000	865	524000	990	598000
15	3-2 × 8	1540	1275000	1800	1488000	2055	1699000
	2-3 × 8	1390	1148000	1620	1340000	1850	1530000
	1-6 × 8	1180	943000	1375	1100000	1570	1257000
	1-3 × 10	1705	1105000	1990	1289000	2270	1473000
	2-2 × 10	1420	921000	1660	1075000	1895	1228000
	1-4 × 10	1220	789000	1420	921000	1625	1052000
	3-2 × 10	950	614000	1105	717000	1265	818000
	2-3 × 10	850	553000	995	645000	1135	737000
	1-6 × 10	735	464000	855	541000	980	618000
	4-2 × 10	710	461000	830	538000	945	614000
	2-2 × 10	960	512000	1120	597000	1280	682000
16	3-2 × 8	1755	1548000	2045	1806000	2340	2063000
	2-3 × 8	1580	1393000	1840	1626000	2105	1857000
	2-2 × 10	1615	1118000	1890	1305000	2155	1490000
	1-4 × 10	1385	958000	1615	1118000	1845	1277000
	3-2 × 10	1075	745000	1260	869000	1435	993000
	2-3 × 10	970	671000	1130	783000	1290	894000
	1-6 × 10	835	563000	975	657000	1130	750000
	4-2 × 10	810	559000	945	652000	1080	745000
	1-8 × 10	615	413000	715	482000	815	550000
	1-3 × 12	1310	746000	1530	871000	1750	994000
	2-2 × 12	1090	621000	1275	725000	1455	828000

(Continues)

FIGURE 6.7-6 *(Continued)*

Span of Beam, ft.	Nominal Size of Beam	Minimum *f* and *E* (psi) for Beams Spaced: 6 ft.-0 in.		7 ft.-0 in.		8 ft.-0 in.	
		f	*E*	*f*	*E*	*f*	*E*
17	2-2 × 10	1825	1341000	2125	1565000	2430	1781000
	1-4 × 10	1565	1149000	1825	1341000	2085	1532000
	3-2 × 10	1215	894000	1420	1043000	1620	1192000
	2-3 × 10	1095	804000	1280	938000	1460	1072000
	1-6 × 10	945	675000	1100	788000	1260	900000
	4-2 × 10	910	670000	1065	782000	1215	894000
	1-8 × 10	690	495000	805	578000	910	660000
	1-3 × 12	1480	894000	1725	1043000	1975	1192000
	2-2 × 12	1235	745000	1440	869000	1645	993000
	1-4 × 12	1060	639000	1230	746000	1410	852000
	3-2 × 12	820	497000	960	580000	1095	663000
18	2-2 × 10	2045	1592000	2385	1858000	2725	2123000
	1-4 × 10	1755	1364000	2045	1592000	2340	1819000
	3-2 × 10	1365	1061000	1590	1238000	1815	1415000
	2-3 × 10	1270	955000	1480	1114000	1695	1273000
	1-6 × 10	1060	801000	1235	935000	1415	1068000
	4-2 × 10	1020	796000	1195	929000	1365	1062000
	1-8 × 10	780	588000	910	686000	1040	784000
	1-3 × 12	1660	1062000	1935	1239000	2210	1416000
	2-2 × 12	1380	885000	1615	1033000	1845	1180000
	1-4 × 12	1185	758000	1385	885000	1580	1011000
	3-2 × 12	920	590000	1075	688000	1230	786000
	2-3 × 12	830	531000	970	620000	1105	708000
19	3-2 × 10	1520	1248000	1775	1456000	2025	1664000
	2-3 × 10	1365	1123000	1595	1310000	1825	1497000
	1-6 × 10	1170	943000	1365	1100000	1560	1257000
	4-2 × 10	1140	936000	1330	1092000	1520	1248000
	2-4 × 10	975	802000	1140	936000	1300	1070000
	1-8 × 10	860	691000	1005	806000	1145	921000
	1-3 × 12	1850	1249000	2155	1457000	2465	1665000
	2-2 × 12	1540	1041000	1800	1215000	2055	1388000
	1-4 × 12	1320	892000	1540	1041000	1760	1190000
	3-2 × 12	1025	694000	1200	810000	1370	926000
	2-3 × 12	925	624000	1080	728000	1230	832000
	1-6 × 12	805	531000	940	620000	1070	708000
20	3-2 × 10	1685	1456000	1965	1699000	2245	1942000
	2-3 × 10	1515	1310000	1770	1529000	2020	1747000
	1-6 × 10	1300	1099000	1515	1282000	1735	1465000
	4-2 × 10	1260	1092000	1475	1274000	1685	1456000
	2-4 × 10	1080	936000	1265	1092000	1445	1248000
	1-8 × 10	960	806000	1120	941000	1280	1075000
	2-2 × 12	1705	1214000	1990	1417000	2275	1619000
	1-4 × 12	1465	1040000	1710	1214000	1950	1387000
	3-2 × 12	1140	809000	1330	944000	1520	1079000
	2-3 × 12	1025	728000	1195	850000	1365	971000
	1-6 × 12	970	620000	1130	723000	1295	826000
	2-4 × 12	730	520000	855	607000	975	694000
21	3-2 × 10	1855	1685000	2165	1966000	2475	2247000
	2-3 × 10	1670	1516000	1950	1827000	2225	2088000
	1-6 × 10	1430	1273000	1670	1485000	1905	1697000
	4-2 × 10	1390	1264000	1625	1475000	1855	1686000
	2-4 × 10	1195	1083000	1390	1264000	1590	1444000
	1-8 × 10	1050	933000	1225	1089000	1400	1244000
	2-2 × 12	1880	1405000	2195	1640000	2510	1874000
	1-4 × 12	1615	1204000	1880	1405000	2150	1606000
	3-2 × 12	1255	937000	1465	1093000	1670	1249000
	2-3 × 12	1130	843000	1320	984000	1505	1124000
	1-6 × 12	970	717000	1130	837000	1295	956000
	2-4 × 12	805	602000	940	702000	1075	802000
22	1-6 × 10	1580	1463000	1845	1707000	2105	1951000
	4-2 × 10	1525	1453000	1780	1696000	2035	1938000
	2-4 × 10	1310	1245000	1530	1455000	1745	1660000
	1-8 × 10	1160	1073000	1355	1252000	1545	1431000
	1-4 × 12	1770	1384000	2065	1615000	2360	1846000
	3-2 × 12	1375	1077000	1605	1257000	1835	1436000
	2-3 × 12	1240	969000	1445	1130000	1655	1291000
	1-6 × 12	1080	825000	1260	963000	1440	1100000
	2-4 × 12	885	692000	1035	807000	1180	922000
	4-2 × 12	1035	808000	1205	943000	1375	1078000
	5-2 × 12	825	646000	965	754000	1105	862000
	3-3 × 12	825	639000	965	746000	1105	852000

(Courtesy American Forest & Paper Association.)

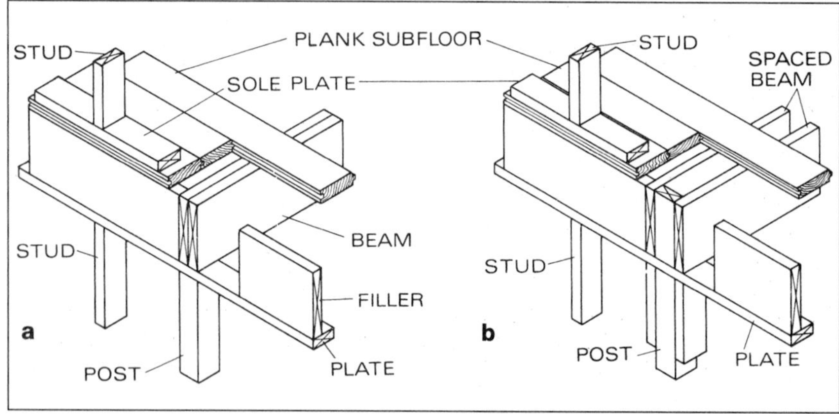

FIGURE 6.7-7 **A plank floor at an upper floor can be supported by (a) a built-up beam or (b) a spaced beam, framed into an exterior wall over a post or top plate.** (Courtesy American Forest & Paper Association.)

arrangements shown in the table are defined as follows:

Type A—Extending over a single span
Type B—Continuous over two equal spans
Type C—Continuous over three equal spans
Type D—A combination of Types A and B

To use this table, determine the plank arrangement type, the span, the supported live load, and the deflection limit, and select from the table the corresponding values for *f* and *E*. The plank used should be of a grade and species that meet or exceed the minimum values listed. The maximum span for a given thickness, grade, and species of plank can be obtained by reversing this procedure. The maximum allowable f and E for most commercial grades and species are tabulated in the *NDS Supplement* (see Section 6.1.3.4).

Regardless of their width, planks should be blind nailed to each support with one 10d nail. In addition, planks 4 in. (100 mm) or 6 in. (150 mm) wide should be face nailed, also with one 16d nail at each support; 8-in. (200 mm) planks should be face nailed with two 16d nails; 10-in. (250 mm) and 12-in. (305 mm) planks with three 16d nails.

A wood strip floor should be laid at right angles to a plank subfloor, using the same procedure followed in conventional construction. Where the underside of the plank is to serve as a ceiling, flooring nails should not penetrate through the plank.

In a plank-and-beam system, interior partitions are usually nonbearing. Where bearing partitions occur, they should be placed over beams and the beams enlarged to carry the added load. If this is not possible, supplementary beams must be

FIGURE 6.7-8 **The resultant deflection under similar loads is less when individual planks are long enough to reach across two spans, than it is in single-span floors.** (Courtesy American Forest & Paper Association.)

FIGURE 6.7-9 Required *f* and *E* Values for 2-in.-Thick Plank

Plank Span, ft.	Live Load, psf	Deflection Limitation	Type A f psi	Type A E psi	Type B f psi	Type B E psi	Type C f psi	Type C E psi	Type D f psi	Type D E psi
6	20	ℓ/240	360	576000	360	239000	288	305000	360	408000
		ℓ/300	360	720000	360	299000	288	318000	360	509000
		ℓ/360	360	864000	360	359000	288	457000	360	611000
	30	ℓ/240	480	864000	480	359000	384	457000	480	611000
		ℓ/300	480	1080000	480	448000	384	571000	480	764000
		ℓ/360	480	1296000	480	538000	384	685000	480	917000
	40	ℓ/240	600	1152000	600	478000	480	609000	600	815000
		ℓ/300	600	1440000	600	598000	480	762000	600	1019000
		ℓ/360	600	1728000	600	717000	480	914000	600	1223000
7	20	ℓ/240	490	915000	490	380000	392	484000	490	647000
		ℓ/300	490	1143000	490	475000	392	605000	490	809000
		ℓ/360	490	1372000	490	570000	392	726000	490	971000
	30	ℓ/240	653	1372000	653	570000	522	726000	653	971000
		ℓ/300	653	1715000	653	712000	522	907000	653	1213000
		ℓ/360	653	2058000	653	854000	522	1088000	653	1456000
	40	ℓ/240	817	1829000	817	759000	653	968000	817	1294000
		ℓ/300	817	2287000	817	949000	653	1209000	817	1618000
		ℓ/360	817	2744000	817	1139000	653	1451000	817	1941000
8	20	ℓ/240	640	1365000	640	567000	512	722000	640	966000
		ℓ/300	640	1707000	640	708000	512	903000	640	1208000
		ℓ/360	640	2048000	640	850000	512	1083000	640	1449000
	30	ℓ/240	853	2048000	853	850000	682	1083000	853	1449000
		ℓ/300	853	2560000	853	1063000	682	1354000	853	1811000
		ℓ/360	853	3072000	853	1275000	682	1625000	853	2174000
	40	ℓ/240	1067	2731000	1067	1134000	853	1444000	1067	1932000
		ℓ/300	1067	3410000	1067	1417000	853	1805000	1067	2415000
		ℓ/360	1067	4096000	1067	1700000	853	2166000	1067	2898000

(Courtesy American Forest & Paper Association.)

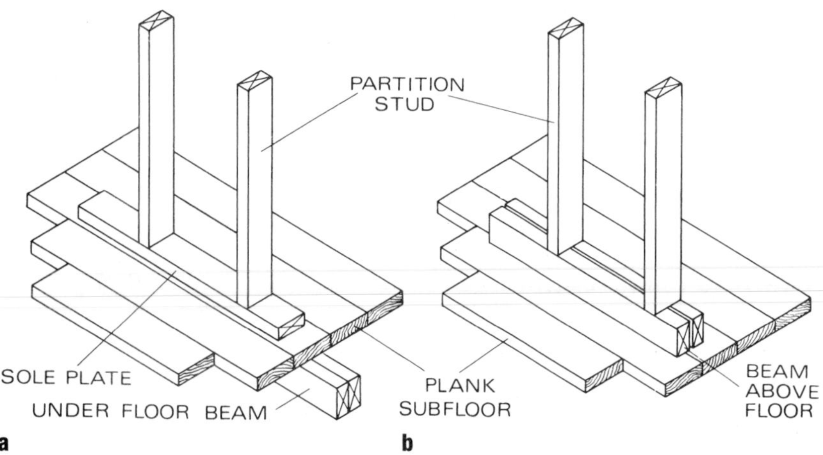

FIGURE 6.7-10 Interior nonbearing partitions running parallel to a plank floor can be supported on (a) a suitable under-floor beam, or (b) a double vertical sole plate. (Courtesy American Forest & Paper Association.)

placed in the floor framing. Nonbearing partitions, which are parallel to the planks, should be supported on a sole plate consisting of double 2 × 4s (50 × 100s) set on edge and spiked together (Fig. 6.7-10b). Where door openings occur, double 2 × 4s (50 × 100s) may be placed under the plank floor and framed into the beams with framing anchors (Fig. 6.7-10a). Nonbearing partitions located at right angles to the planks do not require supplementary framing, because the partition load is distributed across a number of planks.

6.7.2.2 Plank-and-Beam Roofs

Plank-and-beam roofs are typically used on 1-story buildings and for low-slope and flat roofs. These systems are characterized by the use of 2 × 6 (50 × 150) and 2 × 8 (50 × 200) tongue-and-groove (T&G) or splined planks spanning between either longitudinal or transverse beams (see Fig. 6.7-11). Beam spacing and planking thickness are determined by imposed loads, spans, and material strength. Nominal 2-in. (50 mm) planks may be used on spans up to about 8 ft. (2440 mm). The AF&PA publication *Plank and Beam Framing for Residential Buildings* includes tables for the minimum required stress in bending and the modulus of elasticity for beams and planks used in plank-and-beam roofs. Figures 6.7-6 and 6.7-9 reproduce some of these tables. The use of these tables is explained in Section 6.7.2.1.

Longitudinal beams span parallel to the long dimension of a building and are supported on end walls or columns. In buildings up to about 36 ft. (10.97 m) wide, beams typically are placed at quarter points. This divides the width of the building into two equal spans for planking on each side of the center ridge beam. Planks may be supported along the exterior sides by walls or beams. Interior planning flexibility is slightly more restricted than when transverse beams are used, because of the greater number of beam supports.

Transverse beams, generally spaced 6 ft. (1800 mm) to 8 ft. (2440 mm) apart, span perpendicularly to the long dimension of a building. They may be supported by load-bearing walls or beams along the ridge or center of the building.

The system may be used with wood stud exterior walls and is often used with post-and-beam exterior walls.

A plank ceiling and roof deck does not provide space for concealment of ductwork and wiring. If the system requires insulation, it is necessary to use rigid insulation applied to the top of the deck, which in turn may limit the selection of roofing finishes.

Plank-and-beam roof assemblies, in contrast to truss and joist-and-rafter roof and ceiling assemblies, require different design and construction methods and more precision in erection, since the finished appearance is generally an integral part of the system. Planks, beams, and columns are usually left exposed as the finished interior surface.

6.7.3 HEAVY TIMBER

Before the development of balloon frame construction (see Section 6.6) heavy timber framing was the normal method used to build most wood-framed buildings. Timbers for early timber-framed build-

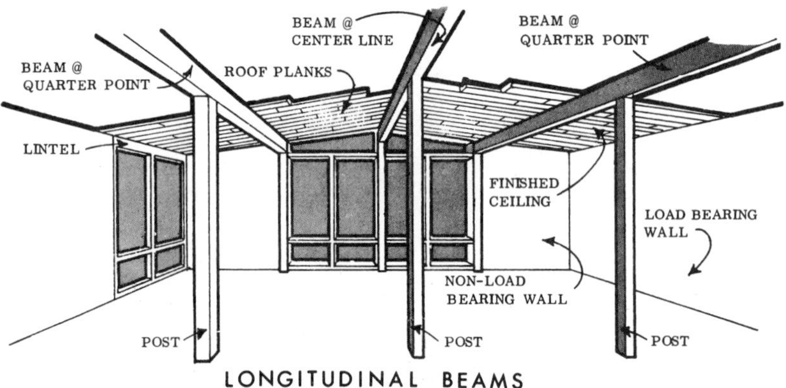

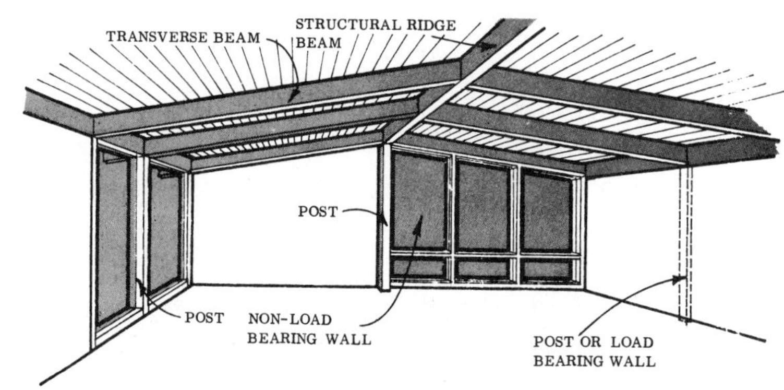

FIGURE 6.7-11 Wood plank-and-beam roof and ceiling assemblies.

ings were hewed from trees by hand using axes and similar cutting tools and two-man pit saws. The development of saws powered by moving water made the production of square and rectangular timbers easier and led to the use of heavy timber framing in most of the industrial buildings of the nineteenth century and many of the homes and farm buildings of that time.

Heavy timber construction using large timbers is rare today, but can be seen in many older buildings. One reason for the demise of this type construction is the development of balloon framing and, later, platform framing, accompanied by the ease of producing lumber for them by modern mechanical means. Another reason is the lack of old-growth trees of the size necessary to produce large timbers. Typical heavy timber columns—for example, 8 × 8 (200 × 200), 12 × 12 (305 × 305), and even 16 × 16 (406 × 406)—are often found in early warehouse, storage, and in factory buildings. Heavy timber beams supporting floor loads are 6 × 10s (150 × 250s) or larger. Trusses may be made using 8 × 8 (200 × 200) members, and decking is often 3 in. (75 mm) thick with a 1-in. (25 mm) wood finish floor.

Even where old-growth timber of the appropriate size exists, harvesting it for construction purposes is often prohibited by law. Where such wood is available, it is usually expensive as compared with other available materials, such as steel, that will do the same job. However, heavy timber construction is acceptable in most codes and still finds a use in expensive residential construction projects.

6.7.4 LAMINATED TIMBER

Although heavy timber buildings using large wood members are seldom built today, people still enjoy the look and feel of wood and many buildings are constructed with laminated timber members according to principles for the use of heavy timber. *Laminated timbers*, which are also called *structural glued laminated members* or *gluelams*, make sizes and shapes possible that are difficult and sometimes impossible to obtain in solid timber. Arches, beams, trusses, and other laminated members are custom fabricated from pieces of stress-graded lumber bonded together with structural glues stronger than the wood itself.

Laminated timbers are used in every type building today. In residences, they appear in utilitarian uses, such as beams to span openings and to produce large open areas. In commercial structures, laminated timbers are used to produce large beams for both utilitarian and decorative purposes. In churches, athletic arenas, and residences they are used to produce arches and domed structures that are at once structural and decorative.

Laminated timbers are used to produce large, rigid space frames and trusses with spans of more than 100 ft. (30 m).

Laminated timbers and their installation should conform to the American Institute of Timber Construction (AITC) publications 190.1, *Wood Products—Structural Glued Laminated Timber*, and *Timber Construction Manual*.

6.7.5 ROOF TRUSSES

Roof trusses for light wood-framed buildings are composed of a number of wood framing members (chords) arranged in a framework of triangles to form a supporting structural element for both a roof and a ceiling (Fig. 6.7-12). In modern trusses, members, generally 2 × 4s (50 × 100s) or 2 × 6s (50 × 150s), are connected by split ring connectors or metal gang-nail plates (Figs. 6.7-13 and 6.7-14) that permit members to be assembled without having to be overlapped at the joints. Older trusses were often connected by wood or plywood gusset plates.

A truss acts as a unit to support roof and ceiling loads and to span the entire width of a building between exterior walls. This clear span allows interior walls to be non–load bearing and permits great flexibility in placing walls (Fig. 6.7-15), while also providing space for insulation and heating, air conditioning, electrical, and plumbing equipment.

In a light truss, individual diagonal

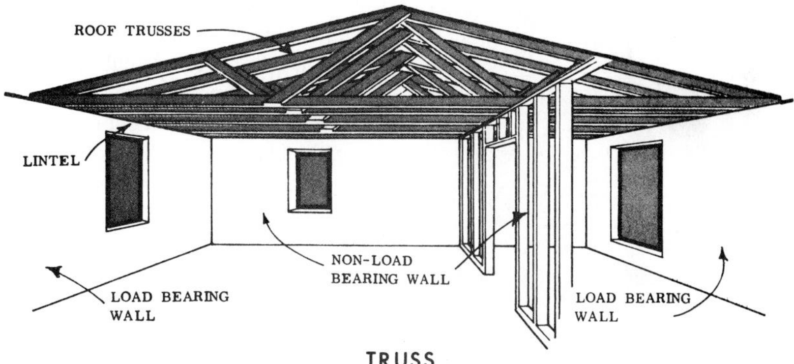

FIGURE 6.7-12 A wood truss roof and ceiling assembly.

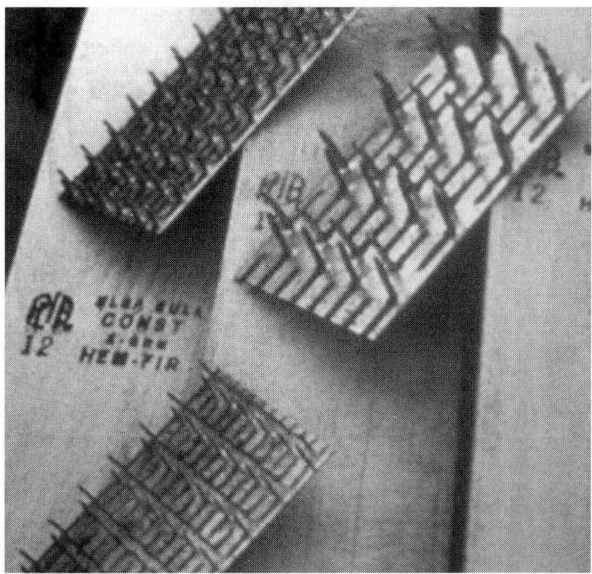

FIGURE 6.7-13 The types of metal plates used to connect wood truss components. (Courtesy Wood Truss Council of America.)

FIGURE 6.7-14 **Spliced joints may have metal plates on one or both sides.** (Courtesy Wood Truss Council of America.)

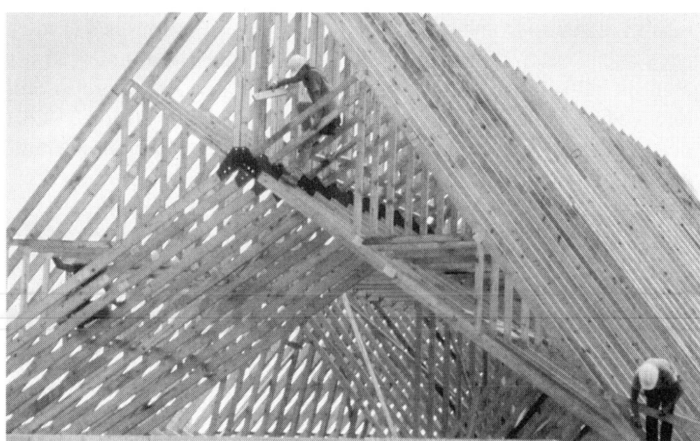

FIGURE 6.7-17 **Trusses with sloped roof and ceiling. Ceilings can also be flat.** (Courtesy Wood Truss Council of America.)

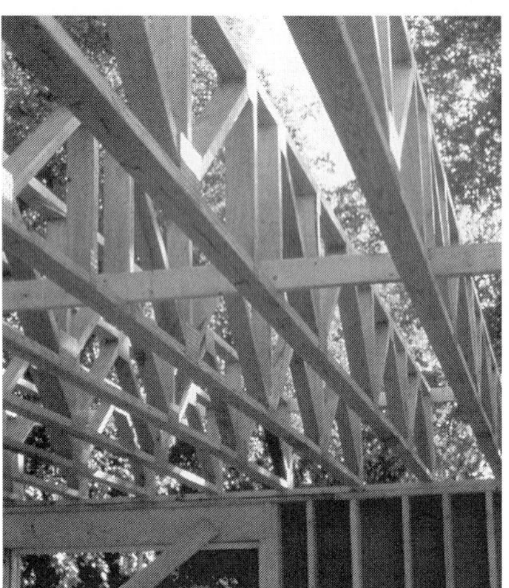

FIGURE 6.7-15 **Trusses provide clear spans and eliminate the need for interior load-bearing walls.** (Courtesy Wood Truss Council of America.)

members resist compression and tension forces. The top and bottom chords also resist bending, because roof and ceiling loads are applied along their entire length. However, since the chords span only between panel points, their spans are much shorter than those of normal rafters and ceiling joists. Therefore, the loads they carry are lighter and the bending stresses they must resist are smaller than those of conventional framing members. As a result, truss top and bottom chords are much smaller than conventional joists and rafters with the same span. Trusses require stress-rated lumber and must be built from engineered designs. They are commonly spaced from 24 in. (610 mm) to 48 in. (1220 mm) on centers, and, like other preassembled components, they can be erected easily, permitting the building to be enclosed quickly (Fig. 6.7-16).

Truss roof systems may be designed for flat (see Fig. 6.7-16) or sloped roofs to receive membrane or shingle roofing finishes (Fig. 6.7-17). Plywood or board sheathing is fastened to trusses to receive roofing. Ceilings can be either flat, sloped (see Fig. 6.7-16), or curved (Fig. 6.7-18).

Plaster bases and gypsum board finishes are commonly applied directly to the lower chord member of a truss, but most any finish may be used. When trusses are

FIGURE 6.7-16 **Individual trusses are positioned easily by hand. Trusses shown will produce a flat roof and ceiling.** (Courtesy Wood Truss Council of America.)

FIGURE 6.7-18 **Sloped roof trusses with segmented bottom chords to permit installation of a curved ceiling.** (Courtesy Wood Truss Council of America.)

FIGURE 6.7-19 **Large open spaces can be created economically using wood trusses.** (Courtesy Wood Truss Council of America.)

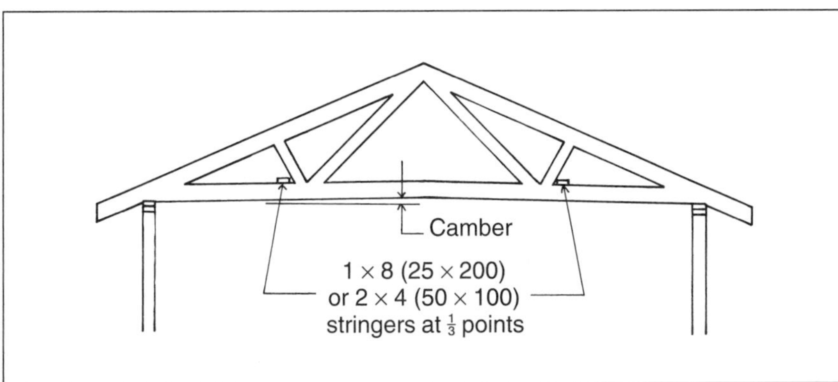

FIGURE 6.7-20 **When cross furring is not used, 1 × 6 (25 × 150) or 2 × 4 (50 × 100) stringers at one-third points may be installed to regulate truss spacing. Additional 2 × 6 (50 × 150) stringers may also be used to level the bottom chords when necessary.**

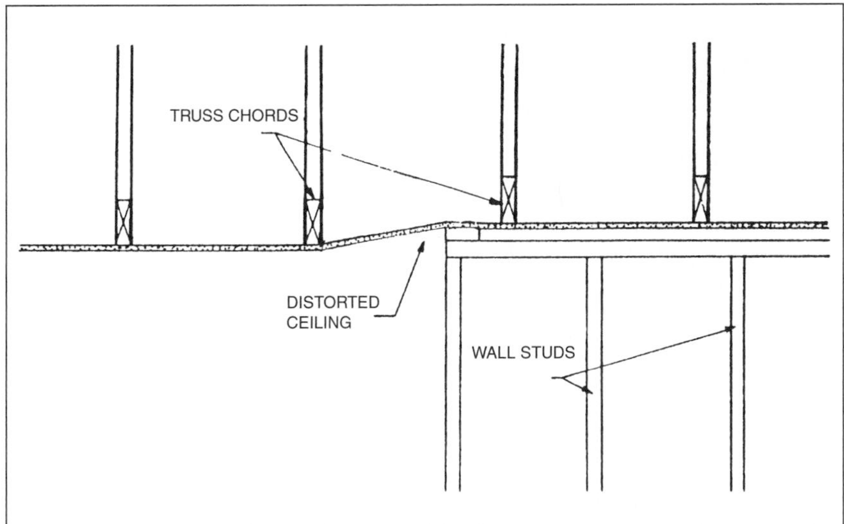

FIGURE 6.7-21 **In truss construction, ceiling distortion may result if substantial loads are introduced after interior partitions have been erected.** (Courtesy Gypsum Association.)

spaced more than 24 in. (610 mm) on centers and when other than board finishes are used, it may be necessary to apply nailing or furring strips across the trusses to receive the interior finish.

Trusses are economical and widely used, even for large structures (Fig. 6.7-19). They provide excellent structural performance, make efficient use of materials, are easy to assemble and to erect, and provide maximum interior flexibility.

Irregularities in spacing and leveling are often magnified in wood truss construction because relatively small members are used for bottom chords and larger spans are involved than in joist construction. Cross furring is generally recommended to remedy these problems. However, when gypsum board is nailed directly to the bottom chords, the chords can be aligned and leveled by using stringers at approximately one-third points continuously across the entire ceiling. Built-in camber to compensate for future deflection of bottom chords is also recommended (Fig. 6.7-20).

With trussed roof construction, exterior walls and the ceiling are sometimes finished before interior partitions are erected and finished. When this method is used, the roofing and other construction elements that would increase roof loads should be installed before interior partitions are erected. If substantial roof loads are introduced after partitions are installed, ceiling distortion may result near interior partitions as trusses deflect (Fig. 6.7-21).

6.7.6 WOOD-POLE CONSTRUCTION

Pole construction has been used for millennia. From prehistoric lake dwellers to twentieth-century farmers, people have been embedding poles in the ground and building simple elevated structures to protect themselves from unfriendly humans, ravaging animals, fierce winds, and engulfing floods.

Over time people have learned to recognize naturally decay-resistant species such as redwood, cypress, and cedar. However, it was not until after World War II, when pressure-treated poles became widely available, that pole construction of farm buildings and beach cottages became commonplace.

The use of chemically treated poles gained impetus from the favorable experience of West Coast builders in the 1950s. They erected houses on steep slopes, on periodically flooded lowlands, and in areas subject to earthquakes, and yet they were able to build them at reasonable costs with a high degree of permanence.

The following information in this section is primarily based on a study by the American Wood Preservers Institute (AWPI), done for the Department of Housing and Urban Development (HUD). Some recommendations and illustrations were also obtained from publications of the Forest Products Laboratory of the U.S. Department of Agriculture and from *Pole Building Design* by Donald Patterson for the AWI.

6.7.6.1 General Description

A pole building is a structure whose contact with the ground is through a series of decay- and termite-resistant round tree trunks (poles) or squared wood members (timbers). Poles are cheaper and longer lasting, are more frequently used, and therefore give their name to the system (Fig. 6.7-22).

Although cedar, redwood, and cypress have inherent decay resistance, their availability in long, thick poles is limited and their decay resistance varies with local growing conditions. Chemical treatment of Douglas fir, southern pine, ponderosa pine, and red pine in accordance with the American Wood Preservers Association's (AWPA) standards ensures long-term resistance to infestation and in-ground durability.

In typical pole construction, poles ranging in diameter from 6 in. (150 mm) to 12 in. (305 mm) are spaced 6 ft. (1800 mm) to 12 ft. (3600 mm) apart to support floor and roof areas up to 144 sq. ft. (13.38 m^2) (Fig. 6.7-23). Most such poles are dropped into holes made either with a power auger or by hand digging. Occasionally, poles are "jetted" with a stream of water or driven with a pile driver. Proper pole embedment and backfilling of the hole are critical to the structural soundness of a pole building. A concrete pad under the pole, or a concrete necklace around the pole, may be required to increase the area of contact with the soil. The pad and necklace help to distribute the load over a larger area.

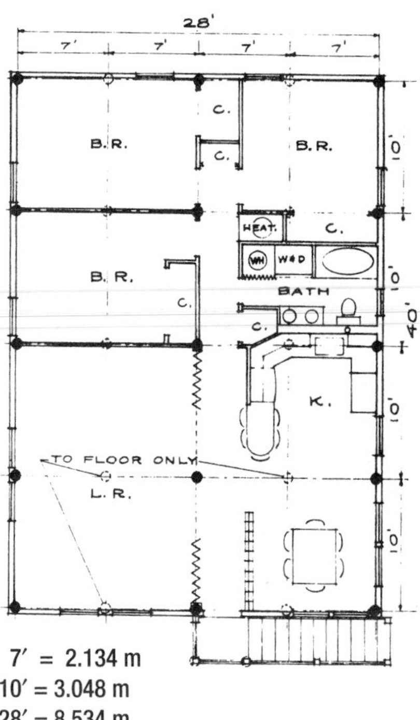

FIGURE 6.7-23 A plan for a small house with all poles inside the house. Short poles could be eliminated and poles placed outside the walls if floor girders were used. (Courtesy American Wood Preservers Institute.)

FIGURE 6.7-22 Treated poles are frequently combined with timbers to form a structural frame and enhance a design. (Courtesy American Wood Preservers Institute.)

Poles may extend substantially above ground to form the load-bearing frame of a superstructure or may terminate at the first-floor level, thus permitting conventional stud-and-joist construction above the pole-framed platform (Fig. 6.7-24).

Embedded poles can support a superstructure either by spreading the loads with a concrete footing or by bearing on rock or very stiff soils. The following recommendations are applicable only where a concrete footing pad, necklace, or backfill will be used to spread the building loads.

All wood-pole structures should be designed by a licensed professional engineer.

6.7.6.2 Advantages and Disadvantages

Pole construction is most advantageous when conventional methods are either uneconomical or do not meet the desired objectives of minimum disturbance to the terrain, freedom from flooding, or resistance to strong winds and earthquakes. Therefore, pole buildings are generally more economical on steep slopes because they use relatively light wood members. Because pole construction does not require heavy equipment for excavation, it

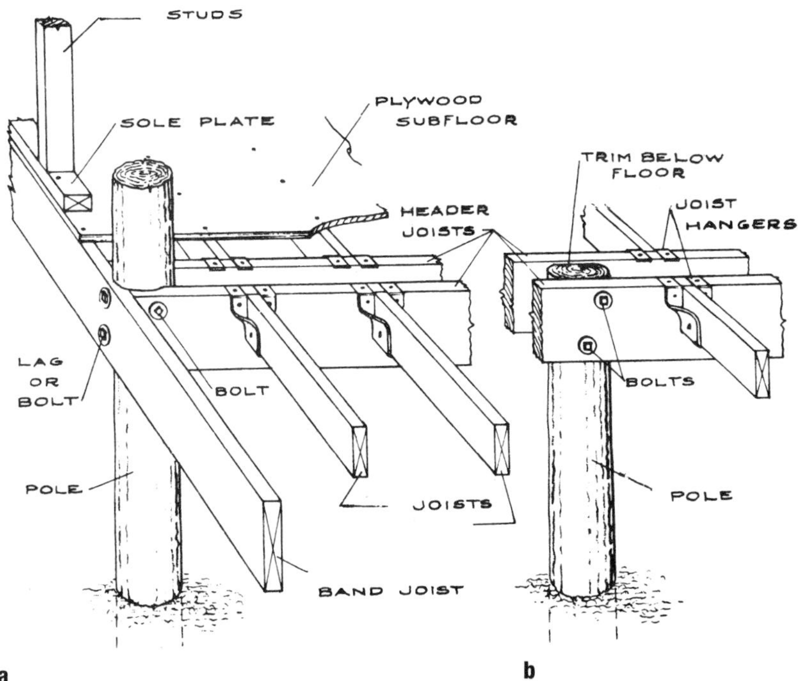

FIGURE 6.7-24 **In a pole-frame building (a), the poles continue up past the main floor. In a pole-platform building (b), the poles stop at the main floor. Lag bolts and joists are used for relatively light loads. Timbers and through-bolted shear connectors are usually required for efficient pole loading.** (Courtesy American Wood Preservers Institute.)

FIGURE 6.7-26 **The pole substructure elevates the house above the natural terrain, permitting unobstructed runoff of storm and flood waters.** (Courtesy American Wood Preservers Institute.)

FIGURE 6.7-25 **A pole house "floats" above the terrain.** (Courtesy American Wood Preservers Institute.)

retains existing drainage patterns and preserves ground cover (Fig. 6.7-25).

On level ground subject to periodic flooding (as in coastal areas from high winds or in river plains from spring floods), pole buildings permit raising the lower floor safely above flood levels. When there are adequate embedment and proper connections, flood waters can flow under the building with minimum damage to the structure (Fig. 6.7-26). However, pole buildings in flood-prone areas should be carefully engineered for specific local conditions, rather than according to the generalized guidelines of this section.

In hurricane-plagued areas, whether on flat or sloping ground, pole construction permits secure anchorage of a structure from roof to footings. Proper connections and adequate embedment are especially important in areas of high wind subject to flooding.

In most parts of the country it is necessary to insulate the first floor and to provide a finish to protect the insulation.

In heavily wooded areas, vegetation and combustible materials should be removed from the area that will be beneath the building when it is raised above grade only enough to create a crawl space. A skirt may be desirable to keep animals out and prevent accumulation of debris under the building.

The underfloor cavity should remain accessible to permit periodic inspections of the poles and the underside of the floor.

6.7.6.3 Site Preparation

The use of pole construction provides a great opportunity to preserve the natural features of a site. Buildings can be fitted to the terrain while observing varying setbacks and sideyards as dictated by codes and good design principles.

Drainage can be crucial on hillside sites. Pavements, roofs, road cuts, and utility trenches may concentrate surface waters into erosive streams. Heavy machinery should be used as little as possible, and necessary roads should be terraced along contour lines. It is generally preferable to dig holes by hand and to remove as little natural vegetation as possible. Hand digging theoretically is possible for holes up to 6 ft. (1800 mm) deep, but may be uneconomical except for shallow poles, as described in Section 6.7.6.6.

On relatively flat sites, machine digging is more practical, but most truck-mounted augers are limited to 18-in. (460-mm)-hole diameters. While these hole sizes are adequate for close pole spacing in good soils, larger diameters are required for most soil conditions.

The number of poles used should be the minimum number that is consistent with other design and structural requirements. The most efficient use of poles usually occurs when the vertical load per pole is about 8000 lb. (3630 kg). Heavier loads require special engineering; one-half the indicated loading is generally inefficient.

Dug holes should be at least 8 in. (200 mm) larger than the diameter of the pole, to permit accurate aligning and plumbing (Fig. 6.7-27). In most cases, the hole is

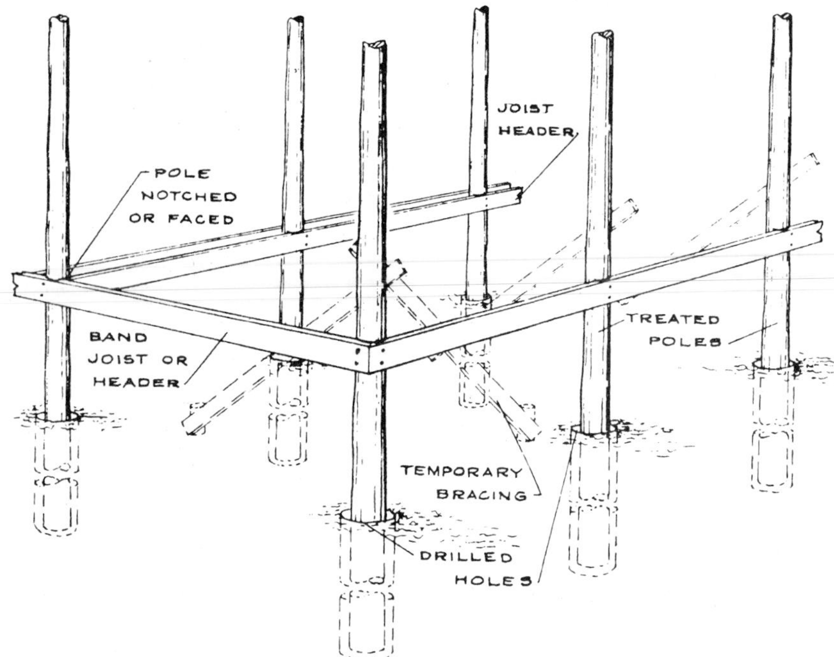

FIGURE 6.7-27 Poles are aligned and plumbed, then braced temporarily, before holes are backfilled. Heavier horizontal members are likely to be used in typical house construction.

wider than this minimum in order to develop a bearing area of sufficient size for the loads.

6.7.6.4 Pole Selection

The main considerations in pole selection are (1) type of treatment and (2) size and length.

Poles should be pressure treated in accordance with AWPA C-23 (see Section 6.1.5), which specifies how much preservative should be retained in each pole and how deep it should penetrate into the wood. Treated poles are frequently branded with the manufacturer's name and location, the year of treatment, pole size, length, wood species, and type of preservative.

6.7.6.5 Construction Methods

There are two principal methods of pole construction: pole frame and pole platform. Footing size and depth of embedment are partly dependent on the method of construction.

In *pole-frame* buildings, the main vertical members extend from below the ground surface to the roof. The poles not only support the vertical weight of the building, but give it considerable lateral resistance to the forces of wind, flood, and earthquake.

In *pole-platform* buildings, treated poles form the substructure, permitting conventional construction methods to be used above. Pole-platform structures generally require greater pole embedment than pole-frame types.

GENERAL RECOMMENDATIONS

Following a few general rules will avoid problems in matching conventional framing methods with pole construction. Poles should not be placed within stud walls, because it is costly to apply finishes around poles that vary in diameter and shape. Beams, girders, and posts should be spaced so that standard, readily available sizes of lumber and panel materials can be used with a minimum of waste.

On level ground and on steep slopes with good soils (6000 psf [29.29 Mg/m^2] and better), noncementitious, granular backfill can be used, provided that a concrete pad or necklace is used (Fig. 6.7-28).

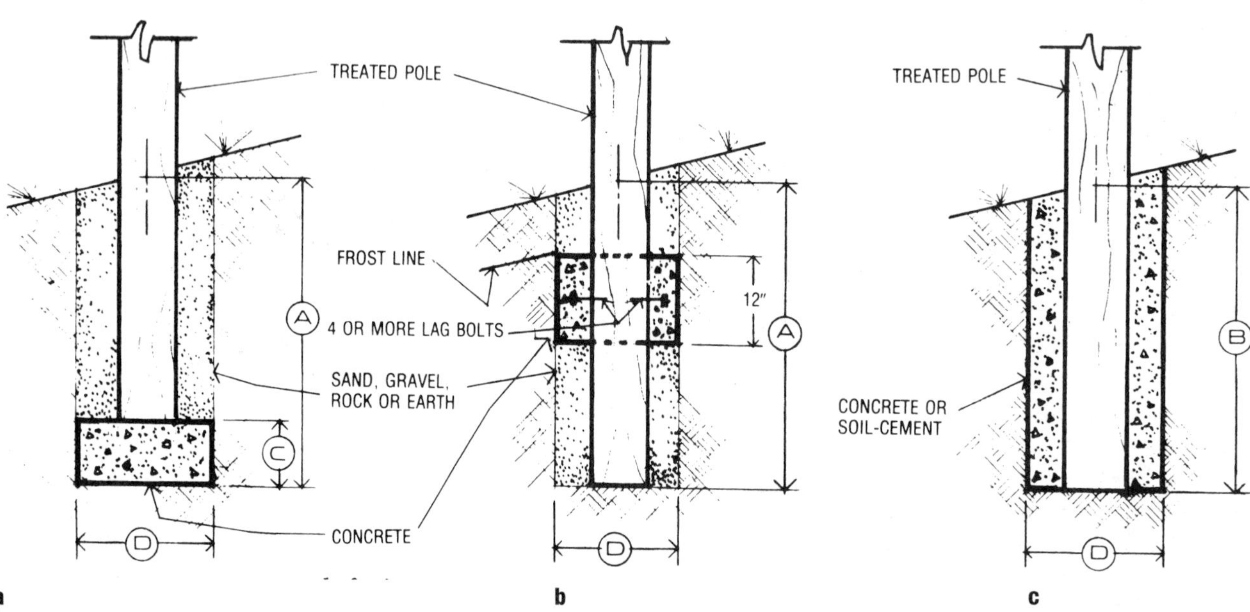

FIGURE 6.7-28 Recommended pole footings: (a) concrete pad, (b) concrete necklace, (c) concrete or soil-cement backfill. Dimensions A, B, and D can be obtained from Figures 6.7-31 through 6.7-35. Pad thickness C should be determined by engineering calculations.

FIGURE 6.7-29 Soil Classification for Pole Construction

Rating	General Description	Soil Groups	Maximum Bearing Value, psf	Lateral Pressure per ft. of Embedment, psf
Good	Compact, well-graded sand and gravel; hard clay or graded fine and coarse sand	GW, GP, SW, and SP	6,000	400
Average	Loose gravel, medium clay, or more compact compositions	GM, GC, SM, SC, and SL	3,000	200
Below average	Sand or clay containing large amounts of silt	CL, OL, MH, CH, and OH	150	100

(Courtesy American Wood Preservers Institute.) (Refer to Figure 1.8-27 for metric conversions.)

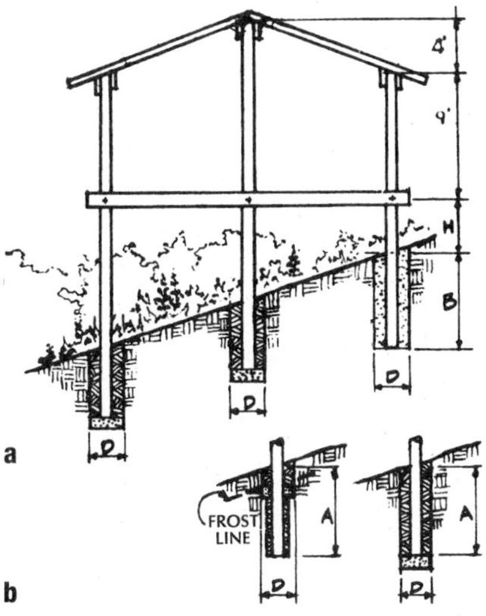

FIGURE 6.7-30 **Building height design assumptions and embedment conditions: (a) concrete pad or necklace and granular backfill, (b) concrete or soil cement backfill.** (Courtesy American Wood Preservers Institute.) (Refer to Figure 1.8-27 for metric conversions.)

Common backfill materials include sand, gravel, and crushed rock. Clean sand is often the least expensive of these and can achieve 100% compaction when properly tamped. Excavated soil from the site also can be used, provided it is free of organic matter and is compacted by tamping.

Under these conditions, backfilling with concrete or with soil-cement (excavated soil mixed with portland cement in a 5 to 1 ratio) will generally reduce the required hole depth by 1.5 ft. (460 mm) to 2 ft. (610 mm) (see Fig. 6.7-28).

On steep slopes with average or below-average soils (3000 psf [14.65 Mg/m^2] and lower), only concrete or soil-cement backfills are suitable. When concrete or soil-cement is used, a separate pad or necklace is not necessary.

DESIGN ASSUMPTIONS

Pole embedment depends on many factors. To reduce the variables to a smaller number that can be tabulated readily, certain design assumptions are made. These are included in the following list. Note that other design assumptions would produce different results from those outlined in the following paragraphs.

Soil bearing values fall in three general categories, as shown in Figure 6.7-29.

Building height of a pole-frame structure is 9 ft. (2743 mm) from floor to top plate; 4 ft. (1220 mm) from top plate to ridge (Fig. 6.7-30).

Embedment Conditions:

Condition A: Each pole either rests on a concrete pad or has a concrete necklace and is backfilled with granular material.

Condition B: Each pole is completely backfilled to grade with either concrete or soil-cement.

POLE-FRAME BUILDINGS

Embedment depths vary according to the slope of the site, as well as the soil, pole spacing, and unsupported pole height.

Flat Slope On sites with a slope of less than 1 in 10, poles should be embedded and supported as recommended in Figure 6.7-31.

Steep Slope On sites with a slope of 1 in 10 up to 1 in 1, poles should be embedded and supported as recommended in Figures 6.7-32 and 6.7-33.

It should be noted that embedment recommendations differ for uphill and downhill poles. Uphill poles, which have a short unsupported height, provide the necessary rigidity for the structure. In order to transfer this rigidity to the longer downhill poles, the floor should be designed as a diaphragm. Properly nailed conventional

FIGURE 6.7-31 Embedment Depths for Pole-Frame Buildings on Sites with Slopes Less Than 1:10

		Good Soil				Average Soil				Below-Average Soil			
		Embedment Depth, ft.				Embedment Depth, ft.				Embedment Depth, ft.			
H, ft.	Pole Spacing, ft.	A	B	D, in.	Tip Size, in.	A	B	D, in.	Tip Size, in.	A	B	D, in.	Tip Size, in.
1½ to 3	8	5.0	4.0	18	6	6.5	5.0	24	6	*	6.0	36	6
	10	5.5	4.0	21	7	7.0	5.0	30	7	*	6.5	42	7
	12	6.0	4.5	24	7	7.5	5.5	36	7	*	7.0	48	7
3 to 8	8	6.0	4.0	18	7	7.5	5.5	24	7	*	7.0	36	7
	10	6.0	4.5	21	8	8.0	6.0	30	8	*	7.5	42	8
	12	6.5	5.0	24	8	*	6.0	36	8	*	8.0	48	8

*Embedment depth is greater than 8 ft., considered excessively expensive.

Note: Where a concrete floor slab is used at grade, embedment depths may be reduced to 70% of those shown.

(Courtesy American Wood Preservers Institute.) (Refer to Figure 1.8-27 for metric conversions.)

joists and ½ in. (12.7 mm) or thicker plywood floors provide such diaphragm action.

POLE-PLATFORM BUILDINGS

Embedment depths of poles vary according to the slope of the site, as well as the other factors tabulated for pole-frame buildings.

FIGURE 6.7-32 Embedment Depths for Downhill Poles of Pole-Frame Buildings on Slopes up to 1:1

	Embedment Depth, ft.		
	Slope of Grade		
Soil Strength	Up to 1:3	Up to 1:2	Up to 1:1
Below average	4.5	6.0	*
Average	4.0	5.0	7.0
Good	4.0	4.0	6.0

(Courtesy American Wood Preservers Institute.)
(Refer to Figure 1.8-27 for metric conversions.)
*Embedment depth is greater than 8 ft., considered excessively expensive.

Flat Slope On sites with a slope of less than 1 in 10, poles should be embedded and supported as recommended in Figure 6.7-34.

Steep Slope On sites with a slope of 1 in 10 up to 1 in 1, uphill poles should be embedded and supported as recommended in Figure 6.7-35. Recommendations for downhill poles are given in Figure 6.7-32. The floor should be designed as a diaphragm to transfer the rigidity of the uphill poles to the rest of the structure.

CONNECTIONS

It is generally better to frame poles with pairs of beams by lag bolting, through bolting, or nailing to the supporting pole (Fig. 6.7-36). To make a stronger connection, poles can be faced or notched (Fig. 6.7-37). The use of a spike grid (Fig. 6.7-38) is preferred to squaring or notching because it minimizes damage to the preservative-treated outside layer and increases the load-bearing value of the connection.

Accepted design practices are outlined in AWPI's *Pole Building Design* and in *Timber Construction Manual* by the American Institute of Timber Construction. Strength values used in designing bolted and nailed connections should conform to the *National Design Specifications for Stress-Grade Lumber and Its Fastenings* by the National Forest Products Association.

6.7.6.6 Special Construction on Rocky Slopes

The following discussion of pole bracing as an alternative to embedment should not be attempted without competent engineering advice.

On rocky slopes, where deep embedment cannot be developed without expensive machine digging or rock drilling, it is possible to use shallow, hand-dug holes,

FIGURE 6.7-33 Embedment Depths for Shorter, Uphill Poles in Pole-Frame Buildings on Slopes up to 1:1

		Good Soil				Average Soil				Below-Average Soil			
		Embedment Depth, ft.				Embedment Depth, ft.				Embedment Depth, ft.			
H, ft.	Pole Spacing, ft.	A	B	D, in.	Tip Size, in.	A	B	D, in.	Tip Size, in.	A	B	D, in.	Tip Size, in.
1½ to 3	6	7.0	5.0	18	8	a	6.5	18	8	a	a	a	a
	8	7.5	5.5	18	9	a	7.0	24	9	a	a	a	a
	10	a	6.0	21	9	a	8.0	30	9	a	a	a	a
	12	a	6.5	24	10[b]	a	a	a	a	a	a	a	a
3 to 8	6	7.5	5.5	18	8	a	7.0	18	8	a	a	a	a
	8	8.0	6.0	18	9	a	8.0	24	9	a	a	a	a
	10	a	7.0	21	10[b]	a	a	a	a	a	a	a	a
	12	a	7.0	24	11[b]	a	a	a	a	a	a	a	a

[a]Embedment depth is greater than 8 ft., considered excessively expensive.
[b]These tip diameters may be decreased 1 in., provided embedment is increased by ½ ft.
(Courtesy American Wood Preservers Institute.) (Refer to Figure 1.8-27 for metric conversions.)

FIGURE 6.7-34 Embedment Depths for Poles in Platform Buildings on Slopes Less Than 1:10

		Good Soil				Average Soil				Below-Average Soil			
		Embedment Depth, ft.				Embedment Depth, ft.				Embedment Depth, ft.			
H, ft.	Pole Spacing, ft.	A	B	D, in.	Tip Size, in.	A	B	D, in.	Tip Size, in.	A	B	D, in.	Tip Size, in.
1½ to 3	8	4.0	4.0	18	5	5.5	4.0	24	5	7.0	5.0	36	5
	10	4.5	4.0	21	5	6.0	4.0	30	5	8.0	5.5	42	5
	12	5.0	4.0	24	5	6.5	4.5	36	5	*	5.5	48	5
3 to 8	8	5.0	4.0	18	6	6.5	4.5	24	6	*	6.0	36	6
	10	5.5	4.0	21	7	7.0	5.0	30	7	*	6.5	42	7
	12	6.0	4.5	24	7	7.5	5.5	36	7	*	7.0	48	7

*Embedment depth is greater than 8 ft., considered excessively expensive.
(Courtesy American Wood Preservers Institute.) (Refer to Figure 1.8-27 for metric conversions.)

FIGURE 6.7-35 Embedment Depths for Uphill Poles in Platform Buildings on Slopes up to 1:1

		Good Soil				Average Soil				Below-Average Soil			
		Embedment Depth, ft.				Embedment Depth, ft.				Embedment Depth, ft.			
H, ft.	Pole Spacing, ft.	A	B	D, in.	Tip Size, in.	A	B	D, in.	Tip Size, in.	A	B	D, in.	Tip Size, in.
1½ to 3	6	5.5	4.0	18	6	7.5	5.0	18	6	[a]	7.0	24	6
	8	6.5	4.5	18	7	[a]	6.0	24	7	[a]	8.0	36	7
	10	7.0	5.0	21	7	[a]	6.5	30	7	[a]	[a]	[a]	[a]
	12	7.5	5.5	24	8	[a]	7.0	36	8	[a]	[a]	[a]	[a]
3 to 8	6	7.0	5.0	18	8	[a]	6.5	18	8	[a]	8.0	24	8
	8	7.5	5.5	18	9	[a]	7.0	24	9	[a]	[a]	[a]	[a]
	10	[a]	6.0	21	10[b]	[a]	7.5	30	10[b]	[a]	[a]	[a]	[a]
	12	[a]	6.5	24	10[b]	[a]	[a]	[a]	[a]	[a]	[a]	[a]	[a]

[a]Embedment depth is greater than 8 ft., considered excessively expensive.

[b]These tip diameters may be decreased 1 in., provided embedment is increased by ½ ft.

(Courtesy American Wood Preservers Institute.) (Refer to Figure 1.8-27 for metric conversions.)

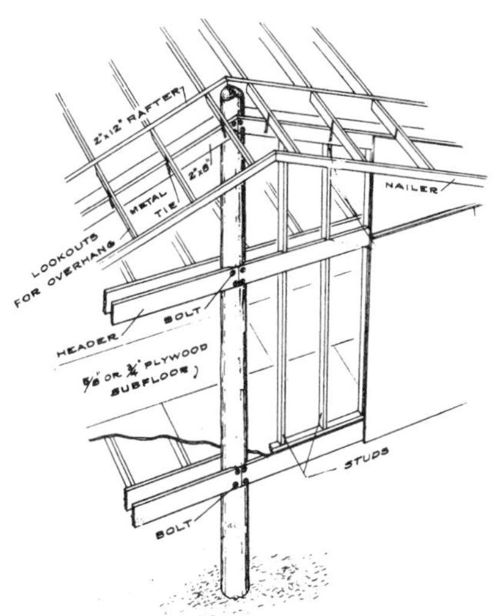

FIGURE 6.7-36 Double headers at exterior end walls frame ridge-high pole and are bolted to it. (Courtesy American Wood Preservers Institute.)

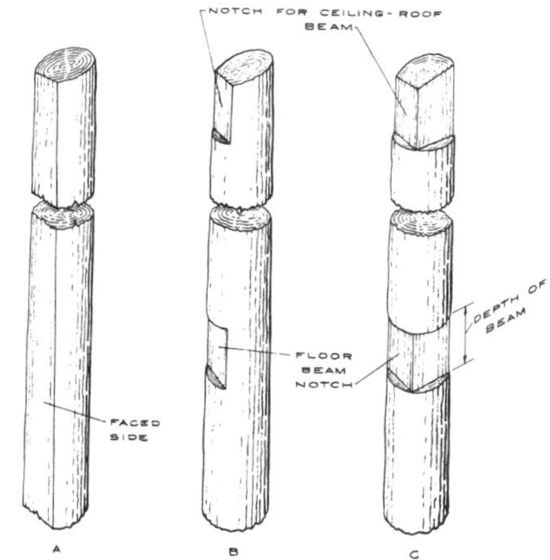

FIGURE 6.7-37 Poles are faced or notched to provide better contact with beams and make suitable connections with shear connectors such as the spike grid shown in Figure 6.7-38. (Courtesy American Wood Preservers Institute.)

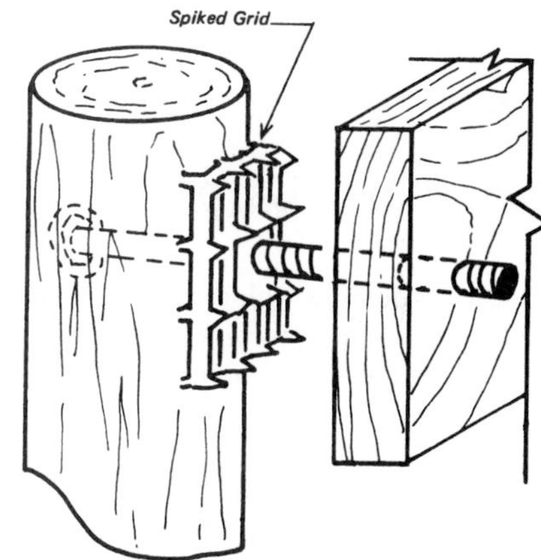

FIGURE 6.7-38 A single curved spike grid with a ¾-in. (19.1 mm) bolt has a carrying capacity of 3800 lb. (1723.6 kg); with a 1-in. (25.4 mm) bolt, 4100 lb. (1859.7 kg). (Courtesy American Wood Preservers Institute.)

provided that the pole structure is adequately braced and anchored.

Alternates to deep embedment include (1) pole seating and cross bracing and (2) shear walls and floor diaphragms.

POLE SEATING AND CROSS BRACING

If the understructure is adequately braced, the poles must be embedded only far enough to prevent slipping downhill. The embedment should be at least 4 ft. (1220 mm), and the poles should be placed in specially dug seats in the rock. Since a rock surface is likely to be irregular, tamped earth or wetted sand should be used to make a flat bearing surface under each seated pole. If a pole seat cannot be dug into the rock, a concrete pad may be used over the rock and the pole anchored to the pad with lag bolts or standard post anchors. The pad should be 8 in. (200 mm) thick and of the same diameter as the hole, which can be 8 in. (200 mm) larger than the pole.

In one cross-bracing method, ⅝-in. (15.9 mm) or ¾-in. (19.1 mm) steel rods are fitted through drilled holes, flooded with preservative, then fastened with nuts, using beveled washers to avoid crushing the wood. Steel-rod cross bracing is not usually needed between poles in every case; several diagonal rods are often adequate.

A pole structure can also be knee braced with 2 × 6 (50 × 150) wood members, nailed or bolted to form a diagonal connection from the poles to the floor beams. The bracing can be left exposed or covered with skirt boards around the perimeter of the house.

SHEAR WALLS AND FLOOR DIAPHRAGMS

A masonry or concrete fireplace block that is supported on concrete footings at least 4 ft. (1220 mm) below grade, or a masonry or concrete wall can act as a shear wall to resist horizontal forces. Similarly, a masonry or concrete utility room below the first floor can act to brace the structure and anchor it.

Key walls can also be used to resist the horizontal forces pushing a structure down a slope (Fig. 6.7-39). For both shear walls and key walls, floors should be de-

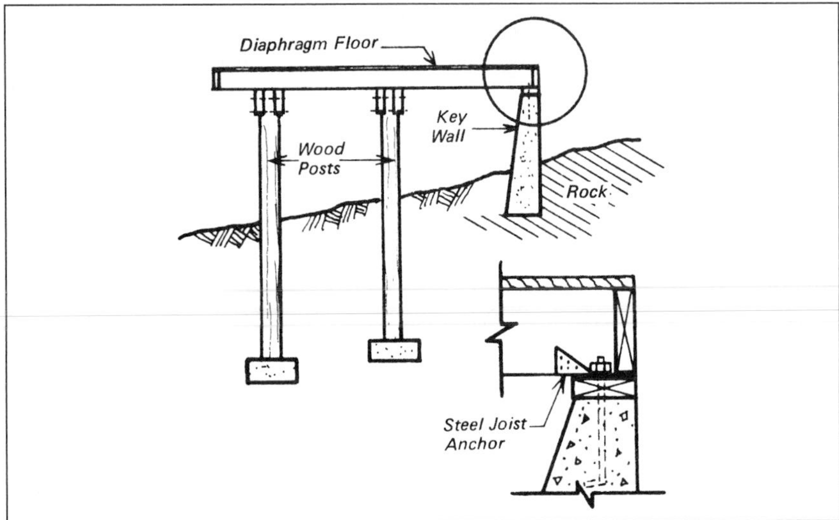

FIGURE 6.7-39 Key wall bracing. (Courtesy American Wood Preservers Institute.)

signed as diaphragms to transfer the lateral rigidity of the walls to the poles. Such floors can be built of conventional wood joists and ½ in. (12.7 mm) or thicker plywood, with nailing as recommended by the APA.

6.8 Finish Carpentry

Finish carpentry, which is sometimes called *architectural woodwork*, traditionally includes exterior siding (see Section 7.7), wood shingles (see Section 7.6), casework and cabinets (see Section 12.1), interior and exterior standing and running trim, interior paneling, and other wood elements addressed in this section. Finish carpentry also traditionally includes wood doors and windows, which are addressed in Chapter 8.

Finish carpentry is generally required to comply with the recommendations of the Architectural Woodwork Institute (AWI), as indicated in its *AWI Quality Standards*, or of the Woodwork Institute of California (WIC) in its *Manual of Millwork*. The AWI standard is applicable throughout most of the United States, but the WIC standard is applicable in some western states. These two standards are similar in many respects but differ in others. For example, they use the same names for material and construction grades (Premium, Custom, and Economy), but the WIC manual adds an additional Laboratory grade for casework and countertops in laboratories where chemical resistance is required. They also use different names for the same finish carpentry components. A designer should obtain the standard applicable in the area of the project and use it as a guide. This section refers to the AWI standards in most cases.

6.8.1 MATERIALS AND GRADES

The materials used in finish carpentry are those discussed in Section 6.1. Both hardwoods and softwoods are used. Plywood is also used for casework and in paneling (see Section 6.3).

The grades in the AWI system are Premium, Custom, and Economy, the Premium grade being of the highest quality. The grade refers to both materials and workmanship. Sometimes a different level of quality is required of the workmanship than of the materials, but most of the time the same grade is applied to both.

6.8.2 STANDING AND RUNNING TRIM AND RAILS

Standing trim is a term used to identify items such as door and window casings, stops, stools, sills, aprons, and the like. *Running trim* includes bases, shoe molding, soffits, fascias, chair rails, cornices, crown molding, and similar items. *Rails* include hospital corridor rails and guardrails.

Figure 6.8-1 shows a few of the many molding shapes available as stock items from wood suppliers and lumber mills that are used to make standing and running trim and rails. In addition, almost any desired shape can be milled at the shop or on the job. Numerous cutter blade shapes are available, and any shape can be ground in a metal shop. Trim shapes and moldings can be combined to form larger units (Figs. 6.8-2 and 6.8-3).

Right-angle intersections of most standing and running trim items are mitered (Fig. 6.8-4a) to make neat corners. Corner joints are either doweled, splined, or held together with plate joiners (biscuits) (Fig. 6.8-4a). Running joints are sometimes back cut and sometimes half lapped (Fig. 6.8-4b). Intersecting joints are sometimes made with mortise-and-tenon jointing (Fig. 6.8-4c).

6.8.3 PANELING

Wood paneling is generally available in either 7⁄16-in. (11 mm) or ¾-in. (19.1 mm) thick panels. Joints in flat paneling may be installed using applied molding (Fig. 6.8-5a) or may be butted and splined to hold them in alignment (Figs. 6.8-5b and 6.8-5c). Articulated joints (Fig. 6.8-6) are also used. These figures show a few of the many possible variations. Raised panels are also often used.

Pattern matching in wood paneling is described in Section 6.3 (see Fig. 6.3-18).

6.8.4 CLOSET AND STORAGE SHELVING

Closet and utility shelving is constructed either of wood boards (Section 6.2) or

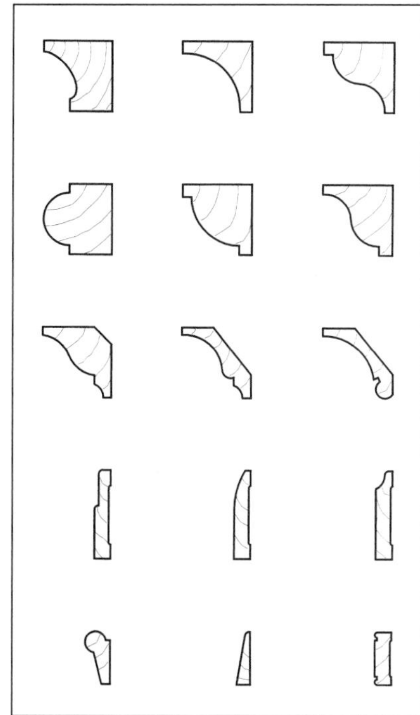

FIGURE 6.8-1 Various molding shapes. (Drawing by HLS.)

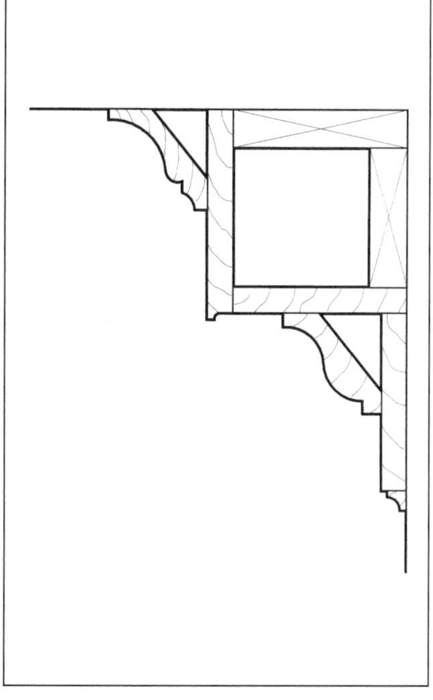

FIGURE 6.8-3 A combination of molding shapes joined to make a complex cornice. (Drawing by HLS.)

panel products (Section 6.3). The edges of shelving made from products other than solid wood are usually edge banded with wood or plastic edging strips.

Shelves are generally required to be ¾ in. (19.1 mm) thick, and those more than 48 in. (1220 mm) long must have intermediate supports. Unsupported shelves more than 48 in. (1220 mm) long must be thicker or must be supported along the front edge by a 2 × (50 ×) wood apron.

6.8.5 STAIRWORK AND HANDRAILS

The AWI *Quality Standards* contain requirements for stairs and handrails. In addition, stair riser heights and tread widths and their ratio to one another, and heights and types of handrails and safety guards are dictated by the applicable code. Freestanding and cantilevered stairs should be designed by a licensed engineer.

Decorative stairs, especially curved and cantilevered stairs, are often constructed in a woodwork shop and delivered to the project in sections for final

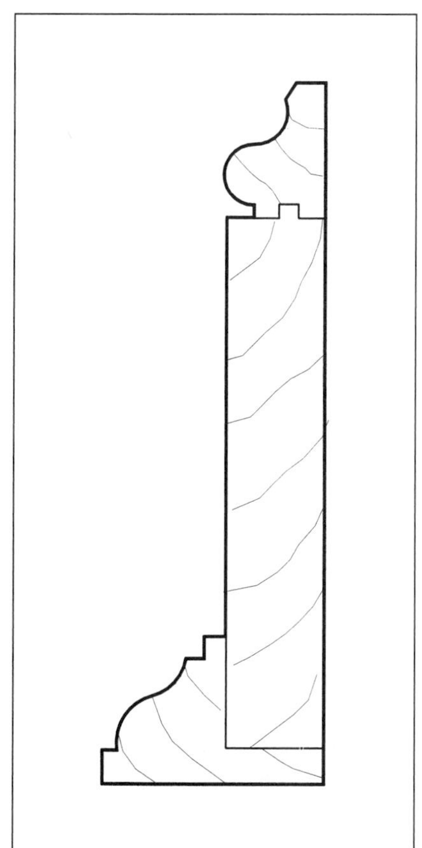

FIGURE 6.8-2 A combination of molding shapes joined to make a wall base. (Drawing by HLS.)

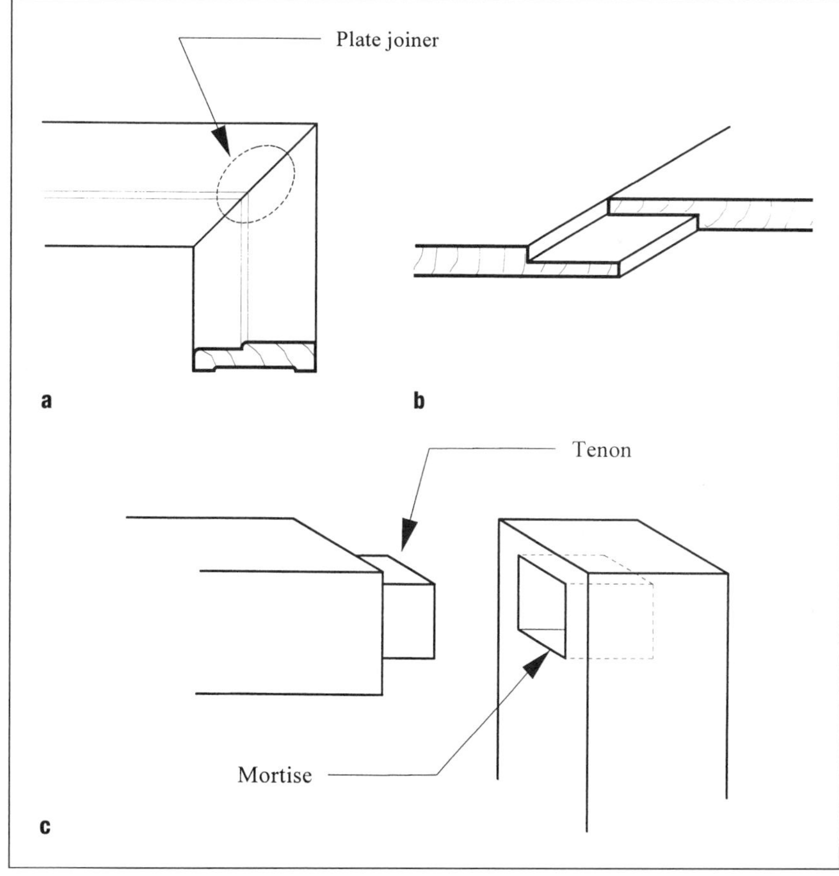

FIGURE 6.8-4 Jointing methods for standing and running trim: (a) mitered corner with a plate joiner, (b) a half-lap joint, (c) a mortise and tenon joint. (Drawing by HLS.)

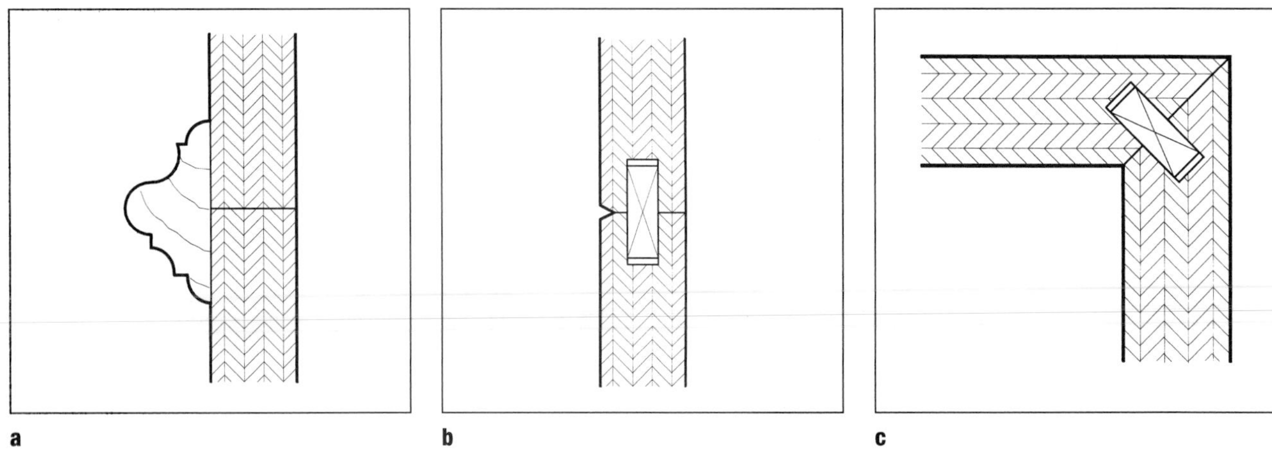

FIGURE 6.8-5 Butt jointing methods for paneling: (a) an applied molding, (b) a butted and splined running joint, (c) a mitered and splined corner joint. (Drawing by HLS.)

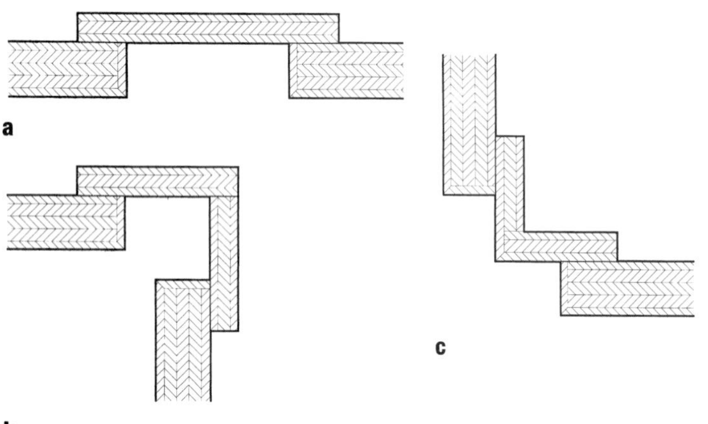

FIGURE 6.8-6 Articulated jointing methods for paneling: (a) a running joint, (b) an inside corner joint, (c) an outside corner joint. (Drawing by HLS.)

field assembly. This process offers better control of the process and often results in a higher-quality finished product than can be achieved when the stair is built completely at the job site.

6.8.6 OTHER FINISH CARPENTRY ITEMS

Many other wood elements are used in modern buildings. Among them are mantles, ornamental grills, pediment heads, pilasters, columns, tracery, and the like. The AWI standards address these as well.

6.9 Plastic Fabrications

Plastics are relative newcomers to the construction field as compared with traditional materials such as wood, glass, and concrete. The first commercial plastic in the United States, cellulose nitrate, was developed just a little more than a century ago. It was created by John Wesley Hyatt to help overcome a shortage of ivory, from which billiard balls were made.

Progress in plastics technology was slow at first. The next commercial plastic to be developed was phenol-formaldehyde, in 1909. Gradually, a few others were introduced, such as casein (1919), alkyds (1926), and polyvinyl chloride (1927). It was not until the 1930s, however, that the modern commercial plastics industry really was started; plastics were not widely used in building construction until after World War II.

Despite their relative newness, plastics have become important materials in the construction field. As a group, they offer a wide range of properties possessed by few other building materials: they are moisture and corrosion resistant, lightweight, tough, and easily molded into complex shapes. Furthermore, the great variety of plastic materials available, each of which offers slightly different advantages, makes plastics well suited for a multitude of purposes.

Sheet and tile flooring is perhaps the most common use for plastics in construction. Vinyl flooring is noted for excellent abrasion and stain resistance and good resilience. Plastic siding is becoming increasingly popular because of its good durability and freedom from maintenance; plastic-covered wood windows and doors offer similar advantages. Plastic laminates are standard materials for counter- and tabletops; they are produced in many decorative designs and bright colors and have good heat and chemical resistance. Plastics also are used for break-resistant glazing, skylights, and roof domes. Plastic films make excellent vapor retarders; plastic foams can be used for a variety of insulation purposes.

Plastics do have limitations, however, and they can create serious problems if used incorrectly. For satisfactory service, a suitable plastic for the intended use should be chosen. The plastic product should be correctly designed and properly integrated with other products.

This section discusses the composition and genetic properties of plastics and describes the chief applications of the most important types. For more detailed information on molecular structure and how it affects the specific properties of materials, see Section 1.10.

6.9.1 COMPOSITION

Twenty to thirty families of materials are classified as plastics. Within each of these families, a multitude of variations and combinations are possible. The complex details of the chemistry involved in the production of these compounds and the terminology used to classify them are beyond the scope of this book. A brief explanation of their composition, molecular structure, and basic types is included, however, to help the reader gain a basic understanding of what plastics are, how they behave, and the ways in which they can be used in construction.

Despite widely varying properties, all plastics have several characteristics in common, which warrants classifying them as a single materials group:

1. At one stage in their formation they exhibit plastic behavior, that is, they will soften and can be formed into desired solid shapes, with or without the use of heat and pressure.
2. They are made up of molecules built around a carbon atom and, therefore, are classed as organic materials. There are some exceptions, however. Some plastics are built around a silicon atom rather than a carbon atom, yet are still considered plastics.
3. They are *synthetic* materials; that is, they are man-made and do not occur in nature.
4. They are high polymers, which are giant molecules made of numerous relatively simple repeating units linked together into large aggregations.

Not all materials that possess the listed characteristics are necessarily plastic. For example, synthetic rubbers have all of these characteristics, but are classed as rubbers.

6.9.1.1 Molecular Structure

Some plastics are soft and flexible; others are hard and brittle. Some will become soft and may melt or even evaporate when exposed to intense heat; others are virtually unaffected by extremes of temperature. An examination of the molecular structure of high polymers can help to explain many of these differences in the properties and behavior of plastics.

HIGH POLYMERS

Most plastics are based on carbon chemistry. This means that they are dependent on the peculiarities and properties of the carbon atom (Fig. 6.9-1). Carbon atoms form bonds with other atoms through covalent bonding (see Section 1.10). They have a valence of 4, meaning that there are 4 points to which other atoms can attach themselves to form covalent bonds.

Atoms of other elements have different valences. For example, hydrogen atoms have a valence of 1, and oxygen atoms have a valence of 2. Therefore, 4 hydrogen atoms can attach themselves to a single carbon atom (one at each valence point) to produce CH_4, the gas known as methane. Two oxygen atoms can attach themselves to a single carbon atom (each oxygen atom taking up two valence points) to produce CO_2, carbon dioxide (see Fig. 6.9-1).

Two adjacent carbon atoms can attach to each other. If this attachment involves only a single valence point, thus creating a single bond between them, and the other valence points are taken up by other atoms, the molecule is said to be *saturated*. In other words, a saturated molecule contains only single bonds between the carbon atoms and has no free (unattached) valence point.

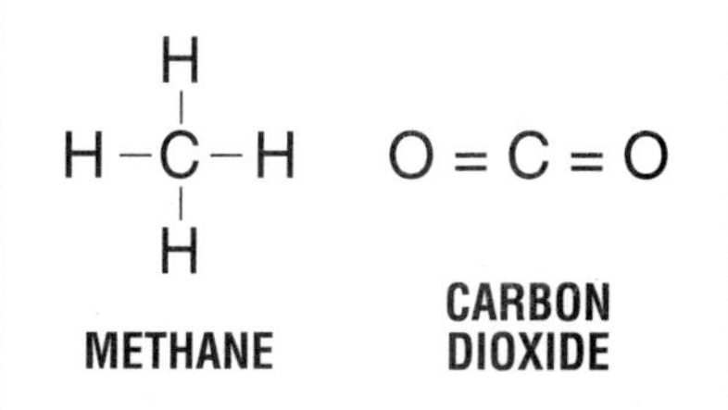

FIGURE 6.9-1 Two molecules based on the carbon atom.

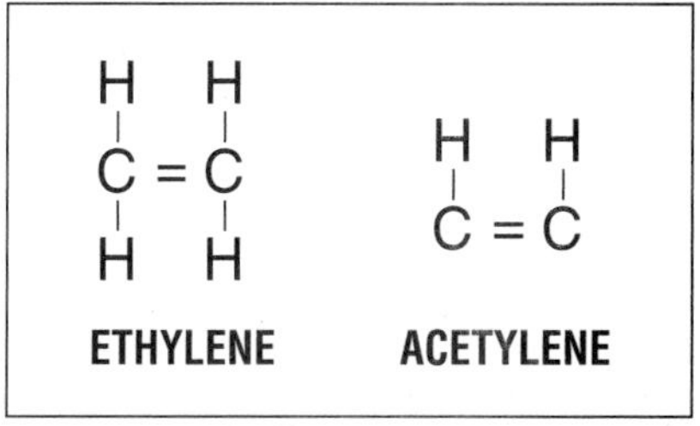

ETHYLENE ACETYLENE

FIGURE 6.9-2 Molecules can have double bonds (ethylene) or triple bonds (acetylene) between the carbon atoms. Such molecules are called unsaturated.

Two carbon atoms may also attach to each other at two valence points, creating a double bond; or at three valence points, creating a triple bond (Fig. 6.9-2). In these cases, the molecules are said to be *unsaturated*. Under certain conditions the double or triple bonds of an unsaturated molecule can be opened up (activated), creating free valence points. Other atoms, or groups of atoms, can then attach themselves to these free valence points. For example, if the double bonds in a large number of ethylene molecules are activated, they can join together to form a chain (Fig. 6.9-3), which is then known as polyethylene. Such chain formation, called *polymerization*, is essential in the production of plastics. Because only unsaturated molecules will react together to form

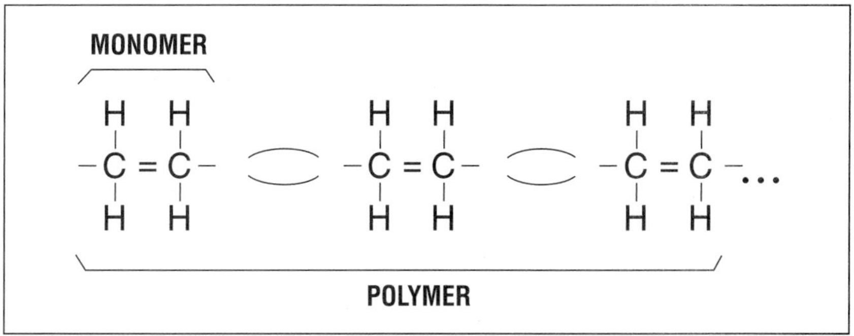

MONOMER

POLYMER

FIGURE 6.9-3 Several ethylene molecules can be linked to form polyethylene, a polymer.

chains, these are the only kinds of molecules that are useful in producing plastics.

A single molecule such as ethylene is known as a *monomer*. Several of these molecules linked together become a *polymer*. When the number of units in a chain is very large (between several hundred and several thousand), it is called a *high polymer*. The number of monomeric units in different chains can vary widely. When chain growth stops, free valence points are left at each end of the chain. These may be taken up by single atoms, such as hydrogen, or by groups of atoms with a single net valence. In some cases, the two ends of a chain loop around and connect, forming a ring.

Plastics are composed of these giant chainlike molecules and are therefore classified as high polymers. This is evidenced by the fact that many of the chemical names for plastics start with the prefix *poly*, which means *many*. Examples are polyethylene, polystyrene, and polyvinyl chloride.

BASIC CATEGORIES

Plastics can be divided into two basic categories based on their molecular structure: (1) *thermoplastic* materials, which can be repeatedly softened and hardened by heating and cooling, and (2) *thermosetting* materials, which are set into permanent shape during forming and cannot be softened again by reheating.

Thermoplastics Thermoplastic materials are made up of long chainlike molecules such as the polyethylene molecule described earlier. These molecules are characterized by being attached to each other. The chemical attraction along the chain is very strong, but the attraction between adjacent chains is relatively weak. Furthermore, these chains are generally saturated, having no double or triple bonds that can be activated. This means that no free valence points can be created to cause a strong attraction to adjacent chains.

At any temperature above absolute zero, molecules are in constant random motion. As a result of the unconnected molecular structure of thermoplastics, the molecules can slide past each other and change shape as in a liquid. At room temperature, this random motion is slight and the plastic holds its shape. However, as the temperature rises, molecular motion increases and the attractive forces between molecules weaken, causing the material to expand and become soft and flexible.

When a plastic reaches a high enough temperature, it will flow and can then be molded into new shapes. The temperature at which this occurs is called the *softening point*, which will vary according to the type of plastic involved. When the material is cooled again, it assumes its former stiffness and holds its new shape at normal-use temperatures. With thermoplastic materials, the process of heating to soften and cooling to harden can be repeated again and again without affecting the properties of the plastic.

The configuration of the molecules in thermoplastics also affects the final properties. Some molecules are arranged in neat, continuous chains (linear molecules); others are highly branched. Highly branched chains do not pack together well, whereas linear molecules can be stacked together closely. As a result, plastics composed of branched molecules tend to be softer, less strong, and lower in density than those composed of linear molecules.

The molecular configuration in a given plastic can be controlled to a certain extent by the method used in its formation. For example, when ethylene is polymerized by high temperature and pressure, the molecules are often highly branched. However, ethylene can also be polymerized through the use of catalysts, which are substances added to the basic chemicals to speed up the chemical reaction. When polyethylene is made by this process, the molecules are generally linear. This means that variations in strength and stiffness can be produced in different batches of the same type of plastic simply by changing the method of polymerization.

There are hundreds of different thermoplastics. A brief description of the most important types is given in Section 6.9.3. However, a few words about the way these different types are developed may be helpful here.

Starting with the ethylene molecule, which is the basic building block of many thermoplastics, substitutions can be made for one or more of the hydrogen atoms involved. Single atoms such as chlorine or fluorine, or whole groups of atoms such as the styrene monomer, can be substituted for the hydrogen atoms (Fig. 6.9-4). For example, if one hydrogen atom is replaced by a chlorine atom, a plastic known as polyvinyl chloride results. If one hydrogen atom is replaced by a phenyl group,

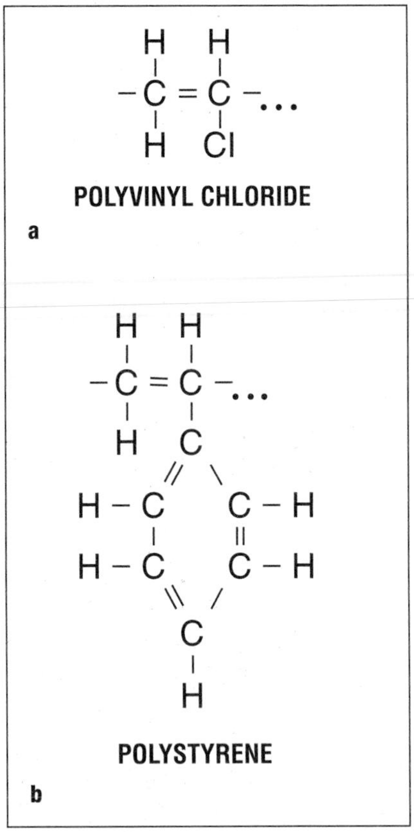

```
   H   H
   |   |
 - C = C -...
   |   |
   H   Cl
```

POLYVINYL CHLORIDE

a

```
     H   H
     |   |
   - C = C -...
     |   |
     H   C
       //  \
  H - C     C - H
      |     ||
  H - C     C - H
       \\  /
         C
         |
         H
```

POLYSTYRENE

b

FIGURE 6.9-4 Starting with an ethylene molecule, substitutions can be made for a hydrogen atom, producing (a) polyvinyl chloride or (b) polystyrene.

polystyrene results. If one hydrogen atom is replaced by a methyl group (CH_3) and a second hydrogen atom is replaced by a methoxycarbonyl group ($COOCH_3$), polymethyl methacrylate (acrylic) results.

Hundreds of such variations are possible, and each variation produces a plastic with slightly different properties. A large group of plastics called *vinyls* are based on the ethylene molecule. Other types are based on different molecules. For example, *cellulosics* are based on modifications of the basic celluloid molecule.

Still other variations are possible by combining different monomeric units into the same molecular chain. These substances then are called *copolymers*. Therefore, instead of having a single repeating monomer along a polymer chain, two different monomers may alternate along the same chain. Numerous variations in the properties of plastics are made possible through such copolymerization. For example, vinyl chloride may be copolymerized with vinyl acetate, producing a substance with some of the properties of each of the original polymers (Fig. 6.9-5).

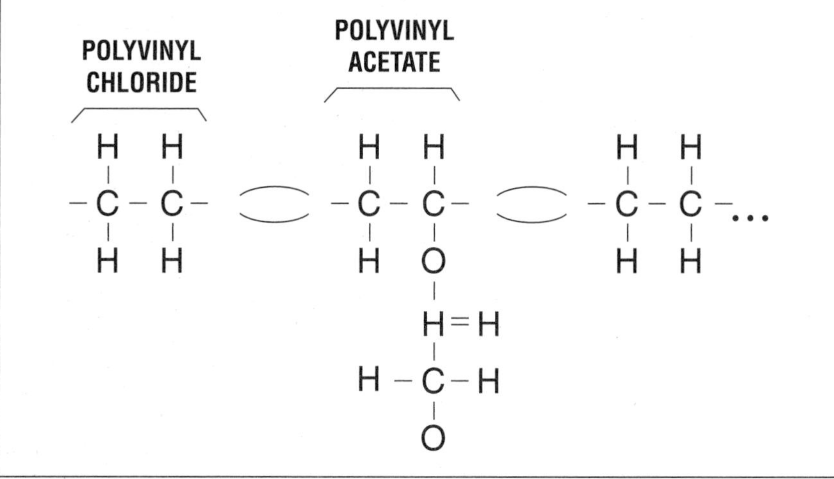

FIGURE 6.9-5 Two different monomers may alternate along a polymer chain, producing a copolymer.

Acrylonitrile-butadiene-styrene (ABS) is another copolymer that has proven to be very useful because of its unique combination of properties.

Thermosets Thermosets differ from thermoplastics in that once they are set (cured), they cannot be softened again. This is because during forming, they go through an irreversible chemical process that develops strong chemical bonds between adjacent molecules. Therefore, the molecules cannot slide past each other as they can in thermoplastic material, but instead form a strongly interconnected molecular structure.

An example of this process is the production of thermosetting plastics from unsaturated polyesters. Being unsaturated, the chainlike polyester molecules contain at least occasional double carbon bonds. These double bonds can be activated, creating free valence points. If other activated molecules are present in the mixture, strong cross links can be developed between them. For example, activated styrene molecules will react with activated polyester molecules, forming a continuous, cross-linked network. The degree of cross-linking that occurs determines the final properties of the material, that is, whether it will be resilient and rubbery, hard and glassy, or somewhere in between.

Another type of thermosetting reaction, called *condensation polymerization*, produces a by-product such as steam. An example of this is the reaction between phenol and formaldehyde, producing the thermoset known as phenol-formaldehyde. During this reaction, the phenol molecules each give up one or more hydrogen atoms and the formaldehyde molecules give up an oxygen atom. This leaves open valence points on both molecules, allowing them to join together to form a strongly interlinked structure. The leftover hydrogen and oxygen atoms join together to form water, which is given off as steam.

Thermosetting reactions are irreversible. Therefore, these plastics go through a pliable stage only once, and after they have set they remain rigid and hard unless actual chemical decomposition (breaking down) occurs due to extreme heat, pressure, or environmental effects.

6.9.1.2 Modifiers

Many plastics in their pure form are difficult to work with, expensive to produce, or lack some property needed for a particular application. These difficulties can sometimes be overcome by the addition of one or more of several types of modifiers.

These modifiers are added to the basic plastic ingredient (the resin) before fabrication, in a process known as *compounding*. It is important that the different ingredients are thoroughly compounded into a homogeneous mass, because the uniformity of the mix contributes greatly to the final properties of the plastic.

The type and amount of modifiers used with different plastics varies greatly. Some mixes consist of 90% to 95% basic resin, with only small amounts of modifiers. Other mixes may contain only 20% to 30% resin, the rest being made up of various plasticizers, fillers, stabilizers, and other additives.

PLASTICIZERS

At ordinary-use temperatures, many thermoplastics are inherently hard and rigid. In order to produce sheets, tubes, films, and other flexible plastic products, plasticizers must be added to these materials (Fig. 6.9-6).

Plasticizers are liquid or solid materials that are blended with plastic resin to impart increased flexibility, impact resistance, resiliency, and moldability. Basically, a plasticizer produces the same effect in a thermoplastic as does high temperature; that is, it lessens the forces of attraction between the long chainlike molecules, allowing them to slide more freely past each other. Unfortunately, this greater flexibility also often reduces the strength of the plastic and lowers its heat resistance and dimensional stability.

To obtain the optimum effects from a plasticizer, it is important that it be compatible with the plastic resin being used. This means that the plasticizer should mix well with the resin and should not separate out during the life of the object. A poor plasticizer may migrate to the surface of the object, creating a greasy film and causing the original plastic to become hard and brittle again. Furthermore, a plasticizer should not deteriorate with age or cause the plastic itself to deteriorate.

FILLERS

Fillers are used most commonly with thermosets, although there are some filled thermoplastics as well. Fillers are used to improve a particular property of a plastic, such as moldability, heat resistance, or toughness. However, some fillers are used

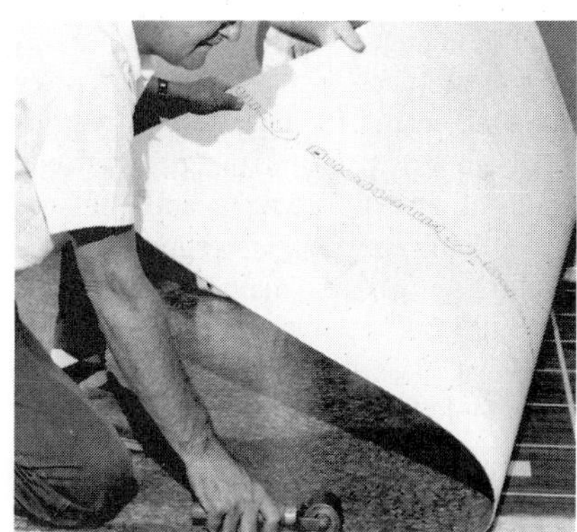

FIGURE 6.9-6 The addition of plasticizers imparts flexibility to vinyl sheet flooring, making it easier to handle and install.

primarily to increase the bulk and therefore decrease the cost of a plastic. These latter materials are called *extenders*.

A wide variety of materials may be used as fillers, depending on the properties desired in the finished product. The amounts used generally vary between 10% and 50% of the total plastic mix. Fillers are used to increase the following properties:

- *Mobility and bulk.* Wood flour, made of finely ground hardwood or nut shells, is commonly added to thermosets to make a plastic that is more easily molded and less expensive to produce.
- *Hardness.* Mineral powders and metallic oxides may be added to a plastic to increase its hardness. However, if too much of these are added, the plastic may become brittle.
- *Toughness.* Fibrous fillers are often added to plastics to decrease their brittleness and increase their impact resistance. Natural fibers such as cotton, sisal, and hemp may be used, as may be synthetic fibers such as rayon, nylon, and polyester.
- *Heat resistance.* Certain kinds of mineral fiber fillers, and other inorganic fillers such as silica, clay, and finely ground limestone, are used to increase the heat resistance of some plastics.
- *Electrical resistance.* Finely ground mica or quartz may be added to increase the electrical resistance of a plastic.

REINFORCING AGENTS

Reinforced plastics are a large and important group in the construction industry. Their strengths and stiffnesses have been significantly increased through the addition of high-strength fillers that act as reinforcing agents. Generally, thermosetting resins such as polyester and epoxy are used in reinforced plastics. Fibrous fillers such as sisal and mineral fibers can be used as reinforcing agents, but the strength of glass fiber is so superior that it is almost always the material selected for this purpose.

Sheet glass is quite brittle, but when drawn into fibers, its tensile strength rises enormously. In fact, pound for pound, glass fiber is the strongest commercially available construction material. Therefore, adding glass fibers to a plastic greatly increases its strength and stiffness. Glass fiber can be used either as chopped strands, yarns, pressed matting, or woven fabrics, depending on the design and properties desired for the finished product.

STABILIZERS

The effects of weathering on construction materials is an important consideration. Many plastics degrade when exposed to the ultraviolet rays of the sun and to oxygen and other gases in the air. As a result, small amounts of *stabilizers* are added to these plastics to act as antioxidants and ultraviolet light absorbers.

Several different substances may be used as stabilizers. For example, carbon black, which is commonly used, changes polyethylene from a quickly degrading material to one that stands up well under extended outdoor exposure.

COLORANTS

The addition of colorants to plastics is a complex subject. Some plastics, such as acrylic and polystyrene, are inherently clear and can be produced in a wide range of colors. Others, such as the phenolics, are dark and opaque and therefore have a more limited range of possible colors.

Both pigments and dyes can be used to color plastics. Pigments produce opaque colors; dyes, transparent ones. The pigment or dye chosen must be compatible with the other ingredients in the plastic mix. It should not bleed out of the plastic to discolor or stain adjacent materials. Furthermore, it should have good light and heat stability so that it will not fade or otherwise change in appearance with age.

6.9.2 PROPERTIES

The range of properties of various plastics is quite broad. They can be flexible or stiff, tough or brittle, transparent or opaque. However, plastics also have certain limitations, which must be considered when choosing them for applications in buildings. These limits can be seen most clearly when the general properties of plastics are compared with those of other construction materials such as steel, concrete, wood, and glass.

6.9.2.1 Strength and Stiffness

Strength can be measured in terms of tension, compression, and bending, with tensile strength being the most important. The tensile strength of most plastics is near that of wood, which is much lower than that of steel. However, some plastic laminates and reinforced plastics approach steel in strength. The highest tensile strengths are found in glass-fiber-reinforced plastics.

Stiffness is more important than strength in most plastics applications. Stiffness is measured by the modulus of elasticity. The higher the stiffness, the greater will be the modulus of elasticity. Generally, plastics are less stiff than steel, glass, and other structural materials, and, therefore, they deflect more than these other materials under a given applied load. Thermosets are somewhat stiffer than thermoplastics. Reinforced plastics have the highest modulus of elasticity (about the same as wood and concrete), but even they are only about 10% as stiff as steel.

Plastics weigh considerably less than other structural materials. Unmodified plastic resins range from a specific gravity of about 0.9 for polypropylene to about 1.5 for polyvinyl chloride. Both filled and reinforced plastics are more dense. In contrast, steel has a specific gravity of about 7.5. Therefore, when strength comparisons are made between plastics and metals on an equal-weight basis, plastics appear to have the advantage.

For this reason, plastics are employed in applications where strength combined with lightness is of prime importance, such as in spacecraft. In other construction, they can provide very strong, lightweight components (such as in sandwich panels), which can appreciably decrease the overall weight of a structure (Fig. 6.9-7).

6.9.2.2 Toughness and Hardness

Toughness, which is strength combined with resilience, is an outstanding characteristic of many plastics. This property is difficult to measure precisely, but it is

FIGURE 6.9-7 Low-density polyurethane foam is injected between two strong, rigid facings to produce a lightweight sandwich panel. (Courtesy Dow USA.)

generally expressed in terms of a material's *impact strength*, or ability to resist impact, such as that from a falling ball or a striking pendulum.

Plastics differ greatly in impact strength, and different formulations of the same plastic also show considerable variation. For example, regular polystyrene is quite brittle, whereas copolymers of styrene, such as acrylonitrile-butadiene-styrene, are noted for toughness.

Compared with other materials such as glass, concrete, and wood, plastics are quite resistant to impact. For example, a ⅛-in. (3.2 mm)-thick sheet of acrylic can withstand 25 times as much impact as a ⅛-in. (3.2 mm)-thick sheet of double-strength window glass. This is illustrated by the widespread use of plastics for break-resistant glazing and lighting fixtures.

Another property of importance is hardness (abrasion resistance). In this respect, plastics do not perform as well as glass or steel. Plastic glazing scratches much more readily than glass, for example. Conversely, many plastics are more resistant to abrasion than wood.

6.9.2.3 Creep

Creep (cold flow) is a permanent change in the dimensions of a material due to prolonged stress. All materials are subject to creep, but in steel the rate of deformation at room temperature is so slight that it is negligible.

Because of their molecular structure, many plastics exhibit creep at room temperature, and the amount of deformation that results can be large. This is particularly true of thermoplastics, whose unconnected molecules begin to slide past each other and assume new forms when an external force is applied. If the force is applied for only a short time, the plastic may recover, but if the force is long term, deformation can be permanent and may eventually result in failure of the material.

Thermosets, laminates, and reinforced plastics have greater resistance to creep, but even they cannot resist prolonged high stresses as can steel and concrete. Nevertheless, plastics can be used in self-supporting and even some load-bearing applications if the proper formulation is chosen.

6.9.2.4 Thermal Properties

The important thermal properties for plastics are expansion and contraction, conductivity, and maximum service temperatures.

EXPANSION AND CONTRACTION

All building materials expand and contract with changing temperatures. Thermal expansion is much higher in plastics than in most building materials. Thermosets exhibit a lower rate of expansion than thermoplastics, and reinforced plastics and laminates generally show the lowest rates among plastics.

Changes in dimension caused by expansion and contraction can cause buckling or cracking of plastic components used in assemblies. These difficulties can be avoided through proper design of the component and through the use of expansion joints to allow for movement. Particular care must be taken when plastics are joined to metals because of the wide difference in rates of expansion of the two materials. Generally, thermosets such as phenolics or reinforced plastics work best in such applications.

CONDUCTIVITY

Plastics are generally good heat insulators. They have a much lower thermal conductivity than metals. Pure, unmodified plastics also compare favorably with concrete, glass, and brick. Wood, however, has a lower conductivity than most plastics.

The exception is foam plastics, which outperform even wood in this respect and are among the best heat insulators available. Polyurethane foam can have a conductivity as low as 0.15, which is lower than even that of mineral wool, which is 0.27. Values for other foams vary according to density, type of foaming agent used, and whether the foam is of the open-cell (interconnecting) or closed-cell variety.

Because of these low conductivity values, plastic foams make excellent insulation materials. However, some problems have been encountered with their use in construction due to their flammability, the toxic fumes they can produce when burned, and the tendency of some of them to reduce in their insulating value over time. Usually, these problems can be overcome with proper material selection and good design. Codes require that some plastic insulations not be left exposed, but rather be covered with more fire-resistive material.

SERVICE TEMPERATURES

The maximum continuous service temperature of a material is the highest prolonged temperature at which it can be used without softening. Service temperatures for plastics are low as compared with those of other building materials. Many thermoplastics begin to soften at temperatures near the boiling point of water. Thermosets have better heat resistance, and some can be used at temperatures up to the wood char point (380°F [193.3°C] to 400°F [204.4°C]). Fluorocarbons have the highest continuous service temperature of unmodified plastics and can be used in non-load-bearing applications at 500°F (260°C).

Because additives can substantially improve the heat resistance of plastics, many filled and reinforced plastics can withstand higher temperatures than the pure, unmodified resins.

Service temperatures are important in determining the suitability of a plastic for a particular application. Most plastics can withstand the temperatures encountered in ordinary building structures; for more demanding applications, such as piping for hot-water supply lines, special heat-resistant formulations may be required.

6.9.2.5 Corrosion Resistance

Plastics have excellent corrosion resistance. They resist attack by most chemicals, with the exception of certain organic solvents. As with other properties, performance varies according to composition. Fluorocarbons are extremely inert and can be attacked only by the most powerful chemical reagents. Other plastics are attacked selectively by certain groups of solvents. Therefore, care must be taken to choose a suitable plastic for the corrosive conditions involved in the application.

6.9.2.6 Fire Resistance

Flammability of materials is an important consideration in construction. Being organic in nature, plastics will burn. Some plastics, such as cellulose nitrate, are extremely flammable and are generally not permitted as construction materials by most building codes. Others are slow burning, and some will stop burning after the source of flame has been removed. Flammability can be somewhat controlled through the use of additives. For example, compounds of chlorine, fluorine, bromine, and phosphorus are commonly added in reinforced plastics to retard flame spread.

The production of smoke and toxic fumes is another element of fire hazard. The amount of smoke and type of fumes produced depend on the composition of the plastic; those plastics rendered flame retardant through the use of additives generally produce the heaviest smoke. Furthermore, if a plastic contains compounds of chlorine, fluorine, nitrogen, or sulfur, they will also be present in the gases given off during combustion. These gases, in combination with carbon monoxide and carbon dioxide, which are by-products of all organic combustion, may prove highly toxic.

In short, plastics are no more flammable than wood, fabrics, or other organic materials. However, because plastics often substitute for traditional nonflammable materials like glass and metals, they should be used with discretion and only after careful consideration of their potential fire hazard.

To deal with these and other problems associated with the use of plastics in construction, most of the major building codes now have chapters devoted specifically to plastics applications. Tests developed or accepted by the American Society for Testing and Materials (ASTM) and published as ASTM standards can be used to rate plastics as to their burning rate, fire endurance, and flame spread. Building codes generally require that plastics suitable for installation in buildings be able to pass the tests in one or more of these ASTM standards. In addition, codes often restrict the use of plastics in certain areas by special requirements, such as requiring that foamed plastics be covered with noncombustible materials, requiring that plastic glazing be spaced with intermittent noncombustible surfaces, and by limiting the amount of plastic materials that may be used as interior finish.

FIGURE 6.9-8 ASTM Abbreviations for Plastics

Term	Abbreviation	Thermoplastic	Thermosetting
Acrylonitrile-butadiene-styrene	ABS	x	
Acrylic			
Poly(methyl methacrylate)	PMMA	x	
Cellulosics			
Cellulose acetate	CA	x	
Cellulose acetate-butyrate	CAB	x	
Cellulose acetate-propionate	CAP	x	
Cellulose nitrate	CN	x	
Ethyl cellulose	EC	x	
Epoxy, epoxite	EP		x
Fiberglass-reinforced plastics	FRP		x
Fluorocarbons			
Polytetrafluoroethylene	PTFE	x	
Fluorinated ethylene propylene	FEP	x	
Melamine-formaldehyde	MF		x
Nylon			
Polyamide	PA	x	
Phenolic			
Phenol formaldehyde	PF		x
Polycarbonate	PC	x	
Polyester	—		x
Polyethylene	PE	x	
Polypropylene	PP	x	
Polyurethane			
Urethane plastics	UP		x
Styrene			
Polystyrene	PS	x	
Styrene-acrylonitrile	SAN	x	
Styrene-butadiene plastics	SBP	x	
Silicone plastics	SI		x
Urea-formaldehyde	UF		x
Vinyls			
Poly(vinyl acetate)	PVAc	x	
Poly(vinyl alcohol)	PVAL	x	
Poly(vinyl butyral)	PVB	x	
Poly(vinyl chloride)	PVC	x	
Poly(vinyl fluoride)	PVF	x	

6.9.3 MAJOR TYPES AND USES

The multitude of commercially available plastics can be broken down into several major groups. This section includes a brief description of the chief characteristics and applications of plastics commonly used in construction. A listing of plastics and their abbreviations compiled by ASTM is given in Figure 6.9-8.

6.9.3.1 Thermoplastics

Thermoplastics are by far the larger group of plastics. Although thermoplastics can be softened by heating, they are dimensionally stable at normal-use temperatures.

ACRYLONITRILE-BUTADIENE-STYRENE (ABS)

ABS copolymers are noted for their unique combination of properties. They are an outgrowth of the high-impact polystyrenes, which are a combination of styrene and butadiene-rubber compounds. Polystyrene plastics alone are very brittle, but their toughness can be greatly improved by adding a butadiene-rubber compound to them. The further addition of acrylonitrile to the mixture improves other properties, such as chemical resistance, rigidity, and tensile strength.

ABS plastics are widely used in the construction industry for piping and pipe fittings (Fig. 6.9-9). They are used for water and gas supply lines and in drain, waste, and vent systems. They are tough and will withstand rough usage and fairly high temperatures for extended periods.

FIGURE 6.9-9 Plastic piping made of ABS is light, easy to handle, and resists corrosion. (Courtesy Plastic Pipe and Fittings Association.)

ABS plastics are also used to make hardware items such as doorknobs, handles, and latches and to manufacture handrails, furniture, appliances, and machine parts.

ACRYLICS

Polymethylmethacrylate (PMMA) is the best known of the acrylic plastics. These transparent plastics are noted for their superior optical properties and can be used readily for glazing. Unlike glass, they also have the ability to *pipe* light. This means that a beam of light can be bent around a corner due to the internal reflections of the light by the outer layer of the plastic.

Acrylics have good resistance to breakage and outstanding weathering properties. Their transparency and physical stability are not adversely affected by long outdoor exposure. However, they cannot withstand high temperatures and begin to soften at 200°F (93.3°C). In addition, their low abrasion resistance makes them more susceptible to scratching than glass.

Acrylics are used for skylights and roof domes and for glazing in public buildings where the likelihood of breakage hazard is high. They also are used in lighting fixtures and as translucent ceiling panels, lamp enclosures, and light diffusers. Their ability to pipe light is utilized in surgical instruments and in lighted outdoor signs. Other applications include clear or translucent corrugated sheets for roofing, films or sheets bonded to wood or metal for exterior finishes, and various molded pieces of hardware.

In addition, acrylics can be dispersed as finely divided particles in a liquid or mastic medium, producing a *latex*. In this form they are used widely for making paints (see Section 9.15) and building sealants.

CELLULOSICS

Cellulosics are based on modifications of the cellulose molecule. Cellulose acetate (CA) and cellulose-acetate-butyrate (CAB) are the most common cellulosics.

Cellulosics are exceedingly tough and will withstand a great deal of rough usage. They can be produced clear or translucent, although their optical properties are not as good as those of the acrylics. CA is not suitable for extended outdoor exposure, but CAB can be formulated so as to be quite resistant to weathering.

Cellulose acetate is known primarily for its use in photographic film and recording tape. Cellulose-acetate-butyrate is used for piping and pipe fittings (primarily for gas and chemicals), as well as for outdoor lighting fixtures. Because of their toughness, both types of cellulosics are also used for a variety of hardware items, for handrails, and for tool handles. Cellulosics also find use as adhesives and coating compounds.

FLUOROCARBONS

Polytetrafluoroethylene (PTFE), which is commonly known by the trade name *Teflon*, is the primary fluorocarbon used in the construction industry. It is remarkably inert to chemical attack. In fact, strong chemicals such as boiling aqua regia, chlorine, and bromine do not affect it at all. Furthermore, it remains serviceable over an extensive range of temperatures (from –450°F [–267.8°C] to +500°F [+260°C]), which is unmatched by most other plastics. It also has one of the lowest coefficients of friction of any known solid, which means that it has superior antistick properties. However, it is quite difficult to form and, consequently, expensive to produce.

PTFE is used in particularly demanding applications such as piping for extremely corrosive chemicals at high temperatures, for low-friction slider pads to permit movement in steam lines, on the surface of bearing pads between beams and columns of steel structures to permit movement, and as nonstick linings for cooking utensils.

NYLONS

Nylon, which was originally a trade name, is now the common name for a group of plastics known as polyamides (PA). These plastics have good all-around toughness, high strength, and good chemical resistance. They are also resistant to abrasion, but they are not recommended for continuous outdoor exposure.

Nylons are perhaps best known for their use in synthetic fibers and filaments. They produce high-strength fabrics that are durable and tear resistant. These fabrics are used for sails, parachutes, and many articles of clothing. In the construction industry, nylon fabrics, which may be coated or reinforced as necessary, have been used in air-supported structures similar to that shown in Figure 6.9-10. Nylon fibers are also used to make carpets.

In addition, because of its toughness, nylon is used to produce molded parts such as locks, latches, rollers for sliding windows and drawers, gears, and cams.

POLYCARBONATES

Polycarbonates (PC) are transparent plastics with exceptionally high impact strength, good heat resistance, and good dimensional stability. They are used in break-resistant safety glazing, lighting

FIGURE 6.9-10 **Air-supported structures are lightweight and portable, making them ideal for such uses as temporary construction enclosures.** (Courtesy Thermo-Flex, Inc.)

fixtures, lighted signs, and various pieces of molded hardware.

POLYETHYLENE

Polyethylene (PE) is a well-known plastic made by polymerization of the ethylene molecule. It is strong, light, and flexible and retains its flexibility even at low temperatures. It is noted for its good water resistance and low moisture-vapor transmission. PE is produced in several different formulations, including linear and branched molecular configurations, and these may vary considerably as to strength, density, and stiffness.

Polyethylene is used widely for vapor retarders in floors and walls (Fig. 6.9-11), for dampproofing of basement walls, including wood foundation walls, and for sealant backer rods. It is also used for some types of piping (cold water, gas, and chemicals), but it is not suitable for use under severe conditions involving high temperatures or extremely corrosive chemicals. Another common application of polyethylene is for wire and cable insulation.

POLYPROPYLENE

Polypropylene (PP) is similar to polyethylene in many respects—both are classed as polyolefins, for example—but it is much stiffer and more heat resistant than polyethylene. This makes it suitable for piping and pipe fittings used under more severe conditions, such as for hot-water supply lines and drainage systems. Although polypropylene lacks the flexibility of polyethylene and is also more expensive to produce, it is used for some films and sheeting, primarily in the packing industry.

FIGURE 6.9-11 A polyethylene vapor retarder on the interior of a wall assembly reduces vapor condensation inside the wall. (Photograph furnished by Owens-Corning Fiberglas Corporation.)

Another common use for polypropylene is in the production of high-strength fibers for carpeting.

POLYSTYRENE

Polystyrene (PS), made from the styrene monomer, is a transparent, water-resistant, and dimensionally stable plastic. It is not affected by extended low temperatures, but it begins to soften at temperatures near that of boiling water (212°F [100°C]). The main drawbacks of polystyrene are that it is quite brittle and not very weather resistant, becoming even more brittle and turning yellow with outdoor exposure.

To improve its properties, polystyrene is often copolymerized with other substances. One such copolymer is high-impact polystyrene produced by adding small amounts of butadiene to the styrene. This results in a much tougher plastic with greater impact resistance and greater flexibility. Another well-known copolymer of styrene is acrylonitrile-butadiene-styrene (ABS), discussed earlier.

Polystyrene is used for lighting fixtures and lighted signs and for various molded pieces of hardware. Expanded (foamed) polystyrene, of which Styrofoam® is a well-known example, is used widely in the construction industry for duct and pipe insulation; insulation for freezers and walk-in refrigerators; and in insulation for walls, floors, and ceilings of residences and commercial buildings (Fig. 6.9-12). Polystyrene foam is also used as the core material in the manufacture of doors and sandwich panels.

VINYLS

Vinyls are a large group of plastics based on modifications of the ethylene molecule. Included in this group are such plastics as polyvinyl acetate, polyvinyl alcohol, polyvinyl butyral, polyvinyl chloride, polyvinyldene chloride, and many others. These plastics exhibit a wide range of properties. They generally are characterized by good strength and toughness, fair chemical resistance, and a slow rate of water absorption. They perform well at normal-use temperatures as well as at relatively low (freezer) temperatures, but cannot withstand prolonged exposure to high temperatures. In fact, some vinyls begin to soften at 130°F (54.4°C). Most types are recommended for indoor use, but some are suitable for outdoor exposure.

The most widely used vinyl in the construction industry is polyvinyl chloride (PVC). This is a very versatile plastic, with high impact resistance, good dimensional stability, abrasion resistance, and good aging characteristics. It is used to produce a large variety of products such as sheet and tile flooring, gutters and downspouts, moldings, siding, window frames, piping, and drainage systems. In sheet form it is used for facings in sandwich construction, and is bonded to wood or metal for exterior building finishes, doors, and window frames.

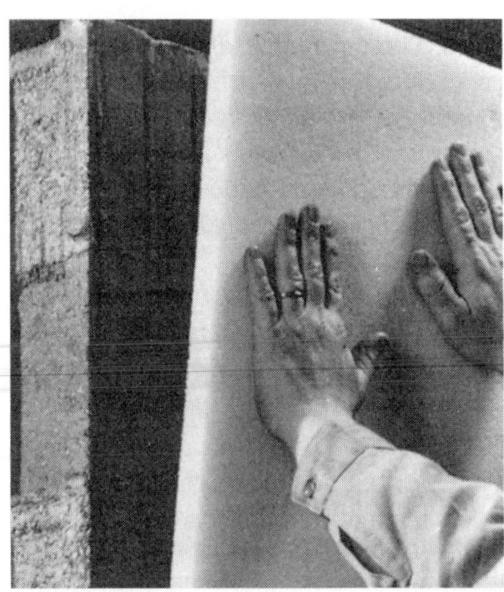

FIGURE 6.9-12 Polystyrene foam insulation is available in blocks or boards, which can be bonded directly to masonry walls. (Courtesy Dow USA.)

PVC is also available as a foam (either rigid or flexible) and is used for the core material in sandwich construction and for cushions in seating. PVC is copolymerized with many other plastics to produce a variety of adhesives, binders for terrazzo, and other floor toppings.

Other vinyls encountered in construction include polyvinyl butyral (PVB), which is used for the interlayer in safety glass and as a coating for upholstery fabrics; polyvinyl acetate (PVAC), which is used in adhesives, mortars, and paints; and vinyl-coated polyester, which is used in demountable air-supported structures.

6.9.3.2 Thermosets

Thermoset plastics are permanently set during forming and cannot be softened again by reheating. In general, they have better heat resistance and greater stiffness than thermoplastics, but are more difficult to form and more brittle. They have a more limited range of applications than do thermoplastics.

EPOXIES

Epoxies (EP) were introduced in the 1950s and were quickly adopted for a wide variety of uses because of their excellent adhesive strength, combined with good chemical and moisture resistance. They will adhere to, and can be used to bond, metals, glass, masonry, other plastics, and most other building materials.

Epoxies are used primarily in coating compounds and adhesives. They are often produced as two-component systems, consisting of a resin and a hardener that must be mixed together just before application. Because they bond well to metals, they are used as protective coatings for appliances, automobiles, and piping. As adhesives, they are used in the construction of sandwich panels; for bonding facings to walls, doors, and other assemblies; and for applying tile finishes. They make excellent high-strength mortars for concrete masonry units and for patching cracked monolithic concrete. Compounds are available for patching deteriorated wood members in existing construction. Epoxy compounds can also be mixed with mineral aggregate or plastic chips to produce terrazzo.

Most glass-fiber-reinforced plastic products are made with polyesters, but sometimes epoxies are used instead. Cured, pure, unmodified epoxy is hard and brittle, but when reinforced with glass fibers, it becomes a strong, tough material that can be used in structural and semi-structural applications.

MELAMINES

Melamines (MF) are produced by the reaction between melamine and formaldehyde. They are classified as amino plastics, as are ureas. Melamines are hard and scratch resistant and can be used at temperatures approaching the boiling point of water. They also stand up well to chemical attack.

The most important application for melamine plastics is in high-pressure laminates, which are used widely for countertops and cabinet finishes. They are also used as adhesives (notably for plywood) and as a protective treatment for fabrics and paper.

As molding compounds, unmodified melamines are somewhat expensive and difficult to work with. Therefore, they are usually mixed with mineral or wood flour fillers. In this form, they are used for molded hardware and for electrical fittings.

PHENOLICS

Phenolics (PF) are produced by the reaction of phenol and formaldehyde. Like many of the thermosets, pure phenolics are quite brittle and hard and are generally mixed with fillers to improve their strength and impact resistance. Many different formulations are available, with varying properties. Phenolics have excellent heat resistance, and glass-fiber-reinforced types will perform well at temperatures up to 400°F (204.4°C). They also have excellent thermal and electrical insulating properties.

Phenolics are among the more widely used thermosetting plastics. As molding compounds, they are used for electrical parts (sockets, switch boxes, circuit breakers), for a great variety of hardware items, and for appliance cabinets and parts.

Phenolic resins are important in the production of high-pressure laminates. They are used to impregnate the sheets of kraft paper that form the base of most laminates. These resins are also used to stiffen paper honeycomb, which is then bonded to a variety of facings to produce lightweight, strong sandwich panels for walls, doors, and room dividers. Adhesives and protective coatings are also made from phenolic resins.

Another major use for phenolics is as foamed insulation. In this form they are used as the core for sandwich panels and for insulation around piping and ducts.

POLYESTERS

Polyester is the generic name for a large group of plastics that are used for a wide variety of purposes. Examples of different types of polyesters include oil-modified polyesters (sometimes referred to as alkyds), used in paints and protective coatings; saturated polyesters, used as plasticizers in other plastic compositions; and saturated polyesters of high molecular weight, used to form fibers and film. The most important polyesters for building purposes are the unsaturated polyesters, which form the basic constituent of reinforced plastics.

Unsaturated polyesters can be cross-linked with a compatible monomer (such as styrene) to form a hard, strong thermosetting material. They have become the most widely used resins for the production of reinforced plastics because of an important combination of properties: they are easy to work with and combine readily with a variety of fibers and reinforcing agents; they cure rapidly at ordinary room temperatures without giving off volatile by-products; and they have excellent dimensional stability, combined with high strength and toughness. Furthermore, they can be specially formulated to provide good weathering properties and good heat and flame resistance.

For maximum strength and stiffness, glass fiber is used as the reinforcing agent in polyester compounds. This produces an exceptionally strong, lightweight material that can be used in many structural building applications. Some of the common uses for glass-fiber-reinforced polyesters are prefabricated sections for roof structures; translucent sheets for roofing and interior partitions; window frames and sash; facings for sandwich panels; and many large molded articles such as bathtubs, shower stalls, sinks, and cabinets. Products such as bathtubs and countertops may be made by molding reinforced polyester within an acrylic shell to give a combination of rigidity, strength, and attractive appearance.

POLYURETHANES

Polyurethanes (UP), including polyisocyanurate, are a class of plastics known primarily for their use as low-density foams, which can be produced with widely varying properties. These range from soft, flexible open-cell types to tough, rigid closed-cell types. Urethane foams have excellent heat and chemical resistance and good tensile strength. They can be formulated to be fire resistant.

Rigid urethane foams make excellent insulation material. They have the lowest thermal conductivity of any building material, including mineral wool. They are used widely in construction for wall, floor, and ceiling insulation, for insulation around pipes and ducts, and as the core material in sandwich panels. Flexible urethane foams are used for cushions, upholstery, and padding.

Urethane resins also are used in the production of elastomers (synthetic rubbers), adhesives, sealing and glazing compounds, and no-seam flooring.

SILICONES

Silicone plastics (SI) are based on the silicon atom rather than the carbon atom. They are therefore an exception to the general rule that plastics are organic compounds.

Silicones are very stable compounds, characterized by excellent corrosion resistance and good electrical insulating properties. They can be used over a wide range of temperatures (from –80°F [–62.2°C] to +500°F [+260°C]). They also stand up extremely well to most weathering conditions.

Silicones, noted for being extremely water repellent, are used as water-resistant coatings on concrete masonry and brick (Fig. 6.9-13), although they are somewhat controversial in this use. They also make excellent sealants and are used in that capacity in metal curtain walls, as joint sealers around windows and doors, in other joints in exterior walls, and in wet locations such as kitchens, bathrooms, showers, and toilets. Small amounts of silicone are frequently added to reinforced plastics to improve the adhesion of the polyester resin to the glass fibers and to increase the strength and moisture resistance of the product.

UREAS

The most common *urea* plastic, urea-formaldehyde (UF), is produced by the reaction between urea and formaldehyde. Ureas are chemically related to melamines and are classified with them as amino plastics.

Ureas have lower heat and chemical resistance than melamines, but they are more scratch resistant and have good electrical resistance. They are used for molded hardware items and electrical fittings. Urea resins provide adhesives for use in sandwich panels, and they are also available as insulating foams.

FIGURE 6.9-13 Silicone water repellent being applied to a masonry wall. (Courtesy Sika Corporation.)

6.10 Additional Reading

More information about the subjects discussed in this chapter can be found in the references listed in Section 6.11 and in the following publications.

Allen, Edward. 1999. *Fundamentals of Building Construction: Materials and Methods*. 3d ed. New York: John Wiley & Sons, Inc.

Ambrose, James E. 1997. *Simplified Design of Wood Structures*. New York: John Wiley & Sons, Inc.

Faherty, Keith F. 1998. *Wood Engineering and Construction Handbook*. New York: McGraw-Hill.

Goldstein, Eliot E. 1998. *Timber Construction for Architects and Builders*. New York: McGraw-Hill.

Halperin, Don A., and Thomas G. Bible. 1994. *Principles of Timber Design for Architects and Builders*. New York: John Wiley & Sons, Inc.

Newman, Morton. 1994. *Design and Construction of Wood Framed Buildings*. New York: McGraw-Hill.

Parker, Harry, and James Ambrose. 1994. *Simplified Design of Wood Structures*. 5th ed. New York: John Wiley & Sons, Inc.

Petree, Jack. 1996. "Recycled Lumber." *The Construction Specifier* 6(June):70.

Wilcox, W. Wayne, Elmer E. Botsai, and Hans Kubler. 1991. *Wood as a Building Material: A Guide for Designers and Builders*. New York: John Wiley & Sons, Inc.

6.11 Acknowledgments and References

ACKNOWLEDGMENTS

We gratefully acknowledge the assistance of the following organizations and individuals in preparing this chapter. We are also indebted to them for permission to use their illustrations when requested and for the use of their publications as references.

Addison Wesley Publishing Company
AIA and ACSA Council on Architectural Research
American Forest & Paper Association (AF&PA)
American National Standards Institute (ANSI)
American Plywood Association (APA)
American Society for Testing and Materials (ASTM)
American Wood Preservers Association (AWPA)
American Wood Preservers Institute (AWPI)
Architectural Woodwork Institute (AWI)
Armstrong World Industries
California Redwood Association (CRA)
Dow USA
Fine Hardwoods—American Walnut Association
Gypsum Association
Hardwood Plywood and Veneer Association (HPVA)
Kawneer Company
National Association of Home Builders (NAHB), Research Foundation
National Hardwood Lumber Association (NHLA)
National Institute of Building Sciences (NIBS)
National Institute of Standards and Technology (NIST)
National Terrazzo and Mosaic Association (NTMA)
Owens Corning Fiberglas Corporation
Plastic Pipe and Fittings Association
Redwood Inspection Service (RIS)
Rohm and Haas Company
Sika Corporation

Society of the Plastics Industry
Southern Forest Products Association (SFPA)
Southern Pine Council (SPC)
Southern Pine Inspection Bureau (SPIB)
Thermo-Flex
United States Gypsum Corporation (USG)
United States Plywood Corporation
U.S. Department of Agriculture (USDA)
U.S. Department of Housing and Urban Development (HUD)
U.S. Forest Products Laboratory (FPL)
West Coast Lumber Inspection Bureau (WCLIB)
Western Electric Company
Western Wood Products Association (WWPA)
Weyerhaeuser Company
Window & Door Manufacturers Association (WDMA)
Wood Truss Council of America (WTCA)
Woodwork Institute of California (WIC).

REFERENCES

We would also like to thank the authors and publishers of the publications in the following list for their contribution to our research for this chapter.

American Forest & Paper Association (AF&PA). *Design Values for Joists and Rafters*. Washington, DC: AF&PA.

———. *National Design Specifications for Wood Construction*. Washington, DC: AF&PA.

———. *Design Values for Wood Construction (NDS Supplement)*. Washington, DC: AF&PA.

———. *Span Tables for Joists and Rafters*. Washington, DC: AF&PA.

———. *Wood Structural Design Data*. Washington, DC: AF&PA.

———. WCD No. 1. *Details for Conventional Wood Frame Construction*. Washington, DC: AF&PA.

———. WCD No. 4. *Plank and Beam Framing for Residential Buildings*. Washington, DC: AF&PA.

———. WCD No. 5. *Heavy Timber Construction Details*. Washington, DC: AF&PA.

———. WCD No. 6. *Design of Wood Frame Structures for Permanence*. Washington, DC: AF&PA.

———. *Wood Frame Construction Manual for One- and Two-Family Dwellings*. Washington, DC: AF&PA.

———. *Load and Resistance Factor Design Manual for Engineered Wood Construction*. Washington, DC: AF&PA.

American Institute of Timber Construction (AITC). *Timber Construction Manual*. Englewood, CO: AITC.

———. AITC 104. *Typical Construction Details*. Englewood, CO: AITC.

———. AITC 108. *Standard for Heavy Timber Construction*. Englewood, CO: AITC.

———. AITC 112. *Standard for Tongue-and-Groove Heavy Timber Roof Decking*. Englewood, CO: AITC.

———. AITC 117. *Standard Specifications for Structural Glued Laminated Timber of Softwood Species, Design Requirements*. Englewood, CO: AITC.

———. AITC 119. *Standard Specifications for Structural Glued Laminated Timber of Hardwood Species*. Englewood, CO: AITC.

American National Standards Institute (ANSI). Standards:

A 58.1-1982, "Minimum Design Loads, Buildings and Other Structures." New York: ANSI.

CC/ANSI A117.1-1986, "Guidelines for Accessible and Usable Buildings and Facilities." New York: ANSI.

A 190.1-1992, " Wood Products—Structural Glued Laminated Timber." New York: ANSI.

A 194.1-1985, "Cellulosic Fiber Board." New York: ANSI.

A 208.1-1999, "Particleboard." New York: ANSI.

APA—The Engineered Wood Association. A310, "Nonresidential Roof Systems." Tacoma, WA: APA.

———. B360, "HDO/MDO Plywood." Tacoma, WA: APA.

———. E30, "Residential and Commercial." Tacoma, WA: APA.

———. E445, "Performance Standards and Policies for Structural-Use Panels." Tacoma, WA: APA.

———. F405, "Performance Rated Panels." Tacoma, WA: APA.

———. K435, "Sanded Plywood." Tacoma, WA: APA.

———. L335, "Installation and Preparation of Plywood Underlayment for Resilient Floor Covering." Tacoma, WA: APA.

———. N335, "Proper Installation of APA Rated Sheathing for Roof Applications." Tacoma, WA: APA.

———. Q220, "Preservative-treated Plywood." Tacoma, WA: APA.

———. Q300, "Builder Tips: Steps to Construct a Solid, Squeak-free Floor System." Tacoma, WA: APA.

———. S811, "PDS Supplement #1: Design and Fabrication of Plywood Curved Panels." Tacoma, WA: APA.

———. S812, "PDS Supplement #2: Design and Fabrication of Plywood-lumber Beams." Tacoma, WA: APA.

———. U813, "PDS Supplement #3: Design and Fabrication of Plywood Stressed-skin Panels." Tacoma, WA: APA.

———. U814, "PDS Supplement #4: Design and Fabrication of Plywood Sandwich Panels." Tacoma, WA: APA.

———. X461, "House Building Basics." Tacoma, WA: APA.

———. X505, "Panel Handbook and Grade Glossary." Tacoma, WA: APA.

American Society for Testing and Materials (ASTM) Standards:

C 79/C79M, "Standard Specification for Treated Core and Nontreated Core Gypsum Sheathing Board." West Conshohocken, PA: ASTM.

D 245, "Standard Practice for Establishing Structural Grades and Related Allowable Properties for Visually Graded Lumber." West Conshohocken, PA: ASTM.

D 1600, "Standard Terminology for Abbreviated Terms Relating to Plastics." West Conshohocken, PA: ASTM.

D 1784, "Standard Specification for Rigid Poly(Vinyl Chloride) (PVC) Compounds and Chlorinated Poly(Vinyl Chloride) (CPVC) Compounds." West Conshohocken, PA: ASTM.

D 2000, "Standard Classification System for Rubber Products in Automotive Applications." West Conshohocken, PA: ASTM.

D 2287, "Standard Specification for Nonrigid Vinyl Chloride Polymer and Copolymer Molding and Extrusion Compounds." West Conshohocken, PA: ASTM.

D 4442-e1, "Standard Test Methods for Direct Moisture Content Measurement of Wood and Wood-Base Materials." West Conshohocken, PA: ASTM.

D 4444-e1, "Standard Test Methods for Use and Calibration of Hand-Held Moisture Meters." West Conshohocken, PA: ASTM.

E 84, "Standard Test Method for Surface Burning Characteristics of Building Materials." West Conshohocken, PA: ASTM.

E 96, "Standard Test Method for Water

Vapor Transmission of Materials." West Conshohocken, PA: ASTM.

E 119, "Standard Test Methods for Fire Tests of Building Construction and Materials." West Conshohocken, PA: ASTM.

E 136, "Standard Test Method for Behavior of Materials in a Vertical Tube Furnace at 750 deg. C." West Conshohocken, PA: ASTM.

American Wood Preservers Association (AWPA). 1999. *Book of Standards*. Granbury, TX: AWPA.

American Wood Preservers Institute (AWPI). *Pole Building Design Manual*. Fairfax, VA: AWPI.

———. *How to Design Pole-type Buildings Manual*. Fairfax, VA: AWPI.

Architectural Graphic Standards. See Ramsey/Sleeper.

Architectural Woodwork Institute (AWI). 1999. *AWI Quality Standards*. 7th ed., version 1.2. Reston, VA: AWI.

Arcom Master Systems. MASTERSPEC®. Basic Sections:

06100, "Rough Carpentry." Salt Lake City, UT: Arcom.

06105, "Miscellaneous Carpentry." Salt Lake City, UT: Arcom.

06130, "Heavy Timber Construction." Salt Lake City, UT: Arcom.

06150, "Wood Decking." Salt Lake City, UT: Arcom.

06185, "Structural Glue Laminated Timber." Salt Lake City, UT: Arcom.

06192, "Metal-Plate-Connected Wood Trusses." Salt Lake City, UT: Arcom.

06200, "Finish Carpentry." Salt Lake City, UT: Arcom.

06401, "Exterior Architectural Woodwork." Salt Lake City, UT: Arcom.

06402, "Interior Architectural Woodwork." Salt Lake City, UT: Arcom.

06420, "Panelwork." Salt Lake City, UT: Arcom.

06605, "Interior Plastic Ornamentation." Salt Lake City, UT: Arcom.

Council of American Building Officials (CABO). *One- and Two-Family Dwelling Code*. Falls Church, VA: CABO. This code is in the process of being transferred to the control of the International Code Council (ICC). (See Section 1.6.3.)

Forest Products Laboratory. 1999. *Wood Handbook—Wood as an Engineering Material*. Gen. Tech. Rep. FPL-GTR-113. Madison, WI: U.S. Department of Agriculture, Forest Service, Forest Products Laboratory.

———. 1998. *Wood-Frame House Construction*. Madison, WI: U.S. Department of Agriculture, Forest Service, Forest Products Laboratory.

Gypsum Association. GA-253, "Application of Gypsum Sheathing." Washington, DC: Gypsum Association.

———. GA-254, "Fire Resistant Gypsum Sheathing." Washington, DC: Gypsum Association.

Hardwood Plywood & Veneer Association (HPVA). 1994. ANSI/HPVA HP-1, *American National Standard for Hardwood and Decorative Plywood*. Reston, VA: HPVA.

Montella, Ralph. 1985. *Plastics in Architecture: A Guide to Acrylic and Polycarbonate*. New York: Marcel Dekker.

National Electrical Manufacturers Association (NEMA). NEMA LD 3-1995, *High Pressure Decorative Laminates*. Rosslyn, VA: NEMA.

National Institute of Building Sciences (NIBS). Catalog No. 4010-1, *Wood Protection Guidelines, Protecting Wood from Decay, Fungi and Termites*. Washington DC: NIBS.

Ramsey/Sleeper, The AIA Committee on Architectural Graphic Standards. 1998. *Architectural Graphic Standards*. 9th ed. New York: John Wiley & Sons, Inc.

Reiff, John, and Larry Foley. 1995. "Glulams." *The Construction Specifier* 9(September):62.

Southern Forest Products Association (SFPA). (See Southern Pine Council.)

Southern Pine Council (SPC). 1998. *Permanent Wood Foundations, Design and Construction Guide*. Kenner, LA: SFPA.

Southern Pine Inspection Bureau (SPIB). 1994. *Grading Rules*. Pensacola, FL: SPIB.

Truss Plate Institute (TPI). *National Design Standard for Metal Plate-Connected Wood Truss Construction and Commentary and Appendix 1*. Madison, WI: TPI.

Underground Space Center. 1988. *Building Foundation Design Handbook*. Minneapolis, MN: Underground Space Center.

U.S. Department of Agriculture. *G 64—Subterranean Termites: Their Prevention and Control in Buildings*. Washington, DC: U.S. Department of Agriculture.

U.S. Department of Commerce. PS 1, "U.S. Product Standard for Construction and Industrial Plywood." Washington, DC: U. S. Department of Commerce.

———. PS 20, "American Softwood Lumber Standard." Washington, DC: U.S. Department of Commerce.

West Coast Lumber Inspection Bureau (WCLIB). *Standard Grading Rules for West Coast Lumber*. Portland, OR: WCLIB.

Western Wood Products Association (WWPA). *Western Lumber Grading Rules, '98*. Portland, OR: WWPA.

———. "WWPA Grade Stamps." Portland, OR: WWPA.

———. A-2, "Specifying Lumber." Portland, OR: WWPA.

———. *Western Woods Use Book*. 4th ed. Portland, OR: WWPA.

Window and Door Manufacturers Association (WDMA). I.S.4-1999, "Industry Standard for Water-repellent Preservative Non-pressure Treatment for Millwork." Des Plaines, IL: WDMA.

Woodwork Institute of California (WIC). 1998. *Manual of Millwork*. 9th ed. West Sacramento, CA: WIC.

CHAPTER 7

Thermal and Moisture Protection

Introduction

Applicable *MasterFormat*™ Sections

Moisture Control

Waterproofing and Dampproofing

Building Insulation

Exterior Insulation and Finish Systems

Low-Slope Roofing

Steep Roofing

Siding

Flashing and Sheet Metal

Metal Roofing

Fireproofing

Joint Sealing

Additional Reading

Acknowledgments and References

Introduction

A building's shell is required to prohibit weather penetration into the building and to maintain a comfortable environment within. Assault from the exterior comes from temperature variations, wind, precipitation, and ground water. This chapter discusses construction methods and materials used in modern construction to help maintain a livable environment within buildings, including the exclusion of water and control of heat migration through building walls and roofs.

Applicable *MasterFormat*™ Sections

The following *MasterFormat* Level 2 sections are applicable to this chapter.

07050 Basic Thermal and Moisture Protection Materials and Methods
07100 Dampproofing and Waterproofing
07200 Thermal Protection
07300 Shingles, Roof Tiles, and Roof Coverings
07400 Roofing and Siding Panels
07500 Membrane Roofing
07600 Flashing and Sheet Metal
07700 Roof Specialties and Accessories
07800 Fire and Smoke Protection
07900 Joint Sealers

7.1 Moisture Control

Moisture in its visible liquid state—water—is relatively simple to understand and to cope with in construction. Roofs shed it, earth is graded to divert it, pipes conduct it, and flashing leads it back to the exterior when it penetrates the outer layer of a building. The technology needed to prevent liquid water penetration into buildings is available. Unfortunately, many designers do not give adequate consideration to preventing this penetration. No competent architect will say that leaks are acceptable, but many will say privately that they do not permit potential maintenance problems to interfere with their design goals. As a result, more than half the lawsuits involving architects are related to water leaks.

Water also is present as an invisible gas—water vapor—in the air all around us and in cavities of construction assemblies, such as stud and joist spaces, in masonry cores, in furred spaces, and in chases containing plumbing or heating, ventilating, and air conditioning ducts. As absorbed water, moisture is present in most construction materials and affects their properties, especially dimensional stability. Moisture may also be present as a solid—ice—within the range of temperatures encountered in most parts of the United States.

Moisture is not damaging to construction in its vapor form. However, it becomes dangerous as it condenses (liquefies) or freezes (solidifies). Over time, hidden condensation can cause organic materials to decay, metal products to corrode, and paint coatings to blister. When it freezes, it can cause concrete and masonry to crack or spall.

Condensation and freezing may occur as a result of either a temperature drop or the migration of water vapor to areas of lower temperature. Therefore, moisture problems occur (1) in cold climates in heated occupied buildings, which generate substantial amounts of water vapor, and (2) in warm, humid climates in cooled buildings.

Buildings built 50 or more years ago were not as well built or sealed against the weather as they are today. This permitted considerable air leakage, dissipating moisture harmlessly through the structure. Economic and comfort considerations require that modern buildings be built tighter and be kept warmer in winter and cooler in summer.

In addition, just as dehumidification is an important aspect of summer comfort conditioning, so is deliberate humidification an integral part of typical winter heating systems. All these factors create conditions conducive to moisture problems, unless provisions to cope with them are included in design and construction.

The properties and behavior of moisture in its various forms are important considerations in many aspects of building construction, ranging from soil mechanics to roofing. This section addresses the principles of water vapor movement, problems associated with its condensation to visible water, and methods of controlling condensation in buildings.

7.1.1 WATER VAPOR PROPERTIES

Moisture can exist in several states: as a liquid (water), as a solid (ice), and as a gas (water vapor). *Condensation* is the change of water vapor to liquid water. *Evaporation* is the change of liquid water to a gas. The transition from ice directly into water vapor is called sublimation.

The change of state from water or ice to vapor requires considerable heat input, called *latent heat*, measured in British thermal units (Btu's) (joules [J]). Therefore, when water evaporates from a surface, that surface is cooled. Perspiring skin, for example, is an effective cooling system for a human body. Like any warm object, the skin gives off heat to the surrounding atmosphere. Heat absorbed by a substance without changing its state is known as *sensible heat* and also is measured in Btu's (J).

When water turns to vapor, it mixes with the air and occupies all available space. In many ways water vapor acts independently of the air because its general properties do not depend on the presence of air. However, when air is moved by wind or a fan, or is heated or cooled, associated water vapor is similarly affected. It is therefore convenient to consider moist air as a mixture of dry air and water vapor.

7.1.1.1 Relative Humidity

The measure of the amount of water vapor in air is known as *relative humidity* (RH), which is stated as a percentage. This percentage is a ratio of the actual amount of water vapor in the air as compared with the maximum amount the same air can contain at a specific temperature and atmospheric pressure. For instance, air with 50% relative humidity contains half the amount of vapor it can hold at that air temperature and barometric pressure. At 100% relative humidity, air contains the maximum amount of water vapor possible. Such air is said to be *saturated.*

An observable law of physics is that the warmer a mass of air, the more water vapor it can hold. Therefore, if the temperature of saturated air is lowered, some vapor must be given up by condensation. Up to the point of condensation, relative humidity increases as air temperature decreases, while the amount of moisture remains the same.

As shown in Figure 7.1-1, air at 70°F (21.1°C) can hold twice as much moisture as the same volume of air at 50°F (10°C) when the air is completely saturated. Therefore, at 70°F (21.1°C) the air is said to be half-saturated or to have a relative humidity of 50%. If the temperature were lowered below the saturation point, say to 40°F (4.4°C), the relative humidity would remain at 100%, but about 18 grains of moisture per pound (0.00257 kg/kg) of dry air would condense into visible water on cool surfaces. Further cooling to 30°F (–1.1°C) would result in the removal of a total of 30 grains/lb. (0.0043 kg/kg) of air (see Fig. 7.1-1).

HUMAN COMFORT

Air movement, relative humidity, and temperature all affect the rate at which perspiration evaporates from the skin and at which a human body loses heat. Either too little or too much humidity can be uncomfortable. Low humidity encourages evaporation and increases the sensation of coolness, which is undesirable in the winter. High humidity reduces skin evaporation, retards body cooling, and makes people feel warmer, which makes them uncomfortable when the air temperature is near the upper limit of a comfortable condition. This can happen in any season, but is more likely to occur in the summer.

High humidities at higher temperatures can contribute to fatigue and reduced working efficiency. Air that is too dry can irritate nasal and sinus passages, contribute to chapped and dry skin, and encourage the generation of annoying and, in locations like computer rooms, damaging static electricity.

The specific humidity or temperature "comfort" level is largely a subjective judgment that depends on many factors, such as prevailing climate, seasonal variations, and an individual's age, health, living habits, clothing, and activities.

As mentioned, cold air cannot hold nearly as much water vapor as warm air. Therefore, when cold outside air with a low moisture content (even though its relative humidity may be high) is brought inside and heated, a low relative humidity results. For instance, as shown in Figure 7.1-1, saturated air at 30°F (–1.1°C) can hold 24 grains of moisture per pound (0.0034 kg/kg) of air, and half-saturated air at 30°F (–1.1°C) has a moisture content of 12 grains/lb. (0.0017 kg/kg) of air. If this air were heated to 70°F (21.1°C), it could hold 108 grains/lb. (0.0154 kg/kg), and its relative humidity would be 100% times 12/108 (0.0017/0.0154), or 11.1% relative humidity. The resulting humidity level is too low for comfort. A humidity level of 25 to 40% is usually considered conducive to good health, but such levels are difficult to maintain in cold weather.

EFFECTIVE TEMPERATURE

The term *effective temperature* (ET) identifies an arbitrary index that combines in a single value the effects of temperature, humidity, and air movement on the sensation of warmth or cold felt by a human body. It has been developed by the American Society of Heating, Refrigerating and Air-Conditioning Engineers (ASHRAE) in cooperation with the U.S. Public Health Service and is based on extensive polls of trained individuals subjected to various temperature and humidity conditions.

Effective temperature is the temperature of saturated air (100% relative humidity) that gives the same physical sensation of warmth as various other combinations of air temperature and relative humidity, with relatively still air movement rates ranging from 15 to 25 feet per minute (fpm) (0.076 to 0.127 m/s).

Winter Comfort It has been found that 98% of people tested were comfortable in the winter season at an effective temperature of 68, which is 100% relative humidity and 68°F (20°C). Humidity and temperature combinations that gave this same feeling of warmth are tabulated in Figure 7.1-2. This table shows that in winter, as the relative humidity is reduced, higher temperatures are necessary to maintain the same sensation of warmth. Conversely, a higher relative humidity

FIGURE 7.1-1 Temperature and Humidity Relationships[a]

Temperature	Actual Moisture Content at Given Temperature, grains/lb. dry air	Maximum Moisture Control at Dew Point, grains/lb. dry air	Relative Humidity, %RH	Condensed Moisture,[b] grains/lb. dry air
70°F	54	108	54/108 = 50	None
60°F	54	77	54/77 = 70	None
50°F	54	54	54/54 = 100	None
40°F	36	36	36/36 = 100	54 – 36 = 18
30°F	24	24	24/24 = 100	54 – 24 = 30

[a]At 29.92 in. Hg pressure.

[b]As air at 50% RH and 70°F drops in temperature (read down in first column).

(Reprinted with permission of the American Society of Heating, Refrigerating and Air-Conditioning Engineers.)

FIGURE 7.1-2 Temperature-Humidity Equivalents of Optimum Summer and Winter Effective Temperatures[a]

Winter Optimum, ET 68°F[b]		Summer Optimum, ET 71°F[b]	
Relative Humidity	Dry Bulb Temperature	Relative Humidity	Dry Bulb Temperature
100%	68°F	100%	71°F
80%	69°F	80%	73.1°F
70%	70°F	70%	74.3°F
60%	71.1°F	60%	75.6°F
50%	72.2°F	50%	76.9°F
40%	73.5°F	40%	78.3°F
30%	74.8°F	30%	79.7°F
20%	76.1°F	20%	81.2°F
10%	77.5°F	10%	83.0°F

[a]Air movement of 15 to 25 fpm.

[b]Considered comfortable by 98% of people tested.

(Reprinted with permission of the American Society of Heating, Refrigerating and Air-Conditioning Engineers.)

produces warmth with lower temperatures. However, the thermal and vapor permeance characteristics of exterior construction impose practical limits on interior humidity, which vary with the severity of winters.

In areas of low winter design temperatures (0°F [–17.8°C] or lower), few buildings can tolerate 40% relative humidity without developing condensation and moisture problems. At this interior humidity and exterior temperature combination, even sealed insulating glass will develop condensation (Fig. 7.1-3). Single glazing and highly conductive surfaces such as metal wall panels and metal door and window frames without thermal break features will show condensation even at 15% relative humidity. To minimize condensation and yet provide for human comfort, about 25% relative humidity should be maintained when the outside temperature is 0°F (–17.8°C) or lower. The humidity can be maintained at 30% when the outside temperature is 10°F (–12.2°C) or lower, at 35% when 20°F (–6.7°C), and at 40% when 30°F (–1.1°C) outside (see Fig. 7.1-3). Relative humidities above 40% are generally not recommended during the winter heating season.

Although a relative humidity of 25% or higher is considered desirable during the winter season for health reasons, the effect of relative humidity on occupant comfort has been overestimated in the past.

Figure 7.1-4 shows the results of both older tests measuring the effect of temperature and humidity on human comfort immediately upon entering a room (curved lines top and bottom) and recent tests on occupants after a 3-hour stay (solid vertical lines). Unlike the optimum summer and winter ET lines derived from earlier studies (see Fig. 7.1-2), the newer data suggest (1) that a dry-bulb temperature of 77.5°F (25.3°C) is most comfortable and (2) that humidity has no perceptible effect on the sensation of warmth within a range of 25 to 60% relative humidity. This is discussed more fully in the following subsection, "Summer Comfort." Section 15.2.1.3 contains an explanation of the terms *dry-bulb temperature* and *wet-bulb temperature*.

Summer Comfort In summer, 98% of those tested in the older test mentioned above were comfortable at an effective temperature of 71 (100% relative humidity and 71°F [21.67°C]). Figure 7.1-2 shows other relative humidity and temperature combinations that give the same physical sensation in summer.

The dashed lines in the ASHRAE comfort chart (see Fig. 7.1-4) represent effective temperatures. Each line represents graphically equivalent temperature and humidity combinations. To select the corresponding dry-bulb temperatures (hori-

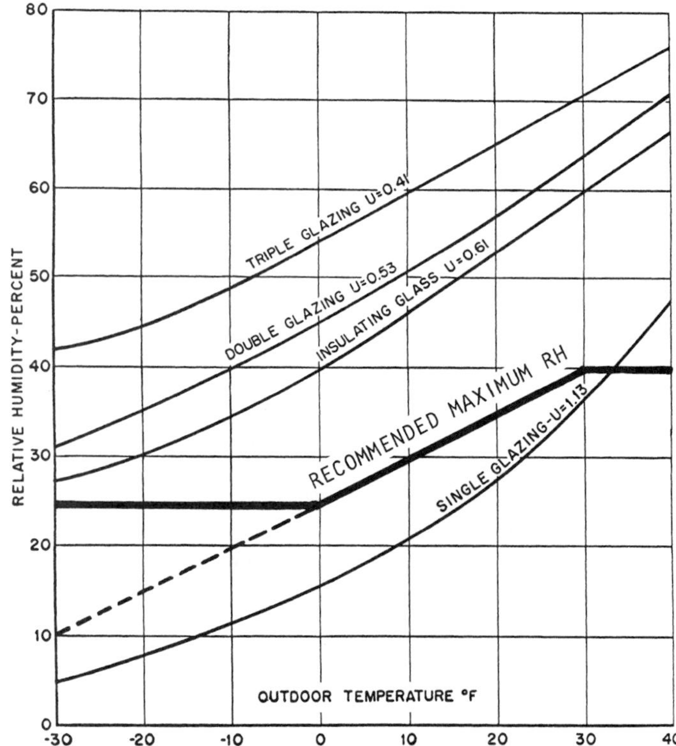

Below 0°F, RH values maintained at 25% to provide acceptable degree of occupant comfort (solid line); however, to minimize condensation on single glazing, use RH values represented by dashed line.

FIGURE 7.1-3 Recommended maximum interior relative humidity to prevent condensation damage. (Reprinted with permission of the American Society of Heating, Refrigerating and Air-Conditioning Engineers.)

zontal coordinate), follow the 68 and 71 effective temperature lines to their intersections with the solid diagonal relative humidity lines. The temperature-humidity equivalents shown in Figure 7.1-2 can thus be derived from this chart.

Note that the optimum summer ET 71 line strikes near the top of the summer curve representing the distribution of responses by the subjects tested. From this curve it can be established that in summer nearly all subjects were comfortable at an effective temperature of 71, and half were comfortable within an effective temperature range of 68 to 75. Similarly, the winter comfort curve indicates that nearly all subjects were comfortable at an effective temperature of 68, and about 86% were comfortable at effective temperatures between 65 and 70.

The effective temperature index, however, is useful only in establishing human response to temperature-humidity conditions immediately upon entering a room. Humidity played a much smaller role in determining the comfort of subjects tested after a 3-hour stay. These findings are represented by solid lines 3, 4, 5, and 6 in Figure 7.1-4. For both summer and winter, a dry-bulb temperature of 77.5°F (25.3°C) was equally comfortable for relative humidities from 25% to 60%. Only above 60% relative humidity (curved portion of line 4) must the temperature be reduced to maintain the same degree of comfort. The lines represent averaged responses for both summer and winter conditions; actually there is a spread of about 1.2°F (0.11°C) between winter and summer responses. The maximum comfort line, therefore, would be at about 77°F (25°C) in winter and 78°F (25.6°C) in summer.

7.1.1.2 Vapor Movement

Water vapor becomes a hazard when it condenses (liquefies) inside the structural cavity of a floor, wall, or roof assembly. Here it can reduce the effectiveness of insulating materials and cause organic materials to decay, metals to corrode, and protective coatings to deteriorate. Wall, floor, and ceiling assemblies should be designed and constructed so as to minimize vapor migration into structural cavities.

Vapor moves largely independently of air movement, from areas of a higher vapor concentration and temperatures to areas of lower concentration and temperatures. Water vapor establishes a pressure proportional to the amount of water vapor present in the air mix. Air with more vapor has a higher vapor pressure. Vapor moves through air by *diffusion* from regions of high vapor pressure to regions of lower pressure, without relying on air circulation to carry it.

Vapor pressure is commonly expressed in inches of mercury (Hg). At 70°F (21.1°C) and 100% relative humidity, vapor pressure is 0.739 in. Hg (0.091 Pa), and at 0°F (–17.8°C) and 100% relative humidity, vapor pressure is only 0.0377 in. Hg (0.128 Pa), or about 1/20 of the saturated condition.

With a winter condition of 0°F (–17.8°C) and 75% percent relative humidity, an outside vapor pressure of 0.027 in. Hg (0.091 Pa) would exist, while inside a building heated to 70°F (21.1°C) and with 35% relative humidity, vapor pressure would equal 0.259 in. Hg (0.877 Pa). The vapor pressure inside would be nearly 10 times the pressure outside. Like other gases, water vapor moves from an area of high pressure to an area of low pressure until equilibrium is established. During cold weather, the difference in pressure between inside and outside causes vapor to move out through every available crack and directly through permeable materials.

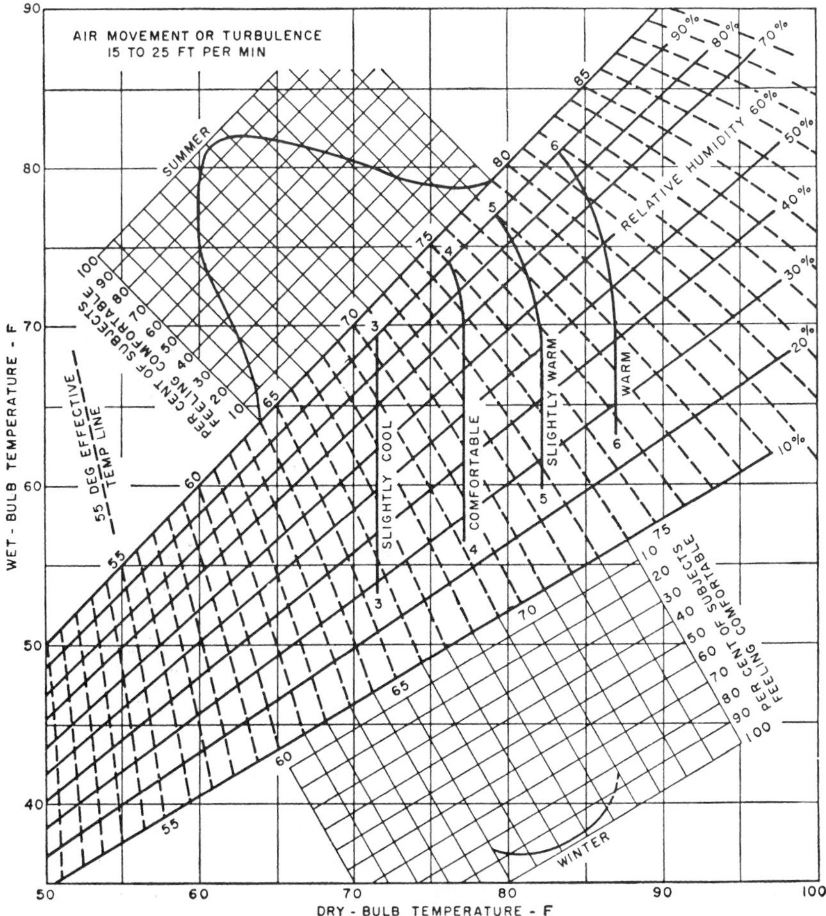

FIGURE 7.1-4 ASHRAE comfort chart showing ETs and subjective responses to temperature and humidity combinations. (Reprinted with permission of the American Society of Heating, Refrigerating and Air-Conditioning Engineers.)

PSYCHROMETRICS

The behavior of moist air (mixture of dry air and water vapor), involving saturation of air and condensation of water vapor at varying temperatures, is generally referred to as *psychrometrics*. Changes in the properties of moist air with heating and cooling can be followed readily with the aid of a psychrometric chart (Fig. 7.1-5). The curved saturation line (100% relative humidity) represents the maximum concentrations of water vapor (pounds of water per pound of dry air) (kg of water per kg of dry air) that can exist as vapor, without condensing, at various temperatures.

Figure 7.1-6 is a simplified psychrometric chart for a specific set of conditions. A possible winter condition inside

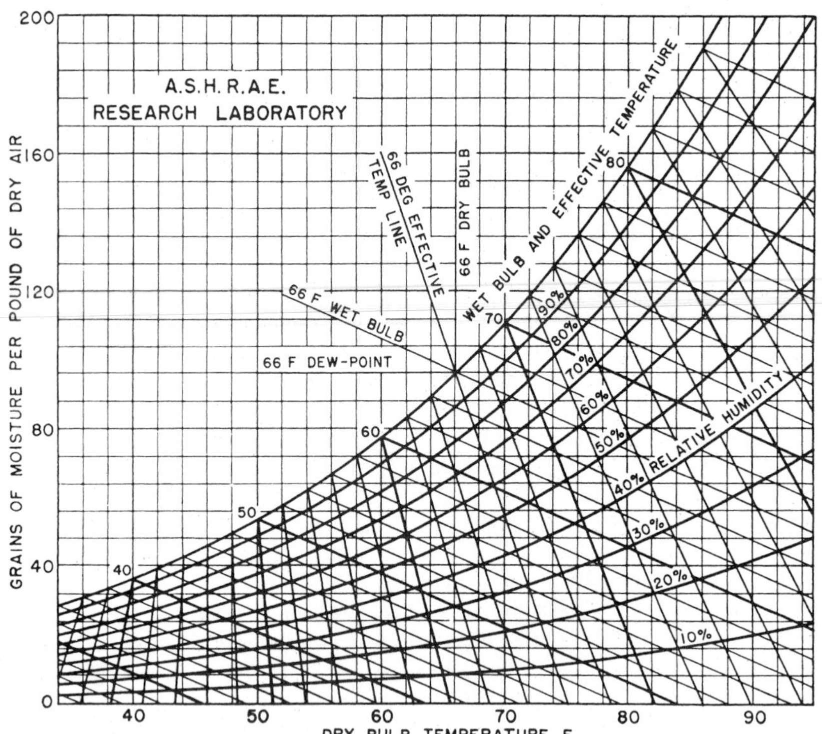

FIGURE 7.1-5 Psychrometric chart. (Reprinted with permission of the American Society of Heating, Refrigerating and Air-Conditioning Engineers.)

buildings, 70°F (21.1°C) and 40% relative humidity, is represented by point A. This is a partial saturation condition (less than 100% relative humidity). As air is cooled from 70°F (21.1°C) (point A) to 44.6°F (7°C) (point B on the saturation curve) the relative humidity keeps increasing until it becomes 100%.

DEW POINT TEMPERATURE

The temperature at which saturation occurs (point B in Fig. 7.1-6) is called the *dew point temperature* (*frost point*, if below 32°F [0°C]). With further cooling to 35°F (1.67°C), the original amount of vapor (44 grains/lb. [0.0063 kg/kg] of dry air) is reduced through condensation to the condition represented by point C (30 grains/lb. [0.0043 kg/kg] of dry air). The progression from point A to B to C illustrates what an air-vapor mixture experiences when it comes in contact with a cool window surface. Cooling from point B to point C results in visible condensation on the glass surface. If point C were below 32°F (0°C), the condensation would be frost.

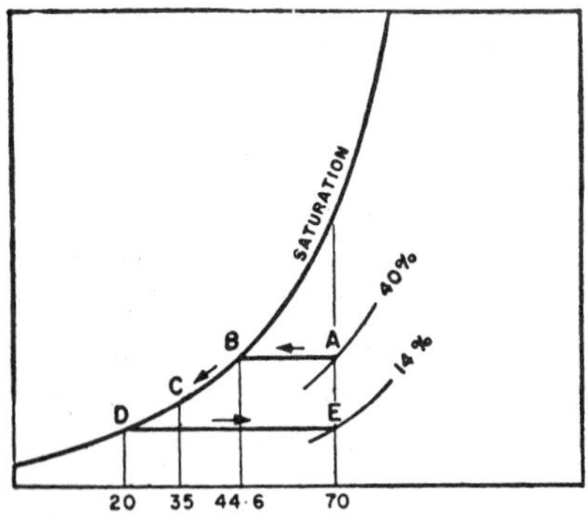

FIGURE 7.1-6 Use of a psychrometric chart illustrated. (Reprinted with permission of the American Society of Heating, Refrigerating and Air-Conditioning Engineers.)

Once the temperature drops below the dew point (or frost point), vapor pressure is also reduced. This reduction in vapor pressure causes diffusion currents within a room, moving water vapor continuously to window surfaces to be condensed as long as the vapor concentration is maintained in the room.

The move from point D to point E (see Fig. 7.1-6) illustrates a common winter phenomenon. When outside air at 20°F (–6.7°C) and 100% relative humidity is heated to 70°F (21.1°C), the relative humidity is lowered to 14%. This is the reason that the relative humidity within a building is greatly reduced when cold outside air enters and is heated to room temperature.

High dew point temperatures are common in the summertime. The psychrometric chart shows that for a dry-bulb temperature of 80°F (26.7°C), the dew point temperature is 60°F (15.6°C) when the relative humidity is 50% (see Fig. 7.1-5). However, raising the relative humidity to 90% increases the dew point to 78°F (25.6°C). As the humidity rises, the dew point gets closer to the dry-bulb temperature but cannot exceed it. If the temperature is raised to 90°F (32.2°C) and the relative humidity is maintained at 90% (by adding moisture to the air), the dew point rises to 87°F (30.6°C).

7.1.1.3 Adsorption and Absorption

The surfaces of most materials have an affinity for water, and a film of water molecules is often held to the surface by molecular forces of attraction. This phenomenon is known as *adsorption*. Surface films at low humidities may be only one molecule thick; at moderate humidities, multimolecular films may be established. The film thickness, and therefore the amount of water held in equilibrium with the surrounding atmosphere, is roughly proportional to the relative humidity. At humidities close to 100%, the films become so relatively thick that small surface pores become completely filled and larger capillaries may be partially filled. At saturation conditions, all voids in the material are completely filled.

Fibrous materials such as wood, plywood, and fiberboards present very large effective surface areas to water molecules, so that the amount of water held on the surface, and subsequently absorbed into these materials through surface pores, is relatively large even at moderate humidities. These materials are said to be *hygroscopic*. Materials that present relatively small surfaces, such as most metals, are not penetrated by water molecules and take on only small amounts of water, as surface film.

DIMENSIONAL INSTABILITY

Changes of moisture content can cause significant dimensional changes in construction materials. For example, when wood is dried from an air-dry condition of 12% or 25% moisture content to an oven-dry condition (0 moisture content), it shrinks about 0.1% longitudinally, 2% radially, and 4% tangentially.

The continuous movement of dimensionally unstable materials must be carefully considered in building design and detailing. For example, millwork elements lap one another to permit movement without developing unsightly cracks; air gaps around wood strip floors allow expansion without buckling in humid weather.

DETERIORATION OF MATERIALS

Moisture can cause or contribute to the breakdown of materials by (1) chemical changes such as the rusting of steel, (2) physical changes such as masonry spalling by frost action, or (3) biological processes such as wood decay. Measures to control liquid water (other than water vapor) include waterproofing and dampproofing (Section 7.2), flashings (Section 7.8), and storm drainage (Section 2.4).

Since air saturation is more likely to take place at low temperatures, when the risk of freezing is also enhanced, condensation and freezing are both major design considerations. Moreover, condensation may occur over a period of time in concealed locations within wall, floor and ceiling, or roof assemblies and not be apparent until a conspicuous failure occurs.

CONDENSATION TOLERANCE OF MATERIALS

Some materials are able to tolerate condensation with little damage; others are not. Each material should be assessed individually. The metal panels in refrigerated storage boxes, for example, almost always contain a frost coating with no resulting damage. Some materials can store condensation with little or no effect. Concrete masonry units, for example, can easily have their saturation level raised from 40% to 50% by absorbing condensation without being noticeable. Metal panels absorb little moisture, but can rapidly release collected moisture or frost through evaporation when weather conditions change.

EFFECT ON HEAT FLOW

Moisture can have a marked effect on the heat transmission through building materials. When present in liquid form, it increases the conductivity of materials by increasing available heat flow paths.

Vegetable fiber products and other hygroscopic insulations, which readily absorb and hold moisture, do not dry out as quickly as nonabsorbent fiber insulations and lose their thermal effectiveness for longer periods. Nonabsorptive products, however, have limited moisture storage capacity and begin "dripping" sooner, possibly damaging interior finishes.

7.1.1.4 Water Vapor Transmission

The water vapor permeance of component materials should be considered when designing exterior walls, floors, and roofs of habitable spaces. The amount and rate of water vapor transmitted through building materials is dependent on (1) the *gradient* (differential) in vapor pressure from one side of the material to the other, (2) the material's area, and (3) its *permeance* (ability to permit vapor passage).

PERMEANCE AND PERMEABILITY

Water vapor *permeance* is a measure of water vapor flow through a material of specific thickness or an assembly of several materials. The unit of permeance, a *perm*, states the amount of vapor flow in grains per hour per square foot of surface, per inch Hg (mercury) (3.386 Pa) vapor pressure gradient. One perm is 57.452 5 nanograms per pascal second square meter (ng/(Pa·s·m^2)).

Water vapor *permeability* is the permeance of a 1-in. (25.4 mm) thickness of a homogeneous substance. Permeability is designated in perm-inches and states the amount of vapor flow through 1 in. (25.4 mm) of material, in grains per hour (ng/Pa·s·m) per square foot (m) of surface, per inch Hg (3.386 Pa) vapor pressure gradient. One perm inch is 1.459,29 nanograms per pascal second meter (ng/Pa·s·m).

Most building materials are permeable to a degree. Some, such as metals and glass, are almost completely impermeable, permitting almost no passage of water vapor. Such materials have a perm rating (permeance) approaching zero. Materials that are either impermeable or that have a permeance of 1 perm (57.45 ng/Pa·s·m^2) or less were formerly referred to as *vapor barriers*. However, in recognition that even the best of these materials permit some vapor transmission, ASHRAE now calls them *vapor retarders*. This text uses this more accurate term throughout. In addition, modern construction methods demand better vapor retarders than those approaching 1 perm (57.45 ng/Pa·s·m^2). Therefore, vapor retarders are frequently required to have a rating of 0.5 perm (28.73 ng/Pa·s·m^2) or less. Permeances for representative construction materials are given in Figure 7.1-7.

Many of the permeance numbers in Figure 7.1-7 were obtained by different test methods, which can produce somewhat varying ratings. Therefore, it is not valid to compare the relative performance of those materials with small variations in their permeances. For instance, it is not meaningful to compare hot melt asphalt, 2 oz./sq. ft. (56.68 g/0.093 m^2) (perm, 0.5 [28.73 ng/(Pa·s·m^2)]) with asphalt-saturated paper (perm, 0.2–0.3 [11.49–17.24 ng/(Pa·s·m^2)]). Vapor retarder materials rated at under 1 perm (57.45 ng/Pa·s·m^2) can be more readily compared to products with permeances substantially higher than 1 perm (57.45 ng/Pa·s·m^2).

The overall permeance (M_t) of an assembly of materials can be calculated from the permeances of the individual components. The calculation is similar to that used in determining overall coefficient of heat transmittance (U) from individual material conductances (C):

$$M_t = \frac{1}{\frac{1}{M_1} + \frac{1}{M_2} + \frac{1}{M_3} + \cdots + \frac{1}{M_n}}$$

where M_t = overall, total permeance and M_1, M_2, M_3, through M_n = the individual permeances of materials in an assembly.

WATER VAPOR TRANSMISSION

Sometimes it is more significant to report the vapor resistance of materials at a specific pressure difference rather than at 1 in. Hg (3.386 Pa), as in permeance and permeability. Water vapor transmission (WVT) states the weight of vapor (grains [kg]) transmitted per unit of time (typically 1 hour) per unit of area (1 sq. ft. [m^2]); values may be high or low depending on the vapor pressure gradient chosen for the test. Water vapor transmission data can be converted to perms by the following formula:

Permeance = *WVT* rating/Δp,

where *WVT rating* equals the weight of vapor transmitted in grains per hour per square foot and Δp equals vapor pressure gradient in the test in inches (Pa) of mercury.

7.1.2 CONDENSATION CAUSES AND PREVENTION

Condensation occurs when vapor-laden air at a high enough humidity contacts cool surfaces. Condensation problems may occur in cooler climates in occupied, heated buildings and in humid, warm climates in air conditioned buildings. Production of condensation-causing moisture

FIGURE 7.1-7 Water Vapor Permeance of Building Materials

Material	Permeance (Perm)
Materials used in construction	
1 in. concrete (1:2:4 mix)	3.2
Brick masonry (4 in. thick)	0.8
Concrete block (8 in. cored, limestone aggregate)	2.4
Plaster on metal lath (¾ in.)	15.0
Plaster on plain gypsum lath (with studs)	20.0
Gypsum wallboard (⅜ in. plain)	50.0
Gypsum sheathing (½ in. asphalt impregnated)	10.0
1-in. structural insulating board (sheathing quality)	20–50
Structural insulating board (interior, uncoated, ½ in.)	50–90
Hardboard (⅛ in. standard)	11.0
Hardboard (⅛ in. tempered)	5.0
Built-up roofing (hot-mopped)	0.0
1-in. wood sugar pine	0.4–5.4
Plywood (douglas-fir, interior, blue, ¼-in. thick)	1.9
Thermal insulations, 1 in. thick	
Air (still)	120.0
Cellular glass	0.0
Corkboard	2.1–9.5
Mineral wool (unprotected)	116.0
Expanded polyurethane (R-11 blown)	0.4–1.6
Expanded polystyrene, extruded	1.2
Expanded polystyrene, bead	2.0–5.8
Unicellular synthetic flexible rubber foam	0.02–0.15
Plastic and metal foils and films	
Aluminum foil (1 mil)	0.0
Aluminum foil (0.35 mil)	0.05
Polyethylene (2 mil)	0.16
Polyethylene (4 mil)	0.08
Polyethylene (6 mil)	0.06
Building papers, felts, roofing papers*	
Duplex sheet, asphalt laminated, aluminum foil one side (43)	0.002
Saturated and coated roll roofing (326)	0.05
Kraft paper and asphalt laminated, reinforced 30-120-30 (34)	0.3
Blanket thermal insulation backup paper, asphalt coated (31)	0.4
Asphalt saturated and coated vapor-barrier paper (43)	0.2–0.3
Asphalt saturated but not coated sheathing paper (22)	3.3
15-lb. asphalt felt (70)	1.0
15-lb. tar felt (70)	4.0
Single-kraft, double infused (16)	31.0
Protective coatings	
Paints, 2 coats	
Aluminum varnish on wood	0.3–0.5
Enamel on smooth plaster	0.5–1.5
Various primers plus 1 coat flat oil paint on plaster	1.6–3.0
Water emulsion on interior insulating board	30.0–85.0
Styrene-butadiene latex coating (2 oz./sq. ft.)	11.0
Polyvinyl acetate latex coating (4 oz./sq. ft.)	5.5
Asphalt cut-back mastic (1/16 in. dry)	0.14
Hot melt asphalt (2 oz./sq. ft.)	0.5

*Numbers in parentheses are weights in lb. per 500 sq. ft.

(Reprinted with permission of the American Society of Heating, Refrigerating and Air-Conditioning Engineers.)

within buildings depends on the types of activities there. Commercial kitchens and shower rooms, for example, can generate enormous quantities of vapor. Four people in a normal home environment generate between 22.5 and 50 lb. (10.21 and 22.68 kg) of vapor in 24 hours. In winter, condensation may collect as water or frost on cold surfaces such as glass, metal wall panels, metal window frames, and metal doors and frames.

Although condensation forming on a surface may enter surface pores as fast as it forms and thus become invisible, any condensation on a visible surface is referred to as surface condensation to distinguish it from concealed condensation, which occurs within a material or assembly.

7.1.2.1 Interior Ventilation

In many locations (see Section 7.1.3), summer cooling does not always create serious concealed condensation problems, because cooled interior air seldom produces conditions within a wall that are conducive to condensation. However, during periods of high humidity, condensation may form on cool basement walls and floors. In humid climates, condensation may occur in concealed spaces (see Section 7.1.3).

Where high water vapor levels are generated indoors, the likelihood of both surface and concealed condensation can be greatly minimized by introducing dry night air or by exhausting excessive water vapor directly at its source, such as in a shower room, kitchen, laundry, or residential bathroom, using exhaust fans. In many buildings a sufficient amount of fresh air is available through infiltration to make up the air lost by exhaust. Where this is not the case, it will be necessary to introduce a fresh air supply to balance this loss.

During cold weather, interior relative humidity often becomes too low for human comfort, and mechanical humidification of occupied spaces is required. Even then, ventilation in areas where high humidity is present, such as those mentioned earlier, is desirable to reduce excessively high local humidities, which could easily result in surface condensation.

The effectiveness of ventilation in reducing humidity is shown in Figure 7.1-8. For simplicity, a 2000 sq. ft. (185.8 m^2) residence is used as an example. Figure 7.1-8 shows the small amount of water vapor escaping into well-insulated walls and ceilings of a building provided with vapor retarders (the floor is not considered). With indoor temperature at 70°F (21.1°C) and a relative humidity of 40%, ventilation at the rate of 2000 cu. ft. (56.6 m^3) per hour removes about 21 lb. (9.53 kg) of moisture per day. Assuming a production of 22.5 lb. (10.21 kg) a day, less

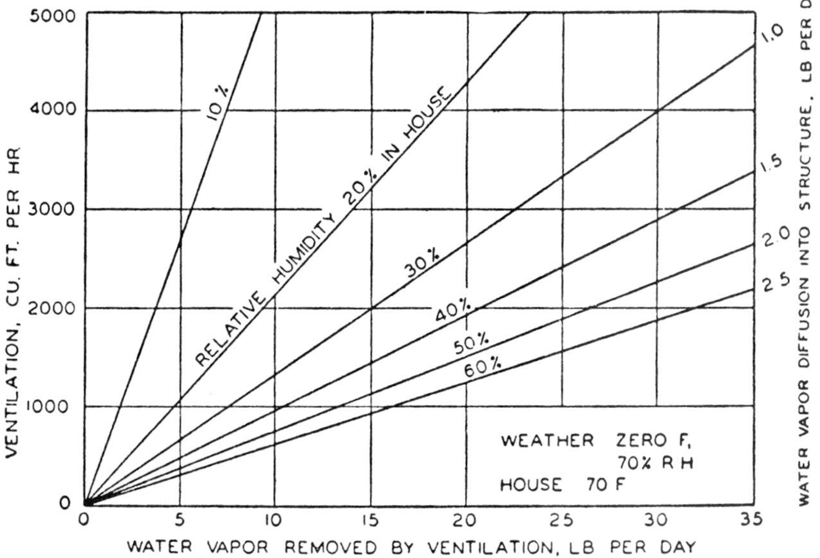

FIGURE 7.1-8 Water vapor balance maintained by ventilation. (Reprinted with permission of the American Society of Heating, Refrigerating and Air-Conditioning Engineers.)

than 1.5 lb. (0.68 kg) escapes into the walls and ceilings of the structure. As mentioned earlier, 22.5 lb. (10.21 kg) in 24 hours is about average for a family of four at home.

In a residence, ventilation at the rate of 2000 cu. ft. (56.6 m^3) per hour is near the minimum for odor control during cooking. Therefore, mechanical ventilation may not be required just to get rid of excess moisture generated in a home. In older buildings of all types, the action of exhaust fans is often supplemented by substantial *exfiltration* (air leakage) through walls, windows, and doors to keep moisture in check. However, tight newer buildings will often require more extensive ventilation systems, particularly in periods of cold weather when the relative humidity must be limited to prevent condensation (see Fig. 7.1-3).

7.1.2.2 Surface Condensation

Condensation occurs when a surface is colder than the dew point of the nearby air. The temperature of interior surfaces such as walls, ceilings, and floors is dependent on (1) air temperatures both inside and outside the building and (2) heat transmittance of the wall, ceiling, or floor assembly. Ratings of overall heat transmittance for an assembly can be derived by formula from individual transmittances of the materials in the assembly.

Surface condensation occurs when a surface loses heat fast enough to reach the dew point (condensation temperature). Adequate insulation reduces heat loss and increases surface temperature, thereby reducing condensation hazard. Therefore, when enough insulation is provided in exterior walls, ceiling and floor assemblies, and roof and ceiling assemblies to keep surfaces that are in contact with moisture at a temperature below the dew point, there will be no surface condensation. Where installing sufficient insulation to accomplish this is not feasible, warm air currents can be induced to increase surface temperature and dissipate moisture. Keeping interior relative humidity within the range recommended in Figure 7.1-3 will also minimize surface condensation.

Figure 7.1-9 shows the relative humidities at which condensation will appear on interior surfaces with various U-values in a room with a 70°F (21.1°C) air temperature. Since condensation can appear at any sufficiently cold spot, values given in Figure 7.1-9 should be used with caution. U-values constitute an average for large surface areas, within which there may be spots where the heat transmittance is higher, such as near structural supports or metal outlet boxes. Also, the inside surface temperature of a stud wall will often be lower at the base due to air stratification inside the stud cavity. As a result, the maximum limit of relative humidity for a nonhomogeneous wall is lower than might be inferred from its average U-value.

Glass areas are normally the most likely surfaces in heated buildings on which condensation may appear; however, an uninsulated wall surface shielded from radiation and convection by furniture may also be quite cold. Over time, surfaces with temperatures near the dew point temperature of the adjacent air may experience swelling, mold, staining, or decay.

Condensation on interior room surfaces can be controlled both by suitable construction and by operating precautions

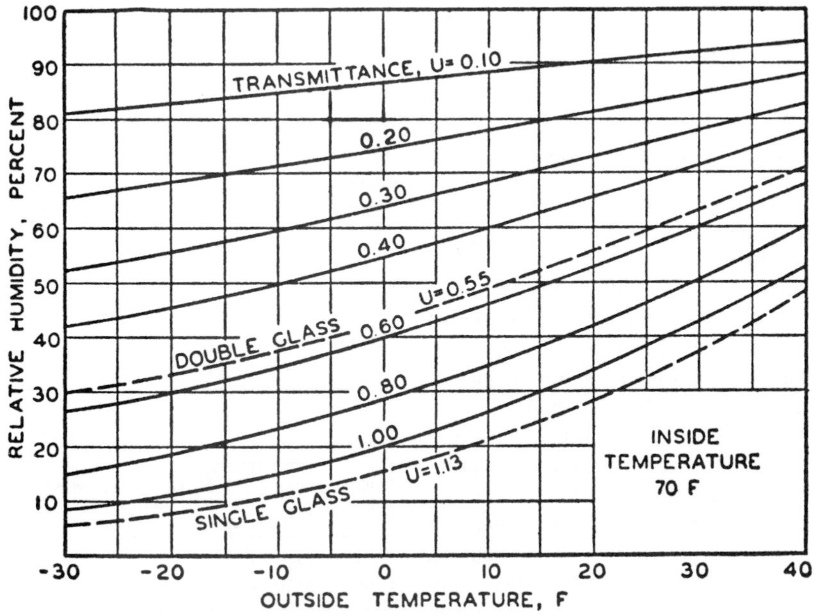

FIGURE 7.1-9 Relative humidity and condensation. (Reprinted with permission of the American Society of Heating, Refrigerating and Air-Conditioning Engineers.)

such as (1) reducing the interior dew point temperature or (2) raising the temperatures of interior surfaces above the dew point.

Interior air dew point temperature can be lowered by removing moisture from the air, through either ventilation or dehumidification. Surface temperatures in winter can be maintained sufficiently high by incorporating adequate thermal insulation, using double glazing, circulating warm air over the surfaces, or heating the surfaces directly.

BELOW-GRADE CONSTRUCTION

Visible surface condensation may occur in summer on concrete basement walls and floors and on exposed cold-water pipes. Because of their large mass and the insulating effect of the adjacent soil, basement floors and walls tend to maintain a relatively cool and constant temperature. If the dew point temperature within the space rises above the temperature of walls, floors, and pipes, condensation occurs. Finish surfaces in below-ground spaces can be seriously damaged by condensation.

Condensation on interior surfaces can be controlled by (1) reducing the air intake from the outside when outdoor dew point temperatures are high, (2) warming the surfaces, or (3) most successfully, by deliberate dehumidification. For instance, pipes to water closets can be warmed by using lukewarm water for flushing; in climates having low temperatures at night, it may be feasible to ventilate only at night when the air contains less moisture and thus reduce the moisture absorption of hygroscopic materials.

SLAB-ON-GRADE CONSTRUCTION

As do basement walls, slabs on grade tend to maintain a relatively constant temperature. Surface condensation can develop when such a slab is sufficiently cold in relation to a warm interior atmosphere that has a high dew point. In climates where high dew point temperatures frequently occur in summer, condensation may damage finish flooring laid on a slab on grade. Thick rugs or carpets aggravate the problem because they insulate the slab from interior warmer air and retain the slab at a low temperature. Because woven, knitted, and tufted floor coverings are often vapor permeable, moisture may condense under the covering and may promote decay, mold, and a musty odor.

Methods of controlling surface condensation are similar to those for above-grade construction: deliberate dehumidification and increasing surface temperature. A slab on grade can be kept warm by insulating the slab from cooler ground temperatures with rigid insulation at the perimeter and with a granular, well-drained base course under the entire slab.

7.1.2.3 Concealed Condensation

In winter in most parts of the country the continual production of vapor inside an occupied building normally raises the vapor pressure above that outside the building. This pressure difference provides the force that causes vapor diffusion into exterior walls and ceilings. The vapor pressure rise in the building is directly proportionate to the amount of vapor produced and inversely proportionate to its ability to escape. In other words, the pressure rises as more vapor is produced, but the pressure is relieved if it is permitted to diffuse freely to areas of lower concentration.

Exterior walls, floors over unheated spaces, and roof and ceiling assemblies should be protected with vapor retarders and provisions for ventilation. In most areas, because moisture sources are usually within a building, a vapor retarder should be placed near the interior finish to prevent moisture migration into concealed structural spaces. However, a suspended floor above a heated crawl space need not be insulated. Since the ground constitutes the chief source of moisture (Fig. 7.1-10), a vapor retarder should be used as a ground cover.

As water vapor moves through a portion of a structure, such as a wall, floor, or roof and ceiling assembly, it may reach a point within the assembly at which it is cooled to the point of condensation. In older frame buildings with little or no insulation, so much heat was lost through the building's shell that the entire structural cavity was often warm enough to be above the dew point (Fig. 7.1-11a). The dew point temperature was reached somewhere in the sheathing or siding, but condensation was rarely a problem because moisture was carried off by air leakage through joints in the sheathing and siding, which were usually of the board type.

Although the theoretical dew point line in an insulated frame wall generally lies well within the cavity or even within the insulation (Fig. 7.1-11b), the vapor does not condense at this plane but moves on through the vapor-permeable fibrous insulation until it reaches the sheathing. Modern sheathing and siding products are frequently made in broad panels rather than narrow boards, thus reducing the opportunity of moisture diffusion by air leakage through joints. In addition, many siding and sheathing materials such as plywood and hardboard are inherently vapor resistant, so only small amounts of vapor can move directly through the material itself. These factors, coupled with the generally higher interior humidity levels maintained by modern heating systems, account for the greater incidence of condensation problems in modern construction if an adequate vapor retarder is not provided.

WALL CONSTRUCTION

When water vapor is allowed to enter a wall assembly and condensation occurs within its cold outer elements, frost or water may develop, depending on the outside temperature (Fig. 7.1-12). In weather that is continuously cold for a long period, the frost may build back into the wall cavity or fibrous insulation, forming an ice mass. When the outside temperature rises, this frost or ice may melt and be

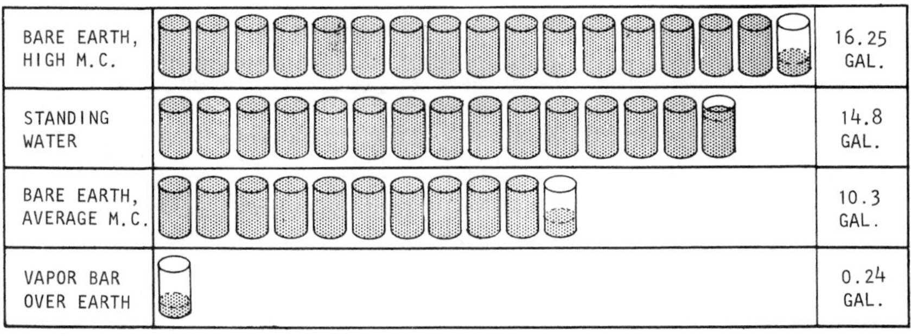

FIGURE 7.1-10 Daily evaporation from a crawl space floor. (U.S. Forest Products Laboratory. Not subject to copyright in the United States.)

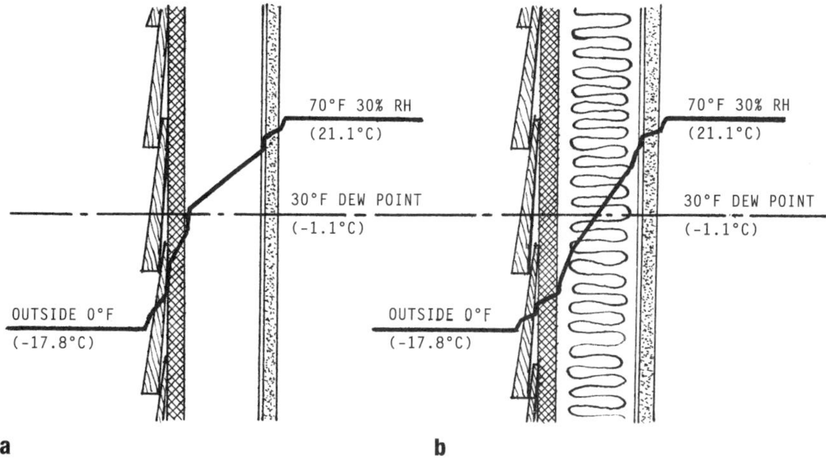

FIGURE 7.1-11 When there is no insulation in a wall cavity (a), so much heat is lost that the temperature in the cavity often remains above the dew point. When insulation is added (b), the dew point will usually occur within the cavity.

FIGURE 7.1-12 Sheathing reversed to show accumulation of frost on inside surface. This cavity was insulated but did not have a vapor retarder. (U.S. Forest Products Laboratory. Not subject to copyright in the United States.)

FIGURE 7.1-13 Condensation formed frost and ice behind this siding during cold weather, then weeped through laps in the siding as the weather moderated, staining the siding. (U.S. Forest Products Laboratory. Not subject to copyright in the United States.)

absorbed by hygroscopic materials such as wood, or it may run down nonabsorbing or already-saturated surfaces (Fig. 7.1-13).

In masonry walls, water seepage to the outside may occur harmlessly when the weather is above freezing, but water seepage into the building must be prevented. The inclusion of weep holes and base flashings encourages runoff to the outside and bars seepage to the interior (see Sections 4.6 and 4.7).

An accumulation of moisture within insulation lowers its thermal efficiency, causing greater heat loss and condensation problems. Trapped moisture in siding and sheathing is a principal cause of paint failure. In seeking to escape from behind vapor-resistant paint films, moisture can cause paint to blister and peel, especially when it is warmed by solar radiation (Fig. 7.1-14). In extreme cases, moisture trapped in wall cavities will eventually cause decay (Fig. 7.1-15).

The installation of an adequate vapor retarder is the most important moisture control measure in exterior wall construction.

ROOF AND CEILING ASSEMBLIES

The principles of vapor movement and control within a roof and ceiling assembly are much the same as for walls. Water vapor moves from the side with the higher vapor pressure toward the side with the lower vapor pressure. When the side with the lower pressure is also colder, the water vapor may condense in concealed spaces. Some roofing materials, such as wood shingles, can "breathe" and are little affected by trapped vapor; others, such as

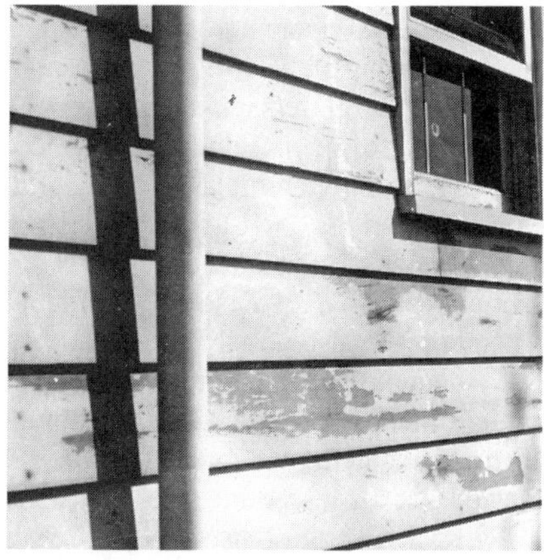

FIGURE 7.1-14 Blistering and deterioration of paint surfaces caused by vapor pressure behind a paint film. (U.S. Forest Products Laboratory. Not subject to copyright in the United States.)

FIGURE 7.1-15 Decay of studs and sheathing due to frequent condensation in the wall cavity. (U.S. Forest Products Laboratory. Not subject to copyright in the United States.)

built-up roofs, are likely to be ruptured by excessive moisture. When a built-up roof blisters and cracks, permitting water to enter the roofing, continuing and rapid deterioration is imminent.

In building spaces with high relative humidities, such as laundries, saunas, and pool enclosures, special consideration must be given to the proper placement of vapor retarders and sufficient insulation.

In unventilated roof and ceiling assemblies, condensation may collect in the form of ice, frost, or water on the framing members, roof deck, and projecting fasteners. Condensation problems seldom exist in roof and ceiling assemblies that have insulation above the ceiling, adequate ventilation, and a properly installed vapor retarder on the warm side of the insulation. But roof constructions with rigid insulation above a wood or metal deck may require a vapor retarder or details that permit venting of the insulation. On sunny days, even at low temperatures, frost or ice may melt and seep to the ceilings below, causing staining and possible damage to the ceilings (Fig. 7.1-16). In extreme cases, rapid deterioration and decay of wood framing and rusting of steel roof structure members are also possible.

Adequate vapor retarders on the warm side of the insulation and the ventilation of roof and ceiling assemblies above the insulation are essential (see "Roof and Ceiling Assemblies" in Section 7.1.3).

CRAWL SPACE CONSTRUCTION

The presence of water in a crawl space may produce considerable amounts of water vapor. This accumulation of water vapor may increase the possibility of decay and termite attack on wood members and rusting of steel members and result in unpleasant odors within the building. Therefore, it is important to ventilate enclosed foundation spaces adequately to provide for the dissipation of water vapor. The amount and methods of ventilation depend on the type of space being ventilated, the degree of water vapor present, and the type of vapor retarder installed.

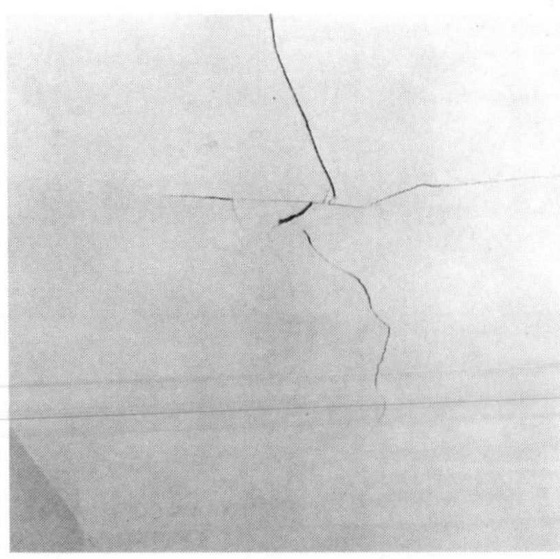

FIGURE 7.1-16 Repeated wetting caused this plaster ceiling to crack and produced a visible stain. (U.S. Forest Products Laboratory. Not subject to copyright in the United States.)

Water vapor rising from the ground below a crawl space is usually a greater problem than vapor coming from inhabited spaces above. Moisture in the soil can rise as much as 11 ft. (3352.8 mm) above the water table. Depending on the level of the groundwater table, the character of the soil, and drainage around the foundation, large amounts of water may be introduced into the crawl space by evaporation from the ground (see Fig. 7.1-17).

Excessive vapor can cause condensation on structural members that have been cooled by heat loss to the outside. Water vapor from a crawl space may also enter walls, move upward within them, and even reach roof construction when the wall construction permits.

These problems can be prevented simply by reducing evaporation from the ground with a vapor retarder and providing adequate ventilation.

7.1.3 DESIGN AND CONSTRUCTION RECOMMENDATIONS

Moisture condensation in walls, floors, and roof and ceiling assemblies can be controlled (1) by providing a vapor retarder on the interior side of the assembly to limit vapor entrance into the construction and (2) by ventilating structural spaces such as attic areas and crawl spaces to dissipate vapor to the outside. This double line of defense is desirable in all areas where severe or even moderate condensa-

FIGURE 7.1-17 Beads of condensation formed on joists in a crawl space not provided with foundation vents or ground cover. (U.S. Forest Products Laboratory. Not subject to copyright in the United States.)

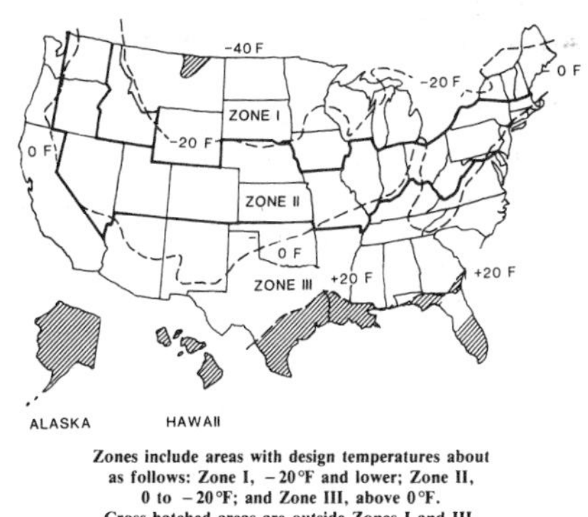

FIGURE 7.1-18 Condensation zones in the United States. (Reprinted with permission of the American Society of Heating, Refrigerating and Air-Conditioning Engineers.)

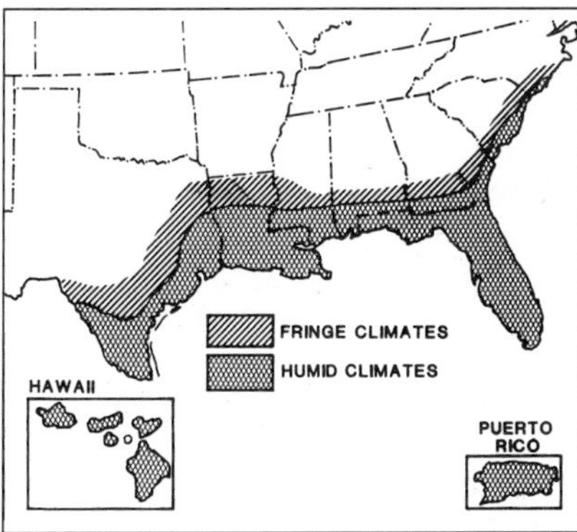

FIGURE 7.1-19 Humid climates in the continental United States. (Reprinted with permission of the American Society of Heating, Refrigerating and Air-Conditioning Engineers.)

tion hazard exists. However, in wall assemblies, reliance is placed on a retarder, and air movement through the cavity is discouraged to minimize heat transfer. The continental United States can be divided into three condensation zones, based mainly on the isotherms of the –20, 0, and 20°F (–28.9, –17.8, and -6.7°C) winter design temperatures (Fig. 7.1-18).

Humid climates present slightly different condensation problems, because in air conditioned buildings located in such climates, the flow of water vapor is from outdoors to indoors. This means that vapor retarders should be located on the exterior side of insulation rather than on the inside. Figure 7.1-19 is a map produced by ASHRAE that shows the location of such areas in the United States.

When construction limitations make it impractical to provide a vapor retarder, as would be the case, for example, in a masonry wall with interior insulation and no applied finish, reliance must be placed on ventilation. Conversely, when adequate ventilation is hard to achieve, vapor-resistant interior construction is essential.

In this discussion, the term *ventilation* is used both in its broad, all-inclusive sense—as any attempt to dissipate moisture by using air as the vehicle—as well as in a more specific sense. In this narrower meaning, *ventilation* describes typical methods of dissipating moisture from structural spaces by positive air movement through continuous air passages linked to the general atmosphere. In this use, the term *ventilation* is often accompanied by a quantitative recommendation, such as a ratio of vent area to total area. This narrower meaning can be contrasted with another type of ventilation, *venting*, which describes the inclusion of air leaks or gaps in an otherwise vapor-resistant material, finish, or membrane, such as vinyl siding, as a means of vapor diffusion. In venting, air spaces are not necessarily interconnected and air movement is nonexistent or sluggish; ventilation infers much more positive air movement and more rapid moisture dissipation.

7.1.3.1 Vapor Retarder Selection

Although a vapor retarder by definition is any material with a permeance of 1 perm (57.45 ng/Pa·s·m^2) or less, not all materials are equally suitable for use in a particular assembly. Some materials are inherently better vapor retarders than others; however, the effectiveness of a retarder depends on the method of installation, as well as its physical properties. For instance, 1-mil-thick aluminum foil has a permeance of 0.00; the permeance of a typical 0.35-mil (0.00889 mm) paper-backed product is likely to be about 0.05 (28.73) when installed in large sheets across the face of studs. The practice of stapling aluminum-faced batts to the sides of wood studs may increase the installed permeance about twentyfold, to as high as 1 perm (57.45 ng/Pa·s·m^2).

Vapor retarder selection depends on a variety of factors, such as prevailing interior and exterior humidities, length and severity of the winter season, and materials and methods of construction. In general, the severity of the winter season, as expressed by the winter design temperature, is a prime factor in determining the degree of vapor protection required (see Fig. 7.1-18).

Figure 7.1-20 summarizes vapor retarder and ventilation recommendations for roof, wall, and floor assemblies in the three zones. Structural space ventilation is recommended for attic spaces and unheated or cooled crawl spaces in all zones. Vapor retarders are recommended for most assemblies in Zones I and II.

FIGURE 7.1-20 Vapor Retarders and Structural Space Ventilation in Different Condensation Zones

	Atticless Joist Roof/Ceiling		Attic-Type Joist Roof/Ceiling		Crawl Space		Floor Over Open Unheated Area, V.R., Perms	Exterior Walls, V.R., Perms	Slab on Ground V.R., Perms
Condensation Zone	Ceiling V.R., Perms	Structural Ventilation[a]	Ceiling V.R., Perms	Structural Ventilation[a]	Ground V.R., Perms	Structural Ventilation[a,b]			
Zone I	0.1	1/300	0.5	1/300	1.0	1/1500	1.0	1.0[c]	3.0
Zone II	0.5	1/300[d]	1.0	1/300[d]	1.0	1/1500	1.0	1.0[c]	3.0
Zone III	1.0	1/300[d]	None	1/300[d]	1.0[e]	1/1500[e]	None	None	3.0

[a]Ratio of net free ventilating area (NFA) to total floor or ceiling area.

[b]Ventilation to atmosphere not required when crawl space is heated or cooled and foundation wall is insulated.

[c]Venting of walls may be desirable where highly impermeable exterior finish is used.

[d]Ratio of 1/150 preferred to relieve high summer solar heat gain.

[e]When crawl space is not heated or cooled, vapor retarder may be omitted provided vent area is increased to 1/150.

(Reprinted with permission of the American Society of Heating, Refrigerating and Air-Conditioning Engineers.)

In Zone III, the warm southern part of the United States, average winter temperatures are high enough that large temperature and vapor pressure differentials between the inside and outside are not developed. Therefore, vapor retarders are often omitted from walls and roof and ceiling assemblies in Zone III; however, a retarder should be installed beneath slabs on grade and in crawl spaces as a general precaution against moisture in this zone, as in Zones I and II.

Figure 7.1-20 shows three levels of retarder efficiency: (I) permeance of 0.1 perm (5.75 ng/Pa·s·m^2) or less, usually aluminum foils or plastic films; (II) 0.5 perm (28.73 ng/Pa·s·m^2) or less, asphalt-coated felts or asphalt-laminated papers; (III) 1.0 perm (57.45 ng/Pa·s·m^2) or less, same as (II), but with less asphalt per sq. ft. (m^2).

Depending on the type, vapor retarders may be incorporated with or installed separately from other materials in the construction. Typical vapor retarders in frame construction include (1) a separate plastic or metallic membrane, job-installed over framing members and under the interior finish material; (2) a metallic foil, factory-applied on the back of an interior finish material such as gypsum wallboard; (3) aluminum-faced paper or asphalt-laminated paper, factory-applied over batt or blanket insulation; and (4) insulations made of inherently vapor-resistant materials, such as layers of reflective metallic foil or closed-cell plastic foam.

RETARDER-FACED INSULATION

Retarder-faced insulation is discussed in Section 7.3.2. Where the vapor retarder material on such insulation is integral with the batts or blankets and the flanges are nailed to stud sides, as recommended by most manufacturers, it is almost impossible to avoid air and vapor leaks. Therefore, this installation method should be restricted to areas of moderate condensation hazard, as in Zone II. In Zone I, the vapor retarder flanges should be secured across the stud faces, or a separate continuous vapor retarder should be used.

FOIL-BACKED BOARDS

Foil-backed gypsum boards are discussed further in Section 9.5. Foil-backed gypsum lath and gypsum wallboard have an aluminum foil retarder laminated to the back of each board (Fig. 7.1-21). However, breaks exist at board joints, which are typically 16 in. (406.4 mm) apart vertically and 48 in. (1219 mm) apart horizontally for lath, and 48 in. (1219 mm) apart horizontally and 96 in. (2438 mm) apart vertically for wallboard. These joints permit some vapor leakage (especially where the joint is not backed up by a structural member), and the efficiency of the retarder is reduced.

Foil-backed gypsum lath boards are generally installed horizontally with the long edges perpendicular to the studs. This method permits a great deal of vapor leakage and is not recommended as the sole means of vapor control in either Zone I or II.

Using foil-backed wallboard results in effective vapor isolation if the boards are installed vertically, with all joints over studs. It is suitable for vapor control in exterior walls in all zones.

SEPARATE SHEETS

The most efficient vapor retarder installation is one in which the vapor retarder is applied separately from the insulation and the interior finish. It is installed in the largest possible sheets and with joints backed up by framing members. Insulation can be provided as pressure-fit blankets or rigid boards. When combined with a suitable insulation, a separate-sheet vapor retarder is particularly suitable for severe condensation conditions, as in Zone I.

Polyethylene film and kraft paper–supported aluminum foil are typical products available in roll widths from 3 to 20 ft. (0.914 to 6.1 m). Best results are obtained when joints are made over a structural member. This "vaporproofing" method is highly recommended for Zone I.

INSULATING BOARDS

A number of synthetic insulating boards, such as foamed glass, polystyrene, and polyurethane, are being marketed; many of these materials have a closed-cell construction and are inherently good vapor retarders. However, their permeability varies with their manufacturing process and should be checked in each case to determine that the required permeance can be achieved in the thickness contemplated. Boards should also be butted

FIGURE 7.1-21 **Vapor control can be provided by foil-backed gypsum board.** (Courtesy USG Corporation.)

tightly together and joints should be sealed with a vapor-resistant mastic.

PAINTS AND OTHER COATINGS

Although many types of paints have proven to be good vapor retarders under laboratory conditions, the uncertainties of field application make reliance on paint coatings somewhat questionable. In new construction, the more reliable factory-produced vapor retarders are recommended. Paints and other surface coatings should not be used as the sole means of vapor control in Zones I and II.

However, paint coatings can sometimes be used correctively in older structures and as an adjunct to other vapor control methods. Several manufacturers market products for specific use as vapor retarders. Alkyd, aluminum, and bituminous coatings generally have low permeability (high resistance) to vapor. Because the vapor-retarding properties of specific products depend on formulation, the manufacturer should be consulted. Most coatings are more vapor-resistant when their finish appearance is shiny or glossy.

7.1.3.2 Vapor Retarder and Ventilation Design

Ventilating openings in exposed locations often are fitted with louvers to keep out wind and rain; screening is usually included to bar insects and vermin. Ventilating requirements are generally given as net free area, which is the space actually available to the moving air, after deducting obstructions such as the screening and louvers. Figure 7.1-22 shows how to calculate the required gross area of various ventilating openings.

FIGURE 7.1-22 Required Gross Area of Vent Openings

Vent Covering	RGA[a] / NFA[b]
Hardware cloth, ¼-in. mesh	1
Screening, ⅛-in. mesh	1¼
Insect screen, 1/16-in. mesh	2
Louvers and hardware cloth ¼-in. mesh	2
Louvers and screening, ⅛-in. mesh	2¼
Louvers and insect screen, 1/16-in. mesh	3

[a]Required gross area.

[b]Net free area (which can be determined from Fig. 7.1-20).

(Reprinted with permission of the American Society of Heating, Refrigerating and Air-Conditioning Engineers.)

WALLS

Ventilating exterior wall cavities may provide adequate vapor control in a moderate climate. However, wall openings designed to move substantial amounts of air and vapor may waste considerable heat in colder climates. Therefore, in walls, as in most other assemblies, vapor retarders are the primary means of vapor control; wall ventilation (more typically venting), when provided, is an adjunct to vapor control.

Venting of an exterior surface should be considered when the siding is a vapor-resistant material in sheet or panel form. Recognizing the high vapor resistance of plastic and metal sidings, manufacturers of these products usually include vent holes. No special venting efforts are required with wood and plywood horizontal board sidings, which are self-venting at the overlap. However, plywood and hardboard panel sidings require more effective interior vapor retarders and may require special venting details.

In older buildings that have no vapor retarders, painted siding may resist vapor flow sufficiently to trap moisture within stud spaces. In Zones I and II, the resulting vapor pressure under the paint film may cause blistering and peeling. In these cases, ventilating stud spaces with individual vents at top and bottom (Fig. 7.1-23) will help dry out the stud cavities. However, if vents become clogged with cobwebs and dust, it will be necessary to provide an interior vapor retarder.

It is essential to form a continuous retarder envelope on the inside of walls that contain any material that would be damaged by moisture or freezing. This includes most thermal insulation, wood and metal studs, and sheathing. The vapor retarder material should have a permeance of 1.0 perm (57.45 ng/Pa·s·m^2) or less. Relatively permeable materials should be used on the outside of a wall. Sheathing paper, for example, should be a "breathing" type, having a permeance of 6 perms (344.72 ng/Pa·s·m^2) or more.

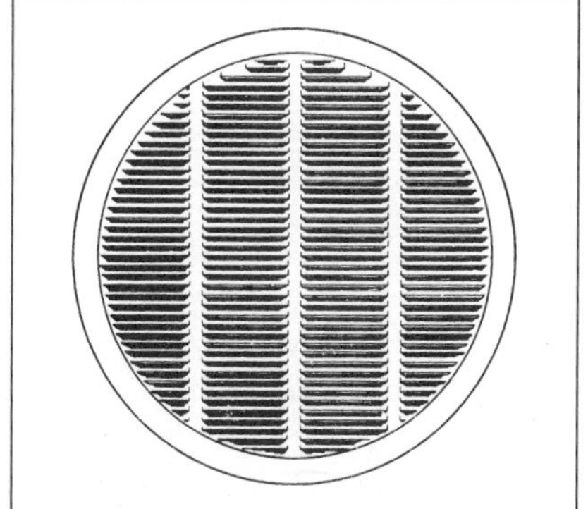

FIGURE 7.1-23 Prefabricated circular vent. (Courtesy Easy-up Home Ventilation from The Solar Group.)

Special emphasis is placed on sheathing paper because the No. 15 asphalt-saturated felt typically used for this purpose is also used for built-up roofing and is often too vapor resistant for use in a wall. Figure 7.1-7 shows No. 15 asphalt felt and tar felt as having a permeance of 1.0 and 4.0 perms (57.45 and 229.81 ng/Pa·s·m^2), respectively. These are average figures. Permeance of a particular product can be higher or lower, depending on the degree of saturation with bitumens.

An often used rule of thumb to ensure safe outward vapor diffusion is that the combination of exterior materials should have a total installed permeance of at least 5 times the total permeance of the interior vapor retarder plus the interior finish materials. Therefore, if the interior retarder-and-finish permeance is 1 perm (57.45 ng/Pa·s·m^2), the total exterior permeance should be 5 perms (287.26 ng/Pa·s·m^2) or more. Conversely, when vapor-resistant finishes such as plywood or hardboard panels will be used on the exterior, a more effective retarder and installation method should be used.

In air conditioned buildings located in the humid areas shown in Figure 7.1-19, water vapor diffuses from the exterior to the interior. Therefore, in these areas the exterior portions of a wall should have a higher vapor resistance than the interior, and when vapor retarders are used they should be on the outside. Water vapor is then directed toward the interior rather than the exterior, as is normal in other climate zones, where it can be removed by the air conditioning system. The air conditioning system must be designed with sufficient dehumidification potential to accomplish this removal. Normal systems employed in other areas of the country may not be appropriate. Reheating of circulating air may be necessary. Ventilating air must be dehumidified as well to prevent introducing high outside humidity levels into a building.

CRAWL SPACES

When a crawl space underlies a heated or air conditioned area, it is generally more economical to limit heat loss (or gain) by insulating the foundation wall rather than the floor above. Under these conditions, there is no need to ventilate the crawl space to the outside, but the crawl space should be heated or cooled. Introducing nominal amounts of air from the supply side (or conducting air to the return side) of an air-handling unit will usually ensure adequate crawl space ventilation.

In closed crawl spaces not equipped with a means of ventilation, moisture rising by capillary action from the ground can cause condensation problems. To stop the moisture at its source, buildings with either heated or cooled crawl spaces in all zones should be provided with a vapor retarder laid directly over the soil as a ground cover.

A vapor retarder for this purpose should be a material that is not subject to decay or insect attack. Typical materials are 4 or 6 mil (0.102 or 0.152 mm) polyethylene film, a coated 30 psf (146.5 kg/m^2) or heavier roofing felt, or roll roofing. This retarder should be installed in the largest possible widths, with edges lapped 6 in. (152.4 mm) and taped or sealed (unless ballast is used), and the sides turned up approximately 6 in. (152.4 mm) against the foundation walls. A pea gravel or sand ballast over the retarder is desirable to keep it in place and to prevent mechanical damage (see Fig. 2.3-1).

Ventilation to the atmosphere is recommended when a foundation wall is not insulated, heated, or cooled. At least four vents (Fig. 7.1-23) should be provided, with one not more than 3 ft. (914 mm) from each corner. They should have a total net free area not less than 1/1500 of the crawl space area. Vents should be placed as high as possible. Vents are not required if the crawl space is completely open on one side to a basement; however, if the crawl space area is greater than the basement, it should be separately ventilated by supplying or returning air.

In older buildings or where the crawl space is not cooled or heated, the ground cover can be omitted, provided the net free area of vents is increased 10 times, to 1/150 of the crawl space area. In the absence of a ground cover, more positive ventilation is required and at least four vents should be provided, with at least one in each wall.

In Zone I, crawl spaces are usually heated and ventilation is limited. When a crawl space will be ventilated in the winter, it is necessary to insulate the floor above it to reduce heat loss and improve occupant comfort. During cold weather, the air in the crawl space will have a low temperature and low moisture content, so there will be vapor pressure outward from the building. To prevent condensation on the framing, a vapor retarder should be installed on the upper, warm side of the insulation. Normal retarder-faced blankets have nailing flanges on the same side as the retarder. If the insulation is installed from below the joists, special reverse flange blankets or retarder-wrapped blankets should be used (see Section 7.3). Wide polyethylene sheets laid over the subfloor or joists are most effective, but retarder-faced or retarder-wrapped blankets are acceptable.

ROOF AND CEILING ASSEMBLIES

For purposes of this discussion, roof and ceiling assemblies can be divided into three categories: *insulated framed systems*, *compact conventional systems*, and *protected membrane systems*. Unfortunately, the industry is not in complete agreement about the use of ventilation and vapor retarders in each of these systems.

Insulated Framed Systems In insulated framed systems, an air space separates the ceiling and insulation from the roof deck. Such assemblies may have the roof deck in contact with the framing (Fig. 7.1-24) or may have an attic space above the ceiling framing. Of the three roof and ceiling assembly categories, this one is the most likely to experience condensation problems. Not only are such roofs more susceptible to condensation, but because many are framed in wood, the result is more damaging.

In most areas of the United States, the ceiling in an insulated framed system should have a continuous vapor retarder membrane on the warm side of the insulation. In the humid areas shown in Figure 7.1-19, however, a vapor retarder should be installed only if local experience so dictates. In these areas, condensation may form on the outside of a vapor retarder whose temperature is below the dew point of the outside air.

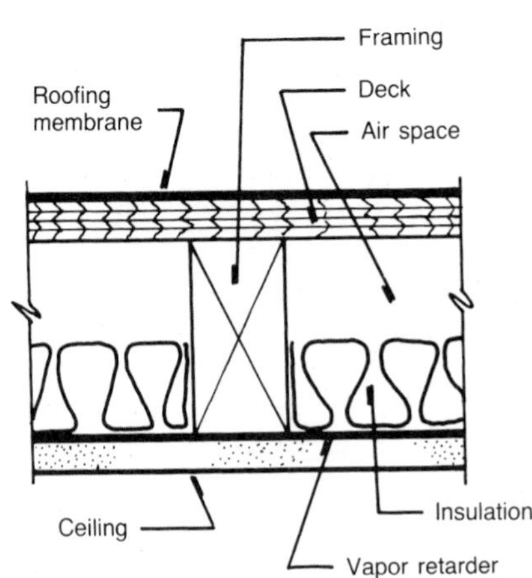

FIGURE 7.1-24 Insulated framed roofing system. (Courtesy H. Leslie Simmons.)

Ventilation of insulated framed systems is particularly important, because the area above their ceilings is a natural depository of moisture from inside the building. Convection currents can carry moisture-laden air up wall cavities and into the roof space through openings inadvertently left at the top of partitions or even through the voids in concrete masonry unit. Light fixtures and other equipment installed in ceilings can also create air and vapor leaks.

Ideally, such vapor leaks should be eliminated; in practice this is not possible, and a fair amount of moisture will penetrate past the ceiling. Fortunately, suitable ventilation details can dissipate this moisture safely to the atmosphere. In fact, ventilation alone can take care of all such moisture, except in the coldest part of Zone I, with winter design temperatures of −10°F (−12.2°C) or lower.

In such systems having attic spaces, wind direction and velocity, as well as the vent type, greatly affect air movement distribution and volume within the attic spaces. "Dead" spots of minimum air flow create potential condensation pockets. Therefore, ventilation based only on the ratio of vent opening to ceiling area may be inadequate. Vents should be selected and located for maximum air flow and uniform air distribution; generally, at least one-half of the required vent area should be located at the ridge or upper half of the roof, the rest at the eaves.

The amount of ventilation required is 1/300 of the ceiling area below when gable end ventilators can be used on both ends of the area. When gable end vents cannot be used, a vent area equal to 1/600 of attic area should be uniformly distributed at the eaves and an additional 1/600 of attic area should be provided at the ridge. All attic spaces should be interconnected.

In older buildings that do not have vapor retarders, vent openings should have a net free area not less than 1/150 of the ceiling area. This larger ventilation volume may also be desirable in moderate and southern zones, regardless of the pres-

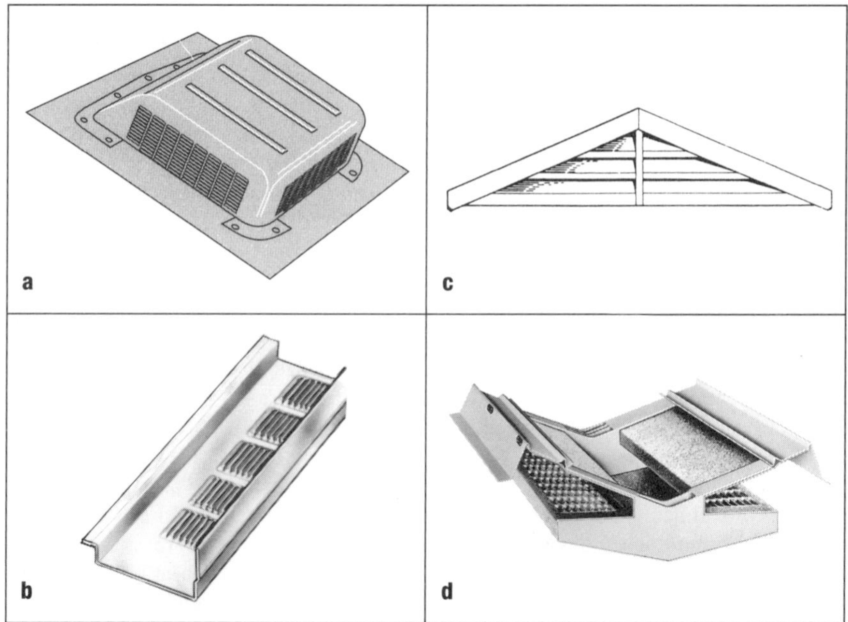

FIGURE 7.1-25 Typical roof ventilators: (a) roof louvers, (b) soffit vent, (c) gable end louver, and (d) continuous ridge vent. (a and c courtesy Easy-up Home Ventilation from The Solar Group); (b and d courtesy Alcoa Building Products.)

ence of retarders. The effectiveness of a ventilation system in removing both heat and moisture depends on the volume of air moved and the extent to which the air flow reaches into all parts of the void space.

Roof louvers (Fig. 7.1-25a) when used alone contribute little to effective air flow through an attic space. However, when placed in the upper half of the attic space and combined with soffit vents (Fig. 7.1-25b) in equal amounts, the air distribution is improved and air flow is almost doubled. The combined system also becomes more responsive to wind, increasing air volume in direct proportion to wind velocity.

Continuous soffit vents (see Fig. 7.1-25b) produce substantial air flow regardless of wind direction, and the flow is directly proportionate to wind velocity. However, most air flow remains near the attic floor, and additional vent openings in the upper roof or ridge are needed for good circulation.

Gable end louvers (Fig. 7.1-25c) are most effective for wind perpendicular or diagonal to the louver face. However, air flow is one-half to one-third this amount when the wind direction is parallel to the louver face, perpendicular to the ridge. Although the addition of soffit vents does not improve air flow for the ideal condition, under adverse wind conditions, the addition of continuous soffit vents increases air flow by 50%.

Continuous ridge vents (Fig. 7.1-25d) are unique in that they provide air flow due to temperature differences with roof slopes as low as 4/12. They provide about the same air flow at zero wind velocity as at 10 mph (16.09 km/h); however, air flow volume is several times larger with wind parallel to the ridge as it is with the wind perpendicular. Continuous ridge vents combined with continuous soffit vents produce the greatest volume and most uniform air flow under most wind conditions.

Ventilation of insulated framed systems with no attic space (see Fig. 7.1-24) should include outlets from every joist space, either with individual vents (see Fig. 7.1-23) or with continuous soffit ventilation on the sides perpendicular to the joists. Soffit ventilation can be provided either with prefabricated vents (see Fig. 7.1-25b) or job-built vent details (Fig. 7.1-26). A continuous ¾-in. (19.1 mm)-wide screened opening is usually adequate for buildings of average width, but the net free area should be verified as discussed earlier.

Compact Conventional Systems Compact conventional systems may have either metal decks, as shown in Figure 7.1-27, or wood plank decks.

Because built-up roofing and single-ply roofing membranes, which are typically used with compact conventional roofing systems, are nearly perfect vapor retarders (0.00 perm), it is impossible for vapor to be diffused through them. A vapor retarder on the underside of insulation beneath such roofing can actually

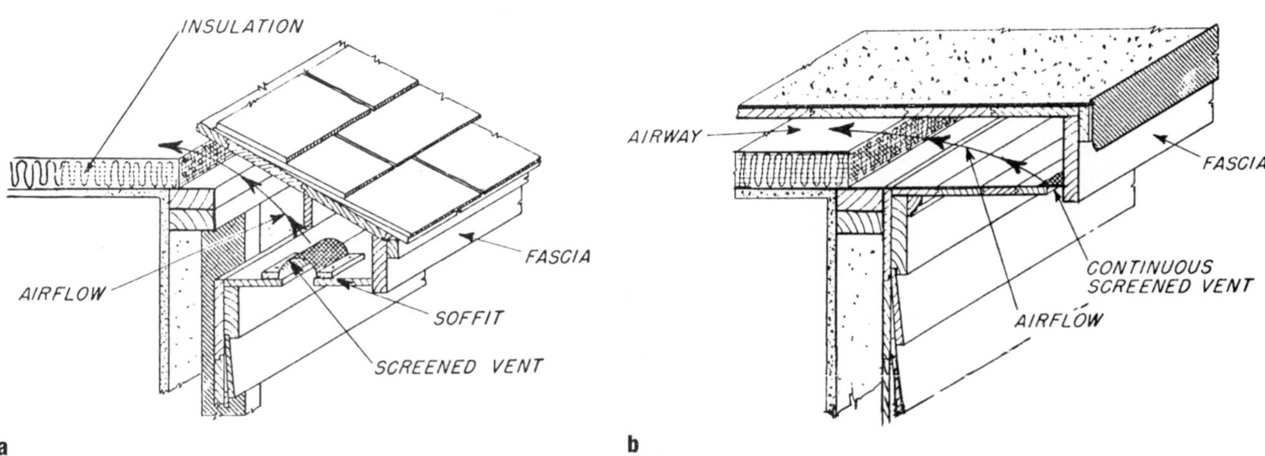

FIGURE 7.1-26 Continuous soffit vents can be used to ventilate both attic-type and atticless insulated framed roofs.

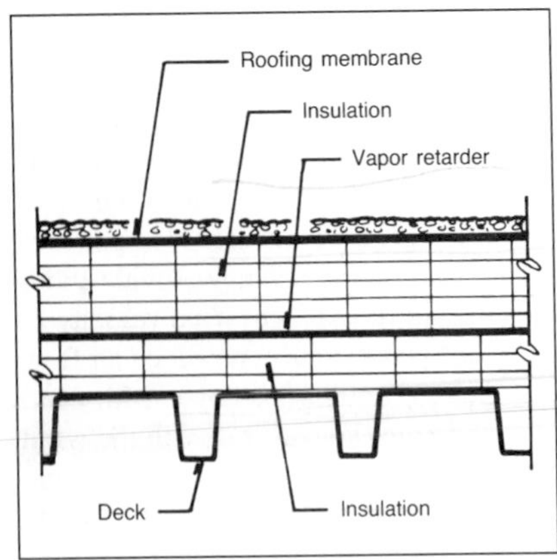

FIGURE 7.1-27 Compact conventional roofing system. (Drawing by HLS.)

cause more problems than it solves, by trapping moisture within the insulation. If a vapor retarder is used in such systems, it should be so located that the calculated dew point will fall on the low-vapor-pressure side of the retarder. At any rate, the circumstances under which a retarder should be used are controversial.

In the past it was believed that a vapor retarder was not needed in a conventional compact system when the interior winter relative humidity averages less than 40%. It was thought that a retarder might be needed when the interior relative humidity exceeded 40% and the average January temperature was 35°F (7.2°C) or lower. Currently, however, vapor drive maps are a better way to determine the need for a vapor retarder.

The 35°F (7.2°C) January isotherm coincides approximately with the 0°F (–17.8°C) isotherm shown in Figure 7.1-18; humidities over 40% are encountered in public shower rooms, kitchens, laundries, and pool areas. In single-family structures, only pool areas merit special consideration; other high-moisture areas, such as baths and kitchens, are usually equipped with exhaust fans to draw off the moisture.

Unless a vapor retarder is used, it is not necessary to dissipate moisture with structural ventilation or venting. The insulation is generally absorptive enough to hold the vapor for short periods without condensation. The roof deck is generally permeable enough to permit dissipation of vapor back into the interior, when the vapor pressure is reversed.

However, if a vapor retarder is provided in cold climates and with high interior relative humidities, the possibility of trapping moisture between the retarder and the roofing is greatly increased. The moisture may be present during construction or may enter the insulation later through leaks from above or below. In any case, moisture cannot readily escape through the roofing or the vapor retarder by diffusion, and peripheral air leaks (venting) must be provided to permit dissipation of moisture.

Edge venting is a method of connecting roof insulation with the exterior atmosphere through built-in vent openings at a roof's perimeter (Fig. 7.1-28a). Edge venting is adequate for roofs not exceeding 40 ft. (12.19 m) in width; for wider roofs, stack venting as well as eave venting is required.

Stack venting consists of simple sheet metal stacks about 3 in. (76.2 mm) in diameter, reaching through the roofing membrane into the insulation (Fig. 7.1-28b). One stack should be provided for each 900 sq. ft. (83.61 m^2) of roof area.

Unfortunately, vents can produce more problems than they solve. They can themselves admit water vapor and even free water from drifting snow into the roofing system. In addition, water leaks can occur where the vents are flashed into the roof. Vents may also act as chimneys carrying moisture-laden air into the roof system from faults in the vapor retarder.

Protected Membrane Systems A protected membrane roofing system is a compact system in which the roofing membrane is below at least a portion of the insulation (Fig. 7.1-29). The roofing lies on the warm side and acts itself as a vapor retarder. In most areas of the country, there is no condensation associated with this type of assembly when it has been properly designed and installed. However, in the humid areas shown in Figure 7.1-19 where the indoor vapor

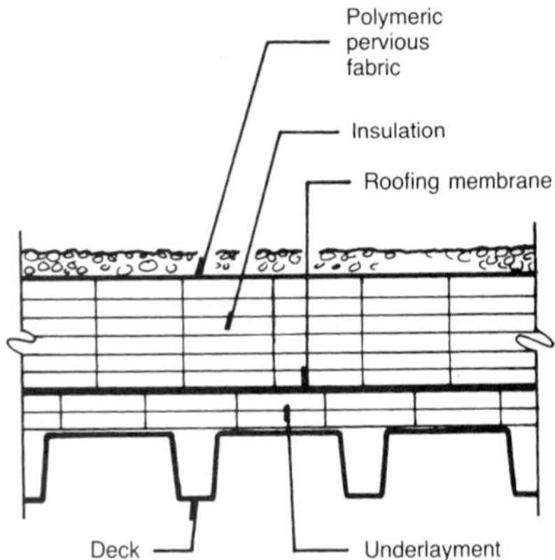

FIGURE 7.1-29 Protected membrane roofing system. (Courtesy H. Leslie Simmons.)

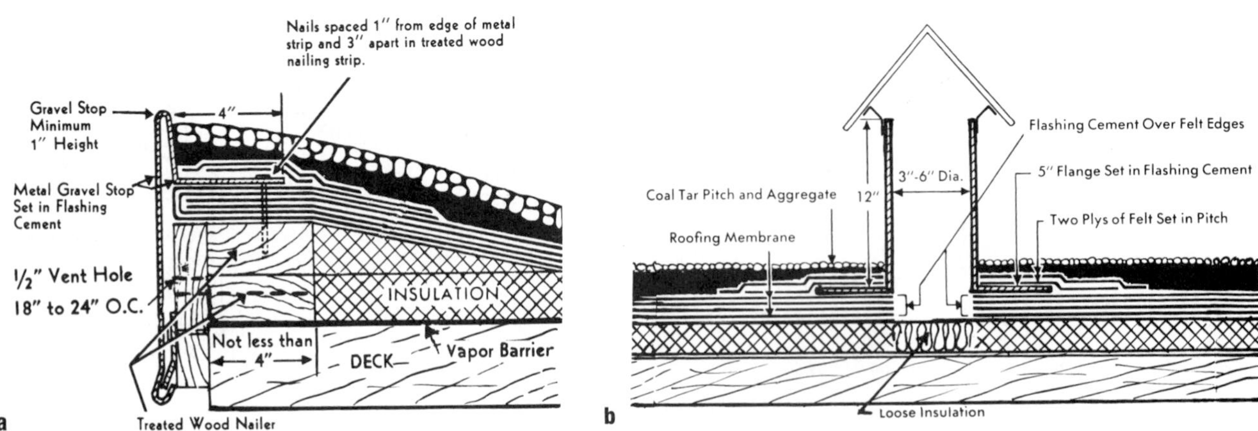

FIGURE 7.1-28 Insulated wood plank deck showing (a) edge venting and (b) stack venting. Insulated steel deck systems are similar. (U.S. Forest Products Laboratory. Not subject to copyright in the United States.)

pressure may be lower than that outdoors, condensation may occur in such roofing assemblies over air conditioned buildings. Calculations are required to determine their suitability.

SLABS ON GRADE

Whether at grade or below grade (basements), slabs on grade are in direct contact with the soil, which often contains large amounts of vapor or free water. Although 4 in. (101.6 mm) of concrete has good resistance to vapor (perm rating of less than 1.0 [57.45 ng/(Pa·s·m^2)]), the vapor pressure from the moist slab underside to the interior may be high enough and sufficiently prolonged to cause large amounts of vapor to be diffused through the slab. Additional moisture, in vapor or liquid form, can seep through cracks and structural joints in the slab. Therefore, a separate vapor retarder should be placed under every building slab on grade. It may seem that a vapor retarder is unnecessary when a slab on grade is underlain by well-drained granular fill and is in a region where irrigation and heavy sprinkling are not practiced. Nevertheless, because a retarder can be installed at nominal cost during construction and because it is costly to correct moisture problems later, a retarder should always be placed under slabs of habitable rooms. The vapor retarder should have a permeance of less than 0.30 perm (17.24 ng/Pa·s·m^2), should be installed with 6-in. (152.4 mm) laps, and should be fitted carefully around utility and other service openings.

Vapor retarder materials suitable for slab-on-ground construction include single-layer membranes such as polyethylene film and multiple-layer membranes such as reinforced waterproof paper with a polyethylene film extrusion coated onto both sides. Polyethylene film used over gravel or crushed stone, or under a reinforced slab, should be at least 6 mil (0.15 mm) in nominal thickness; thinner, 4-mil (0.10 mm) film can be used over sand or firmly compacted soil.

7.1.3.3 Vapor Retarder Installation

Three fundamental principles should be observed when applying vapor retarders: (1) the retarder should be as near as possible to the warm (heated in winter) side of the insulation or the warm face of the assembly, (2) the retarder should be as continuous as possible, with a minimum of joints, breaks, openings, and air leaks, and (3) the retarder should be installed by a method suitable for the specific condensation hazard.

Because a vapor retarder is frequently an integral part of insulation, effective vapor control generally depends on the proper selection and attachment of insulation. Insulation and retarder should be fitted neatly behind piping and equipment. The flanges of retarder-faced insulating blankets should be snugly stapled about 6 in. (152 mm) on centers to framing members, so as to avoid air leaks.

WALLS

Gaps in insulation near wall cavity bottoms or tops encourage convection currents, which carry heat and vapor from the inside to the outside face of the cavity. Therefore, insulation should be pushed tightly against the back of the sheathing and should fill the cavity from top to bottom.

Batts and blankets are made to fit a standard 14½-in. (368.3 mm)-wide cavity resulting from 16 in. (406.4 mm) on center wood stud or joist spacings. Nonstandard spaces less than 14½ in. (368.3 mm) wide should be fitted with blankets cut 1 in. (25.4 mm) oversize to form a stapling flange on the cut size. Nonstandard spaces wider than 14½ in. (368.3 mm) can best be protected by a separate retarder (Fig. 7.1-30) or by foil-faced gypsum board.

If the standard spacing of wood members in a structure is 12 or 24 in. (305 or 610 mm), pressure-fit insulation with a separate retarder should be used. Cracks around window and door frames should be stuffed with loose insulation and covered with a separate retarder.

Metal studs do not lend themselves to the use of retarder-faced blankets, due to the difficulty of maintaining a continuous retarder. When metal studs are used, retarders should be formed either with foil-faced gypsum board or with separate sheet retarders.

FLOORS

The perimeter of a floor assembly just behind the band joist in wood framing or at the ends of steel joists is frequently overlooked when insulation and a vapor retarder are installed. This is particularly true when the floor is between two habitable spaces and no insulation is used in the joist cavities.

When a floor is over an open-air area or an unheated crawl space, insulation is likely to be required in the floor. Here, too, care should be taken to provide thermal and vapor control near the perimeter.

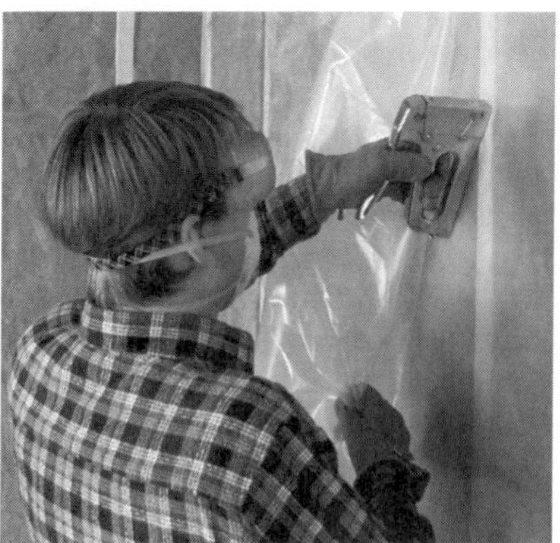

FIGURE 7.1-30 Polyethylene vapor retarder being installed across an entire wall surface. (Photograph furnished by Owens-Corning Fiberglas Corporation.)

The vapor retarder should be on the upper (heated in winter) side of the insulation. When the blankets are installed from above, they can be readily stapled to the joists. When installed from below, either reverse flange insulation should be used, or the blankets should be placed with both retarder and flanges facing up, using wire supports below.

Retarder-faced blankets are usually not the best way to provide a vapor retarder in a metal-framed (joist) floor system. A continuous sheet retarder is likely to be more successful. When retarder-faced blankets are used in metal-framed floors, they can be supported with wire supports or mesh, but providing a continuous retarder using this type of installation is difficult.

ROOF AND CEILING ASSEMBLIES

Blankets in wood ceiling joists should be extended part way over the top plate but not enough to block eave ventilation. When blankets are installed from below with recessed stapled flanges, the gap at the end should be stuffed with loose insulation. Insulation and vapor retarder should extend over dropped soffits in bathrooms and kitchens to provide thermal and vapor control. Tightly wedged glass fiber or rock wool insulation above the soffit also isolates the attic area from the soffit space and prevents the spread of fire. Holes for mechanical equipment through top plates should be filled with loose insulation to minimize air and vapor leaks into the roof space.

A separate vapor retarder should be provided when poured insulation is used between ceiling joists.

7.2 Waterproofing and Dampproofing

The below-grade walls of buildings, tunnels, and pits and horizontal decks below paving or earth are subject to groundwater penetration. *Waterproofing* and *dampproofing* are used to prevent this penetration. Waterproofing is a membrane, coating, or sealer used in concealed locations to prevent water from entering such structures. Waterproofing will work even when the adjacent water is under hydrostatic head. Vertical waterproofing is usually used in conjunction with a foundation drainage system (see Section 2.4). Horizontal waterproofing systems also require a positive means of draining away water that penetrates the wearing surface.

Dampproofing is usually a bituminous coating applied to prevent materials from absorbing adjacent moisture or to prevent absorbed moisture from penetrating farther into an invaded system. Bituminous dampproofing is not appropriate when the adjacent water is under hydrostatic pressure. Bituminous dampproofing may be used on either the interior or exterior, but is usually concealed in either case. Interior uses include interior surfaces of exterior above-grade walls; interior surfaces of the exterior wythe of exterior cavity walls; and for coating structural steel subject to high-moisture conditions.

7.2.1 WATERPROOFING

There are three basic types of waterproofing materials: membrane, clay, and cementitious.

7.2.1.1 Membrane Waterproofing

On vertical surfaces, a membrane waterproofing system usually consists of a membrane, protection course, drainage medium, and a layer of filter fabric (Fig. 7.2-1). Parging may also be required over a masonry substrate.

Several arrangements are possible on horizontal surfaces, but they all have characteristics similar to those of the system shown in Figure 7.2-2. They have a membrane, a protection course, a drainage medium (or space), a layer of filter fabric (when a drainage membrane is used), and a wearing surface. In some systems a layer of insulation may be required. Sometimes this layer of insulation also serves as the protection course. Some systems require a working slab beneath the wearing surface.

Membranes may be either fluid-applied,

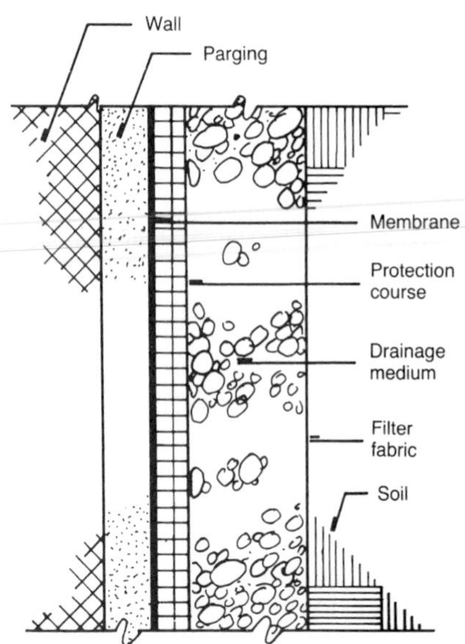

FIGURE 7.2-1 Vertical membrane waterproofing system. (Courtesy H. Leslie Simmons.)

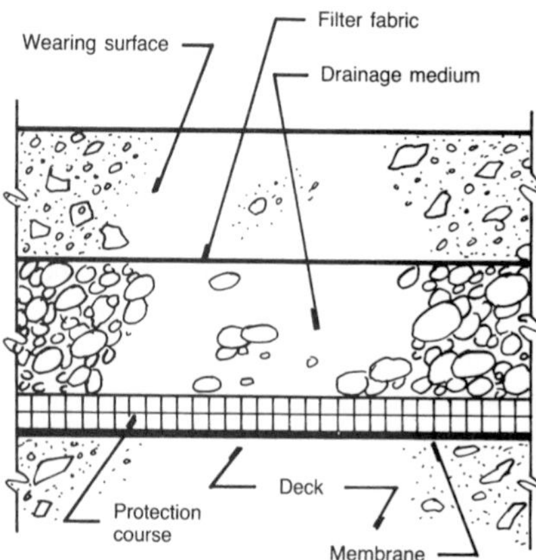

FIGURE 7.2-2 Horizontal membrane waterproofing system. (Courtesy H. Leslie Simmons.)

sheet, or bituminous, which is a combination product. Fluid-applied membranes are usually thicker than sheet membranes and more foolproof when properly installed. In addition, it is easier to locate leaks in them. The materials used in fluid-applied membranes tend to be less expensive than sheet membranes, but they cost more to install because their installation is more difficult. A major advantage of fluid-applied membranes is that they conform easily to irregular surfaces.

Connections between two portions of fluid-applied membrane are often easily made using the same material as in the membrane. Separate sheets of the same or compatible material may be bonded at intersections in sheet membranes. Flashing of the edges of fluid-applied and sheet membranes is essentially the same process. Sheet materials are sometimes finished in a reglet, but usually the edges of both should finish beneath a compatible flashing material. Metal flashing is not appropriate when the two materials must be bonded, because of the differential rates of thermal expansion between the waterproofing materials and the metal. Metal can be used, however, as a cap flashing where appropriate.

A major disadvantage of fluid-applied membranes is that they require considerable skill to obtain a uniform coating thickness. Another is that these materials are incompatible with some substrate materials.

Most bituminous membranes are built-up courses of felts and bitumen. Some bituminous membranes, however, are totally fluid-applied systems consisting of bitumens reinforced with nylon or glass fiber filaments.

FLUID-APPLIED MEMBRANES

Most concealed fluid-applied horizontal waterproofing membranes and many concealed vertical waterproofing membranes used today are either polyurethane or hot rubberized asphalt. Fluid-applied silicone and neoprene systems are also available. These have wearing and weathering abilities beyond those of polyurethane or rubberized asphalt systems, but are also more expensive. They are, therefore, most often used where they will be exposed, such as on roofs and decks. They may, however, also be used as concealed waterproofing where conditions warrant their use.

Polyurethane Fluid-applied polyurethane waterproofing is available as either one- or two-part materials. One-part materials are convenient to use and offer more quality control than two-part materials. The proportions of two-part materials can be adjusted by skilled applicators to suit the peculiarities of a particular project, and they tend to cure faster. The in-place performances of one- and two-part systems are comparable.

Hot Rubberized Asphalt Rubberized asphalt waterproofing is not as elastic as polyurethane and requires a thicker application. This is desirable over some substrates but not over others. The manufacturer's advice should be sought before selecting this system.

SHEET MEMBRANES

Sheet membrane materials include butyl synthetic rubber sheet, ethylene propylene diene monomers (EPDM) sheet, premolded bituminous sheet, rubberized asphalt sheet, and self-adhesive butyl sheet.

Butyl synthetic rubber sheet is uniform flexible sheet material about 60 mils (1.52 mm) thick. EPDM sheet is a flexible sheet material that is available in several thicknesses, the minimum being 1/16 in. (1.6 mm). *Bituminous* sheet waterproofing is a proprietary seven-ply material with a water vapor transmission rating of zero. *Rubberized asphalt* sheet waterproofing is a self-adhering membrane bonded to a polyurethane sheet, with a total thickness of about 56 mils (1.42 mm). *Self-adhering butyl* waterproofing sheet consists of a butyl rubber core bonded to polyethylene plastic to form a sheet about 60 mils (1.52 mm) thick. It requires an adhesive primer.

BITUMINOUS MEMBRANES

Bituminous waterproofing is usually applied on vertical below-grade surfaces, but in some instances may be appropriate for horizontal surfaces as well. It is usually either hot coal-tar pitch or hot asphalt. Its number of plies, overall thickness, and content will vary, depending on hydrostatic head (Fig. 7.2-3). Because it has a proven record of successful use and is highly resistant to deterioration, coal-tar pitch should be used when the melting point can be kept low and there is no incompatibility between that material and adjacent materials. Hot asphalt is necessary when a higher melting point is needed.

Cold asphalt materials are less costly for small applications, and they reduce the danger of air pollution present with hot-applied materials, eliminate the danger of injury due to fire, and produce less objectionable odor. They are not, however, as reliable or foolproof as the hot-applied materials, and they are suitable only when the hydrostatic head is of short duration and never exceeds 3 ft. (914 mm). Cold asphalt waterproofing is usually composed of two plies of glass fiber mat and three coats of emulsion, each applied at a rate of 3 gal./100 sq. ft. (1.147 L/m^2).

Materials used in bituminous waterproofing include:

- Coal-tar–based priming oil conforming with American Society for Testing and Materials (ASTM) D 43
- Coal-tar bitumen conforming with ASTM D 450
- No. 15 coal-tar felt conforming with ASTM D 227
- Cut-back type asphalt primer conforming with ASTM D 41
- Waterproofing asphalt conforming with ASTM D 449 (generally Type I for vertical surfaces, Type II for unexposed horizontal surfaces, and Type C for exposed surfaces)
- No. 15 asphalt-saturated organic felt conforming with ASTM D 226
- Bituminous-saturated organic fabric conforming with ASTM D 173
- Glass fiber fabric conforming with ASTM D 1668
- Glass fiber mat conforming with ASTM D 2178
- Base sheet conforming with ASTM D 2626
- No. 5 rosin-sized building paper as a slip sheet
- Either 1/8-in. (3.2 mm)-thick semirigid asphalt saturated and coated laminations or felt; 1/2-in. (12.7 mm)-thick fiberboard protection board

INSTALLATION

Substrates should be installed and prepared, and fluid-applied, sheet, and bituminous membrane waterproofing materials should be applied in accordance with their manufacturers' recommendations.

Decks to receive waterproofing membranes should be sloped at least 1/8 in./ft. (9.75 mm/m) and preferably 1/4 in./ft. (19.5 mm/m) or more. This slope should be present even after the substrate has deflected. The best surface to receive horizontal waterproofing is a structural concrete slab. Toppings are more likely to move and crack. Precast concrete decks should be covered with a 1½-in. (38.1 mm)-thick or thicker reinforced topping.

Concrete substrates should be cured the number of days required by the waterproofing manufacturer, usually at least 14 days. Substrates should be prepared by chipping or grinding away projections that would interfere with or puncture the waterproofing. Substrates should be cleaned of curing compounds, dirt, dust, and other materials that would interfere with membrane placement or bond.

A primer should be applied to substrates when so recommended by the waterproofing material manufacturer. Primed areas should be covered with waterproofing within 24 hours.

Waterproofing materials should be thoroughly bonded to vertical substrates, and where recommended by the manufacturer, to horizontal substrates as well. Seams and edges should be sealed and flashed to prevent water penetration. Expansion joints should be provided at each expansion joint in the substrate, where the substrate material changes, and in other locations recommended by the manufacturer.

Except over slabs on grade, horizontal waterproofing should be tested for leaks by covering it with 1 in. (25.4 mm) of water for 24 hours and observing the drop in water level, if any. Repairs should be made before a protection course is installed.

PROTECTION COURSE

Most applications of membrane waterproofing should be covered with a protecting material to prevent damage to the waterproofing during subsequent construction operations, including backfilling. This *protection course* is usually a layer of insulating material or other board materials. Each waterproofing material requires a different type of protection course. The waterproofing manufacturer should be consulted to determine the correct type.

FIGURE 7.2-3 Composition of Reinforced Hot Membrane*

Hydrostatic Head, ft. (m)	Number of Fabric Plies	Number of Felt Plies	Bitumen, lb. (kg)	Number of Moppings
3 (0.914)	1	1	75 (34.02)	3
10 (3.05)	1	2	100 (45.36)	4
25 (7.62)	2	2	125 (56.7)	5
50 (15.24)	2	3	150 (68.04)	6
100 (30.48)	3	3	175 (79.38)	7

*Possible to use one ply fewer when a coated base sheet is used as an underlayment.

7.2.1.2 Clay Waterproofing

Clay waterproofing is a layer of an expanding material called *bentonite clay*. When wetted, bentonite expands to between 10 and 15 times its dry volume. It is available loose for mixing with water and spraying in place. It is also available in dry sheets between cardboard, adhered to a drainage board, and adhered to a plastic sheet.

Bentonite is applied below grade to the exterior surfaces of building walls and tunnels. It should not be installed beneath a wearing surface or under earth or any other material that is less than 18 in. thick. A masonry substrate requires a layer of parging before bentonite is installed over it. Cracks wider than ⅛ in. (3.2 mm) must be filled in all substrates, because bentonite cannot bridge wider gaps.

Bentonite is especially useful in tight locations where access is not available to install membrane waterproofing. This is often the case where existing construction occurs.

Bentonite sheets are usually nailed in place or adhered with bitumen. But installation methods vary, depending on hydrostatic pressures, substrate materials, edge conditions, and other factors. Therefore, the manufacturer should be consulted in each instance.

7.2.1.3 Cementitious Waterproofing

Some forms of cementitious materials are primarily decorative, even though they may have some waterproofing characteristics. One type that is specifically used as waterproofing is composed of portland cement, aggregate, and an acrylic or plastic admix. It may also contain pulverized iron fillers, in which case it is known as *iron oxide waterproofing*. An iron oxidizing agent in the mix causes the iron to rust quickly and expand to fill the pores in the plaster. This type of waterproofing is used on either the exterior or the interior of below-grade walls, pits, sumps, and the like.

A second form of cementitious waterproofing is *hydraulic cement*, which is a compound of cement and rapid-setting nonshrinking hydraulic material. It is used to seal holes, cracks, and open joints and will set even in free-flowing water.

Nondecorative cementitious waterproofing tends to be brittle and unable to span even hairline cracks. Therefore, its substrate must be smooth and crack-free.

7.2.2 DAMPPROOFING

Bituminous dampproofing is available in either cold- or hot-applied forms.

7.2.2.1 Hot-Applied Bituminous Dampproofing

Hot-applied bituminous dampproofing consists of either an asphalt primer conforming with ASTM Standard D 41 and dampproofing asphalt conforming with ASTM Standard 449, Type I; or a coal-tar-based primer conforming with ASTM Standard D 43 and coal-tar bitumen conforming with ASTM Standard D 450, Type II or Type III.

Hot-applied bituminous dampproofing is used on the exterior of buildings, usually underground, and especially on large projects when the use of a hot system with solvents is permitted. It is seldom used on interiors, because of the problem of transporting and using hot products indoors. It cannot be used when the dampproofing must be able to transmit water vapor (be able to breathe). Coal-tar systems are almost never used on the interior of buildings, because they have a distinctively unpleasant odor that never completely disappears.

Hot-applied dampproofing may be applied directly on smooth concrete walls or on a coat of parging over masonry.

7.2.2.2 Cold-Applied Dampproofing

Cold-applied dampproofing may be either a coal-tar- or asphalt-based system. It may be used either on the interior or the exterior and is especially useful on small projects where hot products would be difficult or too costly to use. As mentioned earlier, however, coal-tar products are seldom used on the interior of buildings.

Asphalt-based cold-applied bituminous dampproofing systems consist of either cut-back asphalt or asphalt emulsion materials. Both types are available as liquid, semifibrated, or heavy fibrated types. Fibrated types are reinforced with mineral fibers, and most also contain various fillers. Dampproofing that must breathe should be an asphalt emulsion type.

Coal-tar-based cold-applied bituminous dampproofing systems consist of a coal-tar bitumen compound including ASTM Standard D 450, Type II, coal-tar pitch and solvent compound, an inorganic fiber reinforcement, and various fillers.

Cold-applied dampproofing should be applied on exterior wall surfaces over smooth concrete or masonry that has been given a parging coat. Coal-tar products or fibrated asphalt products may be applied over a rougher concrete surface, but it is better to first parge rough concrete surfaces.

Cold-applied dampproofing is sometimes applied over the interior face of exterior above-grade walls or on the interior of the exterior wythe of masonry cavity walls. However, the Brick Institute of America says that mortar coatings are more effective in such locations and cautions against the use of dampproofing there. The problem is that dampproofing applied on an interior surface allows water to enter brick but prohibits its escape, thus contributing to efflorescence and spalling of the masonry when the water in it freezes.

The National Concrete Masonry Association recommends that an impermeable barrier, such as a cut-back asphalt dampproofing, be used on the interior face of concrete masonry unit walls. Under these circumstances, the exterior finish must be completely impermeable. Even oil- or alkyd-based paints may not be used there.

Liquid cold-applied dampproofing materials may be applied with brush or spray; semifibrated materials by brush or spray; and heavy fibrated materials by trowel. In every case, they should be concealed from view.

7.3 Building Insulation

In some climates, using structural or finish materials with high thermal capacities is sufficient to adequately control heat gain or loss through a building's envelope and thereby maintain a healthy and comfortable indoor environment. In most cases, though, a building's structural and finish materials permit excessive heat flow, and insulating products should be added in the roof, walls, and floors to make them more energy efficient.

7.3.1 MATERIALS

The most common insulation products are classified as rigid board, mineral fiber, loose fill, and foamed-in-place. Other materials that provide an insulating function are also sometimes used, although they are not all intended specifically to provide that function. These include insulated sheathing, fiberboard, reflective insulation, and structural insulating roof deck.

The trend today is toward manufacturers listing their thermal insulation products according to their ability to resist heat flow, which is called *thermal resistance* or *R-value*. Thermal resistance is the reciprocal of *thermal conductivity* (k-value). Some manufacturers still list their products according to their k-values, and some even list U-factors. R-values are the preferable listing, however, because they can be added directly to calculate an assembly's overall thermal resistance. The overall R-value is the reciprocal of the assembly's *thermal coefficient* (U-factor), which is finally used to calculate the heat loss through the assembly.

The R-values for some insulation products are given in Figure 7.3-1. Note that in most cases the values listed are averages taken from several competitive products. Values for specific products may vary somewhat from those shown.

Acceptable R-values have increased significantly since the large jump in energy costs that occurred in the 1970s. Figure 7.3-2 shows some typical values. The actual values used for a particular project should be calculated, taking into account such economic factors as insulation costs, increased construction cost that may be necessary to provide more insulation, cost of heating equipment, and building operation cost. In addition, some codes dictate minimum required R-values.

FIGURE 7.3-1 Resistance (R) Values of Some Insulating Products[a]

Product	R-Value per 1-in. (25.4 mm) Thickness
Loose fill	
Mineral fiber	2.20–3.00
Cellulose	2.80–3.70
Perlite	2.70
Vermiculite	2.13
Batts and blankets	
Mineral fiber batts	3.00–3.70
Glass fiber batts	3.00–4.00
Rigid	
Cellular glass	2.78
Expanded polystyrene, extruded	5.00
Expanded polystyrene, molded	3.60–4.40
Expanded polyurethane	6.25
Polyisocyanurate	5.60–7.70
Phenolic	8.30
Organic fiberboard	2.50–2.80
Mineral fiberboard	3.45
Glass fiberboard	3.80–4.80
Reflective	
Aluminum foil[b]	3.48
Foamed-in-place	
Polyurethane	6.25
Polyisocyanurate	7.00–8.00
Phenolic based	4.50

[a]Values are for 75°F (23.9°C) mean temperature.
[b]Thickness of foil not a factor; ¾-in. (19.1 mm) air space on room side.

7.3.1.1 Rigid Insulation

Rigid insulations are usable in all parts of a building. They are made in board form for use in walls, foundations, and roofs, and in special shapes for use in the cores of concrete masonry units. Most rigid insulations are either polystyrene, polyurethane, polyisocyanurate, perlite, phenolic, cellular glass block, organic fiber, or glass fiber. Various composites made from more than one of the listed materials and from other materials are also available.

POLYSTYRENE

Polystyrene is a preformed nonstructural plastic cellular material. It is made in board form for use over structural roof decks, beneath promenade walking surfaces, at foundation perimeters, and in masonry cavities (Fig. 7.3-3). These boards are also used as wall sheathing when wind bracing is provided by other means.

Polystyrene board products are made in two forms: extruded and molded. *Extruded* polystyrene is a closed-cell material made in five densities ranging from 1.35 to 3 lb./cu. ft. (21.62 to 48.06 kg/m^3). *Molded* polystyrene is an open-cell material made in four densities, ranging from 0.9 to 1.8 lb./cu. ft. (14.42 to 28.83 kg/m^3).

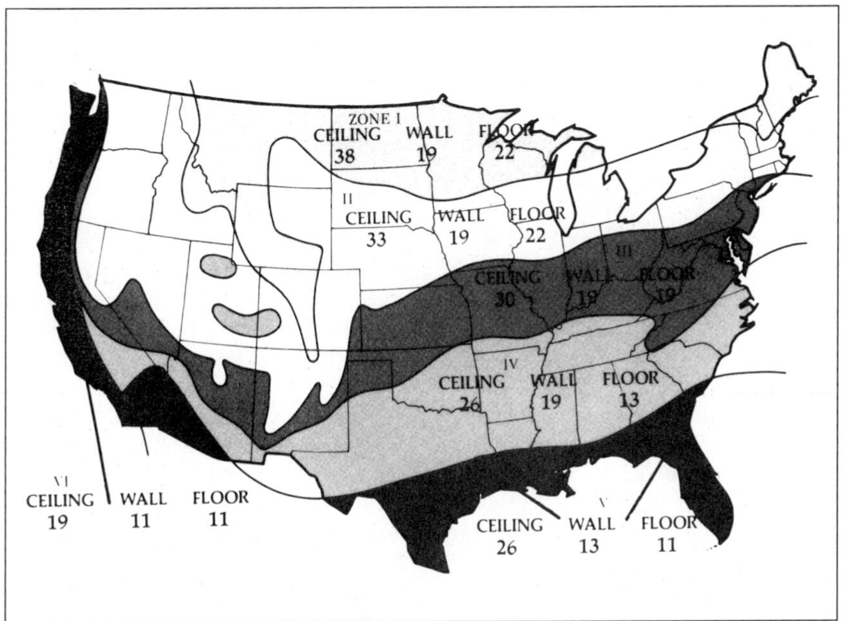

FIGURE 7.3-2 Target insulation R-values.

FIGURE 7.3-3 **Insulation is inserted in cavity as wall is built.** (Courtesy Dow USA.)

The extruded material is then denser and stronger. Its closed-cell structure also means that it is more resistant to moisture absorption and less permeable to water vapor than the molded form. The molded type is commonly called *beadboard*, because it is made up of compressed pellets or beads.

Polystyrene insulations have properties similar to those of some other plastic foam insulations, but lower R-values. They are sensitive to daylight and will deteriorate after prolonged exposure. Like other cellular plastics, they will give off carbon monoxide gas while burning. They do not, however, smolder or add significant fuel to a fire.

Polystyrene is also produced in tapered form. When used under a waterproof membrane, the tapered form permits positive drainage of a roof or promenade.

POLYURETHANE

Urethane foam is a cellular plastic product noted for its excellent resistance to heat transmission and deterioration from chemicals. When formed into rigid boards, its low-density closed-cell structure produces a lightweight and thermally efficient product. Urethane foam has a resistance per in. (25.4 mm) of about 6.25. Polyurethane boards are made with asphalt-impregnated felt bonded to both sides for use in roofing systems.

Urethane will burn and must be protected by some type of finish material if used for interior applications or on roofs. It may also expand after prolonged exposure to moisture.

POLYISOCYANURATE

Polyisocyanurate foam (a form of urethane) is manufactured in rigid boards faced with reflective aluminum foil, foil-craft material, or glass fiber felt. It is used for wall sheathing and in roofing systems. It has a high R-value per inch (7.2). Like other urethanes, it must be protected from flame in interior applications.

PERLITE

Expanded perlite, mineral binders, and waterproofing agents are combined to make boards with weight characteristics similar to those of the plastic insulations. Unlike the cellular plastics, however, perlite boards are inorganic and will not burn or smolder. The R-value per inch is only about 2.78, which is considerably lower than that of the plastics. Perlite should be kept dry because of its tendency to absorb water.

PHENOLIC

Phenolic insulation is a closed-cell foam core material with a density of about 2.5 lb./cu. ft. (40.05 kg/m^3). It is made in board form with a foil-kraft liner or a glass fiber scrim. It can be used in walls and in roofing systems. It has a very high R-value (8.3) and tends to hold that value better than the other foam plastic insulations, but it must be handled carefully because it tends to break easily. It is also reported to corrode metal.

CELLULAR GLASS BLOCK

Cellular glass is also used to produce a noncombustible and lightweight board insulation. Its closed-cell structure makes it impervious to water vapor and moisture absorption. Its R-value per inch (25.4 mm) is about 2.5, which is comparable to that of expanded perlite insulation.

This product is used on roof decks, exterior decks, and promenades, and as perimeter insulation for slabs and foundations. It is available in thicknesses ranging from 1½ to 4 in. (38.1 to 101.6 mm) and can be produced in tapered form on special order. When used under a waterproof membrane, the tapered form permits positive drainage of a roof or promenade. Cellular glass must be handled carefully because it breaks easily even though it has a high compressive strength. It will deteriorate severely if exposed to freezing and thawing and should be protected from this by a coating of bituminous compound. In severe conditions it should be coated on all faces and set with adhesive-filled joints. The manufacturer's advice on this issue should be sought and followed.

ORGANIC FIBER

Organic fiber rigid insulation is composed of wood fibers or other organic fibers and water-resistant binders. These materials are asphalt impregnated and chemically treated to prevent their deterioration and treated on the top surface to reduce asphalt absorption from roofing. They are used primarily in roofing systems and are available tapered to provide a positive roof slope. They are available in both regular and high-density versions.

GLASS FIBER

Rigid glass fiber boards are made from inorganic glass fibers bonded together by a resinous binder. They are used primarily in roofing systems. For this use, they are given an integrally bonded asphalt-impregnated felt or kraft paper top covering. They are available both flat and tapered and in thicknesses varying from ½ to 3 in. (12.7 to 76.2 mm).

Semirigid glass fiber boards and glass fiber batts and blankets are covered in Section 7.3.1.2.

COMPOSITES

Composites of urethane or polyisocyanurate and perlite, waferboard, or fiberboard take advantage of the high R-values of the plastic foams and the stability and higher heat resistance of the other materials. Composites are also available with gypsum board and glass fiber boards.

7.3.1.2 Mineral Fiber Insulation

Mineral fiber insulation includes products made of fibers manufactured from rock, slag, or glass. Mineral fiber insulations produced in the United States are made either from slag or glass. Products produced from slag are sometimes called rock wool or mineral wool, but are more properly termed "semirefractory fiber insulations" because of their natural resistance to high temperatures. Because glass fiber and semirefractory fiber insulations are inorganic, they are naturally resistant to fire, moisture, and vermin.

Mineral fiber insulation is made in rigid and semirigid boards, and in blankets or batts. This section addresses semirigid boards, blankets, and batts. Rigid mineral fiber insulation used mostly for roofing is covered in Section 7.3.1.1.

SEMIRIGID BOARDS

Semirigid mineral fiber insulation boards are available 2 by 4 ft. (609.6 by 1219 mm) and 1 to 3 in. (25.4 to 76.2 mm) thick. They can be cut to size in the field. They are used in curtain walls, between exterior wall furring members, as horizontal support panels for other insulation, and in other locations where a semirigid product is needed.

BATTS AND BLANKETS

Batts and blankets are similar in appearance and composition, but differ in length. Batts are 48 in. (1219 mm) long; blankets are longer and are packaged in large rolls (Fig. 7.3-4). These are used in areas where it is not practical to install loose fill or where the attached foil or kraft paper facing is desired as a vapor retarder. They are typically used as thermal insulation in walls and partitions, below floors in crawl spaces, and above ceilings in attics and the like, and as sound insulation.

Batts and blankets are produced in several types: (1) wrapped with kraft paper on the edges and a vapor retarder on one or both sides, (2) faced with a vapor retarder on one side only, (3) friction-fit without any covering, in which the interlaced fibers have sufficient resilience to remain upright in a cavity. Insulation with kraft paper or foil paper facing should be protected from flame.

Batts and blankets are manufactured to fit between wood studs placed at standard spacings of 16 and 24 in. (400 and 600 mm) on centers. Faced and wrapped products are provided with a stapling flange for secure installation. Product thickness ranges from 3⅓ to 12 in. (86.6 to 304.8 mm). Thickness, however, is not a reliable measure of insulating effectiveness and should be accompanied by an R-number stamped on the wrapper or facing.

7.3.1.3 Loose-Fill Insulation

Fibers, granules, or chips are the usual components of loose-fill insulations. They are either poured or blown into attics or wall cavities. Most loose-fill insulation used in attics and walls is manufactured from mineral wool, glass wool, or cellulosic fibers. Loose perlite and vermiculite are used to fill the voids in concrete masonry units and the spaces between wythes in masonry walls.

Loose-fill insulation that is other than free flowing should not be used in walls or other closed cavities where the large fibers may catch on nails or other obstructions, which could leave uninsulated pockets.

POURING AND BLOWING WOOL

The types of mineral matter included under this heading are slag and glass wool. Both are moisture and vermin resistant. They are less often used in walls than are semirigid boards, batts, or blankets, because the large clumps of fiber tend to catch on nails or other obstructions, leaving unfilled voids. The most frequent application is between ceiling joists in attics, where the wool can be poured from bags or blown by machine (Fig. 7.3-5).

VERMICULITE AND PERLITE

Vermiculite and perlite are used in areas where they can be readily poured or blown, such as in masonry cavities (Fig. 7.3-6). Vermiculite is manufactured by expanding aluminum magnesium silicate, a form of mica. Perlite is an expanded volcanic glass. They are both noncombustible, vermin and moisture resistant, and do not compress in a cavity.

CELLULOSIC FIBER

Cellulosic fiber insulations are composed of recycled newsprint, wood chips, and other organic fibers. They can be installed by either blowing or pouring into cavities and can be combined with adhesives for sprayed application.

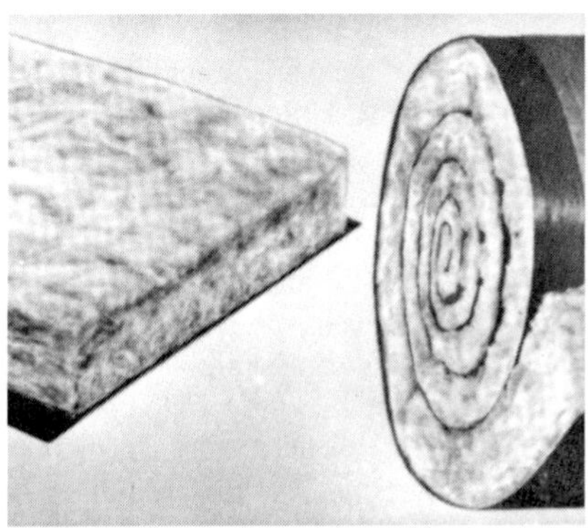

a b

FIGURE 7.3-4 Batt insulation (a) is produced in panels. Blanket insulation (b) is made in rolls that can be cut to size.

FIGURE 7.3-5 Large areas can be quickly insulated with a pressurized hose.

FIGURE 7.3-6 Vermiculite and perlite can be poured directly from the sack into a masonry cavity.

The fibers are chemically treated to resist fire, minimize smoke contribution, and resist vermin. However, certain clearances adjacent to heat-producing fixtures must be maintained. These fibers have a neutral pH to prevent reaction with metal and will not decay after extended exposure to moderate dampness. Cellulosic insulation should be UL listed and labeled to ensure that it is fire resistant.

7.3.1.4 Foamed-in-Place Insulation

Most foamed-in-place insulations are either polyurethane-, polyisocyanurate-, or phenolic-based materials. They are created by a chemical reaction that expands a mixture of components as much as 30 times. When cured, which takes from 24 to 72 hours, they solidify into a cellular plastic.

URETHANE FOAM

Urethane foam has a high resistance per inch (R-6.25 or higher), is resistant to moisture and most chemicals, and has fire-retardant capabilities if chemically treated. It has many applications, including floor and wall cavities, roofs, and structural sandwich panels. The R-value per inch may vary with field conditions at the time of placement.

Urethane foam is particularly effective where a high R-value must be achieved in limited space or on irregularly shaped construction. Care must be taken not to overfill a cavity, because the foam expands to its ultimate volume after injection.

This type of insulating foam should be used in open cavities or where it is possible to accurately compute the amount of foam needed. Even when fire-retardant treated, it will burn and should be protected from flame by a fire-resistant finish.

POLYISOCYANURATE FOAM

Polyisocyanurate foam can be used in the same kinds of locations as can polyurethane foams. It has a higher R-value than polyurethane (7 to 8) and a comparable density.

PHENOLIC-BASED FOAM

Phenolic-based foams may be used in most of the same kinds of locations as polyurethane foams. They have a lower R-value (about 4.5) and a lower density than polyurethane.

7.3.1.5 Other Insulation Materials

Other materials used as insulation are fiberboard, reflective insulation, insulated sheathing (see Section 6.6.5.1), and structural insulating roof deck (see Section 3.17).

FIBERBOARD

Fiberboard is a general-purpose, low-cost insulation board sometimes used as an interior wall finish in cottages and temporary buildings. However, because of their softness, fiberboard wall panels are seldom used in permanent buildings.

Fiberboard for wall use is available with a flame-resistant, predecorated factory finish. It is manufactured ½ and ¾ in. (12.7 and 19.1 mm) thick in tongue-and-grooved plank or panel form. The ½-in. (12.7 mm) thickness can be used when the supporting members are spaced 16 in. (400 mm) on centers; ¾-in. (19.1 mm) thickness is recommended when 24-in. (600 mm) spacing is used.

Fiberboard ceiling tile is stapled to furring strips or applied with mastic to a planed smooth surface. It can also be installed with wood or metal hangers to form a suspended ceiling.

REFLECTIVE INSULATION

Reflective insulation is composed of one or more layers of metallic foil. It is different from other insulations in that its insulating effectiveness depends on how shiny the reflective surface remains over time, whether a ¾-in. (19.1 mm) air space is provided on the warm side, and whether it is intended to reduce heat flow in or out of the structure. Therefore, reflective insulation should face an air space of at least ¾ in. (19.1 mm) and should remain free of dust and other materials that could reduce its reflective qualities.

7.3.1.6 Standards of Quality

Insulating products are manufactured to various standards, including those established by the ASTM and, where applicable, those set by the General Services Administration, known as Federal Specifications. Figure 7.3-7 summarizes current standards of quality.

7.3.2 DESIGN AND INSTALLATION

Heat flow through ceilings, walls, and floors can be controlled by one or more of the following techniques: (1) by constructing the assembly with materials that are inherently good insulators; (2) by using concrete, stone, solid masonry, or other dense materials that have the capacity to store heat and slow heat flow; or (3) by adding thermal insulation to the assembly. The first and second methods are discussed in Section 15.2, the third in this section. Whichever method is employed, correct installation is necessary to achieve the full benefit.

7.3.2.1 General Recommendations

For conventionally heated and cooled buildings, the objective of heat control is to block the flow of heat through the floors, walls, and ceilings that surround habitable areas. In most circumstances, except in areas of high humidity, as discussed in Section 7.1.3.2, to most effectively block heat flow, insulation is placed near the inside of the building envelope. Insulated sheathing, of course, is an exception.

However, in some passive solar-heated buildings, the insulation is placed on the outside face of the wall, rather than on the inside, as in conventional systems. Such systems are designed to absorb solar heat in masonry or concrete walls during sunshine hours and to release (radiate) this heat slowly into the interior during sunless periods. To prevent rapid cooling of the warm masonry during sunless periods, insulation is placed on its outside face.

In exterior insulation and finish systems (see Section 7.4) the insulation is also placed on the exterior of the building.

Some insulation products are made from flammable materials or materials that produce toxic combustion discharges when they are heated or burned. These

FIGURE 7.3-7 Summary of Insulating Products

Material	Standard	Description
Polystyrene	ASTM C 578	"Preformed Cellular Polystyrene"
Polyurethane	ASTM C 591	"Unfaced Preformed Rigid Cellular Polyurethane"
Polyisocyanurate	Fed. Spec. HH-I-1972/1	"Insulation Board, Polyurethane or Polyisocyanurate"
Phenolic	ASTM C 1126	"Faced or Unfaced Rigid Cellular Phenolic Thermal Insulation"
Perlite		
Loose fill	ASTM C 549	"Perlite Loose Fill Insulation"
Board	ASTM C 728	"Perlite Thermal Insulation Board"
Vermiculite	ASTM C 516	"Vermiculite Loose Fill Insulation"
Cellular glass	ASTM C 552	"Cellular Glass Block and Pipe Thermal Insulation"
	ASTM E 136	"Behavior of Materials in a Vertical Tube Furnace"
Organic fiber	ASTM C 208	"Cellulosic Fiber Insulation Board"
	ASTM C 209	"Test for Cellulosic Fiber Insulating Board"
Mineral fiber	ASTM C 553	"Mineral Fiber Blanket and Felt Insulation"
	ASTM C 612	"Mineral Fiber Block and Board Thermal Insulation"
	ASTM C 665	"Mineral Fiber Blanket Thermal Insulation"
	ASTM C 687	"Determination of Thermal Performance of Loose-Fill Building Insulation"
	ASTM C 726	"Mineral Fiber and Mineral Fiber, Rigid Cellular Polyurethane Composite Roof Insulation Boards"
	ASTM C 764	"Mineral Fiber Loose Fill Thermal Insulation"

should be protected by gypsum wallboard or another fire-resistant finish, as most building codes require.

Polyurethane, polyisocyanurate, and polystyrene should not be exposed to daylight, to prevent ultraviolet deterioration. Molded polystyrene and expanded perlite boards must also be protected from contact with water to prevent absorption. As mentioned earlier, cellular glass boards must be protected from freezing.

7.3.2.2 Slabs on Grade

Heat loss related to floor slabs built on ground is more nearly proportional to the perimeter than to the area of the slab. Heat loss to the ground below is minor. Therefore, in areas where heating systems are required, insulation should be installed around the entire slab perimeter so as to prohibit excessive heat loss directly to areas of low outside air or ground temperature. Uninsulated slab perimeters may be uncomfortably cold, may produce condensation, and may result in uneconomical heating efficiency.

This *perimeter* insulation should be (1) virtually nonabsorptive of moisture (noncapillary), (2) not harmed permanently by wetting or contact with wet concrete, (3) sufficiently rigid to resist damage and compression during placing of concrete, and (4) resistant to decay and insect damage.

Perimeter insulation should be at least 1 in. (25.4 mm) thick and preferably 2 in. (50.8 mm) thick. Even greater thicknesses may be common in very cold climates. Insulation should be provided around the entire perimeter of the slab and should extend either vertically 24 in. (600 mm) below grade level (or below the frost line if it is deeper) or horizontally for a distance of 24 in. (600 mm) under the slab. Alternate methods of perimeter slab insulation are shown in Figure 7.3-8. Insulation may also be placed on the exterior of the foundation wall (Fig. 7.3-9).

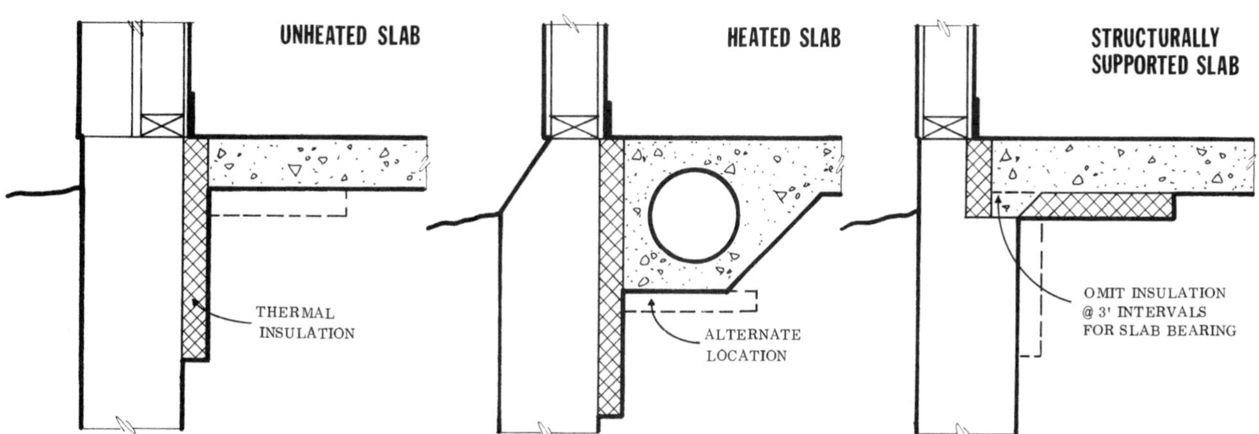

FIGURE 7.3-8 Interior locations for perimeter insulation at slabs.

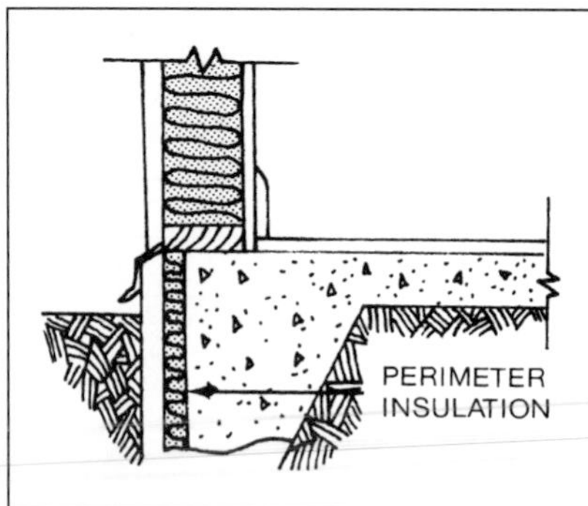

FIGURE 7.3-9 Exterior location for perimeter insulation.

Glass fiber insulation board and other plastic foam boards are sometimes used for perimeter insulation, but most perimeter insulation is polystyrene.

7.3.2.3 Crawl Spaces and Floors

Heat loss through floors over heated crawl spaces can be reduced by installing insulating material against the foundation walls (Fig. 7.3-10). If the space is unheated, the insulation should be placed between the floor joists (Fig. 7.3-11). In either case, the vapor retarder should face the living space.

In basements and heated crawl spaces, sill sealer insulation should be used to prevent air infiltration between sill and foundation or between header joists and foundation (Fig. 7.3-12). In addition, the insulating sheathing should extend to the base of the plate.

The space behind the header joist should be covered with short pieces of flexible insulation. If the wall is insulated, as in Figure 7.3-10, the end of the blanket

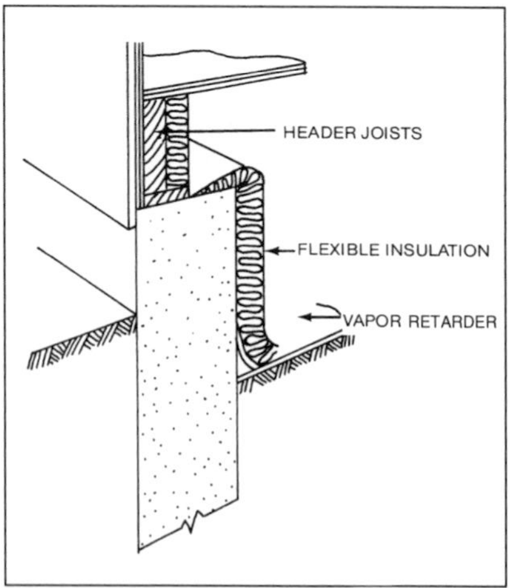

FIGURE 7.3-10 Insulation location on the perimeter walls of a heated crawl space.

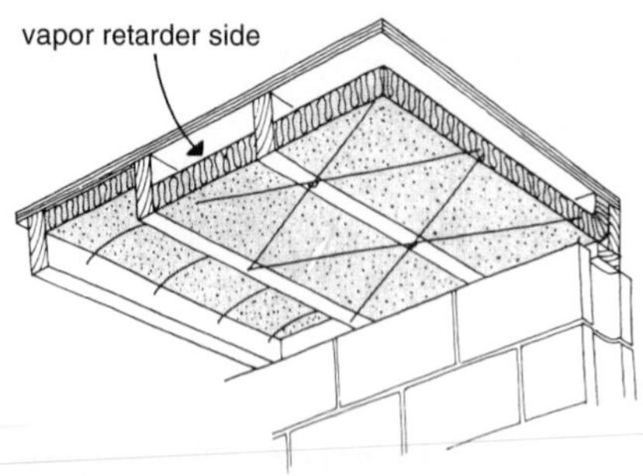

FIGURE 7.3-11 Insulation location over an unheated crawl space. Insulation may be stapled in place if installed from above or if reverse flange blankets are used. It may be held up by wires, as shown here, when installed from below. Wires can also be used when the framing is metal.

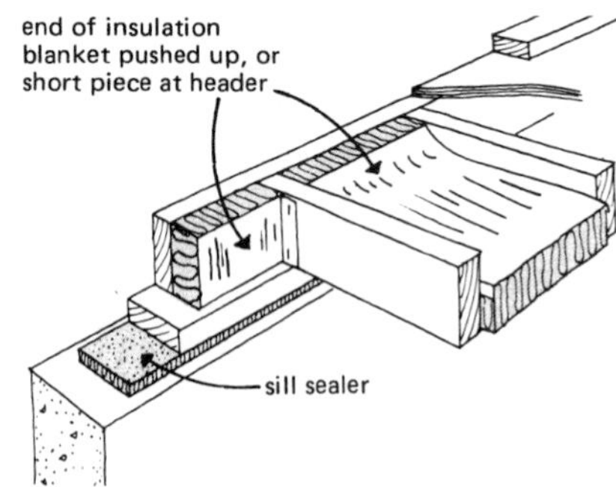

FIGURE 7.3-12 Sill sealer insulation is placed on top of foundation wall. Insulation should be fitted snugly against band joist.

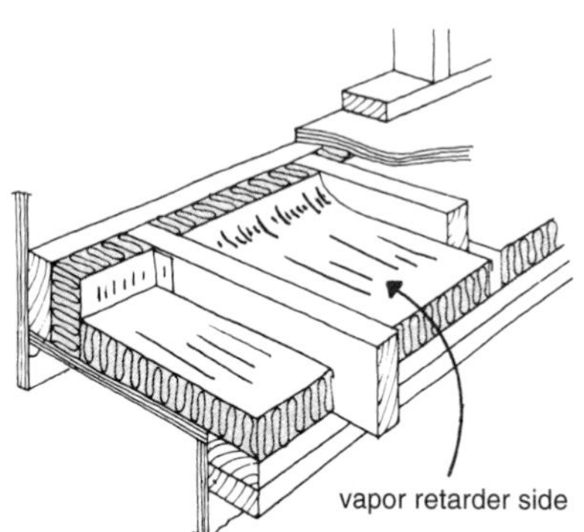

FIGURE 7.3-13 Small cantilevered areas should not be overlooked, but should be carefully insulated, comparably with the rest of the building.

can be extended upward and pushed against the header. Cantilevered floors should be insulated as shown in Figure 7.3-13. In every case, the outside edge of a floor cavity should be insulated with the same insulation value and vapor retarder as the adjacent walls (Fig. 7.3-14).

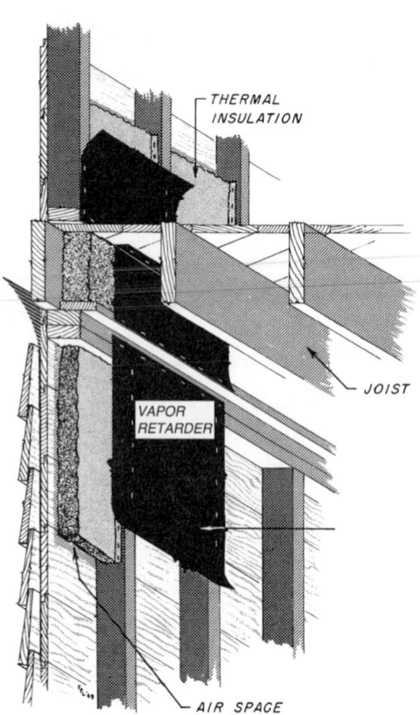

FIGURE 7.3-14 Detail showing insulation and vapor retarder at the outside edge of a floor cavity. (U.S. Forest Products Laboratory. Not subject to copyright in the United States.)

7.3.2.4 Walls

Heat transmission through wall assemblies can be reduced by selecting materials that are inherently good insulators and by adding insulation. The amount of thermal insulation installed in a wall depends on considerations of human comfort, the effect of total heat loss and heat gain, and, therefore, on the cost of heating and cooling the building. Estimated savings in both heating and cooling costs may be used as a guide to determine the amount of insulation that can be installed economically. In some mild climates, insulation may be necessary only to raise or lower the inside surface temperature of a wall to increase human comfort. Cold wall surfaces absorb radiant heat from the body and may produce a feeling of chill even though the inside air temperature is 70°F (21.1°C) or above. In most climates, insulation is added to every exterior wall.

Hourly heat loss or gain through an exterior wall is measured by its U-value. U-value is the amount of heat in British thermal units (Btu's) (W) transmitted in 1 hour per sq. ft. (0.93 m^2) of wall area, for a difference in temperature of 1°F (0.11°C) between the inside and outside air. Figure 7.3-15 gives the U-values of 8-in. (203 mm) masonry walls with vari-

FIGURE 7.3-15 Estimated U-Values for 8-in. (200 mm) Hollow Concrete Masonry Walls

Wall Details	U-Value Based on Density of Concrete Used in Block				
	60 (292.9) pcf (kg/m²)	80 (390.6) pcf (kg/m²)	100 (488.2) pcf (kg/m²)	120 (585.9) pcf (kg/m²)	140 (683.5) pcf (kg/m²)
No insulation	0.32	0.34	0.38	0.43	0.55
No insulation, ½-in. (12.7 mm) gypsum board on furring strips	0.21	0.23	0.25	0.27	0.31
No insulation, ½-in. (12.7 mm) foil-backed gypsum board on furring strips	0.15	0.15	0.16	0.17	0.19
Loose-fill insulation in cores	0.12	0.14	0.18	0.21	0.35
Loose fill in cores, ½-in. (12.7 mm) gypsum board on furring	0.10	0.12	0.14	0.17	0.24
Loose fill in cores, ½-in. (12.7 mm) foil-backed gypsum board, furring	0.08	0.10	0.11	0.12	0.16
1-in. rigid glass fiber, ½-in. (12.7 mm) gypsum board	0.14	0.14	0.15	0.15	0.17
1-in. polystyrene, ½-in. (12.7 mm) gypsum board	0.12	0.12	0.12	0.13	0.14
1-in. polyurethane, ½-in. (12.7 mm) gypsum board	0.10	0.10	0.11	0.11	0.12
Loose fill in cores plus 1-in. (25.4 mm) rigid glass, ½-in. (12.7 mm) gypsum board	0.08	0.09	0.10	0.11	0.14
Loose fill in cores plus 1-in (25.4 mm) polystyrene, ½-in. (12.7 mm) gypsum board	0.07	0.08	0.09	0.10	0.12
Loose fill in cores plus 1-in. (25.4 mm) polyurethane, ½-in. (12.7 mm) gypsum board	0.07	0.07	0.08	0.09	0.11
R-7 blanket insulation, ½-in. (12.7 mm) gypsum board, furring	0.09	0.10	0.10	0.10	0.11

ous combinations of finish and insulation. R-value is the reciprocal of the U-value (R = I/U). R-values are useful for estimating the effect of the individual components of a building section on the total heat flow, because R-values can be directly added. Figure 7.3-16 lists R-values for single-wythe concrete masonry walls. Figure 7.3-17 gives R- and U-values for some typical concrete masonry walls.

The insulation values recommended in most locations today are often difficult to accomplish using a single layer of insulation. For example, the cavity in a 2 × 4 (50 × 100) wood stud wall will permit only 3½ in. (88.9 mm) of blanket insulation with an R-value ranging from about 11 to about 17, depending on the insulation material and its density. A similar situation exists when exterior walls are framed with metal studs. In most areas of the country, total recommended wall R-values are 19 or higher. This leaves a gap between the R-value of the insulation alone and the required value for the entire wall. If insulation in the higher R-value range is selected, the other portions of the wall, such as interior and exterior finishes and sheathing, may produce sufficient total wall R-value. When such is not the case, one way to increase the insulation thickness in wood- or metal-stud-framed structures is to use 6-in. (150 mm) wall studs or staggered studs, but even this is not always enough. Insulations with higher R-values and other construction methods must be used when high R-values are needed. Batt and blanket insulation will most likely not be the proper selection in such cases.

FIGURE 7.3-16 R-Values of Single-Wythe Concrete Masonry Walls[a]

Nominal Wall Thickness, in. (mm)	Insulation in Cells[b]	R-Value Based on Concrete Unit Weight				
		60 (292.9) pcf (kg/m²)	80 (390.6) pcf (kg/m²)	100 (488.2) pcf (kg/m²)	120 (585.9) pcf (kg/m²)	140 (683.5) pcf (kg/m²)
4 (100)	Filled	3.36	2.79	2.33	1.92	1.14
	Empty	2.07	1.68	1.40	1.17	0.77
6 (150)	Filled	5.59	4.59	3.72	2.95	1.59
	Empty	2.25	1.83	1.53	1.29	0.86
8 (200)	Filled	7.46	6.06	4.85	3.79	1.98
	Empty	2.30	2.12	1.75	1.46	0.98
10 (250)	Filled	9.35	7.45	5.92	4.59	2.35
	Empty	3.00	2.40	1.97	1.63	1.08
12 (300)	Filled	10.98	8.70	6.80	5.18	2.59
	Empty	3.29	2.62	2.14	1.81	1.16

[a]R-values do not include the sums of the effect of air film or surface conductance on the inside of the walls ($1/f_i = 0.68$) and on the outside ($1/f_o = 0.17$).

[b]Loose-fill insulation such as perlite, vermiculite, or others of similar density.

FIGURE 7.3-17 R- and U-Values of Some Typical Concrete Masonry Walls

	Heat-Resistance Values, R							
Components	**A**	**B**	**C**	**D1**	**D2**	**E**	**F1**	**F2**
Surface film (outside)	0.17	0.17	0.17	0.17	0.17	0.17	0.17	0.17
8-in. (200 mm) hollow concrete masonry at 100-pcf (1601.9 kg/m^3) density (cores open)	1.75			1.75	1.75			
Cores filled with bulk insulation			4.85				4.85	4.85
Concrete brick at 140-pcf (2242.6 kg/m^3) density		0.44				0.44		
2-in. (50 mm) air cavity		0.97						
4-in. (100 mm) hollow concrete masonry at 100-pcf (1601.9 kg/m^3) density		1.40				1.40		
2-in. (50 mm) bulk insulation in cavity						4.00		
Batt insulation between furring strips: 1 in.				3.70			3.70	
2 or 2¼ in.					7.00			7.00
½-in. (12.7 mm) gypsum board interior finish				0.45	0.45		0.45	0.45
⅝-in. (15.9 mm) plaster, lightweight aggregate		0.39				0.39		
Surface film (inside)	0.68	0.68	0.68	0.68	0.68	0.68	0.68	0.68
TOTAL resistance, R	2.60	4.05	5.70	6.75	10.05	7.08	9.85	13.15
U-value, 1/R	0.384	0.247	0.175	0.148	0.100	0.141	0.102	0.076

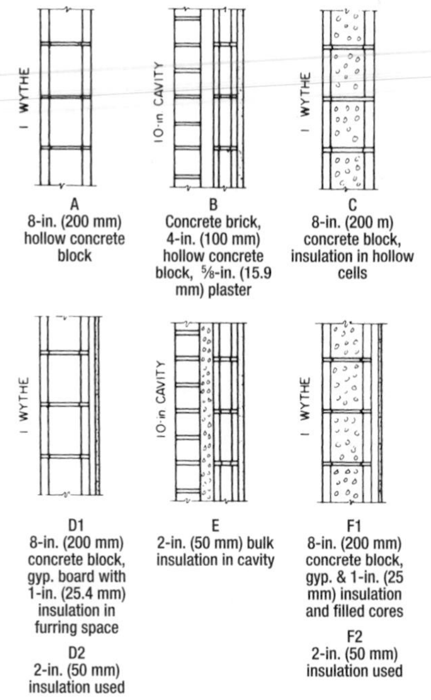

1 WYTHE
A
8-in. (200 mm) hollow concrete block

10-in CAVITY
B
Concrete brick, 4-in. (100 mm) hollow concrete block, ⅝-in. (15.9 mm) plaster

1 WYTHE
C
8-in. (200 m) concrete block, insulation in hollow cells

1 WYTHE
D1
8-in. (200 mm) concrete block, gyp. board with 1-in. (25.4 mm) insulation in furring space
D2
2-in. (50 mm) insulation used

10-in CAVITY
E
2-in. (50 mm) bulk insulation in cavity

1 WYTHE
F1
8-in. (200 mm) concrete block, gyp. & 1-in. (25 mm) insulation and filled cores
F2
2-in. (50 mm) insulation used

RIGID INSULATIONS

When high R-values are required, polyurethane foam insulations may appear to be the proper choice because of their relatively high R-values. However, all plastic foam insulations are good vapor retarders and will prevent moisture that finds its way into a wall from escaping unless certain precautions are taken. In most climates, a continuous polyethylene vapor barrier on the room side of the assembly, and venting the wall cavity, are highly recommended. Specific venting methods vary with the manufacturer, but its recommendations should be followed. Refer to Section 7.1.3.2 for a discussion about the placement of vapor retarders and venting in warm, humid climates.

BATT AND BLANKET INSULATIONS

When doing so will not interfere with installing finish materials, insulation flanges in wood stud framing may be attached to the stud face to make a continuous vapor retarder. For stud spaces narrower than 14½ in. (368.3 mm), standard barrier-faced blankets may be cut down to form a stapling flange. These flanges should then be stapled to the inside faces of the studs. However, facing the studs with an insulation flange will interfere with the installation of board products, such as gypsum board, and this type of application should not be used when such products are to be installed. Batts and blankets in stud spaces should touch the sheathing or siding and fit tightly against the top and bottom plates.

When a stud face must be kept clear, or friction fit insulation or other insulation without a vapor retarder is used, or the stud spacing is less than 16 in. (400 mm), the inside face of the wall should be completely covered with a vapor retarder such as polyethylene film (see Fig. 7.1-30) or aluminum foil. Necessary openings should be cut out later. The retarder material should be installed in the largest possible sheets to minimize vapor leakage. Retarder joints should be made over framing members.

Cracks around window and door openings should be packed with loose insulation and covered with polyethylene to reduce air infiltration (Fig. 7.3-18). Insulation should also be packed around pipes, ducts, and electrical boxes to prevent cold spots in the wall (Fig. 7.3-19).

Gaps in the insulation at the top or bottom of the cavity may allow convection currents to carry heat and vapor from the inside to the outside of the cavity.

INSULATING MASONRY WALLS

Masonry walls can be insulated by installing semirigid, batt, or blanket insulation between wood or metal furring strips (Fig. 7.3-20) or by applying mineral fiber or plastic foam board insulation directly to the wall (Fig. 7.3-21). Board insulation that is not inherently fire resistant should be protected with gypsum board.

Gypsum board may be adhesively attached to a mineral fiber semirigid board,

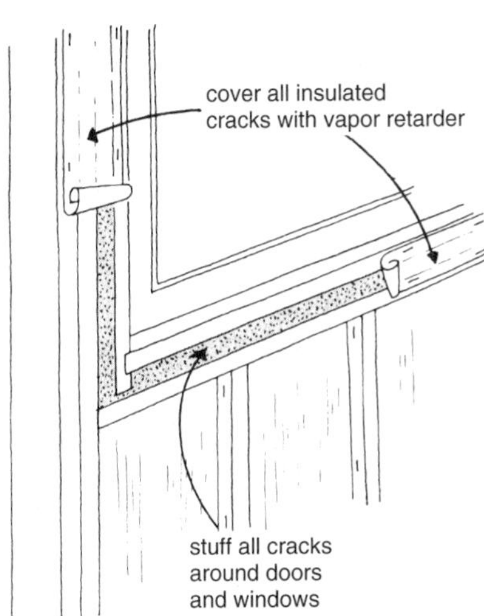

FIGURE 7.3-18 Detail showing stuffing of cracks around a window with insulation and covering it with a vapor retarder.

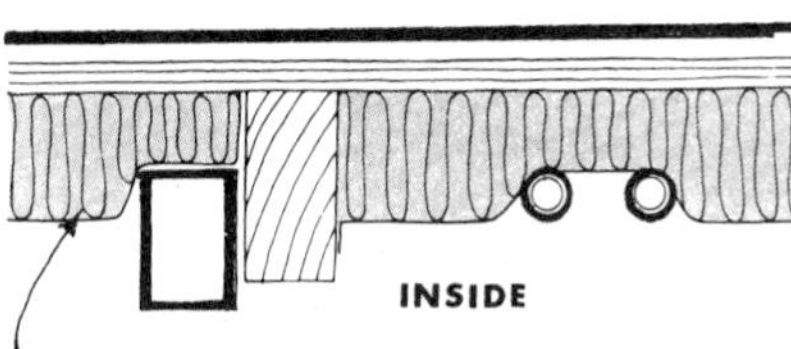

FIGURE 7.3-19 Insulation should be packed behind ducts, pipes, and electrical boxes, filling the spaces between them and the exterior finish or sheathing.

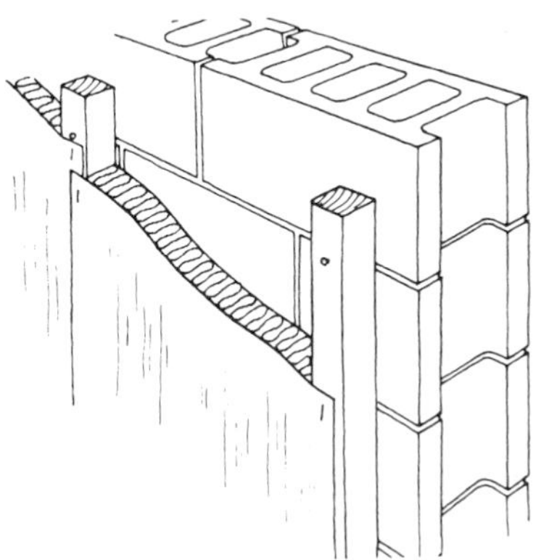

FIGURE 7.3-20 Batts, blankets, or semirigid insulation can be applied between furring strips behind a wall finish. This illustration shows blankets or batts with a nailing flange attached to the face of the furring.

but should not be glued directly to plastic foam materials. If it is, heat will pass through the gypsum board in a fire and melt the plastic, permitting the gypsum board to fall away, leaving the foam board

a

b

FIGURE 7.3-21 (a) Insulation board is bonded to the interior face of masonry using adhesives. (b) Gypsum board is bonded to glass-fiber insulation using adhesives. (Courtesy Dow USA.)

exposed. Instead, such gypsum board should be attached to furring.

Closed-cell plastic boards are highly resistant to moisture and vermin and can be installed on the outer faces of masonry and frame walls. However, polystyrene and polyisocyanurate board insulation must be painted or otherwise protected from exposure to daylight. All plastic boards should also have some type of protection to prevent dents and chips if used in an exposed area.

In general, the methods used to add insulation to masonry walls include (1) batt or blanket insulation placed in a furring space on the interior face of a wall (Fig. 7.3-22; also see Fig. 7.3-20); (2) rigid or semirigid insulation placed in or applied to the face of a wall (Figs. 7.3-23, 7.3-24, and 7.3-25; also see Fig. 7.3-21); and (3) fill insulation poured or blown into the masonry cavity or core space (Fig. 7.3-26; also see Fig. 7.3-6).

Loose-fill insulation used in cavity walls and the cores of hollow masonry should possess the following properties:

- The insulation should not transmit moisture from exterior to interior wythe.
- Insulating efficiency should not be impaired by retained moisture.
- Loose fills should support their own weight without settlement.

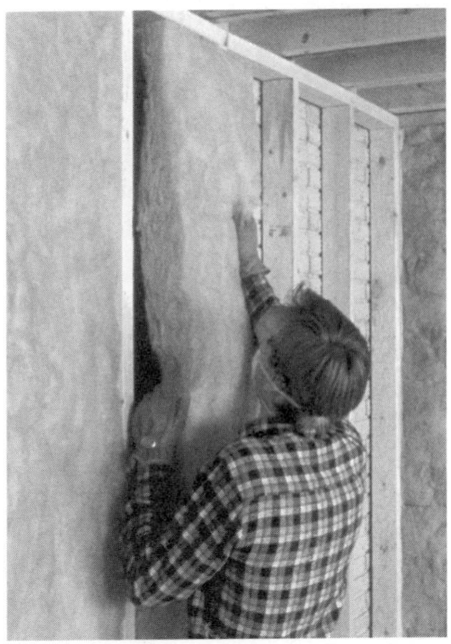

FIGURE 7.3-22 Friction-fit glass-fiber blanket insulation being installed between wood furring at a masonry wall. (Photograph furnished by Owens-Corning Fiberglas Corporation.)

FIGURE 7.3-23 Rigid foam insulation board installed in a cavity wall. (Courtesy Brick Institute of America.)

FIGURE 7.3-24 Glass-fiber insulation being installed in a cavity wall. Brick ties hold insulation against inner masonry wythe, leaving an air space between the outer wythe and the insulation. (Courtesy Brick Institute of America.)

- The insulation should be inorganic or have comparable resistance to decay, fire, and vermin.

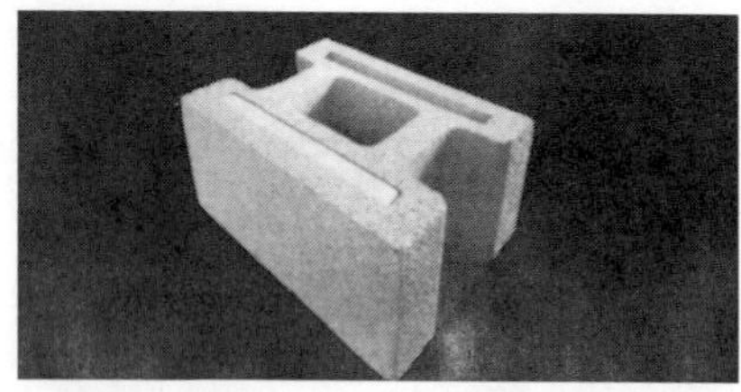

a

b

FIGURE 7.3-25 (a) Insulation inserted in the face shell of specially formed blocks. (b) Specially formed insulation inserted in cores of concrete masonry units. (Courtesy Concrete Block Insulating Systems, Inc.)

FIGURE 7.3-26 Granular insulation fills a brick cavity wall. It can also be used to fill the cores in concrete masonry units. (Courtesy Brick Institute of America.)

- Granular fills should be pourable in lifts of at least 4 ft. (1219 mm).

Types of loose fill meeting these criteria are (1) water-repellent, expanded vermiculite and (2) silicone-treated, expanded perlite.

Rigid boards having the required properties include (1) expanded or molded polystyrene, (2) polyurethane, (3) polyisocyanurate, (4) cellular glass, (5) preformed glass fiber, and (6) perlite board.

MASONRY WALL CAPACITY INSULATION

In warm climates and in the summer months in cooler climates, the heat gain characteristics of walls are important. Heat gain in buildings with masonry walls depends on the *capacity insulation*, which is the wall's heat storage capacity. The greater the capacity insulation, the lower the heat gain and the cooler the inside temperatures. In general, heavier walls have a greater capacity insulation. They absorb and retain heat during the day and dissipate it to the outside during the cooler night hours. However, the wall must be thick enough that its entire thickness does not reach a temperature greater than that of the interior space, in which case, the wall will radiate heat into the cooler space. The more difference there is between the daytime and the nighttime outdoor temperatures, the more efficiently a wall will radiate the heat built up in it during the daytime.

7.3.2.5 Ceilings and Roofs

CEILINGS

Except in the humid areas discussed in Section 7.1.3.2, flexible insulation with vapor retarders should be installed in ceilings so that the retarder faces the living space. Loose-fill or batt or blanket insulation without a vapor retarder should have separate polyethylene film placed below the insulation facing the living space.

For maximum effectiveness against vapor transmission, the kraft vapor retarder facing on the insulation should overlap the faces of the framing members. If such a procedure is not appropriate because it interferes with tight installation of gypsum board or other finish materials, a separate vapor retarder should be installed.

The insulation should extend entirely across the top plate to prevent heat loss at the joint. If necessary, the junction between the insulation and the plate should be stuffed with loose insulation. When eave vents are used, the insulation should not block the air movement between the vent and the space above the insulation.

Eave baffles or special cardboard ducts should be installed to prevent loose-fill insulation from covering the vents and ensure air movement from the eaves to the general attic. Cellulosic insulation should also be kept from contact with potential sources of heat, such as electrical fixtures, by providing at least a 3-in. (76.2 mm) clearance along each face of the fixture.

ROOFS

Heat transmission through a roof can be reduced by (1) using optimum amounts of insulation, (2) using shiny or light-colored roof materials to reflect heat, (3) ventilating attic spaces, (4) increasing the thermal mass to achieve a time lag in the heat flow, (5) using louvered rather than solid overhangs to free heated air trapped beneath eaves, and (6) using steeply pitched roofs on the windward side of the building to deflect cold winds and reduce the roof exposure.

All the listed rigid insulations are used over structural roof decks as a base for roofing as well as to provide thermal resistance.

7.3.3 REINSULATION OF EXISTING BUILDINGS

Although the heat control concepts discussed here apply to existing construction, it is not always possible to insulate an older structure in the same way and to the same level as a new building. Often it is not economical to add insulation to less accessible parts of a completed building, such as walls and floors.

In general, attics with no insulation should be insulated to the same levels as new construction. If the attic already has 6 in. (152 mm) of batt insulation or 8 in. (203 mm) of loose-fill insulation, the benefits of additional insulation should be examined on a case-by-case basis. Other parts of the building envelope, such as floors and walls, should be reinsulated to those levels when it is economically practical.

Whenever insulation is added to an existing structure, proper vapor control is essential. This involves installation of a vapor retarder facing the living space, provision for adequate cavity ventilation, or both. Section 7.1 provides a more detailed description of proper techniques.

7.3.3.1 Foundations

Existing foundations of concrete or masonry can be insulated from either inside or outside the wall. Outside application is most frequently used when the interior already has a wall finish that would be covered by the new insulation.

INSIDE THE WALLS

Mineral wool batts or blankets or rigid plastic foam boards are the typical materials used in inside walls.

Batts or blankets are installed by furring out the wall with wood strips and stapling the insulation to the strips (Fig. 7.3-27). The added wall thickness may not

FIGURE 7.3-27 If 2 × 4 (50 × 100) studs are used, a wall can be insulated to R-11; 2 × 3 (50 × 75) studs will accommodate only R-7 insulation.

be acceptable for inhabited above-ground rooms but can easily be accommodated in basements that are not used as primary living space.

Mineral wool has the advantage of being fireproof if a paper backing is not attached. If the batts are paper-faced, they should be protected from flame by a wall covering (Fig. 7.3-28).

Rigid plastic boards can be installed on the interior wall between traditional 1 × 2 (25 × 50) and 2 × 2 (50 × 50) wood furring strips with the wall finish attached to the strips. There are also various proprietary systems that use mastics, adhesives, or metal channels in place of furring. Adhesives and mastics are thermally more efficient than the furring system and can be installed at a lower cost by a contractor familiar with their use. However, these systems can be used only if the wall is straight and true, which basement walls often are not. In addition, mastics and adhesives can be used only on clean, dry, unpainted surfaces.

A major advantage of plastic boards is that they often require less space to achieve a particular R-value. Extruded polystyrene has an aged R-value per inch of 5 or more. Mineral wool and glass fiber boards have an R-value per inch of about 4, batts of about 3. Thus, a required R-value of 19 would demand 5 in. (127 mm) of mineral wool or glass fiber boards, 6½ in. (165 mm) of mineral wool or glass fiber batts, but only 4 in. (101.6 mm) of polystyrene. Some densities of glass fiber boards have R-values per inch of 4.4, which is close to, but still not as large as, that of extruded polystyrene. The real difference comes when the foam is polyurethane, polyisocyanurate, or closed-cell phenolic. With their R-values of 7 to 8, only 2½ to 3 in. (63.5 to 76.2 mm) are required to reach the 19 or more in our example. A major disadvantage of foamed plastic board is that it must be covered with gypsum board or another fire-resistant material to be safe from fire.

OUTSIDE THE WALLS

Only plastic foam boards, such as urethane, polyisocyanurate, and extruded polystyrene, are suitable for installation on outside walls, because of their resistance to moisture. The mastic or adhesive normally used to secure the board to the masonry requires a sound, unpainted masonry, stucco, or parged surface for a good bond. To be most effective, the insulation should extend below the frost line, but not less than 2 ft. (600 mm) below ground.

Plastic foam boards are soft materials that can be dented and chipped if not protected by some harder material (Fig. 7.3-29). Several commercial systems available involve bonding a layer of glass fiber reinforcement to the insulation and applying a stuccolike finish. Polystyrene and polyisocyanurate are also sensitive to daylight and must be protected to prevent deterioration.

Outside-the-wall reinsulation is often less economical than that inside the wall because of higher installation and material costs, but it has the advantage that installation can proceed without disrupting interior spaces or covering up interior finishes.

7.3.3.2 Crawl Space

An accessible crawl space can be insulated in the same way as new construction. Special attention should be given to the area between the floor joists and behind the header joist—an often ignored or underinsulated spot (Fig. 7.3-30). If the crawl space is not accessible, it can be insulated on the outside of the wall as described for basements.

7.3.3.3 Walls

Three basic types of walls—conventional stud, brick veneer, and masonry cavity—are most economically insulated at the time of construction. They can be effectively insulated later only if there was no insulation installed initially and if there are no major mechanical or electrical obstructions in the wall.

STUD WALLS

The thermal efficiency of stud walls can be improved by adding insulation to the stud cavity or to the exterior wall face before adding new siding.

Stud cavities are typically reinsulated with either blown mineral wool, cellulose, or foamed-in-place plastic foam.

Before insulation is added, a wall should be checked for existing insulation

FIGURE 7.3-28 A wall covering should be applied over paper-faced insulation.

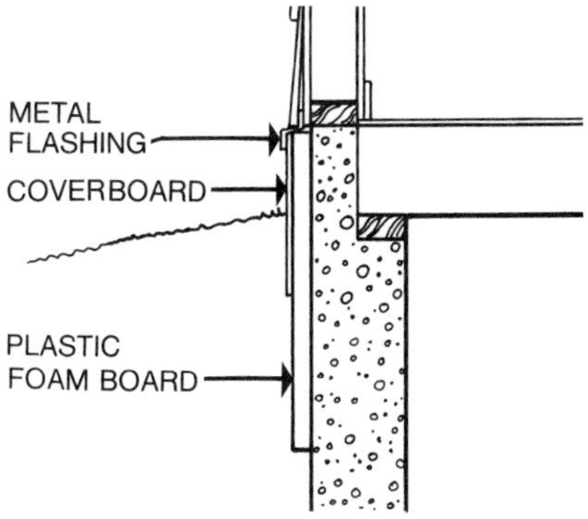

METAL FLASHING
COVERBOARD
PLASTIC FOAM BOARD

FIGURE 7.3-29 Plastic foam board should be protected at the top edge with metal flashing and on its face with a hard, waterproof covering.

FIGURE 7.3-30 Strips of insulation should be stuffed behind header joists.

by removing the face plate of a wall receptacle. If no insulation is present, the cavity should be probed with a plumb-bob or other weight to detect obstructions that may cause blockages. The insulation should be blown or squirted into the wall through holes drilled between the studs (Fig. 7.3-31). Exterior installation is usually preferable, because most often siding can more readily be removed and replaced than can parts of the interior finish.

a

b

c

FIGURE 7.3-31 Installing insulation in an existing wall is a three-step process: (a) open the wall cavity, (b) add insulation, and (c) close the wall cavity.

Exterior re-siding can provide an opportunity for adding plastic foam sheathing board under the new finish. This simple procedure requires only that the insulation be attached to the existing siding before new siding is applied (Fig. 7.3-32). Special metal channels accommodate the extra wall thickness around door and window frames. Although this procedure is not as thermally beneficial as fully insulating the stud cavity, the addition of 1-in. (25.4 mm) rigid polystyrene board, for example, can add 3.57 to the total R-value of the wall.

BRICK VENEER

Insulation can be added to a brick veneer wall by blowing or foaming. It is often difficult, however, to gain access to a stud cavity unless the fascia board or a few bricks near the eaves can be removed to expose an opening into the cavity.

MASONRY CAVITY

Masonry cavity walls are rarely reinsulated because of the difficulty of gaining access to the cavity. However, it may be possible to remove several bricks or a fascia board and thus squirt the insulation into the cavity. In addition to plastic foams and blown mineral wool, vermiculite and perlite can be used for such applications.

7.3.3.4 Attics

Attics are often the most economical areas to reinsulate because they are normally accessible and can account for a substantial part of heat loss or gain. Where the insulation is placed usually depends on whether the attic has been left unfinished or finish materials have been added to floors, walls, and ceilings.

FIGURE 7.3-32 Foam insulation board completely covers the existing walls, thus reducing air infiltration as well as thermal transmission.

UNFINISHED ATTIC

When the attic is unfloored, loose-fill insulation can be blown into the joist cavities or mineral wool batts can be placed between the joists in the same way that new attics are insulated. If the attic already has some insulation, additional loose fill can be blown or poured over the old, or another layer of unfaced batts can be added. If the old insulation was wrapped or if the new insulation is faced, the kraft paper or aluminum foil facing on the top face should be punctured at 12-in. (305 mm) intervals to prevent moisture condensation within the insulation. Condensed moisture could reduce insulating effectiveness and cause structural member decay.

If the attic is floored, it is often possible to remove several boards and blow insulation under the remaining floor before replacing the boards (Fig. 7.3-33). If the space is not used for storage, unfaced batts or blankets can be laid directly on the floor (Fig. 7.3-34). If there is a possibility that the space will be finished and heated in the future, it may be preferable to attach the insulation to the sloping rafters.

FIGURE 7.3-33 Cavities should not be overfilled to the extent that the insulation is compacted when the floor boards are reinstalled.

FIGURE 7.3-34 Insulation over an existing floor should be fitted snugly together to prevent thermal leaks.

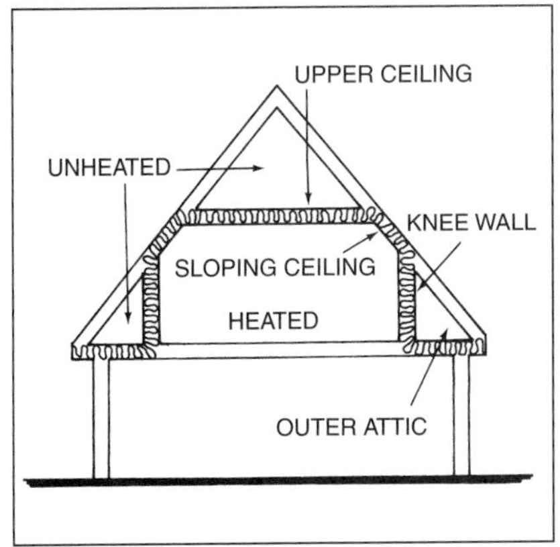

FIGURE 7.3-35 A layer of insulation should be provided between heated or air conditioned spaces and unheated or unconditioned spaces.

FINISHED ATTICS

A finished attic usually has three parts that may need insulation: upper ceiling, sloping ceiling, and outer attic (Fig. 7.3-35).

Upper ceilings can be insulated like unfloored attics, with either blown loose-fill insulation or with flexible batts or blankets if the space is accessible.

Proprietary systems are also available that provide insulation and ceiling finish in one integral panel. In such systems, mineral wool batts or blankets are attached to the back side of fiberboard ceiling panels and are fastened to the ceiling with metal grids. They typically require 3 in. (76.2 mm) of ceiling clearance and provide an R-value of about 12.

Sloping ceilings can be insulated if either the outer attic or upper attic is accessible to permit stuffing the rafter cavities with loose fill or batts or blankets. It may also be possible to install the thermal ceiling panels described above. Such panels can be applied to the face of an existing sloping or cathedral ceiling if the panel thickness does not reduce room dimensions below acceptable levels.

Outer attics can be insulated either at the knee wall and outer attic floor or in the sloping rafters. It is generally preferable to insulate the knee wall and outer attic floor when there are eave vents, or eave vents can be installed, and if the outer attic is sealed from the living space and is not to be used for storage.

7.4 Exterior Insulation and Finish Systems

Exterior insulation and finish systems (EIFS) are often called synthetic stucco, because they look like stucco. Most of them consist of a layer of a reinforced finish (lamina) over a layer of insulation, which is fastened to a layer of sheathing applied to furring or framing. Other configurations are possible, however, and many have been used since EIFS were introduced into this country from Europe in the 1960s. EIFS are available as either field-applied systems or prefabricated panels.

There is no standard for EIFS published by a nationally recognized standards-setting body such as ASTM or ANSI. Therefore, the only overall standards for them are those published by the EIFS Industry Members Association (EIMA). These are listed in Section 7.13.

7.4.1 SYSTEM CLASSES

EIFS are classed by EIMA as either Class PB or Class PM. Systems of both classes consist of a textured protective finish, a reinforced base coat, a layer of insulation, a layer of sheathing, and necessary accessories. The configuration and materials are slightly different, however (Figs. 7.4-1 and 7.4-2).

7.4.1.1 Finish Coats

Both classes use either a polymer-based or polymer-modified finish coat. There are many colors and texture combinations available. Class PB finishes are, however, softer and more flexible than PM finish coatings. For this reason, and because of the difference in their insulation material, PB systems are sometimes called *softcoat* systems and Class PM systems are called *hardcoat* systems. There is no mention in the EIMA documents about the composition of the finish coat. Some coating mixes contain random fibers, others do not. Some contain sand.

Some finish coat materials may re-emulsify if allowed to become saturated or to stand in free water. Therefore, horizontal surfaces should be sloped to provide adequate drainage.

7.4.1.2 Base Coats

In PB systems the base coat varies from $\frac{1}{16}$ in. (1.6 mm) to $\frac{1}{4}$ in. (6.4 mm), depending on the manufacturer. It is typically reinforced with a glass fiber mesh embedded during installation.

In Class PM systems base coats vary from $\frac{1}{4}$ in. (6.4 mm) to $\frac{3}{8}$ in. (9.5 mm), depending on the number of layers of reinforcement. A typical method of reinforcing a base coat for Class PM systems is to fasten a layer of wire lath or glass

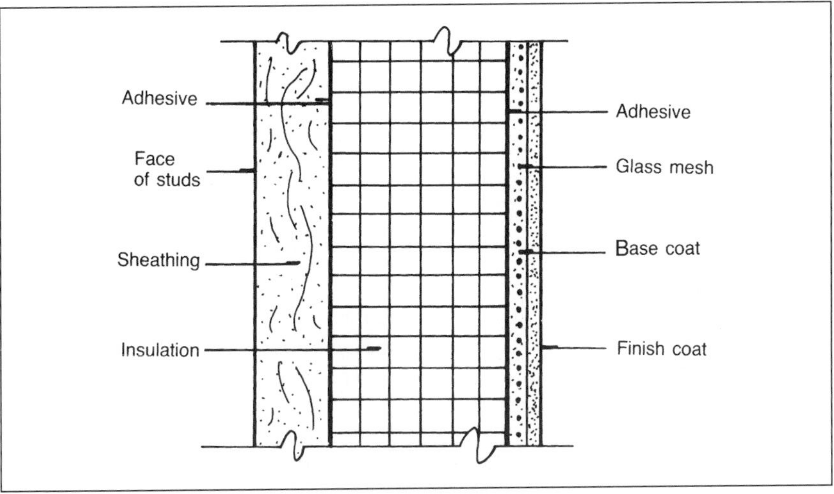

FIGURE 7.4-1 Typical Class PB EIFS. (Courtesy H. Leslie Simmons.)

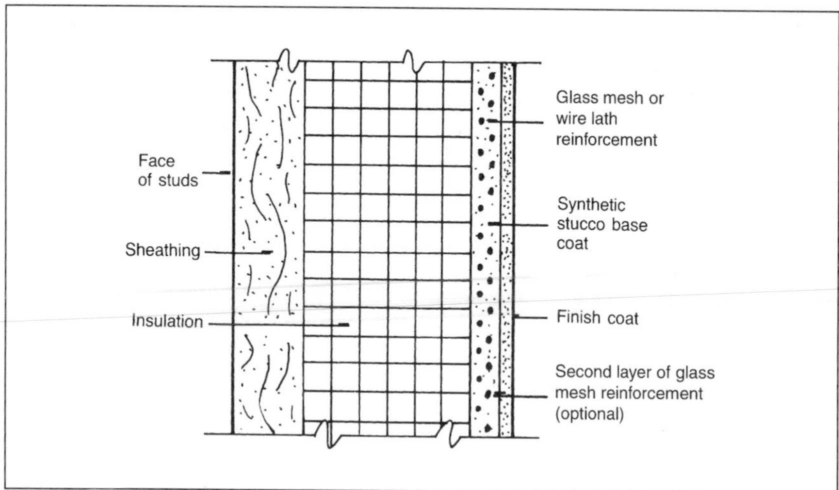

FIGURE 7.4-2 Typical Class PM EIFS. (Courtesy H. Leslie Simmons.)

fiber mesh to the insulation. The base coat is then applied over the reinforcement. In some, but not all, systems, another layer of reinforcement (usually glass fiber mesh) is embedded in the base coat as it is applied. Different weights of glass fiber mesh are selected according to the abuse expected for different segments of an installation.

Additional reinforcement should be provided at the corners of openings, such as windows and doors, in both classes.

There is no mention in the EIMA documents about the composition of a base coat.

7.4.1.3 Insulation

The type of insulation varies with the manufacturer. Most Class PB systems use molded rigid cellular polystyrene. Most Class PM systems contain extruded rigid cellular polystyrene, which is more expensive, less vapor permeable, and denser than the molded product. However, some Class PM systems have molded polystyrene.

The insulation of some Class PB systems is mechanically attached, and with others it is attached both mechanically and with adhesives. However, the insulation in Class PM systems is always mechanically fastened.

7.4.1.4 Sheathing

Failures and fear of failures have created some controversy about the type of sheathing that should be used in EIFS. The traditional sheathing for both Class PB and Class PM systems has been gypsum, because it is relatively inexpensive, fire resistant, light, and thin. Unfortunately, failures have occurred in adhesively applied EIFS that relate to the adhesive interface with the sheathing. EIFS manufacturers claim that these failures are small in number and are due to failures in the adhesives themselves.

Gypsum sheathing represents a small portion of the cost of an EIFS wall system, but some gypsum products manufacturers and the Gypsum Association, apparently fearing that such systems will fail because the paper facing delaminates from the core of their sheathing when used in these systems, have taken the position that they will not be responsible for the use of their products in adhesively attached EIFS systems. One manufacturer has specifically stated that its gypsum sheathing is not endorsed for use in EIFS other than with insulation that is mechanically fastened through the sheathing into the support system. The Gypsum Association's GA 252 states that gypsum sheathing can be used with EIFS having either adhesive or mechanical fastening of the insulation, but that the "performance of these systems and recommendation of the proper method of attachment are the sole responsibility of the EIFS manufacturer."

Other types of sheathing can also be used in EIFS, but they are more expensive than gypsum sheathing and are not likely to replace the gypsum product in spite of the controversy.

The sheathing used should be selected, installed, and protected as recommended by its manufacturer and the manufacturer of the EIFS. Conflicts between the two should be resolved before a system is installed.

7.4.2 CHARACTERISTICS OF EIFS

The two basic EIFS (PB and PM) vary in their appearance and performance characteristics. Failures may occur for several reasons related to failing to recognize the characteristics of the system being used and to accommodate them in the design and installation of the system.

7.4.2.1 Vapor Permeability and Condensation

Both Classes of EIFS have protective coatings that are vapor permeable. However, this permeability is reduced considerably in systems with a second layer of reinforcement added to increase impact resistance. EIFS systems should always have a complete vapor retarder on the interior. Vapor that enters the system in spite of the retarder can cause the coating to separate from the insulation, the insulation to separate from the sheathing, or the sheathing to deteriorate. Condensation can also cause loss of thermal efficiency if it saturates the insulation either in the EIFS or in the wall behind it.

Both types of polystyrene insulation act as a vapor retarder. However, extruded polystyrene is a much better vapor retarder than the molded product. In addition, the extruded material has a much lower rate of absorption, which may help reduce loss of insulation effectiveness in the event that condensation does occur.

To help prevent condensation from forming, the dew point should be calculated to fall within the cellular plastic board insulation in an EIFS.

7.4.2.2 Cracking

The finish coats on Class PM systems are comparatively brittle and the insulation usually employed (extruded polystyrene) more rigid than the materials used in a typical Class PB system. Therefore, Class PM systems must be divided into panels by control joints, much as stucco does. Class PB systems do not require such division and need joints only where expansion or control joints are needed in the substrates.

7.4.3 SELECTION OF A SYSTEM

The building, including its construction, use, environment, and construction budget, should be analyzed to determine the most effective EIFS to use. EIFS are proprietary products, and each is different from the next, even when in the same class. It is necessary to either name a specific product and design around it, or specify the required characteristics. The physical properties used include a system's accelerated weathering characteristics, wind-driven-rain resistance, salt-spray resistance, absorption-freeze resistance, mildew resistance, abrasion resistance, water vapor transmission, impact resistance, and negative wind load performance.

7.5 Low-Slope Roofing

This section discusses three types of roofing used on low-slope roofs: built-up bituminous roofing, bituminous roll roofing, and single-ply roofing. In general, *low-slope roofs* are those whose slope is less than 1½ in./ft. (125 mm/m). However, mineral-surfaced roll roofing can be used on higher slopes.

Both built-up roofing and single-ply roofing form a waterproof barrier. Roll roofing sheds water in a manner analogous to shingles, but employs sheet materials instead of small individual units and incorporates bituminous materials between the sheets. Sheet materials are discussed in this section; shingles in Section 7.6.

Low-slope roofing is a system of compatible components selected to work together to form a waterproof membrane. Each component has a unique function. Possible components include a vapor retarder, underlayment, insulation, waterproof membrane, and ballast.

7.5.1 ROOF CONFIGURATIONS

The three typical roof configurations are called *insulated framed, compact conventional,* and *protected membrane.*

Most low-slope roofing systems are insulated. In a *framed system,* the insulation is separated from the roofing by the roof deck or sheathing. A system in which a ventilated air space is provided between the insulation and the deck or sheathing is called a *ventilated system* (see Fig. 7.1-24).

In *compact systems* (see Fig. 7.1-27) the insulation, membrane, and vapor retarder, if there is one (see Section 7.1), are all in contact. The conventional order of components in a compact system, starting from the exterior, are ballast, membrane, insulation, vapor retarder (if there is one), insulation, and roof deck. In some systems, the ballast is replaced by another surfacing material.

In a *protected membrane* (PMR) *system,* the order is ballast, pervious fabric, insulation, membrane, underlayment, and deck (see Fig. 7.1-29). The membrane serves as a vapor retarder. Ballast must be used to hold the insulation in place.

A system in which there is insulation both above and below the roof deck is called a *combination framed and compact system.*

7.5.2 WARRANTIES AND MAINTENANCE AGREEMENTS

Two types of warranties are issued covering low-slope roofing: a manufacturer's warranty, or guarantee, and a roofing contractor's warranty and maintenance agreement. There is considerable controversy about the value of both.

Some experts believe that some warranties issued by roofing manufacturers limit rights an owner would otherwise have under current law, rather than expand the owner's protection.

The effectiveness of contractors' warranties and maintenance agreements is only as good as the organizations that sign them. In addition, some standard warranty agreements are biased toward the roofing contractor, offering it more protection than is granted to the building owner. A normal period of such warranties is 2 years, because about 65% of roofing problems occur during that period.

The purpose of a maintenance agreement is to ensure that the roof is properly maintained and repaired if a problem appears. Such agreements usually cover about a 5-year period and are usually renewable.

To be more sure of proper protection, a building owner should seek legal council before agreeing to any roofing warranty or maintenance agreement.

7.5.3 SUBSTRATES

Proper functioning of low-slope roofing is dependent not only on the roofing material and its application, but on the roof structure as well. This structure must be sound and smooth, have sufficient slope to prevent ponding of water on the roof, and be designed to support the loads that will be imposed. There must be enough roof drains of the proper size to remove water from the roof.

There are essentially three types of substrates used beneath built-up bituminous and single-ply roofs: nailable, insulated, and concrete.

7.5.3.1 Nailable Substrates

Nailable decks include wood planks, plywood, cement-wood fiber planks, lightweight insulating concrete, poured-in-place gypsum, and precast gypsum planks.

7.5.3.2 Insulated Substrates

Insulated decks are covered with one or more layers of insulating material. The underlying deck may be nailable, metal, or concrete. Insulation may be boards, sprayed-in-place plastic foams, or a thermosetting insulating fill. Steel decks are covered with a layer of insulation (usually boards), even when the insulation is not necessary for thermal purposes, because low-slope roofing materials are not designed to span the flutes in steel decking.

7.5.3.3 Concrete Substrates

Concrete decks may be either poured-in-place or precast.

7.5.4 SELECTING SYSTEMS

Important factors in selecting a low-slope roofing system include the configuration of the system and the components used in it. Two overriding selection criteria are (1) compliance with the applicable codes and (2) compliance with the applicable insurance requirements.

The decision about whether to use a built-up roofing system or a single-ply system is influenced by the advantages of each in the particular circumstances. For

FIGURE 7.5-1 Selection Criteria for Low-Slope Roofing Type

Selection Item	Criteria
Overriding criteria	Code compliance; insurance requirements
Built-up vs. single-ply	Climate; availability of competent installers; presence of existing roofing; condition of the substrates
Configuration	Building's design, construction, and location; energy use; construction budget; aesthetics; code compliance; insurance requirements
Vapor retarder	Climate; roofing configuration; insulation material; roofing membrane material
Insulation type	Code compliance; climate; desired thermal characteristics; economy; roofing system configuration; fire resistance; traffic indentation resistance; presence of a vapor retarder; use of hot-applied or solvent-applied roofing membrane; roofing materials used

example, single-ply roofing can be installed in a much wider range of temperatures than can a built-up bituminous system. The availability of competent applicators is another factor. The success of a single-ply roofing system in preventing leaks and other problems depends to a much larger extent on the ability of the installer than is true for conventional built-up roofing. The sealing of single-ply roofing seams must be perfect. Even a slight error will usually result in a leak. In addition, roof decks that will receive single-ply membranes should be sloped to provide positive drainage, because they are susceptible to damage from standing water and ice. Some single-ply systems can be installed directly over existing roofing.

Figure 7.5-1 is a summary of criteria for selecting low-slope roofing types.

7.5.4.1 Configuration

Which components a roofing system should contain depends on the building's design, construction, and location. Other factors influencing this decision include energy considerations, budget considerations, aesthetics, and codes. For example, the applicable code may dictate the type of insulation that can be used in contact with a steel or wood deck.

Whether a vapor retarder is required depends on local weather conditions and the roof configuration selected. Review Section 7.1 for a discussion of vapor retarders in roofing assemblies.

7.5.5 ROOF INSULATION

7.5.5.1 Materials

The type of roof insulation selected depends on applicable code requirements, the thermal characteristics desired, the selected roofing system configuration, economical considerations, fire resistance, traffic-indentation resistance, weight, and available insulation panel dimensions. Different insulation materials may be selected according to whether the material should be vapor permeable or relatively impermeable. This may depend on whether the selected system contains a vapor retarder. If a vapor retarder is included, the insulation should not be impermeable, because there would then be two vapor retarders, which could trap moisture between them.

Hot-applied built-up roofing used directly over polyisocyanurate, polyurethane, and phenolic foam insulations sometimes leads to delamination and blistering problems. Therefore, the NRCA (National Roofing Contractors Association) recommends that a layer of wood fiber, perlite, or glass fiber board insulation be placed over these plastic foam insulations when a hot-applied built-up roofing system will be used.

Insulation beneath loose-laid single-ply roofing can be installed without adhesives, and the roofing membrane is not attached using adhesives. Thereby, compatibility problems are reduced and the selection of usable insulation types is much wider.

Some roofing materials are not compatible with certain insulation products. Therefore, the roofing membrane manufacturer's recommendations must be followed when selecting roof insulation that will be in contact with roofing membranes. This is especially true for roofing membranes that are installed using the adhered method discussed in Section 7.5.10.

7.5.5.2 Application of Insulation

Most roof insulation should be applied in at least two layers with the joints staggered in the two layers, in order to reduce thermal bridges through the joints, to reduce ridging that sometimes occurs in single-layer applications, and to make mechanical attachment easier (Fig. 7.5-2).

Beneath built-up and adhered single-ply systems, insulation should be mechanically fastened to nailable decks; however, it may be laid in strips of hot bitumen when a vapor retarder has been installed over the deck.

Over a steel deck, single-layer insulation and the first layer of two-layer board insulation beneath built-up or adhered single-ply roofing should be mechanically fastened in place (Figs. 7.5-3 and 7.5-4). When fasteners are not used, the boards should be bedded in strips of hot bitumen laid along the deck flutes (Fig. 7.5-5). Boards should be placed with the long di-

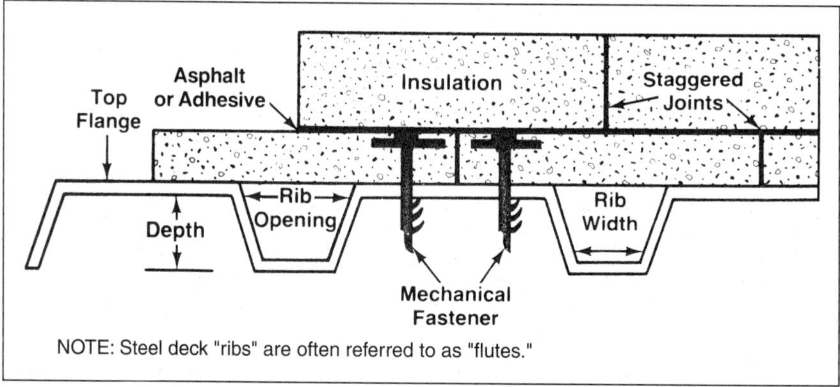

FIGURE 7.5-2 Typical insulation over steel deck. Vapor retarder is not always present. (Reprint courtesy National Roofing Contractors Association.)

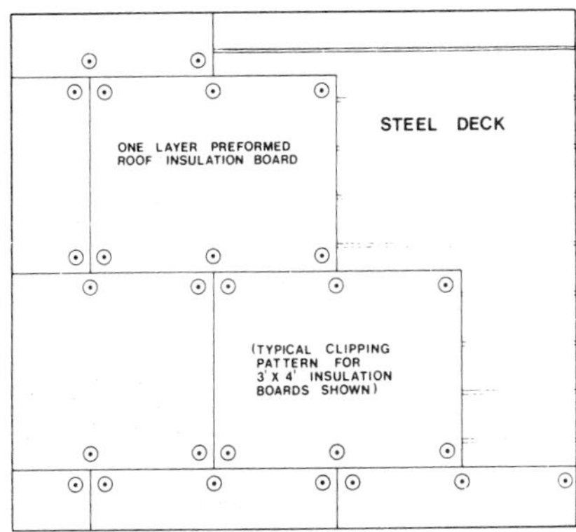

FIGURE 7.5-3 Single-layer insulation over steel deck. (Reprint courtesy National Roofing Contractors Association.)

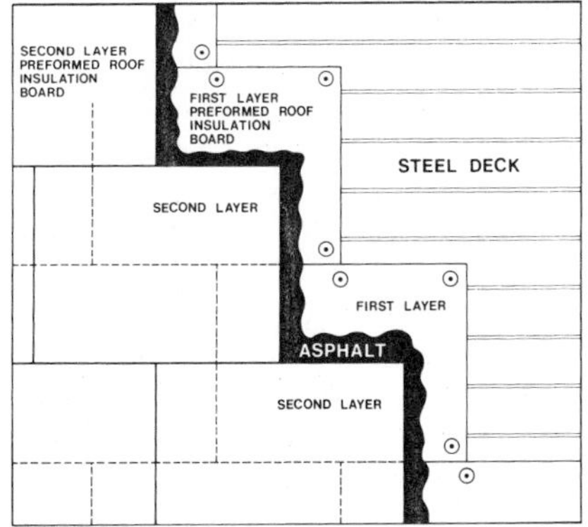

FIGURE 7.5-4 Two-layer insulation over steel deck. (Reprint courtesy National Roofing Contractors Association.)

FIGURE 7.5-5 Placing strips of bitumen on flutes of steel deck. (Reprint courtesy National Roofing Contractors Association.)

mension parallel to the deck flutes and perpendicular to the roof slope (see Fig. 7.5-6).

Board insulation beneath built-up or adhered single-ply roofing should be applied over nonnailable decks in hot bitumen or adhesive.

FIGURE 7.5-6 Laying insulation boards over steel deck in bitumen strips. (Reprint courtesy National Roofing Contractors Association.)

Board insulation may be loose laid beneath loose-laid single-ply systems. The second layer of board insulation should be laid in a full mopping of bitumen (see Fig. 7.5-4).

7.5.6 FLASHING

Flashing is necessary in roofing systems to seal the gaps between the roofing and adjoining materials, such as walls, pipes, and other penetrations. Flashing may be one of several products, including metal, fabric, and plastics.

Flashing associated with low-slope roofing systems should be a part of the roofing. The materials selected should be compatible with the roofing both chemically and in thermal expansion characteristics. This is especially true when there is a large differential in the coefficient of expansion of the roofing material and the flashing. Therefore, metal flashing is not appropriate as a base flashing with most roofing materials. Conditions where metal flashings, such as gravel stops, are built into roofing membranes are particularly susceptible to failure.

Refer to Section 7.8 for a detailed discussion of flashings.

7.5.7 BITUMINOUS MEMBRANE ROOFING PRODUCTS

Four basic materials are used in the manufacture of bituminous roofing products: (1) a reinforcing base, which is either organic dry felt or glass fiber mat, (2) bitumen (asphalt or coal tar), (3) mineral stabilizers, and (4) either fine or coarse surfacings.

In general, the manufacture of organic felt bituminous roofing products includes processing cellulose fibers into dry felt, saturating and coating the felt with asphalt or coal tar, and then, depending on the finished product, surfacing the coated felt with selected mineral aggregates. When glass fiber mat is used as a reinforcing base instead of organic felt, the mat goes directly from a dry looper into a coater, bypassing the saturating process.

7.5.7.1 Raw Materials

The raw materials and some of the processing steps required to manufacture finished products are shown diagrammatically in Figure 7.5-7.

DRY ORGANIC FELT

Dry organic felt, used as a base for making reinforcing felts for built-up roofing and waterproofing, underlayment, smooth and mineral-surfaced roll roofing, and asphalt shingles, is made from combinations of cellulose fibers such as those derived from rags, paper, and wood on a machine similar to the type used for making paper (Fig. 7.5-8). The fibers are prepared by various pulping methods, then blended and proportioned to produce felt with the necessary weight, tensile strength, absorptive capacity, and flexibility required to make a suitable roofing product. The felt must be able to absorb 1½ to 2 times its weight in asphalt saturants and be strong and flexible enough to withstand strains placed on it during the manufacturing process.

As the felt comes off the end of the machine in a continuous sheet, it is cut into specified widths and wound into individual rolls from 4 to 6 ft. (1219 to 1829 mm) in diameter, weighing up to a ton or more.

GLASS FIBER MAT

Dry glass fiber mats are used as a base for reinforcing felts for built-up roofing and waterproofing, coated smooth rolls and mineral-surfaced roll roofing, and asphalt shingles. To make glass fiber mats, sand, soda ash, and limestone are first combined to make chopped strand glass fibers. These fibers are then mixed with a binding agent and cured to produce a mat with

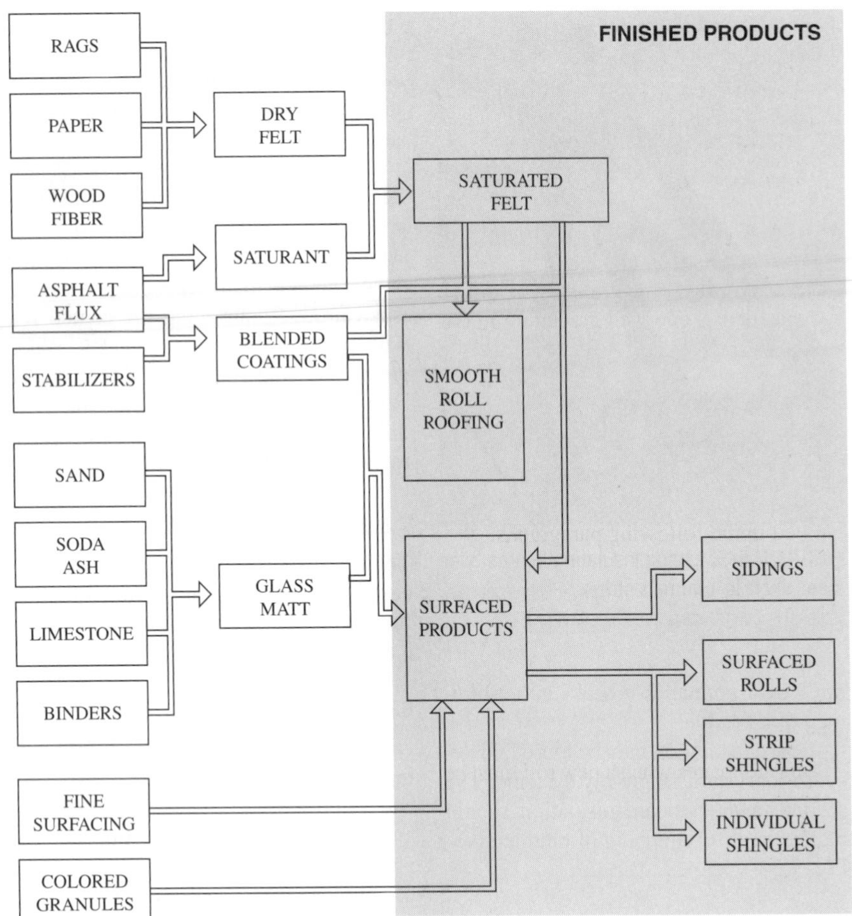

FIGURE 7.5-7 Raw materials and processing steps used to make asphalt roofing products. (Courtesy Asphalt Roofing Manufacturers Association.)

the thickness, tensile strength, tear strength, weight, and flexibility required to produce roofing and waterproofing products.

BITUMENS

The bitumens used in bituminous roofing products are asphalt and coal tar. They are used to saturate and coat dry organic felts and to coat glass fiber mats in the manufacture of roofing felt and to weld or fuse roofing or waterproofing felts together to form a built-up roof or waterproofing membrane. Both asphalt and coal-tar bitumens are thermoplastic, which means that they become more fluid when heated and return to their former solid state when they cool.

A bitumen's equiviscous temperature (EVT) is the optimum temperature at which it should be used. Each asphalt and coal-tar bitumen has its own EVT. The type with the lowest number has the lowest EVT.

Asphalt Asphalt used to make roofing and waterproofing products, known as asphalt flux, is a petroleum product obtained from the fractional distillation of crude oil. Asphalt flux is processed to produce roofing grades of asphalt called saturants or, when combined with mineral stabilizers, coating asphalts. Both forms combine with dry felt to make asphalt roofing or waterproofing membranes.

The four types of asphalt used in built-up roofing are classified in ASTM D 312 as Type I, *dead level asphalt*; Type II, *flat asphalt*; Type III, *steep asphalt*; and Type IV, *special steep asphalt*. Type I is limited to slopes of ½ in./ft. (41.67 mm/m) and lower; Type II may be used on slopes up to 1½ in./ft. (125 mm/m); Type III up to 3 in./ft. (250 mm/m); and Type IV up to 6 in./ft. (500 mm/m). However, steep asphalts are not as durable or as easy to apply as the lower-slope types and should be used only when the slope is steeper.

The preservative and waterproofing characteristics of asphalt reside largely in certain oily constituents. In making roofing products, the body of the highly absorbent felt sheet is first impregnated (saturated) to the greatest possible extent with saturants that are oil-rich asphalts. The saturant is then sealed in with an application of a harder, more viscous coating asphalt, which in turn can be further protected by a covering of opaque mineral granules. A primary difference between saturants and coating asphalts is the temperature at which they soften. The softening point of saturants varies from 100 to 160°F (37.8 to 71.1°C); that of coatings is as high as 260°F (126.7°C). Saturant asphalt is not used with glass fiber mat. Rather, coating mixtures are used to completely surround the glass fibers and to provide a layer of coating on both sides of the mat.

Asphalt flux is also the base material used to make asphalt plastic cement, quick-setting roof adhesives, asphalt primers, other roof coatings, and adhesives.

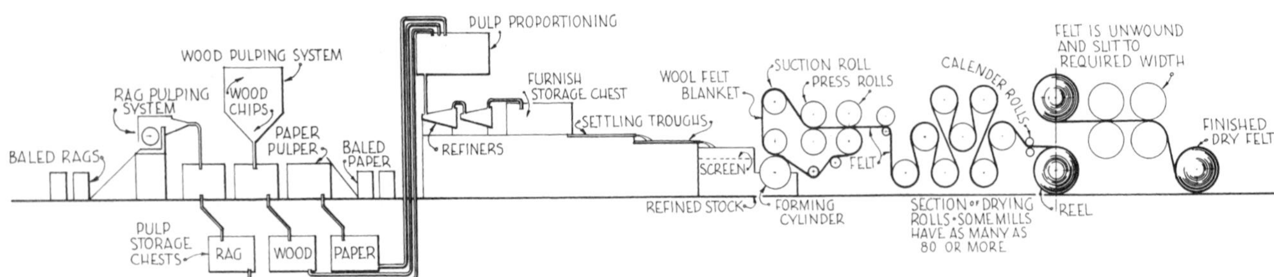

FIGURE 7.5-8 Flow diagram of a typical felt mill. (Courtesy Asphalt Roofing Manufacturers Association.)

Coal Tar Coal tar used to make roofing products is a distillation of bituminous coal. It is considered basically superior to asphalt for use in low-slope roofing. It is most effective at a slope of ¼ in./ft. (20.83 mm/m) and can be used on slopes up to ½ in./ft. (41.67 mm/m), but cannot be used on slopes steeper than ½ in./ft. (41.67 mm/m).

The two types of coal tar recommended for use in built-up roofing are classified in ASTM D 450 as Type I, *coal-tar pitch*, and Type III, *coal-tar bitumen*. Type III has a higher softening point than does Type I.

MINERAL STABILIZERS

Finely ground minerals, such as silica, slate dust, talc, micaceous materials, dolomite, and trap rock, which are called stabilizers, when combined with coating bitumens, control hardness, elasticity, adhesion, and weathering. Coating bitumens containing stabilizers resist weather better, are more shatter- and shockproof in cold weather, and significantly increase roofing product life.

SURFACINGS

Bituminous roofing product surfaces may be either smooth or surfaced with coarse mineral granules.

Finely Ground Minerals Powders, usually talc and mica, are dusted on the surface of smooth roll roofings and the backs of mineral-surfaced roll roofings and shingles to prevent layers in packages from sticking together. They are not a permanent part of the finished product and gradually disappear from exposed surfaces after application.

Coarse Mineral Granules Coarse granules are usually colored slate or rock granules (either natural or colored by a ceramic process) that are used on shingles and on certain roll products. These mineral-surfaced products have increased weather and fire resistance and provide a wide range of colors and color blends. To protect the underlying bitumen from the impact of light rays, the granules should be opaque, dense, and well graded for maximum coverage.

7.5.7.2 Manufacture of Roofing

Raw materials are processed into finished roofing products in a continuous sequence (Fig. 7.5-9). The principal steps are described in the following paragraphs.

DRY LOOPER

A roll of dry felt or glass fiber mat installed on a felt reel is unwound onto a dry looper. This looper provides a felt material reservoir that can be drawn on by the machine as circumstances demand, eliminating stoppages when a new roll must be put on the felt reel or when imperfections in the felt must be cut out.

SATURATION OF FELT

The felt is then subjected to a hot saturating process, which eliminates moisture and fills the felt fibers and intervening spaces as completely as possible with saturant. Glass fiber mat does not undergo this saturating process. Rather, coating bitumen is used to fill the spaces between the glass fibers.

WET LOOPER

Following the saturating process, excess saturant usually remains on the surface of the sheet. Therefore, the sheet is held for a time on a wet looper so that natural shrinkage of the bitumen upon cooling will draw the excess into the felt, resulting in a high degree of saturation. At this point, uncoated products, such as saturated underlayment felt, move directly to the cooling looper and onto the winding mandrel for final packaging.

COATER

Material to be coated continues to the coater, where coating bitumen is applied to both the top and bottom surfaces. Coating thickness is regulated by rollers, which can be brought together to reduce the amount of coating or separated to increase its thickness.

The final weight of the product is controlled by the roofing machine operator. Machines are often equipped with automatic scales that weigh the sheets during manufacture to warn the operator when the material is running over or under weight specifications. When smooth roll roofing is being made, talc or mica is applied to both sides of the sheet by spreading and pressing it through a pressure roll.

MINERAL SURFACING

When shingles and other mineral-surfaced products are made, granules of specified color or color combinations are applied from a hopper and spread thickly on the hot coating bitumen, and talc or mica is applied to the back (smooth) side. The sheet is then run through a series of press and cooling rollers, subjecting the sheet to a controlled pressure that embeds the granules in the coating to the desired depth.

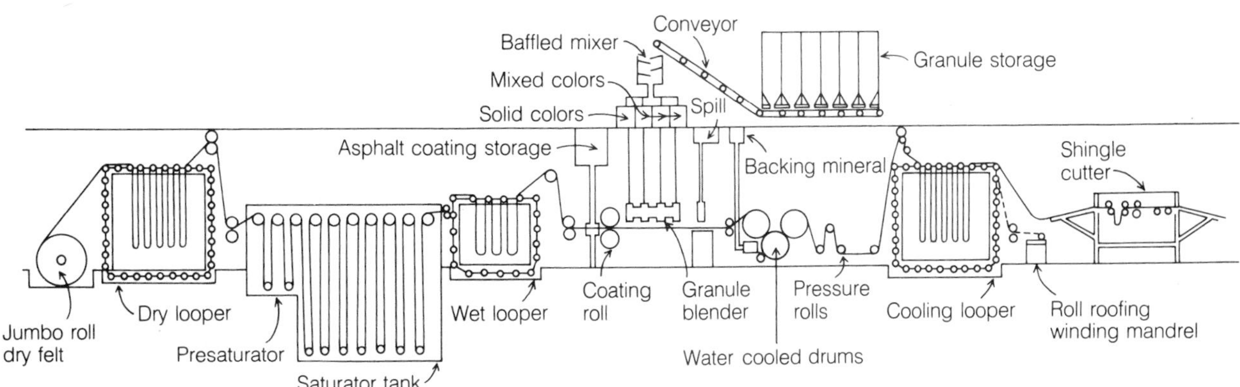

FIGURE 7.5-9 Asphalt roofing products are produced in a continuous process as shown in this diagram. (Courtesy Asphalt Roofing Manufacturers Association.)

TEXTURE

At this stage some products are textured by the pressure of an embossing roller, which forms a pattern in the sheet's surface.

COOLING LOOPER

The sheet then travels into a cooling or finish looper, where it is cooled so it can be cut to length or into shingles and packed.

ROLL ROOFING WINDER

When roll roofing is made, a sheet is drawn directly from the finish looper to a winding mandrel. Here it is wound, and the material is measured to length as it turns. After sufficient material accumulates, it is cut off, removed from the mandrel, and wrapped for warehousing and shipment.

CARE OF MATERIALS

The bituminous products discussed here provide a waterproof membrane after application, but can be harmed if stored outside without protection, particularly if they are applied wet. The following simple rules should be observed in storing finished products in the warehouse and on the job to ensure suitable performance:

- Store roll goods on end and upright; if stored in tiers, place boards between the tiers to prevent damage to the ends of rolls.
- Do not store shingle bundles to a height of more than 4 ft. (1219 mm) to eliminate the possibility of sticking or discoloration.
- Rotate warehouse stock, particularly white and pastel roofing, by moving out of storage first the material held longest.
- Cover bituminous roofing and waterproofing materials stored outside for weather protection.

7.5.7.3 Classification of Products

Bituminous roofing products include (1) bitumens, (2) asphalt coatings and cements, (3) sheet products made on an organic felt base, and (4) sheet products made on a glass fiber mat base.

COATINGS AND CEMENTS

Coatings and cements are bituminous materials combined with special ingredients and either (1) mixed with cut-back type solvents or (2) emulsified in water. These materials are processed to various consistencies depending on their use. They are flammable and should never be warmed over an open fire or placed in direct contact with a hot surface. They may be softened by placing the container, unopened, in hot water or by storage in a warm place.

Cut-back asphalt coatings and cements should be applied to a dry, clean surface and troweled or brushed vigorously to force the material into cracks and openings and to eliminate air bubbles. Emulsions may be applied to damp or wet surfaces, but should never be applied in an exposed location if rain is anticipated within 24 hours.

These materials may be classified into several groups, including plastic roofing cements, flashing cements, lap cements, quick-setting adhesives, roof coatings, asphalt water emulsions, primers, and roofing tape. These may be used in accordance with information provided in Figure 7.5-10.

PLASTIC ROOFING CEMENT (ROOFING MASTIC) AND FLASHING CEMENT

These materials are processed with mineral fibers and cut-back asphalt or coal tar, which are especially plastic and elastic, combined with nonoxidizing minerals, oils, and solvents to produce black plastic repair and flashing cements. They will not flow at summer temperatures nor become brittle at low temperatures. They are sufficiently elastic after setting to compensate for normal roof deck expansion and contraction and movement between the deck and other structural elements. They have excellent acid and alkali resistance, will not absorb water, and provide good adhesion to all surfaces. These materials are similar but not interchangeable. Flashing cement is the stiffer of the two because it has more and longer fibers. It should be used for vertical applications and on cants. Plastic roofing cement is

FIGURE 7.5-10 Uses and Properties of Asphalt Coatings and Cement Products

Product	Uses	Properties
Plastic roofing cement	Flashing assemblies; patching; roof repairing; caulking	Remains elastic after setting; will not flow at summer temperatures or become brittle at low temperatures; excellent resistance to acids and alkalies; will not absorb water; good adhesion to all surfaces
Lap cement	Watertight bond between lapping elements of roll roofing	Not as thick and heavy as plastic cements
Quick-setting adhesive	Cementing down the free tabs of strip shingles and laps of roll roofing	Available in brush, trowel, or gun consistency; same consistency as plastic asphalt cement, but more adhesive; sets quickly
Roof coating	Resurfacing old roll roofing and built-up asphalt roofs	Thin enough to be applied with spray or brush
Asphalt water emulsion	Roof coating	Must be kept from freezing and applied when no rain is forecast for 24 hours
Primer	Preparing masonry surfaces to receive other bituminous products; ensures good bond between the masonry and built-up roofing, plastic cement, or roof coatings	Very fluid; applied with brush or spray

(Courtesy Asphalt Roofing Manufacturers Association.)

suitable for horizontal applications in flashing assemblies, patching, and roof repairing.

LAP CEMENTS

Lap cements are of various consistencies but are not as thick and heavy as plastic cements. They should be used as recommended by the manufacturer. Lap cements are intended to make a watertight bond between lapping elements of roll roofing. Lap cement should be spread over the entire lapped area. Nails used to secure the roofing should pass through the cement so that the shank of the nail will be sealed where it penetrates the deck material.

QUICK-SETTING ASPHALT ADHESIVE CEMENT

This material, available in brush, trowel, and gun consistencies, is used for cementing down the free tabs of strip shingles and for laps of roll roofing.

Quick-setting asphalt adhesive cement is of about the same consistency as plastic asphalt cement but is more adhesive. It contains a solvent that evaporates rapidly on exposure, causing the cement to set quickly,

ROOF COATINGS

Roof coatings are of various types and consistencies but are usually thin enough to be applied with a spray or brush. Coatings are used to resurface existing roll roofing and built-up asphalt roofs.

ASPHALT WATER EMULSIONS

Asphalt water emulsions are a special type of roof coating made with asphalt, sometimes combined with other ingredients, and emulsified with water. Emulsions must be stored where they will not freeze and should be used at times when they will not be rained on for at least 24 hours after application.

PRIMERS

Primers are very fluid bituminous materials that are applied with a brush or spray. They must be thin enough to penetrate rapidly into masonry or concrete surface pores without leaving a continuous surface film. Their purpose is to prepare masonry or concrete to receive other bituminous products and ensure a good bond between the masonry or concrete and built-up roofing, plastic cement, or bituminous coatings.

ROOFING TAPE

Some manufacturers make a roofing tape of cotton fabric or other porous material saturated with asphalt or coal tar. It is available in varying widths from 4 to 36 in. (101.6 to 914.4 mm) and is provided in rolls up to 50 yd. (45.72 m) long. It is used with bituminous coatings and plastic flashing cements, and in patching seams, breaks, and holes in bituminous and metal roofing.

7.5.7.4 Products Made on an Organic-Felt or Glass-Fiber-Felt Base

Bituminous roofing and siding products made on an organic-felt or glass-fiber-felt base may be classified broadly into two groups: (1) shingles and (2) rolls.

Sizes, exposures, and other information covering products typical of these two classifications are given in Figure 7.5-11. Products within each group may differ as to weight, dimension, and design. For example, the weight of a three-tab square butt strip shingle ranges from 200 to 300 lb. per square (90.72 to 136.07 kg per square).

Bituminous roll products are usually referred to by a number (No. 15, No. 30, No. 40, etc.). These numbers represent the approximate weight in pounds as shipped of the amount of that material needed to cover a square (100 sq. ft.) (9.29 m^2) with one ply of the material.

BUILT-UP ROOFING FELTS

Built-up roofing felts are broadly called reinforcing felts. They consist of base sheets, ply sheets, and cap and flashing sheets. They may be saturated, saturated and coated, impregnated, or prepared in some special way for a specific purpose.

Dry felts impregnated with asphalt or coal-tar saturants, but otherwise untreated, are called *saturated felts*. They are also used as underlayment for shingles and as sheathing paper. Saturated felts are made in several weights; the most common are No. 15 and No. 30, weighing about 15 and 30 lb. per square (6.8 and 13.6 kg per square), respectively.

Although glass fiber felts, which are impregnated with bitumen but otherwise untreated, may be classified under this broad category, they are not actually saturated, since glass fibers do not absorb bitumen. In reality, the mat is lightly coated so that all fibers are thoroughly surrounded by bitumen.

Saturated and coated sheets are saturated felts that are then surfaced on one or both sides with a fine mineral material to prevent adhesion before application. These are generally used as base sheets in built-up roofing systems.

Impregnated sheets have less coating material than coated sheets, which makes them lighter and able to ventilate vapors during their application.

Prepared roofing materials are saturated and coated or impregnated with fine decorative mineral granules. Some are combination products made of layers of glass fiber felt, organic fiber felt, or both.

Base Sheets Base sheets are used in some, but not all, built-up roofing systems. Usual materials are No. 30 asphalt-saturated organic felt, No. 40 asphalt-saturated and coated organic felt, asphalt glass fiber mat roll roofing, and venting asphalt-saturated and coated inorganic sheet.

Ply Sheets Ply sheets are usually No. 15 asphalt-saturated organic felt, No. 15 coal-tar-saturated organic felts, asphalt-impregnated glass fiber felts Type III, or asphalt-impregnated glass fiber felts Type IV. Type IV glass fiber impregnated sheets are heavier and have a breaking strength roughly twice that of Type III sheets.

Cap and Flashing Sheets Cap and flashing sheets are usually wide-selvage asphalt roll roofing cap sheet surfaced with granules, No. 90 mineral-surfaced organic roll roofing, No. 50 or No. 60 smooth-surfaced asphalt roll roofing, inorganic cap sheet with mineral granules, or reinforced bituminous flashing material.

ROLL ROOFING MATERIALS

Roll roofing and siding are made by adding a more viscous, weather-resistant asphalt coating to a felt that has first been impregnated with a saturant asphalt or to a glass fiber mat. Some roll roofing and all siding are surfaced with mineral granules providing a wide color range, and some styles are furnished in split rolls designed to give an edge pattern when applied to the roof.

7.5.7.5 Standards of Quality

Asphalt sheet roofing should conform to Underwriters Laboratories, Inc. (UL) standards. The UL specifications constitute the recognized standard for the asphalt roofing industry and include pro-

Typical asphalt rolls								
Product	Appx. shipping weight per roll (lbs.)	Appx. shipping weight per square (lbs.)	Squares per pkg.	Width (in.)	Length (ft.)	Selvage (in.)	Exposure (in.)	ASTM fire and wind ratings
Mineral surface roll 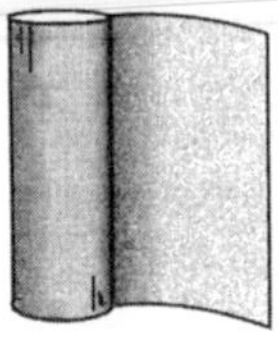	75-90	75-90	1	36-39 3/4	32.7-38	2-4	32-34	Some Class C
Mineral surface roll (double coverage) 	55-70	110-140	1/2	36-39 3/4	32.7-36	19	17	Some Class C
Smooth surface roll 	50-86	40-65	1-2	36-39 3/4	32.7-72	2-4	34-37 3/4	None
Non-perforated felt underlayment 	24-60	6-30	2-8	36	72-288	2-19	17-34	May be a component in a complete fire-rated system. Check with manufacturer for details.
Self-adhered eave and flashing membrane 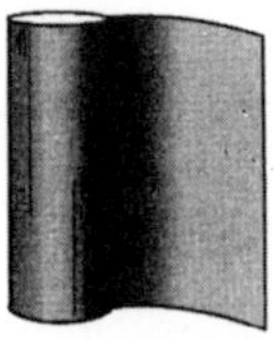	35-82	33-40	1-2 1/4	36	36-75	2-6	34	May be a component in a complete fire-rated system. Check with manufacturer for details.

FIGURE 7.5-11 Summary of typical asphalt roll roofing products. (Courtesy Asphalt Roofing Manufacturers Association.)

visions for (1) fire resistance (Fig. 7.5-12), (2) wind resistance, (3) felt saturation tests to determine saturant quantity and saturation efficiency, (4) coating asphalt thickness and distribution, (5) granule adhesion and distribution, (6) finished product weight, count, size, coloration, and other characteristics before and after packaging, and (7) application instructions.

The UL maintains a check on labeled products by making periodic factory inspections. UL inspectors are afforded the privilege of using the manufacturers' control laboratories and to visit, inspect, and analyze product samples. When necessary, they recommend to the manufacturer measures that are required to maintain the product standards to remain eligible to use the UL label.

Products that conform to UL standards are identified by Underwriters Laboratories' labels.

Application of asphalt roofing systems should follow the specifications of the manufacturer of the system and the recommendations of the National Roofing Contractors Association (NRCA). In addition, metal flashing and accessories should be fabricated and installed in accordance with the recommendations of the Sheet Metal and Air Conditioning Contractors National Association, Inc. (SMACNA). Conflicts in the recommendations of these various parties should be reconciled before installation of a roofing system.

Temperature affects bitumen flow rate. Bitumens that have been heated with too high a temperature flow too readily, which affects the amount of bitumen in a roofing system. They may also separate and leave voids between sheets of membrane, which are potential leak points. Failure to raise the temperature of bitumen high enough can also cause problems. Bitumen that is not sufficiently heated will not flow as it should and may not fuse with and adhere properly to the membranes.

The NRCA recommends that each bitumen package be labeled with its *equiviscous temperature* (EVT) and its *flash point* (FP). A bitumen's EVT is its optimum temperature. A bitumen should never be heated above its flash point, and should be applied at its EVT. In addition, bituminous materials should not be kept at high temperatures or allowed to stand in luggers for prolonged periods.

7.5.8 ROLL ROOFING

7.5.8.1 Materials

Roll roofing may be either mineral surfaced or smooth, but organic felts are not appropriate, because roll roofing is designed to be water-shedding, rather than waterproof, and organic felts tend to absorb water.

One entire side of *standard mineral-surfaced roll roofing* is coated with granules. Most of this material weighs between 75 and 100 lb. per square (100 sq. ft. [9.29 m^2]). This is between 34.02 and 45.36 kg per square. Double-coverage roll roofing is coated with granules for only the first 17 in. (431.8 mm) of a 36-in. (914.4 mm)-wide sheet. The rest is smooth. These sheets weigh between 55 and 70 lb. per square (24.95 and 31.75 kg per square).

7.5.8.2 Application

SHEATHING

Most roll roofing is applied over plywood or wood board sheathing (see Chapter 6), but it can also be applied over most nailable decks.

APPLICATION OF UNDERLAYMENT

Roll roofing requires the application of a continuous underlayment. This is usually No. 15 asphalt-impregnated unperforated roofing felt applied with the long edge parallel with the eaves. On slopes of 4 in./ft. (333.3 mm/m) and steeper, one layer is sufficient. Two layers should be used on lower slopes.

Layers of underlayment should be laid shingle fashion so that they shed water, and they should be nailed using sufficient nails to hold them in place. Ends should be lapped at least 6 in. (152.4 mm). Edge joints should lap at least 2 in. (50.8 mm). The top layer in double-layer applications should lap 19 in. (482.6 mm) over the lower layer. Underlayment should lap over hips and ridges so that at least two layers cover them.

When the mean January temperature is 30°F (−1.1°C) or lower, an ice shield should be applied. When one layer of underlayment is used, this ice shield should consist of an additional layer of No. 15 felt laid and set in mastic from the eave to a line 24 in. (610 mm) inside the inner face of the exterior wall. When two layers of underlayment are used, the ice shield should be formed by applying the underlayment layers in hot asphalt from the eave to a line 24 in. (610 mm) inside the inner face of the exterior wall.

APPLICATION OF DRIP EDGE

A metal drip edge should be placed along rakes and eaves (Fig. 7.5-13). The underlayment should be placed beneath the drip edge at the rakes and above it at the eaves.

APPLICATION OF FLASHING

Flashings at penetrations are usually metal. Base flashings at walls, chimneys, curbs, and the like are usually bituminous, but may also be metal. Valley flashings may be either bituminous or metal. Metal flashings are discussed in Section 7.8.

USE OF FASTENERS

Roll roofing is usually applied using 11- or 12-gauge (3.1 and 2.7 mm) annular or

FIGURE 7.5-12 Classifications Established by Underwriters Laboratories, Inc. Standard UL 790, "Test Methods for Fire Resistance of Roof Covering Materials"

Class A	Includes roof coverings that are effective against severe fire exposures. Under such exposures, roof coverings of this class are not readily flammable and do not carry or communicate fire; afford a fairly high degree of fire protection to the roof deck; do not slip from position; possess no flying brand hazard; and do not require frequent repairs to maintain their fire-resisting properties.
Class B	Includes roof coverings that are effective against moderate fire exposures. Under such exposures, roof coverings of this class are not readily flammable and do not readily carry or communicate fire; afford a moderate degree of fire protection to the roof deck; do not slip from position; possess no flying brand hazard. May require infrequent repairs to maintain their fire-resisting properties.
Class C	Includes roof coverings that are effective against light fire exposure. Under such exposures, roof coverings of this class are not readily flammable and do not readily carry or communicate fire; afford at least a slight degree of fire protection to the roof deck; do not slip from position; possess no flying brand hazard. May require occasional repairs or renewals to maintain their fire-resisting properties.

(Courtesy Asphalt Roofing Manufacturers Association.)

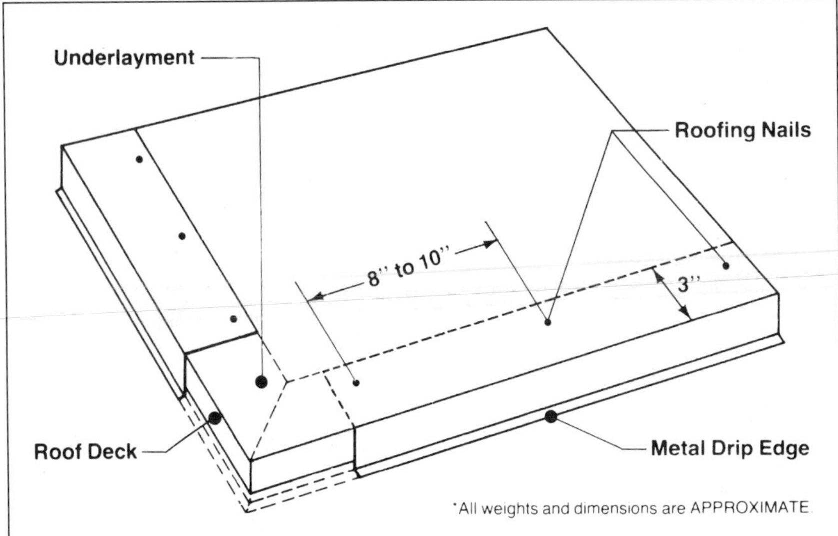

FIGURE 7.5-13 Metal drip edge at eaves and rake beneath roll roofing. (Reprint courtesy National Roofing Contractors Association.)

screw-threaded, galvanized steel or aluminum roofing nails, having $\frac{3}{8}$-in. (9.5 mm)-diameter heads, and lengths between $\frac{7}{8}$ in. (22.2 mm) and $1\frac{1}{4}$ in. (31.8 mm). Nails should penetrate at least $\frac{3}{4}$ in. (19.1 mm) into the underlying nailable deck or sheathing.

APPLICATION OF ROOFING

Roll roofing sheets may be placed either parallel or perpendicular to the eaves. When perpendicular, the sheets should be tilted toward the rake at a rate of $\frac{1}{8}$ in./ft. (10.42 mm/m) of rake.

Standard Mineral-Surfaced Roll Roofing Sheets may be applied with either concealed or exposed nails. The concealed nailing method (Fig. 7.5-14) should be used (1) when the sheets are applied parallel with the eaves and (2) when the slope is 3 in./ft. (250 mm/m) or more, even when the sheets are laid perpendicular to the eaves.

The exposed nailing system is limited to (1) slopes exceeding 2 in./ft. (166.7 mm/m) when the roofing is applied parallel with the eaves (Fig. 7.5-15), and (2) slopes 4 in./ft. (333.3 mm/m) or more when the roofing is applied perpendicular to the eaves (Fig. 7.5-16). End laps are usually 6 in. (152.4 mm); edge laps 2 in. (50.8 mm).

In the concealed nailing system, a 12-in. (304.8 mm)-wide starter strip is applied completely around the perimeter of the roof and subsequently covered with roofing. Roofing cement is applied over the entire area of each lap of roofing material, whether over the starter strip or over other roofing material.

In the exposed nailing system, edges at rakes and eaves and laps of roofing material should be sealed with lap cement before nailing.

Ridges and hips should be covered with 12-in. (304.8 mm)-wide strips of roofing. In concealed nailing systems, these strips are nailed, then cemented, and the next section lapped over the succeeding one to conceal the nails. In exposed nailing systems, the sheets are laid with a 2-in. (50.8 mm)-wide band of roofing cement along each edge and nailed in place. In either case, laps should be fully bedded in roofing cement.

Double-Coverage Mineral-Surfaced Roll Roofing Sheets may be applied with the long dimension either parallel or perpendicular to the eave (Figs. 7.5-17 and 7.5-18). Nailing should be done in the selvage area. Laps should be fully bedded in roofing cement. Sheets should be laid so that only the areas with a granular finish are exposed.

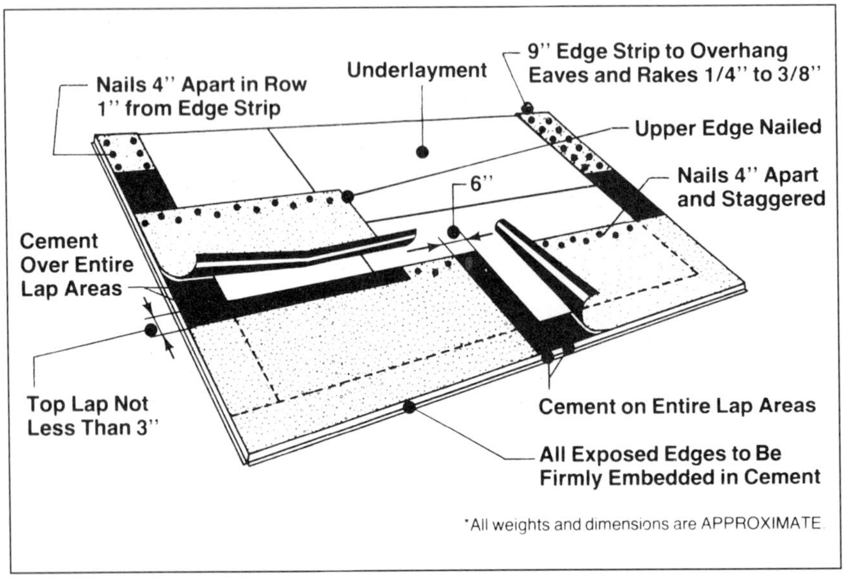

7.5.9 BUILT-UP MEMBRANE ROOFING

7.5.9.1 Materials

The NRCA has developed a four-letter system of generic designations to describe low-slope roofing systems, depending on their makeup. The four letters stand for the deck type, the bitumen used, the felt used, and the surface type (Fig. 7.5-19). For example, an NAGA system has a *n*ailable deck, *a*sphalt bitumen, *g*lass fiber felt, and *a*ggregate surfacing.

FELT SELECTION

Most felts used in built-up roofing today are glass-fiber-base felts, because they are stronger and more resistant to slippage than the organic felts that were favored in the past.

The polymer-modified bitumen sheets discussed in Section 7.5.10.1 may also be used in built-up roofing.

FIGURE 7.5-14 Standard roll roofing installed by the concealed nailing method, with sheets laid parallel with eaves. (Reprint courtesy National Roofing Contractors Association.)

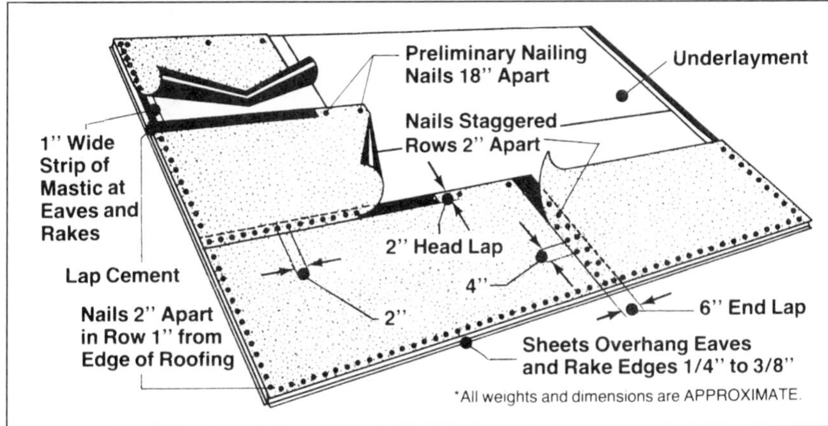

FIGURE 7.5-15 **Standard roll roofing installed by the concealed nailing method, with sheets laid perpendicular to eaves.** (Reprint courtesy National Roofing Contractors Association.)

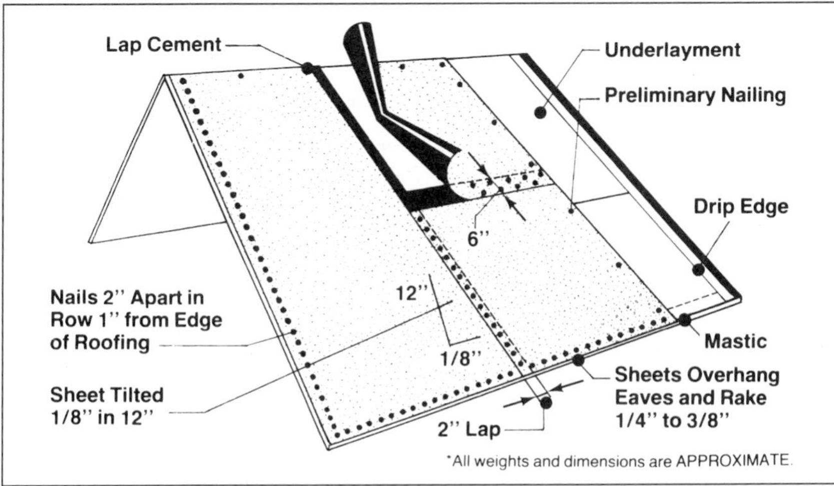

FIGURE 7.5-16 **Standard roll roofing installed by the exposed nailing method.** (Reprint courtesy National Roofing Contractors Association.)

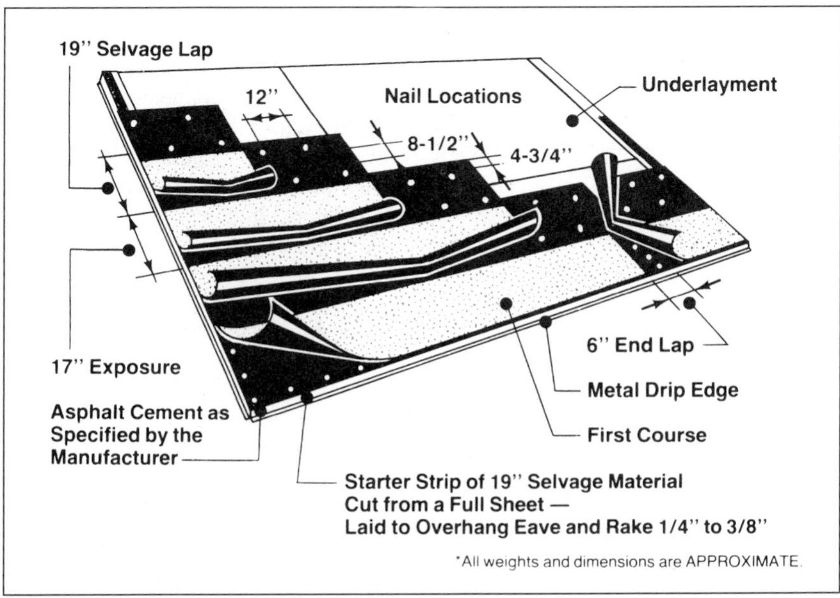

FIGURE 7.5-17 **Double-coverage roll roofing laid parallel with eaves.** (Reprint courtesy National Roofing Contractors Association.)

BITUMEN SELECTION

Coal tar is considered superior to asphalt bitumen, but is limited to use on slopes of ½ in./ft. (41.67 mm/m) or less in the coldest areas of the United States and to even lower slopes in more southern climates. In the hottest areas, the slope should be less than ¼ in./ft. (20.8 mm/m).

Asphalt bitumen can be used over a wide range of slopes and climatic conditions. In general, the lowest-numbered type that will work should be selected for a given slope.

Soft asphalt and coal-tar bitumens are incompatible and should not be used in the same roof system. Some roof systems, however, have successfully combined hard asphalt in the felts with coal-tar bitumen between the felts, but such systems must be applied more carefully than fully compatible systems.

BASE SHEET

Some systems include base sheets and some do not. The supposed function of base sheets is to protect the roofing system against moisture in the substrates and from condensation on the bottom of the membrane. Their effectiveness in this regard is not universally accepted, however. Following local practice is best when deciding whether to include a base sheet in a roofing system. When the base sheet is omitted, an additional ply sheet should be added.

PLY COUNT

A built-up roofing system's ply count is the number of felts in the system. Bitumen coats between plies and topping flood coats and ballast above the top ply do not count as plies. Normal built-up roof systems for commercial projects have either three or four plies.

SURFACING

In most areas of the country, built-up roofing is usually made up of smooth-surfaced felts covered with stone or unit-paver type ballast. Stone ballast, however, is limited to slopes of 3 in./ft. (250 mm/m) or less. In some areas, however, mineral-surfaced felt is often used as the topping sheet over built-up roofing, and no ballast is applied.

When roof slope prohibits ballast use, the top ply is often coated with either bitumen or another coating material such as a fibered or nonfibered aluminum coating.

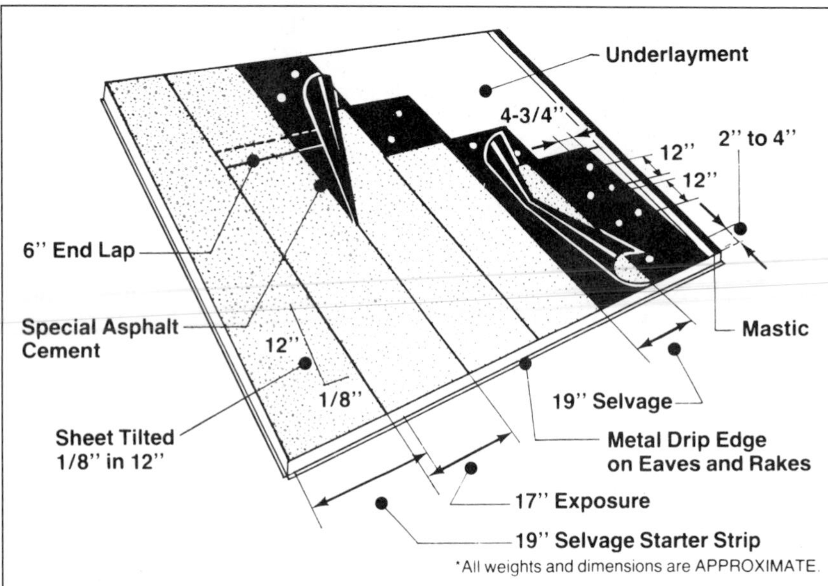

FIGURE 7.5-18 **Double-coverage roll roofing laid perpendicular to eaves.** (Reprint courtesy National Roofing Contractors Association.)

a

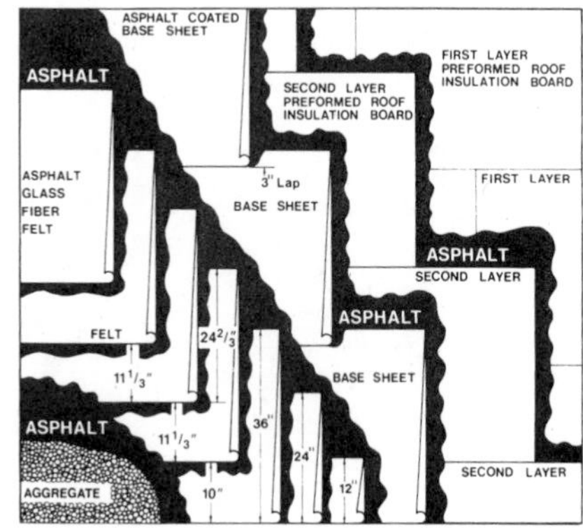

b

FIGURE 7.5-20 **Built-up membrane roofing (a) without a base sheet and (b) with a base sheet.** (Reprint courtesy National Roofing Contractors Association.)

7.5.9.2 Application

A built-up roofing application should be scheduled and carried out so that neither insulation nor roofing is left directly exposed to precipitation or condensation at any time. Roofing plies should be laid shingle-fashion, with laps sufficient to provide the selected number of plies throughout (Fig. 7.5-20). The shingles should be placed in the proper direction to shed water. On slopes in excess of 2 in./ft. (166.7 mm/m), the sheets should run parallel with the slope.

Built-up roofing plies should be nailed to nailable substrates as directed by the system manufacturer and the standards noted earlier, and to built-in nailers where appropriate. Nailers should occur at flashing locations and in other locations as needed. Nails should be concealed in the finished roofing.

Forty-five-degree cant strips and tapered edge strips should be installed where the roofing abuts vertical surfaces and at gravel stops.

In systems where a base sheet will be used, it should be applied first. The base sheet should be mopped and fastened in place.

The selected number of interply sheets should be applied in full hot bitumen beds having the bitumen weight per square specified for the selected system. A glaze-coat top surface of bitumen should then be applied over the top roofing ply. Bitumen should be heated to a temperature that will permit its application at its EVT, but never to its flash point.

Plies should be installed over cants and tapered edge strips and should extend at least 2 in. (50.8 mm) above the top of the cant or tapered edge (Fig. 7.5-21). A folded-back envelope should be built in at the edges of built-up roofing that does not terminate at a cant or at a tapered edge strip, to prevent bitumen from flowing over the edge of the roofing onto adjacent surfaces (Fig. 7.5-22). This envelope should be turned back over the plies and bedded in hot bitumen.

Expansion joints (Fig. 7.5-23), area dividers (Fig. 7.5-24), accessories, and flashings (Fig. 7.5-25) should be built into the roofing according to the standards listed earlier. Counter flashings are discussed in Section 7.8.

As soon as possible after the roofing has been installed, the top should be flood-coated with hot bitumen and the selected aggregate ballast should be laid in it (Figs. 7.5-26 and 7.5-27). Where aggregate ballast will not be used, the selected aggregate-surfaced asphalt felt, cap sheet, emulsion top coating, or aluminum reflective coating should be promptly applied.

FIGURE 7.5-19 **NRCA Four-Letter Designations for Built-Up Roofing Systems**

First Letter (Deck Type)	**Second Letter (Bitumen Type)**
N = nailable	A = asphalt
I = insulated	C = coal tar
C = concrete	
Third Letter (Surfacing)	**Fourth Letter (Ply Felt Type)**
A = aggregate surface	O = organic
E = emulsion coating	G = glass fiber
M = mineral surface	

Example

NAAO = nailable deck type (N), asphalt bitumen (A), aggregate surfaced (A), organic felt (O)

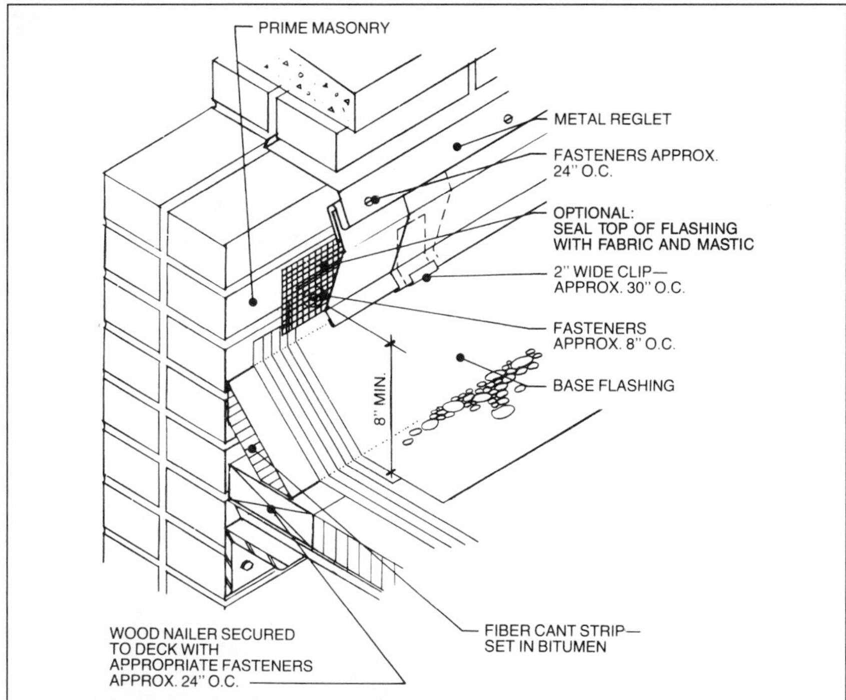

FIGURE 7.5-21 Base flashing for wall-supported deck. When the deck is not supported by a wall, built-up roofing should be applied against an expansion joint to permit wall movement from damaging the roofing membrane. (Reprint courtesy National Roofing Contractors Association.)

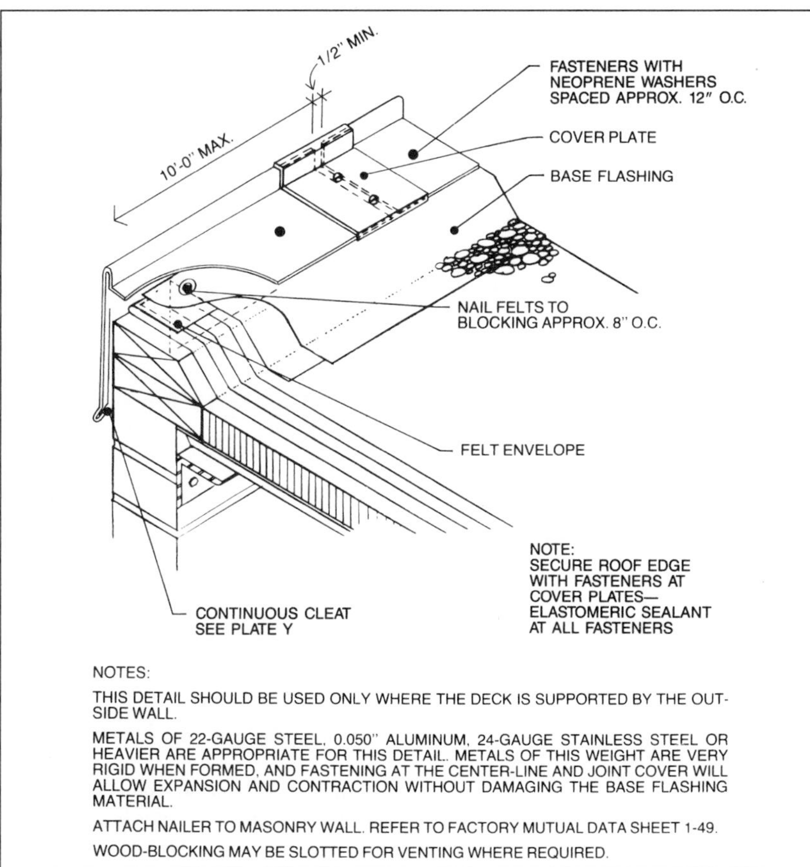

FIGURE 7.5-22 Built-up roof edge at a heavy metal gravel stop. (Reprint courtesy National Roofing Contractors Association.)

7.5.10 SINGLE-PLY MEMBRANES

Single-ply sheet roofing technology was first developed in Europe, and many of the currently available systems first appeared there.

7.5.10.1 Membrane Materials

This section covers single-ply membranes that fall into one of the following categories: vulcanized elastomers, nonvulcanized elastomers, thermoplastics, and polymer-modified bitumen (Fig. 7.5-28). These materials are probably the most popular in use today, but they are not the only ones. Single-ply roofing technology has developed rapidly and continues to develop. Some materials that were available a few years ago have disappeared from the market, and others appear from time to time. In addition, some materials discussed here may also disappear in time and new systems will most surely be developed.

The thicknesses given here are standard today, but the trend is toward heavier membranes.

VULCANIZED ELASTOMERS

Factory cured (vulcanized) elastomers include EPDM and vulcanized neoprene. EPDM is made from ethylene (*E*), propylene (*P*), a diene monomer (*DM*), and plasticizers, accelerators, and fillers. Neoprene (polychloroprene) is a synthetic rubber. Because vulcanized elastomers cannot be molded, reshaped, or welded by heat, they can only be bonded to themselves using adhesives.

Most EPDM is 45 mils (1.143 mm) thick, but 60 mil (1.524 mm) material is also available. Neoprene is usually 60 mils (1.524 mm) thick.

Some formulations of neoprene will accept a chlorosulfonated polyethylene (CSPE) color coating.

NONVULCANIZED ELASTOMERS

Nonvulcanized (uncured) elastomer sheet materials include those made from chlorosulfonated polyethylene (CSPE), chlorinated polyethylene (CPE), polyisobutylene (PIB), and nonvulcanized neoprene. Uncured elastomers can be molded, shaped, and welded by heat. Therefore, seams in these materials can be heat welded or chemically welded. Most of these materials cure in place after they have been installed and are then said to be "thermoset." Thermoset materials can be installed using adhesives in the seams.

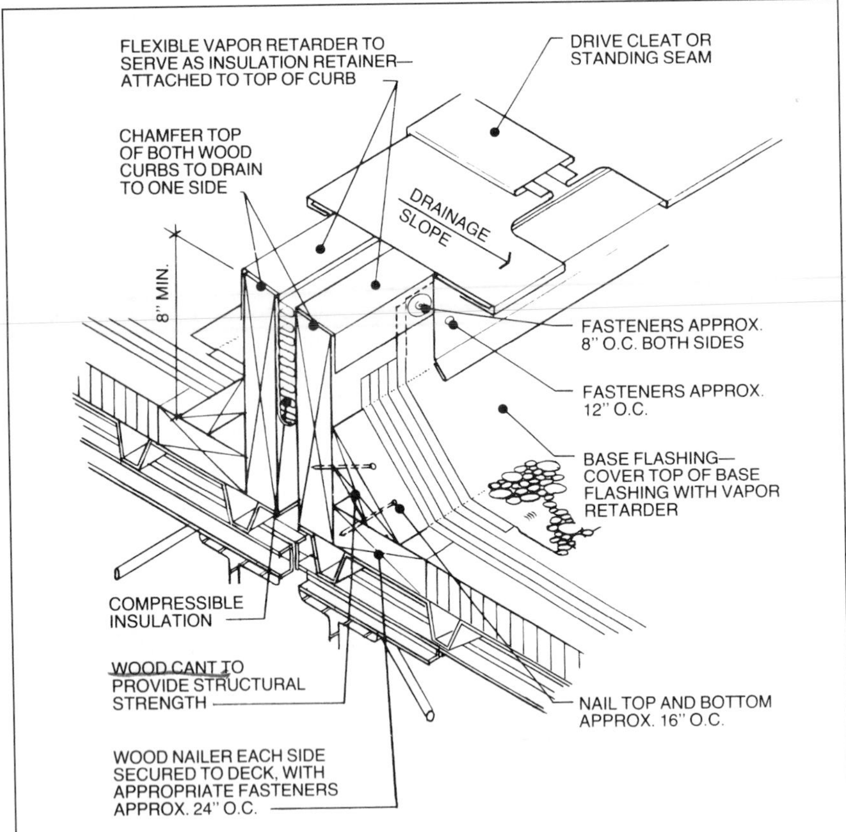

FIGURE 7.5-23 **Expansion joint with high curb in built-up roofing. Expansion joints should be carried through the roofing whenever they occur in the structure. Expansion joints may also be covered with preformed elastomeric or combination metal and elastomeric expansion caps. Expansion joints bonded into built-up roofing without high curbs are not recommended.** (Reprint courtesy National Roofing Contractors Association.)

FIGURE 7.5-25 **Installation of bituminous base flashing at a roof penetration.** (Reprint courtesy National Roofing Contractors Association.)

FIGURE 7.5-26 **Spreading aggregate ballast into hot bitumen adjacent to roof penetrations.** (Reprint courtesy National Roofing Contractors Association.)

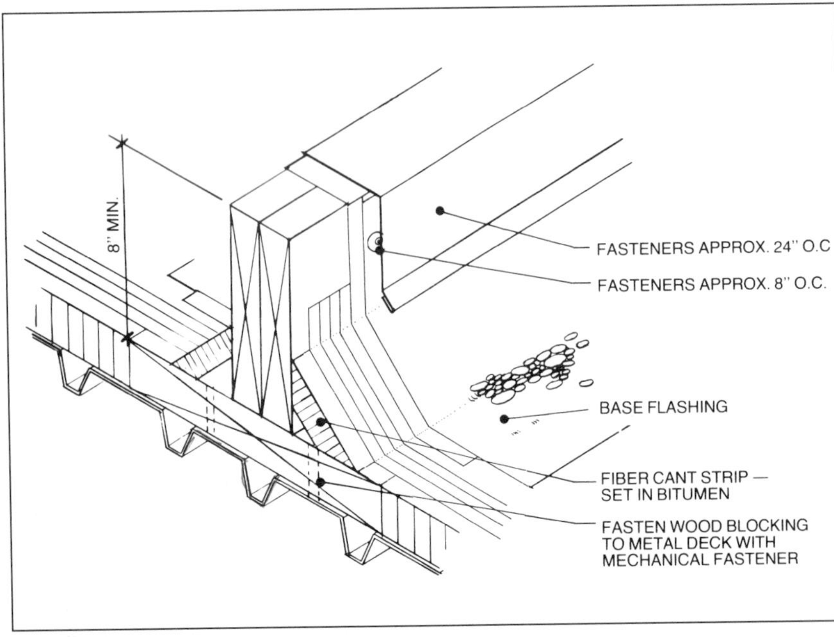

FIGURE 7.5-24 **Area dividers should be located in built-up roofing between expansion joints at 150- to 200-ft. (45.72 to 60.96 m) intervals to provide expansion relief in the membrane. They should not restrict water flow to drains.** (Reprint courtesy National Roofing Contractors Association.)

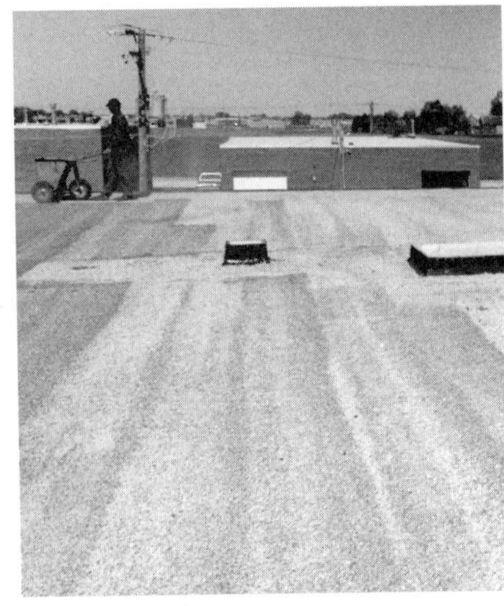

FIGURE 7.5-27 **Placing aggregate ballast over built-up roofing.** (Reprint courtesy National Roofing Contractors Association.)

FIGURE 7.5-28 Summary of Single-Ply Roofing Types

Type	Materials	Thickness, mils (mm)
Vulcanized (cured) elastomers	EPDM	45–60 (1.14–1.52)
	Neoprene[a]	60 (1.52)
Nonvulcanized (uncured) elastomers	CSPE[b]	60 (1.52)
	CPE	40–48 (1.02–1.22)
	PIB	100 (2.54)
	Neoprene[a]	60 (1.52)
Thermoplastics	PVC	40–60 (1.02–1.52)
Polymer-modified bitumen	APP	40–60 (1.02–1.52)
	SBS	40–60 (1.02–1.52)
	SBR	40–60 (1.02–1.52)

[a]Neoprene is available cured and uncured.

[b]CSPE (Hypalon) is also available in liquid form for use as a coating.

CSPE (DuPont Hypalon) CSPE is a synthetic rubber produced in many colors. Sheets are made of CSPE laminated to a mineral fiber backing or a polyester scrim. Sheets are usually 60 mils (1.524 mm) thick. It is also available in liquid form for use as a coating. It is weather, chemical, and pollutant resistant, which makes it usable as an exposed finish.

CPE CPE is an unreinforced or polyester-reinforced sheet material, usually 40 to 48 mils (1.016 to 1.219 mm) in thickness. It can be pigmented with many colors. The raw material used to make CPE is manufactured by the Dow Chemical Company. This material is compatible with bitumen and can, therefore, be installed directly over an existing asphalt or coal-tar pitch roof. It is also resistant to weathering, chemicals, and pollutants and is suitable for exposed locations.

PIB PIB is formulated from isobutylene and other polymers, carbon black, and other additives. A sheet consists of a 60-mil (1.524 mm) PIB membrane laminated to a 40-mil (1.016 mm) polyester backing. Sealing edges are left unbacked. It can be used in exposed locations and in contact with asphalt, but must be protected from petroleum distillates, organic oils and fats, and tar.

THERMOPLASTICS

Thermoplastic materials can be softened and reformed throughout their life. Polyvinyl chloride (PVC) when combined with plasticizers forms a thermoplastic roofing membrane that is usually between 40 and 60 mils (1.016 and 1.524 mm) thick. It is available unreinforced and reinforced with polyester or glass fiber. It can be used as exposed roofing, but is not compatible with bitumens. PVC comes without a finish for field finishing with coatings or with a factory-applied acrylic coating for weather exposure.

POLYMER-MODIFIED BITUMEN

Polymer-modified bitumen sheets are made from asphalt that has been modified. Typical modifiers include atactic polypropylene (APP), styrene-butadiene-styrene (SBS), and styrene-butadiene-rubber (SBR). The sheets are reinforced with glass fiber mats or scrim, polyester mats or scrim, polyethylene sheets, or copper or aluminum foils. Most are between 40 and 60 mils (1.016 and 1.524 mm) thick.

Some polymer-modified bitumen sheets are surfaced at the factory with mineral granules to protect them from the elements and to improve their appearance. APP-modified bitumen sheets must have a coating of granules. Some sheets are not factory-coated, but are suitable for field application of granules.

APP-modified bitumen sheets are not compatible with asphalt. SBS and SBR sheets are compatible with asphalt and can be used in built-up roofing applications. All three can be made suitable for torch application. Some ABS and SBR sheets are self-adhering. In general, APP sheets work better in warm climates. The two other types are better in cold climates.

POLYURETHANE FOAM

Sprayed-in-place, closed-cell polyurethane rigid foam systems provide excellent waterproof roofing for both new buildings and reroofing applications. They have no joints to open up and will not shrink. They are flexible, so they move with the structure. They are ultraviolet (UV), mildew, and fungus resistant. They are also lightweight, which reduces dead load on the structure. These systems are self-flashing by forming a tight bond between the roofing and vertical surfaces.

Sprayed-in-place polyurethane foam is coated with a silicone or acrylic formulation and may be finished with a broadcast application of roofing granules. It may also be finished with a fleeceback membrane layer applied with a polyurethane adhesive.

7.5.10.2 Coatings and Surfacings

Liquid hypalon (CSPE) can be used to coat EPDM and some formulations of neoprene to provide color. Silica sand coatings are also sometimes used on EPDM.

PVC and some modified bituminous membranes are sometimes coated with acrylic emulsions to improve their appearance.

Some modified bitumen membranes that do not have a factory-applied finish require a field-applied protective layer of mineral granules or aggregate to make them fire resistant. Aluminum coatings are also sometimes used over modified bitumen membrane roofing.

7.5.10.3 Application of Single-Ply Membranes

Single-ply membrane roofing application should follow the recommendations of the roofing system manufacturer, the NRCA, and the Single-Ply Roofing Institute (SPRI). Associated metal flashing and accessories should comply with the recommendations of the Sheet Metal and Air Conditioning Contractors National Association, Inc. (SMACNA). Conflicts in these recommendations should be reconciled before roofing system installation.

The methods available for installing single-ply roofing include *loose laid and ballasted*, *mechanically fastened*, and *adhered*. Each manufacturer issues specific instructions concerning application of its product which must be followed if a warranty is to be issued.

Regardless of the system selected, plies should be laid shingle-fashion, starting at low points in the deck so that water drains across the roof without interruption. In all three methods, the membrane should be laid over a rosin-sized building paper separator, unless the manufacturer requires another type of separator.

LOOSE-LAID AND BALLASTED SYSTEMS

EPDM, neoprene, hypalon, PIB, PVC, and some modified bitumen sheets may be laid using the loose-laid and ballasted system. In this system, the roofing is independent of the roof deck, which allows the structure to move without affecting the roofing. This same feature, however, means that such roofing will blow off quite easily due to wind uplift. This tendency can be reduced, but not necessarily eliminated altogether, by a layer of ballast. Most installations include washed river gravel as the ballast, but concrete pads are also sometimes used. Ballast adds weight to the roof and often requires additional structure. Loose-laid roofing with too little ballast or ballast that is too light may suffer damage from wind uplift.

Loose-laid roofing is placed over the substrates with only minimal fastening around the edges and at penetrations (Fig. 7.5-29). Adjoining sheets should be lapped and bonded together using the roofing manufacturer's recommended sealant.

Aggregate ballast should be applied uniformly at a rate recommended by the roofing manufacturer. This weight should be determined according to the location and height of the building, but should not be less than 10 to 12 lb./sq. ft. (48.82 to 58.59 kg/m^2). Precast concrete paver ballast should be applied in accordance with the roofing manufacturer's recommendations. The weight of the ballast must be taken into account in designing the roof structure. Care should be taken to ensure that the membrane is not damaged during installation of ballast.

MECHANICALLY FASTENED SYSTEMS

When reducing the dead load on a roof is desirable, or the roof is more steeply sloped than can be accommodated by ballast, or the building is in a location that is subject to high winds, single-ply roofing can be installed with a mechanically fastened system. Mechanically fastened systems should not be used when a vapor retarder is required, because the fasteners penetrate into the decking and, therefore, through the installed vapor retarder.

Single-ply membranes may be applied using either penetrating or nonpenetrating fasteners. Penetrating fasteners pass through the membrane (Fig. 7.5-30). Nonpenetrating fasteners are anchored to the structural deck, and the membrane is fastened to them using clamps or snap-on caps (Fig. 7.5-31).

The spacing, location, and types of fasteners should be in accordance with the roofing manufacturer's recommendations (Figs. 7.5-32 and 7.5-33).

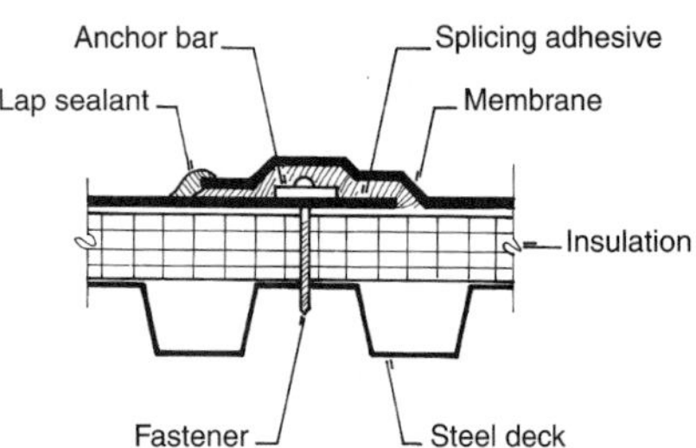

FIGURE 7.5-30 **Penetrating-type fastener for single-ply roofing. Several different types are available, but their principle is similar to the device shown.** (Drawing by HLS.)

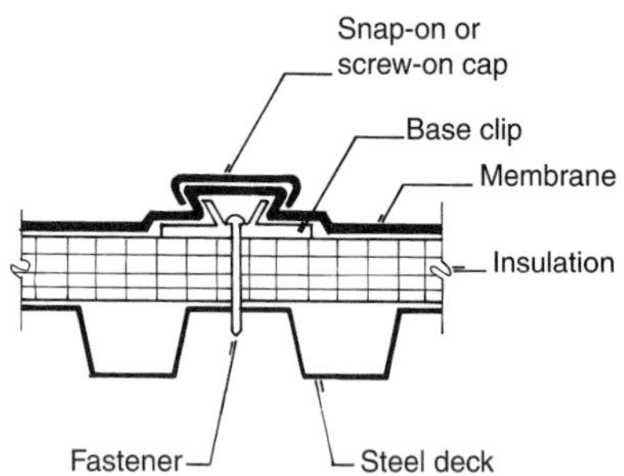

FIGURE 7.5-31 **Nonpenetrating fasteners do not pass through the membrane. These are proprietary systems.** (Drawing by HLS.)

FIGURE 7.5-29 **The home court of the NBA's Minnesota Timberwolves is protected by a ballasted, 45-mil (1.143 mm) EPDM membrane system.** (Courtesy Firestone Building Products Company.)

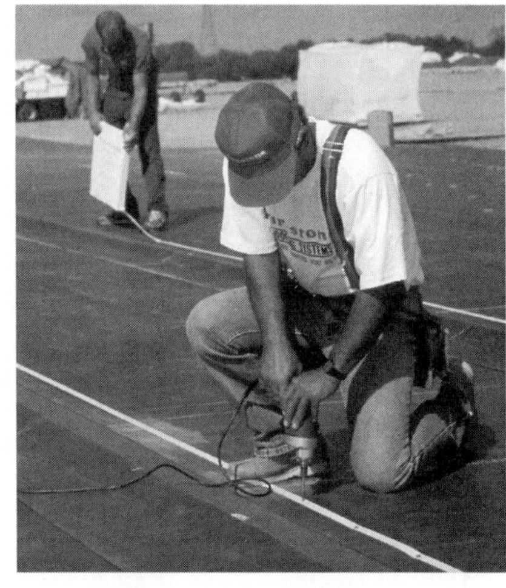

FIGURE 7.5-32 **A polymer batten bar is used to mechanically attach sheets of single-ply EPDM roofing membrane to the substrate atop JVC Company of America's distribution center in Aurora, Illinois.** (Courtesy Firestone Building Products Company.)

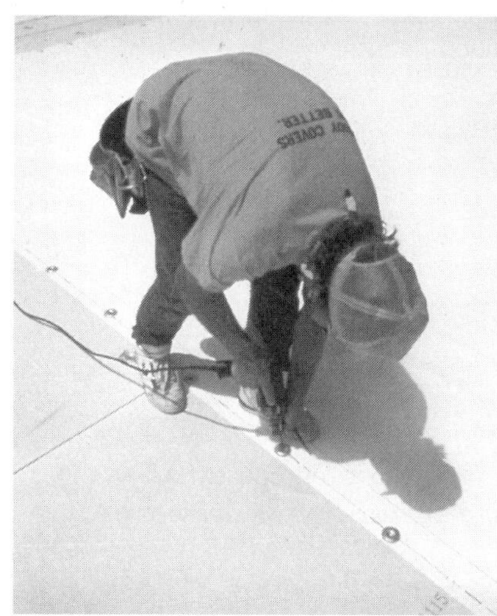

FIGURE 7.5-33 **Fasteners are installed along the leading edge of this white, thermoplastic alloy roofing membrane to attach it securely to a 2¾-in. (69.9 mm)-thick layer of polyisocyanurate insulation.** (Courtesy Firestone Building Products Company.)

ADHERED SYSTEMS

Adhering a single-ply membrane to the roof deck or to insulation permits applications over complex shapes, curves, and steep slopes. Membranes may be either fully or partially adhered.

In a *fully adhered* system, the membrane is completely attached to the substrates using hot- or cold-applied bitumen, cold-applied adhesives, solvents by heating (torching) the back of the membrane, or by pressing a self-adhering membrane in place (Figs. 7.5-34 and 7.5-35).

In a *partially adhered* system, the roofing membrane is laid into beads or strips of bitumen, adhesive, or solvent and rolled, or is adhered by similar materials placed on the top plates of the fasteners that hold down the insulation.

FIGURE 7.5-34 **Fully adhered single-ply roofing systems are particularly useful on complex roof shapes, such as domes, barrels, and parabolas, and, of course, on flat roofs like the one shown here.** (Courtesy the Goodyear Tire and Rubber Company.)

FIGURE 7.5-35 **Installing a fully adhered single-ply roofing system.** (Courtesy the Goodyear Tire and Rubber Company.)

JOINTS

Seams should be made by lapping adjacent sheets and sealing the joints. Different joining methods are required for each roofing type. The manufacturer's recommendations should be followed. Before laps can be sealed, separation coatings, such as talc or mica, must be removed from the area to be joined. This can be done using a solvent.

Joints in nonvulcanized elastomers can be sealed using either heat- or chemical-welding.

Vulcanized and nonvulcanized materials that have been cured must be sealed using adhesives (Fig. 7.5-36). The type of adhesive should be that recommended by the roofing manufacturer. Some sheets require contact adhesives, some require adhesives made from the same material as the membrane, and others will work with generic adhesives. Some sheets are naturally self-adhering. Some are made self-adhering by the addition of an adhesive coating or self-adhering strips bonded to the sheets.

Joints in thermoplastic materials can be either heat- or solvent-welded (Fig. 7.5-37).

Joints in self-adhering modified bitumen sheets should be made by pressing the sheets together and rolling. Joints in non-self-adhering APP sheets should be torch-welded (Fig. 7.5-38). Joints in non-self-adhering SBS and SBR sheets can be either torch-welded or sealed with hot asphalt.

FIGURE 7.5-36 **Applicator ensures that adjoining EPDM sheets are lapped a minimum of 3 in. (76.2 mm), cleaned, and spliced with splice adhesive to form a continuous, watertight membrane atop the Target Center in downtown Minneapolis.** (Courtesy Firestone Building Products Company.)

FLASHING

Base and penetration flashings are usually of the same material as the sheet membrane or of a nonvulcanized compatible elastomeric sheet material (Fig. 7.5-39). Metal base flashings are inappropriate, and joints between elastomeric base flashings and metal cap (counter) flashings should be avoided when possible. Gravel stops may be metal, made from compati-

FIGURE 7.5-37 **This white thermoplastic alloy roofing system uses heat-weldable seams to form a continuous watertight membrane.** (Courtesy Firestone Building Products Company.)

FIGURE 7.5-38 **Torch-down application of APP modified bitumen systems.** (Courtesy Firestone Building Products Company.)

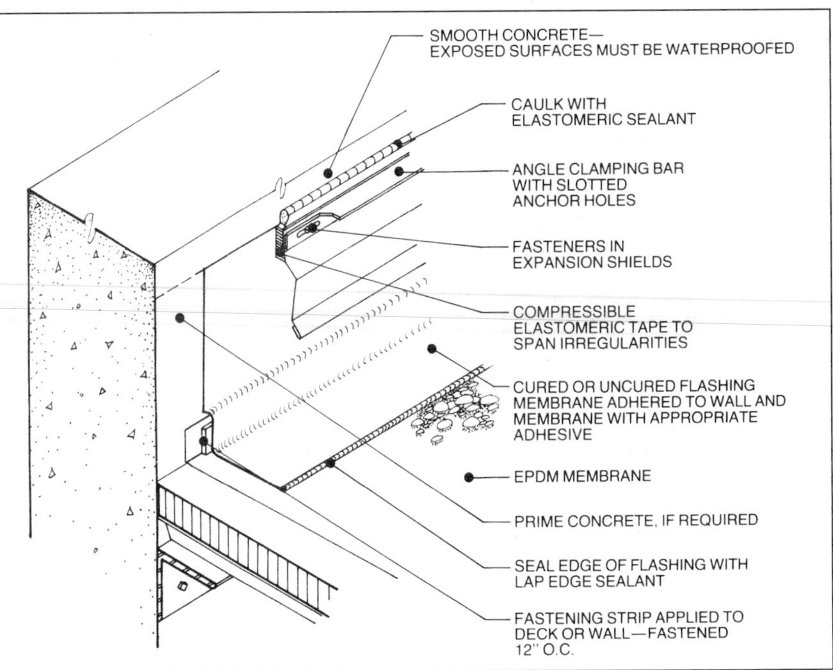

FIGURE 7.5-39 **Wall flashing for a single-ply membrane.** (Reprint courtesy National Roofing Contractors Association.)

ble plastic materials, or coated with the same elastomer used in the roofing membrane to permit proper bonding with the roofing.

Joints between roofing membrane and elastomeric flashings should be sealed with the same material used in the seams in the roofing membrane (Fig. 7.5-40). Expansion joint covers should be made using compatible materials (Fig. 7.5-41).

When the roofing manufacturer so requires, nailers should be provided and elastomeric flashings should be securely fastened in place.

Metal cap flashings are addressed in Section 7.8.

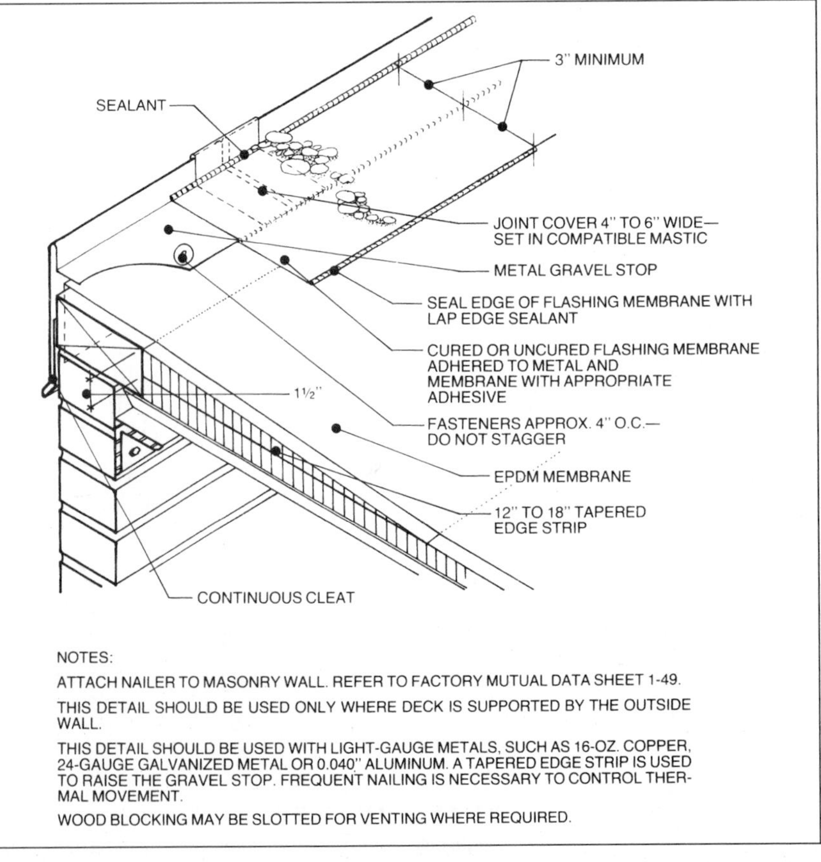

FIGURE 7.5-40 **A light-metal gravel stop at the edge of a single-ply membrane. Some roofing manufacturers recommend that a heavy gravel stop coated with an elastomeric material compatible with the roofing be used in such a condition. When this is not acceptable for aesthetic or other reasons, the detail shown can be used.** (Reprint courtesy National Roofing Contractors Association.)

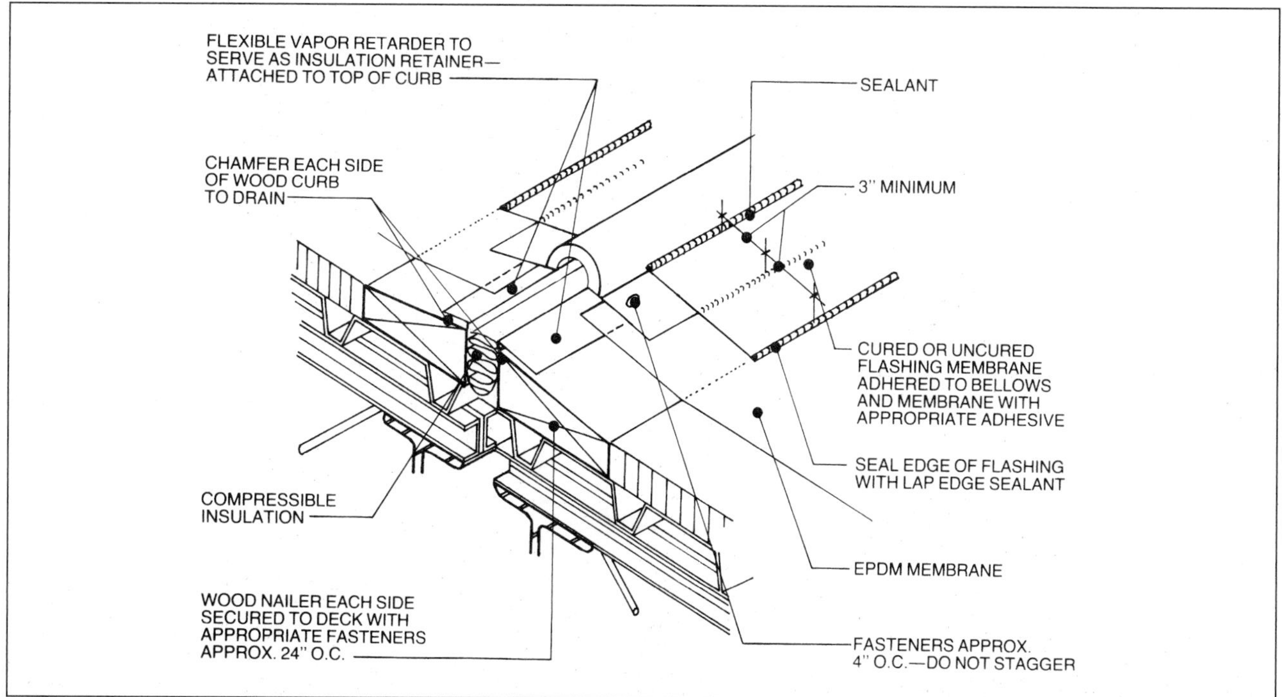

FIGURE 7.5-41 **An expansion joint in an elastomeric single-ply roofing membrane. Some manufacturers recommend other types of expansion joints, but some type of expansion provision should occur in roofing over every expansion joint in the structure. However, separate expansion joints may not be needed in loose laid systems if a metal plate is placed over the joint beneath the insulation and fastened to the deck at one side of the joint only. The roofing manufacturer's advice should be followed.** (Reprint courtesy National Roofing Contractors Association.)

7.6 Steep Roofing

Steep roofs are those whose slope exceeds 1½ in./ft. (125 mm/m). Roofs with slopes less than 1½ in./ft. (125 mm/m) are called *low-sloped* (see Section 7.5). Roofing for steep roofs comes in many forms, including ancient, traditional shapes and newer configurations, and in materials ranging from the stone and clay materials that have been used for centuries to modern-day equivalents made from concrete, mineral-fiber-cement, or metal.

Stone and clay tiles and slate shingles were early hard roofing materials. Concrete and mineral-fiber cement tiles were originally produced, and are still used, primarily to mimic clay tile, slate roofing, or wood shingles or shakes. Those newer materials share several general advantages as compared with the earlier forms. They are often lighter, more readily available, and less susceptible to damage. In many cases, the newer materials cost less than the roofing they mimic. Wood shingles are an exception, usually costing less than clay, concrete, or mineral-fiber-cement substitutes, but wood shingles do not have the fire-resistance characteristics shared by the others.

The newer forms of hard roofing materials share the disadvantage of being machine made. The hand-made forms have somewhat irregular surfaces, which appear more rustic and have greater charm than roofs made with their modern, mass-produced counterparts. Another disadvantage is that, in most cases, the modern materials do not stand up to close visual comparison with the products they are imitating and are immediately apparent when installed alongside the real thing.

When used alone, however, some modern materials are often mistaken for the materials they mimic when viewed from the ground by people who are not experts in recognizing the materials imitated. By contrast, composition shingles that are supposed to look like wood shingles do not fool anybody.

Stone tile, perhaps the earliest heavy roofing, is generally used today only in repairing historic buildings. Stone tile was never used much in the United States and is not found often enough to warrant a detailed discussion here.

This section discusses the materials and methods needed to provide steep roof finishes using composition shingles; wood shakes and shingles; the hard materials, slate, clay and concrete tile, and mineral-fiber-cement shingles.

Steep roofing functions by shedding water. By contrast, low-slope roofing provides a waterproof membrane (see Section 7.5). Mineral-surfaced roll roofing, which functions like steep roofing and can be applied on roofs with slopes of 1 in./ft. or more, is also discussed in Section 7.5. Metal roofing is covered in Section 7.9.

In addition to cost, other criteria to be considered in selecting steep roofing and its application methods include roof slope, expected service life, wind resistance, fire resistance, and local climate. Depending on the color and kind of material selected, the total heat loss or gain of a roof may also be affected.

7.6.1 GENERAL RECOMMENDATIONS

Individual shingle and tile units are applied to steep roofs in an overlapping fashion to shed water and resist weather penetration. They can be applied effectively on most roofs that have sufficient slope to ensure good drainage.

Depending on the material type and application method selected, aesthetic effects range from surfaces having relatively smooth, uniform coursing to those with random coursing, which gives a heavily textured appearance. Color can be achieved by a variety of factory- or field-applied coatings, or by leaving the material in its natural color.

In addition to the shingle or tile units themselves, many accessory materials are required to prepare a roof surface for their application, and to apply them. Depending on the shingles or tile type and application method, accessory materials may include sheathing, underlayment, sheathing paper, backer board, undercoursing, flashing, drip edges, roofing cement, caulking, and nails and other fasteners. Depending on the kinds of shingles or tile, other accessories, such as starter shingles and hip and ridge units, may also be required. Typical shingle and tile materials and application methods for roofs are discussed in the following text, but the shingle or tile manufacturer's instructions and recommendations should always be consulted to ensure the most suitable performance.

In this section, the underlying roof is assumed to be correctly and adequately ventilated, a practice that is essential to the prevention of condensation. Condensation, the actual cause of many roof leakage complaints, can be eliminated as a hazard by providing sufficient ventilation in attic areas or enclosed air spaces beneath the roof. Refer to Section 7.1 for a discussion of vapor retarders and ventilation.

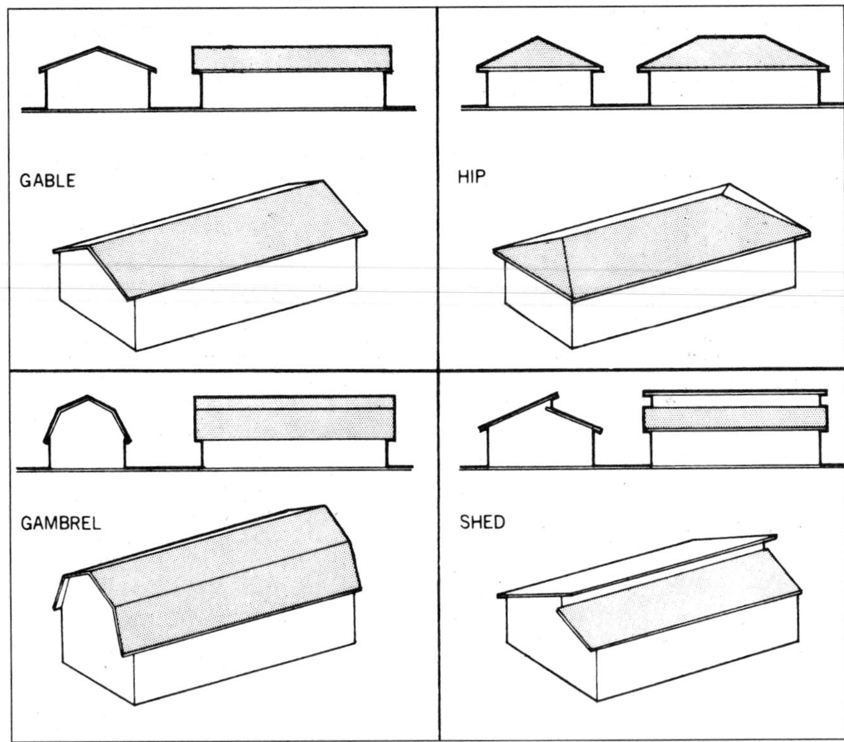

FIGURE 7.6-1 Types of sloped roofs.

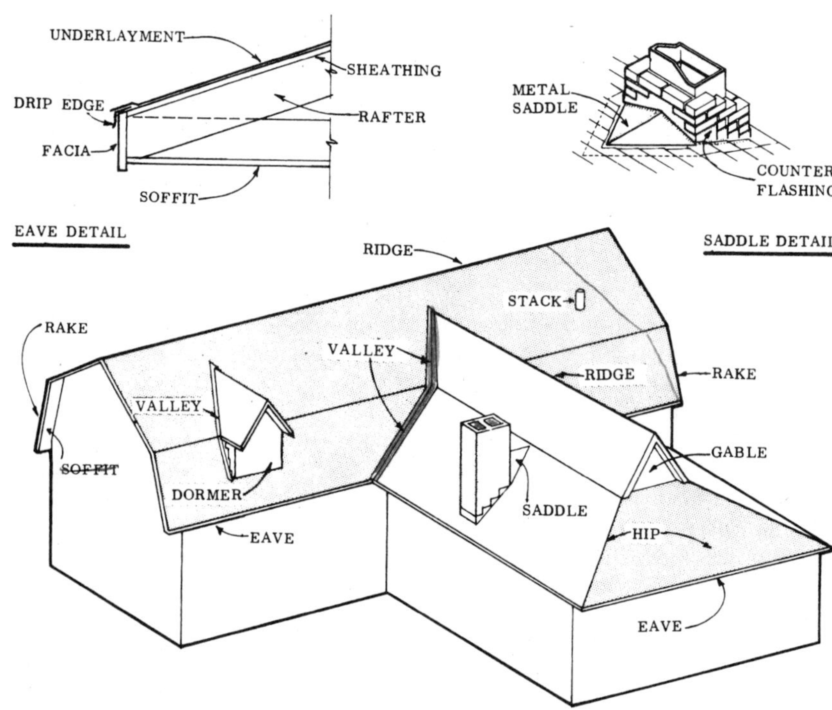

FIGURE 7.6-2 Composite diagram of sloped roof types showing terminology.

7.6.1.1 Nomenclature

Some definitions of roof and roofing terminology are given here to assist in understanding the text and illustrations.

ROOF TERMINOLOGY

Figure 7.6-1 shows several common steep roof types. These simple shapes are often complicated by intersecting walls or roofs, and projections through the roof such as dormers, chimneys, plumbing vents, and roof ventilators.

The composite drawing in Figure 7.6-2 names the elements of a steep roof.

Slope and Pitch The terms *slope* and *pitch* are often incorrectly used synonymously in referring to the incline of a steep roof. Both are defined here and in Figure 7.6-3, which also compares some common roof slopes with corresponding roof pitches. *Roof slope* is the term used throughout this text.

Slope indicates the roof incline as a ratio of vertical rise to horizontal run. It is expressed sometimes as a fraction, but typically as *x in 12*. For example, a roof that rises at the rate of 4 in. (101.6 mm) for each foot (12 in. [304.8 mm]) of run, is designated as having a 4 in 12 slope. The triangular symbol above the roof in Figure 7.6-3 conveys this information.

Pitch indicates the incline of a roof as a ratio of the vertical rise to twice the horizontal run. It is expressed as a fraction. For example, if the rise of a roof is 4 in. (101.6 mm) and the run is 12 in. (304.8 mm), the roof is designated as having a pitch of 1/6.

SHINGLE AND TILE ROOFING TERMINOLOGY

The following terms are peculiar to shingle and tile roofing. Other terms are defined when they first appear in the text or in the Glossary.

Coverage. The amount of weather protection provided by the overlapping of shingles or tile. Depending on the kind of shingle and method of application, shingles may furnish one (single coverage), two (double coverage), or three (triple coverage) thicknesses of material over the surface of a roof.

Shingles providing single coverage are suitable for reroofing over existing roofs, in effect providing a new surface for an old roof that is still serviceable except for isolated trouble spots. Shingles providing double and triple coverage are used for new construction, both having increased weather resistance and a longer service life.

Endlap. The shortest distance in inches (mm) by which adjacent shingles, sheets, or tiles horizontally overlap each other (Fig. 7.6-4).

Exposure. The shortest distance in inches (mm) between exposed edges of overlapping shingles or tiles (see Fig. 7.6-4).

Headlap. The shortest distance in inches (mm) from the lower edges of overlapping shingles to the upper edge of the unit in the second course below (see Fig. 7.6-4).

Shingles. Relatively small individual roofing units designed to overlap other similar units to provide weather protection. Typically, shingles are applied to a nailing base, such as sheathing or horizontal nailing strips, which supports the shingles between structural framing members.

Shingle butt. The lower, exposed edge of a shingle.

Sidelap. See *endlap*.

Square. Roofing is estimated and sold by the *square*, although packages of one-third square are usually available so that individuals or contractors do not have to buy a full square of roofing when the amount needed is not an exact multiple of squares. A square of roofing is the amount required to cover 100 sq. ft. (9.29 m^2) of roof surface.

Toplap. The shortest distance in inches from the lower edge of an overlapping shingle or sheet to the upper edge of the lapped unit in the first course below (that is, the width of the shingle minus the exposure) (see Fig. 7.6-4).

7.6.1.2 Roof Slope Limitations

Free and effective drainage from sloped roof surfaces is essential for long shingle life. As the roof slope decreases, runoff is slower and the susceptibility to leakage from wind-blown rain or snow driven underneath the roofing increases. Accordingly, the roof slope governs, or may impose limitations on, the following: (1) roofing material selection, (2) exposure, (3) underlayment requirements, (4) eave flashing requirements, and (5) methods used in roofing application.

In this text, when a reduction in roof slope requires precautions other than normal practice, the roof is referred to as being *of lower slope than normal*; otherwise, the term *normal slope* is used.

The minimum slope recommended for each shingle and tile type is shown in Figure 7.6-5. Slope limitations for underlayment and eave flashing requirements vary with each shingle and tile type and are discussed later in this section. Application methods are discussed in the following subsections for each shingle type.

7.6.1.3 Roof Framing and Sheathing

Steep roofing performance is dependent on the supporting structure. Roof joists, rafters, headers, and other supports should be adequately sized and spaced to carry the necessary superimposed loads over the required spans.

Sheathing should be smooth, securely attached to supports, and provide an adequate base to receive nails and other fasteners. All shingle types are applied over solid sheathing composed of wood boards

12
X Slope
Rise
Run

Assume:
Rise = 4 ft.; Run = 12 ft.

Slope: 4/12, or 4 in 12

Pitch: $\frac{4}{2 \times 12} = \frac{4}{24} = \frac{1}{6}$

$$Slope = \frac{Rise}{Run}$$

$$Pitch = \frac{Rise}{2 \times Run}$$

Slope	Pitch
2 in 12	1/12
3 in 12	1/8
4 in 12	1/6
5 in 12	5/24
6 in 12	1/4
7 in 12	7/24
8 in 12	1/3
10 in 12	5/12
12 in 12	1/2

FIGURE 7.6-3 Slope and pitch.

E = Exposure
TL = Toplap
HL = Headlap
SL = Sidelap
W = Width for Strip Shingles or Length for Individual Shingle

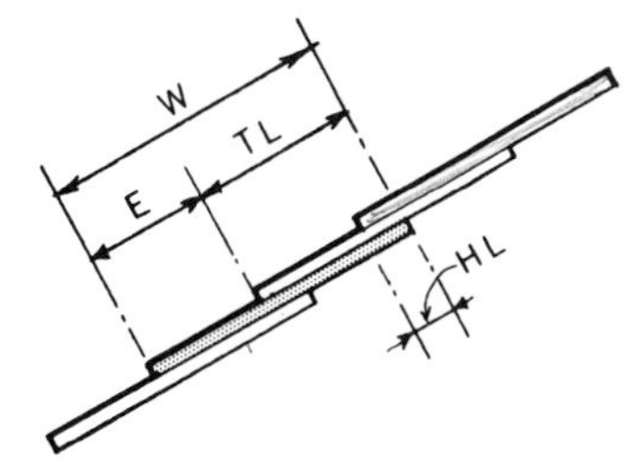

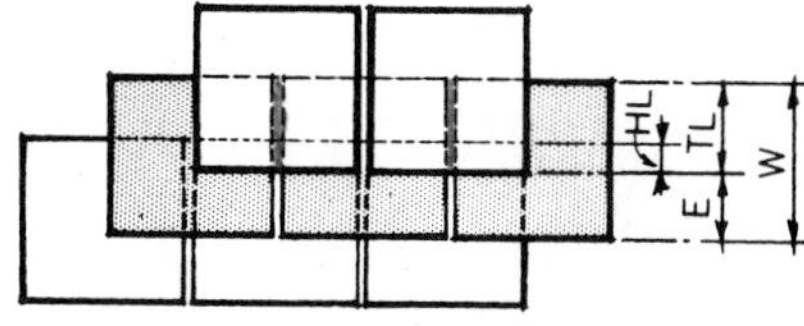

AMERICAN METHOD

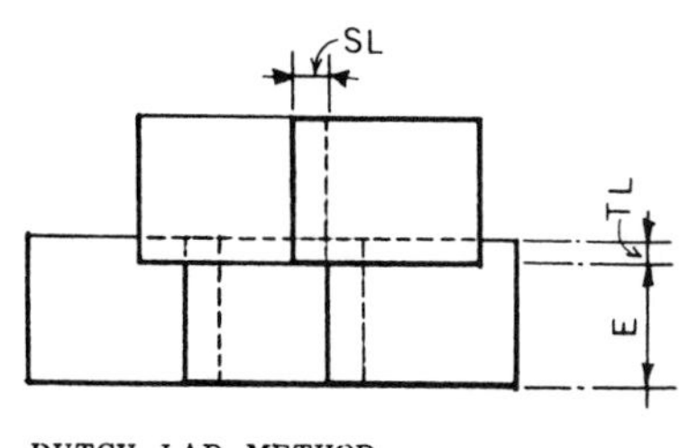

DUTCH LAP METHOD

FIGURE 7.6-4 Shingle roofing terminology.

FIGURE 7.6-5 Summary of Recommended Minimum Roof Slope

Type of Roofing	Minimum Slope[a]
Mineral fiber-cement shingles	3 in 12[b]
Asphalt shingles	2 in 12[c]
Wood shakes	3 in 12[d]
Wood shingles	3 in 12[e]
Slate shingles	4 in 12[b]
Clay tile	
Flat	5 in 12[b]
Other types	3 in 12[b]
Concrete tile	
Flat	5 in 12[b]
Other shapes	4 in 12[b]
Mud set	4 in 12

[a]Where eave flashing is required, it should be cemented.

[b]Requires double underlayment.

[c]Strip shingles only; requires double underlayment and wind-resistant shingles or cemented tabs.

[d]Requires solid sheathing, underlayment in addition to interlayment, and reduced weather exposure.

[e]Requires reduced weather exposure.

or plywood. Wood shingles and shakes are also applied over spaced sheathing.

Roof framing and sheathing are discussed in Chapter 6.

7.6.1.4 Roofing Cements

Roofing cements include plastic asphalt cements, lap cements, quick-setting asphalt adhesives, roof coatings, and primers. They are used for installing eave flashing, for flashing assemblies, for cementing asphalt shingle tabs and laps in sheet materials, and for roof repairing. They are discussed in Section 7.5.

Materials and application methods for steep roofs should be as recommended by the roofing material's manufacturer.

7.6.1.5 Roofing Underlayment

Roof underlayment performs several functions:

- Protecting sheathing from moisture absorption until roofing is applied;
- Providing important additional weather protection by preventing the entrance of wind-driven rain below the roofing onto the sheathing or into the structure;
- Preventing direct contact between asphalt shingles and resinous areas in wood sheathing which, because of chemical incompatibility, may damage the shingles.

Underlayment should be asphalt-saturated felt or another material having a low vapor resistance. Laminated waterproof papers, coated felts, and other materials that act as vapor retarders should not be used. Such materials may permit moisture or frost to accumulate between the underlayment and the surface of the sheathing.

Underlayment requirements for different steep roofing types for various roof slopes are summarized in Figure 7.6-6.

Underlayment should be applied over the entire roof immediately after the roof sheathing application. It should lap at least 2 in. (50.8 mm) at horizontal joints and 4 in. (101.6 mm) at end joints (Fig. 7.6-7), and should be lapped 6 in. (152.4 mm) from both sides over hips and ridges. Only a sufficient number of fasteners need be used to hold the underlayment securely in place until roofing is applied. Roofing should not be applied over wet underlayment.

FIGURE 7.6-6 Summary of Underlayment for Shingle Roofs

	Underlayment Requirements				
Material	**No. 15**	**No. 30**	**1 Layer**	**2 Layers**	**Remarks**
Roll roofing	X		X	X	a, b, c
Composition shingles					
4 in./ft. (333.3 mm/m) and steeper	X		X		b
3 in./ft. (250 mm/m) to 4 in./ft. (333.3 mm/m)	X			X	c
2½ in./ft. (208.3 mm/m) to 3 in./ft. (250 mm/m)	X				d
Slate					
4 in./ft. (333.3 mm/m) and steeper					
¼ in. (6.4 mm) thick and thinner		X	X		b
More than ¼ in. (6.4 mm) thick		X		X	c, e
Less than 4 in./ft. (333.3 mm/m)					i
Graduated, all slopes		X		X	c
Wood shingles		X	X		f
Wood shakes		X	X		f, g
Clay tile		X	X		h
Concrete tile		X	X		h
Mineral fiber-cement tile					
4 in./ft. (333.3 mm/m) and steeper		X	X		f, g
3 in./ft. (250 mm/m) to 4 in./ft. (333.3 mm/m)		X	X		g
Class A roofs					j
Metal tile		X	X		b

[a]Use one layer on slopes 4 in./ft. (333.3 mm/m) and more. Use two layers on flatter slopes.

[b]Where one layer of underlayment is required, install two layers of No. 15 felt or one No. 50 felt from eave line to 24 in. (610 mm) inside wall line as an ice shield when mean January temperature is 30°F (–1.1°C) or lower. At slate roofing and metal tile, use two layers of No. 30 felt. Set ice-shield felt in hot asphalt or mastic. As an alternative, use adhered bituminous membrane manufactured for the purpose as an ice shield.

[c]Where two underlayment layers are required, apply underlayment from eave to 24 in. (610 mm) inside wall using hot asphalt or mastic when mean January temperature is 30°F (–1.1°C) or less.

[d]Entire underlayment applied using hot asphalt or mastic.

[e]Alternative: One layer of No. 55 or No. 65 roll roofing over entire roof, with portion from eave to 24 in. (610 mm) inside wall applied using hot asphalt or mastic when mean January temperature is 30°F (–1.1°C) or less.

[f]Underlayment scheduled is optional except in snow areas (see text).

[g]Felt eave strips and felt interlayers between shakes are essential (see text).

[h]Underlayment scheduled is often used for roof slopes of 4 in./ft. (333.3 mm/m) and higher, but there is disagreement in the industry about underlayment requirements for clay and concrete tile (see text). In applicable areas, the requirements in notes b and c apply.

[i]Slate on slopes less than 4 in./ft. (333.3 mm/m) should be installed over membrane roofing or waterproofing (see text).

[j]See text.

(Courtesy H. Leslie Simmons.)

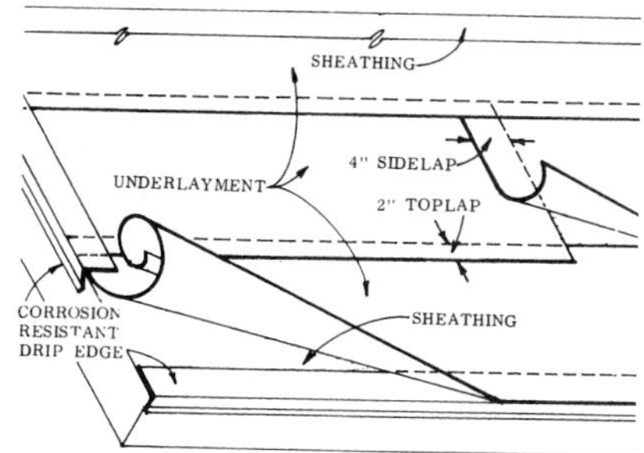

FIGURE 7.6-7 Underlayment requirements for steep roofing.

7.6.1.6 Flashing and Caulking

Roofs are often intersected by other roofs, walls, and projections, creating opportunities for leakage. Flashing must be installed at these locations to weatherseal the roof. Flashing must be watertight and water-shedding.

FLASHING MATERIALS

Corrosion-resistant metal flashing should be at least 26-gauge (0.55 mm) galvanized steel, 0.019-in. (0.4826 mm) aluminum, or 16-oz. (453.6 g) copper. Refer to Section 7.8 for a discussion of flashing materials.

Additional requirements and typical flashing methods are described in the following subsections for each steep roofing type.

ROOF FLASHINGS

Valley Flashing The joint formed by two sloping roofs meeting at an angle is called a *valley*. Because drainage concentrates in a valley, causing runoff toward and along the joint, valleys are especially vulnerable to leakage. An unobstructed drainage way must be provided with enough capacity to carry the water away rapidly.

Chimney Flashing A chimney is usually built on a foundation separate from the foundation that supports the structure and is normally subject to some differential settling. To permit this movement between a chimney and associated roof structure without damage to the waterseal, it is necessary to secure base flashings to the roof deck and secure counter flashing, which is bent down over the base flashing, to the masonry.

A saddle (see Fig. 7.6-2) should be built between a roof and the back face of a chimney that falls below the ridge. A saddle will prevent snow and ice accumulation behind the chimney and deflect water down around the chimney.

Eave Flashing Eave flashing is recommended in areas where the outside design temperature is 30°F (–1.1°C) or colder (Fig. 7.6-8) or wherever there is a possibility of ice forming along the eaves. This protection is important to avoid leaks and moisture penetration caused by water backing up under shingles behind ice dams and snow that collect at the eaves.

Different flashing methods are used to prevent leakage from this cause, depending on roof slope, icing conditions, and roofing type.

For normal slopes, eave flashing is usually formed by a smooth or mineral-surfaced coated roll roofing course applied over the underlayment. It extends up the roof to cover a line at least 12 in. (304.8 mm) inside the interior wall line of the building (Fig. 7.6-9a).

On slopes that are lower than normal, in areas subject to severe icing, a double layer of underlayment should be applied over the entire roof. The two layers should be cemented together with asphalt cement up the roof to a line at least 24 in. (609.6 mm) inside the interior wall line of the building (Figs. 7.6-9b and 7.6-10).

Specific recommendations for eave flashing are described in the following subsections for each kind of steep roofing.

Drip Edge A continuous corrosion-resistant, nonstaining material such as 26-gauge (0.55 mm) galvanized steel, formed to provide a drip and nailed along the

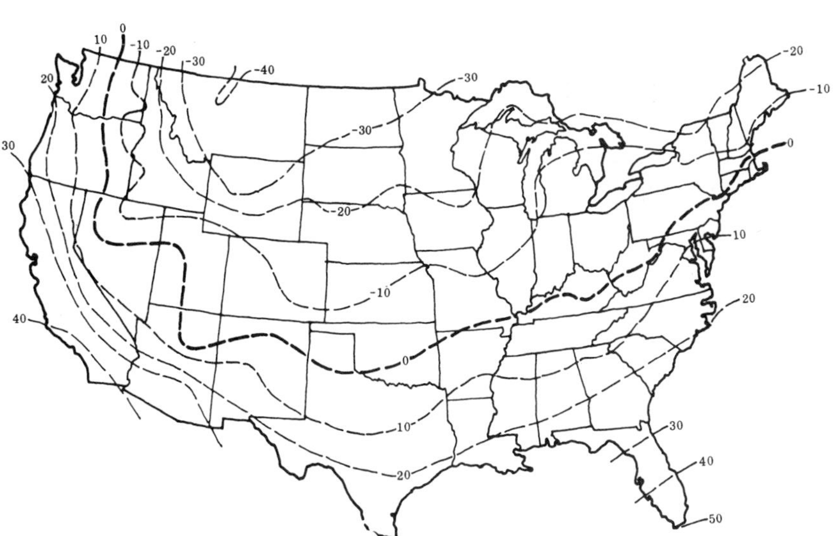

FIGURE 7.6-8 Outside design temperatures. (Reprint courtesy National Weather Service.)

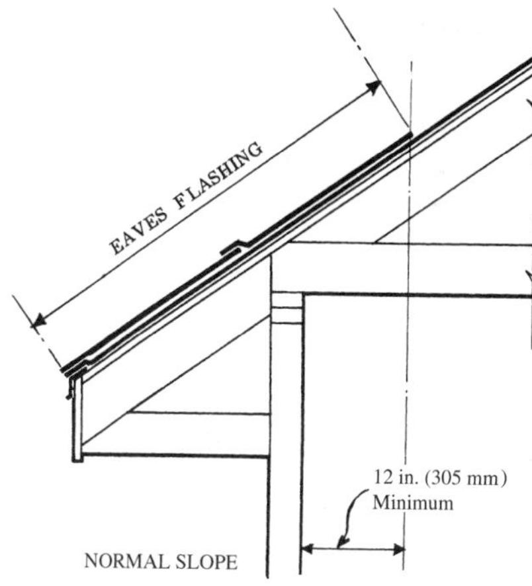

a

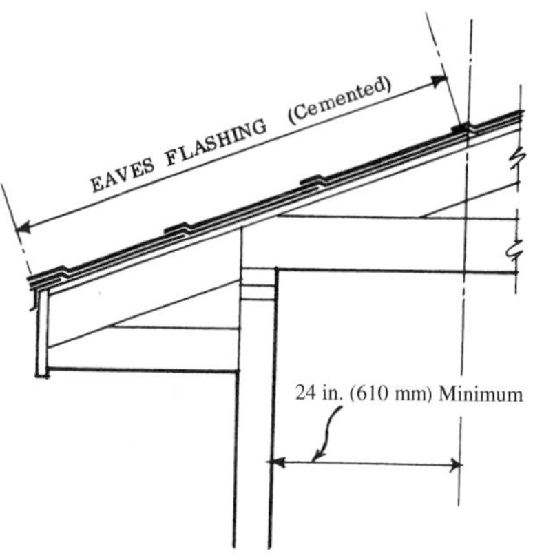

b

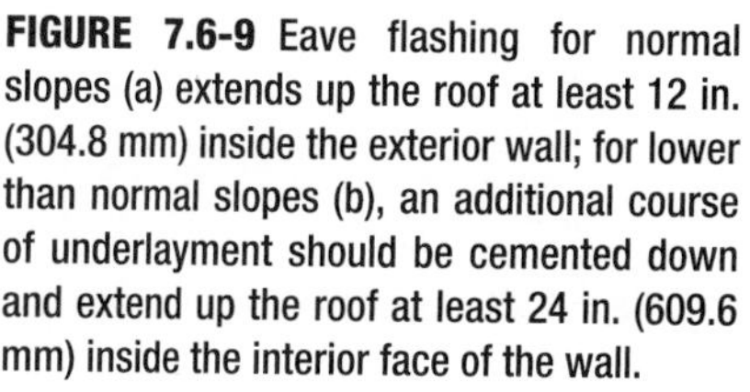

FIGURE 7.6-9 Eave flashing for normal slopes (a) extends up the roof at least 12 in. (304.8 mm) inside the exterior wall; for lower than normal slopes (b), an additional course of underlayment should be cemented down and extend up the roof at least 24 in. (609.6 mm) inside the interior face of the wall.

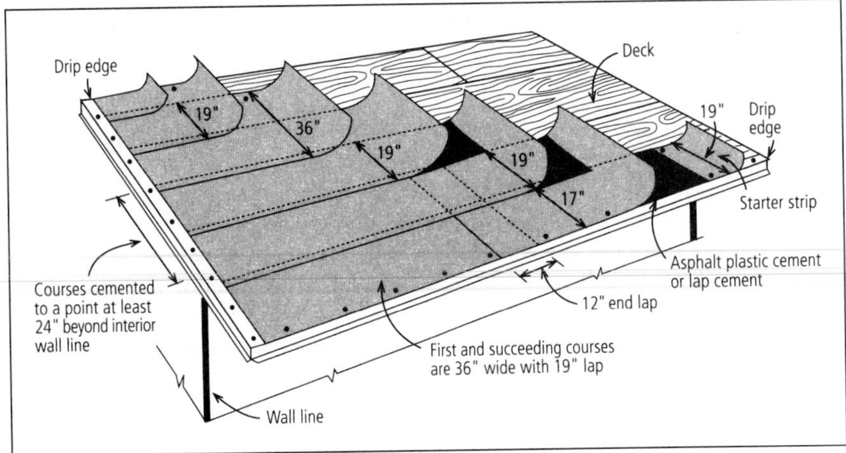

FIGURE 7.6-10 Application of double-layer felt underlayment and eave flashing. (Courtesy Asphalt Roofing Manufacturers Association.)

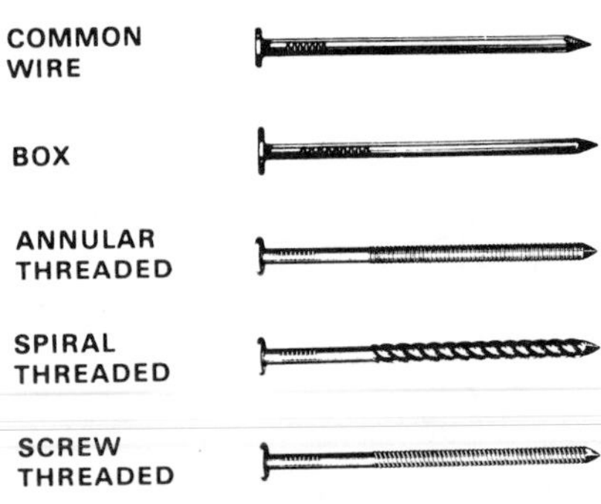

FIGURE 7.6-13 Types of smooth and threaded shank nails used for application of shingle roofing other than for application of asphalt shingles (see Fig. 7.6-32).

eaves and rakes, is recommended for most shingle roofs. Drip edges should be designed and installed to protect deck edges and prevent leaks at this point by allowing water to drip free of underlying eave and cornice construction. Typical preformed drip edge shapes are shown in Figure 7.6-11.

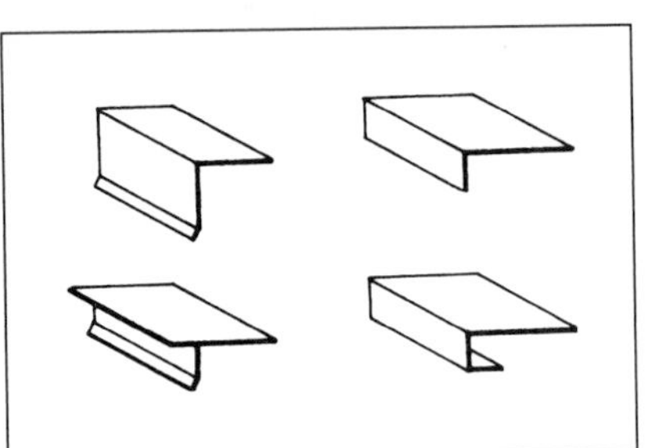

FIGURE 7.6-11 Typical drip-edge shapes.

A drip edge is applied to the sheathing and under the underlayment at the eaves, but over the underlayment up the rake (Fig. 7.6-12).

Flashing at Wall and Roof Intersections Sheet metal flashing should be installed at horizontal intersections and at vertical intersections when the exterior finish material does not provide a self-flashing joint.

7.6.1.7 Nails and Fasteners

No single step in applying steep roofing is more important than proper nailing. Suitability is dependent on (1) selecting the correct nail for the roofing and sheathing type (Fig. 7.6-13), (2) using the correct number of nails, (3) locating them correctly, and (4) choosing a nail metal compatible with the metal used for flashings.

Specific recommendations for the type, size, number, and spacing of roofing nails are given in the following subsections for each roofing material. Some general recommendations follow.

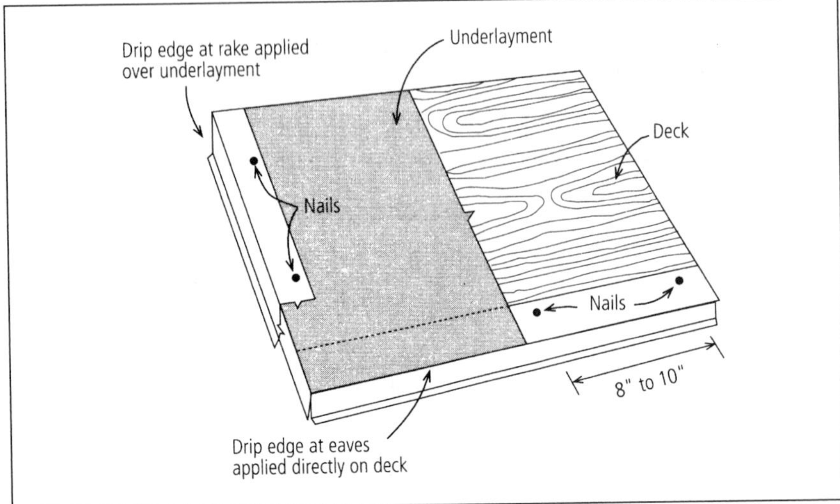

FIGURE 7.6-12 Application of drip edge at rake and eave. (Courtesy Asphalt Roofing Manufacturers Association.)

LUMBER BOARDS, PLANK DECKING, AND PLYWOOD SHEATHING

Roofing nails should be long enough to penetrate through the shingle and through the lumber boards or plywood sheathing. They should penetrate at least 1 in. (25.4 mm) into plank roof decking. Nails used in applying shingles over plywood sheathing should have threaded shanks.

NONWOOD SHEATHING

When gypsum, concrete, fiberboard, or other nonwood sheathing materials are used, special fasteners and special details for fastening are often necessary to provide adequate anchorage. The nonwood material manufacturer's recommendations should be followed to ensure acceptable performance.

7.6.2 ASPHALT SHINGLES

Unless specific exceptions are made in the following text, the provisions of Section 7.6.1 apply to asphalt shingles and their application.

7.6.2.1 Materials

Refer to Section 7.5.7 for a general discussion of the materials and products used in asphalt roofing and their manufacture. Raw materials are processed into finished shingle roofing products in a continuous sequence as described in Section 7.5.7. The data in the next subsection expand the material in Section 7.5.7 to cover asphalt shingles.

MANUFACTURE

The dry organic felt used as a base for making underlayment and asphalt shingles is described in Section 7.5.7.

After the felts have passed through the saturator tank, wet looper, and coater, as described in Section 7.5.7, granules of specified color or color combinations are applied from a hopper and spread thickly on the hot coating bitumen. These coarse granules are usually colored slate or rock granules (either natural or colored by a ceramic process). These granules give shingles increased weather and fire resistance and provide a wide range of colors and color blends. To protect the underlying bitumen from the impact of light rays, the granules should be opaque, dense, and well graded for maximum coverage.

Then a powder, such as talc or mica, is applied to the back (smooth) side of the felt to prevent layers of shingles in packages from sticking together. The powder is not a permanent part of the finished product and gradually disappears from exposed surfaces after application.

The sheet is then run through a series of press and cooling rollers, subjecting the sheet to a controlled pressure that embeds the granules in the coating to the desired depth. At this stage some shingles are also textured by the pressure of an embossing roller, which forms a pattern in the surface.

The sheet then travels into a cooling or finish looper where it is cooled so it can be cut into shingles in a shingle-cutting machine. The sheet is cut from the back (smooth) side as it passes between a cutting cylinder and an anvil roll. Shingles are then separated into stacks of the proper number for packaging. The stacks are moved to packaging equipment, where the bundles are prepared for warehousing and shipment.

In addition to the rules for storing asphalt roofing products in the warehouse and on the job, as outlined in Section 7.5.7, shingle bundles should not be stored to a height of more than 4 ft. (1220 mm), to eliminate the possibility of sticking or discoloration.

7.6.2.2 Shingle Types

There are many weights and patterns of both individual and strip shingles available for both new construction and reroofing projects. Sizes, exposures, and other information covering some shingle types are given in Figure 7.6-14. Available products may differ as to weight, dimension, and design from those shown.

7.6.2.3 Standards of Quality

Asphalt shingles should conform to Underwriters Laboratories, Inc. (UL) standards. The UL specifications constitute the recognized standard for the asphalt roofing industry and include provisions for (1) fire resistance (see Fig. 7.5-12), (2) wind resistance, (3) felt saturation tests to determine saturant quantity and saturation efficiency, (4) coating asphalt thickness and distribution, (5) granule adhesion and distribution, (6) weight, count, size, coloration, and other finished product characteristics before and after packaging, and (7) application instructions.

UL maintains a check on labeled products by making periodic factory inspections. UL inspectors are afforded the privilege of using the manufacturers' control laboratories and visit, inspect, and analyze product samples. When necessary, they recommend to the manufacturer measures that are required to maintain product standards to remain eligible to use the UL label.

Shingles should conform to UL 790 with a minimum of a Class C rating. Many shingles today have a Class A rating. Each shingle bundle should be identified with the UL class label. In areas subject to winds of hurricane force, which the United States Weather Service defines as those of 75 mph (120.7 km/h) or more, *wind-resistant* shingles in accordance with UL 997 Type I are recommended. Such shingles are provided with factory-applied adhesive or integral locking tabs; the bundles carry the UL class label and include the words "Wind Resistant."

Asphalt roof shingles are of four basic types: (1) *strip* shingles, (2) *laminated* (multithickness shingles), (3) *interlocking* shingles, and (4) *giant individual* shingles for application by either the *American* or *Dutch Lap* method. Most strip shingles have cutouts to separate shingle *tabs*, but some are made without cutouts. Three-tab shingles are the most popular type in use today.

Wind-resistant strip shingles have tabs with factory-applied adhesive or integral locking tabs. Lightweight individual shingles applied by the Dutch lap method are intended primarily for reroofing existing roofs.

Each kind of shingle is applied in a distinct way, giving a different appearance to a roof. Starter units, hip and ridge shingles, and valley roll material are also manufactured to match asphalt shingles.

7.6.2.4 Preparation for Asphalt Shingle Installation

Limitations imposed by the slope of a roof are shown in Figure 7.6-3. Asphalt shingles may safely be applied using normal application methods on roofs with slopes of 4 in 12 or greater. Asphalt shingles may be used on slopes as low as 2 in 12 when applied using the lower than normal slope methods described in this section.

ROOF SHEATHING

Sheathing types, thicknesses, spans, and nailing requirements are discussed in Chapter 6. If lumber boards are used, they should not be more than 6 in. (152.4 mm) in nominal width. Wider boards may swell or shrink in width enough to buckle or distort the shingles. Badly warped boards and those containing excessively resinous areas or loose knots should be rejected. If any of these defects appear after the sheathing has been applied, they should be covered with 26-gauge (0.55 mm) galvanized steel patches before the underlayment is placed.

UNDERLAYMENT

Underlayment should be No. 15 asphalt-saturated felt. No. 30 asphalt felt should not be used as general underlayment beneath asphalt shingles, because it is too stiff and may cause wrinkling of the shingles applied over it. However, on lower than normal slope applications (2 in 12 to 4 in 12), two layers of No. 15 felt should be installed over the entire roof.

EAVE FLASHING

Eave flashing is recommended in areas where the outside design temperature is 30°F (–1.1°C) or colder (see Fig. 7.6-8) or where there is a possibility of ice forming along the eaves and causing a water backup.

Normal Slope (4 in 12 or greater) A course of No. 90 mineral-surfaced roll roofing or not less than No. 55 smooth roll roofing should be installed at the eaves over the underlayment. It should overhang the underlayment and drip edge from ¼ to ⅜ in. (6.4 to 9.5 mm) and should extend up the roof to cover a line

Typical asphalt shingles								
Product	**Configuration**	**Appx. shipping weight per square (lbs.)**	**Shingles per square**	**Bundles per square**	**Width (in.)**	**Length (in.)**	**Exposure (in.)**	**ASTM fire and wind ratings**
Laminated self-sealing random tab shingle 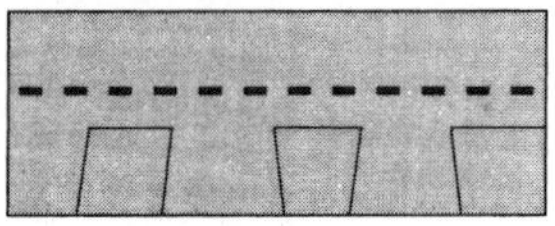	Various edge, surface texture and application treatments	240-360	64-90	3-5	11 1/2 - 14 1/4	36-40	4-6 1/8	Class A or C fire rating. Many wind resistant.
Multi-tab self-sealing square tab strip shingle 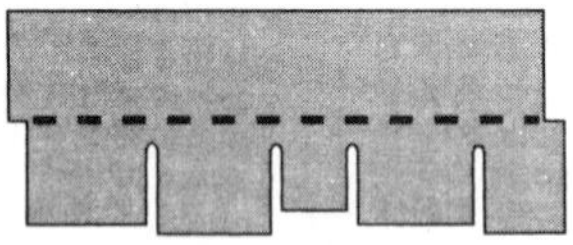	Various edge, surface texture and application treatments	240-300	65-80	3-4	12-17	36-40	4-8	Class A or C fire rating. Many wind resistant.
Multi-tab self-sealing square tab strip shingle 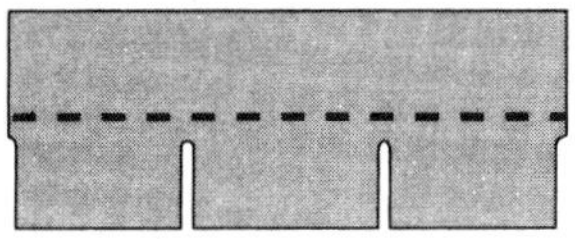	Three-tab or four-tab	200-300	48-80	3-4	12-13 1/4	36-40	5-5 5/8	Class A or C fire rating. All wind resistant.
No-cutout self-sealing square tab strip shingle 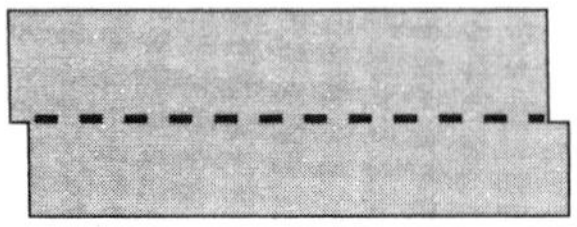	Various edge and surface texture treatments	200-300	65-81	3-4	12-13 1/4	36-40	5-5 5/8	Class A or C fire rating. All wind resistant.
Individual interlocking shingle (basic design) 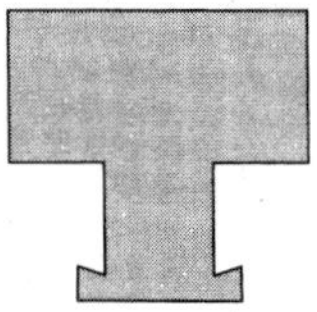	Several design variations	180-250	72-120	3-4	18-22 1/4	20-22 1/2	n/a	Class A or C fire rating. Many wind resistant.

FIGURE 7.6-14 Typical asphalt shingles. (Courtesy Asphalt Roofing Manufacturers Association.)

at least 12 in. (304.8 mm) inside the interior wall line of the building (see Fig. 7.6-9a). When the eave overhang requires the flashing to be wider than 36 in. (914.4 mm), the necessary horizontal lap joint should be cemented and located outside the exterior wall line of the building.

Lower-Than-Normal Slope (as low as 2 in 12) Asphalt shingles may be applied on slopes as low as 2 in 12. For slopes less than 4 in 12 down to 2 in 12, or in areas subject to severe icing, a double layer of underlayment should be used over the entire roof. Eave flashing may be formed by cementing the two layers of underlayment together with asphalt cement up the roof to a line that is at least 24 in. (609.6 mm) inside the interior wall line of the building (see Figs. 7.6-9b and 7.6-10).

This eave flashing should be formed by applying a continuous plastic asphalt cement layer, at the rate of 2 gal./100 sq. ft. (0.815 L/m^2), to the surface of the underlayment starter course before the first full underlayment course is applied. Cement should also be applied to the 19 in. (482.6 mm) underlying portion of each succeeding course that lies within the eave flashing area before the next course is placed. This cement should be applied uniformly with a comb trowel, so that underlayment does not touch underlayment at any point when the application has been completed. The overlying sheet should be pressed firmly into the entire cemented area (see Fig. 7.6-10).

DRIP EDGES

A corrosion-resistant, nonstaining drip edge should be installed along the eaves and rakes (see Fig. 7.6-12).

FLASHING

See Section 7.6.1.6.

Valley Flashing Either the open, closed-cut, or woven method may be used to construct valley flashing.

In constructing *open* valley flashing, a valley underlayment strip of No. 15 asphalt-saturated felt, 36 in. (914.4 mm) wide, is applied first, as shown in Figure 7.6-15, centered in the valley, and secured with enough nails to hold it in place. When laps are necessary, they are formed so that the upper strip overlaps the lower by at least 12 in. (304.8 mm) and are bonded with asphaltic plastic cement. The lower edge is trimmed flush with the eave drip strip. In areas experiencing heavy rainfalls, a layer of self-adhered eave and flashing membrane should be installed beneath the valley underlayment.

The horizontal roof underlayment courses are cut to overlap this valley strip a minimum of 6 in. (152.4 mm). Where eave flashing is required, it is applied over the valley underlayment strip.

A 26-gauge galvanized steel or other corrosion-resistant metal strip is placed over the underlayment strip from ridge to eave and trimmed flush with the eave drip edge (Fig. 7.6-16). The metal is secured, without puncturing it, by means of metal cleats spaced 8 to 12 in. (203.2 to 304.8 mm) apart. Overlaps should be 12 in. (304.8 mm) and cemented.

A chalkline should be snapped on each side of the valley for its full length before shingles are applied. The chalklines should start 6 in. (152.4 mm) apart at the ridge and spread more widely apart at the rate of ⅛ in./ft. (10.42 mm/m) down to the eave. They should serve as a guide in trimming the last shingle units to fit the valley, ensuring a clean, sharp edge. The upper corner of each end shingle should be clipped, as shown in Figure 7.6-17, to direct water into the valley and prevent water penetration between courses. Each shingle should be cemented to the valley lining with asphalt cement to ensure a tight seal. No exposed nails should appear along the valley flashing.

Closed (woven) valleys can be used only with strip shingles. These methods double the shingle coverage throughout the length of the valley, increasing weather resistance at this vulnerable point. Individual shingles cannot be used, because nails would be required at or near the center of the valley lining.

To provide either a closed-cut or woven valley flashing, first a valley lining made from a 36-in. (914.4 mm)-wide strip of No. 50, or heavier, roll roofing should be placed over the valley underlayment and centered in the valley (Fig. 7.6-18).

In a *closed-cut valley flashing* (see Fig. 7.6-18), the shingles on one side of the

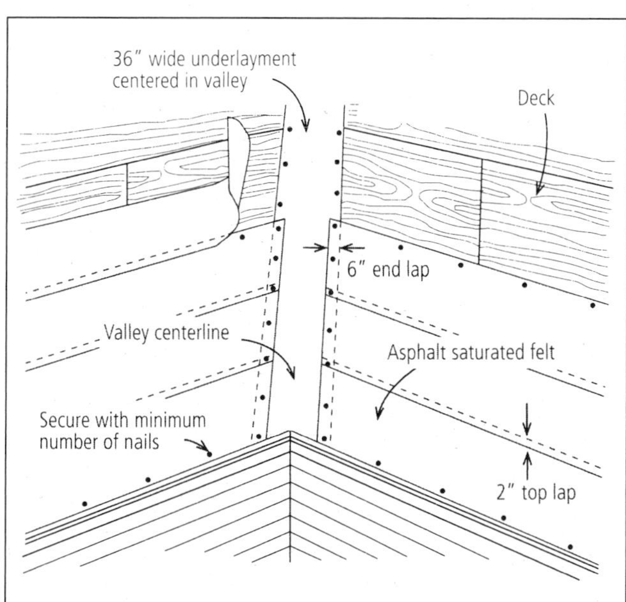

FIGURE 7.6-15 **Flashing underlayment in a valley.** (Courtesy Asphalt Roofing Manufacturers Association.) open

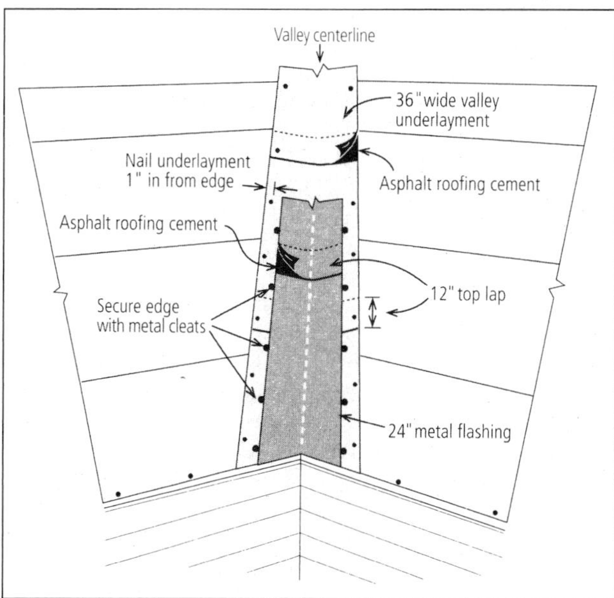

FIGURE 7.6-16 **Open valley flashing using corrosion-resistant metal.** (Courtesy Asphalt Roofing Manufacturers Association.)

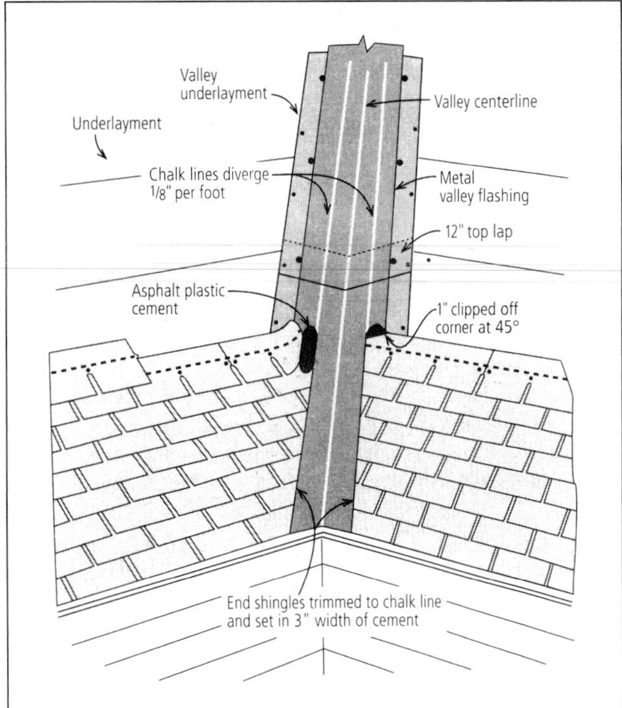

FIGURE 7.6-17 Shingle application in an open valley. (Courtesy Asphalt Roofing Manufacturers Association.)

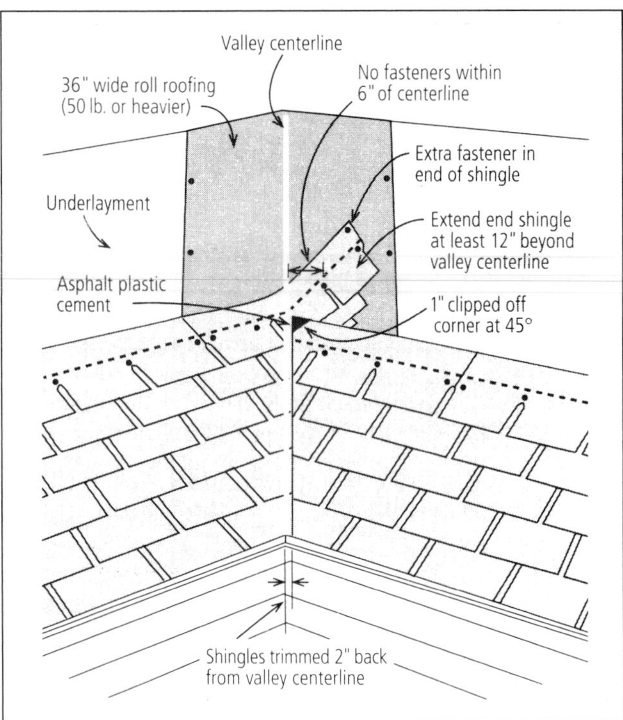

FIGURE 7.6-18 Shingle application in a closed cut valley. (Courtesy Asphalt Roofing Manufacturers Association.)

valley are laid and extended at least 12 in. (304.8 mm) onto the roof surface on the opposite side of the valley (Fig. 7.6-18). A joint is never made in the valley. Normal shingle nailing is used, except that no nail falls within 6 in. (152.4 mm) of the valley center and each shingle crossing the valley is nailed at its end with two nails. A chalkline is placed about 2 in. (50.8 mm) from the valley centerline on the unshingled side. The shingles are then placed on the unshingled side of the valley and trimmed to the chalkline. The upper corner of each shingle on the chalkline is trimmed 1 in. (25.4 mm) to help direct water into the valley. Then each shingle end is embedded in a 3-in. (76.2 mm)-wide strip of asphalt plastic cement.

In a *woven valley flashing* installation (Fig. 7.6-19), valley shingles should be laid over the lining either by (1) applying them on both roof surfaces at the same time, weaving each course in turn over the valley, or (2) covering each surface first to a point approximately 36 in. (914.4 mm) from the center of the valley and weaving the valley shingles in place later.

In either case, the first course at the valley should be laid along the eaves of one roof surface over the valley lining, extending it along the adjoining roof surface for a distance of at least 12 in. (304.8 mm). The first course of the adjoining roof surface should then be carried over the valley on top of the previously applied shingle. Succeeding courses should then be laid alternately, weaving the valley shingles over each other.

The shingles should be pressed tightly into the valley and nailed normally, except that no nail should be located closer than 6 in. (152.4 mm) from the valley centerline and two nails should be used at the end of each terminal strip.

Vertical Sidewall Flashing When the rake of a roof abuts a vertical wall, the most satisfactory method of protecting the joint is to use stepped metal base flashing applied over the end of each shingle course (Fig. 7.6-20).

Individual pieces of this metal flashing are rectangular in shape, approximately 7 in. (177.8 mm) by 10 in. (254 mm). With strip shingles laid 5 in. (127 mm) to the weather, each flashing strip will lap the shingle by 2 in. (50.8 mm). This flashing should be bent, with one end extending 5 in. (127 mm) onto the roof underlayment, the rest extending up the wall surface. Each flashing piece should be placed up-roof from the exposed edge of the shingle that overlaps it and be secured to the wall sheathing with one nail in the top corner. When the shingles are laid 5 in. (127 mm) to the weather, each 7-in. (177.8 mm) width of flashing will lap the next flashing piece by 2 in. (50.8 mm). The wall siding should be finished over the flashing to serve as counter (cap) flashing, but should be held far enough away from the shingles so that the ends of the siding may be painted for weather protection (Fig. 7.6-21).

Chimney Flashing Before flashings are placed, shingles should be applied over the underlayment up to the front face of a chimney and a coat of asphalt primer applied to the masonry to seal the surface at all points where plastic cement will be applied.

Base flashing may be No. 90 mineral-surfaced roll roofing. The chimney front (Fig. 7.6-22) should be applied first by securing the lower section over the shingles in an asphalt plastic cement bed. The vertical upper section should be secured against the masonry with plastic cement and nails driven into the mortar joints; the triangular ends should be bent around the corners of the chimney and cemented into place.

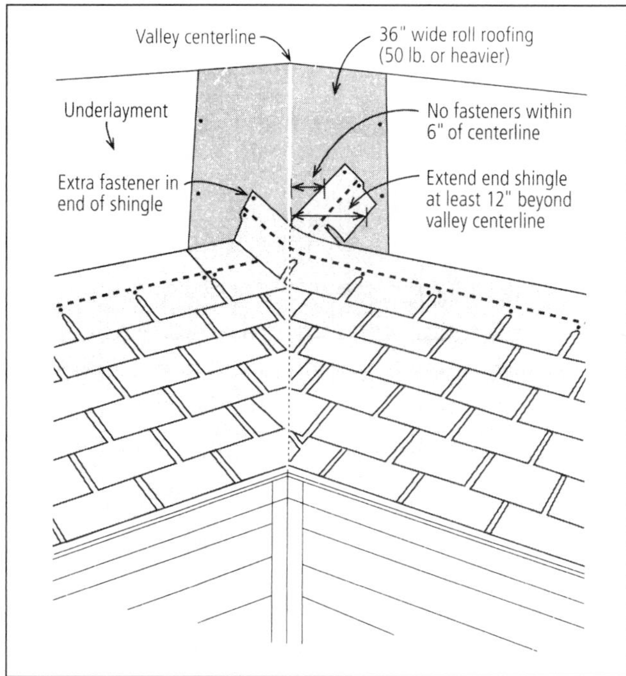

FIGURE 7.6-19 Shingle application in a woven valley. (Courtesy Asphalt Roofing Manufacturers Association.)

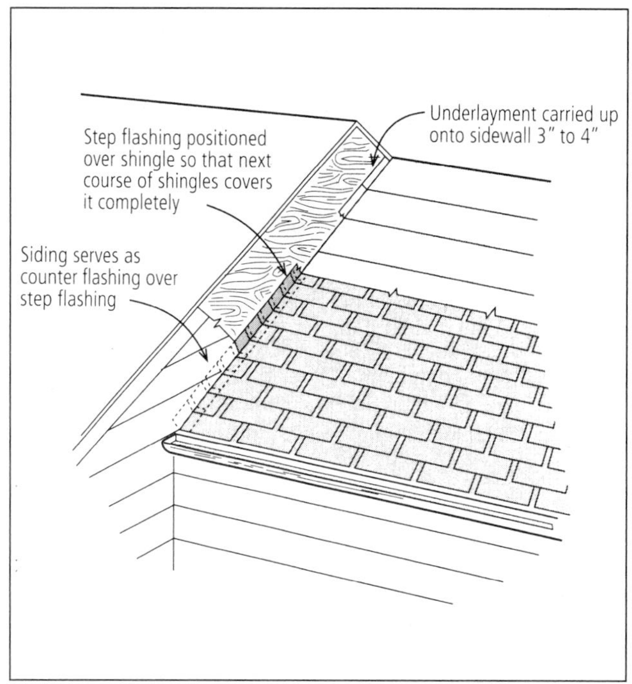

FIGURE 7.6-21 Vertical wall flashing using step flashing. (Courtesy Asphalt Roofing Manufacturers Association.)

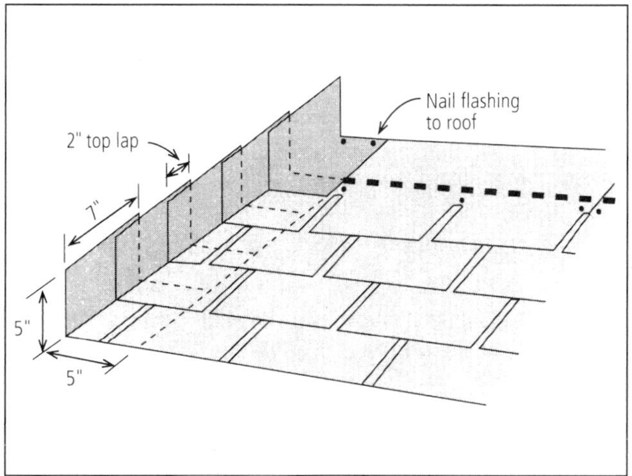

FIGURE 7.6-20 Application of step flashing. (Courtesy Asphalt Roofing Manufacturers Association.)

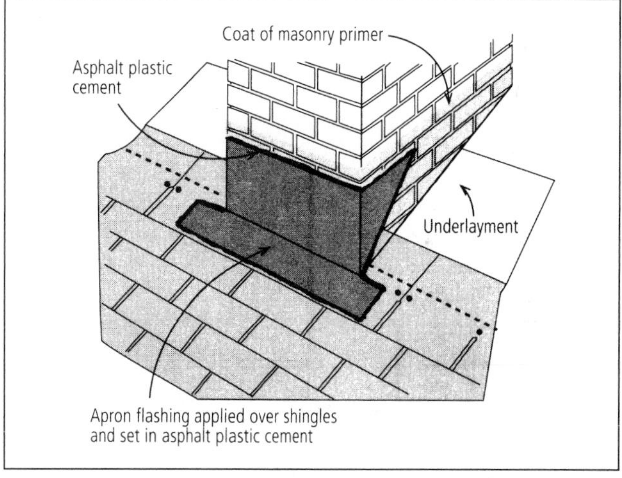

FIGURE 7.6-22 Chimney front apron flashing. (Courtesy Asphalt Roofing Manufacturers Association.)

The base flashing for chimney sides should be applied as shown in Figure 7.6-23. It should be secured to the deck with plastic cement and to the brick with plastic cement and nails; the triangular ends of the upper section should be turned around the chimney corners and cemented in place over the front base flashing.

The base flashing covering a saddle should be secured in plastic cement over the underlayment. The corners should be installed as shown in Figure 7.6-24. The flashing over the cricket should be installed as shown in Figure 7.6-25, with the standing portion extending up the masonry 6 to 12 in. (152.4 to 304.8 mm). This should be capped as shown in Figure 7.6-26. The capping should be bedded tightly in plastic cement, centered over the saddle flashing, and extend up the roof. This provides added protection at the point where the ridge of the saddle joins the roof.

Metal counter flashing should be applied over the base flashing to complete the flashing installation (Figs. 7.6-27 and 7.6-28). One method of securing counter flashing to masonry is to rake out the mortar joint to a depth of 1½ in. (38.1 mm) and insert the bent back edge of the flashing in the raked joint between bricks. The joint should be then filled with portland cement mortar or plastic cement, and the flashing bent down over the base flashing.

The counter flashing on the chimney front should be one continuous piece. On

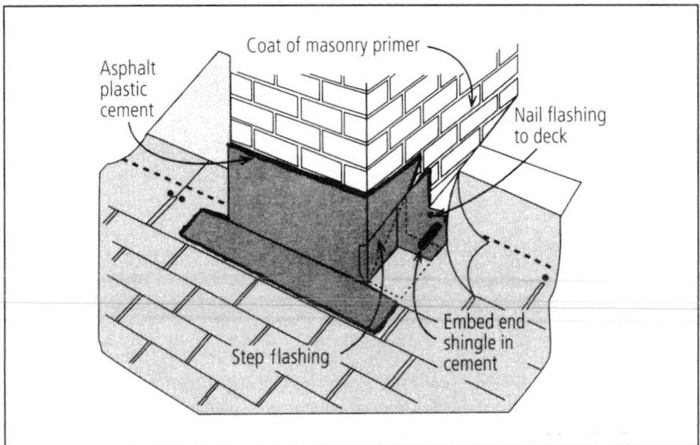

FIGURE 7.6-23 Step flashing at side of chimney. (Courtesy Asphalt Roofing Manufacturers Association.)

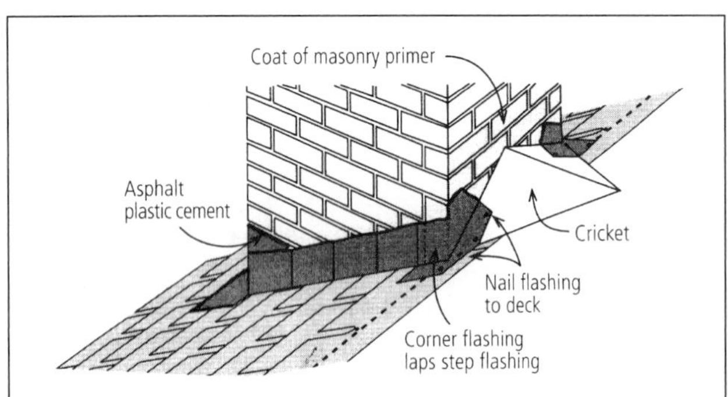

FIGURE 7.6-24 Corner flashing at rear of chimney. (Courtesy Asphalt Roofing Manufacturers Association.)

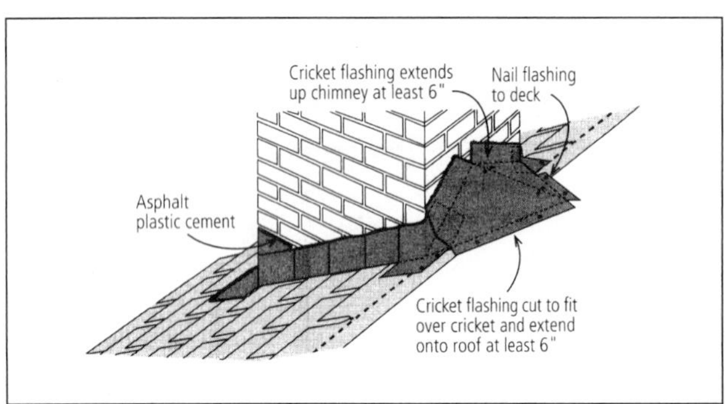

FIGURE 7.6-25 Flashing over cricket. (Courtesy Asphalt Roofing Manufacturers Association.)

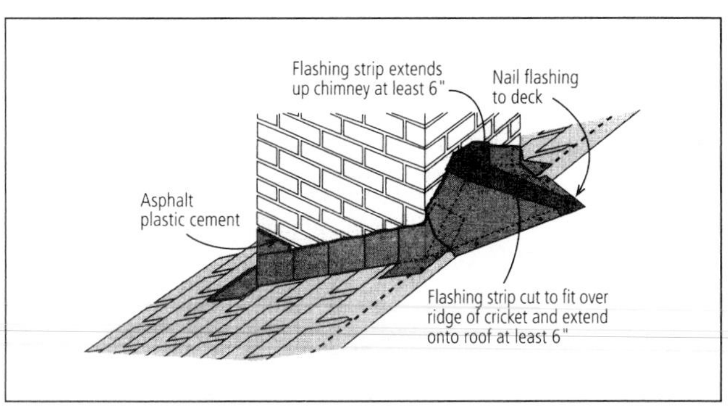

FIGURE 7.6-26 Flashing over ridge of cricket. (Courtesy Asphalt Roofing Manufacturers Association.)

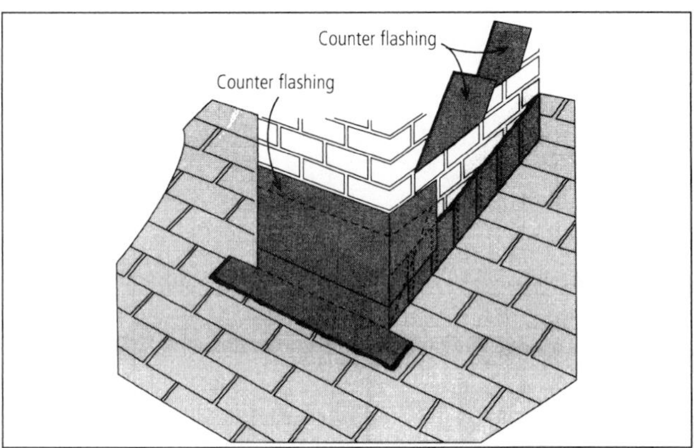

FIGURE 7.6-27 Counter flashing at front and side of chimney. (Courtesy Asphalt Roofing Manufacturers Association.)

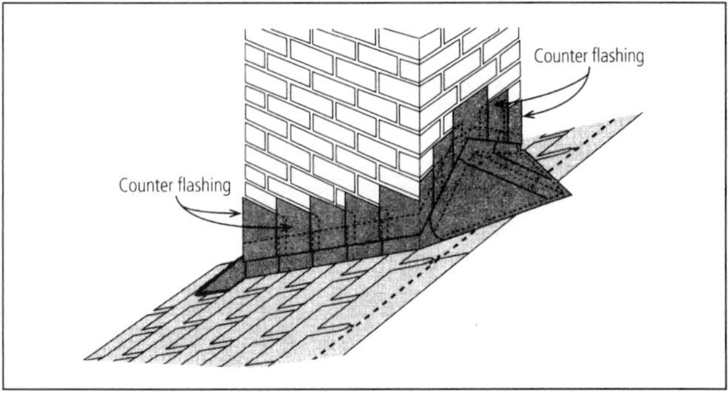

FIGURE 7.6-28 Counter flashing at side and rear of chimney. (Courtesy Asphalt Roofing Manufacturers Association.)

the sides and rear, the sections should be cut to conform to brick joint locations and roof pitch and should lap each other at least 3 in. (76.2 mm).

Stack Flashing Pipes projecting through a pitched roof require special flashing methods. Either corrosion-resistant metal sleeves that slip over stacks and have adjustable flanges to fit any roof slope, or asphalt products may be used with metal stacks. Plastic flanges are available for use with plastic plumbing stacks.

Roofing should be applied up to the point where the stack projects and should be fitted around it (Fig. 7.6-29). A shingle is cut to fit over the pipe and set in asphalt roofing cement. A preformed flange is then fitted snugly over the pipe and sealed to the pipe and to the shingles with asphalt plastic cement (Fig. 7.6-30). Plastic flanges are welded to plastic pipes using adhesives recommended by the pipe manufacturer. The flange should lay flat on the roof. Laying of shingles continues, with shingles being cut to fit around the pipe and embedded in plastic roofing cement. The completed installation should appear as it does in Figure 7.6-31.

7.6.2.5 Shingle Application

Shingles should be laid so they overlap and cover each other to shed water. The

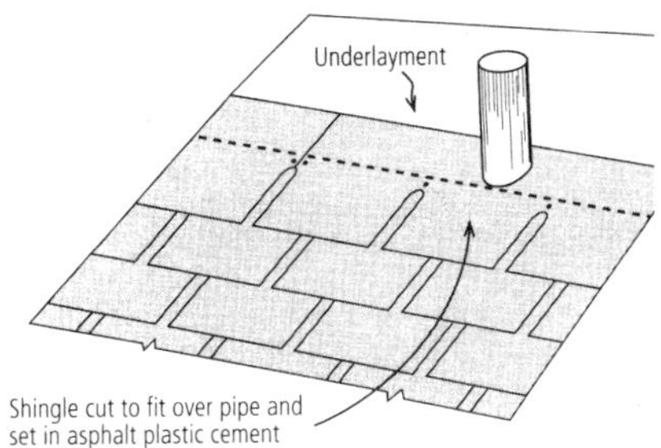

FIGURE 7.6-29 Application of shingle over vent pipe. (Courtesy Asphalt Roofing Manufacturers Association.)

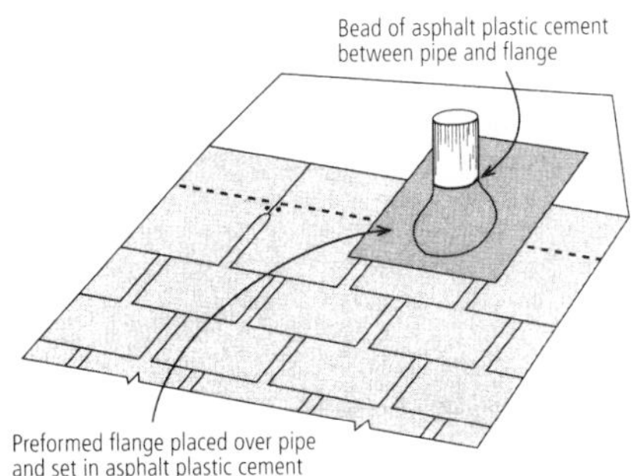

FIGURE 7.6-30 Flashing at vent pipe. (Courtesy Asphalt Roofing Manufacturers Association.)

extent of overlapping, which depends on their shape and method of application, furnishes single, double, and sometimes triple coverage over different roof areas. Strip shingles, some interlocking shingles, and giant individual shingles applied by the American method, providing either double or triple coverage, are suitable for new construction, but as stated earlier, strip shingles are the only ones in general use today.

Strip and individual hexagonal shingles and giant individual shingles, applied by the Dutch lap method, provide only single coverage and are used only to reroof existing construction. Refer to manufacturers' literature for details of their application.

Shingles should be applied to a tight roof deck providing a suitable nailing base, with underlayment, drip edge, and flashings in place. Chimneys should be completed and counter flashing installed. Stacks and other equipment requiring openings through the roof should be in place, with provision for counter flashing where necessary; gutters should be hung.

NAILS AND FASTENERS

Nails (Fig. 7.6-32) for applying asphalt roofing should be hot-dipped galvanized steel, aluminum, or another corrosion-resistant metal and should have sharp points and large, flat heads, 3/8 to 7/16 in. (9.5 to 11 mm) in diameter. Shanks should not exceed 0.135 in. (3.429 mm) or be less than 0.105 in. (2.67 mm) in overall outside diameter (10- to 12-gauge wire). Shanks may be smooth or threaded, but threaded nails are preferred for increased holding power in lumber board sheathing and should definitely be used in plywood sheathing. Aluminum nails should have screw threads with an approximately 12½-degree thread angle. Galvanized steel nails, if threaded, should have annular threads. Nail lengths typically required are given in Figure 7.6-33.

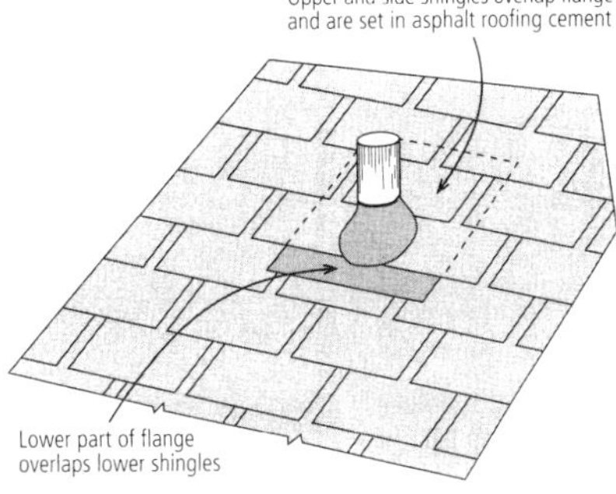

FIGURE 7.6-31 Application of shingle around stack flashing. (Courtesy Asphalt Roofing Manufacturers Association.)

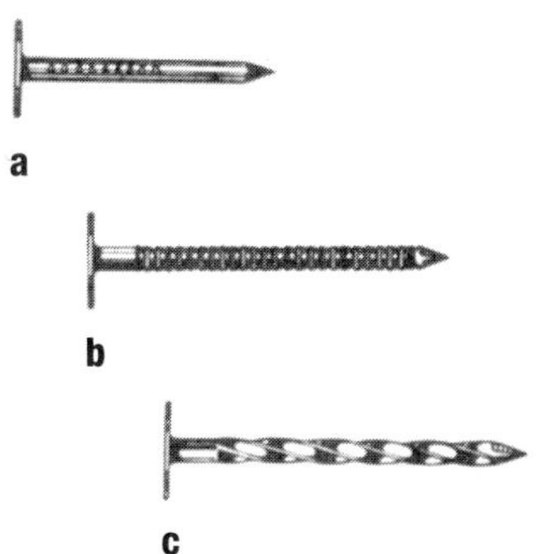

FIGURE 7.6-32 Asphalt shingle nails: (a) smooth, (b) annular threaded, and (c) screw threaded.

STRIP SHINGLES (NORMAL SLOPE)

On small roofs, strip shingles may be laid from either rake. On roofs 30 ft. (9.15 m) and longer, they should be started at the center and worked both ways from a vertical line to assure more accurate vertical alignment and to provide for meeting and matching above a projection such as a dormer or chimney.

FIGURE 7.6-33 Asphalt Shingle Nail Lengths

Application	1-in. (25.4 mm) Sheathing	3/8-in. (9.5 mm) Plywood
Strip or individual shingle (new construction)	1¼ in. (31.8 mm)	7/8 in. (22.2 mm)
Over asphalt roofing (reroofing)	1½ in. (38.1 mm)	1 in. (25.4 mm)
Over wood shingles (reroofing)	1¾ in. (44.45 mm)	—

When a roof surface is broken, shingles should be started from a rake and laid toward a dormer or into a valley. Where unbroken, they should be started at the rake that is most visible. If both rakes are equally visible, shingles should be started in the center and worked both ways.

Minor variations in the dimensions of asphalt shingles are unavoidable. Variations seldom exceed plus or minus ¼ in. (6.4 mm) in either length or width in a 12 by 36 in. (304.8 by 914.4 mm) strip shingle, and usually will be no more than plus or minus 1/8 in. (3.2 mm). However, to control the placement of shingles so that cutouts will be accurately aligned horizontally, vertically, and diagonally, horizontal and vertical chalklines should be used to control alignment.

STARTER COURSE

The purpose of a starter course is to back up the first regular course of shingles and fill in the spaces between tabs. The starter course may be either a 7-in. (177.8 mm)-wide (or wider) mineral-surfaced roll roofing *starter strip* of a color to match the shingles, or a row of inverted *starter course* shingles or strips made from the top lap portion of the shingles with the tabs cut off.

A *starter strip* should be applied over an eave flashing strip, even with its lower edge along the eave. The starter strip should be fastened with roofing nails 3 or 4 in. (76.2 or 101.6 mm) above the eave edge and spaced so that the nail heads will not be exposed at the cutouts between tabs on the first course (Fig. 7.6-34).

Starter course shingles should be applied with tabs facing up the roof over the eave flashing strip. Nails should be spaced so they will not be exposed at the cutouts of the first course (see Fig. 7.6-34).

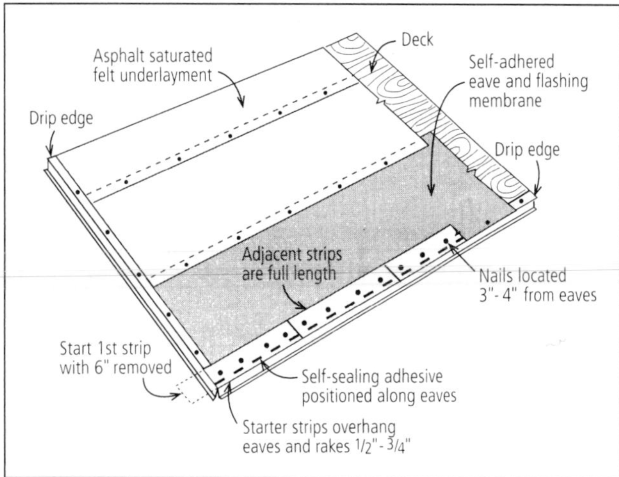

FIGURE 7.6-34 Application of starter strip. (Courtesy Asphalt Roofing Manufacturers Association.)

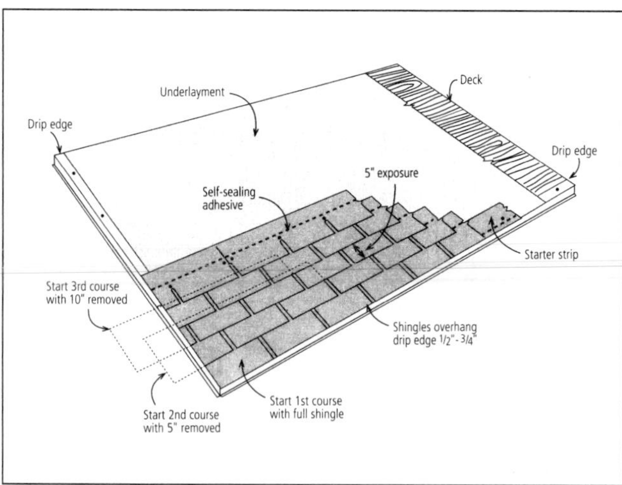

FIGURE 7.6-36 Application of shingles using the 5-in. (127 mm) method. (Courtesy Asphalt Roofing Manufacturers Association.)

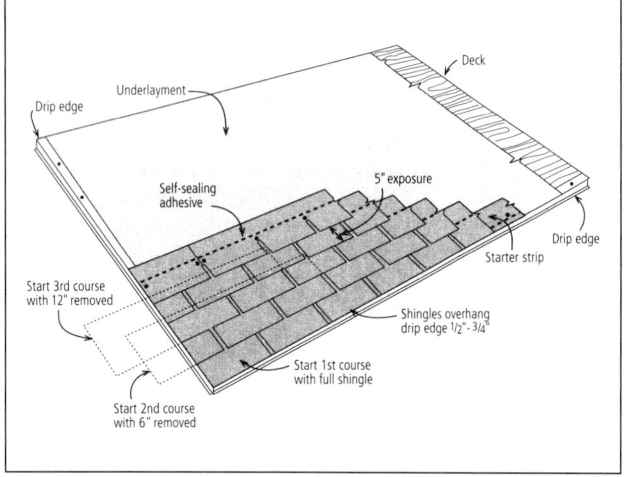

FIGURE 7.6-35 Application of shingles using the 6-in. (152.4 mm) method. (Courtesy Asphalt Roofing Manufacturers Association.)

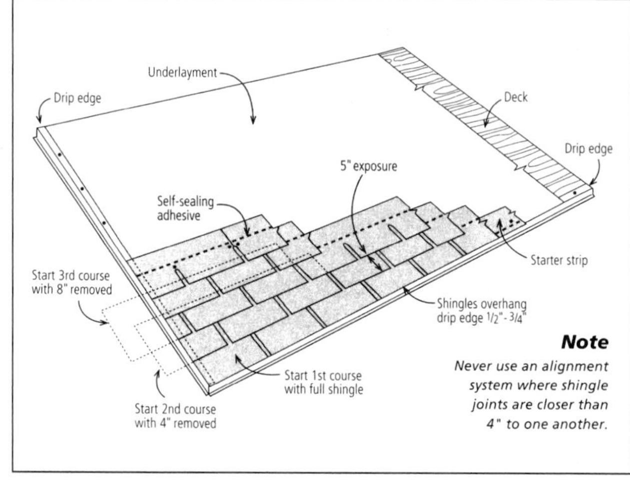

FIGURE 7.6-37 Application of shingles using the 4-in. (101.6 mm) method. (Courtesy Asphalt Roofing Manufacturers Association.)

First and Succeeding Courses The first course should be started with a full shingle. Succeeding courses start with portions removed, depending on the shingle style and the pattern desired. Shingles with no cutouts and those with variable butt lines should be installed according to their manufacturers' directions. There are three methods of applying three-tab shingles: the 6-in. (152.4 mm) method, the 5-in. (127 mm) method, and the 4-in. (101.6 mm) method.

The 6-Inch Method In this method (Fig. 7.6-35), the second course is installed with the first 6 in. (152.4 mm) removed. Each succeeding course through the sixth is started with a shingle having an additional 6 in. (152.4 mm) removed. Adjacent shingles are installed full width. The seventh course then starts with a full shingle, and the pattern described is followed again.

The 5-Inch Method In this method (Fig. 7.6-36), the second course is installed with 5 in. (127 mm) removed. Each succeeding course through the seventh is installed with an additional 5 in. (127 mm) removed. Adjacent shingles are installed full width. Additional courses continue the 5-in. (127 mm) offset pattern. One method is to start the eighth course with 11 in. removed from the first shingle.

The 4-Inch Method In this method (Fig. 7.6-37), the second course is started with a shingle having 4 in. (101.6 mm) removed. Each succeeding course through the ninth is installed with an additional 4 in. (101.6 mm) removed. Adjacent shingles are installed full width. The tenth course then starts with a full shingle, and the pattern described is followed again.

NAILING AND FASTENING STRIP SHINGLES

The number of nails per strip and nail placement are vital for suitable roof application. To prevent buckling, each shingle should be in perfect alignment before nails are driven. Nailing should be started at the end nearest the shingle last applied and proceed to the opposite end, with nails driven straight to avoid cutting the fabric of a shingle with the edge of a nail head. Nail heads should be driven flush, not sunk below the surface of the shingles. When a nail fails to penetrate the sheathing, an additional nail should be driven in a new location.

Nailing Three-Tab Shingles Three-tab shingles installed in locations having normal weather conditions require 4 nails for each three-tab strip. When such shingles are applied with a 5-in. (127 mm) exposure, the nails should be placed ⅝ in. (15.9 mm) above the top of the cutouts and located horizontally with one nail 1 in. (25.4 mm) back from each end and one nail on the centerline of each cutout.

Three-tab shingles installed in high wind regions require 6 nails for each three-tab strip. When such shingles are applied with a 5-in. (127 mm) exposure, the nails should be placed ⅝ in. (15.9 mm) above the top of the cutouts and located horizontally with one nail 1 in. (25.4 mm) back from each end and two nails at each cutout, located 1 in. on each side of the cutout.

Windy locations require additional protection. In areas where wind velocities are 75 mph (120.7 km/h) or greater, shingles manufactured with factory-applied adhesives or integral locking tabs that conform to the UL Standard for Class C Wind Resistant Shingles should be used.

If UL Class C Wind Resistant Shingles are not used, an effective way to obtain protection against high winds is to cement the free tabs in place with an asphaltic plastic roofing cement. A spot of cement, about the size of a quarter, should be applied on each tab with a putty knife or caulking gun. The free tab should then be pressed against the cement. In applying the cement, shingles should not be bent back farther than necessary.

HIPS AND RIDGES

Hips and ridges may be finished by using hip and ridge shingles furnished by the manufacturer or by cutting pieces at least 9 by 12 in. (228.6 by 304.8 mm) either from 12 by 36 in. (304.8 by 914.4 mm) shingle strips, or from mineral-surfaced roll roofing of a color to match the shingles. They should be applied by bending each shingle lengthwise down the center, with equal exposure on each side of the hip or ridge. In cold weather the shingles should be warmed before bending. Application should begin at the bottom of a hip or at one end of a ridge with a 5-in. (127 mm) exposure. Each shingle should be secured with one nail at each side, 5½ in. (139.7 mm) back from the exposed end and 1 in. (25.4 mm) up from the edge.

Metal ridge roll subject to corrosion may discolor the roof and should never be used with asphalt roofing products.

STRIP SHINGLES (LOWER THAN NORMAL SLOPE)

Strip shingles may be used on roof slopes less than 4 in 12 but not less than 2 in 12. Application methods for slopes lower than normal require (1) double underlayment, (2) cemented eave flashing strip, (3) shingles provided with factory-applied adhesives conforming to the UL Standard for Class C Wind Resistant Shingles, or each free tab of strip shingles cemented in place. Any shingle arrangement described for normal slope application methods may be used.

7.6.2.6 Reroofing Existing Construction

When a roof is to be reroofed with asphalt shingles, existing wood shingles, asphalt shingles, roll roofing, built-up asphalt roofing, or slate need not be removed before applying the new shingles, provided that (1) the strength of the existing roof framing is adequate to support the new dead and live loads, (2) the existing roof sheathing is sound and will provide adequate anchorage for nails used to apply new roofing, and (3) not more than two roofs are currently in place (original and one reroof). Tile and cedar shake roofing must be removed. When either the framing or sheathing is inadequate, the old roofing, regardless of type, must be removed to correct the defects.

EXISTING ROOFING TO BE REMOVED

When old roofing must be removed, a roof should be prepared for new roofing by (1) stripping the old roofing down to the roof deck, (2) repairing and reinforcing existing roof framing where required to provide adequate strength, (3) removing existing damaged sheathing and installing new sheathing in its place, (4) filling spaces between boards with securely nailed wood strips of the same thickness, or removing existing sheathing altogether and resheathing, (5) removing protruding nails, and (6) covering large cracks, slivers, knotholes, loose pitchy knots, and excessively resinous areas with sheet metal securely nailed to the sheathing.

EXISTING ROOFING TO REMAIN

When inspection discloses that existing roofing need not be removed, the existing roofing's surface should be prepared to receive new roofing, as described in the following paragraphs for various roofing types.

Existing Wood Shingles Preparation includes:

- Removing loose or protruding nails and nailing the shingles in a new location;
- Nailing down loose shingles;
- Splitting badly curled or warped shingles and nailing down the segments;
- Replacing missing shingles with new ones.

When shingles and trim at eaves and rakes are badly weathered or the location is subject to high winds, the shingles there should be cut back and 1 × 4 (25 × 100) or 1 × 6 (25 × 150) boards nailed, with their outside edges projecting beyond the edges of the deck the same distance as the wood shingles did (Fig. 7.6-38). Beveled wood *feathering strips* should be used along the butts of each course of old shingles to provide a relatively smooth surface to receive new asphalt roofing.

Existing Asphalt Shingles Preparation includes:

- Nailing loose and protruding nails;
- Removing and replacing badly weathered edging strips;
- Removing or renailing loose, curled, lifted, or broken shingles;
- Replacing missing shingles with new ones.

Existing Roll Roofing Preparation includes:

- Splitting buckles and nailing segments down smoothly;
- Removing loose and protruding nails, and placing new fasteners;
- Trimming damaged and torn roofing to have square or rectangular sides;
- Patching tears in the roofing with new roofing.

Existing Built-up Roofs If the slope of the roof is not less than 2 in 12 and if the roof sheathing is sound and can be expected to provide good nail-holding power, new shingles can be applied. Old slag, gravel, and other coarse surfacing materials should first be removed, leaving the surface of the underlying felts smooth and clean. New asphalt shingles should be applied directly over the felts as described under "Strip Shingles (Lower than Normal Slope)" in Section 7.6.2.5.

UNDERLAYMENT (REROOFING)

When existing roofing must be removed before applying new material, underlay-

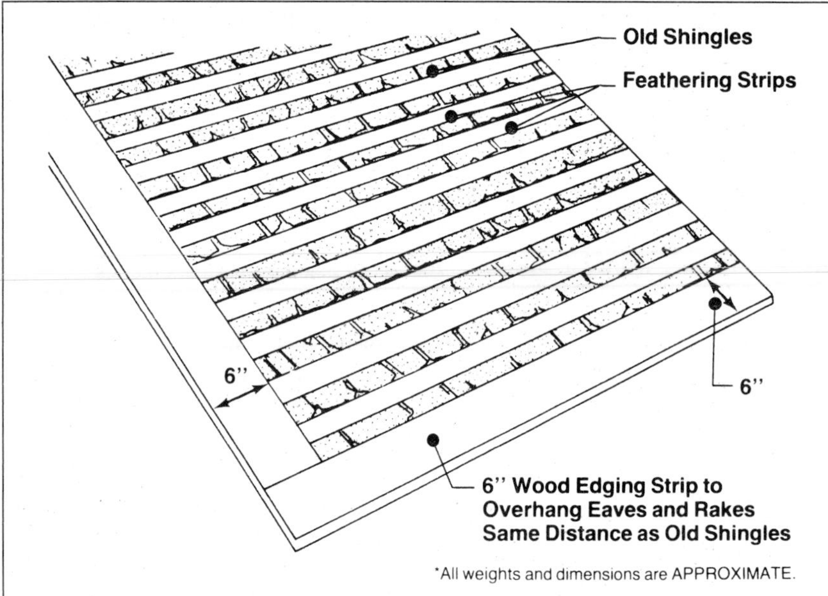

FIGURE 7.6-38 Preparation of existing roof to receive new shingles. (Reprint courtesy National Roofing Contractors Association.)

ment is required over the entire roof as it would be prescribed for new construction.

When new shingles can be applied directly over old roofing, no underlayment is required as the old roofing adequately serves the same purpose. However, underlayment may be required by code over old wood roofing.

DRIP EDGE AND EAVE FLASHING STRIP (REROOFING)

When installing new asphalt shingles over an existing wood shingled roof, drip edges and eave flashing strip should be installed as for new construction (Fig. 7.6-39).

FLASHING (REROOFING)

Existing metal flashing, unless perforated or deteriorated, may be left in place and reused.

Valley Flashing When the existing roof has an open valley, the valley is built up with mineral-surfaced roll roofing and valley flashing is installed in the same way as for new roofing.

Flashing Against a Vertical Wall When new asphalt shingles are being applied over old asphalt shingles, the joint between a vertical wall and the rake of an abutting roof may be treated as shown in Figure 7.6-40.

When asphalt shingles are being installed over old wood shingles, a 6- or 8-in. (152.4 or 203.2 mm) strip of smooth No. 50 roll roofing or a mineral-surfaced roll roofing placed upside down should be applied over shingles abutting a wall surface, using a row of nails placed about 4 in. (101.6 mm) on centers along each edge. As the work proceeds, the strip should be covered with asphalt plastic cement and the end of each course of new shingles should be firmly secured by bedding it in the cement. Improved appearance and a tight joint can be achieved by using a caulking gun to apply a final cement bead between the ends of the shingles and the siding.

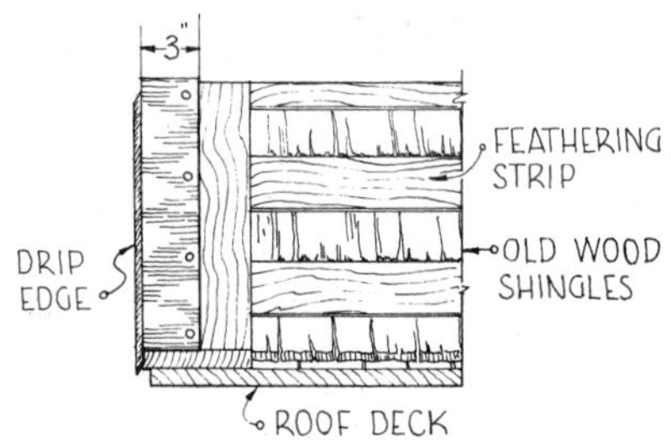

a

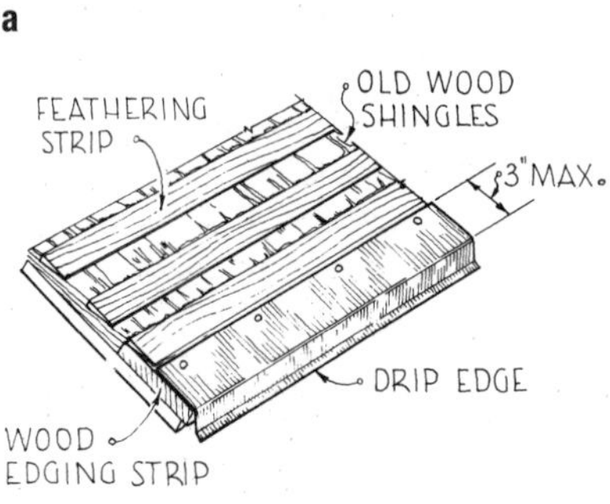

b

FIGURE 7.6-39 Application of drip edge at rake (a) and at eave (b). (Courtesy Asphalt Roofing Manufacturers Association.)

Chimney Flashing When existing flashing is damaged, it should be removed and new flashing installed as described earlier in this section. When existing flashing is in good condition, it may be lifted and shingles installed beneath it. The area of shingle beneath the flashing should be coated with asphaltic plastic cement. At the sides and at metal crickets, the shingles should be trimmed to within ¼ in. (6.4 mm) of the flashing and the last 3 in. (76.2 mm) of the shingles embedded in asphaltic plastic cement. Shingled crickets are finished in the same way as new shingled crickets.

SHINGLE APPLICATION (REROOFING)

After the existing roof has been prepared, the methods of applying the various shingle types are similar to those described in this section for new construction. The major difference is a nesting procedure used to minimize unevenness that may result from new shingles overlapping the butts of the existing shingles (Fig. 7.6-41).

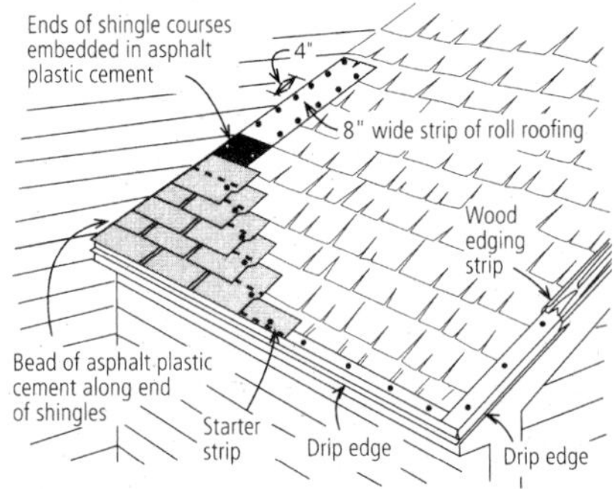

FIGURE 7.6-40 Wall flashing (reroofing). (Courtesy Asphalt Roofing Manufacturers Association.)

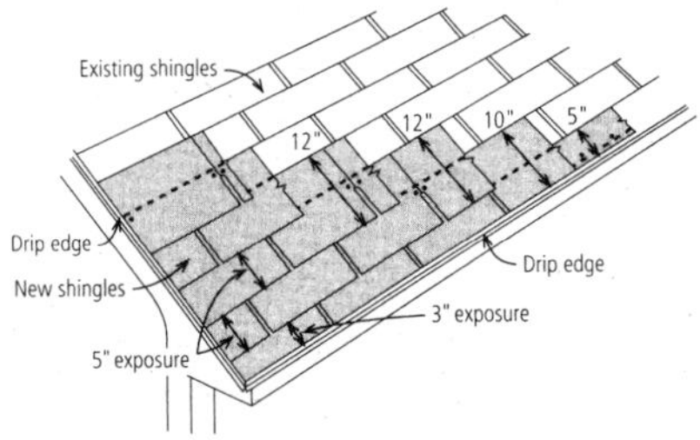

FIGURE 7.6-41 Application of new asphalt shingles over existing asphalt shingles using the nesting method. (Courtesy Asphalt Roofing Manufacturers Association.)

This nesting is achieved by removing the tabs from the starter strip and removing an additional 2 in. from the top so that when installed, the starter course does not overlap the existing course above. In addition, 3 in. (76.2 mm) is removed from the rake end of the starter course.

Two inches (50.8 mm) are also removed from the top of the first course, so that those shingles fit between the butts of the existing third course and the eave edge of the new starter strip without overlapping the existing third course.

Full-width shingles are installed for the second and succeeding courses. This will leave a 3-in. (76.2 mm) exposure on the first course instead of the normal 5 in. (127 mm). Succeeding courses, however, will have the full 5-in. (127 mm) exposure and will not overlap the butts of existing shingles.

7.6.3 SLATE ROOFING

Slate shingles are not widely used today, especially on commercial projects, but so much slate roofing exists that some discussion is appropriate.

7.6.3.1 Materials

SLATE

Slate is hard, practically nonabsorbent rock, characterized by its natural proclivity to split along a single plane. This single-direction cleavage characteristic permits slate to be split into thin sheets. Roofing slate is usually split in the direction of its natural grain.

Roofing slate is classified as either *commercial standard*, *textural*, or *graduated*. Commercial standard roofing slate is smooth and about 3⁄16 in. (4.8 mm) thick. Textural roofing slate is 3⁄8 in. (9.5 mm) thick or thinner. Its color is more variable than that of commercial standard slate, and its surface texture is rougher. Graduated roofing slate varies in thickness from 3⁄16 in. (4.8 mm) to 1½ in. (38.1 mm) or thicker. It also varies more in color and texture than either commercial standard or textural slate. Some slate used in roofing (*ribbon stock*) has strips of other rock embedded in it. Slate without ribbons is called *clear slate*.

Individual slate roofing pieces (*slates*) vary in size from 10 by 6 in. (254 by 152.4 mm) to 24 by 16 in. (609.6 by 406.4 mm). Most slates are square, but corners are sometimes clipped to produce special effects. Most slate is split so that the grain runs with the length of the slate.

Slate is available in several colors and shades, including those classified in the industry as black, blue black, gray, mottled gray, purple, variegated purple (mottled purple and green), green, and red. Other colors (*specials*) may also be used. Existing slate will almost certainly have changed color over time and may be difficult to match with new material. New slate, will, of course, also change color, but there may be a time when new slates are apparent. Slate is classified as *permanent* (or *unfading*) or as *weathering*, denoting the degree of color fading expected.

Before selecting material to match existing slate, it is important to determine the source of the existing slate and to try to find the same or similar slate from the same source as a match. Slate is currently mined in Virginia, New York, Maine, Vermont, and Pennsylvania, but existing slate may have come from another location. Slate from different states, even slate from different quarries in the same state or from a single quarry, may not match. Fortunately, some variation is normal, especially in some colors, so that an acceptable match is often possible.

Every slate should have at least two nail holes machine punched at the quarry. Slates 3⁄4 in. (19.1 mm) or more in thickness and 20 in. (508 mm) or more in length should have four nail holes. Hand punching should be done only where absolutely necessary, such as for fitting hips.

FASTENERS

Zinc nails and copper, brass, yellow metal, and other nails with copper content have all been used to install slate with success. For repairing slate work, however, the best choice is large-head, copper slaters' nails. Under no circumstances should nails intended for any other purpose, such as roofer's nails or galvanized steel nails, be used with slate roofing.

Nails should be at least 1 in. (25.4 mm) longer than the slate thickness, but never less than long enough to penetrate the roof sheathing. Nails should not, however, project on the other side of the sheathing unless they will be concealed in the finished work.

Nail gauge depends on the weight of the slates and their locations on the roof. For example, 3d nails can be used for commercial standard slates up to 18 in. (457.2 mm) long. Longer commercial standard slates require 4d nails. Hips and ridges of roofs covered with commercial standard slate should be installed using 6d nails. Heavier slates require heavier nails.

7.6.3.2 Preparation for Slate Roofing

SHEATHING

In the past, slate was often installed using wood pegs hooked over wood lath. More recently, some slate was hung from wood lath or battens or directly from steel structural elements by heavy wire hooked to the wood or structural elements. Sometimes slate was held to steel sections using long, bent copper nails. Today most slate roofing is installed over continuous nailable substrates, often wood, plywood, or a wood fiber product, sometimes over a nailable portland-cement-based material, and fastened in place with nails.

UNDERLAYMENT

Surfaces to receive slate roofing 1⁄4 in. (6.4 mm) or less in thickness, where the roof slope is 4 in./ft. (333.3 mm/m) or greater, should be covered with an underlayment consisting of at least one layer of No. 30 asphalt-impregnated, unperforated roofing felt.

Surfaces to receive slate roofing with a thickness greater than 1⁄4 in. (6.4 mm), where the roof slope is 4 in. (101.6 mm) or more per ft. (304.8 mm) should receive an underlayment consisting of two layers of No. 30 asphalt-impregnated, unperforated roofing felt or a single layer of No. 55 or 65 asphalt-impregnated, unperforated roofing felt.

Surfaces to receive slate roofing where the roof slope is less than 4 in./ft. (333.3 mm/m) cannot rightly be called steep roofs. In such conditions, slate should be installed as tile (without laps) as a ballast over membrane roofing or as a promenade surface over membrane waterproofing.

Under the conditions outlined earlier in this section, slate roofs should have ice shields. The ice shield should consist of at least two layers of No. 30 felt.

7.6.3.3 Slate Installation

Slates installed using nails should each receive no fewer than two nails. Four nails should be used in larger slates. Nails should be driven so that the head just touches the slate. The slate should not be clenched by the nail, but should hang on the nail.

Slates should be set so that joints fall as close as possible to the midpoint of the length of the slate immediately above or below, but never less than 3 in. (76.2 mm) from the end of those adjacent slates (see

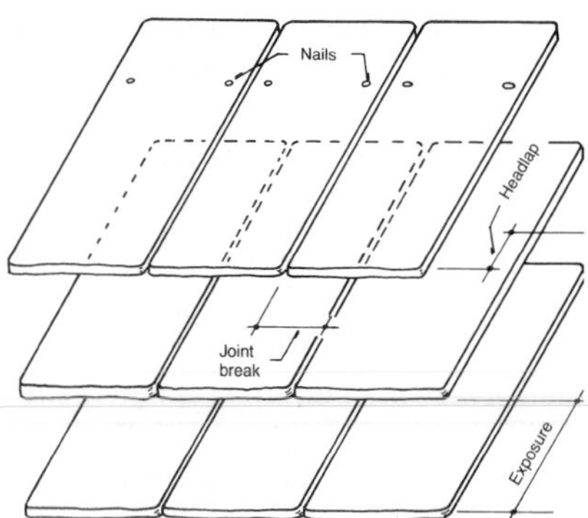

FIGURE 7.6-42 Slate jointing. (Courtesy H. Leslie Simmons.)

"Joint Break" in Fig. 7.6-42). Joints should never be open clear through to the underlayment below.

Slate weather exposure depends on the size of the slate and the headlap (see Fig. 7.6-42). Exposure is the shingle length minus the headlap divided by 2. The exposure for a 24-in. (609.6 mm) shingle with a 3-in. (76.2 mm) headlap is 10½ in. (266.7 mm).

The amount of headlap varies with the roof slope and the region. On slopes from 0 to 4 in./ft. (333.3 mm/m), slate should be laid as tile, with no lap. On slopes from 4 to 8 in./ft. (333.3 to 666.7 mm/m), headlap is often the "standard" 3 in. (76.2 mm), but roofers in some parts of the country consider 4 in. (101.6 mm) necessary. On slopes from 8 to 20 in./ft. (666.7 to 1666.7 mm/m) the standard 3-in. (76.2 mm) headlap is generally accepted as sufficient. In some southern areas, a headlap of 2 in. (50.8 mm) is considered adequate, but over most of the country a 2-in. (50.8 mm) headlap is restricted to mansards and other roofs with slopes of more than 20 in./ft. (1666.7 mm/m).

Slate is sold in squares. A square of slate is the amount needed to cover a square of roof (100 sq. ft. [9.29 m^2] of roof surface) when the headlap is 3 in. (76.2 mm). The actual headlap to be used must be considered when calculating the amount of slate necessary to cover a given area of roof. More headlap will result in less coverage for a square of slate.

Random-appearing slate roofs using standard and textural slate are achieved by varying the slate width while keeping the length equal or by varying both widths and lengths. Equal-width slates are seldom used.

Variation in graduated slate roofing is achieved by varying lengths, widths, and thicknesses. Larger and thicker slates are used in lower courses and graduate to thinner and smaller slates at the ridge. All courses contain various slate thicknesses.

Repaired hips and ridges should match the existing ones. The National Roofing Contractors Association's 1986 *NRCA Steep Roofing Manual* and the National Slate Association's *Slate Roofs* each contain detailed descriptions and drawings showing how several common types of hips and saddles should be formed.

Ridges may be *saddle* type or *comb* type. A saddle ridge is one in which the top slates abut evenly. In a regular saddle ridge, the top slates overlap the adjacent top slates (Fig. 7.6-43a). Two nails, concealed by the next higher slate, are used in each top slate. In a strip saddle ridge, the top slates do not overlap, they abut, and each is fastened down using four nails. Figure 7.6-43b is a section through either a regular or strip saddle ridge.

Comb ridges are formed similarly to strip saddle ridges, except that the tops of the ridge slates do not abut, but overlap by between 1⁄16 in. (1.6 mm) and ⅛ in. (3.2 mm) (Fig. 7.6-44). The projection may occur all on one side, or the projecting slates may alternate from one side to the other in a configuration called a *coxcomb ridge*.

There are several ways to form slate roofing hips. Four common methods are known as *saddle*, *mitered*, *fantail*, and *Boston hips*. *Saddle hips* and *strip saddle hips* are similar to saddle and strip saddle ridges, except that each slate is fastened in place using four nails.

In *mitered hips*, the slates forming the hip are in the same plane as the regular roofing slates and are cut to the slope of the hip (Fig. 7.6-45).

Fantail hips are a variation of mitered hips in which the bottom edge of the hip shingle is cut at an angle to form a fantail (Fig. 7.6-46).

Boston hips are similar in appearance to saddle hips, but are constructed by weaving three shapes of cut shingles together (Fig. 7.6-47).

Flashing beneath the hip types mentioned here is usually not necessary. Some existing conditions, however, may have

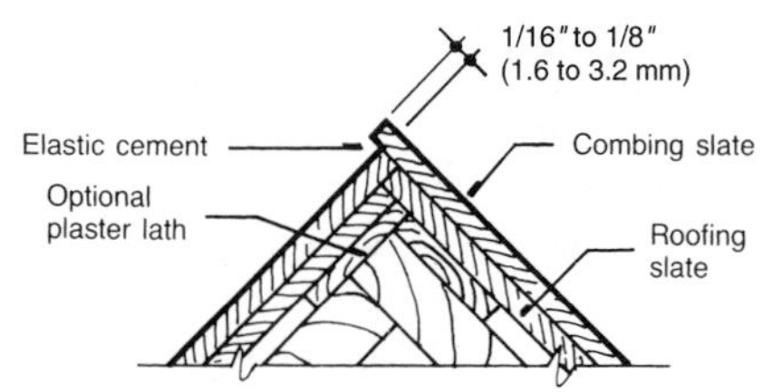

FIGURE 7.6-44 Detail at a comb ridge. (Courtesy H. Leslie Simmons.)

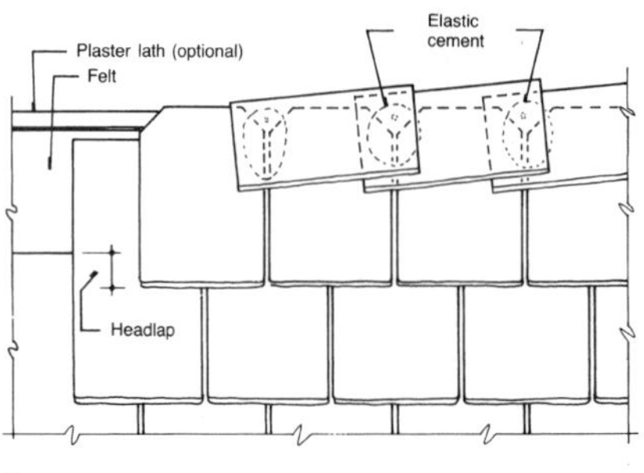

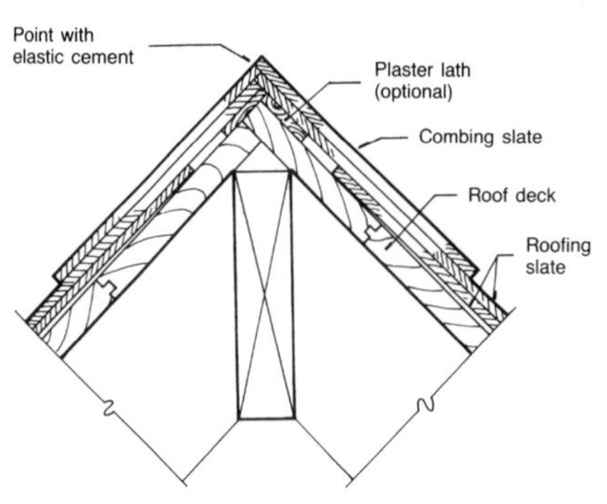

FIGURE 7.6-43 (a) **Regular saddle ridge.** (b) **Section through a saddle ridge.** (Courtesy H. Leslie Simmons.)

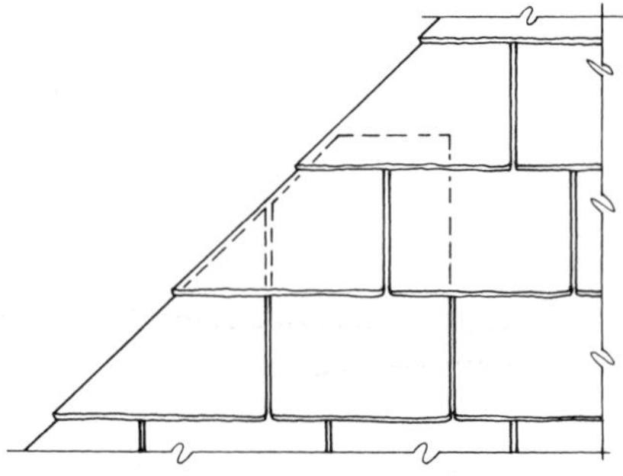

FIGURE 7.6-45 Mitered hip. (Courtesy H. Leslie Simmons.)

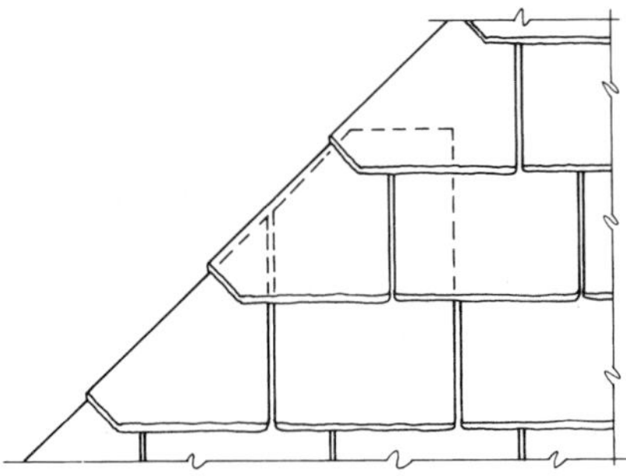

FIGURE 7.6-46 Fantail hip. (Courtesy H. Leslie Simmons.)

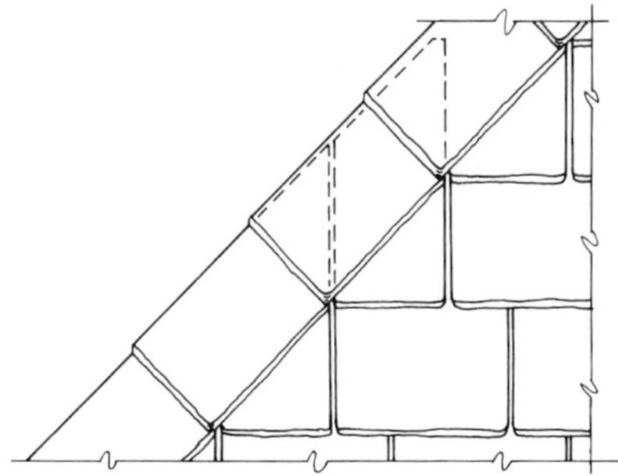

FIGURE 7.6-47 **Boston hip.** (Courtesy H. Leslie Simmons.)

been flashed with metal strips woven in with each course.

Valleys may be either *open*, *closed*, or *round*. Open valleys are protected using metal flashing. They should start with the slates 4 in. (101.6 mm) apart at the top, and widen uniformly from top to bottom at a rate of 1 in. (25.4 mm) in every 8 ft. (2400 mm).

In closed valleys, the slate is cut to abut in the valley and metal flashing is woven in as the slate is installed.

Flashing is not required in properly installed round valleys. Round valleys require considerable underlying support and special slate sizes. Considerable expertise is needed to properly install round valleys. Round valley slates are sometimes bedded in elastic cement. Round valleys built by inexperienced or insufficiently skilled personnel should be flashed. Round valleys susceptible to ice damage should always be flashed. Round-valley flashing may be of metal or bituminous materials.

In all types of ridge and hip conditions, the top regular slate courses should have their edges set in elastic cement to resist wind uplift. The joint between ridge slates and the overlap in comb ridges should be sealed with elastic cement. Some slate roofing is installed with elastic cement in other locations, such as along flashing lines and in the field of the roofing. Some slate roofing is installed with no elastic cement whatever and experiences no leaks as a result. The absence of elastic cement does not, in itself, mean that there is a problem or that the installation in inferior.

In every case, however, exposed nails should be coated with elastic cement. Where slates are nailed near flashing, the nails should not penetrate the flashing.

7.6.4 TILE ROOFING

Although there are industry standards for some other kinds of roofing, such standards for tile roofing are scarce. Sheathing, underlayments, fasteners, and tile should be installed in accordance with the tile producers' recommendations, applicable building code requirements, and the highest standards of the trade. The requirements in this section are a conglomerate of those generally recommended by the producers. Recommendations of industry-recognized sources, such as the National Roofing Contractors Association, have also been taken into account where they were found. Individual producer's and their associations' recommendations and building codes may differ, however.

7.6.4.1 Materials

CLAY TILE

Clay tile is a dense, hard, and nonabsorbent material produced by baking molded clay units in a kiln. It may be either glazed or unglazed.

Clay tile was used before the Golden Age of Greece. Since at least the early part of the twelfth century, some clay tile shapes have remained virtually the same. Clay tile for roofing comes in four general styles: "S" (Spanish), half-round (Mission), pan and cover (Roman), and flat tile. Figure 7.6-48 shows some general clay tile shapes, including three in the flat category. Available shapes may vary from those shown, especially in the flat and pan tile types. It is also possible that an existing roof will have a shape not shown.

Flat shingles may have special butt shapes. These are sometimes called *Persian* tile. Special butt shapes include those with rounded corners, round ends, half-hexagonal ends, half-octagonal ends, single-pointed ends, and many others.

Some flat interlocking tile comes in ½-, ¼-, and 1½-wide units to permit finishing at rakes.

Special shapes are used for finishing hips and ridges, and for trim and closures at ridges, hips, rakes, eaves, and valleys. Special shapes include regular tiles with the ends closed and specially shaped pieces to fit within or between tiles. Shapes vary to work with each tile type.

Round building elements, such as towers and turrets, may be roofed using *graduated* tiles. Graduated tiles are units made to match the regular types, but specially shaped to fit a rounded surface. Tile manufacturers furnish such tile routinely, but matching an existing graduated tile may be a problem. If the particular existing shape is no longer available, it may be necessary to have a match manufactured.

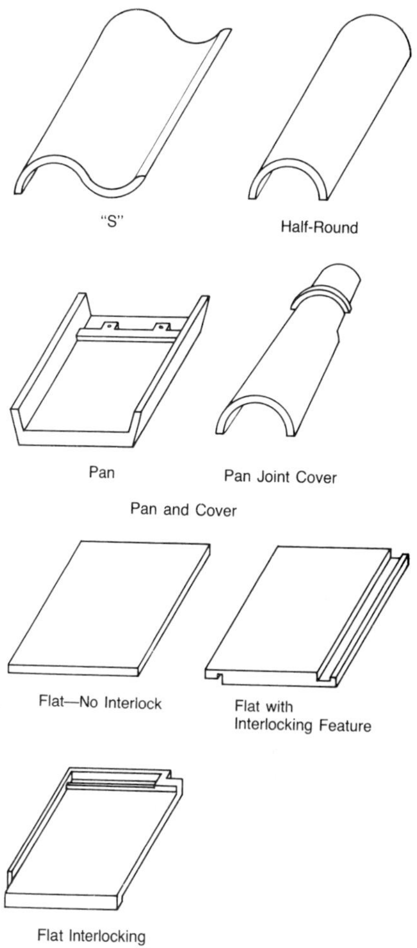

FIGURE 7.6-48 **Clay tile shapes.** (Courtesy H. Leslie Simmons.)

Some clay tile is finished to mimic wood shingles or slate. Broken or cut tiles will show clay's characteristic red color, however, and are easily identified as clay.

CONCRETE TILE

Concrete tile is made from a mixture of portland cement, aggregate, and water, and cured under pressure at a controlled temperature and humidity. Some tile is cast, other tile units are extruded. Concrete mixes vary from manufacturer to manufacturer. Some manufacturers use sand as the aggregate; others use lightweight materials, such as perlite. Some concrete tile is finished on the exposed surface while still wet, using a pigmented cement slurry. Some is colored throughout by pigment added to the concrete mix. Some is the natural concrete color with no added coloration. Concrete tile may be glazed or unglazed.

Some concrete tile is made to mimic wood shingles, clay tile, or slate in shape, color, and texture. Except under close scrutiny, the match with clay tile is fairly successful when the tile is new, differing only in individual tile shape. An attempt to match other materials, however, often fails. For example, some concrete tile is too large to successfully ape wood shingles. Broken edges identify concrete tile made to look like slate or wood. Some concrete tile matches the color of new wood shingles effectively, but, of course, does not age as wood does. In addition, the fungus that grows on wet concrete does not look like the fungus that grows on aging wood. Concrete tile does, however, enjoy the major advantage over wood of being fire resistant. And some concrete tile is colored during manufacture to match closely the appearance of mossy, aged wood. A great advantage of using this concrete tile instead of old wood is that the concrete tile will have the same appearance 50 to 75 years later, whereas wood that far gone will soon need to be replaced. Such colored concrete tile can be made to look more natural by using different colors placed randomly.

All concrete tile absorbs some moisture. Tile made with some lightweight aggregates absorbs more moisture than tile made with sand aggregate. The moisture absorption does not seem to harm the tile or its reinforcement, but is sometimes unsightly and can contribute to rusting metal or rotting wood in the underlying structure. Some concrete tile producers include water-absorption inhibitors in their mixtures to prevent such damage.

Most concrete tile used for roofing is either of the standard barrel tile shape (Fig. 7.6-49) or of a flat interlocking shape similar to the unit called "Flat with Interlocking Feature" in Figure 7.6-48. Shapes used on an existing building may vary slightly from those shown. A common variation, for example, is in the width of the valley in the barrel tile shape. Concrete tile on an existing roof may also be similar to one of the other styles shown in Figure 7.6-48 for clay tiles, or another shape not shown here.

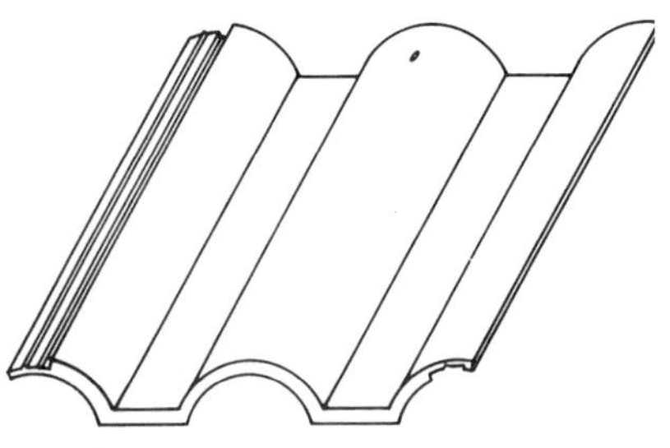

FIGURE 7.6-49 Concrete barrel tile type. (Courtesy H. Leslie Simmons.)

Special shapes are used for finishing hips and ridges and for trim and closures at ridges, hips, rakes, eaves, and valleys. Special shapes include regular tiles with the ends closed and specially shaped pieces to fit within or between tiles. Shapes vary to work with each tile type.

Round building elements, such as towers and turrets, may be roofed using *graduated* tiles, as described for clay tile earlier in this section.

FASTENERS

Tile may be nailed in position or hung from wire hangers, depending on the tile type, substrates, and code requirements.

Fasteners should be noncorrosive, such as stainless steel, aluminum, or a copper-bearing metal such as hard copper, silicon copper, or yellow metal. Nails should be at least 11 gauge (3.1 mm), large-headed (minimum 5⁄16 in. [7.9 mm] diameter), and long enough to penetrate completely through plywood sheathing, not less than 3⁄4 in. (19.1 mm) into other wood nailers, and the distance into other nailable materials recommended by the tile manufacturer. Nails should not, however, penetrate completely through board plank decks. Those for use in plywood should be ring-shank nails. Nails for use in boards should be smooth shank. Nails for use in gypsum plank or nailable concrete should be stainless steel or silicon bronze screw-shank nails long enough to penetrate 1⁄2 and 3⁄4 their length, respectively, into deck. They should never penetrate fully through the deck. For very hard decks, it may be necessary to use smooth-shank nails. Some so-called nailable decks may not have sufficient nail-holding power to support concrete tile.

Wire hangers are sometimes used to hang tile from wood battens, but are more often used on concrete and steel substrates. Hangers are also used, even where the substrate is wood, in locations where driven nails would penetrate flashings and in locations where driving nails is difficult. Wire hangers and associated ties are available in stainless steel, brass, copper, and galvanized steel. One system employs a 1½-in. (38.1 mm) by ½-in. (12.7 mm) wire strip hanger and 14-gauge (1.9 mm) tie wire, but a variety of hanger and tie types are available that may be acceptable to a particular tile manufacturer.

Hurricane clips, also called *storm clips*, are manufactured from brass or galvanized steel to shapes necessary for use with each tile type. Clips should not be visible in the completed roof. Hurricane clips are fastened positively to the roof deck and clip over tile edges to hold the tile in place during high winds.

MORTAR AND CEMENT

Most mortar used in clay tile roofing is composed of one part portland cement, four parts sand, and coloring to produce a mortar that matches the color of the tile.

Concrete tile manufacturers do not agree on the mix for mortar for use with concrete tile, or even on the materials to be used. It may be difficult, or impossible, to determine the actual mortar mix that was used in an existing application. Mortar color usually matches tile color.

In "mud-set" tile applications, concrete tile manufacturers usually recommend using Type M mortar as defined in ASTM Standard C 270.

Plastic cement should be a heavy-bodied asphalt roofing cement. Some tile manufacturers prefer that a silicone sealant be used in lieu of plastic cement. Plastic cement and sealant should be clear or colored to match the tile.

7.6.4.2 Preparation

SUBSTRATES, BATTENS, AND SUPPORT STRIPS

Clay and concrete tile may be installed over wood, metal, or concrete decks. Fastener requirements differ, depending on the tile configuration and deck type.

Wood sheathing for clay or concrete tile should be either exterior-grade plywood or boards. Plywood thickness should be selected to comply with structural requirements and fastener requirements, but should not be less than ½ in. (12.7 mm). Plywood joints should be left open 1⁄16 in. (1.6 mm) to permit expansion.

Board sheathing should be at least 1 × 6s (25 × 150s), spanning not more than 24 in. (609.6 mm). Supporting structure spans wider than 24 in. (609.6 mm) require heavier boards. Spaced sheathing is not generally recommended for clay tile, but may have been used on some existing roofs. Some sources say that spaced sheathing is not recommended in concrete tile installations, but some manufacturers recommend installation over spaced board sheathing, and some concrete tile roofs have been installed in that manner. Concrete tile applied over spaced sheathing should be installed in accordance with the tile manufacturer's recommendations.

Some manufacturers and many tile workers prefer that concrete tile be installed over wood battens on every roof, regardless of slope, because replacing damaged tile is much easier if battens are used. Another reason is the general feeling that nails do not hold in plywood well enough to support heavy concrete tile over an extended period. Many low-slope concrete tile roofs have been installed without battens, of course, but battens should be used on slopes of 7 in./ft. (583.3 mm/m) and steeper, regardless of tile type or manufacturer's recommendations. Battens are usually 1 × 2s (25 × 50s), fastened to the substrate at spacings recommended by the tile manufacturer. Spacing is dependent on tile type and size.

Clay and concrete tiles are seldom installed over cast-in-place concrete decks, because few cast-in-place concrete decks slope as much as 3 in./ft., which is the minimum at which the tile can be nailed to a roof. Such tile applied over cast-in-place concrete is usually fastened in place using wood battens in a configuration known as the *counter batten* system. In a counter batten system, wood battens, usually 1 × 2s (25 × 50s), are nailed to wood nailers cast into the concrete deck from eave to ridge at about 20 in. (508 mm) on center. Underlayment is installed over the deck and cast-in nailers before battens are applied. Batten spacing depends on tile size.

Clay and concrete tile installed over precast concrete decks (and sometimes over cast-in-place concrete) is usually fastened to a counter batten system, in which the wood members running up the slope are through-bolted to the concrete. Sometimes the batten running up the slope is omitted and the horizontal battens are themselves fastened directly to the deck by through-bolting.

A system of nailers and wood battens may also be used over metal decks. Often, however, fire ratings or code requirements will dictate that a fire-retardant sheathing be applied over the metal decking. Nailable fire-retardant sheathing, when the sheathing manufacturer so recommends and the tile manufacturer concurs, can be used the same as a plywood deck.

Battens can also be fastened over underlayment to nailers set flush with nonnailable insulation over any deck. Insulating products that are also nailable can be used like plywood when their manufacturer so recommends and the tile manufacturer concurs.

Manufacturers of some clay tile types recommend that the tile be applied over battens under all, or sometimes just certain selected, conditions. Such recommendations should be followed.

A means should be provided for water that finds its way through the tile to drain out of the roofing system. One way to do this is to separate the end joints between battens about ½ in. (12.7 mm), with end joints about 40 in. (1016 mm) on center. The problem with that method is that since joints must fall over the nailers below, the battens must be in many relatively short pieces. A better way is to raise the battens at least ¼ in. (6.4 mm) above the underlayment, using cut shingles or wood lath strips laid beneath the battens along the line of the nailers and fastened through the underlayment to the nailers.

Tile may also be attached directly to concrete or steel decks using wire hangers.

"S" and half-round clay tiles require nailers at ridges, hips, rakes, and projections. Flat clay and concrete tiles require nailers at ridges and hips. Nailer size varies with tile type and rise, which may be different for different manufacturers and is definitely different for different tile types. Nailers are either 1-in. (25 mm) or 2-in. (50 mm) nominal lumber, depending on the tile configuration and the tile manufacturer's recommendations. Hip and ridge nailers are usually set in cement mortar.

Eave strips are required, except for some types of flat clay tile, for concrete tile where starter course elevation is achieved using a specially made tile strip, and for other tile where special tile accessories are used to elevate the first course. Eave strips are usually 1 × 2 (25 × 50) wood members. Another way to raise the starter course is to elevate the fascia board. In any case, a tapered cant just above the nailer strip or tile starter course, or some other device, is necessary to prevent water from ponding at the eave line.

UNDERLAYMENT

Surfaces to receive clay tile should be completely covered with at least one layer of No. 30 asphalt-impregnated roofing felt. Additional layers of underlayment are required in some circumstances, but industry sources do not all agree on which circumstances dictate additional layers. For example, some sources say that roof slopes of 3 in./ft. (250 mm/m) or steeper can receive any type clay tile when completely covered with two layers of No. 30 asphalt-impregnated felt set in hot asphalt or mastic. The same sources imply that flat clay roofing tile needs a double layer of No. 30 felt underlayment on all slopes less than 5 in./ft. (416.7 mm/m), but can be installed on a single layer of felt for steeper slopes.

Other sources, including many tile manufacturers, however, do not agree on these requirements. Some manufacturers concur that any clay tile shape can be used on any slope exceeding 3 in./ft. (250 mm/m) when there is a double layer of No. 30 felt underlayment. Other manufacturers, however, recommend a double layer of No. 30 felt beneath flat tile on all slopes. Still others see no need for double layers of felt beneath clay tile except for flat clay tile, regardless of slope. Some say that one layer of No. 40 felt can be substituted for the two No. 30 layers. And some say that underlayment on steep slopes should consist of a single No. 40 felt layer, and that underlayment on low slopes should consist of No. 60 felt or two layers of No. 30 felt.

The level of disagreement related to underlayment for concrete tile is similar. Some sources say that, to be waterproof, concrete tile should be installed on slopes of 4 in./ft. (333.3 mm/m) or steeper, that flat shingle tile requires a slope of 5 in./ft. (416.7 mm/m) or more, and that any slope below 4 in./ft. (333.3 mm/m) should have two layers of roofing felt underlayment set in hot asphalt or mastic.

Other sources, however, do not agree with these requirements, nor do they agree with each other. Some say that concrete tile can be nail-applied on slopes of 3 in./ft. (250 mm/m) or steeper. Some manufacturers recommend that a single-ply No. 40 asphalt-impregnated, nonperforated roofing felt underlayment be applied as a minimum. Some recommend that surfaces with slopes between 3 in./ft. (250 mm/m) and 4 in./ft. (333.3 mm/m) to receive nailed-on concrete tile should be completely covered with an underlayment consisting of two layers of No. 15 felt with hot asphalt between the plies, or a waterproofing membrane, or a self-adhering bituminous membrane. Others say that slopes of 4 in./ft. (333.3 mm/m) or steeper should receive a single layer of No. 30 asphalt-impregnated roofing felt.

Consequently, an existing clay or concrete tile roof may have been installed using any one of those underlayment systems or another system entirely.

There is general agreement that both clay and concrete tile on slopes below 3 in./ft. (250 mm/m) must be installed only over built-up roofing or proper membrane waterproofing. Some manufacturers say that concrete tile on slopes of less than 4 in./ft. (333.3 mm/m) should be set in mortar beds over membrane roofing or membrane waterproofing. This method is called *mud set*.

Some manufacturers recommend that mud-set tile applications be made over mineral-surfaced roll roofing weighing not less than 83 lb. (37.65 kg) per square. Others have different requirements.

All sources also agree that valleys should have an additional 36-in. (914.4 mm)-wide layer of No. 30 roofing felt beneath the flashing, and that felts should be carried 6 in. (152.4 mm) up vertical surfaces.

Under the conditions outlined earlier, clay and concrete tile roofs should have ice shields. The ice shield should consist of at least two layers of No. 30 felt.

7.6.4.3 Installation

The following discussion assumes that clay and concrete tile, other than in mud-set applications of concrete tile, will be installed using nails. Similar requirements also apply to tile installed using wire hangers. Wire hanger installation should be done in strict accordance with code requirements and the tile and wire hanger manufacturers' recommendations.

In some areas of the country, on roof slopes of less than 4 in./ft. (333.3 mm/m), concrete tile is sometimes installed without nails in mortar beds laid directly over roll roofing. These mud-set applications should conform to the tile manufacturer's recommendations.

Clay tile should be installed with one, two, or three nails. Nail-applied concrete tile should be hooked over battens (where battens are used) and nailed in place with one, two, or three nails. In each case, the number of nails depends on the type of tile and the tile manufacturer's recommendations. Most field tile will require only one nail. Hip and ridge covers require one or more nails. Tile overlapping flashing should be supported with wire hangers and set in plastic cement. Nail heads should just clear the tile and should never be driven down to the tile. Nails must be driven into nailable material, never into joints in the substrates. Some codes require that every concrete tile be nailed. Other codes permit nailing alternate tile rows, or even every third tile row. Sometimes requirements vary depending on roof slope.

The first course of tile should be elevated to proper height by special tile units manufactured for the purpose, by wood strips, or by a raised fascia. Provisions should be made to permit water to drain from the roof beneath the tile. The first tile course should extend beyond the edge of the sheathing or fascia board by 1½ to 3 in. (38.1 to 76.2 mm), according to the tile manufacturer's recommendations.

Tile is usually installed with butts parallel to eaves and with tile perpendicular to eaves.

Tile should be cut to follow the line of hips and valleys, or special tile may be used. Valleys may be open, closed, or rounded. Most valleys in tile roofs, however, are open. Open ends of tile at valleys should be sealed with special units or mortar, but tile at valleys should not be sealed in a way that would prevent water that has penetrated through the tile to the underlayment surface from draining into the valley. Fan-shaped valley tile should not be sealed at the laps. Battens along valleys should be broken to permit trapped water to escape onto the valley flashing. Tile should lap valley flashing by at least 4 in. (101.6 mm).

Open valleys should be left open 6 in. (152.4 mm) at the top and should taper wider at the rate of 1 in. (25.4 mm) for every 8 ft. (2438 mm) of length.

Roofing tile should be cut and fitted close to and sealed against hip and ridge boards with plastic cement or silicone sealant. Similarly, beads of cement should be placed between tiles at hips and ridges. Hip and ridge covers should be nailed in place and lapped 3 in. (76.2 mm). Laps should be sealed with plastic cement. Open space within hip and ridge covers should not, however, be sealed against air circulation. The space between roofing tiles at ridges should be sealed with special units made for the purpose or with mortar. Additional nailers should be applied if special ridge-filler-tile units are used. Hip and ridge cover ends should be closed with special units designed for the purpose or with cement mortar. Where cement mortar is used, care should be exercised to prevent water intrusion beneath the tile, where it can lead to efflorescence on the face of the tile.

On flat-tile roofs, one-and-one-half-width tiles rather than half-tile units are usually used at gables. The smaller tiles are too light and tend to crack or blow off in windstorms.

Most tile is set with uniform exposure, but some is set with varying exposure. During installation, the spacing of tiles may need adjustment to achieve uniform exposure. Exposure is determined by tile size and headlap (the distance a tile laps over the preceding tile course). Recommended headlap may vary from manufacturer to manufacturer, but most recommend that headlap be not less than 3 in. (76.3 mm).

Hurricane clips are required on the nose end of each eave-course tile in areas classified by code as wind hazard areas and, whether so classified or not, where winds frequently exceed 70 mph (112.65 km/h). On steep roofs, hurricane clips should also be used periodically throughout the roof, or the butt of each tile may be embedded in a plastic cement or silicon sealant bead. Some codes may require additional provisions.

Tile on steep or vertical roofs should be protected against lifting by wind, even when not in a wind hazard area, by means of wind lock clips.

Other precautions are necessary in high-wind areas. For example, tile headlap should be increased to 4 in. (101.6 mm), and a bead of mastic should be applied over nail heads at gable, rake, and ridge tiles.

Open-end tile should be sealed at the eaves with special units made for the purpose or sealed with mortar. Either method must permit trapped water to drain out.

7.6.5 MINERAL-FIBER-CEMENT SHINGLES

Because of possible health hazards, roofing and siding shingles that were earlier made with asbestos fibers are now manufactured using nonasbestos mineral fibers. Therefore, the term used here for such shingles is *mineral-fiber-cement shingles*. Similar materials found in older buildings, however, may contain asbestos. Readers are therefore strongly advised to consult current OSHA (Occupational Safety and Health Administration), EPA (Environmental Protection Agency), state, and local regulations and suggestions with regard to safety precautions and to obtain professional assistance when dealing with such existing materials.

Refer to Section 7.6.1 for (1) definitions of terminology, (2) limitations imposed by the slope of a roof, (3) sheathing guidelines (see also Chapter 6), and (4) general guidelines for typical accessory materials such as underlayment, flashing, roofing cements, eave flashings, drip edge, nails, and caulking for preparation of a roof for application of shingles. Unless specific exceptions are made in the following text, the provisions of Section 7.6.1 apply to mineral-fiber-cement shingles and the methods used to apply them.

FIGURE 7.6-50 Mineral-Fiber-Cement Roof Shingles*

Type	Width, in. (mm)	Height, in. (mm)	Thickness, in. (mm)	Exposure, in. (mm)	Weight per Square, lb. (Mg)
Individual unit	8 (203)	16 (406)	5/32 (3.99)	7 (177.8)	350 (1587.5)
			1/4 (6.4)	7 (177.8)	585 (2653.4)
	9 (229)	18 (457)	1/4 (6.4)	8 (203.2)	570 (2585.4)
Multiple unit	30 (762)	14 (356)	5/32 (3.99)	6 (152.4)	440 (1995.7)
Dutch lap	16 (406)	16 (406)	5/32 (3.99)	12 × 13 (305 × 330)	265 (1202)
Ranch style	24 (610)	12 (305)	5/32 (3.99)	20 × 9 (508 × 229)	255 (1156.6)

*Products of individual manufacturers may vary slightly from the dimensions listed.

7.6.5.1 Materials

Mineral-fiber-cement shingles can be discolored by moisture and dampness while still in packages and until applied. Therefore, they should be kept clean and dry and completely protected from the weather until they have been applied. Shingles should not be applied over wet sheathing, underlayment, or sheathing paper.

ROOF SHINGLES

Mineral-fiber-cement roof shingles are manufactured in four basic types: (1 and 2) *individual units* and *multiple units*, for application by the *American method*; and (3 and 4) *Dutch lap units* and *ranch style units*, for application by the *Dutch* (or *Scotch*) *method*. Each type of shingle should be applied in a distinct way, giving a different appearance. Starter units, hip and ridge shingles, and ridge roll are also available.

Roof shingles are made with prepunched holes correctly located to receive concealed roofing nails. Additional holes are provided in Dutch lap and ranch style units to receive storm anchors.

Mineral-fiber-cement roof shingles should conform to the requirements of ASTM C 222. Data covering the four types are summarized in Figure 7.6-50.

NAILS AND FASTENERS

Roofing nails for applying mineral-fiber-cement shingles should be hot-dip galvanized steel, aluminum, stainless steel, or another corrosion-resistant metal, and should be needle or diamond pointed, with large, flat heads approximately 3/8 in. (9.5 mm) in diameter. Shanks should not exceed 0.140 in. (3.57 mm) in overall outside diameter, so as not to exceed the diameter of prepunched holes in the shingles. Shanks may be smooth or threaded, but threaded nails are preferred to increase holding power. Aluminum nails should have screw threads with approximately a 12½-degree thread angle; galvanized steel nails, if threaded, should have annular threads.

Nails for applying shingles over solid lumber sheathing should be approximately 1¼ in. (31.8 mm) by 0.120 in. (3.048 mm) and should have threaded shanks.

Nails for application of shingles over plywood sheathing should have threaded shanks and should be approximately 1¼ in. (31.8 mm) by 0.120 in. (3.048 mm), or long enough to penetrate through the sheathing.

In addition to nails, *storm anchors* should be used to apply ranch style and Dutch lap units. Storm anchors should be copper, aluminum, or lead, approximately 3/4 in. (19.1 mm) long with a flat base, and have sufficient ductility to permit repeated clinching and bending upward without cracking the shank (Fig. 7.6-51).

7.6.5.2 Preparation for Mineral-Fiber-Cement Shingle Roofing

SLOPE LIMITATIONS

Limitations imposed by roof slopes are shown in Figure 7.6-5. Mineral-fiber-cement shingles are applied by normal slope methods on roofs with slopes of 5 in 12 or greater. Lower-than-normal-slope methods may be used on slopes as low as 3 in 12.

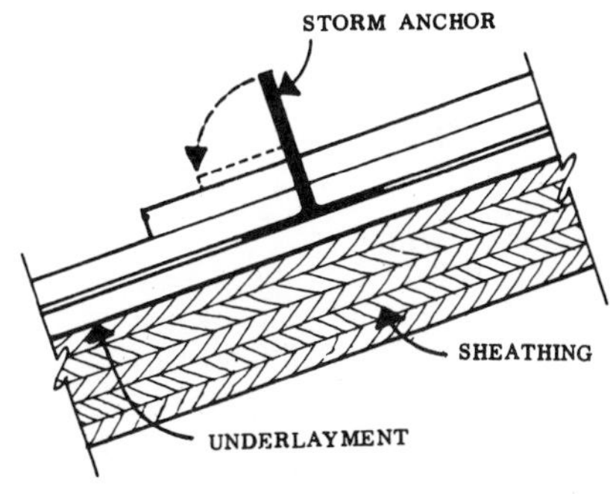

FIGURE 7.6-51 Storm anchors.

Mineral-fiber-cement shingles are not recommended for use on roof slopes of less than 3 in 12.

ROOF SHEATHING

Sheathing types, thicknesses, spans, and nailing requirements are discussed in Chapter 6.

UNDERLAYMENT

Underlayment material should be unperforated No. 15 asphalt-saturated inorganic roofing felt or No. 30 asphalt-saturated organic felt. Coal-tar-saturated felt, which may stain the shingles, should not be used. Underlayment should be applied over the entire roof.

CANT STRIP

When the American method is used with individual-unit and multiple-unit shingles, a ¼ in. (6.4 mm) by 2 in. (50.8 mm) wood strip should be applied along the eaves to pitch (cant) the starter strip at the proper angle (Figs. 7.6-52 and 7.6-53).

DRIP EDGE

A corrosion-resistant nonstaining drip edge should be installed along the eaves and rakes. Where cant strips occur, the drip edge should be applied over the cant strips.

EAVE FLASHING

In areas where the outside design temperature is 30°F (−1.1°C) or colder (see Fig. 7.6-8), or where the possibility of ice forming along the eaves could cause a backup of water, eave flashing should be installed as described below.

Normal Slope (5 in 12 or Greater) Eave flashing should be formed by cementing an additional strip of underlayment over the first underlayment course. A continuous layer of plastic asphalt cement should then be applied, at the rate of 2 gal./ square (0.815 L/m^2), from the eave to cover a line at least 12 in. (304.8 mm) inside the interior wall line of the building (see Fig. 7.6-10), and this should be covered with the additional ply of underlayment.

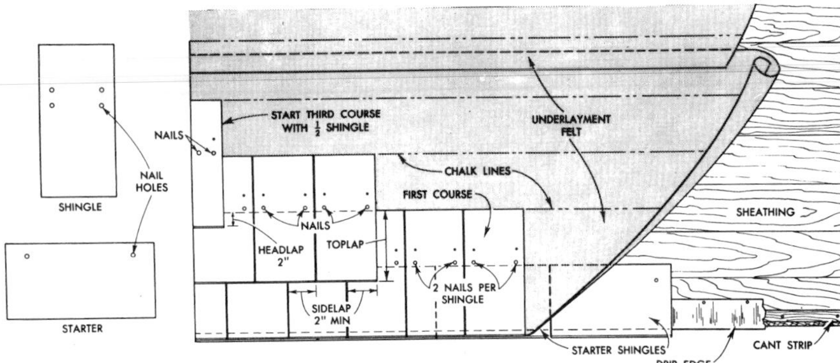

FIGURE 7.6-52 Application of individual shingles with straight butt lines.

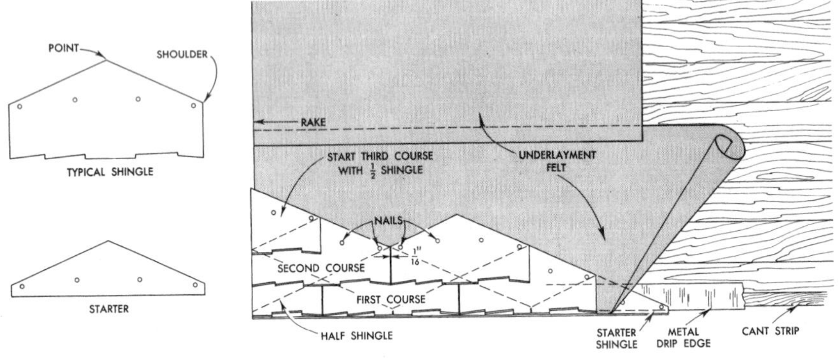

FIGURE 7.6-53 Application of multiple-unit shingles.

Lower-Than-Normal Slopes (3 in 12 to 5 in 12) A cemented eave flashing for lower than normal slopes should be formed as described for normal slopes. However, this flashing should extend from the eave to cover a line at least 24 in. (609.6 mm) inside the interior wall line (Fig. 7.6-10).

FLASHING

The following requirements are in addition to those in Section 7.6.1.6.

Valley Flashing The open or closed method may be used to construct valley flashing, as shown in Figure 7.6-54.

Chimney and Side Wall Flashing When shingles butt against vertical surfaces such as side walls, dormers, and chimneys, individual flashing units of corrosion-resistant metal should be inserted between each course of shingles. These metal flashing units should be folded from a width of 19-in. (482.6 mm) material and have a standing leg 6 in. (152.4 mm) high and a flat leg 4 in. (101.6 mm) wide on the shingle. The length of this flashing should be 3 in. (76.2 mm) greater than the exposure of the shingles. Side wall exterior finish materials should cover the stand-

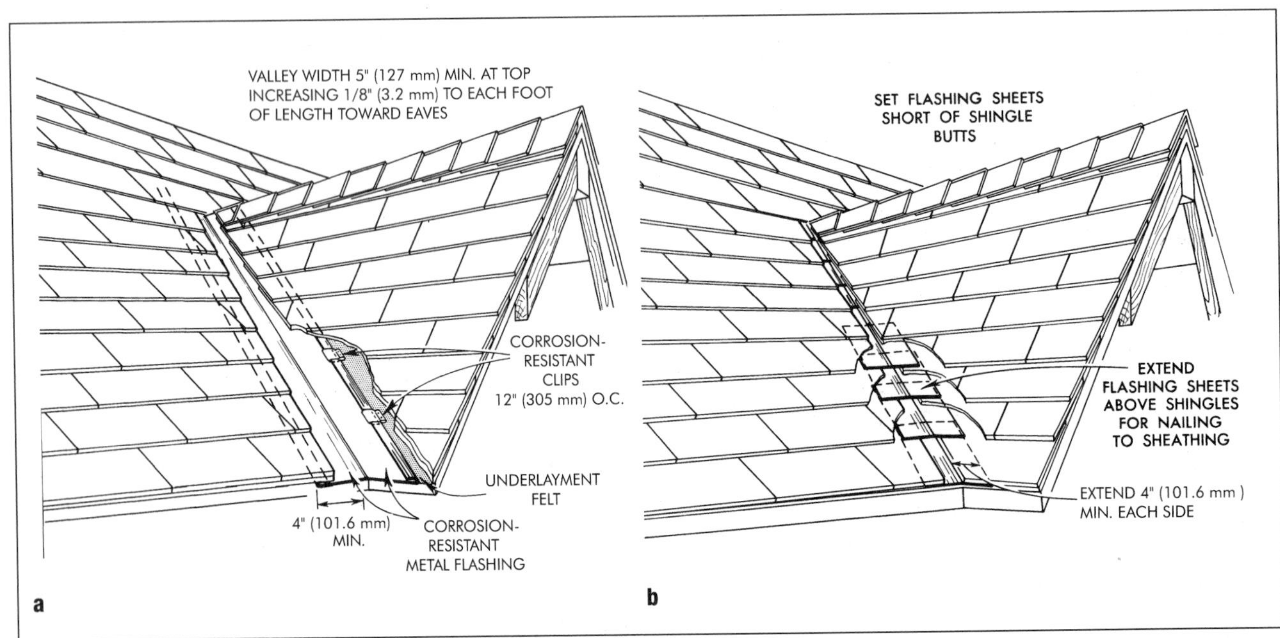

FIGURE 7.6-54 (a) Open valley flashing. (b) Closed valley flashing.

ing leg. At masonry chimneys, a stepped counter flashing of corrosion-resistant metal should cover the standing leg of the flashing units and be trimmed off just above the surface of the shingles.

7.6.5.3 Shingle Roofing Application (New Construction)

The *American method* should be used to apply *individual* and *multiple* shingles without a sidelap. The *Dutch method* with sidelap and toplap should be used to apply Dutch lap and ranch style shingles. In both methods, the underlayment felt should be rolled back and starter shingles applied directly on top of the metal drip edges previously nailed. Starter shingles should overhang the rakes and the eaves about ¾ in. (19.1 mm). When starters match the color of the field shingles and are used without a gutter, they are usually applied with the colored side down for better appearance from below.

AMERICAN METHOD (NORMAL SLOPE)

Individual Shingles (Straight Butt Line) After the underlayment is replaced over the starter shingles, the first shingle course should be applied, starting with a half-shingle overhanging the rake about ¾ in. (19.1 mm) and in line at the butt with the starter shingles. Each shingle should be applied with nails through two of the prepunched holes. The second course should be started with a full shingle, and the third course, like the first course, should start with a half-shingle. For accurate alignment, chalklines should be marked on the underlayment for each shingle course (see Fig. 7.6-52).

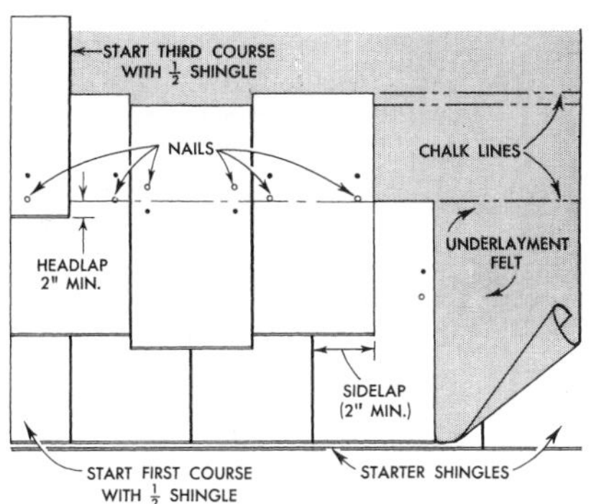

FIGURE 7.6-55 Staggered butt line can be achieved with individual shingles by varying the exposure.

Individual Shingles (Staggered Butt Line) Individual shingles have two sets of prepunched holes to permit a staggered application. The first course should be applied as described for a straight butt line. In starting the second course, the exposure of every second shingle should be reduced by dropping it down and nailing through the upper set of nail holes. Staggering can be varied, depending on the appearance desired, but is limited to the distances between the upper and lower holes. The third course should be started with a half-shingle, and alternating shingles should be staggered as in the previous course.

Individual shingles with either a straight or staggered butt line should have a minimum headlap and sidelap of 2 in. (50.8 mm) (Fig. 7.6-55) (see also Fig. 7.6-52).

Multiple-Unit Shingles After the underlayment is replaced over the starter shingles, the first shingle course should be applied starting with a half-shingle overhanging the rake about ¾ in. (19.1 mm) and in line at the butt with the starter shingles. The second course should be started using a full shingle, and the third course, like the first course, should be started with a half-shingle. Chalklines are not necessary if the visual alignment design of the shingles is properly utilized (see Fig. 7.6-53). Overdriving of nails causes the butts of the shingles to cock up and should be avoided to prevent damage to the shingles.

Because the design of multiple-unit shingles may vary among manufacturers, the particular manufacturer's application instructions for nailing, positioning, and alignment should be followed.

DUTCH METHOD (NORMAL SLOPE)

The Dutch method is used to apply Dutch lap shingles (Fig. 7.6-56a) and ranch style shingles (Fig. 7.6-56b). The following application methods refer to shingles having a 4-in. (101.6 mm) sidelap.

Starter shingles should be applied as shown in Figure 7.6-57a and b. After the felt has been replaced over the starter shingles, a 4-in.(101.6 mm)-wide piece cut from a full shingle should be applied to form a rake starter. As shown in Figure 7.6-57c, storm anchor A should be inserted from below and the rake starter secured in place using two nails. A full-size shingle should be applied over the rake starter by first inserting storm anchor B and positioning the shingles so that the storm anchor A appears through the free corner (Fig. 7.6-57d). Storm anchor A should be clinched and the shingle secured

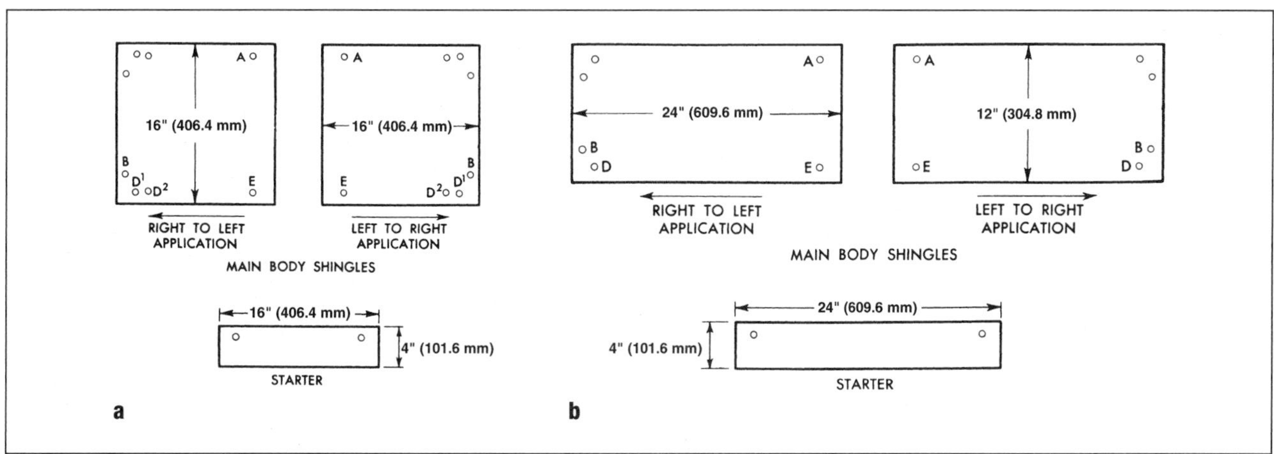

FIGURE 7.6-56 (a) Dutch lap shingles and (b) ranch-style shingles. Holes marked A and B are for nails; holes D and E are for storm anchors.

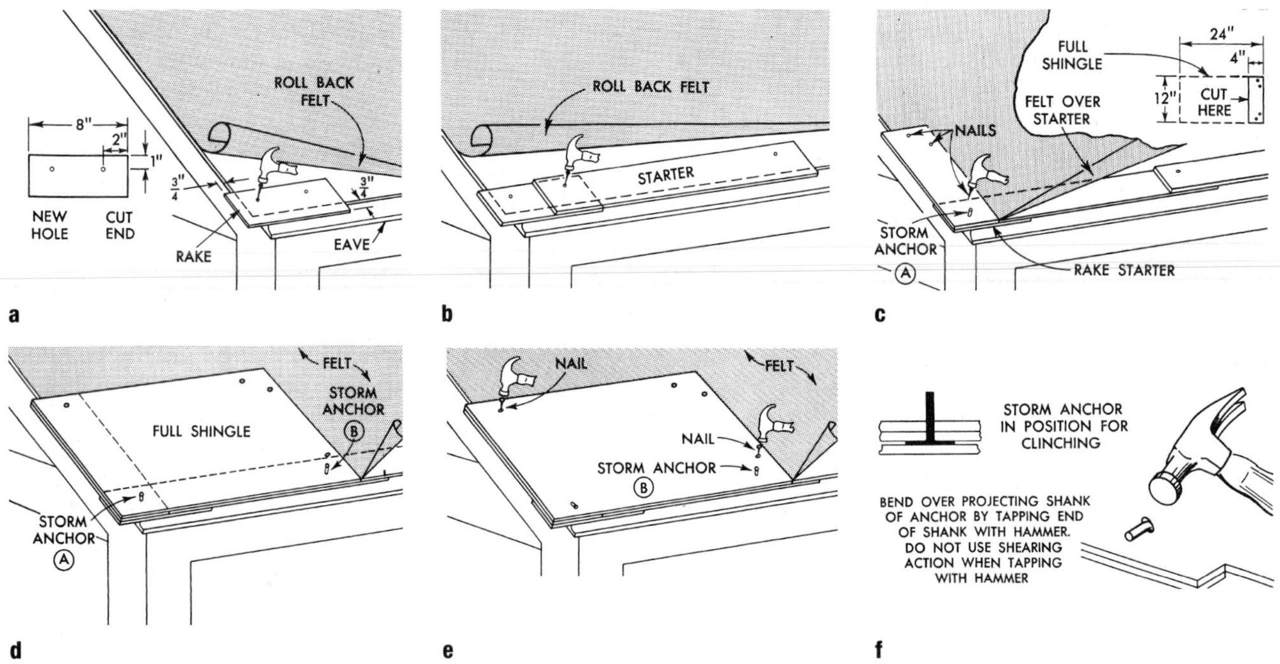

1″ = 25.4 mm; 2″ = 50.8 mm; 8″ = 203.2 mm; 3⁄4″ = 19.1 mm; 4″ = 101.6 mm; 12″ = 304.8 mm; 24″ = 609.6 mm

FIGURE 7.6-57 Application of Dutch lap and ranch-style shingles. (a, b) Apply starter shingles. (c) Insert storm anchor A from below and secure the rake starter in place using nails. (d) Insert storm anchor B and position shingles so that storm anchor A appears through free corner. (e) Clinch storm anchor A and secure shingle with nails.

with two nails (Fig. 7.6-57e). Each succeeding shingle should be applied in the same manner, first inserting a storm anchor in one corner and then engaging the other corner over the previous storm anchor. The second shingle course should be started with a shingle 4 in. (101.6 mm) less than full width, punched with a new hole as required to hold the shingle along the rake, and should be completed using full shingles. The third course should be started with a shingle 8 in. (203.2 mm) less than full width; the fourth course with a shingle 12 in. (304.8 mm) less than full width, and so on up the rake. Automatic alignment of the shingles is usually provided by the location of the prepunched holes.

HIPS AND RIDGES

Before hip and ridge shingles (Fig. 7.6-58) are applied, wood nailing strips equal in thickness to the roof shingle assembly should be applied at the hips and ridges to provide a flat surface. Hips and ridges should then be covered with two thicknesses of underlayment (Fig. 7.6-59), or a double thickness of 8-in. (203.2 mm)-wide felt can be adhered with plastic cement over the nailing strips. Application of hip and ridge shingles should be started with a half-length shingle, covered with a full-length shingle. Thereafter, full-length shingles should be applied with a maximum of 8 in. (203.2 mm) exposure (depending on the design of the shingle used) by nailing each shingle through two nail holes. Alternating shingles should be lapped on either side of the hip or ridge. Finally, the ridge joint should be pointed with plastic asphalt roofing cement.

RIDGE ROLL

A tapered half-round ridge roll is used in some localities (Fig. 7.6-60). Before a ridge roll is applied, a wood nailing strip should be installed along the apex of hips and ridges. The nailing strip should then be covered with a double thickness of underlayment. Whenever possible, the application of a ridge roll should start at the end of the ridge and be laid in a direction opposite to that of the prevailing winds; at hips, a ridge roll should be started at the eaves.

The first ridge roll section should be placed with the large end at the starting point, overhanging the roof shingles 1⁄4 in. (6.4 mm). Each end should be fastened to the nailing strip by nailing at the head end through a corrosion-resistant clip designed for this purpose. The next ridge section should overlay the starter at least 2 in. (50.8 mm). These clips should be bent back to fasten the exposed end of the next ridge roll section. This process should be continued until the entire hip or ridge has been covered. Ridge pieces should be embedded in asphalt plastic cement, and exposed nails should be pointed with plastic cement. The openings at end units may be closed with colored cement mortar or with a closure consisting of a half-round piece cut from a shingle, nailed or screwed into the end of the nailing strip.

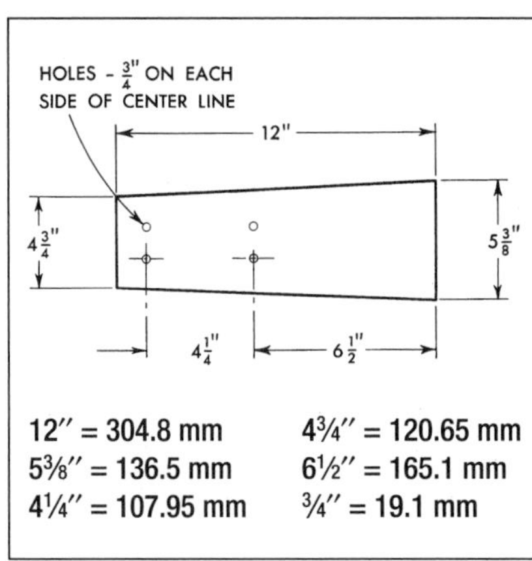

FIGURE 7.6-58 Hip and ridge shingles.

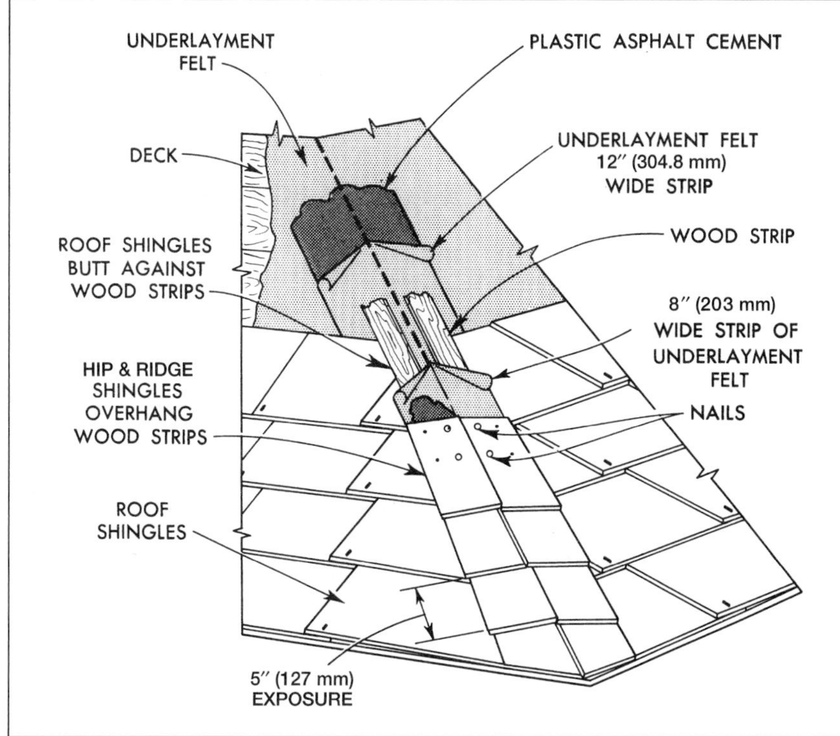

FIGURE 7.6-59 Application method for finishing hips and ridges.

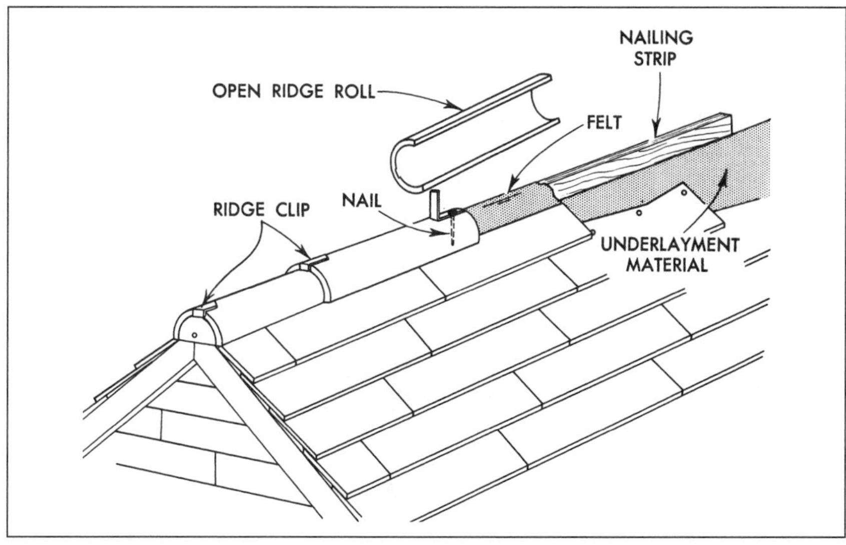

FIGURE 7.6-60 Ridge roll may be used on ridge as alternate to ridge shingles.

LOWER-THAN-NORMAL-SLOPE METHOD (3 IN 12 TO 5 IN 12)

Methods for applying shingles on roof slopes of less than 5 in 12 but not less than 3 in 12 are the same as described for normal slopes of 5 in 12 and over, with certain exceptions in regard to underlayment.

For slopes of 3 in 12 up to 4 in 12, two underlayment layers should be applied over the entire roof. For slopes of 4 in 12 up to 5 in 12, one underlayment layer is sufficient, except that two layers are required in areas where the outside winter design temperature is 30°F (–1.1°C) or lower.

When eave flashing is required on roofs with lower than normal slopes, the additional layer should be cemented at the eaves, as described in Section 7.6.6.2.

7.6.5.4 Reroofing Existing Roofs

When mineral-fiber-cement shingles are applied to an existing roof, the surface must be made as smooth as possible. Mineral-fiber-cement shingles may be applied directly over old asphalt shingles, provided that the sheathing and roof framing members are sound and strong enough to support the increased loads.

Loose or curled wood shingles should be split and nailed down, and missing shingles replaced. Beveled wood strips should be installed below the butts of each course of shingles (Fig. 7.6-61). When existing wood shingles fail to retain nails with a force of at least 40 lb. (18.14 kg) per fastener, the shingles should be removed and the existing sheathing prepared as it would be in new construction. If old wood shingles were applied over spaced sheathing, new sheathing should be fitted between the strips to fill the spaces and provide a solid deck. In this case, it is often simpler to remove the old sheathing strips and construct a new deck.

Nails for applying shingles to existing roofs should be long enough to penetrate through the existing sheathing. Threaded-shank roofing nails provide improved holding power.

REPLACING DAMAGED SHINGLES

Damaged shingles may be removed and replaced as shown in Figure 7.6-62 and as follows.

First, the damaged shingle should be shattered and removed. The nails may then be withdrawn or cut off by hooking a slater's ripper over the nails and striking the offset handle of the ripper with a ham-

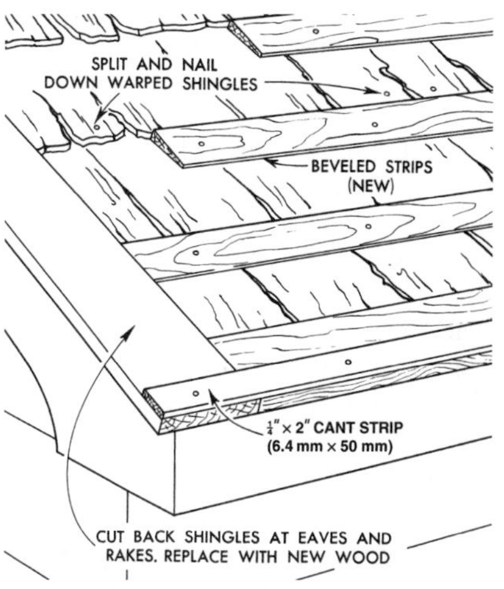

FIGURE 7.6-61 Preparation of existing roof for application of new shingles.

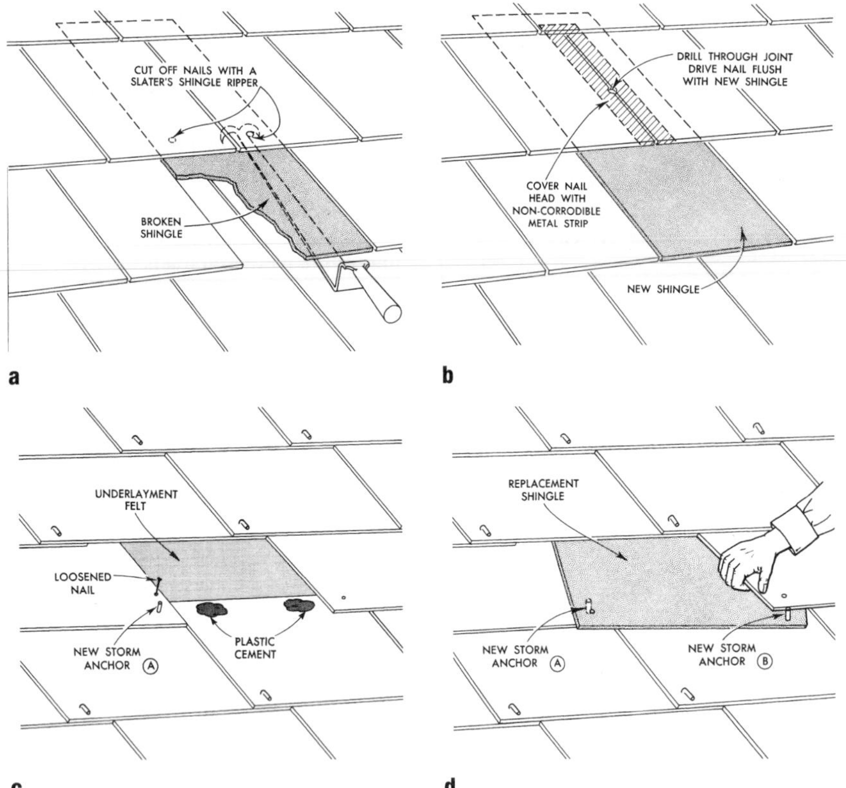

FIGURE 7.6-62 Replace damaged shingles as follows: (a) Remove damaged shingle and nails. (b) Nail new shingle through side joint and cover nail with corrosion-resistant metal strip inserted in lap. (c) In Dutch method, remove damaged shingle and shear off nails and storm anchor by engaging them with the hook end of a slater's ripper and striking offset on handle with hammer. Loosen nail in exposed shingle, remove sheared-off storm anchor, and insert new storm anchor. Drive loosened nail home and apply plastic cement. (d) Place storm anchor B in new shingle. Raise overlapping shingle to allow storm anchor to align with holes. Embed shingle in cement and bend storm anchor down.

mer (Fig. 7.6-62a). A new shingle should then be positioned.

With the American method of application, a hole should be drilled through the joint between the overlying shingles and the head of the new shingle. One nail should be driven through this hole, and a 3-in. (76.2 mm)-wide corrosion-resistant metal flashing strip should be placed over the nail head and under the overlying shingle (Fig. 7.6-62b).

For the Dutch method of application, the new shingles should be anchored in place with two spots of roofing cement placed on the head of the underlying shingle and by engaging the storm anchors. Nailing is not required (Fig. 7.6-62c and d).

7.6.6 WOOD SHINGLES AND SHAKES

Refer to Section 7.6.1 for (1) definitions of terminology, (2) limitations imposed by the slope of a roof, (3) sheathing guidelines (see also Chapter 6), and (4) general guidelines for typical accessory materials, such as underlayment, flashing, roofing cements, eave flashings, drip edge, nails, and caulking for preparation of a roof or wall for application of shingles. Unless specific exceptions are made in the following text, the provisions of Section 7.6.1 apply to wood shingles and shakes and methods of their application.

7.6.6.1 Materials

Most wood shingles and shakes used for roofing are produced from western red cedar trees—giant, slow-growing coniferous trees of the Pacific Northwest—but northern white cedar, bald cypress, and redwood are also used. This section discusses only western red cedar shingles and shakes. Those made with other woods are installed in a similar manner.

Western red cedar has an extremely fine and even grain, exceptional strength in proportion to weight, low expansion and contraction with changes in moisture content, and high impermeability to water. Its outstanding characteristics are great durability and resistance to decay.

Cedar shingles are sawn, have smooth faces and backs, and are produced in four grades and a variety of types. Tapersawn shakes are sawn on both sides, but their thickness is more like that of other shake types. They are produced in two grades. Other shakes have rough split faces and either sawn or split backs and are produced in one grade and in three types.

Both shingles and shakes are made from selected cedar logs cut to proper lengths—16, 18, and 24 in. (406, 457, and 610 mm) for shingles; 18 and 24 in. (457 and 610 mm) for shakes. These lengths are again cut or split into blocks of the proper size for the shingle-cutting machine or to a convenient size for the splitting of shakes. In reducing the large log sections into blocks, each is quartered, split, and requartered to produce blocks having a true edge grain face (Fig. 7.6-63).

Each bundle of wood shingles and shakes should carry the official grade-

a

b

FIGURE 7.6-63 Logs are cut to length (a), then quartered to produce edge-grain blocks for sawing or splitting into shingles or shakes (b). (Courtesy Cedar Shake and Shingle Bureau.)

FIGURE 7.6-64 A sawyer *edges* shingles after they have been cut from cedar blocks by the circular saw at the sawyer's left. (Courtesy Cedar Shake and Shingle Bureau.)

marked label of the Cedar Shake and Shingle Bureau (CSSB).

CEDAR SHINGLES

Roof shingles are cut from cedar blocks on an upright shingle machine (Fig. 7.6-64). This cutting machine carries a block past a thin, razor-sharp circular saw, which slices a shingle smoothly at each stroke. Shingles cut from one block are jointed or edged and graded by a sawyer. The sawyer can adjust the block in the machine from time to time so that the highest possible grade of shingles will be produced.

A sawyer picks up each shingle as it falls from the saw and places its butt firmly against a guide. Using a circular saw set at right angles to the butt, the sawyer clips off one edge of the shingle smoothly. Flipping the shingle over, the sawyer repeats the process, making another smooth edge parallel to the one first cut. If defects are present, the sawyer trims the shingle in such a way that a narrower shingle, or two shingles without defects, can be obtained. Depending on the grade, the sawyer then drops the shingle into a chute that leads to the packing bins.

Shingles are graded again as they are packed into bundles in a packing frame. Bundles of shingles are held together by two bandsticks connected by metal straps. Packaged in this manner, bundled shingles are seasoned and remain perfectly flat in the process.

Grades, Types, and Sizes Shingles are marketed according to grade, thickness, and type. Nominal shingle lengths are 16, 18, and 24 in. (406, 457, and 610 mm) within ¼ in. (6.4 mm). A 1-in. (25.4 mm) tolerance over or under the specified length is permitted in 10% of the shingles. All grades are produced in these lengths, except that Undercoursing Grade is not produced in the 24-in. (610 mm) length. A 16-in. (406 mm) shingle is known as XXXXX or FiveXs; an 18-in. (457 mm) shingle as Perfection; the 24-in. (610 mm) as Royal.

Most cedar shingles conform to one of three grades: No. 1 (Blue Label), No. 2 (Red Label), and No. 3 (Black Label). Other available units include No. 4, undercoursing; No. 1 and No. 2, rebutted-rejointed; No. 1, machine grooved; No. 1 and No. 2, dimension; No. 1 and No. 2, preformed; and factory-built hip and ridge units.

Except for dimension shingles, all shingles are produced in random widths from 3 in. (76.2 mm) up to 14 in. Shingle thickness varies depending on length. A 16-in. (406 mm)-long shingle must be thick enough that the combined butt thicknesses of five shingles will measure at least 2 in. (50.8 mm). Thus, a shingle thickness is designated as 5 in 2 in. (50.8 mm) or 5/2 in. An 18-in. (457 mm) shingle is of sufficient thickness that the combined butt thicknesses of five shingles will measure at least 2¼ in. (5/2¼ in.) (57.2 mm); and 24-in. (610 mm) shingles are of sufficient thickness that four shingles will measure at least 2 in. (4/2 in.) (50.8 mm). These dimensions are for unseasoned shingles measured green. If the shingles are measured after seasoning has occurred, a 3% allowance is made for shrinkage.

Shingles with extreme cross grain are not permitted in any grade.

Grade No. 1 (Blue Label) is the best or premium grade of shingle for roofs. They are of 100% edge grain, consist entirely of heartwood, and contain no knots. On 8 in 12 sloped roofs or steeper, Grade No. 1 shingles are estimated to have a service life of more than 35 years under normal climatic conditions; on roof slopes of 6 in 12, a service life of at least 25 years can be anticipated; and on 4 in 12 slopes, 20 years.

Grade No. 2 (Red Label) is a good grade for roofs on secondary buildings and may be used on major building roofs when exposures are reduced about 20%. Shingles of this grade must be clear, or free from blemishes for 10, 11, and 15 in. (254, 279.4, and 381 mm) for 16-, 18-, and 24-in. (406, 457, and 610 mm) shingles, respectively. A maximum 1-in. (25.4 mm) width of sapwood is permissible in the first 10 in. (254 mm). Mixed edge and flat grains are allowed.

Grade No. 3 (Black Label) is a utility grade to be used where economy is the prime consideration. It may be used for roofs of secondary buildings. These shingles may contain flat grain and sapwood, but must be 6 in. (152.4 mm) clear from the butt for 16- and 18-in. (406 and 457 mm) shingles, and 10 in. (254 mm) clear for 24-in. (610 mm) shingles.

Hip and ridge units are factory cut, mitered, and assembled to fit hips and ridges of various roof slopes. These units, which reduce costs by eliminating on-the-job trimming and fitting (Fig. 7.6-65), are produced in either No. 1 or No. 2 Grades in 16- and 18-in. (406 and 457 mm) lengths for a 6 in 12 roof slope, and are adjustable to fit any slope between 4 in 12 and 8 in 12.

A summary of cedar shingle types, grades, and coverage is given in Figures 7.6-66 and 7.6-67. The labels shown in Figure 7.6-66 are assurance that the shingles meet the quality standards of the CSSB. The maximum exposure for shingles is shown in Figure 7.6-68.

CEDAR SHAKES

Wood shakes are manufactured in four types of units: (1) tapersawn; (2) handsplit and resawn; (3) tapersplit; and (4) straightsplit. Tapersplit shakes are produced in 24-in. (610 mm) lengths. The others are produced in both 18- and 24-in. (457 and 610 mm) lengths. Each is 100% clear wood and 100% heartwood. Both tapersplit and straightsplit shakes, produced largely by hand, are of 100% edgegrain.

Factory-assembled hip and ridge units are also available.

Each bundle of shakes should carry the official grade-marked label of the Cedar Shake and Shingle Bureau.

Both faces of tapersawn shakes are

FIGURE 7.6-65 Preassembled hip and ridge units. (Courtesy Cedar Shake and Shingle Bureau.)

sawn like shingles, but they are thicker and heavier than shingles.

Most shakes are split by machine (Fig. 7.6-69b), although some are still split by hand using a hardwood mallet and steel froe (Fig. 7.6-69a). As a result of being split, no two shakes are exactly alike. The three types of shakes produced are tapersplit, straightsplit, and handsplit and resawn (Fig. 7.6-70).

Tapersplit and straightsplit shakes (Fig. 7.6-70a and b) are made by machine or by hand-riving individual shakes from the block. To obtain the butt-to-tip taper for tapersplit shakes, blocks are turned end-for-end after each split. Straightsplit shakes are split from the same end of the block and therefore are of uniform thickness.

Handsplit and resawn shakes (see Fig. 7.6-70c) are produced by splitting blocks into boards of the desired thickness. The boards are then passed diagonally through a thin bandsaw to form two shakes, each with a split face and a sawn back having thin tips and thick butts and up to 10% flat grain. These shakes are available either as heavy handsplit and resawn (also called *heavy resaw*) or medium handsplit and resawn (also called *medium resaw*). The latter are thinner and lighter.

Packers assemble finished shakes into bundles in a standard-size frame, compressing the bundles slightly and binding them with wooden bandticks and steel strapping (Fig. 7.6-71).

Each type of shake may be used for roof shingles. Handsplit and resawn shakes have the heaviest butt lines and give the most rugged textured appearance to a roof; tapersplit and straightsplit shakes give a more uniform texture.

On 8 in 12 sloped roofs or steeper, shakes are estimated to have a service life of more than 60 years. On lesser slopes, to a minimum of 4 in 12, a life of about 40 years may be expected.

Grades, Types, and Sizes Tapersawn shakes are available in Grades 1 and 2. In Grade 1, no defects are permitted. In

Grade	Size	Bundles or Cartons* Per Square		Description	Labels
		No.	Weight		
No. 1 BLUE LABEL	24″ (Royals) 18″ (Perfections) 16″ (XXXXX)	4 bdls. 4 bdls. 4 bdls.	192 lbs. 158 lbs. 144 lbs.	The premium grade of shingles for roofs and sidewalls. These shingles are 100% heartwood, 100% clear and 100% edge-grain.	CERTIGRADE Red Cedar SHINGLES BLUE 1 LABEL
No. 2 RED LABEL	24″ (Royals) 18″ (Perfections) 16″ (XXXXX)	4 bdls. 4 bdls. 4 bdls.	192 lbs. 158 lbs. 144 lbs.	A good grade for most applications. Not less than 10″ clear on 16″ shingles, 11″ clear on 18″ shingles and 16″ clear on 24″ shingles. Flat grain and limited sapwood are permitted.	CERTIGRADE Red Cedar SHINGLES RED LABEL
No. 3 BLACK LABEL	24″ (Royals) 18″ (Perfections) 16″ (XXXXX)	4 bdls. 4 bdls. 4 bdls.	192 lbs. 158 lbs. 144 lbs.	A utility grade for economy applications and secondary buildings. Guaranteed 6″ clear on 16″ and 18″ shingles, 10″ clear on 24″ shingles.	CERTIGRADE Red Cedar SHINGLES BLACK 3 LABEL
No. 4 UNDER-COURSING	18″ (Perfections) 16″ (XXXXX)	2 bdls. 2 bdls.	60 lbs. 60 lbs.	A low grade for undercoursing on double-coursed sidewall applications.	RED CEDAR SHINGLE BUREAU UNDERCOURSING Red Cedar SHINGLES
No. 1 or No. 2 REBUTTED-REJOINTED	18″ (Perfections) 16″ (XXXXX)	1 carton 1 carton	60 lbs. 60 lbs.	Same specifications as No. 1 and No. 2 Grades above but machine trimmed for exactly parallel edges with butts sawn at precise right angles. Used for sidewall application where tightly fitting joints between shingles are desired. Also available with smooth sanded face.	No. 1 & No. 2 labels (above)
No. 1 MACHINE GROOVED	18″ (Perfections) 16″ (XXXXX)	1 carton 1 carton	60 lbs. 60 lbs.	Same specifications as No. 1 and No. 2 Grades above; these shingles are used at maximum weather exposures and are always applied as the outer course of double-coursed sidewalls.	CERTIGROOVE CEDAR SHAKES NUMBER 1 GRADE RED CEDAR SHINGLE BUREAU
No. 1 or No. 2 DIMENSION	24″ (Royals) 18″ (Perfections) 16″ (XXXXX)	4 bdls. 4 bdls. 4 bdls.	192 lbs. 158 lbs. 144 lbs.	Same specifications as No. 1 and No. 2 Grades above, except they are cut to specific uniform widths and may have butts trimmed to special shapes.	No. 1 & No. 2 labels (above)
No. 1 or No. 2 HIP AND RIDGE	18″ (Perfections) 16″ (XXXXX)	—	—	Same specifications as No. 1 and No. 2 Grades above; factory-cut, mitered and assembled units produced for a 6 in 12 slope and adjust to fit slopes between 4 in 12 and 8 in 12.	No. 1 & No. 2 labels (above)

*Nearly all manufacturers pack 4 bundles to cover 100 sq. ft. when used at maximum exposures for roof construction; Undercoursing, rebutted and rejointed and machine grooved shingles typically are packed one carton to cover 100 sq. ft. when used at maximum exposure for double-coursed sidewalls.

FIGURE 7.6-66 Types of red cedar shingles. (Courtesy Cedar Shake and Shingle Bureau.)

FIGURE 7.6-67 Coverage of Wood Shingles at Varying Exposures

Length and Thickness[a]	Approximate Coverage of 4 Bundles or 1 Carton, sq. ft.[b]																		
	Weather Exposures, in.																		
	3½	4	5	5½	6	7	7½	8	8½	9	10	11	11½	12	13	14	15	15½	16
Random width and dimension																			
16 in. × 5/2 in.	70	80	100[c]	110	120	140	150[d]	160	170	180	200	220	230	240[e]	—	—	—	—	—
18 in. × 5/(2¼) in.	—	72½	90½	100[c]	109	127	136	145½	154½[d]	163½	181½	200	209	218	236	254½[e]	—	—	—
24 in. × 4/2 in.	—	—	—	—	80	93	100[c]	106½	113	120	133	146½	153[d]	160	173	186½	200	206½	213[e]
Rebutted and rejointed																			
16 in.	—	—	—	—	50	59	63[d]	67	72	76	84	93	—	100[e]	—	—	—	—	—
18 in.	—	—	—	—	43	50	54	57	61[d]	64	72	79	—	86	93	100[e]			
Machine-grooved																			
16 in.					(Normally applied at maximum exposure)									100[e]					
18 in.					(Normally applied at maximum exposure)											100[e]			

[a]Sum of the thickness; e.g., 5/2 in. means 5 butts equals 2 in.

[b]Nearly all manufacturers pack 4 bundles to cover 100 sq. ft. when used at maximum exposures for roof construction; rebutted and rejointed and machine-grooved shingles typically are packed one carton to cover 100 sq. ft. when used at maximum exposure for double-coursed sidewalls.

[c]Maximum exposure recommended for roofs.

[d]Maximum exposure recommended for single-coursed sidewalls.

[e]Maximum exposure recommended for double-coursed sidewalls.

(Courtesy Cedar Shake and Shingle Bureau.)

FIGURE 7.6-68 Maximum Exposure for Wood Roof Shingles

	Shingle Type No. 1			Shingle Type No. 2			Shingle Type No. 3		
Shingle Length	**16**	**18**	**24**	**16**	**18**	**24**	**16**	**18**	**24**
Roof Slope	**Exposure in Inches (mm)**								
3½ to 4½	3¾ (95.3)	4¼ (107.9)	5¾ (146.1)	3½ (88.9)	4 (101.6)	5½ (139.7)	3 (76.2)	3½ (88.9)	5 (127)
4½ and Steeper	5 (127)	5½ (139.7)	7½ (190.5)	4 (101.6)	4½ (114.3)	6½ (165.1)	3½ (88.9)	4 (101.6)	5½ (139.7)

FIGURE 7.6-71 Grade labels beneath the bandsticks of both shingles and shakes ensure that they meet or exceed Cedar Shake and Shingle Bureau quality standards. (Courtesy Cedar Shake and Shingle Bureau.)

a

b

FIGURE 7.6-69 Splitting shakes (a) by hand, (b) by machine. (Courtesy Cedar Shake and Shingle Bureau.)

Taper split

a

Straight split

b

Hand split and resawn

c

FIGURE 7.6-70 Three types of shakes: (a) taper split, (b) straight split, (c) hand split and resawn. (Courtesy Cedar Shake and Shingle Bureau.)

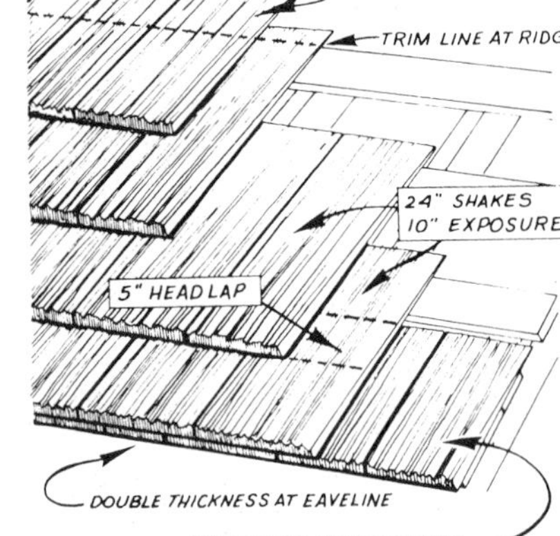

FIGURE 7.6-72 Starter-finish shakes are more economical than regular shakes for use in the initial undercourse at the eave and in the final course at the ridge. (Courtesy Cedar Shake and Shingle Bureau.)

Grade 2, flat and cross grains are allowed and the top half is permitted to have tight knots and other defects.

There is only one grade, No. 1, for other cedar shakes. Shakes of this grade are 100% clear wood, graded from the split face in the case of handsplit and resawn shakes, and from the best face in the case of tapersplit and straightsplit shakes. All are 100% heartwood, free of bark and sapwood. Tapersplit and straightsplit shakes are 100% edgegrain; handsplit and resawn shakes may have up to 10% flat grain.

Shake thickness is determined by measuring the area within ½ in. (12.7 mm) from each edge. If corrugations or valleys exceed ½ in. (12.7 mm) in depth, a minus tolerance of ⅛ in. (3.2 mm) is permitted in the minimum specified thickness.

Shakes of random width, none narrower than 4 in. (101.6 mm), are packed in straight courses in 18- and 20-in. (457.2 and 508 mm)-wide frames. Curvatures in the sawn face of hand split-and-resawn shakes should not exceed 1 in. (25.4 mm) from the level plane in the length of the shake. Excessive grain sweeps on the split face are not permitted.

Starter-finish shakes, produced in 15-in. (381 mm) lengths, are used as a beginning or underlay course at the eave and as the final course at the ridge (Fig. 7.6-72).

Factory-cut, mitered, and assembled hip and ridge units are available to eliminate on-the-job trimming and fitting (Fig.

FIGURE 7.6-73 Hip and ridge shake units. (Courtesy Cedar Shake and Shingle Bureau.)

7.6-73). They are produced in 18- and 24-in. (457.2 and 609.6 mm) lengths for a 6 in 12 roof slope and are adjustable to fit any slope between 4 in 12 and 8 in 12.

A summary of red cedar shake types, grades, and coverage is given in Figures 7.6-74 and 7.6-75. The label shown in Figure 7.6-74 is assurance that the shakes meet the quality standards of the CSSB. The maximum exposure for shakes is shown in Figure 7.6-76.

STANDARDS OF QUALITY

The manufacture of cedar shingles and shakes should conform to the specifications of the CSSB. These specifications constitute the recognized standards of the industry.

In addition to formulating specifications, a primary function of the CSSB is to establish and enforce rigid quality controls covering the manufacture and grading of red cedar shingles and shakes. To ensure compliance with these standards, products made by members of the CSSB are inspected at frequent unannounced intervals by CSSB inspectors. Mills that conform to these standards are privileged to use the *Certigrade* shingle or *Certi-Split* shake labels of the CSSB on their products.

NAILS

Nails for applying shakes and shingles should be corrosion resistant, such as hot-dipped galvanized steel or bright or colored aluminum alloy. Common wire nails or even electrogalvanized nails may shorten the life of a shake or shingle roof or wall and should not be used. Nails may have smooth or threaded shanks and medium or blunt diamond points. Threaded nails provide increased holding power, although smooth-shank nails are used in most applications over wood sheathing and give suitable service. Nails for applying wood shingles or shakes to plywood sheathing should be threaded. If threaded, galvanized steel nails should have annular threads; aluminum nails should have screw threads with approximately a 12½ degree thread angle.

Box nails, having smaller diameters and longer lengths than common nails, are also suitable. For new roofs, nail lengths typically required are 6d (2 in. [50.8 mm]) for shakes with ¾-in. (19.1 mm) to 1¼-in. (31.8 mm) butt thickness, and 5d (1¾ in. [44.45 mm]) for shakes with ½- to ¾-in. (12.7 to 19.1 mm) butt thickness. Nails 2d sizes larger should be used to apply hip and ridge units. Nails should be long enough to penetrate completely through the sheathing. For reroofing, nails should be long enough to penetrate through the existing sheathing.

Nail dimensions typically required for roof installations are given in Figure 7.6-77. Nails of 2d sizes larger should be used to apply hip and ridge units. For overroofing, nails should be long enough to penetrate through the existing sheathing; the 5d size (1¾ in. [44.45 mm]) is normally adequate.

Grade	Length and Thickness	Bundles Per Square*	Weight (lbs. per square)	Description	Label
No. 1 HANDSPLIT & RESAWN	18″ x 1/2″ to 3/4″ 18″ x 3/4″ to 1-1/4″ 24″ x 3/8″ 24″ x 1/2″ to 3/4″ 24″ x 3/4″ to 1-1/4″ 32″ x 3/4″ to 1-1/4″	4 5 4 4 5 6	220 250 260 280 350 450	These shakes have split faces and sawn backs. Cedar blanks or boards are split from logs and then run diagonally through a bandsaw to produce two tapered shakes from each.	
No. 1 TAPERSPLIT	24″ x 1/2″ to 5/8″	4	260	Produced largely by hand, using a sharp-bladed steel froe and a wooden mallet. The natural shingle-like taper is achieved by reversing the block, end-for-end, with each split.	CERTI-SPLIT Handsplit Red Cedar Shakes NUMBER 1 GRADE RED CEDAR SHINGLE & HANDSPLIT SHAKE BUREAU
No. 1 STRAIGHT-SPLIT (BARN)	18″ x 3/8″ (True-Edge) 18″ x 3/8″ 24″ x 3/8″	4 5 5	200 200 260	Produced in the same manner as tapersplit shakes except that by splitting from the same end of the block, the shapes acquire the same thickness throughout.	
No. 1 STARTER-FINISH	15″ x 1/2″ to 1-1/4″	—	—	These shakes are used as the starting or underlay course at the eaves and as the final course at the ridge.	
No. 1 HIP & RIDGE	18″ x 1/2″ to 1-1/4″ 24″ x 1/2″ to 1-1/4″	—	—	Factory cut, mitered and assembled units are produced for a 6 in 12 slope; they adjust to fit slopes between 4 in 12 and 8 in 12.	

* Generally, the number of bundles required to cover 100 sq. ft. when used for roof construction at the maximum recommended weather exposure.

FIGURE 7.6-74 Types of cedar shakes. (Courtesy Cedar Shake and Shingle Bureau.)

FIGURE 7.6-75 Coverage of Shakes at Varying Weather Exposures

Length and Thickness	Number of Bundles	Approximate Coverage, sq. ft. — Weather Exposures, in. 5½	6½	7	7½	8	8½	10	11½	13	14	15	16
Hand-split-and-resawn													
18 in. × ½ in. to ¾ in.	4	55[a]	65	70	75[b]	80	85[c]	—	—	—	—	—	—
18 in. × ¾ in. to 1¼ in.	5	55[a]	65	70	75[b]	80	85[c]	—	—	—	—	—	—
24 in. × ⅜ in.	4	—	65	70	75	80	85	100	115[c]	—	—	—	—
24 in. × ½ in. to ¾ in.	4	—	65	70	75[a]	80	85	100[b]	115[c]	—	—	—	—
24 in. × ¾ in. to 1¼ in.	5	—	65	70	75[a]	80	85	100[b]	115[c]	—	—	—	—
32 in. × ¾ in. to 1¼ in.	6	—	—	—	—	—	—	100[a]	115	130[b]	140	150[c]	—
Taper-split													
24 in. × ½ in. to ⅝ in.	4	—	65	70	75[a]	80	85	100[b]	115[c]	—	—	—	—
Straight-split													
18 in. × ⅜ in. (true-edge)	4	—	—	—	—	—	—	—	—	—	100	106	112[d]
18 in. × ⅜ in.	5	65[a]	75	80	90	95	100[c]	—	—	—	—	—	—
24 in. × ⅜ in.	5	—	65	70	75[c]	80	85	100	115[c]	—	—	—	—
15-in. Starter-finish course		Use supplementary with shakes applied not over 10-in. exposure											

[a]Recommended maximum weather exposure for three-ply roof construction.
[b]Recommended maximum weather exposure for two-ply roof construction.
[c]Recommended maximum weather exposure for single-coursed sidewalls.
[d]Recommended maximum weather exposure for double-coursed sidewalls.
(Courtesy Cedar Shake and Shingle Bureau.)

FIGURE 7.6-76 Maximum Exposure of Wood Shakes

Type	Length, in.	Maximum Exposure — Double Coverage, in.	Triple Coverage, in.
Hand-split-and-resawn shakes	18	7½	5½
	24	10	7½
	32	13	10
Taper-split shakes	24	10	7½
Straight-split shakes	18	—	5½
	24	—	7½

Note: In areas not subject to wind-driven snow, sheathing such as nominal 1 × 4 or 1 × 6 boards may be spaced equal to the exposure of the shake, but not more than 10 in.
(Courtesy Cedar Shake and Shingle Bureau.)

FIGURE 7.6-77 Nail Sizes for Application of Wood Shingles

Size	Nail Size — Length, in.	Gauge	Head, in.	Shingle Length, in.
3d[a]	1¼	14½	7/32	16 and 18
4d[a]	1½	14	7/32	24
5d[b]	1¾	14	7/32	16 and 18
6d[b]	2	13	7/32	24

(Courtesy Cedar Shake and Shingle Bureau.)
[a]3d and 4d nails are used for new construction.
[b]5d and 6d nails are used for overroofing.

7.6.6.2 Preparation for Roof Shingles and Shakes

SLOPE LIMITATIONS AND EXPOSURE

Roof slope limitations are discussed in Section 7.6.1.2. The exposure of wood shingles and shakes on a roof is dependent on the slope of the roof.

Shingles Standard exposures are shown in Figure 7.6-68. Four bundles of shingles will cover 100 sq. ft. (9.29 m^2) (one square) of roof area when the shingles are applied at standard exposures. Shingles are not recommended on slopes of less than 3 in 12.

Shingle exposure should not be larger than the length of the shingle minus 1 in. (25.4 mm), divided by 3. This will provide at least triple coverage (three layers of wood at every point) to ensure complete freedom from leakage in heavy wind-driven rain or snowstorms. Approximate coverage of four bundles of shingles at varying exposures is shown in Figure 7.6-67.

Shakes The maximum recommended exposure for double-coverage roofs is 10 in. (254 mm) for 24-in. (609.6 mm) shakes and 7½ in. (190.5 mm) for 18-in. (457.2 mm) shakes. A triple-coverage roof is achieved by reducing these exposures to 7½ in. (190.5 mm) for 24-in. (609.6 mm) shakes, and 5½ in. (139.7 mm) for 18-in. (457.2 mm) shakes. Approximate roof coverage for shakes using different exposures is shown in Figure 7.6-75. Maximum exposure to provide double and triple coverage is summarized in Figure 7.6-76.

Shakes are recommended only on slopes 4 in 12 or steeper. However, suitable installations have been achieved on slopes as low as 3 in 12 by taking the following extra precautions: (1) reducing exposures to provide triple coverage and (2) using solid sheathing with an underlayment of No. 30 asphalt-saturated felt applied over the entire roof, with No. 30 asphalt-saturated felt interlayment between each course.

ROOF SHEATHING

Sheathing types, thicknesses, spans, and nailing are discussed in Chapter 6.

Shingles Either spaced or solid sheathing may be used with wood shingles. When solid sheathing is preferred, either nominal 1-in. (25.4 mm) boards or plywood may be used. When spaced sheathing, such as nominal 1 × 3 (25 × 75) or 1 × 4 (25 × 100) boards, is used, it should be applied so that the space between boards is not greater than the width of the boards themselves.

Shakes Either spaced or solid sheathing may be used with wood shakes, depending on the climatic conditions of the region.

In areas not subject to wind-driven snow, spaced sheathing, such as nominal 1 × 4 (25 × 100) or 1 × 6 (25 × 150) boards, is suitable (Fig. 7.6-78). The center-to-center spacing of boards should be equal to the exposure of the shake, but not more than 10 in. (254 mm).

In areas subject to wind-driven snow (generally where the outside design temperature is 30°F (–1.1°C) or lower), solid sheathing of nominal 1-in. (25.4 mm) boards or plywood is recommended.

UNDERLAYMENT AND INTERLAYMENT

Shingles Underlayment typically is not required between shingles and spaced or solid sheathing, but may be desirable for the protection of the sheathing and to ensure against air infiltration.

When underlayment is desired, No. 15 asphalt-saturated felt may be used.

Shakes Interlayment strips and the underlayment course used at eaves should be No. 30 asphalt-saturated felt. The installation of interlayment between each shake course, essential to provide a baffle against infiltration of wind-driven rain or snow, is described in Section 7.6.6.4.

EAVE FLASHING

In areas where the outside design temperature is 30°F (–1.1°C) or colder (see Fig. 7.6-8), or where there is a possibility of ice forming along the eaves and causing a backup of water, eave flashing is recommended. Sheathing should be applied solidly above the eave line to cover a line at least 24 in. (609.6 mm) inside the interior wall line of the building.

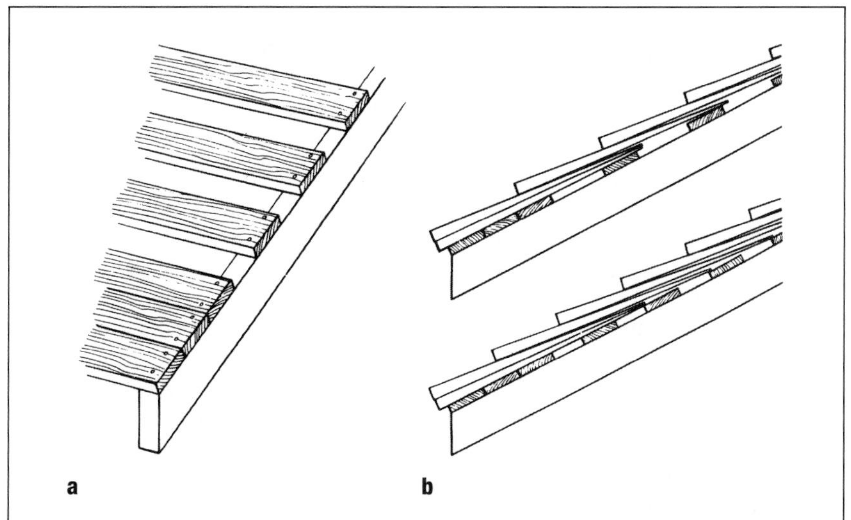

FIGURE 7.6-78 **(a) Spaced sheathing. (b) Maximum recommended exposure produces a two-ply roof; reducing exposure produces a three-ply roof.** (Courtesy Cedar Shake and Shingle Bureau.)

Shingles For 4 in 12 and steeper slopes, eave flashing should be formed by applying a double layer of No. 15 asphalt-saturated felt to cover the solid sheathing. When the eave overhang requires flashing wider than 36 in. (914.4 mm), the horizontal joint between felts should be cemented and located outside the exterior wall line.

For slopes 3 in 12 up to 4 in 12, or in areas subject to severe icing, eave flashing may be formed as described earlier for 4 in 12 slopes, except that the double layer of No. 15 asphalt-saturated underlayment should be cemented. The eave flashing should be formed by applying a continuous layer of plastic asphalt cement, at the rate of 2 gal. per 100 sq. ft. (0.815 L/m^2), to the surface of the underlayment starter course before the second underlayment layer is applied. Cement should also be applied to the 19-in. (482.6 mm) underlying portion of each succeeding course that lies within the eave flashing area, before placing the next course. The cement should be applied uniformly with a comb trowel, so that at no point does underlayment touch underlayment when the application has been completed. The overlying sheet should be pressed firmly into the entire cemented area (see Fig. 7.6-10).

Shakes On slopes 4 in 12 or steeper, eave flashing should be formed by applying an additional course of No. 30 asphalt-saturated felt over the underlayment starting course at the eaves. The eave flashing should extend up the roof to cover a line at least 24 in. (609.6 mm) inside the interior wall line of the building. When the eave overhang requires the flashing to be wider than 36 in. (914.4 mm), the necessary horizontal joint should be cemented and located outside the exterior wall line of the building (see Fig. 7.6-10).

For slopes 3 in 12 up to 4 in 12, or in areas subject to severe icing, eave flashing may be formed as described earlier for 4 in 12 slopes, except that the double layer of No. 30 asphalt-saturated underlayment should be cemented. The eave flashing should be formed by applying a continuous layer of plastic asphalt cement, at the rate of 2 gal. per 100 sq. ft. (0.815 L/m^2), to the surface of the underlayment starter course before the second underlayment layer is applied. Cement should also be applied to the 19-in. (482.6 mm) underlying portion of each succeeding course that lies within the eave flashing area, before the next course is placed. It is important to apply the cement uniformly with a comb trowel, so that at no point does underlayment touch underlayment when the application has been completed. The overlying sheet should be pressed firmly into the entire cemented area.

DRIP EDGE

Wood shingles and shakes should extend out over the eave and rake from 1 in. (25.4 mm) to 1½ in. (38.1 mm) to form a drip. Accordingly, a corrosion-resistant drip edge is not necessary at these points since the shingle starter course and rake shingles serve the same purpose.

FLASHING

Refer to Section 7.6.1.6 for general comments about flashing.

Unless special precautions are taken (see Fig. 7.6-79), copper flashing materials are not recommended for use with red cedar shingles or shakes. Premature deterioration of the copper may occur when the metal and wood are in intimate contact in the presence of moisture.

Valley Flashing for Wood Shingles Only the open method should be used to construct valley flashing for wood shingles. Butting shingles together in the valley at the centerline to form a closed valley is not recommended.

On slopes up to 12 in 12, metal valley sheets should be wide enough to extend at least 10 in. (254 mm) on each side of the valley centerline (Fig. 7.6-80). On roofs of steeper slope, narrower sheets may be used extending on each side of the valley centerline for at least 7 in. (177.8 mm).

To avoid restricting the delivery of a valley and to prevent water splashing from a steep slope onto the shingles of a lower slope, the edges of the shingles on the lower slope should be lined up at least 1 in. (25.4 mm) farther back from the valley centerline than those on the steeper slope. As an alternative, a vertical ridge, or waterstop, can be made by crimping the metal up in the center of the valley. Valley sheets should be edge-crimped by turning the edges up and back approximately ½ in. (12.7 mm) toward the valley centerline. This provides an additional waterstop, directing water down the valley. The open portion of the valley should be at least 4 in. (101.6 mm) wide, but valleys may taper from a width of 2 in. (50.8 mm) where they start and increase at the rate of ½ in. (12.7 mm) per 8 ft. (2438 mm) of length, to become wider as they descend.

In areas where the outside design temperature is 30°F (–1.1°C) or colder, underlayment should be installed under metal valley sheets.

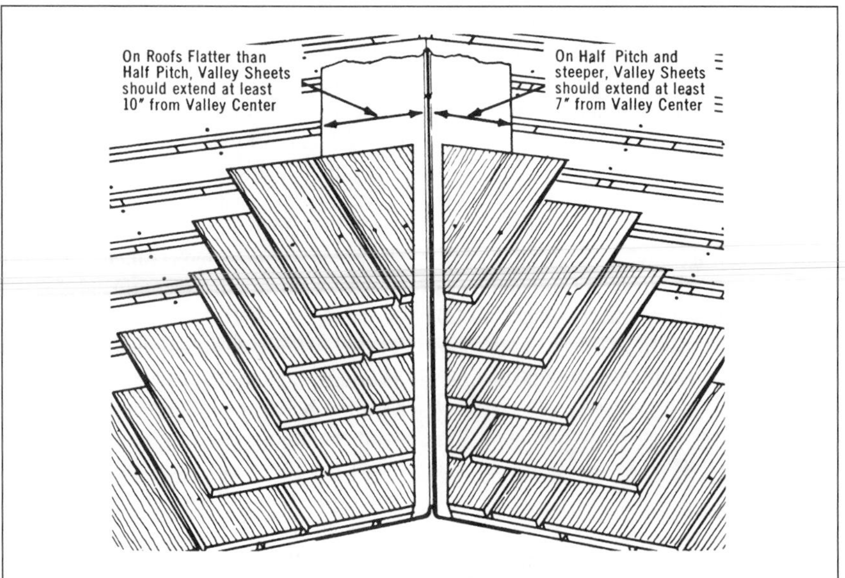

FIGURE 7.6-80 Method for constructing open valley flashing. (Courtesy Cedar Shake and Shingle Bureau.)

FIGURE 7.6-81 Open-valley flashing for shakes. (Courtesy Cedar Shake and Shingle Bureau.)

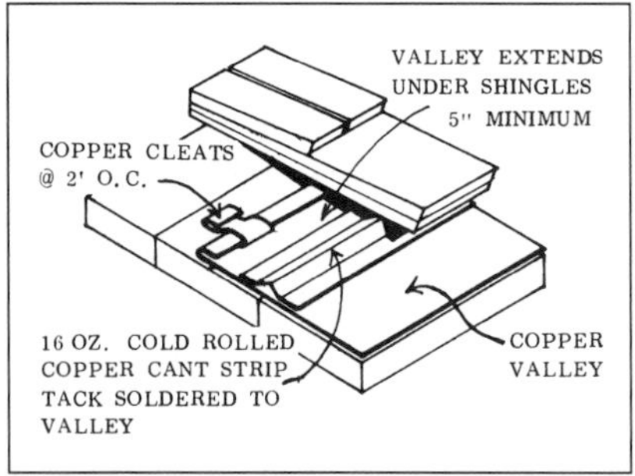

FIGURE 7.6-79 Surface contact of copper with red cedar must be minimized. (Courtesy Cedar Shake and Shingle Bureau.)

Valley Flashing for Wood Shakes The open method, using roofing felt and sheet metal, or the closed method, using hand-fitted shakes, may be used to construct valley flashing for wood shakes. For longest service life, the open method is highly recommended.

Open valleys are first covered with a No. 30 asphalt-saturated felt underlayment strip, at least 20 in. (508 mm) wide, centered in the valley and secured with enough nails to hold it in place (Fig. 7.6-81). Metal flashing strips, 20 in. (508 mm) wide, are then nailed over the valley underlayment. If galvanized steel flashing is used, it should be 18 gauge (1.3 mm) and never less than 26 gauge (0.55 mm), preferably center-crimped, and painted on both surfaces. Valley sheets should be edge-crimped by turning the edges up and back approximately ½ in. (12.7 mm) toward the valley centerline. This provides an additional waterstop, directing water down the valley. Shakes laid to finish at the valley should be trimmed parallel with the valley to form a 6-in. (152.4 mm)-wide gutter.

Closed valleys should first be covered with a 1 × 6 (25 × 150) wood strip, nailed

FIGURE 7.6-82 Flashing at a wall. (Courtesy Cedar Shake and Shingle Bureau.)

FIGURE 7.6-83 Flashing at a chimney. (Courtesy Cedar Shake and Shingle Bureau.)

flat into the saddle and covered with roofing felt, as discussed earlier for open valleys. Shakes in each course should be edge-trimmed to fit into the valley, then laid across the valley with a metal flashing undercourse having a 2-in. (50.8 mm) headlap, and extending 10 in. (254 mm) under the shakes on each side of the saddle.

CHIMNEY AND WALL FLASHING

Individual corrosion-resistant metal flashing units should be inserted between each course of shingles or shakes that butt against vertical surfaces such as walls, dormers, or chimneys. One leg of this flashing should extend at least 6 in. (152.4 mm) under the shingles or shakes with the upward leg covered by the vertical wall finish or by metal counter flashing, as shown in Figure 7.6-82.

On new construction, chimney flashing should be built in by the mason (Fig. 7.6-83). If masonry work has been completed, flashing should be inserted between bricks to a depth of ¾ in. (19.1 mm) by removing mortar and filling over the flashing with bituminous mastic.

Galvanized steel flashing should be painted on both surfaces before it is placed. Flashing strips that must be bent to sharp angles should be painted after bending; if crimped to form a waterstop in the centers of valleys, the crimped surface should be given a double coat of paint.

7.6.6.3 Application of Roof Shingles (New Construction)

The first course of shingles at the eaves should be doubled or tripled and should project 1 to 1½ in. (25 to 38 mm) beyond the eaves to provide a drip. The second layer of shingles in the first course should be nailed over the first layer to provide a minimum sidelap of at least 1½ in. (38 mm) between joints (Fig. 7.6-84). If possible, joints should be "broken" by a greater margin. A triple layer of shingles in the first course provides additional insurance against leaks at the cornice. No. 3 Grade shingles are frequently used for the starter course.

Shingles should be spaced at least ¼ in. (6.4 mm) apart to provide for expansion. Joints between shingles should be separated not less than 1½ in. (38 mm) from joints in adjacent courses above and below, and joints in alternate courses should not be in direct alignment. When shingles are laid with the recommended exposure, triple coverage of the roof results (see Fig. 7.6-84).

Shingles at a valley should be carefully cut to the proper miter at the exposed butts and should be nailed in place first so that the direction of shingle application is away from the valley. This ensures that shingle joints will not break over the valley flashing.

The application of wood shingles can be speeded by the use of a special shingler's or lather's hatchet (Fig. 7.6-85). The

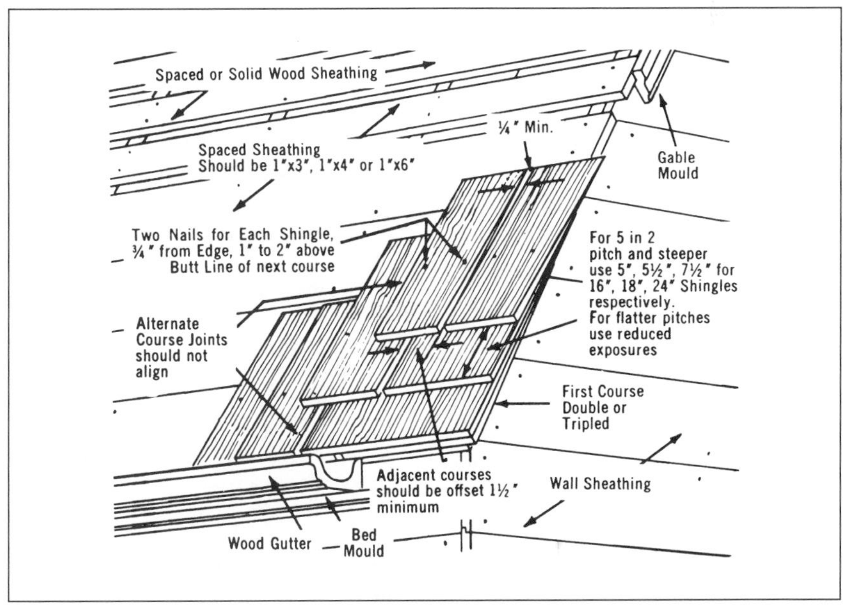

FIGURE 7.6-84 Application of wood shingles over spaced or solid sheathing. (Courtesy Cedar Shake and Shingle Bureau.)

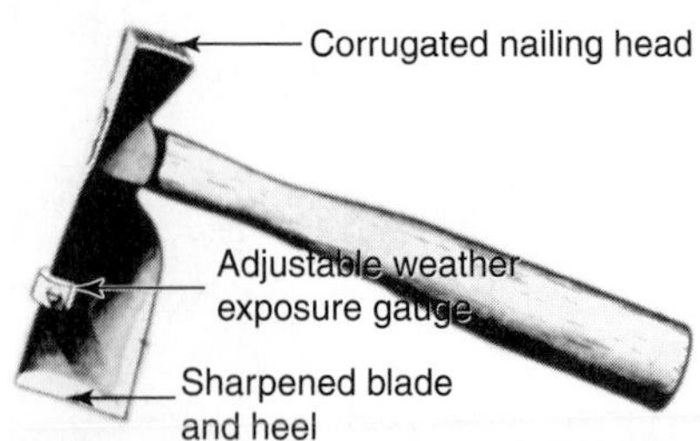

FIGURE 7.6-85 A shingler's hatchet. (Courtesy Cedar Shake and Shingle Bureau.)

hatchet, which is used for nailing, carries an exposure gauge and has a sharpened heel that expedites trimming.

NAILING

To ensure that shingles will lie flat and give maximum service, only two nails should be used to secure each shingle on a roof. Nails should be placed not more than ¾ in. (19.1 mm) from the side edge of shingles, at a distance of not more than 1 in. (25.4 mm) above the exposure line. Nails should be driven flush but not so that the nail head crushes the wood.

HIPS AND RIDGES

Hips and ridges should be of the modified "Boston" type with protected nailing. Either site-applied or preformed factory-constructed hip and ridge units may be used. Nails should be at least two sizes larger than the nails used to apply the shingles.

Construction of hips and ridges is shown in Figure 7.6-86. Hips and ridges should begin with a double starter course.

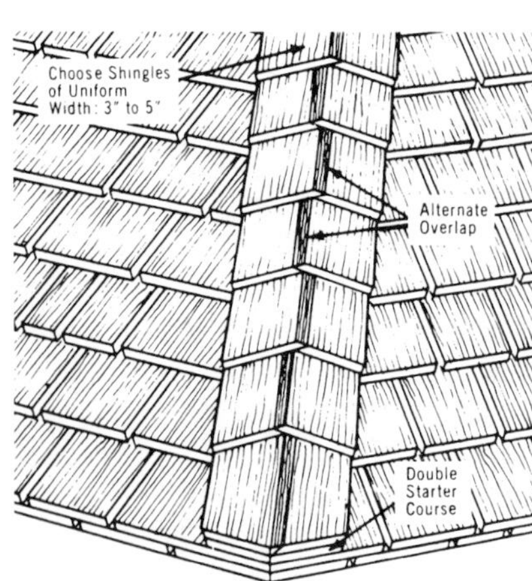

FIGURE 7.6-86 Application of hips and ridges. (Courtesy Cedar Shake and Shingle Bureau.)

GABLE RAKES

Shingles should project 1 to 1½ in. (25 to 38 mm) over the rake. End shingles can be canted to eliminate drips, as shown in Figure 7.6-87.

FIGURE 7.6-87 Beveled strip also provides decorative termination at gables. (Courtesy Cedar Shake and Shingle Bureau.)

ALTERNATE APPLICATION METHODS

Several alternate methods of applying shingles, giving different appearances to a roof, are shown in Figure 7.6-88.

REROOFING EXISTING CONSTRUCTION

It is not necessary to remove existing wood or asphalt shingles in order to install new wood shingles. Existing slate, tile, asbestos-cement, or mineral-fiber-cement shingles, which do not provide an adequate nailing base, must be removed.

However, the first course of existing shingles at the eaves should be removed just below the butts of the shingles in the second course. Old shingles should also be cut back and removed along the gable rakes (Fig. 7.6-89). The removed shingles along the eaves and gable rake edges should be replaced with boards having a thickness approximately equal to the thick-

a Thatch. Shingles are positioned above and below a hypothetical course line, with deviation from the line not to exceed 1″.

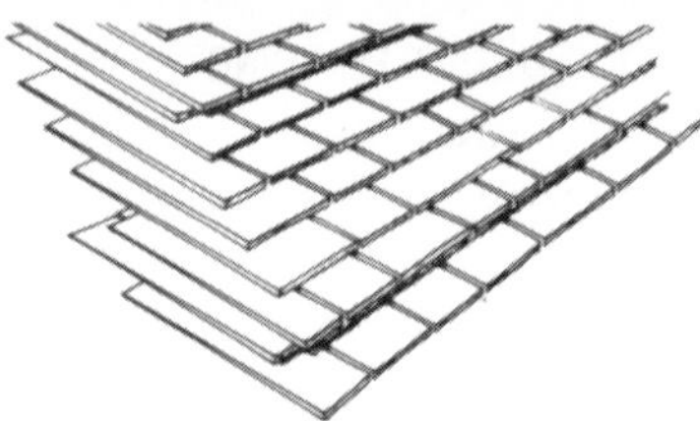

b Serrated. Courses are doubled every third, fourth, fifth, or sixth course. Doubled courses can be laid butt edge flush or with a slight overhang.

c Dutch Weave. Shingles are doubled or superimposed at random throughout the roof area.

d Pyramid. Two extra shingles, narrow shingle over a wide one, are superimposed at random.

FIGURE 7.6-88 Alternate shingle applications. (Courtesy Cedar Shake and Shingle Bureau.)

FIGURE 7.6-89 **Preparation of existing roof to receive new wood shingles or shakes.** (Courtesy Cedar Shake and Shingle Bureau.)

FIGURE 7.6-90 **Application of shakes over spaced sheathing.** (Courtesy Cedar Shake and Shingle Bureau.)

ness of the old roof section. These boards provide a strong base at the perimeter of the new roof, concealing the old roof from view. Ridge shingles should be replaced with strips of bevel siding, laid with the thin edges down the roof.

If necessary, new flashing should be installed over existing flashing.

New shingles should be applied by the same methods used in new construction. If the existing roof sheathing is spaced, it is not necessary for all new nails to strike the sheathing.

7.6.6.4 Application of Roof Shakes (New Construction)

First, a 36-in. (914.4 mm)-wide No. 30 asphalt-saturated felt starter strip should be laid along the eave line over the sheathing. The starter course of shakes at the eave line should be doubled, using an undercourse of 24-, 18-, or 15-in. (609.6, 457.2, or 381 mm) shakes, the latter being made expressly for this purpose. Striking visual effects can be achieved by using a double undercourse at the eave line. After each course has been applied, an 18-in. (457.2 mm)-wide strip of No. 30 asphalt-saturated felt interlayment should be applied over the top portion of the shakes extending onto the sheathing (Fig. 7.6-90). The bottom edge of the interlayment should be positioned at a distance above the butt equal to twice the exposure. For example, if 24-in. (609.6 mm) shakes are being laid with a 10-in. (254 mm) exposure, the bottom edge of the felt should be positioned at a distance 20 in. (508 mm) above the shake butts; the strip will then cover the top 4 in. (101.6 mm) of the shakes and extend 14 in. (355.6 mm) onto the sheathing.

Individual shakes should be spaced approximately ¼ to ⅜ in. (6.4 to 9.5 mm) apart to allow for possible expansion due to moisture absorption. The joints between shakes should be offset at least 1½ in. (38.1 mm) in adjacent courses. The joints in alternate courses should also be kept out of direct alignment when a three-ply roof is being built.

When straight-split shakes, which are of equal thickness throughout, are applied, the froe-end of the shake (the smoother end from which it has been split) should be laid undermost.

The application of shakes can be speeded by the use of a special shingler's or lather's hatchet (see Fig. 7.6-85). This hatchet, used for nailing, carries an exposure gauge and has a sharpened heel that expedites trimming.

NAILING

Only two nails should be used to apply each shake regardless of the width of the shake. Nails should be placed about 1 in. (25.4 mm) in from each edge, and from 1 to 2 in. (25 to 50.8 mm) above the buttline of the succeeding course. Nails should be driven until the heads meet the shake surface, but no farther (Fig. 7.6-91).

HIPS AND RIDGES

The final course at the ridge line, as well as shakes that terminate at hips, should be secured with additional nails and should be composed of smoother-textured shakes. A strip of No. 30 asphalt-saturated felt, at least 12 in. (304.8 mm) wide, should be

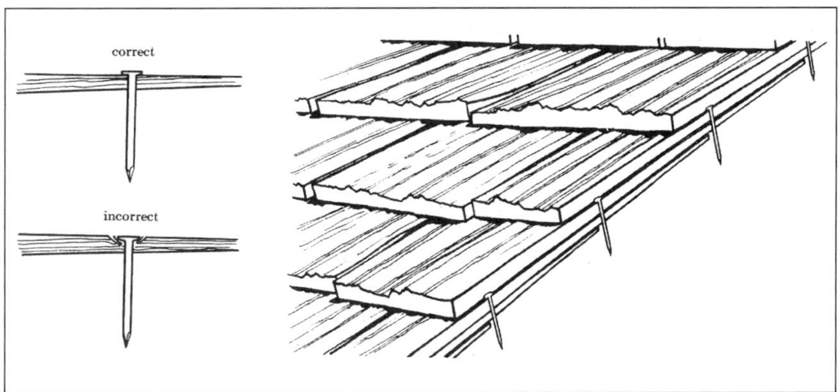

FIGURE 7.6-91 **Nails should be correctly placed, and heads should not break the shake surface.** (Courtesy Cedar Shake and Shingle Bureau.)

applied over the crown of hips and ridges, with an equal exposure of 6 in. (152.4 mm) on each side. Prefabricated hip and ridge units can be used, or the hips and ridges may be cut and applied on the site.

In site-construction of hips, shakes about 6 in. (152.4 mm) wide should be sorted out. Two wooden straightedges should be tacked on the roof, 6 in. (152.4 mm) from the centerline of the hip, one on each side. The starting course of shakes should be doubled. The first shake on the hip should be nailed in place with one edge resting against the guide strip. The edge of the shake projecting over the center of the hip should be cut back on a bevel. The shake on the opposite side should then be applied and the projecting edge cut back to fit. Shakes in the following courses should be applied alternately in reverse order (Fig. 7.6-92). Ridges should be constructed in a similar manner. The exposure of hip and ridge shakes is normally the same as that of the shakes.

Ridge shakes should be laid along an unbroken ridge that terminates in a gable at each end. They should be started at each gable end and terminated in the middle of the ridge. At that point, a small saddle should be face-nailed to splice the two lines. The first shake course should always be doubled at each end of a ridge.

FIGURE 7.6-92 Hips and ridges. Starter course should be doubled as shown. (Courtesy Cedar Shake and Shingle Bureau.)

TILTED STRIP AT GABLES

Dripping water may be eliminated by inserting a single strip of beveled siding the full length of each gable rake, with the thick edge flush with the sheathing edge (see Fig. 7.6-87). The resulting inward pitch of the roof surface diverts water away from the gable edge.

REROOFING EXISTING CONSTRUCTION

Structures that require new roof coverings, either for appearance or serviceability, may be covered with any of the four types of cedar shakes. Except over slate, tile, asbestos-cement, or mineral-fiber-cement shingles, shakes may be applied directly over the existing roof.

When the existing roof is to be overroofed with shakes, a 6-in. (152.4 mm) wide strip of the old roofing should be removed along the eaves and gable rakes and then filled in with boards having a thickness approximately equal to the thickness of the old roof section (see Fig. 7.6-89). These boards provide a strong base at the perimeter of the new roof, concealing the old roof from view. The old ridge covering should be removed and replaced with bevel siding, overlapping the butt edges at the peak. If necessary, new flashing should be installed over existing flashing. Nails for applying a new roof should be long enough to penetrate the underlying existing sheathing.

7.7 Siding

This section discusses the materials and methods needed to provide wall finishes using mineral-fiber-cement shingles, wood shingles and shakes, and aluminum, vinyl, wood, plywood, and hardboard siding.

In addition to cost, other criteria to be considered in selecting exterior siding materials and the method of their application include the expected service life of the siding, wind resistance, fire resistance, and local climate. Depending on the siding color and kind selected, the total wall heat loss or gain may also be affected.

7.7.1 GENERAL

Shingles, tiles, and aluminum, vinyl, wood, plywood, and hardboard siding units are applied to exterior walls in an overlapping or interlocking fashion to shed water and resist weather penetration. They can be applied effectively on most walls. These products are also used on the interior for aesthetic purposes. In either case, they may contribute color and texture to a wall surface, and, when properly applied and maintained, they provide a suitable finish offering long service.

Depending on the siding type and application method selected, aesthetic effects range from wall surfaces having relatively smooth, uniform coursing to those with random coursing, which gives a heavily textured appearance. Color can be achieved by a variety of factory- or field-applied coatings or by leaving the material in its natural color.

In addition to the wall finishing units themselves, many accessory materials are required to prepare a wall surface for application and to apply them. Depending on the type of finishing materials and the method of application, accessory materials may include sheathing, underlayment, sheathing paper, backer board, undercoursing, channel and corner moldings, flashing, caulking, and nails and other fasteners. Although many materials and typical methods of applying the described finishes are discussed in the following text, the siding manufacturers' instructions and recommendations should always be consulted to ensure the most suitable performance.

In this section, the wall construction is assumed to have been correctly installed and structurally adequate, and walls are assumed to have been insulated correctly and adequately where climatic conditions require it. It is further assumed that precautions have been taken to prevent water vapor condensation within walls. Condensation within walls can normally be eliminated by providing a vapor retarder on the warm side of the insulation. Refer to Section 7.1 for a discussion of vapor retarders and ventilation.

7.7.1.1 Nomenclature

Some definitions of siding terminology are given here to assist in understanding the text and illustrations. Other terms are defined when they first appear in the text or are included in the Glossary.

Backer board. An undercoursing material, typically fiberboard, used beneath siding materials such as mineral-fiber-cement

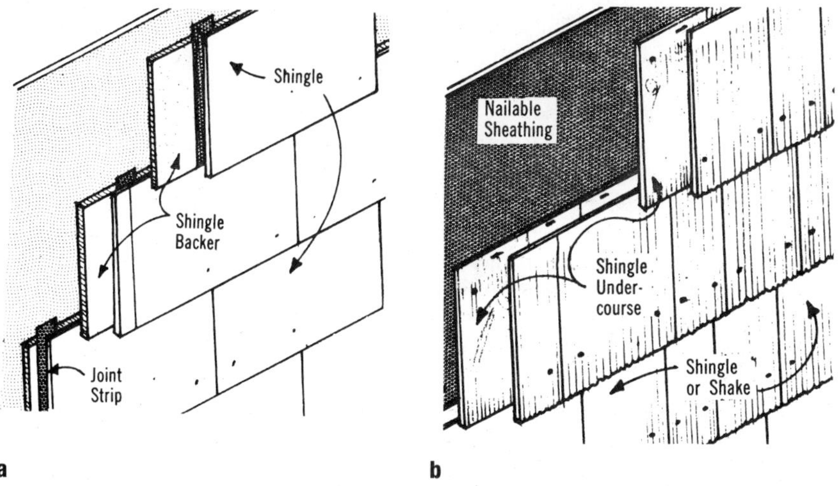

FIGURE 7.7-1 Double-coursed walls with shingles applied over an undercourse: (a) mineral-fiber-cement shingles, (b) wood shingles or shakes.

shingles (Fig. 7.7-1a). It adds insulation value, increases resistance to impact, and provides a heavier shadowline at the butt than shingle siding applied directly without the use of a backing material.

Coverage. The amount of weather protection provided by overlapping shingles. Depending on the shingle type and application method, shingles may furnish one (single coverage) or two (double coverage) material thicknesses over a wall surface.

Double coursing. A method of applying shingles over an undercourse of lower-grade shingles or other suitable material (Fig. 7.7-1b). Nails are exposed as a result of face (butt) nailing and greater shingle exposures are possible than with single coursing. Double coursing usually results in greater wall coverage at lower costs.

Endlap. The shortest distance that adjacent shingles, sheets, or tiles horizontally overlap each other (Fig. 7.7-2).

Exposure. The shortest distance between exposed edges of overlapping shingles or tile (see Fig. 7.7-2).

Headlap. The shortest distance from the lower edges of overlapping shingles to the upper edge of the unit in the second course below (see Fig. 7.7-2).

Shingles. Relatively small individual units designed to overlap other similar units to provide weather protection. Typically, shingles are applied to a nailing base, such as sheathing or horizontal nailing strips, which supports the shingles between structural framing members.

Shingle butt. The lower, exposed edge of a shingle.

Sidelap. See *endlap*.

Single coursing. A method of applying shingles without an undercourse (Fig. 7.7-3). In wood shingles, this method results in concealed nails and a smaller exposure than is found in double-coursed walls.

Toplap. The shortest distance from the lower edge of an overlapping shingle or sheet to the upper edge of the lapped unit in the first course below (that is, the width of the shingle minus the exposure) (see Fig. 7.7-2).

7.7.1.2 Wall Framing and Sheathing

The performance of siding is dependent on the supporting structure. Studs, headers, and other supports should be adequately sized and spaced to carry the necessary superimposed loads over the required spans.

Sheathing should be smooth, securely attached to supports, and provide an adequate base to receive nails and other fasteners. Both mineral-fiber-cement and wood shingles are applied over solid sheathing composed of wood boards or plywood. Wood and hardboard siding are applied over wood board, plywood, or nonwood sheathing, such as gypsum board, rigid foam, and fiberboard.

Wall framing and sheathing are discussed in Chapter 6.

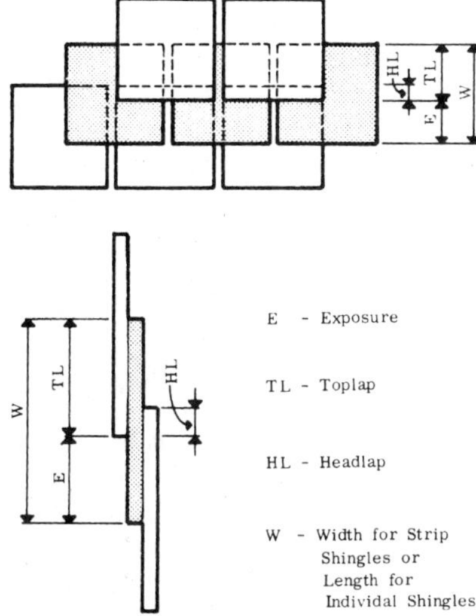

FIGURE 7.7-2 Shingle siding terminology.

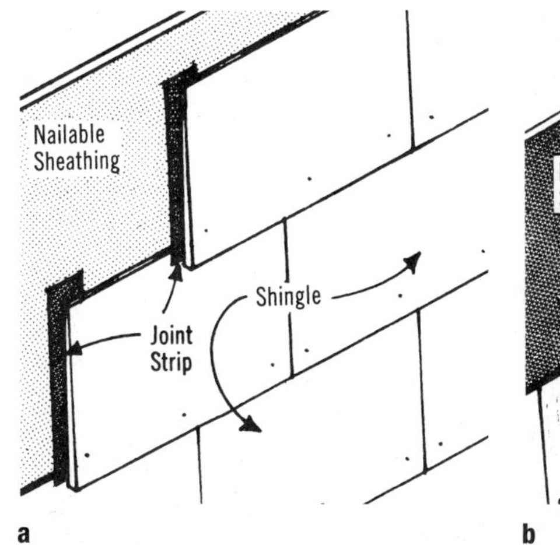

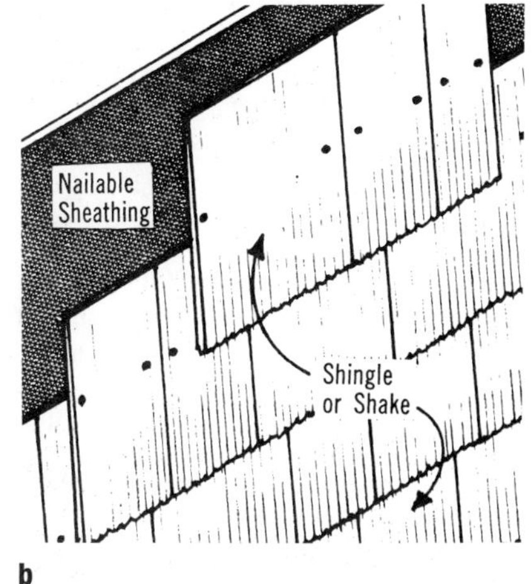

FIGURE 7.7-3 Single-coursed walls have shingles applied without an underlayment: (a) mineral-fiber-cement shingles, (b) wood shingles and shakes.

7.7.1.3 Flashing and Caulking

Walls may contain window or door openings, intersect with sloped roof surfaces, or terminate in exposed locations against other materials, creating opportunities for leakage. Flashing must be installed at these locations and be made watertight and water shedding.

FLASHING MATERIALS

Corrosion-resistant metal flashing should be at least 26-gauge (0.55 mm) galvanized steel, 0.019 in. (0.4826 mm) aluminum, or 16 oz. (453.6 g) copper. Refer to Section 7.8 for a discussion of flashing materials.

Additional requirements and typical methods of flashing are described in the following subsections for each siding type.

WALL FLASHING

Wall flashing is discussed in Section 7.8.

Flashing at Openings Heads of openings should receive drip caps. Bottoms of openings should receive sills. Drip caps and sills (see Fig. 7.8-4) should be flashed.

At heads of openings in wood and metal stud framed walls, metal flashing should be installed, extending from at least 2 in. (50.8 mm) above the trim, and be turned down over the outside edge of drip caps, forming a drip. However, in areas not subject to wind-driven rain, head flashing may be omitted when the vertical height between the top of finish trim of openings and the bottom of eave soffits is equal to or less than one-quarter of the eave overhang.

Flashing should be installed beneath sills and turned out down to form a drip beyond the face of the siding.

Where sheathing paper is omitted, flashing should extend up behind the sheathing.

In unsheathed walls, jambs should be flashed with a 6-in. (152.4 mm)-wide strip of metal, 3 oz. (915.5 g/m^2) copper-coated paper, or 6 mil (0.1524 mm) polyethylene film.

Flashing at Wall and Roof Intersections Sheet metal flashing should be installed at horizontal intersections and at vertical intersections when the exterior finish material does not provide a self-flashing joint.

Flashing at Base of Walls and at Horizontal Material Changes Bottoms of walls and horizontal changes in materials should receive watertables. On these watertables, metal flashing should be installed to a line at least 2 in. (50.8 mm) above the bottom of the siding and turned down over the watertable to form a drip (see Fig 7.8-4).

CAULKING

A nonshrinking caulking compound, either white or a matching color, should be used to weatherseal joints where siding abuts wooden trim, masonry, or other projections, and in unprotected vertical joints in panel siding.

7.7.1.4 Nails and Fasteners

No single step in applying siding materials is more important than proper nailing, which includes:

- Selecting the correct nail for the kind of siding and sheathing
- Using the correct number of nails
- Locating nails correctly
- Choosing a nail metal compatible with the metal used for flashings

Specific recommendations for the type, size, number, and spacing of nails are given in the following subsections for each wall finish material. Some general recommendations are given here.

SHINGLES OVER WOOD AND PLYWOOD SHEATHING

Nails for attaching shingles should be long enough to penetrate through the shingles and through the wood or plywood sheathing. Nails for applying shingles over wood and plywood sheathing should have threaded shanks.

SHINGLES OVER NONWOOD SHEATHING

When gypsum, rigid foam, fiberboard, or other nonwood sheathing materials are used, special fasteners and particular details for fastening shingles are often necessary to provide adequate anchorage. The nonwood material manufacturer's recommendations should be followed.

WOOD, PLYWOOD, AND HARDBOARD SIDING

Nails for attaching wood, plywood, and hardboard board and panel siding should penetrate through the sheathing and at least 1½ in. (38 mm) into the underlying studs.

7.7.2 MINERAL-FIBER-CEMENT SHINGLES

Because of possible health hazards, siding shingles that were earlier made with asbestos fibers are now manufactured using nonasbestos mineral fibers. Therefore, the term used here for such shingles is *mineral-fiber-cement shingles*. Similar materials found in older buildings, however, may contain asbestos. Readers are, therefore, strongly advised to consult current OSHA, EPA, state, and local regulations and suggestions with regard to safety precautions, and to obtain professional assistance when dealing with such existing materials.

Unless specific exceptions are made in the following text, the provisions of Section 7.6.1 apply to mineral-fiber-cement shingles and the methods used to apply them.

7.7.2.1 Materials

Mineral-fiber-cement shingles can be discolored by moisture and dampness while still in packages and until applied. Therefore, they should be kept clean and dry and completely protected from the weather until they are applied. Shingles should not be applied over wet sheathing, underlayment, or sheathing paper.

Mineral-fiber-cement wall shingles are usually supplied in rectangular units having straight or wavy butt edges. These units are available in the sizes shown in Figure 7.7-4, and other sizes may be available. They are designed to be applied with a minimum nominal toplap of 1 in. (25.4 mm). Depending on their size, these units range up to approximately 3⁄16 in. (4.8 mm) in thickness.

Wall shingles are manufactured with prepunched nail holes correctly sized and located to receive exposed face nails. These holes are aligned the proper distance above the lower edge of the units to establish the recommended amount of toplap (Fig. 7.7-5) and to minimize damage to shingles during their installation. For standard application, face nails driven through these holes and resting on the top edge of the next-lower course serve as guides for the correct amount of toplap, exposure, and unit alignment.

Mineral-fiber-cement wall shingles should conform to the requirements of ASTM C 223.

NAILS AND FASTENERS

Face nails should be long enough to enter into and hold securely to the nailing base. Face nails are usually furnished by the siding manufacturer to ensure the use of the proper kind and type. Nails and fasteners should be permanently corrosion resistant and stain resistant. Only those furnished

FIGURE 7.7-4 Mineral-Fiber-Cement Wall Shingles

Sizes,[a] in. (mm)	Exposure, in. (mm)	Pieces per Bundle	Pieces per Square[b]
12 × 24 (305 × 610)	11 (279.4)	18	54
9 × 32 (229 × 813)	8 (203.2)	19	57
9⅝ × 32 (244.5 × 813)	8⅝ (219)	17–18	52
11¾ × 32 (298.5 × 813)	10¾ (273)	14	42
14⅝ × 32 (371.5 × 813)	13⅝ (346)	11	33
14¾ × 32 (374.7 × 813)	13¾ (349)	11	33
8¾ × 48 (222 × 1219)	7¾ (197)	13	39
12 × 48 (292 × 1219)	11 (279)	9	27

[a]Three nail holes per shingle.

[b]Weight per square ranges from 153 to 206 lb. (693.97 to 934.37 Mg).

with the siding or recommended by the siding manufacturer should be used.

Figure 7.7-6 is a summary of face nail types recommended for use in applying mineral-fiber-cement wall shingles.

7.7.2.2 Preparation for Wall Shingles

WALL SHEATHING

Mineral-fiber-cement shingle siding can be applied over lumber, plywood, and other suitable types of rigid sheathing that are smooth and dry and provide adequate support and an adequate nailing base. Sheathing material thicknesses, spans, and nailing requirements are discussed in Chapter 6.

SHEATHING PAPER

Sheathing paper for use over lumber and plywood sheathing should be water repellent yet vapor porous (permeable to water vapor), such as No. 15 asphalt-saturated felt. Coal-tar saturated felt, which may stain the shingles, should not be used. Vapor retarder membranes should not be used for sheathing paper between siding and sheathing.

Sheathing paper for walls is discussed in Section 6.6.5.1.

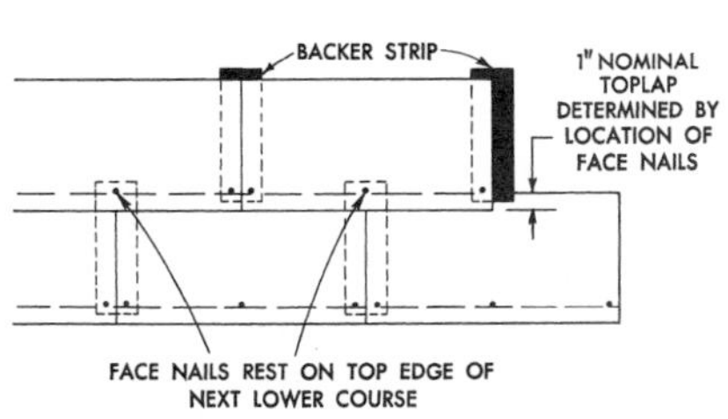

FIGURE 7.7-5 Prepunched nail holes in mineral-fiber-cement siding shingles.

JOINT FLASHING STRIPS

Asphalt (not coal-tar)-saturated and coated, water-repellent joint flashing strips are supplied by the manufacturer with all types of mineral-fiber-cement shingle siding, and should be applied behind every vertical joint between wall shingles (Fig. 7.7-7).

CANT STRIPS

A ¼-in. (6.4 mm) by 1½-in. (38.1 mm) wood strip should be nailed level along the bottom edge of the sheathing so that it overhangs the top of the foundation wall a sufficient amount to seal the joint between the top of the foundation, the wood sill, and the bottom of the sheathing (see Fig. 7.7-7). The bottom edge of the shingles should align with the bottom edge of the cant strip. This strip gives the necessary cant (pitch) to the first shingle course and provides solid bearing to prevent breakage.

FLASHING

Flashing should be done in accordance with Section 7.6.1.6.

FIGURE 7.7-6 Face Nails for Use with Mineral-Fiber-Cement Wall Shingles

	New Construction	
Sheathing Type	**Nailed Directly to Sheathing or Wood Nailing Strips**	**Nailed Through Fiberboard Shingle Undercoursing**
Board lumber	1⅛-in. (28.5 mm) screw thread 1⁷⁄₁₆-in. (36.5 mm) screw thread 1½-in. (38.1 mm) annular thread	2-in. (50.8 mm) annular thread
Plywood	1⅛-in. (28.5 mm) screw thread	1¾-in. (44.5 mm) screw thread
Wood nailing strip	1⅛-in. (28.5 mm) screw thread	1¾-in. (44.5 mm) screw thread
Nailbase fiberboard	1⅜-in. (34.9 mm) special annular thread	2-in. (50.8 mm) special annular thread
	Residing (over existing wood siding or shingles)	
Nailing Base	**Nail Types**	
Sound board lumber	1¾-in. (44.5 mm) or longer annular thread, screw thread, or helical thread	
Wood nailing strip	1⅛-in. (28.5 mm) screw thread	

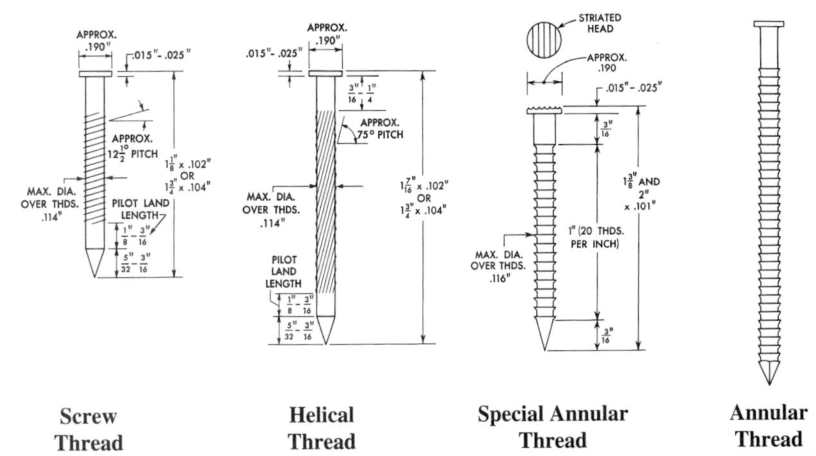

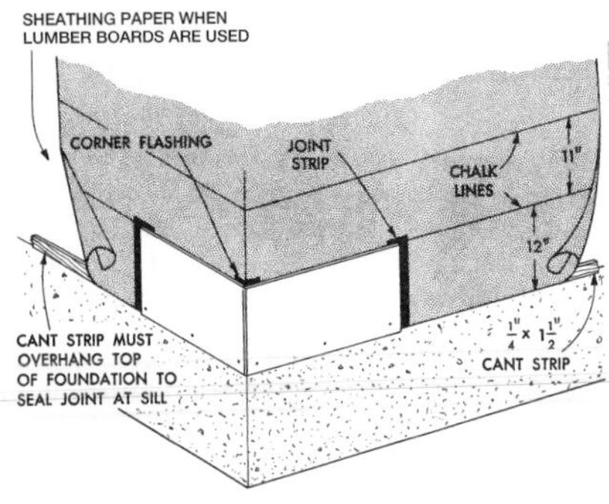

11″ = 279.4 mm
12″ = 304.8 mm
¼″ × 1½″ = 6.4 × 38.1 mm

FIGURE 7.7-7 Direct application of mineral-fiber-cement shingles showing application of joint flashing strips.

7.7.2.3 Wall Shingles Application (New Construction)

Depending on the sheathing type, mineral-fiber-cement shingles may be applied by several methods: (1) *direct application* to nailable sheathing, (2) *shingle backer method*, (3) *channel method* over nailable sheathing or directly over framing members without sheathing, and (4) *wood nailing strip method* for use over any type of sheathing or directly over framing members without sheathing.

The methods described are based on applying 12- by 24-in. (305 by 610 mm) wall shingles from left to right on new construction. The manufacturers' instructions should be followed when applying other shingle sizes.

Careful layout should establish the location of the first horizontal course and ensure that all courses will meet and match at corners and are level all around the building. The total number of horizontal courses to complete one wall should be laid out, and adjustments should be made so that the top courses at the eaves will not be too narrow. This may require uniform adjustments in reducing the exposure of each shingle course. A level chalk line should be marked all around the building to fix the location of the top edge of the first course of shingles. Additional horizontal chalk lines should mark the top of each succeeding course. The lines should be spaced to provide the desired exposure (see Fig. 7.7-7).

DIRECT APPLICATION METHOD

The direct application method is used for sheathing types that provide a nailing base, such as wood boards or plywood.

First and Odd-Numbered Courses A full-size shingle should be started at the left corner of the wall. This unit should be carefully applied plumb, level, and in alignment with a chalk line, because it guides the laying of all other units.

The correct face nails for the sheathing type and application method being used are supplied by the manufacturer, with the shingles as described in Figure 7.7-6. Nails should be driven snug but not too tight.

Before the last nail is driven at the right end of a unit, a felt joint flashing strip should be inserted in place and secured with the last nail. Joint flashing strips should always be used, centered behind the butted joint between wall shingles, and with the lower end overlapping the cant strip or head of the next lower course.

Application of full-size shingles should be continued in the first course with the top edges aligned to the chalk line. The last unit in a course or at an opening should not be less than 6 in. (152.4 mm) wide, and if smaller spaces will remain to be filled, a few inches should be cut from the units applied earlier in the course. Necessary holes for face nails in less-than-full-size pieces of siding should be punched. Unit ends should be butted tightly end to end with no space in between. Wedge-shaped spaces between unit ends indicate incorrect application.

Second and Succeeding Even-Numbered Courses In succeeding courses of 12-in. by 24 in. (305 mm by 610 mm) shingles, vertical joints should break on *halves*; therefore, the second course should be started with a half unit. When 32-in. (812.8 mm)-long units are used, vertical joints should also break on halves. When 48-in. (610 mm)-long units are used, they should break on *thirds* and the second course should be started with a two-third unit. Some bundles of shingles contain the necessary shorter pieces for starting courses.

A partial unit should be started at the left corner with its head edge aligned to the chalkline and its lower (butt) edge overlapping the head of the next-lower course the correct distance to provide the necessary toplap between courses. A nail should be inserted in a face nail hole and the siding positioned so that the shank of the nail is resting on the top edge of the next-lower course, thus establishing the proper amount of toplap. The unit should then be nailed in place with a felt joint flashing strip installed as described earlier. The course should be continued with full-size units.

The face nails described in Figure 7.7-6 should always be used. The nails should penetrate completely through the sheathing. A standard threaded face nail should be used when the siding is applied directly to nominal 1-in. (25.4 mm) wood board sheathing or 5⁄16 in. (7.9 mm) and thicker plywood without wood nailing strips or shingle backer.

SHINGLE BACKER METHOD

The shingle backer method uses rigid insulating fiberboard as a backer for the shingles and may be used over wood boards or plywood, since they provide a nailing base for both the shingle backer and the shingles (Figs. 7.7-8, 7.7-9, and 7.7-10). Special methods may be used to apply shingles directly over framing members.

A rigid, water-resistant, insulating fiberboard shingle backer should be at least 5⁄16 in. (7.9 mm) thick and should underlay the full width and length of each course of shingles completely. The shingle backer should be approximately ¼ in. (6.4 mm) narrower than the shingles. The shingles should be applied with their head edges flush with the head edges of the shingle backer and their butt edges extended down and beyond the butt edges of the underlying shingle backer to provide a drip edge. The shingle backer should be at least 4 ft. (1220 mm) long and be applied with staggered joints in succeeding courses. Vertical joints between shingles should not occur over the ends of the shingle backer. The shingle backer should be applied with its butt edges overlapping the head of the lower course of shingles about ¼ in. (6.4 mm) less than the headlap used between shingle courses.

The shingle backer and the shingles should be secured to the wall with face nails driven through holes in the shingles and into the underlying sheathing. The shingle backer should be held in position before the shingles are applied, by nailing each panel about 3 in. (76.2 mm) from the top edge with two or three 1¼-in. (31.8 mm) galvanized roofing nails. The type and length of the face nails should be as described in Figure 7.7-6.

Succeeding shingle courses should be applied in a similar manner. Felt flashing joint strips are not required if the shingle backer board is water repellent and stain resistant.

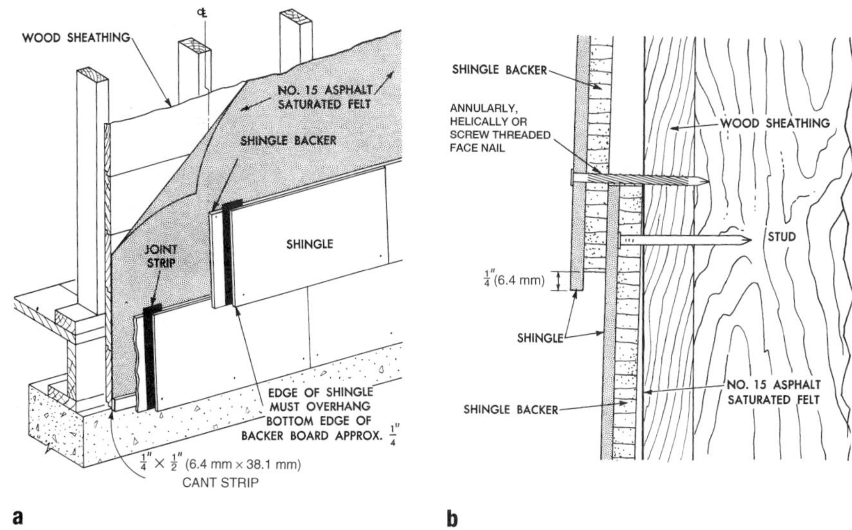

FIGURE 7.7-8 Application of mineral-fiber-cement wall shingles using the shingle backer method.

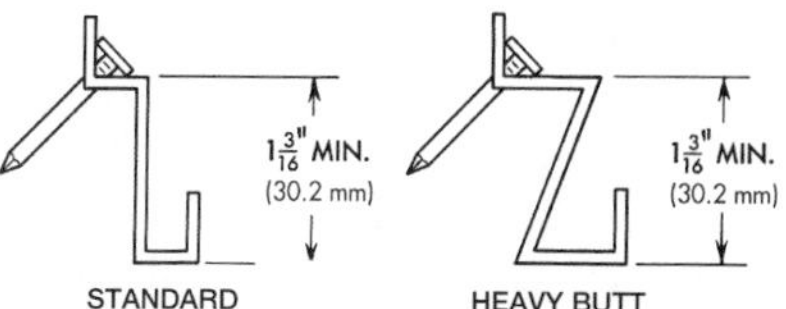

FIGURE 7.7-11 Typical profiles of metal channel moldings for installing mineral-fiber-cement shingles by the channel method.

Direct attachment to framing members is possible using a rabbeted nailing strip with a special shingle backer that also serves as sheathing (see Fig. 7.7-9). This method provides firm backing and a heavy shadowline. A similar method utilizing a grooved shingle backer is shown in Figure 7.7-10.

CHANNEL METHOD

The channel method uses one of several styles of metal channel molding designed for securing mineral fiber siding to nailable sheathing or directly to framing members. Metal channel should be corrosion-resistant and stain-resistant metal of sufficient thickness to provide adequate support and rigidity. Moldings are usually prepainted to match the shingle color and also provide either a standard or heavy butt shadow (Fig. 7.7-11).

A 1³⁄16-in. (30.16 mm) channel permits standard prepunched shingles to be used when units are reversed head to butt (Fig. 7.7-12). Channels should be secured at 16-in. (406.4 mm) intervals, either to the framing member or to nailable sheathing, with 10- (0.135 in.) to 12- (0.106 in.) gauge (2.7 mm), hot-dip galvanized steel, large-head roofing nails, long enough to penetrate and hold securely in the nailing base. Nails should be driven at an angle and set well to provide sufficient pressure on the nailing flange to ensure a tight, vibration-free assembly (Fig. 7.7-13).

Channel molding should have a vertical nailing flange at least ⅜ in. (9.5 mm) high that can bear against the face of the sheathing, and should be strong and rigid enough to secure the siding to the wall sheathing.

Direct attachment to framing members requires shingles wide enough to span from center to center of studs. Face nail

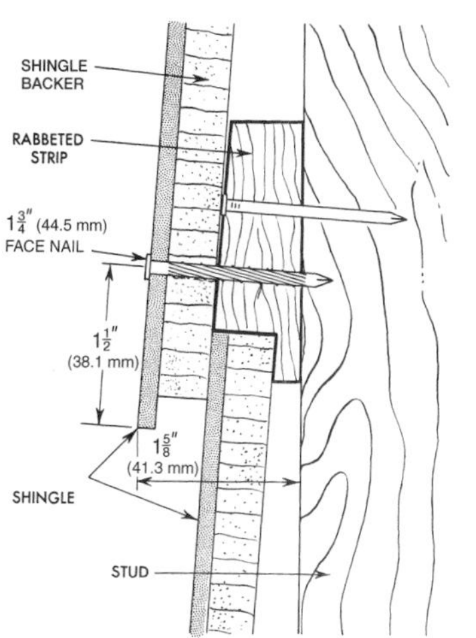

FIGURE 7.7-9 Application of mineral-fiber-cement wall shingles using rabbeted nailing strips.

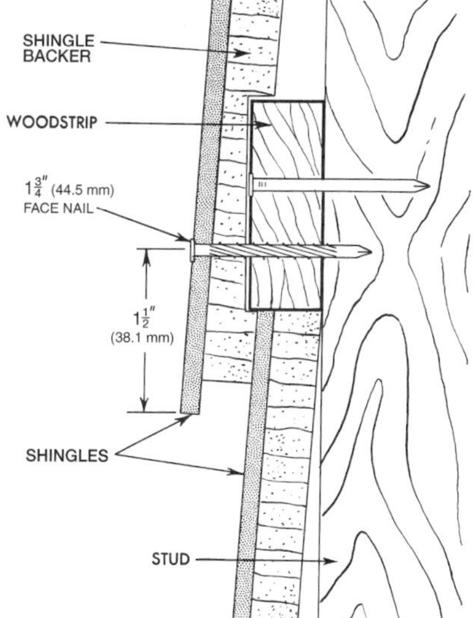

FIGURE 7.7-10 Application of mineral-fiber-cement wall shingles using grooved shingle backer.

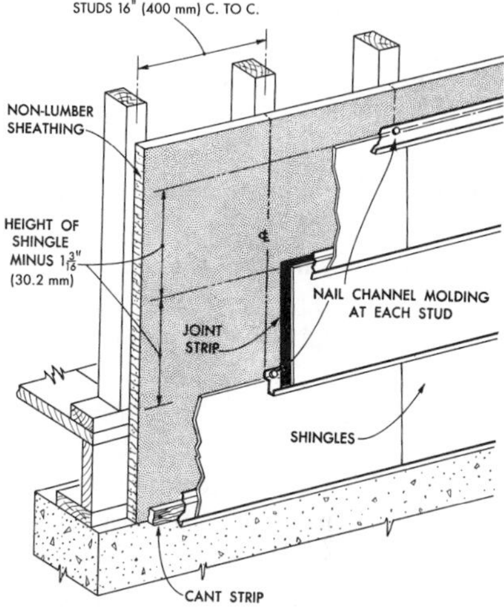

FIGURE 7.7-12 Application of mineral-fiber-cement shingles using the channel method. See also Figure 7.7-13.

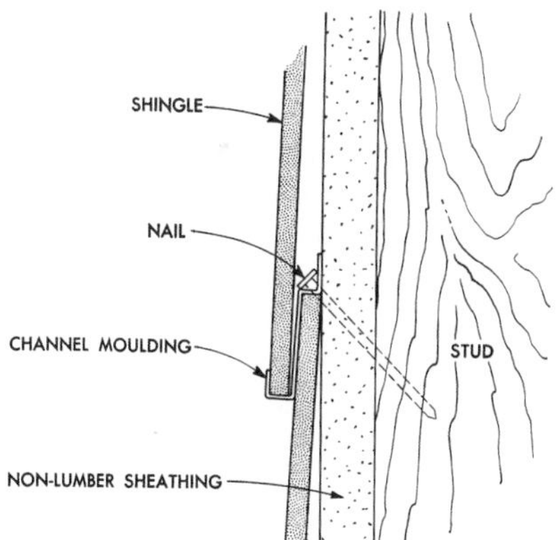

FIGURE 7.7-13 Channels are nailed directly to framing members when sheathing does not provide a nailing base.

holes should be punched within ⅜ to ½ in. (9.5 to 12.7 mm) from the vertical edge of the shingles to ensure nailing into the stud when two shingles are butted together. Intermediate face nail holes should be provided for nailing each stud, and face nails should be long enough to penetrate at least 1 in. (25.4 mm) into the stud.

When this method of nailing is used, studs must be located accurately on 16 in. (406 mm) centers wherever possible. Warped studs should not be used.

When it is not possible to nail directly into a stud, either a *twist nail* or a *self-clinching* nail should be used to attach the corner of the siding to nonlumber sheathing. A twist nail is a slender copper or aluminum nail with flat head and medium needle point, designed for twist clinching, that is, manual twisting of the exposed shank. A self-clinching nail is designed so that its shank clinches automatically when the nail has been fully driven.

The inside of exterior walls should be left exposed for final inspection to ensure that shingle nails have been properly driven into the studs and that twist nails and self-clinching nails are properly clinched.

WOOD NAILING STRIP METHOD

The wood nailing strip method may be used over lumber boards, plywood, and any type or thickness of nonlumber sheathing that does not have adequate nail holding power, because the strips rather than the sheathing serve as the nail base. No special nails, fasteners, or channels are required, and the standard face nails shipped with the siding can be used.

In addition to providing maximum security for the siding and economy in application, this method produces a heavy shadow-line along the butt edge of each course of shingles.

Wood nailing strips for the application of mineral-fiber-cement shingles should be at least nominal ⅜-in. (9.5 mm) by 3½-in. (88.9 mm) boards that are straight, sound, and free of knots or knotholes that may cause splitting or hinder nailing. Strips should be not less than 4 ft. (1220 mm) long, and preferably should be longer, so that they can span continuously over several framing members. Seasoning conditions and face nailing of spaced boards should be the same as described for solid sheathing.

Wood nailing strips should be placed to overlay the head of the lower course of shingles in an amount equal to the desired toplap less ¼ in. (6.4 mm) (Fig. 7.7-14). This firmly clamps the head of the course of shingles to the wall. At each framing member, the wood strip should be nailed through the sheathing and to each underlying stud with two 2-in. (50.8 mm) by 0.113-in. (2.87 mm) (6d) or longer annular threaded nails, penetrating the stud at least 1 in. (25.4 mm). The wood strips should end over framing members; joints in the strips should be staggered in succeeding courses. Vertical joints between shingle units should not occur over the end joints of wood strips.

The next course of shingles should then be applied with its butt edge placed over the wood strip, with the lower edge of the shingles extending down over the lower edge of the strip not more than ¼ in. (6.4 mm), to provide a drip. The siding should then be nailed through the face nail holes to the wood nailing strips with the threaded face nails supplied with the shingles.

CORNER FINISHING

When wall surfaces join at inside corners, the shingle courses are usually butted, although metal molding or wood corner strips can be used. Outside corners can be finished by any one of three methods.

Metal Corner Finish Several styles of individual stain-resistant metal corners and finish moldings are available (Fig. 7.7-15). They should be applied as recommended by the manufacturer.

Wood Corner Board Finish Corners finished with wood strips can be made of previously primed or painted nominal 1-in. (25.4 mm) lumber applied along one corner. A 1 × 4 (25 × 100) should be applied along one corner so that its side edge projects a distance beyond the corner equal to the thickness of the board applied on the adjoining wall. A 1 × 3 (25 × 75) should then be applied with its edge butting against the back of the 1 × 4 (25 × 100) first applied. Therefore, the wood corner strips will appear the same width on each wall (Fig. 7.7-16). When wood nailing strips or shingle backer, or both, are used, wood corner boards thicker than nominal lumber or blocking may be required.

Lapped or Woven Corner Finish This method is appropriate for white or gray shingles only. For lapped adjoining courses at outside corners, a whole shingle should be placed temporarily against each adjoining wall, so that shingles meet and join at the corner. The adjoining vertical edges of the shingles are scribed and cut at the proper angle to form a lapped or woven corner joint, and the edges are dressed with a file. Corner joints should be lapped to the left and right in alternate courses.

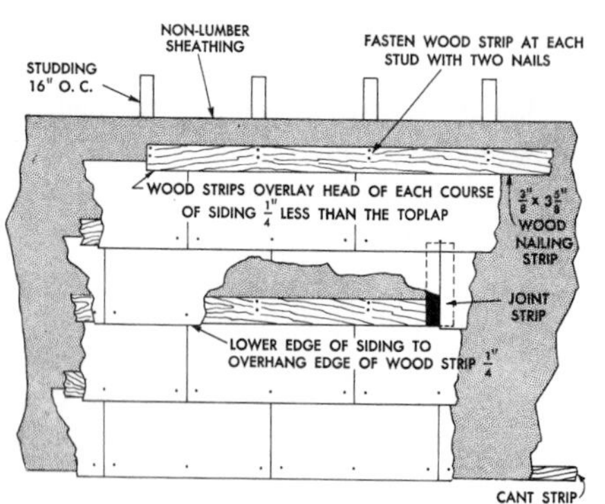

16″ = 400 mm
¼″ = 6.4 mm

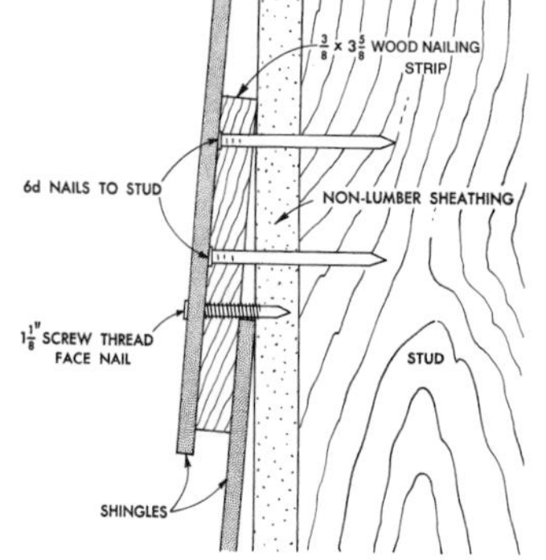

⅜″ × 3⅝″ = 9.5 mm × 92.1 mm
1⅛″ = 28.6 mm

FIGURE 7.7-14 Application details of mineral-fiber-cement shingles using the wood nailing strip method.

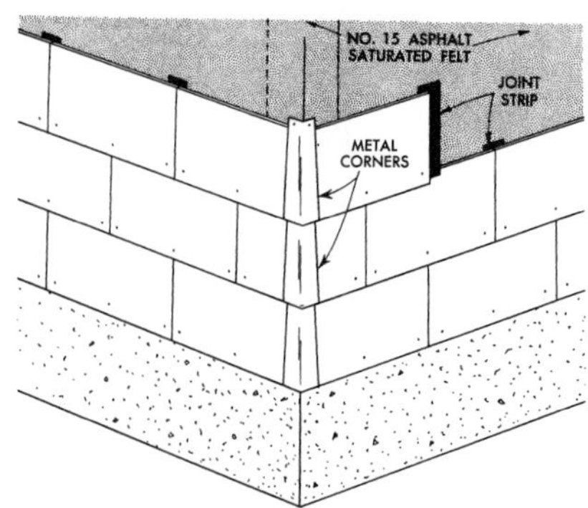

FIGURE 7.7-15 Metal corner for mineral-fiber-cement shingles.

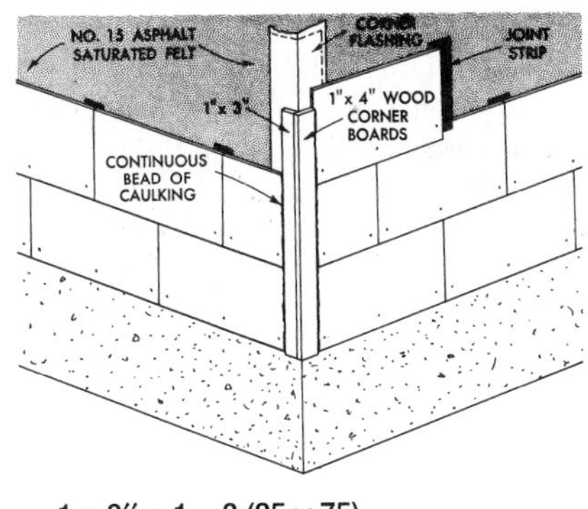

1 × 3″ = 1 × 3 (25 × 75)
1 × 4″ = 1 × 4 (25 × 100)

FIGURE 7.7-16 Wood board corner for mineral-fiber-cement shingles.

Courses on adjoining walls should meet and match at corners.

FLASHING AT OPENINGS AND CORNERS

Metal flashing should be applied over exposed door and window openings (Fig. 7.7-17). On old structures, the flashing must be renewed if not in good condition. Drip edges should be provided at heads of openings, under sills, and at projections to keep water away from the finished wall surface below.

Outside and inside corners should be flashed with a 12-in. (305 mm)-wide strip of underlay material centered in or around the corner when corner boards are used (see Fig. 7.7-16).

In other corner treatments, the underlayment should be carried around the corner of each side wall so that a double thickness of felt results over the corner (see Fig. 7.7-15).

7.7.2.4 Reshingling Existing Walls

An old wall that is to be covered with new mineral-fiber-cement shingles should be repaired so that it has a smooth, substantially true surface with adequate nail-holding capacity and the ability to support the new shingles. When the existing exterior finish consists of relatively thin bevel siding applied directly to widely spaced framing members and does not provide an adequate nailing base for new shingles, beveled wood strips should be placed at nailing locations along the wall. The face nails described in Figure 7.7-6 should be used for application of new shingles.

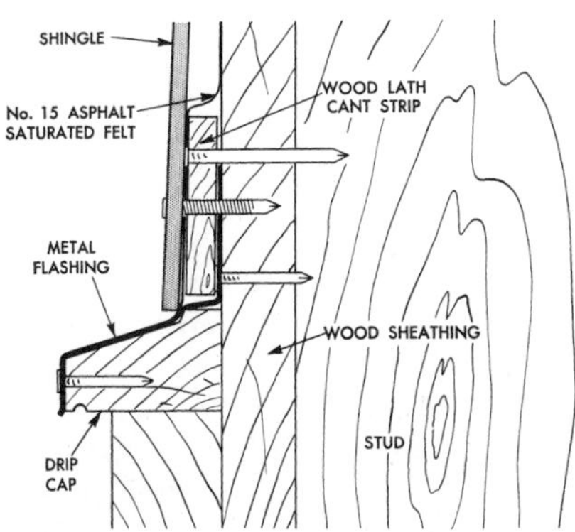

FIGURE 7.7-17 Head flashing over an opening in a mineral-fiber-cement shingle wall.

RESHINGLING OVER NONLUMBER SHEATHING

When mineral-fiber-cement shingles are to be applied to a structure that was originally sheathed with a nonnailable sheathing and the existing siding to be covered will not provide an adequate nailing base, the only recommended attachment method is the wood nailing strip method discussed in Section 7.7.2.3. Using any other method will require that both the existing siding and the existing sheathing must be removed down to the existing framing members and replaced with new sheathing to receive new shingles.

RESHINGLING OVER STUCCO

When new shingles are to be applied over old stucco, the best practice is to remove the old stucco, nails, lath, and protruding elements and apply the new shingles directly over the existing sheathing, if any exists. However, if the stucco is in reasonably good condition or is removed to bare studs, new shingles can be applied by the wood nail strip or metal channel method described in Section 7.7.2.3, modifying the steps as necessary to meet special conditions.

7.7.3 WOOD SHINGLES AND SHAKES

Unless specific exceptions are made in the following text, the provisions of Section 7.7.1 apply to wood shingles and shakes and methods of their application.

7.7.3.1 Materials

Most wood shingles and shakes used for side walls are produced from western red cedar trees—giant, slow-growing coniferous trees of the Pacific Northwest—but northern white cedar, bald cypress, and redwood are also used. This section discusses only western red cedar shingles and shakes. Those made with other woods are similar and are installed in a similar manner.

Western red cedar has an extremely fine and even grain, exceptional strength in proportion to weight, low expansion and contraction with changes in moisture content, and high impermeability to water. Its outstanding characteristics are great durability and resistance to decay.

Cedar shingles are sawn, have smooth faces and backs, and are produced in four grades and a variety of types. Tapersawn shakes are sawn on both sides, but their thickness is more like that of shakes. They are produced in two grades. Other shakes have rough split faces and either sawn or split backs and are produced in one grade and in three types.

Both shingles and shakes are made from selected cedar logs cut to proper lengths—16, 18, and 24 in. (406, 457, and 610 mm) for shingles; 18 and 24 in. (457 and 610 mm) for shakes. These lengths are again cut or split into the proper size blocks for the shingle-cutting machine or to a convenient size for splitting of shakes. In reducing the large log sections into blocks, each is quartered, split, and requartered to produce blocks having a true edge-grain face (see Fig. 7.6-63).

Each bundle of wood shingles and shakes should carry the official grade-marked label of the Cedar Shake and Shingle Bureau (CSSB).

CEDAR SHINGLES

Wall shingles are produced from cedar blocks in the same manner described in Section 7.6.6.1 for roof shingles.

Grades, Types, and Sizes Shingles are marketed according to grade, thickness, type, and size, as described in Section 7.6.6.1.

Grade No. 1 (Blue Label) is the best, or premium, grade of shingles for single-coursed walls and the surface course of double-coursed walls, but are intended primarily for roofing. They are 100% edge grain, consist entirely of heartwood, and contain no knots. On walls they will have an indefinite life, usually greater than the useful life of the structure.

Grade No. 2 (Red Label) is an excellent grade for single-coursed walls and the surface course of double-coursed walls. Shingles of this grade must be clear, or free from blemishes for 10, 11, and 15 in. (254, 279.4, and 381 mm) for 16-, 18-, and 24-in. (406.4, 457.2, and 609.6 mm) shingles, respectively. A maximum width of 1 in. (25.4 mm) of sapwood is permissible in the first 10 in. Mixed edge and flat grain are allowed.

Grade No. 3 (Black Label) is a utility grade to be used where economy is the prime consideration. It may be used as the undercourse in wall double-coursing and for walls of secondary buildings. These shingles may contain flat grain and sapwood, but must be 6 in. (152.4 mm) clear from the butt for 16- and 18-in. (406.4 and 457.2 mm) shingles, and 10 in. (254 mm) clear for 24-in. (609.6 mm) shingles.

Undercoursing shingles, a lower grade than No. 3, are available in 16- and 18-in. (406.4 and 457.2 mm) lengths. They are produced expressly for use in the inner and completely concealed course of double-coursed walls (Fig. 7.7-18a). They should not be used for any other purpose. Some mills mix Grade No. 3 and undercoursing shingles and call them *special undercoursing shingles*.

Rebutted and rejointed shingles are machine trimmed for exacting parallel edges, with butts cut to exact right angles. They are also available with smooth-sanded faces (Fig. 7.7-18b). Produced in No. 1 and No. 2 Grades in 16- and 18-in. (406.4 and 457.2 mm) lengths, they can be used on either single- or double-coursed walls where tight joints between shingles are desired.

Machine-grooved shingles (also known as machine-grooved shakes) are similar to rebutted and rejointed shingles, except that the face is machine striated to give a corrugated appearance (Fig. 7.7-18c). They are produced in No. 1 Grade only, in 16- and 18-in. (406.4 and 457.2 mm) lengths, and either natural, prime-coated, or finished coated. These units can be used for the surface course in double-coursed wall applications over low-grade undercoursing.

Dimension shingles are cut to specific uniform widths where special effects are desired (Fig. 7.7-18d). They are available in 5- or 6-in. (127 or 152.4 mm) widths in 16-in. (406.4 mm) lengths in No. 1 and No. 2 Grades, in 5- or 6-in. (127 or 152.4 mm) widths in 18-in. (457.2 mm) lengths in No. 1 Grade, and 6-in. (152.4 mm) width in 24-in. (609.6 mm) lengths in No. 1 Grade. *Fancy-butt* shingles (Fig. 7.7-18e) are produced from dimension shingles. The butt designs are produced by trimming the shingles on a band saw.

A summary of cedar shingle types and grades is given in Figure 7.6-66. The labels shown are assurance that the shingles meet CSSB quality standards. The coverage of wood shingles is shown in Figure 7.6-67. The maximum exposure for wall shingles is shown in Figure 7.7-20.

FIGURE 7.7-18 (a) A double-coursed wall showing complete concealed undercover shingles; (b) a wall made with rebutted and rejointed shingles; (c) machine-grooved shingles; (d) dimension shingles; (e) *fancy-butt* shingles (many other shapes are available). (Courtesy Cedar Shake and Shingle Bureau.)

CEDAR SHAKES

Wood shakes are manufactured in four types: (1) tapersawn, (2) hand-split-and-resawn, (3) tapersplit, and (4) straightsplit. Tapersplit shakes are produced in 24-in. (609.6 mm) lengths. The others are produced in both 18- and 24-in. (457.2 and 609.6 mm) lengths. Each is 100% clear wood and 100% heartwood. Both tapersplit and straightsplit shakes, produced largely by hand, are of 100% edgegrain. All four types may be used for walls, but tapersawn, tapersplit and straightsplit types are usually preferred because of their more uniform texture

Factory-assembled hip and ridge units are also available.

Sizes, types, grades, method of producing, grade trademarks, and packaging are the same for wall shakes as for roofing shakes, as described in Section 7.6.6.1.

STANDARDS OF QUALITY

The manufacture of cedar shingles and shakes should conform to CSSB specifications. These specifications constitute the recognized standards of the industry. Refer to Section 7.6.6.1 for additional discussion.

NAILS

Nails for applying wall shakes and shingles should be corrosion resistant, such as hot-dipped galvanized steel or bright or colored aluminum alloy. Common wire nails or even electrogalvanized nails may shorten the life of a shake or shingle wall and should not be used. Nails may have smooth or threaded shanks and medium or blunt diamond points. Threaded nails pro-

FIGURE 7.7-19 Minimum Nail Lengths for Wood Shingles and Shakes[a]

	Nail Lengths[b]	
Type	Double-Coursed Sidewalls, in. (mm)	Single-Coursed Sidewalls, in. (mm)
Shingles	1¾ (44.5) (5d)	1¼ (28.6) (3d)
Shakes[c]	2½ (63.5) (8d)	1½ (38.1) (4d)

(Courtesy Cedar Shake and Shingle Bureau.)

[a]For application over plywood sheathing, nails should be threaded.

[b]Nails should be long enough to penetrate through sheathing.

[c]Because thickness and exposure of shakes vary, larger nails may be necessary to obtain suitable penetration in the sheathing.

FIGURE 7.7-20 Maximum Exposure for Wood Shingles and Shakes

Type	Length, in.	Single Coursing, in.	Double Coursing, in.
Hand-split-and-resawn shakes	18	8½	14
	24	11½	20
	32	15	—
Taper-split shakes	24	11½	20
Straight-split shakes	18	8½	16
	24	11½	22
Random-width and dimension shingles	16	7½	12
	18	8½	14
	24	11½	16
Rebutted and rejointed shingles	16	7½	12
	18	8½	14
Machine-grooved shingles	16	—	12
	18	—	14

(Courtesy Cedar Shake and Shingle Bureau.)

vide increased holding power, although smooth-shank nails are used in most applications over wood sheathing and give suitable service. Nails for applying wood shingles or shakes to plywood sheathing should be threaded. If threaded, galvanized steel nails should have annular threads; aluminum nails should have screw threads with approximately a 12½-degree thread angle.

Box nails, having smaller diameters and longer lengths than common nails, are also suitable. Nails should be long enough to penetrate completely through the sheathing.

Nail lengths typically required for wall installations are given in Figure 7.7-19.

7.7.3.2 Preparation for Wall Shingles and Shakes

EXPOSURE

The exposure of shakes and shingles is dependent on their length and whether the walls are to be single or double coursed. For single coursing, shake and shingle weather exposure should never be greater than half the shingle or shake length minus ½ in. (12.7 mm), so that two layers of wood occur over the entire wall.

The maximum recommended weather exposures for single- and double-coursed shingle and shake wall construction are shown in Figure 7.7-20. Approximate wall coverage for shingles using different exposures is shown in Figure 7.6-67.

WALL SHEATHING

Sheathing types, thicknesses, spans, and nailing recommendations are discussed in Chapter 6.

SHEATHING PAPER

To provide moisture protection when lumber boards are used for sheathing, No. 15 or No. 30 asphalt-saturated felt should be applied over the entire wall as described in Chapter 6. When spaced (open) sheathing is used, sheathing paper should be applied first to the framing members (Fig. 7.7-21).

7.7.3.3 Application of Walls Shingles and Shakes

The two basic methods used to apply shakes and shingles are (1) single coursing and (2) double coursing. When single-coursed (Fig. 7.7-22), units can be applied in the same manner as wood roof shakes and shingles, except that greater weather exposures can be used. When double-coursed (Fig. 7.7-23), units can be applied at much greater weather exposures over a course of either No. 3 Grade or undercoursing-grade shingles.

Using either method, at least two layers of shingles (double coverage) occur at

FIGURE 7.7-21 Spaced sheathing being applied over sheathing paper. (Courtesy Cedar Shake and Shingle Bureau.)

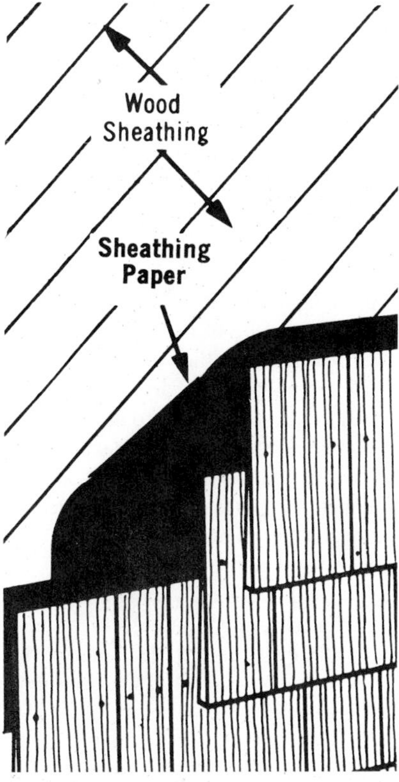

FIGURE 7.7-22 Single coursing of wood shingle or shake wall. (Courtesy Cedar Shake and Shingle Bureau.)

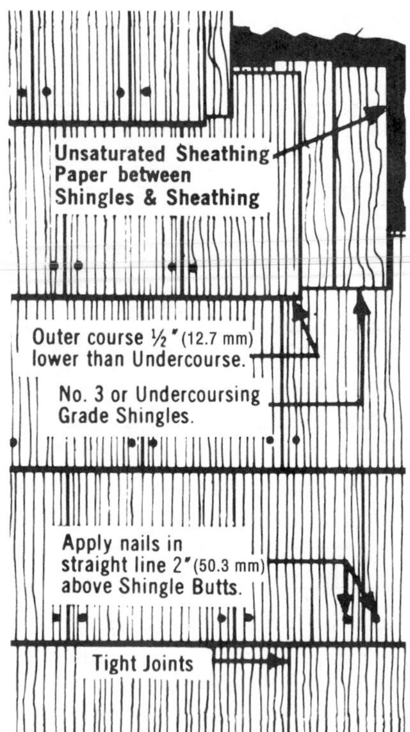

FIGURE 7.7-23 Double coursing of wood shingle or shake wall. (Courtesy Cedar Shake and Shingle Bureau.)

FIGURE 7.7-24 Courses are laid out around the building using (a) a story pole or (b) a chalkline. (Courtesy Cedar Shake and Shingle Bureau.)

FIGURE 7.7-25 Two layers of undercoursing under the first course. (Courtesy Cedar Shake and Shingle Bureau.)

FIGURE 7.7-26 Straightedge forms guide for nailing undercoursing and exposure courses. (Courtesy Cedar Shake and Shingle Bureau.)

every point on the wall surface. Except where specifically noted, the application methods described and illustrated in the following text are applicable to both shakes and shingles.

LAYING OUT

The exposure and number of courses should be determined by measuring the height of the dominant wall from a point 1 in. (25.4 mm) below the top of the foundation to the top of the windows, underside of soffits, or other vertical reference.

The height is then divided into equal parts, as near as possible to the maximum recommended shingle or shake weather exposure. This measurement and the number of courses can be transferred to a story pole or can be established with chalk lines all around the building (Fig. 7.7-24). Course exposure can vary slightly to line up with the bottom and top of openings and at the eave. Story-pole marks or chalk lines should be made after sheathing paper has been applied.

SINGLE COURSING

The first course at the foundation should be doubled. Each shake or shingle should be secured with the correct nails (see Fig. 7.7-19), driven about ¾ in. (19.1 mm) from each edge and not to exceed 1 in. (25.4 mm) above the buttline of the succeeding course. Joints between shingles in one course should be separated at least 1½ in. (38.1 mm) from joints in adjacent courses, and joints in alternate courses should not be in direct alignment.

When fiberboard or gypsum sheathing occurs, shingles and shakes should be applied over 1 × 3 (25 × 75) nailing strips (see Fig. 7.7-21) spaced according to shingle exposure, or special fasteners designed for this purpose should be used.

DOUBLE COURSING

The bottom course should be laid with two undercoursings and one outer course. This triple coursing eliminates the need for a costly drip cap and makes the bottom course as thick as the succeeding courses (Fig. 7.7-25).

A straightedge, such as a piece of wood siding or 1 × 4 (25 × 100) lumber, should be tacked tightly along a chalkline snapped between the bottom story-pole marks. Undercourse shingles, spaced about ⅛ in. (3.2 mm) apart, should be laid on the upper lip of the straightedge and fastened with a single staple near the top (Fig. 7.7-26).

The outer course should then be applied, laid tight without spacing. As each double course is laid, a chalked string should be stretched from opposite corner marks to indicate the level of the next course and relocate the position of the straightedge.

Each wall shingle or shake should be nailed with a minimum of two nails, applied 2 in. (50.8 mm) above the buttline and ¾ in. (19.1 mm) from the edge of each shingle or shake. On shingles wider than 8 in. (203.2 mm), two additional nails should be driven near the center and about 1 in. (25.4 mm) apart.

CORNERS

Outside corners may be lapped alternately. One outer course should protrude slightly past the corner, butting the shake on the other side of it. The protruding corner should then be trimmed with a knife and a plane and touched up with stain (Fig. 7.7-27). Metal corner molding designed for this purpose may also be used.

Inside corners should be fitted with a lumber strip about 1½ in. (38.1 mm) square to receive the undercourse and outer shakes or shingles with a minimum of fitting. The strip should be stained after application.

FIGURE 7.7-27 **Corner treatments for shakes: (a) jointed inside corner, (b) woven outside corner, (c) mitered inside corner, (d) mitered outside corner. Corner treatments for shingles: (e) overlapping outside corner, (f) mitered inside corner using metal flashing in the joint.** (Courtesy Cedar Shake and Shingle Bureau.)

7.7.3.4 Re-siding Existing Walls

Shingles or shakes can be applied directly to existing siding or to new sheathing. Building paper should be applied over old siding, and other steps are the same as described earlier (Fig. 7.7-28). Before applying courses around windows and doors, molding strips, to which shakes and shingles should be joined, should be nailed on the face of old casings, flush with the outer edges. When re-siding over stucco, nailing strips first should be nailed over the wall to provide a good base for nailing double coursing (Fig. 7.7-29). These horizontal strips should be spaced with centers 2 in. (50.8 mm) above the buttlines of each course to receive nails.

The lath strip application shown in Figure 7.7-30 should be used to apply shingles and shakes over insulating sheathing. Lath nailed at the studs provides a nailing base for double-coursing nails. As an alternate for attaching to nonwood sheathing, $\frac{3}{8}$- by $1\frac{1}{2}$-in. (9.5 by 38.1 mm) treated or decay-resistant lath may be used. Lath should be spaced according to shingle exposure, with the butts of the undercoursing shingles resting on the lath. Lath should be nailed to the studs through the sheathing with $2\frac{1}{8}$- by 0.109-in. (7d) (53.98 by 2.78 mm) annular threaded or $2\frac{1}{2}$- by 0.131-in. (8d) (63.5 by 3.4 mm) common wire nails. The outer course of shakes or shingles should be nailed to the lath with small-headed corrosion-resistant annular threaded nails penetrating the lath. The butts of the outer course should project about $\frac{1}{2}$ in. (12.7 mm) below the lath (see Fig. 7.7-30).

FIGURE 7.7-30 **Lath strips are used with nonnailable sheathing for double-coursed walls.** (Courtesy Cedar Shake and Shingle Bureau.)

FIGURE 7.7-28 **Reshingling is similar to new wall construction.** (Courtesy Cedar Shake and Shingle Bureau.)

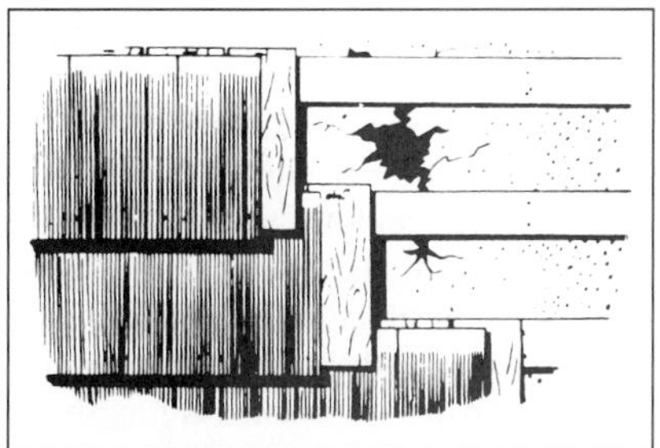

FIGURE 7.7-29 **Horizontal nailing strips must be used over nonnailable sheathing or stucco.** (Courtesy Cedar Shake and Shingle Bureau.)

Joints should not be closer than 1½ in. (38.1 mm) to joints of the undercoursing in the same course. Joints in the following courses should be staggered at least 1½ in. (38.1 mm).

7.7.4 ALUMINUM AND VINYL SIDING

Aluminum siding was first introduced to the American public following World War II. Vinyl siding came along a little later. Early aluminum siding was made in the natural aluminum finish, but soon afterward an enamel finish was applied, making it available in a choice of colors.

The prominent features of both aluminum and vinyl siding include minimal maintenance, good durability, light weight, and high resistance to corrosion. For multiple design possibilities, these sidings are available in a variety of colors, textures, sizes, and shapes. They are used in both new and existing construction (Figs. 7.7-31 and 7.7-32).

7.7.4.1 Aluminum Siding

The recommendations in this subsection are based largely on the American Architectural Manufacturers Association (AAMA) publication 1402, "Standard Specifications for Aluminum Siding, Soffit, and Fascia" and AAMA publication RA-15, "Aluminum Siding Application Manual." These constitute the recognized quality standard for this product.

MATERIAL

Aluminum sheet is fabricated into horizontal and vertical panels by cold rolling. The typical alloy used in aluminum siding is known as 3003H14, a magnesium-containing alloy distinguished by moderate strength and good workability. See Section 5.4 for additional information about aluminum fabrication techniques. The finishes used on aluminum siding are generally organic enamels such as vinyls, acrylics, or alkyds. They are baked on at the factory after proper surface pretreatment.

Siding Types Horizontal aluminum siding is available in several typical widths. The most commonly used siding type is a horizontal panel with 8-in. (203.2 mm) exposure (face). Other types include double 4-in. (101.6 mm) and double 5-in. (127 mm) exposure effects, which are created by grooves bisecting a single 8- or 10-in. (203.2 or 254 mm) horizontal panel.

Aluminum siding can be plain or insulated. Some insulated board has a polystyrene or fiber backer board installed at the factory. Other types are supplied with separate backer boards for field installation. Thicknesses vary from ⅜ to ½ in. (9.5 to 12.7 mm). Insulated siding is available in 8-in. (203.2 mm), double 4, and double 5 styles.

Vertical siding is available in both 12- and 16-in. (304.8 and 406.4 mm) exposures. Both exposures lend a board-and-batten effect, and the 16-in. (406.4 mm) exposure is often bisected by a V-groove for additional interest.

Like other metals, aluminum is an effective vapor barrier. Therefore, aluminum siding should be equipped with vents or weep holes to permit water vapor to escape and prevent moisture accumulation in the wall.

Accessories Accessories for use with aluminum siding include starter strips, J-channels, window head flashing, inside corner posts, outside corner posts, and individual corner caps. Other accessories include aluminum nails, caulking, touch-up paint, aluminum breather (perforated) foil, and adhesive. Adhesives used to bond backer boards to siding should demonstrate sufficient moisture resistance, strength, and flexibility to maintain a permanent bond under service conditions. AAMA specifications outline specific performance tests for the adhesives, as well as for the finish coating, and requires compliance with other industry standards for accessory products employed in siding manufacture and installation.

INSTALLATION

In refinishing projects, the quality of the finished job depends heavily on the initial surface preparation. Loose boards should be nailed down and irregular surfaces leveled. Unbacked siding cannot bridge low spots as readily as the backed type, and furring of low spots may be required to provide adequate support. In some cases, the whole wall must be furred. Good initial preparation permits siding joints to close up and avoids the wavy look of poor siding installations.

In new construction, backed siding can be applied directly to stud framing over sheathing paper. Unbacked siding should be installed only over a solid sheathing surface. When wood board sheathing is used, sheathing paper should be provided; when panel sheathing materials are applied in a way that minimizes air leakage, the sheathing paper can be omitted.

Aluminum siding is often applied over an insulated sheathing to increase a wall's R-value, especially in existing construction. This insulated sheathing may be a faced panel with a core of cellular kraft paper or of a foamed plastic such as polystyrene or polyisocyanurate. Care must be taken to ensure that using an impervious sheathing will not trap moisture in the wall cavity. If it will, a pervious board should be used (see Section 7.1).

Aluminum foil laminated to kraft paper is sometimes used as sheathing paper under aluminum siding to increase a wall's R-value. It may be installed over an insulated sheathing or directly on an existing wall or on furring where there is no insulated sheathing. This foil should be applied with its shiny side facing an adjacent air space. However, aluminum foil is an excellent vapor retarder and will trap moisture unless it is deliberately perforated to allow vapor to escape from the wall cavity. Foil products intended for use on an existing wall should be factory perforated and have a perm rating of at least 10 (632 ng/[Pa·s·m^2]). An efficient vapor retarder with a permeance of 5 perms (287.26 ng/[Pa·s·m^2]) or less should be used on the room side of the wall assembly. Some manufacturers also recommend installing vent cups at the top and bottom of each stud space (see Section 7.1). In new construction, the foil should be perforated or unperforated according to the wall design and vapor retarder location.

Careful application of accessories, such as starter strips and corner trim, is essential to provide a level and square installation. In most cases, starter strips are installed first to ensure proper alignment and fastening of siding.

Continuous inside and outside corner posts are used to receive siding at corners, insets, projections, and wall junctions. Since aluminum expands and contracts with temperature variations, a space should be allowed where siding ends abut a post or any rigidly supported surface. Outside corners can also be completed with the use of outside corner caps, installed after siding is in place.

In installing horizontal siding, extra care should also be taken in the application of the first course, making certain that it is level and securely locked into the starter strip. Each length of siding should be lapped over the preceding length to create the effect of a single panel extended across the building. Factory-finished ends should be exposed and cut ends lapped under the edge of adjoining panels.

To fit siding around areas such as windows, eaves, and doors, panels can be cut to the desired size. Joints should be planned so that, wherever possible, no piece is less than 20-in. (508 mm) long.

Proper nailing is essential to successful installation. Panels should be secured with aluminum nails driven through the slotted holes in the siding at intervals of about 16 in. (406.4 mm) along its length. Siding should be suspended on the nails, not nailed rigidly to the wall, because locked siding will not be able to expand or contract with temperature changes.

Matching gutters, downspouts, fascia boards, and soffit panels are also available. Because the natural corrosion resistance of aluminum and the durability of the baked-on enamel finish, these products are particularly well adapted to their severe exposures.

To prevent water entry and improve appearance, caulking should be used at junctions where siding abuts wood, brick, stone, or other metals. Caulking compounds should be durable synthetic materials, compatible with the aluminum and paint finish.

GROUNDING

Although the possibility is remote, some building codes require that aluminum siding be grounded as a precaution against electrical shock from faulty wiring. This can be done by connecting a No. 8 copper wire to a cold-water pipe or a steel rod embedded in the earth. Grounding allows stray electrical fault currents to be dissipated safely into the ground and eliminates shock hazard to persons touching the siding.

MAINTENANCE

Baked-on siding finishes are generally resistant to the effects of intense sunlight, snow, sleet, hail, ice, and salt spray. In addition to the natural cleaning by normal rainfall, siding surfaces should be kept clean by periodic rinsings with a garden hose. Persistent stains resulting from industrial fallout, tree sap, or insecticides can be removed with a nonabrasive household cleaner or special siding cleaner.

Panels damaged by storm, fire, and other hazards can be replaced individually without disturbing adjacent panels. Aluminum siding may also be repainted if desired or when necessary due to damage. A special paint intended for aluminum surfaces is recommended, although ordinary house paint can also be used.

7.7.4.2 Vinyl Siding

The recommendations in this subsection are based largely on the Vinyl Siding Institute (VSI) publication, "Rigid Vinyl Siding Application Instructions." This publication constitutes the recognized standard for installation of this product. In addition, vinyl siding should comply with ASTM Standard D 3679.

MATERIAL

Vinyl siding and accessories are made from a polyvinyl chloride (PVC) compound and additives such as lubricants, stabilizers, impact modifiers, ultraviolet inhibitors, and colors. PVC is a thermoplastic material, meaning that it can be reformed upon heating. It is flexible and has high impact resistance. The finishes on vinyl siding are integral.

Siding finishes range from almost smooth, through matte, to deeply embossed wood grain surfaces that simulate wood clapboard. Many colors are available.

Siding Types Horizontal and vertical vinyl siding is available in several widths and configurations. Figure 7.7-33 shows some of the more popular sizes and shapes. Special shapes and even vinyl latticework are also available (Fig. 7.7-34) (see also Fig. 7.7-31).

Drop-in backer board is supplied with some vinyl siding, but VSI does not recommend its use, preferring instead a continuous sheathing backup. The manufacturer should be consulted for advice about the use of drop-in backer board.

FIGURE 7.7-31 Vinyl siding on a new building. (Courtesy Wolverine Technologies.)

FIGURE 7.7-32 Vinyl siding on an existing building. (Courtesy Alcoa Building Products.)

Accessories Accessories for use with vinyl siding include starter strips, J-channels, window head flashing, inside corner posts, outside corner posts, and others. Their sizes and shapes vary with the manufacturer. Fasteners for installing vinyl siding should be aluminum, galvanized steel, or other corrosion-resistant nails or staples.

Nails should be long enough to penetrate at least ¾ in. (19.1 mm) into a solid nailable backing surface. Nails heads should be 5/16 in. (7.9 mm) in diameter and 1½ in. (38.1 mm) long for general use. Nails used in re-siding should be longer.

Power-driven staples may be employed to install vinyl siding. A special fitting must be used on power equipment, however. The manufacturer's instructions should be followed.

INSTALLATION

In refinishing projects, the quality of the finished job depends heavily on initial surface preparation. Loose boards should be nailed down and irregular surfaces leveled. Loose caulking should be removed and joints recaulked to make them watertight.

In new construction, vinyl siding should be applied over solid sheathing. Installing a new layer of sheathing over an existing wall surface will also produce the best vinyl siding job. However, furring or strapping can be used over an existing wall to even out the face. This furring or strapping should be shimmed where necessary to produce an even face.

Vinyl siding can be applied over insulated sheathing to increase a wall's R-value. The insulation and siding manufacturers' advice should be sought and followed in making this installation.

Careful application of accessories, such as starter strips and corner trim, is essential to provide a level and square installation. In most cases, starter strips are installed first to ensure proper siding alignment and fastening.

Continuous inside and outside corner posts are used to receive siding at corners, insets, projections, and wall junctions (Fig. 7.7-35a and b). Since vinyl expands and contracts with temperature variations, a space should be allowed where siding-ends abut a post or any rigidly supported surface.

Extra care should also be used in applying the first course of horizontal siding, making certain that it is level and securely locked into the starter strip. Each length of siding should be lapped over the preceding length to create the effect of a single panel extended across the building. Factory-finished ends should be exposed and cut ends lapped under the edges of adjoining panels (Fig. 7.7-36).

To fit siding around areas such as windows, eaves, and doors, panels can be cut to the desired size with saws or snips. Joints should be planned so that, wherever possible, no piece is less than 20 in. (508 mm) long.

Proper nailing is essential to successful installation. Panels should be secured with nails or staples driven through the slotted holes in the siding at intervals of about 16 in. (406.4 mm) along its length. Siding should be suspended on the nails, not nailed rigidly to the wall, because locked siding will not be able to expand or contract with temperature changes (Figs. 7.7-37 and 7.7-38).

Matching gutters, downspouts, fascia boards, and soffit panels are also available. Because of the natural corrosion resistance of vinyl and the durability of vinyl siding, these products are particularly well adapted to their severe exposures.

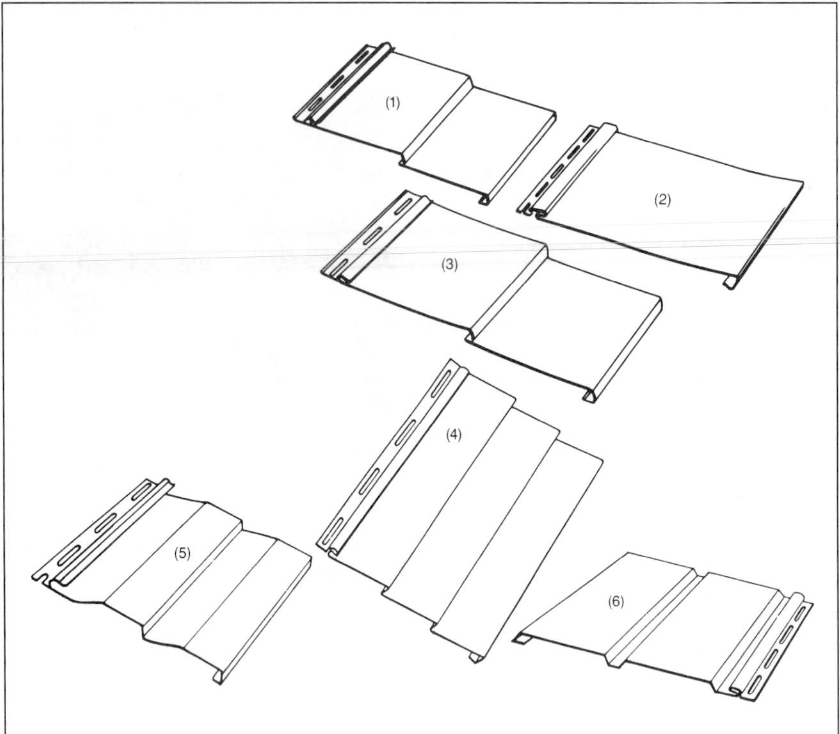

FIGURE 7.7-33 Vinyl siding: (1) 4-in. (101.6 mm) double; (2) 6-, 8-, 9-, or 10-in. (152.4, 203.2, 228.6, or 254 mm) single; (3) 5-in. (127 mm) double; (4) 3-in. (76.2 mm) triple; (5) Dutch lap; (6) vertical is available in 5-in. (127 mm) double, 3-in. (76.2 mm) triple, and 4-in. (101.6-mm) quad. (Courtesy Vinyl Siding Institute.)

FIGURE 7.7-34 Shingle shapes over the entranceway are only one of several shapes available in vinyl siding material. (Courtesy Alcoa Building Products.)

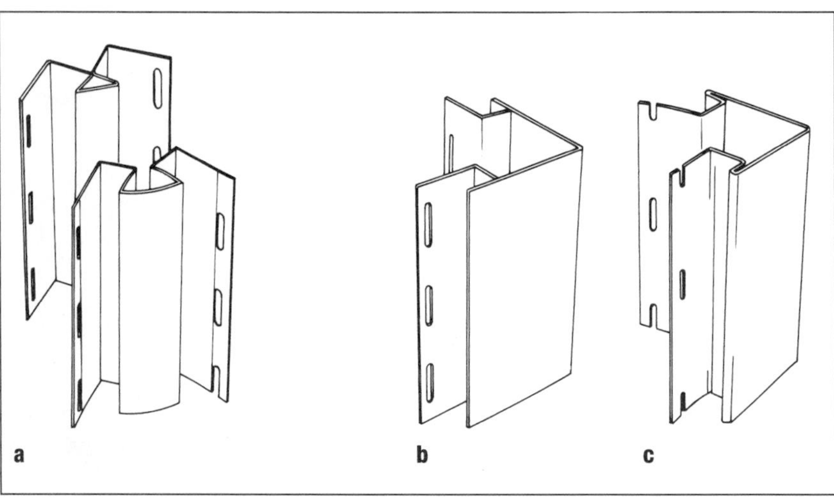

FIGURE 7.7-35 (a) Inside corner posts; (b) and (c) outside corner posts. (Courtesy Vinyl Siding Institute.)

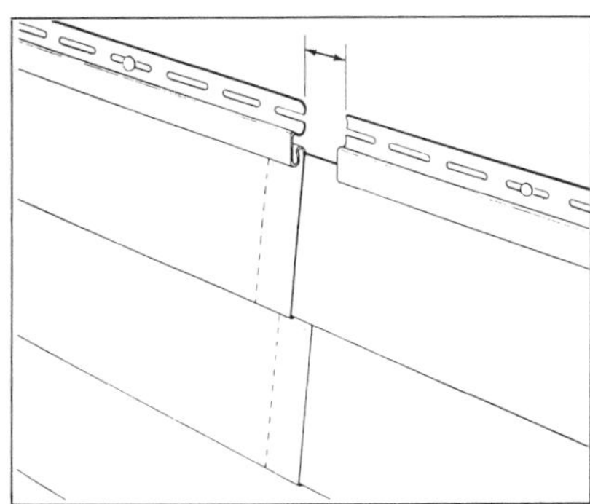

FIGURE 7.7-36 **Lapped end joints.** (Courtesy Vinyl Siding Institute.)

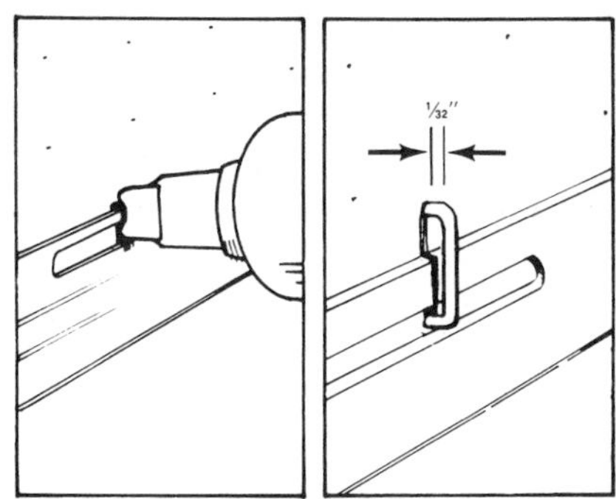

FIGURE 7.7-38 **Staples should be driven so that the panels remain loose and movable.** (Courtesy Alcoa Building Products.)

Siding overlaps should not be caulked, and siding panels should not be caulked where they meet receivers, corners, or J-trim, but should rather be permitted to float free to account for expansion and contraction.

MAINTENANCE

Vinyl siding is generally resistant to the effects of intense sunlight, snow, sleet, hail, ice, and salt spray. In addition to the natural cleaning by normal rainfall, siding surfaces should be kept clean by periodic washings with a long-handled car-washing brush mounted on the end of a water hose. Hard-to-remove soil may be removed with a solution of ⅓ cup (80 mL) of detergent and ⅔ cup (160 mL) of household cleaner in a gallon (3.79 L) of water. Mildew may be removed using the same solution, but substituting 1 quart (.95 L) of liquid laundry bleach for 1 quart (.95 L) of the water. More aggressive means and chemicals such as solvents, cleaning fluids, oxalic acid, and lacquer thinner can be used to remove more stubborn stains, but the advice of the siding manufacturer or VSI should be sought first.

7.7.5 WOOD AND HARDBOARD BOARD SIDING

The following text describes wood and hardboard siding intended for exterior use on building walls. Interior paneling is discussed in Chapter 6. Miscellaneous boards used for decking and other purposes are not included.

Unless specific exceptions are made in the following text, the provisions of Section 7.7.1 apply to wood and hardboard board siding and their methods of application.

7.7.5.1 Materials

WOOD BOARD SIDING

Wood board siding is made from softwood, such as cedar, Douglas fir, various species of pine, cypress, redwood, western hemlock, spruce, yellow poplar, western larch, and other softwoods. The wood should be of a high grade, free from knots, pitch pockets, and waney edges. Moisture content at the time of application should be between 10% and 12%, except in very dry areas, where moisture content should be between 8% and 9%.

There are four basic types: *bevel*, *tongue-and-groove*, *lap*, and *flat board* siding. Figure 7.7-39 shows a few of the many wood board siding profiles available.

Bevel Siding Bevel siding has both a smooth side and a rough-sawn side. When staining is desired, the rough-sawn side may be exposed. For painting, the smooth side is exposed. Widths vary from 3½ in. (88.9 mm) to 12 in. (304.8 mm). The taper is usually 3⁄16 in. (4.8 mm) at the thin edge and varies from ⅝ to 1¼ in. (15.9 to 31.8 mm).

Tongue-and-Groove Siding Tongue-and-groove siding may be used either vertically or horizontally. Nominal board widths are generally 4, 6, or 8 in. (101.6, 152.4, or 203.2 mm). Thickness is usually ¾ in. (19.1 mm).

Lap Siding Lap siding boards include channel siding and other lap siding configurations. Lap siding may be used vertically, horizontally, or diagonally. Various profiles are available in widths for 6, 8, and 10 in. (152.4, 203.2, and 254 mm), and other widths may be available. Overall thickness is usually ¾ in. (19.1 mm).

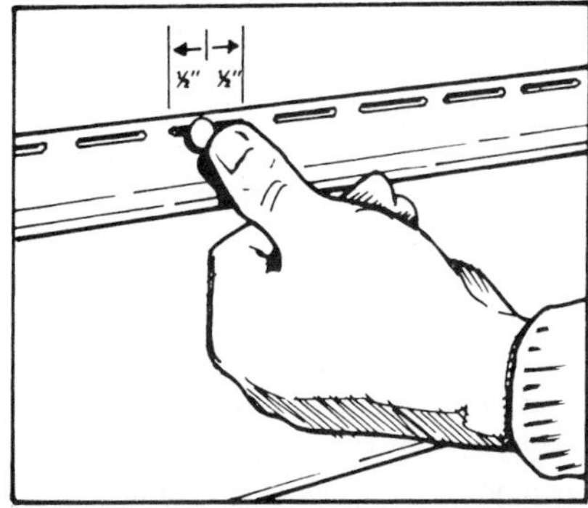

FIGURE 7.7-37 **Nails should be driven into the center of the slot. It should be possible to move the siding ½ in. (12.7 mm) each way after fastening to allow for normal expansion and contraction.** (Courtesy Alcoa Building Products.)

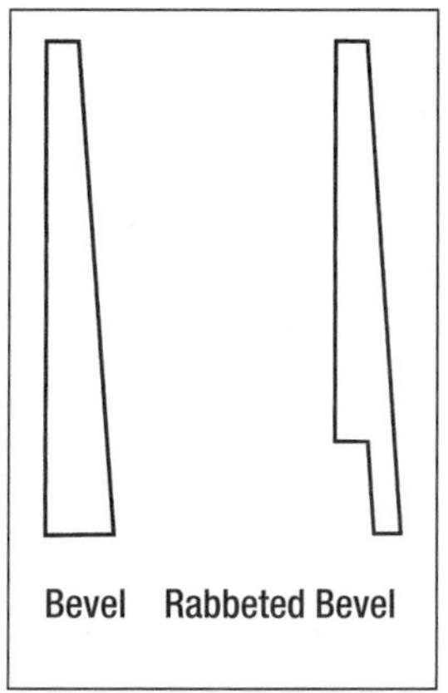

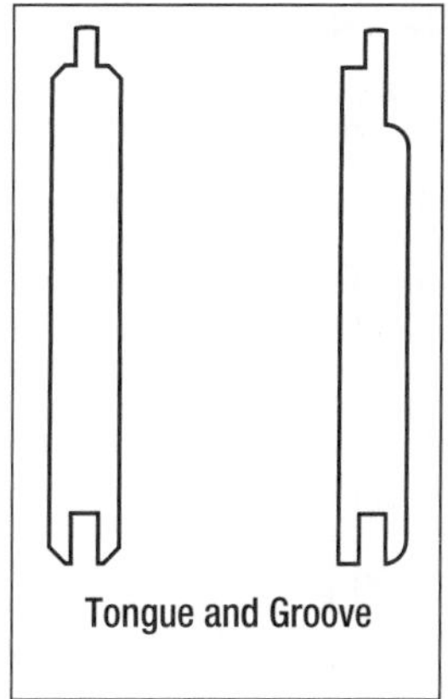

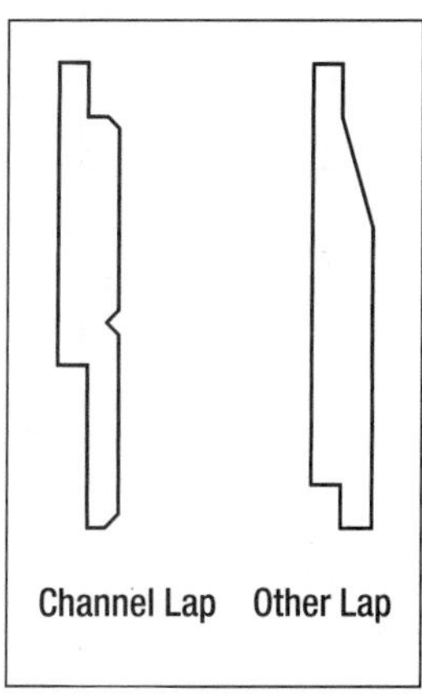

FIGURE 7.7-39 **Wood siding: (a) beveled, (b) tongue-and-groove, (c) lap. Other shapes are available in each category.** (Drawing by HLS.)

Boards for Board-and-Batten Siding Flat boards are available in various nominal widths from 2 in. (50.8 mm) to 12 in. (304.8 mm) and in thicknesses nominally 1, 1¼, and 2 in. (25, 32, and 50 mm). They may be rough, surfaced on one side and two edges, or surfaced on all four sides.

HARDBOARD SIDING

Hardboard board sidings are proprietary products made from thermoset hardboard, usually finished with a fused resin-oil finish. They should comply with all code requirements, should have a flame spread rating of Class III, < 200, should be certified by the American Hardboard Association, and should comply with the requirements of ANSI 135.6.

Types and Size Hardboard board siding is available in various widths, depending on the profile of the siding. Flat lap siding comes in 6-, 8-, 9½-, and 12-in. (152.4, 203.2, 228.6, and 304.8 mm) widths. Most interlocking flat-faced lapped siding comes in nominal 8-in. (203.2 mm) widths. Beveled lap siding usually comes in 11½-in. (292.1 mm) widths. Other widths are also available. Most are available in 7⁄16-in. (11 mm) thicknesses, but some are ½ in. (12.7 mm) thick. Figure 7.7-40 shows some of the available profiles.

Trim boards made from the same material are available in 4-, 5-, 6-, 8-, 10-, and 12-in. (101.6, 127, 152.4, 203.2, 254, and 304.8 mm) widths.

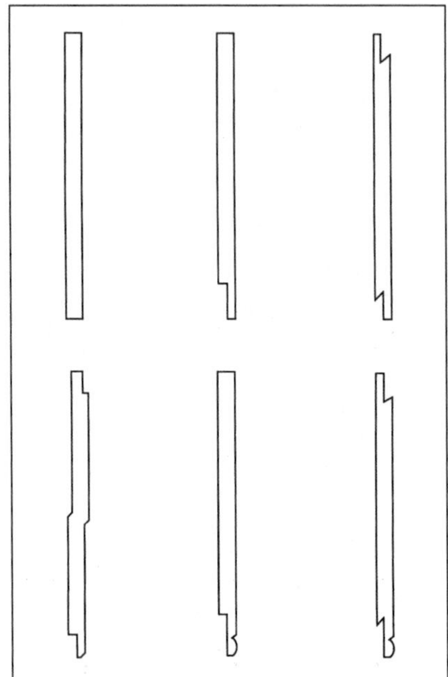

FIGURE 7.7-40 Typical shapes of hardboard siding. Many others are available. (Drawing by HLS.)

7.7.5.2 Preparation for Wood and Other Board Siding

EXPOSURE

Overlap for beveled siding should be not less than 1 in. (25.4 mm). Lap for drop siding and diagonal and vertical siding is dictated by the profile of the boards.

WALL SHEATHING AND SHEATHING PAPER

Wall sheathing and sheathing paper are discussed in Section 7.7.1. Sheathing paper is recommended over lumber and plywood sheathing and, by some sources, over gypsum board sheathing, but it is often omitted from nonwood sheathing.

NAILS

Nails may be aluminum, stainless steel, or hot-dip galvanized. The preferred material is No. 304 stainless steel for most locations, with No. 316 being used in seacoast locations. Other nail materials should not be used.

Nails should be special siding nails with thin shanks and blunt points. They should be ring-shank or spiral type. Nails with textured heads will present an appearance closer to that of the siding when finished.

7.7.5.3 Application of Wood and Hardboard Board Siding

INSTALLATION OF SIDING

Wood Beveled Siding Installation begins at the bottom course with a starter strip (Fig. 7.7-41). Each succeeding course laps the lower course. One nail should occur in each board at each stud.

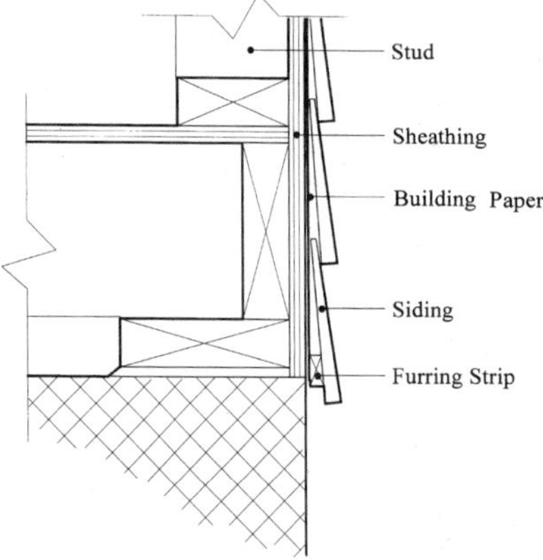

FIGURE 7.7-41 Condition at bottom of beveled siding. (Drawing by HLS.)

Nail sizes should be 7d, 8d, or 10d, depending on the thickness of the siding and the sheathing. For 1-in. (25.4 mm)-thick foam, 10d nails should be used. If ring-shank nails are used, sizes may be smaller, but penetration must be at least 1½ in. (38.1 mm) into the studs, regardless of the nail size. Nails should be located high enough to miss the course below by at least ⅛ in. (3.2 mm) to permit movement in the boards.

Tongue-and-Grove (T&G) Siding T&G siding is installed in a similar manner to beveled siding, except that the exposed face is fixed and always the same from board to board. Boards up to 6-in. (152.4 mm) wide can be blind nailed through the base of each tongue with one nail into each stud. Wider boards should be face nailed with two nails into each stud. Nails should be long enough to penetrate at least 1½ in. (38.1 mm) into the studs.

Lap Siding Lap siding is installed in a manner similar to that of T&G siding, except that a ⅛-in. (3.2 mm) gap should be left at the top of each tongue to provide for expansion. Boards should be face nailed into each stud, with one nail in boards up to 6-in. (152.4 mm) wide and two nails in wider boards. Nails should penetrate at least 1½ in. (38.1 mm) into the studs.

Board and Batten Siding Horizontal blocking or furring strips are required to provide bearing at the proper locations for fastening board and batten siding. A minimum ½-in. (12.7 mm) space should be left between boards for expansion. Battens should overlap boards by at least ½ in. (12.7 mm), with more overlap on wider boards. Boards up to 6-in. (152.4 mm) wide should be fastened with a single nail in the center of each board. Wider boards should be fastened to each support with two nails approximately 3 in. (76.2 mm) apart. Battens should be nailed with one nail into each support. Nails should penetrate at least 1½ in. (38.1 mm) into the blocking.

Hardboard Board Siding Installation, including nail types and sizes, and the methods of starting and finishing should be in accordance with the manufacturer's directions.

END JOINTS

End joints in horizontal board siding should be avoided when possible. Unavoidable end joints should be made over a stud and staggered from course to course. Cut ends should be brushed with a preservative.

End joints in vertical board siding may be made using the configuration of the boards (lap or T&G) or may be made using battens or board on board designs (Fig. 7.7-42).

CORNERS

When wall surfaces join at inside corners, the courses of wood and hardboard board siding are either butted into a wood corner strip or trimmed with a wood corner strip (Fig. 7.7-43a). Outside corners of wood siding are either mitered (Fig. 7.7-43b) or finished with wood corner boards (Fig. 7.7-43c and d). Hardboard board sidings are usually finished at corners with wood corner boards or corner boards made of the same materials as the siding (see Fig. 7.7-43c and d).

Mitered Finish Mitering is appropriate for wood siding only. Mitered joints are cut with a compound miter saw so that the corners finish neatly.

Wood Corner Board Finish Corners finished with wood strips can be made of previously primed or painted nominal 1-in. (25.4 mm) lumber applied along one corner. A 1 × 4 (25 × 100) should be applied along one corner so that its side edge projects a distance beyond the corner equal to the thickness of the board applied on the adjoining wall. A 1 × 3 (25 × 75) should then be applied with its edge butting against the back of the 1 × 4 (25 ×

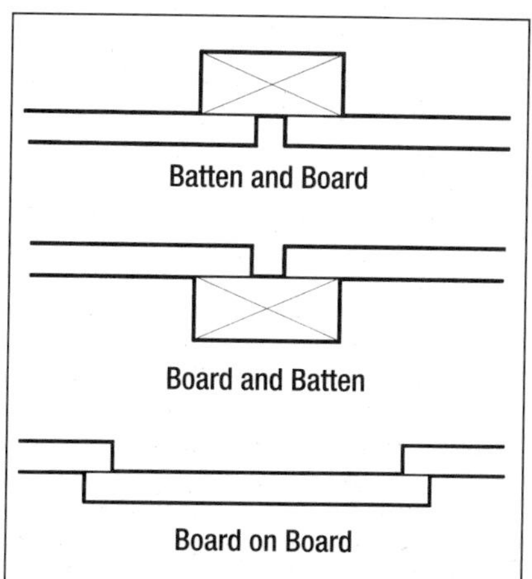

FIGURE 7.7-42 Joints in panel siding. Similar joints are sometimes used in vertical board siding applications. (Drawing by HLS.)

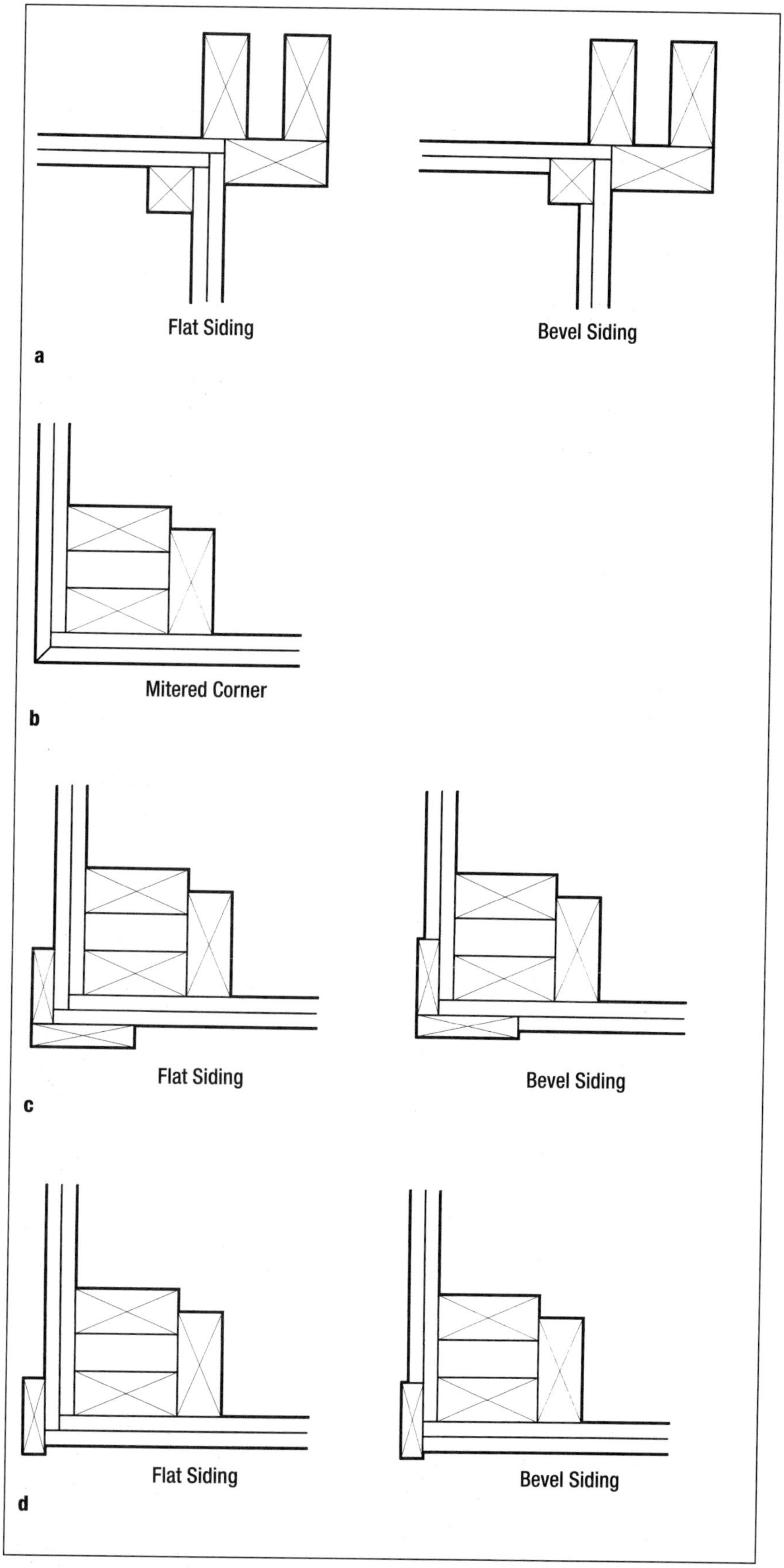

FIGURE 7.7-43 Corners in board and panel siding applications: (a) inside corners, (b) mitered corner, (c) two-board corner, (d) single-board corner. (Drawing by HLS.)

100) first applied. Therefore, the wood corner strips will appear the same width on each wall (see Fig. 7.7-43c). For lower-quality work, a single corner board may be used (see Fig. 7.7-43d).

Hardboard Corner Board Finish Corners may be made with special hardboard products made of the same material as the siding especially for this purpose. The method is similar to that described earlier for wood corner boards.

FLASHING AT OPENINGS AND CORNERS

Metal flashing should be applied over exposed door and window openings. Drip edges should be provided at heads of openings, under sills, and at projections to keep water away from the finished wall surface below.

Outside and inside corners should be flashed with a 12-in. (305 mm)-wide strip of underlay material centered in or around the corner or with metal flashing when corner boards are used.

In other corner treatments, the underlayment should be carried around the corner of each side wall so that a double thickness of felt results over the corner (see Fig. 7.7-15).

7.7.6 PLYWOOD AND HARDBOARD PANEL SIDING

Unless specific exceptions are made in the following text, the provisions of Section 7.7.1 apply to plywood and hardboard siding and their application.

7.7.6.1 Materials

PLYWOOD

Plywood is discussed in Chapter 6. Plywood used for siding includes paper-overlaid and textured panels made under their own trade names. Three available textures are shown in Figure 7.7-44.

HARDBOARD

Hardboard panel siding is made in 4-ft. by 8-ft. (1219.2 by 2438.4 mm) and 4-ft. by 10-ft. (1219.2 by 3048 mm) sizes and is usually 7⁄16 or 5⁄8 in. (11 or 15.9 mm) thick, but other sizes may also be available. It is available with flat surfaces and with vertical grooves of varying widths and spacings. Wood-grain surface textures are also available.

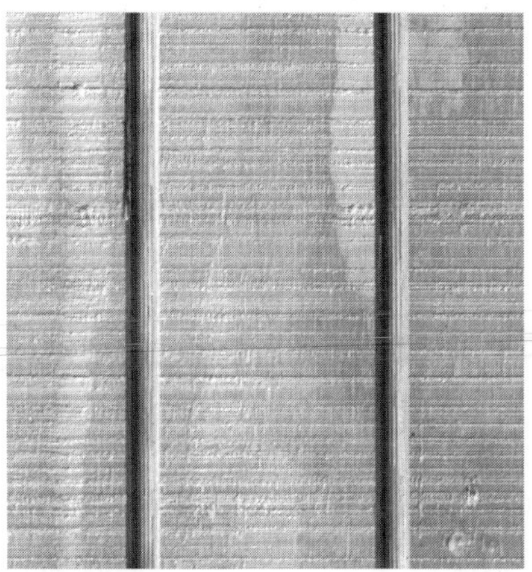

FIGURE 7.7-44 Three available textures of plywood that serves as combination siding and sheathing. (Courtesy American Plywood Association.)

7.7.6.2 Preparation for Plywood and Hardboard Siding

WALL SHEATHING

Plywood paneling can serve as a combination sheathing and siding, or can be applied over another sheathing, such as insulating sheathing. Hardboard panel siding can be applied over wood, plywood, or nonwood sheathing.

SHEATHING PAPER

Sheathing paper should be used over wood, plywood, and gypsum sheathing, but is often omitted over nonwood sheathing.

7.7.6.3 Application of Plywood and Hardboard Siding

PLYWOOD SIDING

Some plywood siding can be installed horizontally, but most is installed vertically. Sheets should be nailed to each stud at 6-in. (152.4 mm) intervals around the perimeter and at 12-in. (304.8 mm) intervals at intermediate members. Nails should penetrate supports by at least 1½ in. (38.1 mm). Joints may butt and be caulked, or may have battens or a board-on-board design (see Fig. 7.7-42).

HARDBOARD SIDING

Hardboard panel siding should be installed in strict accordance with the recommendations of its manufacturer.

CORNERS

Corners in panel siding can be finished as described earlier for wood and hardboard board siding.

7.8 Flashing and Sheet Metal

Flashing is installed at vulnerable locations to help prevent water from entering a building and to collect water that has entered the exterior shell and divert it back to the exterior. Flashings are formed from sheet metal, bituminous fabric membranes, plastic sheet materials, or a combination of these materials.

7.8.1 MATERIALS

Because replacement costs can exceed the initial installed costs, it is prudent to select a durable flashing material for the original installation.

Flashing materials must possess at least four important properties: (1) imperviousness to moisture penetration, (2) resistance to corrosion caused by exposure to the atmosphere or by caustic materials in adjacent materials, such as the alkalis present in mortars, (3) toughness to resist puncture, abrasion, and other damage during installation, and (4) formability so that the flashing can be worked to the correct shape and retain this shape after installation.

7.8.1.1 Sheet Metals

The metal materials generally used in flashing for walls and roofs include copper, lead-coated copper, lead, galvanized steel, aluminum, and stainless steel.

ALUMINUM

Aluminum is a light, bluish silver-white, ductile, and malleable metal. It is resistant to oxidization, but is subject to attack by the caustic alkalis present in fresh, unhardened mortar. Although dry, seasoned mortar will not affect aluminum, corrosion can occur if the adjacent mortar becomes wet. Therefore, aluminum should not be used as a masonry or stone wall flashing. If it must be so used, it should be coated to completely protect it from contact with both fresh and cured mortar.

Aluminum sheet flashing exposed to weather should be at least 0.019 in. (0.48 mm) thick and should have a 16,000 psi (11,249.12 mg/m^2) tensile strength. Aluminum sheet that is completely concealed should be not less than 0.015 in. (0.38 mm) thick and should have a 14,000 psi (9842.98 mg/m^2) tensile strength.

COPPER, PLAIN AND LEAD-COATED

Copper is a reddish ductile and malleable metal. The three types of sheet copper available are cold-rolled copper, soft copper, and lead-coated copper. Cold-rolled copper sheet is less malleable than soft copper, but much stronger. Accordingly, it is used much more often. The copper in lead-coated copper is similar to cold-rolled copper.

Though more costly than most other flashing materials, copper is durable, available in special preformed shapes, and an excellent moisture barrier. Because it is not materially affected by the caustic alkalis present in masonry and stone mortars, copper can safely be embedded in fresh mortar and will not deteriorate in continuously saturated, hardened mortar unless excessive chlorides are present. When using copper flashing, adding calcium chloride or other chloride-based additives to the mortar should be prohibited.

Copper sheet used for flashing should be cold-rolled copper conforming with ASTM Standard B 370 or lead-coated copper conforming with ASTM Standard B 101, Type I or Type II, Class A. Type I has the lead coating applied by hot dipping, Type II by electrodepositing. Class A denotes a lead coating on both sides of the sheet, with the total lead weight on both sides at between 12 and 15 oz. (340 and 425 g) of lead on each 100 sq. ft. (9.29 m^2) of sheet.

Copper thicknesses for various uses should be in accordance with the recommendations of the Copper Development Association's (CDA) publication, "Copper, Brass, Bronze Design Handbook, Sheet Copper Applications." In general, copper flashing that will be completely concealed should weigh not less than 10 oz./sq. ft. (3051.52 g/m^2); when exposed to the weather, not less than 16 oz./sq. ft. (4882.4 g/m^2). Base, valley, gable end, and some other flashings and gravel stops, gutters, and downspouts should weigh between 20 and 24 oz./sq. ft. (7323.7 g/m^2), depending on their size and use.

When copper is exposed to weather, its normal oxidization will wash across adjacent surfaces. This may stain or discolor those surfaces, which is particularly objectionable when copper flashing is installed in light-colored masonry or stone or when the wash-off from copper will pass over other metals, such as aluminum, or flow down walls. Where such copper staining would be objectionable, lead-coated copper can be used.

GALVANIZED STEEL

Galvanized steel is steel that has been coated with a layer of zinc, either electrolytically or by the hot-dip process. It will corrode in fresh mortar and may corrode even when placed in contact with cured mortar. Therefore, it should not be used in masonry or stone where such use is avoidable, and when it is so used it should be protected with a bituminous coating. It will also corrode eventually in the air unless protected with a coating such as a bituminous compound or paint. However, because of its low cost, galvanized steel is widely used for flashing, especially in small commercial and residential buildings. While it can perform adequately in such a use when it is well protected, exposed galvanized steel requires periodic maintenance and should be accessible for inspection and painting.

Galvanized steel for wall flashings should conform with the requirements of ASTM A 653 and should be 28 gauge (0.37 mm) where completely concealed and 26 gauge (0.55 mm) where exposed.

LEAD

Lead is a bluish, soft, heavy, malleable, ductile, plastic (but not elastic) metal. Some lead roofs have been in constant use for hundreds of years. Lead will corrode in fresh mortar and may even corrode when placed in contact with cured mortar. Therefore, it should not be used in masonry or stone where such use is avoidable, and when so used it should be protected with a bituminous coating. For this reason, and because elastomeric flashings are less expensive and easier to work with, lead is seldom used today for wall flashing.

STAINLESS STEEL

Stainless steel is an alloy of steel containing more than 12% chromium and sometimes also nickel, magnesium, or molybdenum. Most stainless steel will not rust and corrodes only under very severe conditions. It is available in either hard or soft temper.

Stainless steel flashing materials are available in several gauges and finishes. They are durable, highly resistant to corrosion, excellent moisture barriers, and surprisingly workable. There are a number of types available, the varieties depending basically on the proportions of chromium, nickel, and steel. These varieties have differing degrees of resistance to corrosion and are priced accordingly. Stainless steel for masonry wall flashing should be ASI Type 304 with a No. 2B finish; it should be specified by number and not by the generic term alone. Stainless steel flashing that will be exposed to the weather should be 28 gauge (0.47 mm); 30 gauge (0.39 mm) when completely concealed. Associated fascias should be formed from 22-gauge (0.85 mm) material, but base, cap, and counter flashings may be 28 gauge (0.47 mm).

In addition to its other favorable properties, stainless steel will not stain adjacent materials, resists rough handling, and can be soldered, welded, or brazed.

SHEET METAL FINISHES

Sheet metal materials associated with prefinished metal roofing are usually finished to match the roofing. Those not associated with prefinished roofing are usually painted when exposed to view or when a protective coating is required. Copings, gravel stops, gutters, fascias, and downspouts are often factory finished. Those used with single-ply roofing are often coated with the same material used in the roofing.

Copper, lead, aluminum, stainless steel, and galvanized steel do not require painting other than for aesthetic purposes. However, if left unpainted, galvanized steel will eventually rust.

Terneplate made with stainless steel does not need painting. That made with normal steel does, because the terne coating contains pinholes that will permit the underlying steel to rust.

Painted sheet metals are usually painted with an oil- or alkyd-based paint. Some are factory primed, others must be primed in the field.

Factory-finished sheet metals may be finished with porcelain enamel, fluoropolymer, acrylic enamel, siliconized polyester, or the same materials used in single-ply roofing membranes. Aluminum may be clear or color anodized.

7.8.1.2 Flexible Sheet Flashings

Flexible sheet flashings include elastic sheet materials made from polyvinyl chloride (PVC) or modified polymers and membranes composed of glass fiber or organic felt fabrics saturated with bitumen. These flashings are used as damp checks and as low-cost substitutes for metal flashing at wall bases, opening heads, windowsills, floor lines, and in other locations where wall flashings are needed. Elastic sheet materials are also used in conjunction with single-ply roofing and as base flashings and penetration flashings in some types of built-up bituminous roofing (Fig. 7.8-1). Elastomeric sheets, usually plain or reinforced chlorinated polyethylene, are used as roof expansion joint covers. Such joint covering is often bonded to a layer of foam insulation.

Some elastic flashings are tough, resilient, and highly resistant to corrosion, acid, and salt attack. This is particularly true of flashing materials, such as EPDM synthetic rubber, recommended by the manufacturers of single-ply roofing membranes for use with their products. Elastomeric products are also able to withstand greater movement without failure in walls and roofs than metal flashings and can be more easily formed into loops or bellows to accommodate movement. Since the chemical compositions of plastics vary so widely, they cannot all be used in every location. Some plastic flashings, for example, cannot withstand the corrosive effects of masonry mortars, and others, such as PVC materials, cannot be exposed to sunlight.

When properly installed, both bituminous and plastic flashings can be effective. However, flashing must permanently exclude the passage of water. Unfortunately, bituminous and plastic fabric flashings are often inadvertently cut by trowels or abraded by masonry units during construction so that they do not provide a continuous membrane. In addition, thin fabrics may sag and not retain the desired shape as a result of changes in temperature or poor workmanship. This is a particular problem when such flashings are concealed, because the changes will not be apparent until the wall begins to leak or the water held in pockets in the deformed flashing freezes and damages the wall.

FIGURE 7.8-1 Elastic flashing is easily installed around a pipe vent.

Since recognized standards have not yet been established for either bituminous or plastic wall flashing materials, it is necessary to rely on material performance and test records, and the manufacturer's reputation to ensure suitable quality and performance.

7.8.1.3 Combination Sheet Flashing

Combination flashings consist of materials combined to utilize the best properties of each. Figure 7.8-2 shows two examples. Other examples are a copper sheet bonded with asphalt between two layers of glass fiber cloth and a rubberized asphalt compound sheet bonded to a polyethylene film. There are many more. The combination of materials can provide a lower-cost flashing by reducing the metal thickness required, or it may permit the use of corrodible metals that may otherwise prove unsatisfactory unless coated.

a

b

FIGURE 7.8-2 Two types of combination wall flashing: (a) a copper sheet bonded to asphalt-saturated kraft paper (aluminum may be used in lieu of copper); (b) a five-layer assembly composed of a copper or aluminum sheet sandwiched between layers of asphalt-saturated cotton fabric, which is then faced with outer layers of ductile mastic.

Some combination flashing materials are made by coating sheet metal with another material. Coating materials commonly used include bitumen, kraft paper, bituminous saturated cotton fabrics, glass fiber fabrics, and combinations of these. The sheet metals commonly used for combination flashing include 3, 4, and 5 oz./sq. ft. (915.5, 1220.6, and 1525.8 g/m^2) copper and 0.004 or 0.005 in. (0.102 or 0.127 mm)-thick aluminum.

Composites consisting of a reinforced insulated chlorinated polyethylene joint cover bonded on each edge to a galvanized steel, stainless steel, or copper flange are used as roof expansion joints.

While no recognized standards exist for combination flashing, the weights of the copper and the thicknesses of the aluminum sheet metal given here, when coated on both sides with any of the coatings listed above and installed with unbroken skins, provide suitable concealed flashings.

7.8.1.4 Fasteners

Aluminum sheets should be fastened in place with screws. Other metals should be fastened in place with either screws or twisted or threaded nails. Screws are often used in roof specialties; nails in flashings. Fasteners should extend not less than 1½ in. (38.1 mm) into solid, pressure-preservative-treated wood.

Cleats in lead roofing should be installed with hard copper wire nails. Most other fasteners in sheet metal flashings, roof accessories, and roofing should be of the same material as the metal being fastened. When that is impracticable, the fasteners should be noncorrosive and compatible with the metal being fastened so that their contact will not produce galvanic corrosion.

Lead sheets, regardless of their size, and other metal sheets more than 18 in. (457.2 mm) wide, should not be nailed. These should rather be held in place with cleats of the same metal and thickness as the sheets, spaced about 12 in. (304.8 mm) on centers (Fig. 7.8-3). However, cleats should be continuous (1) on slopes less than 3 in. (76.2 mm) per foot, (2) where the sheets are subject to high winds, and (3) in other locations where the metal manufacturer so recommends. Cleats should not be less than 2 in. (50.8 mm) wide and 3 in. (76.2 mm) long. They should be nailed in place with at least two nails and lock-seamed into the sheet being held.

Narrow sheets should be nailed on one edge only. In general, nails should be staggered at about 3 in. (76.2 mm) apart and ½ in. (12.7 mm) from the edge of the sheet.

Metal flashing flanges should be nailed to wood members (nailers) at a maximum spacing of 6 in. (152.4 mm).

7.8.2 WALL FLASHING

Refer to Section 4.7 for information about wall flashing in masonry walls.

Except in areas of very slight exposure to precipitation where local experience supports omitting it, flashing should be installed in other types of walls at the following locations (Fig. 7.8-4): (1) at the bottom of walls, (2) under wall opening sills, (3) over wall opening heads, (4) at intermediate floor lines, depending on construction type, (5) where a wall penetrates a roof surface, (6) beneath copings, and (7) at expansion joints.

7.8.2.1 Base of Walls

Flashing and damp checks are used at the base of walls. Flashing diverts to the exterior moisture that enters the wall above the flashing location. A damp check stops the upward capillary travel of ground moisture. If installed properly, metal shields used for termite protection (see Sections 2.3 and 6.5.2) may serve these purposes.

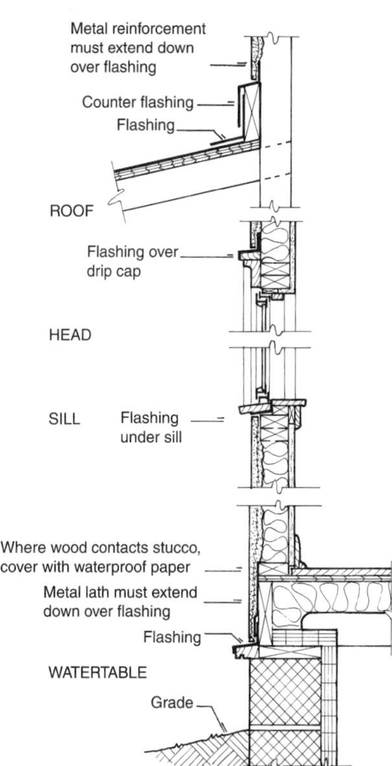

FIGURE 7.8-4 Wall flashing locations in wood stud construction. (Drawing by HLS.)

7.8.2.2 Wall Openings

SILLS

Sill flashing should be placed under and behind every sill. The ends of sill flashing should extend beyond the sides of an opening and turn up at least 1 in. (25.4 mm) into the wall. Sills should slope and project from the wall to drain water away from the building and to prevent staining. When the undersides of sills do not slope, a drip notch should be provided or the flashing may be extended and bent down to form a drip.

HEADS

Head flashing should be placed over all openings except those completely protected by overhanging projections. At steel and wood lintels the flashing should be placed under and behind the facing material with its outer edge bent down over the lintel to form a drip.

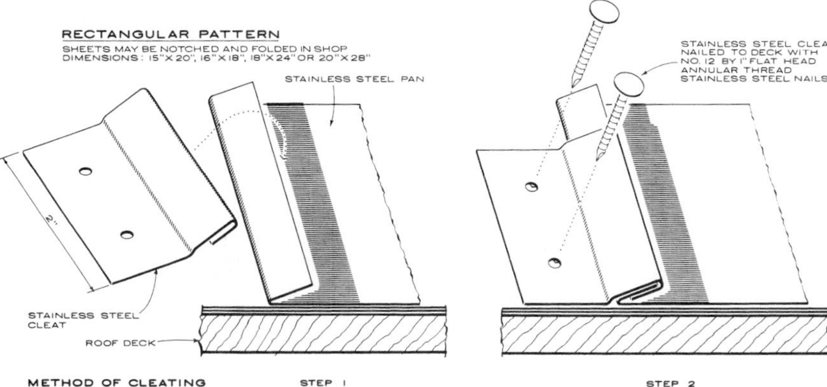

FIGURE 7.8-3 A typical cleat. (Courtesy American Iron and Steel Institute.)

7.8.2.3 Floor Lines

Flashing should be used at floor lines where water entering a wall may be able to find its way into the building through floor edge construction. It should extend from the exterior face of the wall below the floor construction, up the exterior face of the floor construction, pass beneath the wall above, and turn up. This flashing may be any of the materials described in Section 7.8.1.

7.8.2.4 Copings

Flashing should be carried beneath copings where the construction will not otherwise prevent water penetration. Coping flashing may consist of any of the materials listed in Section 7.8.1.

7.8.2.5 Walls Penetrating Roofs

At chimneys and other walls passing through a roof, the base flashing described in Section 7.8.3 should be capped with metal counter (cap) flashing that turns down over the base flashing and is anchored securely in the mortar joints.

7.8.2.6 Expansion Joints

Flashings are required at expansion joints in walls where water may enter the building. These flashings may be preformed of metal or plastic, or they may be formed in the field, usually from metal fashioned into a bellows.

7.8.3 ROOF FLASHING

Refer to Section 7.5 for information about flashing associated with built-up and single-ply roofing, and to Section 7.6 for similar information about flashing associated with shingle and tile roofs.

Flashings should be installed in roofs:

- At eaves, rakes, and valleys of pitched roofs;
- Where walls, pipes, curbs, and other elements penetrate;
- At roof drains.

Except for counter (cap) flashings, flashings associated with field-formed metal roofing are often locked into the roofing with a lock-seam joint.

Flashing for preformed roofing should be made specifically to fit the configuration of the roofing or should be flexible enough to fit the particular roofing's shape. Flashings should be manufactured or supplied by the manufacturer of the preformed roofing.

7.8.3.1 Eave, Rake, and Valley Flashings

Metal eave, rake, and valley flashings are described and illustrated in Section 7.6 for each type of pitched roofing covered there.

7.8.3.2 Roof Penetration Flashings

Roof penetrations include major penetrations, such as walls and chimneys, and minor penetrations, such as pipes, curbs, guy wires, roof drains, and other elements that penetrate a roof. Essentially, flashings at penetrations consist of a base flashing turned up against the penetration and a counter, or cap, flashing turned down over the base flashing. However, several variations of such flashings are used, depending on the roofing and the element to be flashed.

BASE AND CAP FLASHINGS

Separate base and cap flashings are used with walls, chimneys, and curbs. Metal base flashings of the same material and thickness as the roofing are used with metal roofs. Both metal base flashings and bituminous base flashings are used with shingles (see Section 7.6). Sheet metal base flashings are not usually used with built-up or single-ply roofing, because metals have coefficients of expansion that are not compatible with those of most roofing membranes. Base flashings for built-up roofing are usually bituminous or an elastomeric material compatible with the roofing. Base flashings for single-ply roofing are usually of the same materials as the roofing membranes or a compatible elastomeric material. Refer to Section 7.5 for additional discussion about base flashings for use with built-up and single-ply roofing.

Metal base flashings in steep roofing may be either continuous or pieced. Pieced base flashing is woven into the roofing as it is placed, with the pieces overlapped by at least 3 in. (76.2 mm) in the direction of water flow. Sloped-roof base flashing should turn out onto the roofing and up the wall not less than 4 in. (101.6 mm) when there is a cap flashing, 6 in. (152.4 mm) when a wall finish forms the cap flashing, and 8 in. where there is no cap flashing. When the slope is away from the wall, the base flashing should extend up the wall at least 5 in. (127 mm).

Every base flashing requires some kind of counter (cap) flashing. A wall finish may act as a cap flashing, but most cap flashings are metal. The metal material used in cap flashings varies, depending on the base flashing material and the type of penetration being flashed. Figure 7.8-5 shows a typical cap flashing, but many other configurations are possible. A cap flashing may extend beneath a windowsill, for example. Cap flashings at sloped roofs should be in sections, with no section more than 8 in. (203.2 mm) high. Cap flashings should overlap base flashings by at least 4 in. (101.6 mm).

FLASHINGS FOR SMALL PENETRATIONS

Small penetrations such as pipes, conduits, ducts, and columns are sometimes flashed by building a curb around them and providing base and cap flashings. More often, though, such elements are flashed with sleeves, formed metal housings, or collars.

Flashings for some penetrations consist of a factory-formed collar of an elastomeric material shaped to fit the penetrating element. This is especially true in single-ply roofs, where such a collar can be sealed to the roofing membrane with the same materials used to seal seams in the roofing membrane.

Pipes are flashed with metal in several ways. One common way consists of a lead-sheet flange that extends out onto the roofing about 4 in. (101.6 mm) to be flashed in with roofing membranes. A sleeve is soldered or welded to this flange, extends up the pipe for its full height, and is then turned down into the pipe at least 2 in. (50.8 mm).

Penetrations such as shaped steel columns, conduits, and ducts are some-

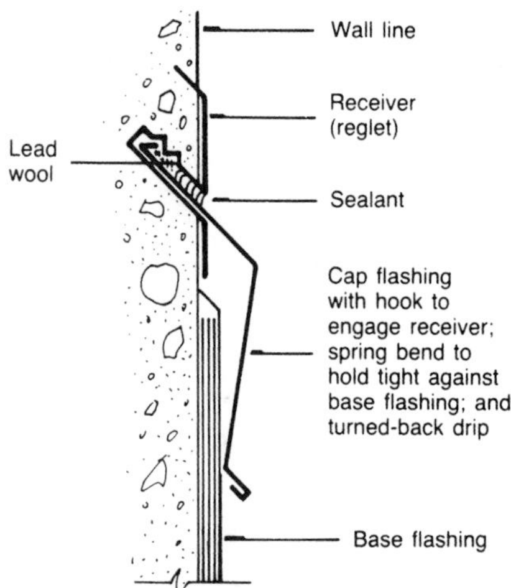

FIGURE 7.8-5 A typical cap flashing in a reglet. (Courtesy H. Leslie Simmons.)

times flashed using a two-piece metal housing. The lower portion of the housing is a sleeve that extends 8 in. (200 mm) up the penetrating element. It may either be fitted to the penetration's shape or be square or rectangular. This sleeve has a 4-in. (101.6 mm)-wide flange that is flashed into the roofing using roofing membrane. The upper portion of the housing is a hood that is attached firmly to the penetrating element and extends down over the sleeve. The joint between the hood and the penetrating element should be filled with elastomeric sealant.

Penetrating items that are difficult to flash, such as guy wires and sloping-roof top-mounted sign supports, are sometimes flashed with a pitch pan. This is a two-piece soldered metal pan set over the roofing membrane and flashed into the roofing by the roofer. The pan is then filled with a layer of mortar or a mixture of roofing cement and sand and topped off with a layer of a polyurethane sealant, coal-tar pitch, or another self-leveling topping. All materials used must be compatible with the adjacent roofing. The use of pitch pans is discouraged by most authoritative sources, but their use persists, nevertheless, because there is sometimes no other way to provide flashing for a penetrating element.

ROOF DRAIN FLASHING

Roof drains in low-slope roofs are usually flashed by extending the roofing membrane into the flange ring of the drain. Then a 30-in. (762 mm) square of 4-lb. (1.8 kg) lead or 16-oz. (4882.4-g/m^2) copper is placed on the roofing membrane, extending into the drain's flashing ring and bedded in roofing cement. This metal pan should be stripped into the roofing by the roofer. In some single-ply roofing systems, the roofing membrane is extended into the drain's flashing flange and the metal sheet is omitted.

Sometimes the roofing insulation is tapered down to the drain. When the flashing material can be exposed to sunlight, a gravel stop is sometimes installed to surround the drain.

7.8.4 ROOFING ACCESSORIES AND SPECIALTIES

7.8.4.1 Roofing Accessories

Roofing accessories include ridge vents, curbs, gravity ventilators, skylights, heat and smoke vents, and roof hatches.

Roof accessories often come with built-in 4-in. (101.6 mm)-wide roof flanges and counter flashings. Base flashing should be extended up the accessory's curb and beneath the built-in cap flashing. Roofing accessory flanges for use with metal roofing should conform with the roofing's profile.

In roof accessories for formed-in-place metal roofing, the portions that contact the roofing are usually made from the same material as the roofing.

Specialties for use with preformed roofing should be specifically designed to fit the roofing profile. They should be made from the same materials as the roofing and be finished to match it. Using roofing specialties with standard curbs not specifically designed for use with the particular preformed roofing product will usually result in leaks.

7.8.4.2 Roofing Specialties

Roof specialties include copings, expansion joint covers, gravel stops, fascias, scuppers, gutters, and downspouts.

COPINGS

Most metal coping covers are either formed sheet metal or extruded aluminum. They may be factory finished or primed for field painting. Most are installed with open joints backed by flashing or with joint covers. Sections are usually 8 or 10 ft. (2438.4 or 3048 mm) long. Most slope down toward the roof side so that water falling on them will not run down the face of the building.

Installation methods vary with the coping, but some characteristics apply to most. Some come complete with their own compatible support system. Others are supported on pressure-preservative-treated wood nailers. Most are installed over a layer of building paper. Outer edges are snapped down in place over continuous cleats. Inner edges are screwed to wood framing or snapped over metal cleats.

EXPANSION JOINT COVERS

Roof expansion joints may be either preformed metal, field-formed metal, preformed elastomeric, or a combination preformed unit of metal and an elastomeric material. See Figures 7.5-23 and 7.5-41.

Except in some single-ply roofing systems, roof expansion joints should be installed over pressure-preservative-treated wood curbs. These wood curbs should be flashed into the roofing with base flashing. The expansion joint cover acts as a cap flashing.

Elastic expansion joints are sometimes installed flush with the roofing, but this is discouraged by most roofing experts. This type of installation should be used only in single-ply roofing where specifically recommended by and warranted by the roofing membrane manufacturer, and where a professionally competent independent roofing expert concurs.

GRAVEL STOPS AND FASCIAS

Gravel stops and fascias are usually prefabricated from either formed sheet metal or extrusions. Sections length is usually 8 ft. (2438.4 mm). Roof flanges should extend 4 in. (101.6 mm) out onto the roof, be securely nailed in place, and be stripped into the roofing by the roofer. Light fascias and gravel stops should have hook strips at the bottom to hold them in place. Heavy extruded items may not require a hook strip.

SCUPPERS

Scuppers are outlets through a parapet or gravel stop that permit water to flow from a roof without traveling through a pipe inside the building. Water flowing through a scupper may fall to the ground, with no connection to a drain system, or fall directly into a conductor head, which is a short section of gutter with the ends sealed. Conductor heads are usually connected to a downspout.

GUTTERS

Gutters collect water from a roof and distribute it to a downspout. They are built in many shapes of formed sheet metal or plastic. Some are built in, but most are hung from a fascia. Most hung gutters are made from aluminum, copper, galvanized steel, or stainless steel in rectangular, half round, or ogee shapes. Built-in gutters are usually made from one of the metals used for flashing and sheet metal roofing. Built-in gutters are seldom made from aluminum, however, because it cannot be soldered and making joints watertight is difficult.

Hung gutters are supported by straps, brackets, or spikes and ferrules of the same metal as the gutter. The size, thickness, and spacing of these hangers should be based on the size and material of the gutter supported. Spacings are usually 30 in. (762 mm) maximum on centers.

Gutters should be set with a slight slope toward the downspouts to permit them to drain. Screens should be used over gutters to prevent them from becoming clogged with leaves.

Expansion should be accounted for in gutter design, especially in built-in gutters. In built-in gutters, the method depends on the design of the gutter. At the least, watertight expansion joints must be installed. Expansion in hung gutters can be accommodated by lapping sections of the gutter with a water barrier between the sections. Expansion can also be accomplished by building separate sections of gutter with ends and closing the gap between them with a cover plate.

DOWNSPOUTS

Downspouts draining built-in gutters are plumbing items. Downspouts for hung gutters are usually built of the same material as the associated gutters. Their size depends on their spacing and the area of roof drained.

Downspouts should be supported at intervals depending on their size, but never more than 5 ft. (1524 mm) apart.

Many downspouts terminate with elbow-type fittings draining out onto splash blocks or pans. However, on many commercial buildings and some residences, downspouts terminate in boots connected to the site's storm sewer.

7.8.5 SHEET METAL INSTALLATION

Additional information about installing flashing in masonry walls is given in Section 4.7, in built-up and single-ply roofing in Section 7.5, in roof shingles and tile in Section 7.6, and in wall shingles and siding in Section 7.7.

7.8.5.1 Preparation

Surfaces to receive flashing should be smooth and free from projections that would puncture the metal material or destroy its effectiveness.

Metal flashings over a wood or cementitious substrate should be installed over a red-rosin-paper slip sheet and an underlayment of polyethylene sheet. The flanges of metal flashing should be bedded in roofing cement.

7.8.5.2 Installation

Sheet metal flashings and roofing should be installed in accordance with the SMACNA *Architectural Sheet Metal Manual*. Elastic flashings should be installed as recommended by the manufacturer.

JOINTS AND EDGES

Seams in metal flashing must be overlapped and thoroughly bonded or sealed to prevent water penetration. There are many variations of such joints. Most sheet metal flashing materials, other than aluminum, can be soldered, but lockslip joints are required at intervals to permit thermal expansion and contraction. Joints in lead sheets are sometimes welded. Most joints in sheet metals are soldered with a tin and lead solder. The exact composition of the solder depends on the metal being soldered. Aluminum sheets are usually lapped, lock-seamed, or butted and back flashed. Aluminum gutters and downspouts are often lapped and riveted. Joints in aluminum should be filled with sealant.

Many plastics can be permanently and effectively joined by heat or an appropriate adhesive. The elastic pliability of plastic flashings eliminates the need for most expansion and contraction seams.

Metal flashings to be sealed with sealants should be overlapped at least 3 in. (76.2 mm). Others should be connected with seams to form a continuous watertight system.

Sheet metal edges that are not attached or locked to adjacent metal should be hemmed for strength.

EXPANSION AND CONTRACTION

There is usually a large differential in the expansion coefficient of most sheet metals and that of their supports. Therefore, it is necessary to place expansion joints in sheet metal. The normal intervals are 32 ft. (9.75 m) for aluminum and 40 ft. (12.2 m) for most other metals. Additional joints are also required when the distance from the last joint to the end of the sheet metal is more than half the normal interval.

Extruded aluminum gravel stops, copings, and fascias should have expansion joints about 12 ft. (3657.6 mm) apart.

FLASHING RECEIVERS

Most metal counter flashings should be set into mortar joints as the masonry is built or later inserted into flashing receivers. Reglets should be used at concrete walls. Flashing receivers are made from the same metal as the cap flashing. Reglets are made from various metals and plastics. The material used should be compatible with the flashing.

Flashing should be snapped or inserted into metal flashing receivers and held in place by the receiver's snaplock feature or by fasteners.

Depending on reglet design, flashing may be snapped into it or inserted and wedged in place with lead wedges or lead wool (see Fig. 7.8-5). After the flashing has been inserted, a reglet should be filled with an elastomeric sealant. Lead flashing, however, is an exception. It should be locked to a separate lead strip, which can then be caulked directly into a reglet.

Cap flashings are sometimes inserted into a *raggle* (cut slot) left by the mason or cut into a finished masonry joint and held in place by lead wedges. The raggle should then be filled with sealant.

7.9 Metal Roofing

Metal roofs have protected buildings for at least 2000 years. The Pantheon in Rome had a lead roof. The 1400-year-old lead roof on the church of Haghia Sophia in Istanbul is still in usable condition. Notre Dame Cathedral in Paris and St. Paul's Cathedral in London both have lead roofs. The balcony of the castle of Fürstenburg, built in the fifteenth century, has a gilded copper roof. The Potala Palace in Lhasa, Tibet, built between 1645 and 1693, also had a gilt copper roof. Many buildings erected during the Revolutionary War period had copper roofs. Copper sheets have been rolled specifically for roofing since about 1740.

Today, galvanized steel, stainless steel, aluminum, lead-coated copper, zinc, and terne-coated stainless steel have been added to the old metal roofing mainstays, copper and lead. Composite panels are also available that combine insulation with a metal skin.

The requirements related to sheet metals discussed in Section 7.8 apply to metal roofing except where they are specifically modified in the following discussion.

Sheet metal roofs are used on every building type. They are made from a variety of sheet metals (Fig. 7.9-1) and in many configurations.

Metal roofing must (1) shed water and prevent its penetration into the building, (2) resist corrosion caused by exposure to the atmosphere and pollutants, (3) resist puncture, abrasion, and other damage during installation and use, and (4) be made from a formable metal that can be worked to the correct shape and retain this shape after installation.

Because replacement costs can exceed the initial installed costs, it is prudent to select a durable metal roofing material for the original installation.

7.9.1 SYSTEMS

7.9.1.1 Preformed Metal Roofing

Preformed metal roofing consists of either composite panels, called metal roof and wall panels, or sheet metal panels. Preformed roofing is ready to install with joint configurations formed as part of the panels when they are shipped to the project site.

Many preformed panels are used also on vertical surfaces.

COMPOSITE PANELS

Composite panels have a sheet metal facing over an insulating material such as polyurethane, polyisocyanurate, or another rigid foam. Some have uninsulated honeycomb cores. The metal facing may be any of the metals described in this section, but are usually stainless steel, galvanized steel, aluminum, galvalume, copper, or zinc. Panel types include standing seam, corrugated, and flat with interlocking joints (Fig. 7.9-2). Some flat panels have ridges or grooves; some have embossed finishes.

Composite panels are available in a variety of widths up to 42 in. (1066.8 mm), and generally in thicknesses from 1¼ in. (45 mm) to 6 in. (152 mm). Wider panels and thicker panels may be available from some manufacturers. R-values range from 4 to 50.

FIGURE 7.9-1 A sheet copper roof. (Courtesy Copper Development Association.)

FIGURE 7.9-2 A typical insulated composite panel. Many configurations are available. (Drawing by HLS.)

INTERLOCKING UNINSULATED PANELS

Uninsulated interlocking panels are formed into various configurations at the factory (Fig. 7.9-3). They are made from stainless steel, galvanized steel, aluminum, and galvalume sheets.

Interlocking panels come in a variety of panel widths, including 10, 12, 16, and 18 in. (254, 304.8, 406.4, and 457.2 mm), and in depths of usually ¾ in. (19.1 mm) to 1½ in. (38.1 mm). Other widths and depths are available from some manufacturers.

PANELS FOR SEAMED ROOFING

Preformed panels for standing-seam, batten-seam, and step-seam roofing are formed into U-shaped configurations at the factory (Fig. 7.9-4). They are made from stainless steel, galvanized steel, aluminum, and galvalume sheets.

Panels for seamed roofing are available in nominal widths of 8, 12, 16, 18, and 24 in. (200, 300, 400, 450, and 600 mm) and with seams finishing 1¾, 2½, 3, and 4 in. (44.5, 63.5, 76.2, and 101.6 mm) high. Actual widths and seam heights vary from manufacturer to manufacturer and from system to system. Structural panels tend to be thicker and have higher ribs than architectural panels. The 16-in. and 18-in. (400 and 450 mm)-wide panels are the most popular. Most are made from 24-gauge (0.70 mm) metal or the equivalent, but some are made from thinner material.

Long panels are more economical than short ones, but length is limited by expansion and contraction to 48 ft. (14 m) for steel pans and 32 ft. (10.5 m) for aluminum pans. Materials producers and their associations may recommend even shorter panels than these.

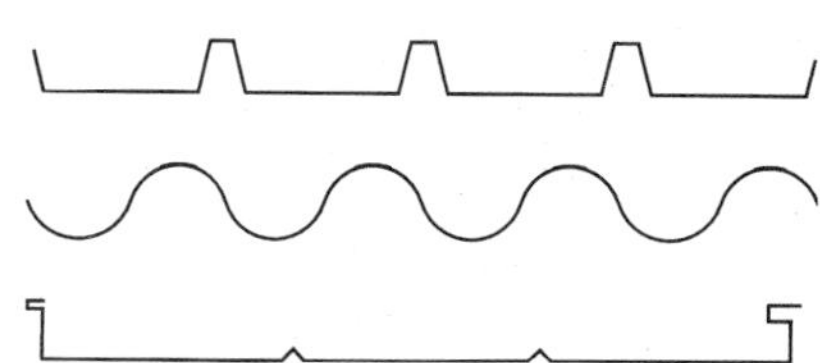

FIGURE 7.9-3 Some typical interlocking and overlapping panels. Many configurations are available. (Drawing by HLS.)

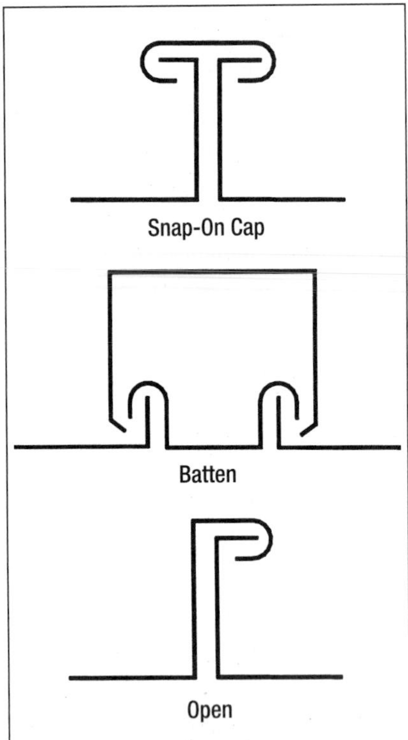

FIGURE 7.9-4 Typical standing seams made with shop-formed panels. Other configurations are available. (Drawing by HLS.)

7.9.1.2 Field-Formed Systems

Field-formed systems are formed to their final U-shaped configurations by roll forming them over anchor clips or cleats at the construction site. They may be made from any of the materials outlined later in this section.

Panel widths for field-formed metal roofing is usually 16 or 18 in. (400 or 450 mm) nominal. Ribs should be at least 1 in. (25.4 mm) high.

7.9.2 TYPES

Sheet metal roofs are either structural or architectural.

7.9.2.1 Structural Sheet Metal Roofing

Structural metal roofing is a structural roof that also forms a weather barrier. It is designed to span between structural roof supports. No roof deck is required. It can be used on slopes as low as ¼ in./ft. (20.83 mm/m). Most such roofs consist of steel or aluminum panels joined by raised seams that provide stiffness. Clips are used to attach such roofing to structural supports, usually purlins. Each manufactured product has its own specific installation methods. The manufacturer's instructions should be followed.

7.9.2.2 Architectural Sheet Metal Roofing

Architectural metal roofing is solely a weather barrier. It has no structural component and must be installed over a supporting roof deck. It is designed to shed water rather than act as a total barrier to it and is, therefore, usually installed on slopes of 3 in./ft. (150 mm/m) or steeper. Some systems can be installed on slopes as low as 1 in./ft. (83.3 mm/m), but these require special precautions to prevent water intrusion. These systems include welding every joint in lead roofing, filling every joint in aluminum roofing with a nonhardening sealant, and soldering every joint in other types of metal.

Architectural sheet metal roofing may be either preformed or formed in place. Installation methods are basically the same for both.

7.9.3 MATERIALS

The metals used in roofing are the same as those used in flashings, and the descriptions of them in Section 7.8 apply. Additional descriptions of these materials can be found in Chapter 5. The descriptions in this section related to materials supplement those in Section 7.8.

7.9.3.1 Aluminum

Sheet aluminum is seldom used to make field-formed roofing. Preformed sheet aluminum roofing, aluminum-coated sheet steel roofing, and aluminum-covered composite panels, however, are quite common. The usual thickness of aluminum roofing is about 0.032 in. (0.813 mm). Roof accessories such as copings, gravel stops, gutters, and downspouts are usually thicker, varying with the manufacturer.

7.9.3.2 Copper

Copper thicknesses for various uses should be in accordance with the recommendations of the Copper Development Association's (CDA) publication "Copper, Brass, Bronze Design Handbook, Sheet Copper Applications."

Most batten-seam, flat-seam, and standing-seam copper roofings should weigh 20 oz./sq. ft. (6103 g/m^2). Some types of roofing may weigh less (12 to 16 oz. [3662 to 4882 g/m^2]) when the pans are 16¾ in. (425.5 mm) wide or smaller.

7.9.3.3 Galvanized Steel

Galvanized steel is steel that has been coated with a layer of zinc, either electrolytically or by the hot-dip process. Because of its low cost, galvanized steel is widely used for roofing, especially in small commercial and residential buildings. While it can perform adequately in such a use when it is well protected, exposed galvanized steel requires periodic maintenance and should be accessible for inspection and painting.

Galvanized steel roofing should be 26 gauge (0.55 mm) for panels up to 24 in. (609.6 mm) wide and 24 gauge (0.70 mm) for wider panels.

7.9.3.4 Lead

Most sources recommend that sheet lead roofing and associated flashings and roof specialties be formed of not lighter than 3-lb. (14.65 kg/m^2) hard lead. Small pieces may be formed from 2½-lb. (12.21 kg/m^2) hard lead. Batten-seam roofing may also be formed of 2½-lb. (12.21 kg/m^2) lead when the battens are kept not more than 18 in. (457.2 mm) apart.

The selection and installation of lead roofing should follow the recommendations of the Lead Industries Association publication "Lead Roofing and Flashing."

7.9.3.5 Stainless Steel

Stainless steel roofing should be 28 gauge (0.47 mm) when the roofing panels do not exceed 18 in. (457.2 mm) and 24 gauge (0.70 mm) for wider panels.

Stainless steel roofing should be selected and installed in accordance with the Specialty Steel Industry of North America publication "Standard Practice for Stainless Steel Roofing, Flashing, Copings."

7.9.3.6 Terneplate

Terne is an alloy of lead and tin, usually in a ratio of 4 parts lead to 1 part tin. Terne-coated steel or stainless steel (terneplate), which is the "terne" used in roofing, is made by coating steel or stainless steel sheet with a layer of terne metal.

Terneplate used for roofing should be 0.015 in. (0.381 mm) thick for panels up to 24 in. (609.6 mm) wide and 0.178 in. (4.5 mm) thick for wider panels.

7.9.3.7 Zinc

Zinc is a bluish-white crystalline metal. It is brittle at ordinary temperatures, but ductile when heated. The zinc sheets used in roofing are an alloy of zinc, copper, and titanium.

Preformed zinc roofing panels are usually 0.027 in. (0.69 mm) thick, but thicker sheets are sometimes available. They carry a guarantee of up to 20 years.

7.9.3.8 Aluminum-Zinc Alloy on Steel

The most durable coated steel products available today have an aluminum-zinc alloy coating applied to them and are referred to as galvalume or zincalume. They carry warranties exceeding 20 years.

7.9.4 FINISHES

Factory-finished sheet metals may be finished with porcelain enamel, fluoropolymer, acrylic enamel, siliconized polyester, or the same materials used in single-ply roofing membranes. Aluminum may be clear- or color-anodized.

Steel and aluminum sheets are also available with flexible and colorful factory-applied coatings that can be shaped on a brake without damaging the coatings.

Sheet metal materials associated with prefinished metal roofing are usually finished to match the roofing. Those not associated with prefinished roofing are usually painted when exposed to view or when a protective coating is required. Copings, gravel stops, gutters, fascias, and downspouts are often factory finished. Those used with single-ply roofing are often coated with the same material used in the roofing.

Copper, lead, aluminum, stainless steel, and galvanized steel do not require painting other than for aesthetic purposes. However, if left unpainted, galvanized steel will eventually rust.

Terneplate made with stainless steel does not need painting. That made with normal steel does, because the terne coating contains pinholes that will permit the underlying steel to rust.

Painted sheet metals are usually painted with an oil- or alkyd-based paint. Some are factory primed, others must be primed in the field.

7.9.5 INSTALLATION

Surfaces to receive roofing should be smooth and free from projections that would puncture the metal material or destroy its effectiveness.

7.9.5.1 Underlayment

An underlayment of not less than No. 15 asphalt-impregnated unperforated roofing felt should be installed beneath metal roofing. Edges should be lapped 2 in. (50.8 mm), ends 6 in. (152.4 mm). Felt should be nailed in place using large-headed roofing nails. In addition to the underlayment, an ice shield should also be provided where the January mean temperature is 30°F (−1.1°C) or less. This ice shield should consist of two layers of at least No. 15 asphalt-impregnated roofing felt, installed with hot asphalt, laid to cover an area from the eave to a line 24 in. (610 mm) inside the interior face of the exterior wall. As an alternative, a single-layer membrane produced specifically for this purpose may be used as an ice shield.

The underlayment beneath metal roofing should be covered by a rosin-sized-building-paper slip sheet to prevent the roofing from sticking to the felts. However, this paper is sometimes omitted from beneath roofing installed with flat seams, because some designers do not demand that such roofing be able to move.

7.9.5.2 Fasteners

FASTENERS IN COPPER ROOFS

Each cleat in a copper roof should be fastened in place with two nails. Nails should be flathead, barbed, wire slating nails, at least 1 in. (25.4 mm) long, of 12-gauge (2.7 mm) hard copper or brass. The required length and gauge may vary with the condition of the substrate and should be verified with the roofing producer.

Screws used in conjunction with a copper roof should be hard copper or brass. They should have round heads and flat seats. A lead washer should be installed between the screw head and the copper. Where screws are exposed and subject to water penetration, copper caps should be installed over the screw heads.

FASTENERS IN ALUMINUM ROOFS

Aluminum sheets should be fastened in place with screws. Other metals should be fastened in place using either screws or twisted or threaded nails. Screws are often used in roof specialties; nails in flashings. Fasteners should extend not less than 1½ in. (38.1 mm) into solid, pressure-preservative-treated wood.

FASTENERS IN LEAD ROOFS

Cleats in lead roofing should be installed with hard copper wire nails. Most other fasteners in sheet metal flashings, roof accessories, and roofing should be of the same material as the metal being fastened. When that is impracticable, the fasteners should be noncorrosive and compatible with the metal being fastened so that their contact will not produce galvanic corrosion.

Lead sheets, regardless of their size, and other metal sheets more than 18 in. (457.2 mm) wide, should not be nailed. Rather, these should be held in place using cleats of the same metal and thickness as the sheets, spaced about 12 in. (304.8 mm) on centers (Fig. 7.8-3). However, cleats should be continuous (1) on slopes less than 3 in./ft. (250 mm/m), (2) where the sheets are subject to high winds, and (3) in other locations where the metal manufacturer so recommends. Cleats should be not less than 2 in. (50.8 mm) wide and 3 in. (76.2 mm) long. They should be nailed in place with at least two nails and lock-seamed into the sheet being held.

FASTENERS IN STAINLESS STEEL ROOFS

Nails for fastening stainless steel clips to secure stainless steel roofing should be flathead, annular-thread, diamond-point stainless steel nails long enough to penetrate the backing by at least 1 in. (25.4 mm).

FASTENERS IN OTHER METAL ROOFING MATERIALS

Fasteners should be compatible with the metal being fastened. They should penetrate the supporting material by at least 1 in. (25.4 mm).

7.9.5.3 Cleats and Clips

Sheet metal roofing should be fastened in place with *cleats* or clips generally of the same material as the roofing sheets. Lead roofing, however, should be installed with 16-oz. (4882.4 g/m^2) hard copper cleats, except that exposed cleats in lead roofing may be of 3 lb. (14.65 kg/m^2) lead.

In general, cleats should be placed about 8 in. (203.2 mm) apart. On very steep slopes, however, cleats should be continuous. Some materials and field conditions may require other spacings. The recommendations of the association representing the roofing material manufacturer should be followed.

Cleats and clips should be fastened to the roof deck and locked into the roofing sheets at the joints. Fasteners should be concealed from view and protected from the weather by the roofing sheets.

On nailable decks, cleats and clips may be nailed or screwed directly to the deck. On insulated decks, cleats should be fastened to wood nailing strips embedded in the insulation, to a layer of plywood or oriented strand board (OSB) attached to the insulation, or to Z-clips attached to the decking. Attaching cleats or clips by driving them through the insulation may void some warranties. Metal decking directly over concrete slabs requires the installation of nailers. On uninsulated steel or concrete decks, cleats or clips can be fastened directly to the decking, but on large surfaces it is better to install Z-shaped purlins over the deck.

7.9.5.4 General Installation Requirements

Sheet metal roofing should be installed in accordance with the SMACNA *Architectural Sheet Metal Manual*. Elastic flashings should be installed as recommended by the manufacturer.

DESIGN CRITERIA

Air Leakage Manufacturers of roofing products vary somewhat in their products' performance relative to air infiltration. A typical system may resist air leakage at not more than 0.06 cfm/ft.^{2} (0.3 L/s/m^2) of roof area at a static pressure of 1.56 psf (0.07 kPa) when tested in accordance with ASTM E 283. Another manufacturer may calculate leakage on the basis of seam length and say it has a maximum leakage of 0.06 cfm per linear ft. (0.33 m^3/h per linear meter) of seam at a static pressure of 6.24 psf (0.3 kPa) when tested in accordance with ASTM E 283. These criteria are both acceptable minimums, depending on the type of system. Slightly more leakage may be tolerable in some systems.

Water Penetration The system should have no water penetration at a positive pressure differential of 6.24 psf (0.3 kPa) or 20% of the design wind pressure, whichever is larger, at a test pressure between 6.24 psf (0.3 kPa) and 12 psf (0.57 kPa) in accordance with ASTM E 1646 or ASTM E 331.

Wind Uplift The system should comply with specific wind uplift resistance criteria, such as UL 90, which is 90 psf (300 Pa), when measured in accordance with UL 580.

JOINTS AND EDGES

Seams in metal roofing must be overlapped and thoroughly bonded or sealed to prevent water penetration. There are many variations of such joints. Although most sheet metal roofing materials can be soldered, lockslip joints are required at intervals to permit thermal expansion and contraction. Joints in lead sheets are sometimes welded. Most joints in sheet metals are soldered with a tin and lead solder. The exact composition of the solder depends on the metal being soldered. Aluminum sheets are usually lapped, lock-seamed, or butted and back flashed. Aluminum gutter and downspouts are often lapped and riveted. Joints in aluminum should be filled with sealant.

The edges of sheet metal that are not attached or locked to adjacent metal should be hemmed for strength.

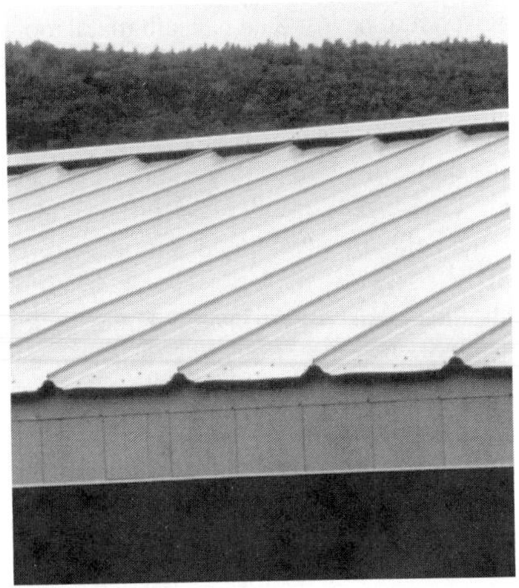

FIGURE 7.9-5 A portion of a roof using one manufacturer's standing seam roof profile. These seams are finished with a mechanical seamer. (Courtesy Metal Sales Manufacturing Corporation.)

EXPANSION AND CONTRACTION

There is usually a large differential in the expansion coefficient of most sheet metals and that of their supports. Therefore, it is necessary to place expansion joints in metal roofing. The spacing of expansion joints in metal roofing depends on the type of roofing and the seaming method. In general, when the ends of roofing are fastened, joints should be at about 15 ft. (4.57 m) on centers; where ends are free to move, joints may be at 30 ft. (9.15 m) intervals. Locations where roofing abuts other materials or penetrating walls should have provisions for expansion. Some metals may require closer or wider spacings. Stainless steel, for example, requires expansion joints between 24 ft. (7.32 m) and 40 ft. (12.19 m) on continuous runs.

7.9.5.5 Installation of Factory-Formed Metal Roofing

Factory-formed roofing, including both composite and formed panel roofing, should be installed in accordance with its

FIGURE 7.9-6 A standing-seam metal roof. (Courtesy Metal Sales Manufacturing Corporation.)

manufacturer's instructions. Methods may differ from manufacturer to manufacturer depending on the design and materials used in a particular roofing system. The general procedures, however, including the use of cleats, are similar to those used to install formed-in-place metal roofing. Different profiles are made by different manufacturers (Fig. 7.9-5). Many profiles are available. The building in Figure 7.9-6 was roofed with factory-formed standing-seam metal roofing.

7.9.5.6 Installation of Formed-in-Place Metal Roofing

The seaming methods most often used for formed-in-place metal roofing are standing seams, flat seams, and batten seams.

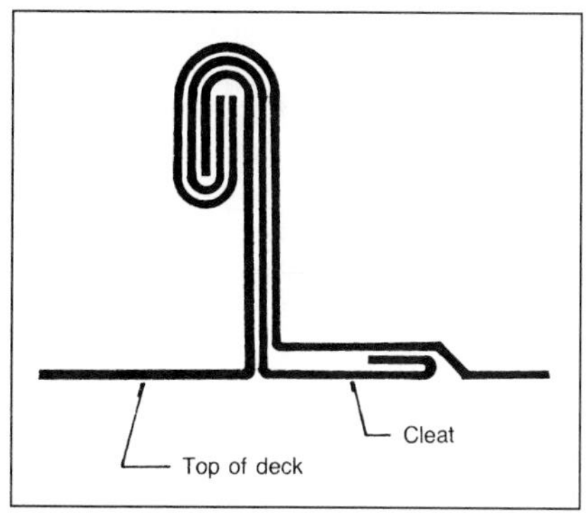

FIGURE 7.9-7 Standing seam. (Courtesy H. Leslie Simmons.)

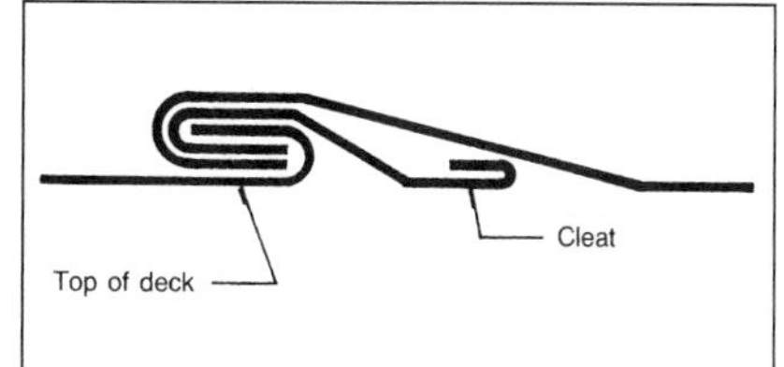

FIGURE 7.9-8 Flat lock seam. (Courtesy H. Leslie Simmons.)

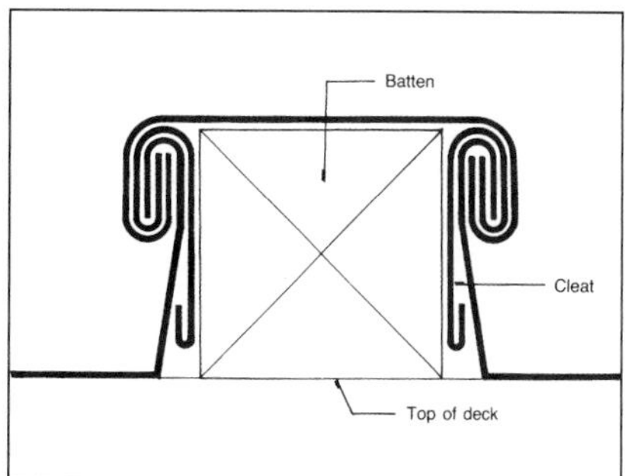

FIGURE 7.9-9 Batten seam. (Courtesy H. Leslie Simmons.)

All three may be used when the roof slope exceeds 3 in./ft. (250 mm/m). Most seams today are made by mechanical seaming machines.

Standing seams (Fig. 7.9-7) should be at least 1 in. (25.4 mm) high.

Flat seams (Fig. 7.9-8) are made by interlocking adjoining sheets. Flat seams are used in both directions in flat-seam roofing and as cross seams in standing-seam and batten-seam roofing. Flat-seam roofing may be laid with the seams parallel and perpendicular to the eaves or in a diamond-shaped pattern with the seams at a 45-degree angle with the eaves.

In batten-seam roofing (Fig. 7.9-9) the battens may run up and down the roof from eaves to ridge or may form ornamental patterns.

7.9.5.7 Oil Canning

Sheet metal roofing characteristically displays a waviness called *oil canning*. This condition is normal, is to be expected, and is almost impossible to eliminate completely. Dark colors increase thermal movement and will increase the amount of oil canning. Low-gloss paints reduce the visual effect of oil canning. Bad workmanship can significantly increase this waviness. Using tensioned, leveled, and resquared metal will reduce this condition; using wedge-shaped panels will make it worse.

7.9.5.8 Flashing and Accessories

Roof flashings, specialties, and accessories are discussed in Sections 7.5, 7.6, and 7.8.

7.10 Fireproofing

The Underwriters Laboratories, Inc. (UL) publication *Fire Resistance Directory* contains a listing and description of fire protection systems and fire ratings for various floor/ceiling assemblies, roof/ceiling assemblies, and structural members.

The UL ratings for assemblies specify the type, size, and spacing of the floor framing (including bar joists), the type and size (material thickness, depth, and configuration) of floor and roof decking, and the methods of fastening framing and decking in place.

The UL ratings for columns, beams, and girders spell out the fireproofing materials and their thicknesses that may be used to accomplish the various hourly ratings. Among the many materials that may be used to fireproof steel are concrete, masonry, gypsum board, plaster, mineral fiber, cementitious materials, and intumescent mastics.

7.10.1 ENCASEMENT

Before the advent of modern lightweight fireproofing materials, structural steel requiring a fire resistance rating was encased in poured concrete or masonry (Fig. 7.10-1). These methods were effective, but they increased building weight considerably and therefore increased framing member sizes, which also increased construction cost.

7.10.1.1 Plaster

Attempts to decrease fireproofing weight led first to the use of plaster as an encasing material (Fig. 7.10-2). The weight was further reduced by a change from sand aggregate to lightweight aggregates such as vermiculite.

7.10.1.2 Gypsum Board

Plaster encasement is seldom used for fireproofing today, having been largely replaced by gypsum board (Fig. 7.10-3). Plaster and gypsum board encasement also serves as a finish.

7.10.1.3 Mineral Fiberboard

When a structure to be fireproofed is concealed from view, other fireproofing types are more often used today. These include mineral fiber and gypsum slabs, which are held against the steel by welded-on attachments (Fig. 7.10-4) and sprayed-on contour materials.

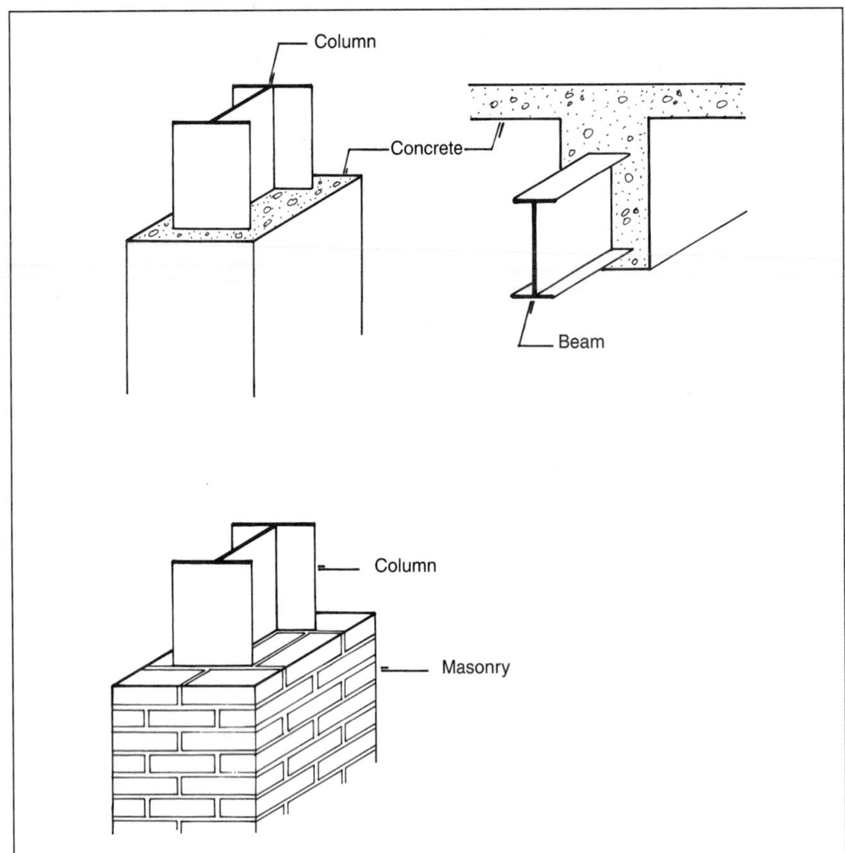

FIGURE 7.10-1 Concrete and masonry fireproofing. (Drawing by HLS.)

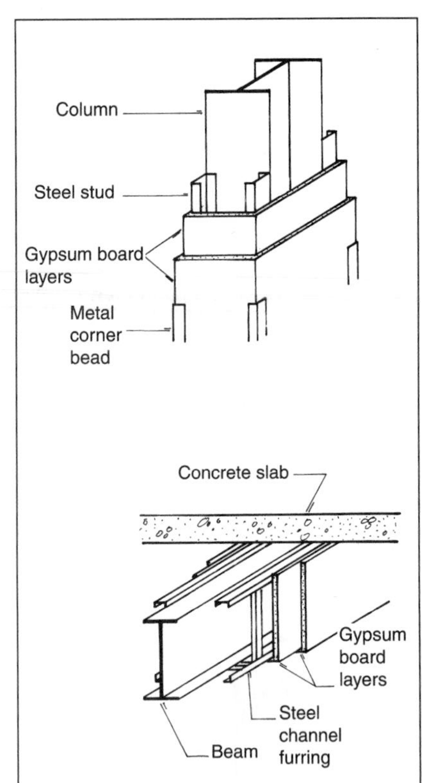

FIGURE 7.10-3 Gypsum board fireproofing. (Drawing by HLS.)

7.10.2 SPRAY-APPLIED FIREPROOFING

Sprayed-on contour fireproofing products (Fig. 7.10-5) are applied following the contour of the structural element being protected. They are also used to protect steel decking and, sometimes, bar joists and trusses. These uses should be approached with caution, however, because thermal and other substrate movement, as may happen in a steel floor or roof deck, can damage sprayed-on fireproofing. Both sprayed-fiber and cementitious products are available. Sprayed-fiber products are made from cement and mineral fiber. Cementitious fireproofing materials include:

- Cement-mineral aggregate
- Mineral fiber
- Foamed magnesium oxychloride
- Magnesium cement

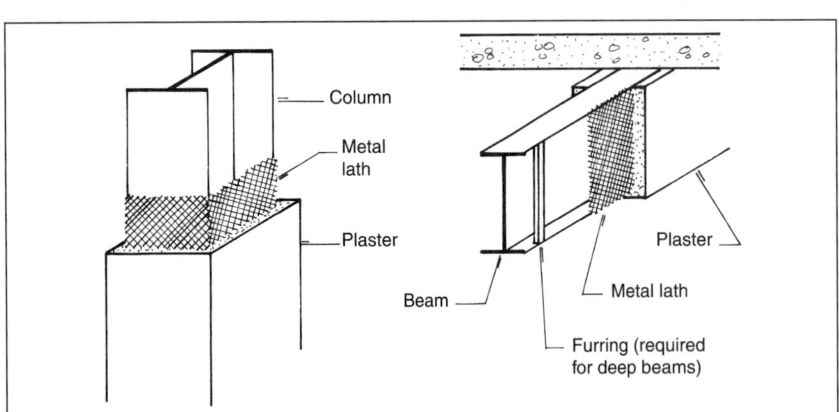

FIGURE 7.10-2 Plaster fireproofing. (Drawing by HLS.)

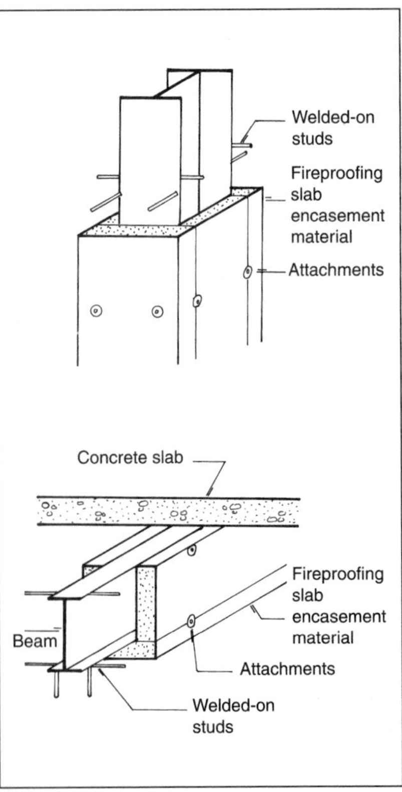

FIGURE 7.10-4 Fireproofing slab enclosure. (Drawing by HLS.)

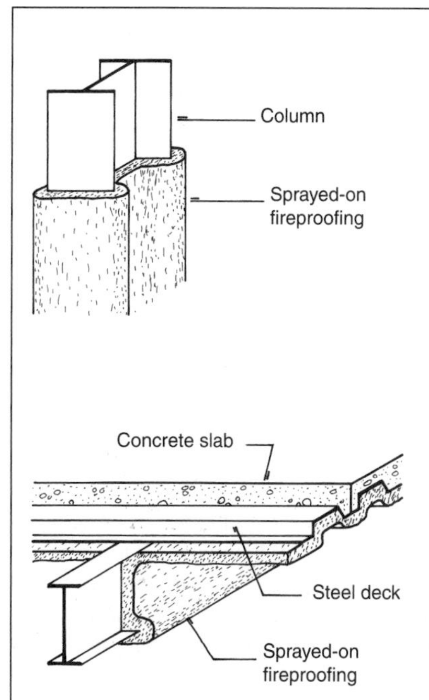

FIGURE 7.10-5 Sprayed-on fireproofing. (Drawing by HLS.)

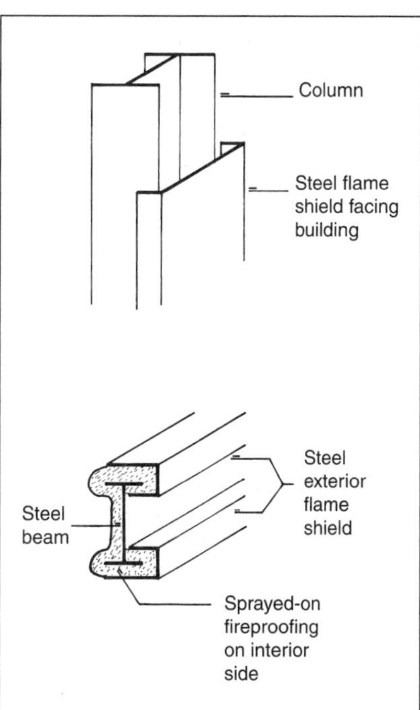

FIGURE 7.10-6 Flame shields. (Drawing by HLS.)

7.10.3 INTUMESCENT COATINGS

Intumescent coatings include both paints and mastics. The paints are intended for locations where the steel is exposed to view. They work by expanding when heated to form a thick carbonaceous foam that thermally insulates the steel and resists further damage from the fire. These coatings are available in various colors, and some can serve as a base coat for other paints.

7.10.4 CEILINGS

Other methods are also used to protect steel from fire. One is to hang a fire-resistant ceiling, such as one of plaster, beneath the structure. Refer to Chapter 9 for a discussion of plaster, gypsum board, and acoustical ceilings.

7.10.5 FLAME SHIELDS

Exterior steel members may be partly protected from contact with a fire by steel *flame shields* (Fig. 7.10-6).

7.10.6 THROUGH-PENETRATION FIREPROOFING

Fire must be prevented from passing from one fire zone to another through openings in fire-rated floor, wall, and other assemblies. Such openings occur where pipes, ducts, and other elements penetrate walls and floors and where exterior curtain walls pass by floor slab edges. This through-penetration fireproofing, which is also called *firestopping* or *fire-safing*, is accomplished by filling open spaces around the penetrations and at the edges of floors with a firestopping material. The following are among the many firestopping materials available today:

- Mineral wool felt
- Mineral fiber blankets
- Nonexpanding cementitious compounds
- Fiber-reinforced foamed cement
- Fireproof sealants
- Intumescent putty, wrap strips, pillows, and spray-applied products
- Endothermic spray that expands when heated
- Fire-rated premanufactured collars

7.11 Joint Sealing

Sealants are materials that are placed in joints in buildings and pavings where a weathertight seal is needed and in other locations to close building and paving joints for aesthetic reasons or as a base for painting. Before the advent of curtain wall construction in the 1950s, most buildings were finished in stone, brick, or terra-cotta. These materials generate relatively small movements across joints in them. In addition, most openings in walls were small. Most windows and doors were wood framed. Oil-based caulking compounds were the only joint sealants around, and they were adequate in most cases.

Curtain walls, however, were made of metals (mostly aluminum) and glass, with their high coefficients of thermal expansion. Movement across joints in and around them was greater than oil-based caulking compounds could accommodate. These materials would harden, crack, and fall out of the moving joints, leaving openings through which wind and water could easily pass. The solution was synthetic rubber (elastomeric) sealants.

The first elastomeric sealants were based on polysulfide polymers. These were quickly followed by elastomeric sealants based on acrylics, silicone, and polyurethane.

Each sealant type has its own characteristics and preparation requirements. A designer must understand these differences and select the proper sealant for the job at hand. The following steps are necessary to ensure proper functioning of a sealant:

- Proper joint design
- Proper joint surface preparation
- Selection of the proper sealant
- Proper sealant installation and finishing

The sealants discussed in this section are *building sealants*, which include those designed for use on horizontal and vertical surfaces in or on buildings, and *paving joint sealants*, which are designed for use in joints in concrete paved roads, curbing, and walkways, and between asphalt and concrete pavements. Glazing sealants are addressed in Section 8.8.

7.11.1 BUILDING SEALANTS

Building sealants include:

- Elastomeric (rubberlike) sealants based on polysulfides, silicone, or polyurethane
- Solvent-release-curing sealants
- Latex emulsion sealants
- Tape sealants
- Acoustical sealants
- Preformed foam sealants
- Oil-based caulking compounds

Until recently, federal specifications have been the standards for most sealants. Today these have been largely replaced by ASTM standards, but some manufacturers still list the old federal specifications as the standards by which their products are manufactured. Fortunately, most manufacturers also list the current ASTM standards. The following descriptions include the applicable standards, which are summarized in Figure 7.11-1.

7.11.1.1 Elastomeric Sealants

Elastomeric sealants are resin-based, rubberlike compounds, intended primarily for use in expansion joints and other moving joints, such as those in curtain walls and between stone panels or precast concrete panels. The polymers used to make most of them are polysulfide, silicone, and polyurethane. A few other polymers are also used, but they appear mostly in proprietary sealants and are not widely used.

STANDARD

The industry standard for elastomeric joint sealants is ASTM C 920. It classifies elastomeric sealants by type, grade, class, and use.

Type Elastomeric sealants are available as single-component (one-part) and multicomponent types. ASTM C 920 classifies one-part sealants as Type S, multicomponents as Type M. Type S sealants are more expensive than those of Type M, because they are prepackaged in small cartridges or other units for direct field application. Type M sealants can be packaged in bulk form. Type M sealants must be mixed in the field, which requires more skill than is necessary for Type S sealants, but Type M sealants cure faster than those of Type S and are often made with better components, which results in a superior finished sealant. In addition, the color of three-or-more-part Type M sealants can be more closely controlled because their coloring is a separate component that is added in the field.

Grade The flow characteristics of a sealant are defined by Grades P and NS. Grade P sealants are also called *self-leveling* because they seek a level and remain level and smooth when applied at temperatures as low as 40°F (4.4°C). They are used on flat horizontal surfaces.

Grade NS sealants are used on vertical surfaces and on sloping horizontal surfaces where a Grade P sealant would flow down the slope. They are also called *nonsag* because they will stay in place even when applied in temperatures as high as 122°F (50°C).

Class There are two Classes in ASTM C 920: Class 12½ and Class 25. The numbers relate to the amount of joint movement a sealant can withstand without cohesive failure. The numbers are the percentage of the joint at the time of sealant application and are plus or minus measurements. A Class 25 sealant, then, would be able to withstand a 25% expansion and a 25% contraction without failure—in other words, a total of 50% of the original joint width.

Some elastomeric sealants withstand greater joint movement than 25% in laboratory tests, and, as a result, their manufacturers claim greater capability. However, most authoritative sources believe that joints should be designed and spaced so that the movement across them does not exceed 25% in each direction, because actual sealant performance can be affected by several imponderables. These factors include the character of joint preparation and sealant application, varying joint widths in the field due to allowable construction tolerances and other factors, and the effect of the temperature at the time of application on later joint movement. A sealant may perform exactly within the rated range when the application temperature is about 70°F (21.1°C), but not perform the same if the application temperature is far below or above that midrange temperature.

Use There are two *use* classifications in ASTM C 920. The first is related to exposure and includes Use T, which covers sealants intended for use on traffic surfaces, and Use NT, which is designed for nontraffic areas.

The second use classification has to do with the ability of a sealant to adhere to various substrates. The classifications are Use M (mortar), Use G (glass), Use A (aluminum), and Use O (everything else).

FIGURE 7.11-1 Standards Governing Sealants

Sealant	Standard
Elastomeric	
Polysulfide	ASTM C 920
Silicone	ASTM C 920
Urethane	ASTM C 920
Solvent-release-curing	
Acrylic	FS A-A-1556A; AAMA 800 Series
Butyl	ASTM C 1805; FS A-A-272A
Small joint	AAMA 800 Series
Latex-emulsion	ASTM C 834
Tape	AAMA 800 Series
Acoustical	None
Oil-based caulking compound	ASTM C 570; FS A-A-272A
Cold-applied paving joint	
Elastomeric	ASTM C 920
Jet fuel–resistant coal-tar-modified polysulfide	FS SS-S-200 E (2)
Jet fuel–resistant urethane	FS SS-S-200 E (2)
Hot-applied paving joint	
Concrete to concrete	ASTM D 3406
Concrete to asphalt	ASTM D 3405
Jet fuel–resistant	
Concrete to concrete	ASTM D 3569
Concrete to tar-concrete	ASTM D 3581

Because the designations M, G, and A refer to specific laboratory material samples, it is advisable to select materials that fall into Use O. Even then, it is necessary to obtain verification that a selected sealant is satisfactory for use with the specific substrates to be found in a given project. This should be evidenced by actual tests of the selected sealant on the joint substrates that will be encountered.

POLYSULFIDE SEALANTS

Polysulfide sealants were the first elastomeric sealants and have a proven record of good performance. Unfortunately, they tend to be more expensive than sealants made with silicone or urethane polymers, and some manufacturers have made them with less polymer to remain competitive. In addition, silicone and urethane sealants are available that have properties of recovery and resistance to ultraviolet light and ozone that are equal or superior to that of polysulfide products. As a result, the percentage of the market for polysulfides is lower today than it once was. Polysulfide sealants can still be made that will give the kind of excellent performance found in earlier polysulfide products, so they should not be disregarded. However, when selecting a polysulfide sealant today it is prudent to compare its polymer content and other properties with those of earlier polysulfide products that have a proven record of dependability.

SILICONE SEALANTS

Silicone sealants perform well in terms of weather resistance, color fading, and durability. They also display low shrinkage and no increase in hardness over time, regardless of the temperatures encountered. Some silicone sealants are marketed for use in traffic-bearing joints, but most manufacturers limit the use of their products to nontraffic joints.

Silicone sealants are available in high-, medium-, and low-modulus formulations. Unfortunately, these designations are useful only in relative terms, as there are no universally recognized standards defining them. In general, high-modulus silicone sealants are to be used where joint movement is 25% (each way) or less. They are intended for use in stopless glazing, curtain wall sealing, and structural glazing (see Section 8.8).

Medium-modulus silicone sealants can be used in the same applications as high-modulus silicone sealants, but they have a higher recovery capability and greater tear resistance. Some of them have demonstrated movement resistance capabilities far above ASTM C 920's Class 25 category. Some can withstand movement as much as 50% each way.

Some low-modulus silicones appear to be able to withstand a joint movement equal to 100% each way without failure.

Both acid-curing and neutral-curing formulations are available. A major difference is that the acid-curing formulations have shown a tendency not to adhere well to certain substrates, such as marble, cement-based materials, copper, galvanized steel, and materials that corrode under the influence of the acids in the sealant.

URETHANE SEALANTS

Urethane sealants rate highly in terms of weather resistance and durability. Their high performance level will persist for a considerable time, but they tend to harden with age, which eventually leads to a diminishing ability to withstand movement in the joints without failure. However, urethanes are available that show a resistance to movement without failure of up to 50% each way. Some formulations are available for use in vertical joints, and others for use in horizontal joints. Some urethane sealants may be used under water.

7.11.1.2 Solvent-Release-Curing Sealants

Solvent-release-curing sealants include acrylic, butyl, and pigmented small-joint sealants. These sealants are used primarily in joints that are fastened together (nonworking joints), such as those in building cladding.

ACRYLIC

Solvent-release-curing acrylic sealants can be used in exterior joints, but are limited to use where the joints will not move more than 12½% either way, and their performance is inferior to that of the elastomeric sealants discussed in Section 7.11.1.1. However, they are able to adhere to a great many surfaces, mostly without application of a primer and with little joint cleaning necessary. There is no current ASTM standard for solvent-release-curing acrylic sealants. The applicable standards are Federal Specification A-A-1556A and American Architectural Manufacturers Association (AAMA) 800 Series, "Voluntary Specifications and Test Methods for Sealants."

BUTYL

Solvent-release-curing butyl sealants are somewhat more limited than solvent-release-curing acrylic sealants in that they are limited to a joint movement of 7½% each way. Butyl sealants are covered by ASTM Standard C 1311 and Federal Specification A-A-272A.

SMALL-JOINT SEALANTS

Solvent-release-curing small-joint sealants are more limited still. They cannot be used in joints wider than 3⁄16 in. (4.8 mm). Movement must be almost nil. Some of these come in spray cans with small nozzles. Others can be applied with needles in very small joints.

7.11.1.3 Latex-Emulsion Sealants

Latex-emulsion sealants are usually used to seal small interior joints having little or no movement. They should comply with ASTM C 834. These sealants are used to fill joints and crevices around doors and windows in wood, concrete, brick, and other masonry surfaces.

7.11.1.4 Tape Sealants

Tape sealants consist of a bead of butyl- or butyl-polyisobutylene-based sealant with a release paper on one side. Tapes may be reinforced with string or unreinforced. They are used for a variety of purposes, the majority being associated with glazing and concealed lap joints.

Tape sealants should conform with the applicable portions of AAMA 800 Series, "Voluntary Specifications and Test Methods for Sealants."

7.11.1.5 Acoustical Sealants

Acoustical sealants are used mostly in concealed locations to seal off sound transmission paths around partitions, electrical outlets, and other openings. They should be nondrying synthetic rubber formulations. No ASTM standard is available for acoustical sealant materials. Acoustical sealant application, however, should be in accordance with ASTM C 919.

7.11.1.6 Preformed Foam Sealants

Preformed polyurethane foam sealants are installed in joints in a compressed state and expand to fill the joint while remaining compressed to a specified degree. There are no current standards covering preformed foam sealants. The manufacturer's instructions should be followed.

7.11.1.7 Oil-Based Caulking Compounds

Oil-based caulking compounds have an oil or oleoresinous base and are used in relatively fixed joints subject to very limited movement. The standards for them are ASTM C570 and Federal Specification A-A-272A.

Oil-based caulking compounds can be used to fill joints and crevices around doors and windows in wood, concrete, brick, and other masonry surfaces that are subject to little or no movement. They are supplied in a gun grade, suitable for use with a caulking gun, and a knife grade, which can be spread with a putty knife. Oil-based caulking compounds tend to dry on the surface but remain soft and tacky inside. The exposed surface of the caulking bead should be painted each time the surrounding area is painted to help extend its life.

In most instances, other sealant types are preferred to oil-based caulking compounds because of their better adhesion and greater elasticity, even at low temperatures. Although most are more expensive than oil-based caulking compounds, their greater reliability and durability is often worth the difference in cost.

7.11.2 PAVING JOINT SEALANTS

Because paving joints are exposed to sun, weather, ice-melting chemicals, and vehicular traffic, they are subject to more severe conditions of expansion, contraction, and wear than most building joints. For this reason, most sealants that will work quite well in building joints will not perform adequately in paving joints. Special sealants have been developed to handle such conditions.

These products fall into two categories: *cold-applied* and *hot-applied* sealants. Selection often depends on the amount of work to be done. Hot-applied sealants require the use of kettles and special installation equipment and, therefore, are usually restricted to those projects where there is a large quantity of joints to be sealed, such as in highway paving.

Several paving sealants are discussed in the following two subsections, but they are not the only ones currently available. In addition, the list will surely grow longer in time and the products available will become better. There is a trend, for example, toward products with higher resistance to joint movement and abrasion and the ability to withstand both higher and lower temperatures without loss of properties.

7.11.2.1 Cold-Applied Sealants

Cold-applied paving joint sealants include pourable elastomeric sealants complying with ASTM C 920, such as a jet fuel–resistant sealant for concrete made with either urethane, coal-tar-modified polymer, or bitumen-modified urethane.

Jet fuel–resistant coal-tar-modified polysulfide and jet fuel–resistant urethane sealants for concrete joints are also available; these should comply with Federal Specification SS-S-200E(2).

Cold-applied silicone paving joint sealants are also available, one for concrete-only joints and another for joints between concrete and asphalt. Both should comply with ASTM C 920.

7.11.2.2 Hot-Applied Sealants

Hot-applied elastomeric paving joint sealants are available for concrete to concrete and for concrete to asphalt joints. Jet fuel–resistant formulations are also available for concrete to concrete and concrete to tar-concrete. Those for concrete to concrete joints should comply with ASTM D 3406, for concrete to asphalt with ASTM D 3405, for jet fuel–resistant concrete to concrete joints to ASTM D 3569, and for jet fuel–resistant concrete to tar-concrete with ASTM D 3581.

7.11.3 ACCESSORY MATERIALS

Accessories for sealed joints include joint backings and bond breakers for building sealants, and joint fillers and bond breakers for paving joints.

Joint backings and joint fillers serve as bond breakers, limit the depth of the sealant to that desired, and support the sealant during installation and tooling. In traffic joints, they also tend to support the sealant under applied loads.

7.11.3.1 Backings and Joint Fillers

Joint backings, also called *backer rods*, include plastic foams and elastomeric tubing. Foams may be of open-cell polyurethane, closed-cell polyethylene, or another plastic material that is acceptable by the sealant manufacturer. Tubing may be neoprene, butyl, EPDM, or silicone complying with ASTM D 1056. Closed-cell foams and elastomeric tubing tend to support a sealant better during installation and tooling. Open-cell foams are better able to accommodate variations in joint width and may help sealants applied over them cure faster. Elastomeric tubing can provide a temporary weather seal until the sealant has been applied, and act as a secondary seal after the sealant has been applied.

Joint fillers for use in joints in traffic surfaces include expanding cork, cork sponge rubber, bituminous-saturated fiberboard, and proprietary products that are composites of cork or fiberboard encased in organic or glass fiber felt.

Joint backing and joint filler materials must be compatible with the sealant and the substrate and should be acceptable to the sealant manufacturer. Backer rods and fillers for use with hot-applied sealants should be those specifically selected by the sealant manufacturer.

7.11.3.2 Bond Breakers

Bond breakers are polyethylene or other plastic tapes designed for installation in joints that have no backer rod or filler to prevent the sealant from bonding to the substrate at the back of the joint. The material used should be acceptable to the sealant manufacturer.

7.11.4 JOINT DESIGN

Proper joint design is an essential element in ensuring that sealants will perform satisfactorily.

7.11.4.1 Joint Types

Two joint types occur in buildings and pavements: *construction joints* and *separation joints*.

CONSTRUCTION JOINTS

Construction joints that require sealants are planned joints that divide a building into buildable segments and prevent damage due to building movement after erection. These include *expansion joints* and *control joints*. Another construction joint type is used to interrupt pours of concrete and break slabs into panels for ease of construction (see Section 3.12), but these are seldom left open and do not usually require a sealant. When they do require a sealant, the principles discussed here apply.

Expansion joints are relatively large joints intended to permit a structure to expand and contract without damage to the structure. These are sometimes called isolation joints. They usually pass completely through the structure and its finishes.

Control joints are relatively small joints used to divide thin materials such as pavements, floors, walls, and finish materials such as ceramic tile into panels of a size that will not be damaged by thermal or other movement.

SEPARATION JOINTS

Separation joints are joints between materials, such as between windows or door frames and adjacent materials, or between two portions of the same material, such as in sheet metals. These joints should be tightly sealed in all climates, whether or not the need for cooling or heating is predominant. To achieve a tight seal around windows and doors, a sealant should be placed at the joint between their frames and the adjoining walls.

7.11.4.2 Joint Movement

Joints can be classified as either dynamic or static. *Dynamic* joints, which are also called *movement joints* or *working joints*, are joints that move with changes in temperature or other building movements. These include all construction joints and some, but not all, separation joints. Dynamic joints in exterior walls are sealed to prevent air, water, and sound from passing through them. Dynamic joints in both interior and exterior horizontal and vertical surfaces are sealed to prevent the passage of light, air, water, sound, and vermin through them and to make them more aesthetically pleasing.

Static joints have little or no movement. They are also called *nonmovement joints* or *nonworking joints*. These include such joints as those between interior door frames and adjacent masonry, plaster, or gypsum board. Static joints are usually sealed to prevent the passage of light, air, and sound through them and to provide a base for painting.

7.11.4.3 Joint Design Procedure

Poor joint design is a major cause of joint sealant failures. The first step in designing dynamic joints is to estimate the amount of movement that will occur across them. To do this, it is necessary to determine the expected temperature extremes and the coefficient of expansion of the materials associated with the joint. Using these data, it is possible to calculate the amount of theoretical movement. Actual movement, however, is very difficult to determine accurately because of many imponderables, such as the exact thermal movement that will occur when more than one material is involved and construction tolerances that result in variable joint widths. Therefore, it is necessary to determine an approximate anticipated movement range and select a joint width that will accommodate this movement. The temperature at the time a sealant will be applied in a joint must be taken into account in joint width selection to ensure that the joint will be of the necessary size at that time. It is usual to assume a temperature range of between 40 and 100°F (4.4 and 37.8°C) as the temperature at the time the sealant will be installed and select a joint width that will work over that entire range.

Two basic joint types are used in buildings: *butt joints* and *shear joints* (Fig. 7.11-2). Shear joints protect the sealant and place less strain on the sealant. However, they are difficult to design and their appearance may not always be desirable. In addition, it is difficult to prepare them to receive sealant and to install the sealant. Shear joints are sometimes used in such applications as sheet metal joints, certain joints in curtain walls, and in articulated stone walls.

For both dynamic and static joints, the minimum joint width at any time during or after sealant application must be wider than the selected sealant's minimum performance width. For most sealants, this is ¼ in. (6.4 mm), but each sealant's minimum performance width should be determined from that sealant's manufacturer and not assumed. Small-joint sealants, for example, cannot be used in joints wider than 3/16 in. (4.8 mm). Their minimum joint widths are sometimes very small, but still significant in the design and sealant selection.

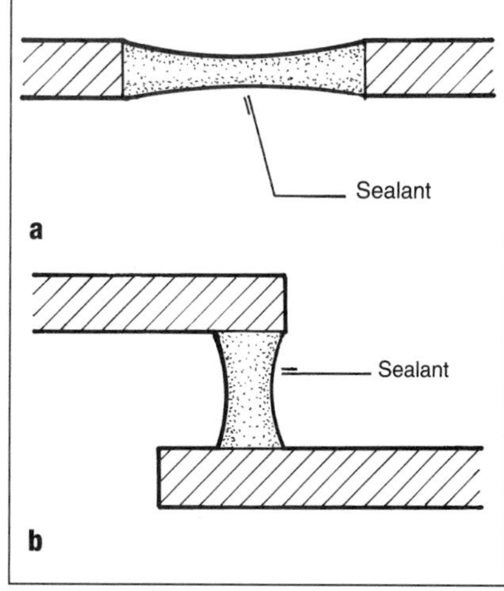

FIGURE 7.11-2 Joint types: (a) butt type, (b) shear type. (Drawing by HLS.)

Joints that are wider than the limitation of a sealant should be avoided. It is better to reduce joint width by providing more joints, but if this is impracticable or undesirable, the joints should be protected by metal plates or preformed elastomeric and metal expansion joints. This is applicable both for joints in traffic surfaces and for vertical joints.

The depth of the space allotted for a sealant must be such that the sealant has the proper shape. If a sealant is too thin, it may tear when it expands to fill a joint. If a sealant is too thick, it may fail to compress properly and may protrude from the joint, or even be squeezed out of it. Under normal conditions, the thickness should be one-half the width, but not less than ¼ in. (6.4 mm) nor more than ½ in. (12.7 mm). This rule of thumb limits the desirable width of joints to be closed with a sealant to 1 in. (25.4 mm).

7.11.5 SEALANT SELECTION

A sealant should be selected that will provide the necessary aesthetic and functional performance. Some sealants are available in many colors, some only in white or black, and others are clear. Joint type and size are prime considerations in selecting a sealant. The sealant must be capable of performing within the confines of the joint in which it is installed. For example, high-modulus silicone sealants should not be used in joints subject to movement of more than ± 25%. Narrow joints require sealants specifically formulated for them.

Sealants must also be selected to suit the location of the joint to be sealed. Traffic joints demand self-leveling sealants; vertical joints, non-sag ones. Traffic joints also require sealants with a higher hardness number. Traffic sealants should have a hardness of at least 25 but not more than 50 when tested in accordance with ASTM Standard C 920.

Special conditions must also be considered in selecting a sealant. Special conditions include the presence of high temperatures, the need for a sealant to be submersed, or the presence of vehicular traffic.

Selecting elastomeric sealants using only the ASTM C 920 classification system is possible, but often impracticable. A perfectly good sealant product may have slightly different characteristics than those required by the ASTM C 920 system. In addition, the ASTM C 920 system is not a completely adequate gauge of sealant performance. Products that exceed its limits should not be rejected on that basis alone. A product that has the desired features for a particular project may be selected even if it does not comply with another aspect of the ASTM C 920 system. The usual practice is to select specific proprietary sealant products. When competition is required, several comparable products may be specified.

7.11.6 INSTALLATION

The sealant manufacturer's instructions should be followed in preparing for and installing sealants.

7.11.6.1 Joint Preparation

Cleaning the surfaces to receive sealants is essential. There must be no contaminants of any kind in joints to receive sealants. These include, but are not limited to, incompatible paint and other coatings, rust and other corrosion, mill scale, water, frost, grease, oil, laitance, and form-release agents. Joints to receive sealant must also be dry and free from dust and grit. These should be removed using dry compressed air or brushes. The sealant manufacturer's instructions should be followed in determining whether a particular substrate is compatible with its sealant and in cleaning and treating such substrates as rubber, plastics, wood, bituminous materials, water-repellent coatings, ceramic materials, glass, and paints and other decorative coatings. Chemical cleaners and solvents must not leave a residue. A corrosion-resistant coating may be used on corrodible joint surfaces, but it must be compatible with the sealant and dry when the sealant is applied.

Joints should have no protrusions into them, such as aggregate projecting from concrete, and should be the same width throughout. Projections should be ground down. Recesses should be filled solidly and permanently.

Joints that are too narrow should be cleaned out or cut wider with power tools to produce the proper width. Under no circumstances should sealant be placed in a joint that is narrower than the minimum performance width of the sealant.

Deep joints should be filled with a joint filler, backer rod, or both to create a sealant pocket of the correct depth (Fig. 7.11-3).

When a backer rod is not installed, a bond breaker tape should be placed over the joint filler or in the back of a joint, as applicable, to prevent the sealant from adhering there (Fig. 7.11-4). If a sealant adheres to the back of a joint, it will not be able to contract properly when it stretches with an expanding joint and will tear apart internally or separate from the sides of the joint.

When the sealant manufacturer recommends doing so, joints to receive a sealant should be primed with a primer recommended by the sealant manufacturer.

Before sealants are applied, adjacent materials should be protected with tape or other means to ensure that sealant does not contaminate them and that the lines of an unconfined sealant run straight and true.

7.11.6.2 Sealant Application

In general, sealants should be applied in accordance with their manufacturer's recommendations in such a manner that they completely fill the joint and completely *wet* the sides of the joint. Hot-pour sealants should be melted within the manufacturers' recommended temperature limits.

Sealants should be confined within a joint when possible. If they finish below the face of the adjoining materials (Fig. 7.11-5), they will be less subject to damage from contact and deterioration by being somewhat shielded from the sun and weathering.

When a sealant must be applied as a triangular fillet seal (Fig. 7.11-6), it should be applied so that the fillet is not less than ⅜ in. (9.5 mm) across the face. The face of a fillet seal should be convex to ensure an adequate volume of sealant.

Mixing procedures for multicomponent sealants should be those recommended by the sealant manufacturer. Care should be taken to prevent raising the temperature of cold-applied sealants during

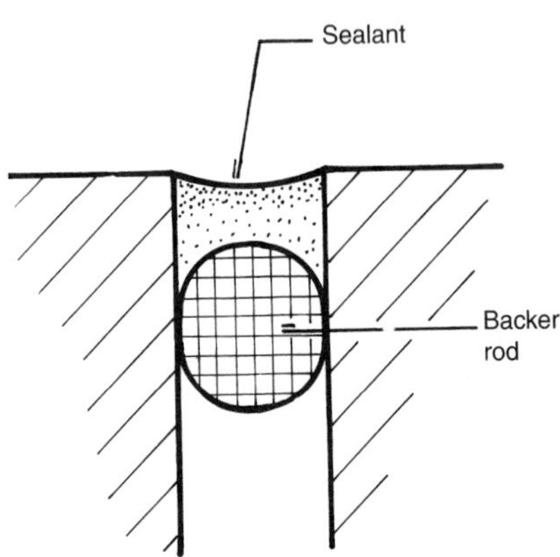

FIGURE 7.11-3 A backer rod should be used to control the joint depth. Sealant depth should not be more than one-half the joint width. (Drawing by HLS.)

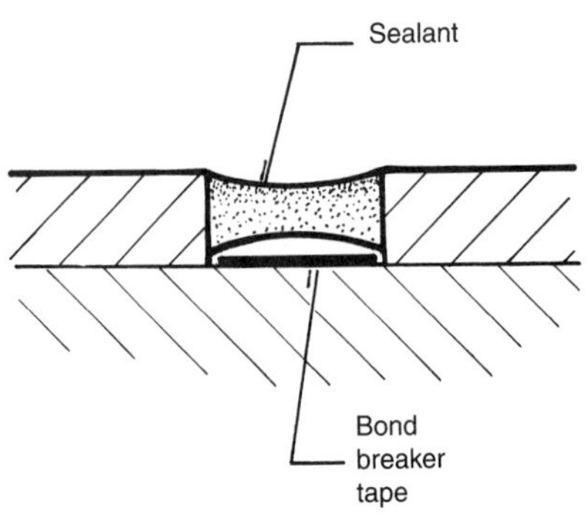

FIGURE 7.11-4 Bond breaker tape is necessary when the joint depth is shallow. (Drawing by HLS.)

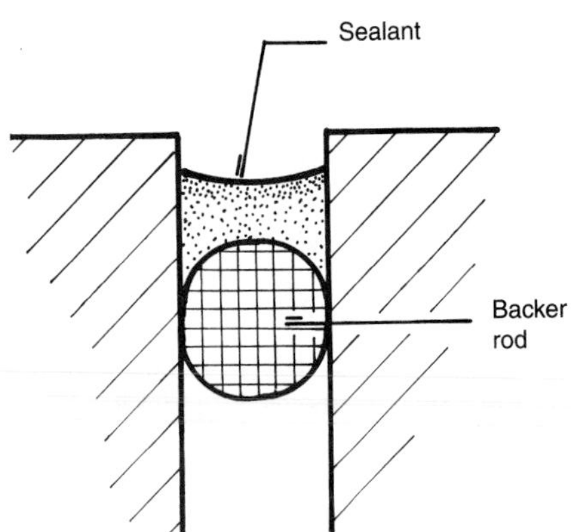

FIGURE 7.11-5 Sealant depressed below the surface. (Drawing by HLS.)

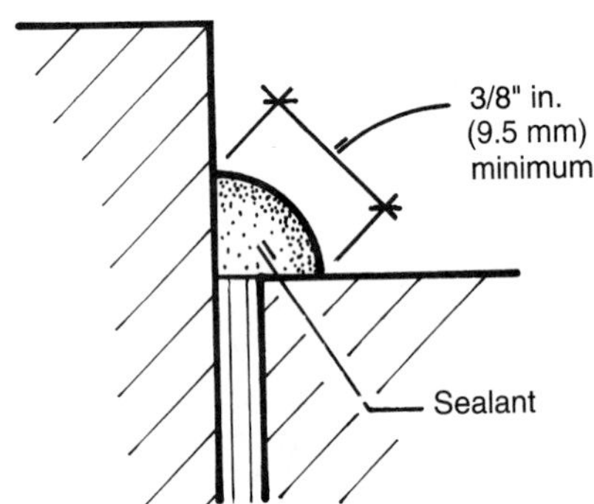

FIGURE 7.11-6 A triangular fillet seal. If the joint exceeds $\frac{3}{16}$ in. (4.8 mm) wide, a bond breaker tape should be used to prevent sealant from adhering to the sides of the joint. (Drawing by HLS.)

mixing and to prevent air from becoming entrapped in the sealant.

Application devices should be those recommended by the sealant manufacturer. Non-sag sealants should be extruded into a joint so that they are forced against the sides. Caulking guns are used to do this with most sealants. Figure 7.11-7 shows a hand-operated caulking gun used with one-part sealants. Similar hand-operated guns (Fig. 7.11-8) and pneumatic guns (Fig. 7.11-9) are used to apply multicomponent bulk sealants.

Hot-pour sealants are placed through a pouring nozzle of a size that just fits in the joint. Cold-pour sealants can be poured directly from the mixing vessel or placed with a special pouring device.

Tape sealants are placed in open joints by first removing the paper slip sheet and laying the tape in the joint. The second layer of substrate is then installed and pressed into the tape sealant.

Preformed foam sealants are forced into a joint and allowed to expand to fill it. Some preformed strips are heated after they have been inserted in a joint to cause them to achieve continuous contact with the joint sides.

Care must be taken with both tape sealants and preformed tape sealants to ensure that they make positive contact where sections join.

7.11.6.3 Tooling

Non-sag sealants should be tooled before *skinning* (curing of the top layers of sealant) occurs. Tooling should be done with the devices recommended by the sealant manufacturer. Tooling should compact the sealant and force it into contact with the sides of the joint. The surface of a tooled joint should be left slightly concave.

7.11.6.4 Protection

Sealants must be protected from contact with people and traffic until they completely cure.

FIGURE 7.11-7 A hand-operated caulking gun for premixed sealants. (Courtesy National Institute of Standards and Technology.)

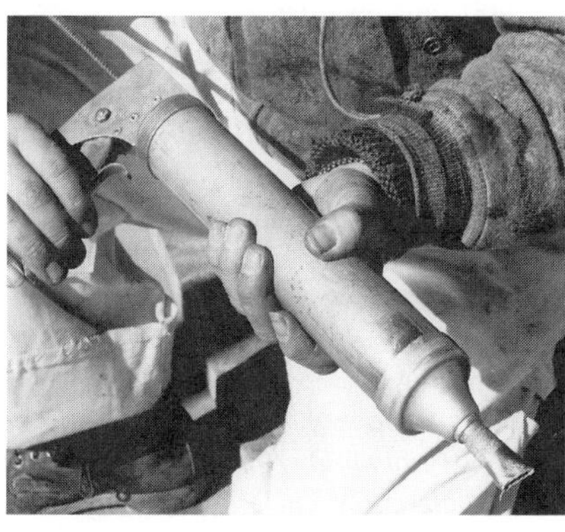

FIGURE 7.11-8 A hand-operated caulking gun for bulk sealants. (Courtesy Sealant Waterproofing and Restoration Institute.)

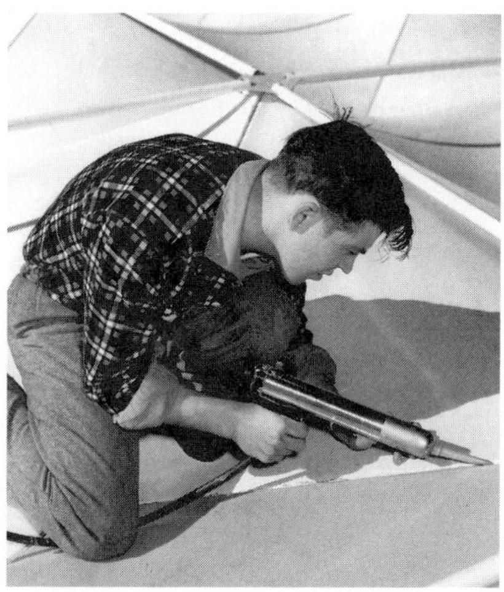

FIGURE 7.11-9 A pneumatic caulking gun for bulk sealants. (Courtesy Sealant Waterproofing and Restoration Institute.

7.12 Additional Reading

More information about the subjects discussed in this chapter can be found in the references listed in Section 7.13 and in the following publications.

Adler, David L. 1989. "Roof Failures and How to Avoid Them." *The Construction Specifier* (January):72.

———. 1994. "Sloped Roof Design Principles." *The Construction Specifier* (July):101.

Adler, David L., and Carl G. Cash. 1995. "Asphalt Shingles: A Primer." *The Construction Specifier* (November): 24.

Ahoph, Arnold J. 1988. "Traditional Building Methods: Moisture Control." *Custom Builder* (February):9–15.

Amstock, Joseph S. 1995. "High-Performance Sealants." *The Construction Specifier* (July):61.

———. 1999. *Construction Sealants and Adhesives Handbook*. New York: McGraw-Hill.

Aulisio, Lorraine M. 1996. "Foam Plastic Insulation: Building Codes and Fire Testing." *The Construction Specifier* (October):68.

Axten, Charles W. 1994. "A Look at Fibrous Insulation." *The Construction Specifier* (March):70.

Beall, Christine. 1992. "Moisture Control." *The Construction Specifier* (October):86.

———. 1998. *Thermal and Moisture Protection Manual: For Architects, Engineers, and Contractors*. New York: McGraw-Hill.

Binsacca, Richard. 1988. "Moisture Condensation Control." *Builder* (December):81–84.

Brouillard, Jeffrey J. 1997. "Flashing Details: A Case Study." *The Construction Specifier* (October):48.

Buchanan, Robert, Jason Pitcole, and Steve Tyler. 1997. "Specifying Firestopping to Save Lives." *The Construction Specifier* (June):40.

Buchinger, Ken. 1994. "Architectural Metal Panels." *The Construction Specifier* (April):92.

Cash, Carl G. 1997. "Asphalt Roof Shingle Performance." *The Construction Specifier* (November):43.

Cash, Kevin B. 1992. "Roofing High Rise Structures." *The Construction Specifier* (September):120.

Ceruti, Jeffry J. 1997. "Metal Roofing Systems: Dos and Don'ts." *The Construction Specifier* (September):64.

Cooper, Martin J., II. 1992. "Specifying Firestopping." *The Construction Specifier* (May):170.

Copper Development Association Inc. 1995. "Sheet Copper Fundamentals." *The Construction Specifier* (January):48.

Coursey, Richard. 1988. "Uplift Resistance, Testing and Design." *Architecture* (January):102–103.

Crosbie, Michael J. 1997. "A Chip Off the Old Slate." *The Construction Specifier* (November):49.

Custom Builder. 1988. "Slate Roofs; State of the Industry." *Custom Builder* (September):29–31.

D'Annunzio, John A. 1993. "Flashing the BUR." *The Construction Specifier* (November):60.

Doelp, Gregory R. 1998. "Rooftop Penetrations: Round Penetrations Are Easier to Flash." *The Construction Specifier* (May):9.

———. 1999. "Wall Leakage Disguised as Roof Problems." *The Construction Specifier* (June):59.

Egan, William F. 1993. "Specifying EIFS." *The Construction Specifier* (December):66.

———. 1995. "Specifying Premium Class PB EIFS." *The Construction Specifier* (February):21.

Fittante, Philip P. 1993. "Asphalt Basics." *The Construction Specifier* (November):76.

Gansen, Neal E. 1988. "Avoiding Roof Ice Dams." *Custom Builder* (November):35–36.

Gordon, Douglas E., and M. Stephanie Stubbs. 1988. "Longevity and Single-Ply Roofing." *Architecture* (January): 95–99.

Grimm, Clayford T. 1994. "The Hidden Flashing Fiasco." *The Construction Specifier* (June):134.

Haddock, Rob. 1994. "Metal Roofing Tips." *The Construction Specifier* (April):50.

Hardy, Steve. 1996. "Metal Roofing Fundamentals." *The Construction Specifier* (November):26.

Haws, Richard B. 1993. "Metal Roofing: The Environmental Choice." *The Construction Specifier* (November):107.

Heineman, Paul. 1987. "Coping with Membrane Roof Penetrations." *The Construction Specifier* (November): 36.

Henshell, Justin. 1995. "Watertight Details: A Checklist." *The Construction Specifier* (May):125.

Hering, Achim. 1995. "The Problem with Intumescents." *The Construction Specifier* (September):80.

Ingalls, Kelly McArthur. 1997. "In Search of Green Building Insulation." *The Construction Specifier* (August):26.

Kalwara, Joseph J., and Chester T. Chmiel. 1996. "Seaming Technology for EPDM Roofing Membranes." *The Construction Specifier* (November):40.

Klamke, Stephan E. 1997. "Exterior Insulation and Finish Systems: Proven Performance." *The Construction Specifier* (December):11.

Kliger, Robert. 1993. "Prefabricated Roof Units." *The Construction Specifier* (November):102.

Kroll, Richard E., and Steven J. Collins. 1993. "Pressure-Equalized EIFS." *The Construction Specifier* (December):69.

Kubal, Michael. 1992. *Waterproofing the Building Envelope*. New York: McGraw-Hill.

Laaly, Heshmat O. 1993. "Flashings and Counterflashings." *The Construction Specifier* (November):56.

———. 1995. "Thermal Insulation." *The Construction Specifier* (March):66.

LaBonté, Mark K. 1997. "Making the Cut." *The Construction Specifier* (February):42.

Levine, Jeffrey S. 1994. "Historic Slate Roofs." *The Construction Specifier* (July):86.

———. 1995. "What Kind of Slate Is That?" *The Construction Specifier* (November):38.

Lstiburek, Joseph, and John Carmody. 1998. *Moisture Control Handbook: Principles and Practices for Residential and Small Commercial Buildings*. New York: John Wiley & Sons, Inc.

Luft, Rene W., and Richard C. Schroter. 1996. "Wall Leaks: Lessons Learned from Sealant Failure Investigations." *The Construction Specifier* (July):84.

Malin, Nadav. 1994. "Beyond the R-Value." *The Construction Specifier* (March):62.

Mautino, Tim. 1995. "Roof Fasteners." *The Construction Specifier* (November):34.

McDonald, Timothy B. 1987. "Flashing for Built-Up Roofs." *Architecture* (October):105–106.

———. 1987. "Standing Seam Metal Roofs." *Architecture* (December):141–143.

McGettigan, Edward. 1994, "Selecting Clear Water Repellents." *The Construction Specifier* (June):120.

Nadal, Michael A. 1993. "Single-Ply Re-Roofing." *The Construction Specifier* (November):72.

Nicastro, David H. 1991, "Water Vapor Transmission." *The Construction Specifier* (June):76.

———. 1994. "12,000 Points of Water Entry." *The Construction Specifier* (July):160.

Nicastro, David, and Joseph Solinski. 1997. "Premature Sealant Failure." *The Construction Specifier* (April):50.

O'Connor, Daniel J. 1993. "Fire Stops and Fire Safety." *The Construction Specifier* (September):136.

O'Connor, Jerome P. 1997. "Exterior Insulation and Finish Systems: Obvious Weaknesses." *The Construction Specifier* (December):11.

Petrie, Edward. 1999. *Handbook of Adhesives and Sealants*. New York: McGraw-Hill.

Pine, Stephen. 1995. "Aluminum Composite Material." *The Construction Specifier* (July):42.

Piper, Richard S. 1993. "Standards for EIFS." *The Construction Specifier* (December):27.

Progressive Architecture. 1988. "Curtain Wall Cautions: Brick Curtain Wall; Exterior Insulation and Finish System; Insulated Metal Panels and Strip Windows." *Progressive Architecture* (June):114–117.

Quirouette, Rick. 1994. "The Rainscreen Roof System." *The Construction Specifier* (October):51.

———. 1997. "The Dynamic Buffer Zone." *The Construction Specifier* (August):50.

Raeber, John A. 1990. "Deciding When to Use Negative-Side Waterproofing." *The Construction Specifier* (December):35.

———. 1991. "Steep Roof Systems." *The Construction Specifier* (November):19.

———. 1992. "Preformed Metal Roofing Systems." *The Construction Specifier* (April):23.

———. 1992. "Roofing and Waterproofing Old or New?" *The Construction Specifier* (June):29.

———. 1992. "Asphalt Shingle Roofing." *The Construction Specifier* (October):25.

———. 1995. "Clay Tile Roofing." *The Construction Specifier* (March):86.
Remmele, Thomas E. 1993. "Detailing Class PB EIFS." *The Construction Specifier* (December):78.
Roe, Richard. 1997. "Membrane Roofing Wind Design: Getting to Know Factory Mutual." *The Construction Specifier* (November):29.
Ruggiero, Stephen S., and Dean A. Rutila. 1991. "Plaza Waterproofing Design Fundamentals." *The Construction Specifier* (January):74.
Sather, Lloyd. 1992. "Firestopping." *The Construction Specifier* (April):111.
Scharff, Robert. 1995. *Roofing Handbook.* New York: McGraw-Hill.
Schroeder, Edward K. 1993. "Regarding Roof Edges." *The Construction Specifier* (November):40.
Schroter, Richard C. 1994. "Metal Roof Specifications." *The Construction Specifier* (August):76.
Seeley, James I. 1997. "The Protocols of White Roofing." *The Construction Specifier* (November):35.
Shrive, Charles A. 1993. "Firestopping Update." *The Construction Specifier* (June):33.
Sievert, Norman. 1988. "Wood Siding Basics." *Custom Builder* (February): 16–20.
Simmons, H. Leslie. 1999. "Architectural Standing Seam Metal Roofing." *The Construction Specifier* (February):49.
Spencer, William. 1996. "Achieving Lasting Impermeability with a Roof." *The Construction Specifier* (November): 32.
Sterling, Walter P. 1996. "Preserving Firestopping Systems." *The Construction Specifier* (July):26.
Subasic, Christine A., Brian E. Trimble, and Janet Peddycord. 1996. "Water Repellents for Brick Walls." *The Construction Specifier* (February):23.
Thomas, Robert, Jr. 1995. "Getting the Best from EIFS." *The Construction Specifier* (February):18.
Truesdell, David E. 1998. "Rooftop Penetrations: Odd-Shape Penetrations Can Be Specified with Confidence." *The Construction Specifier* (May):9.
Tuluca, Adrian. 1996. "Thermal Bridges in Buildings." *The Construction Specifier* (October):38.
Ward, Lonnie. 1992. "Metal Roofing 101." *The Construction Specifier* (November):102.
Warseck, Karen. 1986. "Why Sealant Joints Fail." *Architecture* (December):100–103.
Weaver, Martin E. 1989. "The Basics of Slate Roof Restoration." *The Construction Specifier* (November):49.
———. 1991. "Copper Top." *The Construction Specifier* (November):40.
Wimberly, David. 1995. "Thermoplastic Roofing Systems." *The Construction Specifier* (November):32.

7.13 Acknowledgments and References

ACKNOWLEDGMENTS

We gratefully acknowledge the assistance of the following organizations and individuals in preparing this chapter. We are also indebted to them for permission to use their illustrations when requested and for the use of their publications as references.

ABT Building Products Corporation
Airline Products Company
Alcoa Building Products
American Architectural Manufacturers Association (AAMA)
American Board Products Association
American National Standards Institute (ANSI)
American Society of Heating, Refrigerating and Air Conditioning Engineers (ASHRAE)
American Society for Testing and Materials (ASTM)
Arcom Master Systems
Asphalt Institute
Asphalt Roofing Manufacturers Association
Brick Institute of America
Carlisle Roofing Systems
Cedar Shake and Shingle Bureau
Celotex Corporation
CertainTeed
Coastal Atlantic Associates, Inc.
Concrete Block Insulating Systems, Inc.
Copper Development Association (CDA), Inc.
Dow Chemical Corporation
E/S Products, Inc.
Firestone Building Products Company
Georgia Pacific Corporation
Goodyear Tire and Rubber Company
House and Home (magazine)
Independent Nail Corporation
Koppers Company, Inc.
Lead Industries Association, Inc.
Manville Corporation
W. H. Maze Company
Metal Sales Manufacturing Company
Mineral Fiber Products Bureau
Mineral Insulation Manufacturers Association
National Association of Home Builders (NAHB) Research Foundation
National Institute of Building Sciences (NIBS)
National Institute of Standards and Technology (NIST) of the U.S. Department of Commerce
National Mineral Wool Insulation Association
National Roofing Contractors Association (NRCA)
National Slate Association (NSA)
National Tile Roofing Manufacturers Association (NTRMA)
Nickel Development Institute
Owens-Corning Fiberglas Corporation
Perlite Institute
Portland Cement Association (PCA)
Sealant Engineering and Associated Lines (SEAL), Inc.
Sealant, Waterproofing, and Restoration Institute
Sheet Metal and Air Conditioning Contractors National Association (SMACNA)
Single Ply Roofing Institute (SPRI)
Solar Group
Specialty Steel Industry of North America
Underwriters Laboratories, Inc.
U.S. Department of Agriculture, Forest Service, Forest Products Laboratory
U.S. Department of Housing and Urban Development (HUD)
U.S. Department of the Army, Corps of Engineers
U.S. Weather Service
Vinyl Siding Institute (VSI)
Western Red Cedar Lumber Association
Wolverine Technologies

REFERENCES

We would also like to thank the authors and publishers of the publications in the following list for their contribution to our research for this chapter.

Air-Conditioning and Refrigeration Institute (ARI), Sheet Metal and Air Conditioning Contractors National Association (SMACNA), and National Roofing Contractors Association (NRCA), *Roof Mounted Outdoor Air-Conditioner Installations*. ARI, SMACNA, and NRCA.

Air Movement and Control Association. *Application Manual for Air Louvers*. Arlington Heights, IL: Air Movement and Control Association.

———. *Test Methods for Louvers, Dampers, and Shutters*. Arlington Heights, IL: Air Movement and Control Association.

———. *Laboratory Methods of Testing Louvers for Rating*. Arlington Heights, IL: Air Movement and Control Association.

Alcoa Building Products. 1990. *Vinyl Siding, Soffit, and Designer Accessories Installation Manual*. Sidney, OH: Alcoa Building Products.

Aluminum Association. AA DAF-45, "Designations System for Aluminum Finishes." Washington, DC: AA.

American Architectural Manufacturers Association. 1986. AAMA 1402, "Standard Specifications for Aluminum Siding, Soffit, and Fascia." Des Plaines, IL: AAMA.

———. 1994. AAMA 800, "Voluntary Specifications and Test Methods for Sealants." Des Plaines, IL: AAMA.

American Society for Testing and Materials (ASTM) Standards:

A 167, "Standard Specifications for Stainless and Heat-Resisting Chromium-Nickel Steel Plate, Sheet, and Strip." West Conshohocken, PA: ASTM.

A 463, "Standard Specification for Sheet Steel, Cold-Rolled Aluminum-Coated by the Hot-Dip Process." West Conshohocken, PA: ASTM.

A 591, "Standard Specifications for Steel Sheet, Electrolytic Zinc-Coated for Light Coating Mass Applications." West Conshohocken, PA: ASTM.

A 606, "Standard Specifications for Steel Sheet and Strip, High Strength, Low-Alloy, Hot-Rolled and Cold-Rolled, with Improved Atmospheric Corrosion Resistance." West Conshohocken, PA: ASTM.

A 607, "Standard Specifications for Steel Sheet and Strip, High-Strength, Low-Alloy Columbium or Vanadium, or Both, Hot-Rolled and Cold-Rolled." West Conshohocken, PA: ASTM.

A 611, "Standard Specification for Structural Stainless Steel, Sheet, Carbon, Cold-Rolled." West Conshohocken, PA: ASTM.

A 642, "Standard Specifications for Steel Sheet, Zinc-Coated (Galvanized) by the Hot-Dip Process, Drawing Quality, Special Killed." West Conshohocken, PA: ASTM.

A 653/A 653M, "Standard Specification for Steel Sheet, Zinc-Coated (Galvanized) or Zinc-Iron Alloy-Coated (Galvannealed) by the Hot-Dip Process." West Conshohocken, PA: ASTM.

A 755/A 755M, "Standard Specification for Steel Sheet, Metallic Coated by the Hot-Dip Process and Prepainted by the Coil-Coating Process for Exterior Exposed Building Products." West Conshohocken, PA: ASTM.

A 792/A 792M, "Standard Specification for Steel Sheet, 55% Aluminum-Zinc Alloy-Coated by the Hot-Dip Process." West Conshohocken, PA: ASTM.

A 875, "Standard Specification for Steel Sheet, Zinc 5%, Aluminum Alloy, Coated by the Hot Dip Process." West Conshohocken, PA: ASTM.

B 32, "Specification for Solder Metal." West Conshohocken, PA: ASTM.

B 101, "Standard Specifications for Lead-Coated Copper Sheet and Strip for Buildings." West Conshohocken, PA: ASTM.

B 209, "Standard Specification for Aluminum and Aluminum-Alloy Sheet and Plate." West Conshohocken, PA: ASTM.

B 210, "Standard Specification for Aluminum and Aluminum-Alloy Drawn Seamless Tubes." West Conshohocken, PA: ASTM.

B 211, "Standard Specification for Aluminum and Aluminum-Alloy Bar, Rod, and Wire." West Conshohocken, PA: ASTM.

B 221, "Standard Specification for Aluminum and Aluminum-Alloy Extruded Bars, Rods, Wire, Profiles, and Tubes." West Conshohocken, PA: ASTM.

B 370, "Standard Specifications for Copper Sheet and Strip for Building Construction." West Conshohocken, PA: ASTM.

B 429, "Standard Specification for Aluminum-Alloy Extruded Structural Pipe and Tube." West Conshohocken, PA: ASTM.

B 749, "Standard Specifications for Lead Alloy Strip, Sheet, and Plate Products." West Conshohocken, PA: ASTM.

C 150, "Standard Specification for Portland Cement." West Conshohocken, PA: ASTM.

C 208, "Standard Specification for Cellulosic Fiber Insulating Board." West Conshohocken, PA: ASTM.

C 209, "Standard Test Methods for Cellulosic Fiber Insulating Board." West Conshohocken, PA: ASTM.

C 222, "Standard Specification for Asbestos-Cement Roofing Shingles." West Conshohocken, PA: ASTM.

C 223, "Standard Specification for Asbestos-Cement Siding." West Conshohocken, PA: ASTM.

C 270, "Standard Specification for Mortar for Unit Masonry." West Conshohocken, PA: ASTM.

C 481, "Standard Test Method for Laboratory Aging of Sandwich Construction." West Conshohocken, PA: ASTM.

C 516, "Standard Specification for Vermiculite Loose Fill Thermal Insulation." West Conshohocken, PA: ASTM.

C 549, "Standard Specification for Perlite Loose Fill Insulation." West Conshohocken, PA: ASTM.

C 552, "Standard Specification for Cellular Glass Thermal Insulation." West Conshohocken, PA: ASTM.

C 553, "Standard Specification for Mineral Fiber Blanket Thermal Insulation for Commercial and Industrial Applications." West Conshohocken, PA: ASTM.

C 570, "Standard Specification for Oil- and Resin-Base Caulking Compound for Building Construction." West Conshohocken, PA: ASTM.

C 578, "Standard Specifications for Rigid Cellular Polystyrene Thermal Insulation." West Conshohocken, PA: ASTM.

C 591, "Standard Specification for Unfaced Preformed Rigid Cellular Polyurethane Thermal Insulation." West Conshohocken, PA: ASTM.

C 612, "Standard Specification for Mineral Fiber Block and Board Thermal Insulation." West Conshohocken, PA: ASTM.

C 665, "Standard Specification for Mineral Fiber Blanket Thermal Insulation for Light Frame Construction and Manufactured Housing." West Conshohocken, PA: ASTM.

C 687, "Standard Practice for Determination of Thermal Resistance of Loose-fill Building Insulation." West Conshohocken, PA: ASTM.

C 726, "Standard Specification for Mineral Fiber and Mineral Fiber Roof Insulation Board." West Conshohocken, PA: ASTM.

C 728, "Standard Specification for Perlite Thermal Insulation Board." West Conshohocken, PA: ASTM.

C 764, "Standard Specification for Mineral Fiber Loose-fill Thermal Insulation." West Conshohocken, PA: ASTM.

C 834, "Standard Specification for Latex Sealants." West Conshohocken, PA: ASTM.

C 919, "Standard Practice for Use of Sealants in Acoustical Applications." West Conshohocken, PA: ASTM.

C 920, "Standard Specification for Elastomeric Joint Sealants." West Conshohocken, PA: ASTM.

C 1126, "Standard Specification for Faced or Unfaced Rigid Cellular Phenolic Thermal Insulation." West Conshohocken, PA: ASTM.

C 1193, "Standard Guide for Use of Joint Sealants." West Conshohocken, PA: ASTM.

C 1311, "Standard Specification for Solvent Release Sealants." West Conshohocken, PA: ASTM.

D 41, "Standard Specification for Asphalt Primer Used in Roofing, Dampproofing, and Waterproofing." West Conshohocken, PA: ASTM.

D 43, "Standard Specification for Coal Tar Primer Used in Roofing, Dampproofing, and Waterproofing." West Conshohocken, PA: ASTM.

D 173, "Standard Specification for Bitumen-Saturated Cotton Fabrics Used in Roofing and Waterproofing." West Conshohocken, PA: ASTM.

D 224, "Standard Specification for Smooth Surfaced Asphalt Roll Roofing (Organic Felt)." West Conshohocken, PA: ASTM.

D 225, "Standard Specification for Asphalt Shingles (Organic Felt) Surfaced with Mineral Granules." West Conshohocken, PA: ASTM.

D 226, "Standard Specification for Asphalt-Saturated Organic Felt Used in Roofing and Waterproofing." West Conshohocken, PA: ASTM.

D 227, "Standard Specification for Coal-Tar Saturated Organic Felt Used in Roofing and Waterproofing." West Conshohocken, PA: ASTM.

D 312, "Standard Specification for Asphalt Used in Roofing." West Conshohocken, PA: ASTM.

D 449, "Standard Specification for Asphalt Used in Dampproofing and Waterproofing." West Conshohocken, PA: ASTM.

D 450, "Standard Specification for Coal-Tar Pitch Used in Roofing, Dampproofing, and Waterproofing." West Conshohocken, PA: ASTM.

D 1056, "Standard Specification for Flexible Cellular Material—Sponge or Expanded Rubber." West Conshohocken, PA: ASTM.

D 1079, "Standard Terminology Relating to Roofing, Waterproofing, and Bituminous Materials." West Conshohocken, PA: ASTM.

D 1227, "Standard Specification for Emulsified Asphalt Used as a Protective Coating for Roofing." West Conshohocken, PA: ASTM.

D 1668, "Standard Specification for Glass Fabrics (Woven and Treated) for Roofing and Waterproofing." West Conshohocken, PA: ASTM.

D 1863, "Standard Specification for Mineral Aggregate Used on Built-up Roofs." West Conshohocken, PA: ASTM.

D 1970, " Standard Specification for Self-Adhering Polymer Modified Bituminous Sheet Materials Used as Steep Roofing Underlayment for Ice Dam Protection." West Conshohocken, PA: ASTM.

D 2178, "Standard Specification for Asphalt Glass Felt Used in Roofing and Waterproofing." West Conshohocken, PA: ASTM.

D 2626, "Standard Specification for Asphalt-Saturated and Coated Organic Felt Base Sheet Used in Roofing." West Conshohocken, PA: ASTM.

D 2822, "Standard Specification for Asphalt Roof Cement." West Conshohocken, PA: ASTM.

D 2824, "Standard Specification for Aluminum-Pigmented Asphalt Roof Coatings, Non-fibered, Asbestos Fibered, and Fibered Without Asbestos." West Conshohocken, PA: ASTM.

D 3018, "Standard Specification for Class A Asphalt Shingles Surfaced with Mineral Granules." West Conshohocken, PA: ASTM.

D 3161, "Standard Test Method for Wind Resistance of Asphalt Shingles (Fan-Induced Method)." West Conshohocken, PA: ASTM.

D 3405, "Standard Specification for Joint Sealants, Hot-Applied, for Concrete and Asphalt Pavements." West Conshohocken, PA: ASTM.

D 3406, "Standard Specification for Joint Sealant, Hot-Applied, Elastomeric-Type, for Portland Cement Concrete Pavements." West Conshohocken, PA: ASTM.

D 3462, "Standard Specification for Asphalt Shingles Made from Glass Felt and Surfaced with Mineral Granules." West Conshohocken, PA: ASTM.

D 3569, "Standard Specification for Joint Sealant, Hot-Applied, Elastomeric, Jet-Fuel-Resistant Type for Portland Cement Concrete Pavements." West Conshohocken, PA: ASTM.

D 3581, "Standard Specification for Joint Sealant, Hot-Poured, Jet-Fuel-Resistant Type for Portland Cement Concrete and Tar-Concrete Pavements." West Conshohocken, PA: ASTM.

D 3679, "Standard Specification for Rigid Poly (Vinyl Chloride) (PVC) Siding." West Conshohocken, PA: ASTM.

D 3909, "Standard Specification for Asphalt Roll Roofing (Glass Mat) Surfaced with Mineral Granules." West Conshohocken, PA: ASTM.

D 4434, "Standard Specification for Poly (Vinyl Chloride) Sheet Roofing." West Conshohocken, PA: ASTM.

D 4586, "Standard Specification for Asphalt Roof Cement, Asbestos Free." West Conshohocken, PA: ASTM.

D 4637, "Standard Specifications for EPDM Sheet Used in Single-Ply Roofing Membranes." West Conshohocken, PA: ASTM.

D 4869, "Standard Specification for Asphalt-Saturated Organic Felt Shingle Underlayment Used in Roofing." West Conshohocken, PA: ASTM.

E 84, "Standard Test Method for Surface Burning Characteristics of Building Materials." West Conshohocken, PA: ASTM.

E 96, "Standard Test Method for Water Vapor Transmission of Materials." West Conshohocken, PA: ASTM.

E 108, "Standard Test Methods for Fire Tests of Roof Coverings." West Conshohocken, PA: ASTM.

E 119, "Standard Test Method for Fire Tests of Building Construction and

Materials." West Conshohocken, PA: ASTM.
E 136, "Standard Test Method for Behavior of Materials in a Vertical Tube Furnace at 750°C." West Conshohocken, PA: ASTM.
E 283, "Standard Test Method for Determining the Rate of Air Leakage Through Exterior Windows, Curtain Walls, and Doors Under Specified Pressure Differences Across the Specimen." West Conshohocken, PA: ASTM.
E 331, "Standard Test Method for Water Penetration of Exterior Windows, Curtain Walls, and Doors by Uniform Static Air Pressure Difference." West Conshohocken, PA: ASTM.
E 662, "Standard Test Method for Specific Optical Density of Smoke Generated by Solid Materials." West Conshohocken, PA: ASTM.
E 699, "Practice for Criteria for Evaluation of Agencies Involved in Testing, Quality Assurance, and Evaluating Building Components in Accordance with Test Methods Promulgated by ASTM Committee E-6." West Conshohocken, PA: ASTM.
E 1592, "Standard Test Method for Structural Performance of Sheet Metal Roof and Siding Systems by Uniform Static Air Pressure Difference." West Conshohocken, PA: ASTM.
E 1646, "Standard Test Method for Water Penetration of Exterior Metal Roof Panel Systems by Uniform Static Air Pressure Difference." West Conshohocken, PA: ASTM.
American Society of Heating, Refrigerating and Air-Conditioning Engineers (ASHRAE). 1997. *ASHRAE Handbook: Fundamentals*. Atlanta, GA: ASHRAE.
Anderson, Brent. 1986. "Waterproofing and the Design Professional." *The Construction Specifier* 39(3)(March): 86–97.
Architectural Graphic Standards. See Ramsey/Sleeper.
Architectural Record. 1989. "Metal Roofing: New Versatility." *Architectural Record* (February):120–23.
Arcom Master Systems. MASTERSPEC®. Basic Sections:
07111, "Composite Sheet Waterproofing," 5/95 ed. Salt Lake City, UT: Arcom.
07112, "Polymeric Sheet Waterproofing," 5/95 ed. Salt Lake City, UT: Arcom.
07121, "Fluid-Applied Waterproofing," 2/95 ed. Salt Lake City, UT: Arcom.
07122, "Fluid-Applied Waterproofing," 2/95 ed. Salt Lake City, UT: Arcom.
07160, "Bituminous Dampproofing," 2/94 ed. Salt Lake City, UT: Arcom.
07161, "Modified Cement Waterproofing," 5/97 ed. Salt Lake City, UT: Arcom.
07162, "Crystalline Waterproofing," 5/97 ed. Salt Lake City, UT: Arcom.
07163, "Metal-Oxide Waterproofing," 5/97 ed. Salt Lake City, UT: Arcom.
07170, "Bentonite Waterproofing," 8/96 ed. Salt Lake City, UT: Arcom.
07190, "Water Repellents," 8/96 ed. Salt Lake City, UT: Arcom.
07210, "Building Insulation," 5/95 ed. Salt Lake City, UT: Arcom.
07241, "Exterior Insulation and Finish System—Class PB," 11/96 ed. Salt Lake City, UT: Arcom.
07242, "Exterior Insulation and Finish System—Class PM," 11/96 ed. Salt Lake City, UT: Arcom.
07270, "Firestopping," 5/92 ed. Salt Lake City, UT: Arcom.
07311, "Asphalt Shingles," 2/94 ed. Salt Lake City, UT: Arcom.
07313, "Metal Shingles," 2/97 ed. Salt Lake City, UT: Arcom.
07315, "Slate Shingles," 2/97 ed. Salt Lake City, UT: Arcom.
07317, "Wood Shingles and Shakes," 2/96 ed. Salt Lake City, UT: Arcom.
07320, "Roof Tiles," 2/96 ed. Salt Lake City, UT: Arcom.
07411, "Manufactured Roof Panels," 11/95 ed. Salt Lake City, UT: Arcom.
07411, "Manufactured Wall Panels," 11/95 ed. Salt Lake City, UT: Arcom.
07460, "Siding," 2/96 ed. Salt Lake City, UT: Arcom.
07511, "Built-Up Asphalt Roofing," 5/96 ed. Salt Lake City, UT: Arcom.
07512, "Built-Up Coal Tar Roofing," 5/96 ed. Salt Lake City, UT: Arcom.
07531, "EPDM Single-Ply Membrane Roofing," 11/95 ed. Salt Lake City, UT: Arcom.
07532, "CSPE Single-Ply Membrane Roofing," 11/95 ed. Salt Lake City, UT: Arcom.
07533, "Thermoplastic Single-Ply Membrane Roofing," 11/95 ed. Salt Lake City, UT: Arcom.
07546, "Coated Polyurethane Foam Roofing," 8/95 ed. Salt Lake City, UT: Arcom.
07551, "APP-Modified Bituminous Membrane Roofing," 2/96 ed. Salt Lake City, UT: Arcom.
07552, "SBS-Modified Bituminous Membrane Roofing," 2/96 ed. Salt Lake City, UT: Arcom.
07553, "Self-Adhering Modified Bituminous Membrane Roofing," 11/96 ed. Salt Lake City, UT: Arcom.
07561, "Hot Fluid-Applied Roofing," 11/96 ed. Salt Lake City, UT: Arcom.
07610, "Sheet Metal Roofing," 8/96 ed. Salt Lake City, UT: Arcom.
07620, "Sheet Metal Flashing and Trim, 8/94 ed. Salt Lake City, UT: Arcom.
07710, "Manufactured Roof Specialties," 5/96 ed. Salt Lake City, UT: Arcom.
07716, "Roof Expansion Assemblies," 8/93 ed. Salt Lake City, UT: Arcom.
07720, "Roof Accessories," 8/97 ed. Salt Lake City, UT: Arcom.
07811, "Sprayed Fire-Resistive Materials," 8/96 ed. Salt Lake City, UT: Arcom.
07821, "Board Fire Protection," 8/96 ed. Salt Lake City, UT: Arcom.
07920, "Joint Sealants," 2/97 ed. Salt Lake City, UT: Arcom.
Asphalt Roofing Manufacturers Association (ARMA). 1990. *Built-Up Roofing Systems Design Guide*. Rockville, MD: ARMA.
———. 1990. *Homeowners Guide to Quality Roofing*. Rockville, MD: ARMA.
———. 1997. *Residential Asphalt Roofing Manual*. Rockville, MD: ARMA.
Asphalt Roofing Manufacturers Association (ARMA) and National Roofing Contractors Association (NRCA). 1988. *Quality Control Recommendations for Polymer ModBit Roofing*. Rockville, MD: ARMA and NRCA.
Baumgardner, Gaylorn. 1986. "Modified Bitumens: Versatility in Reroofing." *Exteriors* 4(3)(autumn):62–64.
Baxter, Dick. 1986. "1001 Reasons Not to Roof Over Wet Insulation." *Roofing Spec* 14(8)(August):27–30.
Belles, Don, and Jesse Beitel. 1988. "Fire Performance of Curtainwalls Questioned." *Exteriors* 6(2)(summer):44–50.
Bordenaro, Michael. 1989. "Specifying and Applying Wet Sealants." *Building Design and Construction* 30(5)(April): 72–74.
Bradford, Dane. 1987. "BUR Repairs Made Easy with Torchable Modified Bitumen." *Roofing Spec* 15(3)(March): 22–24.

Brewer, Wilfred B. 1986. "Avoiding Sealant Failure with Preformed Expanded Foam." *The Construction Specifier* 39(12)(December):23–24.

Brick Institute of America. *Technical Notes on Brick Construction.* Reston, VA: BIA.

No. 7C, "Moisture Control in Brick and Tile Walls: Condensation."

No. 7D, "Moisture Control in Brick and Tile Walls: Condensation Analysis."

Cedar Shake and Shingle Bureau. 1987. *Design and Application Manual for New Roof Construction.* Bellevue, WA: Cedar Shake and Shingle Bureau.

———. 1989. *Design and Application Manual for Exterior and Interior Walls.* Bellevue, WA: Cedar Shake and Shingle Bureau.

———. *Design and Application Manual for Roofing, Shake-over-shake.* Bellevue, WA: Cedar Shake and Shingle Bureau.

———. *Certi-Split Manual of Handsplit Red Cedar Shakes.* Bellevue, WA: Cedar Shake and Shingle Bureau.

Commercial Renovation. 1988. "Sprayed On Roofing Prolongs Life." *Commercial Renovation* 10(2)(April):60.

Copper Development Association. *Copper, Brass, Bronze Design Handbook: Sheet Copper Applications.* Greenwich, CT: Copper Development Association.

Council of American Building Officials. 1983. *One- and Two-Family Dwelling Code.* Falls Church, VA. CABO. (In the process of being transferred to the control of the International Code Council (ICC).)

D'Angelo, Charles, and Owen J. Perryman. 1988. "Insulated Panels and the Pre-Engineered Building Market." *Metal Architecture* 4(10)(October):20.

DeMuth, Jerry. 1986. "White vs. Black: Does Membrane Color Matter?" *Roofing Spec* 14(5)(May):25–28.

Doyle, Margaret. 1987. "Extending a Roof's Life." *Building Design and Construction* 28(6)(June):100–102.

———. 1987. "Keeping a Good Roof Down." *Building Design and Construction* 28(6)(June):108–11.

———. 1988. "Trends in Specifying EIFS." *Building Design and Construction* 29(8)(August):58–62.

Dudley, Hubert T. 1987. "Innovations in Vermiculite Roof Insulation Systems." *The Construction Specifier* 40(11)(November): 62–65.

Dunn, William T., and Robert J. Morin. 1988. "An Attractive Alternative, Metal Roofing." *The Construction Specifier* 41(11)(November):106–11.

Dupuis, Rene. 1987. "Choosing the Right Roof Insulation System." *Exteriors* 5(1)(spring):78–81.

———. 1988. "A Renewed BUR Technology is Ready to Take on the 1990s." In *Handbook of Commercial Roofing Systems*, 16–23. Cleveland OH: Edgell Communications, Inc.

EIFS Industry Members Association. "EIMA Guideline Specification for Exterior Insulation and Finish Systems Class PB." Washington, DC: Exterior Insulation Manufacturing Association.

———. "EIMA Guideline Specifications for Exterior Insulation and Finish Systems Class PM." Washington, DC: Exterior Insulation Manufacturing Association.

———. "EIMA Test Method and Standard 101.86, Standard Test Method for Resistance of Exterior Insulation Finish Systems to the Effects of Rapid Deformation (IMPACT)." Washington, DC: Exterior Insulation Manufacturing Association.

Erwin, Gene. 1986. "Exterior Insulation: Specs with Style." *Exteriors* 4(1) (spring):34–40.

———. 1987. "Trouble-Shooting Guide for Exterior Insulations." *Exteriors* 5(1)(spring):83–86.

Exteriors. 1986. "Roof Awareness." *Exteriors* (autumn):44.

———. 1987. "BUR Design Guide." *Exteriors* 5(2)(summer):22–23.

———. 1987. "Certifying Consultants." *Exteriors* (autumn):41–42.

———. 1987. "Skyline: Switchable Coatings." *Exteriors* 5(3)(autumn):42–44.

———. 1987. "Skyline: Roofing Innovations." *Exteriors* 5(4)(winter):19–22.

———. 1988. "Skyline: Roofing Standards." *Exteriors* 6(1)(spring):38–40.

———. 1988. "Skyline: Fiberboard Problems." *Exteriors* 6(1)(spring):41–42.

———. 1988. "Skyline: Modified Bitumen Guide." *Exteriors* 6(1)(spring): 42–43.

———. 1988. "Portfolio: Florida Auditorium Roof Withstands Extreme Temperature Fluctuations." *Exteriors* 6(1)(spring):68.

Federal Specifications, all: *See* U.S. General Services Administration Specifications Unit.

Francis, Geoffrey V. 1987. "A Practical Approach to Sealant and Joint Design." *Exteriors* 5(1)(spring):64–71.

Fricklas, Dick. 1986. "Structural vs. Architectural Metal Roofs." *Exteriors* (summer):8.

———. 1986. "Fire Codes." *Exteriors* 4(3)(autumn):8.

———. 1987. "Better Products Help Prevent Roof Failures." *Building Design and Construction* 28(11)(November): 60–63.

———. 1988. "New Guidelines for Specifying Vapor Retarders." *Exteriors* 6(1)(spring):8.

Georgia Pacific Corporation. *Vinyl Siding Installation Guide.* Atlanta, GA: Georgia Pacific Corporation.

Gordon, Douglas E., and M. Stephany Stubbs. 1988. "Longevity and Single-Ply Roofing." *Architecture* (January): 95–99.

Hasan, Riaz. 1987. "Eliminating Backout and Pullout of Roof Fasteners." *Exteriors* 5(2)(summer):40–50.

Heineman, Paul. 1987. "Coping with Membrane Roof Penetrations." *The Construction Specifier* 40(11)(November):36–43.

Henshell, Justin. 1987. "Specifying Membrane Roofing: A Systematic Approach." *The Construction Specifier* 40(11)(November):102–108.

Hoffman, H. C. 1986. "Sheet Steels for Low Slope Roofing." *The Construction Specifier* 39(11)(November):56–58.

Hoover, Stephen R. 1987. "Wind Uplift on Roofs: One Perspective." *The Construction Specifier* 40(11)(November):76–79.

—. 1987. "Single Ply Roofing: Exploring the Options." *The Construction Specifier* 40(12) (December):92–103.

Huettenrauch, Clarence. 1985. "Building Moisture and Roof System Destruction." *The Construction Specifier* 39(11)(November):15–16.

———. 1987. "Roof Supports and Roof Leaks: Avoiding the Inevitable." *The Construction Specifier* 40(7)(July):27–28.

Israel, Sheldon B. 1986. "Reroofing: Principles for Success." *Exteriors* 4(3)(autumn):50–53.

———. 1987. "The Nuts and Bolts of Peripheral Roofing Components." *The Construction Specifier* 40(11)(November):44–49.

———. 1987. "Roof Renovation: A Vi-able Alternative." *Exteriors* 5(3)(autumn):58–62.

Jentsch, Bryan C. 1988. "Pre-Insulated Panels Providing Excellent Economic Value." *Metal Architecture* 4(10)(October):21.

Lead Industries Association (LIA). *Lead Roofing and Flashing*. New York: LIA.

———. *Sheet Lead: the Protective Metal*. New York: LIA.

Legatski, Leo A. 1987. "Insulating Roof Decks with Cellular Concrete." *The Construction Specifier* 40(11)(November):56–61.

Litvin, Albert. 1968. Portland Cement Association Research and Development Laboratories, Development Department Bulletin D137, "Clear Coatings for Exposed Architectural Concrete." (Reprinted from the May 1968 *Journal of the PCA Research and Development Laboratory* 10(2):49–57.)

Loza, Don. 1985. "Using Exterior Insulating Systems for Renovation." *The Construction Specifier* 38(7)(July):54–56.

Mack, Robert C. 1975. *The Cleaning and Waterproof Coating of Masonry Buildings*. Preservation Brief No. 1. Washington, DC: Technical Preservation Services, U.S. Department of the Interior.

———. 1978. "The Cleaning and Waterproofing of Masonry Buildings." *The Construction Specifier* 31(11)(November):65–70.

McDonald, Timothy B. 1988. "Detailing, Specifying Insulation." *Architecture* (July)137–138.

———. 1988. "Selecting Below-grade Waterproofing." *Architecture* (December):135–137.

Metal Architecture. 1987. "Guide to Metal Wall and Roof Panels." *Metal Architecture* 3(12)(December):20, 22–23, 28–31, 38.

———. 1988. "Guide to Pre-Insulated Panels." *Metal Architecture* 4(4)(April):40–41, 43–46.

———. 1988. "Selecting Fasteners: Manufacturer's Advice." *Metal Architecture* 4(12)(December):26–32.

Miles, J. D., III. 1987. "Tracking the Elusive Leak." *Roofing Spec* 15(3)(March):19–21.

Miller, Michael H. 1984. "Selecting the Right Sealant." *The Construction Specifier* 37(1)(January):72–76.

Nadel, Toby. 1986. "The 25 Percent Solution." *Roofing Spec* 14(5)(May):35–36.

National Concrete Masonry Association (NCMA). *TEK Manual for Concrete Masonry Design and Construction*. Herndon, VA: NCMA.

National Roofing Contractors Association (NRCA). 1985. *Quality Control in the Application of Built-Up Roofs*. Rosemont, IL: NRCA.

———. 1990. *The NRCA Roofing and Waterproofing Manual*. 4th ed. Rosemont, IL: NRCA.

National Slate Association (NSA). 1926 *Slate Roofs*. VT: NSA. (Reprinted in 1977 by Vermont Structural Slate Co., Inc., Fair Haven, VT.)

National Tile Roofing Manufacturers Association (NTRMA). 1987. *Concrete and Clay Tile Installation Manual*. Los Angeles: NTRMA.

———. 1999. *Standard Applications of Clay and Concrete Roof Tile*. Los Angeles: NTRMA.

Nicastro, David H. 1989. "Parameters for Comparing High-Performance Sealants." *The Construction Specifier* 42(4)(April):122–125.

Nimtz, Paul D. 1987. "Comparing Structural and Architectural Metal Roofing." *Exteriors* (winter): 46–49.

———. 1988. "Regular Inspections Can Extend Standing Seam Roof Lifecycle." *Exteriors* 4(9)(September):42.

Perry, Dale C. 1986. "Standing Seam Metal Roofs in Retrofit." *The Construction Specifier* 39(11)(November):50–86.

Petersen, Wayne. 1987. "Foamed Plastic Roof Insulation." *The Construction Specifier* (November):66–74.

Peterson, Arnold. 1987. "Testing Single Ply's Weatherability." *The Construction Specifier* 40(11)(November):21–25.

Poore, Patricia. 1983. "What Most Roofers Don't Tell You About Traditional and Historic Roofs." *The Old-House Journal* (April):56–58.

Porcher, Joel P., Jr. 1988. "What You Should Keep in Mind When Specifying a Modified Bitumen." In *Handbook of Commercial Roofing Systems*, 32–40. Cleveland OH: Edgell Communications.

Portland Cement Association (PCA). 1965. *Moisture Migration—Concrete Slab-on-Ground Construction*. Skokie, IL: PCA.

———. 1986. *Effects of Substances on Concrete and Guide to Protective Treatments*. Skokie, IL: PCA.

Ramsey/Sleeper, The AIA Committee on Architectural Graphic Standards. 1998. *Architectural Graphic Standards*. 8th ed. New York: John Wiley & Sons, Inc.

Rossiter, Walter J., Jr. 1987. "Nondestructive Methods for Inspecting Single Ply Roofing Membrane Seams." *The Construction Specifier* 40(11)(November):92–100.

———. 1988. "Effect of Application Variables on Bond Strength of EPDM Seams." In *Handbook of Commercial Roofing Systems*, 41–47. Cleveland OH: Edgell Communications.

Russo, Michael. 1987. "New Methods Needed to Measure 'Thermal Drift.' " *Exteriors* 5(3)(autumn):68–74.

———. 1987. "A Critical Look at Thermoplastic Roofing." *Exteriors* 5(4)(winter):24–27.

Scharfe, Thomas R. 1987. "Owners, Designers Discover Benefits of Metal Roofs." *Building Design and Construction* 28(6)(June):114–116.

———. 1988. "New Metal Coating Technologies Enhance Design Opportunities." *Building Design and Construction* 29(6)(June):86–89.

———. 1989. "New Technologies Increase Metal Roof Design Options." *Building Design and Construction* 29(6)(June):106–109.

Sealant Engineering and Associated Lines (SEAL). 1987. *SEAL Guide Specification for Sealants and Caulking*. San Diego, CA: SEAL.

Sealant, Waterproofing and Restoration Institute (SWRI). 1995. *Sealants 1995: The Professional's Guide*. Kansas City, MO: SWRI.

Sheet Metal and Air Conditioning Contractors National Association (SMACNA). 1993. *Architectural Sheet Metal Manual*. 5th ed. Chantilly, VA: SMACNA.

Shell, Burt. 1988. "Standing Seam Roof Continues Inroads in Educational Market." *Metal Architecture* 4(4):10–13.

Shmedin, G. 1987. "Bibliography for Slate Roofing." *Association for Preservation Technology Newsletter* 4(2):89.

Simmons, H. Leslie. 1989. *Repairing and Extending Weather Barriers*. New York: Van Nostrand Reinhold.

Single Ply Roofing Institute (SPRI). *Flexible Membrane Roofing: A Pro-*

fessional's Guide to Specifications. Needham, MA: SPRI.

———. *Wind Design Standard for Ballasted Single Ply Roofing Systems.* Needham, MA: SPRI.

———. *Wind Design Standard for Fully-Adhered and Mechanically Fastened Roof Systems.* Needham, MA: SPRI.

Smith, George A. 1987. "Fastener Corrosion to be Judged by New Standard." *Roofing Spec* 15(1)(January):21–22.

Specialty Steel Industry of North America (SSINA). *Stainless Steel: Suggested Practices for Roofing, Flashing, Copings, Fascias, Gravel Stops, and Drainage.* Washington, DC: SSINA.

———. "Stainless Steel Fasteners, A Systematic Approach to Their Selection." Washington, DC: SSINA.

———. "Design Guidelines for the Selection and Use of Stainless Steel." Washington, DC: SSINA.

Stephenson, Fred. 1986. "Clip Design Critical in Standing Seam Roofing." *Exteriors* 4(1)(spring):60–65.

———. 1986. "Specifying Standing Seam Roofing Systems for Conventional Buildings." *The Construction Specifier* 39(11)(November):36–43. (This article was reprinted. from the March 1984 issue of *Exteriors*, which was then called *Roof Design*.)

Sweetser, Sarah M. 1978. *Roofing for Historic Buildings.* Washington, DC: Preservation Brief No. 4. Technical Preservation Services, U.S. Department of the Interior.

Thomas, Robert, Jr. 1991. "Nosweat." *The Construction Specifier* 44(6)(June): 74–79.

Thomsett, Michael C. 1985. "Should You Hire a Consultant?" *The Construction Specifier* 38(10)(October):23–24.

Ting, Raymond. 1986. "Metal Panel Behavior in Exterior Wall Design." *Exteriors* (autumn):65–69.

———. 1989. "Performance Parameters of Composite Foam Panels." *Metal Architecture* 5(9)(September):7–8,73–74.

Tisthammer, Thomas. 1988. "The Case for Specifying Sprayed Polyurethane Systems." In *Handbook of Commercial Roofing Systems*, 51–52. Cleveland OH: Edgell Communications.

Tobiasson, Wayne. 1987. "Vents and Vapor Retarders for Roofs." *Proceedings of the Symposium on Air Infiltration, Ventilation, and Moisture Transfer.* Washington, DC: Building Thermal Envelope Coordinating Council. (Reprinted. in the November 1987 *The Construction Specifier* 40(11):80–90.)

Tobiasson, Wayne, and Marcus Harrington. *Vapor Drive Maps of the U.S.A.* Washington, DC: Cold Regions Research and Engineering Laboratory (CRREL), Corps of Engineers.

Underground Space Center. 1988. *Building Foundation Design Handbook.* Minneapolis, MN: Underground Space Center.

Underwriters Laboratories, Inc. *Building Materials Directory—Class A, B, C: Fire and Wind Related Deck Assemblies.* Northbrook, IL: Underwriters Laboratories, Inc.

———. *Fire Resistance Directory—Time/ Temperature Constructions.* Northbrook IL: Underwriters Laboratories, Inc.

———. 1993. 55A, "Materials for Built-Up Roof Coverings." Northbrook IL: Underwriters Laboratories, Inc.

———. 1994. 580, "Tests for Uplift Resistance of Roof Assemblies." Northbrook IL: Underwriters Laboratories, Inc.

———. 1997. 790, "Tests for Fire Resistance of Roof Covering Materials." Northbrook IL: Underwriters Laboratories, Inc.

———. 1995. 997, "(Type I) Wind Resistance of Prepared Roof Covering Materials." Northbrook IL: Underwriters Laboratories, Inc.

U.S. Department of Agriculture (USDA), Forest Service, Forest Products Laboratory. 1989. *Wood-Frame House Construction.* Washington, DC. USDA. For technical information: Forest Products Laboratory, One Gifford Pinchot Drive, Madison WI 53705 (tel: [608] 231-9223). For copies: U.S. Department of Commerce, National Technical Information Service, 5285 Port Royal Road, Springfield, VA 22161 (tel:[703] 605-6000). Order No. PB90-180662.

U.S. Department of the Navy, Naval Facilities Engineering Command. 1983 (February). Guide Specifications Section 07140, "Metallic Oxide Waterproofing." Washington, DC: Department of the Navy.

———. 1996 (January). Guide Specifications Section 07410, "Metal Roof and Wall Panels." Washington, DC: Department of the Navy.

U.S. General Services Administration Specifications Unit. Federal Specifications:

A-A-272A, "Caulking Compounds." Washington, DC: GSA Specifications Unit.

A-A-1556A, "Sealing Compound (Elastomeric Joint Sealant)." Washington, DC: GSA Specifications Unit.

HH-I-1972, "Insulation Board, Thermal, Polyurethane or Polyisocyanurate, Faced with Aluminum Foil on Both Sides of the Foam." Washington, DC: GSA Specifications Unit.

SS-S-200E(2), "Sealants, Joint, Two-Component, Jet-Blast Resistant, Cold-Applied, for Portland Cement Concrete Pavement." Washington, DC: GSA Specifications Unit.

Vinyl Siding Institute (VSI). *Vinyl Siding Installation: A How-To Guide.* New York: VSI.

Vonier, Thomas. 1988. "Metal Roofing Systems: Scraping Away the Impressions of the Past." In *1988 Handbook of Commercial Roofing Systems*, 48–50. Cleveland OH: Edgell Communications.

Warford, Milan. 1988. "What Aging Does to Roof Materials." *The Construction Specifier* 41(2)(February):29–30.

Western Red Cedar Lumber Association. *Specifiers' Guide.* Vancouver, BC: Western Red Cedar Lumber Association.

CHAPTER 8

Doors and Windows

Introduction

Applicable *MasterFormat*™ Sections

Metal Doors and Frames

Wood Doors

Aluminum Entrances and Storefronts

Aluminum Windows and Sliding Glass Doors

Wood Windows and Sliding Glass Doors

Storm Windows and Doors

Door Hardware

Glazing

Glazed Aluminum Curtain Walls

Additional Reading

Acknowledgments and References

Introduction

Exterior doors and aluminum entrances provide access openings into a building. They minimize heat gain and loss through these access openings and keep the weather outside, and they provide privacy for the building occupants and security for persons and property. They also help control the passage of sound through access openings.

Interior doors control passage between interior spaces. They also provide privacy and security, reduce sound transmission between spaces, and screen storage and utility areas from view.

Windows, storefronts, and curtain walls keep out the weather while admitting light and providing a view. Windows also facilitate ventilation.

Applicable *MasterFormat*™ Sections

The following *MasterFormat* Level 2 sections are applicable to this chapter.

08050 Basic Door and Window Materials and Methods
08110 Metal Doors and Frames
08200 Wood and Plastic Doors
08300 Specialty Doors
08400 Entrances and Storefronts
08500 Windows
08700 Hardware
08800 Glazing
08900 Glazed Curtain Wall

8.1 Metal Doors and Frames

This section discusses interior and exterior doors, frames, and associated sidelights formed from carbon steel, stainless steel, aluminum, and copper alloy sheets. Extruded aluminum and other types of doors and frames are discussed in other sections of this chapter.

8.1.1 GENERAL REQUIREMENTS

The two types of formed door and frame units are *standard* and *custom*. Standard units are mass produced and sold to retail outlets such as lumberyards and home centers. Standard units are also made for a particular project where they have been specified. Custom units are manufactured especially for a particular project to suit an owner's or an architect's design. Both types should meet the same standards of quality. Formed carbon steel, stainless steel, and aluminum doors may be either standard or custom. Copper-alloy doors and frames are custom.

8.1.1.1 Standards

Both standard and custom carbon steel doors and frames should comply with the American National Standards Institute/ Steel Door Institute's publication ANSI A250.8 (formerly ANSI/SDI-100), "Recommended Specifications for Standard Steel Doors and Frames." In addition, a good guide for custom steel door and frame work is the *Hollow Metal Manual*, a collection of National Association of Architectural Metal Manufacturers (NAAMM) publications related to steel doors and frames. It includes the publications of NAAMM's Hollow Metal Manufacturers Association (HMMA) division and the NAAMM *Metal Finishes Manual* series.

In most cases, formed aluminum, stainless steel, and copper-alloy doors and frames should also comply with these standards. In addition, they should conform to the applicable standards of the appropriate organizations, which include:

Aluminum Association (AA)
American Architectural Manufacturers Association (AAMA)
American Iron and Steel Institute (AISI)
American National Standards Institute (ANSI)
American Society for Testing and Materials (ASTM)
Copper Development Association (CDA)
National Association of Architectural Metal Manufacturers (NAAMM)

Formed doors and frames should conform with the following standards:

- Fire doors and frames: National Fire Protection Association (NFPA) 80; tested and labeled by Underwriters Laboratories (UL), Factory Mutual (FM), or another independent testing laboratory
- Sound-rated assemblies: ASTM Standard E 90
- Insulated doors: ASTM Standard C 236
- Door hardware: ANSI Standards A156.1 through A156.21

8.1.1.2 Materials

Refer to Chapter 5 for a general discussion about the metals addressed in this section.

Carbon steel for *interior* door facings and frames should be commercial-quality steel sheet and strip conforming with ASTM Standards A 568 and A 569. It should have a *stretcher-level* standard of flatness.

Carbon steel for *exterior* door facings and frames should be hot-dip, galvanized, commercial-quality carbon steel sheets complying with ASTM Standard A 653. The zinc coating should be not less than class G60. Sheets should be *mill phosphatized* and have a *stretcher-level* standard of flatness.

Sheet aluminum is often used for door faces. Sheet aluminum and bent plate components should comply with ASTM Standard B 209.

Stainless steel is used to make formed doors and frames, sidelights, and storefronts. Door facings are usually flat sheets, but patterned sheets are also used.

Copper-alloy materials are used to make formed doors and frames for both exterior and interior locations. Most door facings on such doors are also copper-alloy sheets.

Concealed components, such as subframes, stiffeners, edge channels, reinforcements, and the like should be fabricated from the same materials as the exposed portions of the door or frame.

8.1.1.3 Finishes

Carbon steel doors are either given a factory finish, such as a baked enamel or a fluorocarbon finish, or are painted in the field (see Section 9.15).

The mill finish is usually left on concealed surfaces and sometimes even on the exposed surfaces of aluminum doors. However, most exposed portions of aluminum entrance doors are anodized (see Section 5.4).

Most exposed stainless steel is polished, but organic coatings are also applied to some assemblies. Most exposed copper alloys are given a patina or architectural-bronze finish. Exposed fasteners are usually finished to match the adjacent metal.

8.1.1.4 Supports, Anchors, Fasteners, and Flashing

Supports, anchors, fasteners, and flashing materials should be compatible with the material being anchored or fastened, and with the substrate, so that galvanic and chemical reactions do not occur between different materials. Carbon steel should be anchored and fastened with galvanized steel, stainless steel, or cadmium-plated steel devices; aluminum and stainless steel with nonmagnetic stainless steel, cadmium-plated steel, or galvanized steel; copper alloy with copper alloy.

Inserts for use in concrete may be cast iron, malleable iron, or hot-dip galvanized steel.

8.1.1.5 Weatherstripping

The function of weatherstripping is to control air infiltration, prevent rain penetration, minimize dust infiltration, and withstand atmospheric conditions. It must be durable under extended use, resist corrosion, and be readily replaceable.

Refer to Section 8.7 for an additional discussion of weatherstripping.

8.1.2 CARBON STEEL DOORS

Carbon steel doors and frames are often called *hollow metal.* Many members of the industry that makes them are represented by the Hollow Metal Manufacturers Association (HMMA), a division of the National Association of Architectural Metal Manufacturers (NAAMM). Its manual of practice is the *Hollow Metal Manual.*

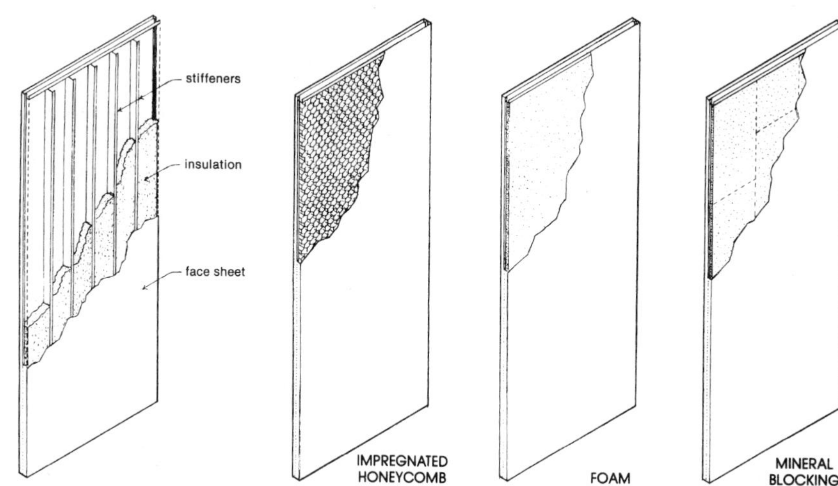

FIGURE 8.1-1 Steel door construction. (Reprinted with permission of NAAMM, the National Association of Architectural Metal Manufacturers.)

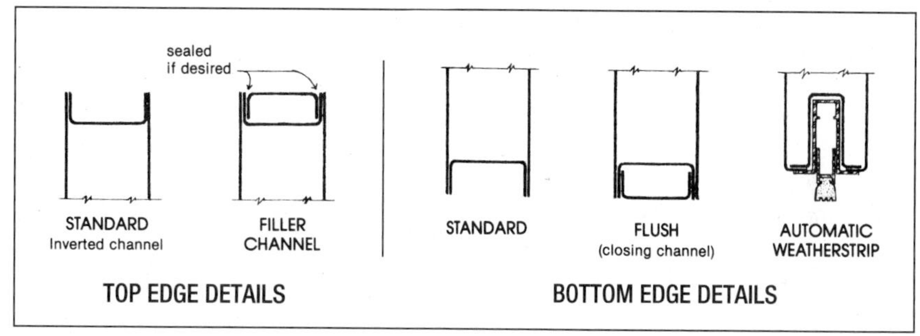

FIGURE 8.1-2 Bottom and top details of a typical steel door. (Reprinted with permission of NAAMM, the National Association of Architectural Metal Manufacturers.)

8.1.2.1 Construction

Carbon steel doors consist of a core, faces, tops, and bottoms. There are two types of core construction. In one type, cores are hollow and braced with steel stiffeners. Most are filled with the manufacturer's standard fireproof, rotproof, sound-deadening material. Insulated doors are filled with glass fiber or plastic foam insulation.

In the second form of core construction, a door's face sheets are laminated to a core of impregnated kraft paper honeycomb, plastic foam insulation, or mineral fiber blocks (Fig. 8.1-1).

The tops and bottoms of standard doors are finished with a recessed, welded-in closure. Exterior doors have tops and bottoms formed flush as an integral part of the door construction or are finished with a welded-in inverted channel. Bottoms may also be fitted with a recess for weatherstripping (Fig. 8.1-2).

Doors should be sized so that the *maximum* clearances between the door and frame, between doors in a pair, and between door and sill are in accordance with the recommendations of the Steel Door Institute's (SDI) publication ANSI/SDI-250.8. These clearances, which are subject to a tolerance of ± 1/32 in. (0.8 mm) are shown in Figure 8.1-3. The security requirements discussed in Chapter 13 may dictate less space between door and frame than that shown in Figure 8.1-3. Codes and the National Fire Protection Association's (NFPA) NFPA 80, which is an applicable requirement in most codes for

FIGURE 8.1-3 Door Clearances

Joint	Clearance, in. (mm)
Door head to frame	1/8 (3.2)
Door jambs to frame	1/8 (3.2)
Between meeting edges of pairs (unrated doors)	1/8–1/4 (3.2–6.4)
Between meeting edges of pairs (fire rated doors)	1/8 (3.2)
Door bottom	3/4 (19.1)
Face of door to stop	1/16 (1.6)

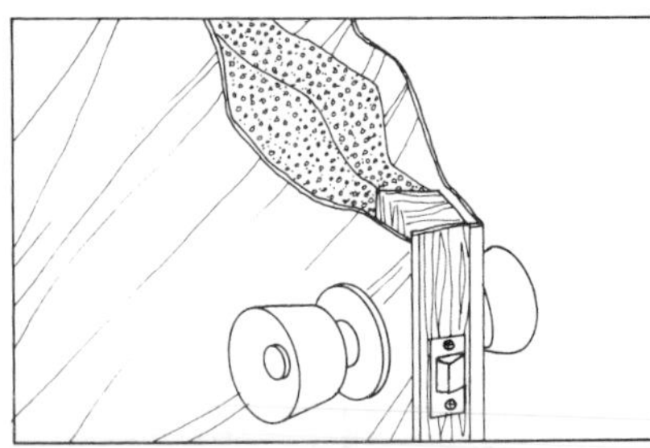

FIGURE 8.1-4 Plastic foam insulation fills the inner core of insulated doors.

fire-rated doors and frames, may require closer spacings and less, or even no, tolerances.

8.1.2.2 Insulated Doors

Insulated doors have an inner core of plastic foam insulation, have high R-values (R-8 to R-15), and are even more effective than the combination of prime and storm doors in reducing heat flow (Fig. 8.1-4). They are often installed with magnetic weatherstrips that are relatively airtight and minimize heat loss by infiltration.

8.1.2.3 Door Types, Styles, and Designs

Most of the steel door types, styles, and designs available today fall into the classification system described in ANSI/SDI 250.8 and SDI 108. SDI classifies doors in three grades:

- Grade I: Standard
- Grade II: Heavy Duty
- Grade III: Extra Heavy Duty

In Grades I and II there are two models:

- Model 1: Full Flush Design
- Model 2: Seamless Design

In Type III, there are five models:

- Models 1 and 1A: Full Flush Design
- Models 2 and 2A: Seamless Design
- Model 3: Stile-and-Rail Flush Panel

Figure 8.1-5, which is derived from SDI 108, shows the uses of the three grades recommended by SDI. It also shows eight basic door designs, their recommended uses, and the abbreviations and acronyms used to identify them. Steel doors that imitate the look of wood stile-and-rail panel doors are also available.

8.1.3 CARBON STEEL FRAMES

Carbon steel frames are brake-formed of steel sheets into one of two basic shapes (Fig. 8.1-6). However, many variations of these basic shapes are made, including cased opening frames that have no stops. Some special frames have tapered or molded shapes. Regardless of their shape or size, frames should be rigid, neat in appearance, and free from defects, warp, and buckle. When possible, they should be fitted and assembled in the manufacturer's plant.

8.1.3.1 General Requirements

The metal thickness of frames is dependent on the door grade and model, as follows:

Grade I: 18 or 16 gauge (1.3 or 1.6 mm)
Grade II: 16 gauge (1.6 mm)
Grade III, Models 1, 2, and 3: 16 or 14 gauge (1.6 or 1.9 mm)
Grade III, Models 1A and 2A: 14 or 12 gauge (1.9 or 2.7 mm)

Except for those that are weatherstripped, or are for double-acting, smoke, or fire doors, frames should be drilled and have three rubber or plastic *silencers* (small resilient bumpers) installed on the strike jambs. Frames for double doors should have silencers on the head.

Some codes and regulations require that interior doors in hospitals, nursing homes, and some other medical buildings have *terminated stops* (also called *cutoff* or *sanitary* stops) (Fig. 8.1-7). These must be omitted, of course, from fire, smoke, lead-lined, and sound-rated doors. The bottom of such a terminated stop is closed with a filler plate.

Plaster guards or mortar boxes are required on frames where plaster or mortar might intrude into the operating hardware. These are sheet metal boxes welded to the frame.

Glazing beads used to hold glass panels in steel frames should be removable. They may be either mitered or butted in the corners, but should be factory fitted. Removable beads should occur on the security side of openings so that they cannot be removed for access.

Frames wider than 49 in. (1245 mm) should be reinforced with a 14-gauge (1.9 mm) steel channel. Lead-lined frames and frames for power-operated doors usually also require reinforcement to support the loads.

8.1.3.2 Frame Types

There are two basic frame types, *welded* and *knockdown*, and several special frame types, including *drywall slip-on*, *storefront*, and *lead-lined* frames.

In *welded* frames, corners and intersections are fully welded or attached together with concealed welded splice plates. Corners should be mitered.

Knockdown frames are made with mechanical joints. Most, but not all, have mitered corners. They are similar in design and appearance to welded frames, except that they are delivered to a project site disassembled for field assembly.

Drywall slip-on frames are designed for installation in drywall partitions after the partitions are in place. Several designs are available.

Storefront-type frames are essentially welded frames. Frame sections that are too large to be shipped in one piece are shipped in several pieces and assembled in the field.

Lead-lined frames are fabricated with lead linings or are of a modified design that will receive lead sheeting in the field. They are used in x-ray rooms and wherever else radiation is a hazard.

8.1.4 FORMED STAINLESS STEEL AND COPPER-ALLOY DOORS AND FRAMES

Most stainless steel and copper-alloy doors and frames are custom-made formed shapes. Therefore, most of the doors and frame designs available for carbon steel doors are also available in stainless steel or copper alloy.

8.1.5 INSTALLATION

The finest door and frame design and the highest quality of manufacture will not compensate for poor installation. Weathertightness, although simple to achieve, can be ensured only if the installation is made by experienced workers following the guidelines and instructions supplied by the manufacturer and the appropriate industry standards. A rough opening should be prepared to receive each unit in such a manner that it can be installed, and will finish out square, plumb, level, straight, and true. The design of most assemblies provides for minor adjustments at the job site, but no unit will operate or weather properly if it is twisted and misaligned during installation.

A properly selected and manufactured door and frame will be weather resistant as a unit. Other things must be done, however, to finish with a completely weather-

FIGURE 8.1-5 Basic Door Grades and Applications

Building Types	Standard Steel Door Grades: Grade I Standard Duty, 1¾ in. (44.5 mm) or 1⅜ in. (35 mm)	Grade II Heavy Duty, 1¾ in. (44.5 mm) only	Grade III Extra Heavy Duty, 1¾ in. (44.5 mm) only	Door Thickness: 1¾ in. (44.5 mm)	1¾ in. (44.5 mm) or 1⅜ in. (35 mm)	Door Design Nomenclature: F	G	V	FG	NL	L
Apartment											
Main entrance			•	•			•		•	•	
Unit entrance	•	•		•		•					
Bedroom	•				•	•					
Bathroom	•				•	•					
Closet	•				•	•					•
Stairwell		•	•	•				•			
Dormitory											
Main entrance			•	•			•		•	•	
Unit entrance	•	•		•		•					
Bedroom	•			•		•					
Bathroom	•			•		•					
Closet	•				•	•					•
Stairwell		•	•	•				•			
Hotel, Motel											
Unit entrance	•	•		•		•					
Bathroom	•				•	•					
Closet	•				•	•					•
Stairwell		•	•	•				•			
Storage and utility	•	•		•		•					•
Hospital, Nursing Home											
Main entrance			•	•			•		•	•	
Patient room		•		•		•					
Stairwell		•	•	•				•			
Operating and exam room		•	•	•		•					
Bathroom	•			•		•					
Closet	•				•	•					•
Recreation		•		•		•		•			
Kitchen		•	•	•				•			
Industrial											
Entrance and exit			•	•			•		•	•	
Office	•	•		•		•	•				
Production			•	•			•				
Toilet		•	•	•		•					•
Tool			•	•		•					
Trucking			•	•			•				
Monorail			•	•		•	•				
Office											
Entrance			•	•			•		•	•	
Individual office	•			•	•	•	•				
Closet	•				•	•					•
Toilet		•	•	•		•					
Stairwell		•	•	•				•			
Equipment		•	•	•		•					
Boiler		•	•	•		•					•
School											
Entrance and exit			•	•			•		•	•	
Classroom		•		•			•			•	
Toilet		•	•	•		•					
Gymnasium		•	•	•		•		•			
Cafeteria		•	•	•			•				
Stairwell		•	•	•				•			
Closet	•			•		•					•

Basic Door Design

Flush (F)

Half glass (G)

Narrow light (NL)

Vision light (V)

Bottom louver (L)

Dutch door (D)

Full glass (FG)

Full louver (FL)

(Courtesy Steel Door Institute.)

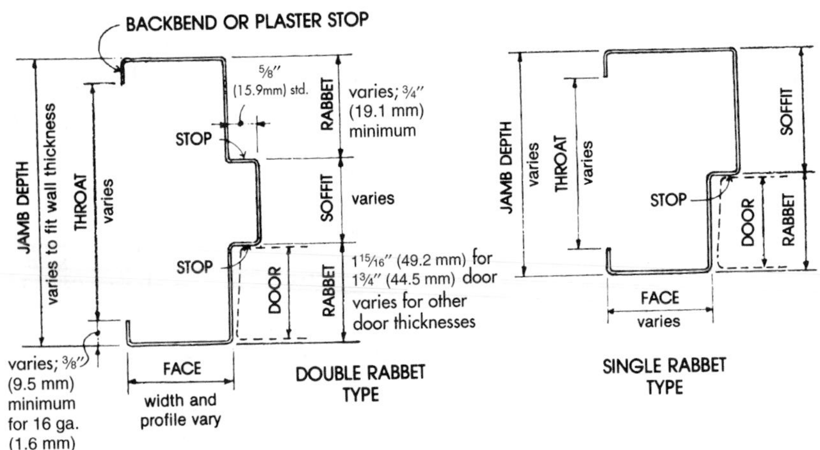

FIGURE 8.1-6 Profiles of basic steel door frames and their parts. (Reprinted with permission of NAAMM, the National Association of Architectural Metal Manufacturers.)

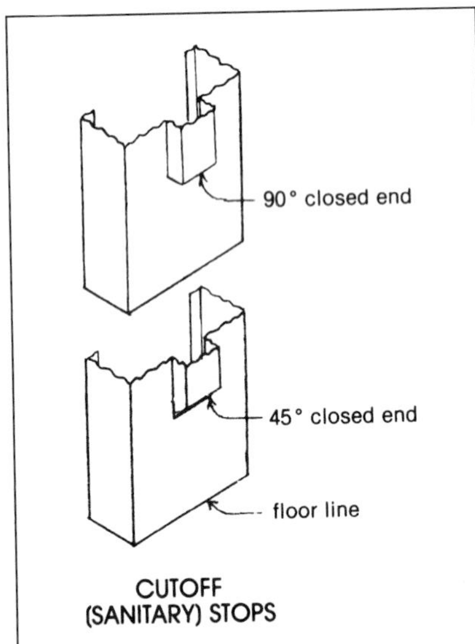

FIGURE 8.1-7 Terminated stops. (Reprinted with permission of NAAMM, the National Association of Architectural Metal Manufacturers.)

tight installation. For example, if the construction is insulated, the space between the rough opening and the unit should be filled with insulation. In addition, the joints between units and adjacent materials must be closed with a sealant, trim members, or both, to prevent the passage of air, dust, and water around the frame.

Door frames are seldom designed to support loads other than themselves. Therefore, unless the frame is specially reinforced to support other loads, a separate lintel or other support is usually required.

Door frames should be supported around the entire perimeter and anchored securely to the supporting construction in a firm and rigid position. They should also be flashed at their heads and sills to provide a path back to the exterior for water that finds its way into the wall.

Some metals, such as aluminum and stainless steel, are compatible and can be used in contact with each other. Other metals, however, react with one another chemically or electrolytically and should be insulated from direct contact with each other by waterproof, nonconductive materials, such as neoprene, waxed paper, or coated felt. Such dissimilar metals that are located where water passing over them may contact another surface should be painted to prevent staining. Refer to Section 5.4 for a discussion about the effect of dissimilar metal materials.

Concealed aluminum in contact with concrete, masonry, or an absorbent material, such as wood, paper, or insulation, should be protected by coating either the aluminum or the adjacent material with a bituminous or aluminum paint, or by coating the aluminum with a zinc chromate primer, to minimize the chance of chemical corrosion of the aluminum by acids, alkalies, and salts leached out of the adjacent materials. Creosote and tar coatings, which may damage the aluminum, should not be used.

Sealants should be used in locations where frames adjoin other materials to

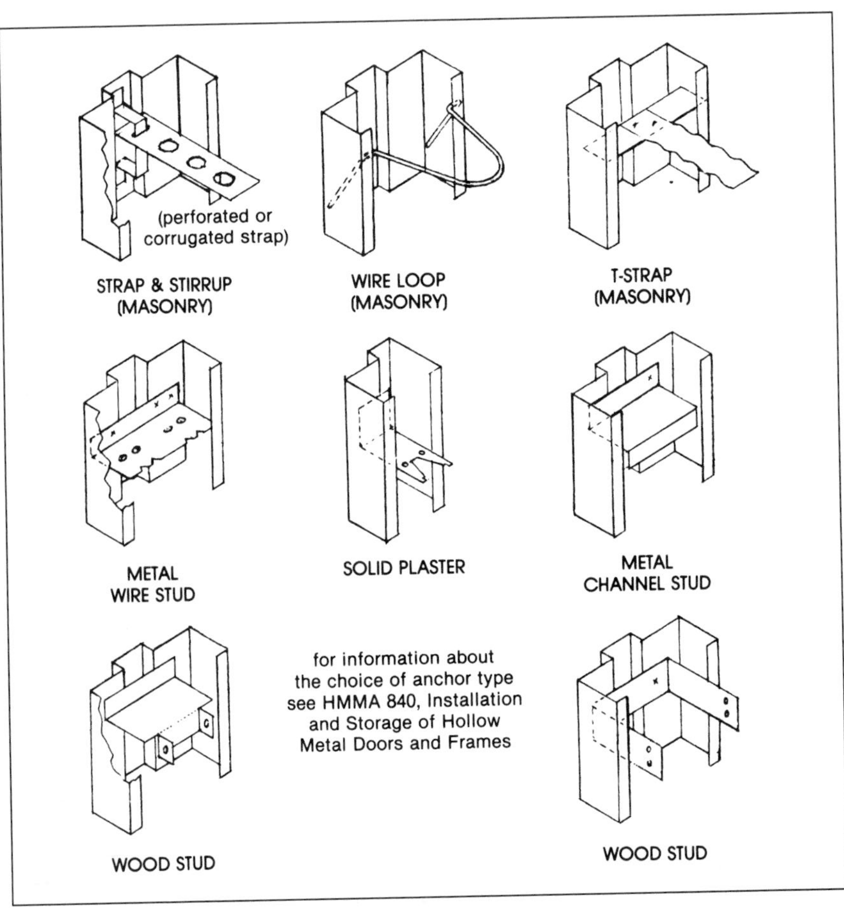

FIGURE 8.1-8 Common types of jamb anchors for conditions where the frames are installed before the wall. (Reprinted with permission of NAAMM, the National Association of Architectural Metal Manufacturers.)

make the joints there airtight and watertight.

8.1.5.1 Formed Frames and Doors

Formed door frames should be installed in accordance with SDI 105. Fire-rated frames should be installed in accordance with NFPA Standard No. 80.

When practicable, frames should be placed before the surrounding construction is built. Each frame should be anchored with jamb anchors near the top and bottom and not more than 24 in. (610 mm) apart in between (Fig. 8.1-8). Jambs should be anchored at the floor with adjustable floor anchors (Fig. 8.1-9).

Anchors in masonry should extend at least 8 in. (203 mm) into the masonry. Anchors for stud walls should be clips designed to attach to the types of studs used. Several different methods are used to anchor frames in previously placed masonry or concrete (Fig. 8.1-10).

Carbon steel and stainless steel frames in masonry should be filled solid with mortar. Copper-alloy and aluminum frames, however, should not be filled with mortar, because the mortar will corrode the metal.

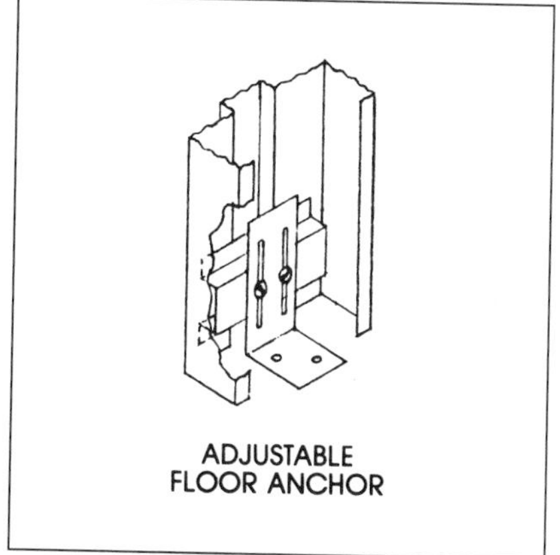

FIGURE 8.1-9 Adjustable floor anchor. (Reprinted with permission of NAAMM, the National Association of Architectural Metal Manufacturers.)

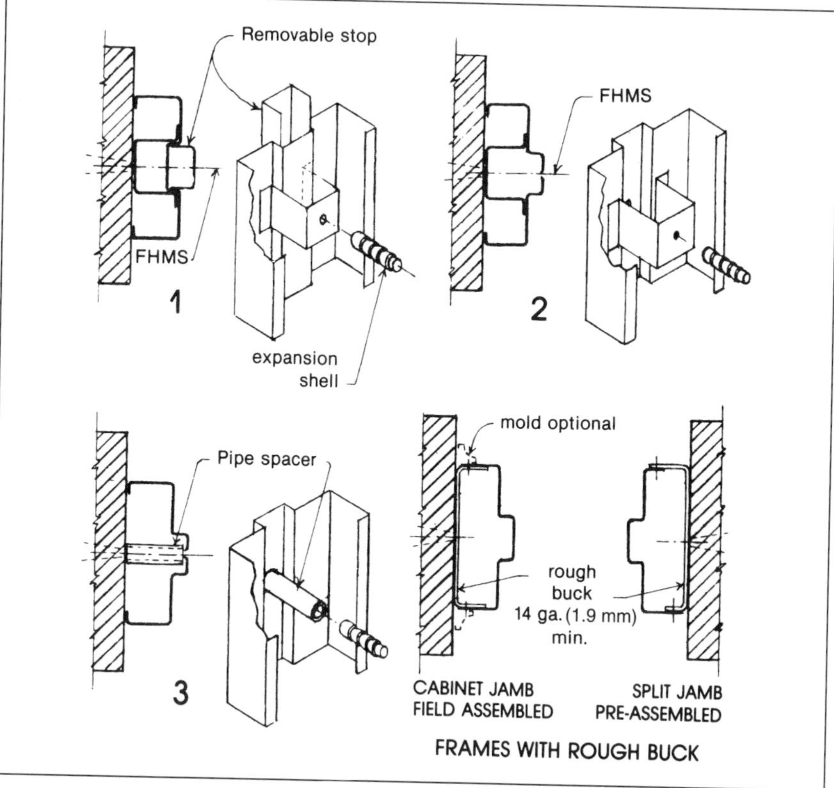

FIGURE 8.1-10 Anchorage of frames in previously placed openings. (Reprinted with permission of NAAMM, the National Association of Architectural Metal Manufacturers.)

8.2 Wood Doors

Because of its ready availability, easy workability, and insulating properties, wood was a very early door material. Before the advent of plywood, doors were assembled from solid wood members. The methods used to conceal joints resulted in the familiar panel treatment of surfaces. Later designs substituted thinner wood or plywood panels in a supporting framework to make interior panel doors, and glass to make sash doors.

Later, the mass production of plywood brought about the widespread use of flush doors, which now account for most wood doors used in all types of construction.

The following discussion outlines the variety of types and uses, methods of manufacture, standards, certification of quality, and installation of wood doors.

8.2.1 TYPES AND USES

Wood doors may be classified according to their construction as follows:

- *Flush:* Plywood, hardboard, or plastic-laminate face panels bonded to solid or hollow cores
- *Stile-and-rail:* Vertical and horizontal members enclosing wood, glass, or louver inserts, creating panel, sash, storm, screen, or louver designs
- *Accordion fold:* Assembled from narrow slats that give a long, drapelike appearance

Figure 8.2-1 summarizes common door types, typical sizes, and applicable industry standards that regulate their construction and manufacture. Standards for various door types are based on construction grades, face material selection, core construction, and vertical and horizontal edge, and core interface. Wood accordion folding doors are made of a variety of woods and are covered with vinyl, plastic laminate, wood veneer, fabrics, and other materials. Plastic accordion fold doors

FIGURE 8.2-1 Summary of Wood Doors

Type and Industry Standard	Typical Sizes	Description
Flush NWWDA I.S.1-A, NWWDA I.S.1-R, AWI, WIC	*Hollow:* 1⅜ in. (35 mm) and 1¾ in. (44.5 mm) thick 6 ft. 8 in. (2032 mm) and 7 ft. 0 in. (2134 mm) high 1 ft. 6 in.–3 ft. 0 in. (457 mm–914 mm) wide *Solid:* 1⅜ in. (35 mm), 1¾ in. (44.5 mm), and 2¼ in. (57.2 mm) thick 6 ft. 8 in. (2032 mm) and 7 ft. 0 in. (2134 mm) high 2 ft. 4 in.–3 ft. 6 in. (812.8 mm–1067 mm) wide	Constructed from *hollow* or *solid* wood or composition cores supporting face panels of wood veneers, hardboard, or plastic laminates. Presents flush smooth appearance unless interrupted by glass lights or louver openings.
Panel NWWDA I.S.6, AWI, WIC	1⅜ in. (35 mm) and 1¾ in. (44.5 mm) thick 6 ft. 8 in. (2032 mm) and 7 ft. 0 in. (2134 mm) high 1 ft. 6 in.–3 ft. 4 in. (457 mm–1016 mm) wide	Assembled from stiles and rails (vertical and horizontal components), which frame and support one or more panels. Has characteristic paneled appearance resulting from raised and recessed areas, patterned molds, etc.
Sash NWWDA I.S.6, AWI, WIC	1⅜ in. (35 mm) and 1¾ in. (44.5 mm) thick 6 ft. 8 in. (2032 mm) and 7 ft. 0 in. (2134 mm) high 2 ft. 0 in.–3 ft. 6 in. (610 mm–1067 mm) wide	Similar in construction and appearance to panel door, except that one or more panels are replaced with glass. Completely glazed doors without panels or *casement* (French) doors.
Storm and Screen NWWDA I.S.6, AWI, WIC	1⅛ in. (28.6 mm) thick 6 ft. 7 in. (2007 mm), 6 ft. 9 in. (2057 mm), and 7 ft. 1 in. (2159 mm) high 2 ft. 6 in.–3 ft. 0 in. (762 mm–914 mm) wide	Lighter, thinner stile-and-rail construction supporting screening (*screen doors*), glass panels (*storm doors*), or interchangeable screen and storm panel inserts (*combination doors*).
Louver NWWDA I.S.6, AWI, WIC	1⅛ (28.6 mm) and 1⅜ in. (35 mm) thick 6 ft. 6 in. (1981 mm), 6 ft. 8 in. (2032 mm), and 7 ft. 0 in. (2134 mm) high 1 ft. 3 in.–3 ft. 0 in. (381 mm–914 mm) wide	Composed of stile-and-rail frame with integral louver construction, mortised into stiles or vertical dividing bars.
Accordion Folding (No industry standard)	6 ft. 8 in.–10 ft. 0 in. (2032 mm–3048 mm) high Practically unlimited length	Assembled from narrow wood strips or single wood slats, 3½ in. (89 mm) to 5 in. (127 mm) wide, with fabric, plastic, or metal hinges, resembling long drapeline doors.

may have a metal or wood frame. Accordion fold doors are not presently covered by industry standards.

8.2.1.1 Flush Doors

In addition to providing the desired surface appearance, face panels of flush doors act as stressed skins to impart strength, rigidity, and stiffness. Face panels may be hardwood veneer, plywood, high-pressure plastic laminate, or hardboard. Wood face panels are usually of two or three plies with a total thickness of at least 1/12 in. (2.13 mm). Door construction is described as 3-, 5-, or 7-ply, by counting the total number of plies in both face panels and counting the core as 1 ply.

Flush doors are available with glazed lights and louver inserts. The U.S. Consumer Product Safety Commission (CPSC) "Safety Standard for Architectural Glazing Materials" requires that all glazing materials in doors be safety glazing material. In addition, industry standards limit the size of the cutout to a maximum of 40% of the door area (Fig. 8.2-2).

SOLID-CORE DOORS

Solid-core construction increases the dimensional stability of a door, provides thermal insulation, and may add sound isolation and fire resistance. Solid-core doors are commonly used as entrance and other exterior doors, but may also be used where sound or fire ratings are required. Doors for which specific fire ratings are required must carry a fire door label from an approved testing agency, such as Underwriters Laboratory (UL). Fire ratings for wood doors range from 20 minutes to 1½ hours, depending on materials and construction. Many regular solid-core doors are capable of a 20-minute fire rating. Applicable codes should be consulted for specific requirements and applications.

Solid-core doors provide sufficient sound isolation for most uses. Where sound isolation is more critical, properly gasketed flush doors with a Sound Transmission Class (STC) rating of up to 42 may be required (see Section 13.1).

Solid cores may be of wood, particleboard, mineral composition, or a combination of these materials. Wood cores are made of strips or blocks of wood forming a glued-block core, framed-block core, or stile-and-rail core (Fig. 8.2-3).

Glued-Block Core Most solid-core doors have a glued-block core construction. They consist of wood blocks bonded together, with the end joints staggered. The stiles and rails are bonded to the core, and the entire assembly is sanded to a smooth, uniform thickness. The face panels are bonded to this assembly and must be a minimum of 1/12 in. (2.13 mm) thick.

Framed-Block Core Framed-block core doors are composed of a stile-and-rail frame surrounding an interior core of wood blocks, which may or may not be bonded to each other, but which are bonded to the face panels. Face panels must be a minimum of 1/12 in. (2.13 mm) thick.

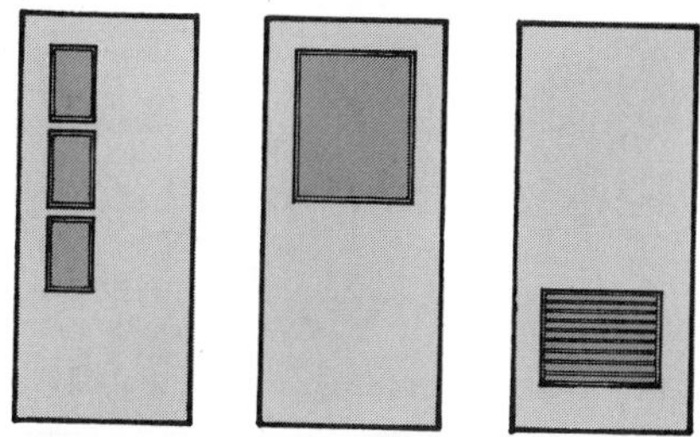

FIGURE 8.2-2 Flush doors with glass and louver inserts. (Courtesy National Wood Window and Door Association [NWWDA].)

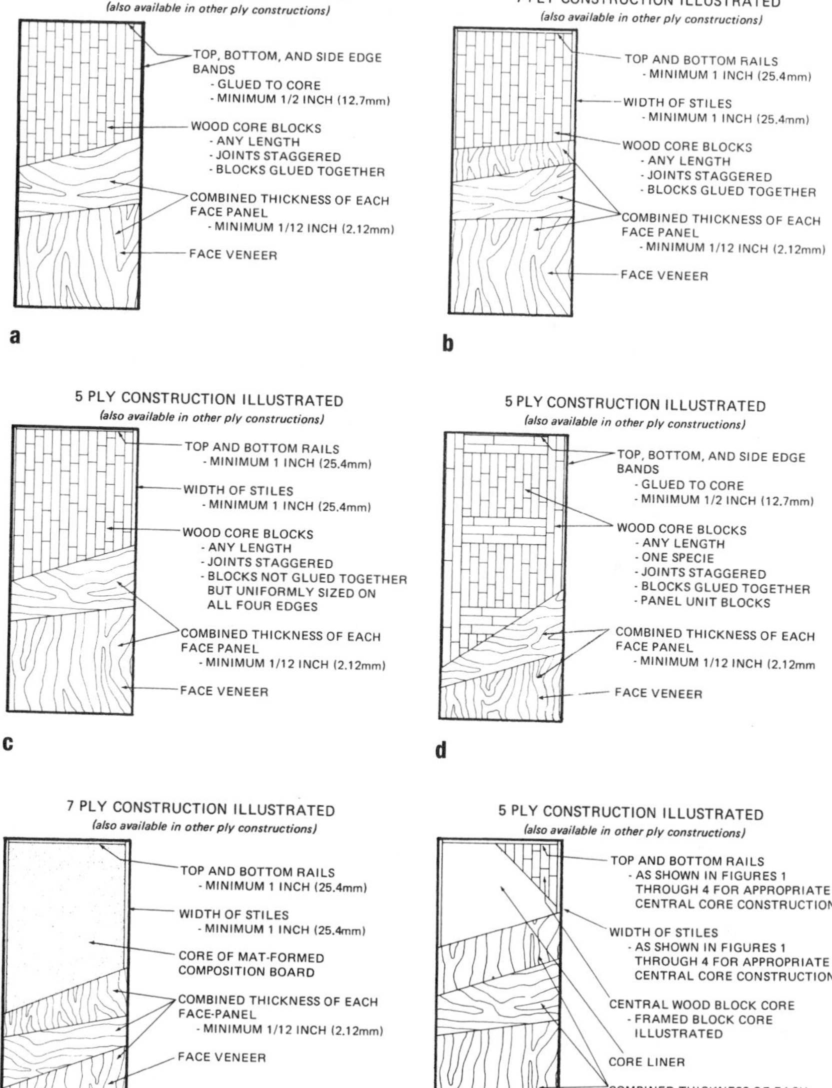

FIGURE 8.2-3 Typical solid-core constructions for flush doors: (a) glued-block core; (b) framed-block glued core; (c) framed-block nonglued core; (d) stile-and-rail core; (e) mat-formed composition-board core; (f) wood-block, lined core. (Courtesy National Wood Window and Door Association [NWWDA].)

Stile-and-Rail Core Stile-and-rail core doors consist of wood blocks bonded into stile-and-rail panel units. These units are assembled to constitute the core of the door and are bonded to the face panels. Again, face panels must be a minimum of 1⁄12 in. (2.13 mm) thick.

Particleboard Core Particleboard core doors are constructed of one or more pieces of particleboard, each the full thickness of the core, which may or may not be edge- or end-glued together. The stiles and rails of the door may or may not be bonded to the core, depending on the type of construction. Open spaces between core pieces or between the core and frame are controlled by industry standards. Face panels must be a minimum of 1⁄12 in. (2.13 mm) thick.

Mineral-Composition Core Mineral-composition cores are normally not available in other than fire-rated doors. These doors are constructed similarly to particleboard core doors, except that the core is made of inorganic, noncombustible materials formed into a rigid slab. Special blocking and framing are normally required in doors of this type to accommodate hardware, glazed openings, and other cutouts.

Wood-Block, Lined Core Wood-block, lined core doors are a combination of the constructions described previously. This type door consists of a central wood-block core, with a 3⁄16-in. (4.8 mm) or thinner liner of particleboard or other material bonded to each face to constitute the full core thickness. Face panels are then bonded to this assembly and to the stile-and-rail frame. These doors may provide added sound isolation, fire resistance, or other performance features, depending on the properties of the liner material.

HOLLOW-CORE DOORS

Flush doors with hollow cores are used for interior locations in residential and some light-commercial buildings. They are occasionally used as exterior doors in residential buildings and can serve satisfactorily in that use if bonded with waterproof adhesives and protected from the weather. They generally do not provide as much heat or sound insulation as solid-core doors, but can be used where these factors are not critical.

Face panels like those used on solid-core doors are available for use on hollow-core doors. Wood face panels must be of at least two plies and must have a total thickness of at least 1⁄10 in. (2.54 mm). The basic hollow-core constructions are mesh (cellular) core and ladder core (Fig. 8.2-4).

Mesh (Cellular) Core Mesh core doors consist of wood or wood-derivative material joined, interlocked, or woven to form a mesh or grid throughout the core area. Honeycomb cores of expanded paper or other material are the most common types of hollow cores. Face panels are bonded to the core and to the stiles, rails, and lock blocks of the door.

Ladder Core Ladder core doors are composed of wood or wood-derivative strips placed horizontally or vertically throughout the core area. Face panels are bonded to these strips and to the stiles, rails, and lock blocks of the door.

Lock blocks are provided in hollow-core doors for lock mortising and backset up to 25⁄8 in. (66.7 mm) Doors with special blocking may be obtained for other hardware mounting.

INSULATED DOORS

Insulated doors have an inner core of plastic foam insulation, have high R-values (R-8 to R-15), and are even more effective than the combination of prime and storm doors in reducing heat flow. Wood-framed insulated doors often have metal skins. Insulated doors are often installed with magnetic weatherstrips that are relatively airtight and minimize heat loss by infiltration.

8.2.1.2 Stile-and-Rail Doors

Stile-and-rail doors derive their strength mainly from a supporting framework of vertical members (stiles) and horizontal members (rails), interlocked and glued together. Stiles and rails may be solid or veneered and are usually of softwoods.

The chief categories of stile-and-rail doors are panel, sash, louver, storm, screen, and combination.

PANEL AND SASH DOORS

Panel doors consist of stile-and-rail frames enclosing flat plywood or raised wood panel fillers. They are available in a variety of designs (Fig. 8.2-5) and may

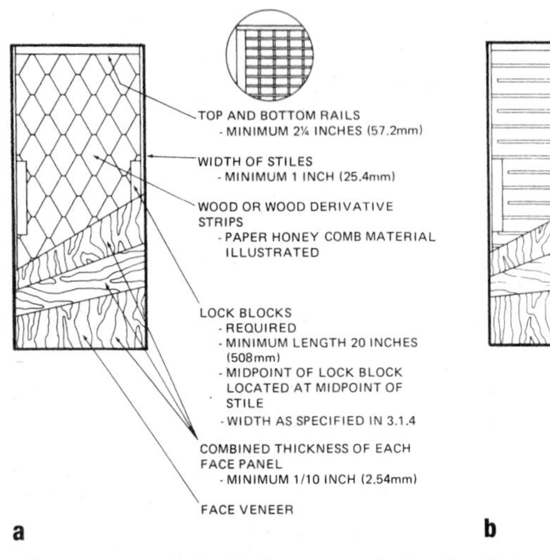

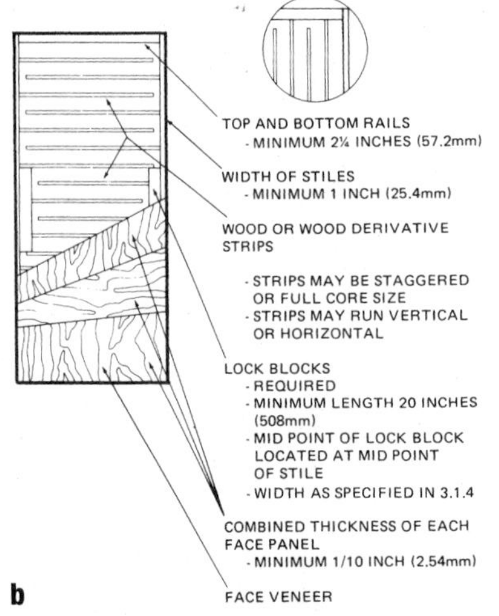

FIGURE 8.2-4 Typical hollow-core construction for seven-ply flush doors: (a) mesh or cellular core; (b) ladder core. (Courtesy National Wood Window and Door Association [NWWDA].)

FIGURE 8.2-5 Panel door designs. (Courtesy National Wood Window and Door Association [NWWDA].)

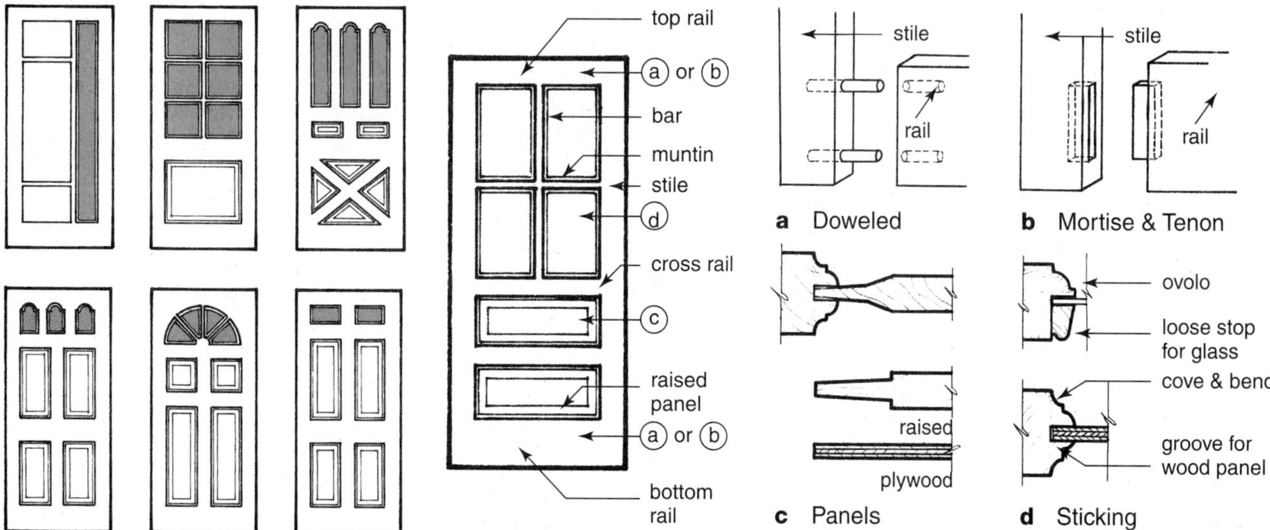

FIGURE 8.2-6 Sash door designs. (Courtesy National Wood Window and Door Association [NWWDA].)

FIGURE 8.2-8 Typical construction of stile-and-rail doors: (a) glued dowel; (b) blind mortise-and-tenon; (c) shown together with mullions, muntins, and bars; (d) edges milled with contoured profiles, grooves, or rabbets.

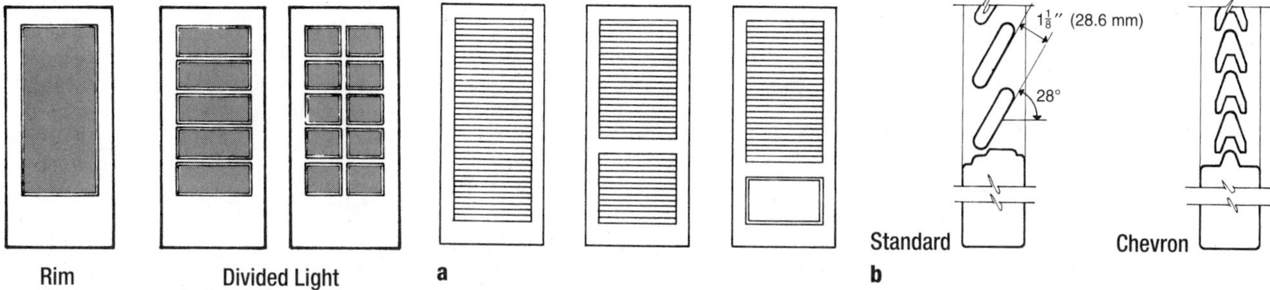

FIGURE 8.2-7 Casement (French) doors. (Courtesy National Wood Window and Door Association [NWWDA].)

FIGURE 8.2-9 Typical louver doors: (a) door designs, (b) slat styles.

be used for either interior or exterior application.

Sash doors are similar to panel doors in appearance and construction, but have one or more glass lights replacing the wood panels. They are used often for main entrance and exterior service doors (Fig. 8.2-6). Casement (French) doors are fully glazed sash doors and are referred to as *rim* doors when they have a single large glass panel, and as *divided light* doors when the glass is divided into smaller lights (Fig. 8.2-7).

Stiles and rails are assembled with glued dowel or blind mortise-and-tenon construction (Fig. 8.2-8a and 8.2-8b). Together with mullions, muntins, and bars, they frame the glass lights or wood panels (Fig. 8.2-8c). The contoured profiles, grooves, or rabbets (sticking) milled on the edges of these members hold the glass or wood panels (Fig. 8.2-8d).

LOUVERED DOORS

Wood louvered doors are made of horizontal slats contained within stile-and-rail frames. The slats usually are blind-mortised. They are available in several common designs and two basic slat styles, chevron and standard (Fig. 8.2-9).

Wood louvered doors are used in interior locations primarily as visual screens and where sound privacy is not required. Appropriate uses include closet doors, space dividers, and closures around residential laundry areas, where ventilation and dissipation of moisture is desirable.

STORM, SCREEN, AND COMBINATION DOORS

Storm, screen, and combination storm and screen doors are lightly constructed glazed doors used with exterior doors to improve weather resistance. Screen doors are of similar construction, but with screening, which provides ventilation while excluding insects. Combination doors combine the functions of both storm and screen doors with interchangeable screen and glass panels (Fig. 8.2-10).

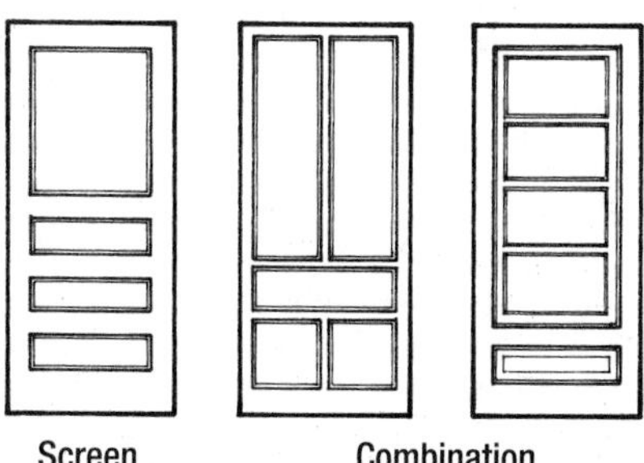

FIGURE 8.2-10 Storm and screen doors give added weather protection and ventilation. (Courtesy National Wood Window and Door Association [NWWDA].)

Some self-storing combination doors are equipped with a screen and a pair of glass panels that operate in a vertical track

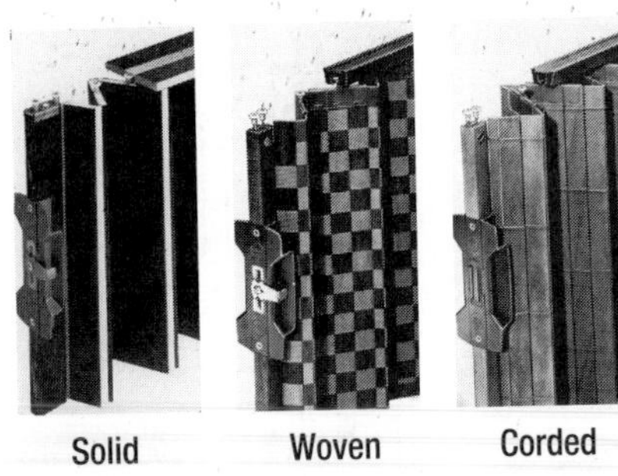

FIGURE 8.2-11 Typical accordion folding doors.

similarly to a double-hung window. Either ventilation or weather protection is possible without removing the panel inserts, but the ventilating area available is limited to one-half that of a full screen door. Glass jalousie inserts in combination doors have similar self-storing advantages with practically no loss of ventilating area. However, even when closed, jalousies permit a high rate of air infiltration and leak water in high-wind storm conditions.

8.2.1.3 Accordion Folding Doors

Wood accordion folding doors are made in several styles. One is made from either wood slats or wood strips, connected with cord or fabric tapes, forming flexible drapelike doors (Fig. 8.2-11). These doors are generally designed to fold to a width less than the thickness of a normal interior partition.

Solid wood slats are connected to each other with continuous metal, vinyl, or nylon fabric hinges. Woven doors are commonly of the basketweave type, in which vertical strips ⅜ to 1 in. (9.5 to 25.4 mm) wide are interwoven with nylon-reinforced vinyl tape. Corded doors are made with similar wood slats connected with cotton cord. Accordion doors are usually suspended from ceiling-mounted tracks and operate on nylon rollers or glides.

These doors are primarily visual screens, for locations such as closets, space dividers, residential laundry enclosures, and other areas where air circulation is desirable. They are not suitable where sound isolation or privacy is a prime requirement. The folding operation also makes them useful where door swings are objectionable.

Other accordion fold doors are made with a framework or core of plastic, metal, or wood and a covering of vinyl, fabric, plastic laminate, or another plastic material.

The folding doors discussed here should not be confused with *folding partitions*, which are similar to folding doors but tend to be much sturdier. Folding doors, for example, are usually limited to heights of about 14 ft. (4267 mm). Folding partitions can be much taller. Some folding partitions are designed to provide both privacy and sound transmission control. These are usually constructed of wood or metal frames covered with plywood or hardboard and a wood, fabric, or plastic-laminate panel. They often have a plastic foam or other solid core. Folding partitions are used where visual and sound privacy are needed, such as to separate meeting rooms or classrooms.

8.2.2 DOOR OPERATION

The operation of wood doors can be classified as swinging, sliding, or folding (Fig. 8.2-12). The use and location of a door determines the most suitable manner of operation. Fully glazed wood sliding glass doors are addressed in Section 8.5.

8.2.2.1 Swinging Doors

Swinging doors typically operate on hinges secured to the side jambs, although they may also operate on pivots supported by the door frame head and the floor. The majority of wood swinging doors are hinged (Fig. 8.2-12a). Either pivot hardware or double-acting hinges are used for double-acting swinging doors (doors that swing in both directions).

Hinged operation provides the greatest degree of security, weather resistance, heat insulation, and sound isolation. Where these features are particularly important, doors may be weatherstripped or gasketed for increased effectiveness.

Swinging door operation is most convenient for frequently used entry and passage doors, but may be used in any location unless door swing into usable space is objectionable.

Door swing is either right-hand, left-hand, right-hand reverse, or left-hand reverse. Assume that the viewer is standing outside a door (on the exterior of an entrance door, in the corridor at a door to a room, or on the side opposite the hinges on a connecting door). A door that swings away from the viewer with the hinges on the right side is a right-hand door; the same door swinging toward the viewer is a right-hand reverse door. When the hinges are on the left and the door swings away from the viewer, it is a left-hand door; if it swings toward the viewer, a left-hand reverse door.

8.2.2.2 Sliding Doors

Interior sliding wood doors are suspended on overhead tracks and operate on rollers attached to the door. These rollers may be nylon or another plastic or metal with plastic bearings. Floor guides are usually provided to prevent sliding doors from swinging laterally. In rare instances, sliding doors are mounted on rollers operating on floor tracks.

Sliding wood doors are used on closets and as room dividers in all kinds of buildings and in storage areas in residential construction, where they are suitable for visual screens. Sliding doors used on closets and other storage areas are usually of the bypass type (Fig. 8.2-12b), but pocket doors are also sometimes used in these locations. Closet sliding doors are often louvered, but may also be flush or panel doors.

Surface sliding and pocket sliding doors are more likely to be found as passage doors (Fig. 8.2-12c and d). Biparting doors (two doors sliding in opposite directions) are used for openings 3 ft. (914 mm) or wider and may be either surface or pocket type. For extremely large openings, multitrack installations permit stacking of several sliding doors at one or both ends. Surface and pocket sliding doors are usually flush or panel doors, but may also be louvered.

Sliding operation does not give a high degree of privacy, sound isolation, or weather resistance, unless custom hardware or preengineered door units are used. The chief advantage of sliding operation is that it eliminates door swings that may interfere with the use of interior space.

8.2.2.3 Folding Doors

Interior folding wood doors are hung on overhead tracks with rollers similar to those used on sliding doors. Some types also require a floor track. Folding operation, like sliding, is generally appropriate for locations requiring primarily visual screening.

Flush, louver, and panel doors are often used for closets in single or paired bifold arrangements (Fig. 8.2-12e), or as space dividers and movable partitions with special hardware for multifolding action and stacking (Fig. 8.2-12f). Accordion folding doors are used as closet doors or as space dividers and may be pocketed or end-stacked within the opening.

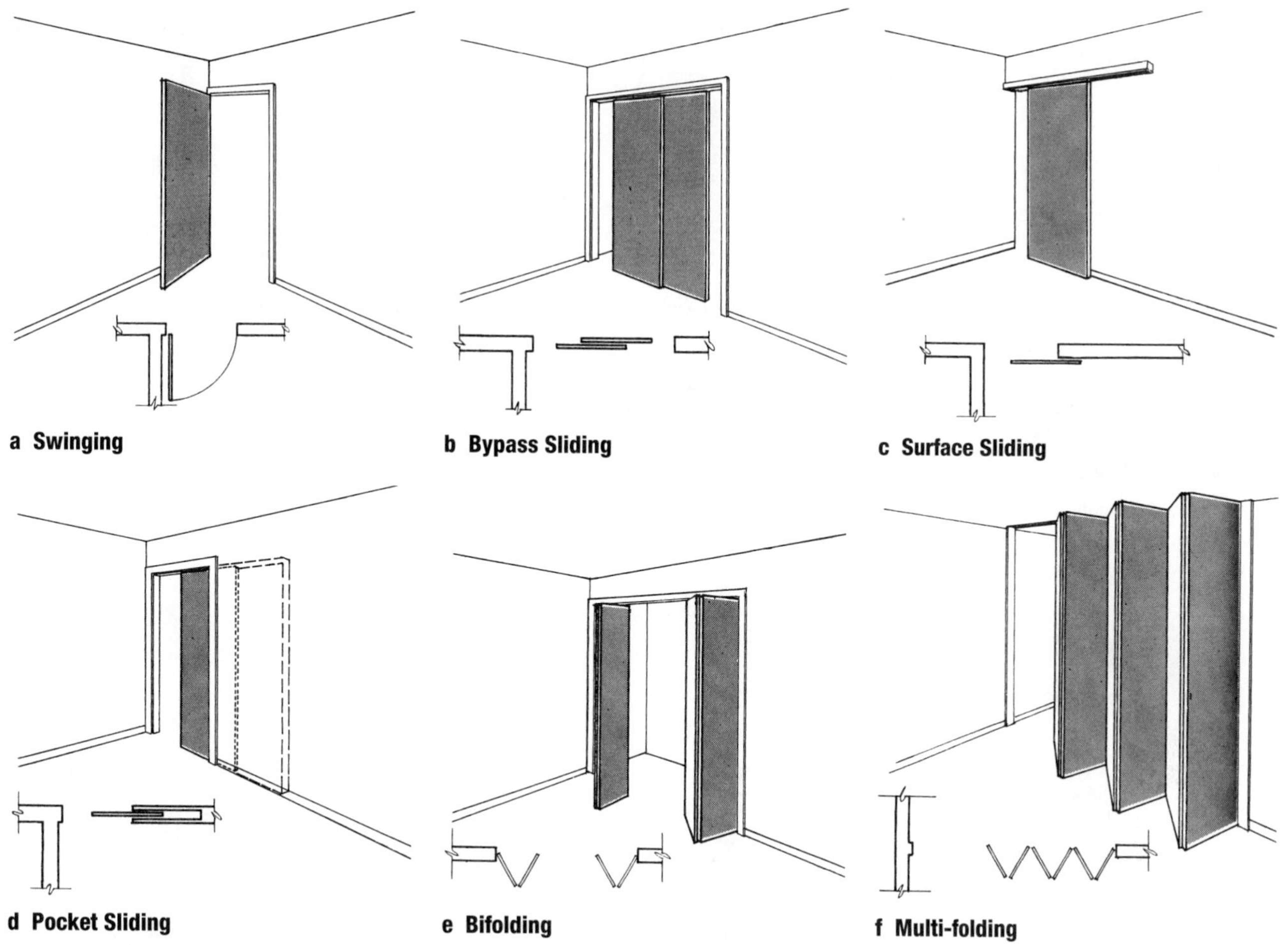

FIGURE 8.2-12 Door operation: (a) swinging; (b, c, d) sliding; and (e, f) folding. Surface sliding doors (c) are used mostly on industrial and agricultural (usually metal prefab) buildings. They are seldom found in modern commercial, institutional, or residential buildings.

8.2.3 MANUFACTURE

The following discussion of door manufacture presumes conformance with industry standards.

8.2.3.1 Materials

Some common species of hardwoods for door manufacture are oak, ash, elm, birch, butternut, cherry, maple, mahogany, walnut, basswood, and lauan. Softwood species are ponderosa (western) pine, Douglas fir, Sitka spruce, and western hemlock.

Most wood-faced flush doors are faced with hardwood veneer. Most stile-and-rail doors are made from ponderosa pine, but some are produced from other softwoods. Hardwood veneered stile-and-rail doors and softwood veneered flush doors, though uncommon, are available. Accordion folding doors are made from both hardwoods and softwoods.

The lumber used in door manufacture is kiln-dried to a moisture content of 12% or less before being assembled into a door.

The solid wood core for a flush door should contain a single species of wood that has a specific gravity of .42 or less. Composition core materials range in density up to about 37 lb./cu. ft. (592.7 kg/m^3) at the same moisture content.

Doors for exterior use are made with waterproof (Type I) adhesives, such as thermosetting melamine urea or phenolic resins. Interior doors are made with water-resistant (Type II) adhesives, such as casein (sometimes fortified with urea resins and formaldehyde), capable of resisting a limited number of wetting cycles.

8.2.3.2 Fabrication and Assembly

Fabrication varies somewhat, depending on the type of door.

FLUSH DOORS

Stiles, end rails, lock rails, and lock blocks for hollow-core flush doors are cut from kiln-dried, surfaced lumber. Stiles and rails are joined with corrugated metal fasteners or with machined dovetail joints. The stile-and-rail frame is laid up in adhesive with one face panel. Cores, lock blocks, and sometimes lock rails, are then set and glued in place. Cores may be ladder, mesh, or implanted blanks (Figs. 8.2-13 and 8.2-14). Assembly is completed by bonding the second skin to the frame and the core.

Most hollow-core flush doors are bonded with water-resistant adhesives that do not require heat to set. These doors are cold pressed (Fig. 8.2-15) and allowed to cure overnight, prior to trimming and further working.

Most solid-core doors are made with *thermosetting* adhesives. Continuous block cores are assembled from wood blocks, glued together, and cured. After additional air drying for 24 hours, face panels are adhesive-bonded to the core and the entire assembly is *hot-pressed* to set the glue line (Fig. 8.2-16).

FIGURE 8.2-13 Implanted-blanks core for hollow-core door is made by feeding wood spirals into an assembled frame.

FIGURE 8.2-16 Hot pressing sets the glue line of solid-core doors bonded with thermosetting adhesives.

FIGURE 8.2-14 Cellular-core construction is made by gluing paper honeycomb into an assembled frame.

FIGURE 8.2-17 Flush doors are routed for glass lights and louvers.

FIGURE 8.2-15 Hollow-core flush doors are cold pressed to ensure proper bonding of face panels to core.

Further operations on both solid- and hollow-core doors may include routing for louver and glass openings and installing muntins and bars prior to glazing (Figs. 8.2-17 and 8.2-18).

Glass and louver openings in flush doors should not be more than 40% of the door area, and openings should not be closer than 5 in. (127 mm) from the door edge. The height of openings in a hollow-core door should not exceed half the door, and face skins should be supported with blocking, glued in place, to frame the opening.

STILE-AND-RAIL DOORS

Kiln-dried lumber is surfaced to exact thickness, and boards are then cut to proper widths and lengths. Pieces for stiles are bored for dowels on a boring machine (Fig. 8.2-19); a molding machine shapes the sticking on stile and rail edges (Fig. 8.2-20); and a double-end tenoner cuts rails to exact lengths, coping both ends to

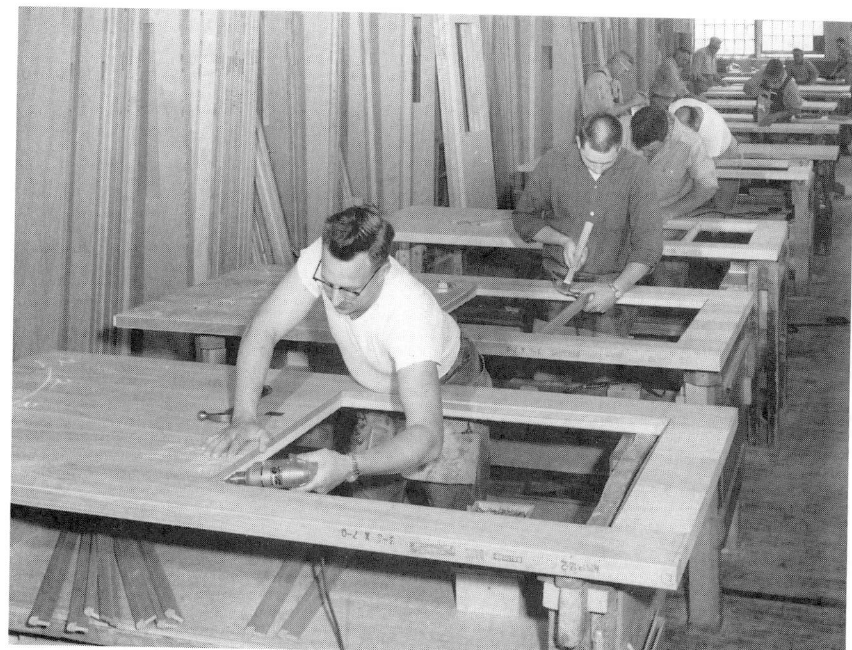

FIGURE 8.2-18 Muntins and bars are fitted and installed before glazing.

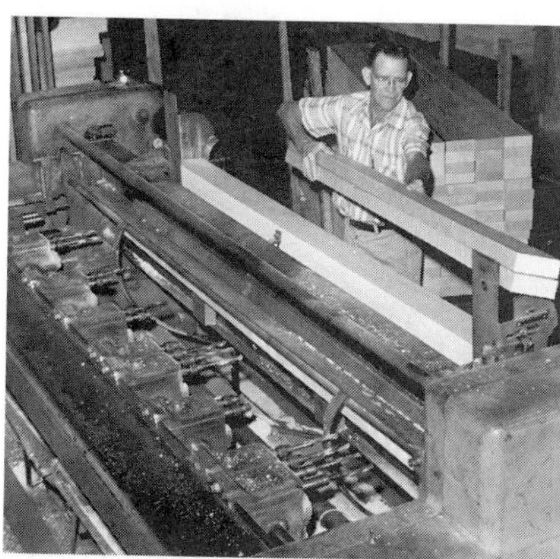

FIGURE 8.2-19 For stile-and-rail doors, holes for dowels are bored in stiles automatically fed from a hopper.

FIGURE 8.2-20 An endless feeder chain carries rails or stiles through a molding machine to form sticking on edges.

fit the stacking on the stiles. Another machine bores holes in the rails, inserts glue, and drives hardwood dowels into the holes (Fig. 8.2-21).

For panel doors, wood or plywood panels are cut to size and may be shaped to form raised panel designs (Fig. 8.2-22). Stiles and rails are assembled with water-resistant adhesives around the panels and clamped in a jig. When the glue has set, the doors are sanded on both sides to a smooth finish (Fig. 8.2-23).

Stile-and-rail frames, and sometimes panels, muntins, and bars for sash doors, are fabricated and assembled in a similar manner. Glass is cut to size and bedded in glazing compound, and stops are accurately mitered and nailed in place (Fig. 8.2-24).

8.2.3.3 Prefabrication

Doors may be further worked and finished in the plant to reduce job-site preparation and installation. They may be seal-coated to prevent moisture absorption and soiling, or may be completely surface-finished. Standard doors and prefinished custom doors are usually prefit to exact opening dimensions so that the finished edges will not be removed by job-fitting. Doors for prefabricated frames are sometimes machined for locks and hinges (Fig. 8.2-25), and bifold doors may have all operating hardware installed.

Plastic laminate–surfaced doors are usually prefabricated and prefit.

Prehung door units are assembled complete with frames, trim, and hardware. The stop may be part of the frame or left loose to be applied during installation. Frames are braced at jamb bottoms and carefully packed to prevent racking and distortion in shipment.

8.2.3.4 Finishes

Door finishes may be either opaque or transparent. *Opaque* finishes are essentially paint. There are several types of *transparent* finish systems used on doors, including lacquers, conversion varnishes, catalyzed vinyls, polyesters, polyurethanes, and oils.

Factory finishes are graded as standard, select, or super.

Standard finishes are simply the manufacturer's standard. They usually include lacquers, conversion varnishes, and catalyzed vinyls.

Select finishes are similar to standard finishes but permit the designer to select

FIGURE 8.2-21 A single machine bores holes, inserts glue, and drives dowels.

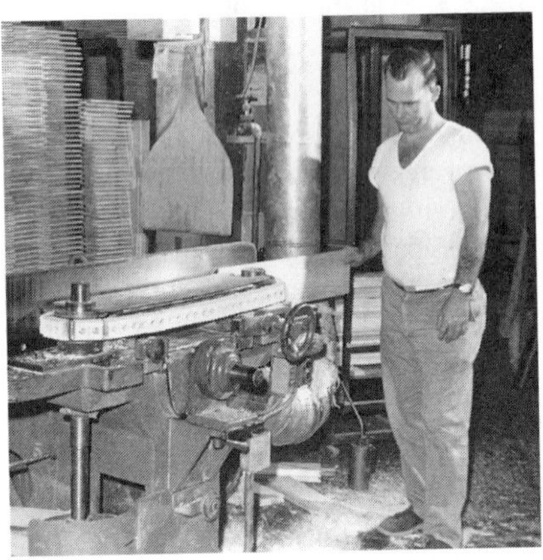

FIGURE 8.2-22 After being cut to size, panels are shaped to make raised designs.

colors and sheens. Select finishes are usually more expensive than standard finishes.

Super finishes are generally superior to both standard and select finishes, but are usually the most expensive of the three. They include polyester, polyurethane, and oil finishes. A designer may select color and sheen.

8.2.4 STANDARDS OF QUALITY

The National Wood Window and Door Association (NWWDA), the Architectural Woodwork Institute (AWI), and the Woodwork Institute of California (WIC) have developed standards applicable to wood doors. The applicable standards are:

Flush Doors

- NWWDA, I.S.1-A, "Industry Standard for Architectural Wood Flush Doors"
- NWWDA, I.S.1-R, "Industry Standard for Residential Flush Doors"
- AWI, *Architectural Woodwork Quality Standards, Guide Specifications, and Quality Certification Program*, Section 1300, "Architectural Flush Doors"
- WIC, *Manual of Millwork*

Stile-and-Rail Doors

- NWWDA, I.S.6, "Industry Standard Wood Stile and Rail Doors"
- AWI, *Architectural Woodwork Quality Standards, Guide Specifications, and Quality Certification Program*, Section 1400, "Stile and Rail Doors"
- WIC, *Manual of Millwork*

These publications are the recognized standards of quality for the wood door industry. They are similar, but vary slightly in detail. The AWI and WIC standards act as supplements to the NWWDA standards. Before making choices related to door selection and design, a designer should examine the requirements of the selected standard and not rely solely on citing the name of the standard. It may be necessary to modify these standards to suit a particular project. For example, one standard may require a particular face veneer thickness for one facing grade, while another standard may require a different minimum thickness or simply be silent on the issue.

In addition to establishing minimum requirements for material, design, construction, and waterproof performance under test conditions, these standards also provide a basis for quality grading of wood doors.

8.2.4.1 Grades

Doors are graded primarily according to their exposed face veneers. However, the standards also include in the grades requirements for vertical edge bands, pair matching, transom matching, beads for light and louver openings, and quality of workmanship. The grades limit the materials that may be used in each grade based on the type of finish to be used (opaque, transparent, or laminate). Flush doors are also limited by core construction.

FIGURE 8.2-23 Stile-and-rail doors are sanded to a smooth finish.

FIGURE 8.2-24 After glazing, accurately mitered stops are nailed in place.

FIGURE 8.2-25 Doors are mortised for locksets on an automatic router.

WOOD FLUSH DOORS

Solid- and hollow-core flush doors are available in *premium*, *custom*, and *economy* grades, depending on the quality of face panels. Flush doors are not recommended for exterior use. Interior doors may be made with either *waterproof* (Type I) or *water-resistant* (Type II) adhesives, depending on their location. Doors in wet or humid areas should be made with Type I adhesives.

Premium Grade Premium-grade face panels are available for either transparent, opaque, or laminate finishes. This is the highest grade available from commercial stock.

For *transparent* finishes, the face veneers are tight and smoothly cut and free from knots, worm holes, splits, shakes, and torn grain. When the face consists of more than one piece, the pieces are of approximately equal width and are matched at the joints for color and grain. Several different matching patterns are used. There are many species used for transparent finishes (see Section 8.2.3.1).

For *opaque* finishes, premium-grade doors have *medium-density overlay* plywood faces.

Laminate faces and edges are nominal 0.050-in. (1.27 mm)-thick high-pressure decorative laminate. The laminate may be solid or may have a grain or texture.

Custom Grade Flush doors with custom-grade face veneers are also intended for a transparent, opaque, or laminate finish. They are intended for use in high-quality projects.

For *transparent* finishes, the face panels for custom-grade flush doors are made of veneers essentially similar to those used in premium-grade doors, but need not be matched at joints for color and grain if sharp contrasts between adjacent pieces are avoided.

For an *opaque* finish, the face veneer may be any close-grained hardwood. Grain pattern, split veneer joints, and other wood characteristics may show through the finish unless special precautions such as filling are taken to prevent this.

Laminate faces are nominal .050-in. (1.27 mm)-thick high-pressure decorative laminate. The laminate may be solid or may have a grain or texture. Edges are close-grained hardwood.

Economy Grade The face veneers in economy-grade flush doors are intended for painting and may contain a variety of color, grain, and surface defects that will not be visible after two coats of paint have been applied. Their selection is generally left up to the mill producing the doors. Some manufacturers make economy-grade doors with hardboard face veneers.

There is no economy grade for laminate-finish doors.

STILE-AND-RAIL DOORS

Grades for stile-and-rail doors are premium, sometimes called *select*, and *standard*, also called *custom*. Both grades are made with water-resistant adhesives and are often treated with water repellents.

Premium Premium-grade stile-and-rail doors are made from stock practically clear of defects. They are intended to receive a transparent finish.

Standard Standard-grade stile-and-rail doors are made from lumber stock that may contain stains and various other minor defects. They are intended to receive a paint finish.

8.2.4.2 Certification of Quality

Doors that conform to the industry standards may be grade marked or labeled by the manufacturer when requested by the buyers. A *grade mark* is usually stamped or branded on the edge of a door; a *label* is a written statement attached to the door, containing information similar to that given by a grade mark.

The NWWDA administers an independent certification program for hardwood-veneered doors based on continuous, impartial inspection and testing of doors. An NWWDA seal ensures that the materials, construction, and manufacture of doors conform to applicable industry standards. A red plastic plug in the top edge of a door indicates a door bonded with waterproof exterior Type I adhesives.

Exterior doors should be treated with a water-repellent preservative for increased dimensional stability and resistance to decay, fungus, and insect attack. The NWWDA administers a certification program for preservative-treated millwork products that includes softwood doors. The NWWDA seal is assurance of adequate preservative treatment in accordance with NWWDA I.S.4, "Industry Standard for Water-repellent Preservative Treatment for Millwork."

8.2.5 INSTALLATION

The satisfactory performance of doors depends on proper handling, storage, hanging, and finishing.

8.2.5.1 Handling and Storage

Unfinished doors should be handled with clean gloves to prevent soiling. They should be stored flat (not on edge) on a level surface in a clean, dry, well-ventilated space. They should not be stored in excessively hot, dry, or humid areas, but should be conditioned for several days to the average prevailing local humidity before hanging or finishing. If doors appear warped, they should be stacked flat under uniformly distributed weights to restore flatness.

8.2.5.2 Hanging

Finished door frames should be square and plumb, and doors should be fitted in the openings with a total clearance of approximately 3/16 in. (4.8 mm) in both horizontal and vertical dimensions to prevent binding under humid conditions. Doors may be trimmed for fitting, but should not be cut to smaller nominal sizes. If a substantial amount of the perimeter is removed, the balance of the door can be destroyed and warping could result.

Installed doors should conform to NFPA 80.

8.2.5.3 Finishing

Unfinished wood doors should receive a seal coat of the finish to be used or another suitable sealer as soon as they are delivered to the job site to prevent undue moisture absorption and soiling. Finishing should be delayed until doors have been trimmed and fitted in the openings. Door surfaces, including both faces and top, bottom, and side edges, should be sealed against moisture with two coats of paint, varnish, or sealer before final hanging.

Door surfaces should be clean and dry before a finish is applied. Soil marks and other surface defects should be removed by light sanding. Doors should not be finished in excessively humid weather. The proposed finish should be tested on a sample of the same species of wood to determine its appearance and suitability.

8.3 Aluminum Entrances and Storefronts

Aluminum entrances and storefronts include both exterior and interior extruded aluminum doors, frames, and storefront systems, as well as revolving doors. Doors and frames formed from steel, stainless steel, copper alloy, and aluminum sheet (as opposed to extruded aluminum) are addressed in Section 8.1.

8.3.1 STANDARDS

Extruded aluminum entrances and storefronts should conform to the applicable standards of the appropriate organization, including:

- Aluminum Association (AA)
- American Architectural Manufacturers Association (AAMA)
- American Society for Testing and Materials (ASTM)
- National Association of Architectural Metal Manufacturers (NAAMM)
- American National Standards Institute (ANSI)

Extruded aluminum components should comply with the AAMA publication *Aluminum Store Front and Entrance Manual.*

Door hardware should conform to ANSI standards A156.1 through A156.21.

8.3.2 MATERIALS

Refer to Chapter 5 for a general discussion about the metals addressed in this section.

A variety of aluminum alloys and tempers are used to make extruded door and frame components. Alloys should conform with ASTM Standard B 221. Common alloys used are 6061, 6063, and 6463. The temper should be suited to the finish to be applied; a common temper is T5.

Sheet aluminum is often used for door faces. Sheet aluminum and bent plate components should comply with ASTM Standard B 209.

Subframes, stiffeners, edge channels, reinforcements, and other concealed components are fabricated from the same materials as the exposed portions of the door or frame.

8.3.3 FINISHES

The mill finish is usually left on concealed surfaces and sometimes even on the exposed surfaces of aluminum doors, but most exposed portions of aluminum doors and storefronts are anodized. Fluoropolymer finishes are also used, but are not recommended for surfaces subject to excessive wear or damage, such as on doors and adjacent frames. Exposed fasteners are usually finished to match the adjacent metal.

8.3.4 SUPPORTS, ANCHORS, FASTENERS, AND FLASHING

Supports, anchors, fasteners, and flashings should be compatible with the material being anchored or fastened and with the substrate so that galvanic and chemical reactions do not occur between different materials. Aluminum should be anchored and fastened with aluminum or with nonmagnetic stainless, cadmium-plated, or galvanized steel.

Inserts for use in concrete may be cast iron, malleable iron, or hot-dip galvanized steel.

8.3.5 WEATHERSTRIPPING

Weatherstripping controls air infiltration, prevents rain penetration, minimizes dust infiltration, and withstands atmospheric conditions. It must be durable under extended usage, resist corrosion, and be readily replaceable. Refer to Section 8.7 for additional discussion of weatherstripping.

8.3.6 EXTRUDED ALUMINUM DOORS, FRAMES, AND STORE FRONTS

Most main entrance doors and their frames, and packaged entrances and storefronts, in commercial and institutional buildings are made from extruded aluminum components.

8.3.6.1 Design Requirements

Extruded aluminum doors, frames, and storefronts should be tested by the manufacturer or by an independent testing agency to ascertain their resistance to deflection due to wind loads, air infiltration, and water penetration. These characteristics are determined by the effect on a door and frame, packaged entrance, or storefront system when tested at specific levels of wind pressure. Since these elements are located on the ground floor, they are not subjected to the high levels of wind pressure that may affect windows and glazed curtain walls. The wind pressures used in the tests, therefore, are lower.

To select a door, frame, packaged entrance, or storefront, first determine the type desired (i.e., stick framing or package, type of door, frame profile, etc.). Then determine the required design pressure.

Design pressure is determined according to the greatest wind speed likely to be encountered at a building's location. Data found in ANSI/ASCE 7-95 can be used to estimate the wind loads likely to be encountered.

The assembly selected should be tested using the expected wind loading and the procedures listed in ASTM Standard E 330. Permanent deflection normal to the plane of the unit should be no more than 0.2% of the clear span, and there should be no glass breakage or permanent damage to fasteners, anchors, or hardware. Deflection during the test should not exceed 1⁄175 of any span. This limit may be decreased to 1⁄360 of any span if a nonflexible finish, such as plaster, is to be used.

The water resistance of an assembly should be determined by testing it in accordance with ASTM Standards E 331 and E 547. The test pressure is usually 6.24 lb./sq. ft. (30.47 kg/m^2). For a unit to pass this test, no water may pass through the unit during the test.

Air infiltration should be tested in accordance with ASTM E 283. The usual pressure is 1.57 lb./sq. ft. (7.67 kg/m^2); the permitted infiltration is 0.06 cu. ft. (0.002 m^3) per minute per sq. ft. (0.09 m^2) of fixed area.

Condensation resistance should be tested for frames having a thermal break feature. Condensation resistance should be determined by procedures outlined in AAMA 1503. An assembly's condensation resistance gives an indication of its ability to resist condensation. The higher the condensation resistance factor, the less likely it is that condensation will occur.

As discussed in Chapter 15, heat will flow quite readily through entrances and storefronts unless some means are used to reduce this flow. The methods for determining conductive heat transmission through these assemblies are included in AAMA 1503.

8.3.6.2 Manufacture

The manufacture of extruded aluminum doors and packaged entrances consists of fabricating aluminum extrusions of the correct sizes and shapes and assembling them together with glass, glazing materials, hardware, weatherstripping, and screws and other fasteners into finished preassembled components.

8.3.6.3 Doors

Each manufacturer has its own standards for its doors, even within the restraints of the standards listed earlier. However, there are only two basic types of swinging doors: (1) stile-and-rail doors and (2) flush-panel doors (Fig. 8.3-1).

STILE-AND-RAIL DOORS

Aluminum stile-and-rail doors consist of a tubular extruded aluminum frame infilled, usually, with glass (Fig. 8.3-1a). Stiles may be *thin* (less than 1¾ in. [44.5 mm] wide), *narrow* (usually 2 in. [50.8 mm] wide), *medium* (4 in. [102 mm] wide), or *wide* (5 in. [127 mm] or more wide). The thickness of most is 1¾ in. (44.5 mm).

Thermal-break doors are available, but most aluminum stile-and-rail doors are standard. The usual glazing is ¼-in. (6.4 mm)-thick glass, but most thermal-break doors are glazed with 1-in. (25.4 mm)-thick insulating glass.

FLUSH-PANEL DOORS

Aluminum flush-panel doors consist of an extruded channel or angle frame, and bracing members, and sheet aluminum facing panels (Fig. 8.3-1b). Most such doors are filled with a core made of a kraft paper honeycomb, rigid plastic foam, or mineral insulation. Faces may be either flat or

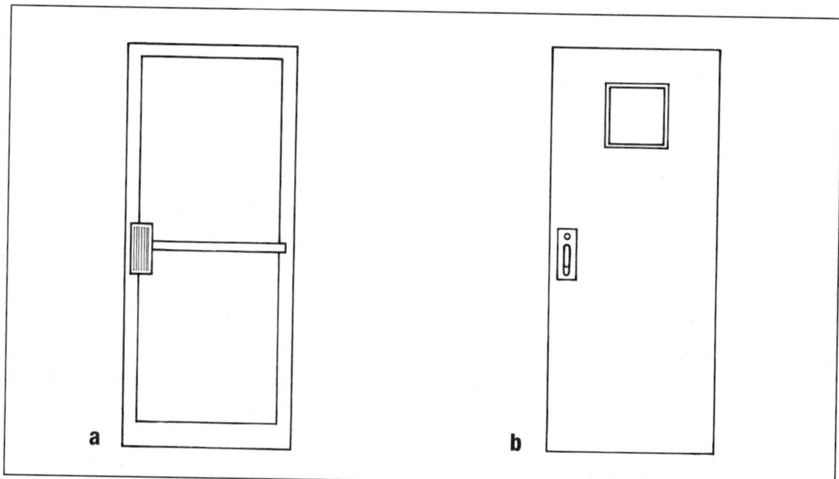

FIGURE 8.3-1 (a) A stile-and-rail door; (b) a flush panel door. (Drawings by HLS.)

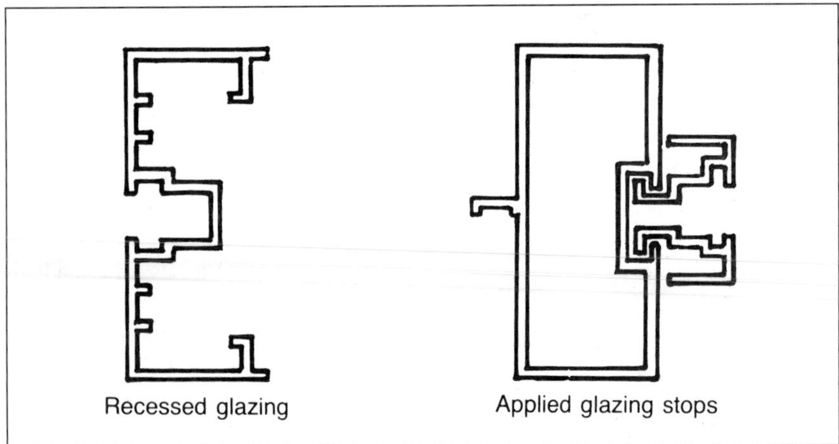

FIGURE 8.3-2 **Typical extruded aluminum entrance and storefront frames. Many other types and shapes are available.** (Courtesy H. Leslie Simmons.)

embossed. Flush-panel doors may have glazed openings or louvers.

8.3.6.4 Frames and Storefronts

Extruded aluminum frames and storefront sections come complete with accessories such as glazing gaskets, glazing stops, clips, trim, fins, flashings, anchors, and fasteners. Incidental components, such as closers, channels, flats, trim, sleeves, and the like, are also usually delivered with the frames and storefronts.

Frame and storefront sections should be accurately cut and fitted to make hairline joints. Where required, joints should be designed and erected to be watertight. Joints between two pieces of metal should be filled with a nonhardening sealant. Welds should be made on the inside so they will not be visible in the finished installation.

Frames and storefronts are usually glazed with ¼-in. (6.4 mm) glass, except where insulating glass is desired. Glazing stops and gaskets should be of the correct size to receive the glass to be used.

Frames and storefront sections are usually either tubes or channels (Fig. 8.3-2). They are typically 1¾ or 2 in. (44.5 or 50.8 mm) thick and vary in depth from 4 to 7 in. (102 to 178 mm). However, other sizes are available.

Frames and storefronts may be either standard (Fig. 8.3-2) or thermal break. They may be either flush glazed or with applied stops (see Figs. 8.3-3 and 8.3-4).

THERMAL BREAK

Where design calculations indicate the need to reduce the heat transmittance of storefronts, significant reduction can be accomplished by incorporating double or triple glazing or insulating glass. Where frame design permits, thermal breaks in the frames will also help. At any rate, when these assemblies are located where the winter design temperature is 10°F (−12.2°C) or less, they should incorporate such devices regardless of calculations.

A metal frame that holds insulating glass should have a *thermal break* to prevent a thermal "short circuit," that is, heat flow being resisted by the insulating glass but flowing freely through the aluminum frame. A thermal break separates the metal parts on the warm side of a frame from the metal on the cold side, using an insulating material such as polyurethane, vinyl, or another material of low conductivity, to keep heat from being conducted through the frame or sash. The insulating separator is also sometimes called a *thermal barrier*.

An aluminum frame with a good thermal break has an R-value of 1.66, which is about equal to that of double-pane insulating glass. A comparable frame with no thermal break would have an R-value of 1.18.

In some areas it may be economically practical to use three layers of glass separated by air spaces. Three layers of sealed glass with two ¼-in. (6.4 mm) spaces, for example, has an R-value of 2.13, as com-

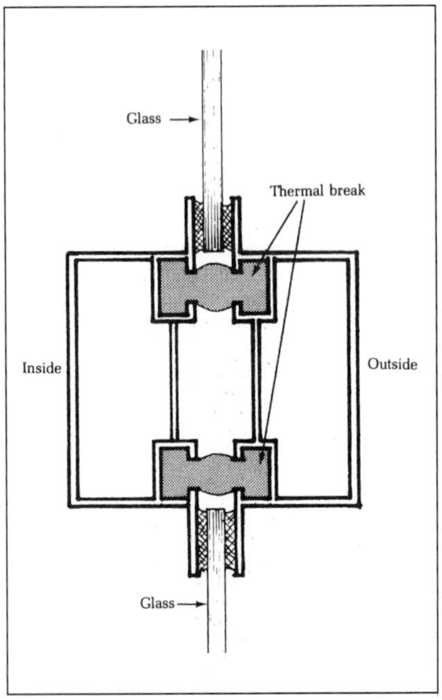

FIGURE 8.3-3 **The shaded thermal break reduces conductive heat transmission through the metal frame.**

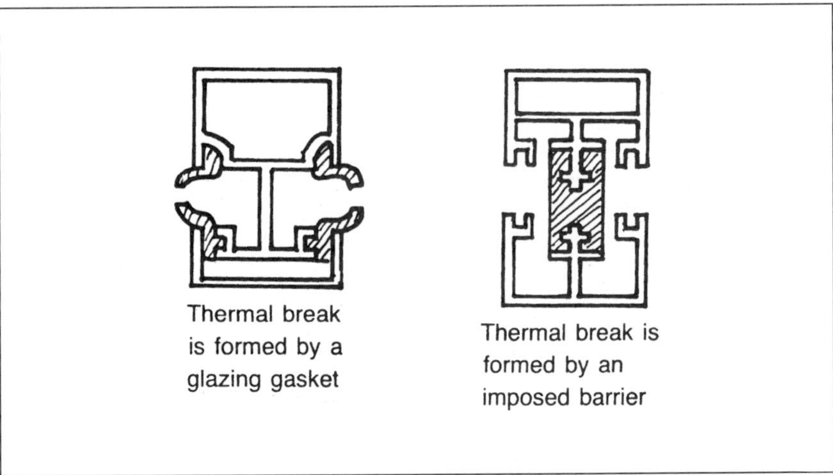

FIGURE 8.3-4 **Two typical frames with a thermal break feature. Many other configurations are also available.** (Courtesy H. Leslie Simmons.)

pared with 1.72 for typical insulating glass.

INTERIOR EXTRUDED FRAMES

Extruded aluminum entrances and storefront components are also used for interior doors and frames. They are available in a number of sizes and shapes to suit various conditions. The requirements for interior frames are similar to those for exterior frames, except that watertightness is not required.

ALL-GLASS ENTRANCES

All-glass entrances feature doors made of glass without metal frames. They are supported on special pivot-type hardware. Such doors are often fitted with automatic operators.

8.3.7 FINISH HARDWARE

Extruded aluminum doors and packaged entrances usually come equipped with all operating hardware except cylinder locks.

Hardware for these entrances may be standard, handicapped-access, automatic, or security type. Standard hardware includes all the devices necessary to operate, lock, and weatherstrip a door, including thresholds.

Handicapped-access hardware includes counterbalanced closers that permit easy operation of a door and specially designed door pulls with extended or oversized grips to permit easy operation by a handicapped person.

Automatic operators open doors when an electromechanical, hydraulic, or pneumatic operator is triggered by an actuating device, such as a floor treadle, motion sensor, sonic sensor, push button, photoelectric cell, or remote switch. Automatic hardware also includes detectors that automatically close doors when smoke or fire is present.

Security hardware consists of high-security cylinders, special control devices such as electromechanical, electromagnetic, or electronic locks, and three-way locks in which a single key operates bolts in the head, jamb, and sill of a door. Security locks may operate by key, push buttons, or card readers.

Refer to Section 8.7 for additional discussion of hardware.

8.3.8 INSTALLATION

The finest entrance or storefront design and the highest quality of manufacture will not compensate for poor installation. Weathertightness, although simple to achieve, can be ensured only if the installation is made by experienced workers following the guidelines and instructions supplied by the manufacturer and the appropriate industry standards. A rough opening should be prepared to receive each unit so that it can be installed and will finish out square, plumb, level, straight, and true. The design of most assemblies provides for minor adjustments at the job site, but no unit will operate or weather properly if it is twisted and misaligned during installation.

A properly selected, manufactured, and installed entrance or storefront will be weather resistant within itself. Other things must be done, however, to provide a weathertight installation. For example, spaces between rough openings and units should be filled with insulation, and joints between units and adjacent materials must be closed with sealant, trim members, or both, to prevent the passage of air, dust, and water.

Entrance frames and storefront sections are seldom designed to support loads other than themselves. Therefore, unless the frame is specially reinforced to support other loads, a separate lintel or other support is required.

Entrance and storefront frames should be supported around their entire perimeters and firmly and rigidly anchored to supporting construction. They should also be flashed at their heads and sills to provide a path back to the exterior for water that finds its way into the wall.

Some metals, such as aluminum and stainless steel, are compatible and can be used in contact with each other. Other metals, however, react with one another chemically or electrolytically and should be insulated from direct contact by waterproof, nonconductive materials such as neoprene, waxed paper, or coated felt. Such dissimilar metals that are located where water passing over them may contact another surface should be painted to prevent staining. Refer to Section 5.4 for a discussion of the effect of dissimilar metal materials.

Concealed aluminum in contact with concrete, masonry, or an absorbent material such as wood, paper, or insulation should be protected by coating either the aluminum or the adjacent material with a bituminous or aluminum paint, or by coating the aluminum with a zinc chromate primer, to minimize the chance or chemical corrosion of the aluminum by acids, alkalies, and salts leached out of the adjacent materials. Creosote and tar coatings, which may damage the aluminum, should not be used.

Sealants should be used in locations where frames, packaged entrances, and storefront sections adjoin other materials to make the joints there airtight and watertight.

Both interior and exterior extruded aluminum entrances and storefronts should be installed in accordance with AAMA's *Aluminum Store Front and Entrance Manual* and the manufacturers' recommendations.

Each interior frame should be anchored with jamb anchors near the top and bottom and not more than 24 in. (610 mm) apart in between. Jambs should be anchored at the floor with adjustable floor anchors.

Anchors in masonry should be long enough to extend at least 8 in. (203 mm) into the masonry. Anchors for stud walls should be clips designed to attach to the types of studs used. Several different methods are used to anchor frames in previously placed masonry or concrete.

8.3.9 REVOLVING DOORS

Revolving doors are both always open and always closed. They prevent passage of cold or warm air and noise while permitting people to pass through easily. They are proprietary packaged units that come complete with hardware (Fig. 8.3-5). They are made from stainless steel, aluminum, and copper alloys. Most revolving doors

FIGURE 8.3-5 Revolving doors usually include a canopy, which can be glazed or solid as shown here. (Courtesy Boonedam Entrance Technology.)

are between 6 and 8 ft. (1800 and 2400 mm) in diameter, but some are larger. Both three- and four-wing designs are available. Many options are available. Custom design is frequent.

Revolving doors operate either manually or with a power assist. They have speed controls for the safety of users and for protection of the door itself. Their optimal speed is 10 revolutions per minute, but slower operation is often desirable in such applications as hospitals, airports, and train stations. Slow-down switches are available to ensure safe passage of persons in wheelchairs (Fig. 8.3-6).

Revolving doors can be fitted with manual or automatic locks. Collapsible units are available so that the doors can provide an open access in the event of fire or other emergencies.

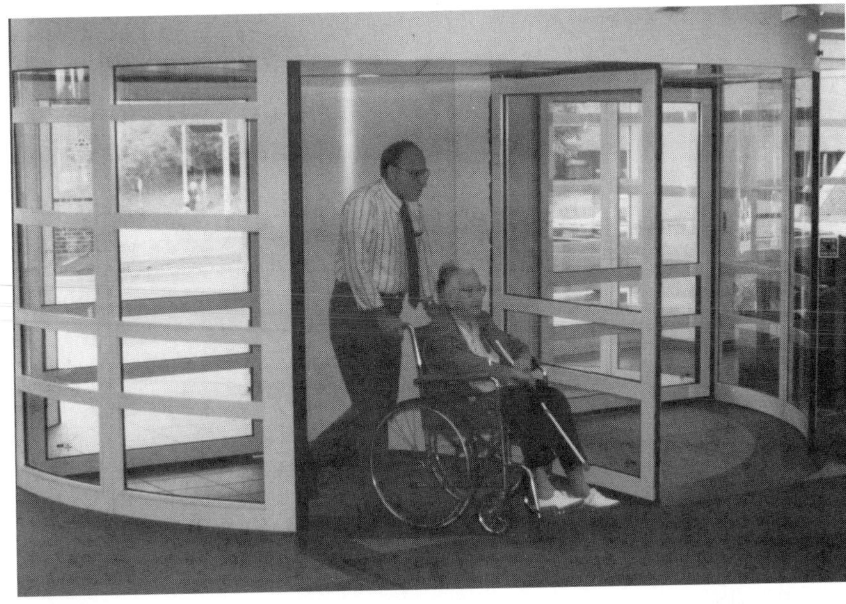

FIGURE 8.3-6 Large doors are available to permit the passage of wheelchairs or of several people at once. (Courtesy Boonedam Entrance Technology.)

8.3.10 ADJUSTMENT, CLEANING, AND PROTECTION

Because aluminum entrances and storefronts are prefinished building components, their installation should be withheld as long as the construction schedule will permit. Once installed, they must be protected against damage to their finish, appearance, and ability to operate properly. Units should be protected from splashing of plaster, mortar, and cleaning solutions.

8.4 Aluminum Windows and Sliding Glass Doors

The use of aluminum for window construction has increased significantly since aluminum windows were introduced for industrial application in the United States about 55 years ago. In contemporary buildings, glass areas have increased—the size of aluminum windows now allows for entire walls of glass and, in the case of aluminum sliding glass doors, movable walls of glass.

8.4.1 MANUFACTURE

The manufacture of aluminum windows and sliding glass doors consists of fabricating wrought aluminum alloys into the correct size and shape, using various techniques of die cutting, punching, notching, piercing, and sawing. The aluminum members thus fabricated are assembled with glass, hardware, weatherstripping, and other components into finished preassembled units.

The workability of aluminum, coupled with the ease and simplicity with which aluminum windows and sliding glass doors can be manufactured, has contributed immeasurably to their widespread availability and use. Modern machinery and production techniques produce a finished product whose parts can be fabricated to extremely close tolerances, resulting in weathertight assemblies having precision operation at relatively low cost.

8.4.1.1 Materials

Pure aluminum has few commercial applications, but when it is combined with other metals, such as manganese, silicon, iron, and magnesium, alloys are formed that are useful in a variety of applications. Refer to Section 5.4 for additional information about aluminum.

The principal materials used in aluminum windows and sliding glass doors are wrought (worked into shape by mechanical means) alloys. The alloys used for windows and sliding glass doors must have strength, be resistant to corrosion, be readily worked and machined for efficient fabrication and simple assembly, and provide surfaces suitable to receive protective and decorative finishes. Although other alloys are also used, the wrought aluminum alloy known as 6063 is a common one for making aluminum extrusions. It contains silicon and magnesium and possesses good formability, corrosion resistance, and medium strength. To make aluminum products, aluminum alloy is formed by the extrusion process and artificially aged by heat treatment. After heat treatment, usually in temper designation T5 (or T6), this alloy, known commercially as 6063-T5 (or 6063-T6), is commonly used for the main frame, sash, and ventilator members in both windows and sliding glass doors. Extruded shapes permit infinite possibilities of cross-sectional configuration to satisfy the requirements of glazing, weatherstripping, condensation control, and assembly.

Window and door locks, hinges, handles, and corners are often made from castings (formed by pouring or forcing molten metal into a mold) of aluminum alloy.

Other aluminum wrought alloys made by roll forming (passing strips of flat aluminum sheet through a series of forming rolls that bend it into a tube or other shape) also have a variety of uses in aluminum windows and sliding glass doors. These include such shapes as screen frames and glazing strips. Aluminum sheet is also cut and formed by stamping in a punch press, where accessories such as corner locks and screen clips can be made.

8.4.1.2 Finishes

The silvery sheen of the natural, mill finish surface of aluminum is a common finish used on low-cost aluminum windows and doors. This natural oxide coating that forms on aluminum provides an adequate degree of protection against corrosion in normal atmospheres and can be readily restored by rubbing with stainless steel wool if damaged or discolored. Normal oxidation over time will convert the silvery sheet to a soft gray. Other finishes often used may be either protective or decorative.

PROTECTIVE FINISHES

Protective finishes are necessary when abnormal atmosphere is present, such as found in heavy industrial and seacoast areas. Protective finishes are also necessary over certain types of decorative finishes, particularly where the surface is etched or polished.

Anodizing, an electrolytic process that provides a heavy oxide film on the surface, is a permanent protective finish against moderate abrasion and atmospheric attack. Clear anodizing, which has a frosty appearance when an etching material is used during the cleaning cycle, is the most economical finish when corrosion protection is a concern.

Certain permanent color tones on aluminum are possible using the anodizing process. These include several shades of gray, bronze, gold, and black. They are produced by varying the alloy constituents, the electric current, the electrolyte, and the time cycle. Practically any color is possible by impregnating the porous surface of an anodic film with dye prior to sealing. But since dye colors are not always permanent, they are not recommended for exterior use. (See Section 5.4 for a discussion of anodizing and other finishing processes.)

Organic coatings, such as paint, that are sometimes applied to aluminum for decorative purposes, may also provide some protection (see Section 9.15).

DECORATIVE FINISHES

Decorative finishes, such as a backed synthetic resin enamel, are colored or opaque coatings. They can be expected to have a life of up to 20 years when the coating is applied and baked on at the factory under controlled conditions. Organic coatings should comply with AAMA 2603.

8.4.1.3 Fabrication

Fabrication of aluminum alloy sections into window or sliding glass door components is accomplished by sawing, drilling, punching, and other traditional methods of metalworking. Small punch presses are often used. Aluminum's machinability permits the use of a variety of intricate punches and dies in a press to cut, form, shape, notch, and pierce the metal. Thus, components can be joined accurately and easily to make weathertight assemblies. A multiple-punch press can cut a long extrusion to the proper length and punch the required notches, holes, and slots, producing a complete framing member in a single operation.

8.4.1.4 Assembly

Materials other than aluminum, such as glass, plastics, and steel, are required to complete the assembly of windows and sliding glass doors.

GLASS AND GLAZING MATERIALS

Many aluminum windows and most sliding glass doors have the glass installed at the factory, which eliminates glazing at the job site and allows controlled methods of glazing not feasible with on-site glazing.

Glass selection by size and grade for windows depends primarily on wind load requirements and is often established by the applicable building code. In the absence of code requirements, the recommendations of AAMA 101/1.S.2 can be used. For sliding glass doors, in addition to wind load requirements, the glass must be of safety-type glass and be in compliance with ANSI Z-97.1 or other similar standards as regulated by codes or laws.

Other functional or other aesthetic requirements often dictate the selection of different types and higher grades of glass. These include double-glazing to minimize heat loss and heat gain, tinted glass for privacy and decorative purposes, tempered glass for additional strength, and tempered, wire, or laminated glass for additional safety.

Glazing materials for both aluminum windows and sliding glass doors include gaskets made from neoprene or polyvinyl chloride and mastic type compounds such as butyl- or rubber-based materials that are compatible with aluminum and do not require painting.

WEATHERSTRIPPING

The function of weatherstripping is to control air infiltration, prevent rain penetration, minimize dust infiltration, and withstand atmospheric conditions. It typically has a resilient part that prevents air penetration when it is pressed between the window and its sash.

Weather stripping must be durable under extended usage, resist corrosion, and be readily replaceable. Many aluminum windows and sliding glass doors are equipped in the factory with fabric-pile weatherstripping (similar to that used in automobiles) or with flexible vinyl plastic neoprene gaskets or fins. In some cases these materials are combined. Thin stainless steel, which compresses to accomplish a seal, is sometimes used.

FASTENERS AND HARDWARE

A variety of fasteners and hardware are required to hold aluminum window and sliding glass door parts together and to make them operable. Although aluminum is used extensively for this purpose, other metals are also used. These other metals must be either compatible with aluminum or insulated from direct metal-to-metal contact with it, to prevent electrolytic corrosion (galvanic action). Therefore, mild steel fasteners must be plated with zinc, cadmium, or another suitable material. Nonmagnetic stainless steel with a chromium content of not less than 16% need not be plated and can be used in exposed locations. Nylon, polypropylene, and other plastic materials are used for slides, bushings, bearings, and similar contact surfaces.

FRAME ASSEMBLY

The main frame, sash, and ventilators of aluminum windows and sliding glass doors are assembled by screwing or welding.

Screwed assemblies are shipped from the factory KD (knocked down) and later assembled by a distributor, dealer, or contractor. A slot to receive the screw threads is often extruded as an integral part of the aluminum member, and holes are prepunched to ensure perfect alignment of the corner joints (see Fig. 5.4-18). Lower corners of the main frame are usually sealed with a mastic to make them watertight. Units that are completely assembled at the factory are either staked or welded together.

Several welding methods are used. *Flash* (butt) welding produces a continuous weld across the entire cross section of a shape. It is made by forcing two members together at their intersection under pressure and applying a voltage that melts the parent metal. Flash welding uses no filler material or flux and forms a rigid, strong, and neat watertight joint.

Inert-gas welding uses filler material and is similar to conventional gas welding used on other metals. It is seldom continuous and is generally used to join members that are of dissimilar shapes. Inert-gas welded joints can be made reasonably watertight, but a mastic sealing material may be required at the lower corners to prevent water leakage.

THERMAL BREAK

A metal frame that holds insulating glass should have a thermal break to prevent a thermal "short circuit," in which heat flow is resisted by the insulating glass but flows freely through the aluminum frame. A thermal break is polyurethane, vinyl, or another material low in conductivity that provides a separation between the outside and inside portions of a frame.

An aluminum frame with a good thermal break has an R-value of 1.66, which is approximately equal to that of double-pane insulating glass. A comparable frame with no thermal break would have an R-value of 1.18.

In some areas it may be economically practical to use three layers of glass separated by air spaces. Three layers of sealed glass with two ¼-in. (6.4 mm) air spaces, for example, has an R-value of 2.13, as compared with 1.72 for typical insulating glass. This technique of triple glazing can also be accomplished by installing storm sash over double-glazed windows.

ACCESSORIES

Window accessories are assembled and attached during all stages of fabrication and assembly. For example, weatherstripping is frequently pulled into retaining grooves in long lengths of extruded frame members before cutting and punching takes place. Operating hardware such as sash balances, pivot arms, guides, and guide followers are usually assembled after the main frames are together. Accessories such as locking hardware and handles are attached by riveting, bolting, screwing, staking, or interlocking with frame members. Manual rotating operators on casement, awning, and jalousie windows are usually the last accessories to be assembled, sometimes being left until after installation of the unit at the job site.

GLAZING

In one method of factory glazing, a glazing compound is applied to the glazing leg of a window as a back-bedding material; the glass is then dropped in and held in place by aluminum or rigid plastic face stops. The stops snap into place to serve as trim molding and to hold tension on the assembly until the glazing compound has set. Another method widely used in the factory to glaze units is to apply a channel-shaped gasket over the edge of the glass and then force the channel shape into the window or door frame around it. Flexible vinyl plastic is most often used for this application.

When windows or sliding glass doors are glazed at the job site, the manufacturer's recommendations should be followed to ensure a satisfactory result.

8.4.2 Standards of Quality

Two types of aluminum windows are covered in AAMA documents: *prime windows* and *architectural windows*. Sliding glass doors are covered in the same documents as prime windows.

8.4.2.1 Prime Windows and Sliding Glass Doors

Most windows used in low-rise buildings are *prime windows*. Their design and construction and that of aluminum sliding glass doors should conform to AAMA 101/I.S.2, which is the recognized standard for the aluminum window and sliding glass door industry. It was developed by AAMA and adopted by ANSI. It includes minimum provisions for frame strength and thickness, corrosion resistance, air infiltration, water resistance, wind load capability, and other requirements necessary to ensure adequate performance.

TYPES

Various types of aluminum windows and doors covered by AAMA 101/I.S.2 are illustrated and described in Figure 8.4-1.

DESIGNATION SYSTEM

Prime aluminum windows and sliding glass doors are given designations that describe the type of window or door, its grade, its design pressure, and supply other information. For example, DH-DW-R-15 conveys the following information: The first letters describe the window or door type—DH means double-hung window (C means casement window, SGD means sliding glass door, etc.); DW signifies a dual window; R means Residential grade (C means Commercial grade, HC means Heavy Commercial grade); and the number designates the design pressure (DP).

PERFORMANCE REQUIREMENTS

AAMA-certified windows and sliding glass doors are required to have certain performance characteristics. These characteristics are determined by the effect on the window or sliding glass door when tested at specific levels of design pressure (DP), structural test pressure (STP), water resistance test pressure (WTP), and air infiltration test pressure and rate of infiltration.

Condensation resistance and heat transmission are two other factors that should be taken into account when selecting windows, depending on a building's location, size, and design.

Selection of a window or sliding glass door begins by determining the type desired (awning, double-hung, projected, sliding glass door, etc.). Then the required *design pressure* (DP) is calculated. *Residential* windows and sliding glass doors are rated at a minimum DP of 15 lb./sq. ft. (psf) (73.24 kg/m^2); *Commercial* windows and sliding glass doors at a minimum of 20 psf (97.65 kg/m^2); and *Heavy Commercial* units at a minimum of 40 psf (195.3 kg/m^2). This rating is called a *performance class*. High-rise structures may require windows of much higher performance classes than these minimums, varying up to as high as 100 psf (488.24 kg/m^2).

Design pressure (DP) is determined by using the fastest wind speed likely to be encountered at a building's location. A map in AAMA 101/I.S.2 can be used to estimate this maximum wind speed. After the expected wind velocity has been determined, the resulting VP (velocity pressure), DP, STP, and WTP are determined from the load tables in AAMA 101/I.S.2. These load tables are based on wind velocities from 70 mph (112.65 km/h) to 110 mph (177 km/h) and the height of a window above the ground from zero (at grade) to 500 ft. (152.4 m). The *velocity pressure* (VP) taken from these load tables

FIGURE 8.4-1 Summary of Aluminum Windows and Sliding Glass Doors

Type	Designation and Application	Description
Double Hung 	DH-R15 Residential DH-DW-R15 Dual Window Residential DH-C20 Commercial DH-DW-C20 Dual Window Commercial DH-HC40 Heavy Commercial DH-DW-HC40 Duel Window Heavy Commercial	*Double-hung windows* have two operable sash. The sash move vertically within the main frame with the assistance of mechanical balancing mechanisms that minimize the effort required to raise the sash. 50% of the window area is available for ventilation.
Casement 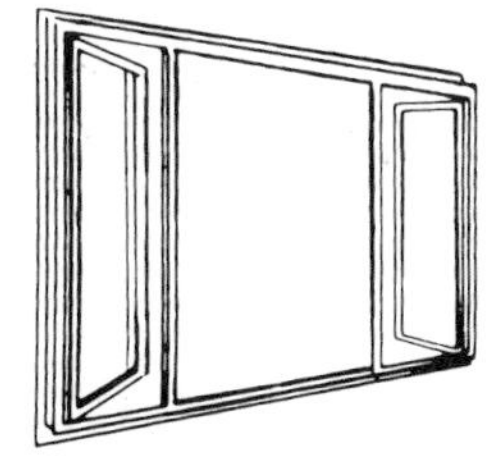	C-R15 Residential C-C20 Commercial C-HC40 Heavy Commercial	*Casement windows* contain side hinged ventilators that operate individually and swing outward. They frequently have nonoperating fixed lights (shown) which the ventilators close to. Casements may be open 100% for ventilation.
Projected 	P-R15 Residential P-C20 Commercial P-HC40 Heavy Commercial	*Projected windows* consist of horizontally mounted ventilators that may project out or in. They differ from awning and hopper windows in that the hinged side of the ventilator moves in a track up or down when the ventilator is operated. Used commonly where large glass areas are desired and where it is desired to deflect incoming air in an upward direction.
Awning 	A-R15 Residential A-C20 Commercial	*Awning windows* consist of a number of top hinged ventilators operated by a single control device that swings the lower edges outward giving an "awning" effect when open. Operation is through a roto operator which also serves as the lock. Ventilating area is 100% of the window area.
Horizontal Sliding 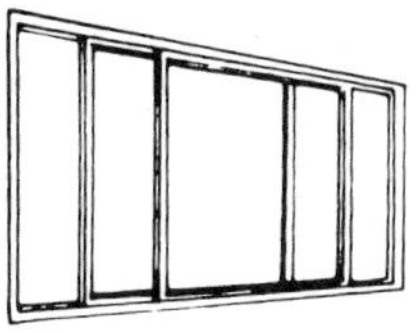	HS-R15 Residential HS-DW-R15 Dual Window Residential HS-C20 Commercial HS-DW-C20 Dual Window Commercial HS-HC40 Heavy Commercial HS-DW-HC40 Duel Window Heavy Commercial	*Horizontal sliding windows* have one or more operable sash arranged to move horizontally within a main frame. Ventilating area is 50% of the window area. This type of window is sometimes combined with fixed lights and constructed in large sizes reaching the proportion of a "window wall."

(Continues)

FIGURE 8.4-1 *(Continued)*

Type	Designation and Application	Description
Jalousie	J-R15 Residential	*Jalousie windows* (louver windows) consist of a series of overlapping horizontal glass louvers that pivot together in a common frame. Ventilating area is 100% of the window area. Jalousies are not as resistant to air and dust infiltration as other types, but are used in moderate climates and as porch enclosures in all climates. They should be used only where large heat losses or heat gains can be tolerated. Insulating effectiveness may be improved by inside storm panels, designed to be interchangeable with screens.
Jal-Awning	JA-R15 Residential	*Jal-awning windows* resemble an awning window in appearance but use different operating and locking mechanisms. These mechanisms are usually separate and require individual operation. Ventilating area is 100% of the window area.
Vertical Sliding	VS-R15 Residential VS-DW-R15 Dual Window Residential	*Vertical sliding windows* resemble double hung windows in appearance but use no sash balancing devices. They are operated manually by lifting and are held in various open positions by mechanical catches that engage in the jamb or hold by friction. Ventilating area is 50% of the window area.
Vertical Pivot	VP-C20 Commercial VP-HC40 Heavy Commercial	*Vertically-pivoted windows* consist of a ventilator sash mounted on pivots located in the center of the top and bottom main frame members. This allows the ventilator to be reversed by rotating and permits the glass to be cleaned from the interior. This window is not used for normal ventilation and is operated solely for cleaning purposes or emergency ventilation. It is used generally in completely air conditioned high rise buildings.
Top-Hinged	TH-C20 Commercial TH-HC40 Heavy Commercial	*Top-hinged, in-swinging windows* consist of a ventilator hinged to the main frame at the top that swings into the room. This permits the glass to be cleaned from the interior. This window is not used for normal ventilation and is operated solely for cleaning purposes and emergency ventilation. Its use is generally in completely air conditioned high-rise buildings.

(Continues)

FIGURE 8.4-1 *(Continued)*

Type	Designation and Application	Description
Sliding-Glass Door 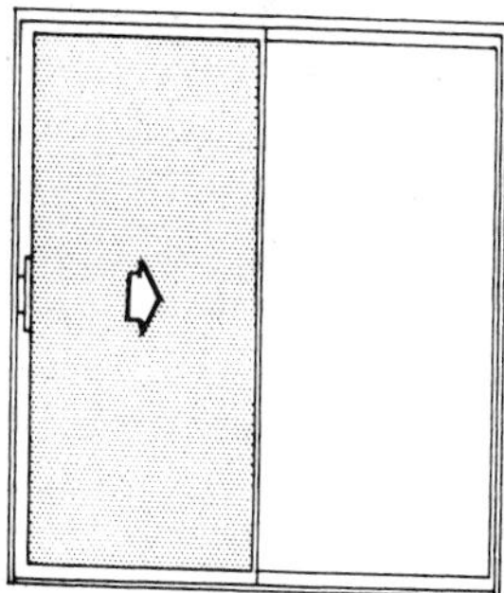	SGD-R15 Residential SGD-C20 Commercial SGD-HC40 Heavy Commercial	*Sliding glass doors* consist of two or more framed glass panels contained in a main frame designed so that one or more of the panels move horizontally in the main frame. The most common arrangement is of 2 panels. "Single-Slide" has one operable panel, a "Double-Slide" has two. The ventilation area of a single and double slide door is 50% of the main frame area. Doors are also made in 3 and 4 panel arrangements: 3 panel arrangement can provide ⅓ or ⅔ of its area for ventilation, and 4 panel can provide ½ or in some cases ¾ of its area for ventilation.
Fixed 	F-R15 Residential F-DW-R15 Dual Window Residential F-C20 Commercial F-DW-C20 Dual Window Commercial F-HC40 Heavy Commercial F-DW-H40 Dual Heavy Commercial	*Fixed windows* may utilize any of the other types. They consist of a frame with no operating sash. Fixed windows may have mullions or muntins.

varies with building height above grade and wind speed. For example, a 70-mph (112.65 km/h) wind will produce a VP of 10 at grade and 17.3 at 100 ft. (30.48 m) up. The VP is determined by the following formula:

$$VP = 0.00256K_Z(IV)^2$$

where

V = the wind velocity in miles per hour taken from the map in AAMA 101/1.S.2.

I = the importance factor as described in AAMA 101/1.S.2.

K_Z = the velocity pressure exposure coefficient as described in AAMA 101/1.S.2. This value depends on the height above ground and the exposure category.

The *design pressure* (DP) is defined as 1.25VP, and the *uniform load structural test pressure* (STP) for aluminum windows and sliding glass doors is 1.25DP.

The *water resistance* of AAMA-rated windows and sliding glass doors is determined by testing them in accordance with ASTM E 331 and E 547. The *water resistance test pressure* (WTP) is established at 15% of DP, but never more than 2.86 psf (13.96 kg/m^2). This is equal to a wind velocity of 33.4 mph (53.75 km/h). At this test pressure, water is applied to the test sample at a rate of 5 gallons (18.93 L) per hour, which is equivalent to 8 in. (203 mm) of rain per hour.

The *air infiltration* of AAMA-certified windows and sliding glass doors is tested in accordance with ASTM E 283. The usual pressure is 1.57 psf (7.67 kg/m^2) for residential and commercial sliding glass doors and windows of all types, 1.57 psf (7.67 kg/m^2) for double-hung and sliding heavy commercial windows, and 6.24 psf (30.47 kg/m^2) for other heavy commercial units. The normal maximum rate of air infiltration is 0.37 cfm (0.01 m^3/minute) per foot (305 mm) of operating ventilator or sash crack for most windows. The rate is 1.50 cfm (0.042 m^3/minute) for jalousie windows, 0.15 cfm (0.004 m^3/minute) for fixed windows, and 0.37 cfm (0.01 m^3/minute) for sliding glass doors, all based on the total area of the unit in square feet (m^2).

The *condensation resistance factor* (CRF) is determined by procedures outlined in AAMA 1503. It gives an indication of a window's ability to resist condensation. The higher the CRF, the less likely it is that condensation will occur.

As discussed in Section 15.2, heat will flow readily through windows and sliding glass doors unless some means are used to reduce this flow. The methods for determining conductive heat transmission through windows and doors are included in AAMA 1503.

Where design calculations indicate the need to reduce the heat transmittance of aluminum windows and sliding glass doors, significant reduction can be accomplished by incorporating double- or triple-glazing and by thermalizing the frames and sash. At any rate, windows and sliding glass doors in areas where the winter design temperature is 10°F (–12.22°C) or less should incorporate such devices regardless of calculations. In fact, double- or triple-glazing is used in most new windows installed today, especially in commercial applications.

Thermalizing involves separating the

aluminum parts on the warm inside portion of a window from those on the cold outside portion through the use of a thermal break, as discussed earlier.

8.4.2.2 Architectural Windows

Architectural windows are covered in a separate AAMA document, GS-001. It includes high-quality windows that were previously known as *monumental windows*.

Since GS-001 covers essentially the same provisions as AAMA 101/1.S.2, much of the discussion in Section 8.4.2.1 also applies to architectural windows. There are some differences, however, in types, grades, and performance criteria.

TYPES

Architectural window types are similar to prime window types, with some exceptions. For example, no awning, jal-awning, jalousie, or greenhouse windows are covered in GS-001 as they are in AAMA 101/1.S.2. Nor are sliding glass doors included. However, GS-001 includes several windows that are not covered in 101/1.S.2. These are horizontally pivoted windows, which are similar to vertically pivoted windows turned on their side, and side-hinged inswinging windows, which are similar to casement windows that swing inward. Some of the features of windows in GS-001 with the same names as those covered by AAMA 101/1.S.2 are slightly different.

GRADES

There are no grades for architectural windows in GS-001. Instead, they are divided into three categories: standard, modified, and custom. *Standard architectural windows* are made entirely from the manufacturer's stock components. In *modified architectural windows*, the manufacturer's stock components may be altered slightly. *Custom architectural windows* are entirely custom-designed units.

PERFORMANCE REQUIREMENTS

The major differences between the requirements of AAMA 101/1.S.2 and AAMA GS-001 are in the performance requirements. For example, the minimum test unit size required by GS-001 is much larger than most of the sizes required by 101/1.S.2. In addition, the static air pressure required in the air infiltration test is 6.24 psf (30.47 kg/m^2) for all units, which is larger than that required for all but a few heavy commercial units in 101/1.S.2. The permitted infiltration rates are also much smaller in GS-001.

The air pressure difference for the water resistance test in GS-001 is 8 psf (39.06 kg/m^2), which is much higher than that required by 101/1.S.2.

Most of the requirements in GS-001 are higher than those in 101/1.S.2, but not all. For example, the minimum structural test pressure (STP) in GS-001 is 30 psf (146.47 kg/m^2); in 101/1.S.2, it is 15 psf (73.24 kg/m^2). However, this test pressure for heavy commercial windows in 101/1.S.2 is 40 psf (195.3 kg/m^2), which is higher than the minimum required by GS-001.

8.4.2.3 Certification of Quality

In addition to developing standard specifications, AAMA sponsors a program of independent testing, evaluating, inspecting, and labeling aluminum products to certify compliance with ANSI/AAMA standards. The certification program is accredited by the American National Standards Institute (ANSI). Products that conform with the requirements of the applicable specifications are identified by the AAMA label.

The manufacturer's code number, appropriate standard, and maximum size tested is indicated on the label. The label must be visible on the metal after installation.

The AAMA certification program is operated in accordance with ANSI Z34.1, which provides for certification to the public on an impartial, independent, and continuing basis. All certifications guarantee compliance with the ANSI specifications.

8.4.3 INSTALLATION

The finest window or door design and the highest quality of manufacture will not compensate for poor installation. Weathertight installation, although simple, can be ensured only if an aluminum window or sliding glass door is installed by an experienced worker following the guidelines and instructions supplied by the manufacturer. A rough opening should be prepared to receive the unit so that it can be installed and will finish out square, plumb, level, straight, and true. The design of most windows and doors provides for minor adjustments at the job site, but no unit will operate or weather properly if it is twisted and misaligned during installation.

A properly selected and manufactured window or sliding glass door will be weather resistant as a unit, but if the building is insulated, the spaces between the rough openings and the windows or sliding glass doors should be filled with insulation. In addition, the spaces between the unit and adjacent materials must be closed with a sealant, trim members, or both, to prevent the passage of air, dust, and water.

Windows and sliding glass doors should be rigidly supported and nailed, screwed, or otherwise anchored securely around the entire frame to the supporting construction. They should also be flashed at their heads and sills to provide a path back to the exterior for water that finds its way into the wall.

Anchoring materials and flashing should be of aluminum or a material compatible with aluminum, such as stainless steel, cadmium-coated steel, or galvanized steel. Dissimilar materials, such as copper and bronze, should be insulated from direct contact with aluminum by waterproof, nonconductive materials, such as neoprene, waxed papers, or coated felts. Dissimilar metals located where water passing over them may contact a window or door should be painted to prevent staining the aluminum. Refer to Section 5.4 for a discussion of the effect dissimilar materials can have on aluminum.

Concealed aluminum in contact with concrete, masonry, or absorbent material such as wood, paper, or insulation should be permanently protected by coating either the aluminum or the adjacent material with a bituminous or aluminum paint, or by coating the aluminum with a zinc chromate primer, to minimize the chance of chemical corrosion of the aluminum by acids, alkalies, and salts leached out of the adjacent materials. Creosote and tar coatings, which may damage the aluminum, should not be used.

Because aluminum windows and sliding glass doors are prefinished building components, they should be handled and treated to prevent damage. Once installed, they should be protected to prevent damage to their finish, appearance, and ability to operate properly. Units should be protected from splashing of plaster, mortar, and cleaning solutions.

8.5 Wood Windows and Sliding Glass Doors

Wood was one of the earliest materials used in the manufacture of windows, and over the years it has proven its suitability for this purpose. Wood's insulating properties, ready availability, ability to take either natural or painted finishes, and ease of fabrication and repair with simple tools all contribute to its widespread use in window manufacture.

Recent technological innovations have further improved wood as a material for window and door construction. Kiln drying of lumber reduces shrinkage, resulting in better operation and weathertightness.

Water-repellent preservative chemical treatments reduce swelling and warping, improve paint retention, and increase wood's resistance to decay and insect attack under all climatic conditions. Windows that cover entire walls consist of combinations of fixed glass for light and view and operating sash for light and ventilation. In some window walls, sliding glass doors also provide access to the outdoors in addition to light and ventilation.

For maximum comfort at minimum heating and air conditioning costs, windows and doors must be weathertight, insulated, and condensation free. In addition, they must be designed and built to permit ready adaptation to varying installation conditions.

This section outlines the manufacture, nomenclature, standards of quality, and installation of stock wood windows and sliding glass doors. Stock windows and doors are manufactured at mills in a variety of common types and sizes.

8.5.1 MANUFACTURE

Most stock wood windows are manufactured in accordance with standards developed by the National Wood Window and Door Association (NWWDA) and adopted by the American National Standards Association (ANSI). The publication containing these standards is ANSI/NWWDA I.S. 2, "Industry Standard for Wood Windows." Sliding door manufacture employs materials and methods generally similar to those used in window manufacture. The manufacture of sliding doors is regulated by NWWDA I.S. 3, "Industry Standard for Sliding Patio Doors." The discussion in this section presumes a level of quality conforming with these standards.

8.5.1.1 Materials

Most stock windows and sliding glass doors are manufactured from ponderosa (western) pine. This species has excellent workability, gluability, nail-holding capacity, and a uniform light color suitable for natural or painted finishes. Other species commonly used are southern yellow pine, Idaho white pine, sugar pine, and Douglas fir. The lumber must be sound, free from defects such as loose knots and excessive checking, and kiln dried to a moisture content between 6% and 12%.

8.5.1.2 Fabrication and Assembly

Kiln-dried lumber is delivered to a millwork shop for fabrication into window and sliding glass door components. There it is cut to specified lengths and widths on ripsaws and crosscut saws, leaving the cut stock substantially clear and without defects. Cut lengths are then sorted and marked to indicate the window or door component for which they are intended. Like pieces are conveyed to molders or lineal milling machines that shape the pieces to the desired profile. The next step is cross-milling, where necessary grooves and channels (dadoes, mortises, and tenons) (Fig. 8.5-1) are cut across each piece to facilitate weathertight joining of units. Milling of components is completed by routing machines that cut notches to receive recessed hardware.

When shaping has been completed, milled components and assembled sash are preservative treated with a solution of toxic, water-repellent agents. Nonpressure treatments, such as the vacuum or immersion processes, are commonly used. Besides imparting water-repellent properties and increasing resistance to decay fungus and insect attack, preservative treatment gives the wood increased dimensional stability, minimizing intermittent swelling and shrinking due to varying atmospheric conditions or contact with water. Preservative treatment should be in accordance with NWWDA I.S. 4, "Industry Standard for Water-Repellent Preservative Treatment for Millwork." Storm sash and exterior components may also be paint-primed or factory finished for additional protection.

Following preservative treatment, weatherstripping is pressure-fit into grooves or surface-tacked, and some hardware is installed. Sash and frames are then assembled from components on automatic nailing machines. A continuous sander gives the assembled sash three separate sandings—coarse, medium, and fine—in one trip through the machine.

The final steps consist of glazing, fitting sash to the frames, and assembly with all operating hardware. Completed window units are usually provided with angle and spacer braces to prevent racking and distortion in shipment.

WEATHERSTRIPPING

Weatherstripping provides a seal against infiltration (leakage of air) and penetration of dust and wind-blown rain. It should be incorporated on all ventilating sash to reduce infiltration substantially below the maximum established by NWWDA standards. The maximum is 0.34 cubic ft. per minute (cfm) (0.0096 m^3 per minute) for Grade 20 (see Section 8.5.3.1) windows and doors; 0.25 cfm (0.007 m^3 per minute) for Grade 40; and 0.10 cfm (0.003 m^3 per minute) for Grade 60. The measurement is per linear foot (305 mm) of sash crack length for operating sash, and per square foot (0.093 m^2) of surface for fixed sash.

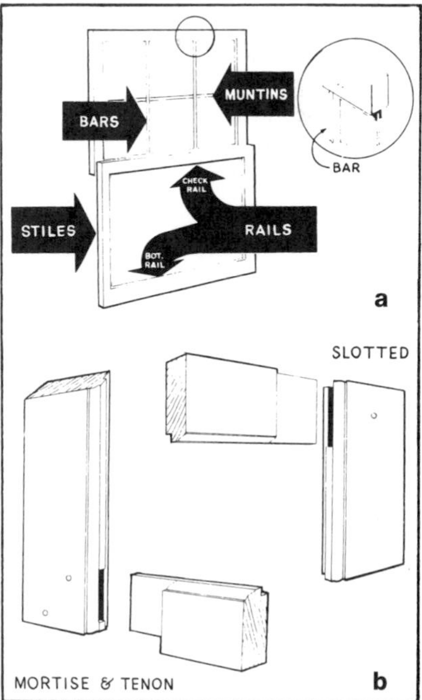

FIGURE 8.5-1 (a) Sash-component terminology; (b) common millwork methods of joining sash corners.

The test pressure is a uniform 1.57 psf ($0.16 m^2$).

Weatherstripping typically has a resilient part that prevents air penetration when it is pressed between the door or window and its sash.

The most common weatherstripping types for wood windows are foam rubber, felt stripping, rolled vinyl, casement stripping, spring metal, interlocking metal, and compressible bulb.

Foam Rubber *Foam rubber weatherstripping* has either an adhesive or wood backing. It is easy to install but lacks durability. The adhesive type can be used wherever there is no friction, such as at the top and bottom of double-hung windows.

Felt Stripping *Felt weatherstripping* has the same advantages and disadvantages as foam rubber, but it also collects dirt more readily. Metal-backed felt, which provides more durability, is also available.

Rolled Vinyl *Rolled vinyl weatherstripping* is available with or without a metal backing. Some types are manufactured with tubular gaskets that can be either hollow or filled with foam (see Fig. 8.7-10c). Rolled vinyl is very durable and can be installed on the exterior of double-hung, sliding, and casement windows.

Casement *Casement window weatherstripping* can be one of several kinds, depending on whether the window is wood or metal. Wood casements can be sealed with felt, adhesive-backed foam, or spring metal.

Spring Metal A common weatherstripping used in wood windows is *spring-tension weatherstripping* of bronze, aluminum, rigid vinyl, stainless, or galvanized steel (Fig. 8.5-2a). Extruded aluminum, cold-formed galvanized steel, and rigid plastic are used as sash guides and integral weatherstripping in many double-hung windows. These may be either fixed to the window frame or cushioned with plastic foam or springs to provide constant pressure against the sash stiles. In the fixed type, *woven felt* in the stiles sometimes engages a projecting fin in the sash guide for improved weather-tightness (Fig. 8.5-2b). The cushioned *compression sash guide*, in addition to affording weather protection, often provides for easy removal of the sash (Fig. 8.5-2c). Metal or plastic sash guides eliminate the need for finishing sash runs and the possibility of their becoming clogged with paint or varnish.

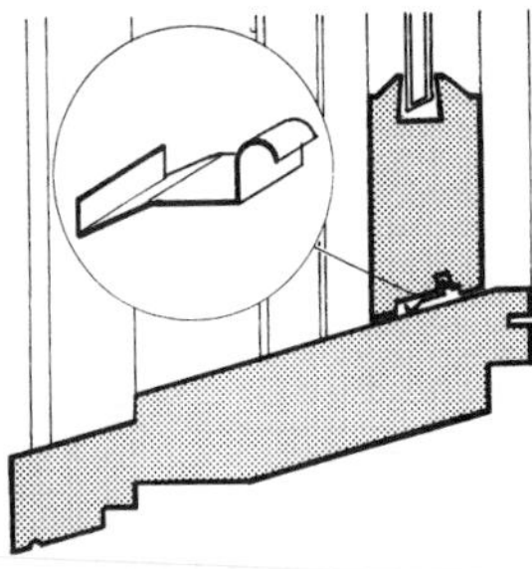

a Spring Tension

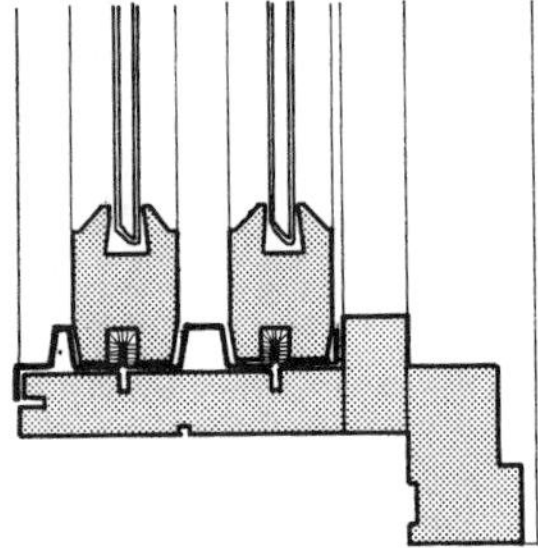

b Woven Felt

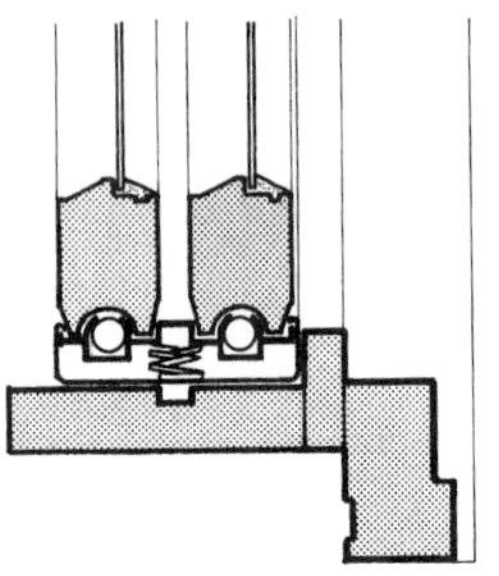

c Compression Sash Guide

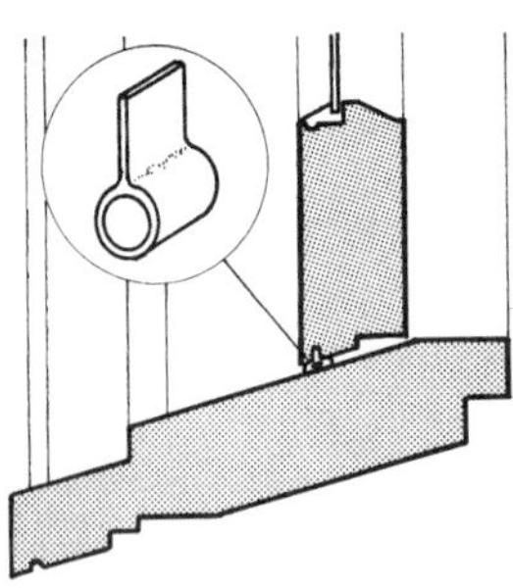

d Compressible Bulb

FIGURE 8.5-2 Common types of weatherstripping: (a) spring-tension; (b) woven felt; (c) compression sash guide; (d) compressible bulb.

Spring metal stripping is available in two types: flat stripping, which is most commonly used for double-hung windows, and V-strip, which is used for doors. Flat stripping is inserted into the narrow side channels of a window sash on the bottom rails of the upper and lower sash and on the top rail of the upper sash.

Interlocking Metal *Interlocking metal stripping* is produced as interlocking channels and interlocking J-strips. Both are difficult to install, because alignments must be exact for a secure interlock. Both also provide excellent, durable seals. The channels are subject to damage, however, because they are exposed. The J-strips are not visible when installed; they are not exposed and, therefore, not subject to damage.

Compressible Bulb Type *Compressible bulb weatherstripping* of vinyl or neoprene is found most often in storm sash, in screens, and in sliding glass doors. It is employed by some manufacturers in main sash (Fig. 8.5-2d). In addition to flexible vinyl, woven fabric weatherstripping is also common in sliding glass doors.

8.5.1.3 Burglar Protection

The construction of windows generally makes them more vulnerable to unauthorized entry than doors. Nine out of ten times a potential intruder will find an open window.

Even when windows are closed and locked, the glass can be broken if an intruder does not mind making noise. When the panes are too small to crawl through, an intruder will be forced to spend more time and make more noise in the process of enlarging the opening. Windows can also be forced open with a small crowbar.

In lieu of regular glass, one of the following security glasses may be used: (1) laminated, consisting of a vinyl panel sandwiched between two glass panels, (2) wired, or (3) tempered. Acrylic or polycarbonate glazing can also be used. Unfortunately, plastics are subject to surface scratches that can detract from their appearance over time.

Exterior ornamental grillwork and interior window grates can be used to protect windows. Gates and grillwork should be installed with nonretractable screws. Grillwork should never be installed on windows needed as fire exits. Interior window grates should have an integral locking knob system, rather than a padlock, so that they do not become an obstacle in case of fire.

Refer to Section 8.7 for additional discussion of burglar protection.

HARDWARE

Hardware provides effective closure, operates a sash, and holds it stationary at the desired degree of openness. Hardware types range from simple hinges and locks on basement windows to complex arrangements on windows that require interior operation without removing screens. Underscreen rotary gear, push bar, and lever-type operators are commonly used for awning windows in combination with sliding friction hinges, which provide projected action (Fig. 8.5-3a). Operators are usually capable of weathertight closure, but on larger sash, additional awning locks

a Awning

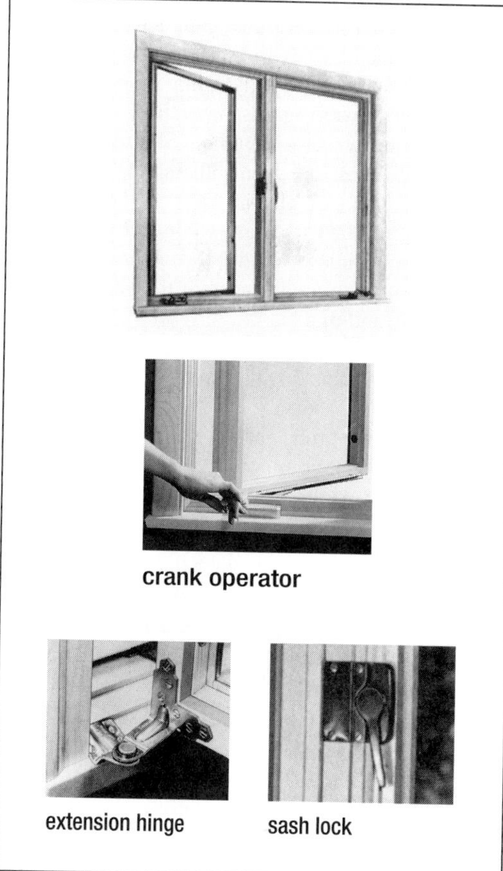

b Casement

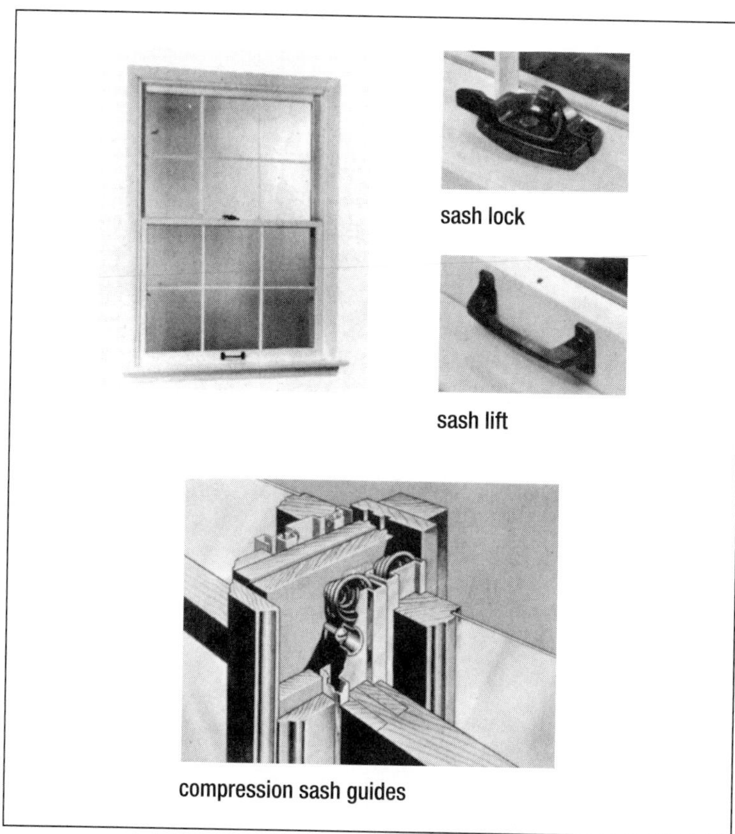

c Vertical Sliding

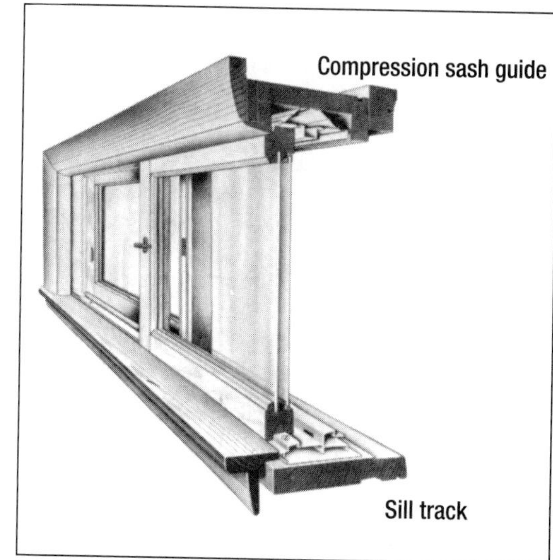

d Horizontal Sliding

FIGURE 8.5-3 Commonly used hardware for various window types. Note in (a) and (b) that sliding friction hinges provide projected action; for example, when awning sash is opened, the top rail moves downward.

may be provided. Outswinging casement windows often utilize similar rotary gear operators, with either sliding hinges or extension hinges (Fig. 8.5-3b); one or more casement fasteners are usually necessary to draw up relatively long sash stiles against the frame.

Double-hung and horizontal sliding windows generally employ cam-action sash locks, which draw the sash together at the meeting stiles (or rails) while pushing the sash outward against the frames to compress the weatherstripping. Sliding windows are often equipped with a spring-cushioned metal or plastic track at the head jamb, similar to the sash guide of the double-hung window; fixed metal or plastic tracks are commonly used at the sill (Fig. 8.5-3c and d).

Double-hung windows are usually equipped with balancing hardware, which assists in raising the sash and holds it in position. Spiral spring, spring, and coiled tape are the most common types of balancing devices (Fig. 8.5-4). Smaller and lighter sash may not require mechanical assistance for raising. In these, the friction fit provided by a compression sash guide may be sufficient to hold sash in place (see Fig. 8.5-3c).

Exposed hardware such as sash locks, sash lifts, and various operators are available in corrosion-resistant standard (zinc, bronze, aluminum, chrome) and proprietary finishes.

LOCK TYPES

Most window locks are vulnerable to prying and are therefore equipped with various locking devices (Fig. 8.5-5).

A friction-type device has a pin and a receptacle into which the pin is screwed. Friction is supposed to keep the window in place, but it takes very little effort to nudge the pin from the receptacle.

A pin-type lock is a keyed device. Its bolt slips into a hole drilled into the upper window frame. Although stronger than the friction type, this lock is vulnerable to prying. Because of the keyed lock, it may be a safety hazard in case of fire.

A wedge-type ventilating lock is self-defeating. It works to hold the window shut by spreading the upper and lower sash apart. A potential burglar can easily slip a knife blade between sashes and cut off the wedge.

A stop-type ventilating lock uses a knob or other protrusion to limit the distance a window can be opened. Although it does not spread the window sashes apart, it too is vulnerable to prying.

A key-operated cam latch replaces an ordinary double-hung window cam latch. Like a pin-type lock, it is vulnerable to prying and may be a safety hazard in case of fire.

An eyebolt, installed according to the recommendations of Consumer's Union, is the most effective device to prevent the forcing of a window.

To accommodate the eyebolts, holes are drilled on the sides of the interior of the window. They should go all the way

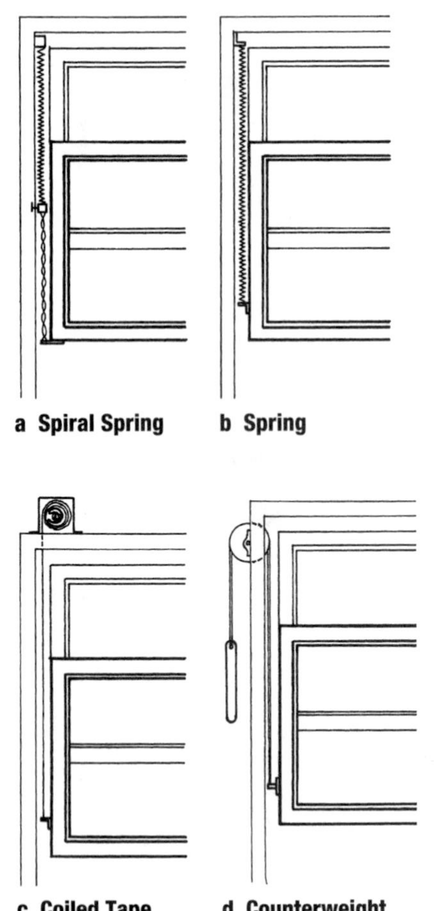

FIGURE 8.5-4 Balancing devices for double-hung windows.

FIGURE 8.5-5 Window locks include (a) friction type, (b) pin type, (c) wedge type, (d) stop type, and (e) key-operated cam latches.

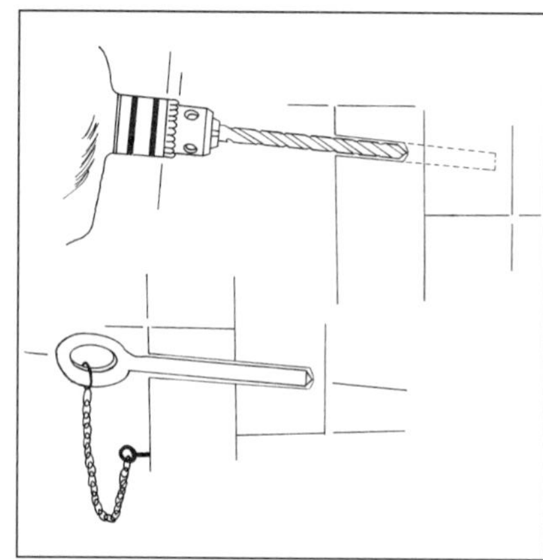

FIGURE 8.5-6 A simple pry-proof window lock: a 5⁄16-in. (23.8 mm) eyebolt is inserted through holes drilled through the inside and outside sashes.

through the inside sash and three-quarters of the way through the outside sash at a slight downward angle, as shown in Figure 8.5-6. Two sets of holes can be drilled in the outside sash so that the window can be partially opened for ventilation. Eyebolts, 5⁄16 in. (7.9 mm) in diameter, inserted into the holes act as locking pins.

GLASS AND GLAZING

Single-strength (3⁄32 in. [2.38 mm]) glass is limited to sizes of less than 76 *united inches* (1930 united mm) (width plus height); and double-strength (1⁄8 in. [3.2 mm]) glass, to sizes of less than 100 united inches (2540 united mm). Additional information on grades and sizes of glass is given in Section 8.8. Most stock windows and sliding glass doors are available with either insulating glass or single glazing; tinted glass and patterned glass are sometimes available on special order. The Consumer Product Safety Commission (CPSC) Safety Standard for Architectural Glazing Materials requires safety glazing material in sliding doors.

Much wood window sash is face-glazed by bedding (back puttying) the glass in a rabbet, securing it with glazing points, and sealing it with a bevel face-putty bead (Fig. 8.5-7a). Sometimes pointing is eliminated by using adhesive bedding materials that cushion the glass, seal it, and secure it in place.

Wood stop glazing replaces both pointing and face puttying (but not bedding) and provides a glass seal of greater strength, durability, and better appearance (Fig. 8.5-7b). Many wood sliding glass doors and large fixed windows are glazed in this manner.

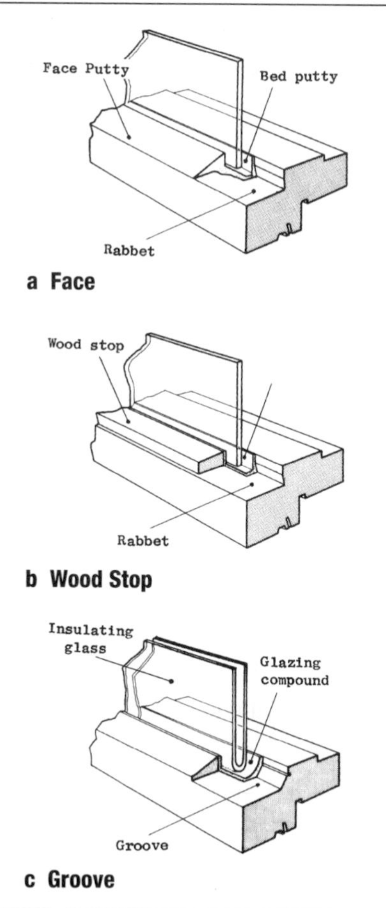

FIGURE 8.5-7 With face glazing (a), glazing compound holds glass in wood seat; with wood stop glazing (b), glass is bedded in compound and secured by wood stop; with groove glazing (c), glass is bedded in compound and set in continuous groove.

Groove glazing, employed by some manufacturers, has the same advantages as wood stop glazing but eliminates the need to handle a separate stop (Fig. 8.5-7c). In this method, grooves in the sash are filled with glazing compound and sash members are assembled around the glass, clamped in a press, and nailed or pinned together. To facilitate reglazing, one of the sash members can be removed by drawing out the pins that join it at the corners. The sash is reassembled after the new glass has been placed in the groove.

SCREENING

Aluminum, galvanized steel, bronze, or vinyl-coated glass fiber insect screening may be used. Mesh size is limited to 18 by 14 or 18 by 16 (number of openings per inch [24.4 mm]).

Screening is secured to wood sash by rolling it into a groove and splining, by tacking, or by stapling. A sash is usually trimmed out with a screen mold to conceal tacks, staples, and splines. In metal-rimmed screens, the screening is rolled into a groove in the rim and held in place by a vinyl or metal spline.

8.5.2 TYPES AND NOMENCLATURE

Wood windows may be classified according to design and manner of operation as *fixed, double- (or single-) hung, awning, hopper, casement, horizontal sliding*, and *basement* types.

Sliding glass doors are essentially large horizontal sliding windows. However, their increased size, weight, and use result in greater demands on frame and sash construction and on sliding hardware. Sliding panels usually are roller supported and operate on metal tracks built into the threshold. Metal parts are covered with wood on the interior to prevent condensation and present a uniform appearance. Door panels are sometimes reinforced with steel to provide necessary strength while retaining a slim profile.

A summary of the most common window and sliding glass door types is given in Figure 8.5-8.

8.5.2.1 Window and Door Components

A window unit consists of a frame, one or more sash, and hardware and weather-stripping necessary to make a complete operating unit. Storm sash and screens may be included as integrated elements of the main frame or contained in a separate combination subframe (Fig. 8.5-9).

SASH COMPONENTS

A wood window sash is made up of horizontal rails, vertical stiles, and sometimes muntins and bars (see Fig. 8.5-1). A wood sliding glass door is composed of a main frame containing one or more operating or fixed door panels. Like window sash, door panels are composed of stiles and rails and may hold either single glass or insulating glass. Most doors provide self-storing sliding screens in a separate track of the door frame.

FRAME COMPONENTS

Figures 8.5-1 and 8.5-10 show the parts of a double-hung window sash. The components of other window types vary somewhat from those shown, of course, but the

FIGURE 8.5-8 Summary of Wood Windows and Sliding Glass Doors

Type and Industry Standard	Description
Fixed ANSI/NWWDA I.S.2	*Fixed windows* usually consist of a frame and glazed stationary sash. They are often flanked with double hung and casement windows, or stacked with awning and hopper units, to make up windows of custom designs. To keep the sight lines (width or height of view area) consistent, fixed sash members are made to same appropriate cross sectional dimensions as adjacent operating sash.
Double Hung ANSI/NWWDA I.S.2 **Single Hung** ANSI/NWWDA I.S.2	*Double hung windows* have two operating sash; single hung have only the lower sash operative. The sash move vertically within the window frame and are maintained in the desired position by friction fit against the frame or with balancing devices. Balancing devices also assist in raising the sash. 50% of the window area is available for ventilation.
Casement ANSI/NWWDA I.S.2	*Casement windows* have side-hinged sash, generally mounted to swing outward. They may contain one or two operating sash and sometimes a fixed light between the pair of sash. When fixed lights are used, a pair of casements may close on a mullion, or against themselves, providing an unobstructed view when open. Operating sash may be opened 100% for ventilation.
Horizontal Sliding ANSI/NWWDA I.S.2	*Horizontal sliding windows* have two or more sash of which at least one moves horizontally within the window frame. In three-sash design, the middle sash is usually fixed; in two-sash units, one or both sash may be ventilating. This type of window sometimes is increased in size to door proportions. Ventilating area is 50% of the window area in most designs.
Awning ANSI/NWWDA I.S.2	*Awning windows* have one or more top-hinged, outswinging sash. Single awning sash often is combined with fixed and other types of sash into larger window units. Several sash may be stacked vertically and may close on themselves or on meeting rails which separate individual sash. When awning windows have sliding friction hinges, which move the top rail down as the sash swings out, they are said to have *projected action*. Ventilating area is considered to be 100% of operating sash area.
Hopper ANSI/NWWDA I.S.2	*Hopper windows* have one or more bottom-hinged, inswinging sash. Hopper sash is similar in design and operation to the awning type, and may actually be an inverted awning sash with minor hardware and weatherstripping modifications. For this reason, windows with hopper sash sometimes are referred to as awning windows. Operating sash provides 100% ventilating area.
Basement ANSI/NWWDA I.S.2	*Basement windows* generally are single sash units of simplified design intended for less demanding installations, particularly in masonry and concrete foundations. The sash may be of awning type, hopper or top-hinged inswinging (shown) type. 100% of the window is available for ventilation.

FIGURE 8.5-8 *(Continued)*

Type and Industry Standard		Description
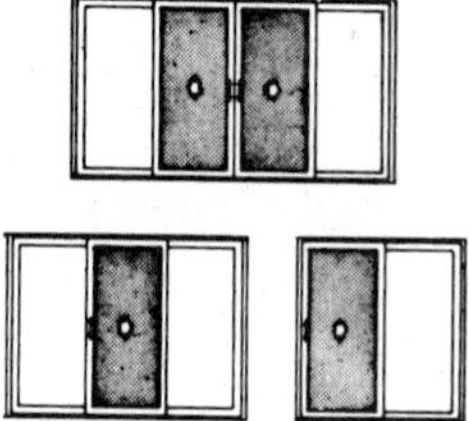	**Sliding Glass Doors** NWWDA I.S.3	*Sliding glass doors* contain at least one fixed and one operating panel in a main frame. Door types are commonly designated by the number and arrangement of panels (as viewed from the outside) using the letter "X" for operating panel, and "O" for fixed panel. The two panel door is most common, but three and four panel designs are also available; two and four panel doors (OX & OXXO) have 50% ventilating area, the three panel unit (OXO) has 33% ventilating area.

basic elements—*sill, side jambs, head jamb, stops,* and *exterior casing*—are present in most window and door designs. *Blind stops* are included in some window frames between jambs and outside casings and may vary in thickness for proper window positioning in wood or metal stud walls with ½- or ¾-in. (12.7 or 19.1 mm) sheathing. *Jamb extenders* adapt windows to various interior finish thicknesses. They are not usually part of the window frame assembly, but may be ordered loose as optional items. *Interior trim* is not usually supplied with stock windows. Typically, it is selected separately and in some installations may be replaced by a gypsum board or plaster reveal.

CLEANING AND MAINTENANCE

To make exterior surfaces accessible for cleaning, double-hung and horizontal sliding windows usually have special hardware that permit easy removal of sash (see Fig. 8.5-3c and d). Awning window sash may be removable, but more often awning windows have sliding friction hinges that create an access space of several inches between sash and frame when the top rail is moved down. Casement windows similarly provide cleaning accessibility with sliding or extension hinges. The inswinging operation of hopper sash makes both surfaces automatically accessible from the inside. Most sash not ordinarily considered removable may be removed with a few simple tools.

STORM SASH AND SCREENS

For greater heat control, any window sash may be glazed with insulating glass (Fig. 8.5-11a) or provided with *storm sash.* Clip-on metal-rimmed *storm panels* are commonly used with hinged sash such as casements, awnings, and hoppers (Fig. 8.5-11b). Sliding and double-hung sash may be equipped with either clip-on storm panels or an external *combination frame* that holds both storm sash and screens (Fig. 8.5-11c).

Combination frames have the advantage of being self-storing and thus eliminate the need to remove external screens for winter storage and provide a year-round choice of storm panel or screen. However, combination frames provide for screening only half the window area, so that only one sash can be opened at a time. When it is desirable to open both top and bottom sash of a double-hung window (or right and left sash of a horizontal sliding window), full screens may be used to cover the entire window.

Outswinging hinged sash (casements and awnings) typically have removable interior screens; hopper sash have exterior screens. Basement windows may have either interior or exterior screens, depending on whether sash swing in or out. Many storm panels and screens are framed with aluminum or stainless steel rims of thinner cross section, which provide greater flexibility in positioning within the limited thickness of a typical frame wall.

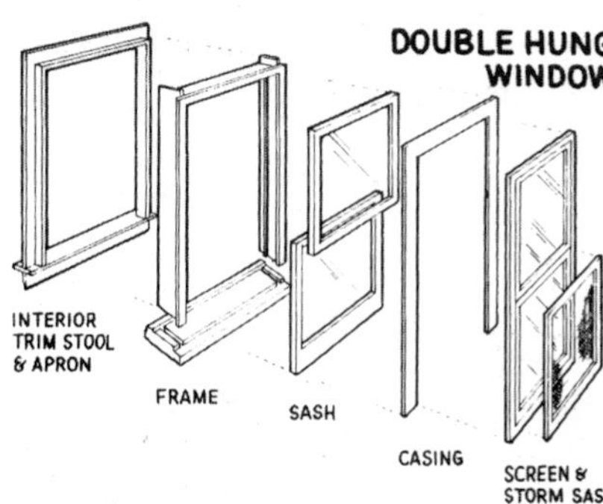

FIGURE 8.5-9 The principal parts of window units, as shown here for a double-hung window.

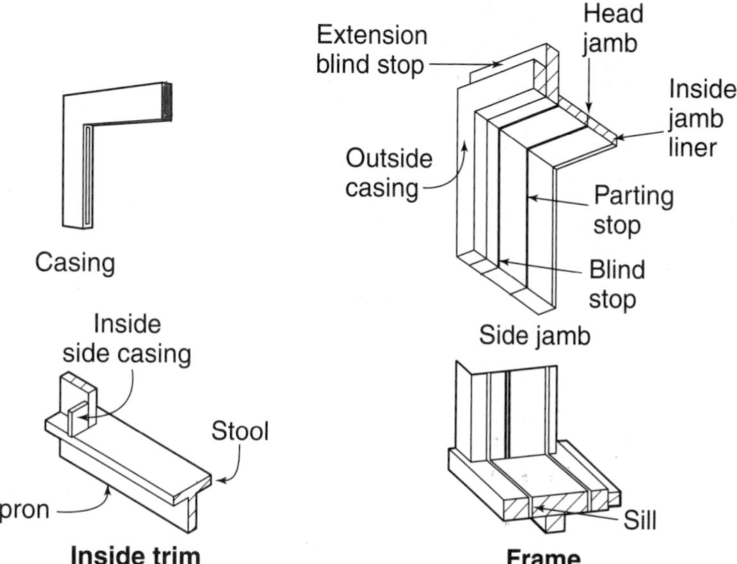

FIGURE 8.5-10 Terminology for typical double-hung windows. Jamb extenders are also called *jamb liners.*

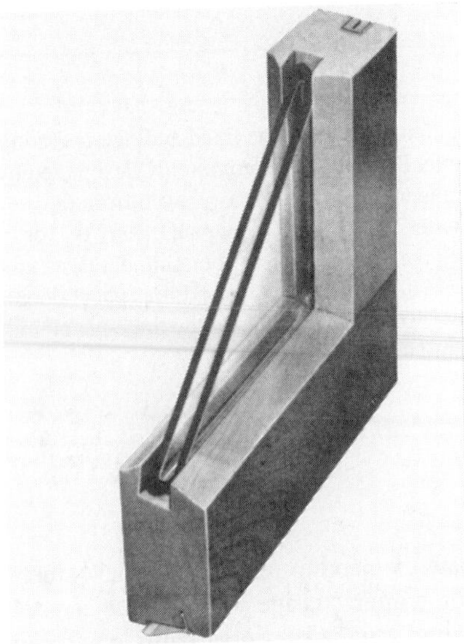

a

b

c

FIGURE 8.5-11 Methods of double glazing: (a) insulating glass shown here groove glazed; (b) clip-on storm panel; (c) combination storm and screen sash.

8.5.3 SELECTION

The selection of windows and sliding glass doors depends on the use and location of the units. The first step is to decide on the types of unit desired (see Section 8.5.2). Then the unit's grade and other required performance criteria should be selected.

8.5.3.1 Performance Criteria

As mentioned previously, ANSI/NWWDA Industry Standard I.S. 2 and NWWDA Industry Standard I.S. 3 constitute the accepted industry standards for wood windows and sliding glass doors. They establish certain minimum performance criteria and rate the products, broadly regulating material, design, and manufacture. For example, both standards recognize three grades of performance: Grade 20, Grade 40, and Grade 60. The numbers represent the minimum structural performance test pressure in pounds per square foot (psf) (4.88 kg/m^2) a unit must be able to sustain without damage. Under ordinary circumstances, Grade 20 units are satisfactory for residential use; Grade 40 for light commercial use; and Grade 60 for heavy commercial construction. However, in high-wind zones (Fig. 8.5-12), higher grades should be selected regardless of the use. The grades also dictate the acceptable operating force, air infiltration rates, and water penetration test pressures. The acceptable levels are different in each grade, becoming more stringent as the grades get higher.

Other items covered by these standards are moisture content of the lumber, preservative treatment, screening, and glass requirements. Appropriate industry standards for wood windows and siding glass doors are shown in Figure 8.5-8.

8.5.3.2 Certification of Quality

The industry standards mentioned do not provide for inspection, testing, or certification of products. These functions are performed by certification programs sponsored by the NWWDA or by independent testing laboratories. Issuance of a label assures that the design, manufacture, and tested performance of the product meets or exceeds the appropriate standard. These certification programs are based on periodic inspection of the manufacturing facilities, quality control procedures, and laboratory testing.

NWWDA offers three certification programs. Wood windows are certified for conformance to NWWDA I.S. 2; wood sliding glass doors are certified for conformance to NWWDA I.S. 3; and water-repellent preservative treating is certified for conformance to NWWDA I.S. 4. Each program consists of initial qualification testing, in-plant inspection, and periodic follow-up testing and inspection to verify that products are manufactured in conformance with the applicable standards. Conformance to the standards results in certification and issuance of a seal. This seal is rubber stamped on various parts of the sash and frame or imprinted on labels affixed to the product. The seal indicates

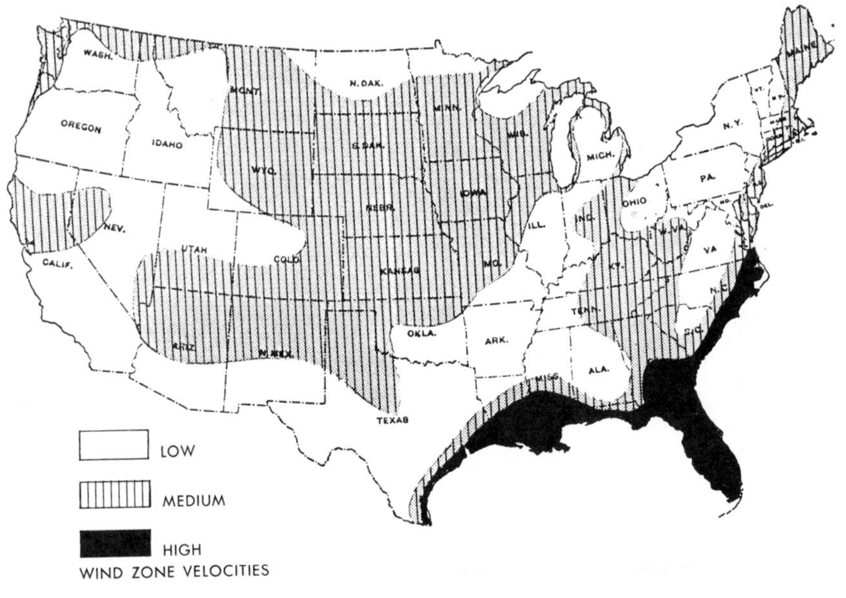

FIGURE 8.5-12 Wind velocity zones.

the fabricator of the product and the industry standard under which the product was produced and is evidence of compliance with the standard.

8.5.4 INSTALLATION

Even the best manufactured windows and sliding glass doors will not perform satisfactorily if they are not installed properly. Although these units are commonly squared and braced before shipping, they may be distorted in transit and should always be checked for squareness and equal spacing of the jambs before they are installed. Braces should be renailed if necessary and left in place until the unit has been installed and is ready for trimming. The frame should be plumbed and fastened securely in the rough opening. Adequate shimming should be provided so as not to distort the frame during nailing. The spaces between finished frame and rough frame should be completely filled with insulation. Drip caps and flashing over the cap or exterior casing are strongly recommended to minimize the possibility of leakage. The entire unit should be well caulked at the juncture of the exterior casing and the wall finish, after both elements have been prime-coated but before finish painting.

Where windows and doors are shipped with loose interior stops, it is essential that these be installed carefully for a snug, weathertight fit against the sash. Weatherstripping and hardware should be adjusted for proper operation and closure according to the manufacturer's instructions. Interior trim should be applied to seal the wall cavity between the rough frame and the finish frame.

Wood windows and sliding glass doors should be handled carefully to prevent damage and protected with a prime coat as soon as possible after arrival at the building site. They should be stored out of the weather in a well-ventilated space. During finishing, it is important to keep paint and varnish off weatherstripping and interior finish hardware.

Exterior hinges and storm sash hardware may be painted for improved corrosion resistance, but the metal or plastic sash guides of double-hung and sliding windows should not be painted. Oil-based glazing compound should be painted, with the paint lapped slightly over the glass on both face putty and back putty runs. Other sealants may or may not be painted, depending on the sealant type and use.

8.6 Storm Windows and Doors

Storm windows and doors are glazed panels or inserts that are sometimes added to windows and doors as an extra line of defense against the elements. Their major use is in residential and small commercial buildings.

Window storm sash are commonly called *storm windows*. They control heat flow by making windows more airtight to reduce infiltration and by providing an insulating layer of air between the window sash and the exterior to reduce conductive transfer.

Storm windows can be of the conventional, combination, or fixed-in-place type. They can be manufactured with wood, metal, or rigid plastic frames. Storm doors are most often made of aluminum, although wood is still sometimes used.

Earlier storm panels were inexpensive sash, made of wood or metal frames with single glazing and hung in place with special clips (Fig. 8.6-1). They were independent of the screens, so they were installed in the fall and taken down in spring. If the building was air conditioned and not properly oriented for natural ventilation, they could be left up year-round to reduce cooling loads. Windows of this type are not often used in new construction.

Contemporary storm windows are usually permanently installed combination types that include both storm sash and screens (Fig. 8.6-2). Frames are usually aluminum and have an anodized or baked enamel finish. They should move smoothly in the frame tracks and seal tightly when closed. They should have tight joints and weatherstripping to ensure a tight seal.

Fixed-in-place interior storm windows are inexpensive temporary installations used almost exclusively in existing residential buildings. They consist of a frameless sheet of rigid plastic held in place with tape or snap-on moldings (Fig. 8.6-3). It is also possible to use polyethylene film taped to the frame. Temporary plastic windows provide an insulating value comparable to that of ordinary storm windows because it is the air films along the glazing surfaces and the air space between that reduce the heat loss, not the material thickness.

8.6.1 GENERAL REQUIREMENTS

Properly installed storm doors and windows provide an effective barrier to the transfer of heat. They are recommended for use in the northern part of the country, above the 4500-winter-degree-day line, and are desirable down to the 2500-degree-day line (Fig. 8.6-4). In these areas they can significantly reduce heating costs and minimize condensation on windows. By preventing cold air from leaking around windows and doors, they also reduce drafts and improve interior comfort. They can also increase comfort in summer, especially in southern areas, by keeping cooled air inside.

Although the term *storm doors and windows* generally refers to units intended for weather protection, it is often used collectively to include screen units as well. Accordingly, the following discussion covers both combination units, containing interchangeable glass and screen inserts, and single-purpose units with fixed glass or screening. The recommendations for storm doors and windows are based largely on standards developed by the American Architectural Manufacturers Association (AAMA), and those for screen doors and screen inserts follow standards established by the Screen Manufacturers Association (SMA).

8.6.2 MANUFACTURE

Most storm doors and windows sold today are made of aluminum. Their manufacture is similar to that of aluminum prime windows and sliding glass doors. They are made in a variety of finishes and colors. A natural mill finish is adequate for many moderate-cost installations where appearance is not a prime consideration. For more demanding projects, where both cor-

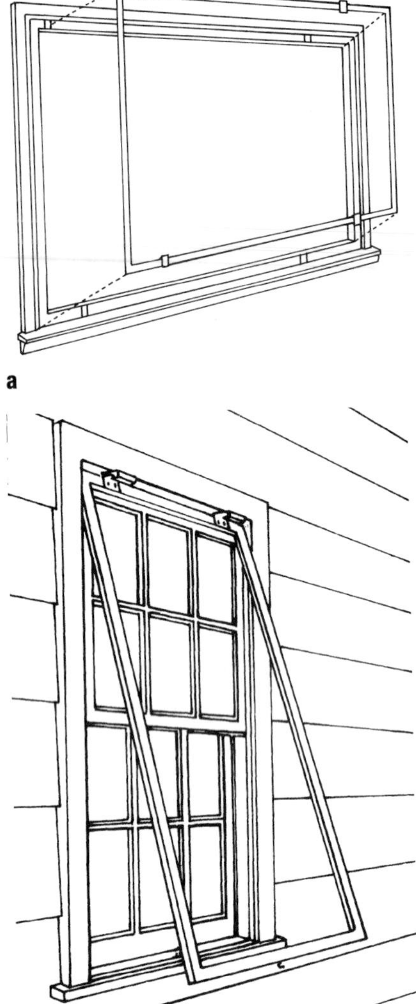

FIGURE 8.6-1 (a) Clip-on metal-rimmed storm panels are typically used with awning or casement windows; (b) storm panels for double-hung windows are attached to the frame.

FIGURE 8.6-2 These windows are also called *triple-track* because of the three panels that move in separate tracks.

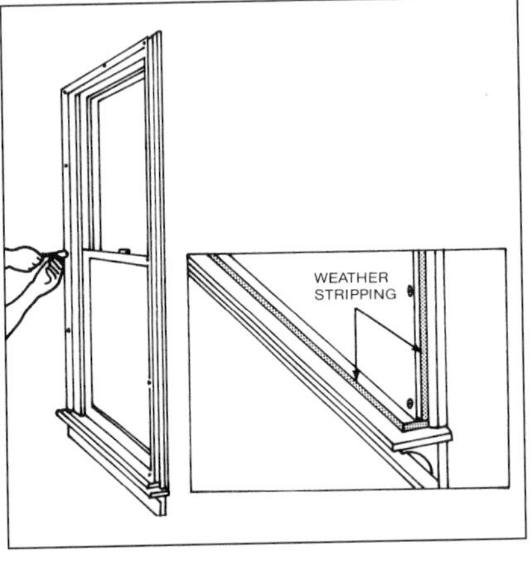

FIGURE 8.6-3 These temporary storm panels attach directly to the window frame.

rosion resistance and appearance are important, an anodized or paint finish is appropriate. Anodized finishes can be of natural aluminum color or deeper shades of bronze or brown-black. Paint finishes are available in a variety of colors.

8.6.3 SELECTION

Both windows and doors can be of the *self-storing* type or may depend on a glazed *storm panel* that is applied and removed seasonally as needed. Self-storing storm windows generally have both glass and screen inserts sliding vertically or horizontally in a frame. In self-storing *combination* (storm and screen) doors, glass and screen inserts slide vertically.

8.6.3.1 Storm Windows and Panels

The most widely used storm windows are of the triple-track and two-track types. When practicable, storm windows should be installed on the outside of the prime sash. Where the prime sash swings out, as in casement and awning windows, the storm windows may have to be on the room side.

To prevent rapid heat transfer and condensation on the interior window frame, an insulating thermal break should be provided between the prime window and storm window frame. A thermal break is recommended regardless of whether the storm window is installed on the room side or outside. Some windows are provided with a plastic or wood liner that serves as the thermal break. An air gap or a small wood molding can be used instead.

Triple-track windows are intended for use mainly with vertically sliding double-hung and horizontal sliding triple-light windows. A typical unit consists of a frame, with each insert riding in its own track. This self-storing feature eliminates the need for switching storms and screens every spring and fall. Instead, selection of glass or screen can be made easily from inside the house.

Two-track windows are generally used with vertically sliding single-hung windows and with two-light horizontal sliders. These units also have three inserts that are stored in the frame. In vertical windows, the bottom light is usually screened; in horizontal sliders, the screen is placed right or left, depending on the prime vent sash location.

Single-track windows have two glass inserts and one screen insert, but only two inserts can be housed in the track at any one time. This makes changing from weather protection to ventilating somewhat less convenient, because one insert must always be stored elsewhere.

Clip-on storm panels are available for outswinging prime windows such as casements, awnings, and jalousies. These consist of single glass panels that are attached either to the individual sash or across the entire prime window, including both fixed and vent sash. When the panel is clipped to the sash only, heat transfer through the glass is reduced, but heat lost through the frame and by air leakage is not affected.

Larger panels covering the entire window control heat loss more effectively but cannot be left in place for the summer.

8.6.3.2 Window Construction

Extrusions used for the main frame and insert frames should be of wrought aluminum alloy with a minimum tensile strength of 22,000 psi (15,468 Mg/m^2) and yield strength of 16,000 psi (11,249 Mg/m^2). Reinforcing members that are not in plain view can be of cadmium, zinc-plated steel, or other noncorrosive materials compatible with aluminum.

Storm windows should be provided with weatherstripping to minimize air infiltration and reduce rattling. Glass inserts generally consist of a light-gauge aluminum frame into which the glass is set with bedding compound, plastic channel gaskets, or glazing tape. These materials

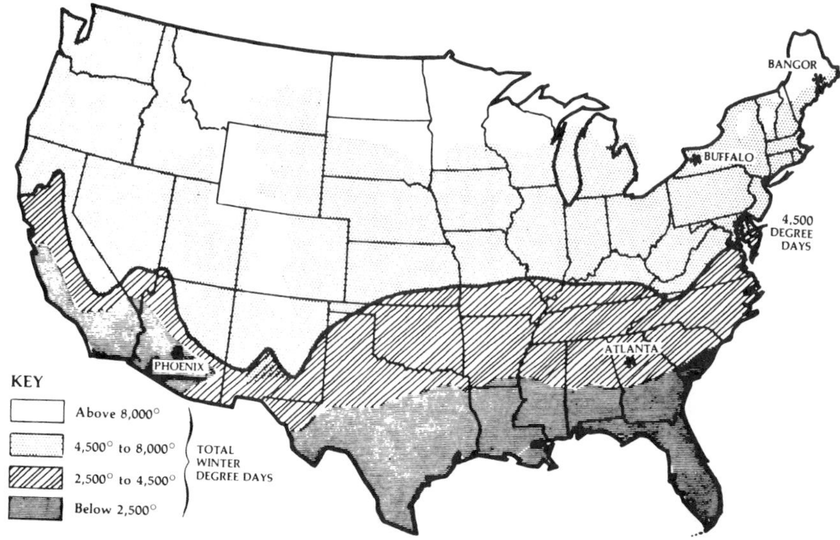

FIGURE 8.6-4 Zones of heating demand (winter degree days).

should cover the glass edge continuously so that there will be no glass-to-metal contact.

Glass used for storm inserts in windows should be not less than B quality and of adequate thickness. Glass inserts in vertically sliding windows should have self-activating latches to hold the insert level in different ventilating positions. Insert frames should be designed to permit reglazing without special tools.

Screen inserts consist of insect screening stretched on an extruded or roll-formed frame. Screening can be of aluminum, vinyl-coated glass fiber, or other synthetic fibers, in 18 by 14 or finer mesh.

8.6.3.3 Storm and Screen Doors

Entry doors are another great source of heat loss. The two most effective means of reducing such heat loss are storm doors and insulated doors. Insulated doors are discussed in Sections 8.1 and 8.2.

Like storm windows, storm doors reduce heat loss by reducing both conduction and infiltration. For example, a typical 1¾-in. (44.5 mm) solid wood exterior door has an R-value of about 2. When a wood storm door is added, the R-value is increased to approximately 3.70 (Fig. 8.6-5).

Storm doors are made either (1) with fixed glass lights, exclusively for weather protection, or (2) with both screen and glass inserts, to permit ventilation and insect control as well. In parts of the country with moderate year-round temperatures, screen doors (without glass inserts) are frequently used. Screen door construction is similar to that of storm doors. Both must be constructed rigidly, to resist sagging, bowing, or rattling from repeated impact.

FIGURE 8.6-5 Door R-Values

Door Type	R-Value
1¾ in. (44.5 mm) exterior door	2.04
1¾ in. (44.5 mm) exterior door and aluminum storm door (50% glass)	3.03
1¾ in. (44.5 mm) exterior door and wood storm door (50% glass)	3.70
Metal-clad insulated door	8.00–15.00

Many storm doors are of the combination type, providing both weather protection and insect control. Such doors are stile-and-rail assemblies, framing a storm window with glass and screen inserts that can be interchanged seasonally. Doors with openings limited to the upper half of the door generally accommodate only one insert at a time.

Self-storing storm doors contain the equivalent of a two-track window, accommodating two inserts in one track and another in the adjacent track. The glass and screen inserts slide up and down just as they do in a vertical storm window.

Both prime doors and storm doors can be of the jalousie type. To provide ventilation and storm protection, interchangeable screen and glass panels can be clipped to the inside of the door. Clip-on glass storm panels not only provide added insulation but also reduce air leakage between the louvers, a common problem with this type of door.

8.6.3.4 Door Construction

Storm and screen doors are opened and closed many times a day and are subject to wind buffeting and other stresses. Therefore, it is important that they be durably constructed and properly installed. Rigid corner connections are particularly important to resist racking forces.

Standard door frames are 1 or 1¼ in. (25.4 or 31.8 mm) thick, with top rail and stiles 2 to 2½ in. (50.8 to 63.5 mm) wide. The bottom rail (kickplate) varies in width depending on the size of the opening. To avoid the hazards of broken glass, storm doors should be safety glazed with transparent plastic or with tempered, laminated, or wired glass. In some locations laws stipulate this requirement.

Doors are currently manufactured in hundreds of types and styles, utilizing many different glass sizes. To minimize waste, storm windows and doors should be required to accommodate standard glass sizes adopted by the industry. Limiting the number of glass sizes not only reduces production costs but also facilitates replacement, particularly of tempered glass.

The construction of glass and screen inserts for doors is similar to that for windows.

8.6.4 STANDARDS OF QUALITY

Aluminum storm doors and windows should conform to standards developed by the Architectural Aluminum Manufacturers Association (AAMA) and adopted as American National Standards by the American National Standards Institute (ANSI), ANSI/AAMA 1002.10, and ANSI/AAMA 1102.7. These constitute the recognized industry standards and establish minimum performance requirements for air infiltration, water leakage, wind load, and structural strength.

Screen inserts should conform to requirements of the Screen Manufacturers Association (SMA), adopted by ANSI as ANSI/SMA 1004.

The AAMA also sponsors an independent certification program of testing, random sampling, and product evaluation to ensure that aluminum storm doors and windows meet ANSI standards. Products that meet all requirements of the standards are identified with an AAMA certification label.

The AAMA certification program is administrated in accordance with ANSI Z34.1, which provides for certification to the public on an impartial, independent, and continuing basis.

8.6.5 INSTALLATION AND MAINTENANCE

In areas above the 4500-degree-day line (see Fig. 8.6-4), storm doors or an entrance vestibule should be provided as part of the initial construction. Storm windows can be omitted in original construction if the prime windows are equipped with insulating glass and a thermal break in the sash and frames.

Most storm windows are sold as complete units, with necessary accessories and hardware. This ensures smooth, convenient operation and adequate security. Installation consists simply of attaching the window frame in a previously prepared opening.

Storm doors are marketed both prehung on aluminum frames, with hardware attached, and unassembled, with frame parts and hardware supplied loose. The unassembled type provides somewhat more flexibility in adapting to special situations, but requires field assembly and attachment of suitable hardware.

A spring-loaded safety chain or other device should be provided to limit door swing and cushion the impact of a sudden opening, as by a gust of wind. A door closer is recommended. A durable latch to maintain the door in a closed position and to permit locking the door from the inside is also desirable.

8.7 Door Hardware

Finish hardware for formed carbon steel, stainless steel, and copper-alloy doors and interior extruded aluminum doors is usually supplied by a hardware supplier. Extruded aluminum exterior doors and packaged entrances usually come equipped with all operating hardware except cylinder locks (see Section 8.3).

8.7.1 FINISH HARDWARE

Hardware may be either standard, handicapped access, automatic, or security type. Standard hardware includes all the devices necessary to operate, lock, and weatherstrip a door, including thresholds.

Door hardware should comply with the following standards listed in Section 8.11:

- *Material:* American National Standards Institute/Builders Hardware Manufacturers Association (ANSI/BHMA) A156 series
- *Installation:* American National Standards Institute/Door and Hardware Institute (ANSI/DHI) A1165 series

8.7.1.1 Locks

Local fire codes should be checked before installing a new lock or upgrading an old lock. Fire codes sometimes restrict the use of locks that may slow escape in the case of fire. Lock types include key-in-knob, key-in-lever, mortise, interconnected, auxiliary, security locks, and padlocks.

KEY-IN-KNOB LOCK

One of the most common and least secure locks is the key-in-knob type (Fig. 8.7-1). Such locks are relatively inexpensive, but have several disadvantages:

- The locking mechanism is in the exterior knob and can be easily stolen.
- The latch is often short and tapered and can easily be pried out of position with a credit card or similar object (Fig. 8.7-2).

However, a trigger bolt attached to the latch bolt can make it impervious to credit card attack. Some models of these locks feature longer latches to make prying less effective, but none deals with the problem of the primary locking mechanism in the exterior knob.

Key-in-knob locks do not provide adequate burglar protection and therefore should be replaced or supplemented with one of the auxiliary locks discussed in the following paragraphs.

KEY-IN-LEVER LOCKS

Key-in-lever locks (Fig. 8.7-3) have the same advantages and disadvantages as key-in-knob locks.

FIGURE 8.7-1 Key-in-knob lock. (Courtesy Falcon Lock, Brea, California.)

MORTISE LOCK

A mortise lock (Fig. 8.7-4) offers more security than a key-in-knob lock because the locking mechanism is inside a metal

FIGURE 8.7-2 Some door locks are easy to pry open.

FIGURE 8.7-3 Key-in-lever lock. (Courtesy Falcon Lock, Brea, California.)

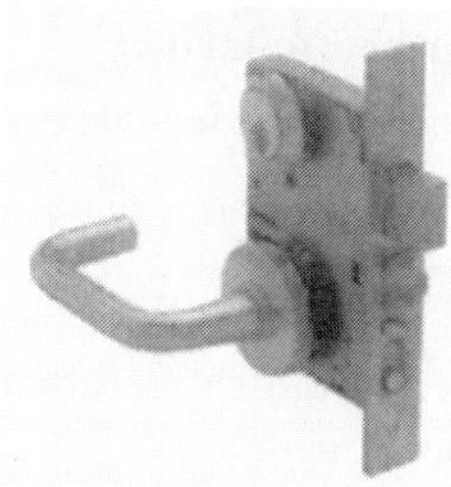

FIGURE 8.7-4 Mortise lock. (Courtesy Falcon Lock, Brea, California.)

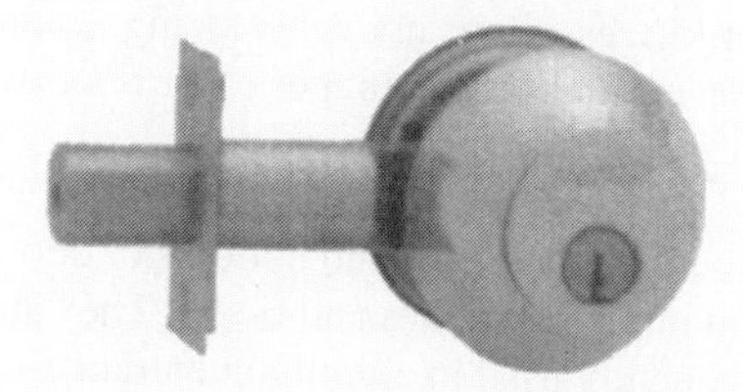

FIGURE 8.7-5 A cylinder deadbolt is an easy and inexpensive means of adding extra security to a door. (Courtesy Falcon Lock, Brea, California.)

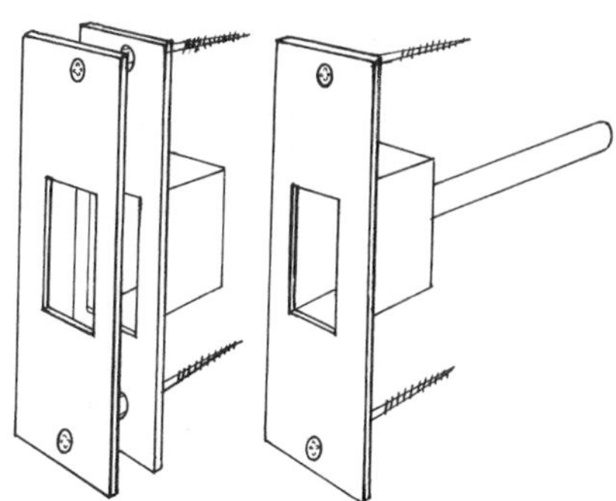

FIGURE 8.7-6 This type of strike fastens securely to the building studs and helps prevent a locked door from being kicked open.

enclosure mounted within the door. It not only has a latch, like the key-in-knob type, but also a deadbolt that cannot be violated by a credit card. Push buttons located under the latch allow the exterior knob to lock in a stationary position or unlock and turn freely.

However, mortise locks have a number of disadvantages:

- A user may fail to engage the deadbolt.
- The large cavity within a door needed to house this lock may impair the structural integrity of the door.
- The keyed cylinder is often poorly anchored and vulnerable to forceful removal.
- The lock may be defeated by loosening the set screw that holds the cylinder in place.

Some locks have faceplates that protect the cylinder. Because of the large cavity occupied by the lock case, the mortise lock is difficult to replace with any other type.

There are also locks that combine elements of both the key-in-knob and the mortise lock types.

INTERCONNECTED LOCK

An interconnected lock has a key-in-knob cylinder, which makes it look like a mortise lock from the outside. However, it lacks the push buttons under the latch that lock and unlock the knob and differs in its method of installation. An interconnected lock requires the boring of holes. It can replace a key-in-knob lock, because the old lock hole can be used and the other necessary holes added.

As with a mortise lock, the user must remember to engage the bolt. Moreover, if the cylinder is unprotected, it is vulnerable to pulling. Since an interconnected lock removes less of the door structure than a mortise lock, the door may stand up to more abuse before breaking.

AUXILIARY LOCK

A second lock is often the easiest and cheapest method of adding security to a door. A cylinder deadbolt fits within a door (Fig. 8.7-5). As with other internal locks, this may weaken the door. Another disadvantage is that two locks may keep out an intruder but may also slow escape in case of emergency.

A vertical deadbolt lock is also called a *rim lock* because it is surface-mounted on the inside rim of the door. This is an obvious, but effective, type of lock. If installed with long enough screws, it resists kicking and picking. Some locks have special shutters that freeze the bolt in place if someone tries to pry it open.

SECURITY LOCK

Security hardware consists of high-security cylinders, special control devices such as electromechanical, electromagnetic, or electronic locks, and three-way locks in which a single key operates bolts in the head, jamb, and sill of a door. Security locks may operate by key, push buttons, or card readers.

PADLOCK

A good padlock consists of (1) a hardened steel shackle, (2) a brass or laminated steel lock case, and (3) a five-pin cylinder. A padlock can add protection to any door, but it is most often used on garage and shed doors. Only the side of the door to which the padlock is attached is accessible.

REINFORCEMENT

Many adequate locks come with inadequate strikes or screws or are installed on inadequate doors.

A *strike* is a jamb fastening that receives the bolt in the locked position. Several kicks can tear away a conventional strike along with its screws and part of the jamb, leaving the lock intact but the door open. A high-security strike consists of a small metal box that fits into the frame's bolt hole and fastens securely to the building framing (Fig. 8.7-6).

The screws that attach the various kinds of locks to the door and the strikes to the jamb are often too short. In general, the longer the screw, the more securely the lock or strike will be fastened. The screws for the average rim lock should be 2 in. (50.8 mm) long. The screws on a strike should be at least 2 in. (50.8 mm) long, and preferably 3 in. (76.2 mm) long, so that they will secure the strike to the framing studs.

On soft-wood doors, a wraparound channel made of metal fits on the door and reinforces the area around the lock, preventing the wood from splitting (Fig. 8.7-7).

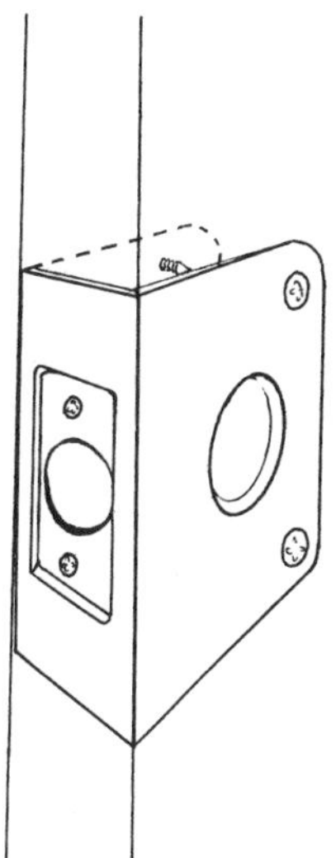

FIGURE 8.7-7 A wraparound channel helps prevent a door from splitting and breaking in a kicking assault.

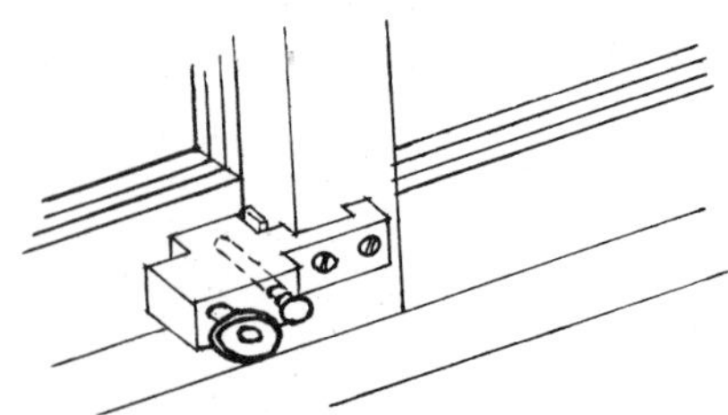

FIGURE 8.7-8 A keyed deadbolt prevents a door from sliding open.

SPECIAL LOCKS

Doors such as sliding glass doors and garage doors need special locks.

Sliding glass doors can be protected by a keyed deadbolt at the door base (Fig. 8.7-8). The disadvantage of this method is that it can slow escape in case of fire.

Some manufacturers make special locks that replace and upgrade the normal locks provided on sliding glass doors. These include double deadbolts operated by both a handle and a key, which permits opening from the outside with a key.

Garage doors can be fitted with a padlock in a hole drilled through the end of a garage door bolt (Fig. 8.7-9). The padlock will secure the door even if the door's regular lock mechanism is destroyed.

8.7.1.2 Hinges

Several types of door hinges are available. Butt hinges are mounted on the door and frame. They may be full-mortise, half-mortise, full-surface, and half-surface types. They may be standard, heavy-duty, security or industrial types. They are available in three-knuckle and five-knuckle types and in standard bearing and ball bearing types. Special hinges are available for sliding and folding doors. Decorative hinges are also available. "Invisible" and piano hinges are available, but are used mostly on cabinetwork. Offset hinges allow doors to swing wider than normal. Spring hinges cause doors to close automatically.

Pivot hinges are a special type that are mounted in the tops and bottoms of doors and into head frames and floors. They are frequently used in aluminum entrances.

Hinges are made from prime coated steel, polished plated steel, chromium-plated metals, brass, bronze, aluminum, and stainless steel.

Hinges are sized and the number used is determined according to the size of any type of door. For example, all exterior, interior solid-core, interior mineral-core doors, and other types as well when they are 7 ft. (2134 mm) or more in height, should be hung with a minimum of three hinges per door, aligned so the door will not twist out of shape. Interior hollow-core doors less than 7 ft. (2134 mm) in height may be hung with two hinges.

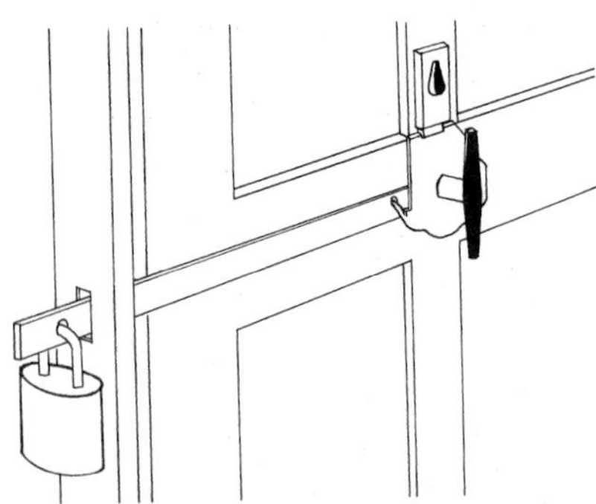

FIGURE 8.7-9 A supplementary padlock can be installed on a garage door.

8.7.1.3 Closers

Closers are devices that close a door using spring or hydraulic action. Types include overhead surface-mounted closers, overhead units that are concealed within the door or frame, and floor-concealed closers.

8.7.1.4 Automatic Operators

Automatic operators open doors when an electromechanical, hydraulic, or pneumatic operator is triggered by an actuating device. These include floor treadles, motion sensors, sonic sensors, push buttons, photoelectric cells, and remote switches. Automatic hardware also includes smoke and fire detectors that automatically close doors when smoke or fire is present.

8.7.1.5 Handicapped-Access Operators

Handicapped-access hardware includes counterbalanced closers that permit easy operation of a door as well as specially designed door knobs with extended or oversized grips to permit easy operation by a handicapped person.

8.7.1.6 Exit Devices

Exit devices consist of a bar across a door, which, when depressed, unlatches the door to permit easy exit even when the door is locked. Many styles and materials are available. Doors with exit devices are equipped with closers and automatically latch upon closing. Exit devices are required by code on exit doors in most public buildings.

8.7.1.7 Other Hardware Devices

Other door hardware devices include:

- Pushes and pulls, which are generally used on entrance doors
- Door stops
- Hold-open devices, which are often, but not necessarily, part of a closer
- Coordinators, which cause closers on two doors to operate in sequence or simultaneously
- Rollers, tracks, and guides for pocket and other sliding doors
- Electronic and key-pad locks
- Finger guards
- Door edges

8.7.2 WEATHER PROTECTION

8.7.2.1 Weatherstripping

Weatherstripping is applied to door frames to prevent air leakage. It typically has a resilient part that prevents air penetration when it is pressed between the door and its frame.

The most common weatherstripping types are:

- Foam rubber
- Felt
- Rolled vinyl
- Spring metal
- Interlocking metal

Foam rubber weatherstripping has either an adhesive or wood backing. It is easy to install but lacks durability. The adhesive type can be used wherever there is no friction, such as at the top and sides of frames (Fig. 8.7-10a).

Wood-backed foam rubber is more conspicuous than the adhesive type (Fig. 8.7-10b).

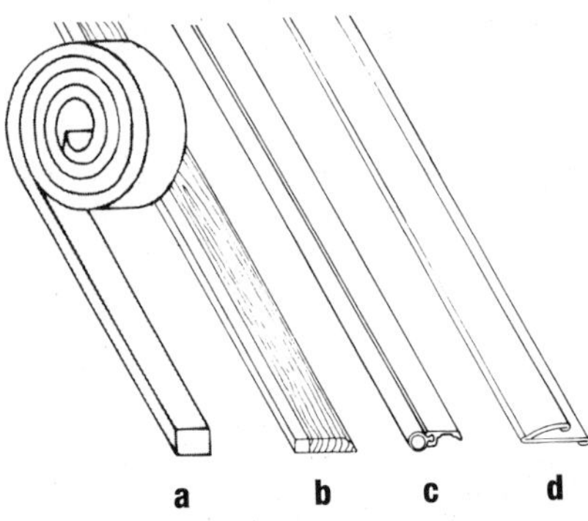

FIGURE 8.7-10 Common types of weatherstripping: (a) adhesive-backed foam rubber, (b) wood-backed foam rubber, (c) rolled vinyl, (d) V-strip.

Felt stripping has the same advantages and disadvantages as foam rubber, but it also collects dirt more readily. Metal-backed felt, which provides more durability, is also available.

Rolled vinyl is available with or without a metal backing. Some types are manufactured with tubular gaskets that can be either hollow or filled with foam (Fig. 8.7-10c). Rolled vinyl is very durable.

Spring metal stripping is available in two types: flat stripping, which is most commonly used for double-hung windows (Section 8.5); and V-strip, which is used for doors. A V-strip is a doubled-over strip of springy metal that fits between the door edge and the jamb (Fig. 8.7-10d). When the door is open, the metal springs apart; when the door is closed, it is forced together to fit tightly between door and jamb. It is durable and inconspicuous after installation, provides an excellent seal, and will last indefinitely.

Interlocking metal stripping is produced as interlocking channels and interlocking J-strips. Both are difficult to install because alignments must be exact for a secure interlock. Both provide excellent, durable seals. The channels are subject to damage, however, because they are exposed. The J-strips are not visible when installed; they are not exposed and, therefore, not subject to damage.

8.7.2.2 Door Bottoms

Air leakage through cracks at door bottoms can be stopped with (1) sweeps attached to the door bottom, (2) door shoes, or (3) weatherproof thresholds.

The most common types of sweeps consist of a felt or rubber flap mounted in a metal channel (Fig. 8.7-11a). They seal most effectively on flat thresholds. They drag on carpets, however, and will eventually wear down from friction. To avoid carpet drag, automatic door sweeps can be used. These are spring-loaded and flip up when the door is opened, down when it is closed.

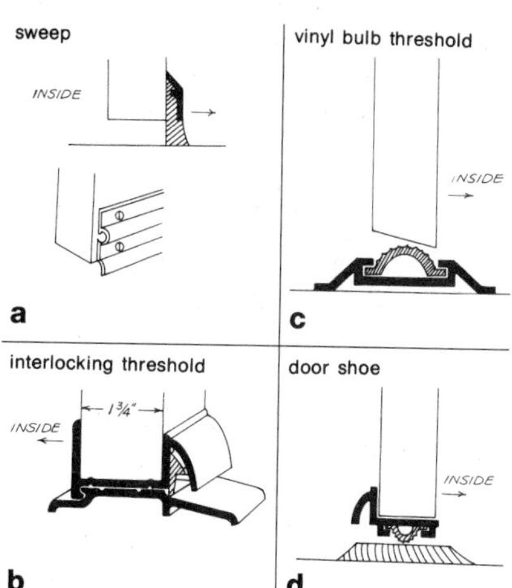

FIGURE 8.7-11 Door bottoms: (a) sweeps, (b) interlocking threshold, (c) vinyl bulb threshold, (d) door shoe.

An interlocking threshold consists of a metal sill that is shaped to interlock tightly with a metal channel attached to the door bottom (Fig. 8.7-11b). It is durable and provides an excellent weather seal but is difficult to install because proper alignment is critical.

Weatherproof thresholds can be either the vinyl-bulb or the interlocking type. These thresholds are difficult to install, because a good fit requires that the door bottom be trimmed at or manufactured with an ⅛ in. (3.2 mm) bevel. A vinyl-bulb threshold contains a flexible vinyl strip that is compressed when the door is closed (Fig. 8.7-11c). The threshold itself is metal and very durable, and, although the vinyl bulb will wear out, it can be replaced.

Door shoes are metal frames with attached vinyl bulbs that press against the threshold to form a tight seal (Fig. 8.7-11d). They are very durable because the bulbs can be replaced when worn. Door shoes are best suited for flat thresholds.

8.7.3 BURGLARY PROTECTION

In most localities today it is necessary to make a special effort to reduce the vulnerability of buildings to unauthorized entry by thieves, vandals, and pranksters, which is called burglary. Burglars do not like challenges. They prefer to work quickly, quietly, and in the dark. If more than 4 minutes are required to gain entry, they will often give up the attempt. No lock will stop a determined professional burglar, but good burglary protection hardware on doors and windows and design of doors and windows to resist intrusion can slow a burglar down. Care should be taken, however, to ensure that devices intended to deter intrusion do not endanger occupants by preventing their exit in an emergency.

To be effective, the devices discussed in this section need to be combined with proper site planning, warning devices, and adequate lighting. Site planning measures to deter unwanted intruders and devices intended to make noise when triggered so that a burglar will not be able to work quietly and to warn occupants and neighbors of an attempted intrusion are discussed in Section 13.3. Outdoor lighting designed to deny a burglar a dark place to work is discussed in Section 16.4.

8.7.3.1 Unauthorized Entry

Before it is possible to design effective measures to prevent intrusion, it is necessary to know how burglars gain entry. Even locked doors and windows offer only a slight challenge to experienced burglars.

Simple lightweight tools can be used to pry open doors equipped with automatic latching devices of the type commonly used in many cylinder locks. Deadbolts are somewhat harder to pry open and require more force and heavier tools.

A credit card or a similar thin piece of plastic can be slipped between a door and its frame to push back the latch and open the door (see Fig. 8.7-2).

A pair of pliers can be used to grip a lock cylinder and pull it out of the lock, making the mechanism accessible. Burglars sometimes use a screwdriver to twist and wrench the core of the locking pins away from the cylinder.

A crowbar may be used to jimmy doors. By inserting the crowbar between the sash and frame, a burglar can pop the bolt out of place. Even if the lock holds, the frame may break.

8.7.3.2 Doors and Frames

Adequate locks are frequently installed on inadequate doors. A lock can only be as good as the door on which it is installed. In general, exterior entry doors should be carefully selected and properly installed. Selection of door and frame types and materials and the types of locks and other security devices used depends on the project budget and the level of security required. In some cases the company insuring a building's contents may dictate the level of security required.

HINGED DOORS

Flush wood solid-core or metal doors, 1¾ to 2 in. (44.5 to 50.8 mm) thick, are the least vulnerable to attack. Wood hollow-core doors and stile-and-rail doors with wood or glass panels are more vulnerable. A burglar can easily break through a hollow door or panels and reach the lock from inside. Refer to Section 8.2 for a discussion of wood doors.

Exterior metal doors consist of a core and a metal frame surrounded by steel skins. Most security experts prefer the use of metal doors over any type of wood door. Doors should fit in their frames snugly, with only enough clearance to ensure easy operation.

SLIDING DOORS

Security problems with sliding glass doors include (1) inadequate locks, (2) flimsy frames, (3) door panels that can be lifted out of their frames, and (4) vulnerability of the glass.

A series of screws along the upper track of a sliding glass door can prevent the door from being lifted out of its track (Fig. 8.7-12). A piece of wood or metal slightly shorter than the door width can be laid in the lower track (Fig. 8.7-13) to supplement the lock and prevent the door from being forced open. To keep the wood or metal bar from being lost, it can be hinged at the frame and leaned against the side jamb when not in use. Commercial devices are available to serve this purpose.

VIEWING DEVICES

One-way viewing devices (peepholes) in doors allow occupants inside a building to see outside without having to open the door.

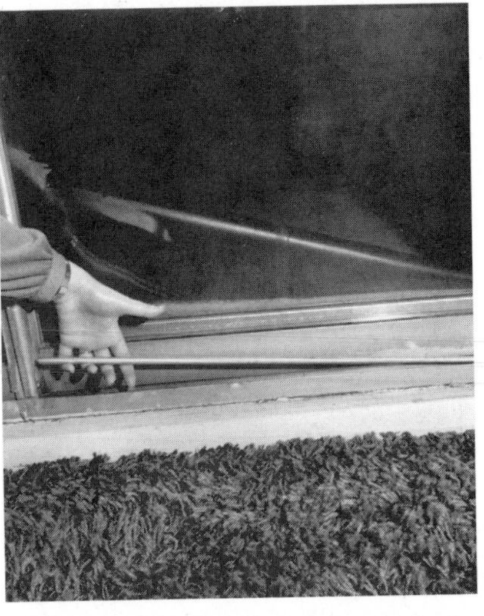

FIGURE 8.7-13 A metal or wood bar in the track prevents a sliding glass door from being forced open from outside. The bar can be hinged to prevent its loss.

DOOR FRAMES

Even a strong door with a good lock and proper hinges can be opened easily by a burglar if the door's frame is inadequate. If the rough opening is oversized and the door frame is inadequately shimmed, a burglar can often pry the frame from the lock to release the door. A space as small as 1/16 in. (1.6 mm) may be large enough to insert a crowbar, screwdriver, or hacksaw. The space between the door and its frame should be 1/64 in. (0.4 mm) or smaller. The same tolerance should be observed at the side jambs and at the top of the door, because an automobile jack can be used to spring the frame at a point just above the lock.

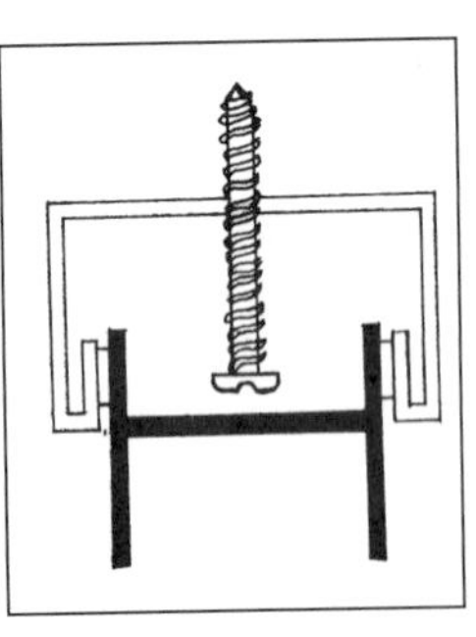

FIGURE 8.7-12 Screws at periodic intervals along the upper track will prevent a would-be intruder from lifting a sliding glass door off its floor track.

8.7.4 HANDICAPPED PROVISIONS

Requirements for compliance with the Americans with Disabilities Act (ADA) are discussed in Section 1.7. The requirements here are supplementary to those in Section 1.7 specifically related to door hardware.

The general principle is that all places of public accommodation must be accessible to persons with disabilities. The major concerns are the ability of a person with a disability to:

- Grasp and operate the pull side of a door exit device
- Avoid contact with obstructions on the surface of a door where a wheelchair footrest is used to push open the door
- Avoid obstructions that protrude into an exitway
- Open and close a door

A solution to the grasping requirement is lever handle devices. Unfortunately, lever devices are much more easily damaged by vandals than are standard knob devices. Freewheeling and breakaway lever designs are available, which are more resistant to deliberate damage than rigid-type lever devices.

Contact at the bottom of a door can be avoided by not specifying surface-mounted vertical operating rods for closers, but instead using concealed rods. Several proprietary devices are also available to solve this problem.

The clear area required at doors is specifically delineated in ANSI A117.1 (see Section 1.7).

Door closers must have sufficient force to close doors completely and to permit latching hardware to operate properly, especially on fire-rated doors. The pressure needed is often more than a person with a disability can exert. Automatic operators and handicapped-access hardware, as described earlier, can solve these problems.

8.8 Glazing

This section discusses glass and plastic glazing materials and their application (glazing). The term *glazing* refers to the installation of vision panels of glass or plastic in a door, formed frame, window, packaged entrance, storefront, or glazed curtain wall. Spandrel glass in curtain walls is also glazing. Other types of opaque panels in a curtain wall are also sometimes said to be *glazed* into the system, but are not generally regarded as glazing materials.

8.8.1 GLASS

Glass results from the fusion of certain common minerals at high temperatures and subsequent controlled cooling. This process causes solidification without crystallization. The chief ingredients of glass are silica, sodium oxide, and calcium oxide (sand, soda, and lime), all of which are abundant minerals throughout the United States.

Glasslike mineral formations resulting from the accidental fusion of silica sand and metallic oxides by the heat of a volcano or lightning (obsidian and lightning stone) can be readily found in nature and may have served as the impetus for the early invention of glass. Glass has been made artificially for more than 4500 years.

The first window glass we know of was made by casting and grinding during Roman times. In eleventh-century Germany, glass was handblown in spheres, elongated into cylinders by a swinging process, then cut apart and flattened to make a product called *broad glass*. *Crown glass* was produced in France during the fourteenth century by a blowing and twirling process. Thomas Jefferson's Monticello contains Bohemian broad glass, and there is crown glass in Williamsburg. Plate glass was produced in the seventeenth century, mainly to provide the smooth surfaces needed to make mirrors. It was cast and then ground and polished. In the nineteenth century a product called *cylinder glass* was made by a blowing and swinging process. It could achieve a maximum size of 10 by 49 in. (254 by 1245 mm). It was this size restriction that governed the building module (48 in. [1219 mm]) of the Crystal Palace built in London in 1851, which contained 900,000 sq. ft. (83,613 m^2) of cylinder glass. Later innovations led to the drawing of molten glass cylinders vertically to a length of up to 50 ft. (15.24 m).

The real revolution in glassmaking dates from early in the twentieth century when the Fourcault and Colburn processes for mass producing large sheets of flat glass were developed.

Although they were all known earlier, glass ranks with concrete, steel, aluminum, and plaster as a true material of the twentieth century. Glass has exerted a tremendous influence on our technology, our modern way of life, and on architectural expression. From the divided-light window of colonial days to modern floor-to-ceiling glazing, glass technology has progressed with building design needs and has, in turn, influenced them.

Just as colonial builders limited the size of windows because of the high cost of glass and the shelter needs of the time, today's designers continually increase glass areas as the cost of glass declines and our ability to control the interior environment improves (Fig. 8.8-1). Largely glass or all-glass buildings are commonplace and create psychological impacts on their inhabitants (Fig. 8.8-2). The ability to view the ever changing panorama of the natural outdoor environment is psychologically satisfying to most people. Possible adverse effects from overexposure to the elements can be minimized by proper orientation and judicious selection of special glass types such as glare-reducing, heat-absorbing, reflective-coated, or insulating glass.

8.8.1.1 Properties of Glass

Glass is unusual in terms of its internal structure. It is mechanically rigid and thus behaves like a solid. However, most mineral solids have a crystalline structure, with the atoms arranged in a definite geometric pattern. In glass, the atoms are arranged in a random or disordered fashion, as is characteristic of a liquid. But the atoms, though arranged at random, are "frozen" in position by the rapid cooling process used to make glass. Therefore, glass combines some of the aspects of a solid and some of the aspects of a liquid. For this reason, it is sometimes called a *supercooled* liquid.

The physical properties of glass can be varied over an extensive range by modifying the composition of the glass, the production techniques used, or both. In selecting an individual glass for a particular product, a combination of properties must be taken into consideration, because one property usually cannot be changed without causing a change in other properties. The following is a brief discussion of some of the properties of glass that are important in architectural applications.

MECHANICAL PROPERTIES

Mechanical properties deal with the action of forces on a material and the effects these forces produce within the material.

An important mechanical property of glass is that it is *fatigue resistant*. This means that if pressure is applied to glass, causing it to bend or stretch, the glass will return exactly to its original shape when the bending or stretching force is removed. Of course, if increasing force is applied, the glass will eventually break, but at any point short of breakage it will not be permanently deformed.

FIGURE 8.8-1 University Tower in Durham, North Carolina, is an example of the kind of glass use that influences architecture today. (Architect Pickle/Thomas; photograph courtesy Guardian Industries.)

The mechanical strength of glass is determined by its ability to withstand forces that cause breakage. Glass generally breaks from stretching or bending (tensile forces), and thus tensile strength is the chief factor used in determining its mechanical strength.

Unlike most metals, glass does not have a clearly defined tensile strength. The *theoretical* tensile strength of glass has been estimated to be as high as 4,000,000 psi (2,812,280 Mg/m^2). The actual tensile strength of annealed glass ranges between 3000 and 6000 psi (2109.2 and 4218.42 Mg/m^2).

The main reason for the wide variation between theoretical and actual strength in glass is that actual strength is dependent on the surface condition of the glass. No matter how carefully it is handled, glass will acquire some small nicks and scratches in the course of manufacture and, later, in use. These surface defects then create weak planes in the material. When a force is applied to glass, stresses tend to concentrate at the weak planes rather than distribute themselves uniformly, as is assumed in theoretical strength calculations. Because of variations in the surface condition of different lights of glass, a group of seemingly identical glass specimens will exhibit a wide range of breaking strengths when tested. Thus, the mechanical strength of glass cannot be stated as a clearly defined figure; it must always be calculated on a probability/stress curve.

The mechanical strength of glass can be improved by both heat and chemical treatments. These processes are discussed in Section 8.8.1.2.

OPTICAL PROPERTIES

An obvious and important property of glass is its transparency. Despite appearances, however, no glass is completely transparent. When light falls on a piece of glass, some of that light is reflected and some is absorbed. The amount of light that does pass through the glass is called *transmittance.* Commonly used clear commercial glass has a transmittance of about 85% to 90% of visible light.

Glass may be treated in many different ways to reflect or absorb varying degrees of light. Glass that has been tinted absorbs a higher percentage of light than clear glass and, depending on the color and thickness of the glass, may transmit anywhere from 21 to 75% of visible light. Glass that has been treated with a thin reflective coating reflects a high proportion of light and may transmit from 8 to 50% percent of visible light, depending on the nature of the coating. The metallic coating on the back of mirrors produces the maximum amount of reflectance possible, allowing no visible transmittance.

Another important optical property of glass is its *fidelity* (lack of distortion). When the two surfaces of a sheet of glass are perfectly parallel, the image that is seen through, or is reflected from, the glass will be free of waves or distortion (Fig. 8.8-3). However, small variations in the thickness of a sheet of glass can cause

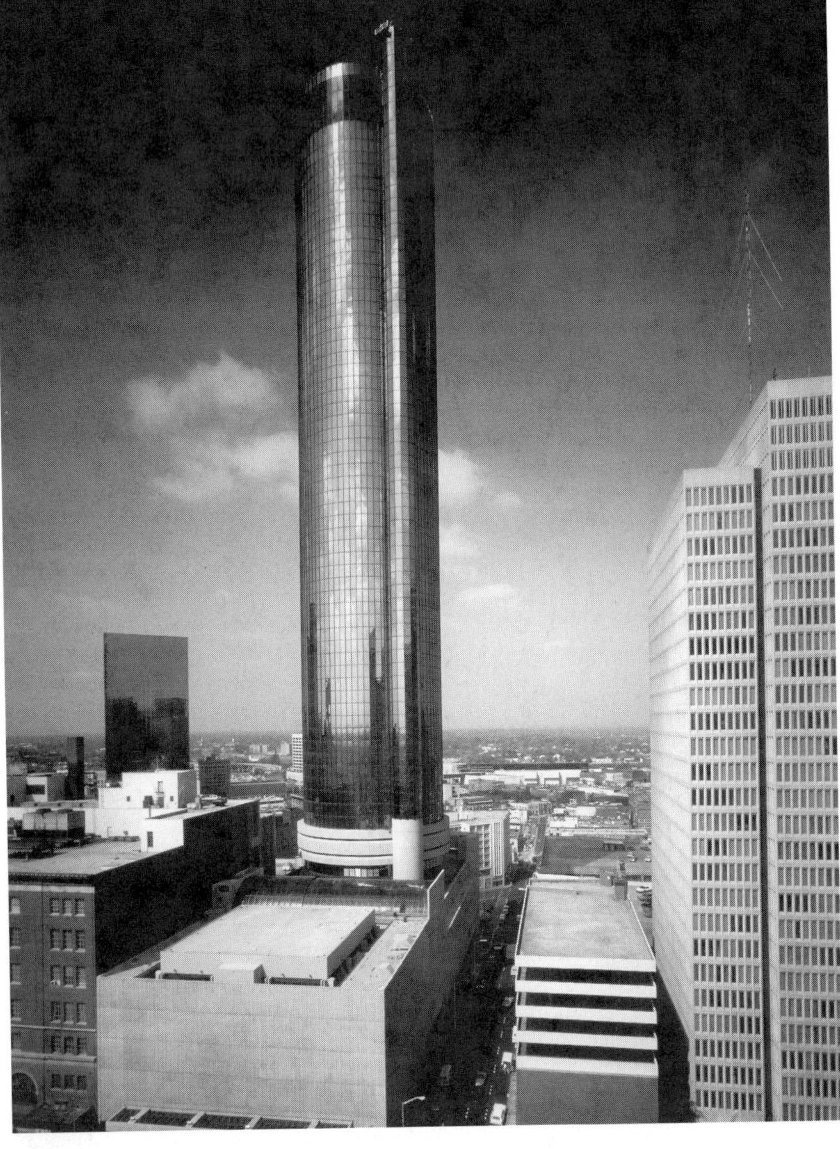

FIGURE 8.8-2 Peachtree Hotel, Atlanta, Georgia. (Architects: John Portman and Associates; photograph courtesy Bruce Wall Systems Corporation.)

FIGURE 8.8-3 The uniform thickness and flatness characteristics of float glass ensure distortion-free views.

considerable distortion in the image seen. For example, in a large area of glass, a variation in thickness of only a few hundred thousandths of an inch (mm) can be quite apparent, because the variation acts as a lens to magnify or diminish portions of the reflection.

THERMAL PROPERTIES

The amount of heat that passes through the glass areas of a building is an important consideration in determining heating and air conditioning loads. Heat can be transferred by conduction, convection, and radiation. Glass has a lower thermal conductivity than most metals, but a great deal of solar energy can be transferred directly through glass by radiation, and air currents that come in contact with glass surfaces can transfer heat by convection. Therefore, overall heat loss in winter and heat gain in summer is generally greater through the glass areas of a building than through ordinary masonry or metal- or wood-framed walls.

However, heat gain (or loss) through windows can be controlled by insulation, reflection, absorption, and shading. Special glass products such as insulating glass, reflective-coated glass, and tinted glass make use of these principles and can substantially reduce heat gain. In addition, shading devices such as blinds, draperies, and awnings can be used with glass to further reduce heat gain (see Section 12.2).

The thermal expansion of glass is another important consideration in the design of buildings. This factor is particularly important when using tinted or reflective-coated glasses, as these glasses absorb a higher percentage of solar energy than ordinary clear glass and therefore will expand and contract more.

When different portions of a single pane of glass heat up at different rates, they expand at different rates, causing thermal stresses to develop within the pane. What often happens under ordinary glazing conditions is that the center portion of the pane heats up faster than the edges (which are shaded by the edge framing), causing the center portion to expand more than the cooler edge regions. This produces stress differentials within the glass, which tend to concentrate at weak points in the glass and may eventually cause breakage. Most thermal failures can be traced to minor edge damage or imperfections, as this is where weak points in a sheet of glass are most likely to occur. Therefore, the best way to ensure against such thermal breakage is to be certain that glass edges are strong and clean-cut.

The thermal expansion of glass is also important when two glasses or a glass and a metal are to be sealed together, as in insulating glass and some types of coated glass. The rates of expansion of the two materials must be closely matched to avoid damaging differential stresses in the finished product.

CHEMICAL PROPERTIES

Glass is much more resistant to corrosion than most metals and other common materials. It is nonporous and nonabsorptive and thus is impervious to many common elements that affect other materials. For this reason, glass windows, after many years of exposure to the elements, remain clear and apparently unchanged.

Nevertheless, glass will corrode and even dissolve under some conditions. For example, hydrofluoric acid will attack soda-lime glass, causing extreme corrosion; similarly, hot concentrated alkali solutions and superheated water will cause soda-lime glass to dissolve. However, these chemicals are usually encountered only in laboratory situations; under ordinary environmental conditions, glass is an extremely inert and durable substance.

ELECTRICAL PROPERTIES

Because it offers high resistance to the passage of electricity, glass is a good electrical insulator. Therefore, glass is used in making many electrical products.

8.8.1.2 Manufacture of Glass

The progress in glass manufacture and the development of glass technology has been outstanding since the turn of the century. Today glass is manufactured in a wide variety of compositions and can be fabricated by several different techniques into a tremendous array of products.

RAW MATERIALS

Glass is manufactured in several generic types, which vary according to chemical composition (Fig. 8.8-4). The most common and least expensive type is *soda-lime glass*, made from the oxides of silicon, sodium, and calcium. This type glass is used for virtually all glazing purposes in both commercial and residential buildings and in automobiles.

Soda-lime glass is composed of approximately 72% silica, 15% sodium, 9% calcium, and 4% various minor ingredients. Other glass types, which have been derived from this basic glass, have slightly different ingredients. These are mentioned only briefly here, as they are of minor importance to the construction industry.

Silica Silica, the principal ingredient of glass, is obtained from sand. It is one of the most common elements of the earth's crust. Unlike the sand found at the seashore, sand for making glass must be more than 99% silica, perfectly white, and not too fine. Sand of this quality is usually obtained from sandstone deposits, which are abundant in the United States, particularly in Pennsylvania, West Virginia, Illinois, and Missouri.

Silica is the glass former in the glassmaking process. It can be used alone to produce glass, but it requires a very high melting temperature (3100°F [1704°C]) and is very difficult to form. It is usually combined with other ingredients to make a more workable glass.

Sodium Sodium for glassmaking is usually obtained from soda ash made from salt, but sodium nitrate and sodium sulfate may also be used. Sodium is added in glassmaking as a flux, to lower the melting point and to make a more workable mix. When sodium is added to silica, it forms a mixture that will melt at 1460°F (793°C), a considerably lower temperature than for silica alone.

Silica and sodium alone form a glass that is not very durable. In fact, it slowly dissolves when placed in contact with water. It also has low resistance to chemical attack. Therefore, an ingredient that will act as a stabilizer must be added to every glass batch.

Calcium The stabilizer usually added to glass is calcium. It is obtained from either limestone or dolomite. When added to silica and sodium, it produces a glass that is both durable and easily worked.

Additives Special elements may be added to, or substituted for, the basic ingredients in glassmaking in order to produce glass with specific properties or characteristics.

Potash, a form of potassium, was used instead of soda ash in the early days of glass manufacture in the United States. At that time wood was the principal fuel for glassmaking, and potash was a by-product

FIGURE 8.8-4 Representative Glass Types, Their Composition, and Their Uses

	Composition[a]								
	Sand	Soda Ash or Salt Cake	Limestone	Dolomite	Potash	Boric Acid	Litharge	Feldspar	
Type of Glass	SiO_2	Na_2O	CaO	MgO	K_2O	B_2O_3	PbO	Al_2O_3	Applications or Uses
100% silica	100%[b]	—	—	—	—	—	—	—	Glass laboratory ware
Soda lime silica[c]	68–74	11–16	8–12	1–4	—	—	—	0–4	Float glass
Borosilicate	65–80	2–4	3	—	0–4	13–28	—	1–3	Heat-resistant glasses
Lead alkali-silicate	35–63	0–10	0–1	—	10–13	0–1	15–58	0–1	Radiation-shielding products
Optical (flint)[d]	35	—	—	—	7	—	58	—	Opthalmic lenses
Aluminosilicate	55–57	0–1	5	8–12	4	4–8	—	20–23	High-thermal-impact glasses

[a]In addition to raw materials shown, small amounts of other elements may be added to help oxidizing, coloring, reducing, etc. Cullet (crushed glass) is also used to help melting.

[b]All figures are percentages of the ultimate oxides in glass compositions.

[c]Most important type based on volume and value.

[d]Optical glass often contains oxides of zinc, titanium, zirconium, and boron in addition to oxides shown.

of wood ash. Potassium is used infrequently today because it is much more costly than sodium. Glass made with potassium is finer, harder, and more brilliant than that made with sodium.

In soda-lime glass, lead is sometimes used to replace the calcium and sometimes to replace part of the silica. It produces a soft, easy-to-melt glass, and when used in quantity, imparts a canary yellow color to the glass. *Lead glass* is relatively expensive, but it has excellent electrical insulating and radiation-shielding properties and is more brilliant in appearance than soda-lime glass. It is used primarily in making electrical products, objects of art, and fine tableware.

Boron and aluminum are used to produce types of glass that are both heat- and shock-resistant. *Borosilicate* glass and *aluminosilicate* glass, as they are called, are used primarily for laboratory ware and oven ware.

There are several additives that are used specifically to produce color in glass (Fig. 8.8-5). For example, nickel oxide produces a tint ranging from yellow to purple, depending on the glass base; cobalt oxide gives an intense blue, and cobalt in combination with nickel produces a gray tone. Red glass is made with gold, copper, or selenium oxides, and many other colors may be produced with other chemicals. Arsenic is sometimes added to glass in small quantities to counteract the greenish color caused by iron impurities in the glass. Some of these colored and tinted glasses are used widely in architectural applications.

MIXING AND MELTING

The first major step in glassmaking is mixing and melting the raw materials. They are granulated to nearly uniform particle size, then carefully weighed and mixed in the proper proportions. Some *cullet* (scrap glass) from a previous melt of the same kind of glass is added to the mixture to assist in the melting process and to conserve materials. The mixed batch is then loaded into a melting unit. The type of unit used depends on the type and quantity of glass needed.

Some glasses that are required in small quantities are melted in *refractory pots*, which are clay containers that can hold up to 3000 lb. (1361 kg) of glass. As many as 20 of these pots can be heated in a single furnace, thus allowing several different types of glass to be melted at the same time.

For somewhat larger quantities of

FIGURE 8.8-5 Additives for Coloring Glass

Additive	% in Glass	Resulting Color
Cadmium sulfide	0.03–0.1	Yellow
Cadmium sulfoselenide	0.03–0.1	Ruby and orange
Carbon and sulfur compounds		Amber
Chromium oxide	0.05–0.2	Green to yellow-green
Cobalt oxide	0.001–0.1	Blue
Copper	0.03–0.1	Ruby
Copper oxide	0.2–2.0	Blue-green
Ferric oxide	Up to 4.0	Yellow-green
Ferrous oxide		Blue-green
Gold	0.01–0.03	Ruby
Iron oxide	(1.0–2.0%) with manganese oxide (2.0–4.0%)	Amber
Manganese oxide	0.5–3.0	Pink-purple
Neodymium oxide	Up to 2.0	Pink
Nickel oxide	0.05–0.5	Brown and purple
Selenides		Amber
Selenium		Bronze
Uranium oxide	0.1–1.0	Yellow with green fluorescence

glass, *day tanks* are used. Day tanks hold enough glass for a single day's operation—up to 20,000 lb. (9072 kg). The tank is filled with raw materials at the beginning of the operation, the glass is melted, and then all of the glass is used before the furnace is refilled.

The units most commonly used in commercial glassmaking operations are continuous tanks, holding 50 to 1000 tons of molten glass. These operate continuously, with raw materials being fed into the loading end as rapidly as the molten glass is removed from the working end. This continuous operation can go on for years, depending on how long it takes for the heat-resistant clay walls of the furnace to be worn away by the hot glass.

GLASSMAKING PROCESSES

After molten glass leaves a melting tank, different manufacturing processes are employed to make various glass products. Processes currently used to make flat glass include (1) *rolling*, for making patterned and wired glass, and (2) *floating*, for making float glass.

Other processes, such as *pressing* and *foaming*, are used to make such products as glass block and cellular glass insulation, respectively. Blowing is, of course, the oldest and best-known glassmaking process, but today it is primarily employed in the manufacture of containers and art objects rather than for making architectural products.

Drawing, an older process for making sheet glass, is no longer used, but is discussed here from a historical perspective.

Drawing In the early days of glass manufacture, sheet glass was made by the blowing process. The molten glass was blown into the shape of a cylinder; the cylinder was then split open, reheated, and flattened to form a small sheet of glass. Glass made by this process could not be produced in large sheets and was of poor quality. It was characterized by uneven surfaces and longitudinal wrinkles that distorted a person's vision.

The next step in the evolution of sheet glassmaking was the drawing process (Fig. 8.8-6). There were three different methods used: the *Fourcault*, the *Colburn*, and the *Pittsburgh* methods. All three utilized the same basic procedure.

The raw materials were first melted in a continuous melting tank. The drawing operation was then started by inserting a *bait* (steel rod) into the molten glass.

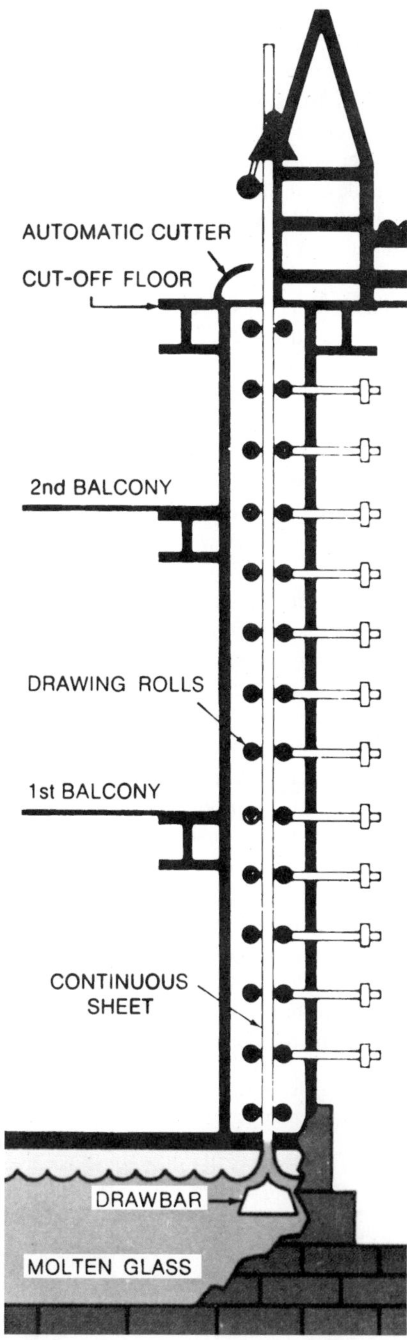

FIGURE 8.8-6 The Pittsburgh process for drawing sheet glass vertically.

Molten glass has a very high viscosity, making it sticky and highly resistant to flow. Therefore, the glass adhered to the bait and was drawn up with the bait as it was raised. The molten glass was drawn directly upward between jets of flame, which fire-polished the surfaces and made them transparent.

Pulling rollers then kept the drawing process going. A continuous sheet was drawn from the molten glass and was pulled slowly through an *annealing lehr*, a continuous oven. Inside the lehr, temperatures were regulated so that the glass was cooled gradually, thus avoiding excessive amounts of strain in the finished product.

After the sheet glass emerged from the annealing lehr, no further finishing operations were required other than cutting to size and packing.

There were some minor differences in technique between the various methods of drawing sheet glass.

The *Fourcault* process was the first process to be developed, and it was the simplest one. The glass sheet was drawn vertically from the tank to a height of about 40 ft. (12.19 m).

The *Colburn process* was similar to the Fourcault process, except that the glass sheet was drawn vertically for only about 3 ft. (914.4 mm) and was then bent over horizontal rollers. The chief advantage of this process was that the glass sheet could be drawn through a horizontal rather than a vertical annealing lehr. The length of a vertical lehr was limited by the height to which it was practical to draw glass vertically—that is, about 30 to 40 ft. (9.15 to 12.19 m). The horizontal lehr used in the Colburn process, however, could be up to 200 ft. (60.96 m) long, which allowed the glass to be cooled much more slowly, producing a glass that was less brittle and easier to cut.

In the *Pittsburgh* process, the glass was drawn vertically, as in the Fourcault process. It differed from the other two processes mainly in that the drawing machine was constructed so that the sheet of glass was not contacted by center rolls until it was hard enough to resist being marred by the rolls.

There were several problems encountered in the drawing process, no matter which method was used. One problem was that the drawing process stretched the sheet, causing "waisting-in" (variations in the width) of the glass. This was overcome by water-cooled rollers at each edge, which maintained sufficient side pull to keep the width constant.

In addition, the drawing process caused waves and ripples to form in the rising sheet of glass before it had cooled enough to become solid, which left an inherent wave or distortion, usually running in one direction. For most purposes this wave was not sufficiently noticeable to cause a problem; however, when large pieces of glass were required or when an oblique angle of view was likely, these imperfections could be quite apparent. This problem largely excluded sheet glass from use for large commercial mirrors, automobile windshields, and display windows, for which the superior qualities of plate or float glass were required.

Rolling Plate glass was originally made by *casting*, which was a very slow and laborious process. The glass was poured onto a casting table, then rolled flat, cooled, and finally ground and polished.

Between 1922 and 1924, however, the major glass manufacturers began using a continuous rolling process for making both plate and patterned glass. In this process, the glass was first mixed and melted in a continuous melting tank. The molten glass then flowed through an opening in the tank into the forming rolls. These rolls were water-cooled, electrically driven, and had precision-ground steel surfaces that could not be harmed by the extremely high temperatures of the glass. They rolled the molten glass to form a semisolid continuous ribbon of glass. The rolls could be adjusted for various thicknesses of glass, in widths up to 11 ft. (3352 mm).

From the forming rolls, the ribbon of glass moved over a series of other steel rolls until it became solidified enough to enter the annealing lehr. There it was cooled gradually to remove internal strains. After leaving the lehr, the glass was cooled still further until it reached room temperature. At this point, the glass was called rough-rolled glass.

If a translucent effect was desired, rough-rolled glass could be used without further processing. When patterned glass was desired, the forming rolls were embossed with various decorative designs; this glass, like rough-rolled glass, did not have to be ground and polished. However, if plate glass was to be made, both surfaces still needed to be ground and polished.

Most plate glass was made by the twin-grinding method, in which the glass was ground on both sides simultaneously. This method assured smooth, parallel surfaces and, thus, distortion-free glass. A mixture of silica sand with water was used in the grinding process.

The sand was graded into several different sizes; the largest particles were used in the first grinding units, and then the size of the particles gradually was reduced. A considerable quantity of glass was removed in the grinding process. Thus, the rough-rolled glass that was used to make plate glass was produced in a thickness greater than that desired in the finished product.

After grinding, the glass was moved to polishing tables, where it was passed along on a conveyor belt between a series of polishing wheels. The wheels were covered with buffs of felt, leather, or a soft metal, and the polishing agent most often used was rouge (ferric oxide) mixed in a solution of copperas water (ferrous sulfate and water). After polishing, the glass was cut to size and packed for shipment.

Rolling is no longer used to make polished plate glass by any manufacturer in this country. Rolling is still used, however, to make patterned and wired glass.

Floating Ground and polished plate glass has been replaced by *float glass* because the latter is substantially cheaper to produce. The float process was introduced in 1959 by Pilkington Brothers, Ltd., of England in an attempt to combine the fire polish and low cost of sheet glass with the flatness and low distortion properties of plate glass.

The first stage in the production of float glass is the same as that once used to produce sheet and plate glass. The batch is melted in a huge continuous tank. From there on, however, the process is quite different.

The molten glass leaves the tank and is floated on a bed of molten metal composed mostly of tin. When a liquid floats on another liquid of higher density, both surfaces of the upper liquid must become flat and parallel. Therefore, when the liquid glass moves across a float bath, it forms into a thin layer with flat parallel surfaces.

The float bath is divided into three sections (Fig. 8.8-7). In the first section, heat from the molten tin and a top heater causes

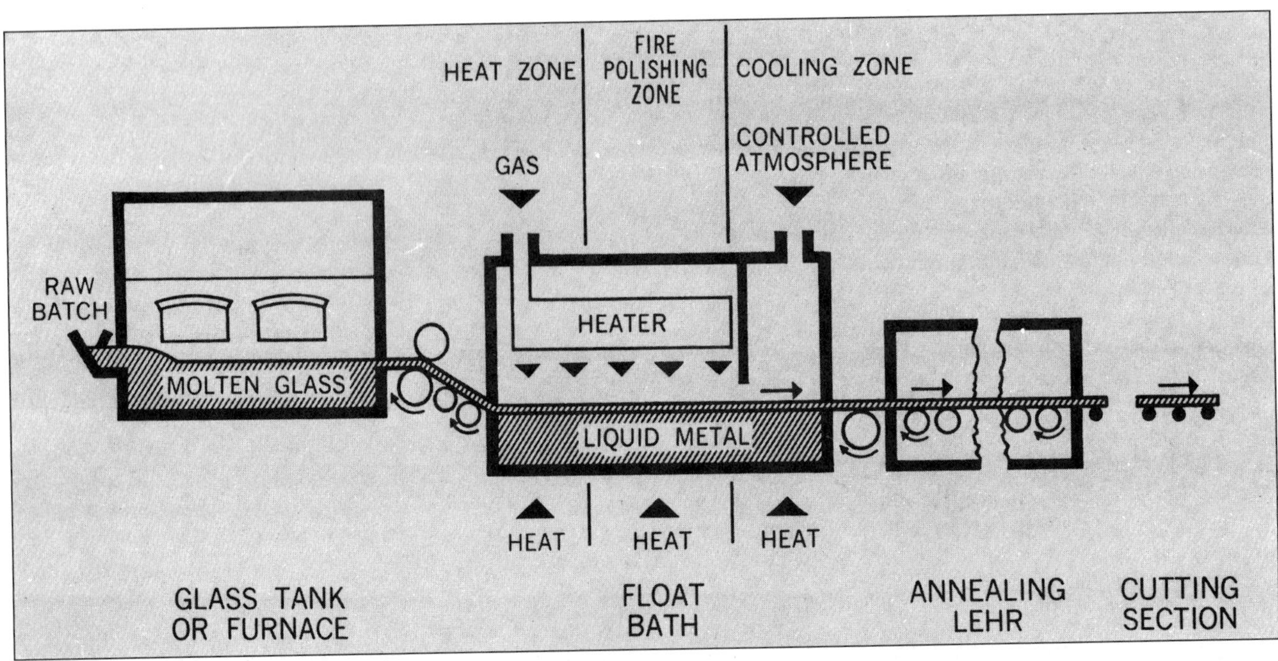

FIGURE 8.8-7 The float process for making glass.

the glass to float uniformly over the flat surface of the tin. In the second section, the glass receives a brilliant fire-polished surface. In the third section, the glass is cooled until it becomes hard enough to be conveyed to rollers in the lehr. Very precise control of heat all along the length of the float bath is required.

After the glass leaves the float bath, it enters an annealing lehr, where it is cooled gradually to ensure that the glass will not contain excessive amounts of strain. It is then washed and moved to a warehouse for inspection, cutting, and shipping.

HEAT AND CHEMICAL TREATMENTS

Several heat and chemical treatments are used to increase the strength and usefulness of flat glass products. *Annealing* relieves internal strain and facilitates fabrication. *Tempering* and *heat strengthening* cause physical characteristics to be developed in the glass so that it cannot be fabricated afterward. Fabrication, such as cutting, grinding, sandblasting, drilling, and notching must be performed before these heat treatments.

Annealing All flat glass, after forming, is subjected to a controlled cooling process called *annealing* to relieve internal stress. If a piece of glass is not annealed, parts of it will cool and contract at different rates, and when the glass approaches room temperature, it may fracture due to differential stresses throughout the sheet.

Annealing is performed by moving the glass on a belt through a continuous oven known as a *lehr*, in which temperatures are carefully regulated. The glass is first raised to a sufficiently high temperature to relieve strain developed in the forming process. Then the temperature is brought down slowly so that all parts of the sheet will cool at the same rate. Therefore, when the glass reaches room temperature, it is free of internal stress and can be easily cut and processed.

Tempering *Tempering* is a heat treatment used to increase the strength of flat glass. The process involves heating the glass to just below the softening point and then chilling it suddenly by subjecting both surfaces to jets of cool air. This causes the surfaces to shrink and harden quickly while the interior is still fluid. The interior begins to cool and shrink next, but since the surfaces are already hardened they cannot flow and adjust to the shrinkage of the interior. This causes the surfaces and edges to be placed in a state of compression while the interior of the glass is in a state of tension. In properly tempered glass the opposing stresses of the surfaces and the interior balance each other out. The resultant glass is much stronger than ordinary flat glass, because the built-in compression of the surface must be overcome before breakage can occur.

This type of glass usually is referred to as *fully tempered* glass to distinguish it from *heat-strengthened* glass, which is only partially tempered. Fully tempered glass is three to five times more resistant to wind loads, impact, and thermal stresses than ordinary annealed glass of the same thickness.

Heat Strengthening The heat treatment known as *heat strengthening* is also used to increase the strength of flat glass. It involves heating the glass and then cooling it in a manner similar to that used in tempering. However, heat strengthening produces lower surface compression stresses in the glass than the tempering process. Therefore, heat-strengthened glass is only about twice as strong as annealed glass of the same thickness.

Chemical Strengthening Glass can be strengthened by chemical treatment as well as by heat treatment. The most commonly used chemical treatment involves an exchange of ions in the surface of the glass. The glass is immersed in a molten salt bath, and the large potassium ions in the salt migrate into the glass surface, replacing the smaller sodium ions there. This action crowds the surface, resulting in compression stresses similar to those produced in glass by thermal tempering.

Full temper in soda-lime glass is difficult to achieve by chemical tempering methods. A high degree of temper by chemical means generally requires accurate formulation of ingredients selected to ensure that glass ions interchange readily with ions in the molten salt bath. Furthermore, in chemically strengthened glass, the compressed surface layer is appreciably thinner than that in thermally tempered glass. Therefore, it is more readily weakened by surface abrasions such as cuts and scratches.

SURFACE FINISHES

The surface finish of glass depends on both the method of manufacture and the subsequent finishing processes to which it is subjected.

Float glass emerges from the annealing lehr with a smooth, glossy, and transparent finish known as a *fire finish*. This type of finish is characteristic of the glassmaking process, and no further finishing operations are required. One of the reasons that plate glass was more expensive to produce than float glass was that it emerged from the annealing lehr with a rough, translucent surface that required grinding and polishing to produce the smooth, transparent finish desired for most flat glass products.

There are special cases, however, where a smooth, transparent glass is not essential. A translucent glass may be desired to provide privacy or to limit the amount of light transmitted into a room; a specially coated or tinted glass may be needed to reduce glare or to control the amount of solar heat transmitted into a room; a patterned or specially treated glass may be preferred just to produce a decorative effect. Some of the surface finishes associated with these special types of glass are discussed in the following paragraphs.

Etching Many different degrees of transparency can be produced in glass by etching. These range from a frosted, almost opaque quality to a semipolished, translucent appearance.

Hydrofluoric acid or one of its compounds is used to etch glass, as it is the only chemical that interacts actively with glass. Either the glass is dipped in the chemical or it is sprayed on; the type of surface produced depends on the composition of the glass, the concentration of fluorides, and the time involved. If an etched design is desired, the glass is first painted with an acid-resistant chemical (the *resist*) to protect the parts of the glass outside the desired pattern. The acid eats away the unpainted surface of the glass, leaving the design.

Sandblasting The translucent finish obtained by sandblasting is usually rougher than that obtained by etching. Sandblasting is done by using compressed air to blow coarse, rough-grained sand against the glass surface. This is often done through a rubber stencil to produce a decorative design.

Both etching and sandblasting reduce the strength of glass, because these processes introduce surface flaws. Sandblasting, for example, reduces the strength of glass by about 50%.

Enameling Translucent and solid colors can be produced on glass surfaces with vitreous enamels fired on at high temperatures. The firing process also imparts a partial or full temper to the glass. This type of glass is used primarily for spandrel areas and other parts of curtain wall construction.

Silvering In the early days of mirror manufacture, silvering was done by hand, the silvering solutions being poured from a pitcher onto the flat glass surface. Some fine mirrors are still produced in this way today.

Most mirrors, however, are manufactured by an automatic process in which the silver is deposited directly on the glass as it passes on a conveyor belt under sprays of silver nitrate and tin chloride. The silvering can then be protected with a coating of shellac, varnish, or paint. For almost permanent protection, an electroplated layer of copper can be applied on the silver.

Reflective Coatings With the advent of the glass-enclosed high-rise, an entire new family of architectural glasses has been developed to control the transmission of solar energy into buildings. These products have a thin reflective coating applied either (1) to one surface of a single sheet of glass or (2) to an inside surface of a laminated or insulating glass unit.

Two processes are used to apply reflective coatings to glass: (1) the pyrolytic process, in which the coating material is sprayed onto heated glass and becomes fused to the glass surface, and (2) vacuum deposition, in which the glass is bombarded with metallic ions, which form an integral molecular bond with the glass.

Under appropriate lighting conditions, reflective-coated glass can become a type of one-way mirror, permitting vision only to the side with the greater amount of light. During the daytime, it creates a mirrorlike effect from the outside of a building, while from the inside the glass is transparent. At night, when the interior lights are on, the effect is reversed, although to a lesser degree. Reflective coatings can be made to reflect varying degrees of light and heat, providing control over the solar energy transmitted into a building.

Tinting Tinting is not really a type of surface finish, because it involves adding special ingredients to the basic glass batch. However, it is used widely to produce glass with heat-absorbing and glare-reducing qualities.

Many different ingredients can be added to glass to produce various tints. Ferrous iron, for example, imparts a bluish-green tint to glass, cobalt oxide in combination with nickel imparts a grayish tint, and selenium is used to make bronze-tinted glass. These different ingredients are added to the glass mixture at the very beginning of the melting process. Float glass can be tinted in this manner.

Tinted glass expands more than untinted glass due to its greater absorption of heat, which must be taken into account when designing the surrounds of the glass.

8.8.1.3 Flat-Glass Products

The multitude of available glazing products makes possible a high degree of specialization in meeting specific design requirements. The method of glass manufacture determines the basic type of the glass. By modifying the ingredients, any type of flat glass can be made clear to provide true vision, or tinted to reduce glare and absorb solar energy. Clear or tinted glass may be coated with a reflective surface to further reduce light and heat transmission.

Where resistance to impact and thermal stresses is important, these glasses can be strengthened by heat treatments, such as tempering or heat strengthening. Glass, either clear or tinted, can be rolled with a wire mesh reinforcement to make wired glass. Moreover, almost any two or more of these glasses can be combined to form a composite multilayer glass. Different types of composite glass are designed to meet specific construction requirements, such as laminated glass for safety and security, and insulating glass for thermal and sound insulation.

STANDARDS OF QUALITY

In the United States, primary float glass, wired glass, and patterned glass are manufactured to meet the requirements of ASTM C 1036; heat-treated glass to meet the requirements of ASTM C 1048; and laminated glass to meet the requirements of ASTM C 1172. Figure 8.8-8 gives a summary of the standards governing some typical glass types. Both ASTM C 1048 and C 1172 reference C 1036 for basic glass used in the glass types covered by these standards. The ASTM standards establish the dimensional, optical, strength, allowable bow (warpage), and other requirements for these glazing products. Figure 8.8-9 gives the typical glass classifications found in ASTM C 1036 and C 1048.

FIGURE 8.8-8 Summary of Glass Standards

Glass Type	Standard
Primary float	ASTM C 1036
Tinted float	
Heat-absorbing and light-reducing	ASTM C 1036
Light-reducing only	ASTM C 1036
Patterned	
Heat-strengthened	ASTM C 1048
Fully tempered	ASTM C 1048
Not heat-treated	ASTM C 1036
Wired	ASTM C 1036
Heat-strengthened	
Coated and uncoated	ASTM C 1048
Fully tempered	
Coated and uncoated	ASTM C 1048
Laminated glass	ASTM C 1172
Insulating glass	ASTM E 774

Sealed insulating glass products should comply with ASTM E 774.

Another important standard for glass is ANSI Z97.1, which covers the safety requirements for fully tempered, laminated, and wired glass, as well as for rigid plastics. It is directed primarily toward ensuring that safety glazings have fail-safe characteristics when broken by human impact.

The use of safety glazing in hazardous locations, such as door sidelights, sliding glass doors, storm doors, and tub-shower enclosures, is required by the Federal Housing Administration (FHA) Minimum Property Standards for Housing, the major model building codes, many municipal codes, the Council of American Building Official's (CABO) *One- and Two-Family Dwelling Code*, and the Consumer Product Safety Commission, Code of Federal Regulations (CFR) 16. In addition, safety glazing requirements have been adopted by a number of states and are being considered by many others.

Certification of Sealed Insulating Glass The Insulating Glass Certification Council (IGCC), founded in 1977, provides independent laboratory testing and periodic

FIGURE 8.8-9 Summary of ASTM C 1036 and ASTM C 1048 Glass Classifications

Types	I	Transparent glass
	II	Patterned or wired glass
Classes	1	Clear
	2	Tinted, heat-absorbing, light-reducing
	3	Tinted, light-reducing
Quality	q^1	Mirror select
	q^2	Mirror
	q^3	Glazing select
	q^4	Glazing A
	q^5	Glazing B
	q^6	Greenhouse
	q^7	Decorative
	q^8	Glazing
Form	1	Wired
	2	Patterned and wired
	3	Patterned
Finish	f^1	Patterned one side
	f^2	Patterned both sides
Mesh	m^1	Diamond
	m^2	Square
	m^3	Parallel strand
	m^4	Special
Pattern	p^1	Linear
	p^2	Geometric
	p^3	Random
	p^4	Special
Kind	HS	Heat-strengthened
	FT	Fully tempered
Condition	A	Uncoated surfaces
	B	Spandrel glass with one surface ceramic-coated
	C	Other coated glass

unannounced on-site plant inspections of participating manufacturers. The Associated Laboratories, Inc. (ALI) and the National Certified Testing Laboratories also provide certification and testing programs. These certification programs are conducted in compliance with ASTM E 774.

Certification of Quality-Safety Glazing The Safety Glazing Certification Council (SGCC) administers a program of testing, random sampling, product evaluation, and certification to ensure that safety glazing materials comply with ANSI Z97.1. This program is administered in accordance with ANSI Z34.1, which provides for certification on an impartial, independent, and continuing basis. Products meeting the requirements of Z97.1 are given a certification label. There is no one universal SGCC label. Instead, each manufacturer submits its own label to SGCC for approval. This label must include reference to ANSI Z97.1, plus an SGCC identification number that indicates the plant in which the product was made. The label must be permanently affixed to the certified product at the time and place of manufacture.

Manufacturers' Warranties Specialty products such as laminated, reflective, and insulating glass are warranted by their manufacturers against manufacturing defects. Laminated glass is typically covered for a period of 5 years against edge separation or defects that obstruct vision. Reflective glass is warranted for 10 years against peeling or deterioration of the reflective coating. Insulating glass usually carries a warranty against any material obstruction inside the glass unit that impairs vision, such as condensation, dust, or foreign particles. The rainbow "oil slick" effect observed occasionally at certain angles is not a manufacturing defect, but the result of light refraction between accurately machined parallel glass surfaces.

Insulating glass carries either a 5-year or 10-year warranty. Stock windows and sliding doors containing insulating glass are normally preglazed with 5-year quality units.

In small window units enclosed by an accurately manufactured sash, the likelihood of the seal's breaking is remote, and the optical quality is not critical if the windows are used primarily for daylighting rather than viewing. For these uses insulating units with a 5-year warranty are generally acceptable.

However, larger insulating units (often used in sliding glass doors and in field-glazed fixed sash) that are intended for viewing should carry long-term assurance against annoying visual obstructions such as internal fogging. These units are generally subject to more mechanical stresses and are larger and more expensive to replace. They should be covered, therefore, by a standard 10-year warranty.

Most glass warranties provide protection against manufacturing defects, but specifically exclude damage due to faulty installation or use in other than ordinary buildings. For instance, the warranty does not apply for applications in ships, vehicles, or commercial refrigeration. The manufacturer usually agrees to deliver a replacement unit without charge to the shipping point nearest the installation, but the cost of forwarding the unit and actual installation must be borne by others.

FLOAT GLASS

Until recently there were three basic types of flat glass in general use in the United States: (1) sheet glass, (2) plate glass, and (3) float glass. Today, float glass has completely replaced the other two and is the only type discussed here.

Float glass has become the most widely used glass product. According to the Flat Glass Marketing Association (FGMA), float glass constitutes as much as 98% of the glass manufactured in the United States. It is used today for many of the purposes once served only by plate glass (high-quality mirrors, display windows, and automobile glazing) and in thinner sheets for the purposes formerly served by sheet glass.

Most of the other types of glass discussed in the following paragraphs are variations of float glass.

According to ASTM Standard C 1036, clear float glass (Type I, Class 1) is available in six qualities, as shown in Figure 8.8-10. It is usually available in thicknesses from 3/32 to 7/8 in. (2.38 to 22.2 mm), but some qualities may be available only in certain thicknesses. For example, q^5 (see Fig. 8.8-9) is usually available only in 3/32-in. (2.38 mm) thickness and q^4 only in 1/8 in. (3.2 mm).

PATTERNED GLASS

Patterned glass is translucent, with a linear or geometric pattern embossed on one or both sides. It is produced by the same method used for plate glass (rolling) but is not ground and polished. Patterned glass is classified in ASTM Standard C 1048 when it is heat-strengthened or fully tempered, and in ASTM Standard C 1036 when it is neither of these. Its normal classifications in all three cases are shown in Figure 8.8-10.

Patterned glass is available in a wide range of textures and patterns, which vary according to manufacturer. These patterns offer different degrees of light transmission and obscurity, depending on the nature of the design and whether it is rolled into one or both surfaces of the glass. Additional obscurity may be obtained by etching or sandblasting the patterned glass. These processes also reduce the strength of the glass, which should be considered in determining suitable sizes and uses.

Patterned glass is available mainly in thicknesses of 1/8 and 7/32 in. (3.2 and 5.56 mm), although some patterns are pro-

FIGURE 8.8-10 Usual Classifications for Some Types of Flat Glass

Product	Classifications*
Clear float	Type I; Class 1; Quality q^1, q^2, q^3, q^4, q^5, or q^6
Tinted heat-absorbing, light-reducing float	Type II; Class 2; Quality q^3, q^4, q^5, or q^6
Tinted light-reducing float	Type I; Class 3; Quality q^3, q^4, q^5, or q^6
Patterned	Type II; Class 1, 2, or 3; Quality q^7 or q^8; Form 3; Finish f^1 or f^2; Pattern p^1, p^2, p^3, or p^4
Wired	Type II; Class 1, 2, or 3; Quality q^7 or q^8; Form 1; Mesh m^1, m^2, m^3, or m^4
Wired/patterned	Type II; Class 1, 2, or 3; Quality q^7 or q^8; Form 2; Finish f^1 or f^2; Mesh m^1, m^2, m^3, or m^4; Pattern p^1, p^2, p^3, or p^4
Heat-strengthened	Type I; Class 1, 2, or 3; Quality q^3, q^4, or q^5; Kind HS; Condition A, B, or C
Heat-strengthened patterned	Type II; Class 1, 2, or 3; Quality q^7 or q^8; Form 3; Finish f^1 or f^2; Pattern p^1, p^2, p^3, or p^4; Kind HS; Condition A, B, or C
Fully tempered	Type I; Class 1, 2, or 3; Quality q^3, q^4, or q^5; Kind FT; Condition A, B, or C
Fully tempered patterned	Type II; Class 1, 2, or 3; Quality q^7 or q^8; Form 3; Finish f^1 or f^2; Pattern p^1, p^2, p^3, or p^4; Kind FT; Condition A, B, or C

*Refer to Figure 8.8-9 for meaning of classification symbols.

duced in other thicknesses on a limited basis (Fig. 8.8-11). Some patterns also are available heat treated.

Patterned glass can provide visual privacy while at the same time allowing diffused light transmission. It has a decorative appearance and is often used for this reason alone. It is used for interior partitions and exterior windows, such as in toilet and shower rooms, where obscured vision is desired. It is also used in tub and shower enclosures; glass in these locations should be tempered for safety.

WIRED GLASS

Wired glass has a wire mesh or parallel wires rolled into the center of the glass thickness. It is available either polished or patterned. Most of it is ¼ in. (6.4 mm) thick, but some manufacturers provide other thicknesses (see Fig. 8.8-11). It should comply with ASTM Standard C 1036 (see Fig. 8.8-10).

The wire in this type of glass holds it together under low levels of impact, thereby reducing injuries from broken glass. It

FIGURE 8.8-11 Summary of Patterned and Wired Glass Types and Sizes

Product	Type	Nominal Thickness in.	Nominal Thickness mm	Maximum Area, in. (mm)[a]	Weight, lb./sq. ft. (kg/m²)	Visible Light Transmission, %
Patterned Glass	Floral Hammered Stippled	⅛	3	60 × 132 (1524 × 3353)	1.60–2.10 (7.81–10.25)	80–90
	Ribbed Fluted[a] Striped[b]	7⁄32	5.5	60 × 132 (1524 × 3353)	2.40–3.00 (11.72–14.65)	80–90
Wired Glass	Polished (square or diamond mesh)	¼	6	60 × 144 (1524 × 3658)	3.50 (17.09)	80–85
	Patterned (square or diamond mesh)	¼	6	60 × 144 (1524 × 3658)	3.50 (17.09)	80–85
	Parallel wired[c]	7⁄32	5.5	54 × 120 (1372 × 3048)	2.82 (13.77)	80–85
		¼	6	60 × 144 (1524 × 3658)	3.50 (17.09)	80–85
		⅜	10	60 × 144 (1524 × 3658)	4.45 (21.73)	80–85

[a]Maximum area varies according to producer; larger sizes may be available from some producers.

[b]These are just a few of the most common patterns available; many patterns are patented and made by one producer only.

[c]This type of wired glass does not carry Underwriter's Laboratory, Inc. fire-retardant rating; it is used mainly for decorative partitions.

is considered a safety glazing material and, when properly made, meets the requirements of ANSI Z97.1. It is used widely for corridor and entrance doors, partitions, skylights, and other potentially hazardous locations. It is widely used in fire doors and windows because of its ability to remain in place even though cracked by excessive heat. When used as a fire-resistant material, it should conform to Underwriters Laboratories, Inc. (UL) standards.

TINTED GLASS

To make tinted glass, an admixture is included in the glass batch to impart glare-reducing and heat-absorbing qualities. Float glass is available tinted. Tinted glass is covered by ASTM C 1036. It may be either simply light-reducing (Class 3) or light-reducing and heat-absorbing (Class 2) (see Fig. 8.8-10).

Most gray- and bronze-tinted glasses are produced in thicknesses up to ½ in. (12.7 mm), and in select glazing quality (q^3) only. Green-tinted glass is generally available only in thicknesses up to ¼ in. (6.4 mm). Equal thicknesses of gray-, bronze-, and green-tinted glass absorb the same percentage of solar energy. Tinted glasses offer a wide range of visible light transmittance, varying from 20% to 78%, depending on the thickness and color of the glass (Fig. 8.8-12).

This type of glass is generally used where reduced light or heat transmission is desired for indoor visual comfort or where the color of the glass can contribute to the design of the building. Because of its heat-absorbing characteristics, special precautions should be taken when using tinted glass to avoid damage to the edges and consequent breakage from thermal stresses.

HEAT-ABSORBING GLASS

If a metallic oxide is added to float glass during the manufacturing process, the glass can absorb significantly more sunlight, especially ultraviolet light. This characteristic distinguishes heat-absorbing glass from tinted glass that merely excludes the visible light useful for illumination (light-reducing).

The sunlight absorbed by glass is reradiated and convected as heat to the inside or outside, depending on which side is cooler. The result is that in summer, more heat is dissipated to an air conditioned interior, because of its cooler temperature. In winter, more heat is dissipated to the outside, because it is then cooler.

However, the net effect is still beneficial, because 17% less heat-producing sunlight is admitted through heat-absorbing glass than through clear glass.

The performance of heat-absorbing glass is substantially improved when it is used as the outer sheet of an insulating unit. In an insulating unit, the absorbed heat must first bridge the air space to the inner sheet of glass before it can be radiated and convected into the building interior. Because heat transfer is slowed by the insulating air space, more heat is dissipated to the outside air, which is in direct contact with the heat-absorbing glass. Furthermore, the outward rate of heat dissipation is greatly accelerated if there is a wind.

COATED GLASS

When sunlight strikes a pane of glass, it is either transmitted, reflected, or absorbed

FIGURE 8.8-12 Summary of Tinted Glass Types and Sizes

Product	Quality	Nominal Thickness in.	Nominal Thickness mm	Maximum Area, in. (mm)[a]	Weight, lb./sq. ft. (kg/m²)	Visible Light Transmission, %	Solar Energy Transmission, %	Principal Uses
Bronze	Glazing	⅛	3	35 sq. ft. (3.25 m²)	1.64 (8.01)	68	65	—
		3⁄16	5	120 × 144 (3048 × 3658)	2.45 (11.96)	58	55	
		¼	6	128 × 204 (3251 × 5182)	3.27 (15.97)	50	46	
		⅜	10	124 × 200 (3150 × 5080)	4.90 (23.9)	37	33	
		½	12	120 × 200 (3048 × 5080)	6.54 (31.93)	28	24	
Gray	Glazing	⅛	3	35 sq. ft. (3.25 m²)	1.64 (8.01)	62	63	Where reduced heat and light transmission is desired
		3⁄16	5	120 × 144 (3048 × 3658)	2.45 (11.96)	51	53	
		¼	6	128 × 204 (3251 × 5182)	3.27 (15.97)	42	44	
		⅜	10	124 × 200 (3150 × 5080)	4.90 (23.9)	28	31	
		½	12	120 × 200 (3048 × 5080)	6.54 (31.93)	20	24	
Green	Glazing	3⁄16	5	96 × 128 (2438 × 3251)	2.45 (11.96)	78	55	—
		¼	6	96 × 128 (2438 × 3251)	3.27 (15.97)	74	48	

[a]Maximum area varies according to producer; larger sizes may be available.

Transmission values vary slightly from one producer to another.

and radiated as heat. By increasing the amount of reflected sunlight, the amount that can be absorbed and transmitted or radiated is reduced.

The category of coated glass called *high-performance glass* has on one side a transparent film of metal or metal oxide. This coating is applied either by *pyrolytic deposition* or *vacuum sputtering*. Those applied by pyrolytic deposition are called *hard-coat* coatings. Some producers claim that these coatings can be exposed to wear. Others do not agree. Coatings applied by vacuum sputtering are called *soft-coat* coatings. Again, some producers recommend them for use where they will be exposed to contact. Others claim that they cannot be used that way and must be applied only on the protected surfaces of panes, such as those in insulated glass or spandrel glass.

Because such coated glass appears mirrorlike when viewed from the exterior, it is called *reflective glass*. Some designers select it as much for this aesthetic property as for its ability to reduce the solar energy transmitted into a building. In insulating units it also minimizes ambient heat flow in and out of the building. Several of the more durable coatings for this glass, such as cobalt oxide, chromium, stainless steel, and titanium, are available on single glass as well as on insulating and laminated glasses. Other available coatings, such as copper, aluminum, nickel, and gold, are less durable and are mostly furnished on an inside surface of insulating or laminated glasses, where the coating is protected from weather and accidental abrasion.

A wide range of metallic colors is available with differing visual and thermal properties. Daylight reflectances range from about 10% for the more subdued coatings to about 45%, and visible light transmittance from 8 to 55%. The more common colors are neutral, gray, gold, bronze, blue, silver, green, and copper. Many coatings are also available on tinted, heat-absorbing glass. When a coating is applied to such glass, it often must be heat-strengthened or fully tempered.

Coated glass products vary widely from one manufacturer to another, and certain colors and tones are offered by only one company. Consequently, individual producers should be consulted for specific product information.

Reflective glass is more effective as the outer sheet of insulating glass than as single glazing. The reason is that, despite its name, reflective glass absorbs more sunlight than clear glass, which permits most of the sunlight that strikes it to pass through. There is, therefore, more heat build-up in reflective glass. When that heat is concentrated in the outer sheet of insulating glass, it is dissipated to the outside air because the air space between glass layers impedes its flow inward.

There are some rather important disadvantages of reflective glass:

- In winter, beneficial heat gain is reduced by it.
- Transmission of visible light for general illumination is severely reduced year-round, making occupants feel as though they are looking through dark sunglasses.
- In spite of its large reduction in light transmission, it does not significantly increase the insulating value of the glass.
- Its high reflectivity can create problems for neighbors and even for traffic on the streets below. In fact, some cities now restrict the amount of reflectivity on buildings.

Most of these disadvantages have been overcome, however, by advances in coating technology. New metallic coatings are of a different type than those used in earlier versions. Some of these coatings produce a more subdued mirror effect, reducing the amount of reflected light. These are called *low-reflectance* glasses. Others are almost transparent. Much light passes through them, but their *emissivity* is reduced sharply. Emissivity refers to the amount of heat absorbed by a material and then reradiated by it. This type of high-performance glass is called *low-emissivity* or simply *Low-E* glass. It can reflect as much as 90% of the radiant energy that strikes it while admitting most of the light. Low-E glass is available in many colors and with many levels of performance characteristics.

Spandrel glass is used most often to cover the spandrel beams and floor construction in curtain walls. It is also sometimes used in other locations on the exterior of a building. It is often coated with ceramic materials. It may also be coated with a high-performance coating resembling that used on the vision panels in the same building. Sometimes the metal oxide coating on spandrel glass is protected with a thin sheet of plastic called an *opacifier*.

HEAT-TREATED GLASS

There are two types of heat-treated glass: fully tempered and heat-strengthened. When either of these types of glass is to be used, the exact sizes desired must be specified, as the glass cannot be cut, drilled, or notched after the heat treatment. These types of glass are governed by ASTM C 1048.

Fully Tempered Glass Except for wired glass and patterned glass with deep patterns, any glass ⅛ in. (3.2 mm) or thicker may be tempered. Tempered glass may be incorporated into insulating glass or laminated glass units; heat-absorbing and heat-reflective glasses may also be tempered.

An inherent characteristic of all tempered glass is a slight deviation from flatness, particularly near the edges. The degree of deviation depends on thickness, width, length, and other factors. Greater thicknesses usually yield flatter products.

Fully tempered glass has three to five times the resistance of annealed glass to uniform loading, thermal stresses, and most impact loads. When fractured, it breaks into a safe pattern of relatively small, harmless particles (Fig. 8.8-13). It is considered a safety glazing material and, when properly made, meets the requirements of ANSI Z97.1. It is frequently required by state or local codes in areas where human contact with glass is probable, such as entrances and sidelights, tub and shower enclosures, storm doors, and sliding glass doors.

Heat-Strengthened Glass Heat-strengthened glass is similar to fully tempered glass, except that it has only a partial heat temper. It is twice as strong as ordinary

FIGURE 8.8-13 Broken fully tempered glass.

annealed glass of the same thickness. When it fails, it breaks into fragments much larger than those of fully tempered glass, and therefore it is not considered a safety glazing material. Heat-strengthened glass is generally used in applications that require a stronger product than regular annealed glass but do not require the full strength of tempered glass.

Any type of ¼-in. (6.4-mm)- or 5⁄16-in. (7.9 mm)-thick glass that can be tempered can be heat-strengthened. The most widely used form of heat-strengthened glass is spandrel glass. This glass has a fired-on ceramic frit on the interior surface, and it is used commonly in spandrel areas in curtain wall construction.

DECORATIVE GLASS

Patterned, articulated, and other forms of decorative glass are available. These may be etched glass or layered glass with different colors in different layers or raised portions of clear or colored glass. Leaded or "stained" glass is another form of decorative glass. Most decorative glass is more expensive than other forms of glass and is, therefore, used primarily in special locations, such as chapels, where the aesthetic effect outweighs the additional expense.

COMPOSITE GLASS

The two main types of fabricated glass, *laminated* and *insulating* glass, are products made by combining two or more layers of glass into a single unit.

Laminated Glass Laminated glass consists of two or more layers of glass with interlayers of *polyvinyl butyral plastic* sandwiched between them, bonded together under heat and pressure to form a single unit. The thickness of the glass layers may range from 3⁄16-in. to ⅞-in. (4.8 to 22.2 mm) float glass. The plastic interlayer is usually .015 or .030 in. (0.381 or 0.762 mm), although it may be as thick as .090 in. (2.29 mm). The face of the glass that is to be bonded to the plastic cannot be patterned or otherwise irregular. Heat-absorbing, heat-reflecting, fully tempered, heat-strengthened, and wired glasses may all be laminated with themselves and in combination with other types.

When laminated glass is broken, the fragments of glass generally adhere safely to the plastic interlayer and do not evacuate the opening. This reduces the potential hazards of flying glass and minimizes cutting injuries. All varieties of laminated glass are considered safety glazing materials and, when properly made, meet the requirements of ANSI Standard Z97.1 (Figs. 8.8-14 and 8.8-15).

FIGURE 8.8-14 Laminated glass was used in this sloped glazing application at the IBM facility in Charlotte, North Carolina. (Courtesy Laminated Glass Corporation.)

Most laminated glass is made of two lights of glass of varying thicknesses with a single plastic interlayer. It is used commonly for interior partitions, doors, and automotive glazing. There are also several special types of laminated glass that are made for particularly demanding applications (Fig. 8.8-16).

Burglar-resistant glass is a laminated glass usually consisting of two layers of ⅛-in. (3.2 mm) glass with a .060- to .090-in. (1.52 to 2.29 mm) plastic interlayer. It is made primarily for use in the protection of show windows displaying valuable merchandise. It has superior resistance and serves as a major deterrent to the "smash-and-run" burglar.

Bullet-resisting glass is a laminated glass having four or more layers totaling ¾ in. (19.1 mm) to 3 in. (76.2 mm) in thickness. Most manufacturers have 1 3⁄16-, 1 9⁄16-, 1 ¾-, and 2-in. (30.16, 49.69, 44.5, and 50.8 mm) thicknesses listed by UL for resistance to medium-, high-, and super-power small arms and high-power rifle arms, respectively. Thicknesses up to 7 in. (177.8 mm) are available on special order. This type of glass is used for bank teller windows, payroll booths, and other security applications.

FIGURE 8.8-15 Laminated security glass protects the users of the CN Tower in Toronto. (Courtesy Laminated Glass Corporation.)

FIGURE 8.8-16 Summary of Laminated Glass Products

Product	Glass Type	Plastic Interlayer, in. (mm)	Nominal Thickness in.	mm	Maximum Area, in. (mm)[a]	Weight, lb./sq. ft. (kg/m²)	Principal Uses[b]
Laminated Glass	SS/SS[c]	.015–.030 (0.381–0.762)	13/64	5	48 × 80 (1219 × 2032)	2.45 (11.96)	Partitions, doors, and automotive glazing
	Lam/Lam[c]	.015–.030 (0.381–0.762)	15/64	6	48 × 100 (1219 × 2540)	2.90 (14.16)	
	DS/DS[c]	.015–.030 (0.381–0.762)	1/4	6	48 × 100 (1219 × 2540)	3.30 (16.1)	
	2 lights 1/8 in.	.015–.030 (0.381–0.762)	1/4	6	72 × 120 (1829 × 3048)	3.30 (16.1)	
	2 lights	.015–.030 (0.381–0.762)	3/8	10	72 × 120 (1829 × 3048)	4.80 (23.44)	
			1/2	12	72 × 120 (1829 × 3048)	6.35 (31)	
			5/8	16	72 × 120 (1829 × 3048)	8.00 (39.1)	
			3/4	19	72 × 120 (1829 × 3048)	9.70 (47.36)	
			7/8	22	72 × 120 (1829 × 3048)	11.40 (55.66)	
			1	25	72 × 120 (1829 × 3048)	12.95 (63.73)	
Burglar resistant	2 lights 1/8 in.	.060–.090 (1.524–2.286)	5/16	8	60 × 120 (1524 × 3048)	3.50 (17.1)	Display windows
Bullet resistant	At least 4 lights	At least 3 plies of .015 (0.381)	1 3/16	30	40 sq. ft. (3.72 m²)	15.50 (61)	Bank teller windows and payroll booths
			1 9/16	40	30 sq. ft. (2.79 m²)	20.00 (97.65)	
			1 3/4	45	25 sq. ft. (2.3 m²)	23.00 (112.3)	
			2	50	22 sq. ft. (2.04 m²)	26.20 (127.9)	
Acoustical[d] STC-36	2 lights	One or more plies of 0.45 (1.143)	9/32	7	60 × 120 (1524 × 3048)	3.40 (16.6)	Office partitions, radio and TV studios, airport glazing
STC-40			1/2	12	60 × 120 (1524 × 3048)	6.36 (31.05)	
STC-43			3/4	19	60 × 120 (1524 × 3048)	9.74 (47.55)	

[a]Maximum area varies according to producer. Larger sizes may be available from some producers.

[b]Double layer products using thicker glass and interlayers and multilayered laminated glass products are also used in other demanding installations, such as zoological animal enclosures, observation platforms (see Figs. 8.8-14 and 8.8-15), and incarceration facilities.

[c]Lam indicates 7/64 in. (2.78 mm) float glass; SS, 3/32 in. (2.38 mm) float glass; DS, 1/8-in. (3.175 mm).

[d]Refers to sound transmission class ratings.

Acoustical glass is a laminated glass with one or more plastic interlayers of 0.045-in. (1.14 mm) thickness. It is particularly effective in reducing sound transmission in the frequency ranges of speech, radio, and TV (250 to 4000 cycles per second range). It is widely used for office partitions and in radio and TV studios and similar applications.

Insulating Glass Insulating glass is a factory-produced, airtight unit, consisting of two layers (lights) of glass, with a dehydrated air space between. Insulating units are produced with two different edge types: spacer-bar edge and fused-glass edge (Fig. 8.8-17). Insulating glass is governed by ASTM Standard E 774.

Any glass type with at least one smooth surface may be used in an insulating glass unit. When heat-absorbing or reflective-coated glass is used in an insulating glass unit, and it is known that the glass will be subjected to severe thermal stresses or high wind loads, one or both lights of the unit should be heat-strengthened or fully tempered.

Insulating glass is designed primarily to protect glazed areas from excessive heat loss in winter and heat gain in summer. Use of insulating glass can substantially reduce both heating and air conditioning costs.

Insulating glass has the added advantage of allowing a much higher interior relative humidity than single glass be-

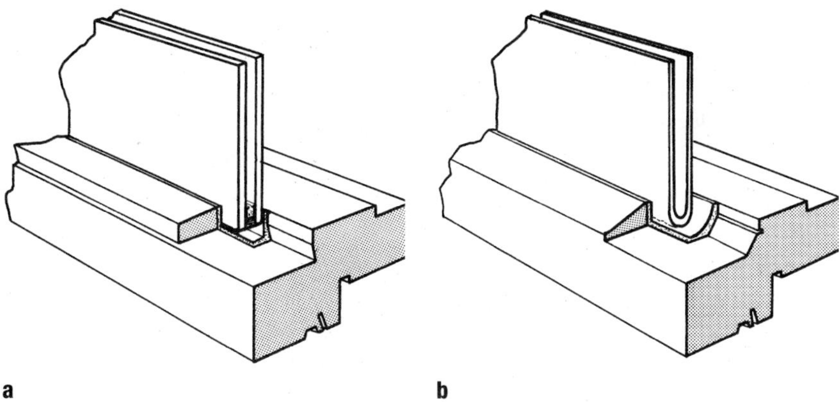

FIGURE 8.8-17 Insulating glass types: (a) spacer bar edge and (b) fused glass edge.

cause it minimizes condensation problems during cold weather (Fig. 8.8-18).

Spacer-edged insulating units consist of two lights of glass separated by a hollow metal or rubber spacer around the edges and sealed airtight with an organic sealant, such as polysulfide or butyl. The entrapped air is kept dehydrated by a desiccant (chemical dehumidifier) located in the spacer. Until recently the edge construction and glass in many units were held together by a metal channel that wrapped around the edge of the unit, encasing both panes of glass. This was designed to provide edge protection and in some cases, to hold the assembly together. This type of metal-edged insulating glass units may be found in some existing buildings, but modern insulating glass units do not have this feature.

Spacer-edged insulating units are most commonly used in commercial construction. Sealed insulating units are manufactured and stocked in a limited number of standard sizes and thicknesses. For example, units with ¼- and ½-in. (6.4 and 12.7 mm) air spaces and glass thicknesses ranging from ⅛ to ⅜ in. (3.2 to 9.5 mm) are available on that basis (Fig. 8.8-19). Special sizes can be provided by some manufacturers at a slight additional cost and longer delivery time.

Fused glass-edge insulating units consist of two lights of SS (3⁄32-in. [2.38 mm]) or DS (⅛-in. [3.2 mm]) float glass with a 3⁄16-in. (4.8 mm) air space between. The two sheets are fused together to form a glass-to-glass sealed edge. The air space in glass-edge units may be filled with dehydrated air or inert gas.

Fused glass-edge units are available in a more limited range of sizes and glass combinations than spacer-edged units (see Fig. 8.8-19). They were developed mainly to provide economical insulating glass for smaller lights in residential and commercial buildings.

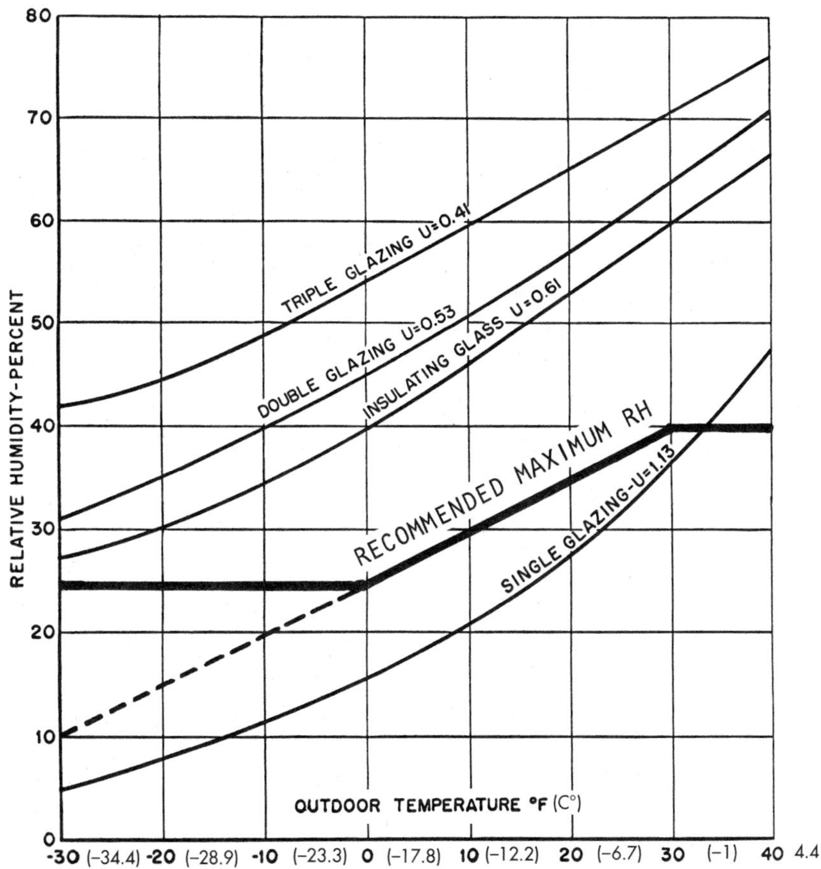

FIGURE 8.8-18 Maximum relative humidity causing condensation on insulating and other glass.

8.8.2 PLASTIC GLAZING MATERIALS

Plastic glazing materials are either polycarbonate or acrylic plastic. Both are tough, breakproof, shatterproof, and crack-resistant thermoplastics. They have a higher coefficient of expansion than glass, are lighter, and resist impacts better. They are, however, much more susceptible to scratching and are flammable, which glass is not.

Acrylics are harder and more impact resistant than polycarbonates, and they wear better. Both can be cold-formed into curves. Polycarbonates are the more heat resistant of the two. Both can be coated with an abrasion-resistant silicone-based material to increase their resistance to weather and chemicals. Polycarbonate sheets can also be coated with polymethyl methacrylate (acrylic) material to reduce the damage normally caused to it by sunlight.

Refer to Chapter 6 for additional discussion of acrylics and polycarbonates.

FIGURE 8.8-19 Summary of Insulating Glass Products

Product	Glass Type and Thickness, in.	Nominal Air Space in.	Nominal Air Space mm	Nominal Unit Thickness in.	Nominal Unit Thickness mm	Maximum Area, sq. ft. (m²)*	Weight, lb./sq. ft. (kg/m²)	Principal Uses
Spacer edge units	2 lights DS, ⅛	¼	6	½	12	22.2 (2.06)	3.27 (15.97)	Residential and commercial glazing
		½	12	¾	19			
	2 lights, 3⁄16	¼	6	⅝	16	34 (3.16)	4.90 (23.9)	
		½	12	⅞	22	41.7 (3.87)		
	2 lights, ¼	¼	6	¾	19	50–70 (4.6–6.5)	6.54 (31.9)	
		½	12	1	25			
Glass edge units	2 lights SS, 3⁄32	3⁄16	5	⅜	10	10 (0.93)	2.40 (11.72)	Smaller lights in residential and commercial glazing
	2 lights DS, ⅛	3⁄16	5	7⁄16	11	24 (2.23)	3.20 (15.6)	

*Maximum area given is for units constructed with clear glass; when tinted or reflective coated glass is used, either the maximum area must be reduced, or special fabrication is required.

8.8.3 GLAZING ACCESSORY MATERIALS

8.8.3.1 Glazing Sealants

Glazing sealants should be selected for their intended use and for their compatibility with the glazing materials and other surfaces they will contact. For example, silicones may become discolored and even fail to function as a sealant if they come in contact with neoprene gaskets or neoprene or EPDM setting blocks or spacers. A particular sealant should not be selected unless its compatibility with adjacent materials has been tested. The results of such tests are generally available from sealant manufacturers.

FACE-GLAZING COMPOUNDS

Oil-based glazing putty is still occasionally used to install small lights in wood and metal sash, but most such glazing today is done with elastic synthetic materials specifically formulated for the purpose. The materials should be suitable for painting.

CHANNEL-GLAZING COMPOUNDS

Channel-glazing compounds may be either elastomeric or elastic oil- and resin-based materials. They are used both for bedding and face glazing into wood or metal. Glazing compounds set firmly but retain a limited flexibility, generally greater than that of putty. Glazing compounds have a life expectancy of 5 to 20 years when protected by periodic painting over the exposed surface.

TAPES

Glazing tape is polybutene or polyisobutylene-butyl tape. It should be 1⁄16 in. (1.6 mm) thicker than the finished seal and should be compressed during application.

SILICONE SEALANTS

Silicone formulations are used as elastic glazing compounds, especially, but not exclusively, in butt-glazing applications. Special formulations of chemically curing, one-part silicone are used in structural glazing applications. These must be specifically formulated for this purpose.

8.8.3.2 Gaskets

Glazing gaskets should be the standard types and materials of the manufacturer of the unit being glazed. Gaskets may be vinyl, neoprene, or another elastomeric material. Lock-strip gaskets are rubber.

8.8.3.3 Miscellaneous Glazing Materials

Setting blocks and spacers should be neoprene, EPDM, silicone, or rubber-extruded blocks. These should be of the size, shape, and hardness recommended by the glass and sealant manufacturers.

Compressible filler rods should be closed-cell or waterproof-jacked rods of synthetic rubber or plastic foam.

Glazing clips should be as required by the conditions.

8.8.4 GLAZING

Glass selection depends on wind-load requirements, thermal requirements, aesthetics, and safety. The type and thickness are often established by the applicable building code. For example, some codes require that glass in entrance doors and the lower portions of sidelights must meet the requirements for "safety type" and be in compliance with ANSI Z-97.1 or another similar standard.

Watertight and airtight installation of glazing materials is the goal of glazing. Installed glazing materials must be able to withstand the effects of temperature changes, wind, and impact with no breakage of glazing material, failure of sealants or gaskets, deterioration of glazing materials, or other defects.

The installation should comply with the requirements of the appropriate standards and codes. It should also comply with the recommendations of the manufacturers of the glazing material; of the sealant, gaskets, and other glazing materials; and of the window, door, frame, entrance, storefront, or curtain wall.

The locations and conditions where glazing materials will be installed should be inspected and unsatisfactory conditions corrected before glazing is started. Surfaces to be glazed should be cleaned and removable coatings removed. Ventilating sash should be properly adjusted before they are glazed.

8.8.4.1 General Requirements for Glazing

Glazing materials vary, depending on the component being glazed. Formed door and frame units are usually glazed using stops, and channel glazing compound or tape sealant, or with gaskets. Most extruded aluminum storefront sections and doors are glazed using compressible gaskets and tape.

When windows or sliding glass doors are glazed at the job site, the manufac-

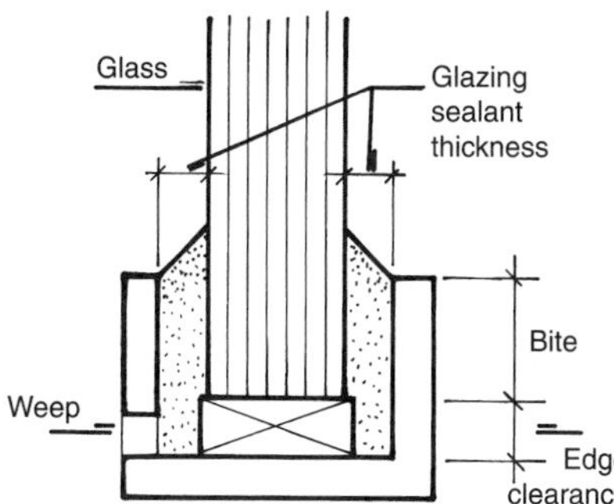

FIGURE 8.8-20 Typical glazing channel terminology. (Drawing by HLS.)

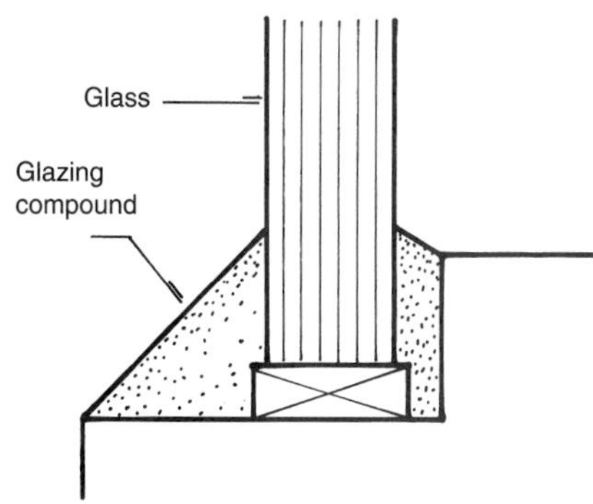

FIGURE 8.8-21 Face glazing for small lights in wood and metal. (Drawing by HLS.)

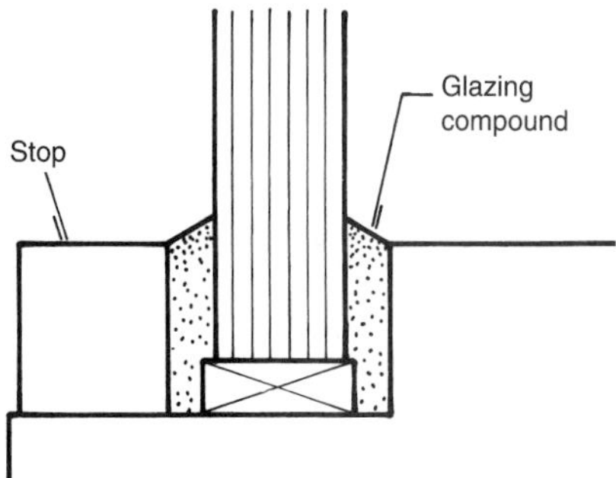

FIGURE 8.8-22 Stop glazing. (Drawing by HLS.)

turer's recommendations should be followed to ensure a satisfactory result.

Glazing includes the installation of glazing beads and gaskets that are furnished by the manufacturer of the unit being glazed. Wood and metal doors and some portions of curtain walls may be either factory preglazed or glazed in the field. Most windows are glazed in the factory. Frames, sidelights, storefronts, and most curtain walls are glazed in the field.

The design of a glazing channel (Fig. 8.8-20) must provide the necessary minimum bite and edge clearance and allow for the proper glazing sealant thickness. The person who installs glazing materials is called a *glazier*. The glazier is responsible for providing the correct glass size for each opening, but, realistically, many of today's glazing materials leave little room for adjustment. For example, heat-treated and preassembled insulating glass cannot be cut in the field. Therefore, the sizes of openings and glazing materials must be coordinated in the shop. Panes of glass and plastic should be fabricated in the shop with the proper edge clearances. Glass clearances should be in accordance with the Flat Glass Marketing Association's (FGMA) *Glazing Manual* or the requirements of the glass manufacturer.

Glazing systems must have provisions to prevent moisture from accumulating in them. This requires that they have weep holes open to the exterior (see Fig. 8.8-20).

Many glazing systems are in use today. Small openings in wood and steel are sometimes glazed by placing the glass in a bed of glazing compound (Fig. 8.8-21). Stop glazing is also used (Fig. 8.8-22). Glazing in metal is often done with stops as well. There, the exterior glazing material may be a liquid or tape sealant but is sometimes a gasket. The interior glazing material may be a tape or liquid sealant, but is more often a gasket (Fig. 8.8-23). In high-performance systems the interior and exterior systems may consist of more than one sealant. For example, there may be a tape topped off with a liquid silicone sealant, or a gasket topped off with an acrylic sealant. In lock-seam gasket glazing, there are no liquid glazing materials (Fig. 8.8-24). The glass is set into a groove on the gasket and the gasket is squeezed against the glass by insertion of the lock-strip with a special tool.

Insulating glass should be installed with the tinted glass toward the exterior, with coated glass, where used, as the outer light. Opaque glass should have its coated side toward the interior.

Insulating glass, laminated glass, and transparent mirrors should not be allowed to come in contact with chlorinated solvents or benzine-related compounds, such as toluene, nor should glazing compounds used with these glass types be thinned with these compounds. They will damage the interlayer in such glasses.

The use of setting blocks, spacers, and edge blocking should be in accordance with the FGMA's *Glazing Manual.*

Wire glass should be installed in accordance with the requirements of NFPA 80.

Two pieces of glass that abut end to end or at a corner should be sealed with beads of silicone sealant, following the manufacturer's recommendations.

8.8.4.2 Structural Silicone-Sealant Glazing

In a structural silicone-sealant glazing system, the glass is bonded to metal or glass supports with a structural-grade silicone sealant.

There are three basic types of structural silicone-sealant glazing systems: (1) those with silicone sealant supporting two sides of the glass and with the other two sides of the glass supported by a metal frame (Fig. 8.8-25); (2) those with structural silicone sealant supporting four sides of the glass; and (3) those with glass mullions called *fins* (Fig. 8.8-26).

The butt joints of glass in these systems

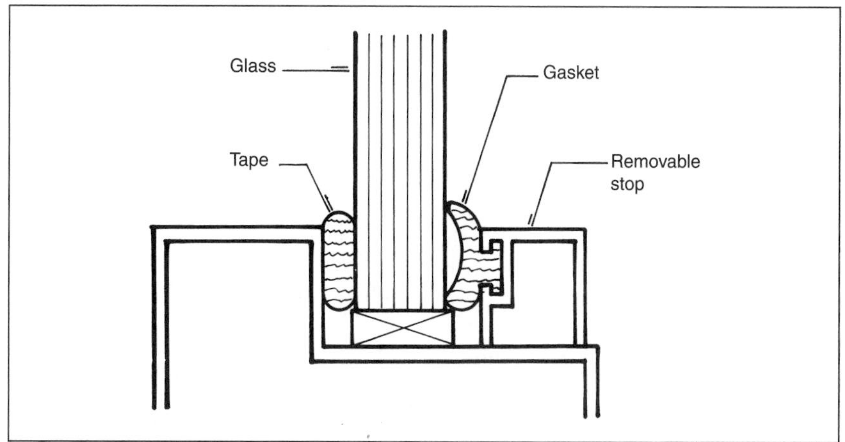

FIGURE 8.8-23 A typical glazing into metal system. Many others are also used. (Drawing by HLS.)

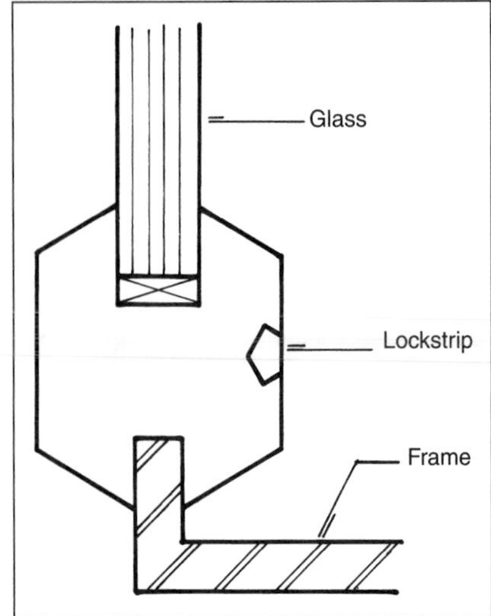

FIGURE 8.8-24 A lockstrip glazing gasket. (Drawing by HLS.)

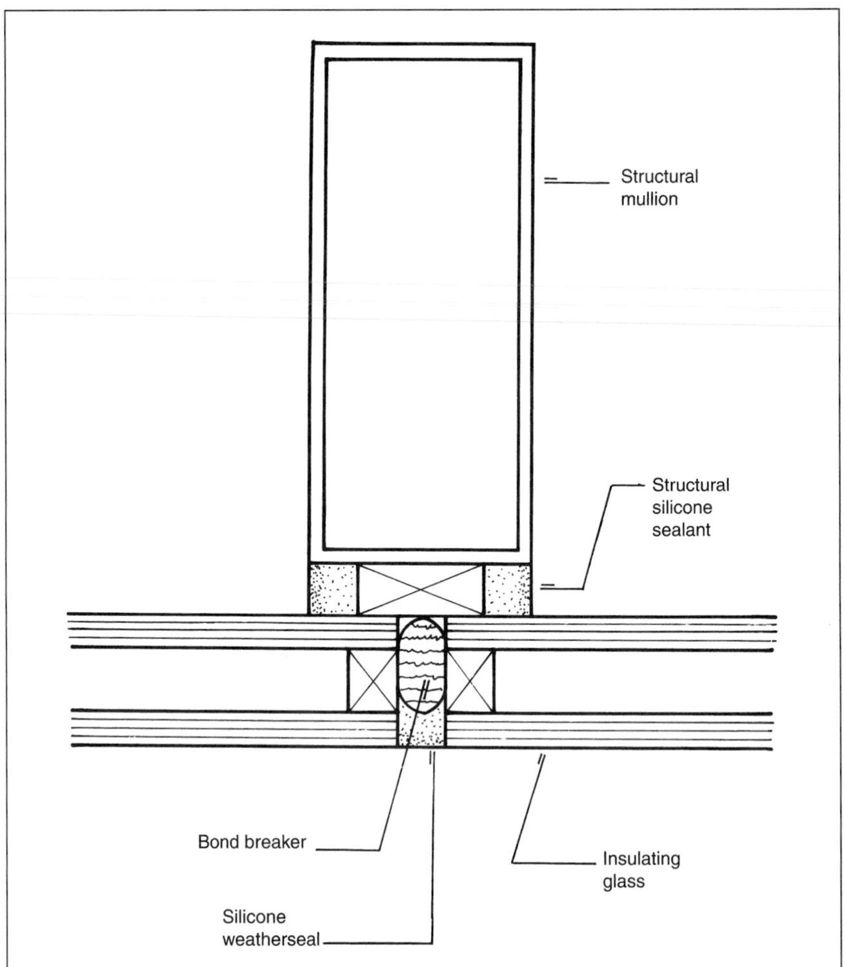

FIGURE 8.8-25 Detail of a typical structural glass. The other two sides of the glass may be similarly supported or supported by a metal frame in the conventional manner (see Figs. 8.9-9 and 8.9-10). (Drawing by HLS.)

are sealed with a silicone sealant, but it is not necessarily a structural sealant.

8.8.4.3 Multiple Glazing

Because of its thinness, a single sheet of glass is a good conductor of heat. One square foot (0.093 m^2) of ¼-in. (6.4 mm) clear glass can conduct 6 to 10 times more Btu's per hour (watts) than a square foot (0.093 m^2) of a typical frame wall. The rate of heat transfer through the glass is so rapid that added thickness is of almost no value. However, if layers of glass are separated by an air space, the path of conduction is interrupted and the rate of heat flow is reduced. Heat can then pass through the air space primarily by radiation and convection and only minimally by conduction.

Storm windows can be effective in reducing conduction and infiltration with an air separation between panes of glass of as much as 6 in. (152 mm). However, an air space is most effective in improving thermal performance when it is between 3⁄16 in. (4.8 mm) and ⅝ in. (15.9 mm) wide. Within this range, the heat flow is reduced markedly as the width is increased from 3⁄16 in. (4.8 mm) toward ⅝ in. (15.9 mm). Increasing the air space beyond ⅝ in. (15.9 mm) does not reduce heat flow to the same extent, because a wider space allows the air in it to circulate more freely and develop convection currents. These currents transport heat from the warmer glass to the colder glass. This increased convection heat loss can more than offset the slight reduction in conductive heat loss through the air. Conversely, a space of less than 3⁄16 in. (4.8 mm) is of little value because heat is readily conducted by the air across such a short distance.

8.8.4.4 Glazing with Plastic

Openings for plastic glazing should be accurately measured and the plastic sheets cut to fit, leaving the appropriate room for expansion.

The size of the sheets should be determined in accordance with the manufacturer's recommendations for edge clearances.

Plastic sheets should be set using the sealants recommended by the plastics manufacturer.

8.8.4.5 Cleaning and Protection

Labels should be removed from insulating glass to prevent thermobreakage or staining from occurring due to the labels. La-

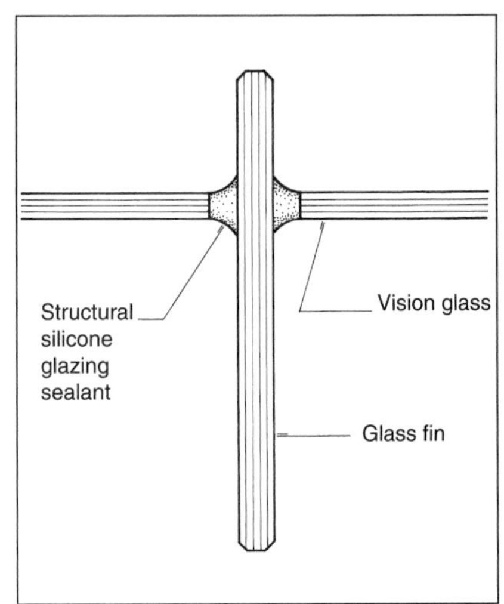

FIGURE 8.8-26 A typical glass mullion. Other types place the glass mullion behind the vision glass and attach the vision glass to the mullions with metal connectors. (Drawing by HLS.)

bels may be left on other types of glass until final cleaning unless the manufacturer of the glass recommends removal.

Glass should be protected from breakage immediately after installation by barriers and warning tapes. Warning tapes should be attached to the frames and not to the glass to prevent their damaging the glass. Glazing materials should also be protected from contaminants, such as masonry cleaning material, that would harm the glass.

Glazing materials should be thoroughly cleaned of foreign materials and polished on both sides. Cleaning procedures should be as recommended by the glazing materials' manufacturers.

8.9 Glazed Aluminum Curtain Walls

A building wall that carries no superimposed loads is called a *curtain wall.* A glazed aluminum curtain wall is an exterior curtain wall that consists essentially of an aluminum (usually extruded) framework infilled with vision panels (fixed or windows) or opaque panels such as spandrels and column covers. Curtain walls are different from *storefronts* in that they generally do not contain entrances and usually occur above the first floor (Fig. 8.9-1). Curtain wall manufacturers call their products *curtain walls* and storefront manufacturers call theirs *storefronts.* Metal doors, frames, entrances, and storefronts are discussed in Section 8.3, windows used in glazed curtain walls in Section 8.4; glazing in metal and wood curtain walls and structural glass curtain walls are covered in Section 8.8.

This section discusses only aluminum-framed glazed curtain walls, because they are the only type specifically manufactured and sold for use as glazed curtain walls. Their components are manufactured to fit together. Other metals and wood are also used in glazed curtain walls, and some glazed curtain walls are made up of unframed glass with or without metal supporting angles or clips. The principles here apply to curtain walls made from other materials, even though they may contain formed rather than extruded members.

8.9.1 SYSTEMS

Most glazed aluminum curtain walls consist of a manufactured system composed of standard extruded components with metal fill in panels where glass is not used. Some designers, however, choose to create *custom* systems. Custom systems may vary in such relatively simple ways as by substituting another material, such as stone, for the metal panels in standard systems. Some custom systems, however, have mullions or other components of configurations and sizes different from those of any standard system. It is important for a designer choosing to introduce a custom aluminum curtain wall system to recognize that such systems are inherently more costly than standard systems, that costly testing of such systems is advisable, and that regardless of the results of such tests the designer will ultimately be held accountable for the performance of such a system.

The American Architectural Manufacturers Association (AAMA) classifies glazed curtain walls by their installation methods in five systems: *stick, unit, unit-and-mullion, panel,* and *column-cover-and-spandrel.*

FIGURE 8.9-1 One Skyline Tower, Baileys Crossroads, Virginia; architect The Wiehe Partnership. (Photograph courtesy Bruce Wall Systems Corporation.)

8.9.1.1 Stick Systems

Most glazed aluminum curtain walls are stick systems. They consist of mullions, the extruded vertical elements that provide the major support for the system; vision panels that may be either fixed glass or plastic sheets or windows; and opaque panels, which are usually insulated (Fig. 8.9-2). They may also include horizontal rails, sills, fascias, copings, and trim. The components in most stick systems are stock off-the-shelf pieces used without modification. Some stick systems, however, use custom-made components. Others are a mixture of stock and custom-made pieces.

Stick systems are usually glazed in the field from the interior, using gaskets or a glazing compound. Systems with lock-strip glazing from the exterior are also available. In these, a glazing gasket is inserted into the frame and the glazing is installed into it with a special zip-in strip gasket. Special tools are needed to install this type of glazing system.

Sloped glazing is a special type of stick system in which portions are either sloped or curved (Fig. 8.9-3).

8.9.1.2 Unit Systems

Unit curtain wall systems are modular panels that are preassembled in the factory and shipped to a construction site in a single piece (Fig. 8.9-4). These panels are usually glazed at the factory and come with spandrel panels already in place. In some designs, field-installed joint covers are required.

8.9.1.3 Unit-and-Mullion Systems

Unit-and-mullion systems are a combination of stick and unit systems (Fig. 8.9-5). The framing (vertical) mullions are shipped loose and installed in the field. The infill panels are factory-made units that are inserted between the mullions in the field. In some unit-and-mullion sys-

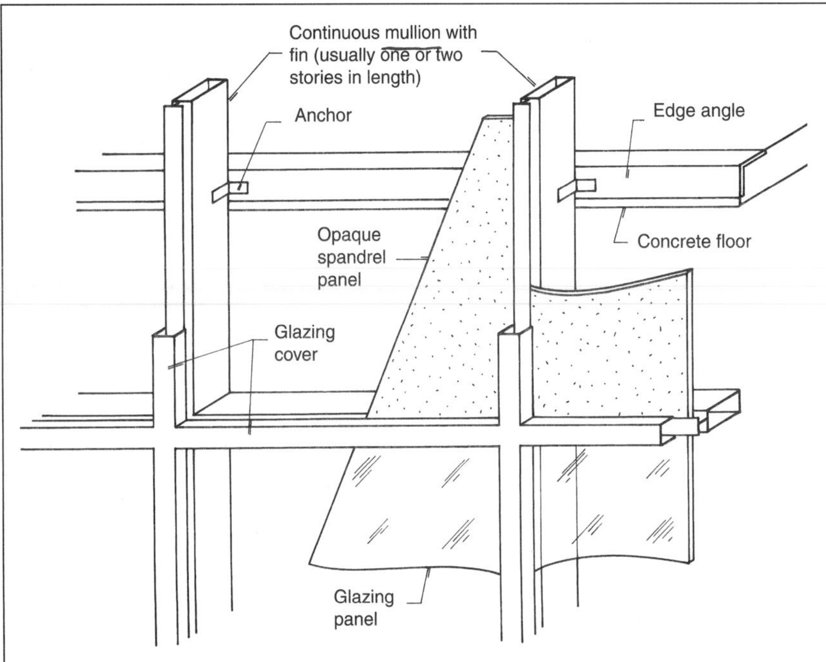

FIGURE 8.9-2 **Details of a stick-type glazed curtain wall system.** (Drawing by HLS.)

tems, the infill panels contain spandrels and vision panels in a single unit. In others, the spandrels and vision panels are separate units, shipped and installed separately in the space between the mullions.

8.9.1.4 Panel Systems

A panel curtain wall system is similar to a unit system, except that a panel system has homogeneous sheet or cast panels with few joints and may not have separate mullions. Unit systems are made up of smaller components. There are two types of panel systems: (1) architectural and (2) industrial.

Architectural panels are custom-designed systems that tend to be expensive (Fig. 8.9-6). They are used mostly on very large projects. Industrial systems are made from sheet materials and in large quantities. They are, therefore, relatively inexpensive. These are the types of panels often used on metal buildings (Fig. 8.9-7).

FIGURE 8.9-3 **A sloped glazing system.** (Courtesy Tubelite Indal Architectural Systems.)

8.9.1.5 Column-Cover-and-Spandrel Systems

Column-cover-and-spandrel systems consist of aluminum column covers and spandrels that fit between them. They are similar to a unit-and-mullion system, except that the building's structure is expressed by the column covers. The space between columns is filled with glazing panels, with or without separate mullions, and spandrel panels (Fig. 8.9-8).

8.9.2 STANDARDS

Glazed aluminum curtain walls should conform to the applicable standards of:

- Aluminum Association (AA)
- American Architectural Manufacturers Association (AAMA)
- American Society for Testing and Materials (ASTM)
- National Association of Architectural Metal Manufacturers (NAAMM)
- American National Standards Institute (ANSI)

In general, glazed aluminum curtain walls should comply with the following AAMA publications:

- 501, "Methods of Test for Exterior Walls"
- 1503, "Voluntary Test Method for Thermal Transmittance and Condensation Resistance of Windows, Doors, and Glazed Wall Sections"
- CW-10, *Curtain Wall Manual*, Volume 10: *Care and Handling of Architectural Aluminum from Shop to Site*
- CW-11, *Curtain Wall Manual*, Volume 11: *Design Windloads for Building and Boundary Layer Wind Tunnel Testing*
- CW-12, *Curtain Wall Design Manual*, Volume 12: *Structural Properties of Glass*
- CW-13, *Curtain Wall Design Manual*, Volume 13: *Structural Sealant Glazing Systems*
- CW-DG-1, *Aluminum Curtain Wall Design Guide Manual*
- TIR-A1, "Sound Control for Aluminum Curtain Walls and Windows"
- TIR-A7, "Sloped Glazing Guidelines"

Other standards are mentioned elsewhere in this text and listed in Section 8.11.

FIGURE 8.9-4 **A panel-unit glazed curtain wall system being installed.** (Courtesy Kalwall Corporation, Manchester, NH, USA.)

8.9.3 MATERIALS AND FINISHES

Refer to Section 5.4 for a general discussion about the metals and finishes addressed in this section.

8.9.3.1 Materials

A variety of aluminum alloys and tempers are used to make extruded curtain wall components. Alloys should conform with ASTM Standard B 221. A common alloy used is 6063. Common tempers are T5 and T6. Other tempers and alloys may be used, however, depending on the design and finish.

Sheet aluminum is often used for closures and trim; bent plate for structural support and trim. Sheet aluminum and bent plate components should comply with ASTM Standard B 209.

Concealed components, such as subframes, stiffeners, edge channels, reinforcements, and the like should be fabricated from the same materials as the exposed portions of the curtain wall.

Many types of opaque panels are used in glazed aluminum curtain walls. Spandrel glass is a common one. Others include metal-faced, ceramic tile–faced, stone, brick, and fiber-reinforced cementitious color-coated materials. Many stock systems come with aluminum-faced panels. Some systems will accommodate only certain types of panels. Therefore, the selection of panel types must often be made before the major framing components can be selected. For example, many stock sections will not accept stone panels. They must either be adapted or a different section must be selected.

Louvers and grills are also common. These may be a part of a system's opaque panels or be themselves infill between the mullions. They may be glazed into a system using glazing beads or gaskets or may be mounted mechanically into the frames.

8.9.3.2 Finishes

The mill finish is usually left on concealed surfaces and sometimes even on the exposed surfaces of glazed aluminum curtain wall components. However, most exposed portions are either anodized or finished with a fluoropolymer (see Section 5.4).

Exposed fasteners should be finished to match the adjacent metal.

8.9.3.3 Supports, Anchors, Fasteners, and Flashing

Supports, anchors, fasteners, window cleaners' bolts, and flashing materials should be of materials that are compatible with the material being anchored or fastened and with the substrate. Aluminum should be anchored, fastened, and flashed with aluminum, nonmagnetic stainless steel, cadmium-plated steel, or galvanized steel. Occasionally a manufacturer's design will require the use of aluminum castings as structural elements. This can be acceptable when the design takes the strength characteristics of the casting into account. However, aluminum castings are not generally used in this way, because

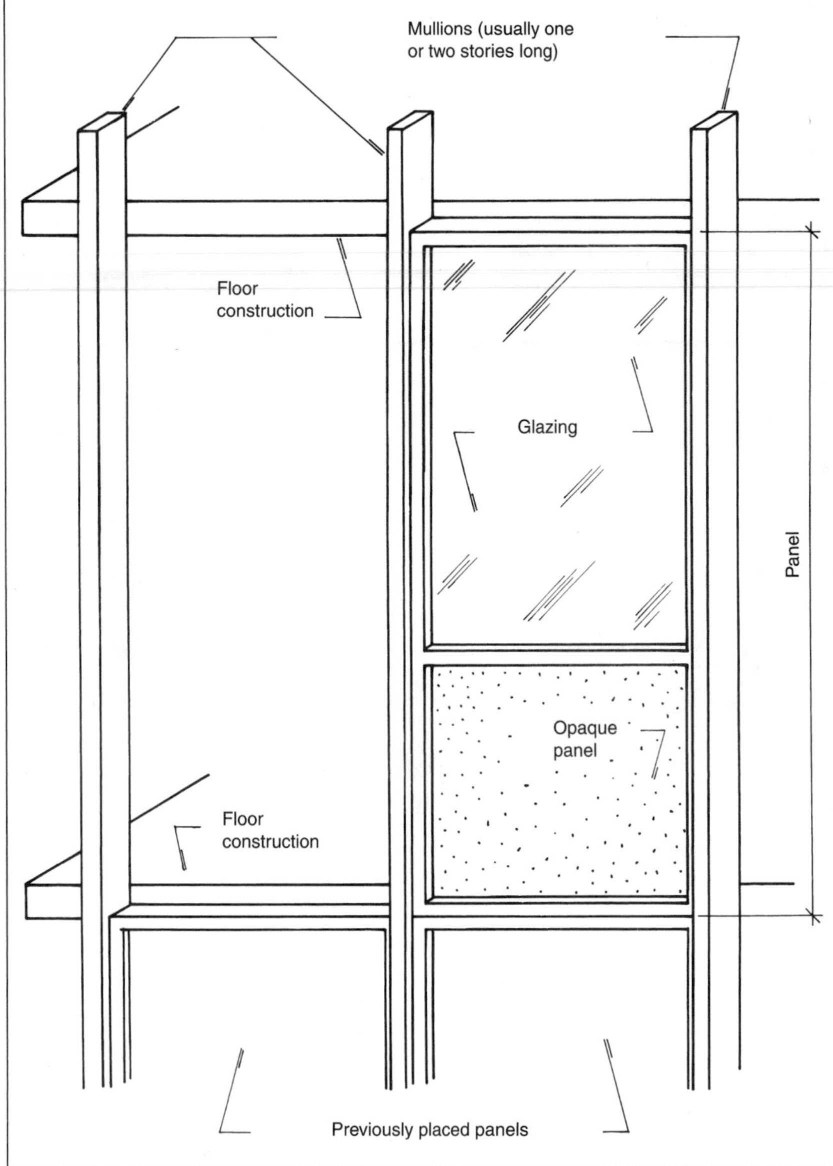

FIGURE 8.9-5 **A unit-and-mullion glazed curtain wall system.** (Drawing by HLS.)

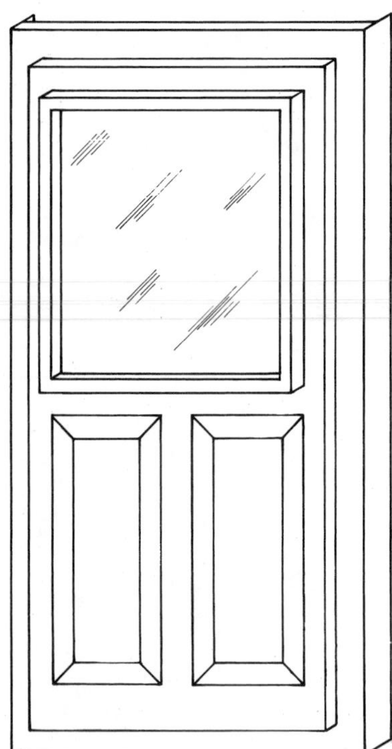

FIGURE 8.9-6 **Architectural panels may take almost any form that can be made in a metal shop, cast in concrete, or molded in plastic.** (Drawing by HLS.)

they sometimes lack sufficient shear strength.

Inserts for use in concrete may be cast iron, malleable iron, or hot-dip galvanized steel.

8.9.4 PERFORMANCE REQUIREMENTS AND TESTING

Standard glazed extruded aluminum curtain wall systems should have been tested by their manufacturer or by an independent testing agency to ascertain their resistance to deflection due to wind loads, water penetration, air infiltration, condensation resistance, thermal transmittance, and, where applicable, sound transmission. The results of these tests should be obtained from the manufacturer and taken into consideration in selecting a system for a particular project. The characteristics of the selected system should be compatible with conditions that are expected to exist at the project site, applicable code requirements, and the demands of the occupancy.

Testing the installed system is advisable for custom-designed systems and often also for standard systems with special requirements, such as those in very tall structures. Such additional testing is expensive, however, and is seldom justified for standard systems or for relatively small projects. Such additional testing should be done by an independent testing laboratory (one not affiliated with the manufacturer or the contractor). Testing may be done on a mockup of the actual system constructed in the laboratory's testing facilities before work on the curtain wall at the building begins. This mockup should be built exactly as the actual work on the building will be. As an alternative, such tests may be done on a mockup of the curtain wall constructed at the site. This mockup is sometimes separate from the building and sometimes a portion of the actual curtain wall. When the designer approves, a satisfactory field-constructed mockup is sometimes incorporated into the building.

The performance characteristics of a system are determined by the effect on the system when tested at specific levels of wind pressure. The test pressure to use in these tests is determined using the fastest wind speed likely to be encountered at a building's location. Data in ANSI/ASCE "Minimum Design Loads for Buildings and Other Structures" can be used to estimate this maximum wind speed. After the

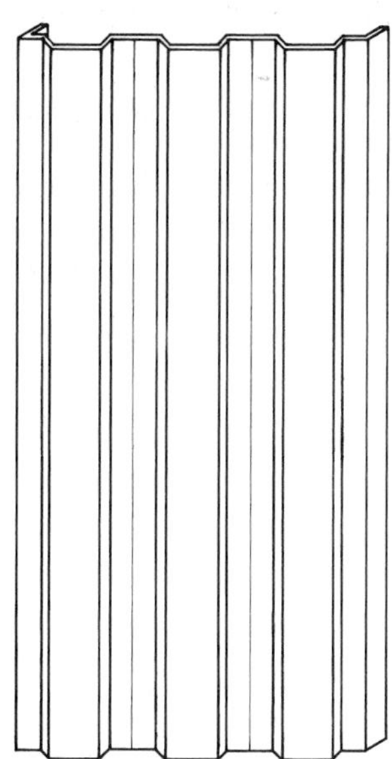

FIGURE 8.9-7 Industrial panels take many shapes, including corrugated ones. (Drawing by HLS.)

expected wind velocity has been determined, the resulting wind loads to be used are determined from load tables. Such load tables are based on wind velocities from 70 mph (112.65 km/h) to 110 mph (177 km/h) and the height of the curtain wall above ground from zero (at grade) to 500 ft. (152.4 m).

The test pressures given in this section are those recommended by the listed standards. Higher pressures may be required for a given project.

Sometimes it is necessary to reinforce aluminum mullions and other components to make them comply with performance requirements.

8.9.4.1 Structural Performance

A glazed aluminum curtain wall system should be tested using the expected wind loading and the procedures listed in ASTM Standard E 330. The test method in ASTM Standard E 1233 may also be used.

Vertical systems should be tested both perpendicular and parallel to the plane of the wall and in both the inward-acting and outward-acting directions. Sloped glazing should also be tested for upward- and downward-acting loads, and in addition should be tested for a uniform load, usually of 200 lb. (90.72 kg) acting on the system at any point. This test is to ensure that a person walking on the system will not damage it. This load may be higher if a larger concentrated load will be superimposed on the system.

The test pressures for perpendicular-to-the-wall tests should be 150% of the wind pressures to be encountered. Pressures may vary with the system's elevation above ground and proximity (within 15 ft. [4572 mm] or so) to corners, where pressures are higher. The actual loads to be used should be established by a registered structural engineer. Some building codes dictate the minimum pressures to be used.

The maximum allowable deflection under these loads is usually 1⁄175 of a span, but project conditions and the system manufacturer's requirements may require limits of as much as 1⁄1000 of a span. In any case, the maximum deflection should not exceed ¾ in. (19.1 mm) over the total length of any member. In addition, under the selected loads:

- No panels should be dislodged.
- No permanent deflection should occur.
- Sealant joints should not be reduced to less than half of the designed width.

The allowable movement parallel to the wall is dictated by panel edge conditions. The bite on glazing and opaque panels should not be reduced to less than 75% of the design value. In addition, the edge clearance between panels and other elements should not be reduced to less than ⅛ in. (3.2 mm). Greater restrictions may be imposed by the method used in assembling the system.

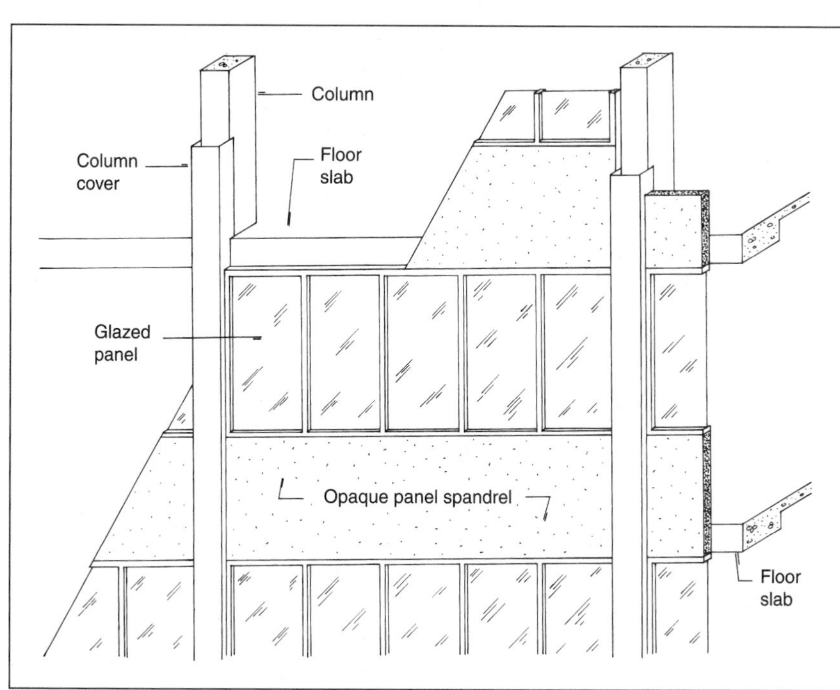

FIGURE 8.9-8 A column-cover-and-spandrel system. The glazed panels may contain windows or may be made up of strip windows. (Drawing by HLS.)

8.9.4.2 Water Penetration

The water penetration resistance of a glazed aluminum curtain wall system is determined by testing it in accordance with ASTM Standard E 331. The test pressure suggested by AAMA 501 is 20% of the inward design wind load pressure but not less than 6.24 psf (30.47 kg/m^2) or more than 12 psf (58.59 kg/m^2). For a system to pass this test, no water may pass through it during the test.

8.9.4.3 Air Infiltration

Air infiltration is tested in accordance with ASTM E 283. The usual pressure is 1.57 psf (7.67 kg/m^2); the permitted infiltration is 0.06 cfm (0.002 m^3/min) per sq. ft. (0.09 m^2) of fixed area.

8.9.4.4 Condensation Resistance

A metal frame that holds insulating glass should have a *thermal break* (thermal barrier) to prevent a thermal "short circuit," that is, heat flow being resisted by the insulating glass, but flowing freely through the aluminum frame. This is done by separating the aluminum parts on the warm inside portion of a frame from those on the

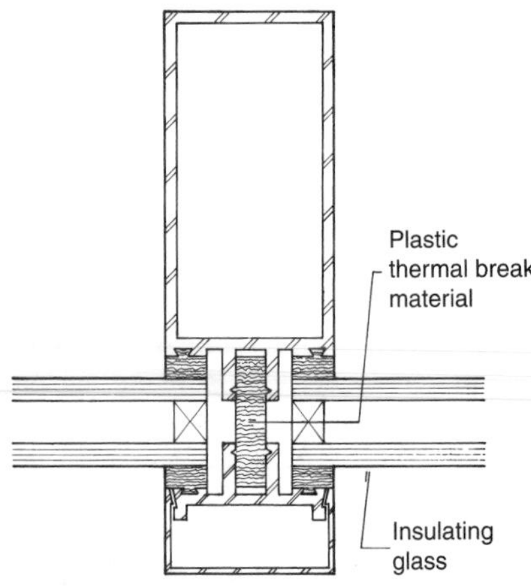

FIGURE 8.9-9 A typical glazed curtain wall mullion containing a thermal break feature. (Drawing by HLS.)

cold outside portion through the use of an insulating material such as polyurethane, vinyl, or another material low in conductivity (Fig. 8.9-9).

An aluminum frame with a good thermal break has an R-value of 1.66, which is approximately equal to that of double-pane insulating glass. A comparable frame with no thermal break would have an R-value of 1.18.

Condensation resistance is tested for frames having a thermal break feature. A system's condensation resistance is an indication of that system's ability to resist condensation. It is determined by procedures outlined in AAMA 1503. The higher the value, the less likely it is that condensation will occur.

8.9.4.5 Thermal Transmittance

As discussed in Section 15.2, heat will flow quite readily through aluminum curtain walls unless some means are taken to reduce this flow. The methods for determining conductive heat transmission through these assemblies are included in AAMA 1503.

In addition, it is usual to require that an glazed aluminum curtain wall be able to withstand the highest ambient temperature differential that it will encounter without damage of any kind. In most parts of the country the ambient temperature differential is about 120°F (48.89°C), can result in a metal surface temperature differential of 180°F (82.2°C). Some extreme southern and northern areas may experience even larger ambient temperature differentials.

8.9.4.6 Sound Transmission

When a specific sound transmission level is required, a glazed aluminum curtain wall should be tested in accordance with ASTM Standard E 90 to establish its sound transmission class (STC). The required STC depends on the project and its location. Buildings adjacent to high noise levels, such as one adjacent to the flight pattern of an airport or near a major highway will require much higher STCs than buildings in other settings. A manufacturing facility will not require as high an STC as a school, office building, or apartment unit.

8.9.5 DESIGN AND FABRICATION

Glazed aluminum curtain wall systems designed to exclude water simply by having tight or sealant filled joints are not always successful, because the pressure within the system is often lower than the pressure on its exterior. This causes water to be sucked into every fault faster than it can run out through weep holes. Water can even be sucked in through weep holes unless they are placed where water cannot reach them.

The answer to this dilemma is to design the system so that minor amounts of water are permitted to infiltrate the system, but are then conducted back to the exterior through a series of water channels and weep holes without entering the building. All sealant-filled joints are located behind the water channels where water cannot reach them. This mechanism works on what is called the *rain screen principle*. A portion of the frame screens the sealants so that water does not come in contact with them. Sealants in curtain walls are called *differential seals*, meaning that the difference in pressure on the two sides of the sealant is almost always more than 1 psf (4.88 kg/m^2). Water will penetrate almost all sealed joints under these circumstances. However, the water channels are open to the air so that the pressure within them is the same as that outside the screen; thus there will be no pressure differential to suck water into the channels. Water that does enter the channels will be quickly led back to the exterior without reaching the sealant joints. Therefore, systems with this type of design are less likely to leak. The concealed sealants should be installed in the shop when possible.

In sloped glazing systems, it is also important to prevent water that runs across the system from building up against horizontal members, where it can cause leaks. This is usually done by a series of gutters and downspouts designed into the system to collect water and lead it quickly to the ground.

Even with these safeguards, glazed curtain walls will sometimes leak through imperfections in glazing seals, incompletely sealed joints, or other inadvertent openings. In addition, even carefully designed systems will sometimes collect condensation. Therefore a water-collection and removal system is advisable on the interior of such systems. This is particularly important on sloped glazing systems, where interior gutters are often advisable.

Glazed aluminum curtain wall systems come complete with accessories such as glazing gaskets, glazing stops, clips, trim, fins, flashings, anchors, and fasteners. Normal glazing gaskets are made from one of several common materials, including, but not limited to, extruded polyvinyl chloride, extruded or molded neoprene, and extruded or molded EPDM. Lock-strip-type gaskets are usually structural rubber. Incidental components, such as closers, channels, flats, trim, sleeves, and the like are also usually delivered with the systems.

Components should be cut and fitted to make hairline joints, except where wider joints are needed to accept sealants or to permit thermal and other movement in the system. Joints in exterior frames should be designed and erected to be watertight. Joints between two pieces of metal should be filled with a nonhardening sealant. Welds should be made on the inside so they will not be visible in the finished installation.

Mullions are usually either tubes or structural shapes, such as T or I shapes. Structural-shape mullions are often, but not always, encased in a rectangular cover. Mullions vary considerably in size and shape. They may be either standard (Fig. 8.9-10) or thermal break (see Fig. 8.9-9). Most systems are flush glazed (see Fig. 8.9-9), but face- (butt-) glazed systems are also used (Fig. 8.9-11).

A system and its anchorages should be designed to permit both horizontal and vertical movement due to thermal changes in the system and building and deflection in the structure.

8.9.6 INSTALLATION

The finest glazed curtain wall design and the highest quality of manufacture will not

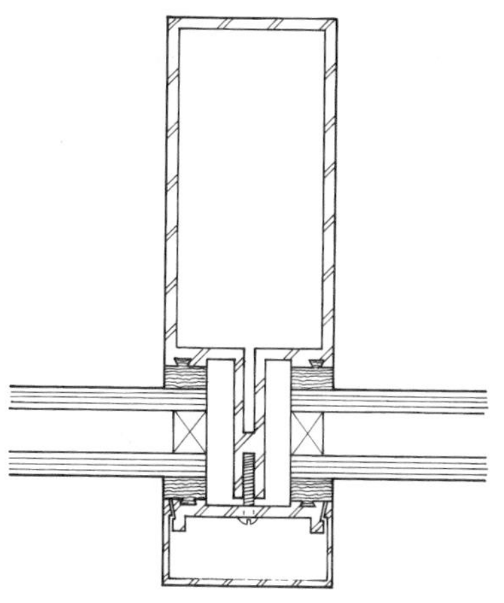

FIGURE 8.9-10 A typical standard glazed curtain wall mullion. (Drawing by HLS.)

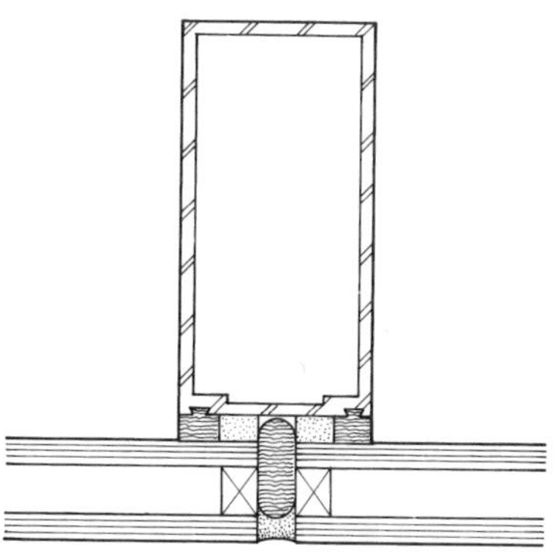

FIGURE 8.9-11 A typical glazed curtain mullion with butt glazing. (Drawing by HLS.)

compensate for poor installation. Proper installation can be achieved only if the work is done by experienced workers following the guidelines and instructions supplied by the manufacturer and the appropriate industry standards. Glazed aluminum curtain walls should be installed in accordance with AAMA's *Aluminum Curtain Wall Design Guide Manual* and the manufacturer's recommendations.

Glazed curtain wall installations include many components and can become very complicated. For example, in addition to the components necessary for a basic system, curtain walls include other elements that would ordinarily be assigned to different trades. These include such items as louvers; copings; fascias; flashings; parapet caps; operable windows; venetian blind tracks; drapery rods; insulation; firestopping; heating, ventilating, and air conditioning unit enclosures; solar shades; masonry infill; and joint sealing. It is, therefore, better that all work directly associated with glazed aluminum curtain walls be done by a single subcontractor to ensure proper coordination of the different elements. This often means that the curtain wall manufacturer is also made responsible for the design and installation of the curtain wall system and all related elements. Unfortunately, not all manufacturers are qualified to offer such a service and union jurisdictions sometimes prevent it as well. Therefore, single subcontractor responsibility is often not possible, except with stock stick systems with few of the additional items mentioned.

Curtain wall components should be protected during shipment and handling and should be installed as soon as possible after they reach the project site to prevent them from being damaged. The substrates should be installed and prepared to receive curtain wall components so they can be properly installed and will finish out square, plumb, level, straight, and true. The design of most systems provides for minor adjustments at the job site, but no system will function properly if its framing members are twisted or misaligned during installation.

A properly designed, manufactured, and installed glazed aluminum curtain wall system should be weather resistant as a unit. Other things must be done, however, to finish with a completely weathertight installation. For example, joints between system components and adjacent materials must be closed with a sealant, trim members, or both to prevent the passage of air, dust, and water around the system.

Glazed aluminum curtain walls should be supported around the entire perimeter and anchored securely to the supporting construction. They should also be flashed where appropriate to provide a path back to the exterior for water that finds its way into the wall.

The installed system should be plumb, level, and in line within specified tolerances. Permissible tolerances vary from project to project and from system to system. However, systems are usually required to be plumb within ⅛ in. (3.2 mm) in any 10-ft. (3048 mm)-long segment, with a maximum of ¼ in. (6.4 mm) in 40 feet; level to within ⅛ in. (3.2 mm) in any 20-ft. (6096 mm)-long segment, with a maximum of ¼ in. (6.4 mm) in 40 feet; and in line within ⅛ in. (3.2 mm), except that the limit should be 1/16 in. (1.6 mm) when the elements are supposed to be flush with each other. Systems should also be within ⅜ in. (9.5 mm) of their planned location.

Curtain wall installation should allow for changes due to thermal movement in the system and in the building, and for building structure deflection.

Metals with dissimilar electrolytic or chemical characteristics should be insulated from direct contact with each other by waterproof, nonconductive materials such as neoprene, waxed paper, or coated felt. Such dissimilar metals located where water passing over them may contact another surface should be painted to prevent staining. Refer to Section 5.4 for a discussion of the effect of dissimilar metal materials.

Concealed aluminum in contact with concrete, masonry, or an absorbent material, such as wood, paper, or insulation, should be permanently protected by coating either the aluminum or the adjacent material with a bituminous or aluminum paint, or by coating the aluminum with a zinc chromate primer, to minimize the chance of chemical corrosion of the aluminum by acids, alkalies, and salts leached out of the adjacent materials. Creosote and tar coatings may damage the aluminum and should not be used.

Sealants should be used within the system, in concealed locations where needed, to make the system airtight and watertight.

Firestopping should be installed in glazed aluminum curtain wall systems where required by code or other governing regulation. It should always be installed at the edges of floor construction where a glazed curtain wall passes. Firestopping in such locations is usually accomplished by filling the space between the floor construction and the curtain wall with a glass fiber or mineral wool insulation specifically made for this purpose.

8.9.7 GLAZING

Glazing material selection depends on wind load requirements, thermal requirements, aesthetics, and safety. The type and thickness are often established by the applicable building code. For example, some codes require that glazing materials close to floors must meet the requirements for safety type and be in compliance with ANSI Z-97.1 or a similar standard.

Aluminum curtain wall systems are usually glazed with ¼-in. (6.4 mm) glass, except where insulating glass is desired. Glazing stops and gaskets should be of the correct size to receive the glass to be used.

Where design calculations indicate the need to reduce heat transmittance, significant reduction can be accomplished by incorporating insulating glass. Thermal breaks in the frames will also help. At any rate, curtain walls located where the winter design temperature is 10°F (–12.2°C) or less should incorporate such devices regardless of calculations.

Some curtain wall components can be glazed at the factory, which is the most satisfactory method. Stick systems, of course, must be glazed in the field. When glazing is done at the job site, the curtain wall manufacturer's recommendations should be followed to ensure a satisfactory result.

8.9.8 ADJUSTMENT, CLEANING, AND PROTECTION

Immediately after glazed aluminum curtain walls have been installed, they should be cleaned. Thereafter, curtain walls should be protected to prevent damage to their finish, appearance, and ability to function properly. Units should be protected from splashing of plaster, mortar, paint, cleaning solutions, and other foreign substances.

8.10 Additional Reading

More information about the subjects discussed in this chapter can be found in the references listed in Section 8.11 and in the following publications.

Agnelli, Philip J. 1993. "Wood Doors Today." *The Construction Specifier* (August):25.

Anderson, Donald L., Maryrose T. McGowan, and James A. Hunt. 1987. "Preventing Unauthorized Access: Doors, Hardware, Windows." *Progressive Architecture* (March):144, 146.

Arasteh, Dariush. 1994. "Energy-Efficient Glazing." *The Construction Specifier* (March):76.

Architectural Record. 1988. "Curtain-Wall Windows in High Rise Office Buildings." *Architectural Record.* (February):104–105.

Beall, Christine. 1990. "Architectural Glass Primer." *The Construction Specifier* (April):48.

Beers, Paul E. 1997. "Security Glazing." *The Construction Specifier.* (October):66.

Berg, James A. 1993. "Recommendations for Architectural-Grade Door Specifications." *The Construction Specifier* (March):135.

Braun, Vince C. 1994. "Steel, Hollow Metal, and Composite Fire Doors." *The Construction Specifier* (April):98.

Bullard, Robert. 1990. "Using Glass in Fire Doors." *The Construction Specifier* (June):63.

Custom Builder. 1988. "Wood Doors: A Review." *Custom Builder* (April):7–10.

Danford, G. Scott, and Edward Steinfeld. 1994. "Automated Doors: Toward Universal Design." *The Construction Specifier* (August):90.

Ellis, William D. 1994. "Hollow Steel Doors." *The Construction Specifier* (April):101.

Fisette, Paul. 1988. "Patio Doors." *Custom Builder* (January):31–34.

Freeman, Craig. 1990. "Revolving Door Options." *The Construction Specifier* (June):70.

George, Louis F. 1990. "Gasketing Doors for Noise Control." *The Construction Specifier* (April):82.

Gordon, Douglas E., and M. Stephanie Stubbs. 1988. "Window Selection Guide." *Architecture* (July):107–109.

Gottwalt, Timothy J. 1994. "Hardware Specification Writing Tips." *The Construction Specifier* (June):89.

Haerer, Donald J. 1990. "Automated Revolving Doors for Security and Barrier-Free Access." *The Construction Specifier* (June):76.

Harpole, Tom. 1995. "Is Tempered Glass Safe?" *The Construction Specifier* (December):36.

———. 1996. "Protective Glazing: Making Credible Choices." *The Construction Specifier* (April):60.

———. 1997. "Security Glazing Options." *The Construction Specifier* (July):52.

Louis, Michael J. 1994. "Drafty Windows." *The Construction Specifier* (November):58.

Makepeace, Chris B., and John O'Connor. 1996. "The Glass Roof: Sloped Glazing Need Not Leak." *The Construction Specifier* (November):50.

Mange, Gary E., and Jose C. Ortega. 1994. "Acoustical Fenestration." *The Construction Specifier* (February):62.

Mayer, Robert Kipp. 1990. "Opening the Door for ANSI Standards." *The Construction Specifier* (June):80.

———. 1991. "Lock Functions: A Digest." *The Construction Specifier* (September):134.

Meneely, Dallas A. 1993. "Of Winds and Windows." *The Construction Specifier* (October):33.

Myers, Donald L. 1996. "Security Locking Systems." *The Construction Specifier* (April):68.

———. 1996. "Security Standards." *The Construction Specifier* (April):70.

Nash, Weldon W., Jr. 1995. "A Systematic Approach to Specifying Door Hardware." *The Construction Specifier* (November):53.

Nashed, Fred. 1996. *Time-Saver Details for Exterior Wall Design*. New York: McGraw-Hill.

Nicastro, David H. 1994. "Buildings That Lose Their Marbles." *The Construction Specifier* (August):160.

———. 1995. "Tempered Glass." *The Construction Specifier* (July):96.

———. 1996. "Glazing System Leaks." *The Construction Specifier* (January):80.

———. 1996. "Falling Snap Covers." *The Construction Specifier* (July):96.

———. 1996. "Structural Sealant Glazing." *The Construction Specifier* (August):112.

———. 1996. "Zipper Gaskets." *The Construction Specifier* (September):96.

———. 1996. "Sloped Glazing Slips." *The Construction Specifier* (December):88.

Olmstead, Patrick R. 1995. "Fire Exit Devices: Companions to the Fire Door." *The Construction Specifier* (October):60.

Raeber, John A. 1989. "Specifying Windows to Code." *The Construction Specifier* (April):21.

———. 1991. "Hardware Finishes: All That Glitters . . ." *The Construction Specifier* (September):31.

———. 1992. "Accessible Door Hardware." *The Construction Specifier* (March):72.

———. 1993. "Exterior Hollow Metal Doors." *The Construction Specifier* (March):39.

———. 1993. "Acoustical Glazing Systems." *The Construction Specifier* (October):27.

———. 1994. "Automatic Pedestrian Doors." *The Construction Specifier* (February):15.

———. 1994. "Exterior Wood Doors." *The Construction Specifier* (June):23.

Rolfe, Ashley R. 1995. "Panic and Fire-Rated Exit Devices." *The Construction Specifier* (August):70.

Rose, Helen. 1997. "Choosing Gasketing Products." *The Construction Specifier* (April):73.

Ross, William J. 1990. "Selecting Fire Door Assemblies." *The Construction Specifier* (June):58.

Schechter, John. 1996. "A Specification Guideline." *The Construction Specifier* (March):70.

———. 1996. "Stainless Steel Doors." *The Construction Specifier* (March): 66.

Spargo, Robert C. 1990. "Planning for Centuries of Use." *The Construction Specifier* (July):88.

Spencer, Greg. 1993. "School Windows." *The Construction Specifier* (October): 33.

Tuluca, Adrian. 1996. *Energy Efficient Design and Construction for Commercial Buildings*. New York: McGraw-Hill.

Voreis, Richard D. 1992. "Architectural Aluminum Windows." *The Construction Specifier* (March):126.

Wilson, Alex. 1996. "What's New in Windows?" *The Construction Specifier* (October):32.

Winandy, Joseph, and John O'Neil. 1996. "Doors, Frames, and Hardware." *The Construction Specifier* (October):54.

Wolfe, Sheldon. 1992. "Window Pains." *The Construction Specifier* (July):106.

Zarghamee, Mehdi S., and Rasko P. Ojdrovic. 1996. "Seismic Design for Curtain Walls." *The Construction Specifier* (September):74.

8.11 Acknowledgments and References

ACKNOWLEDGMENTS

We gratefully acknowledge the assistance of the following organizations and individuals in preparing this chapter. We are also indebted to them for permission to use their illustrations when requested and for the use of their publications as references.

The Adams Rite Manufacturing Company
Aluminum Association (AA)
Amarlite Architectural Products
American Architectural Manufacturers Association (AAMA)
American Iron and Steel Institute
American National Standards Institute (ANSI)
American Society for Testing and Materials (ASTM)
Architectural Woodwork Institute
ASG Industries, Inc.
Boon Edam, Inc.
Bruce Engineering Company
Builders Hardware Manufacturers Association (BHMA)
Cardinal IG
Construction Specifications Institute
Consumer Product Safety Commission
Copper Development Association (CDA)
Door and Hardware Institute (DHI)
Falcon Lock
Glass Tempering Association
Hollow Metal Manufacturers Association (HMMA), Division of National Association of Architectural Metal Manufacturers (NAAMM)
Insulating Glass Certification Council
Kalwall Corporation
Laminated Glass Corporation
Laminated Safety Glass Association
Levelor Lorentzen, Inc.
National Association of Architectural Metal Manufacturers (NAAMM)
National Association of Home Builders (NAHB) Research Foundation
National Fire Protection Association (NFPA)
National Institute of Building Sciences (NIBS)
National Institute of Standards and Technology (NIST)
National Wood Window and Door Association (NWWDA)
Pease Company
Pella Corporation
Screen Manufacturers Association
Steel Door Institute (SDI)
Tubelite Division of Indal Inc.
U.S. Department of Housing and Urban Development (HUD)
Window and Door Manufacturers Association (WDMA)
Woodwork Institute of California

REFERENCES

We would also like to thank the authors and publishers of the publications in the following list for their contribution to our research for this chapter.

Abate, Kenneth. 1985. "Specifying Foam Core Architectural Metal Panel Systems." *The Construction Specifier* 38(9)(September):40–48.

AIA Professional Services Division. MASTERSPEC®. Basic Sections:

08110, "Steel Doors and Frames," 11/94 ed. Washington, DC: AIA.

08114, "Custom Steel Doors and Frames," 11/94 ed. Washington, DC: AIA.

08163, "Aluminum-Framed Sliding Glass Doors," 8/97 ed. Washington, DC: AIA.

08211, "Flush Wood Doors," 11/96 ed. Washington, DC: AIA.

08212, "Stile and Rail Wood Doors," 11/96 ed. Washington, DC: AIA.

08312, "Wood-Framed Sliding Glass Doors," 8/97 ed. Washington, DC: AIA.

08314, "Sliding Metal Fire Doors," 2/95 ed. Washington, DC: AIA.

08351, "Folding Doors," 8/93 ed. Washington, DC: AIA.

08361, "Sectional Overhead Doors," 8/96 ed. Washington, DC: AIA.

08410, "Aluminum Entrances and Storefronts," 8/95 ed. Washington, DC: AIA.

08450, "All-glass Entrances," 8/95 ed. Washington, DC: AIA.

08460, "Automatic Entrance Doors," 8/97 ed. Washington, DC: AIA.

08461, "Sliding Automatic Entrance Doors," 8/97 ed. Washington, DC: AIA.
08462, "Swinging Automatic Entrance Doors," 8/97 ed. Washington, DC: AIA.
08470, "Revolving Entrance Doors," 2/91 ed. Washington, DC: AIA.
08510, "Steel Windows," 11/90 ed. Washington, DC: AIA.
08520, "Aluminum Windows," 2/95 ed. Washington, DC: AIA.
08550, "Wood Windows," 11/96 ed. Washington, DC: AIA.
08710, "Door Hardware," 5/91 ed. Washington, DC: AIA.
08800, "Glazing," 8/97 ed. Washington, DC: AIA.
08920, "Glazed Aluminum Curtain Walls," 5/95 ed. Washington, DC: AIA.

American Architectural Manufacturers Association (AAMA):

AAMA 101/1.S.2-97, "Voluntary Specifications for Aluminum, Poly (Vinyl Chloride) (PVC) and Wood Windows and Sliding Glass Doors." Des Plaines, IL: AAMA.
AAMA 501-94, "Methods of Test for Exterior Walls." Des Plaines, IL: AAMA.
AAMA 604.2-1977, "Voluntary Specifications for Residential Color Anodic Finishes." Des Plaines, IL: AAMA.
AAMA 606.1-1976, "Voluntary Guide Specifications and Inspection Methods for Integral Color Anodic Finishes for Architectural Aluminum." Des Plaines, IL: AAMA.
AAMA 607.1-1977, "Voluntary Guide Specifications and Inspection Methods for Clear Anodic Finishes for Architectural Aluminum." Des Plaines, IL: AAMA.
AAMA 608.1-1977, "Voluntary Guide Specifications and Inspection Methods for Electrolytically Deposited Color Anodic Finishes for Architectural Aluminum." Des Plaines, IL: AAMA.
AAMA 609-93, "Voluntary Guide Specifications and Cleaning and Maintenance of Architectural Anodized Aluminum." Des Plaines, IL: AAMA.
AAMA 610.1-1979, "Voluntary Guide Specifications for Cleaning and Maintenance of Painted Aluminum Extrusions and Curtain Wall Panels." Des Plaines, IL: AAMA.
AAMA 701-92.702-92, "Combined Voluntary Specifications for Pile Weather Strip and Replaceable Fenestration Weatherseals." Des Plaines, IL: AAMA.
AAMA 800-92, "Voluntary Specifications and Test Methods for Sealants." Des Plaines, IL: AAMA.
AAMA 850-91, *Fenestration Sealants Guide Manual*. Des Plaines, IL: AAMA.
AAMA 901-96, "Voluntary Specifications for Rotary Operators in Window Applications." Des Plaines, IL: AAMA.
AAMA 902-98, "Voluntary Specifications for Sash Balances." Des Plaines, IL: AAMA.
AAMA 904-96, "Voluntary Specifications for Friction Hinges in Window Applications." Des Plaines, IL: AAMA.
AAMA 906-96, "Voluntary Specifications for Sliding Glass Door Roller Assemblies." Des Plaines, IL: AAMA.
AAMA 1002.10-93, "Voluntary Specifications for Aluminum Insulating Storm Products for Windows and Sliding Glass Doors." Des Plaines, IL: AAMA.
AAMA 1102.7-89, "Voluntary Guide Specifications for Aluminum Storm Doors." Des Plaines, IL: AAMA.
AAMA 1302-76, "Voluntary Specifications for Forced-Entry Resistant Aluminum Prime Windows." Des Plaines, IL: AAMA.
AAMA 1303.5-1976, "Voluntary Specifications for Forced-Entry Resistant Aluminum Sliding Glass Doors." Des Plaines, IL: AAMA.
AAMA 1402-86, "Standard Specifications for Aluminum Siding, Soffits & Fascia." Des Plaines, IL: AAMA.
AAMA 1503-98, "Voluntary Test Method for Thermal Transmittance and Condensation Resistance of Windows, Doors and Glazed Wall Sections." Des Plaines, IL: AAMA.
AAMA 1504-97, "Voluntary Standard for Thermal Transmittance and Condensation Resistance of Windows, Doors and Glazed Wall Sections." Des Plaines, IL: AAMA.
AAMA 2603-98, "Voluntary Specification, Performance Requirements and Test Procedures for Pigmented Organic Coatings on Aluminum Extrusions and Panels—Series: Components, Coatings and Finishes." Des Plaines, IL: AAMA.
AAMA 2604-98, "Voluntary Specifications for Performance Requirements and Test Reports for High Performance Organic Coatings on Architectural Extrusions and Panels." Des Plaines, IL: AAMA.
AAMA CW-DG-1, *Aluminum Curtain Wall Design Guide Manual*. Des Plaines, IL: AAMA.
AAMA CWG-1-89, "Installation of Aluminum Curtain Walls." Des Plaines, IL: AAMA.
AAMA CW-10, *Curtain Wall Manual*, Volume 10: *Care and Handling of Architectural Aluminum from Shop to Site*. Des Plaines, IL: AAMA.
AAMA CW-11, *Curtain Wall Manual*, Volume 11: *Design Windloads for Buildings and Boundary Layer Wind Tunnel Testing*. Des Plaines, IL: AAMA.
AAMA CW-12, *Curtain Wall Manual*, Volume 12: *Structural Properties of Glass*. Des Plaines, IL: AAMA.
AAMA CW-13, *Curtain Wall Manual*, Volume 13: *Structural Sealant Glazing Systems*. Des Plaines, IL: AAMA.
AAMA CW-RS-1, "The Rain Screen Principle and Pressure-Equalized Wall Design." Des Plaines, IL: AAMA.
AAMA GAG-1, "Glass and Glazing." Des Plaines, IL: AAMA.
AAMA GS-001, "Voluntary Guide Specifications for Aluminum Architectural Windows." Des Plaines, IL: AAMA.
AAMA MCWM-1, *Metal Curtain Wall Manual*. Des Plaines, IL: AAMA.
AAMA SFM-1, *Aluminum Store Front and Entrance Manual*. Des Plaines, IL: AAMA.
AAMA SHDG-1, "Glass Design for Sloped Glazing." Des Plaines, IL: AAMA.
AAMA TIR-A1-1975, "Sound Control for Aluminum Curtain Walls and Windows." Des Plaines, IL: AAMA.
AAMA TIR-A4-97, "Recommended Glazing Guidelines for Reflective Insulating Glass." Des Plaines, IL: AAMA.
AAMA TIR-A7-83, "Sloped Glazing Guidelines." Des Plaines, IL: AAMA.
AAMA WSG-1, "Window Selection Guide." Des Plaines, IL: AAMA.

American National Standards Institute (ANSI) Standards:

A117.1-1998, "Accessible and Usable Buildings." New York: ANSI.
Z34.1, "Certification—Third-Party Certification Programs for Products, Processes, and Services." New York: ANSI.
Z97.1-1984 (R1994), "Glazing Materials Used in Buildings, Safety Performance Specifications and Methods of Test." New York: ANSI.

American National Standards Institute/American Society of Civil Engineers (ANSI/ASCE) Standards:

ANSI/ASCE 7-95, "Minimum Design Loads for Buildings and Other Structures." New York: ANSI.

American National Standards Institute/ Building Hardware Manufacturers Association (ANSI/BHMA) Standards:

A156.1-1997, "Butts and Hinges." New York: ANSI.

A156.2-1996, "Bored and Preassembled Locks and Latches." New York: ANSI.

A156.3-1994, "Exit Devices." New York: ANSI.

A156.4-1992, "Door Controls—Closers." New York: ANSI.

A156.5-1992, "Auxiliary Locks and Associated Products." New York: ANSI.

A156.6-1994, "Architectural Door Trim." New York: ANSI.

A156.7-1988, "Templet Hinge Dimensions." New York: ANSI.

A156.8-1994, "Door Controls—Overhead Stops and Holders." New York: ANSI.

A156.12-1992, "Interconnected Locks and Latches." New York: ANSI.

A156.13-1994, "Locks and Latches, Mortise." New York: ANSI.

A156.14-1997, "Sliding and Folding Door Hardware." New York: ANSI.

A156.15-1995, "Closer Holder Release Devices." New York: ANSI.

A156.16-1997, "Auxiliary Hardware." New York: ANSI.

A156.17-1992, "Self Closing Hinges and Pivots." New York: ANSI.

A156.18-1993, "Hardware—Materials and Finishes." New York: ANSI.

A156.19-1997, "Power Assist and Low-Energy Power-Operated Doors." New York: ANSI.

A156.21-1996, "Thresholds." New York: ANSI.

American National Standards Institute/ Door and Hardware Institute (ANSI/ DHI) Standards:

A115.1-1990, "Preparation for Mortise Locks for 1⅜″ and 1¾″ Doors." New York: ANSI.

A115.1G-1990, "Installation Guide for Doors and Hardware." New York: ANSI.

A115.4-1994, "Specifications for Preparation of 1¾″ Standard Steel Doors and Frames for Manually Operated Lever Extension Flush Bolts." New York: ANSI.

A115.5-1992, "Steel Frame Preparation for Mortise Auxiliary Deadlock Strikes." New York: ANSI.

A115.6-1993, "Preparation of 1¾″ Standard Steel Doors and Steel Frames for Series 2000 Preassembled Door Locks." New York: ANSI.

A115.7-1982, "Preparation for Floor Closers—Light Duty, Center Hung, Single or Double Acting; Center Hung, Single or Double Acting; Offset Hung, Single Acting." New York: ANSI.

A115.12-1994, "Specifications for Preparation of 1¾″ Standard Steel Doors and Steel Frames for Offset Intermediate Pivots." New York: ANSI.

A115.13-1991, "Preparation for Auxiliary Bored Deadlocks and Deadlatches in Standard Steel Doors and Steel Frames." New York: ANSI.

A115.14-1994, "Specifications for Preparation of 1¾″ Standard Steel Doors for Open Back Strikes." New York: ANSI.

A115.15-1994, "Specifications for Preparation of 1¾″ Prehung Insulated Steel Doors and Steel Frames for Series 4000 Bored Locks and Latches." New York: ANSI.

A115.16-1994, "Installation Guide for Doors and Hardware." New York: ANSI.

A115.17-1994, "Specifications for Preparation of 1⅜″ and 1¾″ Standard Steel Doors and Frames for Double Type Locks." New York: ANSI.

A115.18-1994, "Specifications for Standard Steel Door and Steel Frame Preparation for Bored Locks and Latches with Lever Handles for 1⅜″ and 1¾″ Doors." New York: ANSI.

A115.W2-1993, "Preparation of 1¾″ Flush Wood Doors for Series 4000 Bored Locks and Latches." New York: ANSI.

A115.W3-1993, "Preparation of 1⅜″ Flush Wood Doors for Series 4000 Bored Locks and Latches." New York: ANSI.

A115.W6-1993, "Preparation of 1¾″ Flush Wood Doors for Double-Type Locks." New York: ANSI.

A115.W8-1993, "Preparation of 1¾″ Flush Wood Doors for Bored Deadlock and Deadlatches." New York: ANSI.

American National Standards Institute/ Screen Manufacturers Association (ANSI/SMA) Standards:

1004-1987 (R1998), "Aluminum Tubular Frame Screens for Windows, Specifications for." New York: ANSI.

American National Standards Institute/ Steel Door Institute (ANSI/SDI) Standards:

A250.4-94, "Test Procedure and Acceptance Criteria for Physical Endurance for Steel Doors and Hardware Reinforcings." Cleveland, OH: SDI.

A250.5-94, "Accelerated Physical Endurance Test Procedures for Steel Doors, Frames and Frame Anchors." Cleveland, OH: SDI.

A250.6-97, "Hardware on Standard Steel Doors." Cleveland, OH: SDI.

A250.7-97, "Nomenclature for Steel Doors and Steel Door Frames." Cleveland, OH: SDI.

A250.8-98, "Steel Doors and Frames." Cleveland, OH: SDI (formerly SDI 100).

A250.10-98, "Test Procedure and Acceptance Criteria for Prime Painted Steel Surfaces for Steel Doors and Frames." Cleveland, OH: SDI.

American Society of Heating, Refrigerating and Air-Conditioning Engineers (ASHRAE). 1997. *ASHRAE Handbook; 1997 Fundamentals*. Atlanta, GA: ASHRAE.

American Society for Testing and Materials (ASTM) Standards:

A 123, "Standard Specification for Zinc (Hot-Dip Galvanized) Coatings on Iron and Steel Products." West Conshohocken, PA: ASTM.

A 153, "Standard Specification for Zinc Coating (Hot-Dip) on Iron and Steel Hardware." West Conshohocken, PA: ASTM.

A 167, "Standard Specification for Stainless and Heat-Resisting Chromium-Nickel Steel Plate, Sheet, and Strip." West Conshohocken, PA: ASTM.

A 568, "Standard Specification for Steel, Sheet, Carbon, and High-Strength, Low Alloy, Hot-Rolled, and Cold-Rolled, General Requirements." West Conshohocken, PA: ASTM.

A 569, "Standard Specification for Steel, Carbon (0.15 Maximum Percent) Hot-Rolled, and Strip, Commercial Quality." West Conshohocken, PA: ASTM.

A 653, "Standard Specification for Steel Sheet, Zinc Coated (Galvanized) or Zinc-Iron Alloy Coated (Galvannealed) by the Hot-Dip Process." West Conshohocken, PA: ASTM.

B 209, "Standard Specification for Aluminum and Aluminum-Alloy Sheet and Plate." West Conshohocken, PA: ASTM.

B 221, "Standard Specification for Aluminum and Aluminum-Alloy Extruded Bars, Rods, Wire, Profiles, and Tubes." West Conshohocken, PA: ASTM.

C 236, "Standard Test Method for Steady-State Thermal Performance of Building Assemblies by Means of a Guarded Hot Box." West Conshohocken, PA: ASTM.

C 481, "Standard Test Method for Laboratory Aging of Sandwich Construction." West Conshohocken, PA: ASTM.

C 509, "Specification for Elastomeric Cellular Preformed Gasket and Sealing Material." West Conshohocken, PA: ASTM.

C 542, "Standard Specification for Lock-Strip Gaskets." West Conshohocken, PA: ASTM.

C 716, "Standard Specification for Installing Lock-Strip Gaskets and Infill Glazing Materials." West Conshohocken, PA: ASTM.

C 719, "Standard Test Method for Adhesion and Cohesion of Elastomeric Joint Sealants Under Cyclic Movement." West Conshohocken, PA: ASTM.

C 864, "Standard Specification for Dense Elastomeric Compression Seal Gaskets, Setting Blocks, and Spacers." West Conshohocken, PA: ASTM.

C 920, "Standard Specification for Elastomeric Joint Sealants." West Conshohocken, PA: ASTM.

C 964, "Standard Specification Guide for Lock-Strip Glazing." West Conshohocken, PA: ASTM.

C 1036, "Standard Specification for Flat Glass." West Conshohocken, PA: ASTM.

C 1048, "Standard Specification for Heat-Treated Flat Glass—Kind HS, Kind FT Coated and Uncoated Glass." West Conshohocken, PA: ASTM.

C 1115, "Standard Specification for Dense Elastomeric Silicone Rubber Gaskets and Accessories." West Conshohocken, PA: ASTM.

C 1172, "Standard Specification for Laminated Architectural Flat Glass." West Conshohocken, PA: ASTM.

D 2000, "Standard Classification System for Rubber Products in Automotive Applications." West Conshohocken, PA: ASTM.

D 2287, "Standard Specification for Nonrigid Vinyl Chloride Polymer and Copolymer Molding and Extrusion Compounds." West Conshohocken, PA: ASTM.

D 4099. "Standard Specifications for Poly (Vinyl Chloride) Prime Windows." West Conshohocken, PA: ASTM.

E 84, "Standard Test Method for Surface Burning Characteristics of Building Materials." West Conshohocken, PA: ASTM.

E 90, "Standard Method for Laboratory Measurement of Airborne-Sound Transmission Loss of Building Partitions." West Conshohocken, PA: ASTM.

E 96, "Standard Test Method for Water Vapor Transmission of Materials." West Conshohocken, PA: ASTM.

E 283, "Standard Test Method for Determining the Rate of Air Leakage Through Exterior Windows, Curtain Walls, and Doors Under Specified Pressure Differences Across the Specimen." West Conshohocken, PA: ASTM.

E 330, "Standard Test Method for Structural Performance of Exterior Windows, Curtain Walls, and Doors by Uniform Static Air Pressure Differential." West Conshohocken, PA: ASTM.

E 331, "Standard Test Method for Water Penetration of Exterior Windows, Curtain Walls, and Doors by Uniform Static Air Pressure Differential." West Conshohocken, PA: ASTM.

E 413, "Standard Classification for Rating Sound Insulation." West Conshohocken, PA: ASTM.

E 547, "Standard Test Method for Water Penetration of Exterior Windows, Curtain Walls, and Doors by Cyclic Air Pressure Differential." West Conshohocken, PA: ASTM.

E 662, "Standard Test Method for Specific Optical Density of Smoke Generated by Solid Materials." West Conshohocken, PA: ASTM.

E 773, "Standard Test Method for Accelerated Weathering of Sealed Insulating Glass Units." West Conshohocken, PA: ASTM.

E 774, "Standard Specification for the Classification of the Durability of Sealed Insulating Glass Units." West Conshohocken, PA: ASTM.

E 783, "Standard Test Method for Field Measurement of Air Leakage Through Installed Exterior Windows and Doors." West Conshohocken, PA: ASTM.

E 987, "Standard Tests Methods for Deglazing Force of Fenestration Products." West Conshohocken, PA: ASTM.

E 1105, "Standard Test Method for Field Determination of Water Penetration of Installed Exterior Windows, Curtain Walls, and Doors by Uniform or Cyclic Static Air Pressure Difference." West Conshohocken, PA: ASTM.

E 1233-97 "Standard Test Method for Structural Performance of Exterior Windows, Curtain Walls, and Doors by Cyclic Static Air Pressure Differential." West Conshohocken, PA: ASTM.

F 588, "Standard Test Methods for Measuring the Forced Entry Resistance of Window Assemblies, Excluding Glazing Impact." West Conshohocken, PA: ASTM.

F 842, "Standard Test Methods for Measuring the Forced Entry Resistance of Sliding Door Assemblies, Excluding Glazing Impact." West Conshohocken, PA: ASTM.

American Welding Society (AWS). AWS D1.1-98, "Structural Welding Code—Steel." Miami, FL: AWS.

———. AWS D1.2-97, "Structural Welding Code—Aluminum." Miami, FL: AWS.

Architectural Graphic Standards. See Ramsey/Sleeper.

Architectural Woodwork Institute (AWI). 1989. *Architectural Woodwork Quality Standards, Guide Specifications, and Quality Certification Program.*" Centreville, VA: AWI.

Belles, Don, and Jesse Beitel. 1988. "Fire Performance of Curtainwalls Questioned." *Exteriors* 6(2)(summer):44–50.

Benny, James C. 1988. "Ensuring That Wood Doors and Hardware Are Compatible." *Window and Door Specifier* 1(1)(summer):69–70.

Bock, Gordon. 1988. "Glass Notes." *The Old-House Journal* 16(4)(July/August):35–43.

———. 1989. "The Sash Window Balancing Act." *The Old-House Journal* 17(5)(September/October):31–40.

Building Design and Construction. 1987. "Survey Reveals Advantages of Wood Doors and Windows." *Building Design and Construction* 28(6)(June):157–166.

———. 1989. "Roundtable: New Horizons in Curtain Wall Design and Construction." *Building Design and Construction*, 30(1)(January):67–73.

Carter, Roy. 1988. "The Ins and Outs of Access Control Systems." *The Construction Specifier* 41(1)(January):84–87.

Cassidy, Victor M. 1989. "How Windows Get Better." *Modern Metals* 45(1) (February):32–46.

Commercial Renovation. 1987. "Performance Standards for the Wood Door Industry." *Commercial Renovation* 9(4)(August):50.

———. 1987. "Product Focus: Door and Hardware Innovations." *Commercial Renovation* 9(4)(August):52–53.
———. 1988. "New Technology Makes Window Efficiency Shine." *Commercial Renovation* 10(3)(June):48–53.
———. 1988. "Door Security Hinges on Hardware." *Commercial Renovation* 10(5)(October):52–56.
Construction Specifications Institute (CSI). 1988. CSI Monograph 07M411, "Precoated Metal Building Panels." Alexandria, VA: CSI.
———. CSI Monograph 08M710, "Finish Hardware." Alexandria, VA: CSI.
Consumer Product Safety Commission (CPSC). 16 CFR Part 1201, "Safety Standards for Architectural Glazing Materials." Bethesda, MD: CPSC.
Council of American Building Officials (CABO). 1983. *One- and Two-Family Dwelling Code*. Falls Church, VA: CABO. (In the process of being transferred to the control of the International Code Council (ICC).)
Cox, Sarah, and Billy Edwards. 1987. "Keys to Specifying the Right Locking System." *The Construction Specifier* 40(1)(January):25–26.
D'Angelo, Charles, and Owen J. Perryman. 1988. "Insulated Panels and the Pre-Engineered Building Market." *Metal Architecture* October, 4(10):20.
Door and Hardware Institute (DHI). 1976. *Hardware Reinforcements on Steel Doors and Frames*. Chantilly, VA: DHI.
———. 1985. *Basic Architectural Hardware*. Chantilly, VA: DHI.
———. 1990. *Processing Hardware for Custom Aluminum Doors*. Chantilly, VA: DHI.
———. 1990. *Recommended Locations for Architectural Hardware for Standard Steel Doors and Frames*. Chantilly, VA: DHI.
———. 1994. *Installation Guide for Doors and Hardware (ANSI/DHI A115.IG)*. Chantilly, VA: DHI.
———. 1996. *Recommended Locations for Builder's Hardware for Custom Steel Doors/Frames*. Chantilly, VA: DHI.
———. 1996. DHI-WDHS-1, "Templet Book Criteria for Wood Doors." Chantilly, VA: DHI.
———. 1996. DHI-WDHS-2, "Recommended Fasteners for Wood Doors." Chantilly, VA: DHI.
———. 1993. DHI-WDHS-3, "Recommended Locations for Architectural Hardware for Flush Wood Doors." Chantilly, VA: DHI.
Easter, R. Lee. 1985. "Can Glass and Hurricanes Mix?" *The Construction Specifier* 38(5)(May):46–50.
Evans, Lyon D. 1985. "Aluminum Windows and Sliding Glass Doors, the New Specs." *The Construction Specifier* 38(9)(September):56–63.
———. 1986. "Design Techniques for Glazed Curtainwall Retrofit." *Exteriors* 4(3)(autumn):56–60.
Exteriors. 1986. "Skyline: Window Standards." *Exteriors* 4(2)(summer):20.
———. 1986. "Skyline: Glazing Technics." *Exteriors* 4(4)(winter): 24–25.
———. 1987. "Skyline: Curtainwall Insulation." *Exteriors* 5(3)(autumn):44–46.
———. 1987. "Skyline: Low E Glass Use Rising." *Exteriors* 5(4)(winter):18–19.
———. 1988. "Skyline: Glass Strength Debated." *Exteriors* 6(3)(autumn):18–19.
———. 1988. "Skyline: Curtainwall Issues." *Exteriors* 6(4)(winter):75.
Federal Specifications: *See* U.S. General Services Administration Specifications Unit.
Flat Glass Marketing Association (FGMA). *Glazing Manual*. Topeka, KS: FGMA.
———. *Sealant Manual*. Topeka, KS: FGMA.
Francis, Geoffrey V. 1988. "New Developments in Structural Glazing." *Exteriors* 6(1)(spring):50–55.
———. 1989. "Innovation in Structural Sealant Glazing." *The Construction Specifier* 42(3)(March):54–59.
Fulton, Frank. 1988. "High Performance Window Design." *The Construction Specifier* 41(8)(August):80–84.
George, Louis F. 1988. "Specifying Hardware for Schools." *The Construction Specifier* 41(8)(August):52–59.
———. 1989. "Specifying Hardware in Housing for the Elderly." *The Construction Specifier* 42(2)(February):62–66.
———. 1989. "Specifying Systems for Access Control." *The Construction Specifier* 42(6)(June):52–58.
Glass Tempering Association (GTA). 1989. 89-1-6, "Specification for Environmental Durability of Fully Tempered or Heat-Strengthened Spandrel Glass with Applied Opacifiers." Topeka, KS: GTA.
Gregerson, John. 1988. "Metal Panels Have Designs on New Markets." *Building Design and Construction* 29(6)(June):68–71.
———. 1989. "New Glass Technologies Solve Site Challenges." *Building Design and Construction* 30(3) (March):74–77.
Griffiths, Howard. 1987. "Glass: Colors of the City." *The Construction Specifier* 40(8)(August):100–107.
Grossi, Anthony F. 1984. "A Brief on Low E Glass." *The Construction Specifier* 37(9)(September):22–23.
Heerwagen, Judith H. 1987. "Windowscapes—The Psychology of View." *The Construction Specifier* 40(8)(August):31–32.
Heitmann, L. J. 1986. "The Trouble with Shadow Boxes." *Exteriors* 4(2)(summer):10.
———. 1986. "A Closer Look at Structural Glazing." *Exteriors* 4(4)(winter):12.
———. 1987. "The Rain Screen Principle Can Work for You." *Exteriors* 5(1)(spring):12.
———. 1987. "The Trouble with Curtainwall." *Exteriors* 5(2)(summer):12.
———. 1987. "Proper Value Engineering for Curtainwall." *Exteriors* 5(3)(autumn):12.
———. 1987. "Safety First in Overhead Glazing." *Exteriors* 5(4)(winter):12.
———. 1988. "On Designing a Spandrel Cavity." *Exteriors* 6(2)(summer):12.
———. 1988. "A Question of Compatibility." *Exteriors* 6(3)(autumn):12.
Hollow Metal Manufacturers Association. 1987. *Hollow Metal Manual*. Chicago: National Association of Architectural Metal Manufacturers.
Hudnut, Richard. 1989. "The Hardware Metals." *The Construction Specifier* 41(8)(August):31–33.
Insulating Glass Certification Council (IGCC). 1991. *Certified Products Directory—Sealed Insulating Glass*. Cortland, NY: IGCC.
Johnson, Stephen. 1988. "Improvement in Domestic Steel Quality Translating into Better Metal Paneling." *Metal Architecture* 4(9)(September):5, 85.
Johnston, Bob. 1986. "Quality Control Critical with Structural Silicone." *Exteriors* 4(2)(summer):44–50.
Kawneer Company, Inc. 1984. *Architectural Finishes*. Norcross, GA: Kawneer.
Kincaid, Mary. 1983. "Facets of Glass." *The Construction Specifier* 36(3)(March):34–40.
Laminated Safety Glass Association (LSGA). 1987. *LSGA Design Guide*. Topeka, KS: LSGA.
LaTona, Raymond W., Thomas A.

Schwartz, and Glen R. Bell. 1988. "New Standard Permits More Realistic Curtain Wall Testing." *Building Design and Construction* 29(11)(November):42–46.

Lesniak, Joseph G., and Judith P Guy. 1985. "The Architectural Hardware Consultant." *The Construction Specifier* 38(6)(June):34.

Lofgren, Michael. 1983. "Making Building Security Airtight." *The Construction Specifier* 36(11)(November):68–69.

MacDonald, John E. 1978. "The Effects of Light on Wood and Wood Finishing Systems." *The Construction Specifier* 31(3)(March):23–26.

———. 1978. "Architectural Flush Wood Doors." *The Construction Specifier* 31(3)(March):28–32.

Maisel, Murry. 1983. "Custom Hollow Metal Doors, Problems and Solutions." *The Construction Specifier* 36(1)(January):74–77.

Marinelli, Janet. 1988. "Architectural Glass." *The Old-House Journal* 16(4) (July/August):34–42.

Mason, Donald. 1983. "Selecting an Access Control System." *The Construction Specifier* 36(11)(November):71–73.

McAuliffe, William B. 1985. "Finish Hardware, A Complex Specification." *The Construction Specifier* 38(6) (June):30–37.

McInerney, William D. 1987. "Selecting the Proper Hinge and Pivot." *The Construction Specifier* 40(3)(March):116–122.

McKinley, Robert W. 1986. "Saving Energy with Low E Glazing." *Exteriors* 4(1)(March):50–58.

———. 1986. "Windloading of Glazed Curtain Walls." *Exteriors* 4(4)(winter): 32–36.

———. 1989. "Insulating Glass: Designing for the 1990s." *Exteriors* 7(1) (spring):12–16.

Metal Construction News. 1988. *1989 Building Systems Product File and Directory*. Skokie, IL: *Metal Construction News*.

Minor, Joseph E. 1987. "Accommodating Wind Forces in Glazing Design." *The Construction Specifier* 40(8)(August): 25–26.

National Association of Architectural Metal Manufacturers (NAAMM). *Metal Finishes Manual*. Chicago, IL: NAAMM.

National Fire Protection Association (NFPA). 1999. *NFPA 80: Standard for Fire Doors, Fire Windows and Smoke-Control Door Assemblies*. Quincy, MA: NFPA.

———. 1997. *NFPA 101: Life Safety Code*. Quincy, MA: NFPA.

———. 1999. *NFPA 105: Recommended Practice for the Installation of Smoke-Control Door Assemblies*. Quincy, MA: NFPA.

National Glass Association (NGA). *Guide to the Glass and Glazing Requirements of the Model Building Codes*. McLean, VA: NGA.

———. *Glazing Reference Guide*. McLean, VA: NGA.

———. *Glass Standards: A Collection of National and Voluntary Standards Pertaining to Glass and Glazing*." McLean, VA: NGA.

National Institute of Standards and Technology. *See* U.S. Department of Commerce.

National Wood Window and Door Association (NWWDA). *How to Store, Handle, Finish and Maintain Wood Doors*. Park Ridge, IL: NWWDA.

———. *Care and Finishing of Wood Doors*. Park Ridge, IL: NWWDA.

———. ANSI/NWWDA I.S.1-A-91, "Industry Standard for Architectural Wood Flush Doors." Park Ridge, IL: NWWDA.

———. ANSI/NWWDA I.S.1-R-91, "Industry Standard for Residential Wood Flush Doors." Park Ridge, IL: NWWDA.

———. NWWDA I.S.2-87, "Industry Standard for Wood Windows." Park Ridge, IL: NWWDA.

———. NWWDA I.S.3-88, "Industry Standard for Sliding Patio Doors." Park Ridge, IL: NWWDA.

———. NWWDA I.S.4-81, "Industry Standard for Water-Repellent Preservative Treatment for Millwork." Park Ridge, IL: NWWDA.

———. ANSI/NWWDA I.S.6-91, "Industry Standard for Wood Stile and Rail Doors." Park Ridge, IL: NWWDA.

Nicastro, David H. 1988. "Uncovering the Reasons for Curtainwall Failure." *Exteriors* 6(2)(summer):26–32.

Olson, Christopher. 1987. "Heightening Impact with Structural Glazing." *Building Design and Constructions* 28(3)(March):132–136.

Peterson, Charles O., Jr. 1984. "Structural Silicone." *The Construction Specifier* 37(9)(September):56–63.

Raeber, John A. 1985. "Selecting Door Hardware." *The Construction Specifier* 38(6)(June):38–47.

———. 1988. "Selecting Door Contract Hardware." *The Construction Specifier* 41(8)(August):62–71.

———. 1989. "Exposed Metallic Coatings on Glass: A Cautionary Note." *The Construction Specifier* 42(6) (June):31–32.

———. 1989. "Selecting Door-Hanging Hardware." *The Construction Specifier* 42(6)(June):36–42.

Ramsey/Sleeper, The AIA Committee on Architectural Graphic Standards. *Architectural Graphic Standards*. 8th ed. New York: John Wiley & Sons, Inc.

Reynolds, James. 1986. "Specifying Coated Glazing Material." *The Construction Specifier* 39(6)(June):46–59.

Rush, Richard. 1987. "Refining Window Energy Performance." *Building Design and Construction* 28(12)(December):148–153.

Russo, Michael. 1987. "Improving the Art of Glass Design." *Exteriors* 5(3) (autumn):6.

Safety Glazing Certification Council (SGCC). *Certified Products Directory—Safety Glazing Materials Used in Buildings*. Cortland, NY: SGCC.

Saino, Joseph Neuhoff. 1988. "Industrial Doors: The Types and Uses." *The Construction Specifier* 41(3)(March):76–81.

Sanford, A. G. 1987. "Stress Tests and Safety Factors in Structural Glazing." *The Construction Specifier* 40(6) (June):31–37.

Scassellati, Rudy R. 1988. "Architectural Flush Doors." *The Construction Specifier* 41(4)(April):112–120.

Scharfe, Thomas R. 1988. "Meeting the Challenges of Sloped Glazing." *Building Design and Construction* 29(3) (March):140–147.

———. 1988. "New Metal Coating Technologies Enhance Design Opportunities." *Building Design and Construction* 29(6)(June):86–89.

Schulthesis, Joseph A. 1985. "Guidelines for Sloped Glazing." *The Construction Specifier* 38(3)(March):90–95.

———. 1986. "Sloped Glazing with Structural Silicone." *The Construction Specifier* 39(6)(June):36–45.

Screen Manufacturers Association (SMA). *See* American National Standards Institute/Screen Manufacturers Association.

Sealed Insulating Glass Manufacturers Association (SIGMA). 1997. TM-3000,

"Glazing Guidelines for Sealed Insulating Glass Units." Chicago: SIGMA.

———. 1990. TB-3001, "Voluntary Guidelines for Sloped Glazing." Chicago: SIGMA.

Sittnick, Ralph. 1985. "The Evolution of the Electromagnetic Lock." *The Construction Specifier* 38(6)(June):48–49.

Smith, Charles F., Jr. 1989. "Specifying Weatherstripping for Energy, Smoke, and Sound Control." *The Construction Specifier* 42(6)(June):44–51.

Stanbrough, Jerry. 1986. "Energy Efficiency: The Curtainwall as Filter." *Exteriors* 4(2)(summer):34–39.

Steel Door Institute (SDI). SDI-105, "Recommended Erection Instructions for Steel Frames." Cleveland, OH: SDI.

———. SDI-106, "Recommended Standard Door Type Nomenclature." Cleveland, OH: SDI.

———. SDI-108, "Recommended Selection and Usage Guide for Standard Steel Doors." Cleveland, OH: SDI.

———. SDI-109, "Hardware for Standard Steel Doors and Frames." Cleveland, OH: SDI.

———. SDI-110, "Standard Steel Doors and Frames for Modular Masonry Construction." Cleveland, OH: SDI.

———. SDI-111, "Recommended Standard Details for Steel Doors and Frames." Cleveland, OH: SDI.

———. SDI-112, "Zinc-Coated (Galvanized/Gavannealed) Standard Steel Doors and Frames." Cleveland, OH: SDI.

———. SDI-113, "Test Procedure and Acceptance Criteria for Apparent Thermal Performance of Steel Door and Frame Assemblies." Cleveland, OH: SDI.

———. SDI-116, "Standard Test Procedure and Acceptance Criteria for Rate of Air Flow Through Closed Steel Door and Frame Assemblies." Cleveland, OH: SDI.

———. SDI-117, "Manufacturing Tolerances for Standard Steel Doors and Frames." Cleveland, OH: SDI.

———. SDI-118, "Basic Fire Door Requirements." Cleveland, OH: SDI.

———. SDI-128, "Guidelines for Acoustical Performance of Standard Steel Doors and Frames." Cleveland, OH: SDI.

Steel Structures Painting Council (SSPC). 1995. *Good Paint Practice*. Vol. 1 of *Steel Structures Painting Manual*. 3rd ed. Pittsburgh, PA: SSPC.

———. 1995. *Systems and Specifications*. Vol. 2 of *Steel Structures Painting Manual*. 7th ed. Pittsburgh, PA: SSPC.

Stubbs, M. Stephanie. 1986. "Glued-on Glass." *Architectural Technology* (May/June):46–51.

Tatum, Rita. 1984. "The Return of Wood Windows." *The Construction Specifier* 37(9)(September):51–55.

———. 1986. "Metal Building Systems Go High-Rise High-Tech." *The Construction Specifier* 39(3)(March):40–49.

———. 1987. "The Evolution of Energy-Saving Windows." *The Construction Specifier* 40(8)(August):33–34.

Taylor, Paul. 1983. "Glass Skyscrapers Vulnerable." *The Construction Specifier* 36(11)(November):10–12.

Taylor, Tim, and Richard J. Barnhard. 1999. "A Closer Look at Door Width Openings." *The Construction Specifier* (May):61–64.

Technologies Media Corporation. 1989. "Architectural Hardware." *Building Design and Construction* 30(12)(October):115–143.

Ting, Raymond. 1986. "Metal Panel Behavior in Exterior Wall Design." *Exteriors* (autumn):65–69.

———. 1987. "Ensuring a Trouble-Free Curtainwall Renovation." *Exteriors* 5(3)(autumn):64–66.

———. 1988. "Designing a Leak-Free Curtainwall System." *Exteriors* 6(3) (autumn):30–34.

———. 1989. "Performance Parameters of Composite Foam Panels." *Metal Architecture* 5(9)(September):7–8, 73–74.

Trechsel, Heinz. 1988. "Specifying an Energy Efficient Thermal Window." *Exteriors* 6(3)(autumn):36–40.

Umlauf, Elyse. 1989. "Specifying the Right Wood Windows and Doors." *Building Design and Construction* 30(7)(June):100–103.

U.S. Department of the Army. 1995. Corps of Engineers Guide Specifications, Military Construction, CEGS-08110, "Steel Doors and Frames." Office of the Chief of Engineers, Department of the Army.

U.S. Department of Commerce, National Institute of Standards and Technology. 1977. *Window Design Strategies to Conserve Energy*. Washington DC: National Institute of Standards and Technology.

U.S. Department of Housing and Urban Development (HUD). *Minimum Property Standards for Housing*. Washington, DC: HUD.

U.S. Department of the Navy, Naval Facilities Engineering Command. 1995 (July). Guide Specifications Section 08110, "Steel Doors and Frames." Washington, DC: Department of the Navy.

———. 1995 (July) Guide Specifications Section 08120, "Aluminum Doors and Frames." Washington, DC: Department of the Navy.

———. 1995 (July) Guide Specifications Section 08210, "Wood Doors." Washington, DC: Department of the Navy.

———. 1995 (July) Guide Specifications Section 08520, "Aluminum Windows." Washington, DC: Department of the Navy.

———. 1995 (July) Guide Specifications Section 08610, "Wood Windows." Washington, DC: Department of the Navy.

———. 1995 (July) Guide Specifications Section 08710, "Finish Hardware." Washington, DC: Department of the Navy.

———. 1995 (July) Guide Specifications Section 08800, "Glazing." Washington, DC: Department of the Navy.

U.S. General Services Administration (GSA) Specifications Unit. Federal Specification L-S-125, "Screening, Insect, Nonmetallic." Washington, DC: GSA Specifications Unit.

———. Federal Specification RR-W-365, "Wire Fabric (Insect Screening)." Washington, DC: GSA Specifications Unit.

Valdes, Noel. 1988. "Low E Glass." *The Construction Specifier* 41(8)(August): 71–78.

Vild, Donald J. 1986. "Glass and the Building Codes." *Exteriors* 4(1) (March):12.

———. 1986. "Glass: The Energy Saver." *Exteriors* 4(2)(summer):12.

———. 1986. "The Case for Sloped Glazing." *Exteriors* 4(3)(autumn):10.

———. 1986. "Fully Tempered vs. Heat-Strengthened Glass." *Exteriors* 4(4) (winter):10.

———. 1987. "Why Glass Goes Bad." *Exteriors* 5(1)(spring):10.

———. 1987. "The Overhead Glazing Controversy." *Exteriors* 5(3)(autumn): 10.

———. 1987. "Who Selects the Glass." *Exteriors* 5(4)(winter):10.

———. 1988. "Proper Engineering of

Structural Silicone." *Exteriors* 6(1) (spring):10.

———. 1988. "What's Wrong with Glass Spandrels?" *Exteriors* 6(2)(summer): 10.

———. 1988. "How Much Glass Will Break." *Exteriors* 6(3)(autumn):10.

———. 1988. "Some Precautions When Specifying Low-E Glass." *Exteriors* 6(4)(winter):7.

———. 1988. "Clearing Up Building Code, Glass Selection Confusion." *Exteriors* 6(4)(winter):76–80.

———. 1989. "More Problems with Glass Spandrels." *Exteriors* 7(1) (spring):7.

Wherry, Allen P. 1986. "Standard Steel Doors for Customized Design." *The Construction Specifier* 39(7)(July): 106–111.

Wilson, Randy J. 1987. "Sloped Glazing and the Rain Screen Principle." *The Construction Specifier*, 40(8)(August): 94–98.

Woodwork Institute of California (WIC). *Manual of Woodwork*. Fresno, CA: WIC.

Wright, Gordon. 1988. "Curtain Walls Drawing More Critical Attention." *Building Design and Construction* 29(3)(March):162–165.

———. 1989. "Storefront Designs Accommodate Multiple Requirements." *Building Design and Construction* 30(1)(January):60–62.

Zingeser, Joel P. 1988. "Applauding Glass." *The Construction Specifier* 41(8)(August):47–48.

CHAPTER 9

Finishes

Introduction
Applicable *MasterFormat*™ Sections
Plaster Materials
Plaster Support Systems, Bases, and Accessories
Gypsum Plaster Application
Portland Cement Plaster Application
Gypsum Board Systems
Tile
Terrazzo
Acoustical Treatment
Wood Flooring
Dimension Stone Tile Flooring
Resilient Flooring
Carpet
Resinous Flooring
Special Coatings
Paints
Vinyl Wall Covering
Additional Reading
Acknowledgments and References

Introduction

Finishes include materials and systems used to make the visible surfaces of the interior floors, walls, and ceilings of a building. Some of the same products are also used as exterior finishes. There are many products and systems in use today as interior finishes, and there are usually several variations within a particular product or system. Selection of these materials and systems for interior application is influenced by the use of the spaces in which they will be applied and the wear to which they will be subjected. This chapter covers some of the many finishes available.

Applicable *MasterFormat*™ Sections

The following *MasterFormat* Broadscope sections are applicable to this chapter.

09100 Metal Support Assemblies
09200 Plaster and Gypsum Board
09300 Tile
09400 Terrazzo
09500 Ceilings
09600 Flooring
09700 Wall Finishes
09800 Acoustical Treatment
09900 Paint and Coatings

9.1 Plaster Materials

This section discusses the manufacture and use of gypsum materials employed in plaster and board products, of portland cement materials used in plaster, and of plaster products. Other sections of this chapter contain additional information about the manufacture of gypsum board products and other gypsum building products (9.5), support systems and accessories for plaster (9.2), plaster application (9.3 and 9.4), and gypsum board installation (9.5).

9.1.1 GYPSUM

Gypsum, a rocklike mineral known chemically as hydrous calcium sulfate, is found in many parts of the world, usually combined with impurities such as clay, limestone, and iron oxides. In its pure form gypsum is white, but in combination with impurities it may be gray, brown, or pinkish. Relatively pure calcined gypsum is known as *plaster of paris*, after the huge beds of it that underlie the city of Paris where it was mined in abundance in the late nineteenth and early twentieth centuries. Gypsum is quarried or mined throughout the United States where deposits are accessible to transportation (Fig. 9.1-1).

Two properties of gypsum make it useful as a plaster material: (1) it gives up some of its chemically combined water and becomes a fine powder when intensely heated (calcined), and (2) when water is added, it returns chemically to its original rocklike state by forming interlocking crystals (crystallizing). A mixture of calcined gypsum powder and water remains plastic for a short time, during which it can be shaped or molded. After it hardens by crystallizing (sets), it forms an effective fire barrier.

Gypsum has been used in construction as far back as 2000 B.C. Early gypsum plasters, such as plaster of paris, had short, uncontrolled setting times and were difficult to use. Late in the nineteenth century, retarders and accelerators capable of controlling setting time were added to processed gypsum, expanding the use of gypsum plasters for decorative and fire-resistant purposes.

The first large-scale use of gypsum plaster was in the "white palaces" (exhibition buildings) of Chicago's World Columbian Exposition in 1893. Soon after that, gypsum board was invented for use as a plaster base, similar to contemporary gypsum lath. It consisted of layers of gypsum sandwiched between plies of felt paper. Further refinements in ingredients, designs, and manufacturing processes resulted in the modern varieties of plaster, lath, and other gypsum products.

Gypsum provides superior fire resistance. When the temperature of crystalline gypsum is raised to 212°F (100°C), water in the form of steam is released. In addition to absorbing some of the heat and dissipating it, this conversion of water to steam prevents the temperature of the gypsum from rising above 212°F (100°C) as long as steam is present. With continued exposure to heat, this action progresses inward, releasing more water, and the surface becomes calcined, turning to a white, chalky powder. The calcined layers form an insulator, protecting the underlying uncalcined areas and retarding further calcination.

9.1.1.1 Plaster Manufacture

Most gypsum building products can be classed as plaster products or board products. Plaster products consist principally of dry calcined gypsum that requires the addition of water to develop the chemical reaction necessary for setting. Board products are prefabricated at the factory from crystallized gypsum enclosed by paper or vinyl, and are delivered to a job site in immediately usable form (see Section 9.5). Figure 9.1-2 shows a manufacturing sequence for plaster and board products.

About two-thirds of all the crude gypsum mined goes into products for building construction. These products are used extensively where fire resistance or sound control is required and where hard and dense wall surfaces are desired. Many new, engineered gypsum products and assemblies combine the advantages of board and plaster products to meet contemporary construction needs.

After gypsum rock is mined, it is crushed to a 2-in. (50.8 mm) size and passed through a hammer mill where it is further reduced to a ½-in. (12.7 mm) maximum particle size. It is suitable in this form for treating in a rotary calciner, but additional grinding to a fine powder is necessary for treating it in a kettle calciner.

A *rotary calciner* is an inclined steel cylinder lined with fire brick, typically 150 ft. (45.72 m) long and 12 to 15 ft. (3.66 to 4.57 m) in diameter. As a calciner rotates, the rock inside gravitates to the lower end, while heat is injected to produce a temperature of about 350°F (176.67°C). This heat drives off about three-quarters of the combined water, changing the gypsum into a dehydrated form with an affinity for water.

A *kettle calciner* is a vertical cylinder 10 to 12 ft. (3.05 to 3.66 m) in diameter with internal agitators that rotate around a central shaft. As in the rotary calciner, heat is introduced near the bottom, driving off the combined water in the powdered gypsum (see Fig. 9.1-2).

Calcined gypsum from the kettle is further treated in a *tube mill*, a rotary cylinder containing thousands of steel balls. The action of the steel balls produces a finer grind, which contributes to better plasticity and workability.

Calcined gypsum at this point sets too quickly to be suitable for plastering. The type of aggregate used (see Section 9.1.3), local impurities in aggregates and in water, and seasonal variations all affect the setting time. Therefore, the amounts of retarder added vary and are formulated by manufacturers for specific locations and uses to give the desired job set, which is usually a maximum of 4 hours.

Organic fibers are added to fibered gypsum plasters at the mill. Ready-mixed plasters are usually combined with lightweight aggregates before they are packaged and shipped.

FIGURE 9.1-1 **Gypsum is (a) quarried or (b) mined throughout the United States.** (Courtesy Gypsum Association.)

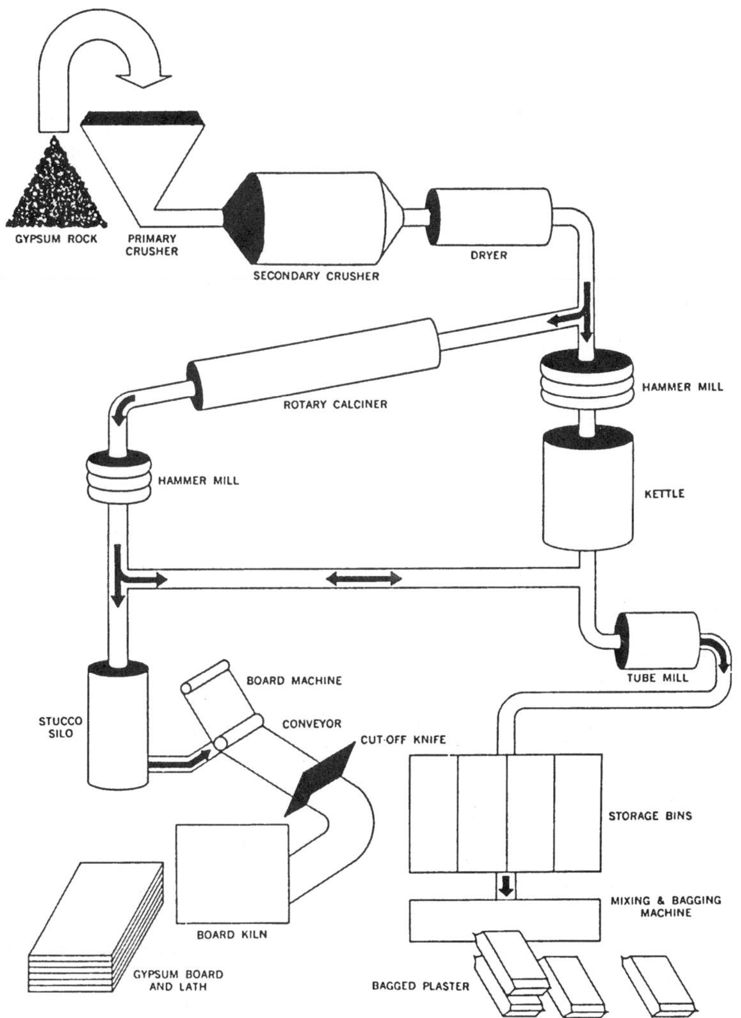

FIGURE 9.1-2 **Flowchart of gypsum board and plaster manufacture.** (Courtesy Gypsum Association.)

9.1.1.2 Plaster Products

Powdered calcined gypsum is the prime raw material in gypsum plaster products and in the cores of board products. It acts as a binder. Others materials used are aggregates (Section 9.1.3), mineral or organic fibers, and lime (Section 9.1.4). Aggregates are fillers that provide dimensional stability, and may also provide increased hardness and fire resistance in some plasters. Wood or other organic fibers are sometimes included in plasters to control the working quality and the keying action of plasters with metal lath. Lime putty is added in finish coat plasters to increase bulk and improve workability. Proprietary chemical retarders and accelerators are usually introduced at the plant to control setting time.

The discussion here is limited to a general description of the properties and uses of packaged plaster products (*bagged goods*) and does not describe the many plasters that may be produced on the job for specific applications. Sections 9.3 and

9.4 discuss application methods and finish types of plasters. Packaged plaster products and job-prepared plaster mixes are both referred to as *plasters* in this text.

Packaged gypsum plaster products consist primarily of powdered gypsum and additives to control setting time and working quality. These products may also contain aggregates and fibers. *Unfibered neat plasters* generally require the addition of aggregates on the job. Ready-mixed basecoat and prepared gypsum finish plasters contain all the necessary ingredients, including aggregates. Adding water to properly proportioned ingredients produces plasters suitable for application over appropriate plaster bases.

Gypsum plasters are suitable for all interior plastering in locations not subject to severe moisture conditions. They can also be used in exterior locations, like the ceilings of open porches and carports or for eave soffits, where they are not directly exposed to water. They are applied over plaster bases, such as gypsum lath and metal lath (sheet, expanded, welded, or woven wire), and masonry or concrete bases.

Plasters used for scratch and brown coats are called *basecoat* plasters; those used for the final coat are *finish coat* plasters. The governing standards of the American Society for Testing and Materials (ASTM) and the physical and working properties of basecoat and finish coat plasters are summarized in Figure 9.1-3.

BASECOAT PLASTERS

Basecoat plasters are classed according to ingredients as *neat*, *wood-fibered*, *ready-mixed*, and *bond* plasters. Basecoat plasters are formulated for hand-trowel or machine applications.

Neat Plaster *Neat gypsum basecoat plaster* does not contain aggregates. It is used where adding aggregates on the job is desired, such as where a specially formulated plaster is required to match an existing plaster or where the typical aggregate in ready-mix plaster is too fine for the desired texture. It may also be obtained lightly fibered with cattle hair, sisal, or other fine organic fibers, or specially formulated for use with lightweight aggregates such as vermiculite or perlite.

Wood-Fibered Plaster Wood-fibered plaster consists of calcined gypsum mixed with selected coarse cellulose fibers, which provide bulk and greater coverage. It is formulated to produce high-strength basecoats for fire-resistant ceiling assemblies.

Ready-Mixed Plaster Ready-mixed plaster is a mill-prepared plaster. It consists of calcined gypsum and an aggregate (usually perlite, but sometimes sand or vermiculite) and requires only the addition of water. This plaster possesses all the features and advantages of neat plaster mixed on the job, but offers mill control of the quality, gradation, and proportioning of aggregates.

FINISH COAT PLASTERS

The finish coat is the final plastering operation. It may serve as a base for further decorating or as a final decorative surface. Finish coat plasters usually consist of calcined gypsum, lime, and sometimes an aggregate. Some finish coat plasters may require the addition of lime or sand on the job.

Gauging Plaster Gauging plaster consists of coarsely ground gypsum of low consistency that soaks up mixing water readily and blends well with lime putty. This plaster controls (gauges) the setting time of plasters when mixed with lime putty (which by itself does not set). Gauging plaster is available as white plaster, which produces a brilliant white finish when gauged with lime putty, and local plaster, which may be somewhat darker in color.

Gypsum Keene's Cement Plaster Gypsum Keene's cement plaster is similar to gypsum gauging plaster, but unlike ordinary gypsum plasters, gypsum Keene's cement is burned in a kiln instead of a calciner or kettle. In this burning process, most of the combined water is removed, producing a *dead-burned* gypsum. The resulting material is denser than ordinary gypsums and has greater resistance to impact, abrasion, and moisture. Select white gypsum rock is used to produce the hardest white gypsum available.

Gypsum Keene's cement plaster has an initial set of about 1½ hours and a final set of 4 to 6 hours. A quick-setting gypsum Keene's cement that provides a final set of about 2 hours also is available.

Gypsum Keene's cement-lime putty-trowel plaster produces a dense, hard finish.

Gypsum Keene's cement-lime-sand float plaster is a highly crack- and spall-resistant plaster.

Prepared Gypsum Finish Plaster Prepared gypsum finish plaster is a ready-mixed plaster that includes all necessary ingredients, except water. It consists of finely ground calcined gypsum with or without fine aggregates. It is available as white or gray *trowel* finish and *sand float*, *colored float*, and *textured* finishes. These finishes are comparable in hardness and appearance to those obtained with job-mixed Keene's cement, but factory blending of ingredients ensures somewhat better results.

Veneer (Thin Coat) Plaster Veneer plaster is a high-strength plaster for application as a 1⁄16-in. (1.6 mm) to ⅛-in. (3.2 mm)-thick one- or two-coat finish over a special plaster base produced specifically for this purpose. Veneer plaster is a dominant type of finish in many parts of the country because it is economical and speeds up job completion.

Molding Plaster Molding plaster is a finely ground gypsum plaster used for casting in molds and for ornamental plasterwork. Its fineness is necessary for the smooth, intricate surfaces and close tolerances common in ornamental work.

Acoustical Plasters Acoustical plasters are proprietary plasters intended to increase sound absorption to reduce sound reverberation. They usually consist of ground gypsum with chemical ingredients that encourage the formation of air bubbles that increase the available surface area so that more sound is absorbed.

Some products touted as acoustical plasters are actually other plaster products with finishes made to simulate the appearance of acoustical plaster. These may offer some sound absorption due to their rough surface, but they are not true acoustical plasters.

9.1.2 PORTLAND CEMENT PLASTER

Portland cement plaster is used in both interior and exterior locations, where it will be exposed to direct wetting or high moisture. In much of the country, exterior portland cement plaster is called *stucco*, while interior portland cement plaster is called simply *plaster*. In some regions, however, especially along the West Coast, both interior and exterior portland cement plaster is referred to as stucco. Portland cement plaster is also used as a parging coat over masonry. *Parging* is often ap-

FIGURE 9.1-3 General Properties of Gypsum Plasters

Basecoat Plasters

Plaster Mix	Standard	Compressive Strength—Dry, psi (g/m²) Average	Compressive Strength—Dry, psi (g/m²) Minimum	Average Tensile Strength—Dry, psi (g/m²)	Setting Time	Characteristics
Neat (with 2 parts sand)	ASTM C 28	900 (633)	750 (527)	175 (123)	2 to 15 hrs.[a]	Greater versatility because aggregates and/or fibers are job-mixed; less control of plaster quality
Ready mixed (with perlite)	ASTM C 28	675 (475)[b]	400 (281)	155 (109)	1½ to 8 hrs.[a]	3½ times better thermal insulator than sanded plaster; closely controlled quality
Wood fibered	ASTM C 28	2100 to 2500 (1476 to 1758)	1200 (844)	400 to 500 (281 to 352)	1½ to 9 hrs.[a]	May be used without aggregate; about 50% greater surface hardness than sanded plasters; generally higher strength and fire resistance
Bond	ASTM C 28	2300 (1617)	—	400 (281)	2 to 9 hrs.	Bonds with nonporous surfaces such as concrete

Finish Coat Plasters

Plaster Mix	Standard	Compressive Strength—Dry, psi (g/m²) Average	Compressive Strength—Dry, psi (g/m²) Minimum	Setting Time	Characteristics
Gauging (with lime putty) Regular	ASTM C 28	2000 to 3000 (1406 to 2109)	1200 (844)	20 to 40 min.	Hard surface; good working properties
Special high strength	ASTM C 28	Variable	5000 (3515)	20 to 40 min.	
Gypsum Keene's cement (with lime putty)	ASTM C 61	4000 to 5000 (2812 to 3515)	2500 (1758)	20 min. to 6 hrs.	Very hard surface; becomes harder with decreased proportions of lime; less susceptible to moisture; good working properties; can be retempered; can be polished
Prepared	None at present	Variable[c]	Variable[c]	Variable[c]	Moderately hard surface; no alkali reaction with paint; not as workable as gypsum–lime putty plasters
Acoustical	None at present	Variable[c]	Variable[c]	Variable[c]	Moderately soft; better sound absorption than conventional plasters

[a]Four-hour maximum for ASTM C 842 Standard Specification for Gypsum Plastering.

[b]Not applicable to ready mix plasters for masonry bases.

[c]Proprietary plaster mixes containing different ingredients, which have variable effects on strength and setting time of plasters.

(Courtesy Gypsum Association.)

plied below ground to increase a wall's resistance to moisture penetration and to serve as a base and leveling coat to receive dampproofing and waterproofing.

9.1.2.1 General Requirements

Portland cement plaster on exteriors and in such interior spaces as shower rooms, steam baths, saunas, laundries, tub-shower enclosures, commercial kitchens, and similar locations is exposed to hazards not normally present in other interior applications. These hazards include water penetration, which may corrode reinforcement and accessories, and stresses due to variations of temperature and humidity. The presence of such hazards requires practices and precautions somewhat different from those needed in less hazardous applications.

A portland cement plaster membrane is, in effect, a thin concrete slab, built up with plastering tools and methods and having many of the desirable properties of concrete. Therefore, the discussion of portland cement in Chapter 3 will be helpful in understanding the data in this section.

Properly applied portland cement plaster membranes are hard, strong, and decorative. They can be formed to almost any shape and can be finished with a wide variety of surface textures and colors. They are resistant to fire, weather, and fungus attack. Although they are used predominantly in the southern, southwestern, and western parts of the country, they have proven durable and economical in both warm and cold climates.

9.1.2.2 Materials

The ingredients of portland cement plaster are essentially the same as those of concrete and portland cement mortar. These include aggregates, which act as inert fillers and provide strength; cementitious materials, which act as binders; water, which, together with the cementitious materials, forms a paste and initiates the reaction that binds the aggregates together into a solid mass; and admixtures, which improve the plasticity or cohesiveness of the mix or retard or accelerate the setting time.

Portland cement plaster requires the use of metal reinforcement and accessories for edge protection and crack control. Exterior portland cement plaster (stucco) also employs backing paper (or felt) and flashings.

CEMENTITIOUS MATERIALS

The cementitious materials used in portland cement plaster are portland cement, masonry cement, plastic cement, and lime. These products should conform to the requirements of the standards listed in Figure 9.1-4. Lime is discussed in Section 9.1.4.

Finishes made with plastic cement or masonry cement, or with portland cement mixed with less than 50% (by volume) of lime, are called *portland cement plaster*. When lime is added to portland cement in amounts greater than 50% by volume, the finish is referred to as lime-cement (or *portland cement-lime plaster*). The principles and methods described here apply to both, whether used on the interior or the exterior of a building.

Portland Cement The types of portland cements suitable for plaster (see Fig. 9.1-4) are also used in concrete; these are discussed in detail in Section 3.1.

FIGURE 9.1-4 Summary of Portland Cement Plaster Materials

Material	Standard Specification
Cementitious Material	
Portland cement Types I, II, III	ASTM C 150
Types IA, IIA, IIIA	ASTM C 150
Masonry cement	ASTM C 91
Plastic cement	ASTM C 150
Lime, hydrated* Special finishing Normal finishing	ASTM C 206
Mason's, Types S and N	ASTM C 207
Quicklime	ASTM C 5
Aggregates	
Sand	ASTM C 144
Lightweight	ASTM C 332
Metal Reinforcement	ASTM C 847
Backing Paper	
Roofing felt	ASTM D 226
Building paper	FS-UU-B-790

*Both special and normal limes are available as air-entraining types (SA and NA).

(Courtesy Gypsum Association.)

Type I (normal) portland cement is used for most portland cement plaster work, except where the special properties of Types II and III are required. Types IA, IIA, and IIIA (air-entraining) portland cement provide improved workability and resistance to bleeding (concentration of water and fines at the surface) and to damage from freezing.

Type I is available in natural gray or white. Truer colors can be obtained in finish coats containing integral coloring or colored aggregate by using a white portland cement. *Waterproof* white portland cement is also available. It contains moisture-resistant additives to increase its inherent moisture resistance. These products must contain lesser amounts of iron oxides than standard portland cement, but in all other respects should conform to the requirements of ASTM Standard C 150 for Type I portland cement.

Type II (moderate heat) portland cement has greater resistance to decomposition from the action of sulfates and lower heat of hydration. Resistance to sulfate attack may be a useful property when portland cement plaster is applied to brick or concrete masonry units containing soluble sulfate materials.

Type III (high early strength) portland cement develops strength faster in the first few days after application. Therefore, Type III may be desirable in cold weather to minimize the hazard of freezing damage and to speed application of successive coats.

Masonry Cement and Plastic Cements Masonry cement and plastic cements consist primarily of portland cement with varying amounts of such materials as natural cements, finely ground limestone, air-entraining agents, plasticizers, and sometimes lime, which makes the field addition of lime unnecessary. They are used extensively for portland cement plaster finishes in some parts of the country.

Plastic cement is produced primarily in the southwestern United States. It is similar in properties and formulations to Types I and II portland cement, but may contain larger amounts of air-entraining agents, insoluble residues, and plasticizing agents not in excess of 12% of the volume. With the exception of these variations, plastic cement should conform to the requirements of ASTM Standard C 150 for Type I and Type II portland cement.

Prepared Finish Coat Factory-prepared mixes (stucco finishes) are available for application over basecoats. These prepared mixes contain all the necessary finish-coat ingredients, except water. They are available in white or a variety of colors and, in general, provide more satisfactory results than those obtained by the addition of pigments in the field.

9.1.3 AGGREGATES

9.1.3.1 Gypsum Plaster Aggregates

The aggregates commonly used in connection with gypsum products are sand, vermiculite, and perlite.

SAND

Most sands used in gypsum plaster are primarily quartz and silica, with small amounts of mica, feldspar, clay, and other minerals. Natural deposits throughout the United States are the main sources of plastering sands. The sand is graded according to particle size and is often washed to remove impurities. Artificially produced sand (stone screenings) of the proper size and type is also suitable.

The natural characteristics of sands vary according to location and must be controlled to ensure their suitability as plaster aggregates. Standards limiting size, gradation, and percentage of impurities, and testing procedures to measure these factors, have been established by the ASTM and the American National Standards Institute (ANSI). In general, rough, sharp, angular sand particles produce stronger plasters than smooth, round ones.

LIGHTWEIGHT AGGREGATES

Lightweight aggregates used in gypsum plaster include vermiculite and perlite.

Vermiculite Vermiculite consists of silica, magnesium oxide, aluminum oxides, and other minerals, combined with 5% to 9% water by weight. In the United States, vermiculite ore is mined primarily in Montana and South Carolina. This micaceous mineral is composed of a series of parallel layers that expand under the pressure of water trapped between them when subjected to intense heat (Fig. 9.1-5). The name originates from the vermicular (wormlike) movement of the layers during expansion.

Processing consists of crushing the raw ore, grading according to particle size, and removing impurities. Ore particles are then heated at 1600 to 2000°F (871 to 1093°C) in a furnace for 4 to 8 seconds to produce expansion, and are cooled rapidly to impart pliability and toughness to the expanded particles. As the material cools, it is air-lifted to remove unexpanded rock particles and to classify particle sizes. The expanded material is soft and pliable, silvery or gold in color, and contains less than 1% water by weight. Expanded vermiculite is classified into five types according to particular size. Type II, the lightest of the standard aggregates, is used in plaster products.

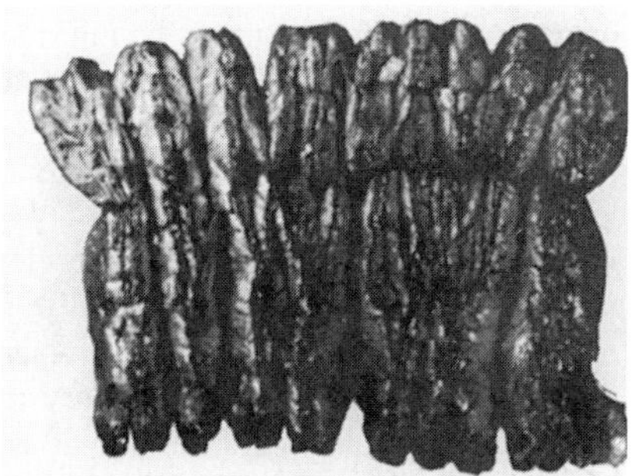

FIGURE 9.1-5 Magnified expanded vermiculite.

FIGURE 9.1-6 Perlite: (a) crude ore, (b) crushed, (c) expanded.

Perlite Raw perlite is a volcanic ore consisting primarily of silica and alumina, with 4% to 6% combined water by weight. It is quarried mostly in the western part of the United States.

Perlite ore is processed by crushing, screening, and grading according to size. It then is expanded in a furnace at 1400 to 2000°F (760 to 1093°C). At this temperature the glassy ore particles approach their melting point and begin to soften. A small amount of the combined water is converted to steam, which pops the particles into frothy glasslike bubbles, up to 20 times the original volume (Fig. 9.1-6). After expansion, an air blast separates and grades the particles according to size.

9.1.3.2 Portland Cement Plaster Aggregates

Aggregates constitute most of the volume of a portland cement plaster mix. One cu. ft. (0.028 m^3) of a typical portland cement plaster mix contains about 0.97 cu. ft. (0.027 m^3) of aggregate and 0.25 to 0.33 cu. ft. (0.007 to 0.009 m^3) of cementitious material. Therefore, the aggregate must be clean, sound, and well graded, with particles ranging from fine up to a maximum size of ⅛ in. (3.18 mm).

The best and most practical mix results when the aggregate size is proportioned so that a minimum of water-cement paste is required to coat the aggregate and fill the voids between particles. An excess of either fine or coarse particles can result in a mix containing too much cement or too much water, either of which can be detrimental to a portland cement plaster finish. Portland cement plaster mixes are discussed in Section 9.4.1.

SAND

Sand aggregate for portland cement plaster should consist of natural or manufactured sand that is clean, granular, and free from deleterious amounts of loam, clay, soluble salts, and vegetable matter. It should be well graded within the limits given in Figure 9.1-7, and, with the exception of grading, should conform to the requirements of ASTM Standard C 144.

LIGHTWEIGHT AGGREGATE

Expanded shale, clay, slate, and slag are sometimes used where sand is not available locally. Vermiculite and perlite have been used in portland cement plaster where lightweight, insulating properties and fire resistance are prime factors. Proportioning of these materials is determined by laboratory test or by making test panels. Thermal insulating properties, gradation, and unit weight for lightweight aggregate should conform to the requirements of ASTM Standard C 332.

9.1.4 LIME

Lime is used in both gypsum and portland cement plasters. It is obtained principally from limestone, a common mineral con-

FIGURE 9.1-7 Grading for Sand Aggregate

Sieve Size	Percent of Aggregate by Weight Retained on Sieve	
	Minimum	Maximum
No. 4	—	0
No. 8	—	10
No. 16	10	40
No. 30	30	65
No. 50	70	90
No. 100	95	100
No. 200	97	100

sisting of calcium carbonate (calcite) and magnesium carbonate (dolomite). To produce building lime, crushed limestone is calcined in a kiln at 2000°F (1093°C). Limestone used to make lime for construction purposes comes from dolomitic deposits in Ohio or calcitic deposits in Arizona, California, Missouri, Pennsylvania, Texas, and Virginia.

The physical and chemical properties of lime depend on the limestone from which it was derived. Building limes may be manufactured from *dolomitic* limestone, which normally contains substantial amounts of magnesium carbonate, or from *calcitic* limestone, naturally high in calcium carbonate. When dolomitic limestone is calcined during manufacture, magnesium oxides result. These oxides produce a plastic putty with better working properties than those of the putty made from the calcium oxides derived from calcitic limestone, but it takes longer to hydrate than the putty made with calcium oxides.

Building limes can be classed as (1) *quicklime* or (2) *hydrated lime*.

9.1.4.1 Quicklime

During the calcining process, limestone gives off carbon dioxide and changes from the carbonate form to calcium oxide or magnesium oxide. *Quicklime* was used as the chief binder in lime plasters for several hundred years, before the introduction of controlled-setting gypsum plasters.

Before quicklime can be used in plaster, it must be *slaked* (combined with water) and soaked for 1 to 2 weeks to hydrate its oxides and make a plastic putty usable in mortar or plaster. Slaking initiates a chemical reaction that transforms an oxide to a stabler hydroxide state. The type of raw material affects the degree and speed of hydration possible during this soaking period. Quicklime identified as calcitic, or high-calcium, can be expected to have a greater hydration of oxides. It is acceptable for use in plaster only if it is identified as *calcitic*, or *high-calcium*, and if it is properly slaked as recommended by its manufacturer. For this reason, some quicklime is sold in the natural pulverized form, but, to eliminate the slaking operation, most is partially hydrated (slaked) at the mill.

9.1.4.2 Hydrated Lime

The material produced by partially hydrating quicklime at the mill is called *hydrated lime*. The addition of water to hydrated lime results in *lime putty*, a plastic form that is combined with gypsum and water to make some finish coat plasters.

If oxides are not completely hydrated during soaking or during the manufacture of hydrated lime, they may be present in the plaster. Under conditions of high humidity these oxides can hydrate there, expand, and crack the plaster containing them. Therefore, dolomitic limes are often autoclaved at the mill to ensure that a minimum of unhydrated oxides remain in the manufactured hydrated lime. To be suitable for use in finish coat plasters, lime must have no more than 8% unhydrated oxides.

Available hydrated limes include *mason's* hydrated lime, intended primarily for masonry mortars, and *finishing* hydrated lime, intended for use in plasters. Both are available in Type S (special), Type SA (special air-entraining), Type N (normal), and Type NA (normal air-entraining).

TYPE N (NORMAL) HYDRATED LIME

Type N hydrated lime is produced in accordance with the requirements of ASTM Standard C 206, except that it has no limitation on unhydrated oxides. However, when calcitic limestone is used in its manufacture, the resulting product (identified as calcitic or high-calcium finishing hydrated lime) consists of at least 95% calcium oxide, which hydrates readily during manufacture. Some oxides may remain unhydrated, but these can be expected to hydrate substantially during the required 12- to 16-hour preuse soaking period for Type N limes. Often, high calcium finishing lime will meet ASTM limitations for unhydrated oxides for Type S lime but may not develop the necessary degree of plasticity within the time limit established by ASTM.

Type N hydrated finishing lime produced in accordance with ASTM Standard C 206 is acceptable for use in plaster if it is identified as calcitic or high-calcium and has been properly soaked as recommended by its manufacturer.

TYPE S (SPECIAL) HYDRATED LIME

Type S (special) finishing hydrated lime conforming with the requirements of ASTM Standard C 206 requires no preliminary soaking and can be mixed with other ingredients before water is added. Because it is easy to use, and because unhydrated oxides are limited during manufacture, Type S lime is most often used and is specifically recommended for plaster mixes.

Type S hydrated lime conforming to ASTM Standard C 206 must contain not more than 8% unhydrated oxides, regardless of the raw materials used.

9.1.5 WATER

Water for mixing and curing should be clean and free from deleterious amounts of oils, acids, alkalies, salts, and organic substances that could adversely affect proper hydration or cause corrosion of metal reinforcement and accessories.

The presence of water promotes hydration of portland cement and carbonation of lime, the chemical processes by which these materials harden. However, more water is generally added to a mix than is necessary to provide for hydration and carbonation, in order to bring the mix to a workable consistency. The amount of water added for workability depends on the characteristics of the cementitious material, gradation of aggregate, suction of the supporting construction, and drying conditions.

9.1.6 ADMIXTURES

Admixtures, called plasticizers, such as mineral, animal, vegetable, or glass fibers, are sometimes added to a plaster mix to improve its plasticity. These plasticizers can be added to the first (*scratch*) coat where greater cohesiveness of the mix is required, particularly for ceilings and soffits, and to reduce the amount of mix lost through openings in metal reinforcement when backing paper is not used. Mineral fibers may also be used for machine-applied basecoats to prevent segregation of the aggregate and to provide lubrication in the hose.

Fibers should be clean and free from harmful amounts of any substance, such as unhydrated oxides, that may be injurious to the plaster ingredients. Animal hair and vegetable fibers should be ½ to 2 in. (12.7 to 50.8 mm) in length and free from grease, oil, dirt, and other impurities.

Other admixtures intended to accelerate or retard setting time are generally not recommended, but may be used where local practice indicates satisfactory results. When used, admixtures should not reduce the compressive strength of the mix more than 15% below the strength of a comparable mix without admixtures.

9.2 Plaster Support Systems, Bases, and Accessories

Before applying a plaster finish, it is necessary to select, install, and prepare the supporting construction, select and apply a plaster base (Fig. 9.2-1), and select and install proper accessories.

9.2.1 SUPPORT SYSTEMS

Both gypsum and portland cement plasters can be applied directly to some concrete and masonry surfaces under certain conditions. In such cases, the concrete or masonry serves as both support system and plaster base (see Section 9.2.2).

Supporting construction must support design loads without excessive deflection and without introducing stresses that may cause finishes to crack. The calculated deflection of ceiling members should not exceed 1⁄360 of the span after the plaster has been applied. Headers or lintels of adequate size should be provided over openings, and all construction should be true and plumb or level.

Supporting construction should be strong and stable, but also capable of accommodating structural movement. Concrete and masonry construction should have control and expansion joints as dictated by good design practice. Control joints in plaster should be placed directly over joints in the supporting construction and in other locations where required to relieve stresses and minimize cracking (see Section 9.4.4).

FIGURE 9.2-1 Applying plaster over a gypsum plaster base. (Courtesy United States Gypsum Company.)

9.2.1.1 Metal Framing and Furring

The thermal insulating and fire-resistant properties of lightweight aggregates, such as perlite and vermiculite, have encouraged the use of portland cement plaster in commercial and institutional structures, especially where fire ratings are required. Lightweight aggregates are not recommended for exterior plaster, however, so they are used primarily for backplaster coats on exterior walls and for fireproofing.

Both gypsum and portland cement plaster are supported by metal framing and furring in interior locations where high-quality construction or fire-resistance ratings are required.

Assemblies consisting of steel studs, furring channels (C-shaped and hat-shaped), and reinforcing rods should be spaced and securely fastened according to the specialty manufacturer's recommendations and the recommendations of the Metal Lath/Steel Framing Association Division of the National Association of Architectural Metal Manufacturers (NAAMM) publication ML/SFA 920, "Guide Specifications for Metal Lathing and Furring." The recommended spacing of metal supports for various types of reinforcements closely parallels that of wood (Fig. 9.2-2). Metal stud framing is as described in Section 9.5 for support of gypsum board.

9.2.1.2 Wood Framing and Furring

Both gypsum and portland cement plaster can be installed over wood framing and furring (see Chapter 6).

Gypsum plaster can be installed over wood framing and furring with either a gypsum lath base or metal reinforcement. Portland cement plaster can be installed over wood framing and furring only using metal reinforcement.

Exterior wood frame construction is either unsheathed (open-frame) or sheathed. Exterior wall studs will usually be as follows:

- Supporting a ceiling and roof only: 2 × 4s (50 × 100s) at 24 in. (610 mm) on centers
- Supporting 1 or 2 floors: 2 × 4s (50 × 100s) at 16 in. (406 mm) on centers
- Supporting more than 2 floors: 2 × 6s (50 × 150s) at 16 in. (406 mm) on centers

UNSHEATHED CONSTRUCTION

The structural frame should be properly braced, by let-in 1 × 4s (25 × 100s) or larger diagonal bracing or by other means, to resist racking forces. Paper-backed

FIGURE 9.2-2 Support Spacing for Metal Reinforcement[a]

Reinforcement	Weight, lb./sq. yd. (kg/m²)	Support Spacing, in. (mm) on centers
Diamond mesh	3.4 (1.8)	16 (406)
Stucco mesh[b]	1.8 and 3.6 (.98 and 1.95)	16 (406)
Flat rib lath	3.4 (1.8)	19 (483)[c]
3⁄8-in. rib lath	3.4 and 4.0 (1.8 and 2.17)	24 (610)
Woven wire	1.7 (.9) (1 in. hex.) (25.4 mm hex.)	16 (406)
	1.4 (.96) (1½ in. hex.) (38 mm hex.)	16 (406)
Welded wire	1.4 (.96) (1 in. × 1 in.) (25.4 × 25.4 mm)	16 (406)
	1.4 (.76) (2 in. × 2 in.) (50.8 × 50.8 mm)	16 (406)
	1.9 (1) (1½ in. × 2 in.) (38 × 50.8 mm)	16 (406)[d]

[a]Support spacing applicable to both wood and metal framing for both vertical and horizontal surfaces, unless otherwise noted.

[b]Not recommended for vertical surfaces over metal framing or for horizontal surfaces.

[c]Spacing of metal framing for horizontal surfaces should not exceed 13½ in. (343 mm) on centers.

[d]Spacing may be increased to 24 in. (610 mm) on centers when stiffener wires are provided 6 in. (152 mm) on centers.

reinforcement can be applied directly. If backing paper is applied separately, a soft annealed 18-gauge (1.3 mm) or heavier steel line wire should be stretched across the faces of the studs in horizontal strands about 6 in. (152.4 mm) apart. The wire should be stretched taut by first nailing or stapling it at every fifth stud, then securing it to the center of intermediate studs and stretching it tightly by raising and lowering the attachments (Fig. 9.2-3a). Backing paper then is nailed over the line wire with edges lapped at least 3 in. (76.2 mm) to prevent the paper from sagging and produce a more uniform thickness of portland cement plaster that is less likely to crack.

Application of flexible reinforcement should start at least one full stud away from corners and should be bent around them to avoid a joint at the corner. Prefabricated corner reinforcement should be used when reinforcement is too rigid to be bent easily. Horizontal laps should be wired at least once between each stud space but not more than 9 in. (229 mm) on centers. Vertical laps should be made over supports. The bottom of a plaster panel should be terminated with a metal drip screed to help support the plaster and permit drainage of water that might otherwise penetrate the plaster (Fig. 9.2-4).

As plaster is applied, it should be forced through the openings in the reinforcement against the paper backing to ensure at least 90% coverage of the reinforcement. An alternate method sometimes used to ensure complete embedment of the reinforcement is to eliminate the backing paper and back plaster the reinforcement between supports. This back-plaster coat should be at least ⅜ in. (9.5 mm) thick and should be applied only after the scratch coat has hardened sufficiently so that pressure will not break the plaster keys. The surface should be uniformly dampened before application and moist-cured for at least 24 hours after application.

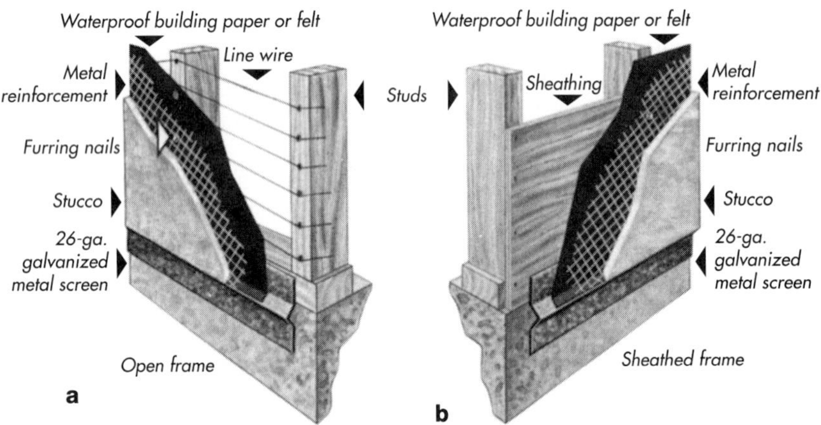

FIGURE 9.2-3 (a) Unsheathed (open-frame) construction; (b) sheathed construction. Line wire is not required with paper-backed reinforcement. (Courtesy Portland Cement Association.)

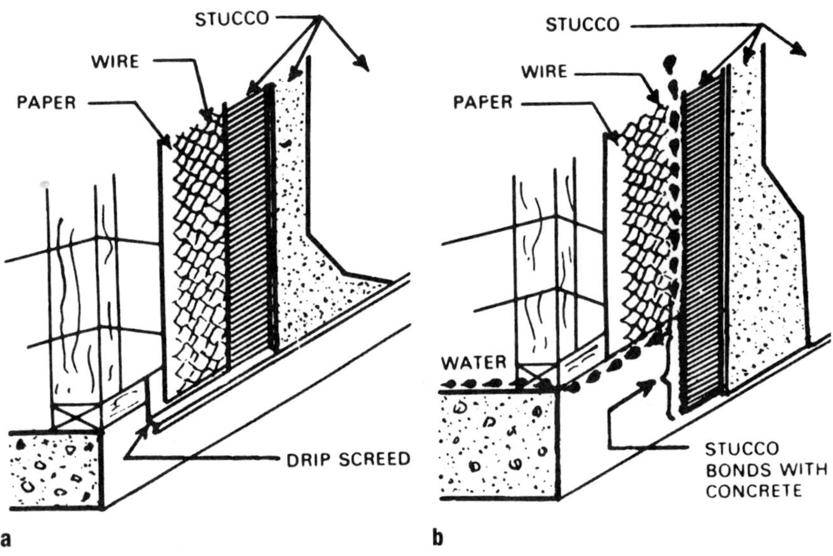

FIGURE 9.2-4 (a) Properly installed drip screed allows drainage and reduces the possibility of water penetration into a wall; (b) if drip screed is omitted, bond between concrete and portland cement plaster (stucco) may result in moisture backup. (Courtesy Portland Cement Association.)

SHEATHED CONSTRUCTION

Wood boards, plywood, or other suitable panel sheathing should be applied horizontally and fastened securely to each stud (see Fig. 9.2-3b). Lumber boards used for sheathing should not be more than 8 in. (203 mm) wide and should be fastened with at least two nails at each support to minimize the possibility of damage from warping. The sheathing should be covered with backing paper, lapped at least 3 in. (76.2 mm), followed by metal reinforcement fastened through the sheathing into the wood supports. Paper-backed reinforcement can be attached directly over the sheathing. The reinforcement should be returned at least 6 in. (152.4 mm) around corners. Vertical laps over supports and horizontal laps should be nailed or stapled at 6 in. (152.4 mm) on centers into the sheathing. Unless the reinforcement is self-furring, nails should be of the furring type; staples can be used for self-furring reinforcement.

9.2.2 PLASTER BASES

Plaster bases are constructions or surfaces that develop adequate bond for the direct application of plaster. Bond can be provided by mechanical key or by suction. Mechanical keying results from the interlocking of a plaster mix with openings or projections in the base surface; suction develops bond by drawing part of the plaster paste into minute pores in the base surface by capillary action. Over metal reinforcement, bond is developed by mechanical keying; over concrete and masonry, bond is developed by mechanical keying, suction, or both.

Both gypsum and portland cement plaster can be applied over metal reinforcement and directly over some concrete, concrete masonry units (CMU), and brick. Gypsum plaster can also be applied directly to gypsum products such as other gypsum plaster, gypsum masonry, and gypsum lath, but these bases do not provide adequate bond for portland cement plaster. Metal lath should be installed over such materials before portland cement plaster is applied; backing paper (see Section 9.2.2.3) should also be used in exterior locations. Wood lath, fiber insulating lath, and sheet metal lath should not be used as a base for portland cement plaster.

Gypsum veneer plaster is applied over *veneer plaster base* (see Section 9.5).

9.2.2.1 Concrete and Masonry

Both gypsum and portland cement plaster can be applied directly over concrete and over CMU and brick that have sufficient bonding capabilities. However, gypsum plaster should not be applied directly to concrete or masonry surfaces that are the interior surfaces of exterior walls or the opposite surfaces of walls or partitions that are exposed to wet conditions, such as would be found in a sauna, shower, or tub and shower enclosure.

CONCRETE

Before plaster is applied directly to concrete, sandblasting should be used to clean the surface of laitance, form oil, and other substances that may impair bond. Smooth or slick spots on fresh concrete caused by form oil can also be removed by scrubbing with strong soap and water and allowing the concrete to dry to restore suction. However, form oil should not be used for concrete forms when direct application of plaster is intended.

Rough lumber can be used for the forms of concrete that are to receive plaster, or the freshly placed concrete can be lightly sandblasted to roughen the surface immediately after the forms have been stripped. Hardened, smooth concrete may require more extensive preparation such as chipping, acid etching, or roughening with a bush hammer. Metal reinforcement should be used when gypsum plaster is to be applied over smooth concrete where roughing is not practical. A bonding agent or a dash coat may also be used where portland cement plaster is to be applied and roughening of the surface is impractical.

Dash Coat When it is doubtful that portland cement plaster can obtain a good bond with concrete or masonry, it is sometimes advisable to use a dash bond coat, consisting of 1 part by volume of portland cement and 1 to 2 parts sand. A small straw broom or long-fibered stiff brush is dipped in the mix, and the material is splattered on the wall with a quick throwing motion (Fig. 9.2-5). A dash coat should not be troweled and should be allowed to set before the plaster base coat is applied. If there is a delay in applying the base coat, the dash coat should be kept moist until the next coat is applied.

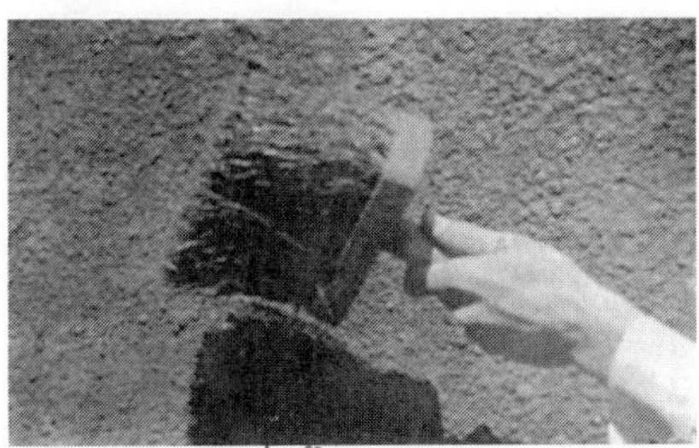

FIGURE 9.2-5 A dash bond coat being applied. (Courtesy Portland Cement Association.)

Bonding Agent Another method of obtaining good bond on concrete and masonry surfaces is to apply a waterproof bonding agent recommended by its manufacturer for exterior use. The supporting surface should be structurally sound and rigid, and free from laitance, dust, grease, and materials that are soluble in water, such as water-based paints or glue.

Acid Etching A solution of 1 part muriatic acid and 6 parts water can be used for acid etching of slick concrete where a dash coat or bonding agent might not adhere well. Before an acid solution is applied, the surface should be wetted so that the acid will act only on the surface. Several applications may be necessary. After treatment, the concrete should be thoroughly washed with water to remove all traces of the acid and be allowed to dry to restore suction.

MASONRY

When plaster is to be applied directly to masonry, mortar joints should be struck flush or slightly raked. The masonry surface should be clean, sound, and firm and should contain no efflorescence, grease, waterproofing compounds, paint, or other substances detrimental to good bond. Efflorescence, oil, and grease can be removed by washing with a 10% solution of muriatic acid and water. Other coatings or substances that will adversely affect bond should be removed with appropriate solvents and rinsed clean.

Medium-hard common and face brick, standard-weight concrete block, and many types of stone generally provide adequate suction and, often, mechanical bond. The surface can be tested for suction by spraying it with clean water to see how quickly the moisture is absorbed. If the water is readily absorbed, good suction is likely to result; if water droplets form, the surface does not provide adequate suction and bond will depend on mechanical key. It can be determined by observation whether the surface is sufficiently rough to ensure mechanical keying with the plaster. Glazed brick will probably be too slick, but standard-weight CMU will usually be rough enough.

On surfaces that absorb moisture too quickly, such as soft common brick and lightweight CMU, portland cement plaster will quickly stiffen and become difficult to apply properly. Excessive suction can be controlled by fine-spraying (not soaking) the masonry with several applications of water immediately before applying the first plaster coat. The surface should appear slightly damp, but there should be no free water present when the plaster is applied. When stucco is machine-applied, it may not be necessary to dampen the surface, because the mix usually contains more water than mixes used in hand application.

Hard-burnt brick, glazed concrete masonry units, and some types of stone may provide poor or nonuniform bond. If suction is not uniform over the entire surface, parts of the wall may draw more moisture than others, and the final finish will be spotty in color. Moreover, nonuniform bonding is likely to cause cracking, because the plaster will move with the supporting base in the areas of good bond and independently of the base in areas of poor bond.

If an adequate or uniform bond cannot be obtained, the surface should be prepared with metal reinforcement, a dash coat, or a bonding agent. When metal reinforcement is used with portland cement plaster, backing paper should also be used to isolate the plaster membrane from the base and prevent dissimilar suction. When plaster is to be applied as a continuous finish over dissimilar rigid bases, a control joint or a strip of metal reinforcement should be placed over the juncture of the dissimilar materials.

Plaster should not be applied directly

a

b

FIGURE 9.2-6 (a) Self-furring wire lath is crimped every 3 in. (75 mm); (b) fiber spacers on furring nails keep plain wire lath ¼ in. (6.4 mm) away from the backing paper.

to a masonry chimney, but should be applied over metal reinforcement. However, backing paper should not be used, because heat may cause deterioration. Flashings should be installed under the chimney cap and at junctures with the roof to prevent water penetration (see Chapter 4).

9.2.2.2 Metal Reinforcement

Metal reinforcement may be used in lieu of gypsum lath for gypsum plaster installed over wood or metal framing or furring. It should also be used in gypsum plaster over concrete or masonry that does not provide sufficient bond for direct plaster application.

Metal reinforcement should also be used when portland cement plaster is applied over either sheathed or unsheathed wood frame construction, or steel frame construction, concrete and masonry providing unsatisfactory bond, and at flashings and chimneys.

Plaster applied over metal reinforcement acts as a thin reinforced slab which, although supported by the underlying construction, is not in sufficiently rigid contact to permit direct transfer of stresses. The reinforcement should be attached securely to supporting construction and stretched tightly between supports to eliminate slack. Loosely attached reinforcement generally results in an uneven thickness of plaster, with the thinner and weaker sections more likely to crack. At least a ¼-in. (6.4 mm) space should be maintained between the reinforcement and the supporting construction by use of either furring fasteners or self-furring reinforcement (Fig. 9.2-6). The metal reinforcement should be of suitable type and weight for the support spacing (see Fig. 9.2-2) and should be attached with fasteners recommended for the particular supporting construction (Fig. 9.2-7).

TYPES

The most common commercially available types of metal reinforcements are *expanded metal lath* and *wire lath* (Figs. 9.2-8 and 9.2-9). Openings in the lath should be large enough to permit the scratch coat to be forced through the openings and to embed the reinforcement, thereby preventing corrosion. Metal reinforcement should conform to the standards listed in Figure 9.1-4 and the weight requirements given in Figure 9.2-9.

Expanded Metal Lath Expanded metal lath is formed from light-gauge steel sheets that have been cut in a regular pattern and expanded (stretched) to form diamond-shaped openings (Fig. 9.2-8). Both expanded zinc-coated (galvanized) steel sheet and cold-rolled steel sheet, coated with corrosion-resistant paint after fabrication, are used.

Plain, *self-furring*, and *paper-backed* types of diamond mesh are produced in sheet form. The plain type generally should be installed with furring fasteners. Diamond mesh is produced in weights of 2.5 and 3.4 lb./sq. yd. (1.36 and 1.84 kg/m^2); only the heavier 3.4 lb. (1.84 kg/m^2) weight is suitable for exterior portland cement plaster use.

The *self-furring* type is dimpled or contains other devices that hold it away from the surface to which it is attached. These devices should keep the back of the lath at least ¼ in. (6.4 mm) away from the supporting construction. Self-furring lath

FIGURE 9.2-7 Fastening of Reinforcement over Wood Supports[a]

	Nails		Staples	
Reinforcement	**Type[b]**	**Maximum Spacing, in. (mm) on centers**	**Type**	**Maximum Spacing, in. (mm) on centers**
Diamond mesh		6 (150)		6 (150)
Stucco mesh[c]		6 (150)	⅞-in. (22.7) leg, 16-gauge (1.6), ¾-in. (19.1) crown	6 (150)
Flat rib lath	1½-in. (38 mm), 11-gauge (3.9), barbed, 7/16-in. (11) head	6 (150)		6 (150)
⅜ in. (9.5 mm) rib lath		6 (150)[d]	1¼-in. (28.6) leg, 16-gauge (1.6), ¾-in. (19.1) crown	4½ (114)
Woven and welded wire		6 (150)	⅞-in. (22.2) leg, 16-gauge (1.6), 7/16-in. (11) crown	6 (150)

[a]Applicable to both vertical and horizontal surfaces, unless otherwise noted.

[b]1-in. roofing nails suitable on vertical surfaces for all but ⅜-in. (9.5 mm) ribbed lath, 1½-in. (38 mm), 12-gauge (2.7 mm), ⅜-in. (9.5 mm) head furring nails suitable to attach welded and woven wire on vertical surfaces.

[c]Not recommended for horizontal surfaces.

[d]Maximum spacing 4½ in. (114 mm) on centers for horizontal surfaces using 2-in. (50.8 mm) nails.

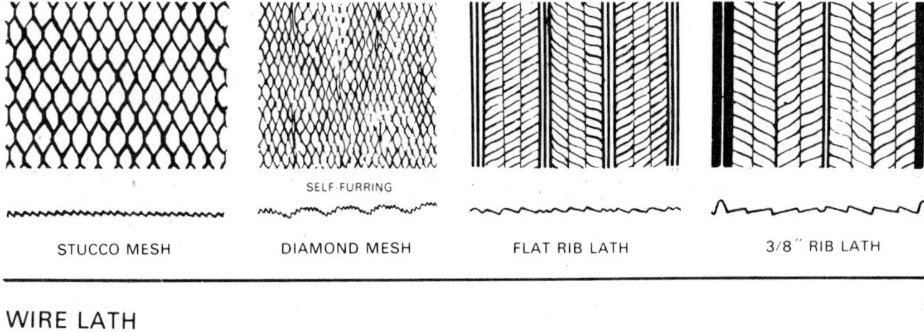

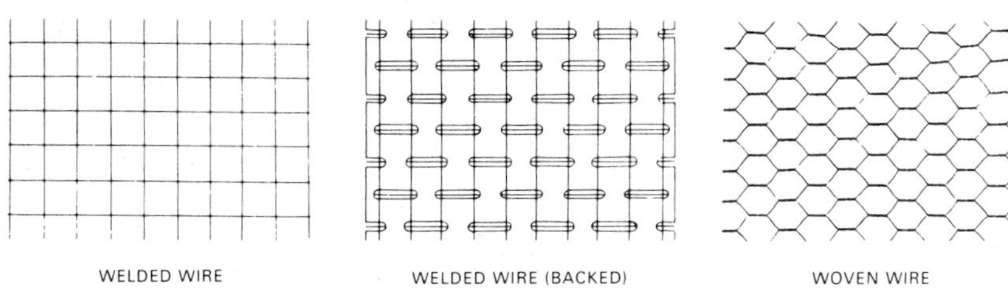

FIGURE 9.2-8 Metal reinforcement suitable for portland cement plaster (stucco) finishes; all but stucco mesh and ⅜-in. (9.5 mm) rib lath are available in plain, self-furring, and paper-backed types.

FIGURE 9.2-9 Metal Reinforcement for Portland Cement Plaster (Stucco) Finishes

Type	Weight, lb./sq. yd. (kg/m²)	Opening Size, in. (mm)	Dimensions
Expanded Metal Lath			
Diamond mesh[a]	3.4 (1.8)	5⁄16 × ⅜ (7.9 × 9.5)	27 in. × 96 in. (686 mm × 2438 mm)
Stucco mesh	1.8 and 3.6 (0.98 and 1.95)	1⅜ × 3⅛ (34.9 × 79.4)	48 in. × 99 in. (1219 mm × 2515 mm)
Flat rib lath	3.4 (1.8)	5⁄16 × ⅜ (7.9 × 9.5)	24 in. × 96 in. (610 mm × 2438 mm) and 27 in. × 96 in. (686 mm × 2438 mm)
⅜-in. (9.5 mm) rib lath	3.4 and 4.0 (1.8 and 2.17)	5⁄16 × ⅜ (7.9 × 9.5)	27 in. × 96 in. (686 mm × 2438 mm) and 27 in. × 99 in. (686 mm × 2515 mm)
Wire Lath			
Woven wire[a]	1.7 (0.92) (18 gauge) (1.3 mm)	1 (25.4) hex.	3 ft. × 150 ft. (.91 m × 45.7 m) (rolls)[b]
	1.4 (0.76) (17 gauge) (1.45 mm)	1½ (38) hex.	
Welded wire[a]	1.4 (0.76) (16 gauge) (1.6 mm)	2 × 2 (50.8 × 50.8)	3 ft. × 150 ft. (.91 m × 45.7 m) (rolls)[b]
	1.4 (0.76) (18 gauge) (1.3 mm)	1 × 1 (25.4 × 25.4)	
	1.9 (1) (16 gauge) (1.6 mm)	1½ × 2 (38 × 50.8)	

[a]Available in plain, self-furring, and paper-backed types.
[b]Paper-backed and self-furring types also available in sheet form.
(Courtesy Portland Cement Association.)

can be nailed directly to smooth reinforced concrete, masonry, and other solid surfaces to receive plaster.

Paper-backed diamond mesh, with paper strips attached to the back surface, is produced primarily for machine application of exterior portland cement plaster. Spaces are provided between the paper strips to permit fastening to studs or other supporting construction.

Stucco mesh is a metal lath similar in pattern to diamond mesh, but with larger openings. It is designed primarily as a base for plaster over solid surfaces. It should be installed with backing paper and furring fasteners.

Rib lath is a type of expanded metal lath that consists of integrally formed V-shaped ribs that provide greater stiffness and permit wider spacing of supports (see Fig. 9.2-2). Laths with ⅛-, ⅜-, and ¾-in. (3.2, 9.5, and 19.1 mm) ribs are produced (see Fig. 9.2-9). The ⅛-in. (3.2 mm) rib lath referred to as *flat rib lath* generally requires furring fasteners and can be used for most plaster applications. It is available in weights of 2.75 and 3.4 lb./sq. yd. (1.49 and 1.84 kg/m²), but only the heavier weight is acceptable as a base for exterior portland cement plaster.

The ⅜-in. (9.5 mm) rib lath does not require furring but is not recommended for severe weather exposure because it is difficult to encase the lath in plaster. It can be used, however, in milder climates and in locations protected from direct wetting, such as at soffits. Its chief advantage is the ability to provide a base for support spacings up to 24 in. (610 mm) on centers (see Fig. 9.2-2).

Wire Lath Wire lath is produced as woven or welded wire fabric. Wire lath should be fabricated from copper-bearing galvanized cold-drawn steel wire of the appropriate gauge (see Fig. 9.2-9). It is used extensively as a base for exterior portland cement plaster.

Woven wire fabric (stucco netting) is a hexagonal woven wire fabric made by interweaving wire into hexagonal openings 1 in. (25.4 mm) to 2¼ in. (57.2 mm) wide (see Fig. 9.2-8). The plain type is nailed to supporting construction with furring nails over backing paper. Woven wire fabric is also available in a self-furring type, with or without backing paper, which eliminates the need for furring fasteners or separate installation of backing paper, depending on the type of lath used. Minimum wire size for 1-in. (25.4 mm) openings should be 18 gauge (1.3 mm); 17 gauge (1.45 mm) for $1\frac{1}{2}$-in. (38 mm) openings (see Fig. 9.2-9).

Welded wire fabric lath consists of a grid formed by welding steel wire at intersections into 1-in. (25.4 mm) by 1-in. (25.4 mm) and 2-in. (50.8 mm) by 2-in. (50.8 mm) squares. It is available with backing paper and stiffener wires and in a self-furring type. This fabric should be galvanized, coated with corrosion-resistant paint after welding, or fabricated from galvanized wire. Wire should be at least 16 gauge (1.6 mm) for fabric with 2-in. (50.8 mm) by 2-in. (50.8 mm) openings, and at least 18 gauge (1.3 mm) for fabric with 1-in. (25.4 mm) by 1-in. (25.4 mm) openings.

INSTALLATION

Reinforcement should be attached to concrete and masonry with suitable fasteners spaced 6 in. (152 mm) on centers vertically and 16 in. (406 mm) on centers horizontally. Horizontal and vertical laps should be secured with fasteners or wire ties 6 in. (150 mm) on centers. Self-furring reinforcement can be attached with powder-actuated fasteners or concrete stub nails; plain reinforcement should be attached with concrete furring nails. These nails should be long enough to provide at least $\frac{3}{8}$-in. (9.5 mm) penetration into the concrete or masonry.

Generally, metal reinforcement should be installed over framing or furring with the long dimension perpendicular to the supports and with end laps staggered, forming a continuous network of metal over the entire surface. Expanded metal lath should be lapped $\frac{1}{2}$ in. (12.7 mm) at the sides and 1 in. (25.4 mm) at the ends. Edge ribs should be nested. Wire lath should be lapped at the ends and sides at least one full mesh but not less than 1 in. (25.4 mm). Generally, lower sheets should be lapped over upper sheets at horizontal laps; however, paper-backed reinforcement should be applied shingle fashion with upper sheets sandwich-lapped over the lower sheets with metal on metal.

Metal reinforcement should be attached to wood framing and furring with corrosion-resistant nails or staples, sized and spaced as shown in Figure 9.2-7. Nails generally should provide at least $\frac{7}{8}$-in. (22.2 mm) penetration into supporting construction on walls and $1\frac{3}{8}$-in. (35 mm) penetration on ceilings. Aluminum nails react chemically with fresh portland cement and should not be used with it. For most types of reinforcement, 1-in. (25.4 mm) or $1\frac{1}{2}$-in. (38 mm) nails are generally suitable. However, $\frac{3}{8}$-in. (9.5 mm) rib-lath is usually nailed at the rib, and $1\frac{1}{2}$-in. (38 mm) or 2-in. (50.8 mm) nails are recommended to ensure the minimum penetration for walls and ceilings.

Metal reinforcement should be attached to metal framing and furring with screws or clips supplemented by screws. At suspended ceilings, metal reinforcement may be wire-tied to furring members.

When sound ratings are required, metal reinforcement may be attached with resilient clips.

External corners can be formed with *cornerbead*, with prefabricated corner reinforcement, or by bending the metal reinforcement around the corner (Fig. 9.2-10). Internal corners can be reinforced with metal corner reinforcement (see Fig. 9.2-13) wired to the reinforcement or the metal reinforcement itself can be bent to conform to the angle (Fig. 9.2-11).

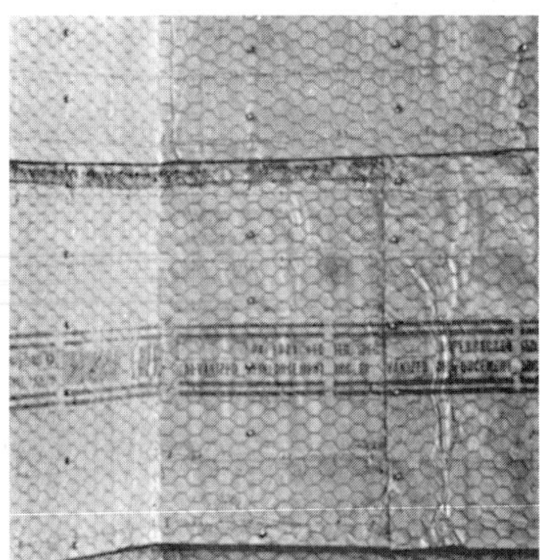

FIGURE 9.2-11 Woven wire lath can be easily bent to form internal corners.

9.2.2.3 Backing Paper

Metal reinforcement for portland cement plaster should have a backing paper in exterior locations, in wet interior areas, and when the plaster will be sprayed on. Backing paper is used to (1) isolate the plaster from the supporting construction, (2) prevent loss of plaster mix through lath openings, (3) facilitate complete embedment of the lath, and (4) resist water penetration through breaks in the plaster membrane. Backing paper is not required for hand-applied plaster on horizontal surfaces protected from direct wetting.

Backing paper should have a vapor permeance of at least 5 perms to permit the plaster to dry and to prevent moisture from becoming trapped within the wall. Waterproof and vapor-permeable products used for sheathing paper, such as 15-lb. (8.14 kg/m^2) asphalt-saturated roofing felt, are suitable. Backing paper may be factory-attached to metal reinforcement or can be installed separately. Absorptive paper attached to some types of metal reinforcement is not water resistant and is not suitable for exterior applications where weather protection is required.

9.2.2.4 Gypsum Lath

Gypsum lath should be fastened to wood wall and ceiling framing and furring with screws in accordance with the lath manufacturer's recommendations (Fig. 9.2-12) or with nails; to metal framing and furring with clips. Where sound ratings are required, gypsum lath may be attached with

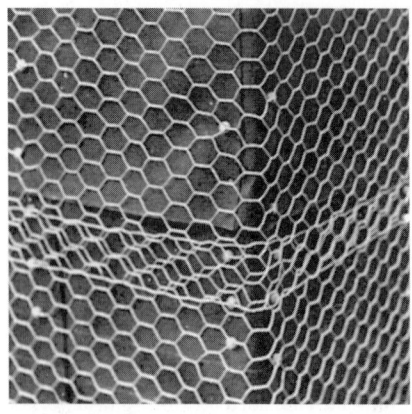

a

b

FIGURE 9.2-10 (a) External corner reinforced by bending lath around the corner; (b) external corner being formed with zinc cornerbead.

FIGURE 9.2-12 Gypsum plaster base applied to metal studs using screws, with clips at corners and end joints. (Courtesy United States Gypsum Company.)

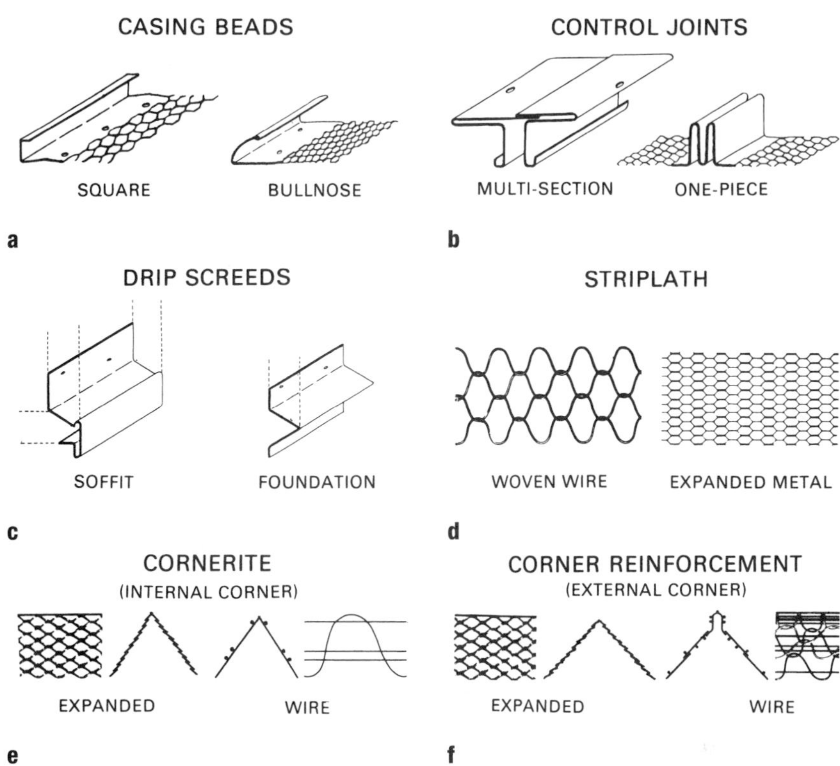

FIGURE 9.2-13 Typical accessories for portland cement plaster (stucco): (a) Casing beads are available with or without expanded flanges. (f) External corner reinforcement is also available with solid metal nosing (cornerbead) in white alloy zinc. Other accessories are produced to meet special conditions.

resilient clips. Refer to Section 9.5 for a detailed discussion about gypsum lath materials and their application.

9.2.3 METAL PLASTER ACCESSORIES

In most residential construction, and in many other buildings as well, wood trim often covers the edges of a plaster finish. When wood trim is not used, exposed edges requiring protection from damage should be trimmed with metal casing beads (Fig. 9.2-13a). Control joints (Fig. 9.2-13b) are incorporated to break large areas into self-contained smaller panels, and drip screeds are used as terminating lines at the bottom of walls (Fig. 9.2-13c). Other accessories such as *strip lath* and *cornerite* (Fig. 9.2-13d and e) are used for reinforcing flat surfaces and internal corners, respectively. External corners are often provided with corner reinforcement (Fig. 9.2-13f) when the metal lath is not flexible enough to be bent around corners.

Corner reinforcement without a solid nosing is designed to permit complete embedment of the metal within the plaster. However, corner beads made with a solid nosing that remains exposed can be used, provided they are made of white alloy zinc (99% pure zinc) where they will be exposed to wetting. Casing beads should be formed from metal of not less than 24 gauge (0.70 mm); other accessories should be at least 26 gauge (0.55 mm). For plaster exposed to wetting, exposed metal accessories (not completely embedded in the plaster) should be as corrosion resistant as white alloy zinc.

9.2.4 FLASHING

Properly installed flashing to prevent water from entering a construction is essential to durable and satisfactory service of plaster surfaces. Flashing (see Section 7.8) should be provided in all locations where there is a possibility of water penetration (see Fig. 7.8-4). Water entering behind a cement plaster finish may cause corrosion of metal reinforcement and accessories and damage to interior finishes. Only corrosion-resistant materials, such as galvanized steel, copper, stainless steel, and suitable plastics, should be used for flashing. Metal reinforcement should be installed over flashing, because these materials do not provide adequate bond for plaster.

9.3 Gypsum Plaster Application

Gypsum plaster mixes and application methods vary with different gypsum plaster products and the bases to which they will be applied.

9.3.1 MIXES

The materials used in gypsum plaster include (1) powdered calcined gypsum, (2) aggregates, (3) mineral or organic fibers, and (4) lime. The last three may be included in a packaged plaster product or added in the field to neat plaster (see Section 9.1.1.2).

9.3.1.1 Basecoat Plaster Mixes

Basecoat plasters include *neat plaster*, *wood-fibered plaster*, and *ready-mixed plaster* (see Section 9.1.1.2).

NEAT PLASTER

Neat gypsum basecoat plaster is usually combined on the job with aggregates in proportions of 1 part plaster to 2 or 3 parts aggregate by weight.

WOOD-FIBERED PLASTER

Although wood-fibered plaster is often used without aggregates, 1 cu. ft. (0.03 m^3) of sand is recommended for each 100 lb. (45.4 kg) of wood-fibered plaster to obtain a basecoat of superior hardness, strength, and added crack resistance, or for use over masonry.

READY-MIXED PLASTER

Ready-mixed plaster is used as it comes from the bag with only the addition of water.

9.3.1.2 Finish Coat Plaster Mixes

Finish coat plasters usually consist of calcined gypsum, lime, and sometimes an aggregate. Some finish coat plasters require the addition of lime or sand on the job.

GAUGING PLASTER

Different gauging plasters provide either a quick set or a slow set. This eliminates the need for adding accelerators or retarders in the field.

GYPSUM-LIME PUTTY-TROWEL PLASTER

Gypsum-lime putty-trowel plaster is a low-cost finish plaster that leaves a smooth surface suitable for most uses. The field mix consists of 100 lb. (45.4 kg) of gauging plaster to 200 lb. (90.7 kg) of dry hydrated lime. Although generally used without aggregates, adding fine silica sand or perlite fines in an amount of at least ½ cu. ft. (0.03 m^3) per 100 lb. (45.4 kg) of dry gauging plaster will increase the crack and crazing resistance of the finish when used over basecoat plasters containing lightweight aggregates.

GYPSUM KEENE'S CEMENT-LIME PUTTY-TROWEL PLASTER

This plaster does not require aggregates except when it is applied over a lightweight aggregate basecoat. In that instance, at least ½ cu. ft. (0.03 m^3) of fine sand or perlite is added to each 100 lb. (45.4 kg) of gypsum Keene's cement and 50 lb. (22.7 kg) of hydrated lime; an extra-hard finish can be obtained by reducing the lime amount to 25 lb. (11.3 kg).

GYPSUM KEENE'S CEMENT-LIME-SAND FLOAT PLASTER

This type of plaster is usually mixed with sand. The resulting textured surface may be painted, or an integral coloring agent may be introduced in the plaster. For a float finish, 150 lb. (68 kg) of gypsum Keene's cement is mixed with 100 lb. (45.4 kg) of hydrated lime and 400 to 600 lb. (181 to 272 kg) of silica sand, depending on the texture desired.

PREPARED GYPSUM FINISH PLASTER

Prepared gypsum finish plaster requires only the addition of water at the job.

MOLDING PLASTER

Most molding plaster mixes consist of 100 lb. (45.4 kg) of molding plaster to 50 lb. (22.7 kg) of dry hydrated lime.

9.3.2 APPLICATION

The minimum thickness and number of coats required for gypsum plaster depend on the plaster base. Three coats are required over metal lath and some gypsum lath installations. Two coats are adequate over most gypsum lath and over masonry and concrete bases.

9.3.2.1 Three-Coat Work

Three-coat work consists of a *scratch coat*, a *brown coat*, and a *finish coat*, each applied separately and allowed to partially set before the next coat is applied. The scratch coat derives its name from cross-raking the wet surface to improve bond with the following brown coat (Fig. 9.3-1).

FIGURE 9.3-1 Scratch coat is raked for three-coat work. (Courtesy Portland Cement Association.)

9.3.2.2 Two-Coat Work

In two-coat work, the scratch coat is applied as in three-coat work, but not cross-raked, and the brown coat is applied immediately, before the scratch coat has set. The finish coat is applied after the basecoat has set and is partially dry (Fig. 9.3-2).

9.3.2.3 Application Methods

Basecoat plasters are formulated for hand-trowel or machine applications. There are three basic methods of applying finish coat gypsum plasters: trowel, float, and spray. The method used determines the surface appearance (Fig. 9.3-3).

Trowel finishes are flat, smooth, and nonporous. They often form the base for further decorating (Fig. 9.3-4a). When an aggregate is used, it is fine silica sand or perlite. Float finishes produce surface textures (Fig. 9.3-4b). The coarseness of the

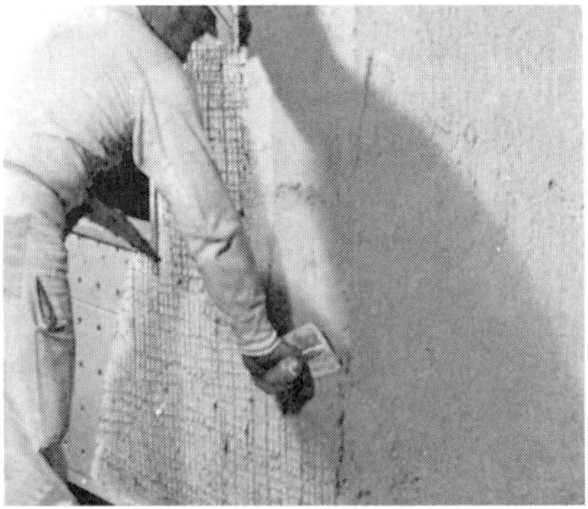

a

b

FIGURE 9.3-2 Brown coat (a) is applied over scratch coat, followed by (b) finish coat in separate operations for three-coat work.

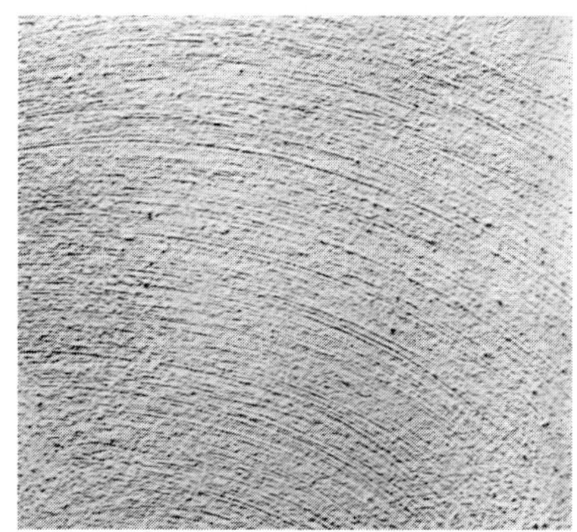

a

b

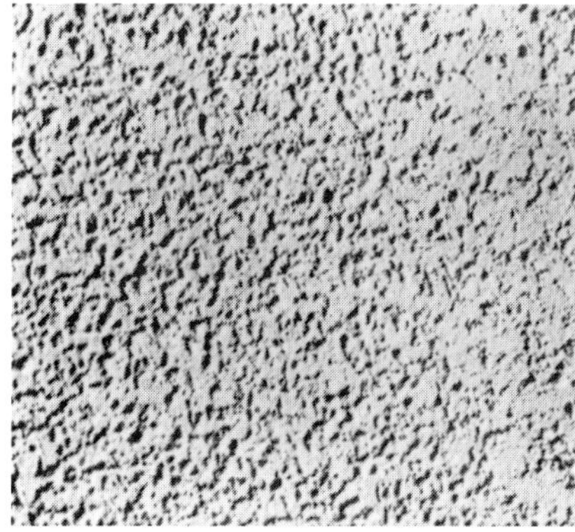

c

FIGURE 9.3-3 Finish coat plasters may be smooth trowel finish or one of the following: (a) float finish, (b) spray finish, (c) texture finish. (Courtesy United States Gypsum Company.)

texture depends on the aggregate, mix proportions, and type of float used.

9.3.2.4 Composition of Coats

The composition of base and finish coats varies considerably. The term *base coat* includes scratch coats and brown coats.

BASE COATS

Gypsum neat plaster is suitable for use as a base coat over metal or gypsum lath, gypsum or clay tile, CMU, and brick of moderate suction (see Section 4.4). Fibered neat plaster is often used as a scratch coat over metal lath because less plaster is lost as mechanical keys are formed behind the lath. When neat plaster is used as a brown coat, sand is usually added to prevent excessive shrinkage.

Wood-fibered plaster may be used over the same bases as neat plaster, but it has greater fire resistance than neat plaster and other sanded base coat plasters.

Ready-mixed plaster of a composition similar to gypsum neat plaster or wood-fibered plaster may be applied where those two plaster types are acceptable.

FINISH COATS

Either trowel or float finishes may be obtained with *gauging plaster*, *gypsum Keene's cement plaster*, prepared *gypsum finish plaster*, or *veneer (thin coat) plaster*. *Acoustical plaster* may be hand applied but is usually machine sprayed (Fig. 9.3-4d).

Gypsum Keene's cement is used for gauging lime finish coat plaster, much the same as gauging plaster, but produces a harder, more moisture-resistant surface. Hardness is proportionate to the amount of gypsum Keene's cement. Both trowel and float finishes are possible with gypsum Keene's cement.

Gypsum Keene's cement-lime-sand float plaster may be applied over all types of base coat plasters but is particularly recommended over lightweight aggregate base coat plasters. It provides greater crack resistance than a trowel finish.

9.3.2.5 Veneer (Thin Coat) Plaster

Veneer plaster may be applied over conventional framing and furring (Sections 9.2 and 9.3.2.7), as part of specialized partition systems and assemblies (Section 9.2.3.7), and in most locations where gypsum board assemblies are applicable (Section 9.5). In assemblies where gypsum board is the outer layer, veneer plaster base is substituted. The remainder of the assembly remains as described. Care must be exercised to ensure that fire or acoustical rated assemblies still comply with these requirements after veneer plaster has

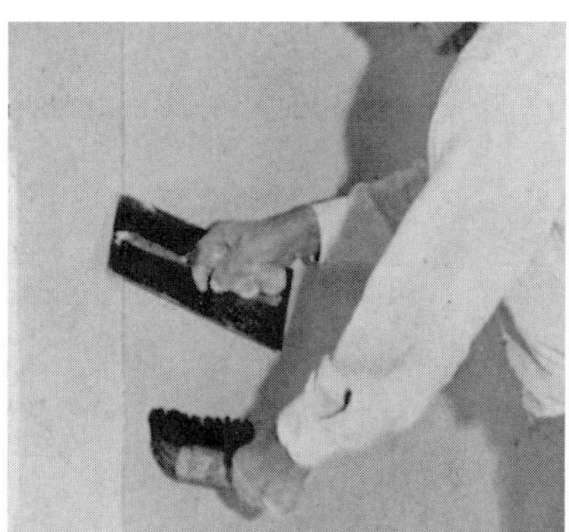

a

b

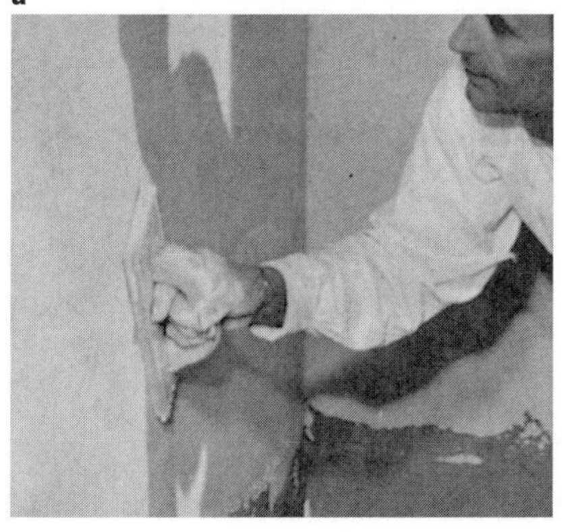

c

d

FIGURE 9.3-4 Methods of applying finish coats: (a) hand trowel, (b) power trowel, (c) hand float, (d) machine spraying.

been substituted for a layer of gypsum board.

Before veneer plaster is applied, joints in the veneer plaster base are treated with a glass fiber tape or reinforced with drywall paper tape and a compatible joint compound, and cornerbeads are installed. Because of the composition and thinness of the plaster coat, it dries rapidly and may be decorated 24 hours after installation. Refer to Section 9.5 for additional information.

Veneer plaster can be applied as a smooth trowel finish or as a textured, sprayed finish. Both are extremely hard and abrasion resistant. Veneer plaster may also be used as a finish coat over bond plaster or as a base coat for gypsum-lime and gypsum Keene's lime finishes where exceptional strength and crack resistance are desired.

9.3.2.6 Acoustical Plasters

Depending on their formulation, acoustical plasters may be hand or machine applied, but most are applied by machine. Noise reduction coefficients ranging from 0.40 to 0.60 may be obtained with these products (see Section 13.1). Surface textures vary from fine to rough, and crack resistance is high. Indentation resistance, however, is much lower than with other plaster products, which limits the use of acoustical plasters to ceiling and upper wall areas not subject to abrasion or impact. Acoustical plasters may be left in their natural off-white or gray color or may be spray painted.

9.3.2.7 Application over Framing and Furring

Gypsum plaster may be applied to either gypsum or metal lath over wood or metal framing or furring.

PARTITIONS

Gypsum plaster finished partitions may be framed with either wood or metal studs 16 to 24 in. (400 to 600 mm) on centers and supported by wood or metal floor and ceiling runners. Wood studs are normally 2 × 4s (50 × 100s) or 2 × 6s (50 × 150s). Metal channel studs are available in 1⅝-in. (41.3 mm), 2½-in. (63.5 mm), 3⅝-in. (92.1 mm), 4-in. (101.6 mm), and 6-in. (152.4 mm) sizes (Fig. 9.3-5a). Greater wall cavities can be developed by using two sets of studs braced with wallboard (Fig. 9.3-5c).

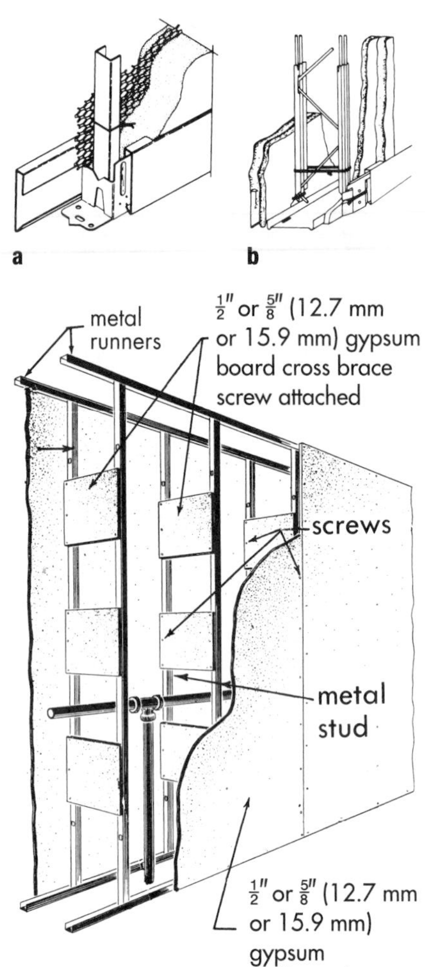

FIGURE 9.3-5 Hollow plaster partitions. (a) With channel studs. (b) With trussed studs. Trussed studs are seldom used in new work today, but may be found in many existing buildings. (c) With braced double studs. Double-stud construction accommodates mechanical equipment and plumbing.

Hollow lath and plaster partitions have gypsum lath screwed, clipped, or nailed to the studs, or metal lath wired, nailed, clipped, or stapled to the studs. Two or three coats of plaster are applied to this plaster base.

CEILINGS

Gypsum plaster is sometimes applied to ceilings over gypsum lath, but most installations are over metal lath (see Section 9.2).

9.3.2.8 Specialized Partition Systems and Other Assemblies

In addition to application over gypsum or metal lath applied to standard framing or furring (Section 9.2), gypsum plaster is also used in proprietary assemblies of framing, lath, board, and plaster with metal accessories designed for each system. A variety of such systems have been developed by gypsum product manufacturers to minimize the number of operations and speed up partition erection. The chief types using gypsum plaster are (1) solid and (2) semisolid partitions. A third type, hollow and demountable (movable) systems, is designed for gypsum board products and does not include gypsum plaster.

SOLID PARTITIONS

Solid partitions are field-erected from plaster on metal or gypsum lath. These include wood or proprietary metal ceiling and floor runners to stabilize the partitions. In lath and plaster partitions, an integral flush metal base acts as a screed. Partitions may also be trimmed out with a resilient or wood base.

Lath and plaster partitions use self-supporting ribbed lath (Fig. 9.3-6a) or ordinary expanded mesh wired to steel bars or channels (Fig. 9.3-6b). To this base, three coats of plaster are applied to each side in thicknesses varying from 1 in. (25.4 mm) to 1¼ in. (31.8 mm). Solid plaster partitions may also be made of ½-in. (12.7 mm)-thick gypsum lath, with two ¾-in. (19.1 mm)-thick plaster faces built up of two or more coats (Fig. 9.3-6c).

Other solid partitions consist of a backing board, a layer of veneer plaster base board, and a veneer plaster finish coat. These are similar to the single and double solid wallboard partitions shown in Figure 9.5-50.

SEMISOLID PARTITIONS

Semisolid partitions (Fig. 9.5-49) are not usually constructed with conventional plaster, but a veneer plaster base may be substituted for the outer layer of gypsum board and finished with veneer plaster.

SPECIAL CEILING ASSEMBLIES

Gypsum plaster may be used in some proprietary electric radiant heating systems. In this case, the heating cables are attached to or through standard gypsum lath and then embedded in the plaster. Such installations should be made strictly in accord with the instructions of the manufacturer of the heating panels (see Section 9.5).

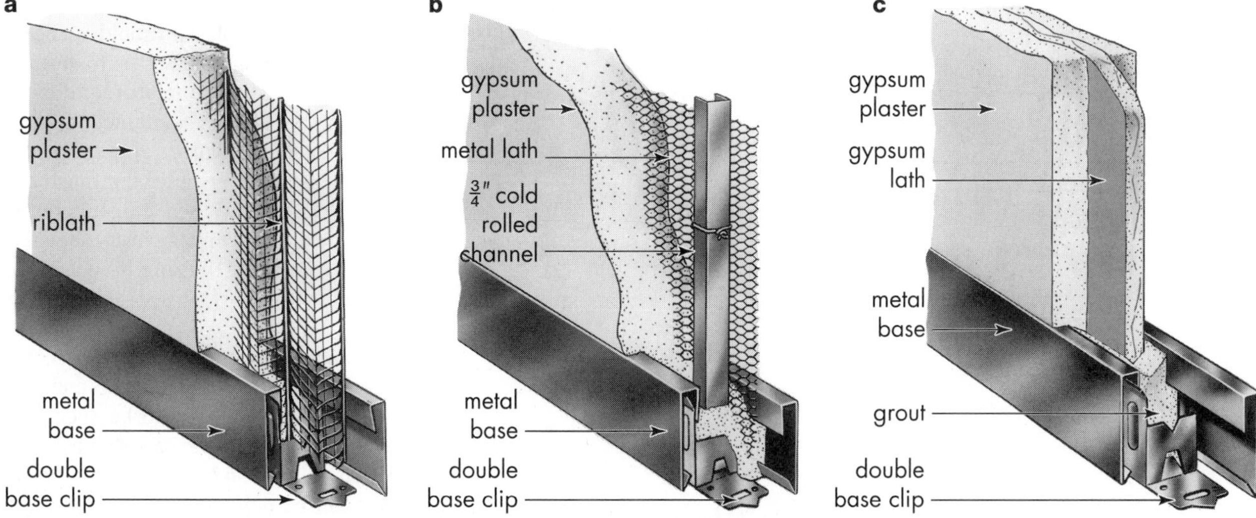

FIGURE 9.3-6 Solid plaster partitions: (a) on self-supporting ribbed lath, (b) on stud-supported woven wire lath, (c) on gypsum lath.

9.4 Portland Cement Plaster Application

Portland cement plaster is used on both interiors and exteriors. It can be used where it will be exposed to wetting.

9.4.1 MIXES

About 0.97 cu. ft. (0.028 m^3) of each cu. ft. (m^3) of a typical portland cement plaster mix is aggregate, and 0.25 to 0.33 cu. ft. (0.007 to 0.009 m^3) is cementitious material. Therefore, the aggregate must be clean, sound, and well graded, with particles ranging from fine up to a maximum size of ⅛ in. (3.3 mm).

In the best mixes, aggregate size is proportioned so that a minimum of water-cement paste is required to coat the aggregate and fill the voids between particles. An excess of fine or coarse particles can result in a mix containing too much cement or too much water, either of which is detrimental to the plaster.

Figure 9.4-1 lists several possible mixes, classified according to plaster groups (C, L, F, and P). The letter designations broadly indicate the type of mix. The selection of a suitable mix for a specific coat depends in part on the bonding properties of the plaster base (Fig. 9.4-2). Mixes containing substantially more portland cement than those shown should not be used.

In general, the proportions should be 1 part cement (including lime when used) to at least 3 and not more than 5 parts by volume of damp, loose sand. Since wide variations in sand quality and gradation exist in different locations, the correct proportion of sand to cement should be determined from trial mixes or previous experience with aggregate from local sources.

Durable finishes depend on proper gradation of aggregate from fine to coarse (see Fig. 9.1-7). When there are too many fine particles, more water-cement paste than is desirable is needed to coat the larger surface area of the particles. When an excess of coarse particles is present, more paste is required to fill the spaces between particles.

A lack of adequate water-cement paste in a mix causes poor workability, but adding either too much water or too much cement to improve workability results in a plaster that is subject to high drying

FIGURE 9.4-1 Stucco Mixes

	Proportions, Parts by Volume				
Group*	Portland Cement	Lime	Masonry Cement Type II	Plastic Cement	Sand
C	1	0 to ¼	—	—	3 to 4
C	1	—	1	—	6 to 7½
C	1	¼ to ½	—	—	4 to 6
L	1	2 to 1¼	—	—	4½ to 9
L	—	—	1	—	3 to 4
F	1	1¼ to 2	—	—	5 to 10
P	1	—	—	1	6 to 10
P	—	—	—	1	3 to 4

*C = (portland) cement; L = lime; F = finish coat; P = plastic cement.

(Courtesy Portland Cement Association.)

FIGURE 9.4-2 Stucco Mix Selection Guide

Stucco Base	Recommended Groups[a] for Stucco Coats		
	Scratch	Brown[b]	Finish[c]
Low absorption (placed concrete, dense clay brick)	C, P	C, L, P	L, F, P
High absorption (concrete unit masonry, clay brick, structural tile)	L, P	L, P	L, F, P
Metal reinforcement[d]	C, P	C, L, P	L, F, P

[a]C = (portland) cement; L = lime; F = finish coat; P = plastic cement.
[b]Use as base coat in two-coat work over concrete and masonry bases.
[c]Finish coat may be factory prepared "stucco finish."
[d]Over any type of supporting construction.
(Courtesy Portland Cement Association.)

shrinkage and cracking. Moreover, sand fines often contain a high percentage of clay particles, which shrink excessively on drying; an excess of coarse particles produces a harsh and difficult-to-apply mix. Therefore, aggregate should be selected carefully and mix proportions should not be relied on to overcome deficiencies in aggregate quality or gradation.

A good mix is easily workable, and easy to trowel. It exhibits good adhesion to the base, and sufficient cohesiveness so that the mix will not sag on vertical surfaces. Once the mix has been determined, uniformity in proportioning of materials from batch to batch will help to ensure uniform suction and color.

Dry plaster ingredients should be mixed to a uniform consistency before water is added. Then only enough water should be added to ensure workability. A power mixer is recommended for uniform blending of materials. It should be operated for at least 5 minutes after all ingredients are in the mixer. Materials should be mixed only in quantities that can be used within 3½ hours. Remixing to restore plasticity is permissible; retempering with additional water is not.

When portland cement plaster is applied by machine, the amount of water should be adjusted to produce a mix of proper consistency for the length of hose, temperature and humidity conditions, and type of plaster base. When measured with a standard 2- by 4- by 6-in. (50.8 by 102 by 152 mm) cone, the slump should be from 2½ to 4 in. (63.5 to 101.6 mm) at the mixer and from 2 to 3¼ in. (50.8 to 82.6 mm) at the nozzle. The hose length from the machine to the working surface should be as straight and as short as possible (not longer than 200 ft. [60.96 m]).

9.4.1.1 Masonry and Plastic Cement Mixes

Masonry cement and plastic cement come from the factory with plasticizing agents in them. Adding lime or plasticizers to them is not necessary. This simplifies field mixing because only sand and water need to be added. Either masonry cement or plastic cement can be combined with portland cement in equal volumes when greater strength is required (see Figs. 9.4-1 and 9.4-2). Plastic and masonry cements alone or in combination with portland cement can be used for any of the three plaster coats.

9.4.1.2 Portland Cement Mixes

Portland cement should be supplemented with lime, plastic cement, or masonry cement to impart suitable workability and plasticity to the mix. For base coats, 1 part portland cement can be mixed with up to 1¼ parts lime, or 1 part masonry or plastic cement. For finish coats, 1 part portland cement can be mixed with 1 part plastic cement or ½ to 2 parts lime (see Figs. 9.4-1 and 9.4-2). Truer colors will be obtained if white portland cement and light-colored sand are used for field-mixed finish coats.

Plasticizers such as mineral, animal, vegetable, or glass fibers can be added to mixes for the first coat, but only in quantities sufficient to obtain the required plasticity or cohesiveness. The kind and exact amount of plasticizing agent will vary with the type of aggregate, plaster base, and manner of application. The amount of a mineral fiber used should not exceed 3% of the cementitious material by weight. The best combination of ingredients should be determined by trial mix.

9.4.2 APPLICATION

Portland cement plaster should not be applied in freezing weather or to surfaces containing frost. The temperature of the plaster should be maintained at 50°F (10°C) or higher during application and for at least 48 hours after the finish coat has been applied. When temperatures are falling, portland cement plaster should be protected against freezing by heated enclosures (Fig. 9.4-3; see also Section 9.4.3).

Hand application of portland cement plaster involves traditional tools and practices that have proven to be successful over centuries. However, machine application enjoys some advantages. The wet mix can be delivered under pressure to scaffolding in multistory buildings; application is speeded; and obvious lap and joint marks, sometimes difficult to avoid with hand application, are eliminated. Machine application results in a more uniform texture and deeper and darker colors than are possible with hand application. Moreover, some textures, such as a uniform rough texture, can be best obtained by spraying. Consequently, finish coats are sometimes applied by machine over a hand-applied base coat.

Generally, portland cement plaster is applied either in two or three coats, depending on the supporting construction.

9.4.2.1 Three-Coat Work

Three-coat applications are required over metal reinforcement and are sometimes also used over a rigid base. They consist of a *scratch coat*, a *brown coat*, and a *fin-*

FIGURE 9.4-3 Protective enclosures in mild climates prevent rapid loss of moisture and facilitate curing. (Courtesy Portland Cement Association.)

a

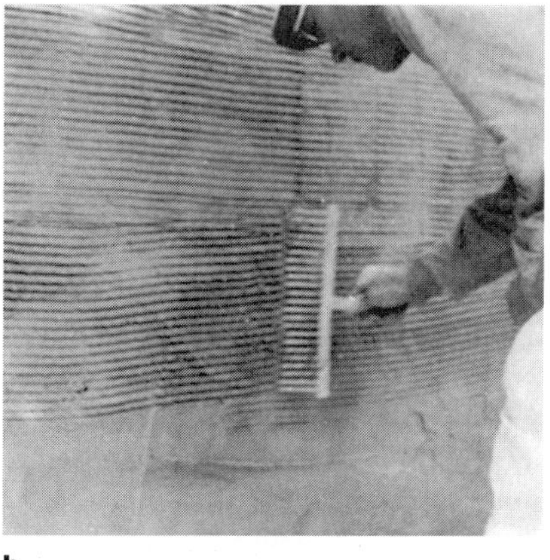

b

FIGURE 9.4-4 The first (scratch) coat (a) is spread on reinforcement with a trowel; then (b) the surface is scored horizontally.

ish coat, totaling ⅞ in. (22.2 mm) in thickness. Each coat is applied separately and allowed to set partially. The scratch coat and brown coat are sometimes referred to as *base coats*.

The first (scratch) coat provides mechanical keying with the reinforcement and a rigid surface for the following coat. Over a rigid base, a *dash bond* coat is sometimes used instead of a scratch coat. The second (brown) coat provides additional thickness and strength and ensures a level surface for the application of the final decorative *finish coat*. Each coat should be moist-cured before the following coat is applied (see Section 9.4.3).

DASH BOND COAT

A dash bond coat's surface is sufficiently rough that cross raking is not required. A dash bond coat should be moist cured for 12 to 24 hours before the brown coat is applied.

SCRATCH COAT

A scratch coat is the first coat over metal reinforcement and can be the first coat over a rigid base that has sufficient bonding capability to make a dash bond coat unnecessary. A scratch coat should be at least ⅜ in. (9.5 mm) thick when measured from a rigid base or backing paper. It should be applied evenly, with sufficient pressure to squeeze the mix through the reinforcement to form mechanical keys (Fig. 9.4-4a). The space between the metal and the backing should be filled thoroughly so that the metal will be sufficiently embedded. The coverage on the surface should be such that no more than 10% of the surface of the reinforcement remains exposed. As soon as the scratch coat has become firm, its entire surface should be cross raked (*scratched*) to provide good mechanical bond for the brown coat (Fig. 9.4-4b). On walls, horizontal cross raking is recommended. Regardless of whether application is by hand or by machine, cross raking is done manually.

BROWN COAT

A brown coat is applied after the scratch coat has been moist-cured and is sufficiently firm to carry the weight of the next coat. If a scratch coat becomes dry, it should be dampened again to control suction before a brown coat is applied. A brown coat should be approximately ⅜ in. (9.5 mm) thick, bringing the thickness of both coats to at least ¾ in. (19.1 mm) (Fig. 9.4-5a).

When possible, a brown coat should be applied over an entire wall panel without stopping. Interruptions in the work should be made at screeds, control joints, corners, pilasters, belt courses, doorways, openings, or other lines where changes in appearance, which may show through the finish coat, will not be obvious.

A brown coat should be *rodded* (struck off) with a straightedge and then darbied to impart a level, uniformly rough texture. Even with machine application, these operations are performed by hand (Fig. 9.4-5b). The same type of darby or float should be used for the entire surface to ensure uniformity of texture. Deep scratches or surface defects should be

a

b

c

FIGURE 9.4-5 Brown coat: (a) mix is spread with a trowel, then (b) leveled with a rod (straightedge). (c) When cornerbead or corner reinforcement is not used, a true plumb corner is formed with a straightedge.

smoothed out, because such irregularities may cause unequal suction or may show through the finish coat. External corners should be formed with a straightedge when corner reinforcement is not used (Fig. 9.4-5c).

After moist curing, the surface should be allowed to dry uniformly before a finish coat is applied (see Section 9.4.3).

FINISH COAT

A brown coat should be uniformly dampened when a finish coat is to be applied by hand. Dampening is not required for machine-applied finishes. A finish coat should be at least ⅛ in. (3.2 mm) thick, bringing the total thickness to at least ⅞ in. (22.2 mm) (Fig. 9.4-6a). Machine-applied sprayed finish coats are often made in two or more consecutive passes (Fig. 9.4-6b). The first pass is made with a thinner mix applied primarily to ensure complete coverage of the brown coat. Successive applications are thicker, to develop the desired finished texture.

Factory-prepared stucco finishes ensure greater uniformity of color in the finish coat. A variety of decorative surface finishes are possible for both machine and hand application (Fig. 9.4-7) (see Section 9.4.5). A plain finish coat that is going to be painted should be moist-cured. Prepared stucco finishes often contain curing agents, so moist curing is not necessary; in fact, it is specifically discouraged by some manufacturers because it may cause streaking of the color surface. The finish coat should be maintained at the proper temperature for at least 48 hours after application (see Section 9.4.3).

9.4.2.2 Two-Coat Work

Sometimes three-coat applications are used over concrete and masonry bases, especially where increased strength or thickness is required. Usually, however, two-coat applications provide sufficient strength and durability over such solid bases. In two-coat applications, a single base coat takes the place of the scratch and brown coats and is followed by a finish coat. The total thickness should be at least ⅝ in. (15.9 mm).

BASE COAT

Except when a chemical bonding agent is used, a surface to be plastered should be dampened to ensure uniform suction. A ⅜-in. (9.5 mm)-thick base coat is then applied. It is not scratched, as in a three-coat application. Mechanical bond for the finish coat is provided by rough floating of the base coat in the same manner as the brown coat in three-coat work (Fig. 9.4-8).

A machine-applied base coat is built up with several applications, with the final application being rodded and darbied manually.

FINISH COAT

A finish coat should be about ¼ in. (6.4 mm) thick. It can be applied in a variety of textures by hand or machine, as in a three-coat application.

a

b

FIGURE 9.4-6 A finish coat can be (a) hand-floated or (b) sprayed by machine.

9.4.3 CURING

Portland cement plaster requires sufficient water and favorable temperatures to facilitate the process of hydration by which the cement gains strength and hardens to a solid mass. The large exposed area of plaster relative to its volume encourages rapid loss of moisture by evaporation. Moist curing of each coat replaces this moisture and maintains an adequate supply for the chemical reaction to continue. An ordinary hose nozzle can be adjusted to provide a fine spray for curing; a steady stream that could erode the surface should not be used.

a

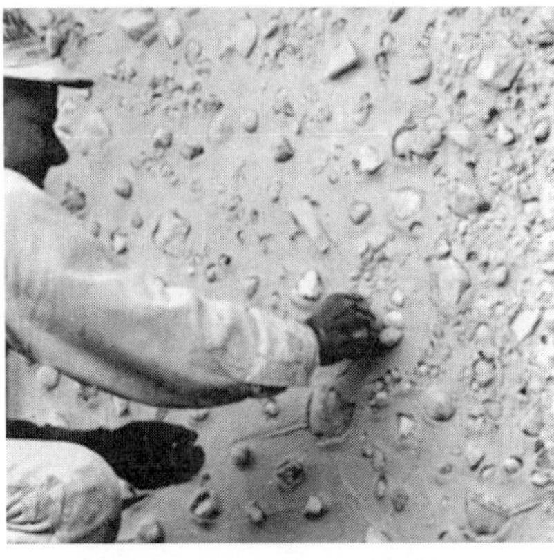

b

FIGURE 9.4-7 Rock dash (marblecrete) finish: (a) small aggregate can be machine-sprayed; (b) larger aggregate is usually placed by hand. (Courtesy Portland Cement Association.)

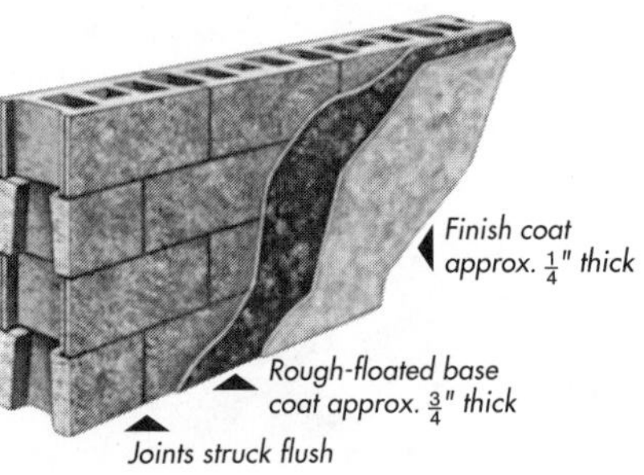

FIGURE 9.4-8 Two coats are adequate over a suitably rough concrete or masonry surface. (Courtesy Portland Cement Association.)

Moist curing does not require spraying at set intervals, but rather periodic observation and spraying when the surface appears to be drying out. Since evaporation increases with increased air circulation and higher temperatures, more frequent spraying is generally required in breezy, warm weather. Moreover, wind breaks or hangings are needed to protect fresh portland cement plaster from hot, dry winds.

In addition to adequate moisture, proper hydration requires that a portland cement plaster mix and each plaster coat be maintained at a temperature of at least 50°F (10°C). A finish coat should be maintained at this temperature for at least 48 hours after it has been applied. In cold weather, it may be necessary to heat the mixing water and provide a heated enclosure around the work area.

While it is generally accepted that adequate water to ensure hydration is essential in producing quality portland cement plaster, there is no general agreement as to exact curing procedures. The curing recommendations given in Figure 9.4-9 are based on the premise that deliberate moist curing ensures a greater measure of success than does dependence on favorable climatic conditions or on water contained in the mix.

In some parts of the country, no deliberate moist curing is performed, and each coat is applied as soon as the previous coat is sufficiently hard to support it. This method, sometimes called *doubling back*, relies on the protective nature of each covering coat to retain the moisture and facilitate curing of the underlying coat.

Another practice involves application of coats at not less than 24-hour intervals, with continuous moist curing between coats and after the finish coat is applied.

The curing method chosen should be based on accepted local practice that has a demonstrated history of satisfactory performance. All curing procedures attempt either to maintain the previous coat uniformly damp or to permit it to become uniformly dry. This is particularly true of the base (or brown) coat, because uniformity of color in the finish coat is dependent on equal suction in the base coat.

The long waiting period between base and finish coats shown in Figure 9.4-9 ensures that the brown coat will be uniformly dry. When shorter intervals are allowed, it is often necessary to keep the base coat damp by continuous moist curing until the finish coat is applied. Unless a rigid schedule for rapid, successive applications can be maintained, the need for continuous moist curing may be more demanding than the recommended short curing period and subsequent drying delay.

9.4.4 CRACK CONTROL

Proper design of supporting construction, selection of materials, and application of portland cement plaster will minimize cracking. Cracks develop in plaster from (1) shrinkage stresses, (2) structural movement, (3) restraints due to penetration of mechanical or electrical equipment or intersecting walls or ceilings, or (4) weak sections in the finish due to thin spots, openings, or changes in materials in the supporting construction.

9.4.4.1 Shrinkage Stresses

As a wet portland cement plaster mix hydrates and dries, it tends to shrink, setting up stresses in the plaster membrane. Cracking occurs when shrinkage stresses exceed the tensile strength of the membrane. It is desirable, therefore, to minimize the degree of shrinkage. Proper moist curing and protection from temperature extremes allow the plaster to dry slowly and uniformly, thus increasing strength and reducing shrinkage. Shrinkage stresses also are minimized by proper proportioning of mixes and by proper placement of control joints.

PROPORTIONING MIXES

The chief factors in proportioning mixes that affect the balance between tensile strength and shrinkage stresses in the finished plaster are the cement content, the water content, and the aggregate quality and gradation.

Rich mixes, high in cement content, generally result in plaster with higher initial shrinkage, balanced to some extent by increased strength. Excessive water in a mix increases the volume of minute voids in the plaster as water evaporates or is used up in hydration, which results in higher shrinkage and lower strength.

Poorly graded sand requires the addition of more water, more cement, or both to obtain suitable workability, which can be detrimental and cause shrinkage cracking (see Section 9.4.1).

CONTROL JOINTS

When portland cement plaster is applied directly to masonry or concrete and no reinforcement is used, control joints in the plaster should coincide with joints in the base. Additional control joints to resist shrinkage cracking in the plaster finish are not usually required, because the underlying base will resist stresses due to shrinkage.

In portland cement plaster applied to metal reinforcement, control joints should be installed in the plaster not more than 18 ft. (5486 mm) apart, forming panels of not more than 150 sq. ft. (13.94 m^2).

Either one-piece or multisection control joints (see Fig. 9.2-13b) can be used for vertical joints in walls and for joints in protected soffits. A multisection control joint is particularly effective in relieving stresses caused by shrinkage and differential structural movement in the supporting construction (Fig. 9.4-10). For horizontal joints, a watertight control joint designed for the purpose should be used (Fig. 9.4-11).

9.4.4.2 Structural Movement

Structural movement may be caused by foundation settlement, deflection of struc-

FIGURE 9.4-9 Curing Schedule for Portland Cement Plaster

Application[a]	Moist Curing[b]	Total Setting Time[c]
Scratch coat	12 to 24 hr.	At least 48 hr. (between coats)
Brown coat	12 to 24 hr.	At least 7 days (between coats)
Finish coat	12 to 24 hr.[d]	At least 48 hr. (after application)

[a]In two-coat work, requirements for base and finish coats are the same as for brown and finish coats, respectively.

[b]Moist curing should be delayed until scratch and brown coats are sufficiently set to prevent erosion and for 12 to 24 hours for finish coat.

[c]At least 50°F (10°C) temperature should be maintained during and after application.

[d]Not recommended for prepared colored "stucco finishes."

(Courtesy Portland Cement Association.)

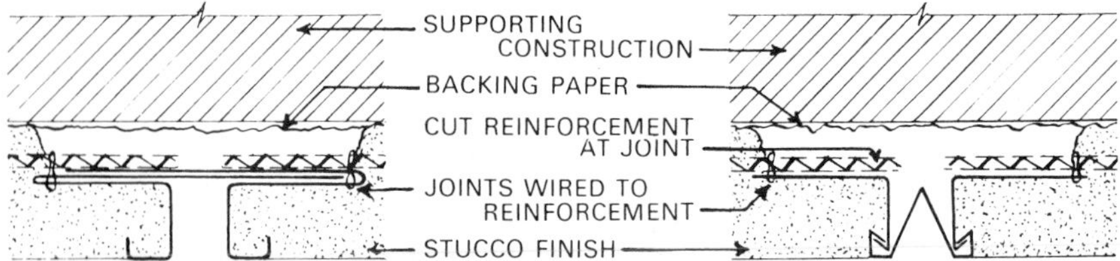

FIGURE 9.4-10 Multisection control joints effectively relieve stress caused by drying shrinkage, structural movement, and temperature variations. (Courtesy Portland Cement Association.)

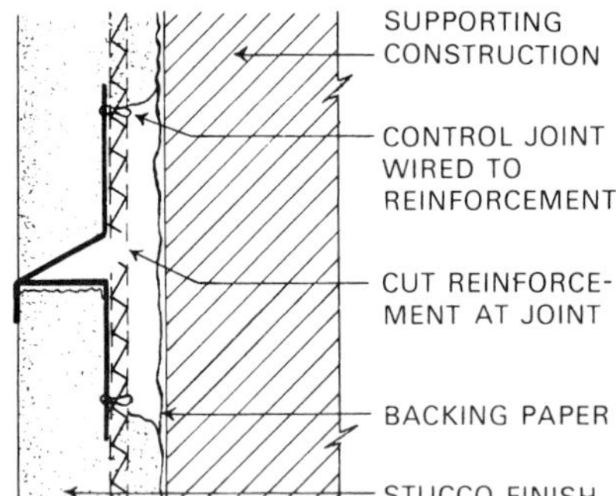

FIGURE 9.4-11 Specially designed horizontal control joints provide weathertightness as well as crack control. (Courtesy Portland Cement Association.)

tural members, racking due to wind forces, and expansion and contraction of materials due to temperature and humidity variations. Good structural design attempts to minimize these movements where possible; specially designed joints are used to localize movements that cannot be reduced to negligible proportions.

Relief joints and control joints are used to localize movement and minimize stresses within a portland cement plaster finish. Relief joints at ceilings and walls should be used where the plaster is intersected (Fig. 9.4-12) or penetrated by structural elements such as beams, columns, or load-bearing walls (Fig. 9.4-13). Location of control joints depends on the type of supporting base.

RIGID BASES

Where portland cement plaster is applied directly to a concrete or masonry base, firm and continuous support of the plaster membrane will tend to resist shrinkage, temperature, and impact stresses in the plaster. However, the finish will be subjected to structural movements in the base and is likely to crack wherever excessive stress concentrations develop. It is particularly important, therefore, in this type of installation to provide proper control joints in the base. Metal control joints should be placed in a plaster finish directly over each underlying joint. Control joints should also be placed in the plaster finish over construction, isolation, or expansion joints that occur in the base and at the juncture of dissimilar materials.

METAL REINFORCEMENT

Portland cement plaster applied over metal reinforcement, with or without backing paper, acts as a thin reinforced concrete slab. Because the plaster is somewhat isolated from the supporting construction, it is not subjected to structural stresses to the same degree as plaster applied directly to rigid bases. However, such a thin reinforced slab is more subject to shrinkage, impact, and temperature stresses because it is not supported and restrained by an underlying rigid base. Of these, shrinkage stresses are the most critical and involve the greatest movement. Therefore, control joints located as recommended for shrinkage purposes will also relieve stresses occasioned by other forces acting on the plaster membrane.

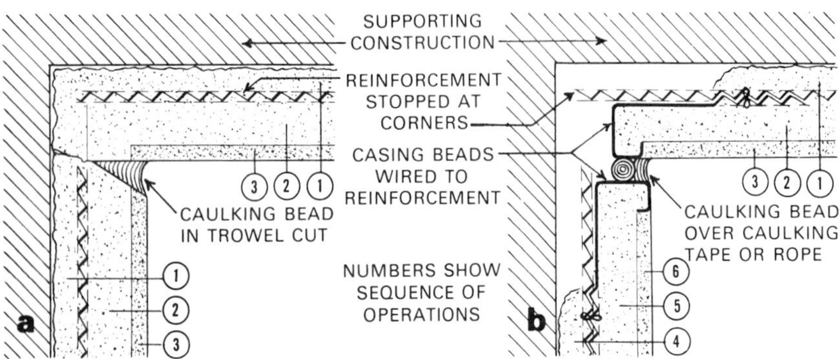

FIGURE 9.4-12 Perimeter relief at internal corner between wall and soffit: (a) using trowel cut; (b) formed with casing beads. (Courtesy Portland Cement Association.)

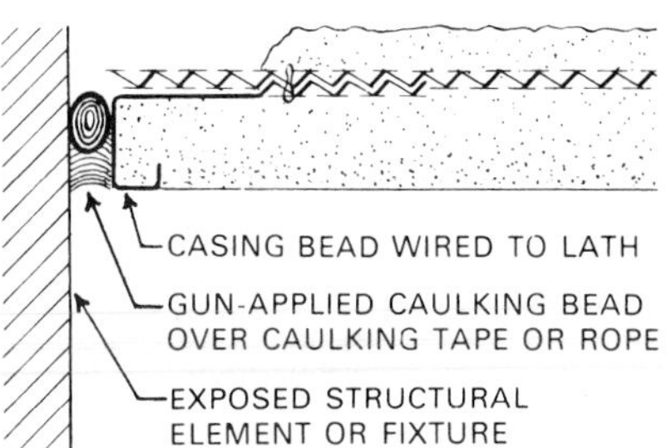

FIGURE 9.4-13 Perimeter relief joint for use where structural elements or fixtures penetrate the surface. (Courtesy Portland Cement Association.)

9.4.4.3 Weakened Sections

Stresses caused by shrinkage, humidity, or temperature variations are evenly distributed in plaster of uniform thickness, but tend to be higher at thin spots. Plaster applied over an irregular, rigid base, or over sagging metal reinforcement, will be of unequal thickness and will contain weakened sections. A concentration of stresses can result in cracking at these weakened points. To minimize this hazard, plaster should be applied over a level base in uniform thickness.

Weakened sections also are created at corners of cutouts and openings in a plaster finish. Metal reinforcement will resist the concentration of stresses at the corners and generally prevent diagonal cracking. When reinforcement is not used and plaster is applied directly to a rigid base, strip reinforcement (*strip lath*) can be installed diagonally at each corner to prevent cracking.

Weakened sections can be introduced deliberately in locations where cracking would not be objectionable. A trowel cut adjacent to a door or window frame will result in a straight, inconspicuous crack rather than random cracking. Often, the trowel cut can be concealed by trim; where trim does not occur, neatly formed cuts can be left exposed and caulked.

9.4.4.4 Restraints

Free movement of a plaster membrane is generally restrained at corners and wher-

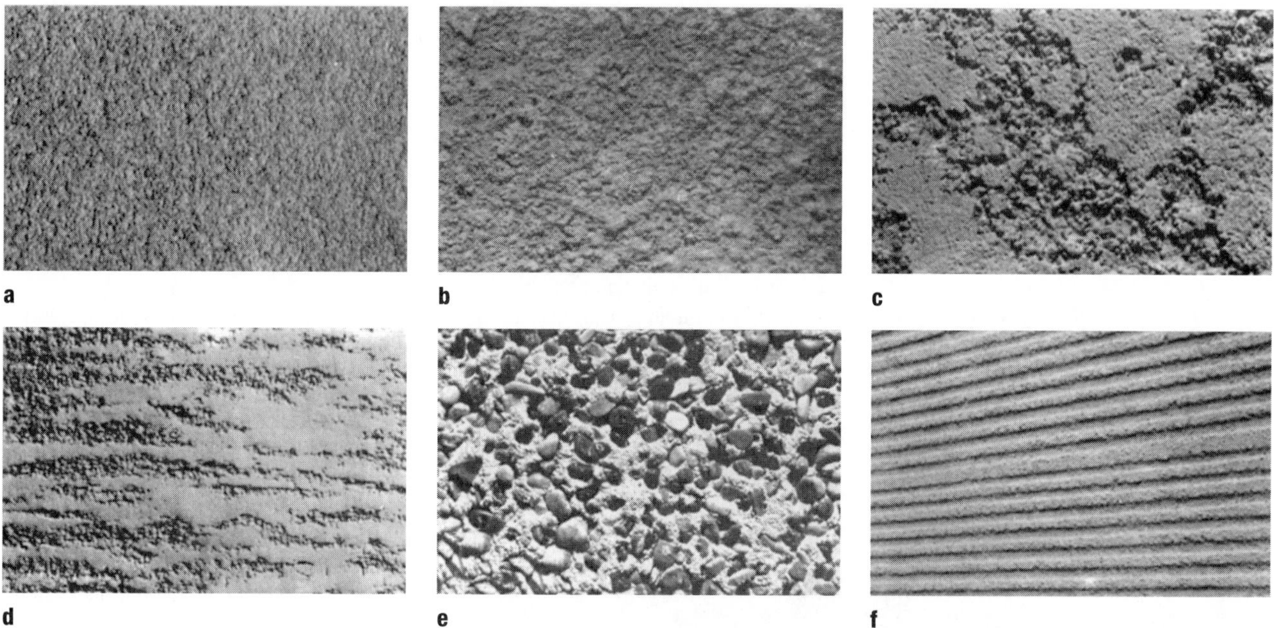

FIGURE 9.4-14 Some typical portland cement plaster (stucco) finishes: (a) *float finish*, made with carpet- or rubber-faced float; (b) *wet-dash*, coarse aggregate mix is dashed on with a brush; (c) *dash troweled*, high spots or dashed surfaces are troweled smooth; (d) *stippled-troweled*, surface is stippled with a broom, then high spots are troweled smooth; (e) *rock-dash*, decorative pebbles are thrown by hand or machine against wet surface; (f) *combed*, surface is formed with a notched templet. (Courtesy Portland Cement Association.)

ever the membrane is penetrated by fixtures or structural elements. Cracking due to restraint is not common at external corners, particularly if control joints are provided in each plane for shrinkage control.

Cracking is more frequent at internal corners formed by the intersection of load-bearing elements (columns, walls, pilasters) and non-load-bearing walls, and at the juncture of walls and suspended or furred ceilings. At such corners, perimeter relief should be provided by omitting cornerite, stopping the metal reinforcement in each plane, and including relief joints that will permit independent movement in each plane.

Relief joints can be formed by a trowel cut (Fig. 9.4-12a) or a pair of casing beads (Fig. 9.4-12b). The casing beads should be wired to the reinforcement, not rigidly fastened to supporting construction. The corner should be caulked with a nonhardening resilient compound after initial shrinkage has taken place.

A relief joint should be provided where plaster abuts or is penetrated by a structural element or a fixture. The plaster should terminate with a casing bead, and the space between the bead and the structural element or fixture should be caulked (see Fig. 9.4-13).

9.4.5 SURFACE FINISHES

In addition to providing the desired decorative surface, the finish selected also affects the weather resistance and maintenance requirements of portland cement plaster. Smooth-troweled plaster finishes tend to make surface irregularities more conspicuous. Rougher textures hide slight color variations, lap joints, uneven dirt accumulation, and streaking.

The range of portland cement plaster textures is practically limitless. Brick, stone, and travertine marble can be simulated, and many other finishes can be produced by hand or machine, depending on the desired texture. The finish may be natural or integrally colored so that painting is not required. Colored finishes can be effected by using a prepared stucco finish, by adding pigment to the mix, or by embedding colored aggregate in the surface.

Figure 9.4-14 shows several common textures. Textures can vary substantially with each applicator. Therefore, a standard sample should be produced at the beginning of the work, and as the work progresses, it should be checked with the sample for conformity.

9.5 Gypsum Board Systems

This section covers the manufacture, properties, and uses of gypsum board products and the materials and methods necessary to provide suitable single-ply and multi-ply gypsum board installations. Gypsum board panels and their supports and accessories are covered, but there are only general recommendations for sound isolation, fire resistance, thermal insulation, and moisture resistance. Additional information about these subjects is given in Chapters 7 and 13.

The use of gypsum board as a finish material started in about 1915, when small, $\frac{3}{8}$-in. (9.5 mm)-thick sheets were developed primarily for remodeling. In the following decades, erection and joint treatment techniques gradually evolved, resulting in improved surface appearance and the use of gypsum board in other types of construction. After World War II, greater emphasis on total system and assembly performance gained gypsum board widespread use (Fig. 9.5-1).

Gypsum board finishes are also called *drywall* finishes, because they have initial low moisture content and require little or no water during application. Plywood, hardboard, and fiberboard are also sometimes called drywall, but the term usually means gypsum board.

Gypsum board finishes consist of one or more plies of gypsum board having incombustible gypsum cores with surfaces of paper or other sheet material. Face (exposed) plies either have plain calendered manila paper, intended for decorating with paint, wall covering, or wall tile, or are predecorated with colored, textured, or simulated wood-grained paper or vinyl coverings.

Most gypsum board installations are made with plain boards, which require fastener and joint treatment and decorating. Predecorated board is applied directly to studs or furring with color-matched fasteners or is laminated to a base layer of gypsum backing board. Joints either are left exposed, treated with special moldings or battens, or concealed with a loose flap of the decorative material. This section refers to plain wallboard finishes except when predecorated panels are specifically mentioned.

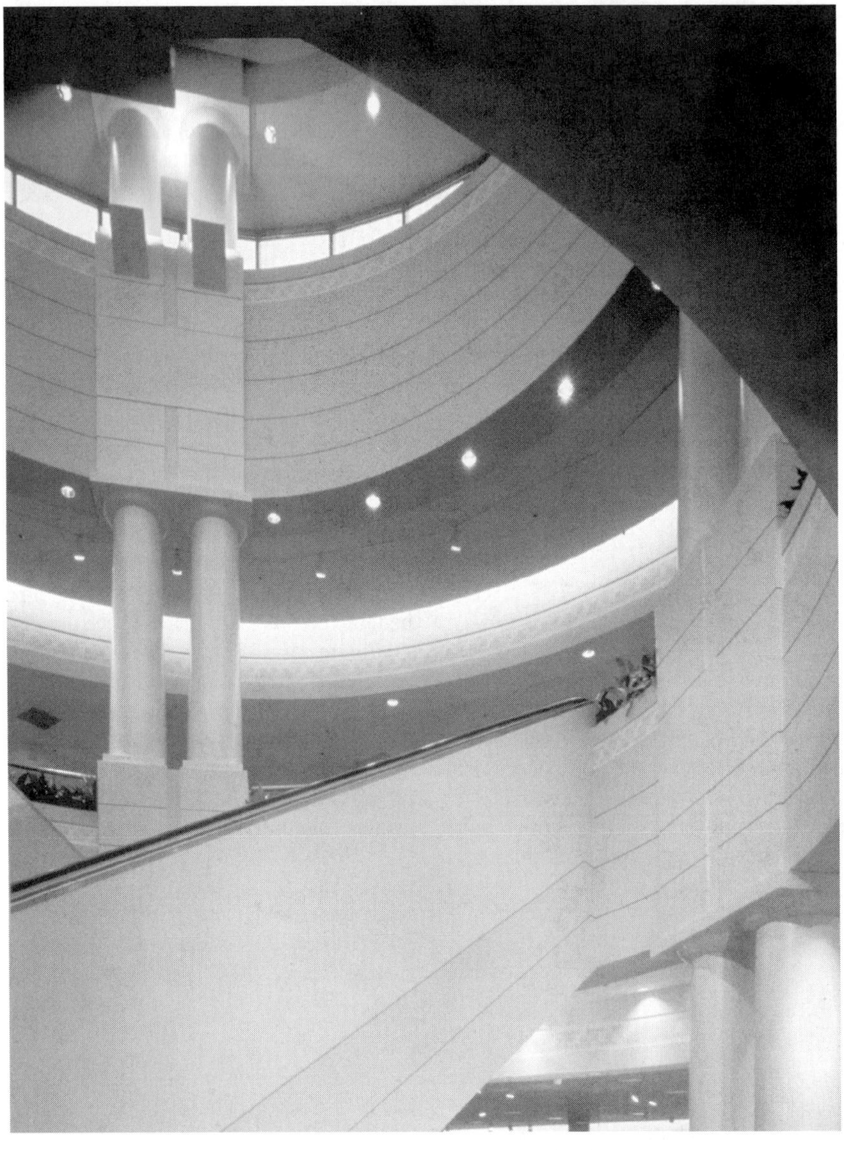

FIGURE 9.5-1 **We usually think of gypsum board being used for flat walls and ceilings. As this photograph shows, gypsum board can also be used in highly imaginative ways.** (Courtesy United States Gypsum Company.)

In addition to their other functions, walls and ceilings are often required to provide sound isolation, fire resistance, thermal insulation, and moisture control. A variety of gypsum boards and engineered systems are available to meet such specific needs. In general, however, performance is closely related to the properties, arrangement, and installation of the wall or ceiling assembly.

For gypsum board finishes to have suitable quality, consideration should be given to:

- Preparation of supporting construction
- Selection of board and accessory materials
- Job conditions
- Board application, including joint and fastener treatment
- Special construction techniques where sound isolation, fire resistance, thermal insulation, or moisture resistance is required

The Gypsum Association publication GA-216, "Recommended Specifications for the Application and Finishing of Gypsum Board," is the currently recognized industry standard and the basis for much of the data in this section.

9.5.1 BOARD MANUFACTURE

As shown in the flowchart in Figure 9.1-2, calcined gypsum without tube milling is used to make gypsum board products. This calcined gypsum is mixed with water to form a slurry. Additives and glass fibers are added for highly fire-resistant gypsum boards. This slurry is fed onto a conveyor belt between two layers of paper. After traveling down the belt for 3 or 4 minutes, the core material has set sufficiently to be cut to length (Fig. 9.5-2), and the boards are then conveyed to kilns for drying. Boards not requiring additional finishes may be bundled and stacked for shipping. Others may receive special backings such as aluminum foil, decorative vinyl paper finishes, or an organic texture coating.

9.5.2 BOARD TYPES

Gypsum board products are panels or slabs consisting of a noncombustible gypsum core, surfaced and edged with covering material specifically designed for various uses with respect to performance, application, location, and appearance.

Gypsum board products are produced for specific uses. They include

FIGURE 9.5-2 Gypsum board on a continuous belt during manufacture. (Courtesy United States Gypsum Company.)

- *Gypsum wallboard:* Board product with either a prefinished surface or a surface suitable for paint finishing after appropriate treatment of fasteners and joints
- *Gypsum sheathing:* Boards used in exterior frame walls as a base for finish materials and for increased fire resistance
- *Gypsum backing:* Boards that provide a base for adhesive application of gypsum wallboards in multi-ply construction and of acoustical tiles in suspended ceilings
- *Gypsum formboards:* Boards that serve as both permanent forms and finished ceilings in poured gypsum concrete roof decks (see Section 3.4)

Board products are made in conformance with ASTM standards in different sizes and thicknesses, with several types of edges (Fig. 9.5-3). The dimensions, edge types, and governing standards of quality of generally used board products are summarized in Figure 9.5-4. Figure 9.5-5 shows some of the terms used in gypsum board work.

9.5.2.1 Gypsum Wallboard

The main types of gypsum wallboard are regular core, predecorated, foil-backed, fire-resistant (Type X), and water-resistant (WR).

REGULAR CORE WALLBOARD

Regular core wallboard is surfaced on the back with a gray liner paper, and on the face and along the longitudinal edges with a calendered manila paper that provides a smooth, even finish suitable for decorating.

Most regular core wallboards are produced with tapered edges, which permit taping and finishing of joints to provide a homogeneous, smooth surface with inconspicuous joints.

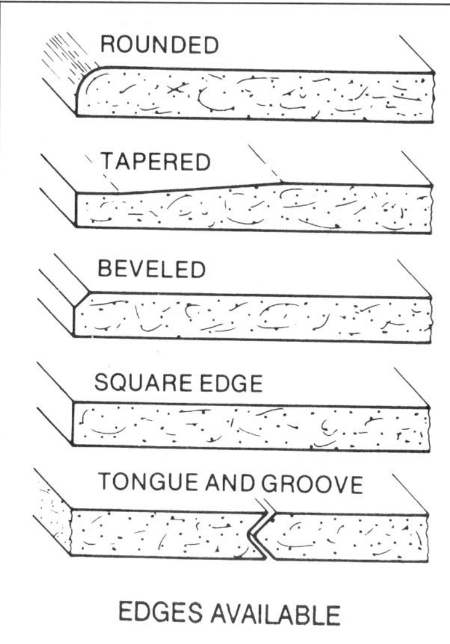

FIGURE 9.5-3 Typical edge designs for gypsum board products. (Courtesy Gypsum Association.)

PREDECORATED WALLBOARD

Predecorated wallboard is regular wallboard with a factory-applied decorative finish on one face. Common finish materials include paper and vinyl in a variety of colors, patterns, and simulated wood finishes (Fig. 9.5-6). These boards usually have accurately formed square or beveled longitudinal edges that do not require further treatment in the field. Bevel-edged and some square-edged boards are simply butted together to create a shadow line that is repeated at regular intervals. Square edges are trimmed with matching moldings or decorative battens to conceal joint irregularities and fasteners. Some manufacturers provide a loose flap of the decorative covering material, which is cemented down after installation to conceal the joint. Predecorated boards may be secured with matching prefinished nails or by adhesive bonding. A lighter-weight, predecorated product, MH Board, is often used in manufactured construction, primarily mobile homes. It is available factory painted or faced with paper or vinyl.

FOIL-BACKED WALLBOARD

Foil-backed wallboard is regular wallboard with a bright-finish aluminum foil bonded to the back to act as a vapor retarder, eliminating the need for a separate vapor retarder in an exterior wall assembly. When the reflective surface faces a ¾-in. (19.1 mm)-wide dead air space, it can also provide increased insulating performance. Installation and edge treatment of foil-backed gypsum board are the same as for regular wallboard.

FIRE-RESISTANT (TYPE X) WALLBOARD

Type X board is similar in exterior covering, appearance, edge treatment, and installation to regular core wallboard, but it has a core specially formulated with additives and glass fibers for greater fire resistance. When prescribed installation methods are followed, fire-resistant assemblies of 45 and 60 minutes are possible with Type X board and frame construction. Fire-resistant wallboard is also available with an aluminum foil vapor retarder.

WATER-RESISTANT WALLBOARD

Water-resistant (WR) boards are used in toilet rooms, bathrooms, kitchens, utility rooms, and other humid areas. They are

FIGURE 9.5-4 Summary of Gypsum Board Products

Type	Thickness in.	Thickness mm	Width, in. (mm)	Length, ft. (mm)	Edge Detail
WALLBOARD (ASTM C 36) Regular	¼	6.4	48 (1200)	8, 10, 12 (2400, 3000, 3600)	Tapered
	5⁄16	7.9	48 (1200)	8, 10, 12, 14 (2400, 3000, 3600, 4000)	
	⅜	9.5	48 (1200)	7, 8, 9, 10, 12 (2000, 2400, 2700, 3000, 3600)	
	½	12.7	48 (1200)	7, 8, 9, 10, 12, 14 (2000, 2400, 2700, 3000, 3600, 4000)	
	⅝	15.9	48 (1200)	8, 10, 12 (2400, 3000, 3600)	
Foil backed	⅜	9.5	48 (1200)	8, 10, 12 (2400, 3000, 3600)	
	½	12.7	48 (1200)	8, 9, 10, 12 (2400, 2700, 3000, 3600)	
Type X	½	12.7	48 (1200)	8, 9, 10, 12 (2400, 2700, 3000, 3600)	
	⅝	15.9	48 (1200)	8, 9, 10, 12 (2400, 2700, 3000, 3600)	
LATH (ASTM C 37) Plain	⅜	9.5	16, 16⅕, 24 (406, 411, 610)	4, 8, 12 (1200, 2400, 3600)	Round
	½	12.7	16, 16⅕, 24 (406, 411, 610)	4 (1200)	
Perforated	⅜	9.5	16 (406)	4 (1200)	
	½	12.7	16 (406)	4 (1200)	
Foil backed	⅜	9.5	16 (406)	4, 8 (1200, 2400)	
	½	12.7	16 (406)	4 (1200)	
Type X	⅜	9.5	16 (406)	4, 8 (1200, 2400)	
BACKING BOARD (ASTM C 442) Regular	¼	6.4	48 (1220)	8 (2400)	Square
	⅜	9.5	48 (1220)	8 (2400)	Square
	½	12.7	24 (610)	8 (2400)	Tongue-and-groove
Foil backed	⅜	9.5	48 (1220)	8 (2400)	Square
Type X	⅝	15.9	24 (610)	7 (2000)	Tongue-and-groove
	1	25.4	24 (610)	7 (2000)	Tongue-and-groove
SHEATHING (ASTM C 79) Regular	⅜	9.5	24 (610)	8 (2400)	Tongue-and-groove
			48 (1220)	8, 9 (2400, 2700)	Square
	4⁄10	10.0	24 (610)	8, 9 (2400, 2700)	Tongue-and-groove
			48 (1220)	8, 9 (2400, 2700)	Square
	½	12.7	24 (610)	8, 9 (2400, 2700)	Tongue-and-groove
			48 (1220)	8, 9 (2400, 2700)	Square
	⅝	15.9	24 (610)	8, 9 (2400, 2700)	Tongue-and-groove
			48 (1220)	8, 9 (2400, 2700)	Square
Type X	⅝	15.9	24 (610)	8 (2400)	Tongue-and-groove
			48 (1220)	8, 9 (2400, 2700)	Square

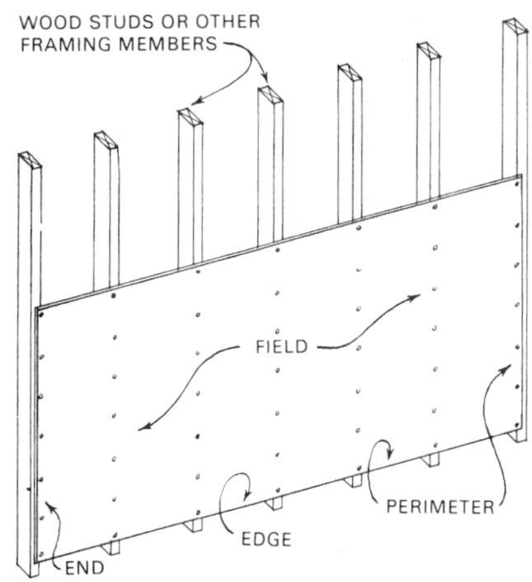

FIGURE 9.5-5 Terms used in gypsum board work. See Glossary for definitions. (Courtesy Gypsum Association.)

FIGURE 9.5-6 Predecorated wallboard.

not recommended for high-moisture areas such as shower rooms, saunas, or steam rooms. They can be painted or can be used as a base for adhesively applied ceramic, metal, and plastic wall tile. Water resistance is achieved by a multilayered covering of chemically treated paper and a gypsum core formulated with asphaltic additives. Edges usually are tapered. Installation is the same as for regular wallboard.

Glass mesh mortar units, a highly water-resistant backing board used as a base for tile in areas subject to direct wetting such as in shower stalls, are described in Section 9.6.

9.5.2.2 Gypsum Backing Board

Gypsum backing boards serve as a base to which wallboards or acoustical ceiling tiles are laminated. Figure 9.5-7 shows two-ply construction of wallboard bonded to backing board. Backing boards differ from regular wallboards mainly in that the paper covering on both faces and edges is a gray liner paper not suitable for decorating. Backing boards may be secured to framing with staples, screws, or nails. The main types are regular, fire-resistant (Type X), and foil-backed. *Coreboard*, a ¾- or 1-in. (19.1 or 25.4 mm)-thick, factory-laminated board, which consists of two layers of regular backing board, is used to build solid gypsum partitions and shaft walls.

A highly water-resistant backing board with treated paper and core is produced as a base for tile in areas subject to direct wetting, such as in shower stalls.

9.5.2.3 Gypsum Sheathing

Gypsum sheathing provides fire resistance, wind bracing, and a base for exterior finishes. It is made of a water-resistant gypsum core completely enclosed with a firmly bonded water-repellent paper, which eliminates the need for sheathing paper. Standard 2- by 8-ft. (609 by 2438 mm) boards have V-shaped tongue-and-groove (T&G) edges to shed water. These are nailed or screwed horizontally across the framing members (Fig. 9.5-8). The 4- by 8-ft. (1219 by 2438 mm) and 4- by 9-ft. (1219 by 2743 mm) boards have square edges and are installed vertically where improved wind bracing and infiltration resistance are desired.

9.5.2.4 Gypsum Formboard

Gypsum formboards are used as permanent forms and as finished ceilings for gypsum concrete roof decking (see Section 3.17). Formboards are surfaced on the exposed face and longitudinal edges with calendered manila paper specially treated to resist fungus growth. The back face is surfaced with gray liner paper as in regular wallboard. A vinyl-faced board, which provides a white, highly reflective and durable surface, is also available.

9.5.2.5 Gypsum Lath

Gypsum lath is used as a base for plastering. It consists of a gypsum core enclosed by a multilayered, fibrous paper covering, designed to ensure good bond with gypsum plaster. A suction bond is created by tiny needlelike crystals of plaster forming in the porous paper covering. Gypsum lath may be stapled, screwed, nailed, or clipped to supporting construction (Fig. 9.5-9). The main types of gypsum lath are

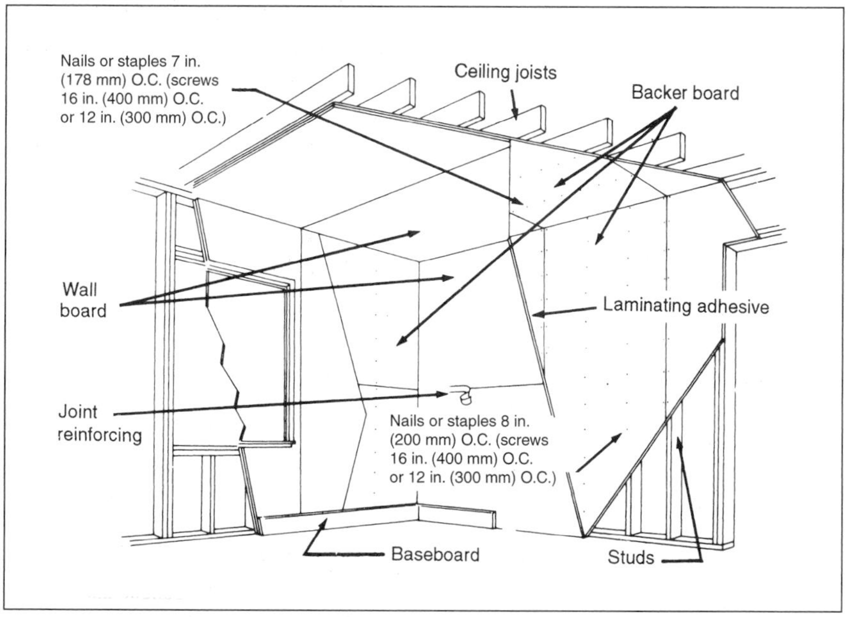

FIGURE 9.5-7 Double-layer (laminated) gypsum board applied to wood construction. Application over metal framing is similar.

FIGURE 9.5-8 Gypsum sheathing applied horizontally to wood construction.

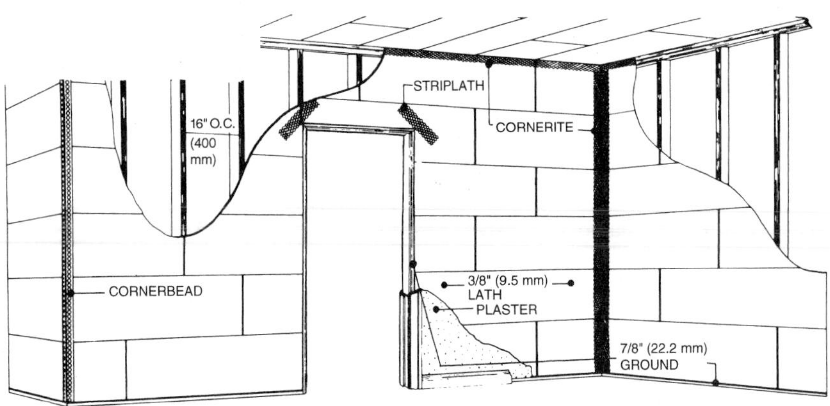

FIGURE 9.5-9 Application of gypsum lath over framing.

plain, perforated, foil-backed, and fire-resistant (Type X).

PLAIN LATH

Plain lath is ordinary gypsum lath consisting of a gypsum core and paper covering, with none of the special features incorporated in the other available types.

PERFORATED LATH

Before the development of Type X lath, perforated lath, with ¾-in. (19.1 mm) holes drilled 4 in. (100 mm) on centers in each direction, was used to improve the lath-to-plaster bond under fire exposure. Covering and core material are the same as for plain lath.

FOIL-BACKED LATH

To make foil-backed lath, a bright-finish aluminum foil is laminated to the back of plain lath. The foil acts as a vapor retarder and can act as a reflective insulation when the reflective surface faces a ¾-in. (19.1 mm)-wide dead air space.

FIRE-RESISTANT (TYPE X) LATH

Special additives and glass fibers in the gypsum core of Type X lath improve its fire-resistant qualities. Covering materials are the same as for plain lath.

9.5.3 SUPPORTING CONSTRUCTION

Gypsum board can be applied directly to masonry, concrete, wood, or metal supporting construction that is capable of supporting the applicable design loads and that will provide a firm, level, plumb, and true base. Generally, the deflection of ceiling members supporting gypsum board finishes should not exceed 1/240 of the span under the full design load. Headers or lintels should be provided over openings to support structural loads, and special construction should be provided to support wall-hung equipment and fixtures.

Wood and metal supporting construction usually consists of:

- *Self-supporting framing:* Studs, joists, and trusses
- *Furring members:* Wood strips and metal channels, supported by underlying construction

Framing and furring should be accurately aligned in a single plane and spaced not to exceed the maximum recommended for the gypsum board thickness being installed. Furring is necessary when the spacing of framing exceeds the maximum allowable span for the gypsum board.

9.5.3.1 Concrete and Masonry

Gypsum board can be applied directly to above-grade interior masonry or concrete if the surface is dry, smooth, plumb, and true. Above-grade exterior cavity walls may also be suitable for direct lamination if the cavity is properly insulated. However, predecorated wallboard with a decorative film that is impermeable to vapor should not be laminated directly to concrete or masonry, because moisture may become trapped within the gypsum core of the board.

Before gypsum board is directly applied, rough or protruding edges in the substrate should be removed and depressions greater than 4 in. (101.6 mm) in diameter or ⅛ in. (3.2 mm) deep should be filled with grout or similar filler. Mortar joints should be cut flush with the face of the wall. Surfaces should be free from form oils, curing compounds, loose particles, dust, grease, and efflorescence to ensure adequate bond. Concrete must be allowed to cure for at least 28 days before gypsum board is applied to it.

Exterior and below-grade walls and horizontal surfaces subject to moisture or excessive heat loss, and surfaces that cannot be prepared readily for adhesive lamination, are not suitable for direct application of gypsum board. These should be furred to separate the gypsum board from possible moisture in the wall and to provide a suitable base for attaching the gypsum board.

9.5.3.2 Metal Framing and Furring

Metal framing and furring are common in commercial, institutional, and high-rise residential buildings where noncombustible construction is required. In such instances, non-load-bearing partitions are typically framed with metal studs; ceilings are usually framed with open-web joists or steel beams.

WALL FRAMING

Metal studs are usually C-shaped, 25-gauge (0.6 mm), cold-formed electrogalvanized steel units, conforming with ASTM Standard C 645. They are available in 20- and 25-gauge (1.0 and 0.6 mm) thicknesses, which are compatible with most drywall sheet metal screws. Proprietary nailable studs (Fig. 9.5-10) were readily available in the past and may be found in some existing construction. They are difficult or impossible to obtain today.

CEILING FRAMING

Floor and roof framing often consists of steel beams or open-web joists. These members are not designed to receive gypsum board directly and are often spaced more than 24 in. (609 mm) apart. Therefore, it is necessary to provide furring or suspension systems for applying gypsum board on the ceilings of metal-framed buildings (Fig. 9.5-11).

FURRING

Wood or metal furring will:

- Provide plumb, true, and properly spaced supports;
- Eliminate capillary transfer of moisture in exterior and below-grade walls;
- Minimize moisture condensation on interior wall surfaces;
- Improve the thermal insulating properties of exterior walls;

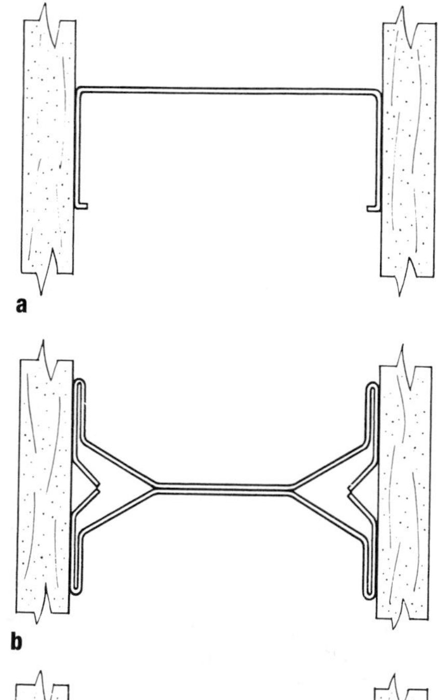

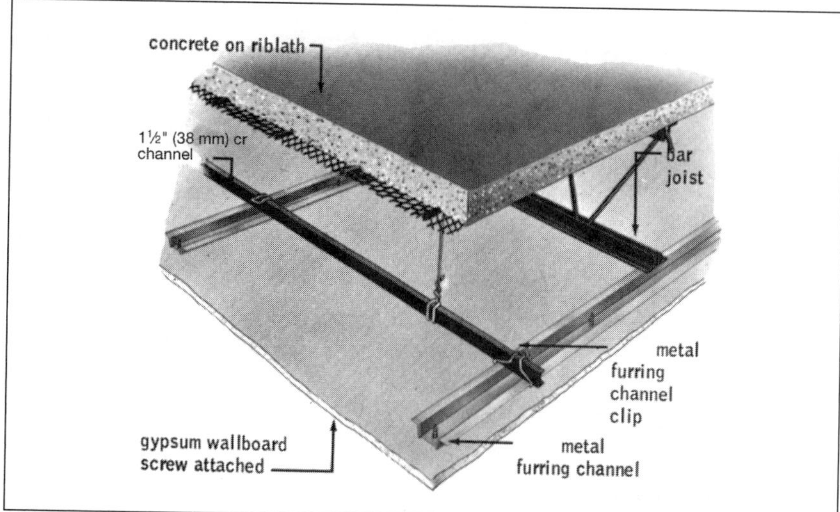

FIGURE 9.5-11 Open-web bar joists usually require a grid of cold-rolled channels and hat-shaped furring channels to support gypsum board.

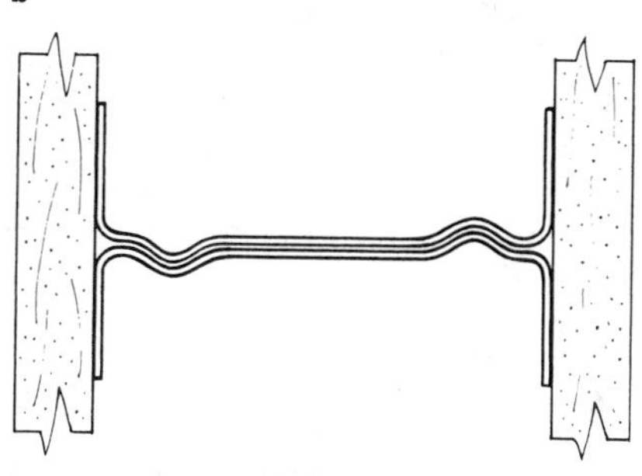

FIGURE 9.5-10 Metal studs: (a) C-shaped stud designed to accept drywall screws (type most used today), (b) nailable stud for special ring-shank nails, (c) nailable stud for smooth-shank nails. Types (b) and (c) are seldom used today and may not be available in many areas.

Plain Channel

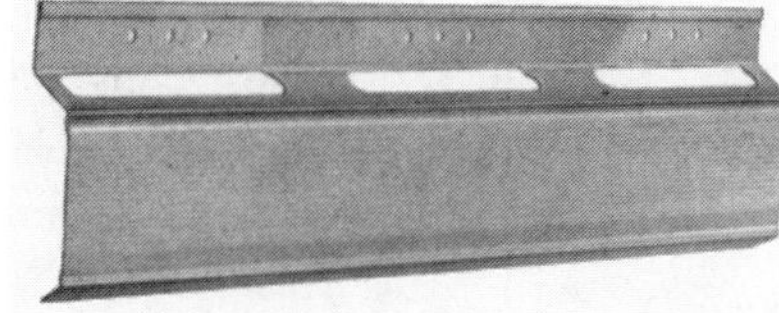

Resilient Channel

Cold Rolled Channel

FIGURE 9.5-12 Metal furring channels. Plain channels are also called hat-shaped channels. (Courtesy Gypsum Association.)

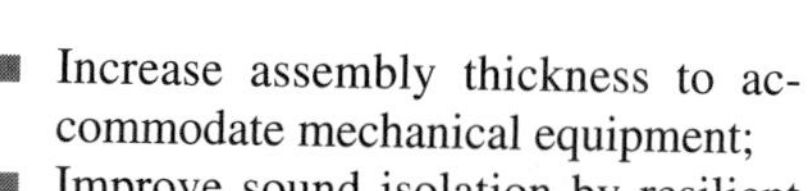

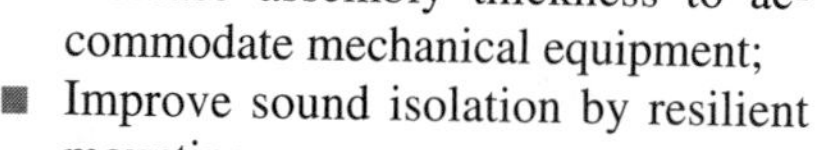

- Increase assembly thickness to accommodate mechanical equipment;
- Improve sound isolation by resilient mounting.

Metal furring can be used with all types of supporting construction and is preferable to wood furring when noncombustible or sound-isolating assemblies are required. Metal furring members may be either cold-rolled channels or drywall channels (Fig. 9.5-12).

Cold-rolled channels are multipurpose members used for both plaster and gypsum board walls and suspended ceiling assemblies. They are usually of 16-gauge (1.6 mm) steel with either a galvanized or a black asphaltum finish. They are ¾ to 2 in. (19.1 to 50.8 mm) wide and 16 to 20 ft. (4880 to 6100 mm) long. Because they are too thick to receive drywall sheet metal screws, these channels are usually wire-tied in place. In gypsum board assemblies, cold-rolled channels are used primarily as a supporting grid for lighter drywall channels to which wallboard can be screw-attached (see Fig. 9.5-11).

Drywall channels are roll-formed of 25-gauge (0.6 mm) electrogalvanized steel, designed for gypsum board attachment with self-drilling sheet metal screws. They are installed perpendicular to wood framing with a 1¼-in. (28.6 mm) screw or a 5d nail at each support and are clipped to cold-rolled furring channels. Over masonry or concrete surfaces, they may be installed horizontally or vertically, whichever is more convenient.

Proprietary clips, runners, and adjustable brackets are available with engineered metal furring systems to facilitate rapid erection over irregular masonry walls (Fig. 9.5-13).

Drywall channels can be either plain (hat-shaped) or resilient. Plain channels are used primarily with unit masonry and concrete wall assemblies that possess inherent sound isolation because of their mass and are not significantly improved by resilient mounting. In suspended ceiling assemblies, a grillage composed of cold-rolled channels and plain drywall channels is used to support a gypsum board finish (Fig. 9.5-14). If sound or vibration isolation is desired, resilient hangers can be wired into the suspension system to isolate the ceiling from the framing.

Resilient channels are used over both metal and wood framing (Fig. 9.5-15). In addition to improving sound isolation, they help to isolate the gypsum board from structural movement, thus reducing joint cracking.

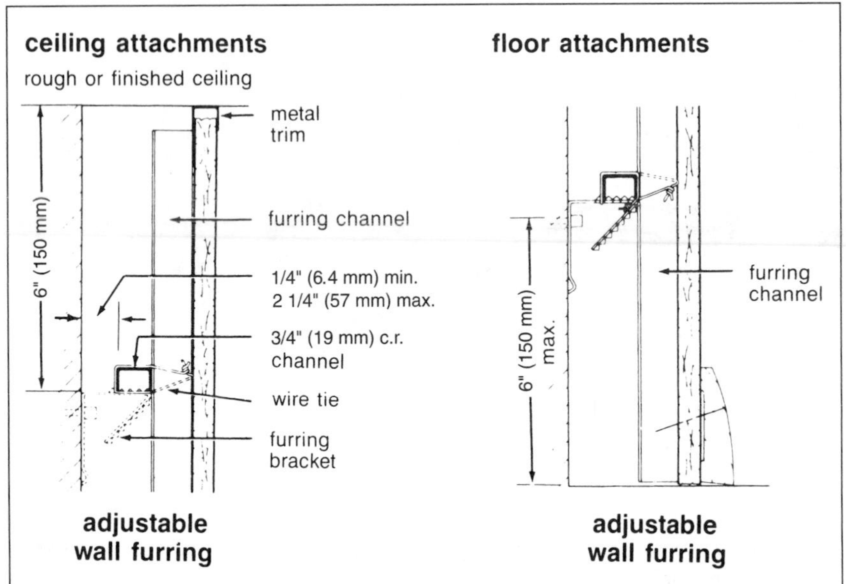

FIGURE 9.5-13 Adjustable brackets have notched surfaces to permit plumb and true installation of channel grid. (Courtesy Gypsum Association.)

FURRING CHANNEL DETAILS

HANGAR SPACING 4'0" C-TO-C MAX.

1-1/2" CHANNELS 4'0" C-TO-C MAX.

FURRING CLIP

FURRING CHANNEL

GYPSUM WALLBOARD

MAX.FURRING CHANNEL SPACING 24" C-TO-C 1/2" AND 5/8" GYPSUM BOARD

4'-0" = 1200 mm
1½" = 38 mm
24" = 300 mm
½" = 12.7 mm
⅝" = 15.9 mm

FIGURE 9.5-14 In suspended ceilings, wire hangers are used to support and level the channel grid. (Courtesy Gypsum Association.)

9.5.4 ACCESSORY MATERIALS

Accessory materials for the application of gypsum board include mechanical fasteners, adhesive, joint compounds, tape, and metal edge and corner trim.

9.5.4.1 Fasteners

Nails and screws are used to apply both single- and multi-ply finishes. Clips and staples are limited to attaching the base ply in multi-ply construction.

NAILS

Nails for applying gypsum board can be bright, coated, or chemically treated. Shanks may be either smooth or annularly threaded, with medium or long diamond points. Nail heads are flat or slightly concave, thin at the rim and not more than 5⁄16 in. (7.9 mm) in diameter. Nail heads of about ¼-in. (6.4 mm) diameter provide adequate holding power without cutting the face paper.

Either annularly threaded nails developed for wallboard (GWB-54) or smooth- or deformed-shank nails suitable for the application of gypsum board should be used. All nails should conform to ASTM C 514. Casing nails and common nails have heads that are too small in relation to the shank or are too thick and should not be used.

Nails should be of the proper length for the wallboard thickness (Fig. 9.5-16). Generally recommended penetration into supporting construction for smooth-shank nails is ⅞ in. (22.2 mm). Annularly threaded nails provide greater withdrawal resistance, require less penetration, and generally minimize nail popping. For fire-rated construction, however, penetration of 1 in. (25.4 mm) or more is usually required and longer smooth-shank nails are used.

SCREWS

Both regular and fire-rated gypsum board can be fastened to both wood and metal supporting construction with *drywall screws*. The usual finish for drywall screws is a zinc phosphate coating with baked-on linseed oil. These screws are typically self-drilling and have self-tapping threads and flat Phillips heads for use with a power screwdriver. Their contour head

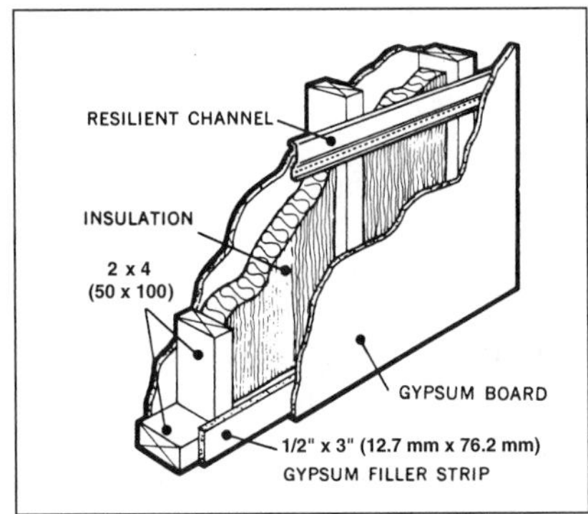

FIGURE 9.5-15 Resilient channel mounting substantially improves sound isolation in wood frame construction. (Courtesy Gypsum Association.)

FIGURE 9.5-16 Nails for Gypsum Board Application

	Recommended Penetration		Board Thickness		Nail Length	
Drywall Nails	**in.**	**mm**	**in.**	**mm**	**in.**	**mm**
Annularly threaded[a] ASTM C 514	¾	19	⅜	9.5	1⅛	29
			½	12.7	1¼	32
			⅝	15.9	1⅜	35
Smooth shank[b] ASTM C 514	⅞	22	⅜	9.5	1¼	32
			½	12.7	1⅜	35
			⅝	15.9	1½	38

Note: For fire-rated constructions, building code and/or manufacturer's specifications should be followed.
[a]Typically, GWB 54: ¼-in. (6.4 mm)-diameter head, 0.098-in. (2.49 mm)-diameter shank.
[b]Typically, ¼-in. (6.4 mm)-diameter head, 0.099-in. (2.51 mm)-diameter shank, with dished head.
(Courtesy Gypsum Association.)

design makes a uniform depression free of ragged edges and fuzz.

Screws pull gypsum board tightly to the framing without damaging the board, thus minimizing fastener surface defects due to loose board attachment. The three basic types of drywall screws for wood, sheet metal, and gypsum construction and the recommended penetration for various uses are shown in Figure 9.5-17. Other specially designed screws are available for attaching wood or metal trim and for wallboard attachment to heavier-gauge load-bearing studs.

Drywall Wood Screws Type W and similar screws are designed for fastening to wood framing or furring. Type W screws are diamond-pointed to provide efficient drilling action through both gypsum and wood and have a specially designed thread for quick penetration and increased holding power. Recommended minimum penetration into supporting construction is ⅝ in. (15.9 mm), but in a two-ply application when the face layer is being screw-attached, the additional holding power developed in the base ply permits reducing the penetration into supports to ½ in. (12.7 mm). Type W screws are available in 1¼-in. (28.6 mm) length. Drywall sheet metal screws, more readily available in longer sizes, may be substituted in two-ply construction.

Drywall Sheet Metal Screws Type S and similar screws are designed for fastening gypsum board to 25-gauge (0.6 mm) metal studs or furring. Type S screws have a self-tapping thread and a mill-slot drill point designed to penetrate sheet metal with little pressure. This is important because steel studs are flexible. The threads should be of adequate depth and should be turned within ¼ in. (6.4 mm) of the head to eliminate stripping. They are available in several lengths, from 1 to 2¼ in. (25.4 to 57.2 mm). Other lengths and head profiles are available for attaching wood trim, metal trim, and metal framing components. Recommended minimum penetration through sheet metal for drywall sheet metal screws is ⅜ in. (9.5 mm).

Drywall Gypsum Screws Type G and similar screws are used for fastening gypsum board to gypsum board. Type G screws are similar to Type W screws, but have a deeper special thread design. They are generally available in 1½-in. (38 mm) length only. Drywall gypsum screws require penetration of at least ½ in. (12.7 mm) of the threaded portion into the supporting gypsum board. Allowing approximately ¼ in. (6.4 mm) for the point, this results in a minimum penetration of ¾ in. (19.1 mm). For this reason, drywall gypsum screws should not be used to attach wallboard to ⅜-in. (9.5 mm)-thick backing board. In two-ply construction with a ⅜-in. (9.5 mm)-thick base ply, nails or longer screws should be used to provide the necessary penetration into supporting wood or metal construction.

FIGURE 9.5-17 Typical Screws for Gypsum Board Application

Drywall Screws*	Recommended Minimum Penetration	Screw Length in.	Screw Length mm	Application
For fastening to wood ASTM C 1002	⅝ in. (15.9 mm)	1¼	32	Single-ply or base-ply application
	⅝ in. (15.9 mm)	1⅝	41	Face-ply application in two-ply construction
For fastening to sheetmetal ASTM C 1002	⅜ in. (9.5 mm)	1	25.4	Single-ply or base-ply application
	⅜ in. (9.5 mm)	1⅝	41	Face-ply application in two-ply construction
For fastening to gypsum board ASTM C 1002	½ in. (12.7 mm); ¾ in. (19 mm)	1½	38	Gypsum board to gypsum board in multi-ply construction
	¾ in. (19 mm); ¾ in. (19 mm)			

Note: Applications are generally acceptable for fire-rated construction.

*Screws shown are Types W, S, and G, respectively; other drywall screws are available.

(Courtesy Gypsum Association.)

STAPLES

Staples are recommended only for attaching the base ply to wood members in multi-ply construction. They should be of 16-gauge (1.6 mm) flattened galvanized wire with a minimum 7⁄16-in. (11 mm)-wide crown and with divergent sheared bevel points. Staples should be long enough to provide a minimum penetration of ⅝ in. (15.9 mm) into supporting construction.

9.5.4.2 Adhesives

Adhesives can be used to attach single-ply wallboard directly to wood framing, masonry, or concrete, or to laminate gypsum board to gypsum board, sound deadening board, or rigid foam insulation. They generally are used in combination with nails or screws, which provide temporary or permanent supplemental support.

Adhesive formulations vary, and manufacturers' recommendations should be consulted in each case. Adhesives for applying wallboard finishes may be classed as *stud adhesives*, *laminating adhesives*, and *contact adhesives*.

STUD ADHESIVES

Stud adhesives are formulated for attaching single-ply gypsum board to wood supporting construction. They generally are rubber or asphaltic based, are of a caulking consistency that will bridge framing irregularities, and are applied with a gun in a continuous bead. Structural stud adhesives should conform to ASTM Standard C 557, which establishes rigid standards for strength, bridging ability, aging, and other qualities necessary for long-term structural performance. Fewer permanent supplemental fasteners are required for structural stud adhesives than are required for adhesive-nail attachment with nonstructural stud adhesives.

LAMINATING ADHESIVES

Most laminating adhesives are casein-based adhesives in powder form to which water is added. They are used to laminate gypsum boards to each other and to suitable masonry or concrete surfaces. They are not suitable for adhesive bonding to wood. In addition to the several adhesives specifically formulated for laminating purposes, some powder embedding compounds (see Section 9.5.4.3) can be used when mixed at the recommended consistency. When specifically recommended by their manufacturers, some stud adhesives and contact adhesives can also be used for laminating purposes.

Only as much adhesive should be mixed as can be used within the working time specified by the manufacturer. Laminating adhesive is applied over the entire board area with a spreader. Installed boards require temporary support or supplemental fasteners until the adhesive develops sufficient bond strength.

CONTACT ADHESIVES

Contact adhesives have a synthetic rubber base, sometimes with resins added, in either solvent or emulsion formulations. They are used to laminate gypsum boards to each other and to bond wallboard to metal studs in demountable partitions. The adhesive is applied by roller, spray gun, or brush in a thin, uniform coating to both surfaces to be bonded. For most contact adhesives some open time is required before surfaces can be joined and bond can be developed. To ensure maximum possible adhesion between mating surfaces, the face board should be impacted over its entire surface with a suitable tool such as a rubber mallet.

Contact adhesives provide an immediate bond, excellent long-term strength, resistance to fatigue, and require no temporary or permanent fasteners in the field of the board. However, they are unable to bridge irregularities in the mating surfaces, thus creating discontinuity in the bond film. In addition, most contact adhesives do not permit repositioning of the boards once contact between surfaces has been made. A slip sheet of polyethylene film or building paper can be used to facilitate gradual bonding of surfaces as the slip sheet is withdrawn.

MODIFIED CONTACT ADHESIVES

Modified contact adhesives combine good long-term strength, bridging ability, and sufficient immediate bond to permit erection with a minimum of temporary fasteners. In addition, they have an open assembly time of about ½ hour before bond develops, during which time boards can be repositioned. Modified contact adhesive is intended for attaching wallboard to all types of supporting construction, including framing, furring, gypsum board, and rigid insulating board.

9.5.4.3 Joint Tape and Joint Compounds

Joint tape and joint compounds are used to produce a smooth, monolithic wallboard finish with inconspicuous joints and no visible fasteners. Joint compounds and tape should conform to ASTM C 475 "Specifications for Joint Treatment Materials for Gypsum Wallboard Construction."

JOINT TAPE

Tape used for joint reinforcement is a strong-fibered tape with chamfered edges. It resists tensile stresses across the joint as well as longitudinally.

JOINT COMPOUNDS

Joint compounds include:

- *Embedding compounds:* For embedding and bonding joint tape
- *Topping compounds:* For final smoothing and leveling over joints and fasteners
- *All-purpose compounds:* Combine the features of the other two

Joint compounds are either premixed or packaged in powder form to be mixed with water on the job. Except for quick-setting embedding compounds, these compounds harden by evaporation of the water and can be kept in a wet-mix state for several days if covered with a moist cloth or a thin layer of water. Both premixed and job-mixed compounds should be protected from freezing.

Embedding Compounds Embedding compounds are casein-based adhesives, similar to laminating adhesives. They are available either premixed or as a powder that is mixed with water. They are used for embedding tape and as a first coat for concealing fasteners and edge and corner trim. Some embedding compounds can also be used for laminating gypsum board in multi-ply construction when mixed to the proper consistency. Usually, embedding compounds require at least overnight drying before following coats can be applied. However, quick-setting compounds that dry in 2 to 4 hours are also available, which permits three-coat application of embedding and topping compounds in one day.

Topping Compounds Topping compounds are surface fillers used to conceal and smooth over embedded tape, fasteners, and trim. These are casein or casein-vinyl formulations, available in either premixed or powder form. They bond well with joint tape and compound, gypsum board, and fasteners. These compounds can be sanded easily and provide a surface with sufficient "tooth" and suction for painting.

All-Purpose Compounds All-purpose compounds work as both adhesives and fillers. They are available in either powder or premixed form, and in either machine or hand-tool consistency. All-purpose compounds can be used for embedding tape, topping over tape, finishing over metal trim, and concealing fasteners. However, they should not be used for laminating gypsum boards unless specifically recommended by their manufacturer.

9.5.4.4 Edge and Corner Trim

Exposed wallboard edges and ends are usually protected by trim. Wood casing trim is often used at window and door frames, and wood or resilient wall base is used to protect wallboard edges at the floor. Metal trim, such as cornerbeads and casing beads, are used in similar locations when inconspicuous edge protection is desired and applied trim is not to be used.

Plastic trim may be used when exposing the trim is acceptable or desired.

Metal trim is available in standard shapes for different conditions and varying board thicknesses (Fig. 9.5-18). On outside corners, metal cornerbeads are used to protect gypsum board from damage and provide neat, slightly rounded corners, which are difficult to obtain with tape and joint compound alone. Casing beads and corner beads with metal flanges are generally nailed or screwed to supporting construction at about 6 in. (150 mm) on centers. Beads with paper flanges are held in place by embedding the flanges in joint compound.

9.5.5 JOB CONDITIONS

Temperature and humidity affect the performance of joint treatment materials, the scheduling of joint finishing operations, the quality of a joint, and sometimes the bonding properties of adhesives.

Interior finishes should not be installed when the outside temperature is less than 55°F (12.8°C), unless the building is completely enclosed and has a controlled heat of 55°F (12.8°C) to 70°F (21°C). This temperature should be maintained for 24 hours before installation, during installation, and until a permanent heating system is in operation. A building need not be completely glazed during summer months, but all materials and installation should be protected from the weather.

Ventilation should be provided to eliminate excessive humidity. If the building has been glazed, windows can be kept partially open to provide air circulation. Temporary air circulators should be used in enclosed areas that have no natural ventilation. Additional drying time between coats of joint compound should be allowed when drying conditions are slow. During hot, dry weather, drafts should be avoided to prevent joint compounds from drying too rapidly.

The delivery of gypsum board should coincide as closely as possible with the installation schedule. Boards should be stored flat, inside, and under cover. Stacking long lengths on short lengths should be avoided to prevent the longer boards from breaking. For short periods of time, boards may be placed vertically against the framing with the long edges of the boards horizontal. Materials should remain in their wrappings or containers until ready for use.

USASI Designation	Trim Profile	Application
CORNERBEAD[1] CB—100 x 100 CB—118 x 118 CB—114 x 114 CB—100 x 114 CB—PF (paper flange steel corner combination bead)	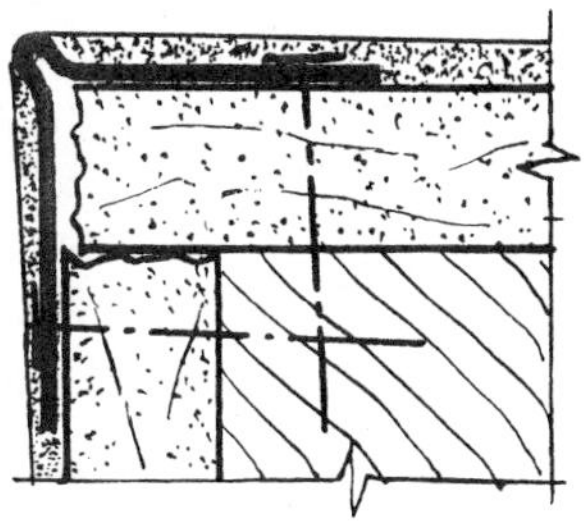	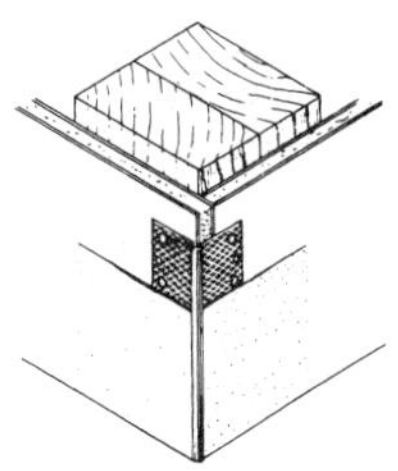
U BEAD[2] U-38 U-12 U-58 U-34	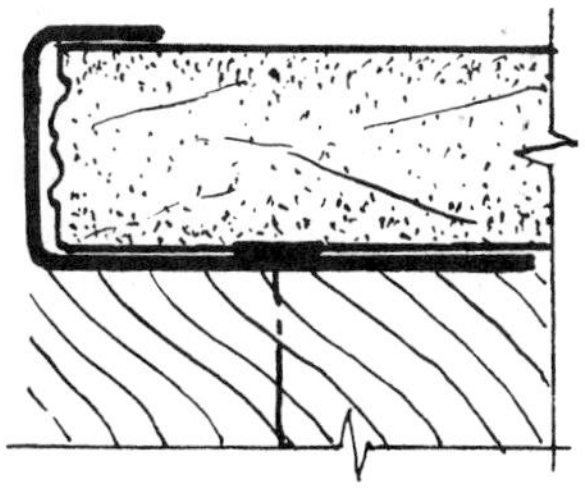	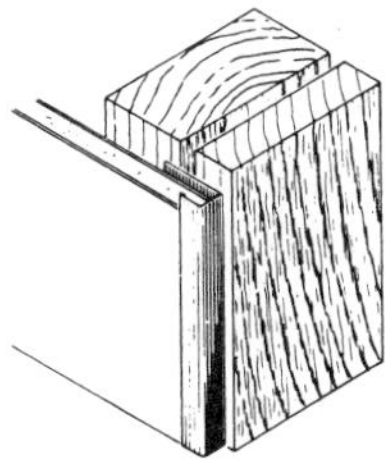
L BEAD[2] LB-38 LB-12 LB-58 LB-34	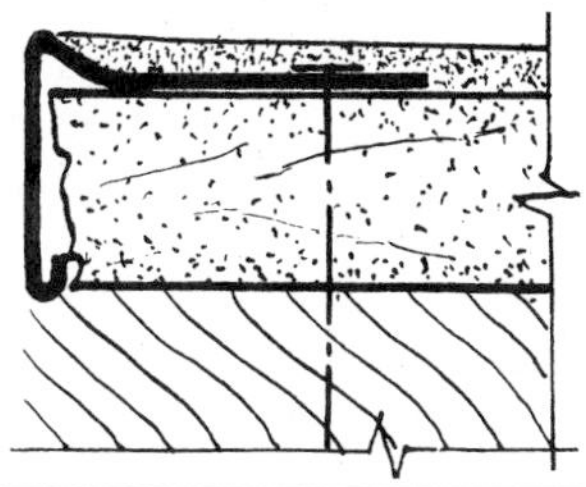	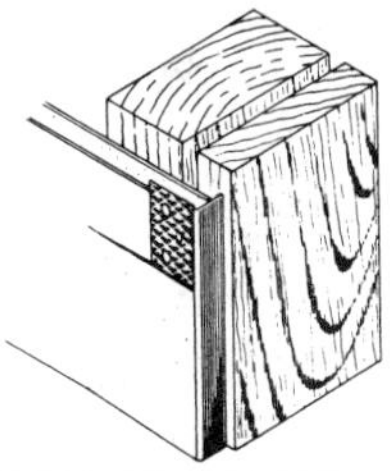
LK BEAD[3,4] LK-S LK-B	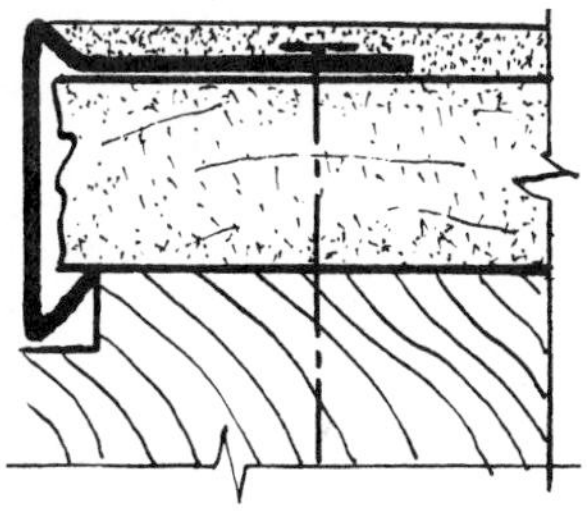	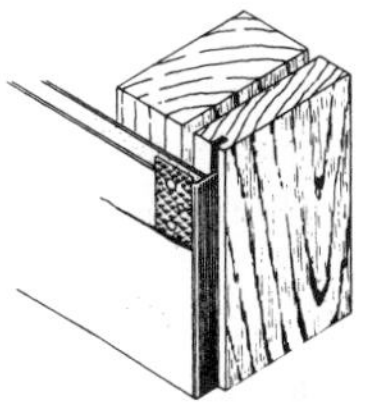
LC BEAD[2] LC-38 LC-12 LC-58 LC-34	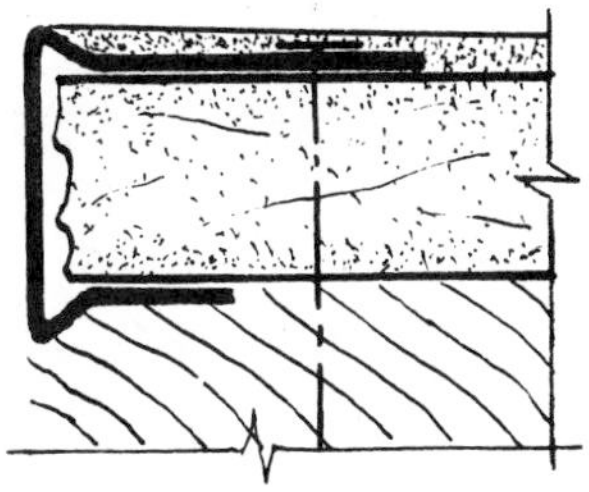	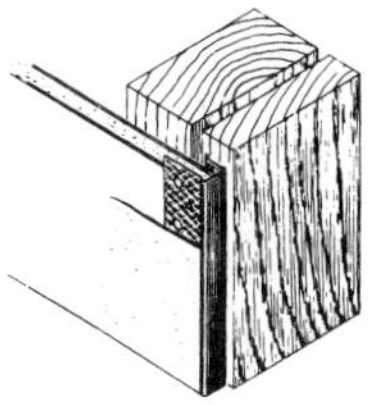

[1]Numbers indicate width of flanges: e.g., 114 × 114 has two 1¼″ (28.6 mm) flanges.

[2]Numbers indicate thickness of board intended: e.g., U-12 is U Bead intended for ½″ (12 mm) wallboard.

[3]Letter (S) indicates square nose; letter (B), bull or round nose.

[4]For use with kerfed jamb.

FIGURE 9.5-18 Standard drywall metal-trim shapes. (Courtesy Gypsum Association.)

9.5.6 BOARD APPLICATION

Gypsum board finishes may be either single-ply using wallboard alone, or multi-ply using wallboard over wallboard, backing board, or sound-deadening board.

Installing wallboard perpendicular to supporting members results in a stiffer assembly and is preferred. Whenever possible, boards should be applied to span ceilings and walls in one piece, to avoid butt-end joints, which are difficult to finish inconspicuously. Unavoidable butt-end joints should be staggered and located as far from the center of a wall or ceiling as possible, so they will be less conspicuous. With the exception of face layers in two-ply construction, board ends and edges that are parallel to supporting members should fall on these members.

Successful application of gypsum board finishes depends on proper measuring, cutting, and fitting; board attachment; and joint and fastener treatment.

9.5.6.1 Measuring, Cutting, and Fitting

Measurements should be taken accurately at the intended location for each edge or end of a board. Accurate measuring will usually indicate irregularities in framing and trimming so that allowances can be made in cutting.

Straight-line cuts across the full width or length of a board are made by scoring through the face paper into the gypsum core (Fig. 9.5-19a). The core of the board is then snapped and the back paper is cut (Fig. 9.5-19b). Cut edges should be smoothed with a rasp or trimmed with a sharp knife (Fig. 9.5-19c). Panels should be cut so that they fit easily in place.

Ceiling panels should be installed first. Joints should be loosely butted without forcing the boards into position. Tapered edges should be placed next to tapered edges, and square-cut ends should be butted to square ends. Square ends should not be placed next to tapered edges, because excessive build-up of joint compound over the tapered edge results in a conspicuous ridge caused by differential shrinkage.

9.5.6.2 Board Attachment

Wallboard and backing board can be attached to supporting construction with mechanical fasteners, adhesive, or a combination of adhesive and mechanical fasteners.

MECHANICAL FASTENERS

Nails and screws are used to attach gypsum board in both single- and multi-ply installations. In two-ply construction, however, staples and clips can be used to attach the base ply. Adhesives, when used, are generally supplemented with mechanical fasteners.

Fasteners should be of a type recommended for the intended method of application. Ordinary wood screws, sheet metal screws, and common nails are not designed to hold gypsum board tightly or to countersink neatly and should not be used. Fasteners should be placed at least 3/8 in. (9.5 mm) from board edges and ends. Application should start in the middle of a board and proceed outward toward the perimeter. Fasteners should be driven as nearly perpendicular as possible while the board is held in firm contact with supporting construction. Nails should be driven with a crown-headed hammer forming a uniform depression around the nail head. Fastener heads should not be seated deeper than about 1/32 in. (0.8 mm) below the board surface, and particular care should be taken not to break the face paper (Fig. 9.5-20).

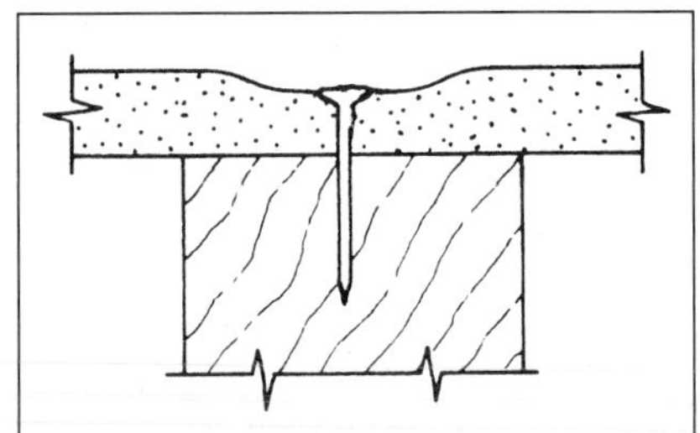

FIGURE 9.5-20 A properly driven drywall nail. (Courtesy Gypsum Association.)

ADHESIVE BONDING

Adhesive can be used to secure single-ply wallboard to wood framing and furring, masonry, and concrete, and to laminate a face ply to a base layer of gypsum board, sound-deadening board, or rigid foam insulation. The surface to be bonded should be free of dirt, grease, oil, and other foreign material.

9.5.6.3 Single-Ply Construction

In single-ply construction, a single layer of gypsum board is permanently nailed, screwed, or adhesive-bonded to supporting wood or metal studs or furring (Fig. 9.5-21). After appropriate fastener and joint treatment, the surface may be painted or finished with wall covering or wall tile.

This subsection discusses assemblies, methods, and precautions needed to obtain suitable finishes with single-ply construction. Refer to Section 9.5.6.6 for a more detailed discussion about provisions for sound isolation, fire resistance, thermal resistance, moisture resistance, and other special construction. Also refer to Chapter 13 for additional information about sound isolation.

Sound ratings ranging from 35 to 50 STC (see Section 13.1) can be achieved with single-ply construction. The higher values require special construction systems such as resilient mounting, separated construction, or sound-absorbing materi-

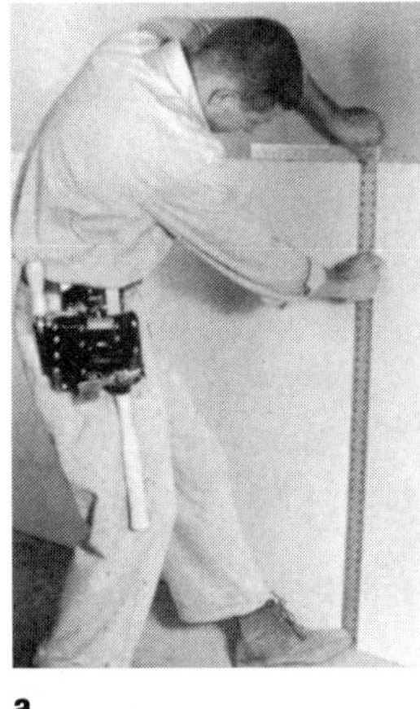

a

b

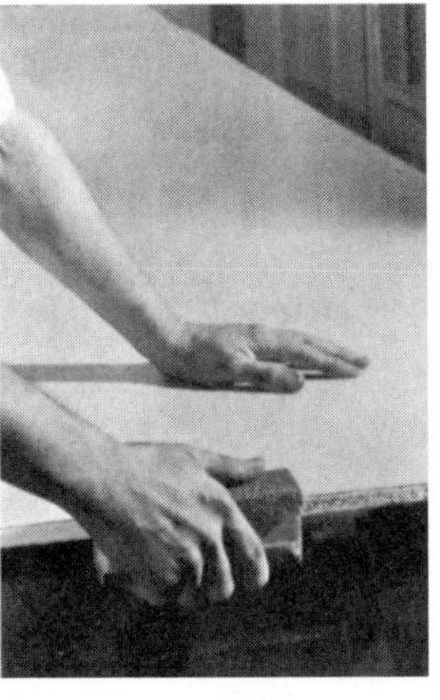

c

FIGURE 9.5-19 (a) Face paper is scored; (b) gypsum core is snapped along score line and back paper is cut; (c) rough edges are smoothed with coarse sandpaper. A rasp or sharp knife can also be used. (Courtesy Gypsum Association.)

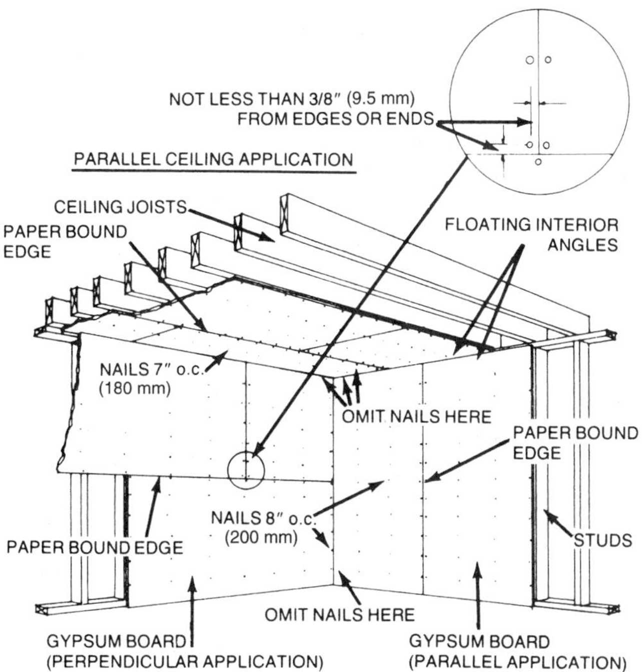

FIGURE 9.5-21 **Gypsum board application over wood-frame construction.** (Courtesy Gypsum Association.)

FIGURE 9.5-22 Support Spacing for Single-Ply Construction

Wallboard Thickness				Maximum Support Spacing, o.c.	
in.	**mm**	**Location**	**Application[a]**	**in.**	**mm**
3/8	**9.5**	Ceilings[b,c]	Perpendicular	16	406
		Walls	Perpendicular or parallel	16	406
1/2 and 5/8	**12.7 and 15.9**	Ceilings[c]	Perpendicular Parallel	24 16	610 406
		Walls	Perpendicular or parallel	24	610

[a]Direction of wallboard edges relative to supports.

[b]3/8-in. (9.5 mm) ceilings should not support insulation.

[c]For ceilings to receive a water-based spray texture finish, use 5/8-in. (15.9 mm) board applied perpendicular to 24-in. (610 mm) supports or 1/2 in. (12.7 mm) perpendicular to 16-in. (406 mm) supports.

(Courtesy Gypsum Association.)

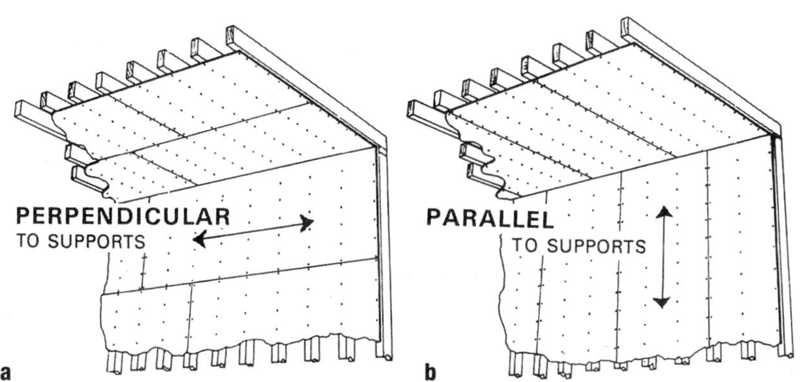

FIGURE 9.5-23 **(a) Horizontal and (b) vertical application of gypsum board to wood frame construction using the single nailing method.** (Courtesy Gypsum Association.)

als. Fire ratings for single-ply framed partitions are generally limited to 1 hour.

Because of the large number of fasteners in the surface of the board and the greater dependence on accurately aligned supporting construction, single-ply finishes are more susceptible to surface problems such as joint ridging, corner cracking, and nail popping. However, with the selection of proper supporting construction and methods of attachment; attention to ridging, cracking, and nail popping precautions; and care in joint treatment, it is possible to produce visually attractive, durable finishes.

Framing or furring members should be spaced as recommended in Figure 9.5-22. Except for 3/8-in. (9.5-mm)-thick wallboard on ceilings, which should be applied perpendicular to the supports, boards can be applied either perpendicular or parallel to supporting members (Fig. 9.5-23). Perpendicular application provides greater strength and shorter joint lengths.

Except in end-floated construction, gypsum board edges and ends should occur over supporting members. Butt-end joints should be staggered with respect to each other on the same side of a partition and with respect to joints on the opposite side of the partition.

WOOD FRAMING AND FURRING

Gypsum board is attached to wood furring and framing with nails or screws or by a combination of adhesives and nails. Resilient metal furring can be used with wood framing where increased sound isolation is desired.

Screw Attachment Fewer fasteners are required when screws are used, and the number of fasteners requiring treatment and possible surface defects is reduced. Drywall wood screws, such as Type W, providing at least 5/8-in. (15.9 mm) penetration into supports should be used, spaced as shown in Figure 9.5-24.

Nail Attachment Nailed gypsum board panels must be drawn tightly against framing and furring members so that the board will not move on the nail shank, which will result in nail popping. Wallboard can be attached by either the single- or double-nailing method, in accordance with GA 216. In general, nail spacing is based on requirements for fire resistance and on intimate board-to-stud contact to reduce possible surface defects, rather than on

FIGURE 9.5-24 Single-Ply Screw Attachment to Wood Supports

Support Spacing, o.c.		Screw Spacing, o.c.* Walls		Ceilings	
in.	mm	in.	mm	in.	mm
16	406	16	406	12	305
24	610	12	305	12	305

*Spacing applicable to both field and perimeter of board.
(Courtesy Gypsum Association.)

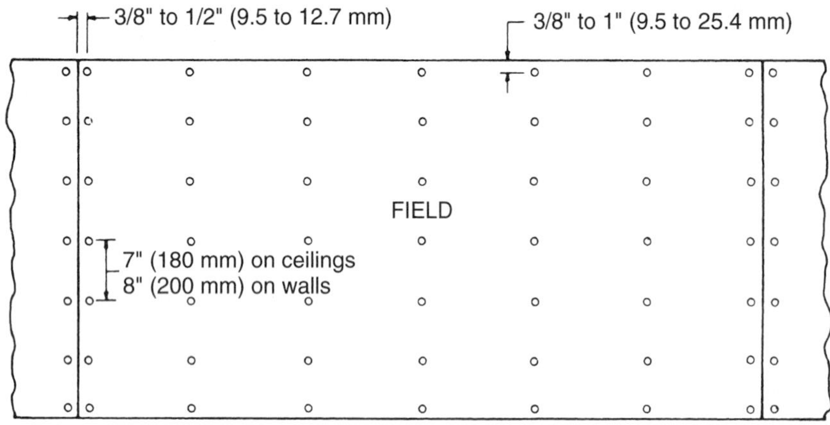

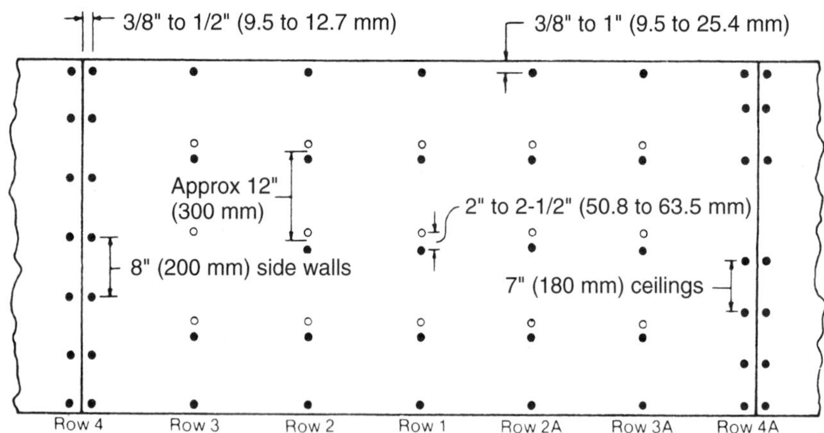

FIGURE 9.5-25 (a) Single nailing method. (b) Double nailing method. First nails (shown by black dots) are installed, starting with Row 1, then Row 2 and 2A, etc., working outward. Second nails (shown by circles) are installed in the same row sequence. (Courtesy Gypsum Association.)

structural support. Depth of nail penetration should be as shown in Figure 9.5-16.

Single nailing can provide satisfactory results if the methods in GA 216 are carefully followed. Nails should be spaced 7 in. (180 mm) on centers on ceilings and 8 in. (200 mm) on centers on walls along supports (Fig. 9.5-25a).

Double nailing ensures consistent board-to-stud contact and fewer defects than single nailing. A set of nails is driven in the field of the panel at 12 in. (300 mm) on centers, followed by a second set of nails 2 in. (50.8 mm) from each of the first nails. Edges and ends falling over supports are nailed at 7 in. (180 mm) on centers on ceilings and 8 in. (200 mm) on centers on walls (Fig. 9.5-25b). The total number of nails per panel is about 10% higher with double nailing, but the nail spots that must be concealed are 20% fewer.

Adhesive-Nail Attachment Adhesive-nail attachment reduces the number of nails in the middle of a panel by at least 50% over the number required for conventional nailing, thereby reducing the amount of fastener treatment. The adhesive between board and supporting members produces stronger, stiffer assemblies and fewer fastener defects. Screws can be used in lieu of nails, but seldom are, because good board-to-stud contact is ensured by the adhesive and the number of face fasteners has already been reduced to the minimum.

A straight bead of *stud adhesive* is applied with a caulking gun to the face of members supporting the field of a panel (Fig. 9.5-26a). Where adjacent panels join over a supporting member, two beads are applied (Fig. 9.5-26b). When a predecorated wallboard joint is to be left untreated, two parallel beads of adhesive should be applied, one near each edge of the member. Generally, the volume of the bead should be such that when the board is applied, the adhesive will spread to an average width of 1 in. (25.4 mm) at least 1/16 in. (1.6 mm) thick.

Care should be taken in applying adhesive to prevent it from being squeezed out through the joints. Adhesive should not be applied to members that are not required for board support, such as diagonal bracing, blocking, and plates.

The number and spacing of supplemental nails (or screws) vary, depending on the adhesive properties.

Nonstructural stud adhesives require fasteners in the field as well as at the

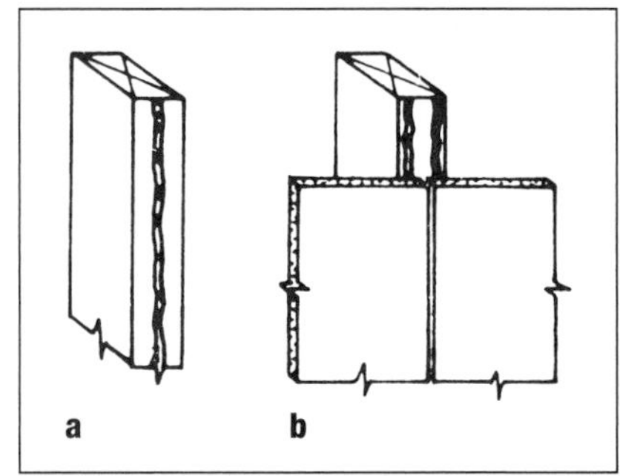

FIGURE 9.5-26 Beads of stud adhesive are applied: (a) single under field of board or (b) parallel under joints. (Courtesy Gypsum Association.)

FIGURE 9.5-27 Fastener Spacing for Nonstructural Adhesive

Location	Support Spacing, o.c.		Fastener Spacing, o.c.*			
			Nails		Screws	
	in.	mm	in.	mm	in.	mm
Ceilings	16	406	16	406	16	406
	24	610	12	305		
Load-bearing walls	16	406	16	406	24	610
	24	610	12	305	16	406
Non-load-bearing walls	16	406	24	610	24	610
	24	610	16	406		

*Applies to both field and perimeter of board.

(Courtesy Gypsum Association.)

perimeter of panels for wall and ceiling installation. Fastener spacing varies with the type of fastener, support spacing, and load-bearing conditions (Fig. 9.5-27).

Structural stud adhesives conforming to ASTM C 557 require only perimeter fasteners for wall application. The fasteners should be spaced 16 in. (400 mm) on centers for edges or ends that fall parallel to supports, and at each support for edges or ends that fall perpendicular to supports. For ceiling application, the perimeter is nailed as for walls, but additional fasteners are required in the field spaced 24 in. (600 mm) on centers (Fig. 9.5-28).

When installing *predecorated wallboard* with beveled or finished edges not requiring joint treatment, it often is desirable to avoid fasteners at the vertical joints. For this type of wallboard and for plain wallboard applied vertically on walls, fasteners may be omitted along panel edges if appropriate means such as prebowing or temporary bracing are used to ensure contact with the adhesive until it develops full bond strength. For plain wallboard, temporary bracing or nailing should be allowed to remain in place and joint or nail treatment should not be started for at least 24 hours after installation.

METAL FRAMING AND FURRING

Resilient channels are nailed or screwed perpendicular to wood framing members through predrilled holes, using a 1¼-in. (28.6 mm) screw or 5d nail at each support. Gypsum board should be attached to metal framing and furring with drywall sheet metal screws, spaced not more than 12 in. (300 mm) on centers along supports for both walls and ceilings. For fire-rated construction, gypsum board is applied parallel to supports with screws spaced 8 in. (203.2 mm) on centers along edges and 12 in. (300 mm) on centers in the field.

CONCRETE AND MASONRY

Concrete and masonry are usually furred before gypsum board is applied. However, gypsum board can be laminated directly to interior and some above-grade exterior masonry (see Section 9.5.3.1) if the surfaces are sufficiently dry, smooth, plumb, and true (Fig. 9.5-29). Some laminating adhesives and stud adhesives are suitable for this purpose, but manufacturers' recommendations vary and should be consulted to ensure proper selection and application. While the adhesive is setting, temporary support is provided with threaded stub nails or cut nails.

9.5.6.4 Multi-Ply Construction

Multi-ply finishes consist of two or more layers of gypsum board and include (1) supported ceiling and wall assemblies and (2) self-supporting all-gypsum partitions. Face layers are laminated to a base layer of wallboard, backing board, or sound-deadening board, with a few fasteners to ensure adhesive bond (see Fig. 9.5-7). Multi-ply construction, therefore, minimizes fastener treatment and surface defects such as shadowing and nail popping. It further provides inherent joint-reinforced construction and reduces the incidence of joint ridging. After appropriate

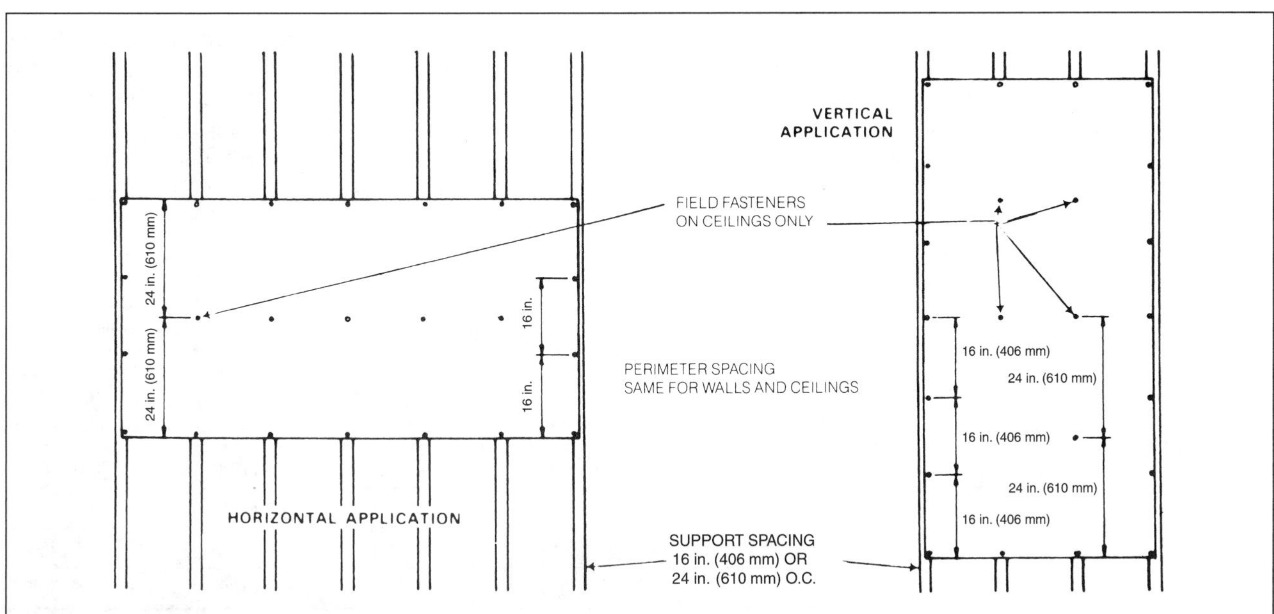

FIGURE 9.5-28 Supplemental fastener spacing for structural adhesives. (Courtesy Gypsum Association.)

a

b

FIGURE 9.5-29 Laminating adhesive is applied with a notched spreader. Stud adhesive is applied in beads with a caulking gun. (Courtesy Gypsum Association.)

FIGURE 9.5-30 Support Spacing for Two-Ply Construction

Thickness of Base Ply in.	Thickness of Base Ply mm	Location	Application of Base Ply[a]	Maximum Support Spacing, o.c. in.	Maximum Support Spacing, o.c. mm
3/8	9.5	Ceilings[b,c]	Perpendicular	16	406
		Walls	Perpendicular or parallel	24[d]	610
1/2 and 5/8	12.7 and 15.9	Ceilings[c]	Perpendicular	24	610
		Walls	Perpendicular or parallel	24	610

Note: Installation with mechanical fasteners with or without adhesive.

[a]Direction of board edges relative to supports.

[b]Two-ply 3/8-in. (9.5 mm) ceilings should not support insulation.

[c]For two-ply ceilings to receive a water-based spray texture finish, use 5/8-in. (15.9 mm) board applied perpendicular to 24-in. (610 mm) supports or 1/2 in. (12.7 mm) perpendicular to 16-in. (406 mm) supports.

[d]16 in. (406 mm) o.c. if no adhesive used between plies.

(Courtesy Gypsum Association.)

fastener and joint treatment, the surface may be painted or finished with wall covering or tile.

Multi-ply finishes are particularly suitable where increased sound isolation or fire resistance is desired and generally provide better surface quality than single-ply construction. Sound ratings up to 60 STC and 2-hour fire ratings are possible with some two-ply framed assemblies and multi-ply all-gypsum partition assemblies. Framed load-bearing wall assemblies with two-ply finishes receive somewhat lower fire ratings than non-load-bearing assemblies.

Two-ply construction on framed ceilings does not provide substantially higher fire or impact-sound ratings than single-ply construction. Therefore, the decision to use two-ply finishes on ceilings is usually based on improved surface quality rather than on fire or sound ratings. Unfortunately, two-ply construction is inherently more expensive than single-ply construction.

The gypsum board finishes discussed here are of two-ply construction supported by wood or metal framing and furring, concrete, or masonry. The spacing for framing and furring depends mainly on the base-ply thickness, whether the base ply is attached perpendicular or parallel to supporting construction, and whether the face ply is adhesively bonded to the base ply (Fig. 9.5-30). Regardless of these recommendations, support spacing in most commercial installations is 16 in. (400 mm) on centers.

BASE-PLY ATTACHMENT

The base ply of supported wall and ceiling finishes may be regular core wallboard, but backing board is used more often because it costs less. Sound-deadening board may be used where increased sound isolation is required. Fastener sizes and spacing for applying sound-deadening board vary for different fire- and sound-rated constructions.

A base ply can be attached to supporting framing or furring with nails, staples, resilient clips, or screws, but resilient clips are limited to walls. Base-ply attachment for self-supporting partitions is discussed in Section 9.5.6.6.

Wood Framing and Furring In wood framing and furring, the fastener spacing for nails, screws, and staples varies primarily with the method of face-ply attachment (Fig. 9.5-31).

Nail attachment of base ply should be

FIGURE 9.5-31 Base-Ply Fastener Spacing for Two-Ply Application in Inches (mm) o.c.

Location	Framing Spacing in. (mm)	Nail Spacing: Laminated Face Ply in. (mm)	Nail Spacing: Nailed Face Ply* in. (mm)	Screw Spacing: Laminated Face Ply in. (mm)	Screw Spacing: Screwed Face Ply* in. (mm)	Staple Spacing: Laminated Face Ply in. (mm)	Staple Spacing: Screwed Face Ply* in. (mm)
Walls	16 (406)	8 (200)	24 (610)	16 (406)	24 (610)	7 (178)	16 (406)
	24 (610)	8 (200)	24 (610)	12 (305)	24 (610)	7 (178)	16 (406)
Ceilings	16 (406)	7 (178)	16 (406)	12 (305)	24 (610)	7 (178)	16 (406)
	24 (610)	7 (178)	16 (406)	12 (305)	24 (610)	7 (178)	16 (406)

*Fastener spacing for face ply should be the same as for single-layer application.

(Courtesy Gypsum Association.)

with drywall nails long enough to provide the recommended penetration into supporting construction. Where the face ply is to be laminated, the base ply should be nailed as recommended for single-ply construction. Double nailing is generally not used, because nail heads will be covered by the face ply and fastener treatment is not a consideration.

When the face ply is to be attached with mechanical fasteners, fewer nails are required in the base ply, because the nails in the face ply will support the base ply as well.

Except for predecorated wallboard, base plies on walls should be installed parallel to framing members. On ceilings, base plies should be installed perpendicular to framing members. At inside corners, only the overlapping base ply should be nailed and fasteners should be omitted from the face ply to create a *floating corner*, which is more capable of resisting structural stresses (Fig. 9.5-32).

Screw attachment of the base ply to wood framing or furring should be with drywall wood screws that provide at least ⅝-in. (15.9 mm) penetration into supporting construction. Spacing of screws depends on the method of face-ply attachment and supporting construction (see Fig. 9.5-31). Inside corners should be floated as described for nailing.

Staple attachment is applicable to wood supporting construction only. Staples should provide at least a ⅝-in. (15.9 mm) penetration into supporting members. Recommended spacing for staples is given in Figure 9.5-31. Staples should be driven with the crown perpendicular to the edges of the board, but where edges fall over supports they should be driven parallel to the edges (Fig. 9.5-33). Staple crowns should bear tightly against the board without breaking the face paper.

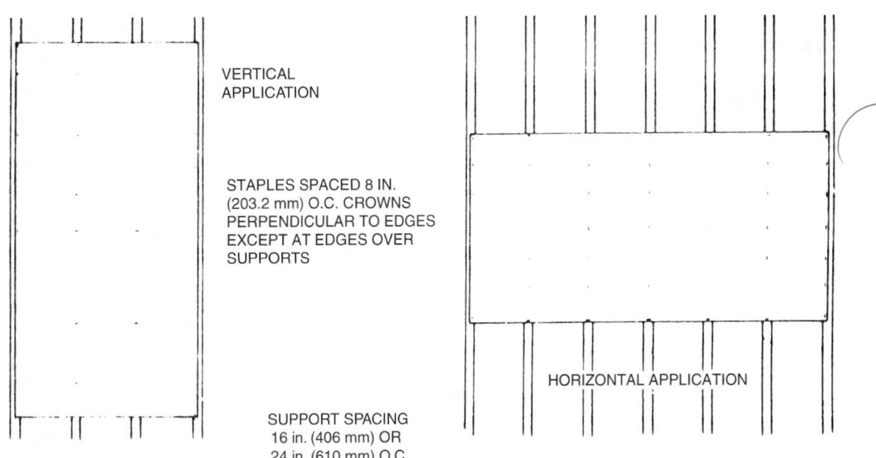

FIGURE 9.5-33 Staple attachment of base ply in wood supporting construction. (Courtesy Gypsum Association.)

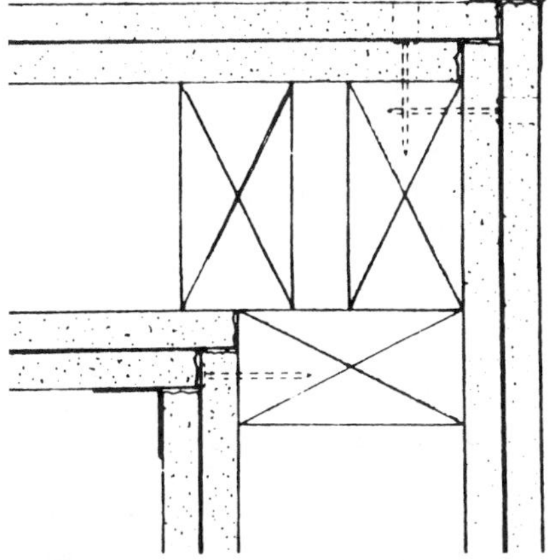

FIGURE 9.5-32 At floating corners, base ply is nailed on only one side of interior corner. (Courtesy Gypsum Association.)

Metal Framing and Furring Base plies should be attached to metal framing or furring with drywall sheet metal screws. Screw spacing on walls should be 12 in. (300 mm) on centers in the field and 8 in. (200 mm) on centers along the edges when the face ply is laminated, and 16 in. (400 mm) on centers throughout the board when the face ply is attached with screws. Screws should be long enough to penetrate metal studs or furring by at least ⅜ in. (9.5 mm).

Two-ply ceiling finishes are not commonly used on metal framing or furring because they do not substantially improve fire resistance or impact-sound isolation. They do provide increased airborne sound isolation and may be used where this is an important consideration.

Concrete and Masonry Concrete and masonry have inherently good fire resistance and sound isolation, which two-ply finishes do not substantially improve. However, when single-ply gypsum board cannot be laminated directly to masonry or concrete and the walls must be furred, two-ply finishes may be desired for superior surface finish. Base plies are applied in the same way they are applied on other furring.

FACE-PLY ATTACHMENT

Joints in the face ply should be offset from joints in the base ply by at least 10 in. (250 mm). Gypsum board can be applied either horizontally or vertically, whichever results in the least waste of materials and shorter joint lengths. Wallboard applied horizontally on walls usually results in shorter and less conspicuous joints and therefore is preferred. However, predecorated wallboard that is to be left with joints untreated and plain wallboard that is to be trimmed with battens are installed vertically.

Nail and Screw Attachment When the face ply is attached with mechanical fasteners with no adhesive, the maximum spacing and minimum penetration recommended for nails and screws in single-ply construction should be observed in accordance with GA 216. However, the specifications for some proprietary sound- and fire-rated partition assemblies may differ.

Adhesive Attachment The face ply is typically attached to gypsum board and sound-deadening board by *sheet lamination* or *strip lamination*. Generally, strip lamination is specified for systems with sound-deadening board.

Sheet lamination involves covering the entire back of the face ply with laminating adhesive using a notched spreader or other suitable tool (Fig. 9.5-34). The sizes and spacing of the notches vary, depending on the adhesive. The wallboard is then erected using moderate pressure. Adhesive squeezed out at joints should be promptly removed. In *strip lamination* the face ply is applied vertically and adhesive is spread in vertical ribbons spaced 16 to 24 in. (400 to 600 mm) on centers, depending on the adhesive and the assembly (Fig. 9.5-35).

FIGURE 9.5-34 Sheet lamination. Notched spreader or laminating box (shown) can be used to apply adhesive. (Courtesy United States Gypsum Company.)

Supplemental mechanical fasteners or temporary wood bracing is required to ensure intimate adhesive bond between base and face plies. When structural ceiling members are used to anchor temporary wood bracing, ceiling installation should be delayed until after the wall finish has been installed.

Temporary fasteners may be used when the face ply is laminated to gypsum board in wood construction and a fire rating is not required. These may be double-headed nails, stucco furring nails, or drywall nails, generally spaced to provide face support approximately 24 in. (600 mm) on centers in each direction. Nails should be long enough to ensure proper penetration into supporting construction. Nails should be driven through gypsum board washers or other rigid material to facilitate removal without damaging the face ply. After nails are removed, nail holes should be *dimpled* (slightly depressed with a curved-face hammer) and finished with compound, as for permanent nails.

FIGURE 9.5-35 Strip lamination. Laminating head applies adhesive beads on the base ply in parallel strips. (Courtesy United States Gypsum Company.)

Some permanent mechanical fasteners are recommended in each laminated gypsum board panel. Usually, these can be spaced at the perimeter of the board where they will be concealed by joint treatment or trim. In the face ply of fire-rated assemblies, permanent fasteners are usually required in the field as well as in the perimeter. Fastener spacing depends on the tested assembly construction and is not related to whether adhesive is used between plies.

In sound-rated partitions where fire resistance is not a consideration, the face ply is generally applied vertically over sound-deadening board, with permanent mechanical fasteners at tops and bottoms of boards. Intermediate fasteners may be omitted along vertical joints and in the field if temporary bracing is provided until the adhesive has developed sufficient bond strength.

9.5.6.5 Cracking and Ridging Precautions

Improper installation and adverse job conditions may cause surface imperfections such as cracking at corners and ridging at joints. *Corner cracking* may result from differential structural stresses at inside corners where adjacent walls or walls and ceilings meet. *Joint ridging* often occurs when joints are treated during cold or humid weather; it occurs more frequently when joint compound is applied too thickly and is feathered over a wide area. These finish defects are not commonly encountered in multi-ply construction.

In single-ply construction, cracking at interior angles can be minimized by *corner floating*; ridging can be substantially reduced by *back blocking* or *strip reinforcing*. However, two-ply finishes also are effective for both these purposes and often may be more practical. A proprietary bullnose-edged wallboard has been developed specifically to overcome the problem of ridging in single-ply construction. It utilizes a special prefill joint compound with nonshrinking and fast-setting properties, which develops high initial bond strength (Fig. 9.5-36).

CORNER FLOATING

Corner floating, which can be used with either screw or nail attachment, consists of omitting some fasteners at interior corners (Fig. 9.5-37). When corners are floated, the gypsum board should be applied to the ceiling first. It should be fitted snugly at both wall-to-wall and ceiling-to-wall intersections. Framing members or blocking should be used at corners to provide solid backup even though they may not receive fasteners.

At ceiling-to-wall intersections, the last row of fasteners should be omitted on the wall so that nails start 8 in. (200 mm) from the intersection. On ceilings, where ceiling joists are parallel to the intersection, nailing should start at the intersection; where the joists are perpendicular,

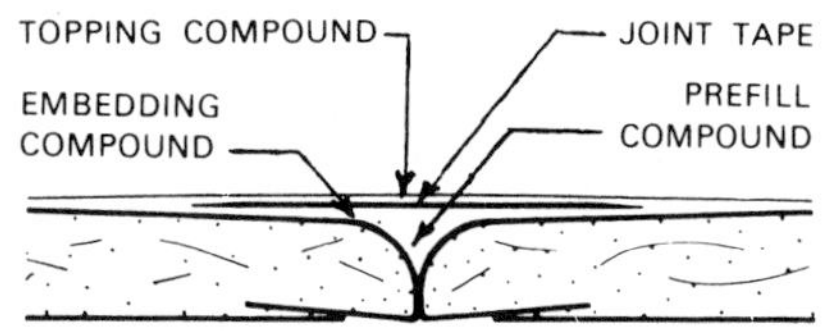

FIGURE 9.5-36 Bullnose-edged wallboard requires prefilling joint channels before standard joint treatment is applied. (Courtesy Gypsum Association.)

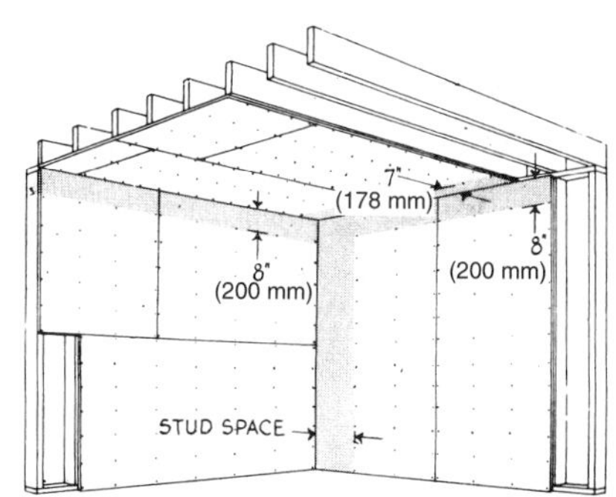

FIGURE 9.5-37 **Corners are floated by omitting fasteners at the intersections of walls and of walls and ceilings, as shown by the shaded areas.** (Courtesy Gypsum Association.)

FIGURE 9.5-38 **Cross section shows how floated end joint is tapered and back-blocked. Brace is temporarily nailed over wood strip, which depresses ends of panels.** (Courtesy United States Gypsum Company.)

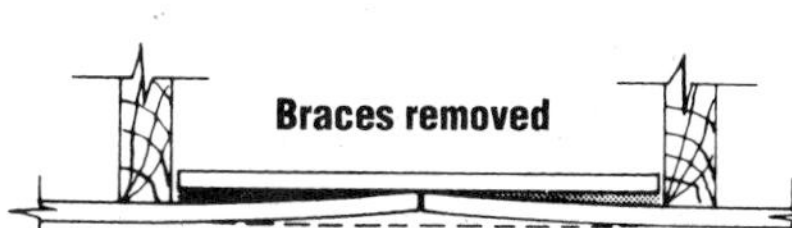

FIGURE 9.5-39 **When the strips shown in Figure 9.5-38 are removed, the tapered formation remains.** (Courtesy United States Gypsum Company.)

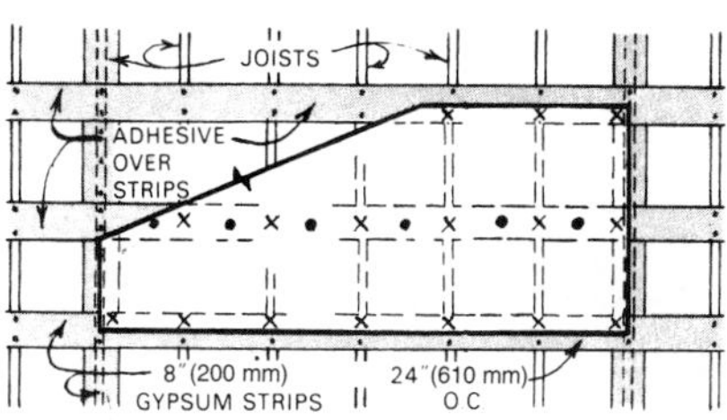

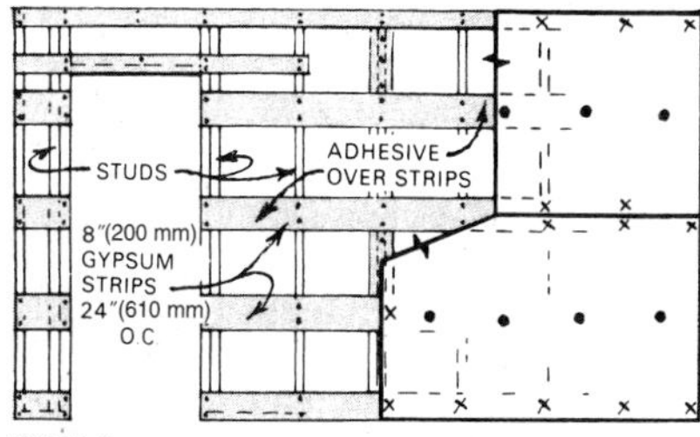

FIGURE 9.5-40 **Strip reinforcing: When support spacing exceeds 16 in. (406 mm) o.c., additional screws (dots) are required midway between supports.** (Courtesy Gypsum Association.)

nails should start 7 in. (180 mm) from the intersection (see Fig. 9.5-37). For interior corners at the meeting of walls, the last row of fasteners should be omitted from the board first applied only, and the overlapped board should be nailed conventionally. When double nailing, the same unrestrained clearances at corners should be maintained as for single nailing, described in Section 9.5.6.3.

BACK BLOCKING

Back blocking is the reinforcing of joints against stresses that cause surface defects, particularly ridging. Simple back blocking consists of laminating wallboard blocks between framing members. To reinforce the wallboard ends by back blocking, it is necessary to float the ends by locating end joints midway between framing members. However, because ends are not tapered, the need for relatively high crowns and wide feathering at such joints may still cause ridging when joint treatment is done under adverse weather conditions. End joint tapering can be used in conjunction with floating to reduce the need for high crowns and wide feathering at joints (Figs. 9.5-38 and 9.5-39).

Back blocking and associated constructions are most often performed on ceilings because imperfections are more conspicuous on such large, unobstructed, obliquely lighted surfaces.

STRIP REINFORCING

Strip reinforcing is a method of reinforcing joints against ridging that offers many of the advantages of two-ply construction at reduced cost. By combining adhesive lamination with mechanical fastening, the number of nails in the face ply is reduced. Although it results in thicker assemblies than back blocking, strip reinforcing can accomplish the same results in possibly less time.

Strips of ⅜-in. (9.5 mm)-thick backing board or scrap wallboard 8 in. (200 mm) wide are installed 24 in. (600 mm) on centers perpendicular to framing members, similar to cross furring. Strips may be 4 in. (100 mm) wide at the base of a partition and at a partition-ceiling intersection. This cross stripping supports the edges of the board and provides intermediate support in the field of the board. To provide support at the butt ends of a board, additional short strips are secured directly over the supporting member between the cross stripping. Strips are nailed about 1 in. (25.4 mm) from the edge. The arrangement of strips, fasteners, and boards is illustrated in Figure 9.5-40.

Laminating adhesive is applied to the strips with a strip laminating head in parallel beads. Wallboard should be ½ in. (12.7 mm) thick or thicker and should be installed perpendicular to the framing members so that long edges fall on the cross stripping and the ends are supported on the short intermediate strips. When supporting members are spaced 16 in. (400 mm) on centers, the nail spacing should be 24 in. (600 mm) on centers at the ends and over each support at the edges for walls, with temporary double-headed nails at every other stud in the field until the adhesive sets. On ceilings, nails should be spaced 24 in. (600 mm) on centers along each support.

When supporting members are spaced more than 16 in. (400 mm) on centers, 1½-in. (38 mm) drywall screws are required in the gypsum board strips midway between framing members for both walls and ceilings, as shown in Figure 9.5-40. Appropriate drywall nails should be used to attach both stripping and gypsum board panels.

9.5.6.6 Special Constructions

Most special constructions designed to improve sound, fire, moisture, and thermal control assume the use of gypsum board finishes with complete joint treatment and panel caulking. Materials used for caulking should be nonhardening, nonshrinking, resilient sealants. In each case, the manufacturer's specific installation recommendations should be followed.

SOUND-ISOLATING CONSTRUCTIONS

The properties of sound and the principles of room acoustics and sound isolation are discussed in Section 13.1. For detailed information on specific assemblies, see Sections 13.1.4 and 13.1.5.

Transmission of airborne and structure-borne sound can be reduced in wall and ceiling assemblies by:

- Eliminating flanking paths
- Separating the supporting structural construction of each finish surface
- Resilient mounting of finishes with channels, clips, or sound-deadening board

- Including sound-absorbing materials in the assembly cavity
- Increasing the weight of the surface construction

Airborne Sound Through Flanking Paths Failing to eliminate airborne sound transmission through flanking paths will seriously impair the effectiveness of other sound-control methods and the performance of the entire assembly. Reduction of flanking paths includes constructing an airtight assembly and avoiding back-to-back location of recessed fixtures such as medicine cabinets, and electrical, telephone, television, intercom, and other devices that perforate a gypsum board surface. In addition, openings for such fixtures and related piping outlets should be caulked (Fig. 9.5-41). The entire perimeter of a sound-isolating partition should be caulked at wallboard edges and under runners to make it airtight.

Assemblies intended for sound isolation should not contain heating or air conditioning ducts or ventilating outlets and should be designed to eliminate flanking paths such as through ceilings or around windows and doors adjacent to the walls or through floors into crawl spaces below.

Separated Construction Separated construction in wood framing can be obtained by using separate sets of studs, each supporting a surface of one or more plies of gypsum board. This is accomplished by staggering the studs on the same top and bottom plates or by placing the studs on separate plates (Fig. 9.5-42). When using separated wall framing systems, cabinets, ceramic tile, lavatories, and other wall-hung fixtures can be installed in the conventional manner, which is not possible on walls with resilient mounting. Separated walls can also be developed with gypsum double-solid and semisolid partitions.

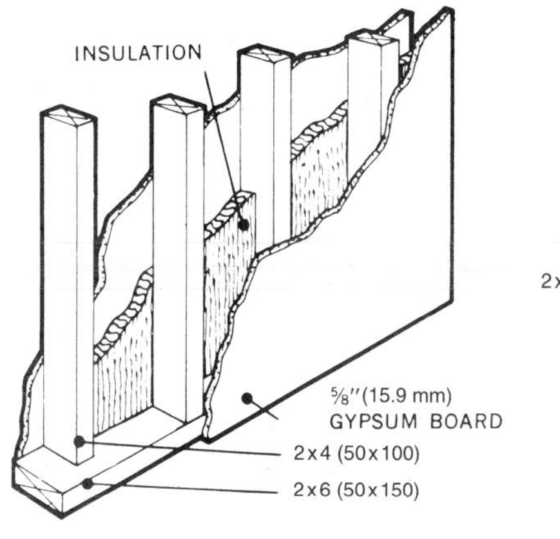

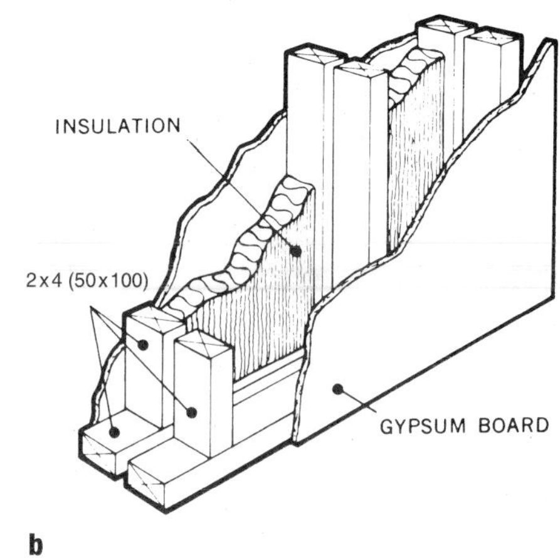

FIGURE 9.5-42 Separated construction: (a) staggered studs on a single plate, (b) double row of studs using separate plates. (Courtesy Gypsum Association.)

Resilient Mounting Resilient-mounting systems and separated structural framing systems are the most effective and practical sound-control means capable of bringing lightweight framed assemblies to acceptable STC and IIC ratings. Gypsum board finishes can be mounted on walls with resilient channels or resilient clips over sound-deadening board. Kitchen cabinets, lavatories, ceramic tile, medicine cabinets, and other fixtures should not be supported on resiliently mounted walls, as their weights may "short out" the construction acoustically.

Resilient channels are used in both wall and ceiling assemblies to isolate gypsum board from structural framing. Channels are screw-attached perpendicular to framing members (see Fig. 9.5-15). In ceiling assemblies, nails used to attach channels to wood joists may work loose as the framing lumber dries. Therefore, screw attachment is preferred.

Resilient clips can be used to attach the base ply of two-ply finishes to wood studs, but are not recommended for ceiling application.

Sound-Deadening Board Sound-deadening board includes (1) wood- or cane-based fiberboard, (2) mineral (including glass) fiberboard, (3) plastic foamboard, and (4) regular core gypsum board. Sound-deadening board is manufactured ¼ to ⅝ in. (3.2 to 15.9 mm) thick, and 4 ft. (1219 mm) wide by 8 to 9 ft. (2438 to 2743 mm) long. It is used as a base layer to which wallboard is attached. It is installed parallel to framing members with end joints staggered on alternate studs (Fig. 9.5-43). Gypsum wallboard can then be laminated to the sound-deadening board, either parallel or perpendicular to supporting construction. Nailing or screwing through the sound-deadening board into the framing is required to satisfy fire ratings, but doing so will reduce sound-isolating effectiveness.

Wood fiberboard and plastic foamboard are used over wood construction or where noncombustible materials are not required. Mineral fiberboard is used with metal framing where noncombustible materials are required.

Sound-Absorbing Materials Mineral fiber insulating batts and blankets may be used in the cavity space of assemblies to improve resistance to sound transmission.

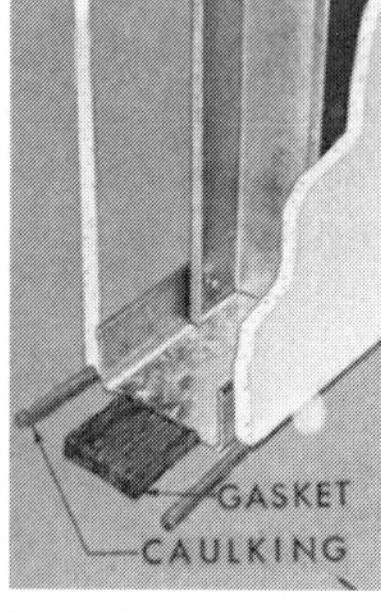

FIGURE 9.5-41 Caulking with acoustical sealant to reduce sound transmission (a) around electrical outlet boxes and (b) beneath partition runners. (Courtesy Gypsum Association.)

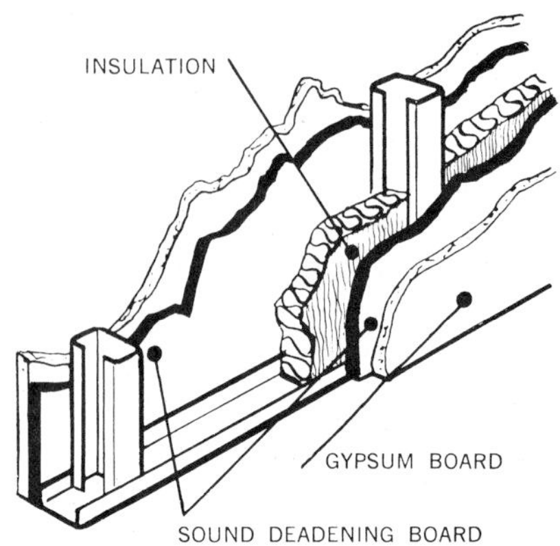

FIGURE 9.5-43 Resilient mounting with sound-deadening board improves sound isolation in metal frame construction. (Courtesy Gypsum Association.)

The full height of the cavity should be filled with batts or blankets carefully fitted behind electrical outlets, blocking, and fixtures, and around cutouts necessary for plumbing lines.

Insulating batts and blankets may be paper-faced or vapor retarder–faced with flanges and are installed by stapling within the stud space. Stapling flanges across the stud face is not recommended, because wrinkled flanges may prevent intimate board-to-stud contact. In metal stud and laminated all-gypsum board partitions, the blankets can be attached to the back of the gypsum board. Batts and blankets without facings required in noncombustible construction are friction-fit within the stud space.

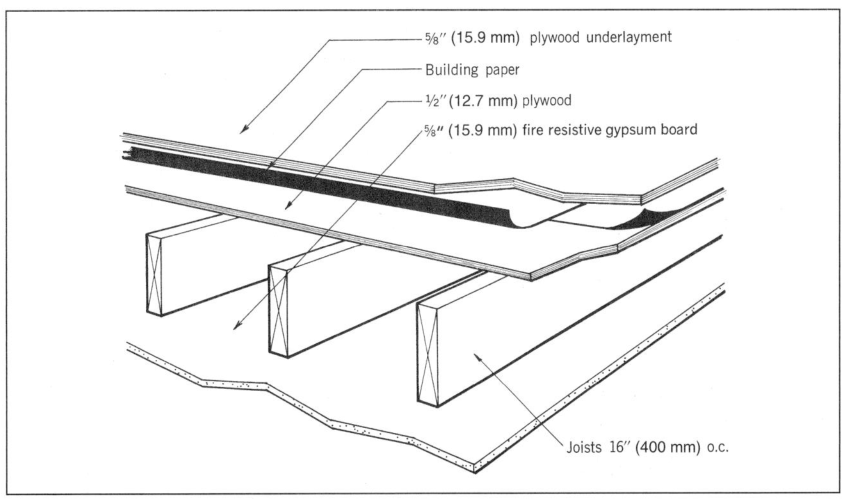

FIGURE 9.5-45 **Properly constructed wood joist floor and ceiling assemblies with ⅝-in. (15.9 mm) fire-resistant gypsum board receive a 1-hour fire rating.** (Courtesy Gypsum Association.)

FIRE-RESISTANT CONSTRUCTION

Fire-resistant construction in gypsum board assemblies requires the use of Type X gypsum board.

Fire-Resistant Ceiling Panels Fire-resistant gypsum board panels are cut into 2-by-2 ft. (609 by 609 mm) and 2-by-4 ft. (609 by 1219 mm) panels for use in some fire-rated ceilings. Common thicknesses are ⅜, ½, and ⅝ in. (9.5, 12.7, and 15.9 mm). Fire ratings up to 2 hours are possible with an approved suspension system (Fig. 9.5-44). Such ceilings also reduce sound transmission but have little sound-absorptive properties. Standard 4- by 8-ft. (1219 by 2438 mm) fire-resistant gypsum board panels may also be used to provide fire-rated floor and ceiling assemblies for steel (see Fig. 9.5-11) and wood (Fig. 9.5-45) construction.

Fire-Resistant Wall Panels Fire-resistance ratings of gypsum board assemblies can be improved by:

- Using fire-resistant Type X board
- Increasing gypsum board thickness
- Filling the stud cavity with mineral wool

Unless all of the materials used in the tested assembly classify as noncombustible, the term *combustible* is often used after the assigned hourly classification. The following ratings are possible with gypsum board construction:

- *1-hour:* Ordinary 2 × 4 (50 × 100) wood stud partitions, when finished with ⅝-in. (15.9 mm)-thick Type X wallboard on both sides and properly installed
- *1-hour:* Wood joist floor and ceiling assemblies with single-layer wallboard
- *2-hour:* Wood and metal stud partitions with two layers of ⅝-in. (15.9 mm)-thick Type X board on each side
- *2-hour:* Metal studs with a ¾-in. (19.1 mm)-thick gypsum board panel on each side.
- *2-hour and more:* Steel-and-concrete assemblies protected by a suspended gypsum board ceiling

Another criterion for fire hazard of exposed surface materials is the rate of surface flame spread. The flame spread rating of gypsum board is under 20, based on a flame spread of zero for mineral-fiber-cement board and 100 for red oak. Gypsum board's low flame spread rating more than satisfies most interior finish requirements.

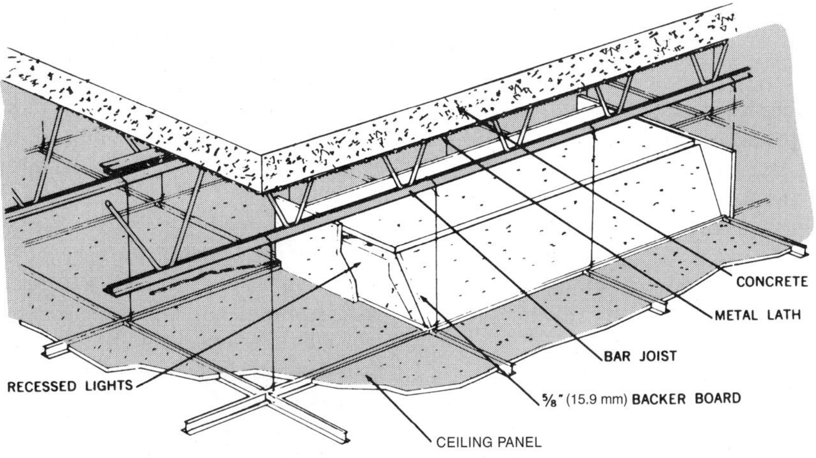

FIGURE 9.5-44 **Fire-resistive ⅝-in. (15.9 mm) ceiling panels in a fire-rated T-grid suspension system provide 2-hour fire protection for steel and concrete construction.** (Courtesy Gypsum Association.)

THERMAL INSULATING CONSTRUCTION

In some parts of the country, exterior wall and ceiling assemblies require thermal insulating construction to minimize heat gain or loss through the assembly. In such areas a vapor retarder is usually required to retard the migration of moisture and condensation within the assembly. For a discussion of moisture and heat control, see Sections 7.1 and 15.2.

Typical materials used to increase the thermal resistance of assemblies with gypsum board finishes include (1) flexible fiber insulation (batts and blankets), (2) loose-fill insulation, (3) rigid plastic insulation, (4) reflective-foil-faced airspace, and (5) air spaces provided by furring over solid materials such as masonry and concrete. For a discussion of thermal insulations, see Section 7.3.

Vapor control can be provided as an integral part of the gypsum board or as part of the insulation. Foil-backed gypsum board will usually provide some insulating value when combined with a dead air space; it is also an effective vapor retarder and can be used with or without other insulating materials. Foil-backed gypsum board should not be used:

- As backing material for ceramic tile;
- As the face ply in laminated two-ply construction;
- For laminating directly to masonry or concrete;
- In connection with electric heating cables in electric panel heating systems.

Rigid plastic foam and reflective foil insulation have inherently low vapor permeability and generally do not require a separate vapor retarder. Flexible insulations such as batts and blankets are available with or without vapor retarders. Loose insulation requires either a separate vapor retarder, such as foil or polyethylene film, or an integral retarder provided by foil-backed gypsum board. Thermal insulation batts and blankets should be attached as recommended for sound-absorbing batts and blankets.

Two important considerations should not be overlooked in insulating design: (1) if reflective foil is to provide effective insulating value, a dead air space of at least ¾ in. (19.1 mm) should be provided, and (2) when a vapor retarder is used, it should be located on the warm side of the insulation. The installation of insulating and vapor-control materials that are independent of a gypsum board finish is essentially the same as for other finishes. Only wood and metal furring and rigid plastic insulation that provide support for a gypsum board finish require special consideration in this section.

Rigid plastic insulation is intended primarily for use on exterior and below-grade concrete or masonry walls. Rigid insulating foamboards are made from expanded polystyrene or polyurethane, in panels ½ to 3 in. (12.7 to 76.2 mm) thick, up to 4 ft. (1219 mm) wide, and 12 ft. (3658 mm) long. These foamboards may be adhesive-bonded to masonry or concrete surfaces. They should not be laminated to masonry or concrete having a surface temperature of less than 45°F (7.2°C). They should be fitted together so that there will be no gaps between the boards that could create cold spots or allow moisture migration. Gypsum board may be bonded to rigid foam insulation with an approved adhesive, or screw attached to special channels provided with the foamboard. At least 24 hours drying time should be allowed before gypsum board is laminated to rigid plastic insulation with an adhesive. In addition, the gypsum board should be held in place by temporary bracing or double-headed nails that can be removed after sufficient bond strength develops. Supplemental fasteners are required.

MOISTURE-RESISTANT CONSTRUCTION

Most gypsum board is sensitive to moisture and should not be used in locations subject to direct wetting or continuous high humidity. However, it can be used in limited exterior locations such as eave soffits and porch and carport ceilings when properly finished and painted and in interior locations such as tub, shower, and laundry areas when protected with wall tile. Such locations, however, require special construction details and precautions.

This subsection covers the use of regular core and moisture-resistant gypsum board. Glass mesh mortar units that can serve as a base for ceramic tile in lieu of moisture-resistant gypsum board are discussed in Section 9.6.

Exterior Locations Regular or water-resistant wallboard used on horizontal surfaces in exterior locations should be at least ½ in. (12.7 mm) thick. Fascia boards at gypsum board soffits should be designed to form a drip edge to prevent water runoff across the board surface and to protect the board edge (Fig. 9.5-46). In addition, edges and ends at the perimeter of the board should be protected by metal edge trim. Unless protected by metal trim or another waterstop, edges should be spaced at least ¼ in. (3.2 mm) from abutting masonry.

Exposed wallboard surfaces and metal trim should be painted with two coats of exterior paint. Adequate ventilation of eaves, ceilings, and attic spaces above gypsum board soffits should be provided.

General Interior Locations Water-resistant gypsum board is recommended for all interior walls subject to high humidity, except for high-moisture spaces like shower rooms, saunas, and steam rooms. Moisture-resistant gypsum board should not be used on ceilings. Foil-backed gypsum board should not be used in interior areas of high humidity because the foil backing may prevent moisture from being dissipated, thus trapping it within the board core. But where a high humidity area is located on an exterior wall, adequate insulation and a vapor retarder on the warm side of the wall are necessary to prevent condensation in the wall cavity.

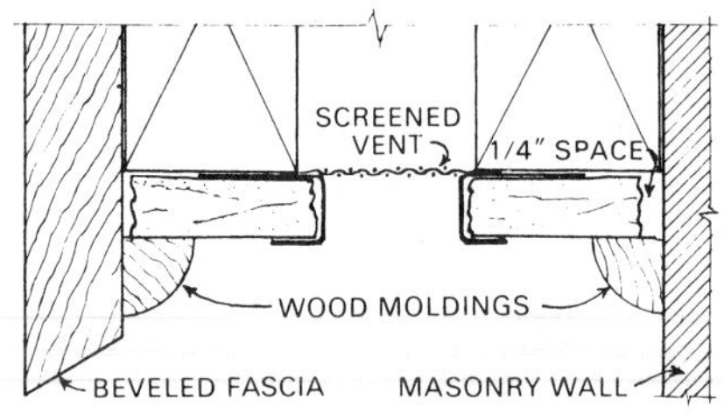

a

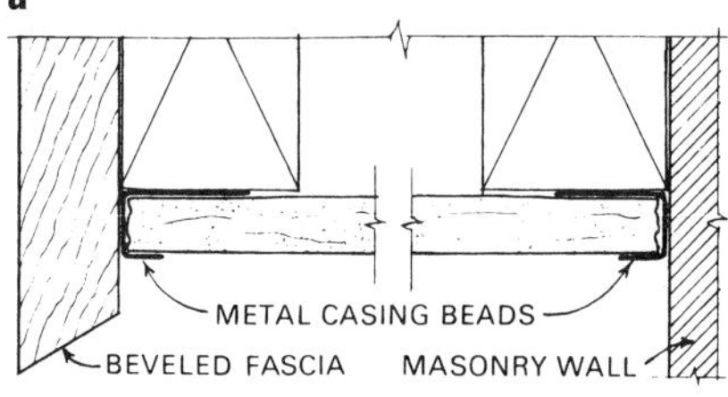

b

FIGURE 9.5-46 Gypsum board can be protected from damp exterior surfaces by (a) air spaces and wood moldings or (b) metal trim. (Courtesy Gypsum Association.)

Tub-Shower Areas Water-resistant gypsum board and proprietary highly water-resistant backing board are recommended as a base for wall tile in tub-shower areas. Gypsum board in tub-shower areas should be at least ½ in. (12.7 mm) thick. Stud spacing should not exceed 24 in. (609 mm) on centers. When the stud spacing exceeds 16 in. (406 mm) on centers, blocking should be installed not more than 4 in. (102 mm) on centers between studs, with the first row of blocking about 1 in. (25.4 mm) above the top of the tub or shower receptor. When a wall is to be finished with tile (see Section 9.6) or handicapped accessories are to be installed (see Section 1.6), closer stud spacings may be required.

Gypsum board should be applied perpendicular to framing members, with paperbound edges parallel to the top of the tub or shower receptor and spaced at least ¼ in. (3.2 mm) above it (Fig. 9.5-47). This space should be filled with a continuous bead of waterproof sealant. Generally, gypsum board should be fastened with nails spaced not more than 8 in. (203 mm) on centers or screws spaced not more than

12 in. (305 mm) on centers. However, when ceramic tile more than 5⁄16 in. (7.9 mm) thick will be installed, nails should be spaced 4 in. (101 mm) on centers and screws 8 in. (203 mm) on centers.

Special water-resistant backing board or regular core gypsum board should be installed as recommended by the manufacturer or the Gypsum Association. When regular core gypsum board is used, joints between adjacent boards, including those at corners, should be taped and treated as for painting, but should not receive finishing coats. Before installing tile, a water-resistant sealer should be applied to the entire wallboard surface, including treated joints and corners. When tile adhesive is used as a sealer, it should be spread separately from the bonding coat in a continuous, uniform coating approximately 1⁄16 in. (1.6 mm) thick, covering all cut ends and all cutouts for fixtures. A Type I, highly water-resistant organic adhesive should be used to apply ceramic wall tiles.

A tile finish should cover all surfaces, as illustrated in Figure 9.5-48. The space between tile and tub rim, shower receptor, or subpan top should be caulked with a continuous bead of sealant (see Fig. 9.5-47). To prevent this caulked joint from opening due to settlement, the tub should be supported at the walls on metal hangers or vertical blocking fastened to the studs. Waterproof shower receptor pans or subpans should have an upstanding leg or flange at least 1 in. (25.4) higher than the water dam or threshold in the entry to the shower.

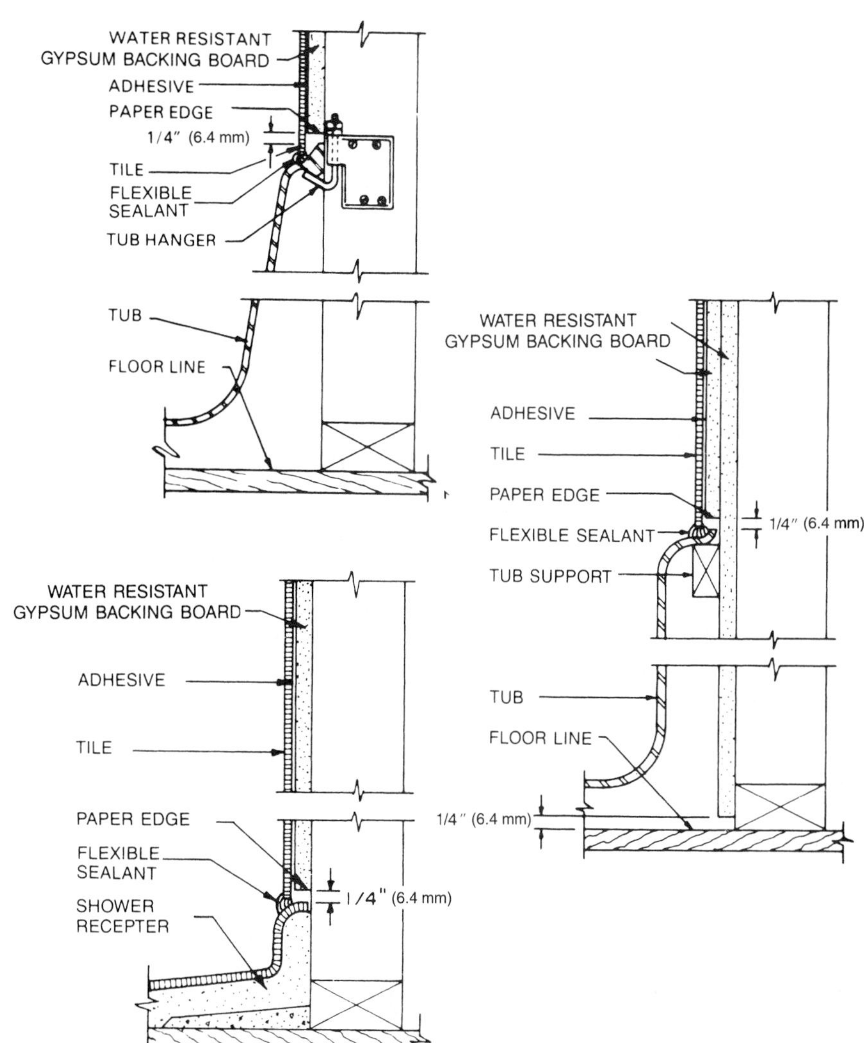

FIGURE 9.5-47 Gypsum board and wall tile installation details for tub-shower areas. (Courtesy Gypsum Association.)

ELECTRIC PANEL HEATING

Electric resistance elements may be incorporated in the gypsum board itself for one-ply application, or in the backing board for two-ply construction. Another type of radiant electric heating ceiling system is a built-up system incorporating a gypsum backing board, heating cables, a filler material, and a gypsum wallboard finish. Standard gypsum lath and plaster products can also be assembled into a radiant heating ceiling system by embedding heating cables in the plaster coats over gypsum lath (see Section 9.3). In each case, the manufacturer's detailed recommendations should be followed.

Heat loss calculations indicate the number of square feet of heating panels required in each space, which is usually less than the total ceiling area. The areas adjacent to heating panels in each space are

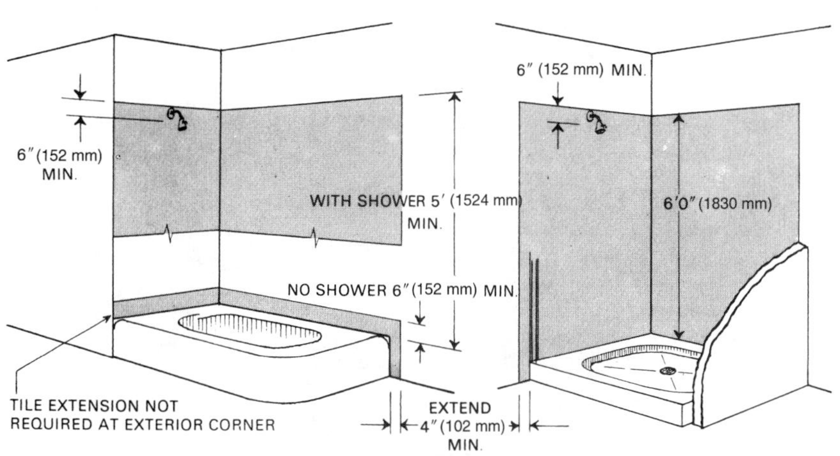

FIGURE 9.5-48 Shaded areas show tile coverage of gypsum board in tub-shower areas. (Courtesy Gypsum Association.)

covered with backing board or wallboard of matching thickness. Wallboard face panels may be finished in the normal manner, but installation of acoustical tile over such a system is not recommended.

Proper insulation design is important to the efficient operation of any electric panel heating installation. Thermal insulating batts used in the ceiling above a panel heating system should be unfaced friction-fit mineral or glass wool.

SELF-SUPPORTING GYPSUM PARTITIONS

Self-supporting gypsum partitions are non-load-bearing, noncombustible multiply partitions, fabricated on the job almost entirely from gypsum boards. Inner plies consist of backing board in rib or sheet form, to which wallboard face plies are laminated. Unlike the ceiling and wall assemblies described elsewhere in this section, which depend on wood, metal, concrete, or masonry construction to support the surface finish, gypsum partitions are entirely self-supporting and require only floor and ceiling runners to stabilize them. Runners should be attached securely to ceiling and floor construction with fasteners spaced 24 in. (609 mm) on centers and should be carefully caulked at partition perimeters and under runners to close air leaks. The maximum height of partitions is limited by the design and dimensions of the components. Therefore, the manufacturer's recommendations should be followed.

A variety of proprietary partition systems have been developed by manufacturers of gypsum products to minimize the number of operations and speed field erection of partitions. The chief types are:

- Semisolid, utilizing backing board ribs
- Solid, with coreboard (thicker backing board) inner plies
- Hollow
- Demountable

All gypsum board components such as ribs, coreboard, backing board, and wallboard are typically installed in the vertical position.

Gypsum partition products are used in self-supporting partition systems, engineered assemblies of board and plaster products with special metal accessories designed for use with each system.

Semisolid Partitions In semisolid partitions, the backing board core is replaced by 6- to 8-in. (150 to 203 mm)-wide gypsum board ribs, laminated at the factory in 1-in. (25.4 mm) and 1⅝-in. (41.3 mm) thicknesses (Fig. 9.5-49a). Ribs are adhesive-bonded in the field to the wallboard faces for single-ply construction or to backing boards for two-ply construction (Fig. 9.5-49b). In two-ply construction, regular or predecorated wallboard is bonded to form the finished surface. Veneer plaster finish may be used over an appropriate plaster base laminated to the ribs. Semisolid partitions generally are not made with standard plaster finishes. Partition thicknesses vary between 2 and 3 in. (50.8 and 76 mm), depending on rib and face panel thicknesses.

The ribs are cut 6 in. (150 mm) shorter than the gypsum board and are laminated to the face ply on one side of the partition either before or after it is erected. Perimeters of door and window openings are reinforced with additional ribs. Adhesive is spread on the erected face ply or on the ribs, keeping it 1 in. (25.4 mm) from the rib edges. These rib-reinforced panels are secured to wood or metal runners at the top and bottom with appropriate drywall screws spaced 12 in. (305 mm) on centers.

The ribs on the erected panels are then spread with adhesive, and facing panels are installed and screw-attached to runners with screws spaced 12 in. (305 mm) on centers. To ensure adequate bond at ribs, 1½-in. (38 mm) drywall screws are provided at each rib 12 in. (305 mm) from the top and bottom and spaced not more than 36 in. (914 mm) on centers.

For partitions with two-ply faces, the ribs and backing board base plies are assembled and installed in the same way as single-ply faces. Face plies are then sheet laminated as for two-ply supported partitions using supplemental drywall screws.

Separated rib construction to improve sound-isolating effectiveness can be provided in either single- or two-ply ribbed partitions. The installation procedure is essentially the same as for ordinary ribbed partitions, but twice as many ribs are required, spaced 12 in. (305 mm) on centers and staggered on opposite sides of the partition (see Fig. 9.5-49b).

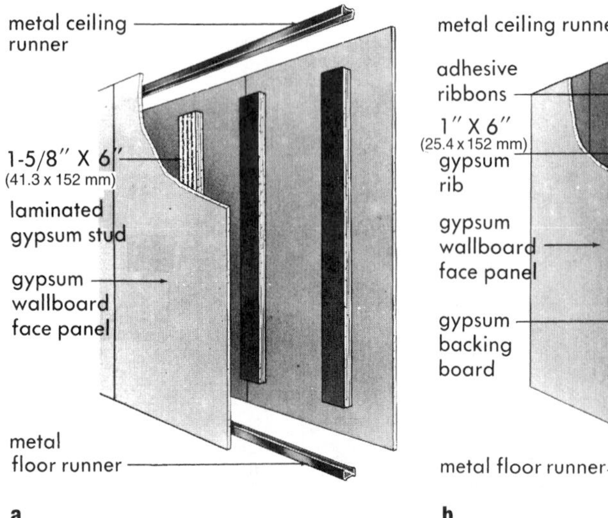

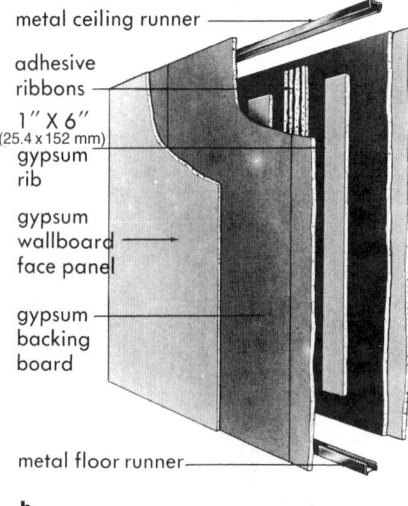

FIGURE 9.5-49 Semisolid gypsum partitions: (a) single-faced with both faces laminated to each rib and (b) double-faced with each two-ply face laminated to alternate ribs, forming separated rib construction.

Solid Partitions Solid gypsum partitions are an extension of the concept of semisolid partitions. They consist of a solid coreboard to which wallboard faces are laminated on each side (Fig. 9.5-50a). The vertical joints in the face ply are offset at least 3 in. (76 mm) from joints in the coreboard. The face ply is attached by sheet lamination with supplemental drywall gypsum screws. Double or triple solid partitions separated by air spaces provide increased fire resistance and sound isolation (Fig. 9.5-50b).

Solid gypsum partitions may be field-erected from combinations of lath and plaster or backing boards and wallboards. Both generally employ wood or proprietary metal ceiling and floor runners to stabilize the partitions. In lath and plaster partitions, an integral flush metal base acts as a screed. Partitions may also be trimmed out with a resilient or wood base.

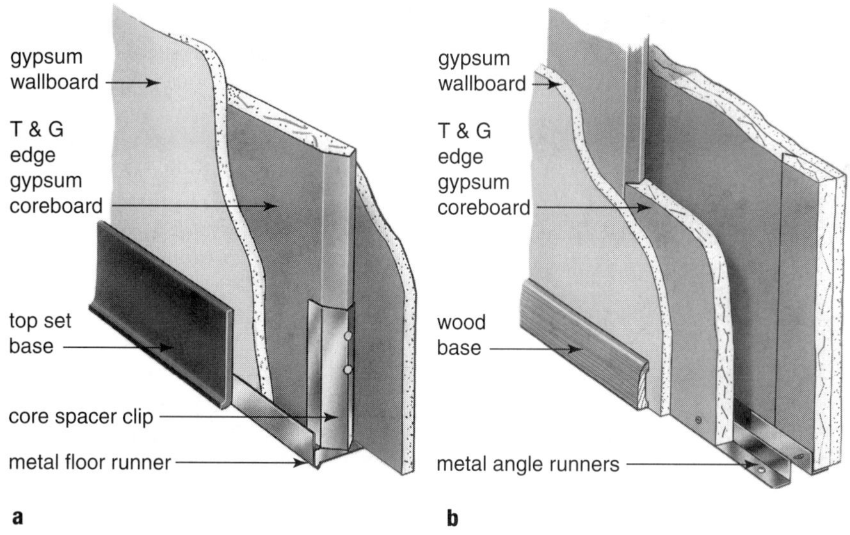

FIGURE 9.5-50 Solid gypsum partitions: (a) single solid partition, (b) double solid partition with air space between the two faces.

Backing board and wallboard partitions consist of a 1-in. (25.4 mm)-thick backing board core to which ½- or ⅝-in. (12.7 or 15.9 mm) wallboard faces are laminated with adhesives. The core may consist of two job-laminated ½-in. (12.7 mm)-thick backing boards or a 1-in. (25.4 mm) factory-laminated product referred to as *coreboard* (see Fig. 9.5-50a). Regular wallboards usually are used, which require joint treatment and decorating, but in some instances predecorated wallboards with V-edges may be installed in a visually attractive repeat pattern. When veneer plaster finish is desired, a veneer plaster base is used in place of wallboard. Double solid partitions are constructed of two separate layers of lath or backing board, separated by an air space, with plaster or wallboard faces on the exterior (Fig. 9.5-50b).

Hollow Partitions Hollow partitions are made with wood or metal studs supporting faces of plaster or wallboard (see Fig. 9.3-5). Studs are usually spaced 16 to 24 in. (406 to 609 mm) on centers and are supported by wood or metal floor and ceiling runners. Wood studs normally are 2 × 4s (50 × 100s) or 2 × 6s (50 × 150s). Metal channel studs are available in 1⅝- (41.4 mm), 2½- (63.5 mm), 3⅝- (92.1 mm), 4- (101.6 mm), and 6-in. (177.8 mm) sizes. Greater wall cavities may be produced with parallel sets of studs braced with wallboard.

Hollow partitions may be of either single-ply or two-ply construction. For single-ply construction, wallboard is nailed or screwed directly to the studs (Fig. 9.5-51a). For two-ply construction, backing board is clipped, nailed, stapled, or screwed to the studs and wallboard is adhesive-bonded to this backing with a minimum of supplemental fastening. Where soundproofing is important, resilient channels may be used in either one- or two-ply construction to mount the board (Fig. 9.5-51b). When regular core wallboard is used as the exterior face, fasteners and joint treatment are required; predecorated wallboard is normally adhesive-bonded to backing board and has V-edge joints, which can be left exposed.

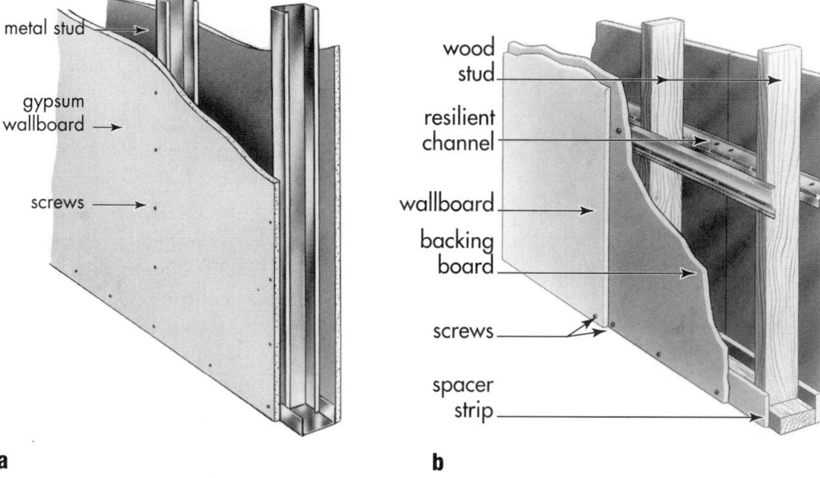

FIGURE 9.5-51 Hollow gypsum board partitions: (a) screwed single-ply, (b) two-ply construction with resilient mounting for improved sound control.

Demountable (Movable) Systems In situations where it may be desirable to rearrange partitions, demountable panel systems are frequently used. They are designed for speedy erection and for minimum loss of components when being demounted and rearranged.

Demountable partition systems use wallboard face panels attached to interior metal studs for stability. Various pressure-fit, clip, or screw systems are used to secure the wallboard panels to the studs. One system uses an H-section stud 24 in. (609 mm) on centers, the flanges of which are pressure-fit into grooves of factory-laminated, predecorated panels. In other systems, predecorated square-edged wallboard is adhesive-bonded directly to studs 24 in. (609 mm) on centers and the edges are secured with nails or screws. Decorative batten trim is then used to conceal fastenings at 24 or 48 in. (609 or 1219 mm) on centers (Fig. 9.5-52).

Shaft Walls Shaft walls are lightweight gypsum board assemblies used for elevator shafts, electrical and mechanical enclosures, and stairwells. Shaft wall assemblies with metal stud framing weigh 10 to 13 psf (48.8 to 63.5 kg/m^2), whereas masonry or masonry-and-plaster assemblies with similar fire ratings weigh between 35 and 65 psf (161 and 317.4 kg/m^2).

Shaft walls built with gypsum board are not only light in weight, but are easier and less expensive to install than conventional masonry construction. They can be erected floor by floor from outside the mechanical or elevator shaft, they do not

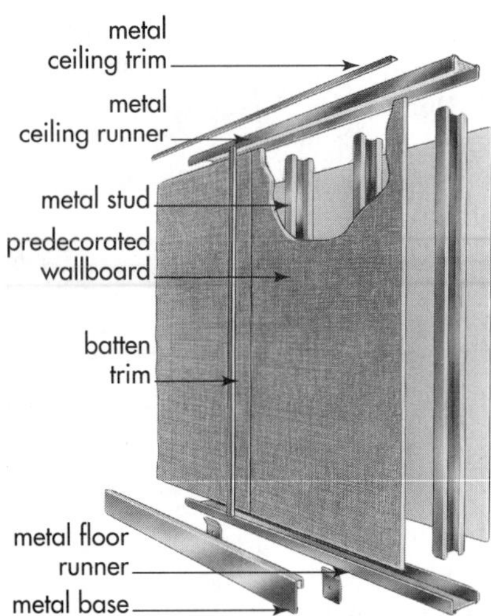

FIGURE 9.5-52 One type of demountable partition.

delay installation of the mechanical equipment inside the shaft due to masonry scaffolding, and they do not require extensive cleanup inside the shaft upon completion.

RESURFACING EXISTING CONSTRUCTION

Gypsum board may be used to provide a new finish on existing walls and ceilings. If the existing surface is sufficiently smooth and true after appropriate preparation and if it provides solid backup for the new wallboard without shimming, ¼-in. (3.2 mm)-thick gypsum board can be applied with an adhesive, nails, or screws. Nails and screws should be long enough to provide recommended penetration into supporting members under the existing finish (see Figs. 9.5-16 and 9.5-17).

Existing surfaces that are not sufficiently smooth to receive gypsum board should be furred with wood or metal furring. Furring members should be shimmed as required to provide a suitably level base. Minimum gypsum board thickness for various support spacings and installation methods should be as recommended for new construction over furring.

Surface trim for mechanical and electrical equipment such as switch plates, outlet covers, and ventilating grilles should be removed and reinstalled over the new gypsum board finish.

VENEER PLASTER BASE

In general, veneer plaster base should be installed in the same manner as other types of gypsum board, and the joints should be taped. Some manufacturers recommend using paper tape; others open-mesh glass fiber tape. Both base coat veneer plaster and joint compounds are used to embed paper tape, depending on the manufacturer of the plaster.

The manner of board placement should produce the least possible number of end joints, which are difficult to conceal.

9.5.7 JOINT AND FASTENER TREATMENT

After gypsum board has been erected, specific areas require treatment to achieve a smooth, homogeneous appearance. Panel joints in the same plane and interior corners require taping. Exterior corners and exposed edges require metal corner and edge trim. Both taped and trimmed areas, as well as fasteners, require finishing with compound (Fig. 9.5-53).

9.5.7.1 Joint Taping and Finishing

A minimum of three coats of joint compound is recommended for taped joints: an embedding coat to bond the tape and two finishing coats over the tape. Each coat should be allowed to dry thoroughly so that the surface can be readily sanded and the compound will obtain maximum shrinkage before the next coat is applied.

Depending on atmospheric conditions, 12 to 24 hours may be required for each successive coat to dry, unless quick-setting compound is used. The second and third coats should be lightly sanded when dry to remove surface irregularities and to blend them evenly with the adjacent board. Care should be taken not to raise the fibers on the board's face paper, as this may result in uneven surface texture and a spotty appearance.

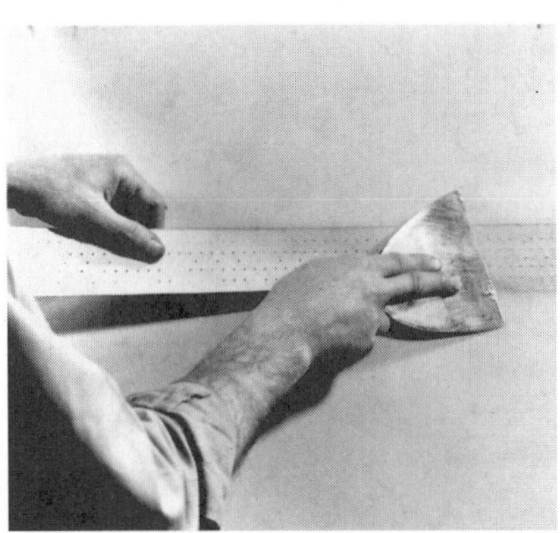

FIGURE 9.5-53 Gypsum board joints are finished by embedding a perforated tape in bedding compound and covering it with several coats of topping compound. (Courtesy Gypsum Association.)

FLUSH JOINTS

Embedding compound should be spread into the depression formed by the tapered edges of adjacent boards (Fig. 9.5-54a) and over all butt-end joints. Then tape is centered over the joint, smoothed to avoid wrinkling, and pressed into the compound by drawing a knife along the joint with enough pressure to squeeze out excess compound (Fig. 9.5-54b). This excess compound may be removed or redistributed as a skim coat over the tape (Fig. 9.5-54c). When the embedding coat is dry, a second coat is applied, with the edges feathered 2 to 4 in. (50.8 to 102 mm) beyond the tape edges (Fig. 9.5-54d). Over the dry second coat, a third coat is spread, with edges feathered 2 to 4 in. (50.8 to 102 mm) beyond the second coat to blend smoothly with the board surface (Fig. 9.5-54e). Joints can be treated in a similar manner by machine application of tape and joint compound (Fig. 9.5-55).

INTERIOR CORNER JOINTS

Compound should be applied to both sides of interior corners. Then tape should be folded along the center crease and embedded snugly in the corner to form a right angle (Fig. 9.5-56). Excess compound is removed, and the surfaces are subsequently finished with additional coats of compound as described for flush joints. Both hand and mechanical tools are available for corner finishing.

9.5.7.2 Edge and Corner Trimming

All trim shapes except metal U-bead are installed after the gypsum board has been fastened in place (see Fig. 9.5-18). Metal cornerbeads and casing beads such as L, LC, and LK are installed with the flange outward and are nailed, crimped, or screwed 5 in. (127 mm) on centers through the gypsum board into the supporting construction. Metal cornerbeads with paper flanges are secured by embedding the flanges in joint compound. All edge and corner trim except U-bead require finishing with joint compound to obtain a smooth surface and to conceal fasteners (Fig. 9.5-54f). Methods and sequence of three-coat application are the same as for fastener treatment.

Unlike other trims, U-bead is fastened through the back flange first, then the gypsum board edges are inserted into the trim. Finishing with compound is not required, because the fasteners are concealed.

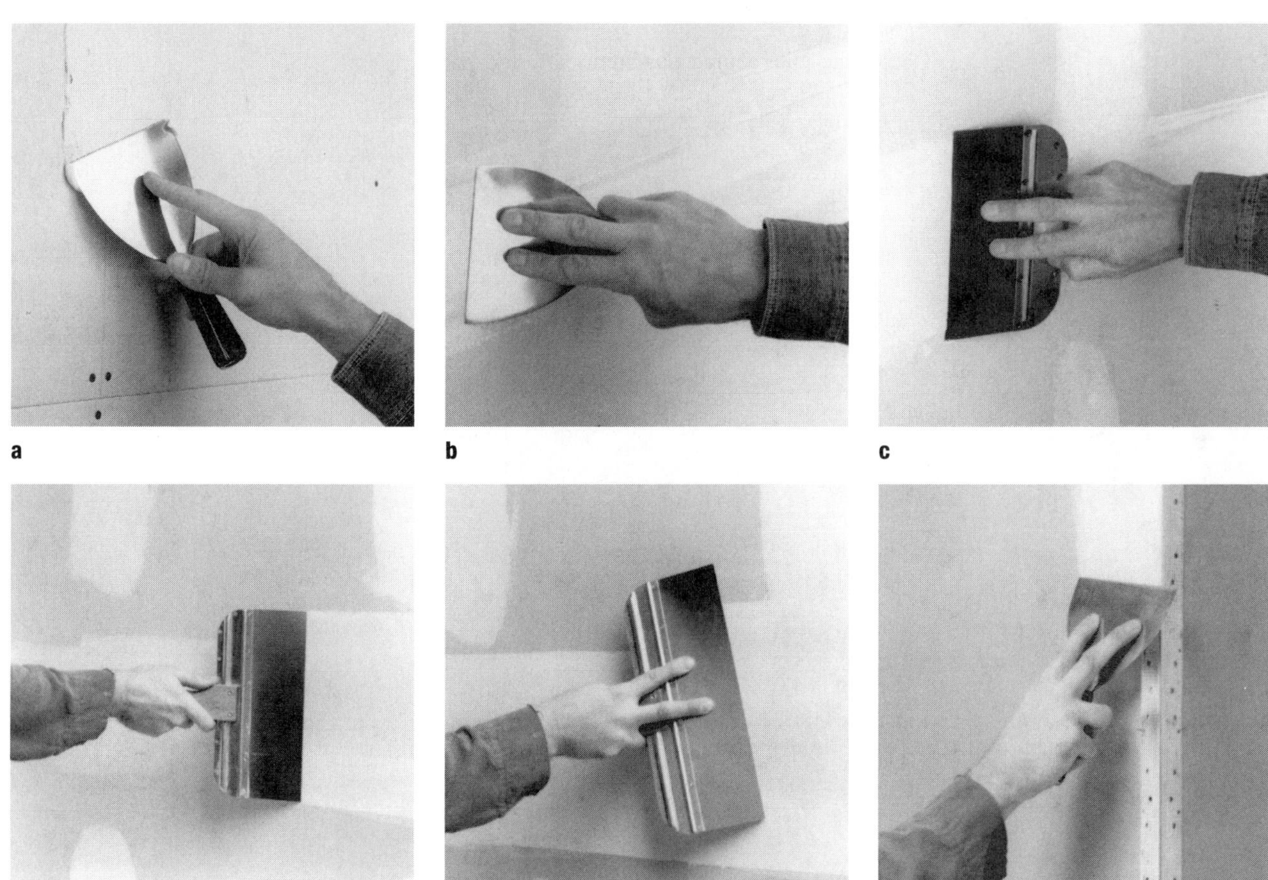

FIGURE 9.5-54 Finishing knives of varying widths are used to spread embedding compound, embed joint tape, and apply successively wider coats of topping compound. (Courtesy United States Gypsum Company.)

9.5.7.3 Fastener Treatment

A first coat of embedding compound is applied over fastener heads at the same time joints are taped. Fasteners may be treated (*spotted*) using an automatic spotter or a finishing knife (Fig. 9.5-57). The knife should be pressed against the surface to finish, level, and remove excess compound. Second and third coats of compound are applied over the fasteners when similar coats are applied to taped joints, edges, and corners. Each successive coat is feathered slightly beyond the one previously applied. Compound is sanded between coats as required. Care should be taken not to raise the fibers in the face paper when sanding. Each coat should be thoroughly dry before the next coat is applied.

9.5.8 FINISH DEFECTS

Improper installation or adverse job conditions may result in surface imperfections such as nail popping, shadowing, cracking at corners, or ridging at joints. Such finish defects can be minimized or completely eliminated if their causes are understood and appropriate precautions are taken.

9.5.8.1 Nail Popping

Nail popping is a surface defect in which compound over a nail head is pushed out beyond the surface of the board. It is due primarily to loose boards caused by wood shrinkage, the use of excessively long nails, or improper nailing.

WOOD SHRINKAGE

Wood shrinks across the grain as its moisture content is reduced (see Section 6.1). As a piece of lumber shrinks, it tends to expose the shank of a nail driven into its edge (Fig. 9.5-58). If shrinkage is substantial and nails are too long, separation between board and framing member can result. A separation as small as $\frac{1}{32}$ in. (0.8 mm) can result in an objectionable nail pop. It is essential that framing and furring lumber have as low an initial moisture content as possible, that lumber be kept dry during storage and installation, and that nail lengths do not substantially exceed those recommended for specific applications. Framing and furring lumber should be at the equilibrium moisture content for the area and should not exceed 15% moisture content. Nail penetration should be as near as possible to that recommended in Figure 9.5-16. Annularly threaded nails require less penetration than smooth-shank nails and tend to minimize nail popping.

IMPROPER NAILING

When gypsum board is improperly nailed, loose boards may result due to loose nails, loose attachment, or punctured face paper.

Loose Nails Nails that either miss the underlying construction or enter it at an

a

b

c

d

FIGURE 9.5-55 **Automatic taping machine applies embedding compound and joint tape on (a) ceiling and (b) wall; (c) broad knife removes excess compound; and (d) finishing tool with interchangeable head applies successively wider coats of topping compound. Occupational Safety and Health Administration (OSHA) standards now require workers to wear hard hats (as in b) when using this type of equipment. Eye protection should also be worn.** (Courtesy United States Gypsum Company.)

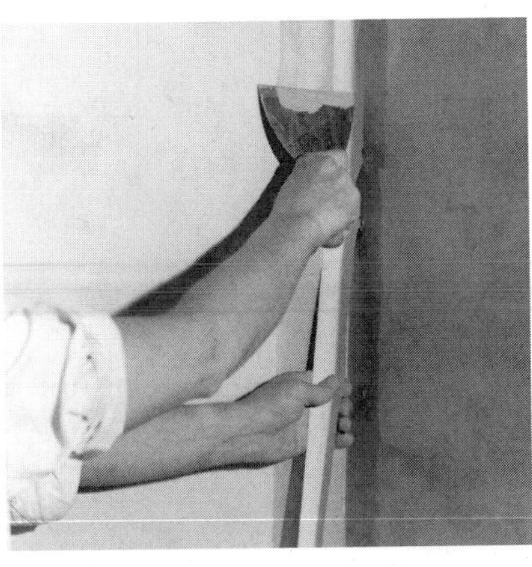

a

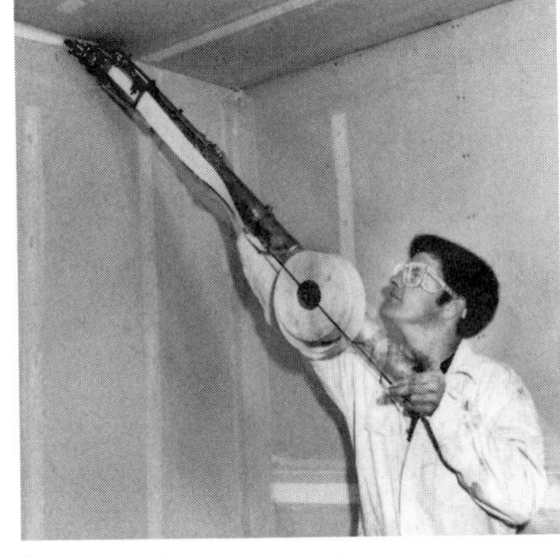

b

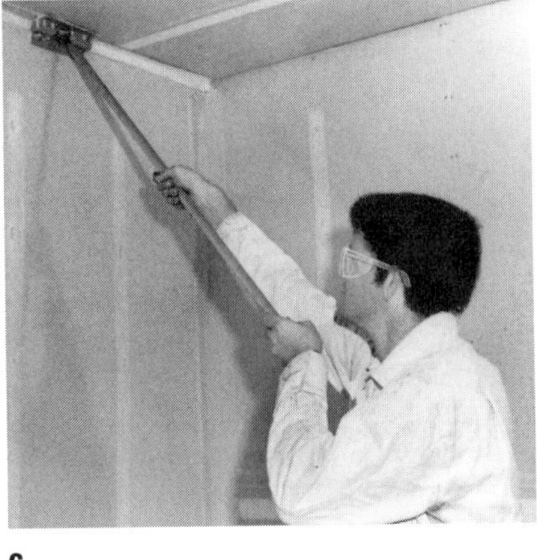

c

FIGURE 9.5-56 **Tape can be embedded in corner (a) by hand or (b) with a tool. Finishing can also be done by hand or (c) with a corner tool.** (Courtesy United States Gypsum Company.)

angle near the edge (Fig. 9.5-59a) can eventually be loosened by vibration until they protrude beyond the finish surface. To guard against such defects, the location of supporting members should be marked carefully on the gypsum board prior to nailing, nails should be driven perpendicular to the surface into the member, and nails that have missed the member should be removed promptly. After nails have been removed, nail holes should be dimpled and treated with compound.

Loose Attachment Loose attachment can occur when too little pressure is applied to a board while it is being nailed (Fig. 9.5-59b). Even when substantial pressure is applied properly, it may be impossible to achieve snug contact with supporting members if the board is cut too large and must be forced into place or if nailing is begun at the perimeter and proceeds inward. To prevent loose board attachment, panels should be cut to fit easily into place, pressure should be applied while nailing, and proper nailing sequence from the middle of the panel outward should be observed. Use of screws and double nailing also tends to minimize loose boards.

Punctured Face Paper Improper nail head design, obliquely driven nails (Fig. 9.5-59c), excessive and repeated impact

a

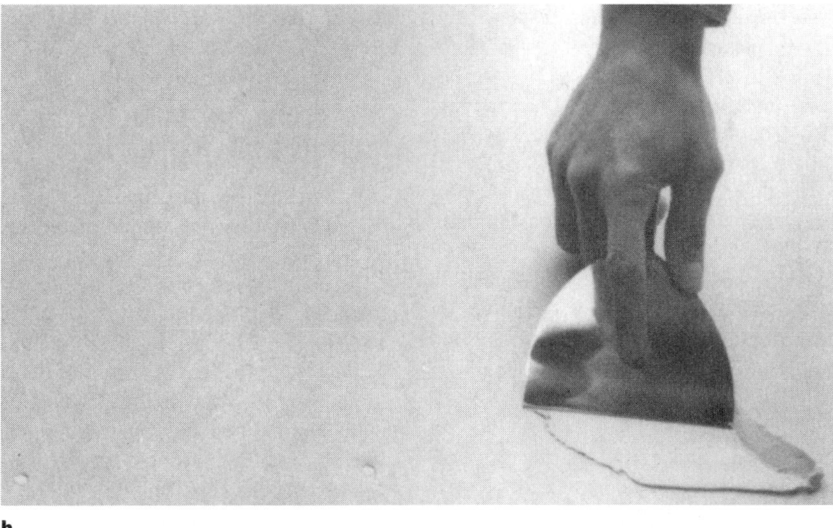

b

FIGURE 9.5-57 **Slightly recessed screw or nail heads are treated with (a) an automatic spotter or (b) a finishing knife. A hard hat and eye protection should be worn.** (Courtesy United States Gypsum Company.)

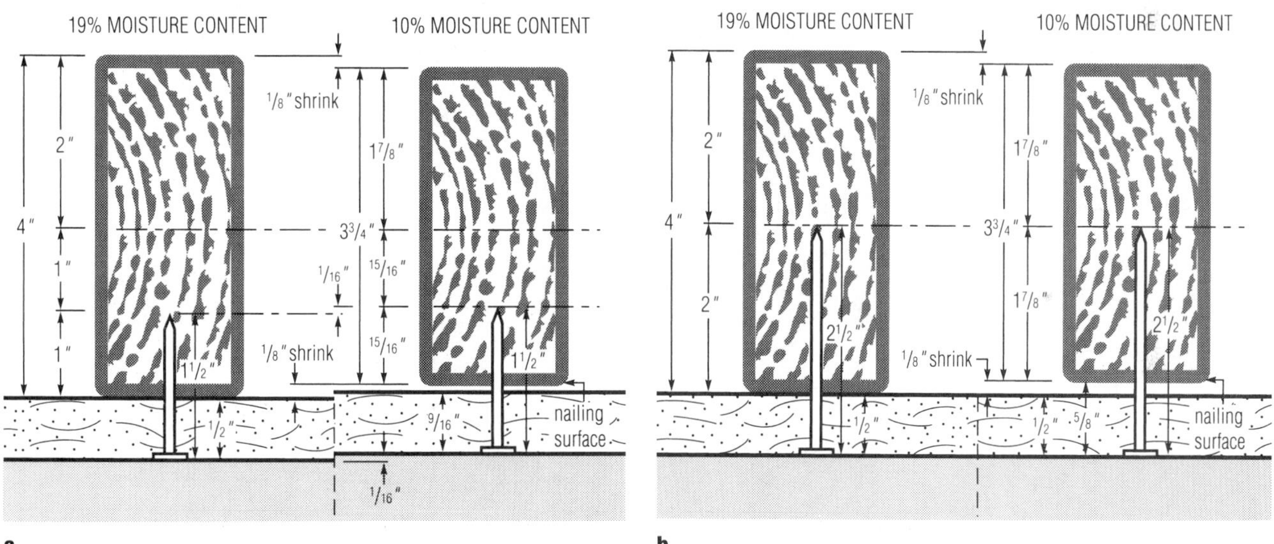

a b

FIGURE 9.5-58 **As a wood stud shrinks, separation may be (a) ¹⁄₁₆ in. (1.6 mm) when nail penetrates 1 in. (25.4 mm) or (b) ⅛ in. (3.2 mm) when nail penetrates 2 in. (50.8 mm).** (Courtesy United States Gypsum Company.)

during nailing, and using boards with extremely dry face paper can all result in the face paper being punctured during board installation. When the paper is punctured, the gypsum core is easily crushed and offers less resistance to loosening of the board from normal impact and vibration. In hot, dry periods, the face paper may become more brittle, increasing the hazard of puncturing. This hazard can be minimized by increasing the air humidity during storage and installation of gypsum board.

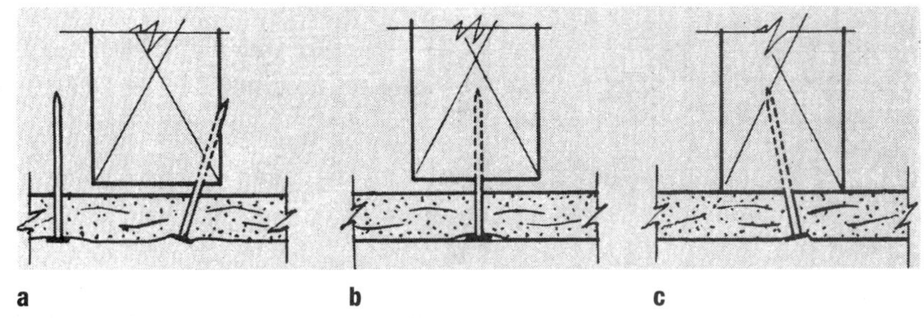

a b c

FIGURE 9.5-59 **Nail popping can result from (a) loose nails that miss framing members, (b) loose boards, or (c) obliquely driven nails that puncture the face paper.**

Other common causes of punctured face paper are misaligned or twisted members and projections caused by improperly installed blocking or bracing (Fig. 9.5-60). Intimate contact between gypsum board and supporting members is impossible in such cases, and hammer impact causes the board to rebound, which often ruptures the paper. Such defects should be corrected before the board is installed. The use of screws, adhesive, or two-ply construction will minimize problems resulting from such defects.

9.5.8.2 Shadowing

Shadowing is actually a surface discoloration, but it may appear similar to nail popping when viewed at a distance. Shadowing is the result of an accumulation of dust or dirt caused by increased thermal conductivity and moisture condensation over a nail head. It occurs commonly on roof and ceiling assemblies and on exterior wall assemblies that are inadequately insulated, and in regions with great indoor-outdoor temperature variations. Shadowing can be remedied by periodic washing or decorating. It can be avoided by proper insulation design and use of laminated two-ply construction. Where a permanent ventilating system is employed, adequate air filtration and humidity control will be helpful.

9.5.9 DECORATING AND REDECORATING

A gypsum board installation should be inspected before it is decorated or redecorated. Special attention should be paid to joints and fasteners. Imperfections should be sanded or repaired.

It is essential that joint compound be thoroughly dry before any form of decoration is applied. Depending on temperature and humidity conditions, 24 hours or more of good drying time is required.

Gypsum board surfaces can be decorated with textured coatings, emulsion or oil-based paints, or with paper, fabric, or vinyl wall coverings. It is necessary to prepare the board surface with a sealer or primer. A good sealer will seal the pores and lay down the fibers in the paper surface so that suction and texture differences between paper and joint compound will be equalized; it also will seal in surface impurities sometimes present in the wallboard surface. A good primer will provide uniform texture and will conceal color or surface variations. Some sealers incorporate the properties of primers as well.

A sealer or primer should be used on gypsum board beneath oil-based paints, lacquers, and enamels, because they tend to raise the fibers in the face paper. Sealers are used under wall coverings primarily to facilitate later removal without marring the board surface and to provide an adequate base for redecorating.

Most emulsion paints such as latex, resin, and casein do not raise the fibers in the face paper. Good latex paints have inherently good sealing and priming properties and require no special preparation of gypsum board surfaces. However, applying a pigmented emulsion sealer as a first coat or using at least two coats of emulsion paint will generally provide a durable, attractive finish if the gypsum board has been properly installed and if joints, fasteners, and trim have been properly treated.

Glue size, shellac, and varnish are not suitable as sealers or primers under paints or wall coverings.

If nail popping and other surface defects occur after decorating, it is advisable to wait until one full heating cycle has been completed before repairing and redecorating.

The information here is general in nature. Refer to Section 9.15 for further discussion about painting gypsum board.

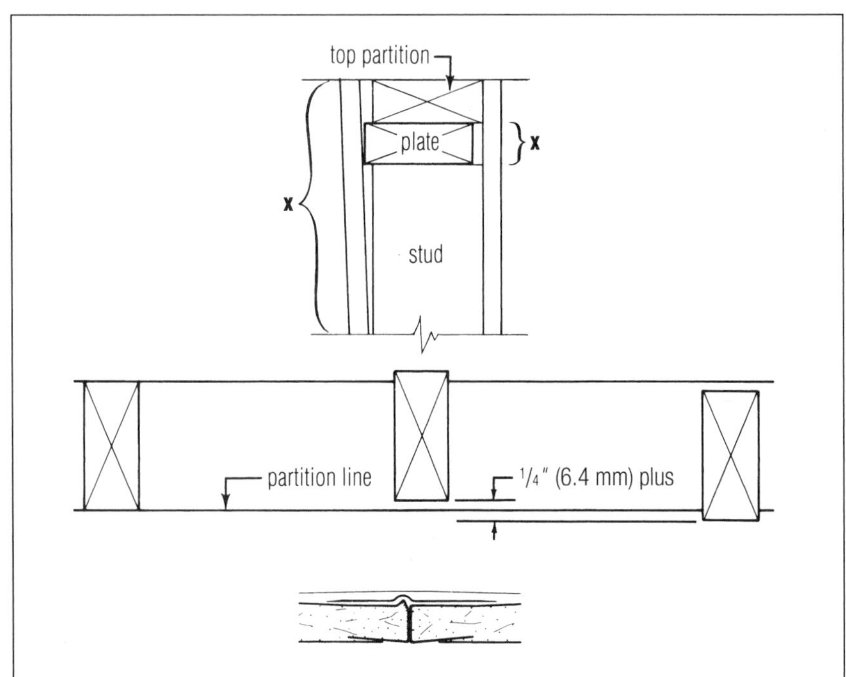

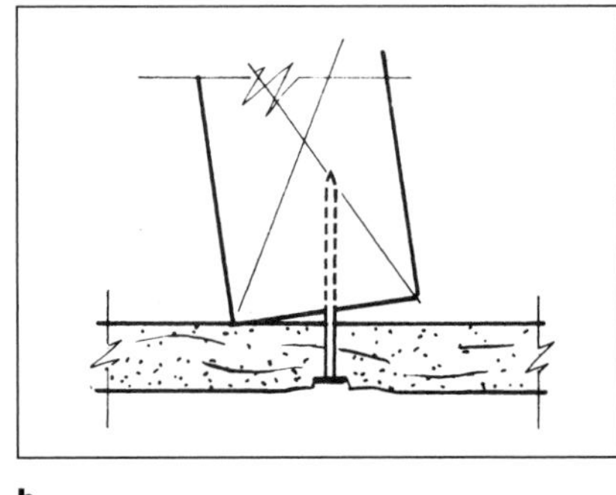

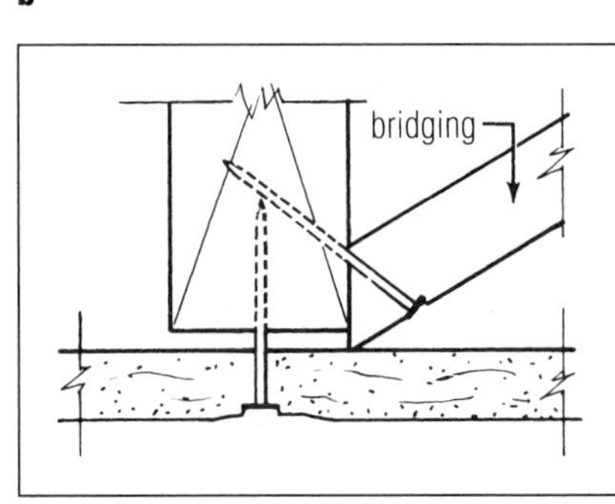

FIGURE 9.5-60 Punctured face paper and loose boards can be caused by (a) misaligned members, (b) twisted members, or (c) projecting blocking. (Courtesy United States Gypsum Company.)

9.6 Tile

The term *ceramic* derives from the Greek *keramos*, which means "potter's clay." Ceramic tile is made from nonmetallic minerals fired at high temperatures and manufactured in modular unit sizes (tiles), which facilitate installation. The term *ceramic tile* includes several products of varying dimensions, properties, and appearance:

- *Glazed wall tile:* 4-in. by 4-in. (101.6 mm by 101.6 mm) or 4¼-in. by 4¼-in. (108 by 108 mm) units used mainly on walls, but some are sufficiently abrasion resistant to be suitable for floors.
- *Ceramic mosaic tile:* 1-in. by 1-in. (25.4 mm by 25.4 mm), 1-in. by 2-in. (25.4 mm by 50.8 mm), or 2-in. by 2-in. (50.8 by 50.8 mm) units used mostly for floors, but used also on walls.
- *Paver tile:* Floor units similar to mosaic tile but generally 4 in. by 4 in. (101.6 by 101.6 mm) or larger.
- *Quarry tile:* 6-in. by 6-in. (177.8 by 177.8 mm) or larger natural clay units for floors. Other sizes and shapes are also available (Fig. 9.6-1).
- *Decorative or commemorative tile:* Ceramic materials made to serve purely decorative or commemorative uses (Fig. 9.6-2).

Ceramic tiles possess good or superior abrasion resistance, a variety of patterns and colors, and a high degree of resistance to moisture. These properties make them uniquely suited for areas exposed to severe foot traffic or in intermittent or continuous contact with water or corrosive chemicals. Ceramic tile is used for most interior areas of residential, commercial, institutional, and industrial buildings. Many types also are suited for exterior use.

The origins of ceramic tile date back 5000 years, as evidenced by burnt natural clay tiles found by archaeologists in early Egyptian and Babylonian excavations. Decorative and functional clay tiles were also an essential part of later Greek and Roman architecture. The Italian town of Faenza is credited with developing, in the fourteenth century, what is now known as *faience tile*, a highly decorative tile with a textured, irregular, handmade appearance. A century later, Holland started production of *delft* tile, characterized by a figure or landscape in blue or violet.

In the United States, production of ceramic tile was spurred by the introduction of the dust-press method at the end of the nineteenth century. Samuel Keys is credited with first producing floor and wall tile in Pittsburgh in 1867. By 1937 there were 52 manufacturers of ceramic tile in the United States. It is generally agreed that more improvements in tile manufacturing and installation techniques have been made by the U.S. ceramics industry during the past 70 years than in the preceding 70 centuries.

Much of that progress is due to the research and educational activities of the Tile Council of America (TCA). In the early 1950s, the TCA sponsored research that resulted in the development of several adhesives and mortars that eliminated the need for thick setting beds, for the laborious presoaking of tile, and for extended curing. To ensure satisfactory tile installations, the TCA:

- Licenses manufacturers of mortars and grout according to Council formulations;
- Promulgates standards for the manufacture and installation of ceramic tile, which has resulted in a series of ANSI Standards;
- Publishes and annually updates the *Handbook for Ceramic Tile Installation*. Much of the discussion and most

FIGURE 9.6-1 Pentagon-shaped tiles lend an elegant appearance to this setting. (Courtesy American Olean [Ceramic] Tile Co.)

FIGURE 9.6-2 **Artist's mosaic of cut tiles. Mosaic murals are also made commercially.** (Courtesy American Olean [Ceramic] Tile Co.)

of the installation details in this section have been abstracted from TCA's handbook.

9.6.1 GENERAL RECOMMENDATIONS

Satisfactory ceramic tile installations depend on a sound, rigid, and dimensionally stable backing surface. Ceramic tile is installed on floors (Fig. 9.6-3) either in an unbonded, isolated, setting bed (thickbed method) or by direct bonding to the backing (thin-set method). Direct bonding is made possible by synthetic resins possessing greater bonding power than conventional portland cement. The following text generally employs terms used in the tile industry. Refer to the glossary for unfamiliar terms.

9.6.1.1 Materials

Materials involved in ceramic tile installation include:

- Tile
- Mortars and adhesives for setting tile
- Grouts for filling joints
- Related materials to complete the installation, such as portland cement, reinforcing metal lath or mesh, and cleavage membrane

CERAMIC TILE

Ceramic tile is made from clay or from clay mixed with other ceramic materials. Oxides may be included for glaze coloring. Ceramic tile products are available in a broad range of sizes (Fig. 9.6-4), appearance characteristics, and functional properties (Fig. 9.6-5), which include perviousness to moisture, abrasion resistance, and other properties. These factors dictate the suitability of a particular tile for a specific purpose.

In discussing the appearance or properties of tile, it is convenient to separate the face from the body of the tile. Some tile faces are glazed, others are not. Glazed

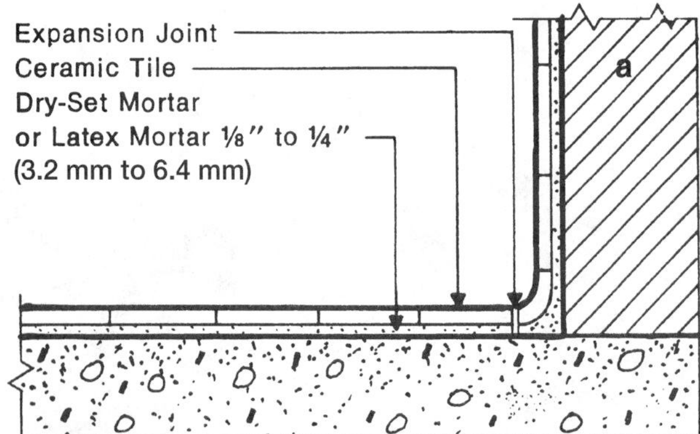

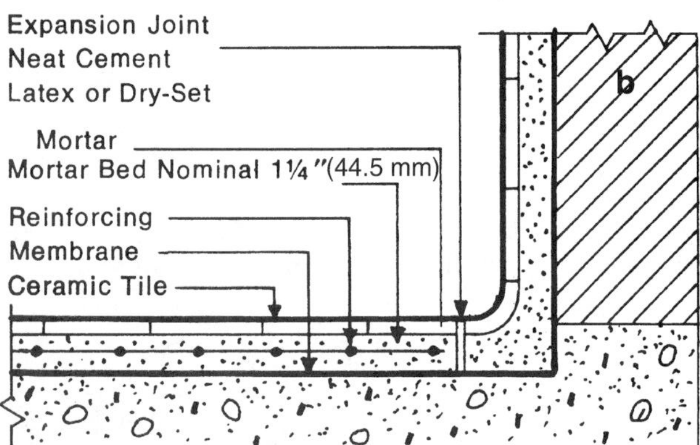

FIGURE 9.6-3 **Basic tile installation methods: (a) thin-set method (tile is bonded to the supporting substrate); and (b) conventional mortar (thick-bed) method (tile and mortar setting bed are isolated from the backing by a membrane).**

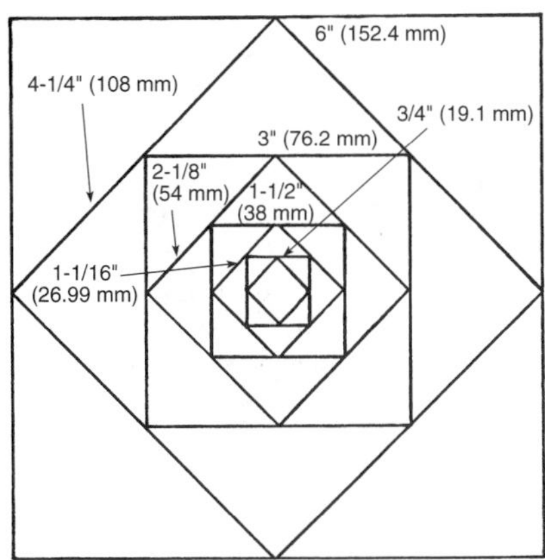

FIGURE 9.6-4 Derivation of normal tile sizes of less than 36 in. (914 mm).

tile may have an opaque decorative glaze over a light-colored body, or a transparent glaze that reveals the natural color of the body. Unglazed tiles, such as some mosaic tiles and all quarry tiles, are hard, dense, units of homogeneous composition throughout, including the face.

Tile bodies have varying degrees of density and porosity, resulting in greater or lesser perviousness to moisture. Tile with a body that has an absorption of less than 0.5% is *impervious*; tile with an absorption of 0.5% to 3% is *vitreous*. The low absorption of impervious and vitreous tiles makes them suitable for exterior installations subject to freezing and to interior installations, such as swimming pools, that are subject to continuous immersion in water. *Semivitreous* tile, with an absorption of 3% to 7%, and *nonvitreous* tile, with an absorption between 7% and 18%, are not suitable for exposure to freezing or immersion in water.

Only two grades of tile are recognized by the TCA and the American National Standards Institute (ANSI): *standard grade* and *seconds*. Standard-grade tile is as perfect as is commercially practicable to manufacture. Individual tiles are harmonious in color, but may vary in shade, and are free from spots and defects visible from a distance of more than 3 ft. (900 mm). *Seconds* may have minor blemishes and defects not permissible in standard grades, but are free from structural defects or cracks.

ANSI Standard 137.1, "Ceramic Tile," is the recognized industry standard for the manufacture, testing, and labeling of ceramic tile. According to this standard, tile should be shipped in sealed cartons with the grade of contents indicated by grade seals, except for quarry tile, which has the grade printed on the carton in lieu of grade seals. A grade seal is a strip of paper, not less than 2 in. (50.8 mm) by 4 in. (101.6 mm), applied over the container in such a way that it cannot be opened without breaking the seal.

Ceramic tile used where appearance is an important factor should be standard grade as defined in ANSI 137.1. Additional evidence of compliance with ANSI 137.1 is recommended, in the form of a manufacturer's written certification, which is called a Master Grade Certificate.

Ceramic tile products are grouped by ANSI 137.1 into four major types:

- Glazed wall tile
- Mosaic tile
- Quarry tile
- Paving tile

Special-purpose tile products combining the properties of the basic type also are manufactured, and those produced by reputable manufacturers have a proven record of success. Figure 9.6-5 summarizes the generally available types, sizes, and uses of ceramic tile. A variety of trim shapes are produced, such as caps, corners, and bases, in matching colors, for both thin-set and conventional installation (Fig. 9.6-6).

Glazed Wall Tile Most glazed wall tile is 1/4 in. (3.2 mm) to 5/16 in. (7.9 mm) thick, and 4 by 4 in. (101.6 by 101.6 mm) or 4 1/4 by 4 1/4 in. (108 by 108 mm), although larger units are also made. It is available in plain and scored surfaces, in bright, matte, and crystalline finishes, and in hand-decorated and sculptured designs (see Fig. 9.6-5). Glazed wall tile will not withstand impact or freezing and thawing conditions. Some glazed tile is suitable for flooring in areas subject to moderate foot traffic, as in most residential and some commercial areas. However, ANSI 137.1 does not cover this type of tile, and the manufacturer's specifications must be relied on to determine suitability for the intended exposure. Glazed tile suitable for flooring usually has a heavier glaze, with a matte or crystalline finish.

Most wall tile is made with an impervious glaze over a *nonvitreous* (absorbent) body and is intended for interior use. Some wall tile, however, is made with an *impervious* or *vitreous* body suitable for exposure to moisture or freezing. All wall tile is glazed, most with a bright (shiny) or *matte* (dull) finish. Some tile is made with other special glazes that are more appropriate for light residential traffic than bright or matte glazes.

Most wall tile is made with a cushioned-edge profile, but square-edged tile is also available. Soap dishes, paper holders, towel bar posts, and other accessories for toilet and bathroom uses are available in matching colors.

Ceramic Mosaic Tile Ceramic mosaic tile is made in two body types, *porcelain* and *natural clay*, in relatively small sizes (Figs. 9.6-7 and 9.6-8). It is generally about 1/4 in. (6.4 mm) thick, with a surface area of less than 6 sq. in. (3871 mm^2). Typical sizes are 1 by 1 in. (25.4 by 25.4 mm), 1 by 2 in. (25.4 by 50.8 mm), and 2 by 2 in. (50.8 by 50.8 mm). This tile may be either glazed or unglazed. Unglazed tiles have a wide range of natural earth colors that extend throughout the body of each tile. Additives can be mixed in with the other components for special effects: abrasives for *slip-resistant* tile and carbon black for *conductive* tile.

Individual tiles generally are face- or back-mounted to form larger units, to facilitate handling, and to speed installation. Back-mounted tile has perforated paper, fiber, resin adhesive, or another suitable bonding material permanently attached to the back and edges of the tile so that a portion of each tile is exposed to the mortar or adhesive. Face-mounted tile has a layer of kraft paper applied to the face of the tile, usually with a water-soluble adhesive, so that it can be easily removed before the joints are grouted.

Available types include:

- *Unglazed porcelain (frostproof) tile:* Made by firing refined ceramic materials; dense, impervious, smooth, and highly stain and wear resistant; clear, luminous colors or granular blends; available with an abrasive surface in a few sizes and colors.
- *Glazed porcelain (frostproof) tile:* Made with the same body as unglazed porcelain tile; glazed with clear (transparent) glaze, ceramic glaze, textured glaze, metallic glaze, or special decorator glaze.
- *Unglazed natural clay (frostproof or nonfrostproof) tile:* Dense and abrasion resistant; made from unwashed clays; rugged and slightly textured surface tough enough to minimize slipping; easily cleaned; also available with a slip-resistant face in 1- by 1-in. (25.4 by 25.4 mm) squares only, in limited colors.

FIGURE 9.6-5 Summary of Ceramic Tile Products

Tile and Description	Suitable Uses	Limitations	Nominal Sizes, in.*
Glazed Wall Tile			
Nonvitreous body, matte or bright glaze	All interior walls in residential, commercial, and institutional buildings	Not suitable for floors or areas subject to freezing	5/16 in. (7.9 mm) thick: 4 1/4 in. × 4 1/4 in. (107.95 × 107.95 mm), 6 in. × 4 1/4 in. (152.4 × 107.95 mm), 6 in. × 6 in. (152.4 × 152.4 mm), scored, octagon, hexagon, and other special shapes
Nonvitreous body crystalline glaze	All interior walls, as well as light-duty residential floors and countertops	Not suitable for areas subject to freezing	
Vitreous body, matte or bright glaze	Interior walls, countertops, refrigerator and chemical-tank linings, and exterior areas subject to freezing	Not suitable for floors	
Ceramic Mosaic Tile Porcelain body, unglazed	All uses, interior and exterior, floors and walls; especially suitable for bathrooms, kitchens, and swimming pools	None except limitations imposed by installation method and grout	1/4 in. (6.4 mm) thick: 1 in. × 1 in. (25.4 × 25.4 mm), 2 in. × 1 in. (50.8 × 25.4 mm), 2 in. × 2 in. (50.8 × 50.8 mm)
Porcelain body, glazed	Interior and exterior walls in residential, commercial, and institutional buildings; decorative inserts with above tile	Not suitable for floors	
Natural clay body, unglazed	Swimming pool runways; floors of porches, entrances, and game rooms in the home; floors and walls of commercial, institutional, and industrial buildings	Exterior use requires special frostproof body	
Natural clay body, glazed	Walls in residential, commercial, and institutional buildings; decorative inserts with above tile	Exterior use requires special frostproof body	
Conductive	Hospital operating rooms and adjoining areas; where required to safely dissipate dangerous charges of static electricity	Requires special installation procedures according to NFPA No. 56A	
Quarry Tile	Exterior and interior; from moderate to extra-heavy-duty floors in all types of construction	Recommended for interior use and exteriors not subject to freezing	1/2 in. (12.7 mm) thick: 3 in. × 3 in. (76.2 × 76.2 mm), 4 in. × 4 in. (101.6 × 101.6 mm), 6 in. × 3 in. (152.4 × 76.2 mm), 6 in. × 6 in. (152.4 × 152.4), 8 in. × 4 in. (203.2 × 101.6 mm), and special shapes
Paver Tile	Same as quarry tile	None	3/8 in. (9.5 mm) thick: 4 in. × 4 in. (101.6 × 101.6 mm), 1/2 in. (12.7 mm) thick: 4 in. × 4 in. (101.6 × 101.6 mm), 6 in. × 6 in. (152.4 × 152.4 mm), 8 in. × 4 in. (152.4 × 101.6 mm)

*Sizes shown are those considered typical by the governing industry standard, ANSI A137.1; many more sizes and shapes are produced by tile manufacturers. Special purpose tiles that combine properties of several types also are produced.

- *Glazed natural clay (frostproof or nonfrostproof) tile:* Same body as unglazed natural clay tile with a clear, ceramic, metallic, or decorator glaze.
- *Conductive tile:* Impervious; with additives that minimize electrostatic build-up and explosion hazard in rooms where explosive gas mixtures are used, such as hospital rooms and laboratories; tile and installation must comply with National Fire Protection Association Bulletin No. 56A.

Quarry Tile Quarry tile is an unglazed tile, with a surface area of 9 sq. in. (58.06 mm^2) or more, 1/2 in. (12.7 mm) thick or thicker, made by the extrusion process from natural clay or shale (Fig. 9.6-9). Typical tile sizes are 3 by 6 in. (76.2 by 152.4 mm), 6 by 6 in. (152.4 by 152.4 mm), 8 by 4 in. (203.2 by 101.6 mm), 9 by 6 in. (228.6 by 152.4 mm), and 9 by 9 in. (241.3 by 241.3 mm) in square, hexagonal, and other special shapes. Tile thicker than 3/4 in. (19.1 mm), made of the same materials by the same process, is called *packing house tile.*

Paver Tile As the name implies, paver tiles are used primarily for floors. They are

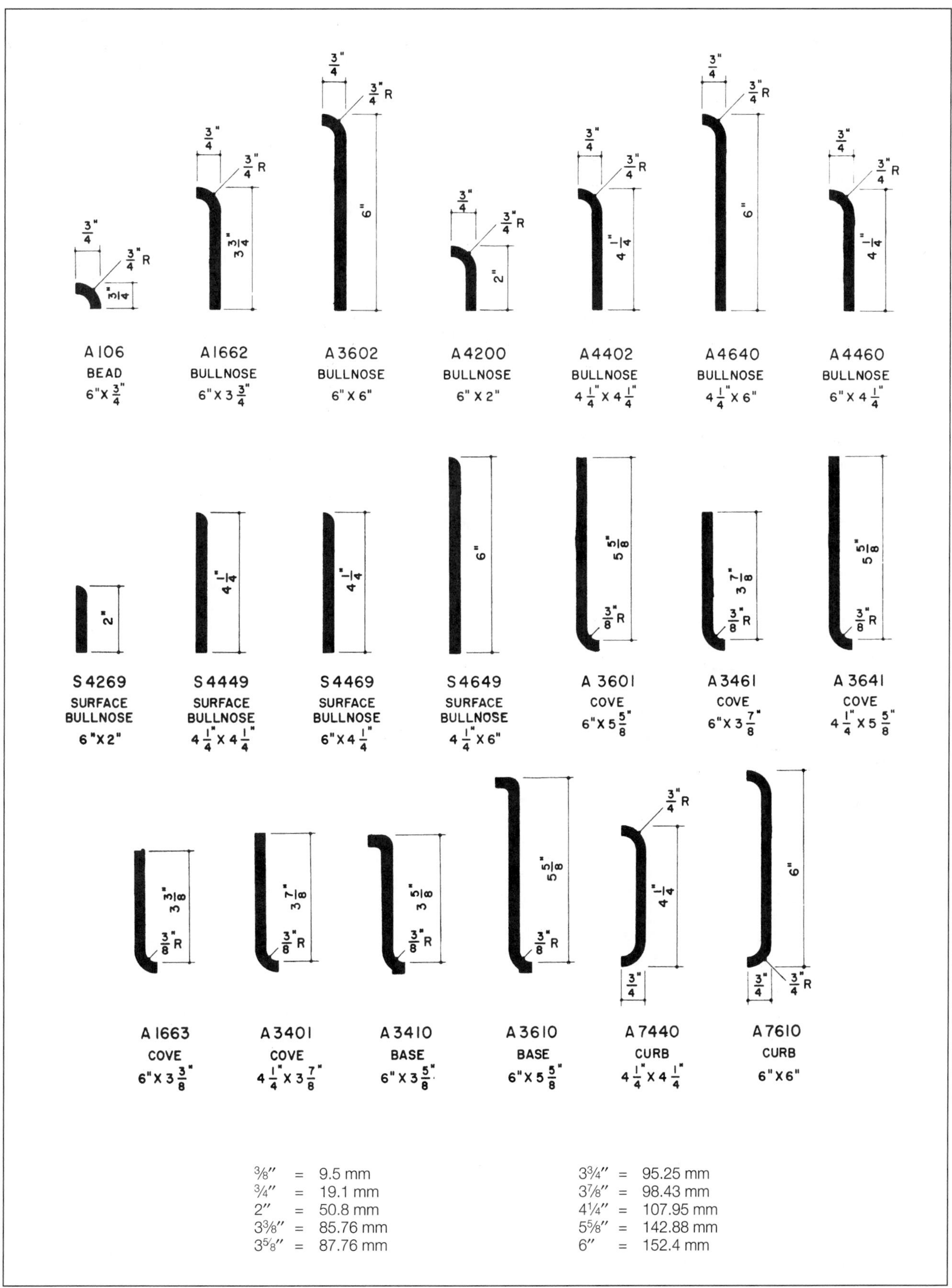

FIGURE 9.6-6 Standard wall tile trim shapes. Similar bullnose, cove, and base shapes are available for other types of tiles (see ANSI A137.1 or manufacturer's literature). (Courtesy American Olean [Ceramic] Tile Co.)

FIGURE 9.6-7 The range of ceramic mosaic tile uses is virtually limitless, including decorative walls and floors indoors and outdoors. The large flat wall on this hotel is finished in ceramic mosaic tile. (Courtesy American Olean [Ceramic] Tile Co.)

FIGURE 9.6-8 This decorative pattern was made with ceramic mosaic tile. (Courtesy American Olean [Ceramic] Tile Co.)

similar in appearance to quarry tile, but more similar to mosaic tile in composition and physical properties, though relatively thicker (½ in. [12.7 mm] typical) and larger (6 sq. in. [3871 mm^2] or more). Typical tile sizes are 4 by 6 in. (101.6 by 152.4 mm), 6 by 6 in. (152.4 by 152.4 mm), and 4 by 8 in. (101.6 by 203.2 mm) (see Fig. 9.6-9).

TILE-SETTING PRODUCTS

Commercially available tile-setting products possess a variety of special properties, each intended to make them suitable for setting tile over certain backings or conditions. Figure 9.6-10 summarizes several common tile-setting products and the governing industry standards. Tile-setting products not covered by industry standards may also be used if they are produced by reputable manufacturers and have a proven record of success.

Portland cement mortar, dry-set mortar, latex–portland cement mortar, epoxy mortar, and organic adhesives, and their installation, should conform to the appropriate industry standards (see Fig. 9.6-10). Products not covered by these standards should conform to the recommendations of their manufacturers and good industry practice.

Neat Portland Cement (Cement-Water Paste) Neat portland cement can be used as a bond coat for tile over a still plastic portland cement mortar setting bed. Neat cement bond coats are considered to be mortars. Dry-set and latex–portland cement mortar can be used to bond ceramic tile to a mortar bed that has been cured for at least 24 hours. Dry-set mortar can also be used as a ¹⁄₁₆-in. (1.6 mm)-thick bond coat over a still-plastic mortar bed. Mortar setting beds to receive organic adhesives must be fully cured and dry. To ensure satisfactory installation in the proper plane, mortar beds should be installed by the tile-setting contractor.

Resinous tile-setting materials, such as epoxies and furans, offer greater bonding strength and chemical resistance than cementitious mortars. However, special tile-setting skills are required, and both labor and materials are more costly than those for cementitious mortars.

Portland Cement Mortar Portland cement mortar is a mixture of portland cement, sand, water, and sometimes hydrated lime. It is the traditional material for installing ceramic tile. The proportions vary, depending on whether the mortar is to be used as a leveling bed or as a setting bed and whether it is to be used on a vertical or a horizontal surface.

Portland cement mortar is suitable for most surfaces and ordinary types of installations over masonry, monolithic concrete, concrete/glass fiber-reinforced backerboard, wood boards, plywood, foam insulation board, gypsum wallboard, and gypsum plaster. Their relatively thick beds, ¾ to 1 in. (19.1 to 25.4 mm) on walls and ¾ to 1¼ in. (19.1 to 28.6 mm) on floors, facilitates accurate slopes and true planes in the tile finish.

FIGURE 9.6-9 Quarry and paver tile provide attractive, low-maintenance floor finishes in heavy-duty traffic areas. (Courtesy American Olean [Ceramic] Tile Co.)

Dry-Set Mortar Dry-set mortar is a mixture of portland cement, sand, and resinous additives that impart water retentivity. Improved retentivity prevents rapid loss of water to the backing surface and improves bond.

Dry-set mortar is suitable for use over masonry, monolithic concrete, concrete/glass fiber-reinforced backerboard, cut-cell expanded polystyrene and rigid closed-cell urethane insulation board, gypsum wallboard, cured portland cement mortar, ceramic tile, and marble. It has excellent impact resistance, is nonflammable, and does not require presoaking of absorptive tile, as do most cementitious mortars. It is the preferred mortar for absorptive glazed tile.

Dry-set mortar is available either unsanded or factory-sanded. Presanded mortars, to which only water must be added, offer greater control over the amount of sand in the mix. Dry-set mortar is not affected by prolonged contact with water and does not form a water barrier. It is limited in truing or leveling the work of other trades to about ¼ in. (6.4 mm) thickness, but can be used in a layer as thin as ¹⁄₁₆ in. (1.6 mm) over a plastic mortar bed.

Latex–Portland Cement Mortar Latex–portland cement mortar is a mixture of portland cement, sand, and a latex additive, which imparts greater flexibility to the mortar. The uses of latex–portland cement mortar are similar to those of dry-set mortar. It should be applied over a cured mortar bed in a coat ⅛ to ¼ in. (3.2 to 6.4 mm) thick. Because latex formulations vary, the manufacturer's directions should be followed.

Epoxy Mortar Epoxy mortar is available in a two-part mortar consisting of an epoxy resin and a hardener, or a less expensive three-part system consisting of an epoxy resin, a hardener, and portland cement powder. When the parts are mixed, a chemical reaction occurs, and the mortar must be used within a limited time (*working time*). Epoxy mortar is suitable for use where chemical resistance or high bond strength is an important consideration. It can be used over concrete, plywood, steel plate, and ceramic tile. It is applied in a layer ¹⁄₁₆ to ⅛ in. (1.6 to 3.2 mm) thick. Working time, bond strength, water-cleanability before cure, and chemical resistance vary with the manufacturer. Epoxy mortar is frequently used in combination with epoxy grout, resulting

FIGURE 9.6-10 Summary of Tile-Setting Products

Type	Description	Industry Standards: Product	Industry Standards: Installation	Limitations
Portland cement mortar	Field mix of portland cement, sand, and water for setting bed; mix portland cement and water for bond coat. Interior or exterior.	ASTM C 150 (Portland cement)	ANSI A108.1	Requires 1¼-in. setting bed, resulting in additional thickness and weight. Requires presoaking of most glazed wall tile.
Dry-set mortar	Prepared mix[a] of portland cement, sand, and certain additives that impart water retentivity and eliminate moist curing. High resistance to moisture, impact, and freezing. Interior and exterior.	ANSI A118.1[b]	ANSI A108.5[c]	Three days of moist curing before grouting recommended for heavy-duty floors.
Latex portland cement mortar	Factory mix of portland cement and sand to which liquid latex resins are added in the field to impart greater bonding strength and a resilience. Interior or exterior.	ANSI A118.4	ANSI A108.5	Not suitable for wood, metal, hardboard, or particleboard backings.
Epoxy mortar	Two-part system of (1) epoxy resin and finely graded sand and (2) hardener. High bond strength and chemical resistance. Interior and exterior.	ANSI A118.3	ANSI A108.6	Requires special installation skills. Material and labor costs relatively high. Not suitable for walls.
Epoxy adhesive	Two-part system of (1) epoxy resin and (2) hardener. Good bond strength and chemical resistance. Interior or exterior. A three-part system containing portland cement powder is also available.	ANSI A118.3	ANSI A108.6	Requires special skills. Installation cost high, but lower than epoxy mortar.
Furan mortar	Two-part system of (1) furan resin and (2) hardener. Highest chemical resistance and installation cost. Interior or exterior.	ANSI A118.5	ANSI A108.8	Same as epoxy mortar.
Organic (mastic adhesive)	Prepared organic material, ready to use without addition of liquid or powder. Adequate bond strength and resilience for light duty on floors and most walls. Interior use only.	ANSI A136.1	ANSI A108.4	Not suitable for exterior use or interior use in continuous contact with water. Applied 1⁄16 in. thick; requires smooth backing.

[a]Requires addition of water in the field.

[b]For conductive dry-set mortar, ANSI A118.2.

[c]For conductive dry-set mortar, ANSI A108.7.

in a highly durable water- and chemical-resistant installation.

Epoxy Adhesive Epoxy adhesive is a two-part system consisting of an epoxy resin and a hardener. It is less costly than epoxy mortar, but also less chemical resistant. It is intended for thin-set application of tile on walls, floors, and counters, where high bond strength and ease of application are overriding considerations. A typical use is in the application of a tile base over a metal refrigeration enclosure in a commercial kitchen. It has a better chemical resistance than organic adhesives.

Furan Mortar Furan mortar is a two-part mortar consisting of a furan resin and a hardener. It is suitable where chemical resistance is the most critical consideration and is therefore used primarily on floors in laboratories and industrial plants. Acceptable subfloors include concrete, steel plate, and ceramic tile. Furan mortar is used in combination with furan grout.

Organic Adhesive (Mastic) Organic adhesives are prepared organic materials that are ready to use with no addition of liquid or powder. They harden by evaporation. Organic adhesives are used in a layer 1⁄16 in. (1.6 mm) thick on floors, walls, and counters. Suitable backings include monolithic concrete, gypsum wallboard, portland cement mortar beds, gypsum plaster, mineral-fiber-cement board, brick, ceramic tile, marble, and plywood.

Organic adhesives eliminate presoaking of tile and are suitable for most residential and commercial uses. They can be used for the walls of residential tub and shower areas, but are not suitable for high-moisture areas, such as shower rooms and swimming pools, or for exterior installations. Their chief advantage is low cost and a flexible bond. Bond strength varies greatly among the numerous available brands. The solvents in some adhesives may be irritating to the skin, and some are flammable.

GROUTS

Grouting materials are available to meet the requirements of the different kinds of ceramic tile and types of exposures. Cementitious grouts consist of portland cement modified to provide specific qualities, such as whiteness, uniformity, hardness, flexibility, and water retentivity. Resinous grouts (epoxies and furans) possess special properties, such as high bond strength and chemical resistance, but are harder and costlier to apply. Figure 9.6-11 summarizes commonly available tile grouts and the governing standards.

Commercial Cement Grout Commercial cement grout is a water-resistant, uniformly colored mixture of portland cement and other ingredients. It is the most commonly specified grout for tile walls subject to ordinary use. Soaking of nonvitreous wall tile is required to prevent moisture loss to the absortive tile body. Moisture in the grout is essential to good bond and full grout strength. Damp curing is needed to ensure an adequate supply of moisture for the hardening process (hydration). Commercial cement grout is usually white, but colored grouts also are available.

Sand–Portland Cement Grout Sand–portland cement grout is an on-the-job mixture of normal portland cement, fine graded sand, and sometimes hydrated lime. It is used with ceramic mosaic tile, quarry tile, and paver tile. Mixes should be 1 part portland cement to 1 part sand for joints up to ⅛ in. (3.2 mm) wide; 1 part to 2 parts for joints up to ½ in. (12.7 mm) wide; and 1 part to 3 parts for joints over ½ in. (12.7 mm) wide. Up to ⅕ part lime may be added for improved workability. Presoaking of nonvitreous wall tile is necessary, and damp curing of the grouted surface is required.

Dry-Set Grout Dry-set grout is a prepared mixture of portland cement and sand, with additives to improve water retentivity. Dry-set grout has the same properties as dry-set mortar and is suitable for grouting walls subject to ordinary use. It eliminates the soaking of tile, although dampening is sometimes desirable under very dry conditions.

Latex Grout Latex grout is a mixture of one of the three preceding products, consisting mainly of portland cement with a latex additive to increase stain resistance and resilience of the joints. Latex grout is suitable for installations subject to ordinary use. It is less rigid and less water permeable than sand–portland cement grout, allowing for more movement and greater exposure to moisture. This grout is therefore particularly suitable in tub and shower areas.

Silicone Rubber Grout Silicone rubber grout is an engineered elastomeric grout system for interior use, employing a sin-

FIGURE 9.6-11 Summary of Tile-Grouting Products

Type	Description	Limitations
Commercial portland cement grout	Prepared mix* of portland cement and other ingredients resulting in a water-resistant, uniformly colored material; usually white, but available in colors. Commonly used for tile walls and floors subject to ordinary use. Suitable for interior and exterior.	Requires presoaking of most glazed wall tile and damp curing of grouted surface.
Sand–portland cement grout	Field mixture of portland cement, fine-graded sand (sometimes lime), and water. Most commonly used grout for ceramic mosaic tile. Suitable for floors and walls, interior and exterior.	Same as for commercial cement grout.
Dry-set grout	Prepared mix* of portland cement, sand, and additives that provide water retentivity and eliminate need for presoaking of glazed wall tile. Damp curing may develop greater strength in portland cement grouts. Generally used on walls, interior and exterior.	Dampening of the tile surface prior to grouting may be required under extremely dry conditions.
Latex–portland cement grout	Prepared mix* of any of the above formulations plus latex additives. It is less rigid and less permeable than regular cement grout. The addition of latex to portland cement grouts eliminates the necessity of damp curing for 72 hours. Can be used on walls and floors, interior and exterior.	Same as for dry-set grout.
Silicon rubber grout	Ready-to-use one-part caulking-type material that cures from moisture vapor in the air. Used mainly for glazed tile walls. Flexible, mildew resistant, stain resistant.	Bonds only to clean, dry surfaces. Must be used with proper adhesive to prevent staining.
Epoxy grout	Ready-to-use two-part system consisting of epoxy resin and hardener. Formulation for wall tile omits coarse fillers, which might scratch glaze; coarse silica fillers generally are included in grout intended for quarry tile and pavers. Highly resistant to chemical staining, hence used for industrial and kitchen floors, countertops.	Requires special skills for proper application. Material cost and labor higher than with other grout system.
Furan resin grout	Ready-to-use two-part system consisting of furan resin and hardener. Highest chemical resistance; used primarily in industrial installations with quarry tile and pavers.	Same as for epoxy grout.

Note: Epoxy grout should comply with ANSI A118.3; other grouts with ANSI A108.10 and A118.6.

*Requires addition of water in the field.

gle-component nonslumping silicone rubber that, upon curing, is resistant to staining, moisture, mildew, cracking, crazing, and shrinking. It adheres tenaciously to ceramic tile, cures rapidly, and withstands exposure to moisture as well as subfreezing temperatures and hot, humid conditions.

Epoxy Grout Epoxy grout is a two-part system consisting of an epoxy resin and a hardener. It is made in several formulations, each intended for a specific ceramic tile. It is highly stain resistant and impervious and is used with epoxy mortar or epoxy adhesive. Its use requires special skills and is usually expensive.

Epoxy grout is formulated without coarse fillers that may scratch glazed wall tile or mosaic tile. It can be used on walls, floors, and counters subject to food staining.

It is also formulated with a coarse silica filler for use with quarry tile and pavers. It is formulated for industrial and commercial installations where chemical resistance is of paramount importance. It has excellent bonding characteristics.

Furan Resin Grout Furan resin grout is a two-part grout consisting of a furan resin and a hardener. It is intended for use with quarry tile and pavers, mainly in industrial areas requiring a maximum of chemical resistance. It is generally used in conjunction with furan mortar.

RELATED MATERIALS AND PRODUCTS

Installation of ceramic tile involves a variety of related materials and products, described below and summarized in Figure 9.6-12.

FIGURE 9.6-12 Summary of Related Materials and Products

Material or Product	Industry Standard
Portland cement, gray or white, Type I	ASTM C 150
Sand for mortar	ASTM C 144
Hydrated lime, Type S	ASTM C 206
Sealants	ASTM C 920
Metal reinforcement	
Welded wire fabric	ASTM A 82 and A 185
Expanded metal lath	ASTM C 847
Cleavage membrane	
Polyethylene film	ASTM D 4397
Asphalt roofing felt	ASTM D 226

Portland Cement Normal (Type I) portland cement is used for mortar setting beds, for leveling beds, for neat cement bond coats, and for field-mixed cement grout. White grout requires the use of white portland cement in the mortar bed and in the grout.

Sand Sand should be clean, free of impurities, and well graded. When fine sand is specified, it should pass through a 16-mesh (1.18 mm) screen.

Hydrated Lime Hydrated lime is used in portland cement–sand mortar to make it more plastic for application on walls, and sometimes on floors. Type S (special) lime should be specified to ensure a dimensionally stable mortar bed.

Sealants Sealants for use in expansion joints and control joints should be suitable for the exposure expected. Conditions that cause sealants to deteriorate include temperature variations, direct sunlight, repeated stretching and compression, frequent wetting and drying, and immersion in water. These sealants can be single-component or two-component types. Sealant color can be selected to match or blend with the tile or adjacent materials. A sealant used on floors must be hard enough to withstand occasional or repeated foot traffic. It should have a Shore A hardness index of not less than 25 for joints subject to occasional foot traffic; not less than 35 in high-traffic areas.

Metal Reinforcement Isolated (unbonded) mortar setting beds should contain a suitable metal reinforcement. Reinforcement for a mortar bed consists of metal lath on walls and welded wire fabric on floors. This welded wire fabric is usually 2- by 2-in. (50.8 by 50.8 mm) mesh, WO.3 by WO.3 (16 AWS gauge or 0.0625-in. [1.6 mm] diameter). Metal lath should be expanded lath weighing not less than 3.4 lb./sq. yd. (1.84 kg/m^2) or sheet lath weighing not less than 4.5 lb./sq. yd. (2.44 kg/m^2).

Cleavage Membrane A cleavage membrane should be installed beneath every unbonded ceramic tile setting bed. It should be one of the following:

- 15 lb. (8.14 kg/m^2) asphalt (or 13 lb. [7.05 kg/m^2] coal-tar)-saturated roofing felt
- Reinforced waterproof duplex asphalt paper
- 4-mil (0.10 mm) or thicker polyethylene film
- Membrane waterproofing

9.6.1.2 Job Conditions

Materials should be delivered in the manufacturers' original sealed containers and stored under cover to prevent damage or contamination. Tile should be set and grouted when the temperature is at least 50°F (10°C) and rising, unless the mortar manufacturer requires a higher temperature or permits a lower temperature.

9.6.1.3 Supporting Construction

A tile installation is only as good as the underlying assembly of structural members and backing (subfloor or wall surface) that support it. If the supporting construction deflects, expands, or contracts, these movements will be passed on to the tile finish.

This is particularly true in bonded installations. Except in small areas, a bonded tile finish cannot absorb these movements and internal stresses are likely to result in cracks. It is essential, therefore, to provide supporting construction with a minimum amount of movement. Where movement is anticipated, expansion or control joints should be provided in the supporting construction and extended into the tile.

Movement is less critical when the mortar bed is isolated from the backing by a cleavage membrane. Although the membrane permits some independent movement in the supporting construction, the amount of movement that can be tolerated still is limited. Therefore, dimensionally stable, rigid construction is desirable.

The *backing* is that part of the supporting construction to which the tile is bonded. It can be a subfloor, wall finish, or mortar leveling coat. In thin-set installations, the backing must be dimensionally stable, sound, true to line, plumb, and sloped or level, as designed. Moreover, the backing must be free of foreign substances and loose materials that may weaken the bond with the tile finish. Trueness of the backing is less critical when a mortar bed is used, but where slopes are required they should be built into the backing, not developed by varying the mortar bed thickness.

FLOORS

Floors should be engineered so that the maximum deflection under full load, including live loads and the weight of the tile, does not exceed 1/360 of the span. For instance, on a span of 15 ft. (180 in.) (4572 mm), the deflection should be not more than 180/360 (4572/360) or ½ in. (12.7 mm). The most common floor assemblies encountered in small buildings involve a concrete or wood subfloor; in large buildings, almost always a concrete subfloor.

Concrete Subfloor Tile can be installed on a concrete subfloor either by the thin-set or the portland cement mortar bed method. For thin-set bonding directly to a floor slab, the surface should be steel troweled and fine broomed, with no variations in the plane or slope exceeding ⅛ in. (3.2 mm) in 10 ft. (3000 mm). For portland cement installations a variation of ¼ in. (6.4 mm) in 10 ft. (3000 mm) is acceptable. Concrete slabs should be thoroughly moist-cured before tile application. Liquid curing compounds and other coatings that may reduce bond should not be used. Slabs should not be saturated with water at the time of installation. To prevent cracking, a slab should be properly reinforced, concrete quality should be controlled, and suitable joints should be provided in the slab.

Wood Subfloor Supporting joists should be spaced not more than 16 in. (400 mm) on centers. For thick-bed installations, either a single-layer subfloor of 1-in. (25.4 mm) nominal boards or ⅝-in. (15.9-mm)-thick plywood is acceptable. For thin-set installations with most tile-setting products (except epoxy mortar) a double floor is required, consisting of a subfloor, as described earlier, and an underlayment. The underlayment should be Exterior Type CC-plugged plywood, ½-in. (12.7 mm) thick for residential use, ⅝-in. (15.9 mm) thick for commercial buildings. When epoxy mortar is used in residential construction, the underlayment can be omitted if a ⅝-in. (15.9 mm)-thick plywood subfloor is used and the longitudinal joints between panels are supported with solid wood blocking 2 × 4 (50 × 100) or larger members.

A ¼-in. (3.2 mm) space should be left between the underlayment and vertical surfaces, such as walls, and projections such as pipes or posts. For organic adhesives, 1/16 in. (1.6 mm) should be left between underlayment ends and edges; for epoxy mortar, ¼ in. (3.2 mm) should be left at ends and edges of panels, and these joints should be filled with epoxy. Adjacent underlayment edges should not be more than 1/32 in. (0.8 mm) above or below each other.

Concrete/glass fiber-reinforced backer board may be used in lieu of a portland cement mortar bed over a ½-in. (12.7 mm)-thick plywood subfloor. This backer board is designed for use with ceramic tile floors and is available in various sizes and approximately ½ in. (12.7 mm) (7/16 in. [11 mm]) thick. It can be nailed or screwed to wood joists through ½-in. (12.7 mm)-thick plywood subflooring. Ceramic tile can be bonded to it with either dry-set or latex–portland cement mortars. Adhesives are not recommended for use with this product.

WALLS

Wall surfaces should be sound, plumb, and true, with square corners. For thin-set application solid backing usually is installed by others (not by the tile installer). The backing surface should not vary from a true plane by more than ⅛ in. (3.2 mm) in 8 ft. (2400 mm). For portland cement mortar bed installations, either a solid backing or open framing is suitable (Fig. 9.6-13). In either case, the tile contractor must attach metal lath (over a membrane) to the framing or backing, then apply a scratch coat and build up a mortar bed ¼ to ¾ in. (6.4 to 19.1 mm) thick, to which the tile is bonded. A similar thickbed installation can be used over irregular masonry or concrete surfaces.

Usually, tile is installed on walls by the thin-set method over gypsum wallboard, gypsum plaster, plywood, concrete/glass fiber-reinforced backing board, or other dimensionally stable solid backing. However, gypsum plaster and plywood should not be used as backing in tub enclosures and shower stalls unless waterproof grout, such as silicone rubber, is used.

Gypsum Wallboard Wallboard should be installed in maximum practical lengths to minimize the number of end joints. Joints should be taped, and both joints and fasteners should be treated with at least two coats of joint compound. In tub and shower areas, water-resistant backing board should be used. Its joints should be taped and filled with one coat (maximum 4 in. [400 mm] wide) of joint compound.

Gypsum Plaster Plaster walls that will receive tile should be steel troweled with a brown coat mix (one part gypsum to three parts sand) or a prepared finish coat plaster. When the tile finish will be flush with an adjacent plastered surface, the finish coat should be omitted and the tile set on the brown coat. Otherwise, a full-thickness three-coat plaster surface should be used (see Section 9.3). The surface to receive tile should be troweled smooth with no trowel marks or ridges more than 1/32 in. (0.8 mm) high.

Plywood A plywood backing should be Exterior Type CC-plugged, ⅜ in. (3.2 mm) thick or thicker. Panels should be applied with the face grain perpendicular to the studs or furring. Solid 2 × 4 (50 × 100) wood blocking should be used under panel edges between studs. Nails should be driven 6 in. (150 mm) on centers at panel perimeters and 12 in. (300 mm) on centers at intermediate supports. Panels should be spaced ⅛ in. (3.2 mm) apart and there should be not more than a 1/32-in. (0.8 mm) variation between the faces of adjacent panels. When ½-in. (12.7 mm)-thick plywood is used, horizontal blocking between studs can be eliminated and panels can be applied parallel to studs. Before in-

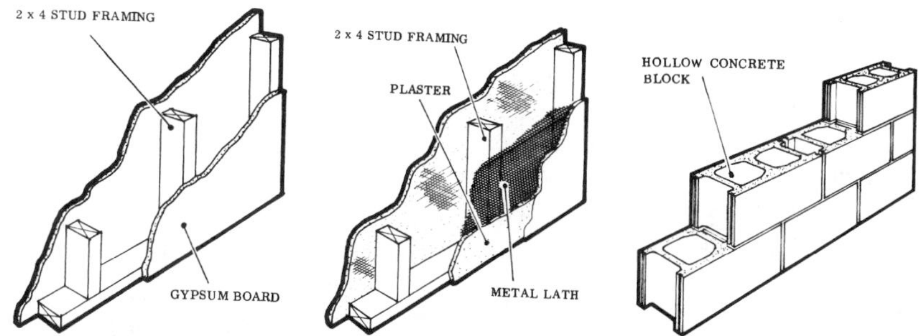

FIGURE 9.6-13 Typical supporting construction for ceramic tile. In areas subjected to wetting, a portland cement mortar bed should be used instead of gypsum plaster. Concrete/glass fiber–reinforced backerboard can also be used where gypsum board is indicated here.

stallation, plywood edges should be sealed with an exterior primer or aluminum paint.

Concrete/Glass Fiber-Reinforced Backing Board Concrete/glass fiber-reinforced backing board is a backer board designed for use with ceramic tile in wet areas around tubs and showers and on floors. Available in various sizes and about ½ in. (12.7 mm) thick, this material can be nailed or screwed in place over wood or metal studs or over a ½-in. (12.7 mm)-thick plywood subfloor. Ceramic tile can be bonded to it with either dry-set or latex–portland cement mortar. It can be used in place of metal lath, portland cement scratch coat, and a mortar bed. It is not suitable as a base for tile installed with adhesives.

Masonry and Cement Masonry and cement backings include monolithic concrete, concrete/glass fiber-reinforced backing board, brick, and concrete masonry units (CMU). For thin-set installations, the wall surface should be clean, sound, and free of paint, efflorescence, and cracked or chipped areas. When latex–portland cement or dry-set mortars are to be used, smooth concrete or brick surfaces should be roughened by sandblasting or bushhammering to ensure good bond.

For mortar bed installations, the surface qualities of the backing are not as critical. If a leveling bed is to be used, the wall surface can be more irregular than if the tile is bonded directly to the wall, but it should be free of surface coatings and loose material that might affect bond. When a cleavage membrane is used, surface contamination or irregularity is not important, but the wall should be structurally sound and dimensionally stable.

9.6.1.4 Tile Layout and Setting

An area to be covered with tile should be checked for squareness, and midlines should be established in both directions. The tile pattern should be laid out so that a tile is either butted to the midline or bisected by it. The most desirable method is the one that produces larger than half-tiles around the perimeter of the area.

Tiles should be cut with a suitable tool rather than split. Wall tile is cut with a tile cutter; mosaics are cut with "nippers"; quarry tile may be cut with a water-cooled masonry saw or heavy-duty cutter. Tiles should be closely fitted to projections, and all ragged edges smoothed with a carborundum stone. Joint lines should be straight, plumb, level, and of even width. Tilework accessories should be evenly spaced, properly aligned with tile joints, and set in the proper plane (Fig. 9.6-14). When the tile is not of the self-spacing type, joint widths should be as shown in Figure 9.6-15, and setting frames or grids are used for setting tile more quickly and accurately.

FIGURE 9.6-14 **Projections make tile layout more difficult and also more important. Tile should be symmetrical with respect to projections and corners. Tiles should be larger than half size.** (Courtesy American Olean [Ceramic] Tile Co.)

Tile is usually set into a previously spread layer of mortar, adhesive, or neat cement (Figs. 9.6-16 and 9.6-17), but it is sometimes more convenient to brush a coat of the tile-setting material on the back of the tile. Alignment of joints should be checked frequently for trueness and squareness (Fig. 9.6-18). After face-mounted tile is set, the mesh or paper is soaked off and removed. Mortar or adhesive is removed from tile faces with a solvent. Strings or ropes used to space tile should be removed as soon as the mortar or adhesive has set sufficiently to hold the tile firmly in place.

FIGURE 9.6-16 **Quarry tile is laid in a bond coat of dry-set mortar and tapped in place with a trowel handle.** (Courtesy Tile Council of America.)

9.6.1.5 Grouting and Curing

Grout should be applied by troweling diagonally across the joints (Fig. 9.6-19). Enough grout should be forced into each joint so that the joint is filled down to the mortar. Joints of square-edged tile should be filled flush with the tile surface; joints of cushion-edged tile should be tooled to the depth of the cushion. The finished grout surface should be uniform in color, smooth, and free of pinholes and low spots. The tile surface should be washed and sponged thoroughly, then polished with a clean, dry cloth or burlap (Fig. 9.6-20).

Glazed wall tile usually requires soaking before it is set with portland cement mortar. If glazed tile has been set with a material other than portland cement mor-

FIGURE 9.6-15 Joint Widths for Ceramic Tile

Tile Description	Joint Width
Ceramic mosaic tile*	
2 in. (50.8 mm) square or smaller	1⁄32 in. to 1⁄8 in. (0.8 mm to 3.2 mm)
Paver and special tile	
2 3⁄16 in. to 4 1⁄4 in. (71.44 mm to 107.95 mm)	1⁄16 in. to 1⁄4 in. (1.6 mm to 6.4 mm)
6 in. (152.4 mm) square and larger	1⁄4 in. to 1⁄2 in. (6.4 mm to 12.7 mm)
Quarry tile	
3 in. to 6 in. (76.2 mm to 152.4 mm)	1⁄8 in. to 3⁄8 in. (3.2 mm to 9.5 mm)
6 in. (152.4 mm) square and larger	1⁄4 in. to 1⁄2 in. (6.4 mm to 12.7 mm)
Glazed tile	
3 in. (76.2 mm) square and larger	1⁄8 in. to 1⁄2 in. (3.2 mm to 12.7 mm)

*Ceramic mosaic tile is mounted, so joint widths are predetermined by manufacturer.

FIGURE 9.6-17 **Spreading a setting bed for wall tile.** (Courtesy American Olean [Ceramic] Tile Co.)

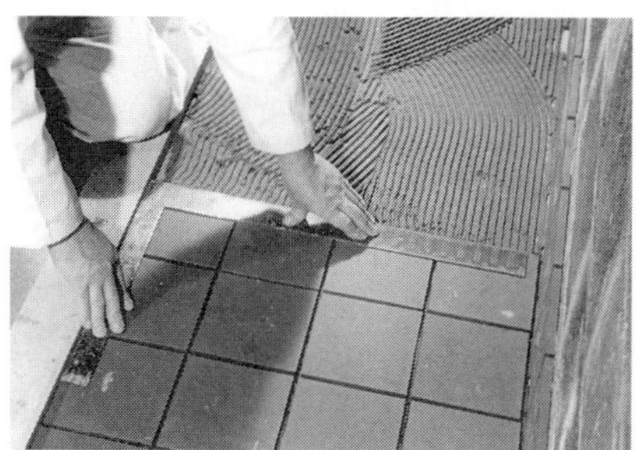

FIGURE 9.6-18 **Quarry tile is spaced to provide ¼- to ½-in. (6.4 to 12.7 mm)-wide joints. A straightedge and metal square are used periodically to check joint alignment and squareness.** (Courtesy Tile Council of America.)

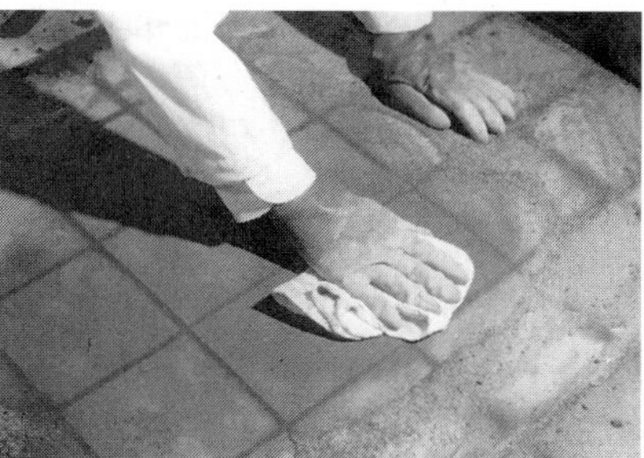

FIGURE 9.6-20 **Excess grout is wiped from the tile surface with burlap or rough-textured cloth.** (Courtesy Tile Council of America.)

FIGURE 9.6-19 **Joints are filled with portland cement grout by troweling diagonally across the joints. Kneeling boards prevent damage to newly-laid tile.** (Courtesy Tile Council of America.)

tar and was not presoaked before it was laid, it should be soaked after setting when a portland cement grout will be used. Soaking is not required for other types of grout.

Commercial cement and portland cement grout should be damp-cured for at least 72 hours. Dry-set grout need not be damp-cured, except when used (1) on floors, (2) on walls exposed to periodic wetting, or (3) in exterior locations. Damp-curing can be accomplished by periodically applying a spray mist to the tile surface or by covering it with polyethylene film after initial and daily spraying.

9.6.1.6 Control and Expansion Joints

To ensure an attractive, crack-free installation, control and expansion joints should be provided in the backing and in the tile finish.

JOINTS IN BACKING

Control joints are provided in cementitious materials such as concrete and stucco to localize shrinkage cracks due to loss of moisture by hydration and evaporation. When control joints are provided, expansion joints usually are not necessary because shrinkage exceeds the likely thermal expansion.

Because shrinkage is minimal in brick masonry, thermal movement is critical and expansion joints must be provided (see Chapter 4). Where differential settlement or movement of adjacent elements is expected, isolation joints should be used. Expansion joints frequently act also as isolation joints or as control joints. To conform to usage in the tile industry, such joints in a tile finish are referred to here as *expansion joints*.

JOINTS IN TILE FINISH

Expansion joints are not required in areas less than 12 ft. (3660 mm) wide or over old, well-cured concrete. However, expansion joints should always be provided in a tile finish (1) directly over expansion, isolation, construction, or control joints in the backing, (2) where the tile finish abuts projecting surfaces, and (3) at proper intervals in floor areas measuring more than 24 ft. (7300 mm) in either direction. Expansion joints in a tile finish should never be narrower than the underlying joint in the backing (Fig. 9.6-21).

Tile edges to which sealant will bond should be clean and dry. Sanding or grinding of these edges is recommended to obtain maximum sealant bond. Some sealant manufacturers recommend priming edges.

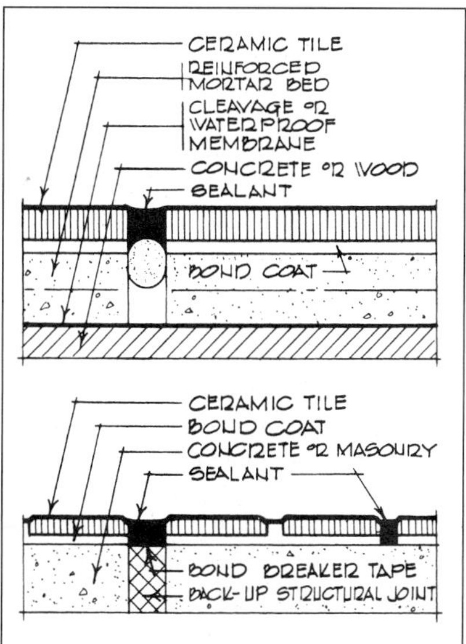

FIGURE 9.6-21 **Expansion joints in tile surfaces.** (Courtesy Tile Council of America, *Handbook for Ceramic Tile Installation.*)

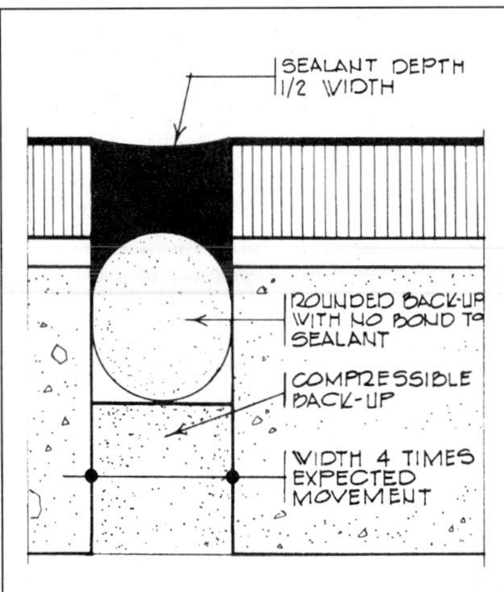

FIGURE 9.6-22 **Proper sealant depth. Good bond at the sides and bond-free contact with backup strip are essential to long-term effective performance.** (Courtesy Tile Council of America, *Handbook for Ceramic Tile Installation.*)

Primers must be kept off the tile face. Sealant depth should be approximately one-half the joint width. Compressible backup strips should be used in a joint to prevent bonding of the sealant to the joint filler and to ensure a proper sealant profile (Fig. 9.6-22).

Exterior Locations Exterior installations should be provided with expansion joints at least ⅜ in. (9.5 mm) wide for joints 12 ft. (3660 mm) on centers, and at least ½ in. (12.7 mm) wide for joints 16 ft. (4880 mm) on centers. These minimum joint widths should be increased ¹⁄₁₆ in. (1.6 mm) for each 15°F (–9.4°C) of temperature differential (range between summer high and winter low) over 100°F (37.8°C). Decks exposed to the sky in the northern United States should have ¾-in. (19.1 mm)-wide joints 12 ft. (3660 mm) on centers.

Interior Locations Thick bed tile installations in interior areas should have expansion joints in the tilework spaced 24 ft. (7320 mm) to 36 ft. (10,973 mm) each way. Joints for interior quarry tile and paver tile should be the same width as the grout joints, but not less than ¼ in. (6.4 mm) wide. Joints for ceramic mosaic tile and glazed wall tile should be ¼ in. (6.4 mm) wide but never less than ⅛ in. (3.2 mm) wide. Large thin-set installations may benefit from appropriately located expansion joints, but no generalizations can be drawn. Joints are always desirable at offsets and changes in backing construction.

9.6.1.7 Thresholds

When it is not practical to depress the subfloor for a conventional mortar bed installation and adjacent finishes are subsequently thinner, a transition is required. A resilient flooring contractor may use rubber or vinyl thresholds; carpenters are likely to use hardwood thresholds in connection with wood flooring; carpeting may be butted to tile with a metal or vinyl edge strip. However, the preferable material for this purpose is a marble or slate threshold located in a doorway or in a cased opening, and set with the same bonding agents as ceramic tile (Fig. 9.6-23).

9.6.1.8 Shower Receptors

Shower receptors can be precast terrazzo, plastic, field-installed ceramic tile, or of a synthetic material. Tile can be installed over wood or concrete subfloors; a mortar bed must be used over wood subfloors. In either case, a waterproof membrane or pan must underlie the tile finish, and it must be sloped ¼ in. per ft. (6.4 mm per 300 mm) to the drain. The sides of the membrane or pan should be turned up the wall 5 in. (127 mm) (Fig. 9.6-24).

Although only the mortar bed method is recommended for shower receptors, the slope should be developed in the backing rather than in the mortar bed. This ensures positive runoff of water that penetrates the tile finish. Crushed stone or tile around the shower drain and weep holes in the drain permit this occasional water to seep into the drainage system (see Fig. 9.6-24).

Concrete subfloors can, themselves, be sloped to the drain. With a wood subfloor, a separate portland cement mortar fill should be used to create the slope. The fill can vary from ½ in. (12.7 mm) to 1¼ in. (28.6 mm) in thickness. It should include a latex additive to improve its bond strength and make it possible to apply it in thin layers.

A waterproof membrane or pan should be installed over the sloped fill, followed by the tile finish set in a mortar bed. Any commercial grout is suitable for grouting the tile joints. Figure 9.6-25 shows proper detailing of tiled tub-shower areas; Figure 9.5-48 shows recommended tile coverage of gypsum board in bathing areas.

9.6.1.9 Tile over Other Finishes

Ideally, existing finishes or coatings should be completely removed, and new tile placed on the original backing. However, if this is impractical, new tile can be installed over existing tile, resilient flooring,

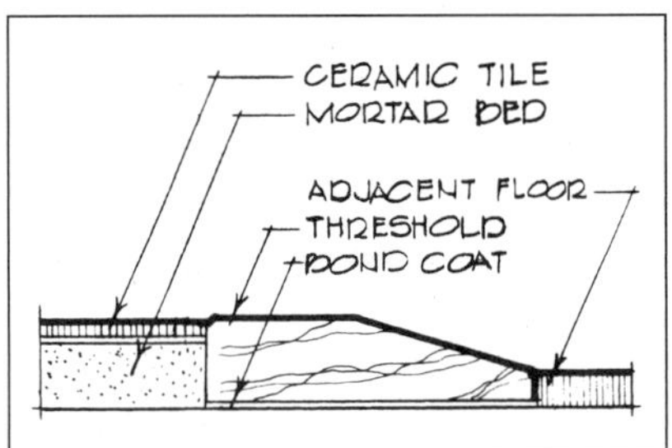

FIGURE 9.6-23 **High end of threshold should be located under door.** (Courtesy Tile Council of America, *Handbook for Ceramic Tile Installation.*)

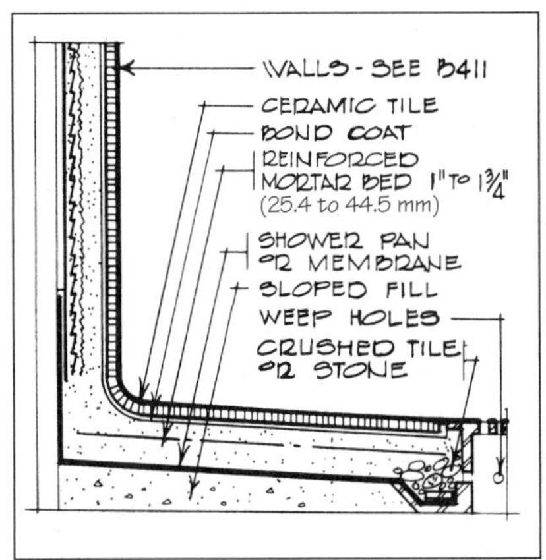

FIGURE 9.6-24 **In a shower stall, including a membrane, crushed stone, and weep holes ensures drainage of water that may seep under the tile finish.** (Courtesy Tile Council of America, *Handbook for Ceramic Tile Installation.*)

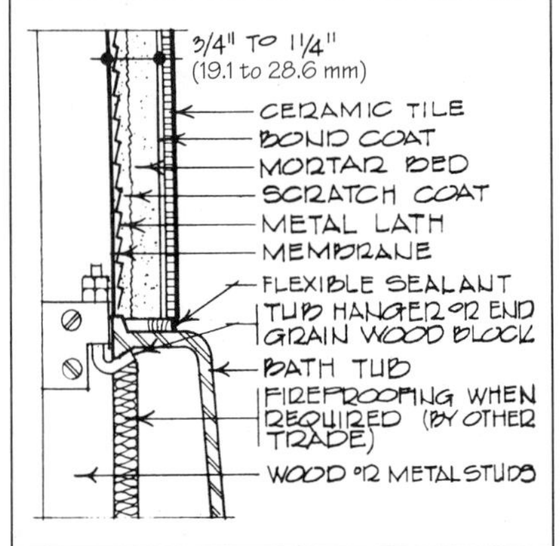

FIGURE 9.6-25 **Applying sealant in the space between tile and tub apron ensures a watertight joint.** (Courtesy Tile Council of America, *Handbook for Ceramic Tile Installation.*)

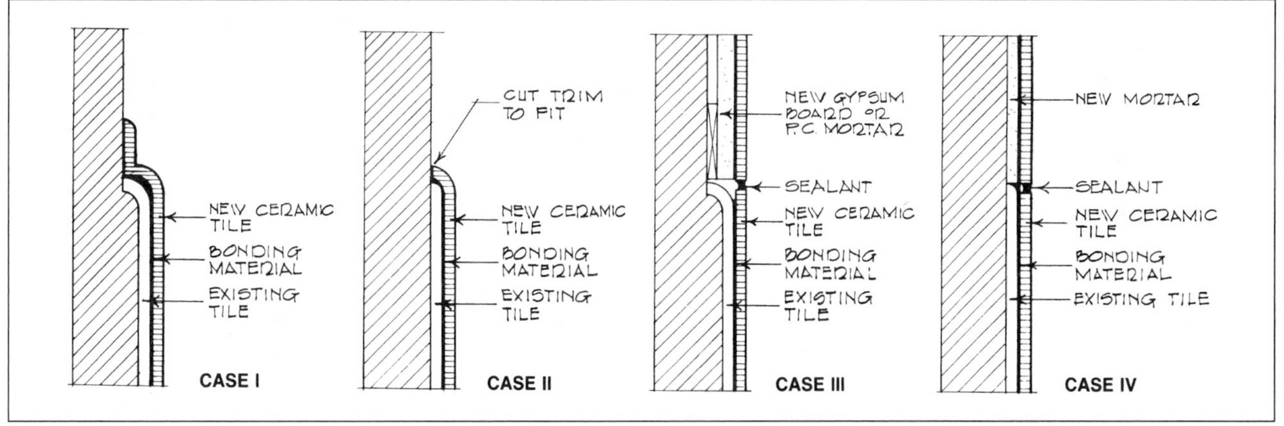

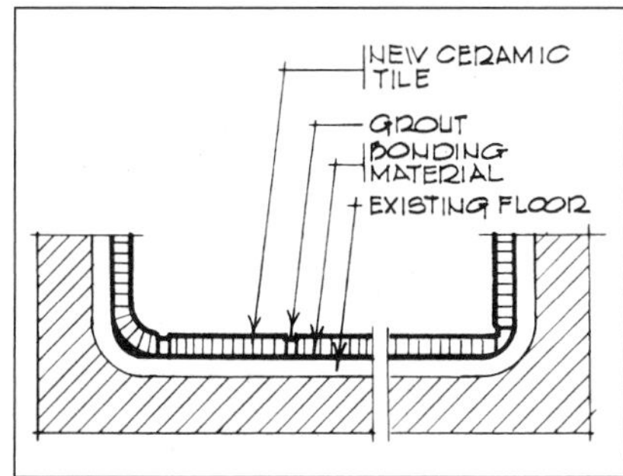

FIGURE 9.6-26 New tile may be bonded directly to existing tile, or a new backing, such as gypsum board, may be imposed between the new and existing tile. (Courtesy Tile Council of America, *Handbook for Ceramic Tile Installation.*)

hardwood flooring, painted concrete, gypsum plaster, and gypsum wallboard, by either installing a new backing surface for thin-set application or installing the new tile using the conventional mortar bed method. Figure 9.6-26 shows several tile-over-tile installation possibilities.

Thickbed installations present no special problems over existing finishes. For thin-set installations, the backing surface must be free of dirt, dust, efflorescence, oil, grease, and other contaminants. The maximum variation in a backing surface should not exceed the dimensions indicated in Section 9.6.1.3. Glossy or scaly painted surfaces should be sandblasted or treated with a chemical paint remover. The adhesive or mortar manufacturer should be consulted to determine whether a sealer or primer is needed. When the structural floor assembly is capable of carrying the weight and the old floor surface is questionable, the more reliable isolated thickbed installation method should be used.

9.6.1.10 Countertops

Either mosaic, quarry, or glazed wall tile can be used as a durable finish for countertops. Tile is typically installed over a wood, plywood, or cementitious board backing by either the mortar bed (thickbed) or the thin-set method (Fig. 9.6-27).

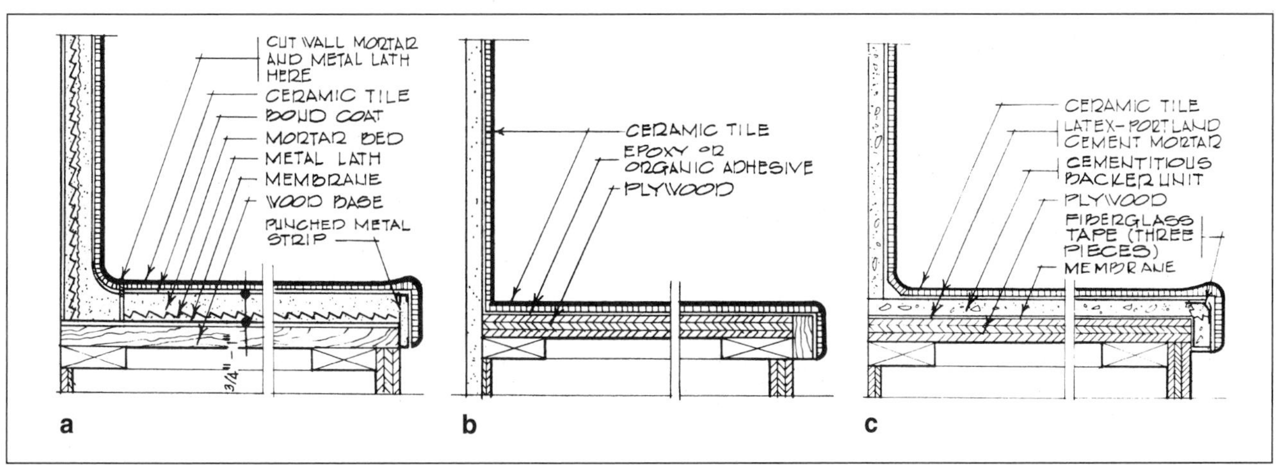

FIGURE 9.6-27 Ceramic tile countertops may be installed by (a) the cement mortar method or (b and c) the thin-bed method. (Courtesy Tile Council of America, *Handbook for Ceramic Tile Installation.*)

The backing for a mortar bed installation should be either 1 × 6 (25 × 150) boards spaced ¼ in. (6.4 mm) apart (Fig. 9.6-27a) or ¾-in. (19.1 mm) exterior plywood. Plywood should be scored with saw cuts 6 to 8 in. on centers to prevent warping. Thin-set installation requires ¾-in. (19.1 mm) exterior plywood, with ¼-in. (6.4 mm) gap panels. This gap should be closed from below with a batten strip (Fig. 9.6-27b).

An epoxy grout or a stain-resistant grout recommended for the purpose by the manufacturer should be used on countertops installed with the thin-set method.

9.6.2 INSTALLATION ON FLOORS

Selection of the most suitable installation method depends on the desired performance level, the supporting construction, and whether the installation is interior or exterior.

9.6.2.1 Installation Performance Levels

The type of traffic (wheeled or foot), impact loads, and contaminants (water, grease, alkalies, chemicals) to which a floor will be subjected are prime considerations in the selection of a suitable performance level (Fig. 9.6-28). Once the performance level has been determined, an appropriate installation method can be selected (Fig. 9.6-29).

In addition to the installation method, consideration should be given to:

- Wear properties of the selected tile
- Fire-resistance properties of the wall or floor assembly
- Acoustical properties (especially impact noise rating) of the floor assembly
- Slip-resistance of the floor tile

Unglazed tile (mosaic, quarry, and paver tile) used for floor surfacing will give satisfactory wear for all the performance levels listed. Decorative glazed tile and soft-body unglazed tile should not be used on floors unless this use is specifically recommended by the tile manufacturer. In determining the suitability of glazed tile for a specific use, color pattern, surface texture, and glaze hardness should be considered. All glazed tile will show traffic wear over time.

9.6.2.2 Interior Concrete

An interior concrete slab on grade is not subjected to the drastic temperature and humidity fluctuations encountered outdoors. In addition, if such a slab is made up of a properly engineered concrete mix and has suitable reinforcement, adequate finishing, proper curing, a well-drained granular base, and proper jointing, the likelihood of excessive movement due to settlement and temperature changes is slight. Therefore, such slabs will accept the complete range of tile installation methods, including a bonded mortar bed, an isolated mortar bed, or direct application with a thin-set mortar or adhesive.

A supported slab, on the other hand, will often deflect enough under loads to cause tile set using other than an isolated mortar bed to crack or even delaminate. Supported slabs may also encounter enough thermal movement to damage tile set using a bonded system.

The specific procedure for installing tile depends mostly on the tile-setting material selected.

PORTLAND CEMENT MORTAR

Neat portland cement can be used as a bond coat only when the concrete surface is still plastic and suitably rough to bond with the neat cement. This condition is difficult to achieve on a structural slab. In addition, an isolated, reinforced mortar setting bed with a cleavage membrane should always be used over a structural (supported) slab, so that stresses induced in the slab are not transmitted to the tile finish.

In a proper concrete slab on grade, as described earlier, the likelihood of excessive movement is minimized and tile can be applied either by the thin-set method directly to the slab or over a bonded setting bed. However, even on slabs on grade, when a mortar bed is used, the bed is often reinforced with wire mesh, and sometimes a cleavage membrane is also installed.

A portland cement mortar bed should not exceed 1¼ in. (28.6 mm) in thickness. After the bed has been installed, a layer of neat cement paste, 1/32 in. (0.8 mm) to 1/16 in. (1.6 mm) thick, should be troweled over it or onto the back of the tile. Alternatively, a layer of dry portland cement, 1/32 in. (0.8 mm) to 1/16 in. (1.6 mm) thick,

FIGURE 9.6-28 Installation Performance Level Ratings

Area Description	Type of Use or Traffic	Performance Level
Office space, commercial reception areas, public spaces in restaurants and stores, corridors, shopping malls	(a) Light high-heel traffic (b) Heavy high-heel traffic	(a) Moderate (b) Heavy
Kitchens	(a) Residential (b) Commercial (c) Institutional	(a) Light or moderate (b) Heavy (c) Extra heavy
Toilets, bathrooms	(a) Residential (b) Commercial (c) Institutional	(a) Light or moderate (b) Moderate or heavy (c) Heavy
Hospitals	(a) General (b) Kitchens (c) Operating rooms	(a) Heavy (b) Extra heavy (c) Heavy
Exterior decks	(a) Roof decks, large on-grade patios (b) Walkways, small patios	(a) Extra heavy (b) Heavy or extra heavy

Note: Performance level ratings are defined as follows: Light: Normal residential foot traffic; occasional 300-lb. (136 kg) loads on soft rubber wheels. Moderate: Better residential, normal commercial, and light institutional use; 300-lb. (136 kg) loads on rubber wheels and occasional 100-lb. (45.4 kg) loads on steel wheels. Heavy: Heavy high-heel traffic; 200-lb. (90.7 kg) loads on steel wheels, 300-lb. (136 kg) loads on rubber wheels. Extra heavy: High-impact service, as in institutional kitchens and industrial work areas; 300-lb. (136 kg) loads on steel wheels.

FIGURE 9.6-29 Suitable Methods for Various Performance Levels

	Installation				
Performance Level	**Subfloor**	**Tile**	**Mortar or Adhesive**	**Grout**	**Remarks**
Light	Wood	Ceramic mosaic or quarry	Organic	Latex–portland cement	Low cost; OK for home bath, entry
	Wood	Ceramic mosaic	Epoxy	Sand–portland cement[a]	Costlier; OK for kitchen, dining room
	Existing tile finish	Ceramic mosaic or quarry	Epoxy	Mastic or latex–portland cement	Residential renovation
Moderate	Wood	Ceramic mosaic	Portland cement mortar bed	Sand–portland cement[a]	OK for depressed wood subfloor
	Wood	Ceramic mosaic or quarry	Epoxy	Epoxy	Very costly; very durable even over wood subfloor
	Concrete	Ceramic mosaic	Dry-set	Sand–portland cement[a]	Moderate cost; smooth surface
	Concrete	Ceramic mosaic	Dry-set	Latex–portland cement	Moderate cost; eliminates damp curing of grout
Heavy[b]	Concrete	Ceramic mosaic or quarry	Dry-set[c]	Epoxy	Costlier, good chemical resistance
	Concrete	Ceramic mosaic or quarry	Portland cement mortar bed	Sand–portland cement[a]	Moderate cost; bed ensures smooth surface
	Concrete	Ceramic mosaic or quarry	Dry-set	Sand–portland cement[a]	Low cost; good all-purpose floor
	Concrete	Ceramic mosaic or quarry	Epoxy	Epoxy	Very costly; most durable and easy to maintain

[a]Damp cured for 3 days after grouting.

[b]Damp cured for 3 days before grouting.

[c]All installations suitable for extra heavy performance when packing house tile is substituted for mosaic tile.

can be dusted over the setting bed as a bond coat and then, immediately before the tile is set, worked lightly with a trowel or brush until damp.

Glazed semivitreous tile should be dampened by placing it on a wet cloth or in a shallow pan. However, no free moisture should remain on the backs of the tile when set. Vitreous and impervious tile, such as mosaic tile and quarry tile, do not require soaking.

When installing face-mounted ceramic mosaic tile, the paper should be wetted within 1 hour of setting, using no more water than necessary for removing the paper and glue (Fig. 9.6-30b). Tiles that are out of line or not level should be adjusted before the mortar assumes initial set.

To obtain maximum contact and a strong bond, tile should be set in position and tamped firmly into the mortar (Fig. 9.6-30a). Surfaces should be brought to a true level at proper elevation. Tamping and leveling should be completed within 1 hour after the tiles have been placed.

Sand–portland cement grout is used with quarry tile, pavers, and ceramic mosaic tile. For quarry tile and pavers, the grout consists of 1 part portland cement to 2 parts sand and ⅕ part hydrated lime by volume. For ceramic mosaic tile, a mix of 1 part portland cement to 1 part finely screened sand is used (Fig. 9.6-30c). Before the joints are grouted, they are filled and pointed with neat cement paste. Grout in paver, quarry, and cushion-edged glazed tile joints should be tooled (compressed) with a pointing tool, so as to firmly bond the grout to the pointing mortar and to the tile.

Excess grout should be removed from the surface with a clean cloth or burlap. Voids and gaps should be filled before the grout sets. When portland cement–based grouts are used, the entire installation should be damp-cured for 72 hours.

The tile surface should be thoroughly cleaned while the grout is still plastic. Muriatic acid may be used to clean glazed tile; sulfamic acid can be used to clean unglazed tile 10 days after setting. Joints should be soaked before acid is applied and flushed thoroughly with water after cleaning. Metal and enameled surfaces should be protected with grease or a strippable coating.

Isolated Setting Bed To create an isolated mortar setting bed, a cleavage membrane is spread over the slab and folded at edges and ends to form a locked joint. Metal reinforcement should be lapped at least one full mesh and should be supported approximately in the middle of the setting bed. The mesh should be cut to fit accurately and kept free of vertical surfaces. Figure 9.6-31 illustrates typical isolated bed construction.

The setting bed mix should consist of 1 part portland cement, 5 parts dry sand (or 6 parts damp sand), and ⅒ part hy-

a

b

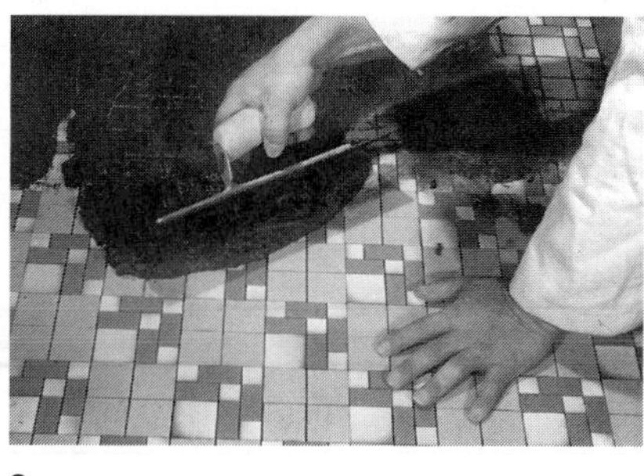

c

FIGURE 9.6-30 Mounted ceramic mosaic tile is (a) laid in a neat cement bond coat; (b) the tile is tamped into place, and the paper is sponged off; (c) grout is troweled into joints, and excess is removed with a clean cloth. (Courtesy American Olean [Ceramic] Tile Company.)

drated lime by volume. Only enough water should be added to the mix to obtain the desired consistency. The setting bed should be applied ¾ in. to 1¼ in. thick, tamped firmly, and screeded to a level surface.

Bonded Setting Bed The bonded bed method should be used infrequently because it does not have the chief advantage of unbonded installations, namely, separation from stresses in the concrete subfloor. Its chief advantages over thin-set methods is its ability to produce a smooth finish over a rougher subfloor and its reliance on familiar, readily available materials such as portland cement and sand. Bonded setting bed construction is similar to isolated construction, except that the cleavage membrane and reinforcement are omitted (Fig. 9.6-32).

The mortar bed can be bonded with a synthetic bonding compound or neat portland cement. For the latter, a fairly rough, wood-floated surface is required. The slab should be thoroughly saturated with water before neat cement and mortar are spread. The mortar mix, grout mix, and installation procedures for a bonded setting bed are the same as those described earlier for an isolated setting bed.

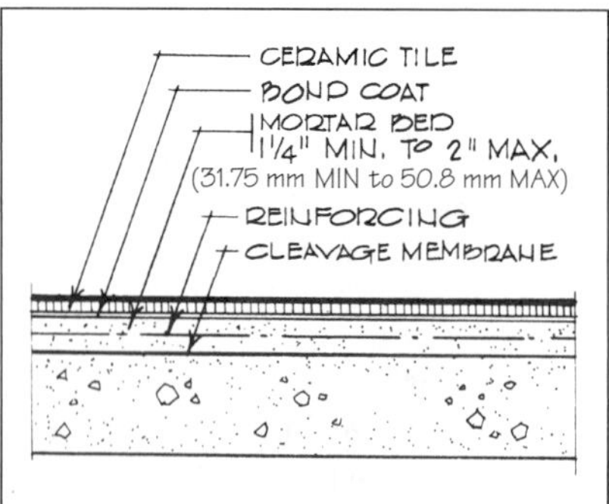

FIGURE 9.6-31 Metal reinforcement should be installed in every isolated thick-bed installation. (Courtesy Tile Council of America, *Handbook for Ceramic Tile Installation.*)

LATEX–PORTLAND CEMENT AND DRY-SET MORTARS

These mortars are generally used to bond tile directly to a concrete slab (Fig. 9.6-33) that has a variation of not more than ¼ in. (6.4 mm) in 10 ft. (3048 mm). On slabs with more variation, tile should be bonded to a cured mortar leveling bed. In addition, the slab should be designed to ensure a minimum of movement. It should be well cured and free of cracks, waxy or oily films, and curing compounds. The slab surface should be steel-troweled and fine-broomed. Dry-set or latex–portland cement mortar should be mixed as recommended by their manufacturer. Mortar should be troweled in a layer about ¼ in. (6.4 mm) thick for mosaic tile and ¼ to ⅜ in. (6.4 to 9.5 mm) thick for quarry and paver tile. The tile should be tamped into the mortar to ensure a level surface and intimate contact.

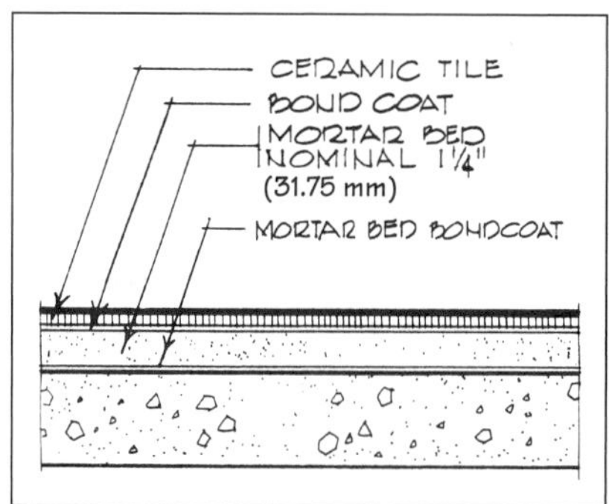

FIGURE 9.6-32 Reinforcement can be omitted when the mortar bed is bonded to the backing. (Courtesy Tile Council of America, *Handbook for Ceramic Tile Installation.*)

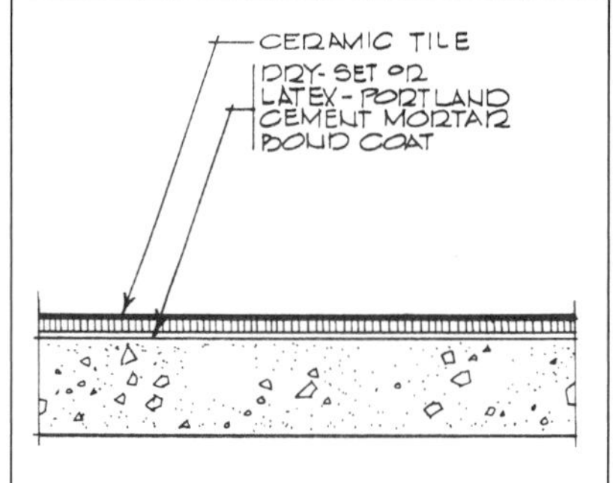

FIGURE 9.6-33 Dry-set and latex–portland cement mortar can be used for thin-set placement of tile on a concrete slab. (Courtesy Tile Council of America, *Handbook for Ceramic Tile Installation.*)

Paper should be removed from face-mounted tile within 1 hour of setting, using a minimum of water. Grouting should be delayed for at least 24 hours, or until the tile is firmly set. Suitable grouts include dry-set grout, latex grout, sand–portland cement grout, and commercial cement grout. Dry-set and portland cement grouts should be damp-cured for 72 hours.

ORGANIC ADHESIVES

Since these products are recommended for use on floors only in light-traffic areas, they are used infrequently over a concrete slab, except in small areas and over relatively smooth surfaces. For organic adhesives, which are applied only ¹⁄₁₆ in. thick, a slab surface should be steel-troweled and fine-broomed and the surface should not vary by more than ¹⁄₁₆ in. (1.6 mm) in 3 ft. (900 mm).

Tile installation with organic adhesives requires the same general precautions as with other thin-setting products (see Fig. 9.6-33). The tile should be tamped into place to ensure maximum contact with the adhesive. Latex–portland cement and epoxy grouts are usually used with organic adhesives, but any grout can be used.

EPOXY MORTAR AND ADHESIVE

The performance and working properties of epoxy adhesives are similar to those of epoxy mortars. Most of the following discussion concerning epoxy mortars applies to adhesives as well.

Epoxy mortar is likely to be used in conjunction with an epoxy grout to provide chemical resistance in laboratories, industrial plants, food preparation areas, and similar clean environments. However, the high bond strength of the mortar, combined with the general stain resistance of the grout, make this also a preferred system for many hard-service areas (such as entries, public toilet rooms, lobbies, and restaurants), particularly when combined with quarry tile. Since the mortar is installed in a layer ¼ to ⅜ in. (6.4 to 9.5 mm) thick, it can absorb minor variation in the subfloor surface, of not more than ¼ in. (6.4 mm) total in 10 ft. (3048 mm). A wood-floated concrete surface is suitable (Fig. 9.6-34).

Epoxy mortar should be the same color as the grout. The tile should be tamped into the setting bed, with care taken not to damage the face of the tile or smear mortar on it. Mortar should be immediately removed from the tile face.

Grouting should be delayed at least 16 hours. The grout should be forced into the joints with a squeegee or rubber trowel so as to completely fill them. The installation should be shaded from direct sun and kept at a relatively even temperature (more than 60°F [33.3°C]) for at least 8 hours after grouting to facilitate curing. Traffic should be kept off the floor for at least 40 hours. Light traffic can be permitted after this period; heavy traffic, after 7 days.

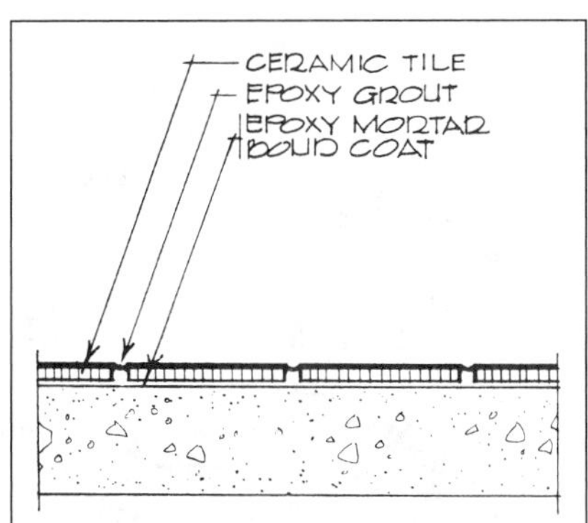

FIGURE 9.6-34 Epoxy mortar and grout possess high bond strength and chemical resistance. Epoxy adhesives can be substituted in less demanding situations. (Courtesy Tile Council of America, *Handbook for Ceramic Tile Installation.*)

Epoxy grout can also be used with other tile-setting mortars. Where chemical resistance at the tile surface is desirable but high bond strength with the backing is not essential, epoxy grout can be used with latex or dry-set mortars. These cementitious mortars generally produce adequate bond and are less expensive than epoxy mortar.

9.6.2.3 Exterior Concrete

Except for installations over built-up roofing or waterproofing membranes, exterior tile installations are likely to be bonded directly to a concrete slab (Fig. 9.6-35). Over a suitably finished concrete surface, dry-set mortar can be used in a coat ¼ to ⅜ in. (6.4 to 9.5 mm) thick to bond tile directly to a slab.

When a structural slab is not suitable for direct application of tile, a mortar setting bed will ensure a suitable substrate (Fig. 9.6-36). Since a mortar bed is installed by the tile contractor, he can follow up with a neat portland cement bond coat while the bed is still plastic.

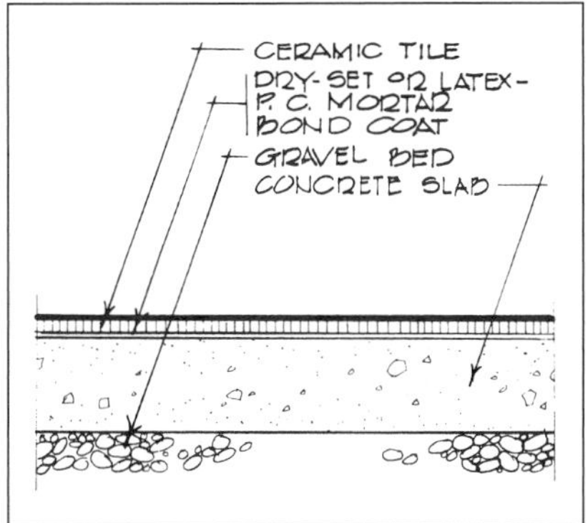

FIGURE 9.6-35 Dry-set mortar can be used for thin-set application of tile in both interior and exterior locations. (Courtesy Tile Council of America, *Handbook for Ceramic Tile Installation.*)

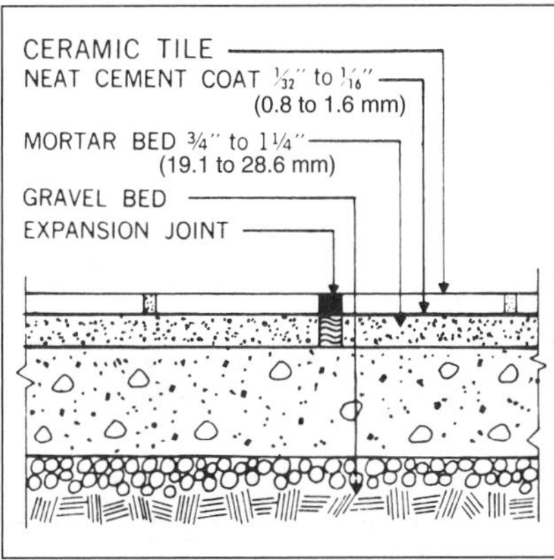

FIGURE 9.6-36 Expansion joint in an exterior mortar bed tile installation.

Expansion joints extending through tile and mortar bed should be provided at intervals of 12 to 16 ft. (3658 to 4877 mm) each way (see Fig. 9.6-36). Expansion joints should also be provided where the tile finish abuts restraining vertical surfaces and where expansion, isolation, or control joints exist in the slab. Both the structural slab and the tile surface should be sloped to drain. See Section 9.6.1.6 for additional discussion of expansion joints.

9.6.2.4 Wood Subfloors

Wood subfloors usually consist of wood joists and plywood or board sheathing. Unless stiffness is carefully considered in the design, joists are likely to deflect and vibrate more than massive structural systems such as concrete or steel. The boards and plywood, being hygroscopic, also are likely to expand and contract with humidity variations.

Therefore, tile should be either completely isolated from the subfloor so as to permit differential movement, or bonded to the subfloor with a tile-setting material that can resist the movement without breaking the bond. When the installation depends on a very strong bond and no setting bed is used, the grout must also be strong enough to resist the internal stresses transmitted to the tile joints.

A typical isolated conventional installation consists of a reinforced mortar setting bed and a latex, dry-set, or neat cement bond coat over a cleavage membrane. A typical bonded thin-set installation utilizes epoxy mortar or epoxy adhesive. In relatively small areas, such as residential bathrooms, tile can also be bonded to a plywood subfloor with organic adhesives. Although this material is not as strong as epoxy, it is adequate to resist the smaller stresses developed in a small, bathroom-sized area.

PORTLAND CEMENT MORTAR

The procedure for installing a mortar setting bed is the same as for concrete subfloors. The grout can be any of the several commercially available formulations or sand–portland cement grout. The complete tile installation, including the mortar bed, should be about 1 in. (25.4 mm) to 1½ in. (38 mm) thick (Fig. 9.6-37).

Since many interior floor finishes are much thinner than 1½ in. (38 mm), a transition problem develops where a tile fin-

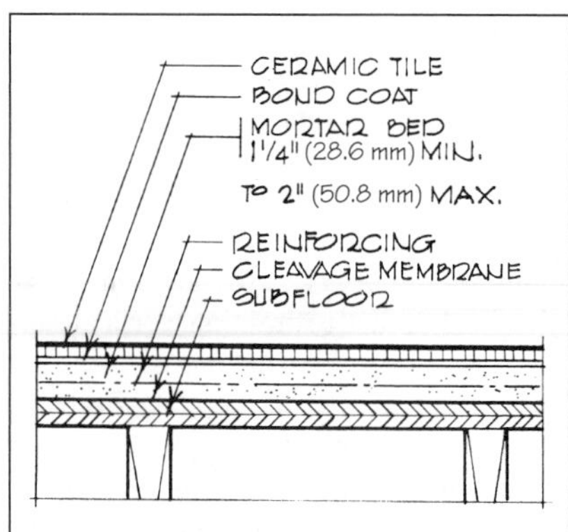

FIGURE 9.6-37 A reinforced isolated mortar bed and tile over a single-layer wood subfloor. (Courtesy Tile Council of America, *Handbook for Ceramic Tile Installation.*)

ish abuts a thinner finish. To minimize this problem, the subfloor can be depressed between the joists, thus gaining ⅝ to ¾ in. (15.9 to 19.1 mm) in height (Fig. 9.6-38). The subfloor, usually plywood, is supported on cleats level with the tops of the joists. This method of tile installation is somewhat more durable in high-moisture areas, such as bathrooms, than the thin-set method. It is also more resistant to impact due to the mass of the mortar bed underlying the tile finish. Depressing the subfloor below the tops of the joists and leveling the joists is not recommended.

ORGANIC ADHESIVE

The organic adhesive method is not suitable for tile installation on the floors of commercial or institutional buildings. It is, however, suitable for all areas in a home, except those subject to excessive moisture, such as in a tiled sunken tub or in a swimming pool. It will not withstand heavy traffic or impact loads. Double-floor construction is generally recommended to achieve a stiffer subfloor (Fig. 9.6-39).

NEAT CEMENT COAT 1/32" to 1/16" (0.8 to 1.6 mm)
MORTAR BED NOM 1/4" (31.75 mm)
REINFORCING
SUB FLOOR
15 LB. (8.14 kg/m2) BLDG FELT
CERAMIC TILE

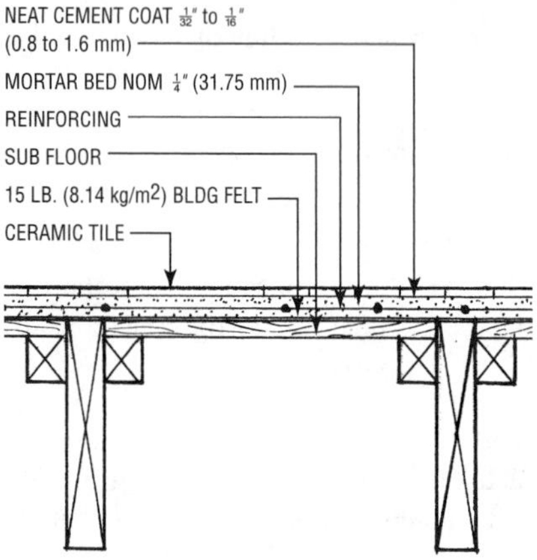

FIGURE 9.6-38 A portland cement mortar bed and tile installation over a depressed subfloor.

The subfloor can be 1-in. (25 mm) nominal boards or ⅝-in. (15.9 mm) plywood. The underlayment should be ⅜-in. (9.5 mm) or thicker plywood. For bathroom floors and other areas subject to occasional moisture, Type I adhesive should be used. This type is capable of withstanding more prolonged exposure to moisture than Type II adhesive. The adhesive should be spread with a notched trowel recommended by the manufacturer, and the tile firmly pressed into the adhesive. Grout can be latex–portland cement or epoxy.

EPOXY MORTAR AND ADHESIVE

Epoxy mortar and adhesive provide a very strong adhesive suitable for bonded thin-set installations over a double wood floor consisting of subfloor and underlayment (Fig. 9.6-40). The subfloor should be ⅝-in. (15.9 mm) plywood or 1-in. (25 mm) nominal boards; the underlayment should be ½-in. (12.7 mm) or thicker plywood (⅝-in. in nonresidential uses). The underlayment should be installed with ¼-in. (6.4 mm) gaps between panels, and these gaps should be completely filled with epoxy mortar. In residential uses, the underlayment may be omitted if the subfloor is ⅝-in. (15.9 mm) or thicker plywood installed perpendicular to the joists and with the edges blocked.

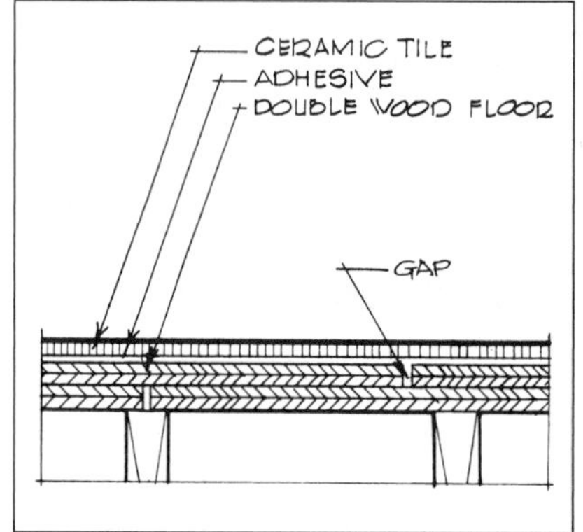

FIGURE 9.6-39 A thin-set tile application over a double wood subfloor. (Courtesy Tile Council of America, *Handbook for Ceramic Tile Installation.*)

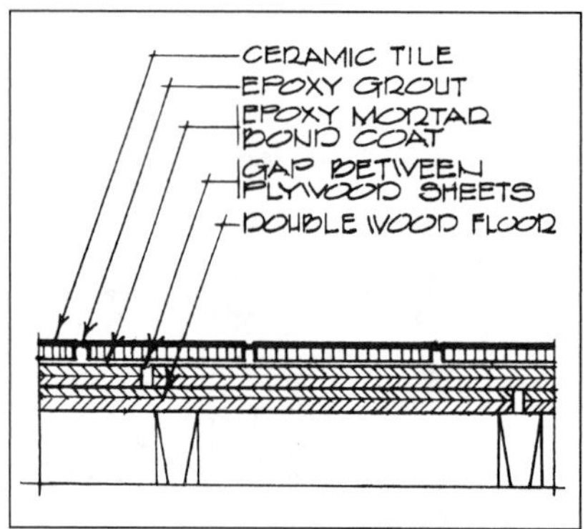

FIGURE 9.6-40 An epoxy mortar-set tile over a double wood subfloor. (Courtesy Tile Council of America, *Handbook for Ceramic Tile Installation.*)

The mortar should be spread with a trowel recommended by the manufacturer. The tile should be tamped into the mortar to ensure a smooth surface and proper bond.

Grouting should be delayed for 16 hours. The installation should be shaded from direct sunlight and kept at a stable temperature of more than 60°F (15.6°C) for 8 hours or more after grouting. Grout and mortar should be cleaned off the face of the tile as soon as possible and before it hardens.

9.6.3 INSTALLATION ON WALLS

The ceramic products most used on walls are glazed wall tile and ceramic mosaic tile. Quarry tile and paver tile are seldom used as wall finishes. Where the wall surface is sufficiently sound, true, clean, and free of coatings, tile can be bonded to the wall by the thin-set method. The surface should not vary more than ⅛ in. (3.2 mm) in 8 ft. (2438 mm) if thin-set materials are to be used; not more than ¼ in. (6.4 mm) in 8 ft. (2438 mm) for portland cement mortar beds.

A mortar setting bed should be used on wall surfaces that are irregular, chipped, cracked, or coated with materials that would impede bond. A mortar setting bed is installed over a base coat of mortar (scratch coat) applied to metal lath over a cleavage membrane.

A *scratch coat* consists of 1 part portland cement, 5 parts damp sand, and ⅕ part hydrated lime. It is normally applied by a plastering contractor, but can also be installed by the tile contractor. A *float coat* is generally applied by the tile contractor to ensure a plastic surface for setting the tile with neat cement.

A solid backing is not essential when tile is to be installed by the mortar bed method. Over wood or metal studs, the backing can be omitted and the membrane and metal lath can be attached directly to

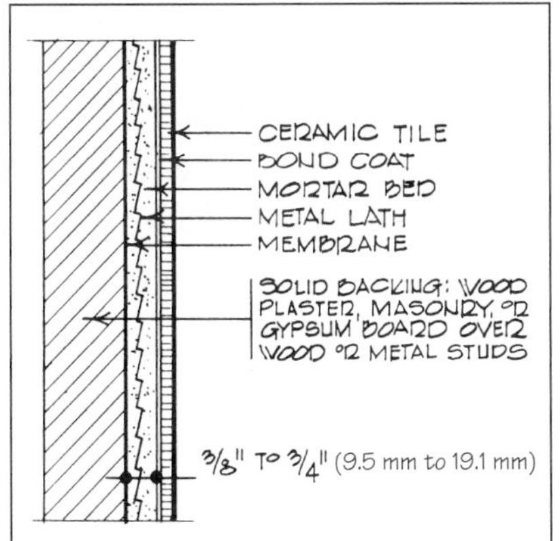

FIGURE 9.6-41 **The one-coat method can be used over a solid backing that is too uneven for thin-set tile.** (Courtesy Tile Council of America, *Handbook for Ceramic Tile Installation.*)

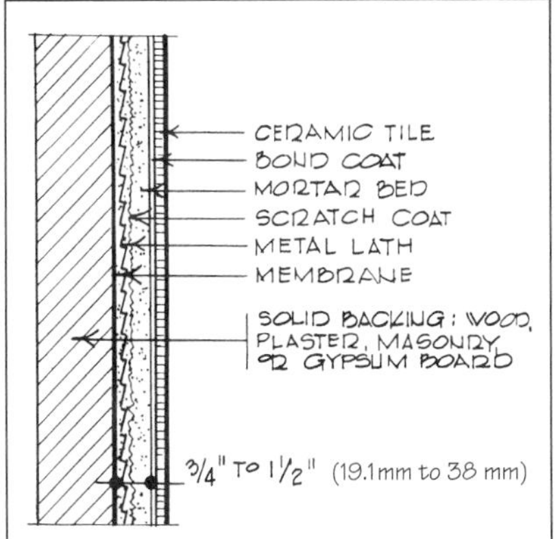

FIGURE 9.6-42 **Two-coat method over open framing or solid backing.** (Courtesy Tile Council of America, *Handbook for Ceramic Tile Installation.*)

the studs. A lath-reinforced mortar bed is generally rigid enough by itself to support tile without excessive deflection between the studs; to ensure adequate vertical rigidity, the metal studs should be at least 2½ in. (63.5 mm) deep at 16 in. (400 mm) on centers, or 3½ in. (88.9 mm) deep when spaced 24 in. (600 mm) on centers.

The materials, installation methods, and suitable backings for tile finishes vary, depending on whether the installation is interior or exterior. Exterior installations are much more critical because of exposure to moisture and temperature variations.

9.6.3.1 Interior Surfaces

Suitable backings for tile finishes include masonry and concrete walls, as well as plywood, gypsum board, and plaster over furring or stud framing. In tub and shower areas, the thin-set method is acceptable, but it is essential that the backing be inherently water resistant or properly prepared to render it so. Dry-set mortar can be used as a bond coat for thin-set application in sunken tubs and pools over a portland cement mortar bed.

PORTLAND CEMENT MORTAR

Traditional installations on walls require a preliminary *scratch coat* followed by a *mortar bed* (*two-coat method*). The scratch coat can be omitted when the wall provides a solid backing for the application of metal lath, mortar, and tile (*one-coat method*). The tile is actually set in a separate bond coat, which is not counted in describing the installation.

The *one-coat method* can be used over existing wall finishes that are not suitable for thin-set application because of surface conditions (Fig. 9.6-41). It is also a preferred method of tiling tub and shower areas in new work. The backing can be masonry, concrete, plaster, wallboard, plywood, or any other solid wall surface. The single bed coat is ⅜ to ¾ in. (9.5 to 19.1 mm) thick and consists of 1 part portland cement, ½ half part hydrated lime, and 5 parts damp sand.

The *two-coat method*, involving both a *scratch coat* and a *bed coat*, is used most frequently over open framing of furring or studs (Fig. 9.6-42). On metal studs, the overall thickness of scratch coat and mortar bed should not exceed 1 in. (25.4 mm). On wood studs, the overall mortar thickness should not be more than 1½ in. (38 mm). The mortar mix is the same as for one-coat installations. The scratch coat should be allowed to cure for at least 24 hours and should be dampened thoroughly before the mortar bed is applied. The mortar bed should be applied in areas no greater than may be covered with tile while the bed is still plastic. Figure 9.6-43

a

b

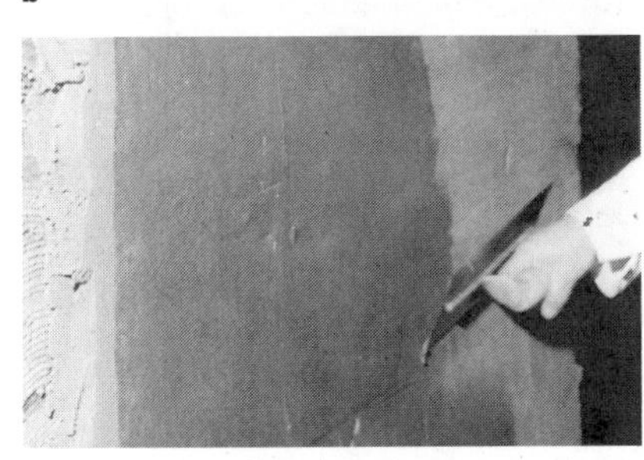
c

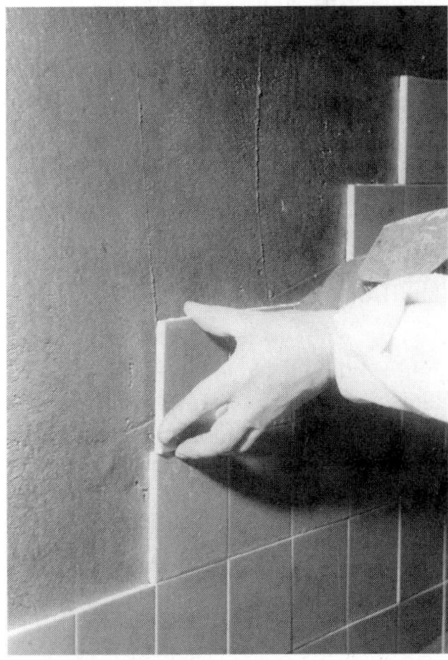
d

FIGURE 9.6-43 **In conventional mortar bed installations (a) a scratch coat is applied to metal lath and temporary screeds are inserted; (b) the float coat is leveled and screeds are removed; (c) a neat cement coat is troweled on; and (d) tile is applied and tamped in place. The glazed tile shown here must be soaked before it is set.** (Courtesy American Olean [Ceramic] Tile Company.)

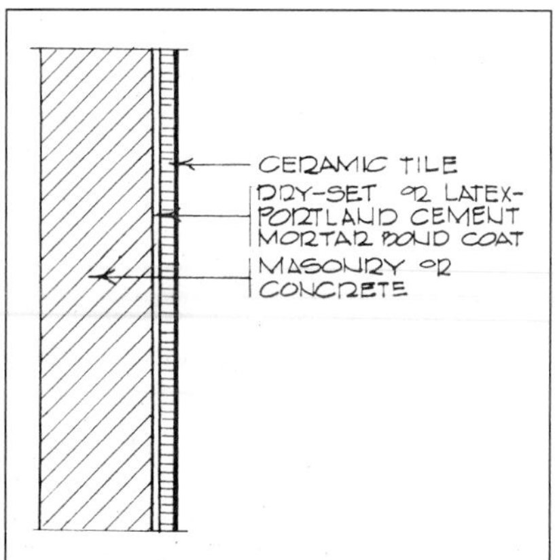

FIGURE 9.6-44 **For thin-set tile applied directly on masonry, the backing surface should be smooth and the joints struck flush. Since exterior thin-set applications directly over masonry are expected to move with the masonry, expansion joints are not as important as in other types of installations.** (Courtesy Tile Council of America, *Handbook for Ceramic Tile Installation*.)

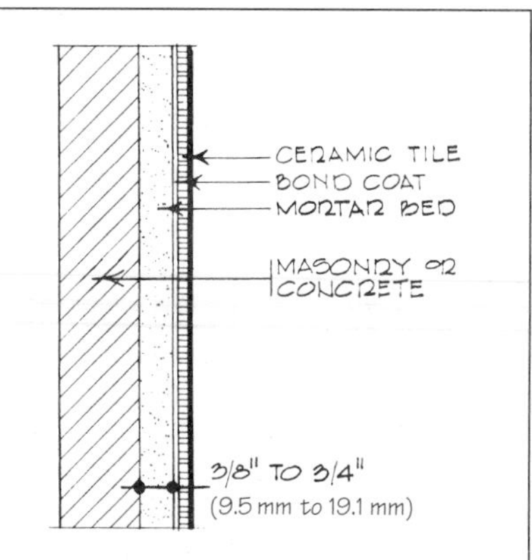

FIGURE 9.6-45 **Before tile is applied over uneven masonry, a leveling bed is placed and cured. The tile is then set with either neat cement or thin-set mortar.** (Courtesy Tile Council of America, *Handbook for Ceramic Tile Installation*.)

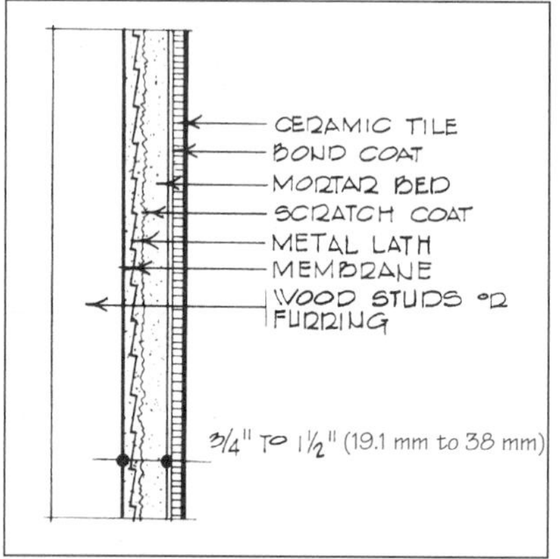

FIGURE 9.6-46 **Open framing requires a two-coat mortar bed over metal lath. Except with dry-set mortar, curing the bed is required for thin-set applications.** (Courtesy Tile Council of America, *Handbook for Ceramic Tile Installation*.)

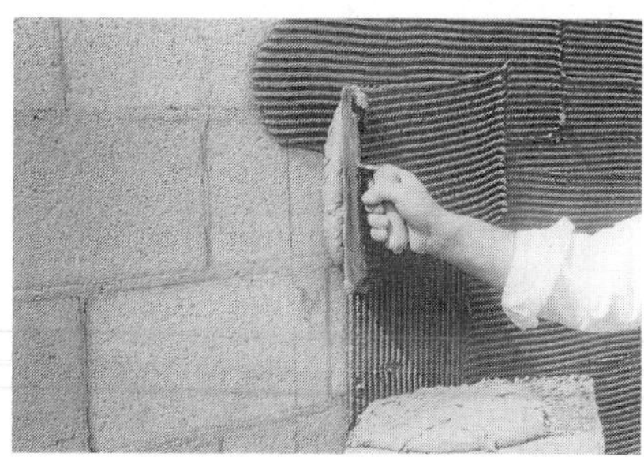

a

b

c

FIGURE 9.6-47 **Thin-set application of glazed wall tile directly to masonry: (a) dry-set mortar is applied with a notched trowel; (b) edge-glued tiles are set and tamped in place; (c) grout is forced into joints and the excess is wiped off.** (Courtesy American Olean [Ceramic] Tile Company.)

illustrates the steps in applying tile over a portland cement setting bed.

A neat cement *bond coat* is used for setting tile with both the one- and two-coat methods. The bond coat should be troweled $^1/_{32}$ to $^1/_{16}$ in. (.08 to 1.6 mm) thick over the float coat, and the tile should be tamped into it. The portland cement for the bond coat should be ordinary gray or white cement, depending on whether the grout is to be gray or white. The grout can be neat portland cement or portland cement mixed with an equal part of sand, latex, or epoxy. Where commercial portland cement or portland cement grout is used, the finish should be cured for at least 72 hours.

DRY-SET AND LATEX–PORTLAND CEMENT MORTARS

Dry-set and latex–portland cement are thin-set mortars suitable for sound, clean, and true surfaces such as gypsum plaster, gypsum board, monolithic concrete, or concrete and brick masonry (Fig. 9.6-44). They can also be used over a bonded mortar bed used mainly for leveling an irregular solid surface (Fig. 9.6-45) or when a full-thickness mortar backing on open framing is already in place (Fig. 9.6-46). When such a mortar backing is used, the backing should be allowed to cure for several days until firm.

The backing should be dampened just before mortar is spread over it. Mortar should be troweled on in appropriate thickness ($^3/_{32}$ to $^1/_4$ in. [2.4 to 6.4 mm] maximum) in an area that can be readily covered with tile before the mortar starts skinning over (Fig. 9.6-47a). The tile generally should be tamped into place to obtain at least 80% coverage; 100% in tubs, showers and other high-moisture areas (Fig. 9.6-47b). Any grout can be used, but in selecting a suitable type, the type of exposure should be considered (Fig. 9.6-47c).

ORGANIC ADHESIVE

The use of organic (mastic) adhesive is limited to light-duty interior areas and over smooth surfaces such as plaster, gypsum board, mineral-fiber-cement board, or plywood (Fig. 9.6-48). It can also be used over concrete or masonry if the surface is relatively smooth (maximum variation $^1/_8$ in. [3.2 mm] in 8 ft. [2438 mm]). Type I adhesive should be used for installations requiring prolonged water resistance, such as in tub and shower areas, and on bathroom floors. Type II adhesive can be used in areas subject to occasional or intermittent wetting. However, neither type is suitable for installations requiring continuous contact with water, such as in sunken tubs or pools. Manufacturers' recommendations should be followed to determine which type is best for the specific use intended and whether a primer is required.

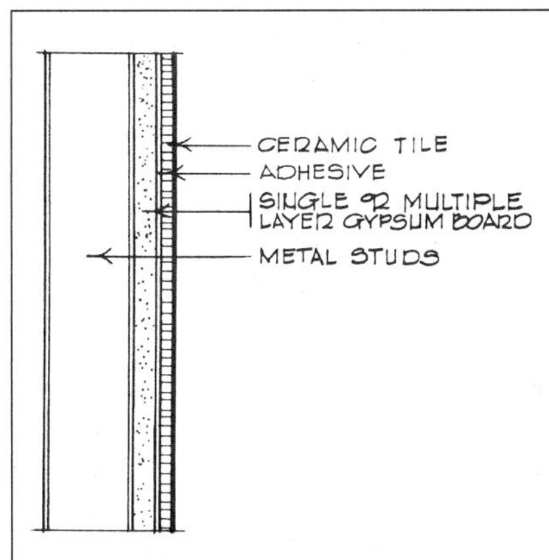

FIGURE 9.6-48 Organic adhesives can tolerate limited moisture exposure (such as in residential shower stalls) if a water-resistant adhesive and backing are used and the board edges are sealed. (Courtesy Tile Council of America, *Handbook for Ceramic Tile Installation.*)

The adhesive should be spread with a notched trowel recommended by the manufacturer, leaving no bare spots (Fig. 9.6-49a). The tile should be tamped into the adhesive to ensure maximum contact (Fig. 9.6-49b). The average contact area of tile removed for inspection should be not less than 75%, and no individual tile or mounted assembly should have less than 40% coverage of adhesive. Adjustments should be made while the adhesive is still plastic. Grouting should be delayed for at least 24 hours to permit solvent evaporation from the adhesive. Joints should be grouted with a product suitable for the conditions. Drywall and latex grouts are commonly used with organic adhesives (Fig. 9.6-49c).

9.6.3.2 Exterior Surfaces

Tile installations on exterior walls require the same precautions regarding movement as exterior floors. The supporting construction, including the backing, must be designed to accommodate movement due to temperature and moisture variations. This means having:

- Suitable control, isolation, and expansion joints;
- Proper metal reinforcement in concrete and masonry;
- Careful control of mortar and concrete mixes;
- Sound engineering design;
- Good workmanship.

Exterior tile can be applied by either the conventional mortar bed method or the thin-set method, depending on:

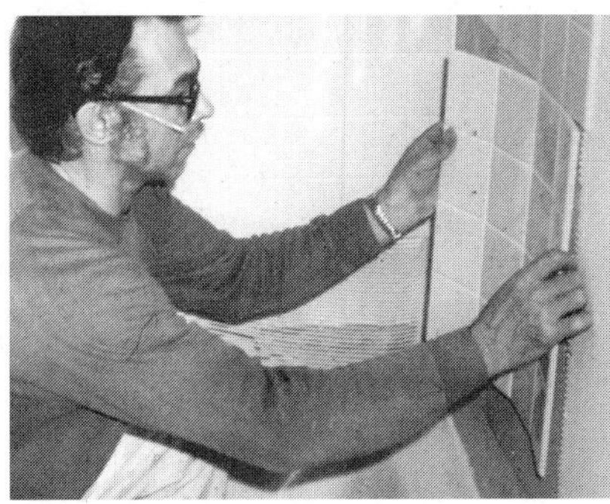

a

b

c

FIGURE 9.6-49 Over previously applied organic adhesive: (a) edge-mounted tile is applied; (b) tile is tamped in place with a "rubber float" trowel; and (c) the same trowel is used to force grout into joints and remove excess grout. (Courtesy American Olean [Ceramic] Tile Company.)

- Whether the backing is suitable for direct bonding
- The dimensional stability of the backing
- Provisions for crack control

On large areas exposed repeatedly to direct wetting, the conventional isolated method usually gives better results. The thin-set bonded method is generally more reliable in covered outdoor areas and when broken into smaller panels by decorative trim, battens, or expansion joints. However, tile can be thin-set over an isolated mortar bed with dry-set mortar.

Regardless of installation methods, expansion joints should be provided in the tile finish over all underlying structural joints, at the tile perimeter, and against any projecting surface. In isolated installations, expansion joints in the tile finish should be provided at 12 to 16 ft. (3660 to 4880 mm) on centers. Where these expansion joints do not coincide with joints in the backing, they should be provided in addition to those over the underlying joints. In a thin-set installation over a suitable stable backing, the additional joints in the tile finish are not as essential because the tile finish is intended to move with the backing.

In exterior tile installations, it is essential to provide suitable flashings, overhangs, or copings to prevent entry of water into, or behind, the tilework. Exterior walls of habitable spaces should include a suitable vapor retarder on the warm side of the wall to prevent condensation in the wall cavity. Additional information on the design and construction of expansion joints is given in Section 9.6.1.6.

PORTLAND CEMENT MORTAR BED

Procedures for installing tile over a mortar bed outdoors is essentially the same as described earlier for interior locations. The entire installation should be backed by cleavage membrane. Metal lath should be firmly attached to the backing and cut at expansion joints (Fig. 9.6-50).

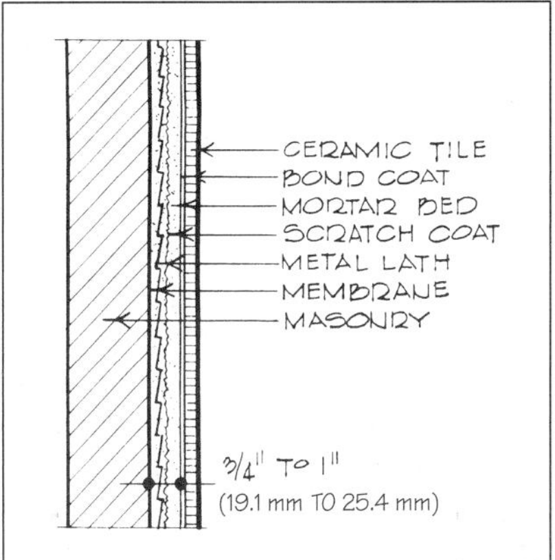

FIGURE 9.6-50 Expansion joints 12 to 16 ft. (3658 to 4877 mm) on centers are essential in exterior isolated thick-bed tile installations. (Courtesy Tile Council of America, *Handbook for Ceramic Tile Installation.*)

A scratch coat consisting of 1 part portland cement, 5 parts damp sand, and ⅕ part lime is usually installed by a plastering contractor. The mortar bed, consisting of 1 part portland cement, 5 parts damp sand, and ½ part hydrated lime, is normally installed by a tile contractor. The overall thickness of both coats should be ¾ to 1 in. (19.1 to 25.4 mm). The mortar bed mix can be varied up to 1 part portland cement, 1 part lime, and 7 parts damp sand. However, extremely rich mixes (too much portland cement) will result in greater drying and shrinkage cracking. Similarly, a total mortar bed thickness in excess of 1 in. should be avoided.

Neat cement can be used for setting tile over a still plastic mortar bed. When dry-set or latex–portland cement mortar will be used, the mortar bed should be cured for at least 24 hours, but longer curing periods of up to 10 days are preferred. Suitable grouts include commercial cement, dry-cure.

THIN-SET INSTALLATION

Waterproof backings suitable for bonding tile with thin-set mortars include monolithic concrete, concrete and brick masonry, structural clay tile, and a cured portland cement mortar bed (Fig. 9.6-44). Latex–portland cement and dry-set mortar are typically used for setting tile. Either latex or dry-set grout can be used for filling tile joints. Damp-curing for at least 72 hours is essential.

9.7 Terrazzo

The term *terrazzo* is derived from the Italian *terrassa* (terrace on the roof) because many early installations were made in upper-level outdoor living areas. The origins of terrazzo date back to the beginning of the Christian era, having evolved from *mosaic*, the art of placing colorful pieces of stone, vitreous enamel, or marble (tesserae) in decorative designs on walls and floors. The earliest terrazzo installations in Roman times were made of a mixture of crushed brick in lime mortar, which was ground and polished after it hardened. Later, the decorative chips salvaged from mosaic work were introduced in outdoor tile terraces to create highly colorful and durable floor finishes.

Terrazzo has been used for centuries as an economical and durable indoor-outdoor finish and has traditionally been defined as a form of mosaic flooring made by embedding small pieces of marble in mortar and polishing. In modern times, the mortar (matrix) for true terrazzo finishes has been portland cement. Historically, the decorative chips were limited to marble. More recently, synthetic resinous binders have been developed and other hard rocks capable of being polished are used.

Because of the expense involved in hand assembly of decorative chips in complex designs for large areas, mosaic work survives today primarily as custom inserts in terrazzo installations. The tesserae are usually assembled in the shop and mounted on paper; then the assembly is installed in a mortar setting bed in the desired location. After the paper is removed, joints are grouted and the entire installation is ground and polished.

Grinding and polishing are distinctive operations that differentiate terrazzo from seamless floorings that also utilize resinous binders. A notable exception to this general rule is *rustic* portland cement terrazzo, which is not ground but washed to reveal the decorative chips. In rustic terrazzo, selected colorful gravels are sometimes used instead of marble chips.

Terrazzo is a durable, low-maintenance floor finish, suitable for both exterior and interior locations. It can be cast in place or precast in a variety of shapes and colors. In precast form, its chief use has been for shower receptors, windowsills, wall bases, and stair treads. More decorative products intended for counters, lavatory tops, and tables are marketed under the names of *art marble* and *cultured marble*. These precast terrazzo products involve installation methods different from those associated with cast-in-place terrazzo floor finishes and are not included in this discussion.

Since many terrazzo floors utilize portland cement mortar and are installed over concrete surfaces, it is recommended that the reader review the requirements discussed in Chapter 3.

9.7.1 MATERIALS

The decorative wear layer of terrazzo finishes is called a *topping*. In some installations, the topping is placed directly on a wood, metal, or concrete subfloor; in others, an intervening portland cement underbed is required. When an underbed is used, it may be rigidly *bonded* to a tile subfloor or it may be *unbonded* by separating it from the subfloor with an *isolation membrane* (Fig. 9.7-1). When no underbed is used, the topping is always bonded to the subfloor. Other terms relating to terrazzo installations are illustrated in Figure 9.7-1 and defined in the glossary.

The basic ingredients of terrazzo toppings are:

- Binders
- Decorative chips
- Pigments
- Divider strips, expansion strips, and other accessories
- Metal reinforcement
- Isolation membranes
- Curing compounds
- Sealing materials

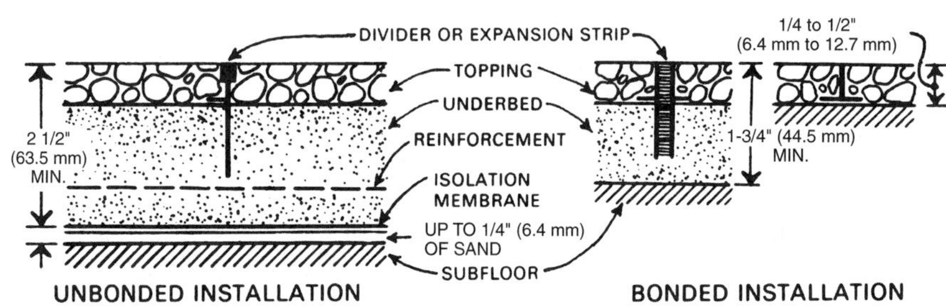

FIGURE 9.7-1 Basic construction elements for unbonded and bonded terrazzo installations.

FIGURE 9.7-2 Summary of Terrazzo Materials

Material	Standard
Cementitious binders	
Portland cement, Type I	ASTM C 150
Resinous matrixes	
Epoxy	NTMA*
Polyester	NTMA*
Polyacrylate	NTMA*
Decorative chips	NTMA*
Aggregate	
Sand	ASTM C 33
Pigments	NTMA*
Metal reinforcement	
Welded wire fabric	ASTM A 185
Isolation membranes	
Roofing felt	ASTM D 226
Polyethylene film	ASTM D 2103
Curing compounds	ASTM C 309

*Specifications of the National Terrazzo and Mosaic Association.

Materials for terrazzo should conform to the industry standards listed in Figure 9.7-2.

9.7.1.1 Binders

Binders in terrazzo are either *cementitious* or *resinous*. Cementitious binders are the older, traditional terrazzo materials; resinous binders are the result of developments in organic chemistry since World War II.

CEMENTITIOUS BINDERS

The primary cementitious binder used today is either white or natural (gray) Type I portland cement. Natural portland cement is less expensive and can be used if a vivid or light-colored matrix is not essential to the color scheme. The National Terrazzo and Mosaic Association, however, cautions that using gray cement may result in an uneven color. White portland cement is more expensive but offers a lighter background for displaying decorative chips and produces clearer, truer colors with mineral pigments. Type IA air-entraining portland cement is recommended for exterior applications subject to freezing and thawing.

RESINOUS BINDERS

Resinous binders possess most of the desirable properties of portland cement binders but have greater strength and chemical and abrasion resistance. In addition, the ability of resinous binders to bond with most subfloor materials and to be applied in ¼- to ½-in. (6.4 to 12.7 mm)-thick toppings makes them suitable for many installations where a portland cement binder would be too thick. However, the high material costs limits resinous terrazzo to uses where its special properties are specifically required.

Synthetic binders for terrazzo finishes include epoxy, polyacrylate, and polyester. When there is a desire to impart special properties, these resins, because of their high cost, are combined with various additives and fillers, which together form the matrix. Epoxy and polyacrylate resins are also combined with portland cement to form *epoxy-modified cement* terrazzo and *polyacrylate-modified cement* terrazzo.

Materials and application for epoxy, polyacrylate, and polyester terrazzo should conform to the requirements of the National Terrazzo and Mosaic Association (NTMA) specifications, details, and product data; other requirements outlined in NTMA's publication *Terrazzo Information Guide*; and the binder manufacturer's recommendations.

Epoxy and polyester matrixes are two-part formulations (a resin and a catalyst) to which a dry mix (fillers, pigment, additives, and decorative chips) is added. These resins are distinguished by good chemical resistance and adhesion, good abrasion resistance, and high impact strength. Epoxy matrixes also exhibit negligible curing shrinkage. In addition to their high cost they are also toxic to the skin until cured.

Polyester matrixes are particularly resistant to acids of high concentration. They are also resistant to water and weather and are strong and durable. However, they have a somewhat lower resistance to alkali than epoxy and a stronger odor during installation and the curing period.

Polyacrylate-modified cement terrazzos have a high bond strength and are resistant to snow-melting salts, foodstuffs, and urine. They are also nontoxic and relatively free of objectionable odors.

9.7.1.2 Decorative Chips

The decorative chips in terrazzo are mainly marble, which is defined to include all calcareous (limestone) rocks such as onyx and travertine, as well as attractive serpentine rocks, capable of being ground and polished. Granite, which is not a marble, is extremely hard and weather resistant and is often used for heavy-duty industrial and rustic terrazzo.

Domestic and imported chips, available in a wide range of colors, are graded by number according to size as shown in Figure 9.7-3. Most producers supply #0, #1, and #2 as separate sizes; some produce #3 and #4 as separate sizes. Larger sizes are frequently grouped as #3-4 mixed, #5-6 mixed, #7-8 mixed, and #4-7 mixed.

The appearance and minimum thickness of toppings are determined largely by the sizes and types of chips used. Toppings are identified as *standard*, *intermediate*, *Venetian*, and *Palladiana* (see

FIGURE 9.7-3 Grading of Decorative Chips for Terrazzo Toppings

Type of Topping and Minimum Thickness	Chip Number	Passes Screen, in. (mm)	Retained on Screen, in. (mm)
¼ in. (6.4 mm) Thin-set	0	⅛ (3.2)	1/16 (1.6)
	1	¼ (6.4)	⅛ (3.2)
⅝ in. (15.9 mm) Standard* **½ in. (12.7 mm) Thin-set**	2	⅜ (9.5)	¼ (6.4)
⅝ in. (15.9 mm) Intermediate	3	½ (12.7)	⅜ (9.5)
	4	⅝ (15.9)	½ (12.7)
¾ in. (19.1 mm) Venetian	5	¾ (19.1)	⅝ (15.9)
	6	⅞ (22.2)	¾ (19.1)
	7	1 (25.4)	⅞ (22.2)
	8	1⅛ (28.6)	1 (25.4)

*Palladiana toppings consist of broken marble slabs set in a standard chip mix and should not be less than ½ in. (12.7 mm) thick.

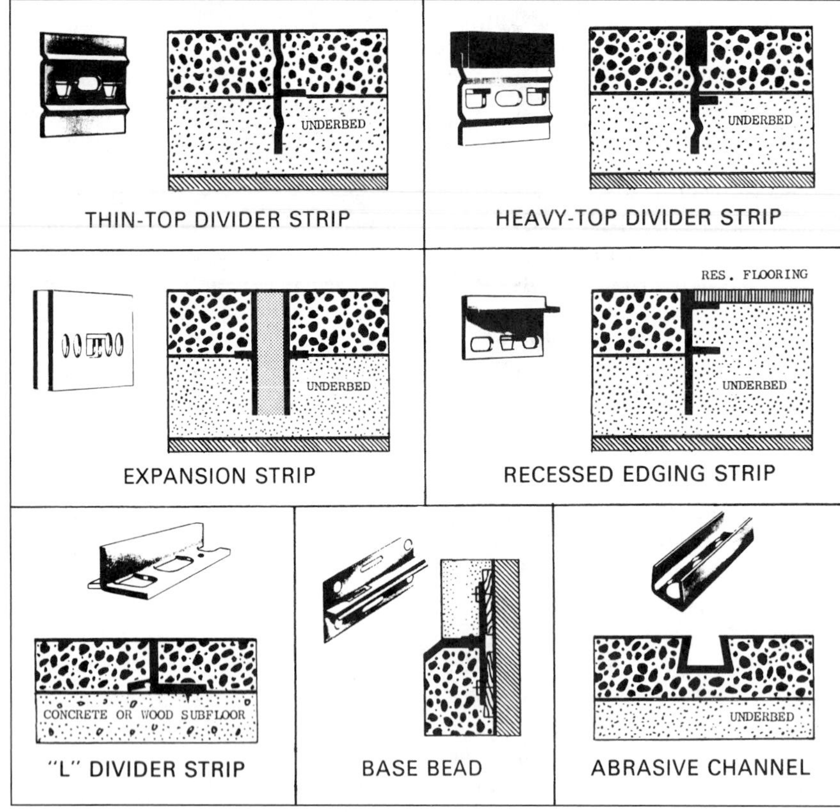

FIGURE 9.7-4 Accessories: shallow T- and L-shaped (angle) strips for installation without underbed are available in thin-top, heavy-top, and edging profiles.

Fig. 9.7-3). The first three use crushed chips, mixed with resinous or portland cement matrix and spread uniformly in the topping. The fourth, Palladiana toppings, are custom finishes similar to mosaics. They contain cut slabs of marble about ½ in. (12.7 mm) thick that are set by hand in the desired pattern. Spaces between these slabs are filled with a standard topping mix.

The NTMA and some producers of chips publish standard color plates as an aid in selecting and specifying decorative chips and color pigments. These marble chips and color combinations are designated by plate number. However, finished terrazzo should not be expected to duplicate the color plate exactly because of slight variations in color and veining between different shipments of chips. Physical samples should be prepared as a standard of comparison at the beginning of each project.

9.7.1.3 Fine Aggregate

The term *aggregate* refers to sand and fillers. It does not include decorative chips. Fine silica sand and fillers are usually included in the proprietary formulations of resinous matrixes. Sand aggregate is not used in a portland cement topping, but is a basic ingredient of a mortar underbed. It should be natural or manufactured, and well graded, clean, granular, and free of deleterious amounts of clay, salts, and organic matter.

9.7.1.4 Pigments

Terrazzo toppings may be the natural color of the binder or may be treated with pigments to produce a wide variety of colors. Pigments should be stable, nonfading mineral or synthetic formulations, compatible with the binder.

9.7.1.5 Accessories

Accessories for terrazzo installations include divider strips, expansion strips, and base beads. Metal accessories are made from half-hard brass or white alloy zinc (about 99% zinc). Plastic-top strips and beads in various colors with brass, zinc, or galvanized steel supports also are available. All-brass, stainless steel, and zinc-supported plastic-top accessories are suitable for exterior use. Aluminum strips are suitable for use in resinous terrazzo, but should not be used with portland cement terrazzo because aluminum is subject to attack by alkalies present in the topping mix. To prevent the accumulation of dangerous electrostatic charges, only plastic-top strips and beads should be used in conductive floors. White alloy zinc accessories are not suitable in exterior locations, because a chemical reaction often causes blooming (a white chalky deposit on the strips). The standard length of strips and beads is 6 ft. (1828 mm), but longer lengths are available.

DIVIDER STRIPS

Divider strips are installed to:

- Localize shrinkage cracking
- Ensure uniformity in topping thickness
- Act as decorative elements
- Permit accurate placement of the different colors forming a floor pattern
- Act as construction joints, subdividing a floor into convenient working areas

A variety of divider strips are available to suit particular installations (Fig. 9.7-4). The thickness of the top of the strip (exposed face) can range from 20 gauge (1.0 mm) up to ½-in. (12.7 mm)-thick bars; 18 gauge (1.3 mm) is the recommended minimum. Divider strips with a ⅛-in. (3.2 mm) or wider face usually have a heavy plastic or metal top section and a thin metal bottom section. For a topping thickness of ⅝ in. (15.9 mm), a 1¼-in. (28.6 mm)-deep divider is recommended; 1½-in. (38 mm)-deep strips are recommended for Venetian toppings ¾ in. (6.4 mm) thick and thicker. For thin-set installations, special angle or T-shaped strips ¼ to ½ in. (6.4 to 12.7 mm) deep are used. Other special strips are available for wall base and for edgings at the juncture with other flooring materials.

EXPANSION STRIPS

Provision for crack control is necessary where substantial structural movement can be expected, particularly over beams and bearing walls and over expansion or isolation joints in the subfloor. Expansion strips for this purpose consist of a pair of divider strips separated by a resilient material such as neoprene (see Fig. 9.7-4). Expansion strips are manufactured in types and sizes similar to those of divider strips.

BASE BEADS

All-metal or plastic-top base beads are used to terminate terrazzo wall base. They are made in several profiles and to accommodate vertical terrazzo thicknesses varying from ½ in. (12.7 mm) to 1½ in. (38 mm).

9.7.1.6 Metal Reinforcement

Metal reinforcement is required in the underbed, except where the underbed is bonded to a concrete subfloor. Underbed reinforcement should be corrosion-resistant welded wire fabric of at least 16 gauge (1.6 mm), with wires spaced not more than 2 in. (50.8 mm) on centers. Thin underlayments for resinous toppings may use welded or woven wire weighing at least 1.4 lb./sq. yd. (0.76 kg/m^2) or expanded metal lath weighing at least 2.5 lb./sq. yd. (1.36 kg/m^2).

9.7.1.7 Isolation Membrane

An isolation membrane is required in all unbonded installations to separate the underbed from the subfloor. The membrane may be 15-lb. (8.14 kg/m^2) asphalt-saturated roofing felt or 4-mil (0.1 mm) polyethylene film.

9.7.1.8 Curing Materials

Curing is necessary for portland cement toppings to develop optimum wear properties. Materials suitable for curing include polyethylene film, nonstaining (nonasphaltic) water-resistant building paper, and clean water. Spray-on curing compounds are sometimes used for curing terrazzo surfaces, but may prevent bond if used to cure concrete slabs that are to receive a thin-set, chemically bonded, or monolithic topping.

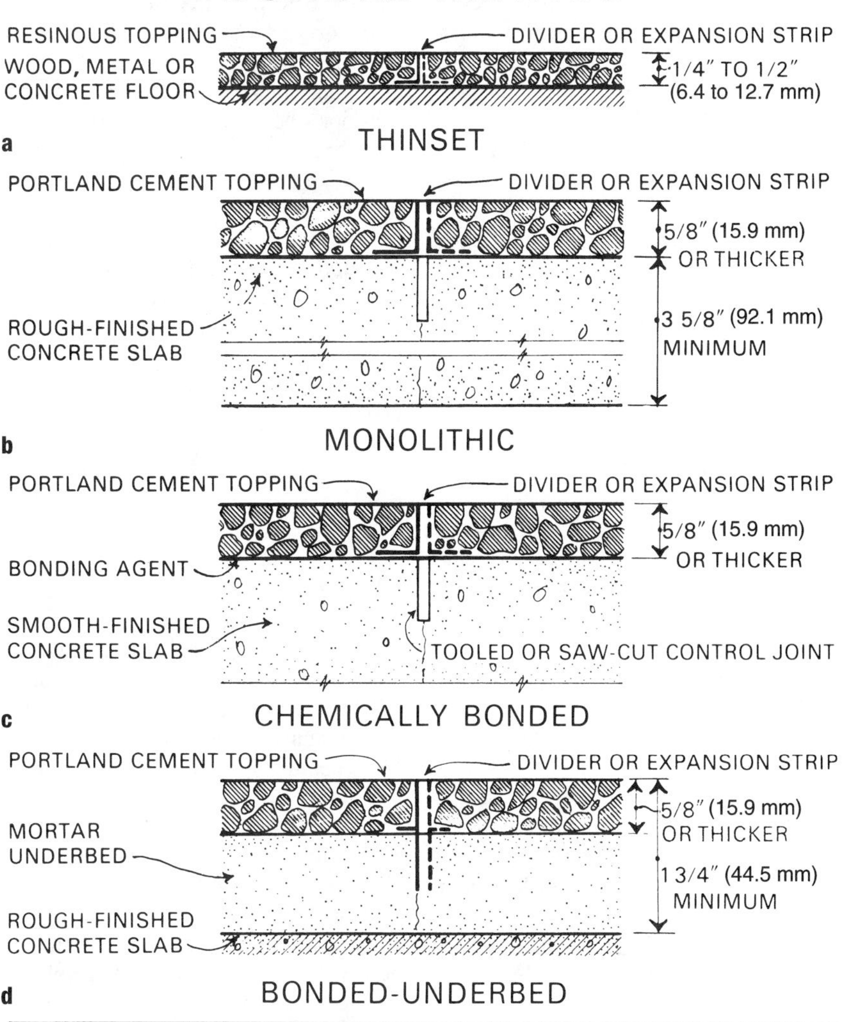

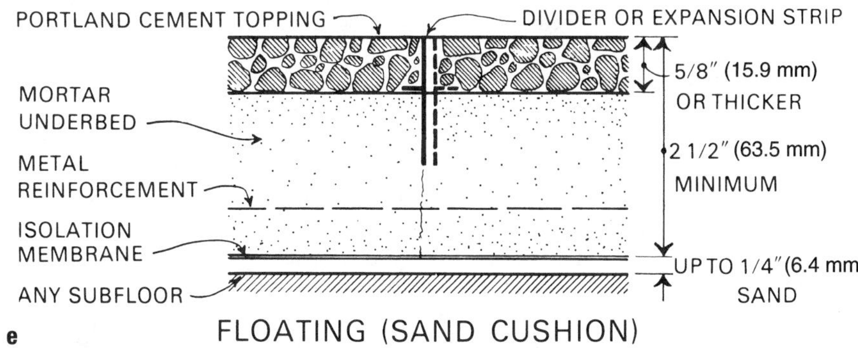

FIGURE 9.7-5 Details for basic terrazzo installation methods.

9.7.2 APPLICATION

Suitable terrazzo finishes require attention to (1) general recommendations common to most installation methods, (2) specific application procedures for particular methods, and (3) various finishing procedures for special surface finishes.

Terrazzo systems can be classed as *bonded* or *unbonded*. Bonded systems include thin-set, monolithic, chemically bonded, and bonded-underbed methods. *Thin-set* installations consist of a ¼- to ½-in. (6.4 to 12.7 mm)-thick resinous topping placed directly over a sound wood, metal, or concrete subfloor (Fig. 9.7-5a). In *monolithic* installations, a somewhat thicker (⅝ in. [15.9 mm] thick or thicker) portland cement topping is applied directly over a rough-finished concrete subfloor (Fig. 9.7-5b). If the concrete subfloor is an existing smooth-troweled concrete slab, a portland cement topping can be *chemically bonded* with a bonding agent (Fig. 9.7-5c). Monolithic and chemically bonded installations depend on the subfloor to resist shrinkage stresses and conspicuous cracking in portland cement toppings. However, *bonded-underbed* installations provide more positive crack control by creating weakened planes at each divider strip (Fig. 9.7-5d).

Unbonded installations provide the

most crack control and are preferred wherever structural movement is expected. Separation between subfloor and underbed can be accomplished by an isolation membrane alone, or with a membrane and sand cushion (Fig. 9.7-5e). When a sand cushion is used, the installation is called a floating installation.

9.7.2.1 General Recommendations

The installation of terrazzo involves (1) preparation of the subfloor and, depending on the method of installation, (2) placement of the topping, strips, and underbed when used.

SUBFLOOR PREPARATION

Regardless of the installation method, the subfloor must be sound, free of loose material, and structurally adequate to support the expected live loads without excessive deflection, as well as the additional dead load imposed by the terrazzo. This dead load will be about 11 lb./sq. ft. for each inch of thickness (53.71 kg/m^2 for each 25.4 mm of thickness) for both resinous and cementitious terrazzo.

For unbonded installations, the surface quality of the subfloor is not critical. In bonded installations, the quality of the installation depends on adequate bond with the subfloor. Therefore, the subfloor must be free of coatings such as paint, wax, and sealing or curing compounds that could prevent proper bond.

For monolithic installations, the surface of the subfloor should not be troweled but should be wood-floated and broomed or otherwise roughened. A chemical bonding agent should be used over a smooth-troweled concrete surface.

Concrete intended for a thin-set resinous installation should be smooth-troweled and be at least 21 days old. Wood floors should consist of a suitable single plywood subfloor or double floor including at least a 3/8-in. (9.5 mm)-thick plywood underlayment (Fig. 9.7-6). Plywood should be Exterior Type CC-plugged grade or better. The recommendations of the manufacturer of the resinous matrix should be followed for subfloor preparation. Over most subfloor materials, the topping can be adhesively bonded, and priming of the subfloor is required to achieve bond. Over plywood, some manufacturers recommend direct adhesive bonding after plywood edges have been bound with glass fiber tape.

The subfloor for monolithic, chemically bonded, and thin-set installations should conform to the profile of the finished surface so that the topping will be of uniform thickness. There should be no more than 1/8-in. (3.2 mm) variation in the subfloor when tested with a 10 ft. (3000 mm) straightedge. A topping mix should not be used to fill voids or develop pitch, because excessive variations in topping thickness cause cracking. Where an existing subfloor is not sufficiently smooth or does not have the desired profile, it should be leveled or pitched with a suitable mastic underlayment such as latex. Over wood strip flooring, these underlayments should be installed at least 3/8 in. (9.5 mm) thick using a 15-lb. (8.14 kg/m^2) felt and expanded metal or wire lath secured to the subfloor at 6 in. (152 mm) on centers along the supports.

UNDERBED

An underbed is necessary in unbonded installations and may be required in some bonded installations. An underbed can be relatively thin, because it is not a structural slab but functions as a surface preparation for a topping. The thickness of an underbed varies, depending on the installation method (see Fig. 9.7-5d and e).

The underbed should be reinforced with welded wire fabric in unbonded installations and in bonded installations over wood or metal subfloors. Unless the underbed is bonded, an isolation membrane should be spread over the subfloor and metal reinforcement placed over the membrane. Except where expansion strips occur, the adjacent sheets of this reinforcing should overlap by at least two mesh spaces. No conduit or pipe should be located within the minimum recommended underbed thickness.

Underbed mortar should consist of 1 part portland cement, 4 to 5 parts sand, and sufficient water to make a workable mortar. The surface over which the underbed is placed should be sufficiently wet to prevent water from being drawn out of the mix. Isolation membranes need only be uniformly dampened; concrete slabs should be thoroughly saturated by overnight soaking.

Underbed mortar should be placed and screeded to a height that will permit installation of the selected topping thickness at the desired finish floor elevation.

STRIP PLACEMENT

When an underbed is used, conventional divider strips (see Fig. 9.7-5) should be installed while the underbed is still semiplastic. Standard 1¼- or 1½-in. (28.6 or 38 mm) strips should be set in the desired pattern, with the top of the strip at the finished floor elevation. The strips should then be troweled firmly along the edge to ensure adequate anchorage in the underbed.

In bonded installations without an underbed, where the terrazzo is applied directly to the subfloor, and in monolithic and chemically bonded installations, it usually is not possible to install strips while the concrete subfloor is still plastic because the subfloor is often installed by a different contractor. For such installations, angle or T-shaped strips (see Fig. 9.7-4) are secured to the concrete with adhesive or mechanical fasteners.

Divider strips should be provided wherever a discontinuity exists in the subfloor, such as at construction, isolation, and control joints, and at changes in materials. Expansion strips may also be used in such locations if desired, and are specifically required at structural expansion joints where a greater degree of movement is anticipated. Divider or expansion strips should be used over beams and girders supporting the subfloor, because tensile stresses are likely to develop due to negative bending moments. Spacing of divider and expansion strips for various methods of installation is given in Figure 9.7-7.

Divider Strips Divider strips are not required in resinous terrazzo installations for control of shrinkage cracking but are often desirable for decorative purposes and for convenience in placing and finishing the topping. In thin-set installations, strips ensure a uniform topping thickness and conserve material. They also minimize the appearance of waviness in large areas.

In portland cement terrazzo installations using an underbed, strips should be spaced not more than 6 ft. (1.83 m) (on centers). When a portland cement topping is placed directly over a concrete slab, the slab should be provided with cut or tooled control joints 15 to 20 ft. (4.57 to 6.1 m) on centers in each direction, forming panels of not more than 300 sq. ft. (27.87 m^2). The length of these panels should not exceed their width by more than 50%. Divider strips should be placed at all such control joints, forming similar panels in the topping (Fig. 9.7-8). In slabs on grade where a minimum of stripping is desired,

FIGURE 9.7-6 Wood Floor Construction for Thin-Set Terrazzo[a]

Single-Floor Construction			
Combination Subfloor/ Underlayment[b,c]	**Maximum Support Spacing, in. (mm) o.c.**	**Nails[d]**	**Nail Spacing**
⅝-in. (15.9 mm) plywood	16 (400)	2 in. (50.8 mm) (6d) threaded	6 in. (150 mm) o.c. around panel edges; 10 in. (250 mm) o.c. at intermediate supports
¾-in. (19.1 mm) plywood	20 (500)		
⅞-in. (22.2 mm) plywood	24 (600)		
1⅛-in. (28.6 mm) plywood (2-4-1)	48 (1200)	2½ in. (63.5 mm) (8d) threaded	6 in. (150 mm) o.c. at panel edges and intermediate supports

Double-Floor Construction			
Subfloor[b]	**Maximum Support Spacing, in. (mm) o.c.**	**Nails**	**Nail Spacing**
½-in. (12.7 mm) plywood	16 (400)	2 in. (50.8 mm) (6d) common	6 in. (150 mm) o.c. around panel edges; 10 in. (250 mm) o.c. at intermediate supports
⅝-in. (15.9 mm) plywood	20 (500)	2½ in. (63.5 mm) (8d) common	
¾-in. (19.1 mm) plywood	24 (600)		
¾-in. (19.1 mm) lumber boards less than 6 in. (152.4 mm) wide	16 (400)	2¼ in. (53.98 mm) (7d) threaded	2 per board at each support
¾-in. (19.1 mm) lumber boards wider than 6 in. (152.4 mm) and less than 8 in. (203.2 mm) wide			3 per board at each support

	Fasteners[f]		
Underlayment[b,e]	**Nails[d]**	**Staples[g]**	**Fastener Spacing**
⅜-in. (9.5 mm) plywood	1¼ in. (28.6 mm) (3d) threaded	1⅛ in. (28.6 mm)	Nails: 8 in. (200 mm) o.c. each direction in middle of panel, 6 in. (150 mm) o.c. around panel perimeter
½-in. (12.7 mm) plywood		1¼ in. (31.8 mm)	
⅝-in. (15.9 mm) plywood	1½ in. (38 mm) (4d) threaded	1½ in. (38 mm)	Staples: 6 in. (150 mm) o.c. each direction in middle of panel, 3 in. (75 mm) o.c. around panel perimeter
¾-in. (19.1 mm) plywood		1⅝ in. (41.3 mm)	

[a]Based on 40 psf (195.3 kg/m^2) live load.

[b]Plywood thicknesses given are for Group 1 softwoods in accordance with U.S. Product Standard PS1; For Groups 2 and 3, increase thickness ⅛ in. (3.2 mm) for Group 4, increase thickness ¼ in. (6.4 mm).

[c]Blocking at longitudinal edges is required unless plywood with T&G edges is used.

[d]Common nails may be substituted for annularly threaded nails by increasing their size ¼ in. (6.4 mm) (1d).

[e]Blocking not required if underlayment joints are staggered with respect to subfloor joints.

[f]Fastener lengths shown are minimum to provide ¾ in. (19.1 mm) penetration for nails and ⅝ in. (15.9 mm) for staples. If longer fasteners are used, penetration should not exceed 1 in. (25.4 mm).

[g]Galvanized divergent chisel, 16 gauge (1.6 mm), ⅜ in. (9.5 mm) crown width.

control joints and divider strips can be placed in doorways and adjacent to partitions where they will be concealed by base trim (Fig. 9.7-9).

Divider strips should be continuous from wall to wall and should form right-angled intersections rather than acute angles. Where a terrazzo wall base is used, strips should be carried up to the base screed. When terrazzo forms a wainscot, strips should be continued from the base in a straight line up the wall. The recommended strip spacing for portland cement terrazzo wainscoting is 3 ft. (914 mm) on centers maximum. Strip spacing for resinous wainscoting is not critical.

Expansion Strips Greater consideration should be given to the need and location of divider and expansion strips in bonded systems because the terrazzo finish will tend to move with the subfloor. In monolithic, chemically bonded, and thin-set installations, expansion strips should be provided about 30 ft. (9.14 m) on centers where unobstructed large areas or long corridors are involved. In monolithic and chemically bonded installations, expansion and divider strips can be alternated 15 ft. (4.57 m) on centers. Expansion strips

FIGURE 9.7-7 Strip Placement

Matrix	Installation Method	Divider Strip Spacing[a]	Expansion Strip Spacing[a]
Portland cement	Unbonded	6 ft. (1800 mm) o.c. maximum	As needed[b]
	Bonded-underbed		
	Chemically bonded	15 ft. to 20 ft. (4500 to 6000 mm) o.c.—panels not over 300 sq. ft.[c]	30 ft. (9000 mm) o.c. maximum[c]
	Monolithic		
Resinous	Thin-set	As desired for decorative or convenience purposes	30 ft. (9000 mm) o.c. maximum[d]

[a]Locate divider strips or expansion strips on beams and girder lines wherever possible.

[b]Provide expansion strips over all expansion joints in the subfloor or at other joints where movement is expected.

[c]Divider strips and expansion strips can be alternated, or expansion strips can be used at 15 ft. (4500 mm) o.c.

[d]Recommended for large areas or long corridors; otherwise, not required.

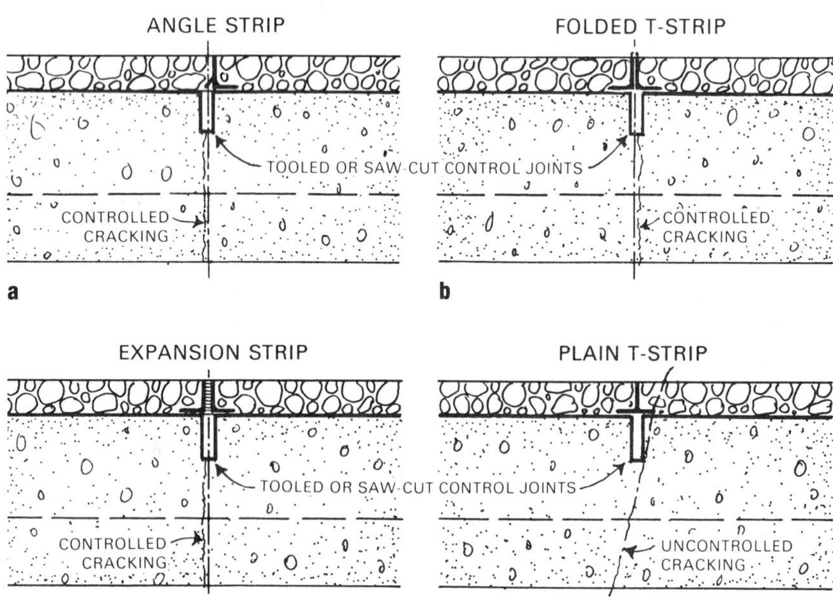

FIGURE 9.7-8 Control joints in monolithic and chemically bonded installations: (a, b, and c) inconspicuous controlled cracking is encouraged adjacent to strip; (d) uncontrolled cracking results from improper installation.

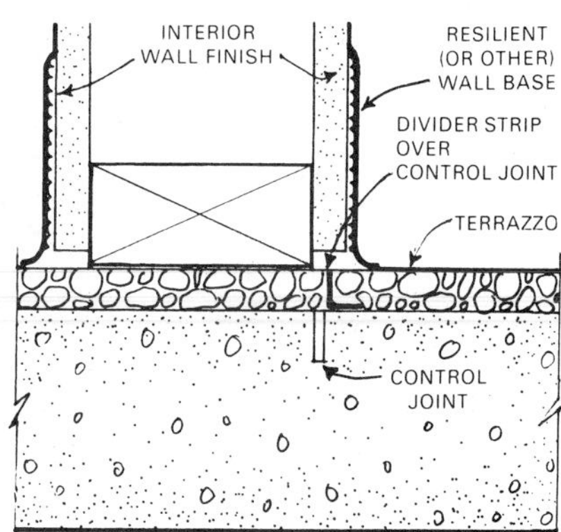

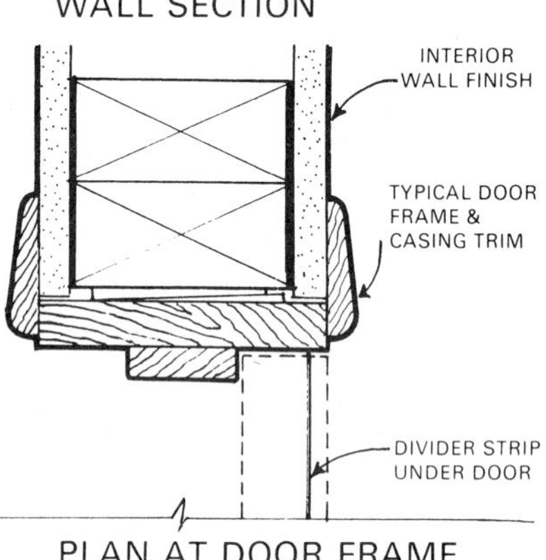

FIGURE 9.7-9 Strip placement in monolithic and chemically bonded terrazzo over a slab on grade when strips are not a design feature.

are not required in such installations if the subfloor is continuous, homogeneous, and 1000 sq. ft. (92.9 m^2) or smaller in area.

Unbonded and bonded-underbed installations will usually have enough divider strips distributed over the floor so that expansion strips are not required except over expansion joints in the subfloor.

TOPPING

This subsection covers portland cement terrazzo with a regular finish. See "Thin-Set Method" in Section 9.7.2.2 for resinous terrazzo and Section 9.7.2.4 for rustic and conductive terrazzo.

Mixing A power mortar mixer should be used to mix a carefully measured and consistently proportioned portland cement topping matrix. A conventional portland cement mix consists of 200 lb. (90.72 kg) of marble chips for each 94-lb. (42.64 kg) bag of cement. Pigment, when used, is measured by weight (not volume) and should be added to the dry mix. After thorough mixing of dry ingredients, water should be added at the ratio of 5 to 5½ gal. (18.93 to 20.82 L) per bag of cement.

Resinous binder formulations vary among manufacturers. Therefore, their proprietary mixing and installation recommendations should be carefully followed.

Placing Installation of a topping over an underbed should be delayed for 12 to 24 hours to permit the underbed to harden. Immediately before a topping is placed, a slush coat of neat cement, pigmented to match the topping mix, is broomed into the underbed (Fig. 9.7-10). The various colored mixes with decorative chips are placed according to the desired pattern. The topping is then troweled even with the tops of divider strips, which act as screeds for leveling the surface (Fig. 9.7-11).

To ensure even distribution and adequate chip content, the surface is seeded with additional chips of the type contained in the mix (Fig. 9.7-12). These chips are wetted, sprinkled over the surface, and embedded by additional troweling. Then

FIGURE 9.7-10 A slush coat of neat cement is broomed into the surface just before the topping is placed.

FIGURE 9.7-11 Topping is spread and troweled even with the divider strips.

FIGURE 9.7-12 Additional chips are seeded and troweled into the surface where necessary for uniform appearance.

the entire area is rolled with a heavy roller until all excess water has been brought to the surface (Fig. 9.7-13). The surface is again troweled smooth and uniform, revealing the tops of the divider strips.

Before finishing operations are started,

FIGURE 9.7-13 Rolling compacts the topping mix, brings up excess water, and creates a harder, more durable surface.

FIGURE 9.7-14 After proper curing, the surface is flooded and is wet-ground to reveal the decorative chips.

a portland cement topping should be wet-cured for 3 to 7 days either by ponding, covering with watertight covers, or spraying with a curing compound.

Preliminary Finishing After proper curing, the preliminary finishing operations of grinding, grouting, and curing are performed.

The surface is ground using a terrazzo grinding machine first with No. 24 or finer abrasive stones, followed by a second grinding with No. 80 or finer stones (Fig. 9.7-14). These grinding operations are performed while the surface is covered with water.

After grinding, the surface is thoroughly rinsed with water to remove dust and loose fine particles. Then a skim coat of neat cement grout of the type used in the topping mix (including pigment to match the topping) is troweled over the entire surface, filling all voids.

The installation is then wet-cured again for at least 3 days using a watertight cover or curing compound. The compound or cover should be allowed to remain in place as protection against physical damage and staining until final finishing is performed.

Final Finishing Final finishing includes polishing, cleaning, and sealing. A floor machine with No. 80 or finer stones is used for polishing. The polishing operation should continue until the grout skim coat is removed (except in voids and depressions) to reveal the decorative chips in 70% to 75% of the area.

A solution of neutral liquid cleaner is used to clean the polished floor. It is then rinsed with clean water and allowed to dry. The cleaned, polished terrazzo surface is then treated with a sealing compound. These compounds are proprietary formulations and should be applied in accordance with their manufacturer's recommendations.

Until all work on the premises is completed, finished terrazzo should be protected with a covering, such as heavy building paper.

9.7.2.2 Installation Methods

The following recommendations are in addition to those in Section 9.7.2.1.

UNBONDED METHOD

Unbonded installations require the use of a reinforced underbed to support the topping. The combined thickness of underbed and topping should be at least 2½ in. (38 mm) (see Fig. 9.7-5e). Separation between subfloor and underbed can be provided by an isolation membrane alone, or by a membrane spread over a 1⁄16- to 1⁄4-in. (1.6 to 6.4 mm)-deep layer of sand. Over a smooth subfloor, a double layer of polyethylene film will provide isolation between surfaces.

Metal reinforcement should be placed over the membrane, lapped at joints at least two mesh spaces, and the pieces wired together. After some mortar has been poured in place, the reinforcement should be lifted to about 1 in. (25.4 mm) above the membrane. Underbed, strips, and topping should be installed as indicated in Section 9.7.2.1.

BONDED METHODS

Bonded installations have a smaller dead load and less finish thickness. Most bonded methods allow the topping to be placed directly over the subfloor without an intervening underbed. However, these methods do not generally develop the degree of crack control obtained when an underbed is used with conventional divider strips. Therefore, the bonded-underbed method (see Fig. 9.7-5d) is often used if the greater dead load and thickness can be accommodated. Since the difference in thickness between bonded-underbed and unbonded methods is only ¾ in. (19.1 mm) (9 lb./sq. ft. [43.94 kg/m^2]), the unbonded method should always be considered because it is most effective in reducing cracking.

In the monolithic, chemically bonded, and thin-set bonded methods, the relatively thin topping by itself cannot resist the various stresses acting on it. Adequate bond between topping and subfloor is necessary so that stresses are resisted across the full depth of the floor construction. Therefore, special care should be taken to remove foreign substances that may impair bond.

Moreover, a topping should be of uniform thickness to prevent concentration of stresses and consequent cracking at thin spots. The subfloor must be level or pitched to conform with the desired finished floor profile.

Bonded-Underbed Method The bonded-underbed method, which can be used over a metal or concrete subfloor, provides a measure of crack control for portland cement terrazzo not available in other bonded installations. The total recommended thickness of underbed and topping is 1¾ in. (44.5 mm) for bonded-underbed, as compared with 2½ in. (63.5 mm) for unbonded installations.

The crack control advantage of this method is derived from the relatively frequent spacing of divider strips. The maximum spacing of 6 ft. (1800 mm) on centers for conventional divider strips induces unnoticeable stress-relieving cracks along the strips. The minimum overall terrazzo thickness results from the normal strip depth (1¼ or 1½ in. [28.6 or 38 mm]) plus a ¼-in. (6.4 mm) tolerance for subfloor irregularity.

The underbed must be bonded to the subfloor. Natural bond will occur between a portland cement underbed and a suitably roughened concrete slab. A chemical bonding agent can be used over a smooth-troweled concrete slab to bond the underbed. Over a metal deck, adequate bond can be expected without special treatment if the surface is free of corrosion, pickling acids, oil, and other foreign substances. Additional bond is gained by the keying action of the underbed with depressions in a cellular metal deck. Corrugated and ribbed metal decks generally depend on composite action of the deck and a reinforced structural concrete slab. In such floor assemblies, the top of the concrete slab constitutes the subfloor, and topping should be installed by the monolithic or chemically bonded method.

Underbed and topping are generally installed as described in Section 9.7.2.1.

Monolithic Method A monolithic bonded terrazzo finish is a portland cement topping installed over a concrete subfloor where natural bond is developed between the portland cement surfaces. Monolithic installations can be used in interior as well as exterior locations.

The concrete subfloor should be reinforced properly and be at least 3⅝ in. (92.1 mm) thick. The surface should be brushed with a wire broom or otherwise roughened to remove laitance and ensure good mechanical bond with the topping. As soon as the concrete is sufficiently hard to resist tearing with a jointing tool or power saw, control joints should be installed in the slab about 15 to 20 ft. (4.57 to 6.1 m) apart. Divider or expansion strips should be installed in the topping over these control joints in the subfloor. The surface should be protected from damage and dirt with suitable coverings until the topping is placed.

A subfloor that has not been adequately protected from dirt, paint, plaster, or other construction residue should be thoroughly scrubbed with detergents and rinsed clean. Then the subfloor should be saturated with water by ponding overnight. Excess water should be removed, and a slush coat of neat portland cement, pigmented to match the topping, should be broomed into the surface immediately before the topping mix is applied. The topping should be placed, finished, and cured as described in Section 9.7.2.1.

Chemically Bonded Method This method of installing portland cement topping is intended for smooth-troweled concrete subfloors and is often suitable for older concrete floors not originally prepared for a terrazzo topping. Where a terrazzo topping is planned in the initial design, a roughened subfloor surface is preferred.

The subfloor should be inspected, and where random cracks exist or control joints are not provided 15 to 20 ft. (4.57 to 6.1 m) on centers, such joints should be installed with a power saw. Divider strips or expansion strips should be placed over all joints in the subfloor.

Chemically bonded installations depend on a bonding agent such as epoxy, epoxy-polysulfide, or polyvinyl-chloride resin. The subfloor surface should be free of coatings such as paint, curing compounds, and wax, and the bonding agent should be applied in accordance with its manufacturer's recommendations.

The topping should be placed, finished, and cured as described in Section 9.7.2.1.

Thin-set Method Using a resinous binder, it is possible to install very thin toppings over a sound, firm subfloor. Epoxy and polyester toppings should be at least ¼ in. (6.4 mm) thick. When they are installed over concrete, the subfloor surface should be smooth-troweled and level to within ⅛ in. (3.2 mm) in every 10 ft. (3000 mm).

Although some manufacturers allow resinous toppings ⅛ in. (3.2 mm) thinner than those recommended here, industry consensus favors the slightly heavier installations. This is particularly true of epoxy and polyester, where a ⅛-in. (3.2 mm) thickness would require decorative aggregate the size of sand, and an acceptable minimum thickness over subfloor irregularities would be difficult to achieve.

Thin-set terrazzo finishes should not be installed over slab-on-grade construction subject to excessive moisture or high alkaline content. To test for moisture, a piece of polyethylene film about 2 ft. (610 mm) square should be taped down in several locations and allowed to remain for 8 hours. Visible moisture accumulation under the mat is an indication of excessive moisture in or under the slab. Under such conditions, the alkalinity of the moisture should also be tested and the thin-set matrix manufacturer should be consulted.

Divider strips are not required with resinous binders for shrinkage control, but should be provided where there is a discontinuity in the subfloor, such as at isolation joints, control joints, and changes in materials. Divider strips may be provided for decorative purposes or as construction joints at the termination of a pour. Since wall base is often installed as a separate operation, a strip should be provided at the toe of the base. In large areas or long corridors, expansion strips should be installed at 30-ft. (9.5 m) intervals. Closer spacing of expansion or divider strips may be preferred for decorative purposes, to minimize the illusion of waviness, or to ensure uniform topping thickness.

The thin-set matrix manufacturer's recommendations should be followed for using primers, mixing ingredients, applying the topping mix, and grinding and sealing the surface. Over wood subfloors, the terrazzo installer's specific recommendations should be consulted for details of taping plywood joints and for use of mastic underlayment and wire or metal reinforcement.

Generally, the dry components such as fillers and chips are mixed first, then added to the liquid resin. The mix is poured onto the subfloor, then spread and leveled with a trowel. The setting time before grinding depends on the specific resin formulations. Epoxy and polyester toppings can be ground the day after application.

In residences, sealing of the surface is often not required, because the resinous matrixes are quite dense and resistant to most staining agents. Marble chips, however, vary in their stain resistance, and, depending on anticipated traffic conditions, a sealer may be desirable. The matrix manufacturer's recommendations should be consulted in each case.

VIBRATION

Traditional terrazzo installations can develop surface imperfections due to portland cement shrinkage, and the appearance of the finished floor sometimes suffers from uneven chip distribution. To overcome these problems, the Terrazzo Industry Research Institute, a branch of NTMA, has developed a special vibrating machine that improves the performance and appearance of most terrazzo floor installations, including rustic and polyacrylate thin-set.

9.7.2.3 Refinishing Existing Floors

New terrazzo finishes can be installed over almost any type of existing floor finish. The unbonded method can be used if the subfloor and structural system are capable of supporting the additional dead load (28 lb./sq. ft. [136.71 kg/m^2] for a 2½-in. [63.5 mm]-thick finish) and the new finished floor height is not objectionable. Since no bond with the existing floor is intended, relatively little preparation of the surface is necessary and the type of existing finish is not critical. An irregular or slightly pitched surface can be leveled with a sand cushion under the isolation membrane.

If existing structural and architectural conditions limit the finish thickness and the dead load, a bonded method should be considered. The chemically bonded and thin-set methods are usually suitable for refinishing existing floors, but the monolithic method, which requires a roughened subfloor for bond, is used infrequently. A chemically bonded topping is thicker, heavier, and requires a concrete subfloor, but it is not limited as to chip size. Lighter-weight resinous thin-set toppings can be used over most subfloors, including wood.

As in new construction, the condition of the subfloor or of the old finish surface is critical for thin-set resinous toppings. Surface coatings such as paint, wax, and sealing compounds must be removed.

A topping can be installed directly over properly cleaned marble tile, quarry tile, slate, and other hard tile products, but resilient flooring and related adhesives must be removed completely. The surface should be ground smooth to expose the subfloor before a terrazzo topping is installed. A chemical bonding agent or resinous binder should be tested first in a small area to ensure proper bond with the prepared surface.

Existing wood strip and lumber board floors should be covered with stable mastic or plywood underlayment at least ⅜ in. (9.5 mm) thick. The matrix manufacturer's recommendations should be consulted to determine the proper type of underlayment and installation procedure for each type of resin matrix.

9.7.2.4 Special Surface Finishes

In addition to other characteristics, discussed earlier, terrazzo floors can be described according to their surface finish (regular, rustic, or conductive) (Fig. 9.7-15).

Terrazzo in homes, apartments, offices, stores, building lobbies, and other locations of normal foot traffic requires no special additives or surface treatment. The usual binder formulations, after normal grinding and polishing, provide adequate resistance to abrasion, staining, discoloration, and slipping hazard.

Special surface finishes have been developed for situations where unusual hazards exist. In exterior locations, a greater degree of protection from slipping hazard is afforded by *rustic terrazzo*. In hospitals and laboratories, where oxygen is used and potentially explosive atmospheres exist, *conductive terrazzo* will prevent electrostatic sparks. For locations subject to damage from acids, such as hospitals, laboratories, and industrial buildings, many resinous matrixes are suitable. Where greater slip resistance than that provided by regular terrazzo is required for interior locations, as on ramps and stairs, regularly spaced abrasive strips are recommended (Fig. 9.7-16). Some resinous matrixes can be formulated to provide inherently good slip-resistant properties. The regular and special surface finishes can be used with most installation methods (see Fig. 9.7-15).

RUSTIC TERRAZZO

Rustic (washed) terrazzo is suitable for nonslip and decorative exterior surfaces. The decorative chips are either weather-resistant marble, quartz, or granite chips, or decorative gravel. Although resinous binders can be used, they generally are limited to wall installations. Portland cement is the common material for rustic terrazzo floors, and in areas subject to freezing, an air-entraining agent is added. Rustic terrazzo should not be installed where the surface will be subjected to deicing salts or chemicals.

After the topping has been placed, rolled, and leveled as described in Section 9.7.2.1, the surface is washed with suffi-

FIGURE 9.7-15 Surface Finishes for Various Installation Methods

Installation Method	Binder	Chip Size	Surface Finish
Thin-set	Polyester and epoxy	#0 and #1	Regular Conductive
	Latex	Standard	
Monolithic Chemically bonded	Portland cement	Standard Intermediate Venetian	Regular Rustic
Bonded-underbed Unbonded	Portland cement	Standard Intermediate Venetian	Regular Rustic Conductive

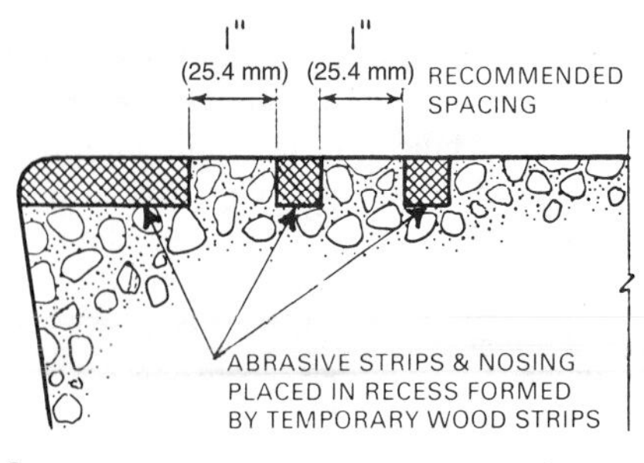

a

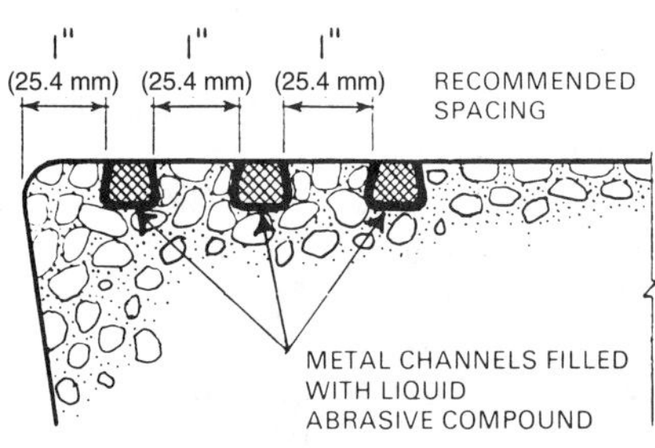

b

FIGURE 9.7-16 Abrasive inserts for nonslip surfaces: (a) prefabricated strips and nosing, (b) abrasive compound poured into channel.

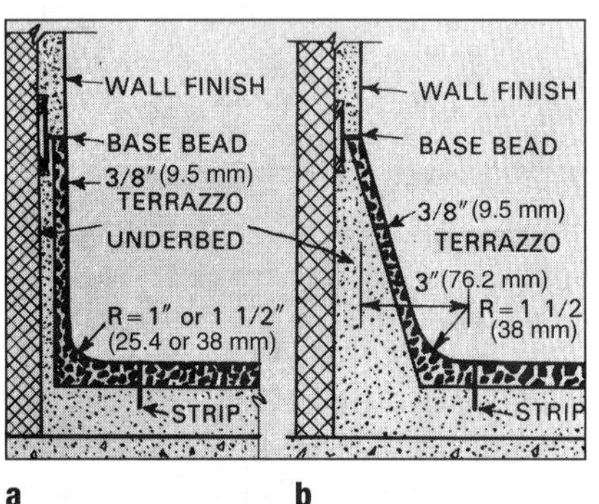

a b

FIGURE 9.7-17 Coved wall base: (a) straight type and (b) splayed type. Underbed thickness may be varied, creating recessed, flush (shown), or projecting base.

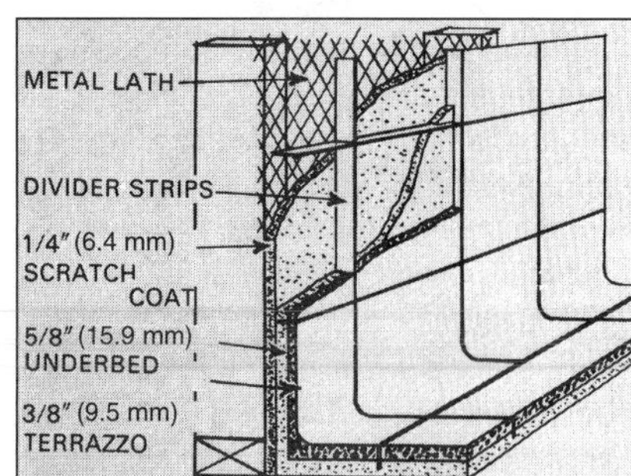

FIGURE 9.7-18 Wainscot can be installed over studs and metal lath (shown) or over any masonry base that will provide adequate bond for portland cement mortar.

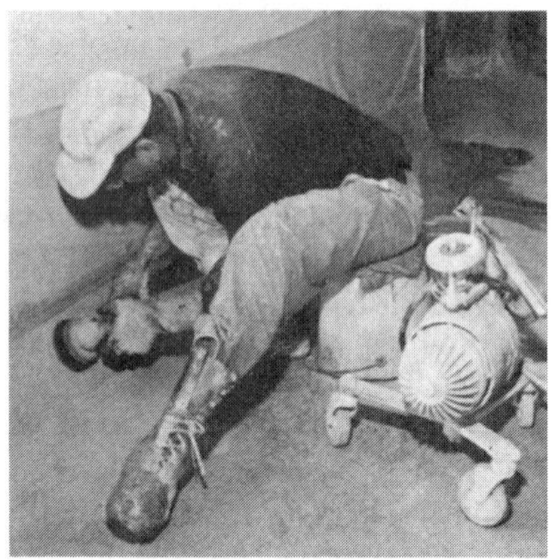

FIGURE 9.7-19 Base grinding: small hand-held wheels are used to grind and polish terrazzo base and wainscoting.

cient water pressure to remove the paste between the chips or gravel to a depth of about 1/16 in. (1.6 mm). The surface is then scrubbed with a solution of 1 part muriatic acid to 10 parts water to remove particles of paste clinging to the aggregate. A final rinsing with clean water removes the acid solution and neutralizes the surface. A light grinding may be desirable for exposed chips in pool areas, and a floor may be sealed, but usually no further operations are necessary.

CONDUCTIVE FLOORS

Conductive floors prevent electrostatic sparks in potentially explosive atmospheres. They consist of resinous matrixes to which a conductive ingredient, usually carbon black, has been added during manufacture. They conduct electricity at levels prescribed in Article 4-6.2 of the National Fire Protection Association's Standard No. 56A. Chips in conductive terrazzo should be no larger than #1 size.

9.7.2.5 Special Elements

Cast-in-place terrazzo can be carried up a wall 4 to 12 in. (100 to 300 mm) to form a coved wall base (Fig. 9.7-17) or can be continued to form a protective wainscot (Fig. 9.7-18). Full-height terrazzo finishes are not practical because of the labor involved in grinding vertical surfaces with relatively small hand-held wheels (Fig. 9.7-19). Cast-in-place terrazzo stairs often are practical and economical (Fig. 9.7-20). Precast treads, risers, and wall base (Fig. 9.7-21) are available.

9.7.2.6 Maintenance

Terrazzo must be protected from damage and soiling and cleaned periodically if it is to maintain its original appearance.

PROTECTION

Portland cement matrixes are porous and require protection from deterioration and staining by various chemical agents. Therefore, portland cement terrazzo requires protection with a proprietary penetrating sealer capable of internally sealing the pores. Resinous matrixes are denser, more stain resistant, and chemically inert. However, all terrazzo finishes consist of at least 70% stone or marble, both of which possess varying degrees of porosity and stain resistance. Resinous terrazzo may require similar protection, depending on the chip type and conditions of use. Sealers should be of a type specifically recommended by the manufacturer for terrazzo surfaces and should be applied in accordance with the manufacturer's recommendations.

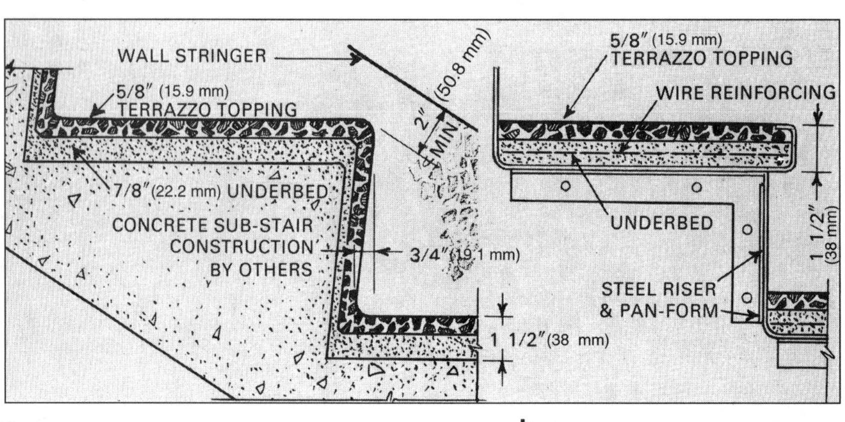

a b

FIGURE 9.7-20 Cast-in-place stairs: (a) continuous treads and risers over concrete substair, (b) treads cast in steel pan forms.

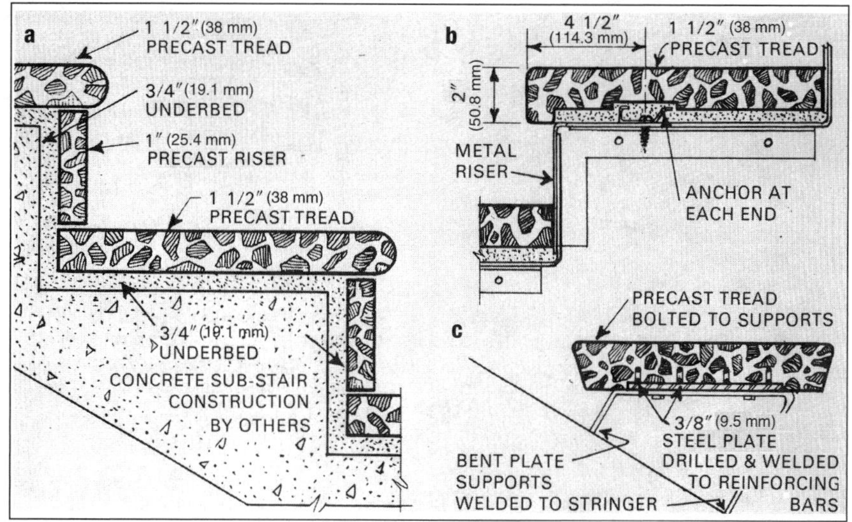

FIGURE 9.7-21 Precast stair treads and risers can be used with (a) concrete or (b and c) metal stair construction.

Surface coatings, such as wax, that provide abrasion protection for softer floor finishes are not required. Foot traffic on unwaxed terrazzo tends to polish and impart further luster to the surface. In addition, surface coatings tend to make the floor slippery when wet and to discolor the surface in low-traffic areas.

CLEANING

Terrazzo should be cleaned periodically by damp mopping and washing with a neutral liquid cleaner recommended for the purpose by the manufacturer. Soaps and scrubbing powders containing detergents, water-soluble inorganic salts, or crystallizing salts should not be used. Sweeping compounds containing oil may stain and permanently discolor a terrazzo surface. Dry buffing after each mopping (and as required between moppings) will remove surface dirt and restore luster.

Terrazzo is susceptible to staining. Stains should be treated as soon as possible because they are more difficult to remove after they have set. The type of stain must be determined and the proper remover selected. There are four general categories of stain removers:

- Solvents, such as carbon tetrachloride, which dissolve grease, chewing gum, and lipstick
- Absorbents, such as chalk, talcum powder, blotting paper, and cotton, which absorb fresh grease and moist stains
- Bleaches, such as household ammonia, hydrogen peroxide, acetic acid, and lemon juice, which discolor stains
- Proprietary stain removers

9.8 Acoustical Treatment

Chapter 13 includes general requirements for and the theory behind sound control in buildings. This section discusses in more detail that part of sound control generally called *acoustical treatment*. This includes acoustical ceilings, integrated ceilings, and acoustical wall treatment and baffles. Two or more of these are often used together (Fig. 9.8-1). Acoustical treatment also includes the furring and suspension systems needed to mount acoustical materials.

Before proceeding it will be helpful to define some of the terms used in this section.

Acoustical baffle. An acoustical panel that is suspended from an edge or that floats cloud-fashion in a space.

Acoustical ceiling. The entire ceiling system including acoustical material and associated insulation, furring, and suspension systems. It does not include auxiliary thermal or acoustical insulation in addition to that normally associated with the acoustical ceiling being discussed. Thus, a layer of acoustical insulation placed above a ceiling and over or adjacent to a partition to achieve increased acoustical performance is not part of an acoustical ceiling. An acoustical blanket or pad that is necessary to achieve the rated acoustical performance of a metal ceiling is, however, a part of an acoustical ceiling. Metal ceilings, as defined below, are also acoustical ceilings.

Acoustical material. A general term that covers both composition and metal acoustical materials as defined in this chapter.

Acoustical wall panels. Acoustical panels that mount on the surface of a wall or partition, regardless of their mounting method or size. Acoustical wall panels may completely cover a wall surface or be mounted as plaques with space between them.

Composition. When referring to acoustical materials, the term *composition* includes cellulose, mineral fiber, glass fiber, asbestos cement, ceramic faced, plastic, gypsum, and wood fiber acoustical materials. Plastic- and metal-faced acoustical ceiling materials in which the plastic or metal is adhered or bonded to a composition material as defined here are also called composition materials.

Integrated ceiling. A ceiling system consisting of acoustical material, suspension system, air distribution outlets, and lighting fixtures.

Metal acoustical ceiling material. Linear metal strips and metal panels and pans are all called metal acoustical materials, even though some of them may be of questionable acoustic value. The term *metal acoustical material* means the ceiling material only. It does not include the associated suspension system or acoustical insulation blankets or pads.

Metal ceiling. The entire ceiling construction, including metal acoustical materials, acoustical pads or blankets, and the suspension system, is called a metal ceiling.

FIGURE 9.8-1 **Acoustical ceiling and wall panels are both used in this auditorium.** (Photograph provided by MBI Products Company.)

9.8.1 ACOUSTICAL CEILINGS

Acoustical ceilings include composition tiles and panels, metal acoustical materials, and associated furring and suspension systems. Integrated ceilings are discussed later in this chapter.

9.8.1.1 Acoustical Materials

The characteristics of acoustical materials are identified according to ASTM Standard E 1264 "Standard Classification for Acoustical Ceiling Products." It classifies acoustical materials according to type, pattern, fire class, acoustical performance, and light reflectance coefficient.

MATERIAL TYPES AND PATTERNS

New acoustical materials will usually be one of the types shown in Figure 9.8-2, with one of the patterns shown in Figure 9.8-3. Other materials are also sometimes available, including ceramic-faced units for use in wet and high-impact areas, units with other abuse-resistant or sanitary plastic faces, gypsum, and wood fiber composition. Acoustical units in existing buildings may contain asbestos. Metal ceiling panels are made from aluminum, steel, and stainless steel. Suspended ceilings may also be of wood, plastic, or sheet metal, but most of these types are more properly classified as decorative rather than acoustical ceiling materials.

ASTM E 1264 lists three fire classes that correspond directly to Classes I, II, and III as required by some building codes. Their flame-spread ratings are as follows:

- *Class A:* 0 to 25
- *Class B:* 26 to 75
- *Class C:* 76 to 200

FINISHES

Most composition acoustical units are painted on the exposed face. Most such painted finishes are considered washable, although gentleness is advised. Some manufacturers list some products as having a "scrubbable" finish, and these are listed in ASTM E 1264, but there is no universally accepted definition of the term.

A variety of finishes are available for

FIGURE 9.8-2 **Acoustical Material Types**

Type I	Cellulose base with painted finish
Type II	Cellulose composition with membrane-faced overlay
Type III	Mineral base with painted finish; type includes: Form 1 Nodulated, cast, or molded Form 2 Water felted Form 3 Dry felted
Type IV	Mineral base with membrane-faced overlay; type includes: Form 1 Nodulated, cast, or molded Form 2 Water felted Form 3 Dry felted
Type V	Perforated steel facing (pan) with glass fiber base backing
Type VI	Perforated stainless steel facing (pan) with mineral or glass fiber base backing
Type VII	Perforated aluminum facing (pan) with mineral or glass fiber base backing
Type VIII	Cellulose base with scrubbable pigmented or clear finish
Type IX	Mineral base with scrubbable pigmented or clear finish; type includes: Form 1 Nodulated, cast, or molded Form 2 Water felted Form 3 Dry felted
Type X	Mineral base with plastic or aluminum membrane-faced overlay or both
Type XI	Mineral base with fabric-faced overlay; type includes: Form 1 Nodulated, cast, or molded Form 2 Water felted Form 3 Dry felted
Type XII	Glass fiber base with membrane-faced overlay; type includes: Form 1 Plastic Form 2 Cloth Form 3 Other
Type XIII	Aluminum or steel strip with mineral or glass fiber base backing; type includes: Form 1 Perforated Form 2 Nonperforated
Type XX	Other types (this category is left for additional types)

FIGURE 9.8-3 Acoustical Material Patterns

A. Perforated, regularly spaced large holes
B. Perforated, randomly spaced large holes
C. Perforated, small holes
D. Fissured
E. Lightly textured
F. Heavily textured
G. Smooth
H. Printed
I. Embossed
J. Embossed-in-register
K. Surface scoured
Z. Other patterns (describe)

metal ceiling panels, including fluoropolymer, baked enamel, anodized aluminum, acrylic lacquer, and even wood veneer.

Light reflectance is rated as one of four coefficients with the following light reflectances:

LR 1: 75% or more
LR 2: 70% to 74%
LR 3: 65% to 69%
LR 4: 60% to 64%

ACOUSTICAL PERFORMANCE

Acoustical products are classified by Sound Transmission Class (STC) and Noise Reduction Coefficient (NRC), by Articulation Class (AC), or by Speech Privacy Isolation Class (NIC′). An acoustical ceiling is rated by Ceiling Sound Transmission Class (CSTC). STC, NRC, and CSTC ratings are used for closed-room situations. AC and NIC′ are used for large spaces that contain several activities, such as occur in an open-plan office space. AC- and NIC′-rated products have been tested for their ability to both control sound reflections and permit white-noise transmission from plenum speakers into a space. White noise is used in open-space designs to mask background noises and help provide speech privacy.

The sound absorption quality of an acoustical material is indicated by NRC, which shows a material's ability to reduce sound levels within a space. An acoustical material's noise isolation control ability is indicated by STC (CSTC for ceilings), which shows a material's ability to prevent sound from passing through it into another space. There is usually an inverse relationship between the two properties. Materials that reduce sound reverberations are usually light and porous. To prevent sound transmission, materials must be heavy and airtight. STC and CSTC are measured by test procedure AMA 1-II "Ceiling Sound Transmission Test by the Two Room Method," so named by the no-longer-existent Acoustical Materials Association. The standard is available from the Ceilings and Interior Systems Contractors Association (CISCA).

Where sound control is an essential element in an existing building, the advice of an acoustical specialist should be sought before changes or repairs are made.

COMPOSITION ACOUSTICAL TILE VERSUS PANELS

Composition acoustical ceiling materials may be either tile or panels, and all the types, patterns, and other characteristics discussed in previous paragraphs are found in both of them. In the past, much tile was used. In recent years, the advent of better acoustical panel materials, more attractive and functional suspension systems, an increased need for access to the space above a ceiling, changes in construction methods, their lower cost, and other factors have led to a much greater use of panels. Some manufacturers have stopped making tile products and concealed grid suspension systems. An advantage of panels over tile is that panels are loosely laid into a grid, which makes them easily removable and the space above them readily accessible. A major disadvantage is that because they are loosely laid into a grid, they are subject to dislocation by wind, pressure applied during cleaning, and air passing through a plenum space above them. Tile and panel edges are square, beveled, or have a special edge such as the reveal edge shown in Figure 9.8-4.

Tile Most modern tile is either 12 or 24 in. (305 or 610 mm) square or 12 by 24 in. (305 by 610 mm). Tile in older buildings is probably 12, 16, or 20 in. (305, 406, or 508 mm) square. Tile joints are flanged, rabbeted, kerfed (splined), or tongue-and-groove. Figure 9.8-5 shows some types, but there are many others in use today and still others have been used in the past.

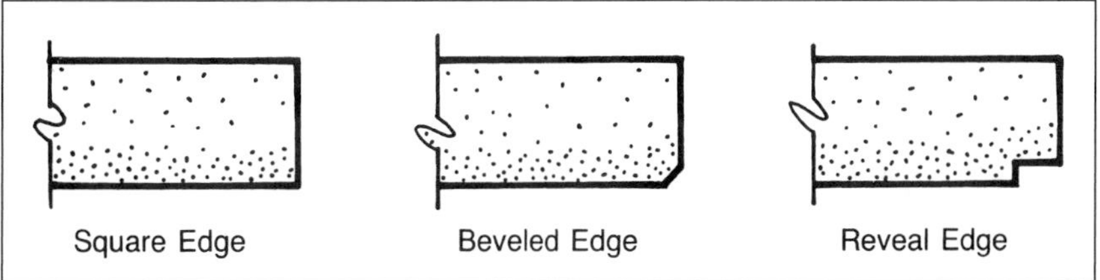

FIGURE 9.8-4 Acoustical tile and panel edges. Many variations of beveled and reveal edges are available. For example, reveal edges may have two, three, or more steps. Bevels may be combined with steps. (Courtesy H. Leslie Simmons.)

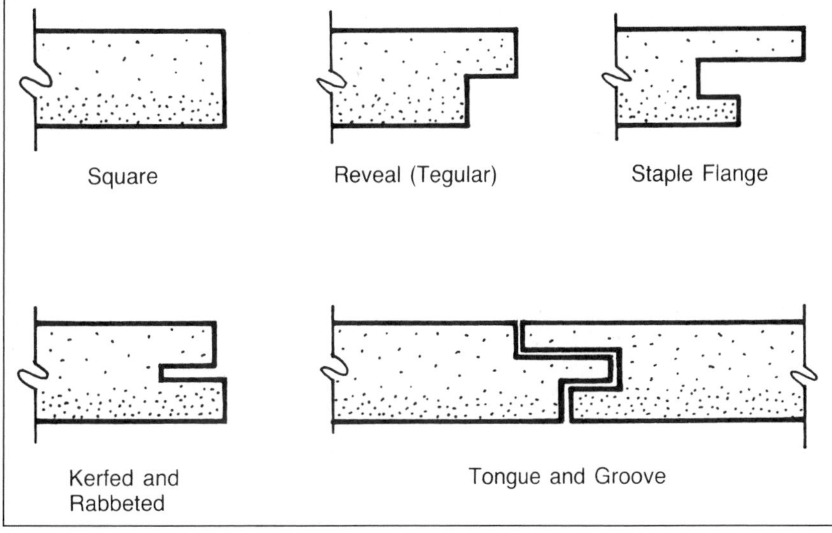

FIGURE 9.8-5 Typical acoustical tile joint conditions. (Courtesy H. Leslie Simmons.)

Sometimes a tile's edges are not all cut the same. The tongue-and-groove joints on two sides of a tile may be quite different from those on the other two sides, so that the tile will nest properly.

Panels Most panels are either 24-in. square or 24 by 48 in. Some are scored so that they appear from the floor to be smaller units. Scoring may produce squares or linear strips, and may be parallel with panel sides or diagonal to them.

Panel joints are usually square or reveal type (see Fig. 9.8-4). In systems where cross suspension members are concealed (semiconcealed systems) the ends of the panels are kerfed and rabbeted to receive the concealed suspension member.

METAL ACOUSTICAL CEILING MATERIALS

In addition to metal-faced composition products, there are many metal acoustical ceiling material styles (Fig. 9.8-6). They include, but are not limited to, linear shapes (strips), flat panels, shaped (corrugated and other shapes) panels, and flat or textured pans. Surfaces may be continuous or perforated, and either flat or embossed. Metal grids and louvers are also available, but whether they are acoustical in nature is debatable. Metal ceiling members may be lay-in type or snap-in type. Snap-in types require special support systems. Some lay-in types also require special support systems; others do not. Some panels come with glass fiber or mineral wool sound-absorbing pads; others do not.

FIGURE 9.8-6 An angled ceiling made from flat, perforated metal acoustical ceiling panels. (Manufactured by Hunter Douglas Architectural Products.)

9.8.1.2 Support Systems

Composition acoustical materials may be directly attached to solid or supported substrates or supported on a suspension system. Metal ceilings usually include a suspension system. The term *suspension system* usually refers to a metal grid that is hung by wires or bars from the structure, but it is also used to refer to a system of wood or metal furring to which acoustical materials, mainly tiles, are screwed, stapled, or nailed. Sheet metal decorative ceilings are often held in place with fasteners, but metal acoustical ceiling materials are seldom so installed.

WOOD FURRING

Except in single-family residences and other small buildings, little acoustical tile is installed directly on wood furring today (Fig. 9.8-7), but in the past such installations were common. The normal installation when wood furring is used today is to install a layer of gypsum board, called backing board, over the furring and fasten the acoustical tile to the gypsum board.

METAL FURRING

In the past, acoustical materials were sometimes screwed directly to metal furring channels (Fig. 9.8-8), but this is seldom done today. Today, when composition acoustical tile is installed over metal furring bars, the method usually used is that described in ASTM Standard C 635 as "Furring Bar Suspension System," in which a layer of gypsum board, called backing board, is applied to suspended nailing or furring bars. The acoustical tile is then fastened to the gypsum board.

Metal acoustical materials are seldom installed over metal furring bars today. They are occasionally installed on snap-in or other grid members that are directly

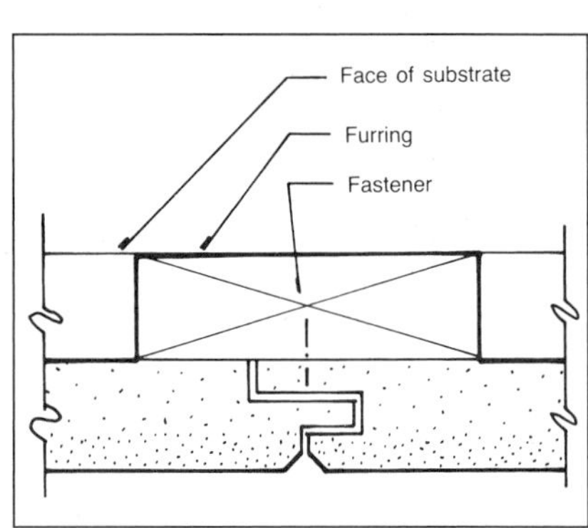

FIGURE 9.8-7 Acoustical tile on wood furring. (Courtesy H. Leslie Simmons.)

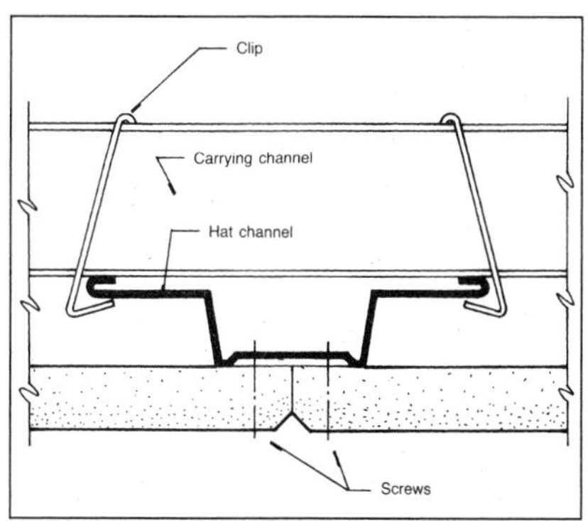

FIGURE 9.8-8 Acoustical tile directly applied to metal furring. (Courtesy H. Leslie Simmons.)

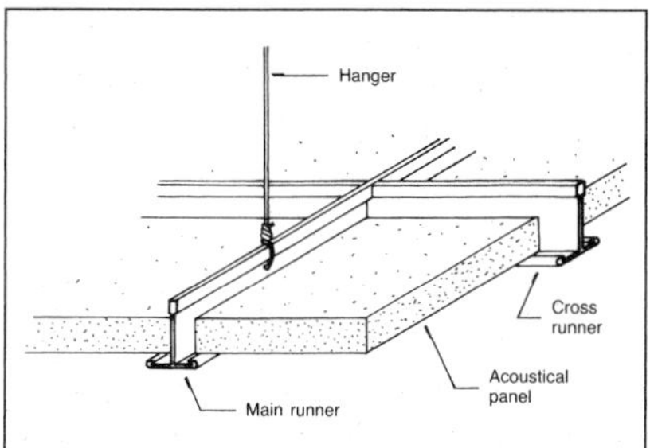

FIGURE 9.8-9 One type of direct-hung acoustical ceiling suspension system. (Courtesy H. Leslie Simmons.)

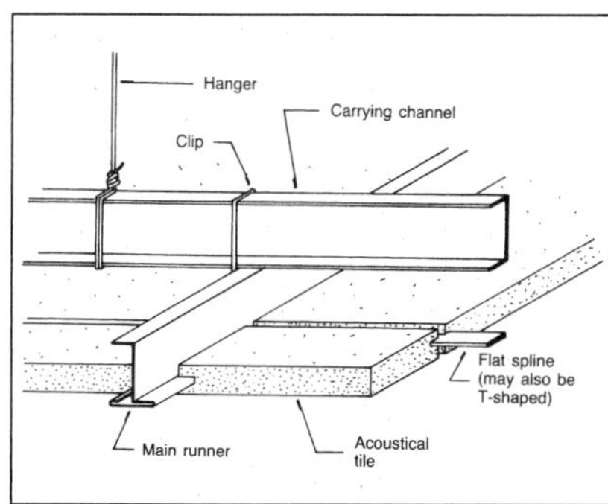

FIGURE 9.8-10 One type of indirect-hung acoustical ceiling suspension system. (Courtesy H. Leslie Simmons.)

clipped to structural elements or inserts in concrete slabs, but this installation is so similar to that described later in this chapter for a metal suspension system that the requirements addressed there apply.

HUNG METAL SUSPENSION SYSTEMS

Metal suspension systems for acoustical ceilings should comply with the applicable requirements of ASTM Standards C 635 and C 636. Acoustical treatment manufacturers may, however, have more stringent requirements.

There are two categories of hung suspension systems for acoustical materials. In *direct-hung* systems, the main runners are suspended from the structure (Fig. 9.8-9). In *indirect-hung* systems, the main runners are suspended from carrying channels (Fig. 9.8-10). The main runners may be any type that is appropriate for the acoustical material being supported. Some common types are shown in Figure 9.8-11, but there are many others.

Lay-in suspension systems may be either direct- or indirect-hung, depending on the system members and the necessary spacing of hangers. *Concealed grid* systems may also be either direct- or indirect-hung, but systems having splines or other less rigid cross members are usually indirect-hung. Except for the carrying channels, most members in both types of systems are proprietary, meaning that they are manufactured or furnished by a single supplier or producer for installation as a system.

Hung suspension systems for composition materials include concealed grid types for tile and exposed or semiexposed grid types for lay-in panels.

Concealed Grid Systems for Composition Tile Most hung suspension systems designed to be concealed in the finished ceiling are H and T systems, which include H-shaped main runners and tee splines similar to those shown in Figure 9.8-11; concealed single- or double-web systems using tee-shaped main runners similar to the standard tee shown in Figure 9.8-11 and spacer bars similar to those shown in the same figure at right angles to the main runners for stiffening; and zee systems (see Fig. 9.8-10) using zees and splines similar to those shown in Figure 9.8-11. Other systems are also used today, and still others have been used in the past, so an existing installation may be different from those noted here.

Concealed suspension systems for composition tile are designed to lock the tile in place so that it cannot be easily removed. To overcome this drawback, manufacturers have devised ways to make sections of tile hinged or removable so that access can be gained where necessary. In direct-hung systems, hinged tile sections, after being released by a special tool, may be moved upward into the plenum or downward into the room. In both types, usually more than one tile is movable. Strips of tile between cross framing members can be either lifted up or pulled down to provide access. The

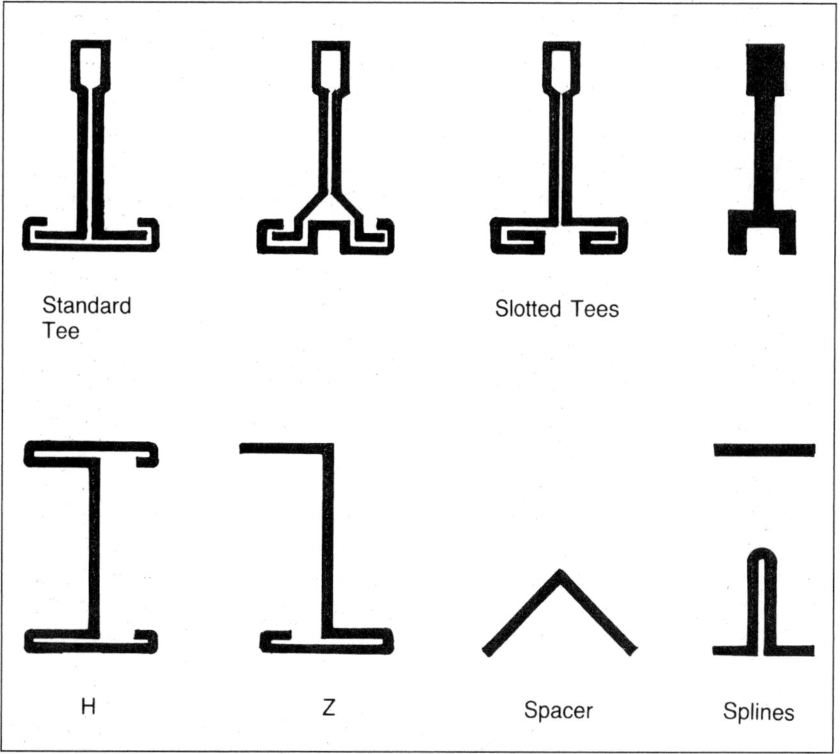

FIGURE 9.8-11 Typical acoustical ceiling suspension system components. (Courtesy H. Leslie Simmons.)

movable panels of tile may be hinged along the cross tees or along the main tees. Once the movable tile has been moved out of the way, the rest of the tile can be removed in successive pieces. Downward access systems are not recommended in some seismic zones.

In indirect-hung systems, access is sometimes gained with special cross-runners or split splines that permit removal of an acoustical unit. Such special runners or splines are identified with unobtrusive permanent markers. Some indirect-hung systems are made accessible with special tools that work only with those systems.

Not all systems, however, have a removable feature. Some older tiles may be removable only by breaking out a tile and then removing the adjacent tile until the necessary opening has been achieved. Where the space above the ceiling is not accessible through the tiles, access doors must be provided.

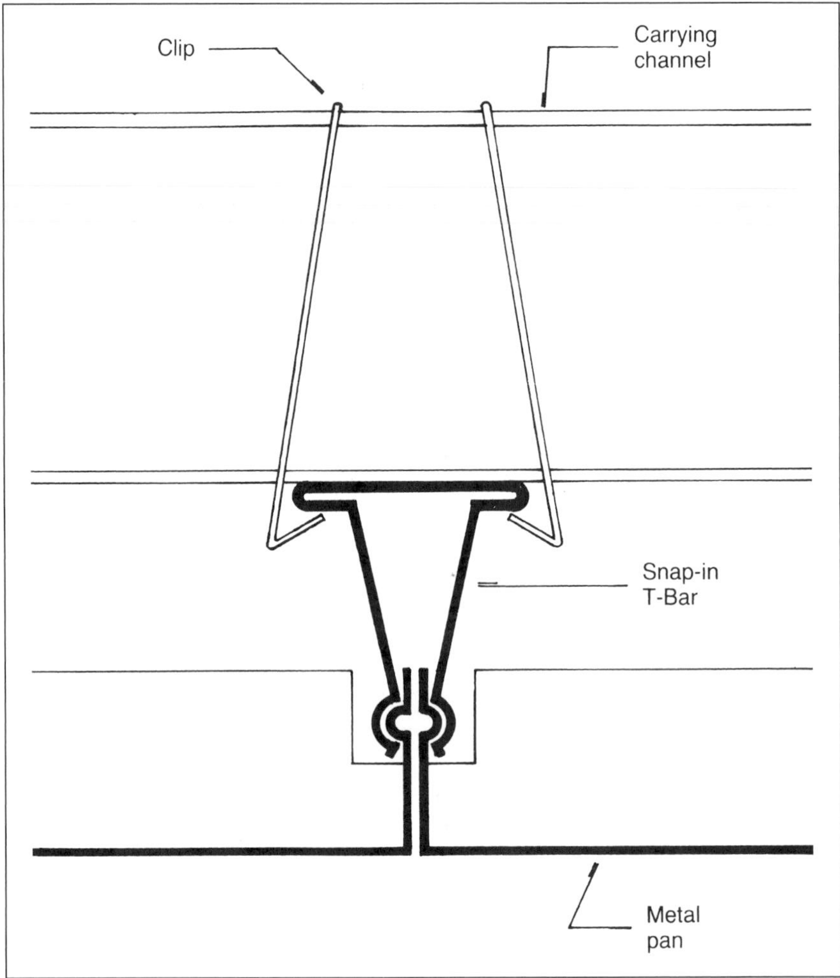

FIGURE 9.8-12 One type of T-bar snap-in metal ceiling system. (Courtesy H. Leslie Simmons.)

Exposed and Semiexposed Grids for Composition Panels Suspension systems designed for lay-in application of acoustical panels usually consist of T-shaped main runners and interlocking cross-tees (see Fig. 9.8-9). The cross-members of semiconcealed systems are usually T-splines. The main runners are usually directly hung from the structure with no carrying channels present. Main runner and cross-runner shapes are proprietary and vary with each system manufacturer. They are available with wide or narrow flat bottom surfaces, roll-formed caps, slotted, with reveals, and in other configurations. A few of the many configurations available are shown in Figure 9.8-11.

Grids for Metal Ceilings Hung suspension systems for metal acoustical ceiling materials include snap-in types and lay-in types. Many metal pans and some panels are installed into snap-in grids. The T-bar snap-in member shown in Figure 9.8-12 is only one of several snap-in-device types in general use. Snap-in grid members are manufactured specifically to fit the edge configuration of the pans or panels being supported. A grid member made by one manufacturer may not correctly fit or support a panel made by a different manufacturer.

Some types of metal ceiling grids are direct hung but do not have cross-furring members because the metal ceiling material itself is stiff enough to stabilize the system.

Some metal pans and panels lay into the same type grid used for composition material. Others hang from similar grids (Fig. 9.8-13).

Support grids for linear metal ceilings, while similar to that shown in Figure 9.8-14, vary from manufacturer to manufacturer to suit their particular ceiling materials.

Structural Classification ASTM Standard C 635 divides hung suspension systems in three structural strength categories: *light duty*, *intermediate duty*, and *heavy duty*. Light-duty systems, which are used in residential and light commercial installations, will support only the acoustical ceiling materials themselves. Intermediate-duty systems, which are used in most commercial structures, will support light fixtures and air diffusers in addition to the acoustical materials. Heavy-duty

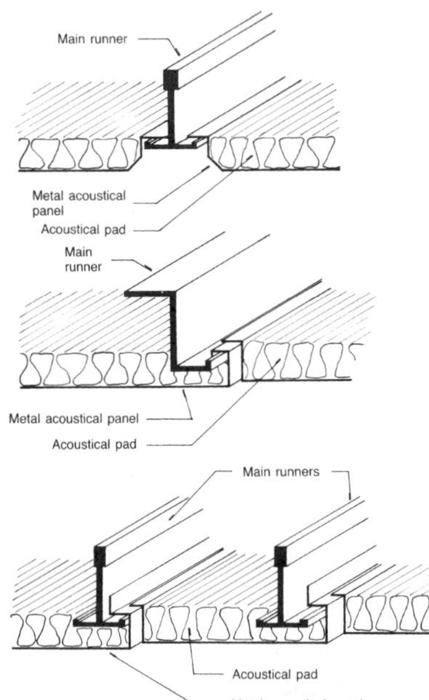

FIGURE 9.8-13 Some typical hung metal ceiling systems. (Courtesy H. Leslie Simmons.)

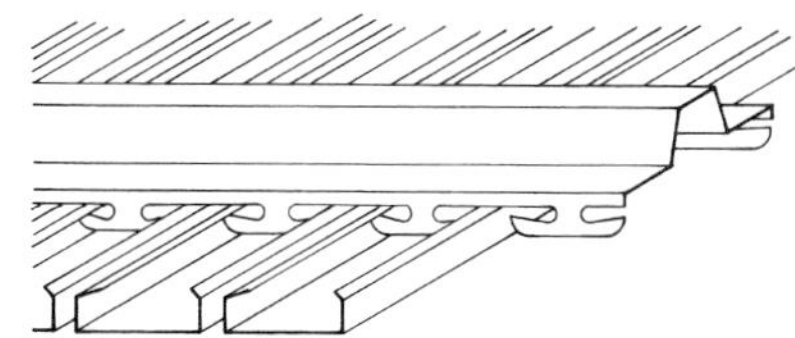

FIGURE 9.8-14 Typical support grid member for a linear metal ceiling. (Courtesy H. Leslie Simmons.)

systems will support heavier loads than the other two classifications.

Edge Moldings and Trim Except for some linear metal ceilings, most acoustical ceiling suspension systems have matching edge moldings at all vertical surfaces to which the acoustical ceiling abuts, including penetrations such as columns. Acoustical tile ceilings applied with fasteners or adhesives should also have edge moldings and trim.

The edge molding and trim type and configuration should be suitable for the suspension system used. Some ceiling types have specialized trim members used only for that type.

Metals and Finishes In exterior installations and in interior installations where high humidity will be present, such as in swimming pool areas or commercial kitchens, concealed metals should be either aluminum, stainless steel, zinc alloy, or hot-dip galvanized steel.

In these locations, exposed grids and edge members should be either aluminum or stainless steel. Aluminum is usually protected by a baked-on paint, fluoropolymer, or anodized finish, but it may also be left natural and protected by a coat of lacquer. Stainless steel may be protected by a coating or left exposed.

The finishes on both concealed and exposed metals in such locations should comply with ASTM C 635 requirements in the paragraph entitled "Coating Classification for Severe Environment Performance."

Concealed metals in grids supporting composition tile and panels in locations other than those discussed in the preceding paragraphs may be the manufacturer's standard steel products with the manufacturer's standard galvanized or painted finish.

Exposed grids and edge members in locations other than those mentioned in the preceding paragraphs are usually steel, finished with the manufacturer's standard paint finish, but may be aluminum or stainless steel where those decorative finishes are selected. Where the acoustical ceiling material has a decorative finish, such as anodized aluminum or stainless steel, exposed suspension system members often match the ceiling material and finish.

Colors of exposed finished metal vary, although white and off-white are the most common.

9.8.1.3 Accessories and Miscellaneous Materials

The types of accessories needed for acoustical ceilings include the following.

Carrying channels. 1½-in. (38 mm) cold-rolled steel channels, weighing a minimum of 0.475 lb./lin. ft. (0.71 kg/m), painted. Where spans require, 2-in. (50.8 mm) cold-rolled steel channels weighing a minimum of 0.590 lb./lin. ft. (0.88 kg/m).

Hanger wire. For typical installation, galvanized carbon steel wire of the size necessary to properly support three times the hanger design loads, but in no case less than 12-gauge (1.6 mm) wire. Some codes may require heavier wire regardless of loads. For support of aluminum or stainless steel grid systems, and in exterior locations and air plenums, stainless steel or aluminum wire of the gauge recommended by the suspension system manufacturer.

Clips. Commercial types acceptable under the building code for support.

Hold-down clips. Suspension system manufacturer's standard types. Hold-down clips are necessary where uplift is a problem from air pressure or to permit cleaning. Clips are also required in some fire-rated assemblies. Security clips are sometimes used to make metal pan ceilings nonremovable.

Tie wire. Galvanized wire, at least 16 gauge (1.6 mm).

Extension hangers for metal ceilings and hanger rods and flat bars. Zinc- or cadmium-coated mild steel. Alternatively, such hangers may be finished with rust-inhibitive paint.

Angle-shaped hangers. Formed from nominal 15- or 16-gauge (1.75 or 1.6 mm) galvanized sheet steel. Legs should not be less than ⅞ in. (22.2 mm) wide.

Hanger anchorage devices. Rust-inhibitive screws, clips, bolts, concrete inserts, expansion anchors, or other devices, sized to support five times the calculated hanger loading.

Screws and other fasteners. Rust-inhibitive and of a type standard with the suspension system manufacturer. Tile fasteners should be stainless steel or cadmium plated of the type recommended by the tile manufacturer and should penetrate the substrate by at least ½ in. (12.7 mm). Exposed fasteners should match the material fastened in appearance.

Acoustical sealant. Permanently elastic, water-based, nondrying, nonbleeding, and nonstaining.

Sound attenuation blankets. Semirigid mineral fiber blankets with no membrane facing, with Class 25 flame spread; 1½ in. (38 mm) or more thick.

Acoustical pads for use in metal pan and panel ceilings. Mineral or glass fiber. Some are wrapped or faced with materials such as polyvinyl chloride. Others are bare.

Adhesives for installing acoustical tile. UL (Underwriters Laboratories, Inc.) labeled with a Class 0 to 25 flame-spread rating.

9.8.1.4 Fire Performance

The ASTM E 1264 Fire Classes refer only to the acoustical material itself. Underwriters Laboratories' (UL) *Building Materials Directory* contains a listing showing the ratings of acoustical ceiling materials by manufacturer and product. Acoustical materials manufacturers usually indicate the flame-spread rating of their products.

Hung suspension systems are classified as fire-resistance rated or non-fire-resistance rated, but sometimes it is hard to tell which is which from the manufacturers' product literature.

Unfortunately, using a rated acoustical material, suspension system, or both together does not alone result in compliance with code or other requirements for fire-rated construction. Fire-rated construction is a function of the entire assembly, including the floor or roof construction, flooring or roofing, insulation, and the installed ceiling. UL's *Fire Resistance Directory* contains a listing of construction systems, which includes the structure (floor, roof, etc.) and finishes, including acoustical ceilings. Ceiling materials are also listed by manufacturer and product.

Fire-resistance ratings are established according to ASTM Standard E 119.

9.8.1.5 Installation

Layout plans should be prepared for spaces where acoustical ceilings will be installed to ensure coordination with lighting fixtures and other ceiling-mounted equipment and devices.

Manufacturers' brochures and installation details and ASTM and CISCA standards should be obtained and followed. Metal suspension systems should be installed in accord with ASTM Standard C 636, unless the acoustical treatment manufacturer has more stringent requirements.

The manufacturers' recommendations for cleaning and refinishing acoustical units and suspension system components should be obtained and stored in a safe place so that they will be available when maintenance and repair are necessary in the future.

Extra material, in the amount of 2% or 3% of the amount installed, is often left at the site for the owner to use in making repairs.

Acoustical materials and suspension system components should be kept dry before and after use. Acoustical materials should be allowed to acclimate to the temperature and humidity of the space before they are installed to prevent dimensional changes that may affect the appearance of the installed material.

Watermarked, rusted, or otherwise damaged units should not be installed.

Acoustical ceiling components should not be installed until the building is weathertight, glazing has been completed, exterior openings have been closed, and mechanical, electrical, fire-fighting, and other work above ceilings has been completed.

Wet work, such as masonry, concrete, and plaster, should be allowed to dry to the moisture content recommended by the acoustical materials manufacturer before acoustical materials are installed. A temperature of at least 60°F (15.6°C) should be maintained in the space during and after installation of acoustical materials.

Ceilings where cleanliness is paramount, such as in hospital operating rooms, often require sealant around the perimeter of the ceiling grid and sometimes even around the perimeter of each panel at its supports. Health department rules and regulations and building codes should be examined to determine the applicable requirements.

Suspension systems for use where fire-resistance ratings are required should comply with governing regulations, and their materials and installations should be identical with assemblies that have been tested and listed by recognized authorities, such as UL.

Suspension system members that abut vertical or overhead building structural elements should be isolated from structural movement sufficiently to prevent transfer of loads into the suspension system. Provisions should also be made in the suspension system to accommodate movement that occurs across control and expansion joints in the substrates.

Suspended acoustical ceilings should have a seismic restraint system in accordance with ASTM Standard E 580 where required by code or usual practice.

Flat surfaces of suspension system members should not be warped, bowed, or out of plumb or level by more than ⅛ in. (3.2 mm) in 12 ft. (3658 mm) in both directions, or by more than $\frac{1}{16}$ in. (1.6 mm) across any joint or ⅛ in. (3.2 mm) total along any single member.

Before acoustical ceiling installation begins, mechanical, electrical, and other work above or within the acoustical ceilings should be completed.

Acoustical units should be handled carefully to prevent damage to edges and faces.

FURRING INSTALLATION

Furring should be installed as is appropriate to support the acoustical ceiling. It should be placed on lines and levels necessary to cause the acoustical material to fall into the proper location. It should correct unevenness in the supporting structure or substrate.

The surface of wood furring to which acoustical material will be fastened should be not less than 1½ in. (38 mm) wide. Generally, 2 × 2 (50 × 50) lumber is used for furring that is attached to a supporting structure, and 1 × 2s (25 × 50s) or 1 × 3s (25 × 75s) where furring is laid directly over solid substrates. Nominal 1-in. (25.4 mm)-thick lumber is too flexible to use over framing. Sometimes acoustical ceilings are furred with 2 × 4s (50 × 100s) hung from structural floor or roof framing by hangers spaced not more than 48 in. (1200 mm) on centers. Frequently, in older buildings the hangers were wood members instead of wire.

Furring should form a complete system adequate to properly support the acoustical tile. Closers should be installed at edges and openings.

Bolts, lag screws, and other anchors are used to anchor wood furring in place. Fasteners are placed near the ends of furring members and not more than 36 in. (900 mm) on centers between. Shorter members, however, should be anchored at 30 in. (750 mm) on centers.

Screws, bolts, clips, and other anchors are used to anchor metal furring in place. Fasteners are placed near the ends of furring members and not more than 24 in. (600 mm) on centers between, but other support spacings may be recommended by the furring bar manufacturer. Some types of furring bars may require closer spacings. The correct spacing should be verified with the manufacturer.

Furring spacings may vary with the substrate and the acoustical material being supported. Minimum spacings are dictated by acoustical material thickness and size and may vary somewhat from manufacturer to manufacturer and system to system. Spacing must be such as to adequately support the acoustical ceiling.

HUNG-SUSPENSION SYSTEM INSTALLATION

Suspension system hangers should be secured to wood, concrete, or steel structural supports by connecting directly to the structure where possible. Where direct connection is not possible, the hangers should be connected to inserts, clips, anchorage devices, or fasteners. Except for hanger attachments installed specifically for the purpose by the steel deck installer, hangers should not be attached to steel decks. Since failures have occurred even when deck tabs acceptable to the deck manufacturer have been used, some sources recommend that ceilings not be supported from steel deck under any circumstances. Hangers should never be attached to pipes, conduit, ducts, or mechanical or electrical devices.

Hangers should be located plumb in relation to primary suspension members. Contact with insulation covering ducts and pipes and other items within the plenum space should be avoided. Hangers should be splayed only where obstructions or conditions prevent plumb, vertical installation. Horizontal forces generated by splayed hangers should be offset by counter-splaying, bracing, or another suitable method.

Frames for grilles, registers, and other items that occur within acoustical ceilings should be properly installed in the correct locations. Additional hangers should be provided for support of the grid systems at lights, grilles, registers, and other items to prevent excessive deflection. At grid systems with a light-duty structural classification, lights, grilles, registers, and other items should be supported separately from the ceiling grid.

Main runner and cross-furring members should be placed in proper positions, leveled, and fastened together in the manner required by the system. T-shaped members are usually locked together without accessories. Some systems, however, require clips.

Except in self-edging systems, edging strips should be provided for all acoustical ceilings, including those with concealed grids, along the entire perimeter and where they abut vertical surfaces. Short lengths should be avoided. Standard moldings should be attached through the web 3 in. (76.2) from each end and not more than 16 in. (406 mm) on centers between. Special moldings may require other spacings. Edging strips are usually of the same material and finish as exposed portions of grid systems.

Light fixtures and other penetrations are often boxed with mineral wool boards to comply with UL rating requirements where ceilings are rated.

Suspension systems should achieve a tight and level ceiling at the spacings and directions recommended by the manufacturer and to comply with UL label requirements.

Installation of Concealed Grid Systems for Composition Acoustical Tile Maximum hanger spacing is usually 48 or 60 in. (1220 or 1524 mm), depending on the system and the components selected. The first hanger on each end of each runner or carrying channel should be no more than 6 in. (150 mm) from the ends. Some systems require closer spacings. Where the hangers must be placed at wider spacings than those recommended for the system, the main runners can be supported by carrying channels. In indirect-hung suspension systems main runners are usually supported by carrying channels.

Installation of Exposed Grid Systems for Composition Acoustical Panels Exposed members should be installed in line, square, level, in one plane, and within the tolerances permitted by ASTM C 636. Tees should be of adequate size to span supporting structure. It may be necessary to support main tees from carrying channels.

Main tees are usually located 48 in. (1220 mm) on centers and all in the same direction. Cross-tees are then installed to span between the main runners with cross-tees at 24 in. (610 mm) on centers. The tees are then locked together using whatever method is applicable to the system being used. The tees should rest on wall moldings at vertical surfaces. Where 24-in. (610 mm)-square panels are used, additional cross-tees are placed from cross tee to cross-tee and securely fastened (or locked) to the cross-tees.

Main tees should be supported by hangers placed not more than 48 in. (1220 mm) on centers. Where the structure will not permit hangers at that spacing, additional carrying channels should be provided at 48 in. (1220 mm) on centers and the main runners should then be supported from the carrying channels. The first hanger on each end of each runner or carrying channel should be no more than 6 in. (150 mm) from the ends.

COMPOSITION ACOUSTICAL TILE INSTALLATION

Tiles should be placed over the entire grid area and cut to fit around fixtures, penetrating hangers, and accessories. Border units should be cut and fit neatly and snugly into the grid, leaving no voids in the finished ceiling.

Splines and other grid members are placed into the kerfed edges of tiles, or tile tongues are inserted into tile grooves so that every joint is fitted tightly and properly supported. The tile field should be held in compression with leaf-type spring steel spacers between the tiles and the edge moldings. Spacers should be placed 12 in. (305 mm) apart.

When a grid system is not used, tiles are held in place with fasteners, adhesive, or both. Before adhesive application, loose dust should be removed from the backs of the tiles and they should be primed with a thin coat of adhesive. Surfaces to receive tiles should be sound, free of foreign materials, clean, and dry. New plaster and concrete should be cured thoroughly. Bare plaster, gypsum board, and concrete should be primed. The alkalinity of surfaces to receive adhesive-applied tiles should be tested with litmus or Hydrian pH paper. Surfaces with pH in excess of 9 should have a neutralizer applied to lower the pH to 8 or less.

Tiles are installed with adhesives by applying spots of adhesive to each tile and pressing the tiles in place with a sliding motion. The thickness of the adhesive after application should be about ⅛ in. (3.2 mm).

Splines are installed in the joints between tiles to keep them in proper alignment and flush with each other. Tiles are scribed to fit at penetrations and edges.

When tiles are fastened in place with fasteners or adhesives, it is necessary to provide access doors in the ceiling where mechanical, electrical, or other equipment must be accessible. Such panels are usually recessed so that they can be faced with tiles like those installed in the ceiling.

COMPOSITION ACOUSTICAL PANEL INSTALLATION

Lay-in panels should be placed into each opening of the exposed grid system. Panels should be cut neatly around fixtures, diffusers, penetrating hangers, accessories, and other items. Border units should be cut to fit snugly into the grid, with no cracks or voids.

METAL ACOUSTICAL MATERIALS INSTALLATION

Linear metal ceiling strips are snapped into supporting grid members to cover the entire ceiling.

Lay-in metal panels and pans are laid into each opening in the grid.

Snap-in panels and pans are placed in each opening in the grid and shoved into place so that the holding feature fully engages and the panels or pans are held securely in place. Where a ceiling is to be nonaccessible, security clips are installed as the tile or pans are placed.

SOUND INSULATION

Sound insulation blankets may be installed over the full area of an acoustical ceiling or only in certain locations, such as close to sound-rated partitions. Unless the ceiling is specifically designed to support it, sound insulation should be separately supported so that its weight does not rest on the acoustical materials or any part of the suspension system.

In metal ceilings, sound insulation pads should fit snugly into each pan, panel, or strip over the entire ceiling area.

Insulation should be tightly butted, with joints in solid contact and without voids. It should be fitted neatly to adjacent surfaces.

ACOUSTICAL SEALANT

Where light leaks, sound leaks, and air movement at the edges of acoustical ceilings are not acceptable, a continuous bead of acoustical sealant is applied on the concealed leg of the ceiling edge molding before it is installed.

9.8.2 INTEGRATED CEILINGS

Integrated ceilings consist of acoustical materials, a suspension system, outlets (grilles, vents, or vanes) for air distribution, and electrical devices, such as lights, combined into a single system, the components of which are compatible visually and mechanically. For the most part, statements made earlier in this section about acoustical ceiling materials and suspension systems and their installation apply as well to materials used in integrated ceilings.

9.8.2.1 System Components

ACOUSTICAL MATERIALS

The acoustical materials used in integrated ceilings include the composition and metal acoustical materials listed in ASTM E 1264 and discussed in Section 9.8.1.1. Besides the types of acoustical materials specifically listed in ASTM E 1264, the other types of materials mentioned earlier that may be used in acoustical ceilings may also be used in integrated ceilings.

SUPPORT SYSTEMS

Unlike suspension systems for normal acoustical ceilings, those for integrated ceilings are usually designed to support all components of the ceiling, including such devices as light fixtures, and the grid members of an exposed grid integrated system are often designed to provide air supply or return, or both.

Grids for integrated ceilings may be either concealed or exposed. In linear metal integrated ceilings and in metal snap-in systems, the supporting members are usually not exposed.

AIR DIFFUSION

Using grid members that are designed to contain the air supply and return outlets of the heating, ventilating, and air conditioning system can result in effective concealment of diffusers and grills. Some integrated systems employ air supply and return slots that cannot be discerned as such from floor level.

Integrated ceilings employ boots or manifolds to connect the integral air supply and return elements to the building's ductwork. Connections from the ceiling to the ductwork are usually flexible.

LIGHTING

Spaces with acoustical ceilings are often lit by standard fixtures not specifically designed for the purpose, which are mounted on the surface of the ceiling or are recessed into the ceiling. Recessed fixtures in acoustical ceilings are often standard units laid into the suspension grid. Some integrated ceilings also use standard recessed or surface-mounted fixtures, but many of them include light fixtures that have been designed specifically for the particular integrated ceiling system.

Lighting design is usually an integral part of the design of an integrated ceiling. Fixture locations and types are sometimes limited, but many integrated ceiling designs result in effective and comfortable lighting. Care must be exercised in some cases, especially where standard fixtures are used, to prevent glare.

ACOUSTICAL PERFORMANCE

Integrated ceilings are often used in open-plan spaces. Such installations are sometimes accompanied by *white noise* generators to mask background noises.

FIRE PERFORMANCE

The fire-resistance classification of an integrated ceiling system must take into account the electrical and mechanical components of the system. When an integrated ceiling must be time-rated for fire resistance, the limitations may be severe. Light fixture and mechanical opening sizes and locations may be restricted. The method of enclosing such devices will probably be dictated by the rating designation. The appearance of the installed system may be affected drastically by such limitations. A fire-rated integrated ceiling may not look much like a nonrated system using the same basic components.

9.8.2.2 Installation

Unlike some acoustical ceilings, integrated ceilings are seldom fastened directly to wood or metal framing or furring. Most integrated ceiling installations, almost by definition, include some sort of metal suspension system.

Suspension systems for integrated ceiling systems, for example, may be either direct-hung or indirect-hung. Grids may be either concealed or exposed, depending on the particular ceiling system. In every case, the installation should follow the integrated ceiling manufacturer's recommendations.

Light fixtures and air distribution components should be correctly installed so that their functions are properly carried out. Their installation must be coordinated with the mechanical and electrical system installations.

9.8.3 ACOUSTICAL WALL PANELS AND BAFFLES

Acoustical wall panels include standard and custom *spline-mounted* and *back-mounted* panels. Baffles are either double or single thickness. Wall panels and baffles are faced with standard, custom, or owner-furnished facing materials.

9.8.3.1 Panel and Baffle Materials and Fabrication

Most acoustical wall panels and baffles consist of a fabric or vinyl facing over mineral, glass, or wood fiberboard. Materials and fabrication may vary, however, between spline-mounted and back-mounted panels.

Metal acoustical wall panels and baffles are also available.

SPLINE-MOUNTED PANELS

The core of most spline-mounted panels is faced on one side only, because the panels are mounted with butted joints. The finish is wrapped around the long edges back to the line of the kerf, however. The short panel ends are unfinished.

Panel Cores The cores of spline-mounted panels are:

- Perforated water-felted mineral fiberboard with a fairly high density, such as 17 to 20 lb./cu. ft. (272 to 320 kg/m^3)
- Low-density mineral fiberboard
- Glass fiberboard with a nominal density between 6 and 8 lb./cu. ft. (96.1 and 128.1 kg/m^3)
- High-density wood fiberboard

The long edges of spline-mounted panels are kerfed and rabbeted to receive mounting splines (Fig. 9.8-15).

Accessories Mounting and trim accessories for spline-mounted panels include H- and J-shaped splines and J-, H-, and angle-shaped moldings, as shown in Figure 9.8-16. Some manufacturers' accessories may vary from those shown. Accessories may be either extruded aluminum or plastic.

Wood battens are sometimes used to cover the joints and sometimes to support panels. Wood battens are usually decora-

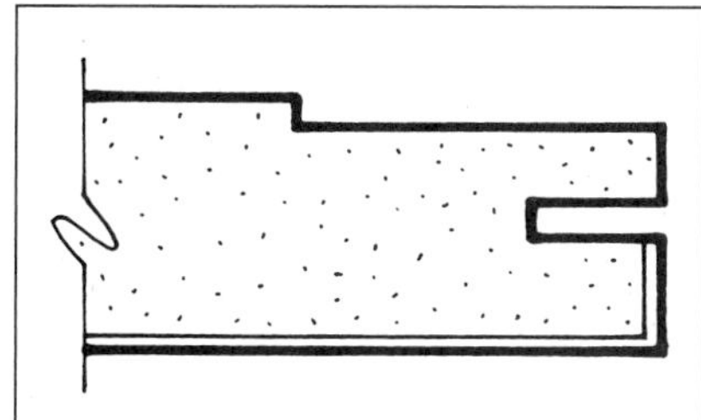

FIGURE 9.8-15 Typical edge detail of a spline-mounted acoustical wall panel. (Courtesy H. Leslie Simmons.)

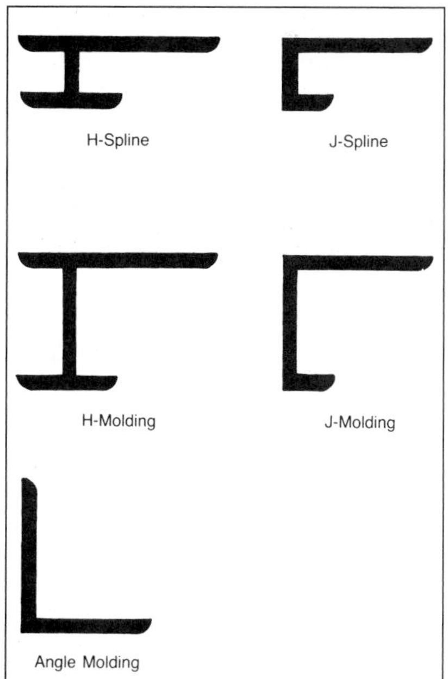

FIGURE 9.8-16 Some typical accessories for spline-mounted acoustical wall panels. (Courtesy H. Leslie Simmons.)

tive hardwood, but may be painted softwood.

Panel Size and Thickness Spline-mounted panels are either 9 or 10 ft. (2743 or 3048 mm) long and 24, 30, or 48 in. (610, 762, or 1220 mm) wide. High-density fiberboard panels vary in thickness from ½ in. (12.7 mm) to 1⅛ in. (28.6 mm). Glass fiberboard and low-density fiberboard panels vary in nominal thickness from 1 in. (25.4 mm) to 1⅛ in. (28.6 mm).

BACK-MOUNTED PANELS

The core face that will be exposed after the panel is installed and the edges of back-mounted panels are usually finished.

Panel Cores The core of back-mounted panels may be glass fiberboard with a nominal density between 4 and 7 lb./cu. ft. (64 and 112 kg/m^3) or high-density wood fiberboards. Glass fiber panels may have impact-resistant faces, which usually consist of a ⅛-in. (3.2 mm)-thick layer of 18-lb./cu. ft. (288 kg/m^3)-density glass fiberboard laminated to the face of the core board.

The edges of the cores of back-mounted panels are either chemically hardened to reinforce the edge and help prevent warpage and damage, or framed with aluminum, zinc-coated steel, or wood.

Panel edges that are not framed are usually wrapped with the facing material. Edge shapes of panels with wrapped edges may be square, beveled, radiused, notched, mitered, or a combination of these (Fig. 9.8-17).

Panel corners may be either square or rounded.

Some panels are faced on the backside with a thin aluminum foil to prevent air passage.

Accessories Mounting accessories for back-mounted panels include:

- Conventional Z-clips
- Strip-clips (Z-clips mounted on a continuous strip)
- Magnetic pads (Fig. 9.8-18)
- Other proprietary clip assemblies
- Hook-and-loop panels (Velcro)
- Adhesives
- Screws (sometimes used to mount high-density wood fiber panels)
- Wood battens (sometimes used to cover joints between panels and sometimes to hold panels in place)

Panel Size and Thickness Back-mounted panels vary in thickness from ½ in. (12.7 mm) to 3 in. (76.2 mm). Panels are available in a large variety of sizes, from 15 in. (381 mm) square up to 60 by 144 in. (13.24 by 3657 mm).

METAL WALL PANELS

Some of the metal ceiling products discussed earlier in this section are sometimes used on walls, but here we will discuss only products that are specifically designed for use on walls. Such panels are available in both aluminum and galvanized steel.

Metal wall panels are available flat or

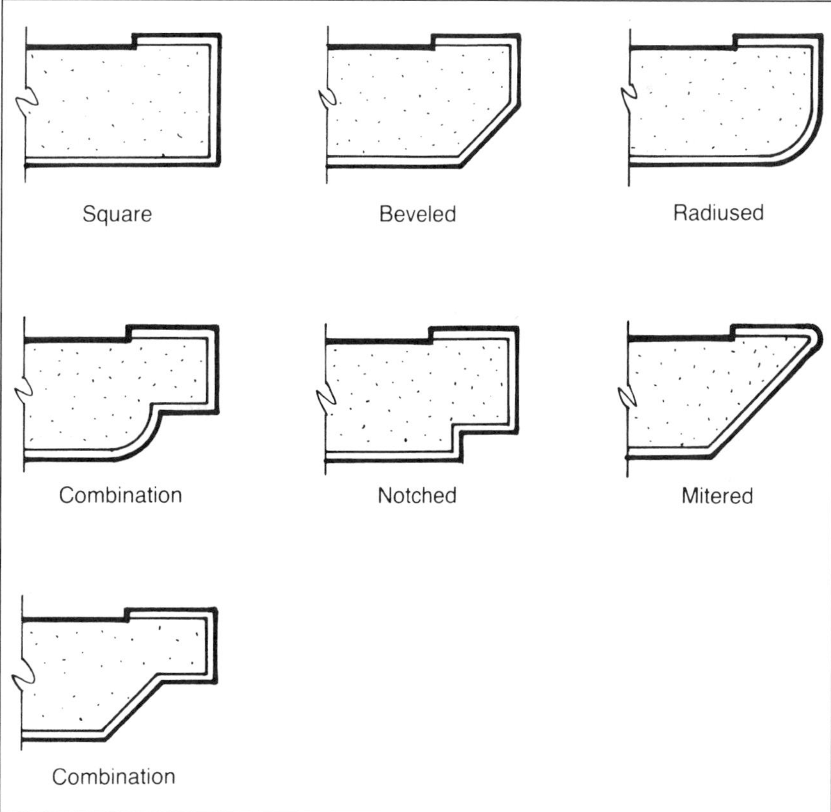

FIGURE 9.8-17 Some typical edge profiles for back-mounted acoustical wall panels. (Courtesy H. Leslie Simmons.)

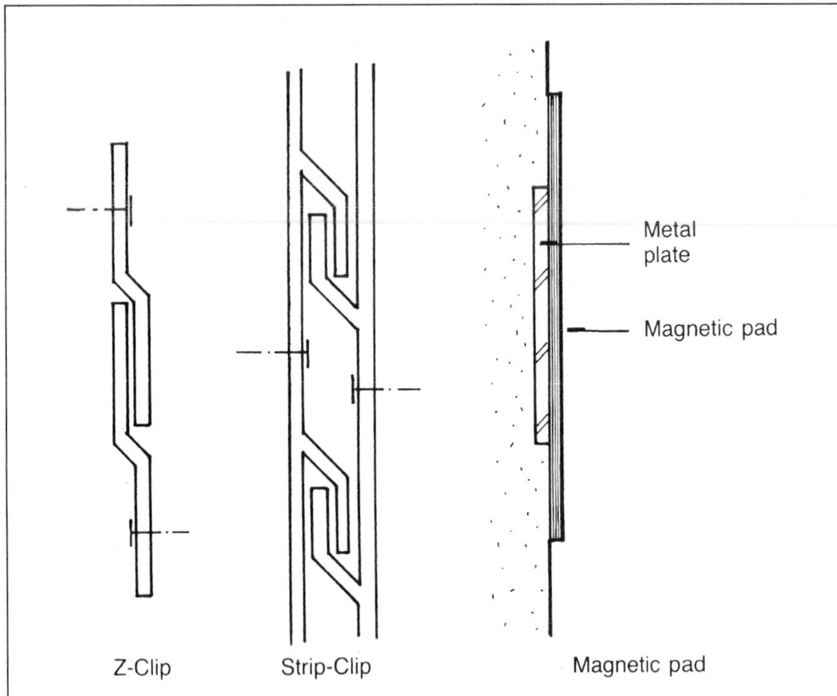

FIGURE 9.8-18 **Typical mounting accessories for back-mounted acoustical wall panels.** (Courtesy H. Leslie Simmons.)

shaped (corrugated, for example), perforated or not perforated, and plain, embossed, or textured. Most panels come with glass fiber or mineral wool sound-absorbing pads. Some pads are wrapped or faced with materials such as polyvinyl chloride. Others are bare.

Special brackets and other accessories are required to install some metal acoustical wall panels.

Most metal wall panel sizes vary from 18 to 48 in. (457 to 1220 mm) wide and from 18 to 144 in. (457 to 3658 mm) long, but some may be smaller or larger.

BAFFLES

A baffle is often nothing more than a standard acoustical wall panel of the back-mounted type that is finished on both faces and all edges. Some baffles are the equivalent of two acoustical wall panels laminated back to back. Some double-thickness baffles are completely encased with finish material. Others have a joint between the two sides of the baffle. Most baffles are flat panels, but cylindrical, octagonal, triangular, and other-shaped baffles are also available. Their basic panel construction is similar to that of flat baffles.

Baffle edges may be finished in the same fabric or vinyl as the face, or framed in wood or metal. Edge configurations are similar to those for back-mounted acoustical wall panels.

Metal baffles may be flat panels similar to the metal acoustical ceiling materials discussed earlier, or more like the metal wall panels discussed in a preceding subsection.

Baffles are made to be mounted or suspended by special hangers either built into or attached to the baffle. Some baffle hangers are nothing more than eye bolts mounted in the narrow edge of the panel or grommets through the panel face. Others use more elaborate hanger systems, which vary from manufacturer to manufacturer. Hangers may be cable, wire, decorative chain, or almost invisible nylon fishing line. Sometimes hangers are solid metal bars. Some baffles are designed to be clipped to overhead framing.

WALL PANEL AND BAFFLE FINISHES

Fabric Standard panels and baffles are available with both woven and nonwoven synthetic fabrics, the majority of which are largely or entirely polyester. Other fabrics are also used. Some manufacturers state that almost any upholstery-weight fabric can be used, but fire-resistance requirements may not permit the use of just any fabric. Some sources state that some wool, nylon, metallic, rayon, and acetate fabrics are not suitable.

Vinyl Perforated vinyl is often used as a facing for acoustical wall panels, especially in spaces such as classrooms, corridors, and the like, where the material is likely to become soiled quickly. It is also sometimes used for facing on baffles. In older installations the vinyl may have had large holes punched in it to provide the proper acoustical performance. Vinyl in newer installations will probably have been punched using a process called microperforating, which eliminates large holes.

Metal Metal panels designed for wall panel or baffle use are often painted with either acrylic or polyester baked-on paint. Other finishes may be available.

Other Finishes More utilitarian baffles may be faced or wrapped with polyethylene or glass fiber cloth.

ACOUSTICAL PERFORMANCE

Acoustical wall panels and baffles are generally classified by their Noise Reduction Coefficient (NRC). Panels and baffles fall into two general categories. The first includes panels that will absorb between 60% and 70% (NRC 0.60 to 0.70) of the sound that strikes them. They are used in normal-sized rooms, such as conference rooms and offices. The second category includes panels that absorb between 80% and 95% (NRC 0.80 to 0.95) of the sound that strikes them. They are used in large spaces, such as open-plan offices, and spaces where high noise levels are generated, such as computer rooms, swimming pools, and gymnasiums (Fig. 9.8-19).

9.8.3.2 Support Systems

Spline-mounted acoustical wall panels may be directly attached to supported substrates or supported by wood or metal furring applied over wood or metal studs or solid substrates. Back-mounted acoustical wall panels are usually directly attached to supported substrates with one of the accessories discussed later in this section. Metal acoustical wall panels may be directly attached to solid or supported substrates, or supported by metal furring applied over studs or solid substrates.

WOOD FURRING

Spline-mounted acoustical wall panels and metal acoustical wall panels are sometimes installed over horizontal wood fur-

FIGURE 9.8-19 Acoustical baffles in an arched ceiling in a large armory. (Photograph provided by MBI Products Company.)

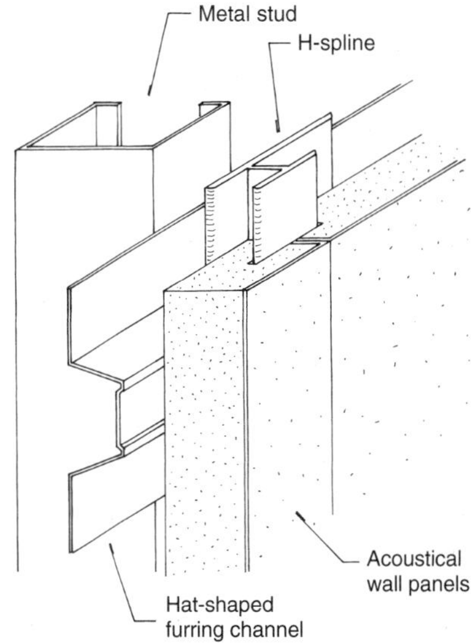

FIGURE 9.8-21 Mounting detail for spline-mounted acoustical wall panels over metal studs. Similar details may be used when the wall panels are installed over solid substrates that are irregular. (Courtesy H. Leslie Simmons.)

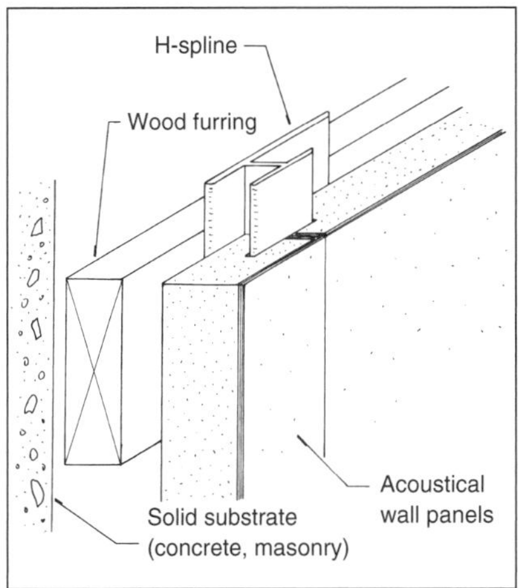

FIGURE 9.8-20 Mounting detail for spline-mounted acoustical wall panels over a solid substrate. (Courtesy H. Leslie Simmons.)

ring (Fig. 9.8-20). Such wood furring is most often used when the substrates are solid ones, such as concrete or masonry, and the substrates are too irregular to directly apply the splines and accomplish a finished surface within acceptable tolerances. Wood furring may also be used over existing wood framing that is not in proper alignment to receive acoustical wall panels, but such use is rare in new buildings. Wood furring may also be used over wood framing for support of acoustical baffles when the framing is not at the proper spacings or locations for the baffle hangers.

Back-mounted acoustical wall panels are usually mounted directly onto a supported substrate, such as plaster or gypsum board. Where wood furring is required in such installations, it is used to fur the supported substrate and not the back-mounted acoustical panels.

METAL FURRING

Spline-mounted acoustical wall panels and metal acoustical wall panels are sometimes installed over horizontal hat-shaped galvanized steel furring (Fig. 9.8-21). Such metal furring is most often used when the substrates are solid ones, such as concrete or masonry, and the substrates are too irregular to directly apply the splines and accomplish a finished surface within acceptable tolerances. Metal furring may also be used over existing metal framing, such as studs, that is not in proper alignment to receive acoustical wall panels, but such use is rare in new buildings. Galvanized or painted steel channel furring may also be used over wood or metal framing for support of acoustical baffles when the framing is not at the proper spacings or locations for the baffle hangers.

Back-mounted acoustical wall panels are usually mounted directly onto a supported substrate, such as plaster or gypsum board. Where metal furring is required in such installations it is used to fur the supported substrate and not the back-mounted acoustical panels.

Metal furring is usually screwed, clipped, or wired to the supporting structure.

9.8.3.3 Fire Performance

Acoustical wall panels and baffles should be tested for flame spread and smoke-developed ratings as complete panels containing the same components they will have when installed. They should rate as Class A according to ASTM Standard E 84. The flame-spread rating of a Class A material ranges from 0 to 25. When the individual components are tested separately, it is difficult to determine a composite value for the entire panel.

Since not all manufacturers have had their acoustical wall panels or baffles tested by independent testing laboratories, such as UL, it is often necessary to obtain specific approval of a particular panel or baffle by local authorities before using it.

9.8.3.4 Installation

Acoustical wall panels should be coordinated with electrical outlets and switches, thermostats, and other wall-mounted equipment and devices to ensure that such items are properly accommodated.

Manufacturers' brochures and installation details should be obtained and followed. The manufacturer's recommendations for cleaning and refinishing wall panels and baffles should be obtained and stored in a safe place so that they will be available when maintenance and repair are necessary in the future. They should include precautions against materials and methods that may be detrimental to finishes and acoustical efficiency.

Extra material is occasionally left at the site for the owner to use in making repairs.

Acoustical wall panels and baffles should be kept dry before and after use. They should be allowed to acclimate to the temperature and humidity of the space before they are installed to prevent dimensional changes that may affect the appearance of the installed panels or baffles.

Watermarked, stained, or otherwise damaged units should not be installed.

Acoustical wall panels and baffles should not be installed until the building is weathertight, glazing has been completed, and exterior openings have been closed.

Wet work, such as masonry, concrete, and plaster, should be allowed to dry to the moisture content recommended by the acoustical wall panel or baffle manufacturer before wall panels or baffles are installed. A temperature of at least 60°F (15.6°C) should be maintained in the space during and after installation.

Furring members should be isolated from building structural elements to prevent transfer of loads into the furring. Provisions should also be made to accommodate movement that occurs across control and expansion joints in the substrates. Furring or panels should not span such joints.

Flat surfaces of furring members and supported substrates should not be warped, bowed, or out of plumb or level by more than ⅛ in. (3.2 mm) in 12 ft. (3658 mm) in both directions, or by more than 1/16 in. (1.6 mm) across any joint or ⅛ in. (3.2 mm) total along any single member.

Acoustical wall units should be handled carefully to prevent damage to edges and faces.

Hangers should be properly designed and sufficiently strong to support the baffles. Where splaying of hangers is necessary to avoid ducts, pipes, conduit, or other items, the forces so generated should be compensated for by counter-splaying or another suitable means.

FURRING INSTALLATION

Furring should be placed on lines and levels necessary to cause the acoustical panels to fall into the proper location and to properly support baffle hangers at the correct spacings. Where the furring will support directly applied panels or baffles, it should correct unevenness in the supporting structure or substrate. The surface of wood furring to which acoustical panel splines will be fastened should be not less than 1½ in. (38 mm) wide. Generally, 2 × 2s (50 × 50s) are used for furring that is attached to the supporting structure and 1 × 2s (25 × 50s) or 1 × 3s (25 × 75s) are used where furring is laid directly over solid substrates. Nominal 1-in. (25.4 mm)-thick lumber is too flexible to be used over framing. Furring to support baffles should be the proper size to span between the framing members without undue deflection in the furring.

Over a solid substrate, furring should be installed over a layer of polyethylene film or other vapor-retarding material as an airflow barrier and vapor retarder.

Furring should form a complete system, adequate to properly support the acoustical wall panels or baffles. Closers should be installed at edges and openings.

Bolts, lag screws, and other anchors are used to anchor wood furring in place. Generally, fasteners should be placed near the ends of furring members and not more than 36 in. (900 mm) on centers between. Shorter members, however, should be anchored at 30 in. (750 mm) on centers. Bolts should have nuts and washers.

Screws, bolts, clips, wires, and other anchors should be used to anchor metal furring in place. Generally, fasteners should be placed near the ends of furring members and not more than 24 in. (600 mm) on centers between, but some types of metal furring members may require closer spacings. The correct spacing should be verified with the furring member manufacturer.

Furring spacings may vary with the substrate and the acoustical panels or baffles being supported. Minimum spacings for wall panels are dictated by the panel thickness and size and may vary somewhat from manufacturer to manufacturer and system to system. Spacing must be such as to adequately support the panels. Spacing of furring for baffle hangers should be such that the baffle hangers fall in the correct locations.

SPLINE-MOUNTED PANELS

The usual method of mounting spline-mounted acoustical wall panels is to place splines vertically and fasten them to a supported substrate such as plaster or gypsum board (Fig. 9.8-22). Then the panels are slipped in place so as to engage the splines in the kerfed and rabbeted panel edges. Spline-mounted panels can also, however, be mounted by fastening the splines to wood or metal furring over masonry, concrete, or another solid substrate (see Figs. 9.8-20 and 9.8-21).

Spline-mounted acoustical wall panels may rest directly on the floor or be supported at the bottom by a base retainer such as a J- or angle-shaped molding (see Fig. 9.8-16).

Panels should not extend into the ceiling plenum. Sometimes the panels are stopped short of the ceiling, leaving a decorative reveal.

When panels must be cut for fitting or other reasons, the panel backs should be resealed with heavy mastic tape or foil to

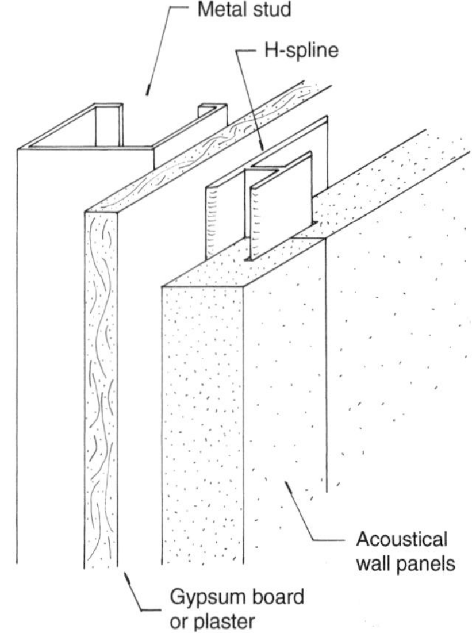

FIGURE 9.8-22 Mounting detail for spline-mounted acoustical wall panels over a supported substrate. (Courtesy H. Leslie Simmons.)

prevent airflow through the panel. Cut edges should be covered by a molding or wrapped with fabric.

Exposed panel edges should be covered with fabric or a molding.

BACK-MOUNTED PANELS

Back-mounted panels are usually mounted directly on a supported substrate such as plaster or gypsum board. Most older panels and many newer panels will have been mounted with a conventional Z-clip, a continuous Z-clip, or a similar proprietary device. Some will have been mounted with adhesives; others will be supported by magnetic pads. Some will have been screw-attached or supported by battens. Many modern installations are supported with hook-and-loop (Velcro) strips.

METAL PANELS

Many metal panels are installed with clips or screws. Some are installed with accessories similar to the moldings shown in Figure 9.8-16 or the clips and magnetic pads shown in Figure 9.8-18.

BAFFLES

Most baffles are suspended vertically by wire, cables, chains, bars, or fishing line from framing or furring. They should be installed in line and vertically or at the proper angle according to the design. Baffles that are suspended in other configurations should be installed according to the design. In every case, hangers should be securely fastened to the supports and to the baffles.

SOUND INSULATION

Where they are used, sound insulation blankets should be installed over the full area occupied by the acoustical wall panels.

Sound insulation pads in metal wall panels should be fitted snugly into each pan, panel, or strip over the entire area of the wall panel installation.

Insulation should be tightly butted, with joints in solid contact and without voids. It should be fitted neatly to adjacent surfaces.

9.9 Wood Flooring

The widespread use of wood flooring can be attributed to its distinctive natural appearance, excellent wearing qualities, moderate cost, underfoot comfort, and ease of installation and maintenance. It can be used in almost all rooms above grade. Improved methods of protecting wood from dampness have also encouraged its use in rooms below grade as well and even in residential kitchens.

Because of its ready availability and the relative ease of working it with hand tools, early flooring was made mostly of softwood. This flooring was often in the form of thick planks $3\frac{1}{2}$ to 10 in. (89 to 254 mm) wide, just as they came from the sawmill. Subsequently, improvements in manufacturing and seasoning methods brought about the development of carefully machined, narrower hardwood strips ($3\frac{1}{4}$ in. [82.6 mm] and narrower), which were easier to install.

Today, hardwood strip flooring is the most popular type (Fig. 9.9-1a), but demand for floors simulating early colonial random-width plank installations has encouraged the production of plank products (Fig. 9.9-1b). Various parquet (pattern) types, such as blocks and uniform-length strips, are also available (Fig. 9.9-2).

Factory-finished (*prefinished*) flooring, suitable for service immediately after installation, is available in strip, plank, and block and in a variety of woods.

A characteristic property of wood that must be given special attention in flooring installations is its tendency to shrink and swell as its moisture content changes. Through modern research and manufacturing methods, products with improved dimensional stability have been developed, but products vary in their sensitivity to moisture and should be selected with a view to the conditions of use.

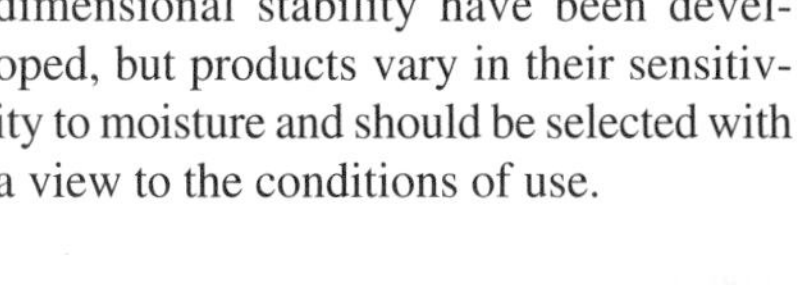

a

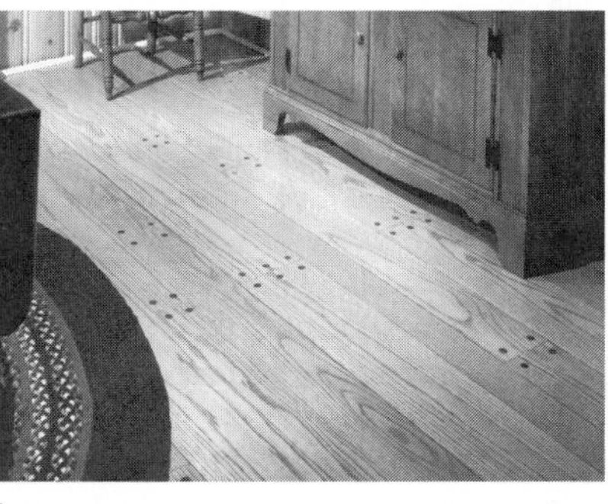

b

FIGURE 9.9-1 Contemporary and traditional interiors are complemented by (a) strip and (b) plank wood floors. (Courtesy Bruce Hardwood Floors.)

a

b

FIGURE 9.9-2 Patterned (parquet) floors may be created with (a) laminated plywood blocks or (b) solid wood slats either preassembled into blocks or individually laid in mastic. (Courtesy Bruce Hardwood Floors.)

9.9.1 MATERIALS

The terms *hardwood* and *softwood*, popularly applied to the two major groups of trees cut for lumber, actually have no bearing on the degree of hardness of the wood. In fact, many softwoods are harder than some hardwoods. These terms are used primarily to distinguish the botanical characteristics of the trees. Arbitrarily, trees having broad leaves are known as hardwoods, and coniferous trees—those bearing needles and cones—are known as softwoods.

About 12 types of woods are regularly manufactured into flooring. Of these, the hardwoods account for about 80%. Their greater popularity can be attributed to their appearance and, in the species used, substantially greater hardness and wear resistance.

9.9.1.1 Hardwoods

Various commercial species of oak constitute more than 90% of all hardwood flooring produced in a typical year. Maple accounts for about 6%. The rest is beech, birch, pecan, and several other hardwoods in limited quantities.

OAK

There are about 20 species of oak in the United States that are considered commercially important in lumber production. Of these, about half are classed as *red oak* and half as *white oak*. As growing trees, the several species within each group are readily distinguishable, but in lumber form the differences are fairly inconspicuous. Therefore, precise separation of the various species within each group of oaks is impractical and unnecessary in flooring manufacture.

Oak is lumbered throughout the southern, eastern, and central states, in forests of the Atlantic Plain and the Appalachian Mountains. In these regions, species of oak grow under a wide range of climatic conditions in many different kinds of soil. There is, accordingly, much variation in the color of the wood, especially the heartwood. The sapwood shades from white to cream color in all species of oak. In the standard grading rules for oak flooring, color is entirely disregarded except in the amount of light-colored sapwood allowed. Sapwood is limited only in *clear grade*, the top grade of flooring.

Red oak and white oak are about equal in mechanical properties. Both make satisfactory floors when properly finished. A special feature of white oak is the prominence of large rays that make an interesting flake pattern in quarter-sawn flooring.

Although grading rules do not differentiate between red oak and white oak, practically all manufacturers supply either all-red or all-white oak flooring except in the lowest grades. Red oak flooring is generally higher in price and more uniform in color than white oak.

MAPLE

Maple flooring is made from sugar maple, logged largely in the Northeast, the Appalachians, and the lake states (Minnesota, Michigan, and Wisconsin). Sugar maple is known as *hard maple* or *rock maple*. It is strong, hard, and abrasion resistant, making it particularly suitable for hard use locations, such as factories and gymnasiums, as well as residences.

The so-called soft maples (silver maple, red maple, and bigleaf maple) are not as hard, heavy, or strong as hard maple and therefore are not used commonly for flooring.

The heartwood of both sugar maple and black maple is light reddish-brown, and the sapwood, which in mature trees is several inches thick, is creamy white, slightly tinged with brown. The contrast in color between heartwood and sapwood in maple is much less pronounced than in oak, and the standard grading rules permit natural color variation in the wood.

BEECH AND BIRCH

Beech and birch are lumbered in the northeastern part of the country and around the Great Lakes. Beech and birch are used only sparingly in the manufacture of flooring. Only two of the almost 20 species of birch that grow in the United States are manufactured into flooring. Of these, *yellow birch* is by far the most abundant and the most important commercially; the other is *sweet birch*. Only one species of beech is native to the United States.

The heartwood of these species is reddish-brown, with a slight variation in color among them. There are similar slight variations in the color of their sapwood, which is of a lighter shade than the heartwood. The varying color of the natural wood is an accepted characteristic in grading beech and birch flooring.

9.9.1.2 Softwoods

In a typical year, more than 50% of the softwood flooring produced is southern pine, more than 40% is Douglas fir, and the rest is western hemlock, eastern white pine, ponderosa pine, western larch, eastern hemlock, redwood, spruce, cypress, and the true firs. Western larch is similar to Douglas fir in strength and is often sold in mixture with Douglas fir of the northern interior region of the western states. Ponderosa pine, eastern white pine, and redwood are softer than desirable where wear is a prime factor. However, the formidable decay resistance of redwood in the all-heartwood grade has prompted its use for porch and deck flooring.

SOUTHERN PINE

Southern pine is a commercial name applied to a group of yellow pines that grow principally in the southeastern states. They include longleaf, shortleaf, loblolly, slash pines, and several other pines of minor importance. Except in dimension lumber and structural timbers, no differentiation between the species is made in marketing the products of this group.

The wood of all southern pines is much alike in appearance. The sapwood and heartwood are often different in color, the former being yellowish-white and the latter a reddish-brown. However, the contrast in color between sapwood and heartwood in southern pine generally is not conspicuous in a finished floor, and the standard grading rules permit sapwood in all grades of southern pine flooring.

When uniform color is desired, the standard flooring specifications can be amended to require all-sap-face stock (for a light color) or all-heart-face material (for a reddish-brown color). However, special selection of stock for color increases the cost somewhat over that of the established grade.

DOUGLAS FIR

Red fir, yellow fir, coast Douglas fir, and Oregon pine are other names by which Douglas fir is known in the western part of the United States and Canada, where it grows. Douglas fir occupies the same important position in the western and Pacific Coast states that southern pine does in the southeastern states.

The sapwood of Douglas fir is creamy white. The heartwood is reddish-brown, and as in southern pine, the contrast in color between the two is not so pronounced as to be objectionable in a finished floor. Pieces containing both heartwood and sapwood are permitted in all grades.

WEST COAST HEMLOCK

Western hemlock grows along the Pacific Coast from northern California to Alaska and as far inland as northern Idaho and northwestern Montana. The bulk of hemlock lumber (West Coast hemlock) comes from Oregon and Washington. Both the heartwood and sapwood of western hemlock are almost white, with a pinkish tinge and very little contrast, although the sapwood is sometimes lighter in color.

Western hemlock has clear color and good finishing qualities, which account for its use in moderate-wear areas, such as bedrooms, where good appearance is the principal requirement. Western hemlock flooring is relatively free from warping and is easy to cut and nail but is not as hard and wear resistant as Douglas fir or larch.

9.9.1.3 Supports

Generally, the joist spacing to support board subflooring should not exceed:

- For ¾-in. (19.1 mm) strip or plank flooring laid parallel to joists: 16 in. (400 mm) on centers
- For parquet flooring: 16 in. (400 mm) on centers
- For ¾-in. finish flooring laid at right angles to joists: 24 in. (600 mm) on centers

9.9.2 MANUFACTURE

Wood flooring is made in three basic styles: *strip*, *plank*, and *block* (Fig. 9.9-3). The following discussion of manufacture is limited to strips and planks, as blocks may be made in many different ways. Methods of manufacture of hardwood flooring differ from those of softwood flooring.

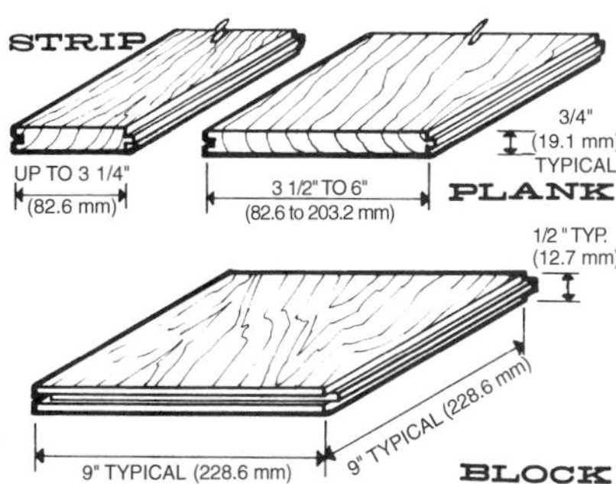

FIGURE 9.9-3 Basic wood flooring types. (Courtesy Bruce Hardwood Floors.)

9.9.2.1 Hardwoods

Hardwood flooring production begins at a sawmill, where logs are rough-sawn into boards, planks, and timbers. Each log is cut to produce the best grade of lumber in each thickness. Thicker pieces containing the highest percentage of clear lumber are reserved for furniture and millwork.

The common nominal thickness of lumber for flooring is 1 in. (25.4 mm), from which the popular ¾-in. (19.1 mm) strips and planks are milled. After air-drying in a yard for several months, rough-sawn lumber is kiln-dried to reduce the moisture content to a range of 6% to 9%. Sawmill operations are described and illustrated in Section 6.2.1.2.

SIZING AND DRESSING

Rough-sawn lumber is often air-dried at a sawmill, but kiln-drying is normally performed at a flooring plant. After cooling for at least 24 hours, the lumber is ready to enter a flooring mill. The first step consists of cutting the boards into the desired widths with a ripsaw (Fig. 9.9- 4). Lumber that is to be tongue-and-grooved (T&G) is cut wider than the face width of the flooring to compensate for the tongue edge and for the material taken off in planing the edges. In some flooring mills, the pieces are then run through a planing machine, which dresses (smooths) the top and bottom surfaces. Then the pieces proceed on conveyor belts to cut-off saws (Fig. 9.9-5), where they are cut to eliminate natural defects that are undesirable in finish flooring. The pieces emerge in random lengths. Then these strips are conveyed to a machine known as a side-matcher.

FIGURE 9.9-4 Rough-sawn lumber is cut lengthwise in a ripsaw. (Courtesy Bruce Hardwood Floors.)

SIDE AND END MATCHING

A side-matching machine smooths the top surface accurately, cuts one or two channels in the bottom side, a tongue on one edge, and a groove in the other (Fig. 9.9-6). The pieces then continue on a belt conveyor to end-matching machines, where one end of each strip is grooved and the other is tongued.

GRADING AND BUNDLING

Next, the pieces go to grading tables, where they are separated by grade, color, and length (Fig. 9.9-7). Each piece is trade- and grademarked, and the pieces of each grade are separated and bundled according to length in multiples of 1 ft. (304.8 mm). An allowance of 6 in. (152.4 mm) under and 6 in. (152.4 mm) over the nominal length is permitted in each bundle, so that a 4-ft. (1219 mm) bundle may contain pieces 3 ft. 6 in. (1067 mm) long to 4 ft. 6 in. (1372 mm) long. However, the most popular flooring size, ¾ in. (19.1

FIGURE 9.9-5 Natural defects are eliminated on a cut-off saw. (Courtesy Bruce Hardwood Floors.)

FIGURE 9.9-6 The top face is smoothed and the sides are tongued-and-grooved on a side-matching machine. (Courtesy Bruce Hardwood Floors.)

FIGURE 9.9-7 Each piece is carefully inspected and graded before it is packaged. (Courtesy Bruce Hardwood Floors.)

mm) by 2¼ in. (57.2 mm), is also bundled in an assortment called *nested flooring*, which includes various lengths ranging from 9 to 102 in. (228.6 to 2591 mm).

Prefinished flooring pieces undergo further treatment, including fine sanding (Fig. 9.9-8), finishing, and final waxing and polishing. Bundles are then shipped to distributors in sealed boxcars or trucks that afford protection from the weather and moisture.

9.9.2.2 Softwoods

Softwood flooring production begins at a sawmill, where green (unseasoned) logs are rough-sawn. The resulting lumber is then seasoned to meet the moisture requirements of the grading rules. Kiln-dried flooring requires a maximum of 12% moisture content. The actual moisture content at the time of installation usually averages about 9% or 10%.

After flooring has been seasoned, it goes directly to a planing mill, where it is manufactured to the desired profile in one operation. Square-end flooring is then trimmed and packaged in equal lengths, six to a bundle. End-matched flooring bundles may contain pieces 1 ft. (304.8 m) to 8 ft. (2438 mm) in length, laid end to end and stacked in one or two tiers of six courses each.

9.9.2.3 Standards of Quality

Of the three basic types of flooring produced, strip flooring and block flooring are regulated by industry-wide specifications. There are no generally accepted standards for plank flooring, but most manufacturers market products in appearance grades similar to those established for strip flooring. The standard for block flooring is American National Standards Institute/American Parquet Association, ANSI/APA 1-1984, "Mosaic-Parquet Hardwood Slat Flooring."

Strip flooring is manufactured to comply with the grading rules of the particular trade association governing the type of wood involved. The National Oak Flooring Manufacturers Association (NOFMA) promulgates and regulates the production of oak flooring through its *Official Flooring Grading Rules*. Since some NOFMA members also produce maple, beech, birch, and pecan flooring, this association also promulgates grading rules for these woods. However, the majority of maple, beech, and birch flooring is produced by members of the Maple Flooring Manufacturers Association (MFMA), which promulgates its own grading rules for these species.

9.9.2.4 Certification of Quality

Strip flooring must be kiln dried, grademarked, and trademarked in order to comply with NOFMA and MFMA grading rules. The two trade associations operate certification programs based on inspections of their members' manufacturing facilities and products, and only qualifying members are permitted to imprint the trademark of the association. This mark is found on the back of each piece of wood and provides visual assurance of conformance with the grading rules. In addition to the grademarks and trademarks, the imprint often contains the mill number and other identifying information.

Softwood strip flooring is manufactured to conform to the grading rules of the Southern Pine Inspection Bureau (SPIB), the Western Wood Products Association (WWPA), and the West Coast Lumber Inspection Bureau (WCLIB). These associations do not require trademarking or grademarking under the grading rules.

9.9.3 STRIP FLOORING

Hardwood flooring is manufactured commonly to a standard pattern (Fig. 9.9-9),

FIGURE 9.9-8 **Disc sanders impart a satin-smooth finish to pieces that will be factory finished.** (Courtesy Bruce Hardwood Floors.)

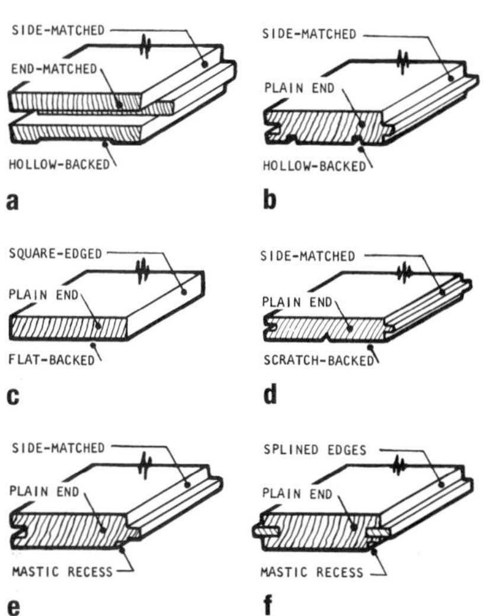

FIGURE 9.9-11 **Flooring patterns: (a, b, c, and d) for installation with nails or screws, (e and f) for mastic installation.** (Courtesy Bruce Hardwood Floors.)

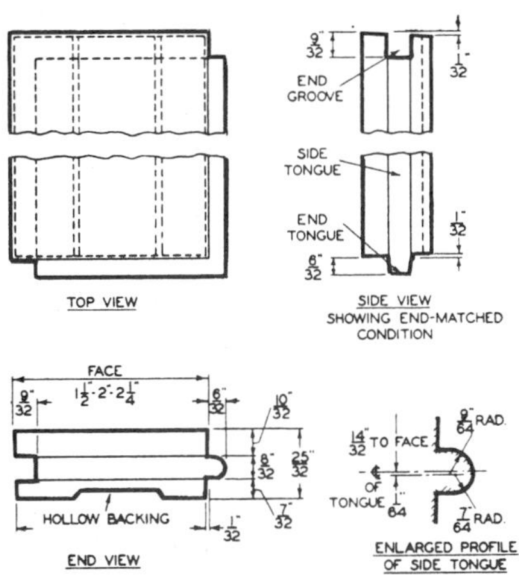

FIGURE 9.9-9 **Detail dimensions of typical ("standard pattern") ¾-in. (19.1 mm) T&G flooring.** (Courtesy Bruce Hardwood Floors.)

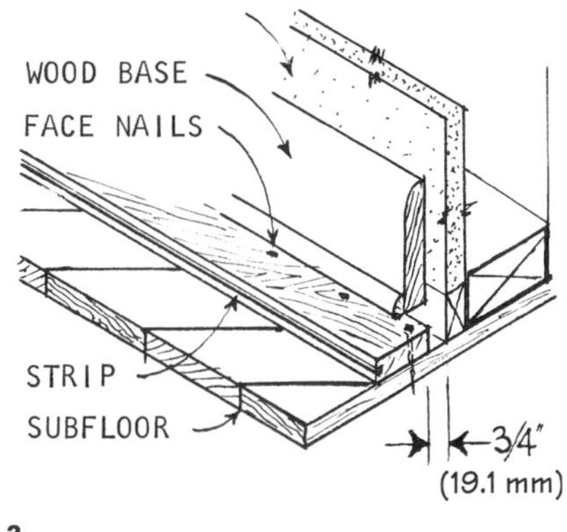

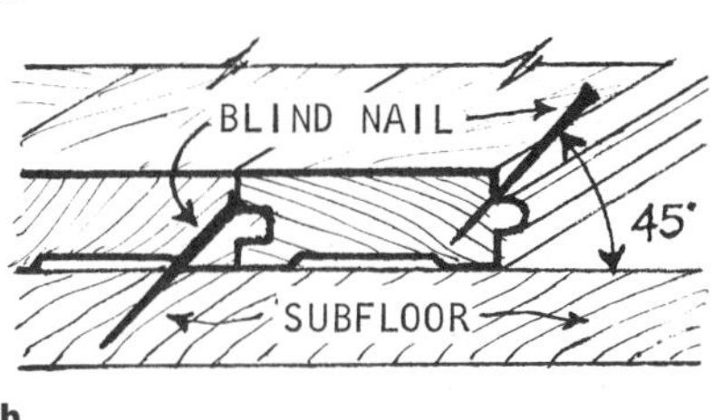

FIGURE 9.9-10 **Strip flooring application: (a) first and last strips are face-nailed; (b) others are blind-nailed.** (Courtesy Bruce Hardwood Floors.)

producing flooring that is side-matched and end-matched. The top face is generally slightly wider than the bottom, so that when the strips are driven tightly together, the upper edges make contact but the lower edges are slightly separated. Standard T&G strip flooring is usually installed over wood subfloors or wood sleepers by *blind* (concealed) nailing at the intersection of the tongue and shoulder (Fig. 9.9-10).

Most strip flooring is either hollow-backed (Fig. 9.9-11a and b) or scratch-backed (Fig. 9.9-11d). The most widely used standard pattern is ¾ in. (19.1 mm) thick and has a face width of 2¼ in. (57.2 mm), but other widths and thicknesses are available. The strips are random length, and the proportion of short pieces depends on the grade.

Other strip flooring, which is used to a limited degree, has square edges and a flat back (Fig. 9.9-11c). It is generally thinner than standard flooring and is usually installed by face nailing. The nails are driven and set so the matching filler used in finishing will fill and conceal the holes.

While standard strip flooring is generally installed by blind nailing to wood subfloors or sleepers (Fig. 9.9-12), uniform short lengths of T&G flooring (*single slat* flooring) can also be installed with *mastic* (paste-like adhesive). Products intended for mastic installation generally have flat backs for greater contact with the adhesive and have a recess milled into the lower surface below the tongue to reduce the possibility of mastic being forced up through the joint (Fig. 9.9-11e and f). Flooring intended for mastic installation may have T&G edges or may be grooved for insertion of a hardwood or metal spline.

Softwood flooring is either end-matched (Fig. 9.9-11a) or has plain ends (Fig. 9.9-11e). It may have a wide hollow back like ¾-in. (19.1 mm) standard-pattern hardwood flooring (Fig. 9.9-11a), a double groove like the thinner standard-pattern flooring (Fig. 9.9-11b), or a single V-shaped groove in the back (Fig. 9.9-11d).

Standard flooring grades are based almost wholly on appearance. They exclude or severely limit such defects as knots, wormholes, and the like in the higher grades and permit increasing sizes and numbers of these characteristics in the lower grades. Natural variations in color generally are not limited, except that in the clear grade of oak the amount of light-colored sapwood is restricted.

Traditionally, hardwood flooring has been packaged and sold in bundles containing pieces of more or less the same length, 6 in. (150 mm) shorter or longer

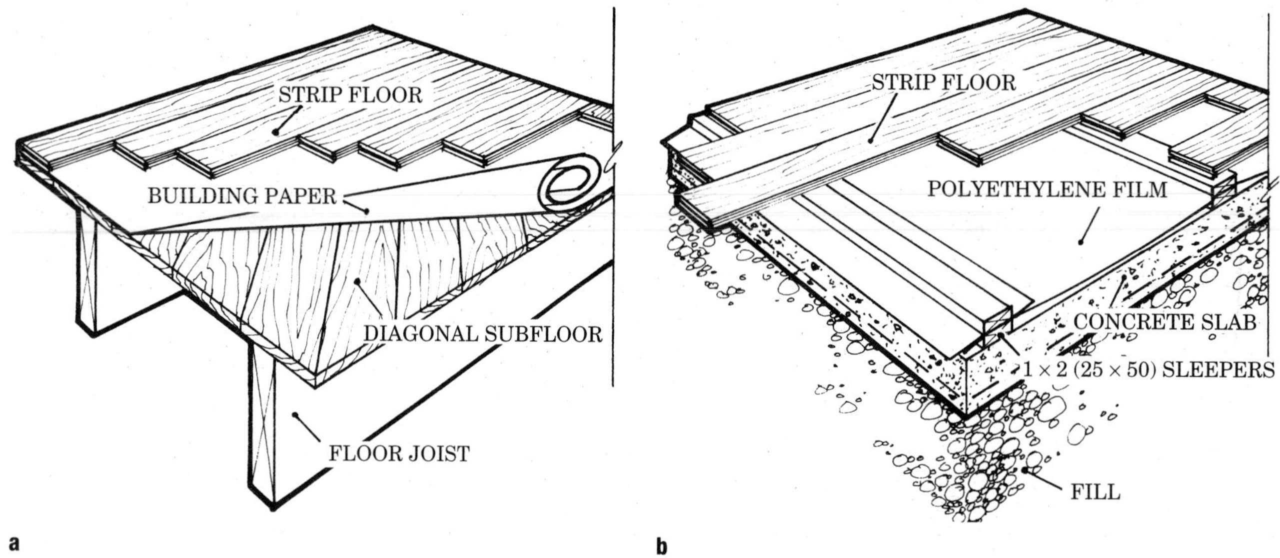

FIGURE 9.9-12 Typical installation methods: (a) wood subfloor and (b) concrete slab; 1 × 2 (25 × 50) sleepers can also be used over a wood subfloor. Polyethylene film and ¾-in. (19.1 mm) plywood can also be installed over a slab as a subfloor. (Courtesy Bruce Hardwood Floors.)

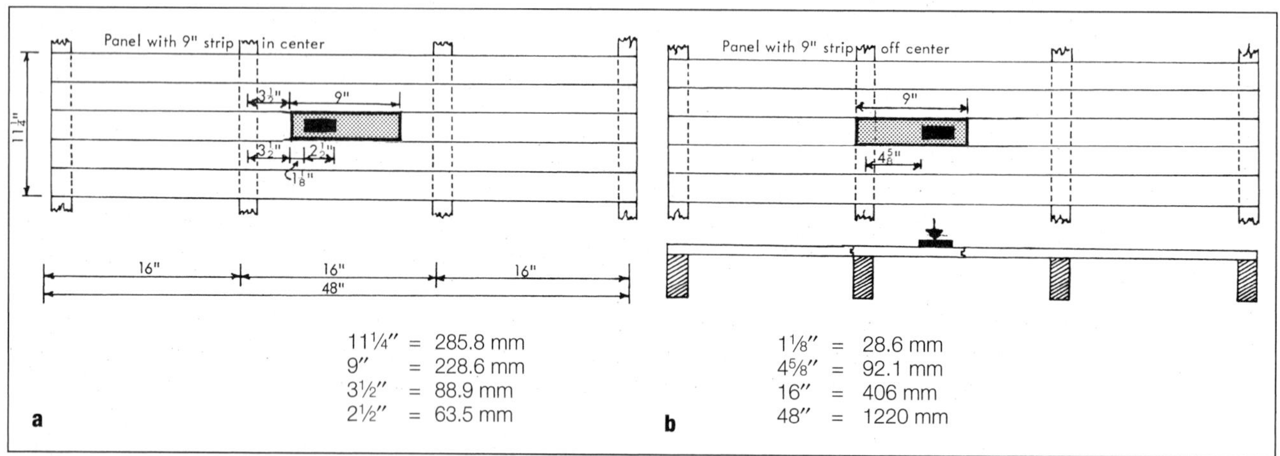

FIGURE 9.9-13 Short pieces of nested flooring (shaded), resting entirely on T&Gs of adjacent strips (a), supported a concentrated load of 845 lb. (383.28 kg), while pieces partly resting on a joist (b) supported more than 915 lb. (415 kg).

than the nominal length. The shortest pieces permitted per bundle and the percentage of short lengths permitted in each grade are regulated by established grading rules of NOFMA and MFMA.

NOFMA grading rules include pieces ¾ in. (19.1 mm) by 1½ in. (38 mm), by 2 in. (50.8 mm), by 2¼ in. (57.2 mm), and by 3¼ in. (82.6 mm) *nested flooring*, which is laid end-to-end continuously to make either 7-ft. (2134 mm)-long or shorter, or 8-ft. (2438 mm)-long bundles. A nested 8-ft. (2438 mm) bundle is four tiers high and three courses wide, or 6 tiers high by two courses wide, each course being 7½ ft. (2286 mm) to 8½ ft. (2591 mm) long. Individual pieces may range from 9 in. (228.6 mm) to 102 in. (2591 mm), but both the proportion of short pieces (9 to 18 in. [228.6 to 457.2 mm]) and the minimum average length are regulated by the grading rules and depend on the grade and species.

The short pieces included in nested flooring are not detrimental to an installation. Experiments have demonstrated that even if the subflooring were omitted and a 9-in. (228.6 mm) piece of strip flooring were placed so that no part of it rested on a support (Fig. 9.9-13), it could support almost 2½ times the concentrated load under the leg of a concert grand piano. This represents an adequate safety factor for what may be assumed to be the most critical loading condition in most uses. Installations over subfloors, or those involving longer pieces, produce lower stresses and certainly are adequate.

9.9.3.1 Oak

Oak is regularly manufactured into plain-sawn and quarter-sawn flooring (Fig. 9.9-14). Most oak flooring is plain-sawn, the lower-priced of the two. Quarter-sawn oak is characterized by a striking grain and flake pattern and by greater dimensional stability, which minimizes shrinking and swelling in width. Oak flooring is graded under the rules of the National Oak Flooring Manufacturers Association (NOFMA).

The most common oak flooring thickness is ¾ in. (19.1 mm). Thinner flooring is available for use over existing floors

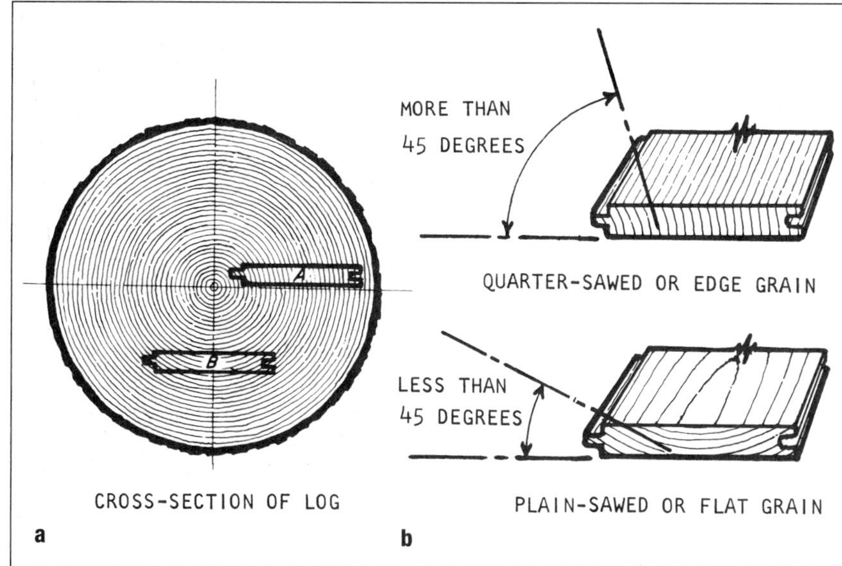

FIGURE 9.9-14 The angle the annual rings make with the surface determines whether wood is (a) quarter-sawn (edge-grained) or (b) plain-sawn (flat-grained).

and for light-service conditions. Flooring thicker than ¾ in. (19.1 mm) is intended for heavy-service conditions.

Although most oak strip flooring is end- and side-matched, some square-edged strip flooring is made. This is intended primarily for industrial, institutional, and commercial use to facilitate replacement of damaged areas, which would be more difficult with tongue-and-groove flooring. Prefinished oak flooring is made in four grades: Prime, Standard, Standard & Better, and Tavern. It is available in all-red and all-white oak selections. Prime grade is available on special order only.

9.9.3.2 Maple, Beech, and Birch

Maple, beech, and birch flooring is graded both under the rules of the Maple Flooring Manufacturers Association (MFMA) and under the rules of NOFMA. Although the grade names are sometimes slightly different, the requirements under the two sets of rules are similar. Both generally disregard normal color variations.

Because these woods are very dense and hard in plain-sawn as well as quarter-sawn stock, the grading rules make no special requirements in this respect. However, quarter-sawn material can be obtained when specified. MFMA defines edge-grain (quarter-sawn) stock as pieces whose annual rings form an angle of more than 30 degrees with the face. This definition is slightly more permissive than the standard definition requiring an angle of 45 degrees or more (Fig. 9.9-14).

9.9.3.3 Southern Pine

Southern pine is graded under the rules of the Southern Pine Inspection Bureau (SPIB). It may be specified in side- and end-matched flooring and in square-end flooring that is side-matched only. Regardless of the pattern to which it is worked, all southern pine flooring is available in both *flat-grain* and *edge-grain* stock. However, edge-grain material is in limited supply, available through special order only, and may be hard to obtain. Edge-grain material in this species must have at least six annual rings per inch across the face, in addition to the standard requirement of grain direction of more than 45 degrees.

A class of southern pine flooring intermediate between edge-grain and flat-grain is known as *near-rift* flooring. This material has annual rings that form an angle of less than 45 degrees with the face, but the rings are so close to this position that each piece shows at least six annual rings anywhere across the face.

9.9.3.4 Douglas Fir and Other Western Softwood Species

With the exception of the B & Better Grade, which is manufactured only from *vertical-grain* material, Douglas fir flooring in the upper grades is regularly manufactured in both vertical-grain (VG) and flat-grain (FG) stock (see Fig. 9.9-14).

The term *vertical-grain* used in these grading rules is almost synonymous with *edge-grain* used for southern pine. However, Douglas fir Grade C requires four rings per inch and only B & Better Grade requires at least six rings per inch (25.4 mm) of face width, as does southern pine. Vertical-grain Douglas fir flooring is harder and possesses better wearing qualities than flat-grain material. Douglas fir flooring is generally T&G but not end-matched. Flooring of this wood is graded under the rules of the West Coast Lumber Inspection Bureau (WCLIB) and the Western Wood Products Association (WWPA).

9.9.4 PLANK FLOORING

Flooring boards wider than 3¼ in. (82.6 mm) are referred to as *planks*. The term *strips* is usually reserved for narrower pieces. However, many random plank installations include both plank- and strip-sized boards, and manufacturers often refer to such assortments as plank flooring. Strip and plank floors are installed in an end-to-end random pattern, with end joints staggered, using nails or screws.

Plank flooring is available end- and side-matched in the same thicknesses as strip flooring. Most plank flooring is white or red oak ¾ in. (19.1 mm) thick and 3½ to 8 in. (88.9 to 203.2 mm) wide. It is available prefinished in dark or light shades and graded according to NOFMA rules for prefinished flooring.

Some plank flooring is designed to simulate the appearance of the random-width planks commonly used in colonial era buildings. These rustic floors had a distinctive appearance resulting from the hardwood pegs used to secure the planks to the subfloor.

Contemporary plank flooring is installed with wood screws in the face of each plank. These screws are then covered with wood plugs to simulate pegs. Narrower, strip-sized boards, which can be secured by blind nailing, sometimes have decorative pegs inserted during their manufacture and are sold as plank flooring.

Because the cross-grain dimension is greater in plank flooring than in strip flooring, it is affected more by variations in atmospheric humidity. To avoid unsightly loose joints, plank edges are often eased or beveled to accentuate rather than conceal the joints.

Plank flooring is sometimes made in three-ply laminated form, with a center cross-ply. This form is more dimensionally stable in width than the solid form, making it possible to install such plank floors with tight joints. Plank flooring is generally sold by the square foot, rather than by the board foot as strip flooring is sold.

9.9.5 PARQUET FLOORING

Parquet (pattern) floors consist of individual strips of wood (*single slats*) or larger units (*blocks*) installed to form a decorative geometric pattern (Fig. 9.9-15).

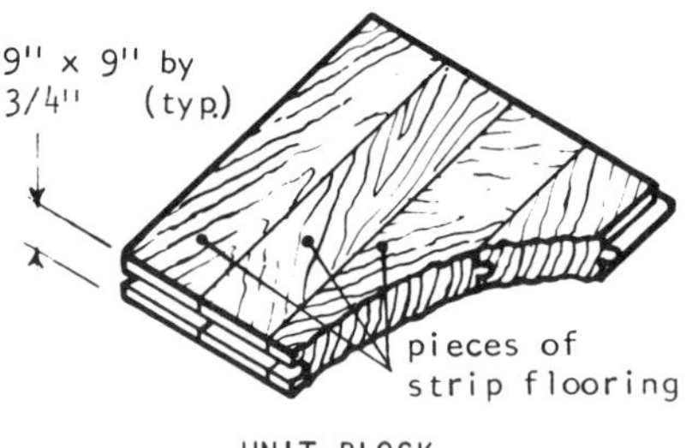

a

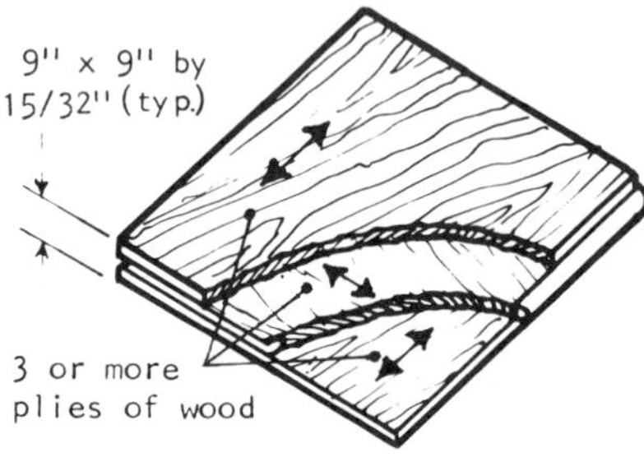

b

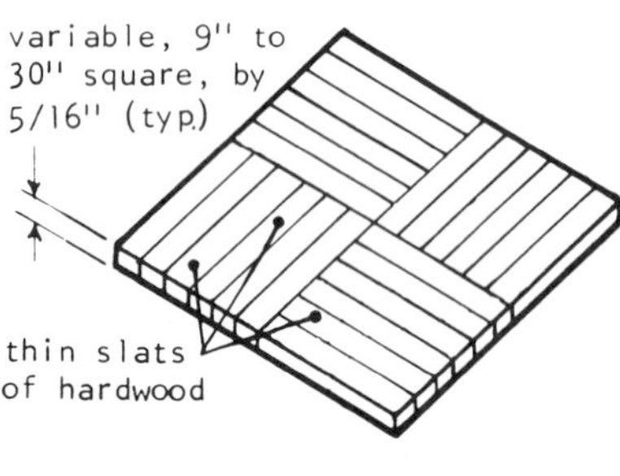

c

9″ × 9″ by 3/4″ = 228.6 × 228.6 × 19.1 mm

9″ × 9″ by 5/32″ = 228.6 × 228.6 × 3.99 mm

9″ = 228.6 mm

30″ = 762 mm

5/16″ = 7.9 mm

FIGURE 9.9-15 Typical construction and dimensions of basic block flooring forms: (a) unit block, (b) laminated block, (c) slat block. Slat blocks often consist of 16 or more basic squares.

Blocks may be made by laminating several hardwood veneers or by gluing a number of solid hardwood pieces into a unit to facilitate installation. Unlike strip and plank flooring, parquet flooring is usually installed with mastic.

Oak is by far the most predominant species used in all types of pattern flooring, but other species, such as maple, walnut, cherry, and East Indian teak, also are available. Sometimes a mixture of hardwoods, such as hickory, ash, elm, pecan, sycamore, beech, and hackberry, are used at random in a single block.

9.9.5.1 Block Flooring

Block flooring is manufactured in several basic types. In *unit block* (also known as *solid unit block*), short lengths of strip flooring are joined together edgewise to form square units (Fig. 9.9-15a). In *laminated* (plywood) block, three or more plies of veneer are bonded with adhesive to obtain the desired thickness (Fig. 9.9-15b). *Slat block* (sometimes called *mosaic parquet hardwood slat*) flooring utilizes narrow slats or "fingers" of wood preassembled into larger units to facilitate installation (Fig. 9.9-15c).

Most unit and laminated block flooring is tongued on two adjoining or opposing edges and grooved on the other two to ensure alignment between adjoining blocks. Some manufacturers produce square-edged blocks, and others include grooves on all four block edges and furnish splines for insertion between adjoining blocks. Both types are designed to be installed with mastic over a wood subfloor or concrete slab. Prefinished blocks usually have eased or beveled edges.

Unit blocks typically consist of several 3/4-in. (19.1 mm) T&G strips, all laid parallel or alternating in each quarter of the block checkerboard fashion (Fig. 9.9-16). Consequently, typical block sizes are multiples of the strip width used (Fig. 9.9-17). Common strip widths are 1 1/2, 2, and 2 1/4 in. (38, 50.8, and 57.2 mm), resulting in typical block dimensions of 6, 6 3/4, 7 1/2, 8, 9, 10, and 11 3/4 in. (152.4, 171.5, 190.5, 203.2, 228.6, 245, and 298.5 mm). The individual strips are held together with wood or metal splines embedded in the lower surface.

LAMINATED BLOCK

This type of block is typically 15/32 in. (11.9 mm) thick in 9-in. (228.6 mm) squares, but other sizes, such as 8 in. (203.2 mm) and 8 1/2 in. (215.9 mm), also are available from some manufacturers. Appearance grades are Prime and Standard.

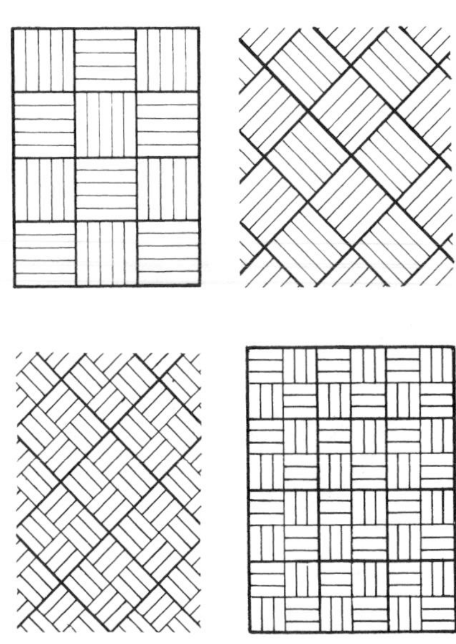

FIGURE 9.9-16 Typical unit block patterns: heavy lines show individual blocks, laid in square or diagonal patterns.

Because of its cross-laminated construction in three to five plies, shrinking and swelling of individual blocks is minimized. This type of wood flooring has good dimensional stability and is often recommended for damp locations, such as slabs on grade.

Adhesives used in the manufacture of laminated block flooring should be capable of resisting the temperature and humidity variations to which the flooring may be subjected. Melamine-urea resin adhesive gives good results at moderate cost and represents the typical adhesive used for most laminated block flooring.

Highly water-resistant adhesives such as phenols, resorcinols, and melamines may also be used if the application warrants it; however, the use of these glues adds to the cost of the product. Adhesives that create high-strength dry bonds but are adversely affected by moisture are not recommended for the manufacture of laminated block flooring.

SLAT BLOCK

Appearance grades are based on grading rules developed by the American National Standards Institute and the American Parquet Association (APA), ANSI/APA 1-1984, "Mosaic-Parquet Hardwood Slat Flooring." APA checks compliance with these rules through periodic inspections

FIGURE 9.9-17 Typical Unit Block Dimensions*

	Strip Thickness, in. (mm)		
Strip Width, in. (mm)	½ in. (12.7 mm)	¾ in. (19.1 mm)	33/32 in. (26.19 mm)
1½ (38)	7½ (190.5) 9 (228.6)	7½ (190.5) 9 (228.6)	9 (228.6)
2 (50.8)	8 (203.2) 10 (254)	8 (203.2) 10 (254)	
2¼ (57.2)		6¾ (171.5) 9 (228.6) 11¼ (285.8)	9 (228.6) 11¼ (285.8)

*All dimensions are for square blocks.

of member plants. These products are suitable for installation in mastic over concrete surfaces, both above and on grade. The basic components of these products are solid, 5/16 in. (7.9 mm) thick, generally square-edged slats of hardwood, ¾ in. (19.1 mm) to 1¼ in. (28.6 mm) wide and 4 to 7 in. (101.6 to 177.8 mm) long, assembled into basic squares. These in turn are factory-assembled checkerboard fashion, with the grain in each adjoining square reversed, into larger flooring blocks up to 30 in. (76.2 mm) long and wide (Fig. 9.9-18).

Several types of slat block flooring are available. Some are assembled into panels held together with a face paper. These are 9½-in. (241.3 mm), 18-in. (457.2 mm), or 19-in. (482.6 mm) squares and are marketed unfinished. Others are made up of single 6-in. (152.4 mm) T&G squares. These are held together by mechanical attachments and are generally factory finished. Still others are assembled into panels held together with a backing material such as asphalt-saturated felt, textile webbing, or another type of felt or nonwoven interfacing. These products generally are 9½, 11, or 12 in. (241.3, 279.4, or 304.8 mm) square; they may be square-edged or grooved and splined, and may be either unfinished or prefinished.

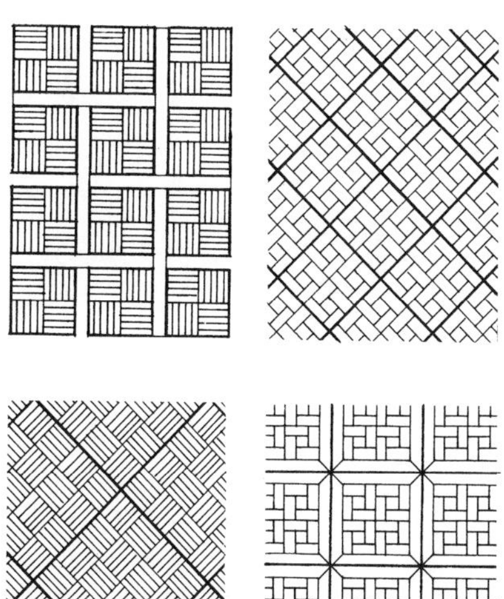

FIGURE 9.9-18 Typical slat block patterns. Pickets forming borders may be part of the block or separate.

9.9.5.2 Single-Slat Parquet

Many decorative motifs can be created with single-slat parquet flooring (Fig. 9.9-19), herringbone being the most common. This type of flooring consists of side- and end-matched strips, machined to uniform lengths ranging from 6 to 18 in. (152.4 to 457.2 mm). Typical thicknesses are ¾ and 33/32 in. (19.1 and 26.2 mm); common face widths are 1½, 2, and 2¼ in. (38, 50.8, and 57.2 mm). Since single-slat parquet flooring is essentially short lengths of strip flooring, it is obtainable in all types of woods, grades, and sizes available in strip flooring.

FIGURE 9.9-19 Popular single-slat patterns achieved with uniform-length precut wood strips.

9.9.6 MATERIALS HANDLING

To maintain proper moisture content, flooring products should not be transported or unloaded in rain, snow, or excessively humid weather. Flooring should not be delivered to a construction site until the building is enclosed, concrete and plaster work has been completed, and all building materials are dry. An interior temperature of 65°F (18.3°C) to 70°F (21°C) should be maintained for at least 5 days before flooring is delivered. Flooring should be stored for several days in the rooms where it will be installed to allow it to become acclimated to local conditions.

In crawl space construction, adequate cross-ventilation should be provided under the floor. The total area of vent openings should equal 1½% of the first-floor area. A ground cover of 4- to 6-mil (0.1 to 0.15 mm) polyethylene film is essential as a moisture retarder. Inadequate moisture control can harm any floor installation by contributing to warping or discoloration of the flooring.

9.9.7 MAINTENANCE

Properly finished wood floors are durable and easy to maintain. It is best to vacuum or dust-mop floors once a week. Wood floors should never be washed or wet mopped, because water seepage between boards can cause staining and warping. Regardless of floor finish type, flooring manufacturers recommend that the finish be protected with a solvent-based liquid buffing wax-cleaner or a paste wax. Spills should be wiped up immediately to prevent damage.

9.10 Dimension Stone Tile Flooring

This section discusses dimension stone used for interior flooring. Most interior flooring stone is either slate, marble, granite, limestone, or soapstone, but some other stones are also used. Basic stone materials and exterior use of dimension stone are described in Section 4.8. It is suggested that the reader review Section 4.8 in conjunction with studying this section.

Stone flooring is available in standard sizes and shapes, depending on the stone material, but stone can be cut to almost any shape using modern cutting tools, including saws and powerful waterjets. Spectacular detail can be achieved in stone projects large and small. Natural stone can be a flexible graphic medium for architectural and artistic expression.

Stone flooring is used today in almost every building type. It occurs in entryways, mudrooms, kitchens, bathrooms, foyers, lounges, dens, offices, retail stores, building lobbies, restaurants, and almost every other type of space (Figs. 9.10-1 and 9.10-2).

All stone types can economically be cut into unique designs. Special inlays are frequently used in lobbies, retail stores, restaurants, foyers, and in other buildings for thematic representations. Designs can be either contemporary or traditional, copies or unique creations. Medallions, insets, border designs, emblems, and logos are easily fabricated with computerized waterjet technology. Also possible are murals, ecclesiastical icons, animal motifs, and representational art. It has been said that "anything that can be drawn and computerized can be realized" (Fig. 9.10-3). Figure 9.10-4 shows a few of the almost unlimited number of possible patterns.

With waterjet technology, metals—ranging from brass to stainless steel—can be combined with marble, granite, and other stones to achieve spectacular results.

9.10.1 MATERIALS

9.10.1.1 General Characteristics

All stone flooring should meet certain standards. The surface abrasion resistance should fall within certain limits. The standard for hardness value, H_a, is ASTM C 241. The Marble Institute of America (MIA) recommends that all floor tile should have a hardness value of at least 10 in residential locations and at least 12 for commercial applications. Different stones in the same area should have hardness values within 5 points of each other.

The MIA does not recommend the use of stone flooring less than 3/4 in. (19.1 mm) thick. This advice is widely ignored, however, and there are many successful applications with thinner tile. There can be

FIGURE 9.10-1 Granite flooring in a lobby. (Photography courtesy of Cold Spring Granite Company.)

FIGURE 9.10-2 **A granite foyer extending into the next room.** (Photography courtesy of Cold Spring Granite Company.)

a problem with thinner tile over substrates that are likely to move, or where a solid mortar bed cannot be achieved, which is sometimes difficult with thin tile. Without a solid bed, tile tends to crack.

The ADA Accessibility Guidelines for Buildings and Facilities, which forms a part of the Americans with Disabilities Act (ADA), encourages the use of flooring materials having a static coefficient of friction of not less than 0.6 for flat surfaces and 0.8 for sloped surfaces. This is also the value required by ASTM D 2047.

9.10.1.2 Slate

Slate has a characteristic cleavage that allows it to be split easily along one plane but not another. Slate is chiefly composed of quartz and illite, with mica, calcite, and minor quantities of other various minerals. Slate has long life and durability. It has high strength and resists chipping, cracking, and abrasion. A few types of slate are capable of withstanding rapid freeze-thaw

FIGURE 9.10-3 **A patterned granite floor showing one of the many designs possible with stone flooring.** (Photography courtesy of Cold Spring Granite Company.)

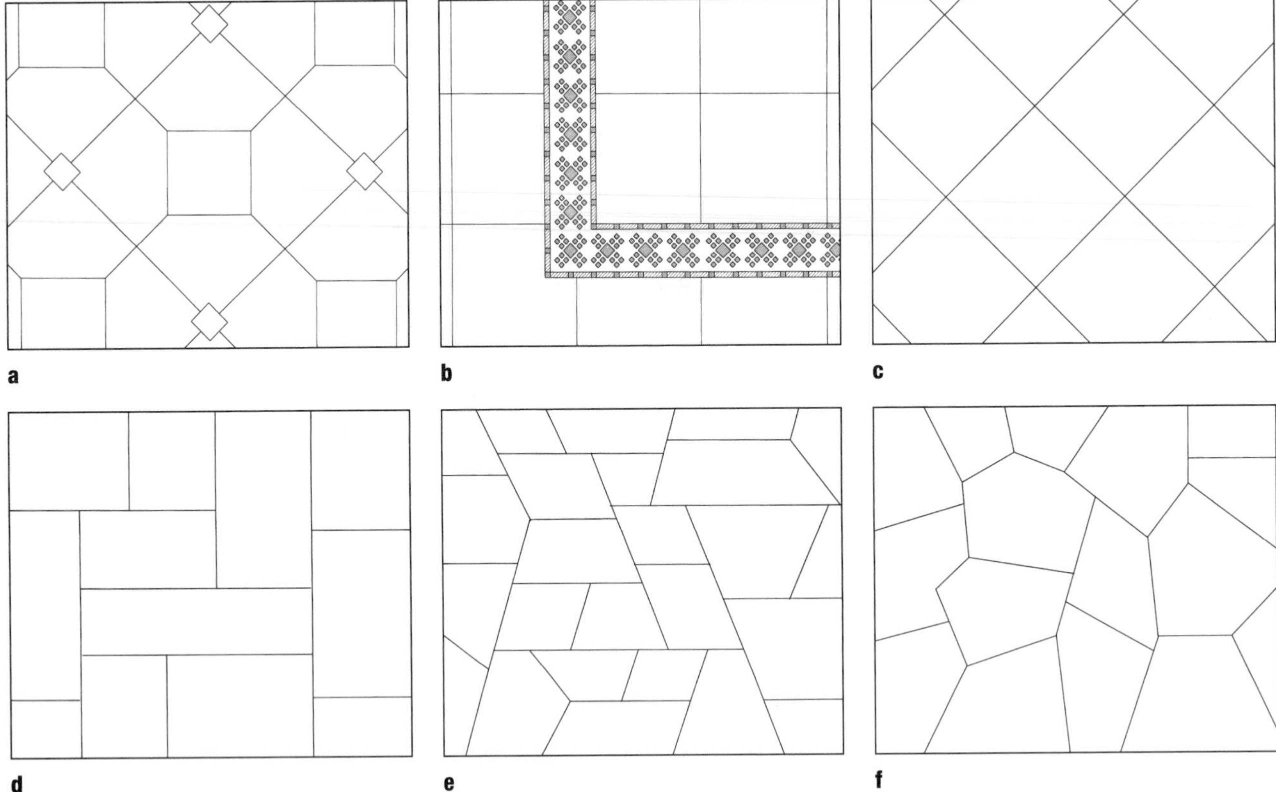

FIGURE 9.10-4 **A few possible flooring patterns: (a) French pattern, (b) decorative border pattern, (c) square pattern, (d) random rectangular pattern, (e) trapezoidal pattern, (f) random shapes pattern.** (Courtesy HLS.)

cycles, but most are not. Slate absorbs only ¼% to 1% when immersed in water for 24 hours. That, together with its dense, compact structure and resistance to chemical acids, gives it a high sanitary value and makes it easy to clean.

Slates come from China, India, Africa, and Brazil, as well as, of course, the United States.

The standard for slate used in flooring is ASTM C 629.

COLOR

Slates vary in color, but are most commonly black, gray, red, purple, green, and multicolored variations of these. A red color is due to the presence of hematite, and green colors are due to chlorite. Gray and black are due to carbon and/or graphite. Shades of red (purple) and tan are related to varying amounts of iron oxides. Slates from a particular area may be available in only one color. Pennsylvania slate, for example, comes only in light or dark gray.

SIZES

Slate tiles, sometimes called "gauge rubbed tiles," are available in slabs and large-format units, but most are stocked only in either 12 × 12 in. (304.8 × 304.8 mm) or 24 × 24 in. (609.6 × 609.6 mm) sizes. In some slates, irregular random shapes are available up to 4 ft. (1.2 m) to a side. Some slates come in sizes from 6 × 6 in. (152.4 × 152.4 mm) up to 12 × 12 in. (304.8 × 304.8 mm) in 3-in. (76.2 mm) increments. Custom sizes are also available. Size range varies with thickness. In some slates, the pieces furnished are roughly rectangular, with about 85% of their edges sawn. Trimming is required to achieve uniform joints and a "broken ice" appearance.

Sawn slate flooring for mortar bed applications are usually rectangles or squares, clean sawn on all four sides. Their nominal sizes range from 6 × 6 in. (152.4 × 152.4 mm) up to 12 × 12 in. (304.8 × 304.8 mm) in 3-in. (76.2 mm) increments in ½-in. (12.7 mm) thicknesses.

The length of stones for stair treads and risers is up to 7 ft. (2.13 m).

Thickness Slate tiles to be set in a thinset bed are ¼ in. (6.4 mm), ⅜ in. (9.5 mm), or ½ in. (12.7 mm) in thickness. For gauged tile, the thickness is usually ¼ in. (6.4 mm). In heavy-use areas, however, thickness is usually increased to ⅜ in. (9.5 mm) for tiles up to 12 × 12 in. (304.8 × 304.8 mm), and to ½ in. (12.7 mm) for larger tiles.

Sawn-slate flooring for mortar bed applications is available in ½-in. (12.7 mm), ⅝-in. (15.9 mm), ¾-in. (19.1 mm), and 1-in. (25.4 mm) thicknesses.

The thickness for stair treads and risers is usually 1 in. (25.4 mm), but 1¼-in. (28.6 mm), 1½-in. (38.1 mm), and 2-in. (50.8 mm) thicknesses are also available.

Weight The nominal weight of slate flooring is 3¾ psi (18.31 kg/m^2) for ¼-in. (6.4 mm)-thick tile, 4¾ psi (23.19 kg/m^2) for ⅜-in. (9.5 mm)-thick tile, 7½ psi (36.62 kg/m^2) for ½-in. (12.7 mm)-thick stone, 11¼ psi (54.93 kg/m^2) for ¾-in. (19.1 mm)-thick stone, and 15 psi (73.24 kg/m^2) for 1-in. (25.4 mm)-thick stone.

Finishes Natural cleft face and bottom are usually specified for interior tile surfaces having random shapes arranged in an irregular pattern. This finish provides a long-wearing, slip-resistant surface. The bottoms should be gauge-rubbed for thinset application.

The finish for sawn slate flooring is

usually a natural cleft surface, but some slates are sand rubbed and others have a machine face or are honed.

The finish for stair treads and risers is usually a sand-rubbed surface, but a honed finish is sometimes used.

9.10.1.3 Granite

Granite floor tile is one of the hardest tile materials available. It offers durability and easy maintenance. It is available in 28 or more colors and several finishes.

The physical characteristics for granite are defined in ASTM C 615.

COLORS

Granite is available in a variety of colors, including white, black, pearl, and pink.

SIZES

Tiles are available in 6 × 6 in. (152.4 × 152.4 mm), 6 × 12 in. (152.4 × 304.8 mm), 12 × 12 in. (304.8 × 304.8 mm), and 18 × 18 in. (457.2 × 457.2 mm) sizes. Special sizes are also available.

Tiles in metric sizes such as 300 mm × 300 mm × 10 mm, and 400 mm × 400 mm × 12 mm, can be ordered, subject to minimum quantities. Custom orders are also available.

THICKNESS

Granite tiles to be set in a thin-set bed are ¼ in. (6.4 mm), ⅜ in. (9.5 mm), ½ in. (12.7 mm), or ⅝ in. (15.9 mm) thick. Standard 12 × 12 in. (304.8 × 304.8 mm) tiles are usually ⅜ in. (9.5 mm) thick; 18 × 18 in. (457.2 × 457.2 mm) tiles are usually ½ in. (12.7 mm) thick. Tolerances are usually less than 1/32 in. (0.8 mm).

WEIGHT

A ⅜-in. (9.5 mm) tile weighs approximately 5.6 psf (27.3 kg/m^2).

FINISHES

Granite finishes include polished, honed, fine-rubbed, thermal, rubbed, sawn, and other special finishes. Tile backs are usually diamond-ground for uniformity in thickness and enhanced bonding.

Granite floors in heavy-traffic areas of shopping malls, airports, and other commercial facilities are typically given a honed finish. Residential floor applications are usually given a polished finish.

9.10.1.4 Marble

On interior and exterior walking surfaces, the quality of marble has been proven many times over. The physical properties of marble give it extraordinary strength and impart a water- and dirtproof surface. Unusually large crystals interlock in such a way as to block the moisture and grime that could otherwise destroy marble. Usually, the more colorful marbles are not as strong as the white varieties.

The physical requirements for marble are defined in ASTM C 503.

COLORS

Marble is available in well over 60 colors, including white, gray, black, white-to-gray with charcoal veining, light gray, dark gray, gray with charcoal veins, pink, pink with charcoal veins, beige, gold, green, brown, and red. Black marble may be streaked with veins of white or gold. Other veining may be red. Veining may be fine and weblike or strong and multicolored. Some marbles have no veining at all, but rather a fine, crystalline texture, with large breccia or small fossils. As a natural product, its colors and textures vary frequently.

SIZES

The standard size of marble tiles is 12 × 12 in. (304.8 × 304.8 mm), but sizes of 6 × 6 in. (152.4 × 152.4 mm), 6 × 12 in. (152.4 × 304.8 mm), 8 × 8 in. (203.2 × 203.2 mm), and 16 × 16 in. (406.4 × 406.4 mm) are also available, as are other custom sizes. The marble dimension stone standard is ASTM C 503.

THICKNESS

The usual marble tile thickness is ⅜ in. (9.5 mm).

FINISH

Standard finishes are *honed* and *polished*. Other finishes include *fine abrasive* and *textured*.

TUMBLED MARBLE

In spite of its appearance, tumbled marble is real marble. It is ideal for use in countertops and fireplace surrounds, but it also can be used as accent pieces in general flooring and as the flooring in small spaces, such as showers.

Tumbled marble is available in a variety of sizes, usually relatively small, such as 2 × 2 in. (50.8 × 50.8 mm), 4 × 4 in. (101.6 × 101.6 mm), and 8 × 8 in. (203.2 × 203.2 mm), but larger sizes, such as 12 × 12 in. (304.8 × 304.8 mm), are also available. The thickness of tumbled marble, regardless of panel size, is ⅜ in. (9.5 mm). It is ideal for mixing with ceramic tile and all types of stone flooring.

Tumbled marble is also available in a variety of precut shapes for use in borders and the like.

9.10.1.5 Limestone

All Indiana limestone meets or exceeds the strength requirements set forth in ASTM C 568 for Type II Dimension Limestone.

When used for flooring, paving, or steps, the abrasion resistance of limestone should be 6 minimum to 17 maximum, when measured according to ASTM C 241.

COLOR

Limestone comes in many colors. Among these are beige, gray, and green.

THICKNESS

Nominal thickness of most limestone flooring tiles is ⅜ in. (9.5 mm).

FINISH

Limestone flooring tile finishes include polished, honed, and special finishes.

9.10.2 INSTALLATION

9.10.2.1 Preparatory Work

Stone flooring should be handled and stored in accordance with the material producer's instructions. Installed stone should not be chipped, broken, or otherwise damaged. It should be stored in the producer's packaging as long as possible to avoid damage.

9.10.2.2 Installation Methods

Dimension stone tiles can be installed thin-set as ceramic tiles on interior surfaces or in accordance with guidelines supplied by the Marble Institute of America, the Terrazzo Tile and Marble Association of Canada, and the Tile Council of America. Complete installation recommendations are available from the manufacturers of the products to be used.

9.10.2.3 Materials

Portland cement mortar, dry-set mortar, latex–portland cement mortar, and epoxy are all acceptable setting materials for stone flooring. However, the material selected should be verified for compatibility by the producer of the stone.

Mortar for interior setting beds should be Type N, ASTM C 270.

APPLICABLE STANDARDS

The following standards are applicable to stone tile installation and provide recognized and accepted technical and installation guidelines:

- American National Standards Institute (ANSI)
  - ANSI A108.1—Thick Bed Mortar Method
  - ANSI A108.4—Thin Bed Mortar Method
- American Society for Testing and Materials (ASTM)—ASTM C 615, "Standard Specification for Granite Dimension Stone"
- International Organization for Standardization (ISO)—*ISO 9002, Quality Systems—Model for Quality Assurance in Production, Installation, and Servicing*
- Tile Council of America—*Handbook for Ceramic Tile Installation*
- Marble Institute of America—*Dimension Stone Design Manual IV*

9.10.2.4 Building Codes

Installation must comply with the requirements of all applicable local, state, and national code jurisdictions.

9.10.2.5 Support Systems

Substrates must be prepared according to stone suppliers' recommendations. Supporting surfaces must be structurally sound, solid, stable, level, plumb, and true. Floors must be free of dust, oil, grease, paint, tar, wax, curing agents, primers, sealers, form release agents and other deleterious substances, and debris. Concrete must be at least 28 days old; 6 weeks is preferable.

Concrete to receive stone flooring should be steel-troweled smooth and then fine-broomed. If deflection will be greater than 1/720 of the spans, the concrete should be finished with a steel trowel and a cleavage membrane should be installed.

Wood, ceramic tiles, and terrazzo, provided they are clean and firmly adhered, can also provide a suitable base for a dimension stone floor.

9.10.3 INSTALLATION

In general, the recommendations of the Tile Council of America for tile setting can be followed when laying stone tile flooring. There is one caveat, however. Always obtain and follow the specific recommendations of the stone provider for materials and methods to be used for each particular type of stone flooring.

Stone flooring should be installed only when the ambient temperature is 50°F (10°C) and remains at that level or above for at least 7 days after installation.

Honed-finish tiles can be laid with narrow joints of ⅛ in. (3.2 mm). Textured tiles should be laid with joints of ¼ in. (6.4 mm) minimum.

9.10.3.1 Installation Methods

TCA-recommended setting methods are shown in figures in Section 9.6, as follows:

- 9.6-31: Concrete slab subject to deflection; portland cement bed; cleavage membrane
- 9.6-32: Concrete slab on grade; portland cement bed
- 9.6-33: Concrete; dry-set and latex–portland cement bed
- 9.6-34: Concrete; epoxy mortar
- 9.6-37: Wood support system; portland cement mortar; cleavage membrane
- 9.6-39: Wood support system; thin-set
- 9.6-40: Wood support system; epoxy mortar

For other possibilities, refer to the publications mentioned earlier in "Applicable Standards."

9.10.3.2 Grouting

Beds must be allowed to set before grouting can begin. Unless recommended otherwise by their producers, honed tiles should be grouted with neat cement and textured tiles with a mixture of 1 part cement to 3 parts sand. Proprietary resin-based and cement-based grouts are also available. Stone set in an epoxy bed are often grouted with an epoxy grout.

9.10.3.3 Sealing

Stone sealers are manufactured specifically to seal and protect stone. Unless the use of a sealer is specifically not recommended by the stone producer, stone flooring should be given a sealer, applied in accordance with the sealer manufacturer's recommendation.

Some slate producers recommend that their stone be sealed. Others say that the use of stone sealer is not required to ensure permanence or durability, but is recommended where a darkening effect or a high-luster appearance is desired.

The application of impregnating coatings on thermal and sandblasted granite tiles can aid in maintenance, but may alter color and texture as well as slip-resistance characteristics. Sample testing on a small and inconspicuous area should always be done before application of a coating on an entire surface.

Despite its appearance, tumbled marble is still just marble and generally needs to have a penetrating sealer applied to it. Any penetrating sealer recommended for marble will work and will leave the stone's appearance unchanged. If a dark tone is desired, a penetrating sealer that contains a color enhancer can be used.

The Indiana Limestone Institute of America suggests that sealers may change the color of limestone or make it shiny. It recommends that sealers be tested on samples of the stone before they are used.

9.10.3.4 Maintenance

Under normal conditions, stone flooring is easily cleaned by regular brushing or vacuuming and washing with warm water and mild detergent, finishing with a chamois leather. Cement and plaster can be removed with a proprietary brick cleaner; paint and stubborn stains can be removed with a proprietary stripper and a scrubbing brush. Cleaning agents should be completely removed with clean water.

9.10.4 OTHER STONE TYPES

Quartz-containing stone and other stone materials are also used in flooring. Some are available from a limited number of quarries, sometimes only a single quarry. Some such stones are selected because of their low cost. Conversely, their uniqueness makes some of these stones very expensive. Some have distinctive colors or graining characteristics. Some are selected because they are local native stone.

In the absence of more detailed data, the weight of such stones can be assumed to be at least 7.4 psf (36.13 kg/m^2).

These materials are generally installed and finished in the same manner as the other stone flooring materials discussed in this section. In addition, the same precautions related to substrate and structural support apply.

9.11 Resilient Flooring

Resilient flooring is a class of flooring products distinguished by resilience and dense, nonabsorbent surfaces. Resilience contributes to indentation resistance and quietness, and the dense surface ensures durability and relative ease of maintenance. A variety of decorative colors, patterns, and textures are available in sheet or tile form.

Availability of self-adhesive tile has made installation of some products easier. No-wax surfaces on both tile and sheet goods have reduced maintenance.

The first resilient flooring material, linoleum, was invented more than 100 years ago in England. After the advent of modern vinyl and vinyl composition materials, linoleum fell into disfavor. Today, it is enjoying a comeback in popularity.

Asphalt tile evolved after World War I from a troweled-on mastic flooring that consisted of a mixture of asphalt, asbestos fibers, and naphtha solvents. Today, asphalt tile has been largely replaced by sheet vinyl and vinyl composition tile products.

Early composition tile products contained asbestos, but since it was discovered to have harmful effects on people, it has been restricted by law and replaced with other, less harmful, fibers. The introduction of vinyl and vinyl composition products, with their excellent wearing properties and decorative potential, substantially increased the range of resilient flooring uses. Today, a variety of resilient flooring products are available for residential and nonresidential construction.

Resilient flooring is used where economical, dense, nonabsorbent wear surfaces are desired. Depending on the product, resilient floor finishes offer a wide range of decorative effects and may provide underfoot comfort, good durability, and relative ease of maintenance. Specific products with enhanced properties are available to suit most functional requirements.

Suitability of a flooring product is related to the product's physical properties and intended use. In addition to cost, ease of maintenance and durability are important considerations in determining suitability. These factors in turn depend on subfloor conditions, proper installation, and maintenance.

9.11.1 FLOORING MATERIALS

Resilient flooring is manufactured in tile and sheet forms, in different thicknesses, and from a variety of ingredients. Tile sizes range from 9 by 9 in. (228.6 by 228.6 mm) up to 20 by 20 in. (508 by 508 mm), depending on material; sheets are available in widths from 4 ft. 6 in. (1350 mm) to 12 ft. (3600 mm), also depending on the material (Fig. 9.11-1). Matching resilient accessories, such as wall base, thresholds, stair treads, and feature strips, are available to complete flooring installations.

Resilient flooring products may be classed according to basic ingredients as (1) *vinyl*, (2) *vinyl composition*, (3) *rubber*, (4) *cork*, and (5) *linoleum*. Many resilient flooring products are manufactured to meet requirements of ASTM standards and federal specifications, which together constitute the industry-recognized standards of quality. In some types, the basic ingredients run the full depth of the flooring; in others, backing material adds desired physical characteristics or reduces cost. Figure 9.11-1 summarizes the ingredients and backing materials, sheet widths, tile sizes, gauges, and applicable standards.

9.11.1.1 Vinyl Composition Tile

Vinyl composition tile is composed of mineral fibers, ground limestone plasticizers, pigments, and polyvinyl chloride (PVC) resin binders. These resins permit light colors and a great variety of designs. Some tile products are available with self-adhesive backs and no-wax surfaces. Older vinyl composition tile contained asbestos fibers, which have been found to cause cancer and other physical problems in humans, especially diseases of the lungs, and are no longer used in resilient flooring products.

ASTM F 1066 classifies vinyl composition tile as either Class 1, which are solid-color tiles, Class 2, in which a pattern extends completely through the tile, or Class 3, in which a pattern is only on the surface of the tile.

Ingredients are mixed under heat and pressure and rolled into blankets to which decorative chips may be added. Additional rolling and calendering produce smooth sheets of desired thickness. After waxing, the sheets are cut into tiles.

9.11.1.2 Vinyl Sheet and Tile

The chief ingredient of vinyl products is polyvinyl chloride (PVC) resin. Other ingredients include mineral fillers, pigments, plasticizers, and stabilizers. Plasticizers provide flexibility; stabilizers fix the mixture to ensure color stability and uniformity. Vinyl products are sometimes referred to as *flexible* vinyl to distinguish them from vinyl composition products, which also are made with PVC resins and are termed *semiflexible* vinyl.

VINYL SHEET

Vinyl sheet products are made with a vinyl wear surface bonded to a backing. The backing may be vinyl, polymer-impregnated mineral fibers, asphalt- or resin-saturated felt, nonfoam plastic, or foamed plastic. Some vinyl products have a layer of vinyl foam bonded either to the backing or between the wear surface and the backing. Backings are classified in ASTM Standard F 1303 as Class A for fibrous formulations; B for nonfoamed plastics; and C for foamed plastics.

In addition to PVC resins, the wear surface of vinyl-surfaced sheets may contain decorative vinyl chips, filler, pigments, and other ingredients. Powdered vinyl resins, fillers, plasticizers, stabilizers, and pigments are mixed, rolled into sheets, and chopped to form these vinyl chips. Vinyl chips of various colors are mixed with additional resins, spread evenly over the backing, and bonded to it under high heat and pressure (Fig. 9.11-2).

The vinyl resins and plasticizers together are called the *binder*.

ASTM F 1303 classifies vinyl sheet flooring as either Type I or Type II. In Type I flooring the binder constitutes at least 90% of the wear layer; in Type II, at least 34%.

ASTM F 1303 designates three grades within each type, to define wear layer thickness:

- *Grade 1:* For commercial, light-commercial, and residential projects; requires a 0.020-in. (0.508 mm) wear layer in Type I and a 0.050-in. (1.27 mm) wear layer in Type II
- *Grade 2:* For light commercial and residential projects; requires a 0.014-in. (0.36 mm) wear layer in Type I and a 0.030-in. (0.76 mm) wear layer in Type II
- *Grade 3:* For residential projects only; requires a 0.010-in. (0.254 mm) wear layer in Type I and a 0.020-in. (0.51 mm) wear layer in Type II

FIGURE 9.11-1 Summary of Resilient Flooring Products

Type	Standards	Backing Materials	Gauges, in. (mm)[a]	Sizes, in. (mm)	Basic Ingredients
Asphalt Tile					
Plain	ASTM F 1066	—	1/8 (3.2), 3/16 (4.8)	9 × 9 (228.6 × 228.6), 12 × 12 (304.8 × 304.8)[b]	Asphaltic and/or resin binders, mineral fibers, and limestone fillers
Greaseproof	ASTM F 1066	—			
Vinyl-Composition Tile	ASTM F 1066	—	1/16 (1.6), .080 (2), 3/32 (2.4), 1/8 (3.2)	9 × 9 (228.6 × 228.6), 12 × 12 (304.8 × 304.8)[b]	PVC resin binders, limestone, mineral fibers
Vinyl Tile					
Solid	ASTM F 1066	Scrap vinyl, mineral or rag felts	1/16 (1.6), .080 (2), 3/32 (2.4), 1/8 (3.2), .050 (1.27) to .095 (2.4)	9 × 9 (228.6 × 228.6), 12 × 12 (304.8 × 304.8)	PVC resin binders and mineral fillers
Backed	ASTM F 1066			4 × 36 (101.6 × 914.4),[b] 9 × 9 (228.6 × 228.6), 12 × 12 (304.8 × 304.8), and sizes up to 36 × 36 (914.4 × 914.4)[b]	
Vinyl Sheet[c]	ASTM F 1303	Scrap vinyl, mineral or rag felts	.065 (1.7) to .160 (4.1) (.010 [.25] to .050 [1.3] wear surface)	72 (1829), 144 (3658), and 54 (1372) wide	PVC resin binders and mineral fillers; clear PVC film for clear vinyl surface
			.065 (1.7) to .160 (4.1) (.010 [.25] to .050 [1.3] wear surface)		
Rubber Tile	ASTM F 1344	—	.080 (2), 3/32 (2.4), 1/8 (3.2), 3/16 (4.8)	9 × 9 (228.6 × 228.6), 12 × 12 (304.8 × 304.8), 18 × 36 (457 × 914.4), 36 × 36 (914.4 × 914.4)[b]	Synthetic or natural rubber, mineral fillers
Cork Tile Protected surface Wood surface	None[d]	—	1/8 (3.2), 3/16 (4.8), 1/4 (6.4), 5/16 (7.9),[b] 1/2 (12.7)[b]	9 × 9 (228.6 × 228.6), 6 × 6 (152.4 × 152.4),[b] 6 × 12 (152.4 × 304.8), 12 × 12 (304.8 × 304.8)[b]	Cork particles and resin binders, wax or resin finish
Vinyl surface			1/8 (3.2), 3/16 (4.8)	9 × 9 (228.6 × 228.6), 12 × 12 (304.8 × 304.8)	Cork particles and resin binders, clear vinyl finish
Linoleum Sheet	None[e]	Burlap	(0.050 [1.27] wear surface), 1/8 (3.2)	72 (1829) wide	Cork and/or wood flour with linseed oil binders
Plain and marbleized Inlaid and molded		Rag felt	.090 (2.3)		
		Burlap	1/8 (3.2)		
		Rag felt	.090 (2.3)		
Battleship		Burlap	1/8 (3.2)		
Linoleum Tile	None[e]		(.050 (1.27) wear surface), 1/8 (3.2)	9 × 9 (228.6 × 228.6), 12 × 12 (304.8 × 304.8)[b]	Cork and/or wood flour with linseed oil binders
		Rag felt	.090 (2.3)		

[a]Thinnest gauges and wear surfaces indicated are minimums suitable for adhesive installation.

[b]Not common, available on special order from some manufacturers.

[c]Also produced with foam backing, self-adhesive back, and no-wax surface; not included in data shown.

[d]Previously covered by Federal Specification LLL-T-00431, now cancelled.

[e]Previously covered by Federal Specification LLL-F-1238A, now cancelled.

FIGURE 9.11-2 Ingredients are keyed in a series of presses to make vinyl sheet flooring.

FIGURE 9.11-3 Felt backing is coated to form a light, smooth background for printed designs on vinyl sheet.

Greater thicknesses are required in Type II than in Type I, because the abrasion resistance and durability per unit of thickness of a wear layer are greater in the binder than in the fillers. The more binder, the thinner the wear layer can be for the same service.

Sheet products with vinyl backings may have a design imprinted on top of the backing or on the underside of the wear surface. They are usually printed with vinyl inks. A clear PVC wear surface is then calendered to the desired thickness and laminated to the backing with heat and pressure (Fig. 9.11-3).

VINYL TILE

Vinyl tile may be of homogeneous solid composition or may be backed with other materials such as organic felts, mineral fibers, or scrap vinyl.

Ingredients for solid vinyl tiles are mixed at a high temperature and hydraulically pressed or calendered into homogeneous sheets of the required thickness. These sheets are then cut into tile sizes. Some tiles are made with a self-adhesive back.

9.11.1.3 Rubber Tile

Natural or synthetic rubber is the basic ingredient of rubber flooring. Clay and fibrous talc or mineral fillers provide the desired degree of reinforcement; oils and resins are added as plasticizers and stiffening agents. Color is achieved by nonfading organic pigments. Chemicals are added to accelerate the curing process.

The ingredients are mixed thoroughly and rolled into sheets. The sheets are calendered to uniform thickness and vulcanized in hydraulic presses under heat and pressure into compact, flexible sheets with a smooth, glossy surface. The backs are then sanded to gauge, ensuring uniform thickness, and the sheets are cut into tiles.

9.11.1.4 Cork Tile

Cork tile is composed chiefly of the granulated bark of the cork oak tree, native to Spain, Portugal, and North Africa. Synthetic resins are added to the granulated cork, and the mixture is pressed into sheets or blocks and baked. Surfaces are finished with a protective coat of wax, lacquer, or resin applied under heat and pressure. Sheets then are cut to tile sizes. Vinyl cork tile has a film of clear PVC vinyl fused to the top surface to improve durability, water resistance, and ease of maintenance.

9.11.1.5 Linoleum

Linoleum consists of a blend of oxidized linseed oil binders, rosins, wood flour, and cork flour fillers, bonded to a burlap, felt, jute, or polyester backing. Linoleum is available in sheet and tile form.

9.11.2 SELECTION CRITERIA

The suitability of a resilient flooring material for a particular location or condition depends on its physical properties. *Moisture resistance* is important in determining whether a material can be used in areas subject to surface or ground moisture. *Grease resistance* and *alkali resistance* have a bearing on suitability for use in kitchens or over concrete, respectively. *Resilience* is related to quietness, indentation resistance, and underfoot comfort.

These and other properties may be combined into ratings for *durability* and *ease of maintenance*. Moisture resistance and alkali resistance must be considered together in determining suitable location with respect to whether the flooring is *suspended*, *on grade*, or *below grade* (Fig. 9.11-4). Installed cost is another factor in flooring selection, but should be considered along with long-term costs, which are affected by physical properties, durability, and ease of maintenance. Selection criteria for resilient flooring products are summarized in Figure 9.11-5.

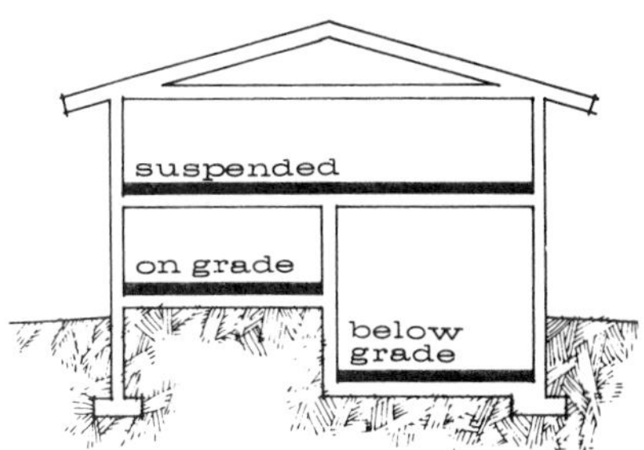

FIGURE 9.11-4 Subfloor locations. Subfloors are considered suspended when built at least 18 in. (457 mm) over a ventilated crawl space.

9.11.2.1 Moisture Resistance and Location

Linoleum contains linseed oil binders and needs sufficiently dry conditions so that it will not be softened from exposure to moisture. Products with organic ingredients or backings are subject to mold growth and decay. Asphalt, vinyl composition, solid vinyl, and sheet vinyl with mineral fiber backings are least affected by moisture and alkalies that may be leached by moisture from concrete floors.

The suitability of some flooring products is shown in the "Location" column in Figure 9.11-5. Consideration has been given to the alkali resistance of the backing, which may not be the same as that of the surface material. In sheet vinyl with asphalt felt backing, for instance, the vinyl surface has excellent alkali resistance, while the backing does not.

In addition to its effect on flooring materials directly, moisture may prevent proper bonding to the subfloor or underlayment. Water-based adhesive will not set up in the presence of water, and asphaltic emulsion adhesives may eventually be displaced under sustained exposure to moisture. Excessive moisture may also deteriorate organic underlayments such as hardboards or plywood.

The main types of moisture affecting resilient floors are (1) *surface moisture*, (2) *subfloor moisture*, and (3) *ground moisture*.

SURFACE MOISTURE

Spilled water, floor moppings, and tracked-in moisture are the primary sources of surface moisture. Spilled water may be common in laundry areas, bathrooms,

FIGURE 9.11-5 Selection Criteria for Resilient Flooring Products[a]

Type of Flooring	Location[b]	Resistance to: Grease	Alkalies	Stain[c]	Cigarette Burns	Indentation	Resilience[d]	Quietness	Ease of Maintenance	Durability
Vinyl-Composition Tile	BOS	2	2	4	2	3	6	6	2	2
Vinyl Tile										
Solid	BOS	1	1	1–2	1–5	1–4	2–5	2–5	3	1
Asphalt felt backed[e]	BOS	1	1	1–3	3–5	2–5	2–4	2–5	3	1
Rag felt backed	S	1	2	1–3	3–5	2–5	2–4	2–5	3	1
Vinyl Sheet										
Asphalt felt backed[e]	BOS	1	1	1–3	3–5	2–5	2–4	2–4	2	1
Rag felt backed[e]	S	1	2	1–3	3–5	2–5	2–4	2–4	2	1
Rubber Tile	BOS	3	3	2	1	4	2	2	3	2
Cork Tile	OS	4	4	5	4	4	1	1	5	5
Vinyl Cork Tile	OS	1	3	1	5	3	3	3	1	4

[a]Numerical rating indicates rank of each product compared with other products. Highest rating is 1 (excellent); 2 = very good; 3 = good; 4 = fair; 5 = poor.

[b]B = below grade, O = on grade, S = suspended.

[c]Varies with staining agent.

[d]Also indicates potential underfoot comfort.

[e]Foam-cushioned products are rated highly for resilience, indentation resistance, ease of maintenance, and quietness within the ranges indicated.

toilet rooms, and kitchens. Entries and corridors may be subjected to tracked-in moisture and excessive wet mopping.

Surface moisture affects flooring materials and adhesives by entering through the seams. Sheet flooring products minimize the number of seams, and when provided with moisture-resistant backings are suitable for use in such areas. The seams of many sheet products may also be sealed to prevent water seepage through seams.

SUBFLOOR MOISTURE

Concrete subfloors and mastic underlayments mixed with water release large amounts of moisture as they cure. With lightweight-aggregate concrete (weighing less than 90 lb./cu. ft.) (1442 kg/m^3), or under conditions that retard the curing process, moisture may be released over long periods of time. Field tests described under "Concrete Subfloors" in Section 9.11.5.2 should be used to establish that the subfloor is sufficiently dry to receive the intended product. Since it originates in the original mix of the concrete, subfloor moisture may be present in below-grade, on-grade, and suspended subfloors.

GROUND MOISTURE

Moisture from the ground is generally limited to below-grade and on-grade locations. Concrete floor slabs in direct contact with the ground, unless protected by a vapor retarder, may transmit moisture by capillary action. A well-drained fill of sand or gravel will retard the migration of moisture.

When resilient flooring is installed on wood subfloors over crawl spaces, a vapor retarder and adequate ventilation of the subfloor space are required. A vapor retarder placed over the ground will prevent transmission of moisture. Ventilation will encourage dissipation of moisture resulting from breaks in the vapor retarder and other sources. Both crawl spaces and concrete slabs on ground require proper surface drainage of exterior finished grade to prevent groundwater from becoming trapped under the floor.

9.11.2.2 Alkali and Grease Resistance

Major sources of grease are oils spilled in kitchens, laboratories, and shops; alkalies may be present from various cleaner residues. If flooring is properly protected with a floor wax or finish and spilled oils or residues are removed promptly before they seep into the seams, such temporary exposure is not a hazard, except for materials with the lowest resistance ratings (see Fig. 9.11-5). Vinyl composition and solid vinyl tile are not affected substantially by either grease or alkalies.

9.11.2.3 Resilience

Resilience is a measure of the instantaneous yielding and recovery of a surface from impact. Indentation resistance, quietness, and underfoot comfort are closely related to resilience.

INDENTATION RESISTANCE

In assessing indentation resistance, the momentary indentation produced by foot traffic and dropped objects is of primary importance. These impact pressures can be quite high and demanding. A 105-lb. (47.63 kg) woman in spike heels, for example, exerts a pressure on a floor of approximately 2000 psi (1,406,140 kg/m^2), while a 225-lb. (102 kg) man with his weight spread over 3-in. by 3-in. (75 mm by 75 mm) heels exerts only 25 psi (17,577 kg/m^2).

Permanent indentation from heavy stationary objects, such as a piano or a desk, may be minimized by using floor protectors to distribute the load (Fig. 9.11-6). Indentation resistance of thinner flooring materials is greatly affected by the subfloor or underlayment and may be increased by selecting harder subsurface materials. Homogeneous vinyl tile and foam-cushioned vinyls have the highest indentation resistance.

Permanent indentation in some flooring types and under certain conditions

WRONG

Narrow surfaces dent floors

RIGHT

Wide surfaces protect floors

Always remove small metal domes from bearing surfaces. Composition furniture cups should be placed under heavy furniture that is only infrequently moved.

WRONG

Hard rollers mark floors

RIGHT

Rubber rollers protect floors

Caster should be used on frequently moved furniture. They should be 2" (50.8 mm) in diameter, with soft rubber treads at least 3/4" (19.1 mm) wide and an easy swiveling ball-bearing action.

WRONG

Remove small metal domes

RIGHT

Use flat flexibl shank glides

Light furniture should have glides with a smooth, flat base and a flexible pin to maintain flat contact with floor. Diameter should be 1" to 2 1/2" (25.4 to 63.5 mm) depending on weight of furniture.

FIGURE 9.11-6 Floor protectors under furniture and equipment reduce indentation.

cannot be entirely prevented. However, these indentations may be less conspicuous in patterned, textured, and low-luster floors.

QUIETNESS

The quietness rating indicates the effectiveness of a floor in reducing sound from foot traffic and other impact noises. Resilient floors soften impact sounds but will not reduce reverberated noises originating from other sources, because they have little sound absorption ability. Cork, linoleum, and foam-cushioned vinyl flooring have high resilience and quietness ratings. Asphalt tile is at the low range of resilience and quietness.

9.11.2.4 Resistance to Sunlight

The actinic rays in sunlight may cause fading, shrinking, or brittleness in some resilient floors. Linoleum and vinyl are resistant to such deterioration. Color pigments are the critical factor in fade-resistant properties. Neutral colors show the best light resistance; pastel tones, especially yellows, blues, and pinks, are least effective in retaining colors under prolonged exposure to sunlight. Cork tile has the same tendency as natural wood to fade under strong sunlight.

9.11.2.5 Ease of Maintenance

The term *ease of maintenance* represents an appraisal of the relative expense (labor and materials) in keeping flooring at an acceptable level of cleanliness and attractiveness. Relative ratings are based on a record of experience with various flooring types as well as laboratory testing. Textured surfaces and darker colors show less scuffing and soiling and generally receive higher ease-of-maintenance ratings. No-wax vinyl products and vinyl cork tile are the easiest to maintain, followed by other vinyl products, vinyl composition, and linoleum.

9.11.2.6 Durability

Its durability is the ability of a flooring product to retain serviceability and attractiveness over time. Durability ratings are based on laboratory tests and are related to the physical properties of the material. Homogeneous vinyl tile is the most durable. Vinyl sheet flooring, rubber tile, and vinyl composition are also rated highly. Foam-cushioned products have good durability because the surfaces of these products absorb impact and resist abrasion.

9.11.2.7 Slip Resistance

The Americans with Disabilities Act of 1990 (see Section 1.7) requires flooring materials to have a static coefficient of friction of not less than 0.60 for level surfaces and 0.80 for ramps. Some manufacturers publish these data for their flooring products, even though there is no consensus in the industry concerning the test methods needed to ensure compliance. It would be prudent to verify the current status of slip-resistance requirements before selecting a resilient flooring for any project.

9.11.3 FLOORING ACCESSORY MATERIALS

The most common resilient accessories used with resilient flooring materials are (1) *wall bases*, (2) *stair treads, risers, and nosings*, and (3) *thresholds, feature strips, and reducing strips* (Figs. 9.11-7, 9.11-8, 9.11-9, and 9.11-10). These accessories are available in more limited color and design variations than the flooring materials.

9.11.3.1 Wall Bases

Rubber or vinyl wall bases are available in 2½-in. (57.2 mm), 4-in. (101.6 mm), and 6-in. (152.4 mm) heights, in straight type and in either set-on or butt types with coved-bottom designs (see Fig. 9.11-8). Vinyl base is available in 50-ft. (15.2 m) coils and in 4-ft. (1200 mm) lengths. Rubber base normally is available in 4-ft. (1200 mm) lengths only. Exterior and interior corners of rubber base are commonly premolded (see Fig. 9.11-9). End stops with a finished edge for terminating

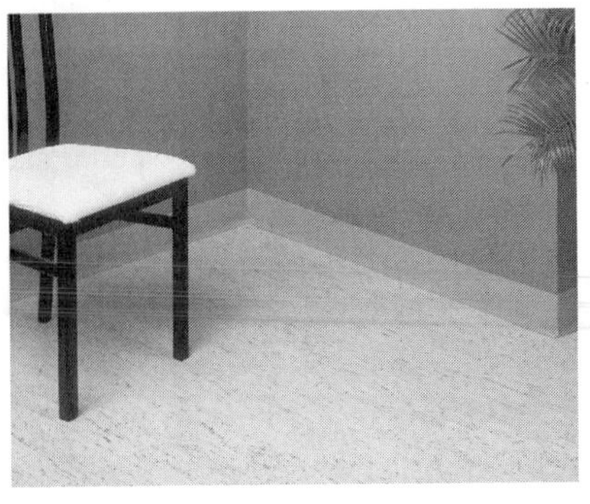

a b

FIGURE 9.11-7 Resilient accessories: (a) wall base (VPI wall base courtesy Vinyl Plastics Incorporated); (b) from left to right, cap for coved sheet goods, tile reducer, vinyl reducer, feature strip, and cove stick (supports sheet flooring for coving at a wall). (Courtesy Mercer Products Company, Incorporated.)

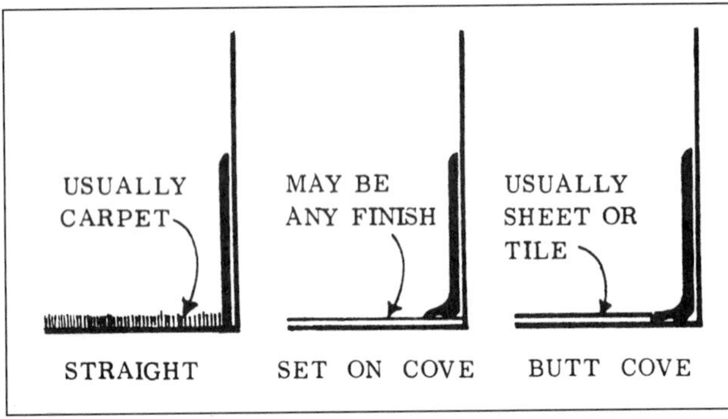

FIGURE 9.11-8 Resilient wall base types.

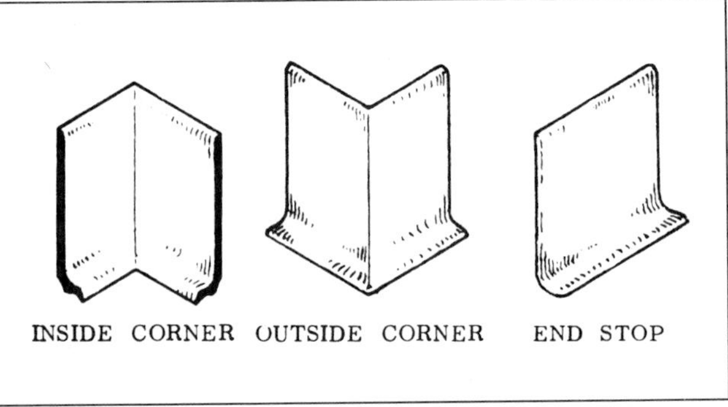

FIGURE 9.11-9 Premolded base accessories.

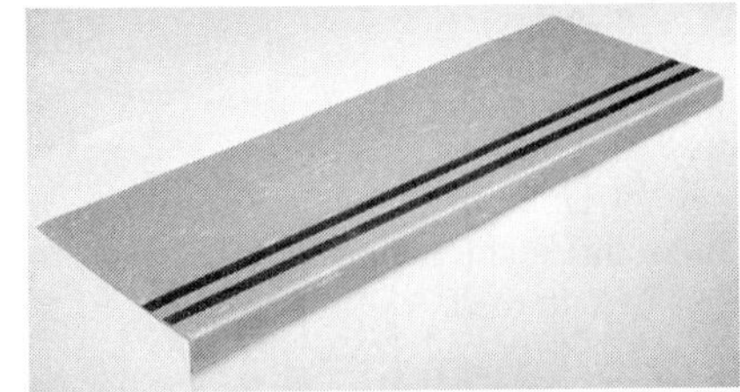

a b c

d

FIGURE 9.11-10 Stair treads and nosings: (a) smooth flat tread; (b) textured (diamond) tread; (c) abrasive strip tread; (d) smooth or ribbed nosings, which may be used with carpet or sheet or tile treads. (a, b, and c courtesy The R.C. Musson Rubber Company; d courtesy Mercer Products Company, Incorporated.)

wall base runs are available in both rubber and vinyl. The greater flexibility of vinyl base permits field-molding around exterior corners (see Fig. 9.11-7a); inside corners are usually scribed and fitted on the job.

9.11.3.2 Stair Treads, Risers, and Nosings

Vinyl and rubber stair treads and nosings are made with smooth or textured surfaces and may have inlaid abrasive grit strips to improve slip resistance (see Fig. 9.11-10). They may be combined with standard 6-in. (152.4 mm)-high wall base or higher stair risers to form complete stair coverings. Lengths vary from 3 to 12 ft. (900 to 3600 mm).

9.11.3.3 Thresholds, Feature Strips, and Reducing Strips

Resilient thresholds are either vinyl or rubber, 2¾ in. (69.9 mm) and 5½ in. (139.7 mm) wide, about ½ in. (12.7 mm) high, and 3 or 4 ft. (900 or 1200 mm) long. *Feature strips* are made of asphalt, vinyl composition, linoleum, and vinyl in widths of ⅛, ¼, ½, and 1 in. (3.2, 6.4, 12.7, and 25.4 mm) and in ½-in. (12.7 mm) increments up to 4 in. (101.6 mm). Thicknesses usually are 1/16 to ⅛ in. (1.6 to 3.2 mm), and lengths are usually 3 and 4 ft. (914 and 1219 mm). Rubber and vinyl beveled *reducing strips*, 1 to 1½ in. (25.4 to 38 mm) wide, 3 to 12 ft. (914 to 3658 mm) long, and 1/16 to ⅜ in. (1.6 to 9.5 mm) thick, permit transition in finished floor heights (Fig. 9.11-7b).

9.11.4 FLOORING INSTALLATION MATERIALS

Resilient flooring installation materials include underlayments, lining felt, and adhesives.

9.11.4.1 Underlayments

Underlayments are used over subfloors to provide a smooth, level, hard, and clean surface to receive flooring products. In general, panel underlayments are used over wood subfloors; mastic underlayments over concrete.

PANEL

Panel underlayments may be used over any structurally adequate suspended subfloor such as plywood or lumber boards. The three common types of panel underlayments are plywood, hardboard, and particleboard.

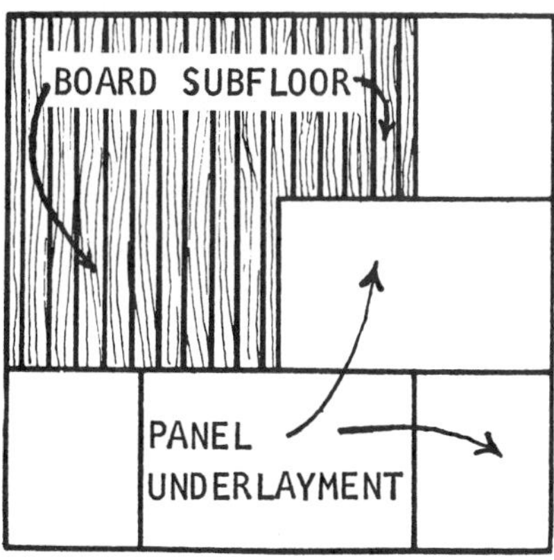

FIGURE 9.11-11 Underlayment layout over a board subfloor.

Stiffer floors result when joints in the underlayment are staggered with respect to joints in a plywood subfloor and blocking can be omitted. Over a board subfloor, underlayment joints should not fall over joints in the floor boards (Fig. 9.11-11). To allow for expansion caused by changes in humidity, underlayment panels should be spaced approximately 1/32 in. (0.8 mm) at joints and about ⅛ in. (3.2 mm) apart at intersections with vertical surfaces. End joints in underlayment panels should be staggered ashlar fashion (see Fig. 9.11-11).

Nailing should start in the middle of the panel and proceed outward to the edges, with nail heads set flush or slightly countersunk. Panel joints should be sanded smooth and flush before the flooring is installed.

Minimum underlayment thicknesses are given below for each material; however, greater thicknesses may be warranted where increased stiffness is desired.

Plywood A sanded Interior Underlayment grade, at least ¼ in. (6.4 mm) thick, is suitable for most plywood underlayments. However, in areas subject to excessive moisture, such as in bathrooms and laundry areas, an Exterior C-C (plugged) or Underlayment grade with exterior adhesive should be used. Underlayment may be installed over a plywood subfloor with the face grain either parallel or perpendicular to the floor joists. However, if the underlayment is installed with the face grain perpendicular to the floor joists, a stiffer floor assembly results. Plywood should be installed over a board or plank subfloor with the face grain perpendicular to the direction of the floor boards (see Fig. 9.11-11), and preferably with long edges over joists. Plywood underlayment should be installed with the fasteners and spacing shown in Figure 9.11-12.

Hardboard Hardboard underlayment will bridge small irregularities and provide a smooth surface over an otherwise structurally sound wood floor. Underlayment grade hardboard is an untempered service type, surfaced on one side. It is manufactured in a nominal ¼-in. (6.4 mm) (0.215 or 0.200 in. [5.46 or 5.08 mm] actual) thickness, in 3-ft. by 4-ft. (914 by 1219 mm), 4-ft. by 4-ft. (1219 by 1219 mm), and 4-ft. by 8-ft. (1219 by 2438 mm) panels, and should conform to the American National Standards Association/American Hardboard Association standard ANSI/AHA A 135.4, Class 4 (service), surface S1S. Installation of panels is the same as for other panel underlayment using fasteners and spacing recommended in Figure 9.11-12.

Particleboard Particleboard underlayment should not be used in areas subject to excessive moisture, and its moisture content as shipped from the mill should not exceed an average of 7%. Particleboard underlayment should be at least ⅜ in. (9.5 mm) thick, should conform to the American National Standards Association Standard ANSI A 208.1, "Wood Particleboard," and should bear the National Forest Products Association grade mark "Floor Underlayment," Grade 1-M-1. Particleboard underlayment is manufactured in 4-ft. (1219 mm) by 8-ft. (2438 mm) panels and in thicknesses ranging from ⅜ to ¾ in. (9.5 to 19.1 mm). Nailing recommendations for various panel thicknesses are given in Figure 9.11-12.

Some resilient flooring manufacturers do not recommend using particleboard as an underlayment for their materials. However, some particleboard manufacturers issue a warranty covering the entire resilient flooring installation if the flooring and underlayment are installed to their specifications.

MASTIC

Mastic underlayments contain a chemical binder such as latex, asphalt, or polyvinyl-acetate resins and portland, gypsum, or aluminous cement. Mixtures consisting of powdered cement and sand to which only water has been added function only as crack fillers; when applied in thin coats, they break down under traffic.

FIGURE 9.11-12 Fastening for Panel Underlayment

Underlayment, in. (mm)	Fastener Length, in. (mm)[a]		Fastener Spacing
	Nails[b]	Staples[c]	
Plywood			
¼ (6.4)	1 (25.4) (2d)	⅞ (22.2)	Nails: 8 in. (203 mm) o.c. each direction in middle of panel, 6 in. (152 mm) o.c. around panel perimeter.
⅜ (9.5)	1¼ (31.8) (3d)	1⅛ (28.6)	
½ (12.7)	1¼ (31.8) (3d)	1¼ (31.8)	
⅝ (15.9)	1½ (38) (4d)	1½ (38)	Staples: 6 in. (152 mm) o.c. each direction in middle of panel, 3 in. (76 mm) o.c. around panel perimeter.
¾ (19.1)	1½ (38) (4d)	1⅝ (41.3)	
Hardboard[d]			
¼ (6.4)	1 (25.4) (2d)	⅞ (22.2)	Nails and staples: 6 in. (152 mm) o.c. each direction in middle of panel, 3 in. (76 mm) o.c. around panel perimeter.
Particleboard[e]			
⅜ (9.5)	1¼ (31.8) (3d)	1⅛ (28.6)	Nails: 10 in. (254 mm) o.c. each direction in middle of panel, 6 in. (152 mm) o.c. around panel perimeter.
⅝ (15.9)	1½ (38) (4d)	1½ (38)	
¾ (19.1)	1½ (38) (4d)	1⅝ (41.3)	Staples: 6 in. (152 mm) o.c. each direction in middle of panel, 3 in. (76 mm) o.c. around panel perimeter.

[a]Fastener lengths shown are minimum to provide ¾-in. (19.1 mm) penetration for nails and ⅝ in. (15.9 mm) for staples. If longer fasteners are used, penetration should not exceed 1 in.

[b]Annularly threaded; common nails may be used by increasing the sizes ¼ in. (6.4 mm) (1d).

[c]Galvanized divergent chisel, 16-gauge (1.6 mm), ⅜-in. (9.5 mm) crown width.

[d]Nominal thickness, actual thickness 0.215 in. (5.5 mm).

[e]Moisture content as shipped from mill should not exceed 7% average.

Latex underlayments are most suitable for applications requiring a thin layer (⅜ in. [9.5 mm] or less), for skim coating, and for patching where the fill must be feather-edged (Fig. 9.11-13).

Skim coating smooths surface irregularities. It does not raise the elevation of the floor.

Asphaltic and polyvinyl-acetate underlayments are used where thicker (more than ⅜-in. [9.5 mm]) underlayments are needed.

Mastic underlayments must be troweled smooth and true, with not more than ⅛-in. (3.2 mm) variation from a straight line in 10 ft. (3000 mm). Subfloors should be free of wax, oil, and surface coatings, such as concrete curing compounds, before mastic underlayment is applied.

FIGURE 9.11-13 Latex underlayment can be feathered to a thin edge when patching subfloor depressions. (Courtesy Kentile Floors.)

Latex Latex underlayments are mixtures containing a synthetic latex emulsion, a cementitious material such as portland cement, fine sand, and water. They are troweled on in layers not more than ⅛ in. (3.2 mm) thick and may be feathered to a thin edge (see Fig. 9.11-13). They are built up in several applications totaling not more than ⅜ in. (9.5 mm). Rough surfaces are machine-sanded after the mastic has dried completely. Because latex underlayment is a quick-setting compound, splashed material should be cleaned off promptly.

Asphaltic solvent-based primers and adhesives used over latex underlayment for installing asphalt or vinyl composition tile should be allowed to set for the time recommended by their manufacturer before the tile is installed.

Asphaltic Asphaltic underlayment is a mixture of an asphaltic emulsion, a cementitious material such as portland cement, fine sand, and water. Proper proportioning varies, depending on the ingredients and the thickness of the underlayment, and the manufacturer. Asphaltic underlayment is not recommended for installations less than ⅜ in. in thickness.

Before placing an asphaltic underlayment, it is necessary to prime the concrete subfloor with a thin solution of the same asphaltic emulsion being used in the underlayment mixture. The mastic is then poured uniformly over the entire surface, with wood screeds used to establish the desired thickness. A straightedge is used to level the mastic over the screeds, after which the surface is troweled to a smooth finish. The setting time of the fill should not be accelerated with excessive heat, as this may cause cracking or crazing. Where it is necessary to trowel the mastic to a featheredge, the floor should be given a brush coat of undiluted asphaltic emulsion to prevent brittleness after the mastic has dried.

Solvent-based asphaltic primers and adhesives should not be applied over asphaltic underlayments; asphaltic emulsion or latex adhesives should be used.

Polyvinyl Acetate Polyvinyl acetate underlayment is formed by mixing portland cement and sand with a specially compounded polyvinyl acetate (PVAc) resin emulsion. The PVAc liquid should also be applied to the concrete subfloor to improve bond with the underlayment. When a fill of ½ in. (12.7 mm) or thicker is required, a portland cement concrete "topping" may be used without PVAc resins, but the PVAc liquid should still be used as a bonding agent on the concrete subfloor.

This type of underlayment may be used with any adhesive and flooring type. Its installation is similar to that described for asphaltic underlayment.

9.11.4.2 Lining Felt

Some manufacturers recommend using lining felt over panel underlayment to minimize the possibility of joints opening and tile splitting due to underlayment movement. However, this is not a significant hazard if the underlayment or subfloor is sufficiently thick and is properly installed. Of greater significance is that the indentation resistance of flooring materials is substantially reduced by lining felt, which imposes limitations on its use.

Lining felt is used in conjunction with some vinyl sheet products over nonporous subfloors, such as old resilient floors, terrazzo, ceramic tile, and concrete that has been treated with curing agents. The liner's purpose is to provide a means for the water in the adhesive to be absorbed, which allows the adhesive to cure.

9.11.4.3 Adhesives

Most adhesives used for the installation of resilient flooring are of a troweling grade for application with a notched trowel (Fig. 9.11-14). Some types, however, are of a more fluid consistency for application with a brush. These often are used with thin-gauge products to eliminate the possibility of trowel marks "telegraphing" to the surface of the finished floor. The major types of adhesives are listed in Figure 9.11-15 and discussed below. Figures 9.11-15 and 9.11-16 summarize the properties and applications of the various adhesives. Adhesives with slightly varying properties are produced in each type, so the recommendations of the flooring and adhesive manufacturers should be followed.

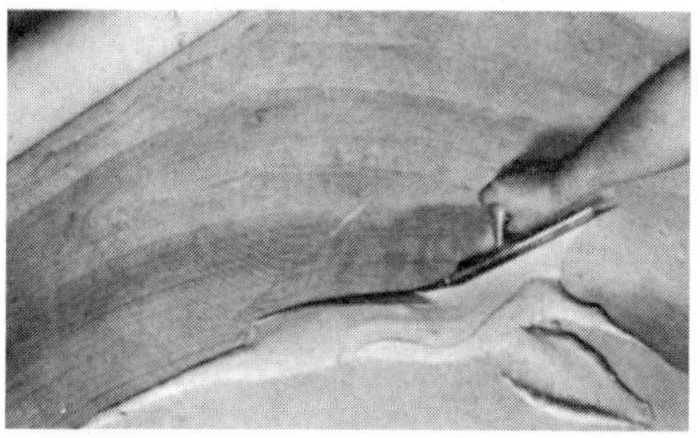

FIGURE 9.11-14 Latex adhesive is gradually spread to a uniform thickness by a swirling motion with a notched trowel. (Courtesy Kentile Floors.)

LINOLEUM PASTE

Linoleum paste is a low-cost adhesive used for installing linoleum, vinyl sheet flooring, and lining felt. It is a water-based adhesive that remains water-soluble even after hardening and should not be used on concrete slabs on or below grade or in other areas subject to moisture.

Flooring should be installed immediately after spreading this adhesive, then rolled with a 100-lb. (45.36 kg) roller to ensure proper bond. The paste dries to a firm film within 24 hours and becomes hard within a few days. The rigidity of the hardened adhesive restrains sheet flooring dimensionally and prevents curling.

ASPHALTIC ADHESIVE

Asphaltic adhesives are low-cost, water-resistant adhesives used exclusively for installing vinyl composition tile. They may be useful on all types of below-grade, on-grade, and suspended subfloors. Asphaltic adhesives work with most underlayments; however, cutback asphalt adhesive and solvent-based primer should not be used over emulsion-based underlayment. Asphaltic adhesives retain their bonding qualities while the tile gradually conforms to minor surface irregularities. They eventually harden to provide a durable water- and alkali-resistant bond. Asphaltic adhesives are either solvent-based or water emulsions.

Solvent-base Solvent-based asphaltic adhesives consist of a solution of asphalt in hydrocarbon solvents. They are available in brushable and thicker formulations. Cutback asphalt adhesive, the most common of the solvent-based asphaltic adhesives, is of a troweling-grade consistency. Because the volatile solvents are flammable and mildly toxic, these adhesives should not be used near an open flame and the area should be well ventilated. They should be allowed to set approximately 30 minutes before tile is applied. The working time ranges from 4 to 18 hours.

Solvent-based adhesives may cause or increase atmospheric damage and are not permitted in some jurisdictions. In time, they may be prohibited altogether.

Water-emulsion Water-emulsion asphaltic adhesives consist essentially of tiny asphalt particles suspended in water. They become water resistant upon drying. They are of troweling-grade consistency and are spread with a notched trowel. The adhesive should be allowed to set for 30 to 60 minutes before tile is installed, but the working time ranges from 8 to 24 hours. This permits application to relatively large areas, overnight setting of adhesive, and installation of tile the following day.

ASPHALT-RUBBER ADHESIVE

An asphalt-rubber adhesive is an asphalt and rubber water emulsion used primarily for the installation of vinyl composition tile. The working characteristics of this adhesive are similar to those of asphaltic adhesive, and it can be used over any type of subfloor or underlayment.

WATERPROOF-RESIN ADHESIVE

Most adhesives of the waterproof-resin type are resins in an alcohol solvent. They have approximately the same application and working characteristics as linoleum paste, but are not soluble in water. However, they are adversely affected by alkaline moisture and are unsuitable for use over concrete in contact with the ground on or below grade. With the exception of asphalt and vinyl composition tile, waterproof-resin cements may be used for applying most resilient flooring in suspended locations requiring waterproof bond. The volatile alcoholic content makes these adhesives flammable, and they should not be used near an open flame or in poorly ventilated areas.

LATEX ADHESIVE

Latex cements consist of rubber resins and water. They are extremely moisture and alkali resistant after they set. On drying, the adhesive turns to a rubbery film, which eventually hardens and becomes insoluble in water. Latex cements generally provide good adhesion and may be used for installing all types of resilient flooring on any type of subfloor or underlayment. They are relatively higher in cost and are not normally used for installing vinyl composition and asphalt tile where cheaper asphaltic adhesives also are suitable. Latex cements are more commonly used for installing rubber tile, vinyl tile, and fiber-backed sheet products on or below grade. Flooring may be

FIGURE 9.11-15 Summary of Adhesives by Type

Adhesive	Type	Flooring	Installation	Precautions
Linoleum Paste	Clay: Water-base Troweling grade	Lining felt All sheet flooring Vinyl tile with backing Rubber tile Cork and vinyl cork tile	Over suspended wood or concrete subfloors, panelboard, and latex underlayments; install flooring immediately.	Do not use on or below grade, over suspended subfloors subject to moisture, or for solid vinyl, asphalt, and vinyl-asbestos tile. Remove with damp cloth.
Asphalt[a] Cutback asphalt	Asphalt: Solvent-base Troweling grade	Vinyl-composition tile	Over all concrete subfloors and panelboard underlayment primed with asphaltic primer. Allow to set 30 min.; install tile within 4 to 18 hr., as recommended.	Do not use under or over lining felt or asphaltic underlayments. Combustible. Remove with fine steel wood and soapy water.
Asphalt emulsion	Clay and asphalt: Water-base Troweling grade	Vinyl-composition tile 1/16 in. and thicker	Over all subfloors and over lining felt. Allow to set 30 to 60 min.; install tile within 8 to 24 hr., as recommended.	Do not use to install lining felt. Keep from freezing. Remove with fine steel wool and soapy water.
Asphalt-Rubber	Asphalt and rubber: Water-base Brush and troweling grades	Vinyl-composition tile	Over all subfloors and over lining felt. Allow to set 30 min. before installing tile or as recommended.	Use brushing grade with 1/16-in. vinyl-asbestos tile; keep from freezing. Remove with fine steel wool and proprietary cleaner.
Waterproof Resin	Resin: Solvent-base Troweling grade	Vinyl with rag felt, vinyl, or rubber backing Rubber tile Cork tile Solid vinyl	Over all suspended subfloors; install flooring within 15 min.	Do not use on or below grade, for asphalt or vinyl-asbestos tile, or to install lining felt. Remove with fine steel wool and proprietary cleaner. Combustible.
Latex[b]	Rubber: Water-base Brush and troweling grades	Vinyl with rubber backing Rubber tile Solid vinyl tile Lining felt All sheet flooring	Over all subfloors; install flooring immediately.	Keep from freezing. Remove with fine steel wool and soapy water.
Epoxy	Resin and catalyst: Troweling grade	Solid vinyl and rubber tile on or below grade	Over all types of subfloors. Mix only as much as can be spread within 30 min. and covered within 2½ hr. or as recommended.	Guard against tile slipping while adhesive sets.
Wall Base Synthetic rubber cement	Synthetic rubber resin: Solvent-base Brush grade	Vinyl cove base Metal nosings and edges All resilient tile except asphalt and vinyl-asbestos to metal surfaces	Apply to both wall and material; install material within 20 min.	Combustible. Remove with proprietary remover.
Standard cove base cement	Solvent-base Troweling grade	Vinyl and rubber cove base	Over dry walls above grade; install base within 15 min.	Combustible. Remove with fine steel wool and soapy water.

Note: Formulations and working properties of adhesives vary and manufacturers' specific recommendations should be followed.

[a]May be used over latex underlayment if sufficient setting time is allowed.

[b]May be used to install asphalt and vinyl-asbestos tile; however, asphaltic adhesives are used more commonly because of lower cost.

installed immediately or as recommended by the manufacturer.

EPOXY ADHESIVE

Epoxy adhesives are extremely moisture-resistant formulations with high bonding strength and excellent durability. However, the relative high cost, difficulty in spreading, and excessive slipping of tiles immediately after installation moderate the use of these adhesives.

Epoxy adhesives consist of two parts, an epoxy resin and a catalyst that activates the mixture. The parts are mixed at the job site. A chemical reaction begins as soon as the components are mixed, so the adhesive must be used within 2 to 3 hours. The actual setting time varies, however, due to variations in the formulations. Therefore, the manufacturer's recommendations should be followed closely.

FIGURE 9.11-16 Summary of Adhesives of Flooring Types and Locations

Flooring	Subfloor and Location: Wood (Suspended)	Concrete (Suspended)	Concrete (On and Below Grade)
SHEET FLOORING			
Vinyl			
Rag felt backed	Waterproof resin	Waterproof resin	Not recommended
Vinyl or rubber backed	Latex Waterproof resin	Latex Waterproof resin	Latex
Mineral fiber backed	Latex	Latex	Latex
TILE FLOORING			
Vinyl			
Solid	Waterproof resin Latex	Waterproof resin Latex	Latex Epoxy
Rag felt backed	Linoleum paste Waterproof resin	Linoleum paste Waterproof resin	Not recommended
Mineral fiber backed	Latex Linoleum paste	Latex Linoleum paste	Latex
Rubber	Linoleum paste Waterproof resin	Waterproof resin Linoleum paste	Latex Epoxy
Vinyl-Composition	Asphalt emulsion Asphalt cutback Asphalt rubber	Asphalt emulsion Asphalt cutback Asphalt rubber	Asphalt emulsion Asphalt cutback Asphalt rubber
Cork	Linoleum paste Waterproof resin	Linoleum paste Waterproof resin	Not recommended

*Foam-cushioned products are installed by specially formulated adhesives recommended by flooring manufacturer.

WALL BASE CEMENT

The two basic adhesives used to secure resilient wall base are standard cove base cement and synthetic rubber cement. Both are solvent-based formulations that set as the volatile solvents escape from the mix; both are flammable and should not be used near an open flame or in poorly ventilated areas.

Standard Cove Base Cement Low-cost standard cove base cement is used for installing resilient wall base and premolded corners. It develops bonding power about 10 minutes after application to either the wall surface or the wall base (Fig. 9.11-17).

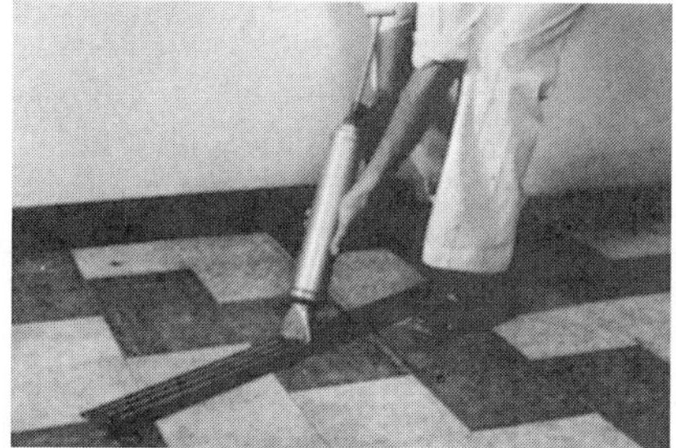

FIGURE 9.11-17 Standard cove base cement may be applied with a notched spreader or by a cartridge-type gun, as shown here.

Synthetic Rubber Cement Synthetic rubber cement is a contact adhesive that must be applied to both the wall and the resilient base. Bond is developed as soon as the two surfaces touch. Synthetic rubber cement also is used for installing metal and resilient nosings, edges, and binding strips. It is particularly useful in bonding resilient accessories in the field where movement after fitting would be undesirable.

9.11.5 GENERAL INSTALLATION RECOMMENDATIONS

Successful installation of resilient flooring depends on (1) job conditions, (2) adequate subfloor construction, and (3) proper selection of underlayment, adhesives, and flooring material.

Because of the variety of proprietary flooring materials, underlayments, and adhesives, the individual manufacturers' recommendations always should be consulted to ensure proper use in each case.

9.11.5.1 Job Conditions

Installation of finish flooring should be delayed until all other interior work, with the possible exception of door trim and touch-up painting, has been completed. If it is necessary to install flooring sooner, the finished floor must be protected with plain, undyed, unsaturated building paper. Door trim may be installed prior to flooring if it is shimmed up with scraps of flooring to permit later installation of the flooring under the trim.

Care should be taken to prevent freezing and damage to materials during delivery, handling, and storage. Materials should be stored on a smooth, level surface in a warm, dry place. Sheet flooring should not be stored for long periods of time, and the rolls should be stored vertically, on end. Tile flooring should be stacked not more than five cartons high.

A temperature of at least 70°F (21°C) and not more than 90°F (32.2°C) should be maintained for 48 hours before installation, during installation, and for 48 hours after installation. A minimum temperature of 55°F (12.8°C) should be maintained thereafter, and adequate ventilation should be provided at all times. Ventilation is particularly important when using solvent-based adhesives and primers, which are highly volatile and flammable.

Figure 9.11-18 shows some of the terms used to describe locations to be covered in a resilient flooring application. The terms used are defined in the body of the text when they occur or in the glossary.

9.11.5.2 Supporting Construction

The two main types of subfloors that occur below resilient flooring are wood and con-

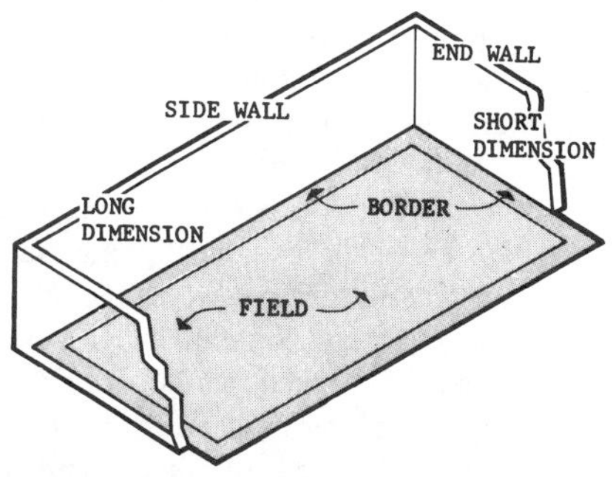

FIGURE 9.11-18 Flooring terms.

crete. All subfloors, whether covered with underlayment or not, should be firmly fastened and structurally adequate to carry the intended loads without excessive deflection.

Both wood joist and concrete slab floor constructions are suitable for the application of resilient flooring. Plank-and-beam construction is similar to wood joist construction, but the subfloor is 2 in. (50.8 mm) thick or thicker and structural members are larger and spaced farther apart. Underlayment is recommended over plank subfloors.

Concrete slab floors may be supported on ground or suspended above grade. Basement floors are concrete slabs supported on ground below grade.

WOOD SUBFLOORS

Suspended wood floors above grade are usually sufficiently removed from excessive external moisture sources, but floors over crawl spaces at or near grade may require that the crawl space be ventilated and the ground inside the crawl space covered with a suitable vapor retarder.

Wood floors may be of single-layer construction consisting of a combination subfloor and underlayment, or of double-layer construction composed of separate subfloor and underlayment. Subfloor and combination subfloor and underlayment materials, minimum subfloor thickness for various joist spacing, and nailing requirements are discussed in Section 6.6.4. Thicker subfloors will result in stiffer floors. Single-lumber board construction is not acceptable for the direct application of resilient flooring.

CONCRETE SUBFLOORS

Concrete subfloors should have a density of at least 90 psf (439.3 kg/m^2). If concrete of lower density has been used, it should be refinished with a 1-in. (25.4 mm)-thick topping of a regular concrete mix, bonded to the existing concrete with a bonding agent such as polyvinyl-acetate compound.

Subfloors on or below grade should be separated from the ground with a suitable vapor retarder, such as polyethylene film (see Section 2.4).

Concrete subfloors generally provide a suitable surface for resilient flooring, but require precaution against excessive moisture. The moisture content of concrete floors should be checked, particularly when lightweight concrete is used, because it cures slowly. In some instances, dampness can be observed under stationary objects on a floor on or below grade, but most often moisture cannot be detected readily and tests should be performed to determine whether the floor is sufficiently dry to receive flooring.

Concrete slabs that have been treated with curing, hardening, or separating compounds should also be tested for adequate bond.

Suitability Tests Several tests may be performed to determine acceptable moisture levels and bonding properties of concrete subfloors. These are a relative humidity test, a mat moisture and bonding test, and a primer test.

A *relative humidity test* is intended for testing new suspended concrete floors for the installation of moisture-sensitive products such as linoleum, cork, and rag felt–backed vinyl. To conduct this test, a relative humidity meter is placed on the concrete next to an interior wall or column. Depending on atmospheric conditions, the meter should read 40% to 65%. The meter is then covered with an 18-in. (450 mm) square of polyethylene film, and the edges are sealed to the floor with adhesive or tape. When substantial temperature fluctuations are expected, the meter should be protected with several layers of burlap or insulating batts. If there are no leaks in the polyethylene film, the reading should stabilize after 24 hours on a normal slab and after 72 hours on thicker-than-normal slabs. A sufficiently dry slab will register a reading of 80% or less. If the reading is more than 80%, the slab should be allowed to cure further and the test should be repeated.

A *mat moisture and bonding test* is designed for checking moisture and surface bonding properties of concrete floors to establish suitability for installation of (1) rubber tile, solid vinyl tile, and vinyl sheet with mineral fiber backing on or below grade and (2) any resilient flooring on slabs from which paint, oil, concrete hardeners, or curing compounds have been removed.

Linoleum and vinyl sheet mats approximately 36 in. (900 mm) square are applied with two bands of adhesives in several locations on the subfloor. One band is a water-soluble adhesive, the other a water-resistant latex adhesive. The mats are then sealed to the floor with adhesive or industrial tape (Fig. 9.11-19). After 72 hours, they are removed and the adhesive examined. If there is too much moisture present, the water-soluble adhesive will be partly or completely dissolved and the water-resistant adhesive will be stringy and will provide little bond. If excessive moisture is still indicated after further curing of the concrete and repeated testing, the above-mentioned flooring types should not be used and consideration given to asphalt or vinyl composition tile, which have greater moisture resistance. However, for these products the primer test also should be performed.

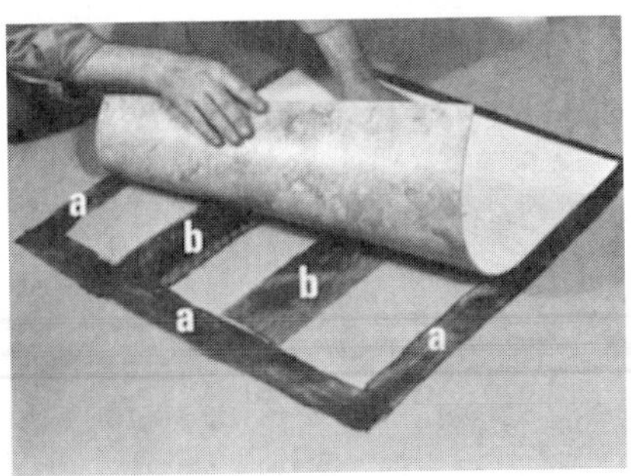

FIGURE 9.11-19 The mat moisture and bonding test for suspended concrete subfloors: (a) water-resistant latex adhesive, (b) water-soluble adhesive.

A *primer test* determines the suitability of a concrete substrate to receive asphalt or vinyl composition tile. Patches of asphaltic primer are applied in several places on the floor and are allowed to dry for 24 hours (Fig. 9.11-20). If the primer does not readily peel and sticks to the floor when scraped with a putty knife, the subfloor is dry enough to receive asphalt or vinyl composition tile.

Surface Preparation Expansion joints, scored joints, cracks, and depressions must be filled with latex floor-patching compound or mastic underlayment to obtain a smooth surface. Concrete with an irregular surface may require grinding or smoothing with mastic underlayment.

FIGURE 9.11-20 The primer test.

Curing or hardening compounds may leave surface films, which will adversely affect adhesive bonding. If a mat bonding test reveals insufficient bonding properties, the film should be removed by power wire brushing or grinding with a terrazzo grinder. An asphaltic primer is required on dusty or porous concrete floors for the installation of asphalt and vinyl composition tile.

9.11.6 SHEET FLOORING INSTALLATION

The following text describes the installation of sheet flooring in new construction. Reflooring existing construction is discussed in Section 9.11.9.

9.11.6.1 Preparation for Flooring

The effectiveness of a flooring installation greatly depends on the proper selection and preparation of the elements that make up the installation. Subfloors that are to receive flooring directly must be firm, smooth, and dense and must possess good bonding properties. When these properties are lacking, it is necessary to prepare the subfloor by grinding or sanding or by installing underlayment. Adhesives must be compatible with all elements, including the flooring itself, and all elements must be suitable for the intended location.

SUBFLOOR

Vinyl sheet products with mineral fiber backing may be installed on below-grade, on-grade, and suspended subfloors. Products with rag felt backs are subject to decay and alkali attack in the presence of moisture and are limited to suspended locations. Linoleum should be installed on suspended floors only.

Concrete subfloors in all locations should be tested for moisture and bonding properties to ensure suitability for the proposed flooring and adhesive (see Section 9.11.5.2).

UNDERLAYMENTS

The need for underlayment is determined by the condition of the subfloor (see Section 9.11.4.1).

ADHESIVES

The adhesives generally recommended for use with sheet products in various locations are summarized in Figures 9.11-15 and 9.11-16.

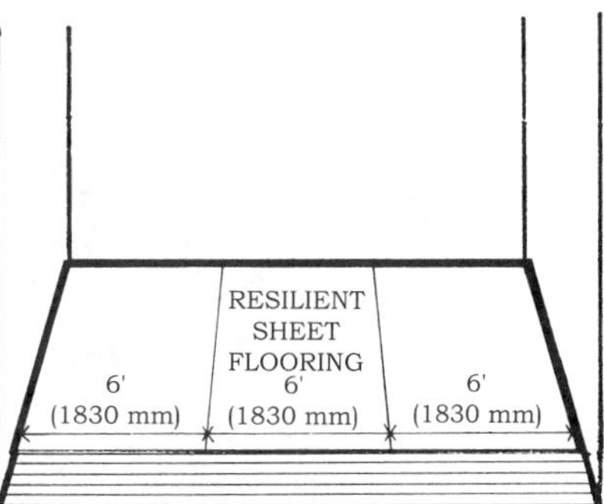

FIGURE 9.11-21 Sheet flooring direction over floor joints.

LINING FELT

Lining felt should be used in the installation of certain sheet flooring products on nonporous floors, as described in Section 9.11.4.2.

9.11.6.2 Application of Flooring

The installation of sheet materials should be planned to minimize the number and total length of seams. Necessary seams should be placed in inconspicuous locations, out of the path of heavy foot traffic. In a rectangular room, running the flooring strips parallel to the side walls generally results in an economical installation with a minimum of seams. However, sheet flooring installed directly over wood strip floors should run perpendicular to the floor joints and may result in seams running the short dimension, parallel to end walls (Fig. 9.11-21).

Installation of sheet flooring generally consists of *fitting and cutting*, *adhesive bonding*, and *seam treating*.

FITTING AND CUTTING

Many sheet flooring products must be kept rolled face out until time of installation. Some manufacturers warn against rolling sheet flooring face in because it may set up internal stresses that can result in residual growth of the sheet flooring long after it is installed. When fitting adjacent strips that have a definite geometric repeat pattern, the pattern should be aligned and matched. Sheet flooring with random overall patterns should be reversed (turned end for end) to minimize differences in graining and shading.

Methods of cutting and fitting are sometimes related to the intended method of bonding the sheet goods. Adhesive application is facilitated by folding the strip back on itself, thus working only half of the strip at a time. The strip is folded back along its length in the *tubing method*, and back along its width in the *lapping method* (Fig. 9.11-22).

Sheet flooring is cut, and fitted to walls and to each other, by *knifing*, *scribing*, and *seam cutting*.

After squaring the end of a roll (Fig. 9.11-23), strips are cut 3 in. (76.2 mm) longer than the distance between end walls to provide excess material for fitting to wall recesses or irregularities. The first strip is placed in position with the side edges butted against the side wall, and the ends of the material are flashed (bent up) against the end walls. The material is scribed along the length of the side wall with dividers or a scribing tool (see "Scribing" below) and cut to fit snugly to the wall surface. Ends that have been flashed against the end walls may also be scribed or may be trimmed by cutting

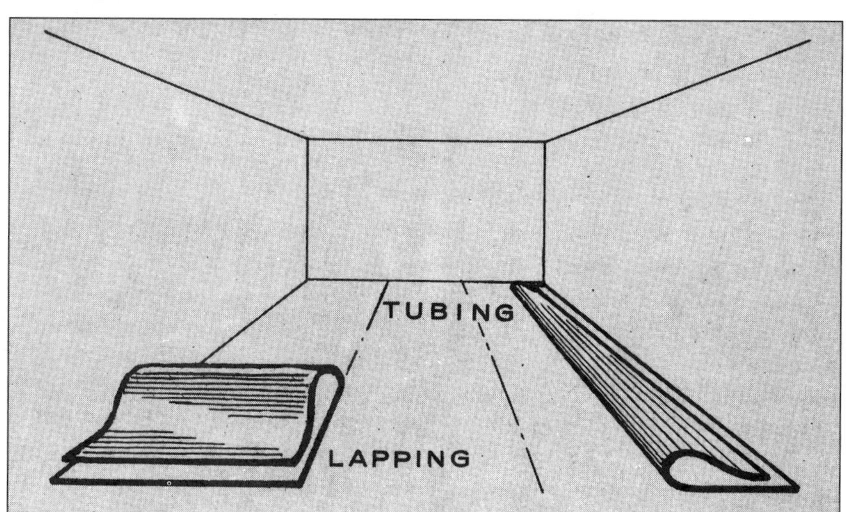

FIGURE 9.11-22 Methods of handling adhesive bonding sheet flooring.

FIGURE 9.11-23 The end of a roll is squared with a straightedge and a square.

away small pieces at a time until the desired fit is achieved.

Knifing Knifing is a method of end fitting that involves some waste. It is used with less expensive felt-backed products. After a strip of flooring is fitted in place against a side wall or adjacent strip, excess material is cut away gradually to fit the end walls and other vertical surfaces.

Safety cuts are made at various points to allow the material to be fitted into difficult locations without cracking. One or more safety cuts should be made first wherever the material tends to buckle as it is flashed against the wall. When the flooring has to be fitted into a corner before it can be knifed at the walls, the *inside corner safety cut* is the simplest (Fig. 9.11-24a). This allows the material to be dropped down snugly into a corner and then flashed up the walls without cracking. An *outside corner safety cut* is used to fit flooring around a curved or cornered projection (Fig. 9.11-24b). A *crossover safety cut* permits bending the flooring under an obstacle such as a radiator.

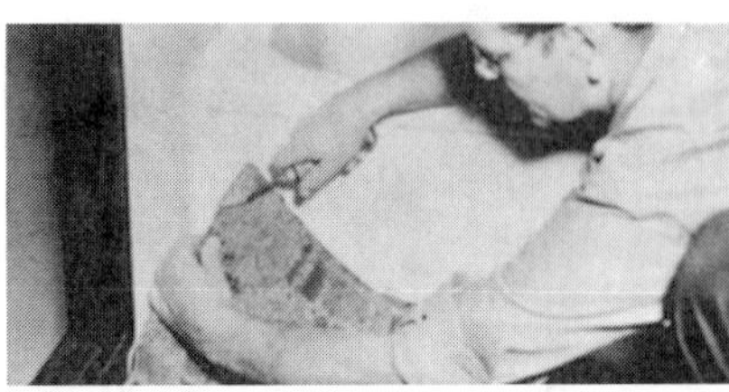

a

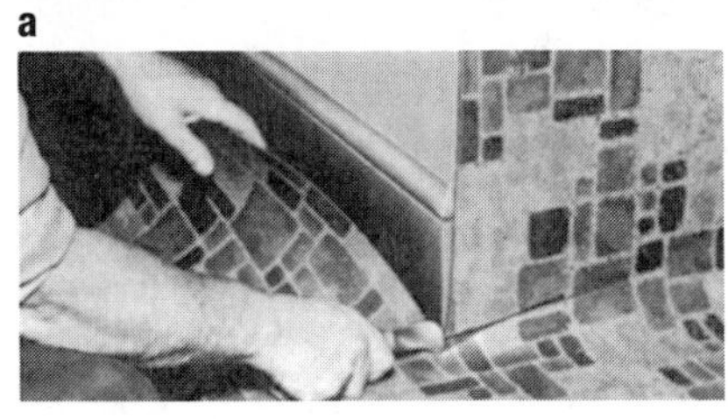

b

FIGURE 9.11-24 Safety cuts facilitate fitting (a) inside corner, (b) outside corner.

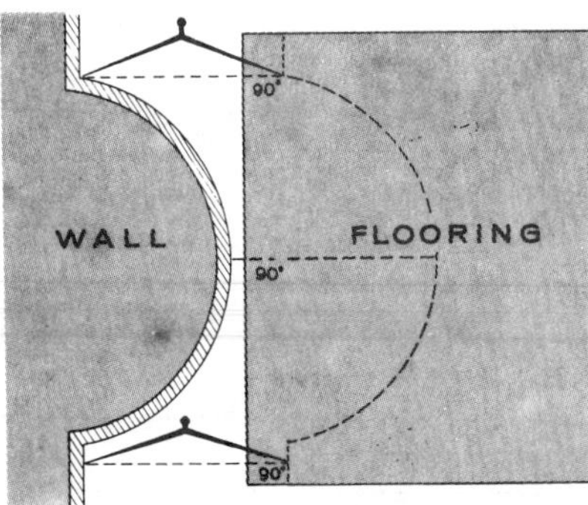

FIGURE 9.11-25 When scribing with dividers, the dividers should be kept perpendicular to the wall surface.

Scribing When flooring must be fitted against a vertical surface such as a wall, column, or stair riser with irregular corners, the exact contour of the vertical surface must be transferred to the flooring material. Scribing is generally used in fitting vinyl flooring to side walls and end walls and is the preferred method for all sheet flooring.

In scribing with dividers (Fig. 9.11-25), the flooring is pushed close to the wall or the projection and the dividers are opened so that one leg touches the wall at the farthest point and the other rests about 1 in. (25.4 mm) inside the edge of the flooring. With the dividers kept perpendicular to the wall, and one point always in contact with it, the profile is transferred to the flooring. The material may then be cut with a linoleum knife or, for vinyl sheet, with a notched-blade knife, and further fitted in place. Cutouts for isolated obstacles such as pipes can also be scribed with a scribing tool in a similar manner (Fig. 9.11-26).

Pattern scribing is used when the size of the room or wall projections make it difficult to scribe the flooring itself. Complex profiles are first scribed to a piece of felt or heavy paper, checked for fit, trimmed, and transferred to the flooring (Fig. 9.11-27). This is particularly desirable when wall base is not used or when the base is a straight type that will not cover minor irregularities in scribing.

End scribing, when the flooring is tubed, is similar to scribing for side walls. However, when the lapping method is used, each strip must be partly cemented before scribing. The unfitted end of the sheet is lapped back and, working toward the end wall from the center of the room, adhesive is applied to within 3 ft. (900 mm) of the end wall. The sheet is laid into the adhesive and rolled. The loose, uncemented end of the material is then scribed to the end wall by drawing a mark perpendicular to the edge of the material 9 to 12 in. (230 to 300 mm) from the wall and extending it from the material onto the subfloor. The material then is buckled and butted against the wall line (Fig. 9.11-28). The perpendicular distance between the line on the subfloor and the line on the material is measured with the scriber. With the scriber set at this distance, the material is scribed to the wall line and cut.

Seam Cutting To make a neat, inconspicuous seam between adjacent strips of flooring, it is necessary to cut away the selvage (factory-cut edge). Sheet widths are generally ½ in. (12.7 mm) greater than the nominal width to permit this excess material to be cut away. On goods with a geometric repeat pattern, the amount of selvage cut away cannot be varied if the pattern at the seam is to match and be aligned properly. In goods with a random overall pattern, the selvage to be trimmed is less critical and may vary (Fig. 9.11-29).

Seams between adjacent strips are made chiefly by *underscribing*, *single cutting*, or *double cutting*. To prevent mis-

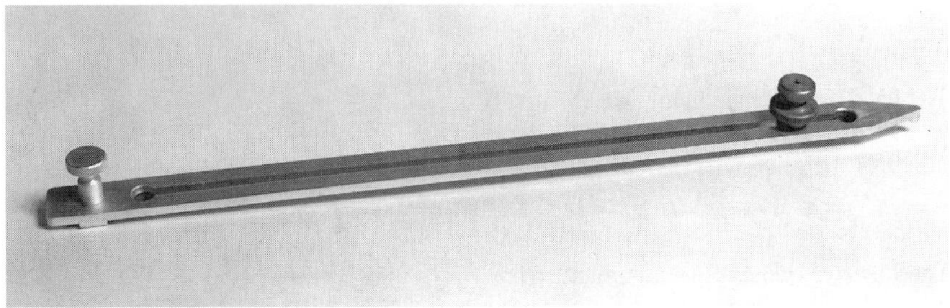

FIGURE 9.11-26 An 18-in. (457 mm) scriber permits scribing under obstacles, which would be impossible with dividers. (Photograph courtesy the makers of Armstrong vinyl flooring.)

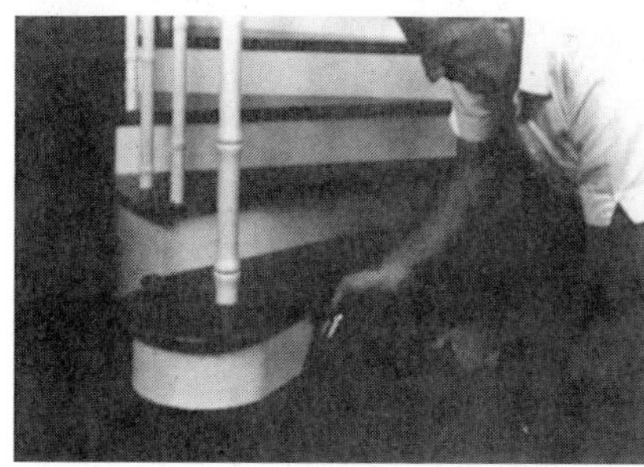

a

b

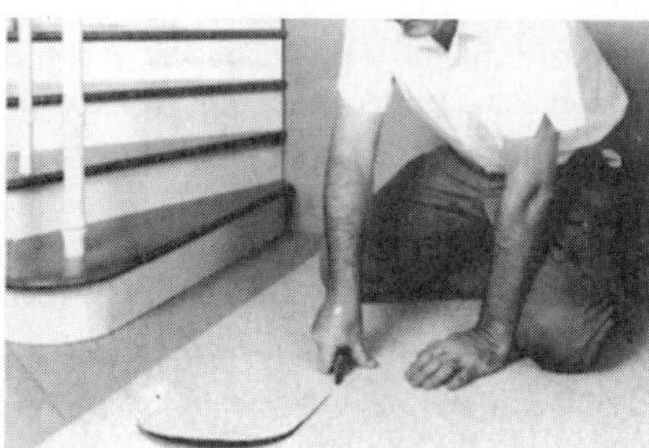

c

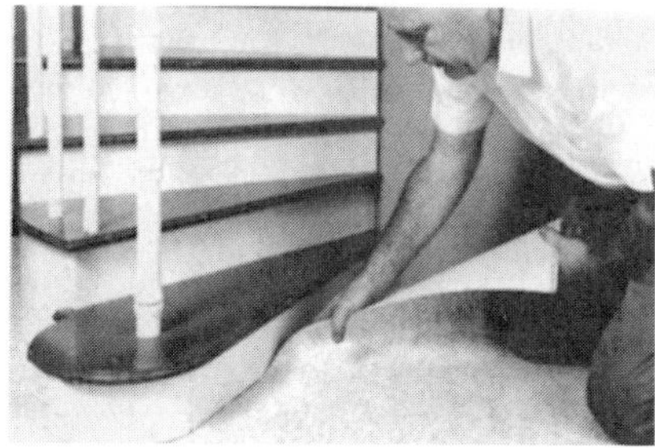

d

FIGURE 9.11-27 Pattern scribing: (a) felt pattern is scribed, (b) pattern is transferred to the flooring, (c) flooring is cut to the pattern, and (d) flooring is checked for fit.

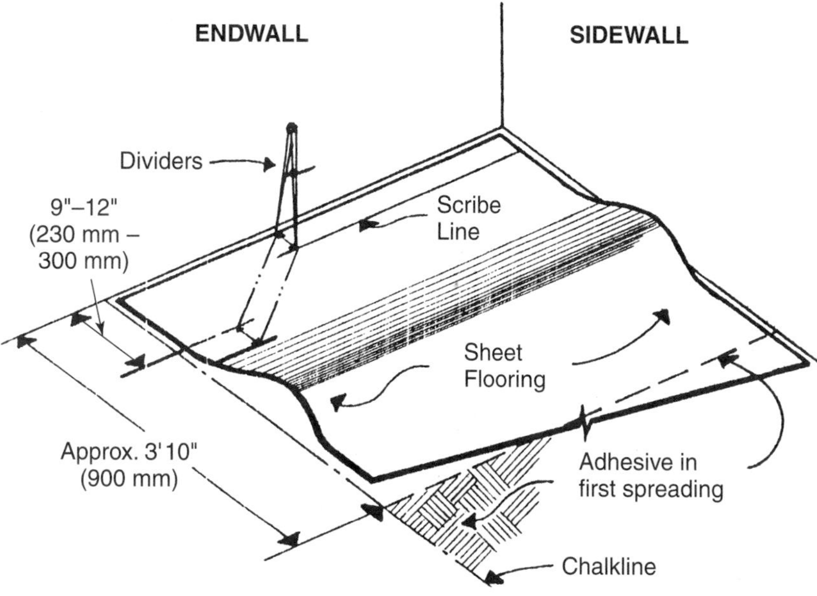

FIGURE 9.11-28 When lapping, each strip is partly cemented before end scribing.

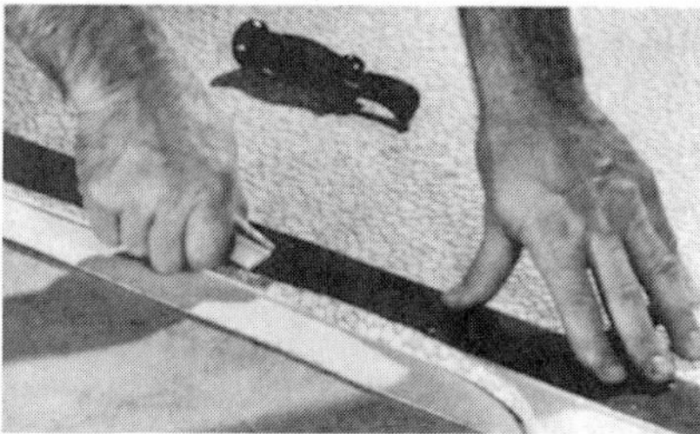

FIGURE 9.11-29 The selvage is trimmed with a straightedge and either a linoleum knife or a straight-blade utility knife.

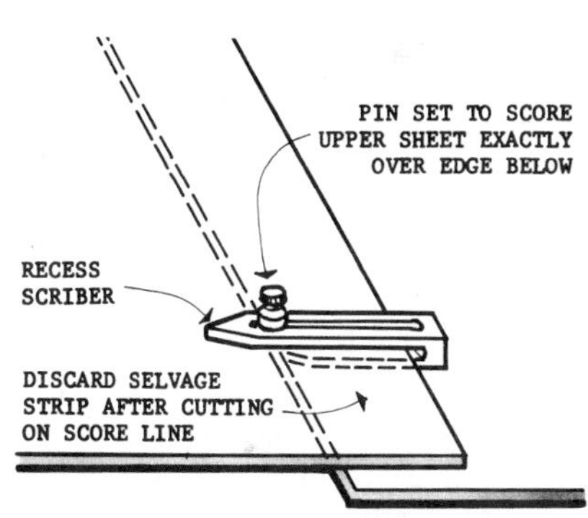

FIGURE 9.11-30 Underscribed seam.

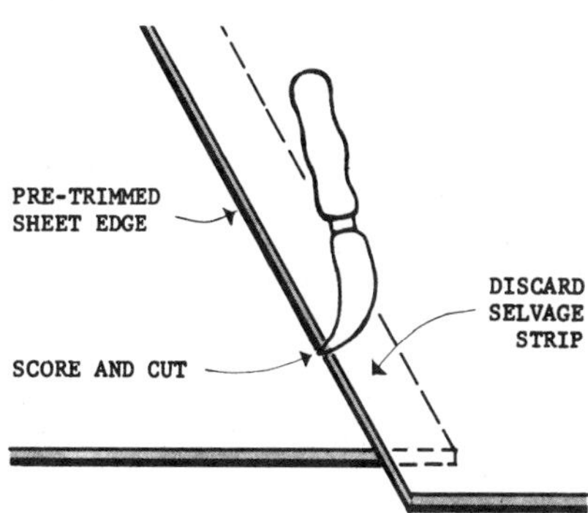

FIGURE 9.11-31 Single-cut seam.

alignment due to shifting, strips are partially or completely bonded to the floor before seams are cut. When using the single- or double-cutting method, adhesive is kept 4 in. (100 mm) or more from the seam until both edges are cut. This prevents adhesive from being pulled up when the lower edge (selvage) is removed. In both straight edging and seam cutting, a piece of scrap flooring is inserted under the cut line to prevent the backing from fraying. The knife is held as low as possible with the blade perpendicular to the floor surface.

In *underscribing*, a recess scriber is used to transfer the exact edge profile from the previously straight-edged strip to the adjacent strip. As it is guided along the finished edge, the tool leaves a score mark on the overlapped adjacent strip directly above. The scribe mark is then used as a guide for cutting (Fig. 9.11-30).

Underscribing is the preferred way of making seams for flooring with filled vinyl wear surfaces. The trimmed edge can be cemented down completely before underscribing, which often minimizes the number of operations involved in fitting and cementing the flooring.

Single cutting (Fig. 9.11-31) is used for making seams in patterned flooring where the pattern must be carefully aligned. The straight-edged strip is laid over the untrimmed strip, overlapping to fit the pattern. The lower sheet is scored with the linoleum knife guided along the upper trimmed edge, and is then cut along the score line.

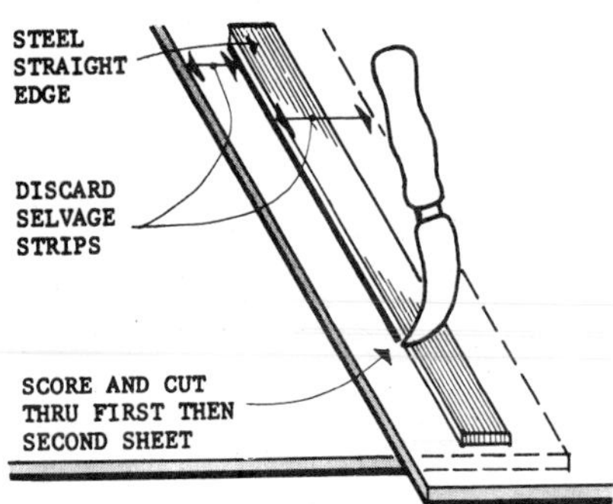

FIGURE 9.11-32 Double-cut seam.

Double cutting (Fig. 9.11-32) is similar to single cutting, except that neither edge is pretrimmed and the cut is made through both strips at the same time. With a straightedge as a guide, the upper sheet is cut in the middle of the overlapped edges, scoring the sheet below. The lower sheet is then cut along the score line. Double cutting is an older method and is not used as commonly as the underscribing and single-cutting methods, but may still be used with some clear (unfilled) vinyl-surfaced sheet, which is not as easily scored with an underscriber.

ADHESIVE BONDING

To facilitate spreading adhesive prior to bonding, the flooring is folded back by either tubing or lapping. The sequence of cutting and fitting, adhesive spreading, and cementing operations depends on which folding method is used. Lapping is the preferred method, but tubing is used where wall projections or the shape of the room make lapping difficult or impossible. Tubing is also recommended for heavy-gauge vinyl sheet and burlap-backed linoleum.

Adhesive is spread with a notched trowel, spreader, or brush, depending on the type of adhesive. As each strip is pasted down, the seam is smoothed with a hand roller and the entire strip is rolled with a 100-lb. (45.36 kg) or heavier roller (Fig. 9.11-33).

a

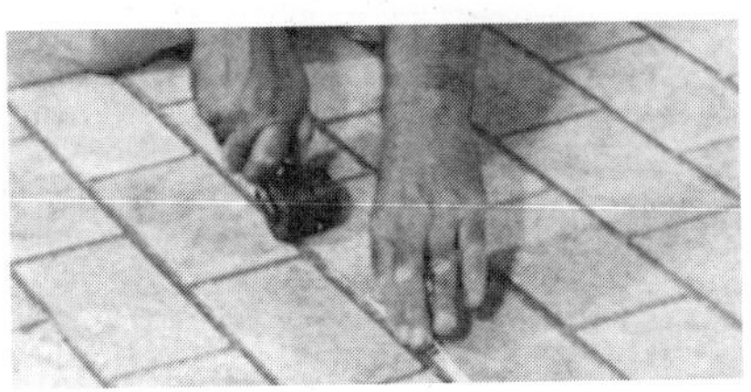

b

FIGURE 9.11-33 Adhesive bonding is completed by (a) rolling the flooring and (b) hand rolling the seam.

SEAM TREATING

Linoleum is softer than vinyl sheet and tends to "flow" around a knife blade without creating ragged edges. In cutting vinyl sheet, however, burrs are often created on the cut edges and must be removed to make the seams inconspicuous. After hand rolling the seams, these burred edges may be removed with the back of a linoleum knife moved in a "chisel" motion along the seam; or a hammer head or scrap of flooring doubled in half (face out) may be rubbed along the burrs while the seam is kept damp to avoid scuffing.

An *electric iron* may also be used to treat seams: a 1-in. (25.4 mm)-wide strip of heavy-duty aluminum foil is laid over the seam (dull side down) and rubbed to make the burrs show through. The foil then is pressed with an electric iron held at an angle and moved in repeated strokes toward the mechanic. When the foil has cooled, it is removed and the seam is cleaned with a damp cloth and a light abrasive cleaner.

TENSION FLOORS

Tension floor installation is a unique method of perimeter bonding recommended for certain sheet flooring products. Flooring may be installed by this method over well-bonded existing resilient floors over concrete and wood subfloors on all grade levels. Special adhesives allow the floor to be bonded only at the room perimeter, seam lines, and at columns or fixtures.

The two types of adhesives used are a two-part epoxy and a special solvent-based mastic. Installation method varies slightly with adhesive type, flooring type, and manufacturer's instructions. As in all resilient floor installations, the subfloor surface must be free of moisture, dust, paint, wax, oil, and grease and must be structurally sound and smooth.

A tension floor installation should be allowed to dry 4 or 5 days before it is cleaned. Floor maintenance is generally the same as for other resilient floors. However, when moving heavy objects such as appliances, a piece of plywood or particleboard should be placed under the object to prevent tearing or buckling the flooring material.

9.11.6.3 Cove-Base Flashing

Sheet flooring may be trimmed out with standard set-on cove base (see Section 9.11.3), or the material may be extended partly up the wall to make an integral flashed cove base (Fig. 9.11-34). To make the cove profile at the meeting of the wall and floor, a ⅞-in. (22.2 mm) wax fillet strip is cemented or a wood cove molding is nailed in place. A binding strip is then nailed or cemented to the wall at the desired height to protect the top of the base. The flooring may be softened with a torch to facilitate molding it against the fillet strip. In heavy traffic areas, inside and outside metal corners may also be desirable

FIGURE 9.11-34 Flashed cove base with (a) outside metal corner, (b) inside metal corner, (c) metal binding strip, and (d) wax cove fillet.

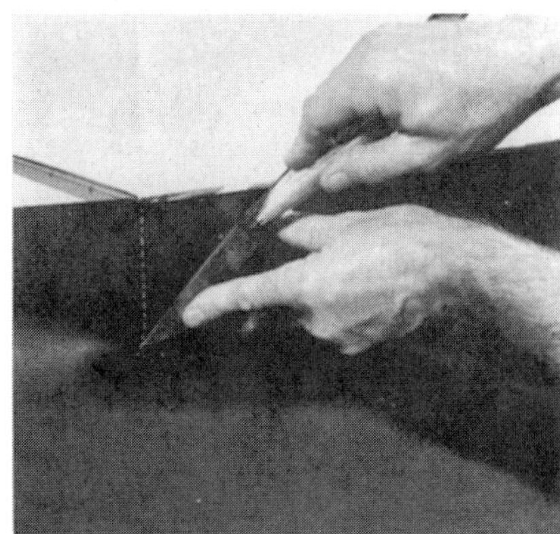

FIGURE 9.11-35 Felt pattern is scribed to the binding strip, and 45-degree V cutout is formed for outside corners of flashed base.

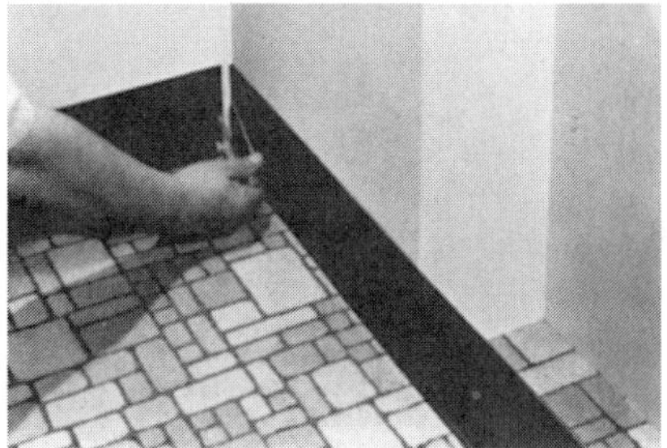

FIGURE 9.11-36 Inside base corner. Right side is butted to corner and bonded; left side is scribed to right side and bonded.

(see Fig. 9.11-34). Similar metal endstops are available for closing base ends at door openings without casing trim.

In rooms requiring a decorative sheet border, the full-width strips may be pasted down in the field first; then the narrower border strips are flashed up the wall, scribed against binding strips, and cemented in place. After the border strip has been cut to proper width, inside and outside corners without metal corner pieces may be scribed and installed.

Where no decorative borders are used, it is necessary to flash the ends and edges of the flooring strips up the wall. When working with such large pieces, it is difficult to cut flooring accurately by fitting and cutting in place. Therefore, it is necessary to make a felt pattern of the entire floor and base, carefully scribing and cutting around projections to the binding strips (Fig. 9.11-35). The pattern is transferred to the sheet flooring, which then can be readily cut, fitted, and cemented in place by either the tubing or lapping method. Inside corners can be made by scribing as shown in Figure 9.11-36 for resilient base. Figure 9.11-37 illustrates the preferred method of making an outside corner. A less durable corner can be made simply by mitering the flashed material at the corner.

A 100-lb. (45.36 kg) roller should be used to roll the floor, but a flashed cove base requires a small hand roller. Care should be taken not to allow the edge of the heavy roller to ride over the fillet, as it may damage the installation.

9.11.7 TILE FLOORING INSTALLATION

The following text describes the installation of resilient tile flooring in new construction.

Resilient floor tile is available in vinyl, vinyl composition, rubber, asphalt, cork, and linoleum. Tile products range in size from 9 in. by 9 in. (28.6 by 28.6 mm) up to 36 in. by 36 in. (914 by 914 mm), in various gauges and in varying colors and patterns. Some tiles are made in no-wax finishes and with self-adhesive backs. Ingredients, gauges, and standards of quality are summarized in Figure 9.11-1; selection criteria in Figure 9.11-5.

9.11.7.1 Preparation for Flooring

The effectiveness of a resilient tile flooring installation greatly depends on the proper selection and preparation of the elements that make up the installation. Subfloors that are to receive the flooring directly must be firm, smooth, and dense and possess good bonding properties. When the subfloor does not have these qualities, it will be necessary to prepare the surface by grinding or sanding or with the use of underlayment. Adhesives should be selected that are compatible with all elements, including the flooring itself. All elements must be suitable for the intended locations.

SUBFLOOR

Solid vinyl, mineral-fiber-backed vinyl, vinyl composition, and rubber tile may be installed on subfloors in suspended, on-grade, and below-grade locations. Cork, vinyl-cork, rag felt–backed vinyl, and linoleum tile should be installed on suspended subfloors only.

Concrete subfloors in all locations should be tested for moisture and bonding properties to ensure suitability for the proposed flooring and adhesive (see Section 9.11.5).

UNDERLAYMENT

The condition of the subfloor will determine the need for underlayment. See Section 9.11.4.1 for a discussion of underlayments and recommended uses. Mastic underlayments should be compatible with the intended adhesives and primers, as discussed below.

ADHESIVES

The adhesives recommended for tile products in various locations are summarized in Figure 9.11-16. For a discussion of

a

b

c

FIGURE 9.11-37 Outside one-piece corner: (a) V cutout is transferred from the pattern to the flooring before installation; (b) corner piece is cut out separately with a square; (c) before matching corner piece is fitted, a narrow strip of backing is removed at the fold.

adhesive types and their properties, see Section 9.11.4.3 and Figure 9.11-15. Solvent-based asphaltic adhesives and primers used for the installation of asphalt and vinyl composition tile should not be used over asphaltic underlayments. For such installations, asphaltic emulsions should be used. When cutback asphalt, which is a solvent-based adhesive, is used over latex underlayment, the amount of time recommended by its manufacturer must be allowed before tile is installed. Porous or dusty concrete floors and panel underlayments intended for the installation of asphalt or vinyl composition tile should be primed with an asphaltic primer before adhesive is applied. Some tiles are made with self-adhesive backs, but these are not generally used in commercial or institutional projects.

9.11.7.2 Installation of Flooring

The first step in the installation of resilient tile flooring is to make a trial design layout of the tiles. The tiles then are cemented down according to the best layout, and walls are trimmed out with a resilient or other type base.

DESIGN LAYOUT

In determining design layout, it is important that tile joints do not fall over joints in the subfloor or in the underlayment below. Because few walls are seldom exactly parallel, it is necessary to center and square off the room. Tiles may be laid in either *square* or *diagonal designs*. Over existing wood strip floors, a diagonal installation is preferred.

Square Designs Tile joints for square designs run parallel to the walls. The trial layout is started by snapping chalklines to divide the room into four equal quadrants (Fig. 9.11-38). If the floor area is extremely large or job conditions prevent working the entire floor area, each quadrant may be subdivided further. Full tiles should be laid starting from the center working toward the walls. The first trial may result in unequal border tiles at side walls and end walls (see Fig. 9.11-38a).

To equalize border tiles at walls, the horizontal row of tiles may be shifted so that it is bisected by the chalklines, as shown in Figure 9.11-38b. In general, layout lines should be shifted so that approximately equal border tiles, at least a half-tile wide or wider, will result.

Checkerboard patterns may be made

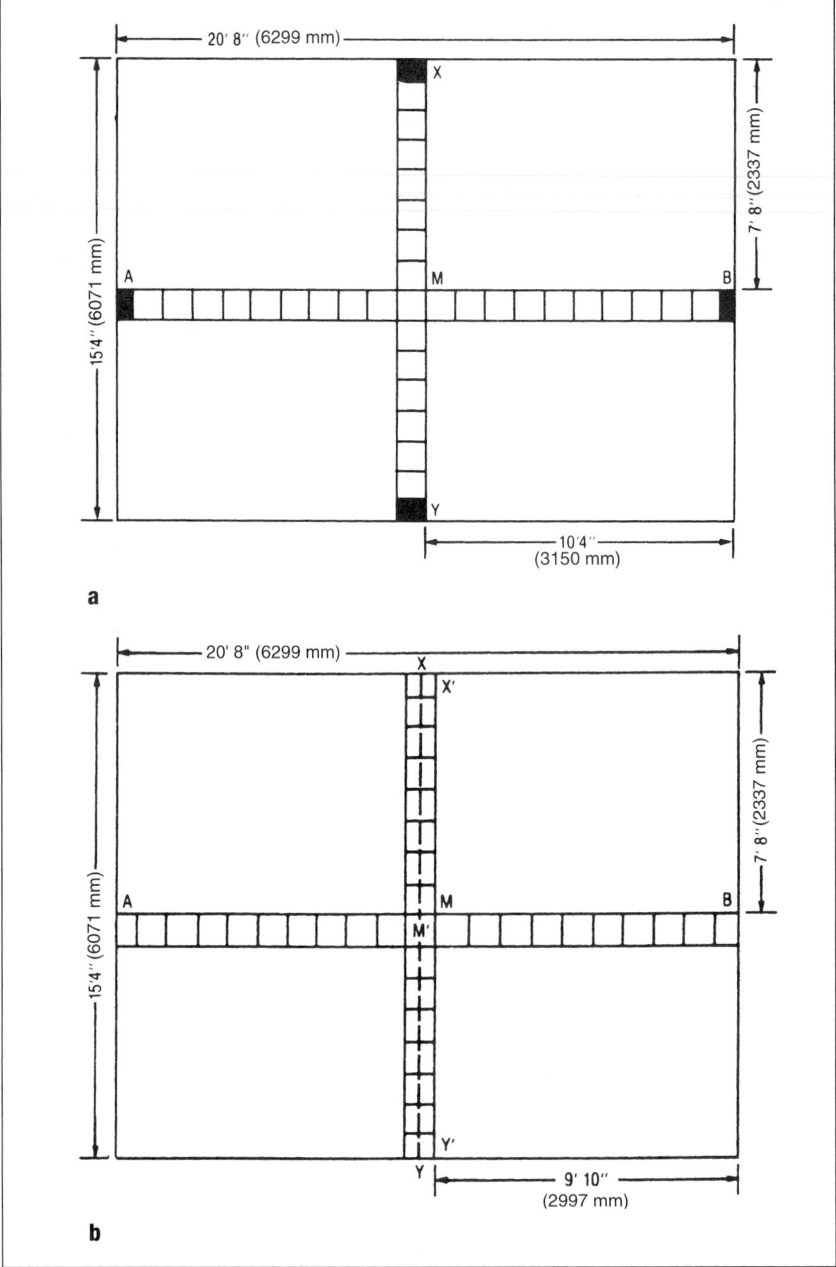

FIGURE 9.11-38 Trial layout, square design: (a) unacceptable layout (size of border tiles differ); (b) acceptable layout (border tiles are equal and wider than a half-tile). (Illustration by the makers of Armstrong vinyl flooring.)

by alternating tiles of two different colors or alternating the grain direction in directional tiles. Either a tile that matches the field or a contrasting one-color sheet or tile can be used for the border.

For a *hollow-square* design, another two-color pattern, it is important to lay out the trial design so that a full repeat of the pattern will occur at all walls (Fig. 9.11-39). Such patterns are commonly used with one-color borders to emphasize the design in the field.

Diagonal Designs Tile joints are run at 45-degree angles to the walls for diagonal designs. A trial row of tiles is laid out along each centerline and diagonally to the corner to determine the resulting condition at the border (Fig. 9.11-40a).

Two-color designs, requiring a border laid on the square, must also be checked to ensure that the half-tiles along the walls will be of the same color. To achieve this, it may be necessary to shift tiles from the trial position shown in Figure 9.11-40a, so

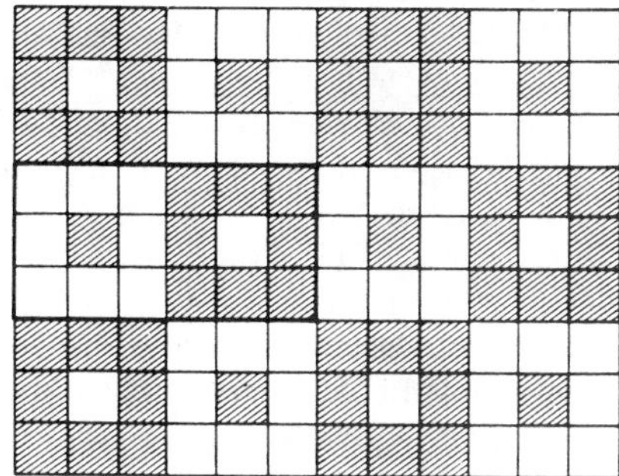

FIGURE 9.11-39 **Hollow-square design; repeated unit indicated by dark outline.** (Illustration by the makers of Armstrong vinyl flooring.)

that tiles will abut rather than be bisected by diagonal lines (Fig. 9.11-40b). The proper starting position of tiles at the center of the room is determined by the dimensions of the room and must be established in each case by trial layout.

If it is determined that the design should start with single tiles at the intersection of diagonals, new diagonals should be struck a half-tile over from the original location to serve as guidelines for the first row of tiles.

One-color designs without a border present no special problem, and the starting point need be adjusted only to ensure that the layout will not result in excessively narrow tiles at the walls. In one-color installations with tiles having a directional grain, it is important that the half-tiles near the border fit properly into a basketweave pattern. If the tiles are cut properly with respect to grain, tiles at both the right and left sides will fit the pattern (Fig. 9.11-41).

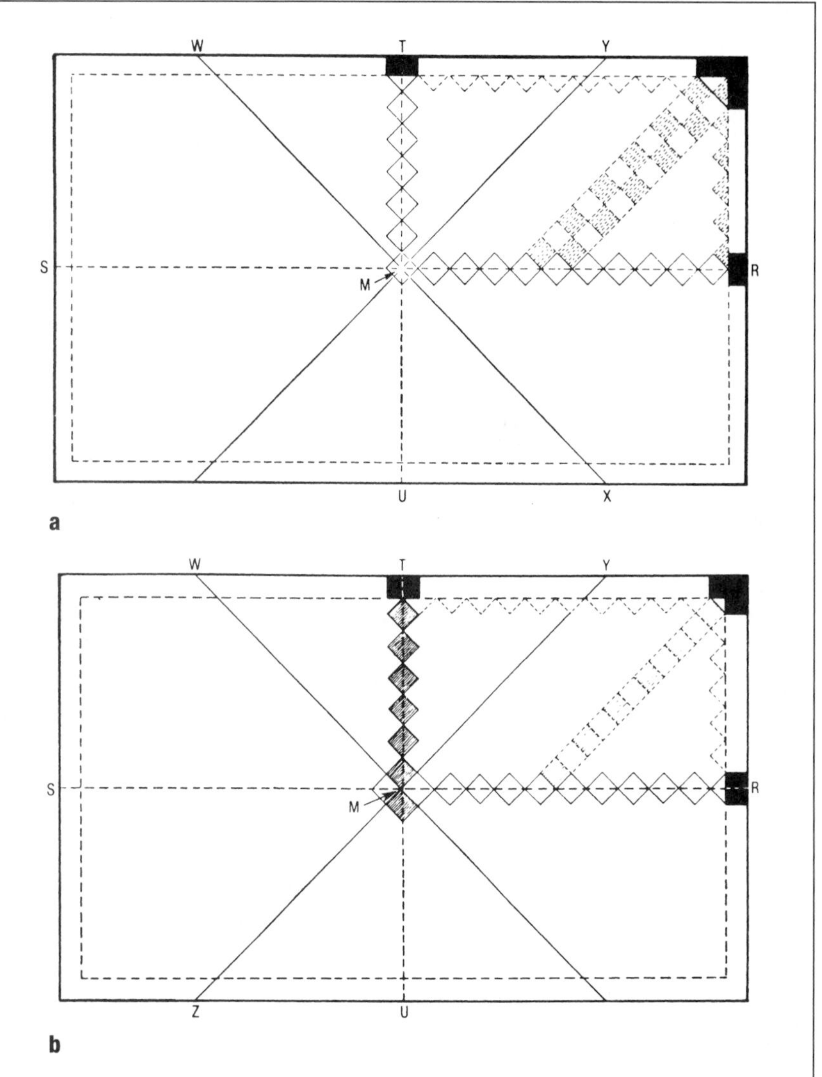

FIGURE 9.11-40 **Diagonal design, trial layout: (a) unacceptable layout, (b) acceptable layout—all half-tiles at the border are of the same color.** (Illustration by the makers of Armstrong vinyl flooring.)

ADHESIVE BONDING

To allow sufficient working space to handle materials, only part of the floor area, depending on the working time of the adhesive, should be covered with adhesive at one time. Care should be taken in applying the adhesive so that chalklines will not be completely concealed. This may be done by leaving ends and intermediate points on the chalkline uncovered.

Adhesive such as cutback asphalt develops holding power by setting before tiles are applied. However, with some adhesives such as waterproof-resin and epoxy, the possibility of tiles slipping may be a problem. Temporary lattice strips nailed at the chalklines will prevent the tiles from sliding out of place during installation.

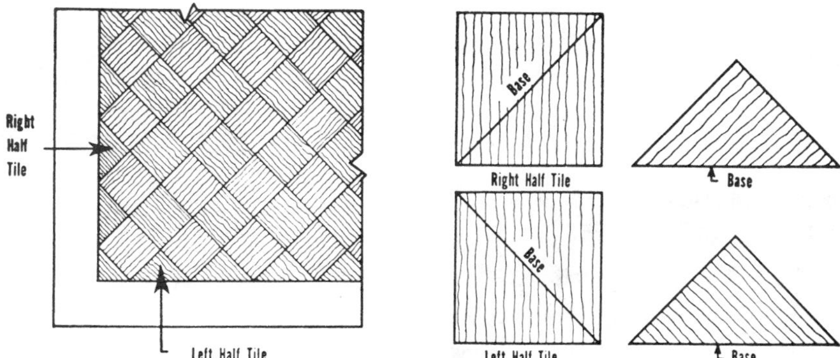

FIGURE 9.11-41 **Tiles must be cut according to grain to obtain half-tiles that will fit into a basketweave pattern.** (Illustration by the makers of Armstrong vinyl flooring.)

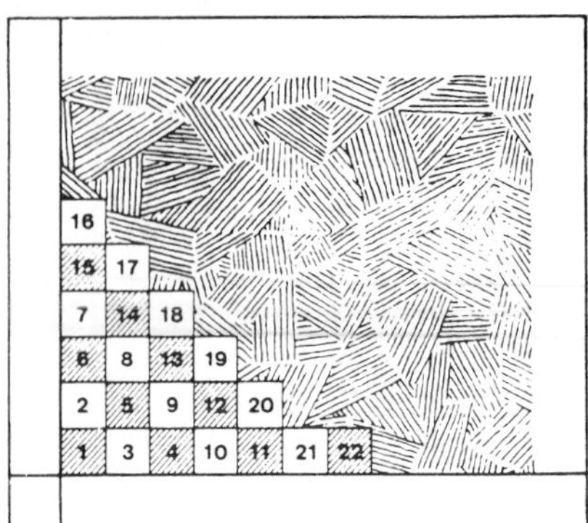

FIGURE 9.11-42 Checkerboard square designs can be installed more quickly by laying tiles of one color in diagonal rows as numbered. (Illustration by the makers of Armstrong vinyl flooring.)

Tiles should be installed with tight, straight, and carefully aligned joints in both directions. Each tile should be butted to the preceding tiles and lowered in place without sliding.

Installation on the square with a two-color checkerboard pattern can be speeded by starting with a row of 10 tiles along each centerline, and laying tiles of one color at a time in diagonal rows (Fig. 9.11-42).

For diagonal designs, tiles may be cemented down in one quadrant at a time, starting at the center, laying the tiles along the diagonal and proceeding clockwise around the room.

FITTING AND CUTTING

Border tiles may require scribing and fitting against walls and obstructions. They are scribed essentially as described for sheet flooring, using dividers or a scribing tool to transfer the contour of the vertical surface (see Fig. 9.11-25).

Most resilient tiles can be cut at room temperature with a linoleum or straight-blade knife. However, it is easier to cut vinyl composition and asphalt tile if the tile and the knife are heated with a blowtorch or a heat gun.

9.11.8 INSTALLATION OF RESILIENT FLOORING ACCESSORIES

9.11.8.1 Wall Base Installation

The two main types of wall base used with resilient tile are set-on and butt-cove base (see Fig. 9.11-8). Most tile installations are made with set-on base because it is easier and more economical to install. This type of base is set on top of the finished floor and requires less accurate fitting of both flooring and base. Butt-cove base provides a more sanitary condition and may be required in spaces where cleanliness is mandated, such as in laboratories and some hospital spaces. Both set-on and butt-cove base are available in rubber and vinyl.

RUBBER BASE

Rubber base is manufactured in straight runs and in premolded outside and inside corners. These facilitate turning corners with a minimum of field labor, the only fitting required being at butted joints. Rubber base is used to match rubber flooring and in cases where premolded corners are desired.

VINYL BASE

The greater pliability of vinyl base permits molding it around corners, and therefore it is manufactured mainly in straight runs. Inside corners are made by notching a V in the back of the base at the fold and heating it to increase pliability. The base may then be formed and pressed into the corner. Inside corners can also be made by butting the first piece squarely in the corner and scribing the second piece around it (see Fig. 9.11-36).

Outside corners are formed by first cutting away about half the thickness from the back of the material at the point of bend. At the back of the cove heel, additional material should be cut away approximately 1 in. (25.4 mm) to each side of the bend. The base can then be molded readily around the corner and set in place with standard base cement or synthetic rubber cement. Synthetic rubber cement is preferred because it provides immediate holding power and prevents movement caused by contraction as bond is developed.

9.11.8.2 Stair Treads, Risers, and Nosings

Resilient sheet materials can be used to provide a covering over wood and concrete stairs. The sheet material can be flashed from the tread up the riser (Fig. 9.11-43a) or used with a set-on coved rubber or vinyl riser (Fig. 9.11-43b). The nosing of a wood tread may be wrapped with the sheet flooring or protected by a separate metal or rubber nosing. Treads on riserless wood stairs may be completely wrapped with sheet materials (Fig. 9.11-43c). Formed concrete stairs can be covered with continuous runs of sheet flooring on both treads and risers (Fig. 9.11-43d). Methods of installing sheet

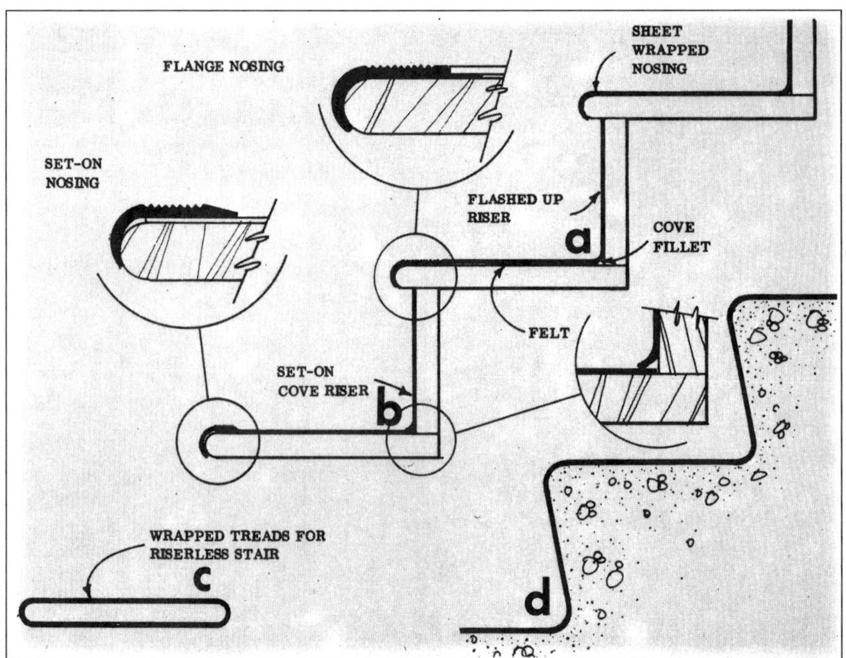

FIGURE 9.11-43 Stair coverings with sheet flooring: (a) one-piece tread and riser, (b) sheet tread and set-on cove riser, (c) wrapped riserless wood tread, (d) continuous risers and treads on concrete.

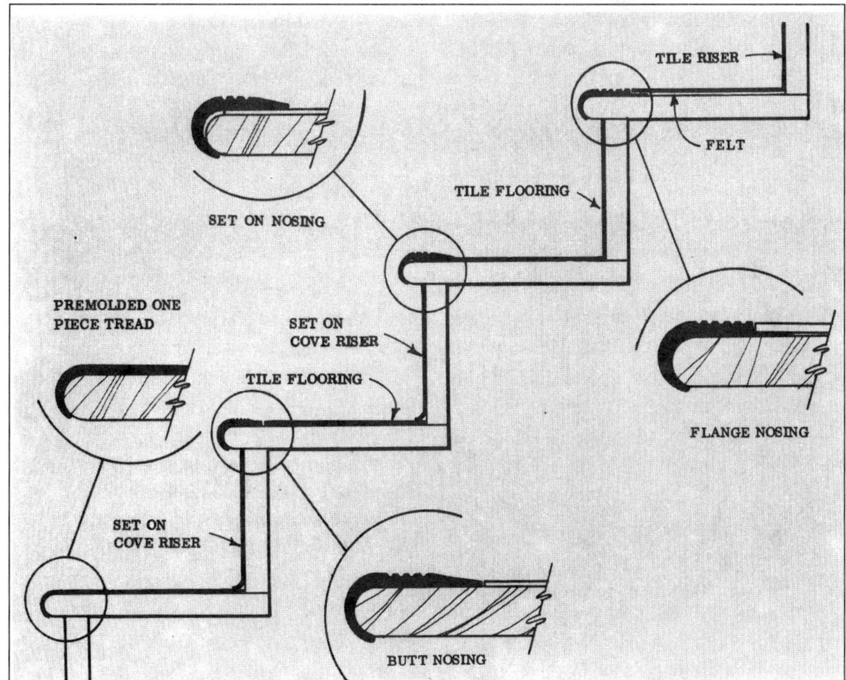

FIGURE 9.11-44 Stair coverings with tile flooring and resilient accessories.

stair covering are similar to those described for sheet flooring. Accurate layout, scribing, fitting, and proper selection of adhesive are essential.

Resilient tiles can also provide an attractive wearing surface on wood and concrete stairs. Tile treads are usually installed with rubber, vinyl, or metal nosings (Fig. 9.11-44). Risers for tile treads may be of matching tile or a one-piece set-on coved riser similar to wall base. Matching one-piece treads with premolded nosings also are available from some manufacturers (see Fig. 9.11-44). Methods of installation are similar to those described for tile floors. Accurate layout, scribing, fitting, and proper selection of adhesives are essential.

9.11.9 REFLOORING EXISTING CONSTRUCTION

Both sheet and tile products may be installed over existing finished floors that have been suitably prepared to provide a firm, smooth, and stable base for the new flooring. The main types of floor finishes encountered in reflooring are (1) wood strip, block, or plank, (2) concrete, (3) ceramic tile, (4) terrazzo, and (5) marble. New flooring may be installed over old resilient flooring, provided that the old flooring is not cushioned or deeply embossed. For specific recommendations, the manufacturer of the new flooring should be consulted.

Existing floors intended for reflooring should be inspected to ensure a firm base free of loose particles. Existing wax, oil, grease films, and most paints should be removed. Surfaces should be smooth and reasonably level, or leveled with an appropriate underlayment when necessary. Recommendations for flooring and adhesive selections with respect to grade locations should be adhered to as in new construction.

9.11.9.1 Wood Floors

A suitable panel underlayment is recommended over existing wood floors. However, sheet products may be applied directly to double-layer floor construction consisting of a lumber board or plywood subfloor and a finish wood strip floor of tongue-and-groove boards, at least ¾ in. (19.1 mm) thick and not wider than 3 in. (76.2 mm), if the existing floor is sufficiently smooth, level, firm, and well seasoned.

When panel underlayment is not used, a protective film on wood floors may be removed with a power sander or a blowtorch. Mild chemical solutions in water should not be used, because the large quantity of water required for scrubbing and rinsing may cause the boards to warp. After a wood floor has been sanded, one brushed coat of floor sizing consisting of equal parts of a wax-free shellac and denatured alcohol should be applied to seal the wood. This prevents excessive moisture absorption from adhesive and other sources and minimizes the possibility of boards warping in the future.

Other double wood floors of block and plank and single-layer plank floors should be covered with underlayment consisting of at least ¼-in. (6.4 mm) plywood, ¼-in. (6.4 mm) hardboard, or ⅜-in. (9.5 mm) particleboard.

9.11.9.2 Concrete Floors

Most paint films can be removed from concrete floors by power grinding, burning with a blowtorch, or scrubbing with a solution of trisodium phosphate and water. After the paint has been removed, the floor should be scrubbed with a strong soapy solution, rinsed with clear water, and allowed to dry thoroughly. Paint removers should not be used, because they react adversely with many adhesives. Because the use of trisodium phosphate and some other paint removers is not permitted in some jurisdictions, it is necessary to check with local authorities before using them.

Paints with a chlorinated rubber or resin base are often tightly bonded to the concrete surface and, if the mat bonding test reveals good adhesion, may be left in place.

9.11.9.3 Ceramic Tile, Terrazzo, and Marble

Ceramic tile, terrazzo, and marble floors should be cleaned with a scrubbing machine using soapy water and clean, sharp sand. A floor grinder used dry with 4½-grit open-coated paper may be preferred to avoid delay due to drying. Badly fitted joints and cracks should be leveled with latex underlayment or a patching compound.

9.11.10 CLEANING

A resilient floor can be damp-mopped immediately and should be polished within 48 hours. However, it should not be given a thorough cleaning until 4 to 5 days after installation. The floor should then be washed with a mild solution of a nonabrasive, nonalkaline cleaner or soap and thoroughly rinsed with cold water. It should be allowed to dry, then waxed with a floor finish recommended by the flooring manufacturer. A soft cotton mop or wax applicator should be used to apply the floor finish as thinly as possible.

No-wax tile and sheet floors should be cleaned the same way as other resilient floors; however, waxing is not necessary. Because they have no renewable surface finish, it is important to protect no-wax floors from severe abrasion, such as caused by gritty dirt and furniture without floor protectors (see Fig. 9.11-6).

9.12 Carpet

The original definition of *carpet* was a rug with a repeating pattern. Wall-to-wall broadloom carpet was developed in the United States, but the art of carpet making has been passed down for many centuries from many cultures. The immediate roots of the American carpet industry are in Western Europe.

The Moorish invasion of Spain in A.D. 711 introduced carpet craft to Europe. The patterns of Spanish carpet intertwined the living forms of humans, animals, or plants and the symbols of church and state with rich Arabesque ornamentation.

When Queen Eleanor of Castille married Edward I of England in 1251, she took many carpets with her to England and thereby introduced the art of carpet making to the English.

Carpet making excelled in France, particularly during the reign of Louis XIV. The patterns of French carpet did not borrow from other cultures. They usually featured floral designs complemented with elaborate scrolls and noble insignias. The secrets of French carpet making were kept primarily in France until 1685, when Louis XIV revoked the Edict of Nantes. This revocation caused many Protestant carpet masters to flee to England. These carpet masters built the first hand looms in England.

The English changed carpet making from a craft to an industry. Before that, carpet weaving, like many other works of art, was primarily commissioned by nobles. As the wealth of Great Britain increased during the eighteenth century, so did the carpet industry. For the first time, people of classes other than the nobility were able to purchase floor coverings.

English inventions during the Industrial Revolution formed the basis for present-day methods of carpet manufacture. As in their needlework, the English incorporated their passion for naturally arranged flowers into their carpet.

The carpet industry in North America began to grow after the American Revolution. The first power loom was invented for weaving carpets in Lowell, Massachusetts, in 1839. For the first time, carpets could be mass-produced. To simplify the process of mass production, carpet makers made carpet without patterns. Carpet was sold primarily for function. However, this monotone product was merely a step in the evolution of carpet.

Today, carpet making has advanced to the point where carpet makers can reincorporate pattern into affordable carpet. As European nobles commissioned carpet to display their insignias, many businesses today order patterned carpet to display their logos. Other businesses order patterned carpet to enhance the aesthetic appeal of their workplaces.

Carpet is often selected for its comfort and decorative values, but it is also effective in reducing impact sound transmission through floor and ceiling assemblies. Therefore, its acoustical performance should be a prime consideration in flooring selection. Many common installations provide a performance considerably higher than the range of –5 to +10 on the Impact Noise Reduction (INR) scale (see Section 13.1). Specially selected combinations of carpet and cushioning may be rated as high as +17 over wood floors, and +29 over concrete slabs.

Carpet is also capable of absorbing sound and reducing sound reflection within a room, much like acoustical ceiling tile. The Noise Reduction Coefficient (NRC) of most carpet ranges between 0.35 and 0.55, which compares favorably with a range of 0.55 to 0.75 for acoustical ceiling tile. Properly selected carpet may achieve absorption coefficients equal to those of acoustical tile. However, carpet is not particularly effective in controlling the transmission of airborne sound, and where this is an important consideration, special sound-isolating construction should be used.

This section describes carpet suitable for long-term wall-to-wall installation, as distinguished from rugs, which usually have bound edges and are laid loose over a finished flooring (Fig. 9.12-1).

FIGURE 9.12-1 Carpet goes from wall to wall; rugs do not. (Courtesy Carpet and Rug Institute.)

Carpet can be identified by its type, which defines how it is made, by the fibers used to make it, and by its face construction.

9.12.1 CARPET CONSTRUCTION

Most carpet consists of pile yarns, which form the wearing surface, and backing yarns, which interlock the pile yarns and hold them in place. Therefore, carpet can be identified according to its construction, which describes the method of interlocking the backing and pile yarns. Comparative factors related to carpet construction, such as pile yarn weight and pile thickness, number of tufts per square inch, and other factors, may be useful in comparing carpets of similar construction.

Its face construction is another factor by which carpet can be identified. The following terms describing face style are defined in the glossary: *level-loop pile*, *cut pile*, *level-tip shear*, *multilevel loop*, *random shear*, *frieze* or *twist*, and *sculptured* or *carved* (Fig. 9.12-2). Some face styles are dictated by the method of manufacture. For example, fusion-bonded goods can only be cut pile. Similar restrictions are identified in the following discussion.

9.12.1.1 Carpet Types

The term *broadloom* is often mistakenly thought to define a type of carpet construction, but it actually refers merely to carpet wider than 6 ft. (1800 mm). With the exception of rubber-backed loomed carpet, and carpet tiles, the carpet types discussed in this section are manufactured in widths of 9, 12, 15, and sometimes 18 ft. (2.74, 3.66, 4.57, and sometimes 5.49 m). Loomed carpet generally is made in a width of 4 ft. 6 in. (1371 mm), with a $\frac{3}{16}$-in. (4.88 mm) attached rubber cushioning.

Carpet can be classed according to the methods of manufacture as woven, tufted, knitted, fusion-bonded, needlepunched, and flocked. Of these, tufted and woven are the methods most often used for making yard goods, and fusion-bonded is the most prevalent method of making carpet tile. In general, weaving and knitting interlock pile and backing yarns in a single operation, and a tufting loom stitches pile yarns onto a premade backing. Tufted carpets can be produced much faster than woven or knitted types, are less expensive to produce, and currently account for about 95% of all carpet manufactured in this country.

WOVEN CARPET

Woven carpet is made on looms that interweave pile yarns and backing yarns in one operation. Backing yarns consist of *weft* (crosswise) yarns and *warp* (lengthwise) yarns. Because the tuft-forming yarns in woven carpet run lengthwise, they too are classed as either *face warp* or *pile warp*. Although more properly applied to woven carpet, the terms *warp* and *weft* are loosely applied to knitted carpet as well.

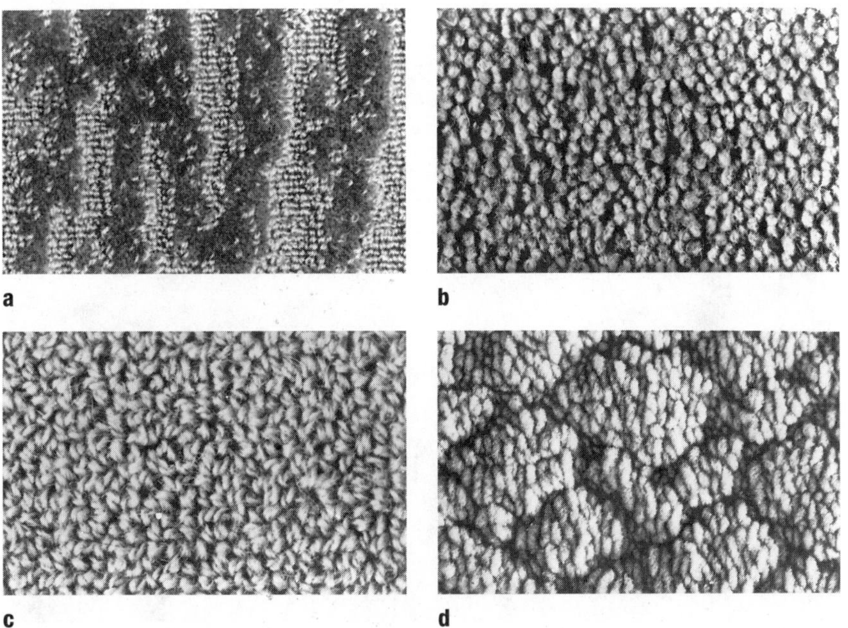

FIGURE 9.12-2 Some carpet texture variations: (a) twist, (b) multilevel loop, (c) random shear, combining cut and loop pile, (d) single-level looped pile. (Courtesy Carpet and Rug Institute.)

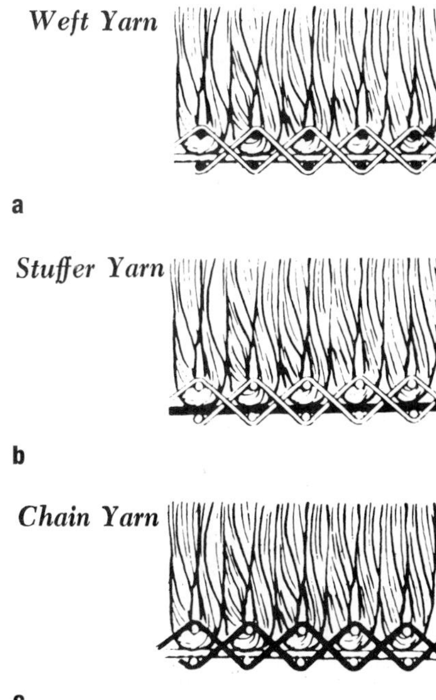

FIGURE 9.12-3 Backing yarns of woven carpet identified by heavy lines: (a) also called weft shots or filling; (b) also called stuffer warp; (c) also called chain warp, binder, or binder warp. (Courtesy Carpet and Rug Institute.)

Weft yarns run crosswise, passing over the pile yarns (Fig. 9.12-3a). *Stuffer yarns*, a type of warp yarns, give strength and dimensional stability to a carpet (Fig. 9.12-3b). Typically, there are two or three stuffer yarns per tuft running the length of the carpet. *Chain yarns*, another type of warp, run the length of the carpet, alternately passing over and under the weft yarns, forming a chain that locks the weft in place and pulls the pile yarns down tightly (Fig. 9.12-3c).

Four weaving processes are used, from which woven carpets derive their names: velvet, Wilton, Axminster, and loomed.

Velvet A velvet weave is the simplest of all weaves (Fig. 9.12-4). Chain, stuffer, and pile yarns are alternately raised and bound together with the weft yarns. The pile is formed by *wires* inserted between pile and warp yarns. The height of the wire determines the depth of the pile. After a row of loops has been formed over the wire and secured by the weft, the wire is withdrawn, leaving a row of loops. Cut pile is created by a razor edge on the wire that cuts the loops as the wire is withdrawn.

Because a velvet loom feeds pile yarn from a single spool, it is best suited for solid colors. However, strands of different color can be twisted together into a single yarn (*moresque*), producing tweeds, salt-and-pepper effects, and stripes. A wide range of textural effects is possible by using uneven wires to form high and low loops.

Velvet weave can be recognized at a cut edge or open selvage where the chain warp is visible, binding all the construction yarns together in a zigzag pattern (see Fig. 9.12-4).

Wilton A Wilton loom (Fig. 9.12-5) is basically a velvet loom fitted with a special *jacquard* mechanism that uses punched cards to select the various colored yarns. These cards, locked together to match the entire surface design of the carpet, are suspended in the loom. Colored pile yarns are fed from spools assembled on frames (racks) in the back of the loom, one frame for each color. As the cards pass, the jacquard mechanism selects and lifts the appropriate pile yarns into position to be looped over the wires, while other yarns remain buried in the body of the carpet. Depending on the number of colors used, for each tuft of yarn showing on the surface, there may be as many as four strands of yarn buried in the backing, adding to the stiffness of the carpet (Fig. 9.12-6).

A carpet woven on a Wilton loom with just one colored pile yarn would have the same construction as a velvet weave and would be designated as a velvet carpet. Wilton carpets are designated as two-, three-, four-, or five-frame Wiltons, depending on the number of colored pile yarns used in their manufacture. Generally speaking, for carpets of the same density and yarn size, the more frames used, the greater the pile yarn weight and, hence, the better the quality.

A Wilton loom can weave sharply delineated sculptured and embossed textures or patterns, which are created by varying the pile height, using high and low loops or combinations of cut and uncut piles. Wilton carpets with uncut loop pile are called *roundwire* Wilton.

In multicolor Wilton weaves, the different colors of the pile fibers are usually visible on the back, and the body of the carpet is thicker than that of other types. As in velvet weaves, the chain warp is also visible at a cut edge.

Axminster Axminster carpet is woven on a loom that draws pile yarns from small spools wound with yarn of various colors. The spools are locked together in a frame equal to the width of the carpet. Since each frame provides one row of tufts across the carpet and each strand of yarn provides a single pile tuft, the sequence of colors determines the finished surface pattern.

An Axminster loom simulates hand weaving because each tuft of yarn is individually inserted, and theoretically each tuft could be of a different color. This flexibility offers unlimited design possibilities, and intricate patterns are produced even in lower-priced carpet. Considerable delicacy and subtlety of design are possible in the higher-priced carpets, in which the rows of tufts are closer together.

The Axminster weave is almost always

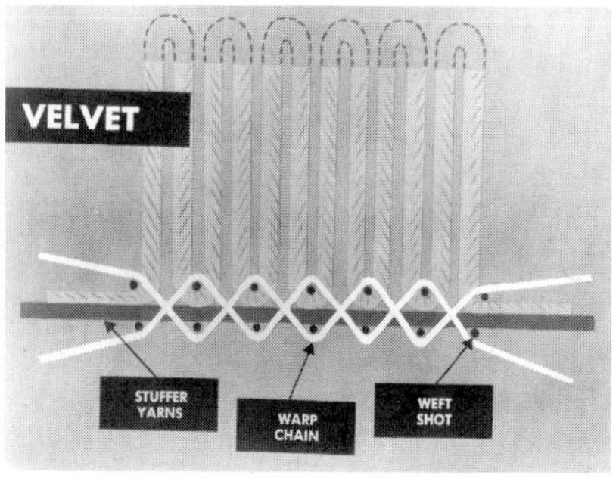

VELVET

STUFFER YARNS

WARP CHAIN

WEFT SHOT

FIGURE 9.12-4 Velvet weave with double chain warp and typically two or three stuffer yarns per pitch. (Courtesy Carpet and Rug Institute.)

FIGURE 9.12-5 A Wilton loom. The punched cards are at the upper left. (Courtesy Carpet and Rug Institute.)

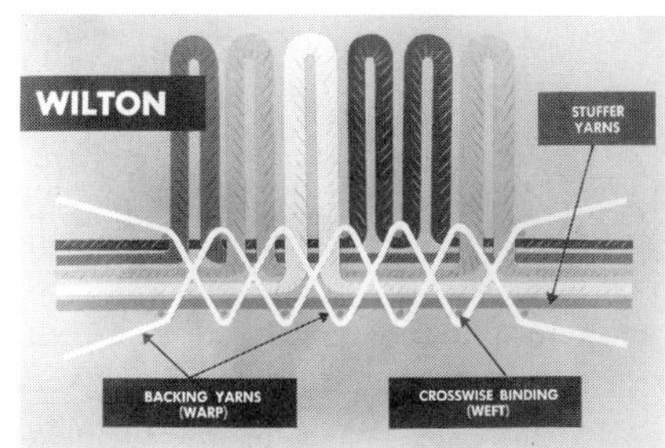

FIGURE 9.12-6 **Wilton weave with four frames (colors), double chain warp, two weft shots per wire, and stuffer yarns.** (Courtesy Carpet and Rug Institute.)

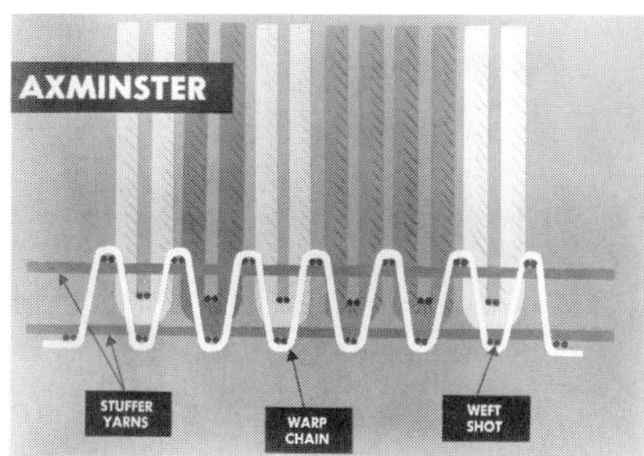

FIGURE 9.12-7 **Axminster weave with single chain warp, three double shots per row, and stuffer yarns.** (Courtesy Carpet and Rug Institute.)

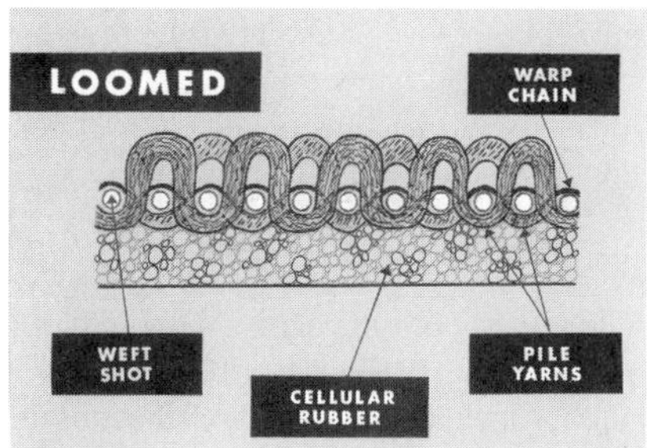

FIGURE 9.12-8 **Loomed weave with double chain warp, single weft shot, and low-level loop pile bonded to rubber cushioning.** (Courtesy Carpet and Rug Institute.)

a cut pile with an even pile height and elaborate patterning with many colors. The backing is so heavily ridged and *sized* (coated) that it usually can be rolled easily only lengthwise. The characteristic chain warp can be seen along a cut edge, as in other woven carpets (Fig. 9.12-7).

Loomed Loomed carpet is intended specifically for use with a bonded rubber cushioning. It is characteristically a high-density nylon with single-level, low-loop pile about ⅛ in. (3.2 mm) high, woven on a modified upholstery loom (Fig. 9.12-8). The pile back is treated with a waterproof adhesive, which serves as a bond coat for the rubber backing. Sponge or foam rubber is applied to the pile fabric, and the backed carpet is heated to cure the rubber. Loomed carpet is generally manufactured in a width of 4 ft. 6 in. (1371 mm) with a rubber backing about $^{3}/_{16}$ in. (4.8 mm) thick.

TUFTED CARPET

Tufted carpet is manufactured by inserting tufts of pile into a premade backing by a machine that resembles a huge sewing machine with thousands of needles operating simultaneously (Fig. 9.12-9). The prime backing is made from jute, kraftcord, cotton, or polypropylene olefin and is coated with a heavy layer of latex to hold the tufts permanently in place (Fig. 9.12-10). On most tufted carpets, a loosely woven, coarse jute fabric (*scrim*) is added as secondary backing to lend additional strength and dimensional stability (Fig. 9.12-11).

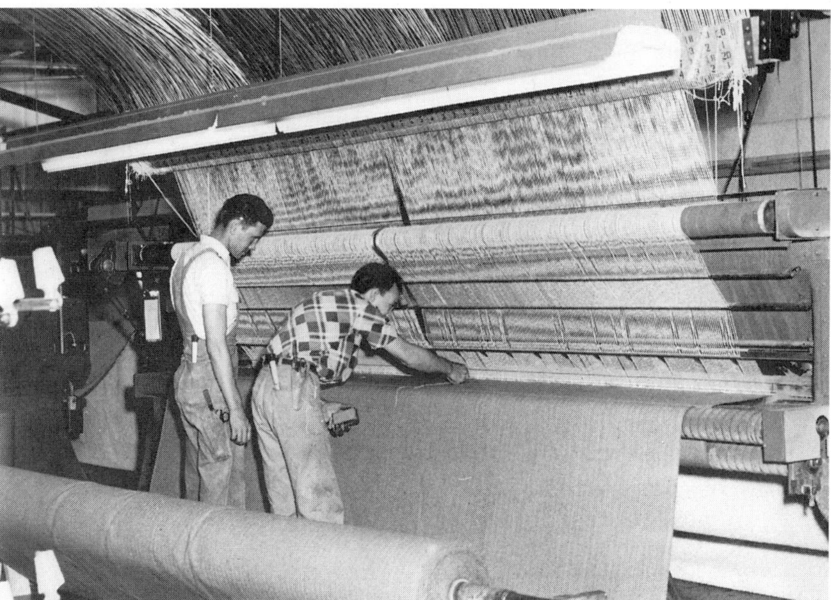

FIGURE 9.12-9 **A tufting machine.** (Courtesy Carpet and Rug Institute.)

FIGURE 9.12-10 **A roll of carpet with a latex coating applied to the back by a special roller machine. Latex is used as a backing on all tufted and knitted carpet, and on many woven carpets. A second backing is also laminated to good-quality tufted carpets.** (Courtesy Carpet and Rug Institute.)

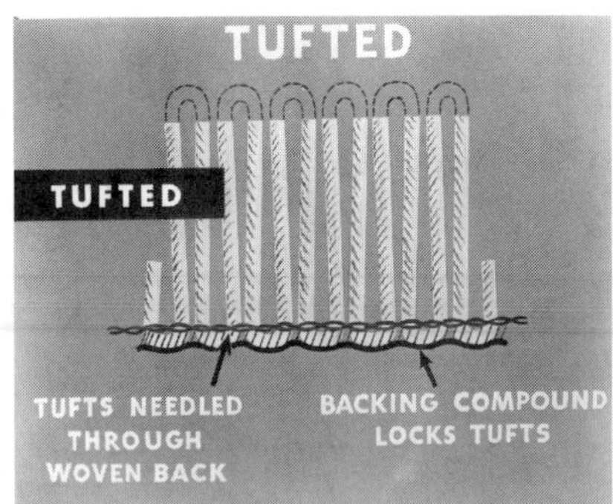

FIGURE 9.12-11 Tufted carpet construction: secondary backing is usually bonded to a prime backing with latex compound. (Courtesy Carpet and Rug Institute.)

Tufted pile can be level-loop, plush, multilevel, cut or uncut, or a combination of cut and uncut. Striated and carved effects can be achieved. Tufted carpets are made in solids, moresques, and multicolor patterns made possible by new dyeing methods.

Tufted carpet can generally be identified on the reverse side, where the even rows of tufts punched through the prewoven backing with punched-in tufts can be seen.

KNITTED CARPET

Knitted carpet is somewhat similar to woven carpet in that the pile and backing yarns are fabricated in one operation. Unlike weaving, the knitting process loops together the backing, stitching, and pile yarns with three sets of needles in much the same way as hand knitting. A coat of latex is applied to the back as a fixative and to give additional body to the fabric (Fig. 9.12-12).

Because a single-pile yarn is used in the knitting operation, as in velvet weaving, knitted carpets are usually solid colors or tweeds, although patterns are also possible. Knitted carpets usually have either single- or multilevel uncut pile, but cut pile can be achieved with modification of the knitting machine.

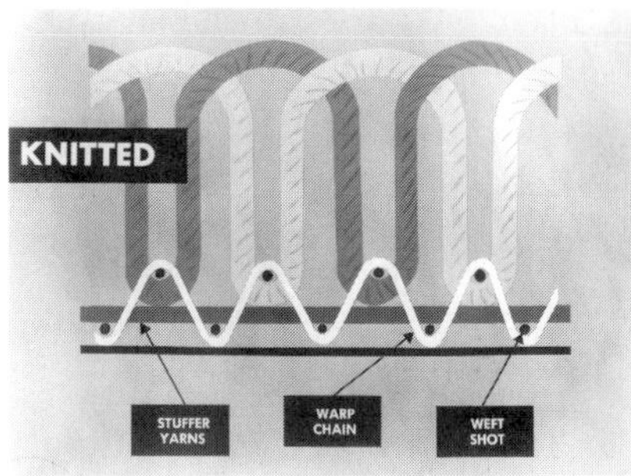

FIGURE 9.12-12 Knitted carpet with typical continuously looped pile, single chain warp, and stuffer yarns, all backed with latex. (Courtesy Carpet and Rug Institute.)

If a knitted carpet is *grinned* (folded back) to expose the backing, it will show the continuous looping of the pile yarn from row to row, held in place by the stitching yarns. The rows of pile run in an irregularly diagonal direction with a random, homespun look, rather than in straight rows as in woven and tufted carpets.

FUSION-BONDED CARPET

Fusion-bonded materials are those most often used for carpet tile. Fusion-bonded carpet is made by sandwiching a yarn bundle between substrates, implanting it there, and heat fusing the two. A blade is then run between the substrates, producing two pieces of carpet. The method of manufacture limits this type of carpet to cut-pile face construction.

NEEDLEPUNCHED CARPET

Needlepunched carpets are made by having hundreds of barbed needles punch loose, unspun fibers (usually polypropylene or nylon) through woven batts or fleeces of fibers to mesh them together, which results in a homogeneous pressed layer of fibers much like an army blanket or heavy felt. Needlepunched carpet is used mostly in the low-cost field.

Since needlepunched carpet is made entirely of synthetic fibers, which are not subject to decay, it can be used on both exterior and interior surfaces. Needlepunched carpet made from solution-dyed polypropylene is often used as outdoor carpet. Needlepunched nylon carpet is often foam-backed for use indoors.

FLOCKED CARPET

Flocked carpet is produced by spraying or electronically depositing short, loose nylon fibers on an adhesive-coated backing, similar to flocked wall coverings. In flocked carpet, polypropylene may be substituted for jute, making it suitable for exterior and high-moisture locations.

9.12.1.2 Comparative Factors

Comparative factors related to carpet construction such as pile yarn weight and pile thickness, number of tufts per square inch, and other factors, may be useful in comparing carpet of similar construction.

The traditional rule of thumb, "the deeper and denser (the pile), the better," is often used in making visual comparisons between carpets of similar construction and fiber type. However, such a generalization is superficial and more detailed measures of carpet construction are generally necessary to judge carpet quality properly. Pile density is actually a combination of pile thickness and pile weight.

Pile thickness (in inches [millimeters]) and pile weight (in ounces per square yard [grams per square meter]) can be established accurately only by laboratory analysis. Pile density (in ounces per cubic yard [grams per cubic meter]) can be derived by a simple formula from established information about pile thickness and weight. All of these criteria are used in determining standards of quality for carpet (see Section 9.12.2.5).

Other comparative factors have been developed as a substitute for, or supplement to, the basic criteria of pile density, weight, and thickness. The number of tufts per square inch; pitch, gauge, and needles; and rows, wires, and stitches are terms that describe the spacing of individual tufts in various carpets. When considered together with yarn size (yarn weight) and number of plies (strands) twisted together to form the pile yarn, these factors provide measures of pile density and other important basic criteria relating to quality.

PITCH, GAUGE, AND NEEDLES

The spacing of tufts across the width of a woven carpet can be described in terms of pitch, gauge, or needles. *Pitch* dictates the number of pile warp yarns (tufts) in a 27-in. (685.8 mm) width of carpet and is a measure of the tightness of crosswise construction (Fig. 9.12-13). Pitch may range between 90 and 256 for velvet and Wilton carpets, and between 189 and 216 for Axminster. The higher the pitch, the tighter the weave and the greater the pile density.

In knitted and tufted carpets where there are no continuous lines present,

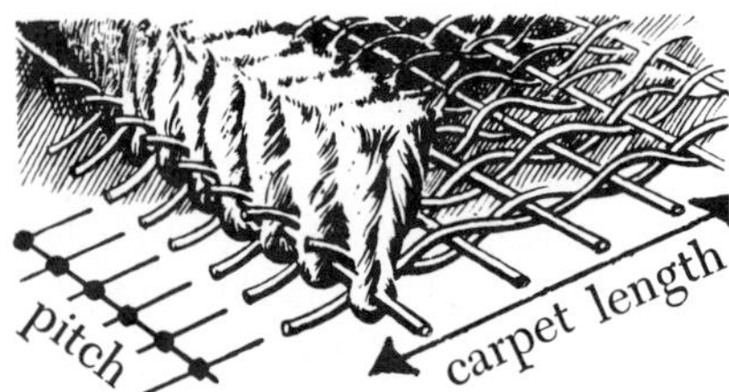

FIGURE 9.12-13 Pitch. (Courtesy Carpet and Rug Institute.)

gauge describes the actual spacing of tufts across the carpet width, expressed as a fraction such as ⅛ in. (3.2 mm), 3⁄16 in. (4.8 mm), or 5⁄32 in. (3.99 mm). For example, ⅛-in. gauge describes carpet with 8 tufts per in. (25.4 mm); 3⁄16-in. gauge indicates 16 tufts in 3 in. (76.2 mm) (a little more than 5 tufts per inch [25.4 mm]).

In tufted carpets, the term *needles* often is used instead of *gauge* to describe the number of tufts per inch across the width. Tufted construction may vary from 4 to 13 needles (tufts) per inch (25.4 mm).

ROWS, WIRES, AND STITCHES

Rows, *wires*, and *stitches* are terms describing the spacing of tufts lengthwise. The number of pile tufts per inch in Axminsters is described in *rows* per inch; for Wiltons and velvets in *wires* per inch (Fig. 9.12-14). Single- or double-weft shots, usually visible on the back of the carpet, correspond to the rows of tufts in the length of the carpet. To give an accurate indication of rows or wires per inch, it is customary to count the visible single- or double-weft shots in a 3-in. (76.2 mm) length of carpet and express this value in rows per inch (25.4 mm), as a fraction where necessary. Thus 17 rows in 3 in. (76.2 mm) would be expressed as 5⅔ rows or wires per inch (25.4 mm). Woven carpets range from 4 to 11 rows or wires per inch (25.4 mm).

In tufted carpets there are no weft yarns that can be related to the spacing of pile tufts. Therefore, lengthwise construction is expressed as the number of tufts per inch. To distinguish this count from tuft count across the carpet (needles), the term *stitches* is used. Tufted carpets may have 6 to 10 stitches per inch.

TUFTS PER SQUARE INCH (mm^2)

The overall density of carpets can be computed in tufts per square inch from information describing both lengthwise and crosswise construction (Fig. 9.12-15) or can be determined by actually counting the tufts in a 1-in. (25.4 mm)-square area. This comparative method is preferred for tufted and knitted carpets, but can also be applied to woven carpet. It is helpful in comparing overall density of dissimilar carpets when different terms are used to describe lengthwise and crosswise construction.

YARN SIZE AND PLIES

Wool fibers and synthetic staple fibers are spun into single-strand yarns. Continuous filament synthetic fibers are not spun, but emerge from the extrusion machine in bundles, which are formed into a single-ply yarn. The strands of yarn may then be further twisted together into thicker yarns, each forming a single pile tuft. Depending on the number of single strands, the pile yarn is identified as *single-*, *two-*, *three-*, or *four-ply* yarn. The size of both single-ply and multi-ply yarns can be measured by *woolen count* (yards [meters] of yarn per ounce [gram]) or denier (grams per 9000 meters of yarn) and other methods, depending on the spinning method employed. Synthetic fibers and yarns are most often measured according to the denier system.

Along with spacing of tufts, the yarn size and number of plies affect carpet density, resilience, compression resistance, and other important performance criteria.

BACKING

In woven and knitted carpets, the backing yarns and pile yarns are interlocked simultaneously during manufacture. The backing yarns are typically jute (sisal fiber), kraftcord (wood pulp fiber), polyester/cotton, nylon, rayon, or polypropylene. Backing yarns hold the pile tufts in place and provide the necessary stiffness so the carpet will lie flat without wrinkling. In addition, they provide sufficient strength to prevent the carpet from tearing when heavy objects are moved across the surface.

Woven carpets such as velvet, Wilton, and Axminster may have single or double chain yarns and several stuffer yarns per tuft running lengthwise; the number of *weft shots per wire* running across the carpet may also vary, resulting in such designations as one-, two-, and three-shot carpet. Increasing the number of backing yarns results in a stiffer, stronger backing and therefore in improved carpet quality. The better Wilton and velvet carpets may be woven through the back by hooking

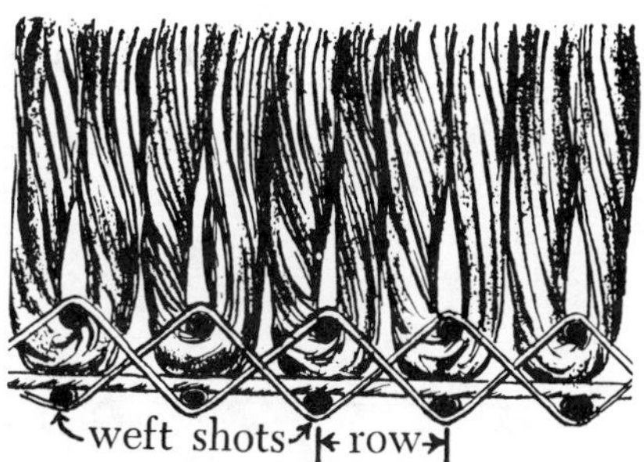
weft shots
row

FIGURE 9.12-14 **Wires or rows per inch (25.4 mm).** (Courtesy Carpet and Rug Institute.)

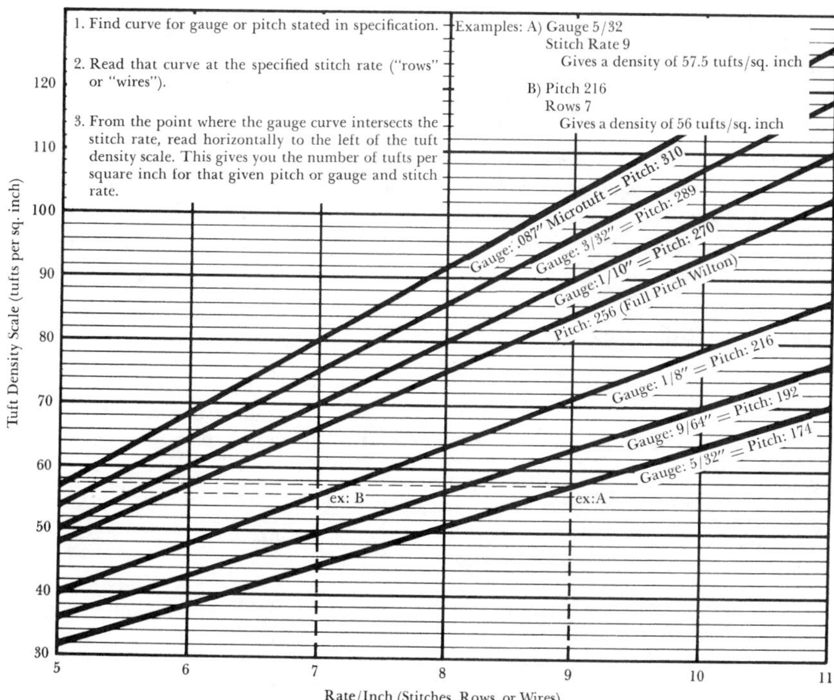
Density Nomograph
1. Find curve for gauge or pitch stated in specification.
2. Read that curve at the specified stitch rate ("rows" or "wires").
3. From the point where the gauge curve intersects the stitch rate, read horizontally to the left of the tuft density scale. This gives you the number of tufts per square inch for that given pitch or gauge and stitch rate.
Examples: A) Gauge 5/32
Stitch Rate 9
Gives a density of 57.5 tufts/sq. inch
B) Pitch 216
Rows 7
Gives a density of 56 tufts/sq. inch
Gauge: .087″ Microtuft = Pitch: 310
Gauge: 3/32″ = Pitch: 289
Gauge: 1/10″ = Pitch: 270
Pitch: 256 (Full Pitch Wilton)
Gauge: 1/8″ = Pitch: 216
Gauge: 9/64″ = Pitch: 192
Gauge: 5/32″ = Pitch: 174
ex: B
ex: A
Tuft Density Scale (tufts per sq. inch)
120
110
100
90
80
70
60
50
40
30
5
6
7
8
9
10
11
Rate/Inch (Stitches, Rows, or Wires)

FIGURE 9.12-15 **The overall density in tufts per square inch is useful in comparing dissimilar carpets, such as tufted (example A) and woven (example B).** (Courtesy Carpet and Rug Institute.)

the pile yarns around the bottom shot yarns, thus providing better anchorage of the pile tufts.

In tufted carpets, pile yarns are stitched into a separately prepared *primary backing*. Most such primary backing today is polypropylene, although jute fabric is still sometimes used. An adhesive such as latex is used to secure the pile yarns in place, and a *secondary backing* of polypropylene or jute is bonded to the primary backing for additional strength, stiffness, and dimensional stability.

Similar *secondary backings* are sometimes used in knitted carpets as well. There is usually no secondary backing on woven carpet.

9.12.1.3 Pile Fibers

Carpet can also be selected according to the material of the pile fibers from which the pile yarns are made. These currently include wool, nylon, acrylics, polyester, and polypropylene.

Although pile fibers differ in their physical properties, all fiber materials included in this discussion will provide adequate service when used in a carpet construction suitable for the traffic exposure.

Wool, the traditional fiber used in carpet manufacture, has proven its performance over the years and remains the classic fiber to which synthetic fibers often are compared. However, synthetic fibers, marketed under a variety of proprietary names, now account for about 90% of all carpet fiber used. The commonly used synthetic fibers can be classed generically as nylon, acrylic, and polypropylene (olefin). Cotton and two other man-made fibers, acetate and rayon, have not proven suitable for long-term wall-to-wall installations and therefore are not considered here. Polyester, one of the newest of the synthetic fibers, shows considerable promise but has not been in use long enough to demonstrate its properties fully.

The type of fiber used does not in itself guarantee carpet quality or performance. The various fibers have unique characteristics that may affect carpet performance and styling. Selected physical properties of pile fibers are given in Figure 9.12-16; other characteristics of fiber related to carpet performance are given in Figure 9.12-17.

In blended carpets, the higher the percentage of a particular fiber, the more the carpet will reflect the characteristics of that fiber. For example, wool may be reinforced by tough nylon in a 70/30 blend, but the carpet will look and feel more like wool. Generally, at least 20% of a fiber must be used for its characteristics to be apparent.

WOOL

The outstanding characteristic of wool is resilience, which, in combination with moderate fiber strength (see "Breaking Tenacity" in Fig. 9.12-16) and good resistance to abrasion, produces excellent appearance retention. Wool's relatively high specific gravity contributes to greater pile density.

Carpet wool is obtained from Syria, Iraq, Argentina, Pakistan, New Zealand, Australia, Scotland, and other countries. Wools from different regions have unique qualities, as various breeds of sheep yield different wools. Some wools are fine and lustrous, others are coarse and springy; in addition, fibers can vary in length from 3½ in. (88.9 mm) to 7 in. (177.8 mm). *Woolen yarn* is made up of interlocked long and short fibers; *worsted yarn* uses only long fibers. Woolen yarns are more irregular, hence are softer and bulkier than worsted yarns. The pile of most wool carpet is made of woolen yarns spun from fibers 3½ in. (88.9 mm) to 6½ in. (165 mm) long (see Fig. 9.12-16).

NYLON

Nylon constitutes about 85% of the commercial carpet market and a large portion of the housing market as well. The extensive use of nylon can be attributed to its lower cost, availability in many bright colors, exceptional resistance to abrasion, and high fiber strength (see "Breaking Tenacity" in Fig. 9.12-16).

Chemically, there are two types of nylon: Type 6, which represents a long-chain polymer, a chemical containing 6 carbon atoms; and Type 6,6, which results from the polymerization of two chemicals, each containing 6 carbon atoms. Although the two types perform similarly to yarn, they differ slightly in their ability to be processed and dyed.

Nylon fiber is produced in two forms: *staple* nylon, composed of 1½-in. (38 mm) to 6-in. (152.4 mm)-length fibers, which are spun into yarn much like wool or other short fibers; and *continuous filament* nylon, consisting of *bundles* of continuous fibers that are formed into yarn without spinning. Both staple nylon and continuous filament nylon yarns are made in round and modified cross sections (Fig. 9.12-18). Staple fibers are mechanically crimped; continuous filament yarns are texturized (Fig. 9.12-19) to impart bulk and covering power to the yarns. Texturizing and crimp give the yarns or fibers an irregular alignment, which further increases their bulk and covering power.

FIGURE 9.12-16 Selected Properties of Pile Fibers

Fiber					
Type	Length	Specific Gravity	Breaking Tenacity, grams/denier	Degrading Temperature[a]	Moisture Absorption, %
Wool	Staple	1.32	1.17	420°–570°F (233–317°C)	15–16
Nylon	Staple and CF[b]	1.14	4–6	428°–482°F (238–268°C)[c]	4–5
Acrylic	Staple	1.18	2–4	400°–500°F (222–277°C)	1.3–2.5
Modacrylic	Staple	1.3–1.4	2.5–3.0	300°F (167°C)	0.4–4.0
Polypropylene	CF[b]	0.91	3–6	333°F (185°C)	up to 0.10

[a]For wool, scorching to charring range; for synthetics, melting temperature.

[b]Continuous filament.

[c]428°F (238°C) for Type 6; 482°F (268°C) for Type 6.6.

(Courtesy Carpet and Rug Institute.)

FIGURE 9.12-17 Selection Criteria for Carpet Pile Fibers[a]

Criteria	Wool	Nylon	Acrylic	Modacrylic	Polypropylene
Resistance to:					
Abrasion	Good	Excellent	Good	Good	Excellent
Alkalies	Fair	Good to Excellent	Good to Excellent	Good to Excellent	Good to Excellent
Acids	Fair	Good to Excellent	Good	Good	Good to Excellent
Insects and fungi	Excellent[b]	Excellent	Excellent	Excellent	Excellent
Burns	Good	Fair	Fair	Fair	Fair
Compression	Good	Good	Good	Good	Fair
Crushing	Excellent	Good	Good to Excellent	Good	Fair to Good
Staining	Good	Good to Excellent	Good to Excellent	Good to Excellent	Excellent
Soiling	Good	Fair to Good	Good to Excellent	Good to Excellent	Excellent
Static buildup	Fair to Good	Fair	Good	Good	Excellent
Texture retention	Excellent	Good to Excellent	Good to Excellent	Good	Fair to Good
Wet cleanability	Fair to Good	Good	Good to Excellent	Good to Excellent	Excellent
Durability	Good to Excellent	Excellent	Good to Excellent	Good to Excellent	Good
Appearance retention	Excellent	Good to Excellent	Good to Excellent	Good to Excellent	Good
Ease of maintenance	Good to Excellent	Good	Good to Excellent	Good to Excellent	Excellent

[a]In each case, criteria vary depending on the properties of the specific fiber type used, carpet construction, and installation procedures.

[b]When chemically treated.

(Courtesy Carpet and Rug Institute.)

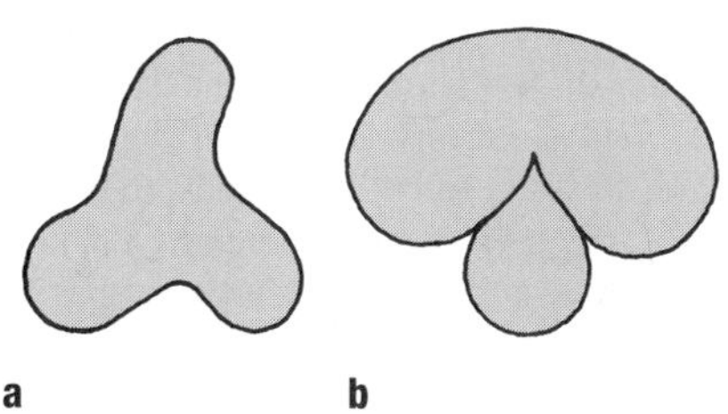

FIGURE 9.12-18 **Modified-fiber cross sections: (a) trilobal nylon and (b) bicomponent acrylic.** (Courtesy Carpet and Rug Institute.)

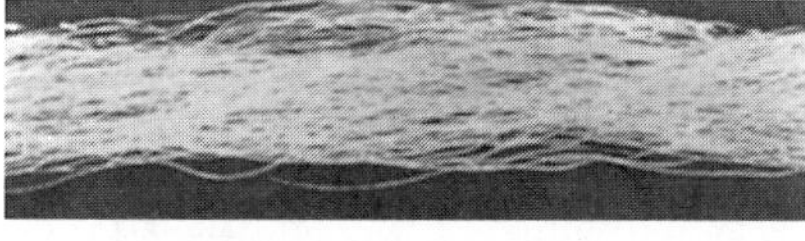

a

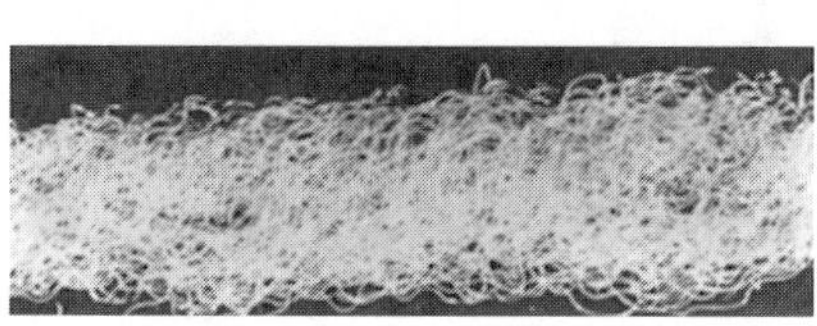

b

FIGURE 9.12-19 **Continuous filament nylon: (a) before texturizing, (b) after texturizing.** (Courtesy Carpet and Rug Institute.)

ACRYLICS AND MODACRYLICS

Acrylics and modacrylics (modified acrylics) are synthetic fibers that closely resemble wool in abrasion resistance and texture.

Chemically, acrylic fibers are composed of 85% or more by weight of acrylonitrile. Modacrylics contain at least 35% of acrylonitrile in chemical combination with modifying materials. Acrylic fiber is produced in round, bean-, and mushroom-shaped cross sections (see Fig. 9.12-18b). The latter is a bicomponent fiber of two filaments welded parallel to each other. Because the shrinking ratios of the two filaments are different, the fiber assumes a coiled shape in its dry equilibrium state, a property that improves texture retention.

Fibers of the acrylic family are now available only in staple form and are characterized by an appearance and high durability comparable to those of wool. Acrylics often are blended with modacrylics in commercial carpet to reduce potential flammability. However, with substantial improvement in acrylic fibers, the trend is toward 100% acrylic fiber.

POLYESTER

Polyester is a fiber-forming thermoplastic polymer. It is made from terephthalic acid and ethylene glycol. Polyester fibers are staple fibers, and polyester yarns are spun yarns.

POLYPROPYLENE OLEFIN

There are two classes of olefin, polyethylene and polypropylene, but only polypropylene has been produced in a fiber suitable for carpet construction. Polypropylene fiber consists of at least 85% propylene by weight; it is generally produced in bulked continuous filament yarn, although staple yarn is also available.

Polypropylene has the lowest moisture absorption rate of any carpet fibers, which gives it superior stain resistance as well as excellent wet cleanability (see Fig. 9.12-16). Its low specific gravity results in superior covering power, and high fiber strength contributes to good wear resistance.

9.12.1.4 Dyeing Methods

Color can be introduced at various stages of carpet manufacture. For both synthetic and wool pile carpets, dyes can be added at either the fiber, yarn, or carpet manufacturing stage. The dyeing method used may vary with the general properties of the fibers, the intended carpet construction, and color effect desired, or may depend on special methods developed for proprietary fibers.

SOLUTION DYEING

Many synthetic fibers are colored by a dye or pigment introduced into the liquid chemical from which fibers are extruded. In this method, the dye becomes chemically a part of the fiber; therefore, solution-dyed yarns are generally colorfast and more resistant to discoloration caused by wet cleaning.

STOCK DYEING

Wool and synthetic staple fibers can be dyed before being spun into yarn by dipping them in kettles of dye. This method ensures uniform distribution of dye and is suitable where large quantities of a single-colored yarn are desired.

SKEIN AND PACKAGE DYEING

When smaller quantities of colored yarn are required, *skein dyeing* is often used. In this method, skeins of natural-colored spun yarn or continuous filament fibers are immersed in vats of hot dye. In *package dyeing*, yarn is wound on a perforated tube and dye is forced through the perforations to soak the yarn.

SPACE DYEING

Both staple and continuous-filament yarns can be space dyed in alternating bands of color, generally for use in multicolored tufted carpet. In the *knit sleeve* method of space dyeing, the yarn is knitted into a fabric, print rollers apply the desired colors, then the fabric is unraveled again into yarn. In the *warp print* method, parallel yarns are fed continuously into a machine where color is applied by rollers.

PIECE DYEING AND RESIST PRINTING

Some woven and many tufted carpets are piece-dyed by immersing the entire carpet in a hot dye bath. However, the carpet need not be dyed only one color; some synthetic fibers can be chemically formulated to absorb certain compatible dyes and repulse others. Thus, it is possible to produce complex color patterns in the same dye bath. A similar technique of selective dyeing can be applied in skein dyeing. *Resist printing*, another method of selective dyeing, involves printing tufted carpet with a dye-resistant agent before piece dyeing.

PRINT DYEING

Print dyeing provides good dye penetration and pattern definition in tufted carpet. With possible color variation of up to six colors, print-dyed tufted carpets compare favorably in appearance with Wiltons and Axminsters but are cheaper to produce. The process involves silk-screening plain tufted carpet with a premetalized dye and applying an electromagnetic charge under the carpet, which drives the dye deeply into the pile. The final pattern is built up by continuously processing the carpet through a series of troughs, each applying a different color.

9.12.2 SELECTION CRITERIA

Carpet performance cannot be attributed to any single element of carpet construction or physical property of its pile fiber. Installed performance is the combined effect of surface texture, construction, backing, cushioning, pile thickness, weight, density, and other variables. Of these, pile weight and pile density are the most uniformly significant, regardless of fiber type or carpet construction.

Although pile fibers differ in their physical properties (see Fig. 9.12-16), all the fibers discussed here will provide adequate service when used in a carpet construction suitable for the traffic exposure (see Section 9.12.2.5). The number of variables in carpet construction is so great that it is difficult to assess their individual effects on performance and selection criteria. Therefore, in the following discussion, selection criteria are outlined for fibers only (Fig. 9.12-17); the effects of carpet construction are described only where appropriate.

Individual selection criteria may be useful in carpet selection for special conditions of traffic or exposure. For example, alkali, acid, and stain resistance may be important in food preparation and serving areas. *Soiling resistance* and *wet cleanability* may be a factor in entry areas. Insect and fungus resistance may require consideration in potentially damp locations. Possible *static build-up* should be considered in low-humidity areas and where such build-up may affect equipment or activities, such as in computer rooms. These and other criteria have been combined into composites such as *durability*, *texture retention*, and *ease of maintenance* for average conditions (see Fig. 9.12-17).

9.12.2.1 Durability

Pile weight and density, backing and cushioning firmness, quality of installation, and maintenance significantly affect carpet durability. Durability is also affected by hazards that tend to destroy pile fibers. The resistance of fibers to such hazards is shown in ratings for resistance to abrasion, alkalies, acids, insects, fungi, and burns. The rating for durability is a composite reflecting overall resistance to destruction or loss of pile fibers and is a measure of the service life of a carpet.

ABRASION RESISTANCE

Abrasion resistance is the resistance of a fiber to wearing away due to foot traffic or other moving loads. It is measured by the actual loss of fiber when exposed to a machine that simulates abrasive forces due to foot traffic. Since loss of fiber occurs as a result of breaking, fiber strength is one of the physical properties that influences abrasion resistance. Nylon and polypropylene have superior fiber strength and, hence, excellent abrasion resistance.

ALKALI AND ACID RESISTANCE

Foods and beverages are generally mildly alkaline or acid in nature; many soaps and detergents are strongly alkaline. If alkaline or acid solutions are not promptly removed when spilled, they may attack the pile fibers. Wool fibers are more vulnerable to such attacks because they have a higher moisture absorption rate and are less inert chemically than the synthetic fibers.

INSECT AND FUNGUS RESISTANCE

The synthetic fibers in modern carpets contain no organic nutritive value and are therefore immune to attack by insects, such as carpet beetles and moths, and to fungi, such as mold and mildew. To render wool carpet resistant to these hazards, most manufacturers treat wool yarn in the dye bath with a chemical preservative that provides long-term protection from insect attack even after wet and dry cleaning. Because many carpets intended for interior use employ at least some organic fibers in the backing, some fungus hazard exists with most carpets, even those with synthetic fiber piles.

In general, the hazard of insect and fungus attack can be reduced be regular vacuuming and exposing the carpet to light and air. This is particularly important where large or heavy pieces of furniture make part of the carpet inaccessible to regular cleaning and ventilation. In such areas, greater insect protection can be provided for wool carpets by periodic spraying with mothproofing agents.

BURN RESISTANCE

Most synthetic carpet fibers are flame resistant to some degree, but they will melt and fuse when exposed to the heat of a burning cigarette or glowing ash, which produces more than 500°F (260°C). Prolonged exposure to such concentrated heat may result in complete local loss of fiber or a slightly fused spot. The fused spot is more resistant to wear and abrasion than the surrounding area, and this causes a visible discoloration in the carpet surface.

Although wool fiber will actually burn by charring, the damage from a lit cigarette is usually less severe and the charred spot will wear off more readily or can be removed with fine sandpaper. A fused spot on synthetic carpet similarly can be snipped off if it is not too severe, and, if followed by brushing, an acceptable appearance usually will result. If the burn is severe, replacement of the burn area may be necessary.

The behavior of a carpet in a fully developed fire is controlled by building codes and laws. All carpet manufactured in the United States must pass a methenamine "pill" test in accordance with ASTM Standard D 2859, "Test Method for Flammability of Finished Textile Floor Covering Materials." In this test, a carpet is exposed to the burning of a methenamine tablet. The medium *flame spread index* or *critical radiant flux* must be 0.04 watts/sq. cm.

Often, carpet will be required by code to have a higher rating than the pill test minimum. The National Institute of Standards and Technology recommends that carpet used in corridors and exitways have a radiant flux of at least 0.22 watts/sq. cm. in commercial occupancies and of 0.45 in institutional occupancies.

A carpet's flame spread during a fully developed fire is measured by the Flooring Radiant Panel Test of ASTM E 648. This flame spread index is called the *critical radiant flux*. It represents the minimum energy necessary to sustain flame. The higher the number, the more resistant the carpet to flame propagation.

Sometimes a carpet's flame spread index is required to be measured by the *Smoke Chamber Test* in Underwriters Laboratories' (UL) 992 "Smoke Chamber Test." This test measures the length of spread and time of travel of a flame. The range is from 0 to 25.

9.12.2.2 Appearance Retention

Appearance retention depends on factors that change original carpet color, texture, or pattern, such as fading, soiling, staining, and compression and crush resistance. Other factors common to many types of fiber and carpet construction that affect appearance without impairing serviceability are pilling, shedding, sprouting, and shading (see glossary). Ratings for appearance retention are composites of these factors.

COMPRESSION RESISTANCE

The extent to which the pile will be compressed under heavy loads or by extensive foot traffic is called compression resistance. For instance, of two samples tested under the same conditions, the one that compresses less is said to have better compression resistance. With the exception of polypropylene, which is rated somewhat lower, all fibers are approximately equal in their ability to resist compression loads. Of greater importance in resisting such forces are the cushioning, pile height, and pile density. Looped pile construction with closely spaced tufts and tightly twisted yarns is more resistant to compression.

CRUSH RESISTANCE

Crush resistance is dependent on the ability of the fibers to recover from short- and long-term compression loads. This property is measured by the extent to which the fiber springs back after a load has been removed, as a percentage of the original height. Wool and acrylic have the highest crush resistance, nylon and modacrylic moderate, and polypropylene somewhat lower. Pile density, spacing of tufts, and tightness of yarns affect crush resistance more than the relative rating of the fibers.

TEXTURE RETENTION

The ability of carpets to retain the surface texture imparted during manufacture is related to the compression and crush resistance of the fibers, as well as the density of the pile and tightness of the yarns. Generally, all fibers have adequate texture retention, with wool rated highest and polypropylene somewhat lower than the other synthetic fibers. Within this range, special fiber types such as heat-set nylon and bicomponent acrylic (mushroom cross section) have improved texture retention.

9.12.2.3 Ease of Maintenance

The ease of maintaining an installed carpet depends mainly on its resistance to soiling and staining, as well as its wet cleanability. Although dry powder cleaning products are useful for moderately soiled areas, wet cleaning (shampooing) remains the most satisfactory method for thorough in-place carpet cleaning under heavier soil conditions.

STAINING RESISTANCE

Staining is a local discoloration of a carpet, as compared with *soiling*, which is more or less uniform over the carpet surface. Many foodstuffs (oil, grease, eggs, chocolate), beverages, and beauty aids may cause staining. If promptly and properly treated, such stains need not be permanent. Many carpet manufacturers publish detailed recommendations for stain removal and indicate the type of solvent or cleaning agent that should be used with each type of stain. Since stains are often moisture-borne, rate of moisture absorption is a factor in staining resistance. Polypropylene and acrylics have low moisture absorption and are therefore rated highly for staining resistance.

SOILING RESISTANCE

Soiling is the result of dirt or soil particles deposited on the carpet surface by foot traffic or dust. Wool and nylon generate higher electrostatic charges in low humidity and therefore tend to attract airborne dust and soil faster than some other fibers. However, the scaly surfaces of wool fibers make soiling less apparent. Polypropylene and acrylics have higher soiling resistance because of low static build-up and low moisture absorption. Bulked fibers with modified cross sections and texturized yarns spun more loosely tend to retain dirt more tenaciously than round, tight, high-density yarns. Moreover, tightly woven, uncut-loop piles will keep soil near the surface where it is easier to remove; multicolored and textured piles of medium shades will conceal soiling better than light or dark smooth plush surfaces.

WET CLEANABILITY

Since it is impractical to remove wall-to-wall carpet for cleaning, the ability of a carpet to respond successfully to in-place wet cleaning is important. Polypropylene, with its low moisture absorption rate, is considered the best in this respect. Generally, synthetic fibers, with their lower moisture absorption rates and inert chemical compositions, clean more readily, dry more rapidly, and are less subject to mildew odors while drying than wool. However, a variety of effective cleaning formulations specifically suited to wool carpet are available.

9.12.2.4 Static Generation

Static electricity is an annoyance in every carpeted space, but can become actually harmful in areas housing computers or other electrically sensitive equipment. Static occurs whenever two different materials come in contact with each other. People usually do not become aware of static until it reaches about 3.5 kilovolts (kv). The normal carpet requirements for solely human occupancies require a level of 3 kv or less. Computers require 2 kv or less.

Of the three factors regarding static electricity—static generation, static dissipation (conductivity), and static decay time—only the first is usually limited in the selection of carpets in nonsensitive locations. Static generation in carpet is measured by the American Association of Textile Colorist and Chemist's "Step Method," which simulates conditions of actual use and measures the charge generated by someone walking across the carpet.

Static dissipation may be also limited in spaces that will house computers. It is measured by a method recommended by the IBM Corporation.

Nylon is inherently high in static buildup. To fight this property, some nylon carpet manufacturers include carbon-loaded nylon fibers that carry a lifetime antistatic guarantee. Some carpet is laced with metal wires that are in contact with a grounded backing. In very sensitive areas, it may be impossible to reduce static sufficiently using antistatic fibers or wires alone. Then, the antistatic properties of such carpet must be supplemented with conductive shoes, furniture, mats, or other devices.

9.12.2.5 Standards of Quality

The Department of Housing and Urban Development (HUD) has issued "Use of Materials Bulletin UM-44C," which covers residential carpet, but it is also referenced in specifying commercial and institutional carpet.

HUD does not permit the use of carpet in baths, kitchens, and service areas such as laundry, utility, and furnace rooms. Carpet is also not appropriate in public toilets or shower rooms. Living areas within a residential unit, such as living, dining, recreation, sleeping, and dressing rooms, are designated as moderate-traffic areas. Public areas such as corridors, entrances, lobbies, waiting rooms, stairways, and elevators are considered heavy-traffic areas. Offices and institutional spaces may be either moderate- or heavy-traffic areas, depending on their actual use.

HUD requirements for carpet in heavy-traffic areas are summarized in Figure 9.12-20. Carpet construction requirements for moderate-traffic areas are discussed below and are illustrated in Figures 9.12-21 and 9.12-22. For both types of traffic exposure, carpet should conform to the following requirements for pile fibers, backing, and cushioning.

CARPET CONSTRUCTION

"Use of Materials Bulletin UM-44C" requirements are based on average values for pile yarn weight, pile thickness, and pile density.

Pile density is expressed as either average pile density (*AD*) or weight density (*WD*). *Pile yarn weight* (*W*) is the average weight of the face pile yarn in ounces per square yard (grams per square meter). *Pile thickness* (*t*) is the height of the tufts above the backing in inches. *Average pile density* (*AD*) is a calculated quantity in ounces per cubic yard (grams per cubic meter) obtained by the formula $AD = (36)(W)/t$. The *AD* is a four-digit number, which should be above 3000 for commercial carpet and 2800 for other carpet. *Weight density* is also a calculated quantity in ounces per cubic yard (grams per cubic meter), obtained by the formula $WD = (AD)(W)$.

UM-44C establishes minimums for pile yarn weight (*W*) and weight density (*WD*), depending on the type of fiber (see Fig. 9.12-22). The minimum weight densities (*WD*) given are always the minimum pile weight (*W*) multiplied by a pile density (*D*) of 2800. The HUD requirements are based on minimum weight-density factors rather than a constant pile density of 2800. If only the pile density were specified, it would establish a simple proportionate relationship between actual pile thickness and pile yarn weight, which would require increasing the minimum pile weight the same extent as the pile thickness. This requirement would be excessive for moderate-traffic areas. Figure 9.12-22 illustrates calculation of the weight-density factor.

PILE FIBERS

Pile yarn should be made of 100% wool, nylon, acrylic, modacrylic, or polypropylene olefin fibers, or blends of these fibers in yarn (exclusive of ornamentation). Not less than 20% of any of these fibers should be used when blended with other fibers, and such blends should not consist of more than two types of fibers. All fibers should be entirely new, not processed or reclaimed from previously knitted, woven, tufted, or felted products. Wool yarn should contain at least 97% wool fibers by dry weight; synthetic fibers should be 15 denier or thicker and should contain not more than 2% chloroform-soluble material.

Carpet pile should be resistant to fading from wet cleaning and exposure to light. Light colors should be rated as satisfactory after laboratory exposure to 20 standard fading hours, and dark colors after 40 fading hours.

BACKING

Backing materials should be those customarily used and accepted in the industry for each type of carpet. For tufted carpet, there should be a secondary backing reinforcement of at least 4 oz./sq. yd. (135.6 g/m^2), exclusive of the adhesive coating.

9.12.2.6 Certification of Quality

There is no industry-wide program for certification of carpet quality at the present time. However, most manufacturers will certify that their product meets or exceeds HUD requirements for moderate- or heavy-traffic areas.

Federal law requires that all textiles be identified with an appropriate label as to fiber content for each fiber type that constitutes 5% or more of the total. In the case of carpets consisting of backing yarns and pile yarns, only the fibers in the pile yarn need be identified. In addition, carpet samples shown to consumers must be similarly identified as to fiber content, name of manufacturer, and country of origin if imported. Several producers of synthetic fibers have agreements with carpet manufacturers that stipulate that only carpet meeting certain minimum

FIGURE 9.12-20 Physical Requirements for Carpet in Heavy-Traffic Areas

	Wilton, Wool or Acrylic	Velvet			Tufted, Wool or Acrylic	Knitted, Wool or Acrylic
		Wool or Acrylic	Nylon*	Wool or Acrylic		
Pile Description	Single-level loop woven through back	Single-level loop pile woven through back	Single-level loop pile woven through back	Multilevel loop woven through back	Single-level loop	Single-level loop
Tufts/sq. in. (mm^2)	52 (0.081)	60 (0.093)	32 (0.049)	34 (0.053)	60 (0.093)	26 (0.040)
Frames	3	—	—	—	—	—
Shots/wire	2	2	2	2	—	—
Weight oz./sq. yd. (g/m^2) Pile	48 (1627.5)	42 (1424)	29 (983.3)	44 (1491.9)	42 (1424)	37 (1254.5)
Total	72 (2441)	60 (2034)	50 (1695.3)	64 (2170)	—	58 (1966.5)
Pile height in. (mm)	Min. 0.250 (6.4) Max. 0.320 (8.1)	Min. 0.210 (5.3) Max. 0.310 (7.9)	Min. 0.200 (5.1) Max. 0.290 (7.4)	Min. 0.190 (4.8) Max. 0.370 (9.4)	Min. 0.250 (6.4) Max. 0.320 (8.1)	Min. 0.230 (5.8) Max. 0.290 (7.4)
Chain	Cotton and/or rayon	Cotton and/or rayon	Cotton and/or rayon	Cotton and/or rayon	—	Cotton, rayon, or nylon
Filling	Cotton and/or rayon, or jute	Cotton and/or rayon, or jute	Cotton and/or rayon, or jute	Cotton and/or rayon, or jute	—	Jute or kraftcord
Stuffer	Cotton, jute, or kraftcord	Cotton, jute, or kraftcord	Cotton, jute, or kraftcord	Cotton, jute, or kraftcord	—	—
Back coating, oz./sq. yd. (g/m^2)	—	8 (271.2)	6 (203.4)	8 (271.2)	—	—
Ply twist turns/in. (turns/mm)	Min. 1.5 (0.06) Max. 3.5 (0.14)	Min. 1.5 (0.06) Max. 3.5 (0.14)	Min. 1.0 (0.04) Max. 3.0 (0.12)	Min. 1.5 (0.06) Max. 3.5 (0.14)	Min. 2.5 (0.09) Max. 4.5 (0.18)	Min. 1.5 (0.06) Max. 3.5 (0.14)

*High bulk or textured continuous filament.

(Courtesy Carpet and Rug Institute.)

FIGURE 9.12-21 Physical Requirements for Carpet in Moderate-Traffic Areas[a]

Pile Fiber	Minimum Pile Yarn Weight (W), oz./sq. yd. (g/m^2)[b]	Minimum Weight-Density Factor (*WD*)[c]
Wool	25 (847.6)	70,000 (25 × 2800)
Acrylic		
Modacrylic		
Nylon (staple and filament)	20 (867)	56,000 (20 × 2800)
Polypropylene olefin		

[a]Based on HUD Use of Materials Bulletin UM-44A.

[b]Computed for blended yarns by multiplying percentage of each fiber by minimum pile weight and adding the two amounts.

[c]Pile density (*D*) = 2800, the minimum density that historically has given satisfactory service for wall-to-wall installations.

(Courtesy Carpet and Rug Institute.)

FIGURE 9.12-22 Weight-Density Calculations for Nylon or Polypropylene Carpet

Weight density (*WD*) calculated for nylon or polypropylene carpet from pile thickness (t) and pile yarn weight (W).

Pile thickness (t) = 0.25 in.

Pile yarn weight (W) = 20 oz./sq. yd.

Calculated pile density (D) = $36W/t$

$$D = \frac{36 \times 20}{0.25} = 2880 \text{ oz./cu. yd.}$$

Calculated weight-density factor (*WD*)

$WD = 20 \times 2880 = 57{,}600$*

*Referring to Figure 9.12-21, pile yarn weight (20 oz./sq. yd.) equals minimum requirements; weight density factor (57,600) exceeds minimum requirement (56,000).

(Courtesy Carpet and Rug Institute.)

construction requirements can display their fiber trademarks.

9.12.2.7 Cushioning

All broadloom carpet should be installed over cushioning to increase resilience and durability. Some carpets, especially those used in direct glue-down installations (Section 9.12.3.2), have an integral secondary backing that serves this function. In loomed carpets, cellular (foam or sponge) rubber cushioning is usually bonded directly to the carpet at the mill. For most carpets, however, cushioning is provided in the form of a separate padding. Currently available paddings include felted hair, rubberized fibers, cellular rubber, and urethane foam.

FELTED HAIR

The most economical traditional type of conventional padding is made of felted animal hair. This padding has a waffle design to provide a skidproof surface and improve resiliency. It is sometimes reinforced with a jute backing or with a burlap center liner. When reinforced with burlap, the hair is punched through the burlap fabric and compressed to a uniform thickness. Sizing (adhesive) is sometimes used to strengthen the bond between the fibers and the burlap core. Many manufacturers also sterilize and mothproof hair padding.

Felted padding of hair or hair and jute may mat down in time or may develop mildew, especially if the fibers become wet, as during cleaning. However, when properly cleaned, sterilized, and treated, hair padding is suitable for reasonably dry floors at all grade levels and on conventional radiant-heated floors.

RUBBERIZED FIBERS

Some padding made of jute or hair is coated with rubber on one or both sides to hold the fibers together and provide additional resilience. Sometimes this padding has an animal hair waffle top and a jute back reinforced with a patterned, rubberized application.

SPONGE RUBBER

In addition to cushioning bonded directly to the carpet, foam or sponge rubber is produced in sheet form with waffle, ripple, grid, or V-shaped rib designs. A scrim of burlap fabric is usually bonded to the rubber sheet to facilitate installation of the carpet. When laid with the fabric side up, this permits a taut and even stretch of the carpet.

Rubber padding is more expensive, but it retains its resilience longer than hair padding. It is highly resistant to decay and mildew and is nonallergenic. It can be used at all grade levels, but the denser cushionings are not recommended for radiant-heated floors.

POLYURETHANE FOAM

Urethane foam padding is a high-density polymeric foam available in two major types: prime and bonded. *Prime* urethane foam padding is available as either prime or densified prime. The latter is modified chemically to make it wear better.

Bonded urethane foam padding is made from heat-fused scraps of urethane, rather than being a homogeneous product as are prime and modified prime urethane padding.

The quality of urethane foam padding is defined by its density or weight per cubic foot (0.028 m³) of foam.

FIGURE 9.12-23 Weight and Thickness of Carpet Padding

Padding Type	Class	Thickness, in. (mm)	Weight
Natural fiber	I	¼ (6.4)	32 oz/sy[a] (1083 g/m²)
	II	5⁄16 (7.9)	40 oz/sy (1356 g/m²)
	III	⅜ (9.5)	50 oz/sy (1695 g/m²)
Synthetic fiber	I	5⁄16 (7.9)	22 oz/sy (746 g/m²)
	II	⅜ (9.5)	28 oz/sy (949 g/m²)
	III	⅜ (9.5)	36 oz/sy (1221 g/m²)
Flat sponge rubber	I	0.225 (5.7)	56 oz/sy (1899 g/m²)
	II	0.250 (6.4)	64 oz/sy (2170 g/m²)
	III	0.250 (6.4)	80 oz/sy (2713 g/m²)
Reinforced rubber	I	0.225 (5.7)	56 oz/sy (1899 g/m²)
	II	0.235 (6)	64 oz/sy (2170 g/m²)
	III	0.250 (6.4)	80 oz/sy (2713 g/m²)
Ripple rubber	I	0.350 (8.9)	64 oz/sy (2170 g/m²)
	II	0.400 (10.2)	80 oz/sy (2713 g/m²)
	III	0.250 (6.4)	80 oz/sy (2713 g/m²)
Bonded urethane	I	⅜ (9.5)	5.0 pcf[b] (80.1 kg/m³)
	II	5⁄16 (7.9)	6.5 pcf (104 kg/m³)
	III	¼ (6.4)	8.0 pcf (128 kg/m³)
Modified prime urethane	I	¼ (6.4)	2.5 pcf (40 kg/m³)
	II	¼ (6.4)	2.7 pcf (43 kg/m³)
	III	3⁄16 (4.8)	3.2 pcf (51 kg/m³)
Densified prime urethane	I	0.350 (8.9)	2.1 pcf (34 kg/m³)
	II	0.265 (6.7)	3.5 pcf (56 kg/m³)
	III	0.265 (6.7)	4.5 pcf (72 kg/m³)

[a]Ounces per square yard.

[b]Pounds per cubic foot.

SELECTION

The density of padding should be as recommended by the Carpet Cushion Council (CCC) in its publication *Selecting the Correct Carpet Cushion for Every Traffic Area*, which recommends the specific density of each available carpet padding type for each type of use, as follows:

- *Class I*, Moderate Traffic: Executive and administrative spaces in such buildings as schools, office buildings, airports, retail establishments, health care facilities, libraries, and museums; sleeping areas in homes, hotels, motels
- *Class II*, Heavy Traffic: The more public areas in the same types of buildings as listed for Class I and in clerical areas, moderate-use corridors, patients' rooms, lounges, dormitories, classrooms, small retail stores, bank lobbies and corridors, and moderate-use public areas
- *Class III*, Extra Heavy Traffic: Heavy-use public corridors, lobbies, locker rooms, dining areas, nurses' stations, cafeterias, major retail stores, restaurant dining rooms, and convention centers

The recommended weights and thicknesses of padding vary within each class, depending on the padding material, and from class to class (Fig. 9.12-23).

The aforementioned CCC publication also recommends carpet padding's Compressed Load Deflection (CLD) and Indentation Load Deflection (ILD), which is called the Indentation Force Deflection (IFD) in ASTM D 3574 (Fig. 9.12-24). The CLD is the force in pounds per square inch (kg/m²) needed to compress padding to a specified percentage of its original thickness. The ILD (IFD) is the number of pounds needed to depress a piece of padding to a specified percentage of its original thickness over an 8-in. (200 mm)-diameter test area.

9.12.2.8 Carpet Tile

Carpet tiles are used in commercial, institutional, and residential buildings. They meet the demands of business environments, including functionality, flexibility, style, value, and performance. Many carpet tiles can be installed with almost invisible seams, so that the installation looks like broadloom. Most are available in 36

FIGURE 9.12-24 Compression Load Deflection (CLD) and Indentation Load (Force) Deflection (ILD) for Urethane Carpet Padding

Padding Material	Class	Load	Percent of Original Thickness
Urethane (CLD)	I	0.75 psi* (3.66 kg/m^2)	25
	II	1.00 psi (4.88 kg/m^2)	25
	III	1.50 psi (7.32 kg/m^2)	25
Bonded urethane (CLD)	I	4.00 psi (19.53 kg/m^2)	65
	II	5.00 psi (24.41 kg/m^2)	65
	III	7.00 psi (34.18 kg/m^2)	65
Densified prime urethane (CLD)	I	0.85 psi (4.15 kg/m^2)	65
	II	1.80 psi (8.79 kg/m^2)	65
	III	2.30 psi (11.23 kg/m^2)	65
Modified prime urethane (ILD)	I	100 lb. (45.36 kg)	25
	II	100 lb. (45.36 kg)	25
	III	120 lb. (54.43 kg)	25

*Pounds per square inch.

× 36 in. (914 × 914 mm) and 18 × 18 in. (457 × 457 mm) sizes. Other sizes may be available from some manufacturers.

Carpet tiles are more adaptable, flexible, and easily installed and removed than broadloom carpet. They also provide unlimited underfloor access to electrical, data, and communications systems. Their backings are designed for easy handling and installation. Most are PVC-free, which means that they can be installed over many existing finishes without complete removal of old adhesives and sealers.

Carpet tiles should be required to comply with the Carpet and Rug Institute's (CRI) Indoor Air Quality Carpet Testing Program. They should have a Class I flammability rating in accordance with ASTM D 648.

9.12.3 INSTALLATION OF CARPET

Carpet can be installed over finished flooring, subflooring of concrete, plywood, particleboard, or any other suitable, smooth surface. The carpet manufacturer's recommendations should be followed with regard to seam locations and direction of carpet. Carpet edge guards should be located where carpet edges would otherwise be exposed. Edge guards should be securely fastened to the substrate.

9.12.3.1 Tacked and Tackless Installation

Carpet that is not rubber-backed can be readily installed over nailable surfaces by tacking the edges or by using *tackless strips* around the perimeter. Tackless strips consist of pieces of plywood of the thickness of the padding, with two or three rows of metal pins that pass through the plywood at an angle and are designed to grip the carpet when it is stretched over the pins (Fig. 9.12-25). Installations on concrete can be made by tacking if wooden plugs are provided 6 to 9 in. (150 to 230 mm) on centers in the concrete slab to receive the tacks. Tackless strips can also be installed on concrete with powder-actuated fasteners or epoxy adhesive. Two-row strips can be used when the carpet is 20 ft. (6.1 m) or less in width, but three-row strips should be used with wider carpet. Strips should be located at the entire perimeter of the carpet.

Padding is placed between the tackless strips and bonded to the substrate with an adhesive. The joints are placed at right angles to the carpet joints. Seams in padding should be taped.

Regardless of the installation method, tacked or tackless carpet should be located, trimmed, stretched sufficiently to eliminate wrinkles and buckles, and attached firmly with tacks or to strips. Seams should be butted tightly and taped or sewn to make them strong enough to resist damage during stretching and to keep them closed without gaps.

9.12.3.2 Direct Glue-Down Installation

In *direct glue-down* applications, carpet is glued to the substrates with no padding between them. Such carpet is fitted in place before adhesive is applied. The adhesive is then applied uniformly to the substrate. Carpet edges are butted tightly so that there are no gaps. The carpet is then rolled lightly to ensure uniform bond to the substrates with no air pockets.

FIGURE 9.12-25 Using a power stretcher, a skilled carpet mechanic can achieve a smooth, taut installation of carpet over tackless strips.

9.13 Resinous Flooring

There are many liquid-applied seamless resinous flooring products on the market for use in laboratories, food-processing facilities, pharmaceutical plants, chemical processing plants, manufacturing and other industrial facilities, high-tech manufacturing plants, chemical-processing facilities, electronics facilities, transportation facilities, food and beverage production facilities, light manufacturing and textile plants, warehouses, clean rooms, commercial kitchens, food stores, medical facilities, parking and bridge decks, stadiums, wastewater treatment plants, automobile facilities, and in other locations where sanitary, dustless, high-performance flooring is needed (Fig. 9.13-1). Some of these products can be applied when the temperature is less than 24°F (–4°C), which makes them ideal for coolers, freezers, and cold weather applications. Resinous flooring systems may have an integral cove base.

9.13.1 SYSTEMS

Resinous flooring systems range from thin-film dustproofing sealers to heavy-duty trowel-applied systems that provide flexible installation options, easy maintenance, and long service life. Antimicrobial, thermal shock–resistant formulations, and low-temperature curing acrylics enhance their performance. Thin-film products range in thickness from 4 to 25 mils (0.12 to 0.63 mm); thick-film products are generally 60 mils (1.52 mm), but some epoxy thick-film coatings are thinner. Most resinous flooring products are U.S. Department of Agriculture (USDA)-approved with 100% solids and zero volatile organic compounds (VOC).

The acrylic reactive resins in resinous flooring chemically bond to each other. Fresh resin dissolves the surface of cured material. The redissolved material cures with the fresh material. The need for surface preparation between layers is eliminated. After years of service, resurfacing or additions without cold joints are simple and fast. This welded bond allows different resinous materials to be combined in monolithic systems for specific performance requirements.

Resinous flooring is usually pigmented in beige, red, green, medium gray, concrete gray, or charcoal gray, but many others colors are possible. Small quantities of vinyl chips in contrasting colors can be randomly broadcast onto fresh resinous flooring for a decorative appearance. A wide variety of premixed quartz color blends are available.

Resinous flooring systems are available in many different forms:

- General service systems having excellent chemical, impact, and abrasion resistance.
- Power-trowelable systems for reduced installation time in larger, open areas.
- Static dissipative and conductive formulations, made to meet exacting electrical specifications. These are designed for electrical manufacturing and assembly areas, computer/control rooms, and clean environments.
- High-build, protective resurfacing systems. These have epoxy or silica aggregate mortar, high-build grout, and seal coats.
- Self-leveling formulations in smooth or skid-inhibiting systems, in an easily applied and durable self-leveling slurry designed to withstand moderate to

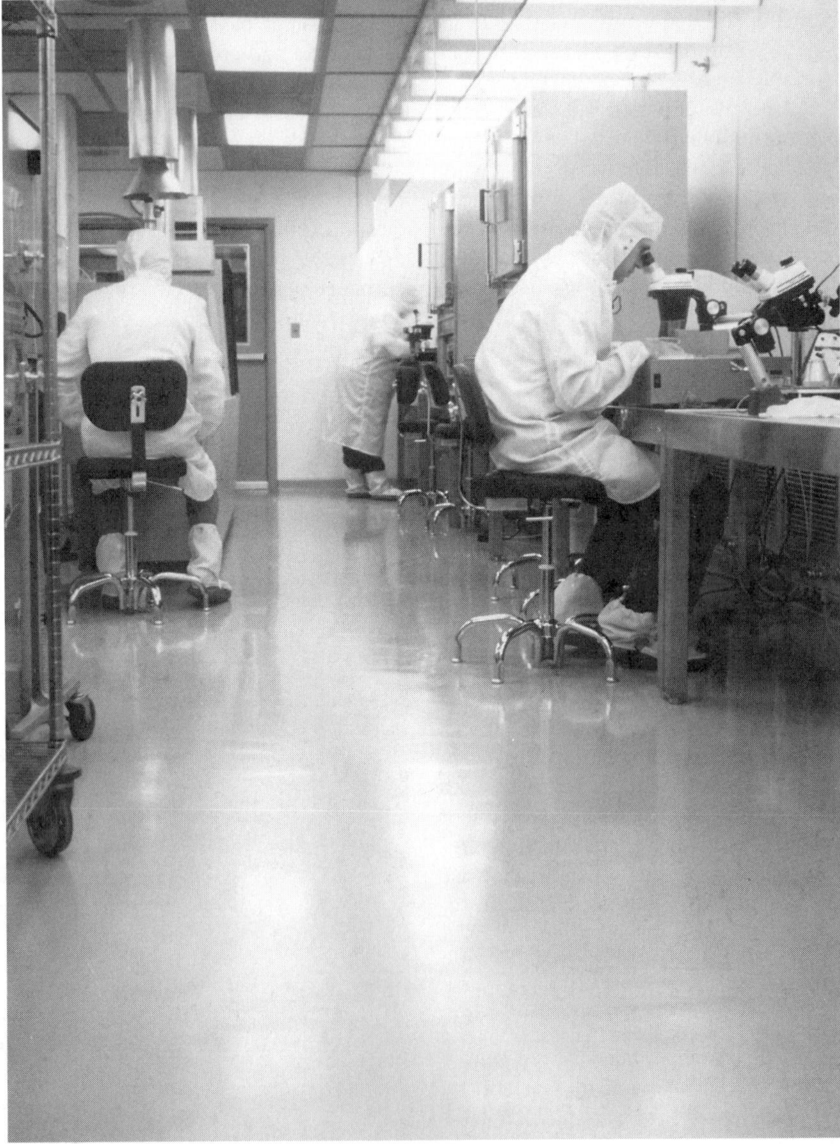

FIGURE 9.13-1 **Free-flowing, static dissipative polymer floor system in a clean room.** (Photo courtesy of Stonhard, Inc.)

heavy traffic, high-impact, and thermal conditions.

- Waterproofing products formulated to eliminate water migration through concrete slabs on parking decks and ramps and mechanical equipment rooms.
- Resurfacers and slurry floor systems, designed to restore old concrete or to provide a protective surface for new concrete in heavy-traffic areas. These are highly abrasion, impact, and chemical resistant. Their excellent adhesion properties make them suitable for use on concrete, brick, quarry tile, and metal surfaces.
- Blended decorative quartz systems that are compacted tightly with a non-yellowing binder and top coat. These systems are available in broadcast and troweled formulations. Their non-porous and monolithic surfaces make them ideal for clean rooms, for medical, pharmaceutical, commercial, and institutional facilities, and for restaurants and kitchens.
- Self-leveling and colored quartz systems for fast-cure, high-strength, monolithic overlays and toppings for industrial and commercial flooring. These systems are suited for heavy-traffic industrial environments where superior toughness, wear, and flexibility are required. Industrial flooring is well suited for heavy metalworking and machining, paper and pulp, high-volume food processing, and other general manufacturing operations that have a need for quick flooring replacement.
- Colored, very smooth, self-leveling overlay products that combine a primer, a matrix layer of filler-enhanced resin, and a standard or chemical-resistant topcoat. Some such systems have an antimicrobial additive that does not support bacterial growth. These products offer exceptional chemical, stain, and impact resistance. Rapid curing ensures short application downtimes.

9.13.2 MATERIALS

Resinous flooring products include methylmethacrylates, polyurethanes, and epoxies.

9.13.2.1 Methylmethacrylate

Methylmethacrylate is a transparent resin to which pigment, aggregate, and fillers are added on-site. Methylmethacrylate floors exhibit excellent resistance to moisture and chemicals, impact abrasion, and thermal shock. They can be placed on concrete, asphalt, and wood. Colored quartz crystals are added as an aggregate to produce colored quartz products. A solvent-free, 100% reactive acrylic resin may also be based on methylmethacrylate. Polymerization (cure) at the job site is effected by adding a hardener and a peroxide initiator. These products feature skid resistance, easy cleaning, and USDA/FDA (Food and Drug Administration) approval for federally inspected food and beverage facilities.

Self-leveling, pigmented products having 100% solids methylmethacrylates provide easy maintenance and high durability for heavy-traffic floors. Their satin finish maintains its appearance for years and features superior stain, chemical, and slip resistance.

9.13.2.2 Polyurethanes

Polyurethane floors exhibit excellent wear, ultraviolet (UV) light resistance, and stain and chemical resistance. Colored quartz can be added to make colored products. Vinyl chips can also be added for color. Formulations are available that offer enhanced chemical resistance. Polyurethane formulations are available in both solvent- and water-based systems.

Trowel-applied, textured polyurethane mortar systems are designed to resist thermal shock/cycling conditions. These are formulated especially for food-processing applications.

Electrostatic discharge polyurethane systems are also available.

9.13.2.3 Epoxies

Epoxy flooring is available in both solvent-based and water-based systems. Its appearance ranges from smooth, high-gloss finishes to orange-peel finishes. Decorative quartz aggregates can be added for color, as can vinyl chips.

Epoxy flooring offers excellent resistance against many chemicals and is easy to clean. These coatings are recommended for assembly and light-manufacturing areas, aircraft hangars, food process areas, laboratories, clean rooms, pedestrian walkways, warehouses, and similar industrial/commercial areas.

Epoxy flooring comes in both thin-film and high-build systems. High-build systems fill in small imperfections in a floor's surface.

Electrostatic discharge control epoxy systems are available in both thin-film and thick-film formulations.

Available epoxy systems include highly chemical-resistant formulations with high organic/inorganic acid resistance.

Solvent-free, flexible epoxy waterproofing membranes are used as crack isolation systems under decorative flooring, as well as stand-alone coating systems for mechanical rooms and high-traffic areas.

There are polymer epoxy industrial flooring systems that combine resins with hard aggregate to provide a tough, traffic-bearing flooring. In these systems, flexibility is achieved without plasticizers or other additives that can separate or migrate as the system ages. Therefore, these products remain flexible and continue to function for many years. Fiberglass scrim may be incorporated into these systems for tensile strength. Hard aggregate provides abrasion resistance, impact resistance, and skid inhibition properties. Various top coats can be specified to provide protection against water, oils, chemicals, and ultraviolet light.

In addition, the following systems are available:

- High-gloss epoxy coating systems with 100% solids.
- High-build, low-odor epoxy coating with 100% solids, either clear or pigmented.
- High-build, chemical-resistant, low-odor, epoxy coatings, having 100% solids, either clear or pigmented. These coatings offer excellent overall resistance to a wide spectrum of chemicals, including aromatic and aliphatic hydrocarbons, acids, alkalies, alcohols, detergents, salts, cutting oils, gasoline, jet fuels, and many solvents.
- High-build, quick-release, low-odor epoxy coatings having 100% solids, which allow for easy removal of build-ups of rubber, wax, mastic, etc.
- High-build, reduced-yellowing, low-odor epoxy coatings having 100% solids.
- Flexible epoxy membrane systems. These specially formulated flexible epoxy coating systems are ideal for applications requiring protection against thermal shock and for areas that use wet processes.
- Two-component, solvent-borne polymer epoxy primer sealers.
- Two-component, waterborne polymer epoxy primer sealers.

- Two-component, 100% solids polymer epoxy coatings.
- Multiple-component seamless polymer epoxy flooring systems.
- Trowel-applied polymer epoxy repair mortar.
- High-performance polymer epoxy top coat products.

9.13.2.4 Limitations

Most resin systems have a very low vapor permeability and act as vapor barriers. Therefore, substrates must be protected against rising moisture.

9.13.2.5 Applicable Standards

The following standards are applicable to resinous flooring:

- American Concrete Institute (ACI)
  - ACI 503.1, "Revised Standards for Epoxy"
  - ACI 503.2, "Standard Specification for Binding Plastic Concrete to Hardened Concrete with a Multi-Component Epoxy Adhesive"
  - ACI 503.3, "Standard Specification for Producing a Skid-Resistant Surface on Concrete by the Use of a Multi-Component Epoxy System"
  - ACI 503.4, "Standard Specification for Repairing Concrete with Epoxy Mortars"
  - ACI 503.5R, "Guide for the Selection of Polymer Adhesives with Concrete"
  - ACI 503.6R, "Guide for the Application of Epoxy and Latex Adhesives for Bonding Freshly Mixed and Hardened Concrete"
- American Society for Testing and Materials (ASTM): C 109, C 307, C 501, C 579, C 580, C 696, D 570, D 635, D 638, D 695, D 790, D 2047, D 2240, D 4541
- U.S. Department of Defense (DOD): Military Standard MIL-D-3134

9.13.3 INSTALLATION

Each resinous flooring and ancillary product requires a different installation procedure. The following information should be considered as generic and generally applicable, but the flooring manufacturer's installation instructions should be strictly adhered to for environmental restrictions, mixing, installation methods, and cure times and methods.

9.13.3.1 Environmental Conditions

Unless the product manufacturer requires other conditions, the minimum ambient and substrate temperatures must be 50°F (10°C) before commencing installation, during installation, and for at least 72 hours after installation has been completed.

Resinous flooring systems should not be applied when the substrate or material temperatures are greater than 104°F (40°C).

9.13.3.2 Precautions

Flooring resin components are flammable, with a 50°F (10°C) flashpoint, which is about the same as that of alcohol. Current regulations controlling flammable liquids must be observed—for example, posting No Smoking signs and taking other precautions. Minutes after placing, the flooring material will no longer support combustion.

Resins can present a minimal dermatological hazard to allergic individuals. Workers should wear protective clothing such as, but not limited to, rubber boots, gloves, and splashproof goggles. Cured materials are completely nontoxic and safe. For further information, consult the manufacturers' Material Safety Data Sheets (MSDOS).

Adequate ventilation must be provided, especially in closed areas. For application in food-processing plants and slaughterhouses, odors of uncured resin require special ventilation or removal of foodstuffs during application.

9.13.3.3 Installer's Qualifications

Some, but not all, manufacturers of resinous flooring products permit their products to be installed only by their authorized representatives or by a contractor acceptable to and certified in writing by the manufacturers.

9.13.3.4 Preparatory Work

Bond tests, as directed by the manufacturer of the flooring, should be conducted to determine adequacy of surface preparation and bond.

SURFACE PREPARATION

The substrate must be clean, durable, level, dry, and free of grease, wax, solvents, curing membranes, oil, and other contaminants. Proper surface preparation requires the following actions.

Decontamination Decontamination is the removal of oils, grease, wax, and other hydrocarbon-based materials. It may require the use of chemical cleaning, microbial decontaminants, high-pressure water, steam cleaning, scrubbing with an industrial-grade degreasing compound or nonionic detergent solution, and mechanical cleaning.

Surfaces to which industrial cleaners or acids have been applied must be scrubbed and flushed with copious amounts of potable water. A moist litmus paper reading should indicate a pH in excess of 10. Material that cannot be chemically removed must be removed mechanically.

Creation of Surface Profile The surface profile of horizontal concrete subjected to traffic should vary less than 5 mils (0.254 mm).

Repair of Surface Irregularities Surface irregularities, including spalls, cracks, deteriorated joints, and the like, must be repaired prior to placement of the resinous flooring, or the system must be designed to offset the thickness irregularities.

Removal and Replacement of Nondurable Concrete Unsound concrete and laitance must be removed by acid etching or appropriate mechanical means. Nondurable, weak, or deteriorated concrete must be removed and new concrete installed before a resinous flooring is installed. Depressions, cracks, holes, and enlarged joints should be filled with a material, usually epoxy, recommended by the flooring manufacturer.

Methods of Surface Preparation The preferred methods of removing dirt, dust, laitance, and curing compounds from concrete slabs is vacuum-grit blasting, sand or grit blasting, or mechanical scarification. Other surfaces, such as cove base (floor and wall intersection) and wall surfaces, should be grit blasted or prepared with other mechanical abrasive techniques.

Resinous flooring systems can be applied to a variety of substrates if such substrates are properly prepared according to the flooring manufacturer's directions. Preparation of surfaces other than concrete will depend on the type of substrate.

Resinous flooring is usually installed directly over nonmoving control joints and cracks that have been treated with the manufacturer's standard crack treatment materials, such as 100% solids epoxy.

Such flooring should not be installed across isolation or expansion joints.

PRIMING

Most resinous flooring systems require priming of the substrate with a primer supplied by the flooring manufacturer. Low-viscosity primers penetrate, seal, and reinforce the substrate, providing a uniform acrylic surface to ensure a deep, strong bond of the topping to the substrate.

9.13.3.5 Installation

MIXING

At the job site flooring resins are mixed with the appropriate quantities of colored quartz sands or pigment and fillers. A hardener is added just before application of the flooring. Mixing is done in small batches with plastic buckets and electric drills with special mixing blades. The pot life of mixed materials is usually 10 to 15 minutes at 68°F (20°C), but may vary with different products.

APPLICATION

After mixing, the flooring material is spread over a properly leveled and primed area, then lightly troweled or gauge-raked to an even thickness dictated by the manufacturer. After troweling, a special roller is used to release trapped air from the topping. For increased skid resistance, a #16 to #25 mesh, clean, dried, angular, quartz sand can be broadcast into the fresh flooring mix. When random scattering of colored vinyl chips is desired, they are also broadcast into the fresh flooring. The chips should rain upon, not be thrown into, the surface.

9.13.3.6 Protection

Resinous floor systems should be protected from damage and wear during other phases of the construction operation, using temporary coverings, as recommended by the manufacturer, if required.

9.13.3.7 Maintenance

CLEANING

Cleaning materials and procedures should be acceptable to the system manufacturer. In general, resinous flooring can be steam cleaned or cleaned with conventional industrial cleaners and wax, but solvent-containing materials should not be used. Mineral spirits may be used to remove rubber marks. Some cleaners will affect the color, gloss, or texture of resinous floor surfaces; therefore, cleaners should be tested in a small, inconspicuous area before they are used.

REPAIR

Chemical bonding allows fast, easy resurfacing when such is necessary.

9.14 Special Coatings

Special coatings are used on both interiors and exteriors of buildings and other structures to waterproof, protect, and beautify. There are several types of coating systems and many coating products available today. This section discusses a number of these.

9.14.1 GENERAL CHARACTERISTICS

Comparing special coatings based on their chemical contents is usually difficult and often impossible. Some manufacturers do not disclose the components of their systems. Others use terms so generic as to be useless in determining the real contents. In the end an architect should base coating selection decisions on achieving a desired result. If the problem is to waterproof a deck or coat a concrete wall, an architect should look for a product that accomplishes that end and make selections based on the reputation of the coating system's manufacturer and the proven performance of the product.

When comparing special coatings, it is necessary to take into account their general characteristics. These properties vary considerably from product class to product class, and from manufacturer to manufacturer even within the same product class. The characteristics of similar products of different manufacturers may vary slightly without significance. If the variation is great, however, it can be concluded that the products are significantly dissimilar and may not be equal in performance.

Most manufacturers list many of the characteristics of their products in their basic catalogs. Other data can usually be obtained directly from the manufacturer or from the manufacturer's Safety Data Sheets. Among the important characteristics are:

- *Recommended use:* This is the first selection criterion, and a critical one.
- *Coating manufacturer:* The manufacturer's reputation is paramount.
- *Product history:* Successful installations are a very good guide.
- *Water vapor transmission:* Perm ratings vary considerably, ranging from 25 to 200, often depending on the coating thickness, but also due to the texture of the coating. Additives that impart texture may affect perm rating.
- *Coating thickness:* Thickness will vary widely with different types of coatings, but the real significance is when thickness varies within the same coating type.
- *Solids content:* Solids content is measured by weight and by volume. These vary considerably from product to product and within a product, depending on texture.
- *Pigment volume concentration:* Such concentrations vary considerably from product to product and within a product due to texture.
- *Abrasion resistance.*
- *Chemical resistance:* Some coatings are more resistant to chemicals than others. This is an important consideration when a coating will come in contact with chemicals, such as ice-melting materials or harsh cleaning agents. Coatings used in high chemical exterior or interior environments must be selected with increased care.
- *Resistance to salt spray.*
- *Resistance to scaling.*
- *Water penetration resistance:* Waterproofing coatings must resist water penetration as measured by the static head of water pressure. For example, a coating might be said to be able to resist penetration of water at up to 4 ft. (1200 mm) of water pressure.
- *Scrub resistance:* Scrub resistance varies somewhat within a class of coatings, depending on thickness.
- *Impact resistance.*
- *Tensile properties.*

- *Surface burning properties.*
- *Chalking resistance:* Most special coatings should have no chalking after 1000 hours of testing.
- *Freeze-thaw properties.*
- *Radon gas penetration:* This characteristic can be important for some below-grade waterproofing installations.
- *Viscosity.*
- *VOC Content:* Usually less than 400 g/L, per ASTM D 5090.
- *Compressive strength:* This is an important characteristic for cementitious coatings.

9.14.2 APPLICABLE STANDARDS

Industry standards are used to test special coatings and identify their characteristics. The following is a partial list of the main standards.

- American Association of State Highway and Transportation Officials: AASHTO T259/T260 National Cooperative Highway Research Program.
- American Concrete Institute: ACI 201.1R, "Making a Condition Survey of Concrete in Service."
- American Society for Testing and Materials (ASTM): Many ASTM standards apply. The following is a partial list, including an abbreviated indication of the use of each standard. Refer to Section 9.18, "Acknowledgments and References," for their full names.
  - B 117: salt spray resistance
  - C 67: sampling and testing brick and structural clay tile
  - C 91: masonry cement
  - C 109: compressive strength of hydraulic cement mortars
  - C 666: resistance of concrete to freezing and thawing
  - C 672: scaling resistance of concrete surfaces exposed to deicing chemicals
  - C 926: application of portland cement–based plaster
  - D 522: bending test of attached organic coatings
  - D 870: water-resistance testing of coatings
  - D 968: abrasion-resistance testing of organic coatings
  - D 1308: test for effect of chemicals on organic coatings
  - D 2243: freeze-thaw test of waterborne coatings
  - D 2370: test of tensile properties of organic coatings
  - D 2486: test for scrub resistance of coatings
  - D 4141: test for accelerated outdoor exposure of coatings
  - D 4258: surface cleaning concrete for coatings
  - D 4259: abrading concrete
  - D 4260: acid etching concrete
  - D 4261: surface cleaning concrete unit masonry for coatings
  - D 5090: ultrafiltration permeate flow performance data
  - E 84: test for surface-burning characteristics of building materials
  - E 96: test for water vapor transmission of materials
  - E 514: test for water penetration and leakage through masonry
  - G 23: operating light- and water-exposure apparatus
- Federal Specifications: TT-C-555b.
- National Cooperative Highway Research Program: NCHRP Report 244.

9.14.3 SPECIAL COATING SYSTEMS

The following classifications were determined by examining the constituents of a wide array of coating products. This section does not mention trade names or manufacturers' names, nor does it suggest which product types are best for a particular situation. It merely displays the variety of coatings available and offers some guidance as to their properties and where they are usually used.

Where the components of a system were identifiable, the system is listed in this section under a heading containing the names of those components (acrylic latex terpolymer coatings, for example). Products for which the components were not identifiable are listed under the generic names (not trade names) used by the manufacturers to identify those systems (polymer modified waterproofing cement coatings, for example). Some of the products listed here under different names may, in fact, be very similar. Other manufacturers may use different nomenclature from that shown here for similar products.

9.14.3.1 Acrylic Emulsion Coatings

Acrylic emulsion coatings are premixed coatings available in smooth, fine, medium, and coarse textures. They are abrasion, mildew, and weather resistant and permit vapor transmission. They bind friable surfaces together and increase the bond strengths of coatings. They are available in tint bases and 50 or more standard colors.

These products are used in the following ways:

- In both vertical and overhead applications as above-grade decorative finishes
- As a primer for water-based coatings, especially on friable, dusty concrete substrates and prepainted chalky surfaces
- As a finish coat over properly prepared concrete, cementitious backer board, CMU, stucco, portland cement plaster, brick, drywall, stone, and painted surfaces
- As deck coatings (some textured acrylic emulsion coatings)

9.14.3.2 Acrylic Emulsion Block Fillers

Block fillers are a special class of acrylic emulsion coatings. They are premixed, smooth, water-resistant, ready-to-use, flat-finish white coatings. They are mildew and weather resistant and may be applied by brush, roller, or spray.

Block fillers are used:

- To fill concrete masonry surface pores and voids prior to coating, and as primers for specialized coatings
- In both vertical and overhead applications on interiors and exteriors
- As a fill coat over concrete, cementitious backer board, block, stucco, plaster, brick, drywall, and stone and as a primer for latex, enamel, epoxy, urethane, and elastomeric coatings

9.14.3.3 Acrylic Polymer Coatings

Acrylic polymer coatings are premixed high-impact coatings available in both elastomeric and synthetic stucco form. They are flexible and crack resistant. They are also alkali, water, mold, and mildew resistant, while permitting vapor transmission. They are available in tint bases and 50 or more standard colors.

They are used as follows:

- On both interiors and exteriors
- As a coating over concrete, cementitious backer board with brown coat, CMU, exterior stucco, brick, exterior insulation and finish systems (EIFS), and gypsum wallboard

9.14.3.4 Silicone Emulsion, Acrylic Polymer Coatings

Silicone emulsion, acrylic polymers are premixed coatings available in both elastomeric and synthetic stucco form. These are flexible coatings that repel water, are recoatable, and are mildew, dirt, algae, and alkali resistant. They are available in tint bases and 50 standard colors.

These coatings are used over concrete, cementitious backer board with brown coat, CMU, exterior stucco, brick, and EIFS.

9.14.3.5 Acrylic Terpolymer Coatings

Acrylic terpolymer coatings are textured, high-build, 100% acrylic terpolymer, water-based coatings with elastomeric properties. They are both decorative and protective. They are available in 16 or more standard colors, and special colors are also available. When roller applied, they produce a smooth, fine texture. Spray application gives them a smooth, fine, coarse, or extra-coarse texture. An embossed texture is also available to help minimize uneven block and mortar joints.

These coatings are used above grade over concrete, stucco, masonry, metal, and other manufacturer-approved surfaces.

9.14.3.6 Acrylic Latex Coatings

Acrylic latex color coating products are 100% acrylic latex exterior finishes. They are weather, moisture, and mildew resistant. They are available in smooth finish only.

These products are used as a top coat over textured primers and over water-based and oil-based textured coatings.

9.14.3.7 Acrylic Latex Terpolymer Coatings

Acrylic latex terpolymer coatings are high-performance coatings with a base of 100% elastomeric resin providing up to 400% elongation. These coatings are flexible in below-freezing temperatures. They resist alkali. They also resist hairline cracking and have excellent adhesion characteristics.

These products are used above grade over concrete, cement plaster, CMU, brick, galvanized metal, cement fiberboard, wood siding, and other manufacturer-approved surfaces.

9.14.3.8 High-Build Acrylic Resin Coatings

High-build acrylic resin coatings are water-based, 100% acrylic resin, textured, waterproof coatings. They are VOC compliant, durable, breathable, and UV, blister, abrasion, wind-driven rain, and mildew resistant. They are available in smooth, fine, and coarse textures, and either with or without algae-resistant additives. They are available for airless spray application, in heavy-textured formulations to simulate stucco, and with a fine texture for use for foot traffic. These coatings can be highly reflective. They are available in 48 or more standard colors. When properly applied, they hide minor construction imperfections and provide a uniformly textured weather-resistant surface.

They are used:

- Over vertical concrete, masonry, and stucco
- On parking garages, stairways, traffic barriers, and in other locations

9.14.3.9 High-Build Latex Coatings

High-build latex coatings are water-based, textured coatings designed for either spray or roller application. They dry rapidly, adhere well, and are flexible.

These products are used above grade only over concrete, cement plaster, block, brick, cement fiberboard, metal, and other manufacturer-approved surfaces.

9.14.3.10 Vinyl Toluene Resin Coatings

Some textured primers and waterproofers are high-build, plasticized, pigmented, solvent systems based on a vinyl toluene resin binder. The cured primer membrane provides fill, texture, and weatherproofing properties over cured or "green" concrete masonry surfaces. These materials have the ability to be applied over damp or highly alkaline concrete. They are available in both smooth and coarse textures. Some of these products can be applied in temperatures as low as 20°F (–6°C). Others require that the temperature be between 40°F (4.4°C) and 100°F (37.8°C).

These coatings are used:

- Above grade over concrete, cement plaster, CMU, brick, and previously painted and other approved surfaces
- For one-coat application on concrete highway abutments and barriers, and on parking garages

These coatings are not recommended for wood or metal surfaces.

9.14.3.11 Vinyl Toluene Acrylic Resin Systems

Some industrial maintenance coatings are high-build, plasticized, vinyl toluene acrylic resin systems, formulated to provide a weatherproof coating over cured or uncured concrete masonry surfaces. They can be placed over damp or high-alkaline surfaces. They are available in both standard and special colors. They may be roller applied for a smooth, fine texture or spray applied to give a smooth, fine, coarse, or extra-coarse texture.

These coatings are used above grade over concrete, cement plaster, CMU, brick, and other manufacturer-approved surfaces.

9.14.3.12 Acrylic Elastomeric Polymer Coatings

Acrylic elastomeric polymer coatings are smooth and premixed.

9.14.3.13 Siliconized Acrylic Elastomeric Polymer Coatings

Siliconized acrylic elastomeric polymer coatings are smooth and premixed.

9.14.3.14 Acrylic Conditioners, Primers, and Sealers

CLEAR CONDITIONERS, PRIMERS, AND SEALERS

Acrylic conditioners, primers, and sealers are ready to use, clear, water-based, alkali-resistant coatings. They reduce the potential for efflorescence while allowing moisture vapor transmission. They consolidate light surface chalk and concrete dust. They are water repellent and stain, abrasion, and mildew resistant. They are VOC compliant and nonflammable and can be cleaned up with water.

These coatings are used:

- As above-grade surface conditioners and primers over concrete, stucco, stone, CMU, brick, plaster, and drywall
- As surface conditioners and primers for elastomeric coatings, synthetic stuccos, water-based textured coatings, acrylic latex paints, and oil-based paints
- As two-coat minimum surface sealers over concrete, stucco, brick, CMU, stone, and plaster

PIGMENTED SEALERS

Pigmented sealers are high-build, VOC-compliant, 100% acrylic waterproofing sealers. They stop wind-driven rain up to 100 mph (161 km/h). They allow the surface to breathe. They span hairline cracks and hide minor surface imperfections. They can be applied in temperatures as low as 20°F (–6°C).

These sealers are used on above-grade masonry, including slick precast, glazed brick, and marginally prepared surfaces, without the use of primers and conditioners.

TINTABLE PRIMERS

Tintable primers are 100% acrylic, ready-mix, opaque, water-based, VOC-compliant primers that dry to a gritty finish to help reduce suction in the substrate, thus providing easier application and better adhesion of various acrylic coatings. They will not degrade in the presence of intermittent moisture. They are breathable and may be applied by brush, roller, or spray. They retard efflorescence. They provide uniform background color prior to top coat applications, and they improve workability and ease of acrylic topcoat applications.

These primers are used on masonry.

9.14.3.15 Polymer-Modified Waterproofing Cement Coatings

Polymer-modified waterproofing cement coatings are used to waterproof masonry.

9.14.3.16 Acrylic Polymer Crack and Surface Repair Materials

Acrylic polymer crack and surface repair materials are premixed, flexible, brush- or knife-grade stucco repair products. They are available either smooth or textured. They are mildew and weather resistant. Brush-grade products are usually white; knife-grade are usually off-white.

These materials are used:

- Above grade in both interior and exterior vertical and overhead applications
- To fill and repair static cracks in concrete, stucco, cementitious backer board, drywall, plaster, block, and brick

9.14.3.17 Penetrating Sealers and Masonry Stains

Penetrating sealers and masonry stains are water-repellent sealers suitable for a wide range of surfaces. They feature long-term water repellence, and resistance to discoloration, efflorescence, dirt, and mildew. They are classed as water repellent and are not intended to serve as waterproofing materials.

These products are used:

- Over concrete, brick, stucco, block, wood, limestone, and aggregate panels
- To protect and maintain bridges, parking garages, airport pavements, industrial plants, buildings, and other concrete and masonry structures

Sealers and stains protect and preserve architectural substrates. Treated surfaces are resistant to staining, spalling, weathering, efflorescence, water intrusion, salt intrusion, fungi and mildews, deterioration, freeze-thaw scaling, and reinforcing steel corrosion. There are several kinds of these sealers.

REGULAR CLEAR SEALERS

Regular clear sealers are penetrating water-repellent sealers. Some contain a hydrocarbon plasticizer that is white, nonglossy, and stain and oxidation resistant.

They are used on both interior and exterior concrete, cement plaster, painted surfaces, natural stone, brick, and wood and other porous surfaces.

HEAVY-DUTY CLEAR SEALERS

Heavy-duty clear sealers are specially formulated for more porous surfaces. Their hydrocarbon plasticizer content is 20%. There are two-coat water-repellent systems that require a 24-hour curing period under proper weather conditions prior to second coat application. Upon drying, the surface is enriched and slightly darkened, with a low sheen.

These sealers are used over concrete masonry, especially split-face block and cinder block, slump stone, adobe block, lightweight concrete, and exposed aggregate systems.

OPAQUE STAIN SEALERS

Opaque stain sealers are solvent-based resin products that penetrate masonry for water repellence and greater depth of color, while allowing moisture vapor to escape. They are available in standard and special colors.

These sealers are used on concrete, CMU, brick, natural stone, and stucco.

SILOXANE AND SILANE SEALERS

Siloxane and silane sealers protect and preserve architectural substrates without altering surface texture or appearance. Treated surfaces retain their vapor permeability and are resistant to staining, spalling, weathering, efflorescence, water intrusion, salt intrusion, fungi and mildews, deterioration, freeze-thaw scaling, and reinforcing steel corrosion.

These sealers are used on flagstones and other pavers, concrete, synthetic stone, CMU, brick, limestone, marble, sandstone, stucco, and synthetic stucco.

Siloxane Sealers Siloxane sealers are one-component penetrating sealers. These products react chemically with the substrate to repel salt and water, while allowing the substrate to retain its permeability. They do not change the substrate's texture, appearance, or skid/slip resistance. They are compatible with most solvent-based surface finishes.

These sealers are used in both traffic-bearing and non-traffic-bearing applications, and on both horizontal and vertical surfaces.

Silane Clear Sealers Silane clear sealers are a chemical treatment that causes concrete, masonry, limestone, and wood to become repellent to salt and water. They are available in two grades for concrete, masonry, and limestone, and one grade for wood. They are claimed to be superior to oligomeric polysiloxanes, stearates, siliconates, silicates, urethanes, butadienes, chlorinated rubber, epoxies, and oils.

Premium-grade silane weatherproofing products contain 40% alkyltrialkoxy silane in mineral spirits. Some will penetrate up to ½ in. (12.7 mm) of the substrate with each application. They are used to protect and maintain bridges, parking garages, airport pavements, industrial plants, buildings, and other concrete and masonry structures in the severest environments.

There are also silane weatherproofing products containing 20% alkyltrialkoxy silane in mineral spirits. These penetrate up to ¼ in. (6.4 mm) with each application. They are used for bridges, parking garages, pavements, plants, and other brick and concrete structures in severe environments.

Silane sealers are usable for traffic conditions. They do not change the substrate's skid/slip resistance, vapor permeability, texture, or appearance. They are usable for both horizontal and vertical applications and are compatible with most solvent-based surface finishes. They are fast acting. Treated substrates, such as bridge decks, are usable in 20 to 45 minutes. These sealers flow into microcracks and penetrate substrates. They contain no hydrocarbons and are not susceptible to UV or chemical degradation.

Microemulsions Microemulsions are VOC-compliant, microemulsifiable concentrates. When silanes and alkoxysilanes are mixed with water, a microemulsion is produced. This diluted form will impart water repellence. These are water-repellent products, They are not intended to serve as waterproofing materials. Top coating these products with other paints or coatings can proceed following a minimum 72-hour curing time in acceptable environmental conditions.

These materials are used above grade over absorbent mineral surfaces.

9.14.3.18 Cementitious Coatings

Cementitious coatings are decorative, water-resistant, portland cement–based finish coats. They are durable and water resistant. They eliminate the need to rub concrete to finish it. They are available in 50 or more factory-blended colors. Building codes may classify cement-based coatings as dampproof rather than waterproof. An architect must verify performance requirements against a product's technical data.

These materials are used as a decorative finish coat over concrete, brick, and stone.

POLYMER-MODIFIED WATERPROOF CEMENT COATINGS

Some polymer-modified waterproof cement coatings are *block fillers*. They are usually one-component systems. They are nontoxic. They reduce radon gas penetration by more than 99%, and waterproof under a maximum of 4 ft. (1200 mm) of water pressure. They are abrasion and moisture resistant and are breathable. They have a white reflective surface.

These coatings are used:

- Both above and below grade and on either interiors or exteriors
- To fill pores and voids in concrete or masonry substrates
- For waterproofing of basements, parapets, cisterns, reservoirs, grain elevators, silos, foundations, and glass-block mortar joints
- For decorating stone, brick, concrete, CMU, planters, fountains, and glass-block mortar joints

POLYMER-MODIFIED WATER-RESISTANT CEMENT COATINGS

Polymer-modified, water-resistant lightweight portland cement–based base coats and finish coats are crack resistant, durable, water resistant, and lightweight.

They are used as a base coat and finish over autoclaved aerated concrete.

CEMENT-BASED WATERPROOF COATINGS

Portland cement–based waterproofing/dampproofing coatings, used in conjunction with an acrylic admixture, become an integral part of the substrate when cured. They resist hydrostatic pressure. They are breathable. They are approved by the National Sanitation Foundation (NSF 61.1).

These coatings are used:

- On both interior and exterior surfaces
- To fill and seal pores and voids in concrete and masonry surfaces
- To correct surface irregularities, thus eliminating the need for concrete rubbing

READY-TO-USE CEMENT-BASED WATERPROOF COATINGS

These are high-build, one-component, water-based, ready-mixed, acrylic waterproofing coatings. They provide a dense surface, which reduces radon infiltration. They should pass a 2000-hour weatherometer test. They clean up with water, are VOC-compliant, and are environmentally friendly.

These coatings are used:

- On interiors in below-grade locations
- For protection of new or existing concrete and masonry surfaces

EXTERIOR BELOW-GRADE WATERPROOF COATINGS FOR FOUNDATIONS

These are portland cement–based waterproofing/dampproofing coatings. When used in conjunction with an acrylic admixture, they become an integral part of the substrate.

They are used on below-grade exterior concrete and masonry surfaces.

9.14.3.19 Flecked Multicolored Interior Finishes

Flecked multicolored interior finishes are spray-applied, textured, decorative coatings that offer a different decorative expression than most coatings, with virtually unlimited color combinations. They hide minor surface irregularities and give surfaces added dimensions in color and texture. Some are finished after application with a clear urethane that offers VOC compliance and features outstanding hardness, durability, abrasion resistance, and gloss retention. They are applied with conventional, airless, or HVLP spray equipment. They are stain and mildew resistant and do not support bacterial growth. Their variegated coloration helps conceal dirt. They are scrubbable with liquid detergents; touch ups are nearly invisible. Their fire resistance is Class A. They are odorless, and meet EPA Clean Air (VOC/VOS) Standards. They dry quickly (2 to 4 hours to the touch). Recoating can be done in 24 hours.

These finishes are used:

- On walls, ceilings, and trim
- Over a wide variety of substrates, including drywall, plaster, masonry, wood, and metals
- Over existing paint or vinyl

9.14.3.20 Antigraffiti Coatings

Antigraffiti coatings are water-based, VOC-compliant coatings that help in easy removal of most kinds of graffiti, such as those made with spray paints and marking pens. They are available in clear or pigmented formulations. The clear formulations do not alter the appearance of the substrates. The pigmented formulations leave a glossy surface.

These coatings are used on metal, concrete, brick, vinyl, stone, stucco, concrete masonry, wood, and EIFS. Some of them are not recommended for glass, polyurethane, or polypropylene surfaces.

9.14.4 INSTALLATION OF SPECIAL COATINGS

The following are some general requirements applicable to most coatings. However, preparation for and installation of coatings should be done only in accordance with the coating manufacturer's specific recommendations, which may vary with the following recommendations.

9.14.4.1 Preparation

Surfaces should be cleaned and repaired in accordance with the requirements of ASTM D 4258, ASTM D 4259, ASTM D 4260, ASTM D 4261, and ACI 201.1R, as applicable.

Surfaces to receive most coatings must be clean, sound, and dry. Some coatings, however, are designed to be applied over damp surfaces. In no case should precipitation be falling or imminent.

Loose, flaking, or oxidized paint should be removed by sandblasting, water blasting, wire brushing, or scraping. Old paint in sound condition can usually be left in place, washed to remove excess chalk, and allowed to dry before coating application.

Most of the time, surface contaminants such as dust, dirt, chalk, mildew, form oils, and other foreign matter must be removed, but for some coatings, some of these may simply be stabilized with an appropriate primer.

Decayed cement plaster must be removed and new stucco installed with a bonding agent.

Cracks, holes, and voids must be filled with cement patching compounds and bonding agents recommended by the coating system manufacturer. The texture of patches should match that of the existing surface.

Joint sealants used must be compatible with the coating and must have cured at least 48 hours prior to priming.

New concrete, masonry, and cement plaster must be allowed to cure for the minimum number of days recommended by the coating manufacturer. Some coatings require a 7-day cure. Others require 28 days.

9.14.4.2 Application

The ambient temperature during installation of most coatings must be between 45°F and 100°F (7°C and 38°C).

Coating application must be made at the rate recommended by the manufacturer. A uniform film of the thickness recommended by the manufacturer should be provided over the entire surface. Material coverage rates can vary due to surface porosity and texture.

To prevent lap marks, a wet edge must be maintained during spraying, brushing, or rolling. Application should not be started or stopped in the middle of a wall or large area. Application should continue to a "natural break," such as a panel edge, seam, or corner. Back rolling while spraying will help minimize spray patterns.

Coatings should be cured in accordance with their manufacturers' recommendations. For most coatings, primers must be cured a minimum of 4 days prior to the application of the color coat. However, for most coatings, primers must not have been exposed to ultraviolet rays for more than 4 weeks.

Subsequent coats should be allowed to dry for the time recommended by the manufacturer. Under ideal conditions, most, but not all, coatings will dry to the touch in about 2 hours and dry hard in 24 hours. Residual matter in the film will then continue to cure for some additional time. Such times will vary if the weather is not favorable, and may be different from those indicated here for some coating systems.

Coatings can damage materials for which they are not intended and can be difficult to remove. It is therefore important to protect adjacent work and surrounding areas from contact with coatings. Adjacent surfaces that become inadvertently contaminated with coatings should be cleaned immediately to help prevent damage.

In their liquid state, some coatings contain chemicals, such as ethylene glycol, that will cause serious illness or death if swallowed, or even if large quantities are absorbed into the body. Therefore, locations where coatings of this type are being applied must be well ventilated, and appliers must be protected with face masks, eye shields, and protective clothing. Some liquid constituents may also be flammable and must be kept away from heat sources.

9.15 Paints

There are no generally accepted definitions in the industry that clearly state the difference between paints and coatings. The two terms are often used interchangeably. Because of the possible confusion created by this double usage, and to clarify some other terminology used in the painting industry, this book defines these terms as follows:

- *Paints:* Air-drying materials; applied as a liquid in the field, except that primers are applied either in the field or in a shop or factory. The complete system of materials needed to produce a paint film is called a *paint system.* The term *paint* includes primers, emulsions, enamels, opaque stains, sealers, fillers, and other applied materials used as prime, intermediate, or finish coats in a paint system.
- *Coatings:* Applied only in the shop or factory; similar in materials and formula to paint; either air-drying or baked-on; include such materials as Kynar coatings on metal, anodizing on aluminum, and baked enamel on lockers or furniture.
- *Special coatings:* Thick high-performance field-applied architectural coatings, such as high-build glaze coatings, fire-retardant coatings, industrial coatings (such as those used in sewage disposal plants), and cementitious coatings.
- *Transparent finish:* A system of materials used to form a protective film on wood that lets the wood color, grain, or both remain visible; applied either in the shop or factory or in the field; sometimes called *clear* wood finishes.

Paints, transparent finishes, and a few special coatings are discussed in this section. Some coatings used on metal are covered in Chapter 5; other special coatings are discussed in Section 9.14.

The comments in this section related to materials have been drawn from data developed by the Scientific Committee of the National Paint and Coatings Association and from recommendations included in the product and use literature of paint manufacturers. The paints and coatings discussed here are not the only ones available. Some manufacturers have other formulations designed for specific purposes, which have advantages in their intended applications over the more commonly used paint and transparent finishes discussed here.

The information in this section related to methods of applying paint and transparent finishes has been developed from data contained in earlier editions of this book. These data were abstracted from the *Paints and Protective Coatings Manual of the Army, Navy, and Air Force* compiled by the National Bureau of Standards (now the National Institute of Standards and Technology) and from data contained in the preparation and application recommendations of paint manufacturers.

Paints and coatings are used to coat, beautify, and protect an underlying surface. They form a protective shield between construction materials and the elements that attack and deteriorate them. Typical causes of failure in paint and transparent finishes are sunlight, temperature variations, salt water, water vapor, decay, mildew, chemicals, and abrasion. When regularly renewed, paint and transparent finishes offer long-term protection that extends the useful life of the materials and the structure. However, as is discussed later, the time between renewals varies considerably, being usually shorter for transparent finishes than for opaque ones.

Paints may serve other functions as well. For example, paint is an essential part of sanitary programs for hospitals, as well as for commercial, institutional, and home kitchens and washrooms. Gloss and semigloss paints provide smooth, nonabsorptive surfaces that can be washed easily and kept free of dirt and germs.

White and light-tinted paints reflect light, brighten rooms, and increase visibility. Flat paints soften and evenly distribute interior illumination. The functional use of color creates comfortable living and working conditions in the home, shop, and office.

Color-coded painting is frequently used to enhance public comfort, safety, and efficiency. Certain colors are stimulating, others are relaxing, and others are universally associated with potentially dangerous situations.

9.15.1 MATERIALS

There is rarely one "best" paint or transparent finish for any surface. Among the factors that must be considered are the condition of the surface, the method of application, the curing conditions, the service expected, and the relative weight of initial cost balanced against appearance and durability. Because the entire paint or transparent finish system must be considered as a unit, it is unwise and often impossible to choose a paint or transparent finish without regard to the surface over which it is to be applied. While it is generally impossible to single out the best paint or transparent finish, it is possible to list commonly encountered surfaces and to recommend the materials most suitable for each.

Paint is a mixture of minute solid particles known as *pigment*, suspended in a liquid medium called the *vehicle*. The pigment provides hiding power and color. The vehicle combines (1) the *solvent* (thinner), which ensures the desired consistency for application by brush, roller, or spray, and (2) the *binder*, which bonds the pigment particles into a cohesive paint film during the drying process. Some paints dry and harden simply by evaporation of the solvent. Other types involve chemical reactions as well as solvent evaporation. Most paints are described according to the type of binder used in them (alkyd, latex, oil, oleoresinous, rubber, urethane).

Materials in any of the types described here may be available in varying qualities. The reputation of the manufacturer is a significant factor in selecting paint and transparent finishes. The reputations of most major paint manufacturers are well known within the construction industry.

Many states, the District of Columbia, and many cities and towns have enacted air pollution regulations that restrict the amounts and types of solvents that may be emitted into the atmosphere by paints, transparent finishes, and coatings, and by the materials used to clean surfaces to receive them and remove them after they have been applied. In addition, many jurisdictions have regulations concerning the use and disposal of even those chemicals that are not prohibited. Therefore, before paint, transparent finish, or coating is applied and before paint removers or chemical cleaning materials are used, it is necessary to verify their acceptability by the authorities having jurisdiction.

9.15.1.1 Interior Primers

Primers are paint or transparent finish materials intended for application over bare surfaces. They serve as a base for succeeding coats. Some topcoat materials also serve as a primer on some surfaces.

WALL PRIMERS AND PRIMER-SEALERS

Interior wall primers and primer-sealers may have a latex, alkyd, or oil binder, depending on the material to be covered and the topcoat that will be applied.

Latex Latex primers and primer-sealers are emulsion paints intended for use on gypsum board surfaces, but they are also used on masonry. They do not raise the fibers on gypsum board as do solvent-thinned primers. They dry quickly, which allows recoating within hours, cleaning equipment with water, and excellent alkali resistance.

Alkyd Alkyd primers and primer-sealers are usually made of odorless alkyd; that is, the binder is an alkyd resin dissolved in odorless mineral spirits. They require overnight drying before recoating. They are suitable for all interior surfaces except paper-surfaced gypsum board and surfaces containing active alkali, such as partially cured or damp plaster or masonry.

Oil The binder for oil primers and primer-sealers is a processed drying oil, often containing some resin. The characteristics of these materials are similar to those of alkyd primers, except that they are softer and dry more slowly. Even so, next-day recoating is possible. They may be thinned with odorless or ordinary mineral spirits.

Alkali-resistant Alkali-resistant primers and primer-sealers are usually based on a butadiene-styrene copolymer or on chlorinated rubber, with one or more softer resins, such as chlorinated paraffin, as plasticizers. Sometimes small amounts of drying oil are added. Aromatic solvents are often required. Application is likely to be more difficult than with other primers. These products are particularly suitable for alkaline surfaces, such as damp or partly cured plaster or masonry. They will not resist hydrostatic pressure, such as may occur below grade, but may be used on above-grade masonry.

WOOD PRIMERS

Interior wood primers include enamel undercoaters, clear wood sealers, and wood fillers.

Enamel Undercoaters Enamel undercoaters may be alkyd, latex, or oil-based, but most are alkyds. Undercoaters are characterized by good hiding ability and low gloss. They are hard, tight films that prevent the penetration of enamel coats applied over them. This property is known as *enamel holdout*. They may be used under any pigmented interior finish, but are particularly suitable under gloss and semigloss enamel where a smooth finish is important. Enamel undercoaters are usually white but may be tinted for use under colored finish coats. They contain no rust-inhibitive pigment, but they may be used as primers on metals where no rusting is expected. They dry sufficiently overnight to be painted the following day. Latex undercoats may be painted in 2 to 4 hours; oil-based undercoats should dry at least 16 hours before recoating.

Clear Wood Sealers Clear wood sealers seal the pores in wood surfaces without impairing the natural appearance. Many sealers contain a transparent pigment to improve sealing and reduce penetration. This reduces the prominence of the grain to some extent, which is usually desirable. These products are generally used under transparent finishes, although they can be used under pigmented materials as well.

Wood Fillers Wood fillers include paste wood fillers and liquid wood fillers. They are used beneath transparent finishes.

Paste wood filler is used to fill the pores of open-grain woods, such as oak or mahogany. A transparent and relatively coarse pigment is combined with sufficient binder to make a stiff paste, which requires thinning with a volatile solvent before application. Stain may be added. When the solvent has evaporated, the filler is rubbed across the grain to force the filler into the pores and to remove excess material. This is followed by a cleanup wipe in the direction of the grain with a coarse cloth such as burlap.

Liquid wood fillers are designed to eliminate thinning. While they generally do not perform as well as paste fillers on many woods, they are adequate when extensive filling is not required. All fillers impart some color to the wood. Fillers containing stearates should not be used under urethane finishes.

MASONRY AND CONCRETE WALL PRIMERS AND FILLERS

Primers Interior masonry and concrete wall primers may be either latex or alkyd. The first coat of an epoxy finish often serves as a primer, but conditions where a heavy-duty finish is needed may require a different epoxy formulation for the primer. Primers are used as sealers on normally smooth masonry surfaces.

Block Fillers When a smoother finish is desired on masonry and concrete, a block filler can be applied beneath the paint primer. Block fillers are relatively thick materials that can be applied by brush, roller, or spray. When applied by spray, they must then be brushed into the surface to ensure that voids are filled. The binder may be latex or a solvent-thinned material, such as an epoxy ester. Latex materials have excellent alkali resistance and may be used on damp surfaces. Solvent-thinned materials may have better chemical resistance.

Cement Grout Instead of a block filler, a thin portland cement mortar may be applied with brush or trowel to give a smooth surface to rough masonry or concrete. These materials are available premixed with latex or in powdery form, which is mixed with a latex emulsion in the field. Mortar made without latex is permeable and not a good substrate for paints.

9.15.1.2 Exterior Primers

The following primers are used on exterior surfaces.

WOOD PRIMERS

Exterior wood primers include both oil and alkyd primers, but most used today are alkyd.

Oil Primers Oil primers contain oil to control penetration. They generally do not contain zinc oxide because it can cause blistering and peeling, particularly under severe moisture conditions. However, paint systems without zinc oxide are more difficult to render mold resistant.

Alkyd Primers Alkyd primers are similar to oil primers, except that they have an alkyd binder. This promotes faster drying; shorter recoating time; resistance to bleeding, mold, and moisture; and controlled penetration.

METAL PRIMERS

Metal primers may have oil, oleoresinous, alkyd, or latex binders. Most metal primers used today have an alkyd binder.

Some of the primers listed in this subsection contain lead. Lead pigment is toxic, and coatings containing more than 0.06% lead should not be used in residential buildings or on structures with which children may come in contact. Many jurisdictions limit the lead content of paint by law, and some prohibit its use altogether.

Red Lead (Red Lead in Oil) Red lead in oil is probably the oldest type of metal primer. It is still in use today because it is very reliable. The oil binder wets the surface well and seeps into cracks and crevices to ensure protection of the metal even on poorly prepared surfaces. However, drying is comparatively slow, and the film may be too soft for some finish coats. The red lead pigment will carbonate and turn white if left exposed for too long. This primer should be covered with a finish coat within 6 months.

Red Lead/Mixed Pigment Red lead/mixed pigment primers are either oil- or alkyd-based primers that depend on the red lead for their corrosion-inhibitive properties. However, they also contain iron oxide and selected extenders, which contribute to the general performance of the coating, adding hiding power, hardness, and general durability. In other respects their properties are similar to those of the red lead in oil primers.

Alkyd-based red lead/mixed pigment primers are similar to their oil-based counterparts but will dry faster and harder, making them suitable for use under harder finish coats. The surface must be cleaned more completely before application than is required for oil-based primers.

Zinc Chromate (Zinc Yellow) Zinc chromate primers are made with zinc potassium chromate and often contain iron oxide as part of the pigment. The binder most used today is alkyd. These primers depend on their zinc chromate component for their anticorrosive effect. They generally dry harder and faster than red lead primers, but require more thorough cleaning of the surface before application. When lead-free, they often are used for priming aluminum and magnesium surfaces.

Zinc-dust/Zinc-oxide Zinc-dust/zinc-oxide primers are available with oil, alkyd, or phenolic resin as the binder. They are especially effective on galvanized steel but may be used on other metallic surfaces as well. The oil type usually is best for old, rusty, or weathered surfaces. The alkyd type is for use on clean surfaces and also as a heat-resistant coating.

Phenolic primers are generally recommended only when the coated surface is to be immersed in water, but they are not recommended for exterior use. When these primers are used, the surface should be given a phosphate treatment before painting.

Latex-Inhibitive Latex-inhibitive primers are water-soluble before they cure, which makes them easy to apply and easy to clean up. They are available in white, permitting coverage with one finish coat, usually within hours. These primers will give good corrosion protection, but may show rust staining if not protected by a moisture-impervious finish coat.

Wash Primers Because wash primers usually contain phosphoric acid, they clean as well as prime a surface. They are useful in promoting adhesion of subsequent coats of paint and are frequently used to protect freshly sandblasted surfaces from further corrosion. Wash primers are furnished as either one- or two-part materials. One-part wash primers have the advantage of longer pot life after thinning, but generally sacrifice some corrosion resistance as compared with the two-part type. Two-part formulations must be used the day they are mixed.

Portland Cement Paints Portland cement paints are either alkyd- or oil-based paints containing portland cement. They are not to be confused with cement grout or dry-powder portland cement paints, which are designed for use on masonry. They are effective primers for galvanized steel. If white portland cement is used, a white primer results, permitting tinting to shade and contributing to the ease with which the primer can be covered by finish coats.

MASONRY PRIMERS

Exterior masonry and concrete wall primers include latex and alkyd primers. The first coat of an epoxy finish often serves as a primer, but conditions where a heavy-duty finish is needed may require a different epoxy formulation for the primer. Masonry primers are used as sealers on normally smooth masonry surfaces. A clear sealer is sometimes recommended to stabilize weathered surfaces. If latex finish coats are to be used, the sealer should be thinned so that it will dry to a low gloss.

Silicone water repellents are also used on exterior masonry surfaces when a water-repellent surface is required without change in surface appearance. They consist of a solution of silicone resin in a solvent, usually an alkyd. Since they are transparent solutions, it is difficult to determine visually when an adequate coating has been obtained. To ensure that the desired water repellence is achieved, the spreading rate should be as recommended by the manufacturer. These materials should not be finished with other coatings until they have weathered for several years.

When a smoother finish is desired on masonry and concrete, block fillers or cement grouts are applied beneath the paint primer, as indicated under "Masonry and Concrete Wall Primers and Fillers" in Section 9.15.1.1.

9.15.1.3 Interior Topcoat Paints

A wide range of interior topcoat paints are available with various surface characteristics.

GLOSS FINISHES

Gloss finishes can be achieved with either paint (opaque) or transparent materials.

Gloss Enamels Most gloss enamels used today are alkyd based. Modern formulations offer easier application than the older oleoresinous gloss paints, and they also have much better gloss retention and resistance to yellowing and alkaline cleaners.

Some latex paints are also available in gloss formulations. They do not produce as high a sheen as alkyd enamels, but their ease of application and cleanup, continued advancements in their quality, and more restrictive VOC requirements will eventually, if they have not already, make them more popular than alkyd gloss enamels.

Floor Enamels Alkyd and oleoresinous enamels are formulated for relatively fast drying and for resistance to abrasion and impact. They are impermeable to water but are not notably resistant to alkali, such as may arise from fresh concrete. These enamels will blister and peel if moisture accumulates behind the film. Most floor enamels used today are alkyd based.

Alkali-resistant Enamels Alkali-resistant enamels are usually based on a solvent-thinned styrene-butadiene resin or chlorinated rubber. They have high resistance to alkali, but relatively poor resistance to solvents. They flow and level well but may sag on vertical surfaces. Bubbling sometimes occurs when they are applied by roller on concrete floors.

Epoxy and Urethane Enamels Epoxy and opaque urethane enamels are available in one- or two-part formulations. The one-part materials are similar to alkyd and oleoresinous enamels but have somewhat improved properties, particularly abrasion resistance. The two-part materials have extremely high adhesion, abrasion resistance, and resistance to water and solvents, but they require great care in surface preparation to develop these properties to the maximum extent.

Dry-fallout Spray Gloss Dry-fallout spray gloss is similar to alkyd gloss enamel but has a much faster-acting solvent and usually a faster-curing resin. It is designed for spray application only. An advantage is that much of the overspray dries quickly before reaching a surface and can therefore be removed easily.

SEMIGLOSS FINISHES

Semigloss paints are available with a range of glosses. Some manufacturers call their semigloss paints *eggshell* or *satin* finishes, and some call different amounts of gloss by different names. Here we refer to all nonflat and non-high-gloss finishes as semigloss.

Semigloss Wall Paint Semigloss wall paint is an oleoresinous enamel with an initial intermediate gloss, which it may lose over time. It is likely to have low resistance to grease and alkaline cleaners and is often too soft for use on horizontal surfaces. Its use is decreasing in favor of other types of semigloss enamels and paints.

Semigloss Enamel Semigloss enamel is an alkyd-based product with better gloss retention and grease and alkali resistance than semigloss oleoresinous wall paint. Application may be somewhat more difficult over large areas. Gloss may be reduced rapidly the first few days after application but will then stabilize.

Semigloss Latex Paint and Enamel Semigloss latex paints and enamels have the usual advantages of latex paints: ease of application and cleanup, rapid drying, little odor, nonflammability, and good film leveling. Its hiding power may not be as good as that of alkyd products, but this disadvantage is rapidly disappearing as improved latex paints are developed.

Dry-fallout Spray Semigloss This material, intended for spray application only, has the properties of its gloss counterpart but at a lower gloss level.

FLAT FINISHES

Flat paint finishes include oil and oleoresinous flats, alkyd flat wall paints, latex flat wall paints, and dry-fallout spray flats.

Oil and Oleoresinous Flats Oil and oleoresinous flat paints are seldom used today. They have a predominantly limed oil vehicle, which reduces penetration and permits easy application and excellent leveling. Finished appearance is excellent, but washability and resistance to yellowing are likely to be poor. Because of their good hiding power in thin films, these flats are sometimes used in repainting acoustical surfaces.

Alkyd Flat Wall Paints Alkyd flat wall paints are usually superior to latex paints in washability and hiding power, although their advantage in hiding power is diminishing as better latex paints are developed. "Odorless" alkyd flats may have very little odor during application, but an afterodor usually occurs as the oil portion of the resin oxidizes. Whether odorless or not, these products should be used only under conditions of good ventilation.

Latex Flat Wall Paint Latex flat wall paint is the most popular kind of interior flat wall paint. High hiding power and good washability may be built into the product, but often one is achieved at the expense of the other. Quick drying, ease of touch-up and cleaning of tools, and absence of flammable solvents and unpleasant odors during and after application are additional advantages. However, ease of application may lead to excessive spreading rates and poor hiding ability. Even quality products may fail to form a coherent film and, hence, show poor washability if applied over a porous surface or under conditions of low humidity or temperature extremes.

Dry-fallout Spray Flats Dry-fallout flats are made with fast-drying solvents and may be either of the alkyd or oleoresinous type. They are used on ceiling areas of industrial or commercial buildings where application is by spray and where it is not practicable to protect all adjacent areas from overspray.

Latex Floor Paints Latex floor paints have most of the application advantages of latex wall paints, as well as fairly good abrasion resistance. The absence of toxic or flammable solvents makes them useful in basements and other areas where ventilation is poor. They may be applied to damp (but not wet) surfaces even if alkalie are present. Continuous exposure to standing water is likely to soften the paint film.

9.15.1.4 Exterior Topcoat Paints

Exterior finish coats include many formulations with oil, alkyd, latex, and other synthetic resins.

A wide range of alkyd- and latex-based house paints are available in flat or glossy sheens. Because of excellent color retention, durability, and ease of cleanup, latex exterior paints are gaining favor over the solvent-thinned alkyd types.

OIL-BASED PAINTS

Oil-based paints include *gloss house paint* and *barn paint.*

Gloss House Paints Traditional house paints were white lead in linseed oil; then various blends with zinc oxide, talc, and other extenders were developed; and, finally, blends with various titanium dioxide pigments were introduced. These paints have been the foundation of the paint industry. Linseed oil is still the binder used in many quality oil-type house paints, although modifications with other materials may provide improved properties. A number of long-oil alkyds (resins with 55% or more oil or fatty acid) are available, which are similar to oleoresinous trim enamels but are softer and more durable.

Lead has been found to be harmful to health and is no longer allowed in residential paints or paints in other locations where children may come into contact with them. Paints containing more than 0.06% lead (nonvolatile content) must carry a warning label and may be prohibited in some jurisdictions.

Zinc oxide is often included to control mold and to aid in the control of chalking; it is also a factor in regulating the hardness of the paint film. Extender pigments are used to control consistency and to develop an optimum blend of other properties.

Titanium dioxide is the most important pigment for providing hiding power in white and tinted paints and contributes to the control of chalking and fading. Some of these factors are of lesser importance with dark-colored paints, which contain little or no white pigment.

Barn Paints Barn paints possess most of the characteristics of oil-based house paints. However, they are more likely to contain vehicle modifications, such as rosin, and they usually employ red iron oxide as the chief hiding pigment. Green, black, and several other colors are also available. Barn paints are formulated to give uniform appearance over poorly prepared and nonuniform surfaces, but lack the long-term protective qualities of house paints. They may also be used on roofs, particularly those of sheet metal.

OLEORESINOUS PAINTS

Oleoresinous paints are rapidly disappearing from the market as alkyd-based and latex paints improve in quality. However, some paints for exterior metal are still made with oleoresinous vehicles, notably those for structural steel. They give good service and protection but usually lose their gloss earlier than similar products made with alkyds.

ALKYD ENAMELS

Alkyd enamels for exterior use include trim enamels, equipment enamels, and masonry paints.

Trim Enamels Trim enamels are designed for use on exterior wood. They are usually made with long oil resins. For this reason and because they are usually designed for easy brush application, they are not fast drying. Drying usually takes from a few hours to overnight. Typically, they are made in high gloss and bright colors and show good retention of color and gloss. They are not suitable where high resistance to acid, alkali, or other chemicals is required. Silicone-alkyds are substantially more durable than conventional alkyd enamels.

Equipment Enamels Equipment enamels are similar to trim enamels, but because they are intended chiefly for metal, a hard-drying resin may be employed and some rust-inhibitive pigment may be added. If designed for spray application, a faster solvent will also be used and drying may be quite rapid. These products are often called *automotive enamels.*

Masonry Enamels Masonry enamels are similar to alkyd trim enamels, except that they contain a higher proportion of pigment. Because very little flexibility is required of a masonry finish, these alkyd-based products perform very well except when excessive moisture and alkali are present.

EXTERIOR LATEX PAINTS

Exterior latex paints include latex house and trim paints, semigloss maintenance finish, masonry paint, and special latex paints.

Latex House and Trim Paints Latex house and trim paints are intended mainly for wood, but some of them can also be used over masonry and metal. They are rapidly gaining in popularity over solvent-thinned alkyd paints because of their excellent color retention, durability, and ease of cleanup. Primers are nearly always required over bare wood, badly weathered surfaces, and bare metal. Where primers are needed, it is best to employ one specifically recommended for use with the intended finish-coat material. These products are similar to other latex paints in their general properties, and they are formulated for good flexibility so that they will accommodate dimensional changes of wood.

Most latex house and trim paints provide a matte finish and offer better color retention than oil and alkyd flats, even in light colors. They are also available with semigloss and gloss finishes.

Semigloss Maintenance Finish Semigloss latex maintenance paints applied over suitably primed metal have shown good performance. However, moisture vapor permeability is fairly high, and they have good resistance to pitting and rusting, but some rust staining may appear unless an impermeable primer is used.

Masonry Paint Paints for masonry were the first large-scale exterior use of latex systems. Because masonry surfaces show less dimensional change than wood and metal and also are rougher, adhesion is less of a problem. For surfaces with efflorescence, a clear or lightly pigmented solvent-thinned sealer is often recommended. Like latex house paints, these products usually give a low-sheen finish.

Special Latex Paints Special latex exterior paints may be used on mineral-fiber-cement and tile roofs. Special latex paints are available for painting roll roofing. They may be used on asphalt shingles, but may curl them.

ALUMINUM PAINTS

Aluminum paints include general-purpose and heat-resistant paints.

General-purpose General-purpose aluminum paints may be made with aluminum powder or paste and a vehicle of drying oil, alkyd resin, or one of a number of varnishes. Quality may vary widely, depending on the amount of aluminum, the ratio of aluminum to binder, and the quality of the binder.

Heat-resistant The most heat resistant of aluminum paints are made with pure silicone resin. When other materials replace part of the silicone vehicle, there will be a reduction in the temperature the paint will withstand and in the paint's durability. Another type intended for high-temperature surfaces uses a small amount of vehicle (usually an oil), which burns off, leaving the aluminum melted and fused to the surface.

ALKALI-RESISTANT PAINTS

Alkali-resistant paints are formulated with styrene-butadiene resin or chlorinated rubber. They dry flat, and ultimate deterioration of the film is by gradual chalking. These materials are used on masonry walls or decks where higher resistance to moisture penetration is required than can be obtained with latex paints. They may be used as primers under less resistant finish coats. Their resistance to water, alkali, and acid is excellent, but their resistance to solvents may be poor.

9.15.1.5 Miscellaneous Opaque Coatings

Miscellaneous opaque coatings include fire-retardant paints, fire-retardant intumescent coatings, portland cement powder paints, zone-marking paints, and roof coatings.

FIRE-RETARDANT PAINTS

Fire-retardant paints are made in a number of types. They can be formulated as strictly functional or as both functional and decorative coatings. Products are available with good fire retardancy after repeated washing and with all the properties of conventional interior coatings. Since some of these paints are sensitive to water and high humidity, suitability for the intended purpose should be determined.

Fire-retardant capability is usually expressed on the basis of flame spread ratings determined and listed by Underwriters Laboratories, Inc. (UL). Flame spread is not a property of paint but depends on the number of coats and their thickness, the surface to which the paint is applied, and other factors. The flame spread rating of a paint may vary under different conditions. In particular, a paint may be noncombustible and have a low flame spread on noncombustible surfaces, but may have little fire retardance on combustible surfaces. However, a paint with a low rating on combustible surfaces will always have a low flame spread on noncombustible surfaces. If coatings are not applied at the listed spreading rates, the flame spread rating will not apply.

FIRE-RETARDANT INTUMESCENT COATINGS

Fire-retardant intumescent coatings slow the progress of a fire by forming a protective foam. In many cases the charred foam may be scraped off and the sound substance repainted.

This material works when heat-sensitive ingredients in it begin to intumesce (swell) at 300°F (149°C). The resulting foam, which is hundreds of times thicker than the original paint film, slows the fire, reduces smoke, and delays the buildup of toxic gases.

Intumescent coatings may be used either in strategic places or throughout a building. These coatings are especially recommended for corridors and areas leading to fire exits. They are available pigmented or clear and are scrubbable.

PORTLAND CEMENT POWDER PAINTS

Portland cement powder paints are made from white portland cement, suitable pigments, and usually small amounts of a water-repellent agent. They are mixed with water just before application. Since hydration of the portland cement requires moisture, painted surfaces should be kept damp as recommended by the manufacturer. These paints are useful, low-cost finishes for rough masonry. Their principal drawback is that they are not good bases for other paints or coatings.

ZONE-MARKING PAINTS

Zone-marking paints, which are sometimes called *traffic paints*, are fast drying, little flow, good hiding ability, resistance to bleeding over asphalt, and usually low gloss. Zone-marking paints for highway use may be reflectorized or nonreflectorized. Nonreflectorized paints are used where there is overhead lighting such as, for example, in many parking lots. Reflectorized paints may be of the drop-in type, to which glass beads or similar material are added after application of the paint, or of the premixed type, with beads incorporated in the paint. The drop-in type is usually more durable and has better visibility early in the life of the paint. The premixed type is easier to apply. A small number of beads may be dropped into the premixed paint after application for early visibility.

Normal zone-marking paints are flat

and are therefore not satisfactory for such interior uses as factory floors or school corridors where dirt collection is a problem. Marking paints for this purpose must have a gloss or semigloss finish.

ROOF COATINGS

Bituminous roof coatings are made of special weather-resistant asphalt dissolved in a solvent. They are intended for use on asphalt roofing surfaces. Mineral fiber and other fillers may be added to prevent sagging on sloping roofs and to permit the application of relatively thick coatings. The addition of aluminum powder enables colors other than black to be obtained. For some roof coatings the asphalt is emulsified in water, permitting application to damp surfaces.

9.15.1.6 Transparent Finishes

Transparent finishes are applied on wood. Transparent finishing materials include wood stains, linseed oil, varnish, shellac, lacquer, and polyurethane.

Transparent finishes are used mostly in interior locations, but some are available for exterior use. In exterior locations, they do not usually perform as well as paints do, because they cannot approach the durability of the same vehicle protected by pigment.

STAINS

Wood stains are made from color pigments suspended in linseed oil or another drying oil. They are then thinned to make them easy to apply. Types of stains include pigmented, dye-type, and shingle stains.

Pigmented Stains Pigmented stains are stains containing insoluble pigments, generally permanent in color. If applied too heavily, they will obscure the texture, giving a painted effect to the wood. They are particularly suitable for subduing the conspicuous irregular grain effect of softwood plywood. These stains will not bleed into varnish or enamel coats applied over them.

Dye-type Stains Dye-type stains employ an organic coloring material that is soluble in the vehicle. They have better clarity and transparency than pigmented stains, and they produce brighter colors. However, they are more likely to bleed into succeeding coats, and the color is less permanent.

Shingle Stains Shingle stains are pigmented products designed to provide color, only partial hiding ability, maximum penetration, and some mold-resistant protection. They may contain creosote or other preservatives; linseed oil and alkyd resin are the usual binders.

LINSEED OIL

Boiled linseed oil is the classic wood finish. It is inexpensive and easy to apply, maintain, and repair. It is normally reduced with turpentine in ratios ranging from equal parts to twice as much oil as turpentine.

VARNISH

Varnish is a homogeneous mixture of resin, drying oil, dryer, and solvent. It can produce a transparent film that is either flat, satin, or high gloss. Varnish is available in colors ranging from clear to dark brown and in several qualities. Most varnishes for interior use include alkyd resins, but epoxy-ester varnishes are also available. The old oleoresinous varnishes have been almost completely replaced today by alkyd varnishes, which are often paler and have better color retention. Epoxy varnishes have good color quality and alkali resistance.

Alkyd varnishes have good initial color and color retention but may crack and peel over time. Some synthetics have good durability but may darken on exposure. A long-oil phenolic, marine spar varnish, has the best history of satisfactory performance, although it too will darken and yellow with time. Penetrating coats of oil or thinned varnish should be used on bare wood, followed by unthinned varnish. Varnishes containing ultraviolet screening agents exhibit less discoloration and greater durability.

Flat and satin varnishes have an added flattening agent such as silica. They do not have the undesirable highlights associated with gloss varnishes, but they are less abrasion resistant.

Floor varnishes are relatively fast drying with good resistance to yellowing and excellent abrasion resistance.

Counter varnishes are hard-drying, printproof films. *Bar top* varnishes should also be resistant to alcohol. Polyurethanes often are used for this purpose.

Paneling and trim varnishes need not be as hard as varnishes for floors or counters but should be resistant to checking and should dry tack-free. Alkyds usually exhibit the best color quality and color retention.

Spar varnish is a varnish in which additives have been included to give the material resistance to salt water.

Clear sealers are solvent-thinned varnishes that penetrate wood to prevent grain rising and to seal porous wood. They are applied liberally, and the surplus is removed by wiping while still wet or by buffing when dry (after about 15 hours). Buffing provides better protection but involves more time and labor. These products give a satin finish with fair abrasion resistance and color retention.

SHELLAC

Shellac produces a finish similar to that of varnish. It is fast drying, light colored, and has excellent color retention. Although its abrasion resistance is poor, the finish is easily repaired. Resistance to water and alcohol is poor, but resistance to petroleum solvents is good. Shellac is often used as a wash coat on wood before final sanding or as a clear primer or stain sealer.

A major disadvantage of shellac is its short shelf life; it is unusable within 4 to 6 months after manufacture.

LACQUER

Lacquer is available in either a flat or glossy finish, but is seldom used for field-applied finishing of building surfaces. It is more likely to appear as a finish for furniture or shop-finished casework.

A major disadvantage of lacquers is that they are not usable over previously finished surfaces. In fact, lacquer can be used as a paint remover.

POLYURETHANE

Some manufacturers classify polyurethanes as varnish. Others call them lacquer. In fact, although they have some characteristics of each, they are not truly either. They are available in both oil-modified and moisture-curing formulations. Oil-modified polyurethanes are transparent materials that are used mostly on interior surfaces. They are water resistant and highly durable. They can produce either matte or gloss finishes, but tend to lose their gloss quickly when used in exterior locations. They cannot be used over shellac, paste wood fillers, and some other materials. The manufacturer's advice should be sought about where they can be used.

Moisture-curing polyurethanes can be either clear or pigmented to produce an opaque finish.

Polyurethanes are often used on bar tops, countertops, and on floors.

Urethanes provide better abrasion resistance than varnishes or lacquers, but generally require more thorough surface preparation prior to application.

9.15.2 SURFACES

This subsection outlines the broad categories of surfaces on which paints, transparent finishes, and coatings are used and describes products suitable for each surface under varying exposure conditions. Unusual surfaces and surfaces that are rarely painted (glass, underground metal, underwater concrete, etc.) require specialized coatings designed for the special conditions of use and are not included in this discussion. Products recommended in the following text are identified by generic name (see Section 9.15.1).

9.15.2.1 Interior Walls and Ceilings

PRIMING

Most surfaces require a primer as a base for a topcoat. The importance of proper selection and application of primers cannot be overemphasized. Regardless of the quality of subsequent finishing, failure of the primer will mean failure of the system.

Plaster New plaster should be primed with a latex, alkyd, or oil-type primer-sealer. A latex flat finish coat may be used as a primer, but if a semigloss or gloss paint is to be used as a finish coat, the primer should be tested for enamel *holdout* (resistance to penetration by the finish coat), which results in uneven gloss. Highly alkaline or fresh, damp plaster requires an alkali-resistant primer or latex system. Wet plaster and plaster having a continuing flow of moisture from the back cannot be painted successfully. Textured or swirl plaster and soft, porous, or powdery plaster should be given an acid wash with a solution of 1 pint (473.2 mL) of white household vinegar in 1 gallon (3.79 L) of water. This procedure should be repeated until the surface is hard; it should then be rinsed well and allowed to dry before painting.

Gypsum Board Latex primer-sealers are recommended for gypsum board and other paper-surfaced materials, because solvent-thinned primers raise the fibers of the paper, giving it an unsightly appearance. A latex finish coat can be used as a primer, but semigloss or gloss paints used as primers should be tested for enamel holdout.

Brick Brick interior walls are usually somewhat porous, but if the mortar is well cured they will not be highly alkaline. They may be sealed with a latex primer-sealer or with an enamel undercoater containing oil or varnish. Latex or solvent-thinned exterior masonry paints may also be used for priming. In every case, efflorescence should be removed by vigorous scrubbing and the brick should be treated with a clear resin sealer.

Concrete Masonry Most concrete masonry, especially the lightweight type, has a very rough and porous surface. Paints and coatings usually cannot bridge the surface voids in these materials. One or two coats of block filler can be used to smooth such surfaces, thus reducing dirt accumulation and improving surface color uniformity. Fillers may be omitted when these considerations are not important, when low-cost finishing is required, and when the sound-absorbing effect of a rough surface is desirable.

Concrete The problems associated with painting concrete are similar in many respects to those encountered in painting concrete masonry. However, the concrete surface is usually denser, less regularly formed, and more likely to be contaminated with oils and other bond-breaking materials. Foreign substances should be removed unless it can be ensured that they will not affect paint adhesion. Rough or honeycombed surfaces can be smoothed with cement grout. This grout is usually thinned with a latex solution, especially when it is to be applied in thin layers. Grout is primed as described earlier for priming plaster. One or two coats of block filler may be used instead of grout, thus eliminating the need for an alkali-resistant primer.

Ferrous Metals For painting ferrous metals, a rust-inhibitive primer is always desirable. It is essential for quality work, for damp locations, when water-thinned paints are used for finishing, and when the finish coats do not produce an impervious film. A primer may be omitted in dry locations and if the finish coats are to be of a solvent-thinned type producing a tight film. Slow-drying oil-based primers wet the surface better and give more positive results, but quick-drying primers are more resistant to moisture penetration.

Zinc-coated (Galvanized) Metals When applied over zinc-coated surfaces such as galvanized steel, many paints and rust-inhibitive primers develop a layer of white, powdery material, which eventually separates the paint film from the metal. This condition can be prevented by using (1) a nonoxidizing coating such as a chemical-resistant coating, or, more commonly, (2) a primer containing zinc dust or portland cement. Certain proprietary primers for galvanized steel (often latex) are very effective for interior use.

Wood In many interior situations, wood may be primed at the same time and with the same sealer used for plaster walls. One or two coats of paint may give a sufficiently smooth base for enamels, but such water-thinned paints will raise the grain and roughen the surface of many woods. Good enamel holdout is necessary under gloss or semigloss finish coats. Solvent-thinned wall primer-sealers provide better adhesion and flexibility. For the best enamel work, an enamel undercoater should be used either over a wall primer-sealer or directly on the wood. A clear wood sealer is recommended for natural finishes, preceded by a paste wood filler, if necessary.

Acoustical Surfaces Acoustical tile and acoustical plaster are rarely painted at the time of their installation but may require finishing later. Since acoustical properties depend on a suitably porous or rough surface, a thin paint should be used so that the pores are not bridged or filled. Paints used on plaster surfaces may be thinned and used on acoustical surfaces. Metal acoustical surfaces should be treated as light-duty areas, using one coat to avoid excessive paint buildup.

FINISHING

Finishing differs slightly, depending on the use of the location.

Light-duty Areas Some building areas have low concentrations of moisture, grease, or dirt in the air and require relatively little maintenance. These include small private offices and similar spaces in commercial and institutional buildings, and living rooms, adult bedrooms, and similar areas in residential facilities. In such spaces a matte finish is both desirable and practical on the ceilings and most walls.

Latex flat wall paints are usually preferred because of their easy application, quick drying, easy cleanup, and lack of odor. Alkyd flats can produce thicker films, have better hiding power, and give a more uniform appearance. These properties are particularly desirable on rough surfaces and when a considerable change is made in color. Both latex and alkyd paints have good resistance to soiling and marring and are relatively easy to clean.

Odorless alkyd flats exhibit little solvent odor during application, but they must be used with adequate ventilation because the solvent fumes, while odor-free, may cause headaches and other physiological symptoms. The curing of these air-drying paints results in odors that can persist for several days, requiring continued ventilation.

Moderate-duty Areas Moderate-duty areas are similar to light-duty areas in most respects, except that the amount of traffic or airborne dirt may dictate a more intensive cleaning schedule. Typical areas are private corridors and large office spaces in commercial and institutional buildings, and children's rooms, entries, halls, and closets in residential occupancies. Finish coats should be chosen from the more washable flat and semigloss paints available. Glossier materials usually have better wear resistance and washability, but the accompanying higher surface shine makes surface irregularities more conspicuous, especially when natural or artificial light strikes a surface at a sharp angle. This consideration becomes less important in smaller spaces or where wall areas are broken up by interior furnishings.

Heavy-duty Areas Heavy-duty areas are those in which there may be large amounts of moisture and grease in the air, or where the amount and type of traffic require a more durable paint film to withstand greater wear or more intensive cleaning. These areas include kitchens, recreation rooms, toilets, public corridors, entrance lobbies, and service areas in institutional, commercial, and apartment buildings, and bathrooms in homes and apartments.

The usual choice is a gloss or semigloss enamel. Where the exposure is particularly severe, hard-gloss coatings can be used. The light reflectivity factors discussed earlier for moderate-duty areas should always be considered in selecting gloss finishes.

9.15.2.2 Interior Trim and Paneling

The selection of primers for interior use on trim and paneling depends primarily on the type of surface material being coated and the desired finish characteristics.

PRIMING

Ferrous Metals Although most interior exposures are not particularly corrosive, a rust-inhibitive primer is usually desirable for ferrous metal surfaces. Any metal primer (oil, alkyd, or latex) may be used. The choice is dictated by the speed of drying, the amount of surface preparation, and the desired color. For very mild exposures, enamel undercoaters may be used, but these products have no corrosion-resistant effect and rust will spread from any break in the film.

Zinc-coated Metals Interior galvanized surfaces that are dry and do not receive heavy wear may be coated with interior latex paints. However, if moisture, abrasion, or repeated cleaning is expected, an inhibitive primer will provide better service.

Wood, Clear Finish Priming will depend on both the wood and type of finish desired. Trim should be primed on all surfaces before installation. Most softwoods will require a clear sealer, especially those, such as pine, that show a marked contrast between springwood and summerwood. A sealer will provide a more uniform surface over hard and soft grain, and thus it is especially recommended if the surface is to be stained.

Dense, close-grained hardwoods (maple, birch, etc.) have less need for a sealer, and the first coat may be a thinned version of the finish coat if so recommended by the manufacturer. In stain systems, the products used and the order in which they are applied will depend on the effect desired. For example, applying the sealer before the stain is applied will minimize both the depth of color and the amount of grain contrast.

Open-grained hardwoods (oak, walnut, mahogany) require a wood filler, which will also serve as a sealer. When the grain is to be subdued, stain may be applied after the filler.

Wood, Opaque Finish Interior wall primers and enamel undercoaters may be used as primers on wood to receive an opaque finish, but water-thinned primers are apt to raise the grain of the wood. An enamel undercoat will give a smoother surface and is preferred under gloss and semigloss enamels, although many semigloss enamels are designed to be self-priming.

FINISHING

Finish coat selection is determined largely by the type of exposure expected.

Light- and Moderate-duty Areas The selection of clear finishes for wood depends on decorative and exposure requirements. For light- and moderate-duty areas, satin or low-luster finishes are often used.

Opaque finishes for wood and metal are usually semigloss enamels or latex enamels. In light- and moderate-duty areas the predominant interior finish used on the adjacent walls may be used.

Heavy-duty Areas Clear finish for wood in heavy-duty areas may be floor varnish or counter varnish. Floor varnish is most resistant to abrasion, and counter varnish has good resistance to water and solvents. The selection should be made on the basis of the exposure anticipated.

Opaque finishes for wood and metal are usually hard-drying enamels. For special exposures, such as in laundries, a special type may be indicated.

9.15.2.3 Interior Floors

The selection of finishes for interior floors depends primarily on the type of surface to be coated.

WOOD FLOORS

Clear Finishes Open-grained woods such as oak may use a filler before coating. Clear finishes include:

- *Shellac:* For light-duty floors where color quality and retention are important; good gloss and color; not water resistant; scratches easily, but is easily patched.
- *Floor varnish:* More resistant to wear and water but somewhat darker in color; may be slippery when wet.
- *Wood sealer:* Lower gloss; usually requires waxing.
- *Specially designed finishes:* For heavy-duty areas such as gymnasium floors.
- *Polyurethane:* For heavy-duty floors.

Opaque Finishes Alkyd enamels and latex floor paints are generally used for opaque floor finishes. Other fast-drying gloss enamels and paints may be used if recommended by the manufacturer. However, products not specifically recommended for floors may be too soft and slow drying, or too hard and brittle.

CONCRETE FLOORS

Enamels suitable for wood floors are generally suitable for concrete floors as well. However, alkali-resistant enamels should be used when a concrete slab is not completely cured or is placed below grade or on the ground without a vapor retarder. Even with alkali-resistant enamels, adhesion problems may be encountered if hydrostatic pressure forces water up through the floor.

Latex floor paints usually dry to a comparatively low gloss. Because they are nonflammable and have little odor, latex paints are preferred for basements and enclosed areas that are difficult to ventilate adequately. Since they are water emulsions and have excellent alkali resistance, the floor need not be completely dry before application. However, latex paints are softened by continued exposure to moisture or solvents and are not suitable for laundry and service areas where solvents are handled. In such exposures, specialized highly resistant finishes should be used.

STEEL FLOORS

A fast-drying, tough enamel applied over a corrosion-resistant primer is recommended for most steel floors. Floors subject to chemical agents, as in industrial installations, should be finished with specialized industrial coatings.

CERAMIC FLOORS

Ceramic floors are seldom painted, but they may be coated with an abrasion-resistant clear coating to seal the surface and facilitate cleaning.

RESILIENT FLOORS

Resilient flooring is rarely coated when new, but coating may be required to restore appearance or change color. The coating chosen will depend on the flooring material, its condition, the adhesive used in the installation, and the effect desired. Specific recommendations should be obtained from the paint manufacturer for each case.

9.15.2.4 Exterior Priming

As with the finishing of interior surfaces, the selection and application of primers are critical to the performance of an exterior coating system.

WOOD

Siding Smooth siding should be primed with a wood primer that is compatible with the intended finish coat. The manufacturer's recommendations for the finish coat should govern the choice of a primer. Nail heads, cracks, and deep surface imperfections should be filled; caulking should be provided at perimeters of doors and windows and wherever required for weather resistance. All filling and caulking should be done after the siding is primed.

For rough siding, shakes, and shingles, a wood primer may be used, or the finish coat may be thinned and used as a primer if it is zinc-free and is recommended by the manufacturer. Flat house paint and shingle stain do not require special primers. Factory-primed siding does not require a field-applied primer unless the surface has been heavily damaged during shipping or application. Minor scratches, dents, and raw edges should be spot-primed.

Plywood After prolonged exterior exposure, plywood surfaces will check and the coating will crack unless an overlaid plywood is used. Medium-density overlaid plywood with a resin-impregnated paper covering is recommended for an opaque paint finish. Painting recommendations for wood siding also are applicable to plywood. Textured plywood, such as grooved, rough-sawn, and striated surface, is not intended for painting and only pigmented stains should be used on it.

Transparent Finishes Wood surfaces intended for a transparent finish are usually primed with a coat of the same material to be used as the finish coat. To improve penetration, some manufacturers recommend that the finish coat be thinned when used for priming. However, some products may be damaged by thinning; therefore, it should be done only when specifically recommended by the manufacturer.

MASONRY

The cost of painting a clay masonry wall in most cases must be charged to appearance alone, not to increased performance or resistance to weather. The only exception is cement-water paint applied to correct a leaky wall condition. In the case of concrete masonry, paint not only provides an attractive finish but may also be necessary to make the masonry surface watertight.

Masonry paints are of four general classifications:

- *Cement-based paints:* Powdered form, mixed with water before application
- *Water-thinned emulsion paints:* Butadiene-styrene, vinyl acrylic, alkyd, and multicolored lacquers
- *Fill coats:* Similar to cement-based paints but contain an emulsion paint in place of some water, giving improved adhesion and a tougher film than unmodified cement paints
- *Solvent-thinned paints:* Oil-based, alkyd, synthetic rubber, chlorinated rubber, and epoxy; except for special-purpose paints and special applications, should be used only on interior masonry walls that are not susceptible to moisture penetration

Masonry surfaces generally do not require a special primer, and the first coat can be the same material as that used for the finish coat. The finish coat material may require thinning, but the manufacturer's recommendations should be followed. Common brick should be sealed with a penetrating type of exterior varnish to minimize spalling and efflorescence, which may cause peeling.

Cement grout thinned with latex may be applied to masonry surfaces that have large voids. Sound but rough surfaces, especially in concrete block, may be surfaced with block filler. Cast-in-place concrete should be cleaned of form oils, and other bond-breaking agents should be removed unless they are guaranteed not to affect paint adhesion. Old surfaces that have become excessively chalky may be bonded or stabilized with a clear or lightly pigmented phenolic or alkyd coating; if the existing surface is soft or crumbly, it should be removed to a sound layer, usually by sandblasting. Where only water repellence is required, a silicone solution may be used above grade; hot melt or cutback asphalt can be used below grade.

METAL

Ferrous Metal Coating materials cannot compensate for inadequate surface preparation. Rust, mill scale, dirt, oils, and old, loose paint must be removed. The standards for surface preparation recommended by the Steel Structures Painting Council should be followed.

An appropriate primer should be used, depending on conditions and the finish materials intended. Oil and oleoresinous materials generally have better bonding properties and are more suitable for hand-cleaned surfaces. Alkyd and other synthetic primers dry more rapidly but require more careful surface preparation. Chemical-resistant coatings and industrial coatings usually are marketed as complete systems of primer and finish coat, and the manufacturer's recommendations as to the appropriate primer should be followed.

Galvanized Steel Zinc-dust/zinc-oxide, latex, and portland cement paints give excellent results, as do certain proprietary paints designed for use on galvanized steel. Conventional exterior paints usually lose adhesion rapidly unless the surface has weathered for a considerable length of time. If the galvanized surface has weathered to the point that substantial amounts of rust are showing, the recommendations for painting ferrous metals should be followed. Chemical pretreatments, usually based on phosphoric acid, improve the adhesion of paints to galvanized metal. However, other acids or chemical solutions sometimes recommended for the preparation of galvanized surfaces rarely contribute to the performance of paints; they may actually be harmful, and their use generally should be avoided.

Aluminum and Magnesium Zinc chromate primers are usually recommended for aluminum and magnesium.

Copper, Brass, and Bronze Copper and copper-bearing metals oxidize rapidly when exposed to air and moisture. To preserve the bright metallic luster of new polished copper and brass, a clear coating should be applied promptly. Old copper and brass should be buffed or polished to a bright color, cleaned with a mild phosphoric acid cleaner, and coated before the surface becomes discolored by oxidation. Specially formulated clear coatings are available that will preserve the bright luster and protect the metal from discoloration by the elements.

Copper gutters and downspouts generally are not painted or coated. The rapid-forming oxide (patina) darkens the surface but protects the underlying metal from progressive deterioration. However, copper may be cleaned and coated either to match the color of adjacent surfaces or to protect them from staining by runoff from the copper. Coating is especially recommended for copper surfaces located above aluminum products to prevent electrolytic deterioration of the aluminum by water wash from above.

9.15.2.5 Exterior Finishing

The choice of finish should be influenced by the type of exposure expected and the performance desired. High temperatures and humidity, particularly in the absence of direct sunlight, encourage the growth of mold (mildew). Therefore, it is important to choose fairly hard-drying finish coats, avoiding those types that may dry to a rough, soft surface, as these finishes tend to trap and hold mold spores. Finish coats containing substantial amounts of zinc oxide are effective in controlling fungus growth. Various other additives are used to provide further mold resistance.

If a high degree of mold resistance is required, the primer as well as the finish coat should be designed for this purpose. However, the use of coatings containing zinc oxide is not recommended on bare wood. The degree of mold resistance is determined by the total formulation rather than by a single component, and the specific product should have adequate resistance under conditions similar to those to be encountered.

Industrial fumes may darken painted surfaces by the reaction of hydrogen sulfide with lead or, occasionally, mercury compounds. Paints used in areas subject to hydrogen sulfide should be free from lead and mercury.

Soot and other airborne materials may settle on and discolor a paint film. Hard-drying films will not retain dirt as tenaciously as softer films. Paints may be formulated to encourage moderate chalking of the surface, which has a self-cleaning effect as the soot and dirt are carried away with the surface layer.

However, for severe weather exposures, resistance to chalking is a desirable property. Repeated exposure to rain and high winds will generally keep the surface clean but will wear away the film too quickly unless a comparatively nonchalking paint is chosen.

WOOD SIDING

Smooth Siding Oil paints and exterior latex paints are preferred for whites and light tints on smooth siding. Oil paints have better hiding ability and will show less staining from wood resins. Latex paints are easier to apply, may be used on slightly damp surfaces, and often have better color retention, particularly in light tints. Where factory-primed siding is used, the recommendations of the siding manufacturer should be followed. Transparent finishes may be used where the natural appearance of the wood is desired, but their durability is much lower than that of pigmented products.

Rough Siding, Shingles, and Shakes Flat house paint, shingle stains, or exterior latex paints are usually used on rough siding, shingles, and shakes.

Latex paints generally give good service if rust-resistant nails have been used. Wood discoloration can be avoided by sealing with a solvent primer or using a latex primer specially designed for resistance to staining. Rough surfaces have substantially greater surface area than smooth surfaces, and more paint must be applied to obtain suitable film thickness.

Shingle stains are often the choice for wood shingles. These stains penetrate the wood but crack rather easily. Impermeable coatings are usually not desirable, because they seal in moisture and are likely to peel. Staining shingles before installation is preferred because many areas are inaccessible after installation.

WOOD TRIM

Alkyd trim enamels should be used where high gloss and good abrasion resistance are desired, as around doors and windows. Any of the finish coats suggested earlier for smooth siding also may be used.

METAL

Oleoresinous and alkyd paints are suitable finishes over compatible primers. Alkyd paints tend to be harder and more abrasion resistant. Paints recommended for wood, such as alkyd enamels, can be used if the surface is properly primed.

CONCRETE AND CONCRETE MASONRY

An alkali-resistant coating is required over concrete and concrete masonry because, even after some aging, they still contain alkali, which will attack oils and oil-based materials. Latex paints or resin-based finishes may also be used. Resin-based finishes are preferred on flat areas, such as ledges and windowsills, and areas subject to foot traffic and standing water, such as patios and roofs. Other types of paints may be used on well-cured concrete, on concrete that is sealed with an impervious primer, and on previously painted surfaces.

BRICK AND STONE MASONRY

Because brick and stone usually do not contain alkali, almost any exterior paint may be used on them with satisfactory results if the mortar is well cured. However, since fresh mortar is usually strongly alkaline, only paints recommended for concrete should be used on new walls. Where the natural appearance of brick or stone is desired, a silicone-resin solution may be used to retard the penetration of water. The solution and spreading rate should be as recommended by the manufacturer. Silicone finishes should not be painted until they have weathered for at least 2 years.

MINERAL-FIBER-CEMENT

Mineral-fiber-cement compositions such as shingles, siding, corrugated boards, and insulating sandwich panels often are supplied with a transparent water-repellent coating, which should be weathered or removed before painting. For most products, trapped moisture is seldom a problem, but alkalinity may be too high for oleoresinous or alkyd paints. However, the insulating core of sandwich panels may absorb moisture, releasing it later to damage the coating. It is recommended that the edges of insulating panels be sealed after cutting and before installation. Exterior latex or solvent-thinned resin paints are generally preferred.

ASPHALT ROOFING

Asphalt shingles, roll roofing, and built-up roofing are rarely coated when new, unless for decorative or heat-reflecting purposes. Most solvent-thinned paints are too brittle to be used over asphalt. The solvent will dissolve the asphalt, and the paint will become stained.

Bituminous roof coatings are most commonly used, and if heat reflection is required, these coatings may be obtained pigmented with aluminum or in a limited range of colors. Where white is required, certain latex paints are sufficiently flexible to give satisfactory performance. However, latex paints are likely to curl shingles and are not recommended for these products. Because severe exposure conditions are encountered on roofs, thick coatings are required to give satisfactory durability.

METAL ROOFING

In general, paints recommended for metal roofs follow the recommendations for other exterior metal. However, because foot traffic is rarely a consideration, softer coatings such as house paints or barn paints may be used for finish coats. As on other roofs, a relatively heavy coating is desirable for resistance to weathering. Terneplate roofs are usually painted with a special iron oxide primer and then finished as suggested here.

9.15.3 APPLICATION

Satisfactory painting and transparent finishing depends on (1) proper selection of paint and transparent finishing materials, (2) adequate preparation of materials and surfaces, and (3) suitable application methods and procedures.

Paints, transparent finishes, and coatings should be standard commercial brands with a history of satisfactory use under conditions equal to or similar to the conditions present in the area concerned. They should be recommended by the manufacturer for the specific use proposed. The printed application instructions should clearly identify the suitability of the material for the type of exposure (exterior or interior), the type of surface to be covered (wood, metal, masonry, concrete, gypsum board, plaster), and the type of service to which the paint will be subjected, such as exposure to moisture, frequent washing, or heavy traffic.

9.15.3.1 Materials Handling

Proper storage conditions and stocking procedures are essential to avoid material damage, explosion hazards, and application problems.

STORAGE

Paint materials should be stored in dry and well-ventilated areas maintained at a temperature between 50°F (10°C) and 90°F (32.2°C). Lower temperatures generally increase material viscosity and make it necessary to temperature-condition materials before use. Freezing temperatures may permanently damage water-based paints.

If a paint material is sensitive to heat, temperatures higher than 100°F (37.78°C) may bring about a chemical reaction within the container. This generally results in increased viscosity to the point of gelation, and the paint becomes unusable. In addition, overheating may cause covers to blow off, creating a serious fire hazard.

Other hazards to be considered are high humidity, which may cause containers to corrode and labels to deteriorate, and poor ventilation, which may allow excessive concentration of solvent vapors that are both toxic and combustible.

STOCKING

Paint materials should be rotated so that oldest stocks are used first. It is desirable to make paints ready for use by conditioning them at the proper temperature and mixing them thoroughly. Using leftover paint should be avoided. When using old paint is necessary, such material should be strained and temperature conditioned again, as recommended below for unopened containers.

Paint material should be stored in full, tightly sealed containers. Required quantities should be carefully estimated so as to have little or no material left at the end of a job. It is generally safer to discard small quantities than to use paint that has skinned. When saved, leftover paint should be placed in smaller containers, filled full and tightly sealed. All information from the original container should be transferred to the new container.

9.15.3.2 Conditioning and Mixing

Unless stored at the recommended temperature, paint materials should be temperature conditioned at 65°F (18.3°C) to 85°F (29.4°C) for at least 24 hours before use. After this time, paints should be mixed, thinned, or tinted as required, and strained if necessary, all according to the manufacturer's recommendations.

Paints consist of two principal components, solid pigment and liquid vehicle. Suspension agents are included in most paints to minimize settling of solids. However, when stored for long periods, separation of the pigment and the vehicle may occur. The purpose of mixing is to reblend settled pigment with the vehicle and to eliminate lumps, skins, and other detriments to proper application.

MIXING

Paint materials should always be mixed prior to issuance or delivery to the job. Mixing can be done by either manual or mechanical methods, but the latter is preferred to ensure maximum uniformity. Some hand stirring should be done just prior to use and periodically during application. Regardless of the procedure used, caution is recommended to avoid the incorporation of an excess of air through overmixing. Figure 9.15-1 outlines the recommended mixing procedures for different coating types.

It is important to mix ready-mixed paints before, as well as after, introducing thinners or other additives. It is best to use the same conditioning procedures for all components of multicomponent paint materials before mixing. Manufacturers' label directions regarding proper mixing should be followed.

FIGURE 9.15-1 Mixing Procedures for Various Paint Types

Coating Type	Mixing Equipment	Remarks
Enamel, semigloss, or flat oil paints	Manual, propeller, or shaker	Mix until homogeneous.
Water-based latex paints	Manual, propeller, or shaker	Use extreme care to avoid air entrapment.
Clear finishes, such as varnishes and lacquers	Manual	Generally require little or no mixing.
Extremely viscous finishes, such as coal tar paints	Propeller	Use extreme care to avoid air entrapment.
Two-package metallic paints, such as aluminum paint*	Propeller	Add small amount of liquid to paste and mix well; slowly add remainder of vehicle, while stirring, until coating is homogeneous.
Two-component systems	Manual, propeller, or shaker	Mix until homogeneous; check label for special instructions.

*With metallic powder, first make into a paste with solvent, then proceed as recommended above.

Mechanical Mixers The two most common mechanical mixers either vibrate the full sealed container or utilize propellers that are inserted into the paint. Vibrating shakers are used for full containers of capacities up to 5 gallons (18.93 L). Propeller mixers are simple hand drill attachments and are generally used on the job for containers holding 1 quart (946.53 mL) or more.

Manual Mixing Manual mixing is more time-consuming than mechanical mixing, but can be used when mechanical equipment is not available and for mixing materials that have recently been well mixed. Paint should be stirred from the bottom up, to dislodge any settled matter on the bottom or lower sides of the container. Properly mixed paint should have a completely blended appearance with no evidence of varicolored swirls or lumps at the top, indicating unmixed pigment solids or foreign matter.

When two or more cans of the same color are to be used in the same area, the paint may be "boxed" by pouring back and forth from one container to the other until uniform. This procedure minimizes the visual difference when switching from one container to the next.

TINTING

Tinting should be done by experienced personnel in a paint shop or at a paint store. On-the-job tinting should be avoided to minimize errors and to eliminate color-matching problems. An exception is the practice of tinting intermediate coats for identification and to ensure complete coverage under the finish coat. Tinting should be done with care, using only colors that are known to be compatible, and preferably adding not more than 4 ounces (118.3 mL) per gallon (3.79 L) of paint, unless otherwise approved by the manufacturer. Tinting of chalking-type exterior paints is not recommended except for identification of intermediate coats.

Many paint manufacturers employ tinting systems for the addition of colorants to their base paints. The colorants are formulated to precise standards of tinting strength and are packaged in tubes or in bulk for use with a colorant dispensing machine at the paint shop. Each color in the system has an identifying number and mixing formula indicating the proportions of base paint and colorants to be combined. Thus, any color can be reproduced accurately at a later time without guesswork.

Tinting Colors There are two types of tinting colors in general use:

Colors-in-oil are limited to use with standard oil-based paints, alkyd resins, chlorinated rubbers, and other solvent systems. They should not be used with the other synthetics or with water-thinned paints.

Universal tinting colors are more compatible with a wide variety of paint materials. Many can be used with both solvent-thinned and water-thinned paints.

Tinting Procedure Paint should be at application consistency before tinting. Ingredients should be compatible and fluid, so as to blend readily. Mechanical agitation is important to ensure uniform color distribution, but overmixing should be avoided. To achieve a paint color not included in a manufacturer's tinting system, a paint sample should be allowed to dry before it is compared with the color chip used for reference. When the desired color has been achieved, it is important to maintain a written record of the tinting formula; the container should be identified by name or number, and a spot of paint applied to the cover for visual reference.

STRAINING

Paint from newly opened containers normally does not require straining. However, a paint showing evidence of skins, lumps, color flecks, or foreign materials should be strained through a fine sieve or commercial paint strainer. Paint that is to be applied by spray gun should be strained to prevent clogging the spray nozzle.

THINNING

Paints are formulated and packaged at the proper viscosity for application by brush or roller without thinning. However, thinning is frequently required and recommended by manufacturers for spray application. The arbitrary addition of thinners should be avoided; unnecessary or excessive thinning may result in an inadequate film thickness and may reduce the durability of the applied coating.

Thinning should be performed by competent personnel using only compatible thinning agents recommended in the label's instructions or the manufacturer's specifications. Thinning should not be used as a means of improving the brushing or rolling properties of cold paint materials; instead, materials should be conditioned to 65°F (18.3°C) to 85°F (29.4°C) before use.

9.15.3.3 Preparation of Surfaces

Proper surface preparation is essential to achieve maximum coating life. Paint will not perform effectively if applied on a poorly prepared surface. The initial cost of adequate surface preparation is rewarded by longer coating life and improved appearance. The selection of a surface preparation method depends on (1) the nature and condition of the surface to be painted, (2) the type of exposure, (3) the surface texture and appearance desired, and (4) material and labor costs.

Surfaces to be painted or to receive a transparent finish or coating often contain contaminants such as dirt, grease, rust, and moisture, which reduce adhesion and cause blistering, peeling, flaking, or cracking of the finished surface. Surface defects that also adversely affect adhesion include metal burrs and irregular weld areas, wood knots and torn grain, powdery concrete and masonry, and old paints in various stages of deterioration. In general, sharp edges, irregular areas, cracks, and holes create conditions conducive to early finish failure.

Because iron and steel corrode readily unless properly prepared and painted, special attention is given in the following discussion to suitable preparation methods for these and other metals. Painting of ferrous metals should be started as soon as possible after the surface has been prepared. The suitability of treatments for various material surfaces is given in Figure 9.15-2.

FIGURE 9.15-2 Preparation of Various Surfaces

Preparation Method	Wood	Metals: Steel	Metals: Other	Concrete and Masonry
Mechanical				
Hand tool cleaning[a]	OK	OK	OK	OK
Power tool cleaning	OK[b]	OK	—	OK
Flame cleaning	—	OK	—	—
Brush-off blast cleaning	—	OK	OK	OK
Other blast cleaning	—	OK	—	—
Chemical and Solvent				
Solvent cleaning	OK	OK	OK	—
Alkali cleaning	—	OK	—	OK
Steam cleaning	—	OK	—	OK
Acid cleaning	—	OK	—	OK
Pickling	—	OK	—	—
Pretreatments				
Hot phosphate	—	OK	—	—
Cold phosphate	—	OK	—	—
Wash primers	—	OK	OK	—
Primers	—	—	—	OK
Sealers	OK	—	—	—
Fillers	OK	—	—	OK

[a]Also suitable for gypsum wallboard and plaster.

[b]Sanding only recommended.

MECHANICAL TREATMENTS IN THE FIELD

Mechanical treatments in the field are intended to remove corrosion, mill scale, and old paint. Solvent cleaning to remove surface oil or grease is recommended prior to other treatments.

Metals can be prepared for painting by cleaning with hand tools, power tools, flame, and abrasive blast. The use of goggles and gloves is essential for blast and flame cleaning and is recommended for the other methods as well. Good ventilation is desirable, and dust respirators may also be required.

Hand Tool Cleaning Hand tool cleaning will remove only superficial contaminants, such as rust and mill scale and loosely adhering paint. It is primarily recommended for spot cleaning where corrosion is not a serious factor. Since manual cleaning removes only surface contamination, it is essential to use primers, which will thoroughly wet the surface and penetrate to the base metal.

Before cleaning is started, the surface should be treated with solvent cleaners to remove oil, grease, dirt, and chemicals. Rust scale and heavy buildup of old coatings should then be removed with impact tools such as chipping hammers, chisels, and scalers. Loose mill scale and nonadhering paint can be removed with wire brushes and scrapers. Caution should be exercised to avoid deep marking or scratching the surface with the tools.

Power Tool Cleaning Power tool cleaning provides faster and more adequate surface preparation than hand tool methods but is less economical than blasting for effective cleaning of large areas. Power tools can be used for removing small amounts of tightly adhering contaminants that are difficult to dislodge with hand tools.

Power tools are driven either electrically or pneumatically and include a variety of attachments for the basic units. Chipping hammers are used for removing tight rust, mill scale, and heavy coats of paint. Rotary and needle scalers are handy for removing rust, mill scale, and old paint from large metallic and masonry areas. Wire brushes can be used for removing loose mill scale, old paint, weld flux, slag, and dirt deposits. Grinders and sanders are useful in removing old paint, rust, or mill scale on small surfaces and for smoothing rough surfaces.

As with hand tools, care is recommended to avoid cutting too deeply into the surface, since this may result in burrs that are difficult to protect satisfactorily. When using wire brushes, care should be taken to avoid polishing the metal surface, which may prevent adequate coating adhesion. Power tool cleaning should be preceded by solvent or chemical treatment; painting should be started and completed as soon after cleaning as possible.

Flame Cleaning Flame cleaning is satisfactory for both new and maintenance work. Oil and grease should be removed prior to flame cleaning both for safety and for adequacy of preparation. Wire brushing normally follows flame cleaning to remove loose matter. A high-velocity oxy-acetylene flame is used in this treatment. Extreme caution and adequate ventilation are necessary to prevent fires.

New paint is generally applied while the surface is still warm, thereby speeding up drying time and making it possible to paint when air temperatures are somewhat below 50°F (10°C). However, when solvent-thinned paints are used, the painting operation should be sufficiently separated from the cleaning operation to avoid a fire hazard.

Blast Cleaning Blast cleaning involves the high-velocity impact of abrasive particles, such as sand or glass beads, on a surface. The abrasive is discharged, either wet or dry, under pressure. In the wet system, water, and sometimes a rust inhibitor, is either mixed with the abrasive in the pressure tank or introduced into the blast stream near the nozzle. Protective clothing, dust masks, and eye protection are needed when blast cleaning to prevent injury.

Metal surfaces that have been blast cleaned should be prime coated the same day, because rust forms rapidly on newly cleaned surfaces. Metal or synthetic shot, grit, glass beads, corn cobs, or similar abrasives are used where recovery of the abrasive is possible; otherwise, sand is generally used. The abrasive should be of a size sufficient to remove surface contamination without pitting the surface or creating extreme peaks and valleys. Pitted surfaces require protection, and high peaks represent areas of possible paint failure.

As in other mechanical cleaning, surface grease or oil should be removed by solvent cleaning prior to blasting. Dry blasting of steel should be avoided if the temperature of the metal is less than 50°F (10°C) above the dew point, to prevent condensation and subsequent rusting.

Blast cleaning is the most effective of the mechanical treatments and is often used on metal structures; it may also be used, with caution, on masonry surfaces. The depth to which the surface is abraded can be varied, depending on the intended paint system and severity of exposure. There are four methods of blast cleaning; the most effective methods, reaching down to, or near, the untarnished "white metal," require more labor, and are therefore costlier.

White metal blast provides for the complete removal of all rust, mill scale, and other contaminants from the surface and is the preferred technique in blast cleaning. It is used for coatings that must withstand exposure to very corrosive atmospheres and where a high initial cost of surface preparation is warranted.

Near-white metal blast surfaces are nearly free of all contaminants but will show streaks and discolorations distributed across the white metal surface. This technique is 10% to 35% cheaper than white metal blasting and has proven adequate for many protective coatings in moderately severe exposures.

Commercial blast describes the removal of superficial contaminants such as loose scale and rust. This method results in a surface quality generally adequate for the majority of paint systems under normal exposure conditions.

Brush-off blasting is a relatively low-cost method, used to remove old finishes, loose rust, and mill scale. Brush-off blasting is not suitable where severe corrosion is prevalent, but is intended as a more efficient substitute for manual and power tool cleaning. Brush-off blasting is also used for the removal of loose or deteriorated paint from masonry.

CHEMICAL AND SOLVENT TREATMENT

Chemical cleaning methods are better suited to shop application, while mechanical methods, other than blast cleaning, are more practical for use in the field. Chemical cleaning is both more efficient and more effective than mechanical methods other than blast cleaning, but chemical cleaning in the field is more difficult than even blast cleaning there because of the potential environmental hazards.

The use of goggles and rubber gloves is recommended, and additional protective clothing is required where acid or alkaline solutions are used. Adequate ventilation is essential during chemical and solvent cleaning. Respirators should be used when operating in confined areas; proper fire precautions should be taken when flammable cleaners are used.

Some chemicals, such as trisodium phosphate, that have been traditionally used in cleaning and the removal of existing paint and coatings are no longer permitted in some areas because of their potential harm to people and the environment. Some jurisdictions permit the use of such chemicals but severely limit the means of their disposal after use. Therefore, before using any chemical for removal of existing paints or coatings or for the cleaning of surfaces before applying paint, transparent finishes, or coatings, it is advisable to check with local authorities to ascertain restrictions on their use and disposal.

Solvent Wiping and Degreasing Solvent cleaning is a procedure used for dissolving and removing surface contaminants such as oil, grease, dirt, and paint remover residues before painting or mechanical treatment. It is best to first remove soil, cement spatter, and other dry materials with a wire brush. The surface should then be scrubbed with brushes or rags saturated with solvent. Other effective methods include immersing the work in a solvent or spraying solvent over the surface.

The solvent quickly becomes contaminated, so it is essential that several clean solvent rinses be applied to the surface. Clean rags should be used for rinsing and wiping dry. Mineral spirits is an effective solvent for cleaning under normal conditions. Toxic solvents and solvents with low flash points constitute health and safety hazards. Use of solvents with flash points below 100°F (37.78°C) or solvents with a maximum safe concentration of less than 100 parts per million is not recommended. Rags should be placed in fireproof containers after use.

Alkali Cleaning Alkali cleaning is more efficient, less costly, and less hazardous than solvent cleaning but is more difficult to carry out. It is suitable for most metals, except stainless steel and aluminum. Alkali cleaner solutions are applied at relatively high temperatures (150°F to 200°F) (65.6°C to 93.3°C), since cleaning efficiency increases with temperature. Alkalies attack oil and grease, converting them into soapy residues that can be washed away with water.

Commercial cleaners contain other active ingredients that aid in removing surface dirt and other contaminants such as mildew. These cleaners also are effective in removing old paint by breaking up the dried vehicle. The most commonly used alkaline cleaners are trisodium phosphate, caustic soda, and silicated alkalies. For steel, these cleaners should contain 0.1% chromic acid or potassium dichromate to prevent corrosion. Cleaners can be applied by brushing, scrubbing, or spraying, or the item to be cleaned can be immersed in a soak tank.

After cleaning, thorough rinsing is necessary to remove the soapy residue as well as all traces of alkali. Rinsing also avoids the possibility of chemical reaction with the applied paint. The rinse water should be hot (175°F) (79.4°C) and applied under pressure. Universal pH test paper should be used to ensure that the surface is neutral and no free alkali is present on the surface after rinsing.

Steam Cleaning Steam cleaning involves the use of steam or hot water under pressure, sometimes with a detergent included for added effectiveness. The steam or hot water removes oil and grease by emulsifying or diluting them, then flushing them away with water. When steam is applied to old paint, the paint's vehicle swells so that it loses its adhesiveness and is easily removed. Under ordinary conditions, steam or hot water alone may be used to remove dirt deposits, soot, and grime. In severe cases, wire brushing or brush-off blast cleaning may be necessary to augment the steam cleaning.

Acid Cleaning Acid cleaning is used to remove oil, grease, dirt, and other surface contaminants and is suitable for cleaning iron, steel, concrete, and masonry. It should not be used on stainless steel or aluminum.

Iron and steel surfaces are treated with solutions of phosphoric acid containing small amounts of solvent, detergent, and wetting agent. Unlike alkali cleaners, acid cleaners also remove light rust and etch the surface to ensure better adhesion of applied coatings. There are many types of metal acid cleaners and rust removers, each formulated to perform a specific cleaning job. The most common formulations are based on phosphoric acid solutions. The procedures usually involve applying the cleaner by brush or rag, allowing time for it to act, then thoroughly rinsing, wiping, and air drying. Smaller objects, particularly in the shop, can be immersed in hot cleaner, followed by several rinses.

Concrete and masonry surfaces can be washed with a solution of muriatic (hydrochloric) acid to remove efflorescence and laitance, to clean the surface, and to improve its bonding properties. Efflorescence is a white, powdery deposit often found on masonry surfaces. Laitance is a fine gray powder that sometimes develops on surfaces of improperly finished concrete. Paints applied over efflorescence or laitance may loosen prematurely and result in early coating failure. It is best to remove as much as possible of these deposits using a stiff fiber or wire brush; putty knives or scrapers may also be used.

Oil and grease should be removed prior to acid cleaning, either by solvent wiping or by steam or alkali cleaning. To start acid cleaning, the surface is first thoroughly wetted with clean water, then scrubbed with a 5% solution (by weight) of muriatic acid and a stiff fiber brush. In extreme cases, up to a 10% muriatic acid solution may be used, and it may be allowed to remain on the surface up to 5 minutes before scrubbing. Work should be done on small areas, not greater than 4 sq. ft. (0.37 m^2) in size. Immediately after the surface has been scrubbed, the acid solution should be rinsed completely from the surface by thoroughly sponging or flushing with clean water. The surface should be rinsed immediately to avoid formation of salts on the surface, which are difficult to remove.

This procedure may also be used to etch hard and dense concrete surfaces with poor surface bonding properties. The etching procedure breaks up the glaze so that paint will adhere to the surface. To determine whether etching is required, a few drops of water are poured on the surface. If the water is quickly absorbed, the surface can be readily painted; if the drops remain on the surface, etching is recommended.

Surface glaze may also be removed by rubbing with an abrasive stone, light sandblasting, or allowing the surface to weather for 6 to 12 months. The surface may also be etched with a solution of 3% zinc chloride and 2% phosphoric acid; this solution should not be washed off but allowed to dry to produce a paintable surface.

Concrete and masonry surfaces are generally alkaline in nature, and it may be necessary to use acid cleaning methods to neutralize these surfaces before applying alkali-sensitive coatings. The presence of free alkali may be detected by testing with pH paper.

Pickling Pickling is used only in the mill or shop to remove mill scale, rust, and rust scale from iron and steel products. Sulfuric, hydrochloric, nitric, hydrofluoric, and phosphoric acids are used individually or in combination. Pickling is usually accomplished by immersing the piece in a tank, followed by several intermediate rinses to remove acids and salts. The final rinse is usually performed with a weak alkali solution to retard rusting and neutralize lingering acid solution.

Paint Removers Paint and varnish removers are used on small areas. Solvent-type removers or solvent mixtures are selected according to the type and condition of the old finish, as well as the nature of the surface. Removers are available as flammable or nonflammable types and in liquid or semipaste consistency. While most paint removers require scraping to remove the softened paint, some removers loosen the paint sufficiently to permit flushing off the debris with steam or hot water.

To retard evaporation, some removers contain paraffin wax, which often remains on the surface after the paint has been removed. It is essential that this residue be removed from the surface prior to painting, to ensure proper bonding of the paint coat. The wax can be removed with mineral spirits or a solvent specifically recommended by the manufacturer. Most removers are toxic to a degree, and some are flammable. Consideration should be given to adequate ventilation and fire control whenever they are used.

METAL PRETREATMENTS

Pretreatment of metal surfaces improves coating adhesion, reduces underfilm corrosion, and extends coating life.

Hot Phosphate Treatments Hot phosphate treatments use zinc or iron phosphate solutions to form crystalline deposits on the surface of a metal. They increase the bond and adhesion of paints while reducing underfilm corrosion. Zinc phosphate usually produces the best results and is most widely used. The use of such treatments is limited to the shop or plant.

When a gloss finish is desired, it is often necessary to apply thicker coats of paint over a zinc phosphate coating, because phosphate coatings absorb considerably more paint. If a higher-gloss finish is desired, iron phosphate is preferred to zinc phosphate because it produces a finer crystalline structure and therefore a thinner film. Hot phosphate treatments are very effective in ensuring good coating adhesion and tight bonds.

Cold Phosphate Treatments Cold phosphate treatments are produced with a mixture of phosphoric acid, a wetting agent, a water-miscible solvent, and water. An acid concentration of about 5% to 7% by weight may be adequate when the area is not exposed during application to high temperatures, direct sunlight, or high wind velocities. Such adverse environmental conditions cause rapid evaporation and consequent high acid concentration.

When a dry, grayish-white powdery surface develops within a few minutes after application, acid has reacted properly and has the proper dilution. If a dark color develops and the surface is somewhat sticky, the acid is probably too concentrated. Then, if the area is small, wiping with damp rags may bring about the desired appearance; otherwise, the surface should be rinsed with water and retreated with a more dilute solution.

Although cold phosphate treatments may produce a crystalline deposit on the metal surface, the density of the deposit is not as great as the hot phosphate treatment; therefore paint adhesion may be less effective. The procedures used for cold phosphating are adaptable to field use on large or small structures.

Wash Primers Wash primers are a form of cold phosphating. They are said to perform more efficiently than standard cold phosphating treatments and have generally replaced them in field use. Wash primers derive their name from the method of applying them in thin "wash" coats. They usually contain polyvinyl butyral resin, phosphoric acid, and a rust-inhibitive pigment such as basic zinc chromate or lead chromate. Wash primers develop extremely good adhesion to blast-cleaned or pickled steel and provide a sound base for top coating. They may also be used to improve adhesion of coatings to hard-to-paint surfaces, such as galvanized steel, stainless steel, and aluminum. Wash primers can be applied in the shop or in the field.

PRIMERS, SEALERS, AND FILLERS

Application of a finish coat should always be preceded by a suitable primer, filler, or sealer or a base coat of the same paint as the finish coat, which performs the same functions.

Primers are often applied on masonry to seal chalky surfaces and to improve adhesion of water-based top coats. Sealers are used on wood to prevent resin exudation or bleeding. Fillers help to produce a smooth finish on open-grained wood and rough masonry.

Primers Ordinary latex paints do not adhere well to chalky masonry surfaces. To overcome this problem, modified latex paints can be used or the surface can be coated with an oil-based primer before latex paint is applied. The entire surface is vigorously wire brushed by hand or power tools, then dusted to remove loose particles and chalk residue. The primer then is brushed on to ensure effective penetration and allowed to dry. A primer should not be used as a finish coat, because it is not as weather resistant.

Knot Sealers Sealers are used on bare wood to prevent resin from bleeding through newly applied paint. Freshly exuded resin, while still soft, may be scraped off with a putty knife and the affected area cleaned with alcohol. Hardened resin may be removed by scraping or sanding. Since a sealer is not intended as a priming coat, it should be used only when necessary and applied only over the affected area. When a previously painted softwood lumber surface becomes discolored over knots, the knots should be covered with sealer before new paint is applied.

Fillers Fillers are used on porous wood, concrete, and masonry to fill the pores and to provide a smoother finish coat.

Wood fillers are used on open-grained hardwoods. In general, hardwoods with pores larger than those in birch (open grain) should be filled (Fig. 9.15-3). When filling is necessary, it should be done 24 hours after staining. If staining is not warranted, natural (uncolored) filler should be applied directly to the bare wood. When desired, the filler may be colored with stain to accentuate the grain pattern of the wood.

Wood filler should be thinned with mineral spirits to a creamy consistency, then liberally brushed across the grain, followed by a light brushing along the grain. It should be allowed to stand 5 to 10 minutes until most of the thinner has evaporated and the finish has lost its glossy appearance. Before it has a chance to set and harden, the filler should be wiped off across the grain, using burlap or other coarse cloth, rubbing the filler into the pores of the wood while removing the excess. The operation is finished by stroking along the grain with clean rags, making sure that all excess filler is removed.

Wiping too soon may pull a filler out of the pores, while allowing the filler to set too long may make it difficult to wipe off. A simple test for dryness consists of rubbing a finger across the surface. If the filler forms into a ball, it is time to wipe; if the

FIGURE 9.15-3 Finishing Characteristics of Major Wood Species

Wood Species	Grain Type	Remarks
Hardwoods		
Ash	Open	Requires filler
Aspen	Closed	Paints well
Basswood	Closed	Paints well
Beech	Closed	Paints poorly; varnishes well
Birch	Closed	Paints and varnishes well
Cherry	Closed	Varnishes well
Chestnut	Open	Requires filler; paints poorly
Cottonwood	Closed	Paints well
Cypress	Closed	Paints and varnishes well
Elm	Open	Requires filler; paints poorly
Gum	Closed	Varnishes well
Hickory	Open	Requires filler
Mahogany	Open	Requires filler
Maple	Closed	Varnishes well
Oak	Open	Requires filler
Teak	Open	Requires filler
Walnut	Open	Requires filler
Softwoods		
Alder	Closed	Stains well
Cedar	Closed	Paints and varnishes well
Fir	Closed	Paints poorly
Hemlock	Closed	Paints fairly well
Pine	Closed	Variable depending on grain
Redwood	Closed	Paints well

(Courtesy Architectural Woodwork Institute.)

filler slips under the pressure of the finger, it is still too wet for wiping. A filler should be allowed to dry for 24 hours before a finish coat is applied.

Masonry fillers can be applied by brush or spray to both old and new bare concrete, stucco, concrete block, and other masonry surfaces. These surfaces should be rough, sound, and free of loose and powdery materials. If surface voids are relatively large, it is preferable to apply two coats of filler rather than one heavy coat. In large areas, filler can be sprayed on, then troweled smooth. One to two hours' drying time should be allowed between filler coats, and painting should be delayed 24 hours after the final filler coat.

9.15.3.4 Repair of Surfaces

Surfaces that will be painted must be clean, sound, and uniform in texture. To achieve this condition:

- Repair or replace damaged materials and surfaces;
- Replace broken windows;
- Remove loose putty and glazing compounds and provide new materials;
- Fill cracks, crevices, and joints with caulking compound or sealants;
- Patch cracks or holes in wood, masonry, and plaster.

Products and materials suitable for this purpose are described in Figure 9.15-4. Refer to Section 7.11 for information on caulking and sealants.

In general, patching of cracks and holes should be done by the installer and finisher of the underlying material. Rarely, the painter might patch minor cracks and fill minor holes. In addition to the materials described in other sections, including Section 7.11, the materials described here may be used for patching holes and cracks.

Cracks, holes, and crevices in masonry, plaster, gypsum board, and wood may be filled with various patching materials. These are supplied either ready for use or as a dry powder to which water is added.

When using patching materials on masonry, plaster, or gypsum board, the crack should first be opened with a putty knife or wall scraper so that weak material is removed and the patching compound can be forced in completely. The compound should be applied with a putty knife or trowel, depending on the size of the hole. Then the surface should be smoothed off slightly convex, to allow for shrinkage.

One of the oldest painter's patching materials is *putty*. It is an oil-based material used to fill nail holes, cracks, and imperfections in wood surfaces. Putty is generally supplied in bulk form and is applied with a putty knife. It should be applied after priming to prevent the wood from drawing the oils out. It dries to a harder surface than caulking compound and should not be used for joints or crevices. Special glazing putty is used to install small lights in wood and metal sash.

Plastic wood is suitable for filling gouges and nail holes in wood. It is applied in a manner similar to putty. After the plastic wood has completely dried, it should be sanded until smooth.

A soupy mixture of portland cement, sand, and water called *grout* can be used to repair cracks in concrete and masonry. Synthetic resins such as latex or epoxy may be added to improve bonding qualities and to permit application in thin layers. Hydrated lime is sometimes added to lengthen working time. This grout should be moist-cured for several days or longer.

Patching plaster is used to repair plaster. It is similar to ordinary plaster except that it hardens more quickly. It is supplied as a powder to be mixed on the job with water.

Spackle is used to fill cracks and small holes in plaster and gypsum board. It is easy to apply and to sand smooth after it hardens. It is supplied both as a paste and as a powder.

Joint compound is intended primarily for sealing joints between gypsum board panels, but it can also be used to repair large cracks. It is supplied as a powder and is used in conjunction with joint tape, which gives it added strength.

9.15.3.5 Paint Application

The most common methods of applying paint are by brush, roller, and spray. Dip-and-flow coat methods are generally limited to shop work. Of the three methods designed for field use, brushing is the slowest and spraying is the fastest. One painter can cover up to 1000 sq. ft. (92.9 m^2) a day by brushing, 2000 to 4000 sq. ft. (186 to 372 m^2) by rolling, and up to 12,000 sq. ft. (1115 m^2) by spraying. Coatings applied by brush may leave brush marks in the dried film, rolling leaves a stippled effect, while spraying yields the smoothest finish if done properly. The choice of method is based on (1) the environmental conditions, (2) the type of surface and coating, and (3) the desired appearance of the finish.

General surroundings may prohibit the use of spray application because of possible fire hazards or potential damage from

FIGURE 9.15-4 Materials for Repair of Surfaces

Material	Where Used	Characteristics
Caulking compound	Filling joints and crevices around doors and windows in concrete, wood, brick, block, and other masonry surfaces	Tends to dry on the surface but remain soft and tacky inside. Exposed surface of caulking; bead should be painted each time surrounding area is painted to help extend its life (5–15 years).
Sealants	Filling joints in locations where movement occurs and a weathertight joint is essential at all times	Better adhesion and greater elasticity than caulking compound, even at low temperatures. Life expectancy 15–30 years.
Putty	Filling nail holes, cracks, and imperfections in wood surfaces; installing small lights in wood and metal sash; not used for joints or crevices	Dries to a harder surface than caulking compound.
Glazing compound	Bedding and face glazing of windows	Sets firmly but retains limited flexibility, generally greater than putty. Life expectancy of 5–15 years when protected by periodic painting over exposed surface.

overspray. Adjacent areas not intended to be coated should be masked when spraying is performed. If the time spent in masking preparations is extensive, it may offset the advantage of the speed of spraying operations.

Roller coating is efficient on large flat surfaces. Corners, edges, and odd shapes, however, generally require brushing. Spraying is especially suitable for large surfaces, but it can be used also for round or irregular shapes. Brushing is ideal for small surfaces or for cutting-in corners and edges. Dip-and-flow coat methods are appropriate for volume production painting of small items in the shop.

Rapid-drying lacquer-type products such as vinyls are usually sprayed. Application of such products by brush or roller is difficult, especially in warm weather or outdoors on breezy days.

JOB CONDITIONS

To obtain optimum performance from a coating, certain basic application procedures should be followed, regardless of the type of equipment used. Cleaned, pretreated surfaces should be prime coated within the specific time limits recommended by the paint manufacturer.

Paint manufacturers generally recommend the following:

- Surface and air temperatures be between 50°F (10°C) and 90°F (32.2°C) for water-thinned coatings, and between 45°F (7.2°C) and 95°F (35°C) for other coatings.
- Paint be maintained at a temperature between 65°F (18.3°C) and 85°F (29.4°C) at all times.
- Paint not be applied when the air temperature is expected to drop to freezing before the paint has dried.
- Wind velocity be below 15 mph (24 km/h).
- Relative humidity be below 80%.

However, some manufacturers may recommend other limits for their products under some circumstances.

Masonry surfaces that are damp (not wet) may be painted with latex or cementitious paints. Other surfaces should be completely dry before painting. Paints should be applied at the spreading rates recommended by their manufacturers. When successive coats of the same paint are used, each coat should be tinted a different color to ensure complete coverage with each coat. Sufficient time should be allowed for each coat to dry thoroughly before the next coat is applied. A finish coat (topcoat) should be allowed to dry for as long as practicable before the space is occupied or put into service.

The line between different colors or finishes should be straight and neat. Previously finished surfaces should be protected by suitable masking or covering. Stain or paint splatter from adjacent surfaces should be removed.

Unless otherwise recommended by the manufacturer, a minimum of two coats should be used. Painted exterior wood surfaces should have a total thickness of 4.0 mils (0.1016 mm) or more (Fig. 9.15-5).

FIGURE 9.15-5 Dry-Film Thickness of Applied Paints

Coverage Rate, sq. ft./gal. (m^2/L)	Approximate Dry-Film Thickness of Finish Coat, mils (mm)
450 (11.05)	2.25 (0.057)
500 (12.27)	2.00 (0.051)
550 (13.5)	1.75 (0.044)
600 (14.73)	1.50 (0.038)
650 (15.95)	1.30 (0.033)
700 (17.18)	1.10 (0.028)
750 (18.41)	1.00 (0.025)

9.16 Vinyl Wall Coverings

Vinyl wall coverings are decorative wall finishes for interior wall surfaces.

9.16.1 APPLICABLE STANDARDS

The industry standards applicable to vinyl wall coverings include the following. Their titles are indicated in Section 9.18.

- American Association of Textile Chemists & Colorists (AATCC): AATCC 90.
- American Society for Testing and Materials (ASTM)
  - D 751
  - D 1308
  - E 84
  - G 21
  - G 22
- Chemical Fabrics and Film Association (CFFA): CFFA-w-101-B
- Federal Specifications (FS): CCC-W-408D

9.16.2 MATERIALS

Vinyl wall coverings are fabric-backed polyvinyl chloride sheets, used in both commercial and residential applications. They are easily cleanable, damage resistant, and decorative finishes. They are also durable, scrubbable, and stain and fade resistant. They hide cracks, gaps, and uneven surfaces. These materials are available in hundreds of colors and many patterns. They are available with antimicrobial additives that inhibit mold, mildew, and fungus growth.

Most vinyl wall coverings are Class A fire rated. In addition, when heated to 300°F (149°C) or higher, some of these materials emit an odorless, harmless vapor that triggers ionization fire detectors. Most are also self-extinguishing; they will not continue to burn when the fire source is removed.

Vinyl wall coverings hide imperfections better and are easier to clean than paint. Their overall lifetime cost is lower than that of paint, due to their long life. Warranties of up to 5 years are available.

These wall coverings are available in several thicknesses, measured by their weight, such as the following:

- 12 oz./lin. yd. (407 g/m^2)
- 15 oz./lin. yd. (509 g/m^2)
- 20 oz./lin. yd. (678 g/m^2)
- 24 oz./lin. yd. (814 g/m^2)

The heavier materials generally have higher resistance to tearing than their thinner counterparts.

They may be applied on new or existing plaster, gypsum wallboard, or CMU. With proper substrate preparation, they may also be applied over existing painted walls, ceramic tile, marble, glass, and other materials. New surfaces must be cured for the length of time recommended by the wall covering manufacturer.

9.16.3 INSTALLATION

Vinyl wall coverings are installed with adhesives of a type recommended by the wall covering manufacturer. Wheat paste should not be used.

Primers are usually required. The types used should be those recommended by the wall covering manufacturer.

Installation methods and cleanup should also follow the manufacturer's recommendations.

Most wall covering manufacturers recommend that substrate moisture content not exceed 5% at the time of application.

The material should be wrapped around inside and outside corners so that joints do not occur there. Vinyl wallcovering should be smoothed with a bristle brush or flexible broad knife to ensure positive adhesion and eliminate bubbles. Excessive adhesive must be cleaned from surfaces immediately to prevent staining.

9.17 Additional Reading

More information about the subjects discussed in this chapter can be found in the references listed in Section 9.18 and in the following publications.

Ayres, Thomas W., and Martin J. Cooper II. 1995. "Acoustical Security Ceiling Systems." *The Construction Specifier* (December):38.

Bartoszek, Edward J. 1993. "PVDF Coatings." *The Construction Specifier* (April):148.

Blackburn, Lane. 1997. "The Waterborne Coatings Revolution." *The Construction Specifier* (August):58.

Blumer, H. Maynard. 1987. "Choosing Stucco Systems; Real vs. Synthetic." *Architecture* (January):110–111.

Bodnar, James. 1986. "Wood Floors a Natural." *The Construction Specifier* (October):62.

———. 1989. "The Revolution of Carpet Tiles." *The Construction Specifier* (April):90.

———. 1990. "When to Use Specialty Flooring." *The Construction Specifier* (October):98.

———. 1991. "Prescription Flooring." *The Construction Specifier* (August): 74.

———. 1993. "Seamless Surfaces." *The Construction Specifier* (April):132.

———. 1993. "Touchdown." *The Construction Specifier* (June):96.

Brungraber, Robert, and John Templer. 1991. "Controlling Slip Resistance." *Progressive Architecture* (March):112–116.

Bucklin, Lisa N. 1995. "Choosing Acoustical Ceilings Without Reservation." *The Construction Specifier* (October):26.

Casper, Douglas P. 1994. "Testing Ceramic Tile for Special Applications." *The Construction Specifier* (May):132.

———. 1994. "Linear Metal Ceilings." *The Construction Specifier* (August): 82.

———. 1994. "Carpet for Health-Care Facilities." *The Construction Specifier* (September):82.

Custom Builder. 1988. "Wall Coverings: Wood Paneling, Fabric Coverings." *Custom Builder* (April):33–38.

Doe, Bruce R., and Susan I. Sherwood. 1985. "Acid Rain and Dimension Stone: A Dangerous Combination?" *The Construction Specifier* (February): 46.

Face, Allen. 1987. "Specifying Floor Flatness and Levelness: The F-Number System." *The Construction Specifier* (April):124.

Factor, David F. 1993. "Calculating VOCs in Construction Coatings." *The Construction Specifier* (December):91.

Fleming, Peter. 1997. *Builder's Guide to Floors*. New York: McGraw-Hill.

Foster, Allan. 1989. "Performance Characteristics of Exterior Plastering Systems." *The Construction Specifier* (August):84.

———. 1990. "Joint Design and Accessories for Portland Cement Plaster." *The Construction Specifier* (October): 60.

Freeman, Katherine. 1992. "Carpet Installation." *The Construction Specifier* (September):39.

Gozdan, Walt. 1994. "Painting Masonry to Last." *The Construction Specifier* (December):33.

Gozdan, Walter J. 1996. "Specifying Quality Architectural Paints." *The Construction Specifier* (July):44.

Grazzini, Edward A. 1992. "The Ground Floor." *The Construction Specifier* (December):86.

Green, Robert. 1993. "Terminal Finishes." *The Construction Specifier* (July):138.

Gypsum Association. 1967. "Gypsum: Its Nature and Use." *The Construction Specifier* (August):30.

Hadfield, Robert W. 1992. "Custom Ceilings." *The Construction Specifier* (March):134.

Harris, Albert Joe. 1993. "Advent of the 'Final Rule.' " *The Construction Specifier* (July):151.

———. 1993. "Corrosion Control by Protective Coatings." *The Construction Specifier* (July):146.

Hill, Russ. 1986. "Investing in a Super Flat Floor." *The Construction Specifier* (March):64.

Houser, James L. 1990. "How to Write Descriptive Finishing Specifications for Gypsum Board." *The Construction Specifier* (October):68.

Hund, Robert. 1986. "Commercial Uses for Marble." *The Construction Specifier* (August):76.

Interiors. 1988. "Textile and Wallcovering Directory Chart." *Interiors* (July): 100–125.

Karofsky, Paul I. 1987. "Focusing on Wall Coverings." *The Construction Specifier* (February):27.

Kent, Keith, and James E. Fell. 1993. "Coating Innovations." *The Construction Specifier* (February):62.

Kohr, Robert L. 1991. "Technics Topics." *Progressive Architecture* (July):45.

Kummer, Robert B. 1996. "The Glass Alternative." *The Construction Specifier* (February):48.

Lavenberg, George N. 1991. "Dimensional Stone?" *Tile & Decorative Surfaces* (March):52–53.

Leary, Michael. 1994. "Carpet Specification: The Balancing Act." *The Construction Specifier* (September):85.

Lore, Gordon. 1993. "New Stones on the Market." *The Construction Specifier* (May):154.

Mahowald, Dave. 1988. "Specifying Paint Coatings for Harsh Environments." *The Construction Specifier* (October):13.

Mayer, Art. 1990. "The Next Step: Coating Selection." *The Construction Specifier* (October):108.

Mazzur, Richard P. 1986. "Choosing the Right Resilient Flooring." *The Construction Specifier* (March):70.

McCaskill, Pete. 1990. "Working with GRG: Problems and Solutions." *The Construction Specifier* (October):80.

McDonald, Timothy B. 1987. "Coatings, Sealants, Consolidants." *Architecture* (March):78–79.

McGowan, Maryrose. 1991. "Specifying Vinyl Wall Covering." *The Construction Specifier* (December):49.

———. 1992. "Specifying Wall Covering When Mildew Is a Problem." *The Construction Specifier* (March):57.

———. 1992. "The Trouble with Vinyl." *The Construction Specifier* (March): 52.

———. 1992. "Vinyl and Hospitality." *The Construction Specifier* (March): 55.

———. 1992. "Fiber Underfoot." *The Construction Specifier* (April):33.

———. 1993. "Stretched Fabric Wall and Ceiling Systems." *The Construction Specifier* (March):53.

———. 1993. "Multicolored Coatings." *The Construction Specifier* (April):47.

———. 1993. "Glass-Reinforced Gypsum." *The Construction Specifier* (June):35.

———. 1995. "Carpet Done Right." *The Construction Specifier* (April):66.

———. 1995. "CRI 104." *The Construction Specifier* (April):71.

McIlvain, Jess. 1994. "Ceramic Tile." *The Construction Specifier* (May):126.

McIntosh, Bruce. 1987. "Stucco, Guidelines, Installation: Trouble Areas." *Professional Builder* (September): 7–14.

Mercolino, Nancy, and Michael Chusid. 1998. "Advanced Technology in Curved Metal Ceilings." *The Construction Specifier* (May):75.

Middleton, George T. 1994. "An Abuse Resistance Standard for Interior Walls." *The Construction Specifier* (October): 91.

Myers, Donald L. 1996. "Abuse-Resistant Ceiling Systems." *The Construction Specifier* (March):56.

Nicastro, David H. 1994. "Of Stress and Stucco." *The Construction Specifier* (April):160.

———. 1995. "Stone Stains." *The Construction Specifier* (November):96.

Parisi, Robert. 1995. "Vinyl Wall Coverings." *The Construction Specifier* (October):34.

Pashina, Keith. 1993. "Coating Concrete." *The Construction Specifier* (December):86.

Professional Builder. 1986. " Metal Lathing, Then and Now." *Professional Builder* (March/April):64–65.

———. 1987. "Skim-Coat Plaster: Better Quality, Yet Nearly Competitive with Drywall." *Professional Builder* (February):29–32.

Raeber, John A. 1989. "Changes in Paints and Coatings: Are We Ready?" *The Construction Specifier* (December):29.

———. 1991. "Codes, Carpets, Fires." *The Construction Specifier* (August): 27.

———. 1993. "Anti-Graffiti Coatings." *The Construction Specifier* (April):27.

———. 1993. "Exterior Ceramic Tile." *The Construction Specifier* (August): 15.

———. 1994. "Stucco Expansion and Control Joints." *The Construction Specifier* (December):13.

Rickert, Stan. 1991. "Industrial Coatings: Specifying for Tough Conditions." *The Construction Specifier* (March):134.

Roessler, David. 1990. " The Trinity Building: Restoring Granite and Limestone." *The Construction Specifier* (July):80.

Roman, Charles A. 1994. "Wallcoverings for Healing Environments." *The Construction Specifier* (September):98.

Rosen, Harold J., and Tom Heineman. 1995. *Architectural Materials for Construction*. New York: McGraw-Hill.

Sacher, Alex. 1994. "Slip Resistance." *The Construction Specifier* (September):90.

Schmid, Edward D., Jr. 1972. "Specifying Indiana Limestone." *The Construction Specifier* (April):38.

Shrive, Charles A. 1993. "Integrated Ceilings." *The Construction Specifier* (April):39.

Simmons, H. Leslie. 1996. "Field-Applied Organic Finish Failures." *The Construction Specifier* (July):54.

Smead, Tom. 1996. "Preventing Masonry Failures with Proper Coating Choices." *The Construction Specifier* (July):48.

Stovall, Gerald L. 1995. "Wood Flooring." *The Construction Specifier* (September):56.

Stover, David. 1995. "If You Can't Hear, It Won't Taste Good." *The Construction Specifier* (April):54.

Swords, Jim. 1992. "Footwork." *The Construction Specifier* (March):66.

Tarver, Joe. 1993. "Setting Stone Tiles." *The Construction Specifier* (May):157.

Thompson, Sheri. 1994. "Better Paint? Better Air?" *The Construction Specifier* (March):25.

———. 1994. "Paint Failures." *The Construction Specifier* (December):21.

Thorp, Erik. 1993. "Concrete Protection." *The Construction Specifier* (December):96.

Weaver, Martin E. 1991. "Covering the Past: Historic Wallpapers." *The Construction Specifier* (March):51.

———. 1992. "Building and Monumental Stones of the United States." *The Construction Specifier* (March):29.

———. 1997. "A Systematic Approach to Graffiti Removal." *The Construction Specifier* (June):24.

Wertz, Sheila. 1990. "Innovative Use of Glass-Reinforced Gypsum." *The Construction Specifier* (October):76.

Wiley, Dave. 1990. "Marble Tile—Understand It!" *Tile & Decorative Surfaces* (May):38.

Wise, Kathryn O. 1996. "The Carpet Industry—Covering the Future." *The Construction Specifier* (June):78.

Woolf, Ken. 1997. "Restoring Acoustical Ceilings." *The Construction Specifier* (June):50.

Young, Parker M. 1985. "Coating Concrete." *The Construction Specifier* (December):52.

Young, Robert T. 1991. "Marble Tile: Pay Me Now—Or Pay Me Later." *Tile & Decorative Surfaces* (May):82.

———. 1992. "Avoiding Problems with Marble Tile Floors." *Dimensional Stone* (February):30.

Zwers, Jim. 1985. "Specifying Vinyl Wallcovering for Long Life." *The Construction Specifier* (April):70.

9.18 Acknowledgments and References

ACKNOWLEDGMENTS

We gratefully acknowledge the assistance of the following organizations and individuals in preparing this chapter. We are also indebted to them for permission to use their illustrations when requested and for the use of their publications as references.

American National Standards Institute (ANSI)
American Olean Tile Company
American Plywood Association
American Slate Company
American Society for Testing and Materials (ASTM)
Ancor Granite Tile Inc.
Architectural Woodwork Institute (AWI)
Armstrong World Industries, Inc.
Athenian Marble
Binks Manufacturing Company
Bruce Hardwood Floors
Buckingham-Virginia Slate Corporation
Burlington Natstone Inc.
Camara Slate, Inc.
Carpet and Rug Institute
Ceiling and Interior Systems Construction Association (CISCA)
Celotex Corporation
Clemco Industries Corporation
Cold Spring Granite Co.
Congoleum Corporation
Coronado Paint Co.
Creative Edge Corporation
Crossfield Products Corporation
Drywall, Lath, and Plaster Bureau
Dur-A-Flex, Inc.
E-Z Painter Corporation
Forbo Vicracoustic Inc.
Garland Floor Company
General Polymers
Georgia Marble Company
Georgia Pacific Corporation
Gold Bond Building Products
Greek Trade Commission
Guard Contract Wallcoverings
Gypsum Association
Harbil Manufacturing Company
Harris Specialty Chemicals, Inc.
Hercules, Incorporate
Hunter Douglas Architectural Products
Indiana Limestone Co., Inc.
International Institute for Lath and Plaster
Johnsonite Division of Duramax, Inc.
Kaylath Division, Georgetown Wire Co., Inc.
Kentile Floors
Kurfees Coatings
MAB Paints
MAC Publishers, Incorporated
Manville Corporation
Maple Flooring Manufacturers Association
Marble Institute of America
Mercer Products Company, Inc.
Metal Building Interior Products Company
Metal Lath/Steel Framing Association Division of the National Association of Architectural Metal Manufacturers
Mid-States Terrazzo and Tile Company
Milliken Carpet
Modac
Mohawk Carpet Mills
Monsanto Company
National Association of Home Builders (NAHB) Research Foundation
National Concrete Masonry Association
National Gypsum Company
National Institute of Building Sciences (NIBS)
National Institute of Standards and Technology (NIST)
National Lime Association
National Oak Flooring Manufacturers Association
National Paint and Coatings Association
National Terrazzo and Mosaic Association, Inc.
Nora Flooring Systems, Inc.
Painting and Decorating Contractors of America
Perlite Institute
Portland Cement Association
Resilient Floor Covering Institute
Ruberoid Company
Selby, Battersby and Company
Sherwin-Williams Company
Sivento Inc., a Hüls Group Company
Solnhofen Natural Stone, Inc.
Southern Pine Inspection Bureau
Steel Structures Painting Council
Stonhard, Inc., an RPM Company
Structural Slate Company
Terrazzo Marble and Supply Company
Textured Coatings of America
Thoro Division of Harris Specialty Chemicals, Inc.
Tile Council of America
Trend Mills, Incorporated
United States Gypsum Company
U.S. Department of Housing and Urban Development (HUD)
Vermiculite Association
Walter E. Selck and Company
West Coast Lumber Inspection Bureau
Western Lath, Plaster, Drywall Industries Association
Western Wood Products Association
Wool Bureau, Inc.
W.R. Bonsal Company
Wunda Weve Carpet Company, Inc.

REFERENCES

We would also like to thank the authors and publishers of the publications in the following list for their contribution to our research for this chapter.

Abate, Kenneth. 1989. "Metal Coatings, Fighting the Elements with Superior Paint Systems." Part I. *Metal Architecture* 5(6)(June):10, 69.

———. 1989. "Metal Coatings, Fighting the Elements with Superior Paint Systems." Part II. *Metal Architecture* 5(7)(July):40–41.

Acoustical Materials Association (AMA). AMA 1-II, "Ceiling Sound Transmission Test by the Two Room Method." Glenview, IL: Ceilings and Interior Systems Contractors Association.

American Association of Colorists and Chemists (ASCC) Standards:

30-88, "Anti-fungal Activity Assessment on Textile Materials, Mildew, and Rot Resistance of Textile Materials." Research Triangle Park, NC: ASCC.

100-88, "Assessment of Anti-bacterial Finishes on Textile Materials." Research Triangle Park, NC: ASCC.

134-88, "Electrostatic Propensity of Carpets." Research Triangle Park, NC: ASCC.

American Association of State Highway and Transportation Officials. AASHTO T259/T260 National Cooperative Highway Research Program. Washington, DC: AASHTO.

American Association of Textile Chemists & Colorists (AATCC). AATCC 90-1965T (R1981), "Detection of Antibacterial Activity of Fabrics: Test Method for Agar Plate Method." Research Triangle Park, NC: AATCC.

American Concrete Institute (ACI) Standards:

ACI 201.1R, "Making a Condition Survey of Concrete in Service." Detroit: ACI.

ACI 503.1, "Revised Standards for Epoxy." Detroit: ACI.

ACI 503.2, "Standard Specification for Binding Plastic Concrete to Hardened Concrete with a Multi-Component Epoxy Adhesive." Detroit: ACI.
ACI 503.3, "Standard Specification for Producing a Skid-Resistant Surface on Concrete by the Use of a Multi-Component Epoxy System." Detroit: ACI.
ACI 503.4, "Standard Specification for Repairing Concrete with Epoxy Mortars." Detroit: ACI.
ACI 503.5R, "Guide for the Selection of Polymer Adhesives with Concrete." Detroit: ACI.
ACI 503.6R, "Guide for the Application of Epoxy and Latex Adhesives for Bonding Freshly Mixed and Hardened Concrete." Detroit: ACI.
American National Standards Institute (ANSI). Standards:
A10.20-1988, "Construction and Demolition—Ceramic Tile, Terrazzo, and Marble Work, Safety Requirements for." New York: ANSI.
A108.1-1992, "Installation of Ceramic Tile." New York: ANSI. Includes A108.1a, A108.1b, A108.1c, A108.4, A108.5, A108.6, A108.7, A108.8, A108.9, A108.10, A108.11, A118.1, A118.2, A118.3, A118.4, A118.5, A118.6, A118.8, A118.9, A136.1.
A137.1-1988, "Specifications for Ceramic Tile." New York: ANSI.
American Olean Tile Company, Inc. *Maintenance Guide Manual*. Landsdale, PA: American Olean Tile Company, Inc.
American Paint Contractor. 1988. "Direct to Rust Coatings." *American Paint Contractor* 65(6)(June):8–19.
American Society for Testing and Materials (ASTM) Standards:
A 82, "Standard Specification for Steel Wire, Plain, for Concrete Reinforcement." West Conshohocken, PA: ASTM.
A 185, "Standard Specification for Steel Welded Wire, Fabric, Plain, for Concrete Reinforcement." West Conshohocken, PA: ASTM.
B 117, "Standard Practice for Operating Salt Spray (Fog) Apparatus." West Conshohocken, PA: ASTM.
C 5, "Standard Specification for Quicklime for Structural Purposes." West Conshohocken, PA: ASTM.
C 11, "Standard Terminology Relating to Gypsum and Related Building Materials and Systems." West Conshohocken, PA: ASTM.
C 28, "Standard Specification for Gypsum Plasters." West Conshohocken, PA: ASTM.
C 33, "Standard Specification for Concrete Aggregate." West Conshohocken, PA: ASTM.
C 35, "Standard Specification for Inorganic Aggregates for Use in Gypsum Plaster." West Conshohocken, PA: ASTM.
C 36, "Standard Specification for Gypsum Wallboard." West Conshohocken, PA: ASTM.
C 37, "Standard Specification for Gypsum Lath." West Conshohocken, PA: ASTM.
C 59, "Standard Specification for Gypsum Plaster Casting and Gypsum Molding Plaster." West Conshohocken, PA: ASTM.
C 61, "Standard Specification for Gypsum Keene's Cement." West Conshohocken, PA: ASTM.
C 67, "Standard Test Methods for Sampling and Testing Brick and Structural Clay Tile." West Conshohocken, PA: ASTM.
C 79, "Standard Specification for Gypsum Sheathing Board." West Conshohocken, PA: ASTM.
C 91, "Standard Specification for Masonry Cement." West Conshohocken, PA: ASTM.
C 97, "Standard Test Method for Absorption and Bulk Specific Gravity of Dimension Stone." West Conshohocken, PA: ASTM.
C 99, "Standard Test Method for Modulus of Rupture of Dimension Stone." West Conshohocken, PA: ASTM.
C 109, "Standard Test Method for Compressive Strength of Hydraulic Cement Mortars (Using 2-in. or [50-mm] Cube Specimens)." West Conshohocken, PA: ASTM.
C 119, "Standard Terminology Relating to Dimension Stone." West Conshohocken, PA: ASTM.
C 144, "Standard Specification for Aggregate for Masonry Mortar." West Conshohocken, PA: ASTM.
C 150, "Standard Specification for Portland Cement." West Conshohocken, PA: ASTM.
C 170, "Standard Test Method for Compressive Strength of Dimension Stone." West Conshohocken, PA: ASTM.
C 206, "Standard Specification for Finishing Hydrated Lime." West Conshohocken, PA: ASTM.
C 207, "Standard Specification for Hydrated Lime for Masonry Purposes." West Conshohocken, PA: ASTM.
C 236, "Standard Test Method for Steady-State Thermal Performance of Building Assemblies by Means of a Guarded Hot Box." West Conshohocken, PA: ASTM.
C 241, "Standard Test Method for Abrasion Resistance of Stone Subject to Foot Traffic." West Conshohocken, PA: ASTM.
C 242, "Standard Terminology of Ceramic Whitewares and Related Products." West Conshohocken, PA: ASTM.
C 270, "Standard Specification for Mortar for Unit Masonry." West Conshohocken, PA: ASTM.
C 281, "Standard Specifications for Gypsum Plasters." West Conshohocken, PA: ASTM.
C 307, "Standard Test Method for Tensile Strength of Chemical-Resistant Mortar, Grouts, and Monolithic Surfacings." West Conshohocken, PA: ASTM.
C 309, "Standard Specification for Liquid Membrane-Forming Compounds for Curing Concrete." West Conshohocken, PA: ASTM.
C 332, "Standard Specification for Lightweight Aggregate for Insulating Concrete." West Conshohocken, PA: ASTM.
C 346, "Standard Test Method for 45-deg Specular Gloss of Ceramic Materials." West Conshohocken, PA: ASTM.
C 423, "Standard Test Method for Sound Absorption and Sound Absorption Coefficients by Reverberation Room Method." West Conshohocken, PA: ASTM.
C 442, "Standard Specifications for Gypsum Backing Board and Coreboard." West Conshohocken, PA: ASTM.
C 472, "Standard Methods for Physical Testing of Gypsum Plasters and Gypsum Concrete." West Conshohocken, PA: ASTM.
C 475, "Standard Specifications for Joint Treatment Materials for Gypsum Wallboard Construction." West Conshohocken, PA: ASTM.
C 481, "Standard Test Method for Laboratory Aging of Sandwich Construction." West Conshohocken, PA: ASTM.
C 501, "Standard Test Method for Relative Resistance to Wear of Unglazed Ceramic Tile by the Taber Abraser." West Conshohocken, PA: ASTM.

C 503, "Standard Specifications for Marble Dimension Stone." West Conshohocken, PA: ASTM.

C 514, "Standard Specifications for Nails for the Application of Gypsum Wallboard." West Conshohocken, PA: ASTM.

C 557, "Standard Specifications for Adhesive for Fastening Gypsum Wallboard to Wood Framing." West Conshohocken, PA: ASTM.

C 568, "Standard Specification for Limestone Building Stone." West Conshohocken, PA: ASTM.

C 579, "Standard Test Method for Compressive Strength of Chemical-Resistant Mortars, Grouts, Monolithic Surfacers and Polymer Concretes." West Conshohocken, PA: ASTM.

C 580, "Standard Test Method for Flexural Strength and Modulus of Elasticity of Chemical-Resistant Mortars, Grouts, Monolithic Surfaces and Polymer Concretes." West Conshohocken, PA: ASTM.

C 587, "Standard Specifications for Gypsum Veneer Plaster." West Conshohocken, PA: ASTM.

C 588, "Standard Specifications for Gypsum Base for Veneer Plasters." West Conshohocken, PA: ASTM.

C 615, "Standard Specification for Granite Dimension Stone." West Conshohocken, PA: ASTM.

C 627, "Standard Test Method for Evaluating Ceramic Floor Tile Installation Systems Using the Robinson-Type Floor Tester." West Conshohocken, PA: ASTM.

C 629, "Standard Specification for Slate Dimension Stone." West Conshohocken, PA: ASTM.

C 630, "Standard Specification for Water-Resistant Gypsum Backing Board." West Conshohocken, PA: ASTM.

C 631, "Standard Specifications for Bonding Compounds for Interior Gypsum Plastering." West Conshohocken, PA: ASTM.

C 635, "Standard Specification for the Manufacture, Performance, and Testing of Metal Suspension Systems for Acoustical Tile and Lay-in Panel Ceilings." West Conshohocken, PA: ASTM.

C 636, "Standard Practice for Installation of Metal Ceiling Suspension Systems for Acoustical Tile and Lay-in Panels." West Conshohocken, PA: ASTM.

C 641, "Standard Test Method for Staining Materials in Lightweight Concrete Aggregates." West Conshohocken, PA: ASTM.

C 645, "Standard Specifications for Nonstructural Steel Framing." West Conshohocken, PA: ASTM.

C 666, "Standard Test Method for Resistance of Concrete to Rapid Freezing and Thawing." West Conshohocken, PA: ASTM.

C 672, "Standard Test Method for Scaling Resistance of Concrete Surfaces Exposed to Deicing Chemicals." West Conshohocken, PA: ASTM.

C 696, "Standard Test Methods for Chemical, Mass Spectrometer, and Spectrochemical Analysis of Nuclear-Grade Uranium Dioxide Powders and Pellets." West Conshohocken, PA: ASTM.

C 754, "Standard Specifications for Installation of Steel Framing Members to Receive Screw-Attached Gypsum Panel Products." West Conshohocken, PA: ASTM.

C 840, "Standard Specifications for Application and Finishing of Gypsum Board." West Conshohocken, PA: ASTM.

C 841, "Standard Specifications for Installation of Interior Lathing and Furring." West Conshohocken, PA: ASTM.

C 842, "Standard Specifications for Application of Interior Gypsum Plaster." West Conshohocken, PA: ASTM.

C 843, "Standard Specifications for Application of Gypsum Veneer Plaster." West Conshohocken, PA: ASTM.

C 844, "Standard Specifications for Application of Gypsum Base to Receive Gypsum Veneer Plaster." West Conshohocken, PA: ASTM.

C 847, "Standard Specifications for Metal Lath." West Conshohocken, PA: ASTM.

C 880, "Standard Test Method for Flexural Strength of Dimension Stone." West Conshohocken, PA: ASTM.

C 897, "Standard Specifications for Aggregate for Job-Mixed Portland Cement-Based Plasters." West Conshohocken, PA: ASTM.

C 919, "Standard Practices for Use of Sealants in Acoustical Applications." West Conshohocken, PA: ASTM.

C 920, "Standard Specifications for Elastomeric Joint Sealants." West Conshohocken, PA: ASTM.

C 926, "Standard Specifications for Application of Portland Cement-Based Plaster." West Conshohocken, PA: ASTM.

C 932, "Standard Specifications for Surface-Applied Bonding Agents for Exterior Plastering." West Conshohocken, PA: ASTM.

C 955, "Standard Specifications for Load-Bearing (Transverse and Axial) Steel Studs, Runners (Track), and Bracing or Bridging, for Screw Application of Gypsum Board and Metal Plaster Bases." West Conshohocken, PA: ASTM.

C 960, "Standard Specifications for Predecorated Gypsum Board." West Conshohocken, PA: ASTM.

C 976, "Standard Test Method for Thermal Performance of Building Assemblies by Means of a Calibrated Hot Box." West Conshohocken, PA: ASTM.

C 1002, "Standard Specifications for Steel Drill Screws for Application of Gypsum Panel Products or Metal Plaster Bases." West Conshohocken, PA: ASTM.

C 1007, "Standard Specifications for Installation of Load-Bearing (Transverse and Axial) Steel Studs and Related Accessories." West Conshohocken, PA: ASTM.

C 1036, "Standard Specification for Flat Glass." West Conshohocken, PA: ASTM.

C 1047, "Standard Specifications for Accessories for Gypsum Wallboard and Gypsum Veneer Base." West Conshohocken, PA: ASTM.

C 1063, "Standard Specifications for Installation of Lathing and Furring to Receive Interior and Exterior Portland Cement-Based Plaster." West Conshohocken, PA: ASTM.

D 16, "Standard Terminology for Paint, Related Coatings, Materials and Applications." West Conshohocken, PA: ASTM.

D 226, "Standard Specification for Asphalt-Saturated Organic Felt Used in Roofing and Waterproofing." West Conshohocken, PA: ASTM.

D 522, "Standard Test Methods for Mandrel Bend Test of Attached Organic Coatings." West Conshohocken, PA: ASTM.

D 570, "Standard Test Method for Water Absorption of Plastics." West Conshohocken, PA: ASTM.

D 578, "Standard Specification for Glass Fiber Yarns." West Conshohocken, PA: ASTM.

D 635, "Standard Test Method for Rate of

Burning and/or Extent and Time of Burning of Self-Supporting Plastics in a Horizontal Position." West Conshohocken, PA: ASTM.

D 638, "Standard Test Method for Tensile Properties of Plastics." West Conshohocken, PA: ASTM.

D 648, "Standard Test Method for Deflection Temperature of Plastics Under Flexural Load." West Conshohocken, PA: ASTM.

D 660, "Standard Test Method for Evaluating Degree of Checking of Exterior Paints." West Conshohocken, PA: ASTM.

D 661, "Standard Test Method for Evaluating Degree of Cracking of Exterior Paints." West Conshohocken, PA: ASTM.

D 662, "Standard Test Method for Evaluating Degree of Erosion of Exterior Paints." West Conshohocken, PA: ASTM.

D 695, "Standard Test Method for Compressive Properties of Rigid Plastic." West Conshohocken, PA: ASTM.

D 751, "Standard Test Method for Coated Fabrics." West Conshohocken, PA: ASTM.

D 790, "Standard Test Methods for Flexural Properties of Unreinforced and Reinforced Plastics and Electrical Insulating Materials." West Conshohocken, PA: ASTM.

D 870, "Standard Practice for Testing Water Resistance of Coatings Using Water Immersion." West Conshohocken, PA: ASTM.

D 968, "Standard Test Method for Abrasion Resistance of Organic Coatings by Falling Abrasive." West Conshohocken, PA: ASTM.

D 1187, "Standard Specification for Asphalt-Base Emulsions for Use as Protective Coatings for Metal." West Conshohocken, PA: ASTM.

D 1308, "Standard Test Method for Effect of Household Chemicals on Clear and Pigmented Organic Finishes." West Conshohocken, PA: ASTM.

D 1653, "Standard Test Method for Water Vapor Transmission of Organic Coating Films." West Conshohocken, PA: ASTM.

D 1730, "Standard Practices for Preparation of Aluminum and Aluminum-Alloy Surfaces for Painting." West Conshohocken, PA: ASTM.

D 1731, "Standard Practices for Preparation of Hot-Dip Aluminum Surfaces for Painting." West Conshohocken, PA: ASTM.

D 1784, "Standard Specification for Rigid Poly (Vinyl Chloride) (PVC) Compounds and Chlorinated Poly (Vinyl Chloride) (CPVC) Compounds." West Conshohocken, PA: ASTM.

D 2000, "Standard Classification System for Rubber Products in Automotive Applications." West Conshohocken, PA: ASTM.

D 2047, "Standard Test Method for Static Coefficient of Friction of Polish-Coated Floor Surfaces as Measured by the James Machine." West Conshohocken, PA: ASTM.

D 2092, "Standard Guide for Preparation of Zinc-Coated (Galvanized) Steel Surfaces for Painting." West Conshohocken, PA: ASTM.

D 2103, "Standard Specification for Polyethylene Film and Sheeting." West Conshohocken, PA: ASTM.

D 2240, "Standard Test Method for Rubber Property-Durometer Hardness." West Conshohocken, PA: ASTM.

D 2243, "Standard Test Method for Freeze-Thaw Resistance of Water-Borne Coatings." West Conshohocken, PA: ASTM.

D 2287, "Standard Specification for Nonrigid Vinyl Chloride Polymer and Copolymer Molding and Extrusion Compounds." West Conshohocken, PA: ASTM.

D 2370, "Standard Test Method for Tensile Properties of Organic Coatings." West Conshohocken, PA: ASTM.

D 2486, "Standard Test Method for Scrub Resistance of Interior Latex Flat Wall Paints." West Conshohocken, PA: ASTM.

D 2859, "Standard Test Method for Flammability of Finished Textile Floor Covering Materials." West Conshohocken, PA: ASTM.

D 3276, "Standard Guide for Painting Inspectors (Metal Substrates)." West Conshohocken, PA: ASTM.

D 3574, "Standard Test Method for Flexible Cellular Materials—Slab, Bonded, and Molded Urethane Foams." West Conshohocken, PA: ASTM.

D 3794, "Standard Guide for Testing Coil Coatings." West Conshohocken, PA: ASTM.

D 4141, "Standard Practice for Conducting Accelerated Outdoor Exposure Tests of Coatings." West Conshohocken, PA: ASTM.

D 4214, "Standard Test Method for Evaluating Degree of Chalking of Exterior Paint Films." West Conshohocken, PA: ASTM.

D 4258, "Standard Practice for Surface Cleaning Concrete for Coating." West Conshohocken, PA: ASTM.

D 4259, "Standard Practice for Abrading Concrete." West Conshohocken, PA: ASTM.

D 4260, "Standard Practice for Acid Etching Concrete." West Conshohocken, PA: ASTM.

D 4261, "Standard Practice for Surface Cleaning Concrete Unit Masonry for Coating." West Conshohocken, PA: ASTM.

D 4397, "Standard Specification for Polyethylene Sheeting for Construction, Industrial, and Agricultural Applications." West Conshohocken, PA: ASTM.

D 4541, "Standard Test Method for Pull-out Strength of Coatings Using Portable Adhesion Testers." West Conshohocken, PA: ASTM.

D 5090, "Standard Practice for Standardizing Ultrafiltration Permeate Flow Performance Data." West Conshohocken, PA: ASTM.

D 5324, "Standard Guide for Testing Water-Borne Architectural Coatings." West Conshohocken, PA: ASTM.

E 84, "Standard Test Method for Surface Burning Characteristics of Building Materials." West Conshohocken, PA: ASTM.

E 90, "Standard Test Method for Laboratory Measurement of Airborne-Sound Transmission Loss of Building Partitions and Elements." West Conshohocken, PA: ASTM.

E 96, "Standard Test Methods for Water Vapor Transmission of Materials." West Conshohocken, PA: ASTM.

E 119, "Standard Test Methods for Fire Tests for Building Construction and Materials." West Conshohocken, PA: ASTM.

E 136, "Standard Test Method for Behavior of Materials in a Vertical Tube Furnace at 750 deg C." West Conshohocken, PA: ASTM.

E 413, "Standard Classification for Rating Sound Insulation." West Conshohocken, PA: ASTM.

E 488, "Standard Test Method for Strength of Anchors in Concrete or Masonry Elements." West Conshohocken, PA: ASTM.

E 514, "Standard Test Method for Water Penetration and Leakage Through Masonry." West Conshohocken, PA: ASTM.

E 580, "Standard Practice for Application of Ceiling Suspension Systems for Acoustical Tile and Lay-In Panels in Areas Requiring Seismic Restraint." West Conshohocken, PA: ASTM.

E 648, "Standard Test Method for Critical Radiant Flux of Floor Covering Systems Using a Radiant Heat Energy Source." West Conshohocken, PA: ASTM.

E 662, "Standard Test Method for Specific Optical Density of Smoke Generated by Solid Materials." West Conshohocken, PA: ASTM.

E 1264, "Standard Classification for Acoustical Ceiling Products." West Conshohocken, PA: ASTM.

E 1347, "Standard Test Method for Color and Color-Difference Measurement by Trist Mulus (Filter) Colorimeter." West Conshohocken, PA: ASTM.

F 1066, "Standard Specification for Vinyl Composition Floor Tile." West Conshohocken, PA: ASTM.

F 1303, "Standard Specification for Sheet Vinyl Floor Covering with Backing." West Conshohocken, PA: ASTM.

F 1344, "Standard Specification for Rubber Floor Tile." West Conshohocken, PA: ASTM.

G 21, "Standard Practice for Determining Resistance of Synthetic Polymeric Materials to Fungi." West Conshohocken, PA: ASTM.

G 22, " Standard Practice for Determining Resistance of Plastics to Bacteria." West Conshohocken, PA: ASTM.

G 23, "Standard Practice for Operating Light-Exposure Apparatus (Carbon-Arc Type) with and without Water for Exposure of Nonmetallic Materials." West Conshohocken, PA: ASTM.

Appleman, Bernard R. 1989. "Coatings for Steel Structures." *The Construction Specifier* 42(3)(March):88–93.

Architectural Graphic Standards. See Ramsey/Sleeper.

Architectural Technology. 1985. "Technical Tips: Write Good Terrazzo Specs." *Architectural Technology* (fall): 54.

———. 1986. "Technical Tips: Metal Lathing—Then and Now." *Architectural Technology* (March/April):64–65.

———. 1986. "Technical Tips: Paints and Coatings Primer." *Architectural Technology* (July/August):64–65.

Arcom Master Systems, Salt Lake City, UT. MASTERSPEC®. Basic Sections:

09210, "Gypsum Plaster," 5/95 ed. Washington, DC: AIA.

09215, "Gypsum Veneer Plaster," 5/96 ed. Washington, DC: AIA.

09220, "Portland Cement Plaster," 5/95 ed. Washington, DC: AIA.

09253, "Gypsum Sheathing," 8/96 ed. Washington, DC: AIA.

09255, "Gypsum Board Assemblies," 2/95 ed. Washington, DC: AIA.

09261, "Predecorated Gypsum Board," 5/95 ed. Washington, DC: AIA.

09265, "Gypsum Board Shaft Wall Assemblies," 5/96 ed. Washington, DC: AIA.

09310, "Ceramic Tile," 11/95 ed. Washington, DC: AIA.

09385, "Dimension Stone Tile," 11/97 ed. Washington, DC: AIA.

09400, "Terrazzo," 8/95 ed. Washington, DC: AIA.

09511, "Acoustical Panel Ceilings," 11/96 ed. Washington, DC: AIA.

09512, "Acoustical Tile Ceilings," 11/96 ed. Washington, DC: AIA.

09513, "Acoustical Snap-in Metal Pan Ceilings," 11/96 ed. Washington, DC: AIA.

09547, "Linear Metal Ceilings," 8/97 ed. Washington, DC: AIA.

09600, "Stone Paving and Flooring," 8/95 ed. Washington, DC: AIA.

09640, "Wood Flooring," 5/97 ed. Washington, DC: AIA.

09651, "Resilient Tile Flooring," 5/96 ed. Washington, DC: AIA.

09652, "Sheet Vinyl Floor Coverings," 5/96 ed. Washington, DC: AIA.

09653, "Resilient Wall Base and Accessories," 6/96 ed. Washington, DC: AIA.

09654, "Linoleum Floor Coverings," 8/96 ed. Washington, DC: AIA.

09671, "Resinous Flooring," 8/96 ed. Washington, DC: AIA.

09680, "Carpet," 8/94 ed. Washington, DC: AIA.

09690, "Carpet Tile," 8/94 ed. Washington, DC: AIA.

09800, "Special Coatings," 5/92 ed. Washington, DC: AIA.

09841, "Acoustical Wall Panels," 2/97 ed. Washington, DC: AIA.

09900, "Painting," 8/95 ed. Washington, DC: AIA.

09950, "Wall Coverings," 5/95 ed. Washington, DC: AIA.

09963, "Intumescent Paints," 8/97 ed. Washington, DC: AIA.

09981, "Cementitious Coatings," 11/96 ed. Washington, DC: AIA.

Association of Tile, Terrazzo, Marble Contractors and Affiliates (ATTMCA). 1983. *Guide for Grouting and Cleaning Ceramic Floors with Latex Grout*. Jackson, MS: ATTMCA.

Association of The Wall and Ceiling Industries—International. *Veneer Plaster Manual*. Washington, DC: Association of The Wall and Ceiling Industries—International.

Bakhalov, G. T., and A. V. Turkovakaya. 1965. *Corrosion and Protection of Metals*. Elmsford, NY: Pergamon Press.

Banov, Able. 1973. *Paints and Coatings Handbook for Contractors Architects and Builders*. Farmington, MI: Structures Publishing Co.

Bartlett, Thomas L. 1983. "Cleaning with Corncobs." *The Construction Specifier* 36(2)(February):6–7.

Bennett, C. R. 1987. "Paints and Coatings: Getting Beneath the Surface." *The Construction Specifier* 40(2)(February):36–41.

Blatterman, Joan F. 1988. "Details Underfoot." *Architectural Record* (July): 118–121.

Blumer, H. Maynard, and Stephanie Stubbs. 1987. "Technical Tips: Choosing Stucco Systems." *Architecture* (January):110–111.

Bocchi, Greg. 1986. "Powder Coatings: A New Technology Takes Off." *The Construction Specifier* 39(9)(September):102–105.

———. 1988. "Powder Coatings Making Inroads in Metal Construction Market." *Metal Architecture* 4(4)(April): 8–9.

Bodner, James. 1985. "Ceramic Tile Makes a Comeback." *The Construction Specifier* 38(10)(October):58–63.

———. 1988. "From the Bottom Up: Good Beginnings for Floorcovering Installations." *The Construction Specifier* 41(4)(April):98–105.

Bower, Norman F. 1985. "Insurance by the Gallon." *The Construction Specifier* 38(4)(April):96–99.

Carpet and Rug Institute (CRI). 1987. *Carpet Specifier's Handbook*. Dalton, GA: CRI.

———. 1987. *Flammability of Carpets*. Dalton, GA: CRI.

Carpet Cushion Council (CCC). 1990. *Selecting the Correct Contract Carpet*

Cushion for Every Traffic Area. Riverside, CT: CCC.

Ceilings and Interior Systems Contractors Association (CISCA). *Acoustical Ceilings—Use and Practice*. Deerfield, IL: CISCA.

Ceramic Tile Institute. *Ceramic Tile Manual*. Los Angeles, CA: Ceramic Tile Institute.

Chalmers, Ray. 1985. "Selecting the Proper Tile Application Method." *Building Design and Construction* 26(9)(September):118–120.

———. 1985. "A Review of Related Products for Ceramic Tile." *Building Design and Construction* 26(9)(September):124–125.

Chemical Fabrics and Film Association (CFFA). CFFA-w-101-B, *Quality Standard for Vinyl Coated Fabric Wallcovering*. CFFA.

Commercial Remodeling. 1981. "Technique: Ion Painting." *Commercial Remodeling* 3(1)(February):24.

Commercial Renovation. 1987. "Product Focus: Ceiling Systems Reach New Heights in Performance and Design." *Commercial Renovation* 9(6)(December):48–53.

Construction Specifications Institute. 1988. *Specguide 09900. Painting*. Alexandria, VA: Construction Specifications Institute.

Construction Specifier, The. 1985. "Painting to Protect." *The Construction Specifier* 38(4)(April):92–94, 123.

Consumer Reports. 1985. "Concrete-Floor Paints." *Consumer Reports* 50(3) (March):166–169.

Council of American Building Officials (CABO). 1983. *One- and Two-Family Dwelling Code*. Falls Church, VA: CABO. This code is in the process of being transferred to the control of the International Code Council (ICC).

Dean, Sheldon W., and T. S. Lee, eds. 1988. *Degradation of Metals in the Atmosphere*. West Conshohocken, PA: ASTM.

Diehl, John R. 1965. *Manual of Lathing and Plastering, Gypsum Association Edition*. New York: MAC Publishers Association.

Domtar Gypsum America, Inc. 1983. *620 Series Shaftwalls*. Oakland, CA: Domtar Gypsum America, Inc.

Doyle, Margaret. 1985. "Ceramic Tile Suits a Variety of Uses." *Building Design and Construction* 26(9)(September):114–117.

Factory Mutual Systems. *Approval Guide*. Norwood, MA: Factory Mutual Systems.

———. *Loss Prevention Data Sheets*. Norwood, MA: Factory Mutual Systems.

Federal Specifications, all: *See* U.S. General Services Administration Specifications Unit.

Foster, Al. 1988. "Textured Coatings: Acoustical in the Name Doesn't Necessarily Mean the Product Absorbs Noise." *PWC Magazine* 50(3)(May/June):34–40.

Frane, James T. 1987. *Drywall Contracting*. Carlsbad, CA: Craftsman Book Company.

Garrison, John Mark. 1985. "Casting Decorative Plaster." *The Old-House Journal* 13(9)(November):186–189.

Gorman, J. R. 1982. "Smooth Specifying for Architectural Textures in Plaster." *The Construction Specifier* 35(5) (July):76–82.

Gypsum Association. GA-216, *Application and Finishing of Gypsum Board*. Washington, DC: Gypsum Association.

———. GA-223, *Gypsum Panel Products Types, Uses, Sizes and Standards*. Washington, DC: Gypsum Association.

———. GA-254-86, *Fire Resistant Gypsum Sheathing*. Washington, DC: Gypsum Association.

———. GA-505, *Glossary—Gypsum Board Terminology*. Washington, DC: Gypsum Association.

———. GA 530, *Design Data: Gypsum Board*. Washington, DC: Gypsum Association.

———. GA-600, *Fire Resistance Design Manual: Fifteenth Edition*. Washington, DC: Gypsum Association.

———. GA-650, *Recommendations for Covering Existing Interior Walls and Ceilings with Gypsum Board*. Washington, DC: Gypsum Association.

Hardingham, David. 1980. "Preparing for Painting." *The Old-House Journal* (October):133–136.

Harris, D. A. 1983. "Ceilings for the Office of the Future." *The Construction Specifier* 36(7)(July):58–65.

Herman, Frederick. 1981. "Refinishing Floors: Think Twice Before Sanding." *The Old-House Journal* (February):27, 44–45.

Italian Tile Center. *The Designer's Guide to Italian Tiles and Their Installation*. New York: Italian Tile Center.

Johnston, Clay M. 1987. "Lath and Plaster: A Capsule Look." *The Construction Specifier* 40(2)(February):54–58.

Jones, Larry. 1984. "Painting Galvanized Metal." *The Old-House Journal* 12(1) (January/February):10–11.

Jowers, Walter. 1986. "Textured Plaster Finishes." *The Old-House Journal* 14(2)(March):75–77.

Kawneer Company, Inc. 1984. *Architectural Finishes*. Norcross, GA: Kawneer.

Kincaid, Mary. 1982. "What Paint Experts Say." *The Construction Specifier* 35(5) (July):54–61.

———. 1983. "Color in the Squares." *The Construction Specifier* 36(6)(June): 70–76.

Kneemiller, Bill. 1982. "Using Exterior Textured Coatings." *The Construction Specifier* 35(5)(July):63–68.

Kosmatka, Steven H. 1987. "Floor-Covering Materials and Moisture in Concrete." *The Construction Specifier* 40(5)(May):35–36.

Labine, Clem. 1982. "Restoring Clear Finishes." *The Old-House Journal* 10(11)(November):221, 238–241.

Lowes, Robert. 1988. "Abrasive Blasting." *PWC Magazine* 50(3)(May/June):26–30.

MacDonald, John E. 1978. "The Effects of Light on Wood and Wood Finishing Systems." *The Construction Specifier* 31(3)(March):23–26.

Mahowald, Dave. 1988. "Specifying Paint Coatings for Harsh Environments." *The Construction Specifier* 41(10) (October):13–16.

Maple Flooring Manufacturers Association (MFMA). 1988. *Grading Rules for Hard Maple*. Northbrook, IL: MFMA.

———. *Sanding, Sealing, Court Lining and Finishing Maple Gym Floors*. Northbrook, IL: MFMA.

———. *Wood Flooring, Maple*. Northbrook, IL: MFMA.

Marble Institute of America, Inc. *Dimension Stone—Design Manual IV*. Farmington, MI: Marble Institute of America.

Martens, Charles R., and The Sherwin-Williams Company. 1974. *Technology of Paints and Lacquers*. New York: Robert E. Krieger Publishing Company.

Mazzur, Richard P. 1986. "Resilient Flooring: Choose Carefully." *The Construction Specifier* 39(3)(March): 70–75.

McDonald, Timothy B. 1987. "Technical Tips: Coatings That Protect Against the Corrosion of Steel." *Architecture* (July):101–102.

McIlvain, Jess. 1982. "The Problem with Dry-Set Mortars." *Tile and Decorative Surfaces* (September):17–20.

———. 1982. "Ceramic Tile Maintenance." *Tile and Decorative Surfaces* (October/November):48–57.

———. 1982. "The Preconstruction Conference." *Tile and Decorative Surfaces* (December):102–103.

———. 1983. "Caution: Latex-Portland Cement Mortar over Plywood." *Tile and Decorative Surfaces* (January/February).

———. 1983. "Plaster vs. Mortar Beds." *Tile and Decorative Surfaces* (April): 41–56.

———. 1983. "Never Again." *Tile and Decorative Surfaces* (July):138–140.

———. 1983. "Dodging the Dry-Set Mortar Gremlins." *The Construction Specifier* 36(5)(June):77–82.

———. 1984. "Prefabricated Exterior Ceramic Tile Cladding." *The Construction Specifier* 37(6)(June):76–91.

———. 1984. "Prefabricated Exterior Ceramic Tile Cladding, Part 1." *Tile and Decorative Surfaces* (January/February):80–85.

———. 1984. "Prefabricated Exterior Ceramic Tile Cladding, Part 2." *Tile and Decorative Surfaces* (March):41–53.

———. 1984. "Prefabricated Exterior Ceramic Tile Cladding, Part 3." *Tile and Decorative Surfaces* (April):50–58.

———. 1984. "Let's Talk Tile." *Tile and Decorative Surfaces* (June):69–74.

———. 1984. "Mortar Beds." *Tile and Decorative Surfaces* (September):118–120.

———. 1984. "The Preconstruction Conference." *Tile Letter* (October):32–33.

———. 1985. "Floor Renovation." *Tile and Decorative Surfaces* (April):48–49.

———. 1985. "Moisture Expansion." *Tile and Decorative Surfaces* (June): 72–86.

———. 1985. "The Goose That Laid the Rotten Egg." *Tile Letter* (August): 42–46.

———. 1986. "Writing Ceramic Tile Specifications." *Tile and Decorative Surfaces* (October/November):81–82.

———. 1986. "Lights, Action." *Tile Letter* (July):36–40.

———. 1987. "Selecting Ceramic Tile Floor Installation Methods. *Tile and Decorative Surfaces* (August):38–42.

———. 1988. "Grout: It Doesn't Have to Be a Problem." *Tile and Decorative Surfaces* (June):42–45.

Metal Architecture. 1987. "Building Repainting System Duplicates Coil Coatings in Appearance, Life Expectancy." *Metal Architecture* 3(11)(November): 24.

Metal Lath/Steel Framing Association (ML/SFA). 1987. *Lightweight Steel Framing Systems Manual*. Chicago, IL: ML/SFA.

———. 1991. *Specifications for Metal Lathing and Furring*. Chicago, IL: ML/SFA.

Minnery, Catherine, and Donald Minnery. 1979. "Repairing Stucco." *The Old-House Journal* 7(7)(July):73, 77–79.

Moit, Dan. 1988. "Coatings for Metals: Preplan the Selection." *Metal Architecture* 4(9)(September):8.

Moormann, Ambrose F., Jr. 1982. "Paint and the Prudent Specifier." *The Construction Specifier* 35(5)(July):69–71.

———. 1982. "Paint and the Prudent Specifier: If You Must Paint Concrete Floors." *The Construction Specifier* 35(9)(September):60–62.

———. 1983. "Paint and the Prudent Specifier: Working Hard to Look Natural." *The Construction Specifier* 36(1)(January):84–87.

Munger, Charles G. 1984. *Corrosion Prevention by Protective Coatings*. Houston, TX: National Association of Corrosion Engineers.

National Association of Tile Contractors. *Reference Manual and Specifications*. Jackson, MS: National Association of Tile Contractors.

National Concrete Masonry Association. 1985. NCMA-TEK 149, "Portland Cement Plaster (Stucco) for Concrete Masonry." National Concrete Masonry Association.

National Cooperative Highway Research Program. NCHRP Report 244, "Concrete Sealers for Protection of Bridge Structures." National Cooperative Highway Research Program.

National Decorating Products Association. 1988. *Paint Problem Solver*. St. Louis, MO: National Decorating Products Association. (Also available from the Painting and Decorating Contractors of America.)

National Fire Protection Association (NFPA). NFPA 253-95, "Standard Test Method for Critical Radiant Flux of Floor Covering Systems Using a Radiant Heat Energy Source." Quincy, MA: NFPA.

———. NFPA 255-96, "Standard Method of Test of Surface Burning Characteristics of Building Materials." Quincy, MA: NFPA.

National Oak Flooring Manufacturers Association (NOFMA). *Hardwood Flooring Installation Manual*. Memphis, TN: NOFMA.

———. *Hardwood Flooring Finishing/Refinishing Manual*. Memphis, TN: NOFMA.

———. *NOFMA Official Flooring Grading Rules*. Memphis, TN: NOFMA.

National Paint and Coatings Association (NPCA). *Finishing Hardwood Floors*. Washington, DC: NPCA.

National Terrazzo and Mosaic Association (NTMA). 1992. *Terrazzo Information Guide*. Des Plaines, IL: NTMA.

O'Donnell, Bill. 1985. "Reconditioning Floors." *The Old-House Journal* 18(10) (December):201, 218–219.

———. 1986. "Unwanted Textured Finish: How to Get Rid of It." *The Old-House Journal* (October):374–377.

Old-House Journal, The. 1982. "Stripping Paint." *The Old-House Journal* (December):249–252.

———. 1983. "48 Paint Stripping Tips." *The Old-House Journal* 11(2)(March): 44–45.

———. 1983. "Our Opinion of 'Peel Away.' " *The Old-House Journal* 11(4) (May):80.

———. 1983. "Ask OHJ: Paint on Paint." *The Old-House Journal* 11(9)(November):202.

———. 1984. "Tips on Mixing Paint." *The Old-House Journal* 12(1)(January/February):8–9.

———. 1985. "Reconditioning Floors." *The Old-House Journal* 13(10)(December):203, 218–219.

———. 1985. "Stripping Clinic." *The Old-House Journal* 13(10)(December):212B.

———. 1987. "Exterior Painting: Problems and Solutions." *The Old-House Journal* (September/October):35–39.

———. 1988. "Commercial Paint Stripping." *The Old-House Journal* 16(4) (July/August):29–33.

Olson, Christopher. 1987. "Improved Products Broaden Tile Applications." *Building Design and Constructions* 28(4)(April):98–101.

O'Neil, Edward F., and James E. McDonald. February 1976. *An Evaluation of Selected Instruments Used to Measure the Moisture Content of Hardened Concrete, Technical Report C-76-1*.

Vicksburg, MS: Concrete Laboratory, U.S. Army Engineer Waterways Experiment Station.

Painting and Decorating Contractors of America (PDCA). 1975. *Painting and Decorating Craftsman's Manual and Textbook.* 5th ed. Falls Church, VA: PDCA.

———. 1982. *Painting and Decorating Encyclopedia.* Falls Church, VA: PDCA.

———. 1984. *Painting and Wallcovering: A Century of Excellence.* Falls Church, VA: PDCA.

———. 1986 *Architectural Specification Manual, Painting, Repainting, Wallcovering and Gypsum Wallboard Finishing.* 3d ed. Kent, WA: Specifications Services, Washington State Council of PDCA.

———. 1988. *Hazardous Waste Handbook.* Falls Church, VA: PDCA.

———. 1988. *The Master Painters Glossary.* Falls Church, VA: PDCA.

———. 1988. *PDCA Estimating Guide.* Falls Church, VA: PDCA.

Petersen, Maurice R. 1984. "Finishes on Metals: A View from the Field—Part 1." *The Construction Specifier* 37(12) (December):36–39.

———. 1985. "Finishes on Metals: A View from the Field—Part 2." *The Construction Specifier* 38(1)(January): 70–73.

Poore, Patricia. 1980. "Sanding a Parquet Floor." *The Old-House Journal* (November):168–171.

———. 1981. "Picking a Floor Finish." *The Old-House Journal* (May):107–113.

———. 1985. "Stripping Paint from Exterior Wood." *The Old-House Journal* 13(10)(December):207–211.

Portland Cement Association (PCA). *Portland Cement Plaster (Stucco) Manual.* Skokie, IL: PCA.

———. *Bonding Concrete or Plaster to Concrete.* Skokie, IL: PCA.

———. 1977. *Painting Concrete, IS134T.* Skokie, IL: PCA.

———. 1982. *Removing Stains and Cleaning Concrete Surfaces.* Skokie, IL: PCA.

———. 1986. *Effects of Substances on Concrete and Guide to Protective Treatments.* Skokie, IL: PCA.

Raeber, John A. 1989. "Composite Plaster and Masonry Construction: A Closer Look." *The Construction Specifier* 42(9)(September):31–32.

Ramsey/Sleeper, The AIA Committee on Architectural Graphic Standards. 1981. *Architectural Graphic Standards.* 8th ed. New York: John Wiley & Sons, Inc.

Resilient Floor Covering Institute (RFCI). *Maintenance of Solid Vinyl Tile for Commercial, Institutional, and Industrial Buildings.* Rockville, MD: RFCI.

———. *Recommended Installation Specifications for Resilient Tile as a Wall Covering.* Rockville, MD: RFCI.

———. *Recommended Installation Specifications for Self-Adhering Tile on Floors.* Rockville, MD: RFCI.

———. *Recommended Installation Specifications for Sheet Vinyl Flooring, Flat-layed, Fully Adhered to the Floor.*" Rockville, MD: RFCI.

———. *Recommended Installation Specifications for Vinyl Composition, Solid Vinyl, and Asphalt Tile Floorings.* Rockville, MD: RFCI.

———. RFCI Specification ADH-1, "Recommended Specification for Vinyl Composition Tile Adhesive." Rockville, MD: RFCI.

———. RFCI Specification ADH-2, "Recommended Specification for Wall Base Adhesive." Rockville, MD: RFCI.

———. RFCI Specification CL-1, "Recommended Specification for Cleaners for Use on Resilient Floor Coverings." Rockville, MD: RFCI.

———. RFCI Specification FP-1, "Recommended Specification for Water Emulsion Floor Polish for Use on Resilient Flooring." Rockville, MD: RFCI.

———. RFCI Specification SV-1, "Recommended Specification for Resilient Floor Covering—Vinyl Plastic Sheet." Rockville, MD: RFCI.

———. RFCI Specification VCT-1, "Recommended Specification for Vinyl Composition Floor Tile." Rockville, MD: RFCI.

———. RFCI Specification VT-100, "Recommended Specification for Solid Vinyl Floor Tile." Rockville, MD: RFCI.

———. RFCI Specification VT-200, "Recommended Specification for Self-Adhering Resilient Floor Tile." Rockville, MD: RFCI.

———. RFCI Specification WB-1, "Recommended Specification for Vinyl Wall Base." Rockville, MD: RFCI.

Sherwin-Williams. 1991. *One-Hundred Twenty-Five Years of Providing Painting and Coating Systems for Specifiers and Applicators.* Cleveland, OH: Sherwin-Williams.

Southern Pine Inspection Bureau (SPIB). *Grading Rules.* Pensacola, FL: SPIB.

Steel Structures Painting Council (SSPC). 1982. *Steel Structures Painting Manual.* Vols. I and II. 2d ed. Pittsburgh, PA: SSPC.

———. SSPC-6, "Surface Preparation Specification No. 6—Commercial Blast Cleaning." Pittsburgh, PA: SSPC.

———. SSPC-10, "Surface Preparation Specification No. 10—Near-White Blast Cleaning." Pittsburgh, PA: SSPC.

———. SSPC-16, "Paint Specification 16—Coal Tar Epoxy-Polyamide Black (or Dark Red) Paint." Pittsburgh, PA: SSPC.

Stubbs, M. Stephanie. 1986. "Technical Tips: Paints and Coatings Primer." *Architectural Technology* (July/August): 64–65.

Tile Contractors Association of America (TCAA). *Products and Materials Guide.* Alexandria, VA: TCAA.

Tile Council of America (TCA). *Handbook for Ceramic Tile Installation.* Princeton, NJ: TCA.

Underwriters Laboratories, Inc. (UL). *Building Materials Directory—Class A, B, C: Fire and Wind Related Deck Assemblies.* Northbrook, IL: UL.

———. *Fire Resistance Directory—Time/Temperature Constructions.* Northbrook IL: UL.

———. UL 992-76, "Smoke Chamber Test." Northbrook IL: UL.

United States Gypsum Company (USG). 1972. *Red Book: Lathing and Plastering Handbook.* 28th ed. Chicago: USG.

———. 1987. *Gypsum Construction Handbook.* 3d ed. Chicago: USG.

U.S. Departments of the Army, the Navy, and the Air Force. 1969. *Technical Manual TM 5-618, Paints and Protective Coatings.* Washington, DC: U.S. Government Printing Office.

U.S. General Services Administration (GSA) Specifications Unit. Federal Specifications:

A-A-8, "Shellac, Varnish." Washington, DC: GSA Specifications Unit.

A-A-378, "Putty: Linseed-Oil Type (for Wood-Sash Glazing)." Washington, DC: GSA Specifications Unit.

A-A-1546A, "Rubbing Varnish." Washington, DC: GSA Specifications Unit.

A-A-2335, "Sealer, Surface (Varnish Type, Wood and Cork Floors)."

Washington, DC: GSA Specifications Unit.

A-A-2336A, "Primer Coating (Alkyd Exterior Wood, White and Tints)." Washington, DC: GSA Specifications Unit.

A-A-2904, "Thinner, Paint, Mineral Spirits, Regular and Odorless." Washington, DC: GSA Specifications Unit.

A-A-2962A, "Enamel, Alkyd, Exterior, Solvent Based, Low VOC." Washington, DC: GSA Specifications Unit.

A-A-2994, "Primer Coating, Interior for Walls and Wood." Washington, DC: GSA Specifications Unit.

A-A-3003, "Lacquer, Spraying, Clear and Pigmented for Interior Use." Washington, DC: GSA Specifications Unit.

A-A-3054, "Paint: Heat-Resisting (204 Deg C)." Washington, DC: GSA Specifications Unit.

A-A-3067, "Paint: Alkyd, Exterior, Low VOC." Washington, DC: GSA Specifications Unit.

A-A-3126, "Coating Systems: Tile-like High Performance Architectural." Washington, DC: GSA Specifications Unit.

A-A-50574, "Enamel, Odorless, Alkyd, Interior, Semigloss, White and Tints." Washington, DC: GSA Specifications Unit.

RR-T-650E, "Treads, Metallic and Nonmetallic Skid Resistant." Washington, DC: GSA Specifications Unit.

TT-C-542E, "Coating, Polyurethane, Oil-Free, Moisture Curing." Washington, DC: GSA Specifications Unit.

TT-C-555B(1), "Coating, Textured (for Interior and Exterior Masonry Surfaces)." Washington, DC: GSA Specifications Unit.

TT-E-489J, "Enamel, Alkyd, Gloss, Low VOC Content." Washington, DC: GSA Specifications Unit.

TT-E-527D, "Enamel, Alkyd, Lusterless, Low VOC Content." Washington, DC: GSA Specifications Unit.

TT-E-1593B, "Enamel, Silicone Alkyd Copolymer, Gloss (for Exterior and Interior Use)." Washington, DC: GSA Specifications Unit.

TT-L-201A, "Linseed Oil, Heat Polymerized." Washington, DC: GSA Specifications Unit.

TT-P-19D, "Paint (Latex Acrylic Emulsion): Exterior." Washington, DC: GSA Specifications Unit.

TT-P-29K, "Paint, Latex." Washington, DC: GSA Specifications Unit.

TT-P-47G, "Paint, Oil, Nonpenetrating-Flat, Ready-Mixed Tints and White (for Interior Use)." Washington, DC: GSA Specifications Unit.

TT-P-645B, "Primer, Paint, Zinc-Molybdate, Alkyd Type." Washington, DC: GSA Specifications Unit.

TT-P-650D, "Primer Coating, Latex Base, Interior, White (for Gypsum Wallboard or Plaster)." Washington, DC: GSA Specifications Unit.

TT-P-664D, "Primer Coating, Alkyd, Corrosion-Inhibiting, Lead and Chromate Free, VOC Compliant." Washington, DC: GSA Specifications Unit.

TT-S-711C, "Stain: Oil Type, Wood, Interior." Washington, DC: GSA Specifications Unit.

UU-B-790A, "Building Paper, Vegetable Fiber (Kraft, Waterproofed, Water-Resistant, and Fire-Resistant)." Washington, DC: GSA Specifications Unit.

CCC-W-408D "Wallcovering, Vinyl Coated." Washington, DC: GSA Specifications Unit.

Weaver, Martin E. 1989. "Fighting Rust, Part I: A Backgrounder." *The Construction Specifier* 42(5)(May):143–145.

———. 1989. "Fighting Rust, Part II: Remedies." *The Construction Specifier* 42(6)(June):129–130.

Weinstein, Nat. 1984. "How to Match Paint Colors." *The Old-House Journal* 12(1)(January/February):7–9.

Weismantel, Guy E., ed. 1981. *Paint Handbook*. New York: McGraw-Hill.

West Coast Lumber Inspection Bureau (WCLIB). *Standard Grading Rules for West Coast Lumber*. Portland, OR: WCLIB.

Western Lath, Plaster, Drywall Industries Association. 1988. *Plaster/Metal Framing System/Lath Manual*. New York: McGraw-Hill.

Wood and Synthetic Flooring Institute (WSFI). *Wood and Synthetic Flooring Specifications, Details, Product Data*. Hillside, IL: WSFI.

Wright, Gordon. 1987. "Trends in Specifying Architectural Coatings." *Building Design and Construction* 28(6) (June):188–192.

Zulandt, David F. 1985. "Consider an Alternative: Veneer Plaster." *The Construction Specifier* 38(7)(July):84–85.

CHAPTER 10

Specialties

Applicable *MasterFormat*™ Section

Fire Safety

Fire Protection Specialties

Additional Reading

Acknowledgments and References

Applicable *MasterFormat*™ Section

The following *MasterFormat* Level 2 section is applicable to this chapter.

10520 Fire Protection Specialties

10.1 Fire Safety

Fires cause loss of property and lives. The primary consideration in a fire is to evacuate all occupants from the building as quickly as possible.

The National Fire Protection Association's NFPA 550, "Guide to the Fire Safety Concepts Tree," identifies the various components of fire safety and shows their relationship to each other. The major components are prevention, detection, containment, control, and extinguishment.

The tree model of NFPA 550 demonstrates that no one of the components of fire safety is sufficient by itself. Rather, a balanced approach is necessary, employing several or preferably all of them. The NFPA *Life Safety Code* also recognizes this need for a balanced approach to fire safety. This section discusses some of the measures that can be taken to afford fire safety. For a more detailed review of the tree concept, refer to NFPA 550.

10.1.1 FIRE PREVENTION

The only way to prevent damaging fires from occurring is to eliminate or control potential sources of fire, to separate them from combustible materials, or to eliminate the combustible materials.

10.1.1.1 Fire Sources

A faulty electrical system is a potential source of fire and a potential source of injury to individuals through electrocution. Methods of controlling this source of fire are discussed in Chapter 16.

A lightning strike can also be a source of fire. Systems to eliminate that danger are covered in Section 13.2.

Kitchens are among the most fire-prone spaces in both dwellings and other building types. Fire-producing devices, such as gas ranges and stoves, should not be located near flammable materials. In a residence, for example, such devices should be kept away from windows because of the risk of curtains catching fire. Flammable wall coverings and other flammable finishes should not be used adjacent to such devices.

Careless smoking is a major cause of fires in both residential and other buildings. Many commercial, institutional, and industrial buildings prohibit smoking indoors.

10.1.1.2 Fire-Retardant Products

Current consumer protection laws require that mattresses and carpeting be treated for flame resistance. Some states have adopted restrictions on upholstered furniture to ensure that the room a fire starts in will remain at temperatures from which occupants can escape for 5 minutes after the fire starts and remain safe from toxic levels of carbon monoxide for 30 minutes. Older mattresses, carpeting, curtains, and drapes can be treated with homemade fireproofing chemicals (Fig. 10.1-1).

Refer to Section 7.10 for a discussion of the fireproofing of building structural elements. The pigmented or clear *intumescent* coatings discussed there slow the progress of a fire by swelling to form a protective foam. This foam slows the progress of a fire, reduces smoke, and delays the buildup of toxic gases. In addition to protecting structural elements, these coatings can be used as general surface coatings in areas subject to high fire hazards and in portions of exitways, such as corridors.

10.1.1.3 Fire-Resistant Plants

In some parts of the country, forest and brush fires can be hazardous to buildings, especially dwellings. In areas where such plants grow, fire-resistant ground cover should be planted within 100 ft. (30.48 m) of the building, and shrubs and trees should not be closer than 30 ft. (9.14 m) from the building and not less than 18 ft. (5.49 m) apart (Fig. 10.1-2). Under these conditions a fire cannot easily jump from one to the next and then to the building.

10.1.2 FIRE DETECTION AND ESCAPE

Most residential fires take place between midnight and 6:00 A.M., when people are usually asleep. More than half of the Americans who lose their lives in residential fires yearly could have been saved if

FIGURE 10.1-1 Flame-Proofing Recipes

Material to Be Treated	Recipe*
Permanent-press fabrics (blends of vegetable and synthetic fibers)	12 ounces (354.9 mL) diammonium phosphate 2 quarts (1.89 L) hot water
Untreated vegetable-fiber fabrics	7 ounces (207 mL) borax 3 ounces (88.7 mL) boric acid 2 quarts (1.89 L) hot water
Paper and cardboard	7 ounces (207 mL) borax 3 ounces (88.7 mL) boric acid 5 ounces (147.9 mL) diammonium phosphate 1 teaspoon (5 mL) liquid dishwashing detergent 3½ quarts (3.3 L) hot water
Christmas tree	4½ gallons (17 L) sodium silicate (water glass) 2 quarts (1.89 L) hot water 2 teaspoons (10 mL) liquid dishwashing detergent

*To treat fabrics and paper products: (1) dissolve the recipe ingredients in hot water, (2) spray or dip the product in the appropriate solution, (3) allow to dry. Treat Christmas trees twice.

FIGURE 10.1-2 Fire-Resistant Plants

Type of Plant	Plant Name
Ground covers	Rosea ice plant (*Drosanthemum hispidum*) Jelly beans (*Sedum rubrotinctum*) Trailing gazania (*Gazania uniflora*) Trailing African daisy (*Osteospermum fruticosum*) Kentucky bluegrass (*Poa pratensis*) Rye grass (*Lolium*, several species) White clover (*Trifolium repens*)
Shrubs	Oleander (*Nerium oleander*) Elephant bush (*Portulacaria afra 'variegata'*) Bluechalksticks (*Senecio serpens*) Toyan (*Heteromeles arbutifolia*) Common lilac (*Syringa vulgaris*)
Trees	Gum trees (*Eucalyptus*, several species) Carob (*Ceratonia siliqua*) California pepper (*Schinus molle*) California laurel (*Umbellularia californica*) Cottonwood (*Populus deltoides*)

they had had smoke and fire alarms and a ready means of escape.

People who die in large buildings usually succumb to smoke and gases rather than the fire itself. These deaths are due to lack of sufficient warning and inadequate means of escape. Current laws have reduced these causes of death, but many older buildings are still inadequately protected.

Fire has four distinct stages: (1) incipient, (2) smoldering, (3) flame, and (4) high heat.

10.1.2.1 Smoke and Fire Alarm Systems

Long before flames actually break out, occupants may be killed or rendered unconscious by smoke, noxious gases, or superheated air. Smoke and heat detection, therefore, are an essential part of any fire alarm system. From the time flames break out, a person may have less than 4 minutes in which to escape.

Smoke signals the presence of fire far sooner than heat; therefore, in most cases, a device sensitive to smoke is more important than one that is sensitive to heat. Smoke detectors are typically of the ionization or photoelectric type. Heat detectors generally have a metal strip that is either melted or distorted by heat to trip the alarm.

In a residential installation, cost should not be a problem in smoke detector selection. Battery-powered ionization detectors cost as little as $10.00; battery-powered photoelectric types may cost a little more. Batteries must be replaced approximately once a year. Many insurance companies provide rate discounts to residents who install the more expensive, permanently wired, combined fire and security protection systems.

Wired-in systems are more expensive but are required by most codes for buildings other than single-family residences.

IONIZATION DETECTORS

In an ionization detector a tiny amount of radioactive material causes the air between two electrodes to conduct an electric current (Fig. 10.1-3). When smoke particles are present, current flow is reduced and the alarm is set off.

The radioactive material in these detectors should not be of concern to building occupants. According to the U.S. Nuclear Regulatory Commission, holding an ionization smoke detector close to the body for 8 hours a day for an entire year would expose a person to 1/10 as much radiation as would one transcontinental round-trip airline flight.

As compared with photoelectric detectors, ionization detectors respond more rapidly to the small, wispy smoke particles of a fast-flaming fire but react more slowly to the large smoke particles of a slow-smoldering fire. The average ionization detector is likely to give 20 to 30 seconds extra notice of a fast fire.

PHOTOELECTRIC DETECTORS

A photoelectric detector aims a light beam into a chamber that contains a light-sensitive photocell. The cell detects the light only when smoke enters the chamber and scatters the beam. When enough light reaches the cell, the alarm goes off (Fig. 10.1-4).

A slow-smoldering fire, as would occur in the case of a cigarette dropped on bedding or upholstered furniture, is the most common type of home fire. A photoelectric detector is likely to give 20 to 25 minutes quicker notice of a slow-smoldering fire than an ionization detector, but is less than a minute slower in detecting a fast fire. For these reasons, many experts prefer the photoelectric type of detector.

HEAT DETECTOR

Some manufacturers offer a heat-sensing device as an optional part of their smoke

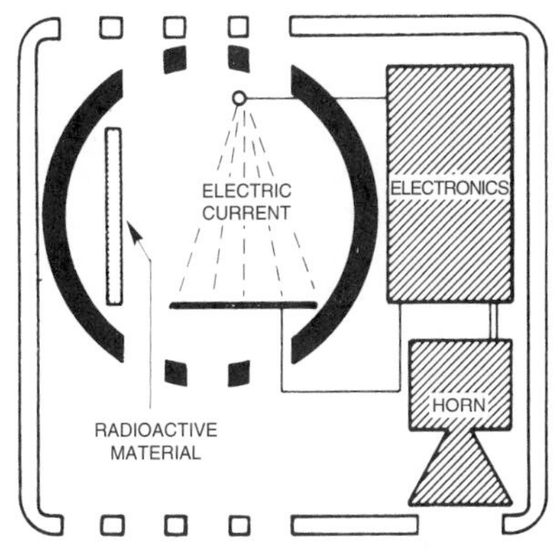

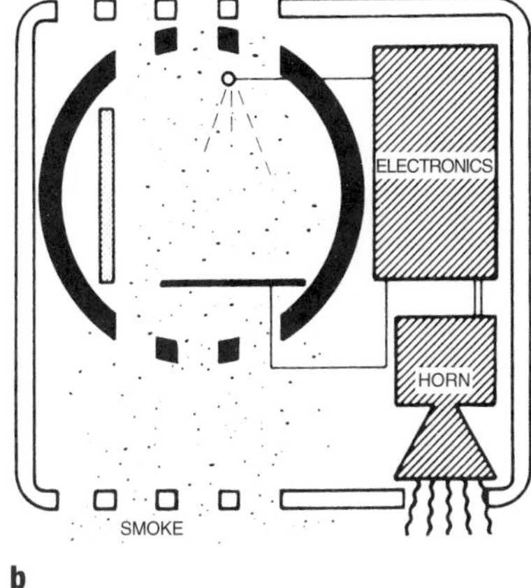

FIGURE 10.1-3 In an ionization detector: (a) a piece of radioactive material "ionizes" the air to cause an electric current to flow across an air gap; (b) smoke reduces the current flow, tripping the alarm.

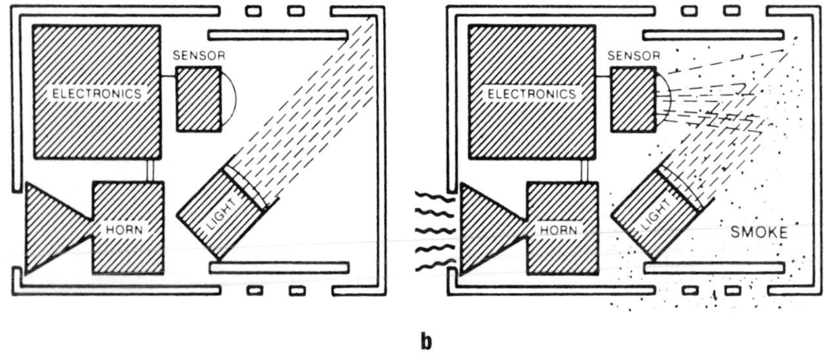

FIGURE 10.1-4 In a photoelectric detector, (a) the sensor does not ordinarily see the light; (b) when smoke is present light rays hit the sensor and sound the alarm.

detectors or as a separate product. Most heat detectors use a metal strip that is either melted or distorted by excessive heat.

Heat detectors may be of either the fixed-temperature or rate-of-rise type. Heat detectors built into smoke detectors are usually of the fixed-temperature type, which trips the alarm when a certain temperature is exceeded. Similar in appearance, separate detector devices may also be of the fixed-temperature or rate-of-rise type; they sound their own alarms or send electrical signals to a central alarm system.

To be effective, heat detectors must be close to the source of a fire. They are especially useful in environments that could fool or disable a smoke detector—such as in a kitchen, where grease particles in the air may cause a smoke sensor to give false alarms. Heat detectors may also be useful in other areas that are unsuitable for smoke detectors, such as furnace rooms, attics, and garages.

However, a heat sensor should not be the only fire protection in a building. It is more often smoke than heat that causes injury and death. A heat detector may ignore a smoldering fire that is emitting lethal amounts of smoke, carbon monoxide, and other toxic gases.

DETECTOR PLACEMENT

The National Fire Protection Association's (NFPA) *Life Safety Code* recommends the installation of at least one smoke detector in each dwelling unit. In some jurisdictions, such a detector is mandated by law. The detector should be powered by the dwelling's electric service and placed in the hallway near bedrooms. In multilevel dwellings, the NFPA recommends that a detector be placed at the top of the stairs. The Council of American Building Officials' (CABO) *One- and Two-Family Dwelling Code* also requires a hard-wired smoke detection system in new residential buildings.

Codes dictate detector types and locations in commercial, industrial, institutional, and multifamily residential buildings.

Many electrically powered smoke detectors are part of security systems wired to a central station. These are required by code in most commercial, institutional, industrial, and multifamily residential buildings. Although such systems are expensive to install in single-family residences, some insurance companies offer discounts of 2.5% to 5% or more on the dwelling's fire insurance rate when these detectors have been installed. When central systems are used in a residence, the alarm bell and test button should be installed in the master bedroom or another suitable location in the sleeping area. In other building types, codes dictate the location of alarm bells and may require that a signal be forwarded electronically to a fire station or alarm-monitoring service. Because of the possibility of failure due to causes originating in another system, the operating power supply should be from a separate and independent transformer, rather than from the entrance signal or other multiple-use transformer.

Model codes and some state and municipal building codes require smoke detectors in residential bedroom hallways. Many codes permit the less expensive, battery-powered models. Some communities require photoelectric detectors. Residents should check local codes before buying and installing smoke detectors.

The Consumer's Union suggests that both photoelectric and ionization detector types may be useful. Although some fire alarm systems combine ionization and photoelectric capacity in one unit, two separate detectors may be less expensive and offer better protection.

For a basic two-detector system in a residence, an ionization detector should be placed in the hallway just outside the bedroom doors and a photoelectric detector should be located in the general living area. The photoelectric detector should not be located where cooking might set it off accidentally. In a larger house, a heat detector may be appropriate in the kitchen. In a two-floor house, a detector should also be placed at the top of the stairs.

Ideally, a smoke detector should be mounted in the middle of a ceiling. If this conspicuous location is objectionable, it may be mounted on a wall between 6 and 12 in. (150 and 300 mm) below the ceiling or on the ceiling not closer than 6 in. (150 mm) to the wall or the corner. Air currents sometimes do not reach the corners of rooms, thus possibly delaying smoke detection.

A smoke detector should not be mounted in front of an air supply duct or near a furnace cold-air return, because excessive airflow prevents prompt detection of smoke. Such locations, however, are acceptable for heat detectors.

A smoke detector should not be located on a ceiling or wall that is substantially warmer or colder than the rest of the room. In this case, an invisible thermal barrier near the wall or ceiling surface can keep smoke from reaching the detector. This thermal barrier may be a problem in poorly insulated buildings and in mobile homes.

When a detector must be mounted under any of the adverse conditions discussed above, a heat detector rather than a smoke detector should be used. A heat detector is preferred in such instances because it is activated by temperature change rather than smoke delivered by airflow.

For added protection a dwelling may be equipped with extra detectors. A photoelectric or heat detector can be placed in the basement near a furnace or boiler. Extra detectors may also be placed in storage areas. A heavy smoker may want an extra photoelectric detector in the bedroom. Both an ionization and a photoelectric smoke detector may be placed on each level of a multilevel house.

Detectors operate on either batteries or electric current. The installation of detectors that use batteries does not require professional training. Those that use electricity have to be wired into the building's electrical system, which requires knowledge of such installations. Unless provided with backup batteries or a separate power source, a wired detector system will become inoperable if there is a power failure. Smoke detectors with batteries should be models that beep when a battery becomes weak.

10.1.2.2 Escape from Fires

Building codes mandate the number and size of exits required. All public spaces and most other spaces require at least two means of egress. Fire escapes are not usually required in one- and two-family dwellings.

10.1.3 FIRE CONTAINMENT AND CONTROL

Fire containment and control measures are dictated by building codes. They include fire walls and fire and smoke doors to limit the spread of fire until inhabitants can escape.

10.1.4 FIRE EXTINGUISHMENT

Fire sprinkler systems are discussed in Section 15.3. Portable fire extinguishers are discussed in Section 10.2.

10.2 Fire Protection Specialties

Fire extinguishing specialties include both the portable fire extinguishers and their housings discussed in this section, and fixed chemical fire extinguishing equipment and systems, which are beyond the scope of this book. Fire sprinkler systems are covered in Chapter 15.

10.2.1 EXTINGUISHING FIRES

Most fires start out small. Often a handheld extinguisher or a sprinkler system can keep them small or put them out entirely.

Every residential kitchen should be equipped with a fire extinguisher. An extinguisher should also be put in a garage, a basement, near bedrooms, and any other place a fire hazard exists. Many insurance companies offer home owners insurance discounts when fire extinguishers are present.

Most codes require portable fire extinguishers in other types of buildings as well, and stipulate their locations. Extinguisher units are usually required in corridors, storage areas, mechanical and electrical equipment rooms, and in other areas where a fire hazard may exist, such as kitchens, laboratories, and manufacturing facilities.

10.2.2 FIRE EXTINGUISHERS

Fire extinguishers smother a fire by depriving it of oxygen. They are made with a variety of fire-fighting agents, including:

- A dry chemical (primarily fluidized and siliconized mono ammonium phosphate, and siliconized sodium bicarbonate)
- Pressurized water
- Liquid carbon dioxide under pressure
- Sodium chloride
- Potassium acetate
- Halon replacement chemicals (Halotron 1, DuPont FE-36, and others)

Each of the extinguisher types described above is effective against one or more classes of fires. Fires can be classed as follows:

A—fire fed by a solid fuel, such as wood, paper, textiles, plastic, rubber, or other nonliquid combustibles
B—fire fed by flammable liquids such as grease, cooking oils, gasoline, and paint solvents
C—fire in a live electrical circuit
D—fire in combustible metals, such as magnesium
K—fire in liquid cooking media

Figure 10.2-1 shows the types of extinguishers required for each class of fire.

For complete protection, it may be necessary to have more than one type of extinguisher. An extinguisher used against a fire for which it is not labeled can actually increase the intensity of the fire. Extinguishers should be rated by the Underwriters Laboratories (UL) and marked with its classification. The markings for extinguishers rated for Classes A and B fires include a letter designating the class of fire for which the extinguisher is effective and a number indicating its relative effectiveness. Extinguishers for use in Class C or D fires are not required to have the number designation. Codes usually specify the class of extinguisher required. Selection is then a matter of comparing the code requirements with the manufacturer's listing. In its NFPA 10, "Portable Fire Extinguishers," the National Fire Protection Association spells out detailed requirements for extinguisher selection.

Fire extinguishers should be located where required by code. There should be one near each doorway and as far as permitted by code from the likely source of a fire. NFPA guidelines require that the top of a fire extinguisher be no more than 5 ft. (1524 mm) above the floor, but within that limit, a fire extinguisher's mounting height is dictated by Americans with Disabilities Act (ADA) requirements and depends on the reaching distance to the extinguisher. For example, if a fire extinguisher handle is 48 in. (1219.2 mm) above the floor, the maximum side reach is 54 in. (1371.6 mm).

Extinguishers should be checked once a month to ensure that they are fully charged. Once an extinguisher has been used, no matter how briefly, it must be recharged. Otherwise, the material left within the discharge valve will gradually allow the pressure to bleed off.

FIGURE 10.2-1 Types of Extinguishers Used for Each Class of Fire

Extinguisher Type	Fire Class				
	A	B	C	D	K
Dry Chemical	X	X	X		
Carbon Dioxide	X				
Sodium Chloride				X	
Potassium Acetate					X
Halon Replacement	X	X	X		

10.2.3 FIRE EXTINGUISHER CABINETS

Fire extinguishers may be mounted on wall brackets, but in most visible loca-

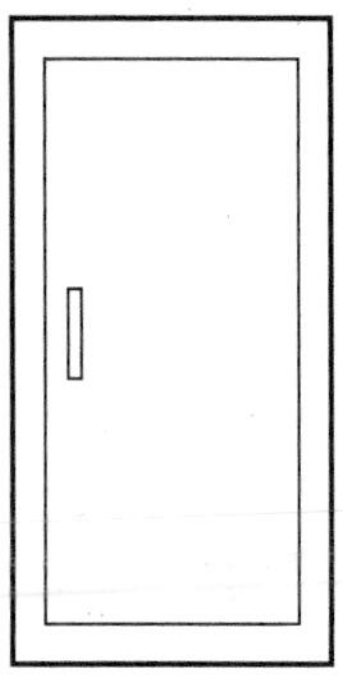

Solid Doors

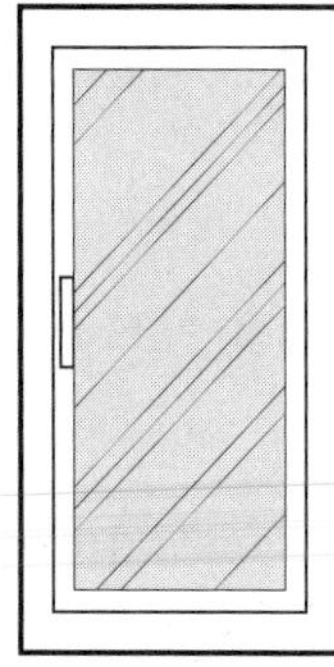

Full Glass Doors

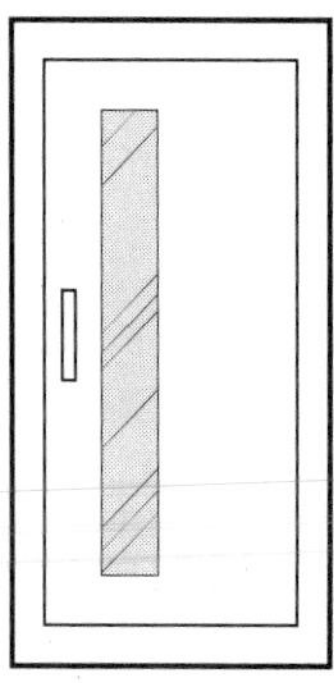

Glass Slot Doors

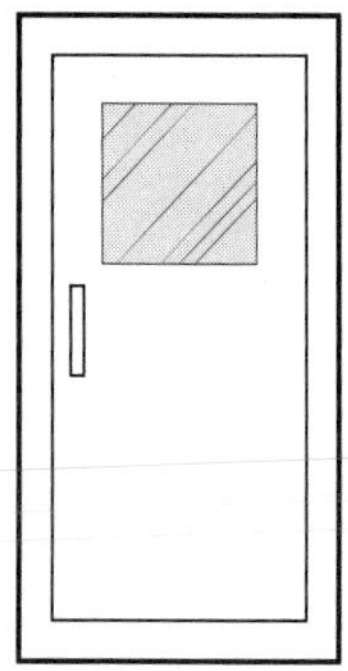

Glass Panel Doors

FIGURE 10.2-2 Typical cabinet front designs.

tions aesthetic considerations dictate enclosing them in cabinets. Fire extinguisher cabinets are made from steel, stainless steel, aluminum, and brass in several types and styles. Basic door styles are shown in Figure 10.2-2. Combination units are also available that hold both extinguishers and fire hoses, or both extinguishers and valves.

Most fire extinguisher cabinets are made of a single wall thickness of steel or aluminum. Insulated fire-rated units are also available, however, for those locations where they are required.

Cabinet sizes are dictated by the fire extinguishers they will hold.

Cabinets may be surface mounted, fully recessed, or partially recessed. The amount they may project from a wall, including their handles, in a space controlled by ADA Accessibility Guidelines is 4 in. (101.6 mm) when their leading edge is between 27 and 80 in. (685.8 and 2032 mm) above the floor.

The ADA guidelines do not require fire extinguisher handles or signage to be in compliance with the guidelines for other doors and signs.

10.3 Additional Reading

More information about the subjects discussed in this chapter can be found in the references listed in Section 10.4 and in the following publications.

Council on Tall Buildings and Urban Habitat. 1992. *Fire Safety in Tall Buildings*. New York: McGraw-Hill.

McGowan, Maryrose. 1993. "Specifying Portable Fire Extinguishers." *The Construction Specifier* (January):25.

Shrive, Charles A. 1993. "Writing Fire Alarm Specifications." *The Construction Specifier* (July):37.

10.4 Acknowledgments and References

ACKNOWLEDGMENTS

We gratefully acknowledge the assistance of the following organizations and individuals in preparing this chapter. We are also indebted to them for permission to use their illustrations when requested and for the use of their publications as references.

American National Standards Institute (ANSI)
Consumer Product Safety Commission (CPSC)
Intermatic Incorporated
J.L. Industries
Larsen's Manufacturing Company
National Fire Protection Association (NFPA)
U.S. Department of Agriculture (USDA)
U.S. Department of Housing and Urban Development (HUD)

REFERENCES

We would also like to thank the authors and publishers of the publications in the following list for their contribution to our research for this chapter.

American Society for Testing and Materials Standard E 814, "Standard Test Method for Fire Tests of Through-Penetration Fire Stops." West Conshohocken, PA: ASTM.

Arcom Master Systems. MASTERSPEC®. Basic Section 10520, "Fire-Protection Specialties." 2/97 ed. Salt Lake City, UT: Arcom.

Bukowski, Richard W. 1991. "Improving the Fire Performance of Building Contents." *The Construction Specifier* 44(2) (February):42–46.

Council of American Building Officials (CABO). 1983. *One- and Two-Family Dwelling Code*. Falls Church, VA: CABO. (In the process of being transferred to the control of the International Code Council (ICC).)

National Fire Protection Association (NFPA). 1995. NFPA 550, "Guide to the Firesafety Concepts Tree." Quincy, MA: NFPA.

———. 1997. NFPA 101, "Life Safety Code." Quincy, MA: NFPA.

———. 1998. NFPA 10, "Standard for Portable Fire Extinguishers." Quincy, MA: NFPA.

Raeber, John A. 1991. "Redundancy in Fire Protection." *The Construction Specifier* 44(2)(February):17–19.

Reid, Ed, and Bob Schifiliti. 1991. "A Balanced Approach to Fire Protection Detection and Suppression Systems." *The Construction Specifier* 44(2)(February):30–41.

Underwriters Laboratories, Inc. UL 8, "UL Standard for Safety for Foam Fire Extinguishers." Northbrook, IL: Underwriters Laboratories, Inc.

———. UL 154, "Standard for Safety for Carbon Dioxide Fire Extinguishers." Northbrook, IL: Underwriters Laboratories, Inc.

———. UL 299, "UL Standard for Safety for Dry Chemical Fire Extinguishers." Northbrook, IL: Underwriters Laboratories, Inc.

———. UL 626, "Standard for Safety for 2½-Gallon Stored-Pressure Water-Type Fire Extinguishers." Northbrook, IL: Underwriters Laboratories, Inc.

———. UL 711, "Standard for Safety for Fire Extinguishers, Rating and Fire Testing of." Northbrook, IL: Underwriters Laboratories, Inc.

CHAPTER

11

Equipment

Introduction

Applicable *MasterFormat*™ Sections

Residential Appliances

Unit Kitchens

Additional Reading

Acknowledgments and References

Introduction

Equipment, as discussed in this chapter, includes devices and materials used for many purposes:

- Building maintenance equipment, including vacuum cleaning systems, window washing equipment, and floor and wall cleaning equipment
- Bank equipment, including vault doors and gates, security and emergency equipment, safes, safe deposit boxes, teller windows and associated equipment, package transfer units, and automatic banking systems
- Ecclesiastical equipment, including baptistries and chancel fixtures
- Library equipment, including book stacks, carrels, theft protection equipment, and book depository equipment
- Theater and stage equipment, including curtains, rigging, and acoustical shells
- Organs, carillons, and bells
- Hotel registration equipment
- Checkroom equipment
- Mercantile equipment, including barber and beauty shop equipment, cash registers, and display cases
- Commercial laundry and dry cleaning equipment, including washers, dry cleaning equipment, extractors, drying and finishing equipment
- Vending equipment, including vending machines and money changers
- Audiovisual equipment, including projectors and screens
- Vehicle service equipment, including lifts and car washing stations
- Parking control equipment, including gates, ticket dispensers, and control units
- Loading dock equipment, including dock levelers, lifts, ramps, bridges, and bumpers
- Solid waste handling equipment, including incinerators, compactors, chutes, and pneumatic waste systems
- Detention equipment
- Water supply and treatment equipment
- Hydraulic gates and valves
- Fluid waste treatment and disposal equipment
- Food service equipment, including food storage, preparation, and dispensing equipment and cleaning and disposal equipment, and bar and soda fountain equipment
- Residential appliances
- Unit kitchens
- Darkroom equipment, including transfer cabinets, film processing equipment, and revolving darkroom doors
- Athletic, recreational, and therapeutic equipment, including scoreboards, backstops, bowling alleys, shooting ranges, and gymnasium, exercise, and therapy equipment
- Industrial equipment for manufacturing, processing, and assembly plants
- Laboratory equipment for scientific purposes, including cabinets, fume hoods, incubators, sterilizers, refrigerators, and safety devices
- Planetarium equipment
- Observatory equipment, including telescopes
- Office equipment
- Medical, dental, and optical equipment, including sterilizers, and examination, treatment, and patient care equipment for the treatment of humans or animals
- Mortuary equipment, including refrigerators, crematorium equipment, and lifts
- Navigation equipment for airports and waterways
- Agricultural equipment
- Exhibit booths and display panels

Many of these types of equipment would require a book of their own to demonstrate them adequately. They are, therefore, beyond the scope of this book. Two of them, however—residential appliances and unit kitchens—which architects encounter frequently, are covered here.

Applicable *MasterFormat*™ Sections

The following *MasterFormat* Level 2 sections are applicable to this chapter.

11450 Residential Equipment
11460 Unit Kitchens

11.1 Residential Appliances

This section contains data related to *residential appliances*, which is the industry term for kitchen and laundry appliances. The appliances discussed in this section include ranges, cooktops, ovens, refrigerator/freezers, freezers, dishwashers, clothes washers, clothes dryers, and room air conditioners. Plumbing for this equipment is covered in Chapter 15; electrical supply and connections in Chapter 16. Central heating and air conditioning systems are addressed in Section 15.2.

The general requirements in this section are also applicable to other residential equipment, such as food waste disposal units, trash compactors, and instant hot water dispensers.

11.1.1 ENERGY CONSERVATION

Standards for major home appliances are included in the National Appliance and Conservation Act of 1987 (NAECA), which authorizes the Department of Energy (DOE) to set energy efficiency standards for major home appliances according to a statutory time schedule. DOE regulations issued under NAECA apply to refrigerator/freezers, clothes dryers, clothes washers, dishwashers, range/ovens, microwave ovens, and room air conditioners.

NAECA amends the Federal Energy Policy and Conservation Act. In turn, NAECA has been amended by the National Appliance Energy Conservation

Amendments of 1988 and by the Energy Policy Act of 1992.

Federal law requires that appliances have attached to them an Energy Guide card. For each room air conditioning unit, the card should show its Energy Efficiency Rating (EER) and an indication of the ratings for the least efficient and most efficient models of similar units. The law further requires minimum ratios, ranging from 8 to 10, depending on the type of unit. An EER of 11 is generally considered the lowest a consumer should accept; some sources say that 12 should be the minimum. The higher the EER, the more energy efficient the unit. For other appliances, the Energy Guide card should show the estimated yearly operating cost for the particular unit and the same data for the highest-cost and lowest-cost similar units. These cards should also show the estimated yearly operating cost for the particular unit for various energy costs.

11.1.2 USE BY HANDICAPPED PERSONS

The Americans with Disabilities Act (ADA) and ANSI A117.1 of the American National Standards Institute (ANSI) (see Section 1.7) impose certain restrictions on residential appliances. The requirements in this section (11.1) are supplementary to those addressed in Section 1.7.

Many wheelchair users have problems in using appliances designed for standing rather than seated users. An ideal cooking arrangement for a wheelchair user consists of the following:

1. A lowered, front-controlled cooktop
2. A lowered sink
3. A lowered work counter at least 30 in. (760 mm) wide
4. An accessible oven, installed as described in the following section

11.1.2.1 Cooktops and Ranges

More accidents are related to *cooktops* and *ranges* than to any other kitchen appliance. To prevent such injury, cooktop and range controls should be at the front of the appliance so that a cook does not have to reach over hot burners to adjust them. People with limited arm coordination must drag rather than lift pots and pans. A cooktop, if separate from the oven, should be adjustable, with the same knee space and height dimensions as sinks and work counters. The knee space should be finished like the rest of the floor. To protect a user from injury, the underside of a cooktop should be smooth, insulated, and free of sharp projections. An electric cooktop should be equipped with a light to indicate that a burner is on.

When residents include small children, controls should be out of sight on a horizontal surface and should either have a protective lid or require an extra movement, such as pushing, before a control can be turned to activate a burner.

Ovens should have side-hinged doors with a pull-out shelf beneath and should be self-cleaning. Ovens without a self-cleaning feature, and the latch side of those that are side-hinged, should be located next to a lowered counter area with accessible knee space. Side-hinged ovens should have a heatproof pull-out shelf below the oven, running the full width of the oven; the depth of the shelf should be not less than 10 in. (255 mm).

11.1.2.2 Refrigerators and Freezers

Many refrigerators with freezers (side-by-side and some over-under styles) are accessible for wheelchair users. *Over-under* refrigerators should have it least 50% of the freezer, most of the refrigerator space, and the controls at a height lower than 54 in. (1370 mm). Freezers that have any of their storage volume above the 54-in. (1370 mm) reach limit should be self-defrosting.

11.1.2.3 Dishwashers

Dishwashers should open in the front, and the racks should be within easy reach. Pull-out racks are the easiest to use.

11.1.2.4 Washers and Dryers

Washers and dryers should be the front-loading type and should be hinged on the side rather than on the bottom.

11.1.3 STANDARDS

Residential equipment should be listed by the Underwriters Laboratories (UL) and should bear UL labels.

Gas-burning residential equipment should bear the American Gas Association's (AGA) Seal of Approval and should comply with the American National Standards Institute's (ANSI) Series Z21 standards.

Appliances that bear the applicable seal of the Association of Home Appliance Manufacturers (AHAM) comply with AHAM standards. AHAM standards cover such items as shelf area ratings for refrigerators and freezers, refrigeration capacity, and amperage ratings. Refer to Section 11.4 for a listing of some AHAM standards. Note, however, that some manufacturers do not belong to AHAM. This does not mean that their products are inferior or would not meet AHAM standards were they eligible for that distinction.

Each appliance should bear the manufacturer's standard warranty. This warranty should cover all factory defects for not less than 1 year. Some states require longer or more complete warranties.

11.1.4 SPECIFIC APPLIANCES

Appliances are available with such a wide range of features and options that it is impossible to list them all here. In addition, the available features and options change regularly as competition increases and technological advances occur. Therefore, the features included here should be considered basic and not fully representative of the many product features available. Appliances should be selected from up-to-date manufacturers' catalogs.

In selecting appliances, great care should be taken to determine the features available and the energy efficiency of the particular unit under consideration, and to compare the units of several reputable manufacturers. Although there are few generally accepted industry-wide standards for residential appliances, units produced by most major manufacturers are usually, but not always, comparable. In addition, some lesser-known manufacturers produce products that compare favorably with those of the major companies, but greater care is advisable in considering their products. While not a guarantee of quality, a manufacturer's reputation can become very important when something goes wrong. Those with good reputations keep those reputations by servicing the products they sell and by upholding warranty agreements.

When practicable, all the appliances in a particular residence or other unit should be obtained from the same manufacturer. This will ensure compatibility of color and design and make service contracts less complicated and, often, less costly. Many manufacturers and vendors discount service contracts when more than one appliance is insured.

11.1.4.1 Ranges

Ranges may be either gas or electric units, and either freestanding (Fig. 11.1-1a) or

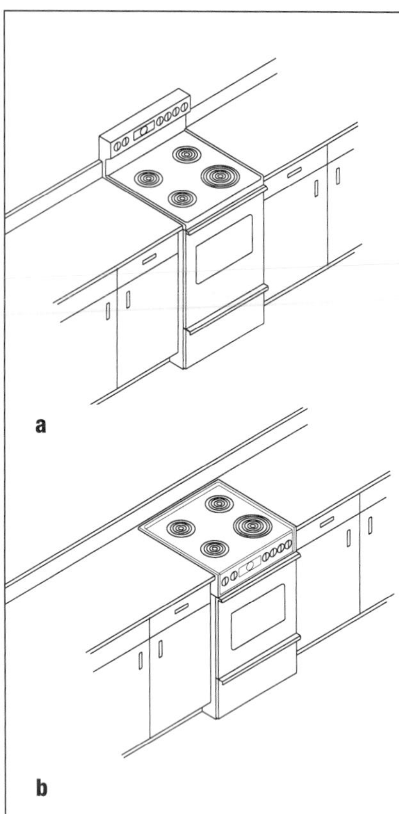

FIGURE 11.1-1 Ranges: (a) freestanding type, (b) drop-in type. (Drawing by HLS.)

drop-in types (Fig. 11.1-1b). Typical units have four plug-in burners, a self-cleaning oven, chrome oven racks, a timer, a clock, and a storage drawer below the oven. Some have glass panels in the oven door, automatic rotisseries, and other features. Some units have built-in microwave ovens. Some gas units have pilot-free electronic ignition systems.

11.1.4.2 Cooktops

Cooktops may be either gas or electric units. Typical units have four plug-in heating elements (Fig. 11.1-2). Some are available with built-in griddles or grills; others have flush heating elements or either glass or ceramic tops. Some gas units have pilot-free electronic ignition systems.

11.1.4.3 Wall Ovens

Wall ovens (see Fig. 11.1-2) may be either gas or electric units. They have the same features generally as the oven portion of freestanding and drop-in ranges.

11.1.4.4 Exhaust Hoods

Exhaust hoods are available in a variety of types. Figure 11.1-2 shows a standard undercounter unit. These come in 24-in. (610 mm), 30-in. (762 mm), and 36-in. (889 mm) widths. Special units include 36-in. (889 mm)-wide by 30-in. (762 mm)-deep, 24-in. (610 mm)-tall island hoods and 30-in. (762 mm)-wide, 24-in. (610 mm)-high wall-mounted hoods.

11.1.4.5 Microwave Ovens

Microwave ovens (Fig. 11.1-3) are available as either freestanding countertop units or undercounter-mounted units. Most have electronic digital controls, but a few have manual dial controls. Wattages vary from 500 to 800 or even more. Units are available with rotisseries, convection cooking, convection broiling, turntables, and recirculating charcoal filters.

11.1.4.6 Refrigerator/Freezers

Refrigerator/freezers are combination units housing a freezer compartment and a refrigerator compartment. They are available as side-by-side types (Fig. 11.1-4a) and top-freezer types (Fig. 11.1-4b).

Units may be either manual-defrost type or automatic-defrost type. Units are also available with bottom-mounted freezers. Some have dispensers on the door for water, cubed ice, and crushed ice. Some have separate small doors to hold high-use items, such as milk and soft drinks. Some manual-defrost units have a single door. Automatic-defrost types typically have separate doors for the freezer and refrigerator compartments.

11.1.4.7 Freezers

Freezers are available as either upright type (Fig. 11.1-5a) or chest type (Fig. 11.1-5b). Capacities range from 4 cu. ft. (0.11 m^3) to 26 cu. ft. (0.74 m^3). Both manual- and automatic-defrost models are available.

11.1.4.8 Dishwashers

Dishwashers are either built-in (Fig. 11.1-6a), freestanding (Fig. 11.1-6b), or convertible (can be used as either an undercounter or a freestanding unit). Standard features include at least two wash cycles, two wash levels, and energy-saving drying cycles. Standard units have porcelain enamel interior liners and insulated cabinets. Racks should extend fully outside the cabinet for loading. Optional features include a forced-air drying cycle, a built-in auxiliary hot water heater, addi-

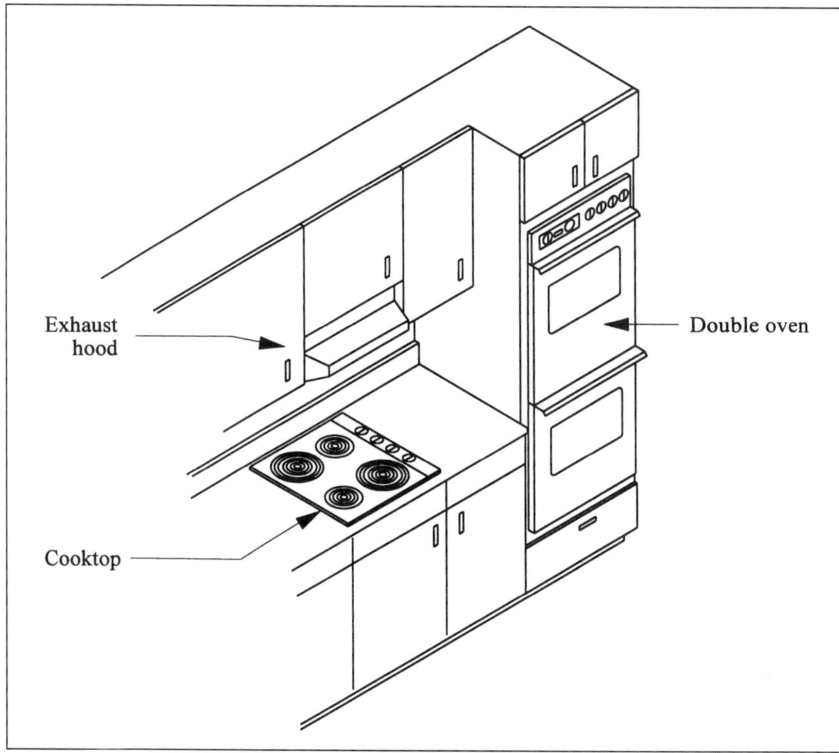

FIGURE 11.1-2 Cooktop, double wall oven, and exhaust hood. Single ovens are also available. Some manufacturers make a unit with a conventional oven and a microwave oven as a unit. Exhaust hoods may be either ductless or with a duct connection to the outdoors. (Drawing by HLS.)

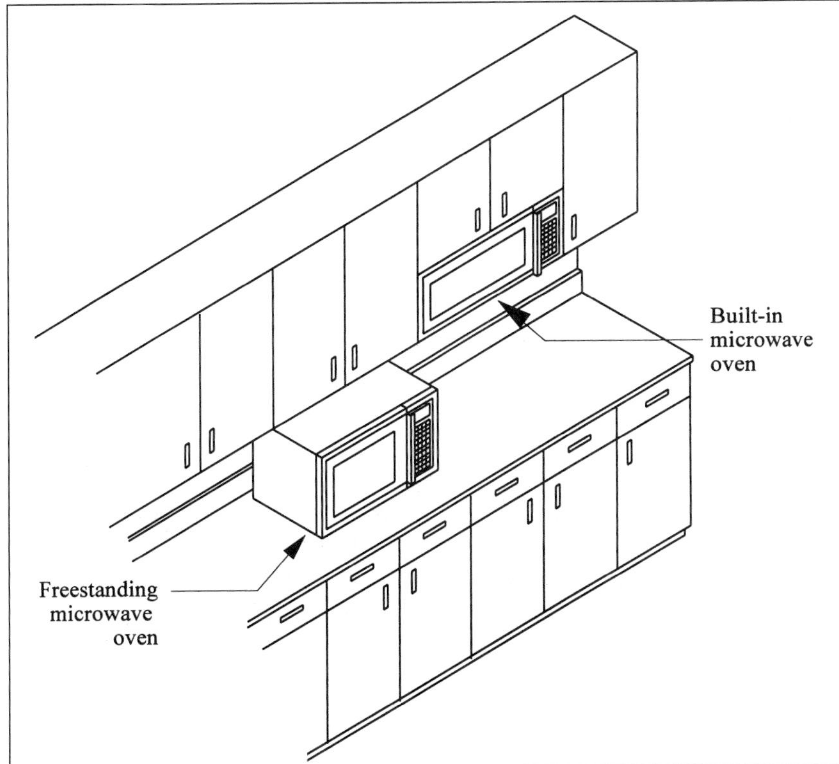

FIGURE 11.1-3 Microwave ovens may be either freestanding or built in. Some can be hung beneath cabinets; others may be wall mounted. Some built-ins contain an exhaust system, which may be either ducted or ductless. (Drawing by HLS.)

tional wash cycles, a built-in soft-food disposer, a plastic liner, delay-start operation, and an automatic rinsing aid dispenser.

11.1.4.9 Clothes Washers

Most clothes washers today are of the automatic top-loading type (Fig. 11.1-7). However, units that must be ADA accessible must be front-loading; these are available in undercounter and space-saving stacked units. Most clothes washers have metered fill-level controls, presettable wash and rinse water temperatures, variable speed and cycle fabric selectors, removable lint filters, and bleach and fabric softener dispensers. Some units have a removable tub for washing small and delicate fabric loads.

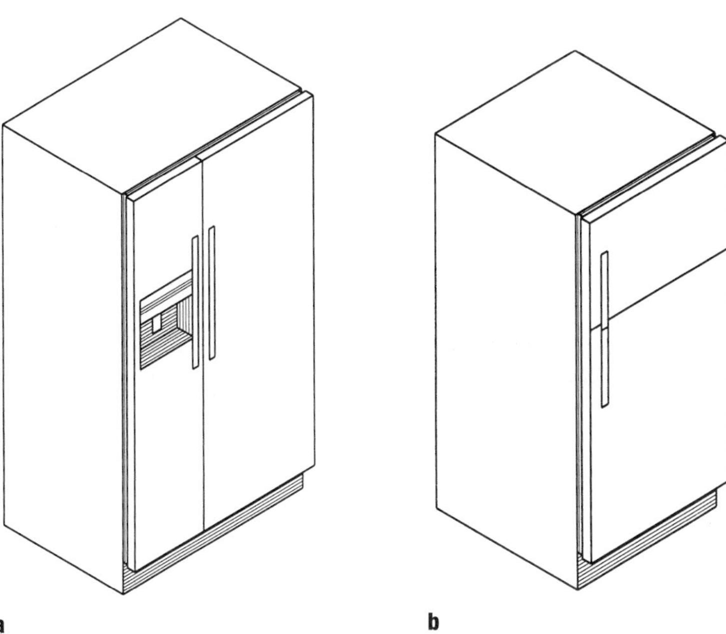

FIGURE 11.1-4 Refrigerators with freezers. The side-by-side type (a) may be as much as 40 in. (1016 mm) wide and slightly taller than the top-freezer type. Some top-freezer types (b) fit into a nominal 36 in. (914 mm) counter opening, but others are narrower. Many combination refrigerator/freezer units are self-defrosting. Upright, single-door refrigerators are also available, with a smaller freezer inside the refrigerator compartment. These are usually not self-defrosting. Single-door, undercounter combination refrigerator/freezer units are also available. (Drawing by HLS.)

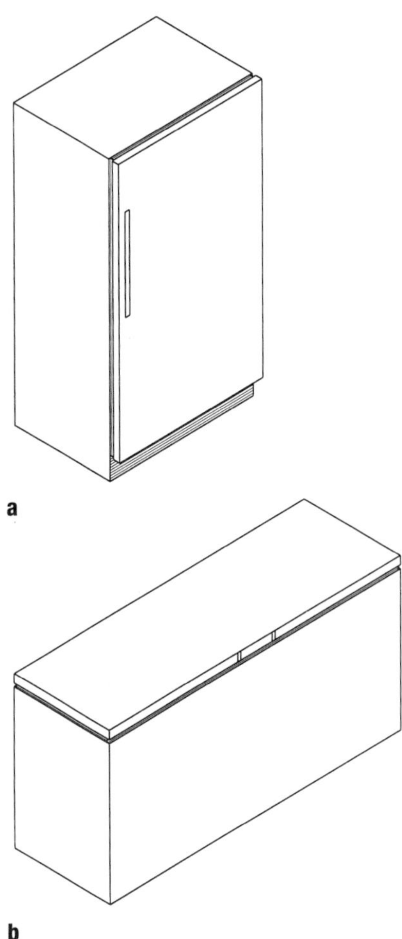

FIGURE 11.1-5 Freezers. The upright type (a) may be 36 in. (914 mm) wide and as much as 72 in. (1830 mm) tall. Most are self-defrosting. Chest-type freezers (b) are a nominal 36 in. (914 mm) high and range from 20 in. (508 mm) to 75 in. (1905 mm) wide. Most chest-type freezers must be manually defrosted. (Drawing by HLS.)

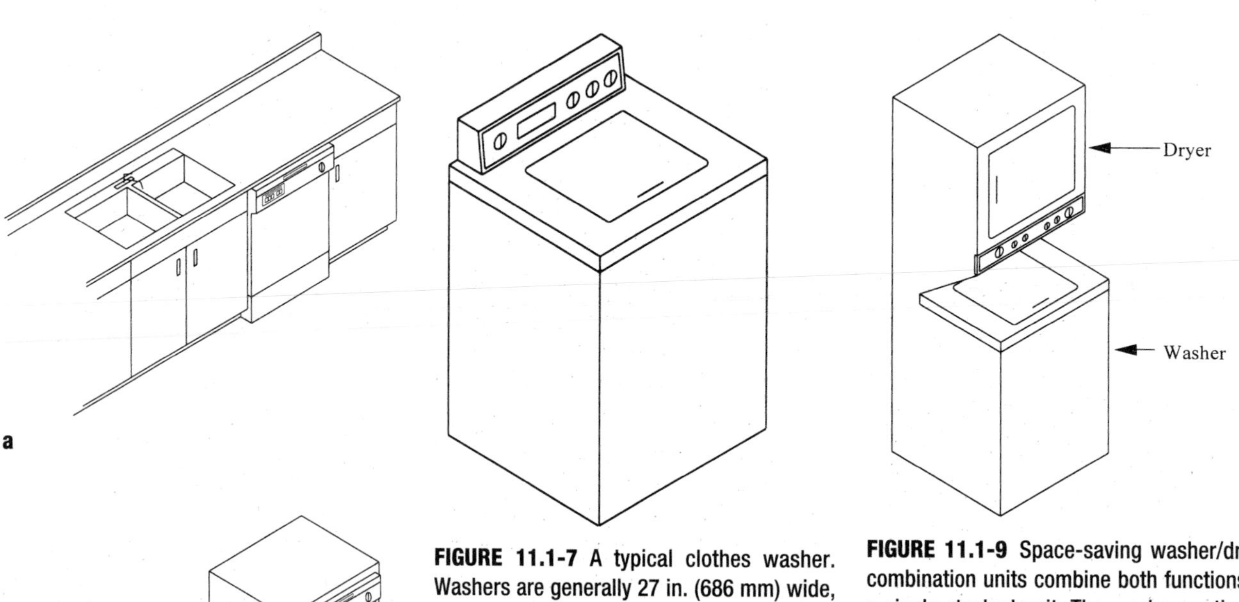

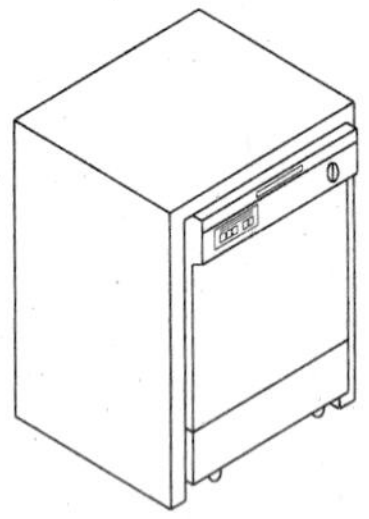

b

FIGURE 11.1-6 Dishwashers. (a) Undercounter (built-in) type, (b) freestanding type. Most built-in units are designed to fit into a 24-in. (610 mm) counter opening. Freestanding units are generally the same size as the built-in type. Some manufacturers make dishwashers that can be either freestanding or rolled into an opening beneath a counter. (Drawing by HLS.)

FIGURE 11.1-7 A typical clothes washer. Washers are generally 27 in. (686 mm) wide, 25 in. (635 mm) deep, and 43 in. (1092 mm) to the top of the control panel. Front-loading undercounter units are also available. (Drawing by HLS.)

FIGURE 11.1-9 Space-saving washer/dryer combination units combine both functions in a single stacked unit. The washer portion is about the same size as a freestanding washer. The overall height is about 66 in. (1676 mm). Units with front-loading washers and large dryers stacked directly on top of the washers are also available. (Drawing by HLS.)

11.1.4.10 Clothes Dryers

Clothes dryers (Fig. 11.1-8) may be either gas or electric units. Standard features include timed cycle selection, a fabric selector, a cycle-end signal, a safety starting control, a removable lint filter, and a no-heat drying cycle. Some gas units have an automatic electric ignition (no pilot light).

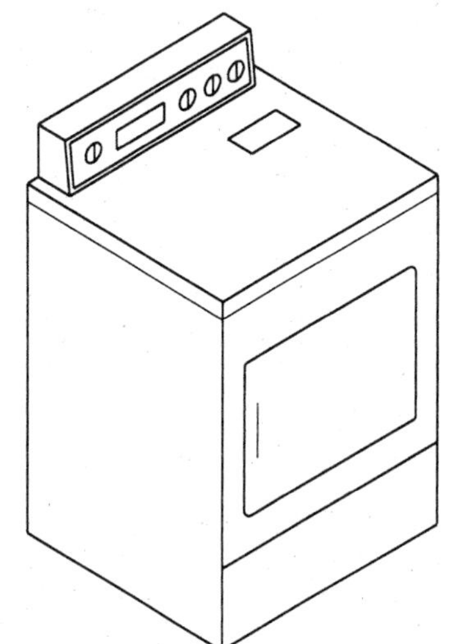

FIGURE 11.1-8 A typical clothes dryer. Dryers are generally 27 in. (686 mm) wide, 27 in. (686 mm) deep, and 43 in. (1092 mm) to the top of the control panel. Undercounter units are also available. (Drawing by HLS.)

11.1.4.11 Space-Saving Washers and Dryers

Washers and dryers are available as sets designed to save space for use in apartments and other small home laundry areas (Fig. 11.1-9). Their features are essentially as described for separate washers and dryers.

11.1.4.12 Air Conditioning Units

Room air conditioners are available in sizes ranging from 4000 Btu (4220.24 kJ) to 23,000 Btu (24,266.4 kJ) or even more. Some have heating coils, and some are actually heat pumps. Most units may be mounted either in a window or through a wall. Large units, units with heating coils, and heat pumps usually require 240-volt service. Most smaller units work on 120-volt standard house power.

11.2 Unit Kitchens

Unit kitchens, sometimes called *kitchenettes*, are self-contained units containing a countertop with cabinets under it, a sink, and wall cabinets (Figs. 11.2-1, 11.2-2, and 11.2-3). Some unit kitchens also contain a range or a burner top and a range hood. Some have an undercounter refrigerator built in. Some are designed for use by people in wheelchairs. Special units provide refrigeration and water along with storage cabinets for use as wet bars. Other appliances sometimes available are ice makers, dishwashers, garbage disposals, and microwave ovens.

Unit kitchens are used in offices, in housing facilities for elderly people, and as nourishment centers in many types of facilities.

Unit kitchens are made of metal or wood; many finishes are available, including natural wood, stainless steel, baked enamel on steel, and plastic laminate.

Unit kitchens that comply with the Americans with Disabilities Act (ADA) are available. These are 34 in. (863.6 mm) high (even above undercounter refrigerators and ovens); handles and controls are designed for easy visual and mechanical operation; and they meet or exceed all "reach" restrictions. Many models can be specified to meet additional state or local accessibility/adaptability requirements as well.

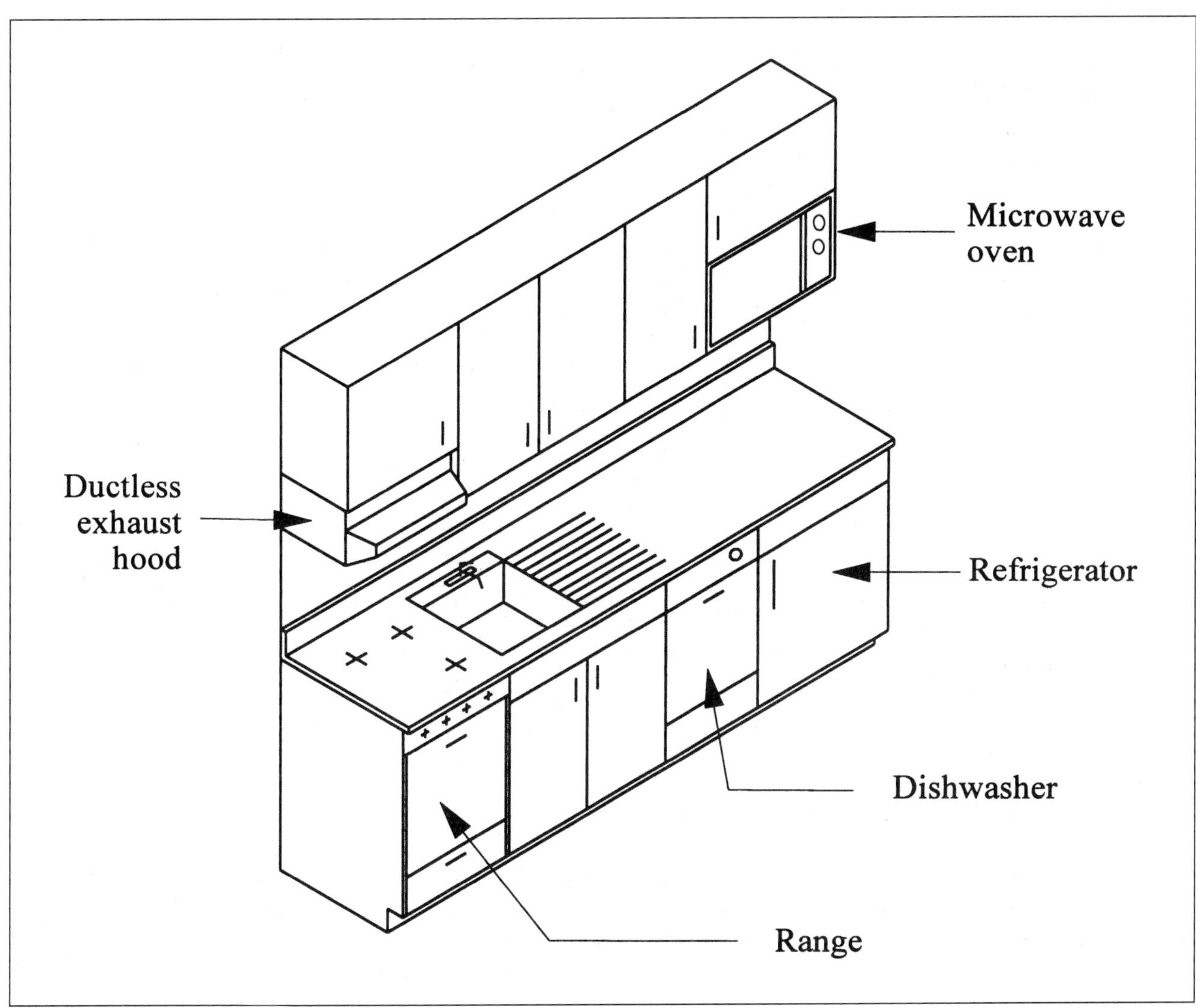

FIGURE 11.2-1 Large unit kitchen containing a sink in a countertop with a drain board, a range containing a conventional oven and surface cooking units, a ductless exhaust hood, a refrigerator, a dishwasher, wall and base cabinets, a continuous backsplash, and a microwave oven. (Drawing by HLS.)

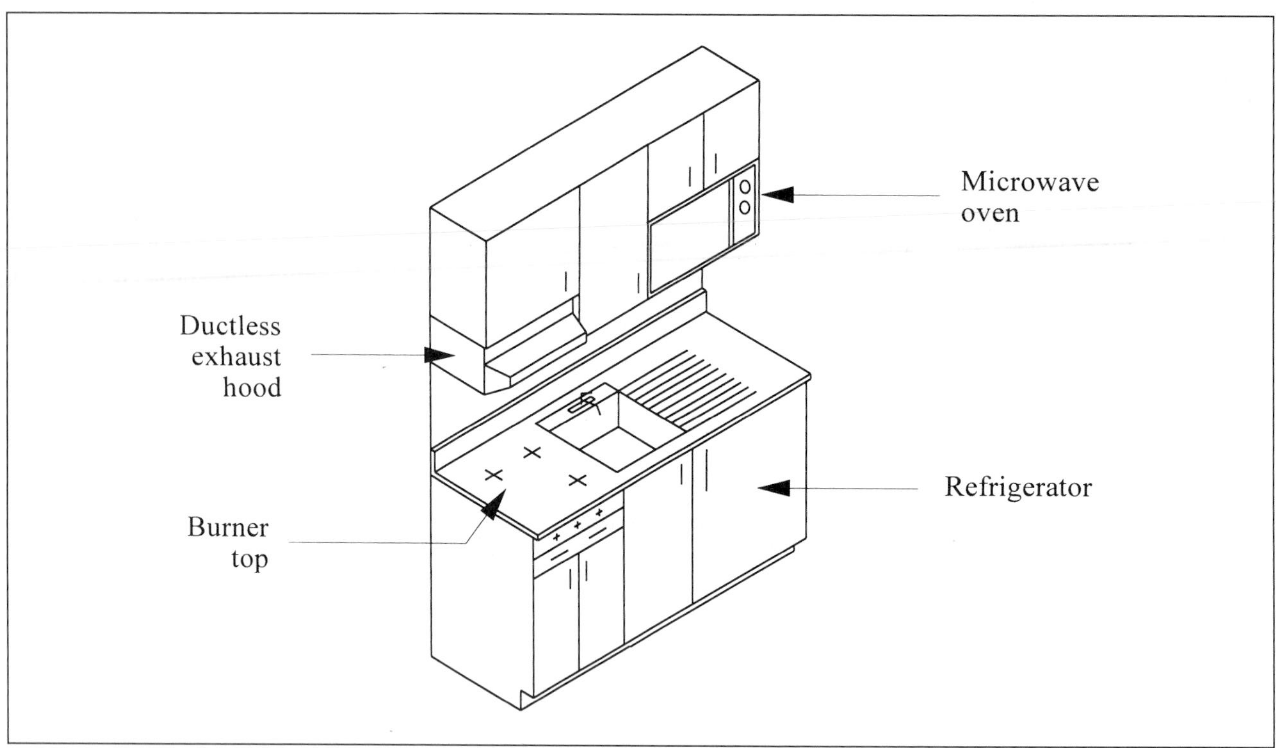

FIGURE 11.2-2 A smaller unit kitchen of a type generally used in offices, conference rooms, and similar spaces. It has a sink in a countertop with a drain board, surface cooking units, a ductless exhaust hood, a refrigerator, wall and base cabinets, a continuous backsplash, and a microwave oven. (Drawing by HLS.)

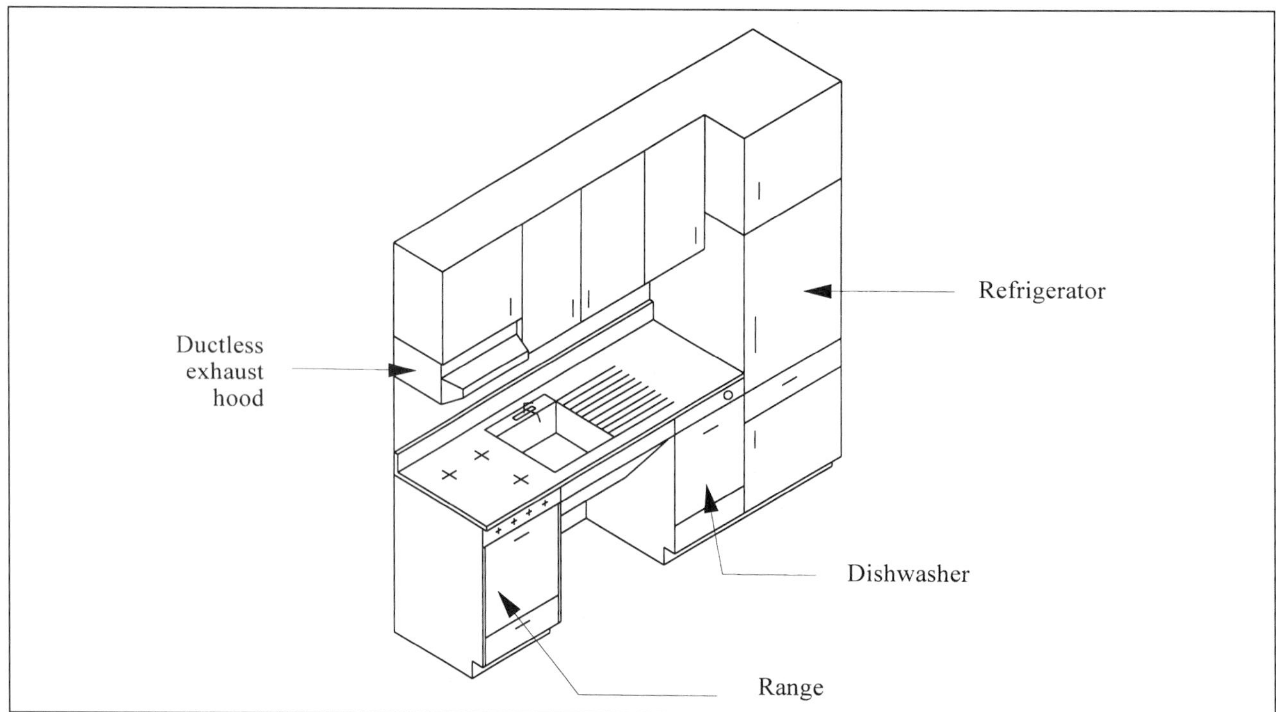

FIGURE 11.2-3 A unit kitchen designed specifically for use by someone in a wheelchair. The unit shown contains a sink in a countertop with a drain board, a range containing a conventional oven and surface cooking units, a ductless exhaust hood, a refrigerator, a dishwasher, wall and base cabinets, and a continuous backsplash. A microwave oven could be easily substituted for the conventional one. Some manufacturers make units of this type that disassemble easily so that they can be converted for conventional use. On such units, the countertop can be raised or lowered, and doors can be easily installed to provide an enclosed undersink storage space. (Drawing by HLS.)

11.3 Additional Reading

More information about the subjects discussed in this chapter can be found in the references listed in Section 11.4, and in the following publications.

Association of Home Appliance Manufacturers (AHAM). 1997. *AHAM Major Appliance Industry Fact Book.* Washington, DC: AHAM.

———. "AHAM Guide to NAECA." Washington, DC: AHAM.

Donohue, Kathleen. 1998. *Kitchen & Bath Theme Design; An Architectural Styling Guide.* New York: McGraw-Hill.

National Kitchen and Bath Association (NKBA). *The Essential Kitchen Design Guide.* Hackettstown, NJ: NKBA.

11.4 Acknowledgments and References

ACKNOWLEDGMENTS

We gratefully acknowledge the assistance of the following organizations and individuals in preparing this chapter. We are also indebted to them for permission to use their illustrations when requested and for the use of their publications as references.

American National Standards Institute (ANSI)

Association of Home Appliance Manufacturers

Dwyer Products Corporation

Hotpoint Division of the General Electric Company

Whirlpool Corporation

REFERENCES

We would also like to thank the authors and publishers of the publications in the following list for their contribution to our research for this chapter.

American National Standards Institute (ANSI). Z21 Series Documents. New York: ANSI.

Architectural Graphic Standards. See Ramsey/Sleeper.

Arcom Master Systems. MASTERSPEC®. Basic Section 11400, "Food Service Equipment." 8/95 ed. Salt Lake City, UT: Arcom.

———. Section 11451, "Residential Appliances." 2/97 ed. Salt Lake City, UT: Arcom.

———. Section 11460, "Unit Kitchens." 5/93 ed. Salt Lake City, UT: Arcom.

Association of Home Appliance Manufacturers (AHAM). ANSI/AHAM DH-1, "Dehumidifiers." Washington, DC: AHAM.

———. AHAM DW-1, " Dishwashers." Washington, DC: AHAM.

———. ANSI/ASSE 1006/AHAM DW-2PR, "Dishwasher Plumbing Requirements." Washington, DC: AHAM.

———. ANSI/AHAM ER-1, "Electric Ranges." Washington, DC: AHAM.

———. ANSI/AHAM FWD-1, "Food Waste Disposers." Washington, DC: AHAM.

———. ANSI/ASSE 1008/AHAM FWD-2PR, "Food Waste Disposer Plumbing Requirements." Washington, DC: AHAM.

———. ANSI/AHAM HLD-1, "Clothes Dryers." Washington, DC: AHAM.

———. ANSI/ASSE 1007/AHAM HLW-2PR, "Home Laundry Plumbing Requirements." Washington, DC: AHAM.

———. ANSI/AHAM RAC-1, "Room Air Conditioners." Washington, DC: AHAM.

———. ANSI/AHAM RF-1, "Household Refrigerators/Household Freezers." Washington, DC: AHAM.

———. ANSI/AHAM TC-1, "Trash Compactors." Washington, DC: AHAM.

———. "Clothes Washers." Washington, DC: AHAM.

———. "Metric Practice Guide for the Appliance Industry (SI)." Washington, DC: AHAM.

Ramsey/Sleeper, The AIA Committee on Architectural Graphic Standards. *Architectural Graphic Standards.* 8th ed. New York: John Wiley & Sons, Inc.

Raschko, Bettyann. 1990. "Specifying Kitchens for the Disabled." *The Construction Specifier* (8)(August):46.

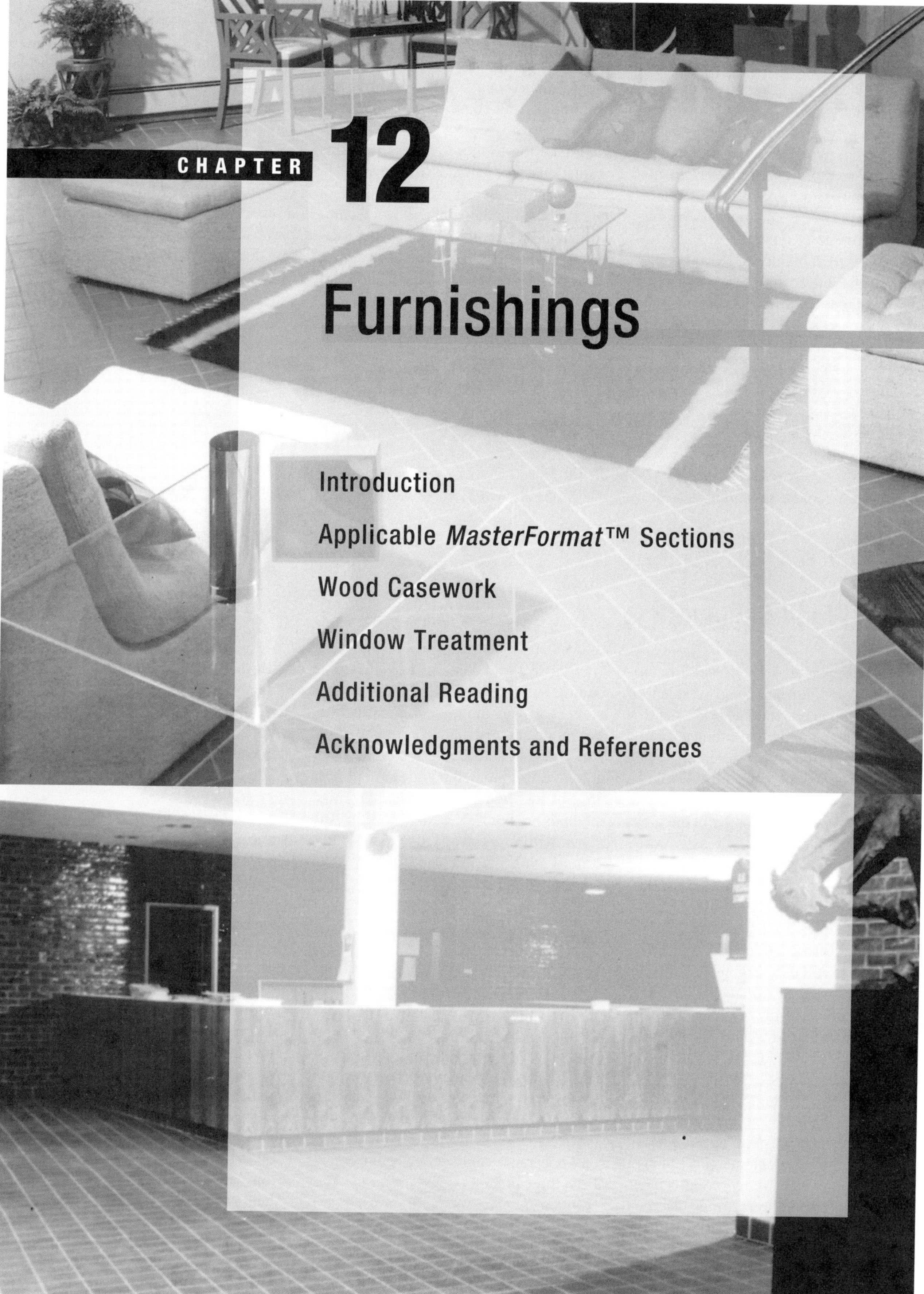

CHAPTER 12

Furnishings

Introduction

Applicable *MasterFormat*™ Sections

Wood Casework

Window Treatment

Additional Reading

Acknowledgments and References

Introduction

MasterFormat Division 12, "Furnishings," covers a broad range of materials and products used in the furnishing of buildings. These include artwork, manufactured casework, window treatments, furniture, accessories, rugs, mats, theater and stadium seating, and interior plants and planters. Space limitations prevent detailed coverage of most of the many furnishings used in buildings. However, two types of these furnishings are sufficiently important to include here. These are manufactured wood casework, including both wood-finished and laminated-plastic-faced units, and some types of window treatment.

In addition to those addressed in this chapter, many other kinds of casework are used in buildings. These include metal casework and both metal and wood casework units that are used in banks, libraries, restaurants, cafeterias, medical facilities, veterinary facilities, ecclesiastical facilities, and as display units in retail establishments. A reader requiring specific data for one of these should first contact the various manufacturers of these products and obtain the necessary catalog data.

The types of window treatment discussed in this chapter are those usually designed and specified by architects. Window treatments used solely for decorative purposes are generally selected by interior design professionals and are not addressed in this book.

Applicable *MasterFormat*™ Sections

The following *MasterFormat* Broadscope sections are applicable to this chapter.

06200 Finish Carpentry
06400 Architectural Woodwork
09300 Tile
12300 Manufactured Casework

12.1 Wood Casework

Wood casework has been used in commercial, institutional, and residential buildings for centuries. The origins of residential bath and kitchen cabinets can be traced back to the sixteenth century, when *ambries* (cupboards) were used to keep rats and other vermin away from food. Ambries were closed food storage areas above open shelving for utensils. More extensive built-in cabinets emerged in the seventeenth century, when the space between the chimney and the side walls was filled with open or closed shelves, giving the effect of built-in kitchen cabinets.

Wood is still used extensively for cabinets. However, increased cost has led to the substitution of plywood and particleboard in cabinets and more use of metal cabinets, especially in commercial and institutional buildings.

Because of the high cost of manufacturing custom cabinets, they are seldom used in residential buildings today, but they still appear in commercial and institutional buildings. Many wood cabinets are mass-produced in factories that manufacture thousands of units a day, using sophisticated milling machines, presses, sanders, laminating devices, conveyors, and finishing equipment.

This section addresses the applicable quality standards and certification programs for, and the typical construction, assembly, and installation of, some of the many cabinet and countertop types available today. The methods described here are not the only ones in use, and other methods are not necessarily inferior. Any construction that produces cabinets that comply with the indicated quality standards is satisfactory.

12.1.1 QUALITY STANDARDS

Wood cabinets in commercial and institutional buildings and in higher-priced residential buildings are usually required to comply with *Architectural Woodwork Quality Standards, Guide Specifications and Quality Certification Program*, published by the Architectural Woodwork Institute (AWI). This standard requires that the designer select the desired grade based on appearance and service requirements from three grade options, *Premium*, *Custom*, and *Economy*. The highest grade is the strongest and most durable, but is usually the most expensive.

In parts of the western United States, designers sometimes require compliance with the standard, *Manual of Millwork*, published by the Woodwork Institute of California (WIC) in lieu of compliance with the AWI standard. The WIC standard has the same grade listings as the AWI standard but adds a fourth, *Laboratory* grade.

The AWI and WIC standards differ in some aspects and use different names for the same cabinet components. A designer should obtain the one of these two standards that will be applied and ascertain its requirements for each aspect of cabinets.

Often, cabinets for laboratories, hospitals, and schools are selected by specifying one or more acceptable manufacturers and product lines. Occasionally, manufacturers are required to modify their stock units to comply with either the AWI or WIC standard, or with even more stringent requirements, such as thicker materials.

The standard usually applied to residential kitchen and bath vanity cabinets is ANSI/KCMA A161.1, "Performance and Construction Standard for Kitchen and Vanity Cabinets." This is a publication of the Kitchen Cabinet Manufacturers Association (KCMA) that has been adopted by the American National Standards Institute (ANSI). It applies to manufactured, factory-finished bath and kitchen cabinets of metal, wood, or other nonmetallic materials, such as plastic laminate, plastics, hardboard, and particleboard.

To ensure high quality, bath and kitchen cabinets in residences should also comply with the requirements of the U.S. Department of Housing and Urban Development (HUD) 4910.1, "Minimum Property Standards for Housing," whether or not the housing will be financed with a Federal Housing Administration (FHA) guaranteed loan. HUD 4910.1 requires compliance with ANSI/KCMA A161.1.

ANSI/KCMA A161.1 is a performance standard. It does not prescribe specific requirements for cabinet design or materials. This allows considerable freedom in the selection of materials and methods of fabrication as long as the final product meets stringent testing procedures for endurance and quality. A161.1 does, however, require that:

- Lumber and plywood parts be kiln dried to a moisture content of 10% or less at time of fabrication.
- Metal cabinets be rust resistant.
- All materials be suitable for use in kitchen and bath environments.
- Cabinets be enclosed with sides, backs, and bottoms, except that backs are not required on vanity units, kitchen sink fronts, sink bases, oven cabinets, or refrigerator cabinets; bottoms are not required on vanity bowl or drawer fronts, or on kitchen drawer units, oven cabinets, refrigerator cabinets, or sink fronts.
- Countertops enclose the tops of base cabinets.
- Wall cabinets have individual tops.
- Swinging doors have a catch or other device to hold them closed.

FIGURE 12.1-1 **The KCMA Certification Seal is placed on the inside of a cabinet door or drawer.** (Courtesy Kitchen Cabinet Manufacturers Association.)

- Adjacent cabinets and doors be in proper alignment with each other.
- Doors be of balanced (warp-resistant) construction and operate freely.
- Drawer slides, shelf standards, brackets, rotating shelf hardware, and other miscellaneous hardware, support the design loads and operational functions described in ANSI/KCMA A161.1.
- Hardware comply with the Builders Hardware Manufacturers Association ANSI/BHMA A156.9, "Cabinet Hardware."

The KCMA Certification Seal (Fig. 12.1-1) offers assurance that cabinets bearing it have been manufactured in conformance with KCMA standards. Any manufacturer of prefinished, factory-engineered kitchen cabinets may apply for a license to participate in the KCMA's Certification Program. KCMA membership is not required. To become a participant, a manufacturer must enter into an agreement with the KCMA, the agreement spelling out the conditions under which the manufacturer may affix a KCMA Certification Seal to its cabinets. However, some manufacturers of high-quality cabinets, even those who are KCMA members, do not participate, because a KCMA Certification Seal represents a *minimum* standard and they do not want to have their products identified with it. Cabinets bearing a Certification Seal must meet performance tests that simulate many years of normal use. These include tests for structural integrity, door operation, drawer operation, and exterior cabinet finish.

12.1.2 CABINETS

Figure 12.1-2 shows normal work heights, mounting heights, and clearances for cabinets. These may vary, especially in commercial and institutional uses and where accommodations for handicapped persons are required (see Section 1.7).

Figure 12.1-3 shows some of the types and sizes of stock cabinets available for residential kitchen use. Cabinets for laboratories, hospitals, and schools are often specifically designed for their applications. These include laboratory fume hoods, specimen cases, workbenches with racks, balance tables, chemistry and physics tables, specialty storage cabinets, reagent racks, glass-front cabinets, and many more. Base and wall cabinets, in general appearance not unlike those shown in Figure 12.1-3, are also used in such applications, but these are usually quite different in construction. For example, they are likely to have heavier and thicker sides, backs, fronts, and tops and stronger joints.

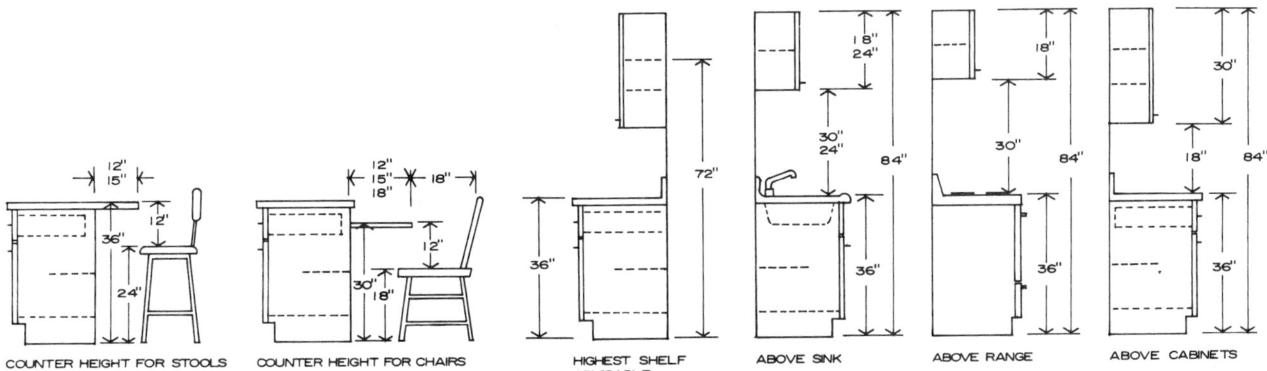

FIGURE 12.1-2 **Typical work heights and cabinet clearances. Refer to Figure 1.8-27 for metric conversions.**

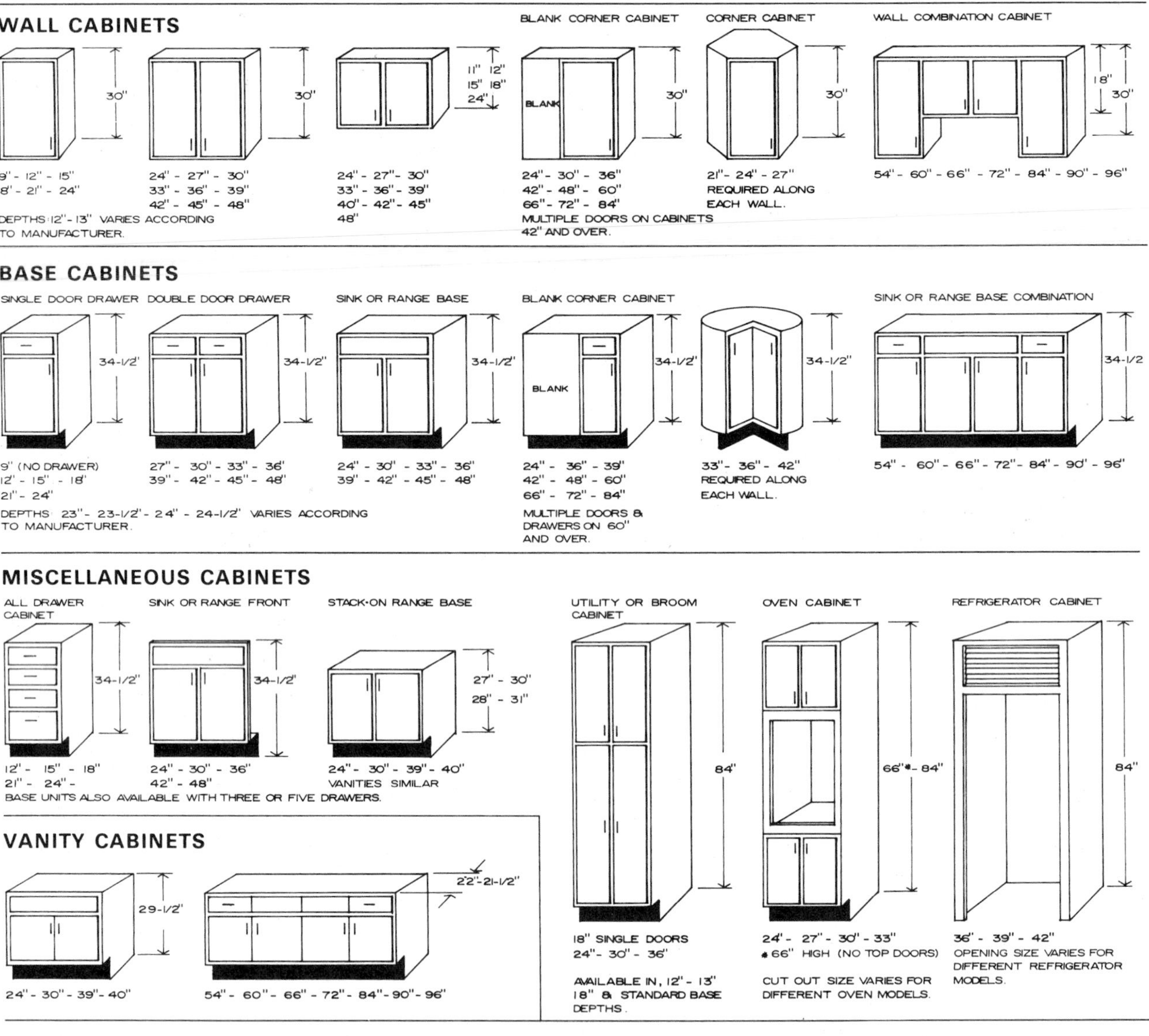

FIGURE 12.1-3 Summary of kitchen and bath cabinet types and sizes. Refer to Figure 1.8-27 for metric conversions.

12.1.3 TYPES AND STYLES

Cabinets are classed generally according to function as *base cabinets*, *wall cabinets*, and *miscellaneous cabinets*. The third category includes laboratory cabinets, school science tables, medicine cabinets, broom cabinets, oven cabinets, and other special pieces.

Cabinets are also classified as to style, which refers to either *design style* or cabinet *face style*. Cabinets are manufactured in many decorative design styles, some with glass doors, others with raised panels, applied panel molds, and other articulations.

In standard manufactured kitchen and vanity cabinets there are three basic face styles, *reveal overlay*, *flush overlay*, and *flush* (Fig. 12.1-4). Other styles may be used in custom cabinets.

12.1.4 MATERIALS AND FINISHES

Cabinets may be made from wood and plywood with a wood finish, from various materials with a plastic laminate finish, or from metal. Hardboard and particleboard are used for some components in some cabinets. Common species of hardwood used are ash, oak, birch, maple, and walnut; common softwood species are ponderosa (western) pine, Douglas fir, and western hemlock. Other wood species are also used, however.

The AWI standard lists eight transparent finish systems and five opaque finish systems for the wood portions of custom cabinets. The same finishes can be specified for standard fabricated units, but not all manufacturers will be able to supply them. ANSI/KCMA A161.1 lists no specific finishes. It rather requires that the finish meet certain performance requirements, leaving the finish up to the manufacturer. The designer must select from the manufacturer's standard finishes. These include various wood grains, stains,

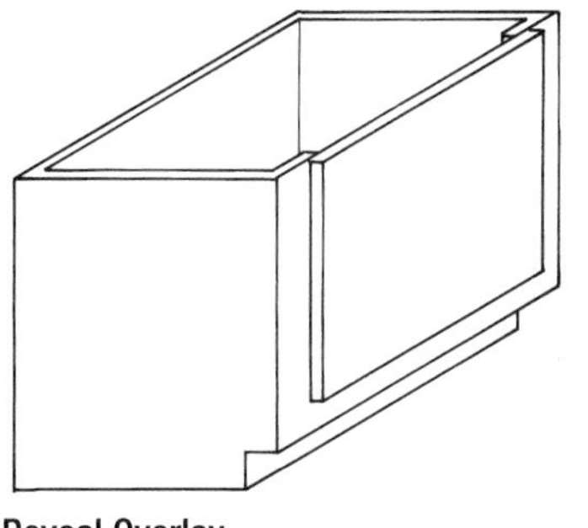
Reveal Overlay

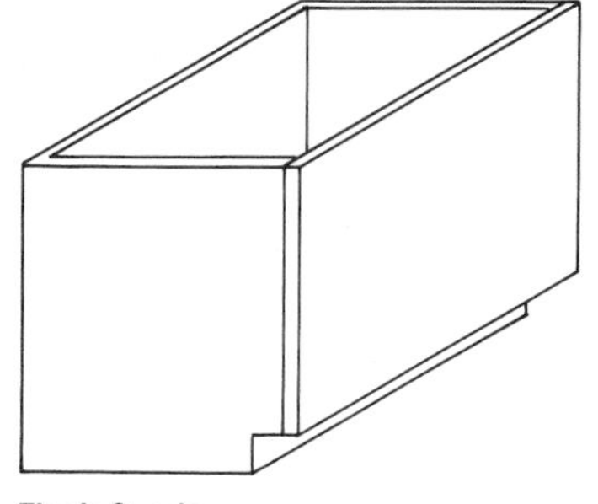
Flush Overlay

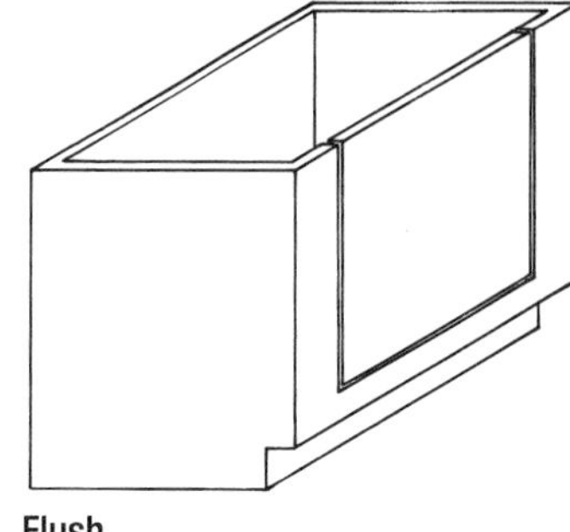
Flush

FIGURE 12.1-4 Cabinet face styles. (Drawing by HLS.)

and opaque colors, reflecting the popularity of a particular color or finish at the time.

12.1.4.1 Construction—General

Performance standards for metal cabinets are basically the same as those discussed here; the construction details, of course, differ. There are no nationally recognized organizations representing metal cabinet manufacturers and no coordinated consensus for construction criteria. Therefore, in specifying metal cabinets, it is necessary to rely on the manufacturer's reputation.

Wood cabinets may have either *face frame* or *frameless* construction. Most cabinets have *face frame* construction, such as that shown in Figure 12.1-5. In frameless construction, the face frame is omitted and the cabinet ends are finished on their exposed edges. Other components are similar to those in cabinets that have face frames.

Wood cabinets are further classified as either *wood-finished* or *plastic-laminate-finished.*

12.1.4.2 Wood-Finished Cabinets

The main components of *wood-finished* cabinets are *face frames*; *back, top, and bottom rails*; *backs*; *tops and bottoms*; *end panels*; *doors*; *drawers*; *shelves*; and *hardware* (see Fig. 12.1-5).

Some manufacturers use panel components similar to those in plastic-laminate-finished cabinets in concealed or semiexposed (visible only with doors or drawers open) locations in wood-finished cabinets. Typical locations are end panels, back panels, and bottoms. Sometimes plastic laminate components are used for shelves in wood-finished cabinets. With that exception, the materials generally used and their nominal thicknesses are shown in Figure 12.1-6. Note that the AWI standard requires that some components be *panel products*. A panel product is defined as plywood, particleboard, fiberboard, or hardboard faced with a hardwood plywood veneer.

Face frame rails, stiles, and *mullions* are doweled (or mortised and tenoned), as well as glued and stapled or nailed for rigidity (Fig. 12.1-7). Side frames are often tongued and grooved into the face frames. Rails are rabbeted to receive top and bottom panels. Back rails are screwed and glued to secure them in place.

Backs are generally fastened to end panels and top and bottom rails.

The *bottoms* of base cabinets and the tops of wall and special cabinets are usually *dadoed* (let) into the sides and interlocked into the hanging rails of wall cabinets.

End panels are typically glued and mechanically fastened to side frames.

Door construction varies with the face style of the cabinet. Doors may be flush overlay style, or they may have solid wood stiles and rails. In some cabinets the doors are hollow; in others they have particleboard, wood, or high-density fiber cores with hardwood or plastic veneer. Sometimes the core is plastic with hardwood veneer overlays. Other doors have

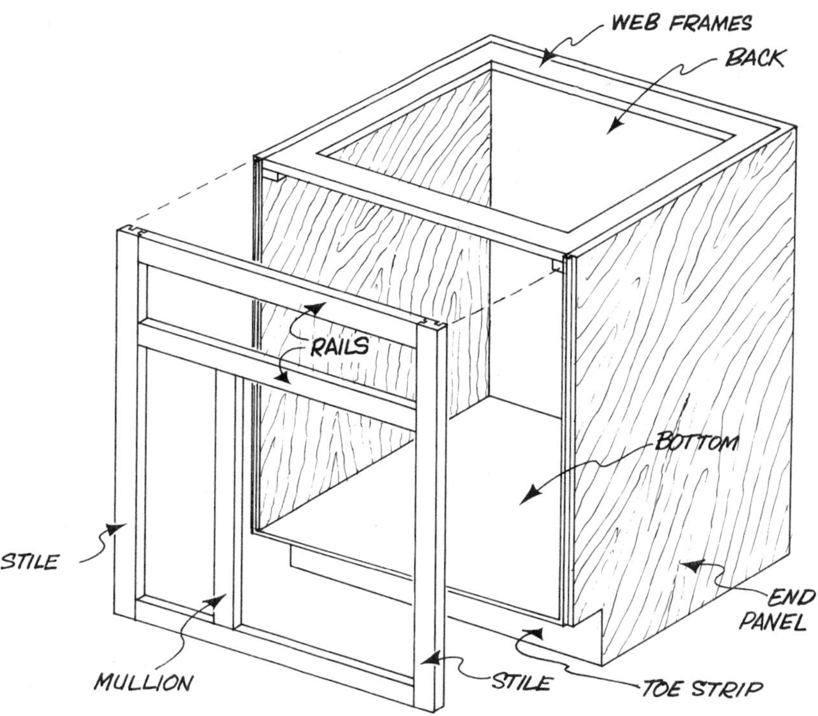

FIGURE 12.1-5 Typical base cabinet construction, showing component parts. Cabinets may have two end panels (as shown), no end panels for use within a series of cabinets, or one end panel to be placed at the end of a series.

FIGURE 12.1-6 Wood-Finished Cabinet Criteria

Component	KCMA Standard	AWI Standard
Face Frames	½- to ¾-in. (12.7 to 19 mm)-thick hardwood	Same
Rails: Back, Top, and Bottom	3-in. (76.2 mm)-wide, ¾-in. (19 mm)-thick solid wood or panel product	Same
Backs	⅛-in. (3.2 mm)-thick hardboard or ⅜-in. (9.5 mm)-thick hardwood plywood	¼-in. (6.4 mm)-thick panel product
Tops and Bottoms of Base Cabinets; Tops of Wall and Special Cabinets	½-in. (12.7 mm)-thick particleboard, or ¼-in. (6.4 mm)-thick hardboard, or ⅜-in. (9.5 mm)-thick hardwood plywood	¾-in. (19 mm)-thick lumber or panel product
End Panels	½-in. (12.7 mm)-thick hardwood plywood	¾-in. (19 mm)-thick panel product
Flush Overlay Doors	½-in. (12.7 mm)-thick plywood with hardwood face veneers	From ¾-in. (19 mm) to 1¼-in. (31.8 mm)-thick for particleboard core doors, and ¾-in. (19 mm)-thick for veneer core doors
Stile and Rail Doors	⅝-in. (15.9 mm)-thick with ¾-in. (19 mm)-thick solid wood center panels, or ¾-in. (19 mm)-thick with applied ¼-in. (6.4 mm)-thick center panels	From ¾-in. (19 mm) to 1¼-in. (31.8 mm)-thick for particleboard core doors, and ¾-in. (19 mm)-thick for veneer core doors
Shelves	From ½- to ¾-in. (12.7 to 19 mm)-thick, lumber, plywood, or particleboard	¾-in. (19 mm)-thick for short spans (32 to 36 in. [812.8 to 900 mm] and 1 in. to 1¹⁄₁₆ in. [25.4 to 27 mm]) for larger spans (42 to 48 in. [1066.8 to 1219.2 mm])

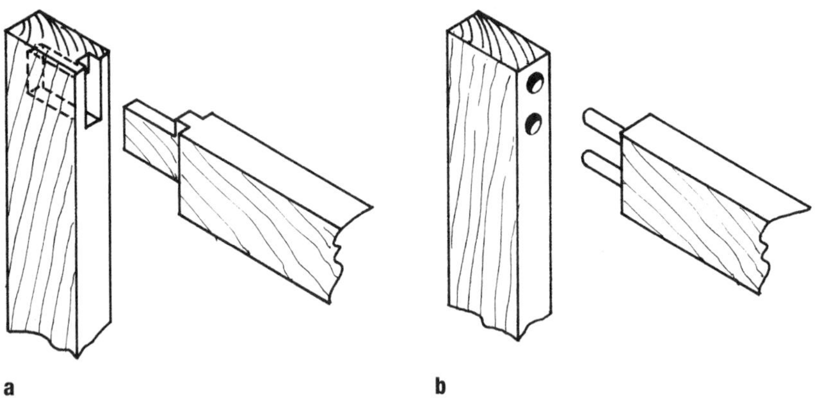

FIGURE 12.1-7 Rails and stiles of frames can be joined by (a) mortise and tenon or (b) dowel joint.

particleboard or plywood cores with plastic faces, and in still others the entire door is plastic. The required thicknesses for doors for cabinets complying with the AWI standard vary with the composition of the core of the door and its grade, width, and height. Face veneers are hardwood plywood.

Drawers often have hardwood lumber sides and backs and plywood bottoms, but the sides may also be plywood or particleboard. Bottoms are often particleboard. Drawer construction varies with cabinet quality. In high-quality cabinets the sides are connected to the front and back with multiple *dovetail* joints; in others, with *lock shouldered* or *square shouldered* joints (Fig. 12.1-8). Drawer bottoms are *dadoed* into the sides, front, and back for more rigid construction. Some drawer units are molded plastic (except for the front), with rounded inside corners for ease of cleaning.

Drawer fronts usually match the *door fronts* and are of the same or similar construction. They should be positively fastened to a wood *subfront* with screws. The *subfront* should be *rabbeted* into the sides and fastened to them with mechanical fasteners and glue. There is as wide a variation in drawer front styling as there is in door design.

Most *shelves* are lumber, plywood, or particleboard, with rounded front edges. Plywood or particleboard shelves may be banded with hardwood front edges.

12.1.4.3 Plastic-Laminate-Clad Cabinets

Many *plastic-laminate-finished* cabinets are constructed in the same way as *wood-finished* cabinets, except that the finish veneers are made from a suitable grade of sound hardwood, or in some cases from softwood, and are covered with plastic laminate.

A second type of plastic-laminate-finished cabinet is made from panels prefinished with plastic laminate and assembled into a finished cabinet. The main components of this type of plastic-laminate-clad cabinets are similar to those of wood-finished cabinets.

Frameless type construction is more likely to be used in plastic-laminate-clad cabinets. In *reveal overlay style* plastic-laminate-clad cabinets, face frames are usually not required.

The *concealed* (never visible) surfaces of plastic-laminate-clad cabinets are often plywood or particleboard. The *exposed* and *semiexposed* faces of components are finished in *high-pressure decorative laminate* (HPDL) or *thermoset decorative panels*. The HPDL faces should comply with the requirements of the National Electrical Manufacturers Association (NEMA) standard, "High Pressure Decorative Laminates," which has been adopted by ANSI as ANSI/NEMA LD-3. LD-3 designates laminate types according to the International Organization for Standard-

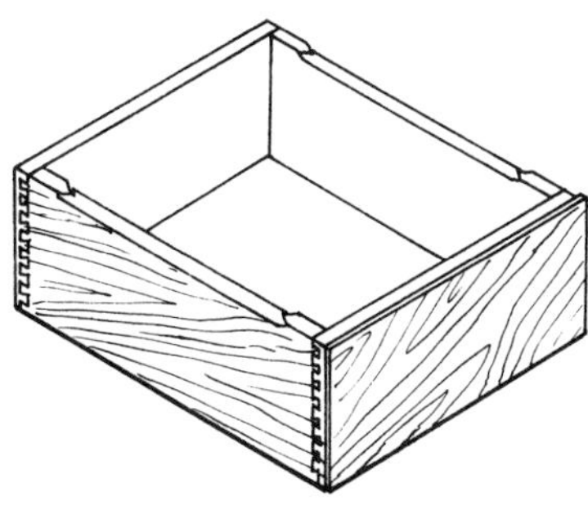

a

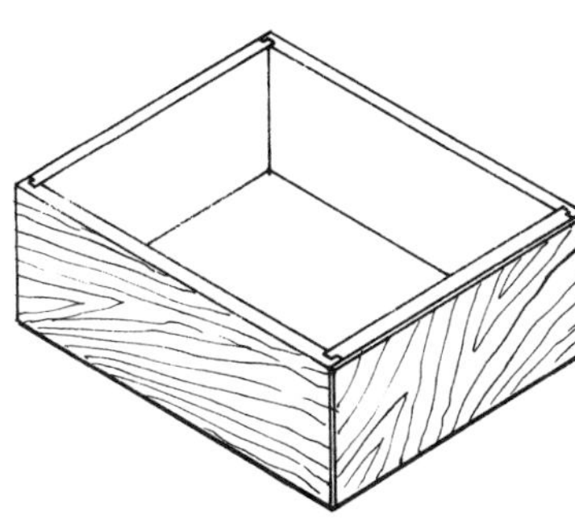

b

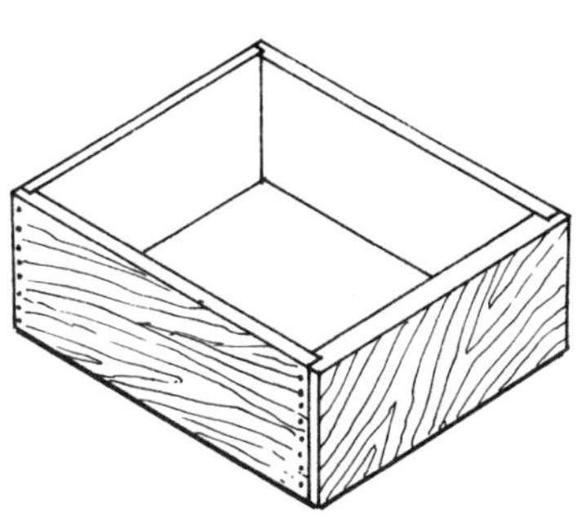

c

FIGURE 12.1-8 Drawer construction: (a) multiple dovetail, (b) lock shouldered, and (c) square shouldered joints.

ization (ISO) nomenclature and thicknesses, such as:

HGS: Horizontal General Purpose, Standard—0.048 in. (1.2 mm)

HGP: Horizontal General Purpose, Postformable—0.038 in. (1.0 mm)

VGP: Vertical General Purpose, Postformable—0.028 in. (0.7 mm)

HGF: Horizontal General Purpose, Fire-Rated—0.048 in. (1.2 mm)

VGF: Vertical General Purpose, Fire-rated—0.028 in. (0.7 mm)

CLS: Cabinet Liner, Standard—0.020 in. (0.5 mm)

The uses of the various types are mostly self-explanatory. For example, materials designated HG are used in horizontal applications, VG in vertical applications, CL for the insides of cabinets where subjected to little or no wear. Postformable sheets can be bent to form curved surfaces, such as a rounded front edge or a coved backsplash.

The panel thicknesses listed are those most used in manufactured cabinets, but thicker materials are readily available and are often used in commercial and institutional casework.

Edges in semiexposed locations are often finished with *high-impact plastic edging*. Exposed edges are usually finished with the same plastic laminate used on the adjacent surface.

Thermoset, or *low-pressure*, decorative laminate panels, such as melamine, are used in semiexposed locations where they are not subjected to wear. They should comply with the American Laminators Association (ALA) standard 1985, "Performance Standard for Thermoset Decorative Panels," which is roughly equivalent to CLS as defined by ISO.

The core materials generally used in plastic laminate cabinets and their nominal thicknesses are shown in Figure 12.1-9.

Face frames are made with mortise and tenon joints.

Stretchers are needed at the top of base cabinets to support the countertops. Sides, dividers, tops, bottoms, shelves, and stretchers are covered on both sides with CLS HPDL.

Backs generally have a *thermoset* decorative laminate on the surface that will be inside the cabinet. Backs are fastened to end panels and top and bottom rails.

Doors usually have a VGP plastic laminate finish on both sides. The thicknesses of doors for cabinets complying with the AWI standard vary, depending on the core type and the height and width of the door. Core types permissible are particleboard and plywood veneer.

Drawer fronts, backs, and bottoms should be *rabbeted* into the sides and fastened with screws and glue. Some drawer units are molded plastic (except for the front), with rounded inside corners for ease of cleaning.

Drawer fronts usually match the door fronts and are of the same construction. They should be positively fastened to a particleboard subfront with glue and fasteners. The subfront should be rabbeted into the sides and fastened to them with mechanical fasteners and glue.

12.1.4.4 Hardware

Cabinet hardware includes *pulls*, *catches*, *drawer suspension devices*, and *adjustable shelf supports*. Many types of each are available. Hardware should comply with ANSI/BHMA A156.9.

Drawer suspension devices should provide for a minimum capacity of 50 lb. (22.7 kg), and should have nylon rollers.

Shelves may be either fixed or adjustable, with metal hardware of various types to permit adjustment.

12.1.4.5 Manufacture

There are two basic kinds of cabinet manufacturing operations: *integrated plants* and *assembly plants*. In an *integrated*

FIGURE 12.1-9 Plastic-Laminate-Finished Cabinet Criteria

Component	KCMA Standard	AWI Standard
Face Frames	1-in. by ⅝-in. (25.4 mm by 15.9 mm) solid wood	
Sides, Dividers, Bottoms, Tops, Shelves, and Stretchers	½-in. (12.7 mm)-thick particleboard	¾-in. (19 mm)-thick with shelves being 1-in. thick for spans of more than 36 in. (914 mm)
Backs	⅛-in. (3.2 mm)-thick hardboard	¼-in. (6.4 mm)-thick panel product
Doors	⅝-in. (15.9 mm)-thick particleboard	¾- to 1¼-in. (19 to 31 mm) thick
Drawers	⅜-in. (9.5 mm)-thick particleboard sides and backs	½-in. (12.7 mm)-thick lumber, plywood, or particleboard; bottoms from ¼-in. (6.4 mm)-thick panel product
Drawer Fronts	Match door fronts with a ⅜-in. (9.5 mm)-thick particleboard subfront	Match door fronts with ½-in.-thick lumber, plywood, or particleboard subfront

wood-finished cabinet plant, wood is processed from raw material (logs) to the finished products (cabinets), with minor purchases of plywood, particleboard, plastic components, and hardware. Such facilities have vast milling operations for the many wood-processing steps needed to produce the final products.

In an *integrated plastic-laminate-finished cabinet plant*, plastic laminates are manufactured and processed on-site. Wood may either be milled on-site or purchased already milled. Plastic components, plywood, particleboard, and hardware are usually purchased.

In an *assembly plant*, raw wood materials have already been completely or partially milled and plastic laminate and other materials are purchased from their manufacturing source. Production is mainly the assembly of components produced elsewhere.

12.1.5 COUNTERTOPS

Many materials are used as *countertops*. These include, but are not necessarily limited to, plastic laminate, filled polymer, cultured marble, ceramic tile, wood, natural stone—such as granite and marble—stainless steel, and the special tops used in laboratories and the like, such as soapstone, synthetic stone, and epoxy resin.

12.1.5.1 Plastic Laminate

The most common type of countertops in residential and general commercial and institutional buildings is finished with horizontal grade plastic laminate. In a plastic laminate countertop, the backsplash, top, and front piece are assembled first onto a wood or particleboard blank. Then adhesive is applied to the wood or particleboard surface, and a plastic laminate sheet is laid over the top and pressed into intimate contact with the surface. If a countertop blank has curved surfaces, the plastic sheet is first heated, then *postformed* (molded) to the blank surface by a special press. Openings for sinks are cut and holes for pipes are drilled at the factory or shop.

12.1.5.2 Filled Polymer

Recently, *filled polymer*, or *acrylic-modified polyester*, tops have gained favor. This material is a rigid-cast acrylic plastic that is homogeneous, solid, and nonporous. It can be cut and shaped with ordinary woodworking tools and sanded to remove burns and scratches. Probably the best-known example of this material is *Corian* by the DuPont Company, but similar materials are made by other manufacturers.

12.1.5.3 Other Materials

Countertops made from synthetic materials called *cultured marble*, *synthetic stone*, and various other names are manufactured by proprietary methods and shipped to a project site ready for cutting and assembly. These materials may or may not have marble dust in their mix and may either be homogeneous or just have a decorative surface applied.

Marble, granite, and other real stone countertops are cut to fit in a shop and shipped to a project site ready for installation. In both real stone and synthetic tops, cutouts for plumbing fixtures and pipes are usually made at a factory or shop.

Ceramic tile countertops are normally built at a construction site from standard tile components. Tops supporting the ceramic tile are shop cut and field fitted.

Wood tops are usually manufactured to fit, but some cutting may be done in the field.

Special tops for use in laboratories are manufactured or cut to fit at the factory.

12.1.6 INSTALLATION

Cabinets, countertops, and devices built into them should be installed level, plumb, and true, in accordance with the manufacturers' instructions and with the requirements of the standards referenced earlier.

12.1.6.1 Cabinets

Attach wall and base cabinets with screws (not nails) to studs or other framing members.

To ensure a plumb and level installation, remove high spots and shim low spots in walls and floors. Remove wall base and chair rail from behind cabinets to ensure a flush fit.

Install wall cabinets first, beginning with a corner unit. Insert screws through *hanging rails* built into the backs of cabinets at both top and bottom. After the cabinets have been leveled and plumbed and doors perfectly aligned, tighten the screws.

Start installing base cabinets in a corner. After installing the base corner unit, use C-clamps to hold adjacent cabinets in alignment. Then, use T-nuts to fasten them together. Check each cabinet front to back and across the front edge with a level, as it is attached to the wall. The front frame should be plumb, and the top should be level. Attach base cabinets with screws through the back frame into wall studs. After base cabinets have been secured in place, install the countertop and finish the toe strip with a tile, resilient, or wood base.

12.1.6.2 Countertops

Install *plastic laminate* and *filled polymer* countertops, using screws through *corner blocks* in the base unit into the top. Spline and glue joints in plastic laminate countertops, then pull them tight with mechanical fasteners. Finish seams in *filled polymer* tops with the manufacturer's joint adhesive.

Install and finish other types of tops in accordance with their manufacturers' instructions. Refer to Section 9.6 for information about installing ceramic tile.

12.2 Window Treatment

Heat flow through windows can be reduced by means of *window treatment*, including interior and exterior shading devices, and interior blinds, shades, and draperies. This section addresses window treatment intended to control heat loss or gain. Shutters, blinds, and draperies used primarily for decoration are beyond the scope of this book, even though some of them—lined draperies, for example—may incorporate features that reduce heat transfer.

Measures often imposed to reduce heat gain and loss through glass include tightly sealed openings, multiple glazing, insulating glass, and storm windows. These measures do result in less heat loss and gain. Unfortunately, even double-glazed windows lose heat 6 to 10 times faster than typical insulated wood-framed walls. Therefore, additional thermal protection is often desirable.

The most effective form of additional protection is provided by *exterior* shading devices, such as *shutters*, *blinds*, *shades*, *awnings*, and *architectural projections*. Though less effective, *interior* devices such as *blinds*, *roll shades*, *thermal shutters*, and *draperies* are also worthwhile.

12.2.1 EXTERIOR SHADING DEVICES

Shading devices are most effective when installed on the exterior of a building where their shade can prevent sunlight from entering. This is critical, because only a portion of the solar radiation that enters a building can be reflected back to the exterior. The rest is absorbed and then radiated to the interior (see Section 15.2).

12.2.1.1 Shutters

A shutter that blocks out all direct sunlight can reduce solar heat gain through a window by 80%. Shading performance, however, depends on how well heat absorbed into the shutter is dissipated to the outside air. Shutters that absorb large amounts of heat radiate a significant percentage of that heat into the building. Therefore, light-colored shutters, which reflect most sunlight before it can be absorbed, are superior to dark shutters.

The most effective shutters have operable-tilt slats that can be adjusted to block the sun and also let air circulate through them. Nonoperable louvered shutters will also permit natural airflow but cannot be adjusted to efficiently track the sun's angle. Air should be allowed to circulate between windows and shutters to dissipate the heat absorbed by the shutters. Shutters used for reduction of heat gain or loss are usually manufactured of wood or aluminum. Decorative shutters are often vinyl. Shutters having a finished surface on one side only are not suitable for use as operable shutters. Several common shutter types are shown in Figure 12.2-1.

Because they reduce airflow, tight-fitting shutters provide more resistance to heat loss than loose-fitting ones. However, even when the fit is not tight or the louvers cannot be completely closed, shutters will reduce heat loss by reducing infiltration and preventing the scouring action of the wind from disturbing the insulating air film at the outer surface of the glass.

a

b

c

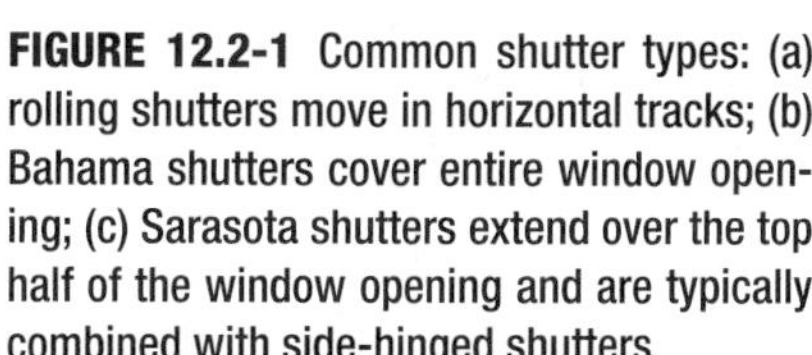

FIGURE 12.2-1 **Common shutter types: (a) rolling shutters move in horizontal tracks; (b) Bahama shutters cover entire window opening; (c) Sarasota shutters extend over the top half of the window opening and are typically combined with side-hinged shutters.**

12.2.1.2 Roll Blinds and Shades

Roll blinds and shades range from simple bamboo or canvas shades to motorized roll blinds that can be operated from the interior. Such devices are highly effective in stopping sunlight from entering a building, even low-angle sun at west elevations. Light colors are most effective because they reflect heat-producing sunlight.

Motorized roll blinds are usually made of aluminum, vinyl, or wood, producing the added benefit of year-round use and durability (Fig. 12.2-2). Acting like tight-fitting storm windows, they can also reduce heat loss in winter by as much as 50%. Independent shade control of each window and ease of operation are other advantages. Their prime disadvantage is

a

b

FIGURE 12.2-2 **(a) Soffits above the windows house the motorized roller that stores the blinds. (b) Even larger windows can be protected, such as on this screened porch.** (Courtesy Levelor.)

that exterior view and natural ventilation are reduced. Blinds are available, however, that can be projected away from the window, like an awning, to provide view and ventilation.

12.2.1.3 Awnings

How well an *awning* shades a window depends on how opaque the awning material is. The more opaque, the more complete is the shade provided.

The exposed surfaces of an awning should be a light color to minimize the amount of sunlight absorbed. When sunlight is absorbed, an awning's temperature is raised. This heat is then radiated through the shaded window.

Clean, white canvas or slatted, light aluminum awnings reflect 80% to 91% of the sunlight that strikes their surfaces. A dark green canvas awning reflects only 21%, and a dark green plastic awning reflects 27% percent of the sunlight.

A light-colored awning can reduce heat gain by 64% on south-facing windows and 77% on west-facing windows. Figure 12.2-3 lists the net effectiveness of several types of awnings.

Heat from sunlight absorbed by an awning will build up under the awning and be transferred to the window unless

FIGURE 12.2-3 Heat Gain Through Single-Glazed Windows with Awnings

Window Orientation	Awning Type	Heat Gain/ 100 sq. ft. Glass Surface, Btu/day (W/m²)	Heat Excluded Total, Btu/day (W/m²)	Heat Excluded % Reduction
South	No awning	62,200 (82)	—	—
	White canvas	22,500 (30)	39,700 (52)	64
	Dark green canvas	27,700 (36)	34,500 (46)	55
	Dark green plastic	35,600 (47)	26,600 (35)	43
West	No awning	84,200 (111)	—	—
	White canvas	19,500 (26)	64,700 (85)	77
	Dark green canvas	23,900 (32)	60,300 (79)	72
	Dark green plastic	34,800 (46)	49,400 (65)	59

FIGURE 12.2-4 Roof Overhang for Complete Summer Window Shading*

Latitude • North, Degrees	Overhang Width: South Exposure	Overhang Width: Southeast/Southwest Exposure
28	1 ft. 5 in. (432 mm)	3 ft. 10 in. (1168 mm)
32	1 ft. 11 in. (584 mm)	4 ft. 2 in. (1270 mm)
36	2 ft. 5 in. (737 mm)	4 ft. 8 in. (1422 mm)
40	3 ft. 0 in. (914 mm)	5 ft. 4 in. (1626 mm)
44	3 ft. 6 in. (1067 mm)	6 ft. 0 in. (1829 mm)

*Assumes window sill is 2 ft. from floor.

air is permitted to escape. Therefore, fabric awnings are often installed with a continuous gap between the top of the awning and the wall. Slatted aluminum awnings inherently provide air circulation between the horizontal slats.

Awnings should provide adequate coverage of a window for its orientation. South-facing windows require only a minimal horizontal projection to be completely shaded, but the sides of such awnings should be closed to prevent sunlight from angling in behind them. Awnings on east- or west-facing windows should extend down a substantial distance to block out early morning or late afternoon sun, as applicable.

12.2.1.4 Architectural Projections

Windows can be shaded by horizontal or vertical planes that extend beyond the face of the glass to intercept the summer sun. If properly designed, these projections will not interfere with the penetration of the beneficial winter sun or block the view.

Roof overhangs are one of the most common ways to shield windows in small buildings, especially houses, from sunlight. Figure 12.2-4 lists the horizontal projections necessary to provide complete summer window shading with a roof overhang.

A deeply recessed window, or one with closely spaced horizontal or vertical fins, can also be effective. In general, east- and west-facing windows are more effectively shaded by vertical projecting planes; south-facing windows by horizontal planes.

The underside of an overhang should be dark so that it will transmit less ground-reflected sunlight through the glass. The net result of such shading is that heat gain through a completely protected window can be reduced by 80%.

12.2.2 INTERIOR HEAT GAIN/LOSS DEVICES

Although not as effective as exterior shading devices, interior heat gain/loss devices have the advantage of greater accessibility and ease of operation. Interior devices used to reduce heat gain in the summer and heat loss in the winter include *blinds*, *shades*, *thermal shutters*, and *thermal draperies*.

12.2.2.1 Blinds

The most common of such devices are *venetian blinds* (Fig. 12.2-5). These are slatted horizontally and can be tilted to reflect a maximum amount of sunlight back to the outdoors or to direct daylight toward the ceiling for glare-free uniform

FIGURE 12.2-5 Venetian blinds can be made to cover large glazed areas such as sliding glass doors. (Courtesy Levelor.)

lighting. Blinds, however, provide only minimal reduction in heat loss in winter because they do not effectively entrap a layer of air.

Vertical blinds offer similar protection against heat gain. They more efficiently reduce heat loss in winter, however, because they trap an insulating layer of air between themselves and the glass (Fig. 12.2-6).

Some proprietary systems encase venetian blinds between layers of glass (Fig. 12.2-7). When fully closed, this system can provide an R-value of 2.50, as

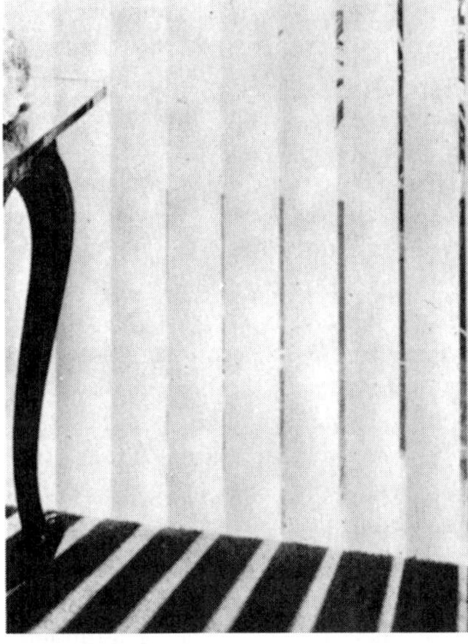

FIGURE 12.2-6 Vertical blinds extend to the floor, thus reducing heat-robbing convection currents. (Courtesy Levelor.)

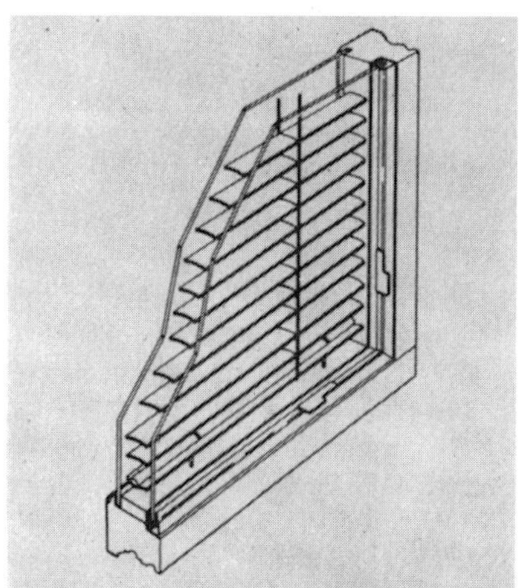

FIGURE 12.2-7 With blinds fully closed, solar heat gain is reduced by about 82%. (Courtesy Pella Corporation.)

FIGURE 12.2-8 Thermal Shutter Materials

Material	R-Value/ 1 in. (25.4 mm)	Window R-Value: With 1 in. (25.4 mm) Insulation	Window R-Value: With 2 in. (50.8 mm) Insulation
Expanded polystyrene (extruded)	4.00	5.00	9.00
Expanded polystyrene (molded)	3.57	4.54	8.11
Polyurethane	6.25	7.14	13.39

compared with an R-value of 2.04 for a similar window with no blinds.

12.2.2.2 Roll Shades

The most effective type of interior protection against window heat gain is an opaque or translucent *roll shade*. Such shades can be easily installed with simple brackets on a window frame and, when not in use, rolled up for unobtrusive storage. A completely opaque roll shade with a white outer face can reflect 80% of the sunlight back through the glass.

A roll shade can have a dark color on one side that effectively absorbs sunlight and a white surface on the reverse side that effectively reflects it. Reversing such a shade, from dark side facing out in winter to reflective side out in summer, will permit it to perform both as a solar collector and a shading device.

12.2.2.3 Thermal Shutters

Covering windows with a rigid insulating material such as polyurethane or polystyrene is an effective way to control heat loss. Figure 12.2-8 lists several common insulating materials and the resulting R-values when they are combined with ¼-in. (6.35 mm)-thick plate glass. They are usually installed on the interior, but they can be placed on the exterior if the insulation is protected from the weather.

A common way to achieve protection from the elements is to face the insulation with aluminum, galvanized steel, or exterior-grade plywood. Foam plastic insulation must also be protected from fire; plywood or metal cladding can serve this function.

Thermal shutters can be hinged, tracked, or mounted with clips. Removable shutters can be taken down and stored during warm weather. Hinged or tracked shutters can be folded or pushed back out of the way when not needed.

When an insulating material is placed in direct contact with glass, heat loss is reduced only in proportion to the R-value of that material. A space left between a thermal shutter and a sash creates an insulating dead air space, which more effectively reduces conductive heat loss. Tightly sealing cracks around a frame will reduce infiltration. Loosely fitting shutters permit air to circulate between the glass and the shutter, carrying away heat when the air passes over the glass and resulting in a dramatic reduction in overall insulating effectiveness.

12.2.2.4 Thermal Drapes

Drapes offer a more traditional way of controlling window heat flow. Because the surface temperature of a drape is much higher than that of window glass, drapes reduce the radiant heat loss from occupants and, therefore, improve thermal comfort.

For maximum benefit from window coverings, room air must not pass between the glass and the covering material. When air is allowed to pass over the glass, the improvement from a loose-fitting drape amounts to an increase from the R-0.85 of an undraped window to only R-0.94. If the heated air from a wall or floor register passes between the covering and the glass, heat loss is actually more rapid than if no covering were provided.

By contrast, when warm room air is prevented from circulating behind the drapes, the R-value of the window and drape can increase to 1.13. To achieve this value, drapes must be tightly sealed at the bottom and sides. Thermal drapes can be fastened to a window frame with hooks, snaps, or Velcro fasteners. Long drapes should touch the floor.

Thermally efficient drapes are usually made of tightly woven fabric. Drapes can be further improved by adding a reflective liner, foam backing, or additional layers of fabric.

12.3 Additional Reading

More information about the subjects discussed in this chapter can be found in the references listed in Section 12.4 and in the following publications.

Bianco, John. 1995. "Commercial Window Treatments." *The Construction Specifier* (July):78.

McGowan, Maryrose. 1992. "Horizontal Louver Blinds." *The Construction Specifier* (June):47.

Pearce, John. 1984. "Sunlight Solutions: Operable Woven Sunscreens." *The Construction Specifier* (March):44.

12.4 Acknowledgments and References

We gratefully acknowledge the assistance of the following organizations and individuals in preparing this chapter. We are also indebted to them for permission to use their illustrations when requested and for the use of their publications as references.

ACKNOWLEDGMENTS

American National Standards Institute (ANSI)
Formica Corporation
Kitchen Cabinet Manufacturers Association
Levelor Lorentzen, Inc.
Nevamar—International Paper, Decorative Products Division
Pella Corporation
U.S. Department of Housing and Urban Development (HUD)
Wilsonart International
Woodwork Institute of California

REFERENCES

We would also like to thank the authors and publishers of the publications in the following list for their contributions to our research for this chapter.

American National Standards Institute (ANSI) Standards:

A117.1-1986, "American National Standards for Buildings and Facilities—Providing Accessibility and Usability for Physically Handicapped People." New York: ANSI.

A208.1-1979, "Particleboard, Mat-Formed Wood." New York: ANSI.

Architectural Graphic Standards. See Ramsey/Sleeper.

Architectural Woodwork Institute. 1988. *Architectural Woodwork Quality Standards, Guide Specifications and Quality Certification Program.* 5th ed. Centreville, VA: AWI.

Arcom Master Systems. MASTERSPEC®. Basic Sections:

06200, "Finish Carpentry." Salt Lake City, UT: Arcom.

06402, "Interior Architectural Woodwork." Salt Lake City, UT: Arcom.

12500, "Window Treatment." Salt Lake City, UT: Arcom.

Council of American Building Officials (CABO). 1983. *One- and Two-Family Dwelling Code.* Falls Church, VA: CABO.

Kitchen Cabinet Manufacturers Association (KCMA). ANSI/KCMA A161.1, "Performance Standard for Kitchen and Vanity Cabinets." Falls Church, VA: ANSI/KCMA.

National Electrical Manufacturers Association (NEMA). NEMA LD 3-1985, "High-Pressure Decorative Laminates." Washington, DC: NEMA.

Ramsey/Sleeper, The AIA Committee on Architectural Graphic Standards. 1988. *Architectural Graphic Standards.* 8th ed. New York: John Wiley & Sons, Inc.

Woodwork Institute of California (WIC). *Manual of Woodwork.* Fresno, CA: WIC.

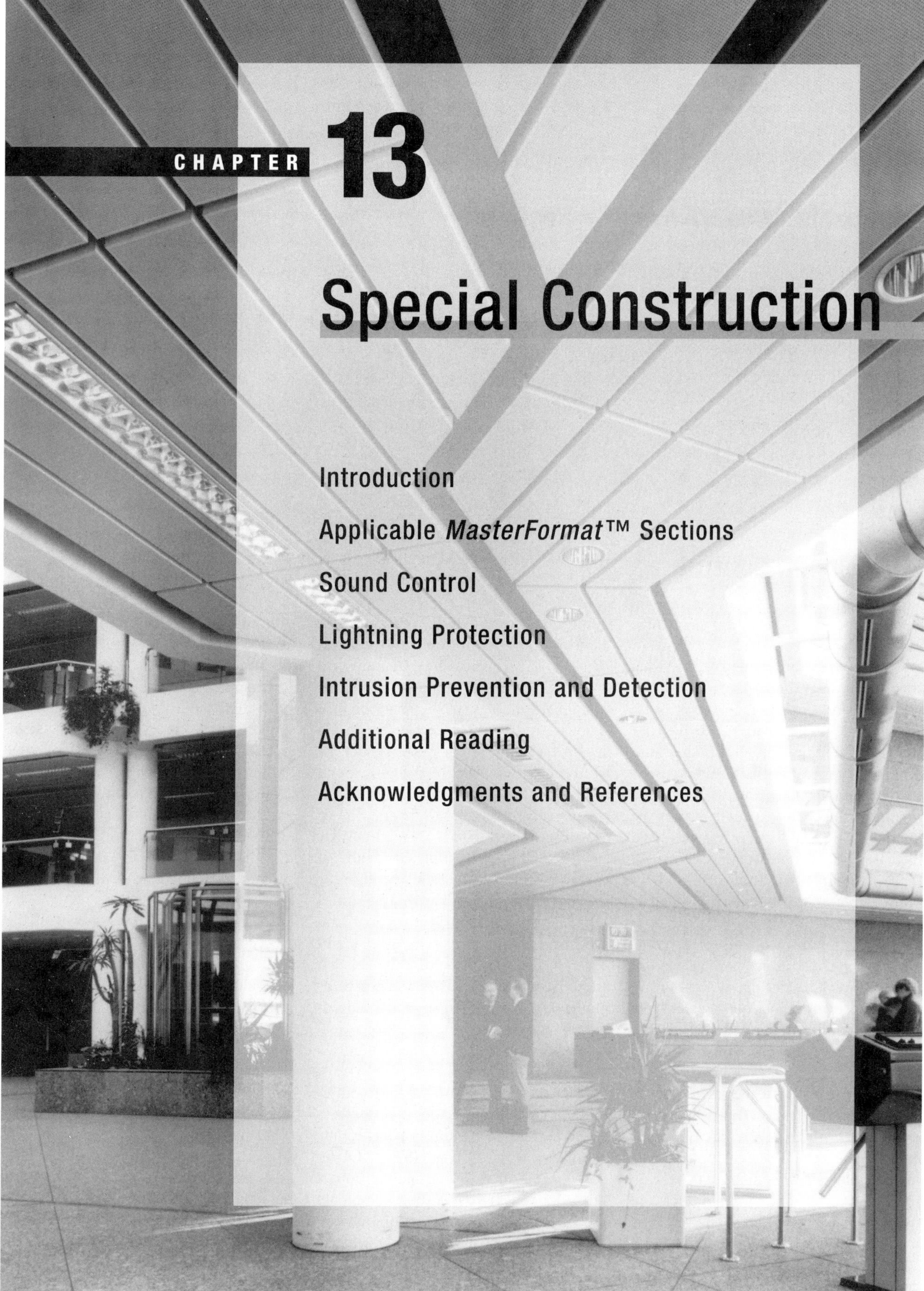

CHAPTER 13

Special Construction

Introduction

Applicable *MasterFormat*™ Sections

Sound Control

Lightning Protection

Intrusion Prevention and Detection

Additional Reading

Acknowledgments and References

Introduction

The types of special construction addressed in this chapter include sound control, lightning protection, and intrusion prevention and detection. *MasterFormat* Division 13 includes other types of special construction that are beyond the scope of this book. These include, but are not limited to, air-supported structures, radiation protection, kennels and animal shelters, preengineered structures, incinerators, and solar and wind energy equipment.

Applicable *MasterFormat*™ Sections

The following *MasterFormat* Broadscope sections are applicable to this chapter.

06100 Rough Carpentry
07200 Thermal Protection
08100 Metal Doors and Frames
08200 Wood and Plastic Doors
08500 Windows
08800 Glazing
09100 Metal Support Assemblies
09200 Plaster and Gypsum Board
09300 Tile
09500 Ceilings
09600 Flooring
09700 Wall Finishes
09800 Acoustical Treatment
12400 Furnishings and Accessories
13080 Sound, Vibration, and Seismic Control
15050 Basic Mechanical Materials and Methods
15400 Plumbing Fixtures and Equipment
15500 Heat-Generation Equipment
15700 Heating, Ventilating, and Air Conditioning Equipment
15800 Air Distribution
16050 Basic Electrical Materials and Methods
16800 Sound and Video

13.1 Sound Control

Computer printers, typewriters, calculators, intercom systems, background music systems, clothes washers and dryers, dishwashers, garbage disposals, refrigerators, vacuum cleaners, radios, television sets, and other electrical devices make life easier and more enjoyable, but at the same time they create noise. Heating, ventilating, and air conditioning equipment produces noise and carries it into every interior space. Outdoor noise levels have increased also. People work and live close to their neighbors. Apartments and townhouses continue to increase in number. Sounds created by high-speed roads, fast trains, and jet airliners add to the problem. Open offices allow workplace noise to travel into adjacent spaces.

Many construction materials exacerbate our noise problems. Glass, plaster, gypsum board, wood paneling, masonry, tile, metal panels, and other hard, relatively flat surfaces reflect up to 98% of the sound that strikes them. Thin, lightweight materials, including some preassembled components, in walls and in floor and ceiling assemblies permit noise to travel easily from room to room, floor to floor, and from one living or working unit to another.

The first major use of the science of sound control was applied earlier in this century in churches, theaters, and auditoriums to make the living environment more pleasant. Its use quickly spread to schools, offices, hospitals, restaurants, retail stores, and other commercial and institutional buildings. Today, sound control is considered an essential element in the design of all buildings. When incorporated into a building's design, sound control can usually be accomplished inexpensively. Conversely, alterations to control unwanted sound transmission after a building has been completed are almost always costly.

This section addresses the properties and sources of sound, identifies common acoustical problems, and discusses methods of measuring and controlling noise. It also covers design and construction techniques and the criteria essential to providing and maintaining the acoustical performance of walls, floor and ceiling assemblies, doors, and windows and to resist transmission of airborne and structure-borne sound through them. It contains sound performance ratings, including the Sound Transmission Class (STC), the Impact Noise Rating (INR), and the Impact Insulation Class (IIC) for some commonly used walls and floor and ceiling assemblies. It also discusses *room acoustics*, in which the goal is to provide good hearing conditions within a space, and *sound transmission*, in which the goal is to control sound transmission, by airborne sound or structural vibrations, between spaces.

Chapter 15 addresses control of the generation and transmission of unwanted sound by and through plumbing systems and fixtures, and the role of heating, ventilating, and air conditioning systems in noise generation and transmission. It also describes methods to reduce both. Chapter 16 discusses methods to reduce sound transmission through and around electrical system devices, as well as noise generation by appliances.

13.1.1 PROPERTIES OF SOUND

Sound is the vibration of an elastic medium. In air, it is the movement of air molecules in a wave motion. When the frequency and intensity of this motion are within certain ranges, they produce the sensation of hearing.

A vibrating body causes the layer of air particles next to it to vibrate. These air particles in turn transmit their motion to the next layer of particles. Because air is compressible and has weight, a small interval of time is required for the first air layer to set the second layer in motion. Succeeding air layers are similarly set in

FIGURE 13.1-1 Sound is a wave motion in air caused by a vibrating source. Wave motion in water is analogous.

FIGURE 13.1-2 Light travels almost instantly, but sound requires about 5 seconds for each mile (1.609 km) it travels.

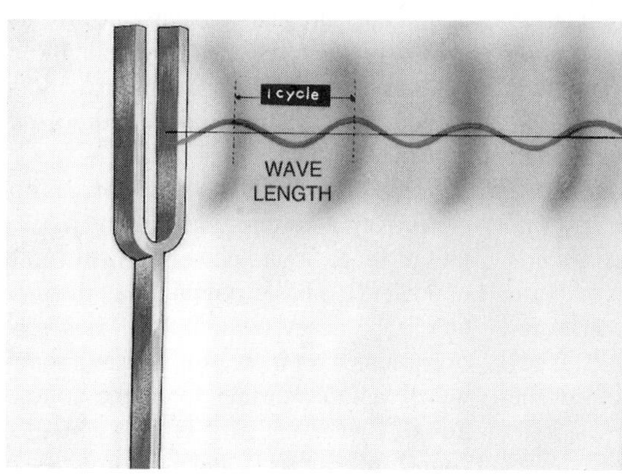

FIGURE 13.1-3 Wavelength is the distance a sound wave travels during one vibration (cycle).

motion after time intervals that are directly proportional to their distance from the source of the sound. This action constitutes a sound wave. The waves produced when a stone is dropped into water are analogous (Fig. 13.1-1).

13.1.1.1 Wave Motion

Water waves travel along the surface and are circular in shape. Sound waves spread out in three dimensions spherically. Each air particle vibrates only in the direction in which the wave is traveling; that is, along a straight line drawn between the particle and the source of the sound wave. Sound travels in air at a speed of about 1130 ft./sec. (344.42 m/s), or about 5 seconds for each mile (1.608 344 km) (Fig. 13.1-2). In water, sound travels at about 4500 ft./sec. (1371.6 m/s), and in steel at about 15,000 ft./sec. (4572 m/s). The speed in a given medium is always the same, regardless of the pitch or loudness of the sound. We hear echoes because we hear the sound from the source and then a reflected sound, which has to travel farther, a few moments later. The same property of sound also causes the unpleasant reverberations we sometimes experience indoors.

13.1.1.2 Frequency

One complete round trip of a vibrating body, starting from its neutral position, moving to one side then to the other side and back to neutral, is called a *cycle* (Fig. 13.1-3). The number of cycles performed in 1 second is called *frequency*. Frequency, then, is measured in cycles per second (cps) and may be expressed as *hertz* (*Hz*) (1 cps). Each air particle in a sound wave has the same frequency as the source. For example, if the sound from a loudspeaker has a frequency of 1000 cps, then the loudspeaker diaphragm and every air particle through which the sound wave passes are vibrating 1000 times each second. The performance and efficiency of acoustical materials in absorbing sound and the ability of materials to transmit sound vary, depending on the frequency of the sound.

The audible frequency range is from about 20 to 20,000 cps, but the range narrows as people grow older. Music covers most of this range, and speech extends from about 100 to 5000 cps. The most important range of frequencies for speech is shown in Figure 13.1-4. However, the human ear is more sensitive to sounds with frequencies between 1000 and 4000 cps, shown by the low dip on the Threshold of Hearing curve in Figure 13.1-4. Sounds between 250 and 1000 cps or between 4000 and 8000 cps must be more intense, and those below 250 or above 8000 cps must be considerably more intense in order to be heard. Most sounds consist of a mixture of numerous frequencies at different relative intensities.

WAVELENGTH

The length of a sound wave is calculated by dividing the velocity of sound by the frequency:

$$\text{wavelength} = \frac{\text{velocity}}{\text{frequency}}$$

Wavelength is the distance a sound travels during a single vibration. Since the velocity of sound is constant, each frequency has a different wavelength. A sound with a frequency of 1000 cps has a wavelength equal to about 1 ft. (0.304 8 m). To reflect a sound effectively, a reflecting surface must be equal to or greater than the wavelength in size. A 1-sq.-ft. (0.092 903 m^2) surface will reflect sound 1000 cps and higher, but sound below 1000 cps will bend around such a surface and pass beyond. A sound with a frequency of 100

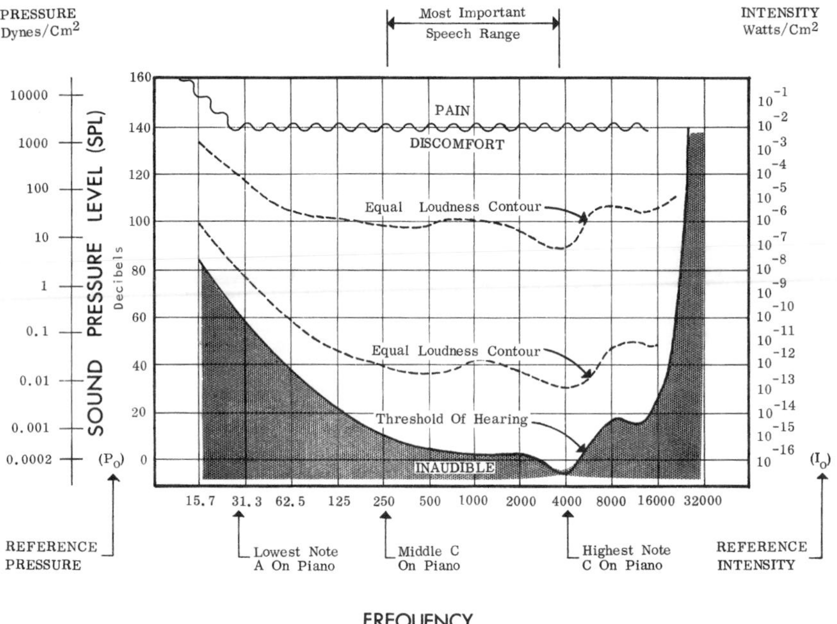

FIGURE 13.1-4 Audible frequency range. Two sounds of the same intensity but of different frequencies do not necessarily sound equally loud. Equal loudness contours represent sounds of different frequencies that seem to the average human ear to be equally loud.

cps has a wavelength of 11.3 ft. (3.45 m). This dimension usually exceeds the ceiling height and often the wall length of residential rooms; thus, the sound wave does not fit within the room, and the sound is not reproduced within the room with fidelity. This accounts for the poor reproduction in small rooms of low-frequency sounds from radios, tape and disk players, televisions, and similar sources.

Octaves and Tones The frequency scale is laid out in octaves in the same manner as a piano keyboard. The term *octave*, taken from musical terminology, is the interval between two sounds having a frequency ratio of 2:1. For example, frequencies between 125 and 250 cps or between 250 and 500 cps constitute an octave. A pure tone is a sound having only one frequency, whereas a complex tone is a sound composed of a number of different frequencies.

Pitch The hearing sensation produced by the frequency of a sound is called *pitch*. The greater number of vibrations per second, the higher the pitch. High-frequency sounds such as squeaks and whistles are heard as high pitches, and low-frequency sounds such as rumbles and roars are heard as low pitches. Pitch is expressed in tones of the musical scale.

Quality The *quality* of a tone is determined by the distribution of its numerous frequencies. Higher-frequency components are called *harmonics* or *overtones* when they are exact multiples of the fundamental frequency. The distribution and relative intensity of the overtones of a musical sound determine the quality (*timbre*) of the sound. Each of the two sounds illustrated in Figure 13.1-5 are complete tones composed of a combination of related frequencies. The vertical distances of the graphs are proportional to the loudness.

SOUND LEVELS

The intensity of the vibration of a sound source depends on the amount of force that sets it in vibration. Sound levels are expressed in *decibels* (dB). The decibel scale extends from 0 dB (threshold of hearing) to more than 120 dB (threshold of feeling). Zero decibel represents a fixed value of sound intensity that is slightly less than that of the faintest audible sound. Sound at 120 dB will cause the sensation of tickling in the ear. Sound at 180 dB, generated by some rocket engines, can cause structural damage to buildings and can be fatal to humans.

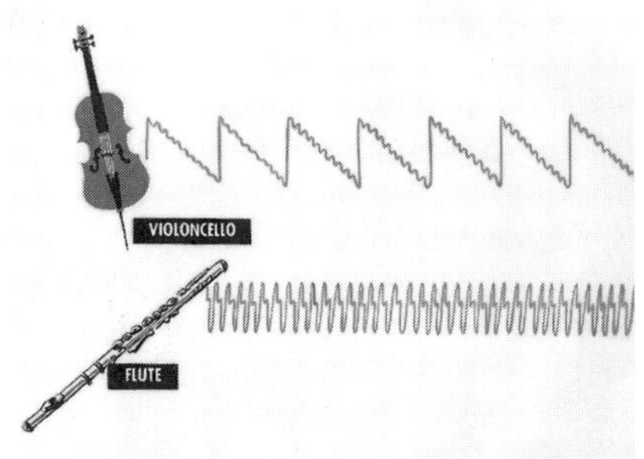

FIGURE 13.1-5 Musical instruments produce sounds of many related frequencies.

A decibel, which cannot be measured directly and has no fixed, absolute value, is the smallest change in sound intensity that can be detected by the average human ear. It is a logarithmic unit that expresses the ratio between a sound level being measured and a reference point. A decibel tells by what proportion one sound level is greater or less than another. The greater the sound level, the higher the rating on the decibel scale.

Sound levels in decibels are calculated from either a ratio of *sound pressures* (*P*) or *sound intensities* (*I*), using a reference point that corresponds to the faintest audible sound.

Pressure is a force per unit of area. Sound pressure at a given distance from a source is the force of moving air molecules spread over a spherical area and is expressed in *dynes* (units of force) per square centimeter.

Power is expended when a force is applied over a certain distance in a given unit of time. Sound power in *watts* (units of power) is a basic expression of the flow of sound energy. Sound power is spread over a sphere of increasing area as a sound wave radiates farther from the source. Sound power per unit of area is called sound intensity (*I*) and is expressed in watts per square centimeter (W/cm^2).

The reference pressure, marked (P_o) in Figure 13.1-4, is 0.0002 dyne/cm^2 and is roughly equivalent to the smallest amount of pressure that will cause an eardrum to vibrate. The corresponding reference intensity, marked (I_o) in Figure 13.1-4, is 10^{-16} W/cm^2.

The ratio in decibels between any measured pressure (*P*) and the reference pressure (P_o) is called the *sound pressure level* (*SPL*). It is found by formula (a) in Figure 13.1-6. The ratio in decibels between any intensity (*I*) and the reference intensity (I_o) is called the *sound intensity level* (*SIL*). It is given by formula (b) in Figure 13.1-6.

The decibel scale makes values com-

(a) Sound Pressure Level

$$SPL = 20 \log_{10} \frac{P}{P_o}$$

(b) Sound Intensity Level

$$IL = 10 \log_{10} \frac{I}{I_o}$$

FIGURE 13.1-6 Formulas for computing the ratio in decibels between (a) the pressure of two sounds and (b) the density of two sounds.

puted for either pressure or intensity directly comparable. Although intensity levels are convenient for discussion, they cannot be measured directly. Therefore, in practice, sound levels are determined by measuring pressures as read on a sound level meter and expressing the measurement as SPL on a decibel scale.

The easiest way to understand sound levels on the decibel scale is to compare them with common, easily recognized sounds (Fig. 13.1-7).

An increment in sound level of 1 dB (say, the difference between 50 and 51 dB) is barely perceptible. A 2- to 3-dB increment is insignificant, a 5-dB increment is clearly noticeable, and a 10-dB increment is quite perceptible. It corresponds to a subjective reaction of doubling (or halving) the loudness.

FIGURE 13.1-7 Levels of Typical Sounds in Decibels

Loudness	Sound Level in Decibels (dB)	
	160	Near jet engine
	130	Threshold of painful sounds; limit of ear's endurance
Deafening	120	Threshold of feeling (varies with frequency) 18 ft. (5.48 m) from airplane propeller
	110	
		Express train passing at high speed
	100	Loud automobile horn 23 ft. (7 m) away
		Noisy factory
Very loud	90	
		Loud hi-fi
	80	Subway train
		Motor trucks 15 to 50 ft. (4.572 to 15.24 m) away
Loud	70	Stenographic room, average TV, loud conversation
	60	Average busy street
		Noisy office or department store
Moderate	50	Moderate restaurant clatter
		Average office, noisy room, average conversation
	40	
		Soft radio music in apartment, average residence
Faint	30	
	20	Average whisper 4 ft. (1.22 m) away
Very faint	10	Rustle of leaves in gentle breeze
	0	Threshold of audibility

Loudness The *loudness* of a sound depends on the intensity and frequency of the sound and on the characteristics of the human ear. People hear sound over a wide range of intensity, extending from a low whisper to a deafening roar. Beyond a certain point, pain in the ears becomes more evident than the sensation of increasing loudness.

There are limitations in the frequency response of the human ear, as well as a definable range of intensities to which it will respond. The psychological sensation of loudness is closely related to the intensity of a sound. The greater the intensity, the louder a sound appears to a listener, but a given sound may appear to be much louder when it is in a middle frequency range than when it is in a high or low range. In fact, Figure 13.1-4 shows that a 15-dB tone, heard easily at 1000 or 2000 cps, is inaudible at 200 and 10,000 cps. The *Equal Loudness Contours* shown in Figure 13.1-4 connect the tones of different frequencies that sound equally loud. The loudness of sound also depends on the characteristics of the listener's ear. Thus, when the intensity of a given sound striking the ear of a normal-hearing person and the ear of a hard-of-hearing person is exactly the same, the sensation of loudness is quite different.

Noise Unwanted sound is called *noise*. Usually, a sound becomes noise when it is loud enough to be a nuisance. Certain sounds, such as scratching, are noise because they are nonuniform vibrations. Even pleasant or otherwise acceptable sound becomes noise when it interferes with concentration, conversation, or other tasks. How loud a sound must be before it becomes objectionable depends on the kind of sound and the person who hears it.

13.1.2 ROOM ACOUSTICS

Room acoustics is a branch of architectural acoustics that deals with both *acoustical correction* and *noise reduction*.

13.1.2.1 Acoustical Correction

Acoustical correction is the physical shaping of a space to produce the best possible hearing conditions. It is usually done only for public spaces, such as auditoriums, concert halls, and churches, but is sometimes done for other types of spaces as well. The objective is to shape an environment where sounds are heard distinctly, in ample loudness, and with as faithful a replication of the source as possible, everywhere in the space. Acoustical correction requires a knowledge of the principles of acoustics, engineering calculations, design of the space to take best advantage of the properties of sound, and the selection and location of acoustical and other finish materials. Sound amplification systems are often involved. Such design requires the skill and judgment of an acoustical specialist and is therefore beyond the scope of this text. In this section, room acoustics deals only with noise reduction.

13.1.2.2 Noise Reduction

Noise reduction is the treatment of interior surfaces with acoustical materials to alleviate the discomfort and distraction caused by the reflection of unwanted sound.

REFLECTION OF SOUND

When a sound source is operating, sound waves travel outward in all directions radially from the source. When the sound waves encounter an obstacle or surface, such as a wall, their direction of travel is changed, or reflected.

The reflection of sound follows the same laws as the reflection of light from a mirror: (1) the direction of travel of the reflected sound always makes the same angle with the surface as that of the incident sound, that is, the angle of incidence equals the angle of reflection; and (2) the reflected sound waves travel exactly in the same manner they would if they had originated at an "image" of the sound source, located the same distance behind the wall as the real source is in front of the wall (Fig. 13.1-8).

Multiple Reflection When a sound is produced in a room having reflective surfaces, sound waves are reflected back and forth by these surfaces and almost immediately the room is filled with sound waves traveling in every direction. This is called *multiple reflection* (Fig. 13.1-9). Multiple reflection causes a buildup of sound intensity to a level greater than there would be if no reflection occurred. It makes noises unnaturally and unnecessarily loud and makes noise last longer.

Aside from causing annoyance and fatigue, excessive reflection interferes with telephoning, conversation, detection of alarms and sound signals, and other types of work or play that depend on accurate hearing.

Reverberation When a sound source is stopped, the reflected waves do not cease to exist at that instant but continue to travel back and forth, *reverberating* (bouncing) between room surfaces. The total sound energy in the room gradually diminishes due to the partial absorption of the sound at each contact with a sound-absorbing surface (Fig. 13.1-10).

Reflected sound waves strike the ear of a listener in rapid succession so that the listener does not hear them as distinct repetitions of the original sound. Instead, the listener hears the original sound being drawn out or prolonged after the source is stopped, and steadily dying out until it becomes inaudible.

Reverberation Time The *reverberation time* of a room is the time in seconds required for a sound to die out to an intensity one millionth the intensity it had at the moment the source was stopped. On the decibel scale this is a drop of 60 dB (Fig. 13.1-11).

The reverberation time (T) of a room depends only on its size (volume, V) and sound-absorbing properties (absorption units, a), not on the power of the source (Fig. 13.1-12).

Values of reverberation time have been established that give the best hearing conditions for speech and music (Fig. 13.1-13). Speech intelligibility becomes

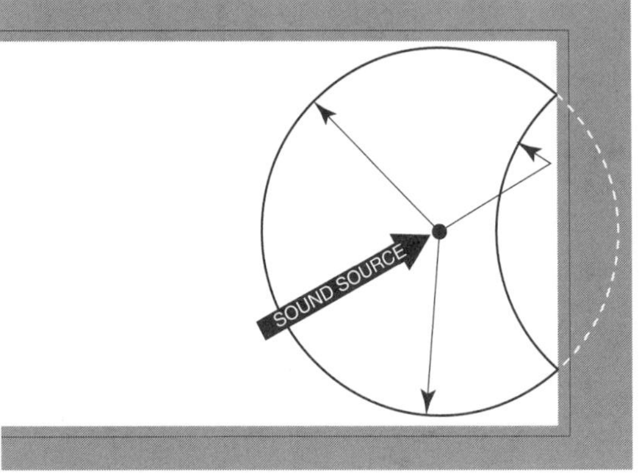

FIGURE 13.1-8 Sound travels radially in all directions and is reflected by room surfaces.

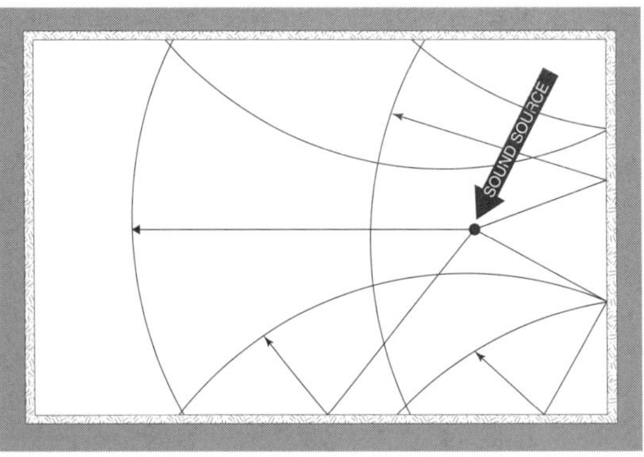

FIGURE 13.1-9 Multiple reflection of sound waves causes sound levels to build up, and sounds to reverberate longer.

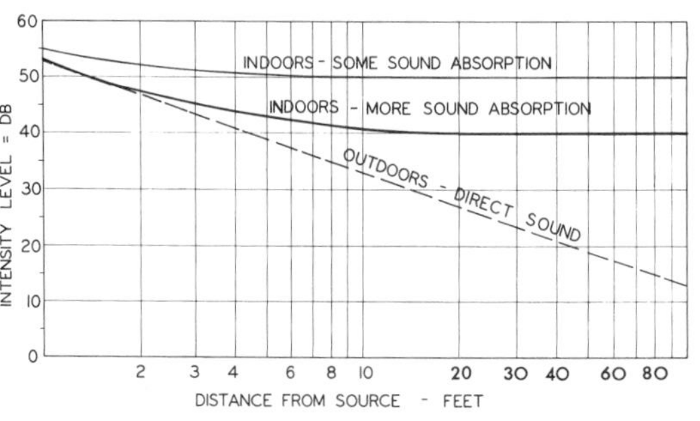

FIGURE 13.1-10 Relation of distance from the source to the resulting sound level in decibels.

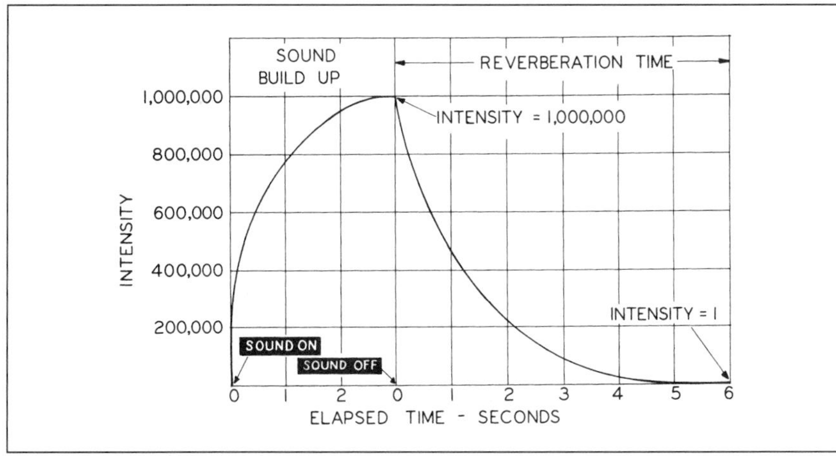

FIGURE 13.1-11 Curve showing building up and dying out of sound in a room.

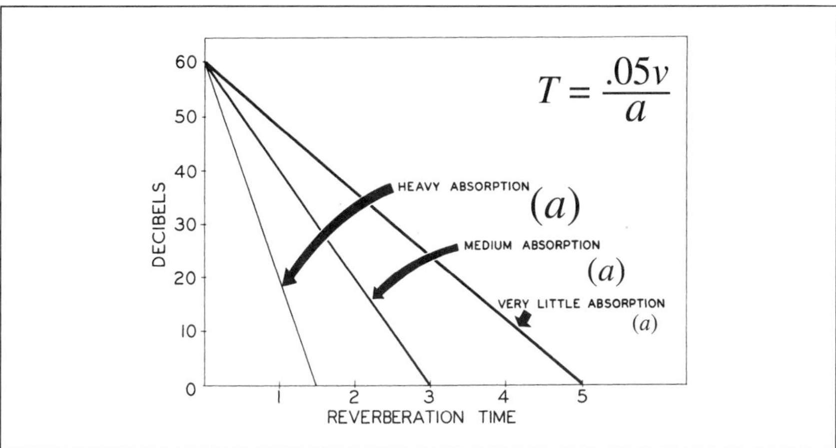

FIGURE 13.1-12 Reverberation time is important for speech intelligibility and quality reproduction of music. T = reverberation time in seconds, V = room volume in cubic feet, and a = room absorption in sabins.

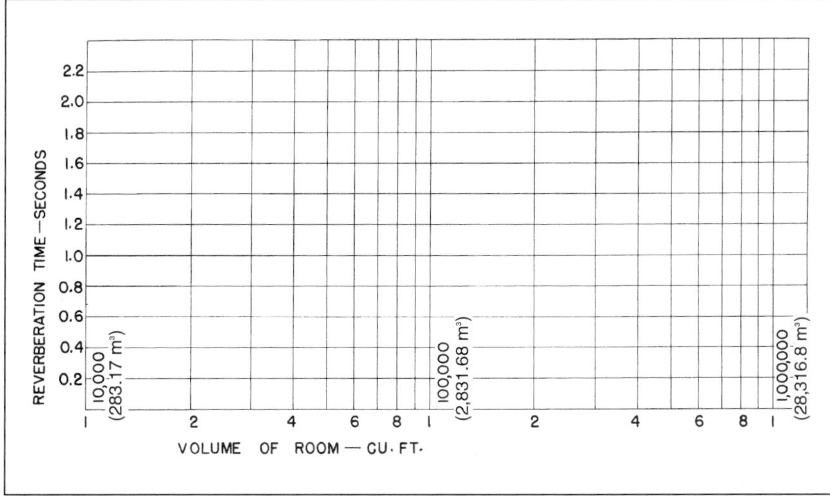

FIGURE 13.1-13 Range of acceptable reverberation times at a frequency of 500 cycles per second (cps) for rooms of various sizes.

poorer as loudness decreases or as reverberation time increases.

Echoes An *echo* is a reflection that can be heard as a distinct repetition of the original sound. Echoes are likely to occur in large rooms where there is an appreciable time lag between the original sound and the reflected sound. To be heard as an echo, the reflected sound must be heard at least 1/17 second later than the direct sound. Therefore, reflected sound must travel a path from source to listener at least 65 ft. (19.81 m) longer than the path of the direct sound. Large rooms make possible the time lag for hearing an echo, and when reverberation time is short the echo is heard more easily above the general reverberation.

Echoes produced by concave surfaces are louder than those produced by flat surfaces because a curved surface converges and focuses the reflected sound, increasing its intensity in the same manner as the curved reflector in a flashlight focuses light (Fig. 13.1-14). A ceiling having a center of curvature near the floor line is the most objectionable type of curved ceiling from the standpoint of focused echoes, since the focal point is brought closer to the audience than when the center of curvature is higher or lower.

Focusing is best corrected by breaking up the curve or changing its shape. Treatment with acoustical materials is rarely enough.

ABSORPTION OF SOUND

Smooth, hard building surfaces reflect up to 98% of the sound that strikes them. This reflection causes the sound to build up to a high level and to die out slowly.

Acoustical materials are designed to absorb most of the sound striking them and reflect less than 50% back into a room. They owe their absorption efficiency to their high porosity.

The amount of absorption of a porous material depends on its thickness, the size and number of its pores, and the frequency of the sound. Air particles moving in and out of the pores cause friction, and sound energy is dissipated as heat. Various materials designed for this purpose are discussed in Chapter 9.

When sound-absorbing materials are placed over the hard surfaces in a room, sound does not build up as much and will die out faster. Although the sound level goes down, no matter how much sound-absorbing material is introduced into the

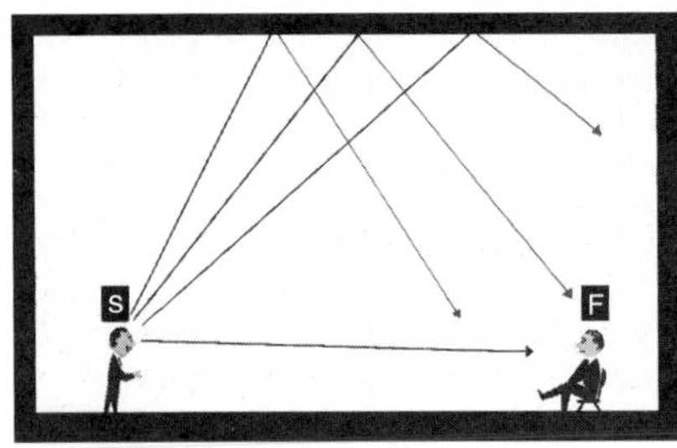

a

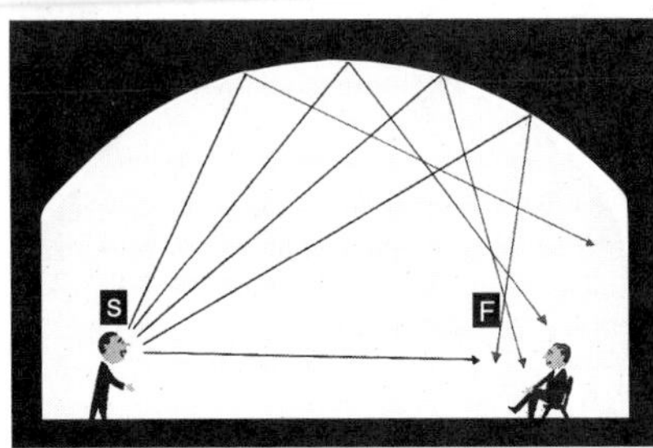

b

FIGURE 13.1-14 (a) Reflections from a flat surface are divergent. (b) Concave surfaces focus sound at points (F) that may be louder than the source (S).

room, the level cannot drop below the level of the direct sound from the source to the listener. Sound-absorbing materials cannot "suck up" the sound from the source; they can only absorb what reaches them. This is analogous to light in a room. When the walls are white (reflective), the whole room is bright, but if the walls are painted black (absorptive), the whole room becomes dark, even though the same source at the same brightness level continues to emit the same amount of light. Acoustical materials act essentially the same with sound as black paint acts with light.

In practice, it is impractical and uneconomical to obtain sound-level reductions of much more than 10 dB by using sound-absorbing materials. For example, 5 to 7 dB is the reduction usually accomplished when the entire ceiling of a normal office is covered with acoustical tile. However, such a reduction is obvious to the user of the space, and the improvement in communications and comfort is usually conspicuous.

Measuring Sound Absorption One unit of absorption (called a *sabin*) is equal to 1 sq. ft. (0.092 903 m^2) of totally absorbent surface. An open window 1 sq. ft. (0.092 903 m^2) in area is equivalent to 1 sound-absorbing unit, because, theoretically, all sound that strikes it passes through; none is reflected. Therefore, an open window has 100% absorption.

An *absorption coefficient* (a) is assigned to materials to measure the percentage of incident sound energy absorbed by them. These vary theoretically from 0 to 1. A perfectly reflective material has $a = 0$, and a perfectly absorptive material has $a = 1$. A material with an absorption coefficient of $a = 0.50$ will absorb 50% of the sound that strikes it and reflect the other 50%. Because absorption coefficients vary with the frequency of the sound, it is necessary to test materials at more than one frequency in order to cover the complete range of sound frequencies. Materials are normally tested at six frequencies: 125, 250, 500, 1000, 2000, and 4000 cps. The test results are plotted as a curve (Fig. 13.1-15). The absorption coefficients for typical construction surfacing materials are given in Figure 13.1-16.

The numerical average of the absorption coefficients at the four middle frequencies, 250, 500, 1000, and 2000 cps, rounded off to the closest 5% is called the *noise reduction coefficient* (NRC). This is a single-number rating used to calculate the required amount of sound-absorbing materials needed. It permits comparison of such materials according to their acoustical performance.

Because small differences in acoustical efficiency cannot be detected by the human ear, and since tests that determine this efficiency cannot be carried out with pinpoint accuracy, the selection of an acoustical material is usually based on a 10-point range (.10) of NRCs (Fig. 13.11-17). A difference of less than 10 points is insignificant because it is seldom detectable in a completed installation.

In addition to the sound-absorption rating of a material, such other properties as light reflectivity, flame resistance, and appearance are also generally considered. Only in special cases where high absorption is required does NRC alone determine the selection. The sound-absorption and noise-reduction coefficients of acoustical materials are published by their manufacturers.

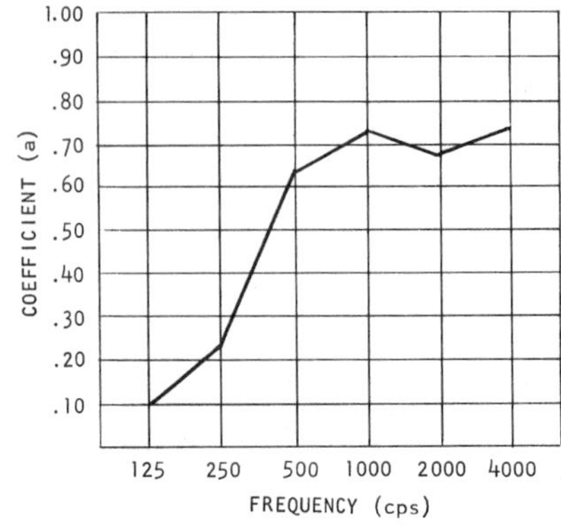

FIGURE 13.1-15 Absorption coefficients and other information are published by the manufacturers of acoustical materials.

Controlling Noise with Sound-Absorbing Materials Acoustical treatment controls noise by reducing multiple reflection of sound waves, thereby reducing the loudness level of reflected sound, and shortening the reverberation time, so that the noise is not prolonged unnecessarily.

There are two basic noise-reduction principles that serve as rough guides in determining the amount of treatment within a given room for satisfactory noise control: (1) the total number of absorption units in spaces with low ceilings should be between 20% and 30% of the spaces' total area of interior surface in square feet; and (2) to produce a satisfactory improvement in an existing space, the total absorption after treatment must be between 3 and 10 times the absorption before treatment.

The following outlines one procedure for calculating the percentage of reduction in loudness of reflected sound produced by acoustical treatment.

1. Determine the number of absorption units in the space before treatment (a_1). To do this, multiply the absorption coefficients of each material in the space (see Fig. 13.1-16) by the area occupied by that material in square feet. This is the number of absorption units contributed by that material. It is usually satisfactory to use the material's absorption coefficient at 500 Hz. Add all the resultant units together. To these add an estimated number of units for furniture at about 0.03 times the area covered by furniture. Then add about 4 units per occupant.
2. Determine the number of units provided by the treatment. To do this, obtain the manufacturer's stated noise-reduction coefficient (NRC) of the selected treatment. This may be .70, for example. Multiply this by the area covered by the treatment in square feet.
3. From a_1 subtract the number of units originally calculated for the material to be covered by the treatment. Then add the number of units provided by the treatment. This is the total absorption in the room after treatment (a_2).
4. Divide a_2 by a_1, to obtain the absorption ratio a_2/a_1.

FIGURE 13.1-16 Absorption Coefficients for Typical Construction Materials

Material	Absorption Coefficients 125 Hz	250 Hz	500 Hz	1000 Hz	2000 Hz	4000 Hz
Brick						
Unglazed	.03	.03	.03	.04	.05	.07
Unglazed, painted	.01	.01	.02	.02	.02	.03
Carpet (heavy) on concrete slab	.02	.06	.14	.37	.60	.65
With 40-oz. (1134 g) hair-felt or foam-rubber pad	.08	.24	.57	.69	.71	.73
With impermeable latex backing and 40-oz. (1134 g) hair-felt or foam pad	.08	.27	.39	.34	.48	.63
Concrete Block						
Porous	.36	.44	.31	.29	.39	.25
Painted	.10	.05	.06	.07	.09	.08
Fabrics						
Light velour (10 oz./sq. yd.) (339.06 g/m^2) hung straight in contact with wall	.03	.04	.11	.17	.24	.35
Medium velour (14 oz./sq. yd.) (474.68 g/m^2) draped to half the area	.07	.31	.49	.75	.70	.60
Floors						
Concrete or terrazzo	.01	.01	.015	.02	.02	.02
Resilient flooring on concrete	.02	.03	.03	.03	.03	.02
Wood	.15	.11	.10	.07	.06	.07
Parquet in asphalt on concrete	.04	.04	.07	.06	.06	.07
Glass						
Large panes heavy plate glass	.18	.06	.04	.03	.02	.02
Standard window glass	.35	.25	.18	.12	.07	.04
Gypsum board, ½ in. (12.7 mm) nailed to 2-in. (50.8 mm) × 4-in. (101.6 mm) studs 16 in. (406.4 mm) o.c.	.29	.10	.05	.04	.07	.09
Marble or glazed tile	.01	.01	.01	.01	.02	.02
Plaster						
Smooth finish on brick or tile	.013	.015	.02	.03	.04	.05
Rough finish on lath	.02	.03	.04	.05	.04	.03
Smooth finish on lath	.02	.02	.03	.04	.04	.03
Plywood paneling, ⅜ in. (9.53 mm) thick	.28	.22	.17	.09	.10	.11

5. Determine the percentage of loudness reduction from Figure 13.1-18.

It is desirable to reduce the reverberation time in small spaces to about 0.5 sec., and in large spaces, especially those used for music as well as speech, to not more than 1 sec. In most spaces, the problem is usually one of reducing reverberation enough to lower noise levels about 5 to 8 dB. Unless the room shape presents special problems, covering the ceiling with a good sound-absorbing material is usually adequate treatment.

Spaces such as corridors generally need sound absorbing materials on the floor as well as the ceiling to reduce the *piping* of sound along the corridor to other spaces.

In small rooms, especially those containing upholstered furniture and heavily carpeted, additional absorption in the form of acoustical ceilings will probably be unnecessary.

Transmission Through Sound-Absorbing Materials Acoustical materials are used to absorb sound. Although some materials

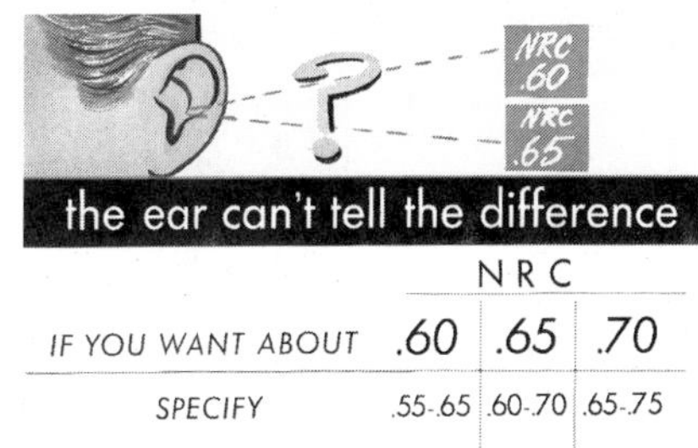

FIGURE 13.1-17 NRC rates the efficiency of acoustical materials.

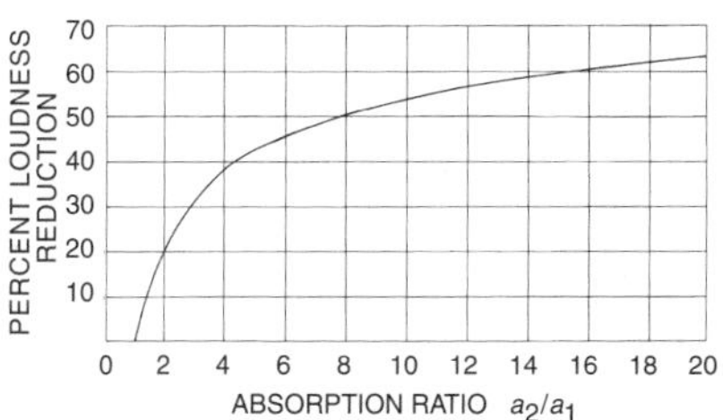

FIGURE 13.1-18 Relation of reduction in loudness (in %) to the absorption ratio.

may be effective in preventing sound transmission, most acoustical surfacing materials are not. Acoustical tile and panels, carpet and pad, and other sound-absorbing surfacing materials placed in a room where noise originates act indirectly in reducing sound transmission to adjoining rooms by lowering the noise level in the room where the sound originates. Used in an adjoining room, they will similarly lower the level of the transmitted noise. These reductions, however, usually amount to only a few decibels. Acoustical surface treatment will supplement but will not take the place of sound isolation. *Sound isolation* is the control of sound transmission through walls and floor and ceiling assemblies.

13.1.3 SOUND TRANSMISSION

Sound is transmitted through buildings when wall, floor, or ceiling surfaces vibrate. This vibration may be caused by (1) the alternating air pressure of incident sound waves, called *airborne sound transmission*, or (2) direct mechanical contact or impact such as may be caused by vibrating equipment, a footstep, an object dropped on a floor, or a slamming door. Transmission as a result of direct mechanical contact or impact is called *structure-borne sound transmission*. In either case, vibrating surfaces generate new sound waves of reduced intensity in the space where the sound originates, as well

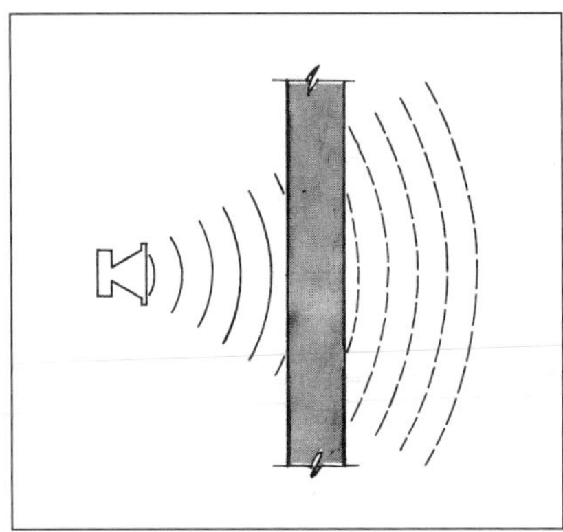

a Airborne

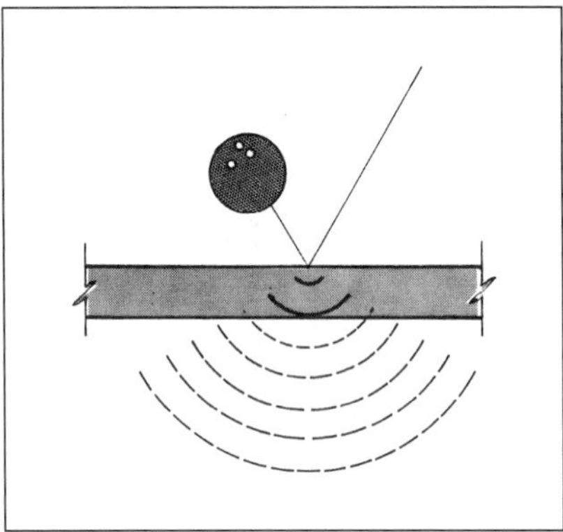

b Direct Impact or Mechanical Contact

FIGURE 13.1-19 Sound is transmitted by (a) alternating air pressures of airborne sound and (b) impact or direct mechanical contact.

as in spaces adjacent to or above or below (Fig. 13.1-19).

13.1.3.1 Airborne Sound Transmission

Airborne sound is produced by sources that radiate sound directly into the air, such as people talking, a TV on or a CD playing, mechanical equipment humming, or traffic rumbling. Airborne sound may be transmitted to adjoining rooms or spaces (1) directly through the intervening constructions or (2) by way of *flanking paths* such as open doors or along interconnecting structures (Fig. 13.1-20). Airborne sound can be isolated by walls and floor and ceiling assemblies that reduce the level of transmission to the point where it is either inaudible or not annoying or interfering.

MEASURING AIRBORNE SOUND TRANSMISSION

To determine the effectiveness of walls and floor and ceiling assemblies in isolating airborne sound transmission, the *transmission loss* of the wall or assembly must be measured. Transmission loss (*TL*) is the number of decibels a sound loses while passing through a wall or floor and ceiling assembly when tested in a laboratory in accordance with American Society for Testing and Materials (ASTM) Standard E 90. In both laboratory and field tests of walls for airborne transmission, the procedure is to generate a steady sound on one side of the wall and measure the sound level in decibels on both sides. The TL of the wall is the measured difference in sound level of the two sides. Therefore, a sound of a given frequency with an intensity level of 70 dB on one side of a wall will have a reduction in intensity to 40 dB on the other side when the wall has a TL of 30 dB at this frequency (Fig. 13.1-21). The higher the TL, the more efficient the wall or assembly.

TL is a physical property of a wall or floor and ceiling assembly. It depends on the materials and methods used in construction, not on the loudness of sound striking the surface. However, because TL varies considerably depending on sound frequency, assemblies are tested at numerous frequencies and plotted as a curve (Fig. 13.1-22). Most assemblies are more efficient at reducing high-frequency sounds than at blocking low-frequency sounds.

It is seldom necessary or economically feasible to build walls or floor and ceiling assemblies that are capable of completely blocking sound transmission. Sound transmitted through a wall or floor will be inaudible or "masked" if its intensity level is below the level of the background noise in the receiving room (Fig. 13.1-23).

Single-Number Rating Systems It is generally impractical to use the TL values at each frequency tested, because it is difficult to identify the frequency of a noise source in sufficient detail. For convenience in selecting or comparing walls or floor and ceiling assemblies, two single-

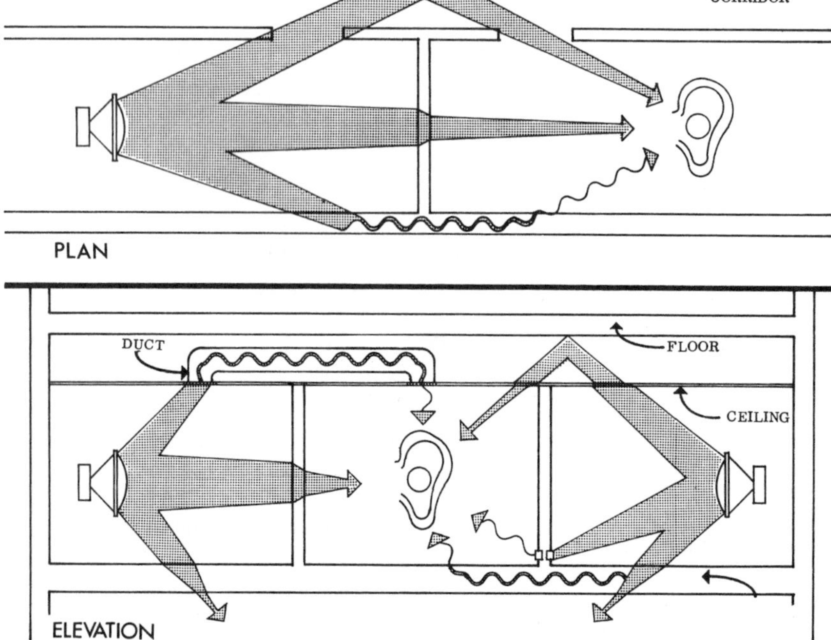

FIGURE 13.1-20 Sound may be transmitted (1) through the intervening construction, (2) through open air paths, or (3) along the interconnecting structure.

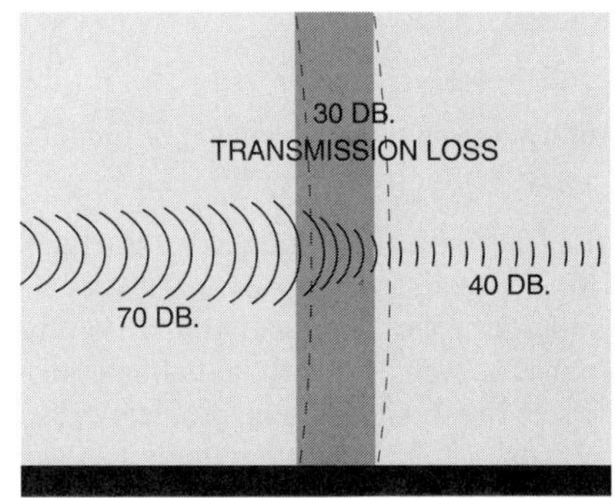

FIGURE 13.1-21 Transmission loss is the difference in sound level on opposite sides of a construction. The wall here has a transmission loss of 30 dB.

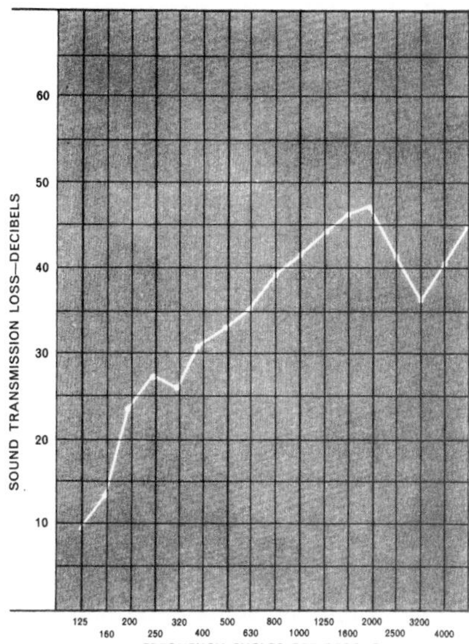

FIGURE 13.1-22 The transmission loss of a wall consisting of ½-in. (12.7 mm)-thick gypsum board nailed directly to both sides of wood studs placed at 16 in. (400 mm) on centers.

number rating systems have been developed, the *average TL* system and the *Sound Transmission Class* (*STC*) system. Either will indicate whether a wall or floor and ceiling assembly will offer sufficient resistance to sound transmission, but the STC system is the one generally used today.

The average TL is the numerical average of the transmission loss values measured at 9 or more frequencies, including 125, 175, 250, 350, 500, 700, 1000, 2000, and 4000 cps. Average TL was used in the past by the National Bureau of Standards (National Institute of Standards and Technology), Riverbank Acoustical Laboratories, and other laboratories for rating airborne sound transmission through walls and floor and ceiling assemblies. It was also once the most commonly used basis for comparison of published data.

To estimate the minimum average TL required of a wall or floor and ceiling assembly to obtain speech privacy, subtract the level of the background noise on one side of the construction from the level of sound that strikes the other side. Hearing conditions on the listening side of walls with varying average TL are shown in Figure 13.1-24. Based on the curve shown in Figure 13.1-22, the average TL of a commonly used wood stud partition is 31 dB, which is not much resistance to sound transmission.

The average TL system gives a fair indication of sound isolation value if there are no sharp dips on the test curve and if the values being compared are based on identical tests performed by the same laboratory under identical conditions. Sharp dips on the curve in a narrow frequency band, which are normal for partitions, are harmful, but the numerical averaging in this system misses them, making it inaccurate for use with partitions.

The Sound Transmission Class (STC) system of rating airborne-sound isolating performance is preferable to the average TL system because it compensates for dips in the test curve by lowering the overall evaluation when they are present. This gives a more reliable indication of speech or noise transmission, especially by partitions.

The STC is a single-number rating for airborne sound transmission that represents the TL of a wall or floor and ceiling assembly at all test frequencies. The higher the STC rating, the more efficient the construction.

FIGURE 13.1-24 Sound-Insulating Properties of Construction According to Transmission Loss (TL)*

Transmission Loss (TL)	Hearing Conditions
30 dB or less	Normal speech can be understood quite easily and distinctly through the wall.
30 to 35 dB	Loud speech can be understood fairly well. Normal speech can be heard but not easily understood.
35 to 45 dB	Loud speech can be heard, but is not easily intelligible. Normal speech can be heard only faintly, if at all.
40 to 45 dB	Loud speech can be faintly heard but not understood. Normal speech is inaudible.
45 dB or greater	Very loud sounds, such as loud singing, brass musical instruments, or a radio at full volume can be heard only faintly or not at all.

*Background noise level on the listening side assumed at 30 dB.

An STC rating is derived by plotting the laboratory TL test curve for a given assembly, tested at 16 frequencies in accordance with ASTM Standard E 90, and comparing this curve with a standard frequency curve (*standard contour*) (Fig. 13.1-25). The standard frequency curve is drawn on a transparent overlay, which is placed over the test curve and moved up and down until the test curve remains above the standard curve at every frequency, within the following tolerances: (1) no single deviation where the sound transmission loss value falls below the standard curve may exceed 8 dB, and (2) the sum of these unfavorable deviations may not exceed 32 dB. The STC rating is the numerical value that corresponds to the sound transmission loss value at 500 Hz (see Fig. 13.1-25).

The ASTM test procedure was first issued in 1950 and has been extensively revised since that time. Some currently published STC ratings for assemblies are based on tests performed before the current revision of ASTM E 90. STC values derived from earlier tests may vary somewhat from tests on similar assemblies performed in accordance with ASTM E 90.

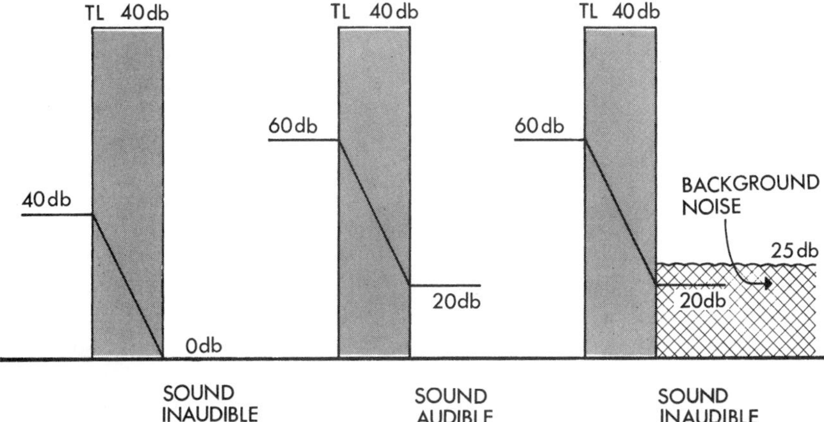

FIGURE 13.1-23 Normal background noise on the listening side of a wall has the effect of masking transmitted sound that may otherwise be objectionable.

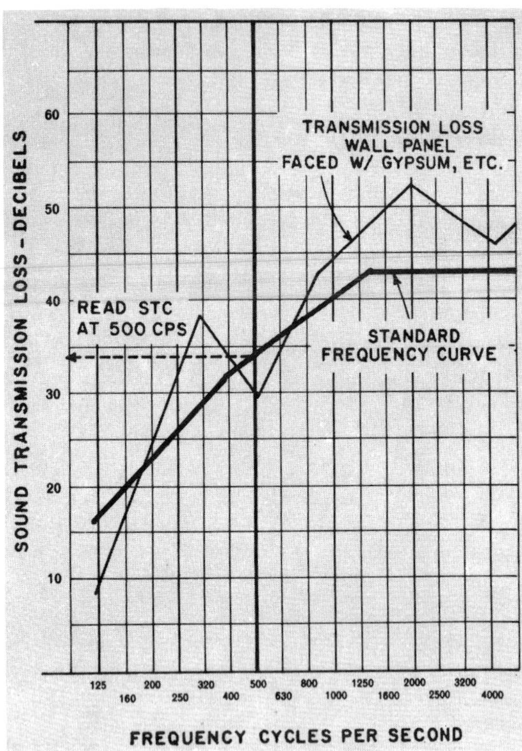

FIGURE 13.1-25 Method for obtaining an STC rating.

FIGURE 13.1-26 Sound Isolation of Walls in Terms of STC Values

Sound Transmission Class (STC)				
	Background Noise[a]			
None	20 dB	30 dB	40 dB	Speech Audibility[b]
40 to 50	25 to 35	15 to 25		Normal speech easily understood.
50 to 60	35 to 45	25 to 35	15 to 25	Loud speech easily understood. Normal speech 50% understood.
60 to 70	45 to 55	35 to 45	25 to 35	Loud speech 50% understood. Normal speech faintly heard but not understood.
70 to 80	55 to 65	45 to 55	35 to 45	Loud speech faintly heard but not understood. Normal speech usually inaudible.
	65 to 75	55 to 65	45 to 55	Loud speech usually inaudible.

[a]Background noise levels on *A* scale of sound level meter.

[b]Loud speech taken as 10 dB higher than normal speech; intelligibility taken as percent of sentences understood; normal source and receiving room absorption considered.

Very old references may vary considerably. Variables that may influence test results include the number of test frequencies used, the size of the panel tested, and the application of the standard frequency curve.

In general, a wall or floor and ceiling assembly provides acceptable privacy when the level of conversational speech in one room is reduced to a level at which it cannot be understood in another room. To determine the minimum STC required to obtain speech privacy, subtract the level of the background noise on one side of the wall or assembly from the level of sound that strikes the other side. The examples in Figure 13.1-26 illustrate the approximate effectiveness of walls with varying STC numbers as influenced by the background noise. Values of STC for some assemblies are given in Section 13.1.5.

CONTROLLING AIRBORNE SOUND TRANSMISSION

Variables that affect the control of airborne sound transmission to provide a suitable acoustical environment include (1) the frequency and sound level of the sources, (2) the sound level that will be acceptable within the space, and (3) the sound-reducing characteristics of the intervening construction.

Methods to control airborne sound transmission include (1) selecting walls and floor and ceiling assemblies that reduce transmission by vibration, (2) closing paths of direct air transmission, and (3) installing sound-absorbing materials within the free air space of frame construction. Usually, a combination of these control measures is necessary to achieve suitable sound privacy. Methods of controlling airborne sound transmitted through walls and through floor and ceiling assemblies are discussed in Section 13.1.4, "Design and Construction."

The sound-reducing characteristics of a wall or floor and ceiling assembly directly influence its cost. Therefore, it is important to minimize or eliminate the need for special construction techniques by proper planning using some basic sound-control principles and common sense. These are discussed in Section 13.1.4, "Design and Construction."

13.1.3.2 Structure-Borne Sound Transmission

Structure-borne sound transmission occurs when walls or floor and ceiling assemblies are set in vibration by direct impact or direct mechanical contact. The vibrating surface generates new sound waves on both sides of the construction. The new sound waves may even occur in different parts of the building. Floors are especially critical in this type of noise transmission. Some structure-borne noise is caused by objects striking or sliding on a floor, such as footsteps, furniture, and dropped objects; objects that bump against a wall, such as furniture; and by the slamming of doors. Other such noise is generated by mechanical equipment, fixtures, and appliances that transmit their own vibrations to adjacent construction. These include motors and fans on heating, cooling, and ventilating equipment; vacuum cleaners; water closets; bathtubs; and showers.

Isolating structure-borne sound requires providing wall and floor and ceiling assemblies that will reduce the level of transmission to the point where the sound is either inaudible or not annoying or interfering.

MEASURING IMPACT SOUND TRANSMISSION

Resistance to impact sound transmission is determined using a standard test method described in ASTM E 492, which is based on the International Organization for Standardization's (ISO) Recommendation 140, "Field and Laboratory Measurements of Airborne and Impact Sound Transmission." This test is performed by placing on the floor to be tested a standard tapping machine, which produces a series of uniform impacts at a uniform rate. The *impact sound pressure level* (ISPL) (measured in decibels) is then recorded in the receiving room below and electronically separated into different frequency bands. A curve can then be plotted, showing how the sound energy in the receiving room is distributed over the audible frequency range (curve 1 in Fig. 13.1-27).

To determine airborne sound transmission resistance, the difference between the sound levels in the sending and receiving rooms is measured and is expressed as a sound transmission loss. Conversely, ISPL curves measure the actual sound levels from a standard noise source that are transmitted through the construction. The lower the transmitted level, the better the acoustical performance of the construction.

ISPL determinations assume that an assembly that will transmit little sound generated by a standard tapping machine will perform similarly with other types of impact sound. Unfortunately, the noise produced by a standard tapping machine does not duplicate the sound produced by many types of impact. For example, much footfall noise is in a frequency below the lowest of the standard test frequencies. Because of the problems with duplicating actual conditions (and other limitations of the test used), model building codes often require that specific construction assembly types be used rather than relying on test performance criteria.

Impact Noise Rating (INR) An *impact noise rating* (INR) is a single-number rating in decibels for evaluating the impact sound isolation performance of a floor and ceiling system. It is determined by analyzing tests performed in accordance with ISO R140. This method has been superseded by the *impact-insulation class* (IIC) system discussed later in this section. The INR rating system is discussed here because some still-available published material uses it to compare the performances of floor and ceiling assemblies.

The INR for an assembly is determined by first measuring the assembly's ISPL, plotting it on a transparent overlay, and placing the overlay over a standard criterion curve (curve 2 in Fig. 13.1-27). The standard curve is then moved up or down until the ISPL test curve falls over the standard criterion curve within the following tolerances: (1) the mean amount by which the test curve exceeds the standard curve must be 2 dB or less, as averaged over the 16 one-third octave frequency bands between 100 and 3200 Hz; (2) the test curve must not exceed the criterion curve by more than 8 dB at any one frequency.

The INR value for a given assembly is the distance the ISPL curve falls below or above the standard criterion curve. Plus (+) INR values indicate better than standard criterion performance; minus (–) INR values indicate less than standard criterion performance. When the test curve meets the standard curve within the allowable tolerance, the INR is 0. If the standard curve must be moved down 5 dB for the test curve to conform, the construction is given a +5 rating, indicating that the assembly is 5 dB better than the standard; if the standard curve must be moved up 5 dB for the test curve to conform, the assembly is given a –5 rating, indicating that it is 5 dB poorer than the standard.

The criterion curve is based on average background noises that are typical of apartments in moderately quiet suburban neighborhoods. For quieter areas, floor constructions with an INR of +5 to + 10 should be used. For noisy urban districts, a construction with a –5 INR may suffice. INR values for various floor and ceiling constructions are given in later in this section.

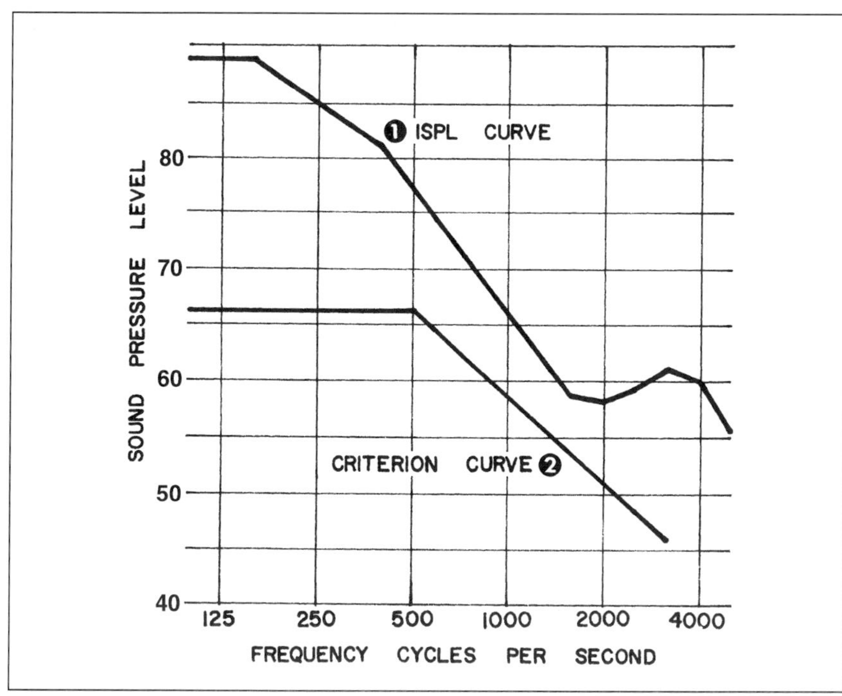

FIGURE 13.1-27 Impact noise rating (INR) is obtained by comparing the ISPL test curve (1) with a standard criterion curve (2).

Impact Insulation Class (IIC) In an effort to reconcile and correlate the various rating systems for airborne sound transmission and impact sound transmission, the Federal Housing Administration (FHA) devised a single-number rating system for impact sound transmission called the *impact insulation class* (*IIC*). The IIC system supersedes the INR system described earlier.

On the average, an IIC rating for a given construction is approximately 51 dB higher than the INR rating. For example, an INR rating of +4 would be comparable to an IIC rating of about 55. However, exact values for IIC ratings should be determined on the basis of individual test data.

To develop the IIC of an assembly, the ISPL is measured and a curve plotted as discussed earlier (see Fig. 13.1-27). A standardized reference contour is plotted on a transparent overlay, placed over the ISPL curve, and moved up and down until only the following allowable tolerances exist: (1) a single unfavorable deviation in which an ISPL value lies above the reference contour may not exceed 8 dB, and (2) the sum of the unfavorable deviations may not be greater than 32 dB. The IIC rating is the value on the right scale corresponding to the ISPL value at 500 Hz (Fig. 13.1-28).

The HUD publication "A Guide to Airborne, Impact, and Structure-Borne Noise Control in Multifamily Dwellings" presents a three-segment standard contour that offers three options, depending on the outdoor nighttime background noise level expected. The contours presented are labeled Grade I (IIC 55), Grade II (IIC 52), and Grade III (IIC 48). The higher the outdoor level, the greater the masking will be and the lower the insulation needed. The most used of the three is the Grade II (IIC 52) contour.

The IIC rating system parallels the STC rating system in that performance of a floor and ceiling assembly is expressed with positive values in ascending degrees of impact sound isolation. The larger the IIC rating, the better the assembly.

CONTROLLING STRUCTURE-BORNE SOUND TRANSMISSION

Most building materials, especially hard or dense ones, transmit vibrations readily. A building's steel frame, for example, can telegraph a vibration throughout a large structure. Clicking of high heels on a concrete floor can be heard distinctly in the room below if no other construction intervenes. The vibration of a structure sets up sound waves that can be heard distinctly at great distances from the source. Impact noise caused by a floor or wall being vibrated by direct mechanical contact is radiated from both sides. This vibration may also be transmitted throughout the structure to walls and reradiated as sound to adjoining spaces. Since footsteps and furniture being moved on floors constitute a major impact problem, control of impact noise is treated here as primarily a floor problem.

Impact isolation methods in general use are more effective in reducing the high-frequency components of impact noise than the low-frequency "thumping" noises caused by heavy impacts. Design requirements for low-frequency isolation are more stringent for wood joist floors than for concrete floors because of the inherent lighter weight and greater flexibility of wood floor joist construction. Methods to control structure-borne sound transmission are discussed in Section 13.1.4, "Design and Construction."

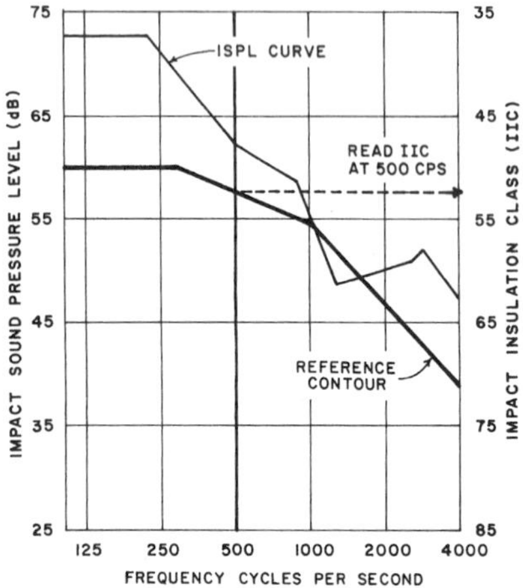

FIGURE 13.1-28 Impact isolation class (IIC) is obtained by comparing the ISPL curve with a standard reference contour and is read as a positive rating.

13.1.4 DESIGN AND CONSTRUCTION

This section deals with (1) criteria and planning principles to aid in the design and selection of walls and floor and ceiling assemblies to control noise within buildings, (2) construction recommendations, including precautions essential to obtaining and maintaining suitable sound isolation between rooms and between living units, and (3) selection criteria for obtaining suitable sound isolation. It also includes the design of doors and windows and selection of appropriate hardware to control sound transmission. Criteria and planning principles that will reduce sound generation by and transmission through plumbing systems are covered in Section 15.1; heating, ventilating, and air conditioning systems in Section 15.2; and electrical systems and appliances in Section 16.3.8.

13.1.4.1 Design and Planning

Sound control involves (1) reducing loud noises at their source, (2) orienting quiet spaces away from noise sources, thus separating the source from the listener by distance, (3) careful interior design and building layout, especially with respect to building shape and the location of door and window openings and other sound leaks, and (4) reducing the amount of noise that reaches a listener by selecting walls and floor and ceiling assemblies that have improved resistance to sound transmission.

BACKGROUND NOISE SOURCES AND LEVELS

The source and intensity of outdoor noise at the site affect the noise-reducing effectiveness of exterior walls and roofs and also have an important effect on the selection of interior wall construction and floor and ceiling assemblies. The noise characteristics of the site should be studied to determine existing and potential noise levels (Fig. 13.1-29).

Theoretically, airborne sound travels from a *point noise source* in a spherical pattern (see Fig. 13.1-1). The sound level of a point noise source is reduced 6 dB each time the distance from the source is doubled (see Fig. 13.1-10). But fast-moving traffic is a *line noise source* rather than a point source. Sound from a line noise source travels in a cylindrical rather than a spherical pattern. To reduce the noise level from a line source 6 dB, the distance between noise source and listener must be increased four times.

The direction and speed of air movement, air temperature, and obstructions such as buildings, trees, and terrain all affect sound level. For example, sites upwind of major noise sources will be quieter than sites an equal distance away but downwind from the noise

Hard exterior surfaces of buildings reflect a large proportion of the sound energy that strikes them (Fig. 13.1-30). Multiple reflection of sound striking a large building can increase the noise problem at a site. Curved buildings, or buildings designed to form courtyards, having

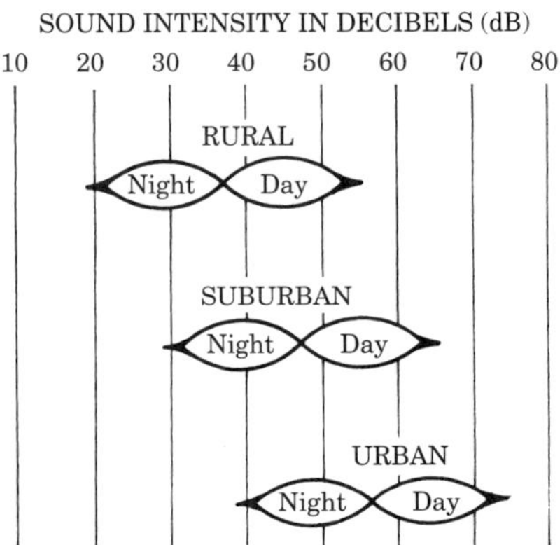

FIGURE 13.1-29 Typical outdoor background noise levels.

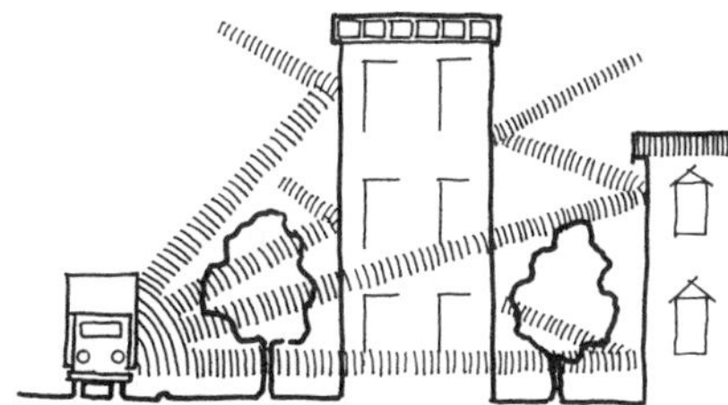

FIGURE 13.1-30 Hard surfaces reflect sound; soft, porous ones, such as are found in plants, absorb some of it.

the concave side facing a noise source will also result in an especially difficult noise problem. Future development of nearby vacant land should be anticipated, and the type and level of probable or possible noise sources should be investigated.

When a proposed site is near high-level noise sources, such as a business district, highway, expressway, railroad line, or airport, both daytime and nighttime outdoor noise levels should be surveyed. When the average background noise level is unusually high or if there are infrequent but intermittent unusually loud or annoying noises, expert acoustical advice should be sought.

Ground Vehicular Traffic Ground vehicular traffic (by automobiles and trucks) is a primary source of outdoor background noise. Background noise levels are affected by topography, landscaping, and the masking or magnifying effect of other buildings. Typical outdoor noise levels are shown in Figure 13.1-31. High-speed highways and dense city traffic generally produce higher noise levels than moderate-speed suburban roads carrying an equivalent number of cars. Trucks increase highway noise about 15 dB, and trucks going uphill will produce the highest highway noise levels. Unless the highway is already saturated with traffic, a future increase in traffic can generally be expected, with a resultant increase in noise.

FIGURE 13.1-31 Approximate Sound Levels for Automobile Traffic at Ground Level

	Distance from Traffic Lane		
Cars/ Minute*	**40 ft. (12.2 m)**	**100 ft. (30.5 m)**	**500 ft. (152.4 m)**
1	50 dB	43 dB	29 dB
10	59 dB	52 dB	38 dB
100	63 dB	60 dB	47 dB

*The maximum capacity of one lane of cars moving freely at approximately 30 mph (48.28 km/h) is 2000 cars per hour, or 34 cars per lane per minute.

Aircraft Traffic The proliferation of jet passenger and transport operations and the rapid expansion of small commercial and business jet aircraft create noise problems even in small communities. Methods for minimizing aircraft noise include techniques for quieting engines, reducing noise exposure by takeoff and landing operational procedures, and reducing the number of people affected by controlling land use around airports through zoning regulations.

Two types of aircraft noise are of concern: (1) noise generated by revving engines on the ground and (2) flyover noise. Figure 13.1-32a illustrates that the noise level during flyover changes very rapidly with time, with the maximum noise sustained for perhaps a second or less, and that sound waves during this short time interval strike a building from all directions.

Figure 13.1-32b illustrates the noise produced by an aircraft revving its engines while stationary on the ground. Here, the relationship between aircraft and building is fixed; only one or two exterior surfaces of the building are directly exposed to aircraft noise, and other surfaces are partially shielded. Intervening obstacles, such as hills, houses, or other terrain features, will partially block the sound.

Figure 13.1-33a illustrates typical noise level contours of jet aircraft during takeoffs; Figure 13.1-33b is an example of noise level contours for the revving of one engine of a jet aircraft.

Of all exterior wall and roof surfaces, windows offer the least resistance to noise transmission. The difference in perceived noise level (PNdB, Fig. 13.1-33) between exterior and interior noise levels from aircraft is illustrated in Figure 13.1-34.

The lower values in each category will be for buildings having window areas comprising more than 20% of the total exposed wall area, with relatively lightweight wood frame walls and roofs (weighing less than 8 psf). The higher values will occur in buildings with relatively small, airtight windows and airtight storm windows of good quality. Examples of aircraft noise reduction through various exterior wall and roof construction assemblies are summarized in Figure 13.1-35.

SITE PLANNING

Site planning can reduce problems caused by exterior noise by (1) increasing the distance between the noise source and buildings, (2) siting buildings to minimize the effect of direct sound transmission, reflection, and reverberation from existing or possible future buildings, and (3) using natural topography, other buildings, barrier walls, and landscaping to shield buildings from major noise sources.

Building Location Buildings should be located as far as possible from noise sources. Buildings reflect noise readily. An orientation that places rows of buildings parallel to each other results in a

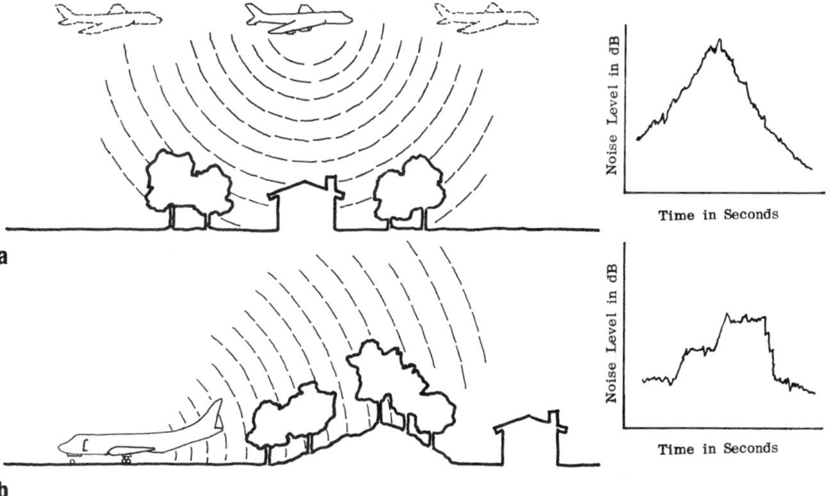

FIGURE 13.1-32 (a) Flyover noise strikes buildings from all angles, but its maximum intensity lasts only a second or less; (b) noise from revving engines on the ground strikes fewer building surfaces but lasts longer.

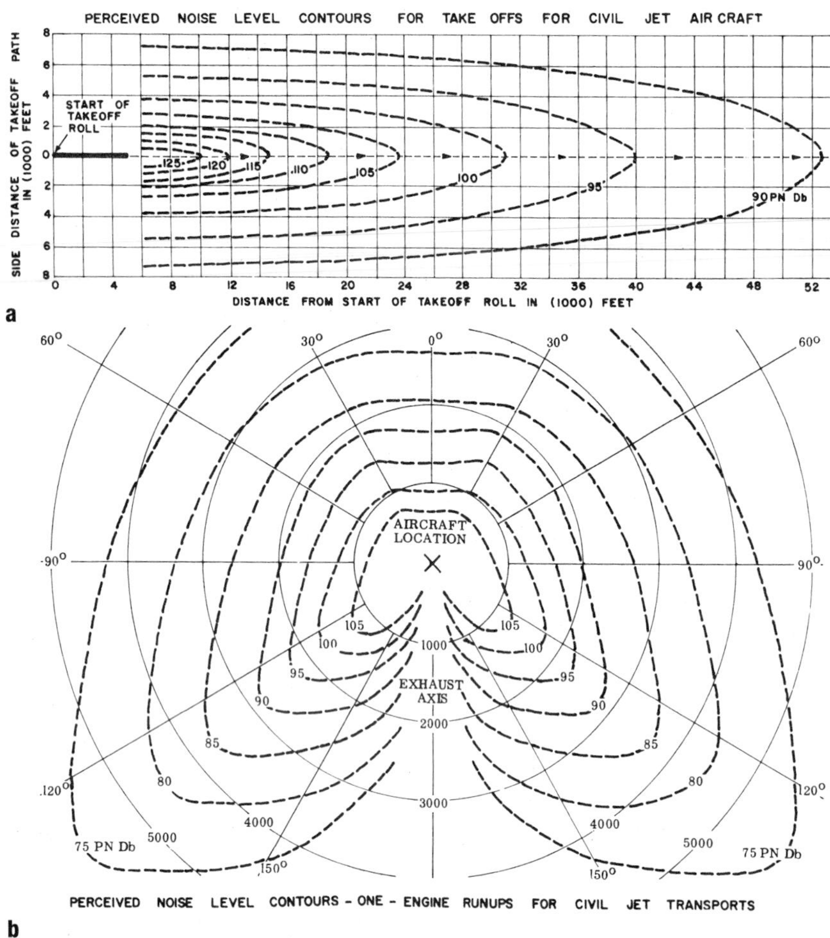

FIGURE 13.1-33 Aircraft noise level contours expressed in perceived noise level (PNdB).*

FIGURE 13.1-34 Perceived Noise Levels from Aircraft

Window Condition	Difference Between Interior and Exterior Noise Levels
Windows open	10 to 20 PNdB
Windows closed*	20 to 30 PNdB
Storm sash installed*	25 to 35 PNdB

*For maximum performance, glass should be sealed airtight within frames; storm sash should be separated by at least 2-in. (50.8 mm) airspace between panes.

buildup or focusing of the noise by multiple reflection of the sound. Acoustically, the best orientation for a building occurs when its long axis is perpendicular to noise sources (Fig. 13.1-36a).

U-shaped courtyards are especially noisy when the open end of the U faces a major noise source (Fig. 13.1-36b). It is acoustically better to have the end of such a courtyard facing away from noise sources. Landscaping in such a courtyard will help to reduce reverberation of sound within the courtyard.

Noise Barriers A building's walls, landscaping, and topography can be used to shield against an exterior noise source. A barrier's effectiveness depends on the frequency of the sound, the angle of the sound shadow, and the height of the barrier (Fig. 13.1-37). Higher-frequency sounds are controlled more effectively by barriers than low-frequency sounds. For greatest effectiveness, barriers should be as high as possible, located as close as possible to the noise source, and impervious to airflow. Barriers should also have some mass, but great weight will not appreciably increase their effectiveness.

Plants absorb some sound and can help to reduce reverberation noise between buildings, but trees and shrubs reduce noise levels only slightly. A 2-ft. (610 mm)-thick, dense cypress hedge will reduce high-frequency noise about 4 dB (Fig. 13.1-38). A single row of trees does little to reduce noise levels. Even dense tropical jungle will reduce levels only about 6 or 7 dB per 100 ft. (30.48 m) of depth, which is only 1 or 2 decibels more than can be achieved by the distance alone. Since noise has a psychological effect, plants that provide a visual barrier to noise sources such as traffic may help to reduce sensitivity to the noise.

Nonresidential buildings parallel and adjacent to a highway can serve as a noise barrier between the highway and residential buildings farther away (Fig. 13.1-39a). Hills and highway cuts can provide natural barriers to major noise sources if the buildings are located to take advantage of these sound shadow barriers (Fig. 13.1-39b).

INTERIOR PLANNING

It is difficult and expensive to correct sound transmission problems caused by faulty interior design after a building has been built. Therefore, it is essential that acoustical principles be recognized in the initial design stages. In addition, design of interior spaces with sound privacy in mind can minimize or eliminate the need for costly construction techniques to control noise. Sound transmission through walls and floor and ceiling assemblies can be reduced by selecting suitable sound-insulating systems and controlling the methods used in the field, as discussed in Section 13.1.4.2.

Sound control should create an environment that is acoustically comfortable, not render rooms soundproof. The basic principles of such planning include (1) locating occupied spaces as far as possible from noise sources, (2) separating noisy areas from quiet areas by isolating the noise source or using buffer spaces, (3) eliminating air paths that permit sound to travel between spaces, especially through the design, sizing, and placement of doors and windows, and (4) reducing the noise level at the source by selecting mechanical equipment and appliances that operate quietly.

In planning for sound control in homes and apartments most of the emphasis is usually placed on sound originating in adjacent homes or apartments or from outdoors. This is because people can control the intensity of noise sources within their own homes and tend to be more tolerant of noises created by their own families than of those that come from other sources.

FIGURE 13.1-35 Summary of Noise Reduction Through Exterior Materials and Assemblies

Typical Constructions	0–10	10–20	20–30	30–40	40–50
	Difference Between Outside and Inside Noise Levels (PNdB)				
Windows					
⅛-in. (3.18 mm) glass in double-hung wood frame		■			
¼-in. (6.35 mm) plate glass, sealed in place			■		
Aluminum frame window, glass panels set in neoprene gaskets, ¼-in. (6.35 mm) and 7/32-in. (5.56 mm) panes, 3¾-in. (95.25 mm) airspace				■	■
Exterior walls					
Wood sheathing (or stucco), 2 × 4 (50 × 100) studs, ½-in. (12.7 mm) plaster board interior surface			■		
Wood sheathing (or stucco), 2 × 4 (50 × 100) studs, ⅞-in. (22.2 mm) plaster interior surface			■	■	
4½-in. (114.3 mm) brick or 6-in. to 8-in. (140 mm to 200 mm) lightweight concrete block				■	
9-in. (230 mm) brick wall					■
Roofs					
Built-up roofing on 1-in. (25.4 mm) wood decking, ½-in. (12.7 mm) gypsum board on 2 × 8 (50 × 200) joists		■	■		
Built-up roofing on 1-in. (25.4 mm) wood decking, ⅞-in. (22.2 mm) plaster on 2 × 8 (50 × 200) joists			■	■	
Built-up roofing or shingles on wood siding, ventilated attic space, ½-in. (12.7 mm) gypsum board ceiling				■	■
Built-up roofing or shingles on wood sheathing, ventilated attic space, ⅞-in. (22.2 mm) plaster ceiling					■

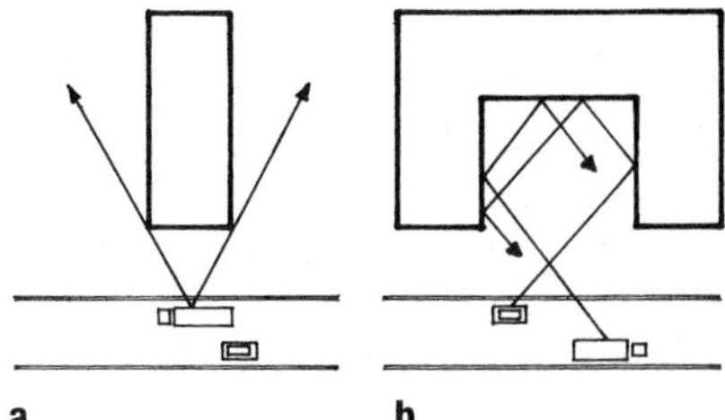

FIGURE 13.1-36 (a) Perpendicular orientation exposes less wall area to noise source; (b) sound reverberates in U-shaped courtyard.

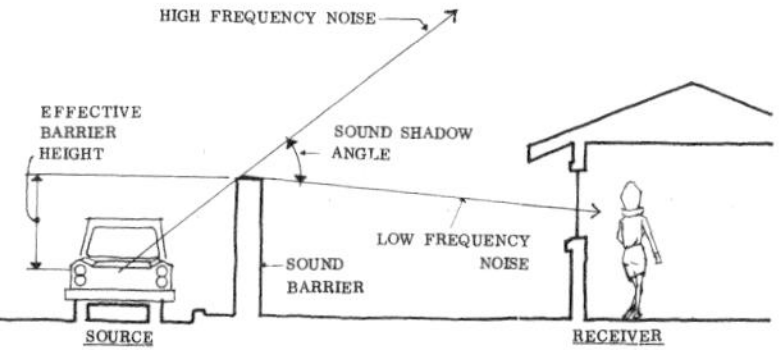

FIGURE 13.1-37 Buildings and walls serving as barriers should be as high as possible and as close as possible to the noise source.

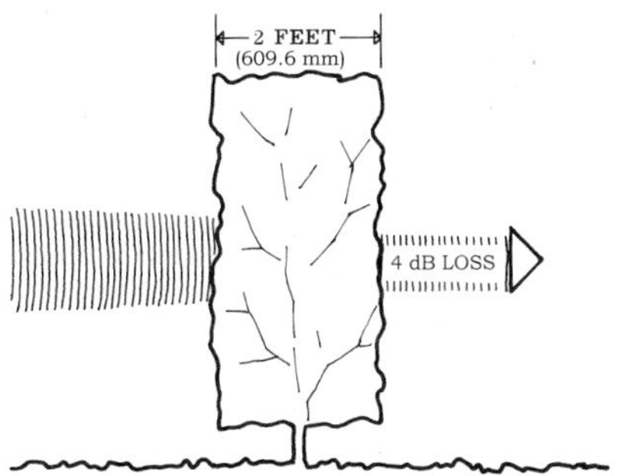

FIGURE 13.1-38 Plants reduce noise levels only slightly, but may provide beneficial visual and psychological barriers.

The following discussion uses apartment living units to illustrate design principles, but most of the principles used are equally applicable to commercial and institutional buildings.

Space Layout Spaces can be zoned acoustically into three general areas: (1) areas that are relatively noisy, (2) areas that should be quiet, and (3) areas that fall between the two extremes.

People and their activities establish acceptable noise levels for different areas. Noisy areas should be grouped together. Buffer spaces such as corridors, stairways, and other spaces can be used to provide acoustical buffers against airborne sound transmission between spaces. Closets, storage rooms, and bookcases backed up to common walls help to reduce airborne sound transmission between rooms, and in addition force the placing of furniture away from the common walls, thereby increasing the distance between a noise source and the occupants of adjacent rooms (Fig. 13.1-40).

In apartment design, it is acoustically preferable that a plan be mirrored horizontally and matched vertically (Fig. 13.1-41). This places noisy areas adjacent to other noisy areas and quiet areas adjacent to other quiet areas so that the sound control requirements can be as low as possible, thereby reducing cost. For example,

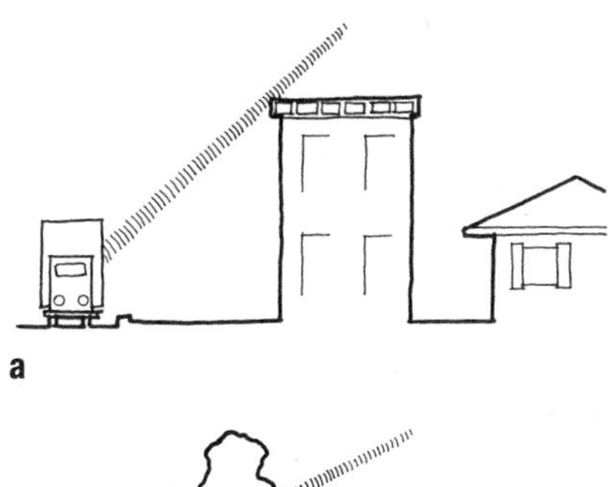

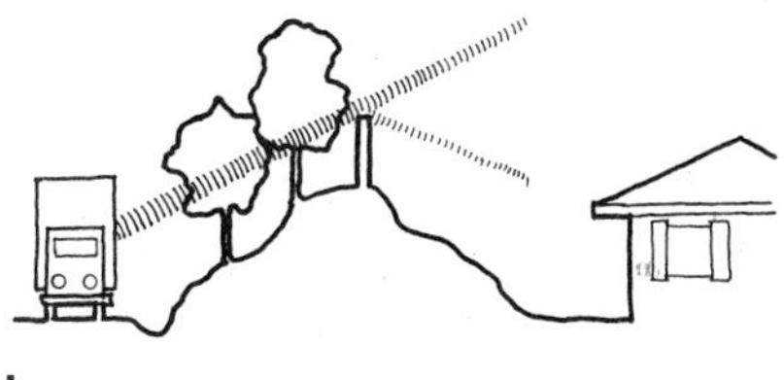

FIGURE 13.1-39 (a) Nonresidential buildings and (b) natural terrain and man-made road cuts and fills can serve as noise barriers.

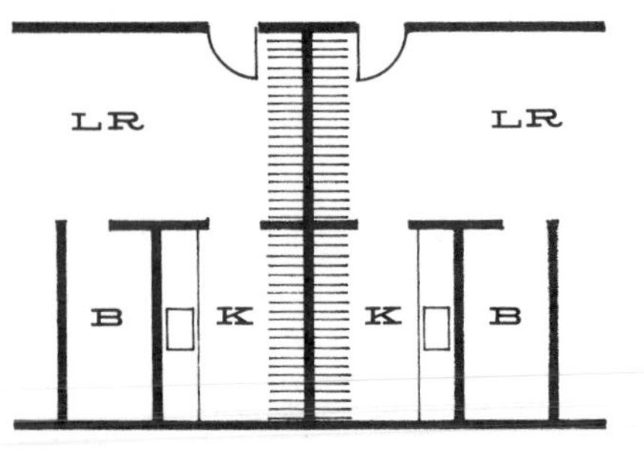

FIGURE 13.1-40 Closets and bookshelves can serve as buffers to separate listeners from a noise source.

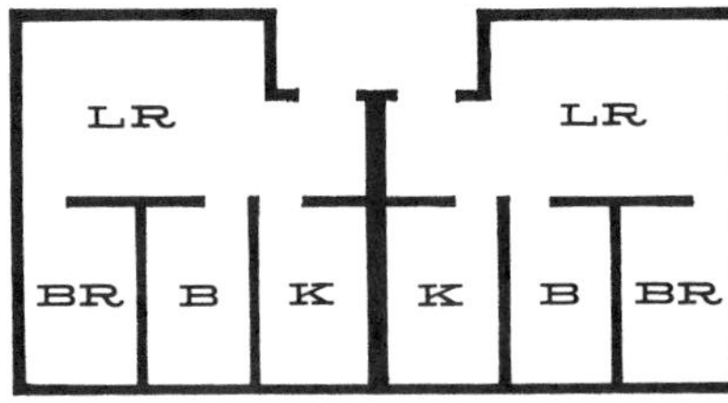

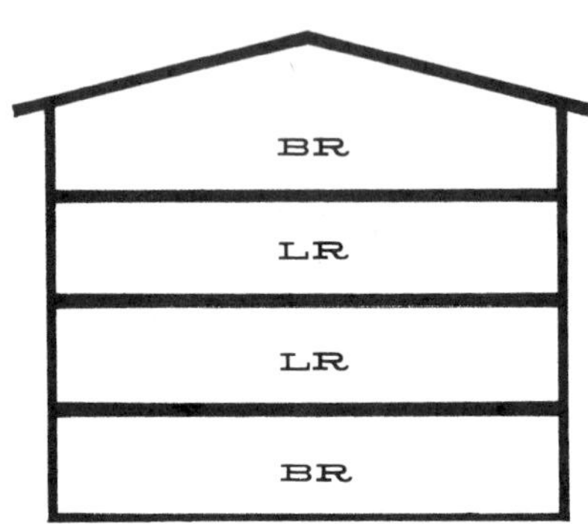

FIGURE 13.1-41 Mirroring the plan results in about equal noise level requirements on both sides of common walls and floor and ceiling assemblies.

kitchens and bathrooms, which have similar noise levels, are located adjacent to and above and below kitchens and bathrooms in adjoining apartments. Similarly, quieter areas such as bedrooms and living rooms should be located adjacent to and above and below quiet areas in adjoining apartments. The noise level requirements on both sides of walls and floor and ceiling assemblies will be the same, and the full effect and cost of improved sound control measures will benefit all apartments. Some additional precautions that should be observed in mirror planning are illustrated in Figure 13.1-42.

Row plans where apartments are arranged in lines, provide better acoustical performance than block plans where each apartment may have as many as three

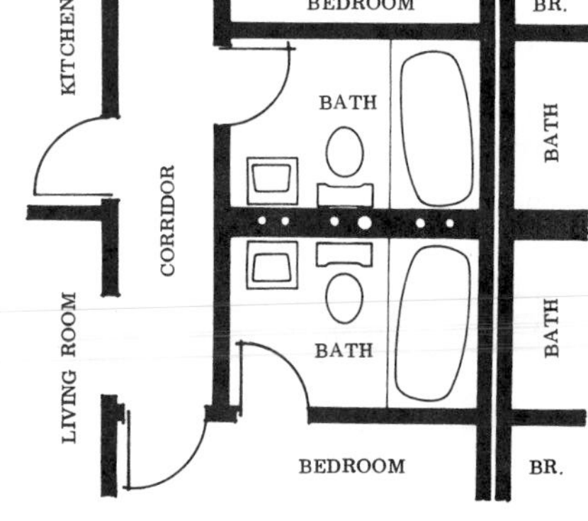

FIGURE 13.1-42 Plumbing should be located in walls within each living unit, rather than in common walls between units.

common walls (Fig. 13.1-43b). In row planning, with halls between every other apartment, each apartment has only one common wall (Fig. 13.1-43a).

Room proportions can be planned to reduce sound transmission. The higher the ratio of common wall to floor area, the more sound the common wall will transmit (Fig. 13.1-44).

Adding large absorptive surfaces, such as acoustical tile, carpeting, or heavy drapes, to rooms will slightly reduce the amount of sound available that can be transmitted. In a room with large hard-surfaced areas, a large portion of the sound is reflected, whereas large soft-surfaced areas will absorb sound.

DOORS AND WINDOWS

Doors and windows are acoustically weak. Maximum acoustical performance requires that they be properly placed and that their edges and frames be sealed airtight. Packing and caulking around their frames is necessary to reduce sound transmission through cracks.

Door Construction and Installation Solid-core and mineral-filled doors are generally better sound isolators than hollow-core doors (Fig. 13.1-45). However, the amount of air space around the edges of a door is usually the controlling acoustical factor. Doors should be gasketed if sound privacy is required (see Fig. 13.1-45). The relative performances of solid- and hollow-core doors, gasketed and ungasketed, are given in Figure 13.1-46.

Soft weatherstripping provides a good seal at the top and sides of a door, but automatic threshold closers are needed to seal the bottom. While it may be costly to

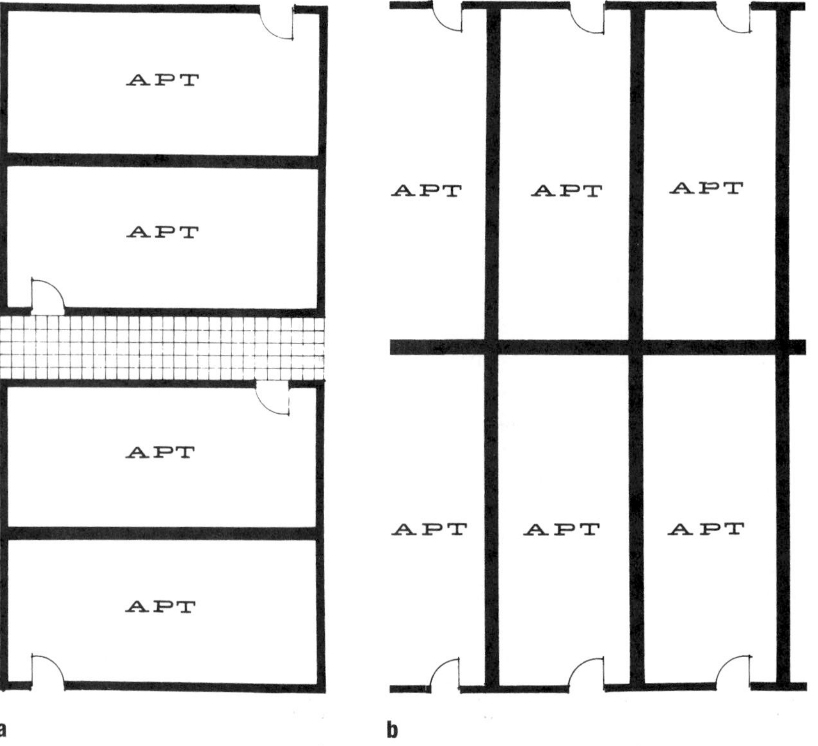

FIGURE 13.1-43 (a) Row planning reduces the number of common walls between units. (b) Block planning can result in as many as three common walls in each unit.

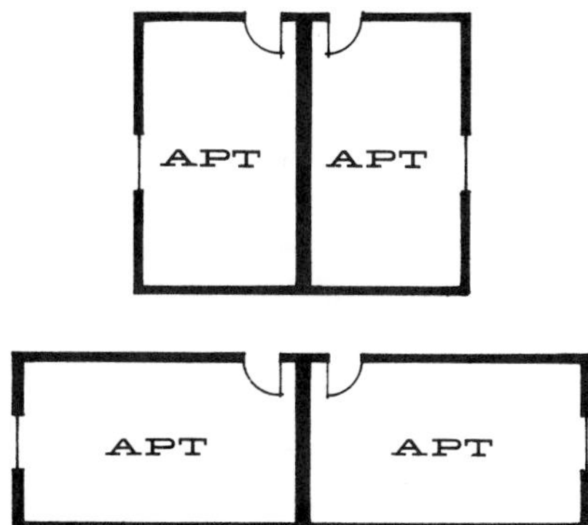

FIGURE 13.1-44 Room proportions can reduce the ratio of common-wall-to-floor area. Hard-surfaced areas should be minimized; soft, absorbent areas, increased.

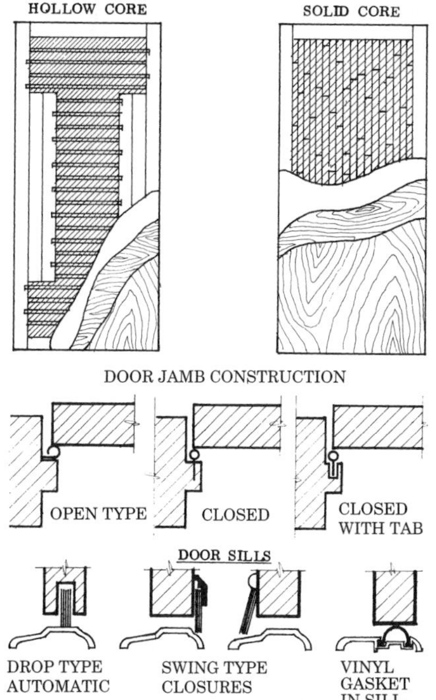

FIGURE 13.1-45 Solid-core doors should be used in sound-conditioned walls, with each door sealed at the jambs, head, and sill.

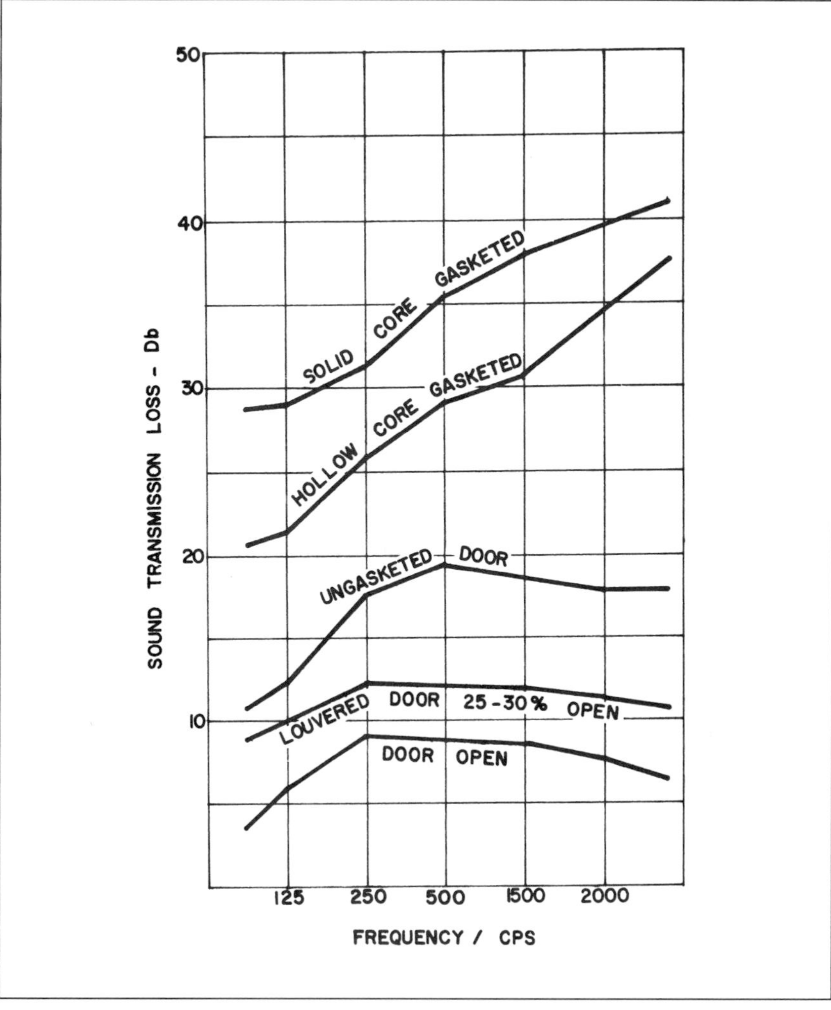

FIGURE 13.1-46 Relative acoustical performance of hollow- and solid-core wood doors.

use gasketed doors, the provision of a cushion such as sponge rubber between the door and its jamb will reduce the impact energy of a door slam and make a door closure quieter. Unfortunately, properly sealed doors, because they are tight fitting, are harder to close and more prone to warping.

Undercutting doors to allow for rugs or return air movement in forced-air ventilation systems essentially destroys their acoustical effectiveness.

Sliding doors provide little sound privacy, because a tight air seal cannot be maintained.

Heavy-duty doors made of metal or metal clad over very heavy core materials, such as fire-rated metal doors, will not necessarily provide good acoustical performance in spite of their mass, even when two doors are installed, unless they are sealed with gaskets and doorsills (Fig. 13.1-47a). Where such doors are installed in locations where loud noises occur, such as in mechanical equipment rooms, a second door with better acoustical performance can be installed (Fig. 13.1-47b).

Window Construction and Installation About 80% of the acoustical problems occurring with windows results from the airtightness of construction and about 20% from the selection of the glass. Sturdy, well-weatherstripped, and tightly closing windows are better sound insulators than loosely fitting ones; a fixed sash performs

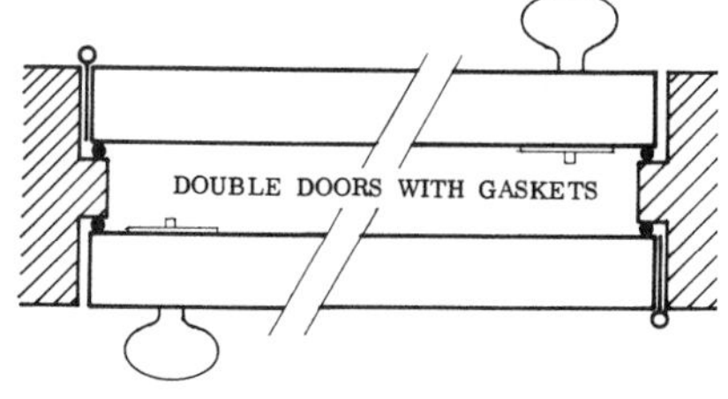

a

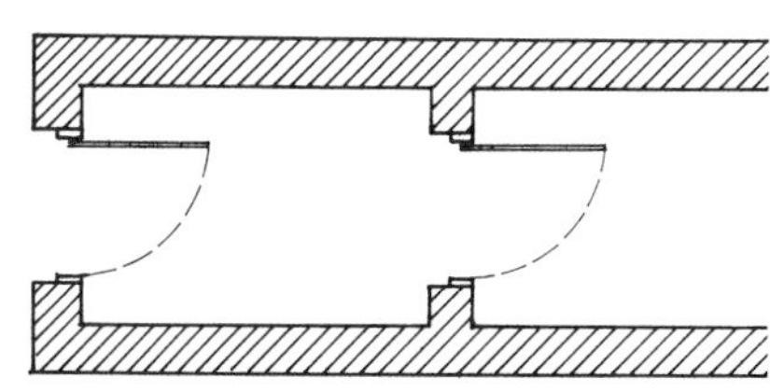

b

FIGURE 13.1-47 "Acoustical traps" can be used to reduce sound transmission from sources of high noise levels such as service rooms and mechanical equipment rooms.

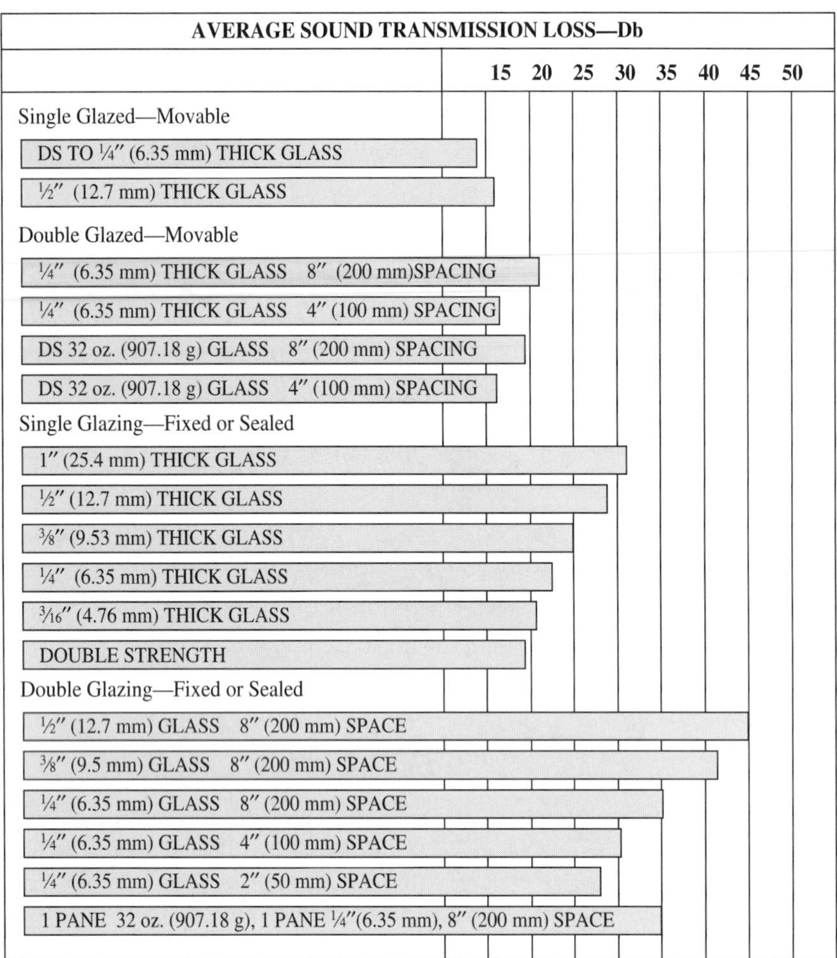

FIGURE 13.1-48 Relative acoustical performance of various types of windows.

better than an operating sash. Refer to Figure 13.1-48 for the relative acoustical performance of various types of windows.

Operating single-glazed wood windows in grooved wooden tracks with weatherstripping, which are pressure sealed when closed, perform well. The performance of wood windows in metallic or plastic slides is fair against air infiltration, but poor against resonant frequencies. Metal windows are available in many grades and quality levels. The better grades will perform acceptably if tight fitting, properly sealed, correctly installed, and sufficiently rigid to reduce air infiltration and resonant vibration.

In operating double-glazed windows (storm windows), the greater the separation between two panes of glass, the more efficient the sound isolation, but to be significant the separation must be relatively large. For example, a separation of 2 to 4 in. (50.8 to 101.6 mm) increases the STC from 33 to only 34. The minimum useful distance is 4 in. (101.6 mm); 8-in. (203.2 mm) spacing is effective but not always practical, and 12-in. (304.8 mm) spacing is required to isolate severe noise. If only a 2-in. (50.8 mm) space is possible, it is more economical to double the weight of the single glass. Storm windows attached to casements are credited for added weight of glass only.

Each glass pane should be mounted in a heavy frame and securely fastened in place. Using flexible or preextruded sealant improves the damping characteristics. Each glass pane should be of a different thickness or weight per sq. ft. (m^2) so that one can offset the deficiency of the other and they will not vibrate together or resonate.

A fixed window having a single pane of plate glass is a fair sound barrier, as compared with an operating window, but a window is still the weakest part of a wall construction. Year-round air conditioning allows glass to be sealed in less expensive frames and reduces air infiltration.

Glass thickness should meet a project's decibel loss requirements. Two layers of insulating glass with a ½-in. (12.7 mm) air space should be considered for fuel savings and prevention of condensation. Acoustically, two layers of closely spaced ¼-in. (6.4 mm) glass is approximately equal to one layer of ½-in. (12.7 mm) glass.

Windows should be installed with a flexible mounting, making certain that the mounting is installed in a firm frame and is airtight to prevent flanking sound from bypassing the glass.

Fixed double-glazed sash should be used for the most demanding noise problems. This consists of two panes of glass separated by at least 8 in. (203 mm) of enclosed air space. The addition of a sound absorber, such as a 1-in. (25.4 mm)-thick resilient material around the perimeter between the two glass panes, will generally improve performance by absorbing some of the sound energy in the air space.

Location Doors, windows, and other openings are direct air-to-air paths that transmit sound readily. Exterior doors, windows, and other openings should be placed on the quiet side of a building when possible. Where they are necessary on the noisy side, their number and size should be minimized when sound control has a higher priority than light, view, or ventilation. Exterior doors and windows should be located as far as possible from, or in positions that do not interfere with, doors and windows in adjacent apartments (Fig. 13.1-49).

When possible, interior doors should be staggered on opposite sides of corridors, so that the least amount of sound is radiated directly from one door across to another. Offsetting doors allows the greatest amount of diffusion and dissipation of sound before it reaches adjoining units (Fig. 13.1-49b).

Noise is transmitted in both directions from open windows. To minimize noise from one apartment to another, windows should be placed as far as possible from the common wall. Casement windows should be arranged to operate in the same direction so that the sound from one room is not reflected into another (Fig. 13.1-50).

PLUMBING, MECHANICAL, AND ELECTRICAL EQUIPMENT AND SYSTEMS

Plumbing and heating, ventilating, and air conditioning equipment and systems produce airborne noise as well as structure-borne noise. It may be that the noise at the

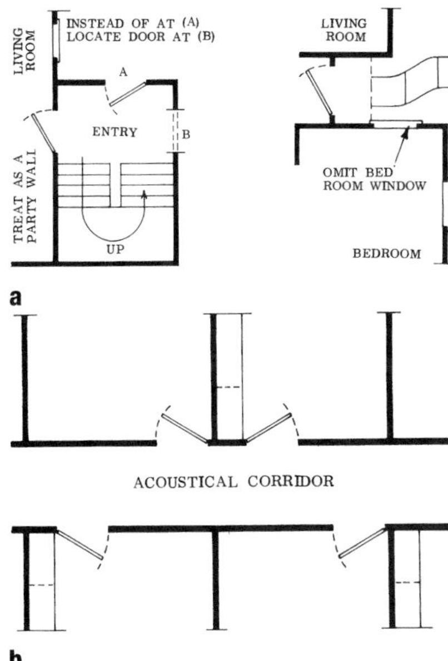

FIGURE 13.1-49 (a and b) Doors and windows should be placed so that sound will not travel through them from one unit to another. (b) Interior doors should be staggered on opposite sides of a hallway.

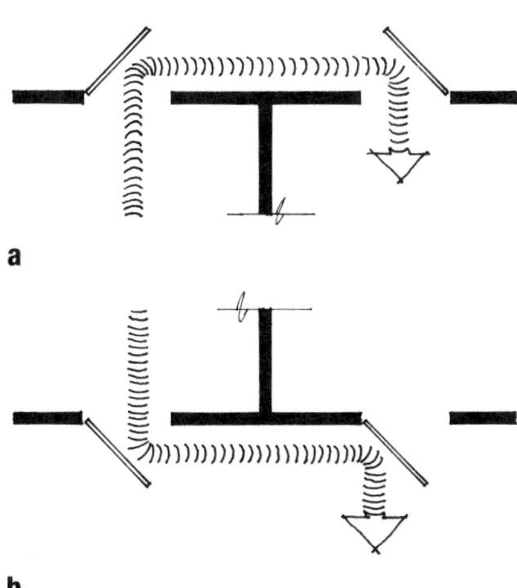

FIGURE 13.1-50 Casement windows should not reflect sound from one unit to another. (a) Improper arrangement; (b) proper arrangement.

source cannot be eliminated, but it can be minimized by selection of equipment, design layout, and construction precautions.

Precautions to avoid noise associated with plumbing systems are discussed in Section 15.1; those associated with heating, ventilating, and air conditioning equipment and systems in Section 15.2; and those associated with electrical systems in Section 16.3.8.

13.1.4.2 Construction Recommendations

The ability of materials and assemblies to reduce noise transmission varies with the frequency of the sound. Some assemblies are efficient at high frequencies, others at low frequencies. A wall or floor and ceiling assembly that will adequately reduce the transmission of conversational speech may not be able to block the low-frequency thump of an air compressor.

Acceptable sound conditioning requires proper selection and construction of sound-isolating walls and floor and ceiling assemblies. It also requires proper selection and installation of plumbing, heating, ventilating, air conditioning, and electrical systems, equipment, and appliances of the types that will reduce noise generation and transmission to acceptable levels.

The sound-isolating efficiency of a wall or floor and ceiling assembly depends on (1) its weight or mass, (2) effective separation of the surfaces of the assembly, and (3) the addition of sound-absorbing materials within the assembly.

To reduce sound transmission through an assembly, do one or more of the following: (1) increase the weight of the assembly; (2) devise discontinuous construction methods that will permit the surfaces of an assembly to be independent of each other; (3) add sound-absorbing materials in the cavities within the assembly.

The ability of a wall or floor and ceiling assembly to transmit sound is inversely dependent on its mass. The heavier or more massive it is, the greater will be its resistance to sound transmission. A wall or floor and ceiling assembly with an infinite mass could not be set in vibration by airborne sound pressures, regardless of their intensity. If the mass was extremely small, even faint sound pressures would move it. However, massiveness soon reaches a point of diminishing returns, since doubling the mass of a wall or floor and ceiling assembly will improve its resistance to sound transmission by only about 5 dB. For example, to achieve a 20-dB reduction in sound transmission in a wall having a single layer of ½-in. (12.7 mm)-thick gypsum board would require adding 16 layers of the same board.

Separating opposite surfaces by discontinuous construction in such a manner that the coupling between the surfaces is very ineffective reduces transmission, because the surface opposite the source is not set in as vigorous vibration as the surface on the source side.

One method is to build the opposite faces of a wall independent of each other with no structural connections between them (Figs. 13.1-51 and 13.1-52). In wood frame constructions such as those with staggered studs or with two rows of studs, the framing provides vibration isolation

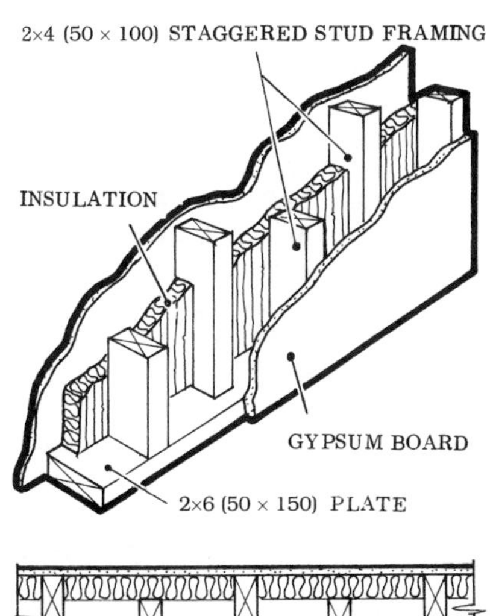

FIGURE 13.1-51 Staggered studs provide discontinuous construction by effectively separating opposite surfaces. Adding insulation improves performance.

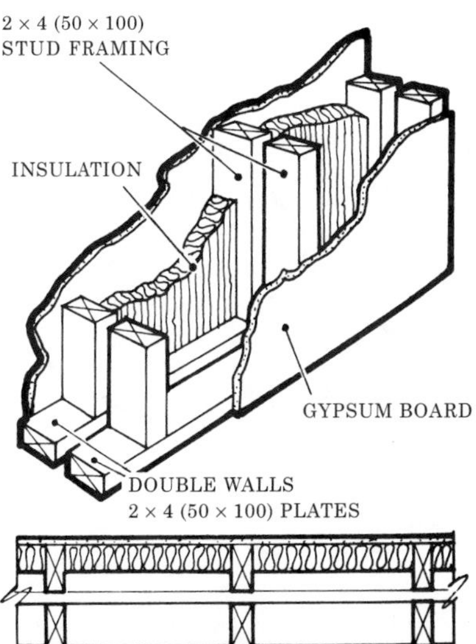

FIGURE 13.1-52 A double row of studs, each row on separate plates, provides discontinuous construction. Adding insulation improves performance, and performance increases as insulation thickness is increased.

between the two wall surfaces. The surfaces can also be isolated with resilient underlayment materials between the outer wall surfaces and the structural framing (Fig. 13.1-53), thus reducing transmission by isolating the surfaces, as an automobile spring isolates the body of the auto from the road. Sound striking one side of a wall causes it to vibrate, but, because of mechanical separation, the vibration of the other side is greatly reduced. Double walls are much more effective than single walls of the same overall weight.

Another method is to attach each face to the structural support, yet still have them mechanically separated with flexible connections or resilient mountings so that the opposite faces do not vibrate sympathetically (Fig. 13.1-54). Then, much of the sound energy is dissipated by vibrating one surface and its mountings, instead of being transmitted to the opposite surface. The improvement in performance due to discontinuous construction or resilient mountings can be increased significantly by placing sound-absorbing insulation blankets within the wall or floor and ceiling assembly to absorb sound energy in the cavity, as shown in Figure 13.1-51.

Discontinuous construction can be obtained in floor and ceiling assemblies by floating the finish flooring and separating the ceiling construction from the floor construction.

Attempting to isolate airborne sound transmission by separating opposite surfaces alone has limitations. Doubling the free air space between the opposite surfaces of a wall or floor and ceiling assembly improves its resistance to sound transmission by only about 5 dB, assuming a constant mass. Within practical limits of conventional building materials, both mass and discontinuous construction are required in order for a wall to reach a 50 to 55 dB STC. STC values greater than that are not practical, regardless of mass and isolation, without construction methods that will control transmission through flanking paths.

Presuming that walls and floor and ceiling assemblies that are selected on the basis of test data will actually perform according to their ratings, the greatest single problem in achieving satisfactory sound conditioning is improper installation. Only when every detail has been correctly taken care of will the actual walls and floor and ceiling assemblies live up to the ratings of the tested assemblies. When possible, workers and other key personnel should be taught the reasons for care in building and applying sound-control measures. Supervision to ensure that the proper methods and techniques are used and coordination between the various trades are essential. Seemingly insignificant variations from a correct technique can nullify the effectiveness of costly sound-control measures. For example, nails that are too long can short-circuit resilient mountings and make them useless. Failing to seal small cracks and holes will create a flanking path and negate the effectiveness of a wall or floor and ceiling assembly. The manufacturer's instructions for installing sound-control devices, such as resilient clips and sound insulation, must be followed precisely. The actual walls and assemblies must be built exactly as were the tested ones.

It is normal to construct walls and floor and ceiling assemblies as continuous surfaces to divide living units into rooms. Windows, doors, and other openings in such surfaces are required. These horizontal and vertical surfaces and their openings, as well as construction imperfections such as cracks and holes, create flanking paths that transmit sound. Just as snapping a rope will cause it to snake along its entire length, sound impulses in a wall or floor and ceiling assembly create vibrations that may transmit sound around, over, or under a partition even though the partition itself has a high resistance to transmission (Fig. 13.1-55).

Open-air flanking paths are openings through which sound travels virtually undiminished. A doorway or window is an enormous hole through a wall. It is an ideal sound leak, and unless properly sealed, can render a wall almost acoustically ineffective.

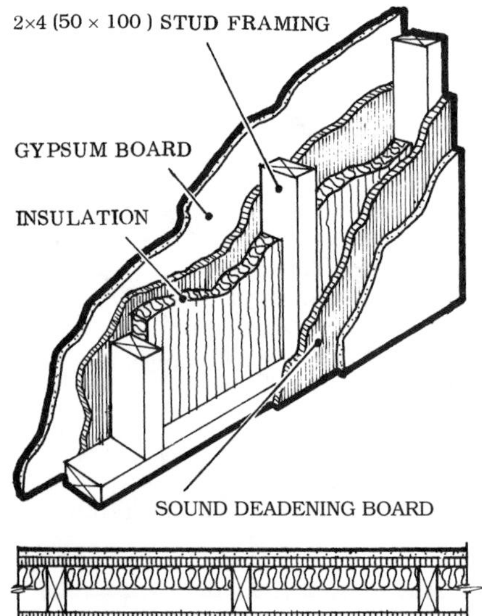

FIGURE 13.1-53 Sound-deadening board reduces transmission by resilient mounting of surfaces and can be used with single-, double-, or staggered-stud walls.

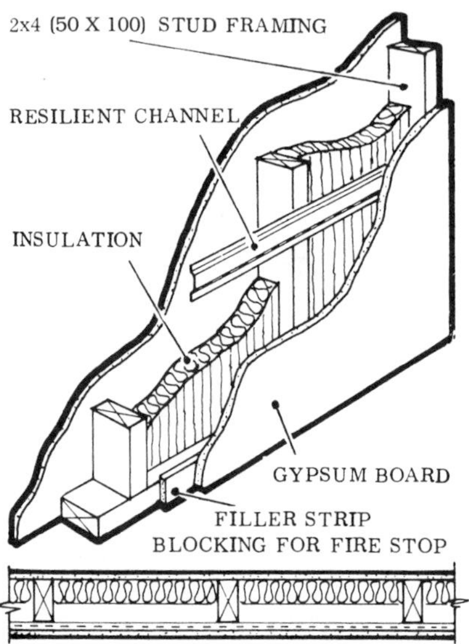

FIGURE 13.1-54 Resilient channels (on one side of the wall) flexibly mount the surface. The performance increases when insulation is added and continues to increase as the thickness of the insulation is increased.

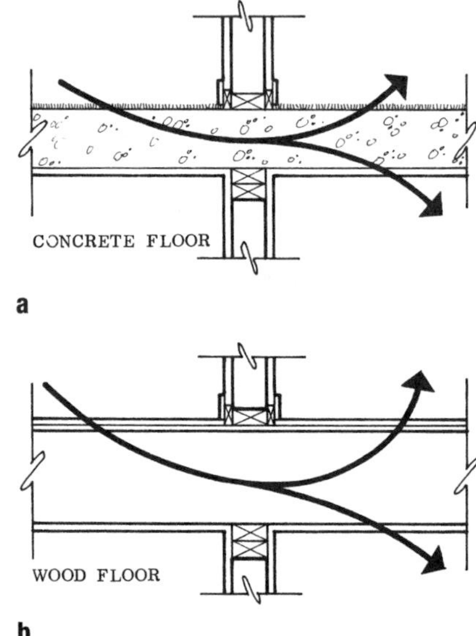

FIGURE 13.1-55 Sound can be transmitted around, over, and under partitions through (a) structure-borne flanking paths and (b) airborne flanking paths.

To reduce the number of flanking paths, doors and windows should be airtight, heavy enough to minimize transmission through them, and installed to maintain their airtightness.

Flanking paths through construction imperfections are illustrated in Figure 13.1-20. Walls, ceilings, and floors are pierced by pipes, ducts, conduits, lighting fixtures, electrical boxes, and similar equipment. Because airborne sound will travel wherever there is an air passage, they all present potential sound leaks. Often they are inadvertently arranged to become ideal "speaking tubes." Air conditioning and ventilating ducts are frequently offenders of this type. Medicine cabinets or outlet boxes placed back-to-back are common sites of sound leaks. Such air leaks must be closed, openings caulked or sealed, fixtures staggered, and ducts lined with absorbent materials or equipped with baffles that interrupt the direct path of the sound and assist in absorbing it.

Airborne sound transmission cannot be controlled satisfactorily solely by acoustical surfacing materials. Covering the surfaces of walls or ceilings with acoustical materials, or a floor with carpeting, adds little to the STC of walls or floor and ceiling assemblies. It is true that the amount of sound available to be transmitted will be reduced somewhat by being absorbed by acoustically treated surfaces, but that reduction is almost never enough.

However, flexible insulation placed in the free air space of frame walls and floor and ceiling assemblies improves performance by absorbing sound within the cavity and can result in as much as a 10-dB improvement when the surfaces are effectively separated.

WALL SYSTEMS

Sound-isolating wall systems may be constructed of wood or metal framing, concrete, or unit masonry. Regardless of the system selected, it is essential that its surfacing materials be installed over the entire wall area, including areas behind soffits, cabinets, and duct enclosures and above suspended ceilings. This precaution is particularly critical for common walls between units to ensure that sound will not be transmitted through an unfinished space (Fig. 13.1-56). Resilient, nonhardening caulking must be used at the floor and ceiling in partitions, where partitions abut other walls and partitions, and around electrical outlets and switches, plumbing pipes, and other penetrations to make the wall airtight (Fig. 13.1-57).

Most sound-transmission test data for wood and metal frame construction have been obtained with gypsum board as the surfacing material because of its extensive use and ease of installation in the laboratory, without the necessity of drying and curing. The following discussion is based primarily on the evaluation of such testing, but the principles and conclusions involved are applicable to lath and plaster and other surfacing materials of comparable weight applied to similar construction. Representative STC ratings for lath and plaster, as well as other typical constructions, are given in Section 13.1.5.

Wood Framing Systems Conventional single-stud walls finished with one layer of gypsum board on each side (Fig. 13.1-58) have an STC of about 34, which represents poor resistance to sound transmission. Sound-absorbing blankets placed in the stud cavity will not appreciably increase the sound-isolation performance of this kind of wall. Either the weight of the covering material must be doubled, using multiple layers of gypsum board (Fig. 13.1-59), or the two wall surfaces must be mechanically separated to achieve an STC rating of more than 40. A summary of the acoustical performance of wood frame walls is given in Figure 13.1-60.

Resilient mounting of the surface finish of a single-stud wall with resilient channels on one side only (see Fig. 13.1-54) will increase the performance of the wall about 6 dB. Adding channels to the other side will not further improve performance because of the resiliency of the wall through sympathetic vibration of both surfaces.

Installing insulation in a resiliently

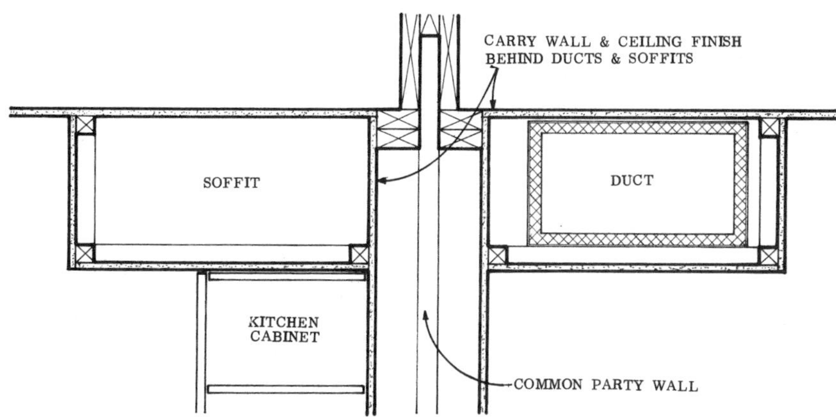

FIGURE 13.1-56 To prevent sound transmission through unfinished spaces, wall and ceiling finishes should cover all common wall and ceiling areas, including those inside furred areas and above ceilings, and those concealed by stairs or other construction.

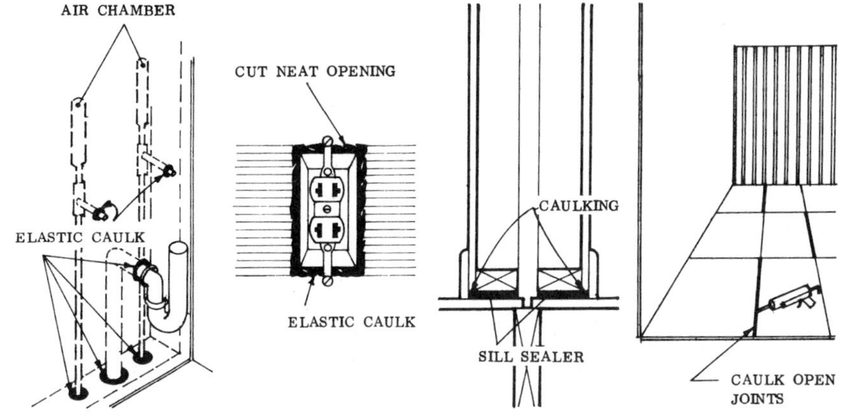

FIGURE 13.1-57 To prevent sound transmission through them, openings in sound-insulated walls, floors, and ceilings must be sealed with nonhardening, resilient sealants or a resilient material.

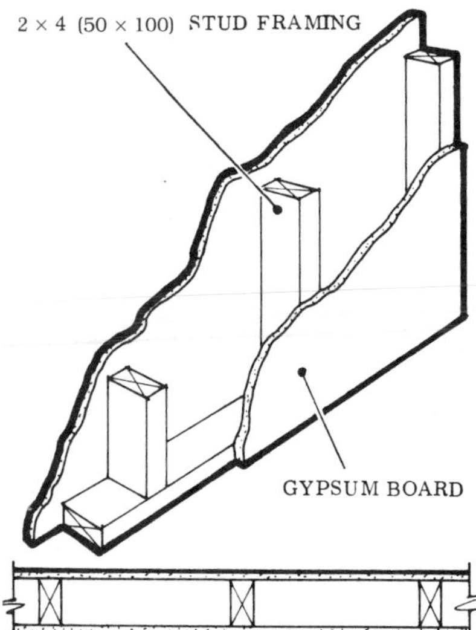

FIGURE 13.1-58 Conventional single-stud walls permit loud speech and other noises on one side to be heard clearly on the other.

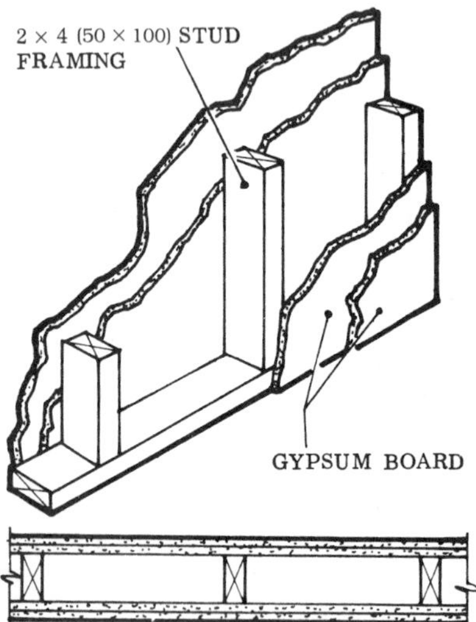

FIGURE 13.1-59 An additional layer of gypsum board on each side slightly improves the performance of walls like the one shown in Figure 13.1-58.

FIGURE 13.1-60 Acoustical Performance of Typical Wood-Framed Walls[a]

Wall Construction	Approximate STC Rating: Without Insulation	With Insulation[b]	Sound-Deadening Board Both Sides[c]
Single-stud wall			
Single-layer surface	34	36	46
Double-layer surfaces	40	—	—
Single-stud wall with resilient channel mounting one side	40	50[d]	—
Staggered-stud wall	41	49	50
Double-stud wall	43	53	50

[a]As illustrated in Figures 13.1-51, 13.1-52, 13.1-53, 13.1-54, 13.1-58, and 13.1-59.

[b]Minimum thickness of insulation (2 or 3 in.) (50 or 75 mm).

[c]Gypsum board surfaces laminated, with exception of double-stud wall, which was laminated and nailed 8 in. (200 mm) o.c. into studs and floor and ceiling plates.

[d]Resilient channels installed throughout; fire-rated wall with firestopping top and middle heights or backer strips for attaching wall base, STC 46.

mounted wall increases its acoustical performance in proportion to the thickness of the insulation and can result in as much as a 10-dB improvement. Noise isolation performance depends on the surfacing materials and how they are installed. Backer strips for attaching baseboards or horizontal blocking at the top and middle height for firestopping decrease the performance. Resilient channels must be handled and installed carefully to avoid permanent deflection. The wall should contain no openings, and heavy objects such as fixtures and cabinets (which would destroy resiliency) should not be mounted on the resilient channel side of the wall.

To obtain an STC of 50 or more, without regard for fire ratings, channels must be used throughout, instead of gypsum backing strips, to totally float the surfacing materials. Resilient sealants must also be installed around the entire perimeter and around every penetration.

A wall with resilient channels on one side generally results in the thinnest possible sound-isolating system in wood frame construction and can be constructed with proper blocking to give a 1-hour fire rating.

Sound-deadening board applied to both sides of a single-stud wall will provide resilient mounting if the surfacing materials are correctly laminated to the sound-deadening board (Fig. 13.1-53) and can result in a wall with an STC of about 46. However, if the surfacing materials are nailed in order to meet fire rating requirements, the STC will be reduced about 10 dB when nails are used at spacings of 8 in. (203.2 mm) on centers, 8 dB with nails spaced 16 in. (406.4 mm) on centers, and 4 dB with nails spaced 24 in. (609.6 mm) on centers.

Staggered-stud walls, consisting of 2 × 4 (50 × 100) studs staggered on a single 2 × 6 (50 × 150) plate (Fig. 13.1-51), with one layer of gypsum board on each side, have an STC of about 42. The addition of insulation in the stud cavity or sound-deadening board to both sides increases the performance of the wall to a range of 45 to 49 STC. The STC rating does not increase appreciably as the thickness of the insulation is increased, because sound will travel through the single plate. Therefore, 2 or 3 in. (50.8 to 76.2 mm) of mineral-fiber blanket insulation is all that is necessary.

One advantage of using staggered-stud construction is that cabinets can be surface mounted on the wall framing without impairing the acoustical effectiveness of the wall, because they are mounted on framing members that do not contact both sides of the wall.

For maximum performance, a wall base should not be nailed to the bottom plate of a staggered-stud wall.

A double-stud wall consists of two separate 2 × 4 (50 × 100) stud walls on two separate 2 × 4 (50 × 100) plates spaced ¼ to 1 in. (19 to 25.4 mm) apart (Fig. 13.1-52). This kind of wall has an STC of about 43, which is 9 dB higher than that of a conventional single-stud wall. Adding sound-deadening board to both sides will increase the rating to about 50. The isolation performance of a double-stud wall improves when the thickness of insulation in the cavity is increased. Ratings as high as 59 STC can be achieved.

A double-stud wall can accommodate surface-mounted cabinets without reduced effectiveness. It also has fewer performance variations due to bad materials or workmanship. However, to perform acoustically, the separate walls should not be tied together in any way when plumbing or wiring is installed.

Metal Framing Systems Because metal studs are themselves resilient, the two sur-

FIGURE 13.1-61 Acoustical Performance of Typical Metal-Channel-Stud Walls[a]

Gypsum Board Surfacing	Approximate STC Rating					
			Sound-Deadening Board One Side		Sound-Deadening Board Two Sides	
	Without Insulation	With Insulation[b]	Without Insulation	With Insulation[b]	Without Insulation	With Insulation[b]
Single layer each side	38	45	43	49	47	54
Double layer one side, single layer one side	44	50	46	51	52	54
Double layer each side	49	57	53	57	54	57

[a]As illustrated in Figures 13.1-62, 13.1-63, and 13.1-64.

[b]Minimum thickness of insulation (2 or 3 in.) (50 or 75 mm).

faces of a wall built with them vibrate independently of each other and may be considered as resiliently mounted. Therefore, the acoustical performance of non-load-bearing metal stud walls differs from that of wood stud walls. For example, the addition of resilient channels, which can be effective with wood studs, does not appreciably improve the performance of metal stud walls.

When resistance to sound transmission is required in metal-framed walls, the measures to be used should be selected on the basis of individual test data for the system and should be installed in accordance with specified procedures.

The precautions recommended for resiliently mounted wood-framed walls also apply to metal-framed walls. For maximum performance, the wall should contain no openings, and heavy objects such as cabinets or fixtures should not be mounted on either side of the wall.

The difference between the acoustical performance of walls constructed with 1⅝- and 2½-in. (41.3 and 63.5 mm)-deep metal studs, or using ½- or ⅝-in. (12.7 or 15.9 mm)-thick gypsum board, is negligible. A summary of the acoustical performance of metal-channel-stud walls is given in Figure 13.1-61. The effects of adding gypsum board, insulation, and sound-deadening board in metal-channel-stud construction are cumulative.

Single metal-channel-stud walls with single-layer surfaces on both sides have an STC of about 37. Adding 2 or 3 in. (50.8 or 76.2 mm) of insulation in the cavity (Fig. 13.1-62) will improve the acoustical performance to an STC of about 46. Adding one layer of mineral-fiber sound-deadening board to one side of the wall will improve the STC to 43 without insulation within the cavity, and to 49 with insulation. Adding sound-deadening board to both sides, and insulation in the cavity (Fig. 13.1-63), can produce an STC of 52 to 54.

Adding layers of surfacing material to one or both sides will increase the STC rating of a wall by 5 or 6 dB for each layer added. Adding insulation to a wall with two layers of gypsum board on one side will improve performance an additional 6 to 8 dB. A layer of sound-deadening board will increase the STC by about 5 dB, which is about the same as adding a single layer of gypsum board.

The effects of adding gypsum board, insulation, and sound-deadening board in metal-stud construction are cumulative (see Fig. 13.1-63). However, the improvement is small when sound-deadening board is added to both sides of a wall that already has two layers of gypsum board on both sides. This is probably because the added mass of the various layers of finish begins to overshadow the resiliency of the studs.

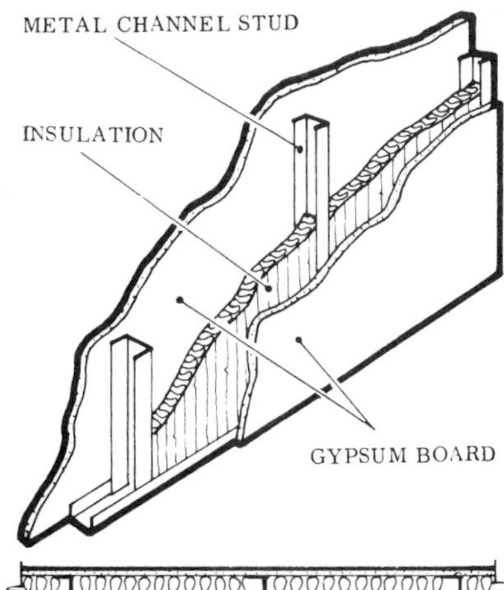

FIGURE 13.1-62 A metal-channel-stud wall acts as a resilient system without the need for resilient mountings.

Double-channel-stud walls consisting of 1⅝-in. (41.3 mm)-deep studs tied together with bracing ties (Fig. 13.1-64) to accommodate plumbing, wiring, and other systems will achieve a rating of 42 STC when surfaced with one layer of gypsum board on each side. This rating can be improved to 50 by adding a layer of sound-deadening board to both sides, and to 52 by adding 2 to 3 in. (50.8 to 76.2 mm) of insulation in the cavity. Double-layer gypsum board and sound-deadening board added to both sides results in an STC of 53. Increasing the cavity to 12 in. (304.8 mm) and adding three layers of insulation (9 in.) (228.6 mm) to a wall with single-layer gypsum board on each side can result in an STC of 55; another layer of gypsum board added to one or both sides of this construction can result in a 59 or 60 STC.

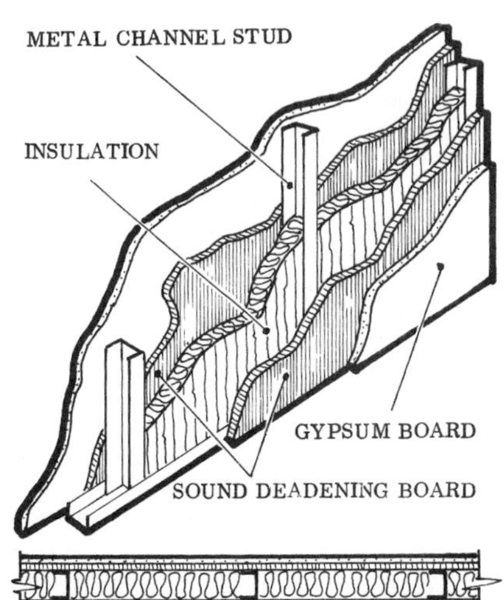

FIGURE 13.1-63 The sound-isolating properties of metal-channel-stud walls are measurably improved when insulation, sound-deadening board, or additional layers of gypsum board are added.

Concrete and Unit Masonry Walls The sound-isolating performance of walls with sealed surface pores is closely correlated with the weight of the wall, and the performance of such walls parallels what is known as the *limp mass law relationship*. The mass law relationship holds true only for limp homogeneous partitions that are nonporous and have uniform physical properties throughout the entire wall panel. This relationship between transmission loss and mass can be expressed by the formula

$$TL = 20 \log m + 20 \log f - 33$$

where m is the mass of the wall stated in lb./sq. ft. and f is the frequency in cps.

Heavy walls tend to reduce the transmission of sound more than light walls, but doubling the mass of a wall increases the transmission loss only 6 dB. As shown in Figure 13.1-65, to achieve a transmission loss of greater magnitude, the increase in mass must be comparatively large.

Nonporous concrete and masonry walls do not permit sound to pass through them, but rather transmit sound by acting as a diaphragm. In a porous wall, the pore structure is the principal transmission path, particularly at low frequencies. When a porous wall's surface is completely sealed with paint or plaster, transmission loss no longer depends on pore structure but instead is determined by the weight and stiffness of the wall. Therefore, the acoustical performance of a painted or plastered porous wall will be essentially the same as that of a nonporous wall of the same weight and stiffness.

It is possible to predict the STC of concrete and masonry walls that are nonporous to the extent that sound transmission through them is due to vibration rather than the passage of airborne sound. This prediction, which is based on the weight of the wall, can be reasonably accurate when there are no openings in the wall and all joints and cracks are sealed. Figure 13.1-65 shows that STC test values are consistently 5 or 6 dB below STC values calculated according to the mass law formula.

Concrete masonry units are produced in hollow and solid units manufactured from either lightweight aggregate or normal-weight aggregates, such as sand and gravel (see Section 4.5). Normal-weight, or dense, aggregate units have a relatively tight pore structure, are considerably heavier than lightweight aggregate units, and will perform acoustically according to their mass.

Lightweight concrete blocks typically contain a larger number of pores (air spaces between aggregate particles and cement matrix) than dense units. For this reason they excel in sound absorption but are not effective in resisting sound transmission. Lightweight blocks are subject to a wide variation in transmission loss when not sealed, because of large differences in the pore structure of blocks made by different manufacturers. Completely sealing their surface pores increases the performance of lightweight units about 7 dB when painted, and about 10 dB when plastered (Fig. 13.1-66). Conversely, the increase gained from painting or plastering dense units is only about 2 dB. But, because of their greater mass, dense units perform considerably better than lightweight units in the first place.

Cavity wall construction and combinations of dense and porous units improve acoustical performance and can provide STCs of 50 or better (see Fig. 13.1-66). Using a combination of two 4-in. (101.6 mm) lightweight block withes with a ⅜-in. (9.5 mm) or wider air space between them and parging one of the withes on the cavity side eliminates the need to seal the faces of the units, thereby retaining intact the sound-absorbing porosity of the lightweight units.

Filling the cores of hollow masonry units with sand or another aggregate will increase the acoustical performance of a well-built wall by only 1 or 2 dB. Filling the cores in a poorly built wall will im-

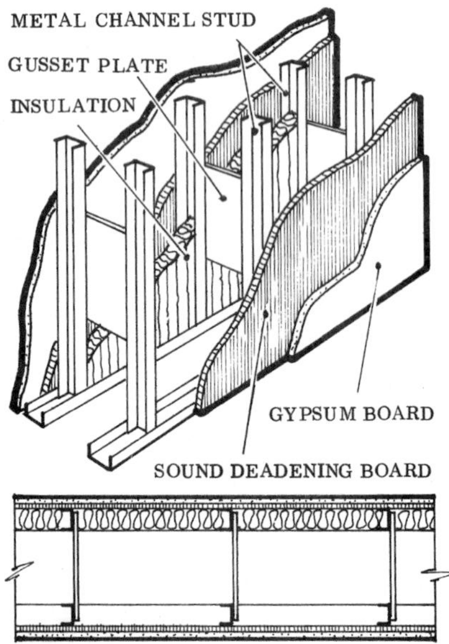

FIGURE 13.1-64 When the cavity depth or insulation thickness is increased and sound-deadening board is used, a metal-channel-stud wall can achieve excellent resistance to sound transmission.

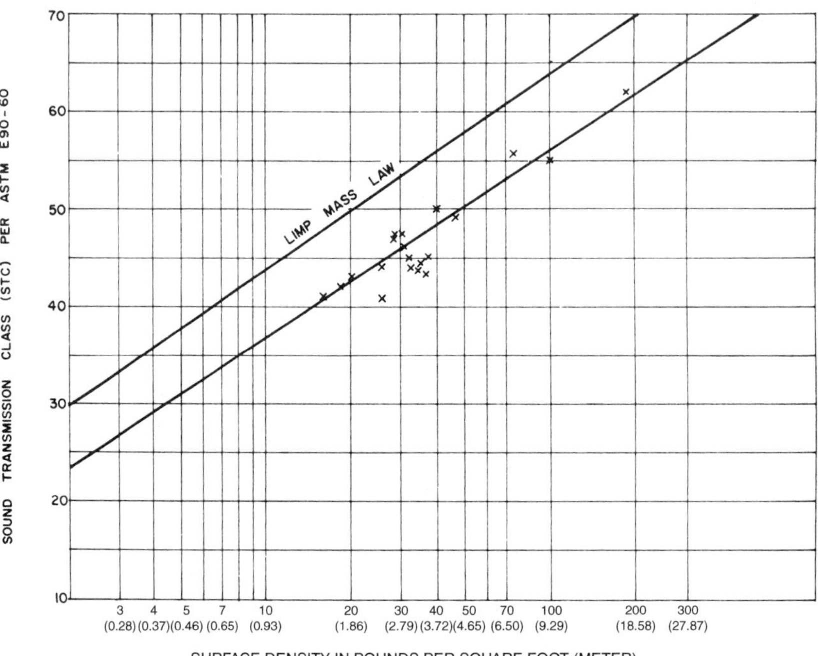

FIGURE 13.1-65 The performance of well-built, nonporous masonry walls can be predicted from their weight according to a mass law formula.

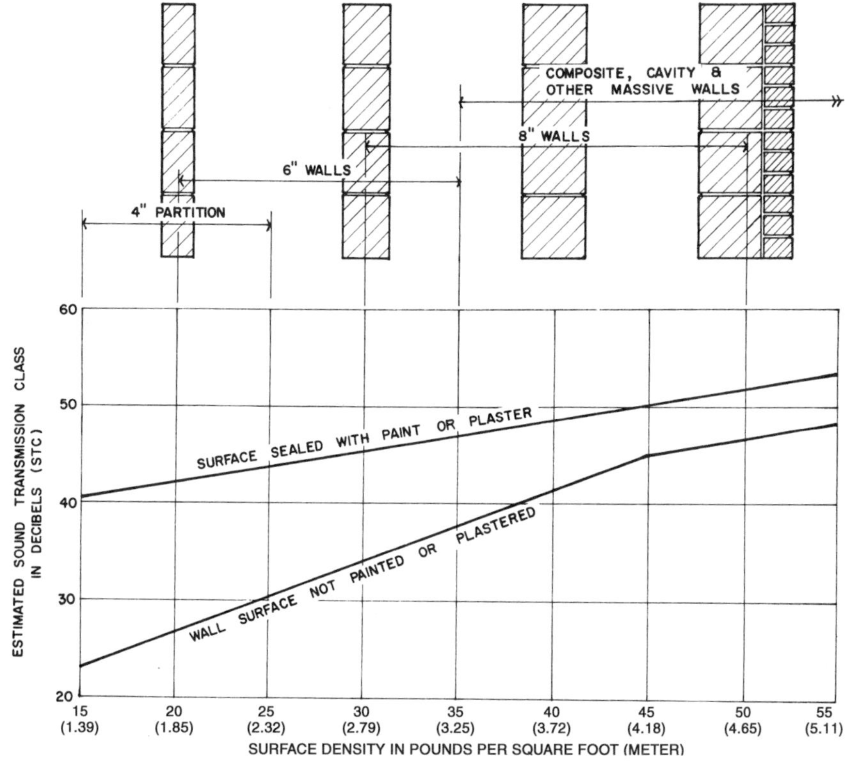

FIGURE 13.1-66 Estimated STC values for sealed and porous concrete masonry walls according to their weight in pounds per square foot.

prove its performance more, but it will never be as good as a properly constructed wall.

There is evidence that hollow units may be better sound insulators than solid units of the same thickness and made with the same aggregate type. A suspected reason is that because the core surfaces are denser than the rest of the block due to the manufacturing process, a *skin effect* occurs, producing a greater-than-expected resistance to sound transmission.

Because clay masonry units are relatively dense and nonporous, the acoustical performance of tightly constructed walls made with them can be estimated according to their mass. Cavity wall construction will provide improved sound isolation with brick masonry, just as it does with concrete masonry.

Regardless of the material, the sound-isolating performance of a masonry wall will depend largely on the extent to which it is airtight. Even relatively fine cracks and open joints such as may occur when a wall is poorly built or as occur at a wall's perimeter will destroy the acoustical effectiveness of a masonry wall.

Because of the great number of mortar joints (and therefore possible acoustical leaks), it is recommended that brick masonry walls be sealed with plaster when acoustical privacy is desired.

Concrete and masonry wall surfaces that have to be sealed should be completely sealed over their entire surfaces, including areas behind cabinets and surfacing materials, in closets, in furred spaces, and above ceilings, to eliminate flanking paths through these areas. Porous masonry walls that are relied on to act as a sound barrier should be plastered or painted even when they are to be furred and surfaced with gypsum board or paneling as the interior finish. A sheet of polyethylene film over a masonry wall does not seal the surface acoustically.

There are often openings at the top and ends of a masonry wall that allow sound to pass to the other side of the wall. Sometimes such openings will occur even at the bottom of a masonry wall. These openings must be closed with acoustical sealant, resilient sound barriers, or lead sheeting.

FLOOR AND CEILING SYSTEMS

Airborne sound transmission is a consideration in selecting floor and ceiling assemblies, but structure-borne (impact) sound transmission is the major problem. Most conventional wood, concrete, and metal floor and ceiling assemblies provide negligible resistance to impact sound transmission. Therefore, some sound-isolating provisions must be added when acoustical privacy is required. Most measures that will control impact sound will also provide a good degree of resistance to airborne sound, providing that essential construction practices are followed.

Measures used to control structure-borne sound include (1) cushioning impact with soft floor-surfacing materials, (2) providing discontinuous construction by floating the finish flooring over the structural framing or subflooring, (3) providing discontinuous construction between the structural floor and the finished ceiling by mounting the ceiling on resilient devices or on separate ceiling joists, and (4) installing sound-absorbing insulation in the cavity of frame construction. Singly and in combination, these construction measures reduce impact sound transmission by breaking the direct structural connections that could carry vibration into other rooms and by absorbing some of the sound within cavities.

Impact cushioning and floated floors are effective with any type of structural floor system. Discontinuous ceilings are generally used only with wood- or metal-framed construction or for critical areas of reinforced concrete, such as for ceilings under equipment rooms and service areas. Installing sound-absorbing materials in the cavity of frame construction is significantly effective only when a resiliently mounted ceiling is also installed.

There is little difference in the acoustical performance of the various hard floor surfacing materials such as wood strip, wood block, and resilient tile and sheet flooring. Impact insulation classes (IIC) for representative floor and ceiling assemblies are given in Section 13.1.5.

Cushioning the Impact The most effective method for controlling impact sound transmission is to cushion the impact with soft resilient floor finish materials, such as carpet and pad. By dissipating a substantial amount of impact energy as it strikes the surface, the amount of sound transmitted to the supporting structural system is substantially reduced. The impact insulation class (IIC) of the carpet and pad combination varies with the thickness of the carpet, the degree of matting of the

FIGURE 13.1-67 Effect of Carpet in Controlling Impact Sound[a]

Carpet[b]	Pad	Floor/Ceiling Construction: Concrete Slab[c] IIC	Floor/Ceiling Construction: Wood Joist[d] IIC[e]
20-oz. (678.11 g/m²) wool	(none)	53	—
40-oz. (1356.23 g/m²) wool	(none)	57	—
60-oz. (2034.34 g/m²) wool	(none)	65	—
44-oz. (1491.85 g/m²) wool	44-oz. (1491.85 g/m²) sponge rubber bonded to carpet	68	54
20-oz. (678.11 g/m²) wool	40-oz. (1356.23 g/m²) felt	69	—
44-oz. (1491.85 g/m²) wool	40-oz. (1356.23 g/m²) hair and jute covered with foam-rubber face and back	70	59
40-oz. (1356.23 g/m²) wool	40-oz. (1356.23 g/m²)	72	—
44-oz. (1491.85 g/m²) wool	40-oz. (1356.23 g/m²) all hair	72	59
44-oz. (1491.85 g/m²) wool	66-oz. (2237.78 g/m²)	72	63
60-oz. (2034.34 g/m²) wool	40-oz. (1356.23 g/m²) felt	73	—
44-oz. (1491.85 g/m²) wool	40-oz. (1356.23 g/m²) hair and jute	73	61
44-oz. (1491.85 g/m²) wool	20-oz. (678.11 g/m²) combination hair and jute, foam-rubber face, latex back	73	61
44-oz. (1491.85 g/m²) wool	Urethane foam	75	—
44-oz. (1491.85 g/m²) wool	40-oz. (1356.23 g/m²) sponge rubber	76	65
44-oz. (1491.85 g/m²) wool	31-oz. (1051.08 g/m²) foam rubber on burlap (⅜ in. [9.53 mm] thick)	79	67
44-oz. (1491.85 g/m²) wool	80-oz. (2712.46 g/m²) sponge rubber	80	68

[a]Based on tests performed by Kodaras Acoustical Laboratories, Inc. for the American Carpet Institute. IIC values approximated by adding 51 to original INR test values.

[b]Carpet pile weight given in oz./sq. yd. (g/m²).

[c]5-in. (127 mm) concrete slab (sand-gravel aggregate), no ceiling finish.

[d]2 × 8 (50 × 200) wood joists, 16 in. (400 mm) o.c.; ⅝-in. (15.88 mm) T&G plywood nailed to joists as subfloor. ⅝-in. (15.88 mm) gypsum board nailed to joists as ceiling.

[e]Values given only where test data are comparable to tests performed on concrete slab construction.

pile, and the weight of both carpet and pad in ounces per square yard (oz./sq. yd. [g/m²]) (Fig. 13.1-67). Unfortunately, there is no appreciable gain in airborne sound isolation by carpet and pad or by acoustical surfacing materials adhered to the ceiling below. Therefore, when cushioning is the only measure to be used to control impact sound transmission, the basic floor and ceiling assembly should be designed to control airborne sound transmission.

Cushion-backed vinyl sheet flooring can be effective in controlling impact sound transmission when carpet is not desired, as may be the case in kitchens, laboratories, toilets, and corridors, for example. Test data for a representative cushion-backed vinyl sheet product show that an IIC of 54 was achieved when this material was installed on a 6-in. (152.4 mm)-thick precast hollow concrete slab. When the ceiling beneath the slab was resiliently mounted, an IIC of 59 was recorded. Adding a 2-in. (50.8 mm)-thick concrete topping to the slab increased these values by 4 dB.

Wood-Framed Construction A conventional wood joist floor and ceiling assembly, finished with hard-surfaced flooring and with a gypsum board ceiling attached directly to the joists (Fig. 13.1-68), will have an IIC of approximately 26. Adding plywood or other subflooring will not appreciably affect this rating. Where an IIC of 51 or better is required, one of the following methods should be employed, depending on the degree of sound isolation required: (1) apply carpet and padding, either alone or in combination with a floated floor or separated ceiling; (2) use separated ceiling construction, by resiliently mounting either the ceiling material or using separate ceiling joists, in conjunction with a floated floor.

The previously mentioned methods for reducing impact sound transmission can also be used in wood-framed construction.

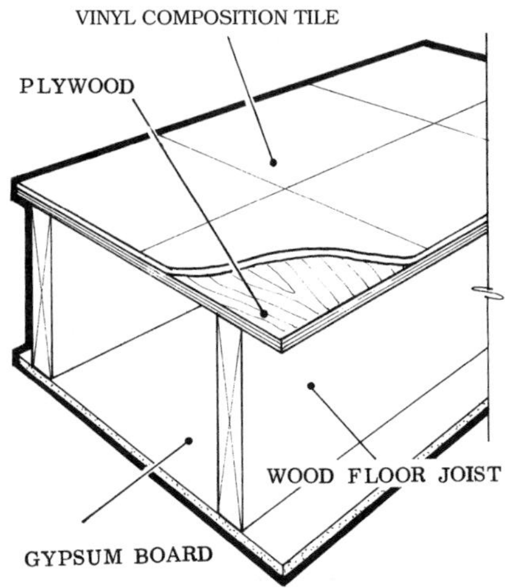

FIGURE 13.1-68 Conventional wood-joist floor construction provides poor resistance to impact sound transmission.

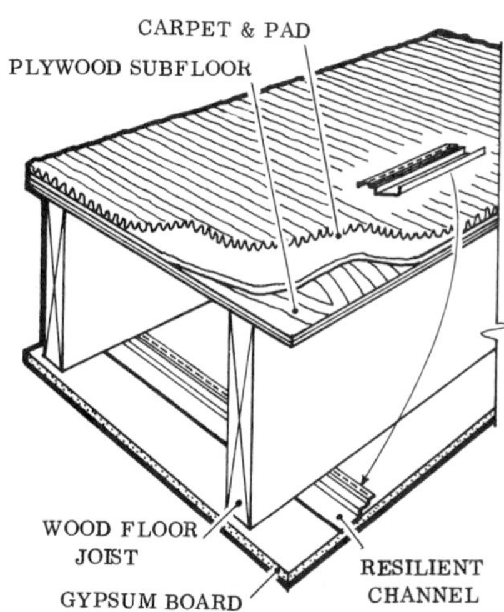

FIGURE 13.1-69 Resilient channels separate the finished ceiling from the structural framing and reduce both airborne and structure-borne sound transmission.

As with other floor and ceiling assemblies, cushioning the impact with carpet and pad of sufficient thickness and density is the most effective method. Carpet and pad used alone or with any of the following tested sound-isolating systems will result in IICs ranging from 42 to 85.

Separating a finished ceiling from an associated structural floor system can be done by either applying the finished ceiling with resilient channels (Fig. 13.1-69) or attaching the finished ceiling to separate ceiling joists (Fig. 13.1-70). Either of these discontinuous construction methods can effectively reduce both impact sound from above and airborne sound from either direction. As is true with every sound-control measure, separating a ceiling from a structural system will not produce an effective acoustical barrier unless the entire floor and ceiling assembly is built correctly, with careful attention to the blocking of flanking paths, and measures are taken to prevent sound from bypassing the ceiling through walls and other structural connections. As with resiliently mounted walls, heavy objects, such as lighting fixtures, should not be mounted on or recessed in the ceiling, and airtight construction should be maintained.

Resilient mounting of a finished ceiling or attaching it to separate joists can improve the IIC of an assembly as much as 7 dB; sound-absorbing material in the cavity can provide an additional 7 dB. But separated ceiling construction alone will not produce a floor and ceiling assembly with much more than 45 IIC when the floor above is hard-surfaced, even though an acceptable STC of 50 is obtained. However, if a floated floor is used instead, even when the flooring has a hard surface the IIC can jump to 57 and the STC will still be 50 or more.

Floated-floor construction over wood joists and subflooring can be accomplished with a resilient underlayment separating the finished flooring and underlayment from the structural floor system (Fig. 13.1-71a), or by placing lightweight (cellular) concrete over the resilient underlayment (Fig. 13.1-71b). Section 13.1.5 shows other possible methods. When resilient underlayment is used (Fig. 13.1-72), the resilient material must be capable of resisting major deformation from su-

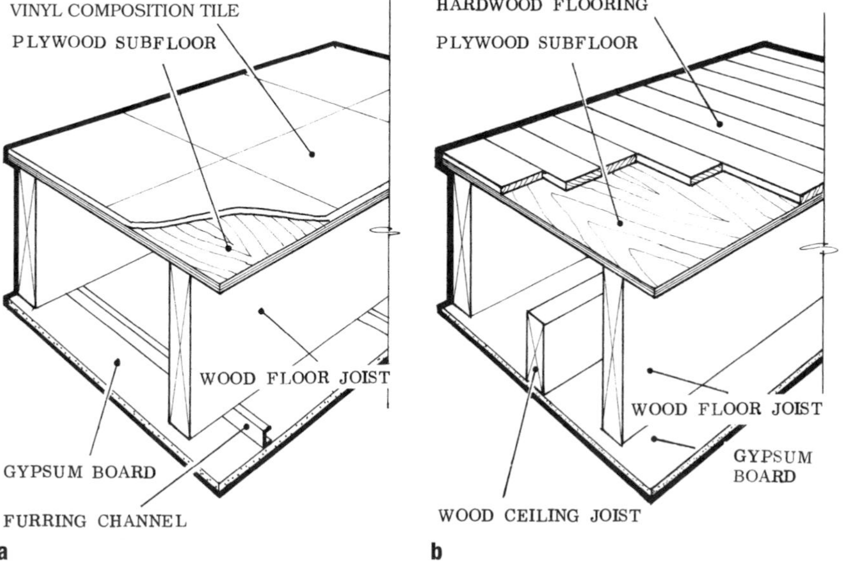

FIGURE 13.1-70 The impact-sound–isolating performance of floor and ceiling assemblies like that shown in Figure 13.1-68 can be improved by either (a) resilient mounting of the finished ceiling or (b) carrying the finished ceiling on independent ceiling joists.

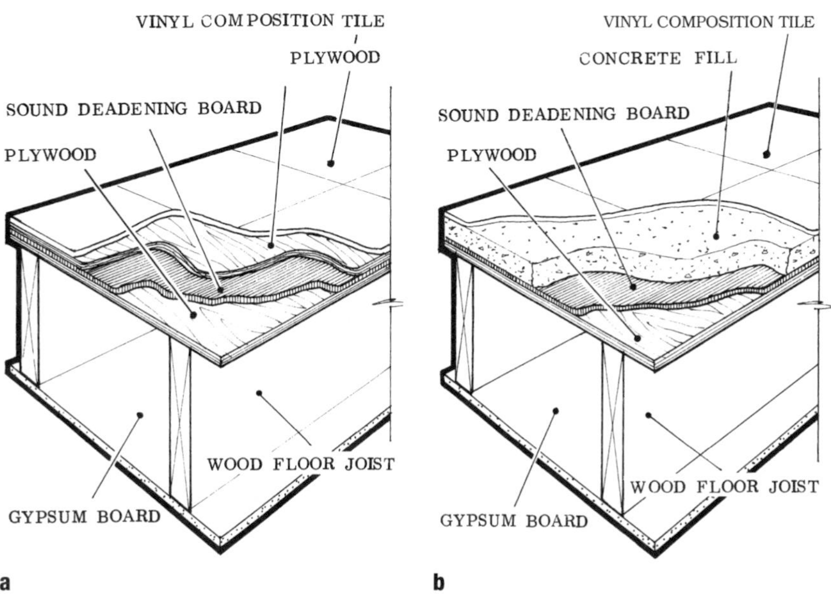

FIGURE 13.1-71 Two methods of making a floated floor. In (a) a resilient underlayment separates the finished flooring from the structural floor. In (b) lightweight cellular concrete has been placed over a resilient underlayment to form a floating floor system.

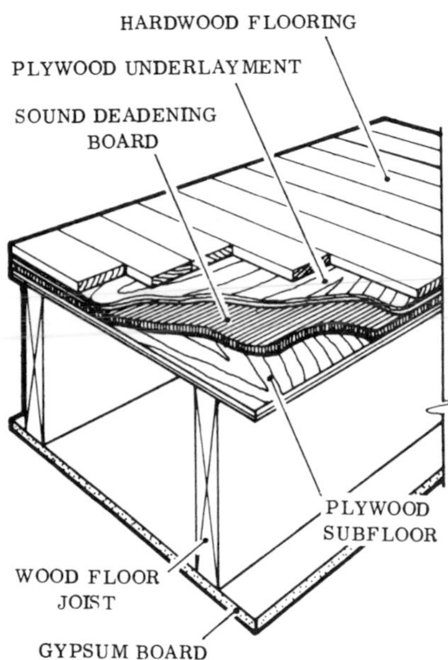

FIGURE 13.1-72 Resilient underlayment (shaded) reduces impact sound transmission by "floating" the finished flooring above the subfloor or structural floor.

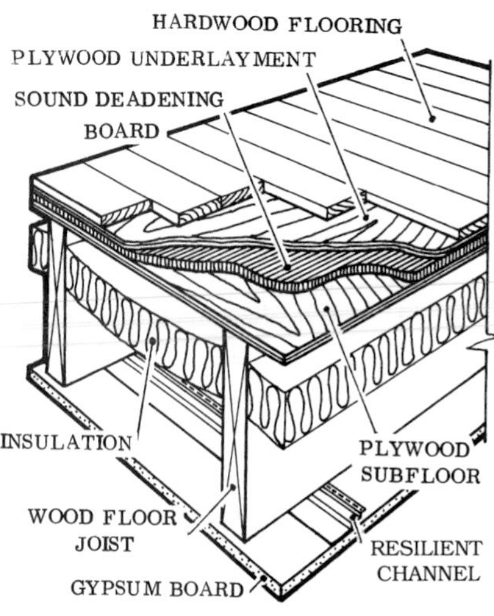

FIGURE 13.1-73 A resiliently mounted ceiling combined with a floated floor effectively resists impact sound transmission.

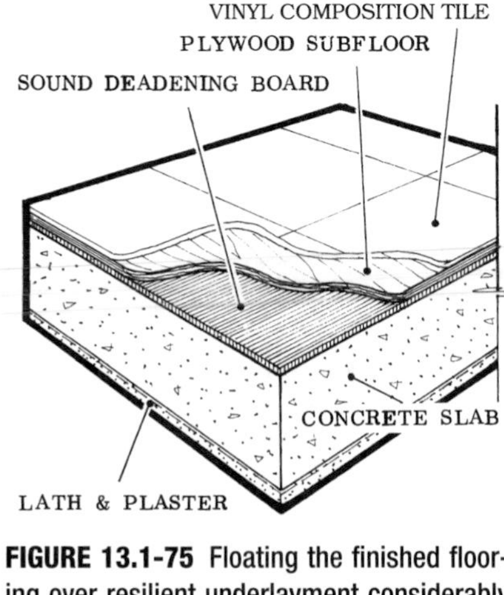

FIGURE 13.1-75 Floating the finished flooring over resilient underlayment considerably improves the acoustical performance of a concrete slab assembly like that shown in Figure 13.1-74.

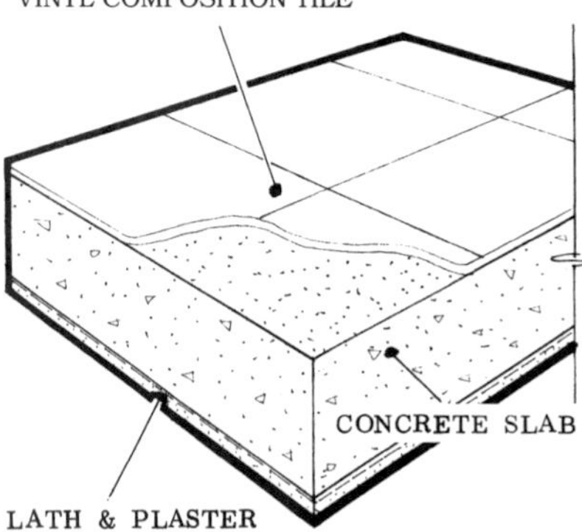

FIGURE 13.1-74 A typical concrete slab with hard-surfaced flooring and a directly applied plaster ceiling offers poor resistance to impact sound transmission.

perimposed loading. Most such material is designed to carry only the relatively light loading of the finished floor, people, and furniture.

In general, the greater the resilience of the underlayment material, the greater the isolation. Like carpeting, resilient underlayment works primarily as a cushion and only secondarily to resist airborne sound transmission through the assembly.

The performance of a floated floor depends on effectively separating the finished floor from the structural elements. The floating part should not be rigidly connected to other elements. Adhesive is generally used to attach resilient underlayment. Nailing or stapling the subflooring or finished flooring through the resilient underlayment into the supporting construction will nullify the sound-isolating properties of the floated floor.

A basic wood-joist construction can be improved approximately 7 dB with a floated-floor system using resilient underlayment. To achieve a rating of more than 51 IIC, the floated floor can be supplemented by a separated finished ceiling. With the addition of sound-absorbing material in the joist space (Fig. 13.1-73), an IIC of 57 can be achieved. The addition of a mineral-fiber insulating blanket in the space between the floor joists absorbs approximately 7 dB of the transmitted noise. Unfortunately, the addition of sound-absorbing materials within the free airspace of a framed floor and ceiling assembly is only significantly effective in controlling impact sound transmission when used with a resiliently mounted ceiling. However, it is effective in reducing airborne sound with all types of discontinuous construction.

Lightweight concrete placed directly over a plywood subfloor does not appreciably improve the impact sound resistance of a basic wood-joist floor and ceiling assembly. When the concrete is resiliently floated, however, the improvement is about 10 dB, but the IIC will still be less than 51 unless carpet and pad are added or the ceiling is resiliently mounted.

Concrete Slab Construction A plain concrete slab with resilient tile flooring and a plaster ceiling applied directly to it (Fig. 13.1-74) will have an IIC of about 36. Increasing the weight or thickness of the slab will have little effect on the impact sound transmission loss produced by it. Sound-control measures that will improve the acoustical performance of a concrete slab are limited to floating the flooring and cushioning impacts with carpet and pad. Because doing so is impracticable and costly, the ceiling under a concrete slab is seldom resiliently mounted and test data for that are limited.

Adding carpet and pad to a concrete slab provides somewhat better performance than adding carpet and pad on a wood-framed construction, with IICs ranging from 53 to 88 or even more, depending on the weight and density of the carpet and pad (see Fig. 13.1-67).

Floating to isolate a finished floor from a concrete slab is accomplished in much the same manner as in a wood frame (Fig. 13.1-75). Provisions should be made to keep the floated floor from rigid contact with structural elements. When possible, floating floors should not touch partition edges. One way to prevent such contact is to install a cushion between the edge of the partition and the floor.

A hard-surfaced floor floated on a resilient underlayment over a 6-in. (150 mm)-thick concrete slab with a directly applied plaster ceiling will have an IIC of about 55.

The performance of precast concrete floor and ceiling assemblies is similar to that of cast-in-place concrete assemblies, but available test data on such assemblies are limited. Data for representative precast floor systems are given in Section 13.1.5.

Steel bar joist and other metal-framed floor systems can be improved by measures similar to those used to improve wood-framed assemblies. Methods for floating finished flooring and resiliently mounting finished ceilings closely parallel those used in wood-framed assemblies. The acoustical effectiveness of steel bar joist construction, as well as metal decks, concrete floor joists, and similar systems, depends primarily on airtight sealing of wall and floor intersections. Walls must be extended to the bottom of the slab or deck above, and the space at their tops must be made airtight. When walls run perpendicular to joists, proper sealing is difficult and expensive to achieve.

Concrete slabs over steel bar joists are normally about 2 in. (50.8 mm) thick, which gives them sufficient mass to produce an STC only in the high 40s. Therefore, when a higher STC is required for either walls or floor and ceiling assemblies, this type of construction is not effective. Even when a wall itself has a high STC, sound will bypass it through the flanking path provided by a relatively lightweight floor system.

Current test data for evaluating metal-framed floor and ceiling assemblies are not as extensive as data for other construction systems. Representative steel bar joist systems and some available test data are given in Section 13.1.5.

WALL AND FLOOR INTERSECTIONS

Noise problems from flanking paths are encountered where walls and floors intersect. Unless precautions are taken, both airborne and structure-borne sound can penetrate around and through the walls and floors, even though the wall and floor constructions are structurally adequate.

Walls and associated floor and ceiling assemblies should have comparable sound-isolating abilities so that a weakness in one will not diminish the effectiveness of the other.

The spaces between floor joists should be blocked where walls and floor and ceiling assemblies join, to reduce airborne sound transmission through such flanking paths (Fig. 13.1-76). Joints at the top and bottom of partitions should be sealed airtight with a resilient, nonhardening acoustical sealant or a resilient material (Fig. 13.1-77).

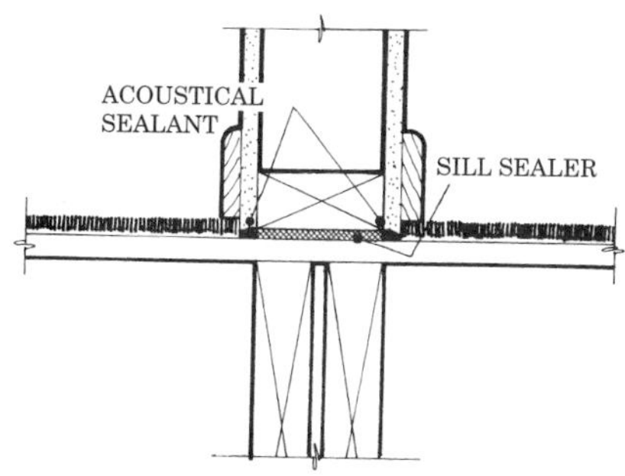

FIGURE 13.1-77 Acoustical sealant should be used to seal small open air paths. This figure shows the sealing of paths at the bottom of a wall. Paths at the top and ends of sound-isolating walls should be similarly sealed.

Loads applied to a floated floor by a structural wall will cause the resilient underlayment to collapse, damaging the floor and maybe even causing the structure to settle. An underlayment material that is sufficiently rigid to support structural loading is too rigid to provide resilient isolation in a floated floor system. Therefore, the bottom plate of a bearing partition should not be supported on a floating floor, but rather should sit directly on and be fastened to the subfloor or supporting structure.

A break in the subflooring under partitions will help to reduce impact noise transmission between adjacent rooms by breaking direct contact of the structural elements (Fig. 13.1-78a).

When possible, a cushion should be provided between a partition and the edge of a floating floor so that they do not touch (Fig. 13.1-78b).

Floated concrete floors on wood-joist construction should be confined to individual rooms to prevent transmission of impact noise along the floor (see Figs. 13.1-78b and 13.1-79). Isolation joints around the perimeter of rooms and across doorways are essential. Before the concrete is placed, walls should be supported on a floor plate, polyethylene film should be installed over the resilient underlayment, and isolation joints should be installed around the perimeter of each room.

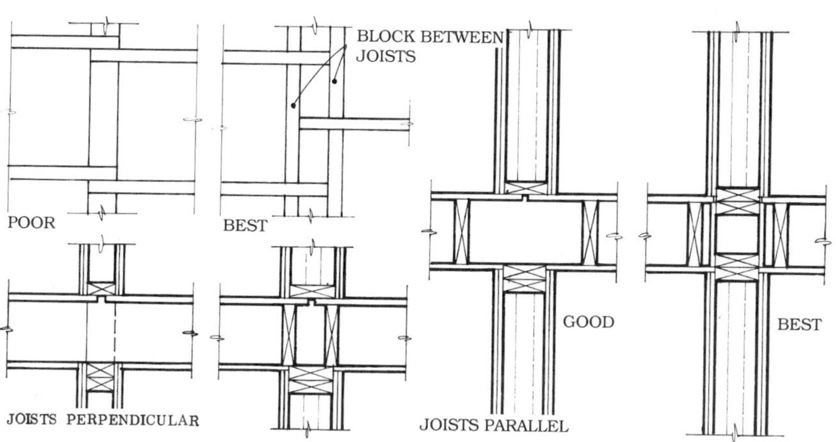

FIGURE 13.1-76 Openings must be closed to eliminate airborne sound flanking paths. The spaces between joists must be blocked, especially when perpendicular to walls.

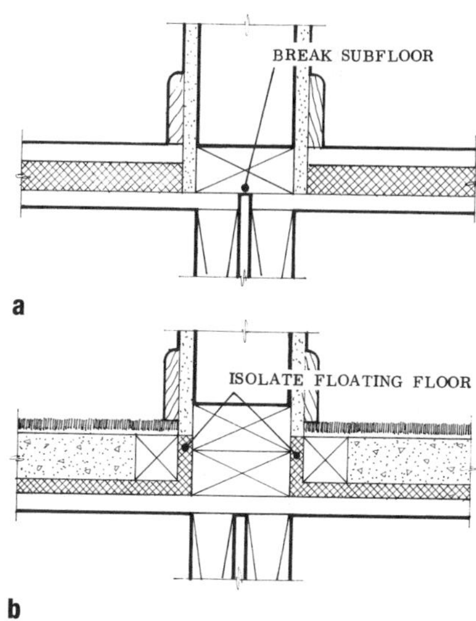

FIGURE 13.1-78 Flanking structure-borne noise can be reduced by breaking direct contact between structural elements, such as the subflooring shown in (a), and by isolating floated floors from walls (b).

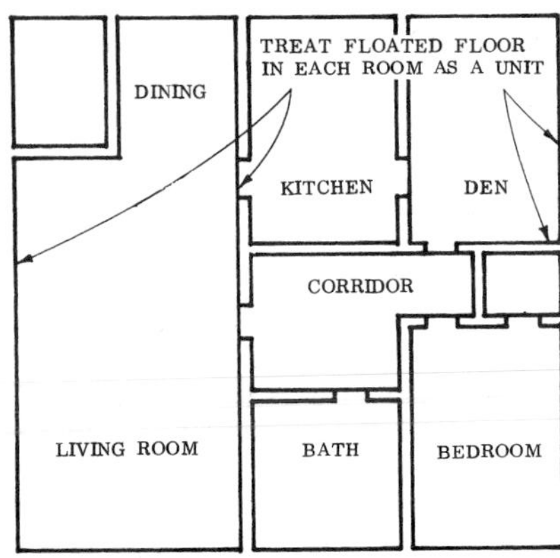

FIGURE 13.1-79 A floated floor should be confined to a single room by separating it from adjacent construction at doors and walls.

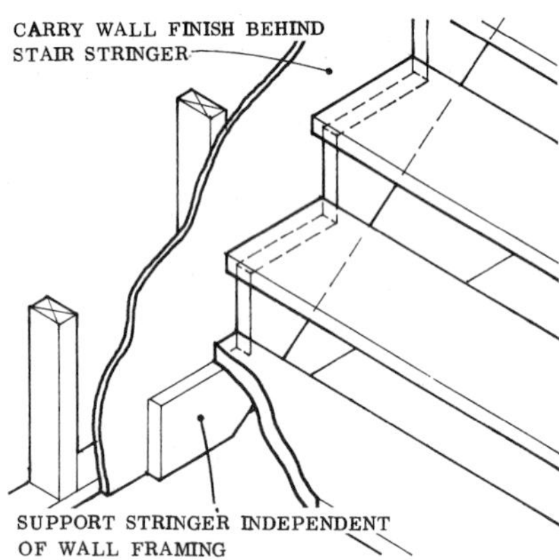

FIGURE 13.1-80 Stair stringers should not be nailed or otherwise fastened to adjacent walls. If they are, footfalls will be transmitted into the walls.

The concrete should not be directly in contact with the walls. Moldings and quarter-round trim can be used to cover isolation joints.

Wall surfacing material should not be tightly forced against resiliently mounted ceilings. Doing so may collapse resilient hangers, channels, or clips and short-circuit the resilient mounting.

STAIRS

Stair stringers should be fastened only at the top and bottom and should not be nailed to adjacent studs. Header joists should be used at landings to support floors instead of nailing the floor joists to the studs (Fig. 13.1-80).

When continuous wall surfacing material is installed on the entire wall adjacent to a stairwell before the stairs are installed, sound transmission between the stairs and adjacent rooms will be reduced.

When stairs are on both sides of a common wall, such as in townhouses, duplexes, and other attached housing units, staggered- or double-stud walls with insulation in the cavity should be used. Stair runners on the wall side should be held away from the studs to accommodate later installation of a finishing material behind the runners. If the walls are to be plastered, temporary construction ladders, as are used in fire-resistant construction, should be used to permit complete plastering and to eliminate breaks in the wall.

Finishing them with carpet or a resilient material, such as a rubber mat, will substantially reduce impact noise transmission from stairs to adjacent areas.

CORRIDORS

Air ducts often are installed before ceiling and wall finishes, which will be behind them, are in place, making it difficult or impossible to install these finishes later. This permits impact noise from a corridor to pass through the ceiling and down the open stud spaces in the walls of adjoining spaces. This problem can be solved by installing only register outlets and the first duct section in a corridor, thus deferring the rest of the duct installation until the interior finishes of the walls and ceilings have been completed. Ducts can then be installed and a furred ceiling hung below them (Fig. 13.1-81).

13.1.4.3 Selection Criteria

The following recommended criteria are suggested for selecting walls and floor and ceiling assemblies to provide resistance to airborne and structure-borne sound transmission.

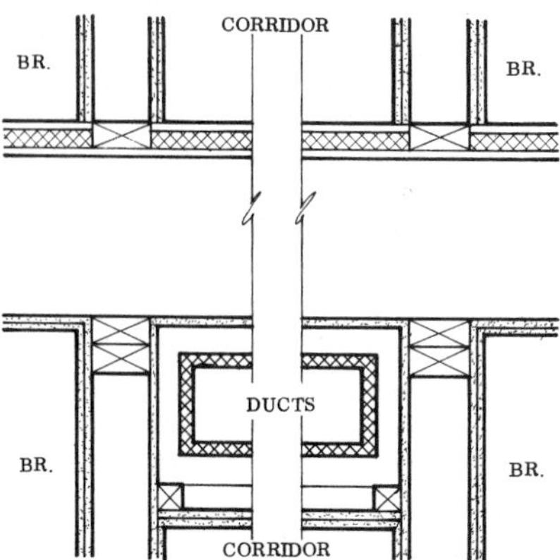

FIGURE 13.1-81 Unless finishes are taken above them, furred spaces become sound transmission paths. Ducts should be installed after the finishes have been placed.

Generally, acceptable overall airborne sound privacy is said to exist when the level of conversational speech is reduced by a wall or floor and ceiling assembly to the point where it cannot be understood in an adjoining room. To estimate the minimum STC required to obtain privacy, the level of background noise on one side of a construction is subtracted from the level of sound that strikes the other side. Figure 13.1-26 illustrates the approximate effectiveness of walls with varying STC values at varying background noise levels.

General recommendations for common walls and floor and ceiling assemblies separating living units in terms of STC and IIC ratings are given in Figures 13.1-82, 13.1-83, and 13.1-84. Exact values for STC and IIC ratings should be determined on the basis of individual test data in accordance with prescribed procedures (see Section 13.1.3, "Sound Transmission").

The criteria recommended by the Federal Housing Administration (FHA) of the U.S. Department of Housing and Urban Development (HUD) for noise control are often used as a basis in the design and construction of multifamily housing units. The following recommendations are included in the "FHA Guide to Airborne, Impact, Structure-borne Noise Control in Multifamily Dwellings." The criteria are based on (1) the level of ambient background noise represented by urban, suburban, and rural locations, (2) minimum-, average-, and high-income housing, and (3) the specific function of the wall or floor and ceiling assembly within the building.

The FHA uses three grades of acoustic environment, which allows its criteria to be applied to a wide range of urban developments, geographic locations, economic conditions, and other factors. Constructions that meet the criteria will provide good sound isolation and will satisfy a majority of the occupants in a building that fits the conditions of the particular grade. Grade II is applicable to the largest percentage of multifamily construction and, therefore, can be considered as a basic guide.

The Grade I category includes primarily suburban and peripheral suburban residential areas, considered to have a quiet background noise level. The nighttime exterior background noise level may be between 35 and 40 dB or lower. In addition, the sound isolation criteria of this grade are applicable in certain special cases, such as dwelling units above the eighth floor in high-rise buildings and luxury apartments, regardless of location.

The Grade II category is applicable primarily in residential urban and suburban areas considered to have an average noise environment. The nighttime exterior background noise levels may be between 40 and 45 dB.

The Grade III criteria represent minimal recommendations. They are applicable in some urban areas that are generally considered to be noisy locations. The nighttime exterior background noise levels may be about 55 dB or higher.

Walls and floor and ceiling assemblies should have STC and IIC ratings equal to or greater than the recommended criterion figures. A floor and ceiling assembly that may provide adequate impact sound isolation may not provide suitable airborne sound isolation. The criteria for both airborne and impact sound isolation of floor and ceiling assemblies must be met.

Recommended basic criteria for airborne and impact sound isolation of common wall and floor and ceiling assemblies that separate dwelling units of equivalent functions are given in Figure 13.1-82. These criteria are based on STC and IIC ratings derived from laboratory measurements. Standard methods of testing for field measurements have not yet been formally adopted.

Figure 13.1-83 gives the recommended criterion values for common walls, related to room or area function. Most of the typical conditions encountered for common walls separating dwelling units in multifamily construction are included. Some of these conditions are clearly undesirable and should be prevented during the initial design of the building. The purpose of this detail is to illustrate the importance of the acoustical separation between sensitive and nonsensitive areas. A partition between dwelling units that is common to several functional spaces should meet the highest criterion value.

Figure 13.1-84 includes criteria for most of the conditions encountered in common floor and ceiling assemblies that separate units in multifamily dwellings. Some of these conditions are also clearly undesirable and should be prevented during building design. Additional precautions and qualifying recommendations are given in the footnotes to Figure 13.1-84.

When the placement of living units vertically or horizontally adjacent to mechanical equipment rooms cannot be avoided, special consideration should be given to provide suitable sound-isolating construction. The recommended criteria for airborne sound between mechanical equipment rooms and sensitive areas in dwelling units are:

Grade I STC > 65
Grade II STC > 62
Grade III STC > 58

Mechanical equipment rooms include furnace-boiler rooms, elevator shafts, trash chutes, cooling towers, garages, and similar areas. The recommended criteria between mechanical equipment rooms and less sensitive areas, such as kitchens and family and recreation rooms, are:

Grade I STC > 60
Grade II STC > 58
Grade III STC > 54

FIGURE 13.1-82 General Criteria for Airborne and Impact Sound Isolation Between Dwelling Units*

Construction	Grade I	Grade II	Grade III
Wall partitions	STC >55	STC >52	STC >48
Floor/ceilings	STC >55 IIC >55	STC >52 IIC >52	STC >48 IIC >48

*Adapted from "FHA Guide to Airborne, Impact, and Structure-Borne Noise Control in Multifamily Dwellings."

FIGURE 13.1-83 Sound Isolation Criteria for Walls Between Dwelling Units[a]

Common Walls Separating Dwelling Units[b]			Sound Transmission Class (STC)		
			Grade I	Grade II	Grade III
Bedroom	and	bedroom	55	52	48
Living room	and	bedroom	57	54	50
Kitchen[c]	and	bedroom	58	55	52
Bathroom	and	bedroom	59	56	52
Corridor	and	bedroom[d]	55	52	48
Living room	and	living room	55	52	48
Kitchen[c]	and	living room	55	52	48
Bathroom	and	living room	57	54	50
Corridor	and	living room[d,e]	55	52	48
Kitchen[c]	and	kitchen[f]	52	50	46
Bathroom	and	kitchen	55	52	48
Corridor	and	kitchen[d,e]	55	52	48
Bathroom	and	bathroom	52	50	46
Corridor	and	bathroom[d]	50	48	46

[a]Adapted from "FHA Guide to Airborne, Impact, and Structure-Borne Noise Control in Multifamily Dwellings."

[b]Acoustically, mirror planning is desirable where common walls separate rooms of similar function; when not possible, walls separating bedrooms or living rooms from dissimilar areas must have greater sound-isolating properties.

[c]Or dining, family, or recreation room.

[d]No entrance door leading from corridor to living unit assumed.

[e]It is commonly and incorrectly assumed that walls containing entrance doors need be no better acoustically than the door. However, the basic corridor wall may separate sensitive living areas without entrance doors and therefore must have adequate sound-isolating properties to ensure privacy in these areas.

[f]Double wall construction is recommended to provide isolation from impact noise resulting from vibration of mechanical equipment and systems.

FIGURE 13.1-84 Sound Isolation Criteria for Floor and Ceiling Assemblies Between Dwelling Units[a]

Common Floor and Ceiling Assemblies Separating DwellingUnits[b]	Grade I		Grade II		Grade III	
	STC	IIC	STC	IIC	STC	IIC
Bedroom above bedroom	55	55	52	52	48	48
Living room above bedroom	57	60	54	57	50	53
Kitchen[c] above bedroom	58	65	55	62	52	58
Family room above bedroom[d]	60	65	56	62	52	58
Corridor above bedroom	55	65	52	62	48	58
Bedroom above living room[e]	57	55	54	52	50	48
Living room above living room	55	55	52	52	48	48
Kitchen above living room	55	60	52	57	48	53
Family room above living room[d]	58	62	54	60	52	56
Corridor above living room	55	60	52	57	48	53
Bedroom above kitchen[e]	58	52	55	50	52	46
Living room above kitchen[e]	55	55	52	52	48	48
Kitchen above kitchen	52	55	50	52	46	48
Bathroom above kitchen	55	55	52	52	48	48
Family room above kitchen[d]	55	60	52	58	48	54
Corridor above kitchen	50	55	48	52	46	48
Bedroom above family room[e]	60	50	56	48	52	46
Living room above family room[e]	58	52	54	50	52	48
Kitchen above family room[e]	55	55	52	52	48	50
Bathroom above bathroom	52	52	50	50	48	48
Corridor above corridor	50	50	48	48	46	46

[a]Adapted from "FHA Guide to Airborne, Impact, and Structure-Borne Noise Control in Multifamily Dwellings."

[b]Acoustically, mirror planning, where common floor and ceiling assemblies separate rooms of similar function, is desirable; when not possible, floor and ceiling assemblies separating bedrooms or living rooms from dissimilar areas must have greater sound-isolating properties. When bedrooms or living rooms are located below less sensitive living spaces, greater impact isolation is required.

[c]Or dining, family, or recreation room.

[d]STC criteria also apply to vertical partitions between these two spaces.

[e]Equivalent airborne sound isolation, but less impact sound isolation generally required for the converse situation.

FIGURE 13.1-85 Sound Isolation Criteria for Walls Within Dwelling Units[a]

Walls Separating Rooms[b]	Sound Transmission Class (STC)		
	Grade I	Grade II	Grade III
Bedroom and bedroom	48	44	40
Living room and bedroom	50	46	42
Bathroom and bedroom	52	48	45
Kitchen and bedroom	52	48	45
Bathroom and living room	52	48	45

[a]Adapted from "FHA Guide to Airborne, Impact, and Structure-Borne Noise Control in Multifamily Dwellings."

[b]Doors leading to bedrooms and bathrooms preferably should be gasketed, solid-core doors to provide maximum privacy.

Double walls are usually necessary to achieve adequate acoustical privacy. When living units are above noisy areas, airborne sound isolation is important, but impact sound isolation is not significant when structure-borne vibration is minimal. However, when mechanical equipment rooms are above living areas, airborne sound isolation must be maintained, impact sound isolation becomes critical, and specially designed sound-isolating construction is required to ensure quiet living areas.

Placing dwelling units vertically or horizontally adjacent to commercial establishments such as restaurants, bars, community laundries, and similar businesses should be avoided when possible. If such situations cannot be avoided, the recommended airborne isolation criteria between business areas and sensitive living areas are:

Grade I STC > 60
Grade II STC > 58
Grade III STC > 56

If the living areas are situated above such areas, adequate impact isolation criteria should be:

Grade I IIC > 60
Grade II IIC > 58
Grade III IIC > 56

However, if the business areas are above living areas, the impact isolation criteria values should be increased by at least 5 points.

If noise levels in mechanical equipment rooms or business areas exceed 100 decibels, as measured using the C scale of a standard sound-level meter, the airborne-sound isolation criteria given in the preceding lists should be raised 5 points.

Figure 13.1-85 lists suggested criteria for airborne sound isolation for walls separating rooms within a given living unit.

Townhouses and rowhouses in which the living unit occupies more than 1 story should be separated by a double-wall construction with an STC rating of 60 or greater. Suggested criteria for sound isolation between rooms in a given dwelling are the same as listed in Figure 13.1-83 and, in addition, the floor and ceiling assemblies should have IIC ratings that are at least numerically equivalent to or greater than the listed STC criteria.

Rooftop and indoor swimming pools, bowling alleys, ballrooms, tennis courts, and gymnasiums require constructions specifically designed in accordance with acoustical analysis.

13.1.5 SOUND PERFORMANCE RATINGS FOR SELECTED ASSEMBLIES

The following data and the accompanying figures include sound performance data for walls and floor and ceiling assemblies. Data on systems that provide poor sound isolation are included to indicate the degree of improvement required to render such systems sufficiently resistant to sound transmission.

Unless otherwise noted, the data presented in the figures are based on tests performed by nationally recognized acoustical laboratories in accordance with the test procedures in ASTM E-90, "Method for Laboratory Measurement of Airborne-Sound Transmission Loss of Building Partitions" and the International Organization for Standardization (ISO) Recommendation R-140, "Field and Laboratory Measurements of Airborne and Impact Sound Transmission." The data include tests performed by Cedar Knolls Acoustical Laboratories, Geiger and Hamme, Kodaras Acoustical Laboratories, the National Bureau of Standards (now the National Institute of Standards and Technology), National Gypsum Company, Ohio Research Corporation, Owens-Corning Fiberglas Sound Laboratory, and Riverbank Acoustical Laboratories, IIT Research Institute.

The STC, INR, and IIC ratings shown have been drawn from a variety of manufacturers' literature and testing laboratory reports. In some instances, only very old test results were available. In others, few correlating test results could be found. Therefore, some ratings have been estimated. Before relying on the data shown for the final design of an actual project, it would be wise to ask the manufacturers of the proposed materials or systems proposed to furnish the results of their latest sound performance tests.

When possible, a range of values is given to account for (1) variations in construction materials, (2) variation in application procedures, and (3) variation in testing facilities.

The reproducibility of laboratory tests on different samples can range up to 5 dB at any one frequency. However, overall STC ratings should be within 2 dB.

13.1.5.1 Wall Assemblies

Refer to Section 13.1.4 for a discussion of the parameters dictating the use of the various materials and construction in the assemblies illustrated in Figure 13.1-86.

The STC ratings given for clay masonry walls have been interpolated from Figure 13.1-66.

13.1.5.2 Floor and Ceiling Assemblies

Whether to use plywood, fiberboard, hardboard, or another underlayment is generally determined by the type of finish flooring or other practical construction techniques. Except for resilient underlayment in floated floor systems, underlayments contribute insignificant impact sound isolation. Unless otherwise noted, the ranges given in Figure 13.1-87 for INR and IIC include assemblies both with and without underlayment. Refer to Section 13.1.4 for additional discussion.

Because floor and ceiling assemblies that control impact sound isolation will also provide a good degree of resistance to airborne sound transmission, much of the test data for floor and ceiling assemblies are given in terms of INR and IIC only. However, the STC values for a particular assembly can be estimated, providing that the essential precautions of construction practices are followed. Where test data for STC ratings were not available, the STC values have been estimated.

FIGURE 13.1-86 Wall Systems

Wood Frame Load-Bearing Walls—Gypsum Board Finish[1]

SINGLE STUD WALLS

Single Layer Finish[2]

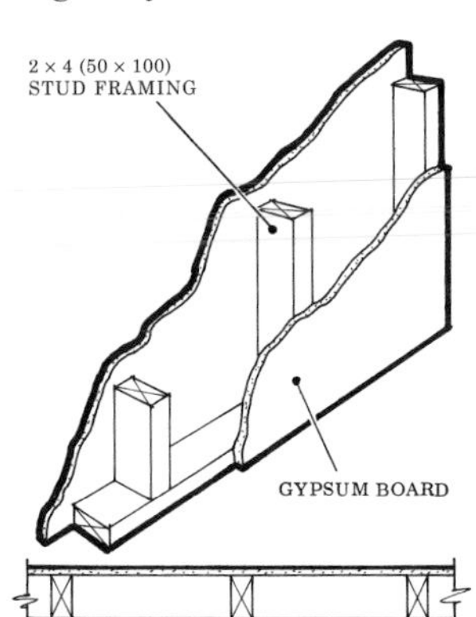

STC 32 to 36

Multilayer Finish[2]

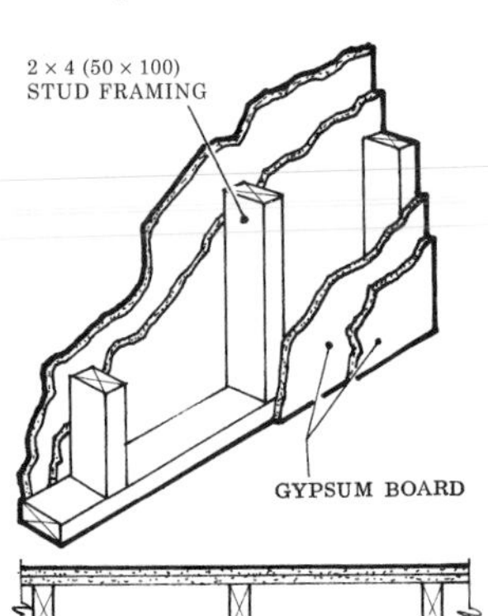

STC 38 to 41

SINGLE STUD WALLS

Single Layer Finish

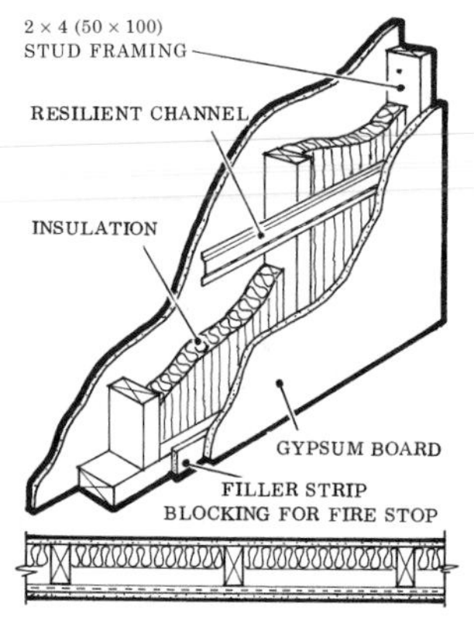

STC 40 to 42

With insulation[7]

STC 47 to 51[4]

Sound Deadening Board[5]

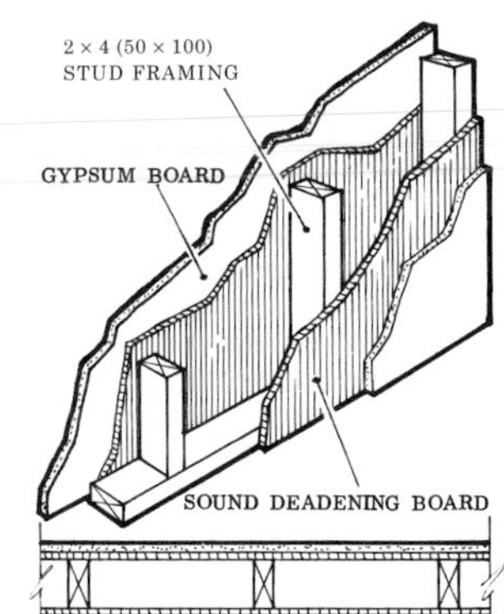

STC 47 to 51[4]

STAGGERED STUD WALLS

Resilient Finish[3]

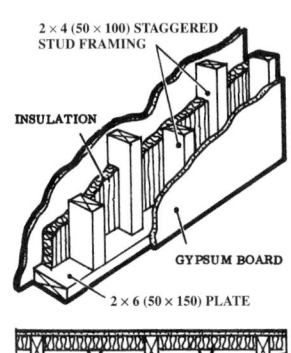

2 × 4 (50 × 100) STAGGERED STUD FRAMING

INSULATION

GYPSUM BOARD

2 × 6 (50 × 150) PLATE

STC 38 to 42[4]

With insulation[8]

STC 46 to 48[4]

Sound Deadening Board[5]

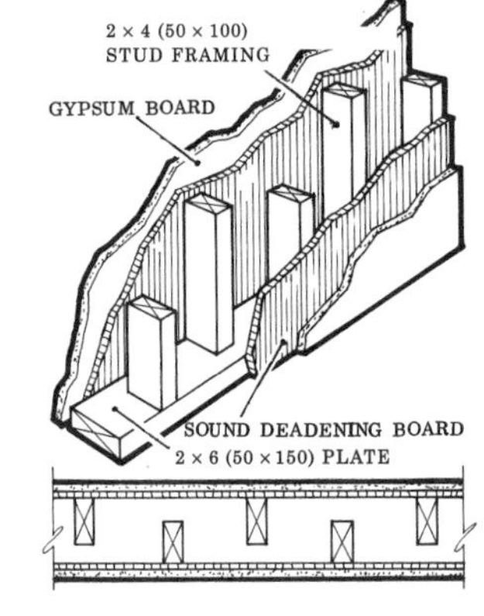

STC 46 to 50[4]

DOUBLE STUD WALLS

Single Layer Finish

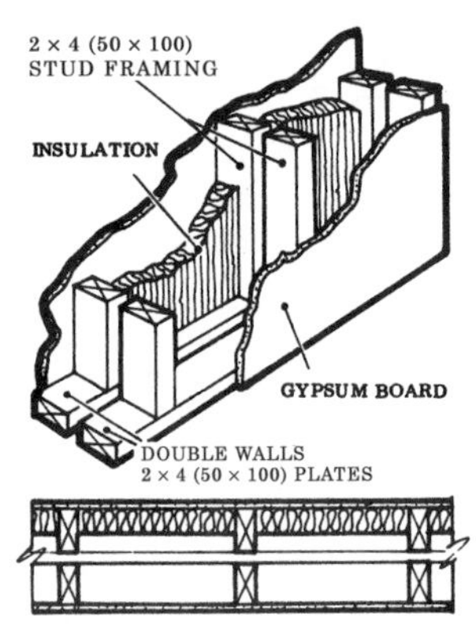

STC 42 to 44

With insulation[8]

STC 50 to 53[4]

Sound Deadening Board[5]

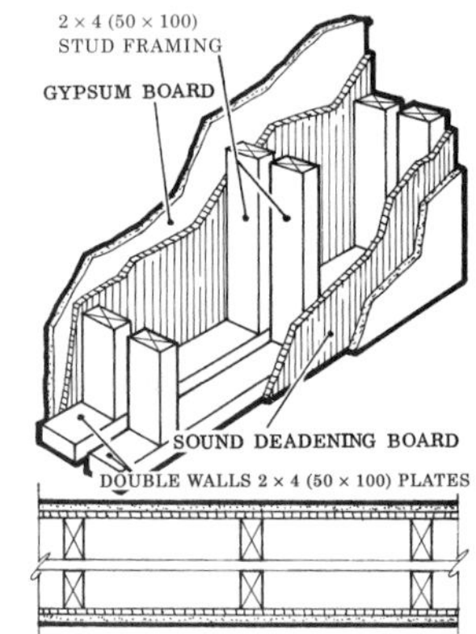

STC 50 to 53[9]

With insulation[8]

STC 54 to 58

1. Ranges include values for 3/8 in. (9.53 mm), 1/2 in. (12.7 mm), and 5/8 (15.33 mm) thicknesses of gypsum board finish.
2. Addition of insulation does not appreciably improve this construction; an increase of approximately 2 to 4 dB can be expected regardless of thickness or density of insulation.
3. Resilient channels or clips may be used one side only to totally float the finish; use of blocking for fire stopping and use of backer strips to attach wall base reduces performance approximately 4 to 5 dB.
4. Range includes values derived from ASTM E90-61T, ASTM E90-66T, and field test procedures.
5. Gypsum board finish laminated to sound deadening board; if gypsum board is nailed or screwed to meet fire ratings, performance decreases approximately 10 dB for fasteners spaced 8 in. (200 mm) o.c., approximately 8 dB for fastener spacing 16 in. (400 mm) o.c., and approximately 4 dB with fastener spacing 24 in. (600 mm) o.c.
6. Ranges include values for 2 × 3 (50 × 75) studs staggered on single 2 × 4 (50 × 100) plate and 2 × 4 (50 × 100) double stud walls on separate 2 × 3 (50 × 75) plates.
7. A minimum thickness of 2 to 3 in. (50 to 75 mm) of insulation can be used effectively as performance does not increase proportionately with increased thickness of insulation.
8. Ranges include 1 1/2 in. (38 mm) to 3 in. (75 mm) thickness of insulation; performance increases with increased thickness. On double stud walls, an STC of 58 to 60 can result with up to 9 in. (230 mm) of insulation.
9. Based on ASTM E90-61T test procedures.

FIGURE 13.1-86 *(Continued)*

Wood Frame Load-Bearing Walls—Plaster Finish[11]

SINGLE STUD WALLS		SINGLE STUD WALLS	
Metal Lath & Plaster[2]	**Resilient Gypsum Lath & Plaster**[3]	**Plaster Base, Veneer Plaster**[2]	**Resilient Plaster Base, Veneer Plaster**[3]
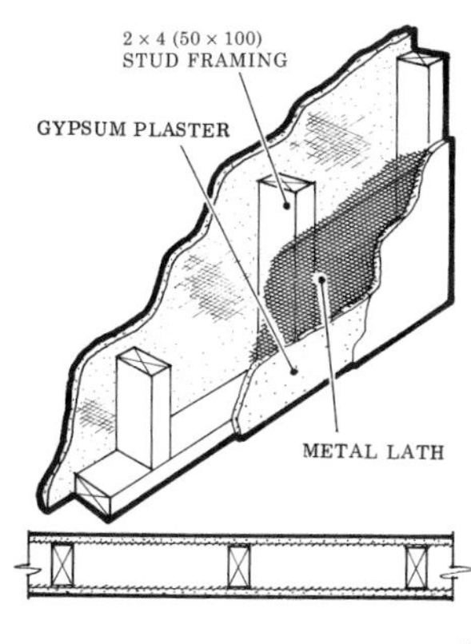	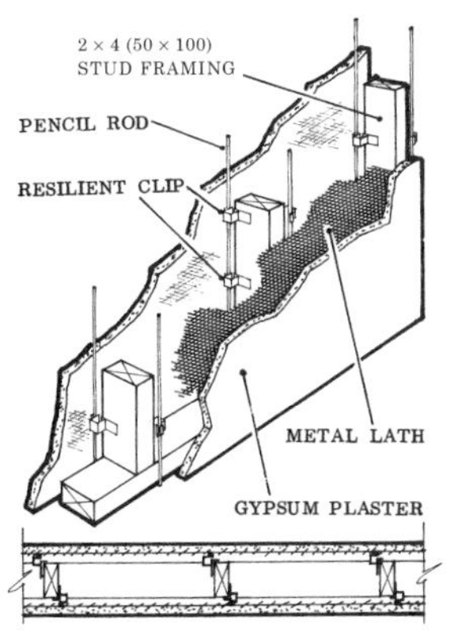	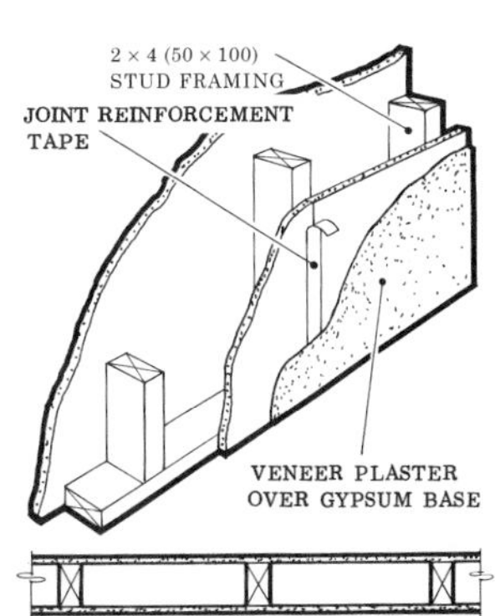	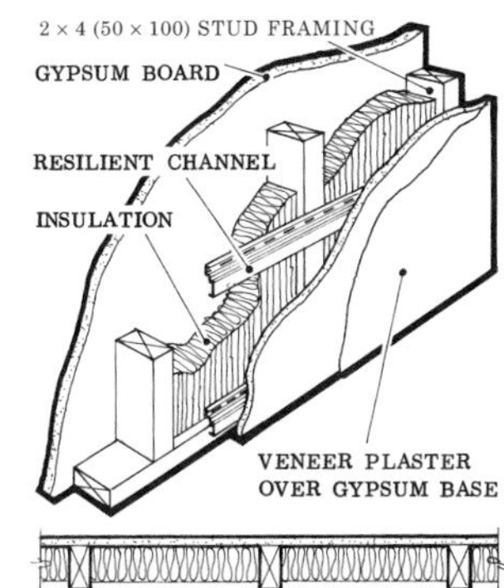
STC 41[10]	**STC 43**[9]	**STC (est) 36** 2 layer base 2 sides **STC (est) 42**	With insulation[8] **STC 49 to 50**[4]

SINGLE STUD WALLS		SINGLE STUD WALLS	
Gypsum Lath & Plaster[3]	**Resilient Metal Lath & Plaster**[3]	**Resilient Plaster Base & Multilayer Base, Veneer Plaster**[3]	**Gypsum Lath & Plaster**
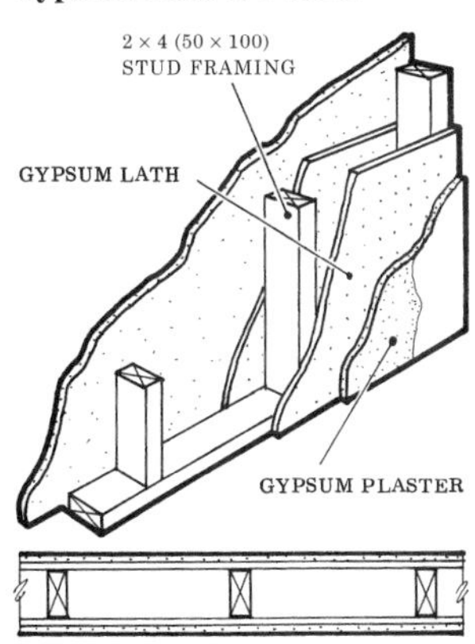	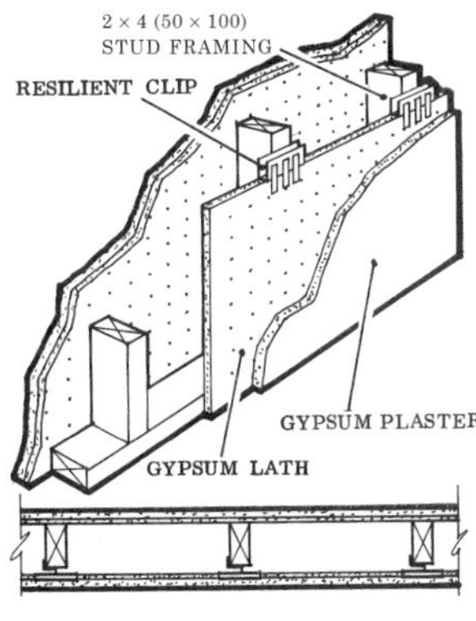	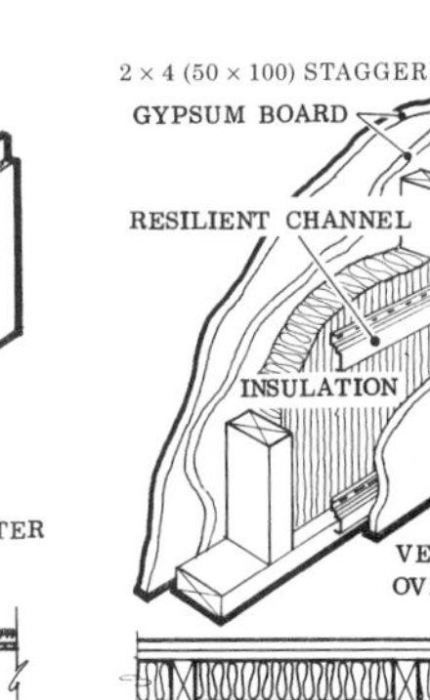 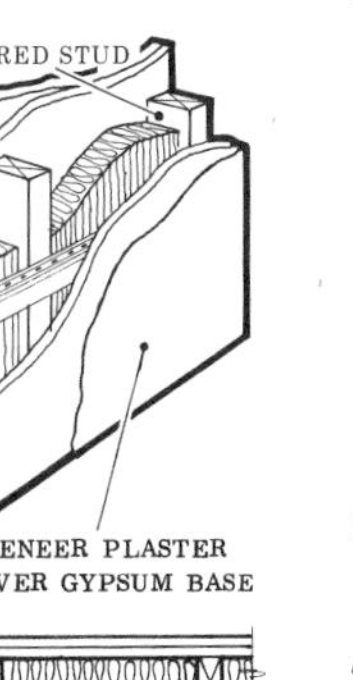	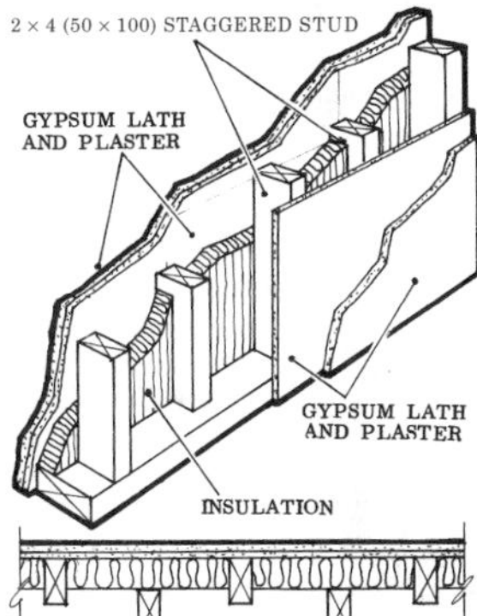
STC 41[10]	**STC 45 to 47**[10] With insulation[8] **STC 50 to 54**[4]	With insulation[8] **STC 53**	**STC 42 to 43**[9] With insulation[7] **STC 50**[10]

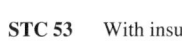

10. Based on tests performed prior to ASTM E90-61T.

(Continues)

11. Test data can vary widely depending on weight and thickness of plaster, and water content (due to length of curing) of plaster.
12. With exception of double stud walls, ranges include values for ½ in. (12.7 mm) or ⅝ in. (15.88 mm) gypsum board thickness of 3 in. (75 mm) or 4 in. (100 mm) studs.
13. Ranges include values for gypsum board screwed, or laminated and screwed, to sound deadening board or base layer.
14. Unbalanced Construction: single layer gypsum board one side, double layer gypsum board one side.
15. Ranges include values for ½ in. (12.7 mm) and ⅝ in. (15.88 mm) board thickness on 2 in. (50 mm) studs.
16. Up to 60 STC with 9 in. (230 mm) of insulation; lower values result with lesser thicknesses of insulation.
17. Performance of concrete and unit masonry systems is highly dependent on minimizing the number of openings through the construction (to accommodate piping, wiring, and other equipment) and on all openings being sealed airtight.
18. Sealed both sides unless otherwise indicated. Sealing with plaster is most effective; when surfaces are to be painted, use of block sealer and/or 2 coats of cement-based paint is more effective than 2 coats of resin-emulsion paint to seal against airborne sound transmission.

FIGURE 13.1-86 *(Continued)*

Metal Frame Non-Load-Bearing Walls—Gypsum Board Finish[12]

SINGLE STUD WALLS

Single Layer Finish

	STC 37 to 41[4]
With insulation[8]	
	STC 44 to 48

Single Layer Finish Sound Deadening Board[13]

SDB 1 side	**STC 43**
SDB 2 sides	**STC 46 to 50[4]**
With insulation[8]	
SDB 1 side	**STC 47 to 50[4]**
SDB 2 sides	**STC 50 to 53[4]**

SINGLE STUD WALLS

Unbalanced Finish[14]

	STC 44 to 47
With insulation[8]	
	STC 49 to 52

Unbalanced Finish[14] Sound Deadening Board[13]

SDB 1 side	**STC 48[9]**
SDB 2 sides	**STC 52[9]**
With insulation[8]	
SDB 1 side	**STC 49[9]**
SDB 2 sides	**STC 54[9]**

SINGLE STUD WALLS

DOUBLE STUD (CHASE) WALLS

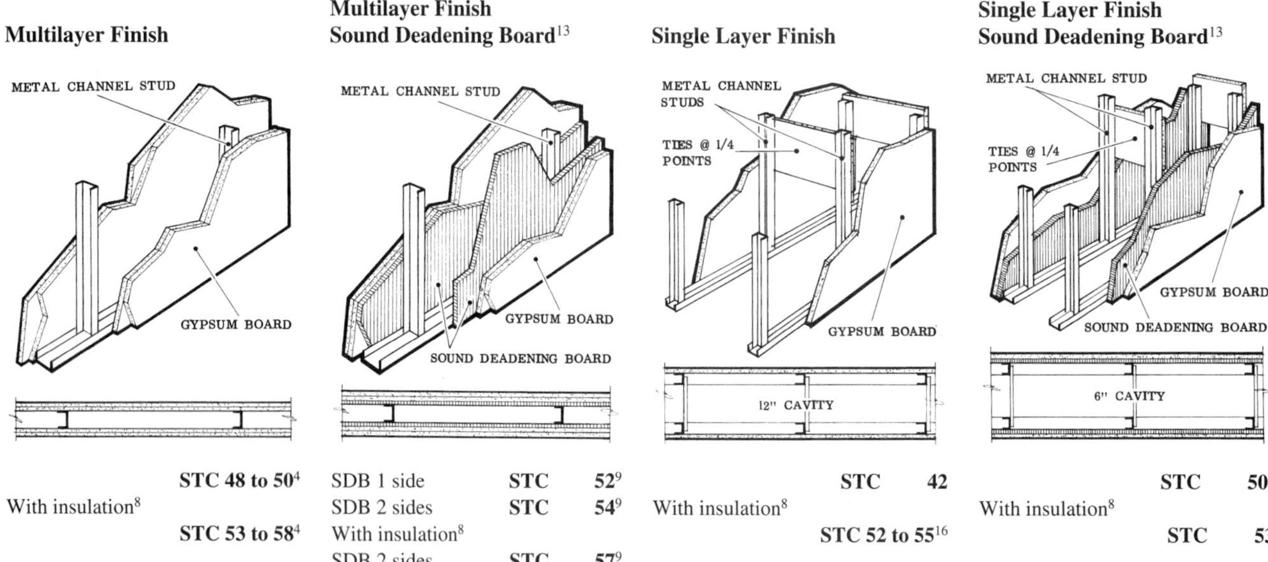

Multilayer Finish (Single Stud)

	STC 48 to 50[4]
With insulation[8]	
	STC 53 to 58[4]

Multilayer Finish Sound Deadening Board[13] (Single Stud)

SDB 1 side	**STC 52[9]**
SDB 2 sides	**STC 54[9]**
With insulation[8]	
SDB 2 sides	**STC 57[9]**

Single Layer Finish (Double Stud)

	STC 42
With insulation[8]	
	STC 52 to 55[16]

Single Layer Finish Sound Deadening Board[13] (Double Stud)

	STC 50[9]
With insulation[8]	
	STC 53

1. Ranges include values for ⅜ in. (9.53 mm), ½ in. (12.7 mm), and ⅝ (15.33 mm) thicknesses of gypsum board finish.
2. Addition of insulation does not appreciably improve this construction; an increase of approximately 2 to 4 dB can be expected regardless of thickness or density of insulation.
3. Resilient channels or clips may be used one side only to totally float the finish; use of blocking for fire stopping and use of backer strips to attach wall base reduces performance approximately 4 to 5 dB.
4. Range includes values derived from ASTM E90-61T, ASTM E90-66T, and field test procedures.
5. Gypsum board finish laminated to sound deadening board; if gypsum board is nailed or screwed to meet fire ratings, performance decreases approximately 10 dB for fasteners spaced 8 in. (200 mm) o.c., approximately 8 dB for fastener spacing 16 in. (400 mm) o.c., and approximately 4 dB with fastener spacing 24 in. (600 mm) o.c.
6. Ranges include values for 2 × 3 (50 × 75) studs staggered on single 2 × 4 (50 × 100) plate and 2 × 4 (50 × 100) double stud walls on separate 2 × 3 (50 × 75) plates.
7. A minimum thickness of 2 to 3 in. (50 to 75 mm) of insulation can be used effectively as performance does not increase proportionately with increased thickness of insulation.
8. Ranges include 1½ in. (38 mm) to 3 in. (75 mm) thickness of insulation; performance increases with increased thickness. On double stud walls, an STC of 58 to 60 can result with up to 9 in. (230 mm) of insulation.

FIGURE 13.1-86 *(Continued)*

Metal Frame Non-Load-Bearing Walls—Gypsum Board Finish[12] *(Cont.)*

DOUBLE STUD (CHASE) WALLS[15]

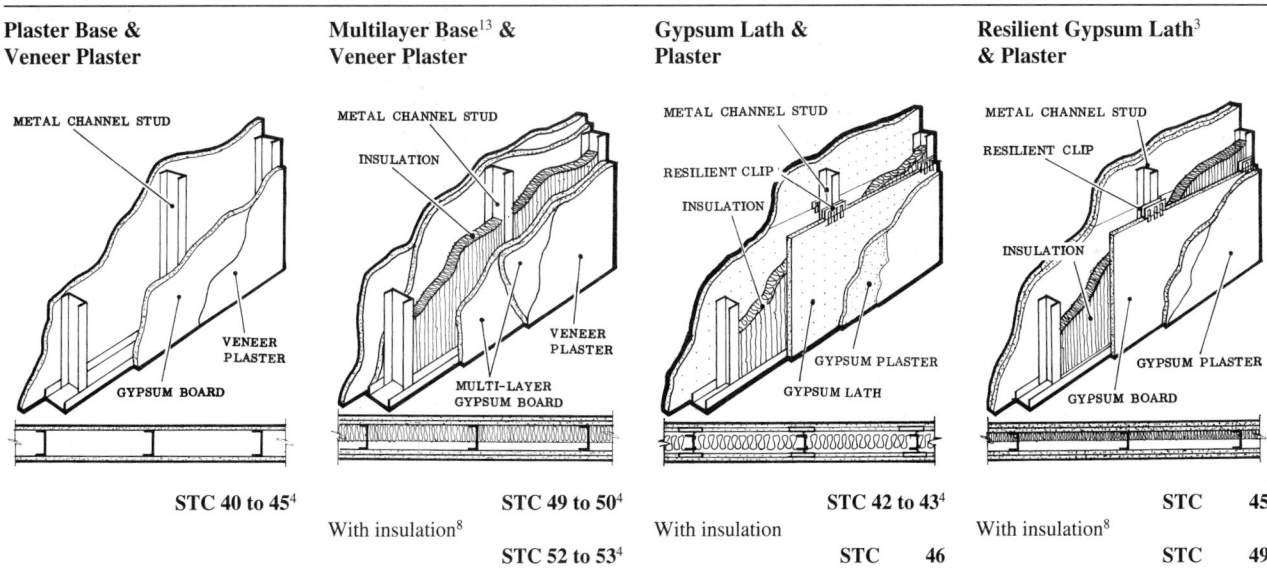

Multilayer Finish

With insulation[8]
Unbalanced Finish[14]

STC 59[16]

Multilayer finish[13]

STC 55 to 60[4 16]

Multilayer Finish
Sound Deadening Board[13]

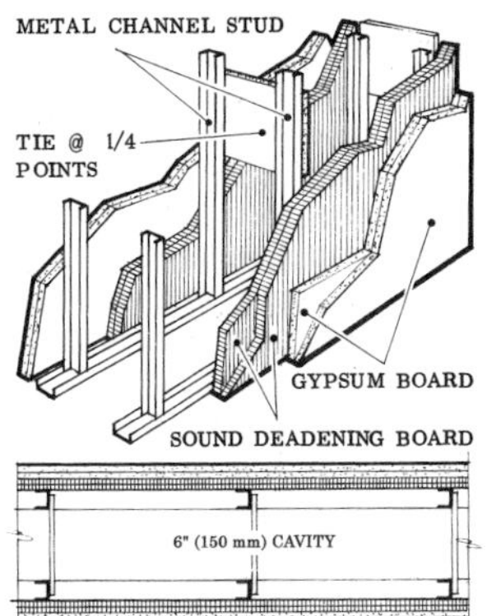

Unbalanced finish[14]

STC 51[9]

Multilayer finish[13]

STC 53[9]

Metal Frame Non-Load-Bearing Walls—Plaster Finish[11]

SINGLE CHANNEL STUD WALLS

Plaster Base & Veneer Plaster | **Multilayer Base[13] & Veneer Plaster**

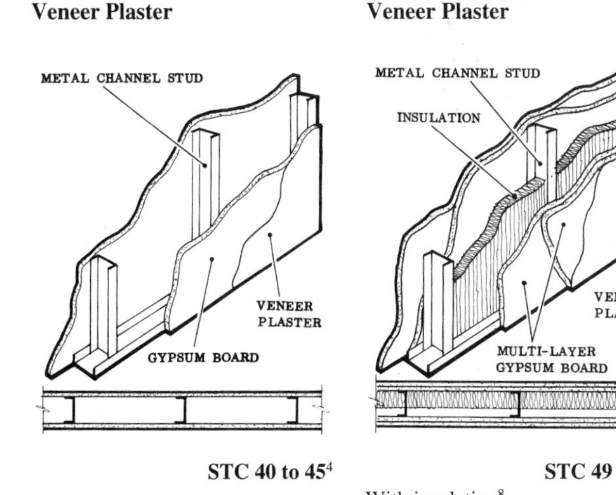

Plaster Base & Veneer Plaster: **STC 40 to 45[4]**

Multilayer Base & Veneer Plaster: **STC 49 to 50[4]**
With insulation[8]
STC 52 to 53[4]

SINGLE CHANNEL STUD WALLS

Gypsum Lath & Plaster

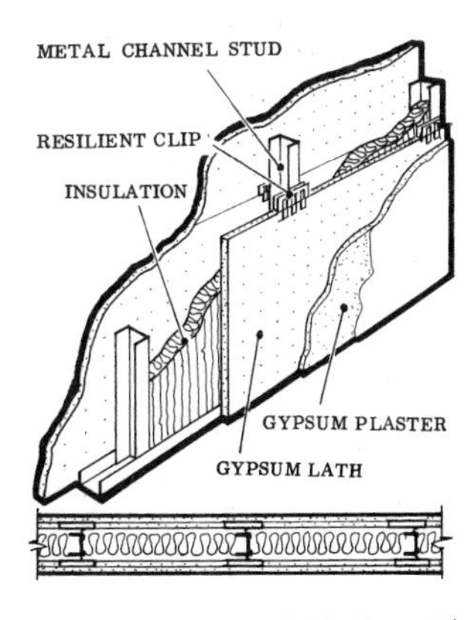

STC 42 to 43[4]
With insulation
STC 46

Resilient Gypsum Lath[3] & Plaster

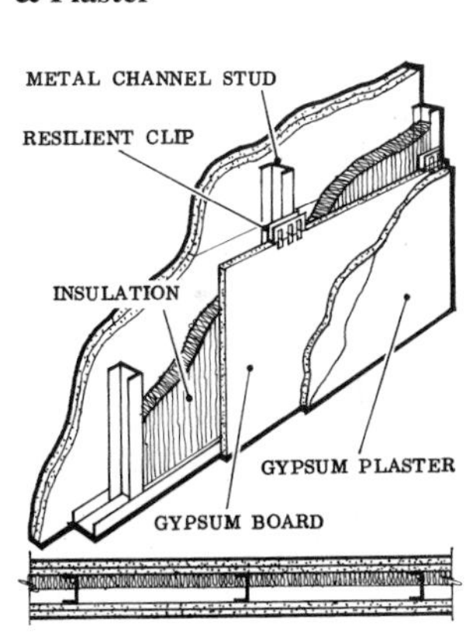

STC 45
With insulation[8]
STC 49

9. Based on ASTM E90-61T test procedures. *(Continues)*
10. Based on tests performed prior to ASTM E90-61T.
11. Test data can vary widely depending on weight and thickness of plaster, and water content (due to length of curing) of plaster.
12. With exception of double stud walls, ranges include values for ½ in. (12.7 mm) or ⅝ in. (15.88 mm) gypsum board thickness of 3 in. (75 mm) or 4 in. (100 mm) studs.
13. Ranges include values for gypsum board screwed, or laminated and screwed, to sound deadening board or base layer.
14. Unbalanced Construction: single layer gypsum board one side, double layer gypsum board one side.
15. Ranges include values for ½ in. (12.7 mm) and ⅝ in. (15.88 mm) board thickness on 2 in. (50 mm) studs.
16. Up to 60 STC with 9 in. (230 mm) of insulation; lower values result with lesser thicknesses of insulation.
17. Performance of concrete and unit masonry systems is highly dependent on minimizing the number of openings through the construction (to accommodate piping, wiring, and other equipment) and on all openings being sealed airtight.
18. Sealed both sides unless otherwise indicated. Sealing with plaster is most effective; when surfaces are to be painted, use of block sealer and/or 2 coats of cement-based paint is more effective than 2 coats of resin-emulsion paint to seal against airborne sound transmission.

FIGURE 13.1-86 *(Continued)*

Metal Frame Non-Load-Bearing Walls—Plaster Finish[11] *(Cont.)*

SINGLE TRUSS STUD WALLS

Metal Lath & Plaster

STC 38 to 41[9]

Resilient Metal Lath[3] & Plaster

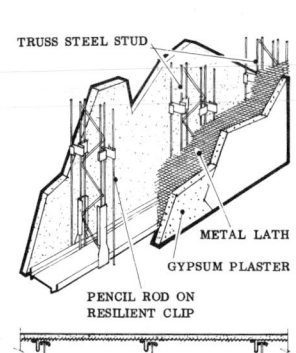

STC 46 to 50[4]

SINGLE TRUSS STUD WALLS

Gypsum Lath & Plaster

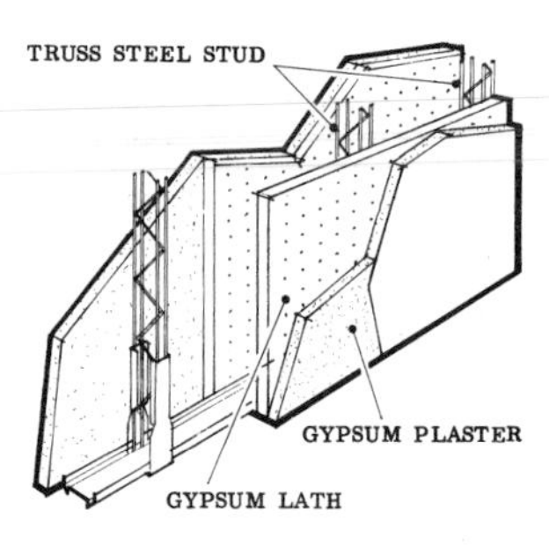

STC 41 to 45[4]

Resilient Gypsum Lath[3] & Plaster

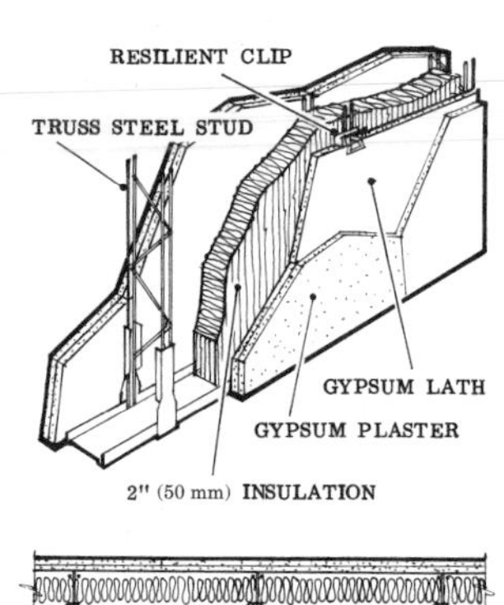

STC 46 to 50

With insulation[8]

STC 46 to 52[4]

Concrete and Concrete Masonry Walls[17]

NONPOROUS (NORMAL WEIGHT) HOLLOW CONCRETE BLOCK WALLS

Unsealed

4″ (100 mm) Block	STC 42 to 45[4]
6″ (150 mm) Block	STC 43 to 46[4]
8″ (200 mm) Block	STC 46 to 49[9]

Sealed[18]

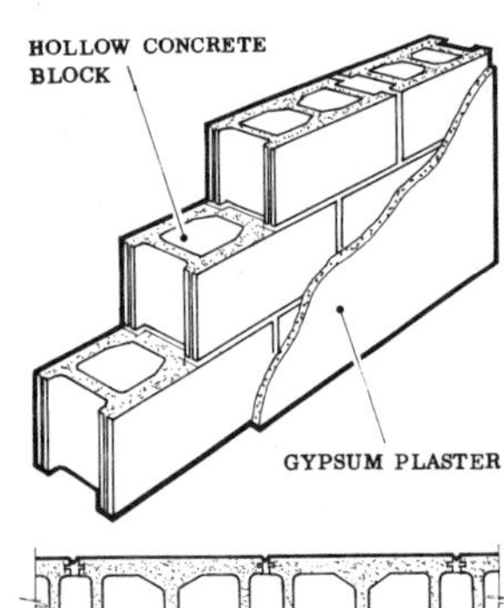

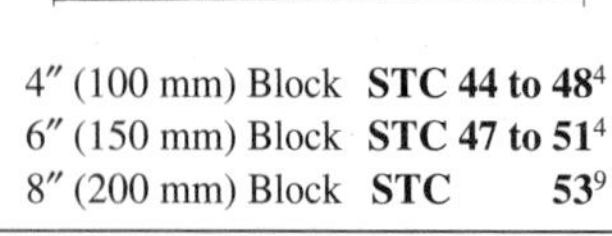

4″ (100 mm) Block	STC 44 to 48[4]
6″ (150 mm) Block	STC 47 to 51[4]
8″ (200 mm) Block	STC 53[9]

POROUS (LIGHTWEIGHT) HOLLOW CONCRETE BLOCK WALLS

Unsealed

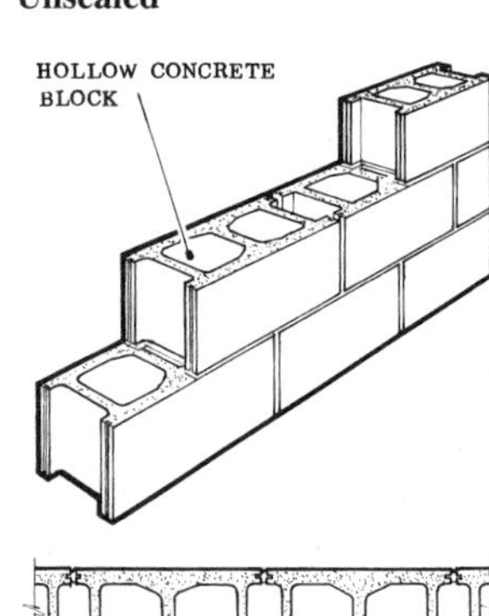

4″ (100 mm) Block	STC 26 to 34[10]
6″ (150 mm) Block	STC 33 to 37[10]
8″ (200 mm) Block	STC 34 to 38[10]

Sealed[18]

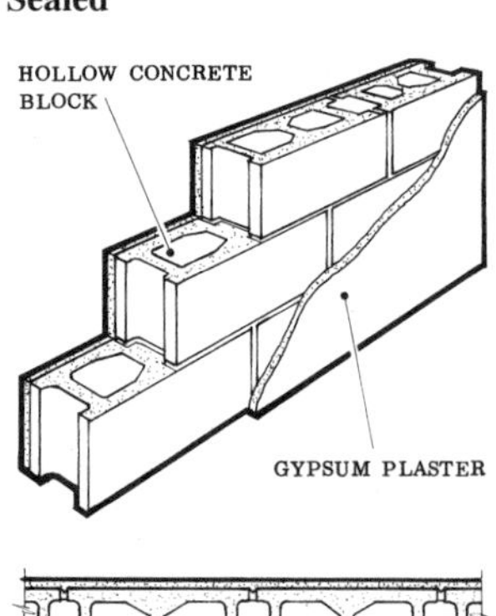

4″ (100 mm) Block	STC 41 to 44[4]
6″ (150 mm) Block	STC 44 to 49[4]
8″ (200 mm) Block	STC 44 to 49[4]

1. Ranges include values for ⅜ in. (9.53 mm), ½ in. (12.7 mm), and ⅝ (15.33 mm) thicknesses of gypsum board finish.
2. Addition of insulation does not appreciably improve this construction; an increase of approximately 2 to 4 dB can be expected regardless of thickness or density of insulation.
3. Resilient channels or clips may be used one side only to totally float the finish; use of blocking for fire stopping and use of backer strips to attach wall base reduces performance approximately 4 to 5 dB.
4. Range includes values derived from ASTM E90-61T, ASTM E90-66T, and field test procedures.
5. Gypsum board finish laminated to sound deadening board; if gypsum board is nailed or screwed to meet fire ratings, performance decreases approximately 10 dB for fasteners spaced 8 in. (200 mm) o.c., approximately 8 dB for fastener spacing 16 in. (400 mm) o.c., and approximately 4 dB with fastener spacing 24 in. (600 mm) o.c.
6. Ranges include values for 2 × 3 (50 × 75) studs staggered on single 2 × 4 (50 × 100) plate and 2 × 4 (50 × 100) double stud walls on separate 2 × 3 (50 × 75) plates.
7. A minimum thickness of 2 to 3 in. (50 to 75 mm) of insulation can be used effectively as performance does not increase proportionately with increased thickness of insulation.
8. Ranges include 1½ in. (38 mm) to 3 in. (75 mm) thickness of insulation; performance increases with increased thickness. On double stud walls, an STC of 58 to 60 can result with up to 9 in. (230 mm) of insulation.
9. Based on ASTM E90-61T test procedures.

FIGURE 13.1-86 *(Continued)*

Concrete and Concrete Masonry Walls[17] *(Cont.)*

POROUS (LIGHTWEIGHT) HOLLOW CONCRETE BLOCK WALLS

Gypsum Board Laminated Both Sides

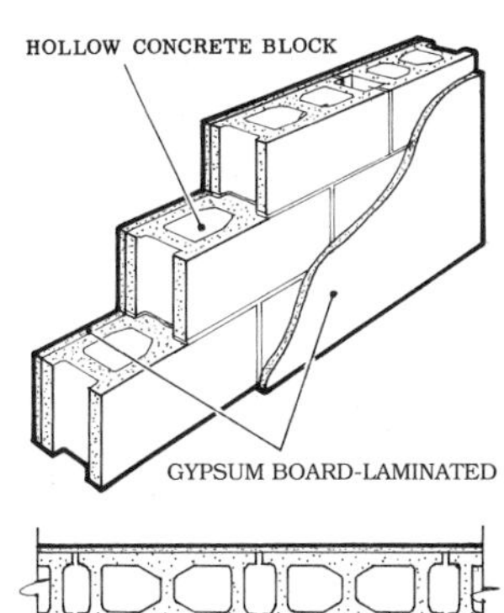

6″ (150 mm) Block **STC** **49**

Resilient Gypsum Board 1 Side, Sealed[18] **1 Side**

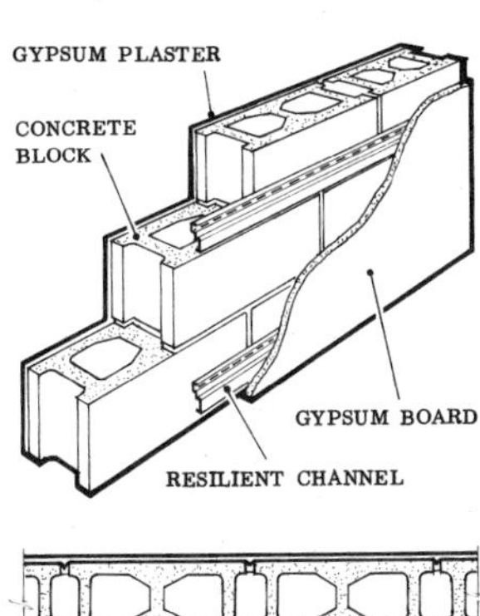

6″ (150 mm) Block **STC** **53**
8″ (200 mm) Block **STC** **48**[9]

POROUS (LIGHTWEIGHT) HOLLOW CONCRETE BLOCK WALLS

Resilient Gypsum Board Both Sides

RESILIENT CHANNEL
GYPSUM BOARD
CONCRETE BLOCK

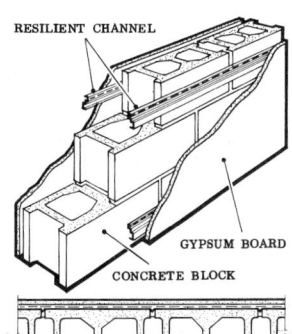

8″ (200 mm) Block **STC** **49**[9]

Cavity Wall Unsealed

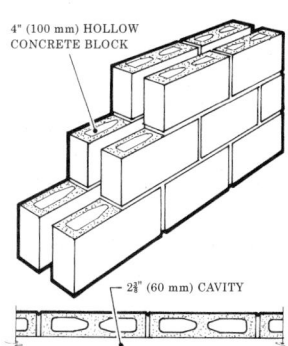

4″ (100 mm) HOLLOW CONCRETE BLOCK
2⅜″ (60 mm) CAVITY

STC **38**[10]

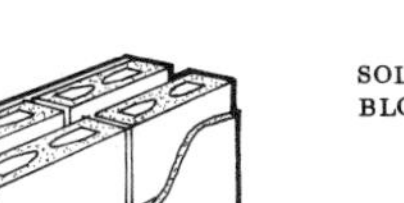

HOLLOW CONCRETE BLOCK WALLS

Porous Cavity Wall Sealed[18]

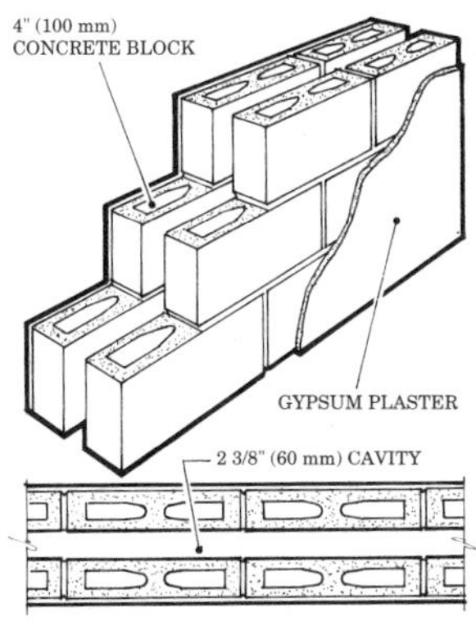

STC 49[10]

Porous/Nonporous Cavity Wall Sealed[18]

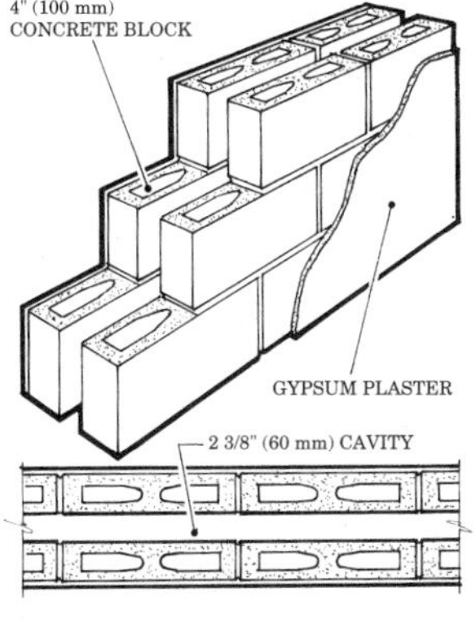

STC 51 to 52[10]

NONPOROUS (NORMAL WEIGHT) SOLID CONCRETE MASONRY WALLS

Unsealed

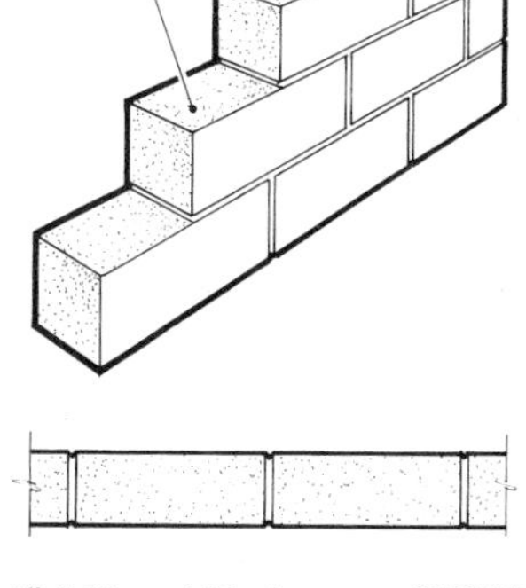

4″ (100 mm) Block	**STC 30**
6″ (150 mm) Block	**STC 38**
8″ (200 mm) Block	**STC 47**

Sealed[18]

SOLID CONCRETE BLOCK
GYPSUM PLASTER

4″ (100 mm) Block	**STC 44**
6″ (150 mm) Block	**STC 48**
8″ (200 mm) Block	**STC 53**

10. Based on tests performed prior to ASTM E90-61T.

(Continues)

11. Test data can vary widely depending on weight and thickness of plaster, and water content (due to length of curing) of plaster.
12. With exception of double stud walls, ranges include values for ½ in. (12.7 mm) or ⅝ in. (15.88 mm) gypsum board thickness of 3 in. (75 mm) or 4 in. (100 mm) studs.
13. Ranges include values for gypsum board screwed, or laminated and screwed, to sound deadening board or base layer.
14. Unbalanced Construction: single layer gypsum board one side, double layer gypsum board one side.
15. Ranges include values for ½ in. (12.7 mm) and ⅝ in. (15.88 mm) board thickness on 2 in. (50 mm) studs.
16. Up to 60 STC with 9 in. (230 mm) of insulation; lower values result with lesser thicknesses of insulation.
17. Performance of concrete and unit masonry systems is highly dependent on minimizing the number of openings through the construction (to accommodate piping, wiring, and other equipment) and on all openings being sealed airtight.
18. Sealed both sides unless otherwise indicated. Sealing with plaster is most effective; when surfaces are to be painted, use of block sealer and/or 2 coats of cement-based paint is more effective than 2 coats of resin-emulsion paint to seal against airborne sound transmission.

FIGURE 13.1-86 *(Continued)*

Concrete and Concrete Masonry Walls[17] *(Cont.)*

POROUS (LIGHTWEIGHT) SOLID CONCRETE MASONRY WALLS

Unsealed

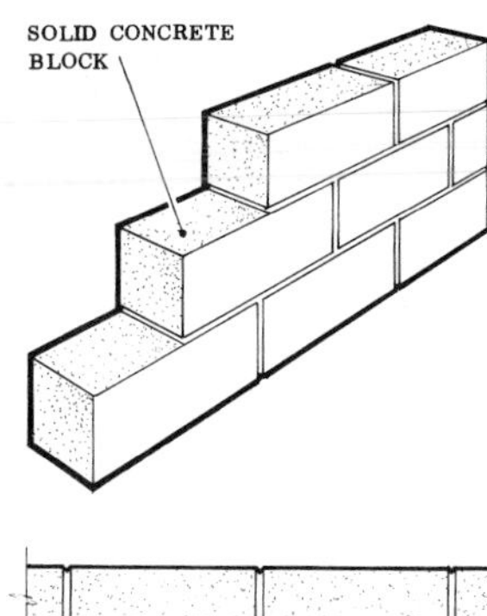

4″ (100 mm) Block	**STC**	**27**
6″ (150 mm) Block	**STC**	**36**
8″ (200 mm) Block	**STC**	**45**

Sealed[18]

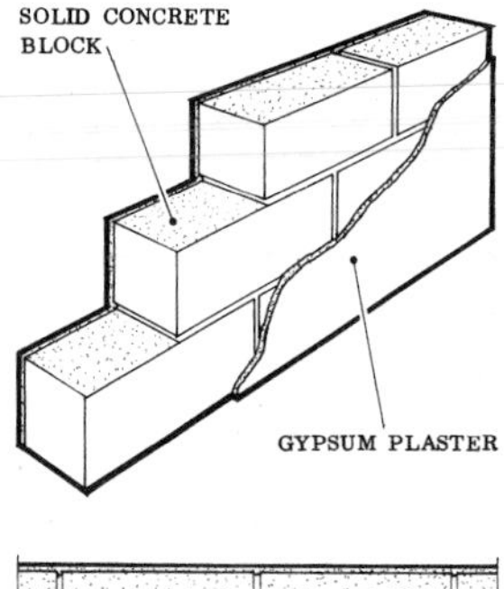

4″ (100 mm) Block	**STC**	**42**
6″ (150 mm) Block	**STC**	**46**
8″ (200 mm) Block	**STC**	**50**

POURED-IN-PLACE SOLID CONCRETE WALLS

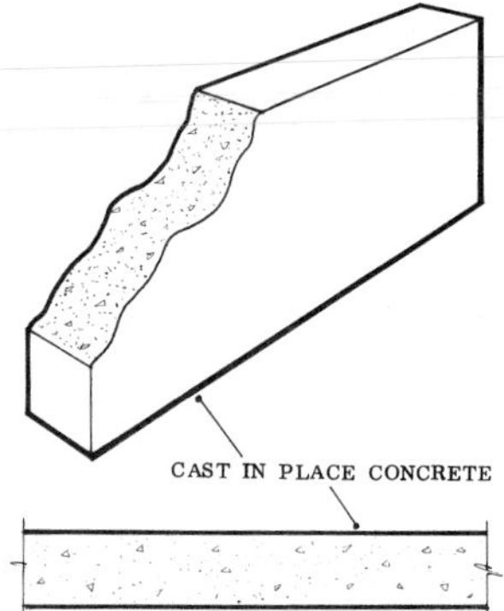

4″ (100 mm) Thick	**STC**	**46**
6″ (150 mm) Thick	**STC**	**50**
8″ (200 mm) Thick	**STC**	**55**

Clay Masonry Walls[17]

SOLID WALLS

Brick Wall Sealed[18]

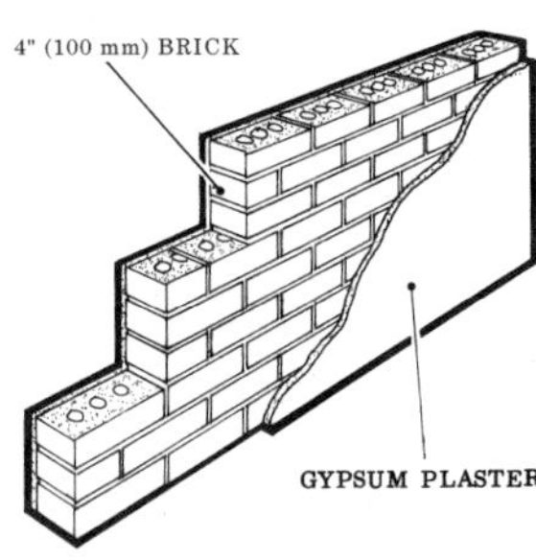

6″ (150 mm) Brick	**STC 45 to 50**
8″ (200 mm) Brick	**STC 50 to 52**
12″ (300 mm) Brick	**STC 55**

CAVITY WALL

Brick Cavity Wall Sealed[18]

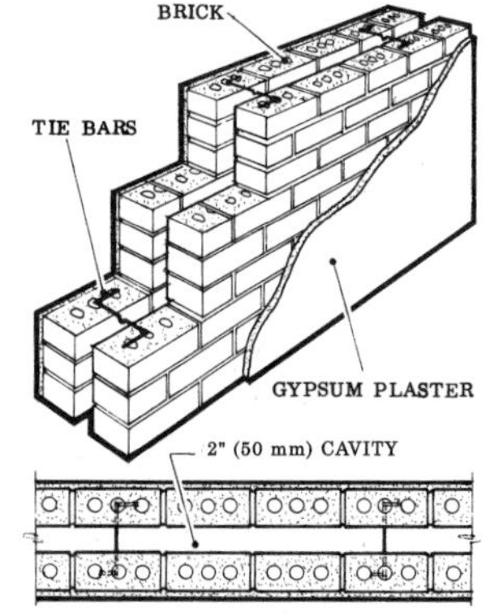

STC 45 to 50

COMPOSITE WALL

Composite Brick, bonded to Concrete Block

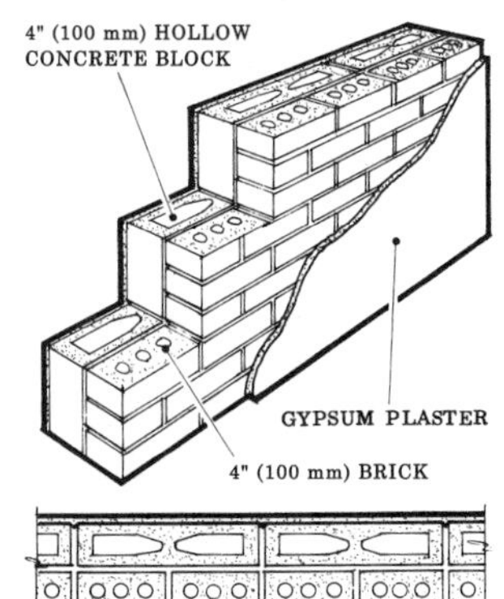

STC 45 to 50

FIGURE 13.1-87 Floor and Ceiling Assemblies

Wood Joist Floors—Gypsum Board Ceiling Finish[2]

HARD SURFACE FLOORING	CARPETING	HARD SURFACE FLOORING[3] RESILIENT CEILING	CARPETING RESILIENT CEILING
STC 38 to 42 (No Ceiling STC 25)	STC 38 to 42	STC 46 to 51	STC 46 to 48
INR −19 to −16 IIC 32 to 35 (No Ceiling) INR −28 IIC 23	Carpet Only INR −10 to −9 IIC 41 to 42 Lightweight Carpet & Pad[4] INR −5 IIC 46 Standard Carpet & Pad[5] INR +8 to +10 IIC 59 to 61	INR −13 to −9 IIC 38 to 42 With Insulation[6] Vinyl Flooring INR −5 to −3 IIC 46 to 48 Wood Block Flooring INR −1 IIC 50	INR +15 to +16 IIC 66 to 67 With insulation[6] Lightweight Carpet & Pad[4] INR +7 IIC 58 Standard Carpet & Pad[5] INR +18 to +20 IIC 69 to 71

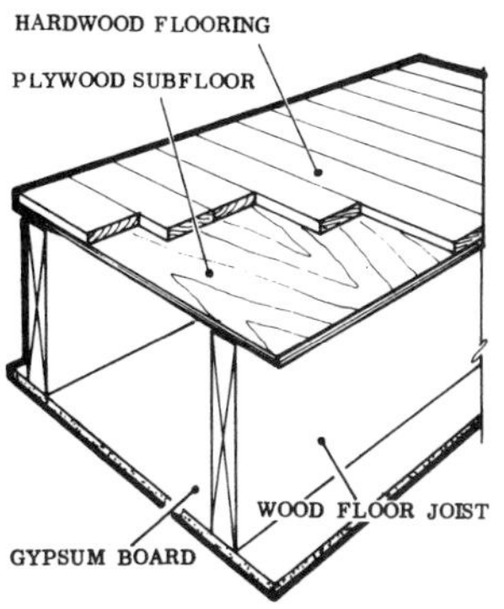

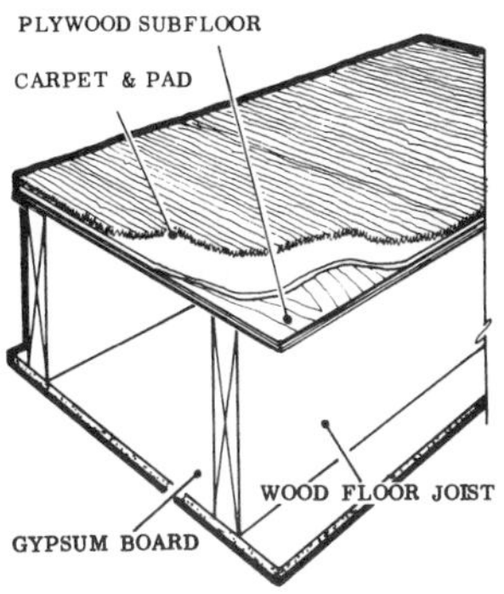

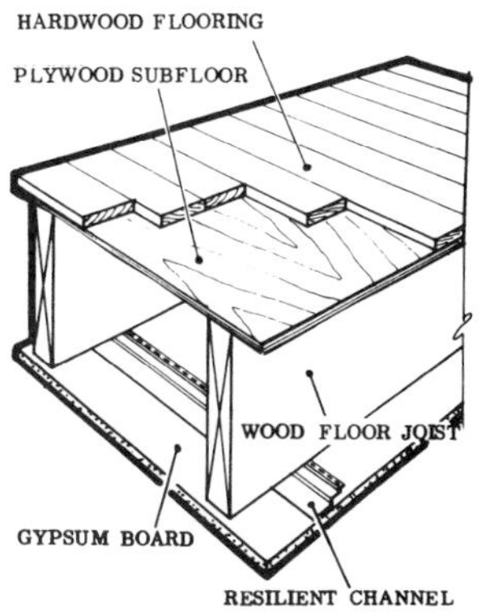

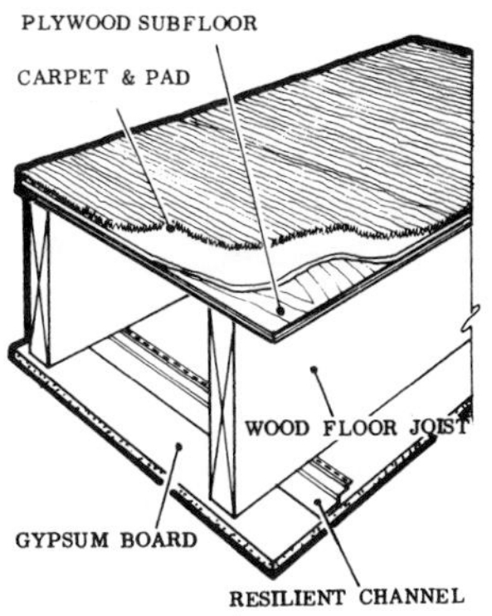

HARD SURFACE FLOORING[3] SEPARATE CEILING JOISTS	CARPETING[5] SEPARATE CEILING JOISTS	FLOATED HARD SURFACE FLOORING	FLOATED HARD SURFACE FLOORING, RESILIENT CEILING
STC 50 to 53	STC 51	STC 46	STC 50 to 52 With insulation[6] STC 51 to 53
With insulation[5] INR −6 to −2 IIC 45 to 49	With insulation[6] INR +29 IIC 80	INR −13 IIC 38	INR −3 to −2 IIC 48 to 49 With insulation[6] INR +1 to +2 IIC 52 to 53

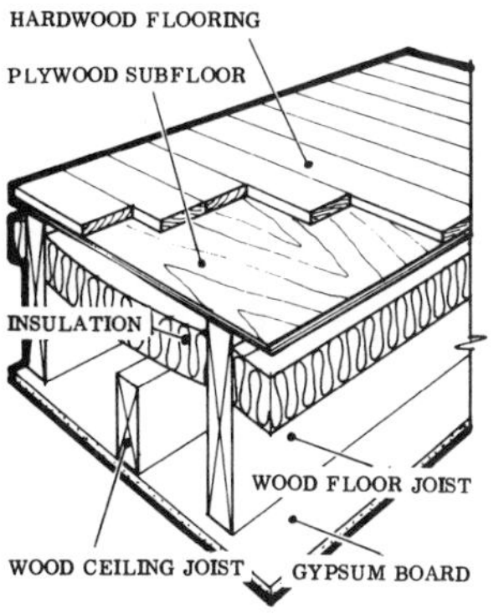

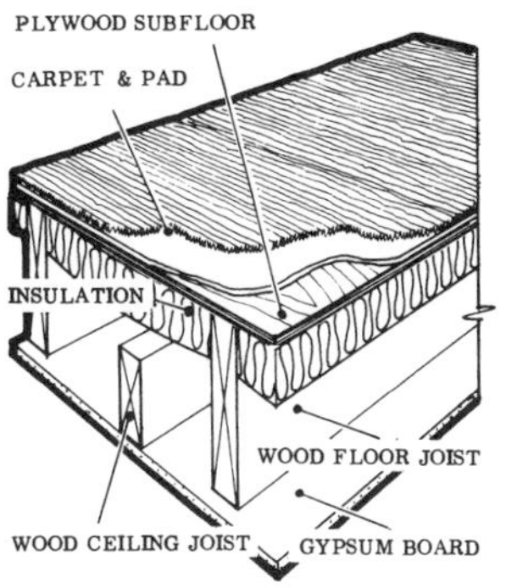

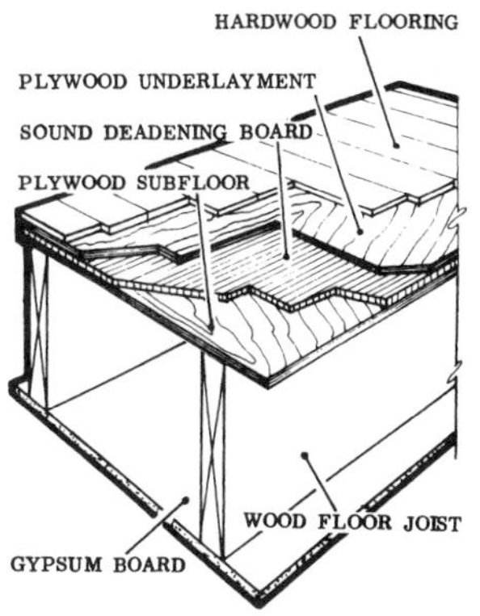

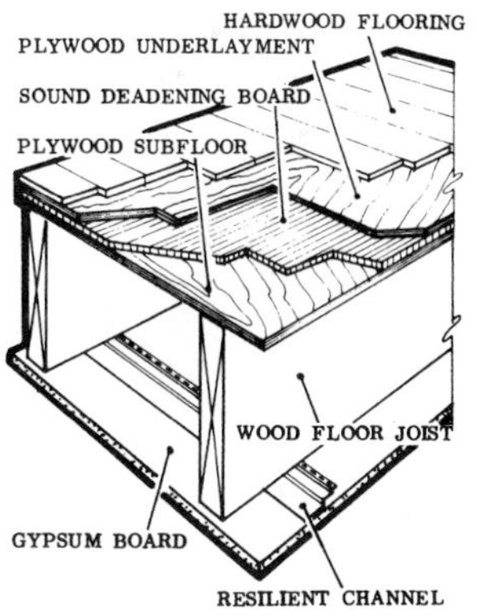

(Continues)

1. INR values are based on tests performed in accordance with ISO Recommendation R-140, "Field and Laboratory Measurements of Airborne and Impact Sound Transmission," January, 1960, IIC values approximated by adding 51 to original INR tests values.
2. Ranges include value for ⅜ in. (9.53 mm), ½ in. (12.7 mm), and ⅝ in. (15.88 mm) thicknesses of gypsum board finish.
3. Hard surface finish flooring (i.e., stripwood, wood block, vinyl, and vinyl asbestos) generally perform similarly in isolating impact sound. Unless otherwise indicated, ranges given include values for construction with any typical hard surface finish flooring.
4. Lightweight carpet and pad having pile weights of 20 oz. per sq. yd. (678.11 g/m^2) and 28 oz. per sq. yd. (949.36 g/m^2) respectively.
5. Standard weight carpet and pad having pile weights of 44 oz. per sq. yd. (1491.85 g/m^2) and 40 oz. per sq. yd. (1356.23 g/m^2) respectively. Ratings shown are based on standard weight combinations unless otherwise noted.
6. A minimum thickness of 2 to 3 in. (50 to 150 mm) of insulation can be used effectively as performance does not increase with increased thickness in floor and ceiling assemblies.
7. Ratings vary with thickness of concrete slabs. The low end of the range indicates value for 6 in. (150 mm) thickness; high end, 8 in. (200 mm) thickness.

FIGURE 13.1-87 *(Continued)*

Wood Joist Floors—Gypsum Board Ceiling Finish[2] *(Cont.)*

CARPETING ON FLOATED FLOOR RESILIENT CEILING	FLOATED HARD SURFACE FLOORING[3], ON FURRING RESILIENT CEILING	CARPETING[5], ON FLOATED FLOOR, FURRING RESILIENT CEILING	FLOATED HARD SURFACE FLOORING[3] SEPARATE CEILING JOISTS
STC 50	STC 48 to 53	STC 52	STC 53
With insulation[6] Lightweight Carpet & Pad[4] INR +8 IIC 59 Standard Carpet & Pad[5] INR +18 IIC 69	INR −7 IIC 44 With insulation[6] INR −2 to +7 IIC 49 to 58	With insulation[6] INR +27 IIC 78	With insulation[6] INR +3 IIC 54

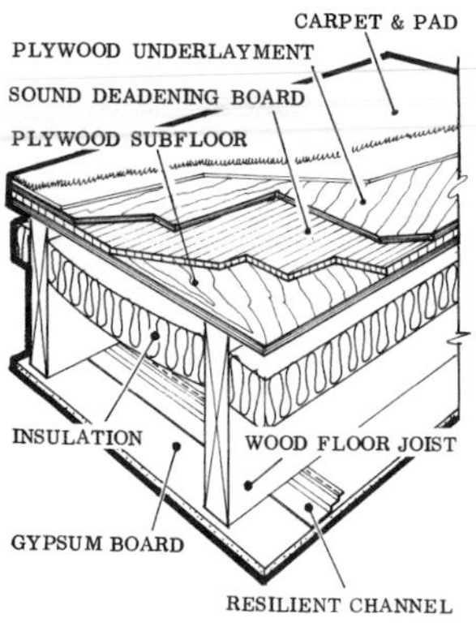

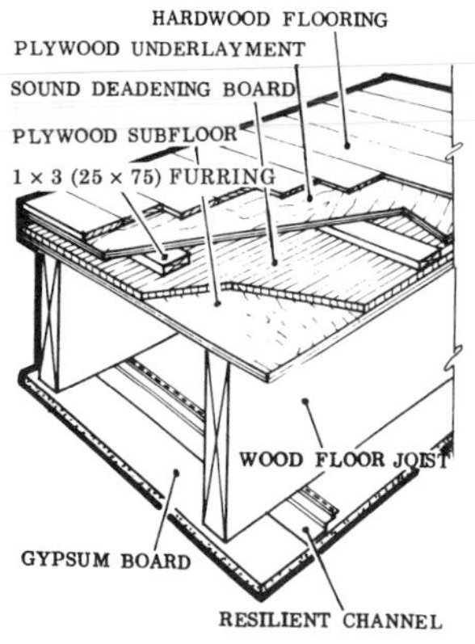

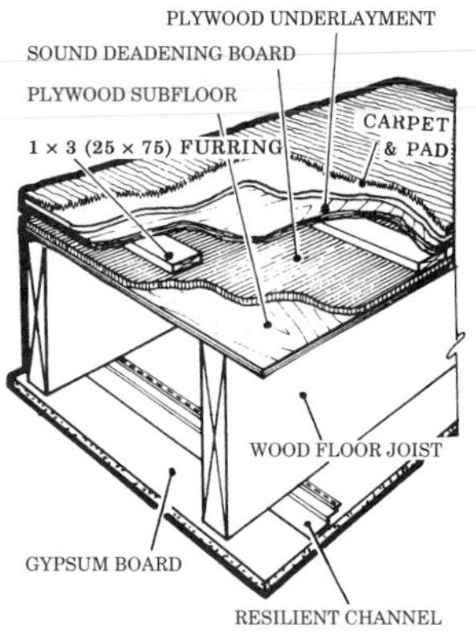

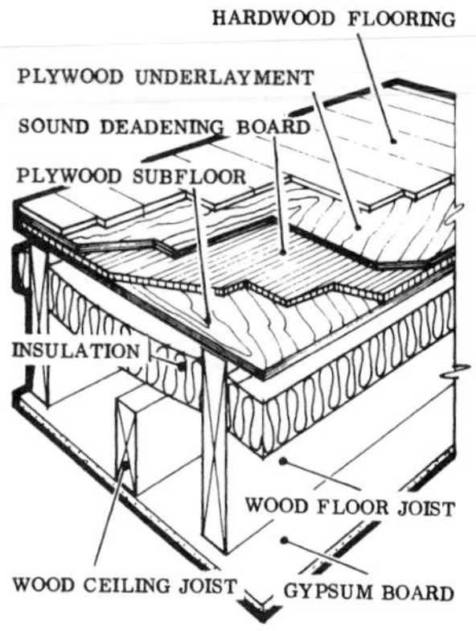

HARD SURFACE FLOORING ON CONCRETE FILL	CARPETING ON CONCRETE FILL	HARD SURFACE FLOORING ON CONCRETE FILL RESILIENT CEILING	CARPETING ON CONCRETE FILL RESILIENT CEILING
STC 46 to 48	STC 46 to 48	STC 50 to 53	STC 50 to 53
INR −14 to −10 IIC 37 to 41	Lightweight Carpet & Pad[4] INR −2 IIC 49 Standard Carpet & Pad[5] INR +10 IIC 61	INR −5 IIC 46 With Insulation[6] Wood Block Flooring INR +1 IIC 52 Cushion Backed Vinyl INR +5 IIC 56	With insulation[6] Lightweight Carpet & Pad[4] INR +13 IIC 64 Standard Carpet & Pad[5] INR +23 IIC 74

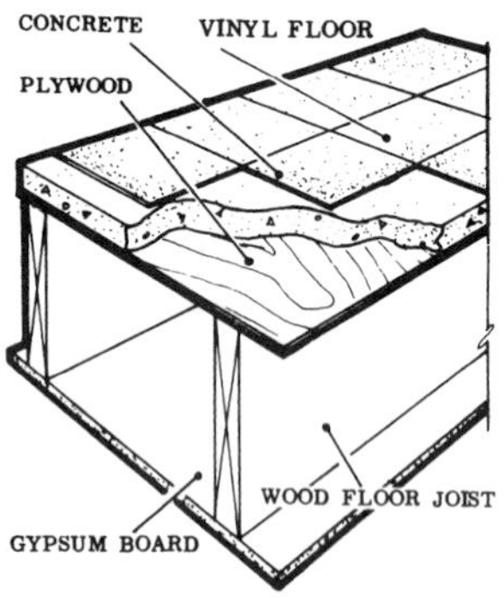

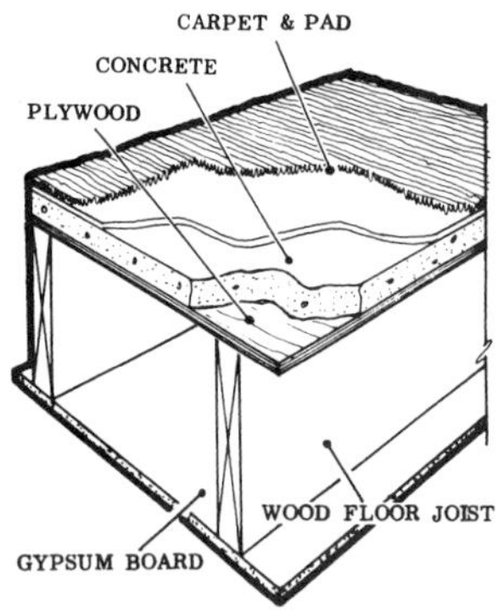

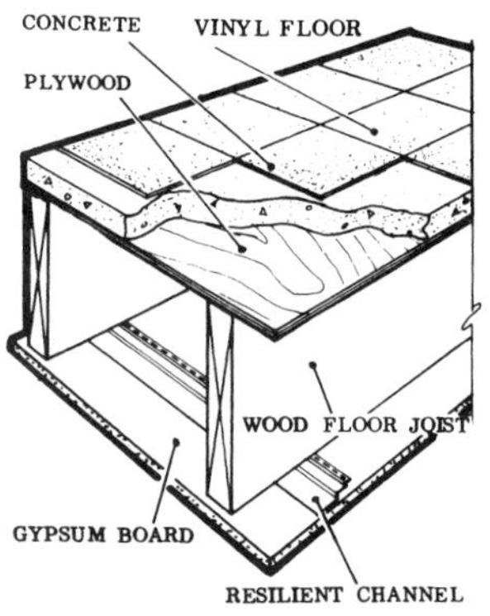

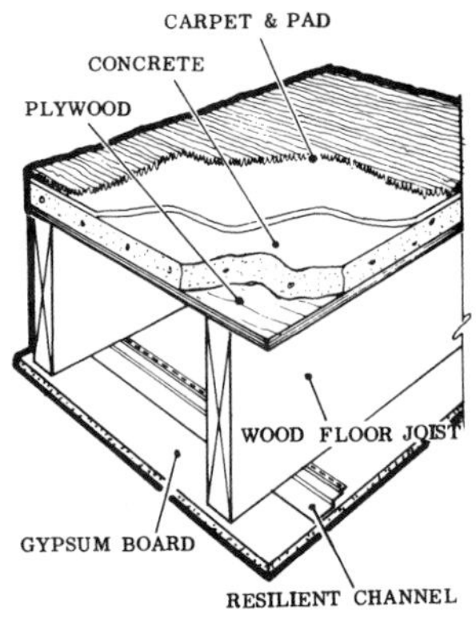

FIGURE 13.1-87 *(Continued)*

Wood/Steel Truss Floors—Gypsum Board Ceiling Finish[2]

HARD SURFACE FLOORING[3]	CARPETING[5]	HARD SURFACE FLOORING[3] RESILIENT CEILING	CARPETING[5] RESILIENT CEILING
STC 45	STC 45	STC 47	STC 47
HARDWOOD FLOORING PLYWOOD SUBFLOOR WOOD / STEEL JOIST GYPSUM BOARD	PLYWOOD SUBFLOOR CARPET & PAD WOOD / STEEL JOIST GYPSUM BOARD	HARDWOOD FLOORING PLYWOOD SUBFLOOR WOOD/STEEL JOIST RESILIENT CHANNEL GYPSUM BOARD	CARPET & PAD 1⅛ (28.12 mm) T. & G. PLYWOOD SUBFLOOR WOOD / STEEL JOIST GYPSUM BOARD RESILIENT CHANNEL
INR −18 IIC 33	INR +5 IIC 56	INR −15 IIC 36	INR +15 IIC 66 With insulation[6] INR +18 IIC 69

HARD SURFACE FLOORING[3] ON CONCRETE FILL | **CARPETING[5] ON CONCRETE FILL**

STC 46 | STC 46

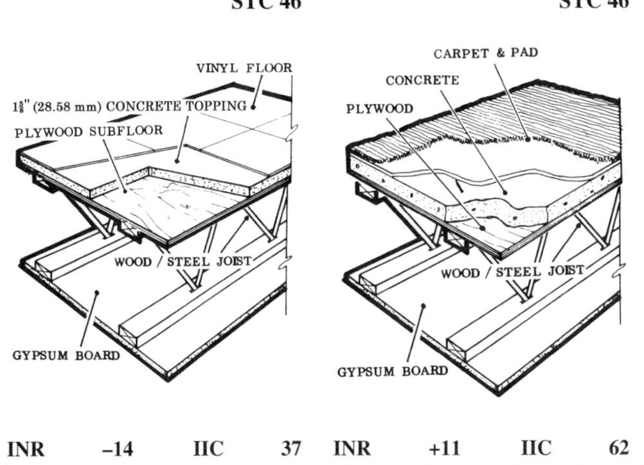

HARD SURFACE FLOORING[3] ON CONCRETE FILL	CARPETING[5] ON CONCRETE FILL
INR −14 IIC 37	INR +11 IIC 62

(Continues)

1. INR values are based on tests performed in accordance with ISO Recommendation R-140, "Field and Laboratory Measurements of Airborne and Impact Sound Transmission," January, 1960, IIC values approximated by adding 51 to original INR tests values.
2. Ranges include value for ⅜ in. (9.53 mm), ½ in. (12.7 mm), and ⅝ in. (15.88 mm) thicknesses of gypsum board finish.
3. Hard surface finish flooring (i.e., stripwood, wood block, vinyl, and vinyl asbestos) generally perform similarly in isolating impact sound. Unless otherwise indicated, ranges given include values for construction with any typical hard surface finish flooring.
4. Lightweight carpet and pad having pile weights of 20 oz. per sq. yd. (678.11 g/m^2) and 28 oz. per sq. yd. (949.36 g/m^2) respectively.
5. Standard weight carpet and pad having pile weights of 44 oz. per sq. yd. (1491.85 g/m^2) and 40 oz. per sq. yd. (1356.23 g/m^2) respectively. Ratings shown are based on standard weight combinations unless otherwise noted.
6. A minimum thickness of 2 to 3 in. (50 to 150 mm) of insulation can be used effectively as performance does not increase with increased thickness in floor and ceiling assemblies.
7. Ratings vary with thickness of concrete slabs. The low end of the range indicates value for 6 in. (150 mm) thickness; high end, 8 in. (200 mm) thickness.

FIGURE 13.1-87 *(Continued)*

Poured-in-Place Reinforced Concrete Slabs		
HARD SURFACE FLOORING[3] WITH OR WITHOUT PLASTER CEILING	FLOATED HARD SURFACE FLOORING[3], WITH OR WITHOUT PLASTER CEILING	CARPETING WITH OR WITHOUT PLASTER CEILING
STC 44 to 48	STC 48 to 50	STC 44 to 48
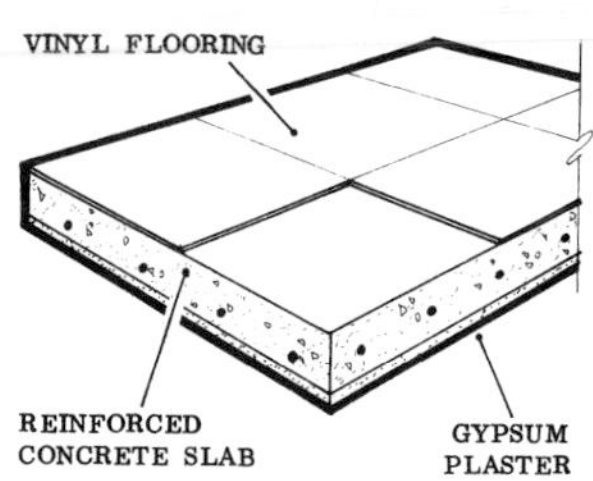	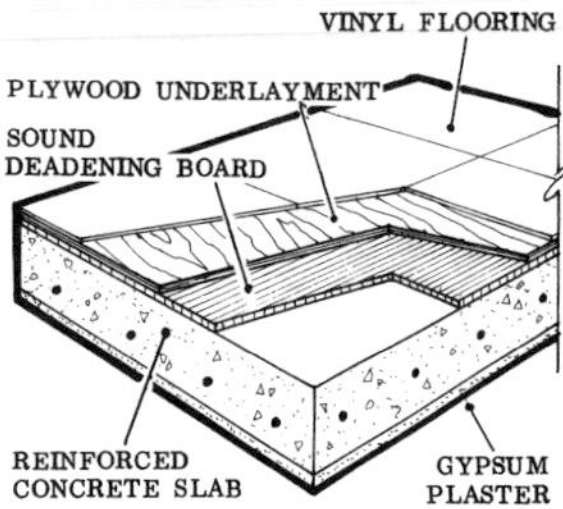	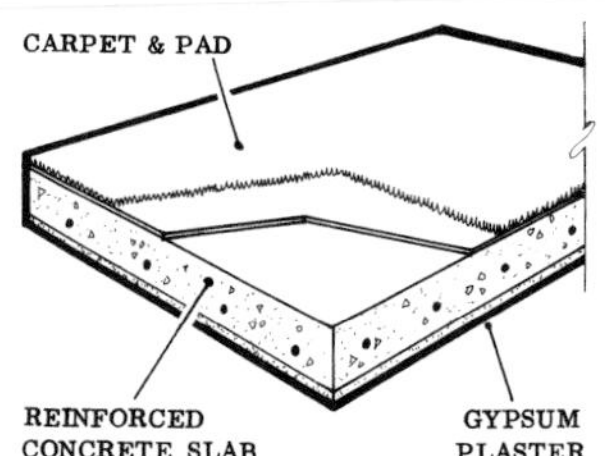
No Flooring **INR** **−17** **IIC** 34 Vinyl Flooring **INR** **−15** **IIC** 36 Wood Block Flooring **INR −9 to −6** **IIC 42 to 45**	**INR −7 to +4** **IIC 44 to 55**	Lightweight Carpet Only **INR** **+2** **IIC** 53 Standard Carpet Only **INR** **+6** **IIC** 57 Standard Carpet & Pad[5] **INR +17 to +22** **IIC 68 to 73**

FIGURE 13.1-87 *(Continued)*

Precast Concrete Slabs[7]

HARD SURFACE FLOORING[3] INTEGRAL CEILING		CARPETING[5] INTEGRAL CEILING		FLOATED HARD SURFACE FLOORING[3] INTEGRAL CEILING	
STC 45 to 47		STC 45 to 47		STC 47 to 51	
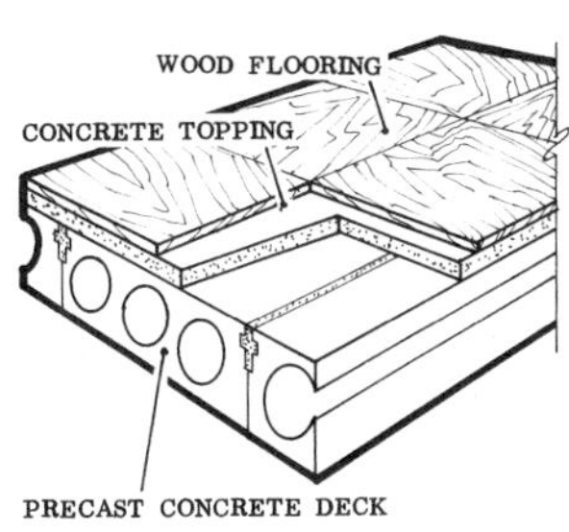		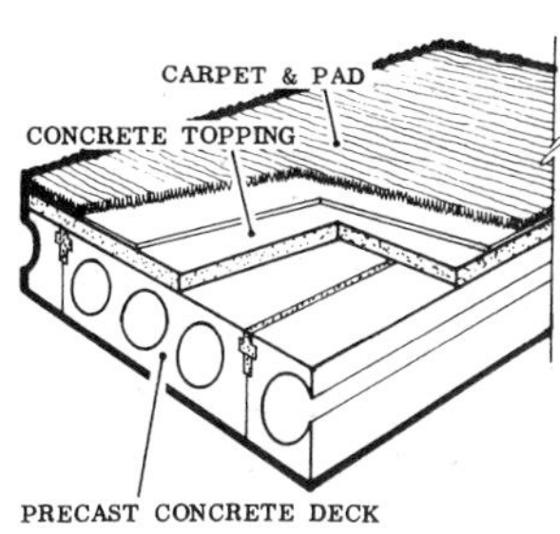		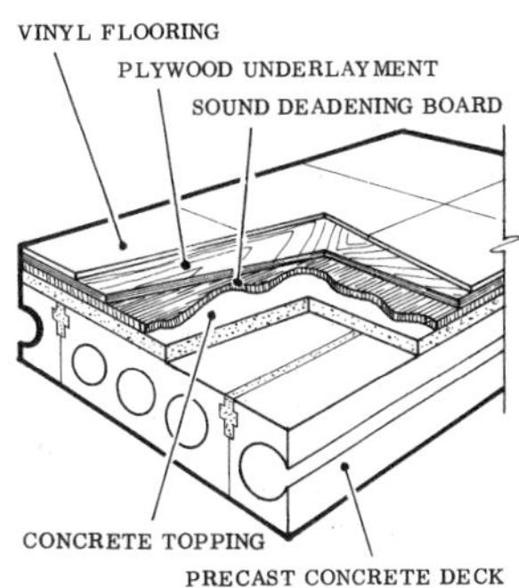	
Vinyl Flooring					
INR –28 to –23	**IIC 23 to 28**	**INR +18 to +22**	**IIC 69 to 73**	**INR +2 to +4**	**IIC 53 to 55**
Wood Block Flooring					
INR –4 to –3	**IIC 47 to 48**				

HARD SURFACE FLOORING[3] ON 1½″ (12.7 mm) CONCRETE INTEGRAL CEILING		CARPETING[5] ON 1½″ (12.7 mm) CONCRETE INTEGRAL CEILING	
STC 49		STC 45	
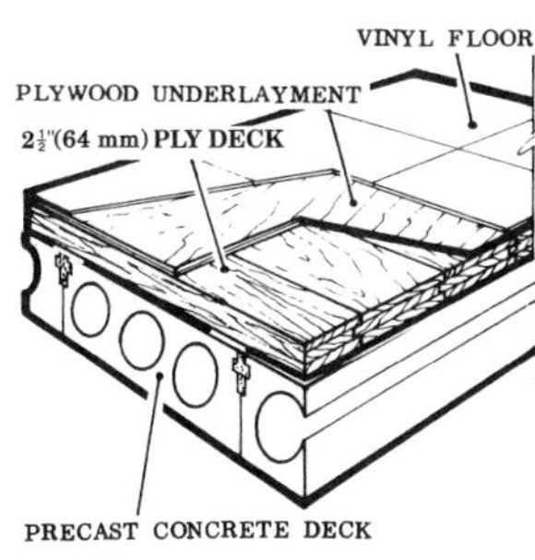		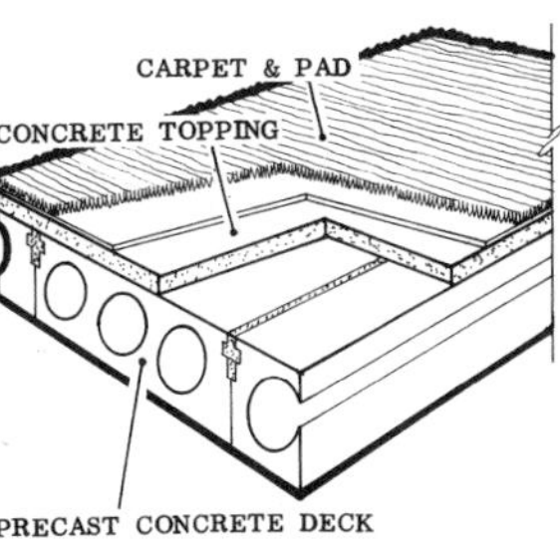	
INR +2 to +4	**IIC 53 to 55**	**INR +25**	**IIC 76**

(Continues)

1. INR values are based on tests performed in accordance with ISO Recommendation R-140, "Field and Laboratory Measurements of Airborne and Impact Sound Transmission," January, 1960, IIC values approximated by adding 51 to original INR tests values.
2. Ranges include value for ⅜ in. (9.53 mm), ½ in. (12.7 mm), and ⅝ in. (15.88 mm) thicknesses of gypsum board finish.
3. Hard surface finish flooring (i.e., stripwood, wood block, vinyl, and vinyl asbestos) generally perform similarly in isolating impact sound. Unless otherwise indicated, ranges given include values for construction with any typical hard surface finish flooring.
4. Lightweight carpet and pad having pile weights of 20 oz. per sq. yd. (678.11 g/m^2) and 28 oz. per sq. yd. (949.36 g/m^2) respectively.
5. Standard weight carpet and pad having pile weights of 44 oz. per sq. yd. (1491.85 g/m^2) and 40 oz. per sq. yd. (1356.23 g/m^2) respectively. Ratings shown are based on standard weight combinations unless otherwise noted.
6. A minimum thickness of 2 to 3 in. (50 to 150 mm) of insulation can be used effectively as performance does not increase with increased thickness in floor and ceiling assemblies.
7. Ratings vary with thickness of concrete slabs. The low end of the range indicates value for 6 in. (150 mm) thickness; high end, 8 in. (200 mm) thickness.

FIGURE 13.1-87 *(Continued)*

Steel Bar Joist Floors

CARPETING[5] ON 2½″ (63.5 mm) CONCRETE LATH & PLASTER CEILING	CARPETING[5] ON 1⅝″ (41.3 mm) CONCRETE GYPSUM BOARD CEILING	CARPETING[5] RESILIENT GYPSUM BOARD CEILING
STC 49	STC 46	STC 47

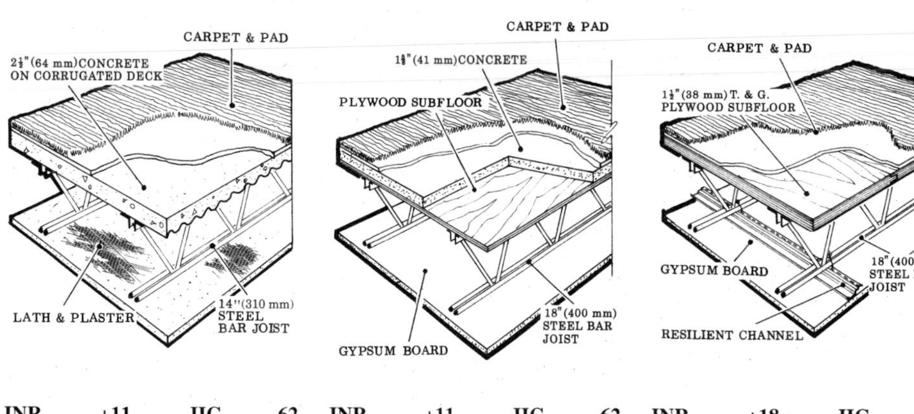

INR	+11	IIC	62	INR	+11	IIC	62	INR	+18	IIC	69

CARPETING[5] ON 2½″ (63.5 mm) CONCRETE LATH & PLASTER CEILING	CARPETING[5] ON 2″ (50.8 mm) CONCRETE RESILIENT GYPSUM PLASTER CEILING
STC 47 to 50	STC 49

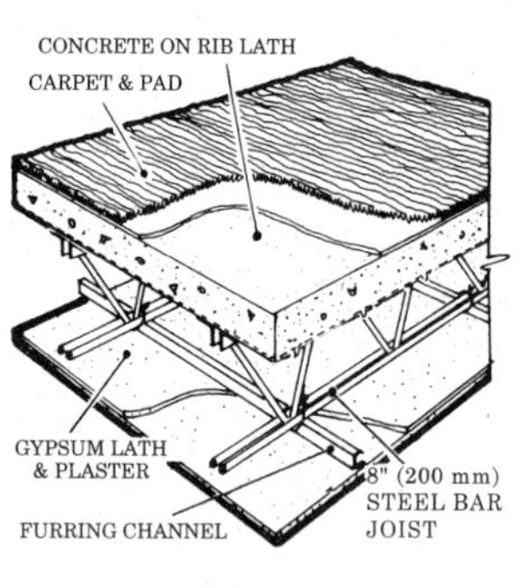

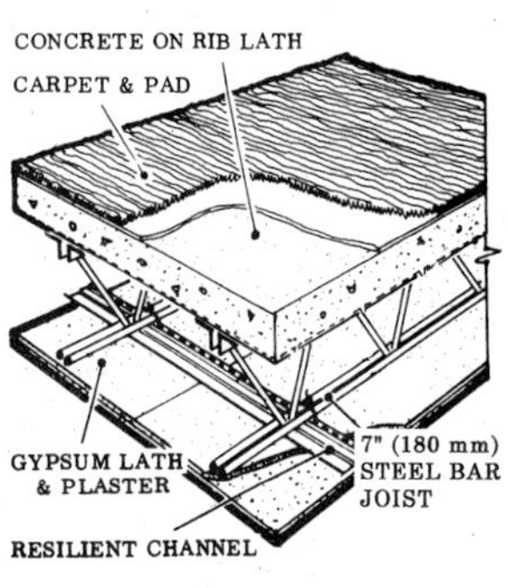

INR	+22	IIC	73	INR	+23	IIC	74
With insulation[6]							
INR	+36	IIC	87				

1. INR values are based on tests performed in accordance with ISO Recommendation R-140, "Field and Laboratory Measurements of Airborne and Impact Sound Transmission," January, 1960, IIC values approximated by adding 51 to original INR tests values.
2. Ranges include value for ⅜ in. (9.53 mm), ½ in. (12.7 mm), and ⅝ in. (15.88 mm) thicknesses of gypsum board finish.
3. Hard surface finish flooring (i.e., stripwood, wood block, vinyl, and vinyl asbestos) generally perform similarly in isolating impact sound. Unless otherwise indicated, ranges given include values for construction with any typical hard surface finish flooring.
4. Lightweight carpet and pad having pile weights of 20 oz. per sq. yd. (678.11 g/m²) and 28 oz. per sq. yd. (949.36 g/m²) respectively.
5. Standard weight carpet and pad having pile weights of 44 oz. per sq. yd. (1491.85 g/m²) and 40 oz. per sq. yd. (1356.23 g/m²) respectively. Ratings shown are based on standard weight combinations unless otherwise noted.
6. A minimum thickness of 2 to 3 in. (50 to 150 mm) of insulation can be used effectively as performance does not increase with increased thickness in floor and ceiling assemblies.
7. Ratings vary with thickness of concrete slabs. The low end of the range indicates value for 6 in. (150 mm) thickness; high end, 8 in. (200 mm) thickness.

FIGURE 13.1-87 *(Continued)*

Steel Bar Joist Floors

**HARD SURFACE FLOORING[3]
ON 2½″ (63.5 mm) CONCRETE
LATH & PLASTER CEILING**

STC 49

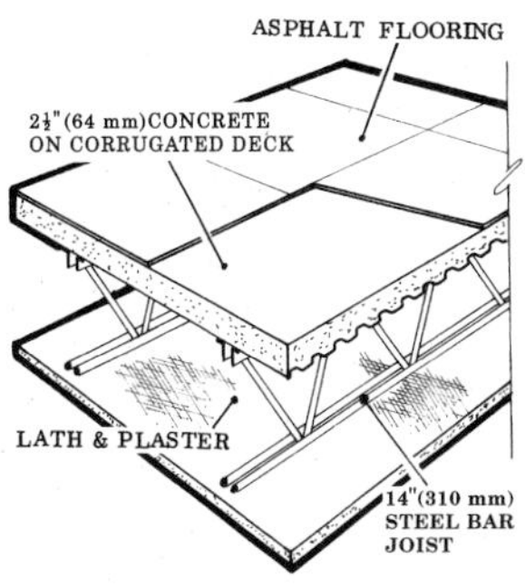

INR **−16** **IIC** **35**

HARD SURFACE FLOORING[3] ON 2½″ (63.5 mm) CONCRETE LATH & PLASTER CEILING	HARD SURFACE FLOORING[3] ON 2″ (50.8 mm) CONCRETE RESILIENT GYPSUM BOARD CEILING
STC 47	**STC 49**
With insulation[6] **STC 50**	

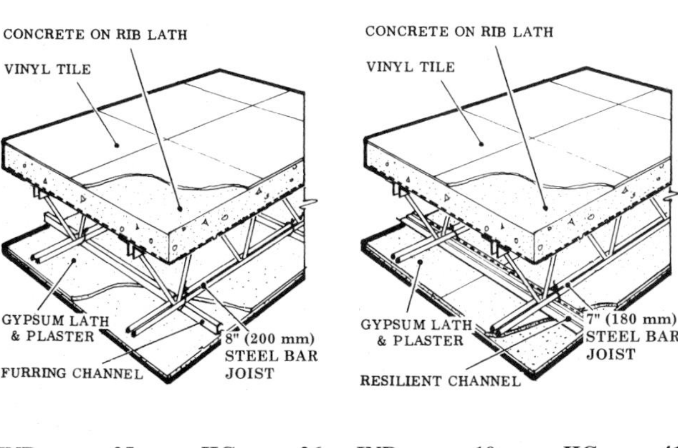

INR −25 IIC 26	**INR −10 IIC 41**
With insulation[6]	
INR −22 IIC 29	

13.2 Lightning Protection

Lightning strikes cause fires in unprotected buildings. This section discusses lightning protection systems intended to eliminate this hazard. Sprinkler systems are covered in Section 15.3, fire detection devices in Section 10.1.

13.2.1 LIGHTNING

Lightning is a discharge of static electricity, a giant spark that jumps between two charged bodies, such as two clouds or a cloud and an object on earth. A study by the Lightning Protection Institute shows that dwellings are the most frequent targets of damaging lightning, and lightning is also a major cause of rural fires in other types of structures. Lightning often strikes large buildings as well, especially tall ones, but most remain undamaged because they have lightning protection systems.

Lightning can enter a building by (1) striking the building directly, (2) striking a metal object such as a TV antenna or radio tower and then traveling down into the building, (3) striking a nearby tree and leaping over to the building, (4) following a power line or ungrounded fence.

On its way to the ground lightning follows the path of lowest electrical resistance. Hence, it frequently follows wiring or plumbing lines, causing fires and damaging electrical appliances.

Television antennas resemble lightning rods in that they are targets for atmospheric discharges, but they cannot safely handle the current of a bolt of lightning even when they are grounded and equipped with a lightning arrester. Lightning can jump from a television antenna to its circuit, to plumbing, or to other electrical wiring, and cause damage.

13.2.2 LIGHTNING PROTECTION SYSTEMS

A good lightning protection system provides an easy, direct path for an electrical current to follow to reach the ground. It must prevent injury, damage, or death as a lightning bolt travels that path. Many insurance companies offer a reduction in premiums for buildings with properly installed lightning protection systems.

A lightning protection system includes (1) air terminals (lightning rods), (2) main conductors, (3) branch conductors, (4) lightning arresters, and (5) ground connections (Fig. 13.2-1).

Air terminals, which are pointed copper or aluminum rods (Fig. 13.2-2), are placed on the highest points of roofs and on every projection, such as a chimney or dormer on a house (see Fig. 13.2-1) and along parapet walls on other types of buildings. Terminals should be sized, spaced, and anchored according to the requirements of the National Fire Protection Association's (NFPA) *Lightning Protection Code*. They are usually between 10 in. (254 mm) and 24 in. (610 mm) long but may be longer. Those less than 24 in. (610 mm) long should be spaced not more than 20 ft. (6.1 m) apart. Longer ones may be placed up to 25 ft. (7.6 m) apart. In any case, there should be a terminal within 2 ft. (610 mm) of each roof end.

Ground connections can be made by (1) clamping the main conductor cable to a copper-clad or galvanized steel rod that has been driven at least 10 ft. (3000 mm) into the ground, (2) burying a stranded copper conducting cable in a trench, (3) clamping copper conducting cable to a sheet metal plate and burying the assembly, or (4) clamping the conductor to a metal water pipe.

The chief requirement for an efficient lightning protection system is a solid connection between the system and the earth. Ground connections of galvanized steel, copper, or copper-clad steel are preferred. Aluminum should not be used for ground connections because it corrodes when in contact with soil.

Every lightning protection system needs at least two ground rods placed as far apart as possible. Usually, they are placed at opposite corners of a building. They should reach below and away from the building to avoid lightning damage to exterior walls.

Lightning systems can be added to an existing structure, but if they are installed

FIGURE 13.2-1 **The components of a lightning protection system include (1) air terminals, (2) down-lead conductors, (3) at least two grounds, (4) roof projections tied into the conductor system, (5) tree protection, (6) at least two terminals on chimneys, (7) dormers rodded, (8) arresters on antennas, (9) tie-in to conductor system of gutters within 6 ft. (1800 mm) of conductor, and (10) arrester on overhead power lines.** (From U.S. Department of Agriculture, Farmers Bulletin Number 2136, "Lightning Protection for the Farm.")

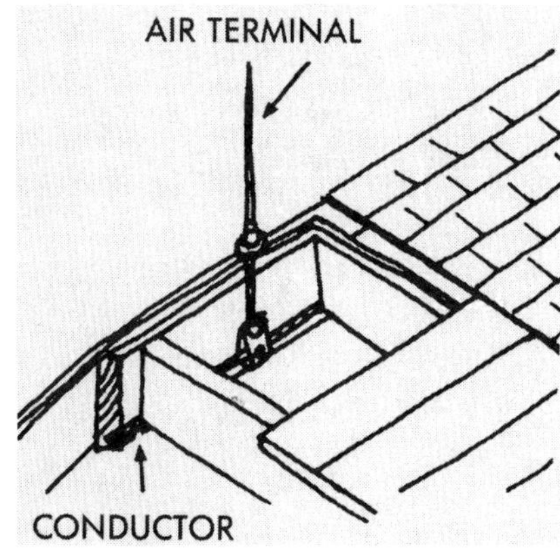

FIGURE 13.2-2 **Air terminals are the only visible parts of a concealed lightning protection system.** (From U.S. Department of Agriculture, Farmers Bulletin Number 2136, "Lightning Protection for the Farm.")

as part of new construction, (1) protection is effective during the construction process, (2) system parts can be more readily protected and concealed, and (3) installation costs may be lower.

Lightning protection system components, such as rods, cables, and connectors, should be Underwriters Laboratory (UL) labeled.

13.2.2.1 Personal Safety

For personal safety during lightning storms, people should stay away from metal objects such as prefabricated fireplaces, stoves, water faucets, appliances, telephones, and metal windows.

13.2.2.2 Tree Protection

Any tree taller than an adjacent building and within 10 ft. (3048 mm) of it should have lightning protection.

When a building with a lightning protection system is within 25 ft. (7620 mm) of a protected tree, the two systems may be interconnected. A common grounding may be used for several large trees, as long as the trees are not more than 80 ft. (24.4 m) apart.

A tree's lightning protection system (Fig. 13.2-3) consists of:

1. Main terminal (special tree) point placed as high as possible within the tree
2. Miniature terminal points clamped on main branches
3. Main conductor (32-strand, 17-gauge (1.45 mm) copper cable, run from the main terminal down)
4. Branch conductors (14-strand, 17-gauge (1.45 mm) copper, run from miniature terminals to main conductor)
5. Grounds (two or more ½- to ¾-in. (12.7 to 19.1 mm)-diameter rods, driven to a 10-ft. (3000 mm) depth at points away from the main root system)

Trees having trunks more than 3 in. (76 mm) in diameter should have two main conductors, one on each side of the trunk. Cables should be copper because the color blends unobtrusively with a tree and because other metals may corrode and harm the tree. Cables should be attached with fasteners that hold them away from the trunk at intervals 3 to 4 ft. (900 to 1200 mm) apart (see Fig. 13.2-3).

13.2.2.3 System Maintenance

Lightning systems should be inspected yearly for (1) bent, loose, or missing air terminals, (2) broken conductor cables, and (3) loose connecting clamps.

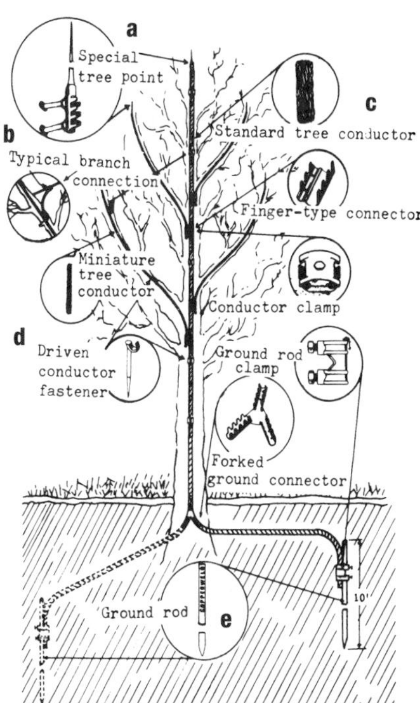

FIGURE 13.2-3 A tree's lightning protection system consists of (a) the main terminal (special tree point), (b) miniature terminals (branch connection), (c) main conductor, (d) branch conductor, and (e) grounds. (From U.S. Department of Agriculture, Farmers Bulletin Number 2136, "Lightning Protection for the Farm.")

13.3 Intrusion Prevention and Detection

Security measures are often necessary to make a facility less vulnerable to unauthorized entry by thieves, vandals, and pranksters.

Burglary, a frequent problem in both urban and rural areas, is a primary threat to security. Nationwide, a burglary occurs about every 10 seconds. About half of all burglars gain entry through an unlocked window or door.

Would-be intruders prefer to work quickly, quietly, and in the dark. If more than 4 minutes are required to gain entry, most burglars will usually give up the attempt. Good burglary protection devices will either slow down or frighten away a burglar. The intent of security devices is to save lives and property by warning residents and neighbors when someone tries to enter a property without authorization.

This section discusses site planning and intruder detector systems. Locking devices and the design of doors and windows to deter intrusion are discussed in Chapter 8; exterior security lighting in Section 16.4.3.2.

13.3.1 SITE PLANNING

The first line of defense against intruders is the area between a building and its property lines. Building location, terrain, landscaping, lighting, and structural barriers are important considerations in creating a sense of physical security. The location of public places, and their convenience and visibility, can encourage the gathering of sociable people and discourage potential criminals.

Barriers that define and protect space may be physical or psychological. Physical barriers include building shape, fences, gates, and doors. The shape of a building can act as a deterrent when all areas are visually accessible and no areas are hidden from view. Psychological barriers do not necessarily bar movement, but rather symbolize boundaries. They include building elements such as open gateways, light standards, short runs of steps, and textural changes in walking surfaces.

13.3.1.1 Site Choice

Where a dwelling is located on the lot, as well as where it is in comparison to other buildings, affects whether the building discourages criminal activity. For example, placing a building far back from the street can increase its vulnerability to crime. A building should be placed so that its entrances and windows are clearly visible to passersby, neighbors, and police patrols.

13.3.1.2 Fences

Fences of wood, chain links, shrubbery, or masonry serve as both psychological and physical deterrents. In many jurisdictions it is difficult to get a trespassing conviction unless property boundaries are clearly marked. A fence should hinder a burglar's attempt to get onto a property, while not providing cover for a break-in.

A 6-ft. (1829 mm)-high brick wall gives a false sense of security. It presents a major barrier, but also hides an intruder's preliminary activities. Fences that protect but do not obscure are better security choices.

A wooden picket fence may be a psychological deterrent but is easy to climb over. A thick hedge, particularly a thorny one, is harder to climb through and over. If either one is no more than 40 in. (762 mm) high, it is easy for neighbors or passersby to see an intruder.

A 6-ft. (1829 mm)-high chain link fence is a good physical barrier that does not obscure vision. It should be installed as close to the ground as possible so that a potential intruder cannot crawl under it. A hedge no more than 40 in. (762 mm) high on one or both sides of a chain link fence can soften its appearance.

Access into the protected area should be through a gate. A simple bell arrangement can sound a signal whenever the gate is opened or closed.

13.3.1.3 Landscaping

Plants that hide a building can also hide an intruder. Trees should be far enough away from a building that a burglar cannot climb them and enter the second story. Bushes should be low enough that they do not obscure windows or doors. Hedges used as fences should be no more than 40 in. (762 mm) high. Landscaping should not obscure a building from the street. Small plants and floral borders can serve to define private and public spaces.

13.3.2 INTRUDER DETECTION SYSTEMS

Various alarm systems are available to warn against intruders. Alarm systems predate the telephone. Alexander Graham Bell's assistant, Thomas Watson, acquired much of his electrician's skill by installing burglar alarms during the 1860s; Alexander Pope patented one of the earliest of these systems in 1853.

Intruder detection devices should be used in conjunction with adequate structural protection such as strong doors and windows and secure door and window hardware (see Section 8.7), adequate site planning, and strategically placed lights (see Section 16.4).

13.3.2.1 Perimeter Intrusion Detection

Exterior doors, windows, and walls make up the perimeter of a building. Perimeter detectors work by setting up an electronic boundary line around this perimeter; breaching the boundary activates the detector's signaling device.

DETECTOR ELEMENTS

Most detectors consist of a protective circuit, a control instrument, and a signaling device. The protective circuit may be made up of light beams, sound waves, or wires that detect the presence of an intruder.

A control instrument monitors a protective circuit. When a circuit is breached, the control instrument activates the signaling device.

A control unit may be mounted indoors or out, depending on the system. Many units have a time delay so that occupants may arm or disarm the system without triggering the alarm. A good lock being as important on a control unit as it is on a door, key-operated systems are preferable. Some control units will sound an alarm if tampered with.

A signaling device issues the actual warning. It may be a light, bell, buzzer, horn, or siren. The signaling device may turn on exterior flood lights. It may also transmit a warning over the telephone. The telephone warning may go directly to a police station or to the police through a contracted service, which is often the same company that installed the system.

EFFECTIVENESS

A perimeter detector's alarm or flood lights will frighten a potential burglar away only if it instills a fear of capture. For a burglar to be caught, someone else must be able to hear the alarm.

Neighbors or occupants should respond to the warning by calling police. Trying to capture a burglar is very dangerous and should be left to professionals.

Police response time is the most important factor in capturing a potential burglar. If the response is within 30 seconds of the alarm, the thief will probably be captured (Fig. 13.3-1). Silent alarm devices that alert the police are available. They may make capture more sure, but they harbor the inherent danger of permitting an intruder to enter the building thinking he or she has not been detected. This can present a heightened danger to an

FIGURE 13.3-1 Police Response Time and Probability of Capture of Burglar

Response Time after Alarm Sounds	Probability of Capture (%)
30 sec.	100
60 sec.	90
2 min.	75
4 min.	50
10 min.	20

occupant who knows the alarm has been sounded and attempts to interfere with the intruder, or who simply stumbles on the intruder. It is better to frighten a burglar away than to face him or her in the dark.

Unfortunately, both noisy and silent intruder alarm systems are subject to false alarms, which diminish their credibility. If false alarms are common, neighbors and police may not respond when a real alarm is sounded, and even occupants may assume that another false alarm has been sounded when there is a real intruder.

SIGNALING DEVICES

Signaling devices may be selected to alert occupants and neighbors, a private security company, the police, or all of these at the same time.

A local alarm is usually a loud bell or siren mounted on an exterior wall of the building so it can be heard by most neighbors. Some alarms continue to ring until they are manually shut off; others sound for a fixed period of time, usually 5 minutes, or are adjustable within a range of 4 to 12 minutes.

A central station alarm is a system in which the protective circuit, or a special dialing apparatus on a regular telephone line, is connected to a private company that monitors it. This is often the same company that installed the system.

When the alarm signals, the company will either relay the information to the police or have their own guards check the alarm. Their cost includes initial installation and a monthly maintenance fee.

In some jurisdictions an intruder alarm may be connected directly to a police station. In those locations extra care should be taken in selecting a detection system, because some of them are very sensitive and prone to signaling false alarms. Some police departments issue fines if they have to respond to too many false alarms. Neighbors may also be annoyed by frequent soundings of a burglar siren.

A telephone dialing system uses an automatic telephone dialing unit with recorded tapes as its signaling device. When activated, the telephone dialer calls a preprogrammed number. Some units can be programmed to dial another number if the first is busy; other units will keep dialing the same number until someone answers.

Many units come with their message on a continuous loop, which is a good idea because of the possibility that an alarm call may be put on hold. These units are called *central station systems* when the programmed number is that of a police department or monitoring company.

DETECTOR TYPES

Detectors may have their protective circuit, control instrument, and signaling alarm all in one device, or each may be in a separate device. They may be either wired or wireless. The NO and NC systems discussed later in this subsection are called wired perimeter detection systems, because their detectors are directly wired to their control and signaling device. The carrier-current and radio-frequency systems, also discussed, are called wireless, because their signals are sent to their control and signaling device by a transmitter that is not hardwired to them. While spot detectors are, in fact, wireless, they actually fit into neither category because they have no separate signaling device.

Some wired systems are versatile and accept both NO and NC intrusion detectors. These combined NO and NC wired systems have the advantage that many forms of protection can be combined into one complete security system.

Both wired and wireless systems have centrally located control and signaling devices, make the same provisions to permit ventilation through windows, and have the same number and type of wired intrusion detectors (Fig. 13.3-2), which attach to doors and windows. Opening the door or window will affect the control device and activate the signaling alarm. They may also have window foil or other glass-breakage detectors. Breaking glass that has a glass-breakage detector attached will cause the control instrument to activate the signaling alarm. When these detectors are used with operable windows, contact switches are provided to permit opening the windows. Both wired and wireless intrusion detector systems may have photoelectric detectors that send a beam of light between a sending and a receiving station. Stopping the path of the light beam breaks the circuit and causes activation of the signaling alarm.

Installing a wired system can be tedious. Wiring must be run from room to room, connecting each door and window back to the central unit. This wiring may be difficult to conceal and, if left exposed, is unsightly. Wireless systems are easier and more convenient to install because they use transmitters rather than extensive wiring to relay a signal to their signaling alarm. Even so, NO and NC wired systems are usually less expensive than other systems.

NO Wired Systems A normally open (NO) wired system has switches in the open position, which is the equivalent of a light switch in the off position. Opening a protected door or window turns the system on. Because current flows through the system and activates the alarm only when electricity completes a loop, removing any switch or breaking any wire may disable the entire system. Accessories such as a smoke detector can be added to an NO system.

NC Wired Systems A normally closed (NC) wired system is less susceptible to defeat than an NO system. Electricity flows through an NC system in a constant loop. An interruption in the current, including breaking or cutting a wire, triggers the alarm. This feature can be an advantage, because a broken wire or sensor will set off the alarm. Conversely, a system failure in an NO system would go unnoticed until a routine check was made.

Carrier-Current Wireless Systems A carrier-current wireless system sends its signal to the control device via regular 120-volt *ac* house wiring. Its intrusion devices are connected to a transmitter that plugs into a wall outlet. The transmitter is the equivalent of a protective circuit. The receiver is a control device that plugs into an outlet in a central area such as a hall. It houses the signaling device.

Radio-Frequency Wireless Systems A radio-frequency wireless system sends its signal from intrusion device transmitters to a control device receiver via radio frequencies. The transmitters operate on batteries. The receiver usually runs on normal 120-volt current with battery backup. In some radio-frequency systems, the battery may sound an alarm when there is a power failure. In others, the battery operates the system on a standby basis only during a power outage. Batteries usually last a year but should be checked occasionally to make sure the system operates properly.

Spot Detectors In a spot detector all of the detector elements are in a single unit designed to be attached to a door or window. The unit buzzes loudly when the door or window is opened. This type of detector is useful in apartments and other areas with few entrance points, but an intruder who takes the time to see where the noise is coming from can often shut it off.

13.3.2.2 Area Controls

Area controls protect the space within a building. They consist of devices either to

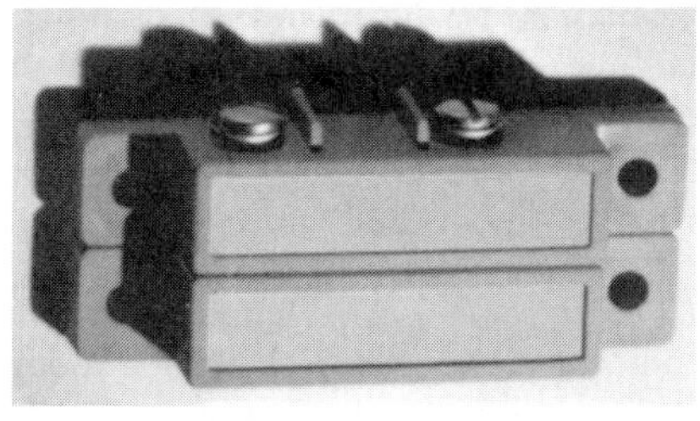

a

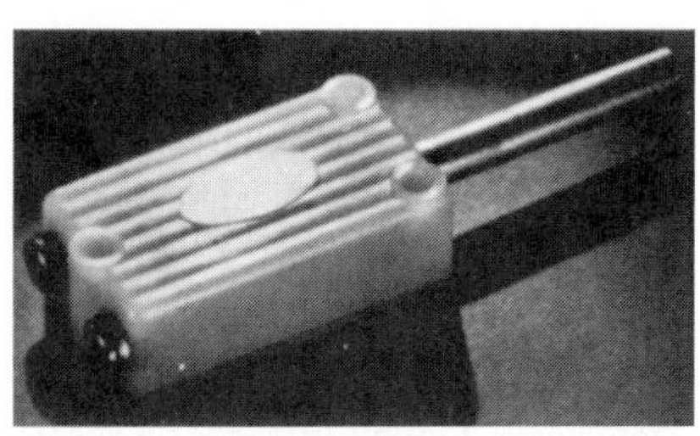

b

c

FIGURE 13.3-2 (a) Separating the housing of a magnetic switch activates the signaling device. Opening a door or window releases the plunger on (b) a surface-mounted and (c) a recessed-mounted detector and activates a signaling device.

fool an intruder into thinking a building is occupied or to detect intrusion within the space.

UNAUTHORIZED ENTRY PREVENTION

Most of the time, a burglar who thinks someone is inside will not break into a building. A burglar can be discouraged by timer-controlled lights and sounds and by decals and labels.

FIGURE 13.3-4 Typical Living Patterns with Four Timers

	A		B		C	
Room	On	Off	On	Off	On	Off
Kitchen	6:00 P.M.	10:00 P.M.	7:00 P.M.	12:00 A.M.	8:00 P.M.	1:00 A.M.
Living room	7:30 P.M.	11:00 P.M.	8:00 P.M.	1:00 A.M.	8:30 P.M.	12:00 P.M.
Bedroom	10:30 P.M.	1:00 A.M.	12:00 A.M.	3:00 A.M.	10:00 P.M.	2:00 A.M.
Bathroom	11:00 P.M.	6:00 A.M.	11:30 P.M.	6:00 A.M.	9:00 P.M.	6:00 A.M.

(Courtesy Intermatic, Inc.)

Timer Controlled Lights and Sounds A cycling timer is a 24-hour electric clock that turns on lights or other appliances at designated times and leaves them on for specified intervals. Some models have a cycle override feature that allows both timed and normal use. Others have multiple cycles that switch lights and appliances on and off, from 1 to 12 times each day. Still others feature variable cycles that switch on at random times (Fig. 13.3-3).

Lights plugged into four timers set in various rooms in a dwelling will give the impression that someone is home. Figure 13.3-4 shows three possible timer use configurations that will suggest various living patterns. If other living patterns are used, it is important to remember to overlap the times that lights are on in at least two rooms.

The sounds produced by radios and television sets can give the illusion that someone is in a building when no one is there. Radios or television sets can simply be left on or can be plugged into a timer.

Decals and Labels Some security systems supply decals for doors and windows that indicate that the property is protected. Protective foil tape on windows also serves to label a building as having security protection.

FIGURE 13.3-3 This type of variable timer plugs into a wall receptacle. The equipment to be timed is then plugged into the bottom of the timer. (Courtesy Intermatic, Inc.)

Some police departments and other organizations lend diamond-tipped engraving pens to mark valuable property with the owner's name and identification. Others provide warning decals for the people who have marked their valuables to place in doors and windows to discourage burglary (Fig. 13.3-5).

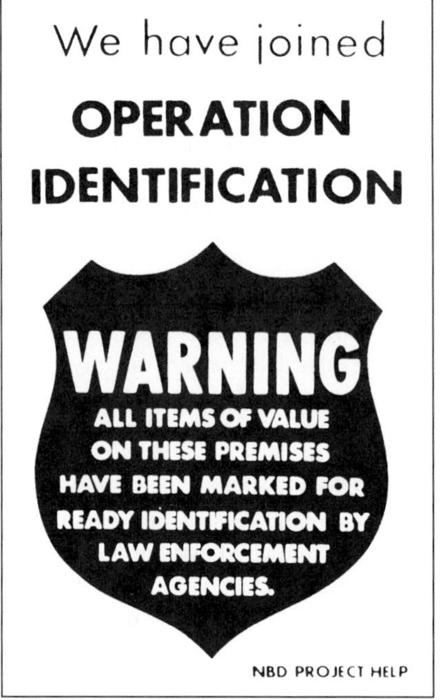

FIGURE 13.3-5 A decal may discourage a burglar because labeled property is hard to dispose of.

INTERIOR AREA DETECTORS

Interior detectors protect the space within a building. In order to activate these detectors, an intruder must have breached the perimeter and be inside the building. Interior area detectors can be used instead of, or in addition to, perimeter detectors. Some types of area detectors cannot be used in dwellings with pets or other buildings with guard dogs, because they are unable to distinguish between the movement of animals and those of human intruders. Combination protection systems utilizing various types of perimeter and interior area detector elements are available.

As is true for perimeter detectors, the effectiveness of area control detectors depends on the burglar's fear of capture, people hearing the alarm and calling the police, and police response time.

Area control detectors have a protective circuit, a control instrument, and a signaling device.

Sound waves, radio frequency waves, and light beams are the equivalents of the protective circuit in a wired system. The control instrument and signaling device are usually relatively small and unobtrusive.

Photoelectric and Infrared Systems Photoelectric and infrared systems can be either perimeter or area detectors and can be connected to either NC or NO systems. A transmitter projects either visible light, electromagnetic waves, or an infrared beam to a receiver. The light can be reflected by mirrors to weave a kind of electronic spider web to increase the area protected. If the beam is broken by an intruder, the circuit is broken and the alarm sounds. Some infrared systems detect the presence of body heat and will sound an alarm even if an intruder is wearing heavy clothing and standing 30 ft. (9000 mm) away.

Ultrasonic Motion Detector An ultrasonic motion detector blankets a room with inaudible sound waves in the 18,000 to 45,000 Hz range. The waves set up an energy field. Some detectors set up an elliptical pattern ranging from 12 × 35 ft.

(3.66 × 10.67 m) to 5 × 20 ft. (1.5 × 6.1 m). A person entering the energy field, also known as the trap zone, interrupts the pattern, and an alarm sounds. Most of these systems have an adjustable sensitivity control to guard against setting off the alarm with normal movement or with ordinary noise, such as that from fans.

An ultrasonic detector, which is about the size of a table radio, is sometimes designed to look like a set of books or a small stereo speaker. Some units require no wiring and are portable. If a detector of this type is used while the building is occupied, people and animals must stay out of the guarded area to avoid tripping the alarm.

Microwave System A microwave detector is a type of radar. Its electromagnetic waves penetrate through wood and plaster but not brick or concrete. This system can protect several rooms, using a device that both transmits and receives a signal. This type transmitter/receiver is placed in a central location in a building. Unfortunately, a microwave system can be a source of false alarms if the energy field is too large. Movement in an adjacent room or apartment, or pedestrian traffic in front of the building, can trigger the alarm. Care in the adjustment and placement of the system should solve these problems.

Sonic System A sonic system resembles an ultrasonic system, except that audible sound waves set off the system. A receiver detects echoes coming from every object in the room. When an intruder enters, the echo changes, triggering the alarm.

Audio Detector An audio detector registers sharp sounds that would indicate an intrusion, such as wood splintering or glass breaking.

Floor Mat Detector A floor mat detector is part of a wired perimeter system. It is included here under area controls because such a device cannot be activated until the intruder enters the building.

13.4 Additional Reading

Additional information about the subjects discussed in this chapter can be found in the references listed in Section 13.5 and in the following publications.

Ambrose, James, and Jeffrey Ollswang. 1995. *Simplified Design for Building Sound Control*. New York: John Wiley & Sons, Inc.

Architecture. 1987. "Transmission Through Floor Assemblies." *Architecture* 11(November):107–109.

Builder. 1983. "Sound Proofing Apartment Walls." *Builder* 12(December): 94.

———. 1983. "Sound Proofing Floors." *Builder* 9(September):96.

———. 1986. "Sound Insulation for Walls and Floors." *Builder* 2(February):138.

———. 1988. "Soundproofing Interior Spaces." *Builder* 8(August):90.

Cavanaugh Tocci Associates. 1999. Edited by Joseph A. Wilkes and William J. Cavanaugh. *Architectural Acoustics, Principles and Application*. New York: John Wiley & Sons, Inc.

Egan, M. David. 1998. "Sound Isolation for Wall Openings: Doors, Windows." *Architecture* 11(November):89–90.

Haas, Steve. 1998. "Five Ways to Reduce HVAC Duct Noise." *The Construction Specifier* 2(February):63.

———. 1998. "Noisy, Rumbling, Hissing." *The Construction Specifier* 2(February):64.

Hansen, Robert A. 1988. "Acoustic Design for Office Tower's Offices, Physical-Fitness Center, Auditorium." *Architectural Record* 8(August):108–113.

Hawkins, Harold D. 1993. "Reverberation and Attenuation." *The Construction Specifier* 9(September):74.

Hughes, Richard L., and Martin F. Johnson. 1983. "Acoustic Design in Natatoriums." *The Construction Specifier* 36(7)(July):67–70.

Interiors. 1987. "Basics of Sound Reflection Problems." *Interiors* 11(November):72.

Irvine, L. K. 1987. "Wall Sound Transmission Comments." *Professional Builder* 5(May):2.

Kincaid, Mary. 1983. "Making Open Office Acoustics Work." *The Construction Specifier* 36(7)(July):70–76.

Mintz, Martin M. 1986. "Sound Control in Stud Walls and Wood Floors." *Builder* 3(March):136.

Patil, Pandit G. 1974. "Flat Glass for Sound Control." *The Construction Specifier* 12(December):22.

Professional Builder. 1987. "Ratings of Various Stud Walls." *Professional Builder* 3(March):48.

Progressive Architecture. 1988. "Noise Insulation Project: Homes, Churches and Schools Near Airport—Denver, Colorado." *Progressive Architecture* 11(November):127.

Rosenberg, Carl J. 1987. "Sound Insulation Between Multi-Family Dwelling Units: Floors, Ceilings, Wall Construction, Pipes." *Architecture* 11(November):85–88.

———. 1988. "Affecting Sound Performance in Public Spaces." *Architecture* 10(October):93–95.

Smith, Homer A. 1976. "Floating Floors and Noise Control." *The Construction Specifier* 2(February):42.

Stanwood, Les. 1983. "Taking a Look at Noise." *The Construction Specifier* 36(7)(July):50–55.

Wagner, Michael. 1987. "Basics of Sound Absorption." *Interiors* 5(May):100.

———. 1988. "Controlling Noise from Air Conditioning Equipment." *Interiors* 8(August):42.

Wagner, Michael, and Nicholas Edwards. "Isolating Outside Sound from a Space." *Interiors* 2(February):42, 46.

Yerges, James F., and John R. Yerges. 1988. "Eight Steps to Noise and Vibration Control in Industrial Facilities." *The Construction Specifier* 3(March):62.

13.5 Acknowledgments and References

We gratefully acknowledge the assistance of the following organizations and individuals in preparing this chapter. We are also indebted to them for permission to use their illustrations when requested and for the use of their publications as references.

ACKNOWLEDGMENTS

American National Standards Institute (ANSI)
American Society for Testing and Materials (ASTM)
Architectural Woodwork Institute
Brick Institute of America (BIA)
Carpet and Rug Institute (CRI)
Cedar Knolls Acoustical Laboratories
Celotex Corporation
Construction Specifications Institute
Consumer Product Safety Commission
Flexicore Manufacturers Association
Geiger and Hamme
Gypsum Association (GA)
IIT Research Corporation
International Organization for Standardization (ISO)
Kodaras Acoustical Laboratories
Lightning Protection Institute
National Association of Home Builders (NAHB) Research Foundation
National Bureau of Standards (now the National Institute of Standards and Technology)
National Concrete Masonry Association (NCMA)
National Fire Protection Association
National Gypsum Company
National Wood Window and Door Association
Ohio Research Corporation
Owens-Corning Fiberglas Corporation
Owens-Corning Fiberglas Sound Laboratory
Portland Cement Association (PCA)
Riverbank Acoustical Laboratories
Rodale Publishing Company
United States Gypsum Company (USG)
U.S. Department of Agriculture
U.S. Department of Housing and Urban Development (HUD)

REFERENCES

We would also like to thank the authors and publishers of the publications in the following list for their contributions to our research for this chapter.

Acoustical Materials Association (AMA). AMA 1-II, "Ceiling Sound Transmission Test by the Two Room Method." Glenview, IL: Ceilings and Interior Systems Contractors Association.

American Architectural Manufacturers Association (AAMA). AAMA TIR-A1-1975, "Sound Control for Aluminum Curtain Walls and Windows." Des Plaines, IL: AAMA.

American Society for Testing and Materials (ASTM) Standards.

C 919, "Practices for Use of Sealants in Acoustical Applications." West Conshohocken, PA: ASTM.

C 920, "Specifications for Elastomeric Joint Sealants." West Conshohocken, PA: ASTM.

E 90, "Laboratory Measurement of Airborne-Sound Transmission Loss of Building Partitions and Elements." West Conshohocken, PA: ASTM.

E 413, "Classification for Rating Sound Insulation." West Conshohocken, PA: ASTM.

E 492, "Method for Laboratory Measurement of Impact Sound Transmission Through Floor-Ceiling Assemblies Using the Tapping Machine." West Conshohocken, PA: ASTM.

American Society of Heating, Refrigerating and Air-Conditioning Engineers (ASHRAE). *1997 ASHRAE Handbook: Fundamentals.* Atlanta, GA: ASHRAE.

Construction Specifications Institute (CSI). 1995. Spec Data 09800, "Acoustical Treatment/Acoustical Insulation and Sealants." Alexandria, VA: CSI.

———. 1999. Spec Data 07920, "Joint Sealants." Alexandria, VA: CSI.

Council of American Building Officials (CABO). *Model Energy Code.* Falls Church, VA: CABO. (In the process of being transferred to the control of the International Code Council (ICC)).

———. 1983. *One- and Two-Family Building Code.* Falls Church, VA: CABO. (In the process of being transferred to the control of the International Code Council (ICC)).

International Organization for Standardization (ISO). 1960. *Field and Laboratory Measurements of Airborne and Impact Sound Transmission, ISO Recommendations #140-1960.* Falls Church, VA: ISO.

Lightning Protection Institute (LPI). L.P.I-175, "Installation Code." Woodstock, IL: LPI.

National Fire Protection Association (NFPA). 1996. NFPA 70, "National Electrical Code." Quincy, MA: NFPA.

———. 1990. NFPA 70A, "Electrical Code for One- and Two-family Dwellings." Quincy, MA: NFPA.

———. 1986. NFPA 78, "Lightning Protection Code." Quincy, MA: NFPA.

———. 1997. NFPA 101, "Life Safety Code." Quincy, MA: NFPA.

———. 1995. NFPA 550, "Guide to the Firesafety Concepts Tree." Quincy, MA: NFPA.

Raeber, John A. 1991. "Redundancy in Fire Protection." *The Construction Specifier* 44(2)(February):17–19.

Ramsey/Sleeper, The AIA Committee on Architectural Graphic Standards. *Architectural Graphic Standards.* 8th ed. New York: John Wiley & Sons, Inc.

Reid, Ed, and Bob Schifiliti. 1991. "A Balanced Approach to Fire Protection Detection and Suppression Systems." *The Construction Specifier* 44(2)(February):30–41.

Underwriters Laboratories, Inc. (UL). 1982 (rev. 1990). *Installation Requirements for Lightning Protection Systems.* Northbrook, IL: UL.

U.S. Department of Agriculture (USDA). *Farmers Bulletin Number 2136.* Washington, DC: USDA.

U.S. Department of Housing and Urban Development (HUD), Federal Housing Administration. 1974. *A Guide to Airborne, Impact, and Structure-borne Noise Control in Multifamily Dwellings.* Washington, DC: HUD.

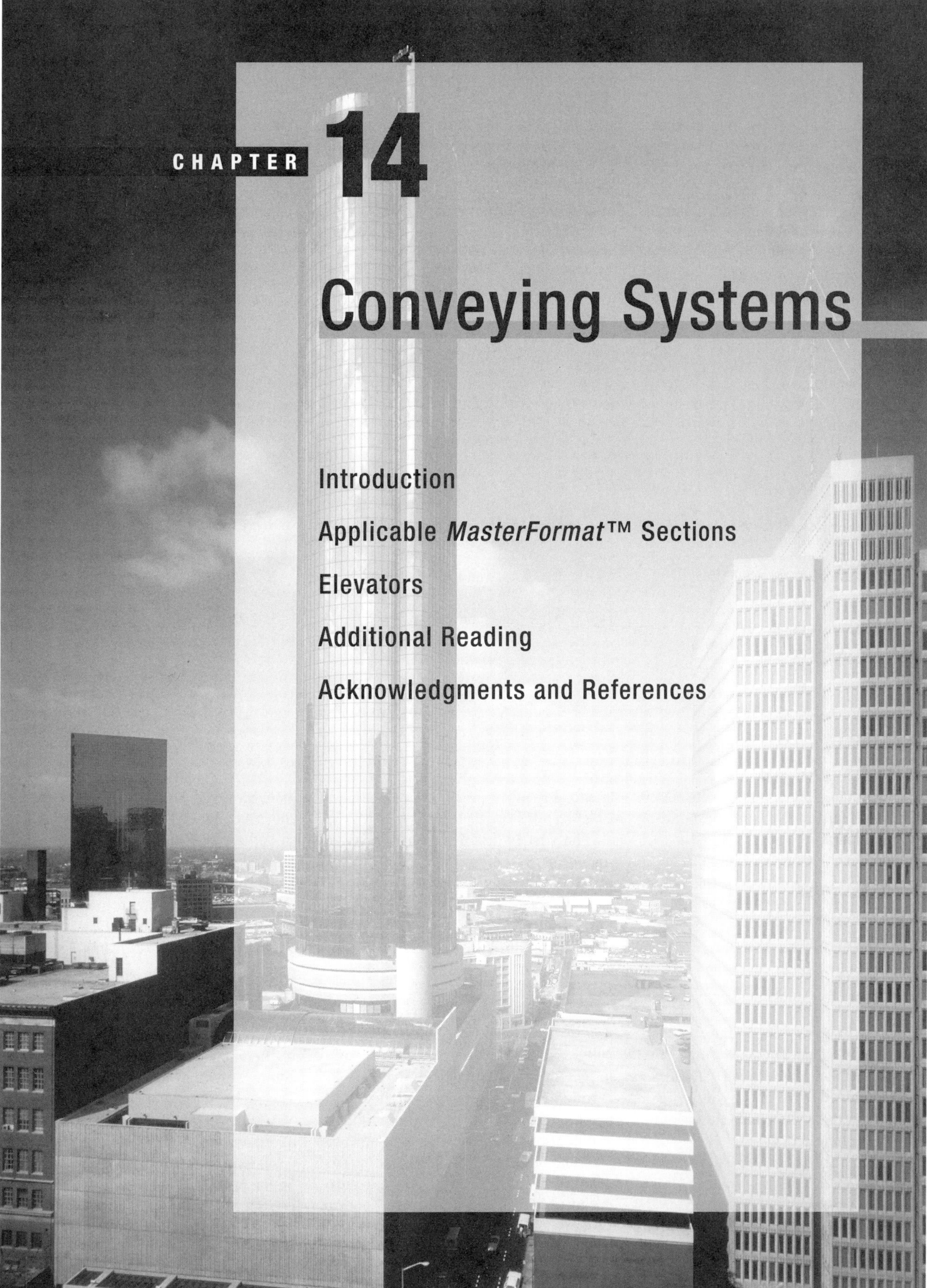

CHAPTER 14

Conveying Systems

Introduction

Applicable *MasterFormat*™ Sections

Elevators

Additional Reading

Acknowledgments and References

Introduction

Many types of conveying systems are used in and around buildings. Among them are dumbwaiters, elevators, escalators, moving walks, lifts, material handling systems, hoists, cranes, turntables, scaffolding, and transportation systems.

Dumbwaiters are essentially small elevators intended for moving materials instead of people. They may be packaged or field assembled and may be motorized or hand operated. Motorized units may operate either electrically or hydraulically.

Elevators are mechanisms designed to move people from one building level to another. They are addressed in more detail in this section.

Escalators are moving stairs designed to convey passengers.

Moving walks consist of moving belts installed either horizontally or on a slant. They are intended to move passengers.

Lifts, commonly known as conveying systems, include people lifts, wheelchair lifts, platform lifts (including stage and orchestra lifts), sidewalk lifts, and vertical lifts. They are designed to lift either people or equipment from one level to another. Conveying systems do not include other lifting devices, such as ski lifts, loading dock levelers, service station equipment, or book and library cart lifts.

Material handling systems include chutes, conveyors (baggage, postal, bucket, screw, vertical, roller, belt, etc.), tube systems (pneumatic tube, etc.), chutes, and similar equipment.

Hoists and *cranes* include fixed and trolley hoists; bridge, gantry, jib, tower, and mobile cranes; and derricks that are part of a completed building.

Turntables are devices designed to rotate sections of a building. They are used in restaurants, stages, exhibit halls, and in rail and other transportation facilities.

Scaffolding classified as vertical transportation includes suspended scaffolding, rope climbers, and telescoping platforms that are part of a completed building.

Transportation systems include monorail systems, people movers, passenger loading bridges, and similar devices used in railroad, subway, monorail, and aircraft facilities.

Applicable *MasterFormat*™ Sections

The following *MasterFormat* Level 2 sections are applicable to this chapter.

14100 Dumbwaiters
14200 Elevators
14300 Escalators and Moving Walks
14400 Lifts
14500 Material Handling
14600 Hoists and Cranes
14700 Turntables
14800 Scaffolding
14900 Transportation

14.1 Elevators

An elevator is a device for moving people or freight vertically from one building level to another.

14.1.1 HISTORY

Since ancient times, people have been using hoists powered by water, people, or animals to move loads vertically. Such devices were first powered by steam around 1800, but even the steam-powered devices had no safety features to prevent them from falling when a rope or some other part of the mechanism failed. Thus, their use was limited to freight hauling only. In addition, early commercial elevators ran only between two floors.

In 1852, Elisha Graves Otis, a master mechanic in a bedstead factory, introduced the first practical safety device for passenger elevators, a system of spring-operated cams that would engage an elevator's guide rails in the event that the cable broke. The first elevator specifically intended to move people was installed in a New York store in 1857.

The first hydraulic elevator was installed in 1878. The hydraulic fluid was water; otherwise, the operation was similar to that of a modern hydraulic elevator. Early hydraulic elevators were of two types. In one, the cylinder was directly beneath the car. In the other, the cylinder pulled ropes that in turn lifted the car.

The first electrically driven elevator was installed in a New York City office building in 1889. Their use made the building of skyscrapers practical.

Early elevator controls were not automatic. A human operator was required. Automatic elevators were available for use in homes as early as 1894, but the first gearless traction elevator was not introduced until 1903, and advanced control systems not until 1924. Fully automatic elevators for use in tall buildings came into general use in 1949.

Improvements have come rapidly since then. Speeds have increased, acceleration and deceleration functions have become smoother, leveling at floor landings has improved, and the combination of developing technology and stringent laws has increased safety immensely.

14.1.2 ELECTRIC TRACTION ELEVATORS

Electric traction elevators consist of a car traveling in a shaft (Fig. 14.1-1). The car is lifted by an electric motor housed in an equipment room at the top of the shaft. The motor turns a grooved wheel (*drive sheave*), over which pass multiple steel cables that are connected to a cast iron counterweight on one end and the elevator car on the other end. As the sheave turns, the cable moves, raising or lowering the car. The combined weight of the car and counterweight keeps the cables taut. There is usually also a second sheave for added traction.

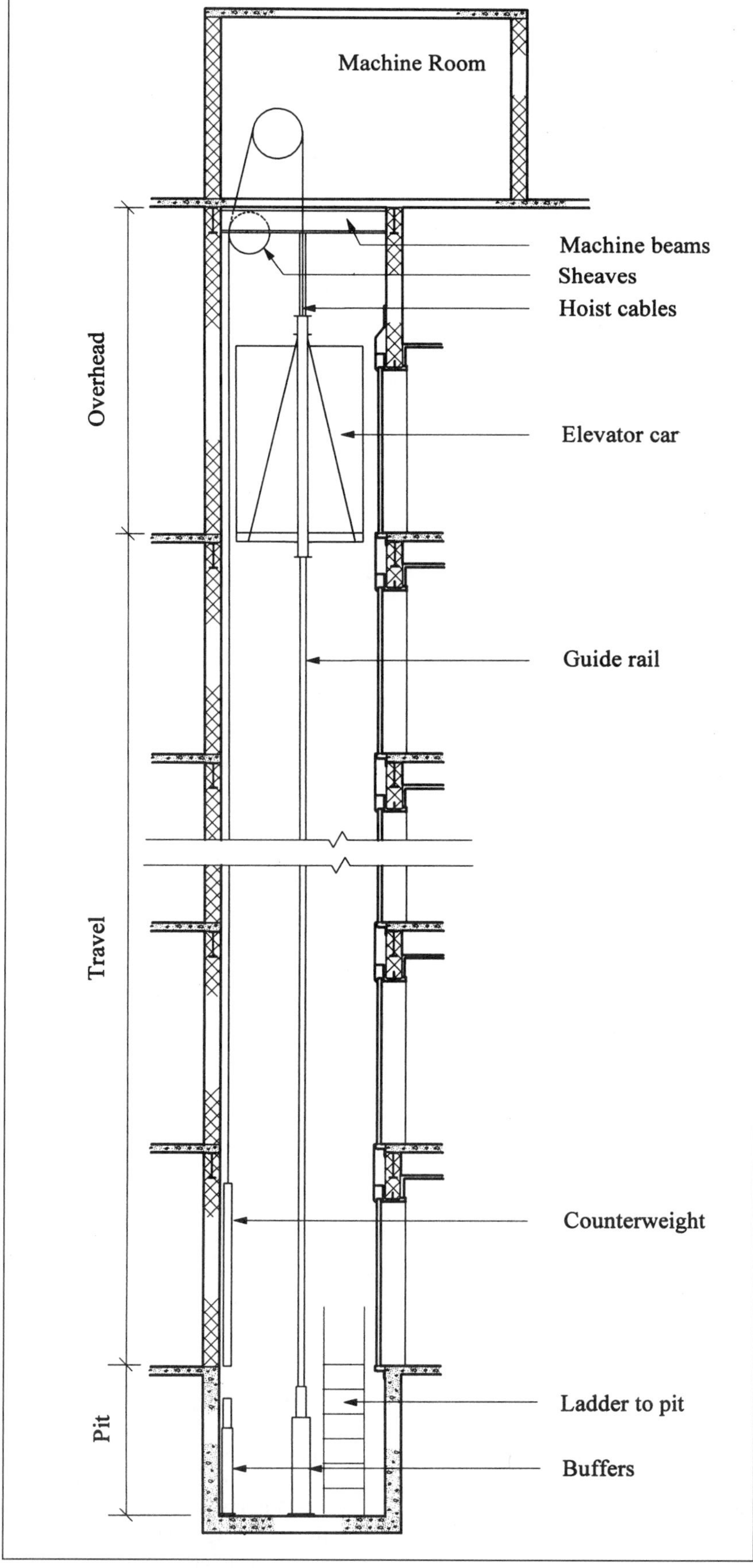

FIGURE 14.1-1 Electric traction elevator. (Drawing by HLS.)

In an *underslung* traction elevator, also called a *basement traction* elevator, the motor is placed in a room adjacent to the bottom of the shaft. The drive wheels on the motor project into the shaft. Cables pass under the drive wheels and extend vertically up the shaft, where they pass over the sheaves. One end is connected to the car, and the other end is connected to a counterweight. Underslung elevators are used in existing buildings and in low-rise new buildings where a penthouse would be objectionable or illegal and where overhead is limited.

In traction elevator installations, both the car and the counterweights have guide rails along the side of the shaft to keep them in proper alignment.

Traction elevators are classified as either geared or gearless. Geared elevators are used in low-rise buildings where speeds between 200 ft. (61 m) and 450 ft. (137 m) per minute are sufficient. At speeds of 500 ft. (152 m) to 2000 ft. (610 m) per minute, which are required for tall buildings, geared machines generate excessive noise and are inefficient. Gearless machines are smoother, quieter, and more efficient than the geared type.

Long, flexible power and control cables run down the shaft to the car to provide the required power for the car. They supply electricity to operate lights, doors, fans, and control systems and connect the control systems in the car with those in the machine room.

14.1.2.1 Safety Features

There are several safety systems protecting elevator cars. A car is connected by cable to an electric governor at the top of the shaft. If the car descends faster than a preset limit, the governor slows down the drive sheaves. If the car then fails to stop, the governor grips the governor cable, which activates safety clamps that provide a wedge between the car and the guide rails, stopping the car.

Elevators also have electrical controls that prevent the car from operating when a door is open.

14.1.2.2 Operation and Performance

Until recently, most traction elevators were powered by direct current (dc) power. Converting alternating current (ac) building power to dc for elevator use was done by a motor generator (MG) set. MG sets consume power while idling, generate copious amounts of heat, and are noisy. Later dc-powered elevators have solid-state motor drives, which use silicon-controlled rectifiers (SCR) units instead of MG sets to convert house current to dc. SCR units use power only when the elevator is running. They are also quieter and produce less heat than MG sets. While dc-powered elevators are still available, the trend today is toward elevator drive units powered by variable frequency alternating current (ACVF).

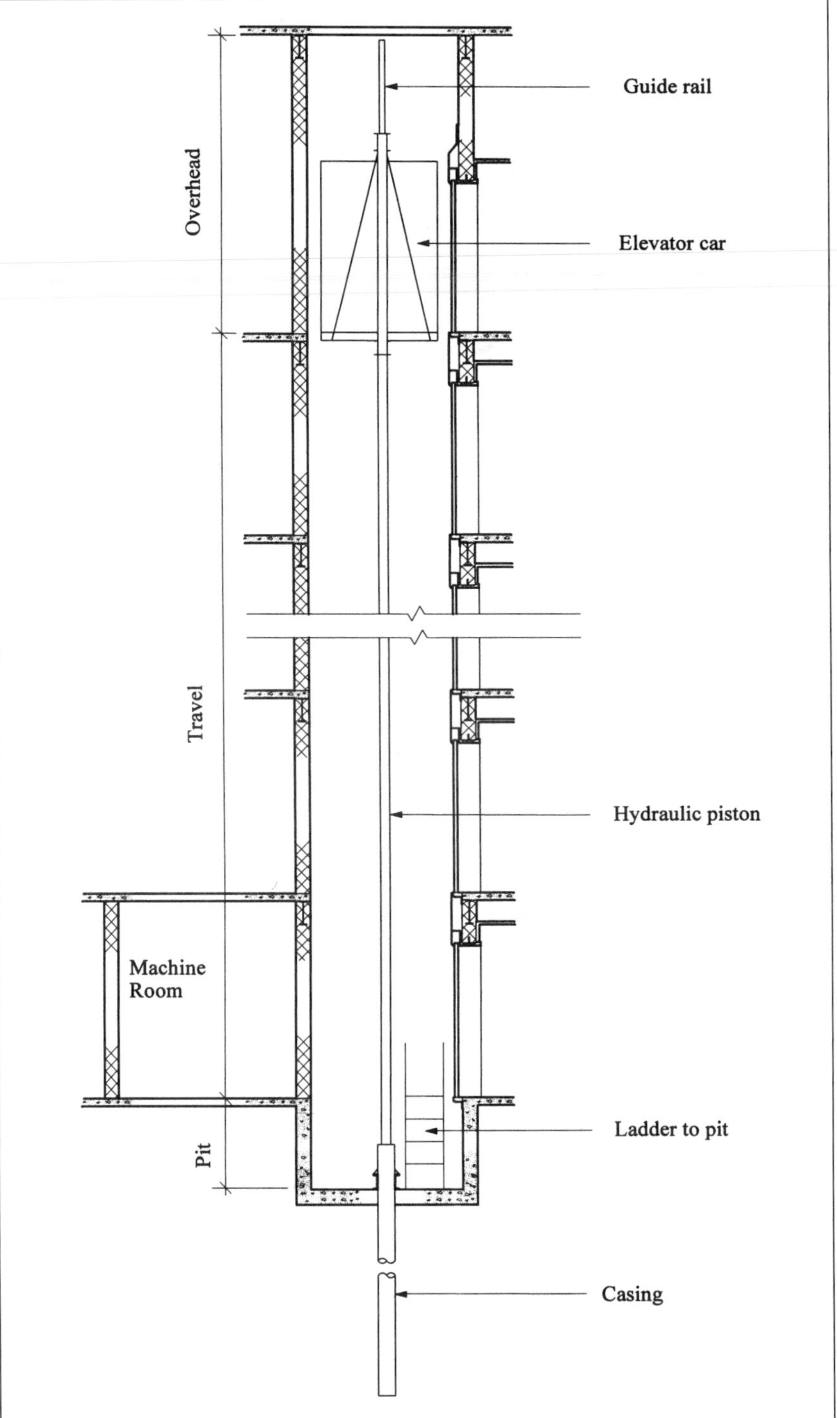

FIGURE 14.1-2 Hydraulic elevator. (Drawing by HLS.)

14.1.3 OIL HYDRAULIC ELEVATORS

Hydraulic elevators are classified as either *inground*, *holeless*, or *roped*.

14.1.3.1 Inground Oil Hydraulic Elevators

Inground hydraulic elevators are supported from below by a steel plunger that moves in a vertical cylinder that is sunk into a shaft drilled below the elevator shaft (Fig. 14.1-2). The cylinder extends below ground to a distance equal to the rise of the elevator, or a little more. For this reason most hydraulic passenger elevators are not practicable for buildings more than 8 stories high. Most are in shorter buildings. The maximum travel of most hydraulic freight elevators is between 40 and 50 ft. (12 and 15 m).

The basic operation of an inground hydraulic elevator is as follows:

- To raise the car, a pump driven by an electric motor forces oil under pressure into the casing, which raises the plunger, lifting the car.
- To lower the car, electrically controlled valves open, permitting the oil to return slowly to a storage tank, thus lowering the plunger and, in turn, the car. The motor and the pump are idle when the car is descending.

A hydraulic elevator can fall if oil leaks from the cylinder. When this rare event occurs, the oil leaks out very slowly, lowering the car at a safe speed onto a buffer at the bottom of the shaft. Therefore, safety clamping systems, such as those used on traction elevators, are not required to prevent hydraulic elevators from falling.

14.1.3.2 Holeless Oil Hydraulic Elevators

Holeless oil hydraulic elevators are driven by a mechanism similar to that used in an inground hydraulic elevator. The cylinder, however, telescopes, making a deep hole unnecessary. Some telescoping cylinders are in shafts that extend into the ground, but in such cases the shaft extends a shorter distance into the ground than the shaft of an inground unit. Holeless units may have either single-stage or two-stage operation. The basic difference between the two is the height of building the elevator can serve. A single-stage holeless installation may be limited to 15 ft. (4.57 m) travel distance. A partially inground machine may be able to travel 20 ft. (6 m). A two-stage machine may be able to travel about 35 ft. (10.7 m), or for four stacked stops. Two-stage machines with additional pit depth and overhead height can travel as much as 41 ft. (12.5 m).

14.1.3.3 Roped Oil Hydraulic Elevators

A *roped* hydraulic elevator is driven by an oil hydraulic mechanism. Plungers (which are in turn supported by devices called jackstands, sliding in above-ground cylinders) are located in the shaft beside the car. Cables are attached to the top of the jackstands, pass over a sheave, and are fastened to the bottom of the car frame. As the plunger is forced up by hydraulic pressure, it raises the sheave, shortening the cable length from the sheave to the car, and thus raising the car. The other end of the cables, being fixed to the jackstands, cannot move.

Roped systems have a traction governor and other safety features similar to those of an electric traction elevator.

Roped machines work in applications of up to 60 ft. (18.3 m).

14.1.4 PASSENGER ELEVATORS

Passenger elevators primarily move passengers. Passenger elevators that are used to move supplies, furniture, and other items, in addition to passengers, are generally designed and finished in the same manner as other passenger elevators.

14.1.4.1 Passenger Elevator Cars

FEATURES

A passenger car should include the following features:

- An emergency light with battery and charger.
- An emergency exit.
- Finished flooring.
- A suspended ceiling. Various translucent plastic, metal, aluminum eggcrate, and other designs are available. Fluorescent light fixtures are often provided above the suspended ceiling, but incandescent lights are also used.
- A handrail. Many types are available, including wood rails and metal flat bars, tubes, and cylindrical shapes.
- A single-speed or two-speed fan for power ventilation.
- As an option, protective pads and pad hooks for side and rear walls.

See also Sections 14.1.5 and 14.1.6.

FINISHES

Many finishes are available for use in passenger elevator cars. Doors, articulation, wall bases, ceilings, and operating panels may be wood, plastic laminate, or metal. Metals typically used are stainless steel, aluminum, Muntz metal, and bronze. Metal panels may be finished in baked enamel in almost any color.

Floor finishes in passenger elevators are usually carpet, but some are resilient flooring.

CAR AND HOISTWAY DOORS

Passenger-type hoistway entrances are usually required to be Underwriters Laboratories, Inc. (UL) "B" labeled. They are usually hollow metal but may be finished in any noncombustible material, such as stainless steel or bronze, or may receive a baked enamel coating.

Opening sizes vary with use and elevator size and capacity. For example, doors for 2000-lb. (907 kg) elevators may be 3 ft. (914 mm) wide; for 2500-lb. (1134 kg) to 3500-lb. (1588 kg) elevators, 3½ ft. (1067 mm) wide; 4000-lb. (1814 kg) to 5000-lb. (2268 kg) elevators, 4 ft. (1220 mm) wide.

DOOR OPERATION

Passenger car doors are horizontal sliding and may be *center-opening*, *single-slide*, or *two-speed* doors (Figs. 14.1-3, 14.1-4, and 14.1-5). Although other door sizes are sometimes used, elevators with small capacities (2000 lb. [907 kg] to 2500 lb. [1134 kg]) generally have single-slide doors; midrange capacities (3000 lb. [1360 kg] to 3500 lb. [1588 kg]), center-opening doors; and higher capacities (4000 lb. [1814 kg] to 5000 lb. [2268 kg]), two-speed doors.

DOOR OPERATOR EQUIPMENT

A door operator opens and closes car and hoistway doors simultaneously. Door movement is cushioned at both limits of travel. An electric contact on each car entrance prevents the elevator from operating unless the car door is closed. An electromechanical interlock prevents elevator operation until all its doors are closed and locked.

Doors are designed so that if the electric power is interrupted, the doors can be opened from within the car by hand. Emergency devices and keys for opening doors from the landing are required by code.

Doors open automatically when the car has arrived at a landing. They close after a predetermined time interval or can be closed immediately by pressing a *door close* button in the car. Momentary press-

FIGURE 14.1-4 Elevator with two-speed doors. The doors in each pair operate at different speeds so that the two hoistway doors and the two cab doors are fully open at the same time and fully closed also at the same time. All four doors stack on the same side of the hoistway opening. (Drawing by HLS.)

FIGURE 14.1-5 Elevator with single-slide doors. Cab and hoistway doors operate simultaneously. Both car and hoistway doors stack on the same side of the jamb opening and between it and the hoistway wall. (Drawing by HLS.)

FIGURE 14.1-3 Elevator with center-opening doors. Cab and hoistway doors operate simultaneously. Pairs of doors stack, with one on each side of the jamb opening and between it and the hoistway wall. (Drawing by HLS.)

ing of a *door open* button in the car reopens the doors and resets the time interval.

Each elevator car door has an electronic sensing device that prevents the doors from closing when an obstruction is present and will cause them to stop and reopen if they encounter an obstruction while they are closing.

14.1.4.2 Passenger Elevator Dispatch

Overall elevator dispatch is governed by a logic control system. For passenger elevators, two of the logic control systems used are *selective collective* and *car group*. Others may also be used.

In a selective collective control system, multiple hall and car calls may be registered at the same time. The control system remembers each call and answers the calls in the order in which the landings are reached, based on the direction of travel of a particular elevator, unrelated to the order in which the calls were registered.

Modern, group logic systems are microprocessor controlled. A processor in each car controls the operations of that car, such as door openings and closings, safety and stop circuits, load weighing, and timing devices. The in-car processors are tied together through a *group controller*, which receives and processes elevator demands from car and hall calls and dispatches individual cars to answer them in the most efficient manner.

14.1.5 FREIGHT ELEVATORS

Freight elevators are intended primarily to move freight and, as such, are utilitarian in nature. Freight elevators are designed according to their load classes, as defined in the American Society of Mechanical Engineers' ASME A17.1:

A. General Freight Loading
B. Motor Vehicle Loading
C1. Industrial Truck Loading (This class permits a forklift to enter the elevator.)
C2. Industrial Truck Loading (Forklifts do not usually enter the elevator.)
C3. Other Loading with Heavy Concentrations

14.1.5.1 Freight Elevator Cars

The interior walls and ceilings of freight elevators are usually of 14-gauge (1.9 mm) cold-rolled steel with no applied finishes. Perimeter bumper guards are often installed to reduce potential wall damage caused by forklifts and loads contacting the walls. Depending on capacity and class of load, flooring in freight elevators may be checkered steel plate, galvanized steel, or high-density wood.

CAR AND HOISTWAY DOORS

A freight elevator car typically has a gate. In most cases, this is a vertically rising device. Freight elevator hoistway doors are usually biparting units in which the upper panel travels upward and the lower panel travels downward. Some freight elevator doors are one-piece units that travel upward. The car riding gate is usually power operated. Hoistway doors may be either manually operated or power operated.

Power-operated doors are opened and closed by means of push buttons located at each landing and within the elevator. In a normal operation, when the car arrives at a landing the doors and gate open either automatically or by momentary pressure on a door-open push button. Closing the doors and gate requires constant pressure on the door-close push button. Door closing may also be done automatically.

When passengers will be permitted to use a freight elevator, special capacity calculations must be made, special doors are required, and additional door operation features are required. For example, a safety edge that automatically opens the door if it meets an obstruction is required on the lower edge of the car riding gate, and an audible signal must be added to sound 5 seconds before the automatic door–closing sequence begins.

14.1.5.2 Freight Elevator Operation

Overall elevator operation is governed by a logic control system. The two most frequently chosen elevator logic controls for freight elevators are *selective collective* (S/C) control and *full automatic push button* (FAPB) control.

Selective collective control systems are described in Section 14.1.4.2. In an FAPB logic control system, only one hall call is registered at a time, so that the elevator may be used without interruption. When not in use, the elevator will answer a registered hall call, but hall calls will not be registered while the elevator is in use.

14.1.6 AUXILIARY EQUIPMENT

Many items of auxiliary equipment and devices are necessary for proper elevator operation. Some of this equipment and these devices are located in the car; some are in the hall or lobby, some on the outside of the car, and others in locations remote from the car.

14.1.6.1 In-Car Equipment and Devices

CAR OPERATION STATION

Each elevator is equipped with a main car operating station, located integrally in a vertical swing panel. This station contains call registration buttons in accordance with the logic operation selected, an emergency stop switch, an alarm button, tactile plates, and a light switch, as well as any other devices required by the applicable code.

The American with Disabilities Act (ADA) requires that devices operable by the general public and mounted in an elevator car operating panel be identified with braille or tactile symbols. In addition, the car operating panel must be located vertically in accordance with ADA requirements.

ADA also requires that an emergency telephone be installed in each elevator cab. It can be a separate unit or can be a part of the main car operating station. Such telephones require specialized controls for the use of handicapped persons.

CAR POSITION INDICATOR

A car position indicator must be provided in each car. This indicator must show the car's position by floor designation.

CAR DIRECTION SIGNS

Each car must contain a car direction sign that indicates the direction in which the car is moving. The direction sign must have an associated audible signal sounding once for *up* and twice for *down*.

14.1.6.2 Hall Operating Equipment and Devices

HALL PUSH-BUTTON STATIONS

A hall push-button station must be provided at each landing. The push buttons in the station must be illuminated to notify waiting passengers that the call has been registered, and remain illuminated until the call has been answered. ADA requires that these push buttons be identified with braille and tactile symbols. In addition, hall push-button stations must be located vertically in accordance with ADA requirements.

In a FAPB logic control system, a single call button is supplied at each landing. When the elevator is already in use, an illuminated signal is activated, indicating that no additional hall call can be registered until the light is extinguished.

HALL POSITION INDICATOR

Hall position indicators are sometimes provided. These show the position of the car in the hoistway with floor numbers or letters.

CAR DIRECTION SIGNS AND HALL LANTERNS

Car direction signs and hall lanterns include directional indications and an audible signal. The appropriate arrow is illuminated to show the direction in which the car is set to travel. An audible signal alerts passengers on the landing, sounding once for *up* and twice for *down*.

EMERGENCY PANEL

An emergency status panel may be provided, and, especially in large buildings, in hospitals, and in some other building types, may be required by code. These panels contain the following items:

- Digital position indicator for each elevator within the group
- Controller indicator lights
- Controller key switches
- Emergency power indicator lights and key switches
- Fire service indicator lights
- Fire recall key switches
- Emergency telephone jack

14.1.6.3 Outside-the-Car Equipment and Devices

Each elevator must have an inspector's operating station mounted on top of the car. It must contain the following items:

- Up and down constant-pressure buttons
- An emergency stop switch
- A light with guard and switch
- A duplex 120 VAC outlet

In addition, a light with a guard and a switch, and a duplex 120 VAC outlet must be mounted on the bottom of each car.

14.1.6.4 Remote Devices

An emergency stop switch is required in each elevator pit. It must cut off the power supply to the motor and bring the car to rest independent of other devices.

An alarm bell must be located in or adjacent to the hoistway. It must be connected to an alarm button mounted in the main operating panel in each car.

14.1.7 SELECTION CRITERIA

Elevators are generally packaged units or are constructed with packaged components. The following criteria should be addressed in determining which type elevator is needed and the types of features it should have:

1. The type of unit (passenger, hospital, service, freight, residential). This influences the type of controls required, the size of the car, the types of doors, and the load-carrying capacity needed. Passenger-use loads generally range in 500-lb. (226.8 kg) increments from 2000 lbs. (907 kg) to 3500 lbs. (1588 kg). For hospital or service use, capacity is usually 4000 lbs. (1814 kg), 4500 lbs. (2041 kg), or 5000 lbs. (2268 kg). Freight elevator capacities range from 2000 lb. (907 kg) up to 100,000 lb. (45,359 kg).

An *observation elevator* is a special type, basically made up of one of the other types but arranged to ride up the exterior of a building or in a multistory atrium space within a glass hoistway.

2. For a freight elevator it is necessary to determine:

- The type of freight that is to be carried, which can range from general freight to loaded semi-trucks
- How the elevator will be loaded and unloaded (e.g., by hand carrying freight or loading material with a forklift)
- The maximum weight of any single item to be moved
- The maximum size of any single item to be moved (width, length, height)
- Whether passengers will use the elevator

3. The number of stops. This influences whether a traction elevator or a hydraulic elevator should be selected and which type of machine should be used (geared or gearless for a traction elevator, and inground or holeless for a hydraulic elevator).

4. Speed. The speed at which an elevator will ascend and descend influences the type of machine (geared or gearless) selected and is in turn influenced by the number of stops and height of the building. Buildings up to about 6 stories can operate well at speeds between 200 ft. (61 m) per minute and 450 ft. (137.4 m) per minute. Taller buildings need greater speed. Hydraulic elevators generally operate at speeds up to 150 ft. (45 m) per minute. Geared traction elevators operate up to about 450 ft. (137.4 m) per minute. Higher speed requirements dictate the use of gearless traction machines. Very tall buildings require very high speeds. The elevators in the John Hancock Center in Chicago move at speeds up to 1800 ft. (550 m) per minute. Freight elevators, on the other hand, have little use for high speeds. They can operate at 50 ft. (15 m) to 350 ft. (107 m) per minute, but most operate in the range of 80 ft. (24.4 m) to 100 ft. (30.5 m) per minute.

5. Travel. The total travel distance of an elevator influences the type of drive and whether an elevator should be of the traction or hydraulic type. Hydraulic elevators work well up to about 5 stories and can be used up to 8 floors. Buildings higher than that should have traction elevators.

6. Operation. The type of operation is dictated by the height of the building and the type and configuration of the elevators.

7. Number of passengers and wait time. The number of passengers and their distribution within the building controls the number and size of the elevators required in a building. Generally, there should be one elevator for every 250 persons in a building of 6 or more floors.

The number and distribution of elevators are also influenced by the length of time passengers will have to wait for an elevator. People become annoyed when made to wait as little as 20 seconds in an office building, but may be willing to wait for about a minute in an apartment building.

When the number of passengers is large, elevators are arranged in groups (*banks*) of as many as eight cars.

In buildings exceeding 15 floors, there may be one or more banks of high-speed *express* elevators that travel directly from the lobby level to an upper floor, bypassing lower floors. These lower floors are served by banks of *local* elevators.

Double-deck elevators are sometimes used to increase capacity without reducing floor space. Similarly, in a *sky-lobby* system, express elevators take passengers to a level high in the building. There, they leave the express elevator and take local elevators to their destination floors. These local elevators start at the sky-lobby floor, which leaves space below them open for rental use. There may be several levels used as sky-lobbies.

8. Platform size. The platform size is dictated by use of the elevator and number of passengers to be accommodated or the amount of freight to be carried. Cars for hospital elevators are deeper front-to-back than passenger cars so as to accommodate patient beds and stretchers. Platform size can be limited by the load capacity of the elevator.

9. Passenger car and hoistway entrance size. The car and hoistway entrance size is dictated by the size and type of elevator.

10. Door operation. Passenger car doors are either single-slide, center-opening (may not be available for 2000-lb. [907 kg] car), or two-speed.

11. For a freight elevator, the following steps are necessary:

- Determine the minimum door size (finished opening dimensions—width and height).
- Determine the type of doors (i.e., manually or power operated, biparting or other).
- Determine the material, configuration, and construction of the doors.
- Determine the load class.

12. Machine room location. For an overhead traction elevator, the machine room is located above the shaft in a penthouse. The machine room for an underslung traction elevator must be adjacent to the shaft and at a lower level. A machine room for a hydraulic elevator can be in any location, but the preferred location is at the lowest level and as close to the shaft as possible, preferably immediately adjacent to the shaft.

13. Weight. The weight of the elevator and associated machinery is a major factor in selecting the type of elevator to use in freight-handling applications. For the same capacity and size, traction elevators tend to be heavier than hydraulic elevators. In addition, the heavy load capacity associated with some freight elevators adds considerable weight and increases the structural requirements of the building substantially. This factor may be an influencing one when the building under consideration does not exceed the height limit of hydraulic elevators. The shaft depth for heavy-load-carrying in-ground hydraulic elevators may have to be greater than the travel distance of the elevator. If there are underground water problems or if drilling is difficult because of underground conditions, the cost of drilling might exceed the cost of the additional framing required for the use of a traction elevator. In general, however, within the practical limits of hydraulic elevators, they are more economical for freight-carrying applications than traction elevators.

14.1.8 INSTALLATION

14.1.8.1 Additional Work Required

To complete an elevator installation, the following additional work must be performed according to the governing codes.

There must be a legal hoistway constructed with the required fire rating, with properly sized rough openings, adequate fastening for hoistway entrance assemblies, and a clear overhead of adequate height including suitably sized support for necessary hoisting. The front entrance wall at the main landing is not constructed until all elevator material has been located in the hoistway. The other front entrance walls are not constructed until door frames and sills are in place.

Supports for guide rails must be installed. Separator beams must also be installed where required.

There must be a properly located, ventilated, heated, lighted, and sound-isolated room large enough to house the elevator machinery. Temperature must be maintained between 55°F (12.7°C) and 90°F (32.2°C), with relative humidity not to exceed 95%. The entrance door to the machine room must be large enough to allow placing of the elevator equipment.

A dry pit, reinforced to sustain impact loads on guide rails and buffers, must be provided.

Landings must be configured to receive entrance sills.

A fixed vertical metal ladder must be installed in the pit. It must extend at least 42 in. (1066 mm) above the sill of the lowest terminal entrance.

The building structure must be designed to support the loads that will be imposed by the elevator and its equipment.

If sprinklers are installed in the hoistway, machine room, or machinery spaces, the electric power must automatically be disconnected before water is applied. Smoke detectors may not be used to activate these sprinklers or to disconnect the power supply.

The proper type of electric power must be brought to the machine room and connected to controller terminals in the machine room.

Electric power with fused disconnect switches must also be installed for car lights and other devices requiring power.

Switched light fixtures and convenience outlets with ground fault circuit interrupters must be installed in the machine room.

A suitable switched light fixture and a convenience outlet with a ground fault circuit interrupter must be provided in the pit.

Where required, a means of standby power must be provided to operate the elevators in the event of a general or building power failure.

14.1.8.2 Installation

Installation is done in accordance with the manufacturer's shop drawings and following the manufacturer's instructions.

14.2 Additional Reading

More information about the subjects discussed in this chapter can be found in the references listed in Section 14.3, and in the following publications.

Barker, Frederick. 1997. "Is 2,000 Feet per Minute Enough?" in *Elevator World.* 3(March). ©1995 Otis Elevator Company, Farmington, CT. All rights reserved. Paper originally presented at International Conference on High Technology Buildings, Council on Tall Buildings and Urban Habitat, São Paulo, Brazil, October 25–26, 1995.

Bertz, Rick. 1992. "Elevator Technology." *The Construction Specifier* 9(September):27.

Fortune, James W. 1992. "Wright to the Top." *The Construction Specifier* 9(September):86.

———. 1995. "Elevator Modernizations." *The Construction Specifier* 11(November):60.

Knorr, Janis. 1989. "The Ups and Downs of Elevator Planning." *The Construction Specifier* 4(April):96.

Lacob, Miriam. 1997. "Elevators on the Move." *Scientific American* 10(October).

Strakosch, George R., ed. 1998. *The Vertical Transportation Handbook.* 3d ed. New York: John Wiley & Sons, Inc.

14.3 Acknowledgments and References

ACKNOWLEDGMENTS

We gratefully acknowledge the assistance of the following organizations and individuals in preparing this chapter. We are also indebted to them for permission to use their illustrations when requested and for the use of their publications as references.

Dover Elevator Systems, Inc.
Montgomery-Kone
Otis Elevator Company

REFERENCES

We would also like to thank the authors and publishers of the publications in the following list for their contributions to our research for this chapter.

ADA Title III, *Americans with Disabilities Act—Buildings and Facilities.* Washington, DC: ATBCB.

American National Standards Institute (ANSI) Standards.

ANSI A17.1, *Safety Code/Elevators, Escalators, Moving Walks.* New York: ANSI.

ANSI A117.1, *Buildings and Facilities, Providing Accessibility and Usability for Physically Handicapped People.* New York: ANSI.

American Society of Mechanical Engineers (ASME). A17.1, *Handbook on Safety Code for Elevators and Escalators.* New York: ASME.

Architectural and Transportation Barriers Compliance Board (ATBCB). 1990.

Arcom Master Systems. MASTERSPEC® Basic Sections.

14210, "Electric Traction Elevators." 11/94 ed. Salt Lake City, UT: Arcom.

14240, "Hydraulic Elevators." 11/94 ed. Salt Lake City, UT: Arcom.

14420, "Wheelchair Lifts." 8/95 ed. Salt Lake City, UT: Arcom.

National Fire Protection Association (NFPA). NFPA 70, *National Electrical Code.* Quincy, MA: NFPA.

———. NFPA 80, "Fire Doors and Fire Windows." Quincy, MA: NFPA.

———. 1997. NFPA 101, "Life Safety Code." Quincy, MA: NFPA.

Underwriters Laboratories, Inc. (UL). 10B, "Fire Tests of Door Assemblies." Northbrook, IL: UL.

CHAPTER 15

Mechanical

Introduction

Applicable *MasterFormat*™ Sections

Plumbing

Heating, Ventilating, and Air Conditioning (HVAC)

Fire Sprinkler Systems

Additional Reading

Acknowledgments and References

Introduction

Building mechanical systems include plumbing systems, heating, ventilating, and air conditioning systems, and fire sprinkler systems. *MasterFormat* covers all of these in its Division 15.

Applicable *MasterFormat*™ Sections

The following *MasterFormat* Level 2 sections are applicable to this chapter.

02500 Utility Services
02600 Drainage and Containment
15050 Basic Mechanical Materials and Methods
15100 Building Services Piping
15200 Process Piping
15300 Fire Protection Piping
15400 Plumbing Fixtures and Equipment
15500 Heat-Generation Equipment
15600 Refrigeration Equipment
15700 Heating, Ventilating, and Air Conditioning Equipment
15800 Air Distribution
15900 HVAC Instrumentation and Controls
15950 Testing, Adjusting, and Balancing

15.1 Plumbing

The earliest discovered plumbing systems were found in the ruins of Minoan palaces on the island of Crete. Four thousand years ago the Minoans produced plumbing systems that incorporated most of the features of systems used today, including flush toilets and vented drainage systems. It was the Romans, however, who gave plumbing its name, based on *plumbum*, the Latin word for lead, which they used for piping. Although famed more for their aqueducts, the Romans also gave their buildings both hot and cold water supply systems. Nineteenth-century England was the setting in which the flushing water closet was perfected, and Thomas Crapper was knighted for this accomplishment.

Versions of this section in earlier editions, and the illustrations in them, were originally prepared for the United States League of Savings Institutions under a research grant by the staff of the Building Research Council, University of Illinois at Urbana/Champaign, Illinois. This section and many of its illustrations have been updated for this edition, but much of the basic data remains essentially the same as it was in earlier editions.

Residential and commercial plumbing installations generally consist of a water supply system, a drainage and venting system to conduct away wastes, and plumbing fixtures and equipment. Figure 15.1-1 shows these basic systems for a residential building. Such systems for larger buildings and other building types will be much more complicated, but their basic components are the same.

15.1.1 CODES, LAWS, AND INDUSTRY STANDARDS

The design of plumbing systems is generally governed by local or statewide plumbing codes. Most of these are based on one of the following model plumbing codes:

1. The National Plumbing Code of the Building Officials and Code Administrators International (BOCAI)
2. The Standard Plumbing Code of the Southern Building Code Congress International (SBCC)
3. The Uniform Plumbing Code of the International Association of Plumbing and Mechanical Officials (IAPMO)
4. The National Standard Plumbing Code of the Plumbing-Heating-Cooling Contractors Association (PHCC)

International plumbing codes are produced by the International Code Council (ICC) of the International Conference of Building Officials (ICBO). Refer to Section 15.5 for a list and to Section 1.6 for a

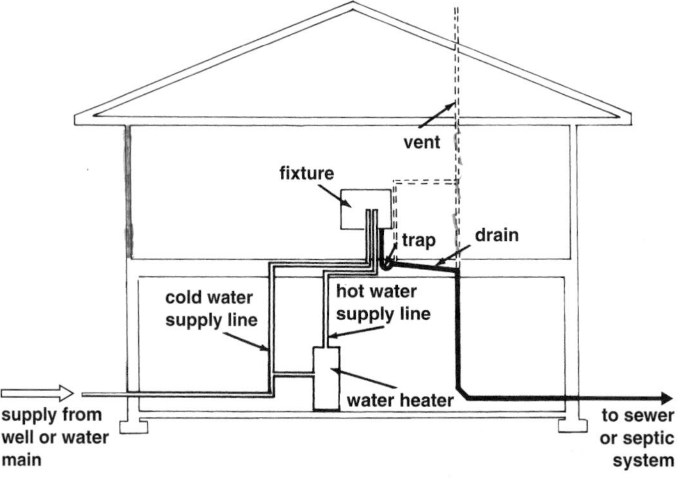

FIGURE 15.1-1 Components of a residential plumbing installation. (Courtesy Small Homes Council, Building Research Council, University of Illinois, Urbana-Champaign.)

discussion of these codes. International plumbing codes may have been adopted in some jurisdictions by the time you read this section.

The power to enforce water quality standards rests with state governments, which in turn generally adopt U.S. Environmental Protection Agency (EPA) standards. The current EPA standard is National Primary Drinking Water Standards. These regulations set maximum limits on trace metal and organic chemical content, turbidity, and chlorine residuals and establish sampling and monitoring procedures. In addition to federal regulations, some local codes impose their own limitations on impurities and in some cases mandate water conserving practices or devices.

Many industry standards have evolved to govern the manufacture and installation of plumbing pipe, fixtures, and equipment. Figure 15.1-2 summarizes the most generally accepted standards. Figure 15.1-3 lists the names of standard-setting organizations and their abbreviations.

Some types of equipment are either *listed* (placed on a listing of acceptable products) or *labeled* by a nationally recognized organization to indicate compliance with certain safety standards. Gas water heaters and pressure relief valves, for instance, should be listed and labeled by the American Gas Association (AGA). Electric and oil-fired water heaters should be listed by the Underwriters Laboratories, Inc. (UL).

Listing means that the manufacturer has agreed to comply with standards that the UL or AGA prescribes and monitors. Lists of approved products and manufacturers are published periodically by AGA, UL, and other testing organizations and are available to the public. Labels are affixed to the equipment to give graphic evidence of compliance, but labeling is not an automatic adjunct to listing. Some products may be listed but not labeled.

FIGURE 15.1-2 Industry Standards for Plumbing Pipe and Equipment

Subject or Title of Standard	Standard Designation
Water Piping Materials Standards	
ABS pipe	ASTM D 2751
Cast-iron soil pipe	ASTM A 74, C 564
Cast-iron water pipe	ASTM A 377
Clay pipe	ASTM C 700, ASTM C 425
Copper drainage tube	ASTM B 306
Copper water tube and pipe, K, L, M	ASTM B 88
Brass pipe	ASTM B 43
CPVC pipe	ASTM F 891
Galvanized steel threaded pipe	ASTM A 53
No-hub cast-iron pipe	CISPI 301
Plumbing fixtures	FS WW-P-541, ANSI A112
Polybutylene water distribution pipe	ASTM D 3309
PVC thin wall	ASTM D 2949
Stainless steel—grade TP-409	ASTM A 312
Conditioning of Domestic Water Supply	
Water softeners	WQA S-100
Water filters	WQA S-200
Sewage Disposal System Standards	
Minimum design standards for community sewerage systems	HUD 4940.3
Tanks, septic, bituminous-coated metal	UL 70
Installing vitrified clay pipe sewers	ASTM C 12
Sewage treatment plant design	ASCE manuals
Plumbing System Standards (General)	
Joints and connections	ASTM C 564, CISPI 301
Pipe protection	AWWA-C-203
Water softening	WQA S-100
Sump pumps	SSPM A
Plumbing fixtures	
Plastic bathtubs, shower receptors, and shower stalls	ANSI Z124.1 and Z124.2, NAHB Research Foundation or U.S. Testing Company, Inc., label
Domestic Water Heater Standards	
Electric hot water heaters	EEI Stds., UL 174
Gas water heaters	AGA Stds. and listed
Oil water heaters	UL 732
Water heater controls	ANSI Z21.22
Pressure-relief valves—water heaters	NBBPVI and AGA Stds.
Hot water tank construction	ANSI Z21.10
Indirect heaters	IBR Code by Hydronics Institute
Safety devices	
Energy shutoff	UL or AGA* listed and labeled
Pressure relief valves	AGA* or NBBPVI listed and labeled by AGA or ASME

*AGA listed shutoff devices and relief valves shall comply with ANSI Z21.22 and shall be so labeled.

(Excerpted from a table produced by and reprinted by courtesy of Small Homes Council, Building Research Council, University of Illinois, Urbana-Champaign.)

15.1.2 PLUMBING FIXTURES

Plumbing fixtures fulfill the basic purpose of plumbing systems; that is, they provide clean water and dispose of wastes. They also play a vital role in ensuring that a water supply remains clean by providing a physical separation between supply and drainage piping. If there were no such separation and the positive pressure in a supply line were lost, dirty water might be sucked back into the supply system (see Section 15.1.2.8).

When selecting fixtures one should consider convenience, cost, durability, ease of repair and maintenance, and noise generated. These considerations are often reflected in the type and grade of fixture. In addition to their obvious cosmetic contributions, colors and finishes may also contribute to ease of maintenance.

Plumbing, health, and handicapped-use codes have specific restrictions related to plumbing fixtures, especially, but not limited to, those in some institutional buildings, such as medical care facilities, and in other building types where access and safety are involved. Be sure to check all applicable codes carefully when selecting fixtures and designing rooms to house them. Specific degrees of floor slope may be required in a shower, for example, and curbs may be prohibited. Floor drains may

FIGURE 15.1-3 Abbreviations for Standards-Writing Organizations

Abbreviation	Organization
AGA	American Gas Association
AHAM	Association of Home Appliance Manufacturers
ANSI	American National Standards Institute
ARI	Air-Conditioning and Refrigeration Institute
ASME	American Society of Mechanical Engineers
ASTM	American Society for Testing and Materials
AWWA	American Water Works Association
CABO	Council of American Building Officials
CISPI	Cast Iron Soil Pipe Institute
EEI	Edison Electric Institute
FHA	Federal Housing Administration (now HUD)
FS	Federal Specification, General Services Administration
HUD	U.S. Department of Housing and Urban Development
NAHB	National Association of Home Builders
NBBPVI	National Board of Boiler and Pressure Vessels Inspectors
NFPA	National Fire Protection Association
SSPMA	Sump and Sewage Pump Manufacturers Association
UL	Underwriters' Laboratories, Inc.
WQA	Water Quality Association

(Courtesy Small Homes Council, Building Research Council, University of Illinois, Urbana-Champaign.)

be required in rooms housing fixtures. Specific requirements for the fixtures themselves, their hardware, and their installation will be stated in the applicable codes.

15.1.2.1 Lavatories

Lavatories come in three basic types: (1) wall-hung units with a ledge at each side, (2) units installed in or as a part of countertops, and (3) pedestal types. Figure 15.1-4 shows several typical lavatories.

Wall-hung lavatories often have a raised back ledge that serves as a backsplash. Their back and side ledges vary from 1 to 10 in. (25.4 to 254 mm) in width; at least 4 in. (100 mm) is desirable. Inside bowl dimensions vary from 10 to 16 in. (254 to 406 mm) front to back and from 13 to 17 in. (330 to 432 mm) side to side. Bowl depth varies from about $5\frac{1}{2}$ to $7\frac{1}{2}$ in. (140 to 190 mm).

Wall-hung units are mounted on a metal bracket attached to the wall. For a secure attachment to wood studs, this bracket should be bolted to a horizontal wood support mortised into the wall framing. When the walls are framed with metal studs, a lavatory's metal bracket can be bolted to a metal plate that spans several studs and is fastened securely to them. Wood or metal furring should be used as lavatory supports at masonry walls. Additional support can be provided by chrome legs under the front corners of a lavatory.

There are four kinds of *counter lavatories:* (1) a drop-in type with a raised self-rim, (2) a drop-in type with a thin-edge metal trim over the joints, (3) a type that is fastened to the bottom of the counter with the counter edge exposed and finished, and (4) a one-piece molded lavatory-counter unit.

Sometimes a single-bowl kitchen sink is installed in a bathroom instead of a lavatory for bathing infants and small children, washing hair, and doing hand laundry. Such a larger basin controls splashing. A hose spray can be added for extra convenience.

Most lavatories are made of vitreous china. Other common materials are porcelain over cast iron or steel, cultured marble (ground marble powder with plastic binder), thermoformed plastic sheet, and filled polymer (acrylic-modified polyester). Materials differ widely in resistance to impact and abrasion, to staining or charring by cigarettes, and to the spillage of cosmetics. Vitreous china and enameled metal have the most chemical resistance.

Molded counter lavatory units can be formed of cultured marble, thermoformed plastics, or filled polymer. Their characteristics vary with the types of binders or plastics used.

Some lavatories come with faucets and drain fittings; some provide appropriate holes for their insertion. Modern lavatories generally have only one spout but may have separate hot- and cold-water valves or may have a single-lever mixing valve. Lavatory faucet-drain assembles usually include a pop-up drain plug similar to those on bathtubs.

15.1.2.2 Sinks and Laundry Trays

Kitchen sinks are available with one, two, or three bowls to permit simultaneous uses. Most new residences include a two-compartment sink, having the spout high enough and one compartment large enough (usually 15 by 18 in. [381 to 457 mm]) to accommodate large cooking utensils.

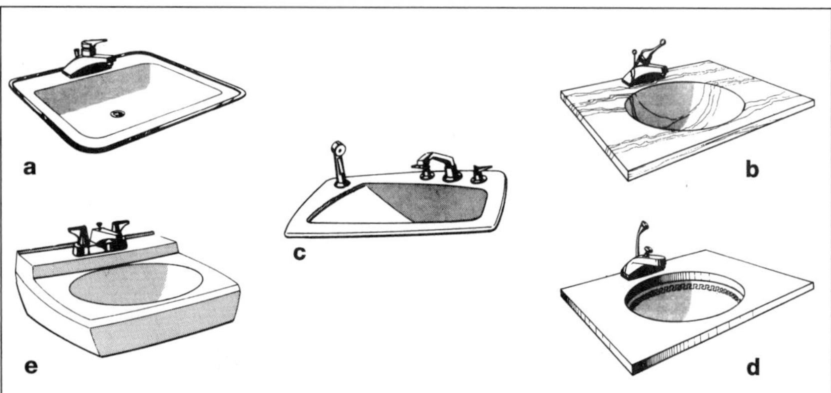

FIGURE 15.1-4 Lavatory types: (a) drop-in with stainless steel rim, (b) molded lavatory counter, (c) drop-in, self-rimming, (d) undercounter rimless, (e) wall hung.

Typical materials for kitchen sinks are porcelain-enameled cast iron or steel, stainless steel, and plastics. These units are generally designed to drop into a countertop and are supported by their rims resting on the counter. Some have self-sealing rims; others require a separate stainless steel sealing rim.

A variety of faucet types are available, either with separate hot and cold controls or with a single control and mixing valve. A diverter valve similar to those on tub-shower combinations is often installed to serve a hose-mounted spray accessory. Several appliances are available as attachments to kitchen sinks, including garbage disposal units, instant water heaters with tap, and small water coolers with tap.

Laundry sinks (laundry trays) are available in the same materials as kitchen sinks, but are larger and deeper. Some are wall-hung, others rest on legs or a trap support.

15.1.2.3 Faucets

Faucets are valves (see Section 15.1.3.3) with spouts attached. They are designed to deliver the water flowing through the valve into a fixture. Modern faucets have one spout and either one or two handles.

Faucets should be selected on the basis of compatibility with the fixture served, ease of operation, durability of finish and operating mechanism, and ease of cleaning. In choosing a faucet for a particular fixture, the spout design should be checked to ensure that when it is installed, an air gap will exist between the tip of the spout and the maximum fill level of the fixture. The height of the air gap should be at least twice the diameter of the spout opening.

A faucet handle that requires a relatively large amount of movement provides easier adjustment of water temperature and flow rate. Round knobs require finger pressure to operate, which is difficult for some people. For example, persons with arthritis in their hands and wrists prefer handles that can be operated by pressure of the palms or sides of their hands. The simplest to operate is a single-lever type that is moved from side to side for temperature control and up and down for flow rate control. Knobs that must be pulled or pushed are the most difficult to adjust.

Chrome on brass is the most durable finish for faucets and trim. Other finishes may be quite thin and subject to corrosion or wear. This is particularly true of the gold-colored finish. Finely detailed decorative handles are hard to keep clean.

Lavatory faucets can be obtained with a spray head similar to a shower or kitchen sink spray, which is useful for washing hair, pets, and children.

STEM FAUCETS

A two-handled faucet has one handle for hot water, the other for cold, and is known as a *stem* or *compression* faucet. Each handle controls a globe valve, or some variation of it, which makes this type of faucet subject to leakage problems. A frostproof outdoor faucet is another type of stem faucet.

SINGLE-LEVER FAUCETS

Single-lever faucets are often called *mixing faucets*, because both hot- and cold-water supplies are controlled by the same handle. A single-handle faucet controls the flow of hot and cold water into the spout by the matching of holes in an alignment valve. By varying the size of the matched opening on the hot- or cold-water side, the desired temperature is achieved.

ANTISCALD FAUCETS

Faucets are available with thermostatically controlled valves to maintain the desired temperature regardless of pressure, flow, or temperature variations (Fig. 15.1-5) (see also Section 1.7). These faucets may also have a safety stop so that they cannot be set hotter than 100°F (37.7°C) without first depressing a warning stop. Less expensive temperature control valves compensate for changes in relative pressure between hot and cold supplies, but not for temperature drift in either supply. Antiscald valves are available for lavatory and tub use, but are most common in shower installations.

15.1.2.4 Water Closets

Water closets have two basic parts: (1) a bowl that provides a receptacle for wastes and also forms the fixture drain trap and (2) a flushing mechanism. They are generally classified according to their flushing mechanism, their bowl type, and their mounting method. Most water closets are made of vitreous china, although some are plastic.

FLUSHING MECHANISM

Water closets are classified by their flushing mechanisms as either *flush tank* or *pressure valve* units.

Flush Tank Units Flush tank water closets are so called because they contain a tank that holds a reserve of water that drains into the bowl when called for to assist flushing. The most common tank-type water closets are (1) one-piece floor-supported, (2) two-piece floor-supported, and (3) two-piece wall-hung (Fig. 15.1-6). One-piece wall-hung models are also available.

Some one-piece models similar to that shown offer a nonoverflow feature and a cleaner appearance, but they are usually

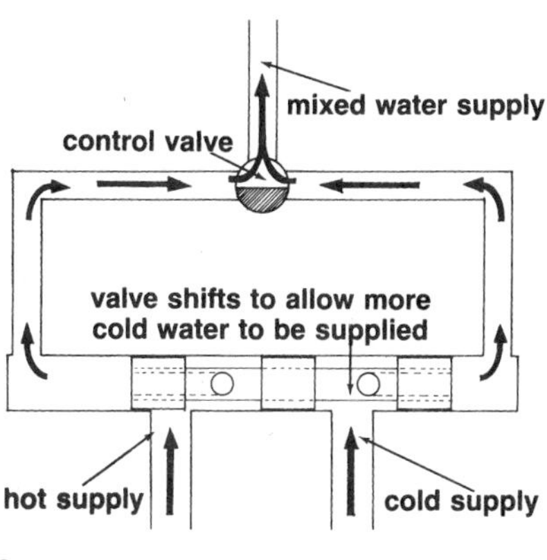

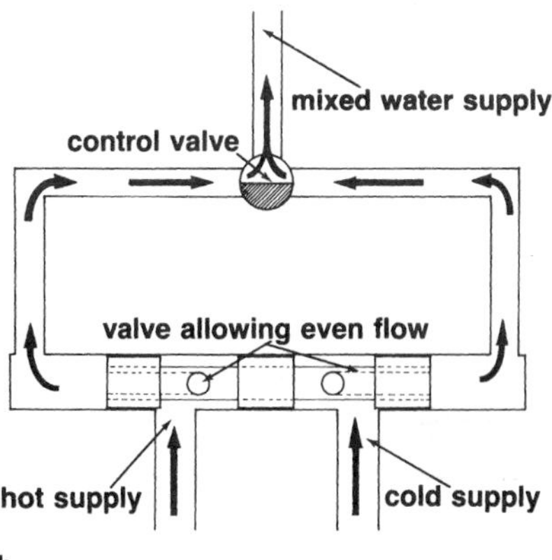

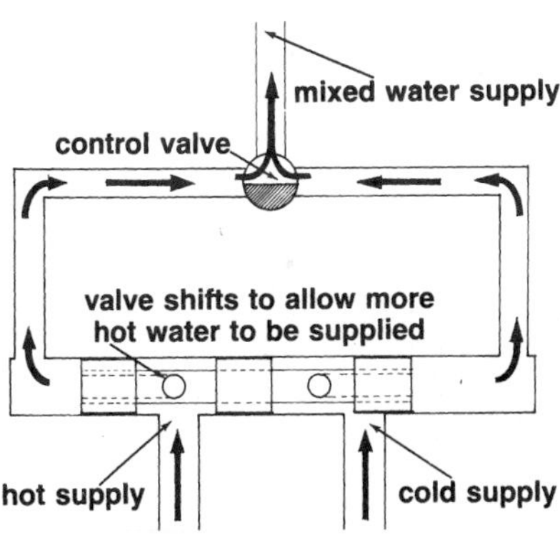

FIGURE 15.1-5 Antiscald valves maintain selected hot- and cold-water flow ratio. (Courtesy Small Homes Council, Building Research Council, University of Illinois, Urbana-Champaign.)

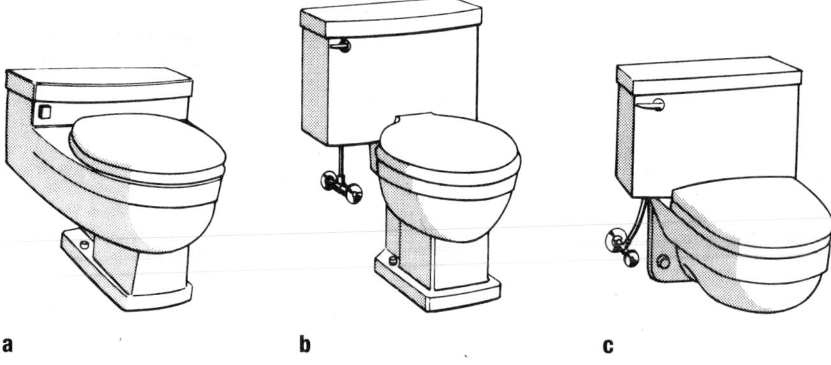

FIGURE 15.1-6 **Water closet types: (a) one-piece floor-mounted, (b) two-piece floor-mounted, (c) two-piece wall-hung.** (Courtesy Small Homes Council, Building Research Council, University of Illinois, Urbana-Champaign.)

more expensive to purchase and install than standard two-piece units. Standard two-piece units have the tank supported on the bowl. In older water closets, the bowl is generally floor-supported and the tank hangs on the wall.

Submerged in the tank of a flush tank water closet are a number of control mechanisms: a flushing valve, a water supply valve, and a combination bowl-refill and tank overflow device (Fig. 15.1-7). The flushing valve is essentially a stopper, or flapper, at the bottom of the tank, which is lifted when the flush lever on the outside of the tank is depressed. This allows the water in the tank to drain into the bowl to promote flushing.

TANK FLUSHING t = 10 sec.
flush handle
overflow pipe
fill valve
water level
flush ball
float
water supply on
water flow out
TANK FILLING t = 5 sec.
flush handle
water level
float
water supply on
water flow stopped
t = 2 min.
flush handle
float
water level
water supply off
water flow stopped
BETWEEN FLUSHES

FIGURE 15.1-7 **Flush tank water closets are favored for homes because they are simple and quiet. Flushing of a tank is shown, starting from the top.** (Courtesy Small Homes Council, Building Research Council, University of Illinois, Urbana-Champaign.)

Flush tank supply valves are available in several designs, all of which have a means of sensing when the water reserve is being lost due to action of the flushing valve. The *ballcock valve* is controlled by the position of a float ball riding on the surface of the water in the tank (see Fig. 15.1-7). As the water flows out, the float falls, thereby opening the ballcock supply valve. As the tank refills, the float rises and eventually shuts off the valve.

A *fluid-level control valve* works similarly to a ballcock valve, except that it has a float mounted on the supply-inlet post, rather than on an arm extending from it (Fig. 15.1-8). It is simpler and quieter than a ballcock valve.

A small tube extending from the high-water level in a flush tank down into the bowl acts as an overflow and refill device. A spur from the water supply line that fills the tank feeds into this tube, refilling the bowl, thereby restoring the trap seal after every flush. Should the water supply valve fail to shut off properly, this tube acts as an overflow drain for the tank.

bowl fill line
overflow pipe
float
fill valve
tank fill line

FIGURE 15.1-8 **Fluid-level control valve is quieter than ballcock mechanism.** (Courtesy Small Homes Council, Building Research Council, University of Illinois, Urbana-Champaign.)

Pressure Valve Units The flushing of a *pressure valve* water closet (also called a *flush valve* unit) is controlled by a *pressure control fill valve* (sometimes called a *flush valve*), which operates on pressure in the serving water line rather than on the pressure of water in a tank. These units are rarely installed in residences due to their noisiness but are sometimes used in areas with limited water supplies (Fig. 15.1-9). Because of their ease of maintenance and low water consumption, pressure valve water closets are frequently installed in commercial installations where quietness is not critical and sufficient water pressure is available. Pressure valve units are available in both floor and wall-hung models.

TANKS

Traditional flush tank units discharge 5 to 7 gal. (18.93 to 26.5 L) of water at each flushing action. Federal law now requires that all new units function with no more than 1.6 gal (6 L) per flush.

In summer, moisture can condense on the outside of a tank when the room humidity is high and the supply water is cold. Tanks are available with factory-installed insulation inside them to minimize condensation. Other remedies for condensation include using a tempering valve to warm the cold water entering the tank, installing interior insulation in the field, providing a decorative fabric jacket around the outside of the tank, and installing a tray at the bottom of the tank to collect the condensate and drain it into the bowl.

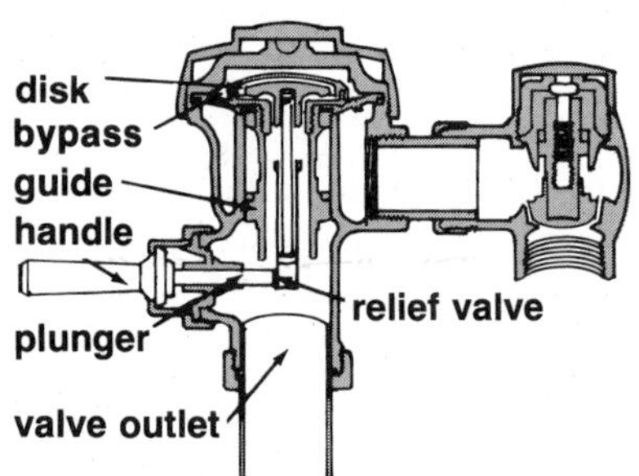

FIGURE 15.1-9 **Pressure valves are noisy but use little water. They are common in commercial buildings.** (Courtesy Small Homes Council, Building Research Council, University of Illinois, Urbana-Champaign.)

BOWLS

There are several types of bowls in general use today:

1. A *siphon jet bowl* is most frequently used in residential construction (Fig. 15.1-10). In this bowl, a jet of water in the trapway initiates the siphon action that empties the bowl. Its advantages are quietness and effectiveness.
2. A *siphon action bowl* is similar to a siphon jet bowl but holds more water and thus reduces where water does not cover the bowl (the dry-fouling area).
3. A *reverse trap bowl* works on the same principle as a siphon jet one; it uses less water and is less expensive, but is noisier (Fig. 15.1-11).
4. A *blowout bowl* flushes by the force of a high-pressure jet in the trapway rather than by siphonage. Blowout bowls are used primarily in commercial buildings due to their noisiness and requirements for high water pressure.

MOUNTING

Water closets are either floor supported or wall hung. Some flush tank units have floor-supported bowls and wall-hung tanks.

Wall-hung units are supported by a metal carrier concealed in the wall. An advantage of wall-hung water closets is that the floor under the tank is more accessible for cleaning.

The height of floor-supported seats normally ranges from 14 to 18 in. (356 to 457 mm), but a 20-in. (508 mm) seat height may be desirable for elderly or handicapped persons. This can be achieved by use of an extension seat on floor-supported fixtures or adjustment of the carrier height of wall-hung fixtures.

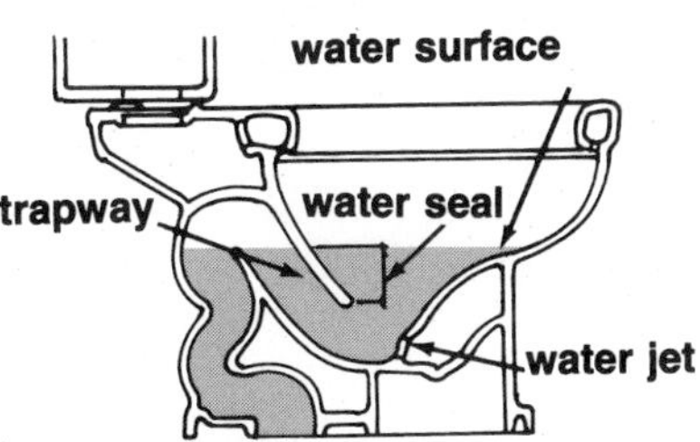

FIGURE 15.1-10 **A siphon jet bowl is relatively quiet and has an effective scouring action.** (Courtesy Small Homes Council, Building Research Council, University of Illinois, Urbana-Champaign.)

15.1.2.5 Bathtubs

Bathtubs are made in various shapes, sizes, and materials.

SHAPES AND TYPES

Three common shapes are available, as shown in Figure 15.1-12. The square and sunken types are more difficult to clean than rectangular and above-floor types. Typical overall dimensions for rectangular tubs are 14 to 16 in. (356 to 457 mm) high, 28½ to 32 in. (724 to 813 mm) front-to-back width, and 60 in. (1524 mm) long. Longer, shorter, and wider tubs are available on special order and at additional cost.

Most residential bathrooms employ a recessed tub with walls on three sides. When walls are available on two sides, a corner tub is used. In custom houses, freestanding tubs are sometimes used; many of these are either surrounded by a raised platform or sunk into the floor. One-piece panelized tub-shower assemblies are available in several types of plastic. They provide low-maintenance shower enclosures with built-in conveniences.

MATERIALS AND FINISHES

The most common and durable tub material is cast iron with a thick porcelain-enamel finish. Tubs of this type weigh approximately 200 lb. (90.72 kg) and require additional floor framing members beneath them when they are placed on a wood-framed floor.

Tubs also are made of steel with a porcelain-enamel finish. These are more prone to enamel chipping and rusting than the cast-iron variety. Because they are lighter, they are also noisier when water strikes them. Some models are available with an acoustical treatment on the underside to reduce water-impact noise. Steel and plastic tubs are lighter and less rigid than cast-iron ones.

Some tubs are made from synthetic marble, acrylic-modified polyester, or similar homogeneous materials. Special shapes and sizes can be made using these materials, including such unusual shapes as round and octagonal ones. Such tubs frequently have whirlpool or hot-tub devices attached to or built into them.

Plastic tub assemblies are made (1) of glass fiber–reinforced polyester (FRP) with a smooth gel-coat surface or (2) of a molded acrylic sheet. Built-in ledges are provided in most plastic molded assemblies. Ledges are handy for storing items used during bathing, but the rim to be stepped over should be narrow to make entry into the tub as easy as possible.

Tubs commonly have two openings for plumbing connections: one for the bottom drain and one for an overflow drain. The overflow opening is usually an inconspicuous slot under the edge of a chrome plate mounted high on the tub wall. The overflow drain protects against two hazards associated with overflowing fixtures: water damage to the building and siphonage of the trap seal. Tubs with whirlpool action devices will, of course, require additional openings to accommodate those devices and the circulation of water.

TUB FITTINGS

Tub drains are often bothersome because they either do not completely seal or do not drain quickly. Slow drainage is often

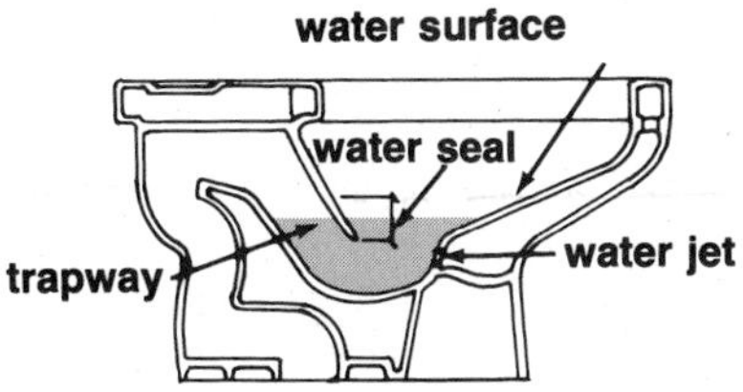

FIGURE 15.1-11 **Reverse trap bowls use less water and are less expensive than siphon jet bowls, but are also noisier.** (Courtesy Small Homes Council, Building Research Council, University of Illinois, Urbana-Champaign.)

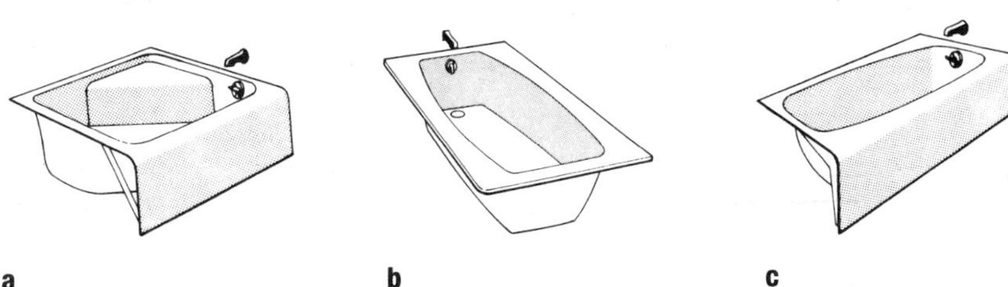

FIGURE 15.1-12 **Bathtub types: (a) recessed square, with walls on three sides, (b) sunken rectangular, and (c) recessed rectangular, the most common type.** (Courtesy Small Homes Council, Building Research Council, University of Illinois, Urbana-Champaign.)

caused by hair collecting on a drain plug mechanism of the type that is controlled remotely from a lever mounted in the overflow plate. In older versions of this mechanism, the lever raises and lowers a brass cylinder which, when lowered, blocks off the fixture drain (Fig. 15.1-13). In newer versions, the lever raises and lowers a vertical arm, which in turn controls a horizontal arm with a pop-up drain plug at the end (Fig. 15.1-14).

15.1.2.6 Showers

A shower unit consists of a showerhead mounted either in an enclosed bathtub or in a shower stall. Showerheads are sprayers that either have a preset spray pattern or can be adjusted from fine mist to heavy spray. Some models offer built-in flow restrictors, which conserve water. Some units also have a shutoff valve on the head, so the water can be turned off during soaping (Fig. 15.1-15a). This is an advantage also in that the temperature setting on the main supply valve need not be disturbed.

A hand-held spray head on a flexible metal-covered hose can be used instead of a fixed showerhead (Fig. 15.1-15b). It can be hung on a permanent bracket, on a sliding bracket, on a support bar to allow adjustment for various heights, or can be hand-held. When such heads are installed over tubs or lavatories, a vacuum breaker should be provided or the showerhead should be hung in such a way that at least a 1-in. (25.4 mm) air gap will be ensured when it is hanging fully extended. This prevents back-siphonage of bath water and contamination of the water supply, should the spray head fall into the water.

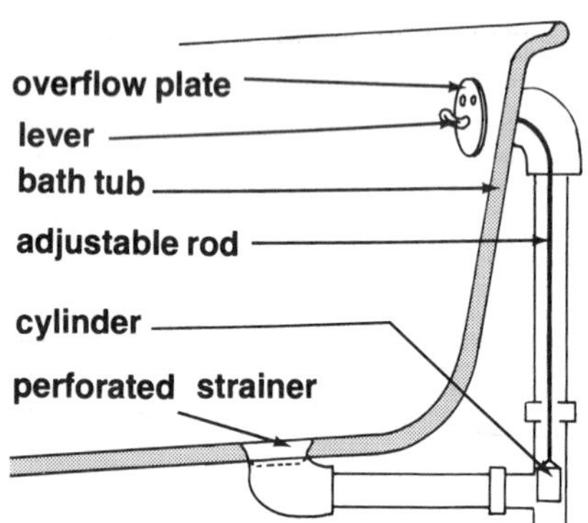

FIGURE 15.1-13 Trip-lever drain uses a cylinder on an adjustable rod to open or close the drain opening. (Courtesy Small Homes Council, Building Research Council, University of Illinois, Urbana-Champaign.)

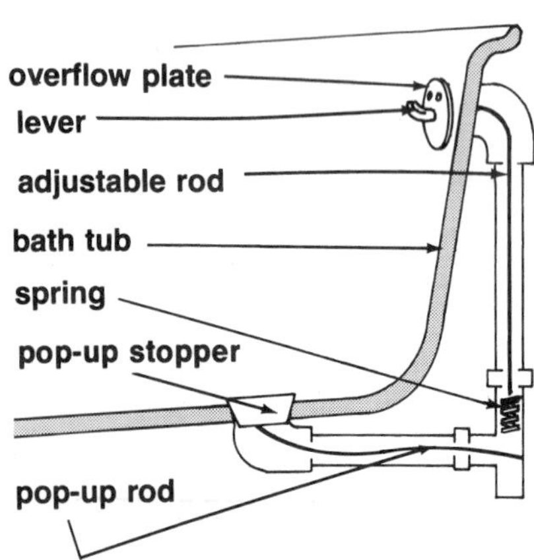

FIGURE 15.1-14 Pop-up drain opens when spring-loaded lever pushes on end of pop-up rod. (Courtesy Small Homes Council, Building Research Council, University of Illinois, Urbana-Champaign.)

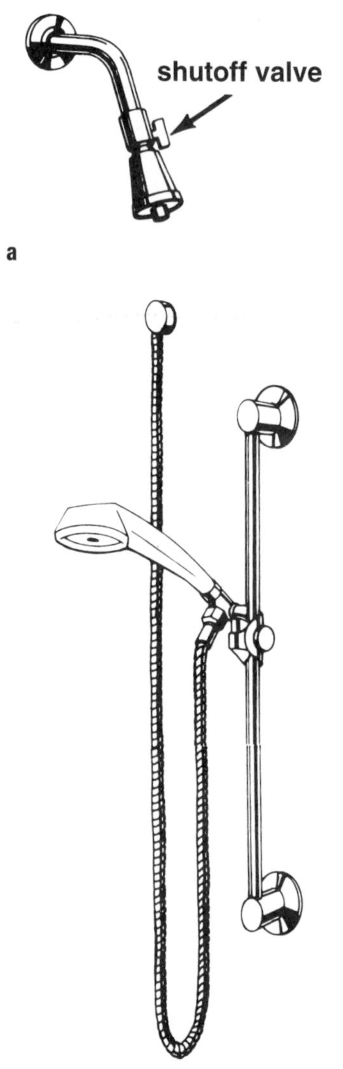

FIGURE 15.1-15 (a) Shutoff valve on showerhead; (b) phone-type adjustable showerhead. (Courtesy Small Homes Council, Building Research Council, University of Illinois, Urbana-Champaign.)

TUB-SHOWERS

Showers installed over a bathtub usually share the tub's water supply valve. The *tub-filler* spout includes a *diverter valve* that redirects water coming to the spout to the shower head instead. Adding a shower to a bathtub also requires that a waterproof enclosure be provided up to the height of the showerhead so that splatter from the shower is directed into the tub and down the drain. On the three walls surrounding the tub, this can be accomplished with metal or plastic panels specially made for this purpose, or with ceramic tile. The risk of leakage is reduced if panel systems without corner joints are used.

In ceramic tile tub enclosures, either cement backer board or water-resistant gypsum board should be used as the backing (see Section 9.6). Care must be taken that grouting between tiles is sound, so as to avoid leaks.

An alternative that eliminates joints and potential leaks from a tub enclosure is a one-piece molded plastic tub-shower unit (Fig. 15.1-16a). The size of such units dictates that they be placed before the bathroom walls are framed. The fourth side of the tub must also be enclosed, either with a shower curtain or with a safety-glazed door.

SHOWER STALLS

Showers not installed in conjunction with a bathtub have their own stalls built around them. Most shower stalls consist of a floor receptor topped by water-resistant walls, either self-supporting or attached to the wall framing. Prefabricated receptors are made of precast concrete, terrazzo, preformed metal, or plastic. Job-built receptors of ceramic tile can also be used if a sloping base and a waterproofing pan are laid beneath them. Complete plastic receptor and wall combinations are available, both as components and as one-piece units (Fig. 15.1-16b).

The fourth side of the shower stall can be enclosed with a curtain, but a higher-than-normal curb is necessary to prevent it from blowing out and allowing water to run onto the floor. Sliding or bifold doors with safety glazing are available for wide stalls. Hinged doors are common, but if they open outward, they should have a

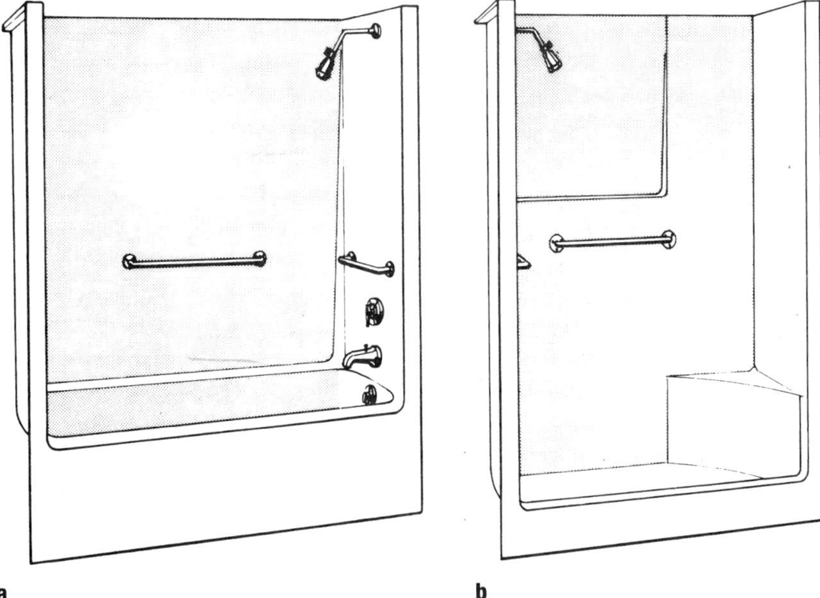

FIGURE 15.1-16 (a) One-piece molded plastic tub shower; (b) one-piece plastic shower. Construction permits inclusion of conveniences, such as soap dishes and seats. (Courtesy Small Homes Council, Building Research Council, University of Illinois, Urbana-Champaign.)

provision to minimize water dripping onto the floor.

A common small shower stall size is 30 in. (762 mm) square. A more convenient stall size is 36 in. (914 mm) square or 36 by 48 in. (914 by 1067 mm). Complete assemblies are about 84 in. (2134 mm) high.

A shower stall offers easier entry and exit and more secure footing than a shower in a tub, especially when provided with grab bars. Portable suction mats or slip-resisting tapes can be used if the floor finish is not slip resistant.

15.1.2.7 Drinking Fountains

Drinking fountains may be simply a type of faucet arranged for ease in drinking, but most deliver chilled water. The water may be chilled in a self-contained unit (Fig. 15.1-17), or it may be cooled by a remote chiller. Units are available either freestanding or wall mounted. Recessed and semirecessed units are available, as are pedestal units and other special designs. Both indoor and outdoor units are available.

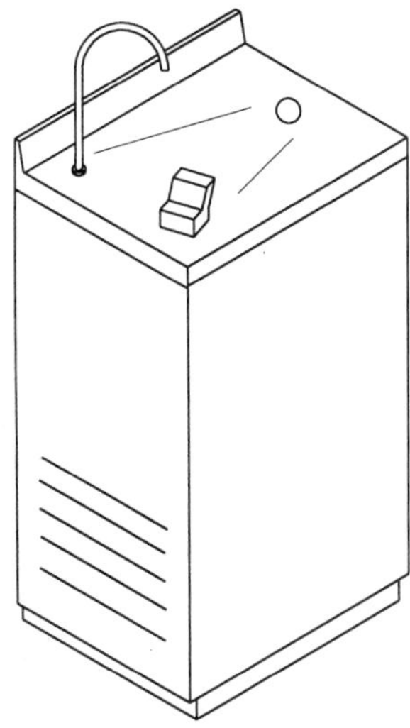

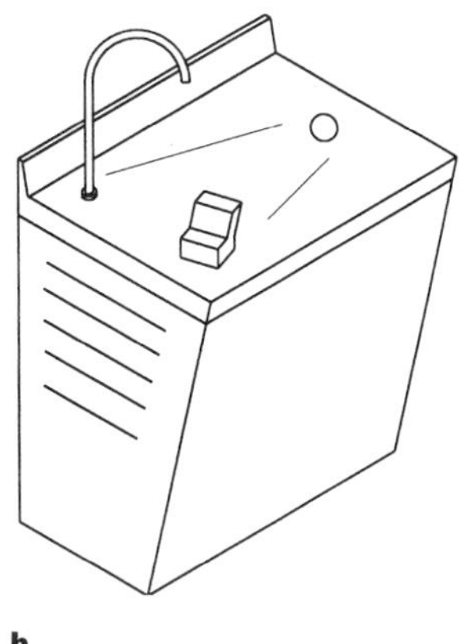

a b

FIGURE 15.1-17 Water coolers. (a) Floor model; (b) wall mounted. (Drawing by H. Leslie Simmons.)

15.1.2.8 Separation of Systems

Most codes demand that plumbing supply and drainage systems be separated to prevent the cross flow of waste fluids into the supply system and vice versa. Contamination of a supply system can have severe effects on the health of occupants.

Water flows into a fixture from a supply system, is used by humans, and then drains out of the fixture into a drainage system. Fixtures are designed to ensure that the flow of water through them is in one direction only, so that drain water is never allowed back into the supply pipes. The unintentional flow of contaminated water into a supply source is called *backflow* or *back siphonage*. This can occur when the pressure in a supply pipe drops below that of the fluid, be it air or water, in which the supply pipe terminates. This pressure reversal cannot occur when the supply water pressure is between 40 and 50 psi (276 to 345 kPa), as is usual, and each supply line ends in air, which has a pressure of about 14 to 15 psi (97 to 100 kPa).

Therefore, plumbing codes require that fixtures be designed so that an air gap separates the supply lines from the highest point at which contaminated water can collect (Fig. 15.1-18). An *air gap* is an unobstructed vertical distance between the

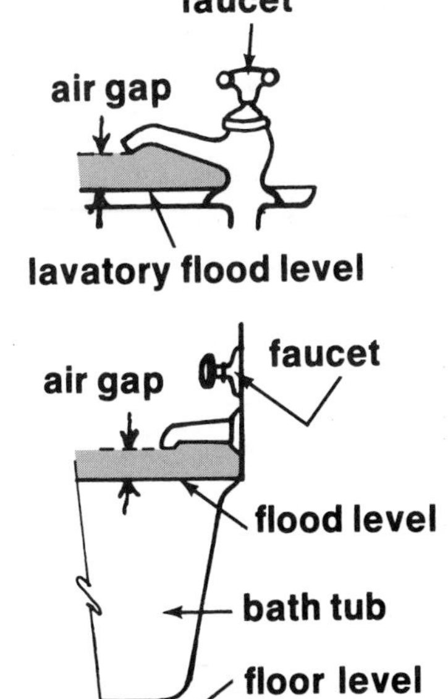

FIGURE 15.1-18 The air gap between overflow level and spout prevents contamination of the water supply. (Courtesy Small Homes Council, Building Research Council, University of Illinois, Urbana-Champaign.)

mouth of a water outlet and the flood-level rim of the water receptacle. The water outlet may be a faucet, spout, or other outlet; the receptacle may be a plumbing fixture, tank, or other receptacle. In some cases, the highest point that contaminated water can reach is permitted to be the level of the overflow drain of a fixture. To ensure that the suction created in the pipe during reversal of flow is not strong enough to draw contaminated water across the air gap between the end of the supply line and the water surface, the required minimum air gap is usually two times the diameter of the faucet or a minimum height of 1 in. (25.4 mm) for lavatories, 1½ in. (38 mm) for sinks and fixtures having ¾-in. (19 mm) bathroom faucets, and 2 in. (50.8 mm) for fixtures having 1-in. (25.4 mm) bath faucets.

Where provision of an air gap is not feasible in fixture design, a device called an *atmospheric vacuum breaker or siphon breaker* is installed. This is a simple mechanical device consisting essentially of a check valve in a supply line and a valve member (on the discharge side of the check valve) opening to the atmosphere when the pressure in the line drops to atmospheric pressure (Fig. 15.1-19). This device acts as a one-way valve, allowing water to flow through it in one direction, but if flow is reversed, it opens, allowing air to enter instead of permitting the reverse flow of water. Vacuum breakers are generally required on dishwashers and other water-using appliances that otherwise lack an air gap.

15.1.3 WATER SUPPLY SYSTEMS

A *water supply system* includes everything used to obtain water from a source, treat it to make it drinkable, and deliver it to its point of use. This section discusses available sources of water and its distribution both to and within buildings.

15.1.3.1 Conservation

Until recently, water was used without regard for future supplies. Today, as groundwater supplies in many areas dwindle, water is regarded as a limited natural resource. Using it wastefully in plumbing systems not only incurs municipal costs in purifying it for consumption, in pumping it to the point of use, and in treating it before discharge into a body of water, but reduces the amount available in underground aquifers. *Aquifers* are large underground stores of water, contained either in large cavities or distributed in smaller voids of porous, granular soils.

Although the earth has an enormous reserve of water, 97% of it is salty and more than 2% is trapped in polar icecaps and glaciers. Therefore, less than 1% of all the earth's water is available to us in lakes, rivers, and underground aquifers. The rapid urbanization and industrialization of many areas increase demand on the aquifers and slow the replenishment process, wherein melting snow and rainwater soak back into the ground. Surface waters are rushed from streets and parking lots to storm sewers and thence to salty oceans or polluted freshwater lakes. This water is lost as a valuable resource until it comes back in the form of rainwater or is purified for reuse.

Spurred by recurring rainfall shortages, groundwater depletion, and the high cost of water treatment, water conservation has become a regional and national issue. Although the design of facilities and selection of water-saving equipment are essential, conservation depends largely on the habits and decisions of the people using the water.

Some local building codes require the use of water-conserving devices or fixtures. Flow restrictors installed on existing showerheads, for instance, can cut the flow rate by about 50%. New energy-saving showerheads are fairly nominal in cost, however, and can cut water flow from the usual 5 gpm (18.9 L/min) to between 1.8 and 2.5 gpm (6.8 to 9.5 L/min).

Older water closets required between 5 and 7 gal. (18.93 and 26.5 L) per flush. The U.S. government currently restricts new water closets to those that use no more than 1.6 gal. (6 L) of water per flush. Plastic bottles or dams and other devices specifically manufactured for this purpose inserted into older water closet tanks can reduce water consumption, but energy-saving water closets are generally much more effective than modified older fixtures because they are designed to provide equivalent cleansing action while using less water.

15.1.3.2 Sources

Typically, water originating in lakes, streams, or underground sources is piped to treatment plants, where it is purified. From there it is conducted through a network of water mains to end users such as residences, businesses, and industries (Fig. 15.1-20).

Modern water supply systems provide potable water for drinking, cooking, and bathing, and for use in conducting away wastes. *Potable water* is water of sufficient quality to be drinkable. Water may be naturally potable or may be made potable by treatment.

Water entering a building's plumbing system generally comes from either a municipal or community water system or a private well. Water from these sources may differ substantially as to quality, pressure, and reliability of supply.

MUNICIPAL AND COMMUNITY SYSTEMS

Water for municipal and community systems comes from rivers, reservoirs, wells, or a combination of these. Federal, state, and local health standards are concerned with the safety of water for human consumption, not its usefulness for other purposes or its interaction with common plumbing materials. Many municipalities, however, treat their water supply not only to make it safe to drink but also to clarify it, soften it, and reduce its corrosiveness.

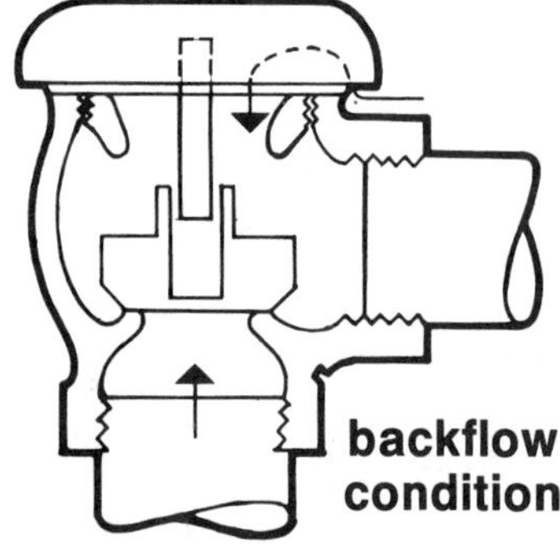

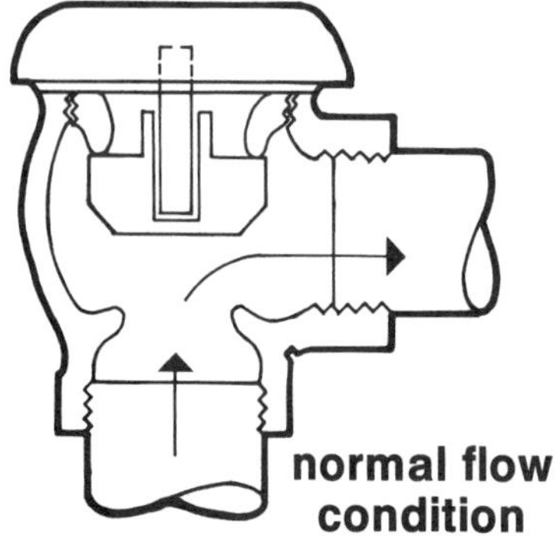

FIGURE 15.1-19 A vacuum breaker uses atmospheric pressure to close the valve when the pressure is reduced on the left side and to reopen it when higher-than-atmospheric pressure is restored on the left side. (Courtesy Small Homes Council, Building Research Council, University of Illinois, Urbana-Champaign.)

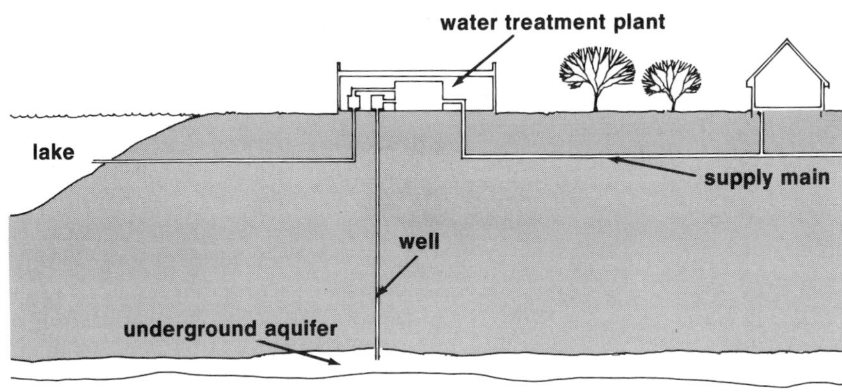

FIGURE 15.1-20 A water treatment plant must be strategically located between the users and the source, which may be an underground pool (aquifer) or lake.

Alum, activated carbon, and coagulant aids are added to remove suspended materials from surface water supplies and to control taste and odor problems. In many localities, fluoride is added to reduce the incidence of tooth decay, and chlorine is added to control bacterial contamination. The water may also be aerated to remove odors and dissolved gases. Equipment is available to perform similar water treatment procedures on a small scale for private well water.

PRIVATE SYSTEMS

Private wells are the sources of domestic water in locations where community or municipal water is not available. Private wells are also used to supply water for irrigation even when community or municipal systems are available. Well water must be tested and treated if not potable. Treatment for other problems is addressed in Section 15.1.3.5.

15.1.3.3 Building Water Supply System Design

A *water supply system* within a building consists of service piping, distribution piping, branches to plumbing fixtures and appliances, and the necessary fittings and control valves. Section 15.1.2 discusses plumbing fixtures. Chapter 11 discusses appliances that receive water and may or may not discharge liquid or solid waste, including, but not limited to, dishwashers, clothes washers, and some refrigerators.

The design of a building's water supply system includes laying out the system and selecting and sizing pipes, flow-control and pressure-regulating devices, and valves.

A water supply system is always filled with water at greater-than-atmospheric pressure. Once in a building, it is separated into two parts. The first forms a cold-water supply tree. The second goes through a water heater and becomes a hot-water supply tree. Each of these trees then feeds the appropriate plumbing fixtures and appliances. The trees are composed of vertical supply pipes called *risers* and horizontal pipe runs called *branches* that are connected to the risers. Plumbing fixtures may be connected to risers, but more frequently are connected through branch runs.

The basic goals in the design of a water supply system are to:

1. Keep the water from becoming contaminated;
2. Maintain adequate pressure and flow rate for operation of plumbing fixtures and appliances;
3. Prevent excessive flow and consequent waste of valuable water;
4. Prevent excessive pressures, velocities, corrosion, and scale buildup;
5. Provide water without noise, fixture staining, or odor.

The efficient functioning of a water supply system depends on pipe sizing, fixture and materials selection, and installation.

CONSUMPTION

Consumption levels for residential and small commercial buildings are generally estimated from past experience with similar occupancies. The number, frequency, and pattern of use of various plumbing fixtures are estimated, based on the number and type of occupants. From this estimate, the total building demand may be determined (Fig. 15.1-21).

For larger installations, a probability method is used, which takes into account the fact that not all fixtures will be in use at any one time. Consequently, the probability method assigns a weighted mathematical factor, known as a *water supply fixture unit* (w.s.f.u.) rating, to each fixture in proportion to the load it imposes on the water supply when in use (Fig. 15.1-22). The w.s.f.u. values for all fixtures in the building are summed and converted to a water demand figure by a graph that expresses the probability factor of simultaneous use of various fixtures (Fig. 15.1-23), taking into account volume, duration of flow, and intervals between operations.

FIGURE 15.1-21 Water Consumption Estimating Guidelines

For estimating domestic water consumption, demand rates, and hot water storage capacity, the following rules of thumb are often used for single-family dwellings:

- C_{av}*/day/per*, cold water consumption per day per person = 100 gal. (378.54 L)
- H_{av}*/day/per*, hot water consumption per day per person (in large house) = 50 gal. (189.27 L)
- H_{max}*/hr/per*, maximum hourly hot water demand per person = 15% × H_{av}*/day/per* = 7.5 gal. (28.39 L)
- H_{av}*/hr/per*, average hourly hot water demand per person = H_{av}*/day/per* ÷ 24 = 2.1 gal. (7.95 L)
- $H_{stor'}$, hot water storage capacity (includes 20% of tank volume as air space) = 1.6 (H_{max}*/hr/per* – H_{av}*/hr/per*) = 8.6 gal. (32.55 L)

Using these rules of thumb, the estimated consumption of a household containing 5 persons would be 500 gal. (1892.71 L) of cold water and 250 gal. (946.35 L) of hot water per day. The maximum hourly hot water demand would be estimated at 37.5 gal. (141.95 L); and the average hot water demand at 10.4 gal. (39.37 L). The hot water heater storage tank capacity would be estimated at 43.3 gal. (163.91 L), so a 50-gal. (189.27 L) tank would be specified. Manufacturer's information on specific water heaters should be used for exact sizing.

(Courtesy Small Homes Council, Building Research Council, University of Illinois, Urbana-Champaign.) Metric added by author.

FIGURE 15.1-22 Fixture Unit Ratings and Water Flow

Fixtures	Fixture Units	Water Flow, gpm (L/m) Cold	Hot
Water closet, flush valve	10	30–45 (113.56)	0
Water closet, tank type	5	18 (68.14)	0
Kitchen sink	2	5 (18.93)	5 (18.93)
Lavatory	1	5 (18.93)	5 (18.93)
Bathtub	2	6 (22.71)	6 (22.71)
Shower stall	2	6 (22.71)	5 (18.93)
Laundry tray	3	10 (37.85)	10 (37.85)
Clothes washer	3	5 (18.93)	5 (18.93)
Dishwasher	3	5 (18.93)	5 (18.93)

(Courtesy Small Homes Council, Building Research Council, University of Illinois, Urbana-Champaign.) Metric added by author.

SYSTEM LAYOUT

The physical configuration of a water supply system should take into consideration the efficiency of the materials used, simplicity of construction, protection from freezing, and ease of maintenance and repair.

A primary principle in efficient layout of water supply systems is that all runs should be as short, straight, and as direct as possible and have as few fittings as possible. Following this principle not only conserves materials, but also reduces internal friction in the pipes. This in turn keeps to a minimum pressure loss associated with flow from the water main to the fixtures.

In the schematic planning stage of a building, an effort should be made to locate rooms that will contain plumbing fixtures as near as possible to the point of entry of the water supply to the building and the point of exit to the sewer line. If possible, rooms that will contain plumbing fixtures should be clustered so that plumbing fixtures in more than one room share the same wall for pipe distribution (Fig. 15.1-24). This can be done by arranging the rooms so that fixtures in two or more rooms are placed back-to-back against the same wall or are vertically aligned or stacked on different floors (Fig. 15.1-25).

Jogs and offsets in plumbing walls containing horizontal lines should be avoided so as to simplify the plumbing lines. Fixtures in a given room should be arranged so that all of their supply, drain, and vent lines fall in one wall. This practice shortens the piping runs and simplifies framing of the building by reducing the number of penetrations through floors and eliminating the need to furr or double frame more than one wall (Fig. 15.1-26).

Walls used in this way for vertical distribution of plumbing are known as *chase* walls, *wet* walls, or *plumbing* walls (Fig. 15.1-27). They are usually thicker than typical interior walls, allowing sufficient space to accommodate the drainpipe plus one supply pipe crossing behind the drainpipe. In most wood-framed construction, chase walls are built with 2 × 6s (50 × 150s) rather than the typical 2 × 4s (50 × 100s).

When more space is needed, a double wall, each wall built with wood 2 × 4s (50 × 100s) or metal studs, is used. Sometimes it is necessary to separate the two walls of a double-walled chase to provide even more space. The double wall method is usually used when the partitions are framed with metal studs. The two metal stud walls forming a plumbing chase are required to be tied together and braced, either with sections of stud or with pieces of gypsum board (see Section 9.5).

Future expansion of a supply system can be facilitated by including the appropriate tees with plugs in the original installation or *roughing in* the plumbing for fixtures to be added at a later time. Roughing in means installing those parts of a plumbing system that can be completed before plumbing fixtures or appliances are installed. Such parts include water supply, drainage, and vent piping and the necessary fixture supports.

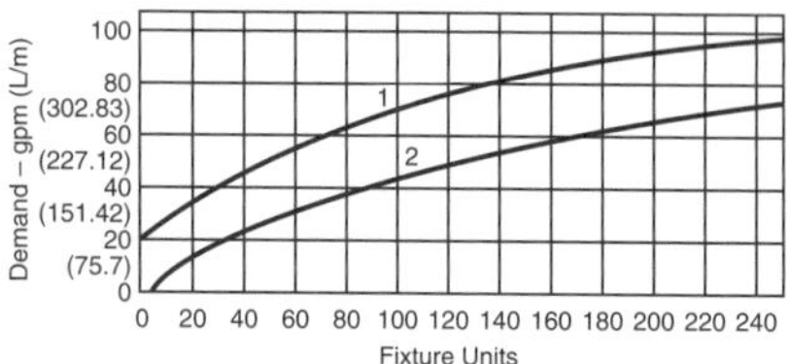

FIGURE 15.1-23 Water demand curves. Locate fixture units on curve and read demand on left scale. Curve 1 is for buildings with flush-valve water closets; 2 is for tank type. (Courtesy Small Homes Council, Building Research Council, University of Illinois, Urbana-Champaign.) Metric added by author.

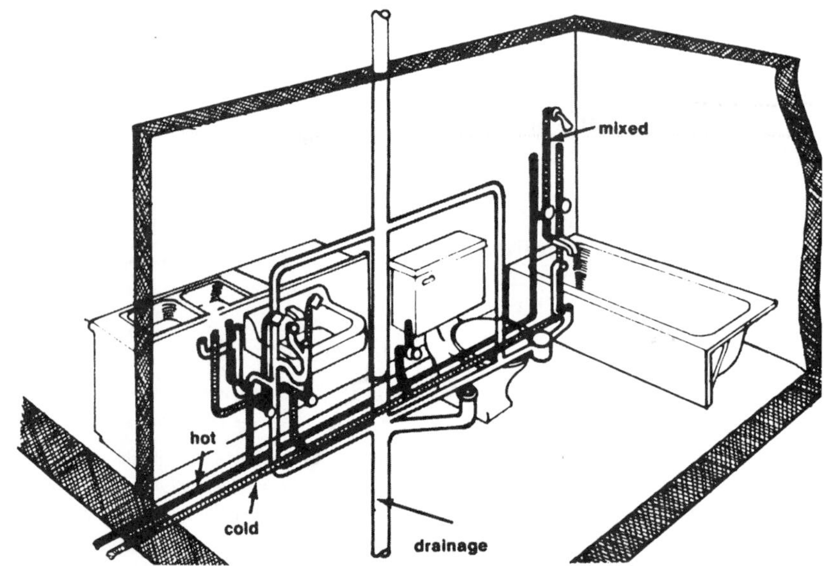

FIGURE 15.1-24 Back-to-back plumbing arrangement is economical and also separates noise-producing equipment from living spaces. (Courtesy Small Homes Council, Building Research Council, University of Illinois, Urbana-Champaign.)

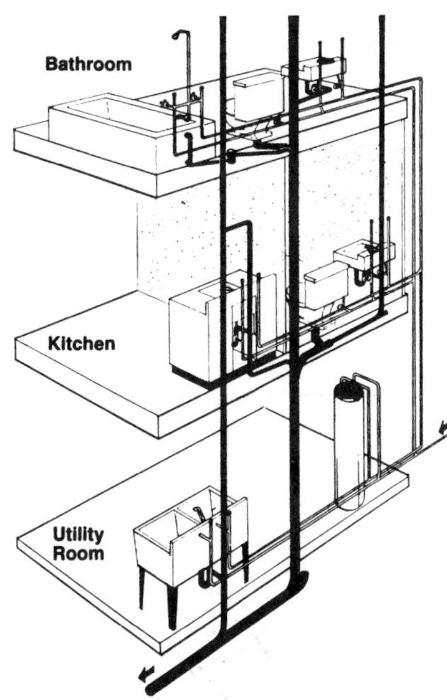

FIGURE 15.1-25 Vertical alignment of fixtures conserves piping and framing. (Courtesy Small Homes Council, Building Research Council, University of Illinois, Urbana-Champaign.)

To reduce the probability of pipes freezing, supply lines should not be located in exterior walls. If such location is unavoidable, the supply pipes should be run on the inside face of the wall with at least 3½ in. (90 mm) of mineral-fiber, or equivalent, insulation separating them from the outside (cold) face of the wall and with no insulation on the warm side. This practice keeps pipes near the heated interior of the building and slows heat loss from the pipes to the exterior. Every effort should be made to keep water-carrying pipes out of areas subject to freezing. In practice, this means providing freeze-proof valves for outdoor sillcocks, providing drain valves for seasonal drainage of outdoor piping, and locating supply pipes away from exterior walls.

When laying out a water supply system and fixture locations, the location of the resulting drainage system must also be considered. Care should be taken in locating framing so that a plumbing wall will not fall in the same plane as roof trusses or floor or ceiling joists, which would obstruct the passage of supply, drain, and vent lines in and out of the plumbing wall. Likewise, if major framing elements run perpendicular to the plumbing wall, care should be taken that the fixtures requiring below-floor drains are located so that the drains do not conflict with the floor framing. Wood joists should not be notched in the middle third of the span, and holes should not be larger than 2½ in. (63.5 mm) in diameter.

During the design process, thought should also be given to the future accessibility of plumbing for repairs. In buildings with crawl spaces, sufficient height should be provided for a person to conveniently access the piping. The minimum crawl space depth required by many codes is 18 in. (457 mm); however, this is the depth required for moisture protection, rather than for access to equipment in the crawl space. For easy access, a crawl space should be at least 2 ft. (610 mm), and preferably 3 ft. (914 mm), high. Access panels into plumbing walls simplify repairs and are usually required by code for access to valves.

Cold-water systems are generally noncirculating, meaning that the water travels to the fixtures and remains stationary in the pipes when no fixture is being used. Hot-water systems may be either noncirculating or circulating. In a noncirculating hot-water system, water stays in the pipes and cools off. When a fixture is opened, the cold water in the line must be drained off before hot water arrives. In a circulating system, a complete loop is made from the heater to each fixture and back to the heater. As hot water rises in the system, colder water is forced downward and back to the heater by gravity. Water stays in the branch piping and may cool off, but cold water in the risers flows back to the heater. Therefore, more hot water is available in the piping and much less water is wasted while waiting for the hot water in the system to arrive. Pumps are used in that portion of such a system where cold water cannot flow by gravity back to the heater.

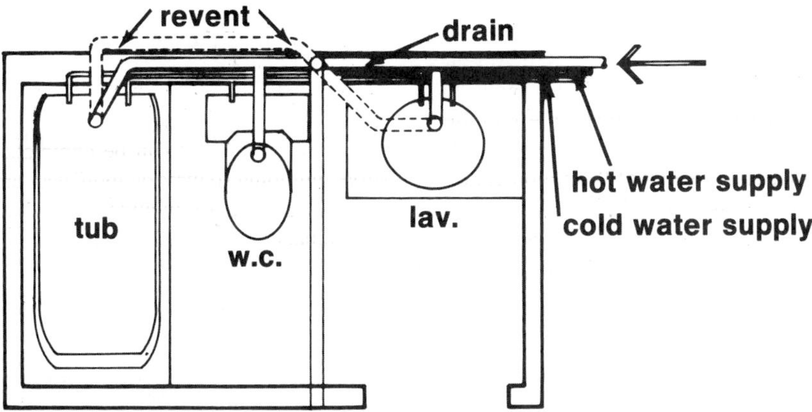

FIGURE 15.1-26 In-line arrangement of plumbing fixtures simplifies piping and reduces costs. (Courtesy Small Homes Council, Building Research Council, University of Illinois, Urbana-Champaign.)

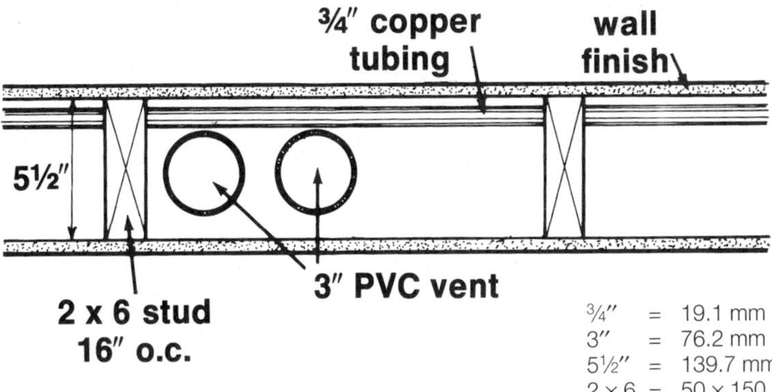

FIGURE 15.1-27 A 6-in. (150 mm) stud wall is wide enough to permit a ¾-in. (19.1 mm) water pipe to cross a 3-in. (76.2 mm) plastic vent stack. (Courtesy Small Homes Council, Building Research Council, University of Illinois, Urbana-Champaign.) Metric added by author.

SYSTEM CONFIGURATION

Both hot- and cold-water supply systems for small buildings are usually of the upfeed type (see Fig. 15.1-1). For larger buildings, cold- and hot-water systems may be either upfeed or downfeed systems, or a combination of the two (Figs. 15.1-28 and 15.1-29). Upfeed systems are simpler and less expensive and are preferred where appropriate.

To determine whether an upfeed system can be used effectively, it is necessary to ascertain the height of the topmost fixture that can be served by the available water service. Upfeed systems are limited in height by service water pressure, static head, friction caused by piping and meters, and the demand of the fixtures. To determine the permissible height of an upfeed system, first make a general estimate, as follows:

1. Determine the pressure in the service main. Many municipal systems deliver water at 50 to 60 psi (345 to 413.7 kPa). Private wells usually deliver water at 40 to 50 psi (275.8 to 345 kPa).

2. Obtain the required pressure for the fixture that requires the most pressure at the topmost building floor. A flush-valve water closet, for example, may require 20 psi (137.9 kPa) of water pressure to operate. Subtract this from the street pressure. If the street pressure is 50 psi (345 kPa), the pressure remaining to overcome pressure losses in the system is 30 psi (206.8 kPa); for a 60 psi (413.7 kPa) service, 40 psi (275.8 kPa) would remain.

3. Determine the pressure loss that will occur when the water flows upward to the topmost fixture (static head). The pressure drop required to raise water in a pipe is 0.434 psi per ft. (9.8 kPa/m) of rise. Divide the pressure available by the static head per foot. If there were no fixtures attached and no friction loss to contend with, the maximum rise for water at 30 psi (206.8 kPa) would be 30 divided by 0.434, or about 69 ft. (21 m). Water at 40 psi (275.8 kPa) could rise about 92 ft. (28 m).

4. Estimate the friction loss for pipe. The loss for a 2-in. (50.8 mm) pipe is about 3.2 psi (22.1 kPa) for each 100 ft. (30.48 m) of length. Therefore, the static head for a 69-ft. (21 m) pipe riser would be about 2.2 psi (15.2 kPa). If the distance from the street main to the riser were 100 ft. (30.48 m), its loss would be 3.2 psi (22.1 kPa). The pressure loss in a meter may be 4 psi (27.6 kPa). The loss for all three would then be 9.4 psi (64.8 kPa). This would lower the permissible height by 21.6 ft. (6.6 m) to a height of 47.4 ft. (14.4 m). Friction loss in pipe fittings lower this height even further. The practical limit is about 40 ft. (12.19 m) for 50 psi (345 kPa) services and 60 ft. (18.3 m) for 60 psi (414 kPa) services.

If the building is higher than the limits determined, an upfeed or, more usually, a combination system is used.

Upfeed Systems The steps in designing an upfeed water supply system and the upfeed portion of a combination system (see Figs. 15.1-28 and 15.1-29) are as follows:

1. Determine the pressure of the incoming service.

2. Determine the pressure needed to raise the water to the height of the topmost fixture (*static head*).

3. Subtract the pressure loss due to static head from the street main pressure. The result is the pressure available to overcome friction and serve the fixtures.

4. Determine the pressure required by the fixtures on the uppermost floor. To do this, find the total number of w.s.f.u. of the fixtures to be served and the amount of water they will require. Figure 15.1-22 shows the number of w.s.f.u. for various fixtures. The water demand curve in Figure 15.1-23 can be used to determine demand, but tables in the applicable plumbing code will give a more accurate estimation. For example, the tables may show that a riser containing 30 w.s.f.u. will demand 20 gpm; for 70 w.s.f.u., the demand is 35 gpm; for 100 w.s.f.u., 53.5 gpm. Be sure to use the hot-water w.s.f.u. when designing the hot-water system and the cold-water units for a cold-water system.

5. Subtract the pressure loss by the fixtures from the pressure left after taking off for static head. This is the pressure left to overcome friction.

6. Determine the friction loss in the system. This is a combination of losses from pipe, fittings, and a meter. First determine the size and type of pipe and meter to be used. Experience has shown that for one- and two-family dwellings of average size, the pipe sizes listed in Figure 15.1-30 will provide adequate service. Selecting pipes for other building types is initially done by trial and error. An educated guess is made, the calculations run, and then the pipe and meter sizes are refined and the calculations run again. The objective in sizing pipe is to ensure that there is adequate pressure and flow at plumbing fixtures and appliances. Losses due to internal friction in pipes are affected by:

- The type of pipe and its internal surface characteristics (rougher surfaces generate more friction)
- The length of the pipe (more length, more total friction)
- The number of turns, fittings, and valves (these each generate friction loss)
- The velocity of the water

The friction loss of the selected pipe, its fittings, and the meter are determined from tables in the plumbing codes, and this total is multiplied by the length of run. Different pipe sizes may be appropriate for different parts of the system. The horizontal pipe from the street main to the riser may be larger than an individual riser. If the total of all friction losses does not exceed the pressure available after subtracting the loss for static head and fixtures, then the system is satisfactory. If not, different sizes can be selected for the pipe and meter, and the calculations rerun until a satisfactory result is obtained. If necessary, fixtures with lower pressure requirements may be selected.

7. Go through the same process for each floor to determine pipe sizes at each level.

Downfeed Systems The design of a downfeed system is similar to that of the upfeed system described earlier, but the configuration differs. The major difference is that the water is pumped up into storage tanks and serves the fixtures by gravity flow (see Fig. 15.1-28). In the usual design for a hot-water downfeed system, the water from the tank is allowed to build up sufficient pressure so that it can serve the fixtures in an upfeed system (see Fig. 15.1-29). The steps in the design of a downfeed system are as follows:

1. Determine the number of w.s.f.u. in each zone and the demand load for them.

2. Verify the height of the storage tank in each zone above the uppermost fixture served by that tank. The uppermost fixtures must be far enough below the tank for the water to build up the pressure required to operate the fixtures. For example, a tank-type water closet may require a pressure of 15 psi (100 kPa) to operate properly. The pressure of falling water increases at 0.434 psi per foot (9.8 kPa/m) of drop. This water will also lose pressure due to friction in the pipe. Therefore, 34.6 ft. (10.55 m) of drop is required to serve the fixture and an additional 5 ft. (1.83 m) is required to overcome friction. Flush-valve water closets require a higher pressure (20 psi or so). Therefore, flush-tank

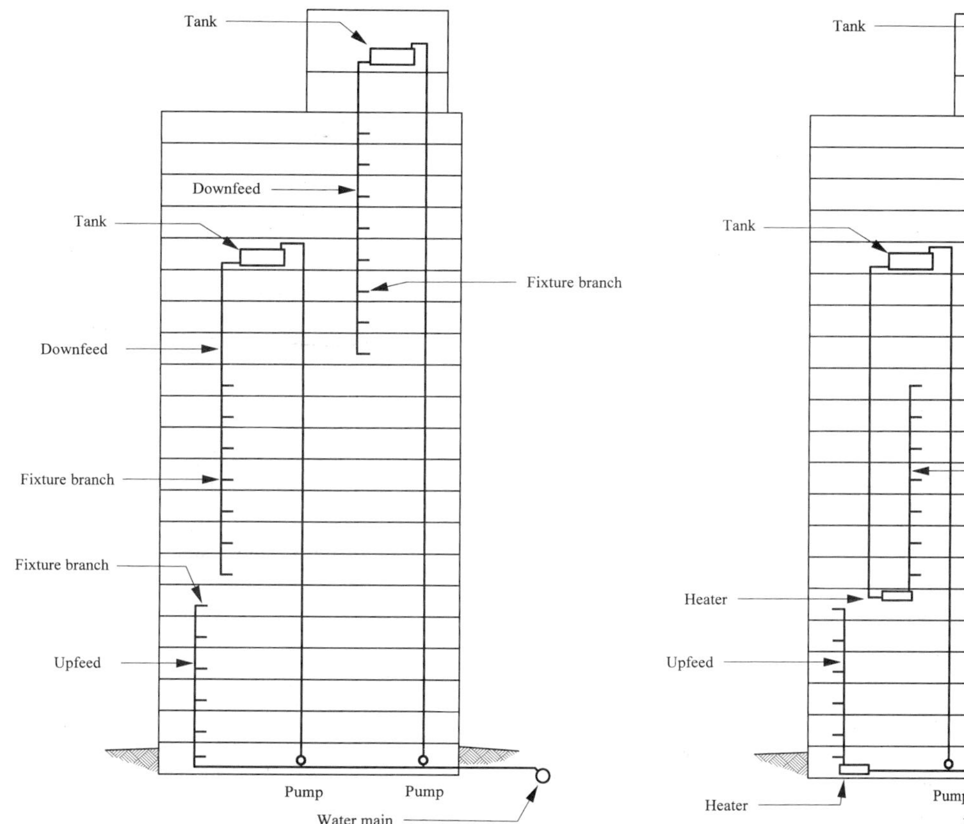

FIGURE 15.1-28 **Combination cold-water supply system.** (Drawing by H. Leslie Simmons.)

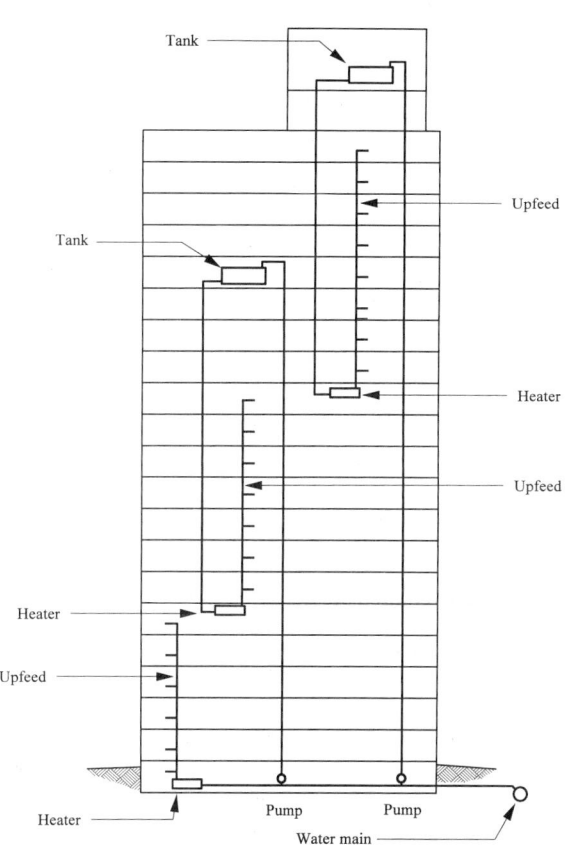

FIGURE 15.1-29 **Combination hot-water supply system.** (Drawing by H. Leslie Simmons.)

fixtures are often selected in high buildings to reduce the height required above the topmost fixture for the holding tank.

3. Determine the maximum height of each downfeed zone. To do this, first obtain the maximum pressure of each fixture from the fixture manufacturer. This pressure may vary from 40 to 60 psi (275 to 415 kPa). Then determine the pressure of the water. Remember that there will be a small loss in pressure due to friction. If the water pressure starts at zero and increases 0.434 psi for each foot (9.8 kPa/m) of fall, the maximum zone height for a fixture requiring a maximum pressure of 40 psi (275 kPa) would be 103.7 ft. (31.61 m). It can be slightly more, due to friction loss.

4. Each riser is then sized in a manner similar to that described for an upfeed system, except that loss due to w.s.f.u. is not a factor. If street main pressure is insufficient to fill the tanks, pumps are employed to raise the water to the tanks.

The upfeed portion of a combination water supply system is designed as described earlier for an upfeed cold-water system.

PIPE MATERIALS

The water supply pipe material to be used for a particular installation is influenced by several criteria. Plumbing code requirements are, of course, paramount. Within the applicable code parameters, the characteristics of the local water supply—that is, its acidity and its air, carbon dioxide, and mineral contents—affect the selection. The original cost is also important, as are the ease of installation and of repair of a particular piping material. Water supply piping should bear the stamp of approval of the National Sanitation Foundation (NSF). Acceptable piping materials are listed in Figure 15.1-31.

Galvanized Steel Galvanizing is a process that deposits a coating of zinc on the surface of a steel pipe to protect it against corrosion. Galvanized steel (also called galvanized iron) pipes are low in cost and high in strength. They are resistant to high pressures and to water hammer (see Section 15.1.5). Their major disadvantage is that joints are typically threaded, although other methods of joining are sometimes used. Threading the initial installation is

FIGURE 15.1-30 **Recommended Water Supply Pipe Sizes**[a]

Fixture	Pipe Size, in. (mm)[b]
House main	1 (25.4)
House service	¾ (19.1)
Supply riser	¾ (19.1)
Water closet	½ (12.7)
Bathtub	½ (12.7)
Lavatory	½ (12.7)
Kitchen sink	½ (12.7)
Laundry tray	½ (12.7)
Clothes washer	½ (12.7)
Dishwasher	½ (12.7)
Shower stall	½ (12.7)

[a]For one- and two-family dwellings of average size.
[b]All pipe sizes except main and service are for either hot or cold.

FIGURE 15.1-31 Acceptable and Unacceptable Water Supply Piping Materials[a]

Material	Service	Distribution Piping: Under Slab	Distribution Piping: Other
Cast Iron			
Water pipe	A		
No-hub pipe and fittings	NA	NA	NA
Copper Tubing			
Type K — Hard temper	A	A[e]	A
Soft temper	A	A[b]	A
Type L — Hard temper	A	A[c]	A
Soft temper	A	A[c]	A
Type M — Hard temper	A	NA	A
DWV — Hard temper	NA	NA	NA
Galvanized Open Hearth or Wrought Iron	A[c]	A	A
Galvanized Steel Pipe	A[c]	A	A
Black Steel Pipe	A[c]	NA	A
Plastic			
ABS	A[d]	NA	A[d]
CPVC-41	NA	A[b]	A
PVC	A[d]	NA	A[d]
PVC thin wall	NA	NA	NA
Polybutylene	A	A[b]	A
Polyethylene	A[d]	NA	A[d]
Styrene rubber	NA	NA	NA

[a]A = Acceptable; NA = Not acceptable.
[b]Install without joints in or under slab floors.
[c]Protected when installed underground.
[d]For cold water only, outside dwelling or in unfinished basement or crawl space.
[e]Brazed joint only.

laborious, and the need to thread new joints in place makes later tapped-in connections to existing systems difficult.

Galvanized pipe has a tendency to corrode or form scale on the pipe interior to a degree that depends on local water chemistry. Galvanized pipe performs best in areas with moderately hard, alkaline water. Naturally soft water should be treated with a silicate inhibitor to produce a protective film on steel pipe, and with caustic soda or lime to raise its pH (see Section 15.1.3.5). Very hard water should be softened to control scarring on steel pipes. When corrosion or scaling occurs, the interior pipe diameter can become greatly reduced, changing the carrying capacity of the pipe. Because it will react chemically, galvanized steel pipe should not be laid in concrete.

There are three available means of joining galvanized steel pipe:

1. *Threaded connections* are the most common way of joining galvanized pipe (Fig. 15.1-32a). Generally, the outside of the pipe run is threaded to match a female galvanized steel or malleable cast-iron fitting with the desired geometry. The threads should be coated with pipe joint compound or wound with plastic joint tape before the fitting is attached.

2. *Unions* (Fig. 15.1-32b) are a means of joining galvanized steel pipe that can be disconnected and reconnected at future times, but they are more costly than conventional fittings. A union consists of two *union nuts* with machined brass mating surfaces, each of which is threaded onto the respective pipe end, and a *ring nut*, which is tightened, drawing the two union nuts together. The brass components of unions, however, can cause localized corrosion in steel pipes, which restricts flow.

3. *Slip couplings* are a third type of connector available for galvanized steel pipes (Fig. 15.1-33). These couplings consist of a sleeve and compression nuts. The sleeve is slipped over the end of one pipe. The two pipes are then aligned, and the sleeve is positioned to cover the joint between them. *Compression nuts* at each end of the sleeve seal it to the pipes with rubber gaskets. Some types of slip couplings also have a screw connection in the center for easier installation.

Where steel pipe is being joined to a dissimilar metal such as brass and the water supply is corrosive or has a high solids content, a *dielectric* (nonconductor of direct electric current) fitting must be used to prevent galvanic corrosion. Where the water supply is sufficiently hard to produce a protective carbonate or silicate film on the pipe, a dielectric fitting is unnecessary.

Copper Although relatively costly, copper pipe is convenient to work, easy to join, and sufficiently corrosion resistant for use with most water supplies. Water that is very soft, very high in carbon dioxide, or very aggressive and hard should be treated before entering a copper plumbing system. Red brass (85% copper, 15% zinc), and 90-10 copper-nickel pipes are good choices where the water supply is corrosive. Copper should not be used in systems designed for velocities greater than 4 ft./sec. (1.22 m/s). While the piping itself is durable, the joints are weaker

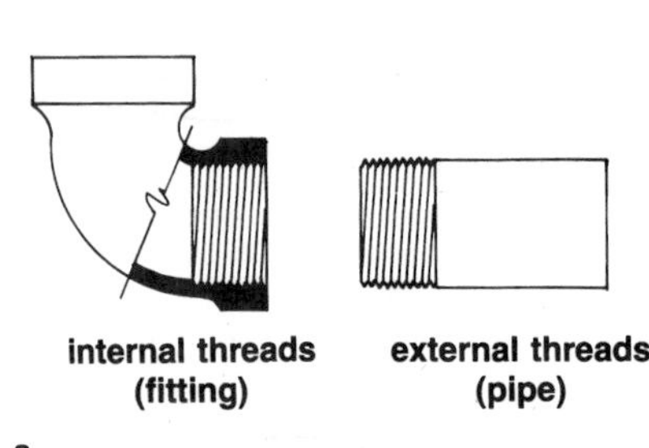

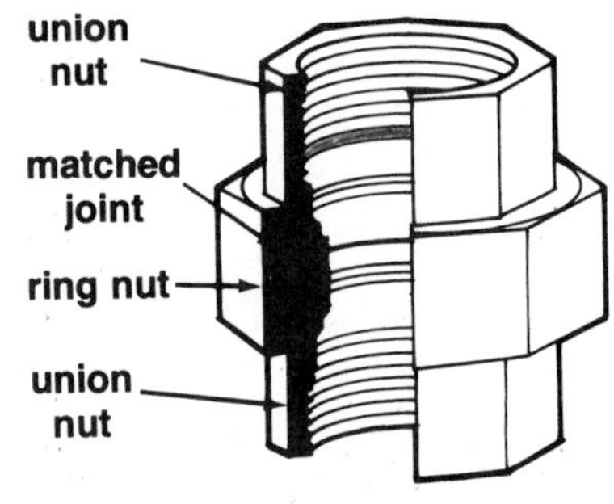

FIGURE 15.1-32 Galvanized pipe joints: (a) threaded fitting receives field-threaded pipe; (b) union fitting, to be threaded onto pipe ends.

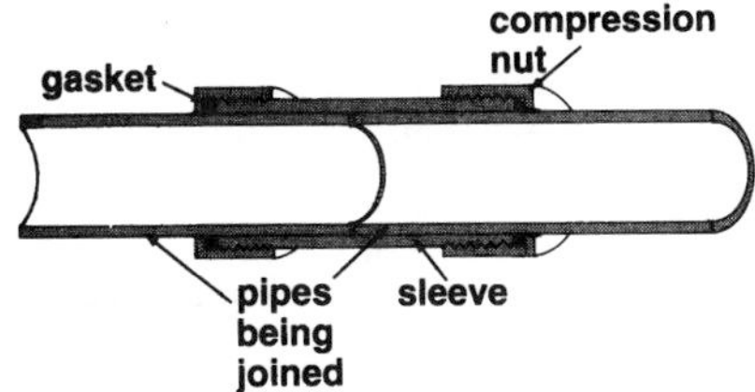

FIGURE 15.1-33 A section through a slip coupling.

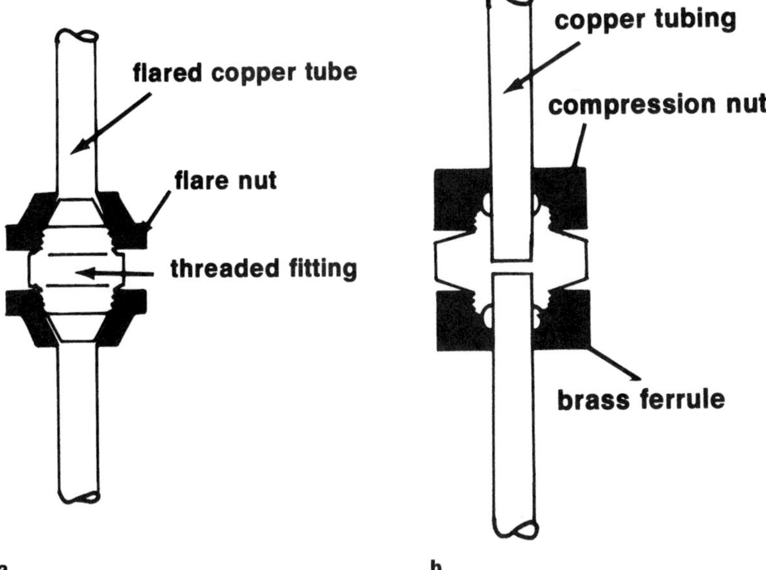

FIGURE 15.1-35 Unsoldered fittings for flexible copper tubing: (a) flared fitting; (b) compression fitting.

than the piping and are subject to damage from water hammer, which necessitates having air chambers at the ends of supply lines (see Section 15.1.5).

Copper supply piping falls into two major categories: rigid pipe and flexible tubing.

Rigid pipe, which is hard tempered, is generally connected using copper fittings that are sealed with tin-lead solder and a noncorrosive flux (Fig. 15.1-34). Excessive use of self-cleaning flux can lead to later corrosion problems. Rigid copper pipe comes in three wall thicknesses—(1) type K (thick), (2) type L (medium), and (3) type M (thin)—and in 10- to 20-ft. (3.05 to 6.1 m)-long straight lengths. The three types are equal in exterior diameter and are designated by the same nominal sizes, but the interior diameters vary due to the differences in wall thickness.

Flexible tubing, which comes in 30-, 60-, and 100-ft. (9.14, 18.3, and 30.5 m) coils, has several advantages over rigid copper pipe. Flexible tubing

- Is easier to install, especially in cramped areas;
- Can be bent into smooth turns, which reduces friction and sometimes permits using smaller pipes;
- Can withstand several freezing cycles without cracking, although with each freeze the pipe expands and the pipe wall becomes thinner.

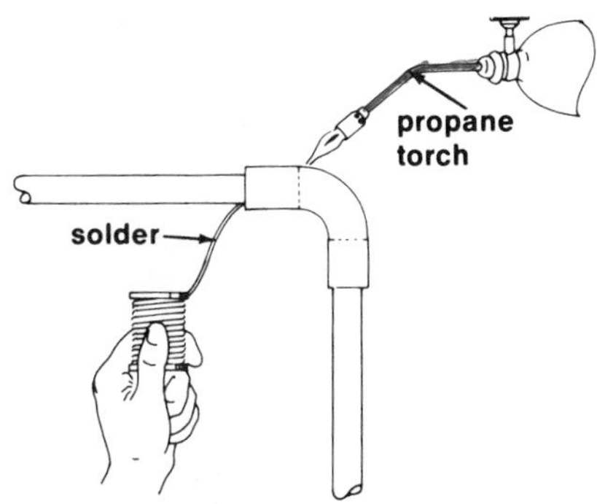

FIGURE 15.1-34 Rigid copper pipe being joined with a soldered fitting.

Flexible tubing also provides more options for connections. Soldered fittings, which require the use of a torch, and flared and compression fittings can all be used. A *flared fitting* (Fig. 15.1-35a) is made by slipping a *flare nut* over the end of the copper tube, re-forming the end of the tube so that it has a flared tip, then screwing the flare nut onto a threaded fitting. A *compression fitting* (Fig. 15.1-35b) is similar, except that the nut bears against a brass ferrule as it is screwed into the fitting rather than against the flared end of the tube. A push-in connector for copper tubing is also available. It consists of stainless steel and neoprene grabbers mounted in a nylon body.

Flexible copper tubing is also available in Type K or Type L wall thickness. Type K is typically used for underground installations, and Type L is used for indoor plumbing.

Plastic Not all of the many varieties of plastic pipe are suitable for carrying drinking water. Those that are suitable bear the stamp of the National Sanitation Foundation (NSF). Plastic pipe is lightweight, being about 1/20 of the weight of equal-sized steel pipe. It is also easily joined, produces low friction, and does not corrode. Joints are typically made using a solvent and special fittings (Fig. 15.1-36). Although these joints are easy to make, the solvent acts quickly and the joint becomes permanent. The only way to correct mistakes is to saw the joint out and start over. The corrosion resistance of plastic allows its internal surface to remain smooth, which maintains the carrying capacity of the pipe as the system ages.

Plastic is much less thermally conductive than metal pipe. Therefore, it is a better insulator, which reduces heat loss from hot-water lines and condensation on cold-water lines.

Plastic and metal piping differ greatly in their rates of thermal expansion, amount of softening at the same temperature level, electrical conductivity, flammability, and weathering characteristics. Plastic has a much larger coefficient of

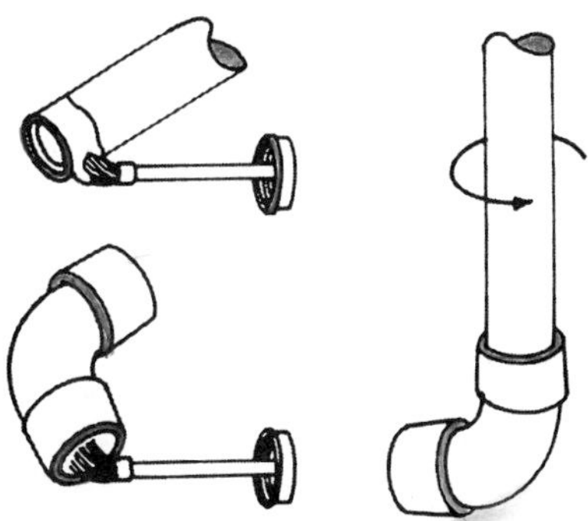

FIGURE 15.1-36 Plastic pipe joint. First cleaner, then solvent, is brushed on the outside of the pipe and inside of the fittings. Then the two are joined, and the pipe is turned about one-quarter turn. A good joint will show a fused solvent bead on the fitting end.

thermal expansion than metal. Because it tends to soften more and often has less strength and stiffness than metal pipe, it is necessary to support plastic pipe more frequently than metal pipe to prevent sagging. Because of its high rate of thermal expansion, plastic pipe is often snaked (placed in gentle undulating curves rather than straight) when it is laid in underground trenches. The upper limit of operating temperature for plastic pipes ranges from 120 to 300°F (48.9 to 148.9°C), depending on the type of plastic.

Plastic piping should be kept away from high-temperature areas such as those near heating ducts. Unlike metal piping, it is not a good electrical conductor. This has the advantage of preventing galvanic or electrolytic corrosion, but it also prevents the use of a cold-water pipe as a ground for a building's electrical system. Depending on the particular material's formulation, plastic pipe varies in its degree of flammability. Plastics weather differently than metal piping; instead of corroding or oxidizing, they chemically degrade when exposed to ultraviolet radiation and become brittle. This action can be controlled by using plastics that are formulated with ultraviolet stabilizers.

There are four major types of plastic pipe used in small buildings: *polyvinyl chloride* (PVC), *polyethylene* (PE), *chlorinated polyvinyl chloride* (CPVC), and *polybutylene* (PB). All four types are suitable for cold-water supplies, but only CPVC and PB are suitable for hot-water supplies.

Polyvinyl chloride (PVC) is most frequently used for cold-water supply lines. It can be used for the entire cold-water supply system within a building. Joints in PVC pipe can be either solvent-welded or made with threaded fittings, which are commonly available in heavier weights of pipe. PVC pipe is manufactured in two types. Type I is unplasticized and has a maximum working temperature of 150°F (65.6°C). Type II is plasticized by the addition of rubber, which makes it less brittle than Type I. This, however, makes its strength and maximum working temperatures lower than those of Type I. In addition to water lines, PVC pipe is used also for natural gas and electrical conduit lines.

Polyethylene (PE) pipe is used for cold-water supply from a municipal water main or a well to a building. It has a colder range of working temperatures than most other plastics: from –67°F (–55°C) to 122°F (50°C). Polyethylene pipe comes in Types I, II, and III, in order of increasing density. Joints in polyethylene pipe are made with tapered, serrated inserts of polystyrene or galvanized steel, which slip into the end of the polyethylene pipe (Fig. 15.1-37). The pipe is then compressed against the insert by means of a metal hose clamp.

Chlorinated polyvinyl chloride (CPVC) is the most prevalent type of plastic for hot-water supply lines. It can be used with water temperatures up to 212°F (100°C), although the water heater relief valve in a system using CPVC pipe should be set not higher than 180°F (82.22°C). Support clamps should be provided on horizontal runs every 3 ft. (914 mm) to prevent sagging. Due to the higher cost of CPVC pipe as compared with PVC, its use is usually restricted to hot-water lines. Joints are solvent welded.

Polybutylene (PB) pipe is suitable for either hot- or cold-water supply pipes. It is so flexible that most turns can be made without fittings. It can be obtained in a high-temperature grade for use requiring up to 221°F (105°C). Polybutylene pipe is similar to Type III polyethylene pipe but is stronger. Like polyethylene pipe, joints are made with tapered inserts and external clamps.

Brass In the past, red brass pipe (85% copper, 15% zinc) was frequently used in areas where the water supply was so corrosive as to make galvanized steel piping unsatisfactory. Today, because of its high cost, brass pipe is seldom used in water supply systems. Modern plastic piping can do the same job at a much lower cost.

Brass piping may still be encountered in old buildings. Its connections will be threaded, similar to those in galvanized steel pipe. While brass is highly resistant to corrosion generally, it is subject to galvanic corrosion when connected to galvanized steel, and there a dielectric coupling or union should be used.

Lead Lead piping was used for water piping in the past and may still be found in some older buildings. It is no longer used for water supply piping inside buildings, however, because minute quantities of lead oxides on the piping dissolve in water, which can result in lead poisoning.

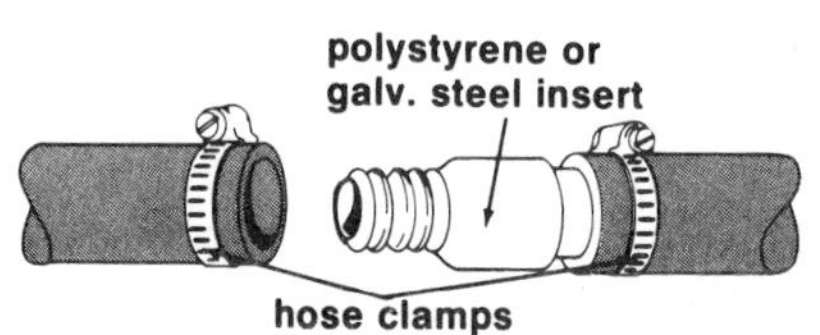

FIGURE 15.1-37 Clamped connection for polyethylene or polybutylene pipe.

FLOW CONTROL

An unsteady rate of water flow from a supply system into a fixture is due to changes in the flow pressure in the supply lines. This happens either because of inconsistent pressure from the water source (main or well) or because of the start-up or shutdown of other fixtures or appliances on the same plumbing tree. If the problem is severe, the inconsistent flow can be corrected with a flow-control device (see Fig. 15.1-38). This is a fitting with a flexible orifice on which the cross-sectional area varies inversely with the flow pressure. As pressure in the line increases, a flexible insert in the fitting is pushed forward against a stop, decreasing the cross-sectional area of the pipe and thereby regulating the flow rate. Flow-control devices are not to be confused with (or used as) pressure-reducing valves.

PRESSURE REGULATION

Most municipal systems deliver water to a building at pressures of 40 to 60 psi (276 to 413.7 kPa). In designing a supply system, it is critical to know the average supply pressure for summer months, when the main pressure is generally lowest due to extensive use of water for lawn watering. The water distribution system in a building is then designed in such a way that its length and friction characteristics will deliver an acceptable flow rate and pressure at the fixtures in the building. If the final pressure at the fixtures is inadequate, which is rarely the case in residential and other small buildings, it can be supplemented by installing a gravity tank

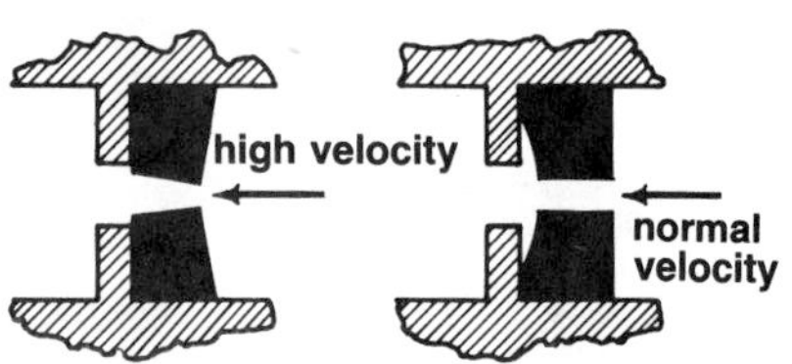

FIGURE 15.1-38 In a flow controller, increasing the velocity deforms a resilient orifice to reduce the opening. When velocity is normal, the orifice increases to normal. (Courtesy Small Homes Council, Building Research Council, University of Illinois, Urbana-Champaign.)

system, hydropneumatic tank system, or a booster pump.

VALVES

Valves used in building water supply systems can be classed according to their operation as *gate valves*, *globe valves*, and *alignment valves*.

Valves of various kinds are located throughout a water supply system. They are, in fact, a system's only moving parts. A *supply main*, *street main*, or *distribution main* is the water line from a treatment plant or well into which individual building supply lines are tapped. The principal, and usually lowest, horizontal artery in a building's water supply system is sometimes called a *main*, but we will not refer to it as such here in the interest of clarity. Where a building supply line leaves the supply main, there is a shutoff (*corporation*) cock. A main shutoff valve is installed just inside the building if the water meter is remotely located. Shutoff valves are installed on either side of in-line appliances such as water meters, water heaters, and water treatment devices. These are typically *gate* or *globe* valves.

Most water valves are made of brass, because of its resistance to corrosion. Valves are made with threaded ends for use with galvanized steel and rigid copper pipe and with solderable ends for use with copper tubing. Plastic-bodied valves are available for use with plastic pipe.

It is also good practice to install a shutoff valve where a supply pipe makes a turn of more than 45 degrees or enters or leaves a wall. To facilitate repairs, a type of globe valve known as an *angle stop valve* (because it is a shutoff valve through which the supply line makes a right-angle turn) should be installed just upstream from each fixture and appliance in a system.

Automatic appliances such as dishwashers and clothes washers have internal valves to cycle the water flow into them. These are usually electrically controlled *solenoid valves*, which open and close very quickly, sometimes causing water hammer (see Section 15.1.5). There are also valves within the faucet assemblies of each plumbing fixture. These are usually globe valves or cocks.

Gate Valves Gate valves are used when full, unobstructed flow is desired when a valve is open and positive closure is needed when it is shut (Fig. 15.1-39a). When the handle on a gate valve is turned, a wedge is lowered into the pipeway, obstructing the flow.

Globe Valves Globe valves contain two chambers connected by a small opening (Fig. 15.1-39b). When the handle is turned, a plunger (*stem*) is lowered, blocking the flow of water from one chamber to the other. The seal over the opening is usually made with a washer mounted on the tip of the stem.

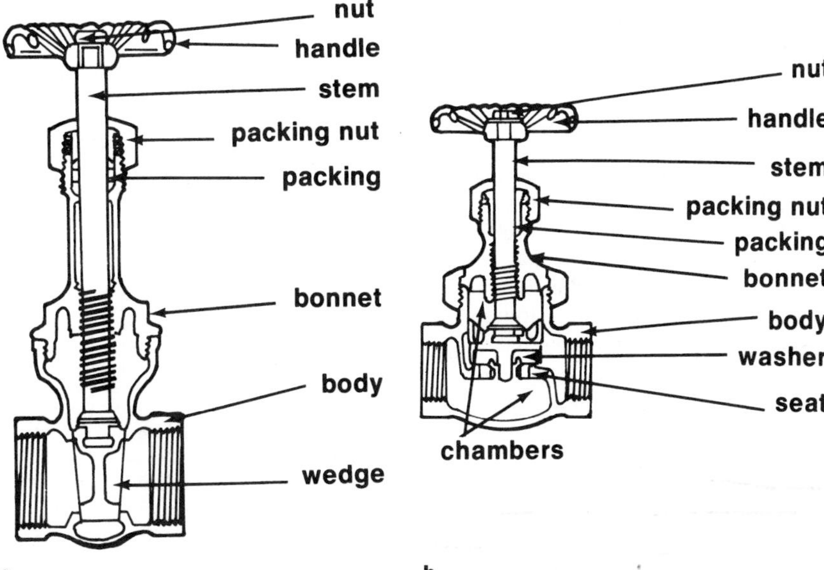

FIGURE 15.1-39 (a) Gate valves are used in key locations to turn water completely on or off. (b) Globe valves can control flow in partly open or closed position.

15.1.3.4 Water Heating

There are two basic systems for heating water for use at plumbing fixtures and appliances. They can be designated as direct or indirect. A solar system is a form of indirect water heater.

DIRECT WATER HEATERS

Direct water heaters are available in several types.

Automatic Storage Type *Automatic storage heaters* heat a volume of water and store it until it is needed. Heaters of this type are usually found in residential buildings, where their capacity generally runs from 20 gal. (75.7 L) to 100 gal. (378.5 L). The automatic storage heaters used in other building types may be identical to those found in a residence, but units with capacities of as much as 4000 gal. (15 141.6 L) are also available.

In automatic storage heaters, the water is maintained at a constant temperature by a thermostatically controlled heat source (Fig. 15.1-40). Generally, the thermostat is set between 120 and 140°F (48.9 and 60°C), but different temperatures may be required in some instances, particularly where use by handicapped persons is required, and for use in institutional buildings. A higher temperature is required for some automatic dishwashers and other appliances and devices. However, modern energy-saving dishwashers and some other appliances are equipped with a booster heater so that 120°F (48.9°C) water can be used.

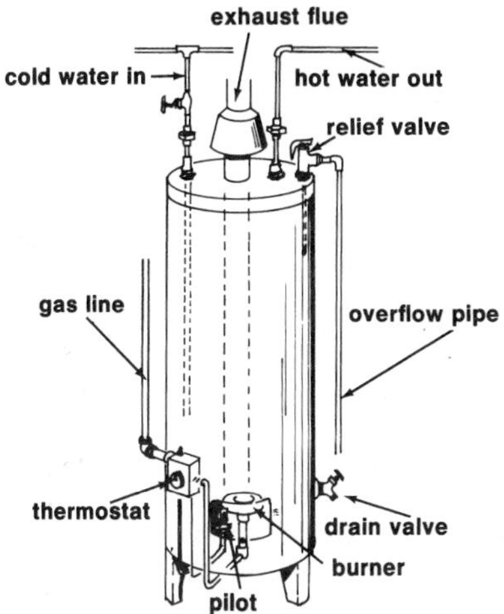

FIGURE 15.1-40 Gas water heater. Heater should be located near floor drain to permit periodic draining of sediment and to handle occasional overflow.

When water is used at a plumbing fixture or appliance, the tank is refilled automatically, as it is connected with an open valve to the pressurized cold-water supply line. This refill lowers the temperature of water in the tank, which the thermostat notes, and calls for heat from the heat source. The thermostatic control also calls for heat to compensate for heat lost through the tank wall when no hot water is being consumed. For this reason, new water heaters should be internally insulated with at least R-7 insulation. Kits are available for applying insulating blankets externally to older models.

The life expectancy of a residential-type automatic storage water heater is limited, by corrosion of the tank wall, to 10 to 15 years. The usual tank material is glass-lined steel. If the local water supply is particularly hard or corrosive, water treatment equipment may be installed to lengthen tank life. An alternative, though expensive, material for tank interiors is monel metal. It forms an oxide coating that resists corrosion. Although aluminum and unlined galvanized steel tanks are available, they fail quickly due to corrosion. Some water heaters are fitted with a sacrificial anode of magnesium, which gives up certain ions to the water more readily than the tank wall. The anode becomes a substitute material for the water to attack corrosively, thus sparing the tank liner material.

Excessively high temperatures and pressures are dangers associated with heating water in closed tanks. High temperatures can result in scalding or burning of persons using the water at fixtures. Pressure buildup in a tank can be an explosion hazard. Under normal operating circumstances, a thermostat cuts off the heat long before either of these two hazards is approached. However, as a safety measure in the event that the thermostatic control fails, UL-approved water heaters are fitted with a *pressure and temperature relief valve* (see Fig. 15.1-40).

A *pressure valve* vents a closed water heater tank in the event that a dangerous level of either pressure or temperature is reached. This valve is located on top of the tank and connects to a drip pipe that directs relief flow from the tank to a safe drainage area. Relief valves used on water heaters should be of the reseating type, which vents the system until an acceptable pressure or temperature level is reached.

Relief valves are specified for water heaters based on the British thermal unit (Btu) rating of the heater. The relief valve must always have a higher Btu rating than that of the heater. Valves are factory-set for the pressure and temperature levels at which they will discharge and should not be altered. The limits set for residential water heaters are approximately 125 psi (862 kPa) of pressure and 200°F (93.3°C) of temperature.

Water heaters are generally chosen on the basis of their volume and their recovery rate. Volume simply refers to the number of gallons the heater storage tank holds. Recovery rate is the number of gallons that the heater can raise 100°F (37.8°C) in 1 hour. This is a measure of a heater's ability to reestablish the thermostat-set temperature after rapid drawdown of the stored hot water. In general, oil-fired water heaters have a high recovery rate, gas heaters have a medium recovery rate, and electric heaters have a low recovery rate. Heaters are also rated as to energy input and output, usually quoted in Btu's. The ratio of the input rating to the output rating is an indicator of the heater's fuel efficiency.

Demand-Type Water Heaters *Demand-type* direct water heaters heat water only when more hot water is called for and do not store a reserve of hot water. Therefore, they are much smaller than storage-type heaters. Because hot water is not stored, standby losses at night and when occupants are away are much lower. Demand heaters are not as popular in this country for residential use as *automatic storage heaters* because of their limited recovery rate, but many demand heaters are used in commercial and institutional facilities.

COMBINATION WATER HEATERS

The second type of water heater is the indirect type, also called *combination heaters*. These use a heat exchanger on a space-heating boiler or furnace and a separate hot-water storage tank. The same boiler or furnace that is used to heat the structure is used to heat domestic water. This works well in cold climates, where heat for the building is required all or most of the year. In milder climates, however, the drawback is that the heater must run even when heat is not required. Sometimes an auxiliary water heater is employed to overcome this difficulty, but operating costs can still be higher and initial cost will certainly be greater.

Sometimes it is economical to use an indirect heater and a direct heater in the same installation. The indirect heater raises the temperature of domestic water to a lower level than is required at the discharge point. Then a secondary, direct water heater is used to elevate the temperature to the desired level.

FUELS

Heat sources for water heaters can be electricity, piped-in fossil fuels (natural gas or heating oil), or sunshine. Electric heaters are limited to small (20 to 100 gal. [75.7 to 378.5 L]) direct water heaters. Gas and oil can be used on every water heater type discussed in this chapter.

Electric water heaters usually have two electrical elements inside a storage tank. They require no supply of combustion air or venting of combustion products (Fig. 15.1-41).

At the base of its storage tank a *gas* water heater has a burner with a flue to exhaust its combustion products. The flue extends up through the center of the tank, acting as a heat exchanger as it does, and is finally connected to a chimney.

The burners on *gas* heaters can be designed either for natural gas or for liquefied propane. Adequate combustion air must be supplied to the burners, and an exhaust flue and chimney must be provided. Requirements similar to those for gas heaters also apply to *oil-fired* heaters, although the burner design is slightly different.

Solar water heaters use the sun as a partial source of energy. The intermittent nature of this heat source makes it neces-

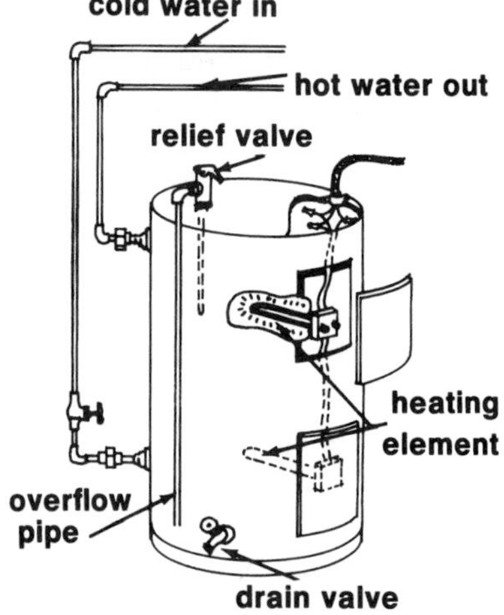

FIGURE 15.1-41 Electric water heaters are generally larger than gas- or oil-fired heaters because their recovery rate is slower.

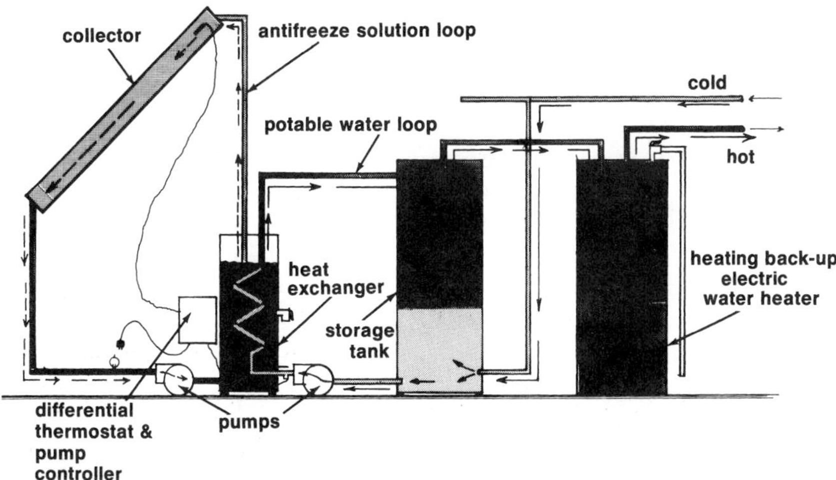

FIGURE 15.1-42 Heated water from a collector is conducted to a heat exchanger, where most of the heat is transferred to a loop conducting heat to a storage tank. From there preheated water is drawn as needed to supply a domestic water heater.

sary to provide extra storage capacity so that hot water will be available when needed at night and on cloudy days (Fig. 15.1-42).

Solar energy is absorbed by a south-facing solar collector mounted outdoors, through which a heat-transfer fluid flows. When heat is available from the collector, the transfer fluid carries it to a heat exchanger coil in a storage tank. The domestic water heater uses this preheated water when replacing water in its own tank.

Except in a few locations and during limited periods, the solar energy available is insufficient to heat water to the desired temperature. Thus, solar heating is usually practical only as an auxiliary heat source and must be backed up with an electric-, gas-, or oil-fired conventional heater. This limitation does not, however, invalidate the potential of solar energy as a means of domestic water heating. In many parts of the country, one-third to two-thirds of needed heating energy can be obtained from the sun.

Sizing The sizing of water heaters is based on a calculation of the maximum hot-water demand per hour. For a dwelling, this is estimated as 7 to 10 gal. (26.5 to 37.9 L) of hot water per person per hour, times the number of persons in the household. For other building types, the requirement is based on calculations made using water supply fixture units (w.s.f.u.). Refer to Section 15.1.3.3 for a discussion about the use of w.s.f.u.

A water heater should be sized so that 70% of the tank capacity plus the heater's recovery rate equals or exceeds the calculated maximum demand per hour. The equation for sizing can be written thus:

7–10 gal./person × number of persons = maximum demand/hr. = 0.7 × (tank capacity/hr.) + (recovery rate).

This requirement can be met with a variety of combinations of tank capacity and recovery rates. These combinations depend largely on the type of fuel being used. In oil-fired heaters, which have a fast recovery rate, a heater with a small tank can be specified and the maximum demand per hour can still be met. In gas-fired heaters, which have a moderate recovery rate, a medium-sized tank would be necessary. In electric systems, with their slow recovery rates, a relatively large tank would be necessary. For example, for a family of four or five people whose maximum demand per hour is in the range of 40 to 50 gal. (151.4 to 189.3 L) of hot water, either of the following systems would meet the size requirement to provide the maximum demand: (1) a 30- to 40-gal. (113.6 to 151.4 L) oil- or gas-fired heater or (2) a 40-gal. (151.4 L) quick-recovery electric heater. For an off-peak electric heater, an 80-gal. (302.8 L) tank is recommended.

TEMPERATURE CONTROLS

Water temperature controls are often advisable, especially where handicapped people, elderly people, or small children will use the plumbing fixtures (see Section 1.7). Two types of temperature control devices are available, primarily for tubs and showers. One type maintains constant relative pressures in the hot and cold supply lines to compensate for sudden draws by nearby fixtures. Consequently, the flow rate, once set, remains reasonably constant. This does not, however compensate for changes in temperature of the water from either supply line, which occurs when the hot water heater is drawn down. The second type of device controls temperature thermostatically and changes the flow rate of the hot and cold supplies to maintain a constant temperature of mixed output.

15.1.3.5 Water Treatment

Water from some sources may have to be processed further to make it suitable for domestic use, even though it meets public health standards for safe drinking water. Problems that may require treatment include excessive hardness, resulting in ineffective soaping; materials in the water that cause corrosion or scarring of piping; and unpleasant odor, color, or taste. Correction of such problems usually involves mechanical filtering or chemical treatment of the water.

Filters require periodic cleaning or replacement to prevent bacterial growth in them. Bacterial contamination of a building's water supply can cause problems more severe than those the filters are intended to correct. Treatment is often confined to the hot-water supply, because higher temperatures accelerate the undesirable chemical reactions and make sensory characteristics more noticeable.

WATER HARDNESS

The most common type of domestic water treatment combats what is known as water hardness. *Hardness* refers to the amount of calcium and magnesium carbonate dissolved in the water; *total hardness* refers to the amount of carbonates plus several other compounds dissolved in the water.

Some degree of hardness is desirable, since it causes a thin film of calcium carbonate to deposit on the interior walls of pipes, forming a protective coating against corrosion. Too much hardness, especially in conjunction with high temperatures, has two undesirable effects. It causes excessive carbonate deposition (scaling) and renders the water ineffective at rinsing soap away. Calcium carbonate scale appears as a cream-colored buildup in piping and fixtures.

Using a *water softener* alleviates both scaling and soaping problems. An *ion-exchange* water softener contains a sodium-zeolite exchange medium, which develops sodium ions. These then interact with calcium (or magnesium) ions in the water. When all sodium ions have been consumed, the filter must be recharged. Some systems are recharged in place by backwashing and soaking with a salt solution; others depend on a service company to supply recharged replacements (Fig. 15.1-43).

Softening water too much raises the sodium content of the treated water, which may make it harmful to persons with heart or kidney disorders. Fortunately, the adverse effects of hardness are more pronounced in the hot-water supply than in the cold. Humans more often drink water from the cold supply than from the hot. Consequently, by installing the water softener on the hot-water system only, human consumption of softened water can be limited.

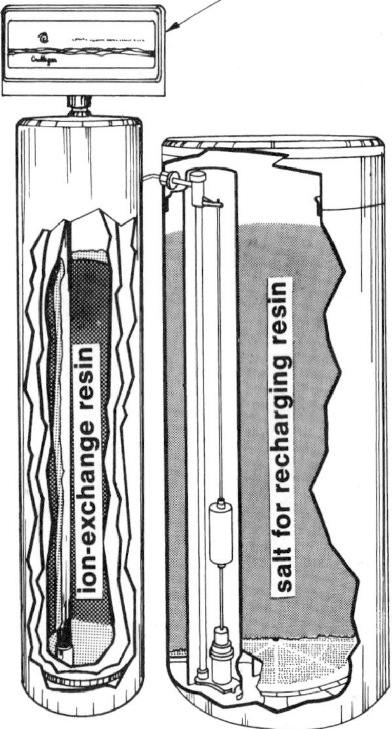

FIGURE 15.1-43 As hard water passes through resin, calcium and magnesium in the water are retained and sodium is given up. Either the salt or the entire tank must be replaced periodically.

In addition, oversoftened water not only fails to provide the protective calcium carbonate film on pipes, but actually makes the water corrosive. For these reasons, it is not advisable to soften water to less than 60 mg of calcium carbonate per liter. If the output of the softening equipment falls below this level, it should be tempered with untreated water from the cold supply before it enters the plumbing system.

CORROSIVENESS

A second major problem that can be treated is corrosiveness. A water supply can be rendered corrosive by high levels of dissolved carbon dioxide (yielding carbonic acid) or by low levels of calcium and magnesium carbonate (excessive softness). The former occurs naturally in water supplies in the east and northwest United States; the latter can occur naturally or can be the result of overzealous water treatment. Corrosion due to either cause is accelerated by high temperatures and therefore is a larger problem in a hot-water system than in a cold-water one.

Symptoms of corrosion are staining of fixtures, reduced flow or pressure resulting from clogging, and leaky pipes. The stains are red if the supply piping is steel, green if copper. Clogging can occur due to accumulated rust particles or due to localized scaling triggered by the changes in pH that accompany corrosion.

There are several solutions to corrosion problems. Mildly corrosive water from the natural supply can be treated with a neutralizing filter containing marble or limestone (calcite) chips. In severe cases, liquid sodium silicate and caustic soda can be added to raise the pH of the water. While this relieves corrosion problems in most of a plumbing system, it does not sufficiently protect low-velocity areas such as storage tanks. Tanks with special protective coatings or sacrificial anodes should be considered.

UNPLEASANT APPEARANCE, ODOR, OR TASTE

Unpleasant appearance, odor, or taste in a water supply can also be treated chemically or mechanically. For example, yellow, cloudy water and rust-stained fixtures are frequently caused by suspended iron particles. In mild cases, these can be removed by an ion-exchange water softener; in more severe cases, an oxidizing filter containing manganese-treated sand is effective.

Particles of organic matter, silt, or algae can degrade the appearance and taste of water. This problem can also cause pipe clogging and reduced water-heater efficiency. Such particles may be inherent in the water supply or may be introduced by the backflow of drainage water into the supply system. If this type of water quality problem persists, fixtures should be checked for air gaps and appliances such as clothes washers, laundry tubs, and dishwashers should be checked to ensure that they have vacuum breakers (backflow preventers) (see Section 15.1.3.3).

If the source is the water supply, a charcoal filter can be installed where the supply piping enters the building. In severe cases, a chemical feeder with a chlorine or alum solution can be used, followed by a sand filter. For small amounts of high-purity potable water, a reverse osmosis unit can be installed on individual fixtures. This device features a membrane through which only pure water can pass.

Rotten egg odor is caused by high levels of dissolved hydrogen sulfide gas. The source of the hydrogen sulfide can be anaerobic decomposition of organic matter somewhere in the supply chain. This problem occurs at many locations in the southwest, midwest, and southeast United States. It can also be produced by magnesium anodes used to protect water heater tanks from corrosion. Water with dissolved hydrogen sulfide also has a black tinge and tarnishes silver. An oxidizing filter with manganese-treated sand can be used to alleviate this problem. Extreme cases call for chlorination followed by sand filtering.

15.1.4 DRAINAGE SYSTEMS

A *drainage system* consists of piping that conveys solid and liquid wastes from their source to a point of disposal. It includes the system portions within a building and on its site, and the mains of a public sewer system or a private septic system.

There are three kinds of drainage systems. A *sanitary drainage system* carries *sewage*, which is liquid and solid waste matter, including fecal matter and food, and industrial waste, as opposed to the kind of material carried off by storm sewers. The portion of a sanitary drainage system that lies within a building and up to a point 3 ft. (914 mm) outside the walls is called a *building drainage system* or, more frequently, a *drain, waste, and vent system*

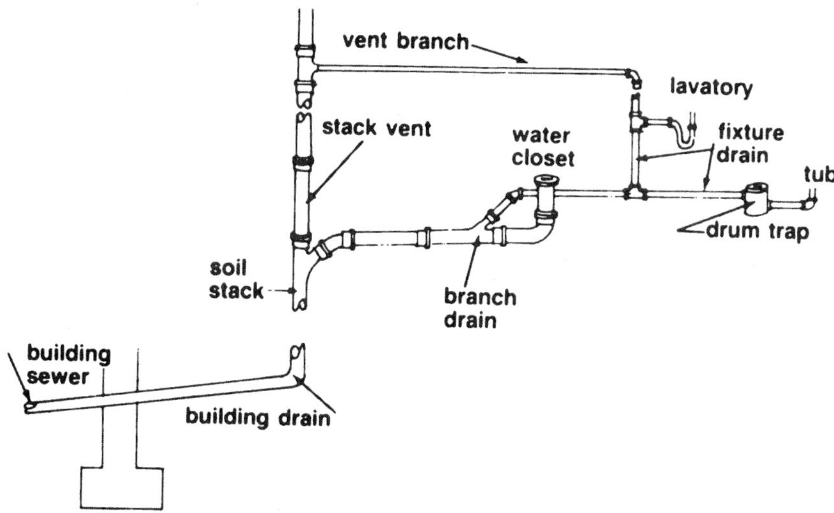

FIGURE 15.1-44 Diagram of major components of a DWV system, showing a loop vent.

(DWV) (Fig. 15.1-44). This collection of pipes facilitates the removal of liquid and solid wastes as well as the dissipation of sewer gases within a building.

The second type, a *storm drainage system*, carries rainwater, surface water, condensate, cooling water, and similar liquid wastes, exclusive of sewage and industrial waste.

The third type of drainage system is a *combined* system that carries both storm water and sewage. Such systems have been responsible for polluting many of our waterways when their capacity was insufficient to handle heavy storms. When excessive amounts of water overwhelm a storm drainage system, water backs up into streets and even buildings. When this happens to a combined system, what backs up is raw sewage. Combined systems, which may still be found in older installations, are not permitted in many jurisdictions and should be prohibited everywhere.

Both storm and sanitary drainage systems include sewers. A *sewer* is an underground conduit used to carry off liquid waste, solid waste, or both. A system of sewers is called *sewerage*. The term *sewerage* is also used to refer to the removal and disposal of liquid and solid wastes by sewers. The horizontal piping that extends from a point 3 ft. (914 mm) outside a building to the public sewer or private individual sewage disposal system is called a *building sewer*. This may be either a *building storm sewer* or a *building sanitary sewer*.

15.1.4.1 Sanitary System Design

A DWV system should be designed and installed so that it (1) maintains separation of drain fluids from the water supply, (2) contains and discharges sewer gases to prevent fouling the interior atmosphere, and (3) effectively moves fluids and solids through the pipes.

COMPONENTS

DWV systems are larger and often more complex than water supply systems. Their geometry is more critical because the pipes are larger and must be installed at a slope for gravity drainage. They consist of drains (waste and soil pipes), traps, and a vent system.

A *drain* is any pipe that carries sewage in a building drainage system. *Waste pipes* carry water-diluted waste products; *soil* pipes carry the discharge of water closets, urinals, and other fixtures with similar functions to the building drain or building sewer (Fig. 15.1-45). Soil pipes may, in addition, carry wastes from other fixtures.

A *trap* is a fitting or device designed to provide, when properly vented, a liquid seal that will prevent the back passage of air without materially affecting the flow of sewage or wastewater through it (Fig. 15.1-46). A *vent system* is a collection of pipes installed to provide a flow of air to or from a drainage system, or to provide circulation of air within such a system to protect trap seals from siphonage and back pressure.

All the components of a DWV system are interconnected. In fact, a given pipe may serve in one portion of its length as a drain and in another portion as a vent, or under some circumstances may serve as a drain and vent simultaneously at the same location.

At the center of a DWV system is one or more vertical combination *drain and vent* pipes that extend from a point above the roof down to the building drain. A *building drain* is the lowest part of a build-

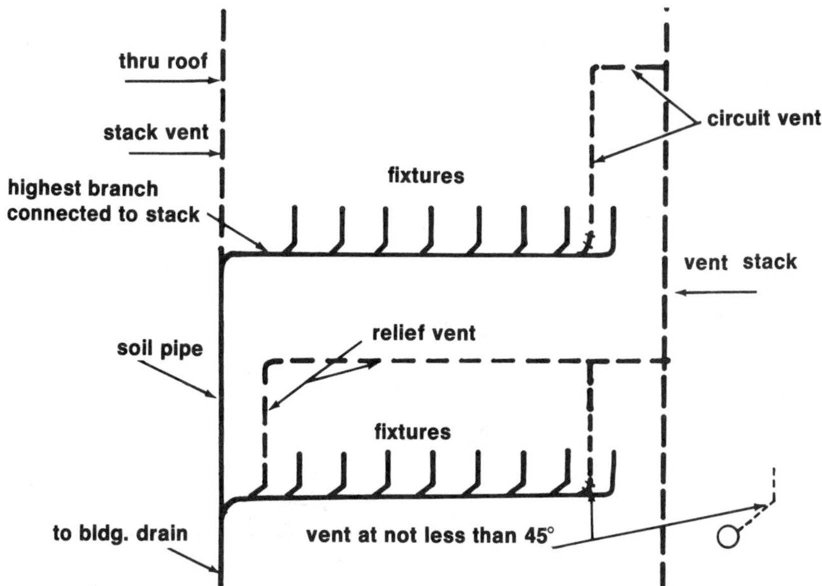

FIGURE 15.1-45 Drainage diagram for larger building showing two ways of venting drain branch: with circuit vent (above) and circuit and relief vent (below). (Courtesy Small Homes Council, Building Research Council, University of Illinois, Urbana-Champaign.)

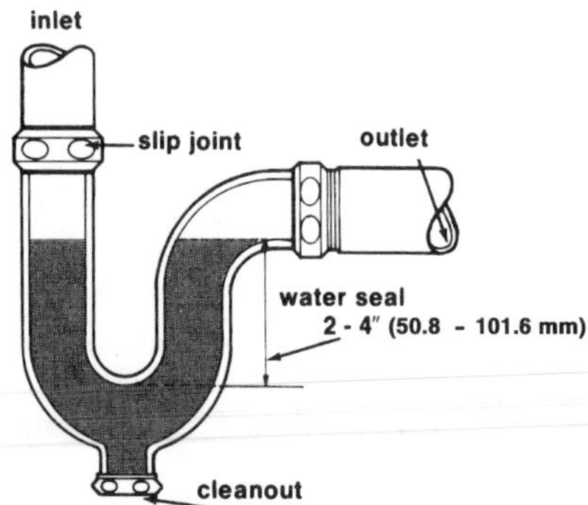

FIGURE 15.1-46 **A P-trap showing the water seal.** (Courtesy Small Homes Council, Building Research Council, University of Illinois, Urbana-Champaign.) Metric added by author.

ing's drainage system. It is the principal artery to which branches may be connected and as such is sometimes called the system's *main*. It receives the discharge from soil, waste, and other drainage pipes inside the walls of a building and conveys it through the foundation walls to a point 3 ft. (914 mm) outside the building, where it joins with the building sewer (Fig. 15.1-47). The *building sewer* is the horizontal piping of a *drainage system* that extends from the end of a *building drain* to a public sewer, private sewer, individual sewage disposal system, or other point of disposal (see Fig. 15.1-47). A building sewer begins at a point 3 ft. (914 mm) outside the building wall, where the building drain ends. A building sewer is sometimes called a *house sewer*.

Soil stacks, or just *stacks*, are the central vertical pipes that are the collection points for the drain and vent lines in the DWV system.

A typical soil stack has connections to one or more branch drains serving several fixtures each (see Fig. 15.1-44). Branch vents serving those same fixtures connect to the stack at points higher than the branch drain connections. Thus, the lower portion of this pipe serves as the system's main vertical drain, or *soil stack*, which leads to the *building drain* and ultimately to the public *sewer*. The portion of the soil stack that lies above the highest horizontal waste-carrying connection to it is called a *stack vent*, sometimes also called a *waste vent* or *soil vent* (see Fig. 15.1-45). A stack vent may collect one or more vent branches before it continues through the roof to the outdoors, providing a source of atmospheric pressure and thus serving, in this case, as the system's *main vent*. A horizontal pipe to which a number of vent stacks are connected before they go through the roof is also sometimes called a *main vent*.

Branch vents and branch drains are fed by fixture drains (Fig. 15.1-48). A *fixture drain* leaves the drainage point of a fixture and runs to the nearest branch drain. It includes a water-filled plumbing trap, which serves to prevent sewer gases from rising through the fixture drain into the building. When the fixture drain line reaches a wall, it intersects with the branch drain/branch vent system. Water flowing out of the fixture drain line falls into the lower portion of the intersected pipe.

A stack vent or branch vent provides an air supply to equalize pressures and smooth the drainage process (see Fig. 15.1-44). The most desirable drainage pattern for water going down a vertical pipe is one in which the water swirls down the pipe, clinging to the walls by surface tension. This process draws air down the pipe with the water. The air is supplied through the vent system. If no air is available through a vent system, or if the flow of water is so heavy that it overcomes the surface tension holding it against the pipe wall, then drainage can be accompanied by gurgling noises, vibration, and loss of the trap seal.

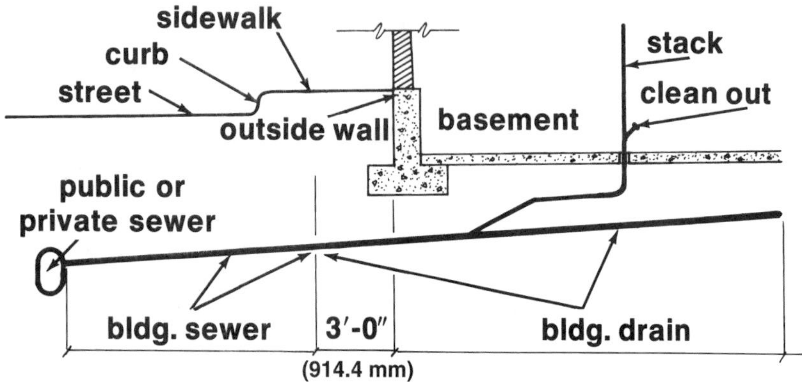

FIGURE 15.1-47 **Diagram showing stack, cleanout, building drain, building sewer, and public or private sewer.** (Courtesy Small Homes Council, Building Research Council, University of Illinois, Urbana-Champaign.) Metric added by author.

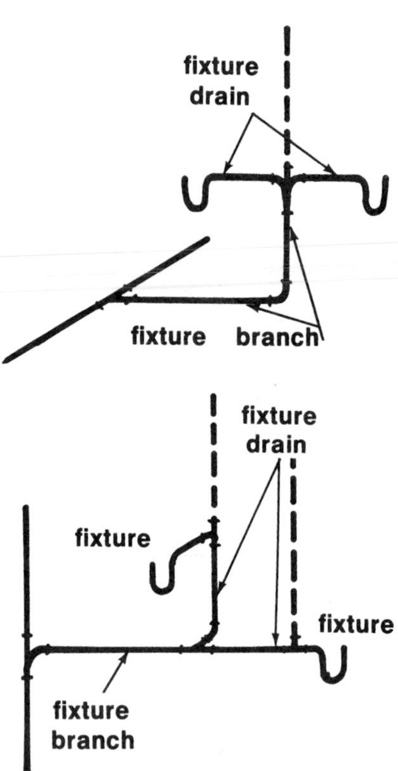

FIGURE 15.1-48 **Fixture drains connect fixtures to fixture branches or branch drains.**

GRAVITY FLOW

In a water supply system, flow occurs because the fluid is pressurized. Drainage system components exist at atmospheric pressure, and flow is due to gravity. Consequently, sanitary drain lines must be sloped and arranged to assist gravity flow.

Sufficient slope should be provided in sanitary drainpipes to maintain a flow velocity of at least 2 ft./sec. (0.61 m/s), which is necessary to scour the pipe properly and keep solids in suspension. For adequate drainage in a horizontal pipe (one with a slope of less than 45 degrees from horizontal), the pipe should (1) slope at least ¼ in. per ft. (20.83 mm/m) of run if it is less than or equal to 3 in. (76 mm) in diameter or (2) slope at least ⅛ in. per ft. (10.42 mm/m) of run if it is more than 3 in. (76 mm) in diameter.

Sufficient pipe supports should be provided to maintain uniform slope. Drains laid underground should have a bed of undisturbed compacted earth or concrete to control differential settlement. Drain lines should not be laid beneath driveways or buildings when avoidable.

Turns and *fittings* should be properly detailed to minimize resistance to water flow. When possible, flow should move in a straight line. If turns are unavoidable, sweeping bends are preferred to sharp turns. In horizontal runs, 45-degree wyes, sweeps, and ½, ⅙, ⅛, and ¹⁄₁₆ bends should be used; for vertical-to-horizontal turns, ¼ and ⅕ bends should be used.

Fittings for use in drain systems are designed differently from those for supply systems, primarily to eliminate shoulders and other impediments to gravity flow (Fig. 15.1-49). Consequently, *fitting* names are preceded by the designation *drain*, as in *drain bend* and *drain wye*.

Connection of branch drains to main drains is always done in the upper half of the horizontal pipe. This reduces the chance of clogging, cuts down on interference in flow at the point of connection, and when there is no flow in the branch, leaves the full diameter of the pipe available to act as a vent.

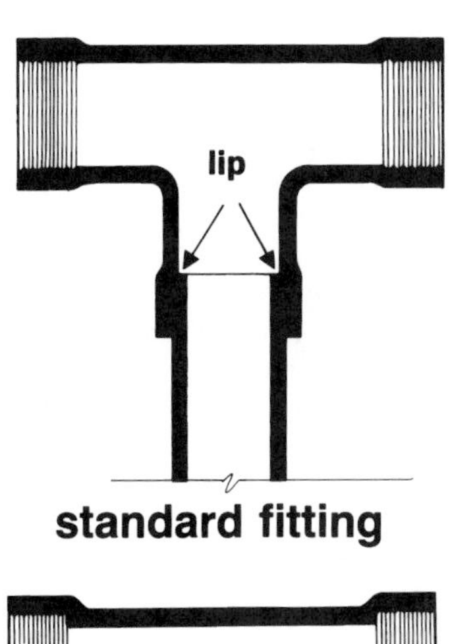

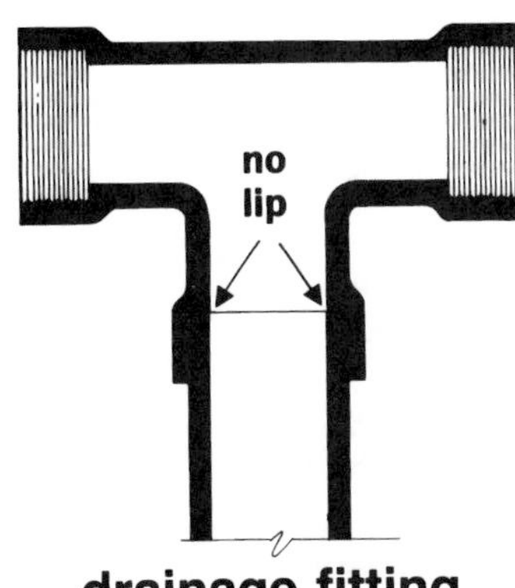

FIGURE 15.1-49 Drainage fittings showing elimination of lips that would impede sewage flow. (Courtesy Small Homes Council, Building Research Council, University of Illinois, Urbana-Champaign.)

When gravity flow of liquid or solid waste is not possible, it may be necessary to mechanically force the material up to the level of the sewer. This is customarily accomplished by means of a sump and a sump pump. A *sump pump* is a mechanical device used for pumping out sewage or liquid industrial wastes from locations below a building's gravity drain. These pumps are located in a *sump*, which is a tank or pit that receives sewage or other liquid waste, located below the normal grade of a building's gravity sewer system, and which must be emptied by mechanical means (a sump pump).

DRAINPIPING

The routing of drainpipes may pose serious problems during construction unless it is carefully considered during the design process. Because a sanitary drainage system drains by gravity, horizontal pipe runs must be sloped downward. This restricts the length of horizontal runs that can be tolerated before the sloping pipe encounters structural members or exceeds the depth of the space in which it is to be concealed.

Sealing and Venting A DWV system must be sealed off from the air to prevent sewer gases from escaping into a building's interior. This is usually accomplished by water seals in fixture traps. A DWV system must, however, be open to the outside air to maintain atmospheric pressure within the system, both to aid in the flow of materials through it and to protect the trap seals from being disturbed. This is usually accomplished through a network of vents.

Traps In addition to waste, a drainage system contains air and foul-smelling sewer gases. For reasons of sanitation and comfort, it is desirable to prevent sewer gases from entering the building through fixture drains. To prevent this from happening, a device known as a *trap* (see Fig. 15.1-46) is installed at each fixture. A trap is simply a length of drainpipe shaped in such a way that it does not drain completely. After each use of the fixture, the trap retains enough water to fill the drainpipe at that point. The trapped water serves as a seal that prevents sewer gas from rising through the fixture drain into the building. Each individual fixture requires its own trap.

Most plumbing fixture traps are shaped like the letter *P* and are called *P-traps* (see Fig. 15.1-46).

A special trap, called a *drum trap*, is usually installed on the drain lines of bathtubs. Such traps offer easier access for cleaning bathtub drains, which are frequently clogged with dirt and hair. A drum trap is constructed so that the drain line from the bathtub enters its lower portion, and the drain line out into the rest of the drainage system leaves the trap 2 to 4 in. (50.8 to 100 mm) higher. Therefore, a water seal of 2 to 4 in. (50.8 to 100 mm) is maintained within the drum. Drum traps are generally 3 to 4 in. (76 to 100 mm) in diameter and have a removable top for easy access. Water closets have a trap built into their internal structure (Fig. 15.1-50).

A *building trap* is a device intended to prevent the circulation of air between the building drainage system and the building sewer (Fig. 15.1-51).

A *grease trap*, unlike other traps, is not designed to prevent gases from rising through the drainpipes, but rather to collect grease particles and prevent them from reaching the sewer or septic system where they can cause blockages. Grease traps generally are not found in small residential buildings but may be used in restaurants, institutional kitchens, and large residential structures.

Ordinary traps are generally maintenance-free, as they are simple devices with no moving parts. Difficulties arise, however, if the water seal in a trap is lost. There are at least five ways in which this

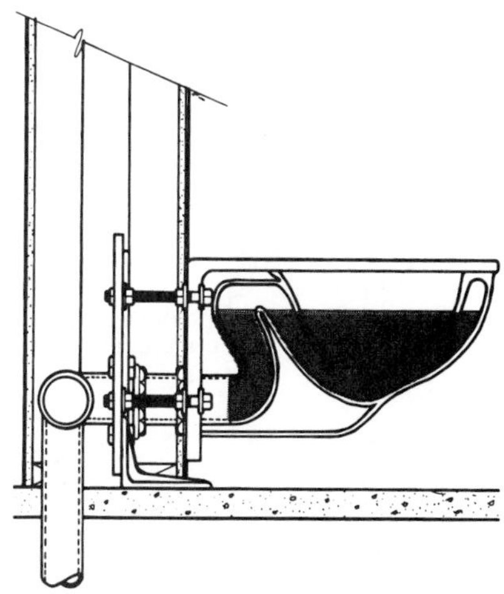

FIGURE 15.1-50 The residual water in the bowl of a water closet forms a water seal similar to that in a sink trap. A wall-hung water closet is shown.

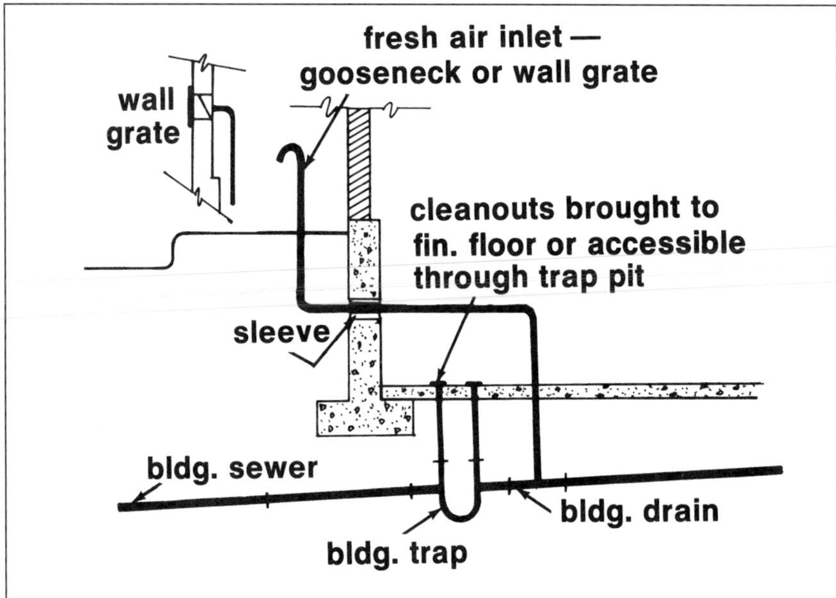

FIGURE 15.1-51 A building trap acts both as a seal and as a cleanout for the building drain. (Courtesy Small Homes Council, Building Research Council, University of Illinois, Urbana-Champaign.)

can happen. The water seal may be lost because of any of the following:

1. Suction (*back siphonage*).
2. Back pressure from a slug of sewage passing through the drain system that forces the water seal up into a lower fixture or sucks it out of a higher trap (Fig. 15.1-52).
3. Evaporation from a rarely used drain.

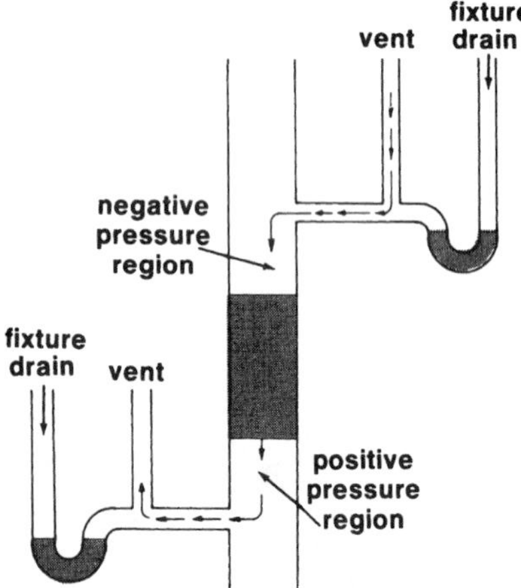

FIGURE 15.1-52 A slug of sewage in a soil stack creates positive pressure in front of it and suction behind. If there were no vents or the vents were clogged, the left-side trap in this figure could be blown out or the right-side one could be sucked out. (Courtesy Small Homes Council, Building Research Council, University of Illinois, Urbana-Champaign.)

4. Capillary attraction through an object caught in the trap (such as a diaper, from which the water can readily evaporate).
5. Outdoor wind conditions that create positive or negative pressures within the drainage system through the vents; positive pressure can push water out of the trap, negative pressure can suck it out of the trap. The latter is much more likely to occur.

Another infrequent maintenance problem with traps is freezing. Freezing of plumbing is generally a greater problem in the supply system than in the drainage system, but it is possible for damage to be caused by freezing of the water seal in a trap. This can result in cracking of the pipes or joints and, consequently, leaks. If a building is to be left unheated during freezing weather, each trap should be individually drained through the fitting at its lowest point or an antifreeze solution should be added to the water in the trap.

Vents Sometimes the pressure developed in front of a column of water falling through a drainpipe is great enough to blow the water seal (and sewer gases with it) through a fixture trap and into the interior of the building (see Fig. 15.1-52). Under other conditions, the pressure behind a falling column of water can be low enough to suck the water plug out of a fixture trap and into the drainage system. To prevent either of these events from occurring, a second set of openings is added to the drainage system. These openings constitute a venting system that provides a free-flowing air connection to the outdoors.

A *vent* is a pipe that connects a drainage system to the atmosphere to permit gases to escape. Vents are usually made of pipe of the same material and size as the lower components that serve as drains.

The vent piping of a drainage system consists essentially of the upper half of the fixture branch pipes, *circuit vents*, *loop vents*, *relief vents*, *stack vents*, and *vent stacks* (see Fig. 15.1-45).

A *circuit vent* is a branch vent that serves two or more traps and extends from a location in front of the last fixture connection of a horizontal branch to a *vent stack*.

A *loop vent* is a branch vent that serves two or more traps and extends from a location in front of the last fixture connection of a horizontal branch to the *stack vent*.

A *relief vent* is a vent whose primary function is to provide circulation of air between *drainage* and *vent systems* (see Fig. 15.1-45).

A *stack vent* is that portion of a *soil stack* above the highest drain line connection.

A *vent stack* is a vertical vent pipe installed primarily for the purpose of providing circulation of air to and from a part of a drainage system.

The maximum distance between a plumbing fixture's trap and its vent is gov-

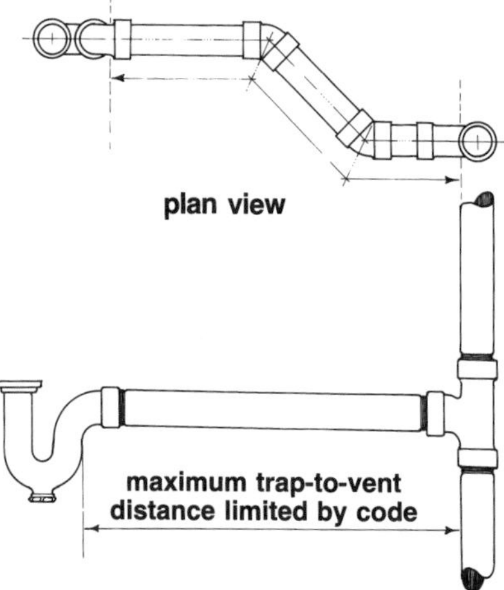

FIGURE 15.1-53 For code compliance, measure trap-to-vent distance as shown in plan view. (Courtesy Small Homes Council, Building Research Council, University of Illinois, Urbana-Champaign.)

erned by the diameter of the drainpipe (Fig. 15.1-53). The larger the diameter of a drainpipe, the farther the vent can be located from the trap and still protect its water seal.

In 1-story buildings or at the top floor of multistory structures, where fixtures are located close to a soil stack, stack venting may be permitted by the applicable codes (Fig. 15.1-54). In this case, the drainpipe is large enough and short enough so that it is never entirely filled with water. Therefore, there is always an air space above the water flow, which acts as a vent passage connecting into the soil stack and ultimately to the outdoors through the top of the soil stack/stack vent.

Sometimes it is inconvenient to run a vent from a fixture that is remote from a wall. In such situations, wet venting can be used. In wet venting, the fixture drain is oversized and used as a vent as well as a drain. Wet venting can be used with fixtures such as sinks and lavatories, but not for venting water closets.

CLEANOUTS

Cleanouts are fittings that provide periodic access to the interior of drainpipes, to permit removal of blockages of lint, hair, grease, and other solids. Cleanouts usually consist of a 45-degree *Y-fitting*, of the same diameter as the drainpipe, with a removable screw-in plug. When the plug is removed, a cleaning tool can be inserted into the main drain line (Fig. 15.1-55).

A cleanout fitting should be located at the base of each soil stack and at the point where the building drain passes through the building's foundation. A cleanout should also be provided at each change of direction of more than 45 degrees, and every 50 ft. (15.24 m) in 4-in. (100 mm) pipe and every 100 ft. (30.5 m) in larger pipes. Cleanouts should be readily accessible, and their height and clearance should be adequate for inserting cleaning tools.

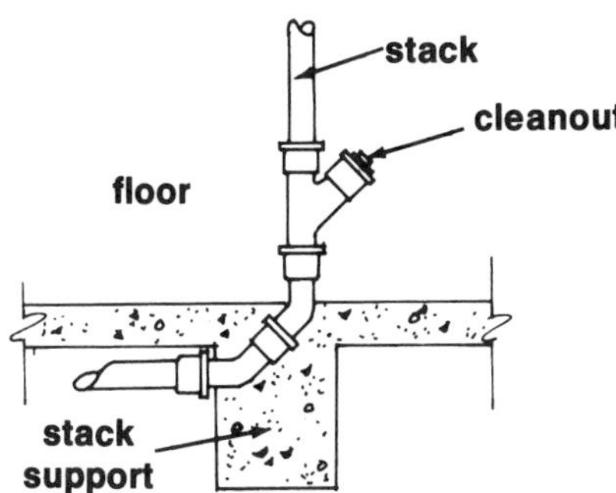

FIGURE 15.1-55 A stack cleanout. (Courtesy Small Homes Council, Building Research Council, University of Illinois, Urbana-Champaign.)

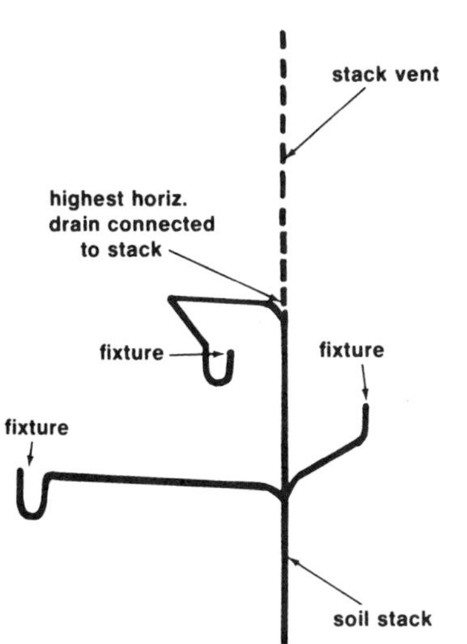

FIGURE 15.1-54 Stack venting permits elimination of fixture vents.

PIPE

Using the correct pipe sizes and materials is essential to the proper functioning of a drainage system.

Pipe Material The most common materials for drainage system piping are cast iron, galvanized wrought iron or steel, and plastic. Copper, brass, and lead can also be used, but their cost is usually prohibitive. See Section 15.1.3.3 for a discussion of these less typical materials. Figure 15.1-56 identifies pipe materials that are acceptable for drain and vent piping.

Cast iron is the long-standing traditional material for drainpiping. It is heavy, durable, and serviceable underground or in concrete. *Hubless* joints and fittings for cast-iron pipe make it possible to run 3-in. (76 mm) cast-iron drain lines in 2 × 4 (50 × 100) wood stud walls (Fig. 15.1-57). Hubless joints consist of a neoprene rubber sleeve gasket that covers the joint, a stainless steel band that fits over the gasket, and worm-screw clamps that compress the band against the pipe. Support must be provided at joints and at 4-ft. (1219 mm) intervals along a cast-iron pipe.

Older installations of cast-iron drainpipe have *bell-and-spigot* or *hub-and-spigot* joints (Fig. 15.1-58). Pipe sections are cast so that one end flares out into a *bell* or *hub*, which fits over the straight end, or *spigot*, of the succeeding section. The gap between the bell and spigot is packed with oakum or tow hemp and filled with molten lead. Such joints have a shoulder on the inner surface and so must be installed directionally to prevent clogging. The direction of flow should always be from the spigot to the bell.

FIGURE 15.1-56 Acceptable Drainage and Vent Piping Materials[a]

Material	Below Grade	Above Grade
Cast Iron		
Soil pipe	A	A
No-hub pipe and fittings	A	A
Copper Tubing		
Type K — Hard temper	A	A
Type L — Hard temper	A	A
DWV Hard temper	A	A
Galvanized Open Hearth or Wrought Iron	A[b]	A
Galvanized Steel Pipe	—	A
Plastic		
ABS[c]	A	A
PVC[c]	A	A
PVC thin wall	NA	A
Polybutylene[c]	A	A
Polyethylene[c]	NA	NA
Styrene rubber	A[c]	NA
VC Pipe	A[c]	—

[a]A = Acceptable; NA = Not acceptable.
[b]Protected when installed underground.
[c]Outside dwelling only.
(Courtesy Small Homes Council, Building Research Council, University of Illinois, Urbana-Champaign.)

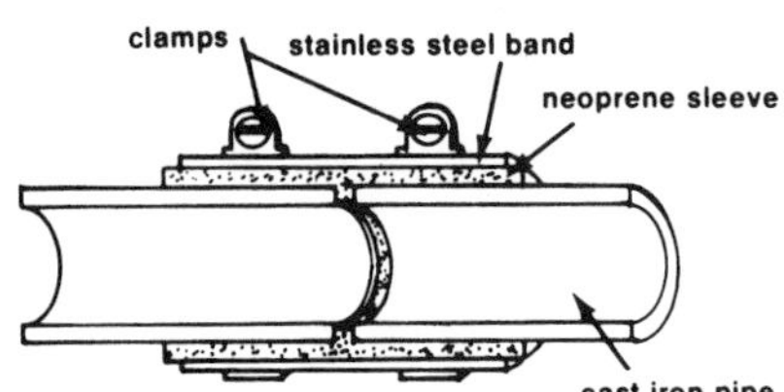

FIGURE 15.1-57 A section through a no-hub cast-iron pipe joint.

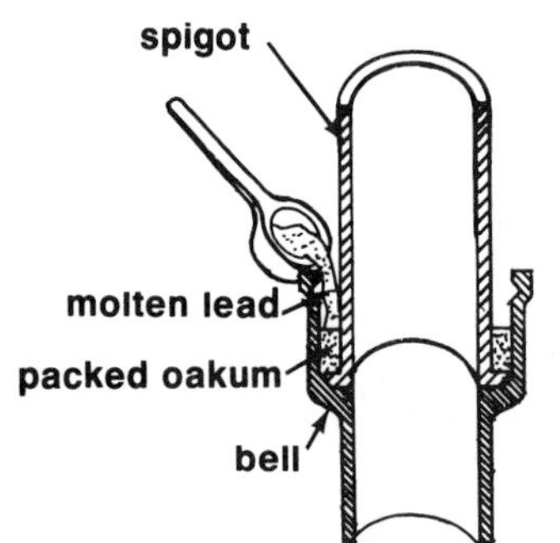

FIGURE 15.1-58 Traditional bell-and-spigot joints in cast-iron pipe are bulky and time-consuming.

Galvanized steel pipe is sometimes called galvanized iron. This pipe is used in drain and vent systems essentially the same way it is used in water supply systems. Threaded joints are typical, but where galvanized pipe connects to cast iron, it is generally fitted into a bell and leaded.

The two types of *plastic pipe* suitable for use in drainage systems are polyvinyl chloride (PVC) and acrylonitrile-butadiene-styrene (ABS). PVC pipe used in drain systems is of the same formulation as that used in water supply systems, but is thicker.

ABS pipe has an operating temperature range of –40 to 80°F (–40 to 26.7°C). It can be obtained with solid walls or with foam-cored walls, the latter being less expensive because it uses less resin. PVC and ABS pipe can be solvent welded at joints with or without a primer, or it can be obtained with threaded fittings (Fig. 15.1-59).

Building and plumbing codes vary in their attitudes toward plastic drainpiping. The applicable codes should always be consulted during the early design of a plumbing system.

Plastic drainpiping has several advantages. Some manufacturers produce a 3-in. (76 mm) drainpipe with joints compact enough to fit within the clearance of a standard 2 × 4 (50 × 100) stud wall (Fig. 15.1-60). This eliminates the need for special framing for plumbing walls. The light weight of plastic pipe means less load on the supporting structure. Since plastic pipe is a better thermal insulator than metal pipe, liquids with suspended fats are less likely to be cooled to the point of depositing fats, which can clog the pipes.

The disadvantages of plastic pipe include its vulnerability to puncture, to damage by rodents, and to some nonhousehold chemicals.

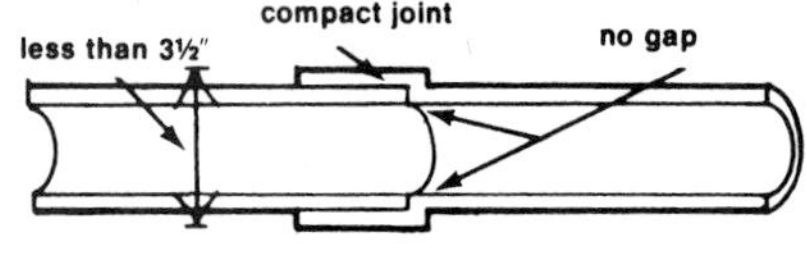

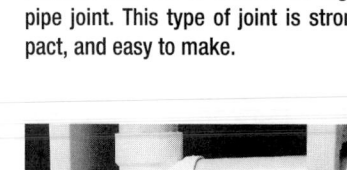

FIGURE 15.1-59 A section through a PVC pipe joint. This type of joint is strong, compact, and easy to make.

FIGURE 15.1-60 A 3-in. (76.2 mm) PVC vent pipe in a 2 × 4 (50 × 100) stud wall. (Courtesy Small Homes Council, Building Research Council, University of Illinois, Urbana-Champaign.)

SIZING

Another factor that should be considered during the design stage is pipe sizing. Expedient selection of pipe size or slope to conform to available floor depth may result in an inefficient and noisy installation. The recommended drain and vent sizes for different parts of a residential system are given in Figure 15.1-61.

Pipe sizing for other building types is designed using water supply fixture units (w.s.f.u.). The procedure is as follows:

1. Make a sketch of the drainage system similar to that shown in Figure 15.1-62. Show all components and all fixtures.
2. Select the minimum trap size for all traps that are not integral to a fixture. In general, every plumbing fixture requires a trap. Select sizes from tables in applicable plumbing codes.
3. Size each fixture branch starting with the section that is farthest from the stack. In Figure 15.1-62, this is the part serving the tub. Determine the number of fixture units for the fixture (for the tub it is 2 w.s.f.u.), using the applicable plumbing code. Figure 15.1-22 is a typical list. Using the w.s.f.u., select the pipe size from tables in the applicable code. In this case, 1½-in. (38 mm) pipe is required. Continue to size each pipe serving a fixture. Then size each successive portion of the branch, making sure to add all the w.s.f.u. together in each case. That is, add the w.s.f.u. of the tub and the lavatory and the sink to size the pipe from the intersection of the tub portion going toward the stack.

FIGURE 15.1-61 Recommended Drain and Vent Pipe Sizes*

Fixture	Drain/Trap Size, in. (mm)	Vent Size, in. (mm)
House sewer	4 (101.6)	—
House drain	3 (76.2)	—
Soil stack	3 (76.2)	—
Branch drain/vent	1½ (38.1)	1¼ (31.8)
Water closet	3½ (88.9)	2 (50.8)
Bathtub	1½ (38.1)	1¼ (31.8)
Lavatory	1¼ (31.8)	1¼ (31.8)
Kitchen sink	1½ (38.1)	1¼ (31.8)
Laundry	1½ (38.1)	1¼ (31.8)
Clothes washer	2 (50.8)	1¼ (31.8)
Dishwasher	1½ (38.1)	1¼ (31.8)
Shower stall	2 (50.8)	1¼ (31.8)

*For one- and two-family dwellings of average size.

(Courtesy Small Homes Council, Building Research Council, University of Illinois, Urbana-Champaign.)

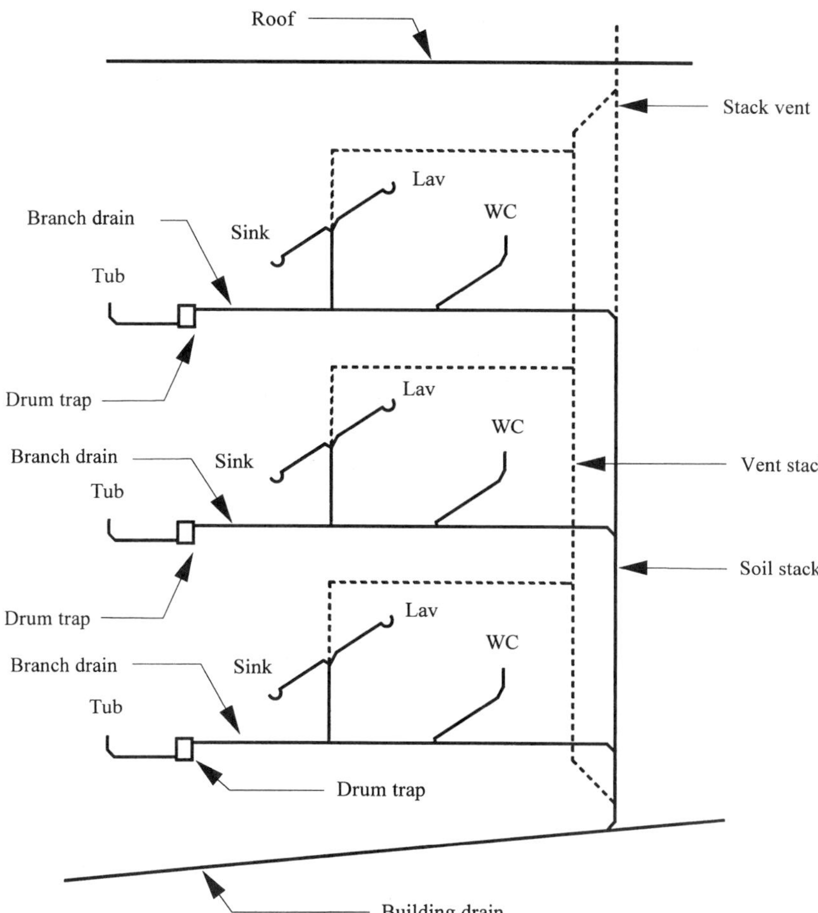

FIGURE 15.1-62 **Drainage riser diagram for an apartment building. A 3-story building is shown, but this kind of diagram is essentially the same regardless of the number of stories. It is necessary, however, to draw every floor, because the pipe sizes differ from floor to floor and a convenient way to keep track of them is to write the sizes on the diagram. Diagrams for other building types are similar.** (Drawing by H. Leslie Simmons.)

4. Next, select the soil stack size. The soil stack is that which carries waste. To do this, add up all the w.s.f.u. above the section being selected and select the size from tables in the applicable code.
5. Then determine the size of the main vent stack. Having determined the length of the vent stack and the size of the soil stack, consult the tables found in the plumbing code and select the vent stack size directly.
6. Finally, size the building drain. Determine the slope of the drain from the position of the street main and soil stacks. Then select the drain size from a table in the plumbing code, using the slope and the number of w.s.f.u. in the building.

In designing a building sanitary drainage system, it is necessary to check carefully the requirements of the applicable plumbing code. Codes often require the drainage component sizes to be determined in accordance with the aforementioned criteria, but with a minimum that may be larger than would be calculated using the required criteria. For example, the code may say that the main soil stack is to be sized in accordance with the tables in the code, but may not be less than 75 mm in diameter.

15.1.4.2 Sanitary Sewage Disposal Systems

Sewage is the name given to liquid waste that contains organic (animal or vegetable) matter in suspension or solution, as well as liquids containing chemicals in suspension. Without treatment to break down such wastes into nontoxic substances, humans increase their amount beyond the ability of the natural environment to absorb them.

PUBLIC SEWAGE DISPOSAL SYSTEM

Large sewage treatment facilities work on the same principle as natural decay and oxidation, except that the action is more concentrated. In a treatment facility, favorable conditions are created for decay bacteria to break down organic solids into chemically simpler water-soluble compounds: phosphates and nitrates. In the natural cycle, limited amounts of phosphates and nitrates are produced and these become the nutrients for other growing things. Growing plants absorb such nutrients through their roots and, with the help of sunlight, recombine them with carbon dioxide from the air to form new living tissue. In sewage treatment, large amounts of phosphates and nitrates are produced and flushed away by the effluent to be absorbed by the soil or discharged into a body of water.

Phosphates and nitrates are relatively harmless when released into the soil or water in limited quantities. When introduced into a lake or stream in large amounts, they spur the growth of small aquatic plants (algae), which use up valuable dissolved oxygen as they proliferate and die off in large quantities. This overenrichment of a body of water with excessive nutrients, proliferation of algae, and reduction of oxygen content is characteristic of the *eutrophication* process, which gradually destroys aquatic life and renders the water useless for recreation and water supply.

Proper disposal of human excreta (body wastes) is of major importance to the health of individuals. Where public sewers are not available, many diseases, such as dysentery, infectious hepatitis, typhoid, and paratyphoid, and various types of diarrhea are transmitted from one person to another through the fecal contamination of food and water, largely due to the improper disposal of human wastes.

Safe disposal of human wastes is necessary to protect the health of the community and to prevent the occurrence of nuisances. To accomplish satisfactory results, such wastes must be disposed of so that they will not:

- Contaminate a drinking water supply;
- Pollute the waters of any bathing beach, shellfish breeding ground, or stream used for water supply or for recreation;
- Be discharged on or near the surface where they would be accessible to insects, rodents, or other small animal

carriers that may come in contact with food or drinking water;

- Be accessible to children and give rise to a nuisance due to foul odor or unsightly appearance;
- Overburden any body of water with excess nutrients or degrade its quality below officially adopted water quality standards.

PRIVATE SEWAGE DISPOSAL SYSTEM

Connection to a public sewerage system is the most satisfactory method of sewage disposal. When this is not feasible, and a when a considerable number of residences or other small buildings are to be served, consideration should be given to the construction of a community sewerage system and treatment plant. Specific information on this matter can generally be obtained from a local health authority.

In spite of the immense growth in public sewage disposal systems in this century, many areas are still beyond their reach. Community systems are not always feasible. In locations with no access to either a municipal or a community system, sewage disposal systems designed to handle a single residence, a small group of dwellings, or a small commercial project continue to be an important method of sewage disposal.

Not long ago, in rural areas human and other domestic wastes were disposed of using cesspools. These are covered and lined underground pits designed to permit liquids to seep into the ground through the bottom and sides, but retain organic matter and other solids, which are then periodically pumped out. Such systems not only present an obnoxious nuisance, they are also a major health hazard and are almost universally prohibited in this country. Comparatively, today's *individual sewage (septic) disposal systems* are safe and nuisance free (Fig. 15.1-63).

Septic Tank Systems These systems (Fig. 15.1-63) consist of a *septic tank*, where the sewage is retained long enough to separate and decompose the solids, coupled with some type of *soil absorption system* (absorption field, absorption bed, or seepage pit) whereby the liquids discharged from the septic tank seep into the soil.

In these systems, sewage is treated both in the tank and in the soil, as anaerobic (oxygen-free) bacteria decompose the organic matter into simpler, harmless compounds. However, because the anaerobic decomposition of organic matter is accompanied by foul odors and pathogenic (disease-producing) organisms, the partially treated liquid sewage flowing from a septic disposal system (*effluent*) cannot be discharged on the ground surface or into a lake or stream without creating a nuisance and a health hazard.

Instead, the effluent from a septic tank is discharged into a properly designed and maintained soil absorption system, which uses the soil for subsurface absorption of treated sewage. It practically eliminates the health and nuisance hazards, but still poses problems in certain areas. A soil absorption system may not work in tightly packed soils, or the land area required may be too large.

The removal of solids and the partial biological treatment that takes place in a septic tank is known as *primary treatment*. Most municipal and community sewage plants provide *secondary treatment*, producing relatively odorless but nutrient-rich effluents. Therefore, sewage plants should be equipped to remove or neutralize the nutrients (a process called *tertiary treatment*) before discharging into lakes or streams.

The discussion and recommendations presented in this section have been abstracted from a publication of the Bureau of Community Environmental Management, U.S. Public Health Service (HEW), entitled *Manual of Septic Tank Practice*, 1967 edition, with some updating to reflect current practice. Codes and health officials should be consulted and their lat-

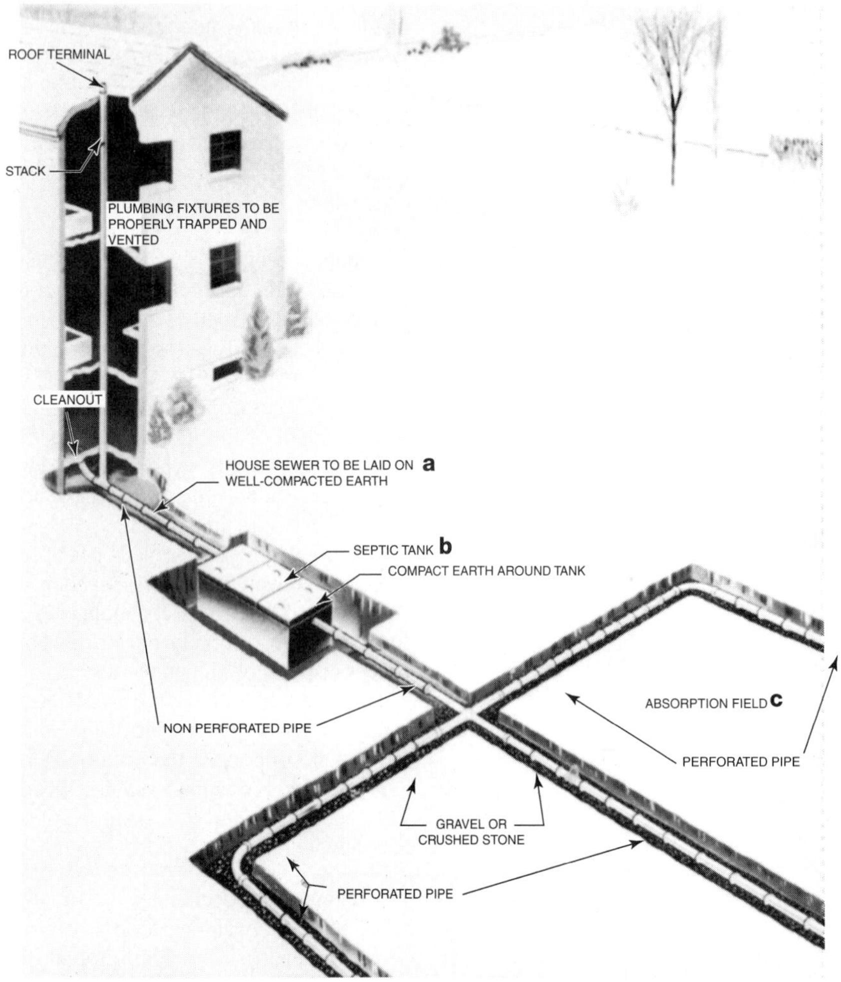

FIGURE 15.1-63 Typical septic-tank–type individual sewage disposal system. Solids are separated and retained in the tank, liquids are passed to the absorption field, and gases are vented through the house sewer and vent stack. (Illustration from the *Manual of Septic Tank Practices*, Bureau of Community Environmental Management, U.S. Public Health Service [now EPA].)

est requirements obtained before beginning design of an individual sewage disposal system.

This discussion of septic tank systems is intended as an aid in understanding the nature of such systems. The recommendations here are made with the full understanding that they may differ in some instances from other authoritative sources such as the standards and manuals referenced in the HUD *Minimum Property Standards for Housing* and the various model plumbing codes. Local, state, and federal laws, codes, and regulations always supersede the recommendations in this section.

Suitability of Soil A sewage disposal system that uses the soil for subsurface absorption of treated sewage is called a *soil absorption system*. A septic tank system is in this category. The first step in the design of a septic tank system is to determine whether the soil is suitable for the absorption of effluent and, if it is, how much absorption area is required. In general, a site is unsuitable for a soil absorption system unless the following conditions can be satisfied:

1. The percolation rate is within the range indicated in Figure 15.1-64, that is, not more than 60 min. for absorption fields and beds, nor more than 30 min. for seepage pits. Percolation time is the time required for the level of water in a standard test pit to drop 1 in. (25.4 mm).
2. The bottom of a trench or seepage pit is at least 4 ft. (1219 mm) above the maximum seasonal elevation of the groundwater table.
3. Rock formations and other impervious strata are more than 4 ft. (1219 mm) below the bottom of a trench, bed, or pit.
4. The area is large enough to provide the required absorption area in trench and bed bottoms or pit sidewalls, as shown in Figure 15.1-64.

FIGURE 15.1-64 Required Absorption Areas in Various Soils

Percolation Rate, Time in min. for Water to Fall 1 in. (25.4 mm)	Minimum Area per Bedroom, sq. ft. (m^2)
2 or less	85 (7.9)
3	100 (9.3)
4	115 (10.7)
5	125 (11.6)
10	165 (15.3)
15	190 (17.7)
30[a]	250 (23.2)
45	300 (27.9)
60[b]	330 (30.7)

[a]Soil not suitable for seepage pit if percolation rate is more than 30 min.

[b]Soil not suitable for any absorption system if percolation rate is more than 60 min.

(Table excerpted from the *Manual of Septic Tank Practices*, Bureau of Community Environmental Management, U.S. Public Health Service [now EPA].) Metric added by author.

The water absorption of soils can be determined best by percolation tests, but a number of clues help in estimating the likely characteristics of a particular soil. These include the soil maps and descriptions of the U.S. Department of Agriculture, on-site visual inspection of soil texture and color, and the record of experience in adjacent areas with similar soils.

For these purposes, soil can be classified as either pervious or impervious. *Pervious soil* is usually granular, such as sand or gravel. It allows water to pass readily through it. *Impervious soil* is a tight, cohesive soil, such as clay, which does not permit the ready passage of water. Most sites contain a mixture of pervious and impervious soils.

Soil particle size governs pore size, which, in turn, influences the rate of water movement. Sandy and gravelly soils generally have large particles and large pores, while clay and silt soils have small particles and small pores (although total void area may be larger). In some soils, especially clay and silt types, there may be clumping of basic particles into larger particles, giving the soil a well-developed structure. A lump of structured soil will break apart with little force along well-defined cleavage planes. Such structured clay soil is likely to have better water absorption characteristics than one with little or no structure.

Another clue to the absorptiveness of a soil is its color. Bright, uniform, reddish-brown to yellow colors throughout a soil profile indicate favorable oxidation conditions, good air and water movement, and desirable absorption properties.

In some cases, examination of road cuts, stream embankments, or building excavations will give useful information. Well drillers' logs can also be used to obtain data on groundwater and subsurface conditions. In some areas, subsoil strata vary widely in short distances, and borings must be made at the site of the system. If the subsoil appears suitable, as judged by the visual characteristics described above, percolation tests should be made at points and elevations selected as typical of the area in which the soil absorption system will be located.

Subsurface exploration is usually necessary to determine soil conditions: the depth of pervious material, the presence of rock formations, and the level of groundwater. Augers with extension handles are often used for making such investigations (Fig. 15.1-65).

Percolation tests help to determine the acceptability of a site and establish the type and size of the required soil absorption system. Figure 15.1-64 shows the area of absorption or seepage area recommended for soils with various percolation rates, as well as the slowest percolation rate acceptable for each type of soil absorption system.

At least six separate test holes should be made. These should be spaced uniformly over the proposed soil absorption area and should reach down to the depth of the proposed absorption system. These holes can be from 4 to 12 in. (100 to 305

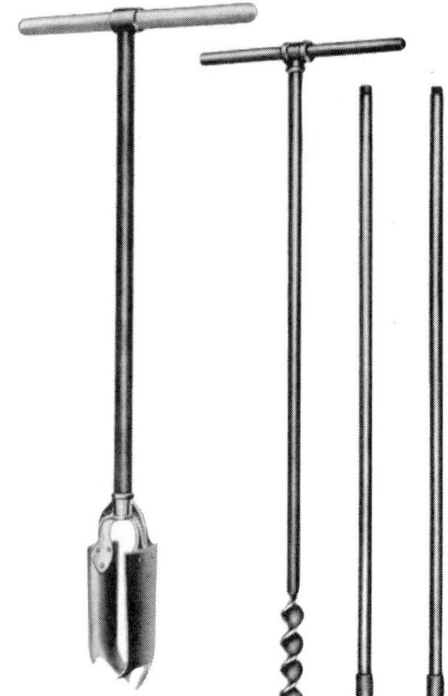

FIGURE 15.1-65 Augers with extension handles can be used to bore deep test holes. (Illustration from the *Manual of Septic Tank Practices*, Bureau of Community Environmental Management, U.S. Public Health Service [now EPA].)

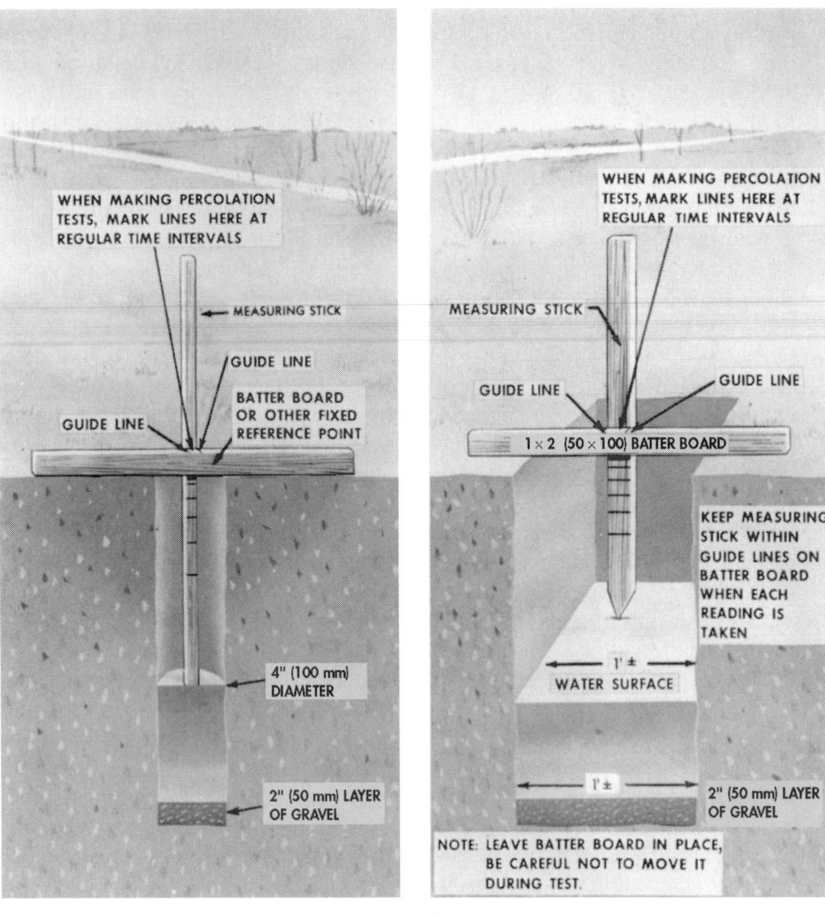

FIGURE 15.1-66 Shallow percolation test holes for absorption trench or bed showing (a) a 4-in. (100 mm) auger-bored hole and (b) a larger test hole dug with a shovel. (Illustration from the *Manual of Septic Tank Practices*, Bureau of Community Environmental Management, U.S. Public Health Service [now EPA].) Metric added by author.

mm) in diameter. To save time, labor, and the volume of water required per test, shallow holes can be bored with a 4-in. (100 mm) auger (Fig. 15.1-66a). Deeper holes may have to be dug by shovel or heavy equipment (Fig. 15.2-66b).

The bottom and sides of a test hole should be scratched with a knife or sharp-pointed instrument to remove smeared soil and to provide a natural soil surface into which water may percolate. Loose material should be removed from the hole, and 2 in. (50.8 mm) of coarse sand or fine gravel should be added to protect the bottom from scouring and sediment.

Saturation means that the void spaces between individual particles of soil are full of water, which can be accomplished in a short period of time. *Swelling* is caused by the intrusion of water into the particles of soil, which is a slow process. It is the chief reason for providing an overnight soaking period, especially in clay.

The suggested test procedures are intended to ensure that the soil is given ample opportunity to swell and to approach the condition that will exist during the wettest season of the year. Thus, the test should give comparable results in the same soil, whether it is performed in a dry or in a wet season. In sandy soils containing little or no clay, the swelling procedure is not essential and the test may be made after the water from a single filling of the hole has completely seeped away.

The length of time required for percolation tests varies, depending on the type of soil. For example, the recommended swelling procedure may not be needed in a sandy soil containing little or no clay, and the test may be made after the water from one filling of the holes has completely seeped away.

To begin a test, clear water should be placed in the test holes until the gravel is covered by at least 12 in. (305 mm) of water. In most soils, it is necessary to refill the hole by supplying a surplus reservoir of water, possibly by means of an automatic siphon, to keep water in the hole for at least 4 hours and preferably overnight (see Figs. 15.1-66 and 15.1-67). The purpose of this preliminary soaking is to obtain the percolation rate that will occur when the soil is soaked and saturated.

If water remains in a hole after the swelling period, more water should be added, or water should be removed, until water covers the gravel by about 6 in. (152 mm). The drop in water level should then be measured from a fixed reference point at 30-min. intervals over a 4-hr. period. The drop that occurs in the final 30-min. period should be used to calculate the percolation rate. For instance, a drop of 2½ in. (63.5 mm) in 30 min. translates to a percolation rate of 12 minutes per in. (30 min. divided by 2.5 in.) (0.47 min. per mm).

If no water remains in a hole after the swelling period, water should be added until water covers the gravel by about 6 in. (152 mm). From a fixed reference point, the drop in water level should be measured at 30-min. intervals for 4 hr., refilling up to 6 in. (152 mm) over the gravel as necessary. The drop that occurs during the final 30 min. should be used to calculate the percolation rate.

In sandy or other permeable soils in which the first 6 in. (152 mm) of water seep away in less than 30 min. after the swelling period, measurements should be taken at 10-min. intervals for 1 hr. The drop that occurs during the final 10 min. should be used to calculate the percolation rate.

Soil Absorption Systems From the results of the percolation tests it is possible to determine whether a septic tank system is feasible and which type of soil absorption system can be used. When a soil absorption system is usable, three types of design may be considered: absorption field, absorption bed, and seepage pit.

The following definitions apply to soil absorption systems.

- *Absorption field:* An arrangement of absorption trenches through which treated sewage is absorbed into the soil. These are also called *disposal fields* and *soil absorption fields.*

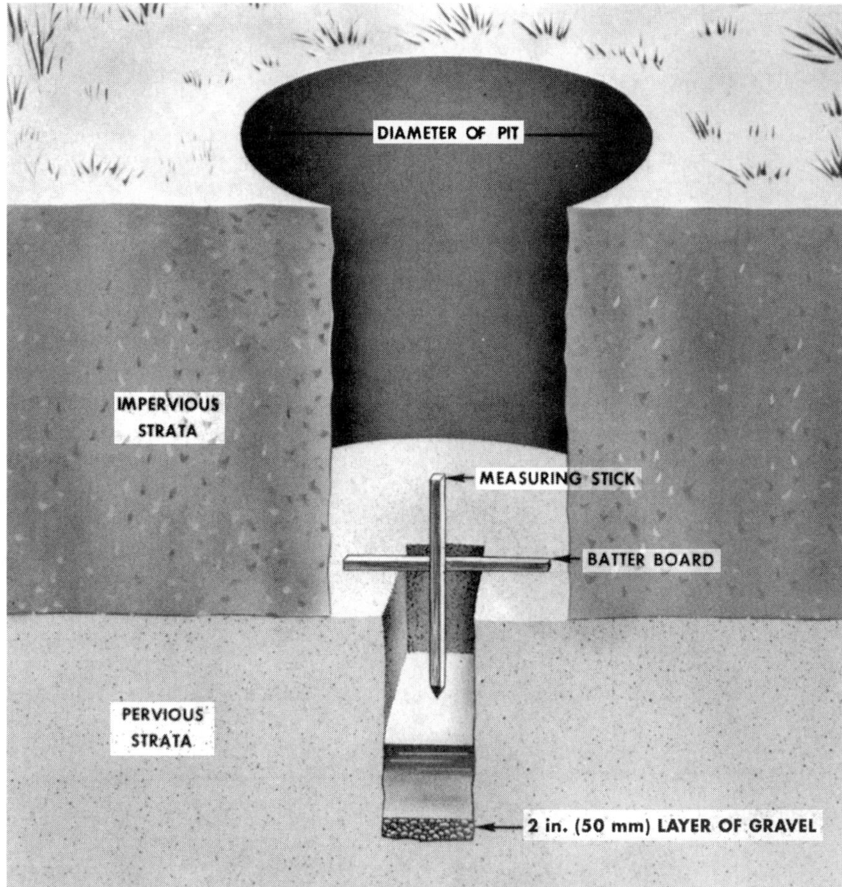

FIGURE 15.1-67 Percolation tests for seepage pits must be made in deep test holes dug down to the level of the proposed bottom of the pit. (Illustration from the *Manual of Septic Tank Practices*, Bureau of Community Environmental Management, U.S. Public Health Service [now EPA].) Metric added by author.

- *Absorption trench:* A trench not more than 36 in. (914 mm) in width, containing at least 12 in. (305 mm) of clean, coarse aggregate and a distribution line, through which treated sewage is allowed to seep into the soil.
- *Absorption bed:* Similar to an absorption trench, except that it is more than 36 in. (914 mm) wide and contains two or more distribution lines. These are also called *seepage beds.*
- *Distribution line:* An open-jointed or perforated pipe that permits effluent to escape into the soil for absorption. It may consist of 1-ft. (305 mm) lengths of 4-in. (100 mm) agricultural drain tile or 2- to 3-ft. (610 to 915 mm) lengths of vitrified clay sewer pipe (both laid with open joints), or perforated plastic pipe. Local experience and applicable codes and ordinances should be reviewed before selecting piping materials.
- *Seepage pit:* A covered underground pit with a concrete or masonry lining. Such pits are designed to permit partially treated *sewage* to seep into the surrounding soil. They are also called *dry wells.* Some jurisdictions do not permit the use of seepage pits.

As Figure 15.1-64 shows, only soils with a percolation rate of 30 min. per in. (25.4 mm) or less are suitable for seepage pits; if the rate is greater than 60 min., the soil is not suitable for any type of soil absorption system. In addition, the selection of a soil absorption system is affected by the location of the system. For example, a safe distance must be maintained between a soil absorption system and a source of water supply. Since the distance pollution will travel underground depends on numerous factors, including the characteristics of the subsoil formations and the quantity of sewage discharged, no specified distance would be absolutely safe in all localities. In general, the horizontal separation between elements of a soil absorption system and other site improvements should not be less than that outlined in Figure 15.1-68.

Seepage pits should not be used in areas where water supplies are obtained from shallow wells, or where there are limestone formations and sinkholes through which effluent may easily contaminate the groundwater.

While *absorption fields*, *absorption beds*, and *seepage pits* should not be in-

FIGURE 15.1-68 Minimum Clearances for Components of a Septic Sewage Disposal System[a]

From	To Well, ft. (m)[b]	To Water Line, ft. (m)[c]	To Foundation Wall, ft. (m)	To Property Line, ft. (m)	To Stream, ft. (m)
Building sewer	50 (15.24)	10 (3.05)	—	—	50 (15.24)
Septic tank	50 (15.24)	10 (3.05)	5 (1.52)	10 (3.05)	50 (15.24)
Absorption field	100 (30.48)	25 (7.62)	20 (6.10)	5 (1.52)	50 (15.24)
Absorption bed	100 (30.48)	25 (7.62)	20 (6.10)	5 (1.52)	50 (15.24)
Seepage pit	100 (30.48)	50 (15.24)	20 (6.10)	10 (3.05)	50 (15.24)

[a]Based on recommendations of U.S. Public Health Service, HEW.

[b]And to suction line between well and pressure tank.

[c]Pressure type.

(Table excerpted from the *Manual of Septic Tank Practices*, Bureau of Community Environmental Management, U.S. Public Health Service [now EPA].) Metric added by author.

termixed, several of any one of them can be arranged in sequence so that each is forced to utilize its total effective absorption area before liquid flows into the succeeding component. This is called *serial distribution*.

The individual branches of an *absorption field* are called *laterals*. These consist of either

1. The length of distribution line between the tee or cross fitting and the farthest point in a closed-loop field (Fig. 15.1-69); or
2. The length of distribution line between overflow pipes (Fig. 15.1-70).

There should be at least two laterals in every system. No lateral should be more than 100 ft. (37.8 m) long. Using more and shorter laterals is preferred because if something should happen to disturb one line, most of the field will still be serviceable. The trench bottom and distribution lines should be level. Laterals should be spaced so as to provide at least 6 ft. (1829 mm) of undisturbed earth between trenches.

Many different designs may be used in laying out subsurface absorption fields. The choice depends on the size and shape of the available disposal area, the capacity required, and the topography of the area (see Figs. 15.1-69 and 15.1-70).

To provide the minimum required gravel depth and earth cover, the trench depth should be at least 24 in. (610 mm) (Fig. 15.1-71). Additional depth may be needed for surface contour adjustment, for extra aggregate under the tile, or for other design purposes.

In considering the depth of absorption trenches, a question may arise concerning the possibility of the lines freezing during prolonged cold periods. Freezing rarely occurs in distribution lines that are adequately surrounded by gravel at the proper depth when the system is kept in continuous operation. However, pipes under driveways or other surfaces that are usually cleared of snow should be close jointed and insulated or should be installed at a greater depth.

The total required absorption area is determined by the results of soil percolation tests. The required area for residences may be obtained from Figure 15.1-64. For example, for a three-bedroom house on a lot where the minimum percolation rate was 15 min. per in. (25.4 mm), the necessary absorption area would be 3 times 190 sq. ft. (17.65 m^2) per bedroom, or 570 sq. ft. (52.95 m^2). For trenches 2 ft. (610 mm) wide, the required total length of trench would be 570 sq. ft. (52.95 m^2) divided by 2, or 285 linear ft. (26.48 m). If three laterals were used on sloping ground (see Fig. 15.1-70), the length of each lateral would be 285 ft. (26.48 m) divided by 3, or 95 ft. (8.83 m). If a closed-loop system on level ground (see Fig. 15.1-69) were used, each of the half-loops could be 100 ft. (9.29 m), and the middle lateral 85 ft. (7.9 m). The spacing of trenches is generally governed by practical construction considerations such as type of equipment, soil, and topography, but the clear space between trenches should not be less than 6 ft. (1.83 m) to prevent short-circuiting (flow between trenches).

In the example cited, trenches are 2 ft. (0.61 m) wide, there are three trenches, and two spaces between trenches at 6 ft. (1.83 m) each. The total width of trenches and spaces is 6 ft. (1.83 m) plus 12 ft. (3.66 m), or 18 ft. (5.49 m). The required area for the absorption field is 18 ft. (5.49 m) times 95 ft. (29 m), which equals 1710 sq. ft. (158.86 m^2), plus the additional land required to provide the clearances recommended in Figure 15.1-68.

Ideally, a house site should be large enough to allow room for an additional system if the first one fails or if bedrooms are added later.

Careful construction is important in obtaining a satisfactory soil absorption system. Attention should be given to protecting the natural absorption properties of the soil. Care must be taken to prevent sealing of the surface on the bottom and sides of the trench. Trenches should not be excavated when the soil is wet enough to smear or compact easily. Soil moisture is right for safe working only when a handful will mold with considerable pressure. Open trenches should be protected from surface runoff to prevent the entrance of silt and debris. If it is necessary to walk in the trench, a temporary board laid on the bottom will reduce compaction. If smearing and compaction do occur, surfaces should be raked to a depth of 1 in. (25.4 mm), and loose material removed, before the gravel is placed in the trench.

A distribution pipe should be laid in a level trench of sufficient width and depth (see Fig. 15.1-71) and should be surrounded by clean, graded gravel or rock, broken hard-burned clay brick, or similar aggregate, ranging in size from ½ to 2½ in. (12.7 to 63.5 mm). Cinders, broken shell, and similar materials are not recommended, because they are usually too fine and may lead to premature clogging. The coarse aggregate should extend from at least 2 in. (50.8 mm) above the top of the pipe to at least 6 in. (152 mm) below the bottom of the pipe. If open-joint drain tile is used, the upper half of each joint opening should be covered with a strip of

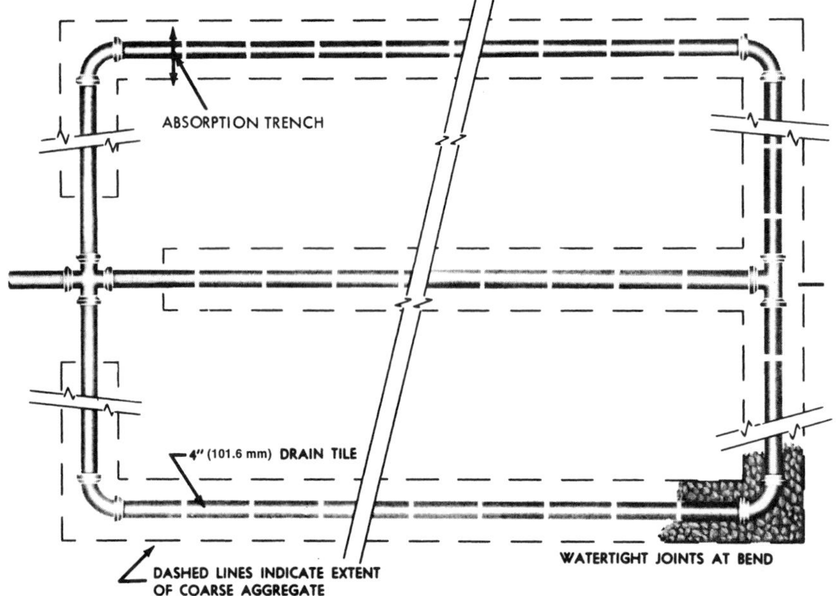

FIGURE 15.1-69 **Typical absorption field layout for level ground. Length of lateral should be figured from start of absorption trench (dashed lines) to farthest point.** (Illustration from the *Manual of Septic Tank Practices*, Bureau of Community Environmental Management, U.S. Public Health Service [now EPA].) Metric added by author.

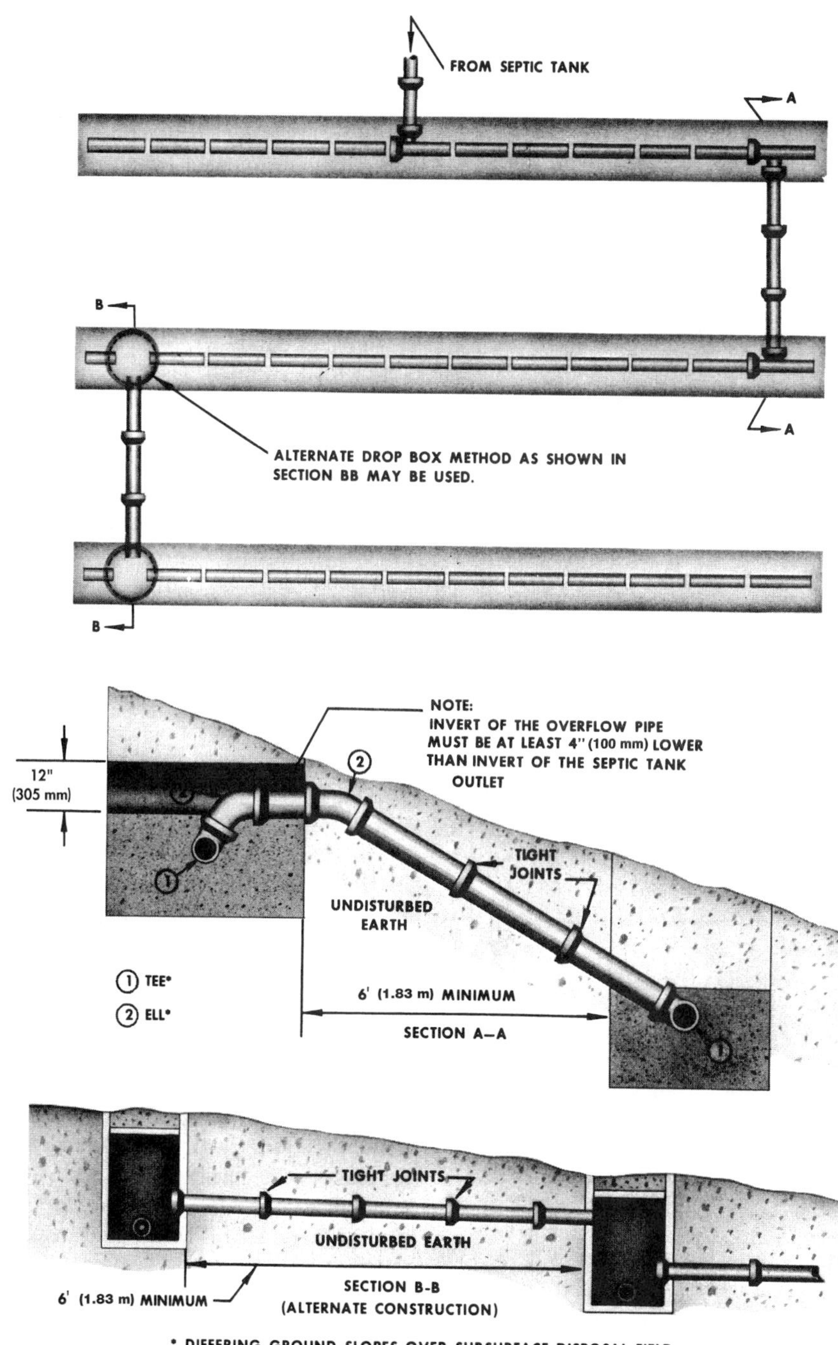

FIGURE 15.1-70 Typical absorption field layout for sloping ground. (Illustration from the *Manual of Septic Tank Practices*, Bureau of Community Environmental Management, U.S. Public Health Service [now EPA].) Metric added by author.

No. 15 roofing paper, as shown in Figure 15.1-71.

The top of the stone should be covered with a vapor-pervious barrier, such as untreated building paper, a 2-in. (50.8 mm) layer of hay or straw, or similar pervious material to permit vapor movement and yet prevent the stone from becoming clogged by the earth backfill. An impervious covering should not be used, as this interferes with *evapotranspiration* (absorption and release of moisture by plants) at the surface. Although generally not figured in the calculations, evapotranspiration is often an important factor in the operation of absorption fields and beds.

Open-joint tile is usually laid with the opening covered by strips of roofing paper, but drain tile connectors, collars, clips, and similar accessories are of value in obtaining uniform spacing and protecting tile joints. Most such accessories are made of galvanized iron, copper, or plastics.

Tree and shrub roots sometimes penetrate into transmission lines and block the flow of effluent. Root problems can be prevented by using a liberal amount of gravel or stone over and under the pipe. Roots generally seek ground moisture; in the few cases in which they become troublesome, there is usually an explanation involving changing moisture conditions. For example, at a residence used only during the summer, roots are most likely to penetrate when the house is uninhabited or when moisture immediately below or around the gravel becomes less plentiful than during the period when the system is in use. In general, trenches constructed within 10 ft. (3.05 m) of large trees or dense shrubbery should have at least 12 in. (305 mm) of gravel or crushed stone beneath the pipe.

The backfill over a new absorption trench should be free of large stones and other deleterious materials and should be hand tamped and overfilled 4 to 6 in. (100 to 152 mm) higher than adjacent grade to compensate for likely settling. Settlement of the fill in a trench to a lower level is very undesirable because it may cause collection of storm water in the trench, which can lead to premature saturation of the absorption field or even complete washout of the trench.

Heavy vehicles will crush the tile in a shallow absorption field and should not be operated there unless special provisions are made to support the weight. Machine tamping or hydraulic backfilling of absorption trenches should not be permitted. Machine grading should be completed before the field is laid.

Where the ground does not exceed a 6-in. (152 mm) height variation in any direction within the absorption area, effluent may be discharged through a system of laterals connected at their extremes to form a continuous closed loop (see Fig. 15.1-69). Such a system permits the effluent to flow into all the tile lines whenever there is a discharge from the septic tank.

Laterals traditionally were connected to a house sewer through a distribution box and the ends were not interconnected. A Public Health Service study for the Federal Housing Administration (FHA) on

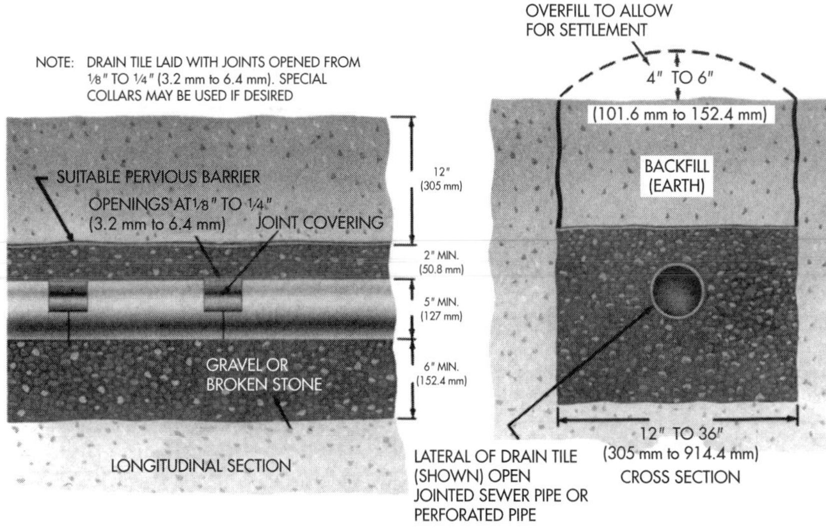

FIGURE 15.1-71 Details of typical absorption trench. (Illustration from the *Manual of Septic Tank Practices*, Bureau of Community Environmental Management, U.S. Public Health Service [now EPA].) Metric added by author.

the effectiveness of distribution boxes reached the conclusion that distribution boxes generally do not ensure equal distribution as intended and therefore are not needed for single-family systems. However, even when local practice includes distribution boxes, the laterals should be interconnected at their ends, as in Figure 15.1-69. Interconnection of laterals ensures access to each point in a lateral from two sides, so that if one part of the lateral becomes damaged, effluent will be able to reach the lateral from the other side.

Where sloping ground is used for an absorption field, it is usually necessary to construct a small temporary dike or surface-water diversion ditch above the field to prevent the absorption field from being eroded by surface runoff. This dike should be maintained or the ditch kept free of obstructions until the field becomes well covered with vegetation.

The maximum ground slope suitable for soil absorption systems is governed by local factors affecting erosion of the ground used for the absorption field. Excessive slopes that are not protected from surface water runoff or do not have adequate vegetation cover to prevent erosion should be avoided. Generally, ground having a slope greater than 1 vertical to 2 horizontal should be investigated carefully to determine whether it is satisfactory from the standpoint of erosion potential. Near banks, the horizontal distance from the side of a trench to the ground surface should be 2 ft. (610 mm) or more to prevent lateral flow of effluent and breakout onto the surface.

In sloping ground each lateral should be connected to the next by a closed overflow line laid on an undisturbed section of ground, as shown in Figure 15.1-70. The arrangement is such that all effluent is discharged to the first trench until it is filled; excess liquid is then carried by means of a closed overflow line to the next succeeding or lower trench. In that manner, each portion of the subsurface system is used in succession.

Design and construction procedures for sloping-ground fields are basically the same as those for level-ground fields. The bottoms of trenches and of distribution lines should be level, and laterals should follow surface contours. Adjacent trenches may be connected with either an overflow line or a drop box arrangement. The *invert* (bottom) of the first overflow line should be at least 4 in. (100 mm) lower than the invert of the septic tank outlet (see Fig. 15.1-70).

Overflow lines should be 4-in. (100 mm) tight-joint pipe with direct connections to the distribution lines in adjacent trenches or to a drop box arrangement. Care must be exercised in constructing overflow lines to ensure at least 6 ft. (1.83 m) of undisturbed earth between trenches. An overflow line should rest on undisturbed earth, and backfill should be carefully tamped.

An *absorption bed* is a variation of the typical trench used in absorption fields. The basic design is essentially the same as for an absorption trench, but it is generally much wider than 36 in. (0.914 m) and contains two or more distribution lines, 4 to 6 ft. (1.22 to 1.83 m) apart. Absorption beds have the advantages of making more efficient use of the land and of modern earth-moving equipment.

The minimum bottom area for absorption beds is the same as for trenches in absorption fields (see Fig. 15.1-64). A bed should have a minimum depth of 24 in. (610 mm) below natural ground level to provide a minimum of 12 in. (305 mm) of gravel and 12 in. (305 mm) of earth backfill. The gravel should extend at least 2 in. (50.8 mm) above and 6 in. (152 mm) below the distribution line. The bottoms of beds and distribution lines should be level. Distribution lines should be spaced not more than 6 ft. (1.83 m) apart and not more than 3 ft. (0.9 m) from the bed sidewall. When more than one bed is used, there should be a minimum of 6 ft. (1.83 m) of undisturbed earth between adjacent beds, and the beds should be connected in series, as illustrated in Figure 15.1-70 for absorption fields in sloping ground.

Seepage pits are deeper than absorption trenches or beds and are, therefore, more likely to contaminate groundwater than other absorption methods. In fact, they are not permitted in some jurisdictions. Because pits handle the same amount of effluent as an absorption field or bed in a smaller land area, the likelihood of clogging and saturation is greater, as is the possibility of effluent breaking through to the surface.

Pits should not be used where absorption fields or beds are feasible, or when the soil has a percolation rate in excess of 30 min. per in. (25.4 mm). Where the top 3 or 4 ft. (0.914 or 1.29 mm) of soil is not porous enough for an absorption field, but is underlaid with suitable granular material, seepage pits constitute an acceptable substitute. Seepage pits may also be used where the land area is too small, too irregular, or too heavily wooded to accommodate absorption field or bed systems.

The required capacity of deep pits should be computed on the basis of percolation tests made in each vertical stratum penetrated (see Fig. 15.1-67). Soil strata in which the percolation rates are in excess of 30 min. per in. (25.4 mm) should not be included in computing the absorption area. The weighted average of the re-

FIGURE 15.1-72 Required Wall Area for Seepage Pits

Symbol*	Character of Soil	Minimum Wall Area per Bedroom, sq. ft. (m^2)
GW	Well-graded gravels, gravel-sand mixtures, little or no fines	50 (4.65)
GP	Poorly graded gravels or gravel-sand mixtures, little or no fines	50 (4.65)
SW	Well-graded sands, gravelly sand, little or no fines	75 (6.97)
SP	Poorly graded sands or gravelly sands, little or no fines	75 (6.97)
SM	Silty sand, sand-silt mixtures	125 (11.6)
GC	Clayey gravels, gravel-sand-clay mixtures	150 (13.94)
SC	Clayey sands, sand-clay mixtures	150 (13.94)
ML	Inorganic silts and very fine sands, rock flour, silty or clayey fine sands or clayey silts with slight plasticity	300 (27.9)
CL	Inorganic clays of low to medium plasticity, gravelly clays, sandy clays, silty clays, lean clays	350 (32.52)
OL	Organic silts and organic silty clays of low plasticity	
MH	Inorganic silts, micaceous or diatomaceous fine sandy or silty soils, elastic silts	
CH	Inorganic clays of high plasticity, fat clays	Unsuitable for Seepage Pits
OH	Organic clays of medium to high plasticity, organic silts	
Pt	Peat and other highly organic silts	

*Refer to Section 2.1 for a discussion on soil classification.

(Table excerpted from the *Manual of Septic Tank Practices*, Bureau of Community Environmental Management, U.S. Public Health Service [now EPA].) Metric added by author.

FIGURE 15.1-73 Effective Wall-Absorption Areas (sq. ft. [m^2]) of Round Seepage Pits

Pit Diameter, ft.	Depth Below Inlet									
	1 ft.	2 ft.	3 ft.	4 ft.	5 ft.	6 ft.	7 ft.	8 ft.	9 ft.	10 ft.
3	9.4	19	28	38	47	57	66	75	85	94
4	12.6	25	38	50	63	75	88	101	113	126
5	15.7	31	47	63	79	94	110	126	141	157
6	18.8	38	57	75	94	113	132	151	170	188
7	22.0	44	66	88	110	132	154	176	198	220
8	25.1	50	75	101	126	151	176	201	226	251
9	28.3	57	85	113	141	170	198	226	254	283
10	31.4	63	94	126	157	188	220	251	283	314
11	34.6	69	104	138	173	207	242	276	311	346
12	37.7	75	113	151	188	226	264	302	339	377

(Table excerpted from the *Manual of Septic Tank Practices*, Bureau of Community Environmental Management, U.S. Public Health Service [now EPA].) Refer to Figure 1.8-27 for metric conversions.

sults should be computed to obtain an average percolation rate. In shallow pits where the soil is relatively uniform, the required absorption area can be established by soil experts through visual analysis of the soil type (Fig. 15.1-72). In other cases, the absorption area should be established by percolation tests as for other absorption methods.

Figure 15.1-64 gives the absorption area requirements per bedroom for soils with different percolation rates. The absorption area of a seepage pit is the vertical wall area of the actual excavated pit (not the pit lining). Only the area of pervious strata below the inlet should be figured. No allowance should be made for impervious strata or for bottom area. After the required area has been established, Figure 15.1-73 may be used to select the desired dimensions of circular or cylindrical seepage pits.

To demonstrate a typical pit capacity calculation, assume that the absorption system is to serve a three-bedroom house on a lot where the percolation rate is 1 in. (25.4 mm) in 15 min. Assume also that the water table does not rise above 17 ft. (5.18 m) below the ground surface, that seepage pits with an effective depth of 10 ft. (3.05 m) can be provided, and that the house is in a locality where it is common practice to install seepage pits of 5-ft. (1.5-m) diameter (i.e., 4 ft. [1.2 m] to the outside walls, which are surrounded by about 6 in. [152 mm] of gravel).

According to Figure 15.1-64, 570 (3 × 190) sq. ft. (52.95 m^2) of absorption area would be needed. Referring to Figure 15.1-73, it can be determined that a 5-ft. (1.52-m) round pit 10 ft. (3.05 m) deep would provide 157 sq. ft. (14.59 m^2) of wall area. Dividing 570 (52.95) by 157 (14.59) yields 3.6, that is, 3 pits 10 ft. (3.05 m) and one pit 6 ft. (1.83 m) deep. If it is desirable to make all pits the same size, four 5-ft. (1.52-m) round pits 9 ft. (2.74 m) deep can be selected from Figure 15.1-73, providing 4 × 141 or 564 sq. ft. (52.4 m^2) of wall area, which is close enough to the total required. Small-diameter, relatively deep pits require the removal of less soil than larger pits with equivalent wall area, but may be harder to excavate. When this is the case, larger pits can be selected: for instance, two 10-ft. (3.05 m) round pits 9 ft. (2.74 m) deep, or two 13-ft. (3.96 m) round pits 7 ft. (2.13 m) deep (see Fig. 15.1-73).

Where more than one pit is provided, they can be connected in series or in parallel. However, they should be separated by a distance equal to three times the diameter of the largest pit. For pits more than 20 ft. (6.1 m) deep, the minimum distance between pits should be 20 ft. (6.1 m) (Fig. 15.1-74). The area of the lot on which a house is to be built should be large enough to maintain this distance between the pits while still allowing room for additional pits if the first ones fail. If this

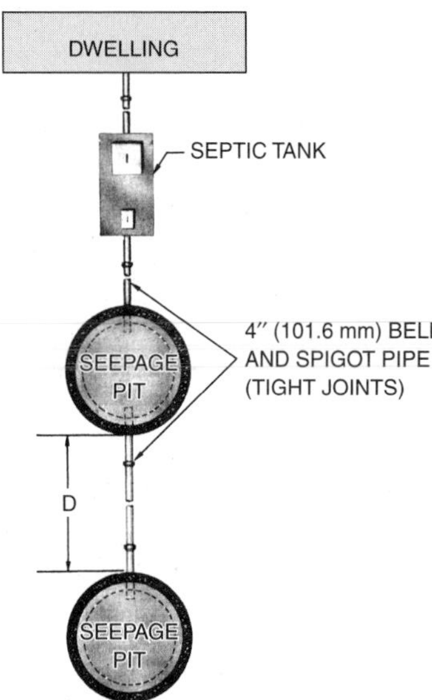

FIGURE 15.1-74 Typical septic tank system with two seepage pits connected in series. (Illustration from the *Manual of Septic Tank Practices*, Bureau of Community Environmental Management, U.S. Public Health Service [now EPA].) Metric added by author.

cannot be done, other sewerage facilities should be explored.

The permeability of the soil may be damaged during excavation. Digging in wet soils should be avoided as much as possible. Cutting teeth on mechanical equipment should be kept sharp. Bucket auger pits should be reamed to a larger diameter than the bucket. Loose material should be removed from the excavation.

Pits should be backfilled with clean gravel to a depth of 1 ft. (305 mm) above the pit bottom or 1 ft. (305 mm) above the reamed ledge to provide a sound foundation for the lining. Lining materials for seepage pits can be clay or concrete masonry units in brick, block, or ring shapes. Rings should have weep holes or openings to provide for seepage. Concrete brick and block should be laid dry (unmortared) with staggered, slightly open joints. Hollow concrete block is frequently laid on its side, with cores horizontal, to facilitate soil absorption. Standard clay brick should be laid flat to form a 4-in. (100 mm)-thick wall. The outside diameter of the lining should be at least 1 ft. (305 mm) smaller than the excavation diameter to provide at least 6 in. (152 mm), and preferably 1 ft. (305 mm), of annular space to be filled with coarse gravel to the top of the lining (Fig. 15.1-75).

Either a corbelled brick dome or a flat concrete cover can be used over a pit. Corbelled bricks should either be laid in cement mortar or have a 2-in. (50.8 mm) covering of concrete. If a flat cover is used, a prefabricated type is preferred. A concrete cover should be reinforced so that it is equivalent in strength to an approved septic tank cover. The cover should rest on undisturbed ground 12 in. (305 mm) beyond the lining. A 9-in. (229 mm) capped opening in the pit cover should be provided for pit inspection.

Connecting lines should be of a sound, durable material, the same as used for the house-to-septic tank connection. All connecting lines should be laid on a firm bed of undisturbed soil throughout their length. The grade of a connecting line should be at least 2% (¼ in. per ft. [30.83 mm/m]). The pit inlet pipe should extend horizontally at least 1 ft. (305 mm) into the pit, with a tee or ell to divert flow downward to prevent washing and eroding of the sidewalls.

Septic Tanks In an *individual sewage disposal system* the building (house) sewer connects the building drain (see Fig. 15.1-51) to a septic tank (see Fig. 15.1-63). A septic tank is a watertight, covered receptacle designed to receive the discharge of sewage from a building sewer; to separate the solids from the liquids, digest the organic matter, and store the digested solids through a period of detention; and to allow the clarified liquids to discharge for final disposal.

Despite their physical position in a system, septic tanks are discussed here—after absorption systems—because it must first be determined as to whether the soil is suit-

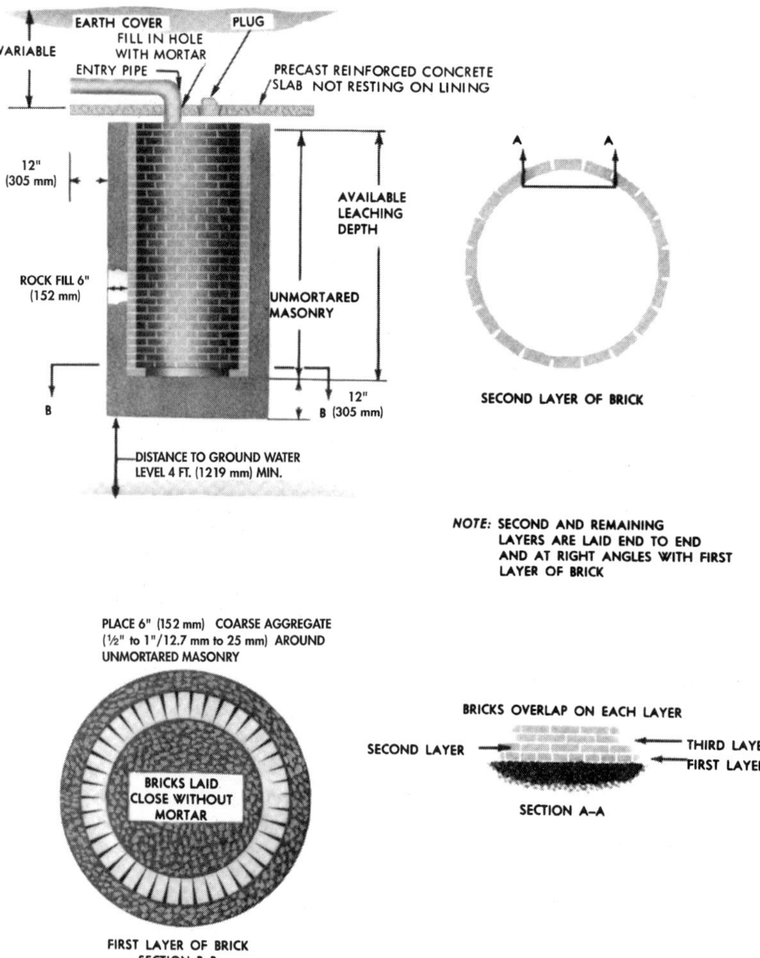

FIGURE 15.1-75 Details of a typical round seepage pit. (Illustration from the *Manual of Septic Tank Practices*, Bureau of Community Environmental Management, U.S. Public Health Service [now EPA].) Metric added by author.

able for a subsurface absorption system and whether adequate area is available for the type of system intended. Except for choosing the tank's capacity and component materials, there are few options in tank design. Because the interior design of a tank is a matter of code compliance and local practice, designs approved by local health authorities can be specified with confidence.

A septic tank performs three basic functions: (1) removal of solids, (2) storage of scum and sludge, and (3) biological treatment. *Scum* is a mass of sewage matter that floats on the surface of the sewage in a septic tank. *Sludge* is the accumulated solids that settle out of the sewage and form a semiliquid mass on the bottom of a septic tank.

A septic tank's chief function is the separation of solids, but its other functions are also essential if the system is to be more than a holding tank, requiring frequent pumping. Biological treatment of the organic matter contained in the scum and sludge ensures that only a small portion of indigestible solid matter remains in the tank permanently between pumpings.

The amount of clogging of the soil by tank effluent varies directly with the amount of suspended solids in the liquid. As sewage from a building sewer enters a septic tank, its rate of flow is reduced so that larger solids sink to the bottom or rise to the surface. These solids are retained in the tank, where most of them are gradually broken down by bacteria into water-soluble components that do not clog the soil.

Heavier sewage solids settle to the bottom of the tank, forming a blanket of sludge. Lighter solids, including fats and greases, rise to the surface and form a layer of scum. A considerable portion of the sludge and scum is liquefied through decomposition or digestion. During this process, gas is liberated from the sludge, carrying a portion of the solids to the surface, where they accumulate with the scum. Ordinarily, they undergo further digestion in the scum layer, and a portion settles again to the sludge blanket on the bottom. The settling action is retarded by grease in the scum layer and by gasification in the sludge blanket.

Solids and liquid in the tank are subjected to decomposition by bacterial and natural processes. The bacteria are of a variety called *anaerobic* (oxygen-free) which thrive in the absence of oxygen. This decomposition or treatment of sewage under oxygen-free conditions is termed *septic*, which explains the name of the tank.

Contrary to popular belief, septic tanks do not accomplish a high degree of bacterial removal. Although sewage undergoes treatment while passing through the tank, *pathogenic* (disease-producing) organisms are not removed. Therefore, septic tank effluents cannot be safely discharged onto the surface or into a ditch, lake, or stream. This, however, does not detract from the value of septic tanks. As previously explained, a tank's primary purpose is to condition the sewage to prevent rapid clogging of the subsurface absorption system.

Further treatment of effluent, including the removal of pathogenic organisms, is effected by percolation through the soil. These organisms slowly die out as their nutrients are used up and as they enter the unfavorable environment afforded by the soil.

Although septic tanks are supposed to be watertight when installed, as they age, some leakage may develop through openings in the inlet and outlet pipes or cracks in the tank. Therefore, septic tanks should be located where they are not likely to contaminate a well, spring, or other source of water supply. Underground contamination may travel for considerable distances, usually in the direction of the water table gradient, that is, from a higher to a lower water table area. Because water tables usually follow ground surface contours, septic tanks should be located downhill from private wells or natural springs.

Unfortunately, however, locating a tank at a lower surface elevation than a well is not a positive guarantee against contamination. Sewage from soil absorption systems occasionally contaminates wells having higher surface elevations. This happens because the elevation of a subsurface absorption system is almost always higher than the level of water in a well that may be located nearby, and contamination may travel downward to the water-bearing stratum (Fig. 15.1-76). It is necessary, therefore, to rely on horizontal as well as vertical distances for protection. Tanks should never be closer than 50 ft. (15.14 m) from any source of water supply, and greater distances are preferred where possible (see Fig. 15.1-68).

A septic tank should not be located within 5 ft. (1.5 m) of a building foundation, because structural damage may result during construction and seepage may enter

FIGURE 15.1-76 A seepage pit can contaminate a private water supply if it is inadequately separated from the well or the water-bearing stratum. (Illustration from the *Manual of Septic Tank Practices*, Bureau of Community Environmental Management, U.S. Public Health Service [now EPA].)

the basement. A tank should not be located in a swampy area, nor in an area subject to flooding. Ideally, tanks should be located where the largest possible area will be available for the absorption system. Consideration should also be given to the location from the standpoint of cleaning and maintenance. A tank should be close enough to a paved area that it can be readily reached by the hose of a septic pumping truck. Where a public sewer may be installed at a future date, provision should be made in the household plumbing system for connection to such a sewer.

Grease interceptors (grease traps) are seldom required in household sewage disposal systems. When one is provided, the discharge from a waste disposer (garbage grinder) should not be passed through it. The septic tank capacities recommended in this discussion are sufficient to receive the grease and solids normally discharged from a typical home.

It is generally advisable to have all sanitary wastes from a household discharge to a single septic disposal system. For household installations, it is also usually more economical to provide a single disposal system than two or more with the same total capacity. Normal household waste, including that from the laundry, bath, and kitchen, should pass into a single system.

Providing adequate capacity is an important consideration in septic tank design. The larger the tank, the longer sewage will be retained in it, and the bet-

FIGURE 15.1-77 Recommended Septic Tank Capacities*

Number of Bedrooms	Minimum Tank Capacity (gal.) (L)
3 or less	1000 (3785)
4	1200 (4543)
Additional bedrooms	Add 250 gal. (946 L)/bedroom

*Provides for use of garbage grinder, dishwasher, automatic clothes washer, and other household appliances.

(Table excerpted from the *Manual of Septic Tank Practices*, Bureau of Community Environmental Management, U.S. Public Health Service [now EPA].) Metric added by author.

ter will be the removal of solids from the effluent and the more complete the biological treatment of the contents. Undersized tanks are more likely to become overloaded, which will result in inadequately treated sewage entering the absorption system and clogging it. The liquid capacities recommended in Figure 15.1-77 allow for the use of all household appliances, including waste disposers, automatic clothes washers, and dishwashers. They also ensure enough excess capacity to retain the sewage a sufficient time to settle out and digest solids. They do not, however, take into account liquids from other sources. For example, water that is free of organic matter should be discharged into a storm sewer or dry well.

A *dry well* is a covered and lined underground pit, similar to a *seepage pit* but intended to receive water free of organic matter, such as from roof drains, floor drains, or laundry tubs. A dry well can be used as an auxiliary to a septic system to avoid overloading the septic system. Dry wells are also called *leaching wells* and *leaching pits*. They should not be confused with *seepage pits*, which are intended to receive septic tank effluent, even though these are also sometimes called dry wells.

Septic tanks should be watertight units constructed of corrosion- and decay-resistant materials such as concrete, coated metal, vitrified clay, heavyweight concrete masonry units, or hard-burned bricks. Precast and cast-in-place reinforced concrete tanks are common. Prefabricated steel tanks meeting Underwriters Laboratories Standard UL 70 are acceptable in some states.

Special attention should be given to job-built tanks to ensure watertightness and prevent contamination. Heavyweight concrete blocks should be laid on a solid foundation, and the mortar joints should be well filled. The interior of a tank should be surfaced with two ¼-in. (6.4 mm)-thick coats of portland cement sand plaster. Some typical septic tanks are illustrated in Figure 15.1-78.

FIGURE 15.1-78 Septic tanks are precast concrete or welded sheet steel in one of several basic shapes. (Illustration from the *Manual of Septic Tank Practices*, Bureau of Community Environmental Management, U.S. Public Health Service [now EPA].)

Precast concrete tanks should have a minimum wall thickness of 3 in. (76 mm) and should be adequately reinforced to facilitate handling. When precast slabs are used as covers, they should be watertight, have a thickness of at least 3 in. (76 mm), and be adequately reinforced so as to be capable of supporting a dead load of 300 psf (1465 kg/m^2).

Backfill around a septic tank should be placed in thin layers and thoroughly tamped in a way that will not produce undue strain on the tank. Backfill may be consolidated using water, provided the backfill is placed and wetted in layers and the tank is first filled with water to prevent floating.

Adequate access to each tank compartment should be provided for inspection and cleaning. Access to each compartment can be provided by either a removable cover or a manhole at least 20 in. (508 mm) wide. When the top of a tank is located more than 18 in. (457 mm) below the finished grade, manholes and inspection holes should extend to approximately 8 in. (203 mm) below the finished grade (Fig. 15.1-79a) or can be extended to finished grade if a *gas seal* is provided to keep odors from escaping. A gas seal is made by filling the top of the manhole with sand or by sealing with a minimum of 1 in. (25.4 mm) of portland cement plaster over an intermediate wood platform (Fig. 15.1-79b). In most instances, an extension can be made using clay or concrete pipe, but attention must be given to the accident hazard involved when manholes are extended close to the ground surface.

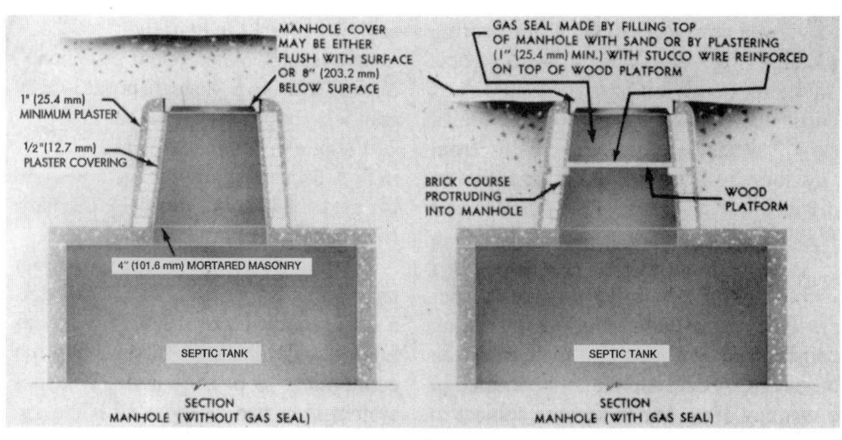

FIGURE 15.1-79 Manholes for septic tank access. (Illustration from the *Manual of Septic Tank Practices*, Bureau of Community Environmental Management, U.S. Public Health Service [now EPA].) Metric added by author.

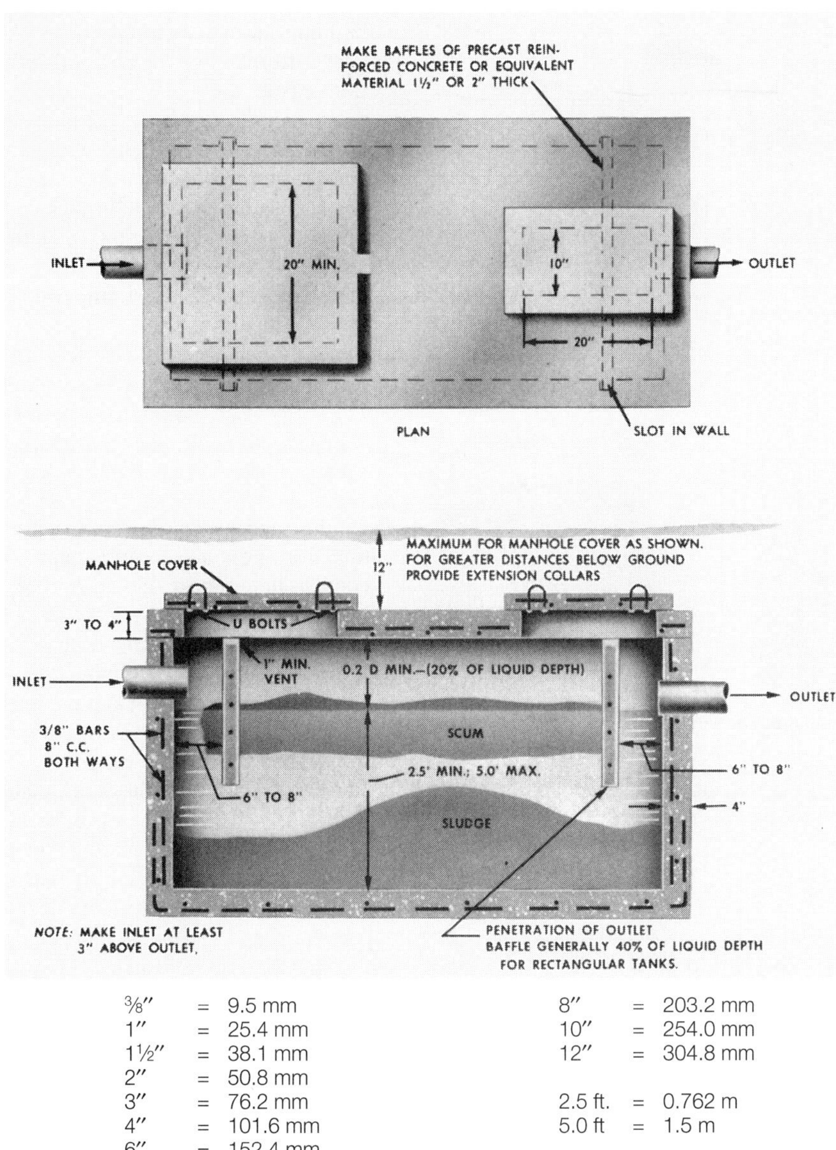

FIGURE 15.1-80 Single-compartment septic tank showing recommended proportions, dimensions, and clearances. (Illustration from the *Manual of Septic Tank Practices*, Bureau of Community Environmental Management, U.S. Public Health Service [now EPA].) Metric added by author.

Typical single- and double-compartment tanks are illustrated in Figures 15.1-80 and 15.1-81. The inlet invert (bottom) should enter the tank at least 3 in. (76 mm) above the outlet invert and the liquid level in the tank to allow for a momentary rise in the liquid level during discharges into the tank. This free drop prevents water from backing up into the house sewer and minimizes stranding of solid material in the sewer.

A vented inlet tee or baffle should be provided to divert the incoming sewage downward. It should penetrate at least 6 in. (152 mm) below the liquid level, but in no case should the penetration be greater than that allowed for the outlet device. A number of arrangements commonly used for inlet and outlet devices are shown in Figure 15.1-82.

The outlet device should penetrate just far enough below the liquid level of the tank to provide a balance between sludge and scum storage volume; otherwise, part of the advantage of capacity is lost. A vertical section of a properly operating tank would show it divided into three distinct layers: scum at the top, a middle zone free of solids, and a bottom layer of sludge. The outlet device retains scum in the tank, while limiting the amount of sludge that can be accommodated without scouring, which results in sludge discharging in the effluent from the tank. The outlet device should generally extend to a distance below the surface equal to 40% of the liquid depth. For horizontal cylindrical tanks, this should be reduced to 35%. For example, in a horizontal cylindrical tank having a liquid depth of 42 in. (1067 mm), the outlet device should penetrate 14.7 in. (42 × .35) (373.4 mm) below the liquid level.

The outlet device should extend above the liquid line to about 1 in. (25.4 mm) from the top of the tank. The space between the top of the tank and the baffle allows gas to pass off from the tank into the house vent stack.

The shape of a septic tank does not affect its operation. Shallow tanks function as well as deep ones. However, the smallest plan dimension should be at least 24 in. (610 mm). Liquid depth may range between 30 and 60 in. (762 and 1524 mm). Tank volume more than 72 in. (1829 mm) below the top of the liquid should not be considered in calculating tank capacity.

Adequate capacity above the liquid line is required to provide for the approximately 30% of the total scum that floats above the liquid. In addition to provision for scum storage, a 1-in. (25.4 mm) clearance between the tops of inlet and outlet tees (or baffles) and the top of a tank usually should be provided to permit free passage of gas back to the inlet and house vent stack.

In tanks with straight vertical sides, the distance between the top of the tank and the liquid line should be about 20% of the liquid depth. In horizontal cylindrical tanks, an area equal to about 15% of the total circle should be provided above the liquid level. This condition is met if the liquid depth (distance from outlet invert to bottom of tank) is equal to 79% of the diameter of the tank.

Although a number of arrangements are possible, the term *compartments* used here refers to a number of units in series. These can be either separate units linked together or sections enclosed in one continuous shell (see Fig. 15.1-81) with watertight partitions separating individual compartments.

A single-compartment tank will give acceptable performance, but a two-compartment tank, with the first compartment equal to one-half to two-thirds of the total volume, generally provides better suspended solids removal. Compartments

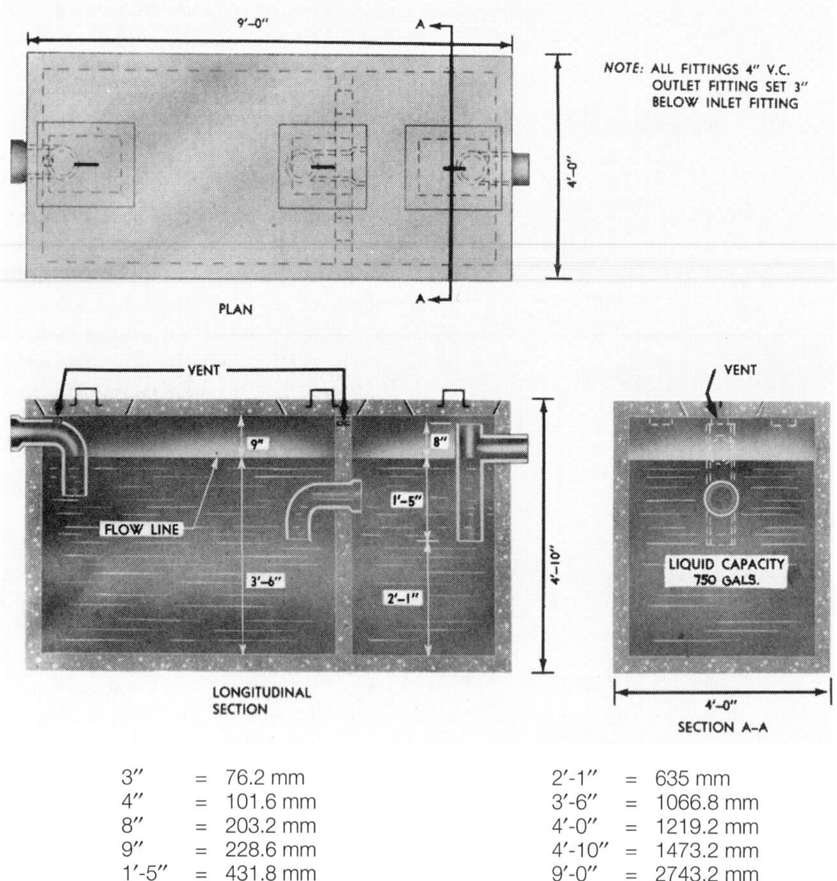

3″ = 76.2 mm		2′-1″ = 635 mm	
4″ = 101.6 mm		3′-6″ = 1066.8 mm	
8″ = 203.2 mm		4′-0″ = 1219.2 mm	
9″ = 228.6 mm		4′-10″ = 1473.2 mm	
1′-5″ = 431.8 mm		9′-0″ = 2743.2 mm	

FIGURE 15.1-81 Typical design of a double-compartment precast concrete septic tank. Dimensions shown are for a 750-gal. (2839 L) tank, which may be found in older homes. Recommended minimum for homes with one to three bedrooms is now 1000 gal. (3785 L). (Illustration from the *Manual of Septic Tank Practices*, Bureau of Community Environmental Management, U.S. Public Health Service [now EPA].) Metric added by author.

should be at least 24 in. (610 mm) in the smallest dimension and have a liquid depth ranging from 30 to 72 in. (762 to 1829 mm).

INSTALLATION INSPECTION

A septic disposal system should be tested and inspected before it is used. Inspection before backfilling is usually required by local regulations, even where plans for the disposal system have been checked before issuance of a building permit. A septic tank should be filled with water and allowed to stand overnight to check for leaks.

A soil absorption system should be inspected before it is covered to be sure that the system is installed according to plans and good trade practice.

15.1.4.3 Storm Drainage Systems

Storm drainage systems consist of gutters and downspouts, roof drains and leaders, underslab and foundation drains, *building storm drains*, *building storm sewers*, *site storm drains*, and *storm sewers*.

Like drain lines in a sanitary sewage disposal system, storm drain lines must be sloped and arranged to assist gravity flow.

Some definitions of terms used in describing storm sewer systems follow:

- *Drain:* A pipe that carries water in a building's storm drainage system
- *Underslab and foundation drains:* Pipes that collect and carry off groundwater and storm water that may seep down into them
- *Building storm drain:* A pipe that picks up water from interior sources and interior roof drains and carries it to a point 3 ft. (914 mm) outside the building
- *Building storm sewer:* A pipe that picks up the water at the end of the building storm drain and carries it to a public storm sewer or disposal location on-site
- *Site storm drain:* A pipe that picks up the overflow from wells and springs, as well as storm water that falls on the site and is conveyed to it through curb inlets, site drains, manholes, and other structures, and carries that water to an outfall on-site, a public storm sewer, or a disposal point
- *Storm sewers:* Pipes that carry collected storm water to a disposal point
- *Disposal point:* A location where storm water is returned to the environment, such as a lake, stream, river, or ocean

Storm drainage systems collect water resulting from groundwater, precipitation, or condensate drainage and carry it to a disposal location. Storm water is collected from a building's roof, from courtyards, and from the surrounding site by a series of inlets and pipes. The roof and site must be sloped or contoured so that water will flow into the system.

Storm drainage systems share many characteristics with DWV systems, but there are some differences as well. For example, because sewer gases are not involved, storm drainage systems do not require traps and usually do not require vents. They do require cleanouts. Many of the principles related to DWV systems apply as well to storm water systems.

SYSTEM TYPE

Storm water is either directed into a community storm drainage system or disposed of in a private system. In a few and dwindling number of jurisdictions, storm water may be directed into a community combined sewer line.

When a community storm sewer is present, storm water may be directed into it by overland flow, as is often the case with single-family residences, or by a system of pipes, as is more usual for larger buildings and extensive sites. The design of parking lots, driveways, and other site conditions to accommodate and collect storm water is beyond the scope of this book.

A private storm water system disposes of water into

- A dry well;
- A low elevation on the property. If the flow is large, riprap or special stone or concrete structures may be required to prevent damage to the surrounding land at the outfall;

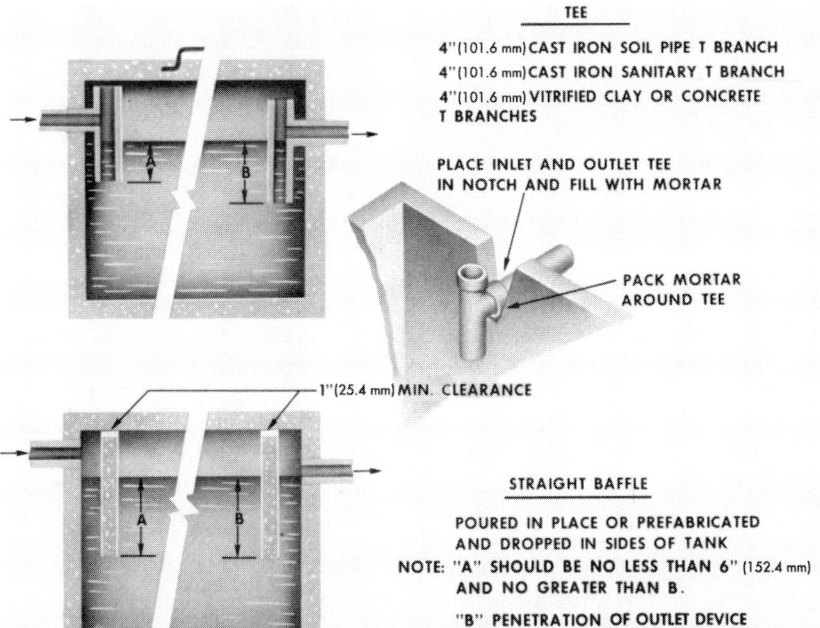

FIGURE 15.1-82 Acceptable inlet and outlet devices for septic tanks. (Illustration from the *Manual of Septic Tank Practices*, Bureau of Community Environmental Management, U.S. Public Health Service [now EPA].) Metric added by author.

- A river, creek, stream, lake, or pond. This may require a permit or may be prohibited by the controlling authority of the waterway.

Footing drains, underslab drains, and other systems for collection of groundwater associated with a building are addressed in Section 2.4.

ROOF DRAINAGE

Drainage from building roofs is a large storm water contributor, and in the case of single-family residences and some other small buildings perhaps the largest. There are several ways to collect roof water.

In buildings with steep roofs (see Section 7.6) roof water may simply be allowed to drain from a roof onto the surrounding landscape or may be collected in a series of gutters and downspouts and deposited either on the ground or into a system of underground pipes leading to a disposal point, as previously described. A low-sloped roof is usually drained by a series of roof drains and pipes, but sometimes these are augmented by overflows called scuppers, at the perimeter of the roof. Scuppers may drain into downspouts or simply allow water to flow over the edge of the roof onto the ground below.

The first step in the design of either of these systems lies in determining the amount of water to be carried away. Plumbing codes indicate the rate of rainfall per hour to use in making these calculations. For example, Maine may have a rate of 4 in. (100 mm) per hour throughout the state, while other states may vary within the state. Virginia, for instance, has a 6-in. (150 mm)-per-hr. rate in its central and eastern parts and a 5-in. (125 mm)-per-hr. rate in its western portion.

Gutters and Downspouts To design a gutter and downspout (leader) system for a steep roof, use the following steps as a guide:

1. Lay out the roof plan and determine the location of downspouts.

2. Size the downspouts to verify that their placement is satisfactory. To do this, multiply the rainfall amount as set forth in the code by the area of the roof to be served by each downspout. Select the downspout size from tables in the applicable plumbing code. For example, in a building where four downspouts will serve a roof area of 3200 sq. ft. (297.68 m^2), each downspout will serve 800 sq. ft. (75 m^2); in a region with a 6-in. (150 mm)-per-hour rainfall, a 3-in. (75 mm) pipe will be sufficient.

3. Size the gutter based on the roof area to be served by each section of gutter. In most designs, one-half of each gutter section spills into each downspout, so the roof area served by each gutter section will be one-half of that served by each downspout—but this may vary, depending on the layout of the system. Whatever the area served, consult the tables in the applicable code for the area to be served, the rainfall amount, and the slope of the gutter, and select the gutter size directly. The slope of a gutter influences the amount of water it can carry in a given time. For example, a 5-in. (125 mm) gutter with a ⅛-in.-per-ft. (10.4 mm/m) slope may serve up to 880 sq. ft. (81 m^2) of roof area where the rainfall per hour is 4 in. (100 mm). Increasing the gutter slope to ¼ in. per ft. (20.9 mm/m) will permit this same pipe to serve 1250 sq. ft. (116 m^2), and a ½-in.-per-ft. (41.7 mm/m) slope up to 1770 sq. ft. (164 m^2).

4. If there is a horizontal element, other than the gutters, in a gutter and downspout system, consult the tables in the applicable code for the area to be served, the rainfall amount, and the slope of the pipe, and select the pipe size directly. As with a gutter, the slope of a horizontal drainpipe influences the amount of water it can carry in a given time.

Low-Slope Roof Drainage To design a roof drainage system for a low-slope roof, use the following steps:

1. Lay out the roof plan and determine the location of the roof drains.
2. From the layout, determine the roof area to be served by each drain.
3. Size the leaders connected to each roof drain. To do this, multiply the rainfall amount given in the code by the area of the roof to be served by each leader. Select the leader size from tables in the applicable plumbing code. For example, in a building where three roof drains will each serve one-third of a roof area of 3200 sq. ft. (297.68 m^2), each roof drain will serve 1067 sq. ft. (99 mm/m^2). In an area with a 6-in. (150 mm)-per-hour rainfall, a 3-in. (75 mm) pipe will be sufficient.
4. Size the horizontal portion of a low-

slope roof system in the same way as described earlier for a steep roof design.

15.1.5 SOUND CONTROL

High water pressure and velocity and quick-closing valves in supply piping systems combine to produce plumbing noises. If a building does not exceed 3 stories and the water pressure at the street exceeds 35 to 45 psi (241 to 310 kPa), a pressure-reducing valve can be installed, preferably as far from the building as feasible. In high-rise buildings, combinations of pressure-reducing valves and pressure-regulating systems may be necessary to provide suitable and uniform pressure levels at the various floor levels.

The noises in a plumbing system can be classified as those associated with the piping and those associated with appliances. The first of these include water hammer, creaking, popping, squealing, cavitation, vibration, and other general noises produced by rushing water. Noises associated with fixtures are usually banging and bumping caused by vibration.

15.1.5.1 Noises Associated with Piping

Water hammer, which is also called hydraulic shock, consists of pounding noises and vibration that sometimes develop in a piping system when its air chambers become filled with water or have been omitted entirely. It may also be due to worn washers in faucets. This unpleasant symptom occurs when there is a sudden change in the velocity of water flowing through a pipe. This causes a shock wave to travel through the piping system, producing a banging as the pipes vibrate against their supports. Such a rapid change in velocity is often caused by the sudden closing of a fast-acting valve, the starting or stopping of certain types of pumps, or improper check valves (Fig. 15.1-83). Continuous water hammer may result in broken pipes, leaky joints, loosened pipe hangers, ruptured tanks, and damage to meters, gauges, regulators, and faucets.

Water hammer can be avoided by using spring-loaded check valves, in lieu of spring or solenoid valves, that do not close fully until the movement of water in the pipe has stopped. It can be minimized by installing air-filled extensions of the supply lines (air chambers) at each fixture or near the service entry (Fig. 15.1-84a). These *air chambers* consist of a piece of pipe about 10 in. (254 mm) long, installed above the hot and cold valves of fixtures such as sinks, lavatories, and clothes washers, to cushion the rush of water as the valve is closed and prevent water hammer. Another way to reduce water hammer is to use plastic piping that is flexible enough to dissipate shock waves.

In older buildings, water hammer may be due to worn-out washers, which cause faucets to close suddenly, or to water-logged air chambers. To restore the air in water-logged chambers, faucets should be opened and water drained from the supply system at its lowest point. In both new and old buildings, water hammer problems can be minimized by installing a shock absorber near the service entrance (Fig. 15.1-84b).

Squealing noises are sometimes emitted by pipes in which water flowing at a high velocity strikes small protrusions on the inside face of the pipe, causing vibration of the pipe. This problem can usually be solved by reducing the velocity of the flow in the pipe. Squealing can also occur when water passes a loose or worn valve washer, causing it to vibrate. This can be stopped by tightening the screw holding the washer in place or replacing the washer. Proper sizing and anchorage of copper water supply piping will also prevent some noises.

Noise resulting from water rushing through piping can be minimized by planning the system with a minimum number of bends and by reducing the number of fittings and valves to a minimum. Valves and right-angle fittings produce substantially more noise than straight runs of pipe due to increased water turbulence in them. The noise produced by four elbows is about 10 decibels higher than the noise produced in a straight run of equivalent-size pipe under conditions of turbulent flow.

Long runs of hot water supply piping produce creaking noises as the piping expands or contracts from hot or cooled water. Long, straight runs of iron or steel piping should be designed with flexibility, with at least one end having a swing arm to permit movement (Fig. 15.1-85a). Flexible L or M piping bent into an S-shaped curve at one end will allow for movement in long runs of copper piping (Fig. 15.1-85b). Supports should be designed to permit the piping to expand and contract without binding.

Supply piping to units that use large amounts of water, such as dishwashers and water closets, should not be undersized.

When pipes are rigidly connected to structures with hangers, clamps, or straps, sound resulting from pipe vibration can be transmitted through the structure and can be amplified from one apartment unit to another. Resilient pipe insulation between the pipe supports and the pipe should be used to help reduce the transmission of these noises (Fig. 15.1-86a). The number of pipe supports should be kept to the minimum required to support the pipe structurally, and supports should

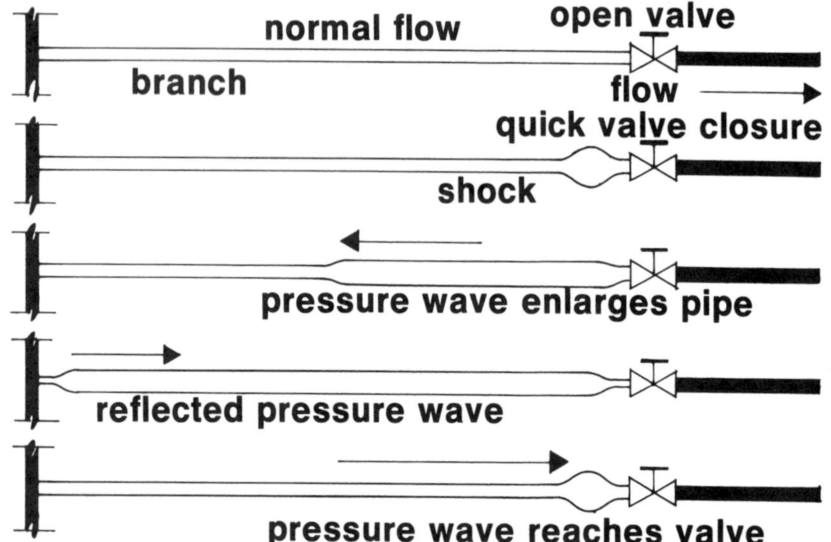

FIGURE 15.1-83 Internal pipe pressures cause water hammer due to a fast-closing valve. (Courtesy Small Homes Council, Building Research Council, University of Illinois, Urbana-Champaign.)

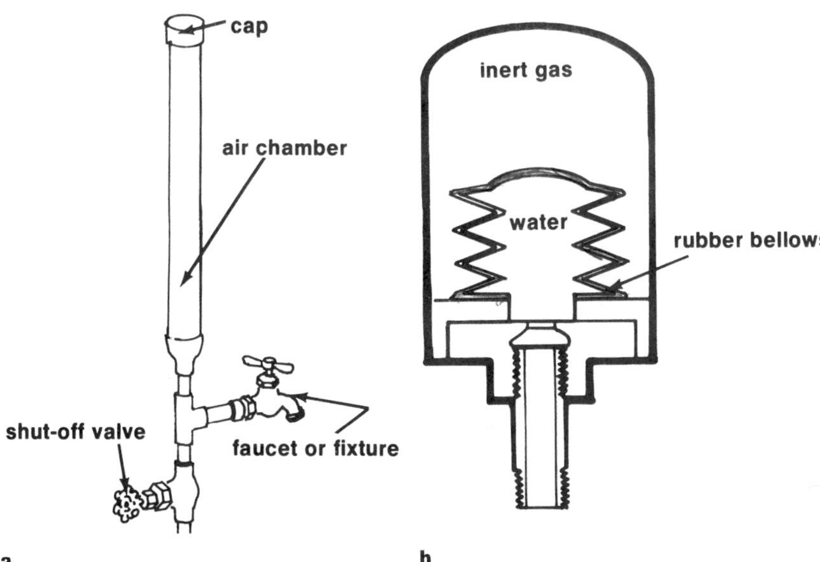

FIGURE 15.1-84 **Water hammer can be minimized by installing (a) an air chamber or (b) a shock absorber near the service entrance. A similar, but smaller, air chamber should be installed at each valved fixture.** (Courtesy Small Homes Council, Building Research Council, University of Illinois, Urbana-Champaign.)

be attached to the most massive structural element available.

The running of drainpiping through the walls of bedrooms or other quiet living areas should be avoided, unless a double sound-isolating wall is provided. Waste dropping down vertical stacks causes piping to vibrate, especially when it hits horizontal runs, unless the pipe is resiliently mounted. When vertical stacks are hung on floor joists, vibrations are transmitted to the structure (Fig. 15.1-86b). Standard pipe clamps and resilient vertical supports or wood brackets should be installed with the end of the support resting on the vertical stack, with a 1-in. (25.4 mm)-thick isolation pad resting on the support (Fig. 15.1-86c). A short section of pipe insulation around a pipe where it passes through a wall sill will keep it in horizontal alignment and isolate it from the structure.

It is essential that holes and openings where piping passes through common walls be sealed with a resilient material, tightly packed with insulation, or both, to isolate vibration of the pipe from the structural framing and to seal against air leaks (Fig. 15.1-87).

Cavitation is a phenomenon that occurs when water flowing at a high velocity makes a rapid change in direction or passes through a small orifice. Cavitation produces a high-pressure region at the outside of a pipe bend and a low-pressure region at the inside of the bend. If the inside pressure is low enough, boiling occurs in the water, allowing air bubbles to form. Farther down the pipe, these bubbles reencounter pressure and are collapsed by it. This produces a noise similar to gravel rattling in the pipe or balloons popping.

Aside from the noise nuisance, cavitation can cause unnecessary wear on pipes and pumps. A solution is to lower the velocity of the water to less than 10 ft./sec. (3.05 m/s). Typical residential installa-

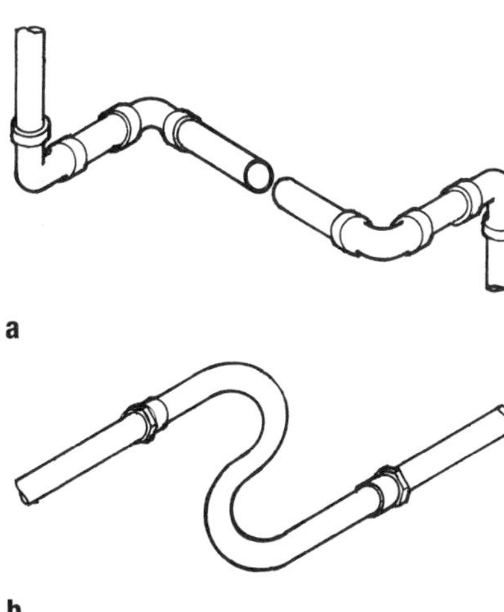

FIGURE 15.1-85 **Flexible connections, using (a) swing arms in piping or (b) S curves in tubing, minimize noise caused by expansion and contraction in long runs of piping.**

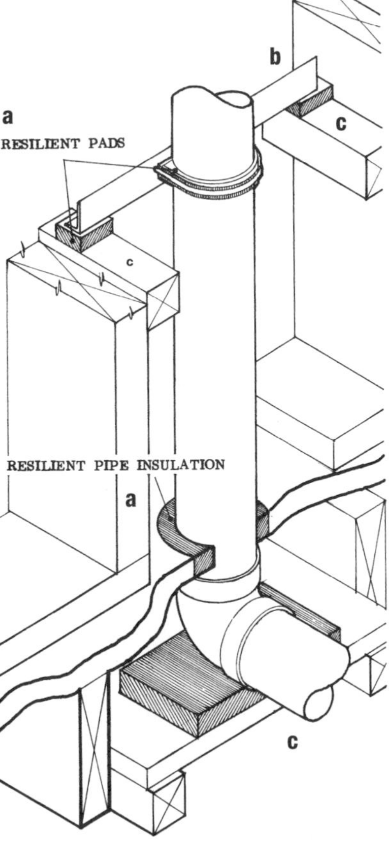

FIGURE 15.1-86 **Vibrations from a plumbing stack can be isolated from the structure by resiliently mounting vertical supports and resiliently supporting horizontal runs.**

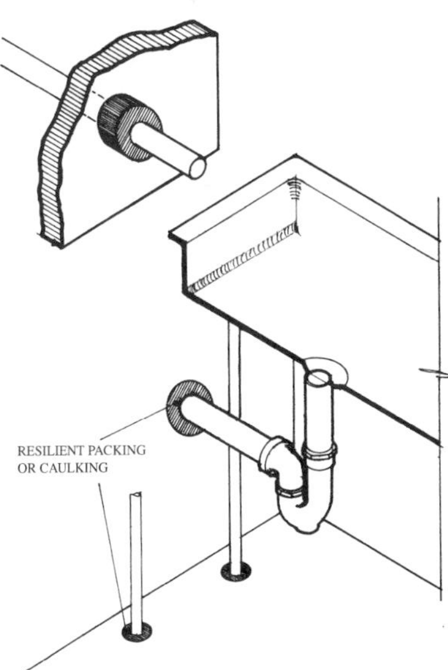

FIGURE 15.1-87 **Where pipes pass through walls, openings must be tightly packed or sealed airtight with resilient materials to isolate vibrations and to close paths for airborne sound transmission.**

tions have water velocities ranging from 4 to 8 ft./sec. (1.2 to 2.4 m/s).

15.1.5.2 Fixtures and Appliances

Plumbing fixtures and appliances can often be selected and installed to reduce the amount of noise. For example, a siphon jet water closet having an adjustable-rate water-supply inlet float valve will operate more quietly. A water closet can be mounted on a floating floor, and the noise transmission from fixtures can be reduced by using a neoprene or rubber gasket between the base of a water closet and the floor (Fig. 15.1-88) or by a resilient underlayment between two layers of floor sheathing. Flexible caulking or tightly packed insulation should be used to seal the space between a toilet bend and the floor sheathing.

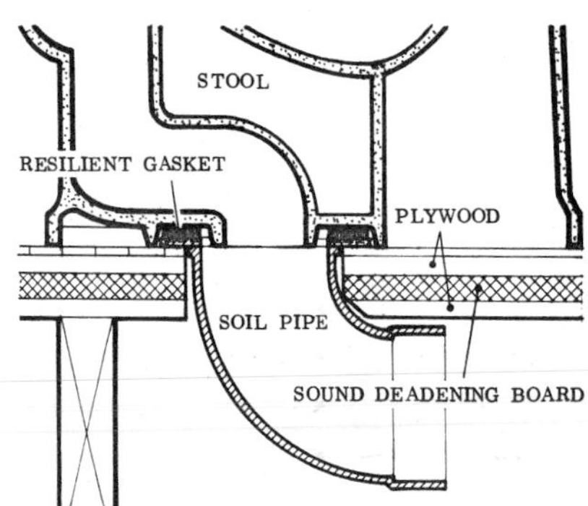

FIGURE 15.1-88 Water closet vibrations can be minimized by resilient mounts and floated floor construction.

Vibrating appliances such as dishwashers, disposers, and clothes washers and dryers should not be rigidly connected to the structure. Resilient or flexible isolation gaskets should be provided between the appliance and the floor, wall, or counter to isolate vibrations, unless a floated floor system is used. Piping connections to appliances should be flexible enough to prevent vibrations from being transmitted to the piping (Fig. 15.1-89). Electrical connections to appliances should also be flexible whenever possible.

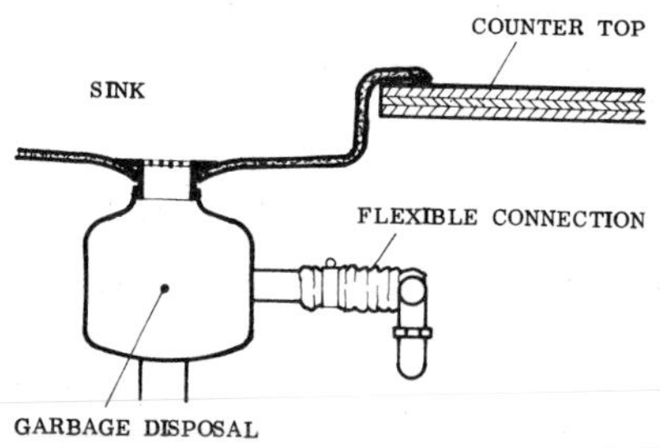

FIGURE 15.1-89 Flexible piping connections prevent vibrations from being transmitted to the piping.

15.2 Heating, Ventilating, and Air Conditioning (HVAC)

Early theories defined heat as a material substance called *caloric*, which was a weightless fluid that had the power to penetrate other materials, expand, and transform them into vapor. It was thought that a substance was hot or cold, depending on the proportion of *caloric* contained in its pores.

The theory of heat was placed on a sound scientific basis in the early nineteenth century by James P. Joule, an English physicist. Joule's experiments led to the principle of conservation of energy, which stated that heat energy was not lost as it was used but was transformed into an equivalent amount of another kind of energy. For instance, heat energy can be converted to kinetic (motion) energy by a steam engine, and from kinetic energy to electric energy by a generator.

Animals, including people, sustain life by converting organic food substances into heat energy. Normally, excess heat energy is produced and the body dissipates this excess to the surrounding atmosphere. Each organism has an optimum rate of body heat loss depending on physical exertion. When the heat loss is at the ideal rate, a person feels comfortable. When heat is lost too rapidly, a person feels cold; when the loss is too slow, the sensation is one of being hot. The objective of interior heat control is to create conditions at which the body loses heat at the optimum rate.

Heat energy generally flows from a warmer region to one that is cooler. Therefore, the flow is to the outside of a building in winter and to the inside in summer. This heat flow affects interior surface and air temperatures.

Cold is the absence of heat. A warm window radiates heat, but a cold window does not radiate cold. A person's sensation of coldness is caused by a rapid loss of body heat to the cold window surface. Likewise, the movement of air over the body produces the sensation of coldness because it carries off heat, not because it brings coldness.

People today expect to find a relatively narrow range of uniform temperatures within a building and want the ability to select that range at will. When energy was less costly, it was relatively easy to force an interior environment into a narrow band of comfort. Today, both architects and building owners are examining the benefits of permitting a broader range of interior temperatures and of adjusting the mode of dress to the temperature.

For instance, putting on a sweater in winter may make a 65°F (18.3°C) interior air temperature acceptable, and removing one's tie and turning on an electric fan may make 80°F (26.7°C) tolerable in the summer. The potential gain from accepting a 5°F (–15°C) variation up or down from traditional thermostat settings can produce a 15 to 30% energy savings over the average costs of heating and cooling.

Until the energy crunch of the 1970s, traditional approaches in interior heat control emphasized reliance on energy-intensive mechanical systems. These systems pumped heat in or out of a building and relied on the interior shell only to minimize heat gain in summer and heat loss in winter. Now, architects tend to design a building's envelope to permit heat flow through the shell when advantageous. For instance, buildings can be designed to make use of sunshine for heating in the winter and of natural breezes for cooling in the summer. Some of the principles of passive solar energy systems are common in many buildings (see Section 15.2.10.2).

This section discusses the principles of heat flow and human comfort, appropriate building design criteria related to heat control, and the design of heating, ventilating and air conditioning systems. Section 7.3 discusses the use of insulation to minimize heat flow through a building's envelope. Chapter 8 contains data about strategies for reducing heat loss and gain through doors and windows. Section 12.2 contains information concerning the use of interior and exterior shading devices and of blinds, shades, draperies, and shutters to control the gain or loss of heat through a building's windows. Together, these techniques help to reduce energy consumption and make buildings more comfortable for occupants, with less reliance on mechanical heating and cooling systems.

15.2.1 HEAT AND COMFORT

Heat is a form of energy. It is produced in a substance by the movement of its molecules—the tiny particles of matter that make up that substance (see Section 1.10). When heat is added, the molecular movement inside that substance is increased. The intensity of movement determines the physical state of the substance, whether solid, liquid, or gas. Temperature, recorded in degrees Fahrenheit (°F) or degrees Celsius (°C), is a measure of that intensity (Fig. 15.2-1). A British thermal unit (Btu) is a measure of heat quantity and is equal to the amount of heat required to raise 1 lb. of water 1°F (–17.2°C). Heat flow rate is measured in Btu's per hour. In the Système International (SI) (metric system), heat quantity is measured in joules (J) and heat flow rate is measured by the watt (W). One watt is equal to one joule/sec., or approximately 0.341 Btu/hr.

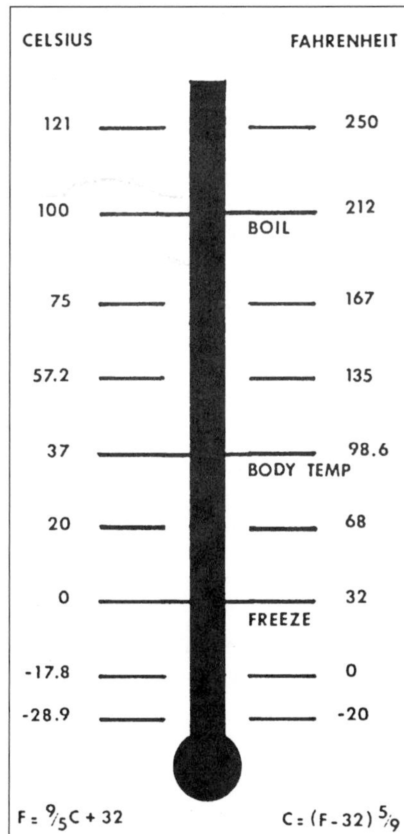

FIGURE 15.2-1 Temperature equivalents on the Celsius (metric) and Fahrenheit (English) scales. Thermometers can be calibrated to a temperature of –273.1°C (–459.6°F), which is known as absolute zero. This is the point at which all molecular movement stops.

15.2.1.1 Sensible and Latent Heat

Energy that causes a rise in temperature but does not alter the physical state of a substance is called *sensible heat*. Heat can be added to a substance without altering its physical state only up to a certain temperature. For example, if a teapot with 1 lb. (0.45 kg) of water in it is placed on a burner, each Btu (1.06 kJ) of heat added raises the temperature 1°F (–17.2°C). If the water started at 62°F (16.67°C), it would take 150 Btu's (158.3 kJ) to bring it to a boil at 212°F (100°C). Heat added beyond the first 150 Btu's (158.3 kJ) would change the water to vapor. It would not raise the temperature of the water above 212°F (100°C).

The energy that would be used to convert the water to steam if the pot were left on the burner is the *latent heat of vaporization*. To change the entire 1 lb. (0.45 kg) of water to steam would require 970 Btu's (1023.4 kJ). To change 1 lb. (0.45 kg) of ice at 32°F (0°C) to water would require 144 Btu's (151.93 kJ), which is the *latent heat of fusion* of water. Only after all of the first 144 Btu's (151.93 kJ) were used and all of the ice had melted would the additional heat be used to raise the water temperature above 32°F (0°C).

15.2.1.2 Methods of Heat Transfer

Heat energy moves from a warmer to a colder area by conduction, radiation, or convection. A person gives up heat by all of these methods, as well as by evaporation of perspiration.

CONDUCTION

The movement of heat through a solid or liquid substance as heated particles transfer their thermal energy to adjacent particles is called *conduction*. Generally, dense materials such as metals can transfer heat faster and are called *conductors*. Less dense substances such as wood and plastic retard heat flow and are called *insulators*.

RADIATION

The transmission of heat by invisible light waves independent of any medium is called *radiation*. The radiant heat from the sun, for example, reaches the earth by traveling through the vacuum of space and the atmosphere. Radiant energy can travel through any nonopaque substance.

An opaque body absorbs radiant energy over the surface struck by the solar radiation and converts it into heat. After absorption, the transfer of heat through the body is by conduction; some heat is also reradiated from the warm body into the air.

Reflectivity Materials vary in their acceptance and rejection of the radiant energy that strikes them. Those that reject much of the radiant energy are called *reflectors*, because of their high degree of reflectivity. Shiny aluminum foil, for instance, reflects 95% of all radiant energy; white paint reflects about 75%. Dark materials, such as asphalt pavement, which absorb more radiant energy, have a high degree of emissivity (Fig. 15.2-2).

Emissivity Emissivity is the converse of reflectivity. Aluminum foil reflects 95% of the radiant heat that strikes it and absorbs the remainder as sensible heat. Therefore, the emissivity of the foil is 5% (see Fig. 15.2-2). The practical significance of such differences is that high-emissivity surfaces absorb and radiate more energy. For example, because asphalt pavements have a high emissivity, air immediately above them will be warmed to as much as 20°F (–6.7°C) higher than air in nearby shaded areas.

CONVECTION

The movement of heat from one region to another through a fluid substance, such as air or water, is *convection* (Fig. 15.2-3). When air in a room contacts a warm baseboard heater, it is heated by conduction. As it is warmed, the air expands, becomes lighter, and rises, thus permitting colder air to take its place. The warmed air often gives up its heat to a cold wall or window and settles to the floor to be reheated. Thus, a *convection loop* is established, which distributes the heat throughout the space.

EVAPORATION

A human body at rest loses heat mainly by convection and radiation. When engaged

FIGURE 15.2-2 Reflectivity/Emissivity Chart

Material	Reflectivity, %	Emissivity, %
Built-up roof, gravel aggregate	7	93
Slate, dark soil, other dark textures	15	85
Grass, dry	30	70
Copper foil		
Tarnished	36	64
New	75	25
Paint		
Light gray	25	75
Red	26	74
Aluminum	46	54
Light green	50	50
Light cream	65	35
White	75	25
Fresh snow	80	20
Aluminum foil	95	5

FIGURE 15.2-5 A sling psychrometer.

in strenuous activity or when the air is too warm, those methods are insufficient to carry off all the excess heat. The body then utilizes perspiration, which cools the body by withdrawing from it the *latent heat of vaporization*. Low humidity aids cooling by perspiration; high humidity inhibits it.

15.2.1.3 Relative Humidity

Most air contains invisible moisture in the form of water vapor. When the moisture becomes visible, it is no longer vapor but a liquid. A cloud or puff of steam is water in an intermediate stage—vapor in the process of becoming liquid, or vice versa. The conversion from a vapor to a liquid is called *condensation*.

Relative humidity (RH) is the ratio of moisture actually present in the air, as compared with the maximum that can exist at a given temperature without condensation. An RH of 50% means that the air holds half the total water vapor it could possibly contain at that temperature. The maximum relative humidity is dependent on the ambient (general) air temperature, because warm air can hold more water vapor than cooler air (Fig. 15.2-4). When all possible moisture is held by the air (100% RH), its *saturation point* or *dew point* has been reached. For additional discussion of relative humidity and saturation see Section 7.1.

Relative humidity is measured by a *sling psychrometer* (Fig. 15.2-5). This device consists of two thermometers mounted side by side on a frame. A wetted cloth sack is attached to one of the thermometers, and both are whirled through the air. The temperature reading given by the uncovered thermometer is the *dry-bulb* temperature (DBT), which is the common way of measuring and stating temperature. Ambient temperatures are usually stated in DBT. The reading shown by the wetted thermometer is the *wet-bulb* temperature (WBT). This is the reading more commonly used in the design of mechanical systems.

Rapid movement through the air causes some of the water to evaporate from the wetted thermometer. The process of evaporation uses up latent heat energy, resulting in a lower temperature reading. The greater the difference between wet- and dry-bulb temperatures, the lower the relative humidity. When the dry-bulb temperature and wet-bulb temperature are known, the relative humidity can be determined from a psychrometric chart (Fig. 15.2-6). For instance, if the DBT is 70°F (21.1°C) and the WBT is 58°F (14.4°C), then the RH is 50%.

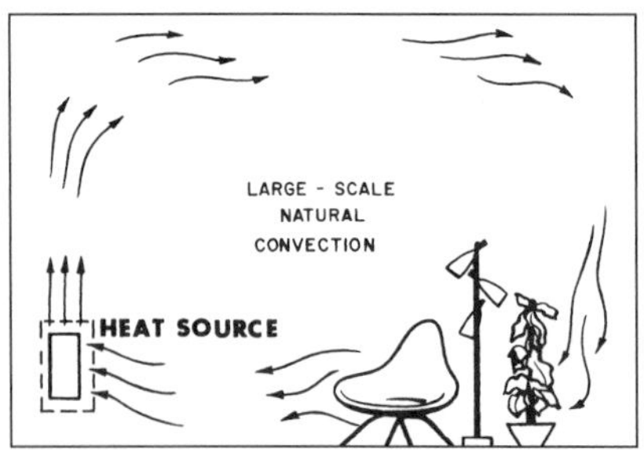

FIGURE 15.2-3 Air warmed by a heater rises, gives up its heat to walls and ceiling, and drops to the floor to be warmed again.

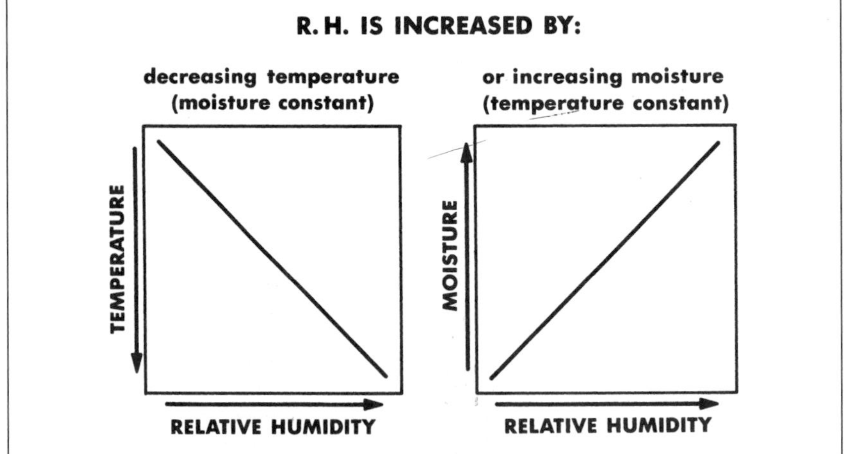

FIGURE 15.2-4 Relative humidity can be increased by decreasing the temperature or by increasing the amount of moisture.

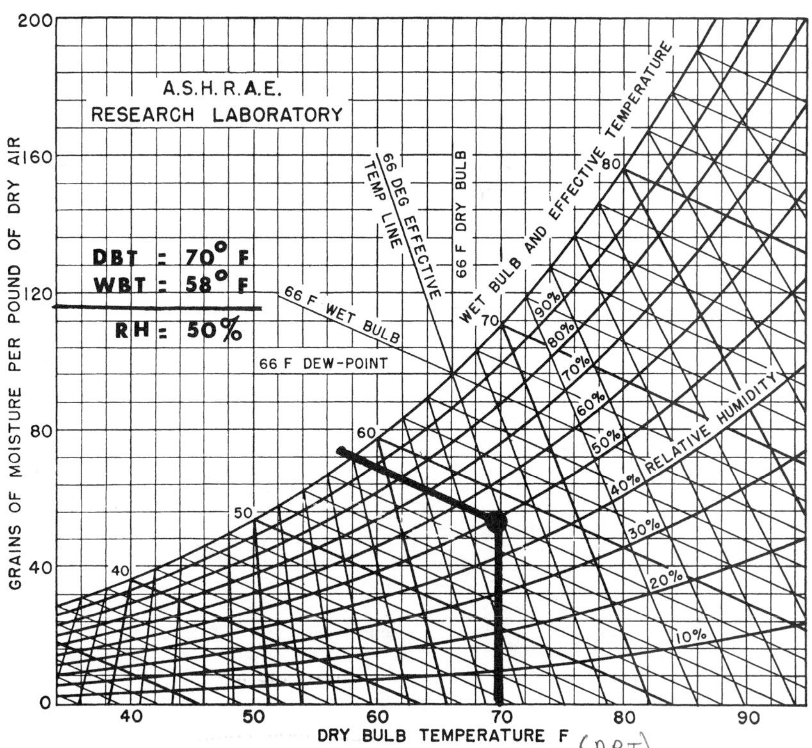

FIGURE 15.2-6 Psychrometric chart. (Reprinted with permission of the American Society of Heating, Refrigerating, and Air-Conditioning Engineers.) (Refer to Figure 1.8-27 for metric equivalents.)

See Section 7.1.1 for an additional discussion of relative humidity and its effect on people and property.

15.2.1.4 Interior Comfort Requirements

The human body is a heat-producing mechanism, generating heat energy from oxidized food. In most cases, it produces more heat than is needed for thermal stability. The amount of extra heat that must be dissipated depends on the activity being performed (Fig. 15.2-7) and ranges between 200 Btu's (211 kJ) and 4500 Btu's (4747.8 kJ) per hour.

ENVIRONMENTAL CONTROL

Control of the interior environment helps the body regulate its temperature and achieve the ideal comfort condition, which is a sensation that is neither too warm nor too cool.

Studies conducted by the American Society of Heating, Refrigerating and Air-Conditioning Engineers, Inc. (ASHRAE) and others have shown a rise in the indoor temperatures people require to be comfortable in the winter from between 65 and 70°F (18.3 and 21.1°C) in 1900 to between 72 and 78°F (22.2 and 25.6°C) today. Therefore, the desire for warmer temperatures may be a result of cultural conditioning rather than biological necessity. The high cost of energy today and a growing awareness of the need to conserve energy may not have changed how people feel about the ideal temperature for comfort, but they have somewhat changed the way people act. There is a greater tendency to turn the thermostat down and wear heavier clothes in winter, and the reverse in summer.

COMFORT FACTORS

Ambient temperature, air movement, relative humidity, and the temperature of a space's surfaces combine to produce the thermal sensations a person experiences within that space. The first three have already been discussed. The fourth is expressed in terms of *mean radiant temperature* (MRT), which represents the collective thermal effect of all the radiant surfaces in a space on an occupant at a given position within the space (Fig. 15.2-8).

EFFECTIVE TEMPERATURE

Effective temperature (ET) is a method of measuring the combined effect of ambient temperature, relative humidity, and air movement. Any combination of temperature and humidity, with an air movement of 15 to 25 ft./min., can be stated as a single-number ET. That number is the dry-bulb temperature at 100% RH (or at 50% RH) that produces the same sensation as the given combination of temperature and humidity. This text uses the old effective temperature definition at 100% RH; the newer 50% RH effective temperature is designated here by an asterisk, ET*.

Early ASHRAE tests suggested that the majority of people were most comfortable in winter at an ET of 68°F (20°C) (ET* 72.2°F [22.3°C]) and in summer at

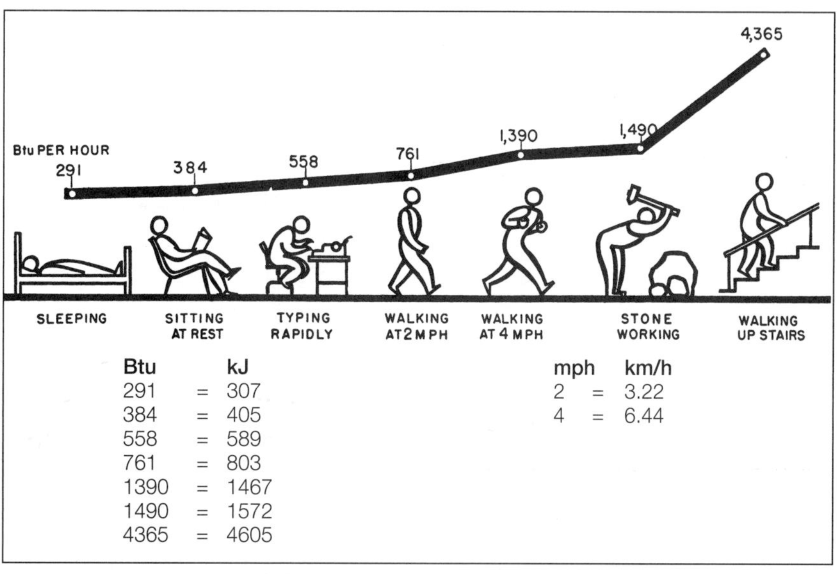

Btu		kJ
291	=	307
384	=	405
558	=	589
761	=	803
1390	=	1467
1490	=	1572
4365	=	4605

mph		km/h
2	=	3.22
4	=	6.44

FIGURE 15.2-7 A body's production of heat depends on how strenuous the activity is.

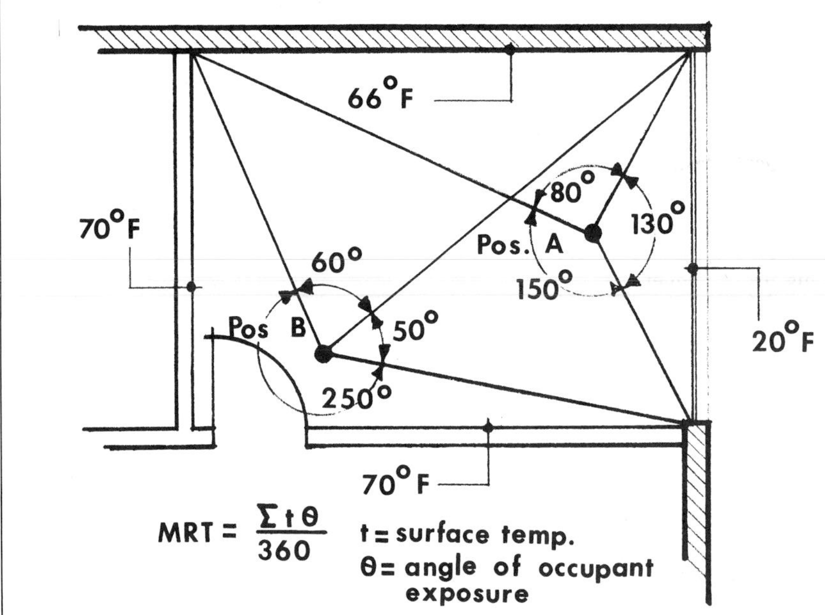

Example: An apartment building has masonry and glass exterior walls with interior surface temperatures of 66° F and 20° F. Interior partitions are at 70° F.

$$\text{Position A: MRT} = \frac{\Sigma t\theta}{360} = \frac{(20 \times 50) + (66 \times 60) + (70 \times 150)}{360} = \frac{18{,}680}{360} = 51^\circ \text{ F}$$

$$\text{Position B: MRT} = \frac{\Sigma t\theta}{360} = \frac{(20 \times 50) + (66 \times 60) + (70 \times 250)}{360} = \frac{22{,}460}{360} = 62^\circ \text{ F}$$

FIGURE 15.2-8 Calculation of mean radiant temperature (MRT). (Reprinted with permission of the American Society of Heating, Refrigerating, and Air-Conditioning Engineers.) (Refer to Figure 1.8-27 for metric equivalents.)

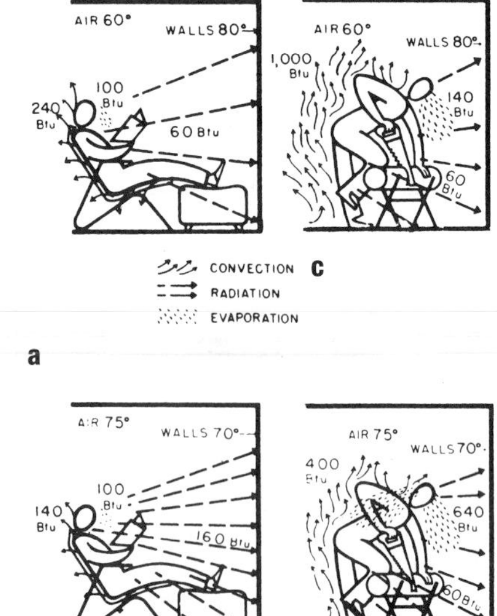

FIGURE 15.2-9 Both ambient (air) and radiant (wall) temperatures are important for comfort. At rest (a and b), thermal balance can be easily maintained with cooler air if walls are warmer; at work (c and d), cool air moves more body heat, requiring less perspiration to maintain balance. (Refer to Figure 1.8-27 for metric equivalents.)

an ET of 71°F (21.7°C) (ET* 76.9°F [24.9°C]). From the psychrometric chart in Figure 15.2-6 it is possible to obtain DBT/RH equivalents at more common seasonal humidities. For example, according to Figure 15.2-6 the winter optimum DBT is near 75°F (23.9°C) at 30% RH, and the summer optimum DBT is near 77°F (25°C) at 50% RH.

A defect in using optimum ETs as a comfort index is that the effect of relative humidity is exaggerated in importance. Later tests concluded that within a temperature range of 60 to 80°F (15.6 to 26.7°C) and a relative humidity range of 25% to 60%, evaporative body cooling is nearly constant. That is, a person at rest loses roughly the same amount of heat by perspiration at 25% RH as he or she does at 60%, so long as the temperature does not exceed about 80°F (26.7°C). However, different temperatures within the 25% to 60% RH range do produce varying sensations of warmth.

Another problem with using optimum ETs as a comfort index is that they ignore the effect of radiant surfaces (Fig. 15.2-9). In summer, cool walls aid body heat dissipation because they absorb radiant body heat that cannot be removed by convection due to high air temperature. A person can therefore lose by radiation any heat he or she would otherwise have to eliminate at greater stress, by perspiration. In winter, warmer wall temperatures compensate for low air temperature by preventing radiant body heat loss.

15.2.1.5 ASHRAE Thermal Comfort Envelope

ASHRAE Standard 55 specifies generally acceptable comfort conditions for sedentary and slightly active, healthy, normally clothed people at altitudes from sea level to 7000 ft. (2133.6 m).

The *comfort envelope* developed in this standard (Fig. 15.2-10) outlines acceptable thermal conditions for 80% of tested Americans and Canadians. It integrates the effect of DBT, MRT, RH, and air movement. In this chart, the effects of DBT and MRT are combined in the *adjusted dry-bulb temperature* (ADBT) scale, and the effect of RH is shown by the water vapor pressure scale. Air movement is assumed at less than 70 ft./min. (21.34 m/min).

Adjusted dry-bulb temperatures and relative humidity levels should be controlled within the ranges shown in Figure 15.2-10. Interior surface temperatures in winter should not be more than 5°F (–15°C) below ambient air temperature, to prevent radiant heat loss from bodies and interior spaces. In summer, walls should be kept slightly cooler than air temperature to aid in body heat dissipation.

15.2.1.6 Thermal Design Criteria

For maximum energy conservation, the latest recommendations of the U.S. Department of Energy (DOE) should be used as interior design temperatures. The DOE is working closely with such organizations as ASHRAE to develop better energy standards.

ASHRAE ENERGY CONSERVATION STANDARD

ASHRAE Standard 90.1 provides performance-oriented criteria for the design of building envelopes and of mechanical and electrical systems. These standards refer to the comfort envelope parameters established in ASHRAE Standard 55 for winter and summer design criteria (see Fig. 15.2-10).

Low relative humidities generally accompany the heating season, and high humidities are characteristic of the cooling season in most of the United States. Therefore, most designers operate near the bottom of the comfort zone in winter and near the top in summer. If energy conservation is a design objective, it can be further assumed that the design ADBT would be near 73°F (22.78°C) in winter and near 78°F (25.56°C) in summer.

HUD STANDARDS

HUD's *Minimum Property Standards for Housing* adheres to the historic interior comfort criteria—70°F (21.1°C) DBT for winter and 75°F (23.9°C) DBT for summer ambient air. This standard does not make any specific requirements for relative humidity or radiant temperature.

15.2.1.7 Climate and Weather Conditions

In order to design a building envelope and mechanical system capable of maintaining interior comfort conditions, it is necessary to predict likely heat gain and loss. To do this, the designer must know not only what indoor temperature is desired, but also the outdoor temperatures that are likely to be encountered.

OUTDOOR DESIGN TEMPERATURES

Outdoor design temperatures are determined from weather records over a 25-year period. They are not absolutely the highest or lowest temperatures for a given area, but rather sustained extremes. Figure 15.2-11 provides a sampling of design temperatures and other related information for some major U.S. cities.

DEGREE DAYS AND COOLING OPERATING HOURS

The severity of winter is expressed in total *degree days heating* (DDH), which is the sum of daily degree days when the temperature drops below 65°F (18.3°C). *Daily degree days* is the numerical difference between 65 and the average recorded temperature of that day. For instance, if there were 10 days in a season when the temperature dropped to 45°F (7.2°C), then the DDH for each day is 20 and the total DDH for the season is 200.

Cooling operating hours is a measure of the intensity of the annual cooling season. These are the hours an air conditioning unit must operate in order to maintain an indoor temperature of 78°F (25.56°C) for 97.5% of the time.

The higher the number of total degree days or annual cooling operating hours for the season, the greater the heating or cooling demand (see Fig. 15.2-11).

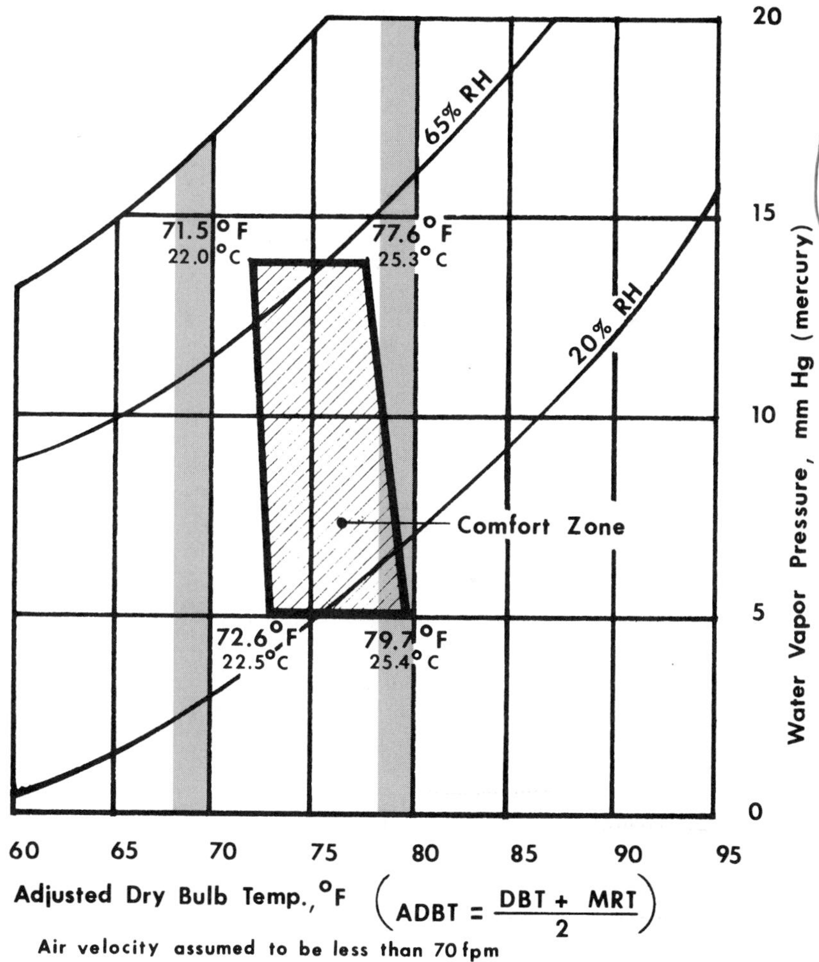

FIGURE 15.2-10 **ASHRAE comfort envelope.** (Reprinted with permission of the American Society of Heating, Refrigerating, and Air-Conditioning Engineers.) (Refer to Figure 1.8-27 for metric equivalents.)

15.2.1.8 Heat Transfer Through Materials

All building materials transfer heat to some extent. Heat transfer can be measured by (1) conductivity (k); (2) conductance (C) (k/t), (3) resistance (R) (1/C), and (4) transmittance (U) (1/R). Figure 15.2-12 shows the heat flow coefficients of some selected materials for conductivity, conductance, and resistance.

Knowledge of how various materials resist the passage of heat enables designers to calculate heat gain and loss and to select materials that minimize heat flow.

CONDUCTIVITY AND CONDUCTANCE

The degree to which a material transfers heat is rated by its conductivity (k), which is the heat flow conductance of 1 in. (25.4 mm) of a substance, or conductance (C), which is the heat flow conductance of the entire thickness of a material. Therefore, C is k divided by the thickness.

Conductivity Conductivity (k) is a measure of the number of Btu's (kJ) per hour that can pass through 1 sq. ft. (0.09 m^2) of a 1-in. (25.4 mm) thickness of a material when its surfaces vary in temperature by 1°F (–17.2°C). For example, a 1 × 1 ft. (304.8 × 304.8 mm) 1 in. (25.4 mm) block of polyurethane foam with a k of 0.16 would permit 0.16 Btu (0.168 kJ) to pass in 1 hour, if one face of the block was at 74°F (23.3°C) and the opposite face was at 75°F (23.9°C).

FIGURE 15.2-11 Representative Outdoor Design Temperatures, Degree Days, Cooling Hours, and Summer Outdoor Daily Range*

City	Winter Design Temperature (97½%, °F) (°C)	Degree Days Heating (DDH)	Summer Design Temperature (2½%, °F) (°C)	Annual Cooling Operating Hours	Summer Outdoor Daily Range
Anchorage, AK	–18 (–27.8)	10,860 (6016)	68 (20)	60	15 (–9.4)
Phoenix, AZ	34 (1.1)	1,680 (916)	107 (41.7)	1,930	27 (–2.8)
Los Angeles, CA	43 (6.1)	1,960 (1071)	80 (26.7)	460	20 (–6.7)
San Francisco, CA	38 (3.3)	3,040 (1671)	77 (25)	170	14 (–10)
Denver, CO	1 (–17.2)	6,150 (3399)	91 (32.8)	700	28 (–2.2)
Washington, DC	17 (–8.3)	4,240 (7338)	91 (32.8)	960	18 (–7.8)
Miami, FL	47 (8.3)	200 (93)	90 (32.2)	2,940	15 (–9.4)
Chicago, IL	–4 (–20)	6,640 (3671)	89 (31.7)	650	20 (–6.7)
Wichita, KS	7 (–13.9)	4,640 (2560)	98 (36.7)	1,030	23 (–5)
Minneapolis, MN	–12 (–24.4)	8,250 (4566)	89 (31.7)	590	22 (–5.6)
New York, NY	15 (–9.4)	4,880 (2693)	87 (30.6)	770	16 (–8.9)
Cincinnati, OH	6 (–14.4)	4,830 (2666)	90 (32.2)	870	21 (–6.1)
Tulsa, OK	13 (–10.6)	3,730 (2054)	98 (36.7)	1,230	24 (–4.4)
Rapid City, SD	–7 (–21.7)	7,370 (4077)	92 (33.3)	610	28 (–2.2)
Houston, TX	32 (0)	1,410 (766)	94 (34.4)	1,880	18 (–7.8)
Salt Lake City, UT	8 (–13.3)	5,990 (3310)	95 (35)	770	32 (0)
Richmond, VA	17 (–8.3)	3,910 (2154)	92 (33.3)	1,010	21 (–6.1)
Spokane, WA	2 (–16.7)	6,770 (3743)	90 (32.2)	460	28 (–2.2)
Green Bay, WI	–9 (–22.8)	8,100 (4482)	85 (29.4)	430	23 (–5)

*Data derived from ASHRAE *Handbook of Fundamentals*, ASHRAE *Bin and Degree-hour Weather Data for Simplified Energy Calculations*, USAF AFM88-29, and NCDC *Climatography of U.S.*

Conductance It is not meaningful to determine heat transfer per 1 in. (25.4 mm) of thickness for products that are not homogeneous or that have air cavities (such as hollow concrete block). Such products are instead given a conductance (C) rating based on the number of Btu's transmitted in 1 hour through 1 sq. ft. (0.09 m^2) of a stated thickness, when the opposite surfaces vary in temperature by 1°F (–17.2°C). For instance, an 8-in. (203.2 mm)-thick hollow concrete masonry unit has a C-value of 0.90. Therefore, 0.90 Btu (0.95 kJ) is transferred in 1 hr. if one face of the unit is at 74°F (23.3°C) and the other is at 75°F (23.9°C).

RESISTANCE

Materials that are normally considered good insulators have low k- and C-values (see Fig. 15.2-12). However, because k and C for these are stated as multidigit decimals smaller than 1.0, they are difficult to understand and compare. As a result, the use of *resistance* (R) numbers has increased as a method of rating building products and thermal insulations. An R-value is computed by taking the reciprocal of k (1/k), or the reciprocal of C (1/C). In the first case, the R-value must be qualified as being for a 1-in. (25.4 mm) thickness; in the second, it is simply the R-value of the product, whatever its thickness.

Its resistance (R) represents the ability of a material to retard heat flow. R-values are also convenient because they are directly proportional to insulating value and can therefore be added or subtracted easily in order to determine the thermal performance of a product or assembly. For example, an 8-in. (203.2 mm) concrete block with a C-value of 0.90 (2.1) would have an R-value of 1.11 (0.48) (see Fig. 15.2-12). If a 1-in. (25.4 mm) slab of polystyrene with an R-value of 5.00 (2.2) were added, the assembly would have a total R-value of 6.11 (2.5), not counting surface films. The molecules of air clinging to the surfaces contribute some insulating value, so R-0.68 (0.29) should be added for still inside air, and R-0.17 (0.07) for outside air at 15 mph (24.14 km/h). The final total for the assembly is R-6.96 (2.86).

Many insulating products are labeled with their R-value on the wrapper. The number given is usually the value of the product only, without the air spaces and films, which must be added to obtain the total installed resistance. Some insulations, such as multilayer foil, however, depend on air spaces and films for performance and can be described only in terms of installed resistance.

TRANSMITTANCE

Transmittance (U) is the number of Btu's transferred in 1 hr. per 1 sq. ft. (0.09 m^2) of a building assembly, for each 1°F (–17.2°C) temperature difference between inside and outside surfaces. The U-value is calculated by adding the Rs for all elements of the building assembly (R_T) and taking the reciprocal of the total ($1/R_T$). If the R_T for an assembly equals 20, then the

FIGURE 15.2-12 Heat Transfer Coefficients of Representative Building Products[a]

Material or Product	Conductivity (k) (Btu/ft · h · °F) (W/m · K)	Conductance (C)[e] (Btu/ft · h · °F · t) (W/m · K · t)	Resistance (R) (ft² · h · °F/Btu)(m² · K/W) Per 1-in. (25.4 mm) Thickness (1/k)	Per Thickness Shown (1/C)
Concrete	5.0–10.5 (8.7–18.2)		0.20–0.10 (0.12–0.1)	
Face brick	9.00 (15.6)		0.11 (0.06)	
Hollow concrete block, 8 in. (203.2 mm)		0.90 (2.1)		1.11 (0.48)
Stucco	5.00 (8.7)		0.20 (0.12)	
Metal lath and plaster, ¾ in. (19.1 mm)		7.70 (17.7)		0.13 (0.06)
Gypsum board, ½ in. (12.7 mm)		2.22 (5.1)		0.45 (0.2)
Plywood, ½ in. (12.7 mm)		1.60 (3.7)		0.62 (0.27)
Pine, fir, other softwoods	0.74–1.12 (1.3–1.9)		1.35–0.89 (0.78–0.52)	
Oak, maple, other hardwoods	1.06–1.25 (1.8–2.2)		0.94–0.80 (0.55–0.46)	
Asphalt shingles		2.27 (5.2)		0.44 (0.19)
Built-up roofing, ⅜ in. (9.5 mm)		3.00 (6.9)		0.33 (0.14)
Wood shingles		1.06 (3.9)		0.94 (0.41)
Vegetable board sheathing, ½ in. (12.7 mm)		0.76 (1.8)		1.32 (0.57)
Mineral wool batts, 3–4 in. (76.2–101.6 mm)		0.09 (0.2)		11.00 (4.8)
5½–6½ in. (139.7 to 165.1 mm)		0.05 (0.1)		19.00 (8.7)
6–7½ in. (152.4 to 190.5 mm)		0.05 (0.1)		22.00 (8.7)
9–10 in. (228.6 to 254 mm)		0.03 (0.07)		30.00 (14.5)
Expanded polystyrene, 1 in. (25.4 mm)		0.20 (0.35)		5.00 (2.2)
Inside surface air film[b]				0.68 (0.29)
Outside surface air film[c]				0.17 (0.07)
Airspace ¾ in. (19.1 mm), nonreflective[d]		1.00 (2.3)		1.01 (0.43)
Airspace ¾ in. (19.1 mm), reflective[d]				3.48

[a]At 75°F (23.9°C) mean temperature, from ASHRAE *Fundamentals Handbook*.

[b]Heat flow horizontal, still air.

[c]Heat flow any direction, 15 mph (24.14 km/h) wind.

[d]Heat flow horizontal.

[e]k multiplied by thickness.

U-value would be 0.05 (1/20). Sample U-value calculations are given in Figure 15.2-13.

HEAT STORAGE CAPABILITY OF MATERIALS

Traditional heat flow calculations are based on the theory that design conditions are sustained for long periods and that heat flows continually from a warmer to a cooler side at a constant rate. This *steady-state* concept of thermodynamics is useful in designing heating or cooling systems where the building shell is not made of massive materials and the day-to-night outdoor temperature variation is not great.

In actual practice, heavy masonry and concrete buildings behave dynamically and do not generally lose or gain heat as rapidly as may be indicated by the steady-state theory. This is especially true in arid, sunny climates where there are relatively warm days in winter and cool nights in summer. Under these conditions, massive buildings respond dynamically, acting as storers of heat, rather than as flow resistors as do insulated lightweight buildings.

In summer, massive structures not only store heat but are able to slow its movement into a building until late afternoon, when the air and structure start cooling off. In winter, heat movement out of the building during the night is slowed by the mass of the structure until the sun warms the walls, and by late afternoon heat actually moves inward. Designers can use this *thermal lag* to help maintain interior comfort conditions with a minimum of man-made equipment.

The principle of using heavy, dense materials as *thermal mass* was well known to the Indian builders of 1000-year-old Arizona cliff communities, as well as to modern adobe home owners throughout the sunbelt states. Despite their limited insulating value, earth-sheltered and adobe homes remain surprisingly cool in summer and warm in winter. The effectiveness of a material as a thermal mass, that is, its ability to retard heat flow, depends on its ability to store heat, which is measured by its *specific heat* and *heat capacity* (Fig. 15.2-14). Refer to Section 15.2.10.2 for a more detailed discussion of this subject.

FIGURE 15.2-13 U-value Computation

Frame Wall Construction	Insulated U and R_T Values R = (ft² · h · °F/Btu)(m² · K/W)		
	Construction	**R**	
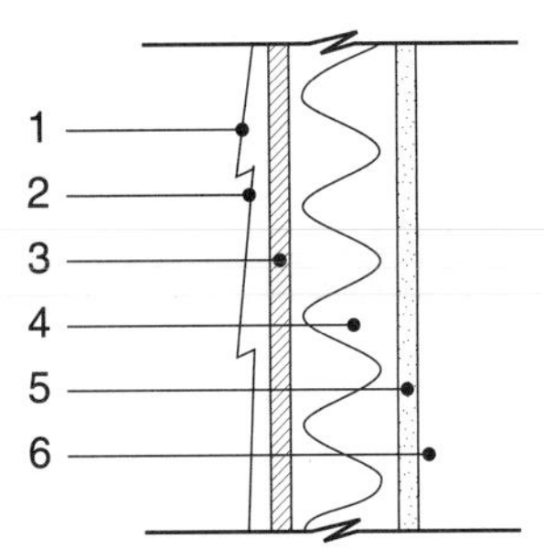	1. Outside Air Film	0.25	0.11
	2. Wood Siding	0.81	0.34
	3. ½″ Insul. Sheathing	1.32	0.56
	4. Insulation	11.00	4.68
	5. ⅝″ Gypsum Board	0.56	0.24
	6. Inside Air Film	0.68	0.29
	R_T =	14.62	6.22
	$1/R_T$ = U =	0.07	0.16

Masonry Wall Construction	Insulated U and R_T Values R = (ft² · h · °F/Btu)(m² · K/W)		
	Construction	**R**	
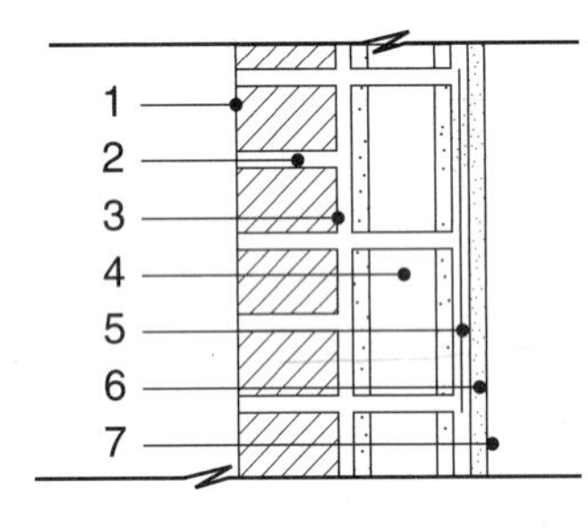	1. Outside Air Film	0.25	0.11
	2. 4″ Common Brick	0.80	0.34
	3. ½″ Cement Mortar	0.10	0.04
	4. 4″ Lt. Wt. Conc. Block	1.50	0.64
	5. ¾″ Air Space	—	—
	6. ⅝″ Gypsum Board	0.56	0.24
	7. Inside Air Film	0.68	0.29
	R_T =	3.89	1.66
	$1/R_T$ = U =	0.07	0.16

FIGURE 15.2-14 Heat Storage Capabilities of Building Materials

Material	Specific Heat, Btu/lb./°F (kJ/kg/°C)	Density, lb./cu. ft. (kg/m³)	Heat Capacity, Btu/cu. ft./°F (kJ/m³/°C)
Water (40°F)	1.000 (4.19)	62.5 (1001)	62.5 (4192)
Steel	0.120 (0.5)	489.0 (7883)	58.7 (3937)
Cast iron	0.120 (0.5)	450.0 (7208)	54.0 (3622)
Copper	0.092 (0.39)	556.0 (8906)	51.2 (3434)
Aluminum	0.214 (0.9)	171.0 (2739)	36.6 (2455)
Basalt	0.200 (0.84)	180.0 (2883)	36.0 (2414)
Marble	0.210 (0.88)	162.0 (2595)	34.0 (2280)
Concrete	0.220 (0.92)	144.0 (2307)	31.7 (2126)
Asphalt	0.220 (0.92)	132.0 (2114)	29.0 (1945)
Ice (32°F)	0.487 (2.04)	57.5 (921)	28.0 (1878)
Glass	0.180 (0.75)	154.0 (2467)	27.7 (1858)
White oak	0.570 (2.39)	47.0 (752.9)	26.8 (1797)
Brick	0.200 (0.84)	123.0 (1970)	24.6 (1650)
Limestone	0.217 (0.91)	103.0 (1650)	22.4 (1502)
Gypsum	0.260 (1.09)	78.0 (1249)	20.3 (1361)
Sand	0.191 (0.8)	94.6 (1515)	18.1 (1214)
White pine	0.670 (2.81)	27.0 (432.5)	18.1 (1214)
White fir	0.650 (2.72)	27.0 (432.5)	17.6 (1180)

The dynamic performance of concrete masonry and other massive shells can be accurately calculated by computers programmed with detailed winter and summer weather information. To perform such calculations manually, utilizing the steady-state approach, is difficult, but a reasonable approximation can be made using a dimensionless correction factor called the *M factor*. The Brick Institute of America can provide more information about the thermal performance of masonry walls.

15.2.2 BUILDING DESIGN RECOMMENDATIONS

Buildings should regulate heat flow in a way that will minimize dependence on mechanical equipment and take advantage of exterior environmental conditions. Energy-saving features can be incorporated economically in buildings if they are considered in the design phase.

15.2.2.1 Site Planning

In locating a building on a site, the following should be considered: (1) general climate and microclimate, (2) terrain and landscape elements, and (3) orientation to the sun.

CLIMATE

Climatic analysis can provide valuable data on which to base energy-efficient design. Such an analysis should include data on the amount and angle of sunshine that strikes the site, prevailing winds, temperature ranges, and precipitation.

The U.S. Weather Service and a number of private meteorologists maintain regional data on these factors. However, natural terrain and proximity to bodies of water may substantially modify the weather information recorded for a particular region. To avoid siting a building in an unsuitable area, the climate variations of each site (microclimate) should be examined. For example, sites in warm climates that act as natural dead-air basins should be avoided. In such areas, temperatures may exceed those of surrounding areas by as much as 35°F (1.67°C). In other areas, prevailing winds and natural terrain can combine to produce unusual wind, snow massing, or rain conditions not typical of a particular climate. Air quality should also be studied to determine whether natural ventilation can be used for cooling.

TERRAIN AND LANDSCAPE ELEMENTS

Terrain configurations and plant material can be used to control heat flow into and out of buildings. Refer to Section 2.5 for a discussion of landscape elements.

ORIENTATION

In cold regions, the area of north-facing windows should be limited to reduce heat loss in winter. The area of north-facing walls can be reduced by berming and sloping roofs (Fig. 15.2-15). In hot climate areas, west wall and glass exposures should be minimized to reduce cooling loads imposed by intense afternoon summer sun.

Buildings in cold regions should be sited so that most windows face south, southeast, or southwest to take advantage of the warming sun. Orientation that will buffer against hot breezes is desirable in hot climates. Prevailing summer breezes that can provide ventilation and natural cooling should be considered in siting a building and locating windows.

15.2.2.2 Building Design

Thermally efficient buildings can be achieved through room organization and effective building envelope design.

ROOM ORGANIZATION

In areas subject to either extreme heat or cold, compact design reduces weather-exposed areas and reduces heat flow. Garages, closets, corridors, and other service spaces can be located against exterior walls to act as insulating buffers.

Inactive living spaces such as kitchens and laundry areas, which contain heat-producing appliances, should be located on the north and east sides. This location permits wintertime use of their excess heat for warming and allows heat dissipation in summer.

Active living spaces should face south or southwest to take advantage of warming afternoon sun in winter. Such spaces should be protected against overheating in summer by external shading devices such as louvers, overhangs, awnings, and trees (see Section 12.2). Interior shading with blinds and drapes is less effective, but acceptable. Rooms likely to remain vacant for long periods should be located where they can be closed off readily. Lower ceiling heights can also reduce the volume of air to be heated or cooled. Built-in furnishings should be located away from supply and return air grilles to prevent obstruction of internal airflows.

FIGURE 15.2-15 Sloping roofs act as wind deflectors.

BUILDING ENVELOPE

For conventionally heated and cooled buildings, the objective of heat control is to block the flow of heat through the ceilings, walls, and floors that surround habitable areas. Such heat flow can be controlled by

1. Constructing wall, floor, and roof assemblies using materials that are inherently good insulators to minimize heat flow out of the building in winter and into the building in summer;
2. Using solid masonry or another dense material that has the capacity to store heat and slow heat flow;
3. Installing loose-fill, flexible, rigid, reflective, or foamed-in-place insulations—in some climates, using materials with high thermal capacities will sufficiently slow heat gain or loss, but in most locations, adequate insulation of the building envelope is essential to controlling heat flow. Refer to Section 7.3 for a discussion and recommendations concerning insulation.

Floors Where solar radiation penetrates interior spaces, floors can act as *heat sinks*. Heat sinks are building components consisting of dense, heavy materials such as concrete or masonry. They are desirable in cold climates because such materials absorb and store heat for long periods before releasing it to the interior by radiation and convection.

The floors of living areas above unheated crawl spaces or basements and slabs on ground should be insulated as recommended in Section 7.3.

Walls Heat transmission through wall assemblies can be reduced by selecting materials that are inherently good insulators or by adding insulation (Section 7.3).

Wind-deflecting devices such as berms, walls, and fences can also be employed to reduce infiltration along exterior walls exposed to cold winds by steering the wind away from the walls.

Roofs Heat transmission through a roof can be reduced by:

1. Insulation (see Section 7.3)
2. Shiny or light-colored roof materials to reflect heat
3. Ventilation of attic spaces
4. Increasing the thermal mass to achieve a time lag in the heat flow
5. Louvered rather than solid overhangs to free heated air trapped beneath eaves
6. Steeply pitched roofs on the windward side of the building to deflect cold winds away from the rest of the roof and reduce the amount of roof exposed to these winds (see Fig. 15.2-15)

15.2.3 HEAT LOSS AND HEAT GAIN

In a typical building, windows and doors lose and gain more heat than any other part of the structure. In some climates this heat loss or gain can account for as much as 70% of the total heating load and 50% of the cooling load, according to a study by the Department of Housing and Urban Development.

Most heat is lost by *infiltration* (air leakage) through cracks around the frames, and *conduction* resulting from warm air coming in contact with the cold window and door surfaces. Most heat is gained through *radiation*, which is transmission of heat by invisible electromagnetic waves through glass. Constructing window and door openings so that unwanted heat loss and heat gain are minimized will conserve energy, create a more comfortable indoor environment, and reduce building operation costs.

Many of the strategies suggested in this section for reducing heat gain and heat loss have been suggested by the National Institute of Standards and Technology publication *Window Design Strategies to Conserve Energy*.

Heat loss through windows can be reduced by tightly sealing openings, using multiple glazing (storm sash or insulating glass), and installing insulating shutters or tightly fitting roll shades.

Besides the other measures discussed here, heat loss through doors is often accomplished by means of vestibules or storm doors.

In most parts of the country these

methods can reduce heating costs, minimize condensation on interior surfaces, and improve interior comfort by reducing drafts and radiant body heat loss. In Zones IV and V of Figure 7.3-2, multiple glazing may not be justified and its use should therefore be decided upon on a case-by-case basis, taking into account fuel and material costs.

Heating and air conditioning equipment is sized on the basis of heat loss and heat gain calculations for the particular area in which the building is situated. Such calculations also provide a measure of the thermal efficiency of a building.

15.2.3.1 Definitions

The following terms are used in performing a heat loss or heat gain analysis:

R-value. A material's resistance to the flow of heat; the larger the R-value, the greater the resistance and the better the insulating value (see Section 7.3).

Heat loss. The amount of heat flow in Btu's (kJ) per hour or per season, from inside a structure to outside, at given design conditions.

Sensible heat gain. The amount of heat gain in Btu's (kJ) per hour that can be recorded as a rise in temperature.

Latent heat gain. Heat introduced into a structure by moisture in the air that must be removed by the cooling system. The latent heat gain is commonly calculated as 30% of the *sensible heat gain.*

Indoor design temperatures. The ideal comfort temperatures that are to be maintained inside a structure in winter and summer. These should correspond with the ASHRAE comfort envelope unless DOE has developed more stringent guidelines.

Outdoor design temperatures. High and low temperatures that can be expected to occur regularly in a particular area (see Fig. 15.2-11).

Design temperature difference. The difference between the outside design temperature (t_o) and the inside design temperature (t_i).

Infiltration. The leakage of heat into or out of a building through cracks in the building structure, especially around windows and doors (Fig. 15.2-16).

Transmittance (U). The number of Btu's (kJ) per hour transferred through 1 sq. ft. (0.09 m^2) of a building assembly (such as a wall) for each 1°F (–17.2°C) temperature difference between inside and outside surfaces. It is calculated by adding the Rs for all elements of the building assembly (R_T) and taking the reciprocal ($1/R_T$) of the total (Fig. 15.2-17).

15.2.3.2 Heat Loss and Heat Gain Formulas

Heat flows into or out of a building:

1. Through exterior walls, roofs, and glass areas;
2. At the perimeter of a building's ground floor;
3. Through floors over unheated spaces;
4. By infiltration around windows and doors.

The following heat loss and heat gain calculation method is intended to be used for small-building design, but the methods are applicable to buildings of all types and sizes. Calculating heat gain and heat loss and determining HVAC system design criteria for commercial installations should be performed by a licensed mechanical engineer.

Further information on calculating heat loss and design criteria for the sizing and location of ducts and registers in small residential buildings may be obtained from the National Association of Home Builders (NAHB) Research Foundation and from the publications of the Air Conditioning Contractors of America (ACCA) and the Sheet Metal and Air Conditioning Contractors National Association (SMACNA).

The following calculation procedures assume use of data and tables in this book, which may not be up to date, for determining building heat loss. In an actual case a designer should obtain and use the latest data available.

15.2.3.3 Heat Loss Calculation

The following steps can be followed to calculate heat loss:

1. Determine the winter outdoor design temperature (t_o) (see Fig. 15.2-11) and the desired indoor design temperature (t_i) from the latest ASHRAE Comfort Envelope or use the HUD guideline temperature. The design temperature difference is equal to the difference between the established indoor and outdoor design temperatures ($t_i - t_o$).
2. Calculate the U-values for roof, wall, and glass areas by summing the Rs for all parts of such an assembly, including air films, (R_T) and taking the reciprocal ($1/R_T$) of that figure (see Fig. 15.2-17).
3. Calculate the surface area (A) of glass and all similarly constructed roof, wall, and floor areas.
4. Calculate building heat loss (H_c) through exterior roof and wall construction and through exterior glass by multiplying the area (A) by its computed U-value, by the design temperature difference ($t_i - t_o$).
5. Calculate the floor heat loss (H_f) by the same method used for walls, roof, and glass, with the following qualifications:
   - If the heating unit is in the basement and ducts or pipes are uninsulated, no floor loss calculation is necessary.

FIGURE 15.2-16 Window and Door Infiltration Values

	Infiltration Airflow, cu. ft./ft. (m^3/m) of Crack*		
Window Type	**10 mph (16.1 km/h)**	**20 mph (32.2 km/h)**	**30 mph (48.3 km/h)**
Double-hung wood frame			
Non-weatherstripped	21 (2)	59 (5.5)	104 (9.7)
Weatherstripped	13 (1.2)	36 (3.3)	63 (5.9)
Double-hung metal frame			
Non-weatherstripped	47 (4.4)	104 (9.7)	170 (15.8)
Weatherstripped	19 (1.8)	46 (4.3)	76 (7.1)
Residential metal casement	18 (1.7)	47 (4.4)	74 (6.9)
Wood or metal door			
Non-weatherstripped	69 (6.4)	154 (14.3)	249 (23.1)
Weatherstripped	19 (1.8)	51 (4.7)	92 (8.5)

*Reduce values by one-third when storm window (or door) is added to weatherstripped construction, by one half when added to non-weatherstripped construction.

FIGURE 15.2-17 U-Value Calculation Procedure

Building Part	Construction Materials	R-Value (ft² · h · °F/Btu)	(m² · K/W)
Roof/ceiling	Outside air film	0.17	0.07
	Shingles	0.44	0.19
	Building paper	0.06	0.03
	Plywood, ½ in.	0.62	0.26
	Attic air film	0.61	0.26
	Insulation	19.00	8.1
	Gypsum board, ½ in.	0.45	0.19
	Inside air film	0.61	0.26
	Total R-Value (R_T)	21.96	9.35
	U-Value ($1/R_T$)	**0.045**	**0.11**
Wall	Outside air film	0.17	0.07
	Siding, wood ½ in. × 8 in. lapped	0.81	0.34
	Sheathing, plywood, ½ in.	0.62	0.26
	Insulation	11.00	4.68
	Interior finish gypsum board, ½ in.	0.45	0.19
	Inside air film	0.68	0.29
	Total R-Value (R_T)	13.73	5.8
	U-Value ($1/R_T$)	**0.073**	**0.17**
Header joist	Outside air film	0.17	0.07
	Siding	0.81	0.34
	Sheathing	0.62	0.26
	Header, wood 1½ in.	1.88	0.8
	Insulation	11.00	4.68
	Inside air film	0.68	0.29
	Total R-Value (R_T)	15.16	6.45
	U-Value ($1/R_T$)	**0.066**	**0.15**
Sill	Outside air film	0.17	0.07
	Siding	0.81	0.34
	Sheathing	0.62	0.26
	Sill, wood 5½ in.	6.88	2.93
	Inside air film	0.68	0.29
	Total R-Value (R_T)	9.16	3.9
	U-Value ($1/R_T$)	**0.109**	**0.26**
Foundation	Outside air film	0.17	0.07
	Concrete block, 8 in.	1.11	0.47
	Insulation	5.00	2.13
	Interior finish gypsum board, ⅜ in.	0.32	0.14
	Inside air film	0.68	0.29
	Total R-Value (R_T)	7.28	3.1
	U-Value ($1/R_T$)	**0.137**	**0.32**

- For floors over unheated basements, use ⅓ of the design temperature difference.
- For floors over unvented crawl spaces with insulated crawl space walls, use ½ of the design temperature difference.
- For floors over vented crawl spaces or other open spaces, use the full design temperature difference.

6. Calculate the building perimeter (P) in feet (meters) if the floor is of slab construction.
7. Calculate the floor slab edge heat loss (H_e) by multiplying a factor of 0.55 by the slab perimeter (P) in feet (meters) if the slab edge is insulated or a factor of 0.8 by (P) in feet (meters) if the slab edge is uninsulated, by the design temperature difference ($t_i - t_o$). Use °F when using feet and °C when using meters.
8. Calculate the length of crack (L) in feet (meters) around window and door openings.
9. Select infiltration airflow volume in cubic feet per hour (cfh) (m³/h) per foot (meter) of crack (see Fig. 15.2-16).
10. Calculate the infiltration heat loss (H_i) for each opening by multiplying the length of crackage (L) by 0.018 times the per-foot (meter) infiltration volume in cfh (m³/h), by the design temperature difference ($t_i - t_o$) in °F for cfh and °C for m³/h.
11. Calculate the total building heat loss (H_T) by summing H_c, H_i, and H_e, or H_f if appropriate.

15.2.3.4 Heat Gain Calculation

The following steps can be followed to calculate heat gain:

1. Determine the winter outdoor design temperature (t_o) (see Fig. 15.2-11) and the desired indoor design temperature (t_i) from the latest ASHRAE Comfort Envelope or the HUD guideline temperature. The design temperature difference is equal to the difference between the established indoor and outdoor design temperatures ($t_i - t_o$).
2. Calculate the U-value for roof and walls by summing the Rs for all parts of each assembly (R_T) and taking the reciprocal ($1/R_T$) of that value (see Fig. 15.2-17).
3. Calculate the surface area (A) of all similarly constructed roof, walls, and floors.
4. Calculate heat gain through unglazed (opaque) roof and wall sections (H_o) by multiplying the calculated area (A) by the calculated U-value, by the design equivalent temperature difference (DETD) (see Fig. 15.2-18).
5. Calculate heat gain through glass areas (H_g) as follows:
   - Locate the glass according to its building face: N, S, E, or W.
   - Calculate the surface area of each particular type of glass (single, double, or heat absorbing).
   - Select the appropriate winter glass load factor from tables in the latest ASHRAE handbook.
   - Multiply the glass area (A) by the heat gain factor for the total window heat gain in Btu's (kJ) per hour.
   - If the glass is partially or completely shaded by a permanent

FIGURE 15.2-18 Design Equivalent Temperature Differences

Design Temperature, °F[a]	85		90			95			100		105	110
Daily Temperature Range[b]	L	M	L	M	H	L	M	H	M	H	H	H
Walls and doors												
Frame and veneer-on-frame	17.6	13.6	22.6	18.6	13.6	27.6	23.6	18.6	28.6	23.6	28.6	33.6
Masonry walls, 8-in. block or brick	10.3	6.3	15.3	11.3	6.3	20.3	16.3	11.3	21.3	16.3	21.3	26.3
Wood doors	17.6	13.6	22.6	18.6	13.6	27.6	23.6	18.6	28.6	23.6	28.6	33.6
Ceilings and roofs												
Ceilings under naturally vented attic or vented flat roof, dark	38.0	34.0	43.0	39.0	34.0	48.0	44.0	39.0	49.0	44.0	49.0	54.0
light	30.0	26.0	35.0	31.0	26.0	40.0	36.0	31.0	41.0	36.0	41.0	46.0
Built-up roof, no ceiling, dark	38.0	34.0	43.0	39.0	34.0	48.0	44.0	39.0	49.0	44.0	49.0	54.0
light	30.0	26.0	35.0	31.0	26.0	40.0	36.0	31.0	41.0	36.0	41.0	46.0
Ceilings under unconditioned rooms	9.0	5.0	14.0	10.0	5.0	19.0	15.0	10.0	20.0	15.0	20.0	25.0
Floors												
Over unconditioned rooms	9.0	5.0	14.0	10.0	5.0	19.0	15.0	10.0	20.0	15.0	20.0	25.0
Over basement, heated crawl space, or concrete slab on ground	0	0	0	0	0	0	0	0	0	0	0	0
Over unheated crawl space	9.0	5.0	14.0	10.0	5.0	19.0	15.0	10.0	20.0	15.0	20.0	25.0

[a]Interpolate for values between those listed.

[b]Daily Temperature Range (see Fig. 15.2-11):

L (Low) calculation value: 12 — Range: less than 15°
M (Medium) calculation value: 20 — Range: 15 to 25°
H (High) calculation value: 30 — Range: more than 25°

(Refer to Figure 1.8-27 for metric equivalents.)

shading device such as an overhang, the shaded portion is treated as north-facing glass. Unshaded portions are considered separately. The shade line is computed by multiplying overhang width by the shade line factor found in the ASHRAE handbook. Glass above the line is in shade, and glass below the line is in sun.

6. Heat gain from infiltration (H_i) is normally insignificant in summer and can be ignored.
7. Estimate the sensible heat gain from people (H_p) at 225 Btu's (237.4 kJ) per hour per person. For residences, estimate the number of occupants at twice the number of bedrooms.
8. Estimate appliance loads (H_a) at 1200 Btu's (1266.1 kJ) per hour.
9. Calculate total sensible heat gain (H_{TS}) by summing H_o, H_g, H_p, and H_a.
10. Estimate total latent heat gain (H_{TL}) as ⅓ H_{TS}.
11. Calculate total heat gain as the sum of total latent heat gain (H_{TL}) and total sensible heat gain (H_{TS}).

15.2.4 HVAC SYSTEMS

Virtually every habitable building in the United States needs at least a minimal mechanical heating system, including those where much of the heating load is carried by a passive or active solar energy system (see Section 15.2.10). Most buildings also need a mechanical cooling system for occupant comfort and acceptable working and living conditions. A cooling system should definitely be considered in locations where the humidity is often high.

In addition, habitable buildings need a method of ventilating toilet rooms, shower rooms, laboratories, bathrooms, and kitchens. Buildings also need a source of fresh air to furnish combustion air for furnaces, boilers, fireplaces, and gas- or oil-burning equipment and devices, such as water heaters, ovens, ranges, and other appliances and devices in which open flames are present. Commercial kitchens and laboratories with fume hoods and the like need a source of makeup air. These ventilation requirements do not, however, always require mechanical ventilation. In a residence, for example, windows or other openings may satisfy most ventilation requirements.

Heating and cooling equipment includes furnaces, boilers, motors and motor-driven compressors, condensing units, cooling towers, heat exchangers, blowers, and pumps.

The heating and cooling equipment for forced-air systems in most single-family residential buildings and other small buildings is usually a packaged furnace or heat pump system selected by the architect or the heating and cooling contractor, based on calculated heating and cooling loads and experience. Professional engineers are seldom employed to design such systems; manuals produced by the Air Conditioning Contractors of America may be used to aid in their design.

More complex systems and those utilizing heavy equipment should be designed by professional structural and mechanical engineers based on design calculations and acoustical analysis. The noise levels and heating and cooling capacities of equipment should be obtained. Many manufacturers can furnish or develop sound-pressure output data based on standard testing for most types of equipment, from which sound-pressure levels can be computed for a specific space.

Selecting better-quality heating and cooling equipment with more than adequate capacity may slightly increase initial cost but will often help to create a quieter installation that will perform more efficiently over the years. Selecting equipment with greater-than-adequate capacity, particularly for fans and pumps, can reduce noise levels considerably, because larger equipment runs more slowly and therefore more quietly.

HVAC systems fall into several categories, depending on the heat medium (air, water, steam, electricity) and the means of delivery. For convenience, HVAC systems can be divided into the categories that follow.

15.2.4.1 Forced-Air Systems

In a forced-air system, air is conditioned in a remote location, such as a mechanical equipment room, and delivered by ductwork to the spaces to be conditioned. All heat and all cooling are delivered by warm or cooled air. In most single-family residences and many other small buildings, the air in a forced-air system is warmed by a gas, oil, or electric furnace and cooled by a unit consisting of cooling coils mounted on the bonnet of the furnace and a condenser located outdoors (see Section 15.2.6) or by a heat pump (see Section 15.2.7). In most large installations in buildings other than residences, the air is warmed by hot water or steam generated in a boiler.

FORCED-AIR DELIVERY

Forced air is delivered through a duct system in one of several possible configurations.

Single-Duct, Single-Zone Systems Most forced-air systems in single-family residences and other small buildings are single-zone, single-duct systems. In these, the plenum at the top of a combined heating and cooling unit is extended through the building. Branch ducts from the extended plenum extend to outlet locations (Fig. 15.2-19). Through these ducts, either heated or cooled air is delivered to each space. The volume of air, and therefore the degree of heating or cooling, delivered to each space is controlled by dampers. In small installations manual dampers are installed at each outlet. In larger systems dampers are controlled by mechanical devices operated by heat sensors.

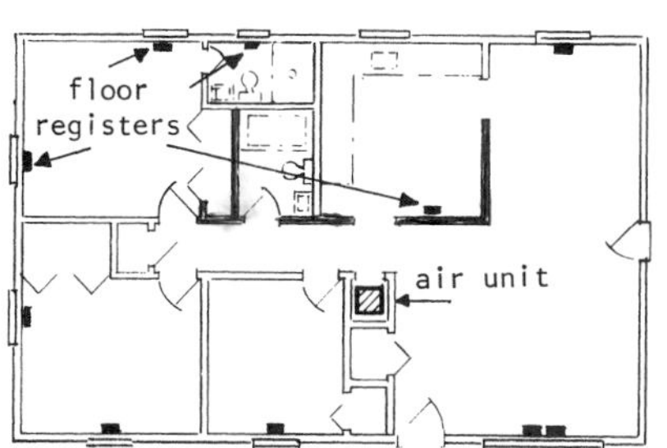

FIGURE 15.2-19 Typical plan showing floor register locations.

Perimeter Systems Two types of perimeter delivery systems are employed in houses and other small buildings. In a *perimeter radial* (*individual supply*) system, individual ducts extend from the top of the furnace to each outlet. In a *perimeter loop* system, a continuous duct loops around the entire building. This duct is fed by lateral ducts extending from the furnace bonnet. In each case, the amount of air delivered is controlled by dampers.

Single-Duct Multizone Larger residences and some medium-sized commercial buildings are divided into several zones. Each floor, or a portion of each floor, may constitute a zone. Each zone is served by a single duct system in which the ducts in all zones emanate from a single air handler. In larger buildings there may be several air handlers, each serving a single zone or group of zones.

Single-Duct Reheat Some single-duct systems in large buildings have hot water or steam reheat coils in the ducts. This permits lower temperatures in the air supply, reducing both duct size and needed fan capacity.

Double-Duct Systems Double-duct systems supply air of different temperature in two separate ducts. The air is mixed at the terminal point to deliver the proper temperature of air into each space. Such systems are used in large buildings where interior and exterior spaces and spaces on opposite sides of the building have different heat gains or losses. This difference may be caused by a particular space's proximity to exterior (solar) heating or cooling loads, the number of people in different spaces, or equipment loads. These loads may vary at different times. In a variation of the double-duct system, one duct may carry high-pressure air that varies in temperature as needed and the second duct may carry air at a constant temperature.

AIR-HANDLING COMPONENTS

Forced-air systems consist of air-handling equipment, ductwork, dampers, and air outlets.

Air-Handling Equipment In single-family residences and other small building systems, the air handler is the furnace. In larger systems, air handlers are located in central locations in mechanical equipment rooms. In very large buildings, more than one air handler will probably be required. Air handlers in large buildings are usually housed in mechanical equipment rooms. Air handlers require access to a supply of fresh air and a means to exhaust excess air to the exterior.

Large air handlers consist of:

1. An outside-air intake that feeds into an outside-air plenum
2. An automatic damper to control the amount of outside air entering the air handler
3. A mixing plenum where fresh air and air from the return-air system, which is also controlled by an automatic damper, are mixed
4. A preheat coil to initially raise the temperature of the mixed fresh and return air
5. A cooling coil to cool the air
6. A heating coil to produce the desired air temperature near the outflow
7. A fan to move the air

All of this equipment is housed in a large metal compartment. A return air fan is also required, but this is housed in the return air duct. When excess air is present in the system, a damper on the return-air duct opens to permit the excess to vent to the exterior.

Ductwork Forced-air heating and cooling system ductwork includes:

- Diffusers and other outlets through which conditioned air is delivered into habitable spaces
- Supply air ducts that carry air from a heating and cooling unit throughout the building to diffusers
- Return air intakes and ducts that pick up heated or cooled air and return it to the heating or cooling unit for recycling
- Fresh-air intakes that provide combustion air and air changes in the building
- Exhaust systems

Supply ducts should be sized to adequately deliver the required amount of air at the desired velocity. Procedures for selecting duct materials and sizes are contained in the Sheet Metal and Air Conditioning Contractors National Association (SMACNA) publication *HVAC Systems—Duct Design*. SMACNA also produces a "Duct System Calculator."

In small buildings, a single-duct system, consisting of a main supply duct with several branch ducts, is usually used. When attic space is available, the main and branch ducts are usually run in that space. When there is no attic, and in intermediate floors in buildings of more than 1 story, the main supply duct can be dropped below the ceiling. The branches can then be either dropped below the ceiling as well or run within the framing space. Ducts in individual supply and perimeter loop systems are often run below the floor in a crawl space or within the floor slab.

Combustion-air and fresh-air intakes may be necessary in buildings that are tightly sealed. A combustion-air intake provides outside air for efficient burning of fossil fuel. Outside air may be simply delivered to the room where the furnace is located if the furnace is in a separate room and a blast of cold winter air is not objectionable. To conserve energy and reduce the discomfort caused by cold air, the outside-air duct may be connected directly to the combustion chamber. This method, in fact, is required by some local utilities and energy conservation codes.

A fresh-air intake improves ventilation and helps to reduce the buildup of indoor odors and other pollutants. ASHRAE Standard 62 specifies a ventilation rate of 15 cfm (.74 m^3/min) to ensure acceptable air quality. Local codes may require a different amount. Fresh-air intakes may be placed in a vented attic or directly outdoors and connected to the return-air duct above the furnace's air filter. Air-to-air heat exchangers can be used to temper the incoming air and save heating and cooling energy.

Intakes should have insect screens and tight-fitting manual or automatic dampers to control the amount of air intake.

In small buildings, a furnace heats air and the associated blower distributes it through the ducts to supply outlets. The return-air system completes the convection loop by returning air to the furnace, where it can be heated and recycled. There may be a return-air duct system or a single return-air grille directly connected to the furnace.

In larger buildings, air is heated and cooled in air-handling units and distributed through ducts to supply outlets. Air is returned to the air handler through ducts.

A return-air system should be sized to equal or exceed the supply volume and should induce the return of air from each conditioned space. A properly designed return-air system helps maintain an even temperature level, with minimal stratification throughout a building.

In small buildings, nonducted volume air returns allow free passage of return air directly to the furnace, usually through a louvered furnace closet door. In nonducted return-air systems doors should be undercut to provide at least 1 in. (25.4 mm) of clearance from floor tile or carpet to ensure proper air circulation. However, a door that must provide acoustical privacy cannot be used as a means of returning air to the furnace. Separate air-return registers can be placed in such rooms and connected to a sound-absorbing duct system. This eliminates blower and furnace noises and reduces cross talk from room to room. Sound-absorbing air inlets can be used to allow air intake from corridors without admitting noise.

Air-return registers should not be located back-to-back for the same reasons supply registers should not be so located (see Section 15.2.11). Similar methods to those suggested for supply-air registers can also be used in return-air openings to reduce sound transmission.

Ducted multiple return-air systems are used in larger homes and other buildings with complex room layouts. Return-air grilles should be placed near the ceiling. The air is returned from grilles in each room through ducts located in the ceiling construction or attic. Smaller rooms such as residential kitchens and baths do not need return air grilles if clear openings, openings fitted with louvers or grilles, or undercut doors permit air to move from the small spaces into an adjacent space that does have a return-air grille.

Most codes require that bathrooms that do not have windows be exhausted directly to the exterior. Some codes have similar requirements for bathrooms with windows and for other spaces as well. Most codes require that toilets, shower rooms, and similar spaces in commercial and institutional buildings be directly exhausted. Whether required or not, separate exhaust systems should be installed in laboratories and in rooms housing cooking appliances, laundry equipment, showers, and other equipment or devices that produce excess heat, high humidity, or undesirable fumes.

Exhaust ducts should be sized to fit the volume of the space they are intended to exhaust with the desired number of air changes per hour. Fans should be selected using the same criteria and should usually be oversized to reduce the noise they generate.

AIR PLENUM SYSTEMS

An air plenum system is a method of distributing heated or cooled air or both to the living area of a dwelling or other small building through a tightly sealed underfloor area, usually a crawl space.

The system may be used with poured concrete or concrete masonry unit (block) foundations but is frequently used with wood foundations (see Section 6.4). Although, theoretically, an air plenum system could be used to heat and cool a 2- or 3-story building, it is presently recommended only for single-story structures.

An air plenum system is cheaper than conventional forced-air systems because it eliminates ductwork.

In a typical 1400 sq. ft. (130 m^2) air plenum house, energy use can be reduced by 10%, as compared with a house with a conventional forced-air heating system and crawl space. Uniform air distribution through well-spaced registers prevents stratification of the convected air. Tests conducted at the University of Florida showed an average temperature differential of 2½°F (16.4°C) between air near the floor and air 60 in. (1524 mm) above the floor. Air quantities can be regulated by adjusting dampers.

The floor surface is warm when the heat is on and cool when the air conditioning is on. A warm floor is particularly pleasant underfoot in winter.

Floor framing members, floor sheathing, underlayment, and finished flooring absorb heat from the system and radiate it into the building. Radiant heat makes occupants feel more comfortable in cold weather, thus allowing system operation at a lower temperature.

In studies by Washington State University, a dwelling in Idaho with an air plenum system used less heat than a dwelling with a conventional forced-air heating system and crawl space. The floor surface temperature ranged between 70 and 82°F (21.1 and 27.8°C); 85°F (29.4°C) is the maximum temperature recommended for floor surfaces.

In tests conducted with air conditioning in Florida, the floor surface temperature ranged between 70 and 73°F (22.2 and 22.8°C).

The information and most of the illustrations in this section were abstracted from a manual prepared by the National Association of Home Builders (NAHB)

Research Foundation, Inc., for the American Plywood Association (APA), American Wood Preservers Institute (AWPI), Southern Forest Products Association (SFPA), and the Western Wood Products Association (WWPA). The data have been updated to agree with the pamphlet *Plenwood System*, issued jointly in 1989 by the American Plywood Association (APA), American Wood Council (AWC), American Forest and Paper Association (AF & PA), Southern Forest Products Association (SFPA), and Western Wood Products Association (WWPA), and some of the illustrations come from that document.

An air plenum system uses a downflow air-handling unit (combination furnace and air conditioner) to create enough air pressure under the floor to deliver heated or cooled air to individual rooms through supply registers in the floor (Fig. 15.2-20). Using the crawl space to distribute conditioned air eliminates costly ductwork and provides both convection and radiant heat (Fig. 15.2-21).

Such a system is simple in design and easy to build. It conserves energy and reduces construction costs. With minor variations, an air plenum system meets the requirements of all the major building code groups.

Air plenum systems are particularly susceptible to the introduction of pollutants, such as radon, into the living environment inside a dwelling. Where radon exists, precautions should be taken to prevent its introduction into the plenum and subsequently into the dwelling.

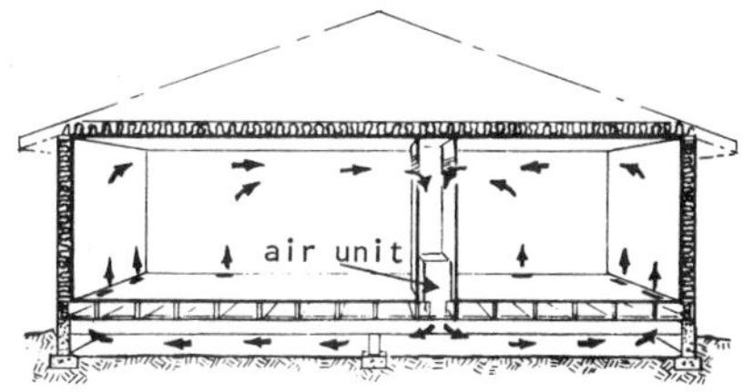

FIGURE 15.2-20 Diagram of an air plenum system.

FIGURE 15.2-21 Entering air heats or cools not only interior air, but floor surfaces as well.

Site Preparation Site clearing and excavation methods should be no different for plenum systems than for conventional foundation systems (see Chapter 2).

No special precautions for drainage are needed when the plenum grade is at or above exterior grade, the plenum space is well drained, the underlying soils are of types that drain well, and the water table height is normal. When the plenum grade is below the exterior grade, the soil is clayey, or the groundwater levels rise close to the plenum floor, foundation drains and sump pumps may be required. Care should also be taken to ensure drainage of rainwater during construction.

When a sloping site results in more of the foundation wall being above grade, thicker insulation should be used on the inside of the wall.

Increased or varying plenum depth may increase air volume and operating cost but does not affect satisfactory performance of the system (Fig. 15.2-22).

Termite Protection In regions subject to termite infestation, the soil should be chemically treated for a distance of 12 in. (304.8 mm) out from the foundation wall. Any wood in the foundation or wood that is below the level of the first floor should be pressure treated and carry the American Wood Preservers Bureau AWPB-FDN mark. Refer to Section 6.5 for additional requirements. If a wood foundation system is used, the preservative treatment methods described in Section 6.4 should be followed.

A shallow-depth wood plenum (less than 18-in. [457 mm] clearance) is not recommended in areas where a termite hazard exists, because soil-treating chemicals should not be used inside a plenum and because visual inspection is difficult.

No building debris or vegetation should be left in a plenum space.

Foundation Construction The basic plenum types are (1) shallow depth, low profile and (2) regular depth, crawl space.

A shallow plenum house looks like a slab-on-grade house from the outside. It is economical because even on sloping sites, very little site grading is necessary. A shallow-depth plenum has less than 18 in. (457 mm) of clearance below the floor joists and can function with as little as 3 in. (76 mm).

A regular-depth plenum is basically the same as a conventional crawl space. A regular-depth crawl space has a clearance of 18 in. (457 mm) or more from the bottom of the floor joists and 12 in. (305 mm) or more from the supporting girders.

A crawl space foundation intended to accommodate an air plenum should be

- Sealable to prevent conditioned air from escaping and water vapor from entering;
- Well drained;
- Fire safe;
- Free of insects and other vermin.

An air plenum system can be constructed with most conventional foundation materials (Fig. 15.2-23). It is also well suited to wood foundations (see Section 6.4), and this combination is common.

Thermal Integrity An air plenum system should be thoroughly caulked and sealed, have a suitable vapor retarder, and be adequately insulated. Foundation vents, typically recommended for crawl spaces, should not be used. Access to an air plenum should be only from inside the dwelling. Materials in a plenum should have a flame-spread rating of 200 or less.

An air plenum system should be sealed to prevent leakage of conditioned air and entrance of ground moisture. A sill sealer should be used between foundation walls

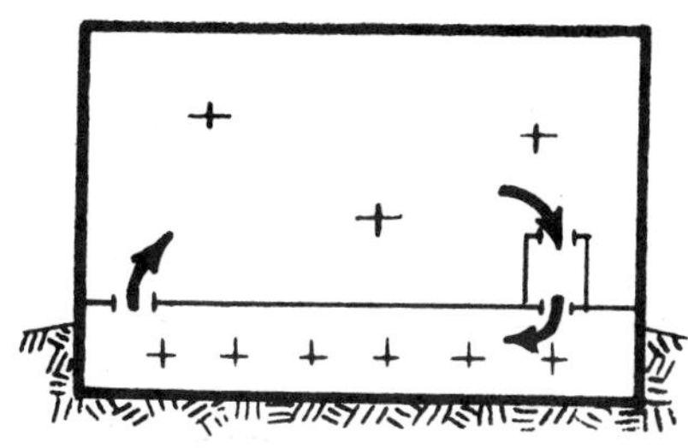

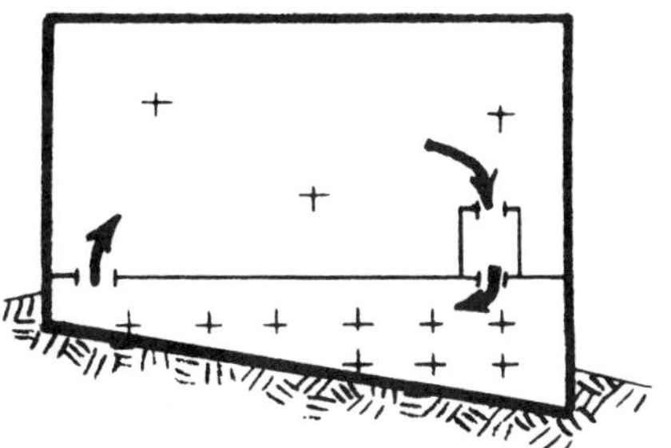

FIGURE 15.2-22 Neither plenum depth nor slope significantly affects system performance.

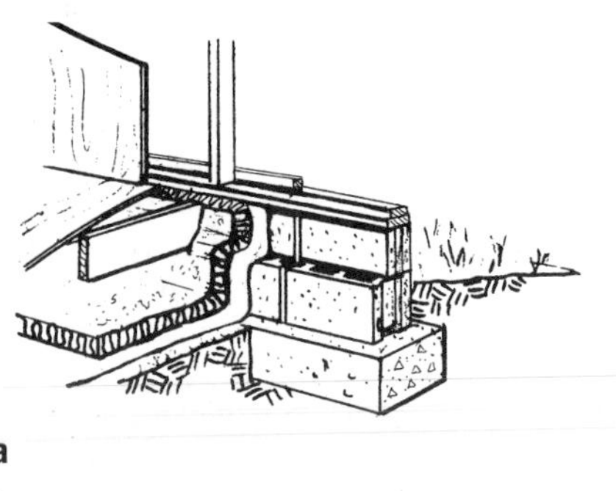

a

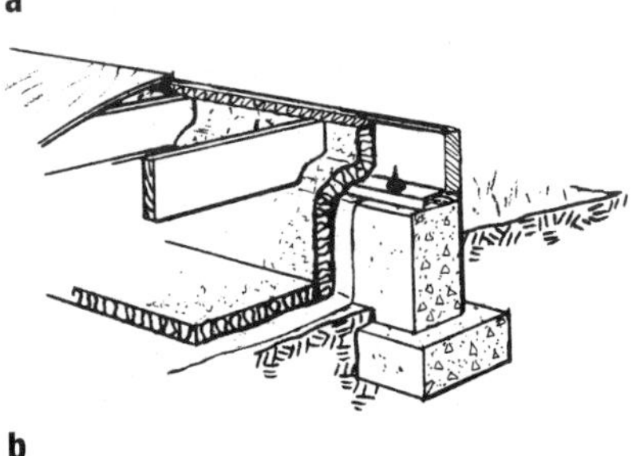

b

c

FIGURE 15.2-23 Air plenum systems may be constructed with foundations of (a) concrete block, (b) cast-in-place concrete, or (c) grade beams on wood posts or concrete piers.

and sole plates in wood-framed construction. Joints should be caulked.

A polyethylene film or similar continuous sheet-material vapor retarder should be used as a ground cover and should extend up and across the inside face of foundation walls to reduce air and vapor leakage and to keep the plenum dry.

The underfloor area should be free of sharp stones and debris that could puncture the vapor retarder film. The film should cover the ground, extend up the foundation walls, and be attached to the top of the sill; a few inches should overhang the sill so that the film may be attached to framing members to form a good seal (Fig. 15.2-24). The film should be applied in the largest practicable sheets. Joints should be sealed, with the exception

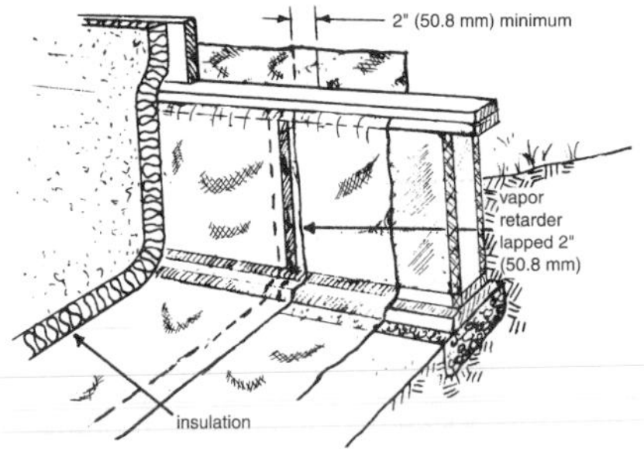

FIGURE 15.2-24 Vapor retarder film in an air plenum.

that joints on the ground should be overlapped 12 in. (305 mm) but not sealed. These unsealed joints provide a means of escape if water should collect in the plenum area. This film may be covered with 2 in. (50.8 mm) of sand or gravel.

Either rigid or batt insulation should be installed in an air plenum system. This insulation should be foil faced and have an R-value equal to or greater than that recommended for the side walls (R-11 or better). Paper-backed and foamed plastic insulations constitute a fire hazard and are not recommended unless they are covered with ½-in. (12.7 mm)-thick gypsum board or ⅝-in. (15.9 mm)-thick plywood. However, covering insulation with such materials increases construction costs.

Insulation should be installed over the vapor retarder on the inside perimeter of the foundation wall and should extend at least 24 in. (610 mm) away from the wall horizontally at the bottom, as shown in Figure 15.2-25. Insulating the entire underfloor area is generally not cost-effective.

Floor Construction The floor over a plenum may be framed conventionally by resting the joists on a wood sill (see Fig. 15.2-25) or in joist hangers. Sometimes joists are supported on a continuous wood ledger mounted on the side of wood foundation walls to lower the dwelling's exterior wall by the depth of the joists. This can reduce the amount of exterior wall sheathing and siding needed by about 9.5%.

Because the moisture content of the air in the plenum is low, the floor system should be built to allow for shrinkage as if it were in a heated basement.

Underfloor Plumbing Seamless copper tubing is recommended for water supply piping to minimize leakage problems.

a

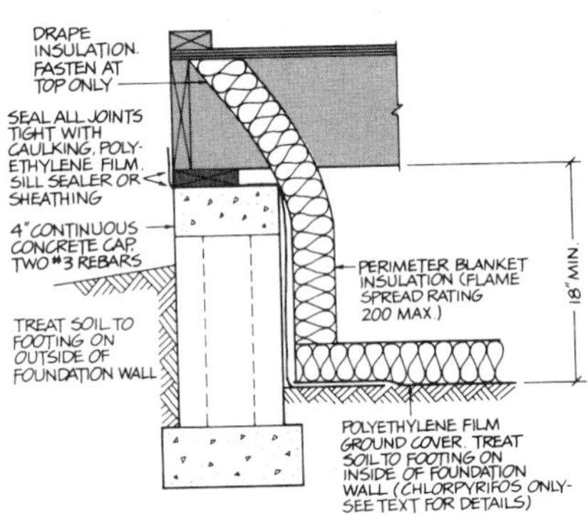

b

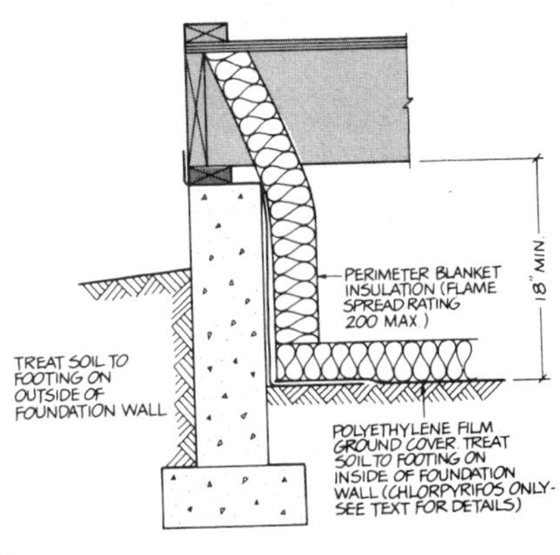

c

4 in.	=	101.6 mm
18 in.	=	457.2 mm

FIGURE 15.2-25 Vapor retarder and insulation in an air plenum at (a) a wood foundation wall, (b) a unit masonry wall, and (c) a concrete foundation wall. (Courtesy American Plywood Association.)

Drain, waste, and vent (DWV) plumbing joints should not be in the plenum, but rather above the floor or below the vapor barrier. DWV cleanouts may be either inside or outside of the dwelling but should not be placed in the plenum space. Underground plumbing should be installed before a vapor barrier is laid on the ground. Refer to Section 15.1 for additional information about DWV systems.

In shallow plenums an access trench is desirable to enable workmen to reach leaky plumbing or faulty equipment.

Underfloor Wiring Electrical wiring within a plenum is restricted by most codes as a potential fire hazard. If permitted, wiring within a plenum should be of approved types of electrical cable or installed in conduit. Some codes require rigid conduit, others permit thin wall.

Mechanical System An air plenum system does not require special equipment. The key to optimum efficiency is the proper sizing of system components.

Usually, a counterflow (down-flow) furnace or heat pump is used. It should be sized according to heat-load calculations; oversizing should be avoided. Blower controls should be set to produce continuous low-volume air circulation at a minimum and higher-volume circulation when heating or cooling is supplied.

The furnace selected should have adequate safety mechanisms to prevent excessive temperature buildups. A downflow furnace should have a safety mechanism to guard against the reversal of airflow should the blower become inoperable.

The heating capacity of the unit and the air volume circulated through the unit by the blower determine the air temperature rise through the furnace. The most efficient underfloor plenum heating units operate at a temperature rise of about 40°F (4.4°C). Figure 15.2-26 shows the recommended furnace output and blower capacities that will result in a 40°F (4.4°C) temperature rise.

Since there are no ducts in an air plenum system, there is little resistance to airflow. Air resistance is mainly due to the air filter and the delivery openings from the plenum into the living or working areas.

The furnace should be located as close to the center of the building as possible. With a central location, air distribution is better and return-air ducting is simpler.

There should be 30 in. (762 mm) of service clearance in front of a furnace or heating unit. The unit should be adequately supported and blocked to prevent vibration. There should be a positive air-

FIGURE 15.2-26 Recommended Furnace Output and Blower Capacities*

Heating Unit Output, Btu (kJ)	Blower Capacity, cfm (m^3/M)
20,000 (21,101)	475 (26.19)
30,000 (31,652)	700 (19.82)
40,000 (42,202)	925 (26.19)
50,000 (52,753)	1175 (33.27)
60,000 (63,304)	1400 (39.64)
70,000 (73,854)	1625 (46.01)
80,000 (84,405)	1850 (52.39)
90,000 (94,955)	2100 (59.47)
100,000 (105,506)	2325 (65.84)

*To obtain 40°F (4.4°C) temperature rise through the furnace.

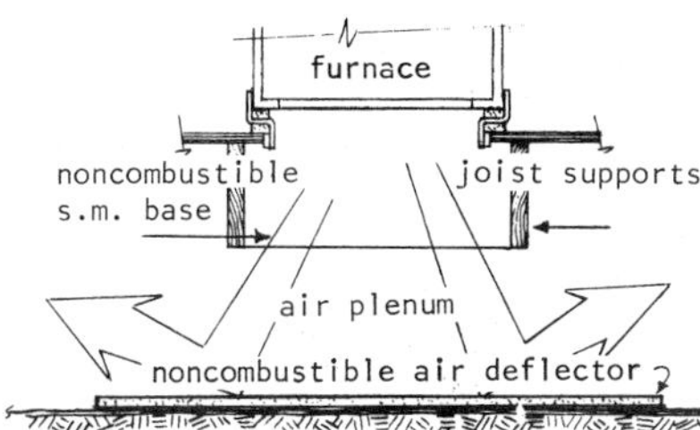

FIGURE 15.2-27 **A noncombustible base and air deflector prevent excessive heat buildup and fires.** (Courtesy American Plywood Association.)

tight seal around both the return-air duct and the furnace base. For fire safety there should be a noncombustible base, usually made of sheet metal. An air deflector pad made of a noncombustible material, such as thin concrete or gypsum board, should be placed directly under the furnace (Fig. 15.2-27).

If air conditioning is included in the system, no furnace base may be needed since the cooling coil housing may serve as the base. There should, however, be provision for collecting condensation from the cooling coil.

If cooling is not provided, the furnace blower should be turned on occasionally in the summer to evacuate humid air from the plenum. Running the blower in hot weather, especially at night, contributes to occupant comfort.

The blower should be wired to allow continuous operation. A continuous blower sound may be less objectionable than intermittent noise. A two-speed blower can provide a continuous low air volume in addition to a higher air volume when heating or cooling equipment is in operation.

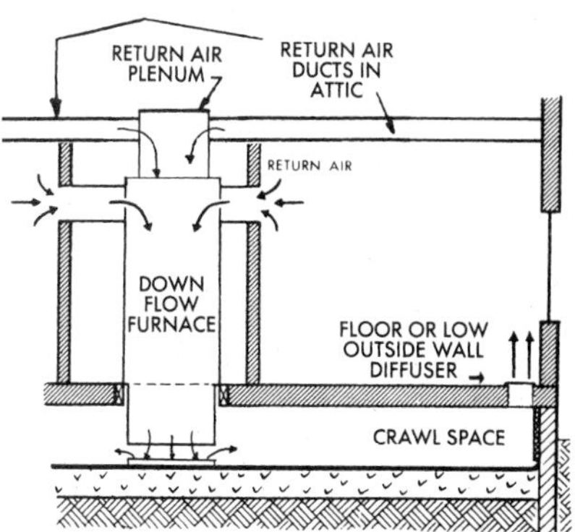

FIGURE 15.2-28 **Return air grille(s) may be provided in furnace room walls, or ducts may be run from each room through the attic to the furnace.**

Blower operation may also be tied in with a separate thermostat.

The furnace heats air, and the blower distributes it through the plenum and supply outlets. The return-air system completes the convection loop by returning air to the furnace, where it can be heated and recycled (Fig. 15.2-28).

The return-air system should be sized to equal or exceed the supply volume and should induce the return of air from each conditioned space. A properly designed return-air system helps maintain an even temperature level with minimal stratification throughout the dwelling.

Nonducted air returns allow free passage of return air directly to a furnace, usually through a louvered furnace room door. Bedroom doors should be undercut to have at least 1 in. (25.4 mm) of clearance from floor tile or carpet to ensure proper air circulation.

Ducted multiple return-air systems are used in larger homes with complex plans. The air is returned from grilles in each room, located preferably near the ceiling, through ducts located in the ceiling construction or attic. Smaller rooms, such as a kitchen or bath, do not need return-air grilles if other means are provided to allow air to return from them to the system.

Combustion-air and fresh-air intakes, as discussed earlier, may also be necessary in buildings with air plenum heating and cooling systems. The same principles apply.

A crawl space cavity can also be used as a return-air plenum rather than as a supply plenum when desired. If it is used as a return-air plenum, an upflow air furnace with overhead ductwork is necessary. Return-air grilles near the outside walls can

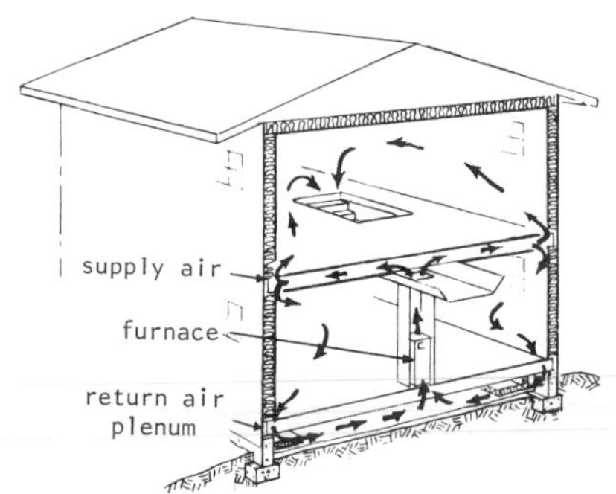

FIGURE 15.2-29 A dropped ceiling above the furnace supplies conditioned air through the joist space between floors; the plenum under the first floor returns the air to the furnace.

be used as the means of routing air to the plenum.

In single-story construction, an overhead main supply duct can be used; the supply duct can either be dropped below the ceiling or located in the attic. In a 2-story dwelling, if the underfloor area of the first floor is used as a return-air plenum, the framing space between stories can serve as a supply plenum (Fig. 15.2-29). A centrally dropped ceiling can provide conditioned air to the space between joists.

Either conventional adjustable floor diffusers, baseboard registers, or a continuous perimeter slot can be used to supply heated and cooled air to living areas (Fig. 15.2-30).

Each *register* should have an adjustable damper to properly balance the airflow. If too much air enters a room, the registers can be partially closed. If there is not enough air, it is easy to replace a small register with a larger one or to add a register.

As a general housekeeping measure, a sheet metal pan should be installed beneath each register opening to catch condensation that may form on the supply outlet and drip into the plenum (Fig. 15.2-31). Such pans may be required by some codes, especially where sand or gravel is not used over the vapor retarder.

In general, register sizing and location are basically the same as with conventional forced-air systems, with a few additional requirements, as outlined here:

1. Heat loss or gain should be calculated for each room and the entire building. Conventional heat loss calculations can be used for living and working areas; a different method of calculating heat loss is used for a plenum system than for conventional ventilated crawl space construction (Fig. 15.2-32).
2. A heating unit with an output capacity equal to 1.5 of the total calculated heat loss should be selected.
3. Generally, at least one register should be located on each exposed exterior wall, except in bedrooms with less than 7000 Btu's (7385 kJ) per hour heat loss. If possible, registers should be located in front of windows (see Fig. 15.2-19). Small rooms near the furnace and interior rooms with no outside walls often need no registers. In cases where kitchen cabinets are on an outside wall, holes can be cut in the floor under the cabinet and conditioned air delivered through a toe-space grille.
4. Register outlet capacity should closely match the air volume of the furnace on a proportional basis. For example, if heat loss for a bedroom is 8400 Btu's (8863 kJ) per hour and total dwelling heat loss is 56,000 Btu's (59,083 kJ) per hour, then that bedroom should receive about 15% of the air volume. If the blower size is 1600 cfm (45.3 m^3/min), 15% of that is 240 cfm (6.8 m^3/min). If two registers are used, each should deliver 120 cfm (3.4 m^3/min). Because the floor acts as a radiant heat source, rooms nearer the mechanical unit will receive a higher proportion of the radiant heat. Registers in those rooms can be slightly undersized, and registers in the rooms farthest from the heat source can be slightly oversized.

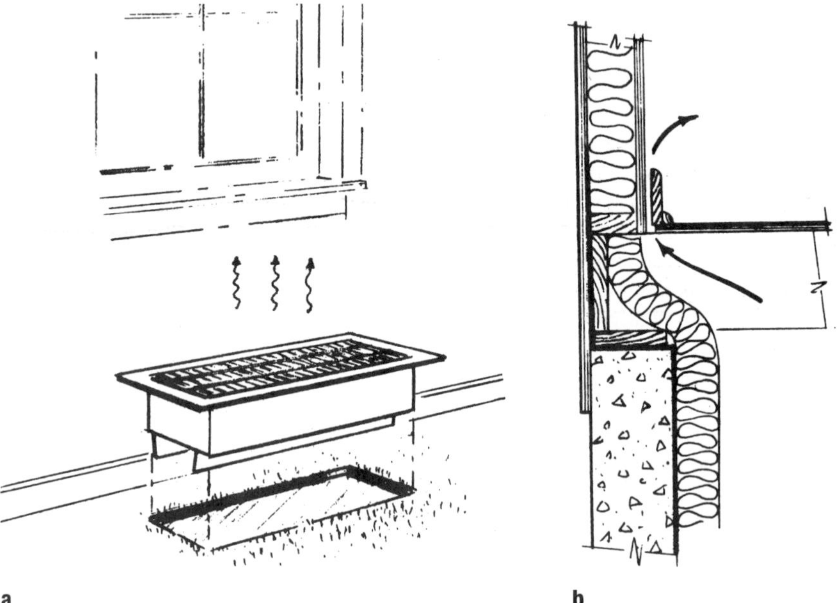

FIGURE 15.2-30 (a) A conventional floor diffuser; (b) a perimeter slot.

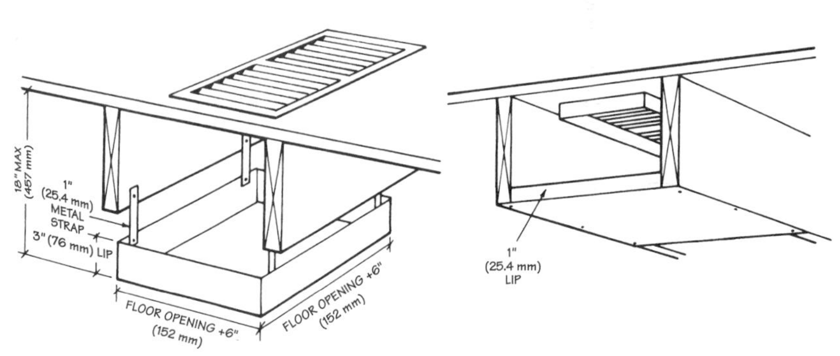

FIGURE 15.2-31 Sheet metal pans below floor registers. (Courtesy American Plywood Association.)

Every square foot of ground under a dwelling can be accounted for by tracing heat flow through circular arc paths to a point below the surface of the exterior grade (a). Heat loss is based on the average conductivity of the soil. Conductivity varies according to soil type, dampness and looseness; usually, dry loose soil is a better insulator than damped packed soil. The average conductivity of soil is 0.8 BTUs per hour per sq. ft. per foot of path length.

The heat loss factors shown below (b) account for arc path length, insulation, heights of exposed foundation wall, foundation construction, surface films and ground temperatures beneath the exterior grade surface of a 50 ft. by 28 ft. dwelling. Though arc lengths vary according to dwelling size, the values shown below may be used as a reasonable estimate for most dwellings. Under severe climatic conditions or with an extremely large dwelling these values will slightly underestimate plenum heat loss.

(a) PLENUM HEAT FLOW PATHS **(b) HEAT LOSS FACTORS, R11/R19 INSULATION**

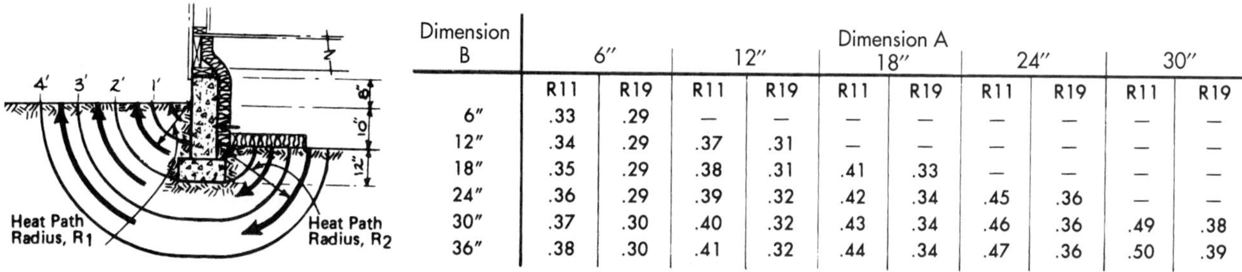

Dimension B	Dimension A 6"		12"		18"		24"		30"	
	R11	R19	R11	R19	R11	R19	R11	R19	R11	R19
6"	.33	.29	—	—	—	—	—	—	—	—
12"	.34	.29	.37	.31	—	—	—	—	—	—
18"	.35	.29	.38	.31	.41	.33	—	—	—	—
24"	.36	.29	.39	.32	.42	.34	.45	.36	—	—
30"	.37	.30	.40	.32	.43	.34	.46	.36	.49	.38
36"	.38	.30	.41	.32	.44	.34	.47	.36	.50	.39

To use (b) Heat Loss Factors:

(1) Select insulation R-value to be used.

(2) Estimate average height of exposed foundation wall and plenum depth from ground to underside of subfloor.

(3) Find appropriate heat loss factor in table by cross-matching exposed foundation in "A" column and plenum depth in "B" column.

(4) From house plan, determine lineal feet of perimeter foundation which encloses the underfloor plenum area.

(5) Determine temperature differential between outside design temperature and plenum perimeter temperature (found to average 85°F).

(6) Multiply heat loss factor times exposed lineal feet of foundation times temperature difference to obtain plenum heat loss: Heat loss factor x exposed lineal feet of foundation wall x temperature difference = Plenum Heat Loss.

(7) Add living area heat loss to plenum heat loss to obtain total estimated heat loss for entire house.

Example: Assume the following characteristics:

- 1400 sq. ft. dwelling.
- Wood foundation, with average 12" exposure to outside.
- Plenum depth (ground to subfloor) about 24".
- R-11 insulation on wall and 24" horizontally around perimeter.
- 156 lineal feet of foundation enclosing the plenum.
- Winter design temperature = 24°F.
- Temperature difference = 85°F minus 24°F = 61°F.

From (b) the heat loss factor is 0.39. Therefore, estimated plenum heat loss is:

0.39 (heat loss factor) x 156 (lineal feet of foundation) x 61°F (temperature difference) = 3,711 BTUs per hour.

Assuming heat loss calculations for the living area of 40,000 BTUs per hour, total heat loss under design conditions for the purpose of sizing equipment will be 40,000 BTUs per hour +3,711 BTUs per hour. Round off to 44,000 BTUs per hour. It is interesting to note that the plenum heat loss in this example is about the same as would have been added for duct heat loss when insulated ducts are located in a conventional ventilated crawl space. Also, over 5,000 additional BTUs per hour would have been lost in such a dwelling through the floor when insulated with R-11 over its entire area. A net energy saving of over 5,000 BTUs per hour under design conditions results by using the underfloor plenum system in this example.

Cooling load calculations for plenum houses can be made in the conventional manner, neglecting heat gains to the underfloor plenum area, unless there is significant exposed foundation wall. In which case, only heat gain through the exposed portion of the foundation wall should be calculated.

(c) WOOD FOUNDATION

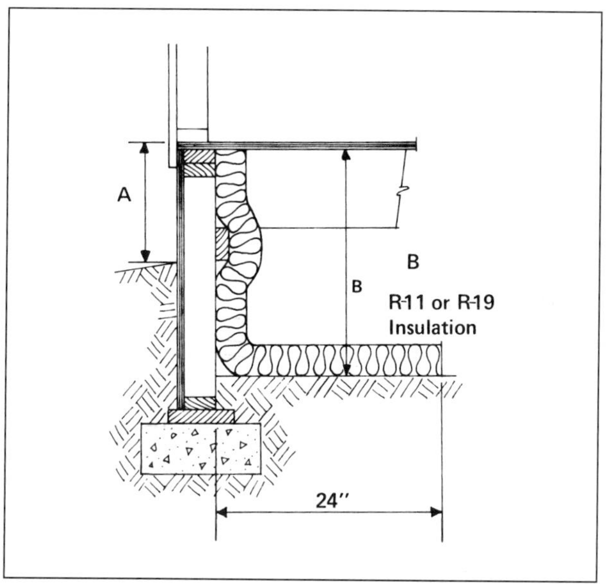

(d) CONCRETE OR BLOCK FOUNDATION

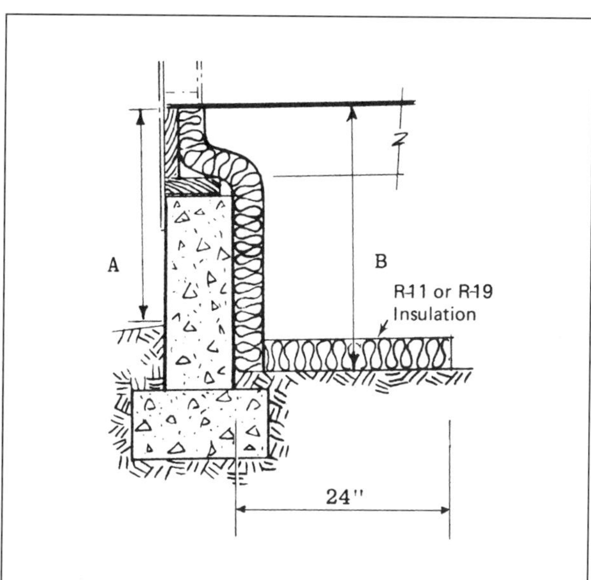

FIGURE 15.2-32 Heat loss calculations for an air plenum system. (Refer to Figure 1.8-27 for metric equivalents.)

FIGURE 15.2-33 A Second Method of Sizing Registers

Assuming a 56,000-Btu/hr. unit with a blower capacity of 1600 cfm and a plenum pressure of 0.1 in. of water, a floor register would deliver about 9 cfm per sq. in. of floor register opening. Therefore, the total register free opening for the house would be 1600/9 = 177 sq. in.; 10 registers each having 18 sq. in. of free area could be distributed uniformly around the house.

a. Register Air Delivery[a]

Free Area of Register	Plenum Pressure, in. of Water				
	0.02	**0.04**	**0.06**	**0.08**	**0.10**
1 sq. in.	4.4	5.8	7.3	8.4	9.4

[a]In cfm per sq. in. of floor or basement register for different plenum pressures.

The same general procedure applies to the perimeter slot method of air distribution. Assuming a 1600-cfm blower and a plenum pressure of 0.1 in. of water, a ¼ in. slot would deliver 19 cfm per lineal foot.

Therefore, total open slot length would be 1600/19 = 84 ft. If a 3/16 in. slot is used, 13 cfm is delivered, so total slot length would be 1600/13 = 123 ft.

b. Perimeter Slot Air Delivery[b]

Slot Width, in.	Plenum Pressure, in. of Water				
	0.02	**0.04**	**0.06**	**0.08**	**0.10**
⅛	3.5	4.7	5.8	6.8	7.8
3/16	5.8	8.4	10.0	12.0	13.0
¼	8.5	12.0	15.0	17.0	19.0
⅜	14.0	20.0	24.0	28.0	31.0

[b]In cfm per sq. in. of slot for different plenum pressures.

(Refer to Figure 1.8-27 for metric equivalents.)

This method of determining register size ignores plenum pressurization and will probably result in slightly oversized registers, which can be partially closed if necessary.

Because conditioned air is fed to the registers directly from the plenum, duct length and size considerations can be ignored.

A low plenum pressure (not more than 0.1 in. [2.54 mm] of water) is desirable to maintain more comfortable velocities through the supply registers and to minimize perimeter air loss. Figure 15.2-33 shows a method for determining register sizes if plenum pressure is not more than 0.1 in. (2.54 mm) of water.

Further information on calculating heat loss and the sizing and locating of registers may be obtained from the National Association of Home Builders (NAHB) Research Foundation.

The perimeter slot method of supplying conditioned air is unobtrusive and avoids the problem of obstructing airflow from a grille or register with furniture. Its primary disadvantage is that it requires some unconventional framing and trimming techniques. It may also be a problem with floor-length drapes, because the heat will be trapped between the drape and wall or window.

The perimeter slot method consists of holding floor sheathing about ½ in. (12.7 mm) from the inside surface of the exterior walls and then blocking the baseboard out uniformly to create an air-delivery gap into the living area (see Fig. 15.2-30b). The width of the slot is determined by heat loss calculations and usually varies between ⅛ in. (3.2 mm) and ⅜ in. (9.5 mm). The amount of air delivered is then fixed by the width of the slot, with no further controls.

Some local codes may require short ducts to be stubbed out from the furnace (Fig. 15.2-34), but they are seldom needed for proper air distribution. Some large dwellings with irregular shapes may find their installation helpful in ensuring adequate air distribution in the far corners of the dwelling (Fig. 15.2-35). Stub ducts often eliminate the need for a separate furnace air deflector (see Fig. 15.2-27).

15.2.4.2 All-Water Systems

In an *all-water* system, hot water and chilled water are produced remotely and delivered to terminal units throughout a building. All-water systems are used mostly in large buildings, but may also be found in larger, more expensive residences and other medium-sized buildings.

HOT WATER SYSTEMS

In hot-water systems, the heating medium (hot water) is generated in a boiler and circulates through the building in pipes. As the pipes pass through heating devices, the medium gives off its heat, which warms the spaces. Three basic configurations are used in hot-water heating systems: *series loop*, *single pipe*, and *two pipe*.

Series Loop Systems In a series loop system, a single pipe carrying hot water from the boiler passes directly through

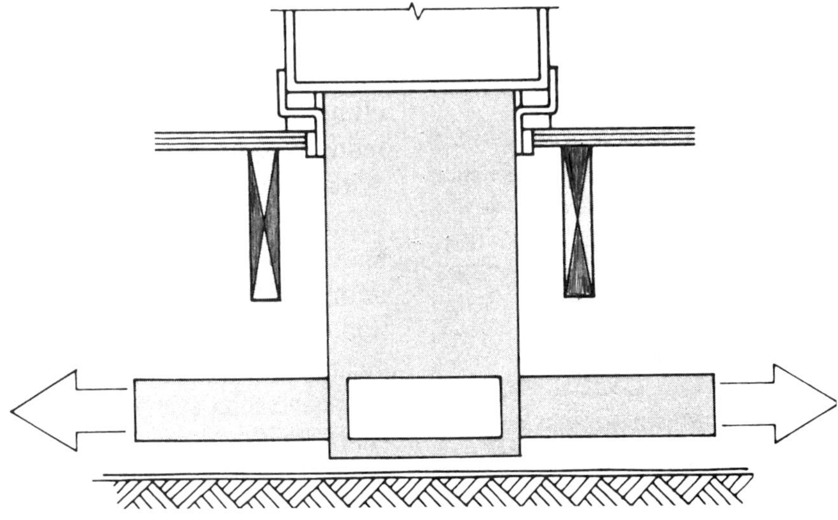

FIGURE 15.2-34 Stub ducts installed under a furnace help direct air to various parts of the dwelling. (Courtesy American Plywood Association.)

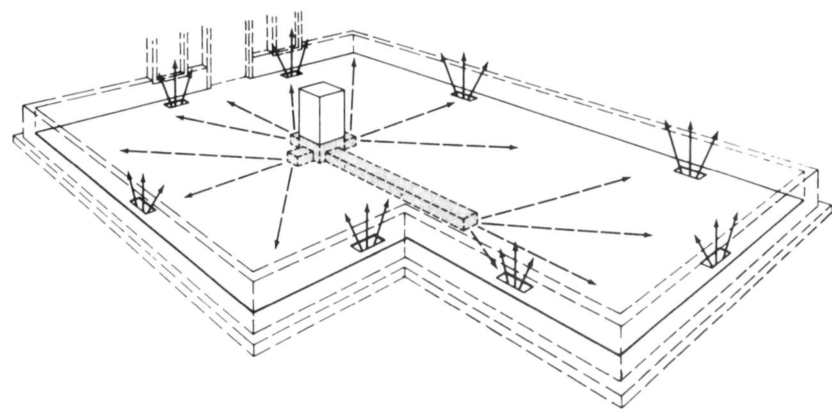

FIGURE 15.2-35 Directional stub ducts in a nonrectangular design. (Courtesy American Plywood Association.)

convectors and returns the now cooled water to the boiler for reheating. Series loop systems are used in small buildings, such as single-family residences. They are economical to install, but the heat produced is difficult to control, because as the water circulates, it cools. Convectors near the end of the loop receive water at a much lower temperature than those at the beginning. Some control can be obtained by dividing the building into zones and installing a separate loop for each zone.

One-Pipe Systems One-pipe systems also have a single pipe running in a loop from the boiler around the building, but each heating device is served by a tee connection and a pipe to that heating device. Therefore, the amount of water to an individual heating device can be controlled. Heating devices at the beginning of the loop are about the same temperature as those farther along. Therefore, there is more heat left at the end of the loop than in a series loop system, and the heat produced by individual heating devices can be regulated. In addition, large buildings can be served, because there is less wasted heat in the piping. Even larger buildings can be served by dividing them into zones, each with its own one-pipe system loop.

Two-Pipe Systems In a *two-pipe* system, one pipe carries heated water from the boiler and a second pipe carries the return water back to the boiler. Heating devices are connected to the supply pipe through a tee. Controls are affected in a manner similar to that used in a one-pipe system but are more effective, since there is no cooled water mixing in the supply pipe. In a *direct return* two-pipe system, there are actually two pipes. The supply pipe connects to the return pipe at the end of a loop. In a *reverse return* system, the supply pipe runs to the end of a loop and then returns to the boiler, passing by each heating device for a second time and picking up water from each as it passes. Like a one-pipe system, a two-pipe system can serve a larger building if the building is divided into zones, with a separate two-pipe system serving each zone from a common boiler.

HOT- AND CHILLED-WATER SYSTEMS

All-water systems that both heat and cool are usually set up in either two-pipe or *four-pipe* systems.

Two-Pipe Systems A two-pipe system for both heating and cooling works in much the same way as a two-pipe system for heating only. The major difference is that the pipes are set up to carry either hot or chilled water. The changeover is controlled by a valve connected into the system's controls (see Section 15.2.13).

Four-Pipe Systems A four-pipe system consists of two separate pipe circuits. One two-pipe circuit is used for circulating hot water, as in a two-pipe system. The second circuit of two pipes is used to circulate chilled water.

TERMINAL UNITS

Various devices are used to translate the heat generated by a boiler into usable temperature rise in building spaces. Of these, radiators, convectors, radiant panels, radiant floors, and unit heaters are capable of heating only, and require that air for ventilation be supplied from another source. Unit ventilators and fan coil units also provide fresh air.

Radiators Air is heated directly by contact with a water-filled cast-iron device—a cast-iron *radiator*, the oldest form of hot-water heat distributor in use today.

Convectors *Convectors* are heat transfer units, usually finned-tube radiators, encased in a metal cabinet. Air passes through the cabinet by gravity, rising as it is heated.

Radiant Panels A *radiant panel* is a surface-mounted or recessed panel containing heating pipes.

Radiant Floors In a radiant floor system, hot water from the boiler passes through pipes embedded in the floor of a space, providing uniform heat throughout the space. It is also possible to embed pipes in a ceiling, but this use is rare because of the possibility of damage due to leaks. In addition, heated air rises, making the placing of a heat source at the ceiling inefficient.

Unit Heaters *Unit heaters* consist of a heat transfer unit, often a finned-tube radiator, and a fan encased in a metal cabinet. Air is forced by the fan to flow across the radiator. Unit heaters are often suspended from the ceiling or a wall, but floor-mounted units are also used. Floor-mounted units may be either surface mounted or recessed.

Unit Ventilators *Unit ventilators* are similar to floor-mounted unit heaters, but can be designed to take in air through openings in the exterior wall at each unit ventilator.

Fan Coil Units Fan coil units are capable of both heating and cooling. They consist of a finned-tube coil, an air blower and motor, and a return-air damper with a filter, all housed in a metal cabinet. When arranged for both heating and cooling, they are supplied either by a four-pipe system, and a double or split coil, or by a two-pipe system with external mixing valves to supply either hot or chilled water as required. When used for heating only, they are supplied by a two-pipe hot water system.

In the cooling cycle, fan coil units both dehumidify and cool air by sucking it into the unit, through a filter, from the air in a room. The air then passes over the coil, which contains circulating chilled water, and is blown back into the room. The condensate generated by the process drips into a pan and is carried away from the unit, either into a condensate drainpiping system or through an adjacent wall to the exterior.

In the heating cycle, air is drawn into the unit through the filter, caused to pass over the coil, which now contains hot water, and is then blown back into the room.

Fan coil units can be arranged to take in ventilation air from a louvered opening in the wall immediately behind the unit or can be used with supplementary air supplied by a second system (see Section 15.2.4.4).

Fan coil units are available in several model types. Standard wall-mounted units are about 25 in. (635 mm) tall, 9 to 12 in. (228.6 to 304.8 mm) deep, and vary in length, according to design requirements, between 30 and 84 in. (762 and 2133.6 mm). High-rise units run from floor to ceiling and vary in size, depending on their capacity. Low-profile units are only about 14 in. (356 mm) high, but are somewhat limited in capacity due to their reduced size. Small units that can be furred into soffits are also available.

15.2.4.3 Steam Systems

Steam heating systems, once extensively used in residences, are now used mostly for commercial installations. In steam systems, the heating medium (steam) is generated in a boiler and circulates through the building in pipes. As the pipes pass through heating devices, the steam gives off its heat, which warms the spaces, and degenerates back to water, which flows back to the boiler for reuse.

Steam systems use one pipe to supply steam to heating devices and a second pipe to carry condensate water back to the boiler. The system may be either a vapor system or a vacuum system. Piping may be arranged in a downfeed layout, in which steam is pumped to a level above the heating devices and is allowed to fall to them in a series of risers, or in an upfeed system by a series of risers. In both cases, condensate is carried back to the boiler in a separate set of risers.

Terminal units for steam heating systems include radiators and unit heaters, as described for hot-water systems.

15.2.4.4 Air-Water Systems

In an *air-water* system, heating or cooling of spaces is done partly by conditioned air delivered from a forced-air system, but mostly by conditioned water delivered to terminal units in the spaces. Air-water systems are used in both small and large buildings. In small buildings, combination systems are generally used, in which hot water provides heat and unit air conditioners provide cooling.

In larger buildings, any of the previously described all-water or steam systems and terminal units may be used to provide heating and cooling. Supplementary heat and ventilation are provided by a forced-air system. The forced-air system usually provides only a small amount of the heating and cooling required in spaces where a water or steam system is being used, such as in zones adjacent to exterior walls, but may supply all of the heating and cooling in interior zones.

15.2.4.5 Direct Refrigerant Systems

Direct refrigerant systems are also called *package* or *unitary* systems. These are self-contained systems that are located within or adjacent to the space being conditioned. Examples include window air conditioning units, rooftop units, and systems that consist of an externally mounted condensing unit and an interior evaporator and air handler. The latter two often provide both heating and cooling. Heat pumps (see Section 15.2.7) are also direct refrigeration units.

15.2.4.6 Electric Heating Systems

Electric heating systems are widely used. They are easy to control and have a low installation cost. Unfortunately, they are costly to operate. None of the heaters described here is capable of providing cooling. In addition to its use in the heaters described in this section, electricity can also be used in centralized systems to power furnaces and boilers (see Section 15.2.5).

CONVECTION-TYPE ELECTRIC HEATERS

Convection-type electric heaters include *baseboard heaters*, which are electric heating elements encased in a low metal cabinet. A variation, *hydronic baseboard heaters*, contain a liquid that stores heat and increases the efficiency of the unit. A *cabinet convector* is an electric heater in a metal cabinet that can be surface mounted or recessed. Cabinet convectors are taller than baseboard units and therefore take a shorter amount of time to deliver the same amount of heat. When convection heaters are the prime source of heat, ventilation must be provided by another system.

RADIANT-TYPE ELECTRIC HEATERS

Radiant heaters include *infrared heaters*, which heat by means of a heat lamp; *radiant panels*, which can be wall or ceiling mounted; and *resistance cable* systems (see Sections 9.3.2.8 and 9.5.6.6). When radiant heaters are the prime source of heat, ventilation must be provided by another system.

FORCED-AIR ELECTRIC HEATERS

Forced-air heaters include unit ventilators, unit heaters, ceiling heaters, wall heaters, and portable heaters. Except for their power source, electrically powered unit heaters and unit ventilators are essentially as described earlier for similar units served by hot water or steam. Both permit ventilation as a part of the unit.

15.2.5 HEAT GENERATION

Heat generation is the elevation of a heating medium (air, water, or steam) to a temperature sufficient to heat a building. Devices that perform this function include furnaces and boilers, which are discussed in this section, and solar energy (see Section 15.2.10).

15.2.5.1 Fuel Types

Furnaces in small buildings are powered with gas, oil, or electricity. Boilers may be fueled by gas, oil, coal, or electricity.

GAS

Furnaces and boilers may be fueled by either natural gas or propane. Most large gas-fired boilers use natural gas.

OIL

Oil-fired furnaces generally run on No. 2 fuel oil. Oil-fired boilers run on either No. 2, No. 4, No. 5, or No. 6 fuel oil. In general, the larger the heating load, the heavier the oil. No. 2 oil will produce about 140,000 Btu's/gal., and No. 6 will produce about 150,000 Btu's/gal.

COAL

Anthracite coal was a major furnace and boiler fuel in the past. It is still used today, but has somewhat fallen out of favor because of the difficulty and cost of preventing the particulates that result from its burning from reaching the atmosphere.

ELECTRICITY

Electricity is used extensively to fuel furnaces, especially in southern regions of the United States. It can also fire boilers, but its use for that purpose is not widespread and is usually limited to small installations.

15.2.5.2 Furnaces

Most types of furnaces for single-family houses, individual apartment or townhouse units, and small commercial establishments are manufactured for heating only or for heating and initial or later installation of air conditioning. Those intended to provide cooling have a larger blower for more volume of air output. Such furnaces are quieter because the larger fan can operate more slowly and still move a larger volume of air.

15.2.5.3 Boilers

Boilers produce heated water or steam by burning fuel adjacent to a tank containing water. Their designed efficiency depends on the number of times the hot gases produced by a burner pass through the water in the boiler (number of passes) and the air-to-fuel ratio in the burner. In large installations more than one boiler is generally used.

Boilers used in small residences could be, and often are, called furnaces, but are actually boilers in the sense that they produce hot water or steam by burning fuel. Such boilers are usually made of steel or iron.

There are two basic boiler types in general use in larger buildings, cast-iron sectional boilers and tube-type boilers.

CAST-IRON SECTIONAL BOILERS

A cast-iron sectional boiler has a water tank made from hollow cast-iron sections bolted together side by side, like donuts stacked in a box. A burner blows hot gases into a combustion chamber either sealed beneath or housed within the water tank. A cast-iron sectional boiler is a single-pass boiler, but a large portion of the tank is exposed to the combustion chamber, which increases its efficiency.

WATER TUBE BOILERS

A water tube boiler consists of a burner and a combustion chamber, and a water tank connected to a series of tubes that contain water. Hot gases produced by the burner pass over the water tubes on their way to the flue. The heated water in the tubes rises into the water tank. A water tube boiler is a single-pass unit.

FIRE TUBE BOILERS

Fire tube boilers consist of a burner and its combustion chamber, and a water tank containing a series of tubes connected to the combustion chamber at one end and to the flue at the other end. Hot gases produced by the burner pass through the tubes on their way to the flue, heating the water in the tank as they pass through. The water is then available to be distributed throughout the heating system.

In a single-pass fire tube boiler, the tubes extend from the combustion chamber directly through the tank to the flue. In a two-pass boiler, some fire tubes extend from the combustion chamber to a secondary chamber on the opposite end of the boiler, and additional tubes extend from the secondary chamber back through the water tank to the flue. In a three-pass boiler, tubes carrying hot gases pass through the tank three times.

WET-BACK SCOTCH MARINE BOILER

A wet-back Scotch marine boiler is similar to a fire tube boiler. The major difference is that the combustion chamber is contained within the water tank. These boilers are available as either two-pass or three-pass units.

OTHER CONFIGURATIONS

Boilers in other configurations are also used. They are often combinations of the types mentioned here.

15.2.5.4 Chimneys and Vent Stacks

The purpose of chimneys and vent stacks is to conduct combustion products to a safe point of disposal: in the atmosphere outside the building, usually above the roof. The minimum height above a roof is dictated by code, but is usually 3 ft. (914 mm) above the roof and 2 ft. (610 mm) above any part of a roof within 10 ft. (3048 mm) of the chimney or vent. Chimneys are designed to create a draft. Vent stacks work with forced-draft designs. Both may be supported on appliances, such as a boiler, when the appliance can carry the load, but offset locations are preferable.

15.2.6 COOLING

15.2.6.1 Refrigeration Cycle

Building cooling systems are made possible by the following physical laws:

1. Compressing a gas releases its *latent* heat and changes it to a liquid.
2. Lowering the pressure of a liquid causes it to absorb latent heat and vaporize.

The cycle of refrigeration consists of compressing a gas to a liquid state and releasing its latent heat, leaving a cooled liquid. The liquid is then caused to absorb heat from the interior air, which vaporizes the liquid and lowers the temperature of the air.

15.2.6.2 Cooling Components

Every building cooling system consists of a compressor, a condenser, and an evaporator. The compressor compresses a gas (refrigerant) until it liquefies. In the condenser, which is located close to the compressor, the latent heat in the refrigerant is released to the surrounding medium, which may be either air or water, but is usually outside air. The refrigerant flows onto the evaporator, where it absorbs latent heat from the surrounding medium (usually air or water), changing the refrigerant back to a gas and cooling the air or water.

The compressor and the condenser are usually located in a single unit on the ground or on the roof. The evaporator is usually inside the building.

Where considerable waste heat is available to augment the system, an absorber and generator sometimes perform the function of the compressor.

Larger systems often employ one or more packaged mechanical refrigeration water-chilling units (*chillers*). Chilled water is circulated from the chillers throughout the building to air-handling units or directly to terminal units, such as fan coil units. There are often one or more cooling towers associated with the condenser in these systems to improve their efficiency. Chilled water systems are complex and vary considerably with the size of the system, type of terminals, and control systems used.

Condenser types include evaporative condensers, water-cooled condensers, dry coolers, evaporative coolers, and cooling towers.

FUEL

Small cooling units are powered by either gas or electricity. Chillers are powered by electric motors or steam turbine drives, or are gas-absorption or gas-engine-driven coolers. Pumps, of course, require electricity.

15.2.7 HEAT PUMPS

Heat pumps draw heat from one medium and transfer it into another medium. The source medium may be exterior air (air exchange), geothermal (ground source), or interior air or water (water loop). The receiving medium may be air or water. Heat pumps have the capacity to either heat or cool.

When cooling, air-exchange and geothermal heat pumps function as direct refrigerant air conditioners (see Section 15.2.6.1), picking up heat from indoor air and depositing it into the outdoor atmosphere (air exchange) or into the ground (geothermal). One major difference between a heat pump and an air conditioner is that the heat pump contains a reversing valve, which allows the refrigerant to pick up heat from a warm exterior medium (air, water, or earth) and deposit it indoors.

Most residential heat pumps function in the heating cycle by absorbing heat from the exterior air. In geothermal heat pumps, which are used more in commercial installations, heat is recovered from relatively high earth temperatures below ground. In the summer, the earth acts as a heat sink to help cool the building.

A form of heat pump called a *water loop* heat pump is used to absorb heat from the air in ducts, or from water in hot or chilled water piping, that is serving building portions that are too warm, and deposit that heat into cooler areas. A water loop heat pump is usually actually a series of heat pumps linked together. Excess heat picked up by the pumps is stored in a water tank until needed. Water loop heat pumps are energy-saving devices that help to balance heat requirements in buildings where heat gain or loss varies considerably from time to time.

Some gas-fired heat pumps are available, but most operate on electricity. Electrically operated heat pumps are not electric heaters. Their efficiency in the heating cycle is directly related to the temperature of the medium from which they pick up heat. As the temperature of the medium drops, the efficiency of a heat pump is reduced and it produces less heat. For this reason, all heat pumps require an auxiliary heater to produce heat when the temperature of the medium is too low. Auxiliary heaters in electrically operated heat pumps are powered by electricity, the least efficient and most costly fuel available. Auxiliary heaters in gas-fired heat pumps, of course, are gas heaters.

Under most circumstances heat pumps become uneconomical to operate when the temperature of the medium falls below about 40°F (4°C). Therefore, air-exchange heat pumps work well and are economical to operate in southern climates, but in northern climates may be less economical than other heating/cooling systems. North of a line stretching from Virginia to northern California, air-exchange heat pumps are not very efficient. Geothermal heat pumps are efficient in more northern climates, of course, since far underground earth temperature is about the same everywhere.

15.2.8 EMBEDDED PIPES AND DUCTS

When ducts and pipes must pass through slabs, care should be taken to prevent them from breaking by isolating them from the slab.

15.2.8.1 Ducts

When galvanized metal ducts are embedded, calcium chloride should not be used in the concrete mixture. Aluminum should not be embedded in concrete.

Heating ducts embedded in a slab on grade may be of metal, fibrous cement, wax-impregnated paper, ceramic tile, concrete, or plastic. Fibrous-cement ducts with watertight joints may be set on a sand leveling bed and backfilled with sand to the underside of the slab. Precautions should be taken to see that the position of ducts is not disturbed during the placing of concrete.

Metal and paper ducts should be completely enclosed in at least 2 in. (50.8 mm) of concrete to prevent moisture from collecting in them; joints may be taped or cement grouted. These ducts must often be weighted so that they will not rise in freshly poured concrete.

When the bottom of the ductwork adjacent to a slab perimeter is below exterior finish grade, the duct system should be of ceramic tile, fibrous cement, or concrete pipe. Joints that are completely or partially below exterior grade should be watertight. In areas where termite control is required, these joints should be resistant to termite penetration and soil-treatment chemicals.

If the slab is not reinforced (Type I slab) (see Section 3.12), welded wire mesh reinforcement (fabric) should be placed to extend 19 in. (482.6 mm) on each side of the centerline of the duct or to the slab edge, whichever is closer (see Fig. 15.2-36a).

15.2.8.2 Pipes

Pipes should pass through a slab vertically; they should be provided with space for expansion and with flexible connections if necessary to accommodate differential movement between the slab and the soil.

When galvanized or copper pipe is embedded, calcium chloride should not be used in the concrete mixture. Aluminum should not be embedded in concrete.

Type II slabs (see Section 3.12) may be

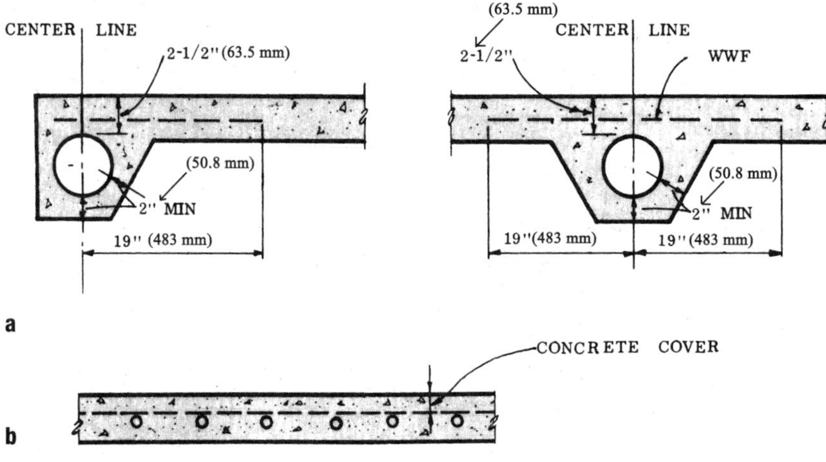

FIGURE 15.2-36 Details for embedding heating sources in a slab.

heated by circulation of heated liquids through embedded pipes. Usually, ferrous or copper pipe is used with approximately 2½ to 3 in. (63.5 to 76.2 mm) of cover over the pipe (Fig. 15.2-36b). Where pipes pass through control joints or construction joints, provision should be made for possible movement across the joints. The piping should be pressure tested before the concrete is placed, and air pressure should be maintained in the pipe during concrete placement; water pressure should not be used for testing. After the concrete is placed, the slab should not be heated until curing of the concrete has been completed. Slabs should be warmed gradually, using lukewarm liquid in the pipe to prevent the cold concrete from cracking.

Using embedded pipes for summer cooling is not recommended with slab-on-ground construction. Under conditions of high humidity, condensation may form on the cooled slab, which may damage flooring and support mold or fungus growth.

15.2.9 HEAT RECOVERY SYSTEMS

Heat recovery systems pick up heat that would otherwise be wasted and use it to provide auxiliary space heating, to preheat outside air intakes, to reheat supply air in air conditioning systems, and to preheat domestic water and boiler feed water. Waste heat is recovered from air, liquids, or gas. Sources include heat generated by people; lights; appliances; motors; generators; refrigeration systems; exhaust air from kitchens, toilets, locker rooms, and swimming pools; and many others.

15.2.10 SOLAR HEATING AND COOLING SYSTEMS

15.2.10.1 Active Solar Heating Systems

Active solar energy systems consist of collectors and storage devices. Such systems are used to heat buildings, heat domestic water, and, rarely, to cool buildings. Solar heating systems used solely to heat domestic water are addressed in Section 15.1.3.4.

COLLECTORS

The most common type of collector is a flat plate collector (Fig. 15.2-37). There are two types, liquid and air. Both have a metal frame with a metal bottom plate and a glass top panel, and the bottom and sides of the box are lined with insulation. On top of the insulation is an absorber plate. In a water system, the space between the top glazing and the absorber plate contains tubes through which water flows. In the air type, the tubes are omitted and air flows through the space between the glazing and the absorber plate. The sun heats the space below the glazing and, subsequently, the air or water passing through this space.

Air systems are generally used in locations where freezing is a problem.

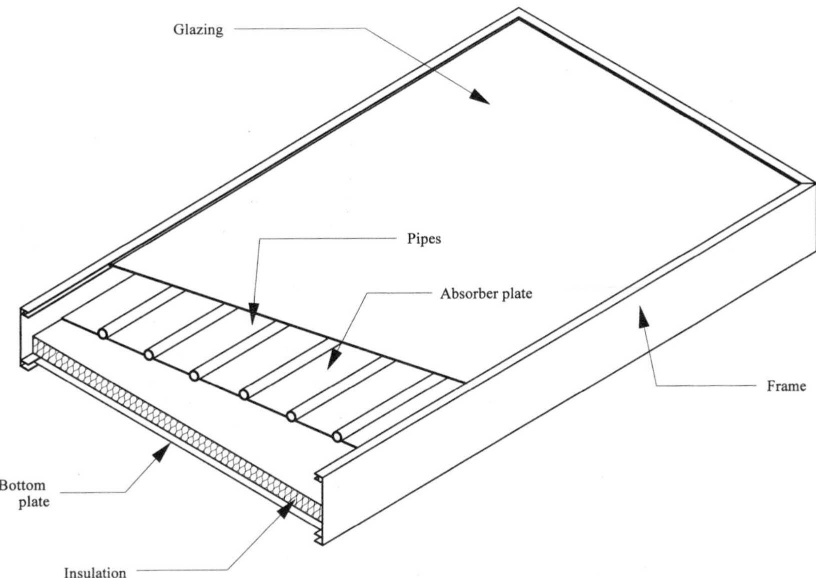

FIGURE 15.2-37 **A typical flat-plate solar energy collector. Many designs are available.** (Drawing by HLS.)

STORAGE DEVICES

Water systems generally use water tanks as their storage devices (Fig. 15.2-38). Air systems generally employ rock beds (Fig. 15.2-39). In a water tank system, the heated water from the collector passes through a heat exchanger in the storage tank, which heats the water in the storage tank. This heated water is then pumped through the heating system to delivery devices, usually fan coil units or space heaters.

In an air system, the heated air flows into a contained bed filled with stone. The air heats the stone and is then pumped back to the collector, where it is again heated. When heat is required, air is caused to flow across the stone bed, picking up the heat stored there and delivering it throughout the supply-air portion of the heating system. Return air flows through the bottom of the stone bed and is returned to the collector, where it is again heated.

15.2.10.2 Passive Solar Heating Systems

HISTORY

Primitive peoples intuitively designed their dwellings to take advantage of the local climate. In each region, a style of building evolved that kept buildings warm in winter and cool in summer with a minimum of energy. This was accomplished passively, that is, through the use of the sun, wind, and other natural phenomena.

In the early Sinaqua and Anasazi Indian cliff dwellings of Arizona, known today as "Montezuma's Castle," an overhanging rock brow protected southern exposures against the high summer sun and kept interiors cool. In winter, the rock brow did not shut out the low afternoon sun, allowing the rock face to warm up. Since rock is a good conductor and storer of heat, it retained the heat and kept the dwellings warm during the night.

The Pueblo Indians of New Mexico also knew how to build energy-efficient dwellings more than 1000 years ago. By facing their structures south and using high, small, deeply cut windows, they were able to delay the entry of heat into the interior of their structures until late in the day when temperatures started dropping. Exterior surfaces were painted a light color to reflect the sun's rays. The builders crowded their dwellings side by side and on top of one another to reduce temperature fluctuations in both summer and winter and cooked in rooms separated

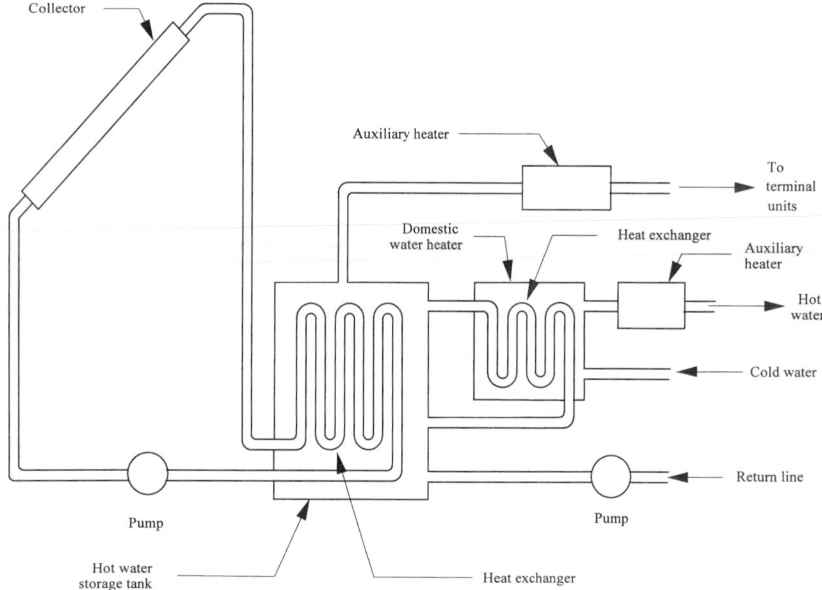

FIGURE 15.2-38 **Diagrammatic illustration of an active solar energy hot-water heating system.** (Drawing by HLS.)

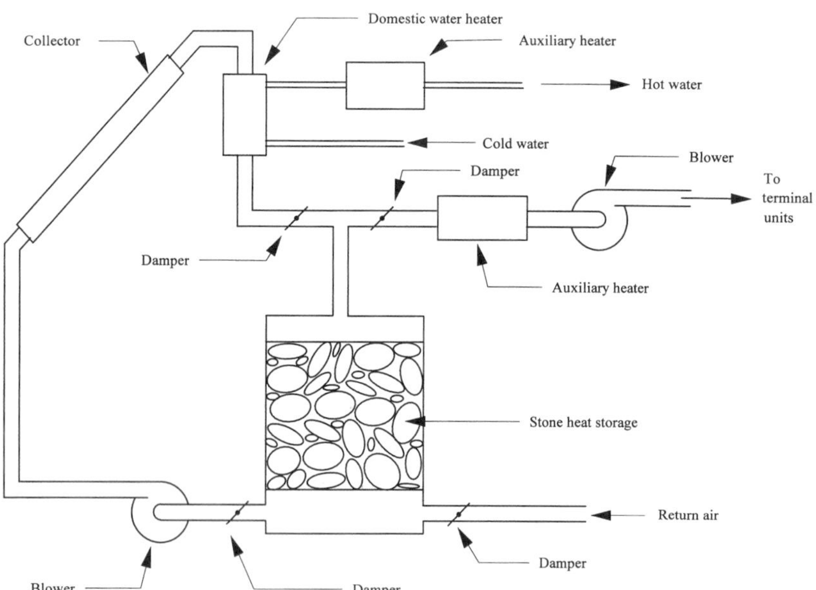

FIGURE 15.2-39 **Diagrammatic illustration of an active solar energy air heating system.** (Drawing by HLS.)

from living spaces to avoid unwanted heat gain in summer.

Descendants of the Pilgrims also evolved a suitable style of building for their northeastern region. In the cold winters of New England, keeping warm meant having efficient heat sources and tight buildings to retain the heat. The typical New England "saltbox" had warming stoves and fireplaces at the center of the house and a minimum of exterior wall surface. A north-sloping roof provided protection from icy winds and allowed snow to build up for added insulation against the cold. Most windows faced south, and the exterior was painted a dark color to absorb as much solar radiation as possible.

Studies of these and other intuitive responses to local climate have laid the groundwork for a more widespread development of similar naturally adaptive heating and cooling techniques. Some architects and engineers have refined and modified these concepts to create innovative and energy-efficient buildings. Their work has resulted in buildings that more effectively collect, store, and distribute solar energy for heating and better utilize natural climatic factors for cooling. These naturally adaptive *passive* heating and cooling systems still require external energy for their operation, but need less of it than do conventional buildings. Solar heating or cooling installations with sophisticated heat collection, transport, control, and storage methods, known as *active* solar systems, are beyond the scope of this book. This section discusses the basics of passive heating and cooling systems.

PASSIVE SOLAR SYSTEM PROPERTIES

Buildings are exposed to the warming rays of the sun and the cooling influences of natural breezes. They can be designed and oriented to take advantage of the local climate for heating and cooling, with reduced reliance on mechanical systems. With such an approach, windows, skylights, walls, roofs, floors, and the natural features of the site regulate heat flow into or out of a building. Conduction, convection, radiation, and gravity provide the means to move the heat through a building.

Passive solar systems are distinguished from active systems in that they use no power-consuming pumps, and because the collection, storage, and distribution facilities are integrated with the building design. In active systems, the collection and storage elements are manufactured products that can be placed in, on, or near the building and tied to it with pipes or ducts. In general, all effective passive system buildings include the following:

1. An envelope that will not leak heat in winter or gain excessive heat in summer
2. A solar collection area, in the form of windows, skylights, glass wall, water bags, or other solar heat traps
3. The capability of storing heat for later use, either in the structural elements or in specially provided facilities

BUILDING ENVELOPE

A passively heated building can collect and store only a limited amount of heat. In all but a few sunny regions, more heat is required for modern comfort than a passive system can deliver. The extra heat is generally supplied by a supplementary source, such as wood stoves, fireplaces, or a conventional heating system. Refer to Section 7.3 and Chapter 8 for information about the design and use of insulation, doors, windows, glazing, and window treatments to minimize overall heat loss. Earlier portions of this chapter discuss the basic principles of heat control in buildings.

An energy-efficient building should be insulated to optimum levels, have tightly caulked and sealed windows, door frames, and sill plates, have double or triple glazing, and have as few nonsolar-collecting windows as possible.

PASSIVE SOLAR COLLECTION

Solar energy reaches the earth as *direct* (parallel) or *diffuse* (nonparallel) rays of the sun. Direct radiation is sunlight that reaches the earth relatively unobstructed. Diffuse radiation is reflected from clouds, dust, and obstructions such as buildings. The relative proportion of direct and diffuse radiation, as well as the total radiation, varies from region to region, making a particular solar energy system more feasible in some areas than in others.

Solar energy that strikes a surface is called *insolation* (incident solar radiation). It is measured in *Langleys* (ly). The amount of energy contained in 1 ly is equal to 1 calorie per square centimeter, or 3.69 Btu's per square foot of surface.

In designing a building to take advantage of passive solar heating, the building shape, orientation, and solar collection area should be given consideration.

A building should be shaped and sited for maximum exposure to sunshine in winter and minimum exposure in summer. In northern latitudes, buildings should expose as much surface to the south as possible. In Pennsylvania, for example, which is at 40 degrees north latitude, a building receives nearly 3 times as much solar radiation on its south face in January as it does in June; due south is therefore the optimum orientation for a collector surface. The east and west sides, on the other hand, receive 2½ times more insolation in summer than in winter, making them inappropriate for heat collection in winter. At latitudes below 35 degrees north, buildings are exposed to even more sun on the south side in winter, and the east and west walls receive 2 to 3 times more sun in summer than the south walls.

A square building is never the optimum form for solar collection at any latitude and should be avoided unless building use and site conditions dictate such a shape. Buildings that are elongated on the north-south axis are even less efficient in winter and summer. The optimum shape is usually a form elongated along the east-west direction.

Solar energy reaches a building in one of two forms, visible light, which is shortwave radiation, and infrared light, which is long-wave radiation. Some shortwave radiation is reflected by glass, but much passes through. Long-wave radiation, on the other hand, is either reflected or absorbed and conducted or radiated away. After passing through glass, shortwave radiation strikes an interior surface and is transformed into an equivalent amount of heat energy. Part of this heat energy is conducted and convected away; the rest is radiated in the form of infrared (long-wave) radiation, sometimes called thermal radiation. Since thermal radiation cannot penetrate glass, it becomes trapped inside the space. This phenomenon is called the *greenhouse effect*. It accounts for a car parked in the sun on a fall day becoming hot and stuffy inside and for an ordinary window's ability to act as a solar collector.

The type of glass used for solar collection can have a significant effect on the efficiency of the system. Clear, heat-absorbing, and reflecting glass all lose about the same amount of heat by conduction, but differ widely in the amount of heat transmitted and trapped under the glass.

Clear single glass is the most effective heat gainer but loses the most by conduction. Clear double glazing has a comparable percentage of gain and loses less by conduction.

When preventing summer heat gain is important, heat-absorbing or -reflecting glass should be considered for south exposures, but only when external shading or screening is not practical. Because reflective glass reduces heat gain year-round, it is desirable only where cooling energy costs are substantially higher than heating costs.

Although skylights and windows are the primary means of passive heat collection, they also are responsible for the greatest winter heat loss and summer heat gain. Passive systems usually employ an insulating or shading technique to prevent unwanted heat loss or gain through the glass collector. Some more traditional approaches to reducing excessive window heat loss and heat gain are described in Chapters 7 and 8.

STORAGE

The more glass exposed to the sun, the more heat will be trapped behind it. Windows used as passive solar collectors have a disadvantage in that even on very cold days, they can quickly overheat the space they serve. In order to soak up the excess collected heat and prevent overheating, and to retain heat for nighttime use, floor and wall surfaces can be utilized as *heat sinks*; that is, they collect and store heat.

Materials such as water, concrete, brick, tile, and stone are ideal heat sinks because they can absorb and store large quantities of heat in a relatively small volume of material. The collected heat is released by radiation to colder surfaces and to the air to maintain a relatively constant temperature. The greater the temperature differential between the mass and its surroundings, the faster the heat is released. And the greater the heat storage capacity (thermal mass) of the materials in the building, the longer it will take for room air to overheat and the longer a room can remain comfortable without receiving heat from an auxiliary source.

A heat-storage material should be located in or near the direct sun. When it is, a thermal storage capacity equivalent to 30 lb. (13.6 kg) of water, or 150 lb. (68 kg) of masonry, should be used for every 1 sq. ft. (0.929 m^2) of south-facing glass. The thermal mass of heat-storage materials in other locations should be increased four times.

Heat Storage Capacity Materials vary in their capacity to store heat. The two most common measures of that ability are specific heat and heat capacity.

A material's *specific heat* is the number of Btu's (kJ) required to raise 1 lb. of that material 1°F (–17.2°C). For example, water has a specific heat of 1.0, which means that 1 Btu (1.06 kJ) is required to raise the temperature of 1 lb. (0.454 kg) of water 1°F (–17.2°C). Since 1 gal. of water weighs 8.4 lb. (3.8 kg), 8.4 Btu's (8.86 kJ) are required for a 1°F (–17.2°C) rise in temperature.

Various materials absorb different amounts of heat while undergoing the same temperature rise. For example, it takes 100 Btu's (105.5 kJ) to heat 100 lb. (45.4 kg) of water 1°F (–17.2°C), but only 22.5 Btu's (23.74 kJ) to heat the same weight of aluminum 1°F (–17.2°C).

Therefore, the specific heat of aluminum is 0.225, while water has a specific heat almost four times as high.

A material's *heat capacity* is the amount of heat needed to raise 1 cu. ft. (0.028 m^3) of that material 1°F (–17.2°C). Dense solid materials are effective as heat storage devices because their high density means that considerable weight is contained in each cubic foot of volume. Such materials therefore are often used as storage devices even though water has a greater specific heat. Concrete, for example, has a specific heat about one-quarter that of water, but because of its density it has a heat capacity of more than half that of water.

PASSIVE SYSTEMS

Passive solar heating and cooling systems rely on adaptive design, not mechanical devices. They cannot be grouped according to engineering specifications or performance standards but must be categorized according to the design approach they utilize. This, in turn, makes generic discussion of passive design difficult, because most designs are the products of their inventor's ingenuity, not of established standards. What has emerged, however, are three distinct design approaches within which most passive systems can be grouped: direct gain, indirect gain, and isolated gain.

Direct Gain Systems A *direct gain* passive solar system is the most common passive solar building solution, having many historic precedents, including the cliff dwellings of the Sinaqua and Anasazi Indians of Arizona. The process can be described as *sun* to *living space* to *thermal mass*. Solar radiation is collected in the living space and then stored in a material having a large thermal mass, such as thick masonry or concrete floors or walls. Thus, the living space is heated by the solar radiation penetrating the windows, and it serves as a "live-in" collector, storage, and distribution facility.

The basic requirements for a building with a direct gain passive solar energy system include the following:

1. A large, south-facing glazed collector area with the living space exposed directly behind
2. A large floor and wall thermal mass exposed to solar radiation
3. A method for protecting the collector windows and storage mass from heat loss to the exterior

Indirect Gain Systems *Indirect gain* passive solar systems differ from direct gain systems in that they deposit collected heat into a thermal storage mass before it is introduced into a living space. The storage mass is usually located between the glazed collector and the living space so that when the living area begins to cool, heat radiates from the mass into the living space.

The principal advantage of such an approach is that collected heat does not travel through the living space to reach the storage mass. As a result, more heat can be collected and stored before the space becomes too warm for the occupants' comfort. In addition, radiant heat flow into living spaces can be controlled more readily in such systems than in direct gain systems. Placing movable insulation on the interior face is one typical regulatory technique. When the space begins to overheat, the insulation can be moved across the wall to reduce heat flow.

The three basic types of indirect heat gain systems are the mass Trombe wall, the water Trombe wall, and the roof pond.

A *mass Trombe wall*, named after its French inventor, Felix Trombe, consists of a glass or plastic panel over a heat-absorbing storage wall. Commonly used materials for the heat storage wall include concrete, brick, stone, and adobe. When possible, the material is painted a dark color to increase its ability to absorb heat. The glazing prevents immediate reradiation of the collected heat to the exterior.

The system operates by permitting solar radiation to pass through the glazed collector and be transmitted directly into the massive storage wall. When the temperature of the wall rises higher than that of adjoining interior surfaces, it begins to radiate its heat to the interior.

Heated air in the space between the glazing and the mass Trombe wall can also be introduced into the living space, and controllable vents in the mass itself can regulate heat supply to the living space and draw cooler air into the mass for heating.

A *water Trombe wall* system uses water as the storage mass. Water is an effective storage medium because it has a higher specific heat than most solid massive materials and can therefore store more heat per cubic foot (m^3).

A disadvantage of water Trombe walls is that containing a liquid is more difficult than containing a solid, such as concrete. Steel drums are the most common containers, but some designers have experimented with concrete block walls in which the voids are filled with vinyl bags containing water. These containers can be left exposed to the interior or embedded in some massive material, such as adobe or concrete.

Convection currents within the water rapidly transfer heat from the collection surface to the entire volume of water. Heat is therefore delivered to the interior spaces at a faster rate than with mass Trombe walls. In addition, the collector surface operates at a lower temperature so that less heat is lost back through the glazed collector before it is absorbed.

A *roof pond*, a variation of the Trombe wall system, utilizes horizontal water containers on the roof of a building to collect and store the sun's energy. Heat is collected by exposing the ponds to the sun by day and retained by covering them with insulation at night to prevent the collected heat from escaping into the cool night air. The solar heat collected in roof ponds is radiated directly to the rooms through a metal ceiling. Thus, a continuous, evenly distributed supply of heat is provided throughout the building. Such a system is capable of delivering heat even when the water temperature drops as low as 70°F (21.1°C).

There are several disadvantages of roof pond collectors. They require a stronger-than-normal structure to support their weight, which increases the cost of the building. The water must be contained in plastic bags or some similar structure. Open ponds have fallen into disfavor for several reasons. They tend to contribute to roof deterioration and leaks, they make roof maintenance more difficult, and they serve as havens for insects and bacteria.

Even contained ponds have disadvantages. Heat from the low-angle winter sun is difficult to collect on a horizontal surface. On a clear December day in Pennsylvania, for example, the sunlight striking a horizontal roof is less than half that hitting an equal area of a south-facing wall. To compensate for this deficiency, more pond area must be provided. Stratification of the heat within a roof pond is another problem, as the warmer, lighter water remains near the top of the pond at night, losing heat to the outside air. The cooler, denser water falls to the bottom, just above the rooms, where heat is needed most. An additional problem is that in many climates, the roof must be protected against snow accumulation, which can block the sun from the water bags.

In general, roof pond collectors are better suited to the warmer south and southwestern states, where the sun climbs higher in the sky on a winter day and snow buildup is not a threat. Roof ponds are also well suited for summer cooling, which is of primary importance in such climates.

Insulated panels should be used to cover a roof pond on summer days to keep the sun from warming the ponds, while the cool water absorbs excess heat from the rooms below. At night the panels must be removed so that the roof pond can radiate heat to the night air. Heat stratification within a roof pond aids this cooling process, as the coldest water lies just above the rooms.

Isolated Gain Systems *Isolated gain* passive systems utilize collectors and storage masses that are thermally isolated from the living space. This concept allows collector and storage to function independently while the building draws heat as required.

Isolated gain systems can generally be grouped in two categories: those that use a greenhouse area or similar structure (often called a *sunspace*) to collect and store heat, and those that employ the *thermosiphon* principle to move collected heat.

Using a sunspace allows maximum solar collection in this space (and consequent overheating) without affecting the comfort within the primary living space. The collected heat can be admitted as needed through doors, windows, or dampers.

Atriums, sunporches, greenhouses, and sunrooms are examples of *sunspaces*. In general, this technique requires a glazed collector space attached to, and yet distinct from, a living space, that has the proper southern exposure. This collector space must contain, or be thermally linked to, a solar storage mass for heat retention and later distribution.

A *thermosiphon* isolated gain system permits heat to build up within a roof, wall, or room-sized collection area before drawing it off to storage or living areas. Thermosiphon (convection), the natural rise of heated air or fluid, is the usual means of transport.

In a typical installation, a glazed collector is located between the direct sun and the living space and is distinct from the building structure. In this space, a convection heat flow occurs when air heated by the sun rises naturally into an appropriately placed living space or storage mass, causing cooler air or liquid to fall again. A continuous heat-gathering circulation is thereby established.

SOLAR COOLING SYSTEMS

Within limits, solar energy can be used to cool buildings.

Active Systems Active solar collectors can be used to produce chilled water that can subsequently cool air. In such a system, heat from water heated in a solar collector operates a generator that makes a lithium bromide water solution. This fluid then passes into an evaporator, where it produces chilled water. The chilled water is then pumped to terminal devices, such as fan coil units. The heat picked up in the chiller is picked up by a condenser and is then dissipated by a cooling tower.

Active solar water chillers are expensive and complicated and are therefore suitable only for large buildings. They are not widely used in the United States.

Other active solar-powered water chilling systems are being developed and may come into general use in the future.

Passive Systems On summer nights, when the outside air is at its lowest temperature, introduction of that air into a house will carry excess heat away from interior floor and wall surfaces. In the process, those surfaces will be cooled and will therefore be able to absorb and store excess heat trapped in the building the following day. This reduces summer heat buildup within the building and aids cooling.

15.2.11 SOUND CONTROL

Heating and cooling equipment and the flowing air and liquids they produce and move about are sources of noise vibration, whether they are part of a small individual package unit or part of a large central system. Heating and cooling equipment can have a wide range of noise levels. It is generally more economical to install a quietly operating (even if more expensive) unit than to attempt to reduce the noise level of a less-expensive unit by acoustical construction.

In residential occupancy, the greatest air conditioning load occurs between 5:00 P.M. and 7:00 P.M. when exterior and interior conditions are at extremes. Some systems are barely adequate for this peak load time and must therefore operate at full capacity late into the night when the surrounding noise level is greatly reduced and equipment noise is generally annoying. Therefore, in residences it is particularly important to select equipment that has greater than adequate capacity, particularly for fans and pumps. Reducing fan speeds only slightly can produce noise level reductions that will more than justify the cost of additional power.

Vibration from mechanical equipment is not usually a significant problem when the equipment is mounted on a concrete slab placed on ground or located in a basement, in a crawl space, in a utility room, or in an attached garage. However, on suspended concrete slab floors, structure-borne vibration from equipment can be transmitted great distances. Equipment placed on wood-framed construction will also cause acoustical problems unless it is isolated from the structural framing.

Whenever possible, mechanical equipment should be mounted resiliently and all related piping, refrigerant lines, condensation lines, electrical connections, and duct systems should be flexibly mounted to the units to prevent transfer of vibrations to walls and floor and ceiling assemblies. Vibration isolation pads should be used between the equipment and the supporting structure. Flues should be isolated from walls by an air gap, especially where they pass through the walls of rooms, such as bedrooms, that are intended to be quiet.

When heating and cooling systems are large or complex, whole machinery rooms can be floated or isolated from the building or special foundations can be constructed to isolate individual pieces of equipment. Floating floor systems are available to isolate entire floor areas acoustically.

There should be the amount of clear space on each side of a heating unit that is recommended by its manufacturer or required by code, whichever is larger. Locating combustible materials too close to furnaces is a potential fire hazard.

The mechanical equipment room should be buffered from the rest of the building by design. Areas containing furnaces, pumps, compressors, laundry, appliances, and similar equipment, as well as areas enclosing elevator shafts, incinerator chutes, and other potential sources of noise, should be located both horizontally and vertically as far as possible from quiet areas such as bedrooms and living rooms.

In single-family dwellings, low-rise

apartments, townhouses, and small commercial buildings, an exterior wall location is usually best acoustically (Fig. 15.2-40a). Unfortunately, in many such buildings, unit forced-air heating and cooling equipment is placed in a closet, an alcove, or a utility room near the center of the living unit or working space to produce shorter and therefore more economical duct runs (Fig. 15.2-40b), thus placing major noise sources in the center of the living or working space. In addition, a single air return is located near the center of the floor plan; often a grille in the wall or door of the furnace enclosure is all that is provided. The need for quiet must be weighed against the economics of central location. For air plenum systems the furnace must be located in the center of the space and other means found to reduce noise levels.

One way to confine mechanical equipment noise in small buildings is to provide three solid walls around the furnace room, with an entry door to the outside for servicing (see Fig. 15.2-40a). In this case, the service door should be a solid-core door and should be gasketed around the perimeter to render it airtight.

Mechanical systems are also a source of exterior noise. While a cooling coil is located inside at the furnace plenum, a water- or air-cooled condensing unit must be located outside. Because ordinances sometimes govern the allowable noise level at property lines, condensers often must be located adjacent to the building in a front or rear location where the distance is greatest to the next building. This creates an exterior noise source at the exterior wall. These units should be as far as possible from windows, porches, patios, and neighboring buildings. An alternate location for the condensing unit is on a vibration-isolated platform on the roof.

Unit air conditioners in the exterior walls in multifamily residences should be located as far as possible from adjacent apartments, so that operation of one unit does not unduly disturb neighbors who do not operate their units at the same time.

Diffusers are a major source of duct noise. Air velocity and the design of diffusers should be considered to produce outlets that do not whistle or cause other annoying noises. Decreasing the air velocity at the outlet diffuser by 50% will reduce the sound level generated by the grille from 15 to 20 decibels, measured 3 to 5 ft. (0.9 to 1.5 m) from the diffuser. Wide-angle directional diffusers generate the most noise; wire mesh or perforated metal diffusers with little air spread capacity are the quietest.

Ducts should be formed of adequate thickness to minimize vibration, and they should be isolated from mechanical equipment with flexible collar connections to prevent transmission of vibrations (Fig. 15.2-41). The space around ducts should be caulked tightly for isolation where they pass through sound-conditioned construction. Mufflers used in ducts near equipment rooms will absorb transmitted noise (Fig. 15.2-42).

Careful planning and installation of duct systems is critical in controlling sound transmission. Ducts act like speaking tubes in transmitting noises. Noise may result from any of the following:

1. Equipment such as fans and motors
2. Sound transmitted through the walls into the ducts and then conducted through the ducts
3. Sound created within the ducts by air rushing through the system or by expansion and contraction of the ducts themselves

Sound does not necessarily have to enter or leave a duct through an opening or diffuser but can be transmitted through walls, particularly where the duct passes vertically through living spaces (Fig. 15.2-43). Planning can minimize the sound transmittance from one room to another through the open air paths of the duct system.

Duct outlets should not be located back-to-back between rooms, because this practice practically eliminates any benefit

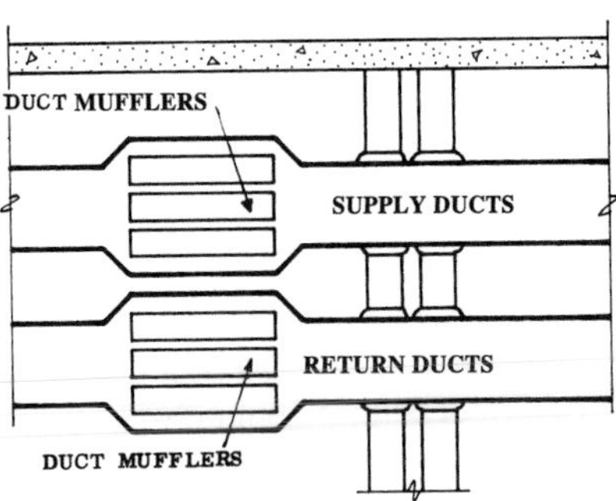

FIGURE 15.2-42 Duct mufflers are effective in reducing noise transmitted from a furnace.

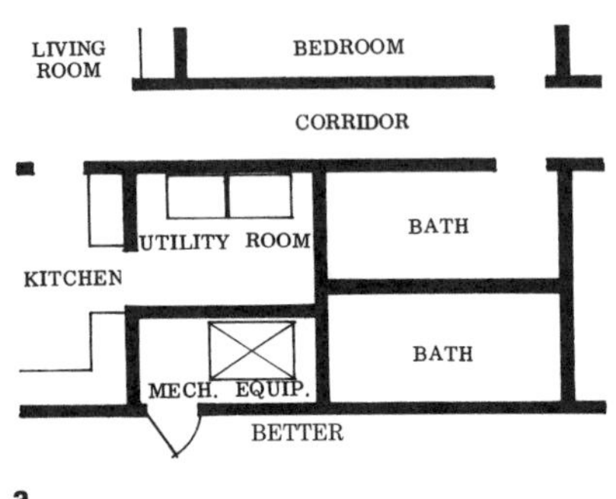

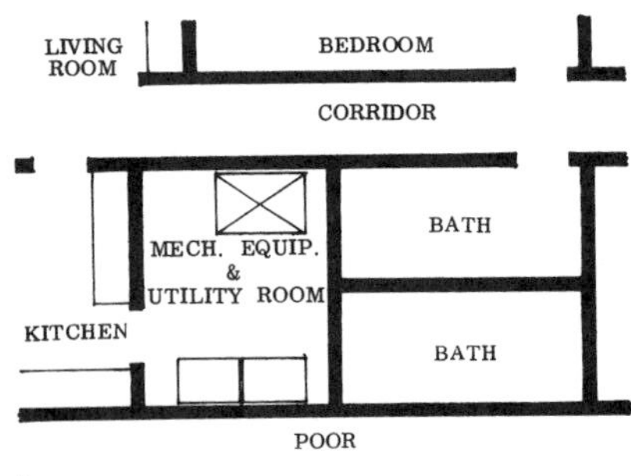

FIGURE 15.2-40 From an acoustical point of view, a mechanical equipment room should be located on an exterior wall (a), rather than in a central location (b).

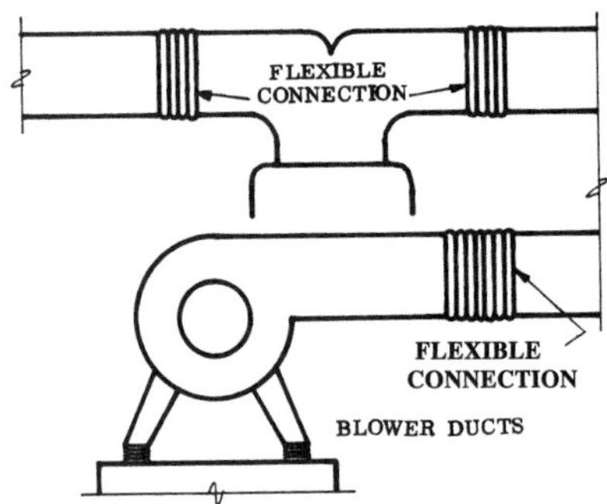

FIGURE 15.2-41 Flexible connections reduce the transmission of vibrations from equipment to the duct system.

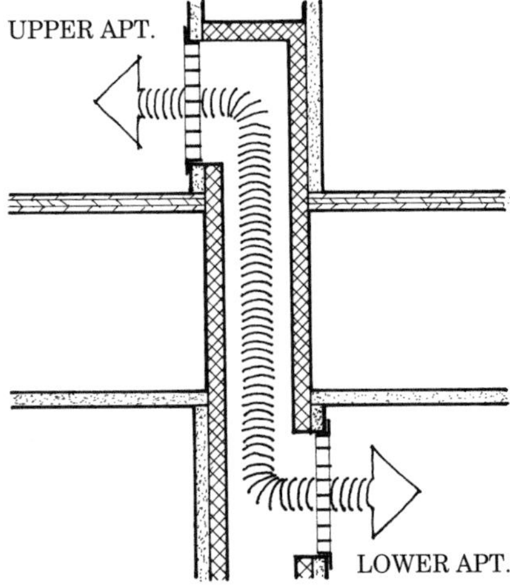

FIGURE 15.2-43 A duct system can conduct noise between units. Ducts should not be housed in common walls or floor and ceiling systems between dwelling units.

obtained from sound-conditioned walls (Fig. 15.2-44). A 6-ft. (1.8 m) length of fibrous duct material adjacent to the outlet generally will provide a reduction of furnace noises and will prevent short-circuiting sound through ductwork between rooms (Fig. 15.2-45).

Duct insulation and duct mufflers will help to reduce furnace noises and to prevent short-circuiting sound through ductwork between rooms.

15.2.12 SMOKE AND FIRE CONTROL

Buildings are required by code to be divided into fire and smoke zones of a size dictated by the code. The size of the zone is controlled by the building type and the construction class of the structure. Refer to Section 1.6 for a discussion of these classifications. Fire-rated or smoke-rated walls, as applicable, are required to surround these zones. Where ducts cross the rated walls, dampers are required to prevent fire or smoke penetration, as applicable, through the rated walls by means of the ductwork in the event of fire or other smoke generation. Where a wall is both a fire and a smoke barrier, combination smoke/fire dampers can be used.

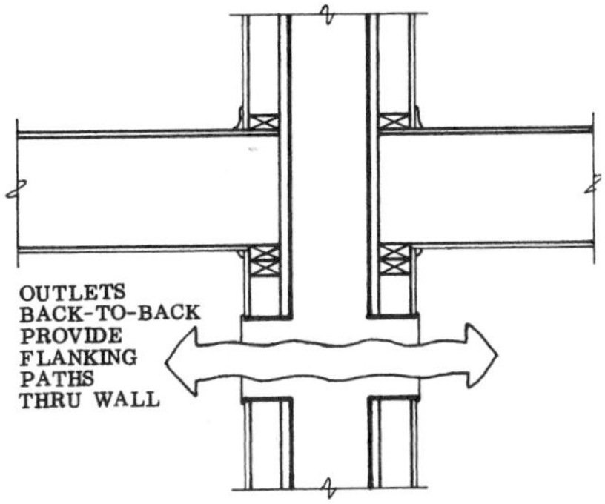

FIGURE 15.2-44 Duct outlets placed back-to-back will nullify the effect of sound-isolating walls.

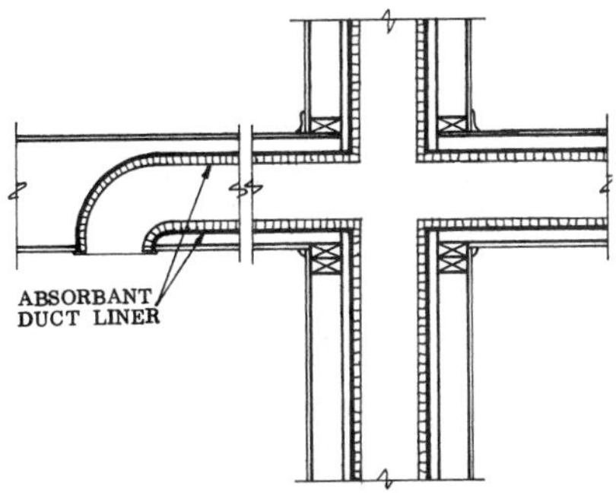

FIGURE 15.2-45 Fibrous insulation in a duct, especially near outlets, helps reduce transmitted noise.

The Underwriters Laboratories, Inc. (UL) dictates the design and type of damper required, and their specific installation requirements, for each specific hazard. Codes mandate compliance with UL assemblies and may have some additional requirements as well.

Fire dampers are connected electrically to a fire alarm system containing fire detectors, or may have fusible links to permit attached springs to close them in the event of fire.

Smoke dampers are tied electrically into a smoke alarm system consisting of smoke detectors.

Dampers similar to smoke dampers can also be used to prohibit passage of other gases or radiation in the event of leaks or accidents.

Some dampers must be manually reset when they have operated. This is an expensive and time-consuming chore. More expensive systems are also available in which the dampers can be opened electrically after an emergency is over. Obviously, dampers containing fusible links cannot be closed electronically; the links must be replaced.

15.2.13 HVAC SYSTEM CONTROLS

A number of separate control systems are required in an HVAC system. Most of these are linked to provide total control of the system.

15.2.13.1 Thermostats

The temperature in a space is measured and controlled by a thermostat of some sort. The thermostat may be actually in the space, in the supply system (duct or water pipe), or in the terminal unit or units serving the space. Thermostats in the space or in an air stream are usually operated by a mercury-type electric switch actuated by a temperature-sensitive spring coil. Thermostats in water lines may be either electric or pneumatic.

Room-type thermostats may be simple devices that have a dial to set the temperature desired, a thermometer to indicate the temperature, and a spring coil and switch to open and close the circuit. Other room-type thermostats are complicated time-clock- or computer-operated devices that can produce different temperatures at different times of the day and even on different days of the year.

Wall-mounted thermostats should be located on an interior partition. They should not be located near lights, cooking or heating appliances, radiators, convection units, or air supply registers, so that the heat and cold air associated with such equipment does not cause a false thermometer reading.

15.2.13.2 Boiler and Furnace Controls

An HVAC system contains operating controls, limit controls, and interlocks, which are designed to provide a preset controlled sequence of operation and to protect furnaces, boilers, chillers, air handlers, and other equipment from damage by taking corrective action when failure occurs or the equipment is not operating in a prescribed manner. *Operating controls* respond to sensors like thermostats and pressure actuated switches and send a signal to *limit controllers*, which are preset and respond to operating control signals to open or close valves, dampers, or relays. *Interlocks* link parts of an HVAC system so that a change in one portion affects the operation of other portions as well. In large modern systems, controls are set and operated by computer.

15.2.14 COMMISSIONING

Commissioning is the process of ensuring that a building's HVAC and electrical systems comply with the contract documents and applicable codes and meet the owner's operational requirements. The process is used for large and small, simple as well as complex, commercial, industrial, and institutional projects.

Commissioning is often specified to be accomplished by the general contractor, but is also sometimes done by mechanical and electrical engineers. Extensive system tests and reports of the conditions observed are usually required.

SMACNA's *HVAC Systems Commissioning Manual* is a guide to the process and contains sample commissioning specifications, reports, and checklists.

15.3 Fire Sprinkler Systems

Sprinkler systems are expensive, but many insurance companies offset some of the costs by offering reductions in premiums for professional installations. In addition, some jurisdictions require sprinklers in most or all commercial and institutional buildings, and some require them in multifamily and even in single-family houses. Some codes permit taller buildings or increased floor areas when sprinklers are used. Installing a sprinkler system throughout an entire existing building can be prohibitively expensive and is seldom justified unless required by law. A sprinkler system should be considered, however, in every new building, even when not required by code.

The purpose of a sprinkler system is to save property and minimize structural damage. There is controversy regarding the life-saving benefits of sprinkler systems as compared with smoke detectors in small buildings. The question is whether the benefits outweigh the higher cost of installation and maintenance of sprinkler systems relative to smoke detectors.

National Fire Protection Association standards NFPA 13D and 13R contain industry standards for the installation of automatic sprinkler systems in residential buildings. Some municipalities have adopted less stringent standards, but require sprinkler systems in all new construction. Other NFPA publications cover sprinkler systems in other building types.

Sprinkler systems are either *wet-pipe* or *dry-pipe* systems. They generally consist of water pipes running at right angles through or under overhead structural framing, and sprinkler heads spaced along the pipes to serve as nozzles.

A *dry-pipe* system's pipes contain air. The loss of air pressure from the opening of a sprinkler head or the detection of a fire condition causes the release of water into the piping system and out of the opened nozzle. Dry-pipe systems are used in areas subject to freezing and in other areas where leaks would cause severe damage.

A *wet-pipe* system's pipes contain water. This system type should be used in areas not subject to freezing, because it tends to deliver water to the fire a little faster than dry-pipe systems.

In case of a fire, a fusible link in a sprinkler head melts, allowing the struts in the center of the head to fall out and release a plug. Emerging water hits a detector, creating a wheel-shaped spray pattern. Sprinkler systems are effective against class A fires (those that can be controlled with water).

15.4 Additional Reading

More information about the subjects discussed in this chapter can be found in the references listed in Section 15.5, and in the following publications.

Kutz, Myer. 1998. *Mechanical Engineers Handbook*. 2d ed. New York: John Wiley & Sons, Inc.

Lechner, Norbert. 1991. *Heating, Cooling, Lighting: Design Methods for Architects*. New York: John Wiley & Sons, Inc.

Long, Jerome A. 1995. "Gas Cooling." *The Construction Specifier* 6(June):52.

McQuiston, Faye C., and Jerald D. Parker. 1994. *Heating, Ventilating, and Air Conditioning: Analysis and Design*. 4th ed. New York: John Wiley & Sons, Inc.

Salas, Marianne, and Carl E. Salas. 1994. "Refrigeration Alternatives." *The Construction Specifier* 3(March):104.

Shrive, Charles A. 1993. "Specifying Electric Heating." *The Construction Specifier* 10(October):35.

Stein, Benjamin. 1997. *Mechanical and Electrical Systems*. 2d ed. New York: John Wiley & Sons, Inc.

Stein, Benjamin, and John S. Reynolds. 1992. *Mechanical and Electrical Systems for Buildings*. 8th ed. New York: John Wiley & Sons, Inc.

Tao, William, and Richard Janis. 1997. *Mechanical and Electrical Systems in Buildings*. Upper Saddle River, NJ: Prentice-Hall.

15.5 Acknowledgments and References

ACKNOWLEDGMENTS

We gratefully acknowledge the assistance of the following organizations and individuals in preparing this chapter. We are also indebted to them for permission to use their illustrations when requested and for the use of their publications as references.

American Forest & Paper Association (AFPA)

American Plywood Association (APA)

American Society of Heating, Refrigerating, and Air-Conditioning Engineers (ASHRAE)

Brick Institute of America (BIA)

Delta Faucet Company

Grinnell Fire Protection Company

National Association of Home Builders (NAHB) Research Foundation

Small Homes Council, Building Research Council

Southern Forest Products Association (SFPA)

U.S. Department of Housing and Urban Development (HUD)

U.S. Environmental Protection Agency (EPA)

Western Wood Products Association (WWPA)

REFERENCES

We would also like to thank the authors and publishers of the publications in the following list for their contributions to our research for this chapter.

Air-Conditioning and Refrigeration Institute (ARI). 1986. 350, *Sound Rating of Non-Ducted Indoor Air-Conditioning Equipment*. Arlington, VA: ARI.

———. 1986. 370, *Sound Rating of Large Outdoor Refrigerating and Air-Conditioning Equipment*. Arlington, VA: ARI.

———. 1991. ANSI/ARI, *Forced-Circulation Air-Cooling and Air-Heating Coils*. Arlington, VA: ARI.

———. 1993. 340/360, *Commercial and Industrial Unitary Air-Conditioning and Heat Pump Equipment*. Arlington, VA: ARI.

———. 1993. 445, *Room Air-Induction Units*. Arlington, VA: ARI.

———. 1993. 850, *Commercial and Industrial Air Filter Equipment*. Arlington, VA: ARI.

———. 1994. 210/240, *Unitary Air-Conditioning and Air-Source Heat Pump Equipment*. Arlington, VA: ARI.

———. 1994. 365, *Commercial and Industrial Unitary Air-Conditioning Condensing Units*. Arlington, VA: ARI.

———. 1994. 1010, *Self-Contained, Mechanically Refrigerated Drinking-Water Coolers*. Arlington, VA: ARI.

———. 1996. Guidelines B, *Roof Mounted Outdoor Air-Conditioner Installations*. Arlington, VA: ARI.

———. 1996. Guidelines D, *Application and Installation of Central Station Air-Handling Units*. Arlington, VA: ARI.

———. 1996. 610, *Central System Humidifiers for Residential Applications*. Arlington, VA: ARI.

———. 1996. 640, *Commercial and Industrial Humidifiers*. Arlington, VA: ARI.

———. 1996. 670, *Fans and Blowers*. Arlington, VA: ARI.

———. 1997. 520, *Positive Displacement Condensing Units*. Arlington, VA: ARI.

———. 1998. 440, *Room Fan-Coils and Unit Ventilators*. Arlington, VA: ARI.

———. 1998. 520, *Ground Water-Source Heat Pumps*. Arlington, VA: ARI.

———. 1998. 840, *Unit Ventilators*. Arlington, VA: ARI.

———. 1999. ARI 430, *Central Station Air-Handling Units*. Arlington, VA: ARI.

———. 1999. ARI 870, *Direct Geoexchange Heat Pumps*. Arlington, VA: ARI.

———. 1999. ARI 450, *Water Cooled Refrigerant Condensers, Remote Type*. Arlington, VA: ARI.

Air Conditioning Contractors of America (ACCA). *Manual C—What Makes a Good Air Conditioning System*. Washington, DC: ACCA.

———. *Manual CS—Commercial Applications, Systems, and Equipment*. Washington, DC: ACCA.

———. *Manual D—Residential Duct Systems*. Washington, DC: ACCA.

———. *Manual G—Selection of Distribution Systems*. Washington, DC: ACCA.

———. *Manual H—Heat Pump Systems: Principles and Applications*. Washington, DC: ACCA.

———. *Manual J—Residential Load Calculations*. Washington, DC: ACCA.

———. *Manual N—Commercial Load Calculations*. Washington, DC: ACCA.

———. *Manual P—Psychrometrics: Theory and Applications*. Washington, DC: ACCA.

———. *Manual Q—Commercial Low Pressure, Low Velocity Duct System Design*. Washington, DC: ACCA.

———. *Manual RS—Comfort Air Quality and Efficiency by Design*. Washington, DC: ACCA.

———. *Manual S—Residential Equipment Selection*. Washington, DC: ACCA.

———. *Manual T—Air Distribution Basics for Residential and Small Commercial Buildings*. Washington, DC: ACCA.

———. *Manual 4—Installation Techniques for Perimeter Heating and Cooling*. Washington, DC: ACCA.

Air Movement and Control Association (AMCA). 1985. 501, *Application Method for Air Louvers*. Arlington Heights, IL: AMCA.

———. 1986. 500, *Test Methods for Louvers, Dampers, and Shutters*. Arlington Heights, IL: AMCA.

———. 1995. B200-3, *AMCA Fan Application Manual*. Arlington Heights, IL: AMCA.

———. 1995. *Standards Handbook*. Arlington Heights, IL: AMCA.

American National Standards Institute (ANSI) Standards:

A112 Series, "Plumbing." New York: ANSI.

Z21.22, "Relief Valves and Automatic Gas Shutoff Devices." New York: ANSI.

Z21.10.1 and Z21.10.3, "Gas Water Heaters." New York: ANSI.

Z124.1, "Plastic Bathtub Units." New York: ANSI.

Z124.2, "Plastic Shower Units." New York: ANSI.

Z124.3, "Plastic Lavatories." New York: ANSI.

Z124.4, "Plastic Water Closet Bowls and Tanks." New York: ANSI.

Z124.6, "Plastic Sinks." New York: ANSI.

American Plywood Association, American Wood Council, National Forest Products Association, Southern Forest Products Association, Western Wood Products Association. *Plen-wood System*. Tacoma, WA: American Plywood Association.

American Society for Testing and Materials (ASTM) Standards:

A 53/A 53M, "Standard Specification for Pipe, Steel, Black and Hot-dipped, Zinc-coated, Welded and Seamless." West Conshohocken, PA: ASTM.

A 74, "Standard Specification for Cast Iron Soil Pipe and Fittings." West Conshohocken, PA: ASTM.

A 312, "Standard Specification for Seamless and Welded Austenitic Stainless Steel Pipe." West Conshohocken, PA: ASTM.

A 377, "Standard Index of Specifications for Ductile Iron Pressure Pipe." West Conshohocken, PA: ASTM.

A 888, "Standard Specification for Hubless Cast Iron Soil Pipe and Fittings for Sanitary and Storm Drain, Waste, and Vent Piping Applications." West Conshohocken, PA: ASTM.

B 43, "Standard Specifications for Seamless Red Brass Pipe, Standard Sizes." West Conshohocken, PA: ASTM.

B 88, "Standard Specification for Seamless Copper Water Tube." West Conshohocken, PA: ASTM.

B 108, "Standard Specification for Aluminum-Alloy Permanent Mold Castings." West Conshohocken, PA: ASTM.

B 134, "Standard Specification for Brass Wire." West Conshohocken, PA: ASTM.

B 135, "Standard Specification for Seamless Brass Tube." West Conshohocken, PA: ASTM.

B 210, "Standard Specification for Aluminum and Aluminum-Alloy Drawn Seamless Tubes." West Conshohocken, PA: ASTM.

B 271, "Standard Specification for Copper-Base Alloy Centrifugal Castings." West Conshohocken, PA: ASTM.

B 306, "Standard Specification for Copper Drainage Tubes (DWV)." West Conshohocken, PA: ASTM.

B 429, "Standard Specification for Aluminum-Alloy Extruded Structural Pipe and Tube." West Conshohocken, PA: ASTM.

B 455, "Standard Specification for Copper-Zinc-Lead Alloy (Leaded Brass) Extruded Shapes." West Conshohocken, PA: ASTM.

B 456, "Standard Specification for Electrodeposited Coatings of Copper Plus Nickel Plus Chromium and Nickel Plus Chromium." West Conshohocken, PA: ASTM.

B 483, "Standard Specification for Aluminum and Aluminum-Alloy Extruded Drawn Tubes for General Purpose Applications." West Conshohocken, PA: ASTM.

B 564, "Standard Specification for Rubber Gaskets for Cast Iron Soil Pipe and Fittings." West Conshohocken, PA: ASTM.

B 584, "Standard Specification for Copper Alloy Sand Castings for General Applications." West Conshohocken, PA: ASTM.

B 587, "Standard Specification for Welded Brass Tube." West Conshohocken, PA: ASTM.

B 650, "Standard Specification for Electrodeposited Engineering Chromium Coatings on Ferrous Substrates." West Conshohocken, PA: ASTM.

C 12, "Standard Practice for Installing Vitrified Clay Pipe Lines." West Conshohocken, PA: ASTM.

C 177, "Standard Test Method for Steady-State Heat Flux Measurements and Thermal Transmission Properties by Means of the Guarded-Hot-Plate Apparatus." West Conshohocken, PA: ASTM.

C 236, "Standard Test Method for Steady-State Thermal Performance of Building Assemblies by Means of a Guarded Hot Box." West Conshohocken, PA: ASTM.

C 425, "Standard Specification for Compression Joints for Vitrified Clay Pipe and Fittings." West Conshohocken, PA: ASTM.

C 564, "Standard Specification for Rubber Gaskets for Cast Iron Soil Pipe and Fittings." West Conshohocken, PA: ASTM.

C 700, "Standard Specification for Vitrified Clay Pipe, Extra Strength, Standard Strength, and Perforated." West Conshohocken, PA: ASTM.

C 976, "Standard Test Method for Thermal Performance of Building Assemblies by Means of a Calibrated Hot Box." West Conshohocken, PA: ASTM.

D 1785, "Standard Specification for Poly(Vinyl Chloride) (PVC) Plastic Pipe, Schedules 40, 80, and 120." West Conshohocken, PA: ASTM.

D 2287, "Standard Specification for Nonrigid Vinyl Chloride Polymer and Copolymer Molding and Extrusion Compounds." West Conshohocken, PA: ASTM.

D 2447, "Standard Specification for Polyethylene (PE) Plastic Pipe, Schedules 40 and 80, Based on Outside Diameter." West Conshohocken, PA: ASTM.

D 2487, "Standard Classification of Soil for Engineering Purposes (Unified Soil Classification System)." West Conshohocken, PA: ASTM.

D 2751, "Standard Specification for Acrylonitrile-Butadine-Styrene (ABS) Sewer Pipe and Fittings." West Conshohocken, PA: ASTM.

D 2949, "Standard Specification for 3.25-in. Outside Diameter Poly(Vinyl Chloride) (PVC) Plastic Drain, Waste, and Vent Pipe and Fittings." West Conshohocken, PA: ASTM.

D 3309, "Standard Specification for Polybutylene (PB) Plastic Hot- and Cold-Water Distribution Systems." West Conshohocken, PA: ASTM.

D 3350, "Standard Specification for Polyethylene Plastics Pipe and Fittings Materials." West Conshohocken, PA: ASTM.

F 891, "Standard Specification for Coextended Poly(Vinyl Chloride) (PVC) Plastic Pipe with a Cellular Core." West Conshohocken, PA: ASTM.

American Society of Civil Engineers (ASCE). "Sewage Treatment Plant Design." *ASCE Manual of Engineering Practice*, No. 36. Reston, VA: ASCE.

———. "Design and Construction of Sanitary and Storm Sewers," *ASCE Manual of Engineering Practice*, No. 37. Reston, VA: ASCE.

American Society of Heating, Refrigerating and Air-Conditioning Engineers (ASHRAE). 1986. *Bin and Degree-hour Weather Data for Simplified Energy Calculations*. Atlanta, GA: ASHRAE.

———. 1996. *Handbook—HVAC Systems and Equipment*. Atlanta, GA: ASHRAE.

———. 1997. *Handbook—Fundamentals*. Atlanta, GA: ASHRAE.

———. 1999. *Handbook—HVAC Applications*. Atlanta, GA: ASHRAE.

———. Standard 55, "Thermal Environmental Conditions for Human Occupancy." Atlanta, GA: ASHRAE.

———. Standard 62, "Ventilation for Acceptable Indoor Air Quality." Atlanta, GA: ASHRAE.

———. Standard 90.1, "Energy Efficient Design of New Buildings Except Low-rise Residential Buildings, and User's Manual." Atlanta, GA: ASHRAE.

———. Standard 90.2, "Energy Efficient Design of New Low-rise Residential Buildings, and User's Manual." Atlanta, GA: ASHRAE.

American Water Works Association (AWWA). C-203-97, "Coal-tar Protective Coatings and Linings for Steel Water Pipelines—Enamel and Tape—Hot-applied." Denver, CO: AWWA.

Architectural Graphic Standards. See Ramsey/Sleeper.

Ashlock, Dennis P. 1994. "Selecting and Specifying Boilers." *The Construction Specifier* 2(February):23.

Bergoust, Don. 1997. "Preventing HVAC Failures." *The Construction Specifier* 6(June):65.

Bobenhausen, William, and Catherine Coombs. 1997. "Lean, Clean, and Green HVAC Systems." *The Construction Specifier* 8(August):44.

BOCA International Inc. (BOCAI).1993. *BOCA National Plumbing Code*. Country Club Hills, IL: BOCAI.

———. 1993. *BOCA National Mechanical Code*. Country Club Hills, IL: BOCAI.

———. 1999. *BOCA National Building Code*. Country Club Hills, IL: BOCAI.

———. 1999. *BOCA National Fire Prevention Code*. Country Club Hills, IL: BOCAI.

Brick Institute of America (BIA). *Technical Notes on Brick Construction*.

No. 4, rev. Jan. 1982. "Heat Transmission Coefficients of Brick Masonry Walls." Reston, VA: BIA.

No. 4A, rev. Feb. 1982 (reissued Sept. 1988). "Heat Gain Through Opaque Walls." Reston, VA: BIA.

No. 4B, rev. Dec. 1992. "Energy Code

Compliance of Brick Masonry Walls." Reston, VA: BIA.
No. 43, rev. May/Jun. 1981. "Passive Solar Heating with Brick Masonry Walls." Reston, VA: BIA.
No. 43B, Sept./Oct. 1979 (reissued Jan. 1988). "Brick Passive Solar Heating Systems, Part III—Performance Calculations." Reston, VA: BIA.
No. 43D, Sept./Oct. 1980 (reissued Sept. 1988). "Brick Passive Solar Heating Systems, Part IV—Material Properties." Reston, VA: BIA.
No. 43G, Mar./Apr. 1981 (reissued Sept. 1986). "Brick Passive Solar Heating Systems, Part VII—Details and Construction." Reston, VA: BIA.
Cast Iron Soil Pipe Institute (CISPI) Standards:
1986. HSN-74, "Specification for Cast Iron Soil Pipe and Fittings for Hub and Spigot Sanitary and Storm Drain, Waste, and Vent Piping Applications." Chattanooga, TN: CISPI.
1997. 301, "Standard Specification for Hubless Cast Iron Soil Pipe and Fittings for Sanitary and Storm Drain, Waste, and Vent Piping Applications." Chattanooga, TN: CISPI.
1997. 310, "Specification for Couplings Used to Connect Hubless Cast Iron Soil Pipe and Fittings for Sanitary and Storm Drain, Waste, and Vent Piping Applications." Chattanooga, TN: CISPI.
History and Uses of Cast Iron Soil Pipe. Vol. I of *CISPI Handbook*. 7th ed. Chattanooga, TN: CISPI.
Champagne, Roger D. 1986. "Planning for Plumbing Efficiency." *The Construction Specifier* 39(6)(June):77–80.
Cornish, Kathy L. 1987. "ASHRAE's Standard 90 Revision: A Split Proposal." *The Construction Specifier* 40(12)(December):111–116.
Council of American Building Officials (CABO). *Model Energy Code*. Falls Church, VA: CABO. This code is in the process of being transferred to the control of the International Code Council (ICC).
———. 1983. *One- and Two-Family Building Code*. Falls Church, VA: CABO. This code is in the process of being transferred to the control of the International Code Council (ICC).
Dagostino, Frank R. 1995. *Mechanical and Electrical Systems in Construction and Architecture*. Upper Saddle River, NJ: Prentice-Hall.
Federal Specifications, all: *See* U.S. General Services Administration Specifications Unit.
Gardner, Robert. 1994. "Air Duct Insulation: Benefits and Options." *The Construction Specifier* 12(December): 86.
Haas, Steve. 1998. "Five Ways to Reduce HVAC Duct Noise." *The Construction Specifier* 2(February):63.
Harrie, David T. 1986. "Clean and Economical Air in Houses." *Architectural Technology* (July/August):33–36.
Harriman, Marc S. 1990. "A Higher Standard." *Architecture* (December):91–93.
Howard, Bion D. 1985. "Masonry Thermal Mass and the Energy Code." *The Construction Specifier* 38(6)(June): 87–95.
Hydronics Institute (HI). *IBR Code, Indirect Water Heater Test Standards*. Berkeley Heights, NJ: HI.
International Association of Plumbing and Mechanical Officials (IAPMO). 1997. *Uniform Plumbing Code*. Walnut, CA: IAPMO.
———. 1997. *Uniform Mechanical Code*. Walnut, CA: IAPMO.
International Code Council (ICC) of International Conference of Building Officials (ICBO).1995. *CABO One- and Two-Family Dwelling Code*. Falls Church, VA: ICBO.
———. 1997. *International Plumbing Code*. Falls Church, VA: ICBO.
———. 1997. *International Private Sewage Disposal Code with 1998 Revisions*. Falls Church, VA: ICBO.
———. 1998. *International Mechanical Code*. Falls Church, VA: ICBO.
———. 1997. *International Fuel Gas Code*. Falls Church, VA: ICBO.
———. 1998. *International Energy Conservation Code*. Falls Church, VA: ICBO.
———. *CABO Model Energy Code*. Falls Church, VA: ICBO.
———. 1998. *International One- and Two-Family Dwelling Code*. Falls Church, VA: ICBO.
Kirsch, Laurence S. 1986. "Indoor Air and the Law." *Architectural Technology* (July/August):37–41.
Lizardos, Evans J. 1995. "HVAC & IAQ." *The Construction Specifier* 10(October):43.
Long, Jerome A. 1995. "Light Commercial/Residential Heat Pump." *The Construction Specifier* 6(June):60.
McNall, Preston E. 1986. "Controlling Air Quality in Office Buildings." *Architectural Technology* (July/August): 29–32.
Mingo, Eric. 1995. "Radiant Floor Heating." *The Construction Specifier* 6(June):58.
Nall, Daniel H. 1998. "Underfloor Air Delivery Systems." *The Construction Specifier* 2(February):45.
National Association of Home Builders (NAHB) Research Foundation. 1983. *NAHB Thermal Performance Guidelines*. Washington, DC: NAHB.
National Climatic Data Center. 1981B, *Climatography of U.S. Number 81, Monthly Normals of Temperature, Precipitation, and Heating and Cooling Degree Days*. Asheville, NC: National Climatic Data Center.
National Fire Protection Association (NFPA). 1999. *Automatic Sprinkler Systems Handbook*. Quincy, MA: NFPA.
———. 1999. NFPA 13, "Installation of Sprinkler Systems." Quincy, MA: NFPA.
———. 1999. NFPA 13D, "Standard for the Installation of Sprinkler Systems in One- and Two-family Dwellings and Mobile Homes." Quincy, MA: NFPA.
———. 1999. NFPA 13R, "Standard for Installation of Sprinkler Systems in Residential Occupancies Up to and Including Four Stories in Height." Quincy, MA: NFPA.
———. 1996. NFPA 15, "Standard for Water Spray Fixed Systems for Fire Protection." Quincy, MA: NFPA.
———. 1990. NFPA 20, "Standard for Installation of Centrifugal Fire Pumps for Fire Protection." Quincy, MA: NFPA.
———. 1997. NFPA 101, "Life Safety Code." Quincy, MA: NFPA.
———. 1999. *National Fire Codes*. Quincy, MA: NFPA.
Plumbing-Heating-Cooling Contractors Association (PHCC). 1996. *National Standard Plumbing Code*. Falls Church, VA: PHCC.
Ramsey/Sleeper, The AIA Committee on Architectural Graphic Standards. *Architectural Graphic Standards*. 8th ed. New York: John Wiley & Sons, Inc.
Reid, Robert N. 1990. "Quality Assurance in Fire Sprinkler Systems." *The Construction Specifier* 43(9)(September): 124–129.
Ross, Robert E. 1985. "Boilers: Low-cost Comfort for Commercial Buildings."

The Construction Specifier 38(3) (March):69–78.

Rudoy, W., and R. S. Douglas. *Effect of Thermal Mass on Heating and Cooling Loads in Residences*. Washington, DC: National Forest Products Association.

Rutkowski, Henry T. 1985. "Designing HVAC Systems That Perform." *The Construction Specifier* 38(3)(March): 50–65.

Sheet Metal and Air Conditioning Contractors National Association (SMACNA).1995. *Thermoplastic Duct (PVC) Construction Manual*. Chantilly, VA: SMACNA.

———. 1991. *Energy Recovery Equipment and Systems*. Chantilly, VA: SMACNA.

———. 1992. *Fibrous Glass Duct Construction Standards*. Chantilly, VA: SMACNA.

———. 1993. *Architectural Sheet Metal Manual*. Chantilly, VA: SMACNA.

———. 1993. *HVAC Systems—Testing, Adjusting, and Balancing*. Chantilly, VA: SMACNA.

———. 1987. *HVAC Systems—Applications*. Chantilly, VA: SMACNA.

———. 1990. *HVAC Systems—Duct Design*. Chantilly, VA: SMACNA.

———. 1992. *Fibrous Glass Duct Construction Standards*. Chantilly, VA: SMACNA.

———. 1992. *Fire, Smoke and Radiation Damper Installation*. Chantilly, VA: SMACNA.

———. 1994. *HVAC Systems Commissioning Manual*. Chantilly, VA: SMACNA.

———. 1995. *HVAC Duct Construction Standard—Metal and Flexible*. Chantilly, VA: SMACNA.

———. 1998. *Indoor Air Quality Manual*. Chantilly, VA: SMACNA.

———. 1998. *Installation Standards for Residential Heating and Air Conditioning Systems*. Chantilly, VA: SMACNA.

———. 1998. *Residential Comfort System Installation Standards Manual*. Chantilly, VA: SMACNA.

Smith, Karen Hass. 1986. "Energy Conservation in Small HVAC Systems." *The Construction Specifier* 39(6) (June):87–92.

Southern Building Code Congress International (SBCCI). 1997. *Standard Plumbing Code*. Birmingham, AL: SBCCI.

———. 1997. *Standard Mechanical Code*. Birmingham, AL: SBCCI.

———. 1999. *Standard Fire Prevention Code*. Birmingham, AL: SBCCI.

———. 1999. *Standard Building Code*. Birmingham, AL: SBCCI.

———. 1999. *National Electric Code*. Birmingham, AL: SBCCI.

———. 1999. *Standard Gas Code*. Birmingham, AL: SBCCI.

Sump and Sewage Pump Manufacturers Association (SSPMA). *Recommended Standards for Sump, Effluent, and Sewage Pump Standards*. Winnetka, IL: SSPMA.

Sussman, Alan. 1988. "Solving HVAC Equipment Vibration Problems." *The Construction Specifier* 41(3)(March): 29–32.

Tilley, Ray Don. 1990. "Reclaimed Resources." *Architecture* (December): 97–102.

Underwriters Laboratories (UL). Standards:

1993. 70, "Septic Tanks, Bituminous Coated Metal." Northbrook, IL: UL.

1996. 174, "Household Electric Storage Tank Water Heaters." Northbrook, IL: UL.

1995. 732, "Oil-fired Storage Tank Water Heaters." Northbrook, IL: UL.

U.S. Department of the Air Force. 1978. AFM 88-29, *Engineering Weather Data*. Washington, DC: Departments of the Air Force, the Army, and the Navy.

U.S. Department of Housing and Urban Development (HUD). *Minimum Property Standards for Housing*. Washington, DC: HUD.

———. Handbook 4940.2, *Minimum Design Standards for Community Water Supply Systems*. Washington, DC: HUD.

———. Handbook 4940.3, *Minimum Design Standards for Community Sewerage Systems*. Washington, DC: HUD.

———. Handbook 4075.12 REV, *Central Water and Sewage Systems*. Washington, DC: HUD.

U.S. Environmental Protection Agency (EPA). *National Primary Drinking Water Standards (NPDWRS)*. Washington, DC: EPA.

———. EPA 625/1-30-012 EPA Design Manual, *On-Site Waste Water Treatment and Disposal Systems*. Washington, DC: EPA.

U.S. General Services Administration Specifications Unit. Federal Specification WW-P-541, "Plumbing Fixtures." Washington, DC: GSA Specifications Unit.

Van Geem, Martha G. 1987. "Thermal Mass, What R-values Neglect." *The Construction Specifier* 40(6)(June): 71–77.

Water Quality Association (WQA) Standards:

S-100, "Voluntary Industry Standards for Household, Commercial, and Portable Exchange Water Softeners." Vol. 1. Lisle, IL: WQA.

S-200, "Recommended Industry Standards for Household and Commercial Water Filters." Vol. 2. Lisle, IL: WQA.

Wilson, Forrest. 1984. *Building Materials Evaluation Handbook*. New York: Van Nostrand Reinhold.

CHAPTER 16

Electrical

Introduction

Applicable *MasterFormat*™ Sections

Fundamentals of Electricity

Electric Power Distribution

Service and Distribution

Lighting

Communications Systems

Conservation of Energy

Additional Reading

Acknowledgments and References

Introduction

Even when an adequate amount of electricity is made available, the extent to which it can be effectively used in a building is determined by the interior wiring. Starting at the service entrance, where electricity enters a building, a wiring system must deliver the correct amount of electricity to every circuit and outlet. Circuits and outlets are either general-purpose ones serving portable devices, or are designed to serve a specific purpose. Switch controls are located with both convenience and safety in mind.

Electrical appliances and devices are constantly being improved to provide higher standards of performance and convenience. The resulting increases in wattage are reflected in greater demands on building wiring systems.

The number of appliances and other electrical devices is also increasing. Work-saving and time-saving devices, such as computers, proliferate in offices, shops, factories, and institutional buildings. Most new homes contain automatic washers, electric clothes dryers, dishwashers, food waste disposers, and freezers. Almost every new commercial, institutional, and residential building is air conditioned. Even though electric heating systems, a mainstay in some parts of the country for single-family houses when electrical rates were lower, are not so common today, they have often been replaced by electric heat pumps, which use somewhat less energy than electric heating systems but nevertheless still use considerable energy.

Such conveniences place large demands on a building's electrical service and wiring system. When loads outstrip the capacities of the wiring system, operating efficiency suffers. When voltage drop becomes excessive in a circuit, poor lighting and inefficient appliance operations result. A 5% voltage loss, for example, produces a 10% loss of heat in an appliance and a 17% loss of light from an incandescent lamp. However, the greatest problems presented by overloading electrical systems are related to safety. Overloaded circuits and undersized circuit breakers can possibly lead to damaged equipment, electric shock, or fires.

Average consumption of electrical power has increased tenfold in the past 50 years, but recent energy consciousness and an energy saving movement, mostly in response to federal action, have tended to abate the increase somewhat. Despite increased generation of electricity by nuclear plants, hydroelectric stations, and other nonfossil means, most of our electrical production today is generated using coal, gas, or oil as a fuel. Spiraling electrical production reduces our nonrenewable fossil fuel reserves and results in large expenditures to prevent the severe adverse environmental damage caused by burning these fuels. The current national commitment to energy conservation and pollution abatement places responsibility on a designer to make electrical systems not only convenient and efficient, but also not wasteful of precious energy.

This chapter includes a primer on the fundamentals of electricity and on electricity distribution and delivery. It also discusses various systems and equipment used in electrical systems in buildings and includes recommendations for planning adequate electrical power and lighting systems. Other related subjects, communications systems and conservation of energy, are covered as well.

Applicable *MasterFormat*™ Sections

The following *MasterFormat* Broadscope sections are applicable to this chapter.

02500 Utility Services

16050 Basic Electrical Materials and Methods
16100 Wiring Methods
16200 Electrical Power
16300 Transmission and Distribution
16500 Lighting
16700 Communications
16800 Sound and Video

16.1 Fundamentals of Electricity

The generally accepted notion of electricity is based on the theoretical structure of an atom. An atom can be visualized as a nucleus orbited by one or more electrons (Fig. 16.1-1). Groups of electrons moving in neighboring orbits can be thought of as belonging to the same orbital shell. Depending on the number of electrons, there can be as many as seven such shells.

The nucleus consists of one or more positively charged protons and one or more electrically neutral particles, called neutrons. Each orbital electron has a negative charge, equal in magnitude and opposite to the positive charge of a nuclear proton.

In a neutral atom, the negative and positive charges balance out. But electrons in the outer shell can be dislodged and become free electrons, separated from the parent atom, which is then left with a net positive charge. A free electron may then attach itself to another neutral atom, lodging on its outer shell and creating a negative charge on the atom (Fig. 16.1-2). Single atoms or clusters of atoms with extra or fewer electrons are called *ions*.

A chemical battery, the simplest source of electricity, is made up of one or more cells. Most chemical batteries deliver small amounts of electricity at a low voltage. A cell usually consists of a carbon rod (positive electrode), a piece of copper or zinc (negative electrode), and a chemical compound (electrolyte). The electrolyte breaks down readily into positively and negatively charged particles (ions), which attach themselves to the electrodes, imparting a negative or positive charge. Connecting the two electrodes with a wire allows the electrons collected on the negative electrode to travel to the positive

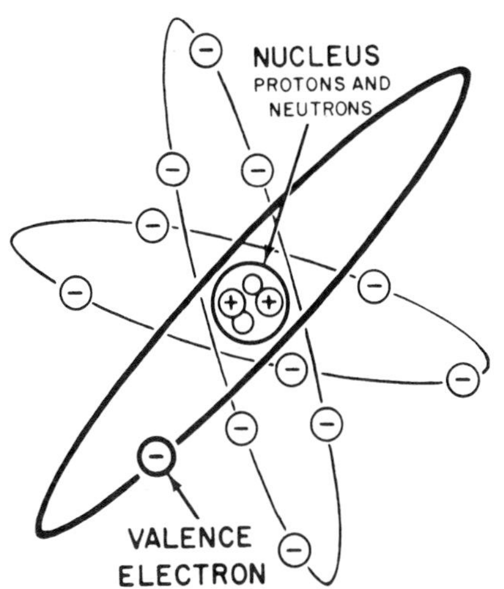

FIGURE 16.1-1 Valence electrons, moving in orbits that are relatively far from the nucleus of an atom, have the least attraction for it and can be dislodged to create electron flow, which is electricity.

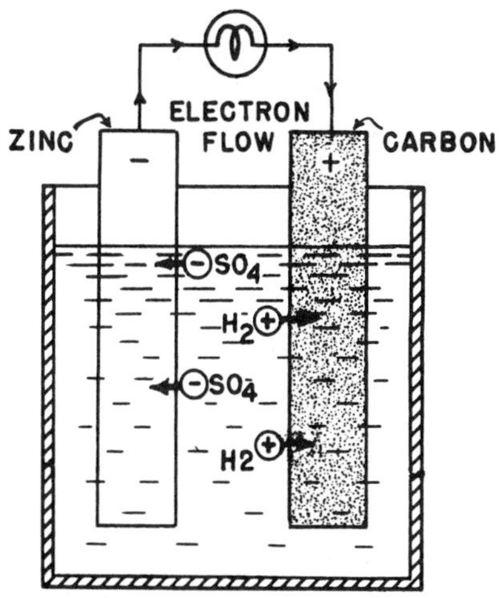

FIGURE 16.1-3 The electrolytic solution (H_2SO_4) in a wet cell breaks down into positive (H_2) and negative (SO_4) ions. Although the SO_4 ions carry electrons to the zinc (Zn) electrode, they also react with it, changing it into $ZnSO_4$.

electrode in a balancing effort. The movement of these electrons through a wire is defined as the flow of electricity, or *current*. This current-generating action can be illustrated by a voltaic wet cell, which uses a solution of sulfuric acid as the electrolyte (Fig. 16.1-3). In a dry cell, such as a flashlight battery, the zinc casing is used as the negative electrode and a special paste acts as the electrolyte between the casing and the carbon rod (positive electrode).

Electricity can also be generated directly from solar radiation by means of silicon photovoltaic cells. This technology was developed by NASA to provide power for satellites and other orbiting and deep space devices. These cells generate low-voltage, direct current power that can then be converted into 120-volt alternating current using an dc-ac inverter. This type of cell may well be used in the future to generate power in large quantities, but it is an expensive way to generate electricity and so far must be considered as experimental, even though a few automobiles and some residential and commercial building projects have made limited use of this technology.

For larger amounts of high-voltage power, an electric generator is the device used today. Unlike batteries, which convert chemical or solar energy into electric power, a generator utilizes kinetic (motion) energy. This kinetic energy usually comes from a steam- or water-driven turbine. Steam is created by burning coal, gas, or oil, or from the heat of a nuclear reactor. Water is caused to drop from a reservoir through a turbine. The turbine then turns a generator armature that has loops of wires wound around it. As the armature turns between the poles of a magnet, the loops cut across the invisible lines of magnetic force between the poles. This movement causes current to flow through the wire loops of the armature (Fig. 16.1-4).

16.1.1 CURRENT FLOW

An electric current can be visualized as a migration of electrons through a conductor from a negative terminal to a positive one. The process can be visualized as follows: a moving electron lodges on the outer shell of an atom, dislodging a second electron, which then migrates to an adjacent atom with the same result. The process is repeated again and again in a chain reaction (see Fig. 16.1-2).

To keep electrons flowing, there must be a continuous uninterrupted circuit (path) from a negative to a positive terminal. If the path is interrupted, electron flow stops, resulting in an open circuit. When a shortcut is inadvertently created, permitting a sudden surge of electrons through an easier path, we call that a short circuit.

16.1.1.1 Voltage

Current is caused to flow by an electromotive force called *voltage*, which is also called *potential difference*. Voltage is the driving force behind electron flow, somewhat like the pressure in a water pipe or an air duct.

Particles with like charges, such as two electrons, repel each other; unlike particles, such as a positive ion and a negative electron, attract each other. These two factors contribute to the force of voltage. The mutual repulsion of electrons tends to make them rush away from each other; the mutual attraction between unlike charges creates a kind of pull that draws electrons toward positively charged atoms.

Electrons actually flow from a negatively charged to a positively charged body, but the older notion that electricity flows from a positive to a negative terminal still appears in textbooks and electrical diagrams as a convention, a kind of accepted fiction or customary code. If electricity is assumed to flow from a positive to a negative terminal, then current must be composed of positively charged

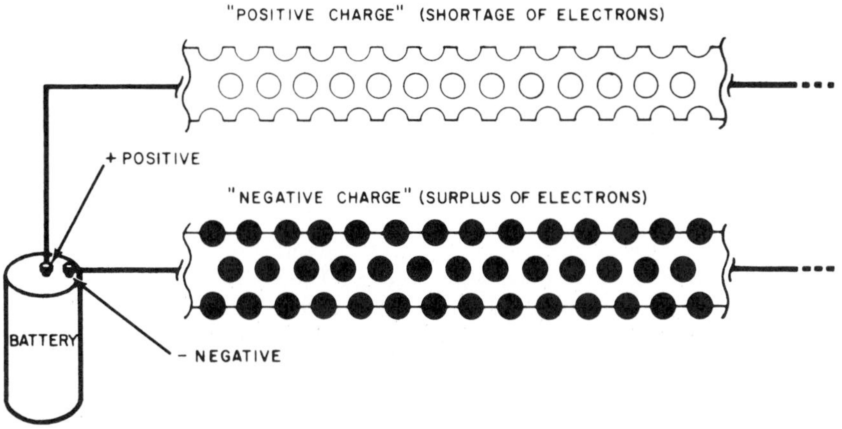

FIGURE 16.1-2 Electrons will flow through a conductive material that connects an area with a surplus of electrons to an area that has a shortage of electrons, such as the two terminals of a battery.

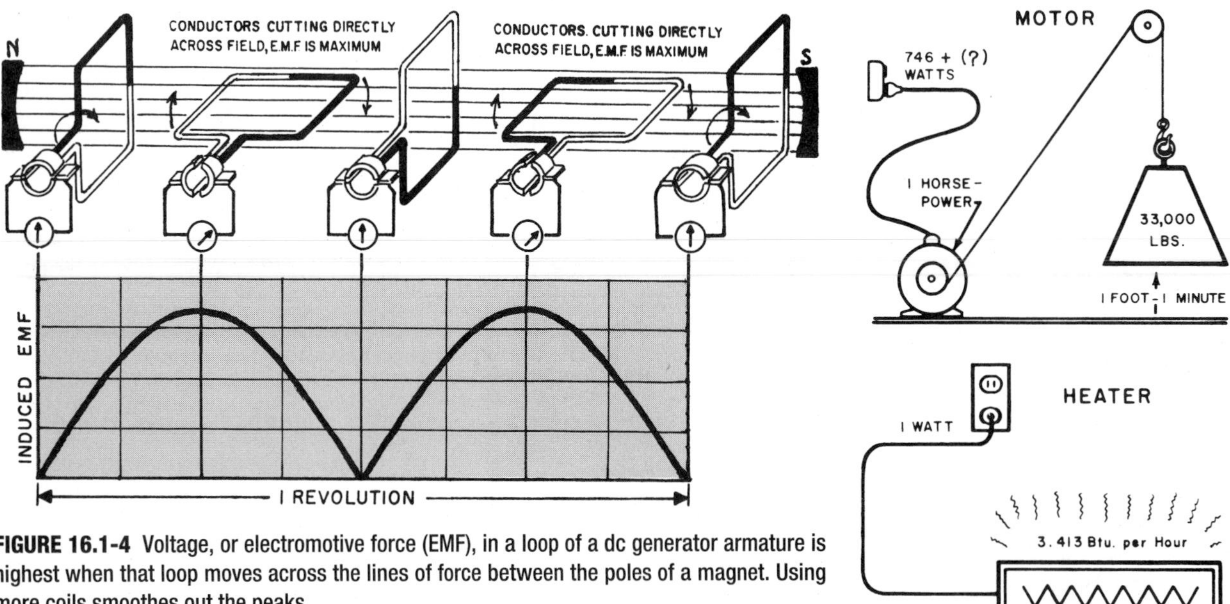

FIGURE 16.1-4 Voltage, or electromotive force (EMF), in a loop of a dc generator armature is highest when that loop moves across the lines of force between the poles of a magnet. Using more coils smoothes out the peaks.

FIGURE 16.1-5 Electric energy can be converted into kinetic (motion) or thermal (heat) energy.

particles. This concept, though contrary to established electron flow theory, is useful in certain calculations.

16.1.1.2 Amperage

The rate at which electrons, or current, flows through a conductor is measured in *amperes* (amps). The speed with which electricity is transmitted is about the same as the speed of light, 186,000 miles (299,338 km) per second. That is, if a current is started at one end of a line 186,000 miles (299,338 km) long, it will be felt at the other end 1 second later. Rate of flow is different from speed because it takes into account *current density*, that is, the number of electrons moving along from atom to atom. Thus, amperage measures the quantity of electricity passing through a circuit per unit of time.

The quantity of electricity without reference to time is measured in coulombs. A *coulomb* is roughly the number of electrons conducted past a point in 1 second. The number is an unimaginable 6,250,000,000,000,000,000 electrons. One ampere equals 1 coulomb of electricity passing a point in the circuit in every second.

16.1.1.3 Resistance

Some metals become *superconductive* at temperatures near absolute zero (–459.6°F [0 K]). Superconductivity is a complete disappearance of electrical resistance. Recently, researchers have developed materials that become superconductive at much higher temperatures. But at normal-use temperatures, the internal structure of the best conductors available resists the flow of electric current and converts some of it into heat. This internal frictionlike effect, which is called *resistance*, is measured in ohms. An *ohm* is that resistance that allows 1 ampere (1 coulomb per second) to flow when pushed by a pressure of 1 volt.

16.1.1.4 Power

The electrical unit of power (energy rate) is the *watt*. In that it measures the rate at which work is done, or energy is used, a watt is similar to British thermal units (Btu's) per hour (Btuh) and horsepower (hp) and can be converted directly into either one. For instance, 1 watt = 3.413 Btuh, 1 hp = 746 watts (Fig. 16.1-5).

An electric motor rated at 1 hp is capable of doing mechanical work at a rate of 33,000 ft.-lb./minute. This means that the motor can lift a 1000-lb. (453.6 kg) weight 33 ft. (10.1 m) every minute. If it were 100% efficient, the motor would then be converting electric energy into mechanical work at a rate of 746 watts. In electric heating, electricity is converted into heat rather than mechanical work (see Fig. 16.1-5).

Stating the relationship in electrical terms, one gets the basic equation for electrical power, $W = EI$, where wattage (W) equals voltage (E) times amperage (I). The equation says, in effect: the heavier the flow rate (amperes) at a given supply pressure (volts), the higher the rate (watts) at which energy is being supplied and used.

An electric utility, for example, may use this equation to find out the load demand on distribution lines in watts or kilowatts (1000 watts). Or a designer may wish to determine the wattage of an electric heater whose rating is 22 amperes. Multiplying 22 amperes times 240 volts gives the answer: 5280 watts (about 5.3 kilowatts).

16.1.2 OHM'S LAW

In an active circuit, voltage, amperage, and resistance are interrelated. The greater the resistance, the greater the voltage required to produce a given amperage. Current and resistance together, then, imply the required voltage pressure, and that is stated in Ohm's law: $E = IR$, where voltage (E) equals amperage (I) times resistance (R). Ohm's law makes it possible to determine any one of these values when the other two are known (Fig. 16.1-6).

By combining the power equation with Ohm's equation, it is possible to derive many equations for the relationship between voltage, wattage, amperage, and resistance (Fig. 16.1-7). The power equation as stated earlier is $W = EI$; substituting IR for E (Ohm's equation), we get $W = (IR)I = I^2R$. In other words, wattage (W) equals amperage (I) squared times resistance (R). This equation helps determine the power used in an electric heater and the power lost in transmitting electricity. In a resistance heater, the power is transformed into useful heat, but in distribution wires it is an undesirable waste of energy.

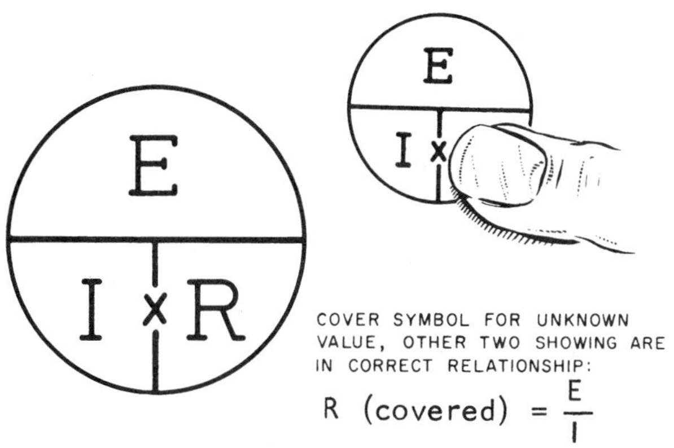

FIGURE 16.1-6 Ohm's law gives the relationship between voltage (E), amperage (I), and resistance (R).

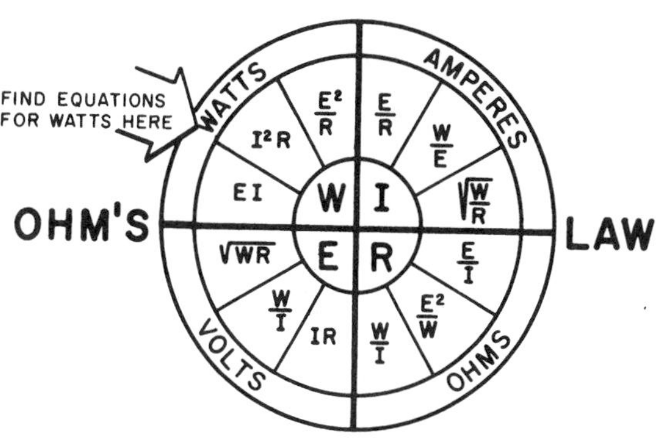

FIGURE 16.1-7 Ohm's law and derivative formulas make it possible to find any of the values in the middle of the wheel when any other two are known.

The usefulness of these equations can be demonstrated by the following example. A 100-ft. (30.48 m)-long No. 14 copper wire has a resistance of about 0.25 ohms. Maintaining a current of 10 amperes at 120 volts, what would be the line loss in the conductors? The line loss (see Fig. 16.1-7) is obtained from the power used or lost equation mentioned in the previous paragraph, $W = I^2R$, or $(10)^2$ amps $\times$ 0.25 ohms = 25 watts. The total power would be $EI = W$, or 120 volts times 10 amps equals 1200 watts. The 25-watt line loss would constitute a little more than 2% of the total power, which is well within the recommended maximum line loss of 3%.

16.1.3 ALTERNATING AND DIRECT CURRENT

So far it has been convenient to discuss electrical principles in terms of direct current rather than the more common commercial form, alternating current. Direct current (dc) always flows in one direction in a circuit: namely, from the negative to the positive terminals of the source. Alternating current (ac) also flows from negative to positive terminals, but the terminals change polarity (whether positive or negative) quickly, causing a reciprocating (back-and-forth) surge of electricity along the wire.

Alternating current is produced by an ac generator (Fig. 16.1-8) designed so that the voltage and current reverse direction periodically. In the commonly used 60-cycle power supply, 1 cycle of a complete

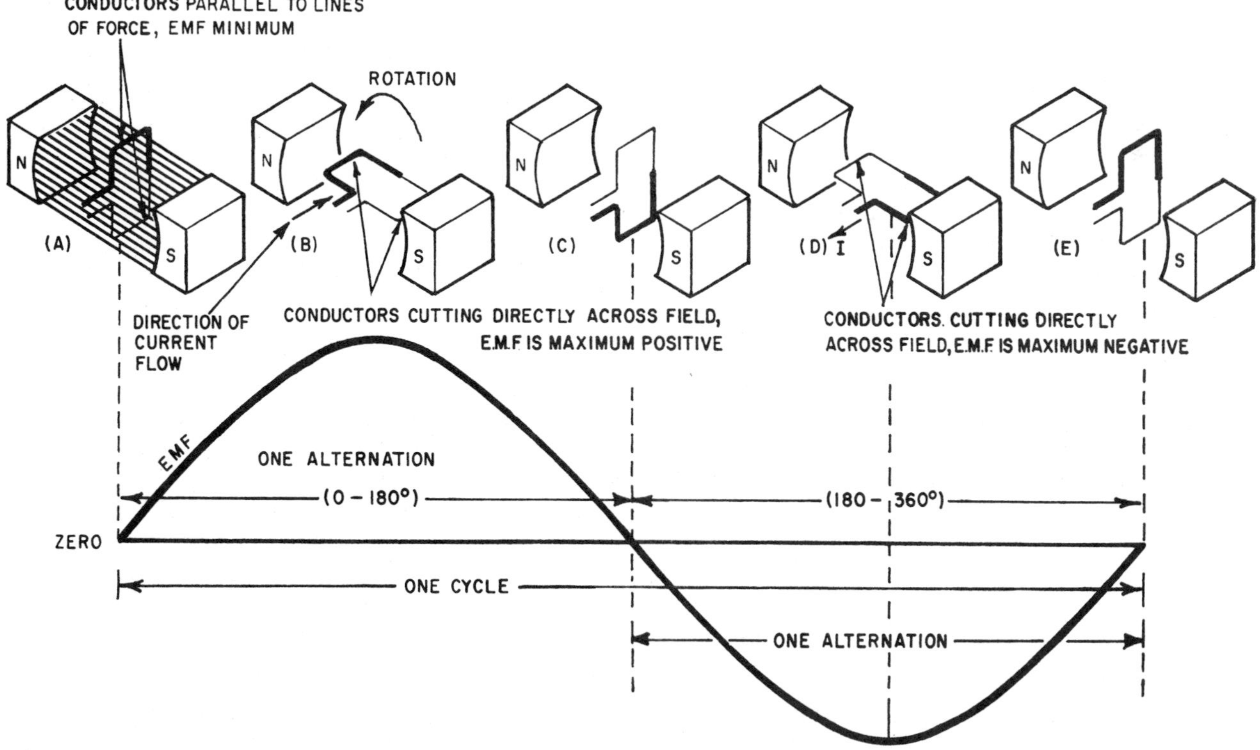

FIGURE 16.1-8 Just as in a dc generator (Fig. 16.1-4), the voltage (EMF) peaks twice in each cycle in an ac generator, but the alternative peaks are of opposite polarity, as current flow reverses in each alternation.

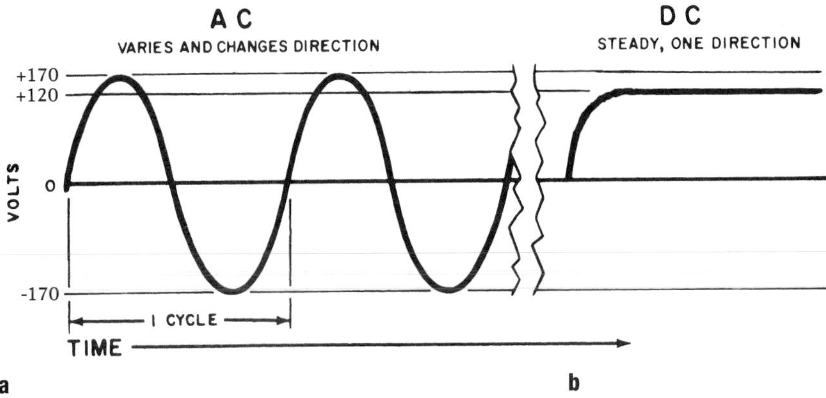

FIGURE 16.1-9 Nominal 120-volt ac current is actually generated with voltage peaks of 170 volts (a), to approximate the effect of a 120-volt dc current (b).

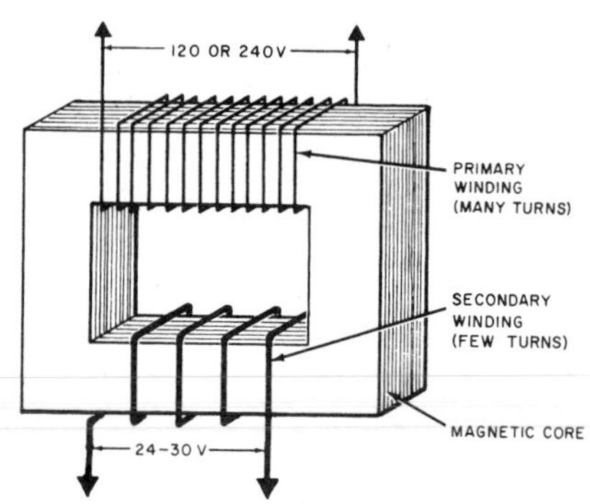

FIGURE 16.1-10 A transformer is used to step up or step down voltage. To reduce voltage from 120 to 30 volts would require four times as many turns in the primary as in the secondary winding.

rise and fall of voltage in each direction takes place in 1/60 second. The voltage of 120-volt direct current is actually 120 volts at the generator and should not drop below 115 volts at the load. However, because alternating current consists of rapidly changing values from 0 to 170 volts, the average or useful voltage appears as 120 volts to the power-consuming device (Fig. 16.1-9). Thus, the effect of both types of current is the same and the preceding equations can be used for both ac and dc calculations.

Alternating current is used almost exclusively in residential and commercial wiring because it provides greater flexibility in voltage selection and simplicity of equipment design. When alternating current is used, the voltage can be stepped up (increased) or stepped down (decreased) through a transformer (Fig. 16.1-10). An alternating current transformer is a very efficient device, wasting only 1% to 4% of the power in heat. By comparison, an electric generator, which is the only means of changing direct current voltage, loses about 25% of its power in waste heat.

Passing current through the primary circuit on one side of the magnet in a transformer generates a current in the secondary circuit. The induced voltage in the secondary circuit has the same relationship to the voltage in the primary circuit as the number of windings in the secondary circuit do to those in the primary circuit. In other words, to double the voltage, it is merely necessary to double the windings in the secondary circuit; to triple the voltage, triple the windings; to reduce the voltage by 50%, use half the number of windings.

16.2 Electric Power Distribution

Electricity from a generating station is transmitted at high voltage to substations, then to transformers near the users. There it is converted to line voltage (120 and 240 volts) for use in small commercial buildings, apartments, and homes, or to 480 or higher voltages for use in commercial, institutional, and industrial facilities. This higher voltage, which may be required for heavy equipment or special devices, is then converted again to line voltage for use in power circuits and lighting. Additional transformers drop line voltage down again where needed for low-voltage circuitry.

16.2.1 PRIMARY DISTRIBUTION

Most power is generated in power plants at 12,000 volts. For transmission over long distances, the voltage is stepped up by transformers to 115,000 volts or more, depending on the distance. Very high voltage is used because more power can be transmitted with a lower power loss in the lines. Utility companies generally use the highest voltage consistent with safety and economy in their transmission lines.

To avoid hazards from high-voltage transmission lines in built-up areas, high voltage is reduced at substations to either 12,000/7200 volts or 4160/2400 volts. Smaller transformers, mounted on poles (Fig. 16.2-1), set on the ground, or placed in underground vaults, further transform current to line voltage (120/240/480 volts).

16.2.2 SINGLE-PHASE SERVICE

Single-phase 120/240-volt service is used almost universally for residences in this country and for many small commercial buildings as well. Small electric heating equipment, heat pumps, and most large, fixed appliances such as water heaters, electric ranges, and clothes dryers, use 240-volt service because they operate more efficiently at the higher voltage.

Power at 480 volts is supplied to many larger commercial, institutional, and industrial facilities, where it is often used for heating, ventilating, and air conditioning systems. Step-down transformers then

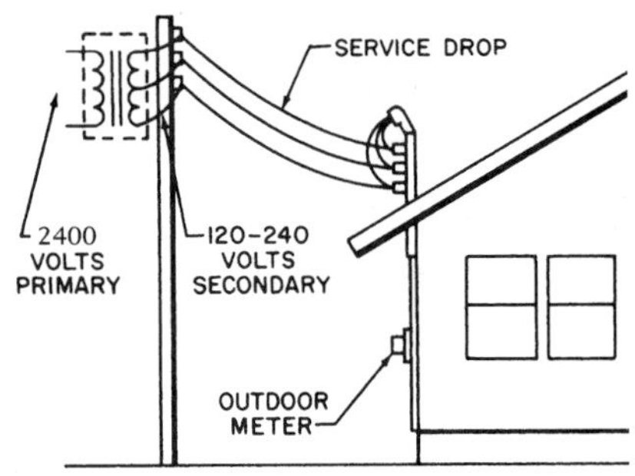

FIGURE 16.2-1 A pole-mounted transformer and service drop. A service drop should be at least 10 ft. (3048 mm) above ground.

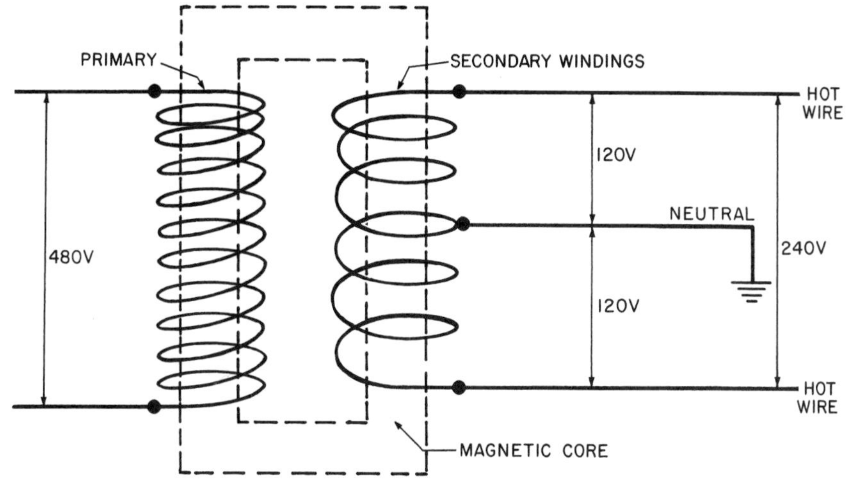

FIGURE 16.2-2 A center-tapped transformer can convert two-wire 480-volt (or higher) current to three-wire 120/240-volt service.

drop the voltage to 240 volts for equipment requiring this voltage and to 120 for general power circuits.

General lighting in residences and many commercial, institutional, and industrial buildings, and smaller, movable appliances, such as refrigerators, mixers, irons, and toasters, operate adequately on 120 volts, as do dishwashers and some other fixed items. Some types of lighting in commercial, institutional, and industrial buildings, however, operate at 240 volts. To satisfy the needs of most buildings, both 120 and 240 volts must be available. This dual voltage is provided by 240-volt, center-tapped secondary transformers (Fig. 16.2-2). This transformer makes both 120-volt and 240-volt power available to a three-wire electric service entrance.

A three-wire service drop runs from a utility pole to a building. A three-wire circuit consists of two hot wires, each carrying 120 volts and a neutral wire having no voltage. Items requiring 240 volts are connected between the two hot wires; loads requiring 120 volts can be connected between either hot wire and the neutral (see Fig. 16.2-2).

The incoming service must be grounded. Refer to Section 16.3.5.5 for a discussion of grounding.

16.2.3 THREE-PHASE SERVICE

Four-wire three-phase service is seldom used in residences in this country, but is frequently used in commercial, institutional, and industrial applications. This system is particularly suitable for electrical heating and cooling installations, and where large motors are involved.

Three-phase service consists of three hot wires and a neutral wire. The voltage in each hot wire is out of phase (out of step) with the others by one-third of a cycle, as if produced by three different generators. Figure 16.1-9 shows single-phase current. In a three-phase diagram, there would be three curves, with the peaks in the second and third curves moved one-third of the diagrammed phase length to the right. Connections made between any of the two hot wires produces nominal 208-volt current; connections made between either of the hot wires and the neutral results in nominal 120-volt current. Although the service has four wires, branch circuits may have two, three, or four wires, depending on the type of equipment served (Fig. 16.2-3).

A three-phase system should be planned so that the total load is about equally divided between the three phases. It should be remembered also that three wires are hot (rather than two, as in three-wire one-phase service) and all three require overload protection.

16.2.4 SERVICE DROP

When the electrical service comes from overhead, a utility service *drop* is usually connected to the customer's service entrance conductors at the building, forming a drip loop, to prevent water from entering the conductor cable. When the service is underground, the service entrance conductors rise in a conduit. From the drip loop or underground cable, conductors go first to a watt-hour meter, which keeps a record of how much power is consumed, and finally to the service equipment. The service equipment is the origin of all the branch circuits that supply a building's electrical needs.

Service equipment provides 120-volt, 240-volt, or 480-volt current for general-purpose, appliance, and special-purpose circuits. In addition, line-voltage circuits can be used to power transformers that supply low-voltage circuits.

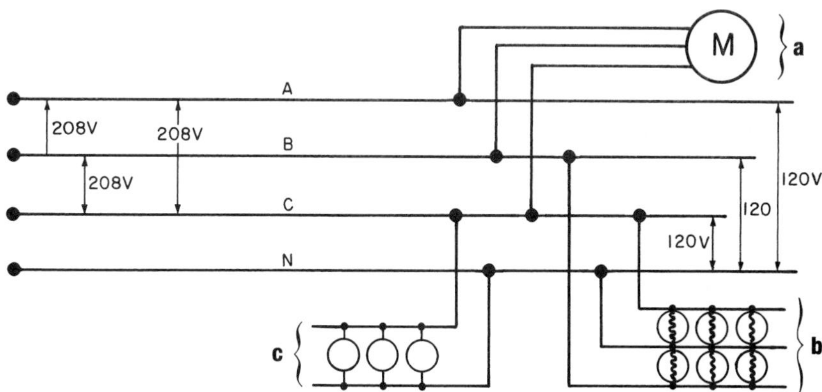

FIGURE 16.2-3 Four-wire, three-phase service provides flexibility in serving different loads. In this figure (a) shows a three-wire, three-phase tap for a motor; (b) a three-wire, single-phase circuit for kitchen appliances; and (c) a two-wire, single-phase circuit for lights and receptacles.

16.3 Service and Distribution

Electric power enters a building through service entrance equipment and from there is distributed to convenience outlets, lights, appliance outlets, and special-purpose outlets. The design and installation of such systems is subject to compliance with codes and standards. Symbols have been developed to help in the indication of requirements on drawings. This section discusses some of these aspects of electric service to buildings.

16.3.1 CODES, STANDARDS, AND SYMBOLS

A familiarity with the applicable codes and standards and the symbols used on drawings is essential to an understanding of the text in this chapter.

Most of the terms used here that have different meanings in this context than their standard dictionary meanings are explained in the body of the text. Those that are not so explained are defined in the glossary.

16.3.1.1 Codes and Standards

Since 1937, the average annual consumption of electrical energy in residences has risen from 805 to more than 8850 kilowatt-hours. In some areas of the country the average consumption is more than 10,000 kilowatt-hours and is still rising. The electrical industry studied this trend and evolved a set of farsighted recommendations, which were first published in 1958 by the American National Standards Institute (ANSI) as USA Standard C91.1-1958, "Requirements for Residential Wiring." Later editions of this standard, now called American National Standard C91.1, are still useful today.

The recommendations in this chapter are based generally on ANSI C91.1, with revisions as necessary to conform with the 1996 *National Electrical Code® (NEC)®** and the HUD *Minimum Property Standards for One- and Two-Family Dwellings*. The purpose of the recommendations here is to suggest design criteria that will produce an efficient and convenient system and will also minimize physical obsolescence. Therefore, they go beyond the requirements of the NEC, whose chief emphasis is on safety.

An adequate wiring installation must conform with applicable safety regulations. The standard for electrical safety is the NEC, developed and published by the National Fire Protection Association (NFPA). Most local ordinances and state and local building codes are based on the NEC or make it a part of their requirements by reference. Therefore, the entire electric system, including its design, layout, equipment, and materials, and the methods used to install it must comply with the requirements of the latest edition of the NEC and local ordinances and laws. It must also be acceptable to the local building authority that has jurisdiction.

Inspection service to determine conformance with safety regulations is available in most communities from the local building official. Where required, a certificate of inspection must be obtained. In the absence of inspection service, an affidavit should be obtained from the electrical contractor attesting to conformity with applicable safety regulations.

16.3.1.2 Symbols

The symbols shown in Figure 16.3-1 are adapted from American National Standards Institute (ANSI) Standard Y32-1972 (reaffirmed 1989), "Graphic Symbols for Electrical Wiring and Layout Diagrams Used in Architecture and Building Construction." These are the standards recommended by the Institute of Electrical and Electronics Engineers (IEEE). They have been made mandatory by some federal agencies, such as the Department of Defense, and should be used on all drawings for electrical systems, whether they are for public or private buildings.

16.3.2 SERVICE POWER REQUIREMENTS

Minimum power requirements are dictated by law, code, or ordinance in many jurisdictions, especially for buildings accessible to the public. Most codes require compliance with the latest edition of the NEC.

16.3.2.1 Residential Buildings

Regardless of law or code requirements, each separate dwelling unit, including apartments, townhouses, and single-family houses, should have at least a three-wire 120/240-volt electric service. Many older residences were served by a two-wire 120-volt service, which is inadequate for today's needs and fails to meet current code requirements.

A 100-amp service will provide a home's needs for lighting, appliances, and fixed equipment totaling not more than 18,000 watts. When the total load is larger than 18,000 watts, as is the case in almost every new home today, the local power company should be consulted. A 150-amp service may be adequate to handle wiring, appliances, and fixed equipment plus 3 to 5 tons of air conditioning. When electric heating, heat pumps, or their equivalent is used, a service of at least 200 amps is called for. In larger homes, especially those having more than one heating system, a 400-amp service is not uncommon.

Inadequate service conductors and service equipment can be serious obstacles to the use of new equipment and appliances in homes. Service entrance conductors should be sized in accordance with the NEC. To provide for probable increases in electrical use, even when electric heating or cooling is not contemplated, service conductors should be oversized to at least No. 1/0 AWG (American Wire Gauge) and service equipment should be of sufficient rating to match the carrying capacity of these wires.

Most portable appliances for residential use are made to work on 110 to 125 volts, but many fixed appliances need 230- to 250-volt circuits. However, some fixed appliances are still manufactured to function on the lower voltages so that they can be used in older residences. When there is a choice, it is more efficient to use the higher voltage.

16.3.2.2 Commercial, Institutional, and Industrial Buildings

Design and selection of adequate wiring and service capacity criteria for commercial, institutional, and industrial buildings should be determined in each instance in consultation with the local power company. Except in the case of very small commercial buildings, such services should be designed by an electrical engineer.

Mechanical equipment, elevators, heavy industrial machinery, and other major equipment for such buildings will usually operate on 240-volt or even 480-

**National Electric Code®* and *NEC®* are registered trademarks of the National Fire Protection Association, Inc., Quincy, MA 02269.

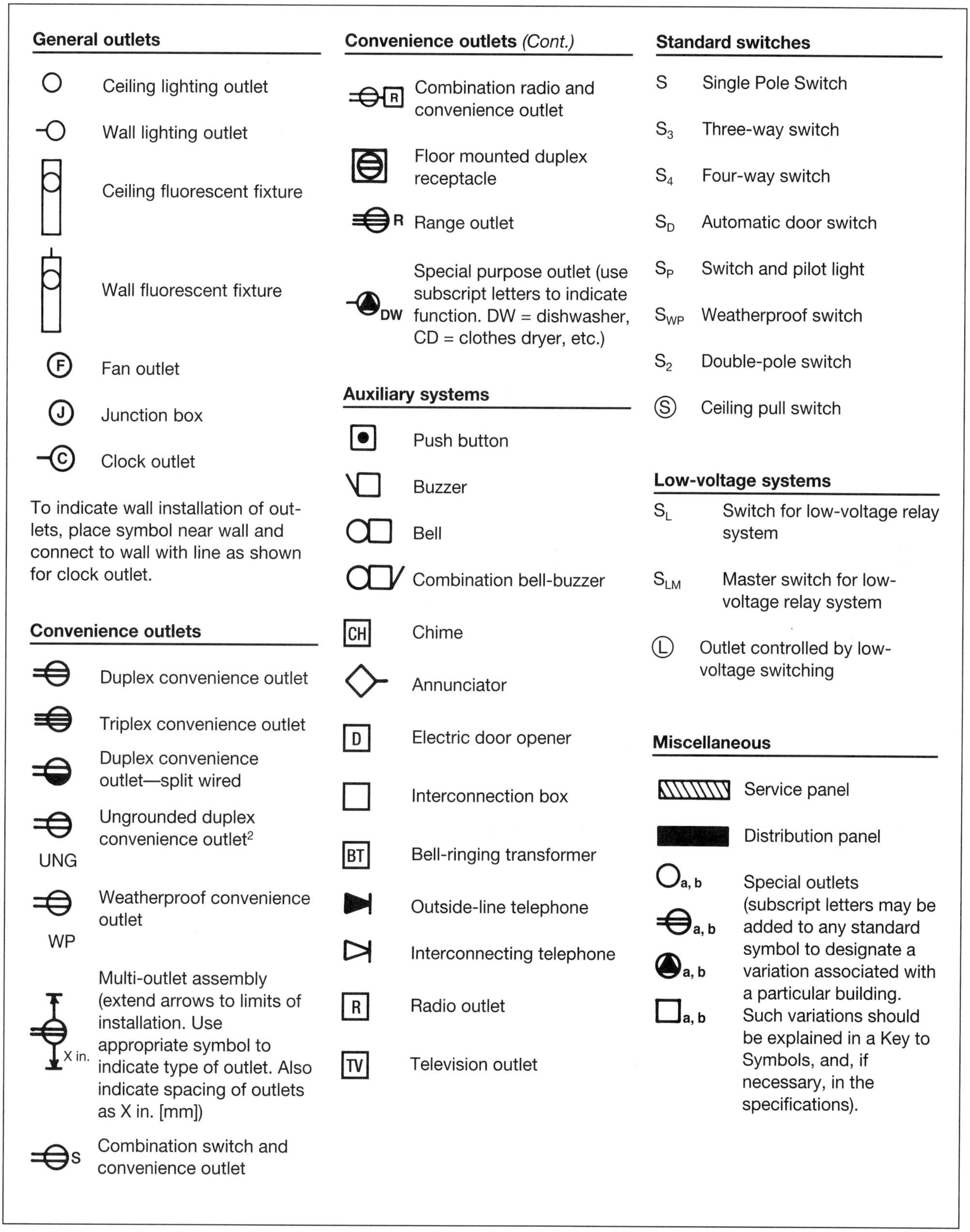

General outlets

Ceiling lighting outlet

Wall lighting outlet

Ceiling fluorescent fixture

Wall fluorescent fixture

F Fan outlet

J Junction box

C Clock outlet

To indicate wall installation of outlets, place symbol near wall and connect to wall with line as shown for clock outlet.

Convenience outlets

Duplex convenience outlet

Triplex convenience outlet

Duplex convenience outlet—split wired

UNG Ungrounded duplex convenience outlet[2]

WP Weatherproof convenience outlet

X in. Multi-outlet assembly (extend arrows to limits of installation. Use appropriate symbol to indicate type of outlet. Also indicate spacing of outlets as X in. [mm])

S Combination switch and convenience outlet

Convenience outlets *(Cont.)*

R Combination radio and convenience outlet

Floor mounted duplex receptacle

R Range outlet

DW Special purpose outlet (use subscript letters to indicate function. DW = dishwasher, CD = clothes dryer, etc.)

Auxiliary systems

Push button

Buzzer

Bell

Combination bell-buzzer

CH Chime

Annunciator

D Electric door opener

Interconnection box

BT Bell-ringing transformer

Outside-line telephone

Interconnecting telephone

R Radio outlet

TV Television outlet

Standard switches

S Single Pole Switch

S_3 Three-way switch

S_4 Four-way switch

S_D Automatic door switch

S_P Switch and pilot light

S_{WP} Weatherproof switch

S_2 Double-pole switch

S Ceiling pull switch

Low-voltage systems

S_L Switch for low-voltage relay system

S_{LM} Master switch for low-voltage relay system

L Outlet controlled by low-voltage switching

Miscellaneous

Service panel

Distribution panel

a, b

a, b

a, b

a, b

Special outlets (subscript letters may be added to any standard symbol to designate a variation associated with a particular building. Such variations should be explained in a Key to Symbols, and, if necessary, in the specifications).

FIGURE 16.3-1 Graphic electrical wiring symbols.[1] (Drawing by HLS.)

[1]Adapted from USA Standard Y32.9-1972 (reaffirmed 1989) "Graphic Symbols for Electrical WIring and Layout Diagrams Used in Architecture and Building Construction."

[2]All outlets of every type are grounded, unless marked UNG.

volt service. Ice makers, food preparation equipment, and other lighter equipment may operate on either 120- or 240-volt service. Lighting and convenience circuits and special-purpose circuits for coffee machines, food dispensing machines, water coolers, and other lighter equipment will often need only 120-volt outlets. Some lighting fixtures will operate at 240 volts; some equipment and some lights will require low-voltage circuits. Exterior lighting may need low voltage, 120-, 240-, or even 480-volt power, depending on fixture type.

As is true of older residences, many older commercial buildings were built with two-wire 120-volt services. Such service is inadequate for today's needs and violates most codes. When major remodeling or renovation is done in such buildings, codes will often require that the service be brought up to code. This can be a major cost item, possibly requiring complete rewiring of the building with all new outlets, and sometimes new conduits, throughout the building.

16.3.3 EQUIPMENT

Electrical equipment includes service entrance equipment, overload protection devices, conductors, raceways, outlets, and switches. This category does not include end-line equipment, such as heating, ventilating, and air conditioning equipment, or portable or built-in appliances.

16.3.3.1 Service Entrance and Distribution Equipment

A *service entrance* is a system of conductors bringing electricity into a building and the associated service entrance and distribution equipment. *Service entrance and distribution equipment* is defined here as all equipment and devices that occur from the outlet side of the watt-hour meter (see Section 16.2.4) to the outlet side of the distribution panels. It includes the service conductors that bring electricity into a building, the service entrance panel, the feeders that connect the service entrance panel to distribution panels, and the distribution panels from which branch circuits distribute electricity to the rest of the building and its grounds. It also includes overcurrent devices mounted in the service entrance and distribution panels. Grounding of the incoming service, also part of the service entrance equipment, is discussed in Section 16.3.5.5. Figure 16.3-2 gives an overall view of service drop, meter, and service entrance equipment.

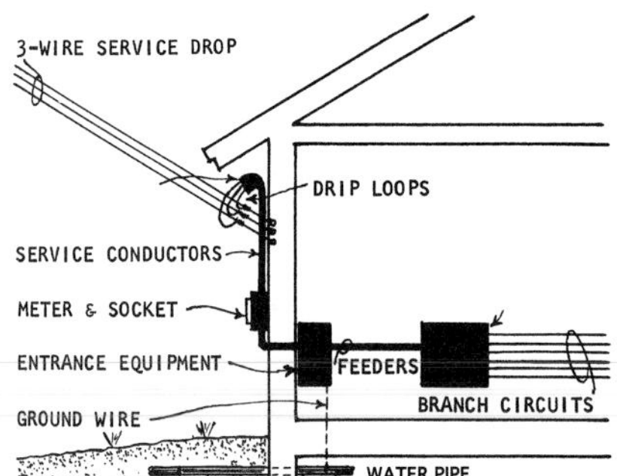

FIGURE 16.3-2 Service entrance equipment.

SERVICE ENTRANCE PANEL

A *service entrance panel*, which in small buildings may also serve as the distribution panel, is an assembly of switches and switchlike devices that permits disconnecting all power or distributing it to various branch circuits through overcurrent devices such as fuses or circuit breakers.

There are several possible arrangements for a service entrance panel. Sometimes, especially in larger systems, a main disconnect switch is housed in a separate panel from the circuit overcurrent protection devices (fuses or circuit breakers). In houses and other small buildings, one panel usually includes both main switch and overcurrent devices.

A main disconnect device provides a means of deliberately deenergizing the entire system, and at the same time protects the service conductors from unexpected overloads. Depending on the load, the code, and the desires of the designer, a main disconnect device could be a knife-blade switch, a pull-out fuse, or a circuit breaker. The overcurrent devices could be fuses or circuit breakers.

Figure 16.3-3 shows a typical circuit breaker type of service entrance panel. This is the type of panel used in most houses and other small buildings today. A main knife-blade switch may be inserted between the main disconnect breaker and the incoming service or may replace the main disconnect breaker. The principles shown are applicable to panels with fuses, but wiring is used in lieu of bus bars. In addition, the principles demonstrated are the same, regardless of the amperage rating of the system.

At the upper left in Figure 16.3-3, the service conductors coming from the meter (not shown) enter the panel. The two hot wires carry the current to the main disconnect breaker, and through it into two hot bus bars. The hot bus bars carry the current to the individual circuit overcurrent protection devices (circuit breakers). Wires are then extended from each breaker to the outlets or devices on a single circuit. When distribution panels are used, the two hot wires run from a breaker on the service entrance panel to hot bus bars in the distribution panel. Breakers are attached to these bus bars, and wires run from the breakers to serve the outlets and devices on a single circuit.

The neutral conductor is neither switched nor fused. It connects to the bonded neutral bus bar, to which the service ground wire also is connected. This grounded bus provides terminals for all load neutrals and ground wires in the entire distribution system.

In residences, because of the many major uses of electricity in a kitchen requiring individual equipment circuits, it is usually best to locate the main service equipment or a subpanel near the kitchen. Such a location is also often close to the laundry, which helps to minimize long circuit runs and wiring costs (Fig. 16.3-4).

For similar reasons, the service entrance in commercial, institutional, and industrial buildings is usually close to the heaviest load center. Where practicable, for example, the service entrance is located close to the boiler plant. In facilities within cities, the service entrance is often in a basement room near a vault under the sidewalk or street adjacent to the building, where the building's main transformer is housed. The idea is to reduce the length of run of the heaviest wiring, thus reducing wiring cost and power loss, which is larger in larger wiring.

FEEDERS

Conductors connecting a service entrance panel or main disconnect switch to distribution panels are called *feeders*.

DISTRIBUTION PANEL

A *distribution panel* is connected to the service entrance panel or main disconnect switch by feeders. In small residences and small commercial buildings, the service entrance panel often serves as the only distribution panel. However, in large residences and in most commercial, institutional, and industrial buildings, there will be several, or many, distribution panels, depending on building size and use. A dis-

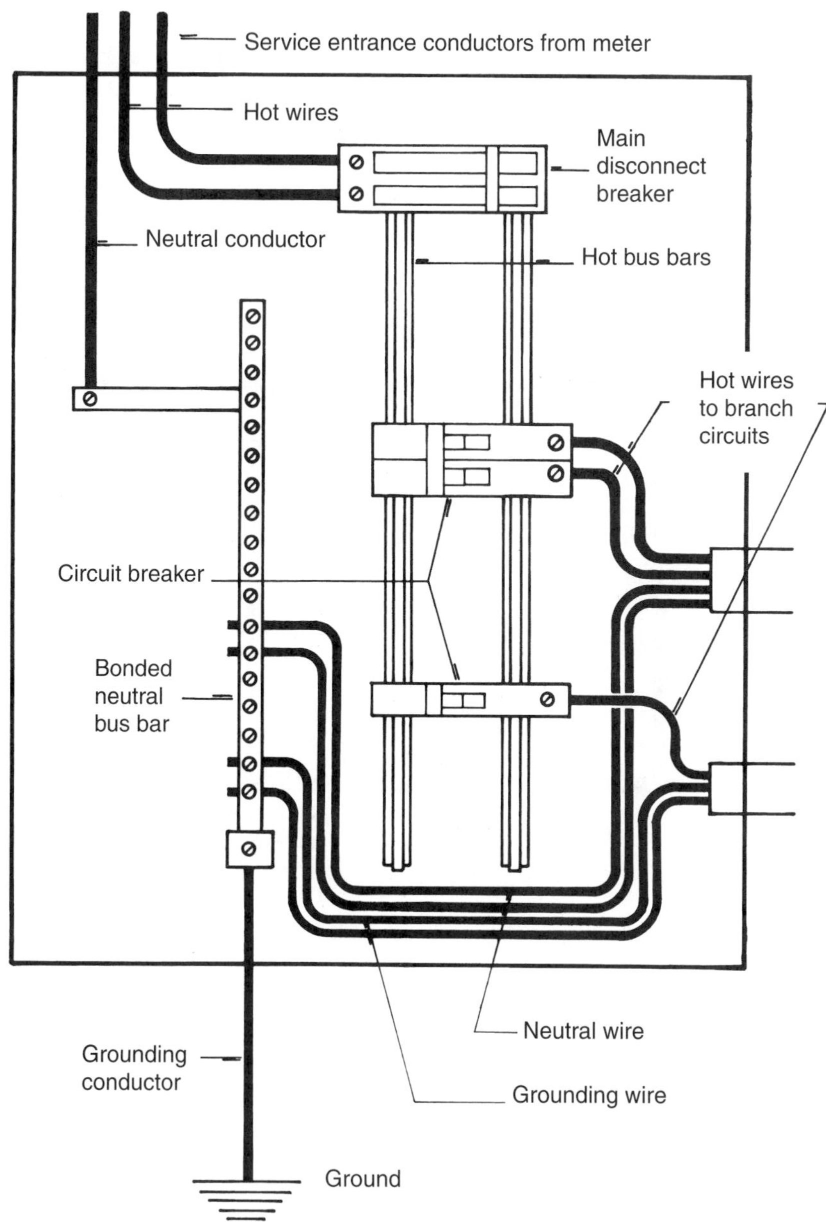

FIGURE 16.3-3 A circuit breaker service entrance panel. To provide a 120-volt branch circuit, insert a circuit breaker by snapping the clip on its back onto one hot bus bar. Then extend to the location of the circuit a hot wire (black or red) from the breaker, a neutral wire (gray or white) from the bonded neutral bus bar, and a grounding wire (bare wire) from the bonded neutral bus bar. To create a 120/240-volt circuit, snap a common-trip two-pole breaker with one handle so that one of the two poles contacts each hot bus bar. Then extend to the location of the circuit, one hot wire from each pole of the breaker, and both a neutral wire and a grounding wire from the bonded bus bar. (Drawing by HLS.)

tribution panel may or may not have a disconnect switch or overload devices like fuses or circuit breakers.

BRANCH CIRCUIT

A *branch circuit* is the portion of a wiring system between an overcurrent device and outlets where loads, such as light fixtures or appliances, can be connected. Refer to Section 16.3.5.1 for recommendations about the allocation of outlets to circuits.

16.3.3.2 Overload Protection Devices

Inadequate or old wiring, faulty wiring installation, and other flaws in wiring or in appliances are major causes of fires. Electricity may also be a source of personal injury, especially in combination with water. When a person who is in contact with the ground comes into contact with a hot wire, current will flow through the person to the ground, resulting in burns at the point where the electricity enters the body as well as where the current leaves the body. Such electricity interferes with the normal operation of a person's nervous system and, in severe cases, can result in convulsions, serious neurological damage, and even death.

Water is a better conductor of electricity than is a human body; it bypasses the protection a person may ordinarily have, such as shoes; and it produces multiple contact points for electricity to enter and leave. Therefore, a person in contact with water is likely to sustain more severe injury when coming in contact with an electrical source than when no water is present.

Overcurrent devices are imposed in electrical circuitry to ensure that in case of a short circuit, current will be interrupted before the conductor overheats and possibly burns up the insulation. The first line of defense is at the main circuit control panel. Older systems, and some large special-purpose systems, may employ fuses. Most new systems use circuit breakers. Several other devices and methods are also used in other parts of a typical system.

FUSES

Fuses have a fusible (meltable) wire, designed to melt and break when the current flow exceeds the rated capacity of the circuit. Thus, the circuit is opened and current flow is stopped before it becomes a hazard. Circuits of 30 amps or less can use plug fuses; for heavier currents, cartridge fuses are required.

Plug fuses are obtainable in sizes from 0 to 30 amperes (Fig. 16.3-5). The NEC requires that type S fuses be used for all plug-fused installations. A type S fuse uses a slightly different adapter for each ampere rating; once the adapter is installed in a fuse box, it cannot be removed readily and only a fuse of a certain rating can be inserted. Existing wiring systems that are being modified should be equipped with type S fuses if the fuse box is retained.

Cartridge fuses are made in all current ratings and voltage ratings from 0 up. Fuses from 0 to 60 amperes use a ferrule-type connection; those above 61 amperes use a knife-blade connection (see Fig. 16.3-5).

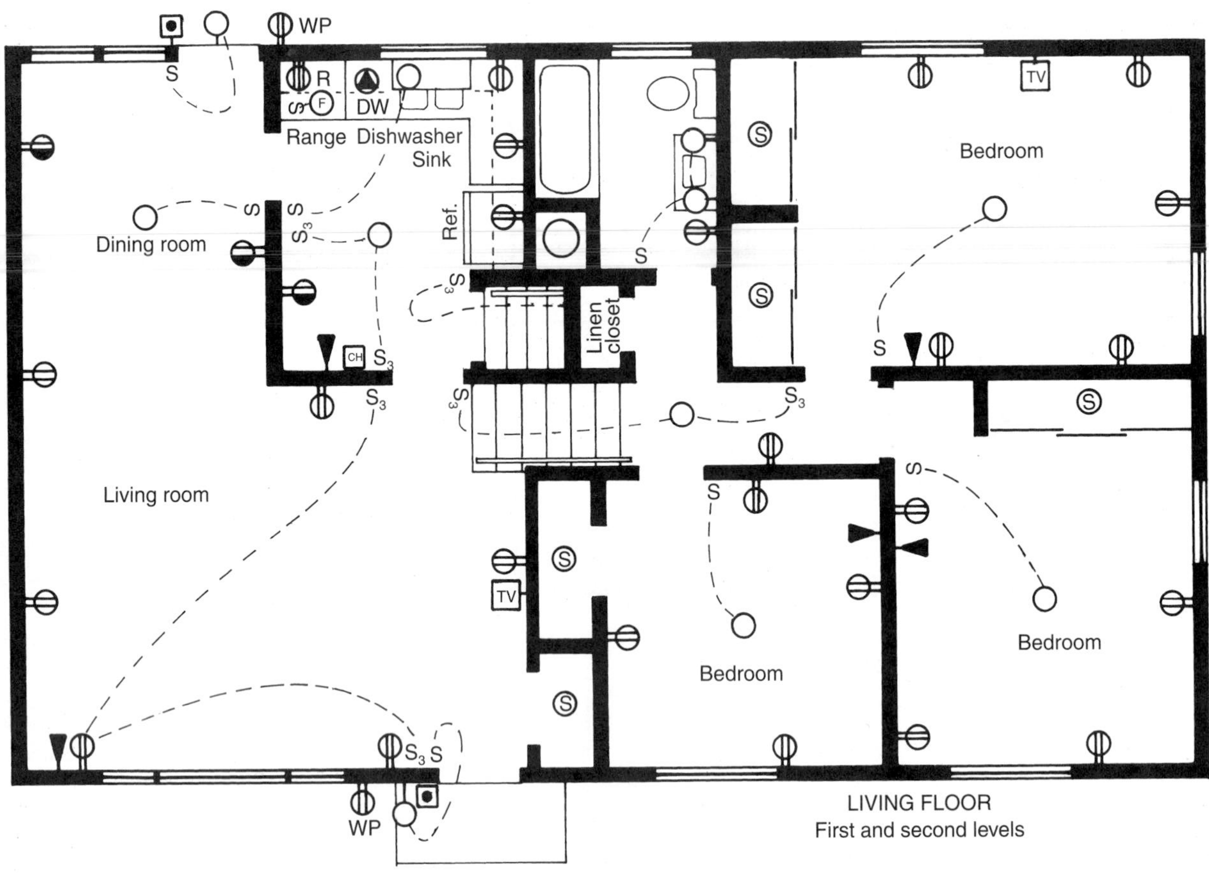

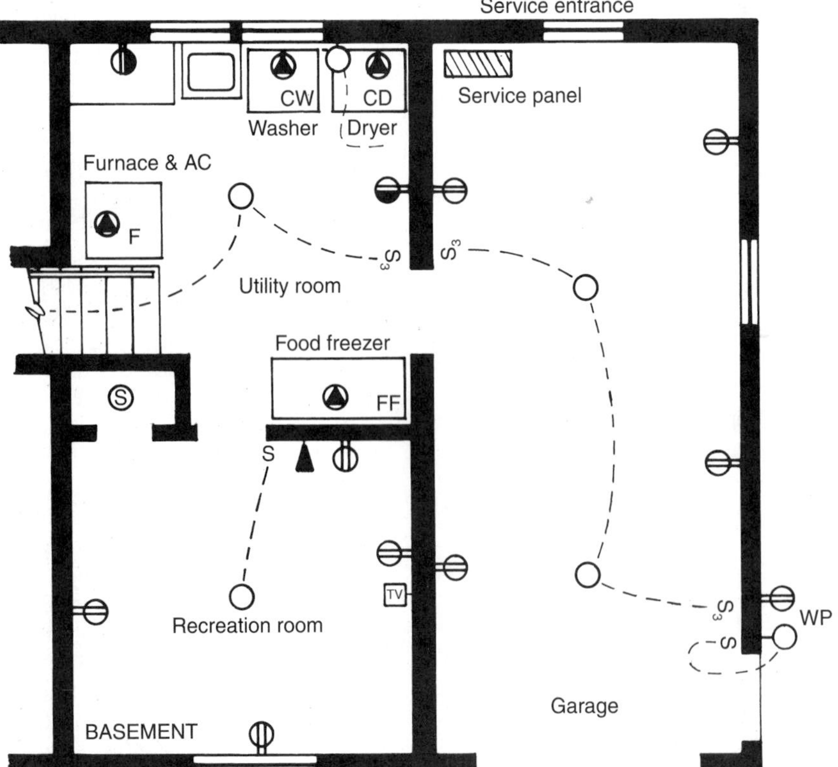

FIGURE 16.3-4 A second distribution panel is not needed here, because the service equipment is centrally located with respect to cooking, laundry, and heating equipment. (Drawing by HLS.)

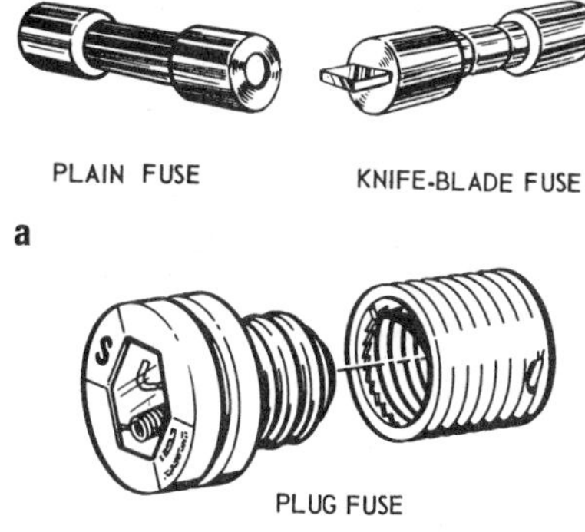

FIGURE 16.3-5 Fuses are either (a) cartridge type (ferrule type at left, knife-blade type at right) or (b) or plug type. A type S adapter makes it impossible to insert a 30-amp fuse in a 20-amp socket, thus avoiding dangerous overloads.

CIRCUIT BREAKERS

Circuit breakers are simple switchlike devices that automatically open a circuit when the rated current is exceeded, such as would result from a short circuit. Unlike plug fuses and most cartridge fuses, circuit breakers are not damaged when they trip (open). To reset (close) the breaker, it is simply necessary to push the toggle into the "on" position. Most small circuit breakers are of the thermal-magnetic type. When such a breaker is tripped, a thermal element in it is heated. It may be necessary, therefore, to wait about a minute for the element to cool before the breaker can be reset.

Circuit breakers are available in a wide range of sizes for many applications. Small breakers for 15-, 20-, 30-, 40-, 60-, and even 100-amp branch circuits are used today in service entrance equipment for residential and commercial buildings. All of these can be combined in a single service panel, along with a larger breaker with a rating equal to the combined ratings of all the circuits to serve as a main switch.

GROUND FAULT CIRCUIT INTERRUPTER (GFCI)

A GFCI is similar to a fuse or circuit breaker in that it stops the flow of electricity when an abnormal electrical situation develops. Unlike circuit breakers and fuses, which keep large current flows from overloading a circuit, a GFCI will be tripped by even very low current flows. GFCIs are usually set to open a circuit when the ground fault current exceeds 5⁄1000 of an ampere. Refer to Section 16.3.4.2 for an indication of locations where GFCIs are appropriate.

POLARIZED PLUGS

Modern portable appliances and light fixtures have polarized plugs with blades of two different sizes so that they can be inserted into an outlet in only one direction (Fig. 16.3-6).

In a lamp with a nonpolarized plug there is a 50% chance that the threaded portion of the bulb socket is live and can give someone a shock. The reason is that one wire in a lamp cord is connected to the contact that touches the bottom of the bulb and another is connected to the lamp socket. One slot in a receptacle is always connected to a hot wire, and the other slot is connected to a neutral wire. When a plug is inserted in the correct orientation, the wire that is connected to a lamp's socket will be plugged into the neutral slot. If the plug is reversed, the socket will be connected to the hot wire. The lamp will work regardless of which way the cord is plugged in. Therefore, there is no way to know whether the socket is hot or safe. Since a polarized plug can be inserted only with the hot connected to the hot and the neutral connected to the neutral, its use will prevent a shock when the socket is touched. However, polarized plugs can prevent shocks only when they are plugged into outlets wired with the neutral wire on the neutral contact and the hot wire on the hot contact.

Modern three-hole grounded receptacles (see Section 16.3.3.4) are made with a large and a small slot to accept a polarized plug, while most old-fashioned nongrounded units are made with two slots of the smaller size and will not accept a polarized plug. Grinding down a new plug to fit an old receptacle is particularly dangerous, because the plug now looks like a modern safe plug but presents the same chance of shock as an old-fashioned nongrounded plug.

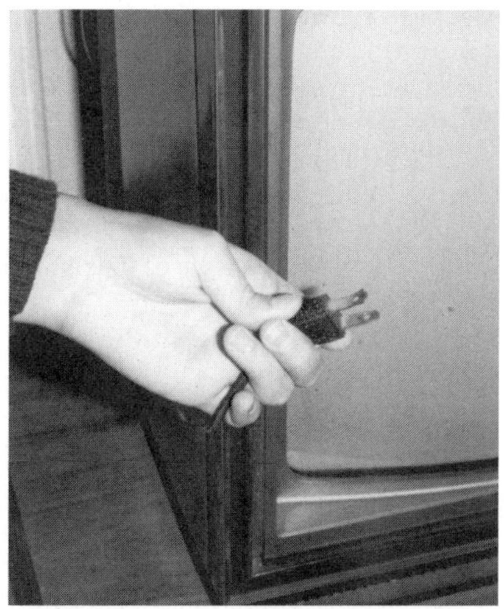

FIGURE 16.3-6 A polarized plug can prevent shocks when it is plugged into a correctly wired outlet.

PLUG COVERS

Many electrical accidents occur when children stick small metal objects into electrical outlets. Simple plastic plug covers fit into unused outlets and serve to prevent mishaps (Fig. 16.3-7).

COPPER/ALUMINUM (CO/ALR) DEVICES

Due to the high cost of copper, aluminum wire was used in many residential installations during the mid-1960s. The potential problems associated with using aluminum wire were not realized until the 1970s.

The basic problem is that aluminum expands when warm and contracts when cold at a greater rate than copper. Over time, repeated expansion and contraction can cause connections in aluminum wire to loosen. In addition, aluminum forms a film of aluminum oxide when exposed to air. This oxide is an insulator rather than a conductor of electricity. The combination of loose wiring and oxide buildup can cause connections to overheat. These

FIGURE 16.3-7 Plug covers not only prevent children from sticking objects into receptacles, they also reduce air infiltration through outlets in exterior walls.

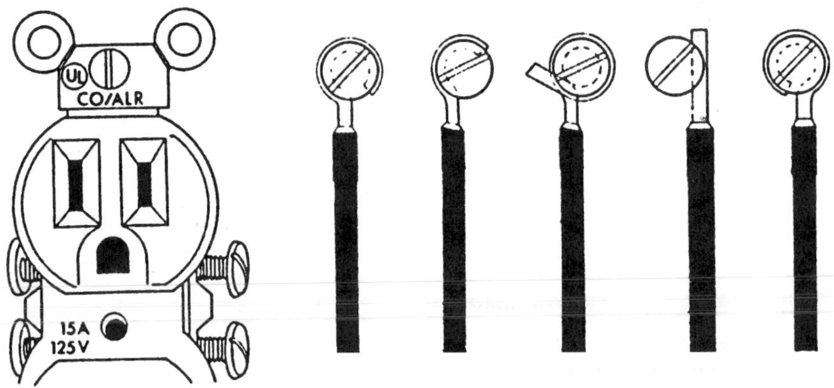

FIGURE 16.3-8 CO/ALR devices have special brass screws with wide heads. Wire should be stripped carefully to avoid nicks and wrapped around a terminal as shown in the drawing of the far left wire. The other wires are incorrectly attached.

problems became apparent when home owners began to report flickering lights, radio and TV static, unpleasant odors, and occasional fires.

In 1972, copper/aluminum (CO/ALR) devices suitable for use with both copper and aluminum wiring came into use. CO/ALR devices have special brass screws with heads wide enough to prevent aluminum wire from working loose. Proper installation is shown in Figure 16.3-8. Only CO/ALR devices should be used with aluminum wire; other types are incompatible.

According to the U.S. Consumer Product Safety Commission (CSPC), about 1.5 million homes were wired with incompatible receptacles between 1965 and 1973, which CSPC believes to be fire hazards.

Owners of dwellings built or rewired during the 1960s can check for aluminum wiring in unfinished spaces such as the basement or attic; the letters *AL* or the word *aluminum* is stamped on the insulation covering the wire.

Signs of trouble for dwellings wired with aluminum include:

- Cover plates or switches that are warm to the touch
- Smoke or sparks at switches or outlets
- Odors, such as the smell of burning plastic
- Lights that flicker periodically
- Receptacles or entire circuits that do not work

When aluminum wiring is discovered, the old switches and receptacles can be replaced with CO/ALR devices. Unfortunately, this replacement will leave other parts of the circuit, such as fixtures and junction boxes, as potential trouble sources.

PIGTAILING

A broader solution to the aluminum/copper mismatch involves attaching short lengths of copper wire into existing aluminum wire with a special crimping tool. These copper-wire pigtails are then connected to the terminals of the electrical device. While CO/ALR devices may be suitable for new installations using aluminum wiring, pigtailing seems more appropriate for retrofitting existing installations.

16.3.3.3 Conductors and Raceways

Electric current is carried and distributed by *conductors*. Technically, conductors are the metallic cores that conduct electricity. In practice, the entire wire, including the insulating cover, is called a conductor. Conductors are protected from mechanical damage by an outer sheath. When a conductor and its protective sheath are fabricated together, the assembly is called a *cable*. Conductors can also be inserted on the job into raceways, which are specially built channels into which wires are placed or pulled.

CONDUCTORS

Conductors consist of an aluminum or copper core, covered by an insulating jacket and an outer coating suitable for the anticipated temperature and moisture conditions.

Conductors for building wiring are available in American Wire Gauge (AWG) sizes, ranging from No. 14 to No. 4/0, and in larger sizes from 250 MCM (thousand circular mils) to 2000 MCM (Fig. 16.3-9). Wiring systems should be designed so that the maximum current carried by a conductor will not exceed the amperages shown in Figure 16.3-9. Copper and aluminum are the typical conductor materials; aluminum is generally limited to larger wire sizes. Due to its lower conductivity, aluminum wire is not made in sizes smaller than No. 12 AWG (rated 15 amps). The same types of insulation are used for both copper and aluminum conductors.

Since its insulation determines the suitability of a conductor to certain exposures, such as moisture, heat, or mechanical damage, conductors are identified by the material and properties of the insulating cover, and sometimes by use. The following are a few of the many designations in the NEC for conductors:

FEP = fluorinated ethylene propylene
H = heat resistant (for use up to 167°F (75°C) maximum temperature)
HH = higher heat resistance (up to 194°F (90°C))
MI = mineral insulation (metal sheathed)
R = rubber
T = thermoplastic
UF = underground feeder
USE = underground service entrance
W = moisture resistance

In addition to the basic letter designations derived from the materials (such as R for rubber), other letters are added to indicate modified or additional characteristics. For example, when an H is added to the basic R designation, the resulting type RH wire may be used in higher temperature conditions. Adding a W to designate an RHW conductor adds a requirement that the conductor also be moisture resistant.

Insulation thickness varies with wire size and may also vary with the type of material and application and the code limitation. The NEC contains tables showing the requirements for each conductor type by its letter designation (MTW, RH, RHH, etc.). These tables also spell out the maximum temperature range for each conductor type and list its required insulation thickness based on wire size. For example, under circumstances of normal use, an RHW (moisture- and heat-resistant rubber) conductor can be used up to 167°F (75°C) operating temperature in dry or wet locations. For voltages up to 600, the following insulation thicknesses apply:

AWG	*Insulation Thickness in mils (mm)*
14–10	45 (1.14)
8–2	60 (1.52)
1–4/0	80 (2.03)
213–500	90 (2.41)
501–1000	110 (2.79)
1001–2000	125 (3.18)

For voltages of about 600, thicker insulation is required.

CABLES

Cables are combinations of two or more conductors, factory encased in a protective covering to safeguard them from mechanical damage. They can have metallic coverings as in armored cable, or plastic or cambric (varnished cotton) jackets as in nonmetallic sheathed cable.

Armored Cable Armored cable, Type AC, which is also sometimes called BX, is a factory assembly of insulated conductors inside a flexible metallic covering intended primarily for use in branch circuits and feeders. It may be run in the voids of masonry block or tile walls when such walls are not exposed to excessive moisture. It should not be run below grade; a transition should be made from armored cable to metal conduit or tubing when supplying a basement appliance.

Armored cable should be securely fastened by approved sleeves, staples, or straps, spaced not more than 4 ft. 6 in. (1372 mm) apart. The cable should be supported within 12 in. (305 mm) of an outlet box or other termination. When the box is mounted on equipment that has to be moved for servicing occasionally, the unsupported cable length can be 24 in. (610 mm).

Armored cable should be cut away to allow at least 6 in. (152 mm) of the conductors into the box. The remaining armored portion should be clamped to the box. When the armor is cut back to provide free conductors, care should be exercised to prevent damaging the conductor insulation. Also, a fiber insulating bushing should be inserted in the end, between the armor and the conductor insulation. This bushing should remain visible within the box.

Armored cable must be handled and bent with care. If it is bent too sharply, the convolutions of the armor can damage the conductor insulation, resulting in possible short circuits or leakage to ground.

Metal-Clad Cable Metal-clad cable, classified as MC, is similar in construction to armored cable but is intended for use in services, feeders, and branch circuits. It is permitted in systems in excess of 600 volts.

Nonmetallic Sheathed Cable This cable, which is also sometimes called Romex, consists of two or more insulated conductors having an outer sheath of moisture-resistant, nonmetallic material (see Fig. 16.3-10). The conductor insulation is rubber, neoprene, thermoplastic, or a moisture-resistant, flame-retardant fibrous material.

Nonmetallic sheathed cable should be securely fastened by approved staples, straps, or other supports, in such a manner that the cable will not be damaged. The supports must not be more than 4 ft. 6 in. (1372 mm) apart. Generally, this cable must be fastened within 12 in. (305 mm) of each termination. In concealed spaces of existing or prefabricated buildings where such support is impractical, this cable may be fished between points of access. When run at an angle to joists, this cable should either be supported on a run-

FIGURE 16.3-9 Properties of Commonly Used Copper Conductors

Conductor			Maximum Current, ampere[a]	
Size: AWG, MCM[b]	One-Half Natural Size	Diameter, in. (mm)[c]	Conductor Type: RUW, T, TW	Conductor Type: RH, RUH, THW, RHW, THWN
14	●	.0641 (1.63)	15	15
12	●	.0808 (2.05)	20	20
10	●	.1019 (2.59)	30	30
8	●	.1285 (3.26)	40	45
6	●	.184 (4.67)	55	65
4	●	.232 (5.97)	70	85
2	●	.292 (7.42)	95	115
1	●	.332 (8.43)	110	130
0	●	.372 (9.4)	125	150
00	●	.418 (10.62)	145	175
000	●	.470 (11.94)	165	200
0000	●	.528 (13.41)	195	230
350 MCM[d]	●	.681 (17.3)	260	310

[a]Based on ambient temperature of 86°F, (30°C); not more than 3 conductors in raceway, cable, or direct burial. Where number of conductors exceeds 3, maximum current given should be derated as follows: 4–6 cond., 80%; 7–24 cond., 70%; 25–42 cond., 60%; 43 or more, 50%.

[b]Sizes 14 to 8 are solid wires. Sizes 6 to 2 are 7-strand cables. Sizes 1 to 0000 are 19-strand cables. Size 350 MCM is a 37-strand cable.

[c]Exclusive of insulation.

[d]MCM is the designation of wire size in thousands of circular mils (350 MCM = 350,000 circular mils).

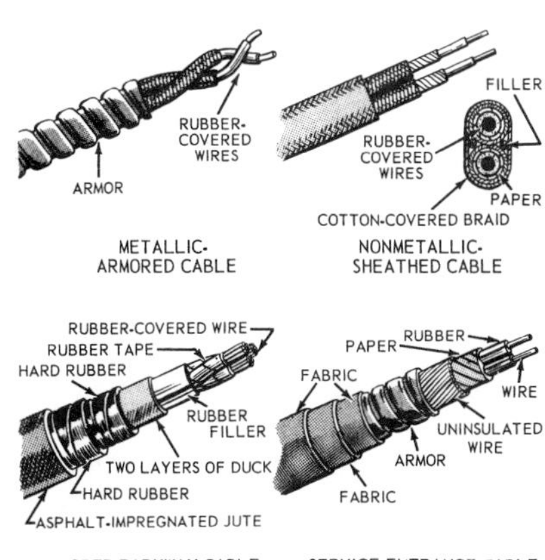

FIGURE 16.3-10 Some common types of electrical cable.

ning board secured to the joists or run through holes bored through the joists.

Care should be taken not to damage the outer sheath of nonmetallic cable during installation. No bend should have a radius less than 5 times the diameter of the cable.

There are two types of nonmetallic sheathed cable: type NM and type NMC.

Type NM cable has conductors that satisfy the requirements for types RH, RW, T, and TW wire, and a sheath that is flame retardant and moisture resistant. Type NM is limited to use in dry locations such as air voids in masonry or tile walls not exposed to excessive moisture.

Type NMC conductors must satisfy the same requirements as type NM, but the sheath must be not only flame retardant and moisture resistant, but also fungus resistant and corrosion resistant. Type NMC may be used in damp or corrosive locations, as well as in dry areas.

RACEWAYS

Raceways are protective enclosures designed for field insertion of conductors. Included in the general term are electrical metallic tubing (EMT), rigid and flexible metal conduit, wireways, cable trays, and underfloor ducts.

Raceways should be securely connected and rigidly supported. Cut ends should be reamed (smoothed) to remove rough edges. Combination couplings should be used to make transitions from flexible conduit to electrical metallic tubing or rigid conduit.

Electrical Metallic Tubing (EMT) This electrical pipe, which is also called *thin-wall conduit*, may be used for both concealed and exposed work (Fig. 16.3-11). It is the most common type of raceway used in single-family, low-rise residential, and commercial construction. With proper joints and connections it can be installed in wet, as well as dry, locations.

EMT is made in ½ to 4 in. (12.7 to 102 mm) nominal sizes. The actual internal diameter is slightly larger than the nominal size. Connectors and couplings for EMT are secured to the tubing by a setscrew or dimples made on the job in the collar of the fitting and tube after assembly. Watertight compression fittings are also available.

EMT is often used in residences and commercial buildings because it is light, easy to handle, and requires no special threading tools. However, bending it,

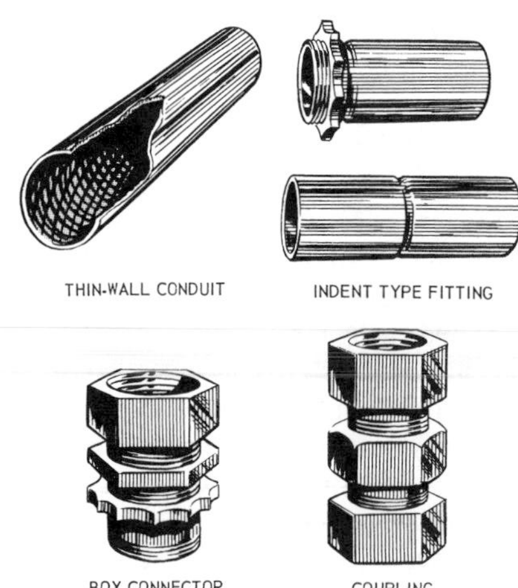

FIGURE 16.3-11 Electrical metallic tubing (EMT), or thinwall conduit, and typical fittings for it.

which should be done with a special type of bender to prevent flattening (Fig. 16.3-12), requires more skill than bending of rigid conduit. The radius of the inner edge of a field bend should not be less than that required in the NEC.

Rigid Metal Conduit This conduit, which is also called *heavywall conduit*, resembles plumbing pipe and is shipped in standard lengths of 10 ft. (3048 mm) (Fig. 16.3-13). A coupling is furnished for one end of each length of rigid conduit. Both ends are reamed and threaded when shipped. If it is cut during installation, the new ends must be reamed and threaded.

Rigid conduit can be joined by screwing it directly into threaded hubs, by locknuts and bushings (see Fig. 16.3-13), or by threadless connectors. When the conduit is connected with a locknut and bushing, the locknut is screwed onto the pipe, the end is inserted through the knockout in a box, and the bushing is screwed on. When the bushing is seated, the outer locknut is screwed tightly against the outer surface

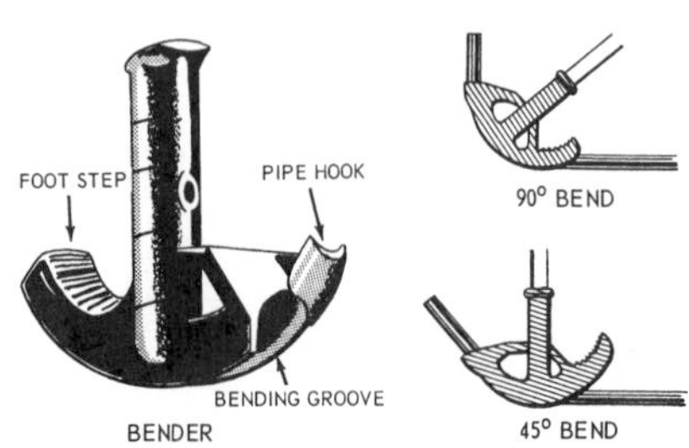

FIGURE 16.3-12 An EMT bender ensures smooth curves, without pinching the interior of a pipe.

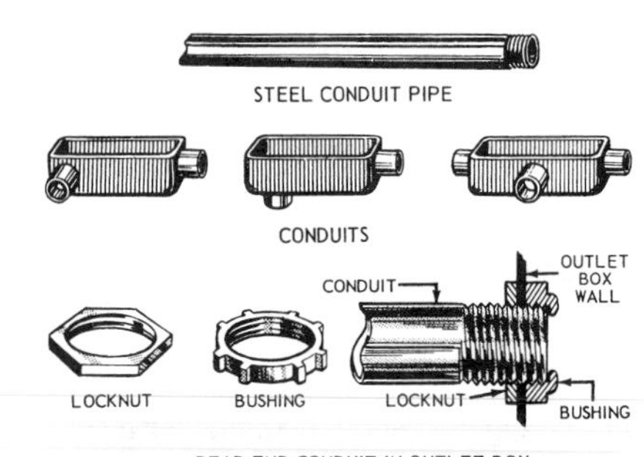

FIGURE 16.3-13 Rigid, or heavywall, conduit, and typical conduits and fittings for it.

of the box. When a conduit bushing is made entirely of insulating material, a locknut must be installed on both sides of the enclosure.

Rigid metal conduit is made in trade sizes of ½ to 6 in. (12.7 to 152 mm). The 6-in. (152 mm) size will accommodate three 2000 MCM (1.63-in. [41.3 mm] diameter) conductors. When greater capacity is required, a bus-type conductor (solid bar) in a bus duct is generally used.

Flexible Metal Conduit Flexible metal conduit, which is also called Greenfield, is similar in appearance to armored cable but does not contain preinserted conductors (Fig. 16.3-14). It cannot be used where it will be subject to corrosive gases or excessive moisture. For the most part, it must conform to the same regulations as rigid conduit with respect to rewiring existing conduit, reaming cut edges, and other code requirements.

Flexible conduit connectors are fastened to the conduit by setscrews or clamps, and to the box by an expanding mechanism, a locknut, or a threaded hub.

Flexible conduit should be installed taut and then fastened securely to facilitate pulling conductors with a fish tape. It should be supported at intervals not greater

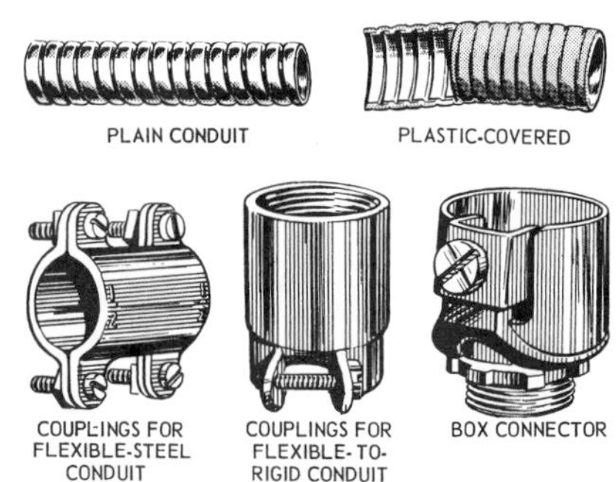

FIGURE 16.3-14 Two types of flexible metal conduit and typical fittings for them.

than 4 ft. 6 in. (1372 mm) and within 12 in. (305 mm) of a box. The radius of bends should not exceed 5 times the conduit diameter.

Nonmetallic Conduit This conduit is made of nonmetallic material that is resistant to moisture and chemical atmospheres. It is flame retardant, resistant to crushing, and resistant to distortion from heat, low temperatures, and sunlight. It can be used underground in floors, walls, and ceilings.

Wireways A *wireway* is a metal trough with a hinged or removable cover, in which conductors are placed after the wireway has been installed. Because they must be accessible throughout their entire length, wireways can be used only for exposed work.

Splices can be made and laid into a wireway if the total conductor and splice cross-sectional area does not exceed 75% of the area of the wireway and is accessible. If the wireway is not fully accessible at the splice location, then the percentage of fill should not exceed 20% of the cross-sectional area of the wireway.

Cable Trays Cable trays are a variety of wireways that are used above ceilings in which the wiring is left exposed. They are used extensively in commercial and institutional buildings, such as hospitals, where a variety of systems are required. Fiber-optic and other communications cabling is often laid in cable trays. Essentially, cable trays are wire or pressed metal troughs with open tops.

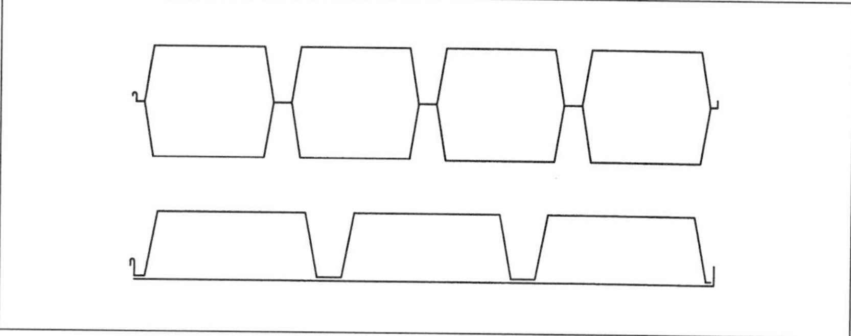

FIGURE 16.3-16 Cellular metal raceways. These raceways replace and serve instead of standard metal deck. Wiring is run through the cells. (Drawing by HLS.)

Underfloor Raceway There are a variety of underfloor duct systems in general use. They are used extensively in commercial and institutional buildings where the framing is steel with bar joists. Essentially, they consist of a metal or fiber header duct that runs parallel to the framing and metal or fiber raceways that cross the steel deck (Fig. 16.3-15). The header duct may be a closed metal or fiber duct or a cellular metal raceway that forms a portion of the steel deck itself (Fig. 16.3-16). A raceway may be a closed duct designed to be covered with concrete (Fig. 16.3-15) or a trench duct with a removable cover (Fig. 16.3-17). The thickness of concrete cover and the circumstances under which a trench duct may be used are dictated by the NEC.

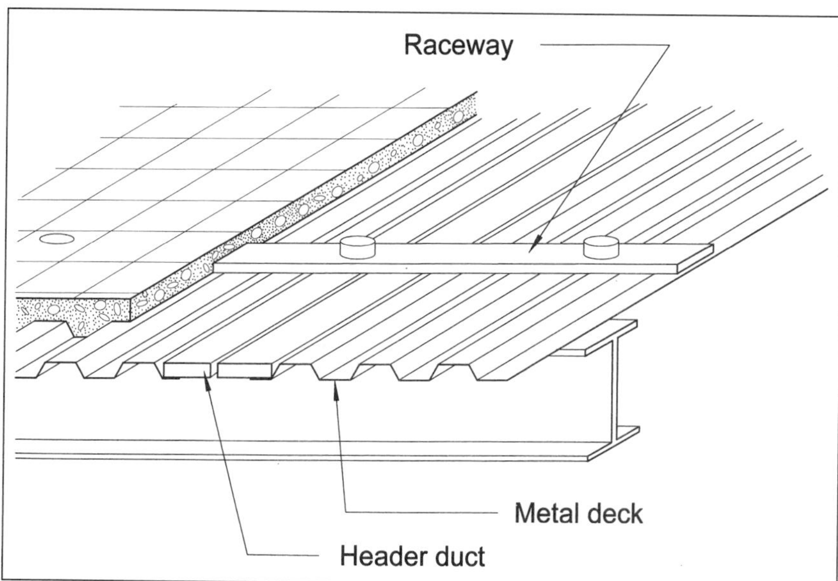

FIGURE 16.3-15 A typical underfloor raceway. (Drawing by HLS.)

16.3.3.4 Outlets and Switches

There are three types of outlets: lighting, convenience, and special purpose.

LIGHTING OUTLETS

A lighting outlet is a metal protective box to which fixed light fixtures, as opposed to portable lights, can be mounted and in which they may be connected to a branch circuit (see Section 16.3.5.1). Lighting outlets must be compatible with the lights selected (see Section 16.4). Figure 16.3-18a shows a standard octagonal protective box of a type that is often used as a lighting outlet. Such a box may also serve as a junction box for pulling wires to remote fixtures in coves, valances, and cornices. Circular and square lighting outlet boxes are also used.

CONVENIENCE OUTLETS

There are two kinds of convenience outlets, standard and multioutlet.

Standard Convenience Outlets A standard convenience outlet is a plug-in receptacle housed in a protective box with plug-in positions for attachment of portable fixtures or minor appliances. The most common type, *duplex*, has two plug-in positions (Fig. 16.3-18b); *triplex* has three positions; a single has only one.

Outlets are sometimes grouped in a single protective box behind a single cover plate. Common groupings are an outlet and a switch, two or three outlets and a switch, and two outlets and two switches.

The NEC requires that outlets on a 15- or 20-amp general-purpose branch circuit should be of the grounding type to minimize the hazard of shock from short circuits (see Section 16.3.3.2). A grounding receptacle is a convenience outlet designed to receive a three-pronged plug.

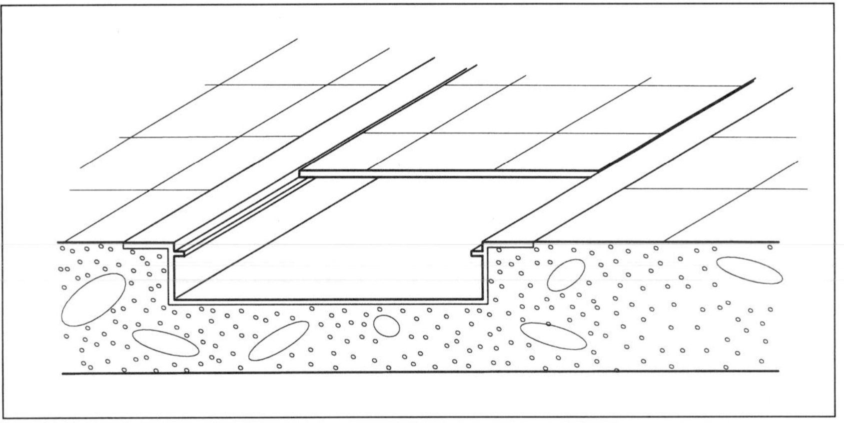

FIGURE 16.3-17 **A typical trench duct. Cover plates are removable for access to wiring. In some ducts of this type the flooring can be run completely across the duct, effectively concealing it from view.** (Drawing by HLS.)

a

b

c

FIGURE 16.3-18 **(a) Box for lighting outlet and junction box use, (b) standard duplex receptacle, (c) special-purpose receptacle.**

The third prong permits an appliance or piece of equipment to be grounded, thus minimizing the shock hazard from faulty wiring.

Regardless of the number of positions, an outlet is considered a single outlet when only one connection has been made to a branch circuit. Other arrangements are sometimes used, however. For example, a duplex receptacle that has been wired so that each plug-in position is on a separate 120-volt circuit or that uses two separate line wires of a single three-wire 240-volt circuit (Figs. 16.3-19 and 16.3-20) is called a *split-receptacle* outlet. Typical uses of such outlets are discussed in Section 16.3.4.

Multioutlet Assembly A multioutlet assembly is a series of plug-in receptacles, usually spaced at regular intervals and contained in a protective raceway (channel). Such assemblies also are called plug-in strips. They may be used in most locations where a standard convenience outlet is needed, especially where several outlets are called for.

SPECIAL-PURPOSE OUTLETS

A special-purpose outlet is a point of connection to the wiring system for a particular piece of equipment, normally reserved for the exclusive use of that equipment. Such outlets may be plug-in receptacles to which the equipment is connected with a flexible cord, as with clothes dryers or ranges (Fig. 16.3-18c), or they may be protective enclosures, called junction boxes, with removable covers containing permanent connections; such junction boxes can be mounted either on the structure or on the equipment.

SWITCHES

A switch is a device for interrupting the flow of electricity. Switches are used in service entrance and distribution panels and in branch circuits. The switches in a panel are either a part of the panel or are in separate metal boxes. Switches in branch circuits are connected to lighting, convenience, and special-purpose outlets.

Switches in branch circuits are mounted in the same kind of protective boxes used for standard convenience outlets. They may be of the push type, toggle type, or slide type. Variable rotary switches are sometimes used to control the power to lights and ceiling fans.

16.3.4 GENERAL RECOMMENDATIONS FOR OUTLET LOCATIONS

16.3.4.1 Commercial, Institutional, and Industrial Buildings

The location of outlets in commercial, institutional, and industrial buildings is dictated by code requirements and the design of the facility. Electrical systems in these kinds of structures must be designed in accordance with sound engineering practices, with outlets and power sufficient to meet the needs of the occupants. The NEC and applicable building codes must also be followed, of course.

16.3.4.2 Residences

This section contains some general recommendations for outlets in both single-family and multifamily residences. These recommendations are necessarily general in nature. Even when unusual conditions make strict compliance impossible, the recommendations here may still be used as guidelines in developing more appropriate provisions.

CONVENIENCE AND SPECIAL-PURPOSE OUTLET NUMBER AND LOCATION

This section contains recommendations for the number and location of convenience and special-purpose outlets. Similar information about lighting outlets is given in Section 16.4.

When these recommendations are based on the number of linear or square feet, the number of outlets should be determined by dividing the total linear or square footage by the recommended linear or square footage per outlet. Resulting numbers that contain a fraction larger than one-half should be increased to the next larger number. The floor area should be computed from the outside dimensions of the house or the centerline of the walls surrounding each room.

Convenience outlets should be located near the ends of a wall space, rather than near the center, thus reducing the likelihood of their being concealed behind large pieces of furniture. In most cases, outlets should be located about 12 in. (300 mm) above the floor line. In every kitchen, living room, dining room, family room, recreation room, den, library, bedroom, and similar habitable space, convenience outlets should be placed so that no point along the floor line of a usable wall space is more than 6 ft. (1800 mm) from an outlet in the same wall space. Where floor-to-ceiling windows prevent meeting this requirement with ordinary convenience outlets, other suitable means, such as floor outlets or multioutlet assemblies, can be used.

Dual-Purpose Rooms Where a space is intended to serve more than one function, such as a combination living and dining room or a kitchen and laundry, the following outlet recommendations are separately applicable to each specific area. In a living and dining space for instance, convenience outlets in the dining area should be of the split-receptacle type in which each receptacle is on a different circuit, and ordinary outlets may be used in the rest of the space. This permits the use of several portable devices, such as a toaster and a coffee maker, in the dining space without overloading a circuit.

Exterior Entrances A weatherproof convenience outlet, located at least 18 in. (457 mm) above the entrance platform near the front entrance, should be provided. This outlet should be controlled by a wall switch inside the building near the entrance for convenient operation of outdoor decorative lighting. Additional waterproof outlets along the exterior of the house are recommended to serve decorative garden treatments and for the use of appliances or electric garden tools, such as lawn mowers and hedge trimmers.

Living and Recreation Rooms When general illumination is to be provided by portable lamps in lieu of fixed lighting, at least two wall-switched separate plug-in positions should be planned. These can be provided with two switched, regular duplex outlets or one switched, plug-in position in each of two split-receptacle outlets (see Fig. 16.3-19).

When regular convenience outlets are wall-switch controlled to accommodate portable lighting, additional unswitched convenience outlets should provided to prevent limiting the use of portable equipment, such as radios and clocks, in that location. In addition, a single convenience outlet should be installed in combination with the wall switch at one or more of the switch locations for the use of a vacuum cleaner or other portable appliances.

One or more outlets for entertainment equipment, such as a television set, radio, or stereo equipment, should be provided at bookcases, shelves, and other suitable locations. In rooms with conventional fireplaces, an outlet may be desirable in or near the mantel shelf.

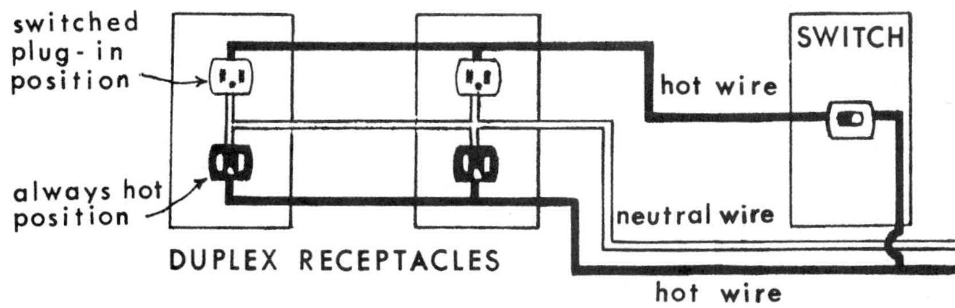

FIGURE 16.3-19 Using two hot wires in a duplex split receptacle permits switching off one receptacle while the other remains hot.

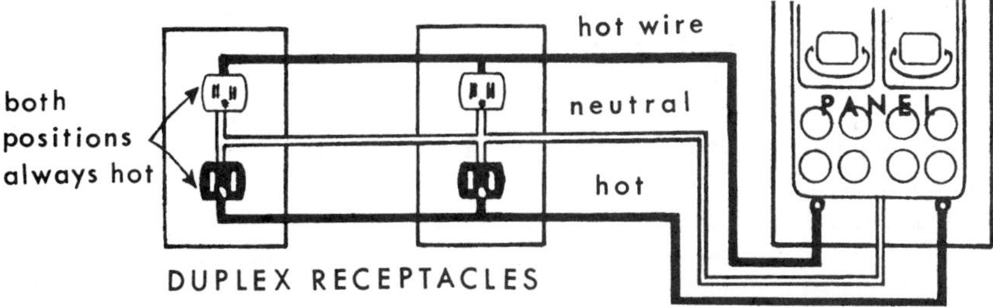

FIGURE 16.3-20 Split receptacles minimize overloads and double the capacity of duplex receptacles by connecting each plug-in position to a separate hot wire. When local codes do not permit this type of installation, and both plugs on a receptacle must be wired to the same circuit, alternate receptacles can be wired to different circuits.

Dining Areas When a dining or breakfast table will be placed against a wall, one of the recommended outlets should be located in that wall, just above table height.

Where open counter space is to be built in, an outlet should be provided above counter height for the use of portable appliances. All convenience outlets in the dining area should be of the split-receptacle type, providing increased capacity at each outlet (see Fig. 16.3-20).

Bedrooms Convenience outlets should be placed so that there is one on each side and within 4 ft. (1220 mm) of the center line of each probable individual bed location. The popularity of bedside radios and clocks, bed lamps, and electric blankets makes increased plug-in positions essential at bed locations. Triplex or quadruplex convenience outlets are recommended at these locations (Fig. 16.3-21c).

Additional outlets should be placed so that no point along the floor line of a wall is more than 6 ft. (1800 mm) from an outlet in that space. It is also recommended that a receptacle be provided at at least one switch location in a bedroom to accommodate the use of a vacuum cleaner, floor polisher, or other portable appliance (see Fig. 16.3-21b).

Bathrooms and Lavatories A convenience outlet should be provided near the mirror, 3 to 5 ft. (900 to 1500 mm) above the floor. An outlet should be installed at each separate mirror or vanity space and at any space that may accommodate an electric towel dryer, razor, hair dryer, or other small electrical equipment. A receptacle that is part of a bathroom lighting fixture is not suitable for this purpose unless it is rated at 15 amps, wired with at least 15-amp-rated wires, and easily accessible. A ground fault circuit interrupter (GFCI) is required by the NEC in every such condition (see Section 16.3.3.2).

A wall-switched, or timer-operated, built-in ventilating fan capable of providing a minimum of eight air changes per hour should be installed in every bathroom. Such a fan is generally required by code where natural ventilation is not provided through windows or skylights. A fan capable of moving 50 to 55 cu. ft. (1.42 to 1.56 m^3) of air per minute (cu. ft./min.) (m^3/min.) will ensure eight air

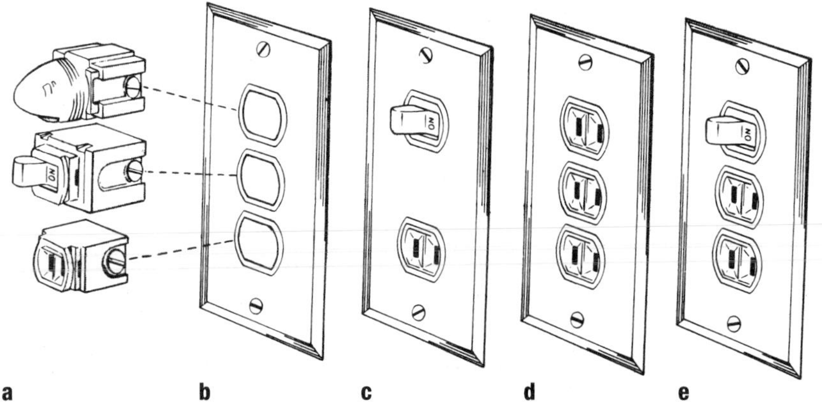

FIGURE 16.3-21 Interchangeable devices can be installed in the same outlet box in various combinations: (a and b) a pilot light, switch, and a receptacle in the same box; (c) a switch and a receptacle; (d) a triplex receptacle; and (e) a switch and a duplex receptacle.

changes per hour in a 5 × 10 ft. (1500 × 3000 mm) bathroom.

In many parts of the country, a wall-switch-controlled electric space heater is desirable for increased comfort during bathing. It may be a part of a ceiling light, fan, heater combination, or a separate unit. Heaters may be of the wall- or ceiling-mounted type and often have a timer switch.

Kitchen A kitchen should have one convenience outlet for the refrigerator and additional outlets placed so that no point along the wall line is more than 24 in. (610 mm) from a convenience outlet, with at least one outlet to serve each separate work surface or planning desk. An outlet should be placed in every wall space that may be used for ironing and at every location that might be used for an electric roaster, microwave oven, toaster oven, can opener, popcorn popper, ice cream maker, wok, or kettle. If a work table or dining table is planned next to a wall, this wall space should have one outlet located just above table level. All kitchen outlets should be protected by a GFCI.

Work-surface outlets should be located about 44 in. (1118 mm) above the finish floor level or about 8 in. (203 mm) above the work surface. Convenience outlets in a kitchen, other than that for a refrigerator, should be of the split-receptacle type (see Fig. 16.3-20).

Possible special-purpose outlets required in a kitchen include those for a range, oven, burner top, ventilating fan, dishwasher, and waste disposer. A special recessed receptacle with a hanger should be installed for an electric clock. A clock should be easily visible from all parts of the kitchen. It is also necessary to provide for a food freezer either in the kitchen or in some other convenient location.

Laundry Area At least one convenience outlet should be provided, preferably near the sorting table if there is one. If properly located, one of the special-purpose outlets (the washer outlet, for example) may also serve as a convenience receptacle for such items as a laundry hot plate or sewing machine. Convenience outlets in the laundry area should be of the split-receptacle type (see Fig. 16.3-20).

A separate special-purpose outlet is generally needed for each of the following pieces of equipment: automatic clothes washer, hand iron or ironer, and clothes dryer. Outlets for a ventilating fan and clock are desirable. Requirements for an electric water heater should be obtained from the local power company.

Halls These recommendations apply to passage halls, reception halls, vestibules, entries, foyers, and similar areas. One convenience outlet should be installed for each 10 linear ft. (3048 mm) of hall, measured along the center line. A hall with more than 25 sq. ft. (2.32 m^2) of floor area should have at least one outlet. In reception halls and foyers, convenience outlets should be placed so that no point along the floor line of a wall is more than 10 ft. (3048 mm) from an outlet in that space.

A receptacle at one of the switch outlets is recommended for convenient connection of maintenance equipment such as a vacuum cleaner, floor polisher, or electric broom.

Stairways Convenience outlets generally are not required, except at intermediate landings of larger-than-normal size, where a decorative lamp, night-light, or portable equipment may be used.

Utility Room At least one convenience outlet should be provided, preferably near the furnace or near a planned workbench location.

A special-purpose outlet is required for each piece of mechanical equipment requiring electrical connection. Such equipment includes a furnace, boiler, water heater, water pump, sump pump, and compressor.

Basement At least two convenience outlets should be provided in an open basement area. If a workbench is planned, one of these outlets should be placed at its specific location.

Additional convenience outlets may be needed for a basement laundry, dark room, or hobby area, and for appliances such as a dehumidifier or portable space heater.

A special-purpose outlet must be provided for each piece of mechanical equipment requiring electrical connections, such as a furnace, boiler, water heater, water pump, sump pump, and compressor.

An outlet for a food freezer also should be considered.

Accessible Attic A convenience outlet is desirable for portable equipment and for trouble lights capable of reaching into distant dark corners. If an open stairway leads to future attic rooms, provide a junction box with direct connection to the distribution panel, for future extension to convenience outlets and lights when the rooms are finished.

A summer cooling fan, multiple-switch controlled from convenient points throughout the house, is often desirable.

Porches and Breezeways At least one GFCI convenience outlet should be provided for each 15 linear ft. (4500 mm) of wall bordering a porch or breezeway. These outlets should be of the weather-proof type if exposed to the elements. They should be controlled by a wall switch inside the house. Since such areas are often used for outdoor dining, at least one outlet should be of the split-receptacle type.

Terraces and Patios At least one weatherproof GFCI outlet should be provided for each 15 linear ft. (4500 mm) of house bordering a terrace or patio. These outlets should be wall-switch controlled from inside the house and located at least 18 in. (457 mm) above patio level.

Garage or Carport At least one GFCI outlet should be provided for one- and two-car garages and carports.

Special-purpose outlets appropriate to the use should be provided when a food freezer, workbench, or automatic door opener is planned. Specific hobbies, such as restoring antique automobiles or ceramic sculpture, may require special-purpose outlets for compressors, welding rigs, or kilns.

16.3.5 WIRING RECOMMENDATIONS

The recommendations given in this section must be supplemented with engineering judgment as to electrical system adequacy in common-use spaces and for general building services. Particular attention must be paid to limiting voltage drop in feeders in order to ensure proper operation of appliances and other devices.

16.3.5.1 Circuit Recommendations

In addition to branch circuits providing power for lights, appliances, and fixed and portable equipment, a system may include feeder circuits serving secondary distribution panels near heavy-use locations.

BRANCH CIRCUITS

A branch circuit is that portion of a wiring system between an overcurrent device and outlets, where loads, such as light fixtures, appliances, or equipment, may be connected. Wiring systems usually include three types of branch circuits: general purpose, small appliance, and special purpose. Each circuit involves consideration as to voltage, amperage, and number and size of conductor wires.

General-Purpose Circuits General-purpose circuits supply lighting and convenience outlets. They should be wired with suitable conductors as required by code, but not smaller than No. 14 AWG (American Wire Gauge) if protected by 15-amp overcurrent device, or No. 12 AWG for 20-amp device. Since No. 14 conductors are more limited in circuit length (30 ft. [9144 mm]) and current-carrying capacity (1725 watts), No. 12 conductors with 20-amp devices are recommended for most general-purpose circuits. Some building codes do not permit the smaller No. 14 wires.

The number of general-purpose outlets required is based on the loads to be encountered and the loads permitted per circuit, both as determined in accordance with the NEC, and the branch circuit amperage to be used. In residences, no less than the following number of general purpose outlets should be provided:

- At least two 20-amp circuits for the first 750 sq. ft. (69.68 m^2) of floor area, plus one circuit for each additional 750 sq. ft. (69.68 m^2) or major fraction thereof, or
- At least two 15-amp circuits for the first 500 sq. ft. (46.45 m^2), plus one circuit for each additional 500 sq. ft. (46.45 m^2) or major fraction thereof.

These floor areas should be calculated from dimensions taken to the outside of exterior walls, including unfinished spaces that are adaptable to future living use; open porches and garages may be excluded from the calculation.

In commercial, institutional, and industrial buildings, the number of general-purpose outlets is determined in accordance with the NEC and the requirements of the facility. The NEC has requirements for health care facilities, hazardous facilities, mobile homes, outdoor areas, show windows, swimming pools, theaters, and other building use types. It also dictates the amperage ratings required in specific locations. Some building designs may require more general-purpose outlets than the NEC requires.

Small Appliance Circuits The actual number and size of small appliance circuits should be determined by the loads to be encountered as calculated in accordance with the NEC. The minimum for residences should be one three-wire 120/240-volt, 20-amp circuit (or two two-wire 120-volt 20-amp circuits), equipped with split-receptacle outlets, in areas regularly intended for cooking, dining, or laundering.

Using three-wire circuits to supply convenience outlets in the locations mentioned is an economical means of increasing the capacity and offers practical operating advantages. Such circuits provide greater capacity at individual outlet locations and lessen voltage drop in the circuit. For maximum effectiveness, the upper half of all receptacles should be connected to the same side of the circuit (see Fig. 16.3-20). A separate 20-amp circuit should be provided for each appliance or piece of equipment rated at more than 1400 watts. However, two separate receptacle circuits may have to be used where codes do not permit split receptacles.

Special-Purpose Circuits Separate circuits should be provided and conductors should be sized appropriately for the specific equipment. Some residential special-purpose circuits and the required conductor sizes are shown in Figure 16.3-22. Conductor sizes for special-purpose outlets and connection points in other types of buildings are determined in accordance with the NEC, based on the load and voltage requirements of the specific equipment being connected.

In residences, spare circuit positions should be provided in each distribution panel for at least two future 20-amp two-wire 120-volt circuits. If the branch circuit or distribution panel is installed in a finished wall, raceways should be extended from the panel to the nearest accessible unfinished space for future use.

In other building types, spare circuits are provided in accordance with the NEC and good design practice for the particular facility. Such requirements may vary in different locations within the same facility. For example, in an office building, where extensive renovation is normal, more spare circuits may be provided than would be required where the use of the facility was expected to remain static.

The majority of portable appliances for residential use are made to operate at 110 to 125 volts. Some more or less fixed appliances, like refrigerators and freezers, also operate in the 110- to 125-volt range. Many fixed appliances, however, operate on 230/250-volt circuits, because of improved operating efficiency. When there is a choice, the higher voltage should be used.

In other building types, both 120-volt and 240-volt devices are common, and some heavy equipment may require 480-volt service.

Most new homes today are air conditioned, and much of the related equipment runs on electricity. Electrical heating systems are not as prevalent today as they were a few years ago, because of their high energy use. However, they are being

FIGURE 16.3-22 Recommended Conductors for Special-Purpose Circuits

Item	Typical Wattage	Recommended Conductor Capacity		
		Amperes	Wires	Volts
Range or electric oven and drop-in cooktop	12,000*	50	3	120/240
	5,000	30	3	120/240
	7,000	40	3	120/240
Automatic washer	700	20	2	120
Electric clothes dryer	5,000*	30	3	120/240
Fuel fired heating equipment	800	20	2	120
Central air conditioning or	5,000*	40	3	120/240
room air conditioner (1 ton)	1,500	20	2	120
Dishwasher and disposer or	2,700	20	3	120/240
dishwasher and waste disposer	1,800	20	2	120
	900	20	2	120
Food freezer	400	20	2	120 or 240
Water pump	500	20	2	120 or 240
Bathroom space heater	2,000	25	2	240
Electric water heater	4,500*	30	3	240

*Typical wattages are shown; verify with manufacturer and/or local power company.

replaced in many instances by electric heat pumps, which use less energy than full electric heating systems but still use considerable electrical energy, especially in colder climates. Electric heating and air conditioning systems and heat pumps are usually custom designed to suit the particular house and local climatic conditions. The equipment used is not custom designed, however. It is selected from a range of standard manufactured products to fit the requirements of each individual house. Adequate wiring and service capacity criteria for such devices should be determined on a case-by-case basis in consultation with the local power company or an electrical engineer.

FEEDER CIRCUITS

Consideration should be given to using subpanels served by appropriately sized feeders to serve heavy-use locations throughout a building. In a residence such areas may be those containing an electric range or clothes dryer.

16.3.5.2 Conductor Size

Conductors should be adequately sized for the design loads and length of run. Figure 16.3-23 lists wire sizes for specific loads, but these should be verified with the latest edition of the NEC. The table gives a direct reading of the maximum distance between distribution panel and load for different conductor sizes that will limit the voltage drop to a maximum of 3%. For example, consider a 3000-watt 120-volt heater that is 75 ft. (22.86 m) from the distribution center. First, watts should be converted to amps: $I = W/E = 3000/120 = 25$ amps (see Section 16.1.2). Checking in Figure 16.3-23 under the 25-amp column heading, one finds that the maximum allowable distance from the source is 138 ft. (42.06 m) for No. 10 AWG wire. This is the maximum distance over which the current will not exceed the permitted 3% voltage drop using that size wire. In selecting a wire size for a known load, the permitted length of run should be greater than the anticipated distance from panel to load.

16.3.5.3 Raceway Size

The number and size of conductors that can be placed in a raceway are limited not only by the size of the interior opening, but also by code restrictions.

The major factor limiting the number of conductors that may be put into race-

FIGURE 16.3-23 Maximum Lengths of Two-Conductor Runs*

Conductor Size, AWG	Maximum Length of Run, ft.																		
	5 amp	10 amp	15 amp	20 amp	25 amp	30 amp	35 amp	40 amp	45 amp	50 amp	55 amp	70 amp	80 amp	95 amp	110 amp	125 amp	145 amp	165 amp	185 amp
14	274	137	91																
12		218	145	109															
10			230	173	138	115													
8					220	182	156	138											
6								219	193	175	159								
4									309	278	253	199							
3										350	319	250	219						
2											402	316	276	232					
1												399	349	294	254				
0												502	439	370	319	280			
00													560	470	403	355	305		
000														588	508	447	385	339	
0000															641	564	486	421	360

*To limit voltage drop to 3% at 120 volt. For 240 volt, multiply length by 2.

ways is *fill*, which is the percentage of the total cross-sectional area of the raceway filled with conductors. The maximum fill is dictated by the NEC. The following percentages are required for non-lead-covered conductors:

1 conductor: 53%
2 conductors: 31%
3 or more conductors: 40%

The following percentages are required for lead-covered conductors:

1 conductor: 53%
2 conductors: 30%
3 conductors: 40%
4 conductors: 38%
more than 4 conductors: 35%

The cross-sectional area of the conductors in a raceway should not exceed the maximum percentage of fill listed here. When using several conductors of unequal size in the same raceway, it is necessary to determine the total cross-sectional area of all the conductors combined, including their insulation, then select the raceway size on the basis of percentage of fill.

The NEC contains extensive tables that list the number of conductors of a given type and size that may be used in each available conduit size so as not to exceed the required maximum fill percentages. For example, it may be possible to install the following number of No. 14 AWG RHW conductors in the listed conduit sizes:

½ in. (12.7 mm): 6
¾ in. (19.1 mm): 10
1 in. (25.4 mm): 16
1¼ in. (31.8 mm): 29
1½ in. (38 mm): 40
2 in. (50.8 mm): 65
2½ in. (63.5 mm): 93
3 in. (76.2 mm): 143

The numbers for No. 12 AWG RHW conductors are:

½ in. (12.7 mm): 4
¾ in. (19.1 mm): 8
1 in. (25.4 mm): 13
1¼ in. (31.8 mm): 24
1½ in. (38 mm): 32
2 in. (50.8 mm): 53
2½ in. (63.5 mm): 76
3 in. (76.2 mm): 117

These numbers may not reflect the NEC requirements at the time you read these data. Therefore, it is always best to consult the latest edition of the NEC when sizing wire.

16.3.5.4 Fitting and Connections

A wiring installation should be adjusted to fit the design of the building. If compliance with a particular recommendation would require alteration of doors, windows, or structural members, alternate provisions should be made. For example, the extensive use of curtain walls may make it necessary to alter the recommended outlet or switch heights, or to resort to surface-mounted raceways.

Conductors should be spliced with approved solderless connectors, or by brazing, welding, or soldering. Splices should be made mechanically secure before soldering. Bare ends of conductors should be covered after splicing with an insulation equal to that on the conductors. Connections should be made in an accessible device such as a junction box, switch box, or fixture base.

Splices should never be made outside a raceway and then pulled into it. They should be made in a wireway that has a removable cover or in an accessible box that can be opened for inspection or repairs. Each conductor run should be complete from one opening to the next before conductors are pulled. Conductors inserted before the raceway is completed can be damaged during the process of making up the raceway.

Terminal connections should be made securely. Terminals for securing more than one conductor should be approved for that purpose.

Outlet, junction, and switch boxes used in wiring installations should provide adequate space for the conductors. Conductors should be sized adequately to minimize line losses and voltage drop. Voltage drop between the distribution panel and the load should be limited to less than 3%.

Both raceway joints and conductor connections should be made strong and secure. A mechanically secure raceway connection is one in which the joint is as strong as the raceway itself. An electrically secure connection will carry a full short circuit, that is, the entire return current, if one of the hot wires should accidentally be grounded. Such an accidental ground will then open the circuit by blowing the fuse or tripping the circuit breaker. An improperly connected system can result in excessive heating at conductor joints and possible fire hazard.

An excessive number of bends in a conduit can result in damage to conductors when they are pulled through it. A run of rigid conduit or electrical metallic tubing should not contain more than the equivalent of four quarter bends (360 degrees total), including those bends located at the fitting outlet.

Because of the different characteristics of copper and aluminum, pressure terminals or pressure connectors should be marked as suitable for mixing copper and aluminum conductors when this combination is used.

16.3.5.5 Grounding

There are two types of grounding involved in a wiring installation: (1) system grounding and (2) equipment grounding. System grounding entails connecting the current-carrying neutral wire to the grounding terminal in the main switch, which in turn is connected to a ground. In a residence, the ground may be a water pipe. In other building types, a system ground is likely to be a pipe or rod driven into the ground or laid in a trench. A neutral wire is also called a grounded wire. Equipment grounding refers to the use of a grounding wire (or the metal raceway) to connect the metal frames of equipment and appliances back to the service cabinet and to ground. The wire connecting a service cabinet to a water pipe or other grounding device is called a ground wire.

SYSTEM GROUNDING

The entire wiring system should be grounded by connecting the neutral wire to the ground, either before it enters the building by connecting it directly to a ground, or, in the case of a house or other small building, to a water pipe inside the building. Ground connection to a water pipe should be made on the street side of a water meter; if connected on the building side, a jumper cable should be used around the meter. The two hot wires are generally color-coded with black or red insulating jackets. The neutral wire (also called the grounded wire) is usually white, but may also be natural gray. The neutral wire is continuous throughout the system, is never interrupted by fuses or switches, and generally is not connected to a hot wire.

For a residential or commercial wiring system, the earth is generally used as the ground. When a metal water supply system terminating in the earth is available, the electrical system is grounded by connecting to a cold-water pipe near the service panel in the building (see Fig. 16.3-24). If a metal water pipe cannot be

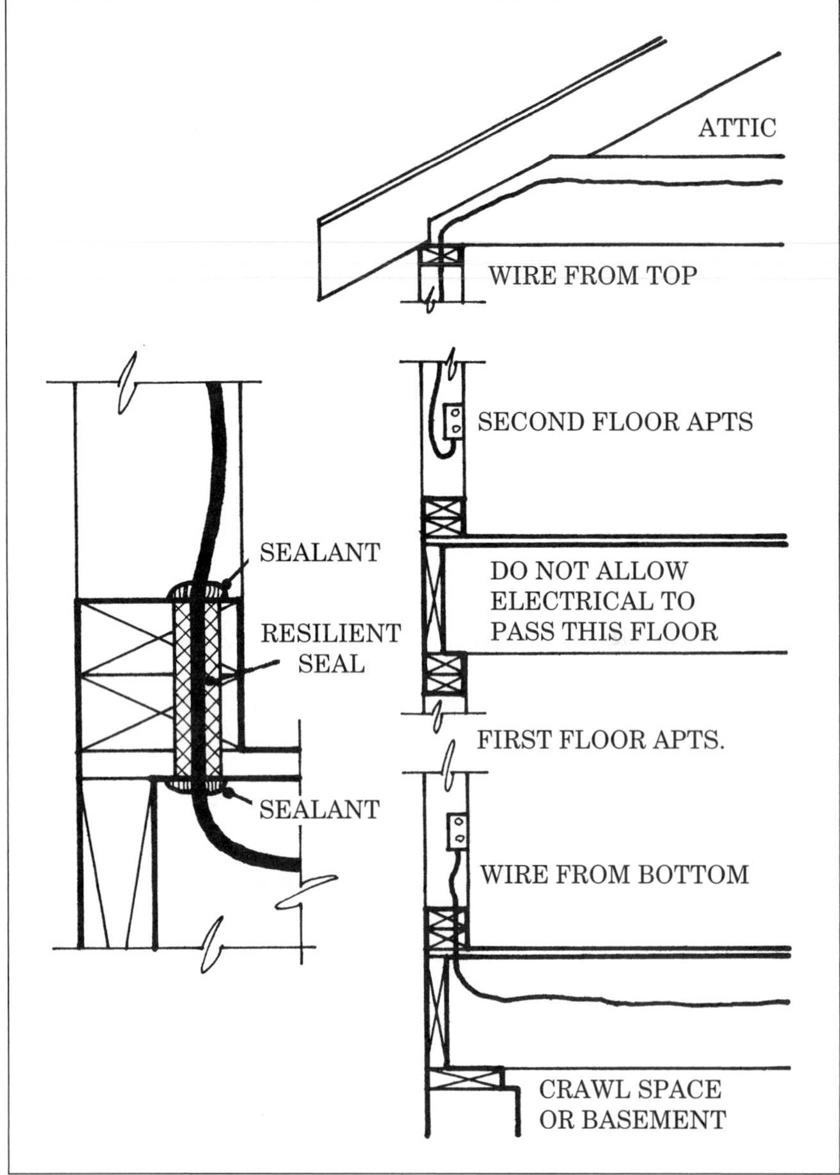

FIGURE 16.3-24 Each apartment should be wired independently, and air passages should be sealed airtight.

used, a metal rod is driven into the ground at least 2 ft. (610 mm) from the building near the service entrance; the neutral wire from the service drop is connected to the rod with a ground wire. The neutral wire of a three-wire service is similarly grounded at the utility pole.

EQUIPMENT GROUNDING

Unlike the grounded (neutral) wire, the grounding wire and a raceway carry no current, except when a short circuit in the equipment or circuit wiring causes a *fault* current to flow. Under these conditions, equipment grounding minimizes the shock hazard to occupants by providing a low-resistance path for the fault current and by opening the circuit overcurrent device in case of excessive flow.

In an electrical system involving conduit, tubing, or metallic cable, metal boxes are used and equipment or fixtures permanently connected at the box become automatically grounded. Portable equipment with cords can be grounded through grounding receptacles and three-pronged grounding plugs. In either case, a grounding wire, color-coded green, is included in the cord or cable connecting the box or receptacle to the frame of the equipment.

When a nonmetallic sheathed cable system is used, a bare grounding wire is often included to tie the system to the ground. Permanently connected equipment has its green grounding wire attached to the bare copper wire in a junction box. Grounding receptacles housed in nonmetallic boxes have a special terminal for the grounding copper wire. Portable equipment cords include a green grounding wire that contacts the grounding system through the round or U-shaped prong of the plug.

Equipment grounding provides adequate safety from shocks, except in high hazard areas such as near pools, in bathrooms and shower rooms, and in outdoor locations. Receptacles should not be installed closer than 10 ft. (3048 mm) from the edge of a pool. Poolside receptacles 10 or 15 ft. (3048 or 4500 mm) from the pool and all outdoor receptacles should be equipped with GFCIs (see Section 16.3.3.2). The use of GFCIs is required by the NEC, because the severity of electrical shock is greatly increased when water is present.

With certain exceptions, the NEC requires GFCIs in new residential construction and in hotels and motels for protection of people. GFCIs are required around swimming pools, in bathrooms, kitchens, garages, crawl spaces, unfinished basements, and in other locations where a damp floor may be expected, and for outdoor receptacles. It may also be desirable to add them in high-risk areas in existing dwellings. Other high-risk areas that need GFCI protection include laundries and workshops. GFCIs should also be considered for any part of a dwelling built on a concrete slab at or below grade.

16.3.6 LOW-VOLTAGE CIRCUITS

Low voltage refers to ac power supplied by step-down transformers from a 120/240-volt line voltage service. Although the NEC treats anything less than 50 volts as *low voltage*, in practice the term usually refers to circuits of less than 30 volts.

Less insulation is required for this type of system, and the NEC, as well as many local building codes, permit low-voltage wiring without protective raceways. Low-voltage wiring is used for bell or buzzer systems, intercoms, control wiring for heating and cooling systems, certain types of lighting fixtures, and other uses requiring relatively small amounts of power.

Because low-voltage wiring is generally used for low-power systems, the conductors are quite small, No. 18 AWG or smaller. In situations requiring multiple and remote switching, low-voltage systems can be used effectively to turn light fixtures and other equipment on and off. The smallness of the wire and elimination of the raceway requirement make this wiring method considerably less costly than 120-volt installations in tubing or conduit. The equipment is fed from the line voltage circuit, as in standard installations, but the switch opening or closing the circuit is activated by a small, low-voltage power relay.

16.3.7 PLANNING FOR FUTURE ADDITIONS

When certain facilities or functions are indicated as future additions, the initial wiring should be arranged so that none of it need be replaced or moved when the ultimate plan is realized. Final wiring for a future addition may be left to a later date if desired.

16.3.8 SOUND CONTROL

Improper electrical installations can reduce the acoustical performance of walls and floor and ceiling assemblies. Causes of reduced performance include (1) open air transmission paths where wires pass through walls or floor and ceiling assemblies, (2) openings provided for outlets, switches, or fixtures, and (3) noise vibration transmitted by wires or fixtures rigidly connecting two otherwise separated structural members or connecting equipment to the structure.

In multifamily housing and other building types having more than one occupancy, each apartment or separate use should be wired as a unit, with wiring kept within the walls of the same unit (see Fig. 16.3-24). Holes should be tightly caulked with resilient, nonhardening compound.

Convenience outlets, switches, telephones, and such other installations as intercoms, should not be placed back-to-back: (1) convenience outlets should be placed at least 36 in. (900 mm) apart on the opposite sides of a wall. In staggered stud walls, installations should be offset horizontally by at least three complete stud spaces (Fig. 16.3-25a); (2) a horizontal distance of 24 in. (610 mm) should be maintained between switches and outlets, and they should not be located in the same stud space (Fig. 16.3-25b); and (3) wall fixtures and equipment should be kept 24 in. (610 mm) apart on opposite sides of a wall (Fig. 16.3-25c).

Where sound control is important, lighting fixtures and other electrical fixtures should be surface mounted (never recessed) and should not be connected to joists above resiliently mounted finished ceilings or connected across separated ceiling joists (Fig. 16.3-25d). If there is no option but to recess a fixture in a sound-controlled wall or ceiling, the fixture should be mounted in an enclosure that provides at least as much sound attenuation as the penetrated wall or ceiling.

Appliances can often be selected and installed to reduce the amount of noise. For example, a clothes dryer can be mounted on a floating floor and the noise transmission from the appliance can be reduced by using a neoprene or rubber gasket or mat between the base of the dryer and the floor or by a resilient underlayment between two layers of floor sheathing.

Appliances should be selected for quiet operation. Vibrating appliances such as dryers should not be rigidly connected to the structure. Unless a floated floor system is used, resilient or flexible isolation gaskets should be provided between the ap-

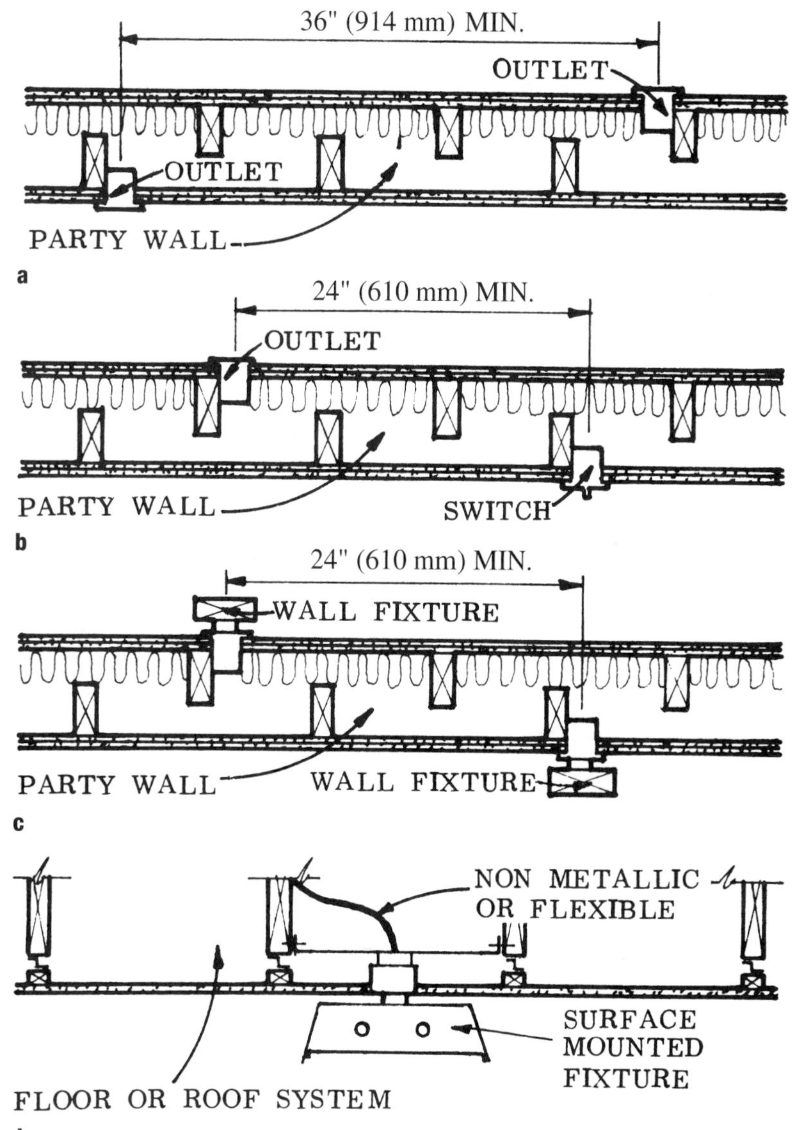

FIGURE 16.3-25 Electrical outlets, switches, fixtures, and equipment should not be mounted back-to-back, should not be located in the same stud space, and should not be spaced closer than shown in (a), (b), and (c) in this figure. Ceiling fixtures should be surface-mounted, as shown in (d), rather than recessed.

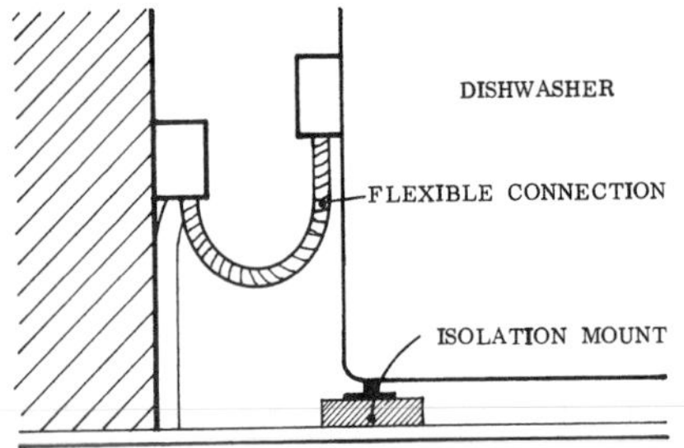

FIGURE 16.3-26 Wiring in rigid conduit can be connected to vibrating equipment with a flexible cable.

pliance and the floor, wall, or counter to isolate vibrations. Electrical connections to appliances should also be flexible whenever possible; a loop as shown in Figure 16.3-26 will reduce vibration transmission. Vents should also be flexible.

Refer to Chapter 13 for additional discussion of sound control.

16.4 Lighting

Lighting outlets should be located where they will produce the most desirable lighting effects. Their types should conform with the lighting fixtures or equipment to be used. The lighting fixtures addressed in this section are fixed rather than portable fixtures.

To understand lighting design and selection, it is necessary to understand some of the special terms used:

Brightness. Luminous intensity of a surface

Candlepower (cp). Luminous intensity of a light source

Candle. Measure for illumination at a surface

Coefficient of utilization. Measure of total number of lumens that will reach the working surface

Footcandle. 1 lumen distributed uniformly over 1 sq. ft. (645 mm^2) of surface 1 ft. (304.8 mm) from a standard candle

Footlambert. 1 lumen per sq. ft. (645 mm^2)

Illumination. Density of light

Lamp. Man-made light source

Light. Radiant energy at a frequency that humans can see

Lumen. The amount of light that falls on a surface that is 1 ft. (304.8 mm) from a standard candle; used to indicate the total output of a light source

Luminaire. A complete lighting unit

Reflector. Portion of a luminaire that reflects light

Transmission factor. The ratio of the light that passes through a material to the amount that strikes its surface

16.4.1 FIXTURE TYPES

Light fixtures are either incandescent or gaseous discharge type.

16.4.1.1 Incandescent Lamps

Incandescent lamps consist of:

- A base, which screws into a socket; several types are available, including the three most common, *medium* (up to 200 watts), *mogul* (more than 200 watts), and candelabra (chandeliers and other decorative fixtures).
- A tungsten filament, which glows to produce the light.
- A clear or frosted glass bulb, which encloses the filament.

Incandescent lamps are used in every type of building, especially in residences. In modern commercial, institutional, and industrial buildings, they are used mostly for spot lighting and special-purpose fixtures.

16.4.1.2 Gaseous Discharge Lamps

Gaseous discharge lamps include fluorescent, mercury vapor, sodium, and neon types.

FLUORESCENT LAMPS

Fluorescent lamps provide more light (lumens) per watt than incandescent bulbs, making them more economical and resulting in fewer power circuits. They also give off less heat, thus reducing air conditioning load. They consist of:

- A cylindrical, sealed glass tube; most are straight, but square, circular, and U shapes are also available.
- Cathodes at the ends of the tube, which produce electrons; these are classed as either rapid start (most fixtures) or instant start (used mostly in cool areas and outdoors).
- A low-pressure mercury vapor within the tube, which is caused to glow by the electrons produced by the cathodes.
- An argon or other inert gas filler.
- A phosphorous coating on the interior of the tubes, which converts invisible ultraviolet radiation produced by the mercury arc into visible light.

Fluorescent fixtures are used in every type of building. They are available in several colors called variously "white," "cool white," "daylight," "warm white," and others.

MERCURY VAPOR, SODIUM, AND NEON FIXTURES

These fixture types are used in some interior locations, such as in arenas, but are mostly used for exterior lighting.

16.4.2 LIGHT CONTROLS

In most cases, fixed lighting outlets are controlled by either single or multiple wall switches. These are usually located at the latch side of doors or at the traffic side of arches and within the room or area where the outlets are located. Some exceptions to this practice are switches for control of

- Exterior lights, such as parking lot and other exterior lights; these may be controlled by manual switches or automatically by timers or light sensors;

- Stairway lights;
- Corridor and other exitway lights;
- Lights in infrequently used areas, such as storage areas;
- Lights in large areas, such as office bays and manufacturing plants; these may be controlled directly from lighting panels, rather than by separate wall switches.

Wall switches are normally mounted about 48 in. (1220 mm) above the floor but may be mounted lower, if desired, to align them with doorknobs or to put them within reach of children. The height of switches in public areas is dictated by considerations for their use by handicapped persons (see Section 1.7).

Spaces that have more than one entrance for which switched lighting is needed are often equipped with a separate switch at each principal entrance. For instance, a door from a living room to a porch is a principal entrance to the porch. However, this door would not necessarily be considered a principal entrance to the living room unless the front entrance to the house is through the porch. Rooms in commercial, institutional, and industrial buildings often have more than one entry. Whether multiple switches are needed is dictated by a room's use and size.

If providing more than one switch in a space would result in the placing of switches controlling the same light within 10 ft. (3000 mm) of each other, one of the switch locations is usually eliminated. Moreover, when a space may be lighted from any one of several sources, as happens when both general and supplementary illumination are provided from fixed sources, multiple switching is required for one set of controls only, usually for the general illumination source.

16.4.3 LIGHTING DESIGN

16.4.3.1 Interior Lighting

The amount and type of illumination required should be fitted to the various tasks carried out in a space. In home living areas, for example, there should be suitable lighting for recreation and entertainment as well as for normal day-to-day living.

In many instances, best results are achieved by a blend of lighting from both fixed and portable luminaires. In an office space or laboratory, for example, the best lighting may be achieved by general lighting alone or by a blend of general lighting and specific task lighting. The type of tasks being performed will dictate which is best. Special lighting will be required where intensive work is being done, such as in surgical operating rooms or automobile manufacturing assembly lines. Good lighting, therefore, requires thoughtful planning and careful selection of permanent fixtures, portable lamps, and other lighting equipment.

COMMERCIAL, INSTITUTIONAL, AND INDUSTRIAL BUILDINGS

Every habitable space should have adequate lighting for the tasks to be performed there. Adequacy is usually determined by the number of footcandles required. This is often a subjective decision, but there are some generally applicable guidelines. Codes sometimes dictate the minimum requirements. Owners will also sometimes have their own requirements, as may be the case, for example, in a hospital or school. The following are some typical footcandle requirements according to use:

- *Schools:* corridors, 20; classrooms, 70 with 150 at chalkboard areas; laboratories, 80
- *Retail stores:* merchandising areas, 150; showcases, 200
- *Industrial facilities:* general areas, 50; fine machinery areas, 80 to 400

After determining the number of footcandles required in each space, then calculate the areas of the spaces. Next, select the luminaires to be used in each space. From manufacturer's data for the selected luminaires, determine the coefficient of utilization (cu) for each. This determination depends on wall, ceiling, and floor reflectance factors, the space's cavity ratios, and the lamp maintenance category.

A reflectance factor is a percentage of light that will be reflected by a surface. These factors are supplied by materials manufacturers. They usually range as follows:

- *Walls:* From 10% to 50%, with most being in the 30% range
- *Ceilings:* From 50% to 80%
- *Floors:* Usually about 10%, but may range up to 30%

Cavity ratios are determined by consulting tables in the Illuminating Engineering Society of North America's (IESNA) *Lighting Handbook*. The data required to enter the tables are the height of the space above the luminaire (ceiling cavity), the height of the luminaire above the work surface (room cavity), the height of the work surface above the floor (floor cavity). The ratios are designated as:

CCR: ceiling cavity ratio
RCR: room cavity ratio
FCR: floor cavity ratio

Effective cavity reflectances are then determined directly from a table in the IESNA *Lighting Handbook* by entering the cavity ratio and the reflectance factor.

The lamp maintenance category is also determined from a table in the IESNA *Lighting Handbook*. It has to do with the configuration of the luminaire.

By entering the data determined, the coefficient of utilization (cu) is then taken either from the manufacturer's data or from a table in the IESNA *Lighting Handbook*. If the actual cavity reflectances do not exactly match those used in the manufacturer's data or the IESNA *Lighting Handbook* tables, it will be necessary to interpolate the final cu based on the actual reflectances.

The next step is to determine the luminaire's lamp lumen depreciation factor (LLD), which is taken from the manufacturer's data. Then determine the luminaire dirt depreciation factor (LDD). This is taken from charts in the IESNA *Lighting Handbook* by entering the LLD, the type of dirt expected in the environment, and the number of months the fixtures are expected to be used without replacement.

The maintenance factor (mf) is the multiple of the fixture's LLD and its LDD.

The total lumens required for the space is determined by multiplying the area of the space by the number of footcandles required, and dividing the result by a multiple of cu and mf.

To determine the number of luminaires needed, divide the total number of lumens required by the lumens output of each luminaire.

Finally, lay out the correct number of luminaires so that the light falls where needed. The layout is often symmetrical, but may be asymmetrical when space shape or lighting needs so dictate.

RESIDENTIAL BUILDINGS

The following requirements comply with the *National Electrical Code* (NEC) as a minimum, but some of them exceed the NEC requirements. When two different requirements demand a lighting outlet at the same location, only one need be installed. However, additional wall-switch controls may be necessary.

For example, a lighting outlet in an upstairs hall may be located at the head of the stairway, thus satisfying the need for both hall lighting and stairway lighting with a single outlet. However, multiple-switch control of this lighting outlet will be required at the head and at the foot of the stairway. In addition, the multiple-switch control rule given in Section 16.4.2 may require a third point of control in the upstairs hall because of its length.

Dual-Purpose Rooms When a room is intended to serve more than one function, such as a combination living and dining room or a kitchen and laundry, the following outlet recommendations are separately applicable to specific areas. Lighting outlet provisions may be combined in a way that will provide both general overall illumination and local illumination of work surfaces. In determining locations for wall switches, the area is considered as a single room.

Living and Recreation Rooms Some means of general illumination is essential in living rooms, recreation rooms, sun rooms, enclosed porches, television rooms, libraries, dens, and similar areas. This lighting may be provided by ceiling or wall fixtures, by lighting in coves, valances, or cornices, or by portable lamps (Fig. 16.4-1). Wall-switch-controlled outlets in locations appropriate to the intended lighting method should be provided. Outlets for decorative accent lighting, such as picture illumination and bookcase lighting, should also be considered.

Dining Areas Each dining room, or dining area combined with another room, or breakfast nook should have at least one wall-switch-controlled lighting outlet. This outlet is normally located over the dining or breakfast table to provide direct illumination for it.

Bedrooms Good general illumination is essential in bedrooms. This may be provided by a fixed ceiling fixture or by lighting in valances, coves, or cornices, similar to that shown in Figure 16.4-1.

Wall-switch-controlled outlets should be placed in locations appropriate to the lighting method selected.

Light sources located over full-length mirrors, or in the bedroom ceiling directly in front of the clothes closet, may serve as general illumination. A master switch in the master bedroom (as well as at other strategic points in the home) controlling selected interior and exterior lights may be desirable.

Bathrooms and Lavatories Illumination of both sides of the face when a person is at a bathroom mirror is essential. This can be accomplished with diffused lighting from a luminous ceiling or with more concentrated light from several sources (Fig. 16.4-2). A single concentrated light source, either on the ceiling or on the wall, generally is not adequate. A 30- or 40-watt fluorescent tube over the mirror may be adequate in a small room with light colored walls. All lighting should be controlled by wall switches not readily accessible while standing in the tub or shower stall.

A ceiling outlet located in line with the front edge of the basin will often provide, in addition to improved lighting at the mirror, general room lighting and safety lighting for a combination shower and tub. When more than one mirror location is planned, equal consideration should be given to the lighting in each case.

A vaporproof fixture should be provided in an enclosed shower stall. It should be controlled by a wall switch that cannot be reached from within the stall.

A switch-controlled night-light is also recommended.

Kitchen Wall-switched outlets for general illumination and for lighting at the sink are essential. Lighting design should provide for illumination of the work areas, sink, range, counters, and tables (Fig. 16.4-3).

Undercabinet lighting fixtures within easy reach may have integral switches or

FIGURE 16.4-1 Valences direct light both up and down for dramatic effect. Coves direct light mostly up; cornices mostly down.

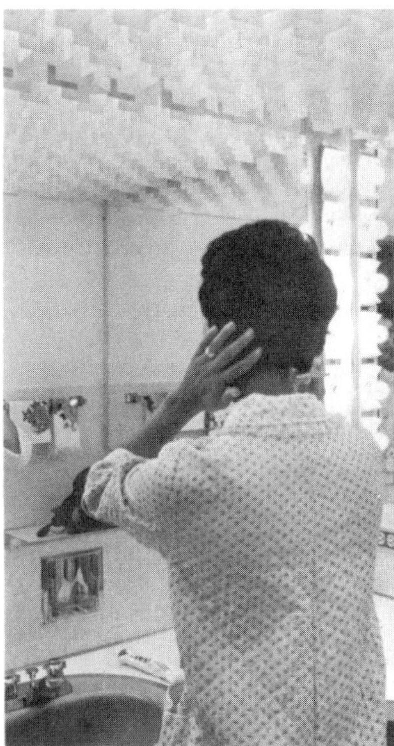

FIGURE 16.4-2 Adequate bathroom mirror lighting may consist of multiple light sources on both sides of the mirror, lights over the mirror, or a luminous ceiling.

FIGURE 16.4-3 In this kitchen, a ceiling luminaire provides overall illumination, while soffit down-lights and undercabinet fluorescent strips highlight the work areas.

may be wall-switch controlled. Consideration should also be given to providing lighting inside deep cabinets.

Laundry Areas Lighting outlets for fixed lights should be installed to provide illumination of work areas, such as laundry tubs, sorting tables, and washing, ironing, and drying centers. At least one outlet in the room should be wall-switch controlled. There should be at least one ceiling lighting outlet centered above laundry trays in unfinished basements or other spaces.

Closets Generally, one lighting outlet should be provided in every closet. Where shelving or other conditions make the installation of lights within a closet ineffective, outlets in the adjoining space should be so located as to provide light within the closet.

Halls Wall-switch-controlled outlets should be installed for proper illumination of the entire area of passage halls, reception halls, vestibules, entries, foyers, and similar areas. Particular attention should be paid to irregularly shaped spaces. A switch-controlled night-light should be installed in each hall that serves bedrooms.

Stairways Fixed wall or ceiling lighting outlets must be installed to provide adequate illumination of each flight of stairs. These outlets should have multiple-switch controls at the head and foot of the stairway. Outlets should be so arranged that the stairs are fully illuminated from either floor, while permitting control of bedroom hall lights independently of ground-floor fixtures. Switches should be grouped whenever possible and should never be located so close to steps that reaching for a switch might result in missing a step.

Utility Room Lighting outlets should be placed to illuminate a furnace area, and work area, if there is one. At least one lighting outlet should be wall-switch controlled.

Basement Lighting outlets should be placed to illuminate designated work areas, the foot of the stairway, enclosed spaces, and specific pieces of equipment, such as a furnace, pump, or water heater, and a workbench if one occurs. Additional outlets for open (unpartitioned) basement space should be provided at the rate of one outlet for each 150 sq. ft. (13.94 m^2).

In unfinished basements, the light at the foot of the stairs should be wall-switch controlled near the head of the stairs. Other lights may be pull-switch controlled.

In basements with finished rooms, with garage space, or with other direct access to the outdoors, the stairway lighting provisions mentioned earlier in this section apply.

A pilot light that lights when the basement light is on should be installed at the switch at the head of the stairs, when the basement is infrequently visited or when the stair layout makes it difficult to determine visually whether the light is on.

Accessible Attic There should be one outlet for general illumination, wall-switch controlled from the foot of the stairs. This switch should have a pilot light to indicate whether it is on. When there are no permanent stairs, the attic light may be pull-switch controlled if located over the access door. When an unfinished attic will later be developed into habitable rooms, the attic-lighting outlet should be multiple-switch controlled at the top and the bottom of the stairs. There should be one outlet for each enclosed space.

Garage or Carport At least one wall-switched ceiling outlet should be installed in one- and two-car garages. If a garage has no covered access from the house, one exterior outlet should be provided and multiple-switch controlled from the garage and the residence.

If the garage is to be used also as a work area or laundry, provisions appropriate to these uses should be made. Additional interior outlets often are desirable even if no specific additional use is planned for the garage. Long driveways need additional illumination, such as would be provided by post lighting. These lights should be wall-switch controlled from the house.

16.4.3.2 Exterior Lighting

Lights are provided on the exterior of buildings to

- Illuminate entrances so that occupants and visitors can find their way about;
- Make exterior areas usable for social and recreational activities;
- Provide for security;
- Make parking areas safer and to illuminate the way from such areas to the building;
- Illuminate driveways and roadways.

Lights are needed on the exterior of buildings in the following locations.

EXTERIOR ENTRANCES

One or more switched lighting outlets should be located at the front (main) and service entrances to every building. Switched lighting outlets should also be installed at other entrances. The principal lighting requirements at entrances are the illumination of walkways and steps leading to the entrance and of faces of people at the door.

In residences, where a single wall outlet is desired, location on the latch side of the door is preferable. Switched outlets in addition to those at the door are often desirable for post lights to illuminate terraced or broken flights of steps or long approach walks.

PORCHES AND BREEZEWAYS

Porches, breezeways, and similar roofed areas of more than 75 sq. ft. (6.97 m^2) in area should have at least one switched lighting outlet. Large or irregularly shaped areas may require more outlets. Multiple-switch controls should be installed at entrances when a porch is used as a passage between a house and a garage.

TERRACES AND PATIOS

One or more fixed outlets on the building wall, or on a post centrally located in the area, is needed for general illumination. These outlets should be switch controlled just inside the door opening onto the area. In commercial and institutional building terraces, much more lighting will be required to provide safety and to illuminate walkways.

GENERAL GROUNDS LIGHTING

Floodlights are often used to illuminate the grounds surrounding buildings. Outlets may be located on the building itself, on the exterior of an adjacent building, such as a garage, or on posts (Fig. 16.4-4). All outlets should be switch controlled from within the building. Multiple and master-switch control from strategic points is also desirable.

PARKING LOTS

Parking lots for all occupancies must be well lit for safety and security and to light the way from parking spaces to the building. Many different kinds of fixtures are used. In large lots, luminaires are usually mounted on high poles. These may be augmented by ground-mounted walkway lights.

EXTERIOR SECURITY LIGHTING

Security demands that a facility be made less vulnerable to unauthorized entry by thieves, vandals, and pranksters. This requires siting to eliminate hidden spots, secure door locks, intruder detection systems, and adequate lighting. Chapter 8 discusses door and window design and hardware that will slow down a burglar. Other security measures are discussed in Section 13.3.

Intruders prefer to work in the dark. Many security experts recommend the installation of all-night security lights to illuminate doorways, yards, public spaces, and other areas subject to intrusion (Fig. 16.4-5). Ordinarily, exterior lights are either manually switched on when they are needed or automatically controlled by a programmed timer or light-sensitive switch.

Security lights are sometimes provided in addition to general site lighting, but more frequently the two functions are combined.

Light Types A variety of lamps and light fixtures are available to suit most security needs. An ordinary incandescent bulb can provide adequate illumination at single-family residential entry doors. It does not have to be bright.

An incandescent flood lamp will light a larger area, such as a side yard or driveway. Flood lamp fixtures hold one to three lamps and allow the lamp to be aimed in whatever direction is necessary.

Mercury vapor lamps cost more than incandescent lamps but emit twice as much light per watt. They also last up to 10 times longer, thus making them the more economical of the two. A standard clear mercury vapor bulb has a blue cast and gives a harsh light; a deluxe white bulb has a pink or orange cast and gives a softer, more balanced light.

Ordinary incandescent fixtures can be adapted for mercury vapor bulbs with components available at an electrical supply store.

Timing Devices Like a deadbolt that is never turned, a light that is never switched on performs no security function. Exterior lights can be automatically switched on by a timer.

A time-of-day device is an electric clock that turns individual fixtures on and off at specified times (Fig. 16.4-6). In single-family residences, exterior lighting can be wired to a timer through its own circuit. The disadvantage of using a timing device is that nights vary in length throughout the year, which means that the timer will have to be periodically adjusted. Some timers can be set to vary the time they go on and off each day, which can give a vacant dwelling a lived-in appearance.

Wired-in timing devices used in commercial, institutional, and industrial facilities are much more sophisticated,

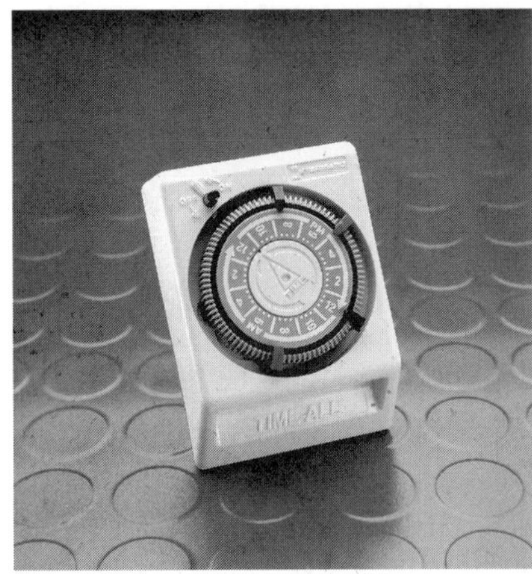

FIGURE 16.4-6 A light may be plugged into a tabletop timer. (Courtesy Intermatic, Inc.)

FIGURE 16.4-4 Exterior lighting can ensure safety and create dramatic views from inside and outside a house.

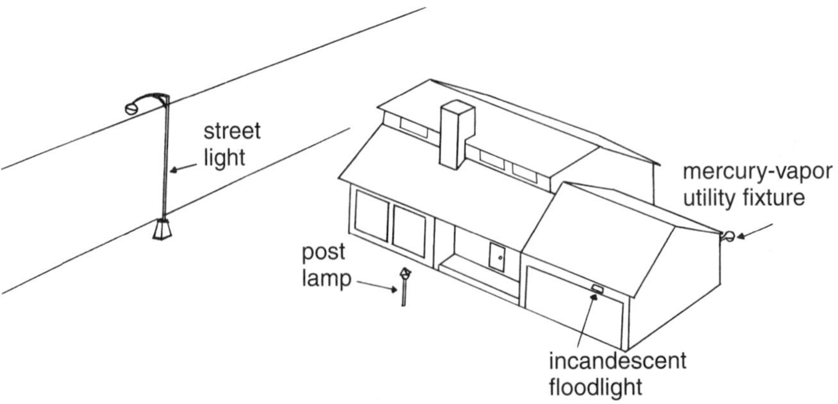

FIGURE 16.4-5 Good security lighting illuminates doors and windows and other intrusion points and reduces shadowed areas where intruders may hide.

overcoming all of the disadvantages mentioned here for residential units.

A photoelectric device senses light. It will automatically turn a light on at dusk and off at dawn. The device can be an integral part of a fixture or a screw-in device that will adapt to a regular fixture.

An automatic sensor (motion detector) can detect anyone walking up to an entrance at night and automatically turn on a light for a few minutes. Such a device is both a convenience for residents and a deterrent to burglars. A photoelectric cell prevents the system from turning on during the day.

16.5 Communications Systems

Additional facilities commonly considered part of a building's wiring system and installed by the electrical trade include entrance signals, radio and TV antenna outlets, telephone outlets, and internal communication (intercom) systems. These systems are discussed in this section.

Some other systems using wiring are often installed by other professionals. These include wiring for nurse call systems in hospitals; computer-controlled systems for equipment and devices such as those in a manufacturing plant; systems for monitoring equipment, such as sensors for patients' vital signs in a hospital, or heating, ventilating, and air conditioning equipment sensors in any building; and many other specialized communications and control systems. These specialized systems are beyond the scope of this book.

16.5.1 ENTRANCE SIGNALS

In a residence, signal push buttons should be provided at each commonly used entrance. In other types of buildings, similar devices are often installed at normally locked or unattended entrances, such as a delivery entrance or a controlled-entry door. In some facilities, such devices are installed also at interior doors where controlled access is desired.

Such devices should be connected to a chime or buzzer that gives a distinctive signal that will not be mistaken as a telephone, smoke alarm, or other signal. Electrical supply for entrance signals is usually obtained from a low-voltage transformer installed near the service equipment.

In smaller homes the chime or buzzer is often installed in the kitchen. In larger homes the chime should be installed in a central location, where it may be heard throughout the house. In a multilevel house, extension signals may be necessary to ensure that they will be heard in quarters located away from the main entry level. Entrance-signal conductors for residential systems should be no smaller than the equivalent of a No. 18 AWG copper wire and, preferably, should be installed in a conduit.

In other types of buildings sounding devices are installed where they will be heard by the person who will answer the signal. Wiring will be dictated by requirements of the device being used.

16.5.2 INTERCOMMUNICATION SYSTEMS

Intercommunication (intercom) systems are quite common even in small houses today and are used extensively in other types of buildings. An intercom system may be a complete sound system with an AM/FM radio, tape deck, or compact disc player, or just a simple push-button two-way talk system. In a residence, an intercom is operated from a master control panel, usually in the master bedroom, recreation room, or kitchen, and connected to remote stations throughout the house. In other types of buildings, the control system is usually in a central location within the system.

A larger house, or one with accommodations for a resident servant, may have additional signal and communication devices. They may consist solely of a dining-room-to-kitchen call signal, with a floor treadle or push-button connected to an annunciator in the kitchen. More extensive systems may include a push button in each bedroom, living room, and other habitable room.

Sometimes an intercom system is tied to the entrance signal or substitutes for it. This permits an occupant to talk with a visitor without opening the door. In apartments, intercom systems are often tied in with electronic locks so that the main entrance door to the building can be operated from each apartment unit. Similar arrangements are often used in other types of buildings, such as at exterior entrances and at interior doors serving restricted areas. Closed-circuit television systems that permit an occupant to see a visitor before opening the door are available. These are expensive and not commonly used in residences, but are used extensively in hospitals and some other types of buildings.

16.5.3 TELEVISION AND RADIO

Provisions for television consist mainly of a nonmetallic outlet box at each television set location, connected to the antenna or incoming CATV source by a suitable transmission line. A convenience outlet should be provided adjacent to each television antenna outlet. Antenna systems with more than two outlets generally require a booster amplifier, except in very strong signal areas.

In areas with a cable TV service, the cable company should be contacted to ascertain its specific requirements for antenna connections.

In most areas, AM radio receivers used purely for the reception of broadcast communication do not require antenna and ground connections. For FM reception, provisions similar to those for television should be included.

16.5.4 TELEPHONE

The number of telephone outlets and the extent of the service in commercial, institutional, and industrial buildings must be determined on a needs basis, depending on the use of the facility. Some types of buildings, such as hospitals, require extensive telephone communications systems. These may be tied into or augment the building's intercom system. Other building types, such as warehouse facilities, may require little more than a single incoming line. Computers require additional telephone lines for Internet access or contact with a company-wide network system.

In residential buildings, at least one

convenient fixed telephone outlet (jack) should be provided in every living unit. Where a living unit occupies more than one floor, at least one outlet should be provided on each floor. Most homes today have an outlet in every habitable room and one in each work space, such as the laundry room, utility room, basement, and garage. Outlets are also often provided on porches and outdoors on patios and the like.

Advance wiring by the telephone company during construction will generally result in an inconspicuous, more efficient installation. The local telephone company should also be consulted for details about service connection before construction, particularly with regard to the installation of protector cabinets and raceways leading to the service entrance equipment.

16.6 Conservation of Energy

The proliferating use of electrical equipment in the home causes increased demands not only on home wiring systems, but on national energy resources as well. Residential use of energy accounts for more than 25% of the total consumption of energy in the country. In addition, the amount of energy used by individual appliances has increased dramatically over the last 45 years. Appliances that were once luxuries are now commonplace. For example, almost every home has a color television set. Home computers are the latest electrical devices to attract the attention of the public. It is difficult to sell a new house today if it does not have all the latest gadgets, unless the price of the house is very low. There are electric dishwashers, clothes dryers, garbage disposals, can openers, built-in popcorn poppers, microwave ovens, woks, frying pans, hotpots, whirlpool bathtubs, hot tubs. The list grows every year.

The increasing use of electronics and computer devices, as well as the perception that the well-being and productive capability of workers are being adversely affected by poor lighting, have also dramatically increased the use of electrical power in buildings other than residences.

Conservation of energy is both a worthwhile national goal and a money-saving consumer objective. By careful design of the wiring system and judicious selection of equipment, it is possible to minimize energy waste without sacrificing comfort and convenience.

16.6.1 LIGHTING

Lighting is responsible for about 40% of the total electricity used in the United States. The use of lighting in all types of facilities is extensive, ranging from necessary illumination to that used to enhance interior and exterior design. However, misplaced or excessive illumination wastes electricity and money.

In residences, seldom does an entire room require a high level of illumination, especially when close-up activity is ordinarily restricted to specific areas. For instance, lighting should be concentrated on areas such as a study desk, kitchen counter, and dining table, but can be diffused and dimmer over the rest of a room.

Proper selection and placement of switches in convenient locations in a residence can reduce wasteful consumption by lights left on unintentionally. Rooms with more than one doorway, such as kitchens, dining rooms, and living rooms, should have switches at each doorway.

Larger bulbs are more energy efficient, producing more light per watt of electric energy. However, increased efficiency should be balanced against the increased overall energy consumption that can be caused by indiscriminate use of larger bulbs. In general, light bulbs should not be larger than required for a specific task or design effect.

Another facet of energy conservation is the saving of related materials. For example, longer-life luminaires require replacement at longer intervals. This saves on the amount of materials and energy used to produce the fixtures.

For hard-to-reach areas such as roof overhangs and post lights, long-life bulbs with a life expectancy of 2500 hours can be used. While ordinary incandescent bulbs have a life span of 750 to 2000 hours, fluorescent tubes are rated for 10,000 to 20,000 hours of life. Moreover, fluorescent tubes produce about 80 lumens per watt, or 4 times as much light as a standard incandescent bulb. However, fluorescent fixtures require more materials than incandescent ones, but the advantages are seen to outweigh this factor.

Newer fluorescent bulbs are smaller (1 in. [25.4 mm] in diameter) than the older ones (1½ in. [38 mm] in diameter), which amounts to some savings in their glass, metal, and chemical contents.

A major drawback of using fluorescent bulbs is the environmental hazard their disposal presents. Fluorescent bulbs contain mercury, and the older electromagnetic ballasts used for them contain polychlorinated biphenyl (PCB). In newer bulbs the amount of mercury has been reduced, but not entirely eliminated. The newer electronic ballasts do not contain PCB. In addition, they use less material, produce almost no heat—which saves on air conditioning load—and weigh less. Unfortunately, replacing the older tubes with the new ones requires disposing of the old ones.

Another way to reduce energy in lighting systems is by the use of electronic timing devices to turn lights off when they are not needed and to prevent turning them on when they are not needed. Such systems can save as much as 50% in lighting energy use.

16.6.2 APPLIANCES

Energy-efficient appliances can reduce operating costs and minimize demands on a wiring system. Before the purchase of any appliance, the life-cycle cost (the long-range, lifetime operating and maintenance cost) of the product should be determined. Buying an energy-efficient appliance may be more costly initially, but significant savings can be realized on subsequent electric bills and energy demands.

The energy efficiency ratio (EER) established by the National Institute of Standards and Technology can be helpful in determining the relative efficiency of certain household appliances. The EER of air conditioners, for example, is determined by dividing the number of Btu's per hour by the number of watts needed for operation. The EER for air conditioners ranges from 5 to 10 for smaller units and up to 12 for larger units. The higher the number, the more efficient the machine, which means less energy and money spent on operation. An EER of 7 or more is considered acceptable for household window air conditioners.

Appliances with extra conveniences, such as self-cleaning ovens and frost-free refrigerators, require more energy and cost more to operate than ordinary models. These special features are largely responsible for the increased energy consumption per appliance recorded in the last 35 years. In selecting appliances, the extra convenience and pleasure provided by the special features should be weighed against a realistic evaluation of the added initial and long-term operating costs. The local power company can be helpful in making an economic life-cycle cost analysis and in providing other energy conserving tips.

Ceiling fans can increase air movement in a space and decrease the need for air conditioning. Except when the humidity is very high, a room with a ceiling fan will remain comfortable even at temperatures above 80°F (26.67°C).

16.7 Additional Reading

More information about the subjects discussed in this chapter can be found in the references listed in Section 16.8 and in the following publications.

Amthor, Geoffrey. 1993. "A Fiber-Optic Backbone." *The Construction Specifier* (October):65.

Bauer, Claude J. 1992. "A New Era in Building Communications." *The Construction Specifier* (September):74.

Breeze, Frank, Steve Eldridge, and Marshall Parsons. 1993. "The Fiber-Optic Infrastructure." *The Construction Specifier* (July):124.

The Construction Specifier. 1993. "Industrial Intercom Systems." *The Construction Specifier* (September):108.

Gordon, Gary, and James L. Nuckolls. 1995. *Interior Lighting for Designers*. 3d ed. New York: John Wiley & Sons, Inc.

Kay, Gersil N. 1995. "Glass Fiber-Optic Lighting." *The Construction Specifier* (August):24.

Loeffler, Mark. 1997. "Energy-Efficient Lighting by Design." *The Construction Specifier* (August):35.

Marsteller, John. 1987. "Philosophy of Light: Nature as a Basis, Examples." *Interior Design* (February):78–80.

Progressive Architecture. 1986. "Floor Systems to House Office Wiring: Advantages and Limitations." *Progressive Architecture* (February):116–121.

Schiler, Mark. 1992. *Simplified Design of Building Lighting*. New York: John Wiley & Sons, Inc.

Stein, Benjamin. 1997. *Mechanical and Electrical Systems*. 2d ed. New York: John Wiley & Sons, Inc.

Stein, Benjamin, and John S. Reynolds. 1992. *Mechanical and Electrical Systems for Buildings*. 8th ed. New York: John Wiley & Sons, Inc.

Willson, David Winfield. 1987. "Light as a Design Tool: Basics to Consider." *Interior Design* (June):98, 200.

16.8 Acknowledgments and References

ACKNOWLEDGMENTS

We gratefully acknowledge the assistance of the following organizations and individuals in preparing this chapter. We are also indebted to them for permission to use their illustrations when requested and for the use of their publications as references.

Acoustical Materials Association
American National Standards Institute
Consumer Product Safety Commission
Illuminating Engineering Society of North America
Intermatic Incorporated
National Fire Protection Association
U.S. Department of Housing and Urban Development (HUD)

REFERENCES

We would also like to thank the authors and publishers of the publications in the following list for their contributions to our research for this chapter.

American National Standards Institute (ANSI). 1958. ANSI C91.1, "Requirements for Residential Wiring." New York: ANSI.

———. 1972 (reaffirmed 1989). ANSI Y32.9, "Graphic Symbols for Electrical Wiring and Layout Diagrams Used in Architecture and Building Construction." New York: Institute of Electrical and Electronics Engineers.

Council of American Building Officials (CABO). *Model Energy Code*. Falls Church, VA: CABO. (This code is in the process of being transferred to the control of the International Code Council [ICC].)

———. 1983. *One- and Two-Family Building Code*. Falls Church, VA: CABO. (This code is in the process of being transferred to the control of the International Code Council [ICC].)

Dagostino, Frank R. 1995. *Mechanical and Electrical Systems in Construction and Architecture*. Englewood Cliffs, NJ: Prentice-Hall.

Illuminating Engineering Society of North America. 1993. *Lighting Handbook*. 8th ed. New York: Illuminating Engineering Society of North America.

Loeffler, Mark. 1998. "Environmentally Responsible Lighting." *The Construction Specifier* (April):57.

National Fire Protection Association (NFPA). 1996. NFPA 70, *National Electrical Code*®. Quincy, MA: NFPA.

———. 1990. NFPA 70A, *Electrical Code for One- and Two-Family Dwellings*®. Quincy, MA: NFPA.

———. 1997. NFPA 101, *Life Safety Code*®. Quincy, MA: NFPA.

Ramsey/Sleeper, The AIA Committee on Architectural Graphic Standards. 1998. *Architectural Graphic Standards*. 8th ed. New York: John Wiley & Sons, Inc.

APPENDIX A

Data Sources

The following organizations contributed to this book by permitting the use of their illustrations or text, by answering the author's queries, or by producing material used in research for the book.

AA Wire Products Company
6100 E. New England Ave.
Chicago, IL 60638
(773) 586-6700
Fax: (773) 586-6710

ABT Building Products Corporation
10115 Kincey Ave., Suite 150
Huntsville, NC 28078
(704) 875-1680
(800) 566-2282
Fax: (704) 875-1680
www.abtco.com

ACI International
P.O. Box 9094
Farmington Hills, MI 48333
(248) 848-3700
webmaster@aci-int.org
www.aci-int.org

Acoustical Society of America (ASA)
American Institute of Physics
One Physics Ellipse
College Park, MD 20740-3843
(301) 209-3100
Fax: (301) 209-0843
www.aip.org

Adams Rite Manufacturing Company
4040 S. Capitol Ave.
City of Industry, CA 91749
(562) 699-0511
(800) 872-3267
Fax: (562) 699-5094

Addison Wesley Longman
(formerly Addison-Wesley Publishing Company)
1 Jacob Way
Reading, MA 01867
(781) 944-3700
Fax: (781) 944-9338
www.awl.com

Adhesive and Sealant Council (ASC)
7979 Old Georgetown Rd., Suite 500
Bethesda, MD 20814
(301) 986-9700
Fax: (301) 986-9795
www.ascouncil.org

Advanstar Communications
(formerly Edgell Communications)
7500 Old Oak Blvd.
Cleveland, OH 44130
(440) 243-8100
Fax: (440) 826-2833
information@advanstar.com
www.advanstar.com

Air-Conditioning & Refrigeration Institute
4301 N. Fairfax Dr., Suite 425
Arlington, VA 22203
(703) 524-8836
Fax: (703) 528-3816
www.ari.org

Air Conditioning Contractors of America Inc. (ACCA)
1712 New Hampshire Ave. NW

Washington, DC 20009
(202) 483-9370
Fax: (202) 588-1217
www.acca.org

Air Diffusion Council (ADC)
104 S. Michigan Ave., Suite 1500
Chicago, IL 60603
(312) 201-0101
Fax: (312) 201-0214
info@flexibleduct.org
www.flexibleduct.org

Airline Products Company
17500 York Rd.
Hagerstown, MD 21740
(301) 582-2500
Fax: (301) 582-2226

Air Movement and Control Association (AMCA)
30 W. University Dr.
Arlington Heights, IL 60004
(312) 394-0150
Fax: (847) 253-0088
www.amca.org

Ajax Magnethermic Corp.
1743 Overland Ave. NE, Dept. DG
Warren, OH 44482
(888) 335-1040
Fax: (330) 372-8608

Alcoa Building Products
1501 Michigan St.
P.O. Box 716
Sidney, OH 45365
(513) 492-1111
(800) 962-6973
Fax: (800) 452-1448
trade.abp@alcoa.com
www.alcoahomes.com

Allied Metal Industries, Inc.
118-144 Harper St.
Newark, NJ 07114
(973) 824-7347
Fax: (973) 624-8404
www.alliedsteel.com

Alside Inc.
P.O. Box 2010
Akron, OH 44309
(800) 922-6009
www.alside.com

Aluminum Association (AA)
900 19th St. NW, Suite 300
Washington, DC 20006
(202) 862-5100
Fax: (202) 862-5164
www.aluminum.org

Amarlite Architectural Products
(*see* Arch Aluminum & Glass Co., Inc.)

American Architectural Manufacturers Association (AAMA)
1827 Walden Office Square, Suite 104
Schaumburg, IL 60173-4268
(847) 303-5664
Fax: (847) 303-5774
webmaster@romepc.com
www.aamanet.org

American Association of State Highway and Transportation Officials (AASHTO)
444 N. Capitol St. NW, Suite 249
Washington, DC 20001
(202) 624-5800
Fax: (202) 624-5806
www.aashto.org

American Association of Textile Chemists and Colorists (AATCC)
P.O. Box 12215
Research Triangle Park, NC 27709-2215
(919) 549-8141
info@aatcc.org
www.aatcc.org

American Bearing Manufacturing Association (ABMA)
(formerly the Anti-Friction Bearing Manufacturers Association)
1200 19th St. NW, Suite 300
Washington, DC 20036
(202) 429-5155
Fax: (202) 828-6042
www.abma-dc.org

American Concrete Pipe Association (ACPA)
222 W. Las Colinas Blvd., Suite 641
Irving, TX 75039-5423
(972) 506-7216
Fax: (972) 506-7682
info@concrete-pipe.org
www.concrete-pipe.org

American Consulting Engineers Council (ACEC)
1015 15th St. NW, Suite 802
Washington, DC 20005
(202) 347-7474
Fax: (202) 898-0068
acec@acec.org
www.acec.org

American Council of Independent Laboratories (ACIL)
1629 K St. NW, Suite 400
Washington, DC 20006
(202) 887-5872
Fax: (202) 887-0021
www.acil.org

American Forest & Paper Association (AF&PA)
1111 19th St. NW, Suite 800
Washington, DC 22036
(202) 463-2700
Fax: (202) 463-2791
www.awc.org

American Galvanizers Association
12200 E. Iliff Ave., Suite 204
Aurora, CO 80014
(800) 468-7732
Fax: (303) 750-2909
aga@galvanizeit.org
www.galvanizeit.org

American Gas Association (AGA)
400 N. Capitol St. NW
Washington, DC 20001
(202) 824-7000
Fax: (202) 824-7115
www.aga.org

American Hardboard Association (AHA)
1210 W. Northwest Hwy.
Palatine, IL 60067
(847) 934-8800

American Hardware Manufacturers Association
801 North Plaza Dr.
Schaumburg, IL 60173-4977
(847) 605-1025
Fax: (847) 605-1093
info@AHMA.org
www.ahma.org

American Industrial Hygiene Association (AIHA)
2700 Prosperity Ave., Suite 250
Fairfax, VA 22031
(703) 849-8888
(703) 207-3561
www.aiha.org

American Institute of Architects (AIA)
1735 New York Ave. NW
Washington, DC 20006
(202) 626-7300
www.aiaonline.com

American Institute of Real Estate Appraisers
The Appraisal Institute
875 N. Michigan Ave., Suite 2400
Chicago, IL 60611-1980
(312) 335-4100
Fax: (312) 335-4400
www.appraisalinstitute.org

American Institute of Steel Construction (AISC)
One E. Wacker Dr., Suite 3100

Chicago, IL 60601
(312) 670-2400
Fax: (312) 670-5403
www.aisc.org

American Institute of Timber Construction (AITC)
7012 S. Revere Pkwy., Suite 140
Englewood, CO 80112
(303) 792-9559
Fax: (303) 792-0669
info@aitc-glulam.org
www.aitc-glulam.org

American Insurance Association (AIA)
1130 Connecticut Ave. NW, Suite 1000
Washington, DC 20036
(202) 828-7100
Fax: (202) 293-1219
www.aiadc.org

American Iron and Steel Institute (AISI)
1101 17th St. NW, Suite 1300
Washington, DC 20036
(202) 452-7100
www.steel.org

American Land Tile Association
1828 L St. NW, Suite 705
Washington, DC 20036
(202) 296-3671
1-800-787-ALTA
Fax: (888) FAX-ALTA
www.alta.org

American Lumber Standards Committee (ALSC)
P.O. Box 210
Germantown, MD 20875
(301) 972-1700

American National Standards Institute (ANSI)
11 W. 42nd St.
New York, NY 10036
(212) 642-4900
Fax: (212) 398-0023
www.ansi.org

American Nursery & Landscape Association (ANLA)
1250 I St. NW, Suite 500
Washington, DC 20005
(202) 789-2900
Fax: (202) 789-1893
www.anla.org

American Olean Tile Company, Inc.
7834 C.F. Hawn Freeway
Dallas, TX 75217
(215) 393-2237
Fax: (215) 393-2784

American Painting Contractor
Douglas Publications, Inc.
2807 N. Parham Rd., Suite 200
Richmond, VA 23294
(804) 762-9600
Fax: (804) 217-8999
www.paintmag.com/apc.html

American Petroleum Institute (API)
1220 L St. NW
Washington, DC 20005
(202) 682-8000
www.api.org

American Planning Association
1776 Massachusetts Ave. NW, Suite 400
Washington, DC 20036
(202) 872-0611
Fax: (202) 872-0643
www.planning.org

American Skylites, Inc.
7451 Dogwood Park
Fort Worth, TX 76118
(817) 589-7811
(800) 772-7401
Fax: (817) 589-5959

American Slate Company
1889 Mt. Diablo Blvd.
Walnut Creek, CA 94596
(925) 977-4880
Fax: (925) 977-4885
http://www.americanslate.com
slatexpert@americanslate.com

American Society for Concrete Construction (ASCC)
3330 Dundee Rd.
Northbrook, IL 60062
(708) 291-0270

American Society for Testing and Materials (ASTM)
100 Barr Harbor Dr.
West Conshohocken, PA 19428-2959
(610) 832-9585
Fax: (610) 832-9555
www.astm.org

American Society of Appraisers
555 Herndon Pkwy., Suite 125
Herndon, VA 20170
(703) 478-2228
Fax: (703) 742-8471
www.apo.com

American Society of Appraisers
5 Revere Dr., #200
Northbrook, IL 60062
(847) 332-0222

American Society of Architectural Hardware Consultants
(*see* Door and Hardware Institute)

American Society of Civil Engineers (ASCE)
1801 Alexander Bell Dr.
Reston, VA 20191-4400
(703) 295-6300
(800) 548-2723
Fax: (703) 295-6222
www.asce.org

American Society of Consulting Planners (ASCP)
1776 Massachusetts Ave. NW, Suite 400
Washington, DC 20036
(202) 872-1498
Fax: (202) 789-2211

American Society of Heating, Refrigerating and Air-Conditioning Engineers, Inc. (ASHRAE)
1791 Tullie Circle NE
Atlanta, GA 30329
(404) 636-8400
(800) 527-4723
Fax: (404) 321-5478
www.ashrae.org

American Society of Landscape Architects (ASLA)
636 Eye St. NW
Washington, DC 20001-3736
(202) 898-2444
Fax: (202) 898-1185
www.asla.org

American Society of Mechanical Engineers (ASME)
Three Park Avenue
New York, NY 10016-5990
(800) THE-ASME
www.asme.org

American Society of Plumbing Engineers (ASPE)
3617 Thousand Oaks Blvd., Suite 210
Westlake, CA 91362
(805) 495-7120
Fax: (805) 495-4861
aspehq@aol.com
www.aspe.org

American Society of Safety Engineers, Inc.
1800 E. Oakton
Des Plaines, IL 60018
(847) 699-2929
Fax: (847) 768-3434
customerservice@asse.org
www.asse.org

American Society of Sanitary Engineering (ASSE)
P.O. Box 40362
Bay Village, OH 44140
(440) 835-3040
Fax: (440) 835-3488

American Solar Energy Society
2400 Central Ave., Unit G-1
Boulder, CO 08301
(303) 443-3130
Fax: (303) 443-3212
ases@ases.org
www.ases.org

American Water Works Association, Inc. (AWWA)
6666 W. Quincy Ave.
Denver, CO 80235
(303) 794-7711
www.awwa.org

American Welding Society (AWS)
550 N.W. Lejune Rd.
P.O. Box 351040
Miami, FL 33126
(305) 443-9353
(800) 443-9353
Fax: (305) 443-7559
www.aws.org

American Wood Council
(*see* American Forest & Paper Association)

American Wood-Preservers' Association (AWPA)
P.O. Box 5690
Granbury, TX 76049
(817) 326-6300
Fax: (817) 326-6306
awpa@itexas.net
www.awpa.com

American Wood Preservers Institute (AWPI)
2750 Prosperity Ave., Suite 550
Fairfax, VA 22031-4312
(703) 204-0500
(800) 356-AWPI
Fax: (703) 204-4610
info@awpi.org
www.awpi.org

Ancor Granite Tile Inc.
435 Port-Royal West
Montreal, Que. H3L 2C3, Canada
(514) 385-9366
Fax: (514) 382-3533
webmaster@ancor.ca
www.ancor.ca

Andersen Corporation
100 4th Ave., N
Bayport, MN 55003
(651) 439-5150
Fax: (651) 430-7246

Anti-Hydro International, Inc.
45 River Rd.
Flemington, NJ 08822
(908) 284-9000
Fax: (908) 284-9464

APA—The Engineered Wood Association
P.O. Box 11700
Tacoma, WA 98411-0700
(253) 565-6600
Fax: (253) 565-7265
www.apawood.org

Arch Aluminum & Glass Co., Inc.
10200 N.W. 67th St.
Tamarac, FL 33321
(954) 724-1775
Fax: (954) 724-9293
webmaster@arch.amarlite.com
www.arch.amarlite.com

Architectural Record
Two Penn Plaza
New York, NY 10121
(212) 904-2594
www.architecturalrecord.com

Architectural Technology
(now a part of *Architecture*)
American Institute of Architects
1735 New York Ave. NW
Washington, DC 20006
(202) 626-7300

Architectural Woodwork Institute (AWI)
1952 Isaac Newton Square West
Reston, VA 20190
(703) 773-0600
Fax: (703) 773-0584
www.awinet.org

Architecture
American Institute of Architects
1735 New York Ave. NW
Washington, DC 20006
(202) 626-7300

Arcom Master Systems
332 East 500 South St.
Salt Lake City, UT 84111
(800) 424-5080
Fax: (801) 521-9163
arcom@arcomnet.com
www.arcomnet.com

Armstrong World Industries
P.O. Box 3001
Lancaster, PA 17604
(717) 397-0611
Fax: (717) 396-4426
www.ceilings.com

Arrow Architectural Hardware
103-00 Foster Ave.
Brooklyn, NY 11236
(718) 257-4700
(800) 221-6529
Fax: (718) 649-9097
www.arrowlock.com

Ashley Corporation
P.O. Box 25522
Richmond, VA 23260
(804) 355-7102

ASME International
Three Park Avenue
New York, NY 10016-5990
800-THE-ASME (U.S./Canada)
www.asme.org

ASM International
9639 Kinsman Rd.
Materials Park, OH 44073
(440) 338-5151
(800) 336-5152
Fax: (440) 338-4634
www.asminternational.org

Asphalt Institute (AI)
Research Park Dr.
P.O. Box 14052
Lexington, KY 40512-4052
(606) 288-4960
Fax: (606) 288-4999
www.asphaltinstitute.org

Asphalt Roofing Manufacturers Association (ARMA)
4041 Powder Mill Rd., Suite 404
Calverton, MD 20705
(301) 348-2002
Fax: (301) 348-2020
www.asphaltroofing.org/contact.html

Associated Air Balance Council (AABC)
1518 K St. NW, Suite 503
Washington, DC 20005
(202) 737-0202
Fax: (202) 638-4833
aabchq@aol.com
www.aabchq.com

Associated Builders and Contractors (ABC)
1300 N. 17th St., 8th Floor
Rosslyn, VA 22209
(703) 812-2000
www.abc.org

Associated General Contractors of America (AGC)
333 John Carlyle St., Suite 200

Alexandria, VA 22314
(703) 548-3118
Fax: (703) 548-3119
info@agc.org
www.agc.org

Association for Preservation Technology
P.O. Box 3511
Williamsburg, VA 23187
(540) 373-1621
Fax: (888) 723-4242
www.apt.org

Association of Home Appliance Manufacturers
1111 19th St. NW, Suite 402
Washington, DC 20036
(202) 872-5955
Fax: (202) 872-9354
www.aham.org

Association of Iron and Steel Engineers
3 Gateway Center, Suite 2350
Pittsburgh, PA 15222
(412) 281-6323
Fax: (412) 281-4657
www.aise.org

Association of Official Seed Analysts (AOSA)
268 Plant Science, 1ANR-UNL, Box 19281
Lincoln, NE 68583
(402) 472-8649
www.zianet.com

Association of the Wall and Ceiling Industries—International (AWCI—International)
803 W. Broad St., Suite 600
Falls Church, VA 22046
(703) 534-8300
Fax: (703) 534-8307
www.AWCI.org

Athenian Marble
369 Syngrou Ave.
P. Phaleron
175 64 Athens, Greece
003 01 94 800 48-9
Fax: 003 01 98 355 43
amarble@acci.gr

Battelle Memorial Institute
Columbus Labs
505 King Ave.
Columbus, OH 43201
(614) 424-6424
www.batell.org

Best Lock Corporation
6161 E. 75th St.
Indianapolis, IN 46250
(317) 849-2250
(317) 595-7620

Bethlehem Steel Corporation
1170 8th Ave.
Bethlehem, PA 18016
(610) 694-2424
Fax: (610) 694-5743
www.bethsteel.com

Binks Manufacturing Company
12705-T Robin Lane
Brookfield, WI 53005-3125
(262) 781-6880

W.R. Bonsal Company
P.O. Box 241148
Charlotte, NC 28224-1448
(704) 525-1621
(800) 334-0784
Fax: (704) 529-5261
http://www.bonsal.com

Boon Edam, Inc.
4050 South 500 West
Salt Lake City, UT 84123
(801) 261-8980
(800) 658-8776
Fax: (801) 261-1612
derek@boonedam.com
http://www.boonedam.com

Brick Institute of America (BIA)
11490 Commerce Park Dr., Suite 300
Reston, VA 22091
(703) 620-0010

Bruce Hardwood Floors
16803 Dallas Pkwy.
Dallas, TX 75248
(972) 506-0480
Fax: (972) 506-7821
www.brucehardwoodfloors.com

Buckingham-Virginia Slate Corporation
1 Main St.
P.O. Box 8
Arvonia, VA 23004-0008
(804) 581-1131
(800) 235-8921
Fax: (804) 581-1130
www.bvslate.com

Builders Hardware Manufacturers Association (BHMA)
355 Lexington Ave., 17th Floor
New York, NY 10017-6603
(212) 661-4261
www.buildershardware.com

Building Design and Construction
1350 E. Touhy Ave.
P.O. Box 5080
Des Plaines, IL 60018-3358
(847) 635-8800
Fax: (847) 390-2152
www.bdcmag.com

Building Officials and Code Administrators International (BOCAI)
4051 W. Flossmoor Rd.
Country Club Hills, IL 60478
(708) 799-2300
Fax: (708) 799-4981
www.bocai.org

Building Owners and Managers Association International
1201 New York Ave., Suite 300
Washington, DC 20005
(202) 408-2662
Fax: (202) 371-0181
www.boma.org

Building Stone Institute (BSI)
420 Lexington Ave.
New York, NY 10017
(212) 490-2530
Note: BSI has no technical personnel. Author recommends contacting individual producers associations or producers rather than BSI.

Building Stone Magazine
(*see* Building Stone Institute)

Burke Company
4810 Dufferin St.
Downsview, Ont. M3H 5S8, Canada
(800) 423-9140

Burlington Natstone Inc.
2701-C W. 15th St., Suite 505
Plano, TX 75075
(972) 985-9182
Fax: (972) 612-0847

Business and Institutional Furniture Manufacturers Association (BIFMA)
2680 Horizon Dr. SE, Suite A-1
Grand Rapids, MI 49546-7500
(616) 285-3963
Fax: (616) 285-3765
email@bifma.org
www.bifma.com

California Lath and Plaster Association
(*see* Western Lath, Plaster, Drywall Contractors Association)

California Redwood Association
405 Enfrente Dr., Suite 200
Novato, CA 94949
(415) 382-0662
(888) 225-7339
Fax: (415) 382-8531
cfgrover@worldnet.att.net
www.calredwood.org

Camara Slate, Inc.
Rte. 22A
Fairhaven, VT 05743
(802) 265-3200
Fax: (802) 265-2211

Capital Development Board, State of Illinois
Third Floor, William G. Stratton Building
401 S. Spring St.
Springfield, IL 62706
(217) 782-2864

Cardinal IG
12301 Whitewater Dr.
Minneapolis, MN 55343
(612) 935-1722
(800) 843-1484
Fax: (612) 935-5538

Carpet and Rug Institute (CRI)
Box 2048
Dalton, GA 30722
(706) 278-3176
Fax: (706) 278-8835
www.carpet-rug.com

Carpet Cushion Council (CCC)
P.O. Box 546
Riverside, CT 06878
(203) 637-1312
www.carpetcushion.org

Cast Iron Soil Pipe Institute (CISPI)
5959 Shallowford Rd., Suite 419
Chattanooga, TN 37421
(615) 892-0137
www.cispi.org

Ceco Concrete Construction, LLC
2900 Brooktree Lane, Suite 200
Kansas City, MO 64119
(816) 459-7000
Fax: (816) 459-7735
gmartin@cecoopctr.com
www.cecoconcrete.com

Ceco Door Products
750 Old Hickory Blvd., Bldg. 1, Suite 150
Brentwood, TN 37027-4504
(615) 661-5030
Fax: (615) 370-5299

Cedar Shake and Shingle Bureau
P.O. Box 1178
Sumas, WA 98295
(604) 462-8961
Fax: (604) 462-9386
www.cedarbureau.org

Ceiling and Interior Systems Construction Association (CISCA)
579 West North Ave., #301
Elmhurst, IL 60126
(708) 833-1919
Fax: (708) 833-1940

Celotex Corporation
4010 Boy Scout Blvd.
Tampa, FL 33607
(813) 873-1700
Fax: (813) 873-4103

Ceramic Tile Institute of America (CTI)
12061 Jefferson Blvd.
Culver City, CA 90230-6219
(310) 574-7800
Fax: (310) 821-4655
ctioa@earthlink.net
www.ctioa.org

Certified Ballast Manufacturers Association (CBM)
1422 Euclid Ave.
Hanna Building, Suite 722
Cleveland, OH 44115
(216) 241-0711

Chamber of Commerce of the United States
Construction, Housing, and Community Development
1615 H St. NW
Washington, DC 20062
(202) 659-6000
Fax: (202) 463-3190
www.chamberbiz.com

Chicago Plastering Institute
6547 N. Avondale
Chicago, IL 60631
(312) 774-4500

Chicago Title and Trust Company
171 N. Clark St.
Chicago, IL 60601
(800) 621-1919
www.ctt.com

Clemco Industries Corporation
One Cable Car Drive
Washington, MO 63090
(636) 239-0300
Fax: (800) 726-7559
www.clemcoindustries.com

Clemson University
Department of Civil Engineering
Lowery Hall
Clemson, SC 29634
(864) 656-3000
Fax: (864) 656-2670
ce@ces.clemson.edu
www.tiger2.ces.clemson.edu/depttest/default.htm

Cold Finished Steel Bar Institute
700 14th St. NW, Suite 900
Washington, DC 20005
(202) 508-1030
Fax: (202) 508-1010
www.cfsbi.com

Cold Regions Research and Engineering Laboratory
72 Lyme Rd.
Hanover, NH 03755-1290
(603) 646-4292
info@crrel.usace.army.mil
www.crrel.usace.army.mil/contactcrrel.htm

Cold Spring Granite Co.
202 South Third Ave.
Cold Spring, MN 56320
(612) 685-3621
Fax: (612) 685-8490
www.coldspringgranite.com

Color Association of the United States (CAUS)
589 Eighth Ave., 12th Floor
New York, NY 10018
(212) 372-8600
www.colorassociation.com

Commercial Renovation
(formerly *Commercial Remodeling*)
20 E. Jackson Blvd.
Chicago, IL 60120
(312) 922-5402

Composite Panel Association (CPA)
18928 Premier Court
Gaithersburg, MD 20879
(301) 670-0604
Fax: (301) 840-1252

Compressed Air and Gas Institute (CAGI)
c/o Thomas Associates, Inc.
1300 Summer Ave.
Cleveland, OH 44115
(216) 241-7333
Fax: (216) 241-0105
cagi@cagi.org
www.cagi.org

Compressed Gas Association (CGA)
1725 Jefferson Davis Hwy., Suite 1004
Arlington, VA 22202
(703) 412-0900
Fax: (703) 412-0128
www.cganet.com

Concrete Block Insulating Systems, Inc.
P.O. Box 1000
Freight House Rd.
West Brookfield, MA 01585
(508) 867-4241
(800) 628-8476
Fax: (508) 867-5702

Concrete Reinforcing Steel Institute (CRSI)
933 Plum Grove Rd.

Schaumburg, IL 60173
(847) 517-1200
Fax: (847) 517-1206
www.crsi.org

Congoleum Corporation
3705 Quakerbridge Rd.
Mercerville, NJ 08619
(609) 584-3000
Fax: (609) 584-3607
www.congoleum.com

Construction Industry Manufacturers Association
111 E. Wisconsin Ave., Suite 940
Milwaukee, WI 53202
(414) 272-0943
Fax: (414) 272-1170
cima@cimanet.com
www.cimanet.com

Construction Specialties, Inc.
49 Meeker Ave.
Cranford, NJ 07016
(908) 272-5200

Construction Specifications Institute (CSI)
99 Canal Center Plaza, Suite 300
Alexandria, VA 22314
(703) 684-0300
(800) 689-2900
Fax: (703) 684-0465
www.csinet.org

The Construction Specifier
(*see* Construction Specifications Institute)

Consumer Reports
(*see* Consumers Union)

Consumers Union
101 Truman Ave.
Yonkers, NY 10703
(914) 378-2000
www.consumer.org

Copper and Brass Fabricators Council
1050 17th St. NW, #440
Washington, DC 20036
(202) 833-8575
www.recycle.new

Copper Development Association (CDA)
260 Madison Ave., 16th Floor
New York, NY 10016
(212) 251-7200
(800) CDA-DATA
Fax: (212) 251-7234
The-Copper-Page@cda.copper.org
www.copper.org

Corning, Inc.
1-T Riverfront Plaza
Corning, NY 14831
(607) 974-9000

Coronado Paint Co., Inc.
P.O. Box 308
308 Old County Rd.
Edgewater, FL 32132
(904) 428-6461
(800) 883-4193
Fax: (904) 427-7130
coronadopt@aol.com
www.coronadopaint.com

Council of American Building Officials (CABO)
(*see* International Code Council [ICC])

Cove, Cooper, Lewis, Inc.
135 E. 55th St.
New York, NY 10022
(212) 421-2822

Craftsman Book Company
6058 Corte del Cedro
Carlsbad, CA 92009
(800) 829-8123
Fax: (760) 438-0398
www.craftsman-book.com

Crane Fulview Door Company
924 Sherwood Dr.
Lake Bluff, IL 60044
(847) 295-2700
Fax: (847) 295-5288

Creative Edge Corporation
601 S. 23rd St.
Fairfield, IA 52556
(515) 472-8145
(800) 394-8145
Fax: (515) 472-2848
sales@cec-waterjet.com
http://www.cec-waterjet.com

Crossfield Products Corporation
Dex-O-Tex Division
140 Valley Rd.
Roselle Park, NJ 07204
(908) 245-2800
Fax: (908) 245-2583

DAP Inc.
2400 Boston St.
Baltimore, MD 21224
(410) 675-2100
(800) 327-1044
Fax: (410) 558-1120

Dekker, Marcel, Inc.
270 Madison Ave.
New York, NY 10016
(212) 696-9000
Fax: (212) 685-4540
www.dekker.com

Delta Faucet Company
P.O. Box 40980
55 E. 111 St.
Indianapolis, IN 46280
(317) 848-1812
Fax: (317) 574-5550
jam@deltafaucet.com
www.deltafaucet.com

Domtar Gypsum America
24 Frank Lloyd Wright Dr.
Ann Arbor, MI 48105-9755
P.O. Box 543
Ann Arbor, MI 48106-0543
(800) 366-8271
Fax: (313) 930-4709
http://search.commerceinc.com/profile/id2046886400.html

Door and Hardware Institute (DHI)
4170 New Brook Dr.
Chantilly, VA 22021-2223
(703) 222-2010
Fax: (703) 222-2410
www.dhi.org

Door Operator and Remote Control Manufacturers Association
655 W. Irving Park, #201 Park Place
Chicago, IL 60613
(312) 525-2644

Dover Elevator Systems, Inc.
6266 Hurt Rd.
Horn Lake, MS 38637
(601) 393-2110
Fax: (601) 342-4309

Dow Corning Corporation
P.O. Box 994
Midland, MI 48686-0994
(517) 496-4000
Fax: (517) 496-4572

Ductile Iron Pipe Research Association
245 Riverchase Pkwy. E, Suite 0
Birmingham, AL 35244
(205) 402-8700
Fax: (205) 402-8730
www.dipra.org

Dur-A-Flex, Inc.
95 Goodwin St.
East Hartford, CT 06108
(800) 253-3539
(860) 528-9838
Fax: (860) 528-2802
afs@dur-a-flex.com
www.dur-a-flex.com

Dwyer Products Corp.
418 N. Calumet Ave.
Michigan City, IN 46360
(219) 874-5236
(800) 348-8508
Fax: (219) 874-2823
Fax: (800) 255-6785
www.dwyerkitchens.com

Edison Electric Institute
701 Pennsylvania Ave. NW
Washington, DC 20004
(202) 508-5000

EIFS Industry Members Association (EIMA)
3000 Corporate Center Dr., Suite 270
Morrow, GA 30260
(770) 968-7945
(800) 294-3462
Fax: (770) 968-5818

Electronic Industries Alliance (EIA)
2500 Wilson Blvd.
Arlington, VA 22201
(703) 907-7500
www.eia.org

ELE International, Inc.
Soil Test Products Division
(formerly Soil Test, Inc.)
86-7 Albrecht Dr.
Lake Bluff, IL 60044
(708) 295-9400
Fax: (847) 295-9414

Ellison Bronze Company, Inc.
125 W. Main St.
Falconer, NY 14733
(716) 665-6522
Fax: (716) 665-5552

ETL Testing Laboratories, Inc. (ETL)
40 Commerce Way #B
Totowa, NJ 07512
(973) 785-9211

Expanded Shale, Clay, and Slate Institute
6218 Montrose Rd.
Rockville, MD 20852
(301) 231-9497
www.escsi.org

Exteriors
1 E. First St.
Duluth, MN 55802
(218) 723-9200

Facing Tile Institute (FTI)
P.O. Box 8880
Canton, OH 44711
(216) 488-1211

Falcon Lock
2650 Orbiter St.
Brea, CA 92821
(800) 266-4456
Fax: (800) 777-8229
http://www.falconlock.com

Federal Housing Administration (FHA)
(*see* U.S. Department of Housing and Urban Development)

Federal Specifications
(*see* U.S. General Services Administration, Specifications Unit)

Federation of Societies for Coatings Technology
492 Norristown Rd.
Blue Bell, PA 19422-2350
(610) 940-0777
Fax: (610) 940-0292
fsct@coatingstech.org
www.coatingstech.org

Ferroalloys Association
900 Second St. NE, Suite 306
Washington, DC 20002
(202) 842-0292
Fax: (202) 842-4840
www.amc.scra.org

Fine Hardwood Veneer Association (FHVA)
(*see* Hardwood Plywood and Veneer Association [HPVA])

Firematic Sprinkler Devices, Inc.
900 Boston Tpk.
Shrewsbury, MA 01545
(508) 845-2121
(800) 225-7288
Fax: (508) 842-3523

Firestone Building Products Company
525 Congressional Blvd.
Carmel, IA 46032
(317) 575-7000
(800) 428-4442
Fax: (317) 575-7100
www.firestonebpco.com

Fitzgerald Formliners
1341 E. Pomona St.
Santa Ana, CA 92705
(714) 547-6710
Fax: (714) 547-7958
www.formliners.com

Flat Glass Marketing Association (FGMA)
(*see* Glass Association of North America)

Flexicore Manufacturers Association
P.O. Box 825
Dayton, OH 45424
(937) 226-8791

FM Global (FM)
1301 Atwood Ave.
Johnston, RI 02919
(401) 275-3000
(401) 275-3029
information@fmglobal.com
www.fmglobal.com

Formica Corporation
10155 Reading Rd.
Cincinnati, OH 45241
(513) 786-3400
(800) 367-6422
Fax: (513) 786-3024

GAF Material Corporation
1361 Alps Rd.
Wayne, NJ 07470
(973) 628-3000
(800) 766-3411
Fax: (973) 628-4954
www.gaf.com

Garland Floor Company
4500 Willow Pkwy.
Cleveland, OH 44125
(216) 883-4100
(800) 321-2395
Fax: (216) 883-9076
www.garlandfloor.com

Gas Appliance Manufacturers Association
1901 N. Moore St., Suite 1100
Arlington, VA 22209
(703) 525-9565
Fax: (703) 525-0718
information@gamanet.org
www.gamanet.org

General Polymers
145 Caldwell Dr.
Cincinnati, OH 45216
(513) 761-0011
(800) 543-7694
Fax: (513) 761-1330
info@generalpolymers.com
www.generalpolymers.com

General Services Administration (GSA)
(*see* U.S. General Services Administration)

Georgia Marble
P.O. Box 238
Georgia Marble Lane
Tate, GA 30177
(770) 735-2611
(800) 334-0122
Fax: (770) 735-4456

Georgia Pacific Corporation
133 Peachtree St., NE
Atlanta, GA 30348
(404) 652-4000
(800) 284-5347
Fax: (404) 827-7006
www.gp.com

Glass Association of North America
310 Miller Ave.
Ann Arbor, MI 48103
(734) 930-9277
(800) 699-9277
Fax: (734) 930-9088
http://www.cssinfo.com/info/gana.html

Glass Tempering Association
(*see* Glass Association of North America)

Gold Bond Building Products
(*see* National Gypsum Company)

Goodyear Tire and Rubber Company
1144 E. Market St., Dept. E.B.
Akron, OH 44305
(330) 796-2121
(888) 899-6354
Fax: (402) 470-4404

Grace Construction Products
W.R. Grace and Company
62 Whittemore Ave.
Cambridge, MA 02140
(617) 876-1400
(800) 354-5414
Fax: (617) 498-4311
www.graceconstruction.com

Graventa (Canada) Ltd.
7505 134 A St.
Surrey, B.C. V3W 7B3, Canada
(604) 594-0422

Greek Trade Commission
168 N. Michigan Ave., #500
Chicago, IL 60601
(312) 332-1716
Fax: (312) 236-5127
info@hepogreektrade.com
http://www.hepogreektrade.com/marble

Greenstreak Plastic Products Co.
3400 Tree Court, Industry Blvd.
St. Louis, MO 63122
(314) 225-9400
Fax: (800) 551-5145

Guard Contract Wallcoverings
1280 N. Grant Ave.
Columbus, OH 43201
(614) 297-6000
(800) 521-5250
Fax: (614) 297-6077

Guardian Industries Corp.
14600 Romine Rd.
Carleton, MI 48117
(734) 654-6264
(800) 521-9040
Fax: (734) 654-4386

Gypsum Association (GA)
810 First St. NE, Suite 510
Washington, DC 20002
(202) 289-5440
Fax: (202) 289-3707
info@gypsum.org
www.gypsum.org

Harbil Manufacturing
1023-T S. Wheeling Rd.
Wheeling, IL 60090
(847) 537-0880
Fax: (847) 537-5530

Hardwood Manufacturers Association (HMA)
400 Penn Center Blvd., Suite 530
Pittsburgh, PA 15235
(412) 829-0770

Hardwood Plywood and Veneer Association (HPVA)
1825 Michael Farraday Dr.
P.O. Box 2789
Reston, VA 22090
(703) 435-2900
hpva@erols.com
www.hpva.org

Harris Specialty Chemicals, Inc.
10245 Centurion Pkwy. N
Jacksonville, FL 32256-0565
(904) 996-6000
(800) 322-7825
Fax: (904) 996-6300

Heat Exchange Institute (HEI)
c/o Thomas Associates, Inc.
1300 Summer Ave.
Cleveland, OH 44115
(216) 241-7333
www.taol.com/hei

Historic Preservation
(*see* National Trust for Historic Preservation)

Hollow Metal Manufacturers Association (HMMA)
8 S. Michigan Ave.
Chicago, IL 60603
(312) 332-0405
Fax: (312) 332-0706
naamm@gss.net
www.naamm.org

Home Ventilating Institute Division of Air Movement and Control Association International, Inc.
30 West University Dr.
Arlington Heights, IL 60004
(847) 394-0150
hvidiv@aol.com
general@hvi.org
www.hvi.org

Honeywell
101 Columbia Rd.
Morristown, NJ 07962
(973) 455-2000
Fax: (973) 455-4807
www.honeywell.com

Hubbell, Inc.
584 Derby-Milford Rd.
P.O. Box 549
Orange, CT 06477
(203) 799-4100
Fax: (203) 799-4254

Hunter Douglas Architectural Products
5015 Oakbrook Pkwy., Suite 100
Norcross, GA 30093
(770) 806-9557
(800) 366-4327
Fax: (770) 806-0214
www.hunterdouglasceilings.com

Hydraulic Institute (HI)
9 Sylvan Way #180
Parsippany, NJ 07054
(973) 267-7772
(888) 786-7744

Hydronics Institute Division of GAMA (HI)
P.O. Box 218
35 Russo Place
Berkeley Heights, NJ 07922
(908) 464-8200

Illuminating Engineering Society of North America (IESNA)
120 Wall St., 17th Floor
New York, NY 10005
(212) 248-5000
Fax: (212) 248-5017
bethbay@aol.com
www.iesna.org

Indiana Limestone Co., Inc.
P.O. Box 1560
Bedford, IN 47421
(812) 275-3341
(800) 457-4026
Fax: (812) 275-3344
www.ilco.com

Indiana Limestone Institute of America, Inc. (ILI)

Stone City Bank Building, Suite 400
Bedford, IN 47421
(812) 275-4426

Industrial Risk Insurers (IRI)
85 Woodland St.
Hartford, CT 06102
(860) 520-7300
(800) 243-8308
www.industrialrisk.com

Inland Steel Company
30 W. Monroe St.
Chicago, IL 60603
(312) 346-0300
(312) 899-3380

Institute of Electrical and Electronics Engineers, Inc. (IEEE)
3 Park Ave., 17th Floor
New York, NY 10016-5997
(212) 419-7900
Fax: (212) 752-4929
www.ieee.org

Institute of Environmental Sciences
940 E. Northwest Hwy.
Mt. Prospect, IL 60056
(708) 255-1561

Institute of Industrial Engineers
25 Technology Park
Norcross, GA 30092
(770) 449-0460
(800) 494-0460
Fax: (770) 441-3295
www.iienet.org

Institute of Real Estate Management
430 N. Michigan Ave.
Chicago, IL 60611
(312) 329-6000
Fax: (312) 329-6039
www.irem.org

Instrument Society of America (ISA)
67 Alexandria Dr.
P.O. Box 12277
Research Triangle Park, NC 27709
(919) 549-8411
Fax: (919) 549-8288
info@isa.org
www.isa.org

Insulated Cable Engineers Association, Inc. (ICEA)
P.O. Box 440
South Yarmouth, MA 02664
(508) 394-4424

Insulated Steel Door Systems Institute
30200 Detroit Rd.
Cleveland, OH 44145
(440) 899-0010
Fax: (440) 892-1404
www.wherryassoc.com

Insulating Glass Certification Council (IGCC)
P.O. Box 9
Henderson Harbor, NY 13651
(315) 938-7444
Fax: (315) 938-7453
jgkent@gisco.net
www.igcc.org

Intermatic, Inc.
Intermatic Plaza
7777 Winn Rd.
Spring Grove, IL 60081
(815) 675-2321
(815) 675-2112

International Association of Plumbing and Mechanical Officials (IAPMO)
20001 E. Walnut Dr., S
Walnut, CA 91789
(909) 595-8449
Fax: (909) 594-3690
iapmo@iapmo.org
www.iapmo.com

International Code Council (ICC)
5203 Leesburg Pike, Suite 708
Falls Church, VA 22041
(703) 931-4533
Fax: (703) 379-1546
www.intlcode.org

International Conference of Building Officials (ICBO)
5360 W. Workman Mill Rd.
Whittier, CA 90601
(562) 699-0541
(800) 284-4406
Fax: (888) 329-4226
www.icbo.org

International Institute of Lath and Plaster
820 Transfer Rd.
St. Paul, MN 55114-1406
(651) 645-0208
Fax: (651) 645-0209

International Lead Zinc Research Organization
P.O. Box 12036
Research Triangle Park
Raleigh, NC 27709-2036
(919) 361-4647
Fax: (919) 361-1957

International Masonry Institute
823 15th St. NW
Washington, DC 20005
(202) 783-3908

International Municipal Signal Association (IMSA)
P.O. Box 539
165 E. Union St.
Newark, NJ 14513
(315) 331-2182
(800) 723-4672
Fax: (315) 331-8205
Info@IMSAsafety.org
www.imsasafety.org

Interprofessional Council on Environmental Design
(*see* National Society of Professional Engineers)

Iron League of Chicago, Inc.
117 N. Jefferson St.
Chicago, IL 60661
(312) 441-9425

Italian Ceramic Tile, Inc.
7521 N. Milwaukee Ave.
Niles, IL 60714
(847) 647-0194

ITW Ramset/Redhead
155 International Blvd.
Glendale Heights, IL 60139
(630) 653-3008
Fax: (630) 350-7985

J.L. Industries
4450 W. 78th St. Circle
Bloomington, MN 55435
(800) 554-6077
Fax: (612) 835-2218
www.jlindustries.com

Johns Manville Corp.
717 17th St.
Denver, CO 80202
(303) 978-2000
(800) 654-3103
Fax: (303) 978-3661
www.jm.com

Johnsonite
Division of Duramax, Inc.
16910 Munn Rd.
Chagrin Falls, OH 44023
(440) 543-8916
(800) 899-8916
Fax: (440) 543-5774

John Wiley & Sons, Inc.
605 Third Ave.
New York, NY 10158
(212) 850-6000
Fax: (212) 850-6088
info@wiley.com
www.wiley.com

Journal of Protective Coatings and Linings
(*see* Society for Protective Coatings)

Kaiser Aluminum Forging and Casting
1015 E. 12th St.
Erie, PA 16503
(814) 454-4571
Fax: (814) 456-9046

Kalwall Corporation
1111 Candia Rd.
Manchester, NH 03109
(603) 627-3861
(800) 258-9777
Fax: (603) 627-7905

Kawneer Company
Department C
Technology Park-Atlanta
555 Guthridge Court
Norcross, GA 30092
(404) 449-5555
Fax: (770) 840-6463
www.kawneer.com

Kentile Operating Company
4532 S. Kolin Ave.
Chicago, IL 60632
(888) 453-6845
www.kentile.com

Kitchen Cabinet Manufacturers Association (KCMA)
1899 Preston White Dr.
Reston, VA 20191-5435
(703) 264-1690
Fax: (703) 620-6530
www.kcma.org

K-Lath Division
Georgetown Wire Co., Inc.
13470 Philadelphia Ave.
P.O. Box 489
Fontana, CA 92335-0489
(909) 360-8288
Fax: (909) 360-6663

Koppers Industries, Inc.
1600 Koppers Building
436 Seventh Ave.
Pittsburgh, PA 15219
(412) 227-2001
Fax: (412) 227-2262

Krieger Publishing Co.
1725 Krieger Lane
Malabar, FL 32950
(407) 724-9542
info@krieger-pub.com

Kurfees Coatings
(formerly Jewel Paint Co.)
201 E. Market St.
Louisville, KY 40201
(502) 584-0151
Fax: (502) 584-1701

Laminated Glass Corporation
P.O. Box 1003
375 E. Church Ave.
Telford, PA 18969
(215) 721-0400
(800) 523-6039
Fax: (215) 721-0402

Laminator's Safety Glass Association
(*see* Glass Association of North America)

Lane Publishing Company
(*see* Sunset Publishing Corp.)

Larsen's Manufacturing Company
7421 Commerce Lane, NE
Minneapolis, MN 55432
(612) 571-1181
(800) 527-7367
Fax: (612) 571-6900
info@larsensmfg.com
www.larsensmfg.com

Lead Industries Association, Inc. (LIA)
13 Main St.
Sparta, NJ 07871
(973) 726-4484
miller@leadinfo.com
www.leadinfo.com

Levelor Corp.
595 Lawrence Expressway
Sunnyvale, CA 94086
(408) 245-4000

Leviton Manufacturing Company, Inc.
5925 Little Neck Pkwy.
Little Neck, NY 11362
(718) 229-4040
(718) 281-6075
Fax: (718) 631-6439
kev@vds.net
http://www.leviton.com/sections/prodinfo/newprod/s5c1p1.htm

Libby-Owens-Ford Building Products
500 Exton Commons
Exton, PA 19341
(610) 363-8994
(800) 523-0133
Fax: (610) 363-9035
http://www.pilkington.com/sites/lof/bpsales.html

Library of Congress
101 Independence Ave. SE
Washington, DC 20540
(202) 287-5000
www.lcweb.loc.gov

Lightning Protection Institute (LPI)
3335 N. Arlington Heights Rd., Suite E
Arlington Heights, IL 60004
(847) 577-7200
(800) 488-6864
Fax: (847) 577-7276
strike@lightning.org
www.lightning.org

Lukens Inc.
(*see* Bethlehem Steel Corporation)

MAB Paints
600 Reed Rd.
Broomall, PA 19008
(610) 353-5100
(800) 622-1899
Fax: (610) 353-8189
www.mabpaints.com

Manufactured Housing Institute
2101 Wilson Blvd., Suite 610
Arlington, VA 22201-3062
(703) 558-0400
Fax: (703) 558-0401
info@mfghome.org
www.mfghome.org

Manufacturers Standardization Society of the Valve and Fittings Industry (MSS)
127 Park St., NE
Vienna, VA 22180
(703) 281-6613
Fax: (703) 281-6671
info@mss-hq.com
www.mss-hq.com

Maple Flooring Manufacturers Association, Inc. (MFMA)
60 Revere Dr., Suite 500
Northbrook, IL 60062
(708) 480-9138
Fax: (847) 480-9282
mfma@maplefloor.org
www.maplefloor.org

Marble Institute of America (MIA)
30 Eden Alley, Suite 301
Columbus, OH 43215
(614) 228-6194
Fax: (614) 461-1497
miaadmin@marble-institute.com
www.marble-institute.com

Martin Fireproofing Corporation
P.O. Box 27, Kenmore Station
2200 Military Rd.
Buffalo, NY 14217
(716) 692-3680
(800) 766-3969
Fax: (716) 693-3402

Marvin Windows & Doors
P.O. Box 100
Warroad, MN 56763
(218) 386-1430

(800) 346-5128
(218) 386-2925

Materials and Methods Standards Association
P.O. Box 350
Grand Haven, MI 49417
(616) 842-7844
Fax: (616) 842-1547

Maxxon Corporation
920 Hamel Rd.
Hamel, MN 55340
(612) 478-9600
(800) 356-7887
Fax: (612) 478-2431
www.maxxon.com

McGraw-Hill Book Company
1221 Avenue of the Americas
New York, NY 10020
(212) 512-4100
(800) 352-3566
Fax: (212) 512-4105
www.booksite.com

McKinley Iron Works Inc.
901 N. Throcmorton
Fort Worth, TX 76106
(817) 335-1268
(800) 433-2303
Fax: (817) 336-0677

Mechanical Contractors Association of America, Inc. (MCAA)
1385 Piccard Dr.
Rockville, MD 20850
(301) 869-5800
Fax: (301) 990-9690

Medeco Security Locks, Inc.
3625 Allegheny Dr.
Salem, VA 24153
(540) 380-5000
Fax: (800) 548-8472

Mercer Products Company, Inc.
37235 SR 19
Umatilla, FL 32784
(352) 357-4119
(800) 447-8442
Fax: (352) 357-9660

Merchant & Evans, Inc.
ZIP-RIB
P.O. Box 1680
100 Connecticut Dr.
Burlington, NJ 08016
(609) 387-3033
(800) 257-6215
Fax: (609) 387-4838

MERO Structures, Inc.
N112 W18810 Mequon Rd.
P.O. Box 610
Germantown, WI 53022
(414) 255-5561
Fax: (414) 255-6932

Metal Architecture
7450 N. Skokie Blvd.
Skokie, IL 60077
(847) 674-2200
Fax: (847) 674-3676
www.moderntrade.com

Metal Building Interior Products Company
5309 Hamilton Ave.
Cleveland, OH 44114
(216) 431-6400
Fax: (216) 431-9000

Metal Building Manufacturers Association (MBMA)
c/o Thomas Associates, Inc.
1300 Summer Ave.
Cleveland, OH 44115
(216) 241-7333
Fax: (216) 241-0105
mbma@mbma.com
www.mbma.com

Metal Construction Association
104 South Michigan Ave., Suite 1500
Chicago, IL 60603
(312) 201-0193
Fax: (312) 201-0214
www.mca1.org

Metal Construction News
7450 N. Skokie Blvd.
Skokie, IL 60077
(847) 674-2200
Fax: (847) 674-3676
www.moderntrade.com

Metal Lath/Steel Framing Association (ML/SFA)
A Division of NAAMM
11 S. La Salle St., #1400
Chicago, IL 60603-1210
(312) 201-0101

Metal Sales Manufacturing Corporation
7800 SR 60
Sellersburg, IN 47172
(812) 246-1935
Fax: (812) 246-1862

Milliken Carpet
P.O. Box 2956
201 Lukken Industrial Dr., W
LaGrange, GA 30241
(706) 880-5511
(800) 241-4826
Fax: (706) 880-5530
www.millikencarpet.com

MODAC
(*see* MAB Paints)

Modern Metals
625 N. Michigan Ave.
Chicago, IL 60611
(312) 654-2300

Mohawk Commercial Carpet
500 Townpark Lane, #400
Kennesaw, GA 30144
(770) 792-6460
(800) 554-6637
Fax: (770) 792-6465

Mohawk Flush Doors, Inc.
U.S. Rte. 11, P.O. Box 112
Northumberland, PA 17857
(717) 473-3557
Fax: (717) 473-3737

Monsanto Company
800 N. Lindbergh Blvd.
St. Louis, MO 63167
(314) 694-1000
Fax: (314) 694-7625
scarlettlfoster@monsanto.com
www.monsanto.com

Montgomery Kone
One Montgomery Court
Moline, IL 61265
(309) 764-6771
(800) 956-5663
Fax: (309) 743-5469

National Academy of Engineering
2101 Constitution Ave. NW
Washington, DC 20418
(202) 334-3200

National Academy of Sciences
Building Research Advisory Board
2101 Constitution Ave. NW
Washington, DC 20418
(202) 334-2000

National Aggregates Association (NAA)
1100 Bonifant St., #400
Silver Spring, MD 20910
Fax: (301) 562-2084
www.nationlaggregates.org

National Asphalt Pavement Association (NAPA)
NAPA Building
5100 Forbes Blvd.
Lanham, MD 20706
(301) 731-4748
(888) 468-6499
Fax: (301) 731-4621
www.hotmix.org

National Association of Architectural Metal Manufacturers (NAAMM)
8 S. Michigan Ave., Suite 1000
Chicago, IL 60603
(312) 332-0405

Fax: (312) 332-0706
naamm@naamm.org
www.naamm.org

National Association of Corrosion Engineers
1440 South Creek
Houston, TX 77084
(281) 228-6200
www.nace.org

National Association of Counties
440 First St. NW, Suite 800
Washington, DC 20001
(202) 393-6226
Fax: (202) 393-2630
www.naco.org

National Association of Floor Covering Distributors
401 N. Michigan Ave.
Chicago, IL 60611
(312) 321-6836
Fax: (312) 245-1085
www.nafcd.org

National Association of Home Builders (NAHB)
1201 15th St. NW
Washington, DC 20005
(202) 822-0200
www.nahb.org

National Association of Home Builders NAHB Research Center
400 Prince Georges Blvd.
Upper Marlboro, MD 20774
(301) 249-4000
www.nahbrc.org

National Association of Housing and Redevelopment Officials
630 Eye St. NW
Washington, DC 20001
(202) 289-3500
Fax: (202) 289-8181
nahro@nahro.org
www.nahro.org

National Association of Housing Cooperatives
1401 New York Ave. NW, Suite 1100
Washington, DC 20005-2160
(202) 737-0797
Fax: (202) 783-7869
www.coophousing.org

National Association of Minority Contractors, Inc.
806 15th St., Suite 340
Washington, DC 20005
(202) 347-8259

National Association of Realtors
430 N. Michigan Ave.
Chicago, IL 60611
(312) 329-8200
www.nal.realtor.com

National Association of Women in Construction
327 S. Adams St.
Fort Worth, TX 76104
(817) 877-5551
Fax: (817) 877-0324
nawic@onramp.net
www.nawic.org

National Board of Boiler and Pressure Vessels Inspectors
1055 Crupper Ave.
Columbus, OH 43229
(614) 888-8320
www.nationalboard.org

National Building Code
(*see* American Insurance Association)

National Building Granite Quarries Association
1220 L St. NW, Suite 100-167
Washington, DC 20005
(800) 557-2848
www.nbgqa.com

National Bureau of Standards (NBS)
(*see* National Institute of Standards and Technology)

National Climatic Data Center
Federal Building
151 Patton Ave.
Asheville, NC 28801-5001
(828) 271-4800
Fax: (828) 271-4876
www.ncdc.noaa.gov

National Coal Association
(*see* National Mining Association)

National Coil Coaters Association
401 N. Michigan Ave.
Chicago, IL 60611
(312) 321-6894
Fax: (312) 527-6705
ncca@sba.com
www.coilcoaters.org

National Concrete Masonry Association (NCMA)
2302 Horse Pen Rd.
Herndon, VA 22071
(703) 713-1900
Fax: (703) 713-1910
www.ncma.org

National Conference of States on Building Codes and Standards
481 Carlisle Dr.
Herndon, VA 22070
(703) 437-0100
www.ncsbcs.org

National Council of Acoustical Consultants, Inc.
66 Morris Ave., Box 359
Springfield, NJ 07081
(973) 564-5859
Fax: (973) 564-7480
info@ncac.com
www.ncac.com

National Council on Radiation Protection and Measurement (NCRPM)
7910 Woodmont Ave., Suite 800
Bethesda, MD 20814
(301) 657-2652
(800) 229-2652
ncrpexec@ncrp.com
www.ncrp.com

National Electrical Code (NEC)
(*see* National Fire Protection Association)

National Electrical Contractors Association, Inc. (NECA)
3 Bethesda Metro Center
Bethesda, MD 20814
(301) 657-3110
Fax: (301) 215-4500
webmaster@necanet.org
www.necanet.org

National Electrical Manufacturers Association (NEMA)
1300 N. 17th St., Suite 1847
Rosslyn, VA 22209
(703) 831-3200
www.nema.org

National Fire Protection Association (NFPA)
One Batterymarch Park
P.O. Box 9101
Quincy, MA 02269
(617) 770-3000
www.nfpa.org

National Forest Products Association (NFPA)
(*see* American Forest & Paper Association)

National Glass Association
8200 Greensboro Dr.
McLean, VA 22102
(703) 442-4890
Fax: (703) 442-0630
www.glass.org

National Guard Products
4985 E. Raines Rd.
Memphis, TN 38118
(800) 647-7874

Fax: (800) 255-7874
www.ngpinc.com

National Gypsum Company
2001 Rexford Rd.
Charlotte, NC 28211
(704) 365-7300
(800) 628-4662
Fax: (800) 329-6421
ng@national-gypsum.com
www.national-gypsum.com

National Hardwood Lumber Association (NHLA)
P.O. Box 34518
Memphis, TN 38184-1818
(901) 377-1818
Fax: (901) 382-6419
info@natlhardwood.org
www.natlhardwood.org

National Housing Conference
815 Fifteenth St. NW, Suite 538
Washington, DC 20005
(202) 393-5772
Fax: (202) 393-5656
nhc@nhc.org
www.nhc.org

National Institute of Building Sciences (NIBS)
1090 Vermont Ave. NW, Suite 700
Washington, DC 20005-4905
(202) 289-7800
Fax: (202) 289-1092
nibs@nibs.org
www.nibs.org

National Institute of Standards and Technology (NIST)
(formerly National Bureau of Standards)
100 Bureau Dr.
Gaithersburg, MD 20899
(301) 975-6478
www.nist.gov

National Kitchen and Bath Association (NKBA)
687 Willow Grove St.
Hackettstown, NJ 07840
(908) 852-0033
Fax: (908) 852-1695
www.nkba.org

National League of Cities
1301 Pennsylvania Ave. NW, 6th Floor
Washington, DC 20004
(202) 626-3000
Fax: (202) 626-3043
www.nlc.org

National Lime Association
200 N. Glebe Rd., Suite 800
Arlington, VA 22203
(703) 243-5463
Fax: (703) 243-5489
www.lime.org

National Lumber Grades Authority (NLGA)
1055 W. Hastings St., Suite 260
Vancouver, B.C. V6E 2E9, Canada
(604) 687-2171

National Mining Association (NMA)
1130 17th St. NW
Washington, DC 20036
(202) 463-6152
http://www.1800miti.com/associations/pagea022f.html

National Oak Flooring Manufacturers Association (NOFMA)
P.O. Box 3009
Memphis, TN 38173
(901) 526-5016
Fax: (901) 526-7022
info@NOFMA.org
www.nofma.org

National Ornamental and Miscellaneous Metals Association (NOMMA)
532 Forest Pkwy., Suite A
Forest Park, GA 30297
(404) 363-4009
Fax: (404) 366-1852
nommainfo@aol.com
www.nomma.org

National Paint and Coatings Association (NPCA)
(formerly National Paint, Varnish, and Lacquer Association)
1500 Rhode Island Ave. NW
Washington, DC 20005
(202) 462-6272
Fax: (202) 462-8549
npca@paint.org
www.paint.org

National Park Service
(*see* U.S. Department of the Interior)

National Particleboard Association (NPA)
(*see* Composite Panel Association)

National Pest Control Association, Inc.
8100 Oak St.
Dunn Loring, VA 22027
(703) 573-8330
Fax: (703) 573-4116
www.pestworld.org

National Precast Concrete Association
10333 N. Meridian St., Suite 272
Indianapolis, IN 46290
(317) 571-9500
(800) 366-7731
Fax: (317) 571-0041
www.precast.org

National Preservation Institute
P.O. Box 1702
Alexandria, VA 22313
(703) 765-0100
info@npi.org
www.npi.org

National Ready Mixed Concrete Association
900 Spring St.
Silver Spring, MD 20910
(301) 587-1400
(888) 846-7622
Fax: (301) 585-4219
www.nrmca.org

National Research Council
2101 Constitution Ave. NW
Washington, DC 20418
(202) 334-3400
www.nas.edu

National Retail Hardware Association
5822 W. 74th St.
Indianapolis, IN 46278
(317) 290-0338

National Roofing Contractors Association (NRCA)
O'Hare International Center
10255 W. Higgins Rd., Suite 600
Rosemont, IL 60018
(847) 299-9070
Fax: (847) 299-1183
www.nrca.net

National Safety Council
1121 Spring Lake Dr.
Itasca, IL 60143-3201
(630) 285-1121
Fax: (630) 285-1315
www.nsc.org

National Sand and Gravel Association
(*see* National Aggregates Association)

National Sanitation Foundation
(*see* NSF International)

National Sash and Door Jobbers Association
10047 Robert Trent Pkwy.
New Port Richey, FL 34655-4649
(727) 372-3665
Fax: (727) 372-2879
www.nsdja.com

National School Supply and Equipment Association (NSSEA)
8300 Colesville Rd., Suite 250
Silver Spring, MD 20910

(301) 495-0240
(800) 395-5550
Fax: (301) 495-3330
NSSEA@nssea.org
www.nssea.org

National Society of Professional Engineers (NSPE)
1420 King St.
Alexandria, VA 22314
(703) 684-2800
www.nspe.org

National Terrazzo and Mosaic Association (NTMA)
110 E. Market St., Suite 200 A
Leesburg, VA 20176
(703) 779-1022
(800) 323-9736
Fax: (703) 779-1026
info@ntma.com
www.ntma.com

National Tile Contractors Association (NTCA)
P.O. Box 13629
Jackson, MS 39236
(601) 939-2071
Fax: (601) 932-6117
ntca@Tile-Assn.com
www.Tile-Assn.com

National Tile Roofing Manufacturers Association, Inc. (NTRMA)
P.O. Box 40337
Eugene, OR 97404
(541) 689-0366
Fax: (541) 689-5530
info@ntrma.org
www.ntrma.org

National Trust for Historic Preservation
1785 Massachusetts Ave. NW
Washington, DC 20036
(202) 588-6000
(800) 944-6847
www.nthp.org

National Woodwork Manufacturers Association (NWMA)
(*see* Window and Door Manufacturers Association)

Nickel Development Institute
15 Toronto St., Suite 402
Toronto, Ont. M5C 2E3, Canada
(416) 362-8850

Nickel Development Institute
214 King St., W, Suite 510
Toronto, Ont. M5H 3S6, Canada
(416) 591-7999
www.nidi.org

North American Association of Mirror Manufacturers
2945 S.W. Wanamaker Dr., Suite A
Topeka, KS 66614-5321
(785) 271-0208
Fax: (785) 271-0166
naamm@glasswebsite.com
http://www.glasswebsite.com

North American Insulation Manufacturers Association (NAIMA)
44 Canal Center Plaza, Suite 310
Alexandria, VA 22314
(703) 684-0084
Fax: (703) 684-0427
insulation@naima.org
www.naima.org

Northeastern Lumber Manufacturers Association, Inc.
272 Tuttle Rd.
P.O. Box 87A
Cumberland Center, ME 04021
(207) 829-6901
Fax: (207) 829-4293
nelma@javanet.com
www.nelma.org

NSF International
P.O. Box 130140
Ann Arbor, MI 48113-0140
(734) 769-8010
(800) NSF-MARK
Fax: (734) 769-0109
info@nsf.org
www.nsf.com

NuTone, Inc.
(formerly Yale and Nutone)
Madison and Red Bank Rds.
Cincinnati, OH 45227
(513) 527-5100
(800) 543-8687
Fax: (513) 527-5177

Occupational Safety and Health Administration (OSHA)
(*see* U.S. Department of Labor)

Otis Elevator Company
One Farm Springs Road
Farmington, CT 06032-2500
(800) 441-6847
Fax: (860) 676-5490
www.nao.otis.com

Owens-Corning Fiberglas Corporation
One Owens Corning Park
Toledo, OH 43659
(419) 248-8000
www.owenscorning.com

Painting and Decorating Contractors of America
3913 Old Lee Hwy., Suite 33B
Fairfax, VA 22030
(703) 359-0826
Fax: (703) 359-2576
www.pdca.org

Painting and Decorating Contractors of America
Washington State Council
1001 S.W. Klickitat Way, Suite 204
Seattle, WA 98134
(206) 346-7322
Fax: (206) 346-6074
contact@wapdca.com
www.wapdca.com

Panel Solutions, Inc.
100 E. Diamond Ave.
Hazleton, PA 18201
(570) 459-3490
(800) 523-6671
Fax: (570) 459-3499
www.panelsolutions.com

Pella Corporation
102 Main St.
Pella, IA 50219
(515) 628-6107
(800) 847-3552
Fax: (515) 621-6754
commercial@pella.com
www.pella.com

Perlite Institute, Inc.
88 New Drop Plaza
Staten Island, NY 10306
(718) 351-5723
Fax: (718) 351-5725
inquiries@perlite.org
www.perlite.org

Perma Tubes, Inc.
3221-N Wellington Court
Raleigh, NC 27615
(919) 872-0820
Fax: (919) 790-0997

Phillips Industries, Inc.
Lasco Products Group
8015 Dixon Dr.
Florence, KY 41042
(606) 371-7720

Pittsburgh Testing Laboratory
850 Poplar St.
Pittsburgh, PA 15220
(412) 922-4000

Plastic Pipe and Fittings Association
800 Roosevelt Rd.
Building C, #20
Glen Ellyn, IL 60137
(630) 858-6540

Plastmo Inc. (East Coast office)
8246 Dandy Court
Jessup, MD 20794
(301) 776-0200
(800) 899-0992
Fax: (410) 792-8047
plastmo@erols.com
www.plastmo-ltd.com

Plumbing and Drainage Institute (PDI)
45 Bristol Dr.
South Easton, MA 02375
(800) 589-8956
Fax: (508) 230-3529
info@PDIonline.org
www.pdionline.org

Plumbing-Heating-Cooling Contractors Association (PHCC)
180 S. Washington St.
P.O. Box 6808
Falls Church, VA 22040
(800) 533-7694
Fax: (703) 237-7442
www.naphcc.org

Plumbing, Heating, Cooling Information Bureau
222 Merchandise Mart Plaza
Chicago, IL 60654
(312) 464-0090
www.phcib.org

Porcelain Enamel Institute (PEI)
P.O. Box 158541
4004 Hillsboro Pike, Suite 224-B
Nashville, TN 37215
(615) 385-5357
Fax: (615) 385-5463
penamel@aol.com
www.porcelainenamel.com

Portland Cement Association (PCA)
5420 Old Orchard Rd.
Skokie, IL 60077
(847) 966-6200
Fax: (847) 966-9781
webmaster@portcement.org
www.portcement.org

PPG Industries, Inc.
One PPG Place, 31N
Pittsburgh, PA 15272
(412) 434-2858
(800) 377-5267
Fax: (412) 434-3675
www.ppg.com

PPG Industries, Inc.
Coil and Extrusion Coatings
151 Colfax St.
Springdale, PA 15144
(724) 274-7900
(800) 258-6398
Fax: (724) 274-2600
coexcoatings@ppg.com
www.ppgcoexcoatings.com

Precast/Prestressed Concrete Institute (PCI)
175 W. Jackson Blvd., Suite 1859
Chicago, IL 60604
(312) 786-0300
Fax: (312) 786-0353
info@pci.org
www.pci.org

Prentice-Hall, Inc.
A Pearson Education Company
1 Lake St.
Upper Saddle River, NJ 07458
(201) 909-6200
www.phptr.com

The Preservation Press
(*see* National Trust for Historic Preservation)

Professional Roofing
(formerly *Roofing Spec*)
(Contact National Roofing Contractors Association.)

PWC Magazine: Painting and Wallcovering Contractor
(*see* Painting and Decorating Contractors of America)

Radio Shack
A Division of Tandy Corporation
700 One Tandy Center
Fort Worth, TX 76102
(817) 390-3011

Raycon Corporation
South Industrial Hwy.
Ann Arbor, MI 48104
(313) 769-2614
www.pace.ch/deutsch/raycon.htm

Redwood Inspection Service (RIS)
California Redwood Association
405 Enfrente Dr., Suite 200
Novato, CA 94949
(888) 225-7339
(415) 382-0662
Fax: (415) 382-8531
cfgrover@worldnet.att.net
www.calredwood.org

Republic Steel Corporation
(*see* Lukens Inc.)

Resilient Floor Covering Institute (RFCI)
966 Hungerford Dr., Suite 12-B
Rockville, MD 20850
(301) 340-8580

Resolite Division
H.H. Robertson Company
P.O. Box 338
Rt. 19 N
Zelienople, PA 16063
(412) 452-6800
(800) 737-6548
Fax: (724) 452-0677

Reynolds Metals Company
P.O. Box 27003
Richmond, VA 23261
(804) 281-2636
(800) 841-7774
Fax: (804) 281-3602

Riesner Vent Brick Corporation
2810 38th Ave.
Long Island City, NY 11101
(718) 392-1570
Fax: (718) 392-1701

Rigidized Metal Corp.
659 Ohio St.
Buffalo, NY 14203
(716) 849-4711
(800) 836-2580
Fax: (716) 849-0401
rmcsales@rigidized.com
www.rigidized.com

Rock of Ages Corp.
560 Graniteville Rd.
Graniteville, VT 05654
(802) 476-3115
(800) 421-0166
Fax: (802) 476-4767

Rodale Publishing Company
33 E. Minor St.
Emmaus, PA 18098
(610) 967-5171

Rohm and Haas Company
Independence Mall West
Philadelphia, PA 19105
(800) 523-7500

Roof Consultants Institute
7424 Chapel Hill Rd.
Raleigh, NC 27607
(919) 859-0742
Fax: (919) 859-1328
www.rci-online.org

Roofing '86
(*see* National Roofing Contractors Association)

The Roofing Industry Education Institute (RIEI)
14 Inverness Drive E., H-110
Englewood, CO 80112-5608
(303) 790-7200
Fax: (303) 790-9006
www.riei.org

Roofing Spec
(*see Professional Roofing*)

Rubber Manufacturers Association (RMA)
1400 K St. NW, Suite 900
Washington, DC 20005
(202) 682-4800
Fax: (202) 682-4854
www.rma.org

Safety Glazing Certification Council (SGCC)
(*see* Insulating Glass Certification Council)

Sargent
100 Sargent Dr.
New Haven, CT 06511
(203) 562-2151
Fax: (203) 776-5992

Schlage Lock Company
1915 Jamboree, Suite 165
Colorado Springs, CO 80920
(800) 847-1864
Fax: (800) 452-0663
www.schlage.com

Screen Manufacturers Association
2850 South Ocean Blvd. #114
Palm Beach, FL 33480
(561) 533-0991
fscottfitzgerald@compuserve.com
www.screenmfgassociation.org

Sealant Engineering and Associated Lines (SEAL)
7867 Convoy Court, Suite 301
San Diego, CA 92111
(619) 569-7906

Sealant, Waterproofing and Restoration Institute (SWRI)
3101 Broadway, Suite 585
Kansas City, MO 64111
(816) 472-7974
Fax: (816) 472-7765
info@swrionline.org
www.swrionline.org

Sealed Insulating Glass Manufacturers Association (SIGMA)
401 N. Michigan Ave.
Chicago, IL 60611-4200
(312) 644-6610
Fax: (312) 527-6783
sigma@sba.com
www.sigmaonline.org

Sears, Roebuck and Company
Sears Tower
Chicago, IL 60684
(312) 875-3000

Selby, Battersby and Company
P.O. Box 1600
Sapulpa, OK 74607
(800) 523-0129

Sentry Group
900 Linden Ave.
Rochester, NY 14625
(716) 381-4900

Sheet Metal and Air Conditioning Contractors National Association (SMACNA)
4201 Lafayette Center Dr.
Chantilly, VA 20151-1209
(703) 803-2980
Fax: (703) 803-3732
www.smacna.org

Sherwin Williams Company
101 Prospect Ave.
Cleveland, OH 44101
(216) 566-2000
www.sherwin.com

Sika Corporation
201 Polito Ave.
Lyndhurst, NJ 07071
(800) 933-7452
Fax: (201) 933-6225
info@sika-corp.com
www.sika-corp.com

Single Ply Roofing Institute (SPRI)
200 Reservoir St., Suite 309A
Needham, MA 02494
(781) 444-0242
Fax: (781) 444-6111
spri@spri.org
www.spri.org

Small Homes Council—Building Research Council
1 E. St. Mary's Rd.
Champaign, IL 61820

Society of American Registered Architects
1245 S. Highland Ave.
Lombard, IL 60148
(708) 932-4622

Society for Protective Coatings (SSPC) (formerly Steel Structures Painting Council)
40 24th St., 6th Floor
Pittsburgh, PA 15222
(412) 281-2331
Fax: (412) 281-9992
research@sspc.org
www.sspc.org

Society of Automotive Engineers (SAE)
400 Commonwealth Dr.
Warrendale, PA 15096-0001
(724) 776-0790
Fax: (724) 776-5760
www.sae.org

Society of Fire Protection Engineers
60 Batterymarch St.
Boston, MA 02110
(617) 482-0686

Society of Plastics Engineers, Inc.
P.O. Box 403
Brookfield, CT 06804-0403
(203) 775-0471
Fax: (203) 775-8490
www.4spe.org

Society of the Plastics Industry, Inc.
1801 K St. NW, Suite 600K
Washington, DC 20006
(202) 974-5200
Fax: (202) 296-7005
www.plasticsindustry.com

Society of Women Engineers
345 E. 47th St., Room 305
New York, NY 10017
(212) 579-9577

Solar Energy Industries Association, Inc.
1111 N. 19th St., Suite 260
Arlington, VA 22209
(703) 248-0702
Fax: (703) 248-0714
plowenth@seia.org
www.seia.org

The Solar Group
P.O. Box 525
2013 Fellowship Rd.
Taylorsville, MS 39168
(601) 785-4711
(800) 647-7063
Fax: (601) 785-8020

Solnhofen Natural Stone, Inc.
1604 17th St.
San Francisco, CA 94107
(415) 552-1500
Fax: (415) 552-3500
http://www.solnhofen.com

Southeastern Lumber Manufacturers Association (SLMA)
P.O. Box 1788
Forest Park, GA 30298
(404) 361-1445
Fax: (404) 361-5963

Southern Building Code Congress International
(*see* Standard Building Code)

Southern Cypress Manufacturers Association

400 Penn Center Blvd. #530
Pittsburgh, PA 15235
(877) 607-SCMA

Southern Forest Products Association
P.O. Box 641700
Kenner, LA 70064-1700
(504) 443-4464
Fax: (504) 443-6612
www.sfpa.org

Southern Hardwood Lumber Manufacturers Association
(*see* Hardwood Manufacturers Association)

Southern Pine Council (SPC)
P.O. Box 641700
Kenner, LA 70064-1700
(504) 443-4464
Fax: (504) 443-6612
www.southernpine.com

Southern Pine Inspection Bureau (SPIB)
4709 Scenic Hwy.
Pensacola, FL 32504
(850) 434-2611
Fax: (850) 433-5594
spib@spib.org
www.spib.org

Southwest Research Institute
6220 Culebra Rd.
San Antonio, TX 78228
(210) 684-5111
www.swri.edu

Specialty Steel Industry of North America
3050 K St. NW
Washington, DC 29551
(202) 342-8360
(800) 982-0355
Fax: (202) 342-8451
ssina@colshan.com
www.ssina.com

Spectrum Glass Company, Inc.
P.O. Box 646
Woodinville, WA 98072
(425) 483-6699
Fax: (425) 483-9007
artglass@spectrumglass.com
www.spectrumglass.com

SRI International
333 Ravenswood Ave.
Menlo Park, CA 94025
(415) 326-6200

Standard Building Code
Southern Building Code Congress International (SBCCI)
900 Montclair Rd.
Birmingham, AL 35213
(205) 591-1853
Fax: (205) 591-0775
info@sbcci.org
www.sbcci.org

Stanlock Division of Standard Products Company
127 W. Perry
Port Clinton, OH 43452
(800) 431-1607
(419) 734-4942
Fax: (419) 734-5246

Steel Deck Institute (SDI)
P.O. Box 25
Fox River Grove, IL 60021-0025
(847) 462-1930
Fax: (847) 462-1940
Steve@sdi.org
www.sdi.org

Steel Door Institute (SDI)
30200 Detroit Rd.
Cleveland, OH 44145-1967
(440) 899-0010
Fax: (440) 892-1404
www.steeldoor.org

Steel Joists Institute (SJI)
3127 10th Ave., North Ext.
Myrtle Beach, SC 29577
(843) 626-1995
Fax: (843) 626-5565
www.steeljoist.org

Steel Structures Painting Council
(*see* Society for Protective Coatings)

Steel Window Institute (SWI)
c/o Thomas Associates, Inc.
1300 Summer Ave.
Cleveland, OH 44115
(216) 241-7333
www.taol.com/swi

Stonhard, Inc.
An RPM Company
One Park Avenue
Maple Shade, NJ 08052
(609) 779-7500
(800) 257-7953
Fax: (609) 321-7525
marketing-services@stonhard.com
www.stonhard.com

Structural Slate Company
222 East Main St.
P.O. Box 187
Pen Argyl, PA 18072
(610) 863-4141
(800) 677-5283
Fax: (610) 863-7016
sscol@ptd.com
http://structuralslate.com

Stucco Manufacturers Association, Inc.
507 Evergreen
Pacific Grove, CA 93950
(408) 649-3466
Fax: (408) 647-1552

Submersible Wastewater Pump Association (SWPA)
1866 Sheridan Rd., Suite 210
Highland Park, IL 60035
(847) 681-1868
Fax: (847) 681-1869
www.swpa.org

Sumner, Rider & Associates, Inc.
355 Lexington Ave.
New York, NY 10017
(212) 661-5300

Sump and Sewage Pump Manufacturers Association (SSPMA)
P.O. Box 647
Northbrook, IL 60065-0647
(847) 559-9233
Fax: (847) 559-9235
www.sspma.org

Sunset Publishing Corp.
80 Willow Rd.
Menlo Park, CA 94025
(415) 321-3600

Superintendent of Documents
(*see* U.S. Government Printing Office)

T&S Brass and Bronze Works, Inc.
P.O. Box 1088
2 Saddleback Cove
Travelers Rest, SC 29690-1088
(864) 834-4102
(800) 476-4103
Fax: (864) 834-3518
tsbrass@tsbrass.com
www.tsbrass.com

Technical Preservation Services
Preservation Assistance Division
National Park Service
(*see* U.S. Department of the Interior, National Park Service)

Textured Coatings of America, Inc.
4101 Ravenswood Rd.
Suite 105A
Fort Lauderdale, FL 33312-5371
(954) 581-0771
Fax: (954) 581-9516
http://www.texcote.com

Thermal Insulation Manufacturers Association (TIMA)
(*see* North American Insulation Manufacturers Association)

Thermo-Flex, Inc.
888 N. Ohio
Salina, KS 67401
(913) 827-7201
(800) 428-2237
Fax: (913) 827-7545

Thoro Division of Harris Specialty Chemicals, Inc.
1367 Elmwood Ave.
Cranston, RI 02910
(904) 996-6000
(800) 322-7825
Technical Service: (800) 327-1570
Fax: (904) 996-6092
http://www.hsc-ss.com

Tile and Decorative Surfaces
20335 Ventura Blvd., Suite 400
Woodland Hills, CA 91364
(818) 704-5555
Fax: (818) 704-6500
www.building.org/texis/db/bix/+YoWYelYBwrmwxTo/profile.html

Tile Contractors Association of America
11501 Georgia Ave., Suite 203
Silver Spring, MD 20902
(301) 949-5995
(800) 655-8453
www.tcaainc.org

Tile Council of America, Inc. (TCA)
100 Clemson Research Blvd.
Anderson, SC 29625
(864) 646-8453
Fax: (864) 646-2821
literature@carol.net
www.tileusa.com

Tile Letter
P.O. Box 13629
Jackson, MS 39215
(601) 939-2071

Transportation Research Board
National Academy of Sciences
2101 Constitution Ave. NW
Washington, DC 20418
(202) 334-2933
Fax: (202) 334-2003
www.nationalacademies.org

Truss Plate Institute (TPI)
583 D'Onofrio Dr., Suite 200
Madison, WI 53719
(608) 833-5900

TRW Nelson Company
7900-T W. Ridge Rd.
Elyria, OH 44036
(800) 321-2005

Tubelite, Inc.
8200 Mackinaw Trail
P.O. Box 118
Reed City, MI 49677
(616) 832-2211
(800) 866-2227
Fax: (877) 299-2414
terrimi@tubeliteinc.com
www.tubelite.com

Underground Space Center
University of Minnesota
500 Pillsbury Dr., SE
Minneapolis, MN 55455
(612) 624-0066

Underwriters Laboratories, Inc. (UL)
333 Pfingsten Rd.
Northbrook, IL 60062
(847) 272-8800
Fax: (847) 272-8129
northbrook@ul.com
www.ul.com

Union Carbide Corporation
39 Old Ridgebury Rd.
Danbury, CT 06817-0001
(203) 794-5300
(800) 335-8550
www.unioncarbide.com

United States Gypsum Company (USG)
125 S. Franklin
Chicago, IL 60606
(312) 606-5566
(800) 874-4968
www.usg.com

United States Plywood Corporation
(*see* Georgia Pacific Corporation)

United States Steel Corporation
(*see* USX)

Urban Land Institute
1025 Thomas Jefferson St. NW
Suite 500 West
Washington, DC 20007
(202) 624-7000
(800) 321-5011
Fax: (202) 624-7140
www.uli.org

U.S. Architectural and Transportation Barriers Compliance Board
1331 F St. NW, Suite 1000
Washington, DC 20004
(800) 872-2253
www.access-board.gov

U.S. Consumer Product Safety Commission
Washington, DC 20207-0001
(301) 504-0990
Fax: (301) 504-0124 and (301) 504-0025
www.cpsc.gov

USDA Forest Service
Forest Products Laboratory
One Gifford Pinchot Drive
Madison, WI 53704-2398
(608) 231-9200
Fax: (608) 231-9592
www.fpl.fs.fed.us/fpldir.htm

U.S. Department of Agriculture
14th St. and Independence Ave. SW
Washington, DC 20250
(202) 720-2791
www.usda.gov

U.S. Department of the Army Corps of Engineers (CE)
Chief of Engineers
20 Massachusetts Ave. NW
Washington, DC 20314
(202) 272-0660

U.S. Department of Commerce (DOC)
14th St. and Constitution Ave. NW
Washington, DC 20230
(202) 377-2000
www.doc.gov

U.S. Department of Commerce, National Technical Information Service
5285 Port Royal Rd.
Springfield, VA 22161
(703) 605-6000

U.S. Department of Defense
Naval Publications and Forms Center
5801 Tabor Ave.
Philadelphia, PA 19120

U.S. Department of Housing and Urban Development (HUD)
451 Seventh St. SW
Washington, DC 20410
(202) 401-0388
www.hud.gov

U.S. Department of the Interior
National Park Service
1849 C St. NW
Washington, DC 20240
(202) 208-6843
www.nps.gov

U.S. Department of Labor
Occupational Safety and Health Administration (OSHA)
Division of Construction Services
200 Constitution Ave. NW
Washington, DC 20210
(202) 693-1707
www.osha.gov

U.S. Department of the Navy
(*see also* U.S. Department of Defense)
Naval Facilities Engineering Command

Washington Naval Yard
Building 212
Washington, DC 20374
(202) 433-5984

U.S. Department of Transportation (DOT)
400 Seventh St. SW
Washington, DC 20590
(202) 366-4000
www.dot.gov

U.S. Environmental Protection Agency (EPA)
401 M St. SW
Washington, DC 20460
(202) 260-2090
www.epa.gov

U.S. General Services Administration
Historic Preservation Office
18th and F Sts. NW
Washington, DC 20405
(202) 655-4000

U.S. General Services Administration (GSA)
Specifications Unit
7th and D Sts. SW
Washington, DC 20407
(202) 619-8925
Fax: (202) 619-8978
www.gsa.gov

U.S. Government Printing Office
Superintendent of Documents
N. Capitol and H Sts. NW
Washington, DC 20405
(202) 512-1991
www.access.gpo.gov

U.S. Postal Service (USPS)
475 L'Enfant Plaza SW
Washington, DC 20260
(202) 268-2000
www.usps.com

U.S. Weather Service
1325 East West Hwy., Room 18444
Silver Spring, MD 20910
(301) 443-8910

USX
(formerly U.S. Steel)
600 Grant St.
Pittsburgh, PA 15219
(412) 433-1121

Vermiculite Association, Inc.
Whitegate Acre, Metheringham Fen
Lincoln LN4 3AL, UK
+44 1526 323990
Fax: +44 1526 323181
tva@vermiculite.org
www.vermiculite.org

Vermont Structural Slate Company
P.O. Box 98
3 Prospect St.
Fair Haven, VT 05743
(802) 265-4933
(800) 343-1900
Fax: (802) 265-3865

Vinyl Siding Institute (VSI)
1801 K St. NW, Suite 600K
Washington, DC 20006
(202) 974-5200
(888) 367-8741
vsi@socplas.org
www.vinylsiding.org

Vulcraft Division of Nucor Corp.
P.O. Box 100520
Florence, SC 29501
(803) 662-0381
www.vulcraft.com

W and P Engineering and Fabrications
7503 Brook Rd.
Richmond, VA 23227
(804) 264-1854

Water Quality Association
4151 Naperville Rd.
Lisle, IL 60532
(630) 505-0160
www.wqa.org

West Coast Lumber Inspection Bureau (WCLIB)
P.O. Box 23145
Portland, OR 97281
(503) 639-0651
Fax: (503) 684-8928
www.wclib.org

Western Lath, Plaster, Drywall Contractors Association
(formerly California Lath and Plaster Association)
8635 Navajo Rd.
San Diego, CA 92119
(619) 466-9070

Western Red Cedar Lumber Association (WRCLA)
1200-555 Burrard St.
Vancouver, B.C. V7X 1S7, Canada
(604) 684-0266
Fax: (604) 687-4930
wrcla@cofiho.cofi.org
www.wrla.org

Western Wood Products Association (WWPA)
522 S.W. 5th Ave., Suite 500
Portland, OR 97204
(503) 224-3930
Fax: (503) 224-3934
info@wwpa.org
www.wwpa.org

Weyerhaeuser Company
Commercial Door Division
P.O. Box 130
Marshfield, WI 54449
(800) 544-6522 (eastern zone)
(800) 826-4003 (western zone)

Weyerhaeuser Company
CH 1K35C
P.O. Box 2999
Tacoma, WA 98477
(253) 924-2345
www.weyerhaeuser.com

Whirlpool Corporation
2000 North M-63
Benton Harbor, MI 49022-2692
(616) 923-3189
Fax: (616) 923-5038
info@whirlpool.com
www.whirlpoolcorp.com

Willard Shutter Company, Inc.
4420 N.W. 35th Court
Miami, FL 33142
(305) 633-0162
(800) 826-4530
Fax: (305) 638-8634
www.willardshutterco.com

Window and Door Manufacturers Association (WDMA)
(formerly National Wood Window and Door Association [NWWDA])
1400 E. Touhy Ave., Suite 470
Des Plaines, IL 60018
(800) 223-2301
Fax: (847) 299-1286
admin@wdma.com
www.nwwda.org

Window and Door Specifier
(*see* Window and Door Manufacturers Association)

Wire Reinforcement Institute (WRI)
1101 Connecticut Ave. NW, #700
Washington, DC 20036
(202) 429-5125

Wolverine Vinyl Siding
750 E. Swedesford Rd.
Valley Forge, PA 19482
(888) 838-8100
Fax: (610) 341-7940
www.vinylsiding.com

Wood and Synthetic Flooring Institute (WSFI)
4415 W. Harrison St., Suite 242-C
Hillside, IL 60162
(708) 449-2933

Wood Truss Council of America
5937 Meadowood Dr., Suite 14
Madison, WI 53711
(608) 274-4849
wtca@woodtruss.com
www.woodtruss.com

Woodwork Institute of California
3164 Industrial Blvd.
West Sacramento, CA 95691
P.O. Box 980247
West Sacramento, CA 95798-0247
(916) 372-9943
Fax: (916) 372-9950
www.wicnet.org

York Manufacturing, Inc.
P.O. Box 1009
Sanford, ME 04073
(207) 324-1300
(800) 551-2828
Fax: (207) 490-2592
yorkmfg@cybertours.com

Zero International, Inc.
415 Concord Ave.
Bronx, NY 10455
(718) 585-3230
(800) 635-5335
Fax: (718) 292-2243

APPENDIX B

Glossary

Boldfaced words in the definitions are also glossary entries.

abrasion Wearing away by friction.

absorption bed A wide trench exceeding 36 in. (910 mm) in width containing a minimum of 12 in. (305 mm) of clean, coarse **aggregate** and a system of two or more distribution pipes through which treated sewage may seep into the surrounding soil. Also called *seepage bed.*

absorption coefficient The ratio of the sound-absorbing effectiveness of 1 sq. ft. (0.09 m^2) of a material to 1 sq. ft. (0.09 m^2) of a perfectly absorptive material at a specific frequency; usually expressed as a decimal value (such as .70) or as a percentage.

absorption field An arrangement of absorption trenches through which treated sewage is absorbed into the soil. Also called *disposal field.*

absorption of sound The ability of a material to absorb rather than reflect sound waves striking it by converting sound energy to heat energy within the material.

absorption of water by clay masonry The weight of water a *brick* or other clay *masonry unit* absorbs when immersed in either cold or boiling water for a stated length of time, expressed as a percentage of the weight of the dry unit.

absorption of water by concrete masonry units (CMU) The weight of water a concrete masonry unit absorbs when immersed in water, expressed in pounds of water per cubic foot of concrete.

absorption rate The weight of water absorbed when a clay *brick* is partially immersed for 1 minute, usually expressed in either grams or ounces per minute. Also called *suction* or initial rate of absorption.

absorption trench A trench of not more than 36 in. (910 mm) in width, containing a minimum of 12 in. (305 mm) of clean, coarse aggregate and a distribution pipe, through which treated sewage is allowed to seep into the soil.

absorption unit *See* **sabin**.

accelerator An **admixture** used in **concrete** to hasten its set and increase the rate of strength gain (the opposite of a **retarder**).

accessible Describes a site, building, facility, or portion thereof that complies with current standards and can be approached, entered, and used by a **physically handicapped person**.

accessible route A continuous unobstructed path connecting all **accessible** elements and spaces in a building or facility that can be negotiated by a person with a severe **disability** using a wheelchair and that is also safe for and usable by people with other disabilities.

acid Corrosive chemical substance that attacks many common building materials, decorative finishes, **coatings**, **paints**, and **transparent finishes**.

acid-resistant brick *See under* **brick masonry unit**.

acoustical correction The planning, shaping, and equipping of a space to establish the best possible hearing conditions for faithful reproduction of wanted sound within the space. *See also* **acoustics**.

acoustics The science of sound; the production, transmission, and effect of sound. *See also* **acoustical correction**, **architectural acoustics**, and **room acoustics**.

acrylic (1) In **carpet**, a generic term including acrylic and modified acrylic (modacrylic) fibers. Acrylic is a **polymer** composed of at least 85% by weight of acrylonitrile; modacrylic is a polymer composed of less than 85% but at least 35% by weight of acrylonitrile. (2) In glazing applications, a transparent plastic material.

actual dimension The actual measured dimension of a **masonry unit**, piece of lumber, or other construction material or assembly. *See also* **nominal dimension**.

actual size *See* **size**.

adaptability The capability of certain building spaces and elements, such as kitchen counters, sinks, and grab bars, to be altered or added so as to accommodate the needs of persons with and without *disabilities*, or to accommodate the needs of persons with different types or degrees of disability.

adhesion The property of a **paint** film that enables it to stick to a surface.

adhesive, drywall Adhesives specifically intended for the application of gypsum board. A *contact adhesive* is an adhesive used to bond layers of gypsum board or for bonding gypsum board to metal studs. A *laminating adhesive* is an adhesive used to bond layers of gypsum board. A *stud adhesive* is an adhesive used to attach gypsum board to wood supports.

adhesive, tile Prepared organic material, ready for use with no further addition of liquid or powder, which cures or sets by evaporation; distinguished from mortars by the absence of siliceous fillers (sand) that are included in mortars either at the plant or in the field.

admixture A material (other than **portland cement**, water, or **aggregate**) used in **concrete** to alter its properties (such as **accelerators**, **retarders**, and **air-entraining agents**).

adobe brick *See under* **brick masonry unit**.

age hardening The continuing increase in strength for long periods of time of aluminum **alloys** after **heat treatment**.

agglomeration Process for increasing the particle size of iron ores to make them suitable for ironworking and steelmaking.

aggregate A hard, inert material mixed with **portland cement** and water to form **concrete**. *Fine aggregate* has pieces ¼ in. (6.4 mm) in diameter and smaller. *Coarse aggregate* has pieces larger than ¼ in. (6.4 mm) in diameter.

aging The period of time in which a **heat-treatable** aluminum **alloy** is allowed to remain at room temperature, after heat treatment (heating and quenching), to reach a stable state of increased strength. *See also* **artificial aging**.

airborne sound transmission *See* **sound transmission**.

air chamber A piece of pipe about 10 in. (250 mm) long, installed above the hot and cold valves of fixtures such as sinks, lavatories, and clothes washers to cushion the rush of water as the valve is closed and prevent **water hammer**.

air drying (1) In **coatings**, **paints**, and **transparent finishes**, capable of forming a solid film when exposed to air at moderate atmospheric temperatures. (2) Of wood, *see* **seasoning**.

air-entrained concrete **Concrete** containing minute bubbles of air up to about 7% by volume.

air-entraining agent An **admixture** used to produce **air-entrained concrete**.

air gap The unobstructed vertical distance between the mouth of a water outlet and the flood level rim of the water receptacle. The water outlet may be a faucet, spout, or other outlet; the receptacle may be a **plumbing fixture**, tank, or other receptacle.

alclad sheet A clad product with an aluminum or aluminum **alloy** coating having a high resistance to **corrosion**. The coating is anodic to the core alloy it covers, thus protecting it physically and electrolytically against corrosion. *See also* **clad alloy**.

alkali A soluble mineral salt present in some soils. Alkalies are chemical substances characterized by their ability to combine with acids to form neutral salts. They are damaging to many **coatings**, **paints**, and **transparent finishes**.

allowable unit stress *See under* **stress**.

alloy The material formed when two or more metals, or metals and nonmetallic substances, are joined by being dissolved into one another while molten. *See also* **brazing alloy** and **clad alloy**.

alloy designation A numerical system used in designating the various **alloys** of aluminum.

alloying element Element added in steelmaking to achieve desired properties.

alloy steel *See under* **steel**.

alumina (Al_2O_3) A hydrated form of aluminum oxide found in **bauxite** and in ordinary clays.

ambient sound A continuous background sound that is a composite of individual sounds coming from exterior sources, such as street traffic, and interior sources, such as ventilating equipment and appliances, none of which can be identified individually by a listener. Also called *background noise*.

ambulatory Able to walk without assistance or difficulty.

amperage Electrical rate of flow, measured in amperes (amps) and comparable to gallons per minute (gpm) in a fluid medium. The strength of a current of electricity.

ampere (amp) A unit of electrical current equivalent to that produced by 1 volt applied across a resistance of 1 ohm. One coulomb of electricity in every second.

anchor A piece or assemblage, usually metal, used to attach parts (e.g., plates, joists, trusses, studs, sills, masonry, windows, doors, and other building elements) to wood, concrete, or masonry.

angular course *See under* **course**.

angular measure The deviation between two lines that meet at a point, expressed in **degrees**, **minutes**, and **seconds**.

annealing *See* **heat treatment**.

annual growth ring The growth layer put on by a tree in a single growth year.

anodic coating A surface coating applied to an aluminum **alloy** by **anodizing**.

anodizing Applying an electrolytic oxide coating to an aluminum **alloy** by building up the natural surface film using an electrical current (usually dc) through an oxygen-yielding electrolyte with the alloy serving as the anode.

antioxidant A compound added to other substances to retard oxidation, which deteriorates plastics.

antique finish A finish usually applied to furniture or woodwork to give the appearance of age.

arch A usually curved compressive structural member, spanning openings or recesses; also built flat. A *back arch* is a concealed arch carrying the backing of a wall where the exterior face is carried by a lintel. A *jack arch* is an arch having horizontal or nearly horizontal upper and lower surfaces. Also called a *flat* or *straight arch*. A *major arch* is an arch with a span greater than 6 ft. (1800 mm) that carries a load that is equivalent to a uniform load greater than 1000 psf (4882 kg/m^2). Typically known as Tudor arch, semicircular arch, gothic arch, or parabolic arch. Has rise-to-span ratio greater than 0.15. A *minor arch* is an arch with a maximum span of 6 ft. (1800 mm) carrying a load that does not exceed 1000 psf (4882 kg/m^2). Typically known as jack arch, segmented arch, or multicentered arch. Has rise-to-span ratio less than or equal to 0.15. A *relieving arch* is an arch that is built over a *lintel*, flat arch, or smaller arch to divert loads, thus relieving the lower member from excessive loading. Also known as *discharging* or *safety arch*. A *trimmer arch* is an arch, usually a low-rise type, of **brick masonry unit** used for supporting a fireplace hearth.

architectural acoustics The **acoustics** of buildings and other structures. *See also* **acoustics** and **room acoustics**.

architectural barrier A physical condition in a building or facility that creates unsafe or confusing conditions or prevents accessibility and free mobility.

architectural terra-cotta Custom-made, hard-burned, glazed or unglazed clay building units, plain or ornamental, that are machine extruded or hand molded. *See also* **ceramic veneer**.

area wall A **retaining wall** around below-grade basement windows.

aromatic solvents Group of organic compounds derived from coal or petroleum, such as benzene and toluene.

artificial aging The heating of an aluminum **alloy** for a controlled time at an elevated temperature to accelerate and increase its strength gain after heat treatment. *See also* **aging**.

ashlar masonry Masonry composed of rectangular units usually larger in size than **brick** with sawed, dressed, or square beds, bonded with **mortar**. Ashlar masonry is also described according to its **pattern bond**, which may be coursed, random, or patterned.

atmospheric pressure steam curing *See* **curing**.

atmospheric vacuum breaker A simple mechanical device consisting essentially of a check valve in a supply line and a valve member (on the discharge side of the check valve) opening to the atmosphere when the pressure in a line drops to atmospheric. Also called a *siphon breaker*.

attenuation Reduction of the energy or intensity of sound.

audible cue *See* **cue**.

austenitic steel *See* **grain structure**.

autoclave *See* **curing**.

average transmission loss (tl) The numerical average of the **transmission loss** values of an assembly measured at nine frequencies. It is a single-number rating for comparing the **airborne sound transmission** through walls and floors.

Axminster carpet *See* **woven carpet**.

back The side opposite the face. The poorer side of a **plywood** panel. The surface of **gypsum board** that will be placed toward the supports.

back arch *See under* **arch**.

back blocking A single-ply gypsum board installation procedure for reinforcing butt-end or edge joints to minimize surface imperfections such as cracking and ridging.

backflow The unintentional flow of water into the supply pipes of a plumbing system from a nonsupply source.

background noise *See* **ambient sound**.

backing The carpet foundation of jute, kraftcord, cotton, rayon, or polypropylene yarn that secures the pile yarns and provides stiffness, strength, and dimensional stability.

backing board *See under* **gypsum board**.

backing, tile A suitable surface for the application of tile, such as a structural subfloor or wall surface.

backparging *See* **parging**.

backplate *See under* **finished mill products**.

back-siphonage A type of **backflow**, usually caused by a temporary occurrence of negative pressure (suction) in pipes.

backup The part of a masonry wall behind the exterior facing.

baking finish A **coating** that is baked at temperatures above 150°F (65.6°C) to dry and develop desired properties.

bark The outer, corky layer of a tree, composed of dry, dead tissue.

bars *See under* **finished mill products**.

base coat The first **plaster** layer in **two-coat work**; or either the **scratch** or **brown coat** in **three-coat plasterwork**.

base course The lowest course of masonry in a wall or pier.

base line A **parallel** of specified **latitude**, used in the **rectangular survey system**, serving as the main east-west reference line, with a **principal meridian** for a particular state or area.

basic stress *See* **stress**.

batter Recessing or sloping a wall back in successive courses; the opposite of **corbel**.

bauxite A raw ore of aluminum consisting of 45% to 60% aluminum oxide, 3% to 25% iron oxide, 2.5% to 18% silicon oxide, 2% to 5% titanium oxide, other impurities, and 12% to 30% water. This ore varies greatly in the proportions of its constituents, color, and consistency.

Bayer process The process generally employed to refine **alumina** from **bauxite**.

beam A structural member transversely supporting a load.

bearding Long-fiber fuzz occurring on some loop pile fabrics, caused by fibers snagging and loosening due to inadequate anchorage.

bearing wall A wall that supports a vertical load in addition to its own weight. *See also* **non-load-bearing wall**.

bed joint Horizontal layer of mortar into which a masonry unit is laid.

beneficiation Concentrating process used to increase the iron content of ores prior to use. *See also* **agglomeration**.

binder (1) In paint and coatings, the **vehicle** ingredient with adhesive qualities (linseed oil, resins, etc.) that binds the **pigment** and other ingredients of a **coating**, **paint**, or **transparent finish** into a cohesive film and facilitates bonding with the underlying surfaces. (2) In **terrazzo**, a cementitious or resinous material that gives the matrix adhesive and other important physical properties.

bleaching (1) The process of lightening raw wood. (2) The process of restoring discolored or stained wood to its normal color or making it lighter.

bleeding (1) In unit masonry, (a) the loss of water from a **masonry unit** having a low **suction** when it comes in contact with mortar, (b) the loss of water from mortar due to a low **water retention** when it contacts a **masonry unit**. Bleeding causes **masonry units** to float (*see* **floating**). (2) In **concrete**, the appearance of excess water rising to the surface shortly after placing of **concrete**. (3) In **coatings**, **paints**, and **transparent finishes**, discoloration of a finish coat by coloring matter from the underlying surface or coat of finishing material. (4) In **plastics**, the diffusion of a colorant out of a plastic part into adjacent materials.

blend to a common level The meeting of two or more surfaces so that there is no abrupt rise or drop in the surface.

blistering The formation of bubbles or pimples on a coated, painted, or transparent finished surface caused by moisture in the underlying material (wood, masonry, concrete, etc.); caused by adding a coat of **coating**, **paint**, or **transparent finish** before the previous coat has dried thoroughly; or caused by excessive heat or grease under a **coating**, **paint**, or **transparent finish**.

blocking (1) In masonry construction, a method of bonding two adjoining or intersecting walls, not built at the same time, by means of offsets whose vertical dimensions are not less than 8 in. (200 mm). (2) In wood construction, short pieces of wood used as nailers, spacers, or fillers between wood members, between wood members and other construction, or between other materials.

blow molding Shaping thermoplastic materials into hollow form by air pressure and heat; usually performed on sheets or tubes.

blushing Describes opaque lacquer that loses its gloss and becomes flat or clear lacquer that turns white or milky.

board *See under* **lumber**.

board foot A measure of lumber. One board foot is the equivalent of a piece of *lumber* whose nominal dimensions are 1 in. (25 mm) thick, 12 in. (305 mm) wide, and 12 in. (305 mm) long.

body Used to indicate thickness or thinness of a liquid **coating**, **paint**, or **transparent finish** material.

bond beam Course or courses of a masonry wall grouted and usually reinforced in the horizontal direction. Alternatively, may be made of reinforced concrete. Serves as horizontal tie of wall bearing courses for structural members, or itself as a flexural member.

bond course A masonry **course** in which the units overlap more than one **wythe** of masonry.

bonded rubber cushioning Rubber or latex cushioning adhered to a carpet at the mill.

bonder *See* **header**.

bonderizing Process to improve **paint** adhesion on steel by dipping lightly galvanized (*see* **galvanizing**) objects in a hot phosphate solution to form a surface film of zinc phosphate.

bond, pattern *See* **pattern bond**.

bond, structural Tying **wythes** of a masonry wall together by lapping units one over another or by connecting them with metal ties.

bond, tensile Adhesion between **mortar** and **masonry units** or reinforcement.

border The resilient flooring at the perimeter of a room adjacent to the walls, which is installed separately from the **field**.

bow *See* **warp**.

boxing Mixing a **coating**, **paint**, or **transparent finish** material by pouring it from one container to another several times. Boxing is not recommended for most such materials.

braille A special raised-touch alphabet for the blind, using a cell of six dots.

brazing A welding process in which the filler metal is a nonferrous metal or **alloy** with a melting point higher than 800°F (426.67°C) but lower than that of the metals joined. A *brazing alloy* is an alloy used as filler metal for brazing. In aluminum the brazing alloy is usually in the 4000 Series of alloys. A *brazing sheet* is an unclad or specially clad sheet for brazing purposes, with the surface of the specially clad sheet having a lower melting point than the core. Brazing sheet of the clad type may be clad on either one or two surfaces.

breathe The ability of a **coating**, **paint**, or **transparent finish** film to permit the passage of moisture vapor without causing **blistering**, **cracking**, or **peeling**.

brick grade Designation for durability of a unit, expressed as SW for severe weathering, MW for moderate weathering, and NW for negligible weathering. See ASTM C 216, C 62, and C 652.

brick masonry unit A unit that is formed into a rectangular prism while plastic. An *acid-resistant brick* is a clay brick suitable for use in contact with chemicals; designed primarily for use in the chemical industry. Usually used with acid-resistant mortars. An *adobe brick* is a large clay brick of varying size, roughly molded and sun dried. A *building (common) brick* is a clay brick for building purposes, not especially treated for texture or color. Formerly called common brick. A *clay brick* is a brick made of clay or shale, formed into a rectangular prism while plastic and burned (fired) in a kiln. A *concrete brick* is a brick made from **portland cement** and an **aggregate** and cured either in normal atmosphere or using either low- or high-pressure steam. It is formed as a rectangular prism, usually not larger than 4 × 4 × 12 in. (100 × 100 × 305 mm). An *economy brick* is a clay brick whose nominal dimensions are 4 × 4 × 8 in. (100 × 100 × 305 mm). An *engineered brick* is a clay brick whose nominal dimensions are 4 × 3.2 × 8 in. (100 × 81.28 × 305 mm). A *facing brick* is a clay brick made especially for facing purposes, often with finished surface texture. Such units are made of selected clays or treated to produce desired color. A *fire brick* is a brick made of refractory ceramic material that will resist high temperatures. A *floor brick* is a smooth, dense brick that is highly resistant to abrasion; used as a finished floor surface. A *hollow brick* is a clay or shale **masonry unit** whose **net cross-sectional area** in any plane parallel to the bearing surface is not less than 60% of its **gross cross-sectional area** measured in the same plane. *Jumbo brick* is a generic term indicating that a clay brick is larger in size than the standard. Some producers use this term to describe oversized brick of specific dimensions manufactured by them. A *Norman brick* is a clay brick whose nominal dimensions are 4 × 2⅔ × 12 in. (100 × 67.7 × 305 mm). A *paving brick* is a clay brick especially suitable for use in pavements where resistance to abrasion is important. A *Roman brick* is a clay brick whose nominal dimensions are 4 × 2 × 12 in. (100 × 50 × 305 mm). An *SCR brick* is a clay brick whose nominal dimensions are 2⅔ × 6 × 12 in. (67.7 × 150 × 305 mm). It lays up three courses to 8 in. (203.2 mm) and produces a nominal 6-in. (150 mm)-thick wall. Developed by the Structural Clay Products Research Foundation. A *sewer brick* is a low-absorption, abrasive-resistant clay brick intended for use in drainage structures.

brick type Designation for **facing brick** that controls tolerance, chippage, and distortion. Expressed as FBS, FBX, and FBA for solid brick and HBS, HBX, HBA, and HBB for hollow brick. See ASTM C 216 and C 652.

British thermal unit (Btu) A measure of heat quantity equal to the amount of heat required to raise 1 pound of water 1 degree Fahrenheit at sea level.

broadloom Carpet woven on a broad loom in widths of 6 ft. (1800 mm) or more.

brown coat The second **plaster** layer in **three-coat work**, which provides additional strength and a suitably true and plane surface for the application of the **finish coat**.

brush A tool composed of bristles set into a handle; often used to apply **paint** and **transparent finish** materials. Sometimes used to apply **coatings**. Bristles may be synthetic (needed for water-thinned paints) or natural, such as hog hair.

brushability Ease with which a *paint* or *transparent finish* material can be brushed.

brush marks Marks of a brush that remain in a dried **coating**, **paint**, or **transparent finish** film.

Btu *See* **British thermal unit**.

buckling Wrinkling or ridging of a carpet after installation, caused by insufficient stretching, dimensional instability, or manufacturing defects.

build The apparent thickness or depth of a **coating**, **paint**, or **transparent finish** film after it has dried.

building brick *See under* **brick masonry unit**.

building drain The lowest part of a building's drainage system. It receives the discharge from soil, waste, and other drainage pipes inside the walls of a building and conveys it to a point 3 ft. (900 mm) outside the building walls, where it joins with the *building (house) sewer*.

building sewer The horizontal piping of a **drainage system**, which extends from the end of a **building drain** to a public sewer, private sewer, **individual sewage disposal system**, or other point of disposal. A **building sewer** begins at a point 3 ft. (900 mm) outside the building wall, where the **building drain** ends. Also called a **house sewer**.

building storm drain A type of **building drain** used for conveying rainwater, surface water, groundwater, subsurface water, cooling condensate; or a combined building sewer, extending to a point not less than 3 ft. (900 mm) outside the building wall, where it joins with a storm sewer or combined storm and sewage sewer.

built environment The collection of buildings, facilities, transportation systems, and other structures and spaces created for the purpose of providing convenient places for work, play, living, and related human activities; contrasted with the natural environment.

bulked continuous filament (BFC) Continuous strands of synthetic fiber made into yarn without spinning; often extruded in modified cross section such as multilocal, mushroom, or bean shape, or textured to increase bulk and covering power.

burling Removing surface defects such as knots, loose threads, and high spots to produce acceptable quality after weaving; also, filling in omissions in weaving.

burnishing Shiny or lustrous spots on a **coating**, **paint**, or **transparent finish** surface caused by rubbing.

butt-end joint Joint in which mill- or job-cut (exposed core) **gypsum board** ends or edges are butted together.

buttering Placing **mortar** on a **masonry unit** with a trowel.

calcimine Water-based **paint** generally consisting of animal glue, zinc white, and calcium carbonate or clay; now seldom used.

calcium chloride An accelerator added to **concrete** to hasten setting (not to be considered an antifreeze).

calendering A process for producing plastic film or sheeting by passing the material between revolving heated rolls.

call In surveying, a statement or mention of a **course** or distance.

cambium A thin layer of tissue that lies between a tree's bark and its wood. The cambium subdivides to form the new wood and bark cells of each year's growth.

cant A log **slabbed** on one or more sides.

cant strip A triangular filler between a roof and a **parapet wall**.

capacity insulation The ability of masonry to store heat as a result of its mass, density, and specific heat.

capillaries (1) Thin-walled tubes or vessels found in wood. (2) In concrete, channels that absorb water and are interrupted by entrained air bubbles.

carbon steel *See under* **steel**.

cardinal points The four major compass headings of north, east, south, and west.

carpet General designation of fabric constructions that serve as soft floor coverings, especially those that cover an entire floor and are fastened to it, as opposed to rugs. *See also* **flocked carpet**, **fusion-bonded carpet**, **knitted carpet**, **needle-punched carpet**, **tufted carpet**, and **woven carpet**.

carved carpet *See* sculptured under **pile**.

case hardening Hardening of the outer skin of an iron-based alloy by promoting surface absorption of carbon, nitrogen, or cyanide, generally accomplished by heating the alloy in contact with materials containing these elements and rapid cooling.

casting (1) In masonry and **concrete**, pouring a mix into a mold and permitting it to set to form the desired shaped object. (2) In metalwork, pouring molten metal into molds to form desired shapes. (3) In **plastics**, the shaping of plastic objects by pouring the material into molds and allowing it to harden without the use of pressure.

cast iron High-carbon iron made by melting **pig iron** with other iron-bearing materials and **casting** in sand or loam molds; characterized by hardness, brittleness, high compressive and low tensile strengths.

catalyst (1) A substance that speeds up or slows down the rate of a chemical reaction without itself undergoing permanent change in composition. (2) A substance used to initiate the polymerization of monomers to form polymers.

caulking compound Semidrying or slow-drying plastic material used to seal joints or fill crevices such as those around windows and chimneys. Either elastomeric or acrylic joint sealants are generally used today where caulking compounds are called for. *See also* **sealant, joint**.

cavity wall A wall built of **masonry units** so arranged as to provide a continuous air space within the wall (with or without insulating material) and in which the inner and outer **wythes** of the wall are tied together with metal **wall ties** or continuous metal **joint reinforcement**.

cavity wall tie *See under* **wall tie**.

C/B ratio The ratio of the weight of water absorbed by a clay **masonry unit** during immersion in cold water to the weight absorbed during immersion in boiling water. An identification of the probable resistance of clay **brick** to the action of freezing and thawing.

cellular plastic A plastic whose apparent density is decreased substantially by the presence of numerous cells disposed throughout its mass. *Closed-cell plastic* is cellular plastic in which there is a predominance of nonconnecting cells. *Open-cell plastic* is a cellular plastic in which there is a predominance of interconnecting cells.

cellulose The principal constituent of wood that forms the framework of the wood cells.

cement A binding agent capable of uniting dissimilar materials into a composite whole. *See also* **portland cement**.

centering Temporary formwork for the support of masonry **arches** or **lintels** during construction. Also called *center(s)*.

ceramic tile A thin surfacing unit having either a glazed or unglazed face, made from clay or a mixture of clay and other ceramic materials and fired to a temperature sufficiently high to produce specific physical properties and characteristics.

ceramic tile finish All of the elements normally installed by a tile contractor, from the backing to the face of the tile. This may include just a tile-setting product, tile, and **grout** or, in addition, a **mortar bed**, reinforcement, and **cleavage membrane**. Also called *tilework*.

ceramic veneer Architectural terra-cotta, characterized by large face dimensions and thick sections. An *adhesion-type ceramic veneer* is a thin section of ceramic veneer held in place without metal **anchors** by adhesion of **mortar** backing. An *anchored-type ceramic veneer* is a thicker section of ceramic veneer held in place by **grout** and wire **anchors** connected to backing.

cesspool A covered and lined underground pit used as a holding tank for domestic **sewage** and designed to retain the organic matter and solids, but to permit the liquids to seep through the bottom and sides. Cesspools are almost universally prohibited in this country and are not acceptable as a means of sewage disposal.

chain warp *See under* **warp**.

chalking Formation of a loose powder on the surface of a **coating** or **paint** after exposure to the elements.

chase A groove or continuous recess built in a masonry or concrete wall to accommodate pipes, ducts, or conduits.

check A lengthwise separation of wood, the greater part of which occurs across the **annual growth rings**. A check that passes entirely through a piece of wood is called a *split*.

checking Type of **coating**, **paint**, or **transparent finish** failure in which many small cracks appear in the surface.

circuit vent A branch vent that serves two or more traps and extends from in front of the last fixture connection of a horizontal branch to a **vent stack**. *See also* **loop vent**.

circulation path An exterior or interior way of passage from one place to another for pedestrians, including, but not limited to, walks, hallways, courtyards, stairways, and stair landings.

clad alloys Alloys having one or both surfaces of a metallurgically bonded coating, the composition of which may or may not be the same as that of the core, and which is applied for such purposes as corrosion protection, surface appearance, or brazing. *See also* **alclad sheet** and **alloy**.

cladding Bonding thin sheets of a coating metal with desirable properties (such as corrosion resistance or chemical inertness) over a less expensive metallic core not possessing these properties. Copper cladding over steel may be applied by hot dipping; stainless steel and aluminum cladding by hot rolling.

clay brick *See under* **brick masonry unit**.

cleavage (isolation) membrane A membrane such as saturated roofing felt, building paper, or 4 mil (0.1 mm) polyethylene film, installed between the **backing** and a **mortar bed** to permit independent movement of a tile finish.

closer The last **masonry unit** laid in a **course**.

CMU *See* **concrete masonry unit**.

coarse aggregate *See* **aggregate**.

coating A mastic or liquid-applied surface finish, regardless of whether a protective film is formed or only decorative treatment results. There are two categories of coatings: coatings and special coatings. Materials called *coatings* in this book are those that are usually applied in the shop or factory on metal, glass, porcelain, wood, and other materials. They include such products as Kynar coatings on metal, anodizing on aluminum, and liquid-applied colored finishes on wood doors and cabinets. Some coatings are virtually identical to **paint**, the major differences being that they are applied in a shop or factory and are usually sprayed on rather than being brushed or rolled on. *Special coatings* are relatively thick, high-performance architectural coatings, such as high-build glaze coatings; fire-retardant coatings; industrial coatings, such as those used in sewage disposal plants; and cementitious coatings. Special coatings are usually applied in the field. These are beyond the scope of this book. *See also* **fire-retardant coatings and paint**, **paints**, and **transparent finishes**.

coke Processed form of bituminous coal used as a fuel, a reducing agent, and a source of carbon in making pig iron.

cold flow Permanent change in dimension due to stress over time, without heat.

cold forming Forming thin sheets and strips to desired shapes at room temperature, generally with little change in mechanical properties of the metal; includes roll, stretch, shear, and brake forming. *See also* **cold working**.

cold rolling *See under* **cold working**.

cold working (1) In aluminum, forming a metal product at room temperature by means of **rolling**, **drawing**, **forging**, or other mechanical methods of forming or shaping. *Cold rolling* is the forming of sheet metal by rolling at room temperature metal that has been previously hot rolled to a thickness of about 0.125 in. (3.2 mm). (2) In steel, shaping by cold drawing, cold reduction, or cold rolling at room temperature; generally accompanied by an increase in strength and hardness. *Cold drawing* is shaping by pulling through a die to reduce cross-sectional area and impart desired shape; generally accompanied by increase in strength, hardness, closer dimensional tolerances, and smoother finish. *Cold finishing* is cold working that results in **finished mill products**. *Cold reduction* is cold rolling that drastically reduces sheet and strip thickness with each pass through the rolls; generally accompanied by increase in hardness, stiffness, and strength and resulting in a smoother finish and improved flatness. *Cold rolling* is a gradual shaping between rolls to reduce cross-sectional area or impart desired shape; generally accompanied by increase in strength and hardness. *See also* **cold forming** and **hot working**.

collar joint Interior longitudinal vertical joint between two **wythes** of masonry.

colorant Concentrated color added to **coatings** and **paints** to make specific colors.

colorfast Fade resistant.

color uniformity Ability of a **coating** or **paint** to maintain a consistent color across its entire surface, particularly during weathering.

column A vertical structural member acting primarily in compression, whose horizontal dimension measured at right angles to its thickness does not exceed 3 times its thickness.

combination process A process used to retrieve additional **alumina** and soda from the red mud impurities of the **Bayer process**.

common alloy An alloy that does not increase in strength when heat-treated (non-*heat-treatable*). Common alloys may be strengthened by **strain hardening**.

common brick *See* **brick masonry unit**.

common wall A wall that separates adjacent dwelling units within an apartment building or adjacent tenants in townhouses and other buildings; also called a party wall.

composite wall A multiple-**wythe** wall in which at least one of the **wythe**s is dissimilar to the other **wythe** or wythes with respect to type or grade of **masonry unit** or **mortar**.

compounding The thorough mixing of a **polymer** or **polymers** with other ingredients such as **fillers**, **plasticizers**, **catalysts**, **pigments**, **dyes**, or curing agents.

compression molding Forming **plastic** in a mold by applying pressure and, usually, heat.

compressive strength A material's ability to resist compressive forces.

concrete A composite material made of **portland cement**, water, and **aggregates**, and sometimes **admixtures**.

concrete block *See* **concrete masonry unit**.

concrete brick *See under* **brick masonry unit**.

concrete masonry unit (CMU) A **masonry unit** having portland cement as its primary cementitious material. A *decorative CMU* is one of various available types of concrete masonry units with beveled face shell recesses or other articulation or texture. A *faced CMU* is one that has a special ceramic, glazed, plastic, polished, or ground face. A *slump block* is a CMU produced so that it will slump or sag before it hardens; for use in masonry wall construction. A *split-face block* is a solid or hollow CMU that is machine fractured (split) lengthwise after hardening to produce a rough, varying surface texture.

condensation The change of water from its gaseous form (water vapor) to liquid water; the liquid water so collected.

condensation polymerization A chemical reaction in which the molecules of two substances combine, giving off water or some other simple substance (*see* **polymerization**).

consistency The relative ability of freshly mixed **concrete** to flow as measured by a **slump test**.

construction The method by which a **carpet** is made (**loom** or machine) and other identifying characteristics including the number of **pile** rows per inch, **pitch**, **wire** height, number of **shots**, yarn count and plies, total **pile** yarn weight, and **pile yarn density**.

construction joint A joint placed in **concrete** to permit practical placement of the work in section with predefined boundaries.

contact adhesive *See* **adhesive, drywall**.

continuous casting A process in which molten metal is used directly to produce semi-finished products such as slabs or billets, bypassing ingot teaming, stripping, soaking, and **rolling**.

control joint A joint placed in **concrete** to form a plane of weakness to prevent random cracks from forming due to drying shrinkage and temperature changes.

control joint A prefabricated metal accessory intended to relieve shrinkage, temperature, or structural stresses in **plaster**, thus minimizing cracking.

coping **Masonry units** forming a finished cap on top of an exposed pier, wall, pilaster, chimney, etc., to protect the masonry below from penetration of water from above.

copolymer A **polymer** formed by the combination of two or more different **monomers**.

copper staining A stain usually caused by the corrosion products of copper screens, gutters, or downspouts washing down on a finished surface. Can be prevented by painting the copper or applying a **transparent finish** to it.

corbel A shelf or ledge formed by projecting successive courses of masonry out from the face of a wall.

core The innermost portion of **plywood**, consisting of either **hardwood** or **softwood** sawed **lumber**, **veneer**, or composition board.

corner cracking Cracks occurring in the apex of inside corners of **gypsum board** surfaces, such as between adjacent walls or at walls and ceilings.

corner floating **Gypsum board** installation procedure that eliminates some mechanical fasteners at interior corners and permits sufficient movement of boards to eliminate **corner cracking**.

correction lines East-west reference lines used in the **rectangular survey system**, located at 24-mile intervals to the north and south of a **base line**.

corrosion Physical deterioration, decomposition, or loss of cross section of a metal due to **weathering**, galvanic action, or **direct chemical attack**. *Galvanic action* is corrosion produced by electrolytic action between two dissimilar metals in the presence of an **electrolyte**. **Direct chemical attack** is corrosion caused by a chemical dissolving of the metal. *Weathering* is galvanic or chemical corrosion produced by atmospheric conditions.

coulomb A unit of electrical charge equal to the number of electrons conducted past a point in 1 second.

count A number identifying yarn size or weight per unit of length (or length per unit of weight), depending on the **carpet** spinning system used (such as **denier**, woolen, worsted, cotton, or jute system).

course (1) Compass direction from one reference point to the next for each leg of a **metes and bounds** survey. An **angular course** is a compass direction in **degrees**, **minutes**, and **seconds**, stated as a deviation eastward or westward from due north or south; used in **metes and bounds** descriptions and **surveys**. (2) One of the continuous horizontal layers of **masonry units**, bonded with **mortar**. One course is equal to the thickness of the **masonry unit** plus the thickness of one **mortar joint**.

coverage Area over which a given amount of **coating**, **paint**, or **transparent finish** will spread and hide the previous surface; usually expressed in sq. ft./gal. (m^2/L).

covering power *See* **hiding power**.

cracking Type of **coating**, **paint**, or **transparent finish** failure characterized by breaks in irregular lines wide enough to expose the underlying surface.

crazing (1) Numerous hairline cracks in the surface of newly hardened **concrete**. (2) Similar cracks in a **paint** or **coating**.

creep *Same as* **cold flow**.

crimping Method of texturing staple and continuous filament yarn to produce irregular alignment of fibers in **carpet** and increase bulk and covering power; also facilitates interlocking of fibers, which is necessary for spinning staple fibers into yarn.

crook *See* **warp**.

crossband A layer of **veneer** in a **plywood** panel whose grain direction is at right angles to that of the face plies.

cross furring Furring members installed perpendicular to framing members.

cross grain *See under* **grain**.

crosslinking A chemical reaction in which adjacent **polymer** molecules unite to form a strong, three-dimensional network; usually occurs during the curing of **thermoset**ting **plastics**.

cross slope The slope of a pedestrian way that is perpendicular to the direction of travel. *See also* **running slope**.

crown, joint The maximum height to which joint compound is applied over a **gypsum board** joint.

cryolite sodium Aluminum fluoride used with **alumina** in the final electrolytic reduction of aluminum. Found naturally in Greenland, generally produced synthetically from alum, soda, and hydrofluoric acid.

cubing The assembling of **CMU**s into cubes after curing for storage and delivery. A cube normally contains 6 layers of 15 to 18 blocks (8 × 8 × 16 in. [200 × 200 × 400 mm]) or an equivalent volume of other size units.

cue A device that alerts a user to an upcoming condition; includes audible, visual, and textural signals. *See also* **detectable warning**. An *audible cue* is a sound or a verbal alert. A *tactile cue* is one that can be detected by touch. A *visual cue* is one that can be seen.

cup *See* **warp**.

curb ramp A short ramp cutting through a curb or built into it.

cure To change the properties of a polymeric system into a final, more stable condition by the use of heat, radiation, or reaction with chemical additives. Sometimes referred to as *set*.

curing (1) The hardening of a **concrete masonry unit (CMU)**. *Atmospheric pressure steam curing* is a method of curing **CMU**s, using steam at atmospheric pressure usually at temperatures of 120°F (48.89°C) to 180°F (82.2°C). Also called low-pressure steam curing. *High-pressure steam curing* is a method of curing **CMU**s, using saturated steam (365°F [185°C]) under pressure, usually 125 to 150 psi (87.88 to 105.46 Mg/m^2). Also referred to as *autoclave curing*. *Moist curing* is a method of curing **CMU**s, using moisture at atmospheric pressure and temperature of approximately 70°F (21.1°C). (2) The process of keeping **concrete** moist for an extended period after placement to ensure proper **hydration** and subsequent strength and quality. (3) Final conversion or drying of a **coating**, **painting**, or **transparent finish** material.

curtain wall An exterior non-load-bearing wall. Such walls may be anchored to columns, spandrel beams, structural walls, or floors. *See also* **panel wall**.

cushioning Soft, resilient layer provided under **carpet** to increase underfoot comfort, to absorb pile-crushing forces, and to reduce impact sound transmission. Also called *underlay* or *lining*; *see also* **padding**.

cut and loop *See* **multilevel loop**.

cut pile A **carpet** face construction in which the pile is cut level so that it stands erect in a low, dense, plush, even surface.

cutting in Painting of an edge, such as wall color at the ceiling line or at the edge of woodwork.

cylinder test A laboratory test for compressive stress of a field sample of **concrete** (6 in. [152.4 mm] in diameter by 12 in. [305 mm] in length).

daily degree day The numerical difference between 65°F (18.3°C) and the average of all recorded temperatures on a given day that are lower than 65°F (18.3°C).

damp course A course or layer of impervious material that prevents capillary entrance of moisture from the ground or a lower course. Often called *damp check*.

darby A tool used to level freshly placed **concrete**.

darbying Smoothing the surface of freshly placed **concrete** with a **darby** to level any raised spots and fill depressions.

dead load The weight of all permanent and stationary construction or equipment included in a building. *See also* **live load**.

decay The decomposition of wood substances by certain fungi.

decibel A logarithmic unit expressing the ratio between a **sound** being measured and a reference point.

decorative CMU *See under* **concrete masonry unit.**

degradation A permanent change in the physical or chemical properties of a **plastic** evidenced by impairment of these properties.

degree A unit of **angular measure** equal to the angle contained within two radii of a circle that describe an arc equal to 1⁄360 of the circumference of the circle; also used to define an arc equal to 1⁄360 of the circumference of a circle.

degree days heating (DDH) The sum of the **daily degree days** when the temperature dropped below 65°F (18.3°C).

delustered nylon Nylon on which the normally high sheen has been reduced by surface treatment.

denier System of yarn count used for synthetic **carpet** fibers: number of grams per 9000 meters of yarn length; one denier equals 4,464,528 yards per pound or 279,033 yards per ounce.

density *See* **pile yarn density**.

detectable warning A standardized surface texture applied to or built into walking surfaces or other elements to warn visually impaired people of hazards in the path of travel. *See also* **cue**.

deterioration *See* **degradation**.

dew point That temperature above freezing at which air becomes **saturated** and **condensation** occurs. *See also* **frost point**.

dimensional stability The ability of a material to retain its dimensions in service.

dimension lumber *See under* **lumber**.

direct chemical attack Corrosion caused by a chemical dissolving a metal.

disability A limitation or loss of use of a physical, mental, or sensory body part or function.

discontinuous construction A construction method used to separate a continuous path through which **sound** may be transmitted. Examples include the use of staggered studs, double walls, and the resilient mounting of surfaces.

dispersion The distribution of a finely divided solid in a liquid or a solid.

disposal field *See* **absorption field**.

distribution line Open joint or perforated pipe intended to permit soil absorption of **effluent**.

divider strips All-metal or plastic-top metal strips provided in a **terrazzo finish** to control cracking due to drying shrinkage, temperature variations, and minor structural movements; are also used for decorative purposes and convenience in placing a **terrazzo** topping.

drain A pipe that carries wastewater or waterborne wastes in a building **drainage system**.

drainage system That piping that conveys **sewage**, rainwater, or other liquid wastes up to a point of disposal, such as the mains of a public **sewer** system or a private **septic disposal system**. *See also* **vent system** and **drain, waste, and vent (DWV) system**.

drain, waste, and vent (DWV) system The collection of pipes that facilitates the removal of liquid and solid wastes and dissipates **sewer** gases.

drawing The process of pulling material through a die to reduce the size, to change the cross section or shape, or to harden the material.

dressed lumber **Lumber** that has been surfaced with a planing machine.

dressed size *See* **size**.

drier An ingredient included to speed the drying of **coatings**, **paints**, and **transparent finishes**.

drip A projection shaped to cause water to flow away from a lower surface, thus preventing it from running down the face of the lower surface.

drying The various stages of **curing** in a **coating**, **paint**, or **transparent finish** film. *Dust-free* is the stage of drying when particles of dust that settle on the surface do not stick to it. *Tack-free* is the stage of drying when the surface no longer feels sticky when lightly touched. *Dry enough to handle* is the stage of drying when the film has hardened sufficiently so the object or surface may be used without marring. *Dry enough to recoat* is that stage of drying when the next coat can be applied. *Dry enough to sand* is the stage of drying when the film can be sanded without the sandpaper sticking or clogging.

drying oil A **coating**, **paint**, or **transparent finish vehicle** ingredient, such as linseed oil, which, when exposed to the air in a thin layer, oxidizes and hardens to a relatively tough elastic film.

dry lumber Under Product Standard PS 20-70, **lumber** with a moisture content of 19% or less.

dry rot A term that is loosely applied to many types of decay, which in the advanced state permit wood to be easily crushed to a dry powder. This term is a misnomer since all fungi require moisture.

drywall nail *See* **nail, drywall**.

drywall screw *See* **screws, drywall**.

dry well A covered and lined underground pit, similar to a *seepage pit* but intended to receive water free of organic matter, such as from roof drains, floor drains, or laundry tubs. Used as an auxiliary to a **septic disposal system** to avoid overloading the **septic tank** absorption system. (Also called *leaching well* or *leaching pit.*)

ductile Capable of being drawn out or hammered, able to undergo cold **plastic deformation** without breaking.

durability The ability of a **coating**, **paint**, or **transparent finish** to retain its desirable properties for a long time under expected service conditions.

dusting The appearance of a powdery material at the surface of a hardened **concrete**.

dye or **dyestuff** Colored material used to change the color of a **coating** or **paint** with little or no hiding of the underlying surface.

dyne A unit of force, which when acting on a mass of 1 gram accelerates it 1 centimeter per second per second.

echo A reflected **sound** loud enough and received late enough to be heard as distinct from the source.

economy brick *See under* **brick masonry unit**.

edge grain *See under* **grain**.

edges, gypsum board *See* **gypsum board**.

edging The finishing operation of rounding off the edge of a **concrete** slab to prevent chipping or damage.

efflorescence Deposit of soluble salts, usually white in color, appearing upon the exposed surface of masonry, **concrete** or **plaster**.

effluent Partially treated liquid sewage flowing from any part of a **septic disposal system**, **septic tank**, or absorption system.

eggshell finish Surface sheen midway between **flat** and **semigloss**.

elastic deformation Deformation caused by a **stress** small enough that when the stress is removed the material returns to its original shape. *See also* **plastic deformation**.

elasticity The ability to recover original size and shape after deformation.

elastic limit The amount of **stress** that, if exceeded, will cause a given material to deform or set permanently.

elastomer (1) A material that at room temperature can be stretched repeatedly to at least twice its original length and that, upon release of the stress, will return instantly and with force to its approximate original length. (2) A rubberlike substance.

electric induction furnace *See under* **furnace, steel**.

electrogalvanizing **Electroplating** with zinc to provide greater corrosion resistance.

electrolysis Also called *galvanic corrosion*. An electrochemical decomposition that results when dissimilar metals are each contacted by the same **electrolyte**, such as water. The process is similar to that which takes place in an automobile battery. One metal acts as a cathode, the other as an anode. When the electrolyte causes an electrical current to flow from one metal to the other, the anodic metal dissolves and hydrogen ions accumulate on the cathodic metal. Electrolysis can also take place in a single metal when one portion of it is cathodic and another portion is anodic if an **electrolyte** makes a bridge between the two portions.

electrolyte A nonmetallic substance in which electricity is conducted by the movement of ions.

electromotive force Something that moves or tends to move electricity.

electroplating A process that employs an electric current to coat a base metal (cathode) with another metal (anode) in an electrolytic solution.

emulsion Mixture of liquids (or a liquid and a solid) not soluble in each other, one liquid (or solid) being dispersed as minute particles in the other, base liquid, with the help of an emulsifying agent.

enamel A **coating** or **paint** capable of forming a very smooth, hard film, sometimes using varnish as the **vehicle**; may be **flat**, **gloss**, or **semigloss**.

enamel holdout Property of producing a tight film that prevents the penetration of subsequent **enamel** coats to underlying surfaces; prevents unequal absorption and uneven gloss.

ends, gypsum board *See* **gypsum board**.

end wall The wall along the short dimension of a room.

engineered brick *See under* **brick masonry unit**.

engineered brick masonry Masonry in which the design is based on a rational, accepted structural engineering analysis.

equilibrium moisture content The **moisture content** at which wood neither gains nor loses moisture when surrounded by air at a given **relative humidity** and temperature.

erosion Wearing away of a **coating**, **paint**, or **transparent finish** film caused by exposure.

etch A surface preparation for a **coating**, **paint**, or **transparent finish** by chemical means to improve adhesion.

evaporation The change of water from a liquid to a gas.

expandable plastic A **plastic** suitable for expansion into cellular form by thermal, chemical, or mechanical means.

expansion strips Double **divider strips** in a **terrazzo finish** separated by resilient material and provided generally for the same purpose as divider strips, but where a greater degree of structural movement is expected.

extender (1) In **plastics**, a low-cost material used to dilute or extend high-cost resins without appreciably lessening the properties of the original resin. (2) In **coatings**, **paints**, and **transparent finishes**, an inexpensive but compatible substance that can be added to a more valuable substance to increase the volume of material without substantially diminishing its desirable properties; in **coatings** and **paints**, extender pigments improve storage and application properties.

exterior wall Any outside wall of a building other than a **party wall**.

external corner A projecting angle formed by abutting walls or vertical surface and soffit (not to be confused with exterior, meaning exposed to the weather).

extrude To form lengths of shaped sections by forcing a plastic material through a shaped hole in a die.

extrusion (1) In aluminum, a product formed by extruding. (a) An *extrusion billet* is a solid, wrought, semifinished product intended for further extrusion into rods, bars, or shapes. (b) An *extrusion ingot* is a solid or hollow cylindrical casting used for extrusion into bars, rods, shapes, or tubes. (2) In plastic, forcing plastic material through a shaped orifice to make rod, tubing, or sheeting.

faced CMU *See* **concrete masonry units**.

faced wall A wall in which facing and backing are of different materials and are bonded together to exert common action under load.

face shell The side wall of a hollow **masonry unit** or clay tile.

face size *See under* **size**.

facing A part of a wall that is used as a finished surface.

facing brick *See under* **brick masonry unit**.

factory and shop lumber *See* **lumber**.

fading Loss of color due to exposure to light, heat, or weathering.

fastener treatment Method of concealing **gypsum board** fasteners by successive applications of compound until a smooth surface is achieved.

featheredging (feathering) Tapering **gypsum board joint compound** to a very thin edge to ensure inconspicuous blending with adjacent gypsum board surfaces.

feather sanding Tapering the edge of a dried **coating**, **paint**, or **transparent finish** film with sandpaper.

ferritic steel *See* **grain structure**.

ferroalloys Iron-based alloys used in steelmaking as a source of desired alloying elements.

ferrous alloys Composite metals whose chief ingredient is iron (ferrum), metallurgically combined with one or more alloying elements.

fiber saturation point The stage in drying (or wetting) of wood at which the cell walls are saturated with water but the cell cavities are free of water, being approximately 30% **moisture content** in most species.

fiber, wood A comparatively long (1/25 in. [1.016 mm] or less to 1/3 in. [8.47 mm]), narrow, tapering unit closed at both ends.

field (1) In **ceramic tile**, the general area of the tile excluding trim. (2) In **gypsum board**, the surface of the board exclusive of the perimeter. (3) In masonry walls, the expanse of wall between openings, corners, etc., principally composed of **stretchers**. (4) In resilient flooring, the area of a floor within the *borders*.

figure The pattern produced in a wood surface by **annual growth rings**, **rays**, **knots**, and deviations from regular **grain**.

fill The sand, gravel, or compacted earth used to bring a **subgrade** up to a desired level.

filler (1) In **plastics**, a relatively inert material added to modify the strength, permanence, or working properties, or to lower the cost of a **resin**. (2) In **coatings**, **paint**, and **transparent finishes**, a pigmented composition for filling the pores or irregularities in a surface in preparation for finishing.

filling *See* **weft**.

film (1) In **plastics**, plastic sheeting having a nominal thickness not greater than 0.010 in. (0.254 mm). (2) In **coatings**, **paints**, and **transparent finishes**, a thin application generally not thicker than 0.010 in. (0.254 mm).

fine aggregate *See* **aggregate**.

fineness modulus A measure of the average size of an **aggregate**, calculated by passing the **aggregate** through a series of screens of decreasing size.

finish coat The final decorative plaster layer in either **two-coat** or **three-coat work**.

finished mill products Steel shapes that can be used directly in construction. *Bars* are hot-rolled or cold-drawn round, square, hexagonal, or multifaceted long shapes, generally larger than wire in cross section. Also hot- or cold-rolled rectangular flat shapes (flats) generally narrower than sheets and strip. *Backplate* is a cold-rolled, flat, carbon-steel product that is thinner than sheet and wider than strip, generally coated with zinc, tin, or terne metal. *Foil* is a cold-rolled flat product less than 0.005 in. (0.127 mm) thick. *Plate* is a hot-rolled flat product generally thicker than sheet and wider than strip. *Sheet* is a hot- or cold-rolled flat product generally thinner than plate and wider than strip. *Strip* is a hot- or cold-rolled flat product generally narrower than sheet and thinner than plate. *Structurals* are hot-rolled steel shapes of special design (such as H beams, I beams, channels, angles, and tees) used in construction. *Terneplate* is backplate or sheet metal that has been coated with *terne* metal. *Tinplate* is backplate that has been coated with tin. *Tubular products* are hollow products of round, oval, square, rectangular, or multifaceted cross sections. In construction, round products are generally called pipe; square or rectangular products with thinner wall sections are called tube or tubing. *Wire* is a cold-finished product of round, square, or multifaceted cross section, generally smaller than bars; round wire is cold drawn, 0.005 in. (0.127 mm) to less than 1 in. (25 mm) in diameter; flat wire is cold rolled, generally narrower than bar.

fire brick *See under* **brick masonry unit**.

fire division wall A wall that subdivides a building so as to resist the spread of fire. It is not necessarily continuous through all stories to and above the roof. *See also* **fire wall**.

fireproofing A material or combination of materials built to protect structural members so as to increase their fire resistance.

fire-resistive material *See* **noncombustible material**.

fire-retardant coating or **paint** A **coating** or **paint** that will significantly (1) reduce the rate of flame spread, (2) resist ignition at high temperatures, and (3) insulate the underlying material so as to prolong the time required for the material to reach ignition, melting, or structural weakening temperature.

fire wall A wall that subdivides a building to resist the spread of fire and that extends continuously from the foundation through the roof. *See also* **fire division wall**.

fixture-unit A mathematical factor used by engineers to estimate the probable demand on a **drainage** or **water supply system** (volume, duration of flow, and intervals between operations) by various **plumbing fixtures**.

flaking Form of **coating**, **paint**, or **transparent finish** failure characterized by the detachment of small pieces of the film from the surface or previous coat; usually preceded by cracking or blistering.

flanking path A wall or floor and ceiling assembly that permits **sound** to be transmitted along its surface. Also an opening that permits the direct transmission of **sound** through the air.

flashing (1) In masonry construction, a thin, relatively impervious sheet material, placed in **mortar joints** and across air spaces in masonry walls to collect water that may penetrate the wall and to direct it to the exterior. (2) In masonry manufacture, the step during the burning process of clay **masonry units** that produces varying shades and colors in the units. (3) In resilient flooring, the bending up of resilient sheet material against a wall or a projection, either temporarily for the purpose of fitting, or permanently so as to form a one-piece resilient base.

flash point Temperature at which a **coating**, **paint**, **transparent finish**, or **solvent** will ignite; the lower the flash point, the greater the hazard.

flash set Undesirable rapid setting of **cement** in **concrete** or **mortar**.

flat Dull, nonreflective; opposite of **gloss**.

flat applicator Rectangular flat pad with an attached handle that is used to paint shingles, shakes, and other special surfaces.

flat finish Finish having no gloss or luster.

flat grain *See under* **grain**.

flats Term applied to **flat coatings** and **paints**.

flatting agent Ingredient added to **coatings** and **paint** to reduce the gloss of the dried film.

flitch (1) A portion of a log sawed on two or more sides and intended for manufacture into **lumber** or sliced or sawed **veneer**. (2) A complete bundle of veneers laid together in sequence as they were sliced or sawed.

floating (1) A **concrete** slab finishing operation that embeds **aggregate**, removes slight imperfections, humps, and voids to produce a level surface, and consolidates mortar at the surface. (2) A condition in which a layer of water occurs between a mortar bed and a masonry unit, usually due to **bleeding**, causing the unit and the mortar to fail to bond with each other. In this condition, the unit is said to *float*. (3) Separation of **pigment** colors on the surface of applied **paint**.

flocked carpet Single-level velvety pile carpet composed of short fibers embedded on an adhesive-coated backing.

floor brick *See under* **brick masonry unit**.

flow The ability of a **coating**, **paint**, or **transparent finish** to level out and spread into a smooth film; materials that have good flow usually level out uniformly and exhibit few brush or roller marks.

fluffing *See* **shedding**.

flux A mineral that, due to its affinity to the impurities in iron ores, is used in ironworking and steelmaking to separate impurities in the form of molten slag. *Basic flux* is a mineral, such as limestone or dolomite, used in basic furnaces to make basic (low-phosphorus) steel. *Neutral flux* is a mineral (fluorspar) used to make slag more fluid.

foamed plastic *See* **cellular plastic**.

foil *See under* **finished mill products**.

footing The base of a foundation, column, or wall used to distribute the load over the subgrade.

forging The working (shaping) of metal parts by forcing between shaped dies. *Press forging* is shaping by applying pressure in a press. *Hammer forging* is shaping by application of repeated blows, as in a forging hammer. *See also* **hot working**.

form A temporary structure erected to contain **concrete** during **placing** and initial hardening.

foundation wall A load-bearing wall below the floor nearest to exterior grade serving as a support for a wall, pier, column, floor, or other structural part of a building.

fractional section *See under* **section**.

frames Racks at the back of a **Jacquard loom**, each holding a different color of pile yarn. In **Wilton carpets**, two to six frames may be used and the number is a measure of quality as well as an indication of the number of colors in the pattern, unless some of the yarns are buried in the **backing**.

freezing cycle day A day in which the air temperature passes either above or below 32°F (0°C). The average number of freezing cycle days in a year equals the difference between the mean number of days the minimum temperature was 32°F (0°C) or below and when the maximum temperature was 32°F (0°C) or below.

frequency The number of complete cycles of a vibration occurring in each second, measured in cycles per second (cps) and expressed in **hertz (Hz).**

frieze carpet *See under* **pile**.

frost point That temperature below freezing at which **condensation** occurs. *See also* **dew point**.

fungicide Agent that helps prevent mold or mildew growth on **paint**.

furnace, blast Tall, cylindrical masonry structure lined with refractory materials, used to smelt iron ores in combination with fluxes, coke, and air into **pig iron**.

furnace, steel Masonry or steel structure lined with refractory materials, used to melt **pig iron**, scrap metal, and sometimes agglomerated ores, ferroalloys, and fluxes into steel. A *basic oxygen furnace* is a suspended, tilting vessel that uses high-purity oxygen to oxidize impurities in hot pig iron and other iron-bearing materials to produce low-phosphorus (basic) steel. An *electric arc suspended furnace* is a kettle that melts scrap metal, ore, and sometimes ferroalloys with the heat of an electric arc to produce steels of controlled chemical composition. An *electric induction furnace* is a steel-encased, insulated magnesia pot in which metal, scrap, and ferroalloys are melted with the heat of an electric current induced by windings of electric tubing; used chiefly to produce small quantities of high-grade steels such as alloy, stainless, and heat-resisting steels. An *open hearth furnace* is a masonry structure with a hearth exposed to the sweep of flames in which hot pig iron, scrap metal, and fluxes are melted and oxidized by a mixture of fuel and air to produce basic or acid steel.

furring A method of finishing the interior face of a **concrete** or masonry wall to provide space for insulation, to prevent moisture penetration, or to provide a level, plumb, and straight surface for finishing. Furring consists of metal channels or studs or of wood strips or studs.

fusion-bonded carpet **Carpet** made by fusing carpet yarn and a backing, then cutting the substrates in two, making two pieces of **carpet**.

fuzzing Temporary condition on new **carpet** consisting of irregular fuzzing appearance caused by slack yarn twist, fibers snagging, or breaking of yarn. Can be remedied by spot **shearing**.

galvanic corrosion *See* **electrolysis**.

galvanizing Zinc coating by electroplating or hot dipping, which produces a characteristic bright spangled finish and protects the base metal from atmospheric **corrosion**.

gang grooved **Plywood** panels produced by passing them under a machine with grooving knives set at certain intervals.

gauge The distance between **tufts** across the width of **knitted** and **tufted carpets**, expressed in fractions of an inch.

glaze Used to describe several types of finishing materials. *Glazing putty* is a compound of creamy consistency applied to fill surface imperfections. A *glazing stain* is very thin, semitransparent, and usually pigmented with a Vandyke brown or burnt sienna, applied over a previously stained, filled, or painted surface to soften or blend the original color without obscuring it. A *glaze coat* is a clear finish applied over previously coated surfaces to create a gloss finish.

glazing compound Doughlike material, consisting of **vehicle** and **pigment**, that retains its plasticity over a wide range of temperatures and for an extended period of time. Either elastomeric or acrylic joint sealants are generally used today where glazing compounds are called for. *See also* **sealant, joint**.

gloss Shiny, reflective surface quality; term sometimes used broadly to include **coatings**, **paints**, and **transparent finishes** with these surface properties.

glueline The line of glue visible on the edge of a **plywood** panel. Also applies to the layer of glue itself.

grade The designation of the quality of wood, steel, and other materials and products made from them, such as **plywood**.

graded aggregate An **aggregate** containing uniformly graduated particle size from the finest fine **aggregate** size to the maximum size of coarse **aggregate**.

graded sand A sand containing uniformly graduated particle sizes from very fine up to ¼ in. (6.4 mm).

grain The direction, size, arrangement, appearance, or quality of the **fibers** in wood. **Cross-grained wood** is sawed with the fibers not parallel with the longitudinal axis of the piece. This grain may be diagonal, in a spiral pattern, or a combination of both. *Diagonal-grained wood* is sawn at an angle with the **bark** of the tree such that the **annual growth rings** are at an angle with the axis of the piece. *Edge-grained wood* is sawn parallel with the **pith** of a log and at nearly right angles to the **annual growth rings**, making an angle of 45 to 90 degrees with the wide surface of the piece. Also called *quarter-sawn* wood and *vertical-grained* wood. *Flat-grained wood* is sawn parallel with the **pith** of the log and nearly tangent to the **annual growth rings**, making an angle of less than 45 degrees with the surface of the piece. Also called *plain-sawn* wood. *Open-grained wood* is a common designation for wood with large pores, such as oak, ash, chestnut, and walnut.

graining Simulating the grain of wood by means of specially prepared colors or stains and the use of graining tools or special brushing techniques.

grain raising Swelling and standing up of wood grain caused by absorbed water or solvents.

grain structure The microscopic internal crystalline structure (size and distribution of particles) of a metal that affects its properties, known as austenitic, ferritic, and martensitic. *Austenitic steels* are tough, strong, and nonmagnetic. Austenitic stainless steels have a chromium content of up to 25%, nickel of up to 22%, and can be hardened by cold working. *Ferritic steels* are soft, ductile, and strongly magnetic. Ferritic stainless steels usually have a chromium content of 12% to 27% and are not hardenable by heat treatment. *Martensitic steels* can be made very hard and tough by heat treatment and rapid cooling. Martensitic stainless steels have a chromium content between 4% and 12%.

great circle A line described on a sphere by a plane bisecting the sphere into equal parts. The **equator** is a great circle, as are pairs of opposing **meridians**.

green concrete Freshly placed **concrete**.

green lumber Under Product Standard PS 20-70, lumber with a moisture content of more than 19%. Unseasoned lumber that has not been exposed to air or kiln drying.

Greenwich meridian *See under* **meridian**.

grin Condition in which the **backing** shows through sparsely spaced **pile tufts**. A **carpet** may be grinned (bent back) deliberately to reveal its **construction**.

gross cross-sectional area In masonry, the total area of a section perpendicular to the direction of the load, including areas within cores, cellular spaces, and other openings in the material.

ground coat Base coat in an antiquing system; applied before graining colors, glazing, or other finish coat.

grounds Nailing strips placed in **concrete** and masonry walls as a means of attaching trim, furring, cabinetry, or equipment.

grout, masonry (1) **Mortar** of a consistency that will flow or pour easily without segregation of the ingredients. (2) A liquid mixture of **cement**, water, and sand of pouring consistency.

grout, tile A formulation used to fill the joints between tiles; may be cementitious, resinous, or a combination of both.

guide meridian *See under* **meridian**.

gusset plate Wood or metal plate used as a means of joining coplanar structural members in trusses. Gusset plates lap the butt joints between members.

gypsum board A panel consisting of a noncombustible core of calcined gypsum, surfaced on both sides with a covering material specifically designed for various uses with respect to performance, application, location, and appearance. *Wallboard* is a class of gypsum board used primarily as an interior finished surface. *Lath* is a class of gypsum board used as a base for gypsum plaster. *Backing boards* are gypsum boards that serve as a base to which gypsum wallboards or tile is applied. *Sheathing* is a class of gypsum board used as a base for exterior finishes. *Edges* are gypsum board extremities that are paperbound and run the long dimension of the board as manufactured. *Ends* are gypsum board extremities that are mill- or job-cut, exposing the gypsum core, and run the short dimension of the board as manufactured.

hardboard A dense panelboard manufactured of wood fibers with the natural lignin in the wood reactivated to serve as a binder for the wood fibers.

hard-burned Clay products that have been fired at high temperatures. They have relatively low absorptions and high compressive strengths.

hardness Cohesion of particles on the surface of a **coating**, **paint**, or **transparent finish** as determined by ability to resist scratching or indentation.

hardwood The botanical group of trees that are broad leaved and deciduous. The term does not refer to the actual hardness of the wood.

header A masonry unit that overlaps two or more adjacent **wythes** of masonry to provide structural bond. Also called *bonder*.

header course A continuous bonding **course** of header **brick**.

head joint The vertical **mortar joint** between ends of **masonry units**.

heartwood That wood that extends from the **pith** to the **sapwood**, the cells of which no longer participate in the growth process of the tree. Heartwood may be impregnated with gums, resins, and other materials, which usually make it darker and more decay resistant than sapwood.

heat capacity The amount of heat required to raise the temperature of 1 cu. ft. of a material 1°F.

heat-set nylon Nylon fiber that has been heat treated to retain a desired shape.

heat sink A material or system that collects and stores heat.

heat-treatable alloys Aluminum alloys capable of gaining strength by being heat treated. The alloying elements show increasing solid solubility in aluminum with increasing temperature resulting in pronounced strengthening.

heat treatment (1) Of aluminum (*see* **solution heat treatment**): *Annealing* is a heat treatment process in which an **alloy** is heated to a temperature between 600°F (315.56°C) and 800°F (426.67°C) and then slowly cooled to relieve internal stresses and return the **alloy** to its softest and most **ductile** condition. (2) Of steel: Controlled heating and cooling of steels in the solid state for the purpose of obtaining certain desirable mechanical or physical properties. *Annealing* is heating metal to high temperatures (1350°F [732.2°C] to 1600°F [871.1°C] for steel), followed by controlled cooling, to make the metal softer or change its ductility and toughness. *Quenching* is rapid cooling by immersion in oil, water, or other cooling medium to increase hardness. *Tempering* is reheating (in steel to less than 1350°F [732.2°C]) after hardening (as by quenching) and slow cooling, to restore ductility.

hertz (Hz) The unit of frequency of a periodic process equal to one cycle per second.

hiding power Ability of a **coating** or **paint** to obscure the surface to which it is applied, generally expressed as number of sq. ft. that can be covered by one gal. of material, or number of gallons required to cover 1000 sq. ft. Also called *covering power*.

high-density plywood *See under* **plywood**.

high-early-strength cement **Cement** used to produce a **concrete** that develops strength more rapidly than normal.

high-low loop *See* **multilevel loop**.

high polymer A **polymer** of high molecular weight.

high-pressure laminate A **laminate** molded and cured at pressures not lower than 1000 psi (703.07 Mg/m^2) and usually between 1200 psi (843.68 Mg/m^2) and 2000 psi (1406.14 Mg/m^2).

high-pressure steam curing *See under* **curing**.

hollow brick *See under* **brick masonry unit**.

hollow masonry unit *See* **masonry unit**.

hollow wall A wall built of **masonry units** arranged to provide an air space within the wall between the facing and backing **wythes**.

horizontal application Applying **gypsum board** with the **edges** perpendicular to supporting members such as studs, joists, channels, or furring strips. *See also* **vertical application**.

hot-dip process Coating a metal with another metal by immersing it in a bath of the molten coating metal. The coating metals most commonly used on steel are zinc, terne metal, and aluminum.

hot rolling *See* **rolling** and **hot working**.

hot working Forming metal at elevated temperatures at which metals can be more easily worked. (1) For aluminum alloys: (a) Hot working temperatures are usually in the 300°F (148.9°C) to 400°F (204.4°C) range. (b) *Hot rolling* is the shaping of aluminum plate by rolling heated slabs of metal. Hot rolling is usually used for work down to about 0.125 in. (3.175 mm) thick. (2) For steel: Shaping hot plastic metal by hot rolling, extruding, or forging. (a) Hot working temperatures are above 1500°F (815.6°C). Hot working is generally accompanied by increases in strength, hardness, and toughness. (b) *Hot forming* is the forming of a hot plastic metal into desired shapes, with little change in the mechanical properties of the metal. (c) *Extruding* is shaping lengths of hot metal by forcing them through a die of the desired profile. (d) *Forging* is shaping hot metal between dies with compression force or impact. (e) *Hot rolling* is gradual shaping by squeezing hot metal between rolls.

house sewer *See* **building sewer**.

hydration The chemical reaction of water and **cement** that produces a hardened **concrete**.

Hz *See* **hertz**.

impact insulation class (IIC) A single-number rating developed by the Federal Housing Administration to estimate the impact **sound isolation** performance of floor and ceiling assemblies.

impact noise rating (INR) A single-number rating used to compare and evaluate the ability of floor and ceiling assemblies to isolate impact sound transmission (*see* **sound isolation**).

impact sound pressure level (ISPL) The **sound** level in **decibels** measured in the receiving room resulting from the transmission of **sound** produced by a standard tapping machine through an adjacent floor and ceiling assembly.

impervious soil A tight, cohesive soil, such as clay, that does not allow the ready passage of water.

incident sound A **sound** striking a surface, as contrasted with a **sound** reflected from the surface. The angle of incidence equals the angle of reflection.

individual sewage disposal system A combination of a **sewage** treatment plant (package plant or **septic tank**) and method of **effluent** disposal (soil absorption system or stream discharge) serving a single dwelling.

inert Having inactive chemical properties.

ingot (1) A mass of aluminum cast into a convenient shape for storage or transportation, to be later remelted for casting or finishing by rolling, forging, or other process. (2) A cast **pig iron** or steel shape made by pouring hot metal into a mold.

inhibitor A substance that prevents or retards a chemical reaction; inhibitors are often added to plastic resins to prolong their storage life.

injection molding Forming **plastic** by fusing it in a chamber with heat and pressure and then forcing part of the mass into a cooler chamber where it solidifies.

insolation (incident solar radiation) The solar radiation that strikes a surface.

intercoat adhesion Adhesion between two coats of **paint**.

internal corner An enclosed angle of less than 180 degrees formed by abutting walls or the juncture of walls and the ceiling (not to be confused with interior, meaning protected from the weather).

isolation *See* **sound isolation**.

isolation joint A joint placed to separate a **concrete** slab into individual panels or from adjacent surfaces.

isolation membrane (1) In tile floors, *see* **cleavage isolation membrane**. (2) In a **terrazzo finish**, a membrane such as asphalt-saturated roofing felt, building paper, or polyethylene film installed between the subfloor and the terrazzo underbed to prevent bond and permit independent movement of each.

jacquard Mechanism for a **Wilton loom** that uses punched cards to produce the desired color design.

Jaspe carpet *See under* **pile**.

joint *See* **mortar joint**.

joint beading *See* **ridging**.

joint compound A material used for finishing joints in **gypsum board**.

joint crown *See* **crown, joint**.

joint reinforcement Steel wire, bar, or fabricated reinforcement that is placed in horizontal **mortar joints**.

joints *See* **control joint**, **construction joint**, and **isolation joint**.

joint tape Paper or paper-faced cotton tape used over joints between **wallboard** to conceal the joints and provide a smooth surface for painting.

joint treatment Method of reinforcing and concealing **gypsum board** joints with tape and successive layers of **joint compound**.

joist One of a series of parallel beams used to support floor, ceiling, and roof loads. Joists are supported in turn by bigger **beams**, girders, or **bearing walls**.

joists and planks *See under* **lumber**.

jumbo brick *See under* **brick masonry unit**.

jute Strong, durable yarn spun from fibers of the jute plant, native to India and the Far East; used in the **backings** of many **carpets**.

kiln drying *See under* **seasoning**.

knitted carpet **Carpet** made on a knitting machine by looping together backing, stitching, and pile yarns with three sets of needles, as in hand knitting.

knot That portion of a branch or limb that has been surrounded by subsequent growth of wood.

kraftcord Tightly twisted yarn made from wood pulp fiber, used as an alternate for cotton or jute in carpet backing.

lacquer Coating that dries quickly by evaporation of its volatile **solvent** and forms a **film** from its nonvolatile constituent, usually nitrocellulose; may be pigmented or clear.

laitance A soft, weak layer of mortar appearing on a horizontal surface of **concrete** due to **segregation** or **bleeding**.

laminate A product made by bonding together two or more layers of materials.

laminated wood A piece of wood built up of laminations that have been joined with glue, mechanical fastenings, or both.

laminating adhesive *See* **adhesive, drywall**.

langley The measure of **isolation**; equal to 1 calorie per square centimeter or 3.69 **Btu**'s per sq. ft.

lap To lay or place one coat so that its edge extends over and covers the edge of a previous coat.

lateral A branch of an **absorption field**, consisting of either (1) the length of distribution line between overflow pipes or (2) the length of distribution line between the tee or cross fitting and the farthest point in a closed-loop field.

lateral support of walls Means whereby walls are braced either horizontally by **columns**, **pilasters**, or cross walls, or vertically by floor and roof construction.

latex A water suspension of fine particles of rubber or rubberlike plastics.

latex paint A paint containing **latex** and thinned with water.

lath, gypsum *See* **gypsum board**.

latitude The position of a point on the earth's surface north or south of the equator, stated as an angular measure (**degrees**, **minutes**, and **seconds**) of the **meridian** arc contained between that point and the equator.

leaching pit *See* **dry well**.

leaching well *See* **dry well**.

lead The section of a masonry wall built up and racked back on successive courses. A line is attached to leads as a guide for building a wall between them.

legal description A written, legally recorded identification of the location and boundaries of a parcel of land. A legal description may be based on a **metes and bounds survey** or the rectangular survey system, or it may make reference to a **recorded plat** or **survey**.

leveling Ability of a **film** to flow out free from ripples, pockmarks, and brush marks after application.

level loop pile A woven or **tufted carpet** style in which the tufts are in loop form and of the same height.

level tip shear Random patterned **cut and loop pile carpet**.

lifting Softening and penetration of a previous film by solvents in the **paint** being applied over it, resulting in raising and wrinkling.

light-fastness Ability to resist **fading**.

light framing lumber *See* **lumber**.

lignin The second most abundant constituent of wood, making up 12% to 28% in most species. It encrusts the cell walls and cements the cells together.

linear polymer A **polymer** that is arranged in a long, continuous chain with a minimum of side chains or branches.

lining felt A mineral fiber or asphalt felt specifically manufactured for use under resilient flooring.

lintel A structural member used to carry the load over an opening in a wall.

lip When related to barrier-free design, an abrupt vertical change in level.

live load The total of all moving and variable loads that may be placed on or in a building. *See also* **dead load**.

longitude The position of a point on the earth's surface east or west of the **Greenwich meridian**, stated as an **angular measure** (**degrees**, **minutes**, and **seconds**) of the arc on the equator contained between a **meridian** passing through that point and the **Greenwich meridian**.

long oil Varnish or **paint** vehicle with more than 55% of the resin consisting of oil or fatty acid.

loom Machine on which **carpet** is woven, as distinguished from other machines on which **carpets** may be tufted, flocked, or punched.

loomed carpet *See* **woven carpet**.

loop pile A woven or tufted carpet style in which the pile surface consists of uncut loops; also called *round wire*.

loop vent A branch vent that serves two or more traps and extends from in front of the last fixture connection of a horizontal branch to the **stack vent**. *See also* **circuit vent**.

loudness A subjective response to **sound**, indicating the magnitude of the hearing sensation, dependent on the listener's ear.

lumber The wood product of a saw and planing mill not further manufactured than by sawing, resawing, and passing lengthwise through a standard planing machine, cross-cut to length, and worked. *Beams and stringers* are large pieces of lumber measuring 5 in. (127 mm) or more in thickness and 8 in. (203.2 mm) or more in width. These are graded with respect to their strength in bending when loaded on the narrow face. *Boards* are lumber up to 2 in. (50 mm) thick and 2 in. (50 mm) or more in width. *Dimension lumber* is lumber from 2 to 4 in. (50 to 100 mm) thick and 2 in. (50 mm) or more in width. *Factory and shop lumber* is intended to be cut again for use in further manufacture. *Joists and planks* are pieces of lumber 2 to 4 in. (50 to 100 mm) in nominal thickness and 4 in. (100 mm) or more in width. These are graded with respect to strength in bending when loaded either on the narrow face as a joist, or on the wide face as a plank. *Light framing* is lumber 2 to 4 in. (50 to 100 mm) in thickness and 2 to 4 in. (50 to 100 mm) in width. *Matched lumber* is lumber that has been edge dressed and shaped to make a close tongue-and-groove joint at the edges or ends when laid edge to edge or end to end. *Patterned lumber* is lumber that is shaped to a pattern or to a molded form in addition to being dressed, matched, shiplapped, or any combination of these. *Posts and timbers* are lumber that is approximately square, 5 in. (127 mm) and thicker, and having a width not more than 2 in. (50.8 mm) greater than its thickness. *Stress grade lumber* is the same as *structural lumber*. *Structural lumber* is lumber that has been machine rated or visually graded into grades with assigned working stresses. Also called stress grade lumber. Includes most yard lumber grades, except boards. *Yard lumber* is intended for general building purposes; includes boards, dimension lumber, and timbers.

machinability The ability to be milled, sawed, tapped, drilled, and reamed without excessive tool wear, with ease of chip metal removal and surface finishing.

main The principal artery of a **drainage** or **water supply system** to which branches may be connected. It is usually the lowest horizontal piping. In a drainage system the main is the building drain; in a water system it is the distributing main, whether located in the basement or on the top floor.

mainstreaming A process that seeks to integrate **physically handicapped persons** into society by (1) equipping them with the personal devices and adaptive skills needed to function effectively in the built environment and (2) removing physical barriers that prevent them from functioning like able-bodied individuals.

main vent The principal artery of a **vent system** to which vent branches may be connected. It may be either a vertical pipe that collects one or more branch vents, or a horizontal pipe to which a number of **vent stacks** are connected before they go through the roof.

major arch *See under* **arch**.

malleability The ability to be shaped without fracture by either hot or cold working.

marine varnish Varnish specially designed for immersion in water and exposure to marine atmosphere.

marl A soil or rock containing calcium carbonate (limestone).

martensitic steel *See under* **grain structure**.

masking (1) The effect produced by **ambient sound** that seems to diminish the loudness of transmitted noise. (2) Temporary covering of areas not to be painted.

masking tape A strip of paper or cloth tape, easily removable and used temporarily to cover areas that are not to be painted.

masonry bonded hollow wall A **hollow wall** in which the **facing** and **backup** are bonded together with solid **masonry units**.

masonry cement A packaged premixture of **portland cement**, lime, and other ingredients to which sand and water are added in the field to make masonry cement **mortar**.

masonry unit A manufactured building unit of burned clay, shale, concrete, stone, glass, gypsum, or other material. A *hollow masonry unit* is one whose **net cross-sectional area** in any plane parallel to the bearing surface is less than 75% of the **gross cross-sectional area**. A *modular masonry unit* is one whose **nominal dimensions** are based on a 4-in. (100 mm) module. A *solid masonry unit* is one whose **net cross-sectional area** in every plane parallel to the bearing surface is 75% or more of the **gross cross-sectional area**.

mastic Heavy-bodied pastelike coating of high build often applied with a trowel.

matched lumber *See under* **lumber**.

matrix, terrazzo Topping mortar consisting of binders, and sometimes **pigments** and inert fillers, that fills the spaces between chips and binds them into a homogeneous mass.

medium-density plywood *See* **plywood**.

meridian Imaginary north-south line on the earth's surface described by a great circle arc from the North Pole to the South Pole. All points on a meridian are of the same longitude. *See also* **great circle**. The **Greenwich meridian**, also called the *prime meridian*, is the meridian passing through the Royal Observatory at Greenwich, England. It is designated as the starting line (zero degrees) for measuring east and west longitude. A *principal meridian* is a meridian of specified longitude, used in the *rectangular survey system*, serving as the main north-south reference lines for a particular state or area. *Guide meridians* are north-to-south reference lines located at 24-mile intervals east and west of a *principal meridian*.

metallics Class of **paints** that include metal flakes.

metes and bounds A system of land **survey** and description based on starting from a known reference point and tracing the boundary lines around an area.

mil One one-thousandth of an inch (0.0254 mm).

mildewcide *See* **fungicide**.

mildew resistance Ability of a **coating**, **paint**, or **transparent finish** to resist the growth of molds and mildew; mildew is particularly prevalent in moist, humid, and warm climates.

mild steel *See under* **steel**.

millwork Lumber that is shaped to a pattern or to a molded form in addition to being **dressed**, **matched**, shiplapped, or any combination of these. Millwork includes most finished wood products such as doors, windows, interior trim, stairways, but not flooring or siding products.

mineral spirits Common **solvent** for **coatings**, **paints**, and **transparent finishes**; derived from the distillation of petroleum.

minor arch *See under* **arch**.

minute A unit of angular measure equal to 1/60 of a **degree**.

modacrylic *See under* **acrylic**.

modifier An ingredient added to a **plastic** to improve or modify its properties.

modular masonry unit *See* **masonry unit**.

modulus of elasticity The ratio of **stresses** lower than the **elastic limit** to their respective **strains**.

moisture content of wood The amount of water contained in wood at the time it is tested, expressed as a percentage of the weight of the wood when **oven-dry**.

monolithic concrete **Concrete** placed in one continuous pour without **construction joints**.

monomer A relatively simple compound that can react to form a **polymer**.

moist curing *See under* **curing**.

moisture content of a concrete masonry unit The amount of water contained in a unit, expressed as a percentage of the total **absorption** (e.g., a concrete masonry unit at 40% moisture content contains 40% of the water it can absorb).

monument A permanent reference point for land **survey**ing whose location is recorded; either a man-made marker or a natural landmark.

moresque Multicolored yarn made by twisting together two or more strands of different shades or colors.

mortar bed, tile A 3/4-in. (19.1 mm) to 1 1/4-in. (31.8 mm)-thick bed of **portland cement** and sand mortar onto which a layer of neat **portland cement** (water-cement paste) is applied to receive tile. Also called *thickbed*.

mortar joint The joint between masonry units. A *tooled joint* is one that is compressed and shaped with a special concave or V-shaped tool. A *troweled joint* is one finished with a trowel to form a struck joint or a weathered joint. In a raked joint the mortar is raked out to a specified depth while the mortar is still green.

mortar, masonry A plastic mixture of one or more cementitious materials, sand, and water.

mortar, tile A mixture of portland cement and other ingredients used to install tile. A *leveling or setting type mortar* is a mixture of portland cement, sand, and sometimes lime, used as a bed into which tile is installed (setting bed), or as a coat to produce a plumb and level surface (leveling coat) so that subsequent coats can be applied in a uniform thickness. A *bonding type mortar* is any one of a variety of formulations used to bond tile to a **backing** or **mortar bed**. Formulations may be mainly cementitious, such as commercial cement mortar; resinous, such as epoxy mortar; or a combination of both, as in latex–portland cement mortar. *Also see* **adhesive, tile**.

multilevel loop A **carpet** style in which the yarns are looped at several levels; also called *high-low loop* and *cut and loop*.

nail, drywall A nail suitable for **gypsum board** application. Such nails are typically bright, coated, or chemically treated low-carbon steel nails with flat, thin, slightly filleted and countersunk heads approximately 1/4 in. (6.4 mm) in diameter, and medium or long diamond points. Annularly threaded nails (GWB-54) and smooth or deformed shank nails should conform to ASTM C 514.

nail gluing A method of gluing wood in which the nails hold the wood members until the glue sets.

nail popping Surface defect in **gypsum board** resulting in a conspicuous protrusion of **joint compound** directly over a nail head.

nail spotting *See* **fastener treatment**.

nap Length of fibers on a paint roller cover.

natural gray yarn Unbleached and undyed yarn spun from a blend of black, brown, or gray wools.

natural resin *See under* **resin**.

neat cement A mixture of **cement** and water (no **aggregates**).

neat portland cement Unsanded mixture of **portland cement** and water used as a bond coat in conventional thickbed installation of tile.

needle-punched carpet **Carpet** made by punching loose, unspun fibers through a woven sheet, which results in a pileless carpet similar to a heavy felt; usually consists entirely of synthetic fibers.

net cross-sectional area The **gross cross-sectional area** of a section minus the area of cores, cellular spaces, and other voids it contains.

noise Unwanted **sound**.

noise reduction Reducing the level of unwanted **sound** by means of acoustical treatment.

noise reduction coefficient (NRC) A single-number index of the noise-reducing efficiency of acoustical materials. Found by averaging a material's sound absorption coefficients at 250, 500, 1000, and 2000 cycles per second.

nominal dimension A dimension greater than the **actual dimension** that is used to identify the size of a material. In masonry, the nominal dimension is greater than the actual dimension by the amount of the thickness of a **mortar joint**, but not more than 1/2 in. (12.7 mm). In wood, the nominal dimension (also called *nominal size*) is greater than the *actual dimension* by an amount roughly equivalent to the size lost due to wood shrinkage during drying, plus the amount planed off to square and smooth the piece. *See also* **size**.

nominal size *See* **nominal dimension**; **size**.

noncombustible material A material that will neither ignite nor actively support combustion when exposed to fire in air at a temperature of 1200°F (648.9°C).

nondrying oil Oil that does not readily oxidize and harden when exposed to air.

non-heat-treatable alloy An **alloy** that is not capable of gaining strength by **heat treatment** and that depends on the initial strength of the **alloy** or **cold working** for additional strength. Also called a *common alloy*.

non-load-bearing wall A wall that supports no vertical load other than its own weight. *See also* **bearing wall**.

nonvolatile vehicle Liquid portion of **coatings**, **paints**, and **transparent finishes** excepting their volatile thinners and water.

normal portland cement **Portland cement**, Type I.

Norman brick *See under* **brick masonry unit**.

octave The interval between two sounds with a frequency ratio of 2 to 1.

ohm A measure of electrical resistance equal to that of a circuit that permits 1 volt to cause 1 ampere to flow through it.

oil color Single **pigment** dispersion in linseed oil used for tinting **coatings** and **paints**.

oil paint *See under* **paint**.

oil stains May be penetrating or nonpenetrating. *Penetrating* oil stains contain dyes and resins that penetrate a surface; *nonpenetrating* oil stains contain larger amounts of pigments and are usually opaque or translucent.

oil varnish *See under* **transparent finish**.

olefins Long-chain synthetic **polymers** composed of at least 85% by weight of ethylene, propylene, or other olefin units.

one-package (one-part) formulation **Coating** or **paint** formulated to contain all the necessary ingredients in one package and generally not requiring any field additions except **pigment** and **thinner**.

opacity Degree of obstruction to the passage of visible light.

opaque coating A **coating** that hides the previous surface or coating.

open-grained *See under* **grain**.

open hearth furnace *See under* **furnace, steel**.

oven-dry wood Wood that has been dried in an oven to a consistent moisture content and, for all practical purposes, no longer holds any water.

overlaid plywood *See under* **plywood**.

oxidation The chemical combination of a substance with oxygen.

package dyeing Placing spun and wound yarn on large perforated forms and forcing the dye through the perforations.

padding Cellular rubber, felted animal hair, jute fibers, or plastic foams in sheet form, used as cushioning under **carpet**. *See also* **cushioning**.

paint A liquid, pigmented material applied in the field and producing an opaque film. **Primers** may, however, be applied by hand or machine in a factory or shop, but are part of the **paint system** and are therefore also called paint. In addition, all parts of a paint system are called paint, collectively and individually. Such parts include **primers**, **emulsions**, **enamels**, opaque stains, **sealers**, **fillers**, and other applied materials used as prime, intermediate, or finish coats. *See also* **coating** and **transparent finish**. An emulsion is a paint with a vehicle consisting of an oil, oleoresinous varnish, or resin binder dispersed in water (see also **emulsion**). A *latex paint* is a paint with latex resin as the chief binder. An *oil paint* is a paint with drying oil or oil varnish as the basic **vehicle** ingredient. A *paste* is a paint with sufficiently concentrated pigment to permit substantial thinning before use. *Water-based paint* is paint with a vehicle that is a water emulsion or water dispersion, or that has ingredients that react chemically with water.

paint remover Compound that softens old **paint** and **transparent finishes** and permits scraping off the loosened material.

paint system Collectively, the several coats that are necessary to produce a complete paint coating. Materials that are used in the various coats are called **paint**.

panel wall An exterior **non-load-bearing wall** in skeleton frame construction, wholly supported at each story. *See also* **curtain wall**.

parallel An imaginary east-west line on the earth's surface, consisting of a circle on which all points are equidistant from one of the poles. All points on a parallel are at the same **latitude**.

parapet wall That part of a wall entirely above the roof line.

parging (1) The application of mortar to the back of the facing material or the face of the backing material within a masonry wall; the mortar so applied. Also called *back-plastering*, *backparging*, or *pargeting*. (2) The application of **portland cement** mortar or **plaster** to the exterior face of masonry, often below or near the ground; the material so applied.

particleboard A composition board consisting of distinct particles of wood bonded together with a synthetic resin or other binder.

partition An interior wall, one story or less in height.

party wall A wall used for joint service by adjoining buildings. *See also* **common wall**.

pattern bond Patterns formed by exposed faces of **masonry units** and their joints.

patterned lumber *See under* **lumber**.

paving brick *See under* **brick masonry unit**.

peeling Detachment of a dried **coating**, **paint**, or **transparent finish** film in relatively large pieces, usually caused by moisture or grease under the finish.

perceived noise level (PNdB) A single-number rating of aircraft **noise** in decibels, used to describe the acceptability or noisiness of aircraft sound. PNdB is calculated from measured interior and exterior noise levels and correlates well with subjective responses to various kinds of aircraft noise.

perforated wall A wall that contains a considerable number of relatively small openings. Often called a *pierced wall* or *screen wall*.

perimeter In **gypsum board** work, the surface (as opposed to the edges) of a **gypsum board** panel near the edges and ends.

pervious soil Usually a granular soil, such as sand or gravel, that allows water to pass readily through it.

physically handicapped person An individual who has a physical impairment, including impaired sensory, manual, or speaking abilities, that results in a functional limitation in gaining access to and using a building or facility.

pickling Removing the oxide scale formed on hot metal as it air cools, by dipping in a solution of sulfuric or hydrochloric acid.

piece dyeing Immersing an entire **carpet** in a dye bath to produce single- or multicolor pattern effects. *See also* **resist printing**.

pier An isolated column of masonry.

pig A **pig iron ingot**.

pig iron High-carbon crude iron from a blast **furnace** used as the main raw material for ironworking and steelmaking. *Basic pig iron* is a high-phosphorus iron used in basic steelmaking furnaces.

pigment Fine, solid particles suspended in the **vehicle** of a **coating** or **paint** that provide color (**hiding power**) as well as other properties.

pilaster A thickened wall section or column built as an integral part of a wall.

pile The raised yarn tufts of **woven**, **tufted**, and **knitted carpets** that provide the wearing surface and desired color, design, or texture. In *flocked carpets*, the upstanding, nonwoven fibers. *Cut loop pile* is a pile surface in which tufts have been cut to reveal the fiber ends. A *frieze* carpet is a rough, nubby-textured carpet using tightly twisted yarns; same as *twist* carpet. *Jaspe* is a carpet surface characterized by irregular stripes produced by varying textures or shades of the same color. *Multilevel* carpet has a texture or design created by different heights of tufts of either cut or uncut loop. *Plush* carpet has a cut pile surface that does not show any yarn texture. A *sculptured (carved)* carpet has surface designs created by combinations of cut and loop pile and variations in pile height. *Shag carpet* has a surface consisting of long twisted loops. *Stria* (*striped*) has a striped surface effect obtained by loosely twisting two strands of one shade of yarn with one strand of a lighter or darker shade. *Twist* carpet is the same as *frieze* carpet. *Uncut loop* is pile in which the yarns are continuous from tuft to tuft, forming visible loops.

pile crushing Bending of pile due to foot traffic or the pressure of furniture.

pile height The height of pile measured from the top surface of the backing to the top surface of the pile. Also called *pile wire height*.

pile setting Brushing after shampooing to restore damp pile to its original height.

pile yarn density The weight of pile yarn per unit of volume in **carpet**, usually stated in ounces per cubic yard (grams per cubic meter).

pilling Appearance defect associated with some staple fibers in which balls of tangled fibers are formed on the carpet surface, which are not removed readily by vacuuming or foot traffic; pills can be removed by periodic clipping.

pipe *See under* **finished mill products**.

pitch The number of *tufts* or pile warp yarns in a 27-in. (685.8 mm) width of woven carpet.

pitch pocket An opening extending parallel to the annual growth rings of a tree. These contain or have contained either solid or liquid pitch.

pith The small, soft core occurring in the center of the growth rings of a tree, branch, twig, or log.

placing The act of putting **concrete** in position (sometimes incorrectly called *pouring*).

plain masonry Masonry without reinforcement, or reinforced only to resist shrinkage or temperature changes.

plain-sawn wood *See under* **grain**.

plaster All elements of a gypsum or **portland cement** plaster membrane, such as the several coats of plaster, metal reinforcement, accessories, and backing paper, when required.

plaster mix Mortar consisting of properly proportioned quantities of cementitious materials, aggregate, water, and sometimes pigments, plasticizers, and other admixtures.

plastic A material that contains as an essential ingredient an organic substance of large molecular weight, or a synthetic or processed substance. A plastic is solid in its finished state, but at some stage in its manufacture can be shaped by flow (*see* **polymer**).

plastic deformation Deformation caused by a **stress** exceeding the **elastic limit** of a material so that it does not return to its original shape when the stress is removed. *See also* **elastic deformation.**

plasticizer (1) In **plastics**, a material added to increase the workability and flexibility of a resin. (2) In **coatings**, **paints**, and **transparent finishes**, a substance added in the liquid state to impart flexibility to a hardened film.

plat A map of **surveyed** land showing the location and the boundaries and dimensions of a parcel. A *recorded plat* is a plat that is recorded at an appropriate governmental office, usually the county recorder's office. In addition to location notes and a boundary line layout, a recorded plat may contain information such as restrictions, easements, approvals by zoning boards and planning commission, and lot and block numbers for a subdivision.

plate *See under* **finished mill products**.

plumbing fixtures Devices or appliances that are supplied with water and receive or discharge liquids or wastes.

ply When preceded by a number, the number of single strands used in a finished yarn.

plywood A crossbanded assembly made of layers of **veneer** or of veneer in combination with a core of lumber, particleboard, or other composition core, all joined with an adhesive. *Standard plywood* is hardwood plywood produced to meet governing industry standards. Also, unsanded interior softwood plywood for sheathing and flooring. *High-density overlaid plywood* is overlaid plywood in which the overlay sheet contains 40% resin by weight. *Medium-density overlaid plywood* is overlaid plywood in which the overlay sheet contains 20% resin by weight. In *overlaid plywood* the face veneer is bonded on one or both sides with paper, resin-impregnated paper, or metal. *Specialty plywood* is hardwood plywood that is not necessarily manufactured to meet industry standards.

P.O.B. (place of beginning) The starting point of a **metes and bounds** description or **survey**.

pointing Troweling mortar into a joint after a **masonry unit** has been laid.

polymer A chemical compound formed by the reaction of simple molecules into more complex molecules of higher molecular weight.

polymerization A chemical reaction in which the molecules of a **monomer** are linked together to form large molecules whose molecular weight is a multiple of that of the original substance.

polypropylene *See* **olefins**.

polyurethane *See* **resin**; **transparent finish**.

polyvinyl acetate *See under* **resin**.

ponding A curing method used on flat **concrete** surfaces whereby a small earth dam or other water-retaining material is placed around the perimeter of the surface and the enclosed area is flooded with water.

portland blast furnace cement **Cement** made by grinding not more than 65% of granulated blast furnace **slag** with at least 35% of **portland cement**.

portland cement A hydraulic cement produced by pulverizing clinker consisting essentially of hydraulic calcium silicates and usually containing one or more of the forms of calcium sulfate as an interground addition. *Types I and IA portland cement* are used in general construction when the special properties of other types are not required. *Types II and IIA portland cement* are used in general construction where moderate heat of hydration is required. *Types III and IIIA portland cement* are used when a high early-strength is required. *Type IV portland cement* is used when low heat of hydration is required. *Type V portland cement* is used when high sulfate resistance is required.

portland-pozzolan cement **Cement** made by blending not more than 50% pozzolan (a material consisting of siliceous or siliceous and aluminous material) with at least 50% of **portland cement**.

postforming The forming of **laminates** that have been cured into simple shapes by heat and pressure.

posts and timbers *See under* **lumber**.

potable water Water of sufficient quality, either through treatment or natural phenomena, to be drinkable.

potential difference The voltage difference between two points.

pot life Time period after mixing reactive components, in a two-component **coating**, **paint**, or **transparent finish** system, during which material can be satisfactorily used.

pots Carbon-lined vessels used in the reduction processing of aluminum. Molten aluminum is collected and siphoned off from the bottom of pots.

precast concrete **Concrete** components that are cast and cured off-site or on-site in a location other than where they will be finally placed.

prefabricated brick masonry Masonry panels fabricated other than in their final location in the structure. Also known as *preassembled*, *panelized*, and *sectionalized brick masonry*.

preframed panels Panels fabricated with precut **lumber** and **plywood**.

preservative A substance that will prevent, for a finite time that varies with quantity, the action of wood-destroying fungi, insects of various kinds, and other destructive life in wood that has been properly impregnated with that substance.

pressure gluing A method of gluing that places the wood members under high pressure until the glue sets.

prestressed concrete **Concrete** that has been subjected to compressive stresses before external loads have been applied by the prestretching (or stressing) of high-strength steel reinforcement within the concrete. This prestressing may be done by pretensioning (stretching the steel before the concrete is placed around it) or by posttensioning (stretching the steel after the concrete has been cured).

prime meridian *See under* **meridian**.

primer, paint First of several coats, intended to prepare a surface for the succeeding coat(s); sometimes a special product, but may be the same as the finish coat.

primer, resilient flooring A brushable **solvent-based** asphaltic preparation recommended as a first coat over porous or dusty concrete floors and panel underlayments; intended to seal the pores and improve the bond with asphaltic adhesives used for the installation of asphalt and vinyl-composition tile.

primer-sealer Product formulated to possess properties of both a **primer** and a **sealer**.

principal meridian *See under* **meridian**.

print dyeing Screen printing a pattern on carpet by successive applications of premetalized dyes, which are driven into the pile construction by an electromagnetic charge.

puddling The compacting or consolidating of **concrete** with a rod or other tool.

putty Doughlike material consisting of **pigment** and **vehicle**; used for sealing glass in sash or frames and for filling imperfections in wood and metal surfaces; does not retain its plasticity for extended period as does a **glazing compound** or joint sealant. Both elastomeric and acrylic joint sealants are superior to putty and are generally used today where putty is called for. *See also* **sealant, joint**.

quarter-sawn wood *See under* **grain**.

quenching *See under* **heat treatment**.

quoin A right-angled masonry corner that is usually projected when of the same material as the surrounding material, but may be flush with the surrounding masonry when of a different material. A typical example of a quoin consists of squared stones set into a rubble stone or brick wall. Quoins are often larger than the surrounding masonry units.

ramp When related to accessibility for handicapped persons, a walking surface in an accessible space that has a running slope greater than 1:20. The maximum slope is limited by the applicable code or law.

random shear A **carpet** style created by shearing only the higher loops of either level loop or multilevel loop carpet.

range lines North-south reference lines used in the **rectangular survey system**, located at 6-mi. intervals between **guide meridians**.

rays Strips of cellulose extending radially within a tree. These store food and transport it horizontally in the tree.

ready-mixed concrete **Concrete** mixed at a plant or in trucks en route to the job and delivered ready for **placing**.

recorded plat *See under* **plat**.

rectangular survey system A land **survey** system based on geographical coordinates of **longitude** and **latitude**; originally established by acts of Congress to survey the lands of public domain and now used in most states. Also called Government Survey System.

red mud Solid impurities collected by either filtering or gravity settling during the **Bayer process** of aluminum refining.

reduction (1) An electrolytic process used to separate aluminum from aluminum oxide. (2) Separation of iron from its oxide by smelting ores in a blast furnace.

refining of steel Melting **pig iron** and other iron-bearing materials in steel furnaces to achieve desired contents of residual and alloying elements.

reflection The return from surfaces of sound not absorbed on contact with the surfaces or transmitted through the material contacted.

refractories When related to steel production, nonmetallic materials with superior heat and impact resistance used for lining furnaces, flues, and vessels for ironworking and steelmaking.

reinforced masonry **Masonry** containing embedded steel so that the two materials act together to resist applied forces.

reinforced plastic A **plastic** material whose strength and, to a lesser degree, stiffness have been upgraded by the addition of high-strength fillers, such as glass fiber.

reinforcing bars Steel placed in **concrete** or *masonry* to take tensile, compressive, and shear **stresses**.

relative humidity The amount of water vapor in air stated as a percentage. The ratio of the amount of water in air at a given temperature and atmospheric pressure to the amount that same air can contain when it is **saturated**.

relief vent A vent whose primary function is to provide circulation of air between **drainage** and **vent systems**.

relieving arch *See under* **arch**.

repeat The distance along the length of a **carpet** from one point in a figure or pattern to the same point at which it again occurs.

residual elements Nonferrous elements (such as carbon, sulfur, phosphorus, manganese, and silicon) that occur naturally in raw materials and are controlled in steelmaking.

resin (1) In **plastics**, the essential ingredients of a plastic mix before final processing and fabrication of the plastic object. (2) In **coatings**, **paints**, and **transparent finishes**, a mixture of organic or synthetic compounds with no sharply defined melting point, no tendency to crystallize, soluble in certain organic solvents but not in water. It is the main ingredient of most coatings, paints, and transparent finishes, binds the other ingredients together, and aids adhesion to the surface. Includes acrylic, alkyd, epoxy, nitrocellulose, polyurethane, polyvinyl acetate, silicone, styrene-butadiene, and vinyl. A *natural resin* may be a fossil of ancient origin, such as amber, or an extract of certain pine trees, such as rosin, copal, and damar. A *synthetic resin* is a man-made substance exhibiting properties similar to those of natural resins; typical synthetic resins used in **coatings, paints**, and **transparent finishes** include alkyd, acrylic, latex, phenolic, urea, and others.

resist printing Placing a dye-resist agent on **carpet** before **piece dyeing** so that the **pile** will absorb color according to a predetermined design.

resonance The sympathetic vibration, resounding, or ringing of such things as enclosures, room surfaces, and panels when they are excited at their natural frequencies.

retaining wall A wall that is subjected to lateral pressure other than wind pressure, such as a wall built to support a bank of earth.

retarder An admixture added to **concrete** to retard its set.

retempering Restoring workability to **mortar** that has stiffened due to evaporation, by adding water and remixing.

retentivity *See* **water retention**.

reverberation The continuing travel of sound waves between reflective surfaces after the original source has stopped transmitting.

reverberation time (T) The time in seconds required for a sound to diminish 60 decibels (dB) after the source has stopped transmitting.

ridging A surface defect in a **gypsum board** surface resulting in conspicuous wrinkling of the **joint tape** at treated joints.

riser In plumbing systems, a water-supply pipe that extends vertically 1 full story or more to convey water to branches or fixtures.

roller Paint applicator having a revolving cylinder covered with lambswool, fabric, foamed plastic, or other material.

rolling Shaping plate and sheet metal, blooms, and billets by passing them through steel rollers. *See also* **cold working** and **hot working**.

Roman brick *See under* **brick masonry unit**.

room acoustics A branch of **architectural acoustics** dealing with both **acoustical correction** and **noise reduction**. *See also* **acoustics**.

rosin Natural resin obtained from various pine trees; an ingredient of varnishes and some man-made resins.

rotary-cut veneer *See under* **veneer**.

rough-in The installation of those parts of a plumbing system that can be completed before the **plumbing fixtures** are installed. This includes drainage, water-supply, and vent piping and the necessary fixture supports. Also, the plumbing system parts so installed.

round wire A pile wire that does not cut the pile loop; or **woven carpet** with an **uncut loop pile**.

rowlock A brick unit laid on its face edge. Usually laid in a wall with its long dimension perpendicular to the wall face. Also spelled *rolok*.

running slope The slope of a pedestrian way that is parallel to the direction of travel. *See also* **cross slope**.

rust preventive paint or primer First coat of **paint** applied directly to iron or steel structures to slow down or prevent rusting.

sabin The measure of sound absorption of a surface, equivalent to 1 sq. ft. (0.093 m^2) of a perfectly absorptive surface.

sags Excessive flow, causing runs or sagging in **coating**, **paint**, or **transparent finish** film during application; usually caused by applying too heavy a coat or thinning too much.

salamander A portable stove used to heat the surrounding air.

sand finish *See* **texture**.

sanding surfacer Heavily pigmented finishing material used for building a surface to a smooth condition; it is sanded after drying.

sandwich panel A composite structural panel made of two thin, strong, hard facings bonded firmly to a core of relatively lightweight, weaker material with insulating properties.

sapwood The living wood of pale color near the outside of a tree.

satin finish *See* **semigloss**.

saturated air Air that contains 100% of the amount of water vapor it can contain. *See also* **relative humidity**.

saturated molecule A molecule that will not unite readily with another element or compound.

saturation coefficient *See* **C/B ratio**.

sawed veneer *See under* **veneer**.

scaling The breaking away of a hardened **concrete** surface (to a depth of about 1⁄16 in. [1.6 mm] to 3⁄16 in. [4.76 mm]), usually occurring at an early age of the concrete.

scarf jointing A joint in which the ends of plywood panels are beveled and glued together.

scoring Partial cutting of **concrete** flat work for the control of shrinkage cracking. Also used to denote the roughening of a slab to develop mechanical bond.

scrap metal Source of iron for ironworking and steelmaking, consisting of rolled product croppings, rejects, and obsolete equipment from steel mills and foundries and waste ferrous material from industrial and consumer products.

scratch coat The first **plaster** layer in **three-coat work**, which embeds the reinforcement and provides a suitably rigid and roughened (scratched) surface for the following coat.

SCR brick *See under* **brick masonry unit**.

screed A wood or metal template to which a **concrete** surface is leveled.

screeding Striking off excess **concrete** in finishing operation of concrete slab work.

screws, drywall Screws developed for **gypsum board** application, usually with self-tapping, self-threading points, special-contour flat heads, and deep Phillips recesses for use with a power screwdriver. *Type S* drywall screws are used for sheet metal studs and furring. *Type W* drywall screws are used for wood framing and furring. *Type G* drywall screws are used for attaching gypsum board to gypsum board.

scribing A method of transferring the profile of an obstruction, projection, or material edge to a piece of material, such as resilient flooring or wood trim, so that it can be accurately cut and fitted.

scrim Rough, loosely woven fabric often used as a secondary **backing** on **tufted carpets**.

scrubbability Ability of a **coating**, **paint**, or **transparent finish** film to withstand scrubbing and cleaning with water, soap, and other household cleaning agents.

sculptured carpet *See under* **pile**.

scum A mass of sewage matter that floats on the surface of the **sewage** in a **septic tank**.

scum clear space In a **septic tank**, distance between the bottom of the scum mat and the bottom of the outlet device (tee, ell, or baffle).

sealant, joint Any one of a number of plastic materials, including rubber, formulated to fill and seal stationary and moving joints. See also **caulking compound**, **glazing compound**, and **putty**, which are types of joint sealants.

sealer (1) In **paint** and **coatings**, a formulation intended to prevent excessive absorption of a finish coat of **coating**, **paint**, or **transparent finish** into a porous surface or to prevent bleeding through the finish coat. (2) In resilient flooring, a solution of equal parts of wax-free **shellac** and denatured alcohol; recommended as a first coat over existing wood strip floors from which finish has been removed; intended to seal wood pores, prevent excessive moisture absorption, and provide a dimensionally stable base for direct application of **lining felt** or resilient flooring.

seasoning Removal of moisture from green wood. *Air drying* is seasoning by exposure to air, usually in a yard. *Kiln drying* is seasoning in a kiln (oven) under controlled conditions of heat, humidity, and air circulation.

second A unit of angular measure equal to 1⁄60 of a minute.

section An area of land used in the **rectangular survey system**, approximately 1 mile square, bounded by **section lines**. The **rectangular survey system** provides for the further subdivision of sections into halves, quarters, and quarter-quarters. A **fractional section** is an adjusted section of land generally containing less (sometimes more) than 1 square mile. The deficiency (or excess) may be the result of the convergence of **meridians**, the presence of bodies of water, or uncertainties in **surveying**.

section lines North-south reference lines used in the **rectangular survey system**, parallel to the nearest **range line** to the east, and east-west lines parallel to the nearest **township line** to the south; these lines divide **townships** into 36 approximately equal squares called **sections**.

seeds Small, undesirable particles or granules other than dust found in a **coating**, **paint**, or **transparent finish**.

seepage bed *See* **absorption bed**.

seepage pit A covered underground pit with concrete or masonry lining designed to permit partially treated **sewage** to seep into the surrounding soil. Also called *dry well* in some parts of the country.

segregation Separation of the heavier coarse **aggregate** from the **mortar**, or of water from the other ingredients of a **concrete** mix, during handling or **placing**.

self-cleaning Controlled chalking of a paint film so that dirt does not adhere to the surface.

selvage The finished long edge of **woven carpet** that will not unravel and will not require binding or **serging**.

semigloss Degree of surface reflectance midway between **gloss** and **eggshell**; also **coatings** and **paints** displaying these properties.

semitransparent Degree of hiding greater than transparent but less than opaque.

separated construction *See* **discontinuous construction**.

septic (sewage) disposal system A system for the treatment and disposal of **sewage** by means of a **septic tank** and a **soil absorption system**.

septic tank A watertight, covered receptacle that receives the discharge of **sewage** from a **building sewer** and is designed and constructed to separate solids from liquids, to digest organic matter during a period of retention, to store digested solids through a period of retention, and to allow the clarified liquids to discharge for final disposal or additional treatment.

serging A method of finishing a cut long edge of **carpet** to prevent unraveling; distinguished from finishing a cut end, which may require binding.

serial distribution A combination of several **absorption trenches**, **seepage**

pits, or **absorption beds** arranged in sequence so that each is forced to utilize the total effective absorption area before liquid flows into the succeeding component.

service temperature The maximum temperature at which a material can be continuously employed without noticeable reduction in strength or other properties.

setting bed *See* **mortar bed, tile**.

settling **Coating**, **paint**, or **transparent finish** material separation in which pigments and other solids accumulate at the bottom of the container.

sewage The liquid and solid waste matter carried off by **sewers**. It contains organic (animal or vegetable) matter in suspension or solution, as well as liquids containing chemicals in solution.

sewer An underground conduit for carrying off **sewage** and rainwater.

sewerage A system of **sewers**; the removal and disposal of liquid and solid wastes by **sewers**.

sewer brick *See under* **brick masonry unit**.

shading Bending or crushing a **pile** surface so that the fibers reflect light unevenly. This is not a defect but rather an inherent characteristic of some pile fabrics.

shake (1) A separation along the grain in a tree, the greater part of which occurs between the annual growth rings. (2) A wood shinglelike roofing material.

shake painter *See* **flat applicator**.

shale Laminated clay or silt compressed by earth overburden. Unlike slate, shale splits along its bedding planes.

shearing A **carpet** finishing operation that removes stray fibers and fuzz from **loop pile** and produces a smooth, level surface on **cut loop pile**.

shear wall A wall that resists horizontal forces applied in the plane of the wall.

sheathing, gypsum *See under* **gypsum board**.

shedding A normal temporary condition of dislodged, loose short fibers in new carpet after initial exposure to traffic and sweeping.

sheen Degree of luster of a dried **coating**, **paint**, or **transparent finish** film.

sheen uniformity Even distribution of luster over the entire surface of an applied finish.

sheet *See under* **finished mill products**.

sheeting A form of **plastic** in which the thickness is very small in proportion to the length and width; usually refers to a product with a thickness greater than 0.010 in. (0.254 mm) (*see also* **film**).

sheet lamination In multi-ply **gypsum board** construction, method of applying adhesive to the entire surface to be bonded. *See also* **strip lamination**.

shellac A **transparent finish** material made from **resins** dissolved in alcohol.

shooting *See* **sprouting**.

short circuit Related to sound control, a bypassing connection or transmission path that tends to nullify or reduce the sound-isolating performance of a construction assembly.

shot *See* **weft**.

shrinkage Decrease in initial volume due to removal of moisture from fresh **concrete** or green wood. May also refer to decrease in volume due to subsequent decreases in temperature or moisture content in concrete.

side wall The wall along the long dimension of a room.

signage Verbal, symbolic, and pictorial information.

silica Silicon dioxide (SiO_2), occurring as quartz, a major constituent of sand, sandstone, and quartzite.

silicone Any of a number of polymeric organic silicon compounds used in **paint**, **sealants**, roof membranes, and water repellents. *See also* **resin**.

siphon breaker *See* **atmospheric vacuum breaker**.

size (1) Water-based formulation, with glue or starch binders, intended as a sealer over existing wall **paint** or **plaster**; now seldom used. (2) The size of lumber. The *actual size* of lumber is its measured size after **seasoning** and **dressing**. The *dressed size* of lumber is its size after surfacing. The *face size* of lumber is the exposed width of a piece of lumber when installed. The *nominal size* of a piece of lumber is its approximate rough-sawn commercial size by which lumber is known and sold.

skein dyeing Immersing batches of yarn (skeins) in vats of hot dye.

skin Tough covering that forms on liquid **coating**, **paint**, and **transparent finish** materials when left exposed to air.

slab (1) In **concrete**, a flat, thin (as compared with its other dimensions) structure. A *structural slab* is a suspended, self-supporting, reinforced concrete floor or roof slab. A *slab-on-grade* is a nonsuspended, ground-supported concrete slab, usually, but not always, having some temperature reinforcement. An *edge-supported slab-on-grade* rests atop the perimeter foundation wall. A *floating slab-on-grade* terminates at the inside face of the perimeter foundation wall and is said to "float" independently of the foundation wall. A *monolithic slab-on-grade* is a combination slab and foundation wall formed into one integral mass of concrete. Also called a thickened-edge slab. (2) In wood, the outside piece cut from a log in squaring it.

slabbed To have removed an outer *slab* from a log.

slag A glasslike waste product, generally from an ironworking or steelmaking furnace. Molten mass composed of fluxes in combination with unwanted elements, which floats to the surface of the hot metal in the furnace and thus can be removed.

slag cement *See* **portland blast furnace cement**.

sliced veneer *See under* **veneer**.

sloshing Attempting to fill vertical joints after units are laid by throwing mortar into the joint with a trowel from above.

sludge The accumulated solids that settle out of **sewage**, forming a semiliquid mass on the bottom of a **septic tank**.

sludge clear space In a **septic tank**, the distance between the top of the **sludge** and the bottom of the outlet device.

slump A measure of the consistency of a **concrete** mix (in inches).

slump block *See under* **concrete masonry units**.

slump test A method of measuring slump by filling a conical mold, removing it, and measuring the sag or **slump** of the sample.

smelting Melting of iron-bearing materials in a blast furnace to separate iron from impurities with which it is chemically combined or mechanically mixed.

soap A brick or tile of normal face dimensions having a nominal 2-in. (50 mm) thickness.

soft-burned clay Clay products that have been fired at low temperature ranges. They have relatively high absorptions and low compressive strengths.

softening range The range of temperature in which a **plastic** changes from a rigid to a soft state. Sometimes referred to as *softening point*.

softness Film property displaying low resistance to scratching or indentation; opposite of **hardness**.

soft temper The state of maximum **workability** of aluminum obtained by **annealing**.

softwood The botanical group of trees that have needle- or scalelike leaves and are evergreen for the most part. Some exceptions are cypress, larch, and tamarack. This term does not relate to the actual hardness of the wood.

soil absorption field *See* **absorption field**.

soil absorption system Any system that utilizes the soil for subsurface absorption of treated **sewage**, such as an **absorption trench**, **absorption bed**, or **seepage pit**.

soil pipe A pipe that conveys the discharge of water closets, urinals, or fixtures having similar functions (with or without the discharge from other fixtures) to a **building drain** or **building sewer**.

solar screen A perforated masonry wall used as a sunshade.

soldier Masonry unit set vertically on end with face showing on the masonry surface.

solid masonry unit *See under* **masonry unit**.

solid masonry wall A wall built of solid or hollow masonry units laid continuously, with joints between units completely filled with **mortar**.

solution dyeing Adding dye or colored pigments to synthetic material while in liquid solution before extrusion into fiber.

solution heat treatment The first temperature-raising step in the thermal treating of a heat-treatable aluminum alloy. Also called *heat treatment*.

solvent The volatile portion of the **vehicle** of a **coating**, **paint**, or **transparent finish**, such as **turpentine** or **mineral spirits**, which evaporates during the drying process.

solvent-based Adhesives and **primers** consisting of cementitious **binders** and **fillers** dissolved in a volatile hydrocarbon carrier such as alcohol or cutback.

solvent-thinned Formulation of a **coating**, **paint**, or **transparent finish** in which the **binder** is dissolved in the **thinner** (as in **oil paint**), rather than emulsified (as in **latex paint**).

sound A vibration in an elastic medium in a frequency range capable of producing the sensation of hearing.

sound attenuation *See* **attenuation**.

sound isolation Materials or methods of construction designed to resist the transmission of airborne and structure-borne sound through walls, floors, and ceilings; the effect of such materials or methods.

sound pressure The instantaneous pressure at a point as a result of the sound vibration minus the static pressure at that point; the change in pressure resulting from sound vibration. It is measured in **dynes** per square centimeter. A sound pressure of 1 dyne per square centimeter is about that of conversational speech at close range and is approximately equal to one millionth of atmospheric pressure.

sound pressure level (SPL) **Sound pressure** measured on the decibel scale; the ratio in decibels between a measured pressure and a reference pressure.

sound transmission The passage of sound through a material or assembly. *Airborne sound transmission* is the transmission of sound through the air as a medium rather than through a solid, such as the structure of a building. Airborne sound is produced when a surface is caused to vibrate, thus producing alternating air pressures adjacent to the surface. The alternating pressures then radiate through the air in waves of higher and lower pressures. *Structure-borne sound transmission* is the transmission of sound through a solid or assembly of solids, such as a wall or the structure of a building. Structure-borne sound occurs when the vibrations from equipment, or the impact of footsteps or dropped objects come in contact with a solid.

sound transmission class (STC) A single-number rating for evaluating the efficiency of assemblies in isolating airborne sound transmission (*see* **sound isolation**). The higher the STC rating, the more efficient the assembly.

space dyeing Alternating bands of color applied to yarn by rollers at predetermined intervals before tufting.

spackling compound Material used as a crack filler for preparing surfaces before painting.

spall A small fragment removed from the face of concrete or of a masonry unit by a blow or by action of the elements.

spandrel wall That part of a panel **curtain wall** above the head of a window; in a multistory building, includes the panel below the sill of the window in the story above.

spar varnish Very durable varnish designed for service on exterior surfaces.

spatter Small particles or drips of liquid **coating**, **paint**, or **transparent finish** materials thrown or expelled when applying these materials.

specialty plywood *See under* **plywood**.

split-face block *See under* **concrete masonry units**.

spot priming Method for protecting localized spots. The only areas primed are those that require additional protection due to rusting or **peeling** of the former coat of **paint** or **transparent finish**.

spreading rate Measure of area that can be covered by a unit volume of **coating**, **paint**, or **transparent finish** material, generally expressed as square feet per gallon (m^2/L).

springwood The portion of the annual growth ring of a tree that is formed during the early part of the season's growth. This is usually less dense and weaker mechanically than summerwood.

sprouting Temporary condition on new **carpets** where strands of yarn work loose and project above the **pile**. Can be remedied by careful clipping or spot **shearing**.

stabilizer A material added to prevent or retard degradation of a plastic when exposed to sunlight or other environmental conditions.

stack A structure or part thereof that contains a flue or flues for the discharge of gases.

stack The vertical main of a system of soil, waste, or vent piping.

stack vent The extension of a soil or waste **stack** above the highest horizontal drain connected to the **stack**. It is sometimes called a *waste vent* or *soil vent*.

stain (1) A discoloration of wood that may be caused by such diverse agencies as microorganisms, metals, or chemicals. (2) A penetrating formulation intended primarily for wood surfaces. An *opaque stain* is classified as **paint**; a *transparent stain* is a **transparent finish** material. An *opaque stain* is a stain that hides the **substrate** and previous surface finish materials. A *transparent stain* changes the color of a wood without obscuring the grain; depending on the amount of pigment, transparent stains may be more or less transparent and may leave little or no surface film.

staple fibers Relatively short natural (wool) or synthetic fibers ranging from about 1½ in. (38 mm) to 7 in. (180 mm) long, which are spun into yarn.

steam curing *See under* **curing**.

steel Iron-based alloy containing manganese, carbon residual, and, often, other alloying elements, characterized by its strength and toughness; distinguished from gray **cast iron** by its ability to be shaped or **hot working** or **cold working** as initially cast. *Alloy steel* is steel in which residual elements exceed limits prescribed for carbon steel, or to which alloying elements are added within specified ranges. *Carbon steel* is steel in which the residual elements are controlled but to which alloying elements are not usually added. *Heat-resisting steel* is a low-chromium steel with at least 4% chromium, which retains its essential mechanical properties at elevated temperatures. *High-strength low-alloy steel* is steel with less than 1% of an alloying element, manufactured to high standards for strength, ductility, and partial chemical specifications. *Mild steel* is carbon steel with a carbon content between 0.15% and 0.25%. *Stainless steel* is steel that contains at least 10% chromium. It has excellent corrosion resistance, strength, and chemical inertness at high and low temperatures.

stock dyeing Dyeing raw fibers before they are carded (combed) or spun.

storm sewer A **sewer** used for conveying rainwater, surface water, condensate, cooling water, or similar liquid wastes, exclusive of **sewage** and industrial waste.

story That portion of a building included between the upper surface of a floor and the upper surface of the floor next above, except that the topmost story is that portion of a building included between the upper surface of the topmost floor and the ceiling or roof above. Where a finished floor level directly above a basement or cellar is more than 6 ft. (1800 mm) above grade, such basement or cellar is considered a story.

story pole Marked pole for measuring vertical masonry courses during construction.

strain The change in cross-sectional area of a body produced by stress. Measured in in. per in. (mm per mm) of length.

strain hardening A method of strengthening **non-heat-treatable** aluminum **alloys** by either **cold rolling** or other physical or mechanical working.

strand caster A machine that continuously casts **steel** slabs and billets.

streaking Irregular occurrence of lines or streaks of various lengths and colors in an applied film; usually caused by some form of contamination.

strength A term used to describe all the properties of wood that enable it to resist forces or loads.

stress The intensity of a mechanical force acting on a body; either tensile, compressive, shear, or the combination of compressive and tension forces known as bending, or the twisting force, torsion. Tensile, compressive, and shear forces are measured by dividing the total force by the area over which the force acts (force per unit of area), such as lb./sq. in. (kg/m^2), for example. In wood, the *allowable unit, or working, stress* is the stress used in designing wood members. It is appropriate to the species and grade of the wood. Values for each type of stress are obtained by multiplying the *basic stress* for that species by the strength ratio assigned to each grade. Most codes include the allowable working stress for each grade. In wood, the *basic stress* is the design stress for a clear wood specimen free from strength-reducing features such as **knots**, **checks**, and **cross grain**. Such a specimen is assumed to have all the factors appropriate to the nature of **structural lumber** and the conditions under which it is used, except those that are accounted for in the strength ratio.

stretcher **Masonry unit** laid with its length horizontal and parallel with the face of the masonry and with its smallest dimension vertical.

stretch forming The stretching of metal for the purpose of flattening it or for forming it to a predetermined shape.

stria (striped) *See under* **pile**.

stringer *See under* **lumber**.

strip *See under* **finished mill products**.

strip lamination In multi-ply **gypsum board** construction, method of applying **adhesive** in parallel strips spaced 16 in. (406 mm) to 24 in. (610 mm) apart. *See also* **sheet lamination**.

strip reinforcing In single-ply **gypsum board** construction, an installation procedure in which strips of gypsum board are applied to the framing members to reinforce gypsum board joints and provide a base for adhesive application.

struck joint *See* **mortar joint**.

structural clay tile Hollow masonry units composed of burned clay, shale, fire clay, or mixtures of these. *End-construction* tile is designed to be laid with the axis of its cells vertical. *Facing tile* is made for exterior and interior use with its face exposed. *Side-construction tile* is intended for placement with the axis of its cells horizontal.

structural lumber *See under* **lumber**.

structurals *See under* **finished mill products**.

stucco **Portland cement**, water, sand, and possibly a small quantity of lime (**portland cement plaster**), along with, perhaps, other **aggregates**; used on exterior surfaces.

stucco finish A factory-prepared mix of **stucco** for application as finish coats.

stud adhesive *See* **adhesive, drywall**.

stuffer yarn *See under* **warp**.

styrene-butadiene *See under* **resin**.

subfloor The structural material or surface that supports a finish floor and floor loads and serves as a working platform during construction.

subgrade An earth surface upon which another material, such as **concrete**, is placed.

sublimation The transition of ice directly into water vapor.

substrate Surface to be covered with a **coating**, **paint**, or **transparent finish**.

suction The initial rate of water **absorption** by a clay **masonry unit**. *See* **absorption rate**.

summerwood The portion of the annual growth ring of a tree that is formed after the **springwood** formation has ceased. It is usually denser and stronger mechanically than springwood.

sump A tank or pit that receives **sewage** or other liquid waste, located below the normal grade of a building's gravity sewer system, and that must be emptied by mechanical means.

sump pump A mechanical device used for pumping out sewage, liquid, or industrial wastes from locations below a building's gravity drain.

surfacer Pigmented formulation for filling minor irregularities before a finish coat of **paint** is applied; usually applied over a **primer** and sanded for smoothness.

survey The measure and marking of land, accompanied by maps and field notes that describe the measures and marks made in the field.

surveying The process of making a **survey**.

swale A low, flat depression in the exterior grade, used to drain away storm water.

system, paint *See* **paint system**.

tack rag Piece of loosely woven cloth that has been dipped into a varnish oil and wrung out. When it becomes tacky or sticky, it is used to wipe a surface to remove small particles of dust.

tactile Describes an object that can be perceived with the sense of touch.

tactile cue *See under* **cue**.

taping In **gypsum board** construction, applying **joint tape** over embedding compound in the process of **joint treatment**.

temper designation Designation (following an **alloy designation** number) that denotes the temper of an aluminum **alloy**.

tempering *See* **heat treatment**.

terne A lead-tin alloy usually consisting of 4 parts lead and 1 part tin.

terneplate *See under* **finished mill products**.

terrazzo A floor topping made of marble or other stone chips set in cement mortar, ground smooth and polished.

terrazzo finish All elements installed by a **terrazzo** contractor from subfloor to finished surface, such as sand bed, **isolation membrane**, **underbed**, and **topping**.

texture In wood, a term often used interchangeably with grain. It refers to the structure of wood.

texture paint A paint that may be manipulated by brush, roller, trowel, or other tool to produce various effects.

thermal mass The heat storage capacity of a material.

thermoplastic A **plastic** that can be repeatedly softened by heating and hardened by cooling.

thermoset A **plastic** that, after curing, forms a permanently hardened product that cannot be softened again by reheating.

thickbed *See* **mortar bed, tile**.

thinners **Solvents** used to thin **coating**, **paint**, or **transparent finish** materials.

thin-set (1) In terrazzo, a method of installing relatively thin **toppings** (¼ in. [6.4 mm] to ½ in. [12.7 mm] thick) directly over a suitable **subfloor**; generally possible with resinous **binders** only. (2) In tile, a method of bonding tile with a thin layer (1/16 in. [1.6 mm] to ¼ in. [6.4 mm] thick) of special mortar or adhesive to a suitable **backing** or to a properly cured **mortar bed**.

thixotropy Property of a material that causes it to change from a thick, pasty consistency to a fluid consistency upon agitation, brushing, or rolling.

three-coat work Application of plaster in three separate layers (**scratch**, **brown**, and **finish coats**), totaling at least ⅞ in. (22.2 mm) in thickness.

tie *See* **wall tie**.

tinplate *See under* **finished mill products**.

tint base Basic **paint** in a custom color system to which **colorants** are added to make a wide range of colors.

tone-on-tone **Carpet** pattern made by using two or more shades of the same hue.

tooled joint *See* **mortar joint**.

toothing Projecting **brick** or **CMU** in alternate courses to provide for bond with adjoining masonry that will be laid later.

topping In **terrazzo finishes**, a decorative wear layer consisting of marble chips embedded in a suitable matrix and requiring grinding, polishing, or washing to form finished **terrazzo**.

touch-up (1) Ability of a **coating**, **paint**, or **transparent finish** film to be spot repaired (usually within a few months of the initial application) without showing color or gloss differences. (2) The repair of a **coating**, **paint**, or **transparent finish** film by selectively adding finishing material to damaged or missed areas after the earlier coats have dried.

toughness Maximum ability of a material to absorb energy without breaking, as from sudden shock or impact.

township An area of land used in the **rectangular survey system**, approximately 6 miles square, bounded by **range lines** and **township lines**.

township lines East-west reference lines used in the **rectangular survey system**, located at 6-mile intervals between **correction lines**.

transfer molding Forming plastic by fusing it in one chamber with heat and then forcing it into another chamber where it solidifies; commonly used with **thermoset**ting **plastics**.

transmission loss The decrease in or attenuation of sound energy, expressed in decibels, of airborne sound as it passes through building construction.

transparent finish A system of materials that are applied as liquids or by hand in the field or in a factory or shop to form a finish through which the substrate or previous finish is visible. *Lacquer* is a transparent finish material that is employed mostly as a shop or factory finish on furniture and casework. It is not usable over existing finish films. In fact, it is an effective paint remover. *Polyurethane* is a hard, highly abrasion- and chemical-resistant transparent finish, often used as a floor finish and as a bar-top finish. Some manufacturers call their polyurethanes varnishes; others call them lacquers. In fact, although they have some characteristics of each, they are neither. Polyurethanes are available in both oil-modified and moist-curing formulations. *See also* **resins**. *Shellac* is a fast-drying transparent finish material consisting of lac resins, produced by the lac insect, dissolved in alcohol. Shellac has a relatively short shelf life. *Spar varnish* is an exterior, weather-resistant transparent finish material, based generally on long-oil phenolic resin; the term originated from the use of this material on spars of ships. *Varnish* is a finish material that dries to a transparent or translucent film when exposed to air. Varnish is a mixture of **resin**, **drying oil**, **drier**, and a **solvent**.

transparent finish system The several coats and materials necessary to produce a **transparent finish**. The various materials may include dyes, bleaches, or stains to change the color of substrates such as wood, and undercoats and finish coats of one of the materials listed under **transparent finish** or another transparent finish material, such as oil.

trap A fitting or device designed to provide, when properly vented, a liquid seal that will prevent the back passage of air without materially affecting the flow of **sewage** or wastewater through it.

trimmer arch *See under* **arch**.

troweled joint *See* **mortar joint**.

troweling A **concrete slab** finishing operation that produces a smooth, hard surface.

tubular steel products *See under* **finished mill products**.

tuck pointing Refilling defective **mortar joints** that have been cut out in existing masonry.

tufted carpet **Carpet** made by inserting the pile yarns through a prewoven fabric **backing** on a machine with hundreds of needles (similar to a huge sewing machine).

tufts Surface loops of pile fabric.

turpentine Colorless liquid used as **thinner** for **oil paints** and **varnishes**, distilled from products of the pine tree.

twist *See* **pile, warp**.

two-coat work Application of **plaster** in two layers (**base coat** and **finish coat**), totaling at least ⅝ in. (15.9 mm) in thickness.

two-package (two-part) formulation **Coating**, **paint**, or **transparent finish** material formulated in two separate packages and requiring that the two ingredients be mixed before characteristic properties can be obtained and the material can be applied.

underbed Layer of nonstructural **portland cement mortar** sometimes used over a **subfloor** to provide a suitable base for **portland cement terrazzo** and to minimize cracking.

undercoat **Primer** or intermediate coat in a multicoat system.

underlay *See* **cushioning**.

underlayment A mastic or panelboard material installed over a **subfloor** to provide a suitable base for resilient flooring when the **subfloor** does not possess the necessary properties for direct application of the flooring.

uniformity In **coating**, **paint**, or **transparent finish**: not varying in gloss, sheen, color, hiding, or other property.

unsaturated molecule A molecule that is capable of uniting with certain other elements or compounds without creating any side products.

vacuum forming Shaping a heated **plastic** sheet by causing it to flow in the direction of reduced air pressure.

vehicle Liquid portion of a **coating**, **paint**, or **transparent finish**, including ingredients dissolved in it.

velvet carpet *See* **woven carpet**.

veneer (1) A single-facing **withe** of **masonry units** or similar materials attached to a wall for the purpose of providing ornamentation, protection, or insulation, but not bonded or attached to intentionally exert common action under load. (2) A thin sheet of wood. *Rotary-cut veneer* is veneer cut in a continuous strip by rotating a log against the edge of a knife in a lathe. *Sawed veneer* is veneer produced by sawing. *Sliced veneer* is veneer that is sliced by moving a log or **flitch** against a large knife.

veneered wall A wall having a face of **masonry units** or other weather-resisting materials attached to the backing, but not so bonded as to intentionally exert common action under load.

veneer wall tie *See* **wall tie**.

vent stack A vertical vent pipe installed primarily for the purpose of providing circulation of air to and from a part of a **drainage system**.

vent system A collection of pipes installed to provide a flow of air to or from a **drainage system**, or to provide circulation of air within such a system to protect trap seals from siphonage and back pressure.

vertical application **Gypsum boards** applied with **edges** parallel to supporting members and borders. *See also* **horizontal application**.

vertical grain *See under* **grain**.

vibrating A mechanical method of compacting **concrete**.

vibrator A mechanical device that vibrates at a speed of 3000 to 10,000 revolutions per minute and is inserted into wet **concrete** or applied to its **forms** to compact **concrete**.

vinyl A **polymer** derived from ethylene; used in **paint**, **coatings**, and fabric. *See also* **resin**.

vinyl foam cushioning **Carpet** cushioning made from a combination of foamed synthetic materials.

visual cue *See under* **cue**.

voids Air spaces between pieces of **aggregate** within a **cement** paste.

volt The **potential difference** between two points in a wire carrying a current of 1 ampere when the power dissipated is 1 watt; equivalent to the potential difference across 1 ohm of resistance when 1 ampere is flowing through it.

voltage Electrical pressure, measured in volts and comparable to pounds per square inch (psi) in a fluid medium. Reference to 120 and 240 nominal voltages includes typical operating ranges of 115 to 125 and 230 to 250 respectively. Same as **potential difference**.

wallboard, gypsum *See under* **gypsum board**.

wall tie A **header** (bonder) or metal anchor that connects **wythe**s of masonry to each other or to other materials. A *cavity wall tie* is a rigid, corrosion-resistant metal tie that bonds two **wythe**s of a cavity wall. It is usually steel, ³⁄₁₆ in. (4.76 mm) in diameter, and formed in a Z-shape or a rectangle. A *veneer wall tie* is a strip or piece of bent metal used to tie a facing veneer to the backing; sometimes in two pieces to permit movement.

wane **Bark** or lack of wood, from any cause, on the edge or corner of a piece of **lumber**.

warp (1) In wood: variation from a true or plane surface. *Bow* is distortion of a board in which the face is convex or concave longitudinally. *Crook* is distortion of a board in which the edge is convex or concave. *Cup* is distortion of a board in which the face is convex or concave transversely. *Twist* is distortion caused by the turning of the edges of a board so that the four corners of any face are no longer in the same plane. (2) In **carpet**: **backing** yarns running lengthwise in the carpet. In *chain warp*, zigzag warp yarn works over and under the **shot** yarns of the carpet, binding the backing yarns together. *Pile warp* is lengthwise **pile** yarns in Wilton **woven carpets** that form part of the backing. *Stuffer warp* is yarn that runs lengthwise in a **carpet** but does not intertwine with any filling (**weft** shot) yarn; serves to give weight, thickness, and stability to the fabric.

washability Ability of a **coating**, **paint**, or **transparent finish** to be easily cleaned without wearing away during cleaning.

water absorption *See* **absorption rate**.

water-cement ratio The ratio, by weight, of water to **cement**, or the amount of water, in gallons, used per 94-lb. sack of **cement** to make **concrete**. It is an index to strength, durability, watertightness, and consistency.

water hammer Pounding noises and vibration that sometimes develop in a piping system when its air chambers become filled with water or have been omitted entirely; the problem may also be due to worn washers in faucets.

waterproofing Prevention of moisture flow through **concrete** or masonry due to water pressure; an impervious liquid or sheet material used to waterproof.

water retention A property of **mortar** that prevents the loss of water to **masonry units** having a high **suction**. It also prevents **bleeding** or water gain when **mortar** is in contact with units having a low suction rate. Also called *water retentivity*.

water spotting **Coating**, **paint**, or **transparent finish** appearance defect caused by water droplets.

water supply system The water supply system of a building consisting of the water service pipe, water distributing pipes, branches to **plumbing fixtures** and appliances, and necessary fittings and control valves.

watt (1) In acoustics, a unit of sound power equal to 1×10^7 dyne per centimeter per second, which is the basic expression of the flow of sound energy. (2) In electricity, a unit of electrical power equal to the work done by a current of 1 ampere under the pressure of 1 volt.

wattage The amount of electrical power measured in **watts**; a single unit combining the effect of both **voltage** (pressure) and rate of flow (**amperage**) by multiplying these quantities (volts times amps equals watts).

watt-hour Unit of energy consumed, consisting of **watts** multiplied by time in hours; the result is often expressed in thousands of watt-hours, called *kilowatt-hours*.

weathering (1) **Corrosion** (galvanic or chemical) produced by atmospheric conditions. (2) Effect of exposure to weather on a **coating**, **paint**, or **transparent finish** film, raw wood, and other materials.

weathering index The product of the average annual number of **freezing cycle days** and the average annual winter rainfall.

weaving Process of forming carpet on a loom by interlacing the **warp** and **weft** yarns.

weep holes Openings placed in **mortar joints** of facing materials at the level of a flashing to divert to the exterior any moisture collected on the flashing.

weft Backing yarns that run across the width of a **carpet**. In **woven carpets**, the weft shot (filling) yarns and the **warp** chain (binder) yarns interlock and bind the **pile tufts** to the **backing**. In **tufted carpets**, pile yarns that run across the carpet are also considered weft yarns.

welding Creating a metallurgical bond between metals with heat and sometimes with the use of pressure and filler metal. *Arc welding* is a welding method employing an electric arc as the source of heat. *Gas welding* is a welding method employing a fuel gas (acetylene, hydrogen) and oxygen as the source of heat. *Shielded welding* is a process in which gases or fusible granular materials are used to shield the weld area from the damaging effects of oxygen and nitrogen in the air. *Shielded metal-arc welding* is arc welding in which a flux-coated metal electrode is consumed to form a pool of filler metal and a gas shield around the weld area. Also known as *manual metal-arc welding* and stick electrode welding. *Inert-gas-shielded arc welding* is arc welding in which shielding is provided by an inert gas envelope (such as argon, helium, a combination of argon and helium, or carbon dioxide) from an external supply. Filler metal is supplied either by a consumable metal electrode, as in inert-gas-shielded metal arc welding (MIG), or by a separate filler rod used with a nonconsumable tungsten electrode, as in inert-gas-shielded tungsten arc welding (TIG). *Submerged arc welding* is arc welding in which the weld area is shielded by a fusible granular material that melts to protect the weld area. Filler metal is obtained from either a consumable electrode or a separate filler rod.

white lead Oldest white pigment, chemically known as lead sulfate or lead carbonate.

Wilton carpet *See* **woven carpet**.

winter rainfall The sum in inches of the mean monthly corrected precipitation occurring between the first killing frost in the fall and the last killing frost in the spring.

wire, steel *See under* **finished mill products**.

wires (pile wire, gauge wire, standing wire) Metal strips over which the **pile tufts** are formed in **woven carpets**. *See also* **round wire**.

woolen yarn Soft, bulky yarn spun from both long and short wool fibers that are not combed straight but lie in all directions so they will interlock to produce a feltlike texture.

workability Relative ease or difficulty with which **concrete** can be placed and worked into its final position within **forms** and around **reinforcing bars**. However, workability is contingent on the absence of segregation of the concrete. If the aggregate segregates, workability is considered to diminish (or terminate) regardless of how easily the concrete flows into place.

working stress *See* **stress**.

worsted yarn Strong, dense yarn made from long **staple fibers** that are combed to align the fibers and remove extremely short fibers.

woven carpet Carpet made by simultaneously interweaving **backing** and **pile** yarns on one of several types of **looms** from which the **carpets** derive their names. *Axminster carpet* is carpet made on an Axminster **loom**, which is capable of intricate color designs, usually with a level cut-pile surface. *Loomed carpet* is carpet made on a modified upholstery **loom** with a characteristic dense, low-level loop **pile**, generally bonded to cellular rubber **cushioning**. *Velvet carpet* is carpet made on a simple **loom**, usually of a solid color or **moresque**, with cut or loop **pile** of either soft or hard-twisted yarns. *Wilton carpet* is carpet made on a **loom** employing a **jacquard** mechanism, which selects two or more colored yarns to create the pile pattern.

wrinkling Development of ridges and furrows in a **coating**, **paint**, or **transparent finish** film when it dries.

wrought aluminum products Products formed by *rolling*, *drawing*, *extruding*, or *forging*. A *bar* is a solid section that is long in relation to its cross-sectional dimensions and has a completely symmetrical cross section that is square or rectangular (excluding flattened wire) with sharp or rounded corners or edges, or is a regular hexagon or octagon. An aluminum bar has a width or greatest distance between parallel faces of ⅜ in. (9.5 mm) or more. An *extruded section* is a *rod*, *bar*, *tube*, or any other shape produced by the *extrusion* process. *Foil* is a solid sheet section rolled to a thickness of less than 0.006 in. (0.1524 mm). *Forging stock* is a *rod*, *bar*, or other section suitable for subsequent change in cross section by *forging*. A *pipe* is a *tube* having certain standardized combinations of outside diameters and wall thicknesses. A *plate* is a solid section rolled to a thickness of 0.250 in. (6.4 mm) or more in rectangular form and with either sheared or sawed edges. A *rod* is a solid round aluminum section ⅜ in. (9.5 mm) or greater in diameter, whose length is greater than its diameter. A *sheet* is a solid section rolled to a thickness ranging from 0.006 in. (0.1524 mm) to 0.249 in. (6.3246 mm), inclusive, supplied with sheared, slit, or sawed edges. *Structural shapes* are solid shapes used as load-bearing members. They include angles, channels, W shapes, tees, zees, and others. A *tube* is a hollow product whose cross section is completely symmetrical and is round, square, rectangular, hexagonal, octagonal, or elliptical, with sharp or rounded corners, and whose wall is of uniform thickness except as affected by corner radii. A *wire* is a solid section that is long in relation to its cross-sectional diameter having a completely symmetrical cross section that is square or rectangular (excluding flattened wire) with sharp or rounded corners or edges, or is round or a rectangular hexagon or octagon, and whose diameter, width, or greatest distance between parallel faces is less than ⅜ in. (9.5 mm).

wrought iron Relatively pure iron, mechanically mixed with a small amount of iron-silicate slag; characterized by good corrosion resistance, weldability, toughness, and high ductility.

wrought steel products Products formed by rolling, drawing, extruding, or forging. *See also* **finished mill products**.

wythe A continuous vertical section of masonry, one unit in thickness. Also called *withe* and *tier*.

yard lumber *See under* **lumber**.

yellowing Development of a yellow color or cast in white, pastel, colored, or clear finishes.

Index

AAMA, *see* American Architectural Manufacturers Association
ABS copolymers, *see* Acrylonitrile-butadiene-styrene copolymers
Absorbed dose, 44
Absorption, 482, 921–923
  of clay masonry units, 219
  measuring, 922
  and noise control, 922–923
  by transmission through materials, 923
Absorption bed, 1013, 1016
Absorption coefficient, 922
Absorption field, 1012–1014
Absorption trench, 1013, 1014
Abutments, 194
ACA (ammoniacal copper arsenate), 377
ACCA, *see* Air Conditioning Contractors of America
Acceleration, conversion factors for, 66
Accelerators (concrete), 128, 230
Accessible curb cuts, 29
Accessible design, 26
Accessible parking spaces, 28–29
Accessible route, 28
Accessible site, building or facility, 28
Accordion folding doors, 628
Accordion fold wood doors, 623–624
ACI, *see* American Concrete Institute
Acid cleaning, 869
Acid etching, 705
Acid fluxes, 299
Acid refractories, 299
Acid resistance (of carpet), 840
Acoustics, 919–923
  correction, acoustical, 919–920
  noise reduction, 920–923
  units for, 59–60
Acoustical baffles, 783, 794
Acoustical ceilings, 784–792
  accessories for, 789
  finishes for, 784–785
  fire performance of, 789
  installation of, 789–791
  materials for, 784–786
  sealants for, 791
  sound insulation blankets for, 791
  support systems for, 786–789
Acoustical correction, 919–920
Acoustical glass, 674
Acoustical plasters, 698, 712
Acoustical sealants, 603
Acoustical surfaces, paints for, 861
Acoustical treatment, 783–797
  for ceilings, 784–792
  integrated ceilings, 792
  wall panels and baffles, 792–797
Acoustical wall panels, 792–797
  back-mounted panels, 793
  baffles, 794
  fire performance of, 795–796
  installation of, 796–797
  metal wall panels, 793–794
  spline-mounted panels, 792–793
  support systems for, 794–795
Acoustic impedance, 60
Acrylic emulsion block fillers, 850
Acrylic emulsion coatings, 850
Acrylic latex coatings, 851
Acrylic latex terpolymer coatings, 851
Acrylic polymer coatings, 850
Acrylics, 471, 839
Acrylic sealants, 603
Acrylic terpolymer coatings, 851
Acrylonitrile-butadiene-styrene (ABS) copolymers, 470–471
Active power, 57
Activity, 44
ADA, *see* Americans with Disabilities Act of 1990
Adaptable buildings, 36–37
Adaptable Housing, 27
ADBT, *see* Adjusted dry-bulb temperature
Additions, electrical systems/planning for future, 1087
Additional services, 2
Adhered roofing installation systems, 528–529
Adhesion type ceramic veneer, 224
Adhesives:
  for gypsum board systems, 727–728, 730, 735–736
  for resilient flooring, 819–821, 823, 826, 827–830
Adhesive bonding:
  aluminum, 349
  steel, 323–324
Adhesive-held fasteners, 321
Adjusted dry-bulb temperature (ADBT), 1030, 1031
Admittance, 57
Admixtures, 122
  concrete, 123, 125–129
  in concrete masonry units, 229–230
  in mortar and grout, 208
Adsorption, 482
AF&PA, *see* American Forest & Paper Association
AGA, *see* American Gas Association
Age-strength relationships (portland cement), 147, 148
Agglomeration (ores), 298
Aggregates, 123, 125–127, 153
  in concrete masonry units, 228–229
  exposed aggregate surfaces, 161–162
  for masonry mortars and grout, 207
  storage of, 259
  water in, 150
A grade (plywood), 397
AHAM (Association of Home Appliance Manufacturers), 895
AIA, *see* American Institute of Architects; American Insurance Association
Air, and sound, 916–917
Airborne sound transmission, 923, 924, 926
Air chambers (in piping), 1024
Air Conditioning Contractors of America (ACCA), 1036, 1038
Air-conditioning units, 898
Aircraft traffic, noise from, 929
Air-dried lumber, 381–382
Air-entraining agents, 128–129, 230
Air-entraining portland cements, 124–125
Air entrainment, 149, 150
Air gap (plumbing systems), 989–990
Air infiltration:
  aluminum windows and sliding glass doors, 643
  barriers to, 438–439
Air plenum systems, 1040–1046
Air supply, fireplaces, 258

Air-supported structures, 471
Air-water HVAC systems, 1048
AISI, *see* American Iron and Steel Institute
Alabaster, 277
Alarms:
  audible, 33
  fire, 887–889
  for intrusion prevention/detection, 966–969
Alclad, 336
Algae, 1009
Alkali resistance:
  of carpet, 840
  resilient flooring, 814
Alkali-resistant enamels, 857
Alkali-resistant paints, 859
Alkali-resistant primers, 855
Alkyd enamels, 858
Alkyd flat wall paints, 858
Alkyd primers, 855, 856
All-glass entrances, 637
Alloys:
  aluminum, 334–337
  copper, 349–353
Alloyed cast iron, 301, 304
Alloying elements, 301, 305
Alloy ores, 299
Alloy steels, 306–307
All-water HVAC systems, 1046–1048
  hot- and chilled-water systems, 1047
  hot-water systems, 1046–1047
  terminal units in, 1047–1048
Alternating current, 1067–1068
Aluminized steel, 319
Aluminizing, 355
Aluminum, 87, 88, 296, 333–349
  alloys, 334–337
  cast products, 340, 341
  corrosion resistance of, 338–340
  for door faces, 618
  fabrication of products, 341–342
  finishes of products, 342–346
  joints and connections, 346–349
  manufacture of, 340–349
  mining and refining of, 334, 335
  production of, 334, 335
  properties of, 337–340
  recycling of, 334, 335
  for roofs, 596
  sheet metal, 589
  strength of, 338
  wrought products, 340–341
Aluminum Association, 338, 343–345
Aluminum entrances and storefronts, 634–635
  anchors for, 634–635
  fasteners for, 634–635
  finishes for, 634
  flashing for, 634–635
  standards for, 634
  supports for, 634–635
  weatherstripping for, 634–635
Aluminum paints, 859
Aluminum siding, 582–583
  grounding of, 583
  installation of, 582–583
  maintenance, 583
  material for, 582
Aluminum sliding glass doors, 638–641, 644
  accessories for, 640
  assembly of, 639
  fabrication of, 639
  fasteners for, 639
  finishes for, 639
  flash (butt) welding, 640
  hardware for, 639
  inert-gas welding, 640
  installation of, 644
  KD (knocked down) frame assemblies, 639–640
  quality standards, 640
  screwed assemblies, 639–640
  thermal break, 640
  weatherstripping for, 639
Aluminum Store Front and Entrance Manual, 637
Aluminum windows, 638–644
  accessories for, 640
  architectural, 640
  assembly of, 639
  awning, 641
  casement (C), 641
  fabrication of, 639
  fasteners for, 639
  finishes for, 639
  flash (butt) welding, 640
  hardware for, 639
  horizontal sliding, 641
  inert-gas welding, 640
  installation of, 644
  jal-awning, 642
  jalousie, 642
  KD (knocked down) frame assemblies, 639–640
  louver, 642
  prime, 640
  quality standards, 640
  screwed assemblies, 639–640
  thermal break, 640
  top-hinged, 642
  vertical pivot, 642
  vertical sliding, 642
  weatherstripping for, 639
Aluminum wiring, 1075–1076
Aluminum-zinc alloy roofs, 597
AMCBO (Associated Major City Building Officials), 25
American Architectural Manufacturers Association (AAMA), 679, 685
American Concrete Institute (ACI), 13, 237, 848
American Forest & Paper Association (AF&PA), 371, 406, 418, 421, 1041
American Gas Association (AGA), 895, 983
American Institute of Architects (AIA), 2, 7–10
American Insurance Association (AIA), 14, 23, 24
American Iron and Steel Institute (AISI), 305–307
American Lumber Standard, 386
American Lumber Standards Committee, 367, 386
American method (of shingle application), 555
American National Metric Council (ANMC), 40
American National Standards Institute (ANSI), 13, 14, 17, 19, 25–27, 40
American Nursery Landscape Association (ANLA), 113
American Plywood Association (APA), 13, 393, 406, 412, 1041
American Society for Testing and Materials (ASTM), 13, 14, 22, 40
American Society of Civil Engineers (ASCE), 14, 237
American Society of Heating, Refrigerating and Air-Conditioning Engineers (ASHRAE), 15, 24, 1029, 1037, 1038
American Society of Mechanical Engineers (ASME), 14
American Softwood Lumber Standard, 380
American Softwood Lumber Standard, 386
American Standard for Nursery Stock, 113
Americans with Disabilities Act of 1990 (ADA), 27, 895
  doors, 660
  elevators, 976
  fire extinguishers, 889, 890
  flooring, 807
  unit kitchens, 899
American Welding Society, 136
American Wire Gauge (AWG), 1076
American Wood Council (AWC), 1041
American Wood Preservers Association (AWPA), 378, 406, 409, 456
American Wood Preservers Institute (AWPI), 456, 1041
Ammoniacal copper arsenate (ACA), 377
Ampere (unit), 44, 1066
Anaerobic bacteria, 1019
Anasazi Indians, 1051, 1054
Anchors:
  for aluminum entrances and storefronts, 634–635
  for concrete, 141
  for metal doors and frames, 619
  for unit masonry, 211–213

Anchorage/anchoring:
- of formed frames and doors, 623
- of masonry walls, 262, 264
- pole construction for, 456
- of stone cladding, 280–283
- for wood-framed buildings, 410

Anchor bolts, 141
Anchored-type ceramic veneer, 224
Anchor plates, 142
Angle clips, 142
Angle draft drawing, 315
Angle stop valves, 999
Angular courses, 76
Angular velocity, units of, 52
ANLA (American Nursery Landscape Association), 113
ANMC (American National Metric Council), 40
Annealed wire, 315
Annealing, 336, 667
Anodes, 87
Anodized finishes, 344–345
ANSI, *see* American National Standards Institute
Antigraffiti coatings, 843
Antiscald faucets, 985
Ants, 107
APA, *see* American Plywood Association
Appliances, residential, *see* Residential appliances
Application for payment, 12
Aquifers, 277, 990
Arches, 121
- laminated timber, 453
- masonry, 253, 254

Architect(s):
- and contract administration, 11–12
- and excavation support systems, 102
- responsibilities of, 11
- role of, 2–3

Architect's First Source for Products, The, 8
Architectural and Transportation Barriers Compliance Board (ATBCB), 26, 27
Architectural barriers, 26
Architectural Barriers Act, 26
Architectural bronze, 349
Architectural custom panels, 405
Architectural drawings, 7–8
Architectural precast concrete, 193–194
Architectural precast concrete units, 190
Architectural terra-cotta, 224, 226–227, 276
Architectural windows, 640, 644
Architectural woodwork, *see* Finish carpentry
Architectural Woodwork Institute (AWI), 13, 390, 462, 632, 904
Architectural Woodwork Quality Standards, 390, 904
Arcom Master Systems, 9
Arc welding, 322, 348
Area, 45
- conversion factors for, 65
- metric conversions for, 71
- SI units for, 61
- units of, 50–51

Area controls, 967–969
Area of absorptive surface, 59
Armored cable, 1077
Art and Architecture Building (Yale University), 122
Articulated joints (paneling), 462, 464
Artificial aging, 336
Asbestos, 811
ASCE, *see* American Society of Civil Engineers
As-fabricated finishes, 351–352
Ashlar stone patterns, 279
Ashlar (stone setting), 280
ASHRAE, *see* American Society of Heating, Refrigerating and Air-Conditioning Engineers
ASHRAE comfort chart, 480–481
ASHRAE Standard 55, 1030, 1031
ASHRAE Standard 62, 1040
ASHRAE Standard 90.1, 1030–1031
ASHRAE standards, 15
ASME (American Society of Mechanical Engineers), 14
Aspdin, Joseph, 121
Aspen, 370
Asphaltic adhesive, 819
Asphaltic underlayments, 818
Asphalt roll roofing products, 520
Asphalt roofing, 516, 865
Asphalt-rubber adhesive, 819
Asphalt-saturate organic felt, 438–439
Asphalt shingles, 536–547
- application of, 542–545
- drip edges for, 539
- eave flashing for, 537, 539
- flashing for, 539–542
- preparation for installation of, 537, 539–543
- roof sheathing for, 537
- standards for, 537
- types of, 537, 538
- underlayment for, 537

Asphalt tile, 811
Asphalt water emulsions, 518, 519
Associated Major City Building Officials (AMCBO), 25
Associates Center (Chicago), 123
Association of Home Appliance Manufacturers (AHAM), 895
ASTM, *see* American Society for Testing and Materials
ASTM's soil classification system, 93
ATBCB, *see* Architectural and Transportation Barriers Compliance Board
ATBCB Minimum Guidelines and Requirements, 26
Atmospheric vacuum breaker, 990
Atomic number, 78
Atomic weight, 78
Atoms, 78, 80
Attics:
- insulation of existing, 510–511
- outlets in, 1082

Audible alarms, 33
Audible cues, 32
Audible warning systems, 33
Audio detectors, 969
Augers, 1011–1012
Austenitic nickel-chromium manganese steels, 308
Austenitic nickel-chromium steels, 308
Austenitic steel, 308
Automatic door hardware, 637
Automatic door operators, 658
Automatic storage water heaters, 999–1000
Auxiliary locks, 657
AWC (American Wood Council), 1041
AWG (American Wire Gauge), 1076
AWI, *see* Architectural Woodwork Institute
Awnings, 911–912
Awning windows, 641, 649
AWPA, *see* American Wood Preservers Association
AWPI, *see* American Wood Preservers Institute
Axminster weave, 834–835

Back blocking, 737
Backer board, 570–571
Backer rods, 604
Backfill(s), 104–105
- compaction of, 106
- nonproblem soils for, 95
- for pole construction, 459

Backflow, 989
Back grade plywood panels, 406
Background noise, 33, 928–929
Backings, carpet, 837–838, 842
Backparging, 268
Back siphonage, 989
Baffles, acoustical, 783, 794
Balancing devices (double-hung windows), 648
Ballcock value (water closets), 986
Balloon framing, 412
B and Better (B&B) grade lumber, 388
Bars, steel, 315
Barn paints, 858
Barrel shells, 121
Barrier-free design, 26–37
- in adaptable buildings, 36–37
- applicable laws and standards, 26–27
- for coordination disabilities, 35

Barrier-free design, (*cont'd*)
doors, 637, 658, 660
for the elderly, 35–36
for general accessibility, 28–32
for hearing impairment, 33
mental and perceptual impairment, 33
residential appliances, 895
for safety, 27–28
for special handicaps, 32–36
for visual impairment, 32–33
wheelchair users, 33–35
Barriers, noise, 930
Baseboard heaters, 1048
Base cabinets, 906
Basecoat plasters, 698, 711, 716
Base coats (EIFS), 511–512
Base flashing, 270, 271, 525
Base line, 74
Basements, 167, 168
lighting in, 1091
outlets in, 1082
in single-story buildings, 4
treated wood foundations for, 407, 408
Basement traction elevators, 973
Basement windows, 649
Base sheets (built-up roofing systems), 519, 523
Base units, 38, 41
Basic fluxes, 299
Basic oxygen furnaces, 302, 303
Basic refractories, 299
Basic services, 2
Basket weave bond pattern, 245, 246
Basket weave pattern, 245, 246
Bathrooms, 27
in barrier-free design, 33, 36, 37
gypsum board in, 740–741
lighting in, 1090
outlets in, 1081–1082
Bathtubs, 987–988
in barrier-free design, 28, 34, 37
finishes for, 987
fittings for, 987–988
materials for, 987
shapes of, 987
showers in, 988
types of, 987
Batteries, chemical, 1064–1065
Batt insulation, 501, 506
Bauxites, 334
Beadboard, 500
Beams:
box beams, 441, 443
cast-in-place construction, 181–184
concrete, 121, 131, 133
girders vs., 414
laminated timber, 453
precast concrete, 189, 190, 191
reshores for, 133
wood, bending in, 373–374
in wood-framed construction, 414
Bearing capacity (of soil), 94–96
Bed coat, 769
Bed joints (masonry), 260, 265–266
Bedrooms:
in barrier-free design, 33
lighting in, 1090
outlets in, 1081
Beech flooring, 798, 803
Bells, 172
Bell-and-spigot joints, 1007, 1008
Belling bucket, 172
Bending:
aluminum, 341
copper alloys, 350
plywood, 399
in wood, 371
in wood beams, 373–374
Bends, 136
Beneficiation (ores), 298
Bentonite clay, 498
Bessemer, Sir Henry, 297
Bessemer process, 297
Bevel siding, 585, 586
B grade (plywood), 397
Bidding, 10–11
documents, bidding, 10–11
process of, 10
requirements, bidding, 9
and selection of bidders, 10
Bilevel buildings, 5–6
Billets (steel), 311
Binary system, 39
Binders, terrazzo, 773
Birch flooring, 798, 803
Bitumens (for roofing), 516–517
Bituminous membrane roofing products, 515–521
classifications of, 518–519
fire resistance of, 521
manufacture of, 517–518
organic-felt or glass-fiber-felt based products, 519
raw materials for, 515–517
standards for, 519, 521
Bituminous membrane waterproofing, 497
Bituminous sheet waterproofing, 497
Blackplate (steel), 312
Black steel pipes, 315, 318
Blanket insulation, 501, 506
Blast cleaning, 868
Blast furnaces, 299–300
Bleeding, 158, 204
Blending to a common level, 30
Blinds:
roll, 911
venetian, 912–913
vertical, 912
Blind headers pattern, 245
Blind stops (windows), 651
Blocks (glass), 284–286
Block fillers, 843, 856
Block flooring, 804–805
Blooms (steel), 311
Blowing insulation, 501
Blowout bowl (water closet), 987
Boards, 379, 388
Board measure (lumber), 383
Board sheathing, 367
Board subflooring, 431
BOCAI, *see* Building Officials and Code Administrators International
BOCA National Building Code, 20–21, 237
Boilers, 1049, 1057
Bolts, 141, 320–321
Bolting, 320–321
Bond breakers, 604
Bond coat, 770
Bonded toppings, 196
Bonderizing, 320
Bonding:
agent, bonding, 705
masonry walls, 262–264
of matter, 81–82
of plywood, 396
steel, 323–324
Bond strengths, 204–205, 261
Book matching (veneer), 402
Borings, soil, 96
Boron, 299
Boston hips, 548, 549
Box beams, 441, 443
Box sill assembly, 417, 419
Bracing:
for excavations, 103
for gypsum board sheathing, 438
plywood, for wall corners, 437
Braille, 31
Brake forming, 317, 350
Branch circuits, 1073, 1083–1084
Branches, 991
Brass, 296, 349
Brass piping, 998
Brass-plated steel, 355
Brazing, 322, 348
Breaking strength, 84
Breezeways:
lighting in, 1091–1092
outlets in, 1082
Brick, 221–222
arches, brick, 253–254
building, 234
cleaning, 275, 276
colors of, 220
designations/dimensions for, 222
exterior finishing of, 864
flooring, 273
grout for, 211
hollow, 222–224, 225, 226
mortar bedding and jointing, 265–267
paints for, 861
pattern bonds, 244–245

paving, brick, 211, 273, 289
sizes of, 221
size variation in, 220
slump, 234
soft-mud process for, 216
sound absorption by, 923
stiff-mud process for, 214, 216
storage/protection of, 259–260
textures of, 217
uses of, 225–228
veneer, brick, 242, 243, 264, 407, 510
Bricklaying, SI units for, 49
Bridging, in wood joist floors, 422–423
Brief dip process (wood preservation), 378
Brightness, 1088
British thermal unit (Btu), 1066
Brittleness, 84
Broad and batten siding, 586
Broadloom carpet, 833
Bronze, 296, 349
Broom finish (concrete), 160
Brown coat, 710, 714–716
Brushed finish, 353
Btu (British thermal unit), 1066
Buffed finishes, 351–353
Buffing, 343
Builders (constructors), 3
Building codes, 18, 19–24
enforcement of, 22
fire safety requirements in, 889
model codes, 22–24
Building (common) brick, 225, 234
Building design, 2–7
and construction systems/methods, 6
construction team, 3
and future of building construction, 6–7
objectives of, 2–3
and shape types, 3–6
and use types, 3
Building Design Systems, Inc., 9
Building drainage system, 1002
Building drains, 1003–1004
Building envelope, 1035, 1052–1053
Building expansion joints, 146
Building module, internationally recommended, 68
Building Officials and Code Administrators International (BOCAI), 22, 23, 24, 25, 982
Building paper, 438–439
Building regulation, 18
Building sewers, 1003, 1004
Building trap, 1006
Built-up girders (wood-framed construction), 414–416
Built-up membrane roofing, 522–526
application of, 524–526
materials in, 522–523
Built-up roofing felts, 519
Built-up roofing systems:
insulation for, 514–515
NRCA designations for, 524
single-ply vs., 513–514
Bullet-resisting glass, 673
Bullnose blocks, 235
Buoyancy, 167
Burglar-resisting glass, 673
Burglary protection, *see* Intrusion prevention and detection
Burning (for site clearing), 99
Burning (masonry), 218
Burnishing clay, 356
Burn resistance (carpet), 841
Bushes, 112
Butt weld process, 315
Butyl sealants, 603
Butyl synthetic rubber sheet waterproofing, 497
Buzzers, entrance, 1093
Byzantine architecture, 202, 203

CAB (cellulose-acetate-butyrate), 471
Cabinets, 904–910
in barrier-free design, 28, 35, 37
for fire extinguishers, 889–890
hardware for, 909
installation of, 910
manufacture of, 909–910
materials/finishes for, 906–910
plastic-laminate clad, 909
quality standards for, 904–905
types/styles of, 906
wood-finished, 907–909
Cabinet convectors, 1048
Cables, 1077–178
Cable trays, 1079
CABO, *see* Council of American Building Officials
CA (cellulose acetate), 471
CADD (computer-aided design and drafting), 8
Cadmium-plated steel, 355
Caisson foundations, 168
Caissons, 172
Calcining, 121
Calcium chloride, 128
Calcium (for glassmaking), 663
Caloric, 1026
Calorific value (mass and volume basis), 67
Candela, 44
Candela, Felix, 122
Candlepower, 1088
Candle (unit), 1088
Cantilevered stairs, 463
Cant strip, 553
Capacitance, 57
Capacity, units of, 51
Capacity insulation, 508
Cap and flashing sheets (built-up roofing systems), 519
Capillarity, 109–110
Carbon:
in steel making, 301
and weldability of steels, 323
Carbonation, 232
Carbon steel doors, 619–620
Carbon steel(s), 304–305, 306, 311
coatings for, 319
for door facings and frames, 618
Carpentry:
finish, 462–464
SI units for, 49
Carpet, 832–845
appearance retention in, 841
in barrier-free design, 32
comparative factors related to, 836–838
construction of, 833–840
cushioning for, 843–844
durability of, 840–841
dyeing methods for, 839–840
ease of maintenance of, 841–842
flocked, 836
fusion-bonded, 836
history of, 832
installation of, 845
knitted, 836
needlepunched, 836
pile fiber materials in, 838–839
pitch of, 836–837
quality standards/certification for, 842–843
selection criteria for, 840–845
sound absorption by, 923
and sound control, 959–962
static generation, 842
subflooring for, 431
tiles, carpet, 843–844
tufted, 835–836
types of, 833–836
Carrier-current wireless alarm systems, 967
Cartridge brass, 349
Casement (C) windows, 640, 641, 646, 649
Casement doors, 627
Casework, wood, *see* Wood casework
Cast aluminum products, 340, 341
Casting(s):
alloys, casting, 350
aluminum, 341
copper alloys, 350
iron, 304
steel, 302
Cast-in-place construction:
cement-based underlayment, 197–198
concrete stairs, 188–189
concrete toppings, 195–196
glass-fiber-reinforced concrete, 195
precast concrete, 189–194
structural insulating roof decks, 196–197
structural slabs, 184–188

Cast-in-place piles, 172
Cast-in-place structural concrete, 180–189
  beams/girders, 181–184
  columns/piers/pilasters, 181, 182
  walls, 180–181
Cast iron, 296, 297, 304
Cast-iron piping, 1007
Cast-iron sectional boilers, 1049
Cast-steel products, 308, 309
Cathodes, 87
Caulking, *see* Sealants
Caulking (siding), 572
Cavitation, 1025–1026
Cavity walls:
  bonding in, 264
  insulation of existing, 510
  masonry, 238, 240–242
  mortar for, 209
  ties for, 212
C&Btr grade lumber, 388
CCA (chromated copper arsenate), 377
CDA (Copper Development Association), 349
Cedar shakes, 559–562, 578
Cedar shingles, 559–562, 577–578
Ceilings. *See also* Roof and ceiling assemblies
  acoustical, 784–792
  distortion in, with truss construction, 455
  and echoes, 921
  fireproofing, 601
  framing of, in gypsum board systems, 724–725
  gypsum plaster, 712
  insulation for, 508
  joist spans, 427–428
  paints for, 861–862
  and sound control, 936, 937, 949, 957–963
Ceiling fans, 1095
Cellular glass block, 500
Cellulose-acetate-butyrate (CAB), 471
Cellulose acetate (CA), 471
Cellulosic fiber insulation, 501–502
Cellulosics, 466, 471
Cement(s), 700
  for roofing products, 518–519, 534
  for tile roofing, 550
Cementation, 355
Cement-based underlayments, 198
Cementitious binders (terrazzo), 773
Cementitious coatings, 843
Cementitious materials, 228
Cementitious waterproofing, 498
CENELEC (Committee for European Electrotechnical Standardization), 17
CEN (European Committee for Standardization), 17
Centering, 131, 134, 187
Central station alarms, 966, 967
Ceramics, 78, 84–87
Ceramic floors, paints for, 863
Ceramic glazing, 217–218
Ceramic materials, ionic bonding of, 81
Ceramic mosaic tile, 751–752
Ceramic tile, 750–755. *See also* Tile
Ceramic veneer, 209–210, 224
Certificate of substantial completion, 12
Certification, 13–14
Cesspools, 1010
CGPM (General Conference on Weights and Measures), 39
C grade (plywood), 397
Chains, 77
Chair-type stair lifts, 33
Change orders, 11–12
Chat-sawed stone finish, 280
Checking (lumber/plywood), 385–386, 401
Chemical batteries, 1064–1065
Chemical brightening, 344
Chemical cleaning, 868–869
Chemical elements, 78–80
Chemical oxidizing, 344
Chemical properties of materials, 87–88
Chemical strengthening (glassmaking), 667
Chilled cast iron, 301
Chillers, 1049, 1050
Chimney blocks, 236
Chimney flashing, 271
  for asphalt shingles, 540–542
  for mineral-fiber-cement shingles, 554
  in reroofing, 546
  for steep roofs, 535
  for wood shingles and shakes, 567
Chimneys, 255–259, 1049
Chlorides, 155
Chlorinated polyvinyl chloride (CPVC), 998
Chromated copper arsenate (CCA), 377
Chromium, 299, 307, 354
Chromium-plated metal, 355
Circuits:
  branch, 1083–1084
  feeder, 1084
  low-voltage, 1086–1087
Circuit breakers, 1075
Circulation path, 28
Circulation space, 35
Cladding, 320, 336
Clay, 92, 94, 95
Clays, 110
Clay caissons, 172
Clay masonry, 214–228. *See also* Architectural terra-cotta; Brick; Clay tile
  architectural terra-cotta, 224
  classifications of, 220–224
  cleaning, 275–276
  hollow masonry units, 222–224
  manufacture of, 214–219
  properties of, 219–220
  raw materials for, 214
  solid masonry units, 221–222
  standards controlling, 224–228
Clay tile:
  hollow, 222–224
  for roofing, 549
  shapes of, 549
  size variation in, 220
  stiff-mud process for, 216
Clay waterproofing, 498
Cleaning:
  carpets, 841–842
  extruded aluminum doors, frames, and storefronts, 638
  glazed aluminum curtain walls, 686
  and glazing, 678–679
  masonry, 275–276
  painting, surface preparation for, 867–868
  resilient floors, 831–832
  windows, 651
Cleanouts, 1007
Clearances (doors), 619
Clear-cutting (forest conservation), 382
Clear floor space (barrier-free design), 34–35, 37
Clearing, site, 99
Clearing and grubbing, 99
Clear sealers, 852, 860
Cleavage membrane, 273
Climate, 1031, 1034
Clinker brick, 220
Clinkers, 121, 123
Clip-on storm panels, 654
Closed-die forging, 308
Closed risers, 31
Closed stirrup-ties, 182
Closeout procedures, 12
Closers, 267, 268, 658
Closets:
  lighting in, 1091
  shelving in, 462–463
Clothes dryers, *see* Dryers
Clothes washers, *see* Washers
CM (construction management) contracts, 7
CMUs, *see* Concrete masonry units
Coal, 1048–1049
CO/ALR (copper/aluminum) devices, 1076
Coal tar, 517
Coal tar-saturated felts, 439
Coated felts, 439
Coated glass, 671–672
Coated stick electrode welding, 322
Coaters, 517
Coatings, 849–854. *See also* Paints
  acrylic emulsion block fillers, 850

acrylic emulsion coatings, 850
acrylic polymer coatings, 850
antigraffiti, 843
cementitious, 843
for copper alloys, 351, 353
for ferrous metals, 317–320
for fire protection, 886
on glass, 668
installation of, 853–854
intumescent, 601
metal, on nonferrous metals, 355
paints vs., 854
for roofing products, 518
for roofs, 518, 519
for single-ply roofing systems, 527
standards for, 850
types and properties of, 850–853
as vapor retarders, 491
Codes, 18. *See also* Standards
building codes, 19–24
for electrical service, 1070
fire safety requirements in, 888, 889
for formwork, 129
for gypsum board sheathing, 438
history of, 18
improvement in, 25
multiplicity of, 22
National Energy Codes, 24
plumbing, 982–983
response to change of, 22
for stairs and handrails, 463
standards vs., 13, 19
and stud spacing, 426
for wood framing anchorage, 410, 411
zoning codes, 18–19
Coefficient of heat transfer:
conversion factors for, 67
units of, 56
Coefficient of linear thermal expansion, 55
Coefficient of utilization, 1088
Cognitive disabilities, 33
Cohesion (of soil), 92, 93
Coke, 298–299
Colburn process (glassmaking), 665
Cold-applied dampproofing, 498
Cold drawing, 310, 341
Cold finishing, 310–312
Cold flow (plastics), 469
Cold-forming (steel), 316
Cold reduction, 310
Cold-rolled copper alloy products, 352
Cold rolling, 310
Cold-weather construction:
concrete curing in, 165
for masonry, 274–275
mortar or grout, 210–211
Collar joints (masonry), 260, 268
Collectors, solar, 1051
Color(s):
anodizing, color, 344–345
in barrier-free design, 35, 36
of clay masonry units, 218, 219–220
in concrete, 162
in concrete masonry units, 234
of glass-reinforced concrete, 195
of granite, 277
of limestone, 277
of mortars, 205, 207–208, 209
in plastics, 468
signage, contrast for, 32
of slate, 277, 547
Columns:
cast-in-place structural concrete, 181, 182
concrete, 131, 133
in light wood-framed buildings, 413
precast concrete, 191, 192–193
Column strips, 187
Combination doors, 627–628
Combination frame (windows), 651
Combination water heaters, 1000
Combined drainage systems, 1003
Comb ridges, 548
Combustion chamber, fireplace, 257
Combustion heat, units of, 55
Comfort, human, 479–481
Comfort envelope, 481, 1037
Comfort levels, 1026, 1029–1030
Commercial brass, 349
Commercial buildings:
lighting in, 1089
outlets in, 1080
service power requirements for, 1070, 1072
Commercial cement grout, 757
Commercial (C) grade windows, 640
Commercial grading rules (lumber), 387
Commercial standard roofing slate, 547
Commercial Standards (CS), 16
Commissioners of Public Works Administrative Offices (Charleston), 123
Commissioning, 1057
Committee for European Electrotechnical Standardization (CENELEC), 17
Common (American) bond pattern, 244–245
Common grades (lumber), 383–384, 388
Common use, 3
Communication systems, 1093–1094
Community water supply systems, 990–991
Compact conventional roofing system, 493–494, 513
Compaction:
ASTM requirements for, 106
of soils, 95
Compartments (septic systems), 1021–1022
Completely fabricated dimension lumber, 390
Components, window and door, 649, 651
Composite glass, 673–675
Composite insulations, 500
Composite piles, 172
Composite walls, 239
Compounding (plastics), 467
Compressibility, units of, 54
Compressible bulb weatherstripping, 646
Compression, 83, 134
in concrete, 121
in wood, 374
Compression faucets, 985
Compression fittings, 997
Compression sash guide, 646
Compression test (concrete), 152
Compressive strength:
of clay masonry units, 219
of concrete masonry units, 232
of masonry walls, 249–250
of mortar or grout, 205
of wood, 369
Computer-aided design and drafting (CADD), 8
Concave joints (masonry), 269
Concealed condensation, 486–488
Concentrated load resistance (plywood), 400
Concentrated loads, 167
Concrete, 120–198, 705. *See also* Concrete masonry
accessories for, 141–143
admixtures in, 127–129
aggregates in, 125–127
anchors for, 141–142
cast-in-place structural concrete, 180–189
consolidating, 157–158
curing/protection of, 164–166
developing technology, 122
fasteners for, 141–142
finishing, 158–164
formwork for, 129–133
foundation systems, concrete, 166–172
furring strips over, 446
handling, 155
history of, 120–122
joint inserts, 142
joints in, 143–146
materials used in, 123–129
mixtures/mixing of, 146–154
paints for, 861
placing, 155–157
portland cement, 123–125
preparation for placement of, 155
reglets, 143
reinforcement of, 134–140
remixing, 154
slabs on grade, 173–180
sleeves, 142
thermal movement in, 143–144
transporting, 155
waterstops, 143

Concrete blocks, 235, 923
Concrete brick, 234
Concrete floors:
paints for, 863
reflooring of, 831
Concrete form plywood panels, 398
Concrete masonry units (CMUs), 228–236, 705, 706
classification of, 234–236
cleaning, 276
foundation walls, 180
manufacture of, 230–231
mortar bedding and jointing, 267–268
pattern bonds for, 245, 246
physical properties of, 232–234
raw materials used in, 228–230
shapes of, 234
standards controlling, 236
storage/protection of, 260
Concrete primers, 857
Concrete Reinforcing Steel Institute, 136
Concrete slab, sound absorption by, 923, 944–945
Concrete substrates, 513
Concrete tile (for roofing), 549–550
Concrete walls, sill plates in, 410
Concreting, SI units for, 49
Condensation, 478, 483–488
concealed condensation, 486–488
and interior ventilation, 484–485
and relative humidity, 484, 485
surface condensation, 485–486
tolerance for, 483
zones of, 489
Condensation polymerization, 467
Condensation resistance factor, 643
Condensers, 1049
Conductance, 57, 1031, 1032
Conduction, 1027, 1035
Conductivity, 86, 87, 469, 1031
Conductors, 1027, 1076–1078, 1084
Conifers, 364
Connections:
in pole construction, 460, 461
for precast concrete columns, 192–193
Conservation:
of energy, 1026. *See also* Energy conservation
in water supply systems, 990
Consistency (of soils), 94, 96, 97
Constitutents, SI units for, 49
Construction joints (concrete), 144, 145
Construction management (CM) contracts, 7
Construction plywood, 391–401
concrete form panels, 398
construction plywood, 391–401
exposure durability of, 396
grades of, 396–397
guide to the use of, 394–395
manufacture of, 391–392
overlaid plywood, 398
paneling, 398
performance -rated panels, 397–398
physical/mechanical properties of, 399–401
quality standards for, 393
sizes of, 399
textured siding panels, 398
tongue-and-groove panels, 398
veneer grades, 396–397
wood species for, 396
Construction Sciences Research Corporation, 9
Construction services, 2
Construction Specifications Canada (CSC), 8
Construction Specifications Institute (CSI), 8, 9
Construction systems/methods, choice of, 6
Constructors (builders), 3
Consumer Product Safety Commission, 25
Contact adhesives (gypsum board systems), 728
Continuous beams, 183
Continuous casting (steel), 302, 304
Continuous footings, 169
Continuous joint reinforcement (masonry walls), 264–265
Continuous ridge vents, 493
Continuous soffit vents, 493
Contracts, 7
Contract administration, 11–12
Contracting requirements, 9
Contraction (plastics), 469
Contractors, 3
Contractors' warranties (low-slope roofing), 513
Controls:
in barrier-free design, 35
lighting, 1088–1089
range, 28
Control joint blocks, 235
Control joints:
in concrete, 142–143, 144, 145
for concrete masonry, 247–248
sealant filled, 269
Controlled fills, 95
Convection, 1027
Convection loop, 1027
Convection-type electric heaters, 1048
Convectors, 1047
Convenience outlets, 1079–1080
Conventional high-strength low-alloy steels, 307
Conversion, metric, 68–72
Conversion coatings, 319, 343, 351, 353, 354
Conversion factors, 65–87
Conveying systems, 972. *See also* Elevators
Cooktops, 28, 895, 896
Cooling loopers, 518
Cooling (masonry), 218–219
Cooling operating hours, 1031, 1032
Cooling systems, 1049–1050
Coordination disabilities, 35
Copings, 593
Copolymers, 466
Copper, 296, 349–353
for roofs, 596
sheet metal, 589
Copper alloys, 618, 620
Copper/aluminum (CO/ALR) devices, 1076
Copper Brass Bronze Design Handbook, 350
Copper-coated steel, 355
Copper Development Association (CDA), 349
Copper piping, 996–997
Copper-plated aluminum, 355
Coreboard, 723
Core construction (carbon steel doors), 619–620
Cork tile, 812, 813
Corner blocks, 235
Corner floating, 736–737
Corners (masonry walls), 261–263
Correction lines, 74
Corridors:
in barrier-free design, 30, 37
and sound control, 946
Corrosion, 87–88
of aluminum, 338–370
coatings and finishes to protect from, 317–320
and plastics, 469
in water supply system, 1002
Cottonwood, base design values for, 370
Couloumb (unit), 1066
Council of American Building Officials (CABO), 25, 888
Counter lavatories, 984
Countertops, 910
in barrier-free design, 35, 37
tile, 763–764
Counter varnishes, 860
Courses, 76, 260–262
for asphalt shingles, 543–544
for wood shingles, 568
Coursed ashlar pattern, 279
Course pole, 261
Courtyards, and sound control, 930
Covalent bonding, 82
Cove-base flashing, 826–827
Coverage (roof), 533
Coxcomb ridges, 548
CPE, 527

C Plugged grade (plywood), 397
CPVC (chlorinated polyvinyl chloride), 998
Crack control:
in concrete masonry, 247
continuous joint reinforcement for, 264–265
in EIFS, 512
in gypsum board systems, 736–737
masonry walls, 147, 246
in masonry walls, 246–249
portland cement plaster, 717–719
in portland cement plaster, 717–719
Cracking:
in ground-supported slabs, 173–176
in gypsum board systems, 736–737
Crane-delivered bucket, 156
Cranes, 972
Crapper, Thomas, 982
Crawl spaces, 167, 168
evaporation from, 486
insulation for, 504
insulation of existing, 509
treated wood foundation, 408
treated wood foundations, 408
ventilation of, 491–492
water vapor in, 488
Crazing, 163–164
Creep (cold flow), 469
Creosote, 378
Cross bracing, 103–104, 461
Cross bridging, 422–423
Cross furring, 446
Cross grain (lumber), 385
Cross slope, 30
Crucible process (steel), 296–297
Crystalline bedrock, 94
CSC (Construction Specifications Canada), 8
CS (Commercial Standards), 16
C Select grade lumber, 388
CSI, *see* Construction Specifications Institute
CSPE (DePont Hypalon), 527
Cubing station, 231
Cue, 32
Cullet, 664
Curbs, 29
Curb ramps, 29
Curing:
of concrete, 164–166
concrete masonry units, 231
portland cement plaster, 716–717
of tile, 760–761
Curing covers, 165
Current, electrical, 43, 56, 1065–1068
Current density, 56, 1066
Curtain walls, 679. *See also* Glazed aluminum curtain walls
Curved lines, 76
Curved roof units, 443, 444
Custom door and frame units, 618
Custom grade (finish carpentry), 462
Cut nails, 412
Cutoff stops, 620, 622

Daily degree days, 1031, 1032
Damp checks, 270
Dampers, 257, 1057
Damping capacity, 85
Dampproofing, 246, 270, 273, 496, 498
Dash bond coat, 715
Dash coat, 705
dB, *see* Decibels
DBT, *see* Dry-bulb temperature
DDH (degree days heating), 1031
Dead level asphalt, 516
Dead loads, 167, 237, 249
De Architectura (Marcus Vitruvius Pollio), 120–121
Decals, security, 968
Decay:
from condensation, 488
in lumber, 384
in plywood, 401
in wood, 376
in wood-framed buildings, 410–411
Decay resistance of wood, 368–369, 375–376
Decibels (dB), 63, 918–919
Deciduous trees, 365
Decimal position notation, 39
Decimal system, 39
Decks:
gypsum concrete, 197
waterproofing membranes for, 497
Decking (lumber), 389
Decoration:
of aluminum windows and sliding glass doors, 639
of gypsum board, 748
Decorative blocks, 235
Decorative glass, 673
Decorative plywood, 401–406
Decorative stairs, 463–464
Deep drawing (steel), 317
Deflection, 82
with floor and ceiling assemblies, 413
and rafter span, 427, 429–430
Deformation, 82–83
De-grade (lumber), 385
Degree days heating (DDH), 1031
Delft tile, 749
Demand-type water heaters, 1000
Demolition, 98–99
Dense aggregates, 229
Density:
conversion factors for, 66
of soils, 94, 96
Derived units, SI, 41, 45
Descaling (steel), 310
Descriptive building codes, 19
Design:
building, *see* Building design
for sound control, 928–935
Design/build contracts, 7
Design equivalent temperature difference (DETD), 1037
Design pressure (DP):
aluminum windows and sliding glass doors, 643
extruded aluminum doors, frames, and storefronts, 635
windows and sliding glass doors, 640
Design professionals, 3
Design services, 2
Design standards, 13
Design temperature difference, 1036
DETD (design equivalent temperature difference), 1037
Detectable warnings, 31, 32
Deterioration of materials, moisture and, 483
Dewatering, 110
Dew point, 482, 1028
D grade (plywood), 397
DH windows, *see* Double-hung windows
Diagonal bond pattern, 245, 246
Diagonal grain (lumber), 385
Diagonal tension, 134
Diamond, 82
Die lines, 342, 352
Differential-acting steam hammers, 171
Differential settlement, 167
Diffusion, 481
Dimensional coordination, 68
Dimensional instability, 482–483
Dimensional stability, 400
Dimension lumber, 386–388, 390, 444
Dimension stock, 379
Dimension stone, 280
Dining areas:
lighting in, 1090
outlets in, 1081
Dip brazing, 348
Direct current, 1067
Direct gain passive solar systems, 1054–1055
Directional textured finishes, 351–353
Direction (in legal descriptions), 76
Direct reduction (iron making), 298, 301
Direct refrigerant systems, 1048
Direct return systems (hot water systems), 1047
Direct water heaters, 999–1000
Dirt resistance, mortar for, 209
Disabilities, 26
Disabled, design for, *see* Barrier-free design
Dishwashers, 895–898
Disposal field, 1012
Distance (in legal descriptions), 76
Distribution line, 1013

Distribution panels, 1072–1073
Distribution rib, 185
Districts, zoning, 19
Diversion ditch, 107
Divided light doors, 627
Divisions (of *MasterFormat),* 8–9
Documents, 7–10
  bidding, 10–11
  drawings, 7–8
  negotiation, 11
  organization of, 7
  project manual, 8–10
DOD, *see* U.S. Department of Defense
DOE, *see* U.S. Department of Energy
Dogleg ramps, 30
Dolomite, 299
Dolomitic limestone, 277
Domes, 6, 121, 187
Doors:
  aluminum entrances and storefronts, 634–638
  in barrier-free design, 30, 33, 34, 36, 37
  barrier-free design for, 27
  carbon steel, 619–620
  finishes, 631–632
  for freight elevators, 976
  in glass-reinforced concrete panels, 195
  hardware for, 656–660
  insulated, 620
  metal doors and frames, 618–623
  for passenger elevators, 975–976
  plastic-laminate clad cabinets, 909
  prefabricated, 631
  revolving, 637–638
  sliding glass doors, 638–653
  solid-core, 625–626
  for sound control, 932–934
  steel, 333
  storm doors, 653–656
  wood doors, 623–634
  wood-finished cabinets, 907, 908
Door clearances, 619
Door grades, 621, 632–633
  flush doors, 633
  stile-and-rail doors, 633
Door-skin plywood, 405
Door swing, 628–629
Double coursing, 571, 580
Double-coverage mineral-surfaced roll roofing, 522
Double cutting, 826
Double doors, 620
Double-duct forced-air systems, 1039
Double-hung (DH) windows, 640, 641, 648, 649
Double tees, 191–192
Douglas fir:
  base design values for, 370–371
  flooring, 798, 803
Dovetail anchors, 212, 213
Dovetail slots, 141
Downfeed water supply systems, 994–995
Downspouts, 594, 1023
Drain, waste, and vent system (DWV), 1002–1005, 1022, 1042–1043
Drainage. *See also* Sanitary drainage systems
  foundation drains, 111
  in-plane drains, 112
  SI units for, 49
  subdrainage systems, 111–112
  underslab, 111, 112
Drainage boards, 112
Drainpipes, 1005–1007
  sealing/venting of, 1005
  traps in, 1005–1006
  vents in, 1006–1007
Drapes, thermal, 913
Drawers:
  plastic-laminate clad cabinets, 909
  wood-finished cabinets, 908
Drawing(s), 7–8, 219, 350
Drawing scales, metric and nonmetric, 61
Dressed lumber, 388
Dressed size (lumber), 383
Drinking fountains, 989
Drip edges, 521
  for asphalt shingles, 539
  for mineral-fiber-cement shingles, 553
  in reroofing, 546
  for steep roofs, 535–536
  for wood shingles and shakes, 566
Driven piles, 170
Drive pin, 142
Drop, service, 1069
Drop hammers, 171, 350
Drop-in lavatories, 984
Drum trap, 1006
Dry-bulb temperature (DBT), 480, 1028
Dryers, 895, 898
Dry-fallout spray gloss, 857
Dry-fallout spray semigloss, 857
Dry loopers, 517
Dry organic felt (for roofing), 515
Dry-pipe fire sprinkler systems, 1058
Dry-press process, 219
Dry-press process (clay), 216
Dry rot, 376
Dry-set grout, 757
Dry-set mortars, 755, 766
Dry-shake colored-concrete surfaces, 162
Drywall finishes, 720
Drywall screws, 727
Drywall slip-in door frames, 620
Dry wells, 1013, 1020
Dry zones, grasses for, 115
D Select grade lumber, 388
Dual window (DW), 640
Ducts:
  embedded, 1050
  in forced-air delivery systems, 1039–1040
  joist ceilings, 431
  and sound control, 1056–1057
Dumbwaiters, 972
Duplex outlets, 1079
Durability:
  of clay masonry units, 219
  of concrete, 147–149
  of mortar, 205
  of plywood, 399
Dusting, 164
Dutch bond pattern, 245
Dutch corners pattern, 245
Dutch method of shingle application, 555–556
DW (dual window), 640
DWV, *see* Drain, waste, and vent system
Dyes:
  for carpet, 839–840
  in plastics, 468
Dye-type stains, 860
Dynamic loads, 167
Dynamic viscosity, units of, 54

Earth, 101
Earth berms, 112
Earthwork, 99–109
  compacting of soil, 106
  erosion control, 106–107
  excavating, 101–104
  excavation support systems, 102–104
  fill/backfill, 104–105
  grading, 105–106
  for slab-on-grade construction, 105, 106
  slope protection, 106–107
  termite control, 107–109
  underpinning, 100–101
Easements, 76
Eastern hemlock, 371
Eastern softwoods, 371
Eave flashing:
  for asphalt shingles, 537, 539
  for mineral-fiber-cement shingles, 553–554
  in reroofing, 546
  for steep roofs, 535
  for wood shingles and shakes, 565–566
EC (European Community), 17
Echoes, 917, 921
Economy grade (finish carpentry), 462
ECPA (Energy Conservation and Production Act), 25
EC standards, 17–18
Edge-grained lumber, 385
Edge venting, 494
Edging (concrete), 159
Edward I, King, 832
EER, *see* Energy Efficiency Rating
Effective temperature (ET), 479–481, 1029–1030
Efflorescence, 275–276

Effluent, 1011
E grade lumber, 388
EIFS, *see* Exterior insulation and finish systems
EIFS Industry Members Association (EIMA), 511
EJCDC (Engineers Joint Contract Documents Committee), 9
Elastic deformation, 83
Elastomeric sealants, 602–603
Elderly people, barrier-free design for, 35–36
Eleanor of Castille, Queen, 832
Electrical conductivity, 58, 87
Electrical equipment, framing for, 423–424
Electrical metallic tubing (EMT), 1078
Electrical outlets, *see* Outlets
Electrical properties of materials, 87
Electrical services, SI units for, 49
Electrical systems, 1068–1095
  codes/standards for, 1070
  communication systems, 1093–1094
  conductors in, 1076–1078
  and energy conservation, 1094–1095
  future additions, planning for, 1087
  and lighting, 1088–1093
  low-voltage circuits in, 1086–1087
  outlets/switches in, 1079–1083
  overload protection devices in, 1073, 1075–1076
  power distribution in, 1068–1069
  and power requirements, 1070, 1072
  raceways in, 1078–1080
  service entrance/distribution equipment in, 1072–1074
  and sound control, 934–935
  sound control with, 1087
  symbols used with, 1070, 1071
  wiring recommendations for, 1083–1086
Electric arc furnaces, 302, 303
Electric capacitance, 44
Electric charge, 57
Electric conductance, 44
Electric cooktops, 28
Electric energy, 1026
Electric field strength, 57
Electric heating systems, 1048
Electric inductance, 44
Electricity, 1064–1068
  concrete that will conduct, 122
  as fuel, 1049
  National Electrical Code, 24
  units for, 56–58
Electric panel heating, 741–742
Electric potential, 57
Electric potential difference, 45
Electric resistance, 45
Electric-resistance weld (ERW) process, 315
Electric traction elevators, 972–974
Electric water heaters, 1000
Electrolytes, 88
Electromotive force, 45, 57
Electrons, 78, 1064–1066
Electron shell, 78
Electroplating, 319–320, 346, 355
Electropolishing, 346
Elements, 78–81
Elevators, 5, 972–978
  auxiliary equipment for, 976–977
  in barrier-free design, 33
  directional lights for, 33
  electric traction, 972–974
  freight, 976
  history of, 972
  installation of, 978
  oil hydraulic, 974–975
  passenger, 975–976
  selection criteria for, 977–978
Elongation, 84
Embankments, nonproblem soils for, 95
Embedding compounds (gypsum board systems), 728
Embossed sheet finish (aluminum), 343
Emergency panels, elevator, 977
Emergency signals, 33
Emissivity, 672, 1027
Empty-cell preservative process, 378
EMT (electrical metallic tubing), 1078
Enameling (of glass), 668
Enamels, gloss, 857
Encasement (for fireproofing), 599
End construction, 222
Endlap, 533, 571
Energy, 45, 63
Energy conservation:
  electrical systems, 1094–1095
  residential appliances for, 894–895
Energy Conservation and Production Act (ECPA), 25
Energy density, 55
Energy efficiency:
  air infiltration barriers for, 438
  of fireplaces, 258
Energy Efficiency Rating (EER), 895, 1094–1095
Energy Policy Act of 1992, 895
Energy Policy and Conservation Act (EPCA), 24–25, 894
Enforcers, 3
Engineered fill, 104, 170
Engineered Wood Association, 406
Engineers Joint Contract Documents Committee (EJCDC), 9
English bond pattern, 245, 263
English corners, 245
English cross pattern, 245
Entrance signals, 1093
Entries, in barrier-free design, 30, 33, 36
Entropy, 55
EPA, *see* U.S. Environmental Protection Agency
EPCA, *see* Energy Policy and Conservation Act
EPDM, *see* Ethylene propylene diene monomers
Epoxies, 473
Epoxy adhesive, 756, 820
Epoxy enamels, 857
Epoxy flooring, 847–848
Epoxy grout, 758
Epoxy mortar, 767, 768
Equal Loudness Contours, 918, 919
Equator, 74
Equilibrium moisture content (wood), 366–367
Equipment, 894
Equipment enamels, 858
Equipment grounding, 1086
Equivalent temperature value, 66
Erosion control, on construction sites, 106–107
ERW (electric-resistance weld) process, 315
Escalators, 5, 972
ET, *see* Effective temperature
Etch finish, 343
Etching, glass, 667
Ethylene propylene diene monomers (EPDM), 497, 525
European Committee for Standardization (CEN), 17
European Community (EC), 17
European standards, 17–18
Eutrophication, 1009
Evaporation, 478, 486, 1027–1028
Evapotranspiration, 1015
Exact conversion of numerical values, 63–64
Excavation(s), 101–104
  demolition as part of, 99
  SI units for, 49
  support systems for, 102–104
Exhaust hoods, 896
Exit devices, 658
Expanded metal lath, 706–707
Expansion anchors, 142–143, 213
Expansion bolts, 321
Expansion joints:
  for clay masonry, 246–249
  in concrete, 144–145, 144–146
  sealant filled, 269
  for single-ply roofing systems, 530
Expansion (of plastics), 469
Experimental aluminum alloys, 336
Explosive forming, 350
Exposure:
  definition of, 270
  roof, 533
  for wood shingles and shakes, 564–565
Exposure 1 and 2 plywood panels, 396

Extenders (in plastics), 468
Exterior casing (windows), 651
Exterior insulation and finish systems (EIFS), 511–513
Exterior lighting, 1091–1093
Exterior plywood, 398
Exterior primers, 856–857
Exterior siding, *see* Siding, exterior
Exterior topcoat paints, 858–859
Exterior type (waterproof) plywood, 396
Exterior window treatments, 911–912
External vibrators (concrete), 157
Extruded aluminum doors, frames, and storefronts, 635–637
  adjustment of, 638
  all-glass entrances, 637
  cleaning of, 638
  finish hardware, 637
  flush-panel doors, 635–636
  installation of, 637
  interior frames, 637
  protection of, 638
  stile-and-rail doors, 635
Extruded polystyrene, 499
Extruding, 309, 310
  aluminum, 341
  copper alloys, 350

Fabrics, sound absorption by, 923
Faced blocks, 235
Faced size (lumber), 383
Faced wall ties, 212
Face grade plywood panels, 406
Face-shell mortar bedding, 267, 268
Facing brick, 225, 226, 227
Facing tile, mortar for, 209
Facing walls, bonding in, 264
Factory lumber, 388
Faience tile, 749
Fan coil units, 1047–1048
Fans, ceiling, 1095
Fantail hips, 548
FAPB, *see* Full automatic push button
Farmers Home Administration, 406
Fasteners:
  for aluminum, 347
  for aluminum entrances and storefronts, 634–635
  for aluminum windows and sliding glass doors, 639
  for asphalt shingles, 543
  for concrete, 142
  for gypsum board sheathing, 438
  for gypsum board systems, 726–727, 730–733
  for metal doors and frames, 619
  for metal roofs, 597–598
  for mineral-fiber-cement shingles, 553
  for roll roofing, 521–522
  for slate roofing, 547
  for steel and iron shapes, 320–322
  for steep roofs, 536
  for tile roofing, 550
Fastening:
  of stone cladding, 280–283
  for strip shingles, 544–545
  of wood frame members, 411
Fast-track delivery systems, 11
Fast tracking, 7
Fast-track projects, 7
Fatigue resistance, 85
Faucets, 34, 985
F-curve, 180
Federal Fair Housing Act of 1988, 27
Federal government standards, 15–17
Federal Housing Administration (FHA), 16–17, 927, 946, 1015–1016
Federal Specifications, 16
Federal Standardization Documents, 16
Feeder circuits, 1084
Feeders, 1072
Felt(s):
  lining, 819
  manufacture of, 517
  for roofing, 515–516
  selection of, for built-up roofing, 522
Felted hair (carpet), 843–844
Felt weatherstripping, 646
Fences, and intrusion prevention/detection, 965–966
Ferritic chromium steels, 308
Ferritic steel, 308
Ferroalloys, 299, 302
Ferromanganese, 302
Ferrous metals, 296, 861, 863–864
FGMA (Flat Glass Marketing Association), 677
FHA, *see* Federal Housing Administration
FHA/HUD Housing Programs, 16–17
Fiberboard, 406
  insulation, 502
  wall sheathing, 437–438
Fiber forms, 129
Fiber saturation point (wood), 366
Field observations, 11
Figure (wood grain), 375, 404
Fill(s), 104–105
  nonproblem soils for, 95
  for subdrainage systems, 111–112
Filled-polymer countertops, 910
Filled shells, 81
Fillers, 467–468, 870–871
Filter fabrics, 107
Final Fair Housing Accessibility Guidelines, 27
Fine-rubbed stone finish, 280
Finger-jointed plywood panels, 399
Finishes, 696–873
  acoustical treatment, 783–797
  air-entrained concrete slabs, 160
  for aluminum, 342–346
  aluminum entrances and storefronts, 634
  for aluminum windows and sliding glass doors, 639
  for bathtubs, 987
  for cabinets, 906–907
  carpet, 832–845
  concrete, 158–164
  of concrete slabs, 161–163
  for copper alloys, 351–353
  dimension stone tile flooring, 806–810
  doors, 631–632
  for ferrous metals, 317–320
  flashing and, 709
  flat, 857–858
  formed concrete surfaces, 163–164
  on glass, 667–668
  for glazed aluminum curtain walls, 681
  gloss, 857
  of granite, 277
  gypsum board systems, 719–748
  gypsum plaster application, 709–713
  joints, masonry, 268–269
  lightweight structural concrete slabs, 161
  on marble, 277
  metal doors and frames, 619
  metallic, 355
  metal plaster accessories, 709
  for metal roofs, 597
  paints, 854–872
  for passenger elevators, 975
  plaster bases, 704–709
  plaster materials, 696–702
  plaster support systems, 703–704
  portland cement plaster application, 713–719
  for preassembled components, 440
  resilient flooring, 811–832
  resinous flooring, 845–849
  semigloss, 857
  sheet metal, 590
  of slate, 277
  special coatings, 849–854
  standard-weight concrete slabs, 158–160
  for stone, 280
  terrazzo, 772–783
  tile, 749–772
  tile over other, 762–763
  transparent, 860–861
  vinyl wall covering, 872–873
  wood doors, 634
  wood flooring, 797–805
Finish carpentry, 462–464
Finish coat plasters, 698
Finish coats, 511, 710, 711, 714–716
Finish grade lumber, 388
Fire alarms, 887–889
Fire clays, 214
Fire codes, 4
Fire control, in HVAC systems, 1057
Fire doors, 620
Fire escapes, 889

Fire extinguishers, 889
Fireplaces, 255–258
Fireproofing, 599–601
Fire rating (of masonry walls), 285–289
Fire resistance:
of acoustical ceilings, 789, 792
of acoustical wall panels, 795–796
of carpet, 841
gypsum board systems, 739
of masonry units, 254–255
of plastics, 469–470
of plywood, 401
of roof covering materials, 521
Fire-resistant plants, 886, 887
Fire-resistant (Type X) lath, 723
Fire-resistant (Type X) wallboard, 721
Fire retardants:
paints, 859
for wood, 378
Fire safety and protection, 24, 886–890
alarm systems, 887–889
containment/control, fire, 889
electrical fires, 1073
escape routes, 889
extinguishers, 889–890
prevention of fires, 886, 887
standards for, 15
Fire-safing, 601
Fire sprinkler systems, 1058
Firestopping, 601
Fire tube boilers, 1049
Fittings (in drainage systems), 1005
Fixed windows, 643, 649
Fixtures:
lighting, 1088
plumbing, 983–990
Fixture drains, 1004
Flakeboard, 406
Flamed stone finish, 280
Flame-proofing recipes, 886
Flammability, *see* Fire resistance
Flanking paths, 924, 937
Flared fittings, 997
Flash (butt) welding, 640
Flashed (clay), 220
Flashing, 589–594
for aluminum entrances and storefronts, 634–635
for asphalt shingles, 539–542
for chimneys, 271
combination sheet, 590–591
cove-base, 826–827
fasteners for, 591
flexible sheet, 590
in low-slope roofing systems, 515
in masonry, 214
for masonry walls, 269–272
materials for, 589–591
for metal doors and frames, 619
for mineral-fiber-cement shingles, 553–555
of plaster surfaces, 709
in reroofing, 546
for roll roofing, 521
roof flashing, 592–594
and siding, 572
for single-ply roofing systems, 529–530
for steep roofs, 535
wall flashing, 591–592
and wall strength, 251
for wood shingles and shakes, 566–567
Flashing cement, 518
Flash welding, 348
Flat asphalt, 516
Flat finishes, 857–858
Flat Glass Marketing Association (FGMA), 677
Flat-glass products, 668–675
coated glass, 671–672
composite glass, 673–675
decorative glass, 673
float glass, 669
heat-absorbing glass, 671
heat-treated glass, 672–673
patterned glass, 669–670
quality standards for, 668–669
tinted glass, 671
wired glass, 670–671
Flat-grained lumber, 385
Flatness, slab, 179–180
Flat plate slabs, 186–188, 191
Flat slicing (veneer), 401–402
Flecked multicolored interior finishes, 843
Flemish bond pattern, 245, 263
Flexible metal conduit, 1078–1079
Flexible tubing, 997
Flexural strength (concrete masonry units), 232
Float glass, 666, 669
Floating, 159, 204
Floating foundations, 167, 170
Flocked carpet, 836
Floors and flooring. *See also* Slab on grade construction
in barrier-free design, 27, 32, 37
brick, 211, 273
cement-based underlayments for, 198
concrete, 131, 134
framing for, in wood-framed buildings, 412
gypsum-based underlayments for, 197–198
as heat sinks, 1035
insulation for, 504
near fireplaces, 255
over air plenum systems, 1042
paints for, 862–863
plank and beam framing systems, 448–452
precast concrete elements for, 191–192
radiant, 1047
resilient, 811–832
resinous, 846–849
sound absorption by, 923
and sound control, 936, 937, 959–963
stone, 211, 806–810
stone for, 280
tile, 758–759, 764–768
in treated wood foundations, 408
vapor retarder installation in, 495
wood, 797–805
wood joist floor systems, 417–422
Floor assemblies:
panelized, 444, 445
in wood-framed buildings, 412–413
Floor decking, 328–330
Floor diaphragms, 461
Floor enamels, 857
Floor joists, 417–422
installation of, 421–422
in light wood-framed buildings, 413
lumber for, 418–419
spans for, 419–421
in wood joist floors, 417–422
Floor mat detectors, 969
Floor projections, wood framing for, 423, 424
Floor sheathing, 439
Floor varnishes, 860
Flow-after-suction, 204
Flow (of a mortar), 204
Flow table, 204
Flue linings, 258
Flues, 257, 258, 259
Fluid-applied waterproofing membranes, 496–497
Fluid capacity:
conversion factors for, 65
SI units for, 61
Fluid-level control valve (water closets), 986
Fluorescent lamps, 1088
Fluorocarbons, 471
Fluorspar, 299
Flush doors, 623–624, 625–626
fabrication/assembly, 629–630
quality standards, 632
Flush joints, 268–269
Flush-panel doors, extruded aluminum, 635–636
Flush tanks, 985–986
Fluxes, 299
Flux of magnetic induction, 58
Flying forms, 122, 131
F-number system, 180
Foamed-in-place insulation, 502
Foam plastics, 469
Foam rubber weatherstripping, 646
Foil-backed wallboard, 490, 721
Folding doors, 628–629
Foot, cushioning of the, 35, 36
Footcandle (unit), 1088

Footings:
forms for, 130–131
in foundations, 169–170
for pole construction, 458
size of, 169–170
for treated wood foundation, 407
for underpinning, 100
Footlambert (unit), 1088
Force, 45
conversion factors for, 66
per unit area, 67
per unit length, 53, 67
SI units for, 62
units of, 53
Forced-air electric heaters, 1048
Forced-air HVAC systems, 1039–1046
air plenum systems, 1040–1046
delivery of air in, 1039
equipment, air-handling, 1039–1040
Forest conservation, 382–383
Forest Products Laboratory, 456
Forging, 308, 309, 350
Forging (aluminum), 342
Formboards, 196
Formed frames and doors, 623
Form liners, 129
Forms:
concrete, 122, 129–133, 155
for gypsum concrete, 196
Foundations:
bracing systems for, 104
high-rise buildings, 5
insulation of existing, 508–509
multi-story buildings, 5
presumptive bearing values, 94
single-story buildings, 4
wood, forces on, 410
Foundation drains, 111
Foundation systems (concrete), 166–172
caissons, 170, 172
design of, 167–168
piles, 170–172
spread foundations, 168–170
Foundation systems (wood), 406–410
Foundation walls:
cast-in-place structural concrete, 180–181
insulation of existing, 508–509
masonry, 252, 253
treated wood foundation, 407–408
Foundries, 300–301
Fourcault process (glassmaking), 665
Four-pipe systems (hot- and chilled-water systems), 1047
Four-way center and butt matching (veneer), 403
Fraction, decimal, and metric equivalents, 64–65
Fractional sections, 74–75
Fracture strength, 84
Frames/framing:
for conventional steel-framed buildings, 325–328
door, 659, 660
for gypsum board sheathing, 438
in gypsum board systems, 724–726
for plaster support, 703–704
and siding, 571
for steep roofing, 533–534
wood, *see* Wood framing
Frame components, 649–650
Framed block core (solid core doors), 625
Framed/frameless glass block systems, 285
Framed roof configurations, 513
Framing lumber, 367
Freezers, 895, 896, 897
Freezing pipes, 993, 1006
Freight elevators, 976
French doors, 627
Frequency, 45
sound, 917–918
units of, 52, 59
Freyssinet, 121
Friction caissons, 172
Friction piles, 171
Frost, 98
Frosted finish, 343
Frost heave, 167
Frost point temperature, 482
Fuels:
for cooling units, 1049
for heating systems, 1048–1049
for iron and steel industry, 298–299
for water heaters, 1000–1001
Full automatic push button (FAPB), 976, 977
Full-cell preservative process, 377–378
Full mortar bedding, 267, 268
Fungus hazards (in wood), 376
Fungus resistance (carpet), 840–841
Furan mortar, 756
Furan resin grout, 758
Furnaces, 301–302, 1043, 1049, 1057
Furnace brazing, 348
Furnishings, 904
Furring, 255, 257, 445–446
Fuses, 1073, 1075
Fusion-bonded carpet, 836
Fusion welding, 355
Future of building construction, 6–7

Gable end louvers, 493
Gable rakes, 568
Gable roofs, 532
Galvanic corrosion (aluminum), 339
Galvanic potentials, 88
Galvanized metals, paints for, 861, 864
Galvanized steel, 319
in drainage systems, 1008
as pipe material, 315, 995–996
for roofs, 596
sheet metal, 589
Galvanizing, 355
Gambrel roofs, 532
Gang-grooved V-grooved panels, 405
Ganister, 299
Garages:
lighting in, 1091
outlets in, 1083
Gas, 1048
Gaseous discharge lamps, 1088
Gaskets, glazing, 676
Gas-shielded metal arc welding, 322, 323
Gate valves, 999
Gauging plasters, 698, 710
General Conference on Weights and Measures (CGPM), 39
General contractors, 3
General Services Administration (GSA), 10, 16, 26
GFCI, *see* Ground fault circuit interrupter
GFRC (glass-fiber-reinforced concrete), 195
Gilders' liquor, 356
Gilders' whiting, 356
Gilding, *see* Gold plating
Girders:
beams vs., 414
cast-in-place construction, 182, 183
forms for concrete, 133
precast concrete, 191
reshores for, 133
in wood-framed construction, 414–416
wood-framed construction, girder-joist connections in, 416
Glare-free lighting, 28
Glass, 78, 661–676. *See also* Glazing
acoustical, 674
for aluminum windows and sliding glass doors, 639
coated, 671–672
composite, 673–675
condensation on, 485
decorative, 673
flat-glass products, 668–675
float glass, 669
heat-absorbing, 671
heat-treated, 672–673
insulating, 674–675
laminated, 673–674
low-reflectance, 672
manufacture of, 663–668
patterned, 669–670
patterned glass, 669–670
properties of, 661–662
selection of, 676
sound absorption by, 923
and sound control, 934
spandrel, 672
standards of quality for, 668–669
tinted, 671
wired, 670–671

for wood windows and sliding glass doors, 649
Glass blocks, 284–286, 500
Glass fiber insulation, 500
Glass fiber mat (for roofing), 515–516
Glass-fiber-reinforced concrete (GFRC), 195
Glass gilding, 356
Glass leaf (gilding), 356
Glass wool insulation, 501
Glazed aluminum curtain walls, 679–686
adjustment, cleaning, and protection of, 686
design and fabrication of, 684
glazing material selection for, 685–686
installation of, 684–685
materials and finishes for, 681–682
performance requirements and testing for, 682–684
standards for, 680
systems of, 679–682
Glazed wall tile, 751
Glazing, 661–686
for aluminum windows and sliding glass doors, 640
ceramic, 217–218
cleaning and protection of, 678–679
general requirements for, 676–678
glass, 661–676
for glass in steel doors, 620
glazed aluminum curtain walls, 679–686
for glazed aluminum curtain walls, 685–686
multiple glazing, 678
with plastic, 678
plastic glazing materials, 675
safety, 27
sealants, 676
SI units for, 49
sloped, 679
structural silicone sealant, 677–678
Globe valves, 999
Gloss enamels, 857
Gloss finishes, 857
Glued block core, 625
Glued floor systems, subflooring in, 434–436
Gluelams, 453
Gluing properties (of woods), 374–375
GMP (guaranteed maximum price), 7
Gold, 354–355
Gold plating (gilding), 355, 356
Government lots, 75
Government system of survey, 74
Grab bars, 28, 37
Grades:
door, 632–633
of wood cabinets, 904
Grade beams, 181
Grade marks, 14, 388, 634
Grade stamps (lumber), 389
Grading, 105–106
Graduated roofing slate, 547
Grain (wood), 375
Granite, 277
Granite flooring, 809
Granular soils, 92
Grasses, 114–116
Gravel, 94, 95, 110, 299
Gravel stops/fascias, 593
Gray cast iron, 301, 304
Grease trap, 1006
Greeks, ancient, 749
Green chain (sawmill), 380
Greenhouse effect, 1053
Green liquor, 334
Green stain (on brick), 275, 276
Greenwich meridian, 74
Grinding, 343
Groined vaults, 121
Ground covers, fire-resistant, 887
Ground fault circuit interrupter (GFCI), 1075, 1081–1083, 1086
Grounding, 1085–1086
of aluminum siding, 583
equipment, 1086
system, 1085–1086
Ground-supported slabs, 173–178
Type I, 173–176
Type II, 175–177
Type III, 177–178
Groundwater, 109–112
Grout, 757–758
for brick paving and flooring, 211
mixing, 210–211
portland cement, 206
portland cement-lime, 206
in reinforced masonry walls, 244
for stone flooring, 211, 810
types, uses/proportions, 209–210
Grout-cleaned finish, 163
GSA, *see* General Services Administration
Guaranteed maximum price (GMP), 7
Guggenheim Museum, 122
Guide meridians, 74
Guide specifications, 9–10
Gutters, 593–594, 1023
Gypsum backing board, 721, 723
Gypsum board, 719–748
adhesives for, 727–728, 730
board attachment, 730
cracking and ridging precautions, 736–737
decorating/redecorating, 745–748
edge and corner trim, 728–729
electric panel heating, 741–742
fasteners for, 726–727, 730
finish defects, 745–748
and fireproofing, 599
fire-resistant construction, 739
foil-backed, 490
job conditions, 729
joint and fastener treatment, 744–745
joint tape and joint compounds, 728
manufacture of, 720
measuring, cutting and fitting, 730
moisture-resistant construction, 740–741
multi-ply construction, 733–736
paints for, 861
resurfacing existing construction, 744
self-supporting gypsum partitions, 742–747
single-ply construction, 730–733
and sound control, 737–739, 923, 935–940
supporting construction for, 724–726
thermal insulating construction, 739–740
and tile, 759
types of, 720–721
veneer plaster base, 744
wall sheathing, 438
Gypsum cementitious underlayment, 197–198
Gypsum concrete:
decks, 197
roof decks, 196–197
Gypsum formboards, 721, 723
Gypsum Keene's cement plaster, 698, 710
Gypsum lath, 708–709, 723–724
Gypsum-lime putty-trowel plaster, 710
Gypsum plaster, 696–698
aggregates, 700–701
application of, 710–713
manufacture of, 696–697
mixes, 709–710
Gypsum sheathing, 721, 723

Half-quarters, 75
Half-round slicing (veneer), 402
Halls/hallways:
in barrier-free design, 30, 37
lighting in, 1091
outlets in, 1082
Hammers, 171
Handicapped access, 5. *See also* Barrier-free design
Handicaps, physical, 26
Handrails:
in barrier-free design, 27, 30
standards and codes for, 463
Hand-rubbed finish, 353
Hand-tooled stone finish, 280
Hanging (of wood doors), 634
Hardboard, as underlayment for resilient flooring, 817
Hardboard panel siding, 588
Hardboard siding, 586–588
Hard conversion of numerical values, 64–65, 67
Hardening (of mortar or grout), 205

Hardness, 85
of plastics, 468–469
water, 1001–1002
Hardpan caissons, 172
Hardware:
for aluminum windows and sliding glass doors, 639
in barrier-free design, 36
for cabinets, 909
for doors, 656–660
for double hung windows, 647–648
for horizontal sliding windows, 647–648
for wood windows and sliding glass doors, 647–648
Hardwoods, 364–365
in decorative plywood, 401–406
glueability of, 375
for paneling plywoods, 398
plain-sawn lumber, 385
quarter-sawn lumber, 385
shrinkage of, 368
visual characteristics of, 402
Hardwood flooring, 798–800
Hardwood lumber, 383, 390
Harmonics, 918
HDO (high density overlaid) plywood, 398
Headers, 260, 423
definition of, 417
in pole construction, 460, 461
in wood joist floors, 422
Header blocks, 235
Head flashing, 270, 271
Head jamb (windows), 651
Head joints (masonry), 260, 266–267
Headlap, 533, 571
Heads, column, 187, 188
Hearing aids, 33
Hearths, 255–257, 257–258
Heartwood, 365, 375–376
Heat:
conversion factors for, 67
metric conversions for, 72
quantity of, 55
and water vapor, 478
Heat-absorbing glass, 671
Heat capacity, 55, 1033–1034, 1054
Heat detectors, 887–889
Heat energy, 1026
Heat flow:
conversion factors for, 67
moisture and, 483
SI units for, 63
units of, 55
Heat flux density:
conversion factors for, 67
units of, 55
Heat gain, 1035–1038
Heat generation, 1048–1049
Heating, ventilation, and air conditioning (HVAC), 1026–1057
air-water systems, 1048
all-water systems, 1046–1048
and ASHRAE thermal comfort envelope, 1030
and climate/weather conditions, 1031
commissioning of, 1057
cooling in, 1049–1050
direct-refrigerant systems, 1048
electric heating systems, 1048
embedded pipes/ducts, 1050–1051
forced-air systems, 1039–1046
and generation of heat, 1048–1049
and heat loss/gain, 1035–1038
heat pumps in, 1050
heat recovery systems, 1051
and heat transfer, 1027–1028, 1031–1034
and interior comfort, 1029–1030
and landscaping, 112, 113
and latent heat, 1027
and relative humidity, 1028–1029
requirements, heat/comfort, 1027–1034
and sensible heat, 1027
site planning for, 1034–1035
smoke/fire controls for, 1057
solar heating/cooling, 1051–1055
sound control for, 1055–1057
standards for, 15
steam systems, 1048
system controls, 1057
systems, HVAC, 1038–1048
and thermal design criteria, 1030–1031, 1035
uniform plumbing and uniform mechanical codes, 23–24
Heat loss, 1026, 1035–1037
Heat recovery systems, 1051
Heat release rate, 56
Heat resistance (of plastics), 468
Heat-resisting steels, 307–308
Heat sinks, 1035
Heat strengthening (glassmaking), 667, 672–673
Heat transfer, 55–56, 1027–1028, 1031–1034
Heat-treatable alloys, 334, 335
Heat-treated constructional alloy steels, 307
Heat-treated glass, 672–673
Heavy commercial (HC) grade windows, 640
Heavy timber framing systems, 452–453
Hedges, 112
High-build acrylic resin coatings, 851
High-build latex coatings, 851
High-capacity rock caissons, 172
High density overlaid (HDO) plywood, 398
High-fired glazing, 218
Highlighting, 343
High polymers, 465–466
High-pressure decorative laminate (HPDL), 909
High-pressure steam curing, 231
High-rise buildings, 5
High silicon bronze, 350
High-strength cast irons, 304
High-strength low-alloy steels, 307
Hindu-Arabic numbering system, 39
Hinges, 658–660
Hips:
in slate roofing, 548–549
wood shakes for, 568–569
wood shingles for, 568
Hip roofs, 532
Hip shingles, 545, 556, 557
HMMA (Hollow Metal Manufacturers Association), 619
Hoists, 972
Holeless oil hydraulic elevators, 974
Hollow bricks, 222, 225, 226, 267–268
Hollow clay tile, 222
Hollow-core doors, 626
Hollow-core flat-slab units, 191, 192
Hollow metal doors, 619
Hollow Metal Manual, 618, 619
Hollow Metal Manufacturers Association (HMMA), 619
Hones, slate, 277
Honed stone finish, 280
Hooks, 136, 140, 182
Hooking, 176
Hopper windows, 649
Horizontal bookleaf matching (veneer), 403
Horizontal cell tile, 222
Horizontal membrane waterproofing system, 496, 497
Horizontal shear (wood beams), 374
Horizontal sliding windows, 641, 649
Hot-applied bituminous dampproofing, 498
Hot-applied built-up roofing, 514
Hot dipping, 319, 355
Hot finishing, 310
Hot-forming (steel), 316
Hot rolling, 309–322
Hot rubberized asphalt waterproofing, 496–497
Hot-water heating systems, 1046–1047
Hot-weather construction, 166
House sewers, 1004
HPDL (high-pressure decorative laminate), 909
Hub-and-spigot joints, 1007
Hubless joints/fittings, 1007, 1008
HUD, *see* U.S. Department of Housing and Urban Development
Human comfort, relative humidity and, 479–481
Human wastes, safe disposal of, 1009–1010

Humidity:
- climates, humid, 489
- relative, 479–481, 1028–1030
- and temperature, 479–481

Huntsman, Benjamin, 296
HVAC, *see* Heating, ventilation, and air conditioning
Hyatt, John Wesley, 464
Hydrated lime, 207, 702, 759
Hydration, 123, 274
Hydraulic cement, 120, 498
Hydraulic shock, 1024
Hydroforming (copper alloys), 350
Hydrogen sulfide, 1002
Hydronic baseboard heaters, 1048
Hygroscopic expansion and contraction, 400
Hygroscopic materials, 482
Hygroscopic properties of wood, 366–369

IAPMO, *see* International Association of Plumbing and Mechanical Officials
IBM Corporation, 842
ICBO, *see* International Conference of Building Officials
ICC, *see* International Code Council
ICC International Codes, 22, 24
IEEE (Institute of Electrical and Electronics Engineers), 1070
IESNA (Illuminating Engineering Society of North America), 1089
IGC (inert gas configuration), 81
IIC, *see* Impact Insulation Class
Illuminance, 45
- conversion factors for, 67
- units of, 59

Illuminating Engineering Society of North America (IESNA), 1089
Impact Insulation Class (IIC), 916, 927–928, 941–944
Impact Noise Rating (INR), 916, 927
Impact resistance (plywood), 400
Impact sound pressure level (ISPL), 927–928
Impact strength, 54, 85–86
Impervious soil, 1011
Improved atmospheric corrosion resistance high-strength low-alloy steels, 307
Improved formability high-strength low-alloy steels, 307
Incandescent flood lamps, 1092
Incandescent lamps, 1088
Independent testing agencies, 14
Index of Federal Specifications, Standards, and Commercial Item Descriptions, 16
Indoor design temperatures, 1036
Induction furnaces, 302, 303
Industry standards, *see* Standards
Inert gas configuration (IGC), 81
Inert-gas-shielded arc welding, 348
Inert-gas welding, 640
Infiltration, 1035, 1036
Infrared detectors, 32, 968
Infrared heaters, 1048
Ingots:
- iron, 300
- steel, 302

Inground oil hydraulic elevators, 974
Initial flow, 204
In-line joist system, 422
Inner ply grade plywood panels, 406
In-place costs, 237
In-plane drains, 112
INR, *see* Impact Noise Rating
Insect damage to wood, 376–377, 385
Insect resistance (of carpet), 840–841
Inserts, joint, 142–143
Insolation, 1053
Instantaneous sound pressure, 59
Institute of Electrical and Electronics Engineers (IEEE), 1070
Insulated doors, 620, 626–627
Insulated framed roof configurations, 513
Insulated framed roofing systems, 482–493
Insulated substrates, 513
Insulated wall sheathing, 438
Insulating boards, 490–491
Insulating glass, 674–675
Insulation, 499–513
- for ceilings and roofs, 508
- crawl spaces and floors, 504
- dew point, 487
- in EIFS, 512
- exterior insulation and finish systems, 511–513
- foamed-in-place insulation, 502
- gaps in, 495
- general recommendations for, 502–503
- loose-fill insulation, 501–502
- for low-slope roofing, 514–515
- materials for, 499–502
- mineral fiber insulation, 500–501
- moisture in, 487
- reinsulation of existing buildings, 508–511
- retarder-faced, 490
- rigid insulation, 499–500
- for slabs on grade, 503–504
- standards of quality for, 502, 503
- and surface condensation, 485
- and vapor retarders, 490
- for walls, 504–508

Insulators, 1027
Intensity, sound, 918
Intercommunication (intercom) systems, 1093
Interconnected locks, 657
Interim Federal Specifications, 16
Interior doors, 618, 620
Interior frames, 637
Interior lighting, 1089–1091
Interior primers, 855–856
Interior topcoat paints, 857–858
Interior trim (windows), 651
Interior type (moisture resistant) plywood, 396
Interior window treatments, 912–913
Interlayment (wood shingles and shakes), 565
Interlocking metal weatherstripping, 646
Internally threaded insert, 141
Internal vibrators (concrete), 157–158
International Association of Plumbing and Mechanical Officials (IAPMO), 23–24, 982
International Building Code, 24
International Code Council (ICC), 24–26, 982
International Conference of Building Officials (ICBO), 22–25, 982
International Fire Code, 26
International Organization for Standardization (ISO), 17–18, 909, 949
International symbol of accessibility, 28
International symbols, 31
International System of Units (SI), 40–72
- adoption of, 40
- conversion to, 63–72
- for design and construction units, 49–63
- preferred dimensions and coordination in, 68
- presentation of numerical values with, 47–48
- presentation of units and symbols in, 46–47
- recommendations for use of, 44, 46–49
- structure of, 40–45
- units in, 40–45

Intrusion prevention and detection, 965–969
- area controls, 967–968
- decals/labels, 968
- and doors, 659–660
- effectiveness, system, 966
- elements, detector, 966
- glass, burglar-resisting, 673
- infrared systems, 968
- interior area detectors, 968–969
- motion detectors, 968–969
- signaling devices, 966–967
- site planning for, 965–966
- timer-controlled lights/sounds, 968
- types of detectors, 967

Intumescent coatings, 601, 886
Investors, 3
Ions, 80
Ion-exchange water softeners, 1002
Ionic bonding, 81–82, 86
Ionization detectors, 887, 888
Iron. *See also* Steel

architectural uses of, 333
cast iron, 304
design and construction, 320–332
history of, 296–297
iron making, 299–301
joints and connections, 320–324
materials and products, 296–320
mining/processing of, 297–299
ornamental iron elements, 333
in plumbing, HVAC, electrical systems, 333
properties of, 304–308
protective coatings/mechanical finishes, 317–320
roofing, siding, cladding, flashing, 333
sitework, 333
Iron founding, 300–301
Iron ores, 298
Iron oxide waterproofing, 498
Ironworkers, 330–332
Irradiance, units of, 55
ISO, *see* International Organization for Standardization
Isolated footings, 169
Isolated gain passive solar systems, 1055
Isolation joints (concrete), 144–146
ISO standards, 17, 18
ISPL, *see* Impact sound pressure level

Jal-awning windows, 642
Jalousie windows, 642
Jamb anchors, 622
Jamb blocks, 235
Jamb extenders (windows), 651
Jamb liners (windows), 651
Jefferson, Thomas, 661
Joinery, SI units for, 49
Joints:
aluminum, 346–349
in concrete, 143–146
in flue walls, 258
gypsum plaster, 744–745
masonry, 260
in masonry, 268–269
mechanically formed, 321–322
in mortared brick paving/flooring, 274
for paneling, 462, 464
sealing of, 601–607
for single-ply roofing systems, 530
in slate roofing, 547–54
for steel and iron shapes, 320–324
in steel-framed buildings, 326–327
tile, 761–762
in treated wood foundations, 408
for trim, 462, 463
Joint compounds (gypsum board systems), 728
Jointing (concrete), 159
Joint inserts (concrete), 142
Joint reinforcement:
in masonry walls, 264–265
for unit masonry, 211–213
Joint tape (gypsum board systems), 728
Joint tooling, 205
Joists, 185, 187, 412
in light wood-framed buildings, 413
steel, 328
Joist-and-rafter roof assemblies, 427, 430
Joist bands, 185
Joist girders:
steel, 328
in wood-framed construction, 416
Joist roofs, 430–431
Joule, James P., 1026

Kalin Associates, 9
KCMA, *see* Kitchen Cabinet Manufacturers Association
Kelly, William, 297
Kelvin, 44
Kettle calciner, 697
Key-in-knob locks, 656
Key-in-lever locks, 656
Keys, Samuel, 749
Keystone finish (concrete), 162–163
Key walls, 461–462
Keyways (in concrete), 142–143
Kickplates, 33, 34
Kiln-dried lumber, 381
Kilns, 218, 219
Kilogram, 44
Kinematic viscosity, 54
Kinetic energy, 1026
Kitchens:
in barrier-free design, 28, 34–35, 36, 37
fire hazards in, 886
lighting in, 1090–1091
outlets in, 1082
unit, 899–900
Kitchen appliances, *see* Residential appliances
Kitchen Cabinet Manufacturers Association (KCMA), 904–906
Kitchenettes, 899
Knifing, 824
Knitted carpet, 836
Knockdown door frames, 620
Knocked down frame assemblies, 639–640
Knot sealers, 870
Knots (in lumber), 365, 384
K-value, *see* Thermal conductivity

Labels, 14, 634, 968
Laboratory grade (finish carpentry), 462
Lacquer, 346, 860
Ladder core, 626
Lagging, 102–104
Laitance, 157
Lally columns, 413
Laminates (plastic), 469
Laminated block flooring, 804
Laminated coatings, 320, 346
Laminated glass, 673–674
Laminated timber, 388
Laminated timber framing systems, 453–455
Laminated waterproof papers, 439
Laminating adhesives, 728
Lamp, 1088
Landfill areas, 99
Landings (in barrier-free design), 27, 30, 31
Landmarks, 33
Landscaping:
in barrier-free design, 30
fire-resistant plants in, 886, 887
and intrusion prevention/detection, 966
lawns, 114–116
planning requirements, 112–113
trees/shrubs, 113–115
Land surveys and descriptions, 73–77
legal descriptions, 75–77
metes and bounds system, 73
rectangular system, 74–75
SI units for, 49
Langley (unit), 1053
Lap cement, 518, 519
Lap siding, 585–586
Large buildings, subsurface investigation for, 96
LAS (Local Acceptable Standards), 17
Latent heat, 478, 1049
Latent heat gain, 1036
Latent heat of fusion, 1027
Latent heat of vaporization, 1027, 1028
Laterals, 1014
Lateral displacement (in soil), 95
Lateral loads, 166, 167
Lateral stability, 250–251
Latex, 471
Latex adhesive, 819–820
Latex flat wall paints, 858
Latex floor paints, 858
Latex grout, 757
Latex-inhibitive primers, 856
Latex paints:
exterior, 858–859
interior, 858
Latex-portland cement, 755, 766, 770
Latex primers, 855
Latex underlayments, 818
Lath:
expanded metal, 706–707
gypsum, 723–724
wire, 707–708
Laundry areas:
lighting in, 1091
outlets in, 1082
Laundry trays, 985
Lavatories, 984
in barrier-free design, 34, 37
lighting in, 1090
outlets in, 1081–1082

Law(s). *See also specific laws*
for accessible design, 26–27
building-related, 18
for fire retardant products, 886
plumbing-related, 983
for smoke/heat detector placement, 888
Lawns, 112, 114–116
Layers (in plywood), 399
Leaching wells, 1020
Lead, 354
for roofs, 596
sheet metal, 589
Lead-coated copper, 356
Leaded nickel silver, 350
Lead-lined door frames, 620
Lead-lined doors, 620
Lead piping, 998
Lean clays, 94
LeCorbusier, 122
Legal descriptions, 75–77
Lenders, 3
Length:
conversion factors for, 65
metric conversions for, 70–71
units of, 43, 50
Leveling (concrete), 159
Levelness, slab, 179–180
Life Safety Code, 24, 27, 886, 888
Lifts, 33, 104, 106, 972
Lift-slab construction, 187
Light, 1088
Light exposure, units of, 59
Light-frame buildings, treated wood foundations for, 406–407
Light framing lumber, 388–389
Lighting, 1088–1093
with acoustical ceilings, 792
along stairways, 1091
in attics, 1091
in barrier-free design, 28, 29–30
in basements, 1091
in bathrooms/lavatories, 1090
in bedrooms, 1090
in closets, 1091
in commercial, institutional, and industrial buildings, 1089
controls for, 1088–1089
in dining areas, 1090
and energy conservation, 1094
exterior, 1091–1093
fixture types, 1088
in garages, 1091
in halls, 1091
interior, 1089–1091
in kitchens, 1090–1091
in laundry areas, 1091
in living/recreation rooms, 1090
outlets for, 1079
for security, 1092
SI units for, 63
timer-controlled, 968
units for, 58–59
in utility rooms, 1091
Lightning protection, 954–955
Lightning Protection Code, 964
Lightning rods, 964
Lightweight aggregates, 229, 233
Light wood framing, 412
Lime:
hydrated, 207, 759
in plaster, 701–702
Lime run, 275, 276
Limestone, 277, 299
Limestone flooring, 809
Limit controllers (furnaces), 1057
Linear acceleration, units of, 52
Linear measurement, SI units for, 50, 60–61
Lining felt, 819
Links, 77
Linoleum, 812, 813
Linoleum paste, 819
Linseed oil, 860
Lintel blocks, 235
Lintels, 253, 254
Live loads, 167, 237, 249
Living rooms, lighting in, 1090
LLD (lumen depreciation factor), 1089
Loads, 167
Local Acceptable Standards (LAS), 17
Locks, 656–658
auxiliary, 657
interconnected, 657
key-in-knob, 656
key-in-lever, 656
mortise, 656–657
padlocks, 657
reinforced, 657
security, 657
special, 658
window, 648–649
for wood windows and sliding glass doors, 648–649
Locker rooms, 27
Lock-seam joining, 322
Loess soils, 95
Logging methods, 379
Loomed carpet, 835
Loose-fill insulation, 501–502
Loose-laid and ballasted roofing installation, 528
Loose leaf (gilding), 356
Loudness, 919
Louis XIV, King, 832
Louvered doors, 627
Louver windows, 642
Louver wood doors, 624
Low-fired glazing, 218
Low-pressure steam curing, 231
Low silicon bronze, 349–350
Low-slope roofing systems, 513–531
bituminous membrane roofing products, 515–521
built-up membrane roofing, 522–526
flashing in, 515
roll roofing, 521–522
roof configurations, 513
roof insulation, 514–515
selecting systems, 513–514
single-ply membranes, 525, 527–531
steep roofs vs., 531
substrates, 513
thermal and moisture protection for, 513–531
warranties and maintenance agreements, 513
Low-voltage circuits, 1086–1087
Lumber, 379–390
base design values for, 370–373
design values for mechanically graded dimension lumber, 372–373
equilibrium moisture content of, 367
for floor joists, 418–419
grades of, 367–368
hardwood, 390
in light wood framing, 413
manufacturing process, 379–383
marketing and measure, 383
metric conversions for, 72
preservative treatment for, 409–410
quality of, 383–386
softwood, 386–390
for wall framing, 425
wall sheathing, 435–436
working stresses of, 371
Lumber core plywood, 403, 404
Lumber sheathing, 439
Lumen depreciation factor (LLD), 1089
Lumen (unit), 1088
Luminaire, 1088
Luminance:
conversion factors for, 67
units of, 59
Luminous efficacy, 59
Luminous flux, 45, 58
Luminous intensity, 43, 58

Maching burn (lumber), 385
Machining (steel), 317
Magnesia, 299
Magnesium stains, 276
Magnetic field strength, 56
Magnetic flux, 45, 58
Magnetic flux density, 45, 58
Magnetic induction, 58
Magnetic potential difference, 56
Magnetic vector potential, 58
Magnetism, units of, 56–58
Magnetization, 56
Magnetomotive force, 56
Magnificent Mile (Chicago), 122
Mailing methods, 411
Mains, water, 999
Main vent, 1004

Malleability (steel), 301, 304
Malleable cast iron, 301, 304
Mandrel, 172
Manganese, 299
Manual metal arc welding, 322
Manual of Millwork, 462, 904
Manual of Standard Practice, 136
Manuals, project, *see* Project manuals
Maple flooring, 798, 803
Maple Flooring Manufacturers Association (MFMA), 800
Marble, 277
Marble flooring, 809
Marble/glazed tile, sound absorption by, 923
Marble Institute of America (MIA), 806
Marine plywood, 398–399
Martensitic chromium steels, 308
Martensitic steels, 307
Masonry, 202–289, 705–706
  arches, 253–254
  backparging, 268
  bonding/anchoring, 262–264
  brick paving/flooring, 273–274
  clay masonry, 259
  clay masonry units, 214–228
  concrete masonry units, 228–236, 260
  exterior finishing of, 864–865
  exterior priming of, 863
  fireplaces and chimneys, 255–259
  fire resistance, 254–255
  furring strips over, 446
  glass unit masonry, 284–286
  grout, 203–211
  history of, 202–203
  joint finishing, 268–269
  laying units, 260–262
  lintels, 253, 254
  mortar, 203–211
  mortar bedding/jointing, 265–268
  paints for, 861
  piers/pilasters, 253–254
  pointing and cleaning, 275
  protection during construction, 274–275
  reinforcement, ties, anchors, flashing for, 211–214
  stone, 276–283
  support for structural members, 265
  unit masonry design, 237–259
  unit masonry erection, 259–276
Masonry bonded multiple-withe walls, 263
Masonry bonded walls, 239
Masonry cement, 207, 700
Masonry cement mortars, 206
Masonry enamels, 858
Masonry paint, 859
Masonry primers, 856, 857
Masonry Society, 237
Masonry veneer construction, 426
Masonry walls, 237–253, 246
  bonding/anchoring, 262–264
  brick pattern bonds, 244–245
  cavity walls, 238, 240–242
  compressive strength of, 249–250
  concrete masonry unit pattern bonds, 245, 246
  continuous joint reinforcement, 264–265
  crack control, 246–249
  dampproofing, 270, 273
  effect of flashing on/strength of, 251
  flashing installation, 269–272
  insulation of, 506–508
  lateral support, 251
  laying, 260–262
  moisture control, 245–246
  parging, 270, 272–273
  pattern bonds, 244–246
  properties of, 285–289
  reinforced walls, 238, 244
  R-values for, 505, 506
  structural design of, 249–253
  thicknesses, heights, lengths of, 251–253
  transverse strength of, 250–251
  U-values for, 505, 506
  veneered walls, 238, 241–244
  waterproofing, 270, 272
  water seepage and, 487
Mass, 45
  conversion factors for, 66
  metric conversions for, 69
  SI units for, 62, 63
  units for, 43
  units of, 52
Mass density, concentration, 53
Mass per unit area:
  conversion factors for, 66
  metric conversions for, 69–70
  units of, 53
Mass per unit length:
  conversion factors for, 66
  metric conversions for, 70
  units of, 53
Mass per unit time, 53, 66
Mass per unit volume, 66, 70
Mass Trombe wall, 1054
MasterFormat, 8–9
MASTERSPEC, 9
Mastic, 756–757
Mastic underlayments, 817–819
Matallizing, 320
Matched lumber, 388
Materials, 77–88
  and bonding of matter, 81–82
  building blocks of matter, 78–81
  chemical properties of, 87–88
  condensation tolerance of, 483
  electrical properties of, 87
  mechanical properties of, 82–86
  stress and strain, 82–86
  thermal properties of, 86–87
  water vapor permeance, 484
Material handling systems, 972
Materials Releases, 17
Material standards, 13
Mat foundations, 170
Mat moisture and bonding test, 822
Matte as-fabricated finishes, 352–353
Matte nondirectional textured finishes, 353
MC 15 lumber, 367
MDO (medium density overlaid) plywood, 398
Mean radiant temperature (MRT), 1029
Measurement, 38–73
  history of, 38–39
  International System of Units, 39–72
  of lumber, 383
  of materials in concrete, 153–154
  metric system, 39
Mechanical equipment, framing for, 423–424
Mechanically fastened roofing installation systems, 528–529
Mechanical properties of materials, 82–86
Mechanical services, SI units for, 49
Mechanical systems, *see specific systems*
Medium density overlaid (MDO) plywood, 398
Melamines, 473
Melting temperature, 86
Membrane waterproofing, 496–497
Merchant pig iron, 300
Mercury vapor lamps, 1092
Meridians of longitude, 74
Metal(s), 78, 82, 84–88, 296–356
  aluminum, 333–349
  chromium, 354
  copper, 349–353
  in door frames, 622
  exterior finishing of, 864
  exterior priming of, 863–864
  gold, 354–355
  iron, *see* Iron
  lead, 354
  metal-coated metal, 355–356
  nickel, 354
  paints for, 861
  pretreatment of, for painting, 869–870
  sheet metal, 589–590
  steel, *see* Steel
  tin, 354
  zinc, 353–354
Metal-clad cable, 1077
Metal conduit:
  flexible, 1078–1079
  rigid, 1078
Metal doors and frames, 618–623
  aluminum entrances/storefronts, 634–638
  basic door grades and applications, 621
  carbon steel doors, 619–620
  carbon steel frames, 620

formed stainless steel and copper-alloy doors/frames, 620
installation, 620, 622–623
sliding glass doors, 638–641, 644
Metal flashing, 270
Metal forms, 129
Metal gauges, metric conversions for, 72
Metal inert-gas shielded arc (MIG) welding, 322, 323, 348
Metallic bonding, 82
Metallic coatings, 319
Metal primers, 856–857
Metal roofing, 594–599, 865
architectural sheet metal, 596
field-formed systems, 596
finishes for, 597
installation of, 597–599
materials used in, 596–597
preformed metal roofing, 595–596
structural sheet metal, 596
Metal studs, and sound control, 938–941, 950–954
Metal-tied, multiple-withe walls, 264
Metal-tied, single-withe walls, 263–264
Metal trim, 728–729
Metes and bounds, 73, 76
Methylmethyacrylate, 847
Metre, 44
Metrication, 38, 40, 63
Metric conversion, 38, 68–72
Metric Conversion Act, 40
MFMA (Maple Flooring Manufacturers Association), 800
MG, *see* Motor generator
MIA (Marble Institute of America), 806
Microemulsions, 853
Microwave detectors, 969
Microwave ovens, 896, 897
Middle strips, 187
MIG welding, *see* Metal inert-gas shielded arc welding
Mild-weather construction, concrete curing in, 165–166
Military agencies, standards of, 16
Military Specifications, 16
Mill finish, 343, 352
Milling, 277.278
Mineral core plywood, 404
Mineral fiberboard, 599
Mineral-fiber-cement shingles, 552–558, 572–577
application of, 555–557, 574–577
materials for, 572–573
materials used in, 553
preparation for, 573
preparation for roofing with, 553–555
reroofing with, 557–558
reshingling with, 577
Mineral fiber insulation, 500–501, 503
Mineral stabilizers (in roofing products), 517
Mineral surfacing, 517
Minimum Property Standards (MPS), 13, 17, 24, 27
Mining:
of aluminum, 334, 335
of clay, 214
of iron materials, 297–298
Minoans, ancient, 982
Mirrors, in barrier-free design, 34, 37
Mismatched lumber, 385
Mitered hips, 548
Mixers (holding furnaces), 300
Mixing faucets, 985
Mobility (of plastics), 468
Model Code for Energy Conservation in New Building Construction, 24
Model codes, 22–24, 27
Moderate exposure, 270
Modifiers (in plastics), 467
Modular coordination, 68, 221
Modular units, 221
Modulus of elasticity, 63, 84
and ceiling joist spans, 428
of concrete masonry units, 232
conversion factors for, 67
and stress grade of lumber, 389
units of, 53, 54
of wood, 374
and wood floor joist spans, 420
Modulus of section:
conversion factors for, 65
units of, 51
Modulus of subgrade reaction, 72
Moisture content:
and dimensional instability, 482–483
of wood, 366
Moisture control, 110, 478–495. *See also* Thermal and moisture protection
concrete masonry units, 245–246
design and construction recommendations, 488–495
and human comfort, 479–481
vapor retarder installation, 495
vapor retarder selection, 489–491
vapor retarder/ventilation design, 491–495
water vapor properties, 478–482
Moisture-controlled concrete units, 236
Moisture penetration, metric conversions for, 72
Moisture resistance:
gypsum board systems, 740–741
resilient flooring, 813–814
Moisture resistant plywood, 396
Molds (in wood), 376, 384
Molded counter lavatories, 984
Molded polystyrene, 499
Molding:
plaster, molding, 698, 710
shapes of, 462, 463
Mole, 44
Molecular materials, 78, 80–81, 82
Molecules, 78, 80–81
Molybdenum, 299
Moment of force, 53, 66
Moment of inertia, 53, 66
Momentum, 53
Monel metal, 354
Monier, F. Joseph, 121
Monomers, 466
Monuments, 73, 76
Mortar, 260
for brick paving and flooring, 211
for brick veneer walls, 243
in cavity walls, 241, 242
dry-set, 755
epoxy, 755–756, 767, 768
on flashed/unflashed walls, 251
furan, 756
for glass blocks, 284
masonry cement, 206
materials used in, 206–208
mixing, 210–211
for parging, 272
portland cement, 206, 754, 755
portland cement-lime, 206
properties of, 204–205
in reinforced masonry walls, 244
removing from masonry, 276
retempering, 211
sand-lime, 206
stains from, 276
for stone/paving flooring, 211
for tile roofing, 550
types, uses/proportions, 208–209
Mortar bedding and jointing, 265–268
brick, 265–267
concrete masonry units, 267–268
hollow masonry units, 267–268
for stone structures, 280
structural clay tile, 267–268
Mortar boxes, 620
Mortared/mortarless brick paving and flooring, 273–274
Mortared paving/flooring, 273
Mortar joints:
finishes for, 268–269
for glass blocks, 284–285
Mortgage loans, 16–17
Mortise locks, 656–657
Mosaic, 772
Motion detectors, ultrasonic, 968–969
Motor generator (MG), 973–974
Moving walks, 972
MPS, *see* Minimum Property Standards
MRT (mean radiant temperature), 1029
Multidirectional loads, 167
Multioutlet assemblies, 1080
Multiple draft drawing, 315
Multiple reflection, 920
Multi-story structures, 4–5
Municipal water supply systems, 990–991

Muntz metal, 349
Muriatic acid, 276
Mutual inductance, 58

NAAMM (National Association of Metal Manufacturers), 619
NAECA, *see* National Appliance and Conservation Act of 1987
NAHB, *see* National Association of Home Builders
NAHB Research Foundation, Inc., 15
Nails/nailing, 321
  for asphalt shingles, 543
  for gypsum board systems, 726, 731–733, 745–746
  for mineral-fiber-cement shingles, 553
  for siding, 572–573, 578–579, 586
  for slate roofing, 547
  for steep roofs, 536
  for strip shingles, 544–545
  of wood frame members, 411
  for wood shingles and shakes, 563–564, 568, 569
Nailable plywood sheathing, 436–437
Nailable substrates, 513
NAPHCC (National Association of Plumbing-Heating-Cooling Contractors), 23
National Academy of Engineering Research Council, 25
National Academy of Science, 25
National Appliance and Conservation Act of 1987 (NAECA), 894–895
National Association of Home Builders (NAHB), 15, 22, 406, 1036, 1040–1041, 1046
National Association of Home Builders Research Foundation, 422
National Association of Metal Manufacturers (NAAMM), 619
National Association of Plumbing-Heating-Cooling Contractors (NAPHCC), 23
National Building Code, 23
National Building Code of Canada, 435
National Bureau of Standards (NBS), 24
National Concrete Masonry Association, 498
National Conference of States on Building Codes and Standards (NCSBCS), 24, 25
National Design Specification for Stress-Grade Lumber and Its Fastenings, 460
National Design Specification for Wood Construction, 371, 388
National Easter Seal Society, 26
National Electrical Code (NEC), 1070, 1076, 1078–1081, 1083–1086, 1089
National Electrical Manufacturers Association (NEMA), 909
National Energy Plan, 24
National Fire Codes, 15
National Fire Protection Association (NFPA), 15, 19, 24, 26, 412, 886–889, 964, 1058, 1070
National Forest Products Association, *see* American Forest and Paper Association (AF&PA)
National Grading Rule Committee (NGRC), 389
National Grading Rule (lumber), 386, 389
National Hardwood Lumber Association (NHLA), 390
National Institute of Building Sciences (NIBS), 25
National Institute of Standards and Technology (NIST), 24, 25
National Oak Flooring Manufacturers Association (NOFMA), 800, 802
National Plumbing Code, 983
National Primary Drinking Water Standards, 983
National Roofing Contractors Association (NRCA), 514, 521, 522
National Sanitation Foundation (NSF), 995, 997
National Standard Plumbing Code, 982
National Terrazzo and Mosaic Association (NTMA), 773, 774
National Wood Window and Door Manufacturers Association (NWWDA), 632
Natural aging, 334
Natural cleft slate, 277
Natural decay resistance (of wood), 375–376
Natural finishes (aluminum), 342–343
NBS (National Bureau of Standards), 24
NC (normally closed) wired systems, 967
NCSBCS, *see* National Conference of States on Building Codes and Standards
Neat plaster, 698, 710
Neat portland cement, 754
NEC, *see* National Electrical Code
Needle beams, 100
Needlepunched carpet, 836
Negative ions, 80
Negotiation, 10, 11
NEMA (National Electrical Manufacturers Association), 909
Nervi, Pier Luigi, 122
Neutrons, 78, 1064
NFPA, *see* National Fire Protection Association
N grade (plywood), 397
NGRC (National Grading Rule Committee), 389
NHLA (National Hardwood Lumber Association), 390
NIBS (National Institute of Building Sciences), 25
Nickel, 299, 354
Nickel-plated metal, 355
Nickel silver, 350, 354
NIST, *see* National Institute of Standards and Technology
Nitrates, 1009
Nodular cast iron, 301, 304
NOFMA, *see* National Oak Flooring Manufacturers Association
Noise, 919. *See also* Sound control
  background, 928–929
  in plumbing systems, 1024–1026
Noise barriers, 930
Noise reduction, 920–923
  and absorption of sound, 921–923
  and reflection of sound, 920–921
Noise reduction coefficient (NRC), 785, 794, 832, 922
Nominal dimensions (lumber), 383, 388
Nondirectional textured finishes, 351–353
Nonetch cleaning, 351, 353
Nonmetallic conduit, 1079
Nonmetallic sheathed cable, 1077–1078
NO (normally open) wired systems, 967
Nonpressure treatments (for wood preservation), 377–378
Nonslip strips, 28
Nonsubterranean termites, 107–108
Nonveneer plywood panels, 406
Nonvitreous tile, 751
Nonvulcanized elastomers, 525, 527
Nonwood roof sheathing, 439
Normal load duration, 373
Normally closed (NC) wired systems, 967
Normally open (NO) wired systems, 967
Normal weight aggregates, 233
Normal-weight aggregates, 229
Nosings (barrier-free design), 27
Notre Dame du Haut, 122
NRC, *see* Noise reduction coefficient
NRCA, *see* National Roofing Contractors Association
NRCA Steep Roofing Manual, 548
NSF, *see* National Sanitation Foundation
NTMA, *see* National Terrazzo and Mosaic Association
Nucleus, 78
Numbering systems, 39
Nuts, 320–321
NWWDA (National Wood Window and Door Manufacturers Association), 632
Nylon, 471, 838

Oak flooring, 798, 802–803
Octaves, 918
Odor (in water supply), 1002
Off bearing, 231
Ohm's law, 1066–1067

Ohm (unit), 1066
Oil, 1048
Oil-fired water heaters, 1000
Oil gilding, 356
Oil hydraulic elevators, 974–975
holeless, 974
inground, 974
roped, 974
Oil primers, 855, 856
Oil size, 356
Olefin, 839
Oloeresinous paints, 858
Omnibus Trade and Competitiveness Act of 1988, 40
One-and-one-half-story buildings, 5
One-and-Two-Family Dwelling Code, 24, 25, 410, 888
100-mm building module, 68
One-pipe systems (hot water heating), 1047
One-side-bright mill finish (aluminum), 343
One-way slabs, 184
Onyx, 277
Oolitic limestone, 277
Open-die forging, 308
Open excavations, 102
Open hearth furnaces, 301–302
Organic adhesives, 756–757, 766, 768, 770–771
Organic-felt or glass-fiber-felt based roofing products, 519
Organic fiber insulating products, 503
Organic fiber insulation, 500
Orientation of building, 1035
Oriented strand board (OSB), 406
Otis, Elisha Graves, 972
Outdoor design temperatures, 1036
Outlets, electrical, 1079–1083
in attics, 1082
in barrier-free design, 32
in basements, 1082
in bathrooms, 1081–1082
in bedrooms, 1081
in commercial/industrial buildings, 1080
convenience, 1079–1080
in dining areas, 1081
in garages, 1083
in halls, 1082
in kitchens, 1082
in laundry areas, 1082
lighting, 1079
in living/recreation rooms, 1081
on porches/breezeways, 1082
in residences, 1080–1083
special-purpose, 1080
on terraces/patios, 1083
in utility rooms, 1082
Ovens:
accessibility of, 895
microwave, 896, 897
wall, 896
Overlaid plywood, 398
Overload protection devices, 1073, 1075–1076
Overtones, 918
Overturning forces, 410
Owners, building, 3

Package dyeing, 840
Padding, carpet, 843–844
Padlocks, 657
Paint(s), 88, 854–872
for aluminum, 345–346
application of, 871–872
coatings vs., 854
conditioning and mixing of, 865–867
exterior finishing, 864–865
exterior primers, 856–857
exterior priming, 863–864
exterior topcoat, 858–859
on interior floors, 862–863
interior primers, 855–856
interior topcoat, 857–858
interior trim and paneling, 862
interior walls and ceilings, 861–862
lead in, 354
opaque coatings, 859–860
primers, 855–857
repair of surfaces for, 871
storage and stocking of, 865
straining, 866
surface preparation for, 867–871
thinning, 866–867
tinting, 866
topcoats, 857–859
transparent finishes, 860–861
vapor pressure behind, 487
as vapor retarders, 491
zinc in, 354
Paintability of wood, 369
Painting, SI units for, 49
Paint removers, 869
Palladiana toppings, 774
Panels, architectural precast concrete, 193–194
Panel doors, 626–627
Panel grades (plywood), 396, 397, 400
Paneling, finish carpentry for, 462, 463
Paneling plywoods, 398
Paneling varnishes, 860
Panelized assemblies, 443–446
Panelized floor and ceiling assemblies, 444, 445
Panelized roof and ceiling assemblies, 445, 446
Panelized wall assemblies, 445, 446
Panel wood doors, 624
Paper, multilayered forms, 129
Parallels of latitude, 74
Parging (masonry walls), 246, 270, 272–273
Parking:
in barrier-free design, 28–29
lighting for, 1092
Parquet flooring, 804–805
Particleboard:
panels, 406
solid core doors, 626
as underlayment for resilient flooring, 817
Particleboard core plywood, 403–404
Partitions:
gypsum plaster finished, 712
self-supporting gypsum, 742–744
Partition blocks, 235
Passage space, 29
Passageways, in barrier-free design, 30, 33
Passenger elevators, 975–976
dispatch system for, 976
doors of, 975–976
features of, 975
finishes for, 975
Passenger loading zones, 29
Passive solar cooling, 1055
Passive solar heating, 1051–1055
Patent leaf (gilding), 356
Paths, access, 29
Patina, 353
Patios, outlets for, 1083
Patterns:
ashlar, 279
brick flooring, 273
wood, 404
Pattern bonds (masonry walls), 244–246
Patterned finishes, 353
Patterned glass, 669–670
Patterned lumber, 388
Paved areas, shading for, 112
Paver tile, 752, 754
Paving:
brick, 227–228, 273, 289
SI units for, 49
stone for, 280
Payment, application for, 12
PB (polybutylene) pipe, 998
PC, *see* Polycarbonates
PCA (Portland Cement Association), 155
PE, *see* Polyethylene
Pedestrian and light-traffic paving brick, 227–228
Pei, I. M., 122
Pelletizing (ores), 298
Penetration tests, soil, 96
Penta (preservative), 377
PE pipe, *see* Polyethylene pipe
Perceptual disabilities, 33
Percolation tests, 1011–1012
Performance codes, 19, 22
Performance-rated panels (plywood), 397–398
Perimeter forced-air systems, 1039
Perimeter insulation, 503–504

Period, units of, 59
Periodic kilns, 218
Periodic number, 80
Periodic table, 80
Periodic time, units of, 59
Perlite, 500, 501, 502, 503, 701
Perm, 483
Permanent forms (concrete), 129
Permanent slate, 547
Permanent Wood Foundations: Design and Construction Guide, 408
Permeability, 483
of EIFS, 512
units of, 58
Permeance, 58, 483, 484
Pervious soil, 1011
PHCC (Plumbing-Heating-Cooling Contractors Association), 982
Phenolic-based foam, 502
Phenolic insulating products, 503
Phenolic insulation, 500
Phenolics, 473
Phosphates, 1009
Photoelectric detectors, 887, 888, 968
Photovoltaic cells, 1065
Physically handicapped persons, 26
PIB, 527
Pickling, 310, 318–319, 869
Piece dyeing, 840
Piers, 171
cast-in-place structural concrete, 181
in light wood-framed buildings, 413
masonry, 253–254
Pig iron, 297, 299, 300, 301, 304
Pigments:
for coloring mortar, 208
in plastics, 468
Pigmented sealers, 852
Pigmented stains, 860
Pigs (ingots), 300
Pigtailing, 1076
Pilasters:
cast-in-place structural concrete, 181
masonry, 253–254
Pilaster blocks, 236
Piles:
for excavation support, 102–103
as foundation support, 170–172
for underpinning, 101
Pile caps, 171
Pile density (of carpet), 842
Pile drivers, 171
Pile foundations, 168
Pine flooring, 798, 803
Pipes/piping:
for drainage systems, 1007–1009
embedded, 1050–1051
freezing of, 993
openings in concrete for, 142
for plumbing systems, 993, 995–998
water, 983
wrought steel, 315
Pipe columns, in light wood-framed buildings, 413
Piping (of sound), 923
PI (plasticity index), 94
Pits, 102
Pitch (of carpet), 836–837
Pitch pockets (in wood), 365, 384
Pitch (roofs), 532–533
Pitch (sound), 918
Pittsburgh process (glassmaking), 665
Place of beginning (P.O.B.), 73
Plain-sawn lumber, 385
Plane angle(s), 45
conversion factors for, 66
SI units for, 61
units of, 51
Planing mills, 382
Plank and beam framing systems, 447–452
floors, 448–452
roofs, 449–452
Planks, in light wood-framed buildings, 413
Plank wood flooring, 803–804
Plants, 112
fire-resistant, 886, 887
hardiness zone map, 113
landscaping considerations, 112–115
for noise control, 930
Plant-cast precast concrete elements, 189–190
Planting strips, 29
Plaster, 696–702
admixtures in, 702
aggregates in, 700–702
backing paper for, 708
bases for, 704–709
on concrete and masonry, 705–706
and fireproofing, 599
flashing and, 709
gypsum lath, 708–709
gypsum plaster products, 696–698
installation of, 708
lime in, 701–702
metal accessories for, 709
metal reinforcement for, 706–708
portland cement plaster, 698–700
sound absorption by, 923
supporting construction, 703–704
water in, 702
Plaster guards, 620
Plastering, SI units for, 49
Plaster of paris, 696
Plastic(s), 84, 85, 86, 87, 464–474
ASTM abbreviations for, 470
composition of, 465–468
corrosion resistance of, 469
creep, 469
fire resistance of, 469–470
glazing with, 678
modifiers in, 467–468
strength and stiffness of, 468
thermal properties of, 469
thermoplastics, 466–467, 470–472
thermosets, 467, 472–474
toughness and hardness of, 468–469
Plastic cement, 700
Plastic concrete, 151
Plastic deformation, 83
Plastic forms, 129
Plastic glazing materials, 675
Plasticity index (PI), 94
Plasticity (of soils), 94
Plasticizers, 467
Plastic-laminate clad cabinets, 909
Plastic-laminate countertops, 910
Plastic piping, 997–998, 1008
Plastic roofing cement, 518–519
Plate aluminum, 340
Platform framing, 412, 417, 418
floor construction in, 422
wood floor joists in, 417
wood stud wall assemblies in, 425–426
Plats, 76
Plolyolefin sheets, 438
Plucked stone finish, 280
Plug covers, 1075
Plugs, polarized, 1075
Plumb, 260
Plumbing and plumbing systems, 982–1026
with air plenum systems, 1042–1043
bathtubs, 987–988
cleanouts, 1007
codes/laws/standards for, 982–983
drainage in, 1002–1024
drinking fountains, 989
faucets, 985
fixtures in, 983–990
framing for, 424, 425
heaters, water, 999–1001
laundry trays, 985
lavatories, 984
National Standard Plumbing Code, 23
pipe materials for, 995–998, 1007–1008
pressure regulation in, 998–999
sanitary drainage, 1002–1009
sewage disposal in, 1009–1022
showers, 988–989
sinks, 984–985
SI units for, 49
sizing for, 1008–1009
and sound control, 934–935
sound control in, 1024–1026
storm drainage, 1003, 1022–1024
supply/drainage separation in, 989–990
treatment, water, 1001–1002
uniform plumbing and uniform mechanical codes, 23–24
valves in, 999
vents, 1006–1007

water closets, 985–987
water supply in, 990–1002
Plumbing-Heating-Cooling Contractors Association (PHCC), 982
Plumbing walls, 992
Ply:
in built-up roofing, 523
in plywood, 399
Ply sheets (built-up roofing systems), 519
Plywood, 391–406
combination subfloor and underlayment panels, 433
concrete form panels, 398
construction plywood, 391–401
for corner bracing in walls, 437
decorative plywood, 401–406
exposure durability of, 396
general classifications of, 404–406
grades of, 394–395, 396–397
guide to the use of, 394–395
hardwood, 401–406
manufacture of, 391–392, 401–404
overlaid plywood, 398
paneling, 398
performance -rated panels, 397–398
physical/mechanical properties of, 399–401
preservative treatment for, 409–410
quality standards for, 393
sizes of, 399
Sturd-I-Floor, 433–435
textured siding panels, 398
tongue-and-groove panels, 398
as underlayment for resilient flooring, 817
veneer grades, 396–397
wall sheathing, 436–437
wood species for, 396
Plywood paneling, sound absorption by, 923
Plywood sheathing, 439
Plywood siding, 588
Plywood subflooring, 431–432
Plywood underlayment, 432–433
Plywood walls, 759–760
PMMA (polymethylmethacrylate), 471
PMR, *see* Protected membrane roof systems
Pneumatic processes (steelmaking), 302
P.O.B. (place of beginning), 73
Point-bearing piles, 171
Pointing (masonry), 275
Point noise source, 928
Polarized plugs, 1075
Pole Building Design (Donald Patterson), 456, 460
Pole construction, *see* Wood-pole construction
Pole-frame buildings, 458, 459–460
Pole-platform buildings, 458, 460–461
Polished stone finish, 280
Polishing, 343
Pollio, Marcus Vitruvius, 120–121
Polybutylene (PB) pipe, 998
Polycarbonates (PC), 471–472
Polyester, 473, 839
Polyethylene (PE), 438, 472, 998
Polyisocyanurate foam, 500, 502
Polyisocyanurate insulating products, 503
Polymers, 466
Polymerization, 465
Polymer-modified bitumen, 527
Polymethylmethacrylate (PMMA), 471
Polypropylene olefin, 839
Polypropylene (PP), 472
Polystyrene insulating products, 503
Polystyrene (PS), 472, 499–500
Polysulfide sealants, 603
Polytetrafluoroethylene (PTFE), 471
Polyurethane foam, 527
Polyurethane insulating products, 503
Polyurethane (UP), 473, 860–861
on floors, 847
as insulation, 500
for waterproofing, 496
Polyvinyl acetate (PVAC), 472
Polyvinyl acetate underlayment, 818–819
Polyvinyl butyral (PVB), 472
Polyvinyl chloride (PVC), 320, 472, 998
for drainage systems, 1008
in vinyl sheet and tile, 811
Polyvinyl fluoride (PVF), 320
Ponding, 165
Pope, Alexander, 966
Porcelain enamels, 320
Porches:
lighting for, 1091–1092
outlets on, 1082
Porosity (of wood), 369
Portable fire extinguishers, 889
Portland blast-furnace slag cement, 125
Portland cement, 122, 759
aggregates, 701
amount of, in concrete mixing, 149–150
in concrete masonry units, 228
lime-mortars and grouts, 206
measuring, 153
mortar, 206, 754, 755, 767–768
neat, 754
origin of, 121
paints, 857
for plaster work, 698–701
ratio of water to, 146
for tile floors, 764–766
types and uses of, 124
types of, 123–125
Portland Cement Association (PCA), 155
Portland cement plaster, 698–700
application of, 714–715
crack control, 717–719
curing, 716
mixes, 713–714
surface finishes, 719
Portland cement powder paints, 859
Portland-pozzolan cement, 125
Positive ions, 80
Posts, in light wood-framed buildings, 413–414
Post and beam framing systems, 447
Postconstruction services, 2
Post-tensioning:
in beams, 183–184
of flat-plate concrete slabs, 188
Potable water, 990
Potash (for glassmaking), 663–664
Potential difference, 57, 1065
Poured-in-place gypsum concrete roof decks, 196–197
Pouring insulation, 501
Powder-actuated fastenings, 142, 321
Power, 45
conversion factors for, 67
SI units for, 63
sound, 918
units of, 54
Power density, 55
Powers of 10, 39
Pozzolan, 120, 125
PP (polypropylene), 472
Preassembled components for wood construction, 439–445
box beams, 441, 443
elements used in, 440–443
floor and ceiling assemblies, 444, 445
panelized assemblies, 443–446
preframed panels, 441, 443
rigid frames, 441
roof and ceiling assemblies, 445, 446
roof units, 443
sandwich panels, 440–442
stressed-skin panels, 440–442
wall assemblies, 445, 446
Precast concrete, 189–194
plant-cast units, 189–190
prestressing/post-tensioning, 194
site-cast units, 189
structural precast concrete units, 190–194
Precast concrete piles, 172
Precipitation, 97, 270
annual rates of, 240
protection of masonry construction from, 274
Precipitation hardening steel, 308
Preconstruction conferences, 11
Predecorated wallboard, 721
Predesign services, 2
Preengineered buildings, 330
Prefabricated doors, 631
Prefabricated formwork, 130
Prefabricated structural insulating roof decks, 197
Prefabrication, 6–7
Preferred building dimensions, 68

Preformed foam sealants, 603
Preformed metal roofing, 595–596
Preframed panels, 441, 443
Preframed panel wall assemblies, 426
Prehydration, 209
Premium grade (finish carpentry), 462
Prequalified bidders, 10
Prescription codes, 19
Preservative treatments (wood), 377–378, 408
President's Committee on Employment of the Handicapped, 26
Pressing (copper alloys), 350
Pressure, 45, 63
  conversion factors for, 67
  sound, 918
  units of, 53, 54
  water, 994, 998–999
Pressure treatments (wood), 377–378
Pressure-valve water closets, 986
Prestressed concrete, admixture in, 128
Prestressing, 121, 122, 184, 190, 194
Primary treatment (septic tank systems), 1011
Prime design professional, 3
Prime grade lumber, 388
Prime meridian, 74
Primers, 855–857, 870
  concrete, 857
  exterior, 856–857
  interior, 855–856
  masonry, 856, 857
  metal, 856–857
  for roofs, 518, 519
  tintable, 852
  wall, 855
  for waterproofing materials, 497
  wood, 855–856
Primer-sealers, 855
Prime windows, 640–644
Priming, exterior, 863–864
Principal meridians, 74
Print dyeing, 840
Private industry standards, 13–15
  standards-setting and testing agencies and, 14–15
  trade associations, 13
Private water supply systems, 991
Procedures for the Development of Voluntary Product Standards, 16
Product descriptions/literature, 12
Product Standards (PS), 13, 15
Projected windows, 641
Projection welding, 323
Project manuals, 8–10
  bidding requirements in, 9
  contents of, 8
  contracting requirements in, 9
  goal of, 8
  guide specifications in, 9–10
  MasterFormat for, 8–9
  specification in, 9–10
Prolongations, 76
Proposal requests, 11–12
Protected membrane (PMR) roof systems, 494–495, 513
Protection course, 497
Protons, 78, 1064
Protruding objects, 28–33
PS, *see* Polystyrene; Product Standards
Psychrometer, 1028
Psychrometrics, 481–482
PTFE (polytetrafluoroethylene), 471
Public sewage disposal systems, 1009–1010
Puddling, 157
Pueblo Indians, 1051–1052
Pugging, 216
Pug mill, 214
Pump hose, 156
Punching shear, 134
Punch list, 12
Pure iron, 296
PVAC (polyvinyl acetate), 472
PVB (polyvinyl butyral), 472
PVC, *see* Polyvinyl chloride
PVF (polyvinyl fluoride), 320

Quality, tone, 918
Quality standards:
  aluminum windows and sliding glass doors, 640
  for flush doors, 632
  for stile-and-rail doors, 632
  for wood doors, 632–634
Quantity, 39
Quarrying, 277.278
Quarry tile, 752
Quarter-quarters, 75
Quarter-sawn lumber, 385
Quarter sections, 75
Quarter slicing (veneer), 402
Quartz rock, 299
Quicklime, 207, 702
Quick-setting asphalt adhesive cement, 518, 519

Raceways, 1078–1080, 1084–1085
Racking resistance (of plywood), 399–400
Radian, 44
Radiant floors, 1047
Radiant panels, 1047, 1048
Radiant-type electric heaters, 1048
Radiation, 1027, 1035
Radiators, 1047
Radio, 1093
Radio-frequency wireless alarm systems, 967
Rafters:
  conversion diagram for, 430
  in joist-and-rafter ceilings, 427–430
  in joist ceilings, 430–431
  spans for, 429
Raft foundations, 170
Rails (finish carpentry), 462
Rain screen principle, 684
Raked joints, 269
Rakers, 103–104
Ramps (in barrier-free design), 30, 35
Random ashlar pattern, 279
Random-matched V-grooved panels, 405
Random matching (veneer), 402-403
Ranges, 28, 895–896
Range lines, 74
Range number, 74
Rationalization, 67–68
Rays (wood), 365
Ready-mixed concrete, 123, 154
Ready-mixed plaster, 698, 710
Recorded plats, 76, 77
Recreational rooms, lighting in, 1090
Rectangular system of survey, 74–76
Recycled aluminum, 334
Red brass, 349
Red lead primers, 856
Redwood Inspection Service, 375–376
Redwood region, 387
Reference numbers, 74
Reference standards, *see* Standards
Reflection, sound, 920–921
Reflective insulation, 502
Reflectivity, 1027
Reflector, 1088
Reflooring, 831
Reforestation, 382–383
Refractories, 299
Refractory pots, 664
Refrigeration, 1049
Refrigerators, 895–897
Refusal, 171
Registers (air plenum systems), 1044
Reglets, 143
Regular core wallboard, 721
Regulation, building, 18
Regulators, 3
Rehabilitation Act of 1973, 26
Reinforced concrete, 121, 122, 134–140
  design/placement, 136, 139–140
  for foundation wells, 180
  materials for, 135–137
Reinforced footings, 170
Reinforced hollow masonry, 244
Reinforced masonry walls, 238, 244, 264
Reinforced plastics, 469
Reinforced solid (or grouted) masonry, 244
Reinforcement:
  SI units for, 49
  structural, 135
  temperature, 135
Reinforcing agents (in plastics), 468
Reinforcing bars, 135–136, 139

Reinforcing deformed bars, 135–136
Reinforcing steel, 136–138, 140, 155
Reinsulation of existing buildings, 508–511
Relative density (of soils), 94, 96, 97
Relative humidity (RH), 63, 479–481, 1028–1030
and condensation, 484, 485
and equilibrium moisture content of wood, 367
Reluctance, 58
Reroofing, 545–547
with mineral-fiber-cement shingles, 557–558
with wood shingles, 568–569
Reshores, 133
Reshoring, 185
Residental (R) grade windows, 640
Residential appliances, 894–898
air-conditioning units, 898
in barrier-free design, 28
circuits for, 1083
clothes dryers, 898
clothes washers, 897, 898
cooktops, 896
dishwashers, 896–898
and energy conservation, 1094–1095
energy efficiency standards for, 894–895
exhaust hoods, 896
freezers, 896, 897
general standards for, 895
microwave ovens, 896, 897
ranges, 895–896
refrigerators, 896, 897
selection of, 895
for use by handicapped persons, 895
wall ovens, 896
Residential buildings:
lighting in, 1089–1091
outlets in, 1080–1083
service power requirements for, 1070
Residual elements, 301
Resilient channels (gypsum board systems), 738
Resilient flooring, 811–832
accessory materials for, 815–817
adhesives for, 819–821
alkali resistance of, 814
cleaning, 831–832
cork tile, 812, 813
durability of, 815
feature strips for, 817
grease resistance of, 814
indentation resistance of, 814–815
job conditions, 821
lining felt, 819
linoleum, 812, 813
maintenance of, 815
moisture resistance of, 813–814
over existing finished floors, 831
paints for, 863
quietness rating of, 815
reducing strips for, 817
rubber tile, 813
selection criteria for, 813–815
sheet flooring installation, 823–827
slip resistance of, 815
stair treads, risers, and nosings, 830–831
stair treads/risers/nosings for, 817
subflooring for, 431, 432
sunlight, resistance to, 815
supporting construction, 821–823
thresholds for, 817
tile flooring installation, 827–830
underlayment for, 433
underlayments, 817–819
vinyl composition tile, 811, 812
vinyl sheet and tile, 811–813
wall bases for, 815–817, 830
Resinous binders (terrazzo), 773
Resinous flooring, 846–849
installation of, 848–849
materials in, 847–848
systems, 846–847
Resistance:
electrical, 1066
to heat flow, 1032
units of, 57, 60
Resistance welding, 322, 348
Resistivity, 57, 58, 87
Resist printing, 840
Restrictions (on plats), 76, 77
Retarder-faced insulation, 490
Retarders (concrete), 128
Retempering, 154, 211
Return (L) corner blocks, 235
Reverberation, sound, 920–921
Reverberation time, 920–921
Reverse return systems (hot water systems), 1047
Revolving doors, 637
RH, *see* Relative humidity
Ribbed slabs, 185–186
Ribbon stock, 547
Rib lath, 707
Ridges:
types of, 548
wood shakes for, 568–569
wood shingles for, 568
Ridge shingles, 545, 556–557
Ridge-supported rafter assemblies, 431
Rift-cut slicing (veneer), 402
Rigid frames, 441
Rigid insulation, 499–500, 506
Rigid metal conduit, 1078
Rigid pipe, 997
Rigid space frames, 453
Rigid steel anchors, 212
Rim doors, 627
Riprap, 107
Risers, 991
Riveting, 320, 346
Rock, 101
Rock caissons, 172
Rock-face stone finish, 280
Rocky slopes, construction on, 460–461
Rod anchors, 213
Roll blinds, 911
Rolled vinyl weatherstripping, 646
Roll forming:
aluminum, 342
copper alloys, 350
Rolling (copper alloys), 350
Roll roofing:
application of, 522–524
materials for, 519, 521–522
winders, 518
Roll shades, 913
Rolok (masonry), 260
Roman numeral system, 39
Romans, ancient, 120, 202–203, 661, 749, 982
Roofs and roofing:
asphalt, 536–547, 865
fire resistance of, 521
flashing for, 592–594
framing, of projections, 423, 424
heat transmission through, 1035
insulation for, 508
for low-slope roofs, 513–531
lumber sheathing for, 439
metal, 594–599, 865
mineral-fiber-cement shingles, 552–558
minimum recommended slope for, 534
multi-story buildings, 4
plank and beam framing systems, 449–452
plywood sheathing for, 439
precast concrete elements for, 191–192
single-story buildings, 4
SI units for, 49
slate, 547–549
for steep roofs, 531–570
in stud systems, 426
terminology associated with, 532–533
terminology for, 532–533, 533
tiles, 549–552
types of sloped roofs, 532
wood shingles and shakes, 558–570
Roof and ceiling assemblies, 426–431. *See also* Roofs and roofing
joist-and-rafter roofs, 427, 430
joist roofs, 430–431
lumber for, 427
panelized, 445, 446
vapor movement and control in, 487–488
vapor retarder installation in, 495
ventilation of, 492–495
Roof boards, 196
Roof coatings, 518, 519, 860

Roof decks:
gypsum concrete, 196–197
prefabricated structural insulating, 197
steel-framed buildings, 328–329
Roof drainage, 1023–1024
Roofing accessories, 593
Roofing mastic, 518
Roofing specialties, 593–594
Roofing tape, 519
Roof louvers, 493
Roof ponds, 1054–1055
Roof sheathing, 435, 439
Roof slope, 532–533
Roof trusses, 453–454
Roof units, preassembled, 443, 444
Rooms:
in barrier-free design, 33, 36–37
heat flow and organization of, 1035
Room temperature precipitation, 334
Roped oil hydraulic elevators, 974
Rotary calciner, 696–697
Rotary cut veneer, 391, 401
Rotational frequency, 52
Rotation forces, 410
Rotten egg odor (in water), 1002
Rough arches, 253
Rough form finish, 163
Rough hardwood dimension, 390
Roughing in, 992
Rough lumber, 388
Roundwire Wilton, 834
Rubbed stone finish, 280
Rubber base, 830
Rubber cement, synthetic, 821
Rubberized asphalt waterproofing, 497
Rubber tile, 812, 813
Rubble (stone setting), 280
Rudolph, Paul, 122
Rugs, and barrier-free design, 27
Running bond pattern, 244, 245, 262
Running slope, 30
Running trim, 462
Rusticated stone finish, 280
R-values, 499, 1032, 1034, 1036

Saarinen, Eero, 122
Saddle ridges, 548
SAE, *see* Society of Automotive Engineers
Safety:
in barrier-free design, 27–28
codes related to, 24
during demolition, 99
elevators, 973
fire safety, 886–890
and formwork for concrete, 129
lightning protection, 965
UL labels for, 14–15
Safety glazing, 27
Safety zones, 29
Salt bath process, 348
Saltbox architecture, 1052
Salt(s), 81, 205, 219, 275
Sand, 92, 94, 110, 299, 759
in gypsum plaster, 700–701
for pole construction, 459
in portland cement, 125, 126
wetness of, 150, 151
Sandblasted stone finish, 280
Sandblasting, 343, 667–668
Sand castings (aluminum), 341
Sand-lime mortar, 206
Sand plates, 136
Sand-portland cement grout, 757
Sand-rubbed slate, 277
Sandstruck bricks, 216, 217
Sandwich panels, 440–442
Sanitary drainage systems, 1002–1009
cleanouts in, 1007
components of, 1003–1004
drainpiping in, 1005–1007
gravity flow in, 1004–1005
piping for, 1007–1008
sizing for, 1008–1009
Sanitary stops, 620, 622
Sapwood, 365
Sash blocks, 235
Sash components, 649
Sash doors, 626–627
Sash wood doors, 624
Satin finishing, 343
Saturated air, 479, 483
Saturation, 517, 1012
Saturation point, 1028
Sawcut stone finish, 280
Sawmillnails, 412
Sawmills, 379–380
SBCCI, *see* Southern Building Code Congress International
Scaffolding, 972
Scaling, 163
Scarf-jointed plywood panels, 399
Schedule of values, 12
Scratch brushing, 343
Scratch coat, 710, 714–715, 768, 769
Screeding, 158–159
Screen doors, 627–628
Screening, 649
Screws (for gypsum board systems), 726–727, 731
Screw ties, 131
Scribing, 824
SCR (silicon-controlled rectifiers), 974
S/C (selective collective) control, 976
Sculptured stone finish, 280
Scum, 1019
Scumming, 275, 276
Scuppers, 593, 1023
S-DRY lumber, 367
Seals, 14
Sealants, 601–607, 759
accessory materials, 604
building sealants, 602–604
in door frames, 622–623
glazing, 676
installation of, 606–607
and joint design, 604
paving joint sealants, 604
selection of, 605–606
standards governing, 602
Sealant-filled joints (masonry), 269
Sealers:
clear, 860
penetrating, 852–853
pigmented, 852
Sealing curing compounds, 165
Seam cutting (resilient flooring), 824–825
Seamless pipes and tubing, 316
Seam welding, 323, 348
Seasoned stone, 279
Seasoning of lumber, 380–382, 385–386
Second, 44
Secondary bonding, 82
Secondary treatment (septic tank systems), 1011
Second moment of area, 51
Sections, townships, 74
Security, *see* Intrusion prevention and detection
Security door hardware, 637
Security lighting, external, 1092
Security locks, 657
Sedimentary rock, 94
Seed-tree cutting (forest conservation), 382–383
Seepage pit, 1013, 1016–1017
Seep holes, 487
Select Decking, 384
Select grade lumber, 383–384, 388
Selective collective (S/C) control, 976
Selective cutting (forest conservation), 382
Select Structural lumber, 384
Self-adhering butyl waterproofing, 497
Self-furring metal lath, 706–707
Self inductance, 58
Semifabricated dimension lumber, 390
Semigloss enamel, 857
Semigloss finishes, 857
Semirigid mineral fiber insulation boards, 501
Semivitreous tile, 751
SEM (Single European Market), 17
Sensible heat, 478, 1027
Sensible heat gain, 1036
Separation joints (concrete), 144–145
Septic tank systems, 1010–1022
inspection of, 1022
soil absorption systems for, 1012–1018
soil suitability for, 1011–1012
tanks for, 1018–1022
Serial distribution, 1014
Series loop systems, 1046–1047
Service drop, 1069
Service entrance (electrical systems), 1072

Service entrance panel, 1072
Service temperature (plastics), 469
Setting:
  concrete, 123
  stone, 280
Settlement, 94–95, 167–168
Severe exposure, 270
Sewage, 1002
Sewage disposal systems, 1009–1022
  public systems, 1009–1010
  septic tank systems, 1010–1022
  standards for, 983
Sewerage, 1003
Sewers, 1003
Sexagesimal system, 39
SFPA, *see* Southern Forest Products Association
S-GRN lumber, 367
Shade (light):
  grasses that grow in, 1114
  trees for, 112
Shades (window), 911
Shadowal, 234
Shadowing (gypsum board systems), 745–746
Shaft walls, 743–744
Shakes, 365, 384. *See also* Wood shingles and shakes
  application of, 579–580
  cedar, 578
  preparation for, 579
Shales, 214
Shapes, building, 3–9
Shear forming (steel), 317
Shear walls (pole construction), 461
Sheathing:
  for asphalt shingles, 537
  condensation concealed behind, 487
  for EIFS, 512
  gypsum, 721
  for mineral-fiber-cement shingles, 553
  for roll roofing, 521
  for slate roofing, 547
  for steep roofing, 533–534
  for tile roofing, 550–551
  for wood shingles and shakes, 565
Sheathing paper, 439, 579, 586
Shed roofs, 532
Sheet aluminum, 340–341
Sheet membrane waterproofing, 497
Sheet Metal and Air Conditioning Contractors National Association (SMACNA), 421, 1036, 1039, 1057
Sheet metal, 589–590, 594
Sheet piles, 102–104
Sheet steel, 307, 311, 318
Shelf-angle insert, 141
Shellac, 860
Shell structures for, 121
Shelter-wood cutting (forest conservation), 382–383
Shelves:
  construction of, 462–463
  wood-finished cabinets, 909
Shielded metal arc welding, 322
Shingles, 533, 571. *See also* Wood shingles and shakes
  cedar, 577–578
  mineral-fiber-cement, 572–577
  terminology related to, 533
  wood, 577–582
Shingle butt, 533, 571
Ship-lapped lumber, 388
Shop lumber, 388
Shores, 131
Shoring, 103
Shot blasting, 343
Shot blast nondirectional textured finishes, 353
Shotcrete, 157
Showers, 27, 988–989
  in barrier-free design, 28, 33, 34, 37
  stalls, shower, 988–989
  tile in, 762
  tub-showers, 988
Shrinkage:
  in clay masonry units, 220
  of concrete, 143–144
  in concrete masonry units, 233
  of framing in wood floor systems, 421–422
  in mortar, 205
  of wood, 366, 368
  and wood girder-joist connections, 416
Shrubs:
  fire-resistant, 887
  planting soil for, 116
  and soil moisture content, 97
Shutoff cock, 999
Shutters, 911, 913
SI, *see* International System of Units; Silicones
Side construction, 222
Side jambs (windows), 651
Sidelap, 533
Sidelights, barrier-free design for, 27
Side wall flashing (mineral-fiber-cement shingles), 554
Siding, exterior, 570–588
  aluminum and vinyl siding, 582–585
  condensation concealed behind, 487
  and equilibrium moisture content of wood, 367
  mineral-fiber-cement shingles, 572–577
  plywood and hardboard panel siding, 588
  wood and hardboard board siding, 585–588
Signage, 31–33
  in barrier-free design, 35
  in elevators, 976
Silane sealers, 852–853
Silencers, 620
Silica, 663
Silicon, 299
Silicon-controlled rectifiers (SCR), 974
Silicon emulsion acrylic polymer coatings, 851
Silicone rubber grout, 757–758
Silicone sealants, 603
Silicones (SI), 473–474
Sills (window), 504, 651
Sill flashing, 270, 271
Sill plates:
  in wood frame construction, 416–417
  of wood-joist floor systems, 410
Siloxane sealers, 852
SIL (sound intensity level), 918
Silt(s), 92, 94, 95, 110
Silvering, 668
Simple beams, 182–183
Simplified Practice Recommendations (SPR), 16
Sinaqua Indians, 1051, 1054
Single coursing, 571, 580
Single cutting, 825
Single-duct forced-air systems, 1039
Single European Market (SEM), 17
Single-family homes:
  foundations for, 180, 252
  soil investigation for, 96
Single-hung windows, 649
Single-lever faucets, 985
Single-phase electrical service, 1068–1069
Single-ply roofing systems, 525, 527–531
  application of, 527–531
  built-up vs., 513–514
  coatings and surfacings for, 527
  insulation for, 514–515
  materials for, 525, 527
Single-story structures, 4, 252
Single tees, 191, 192
Single-track windows, 654
Single-withe concrete masonry walls, 505
Sinks, 35, 37, 984–985
Sintering (ores), 298
Siphon action bowl (water closet), 987
Siphon breaker, 990
Siphon jet bowl (water closet), 987
Site construction, 92–116
  earthwork, 99–109
  lawns/landscaping, 112–116
  site preparation, 98–99
  soils, 92–98
  surface-water/groundwater problems, 109–112
Site planning:
  for barrier-free design, 28
  for sound control, 929–931
Site preparation:
  for pole construction, 457–45
  for treated wood foundation, 407
Skein dyeing, 840

Skin effect, 941
Skip (lumber), 385
Sky-lobby systems, 977–978
Slabs:
  coarse aggregate size in, 126–127
  concrete, 121
  reshores for, 133
  steel, 311
Slab band, 187, 188
Slab beds, 105
Slabs on grade, 134
  control joints for, 145
  forms for, 131
  and foundations, 167, 168
  insulation for, 503–504
  preparation of, 155
  ventilation of, 495
Slab-on-grade construction, 173–180
  base course for, 105, 106
  design of, 173
  flatness/levelness, 179–180
  ground-supported slabs, 173–178
  moisture control, 179
  moisture control for, 110–111
  slab bed for, 174
  structurally supported slabs, 178–179
  and surface condensation, 486
  and termite hazard, 108–109
  thermal control, 179–180
  Type I slabs, 173–176
  Type II slabs, 175–177
  Type III, 177–178
  Type IV slabs, 178–179
Slag, 299
Slag wool insulation, 501
Slaking, 702
Slant nailing, 411
Slash-grained lumber, 385
Slat block flooring, 804–805
Slate, 277
Slate flooring, 807–809
Slate Roofs, 548
Sleeves (concrete), 142
Sliced veneer, 391
Sliding doors, 628–629, 660, 933
Sliding glass doors (SGD), 640, 643, 651
  aluminum, 638–641, 644
  wood, 645–653
Slight exposure, 270
Slip couplings (galvanized pipe), 996
Slip forming, 133
Slip matching (veneer), 402
Slip resistant surfaces, 30
SLMA (Southeastern Lumber Manufacturers Association), 406
Sloped glazing, 679
Sloped roofs, 532–533
Slope limitations:
  pole-frame buildings, 459–462
  for steep roofs, 533–534
  for wood shingles and shakes, 564–565
Slots, anchor, 141
Sludge, 1019
Slump blocks, 235
Slump brick, 234
Slump cone, 152
Slumps, 149–151
Slump test (concrete), 151–152
Slushing (mortar), 266
SMACNA, *see* Sheet Metal and Air Conditioning Contractors National Association
Small buildings:
  foundations for, 180
  plywood use in, 397
  preliminary investigation for, 96
  wood-pole construction for, 456
Smelting, 299
Smoke chamber, fireplace, 257
Smoke control (HVAC systems), 1057
Smoke dampers, 1057
Smoke detectors, 887–889
Smoke doors, 620
Smoke pipes (fireplaces), 258
Smokeshelf, fireplace, 257
Smooth form finish, 163
Smooth rubbed finish, 163
Smooth specular buffed finish, 353
Snap ties, 130
Society of Automotive Engineers (SAE), 306, 349
Socket caissons, 172
Sodium (for glassmaking), 663
Soft conversion of numerical values, 64
Soft-mud process (brick), 216, 219
Softwoods, 364, 383, 386–390
  classifications of, 387–389
  edge-grained lumber, 385
  flat-grained lumber, 385
  glueability of, 375
  major producing regions for, 386–387
  for paneling plywoods, 398
  shrinkage of, 368
  size and grade standardization of, 389–390
  slash-grained lumber, 385
  standards for, 387
  vertical-grained lumber, 385
Softwood flooring, 798–800
Soil(s):
  bearing capacity, 94–96
  classification of, 92–93
  compaction requirements for, 106
  effects of temperature on, 98
  effects of water on, 97–98
  for fill/backfill, 104, 105
  for ground-supported slabs, 174, 175, 177–178
  nonproblem soils, 95
  planting soil, 116
  for pole construction, 459
  problem soils, 95–97
  properties of, 93–94
  and septic tank systems, 1011–1018
  settlement of, 94–95
  for structurally supported slabs, 178
  and subsurface investigation, 95–97
  for Type I slabs, 174
  for Type II slabs, 175
  for Type III slabs, 177–178
  for Type IV slabs, 178–179
Soil absorption systems, 1012–1018
Soil stacks, 1004
Solar cooling systems, 1055
Solar heating systems, 1051–1055
  and building envelope, 1052–1053
  history of, 1051–1052
  and insulation, 502
  and passive solar collection, 1053
  passive systems, 1054–1055
  properties of, 1052
  storage in, 1053–1054
Solar reflective glass blocks, 284
Solar water heaters, 1000–1001
Soldering, 322, 348–349
Solders, 348
Soldiers (masonry), 260
Soldier piles, 102–104
Solenoid valves, 999
Solid angle, 58
Solid blocking, 423
Solid bridging, 423
Solid-core doors, 625–626
Solid flat-slab units, 191, 192
Solid lumber sheathing, 436
Solid masonry walls, 237–240
Solid slabs, 184–185
Solid top blocks, 235
Solution dyeing, 840
Solution heat treatment, 334
Solvent cleaning, 868–869
Sonic detectors, 969
Sound(s), 916–919
  and frequency, 917–918
  levels of, 918–919
  piping of, 923
  timer-controlled, 968
  and wave motion, 917
Sound control, 916–963
  acoustical treatment, 783–797
  background noise, 928–929
  in barrier-free design, 33
  construction recommendations for, 935–949
  in corridors, 946
  design/planning for, 928–935
  doors for, 932–934
  with electrical systems, 1087
  floor/ceiling systems for, 941–945, 949, 957–963
  with gypsum board systems, 737–739
  for HVAC, 1055–1057
  interior planning for, 931–933

plumbing/mechanical/electrical equipment and systems, 934–935
in plumbing systems, 1024–1026
and properties of sound, 916–919
and room acoustics, 919–923
selection criteria for, 946–949
site planning for, 929–931
and sound performance ratings, 949–950
and stairs, 946
and transmission of sound, 923–928
and wall/floor intersections, 945–946
wall systems for, 937–941, 949–956
windows for, 933–934
Sound-deadening board, 738
Sound energy flux, 59
Sounding rods, 96
Sound intensity, units of, 60
Sound intensity level (SIL), 918
Sound isolation, 923
Sound power, units of, 59
Sound pressure level (SPL), 918, 919
Sound-rated doors, 620
Sound transmission:
glazed aluminum curtain walls, 684
of masonry walls, 286–289
Sound Transmission Class (STC), 785, 916, 925–926, 937–940, 945, 949
Southeastern Lumber Manufacturers Association (SLMA), 406
Southern Building Code Congress International (SBCCI), 22–26, 982
Southern Forest Products Association (SFPA), 406, 1041
Southern Pine Council (SPC), 406
Southern Pine Inspection Bureau (SPIB), 367, 800
Southern pine region, 387
Spaced joist construction, 421
Spaced wood girders, 416, 417
Spaced wood sheathing, 436
Space dyeing, 840
Space frames, 330
Space-saving washers/dryers, 898
Spading, 157
Spandrel glass, 672
Span rating (plywood), 397
Span Tables for Joists and Rafters, 418, 427, 431
Spar varnish, 860
SPC (Southern Pine Council), 406
Special construction, 916. *See also* Sound control
intrusion prevention/detection, 965–969
lightning protection, 964–965
Special overlay panels (plywood), 398
Special steep asphalt, 516
Specialty grade plywood panels, 406
Specialty plywoods, 404–405
Specialty Steel Industry of North America (SSINA), 308
Specific acoustic impedance, 60
Specification codes, 19
Specifications, 9–10
definition of, 12
and Federal Specifications, 16
Specific energy, 55
Specific entropy, 55
Specific gravity, 63, 369
Specific heat, 1033–1034, 1053–1054
Specific heat capacity, 55
Specific latent heat, 55
SpecLink, 9
SPECTEXT, 9
Specular as-fabricated finishes, 352
Specular buffed finish, 353
Speed:
conversion factors for, 66
of rotation, 45, 52, 63
units of, 52
SPIB, *see* Southern Pine Inspection Bureau
Spin finishing, 343, 353
Spinning, 350–351
SPL, *see* Sound pressure level
Splits (lumber), 386
Split-face blocks, 235
Split-face stone finish, 280
Split-level buildings, 6
Split-receptacle outlets, 1080
Spot detectors, 967
Spot welding, 323, 348
Spray-applied fireproofing, 600–601
Spread foundations, 168
Spring-tension weatherstripping, 646
Springwood, 365
Sprinkler systems, 1058
Sprinkling (concrete), 165
SPR (Simplified Practice Recommendations), 16
Square (roofing term), 533
SSINA (Specialty Steel Industry of North America), 308
Stabilizers, 468
Stacks, 1004
Stack bond pattern, 245
Stack flashing (asphalt shingles), 542, 543
Stack vents, 494, 1004
Staging, 131
Stainless steel, 88, 307–308, 312
adhesive bonding of, 324
coatings for, 319
doors and frames, 620
for formed doors and frames, 618
mechanical finishes for, 318
for roofs, 596
sheet metal, 589–590
uses of, 333
weldability of, 323
Stains, 376, 384, 860
Stairs, 463–464
in barrier-free design, 31
barrier-free design for, 27
concrete, 188–189
framing openings for, in wood floors, 423
and sound control, 946
Stair nosings, 816, 817, 830–831
Stair treads, 816, 817, 830–831
Stairways:
lighting along, 1091
outlets along, 1082
Staking, 321–322, 347
Stalls, shower, 988–989
Stamping (copper alloys), 350
Standards, 12–18. *See also* Codes
for accessible design, 26–27
aluminum entrances and storefronts, 634
for asphalt shingles, 537
for clay masonry units, 220–222, 224–228
codes vs., 13, 19
for concrete masonry units, 236
for construction plywood, 393
for EIFS, 511
for electrical service, 1070
European, 17–18
federal government, 15–17
for fiberboard sheathing, 437–438
for finish carpentry, 462
for fire resistance of masonry units, 254–255
for insulation, 502, 503
low-slope roofing systems, 519, 520
for lumber quality, 383–384, 386
for metal doors and frames, 618
for moisture content of wood, 367–368
for mortar and grout, 206–211
objectives of, 13
for particleboard, 406
for plastics, 470
plumbing, 982–983
for plywood, 401
for pole preservatives, 458
private industry standards, 13–15
for specialty plywoods, 404–405
for stairs and handrails, 463
steel, 305–307
types of, 13
uses of, 13
for wood preservative treatments, 378
for wood shingles and shakes, 563
Standard bright finish (aluminum), 343
Standard Building Code, 23, 237
Standard contour, 925
Standard door and frame units, 618
Standard hardwood plywood, 405
Standard mineral-surfaced roll roofing, 521, 522, 523
Standard one-side-bright sheet finish (aluminum), 343
Standard Plumbing Code, 982
Standards-setting and testing agencies, 14–15

Standards-setting trade associations, 13
Standing trim, 462
Staples (gypsum board systems), 727
Starter strip (asphalt shingles), 543, 544
States:
  barrier-free design standards, 27
  building codes and standards, 25
Static generation (carpets), 842
Static loads, 167
Statuary bronze, 353
STC, *see* Sound Transmission Class
Steady-state concept, 1033
Steam cleaning, 869
Steam systems, 1048
Steel:
  alloy steels, 306–307
  architectural uses of, 333
  building frame design, 325–330
  carbon steels, 304–305
  cast-steel products, 308, 309
  classification of, 305
  construction systems, 324–332
  design and construction, 320–332
  doors, 333
  erection of steel-framed buildings, 330–332
  fabrication of, 316–317
  fabrication of components, 330
  flat-rolled product classifications, 311–312
  galvanized, 995–996
  high-strength low-alloy steels, 307
  history of, 296–297
  in interior finishes, 333
  joints and connections, 320–324
  manufactured steel products, 308–320
  materials and products, 296–320
  metal coatings on, 355
  in reinforced concrete, 134–135
  stainless and heat-resisting steel, 307–308
  steelmaking, 301–304
  steel structures and construction systems, 324–332
  structural shapes, 312–316
  windows, 333
  wrought-steel products, 308–316
Steel, reinforcing, 136–138
Steel bars, 307
Steel beams, 416
Steel deck, insulation over, 514–515
Steel door frames, 622
Steel doors, 620
Steel floors, paints for, 863
Steel-pipe piles, 172
Steel pipes, 172
Steel posts, in light wood-framed buildings, 413
Steel structures and construction systems, 324–332
  design of, 325–330
  erection of buildings, 330–332
  fabrication for, 330
  joists and joist girders for, 328
  roof and floor decking for, 328–330
  structural frames for, 325–328
  trusses for, 330
Steelworkers, 330
Steelworking, SI units for, 49
Steep asphalt, 516
Steep roof roofing systems, 531–570
  asphalt shingles, 536–547
  cements for, 534
  flashing and caulking for, 535–536
  framing and sheathing, 533–534
  general recommendations, 532–536
  mineral-fiber-cement shingles, 552–558
  nails and fasteners for, 536
  nomenclature, 532–533
  slate roofing, 547–549
  slope limitations, 533
  tile roofing, 549–552
  underlayment for, 534–535
  wood shingles and shakes, 558–570
Stem faucets, 985
Stepped footings, 169
Steps (barrier-free design), 31
Steradian, 44
Stiff-mud process, 214, 216, 219
Stiffness:
  comparison of, 84
  of plastics, 468
  of plywood, 399
Stile-and-rail doors, 626
  extruded aluminum, 635
  fabrication/assembly, 630–631
  quality standards, 632
  wood, 623
Stirrups, 182
Stitching (aluminum), 346–347
Stock dyeing, 840
Stock plywood panels, 405
Stone, 276–283
  anchoring, 281–283
  anchors for, 213
  architectural history of, 276–277
  fastening, 280–283
  finishes for, 280
  granite, 277
  lifting methods for, 278
  limestone, 277
  manufacture of, 277–279
  marble, 277
  patterns for, 279
  paving/flooring, 280
  reinforcements for, 213
  setting, 280
  slate, 277
  ties for, 213
Stone cladding, 280–283
Stone flooring, 806–810
  granite, 807, 809
  grout for, 211
  installation of, 809–810
  limestone, 809
  mortar for, 211
  patterns for, 808
  quartz-containing stone, 810
  slate, 807–809
Stone joint-pointing mortar, 207
Stone-setting mortar, 207
Stops (windows), 651
Storage:
  in barrier-free design, 34, 36
  of concrete masonry units, 231
  of masonry materials, 259
Storefront-type frames, 620
Storm and screen wood doors, 624
Storm doors, 627–628
Storm drainage systems, 1003, 1022–1024
Storm panels, 651
Storm sash, 651
Storm windows and doors, 653–656
  installation/maintenance, 656
  selection of, 654–655
Story pole, 261
STP, *see* Structural test pressure
Strain, 82–86
Strand casting (steel), 302, 304
Strap anchors, 213
Strength:
  of concrete, 146–149
  of concrete masonry units, 236
  and moisture content of wood, 368
  of plastics, 468
Stress, 45, 63, 82–86
  conversion factors for, 67
  units of, 53, 54
Stressed-skin panels, 440–442
Stress grade lumber, 387–388
Stress-rated lumber, 389–390
Stress-strain test, 83–85
Stretchers (masonry), 260
Stretcher level, 618
Stretch forming (aluminum), 341–342
Stretch forming (steel), 316
Stringers:
  definition of, 417
  and truss spacing, 455
Strip lath, 718
Strippable coatings, 346
Strip reinforcing, 737
Strip shingles, 543, 545
Strip steel, 311
Strip wood flooring, 800–803
Struck joints, 269
Structural aluminum shapes, 341
Structural clay facing tile, 226, 227
Structural clay tile, 223, 225–226
  cleaning, 275
  mortar bedding and jointing, 267–268
  storage of, 259
Structural facing tile, 223–224

Structural glued laminated members (gluelams), 453
Structural insulating roof decks, 196
Structural joists and planks, 389
Structural light-framing lumber, 389
Structural lumber, 387–388
Structural panels (plywood), 397
Structural precast concrete units, 190–194
Structural reinforcement, 135
Structural slabs, 184–188
  flat plate slabs, 186–188
  permanent forms, 188
  solid, 184–185
  waffle slabs, 186–188
Structural steels, 87, 307
Structural test pressure (STP), 640, 643
Structural-use panels (plywood), 397
Structure-borne sound transmission, 923–924, 926–928
Stucco, reshingling over, 577
Stucco mesh, 707
Stucco netting, 707–708
Studs, 389
  in balloon framing, 412
  for fiberboard sheathing, 438
  in light wood-framed buildings, 413
  spacing of, 426
Stud adhesives (gypsum board systems), 728
Stud walls:
  construction of, 425–426
  insulation of existing, 509–510
Stud welding, 321
Sturd-I-Floor, 433–435, 440
Subdrainage systems, 111–112
Subflooring, 431–435
  board subflooring, 431
  combination subfloor and underlayment, 433–434
  concrete, 822
  in glued floor systems, 434–436
  near fireplaces, 255
  plywood subflooring, 431–432
  preparation of, for terrazzo, 776
  with resilient flooring, 822–823, 827
  and tile, 759, 767
  underlayment over, 432–433
  in wood joist floor systems, 417, 419
Submerged arc welding, 322, 323
Subpurlins, 196–197
Substantial completion, 12
Substrates:
  concrete, 513
  insulated, 513
  for low-slope roofing, 513
  nailable, 513
  for tile roofing, 550–551
  waterproofing on, 497
Substructure, 166, 167
Subsurface investigations, 95–97
Subterranean termites, 107–108, 376–377
Suction, 219, 259–260
Suction lines, 110
Suction rate, 232
Summerwood, 365
Sump (for poorly drained soils), 408
Sunspaces, 1055
Superconductivity, 1066
Superior grade lumber, 388
Super plasticizers, 128
Superstructure, 166, 412
Superstructure beams, 182
Supplemental services, 2
Supplementary units, SI, 41, 42, 44
Suppliers, 3
Supports:
  for aluminum entrances and storefronts, 634–635
  for metal doors and frames, 619
Support strips, 550–551
Surfaces:
  in barrier-free design, 28, 29, 32, 35
  of roofing products, 517
Surface clays, 214
Surface condensation, 485–486
Surfaced dimension lumber, 390
Surface defects, concrete, 163
Surfaced lumber, 388
Surface finishes (concrete), 161–163
Surface leaf (gilding), 356
Surface repairs, 871
Surface textures:
  aluminum, 343
  concrete masonry units, 233
Surface water, 109–112
Surfacing(s):
  of built-up roofing, 523
  of lumber, 385
  for single-ply roofing systems, 527
Surveys, *see* Land surveys and descriptions
Susceptance, 57
Suspended wood floors, 431
Sweet's Catalog Files, 8
Swelling, 1012
Swinging doors, 628–629
Swirl design (concrete), 162
Switch-back ramp, 30
Switches, 32, 1080, 1089
Symbols, SI, 46–47
Synthetic fiber reinforcements (concrete), 129
Synthetic rubber cement, 821
System grounding, 1085–1086

Tactile objects, 32
Tail joists, 423
TAMAP system, 15
Tanks:
  for septic tank systems, 1018–1022
  water closet, 986
Tape(s), 346
  glazing, 676
  roofing, 519
  sealants, tape, 603
TCA, *see* Tile Council of America
Team, building construction, 3
Teflon, 471
Telephone outlets, 1093–1094
Television, 1093
Temper (aluminum), 336, 337
Temperature(s), 1027
  for concrete mixing/curing, 147, 149
  for curing cement, 165–166
  dew point, 482
  effective temperature, 479–481
  frost point, 482
  and humidity, 479–481
  metric conversions for, 69
  and plant hardiness, 113–115
  service temperature, 469
  SI units for, 62
  units for, 43
Temperature controls (water heaters), 1001
Temperature interval, 45
  conversion factors for, 66
  units of, 55
Temperature reinforcement, 135
Temperature steel, 135
Temperature value, 55
Tempering, 214
  glass, 667
  steel, 310
Temper mill, 318
Temporary forms (concrete), 129
Tendons, 184
Tensile strength, 84
  of aluminum, 338, 339
  of concrete masonry units, 232
  of wood, 369
Tension, 83, 134, 374
Tension floors, 826
Tension parallel to grain, 374
Tension perpendicular to grain, 374
Terminated stops, 620, 622
Termites, 107–109, 411
  air plenum systems, 1041
  damage from, 376–377
  wood-framed buildings, 410–411
Terneplate, 88, 355, 596
Terraces:
  for erosion control, 107
  lighting on, 1092
  outlets for, 1083
Terra-cotta, 214, 224
  cleaning, 275, 276
  colors of, 220
  size variation in, 220
  stiff-mud process for, 216
Terrain, and heat flow, 1035
Terrazzo, 772–783
  application of, 775–783
  binders in, 773

Terrazzo, (*cont'd*)
decorative chips in, 773
existing floors, 781
fine aggregate in, 774
installation methods, 779–781
maintenance of, 782–783
materials in, 772–775
pigments in, 774
portland cement terrazzo with a regular finish, 778–779
strip placement for, 776–778
subflooring for, 431, 432
subfloor preparation, 776
surface finishes for, 781–782
thin-set, 777
underbed for, 776
Testing, soil, 94
Testing and research agencies, 3
Testing standard, 14
Test method standards, 13
Test pits, soil, 96
Textural cues, 32
Textural roofing slate, 547
Texture(s):
aluminum, 343
carpet, 841
of clay masonry units, 220
in concrete masonry units, 233
of roofing products, 518
of wood, 375
Textured plywood siding panels, 398
Thermal and moisture protection, 478–607
building insulation, 499–511
exterior insulation and finish systems, 511–513
fireproofing, 599–601
flashing and sheet metal, 589–594
joint sealing, 601–607
for low-slope roofing, 513–531
metal roofing, 594–599
siding, 570–588
steep roofing, 531–570
waterproofing and dampproofing, 496–498
Thermal break(s):
for aluminum windows and sliding glass doors, 640
extruded aluminum frames and storefronts, 636–637
glazed aluminum curtain walls, 683–684
Thermal capacity (mass and volume basis), 67
Thermal coefficients (U-factor), 499
Thermal comfort envelope, 1030
Thermal conductance, 56
Thermal conductivity, 86, 87, 499
conversion factors for, 67
metric conversions for, 72
units of, 56
of wood, 375
Thermal control (slab-on-grade construction), 179
Thermal drapes, 913
Thermal effects, units for, 55–56
Thermal efficiency (of insulation), 487
Thermal expansion, 87
plywood, 400–401
of wood, 375
Thermal insulance, units of, 56
Thermal insulation (gypsum board systems), 739–740
Thermal lag, 1033
Thermal mass, 1033
Thermal properties of materials, 86–87, 469
Thermal resistance. *See also* R-value
of insulation, 499
units of, 56
Thermal resistivity:
metric conversions for, 72
units of, 56
Thermal shutters, 913
Thermoplastics, 466–467, 527
Thermosets, 467
creep in, 469
decorative panels, 909
Thermosiphon principle, 1055
Thermostats, 472–474, 1057
Threaded connections (galvanized pipe), 996
Threaded fasteners, 321
Three-phase electrical service, 1069
Thresholds:
in barrier-free design, 36
door, 659
resilient flooring, 817
tile at, 762
Threshold of Hearing Curve, 917, 918
Throat, fireplace, 257
Thrust loads, 167
Tiebacks, 103, 104
Tiers, 74
Ties (unit masonry), 211–213
TIG (tungsten inert-gas) arc welding, 348
Tile, 749–772
for acoustical ceilings, 785–786
asphalt, 811
carpet, 844–845
ceramic tile, 750–755
cleaning, 275
cork, 812, 813
countertops, 763–764
floor installation, 764–768, 827–830
grouting and curing, 760–761
grouts for, 757–758
installation materials, 750–758
job conditions for laying, 758
joints, 761–762
layout and setting, 760, 761
origins of, 749
over existing finishes, 762–763
roofing, 533, 549–552
rubber, 812, 813
shower receptors, 762
subflooring for, 431, 432
supporting construction for, 758–760
terra-cotta, 224
at thresholds, 762
tile-setting products, 754–757
vinyl, 812, 813
wall installation, 768–772
Tile Council of America (TCA), 749, 810
Tilt-up panels, 189
Timber Construction Manual, 453, 460
Timber growth areas, 364
Timber piles, 171
Timber(s), 379, 388, 389
heavy timber framing systems, 452–453
laminated timber framing systems, 453–455
Timbre, 918
Time, 43, 45, 51
Time intervals, 51, 61
Timing devices, 1092–1093
Tin, 354
Tinning, 348
Tintable primers, 852
Tinted glass, 668, 671
Titanium dioxide, 858
Title VIII of the 1968 Civil Rights Act, 27
TL, *see* Transmission loss
Toe nailing, 411
Toilets, *see* Water closets
Toilet rooms, 27, 33
Tongue-and-groove plywood panels, 398, 432
Tongue-and-groove siding, 585, 586
Tooled joints (masonry), 269
Topcoat paints, 857–859
Top-hinged windows, 642
Toplap, 533, 571
Topping(s), 192, 776–777
concrete, 195–196
gypsum board systems, 728
Topsoil, removal of, 99
Torn grain (lumber), 385
Torque, conversion factors for, 66
Torsional or bending moment, 53
Toughness, 85–86, 468–469
Townhouses, 5
Townships, 74
Township lines, 74
Township number, 74
Trade associations, 13
Trade contractors, 3
Trade unions, 7
Traffic noise, 929
Translation forces, 410
Transmission, 483
Transmission factor, 1088
Transmission loss (TL), 924–925
Transmittance, 1032–1033, 1036

Transparent finishes, 860–861
Transportation systems, 972
Transverse strength (masonry walls), 250–251
Traps, 1003, 1005–1006
Travertine finish (concrete), 162–163
Travertine marble, 277
Trees, 113–115
coniferous, 112
deciduous, 112
for decorative plywood, 401, 402
and energy needs, 112, 113
fire-resistant, 887
lightning protection for, 965
planting soil for, 116
and site clearing, 99
and soil moisture content, 97
species of, 364–365
Tree growth, 365
Trenches, 102
Trim, metal, 728–729
Trim enamels, 858
Trim (finish carpentry), 462
Trimmers, 423
Trim varnishes, 860
Triple-track windows, 654
Triplex outlets, 1079
Trombe, Felix, 1054
Troweled joints, 268–269
Troweling (concrete), 159–160
Trucking, SI units for, 49
True (placement), 260
Trusses:
laminated timber, 453
roof, 453–454
steel, 330
T-shaped precast elements, 191–192
Tubs, *see* Bathtubs
Tube mill, 697
Tubing, flexible, 997
Tubular steel products, 315
Tuck-pointing mortar, 209
Tufted carpet, 835–836
Tungsten, 299
Tungsten inert-gas (TIG) arc welding, 348
Tunnel kilns, 218
Turns (in drainage systems), 1005
Turntables, 972
TWA Terminal (John F. Kennedy Airport), 122
Two-family homes, foundations for, 180
Two-pipe systems (hot water heating), 1047
Two-track windows, 654
Two-way slabs, 186
Types of construction, 19, 20
Types of hardwood plywood, 405–406
Type M mortar, 208, 209, 211, 272
Type N hydrated lime, 702
Type NM cable, 1078
Type NMC cable, 1078
Type N mortar, 208, 209, 211, 243
Type O mortar, 208, 209
Type S hydrated lime, 702
Type S mortar, 208, 209, 211
Type X lath, 724
Type X wallboard, 721

U-factor, 499
UFAS, *see* Uniform Federal Accessibility Standard
UL, *see* Underwriters Laboratories
Ultimate compressive strength, 249
Ultimate strength, 84, 86
Ultrasonic motion detectors, 968–969
Unbonded toppings, 196
Unclassified materials, 101
Unconfined compressive strength (q_u) of soil, 96
Underbeds (terrazzo), 776
Undercounter rimless lavatories, 984
Underfloor raceways, 1079
Underlayment:
for asphalt shingles, 537
and board subflooring, 431
cement-based, 197–198
combination subfloor and underlayment, 433–434
for metal roofs, 597
for mineral-fiber-cement shingles, 553
over subflooring, 432–433
for reroofing, 545–546
for resilient flooring, 817–819
for roll roofing, 521
for roofing, 534–536
for shingle roofs, 534
for slate roofing, 547
for tile roofing, 551–552
for wood shingles and shakes, 565
Underpinning, 100–101
Underscribing, 825
Underslab drains, 111
Underslung traction elevators, 973
Underwriters Laboratories (UL), 14, 22, 378, 521, 537, 889, 895, 983, 1057
Unfading slate, 547
Unified Numbering System (UNS), 306, 308, 349
Uniform Building Code, 23
Uniform Federal Accessibility Standard (UFAS), 26, 27
Uniform finish, 353
Uniform Fire Code, 23
Uniformity of codes, 22
Uniform loads, 167
Uniform Mechanical Code, 23, 24
Uniform Plumbing Code, 23, 24, 982
Unions (galvanized pipe), 996
Units, 39–47
United States of America Standards Institute (USASI), 14
U.S. Department of Agriculture (USDA), 113–114, 412
U.S. Department of Commerce, 14, 24, 40
Office of European Community Affairs, 18
standards of, 15–16
U.S. Department of Defense (DOD), 26
U.S. Department of Energy (DOE), 894, 1030
U.S. Department of Housing and Urban Development (HUD), 13, 16–17, 24–26, 406, 456, 842, 946–949, 1031
U.S. Department of the Navy, 14
U.S. Department of War, 14
U.S. Environmental Protection Agency (EPA), 983
U.S. Forest Service, 406
U.S. Metric Board, 40
Unit heaters, 1047
Unit kitchens, 899–900
Unit ventilators, 1047
UNS, *see* Unified Numbering System
Unsaturated molecules, 465
Unseasoned lumber, 382
Unspecified as-fabricated finishes, 352
UP, *see* Polyurethanes
Upfeed water supply systems, 994
Uplift, 167
Ureas, 474
Urethane enamels, 857
Urethane foam, 500, 502
Urethane sealants, 603
USASI (United States of America Standards Institute), 14
USDA, *see* U.S. Department of Agriculture
Use groups, 19, 20
Use types, building, 3
U-stirrups, 182
Utilities, underground:
and backfill placement, 105
and excavation preparations, 101
Utility lines, 99
Utility rooms:
lighting in, 1091
outlets in, 1082
Utility shelving, 462–463
U-values, 286–289, 1036–1037

Vacuum process (wood preservation), 378
Valley flashing:
for asphalt shingles, 539–540
for mineral-fiber-cement shingles, 554
in reroofing, 546
for steep roofs, 535
for wood shingles and shakes, 566–567
Valves (water supply systems), 999
Vanadium, 299
Vanity cabinets, 906
Vapor barriers, 483

Vapor permeability, 512
Vapor pressure, 481
Vapor retarders, 110, 111, 155, 246, 483
  and air infiltration, 439
  for cavity walls, 241
  in condensation zones, 488–489
  in floors, 495
  installation of, 495
  for low-slope roofing, 514
  polyethylene, 472
  in roof and ceiling assemblies, 495
  selection of, 489–491
  and ventilation design, 491–495
  in walls, 495
Varnish, 860
Vegetation, and site clearing, 99
Vehicular traffic, noise from, 929
Velocity, 45, 63
  conversion factors for, 66
  units of, 52
Velvet weave, 833–834
Veneers:
  matching of, 402–403
  types of, 401–402
  visual characteristics of, 403
Veneer core hardwood plywood, 403, 404
Veneered masonry walls, 238, 241–244
Veneered wall ties, 212
Veneer grade (plywood), 396–397
Veneer plaster, 698, 711–712, 744
Venetian blinds, 912–913
Vents, 1004–1007
Ventilation, 489
  of attics, 492
  in condensation zones, 488–489
  of crawl spaces, 491–492
  interior, and condensation, 484–485
  of roof and ceiling assemblies, 492–495
  of slabs-on-grade, 495
  vapor retarders and design of, 491–495
  of walls, 491
Vent openings, 491
Vent stacks, 1049
Vermiculite, 501, 502, 503, 701
Vertical butt matching (veneer), 403
Vertical cell tile, 222
Vertical-grained lumber, 385
Vertical loads, 167
Vertical membrane waterproofing system, 496
Vertical pivot windows, 642
Vertical sidewall flashing (asphalt shingles), 540, 541
Vertical sliding windows, 642
V-grooved wall panels, 405
Vibrating (concrete), 157–158
Vinyls, 466, 472
Vinyl base, 830
Vinyl composition tile, 811, 812
Vinyl sheet, 811–813
Vinyl siding, 583–585
  installation of, 583–585
  maintenance of, 585
  material for, 583
Vinyl Siding Institute (VSI), 583
Vinyl tile, 812, 813
Vinyl toluene acrylic resin coatings, 851
Vinyl toluene resin coatings, 851
Vinyl wall coverings, 872–873
Visual cues, 32
Visual impairment, 32–33, 35
Visual warning systems, 33
Vitreous coatings, 320, 346
Vitreous tile, 751
Voids, soil, 94
Voltage, 1065–1066
Volume, 45
  changes of, in concrete masonry units, 232–233
  conversion factors for, 65
  metric conversions for, 71
  SI units for, 61
  units of, 51
Volume rate of flow:
  conversion factors for, 66
  metric conversions for, 72
  units of, 52
V-shaped joints (masonry), 269
VSI (Vinyl Siding Institute), 583
Vulcanized ceoprene, 525
Vulcanized elastomers, 525

Waferboard, 406
Waffle slabs, 186–188
Wales, 103
Walks (barrier-free design), 29, 30
Wall(s):
  acoustical wall panels, 792–797
  capacity insulation in, 508
  cast-in-place concrete, 180–181
  concealed condensation in, 486–487, 488
  concrete, 130–132
  flashing for, 591–592
  heat transmission through, 1035
  insulation for, 504–508
  insulation of existing, 509–510
  masonry, *see* Masonry walls
  paints for interior, 861–862
  plumbing, 992
  and sound control, 935–941, 949, 957–963
  structural precast concrete, 190
  ties for, 212
  and tile, 759–760
  tile installation on, 768–772
  vapor retarder installation in, 495
  ventilation of, 491
  wet, 992
Wall assemblies:
  panelized, 445, 446
  wood framing for, 424–426
Wall base cement, 821
Wallboard, foil-backed, 490
Wall cabinets, 906
Wallcoverings, vinyl, 872–873
Wall flashing, 272
  in reroofing, 546
  for wood shingles and shakes, 567
Wall framing (gypsum board systems), 724
Wall-hung lavatories, 984
Wall ovens, 896
Wall primers, 855
Wall projections, wood framing for, 423, 424
Wall sheathing:
  fiberboard, 437–438
  gypsum board, 438
  insulated, 438
  lumber, 435–436
  near fireplaces, 255
  plywood, 436–437
  wood-framed construction, 435–439
Wane (lumber), 385
Warnings, detectable, 31, 32
Warp, 385, 401, 833
Washers, 895, 897, 898
Wash primers, 857
Water:
  barrier-free design considerations for, 27
  in cement mixtures, 125, 150–151
  in concrete mixtures, 146, 153
  effects of, on soils, 97–98
  as electrical conductor, 1073
  in grouts, 210
  hardness of, 1001–1002
  masonry bonded walls penetration by, 263
  and moisture, 97
  in mortar and grout, 207
  in mortars, 210
  in plaster, 702
  potable, 990
  in steel production, 299
  temperature of, 27–28
Water absorption (concrete masonry units), 232, 233, 236
Water-borne wood preservatives, 378
Water-cement ratio, 146, 149
Water closets, 985–987
  in barrier-free design, 34, 37
  bowls for, 987
  flushing mechanisms in, 985–986
  mounting of, 987
  tanks in, 986
Water gilding, 356
Water hammer, 1024
Water heaters, 999–1001
  combination, 1000
  direct, 999–1000
  fuels for, 1000–1001
  sizing of, 1001

standards for, 983
temperature controls on, 1001
Water mains, 999
Water problems, 109–112
control of, 110–112
grading to control, 106
Waterproofing, 246, 270, 272, 496–498
Waterproof-resin adhesive, 819
Water-reducing admixtures (concrete), 128
Water-reducing retarders, 128
Water-repelling agents, 230
Water resistance test pressure (WTP):
aluminum windows and sliding glass doors, 643
windows and sliding glass doors, 640
Water-resistant wallboard, 721, 723
Water retention (in mortar or grout), 204
Water softeners, 1002
Waterstop(s), 144
in building expansion joint, 146
in concrete joints, 143
in control joint, 145
Waterstruck bricks, 216, 217
Water supply fixture units, 991, 1008
Water supply systems, 989–999
configuration of, 993–995
conservation considerations with, 990
consumption estimation guidelines, 991
corrosion in, 1002
design of, 991–999
flow control in, 998
layout of, 992–993
municipal/community systems, 990–991
pipe materials for, 995–998
pressure regulation in, 994, 998–999
private systems, 991
separation of, from drainage system, 989–990
sources of water, 990–991
valves in, 999
Water table, 109, 110
Watertightness:
of concrete, 149
and mortars, 205
Water Trombe wall, 1054
Water tube boilers, 1049
Water vapor, 478–483
adsorption and absorption, 482–483
condensation causes and prevention, 483–488
in crawl spaces, 488
relative humidity, 479–481
transmission of, 483, 484
vapor movement, 481–482
and ventilation, 484–485
Watson, Thomas, 966
Watt (unit), 1066
Wavelength:
sound, 917–918
units of, 59
Waves, sound, 917–918
Wax coatings, 346
WBT, *see* Wet-bulb temperature
WCLIB, *see* West Coast Lumber Inspection Bureau
WDMA (Window and Door Manufacturers Association), 378
Weather conditions, 1031
Weathered joints, 268, 269
Weathering:
of aluminum, 338–339
clay masonry units, 224–225
of wood, 375
Weathering slate, 547
Weathering steels, 307
Weatherstripping, 658–659
for aluminum entrances and storefronts, 634–635
for aluminum windows and sliding glass doors, 639
for metal doors and frames, 619
for sound control, 932–933
for wood windows and sliding glass doors, 645–646
Weep holes, 270, 275
Weft, 833
Weigh batcher, 230
Weight, units for, 43, 62
Welded door frames, 620
Welded steel pipes, 315
Welded wire fabric, 134–135, 136, 137, 708
Welding, 322–324
aluminum, 347–34
stud welding, 321
symbols used for, 324
Wells, 991
Well points, 110
West coast hemlock flooring, 799
West Coast Lumber Inspection Bureau (WCLIB), 375–376, 800
Western Fire Chiefs Association, 23
Western framing, 412
Western Wood Products Association (WWPA), 375–376, 1041
Western woods region, 387
Wet-back Scotch marine boilers, 1049
Wet-bulb temperature (WBT), 480, 1028
Wet loopers, 517
Wet-pipe fire sprinkler systems, 1058
Wet walls, 992
Wheelchair access. *See also* Barrier-free design
residential appliances, 895
unit kitchens, 900
White bronzes, 349–350
White cast iron, 301, 304
White pocket (or white speck), 376
White portland cement, 125
White scum, 275, 276
WIC, *see* Woodwork Institute of California
Wilton loom, 834
Wind loads, 167
Windows:
aluminum, 638–644
in barrier-free design, 33
cleaning/maintenance of, 651
double-hung (DH), 641
fixed, 643
in glass-reinforced concrete panels, 195
projected, 641
for sound control, 933–934
storm windows, 653–656
wood, 645–653
Window and door components, 649, 651
Window and Door Manufacturers Association (WDMA), 378
Window locks:
eyebolt, 648–649
friction type, 648
key-operated cam latch, 648
pin-type, 648
stop-type, 648
wedge-type, 648
Window treatment, 910–913
exterior, 911–912
interior, 912–913
Wind pressure, 240, 270
Winning, 214
Wire, steel, 315
Wires, 136
Wire bar supports, 138
Wired glass, 670–671
Wire lath, 707–708
Wire mesh ties, 212
Wire nails, common sizes of, 411
Wireways, 1079
Wiring, 1043, 1073, 1075–1076, 1083–1086
Withe (masonry), 260
Wood:
chemical composition of, 366
concrete forms made of, 129
exterior finishing of, 864
exterior priming of, 863
fiberboard, 406
finish carpentry, 462–464
fire-retardant treatment of, 378
fungus/insect hazards, 376–377
general framing requirements, 410–411
gluing properties of, 374–375
hygroscopic properties of, 366–369
lumber, 370–373, 379–390
nonstructural properties of, 375–376
nonveneer panels, 406
oriented strand board, 406
plywood and other panels, 391–406
preservative treatment of, 377–378
properties of, 364–378
specific gravity of, 369
structural properties of, 369–374
treated wood foundations, 406–410

Wood, (*cont'd*)
tree growth, 365
tree species, 364–365
for use in plywood, 396
Wood board siding, 585–588
application of, 586–588
materials for, 585–586
preparation for, 586
Wood casework, 904–910
cabinets, 904–910
countertops, 910
installation of, 910
quality standards for, 904–905
uses of, 904
Wood doors, 623–624
accordion fold, 623–624
finishing, 634
flush, 623–624
handling/storage, 634
hanging, 634
installation of, 634
louver, 624
manufacture of, 629–632
operation of, 628–629
panel, 624
quality standards, 632–634
sash, 624
standards of quality for, 632–634
storm and screen, 624
types and uses of, 623–624
Wood-fibered plaster, 698, 710
Wood fillers, 856, 870–871
Wood-finished cabinets, 907–909
Wood flooring, 797–805
maintenance of, 805
manufacture of, 799–800
materials handling, 805
materials used in, 798–799
paints for, 862–863
parquet flooring, 804–805
plank flooring, 803–804
reflooring of, 831
strip flooring, 800–803
subflooring for, 431
Wood-Frame House Construction, 412
Wood framing, 6, 410–461
anchorage requirements, 410
conventional framing, 412–445
decay and termite control, 410–411
floors/floor and ceiling assemblies, 412–424
furring, 445–446
general requirements for, 410–411
for gypsum board systems, 731–735
heavy timber, 452–453
laminated timber, 453
nailing practices, 411
plank and beam, 447–452
post and beam, 447
preassembled components, 439–445
roof and ceiling assemblies, 426–431
roof sheathing, 435–439
roof trusses, 453–454
and sound control, 937–938, 942–944
subflooring and plywood underlayment, 431–436
wall assemblies, 424–426
wall sheathing, 435–439
wood-pole construction, 455–462
Wood-joist floor systems, 410
Wood-pole construction, 455–462
advantages and disadvantages of, 456–457
construction methods for, 458–461
pole selection for, 458
on rocky slopes, 460–462
site preparation for, 457–458
Wood posts, in light wood-framed buildings, 413–414
Wood primers, 855–856
Wood Products—Structural Glued Laminated Timber, 453
Wood sealers, 855–856
Wood shingles and shakes, 558–570, 577–582
application of, 567–570, 579–580
cedar shakes, 559–562
cedar shingles, 559–562
drip edges for, 566
eave flashing for, 565–566
flashing for, 566–567
materials for, 577–579
maximum exposure of, 564
nails for, 563–564
over existing walls, 581–582
preparation for, 579
sheathing for, 565
standards for, 563
underlayment and interlayment for, 565
Wood Structural Design Data, 413
Wood windows, 645–653
awning, 649, 650
basement, 649, 650
burglar protection with, 646–647
casement, 649, 650
double-hung (DH), 649, 650
fixed, 649, 650
glass and glazing, 649
hardware for, 647–648
hopper, 649, 650
horizontal sliding, 649, 650
installation of, 652–653
lock types for, 648–649
manufacture of, 645–649
screening for, 649
selection of, 652–653
single-hung, 649, 650
summary of, 650–651
types and nomenclature, 649, 650–652
weatherstripping, 645–646
Woodwork Institute of California (WIC), 462, 632, 904
Wool carpets, 838
Workability, 151
concrete masonry units, 230
plywood, 401
Workable mortar, 204
Worked lumber, 388
Work energy:
conversion factors for, 67
units of, 54
Working stress, 371
Workmanship standards, 13
World Columbian Exposition, 696
Woven carpet, 833–835
Axminster weave, 834–835
loomed carpet, 835
velvet weave, 833–834
Wilton, 834
Woven felt weatherstripping, 646
Wright, Frank Lloyd, 122
Wrought alloys (aluminum), 336–338
Wrought aluminum products, 340–341
Wrought iron, 296, 297
Wrought steel pipes, 315
Wrought-steel products, 308–316
WTP, *see* Water resistance test pressure
WWPA, *see* Western Wood Products Association

Yard lumber, 379, 387
Y-fitting, 1007
Yield point, 84
Yield strengths, 84, 85

Zinc, 353–354, 596–597
Zincating, 344
Zinc-based paint pretreatments, 320
Zinc chromate primers, 856
Zinc-dust/zinc-oxide primers, 856
Zinc oxide, 858
Zinc-plated aluminum, 356
Zones:
acoustical, 931
HVAC, 1039
Zone-marking paints, 859–860
Zoning codes, 18–19